Common Functional Groups

General Formula	Example	Name
RH	$CH_3CH_2CH_3$	**Alkane** (Alkanes have no functional group.)
$RCH{=}CH_2$	$CH_3CH_2CH{=}CH_2$	**Alkene** (The functional group is the carbon–carbon double bond.)
$RC{\equiv}CH$	$CH_3C{\equiv}CH$	**Alkyne** (The functional group is the carbon–carbon triple bond.)
ArH		**Arene** (A six-membered ring with three double bonds has different reactions than an alkene so it is given a different name. Arenes or **aromatic rings** can have alkyl or other groups attached to the ring.)
RX	$CH_3CH_2CH_2Cl$	**Alkyl halide** (The functional group is the carbon–halogen bond.)
ROH	CH_3CH_2OH	**Alcohol** (The functional group is the C—O—H.)
ROR	$CH_3CH_2OCH_3$	**Ether** (The functional group is the C—O—C. The alkyl groups on the O can be the same or different.)
RNH_2	$CH_3CH_2CH_2NH_2$	**Amine** (The C—N is the functional group. The other H's on the N can be replaced with alkyl groups.)
$\overset{O}{\overset{\|}{R}}CH$	$CH_3CH_2\overset{O}{\overset{\|}{C}}H$	**Aldehyde** (The functional group is the C=O with at least one H on the C.)
$R\overset{O}{\overset{\|}{C}}R$	$CH_3\overset{O}{\overset{\|}{C}}CH_3$	**Ketone** (The functional group is the C=O with two alkyl groups on the C. The alkyl groups need not be the same.)
$R\overset{O}{\overset{\|}{C}}OH$	$CH_3CH_2\overset{O}{\overset{\|}{C}}OH$	**Carboxylic acid** (The functional group is the $\overset{O}{\overset{\|}{C}}OH$.)
$R\overset{O}{\overset{\|}{C}}Cl$	$CH_3\overset{O}{\overset{\|}{C}}Cl$	**Acyl chloride** (The functional group is the $\overset{O}{\overset{\|}{C}}Cl$.)
$R\overset{O}{\overset{\|}{C}}O\overset{O}{\overset{\|}{C}}R$	$CH_3\overset{O}{\overset{\|}{C}}O\overset{O}{\overset{\|}{C}}CH_3$	**Acid anhydride** (The functional group is the $\overset{O\ \ \ O}{\overset{\|\ \ \ \|}{C}}OC$. The alkyl groups may be different.)
$R\overset{O}{\overset{\|}{C}}OR$	$CH_3\overset{O}{\overset{\|}{C}}OCH_2CH_3$	**Ester** (The functional group is the $\overset{O}{\overset{\|}{C}}OC$.)
$R\overset{O}{\overset{\|}{C}}NH_2$	$CH_3CH_2\overset{O}{\overset{\|}{C}}NHCH_3$	**Amide** (The functional group is the $\overset{O}{\overset{\|}{C}}N$. The other groups on the N may be H's or alkyl groups.)
$RC{\equiv}N$	$CH_3CH_2CH_2C{\equiv}N$	**Nitrile** (The functional group is the carbon–nitrogen triple bond.)

Welcome to your Organic ChemistryNow™ Media Integration Guide!

http://now.brookscole.com/hornback2

The **Media Integration Guide** on the next several pages links each chapter to the wealth of interactive media resources you will find at **Organic ChemistryNow**, a unique web-based, assessment-centered personalized learning system for organic chemistry students.

Media Integration Guide

Media Integration Guide

Media Integration Guide

Chapter	Text Section	Organic ChemistryNow™ Resources: http://now.brookscole.com/hornback2
25 Carbohydrates	25.3 Cyclization of Monosaccharides	**Active Figure** 25.2: The Cyclization of D-Glucose to Form α- and β-D-Glucopyranose (page 1093) **Coached Tutorial Problem** Cyclizations of Carbohydrates (page 1094)
	25.4 Reactions of Monosaccharides	**Coached Tutorial Problem** Reactions of Monosaccharides (page 1102)
	Mastery Goal Quiz	Page 1116
26 Amino Acids, Peptides, and Proteins	26.7 Laboratory Synthesis of Peptides	**Active Figure** 26.4: Mechanism of Amide Formation Using Dicyclohexylcarbodiimide (page 1151) **Coached Tutorial Problem** Reactions Used in Synthesis of Peptides (page 1151)
	26.8 Protein Structure	**Coached Tutorial Problem** α-Helix or β-Sheet (page 1156)
	Mastery Goal Quiz	**Page 1158**
27 Nucleotides and Nucleic Acids	27.2 Structure of DNA and RNA	**Active Figure** 27.1: A Tetranucleotide with the General Structure of DNA (page 1166) **Coached Tutorial Problems** • Complementary Base Pairing (page 1167) • DNA Structure (page 1169)
	Mastery Goal Quiz	Page 1180
28 Other Natural Products	28.5 Steroids	**Active Figure** 28.7: The Cyclization of Squalene Oxide to Lanosterol (page 1199)
	Mastery Goal Quiz	Page 1213

second edition

ORGANIC CHEMISTRY

JOSEPH M. HORNBACK

UNIVERSITY OF DENVER

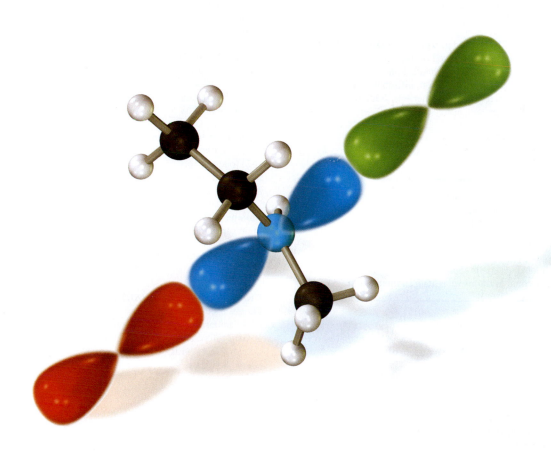

THOMSON

BROOKS/COLE

Australia • Canada • Mexico • Singapore • Spain • United Kingdom • United States

THOMSON
✳
™
BROOKS/COLE

Publisher, Physical Sciences: David Harris
Development Editor: Sandra Kiselica
Assistant Editor: Sarah Lowe
Editorial Assistant: Candace Lum
Technology Project Manager: Donna Kelley
Marketing Manager: Amee Mosley
Marketing Assistant: Michele Colella
Marketing Communications Manager: Nathaniel Bergson-Michelson
Project Manager, Editorial Production: Lisa Weber
Creative Director: Rob Hugel
Print/Media Buyer: Barbara Britton

Permissions Editor: Sarah Harkrader
Production Service: Graphic World Inc.
Text Designer: Gopa & Ted2, Inc.
Copy Editor: Graphic World Inc.
Illustrators: GTS Graphics; Greg Gambino, 2064 Design; and Graphic World Inc.
Cover Designers: Lee Friedman and Lisa Devenish
Cover Image: Greg Gambino, 2064 Design
Cover Printer: Coral Graphic Services, Inc.
Compositor: Graphic World Inc.
Printer: R.R. Donnelley/Willard

Printed in the United States of America
1 2 3 4 5 6 7 09 08 07 06 05

For more information about our products, contact us at:
**Thomson Learning Academic Resource Center
1-800-423-0563**

For permission to use material from this text or product, submit a request online at **http://www.thomsonrights.com**. Any additional questions about permissions can be submitted by email to **thomsonrights@thomson.com**.

Library of Congress Control Number: 2004108121

Student Edition: ISBN 0-534-38951-1
Instructor's Edition: ISBN 0-534-49249-5
International Student Edition: ISBN 0-534-49317-3 (Not for sale in the United States)

Thomson Higher Education
10 Davis Drive
Belmont, CA 94002
USA

Asia (including India)
Thomson Learning
5 Shenton Way
#01-01 UIC Building
Singapore 068808

Australia/New Zealand
Thomson Learning Australia
102 Dodds Street
Southbank, Victoria 3006
Australia

Canada
Thomson Nelson
1120 Birchmount Road
Toronto, Ontario M1K 5G4
Canada

UK/Europe/Middle East/Africa
Thomson Learning
High Holborn House
50/51 Bedford Row
London WC1R 4LR
United Kingdom

Latin America
Thomson Learning
Seneca, 53
Colonia Polanco
11560 Mexico
D.F. Mexico

Spain (includes Portugal)
Thomson Paraninfo
Calle Magallanes, 25
28015 Madrid, Spain

To Melani, Joe, Pat, Jordan, and Cullen,
who bring meaning and joy to my life.

About the Author

Joseph M. Hornback was born and raised in southwestern Ohio. He received a B.S. in Chemistry, magna cum laude, from the University of Notre Dame in 1965. He then attended the Ohio State University on an NSF traineeship and received his Ph.D. in 1968. He next moved to the University of Wisconsin at Madison on an NIH postdoctoral fellowship.

In 1970, he joined the faculty of the Department of Chemistry at the University of Denver, where he is now Professor of Chemistry and Biochemistry. His research interests are in the areas of synthetic organic chemistry and organic photochemistry. He has served in a number of administrative positions, including Associate Dean for Undergraduate Studies; Associate Dean for Natural Sciences, Mathematics, and Engineering; and Director of the Honors Program. But his first love has always been teaching, and he has taught organic chemistry nearly every term, even when he was in administration. He has received the Natural Sciences Award for Excellence in Teaching and the Outstanding Academic Advising Award.

Joe is married and has three children, two sons and a daughter, and one grandson. He enjoys sports and outdoor activities, especially fishing and golf.

Brief Contents

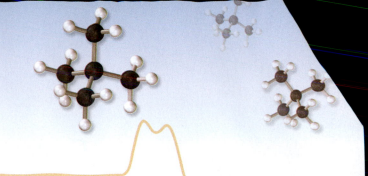

Contents

7 STEREOCHEMISTRY II: CHIRAL MOLECULES   219

8 NUCLEOPHILIC SUBSTITUTION REACTIONS: REACTIONS OF ALKYL HALIDES, ALCOHOLS, AND RELATED COMPOUNDS   257

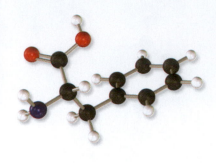

List of Mechanisms

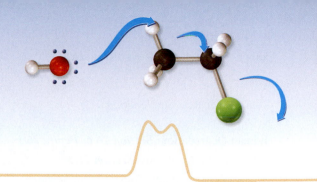

Radical Addition of HBr to an Alkene: *Section 21.9*

Radical Dehalogenation of an Alkyl Halide with Tributyltin Hydride: *Section 21.7; Figure 21.3*

Radical Halogenation of an Alkane: *Section 21.6; Figure 21.1*

Radical Polymerization of an Alkene: *Section 24.1*

Reaction of an Alcohol with Thionyl Chloride: *Section 10.5; Figure 10.3*

Reaction of a Carboxylic Acid with Thionyl Chloride: *Section 19.2; Figure 19.2*

Reaction of an Ester with a Grignard Reagent: *Section 19.9; Figure 19.10*

Reaction of a Nitrile with a Grignard Reagent: *Section 19.10; Figure 19.11*

Reduction of an Alkyne by Sodium: *Section 21.10*

Reduction of an Ester with DIBALH: *Section 19.8*

Reduction of an α,β-Unsaturated Ketone by Lithium: *Section 21.10*

Ring Opening of an Epoxide (Borderline S_N2): *Section 10.10*

Ring Opening of an Epoxide (S_N2): *Section 10.10*

Robinson Annulation: *Section 20.10*

Sigmatropic Rearrangements: *Section 22.8*

Sulfonation of an Arene: *Section 17.6*

Tautomerization of an Enol to a Ketone: *Section 11.5; Figure 11.6*

Unimolecular Elimination (E1): *Section 9.5; Figure 9.6*

Unimolecular Elimination, Conjugate Base (E1cb): *Focus On Box, pages 333–334*

Unimolecular Nucleophilic Substitution (S_N1): *Section 8.6; Figures 8.6, 8.8, and 8.9*

Wolff-Kishner Reduction: *Section 18.8; Figure 18.4*

Preface

This book is intended for use in the first organic chemistry course taken by students majoring in chemistry, biochemistry, biological sciences, and other health-related fields. The unique organization of this book sets it apart from many other texts in this area; specifically, this text is organized according to the mechanisms of the presented organic reactions rather than according to functional groups. It covers all of the same material usually found in a book using the functional group approach but in an order based on mechanistic themes.

The organic chemistry course is a pivotal class taken by science students. In addition to the inherent importance of organic reactions in chemistry and biology, organic chemistry introduces and develops a type of reasoning and logic that is new to many students. A solid understanding of this subject is often critical to subsequent success in a science career.

Organic chemistry often has a reputation among students as being a very difficult course, involving an enormous amount of memorization. For students who approach the course by attempting to memorize all of the important information, it is a daunting subject indeed. The goals of this text are to help students be successful in organizing this vast amount of material, to stimulate their interest by making organic chemistry understandable and relevant, to demonstrate the logic and beauty of the field, and to provide a method to remember all of those many reactions.

Using an organization based on reaction mechanisms rather than the approach based on functional groups has always appealed to me. I believe that the mechanism approach encourages students to develop an understanding of why things occur rather than just memorizing what occurs. In writing this book, my goal throughout has been to present organic chemistry in a way that is clear, understandable, and accessible to students. *Mechanisms are used as the organizing principle that helps students learn organic chemical behavior, not as the major concept to be learned.* Unlike a graduate-level or advanced undergraduate book, this text is not intended to be an encyclopedia of mechanisms. Those mechanisms that are general and that organize a number of reactions are emphasized, whereas those that are uncommon are often not covered at all.

Changes in This Edition

Organization

The first three chapters constitute a review of bonding and an introduction to organic compounds. Functional groups are introduced. Resonance is covered extensively, and numerous examples are provided. Acid–base chemistry is discussed in Chapter 4, and this reaction is used to introduce many of the general features of reactions, including the effect of structure on reactivity. Nomenclature of all of the functional groups is covered in Chapters 5 and 12. In this edition, stereochemistry is covered in two chapters to break up the material: Chapter 6 discusses cis–trans isomers and conformations, and Chapter 7 addresses chiral molecules.

Chapter 8 begins the treatment of organic reactions with a discussion of nucleophilic substitution reactions. Elimination reactions are treated separately in Chapter 9 to make each chapter more manageable. Chapter 10 discusses synthetic uses of substitution and elimination reactions and introduces retrosynthetic analysis. Although this chapter contains many reactions, students have learned to identify the electrophile, leaving group, and nucleophile or base from Chapters 8 and 9, so they do not have to rely as much on memorization. Chapter 11 covers electrophilic additions to alkenes and alkynes. The behavior of carbocations, presented in Chapter 8, is very useful here. An additional section on synthesis has been added to this chapter as well.

IR spectroscopy is covered in Chapter 13, and hydrogen and carbon NMR spectroscopy are covered in Chapter 14. These topics have been separated to provide a more reasonable chapter size and to increase flexibility in order of presentation. Mass spectrometry and UV-visible spectroscopy are covered in Chapter 15, so these topics can be made optional if desired.

Aromatic chemistry is discussed earlier in this edition: Chapter 16 covers aromaticity, and Chapter 17 presents aromatic substitution reactions. Chapters 18 and 19 discuss additions to and substitutions at the carbonyl group. To keep these chapters from being overwhelming, aldol and ester condensations are covered separately in Chapter 20, which deals with reactions of enolate and related nucleophiles. Chapter 21 presents the chemistry of radicals.

All chapters have been designed to be as self-contained as possible, allowing the possibility of presenting them in a different order. For example, it is possible to cover the spectroscopy chapters earlier or later than they are presented in the text. As a further example, those wishing to present carbonyl chemistry earlier can cover Chapters 18, 19, and 20 immediately after Chapter 12.

Chapters 1 through 21 cover the topics that most instructors will include in their courses, with the possible exception of Chapter 15. The remaining chapters offer a choice for the last part of the course. They include chapters on pericyclic reactions, synthesis, and polymers. The chapters on the more biochemical topics—carbohydrates, amino acids and proteins, nucleotides and nucleic acids, and other natural products—concentrate on the organic chemistry of these important biomolecules.

Spectroscopy

Chapters 13 and 14 provide complete coverage of IR and NMR spectroscopy, respectively, and UV-visible spectroscopy and mass spectrometry are discussed in Chapter 15. These chapters can be covered earlier or later in the course as the instructor desires without affecting other chapters. Numerous sample spectra and problems are included, and all spectra have been redrawn for increased clarity. Because all the functional groups have been introduced early, the spectroscopy of all of them is presented at one time. It is not necessary to return to a discussion of spectra each time a new functional group is introduced. Chapters after Chapter 15 have several spectroscopy-based problems in a separate section of the additional problems.

Synthesis

The use of reactions to synthesize organic compounds is introduced early and is an important part of every chapter in which reactions are presented. The first introduction to synthesis occurs in Chapter 8, with problems such as "What reagent and solvent would you use to carry out the following transformation?" and "Show how these compounds could be prepared from alkyl halides." Chapter 10 introduces the concept of synthetic equivalent (acetate for hydroxide and phthalimide for ammonia). Section 10.15 covers the strategy of organic synthesis and introduces retrosynthetic analysis; here the synthesis problems become more complex. A new section on synthesis in Chapter 11 introduces the retrosynthetic arrow and again raises the level of the problems. This process continues through the remaining reaction chapters. Chapter 23, a unique chapter on synthesis, brings together most of the reactions presented in earlier chapters; provides additional discussion of synthetic strategy and the use of protecting groups; and presents some longer, more complex syntheses. Table 23.1 lists the important carbon–carbon bond-forming reactions, and Table 23.2 lists most of the reactions presented in the book according to the functional group that is produced. These tables are very useful in designing syntheses and can also be used as a summary of most of the text.

INTRODUCING ORGANIC CHEMISTRYNOW

http://now.brookscole.com/hornback2

This completely new website is designed to engage students by helping them prepare for examinations. Organic ChemistryNow is an assessment-centered learning tool developed in concert with the approach and pedagogy in the text. Students take a pretest that includes questions that have been authored to reflect the level and approaches discussed in the text. They are then given a personalized learning plan based on their pretest results. The unique personalized learning plan directly links students to Molecular Model Problems, Mechanisms in Motion (animations of organic mechanisms), Coached Tutorial Problems, Building Block Review Problems, and Active Figures.

Key Features

- A brand **new design** has made the book visually appealing and pedagogically easier to read. In addition, color is used to highlight parts of molecules and to follow the course of reactions.

- **Up-to-date information** about reactions, reagents, and mechanisms has been added throughout.

- All **mechanisms** have been examined carefully for inclusion of all steps with arrow pushing, proper reagents, and conditions. Each mechanism is clearly labeled and easily identified by a tan background, and steps are numbered and annotated.

- Each chapter begins with a set of **Mastery Goals.** At the end of the chapter these goals are restated and linked to specific problems so that students can test their command of the material. In addition, **Mastery Goal Quizzes** for each chapter can be found on the website at Organic ChemistryNow.

- More than 1100 problems are included in this book, many with multiple parts. Many new **Practice Problems,** with in-text solutions, have also been added. Many of these include a **Strategy** section to guide students through the thought process involved.

- **Integrated Practice Problems** have been added at the end of many chapters to tie together ideas from the chapter and to demonstrate a process that can be used to decide which reaction in that chapter is occurring.

- Plentiful **problems** appear at the end of each chapter. These problems range from drill to challenging and include applications to the biological sciences, molecular models, and problems using spectroscopy.

- Because Organic Chemistry is a very visual science, considerable effort has gone into the development of chemical **illustrations** to make them both clear and informative. Most figures contain text as well as structures. This decreases the need to refer back and forth between the figure and the text.

- Extensive use of **molecular models,** both in-text and online, helps students visualize the shapes of compounds and how the molecules interact in three dimensions. In addition, **Model-Building Problems** are interspersed throughout the text to give students practice building handheld models. End-of-chapter problems based on online models are also included.

- Students can also assess their understanding of each chapter's topics with additional quizzing, conceptual-based problems, and tutorials at the **OrganicChemistryNow**™ website (see following for details).

- **Electrostatic potential maps** have been added throughout the text to illustrate the important concepts of electrophilicity, nucleophilicity, and resonance.

- The application of organic chemistry to **biological chemistry** is emphasized within the text, in the *Focus On Biological Chemistry* boxes, and within problems designated with the BioLink icon.

- **Focus On and Focus On Biological Chemistry** boxes illustrate applications of organic chemistry to the world around us and to the health sciences, explore topics in more depth, or discuss the history of chemical discoveries. Some topics include *DDT-Resistant Insects, Biological Alkylations and Poisons,* and *Environmentally Friendly Chemistry (Green Chemistry).*

- Animations of key concepts are found on **Organic ChemistryNow** as *Active Figures.* Taken straight from the text, *Active Figures* help students visualize key concepts from the book. The *Active Figures* also include questions so that students can assess their understanding of the concepts.

- Unique to this book are **Tables 23.2 and 23.3** on pages 1031–1043, which summarize the important carbon–carbon bond-forming reactions and most of the reactions presented in the text. Students can use these tables as an aid in designing the synthesis of more complicated problems.

■ All definitions of key terms are collected in a **glossary** at the end of the text. Each glossary listing is keyed to the section of the text where the term is introduced.

■ To improve the **layout and the flow** of the text and to emphasize applications of organic chemistry within the text, this edition has less boxed material. However, if you still wish to use some of the Elaboration boxes from the first edition, that material is available online at **Organic ChemistryNow.**

SUPPORT PACKAGE

For the Student

■ **Student Study Guide,** by Joseph Hornback. This manual contains detailed solutions to all text problems. 0-534-39710-7

■ **Pushing Electrons: A Guide for Students of Organic Chemistry,** third edition, by Daniel P. Weeks, Northwestern University. This paperback workbook is designed to help students learn techniques of electron pushing. Its programmed approach emphasizes repetition and active participation. 0-03-020693-6

ORGANIC
Chemistry ⚛ Now™

■ **Organic ChemistryNow** at **http://www.brookscole.com/hornback2,** developed by Paul R. Young (University of Illinois, Chicago). This web-based assessment-centered learning tool for the Organic Chemistry course was created in concert with this text. Throughout each chapter, icons with captions alert students to media resources that enhance problem-solving skills and improve conceptual understanding. In Organic ChemistryNow, students are provided with a Personalized Learning Plan—based on a diagnostic pretest—that targets their study needs and helps them visualize, organize, practice, and master the material in the text. PIN code access to Organic ChemistryNow is included with every new copy of the text.

■ **OWL (Online Web-based Learning System) for Organic Chemistry**. Developed over the past several years at the University of Massachusetts, Amherst, and class-tested by hundreds of students, Organic OWL is a customizable and flexible web-based homework system and assessment tool. This fully integrated testing, tutorial, and course management system features more than 3000 practice and homework problems. With both numerical and chemical parameterization built in, Organic OWL provides students with instant analysis and feedback to homework problems, modeling questions, molecular structure–building exercises, and animations created specifically for *Organic Chemistry,* second edition. This powerful system maximizes the students' learning experience and at the same time reduces faculty workload and facilitates instruction. A fee-based PIN code is required for access to Organic OWL. To learn more, contact your Thomson Brooks/Cole representative for details. 0-534-42261-6

vMentor

■ **vMentor™ Live Online Tutoring.** One-to-one online tutoring help with a chemistry expert is available with this text. vMentor features two-way audio, an interactive whiteboard for displaying presentation materials, and instant messaging. With vMentor, students interact with the tutor and other students using standard Windows® or Macintosh® microphones and speakers. Inside the vMentor virtual classroom, icons indicate who is in the class and who is speaking, sending a message, or using the whiteboard.

Accessible through Organic ChemistryNow, vMentor lets students interact with experienced chemistry teachers—right from their own computers. For proprietary, college, and university adopters only. For additional information, please consult your local Thomson representative.

For the Instructor

Supporting materials are available to qualified adopters. Please consult your local Thomson Brooks/Cole sales representative for details. Visit the *Organic ChemistryNow* website at **http://now.brookscole.com/hornback2** to see samples of these materials, request a desk copy, locate your sales representative, or purchase a copy online.

- **Ilrn Testing:** Electronic Testing System contains approximately 1000 multiple-choice problems and questions representing every chapter of the text. Available online and on a dual-platform CD-ROM. 0-534-39712-3

- **Test Bank**, by Rainer Glaser of the University of Missouri, Columbia, is a multiple-choice test bank of more than 1000 problems for instructors to use for tests, quizzes, or homework assignments. 0-534-39711-5

- **Multimedia Manager CD-ROM** is a dual-platform digital library and presentation tool that provides text, art, and tables in a variety of electronic formats that are easily exported into other software packages. This enhanced CD-ROM also contains simulations, molecular models, and QuickTime™ movies to supplement your lectures; it also includes PowerPoint™ lecture slides with integrated media by Joseph Hornback. In addition, you can customize your presentations by importing your personal lecture slides or other material you choose. 0-534-49251-7

- **Overhead Transparency Acetates** containing a selection of 125 full-color figures from the text are available. 0-534-39713-1

- **OWL (Online Web-based Learning System) for Organic Chemistry at http://owl.thomsonlearning.com,** by Peter Lillya, Stephen Hixson, and William Vining (University of Massachusetts) and class-tested by hundreds of students, is a fully customizable and flexible web-based homework and course management system and assessment tool. This testing, tutorial, and course management system features practice and homework problems. With both numerical and chemical parameterization built in, Organic OWL provides students with instant analysis and feedback to homework problems, modeling questions, and animations. This powerful system maximizes your students' learning experience and at the same time reduces your workload and facilitates instruction. Utilizing the state-of-the-art MarvinSketch Tool, Organic OWL not only allows you to test your students' understanding of molecular structure by having them build structures online—but even grades those structures for you! 0-534-42261-6

- **WebTutor™ ToolBox on WebCT and Blackboard** is preloaded with content and available free via PIN code when packaged with this text. WebTutor ToolBox pairs all the content of this text's rich Book Companion Website with all the sophisticated course management functionality of a WebCT or Blackboard product. WebTutor ToolBox is ready to use as soon as you log on—or, you can customize its preloaded content by uploading images and other resources, adding web links, or creating your own practice materials. WebCT 0-534-48976-1; Blackboard 0-534-48977-X

ACKNOWLEDGMENTS

It is a pleasure to acknowledge the many people who have made contributions to the development of this book. First I would like to recognize my students. The questions generated by their enthusiasm to learn organic chemistry helped guide the construction of this text. To my team at Brooks/Cole, I am grateful for your guidance throughout the development of this book. My thanks especially to Sandi Kiselica, senior development editor; David Harris, publisher; Lisa Weber, project manager, editorial production; Sarah Lowe, assistant editor; and Donna Kelley, technology project manager. Donald R. Paulson, California State University, Los Angeles, and Neil Allison, University of Arkansas, served as accuracy reviewers, reading page proofs. They have keen eyes, and I am grateful for their expertise. Professor Andrei Kutateladze provided many useful discussions and helped ensure that my explanations are correct. Paul R. Young of University of Illinois, Chicago, put together a pedagogically useful website based on the contents of this book.

I also gratefully acknowledge the reviewers, who made many valuable suggestions:

Neil T. Allison, University of Arkansas
Daniel H. Appella, Northwestern University
Larry Calhoun, University of New Brunswick
Kent Clinger, Libscomb University
David Collard, Georgia Institute of Technology
Robert S. Coleman, Ohio State University
Cathleen Crudden, Queen's University
Stephen R. Daniel, Colorado School of Mines
Ghislain Deslongchamps, University of New Brunswick
Nick Drapela, Oregon State University
Colleen Fried, Hiram College
Rainer Glaser, University of Missouri
Christopher M. Hadad, Ohio State University
Scott T. Handy, State University of New York, Binghamton
Steven A. Hardinger, University of California, Los Angeles
John Haseltine, Université de Sherbrooke

Gene Hiegel, California State University, Fullerton
Robert W. Holman, Western Kentucky University
Christopher Ikediobi, Florida A&M University.
Eric J. Kantorowski, California Polytechnic State University
David Klein, Johns Hopkins University
Susan J. Klein, Manchester College
Devin Latimer, University of Winnipeg
R. Daniel Libby, Moravian College
James Mack, Boston College
Todd L. Lowary, Ohio State University
Donald R. Paulson, California State University, Los Angeles
D. S. Soriano, University of Pittsburgh, Bradford
Laurie Starkey, California State Polytechnic University, Pomona
Jon D. Stewart, University of Florida
James K. Wood, University of Nebraska, Omaha

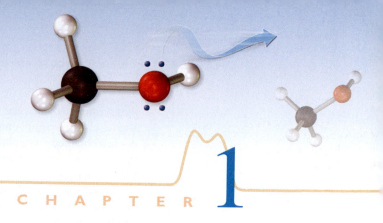

A Simple Model for Chemical Bonds

THIS CHAPTER begins your journey through the subject of organic chemistry. It starts with a brief overview of the field of organic chemistry to show why it is treated as a separate branch of chemistry. Most of the chapter is devoted to the development of a simple picture for the structure of molecules and the bonds that hold their atoms together. Lewis theory is used to explain ionic and covalent bonding. Other topics cover the formal charges of atoms in molecules, a brief introduction to resonance, polar bonds, the shapes of molecules, and polar molecules. Much of this material is a review of topics from general chemistry, but with an emphasis on applications to organic chemistry.

MASTERING ORGANIC CHEMISTRY

▶ Drawing Lewis Structures

▶ Determining Formal Charges

▶ Estimating Stabilities of Structures

▶ Understanding Simple Examples of Resonance

▶ Recognizing Polar Bonds

▶ Determining Shapes of Molecules

▶ Determining Dipole Moments of Molecules

1.1 THE FIELD OF ORGANIC CHEMISTRY

The first question we might ask is "What is organic chemistry, and how did it become a separate branch of chemistry?" A brief survey of the history of organic chemistry will help us understand how the division of chemicals into organic and inorganic originated and why this division persists today.

Some chemical compounds that we today classify as organic, such as sugar, alcohol, and acetic acid (vinegar), have been known since antiquity. However, it was not until the late 1700s that Carl Wilhelm Scheele began his pioneering studies on the isolation of organic compounds. He isolated glycerol from animal fats, tartaric acid from grapes, lactic acid from sour milk, citric acid from lemons, and many other compounds. Neither Scheele nor the other chemists of that time knew the structures of

ORGANIC
Chemistry Now™

Look for this logo in the text and go to OrganicChemistryNow at http://now.brookscole.com/hornback2 to view tutorials and simulations, develop problem-solving skills, and test your conceptual understanding with unique interactive resources.

these compounds or even which elements composed them. Because there was no better way to classify compounds, they were distinguished according to their sources. Compounds obtained from plants and animals were termed *organic* to indicate that their ultimate source was a living organism; those obtained from the mineral kingdom were termed *inorganic.*

This division according to source seemed reasonable, because both the physical and chemical properties of organic compounds are generally different from those of inorganic compounds. Typical organic compounds melt and boil at lower temperatures than typical inorganic compounds (salts). Organic compounds are more easily decomposed by heat than inorganic compounds are. These and other differences led many chemists of that time to believe that organic compounds still contained some of the life force of the organism that made them (the *vital force theory*).

Further progress in organic chemistry was provided by Antoine Lavoisier, who is often called the father of modern chemistry. In 1774 he demonstrated that organic compounds burn to produce carbon dioxide and water, meaning that they are composed predominantly of carbon, hydrogen, and perhaps oxygen. Later he showed that some organic compounds derived from animals also contain nitrogen.

In 1808 John Dalton proposed his atomic theory, making a major contribution to the understanding of inorganic compounds. Essentially, Dalton's theory postulated that the elements are composed of small, indivisible particles called atoms and that these atoms combine in ratios of small whole numbers to form chemical compounds.

Although the atomic theory helped explain the composition of many inorganic compounds, it did not help much in understanding the complexities of organic chemistry. In contrast to the relatively simple inorganic compounds that were known, such as H_2O, NO, and NO_2, there were a vast number of organic compounds, all composed of just a few elements but in a variety of ratios, such as C_2H_2, C_2H_4, C_2H_6, C_3H_8, and many, many more. Difficulties such as this led Jöns Berzelius, who coined the term *organic chemistry* in 1808 and who wrote the first text devoted to the subject in 1827, to wonder whether the atomic theory and other chemical laws even applied to organic compounds.

One tenet of the vital force theory, supported by Berzelius and other chemists of that time, was that only a living organism could make an organic compound. Friedrich Wöhler provided the first indication that this was not the case. In 1828 Wöhler found that urea, an organic compound, was formed when ammonium cyanate, an inorganic compound, was heated. Although this experiment contradicted the vital force theory, more commotion was caused at the time by the fact that both of these compounds have the same formula (CH_4N_2O) and were among the early examples of **isomers** (different compounds that have identical formulas).

Although we know today that there is no vital force that makes organic compounds different from those of the mineral world and that an organization based on source is often not appropriate, we still find it useful to divide compounds into the classes of organic and inorganic. However, today **organic chemistry** is defined as the study of the compounds of carbon. The division is still useful because of the unique chemical behavior of carbon and its compounds. Carbon forms strong single bonds to itself, and this allows chains of carbons to form. This results in an enormous variety of compounds. Organic compounds also behave quite differently from inorganic compounds. Special laboratory techniques have been developed to carry out and study organic reactions, to separate the complex mixtures of very similar compounds that are often produced, and to identify these products.

Although this division into organic and inorganic is useful in organizing the vast subject of chemistry, the division is somewhat arbitrary. For example, a compound that contains both carbon and a metal, such as chlorophyll or hemoglobin, may be considered either organic or inorganic, depending on the interests of the person who is studying it. It is equally difficult (and unnecessary) to draw a line separating organic chemistry and biochemistry.

1.2 SIMPLE ATOMIC STRUCTURE

To understand the structures of organic molecules and how these molecules react, we need a mental picture of the bonds that hold the atoms together. Several different models or pictures are used to describe the chemical bond. Which picture we use depends on what we are trying to accomplish. In this chapter we will learn about the simplest picture, which describes a covalent bond as a shared pair of electrons and uses Lewis structures to represent molecules. Although this model is not complex, it will be adequate for most of our uses. (In Chapter 3 we will look at a more complex model for bonding.) Most of what is covered in this chapter should be a review.

Before looking at molecules, we need to review the structure of atoms. Most of the mass of an atom is concentrated in the **nucleus.** The nucleus consists of **protons,** which are positively charged, and **neutrons,** which are neutral. To counterbalance the charge on the nucleus due to the positive protons, the atom has an equal number of negative **electrons** in **shells** or **orbitals** around the nucleus. Because the electrons in the outermost electron shell (the **valence electrons**) control how the atom bonds, atoms are often represented by their respective atomic symbol surrounded by dots representing the outer-shell electrons. Such representations for some of the elements of interest to us are shown in Figure 1.1. The number of electrons in the **valence shell** of an atom is the same as the group number of that atom in the periodic table.

PROBLEM 1.1
Show Lewis structures for these atoms:
a) Bromine   **b)** Calcium   **c)** Germanium

In the simplest model, bonding can be considered to result from the special stability associated with a filled outer shell of electrons. The noble gases, such as helium, neon, and argon, which already have a filled outer shell of electrons, have little tendency to form bonds. Atoms of the other elements, however, seek to somehow attain a filled outer shell of electrons. The two ways in which they accomplish this goal result in two types of bonding: ionic and covalent.

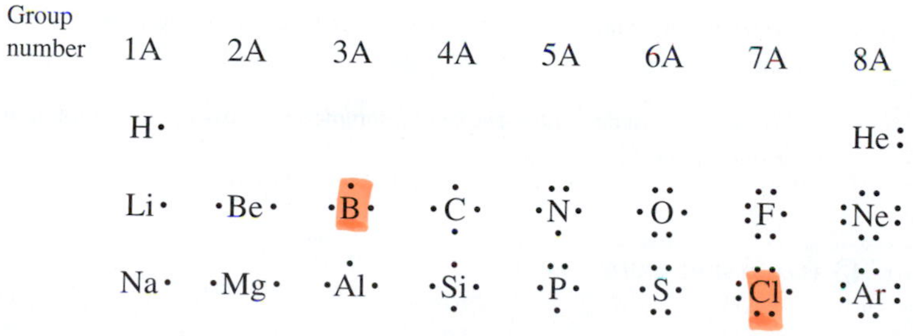

Group number	1A	2A	3A	4A	5A	6A	7A	8A
	H·							He :
	Li·	·Be·	·B·	·C·	·N·	·O·	:F·	:Ne:
	Na·	·Mg·	·Al·	·Si·	·P·	·S·	:Cl·	:Ar:

Figure 1.1

LEWIS STRUCTURES FOR SOME ATOMS OF INTEREST IN ORGANIC CHEMISTRY. The atomic symbol represents the nucleus and the inner-shell electrons. The dots represent the electrons in the outer shell (valence shell).

1.3   IONIC BONDING

In **ionic bonding**, we picture the atoms as gaining or losing electrons to arrive at the same number of electrons as one of the noble gases. Because electrons are charged, this gain or loss results in the formation of charged atoms or ions. This type of bonding is common when metals from the left side of the periodic table combine with nonmetals from the right side. Two examples, the formation of lithium fluoride and magnesium oxide, are shown in Figure 1.2.

PROBLEM 1.2
Show reaction equations using Lewis structures for the formation of these compounds from their elements:
a) Calcium chloride   b) Sodium sulfide

Because ions with opposite charges attract each other, an ionic bond results from the attractive force between the positively charged cation and the negatively charged anion. A typical ionic compound is a high-melting solid. In the solid crystal, several anions surround each cation. Each of these anions is attracted equally to the cation. Likewise, each anion is surrounded by several identical cations that are equally attracted to the anion. Because of these multiple interactions, we cannot say that a particular cation is bonded to a particular anion. Therefore, we do not speak of a molecule of an ionic compound, because this would imply one cation associated with one anion.

The attractive interactions of each cation with a number of anions and vice versa result in a network of strong attractive forces that extend throughout the whole crystal. We can think of these as a three-dimensional net of attractive forces. Disrupting these forces requires a large amount of heat energy; therefore, ionic compounds usually have

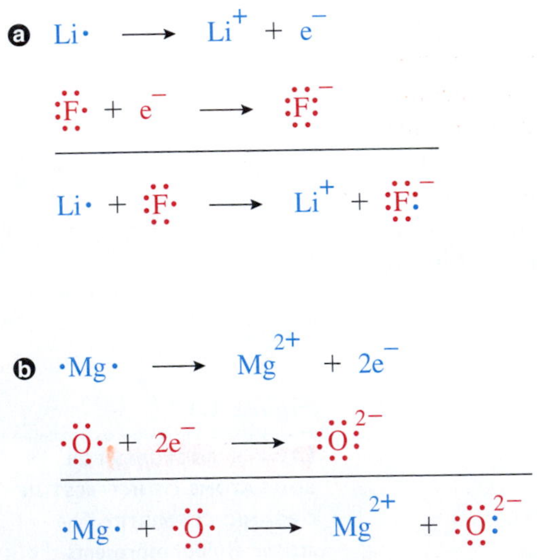

(a) $Li\cdot \longrightarrow Li^+ + e^-$

The easiest way for lithium to attain a noble gas configuration is for it to lose one electron. The resulting lithium cation has the same number of electrons as helium.

$:F\cdot + e^- \longrightarrow :F:^-$

Fluorine prefers to gain one electron. The fluoride anion has the same number of electrons as neon.

$Li\cdot + :F\cdot \longrightarrow Li^+ + :F:^-$

These two reaction equations can be summed (the electrons cancel) to show the formation of lithium fluoride, an ionic compound.

(b) $\cdot Mg\cdot \longrightarrow Mg^{2+} + 2e^-$

Magnesium can lose the two electrons in its valence shell and attain the electronic configuration of neon.

$\cdot\ddot{O}\cdot + 2e^- \longrightarrow :\ddot{O}:^{2-}$

Oxygen can attain the electronic configuration of neon by gaining two electrons.

$\cdot Mg\cdot + \cdot\ddot{O}\cdot \longrightarrow Mg^{2+} + :\ddot{O}:^{2-}$

These two reaction equations can be summed to show the formation of magnesium oxide, an ionic compound.

Figure 1.2

THE FORMATION OF **(a)** LITHIUM FLUORIDE AND **(b)** MAGNESIUM OXIDE.

high melting points and high boiling points. For example, the melting point of magnesium chloride is 708°C, and its boiling point is 1412°C.

For carbon to attain the electronic configuration of a noble gas by this method, it would have to lose or gain four electrons. This would result in a C^{4+} cation (helium configuration) or a C^{4-} anion (neon configuration). The high concentration of charge in a small volume in both of these ions makes them unstable, high-energy species. Because of this high charge-to-volume ratio, carbon prefers to attain the electronic configuration of the noble gases in another manner—that is, by covalent bonding. Many other atoms also attain stable electronic configurations in this manner.

1.4   COVALENT BONDING

In **covalent bonding,** we picture atoms as sharing electrons to arrive at the same number of electrons as a noble gas. The shared electrons are counted as part of the total electrons for both of the atoms involved in the bond. Consider the simple example of the formation of the hydrogen molecule from two hydrogen atoms. Each hydrogen atom has one electron in its valence shell, and each would like to have two (helium configuration). If the two hydrogen atoms come together, each bringing one electron, they can share these electrons. Each hydrogen of the resulting molecule experiences two electrons in its valence shell. The shared pair of electrons in a covalent bond is often represented by a line, as in the middle structure that follows:

$$\text{H} \cdot + \text{H} \cdot \longrightarrow \text{H}{:}\text{H} \quad \text{or} \quad \text{H}—\text{H} \quad \text{or} \quad \text{H}_2$$

In this case, one hydrogen is strongly bonded to the other hydrogen. We can, therefore, speak of these two hydrogen atoms as forming a hydrogen molecule. However, the attractive force between this hydrogen molecule and other hydrogen molecules is quite weak. Because only a small amount of heat energy needs to be added to overcome this weak force and separate individual hydrogen molecules, hydrogen has very low melting and boiling points, near absolute zero (mp = -259°C, bp = -252°C) and is a gas at normal temperatures and pressures. In general, covalent compounds have strong bonds between the atoms of the molecule itself but only weak attractions for other molecules. Therefore, covalent compounds melt and boil at much lower temperatures than do ionic compounds.

Carbon is central to the study of organic chemistry, so let's consider its bonding. Carbon has four electrons in its valence shell. It can attain the electronic configuration of neon (eight electrons in its valence shell) by using each of these electrons, along with four electrons from other atoms, to form four covalent bonds. Thus, the simplest compound of carbon and hydrogen results from the combination of four hydrogen atoms with one carbon atom to produce methane, CH_4. Again, each covalent bond is strong, but the attraction between individual methane molecules is weak. Methane melts at -182°C and boils at -164°C.

$$\cdot \overset{\displaystyle \cdot}{\underset{\displaystyle \cdot}{\text{C}}} \cdot + \ 4\ \text{H} \cdot \longrightarrow \text{H}\overset{\displaystyle \cdot\cdot}{\underset{\displaystyle \cdot\cdot}{:\text{C}:}}\text{H} \quad \text{or} \quad \text{H}—\overset{\displaystyle \text{H}}{\underset{\displaystyle \text{H}}{\overset{|}{\underset{|}{\text{C}}}}}—\text{H} \quad \text{or} \quad CH_4$$

Methane

1.5 LEWIS STRUCTURES

Structures in which all the electrons in the valence shells are shown as dots are called **Lewis structures.** The Lewis structure for H_2O is shown in Figure 1.3. Drawing larger molecules in this fashion can be rather tedious, but there are a number of shorthand ways to represent molecules. For instance, see the previous example for CH_4 and Figure 1.3 for water. All of these representations for water describe the same molecule. Whenever we look at a structure, we should examine it carefully to be sure that we understand exactly what is meant. For example, if we see water represented by the third structure in Figure 1.3, we must realize that oxygen has two unshared pairs of electrons that have not been shown. If these unshared pairs of electrons were not present, the structure would represent a highly unstable and unusual ionic species. *Any unusual feature of a structure must be shown explicitly.* Until you become more familiar with these various ways to represent structures, you should carefully examine each structure that you encounter and picture the complete Lewis structure, including any unshared electron pairs that have not been explicitly shown.

We need to learn the number of bonds that are commonly formed by the atoms that are important in organic chemistry. In the vast majority of cases, hydrogen forms one covalent bond so that it has two electrons in its outer shell. In covalent compounds the atoms of the second period (or row) of the periodic table need eight electrons in their outer shell. (This is called the **octet rule.**) We have already seen that to accomplish this carbon forms four bonds. In a similar fashion we can determine the number of bonds that some other second-period atoms prefer to form. The number of bonds that commonly occur in neutral molecules for the elements of most interest in organic chemistry are listed in Table 1.1.

PROBLEM 1.3
Show Lewis structures for the simplest neutral compounds formed from these elements:
a) Carbon and chlorine   **b)** Hydrogen and bromine

Although the number of bonds shown in Table 1.1 yields neutral structures and is the most common number for each of these elements, other bonding arrangements are also encountered. For example, in some cases, oxygen forms a third bond by using a

Figure 1.3

VARIOUS METHODS USED TO REPRESENT THE STRUCTURE OF WATER.

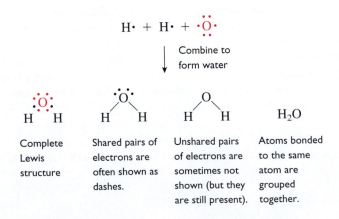

Table 1.1   Number of Bonds Commonly Formed by Some Elements

Atom	Common Number of Bonds	Example		
H	1	H—H	or	H_2
C	4	$\begin{array}{c} H \\ | \\ H-C-H \\ | \\ H \end{array}$	or	CH_4
N	3	$\begin{array}{c} \ddots \\ H-N-H \\ | \\ H \end{array}$	or	NH_3
O	2	$\begin{array}{c} \cdot\ddot{O}\cdot \\ / \quad \backslash \\ H \quad\quad H \end{array}$	or	H_2O
X*	1	H—Br:	or	HBr

*X is used to represent all of the halogens, F, Cl, Br, and I.

pair of its unshared electrons. In such situations the oxygen provides both electrons for the bond that it forms with an electron-poor atom. However, the oxygen still has eight electrons in its outer shell. More examples like this are presented in Chapter 2.

We can also use Table 1.1 to tell us the number of bonds preferred by other atoms that are from the same columns (or groups) of the periodic table as those shown in the table. Elements from the same group have the same number of electrons in their valence shells. Therefore, they form the same number of bonds, and in general, they have similar chemistries. Thus, silicon, being in the same column as carbon, prefers to form four bonds. Phosphorus, like nitrogen, prefers three bonds, whereas sulfur, being related to oxygen, prefers two. All of the halogens (F, Cl, Br, I) form covalent compounds with one bond. However, elements from the third and subsequent periods of the periodic table may have more than 8 electrons in their outer shells (10 and 12 electrons are both common) and may therefore form more than the number of bonds suggested by Table 1.1. These cases will be discussed in more detail later.

PROBLEM 1.4
Show a Lewis structure for the simplest neutral compound formed from hydrogen and sulfur.

It is useful to be able to look at a Lewis structure and judge the stability of the compound represented by that structure. Can the compound be isolated and put in a bottle? Is it more stable than another, similar compound? The most important factor to examine in determining whether a molecule is stable is whether the octet rule is satisfied. Compounds with fewer than eight electrons around a second-period atom are known but are seldom stable. (However, such compounds are encountered as unstable, highly reactive intermediates in some chemical reactions.) Furthermore, compounds with more than eight electrons around a second-period atom are also unstable. In stable molecules, hydrogen strongly prefers to have two electrons. (Later, we will examine some of the more subtle factors that affect stability.)

Let's consider ethane as an example. The formula for ethane, C_2H_6, does not provide enough information for us to judge its stability. We need to examine its structure. The Lewis structure for ethane shows that the octet rule is satisfied for both carbons, and each hydrogen has two electrons. Therefore, ethane is predicted to be a stable compound—and it is.

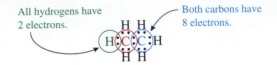

All hydrogens have 2 electrons.   Both carbons have 8 electrons.

PRACTICE PROBLEM 1.1

Discuss the stability of the following species:

$$H\!-\!\overset{\displaystyle\cdot}{C}\!-\!H$$
$$|$$
$$H$$

Solution

This species has a total of seven electrons around the carbon. Because the octet rule is not satisfied, the molecule is predicted to be unstable. (However, we will encounter this and related species as unstable intermediates in some chemical reactions in later chapters.)

PROBLEM 1.5

Discuss the stability of these structures:

a)

$$H$$
$$|$$
$$H\!-\!C\!-\!\overset{\displaystyle\cdot\cdot}{\underset{\displaystyle\cdot\cdot}{O}}\!-\!H$$
$$|$$
$$H$$

b)

$$\quad\quad H$$
$$\quad\quad | \quad\quad H$$
$$H\!-\!N\diagup$$
$$\quad\quad | \quad\diagdown H$$
$$\quad\quad H$$

We need to be able to write a stable Lewis structure for a compound whose formula is provided. As an example, let us write a Lewis structure for CH_4O. First, how many electrons should be used? CH_4O is neutral, so the total electrons shown in the structure must be the same as the total in the valence shells of the neutral atoms involved. In this case the carbon contributes 4 electrons, each hydrogen contributes 1, and the oxygen contributes 6, for a total of 14. This is one way of counting electrons to ensure that the compound is neutral. If the compound were charged, there would be one electron less than this total for each unit of positive charge or one electron more than this total for each unit of negative charge. We then try to assemble the atoms into a structure that satisfies the octet rule. This can often be done by bonding the atoms other than hydrogen together and then filling the valences with the hydrogens. This process is shown in Figure 1.4 for CH_4O.

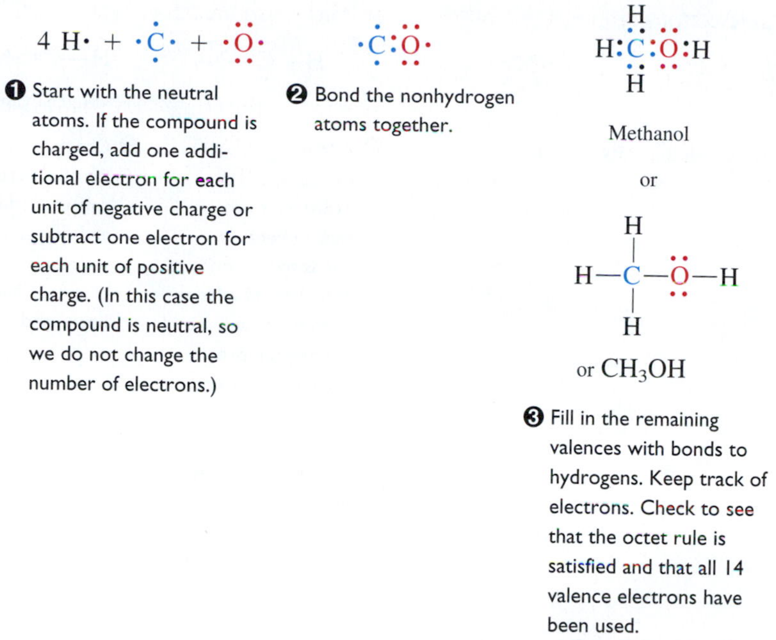

$4 \text{ H·} + \text{·}\overset{\cdot}{\underset{\cdot}{C}}\text{·} + \text{·}\overset{\cdot\cdot}{\underset{\cdot\cdot}{O}}\text{·}$

❶ Start with the neutral atoms. If the compound is charged, add one additional electron for each unit of negative charge or subtract one electron for each unit of positive charge. (In this case the compound is neutral, so we do not change the number of electrons.)

$\text{·}\overset{\cdot\cdot}{\underset{\cdot\cdot}{C}}\text{:}\overset{\cdot\cdot}{\underset{\cdot\cdot}{O}}\text{·}$

❷ Bond the nonhydrogen atoms together.

$$\text{H:}\overset{\overset{\textstyle H}{|}}{\underset{\underset{\textstyle H}{|}}{C}}\text{:}\overset{\cdot\cdot}{\underset{\cdot\cdot}{O}}\text{:H}$$

Methanol

or

$$\text{H}-\overset{\overset{\textstyle H}{|}}{\underset{\underset{\textstyle H}{|}}{C}}-\overset{\cdot\cdot}{\underset{\cdot\cdot}{O}}-\text{H}$$

or CH_3OH

❸ Fill in the remaining valences with bonds to hydrogens. Keep track of electrons. Check to see that the octet rule is satisfied and that all 14 valence electrons have been used.

Figure 1.4

WRITING A LEWIS STRUCTURE FOR METHANOL, CH$_4$O.

Sometimes more than one bond is needed between elements that have a valence greater than 1. If using only single bonds between atoms results in too few electrons to satisfy the octet rule for all of the atoms, then additional electron sharing between atoms must occur, resulting in the formation of double or triple bonds. For each two electrons needed to complete the octets of the atoms in the molecule, one additional bond between two of the atoms must be formed. This process is illustrated in Figures 1.5 and 1.6 for C_2H_4 and HCN, respectively.

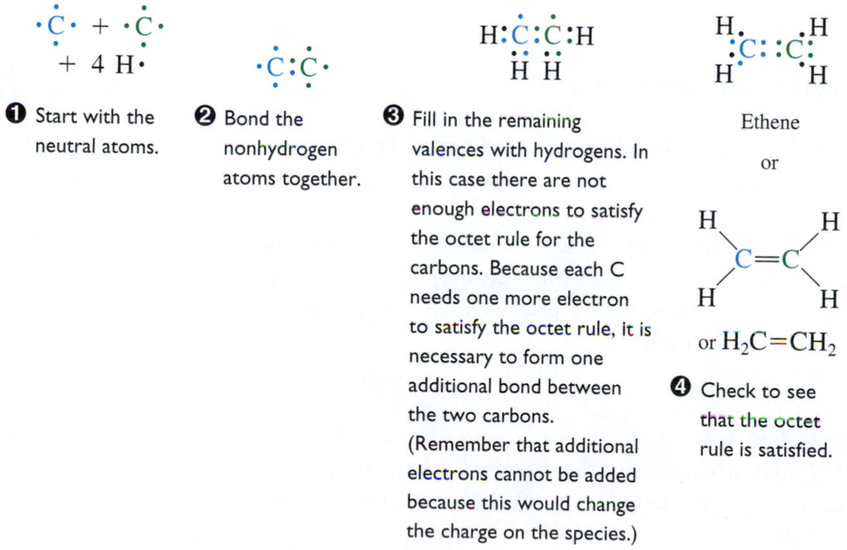

$+ 4 \text{ H·}$

❶ Start with the neutral atoms.

$\text{·}\overset{\cdot}{\underset{\cdot}{C}}\text{:}\overset{\cdot}{\underset{\cdot}{C}}\text{·}$

❷ Bond the nonhydrogen atoms together.

$\text{H:}\overset{\cdot}{\underset{\underset{\textstyle H}{}}{C}}\text{:}\overset{\cdot}{\underset{\underset{\textstyle H}{}}{C}}\text{:H}$

❸ Fill in the remaining valences with hydrogens. In this case there are not enough electrons to satisfy the octet rule for the carbons. Because each C needs one more electron to satisfy the octet rule, it is necessary to form one additional bond between the two carbons. (Remember that additional electrons cannot be added because this would change the charge on the species.)

$$\overset{\text{H}}{\underset{\text{H}}{}}\text{C::C}\overset{\text{H}}{\underset{\text{H}}{}}$$

Ethene

or

$$\overset{\text{H}}{}\underset{\text{H}}{}\text{C=C}\overset{\text{H}}{}\underset{\text{H}}{}$$

or $H_2C{=}CH_2$

❹ Check to see that the octet rule is satisfied.

Figure 1.5

WRITING A LEWIS STRUCTURE FOR ETHENE, C$_2$H$_4$.

Figure 1.6

WRITING A LEWIS STRUCTURE FOR HYDROGEN CYANIDE, HCN.

$\cdot \ddot{C} \cdot$ + $\cdot \ddot{N} \cdot$
+ H $\cdot$

$\cdot \ddot{C} \! : \! \ddot{N} \cdot$

H—$\ddot{C}$—$\ddot{N} \cdot$

H—C≡N:

Hydrogen cyanide

❶ Start with the neutral atoms.

❷ Bond the nonhydrogen atoms together.

❸ Add the hydrogen. In this case, add the hydrogen to the carbon because it needs more bonds to complete its valence. Both the C and the N need two more electrons to satisfy the octet rule. Therefore, form two additional bonds between the C and the N.

❹ Check to see that the octet rule is satisfied and that all the valence electrons have been used.

PRACTICE PROBLEM 1.2

Write a Lewis structure for CH_2O.

Strategy

Follow the same process used in Figures 1.5 and 1.6.

1. Start with the neutral atoms and bond the nonhydrogen atoms together.

2. Add the hydrogens.

3. If the valences of the atoms are not satisfied, form additional bonds between the atoms.

4. Check to see that the octet rule is satisfied at all atoms.

Solution

Bond the C and the O together.

$\cdot \ddot{C} \! : \! \ddot{O} \! :$

Add the hydrogens. In this case, add them to the C because it has more unsatisfied valences.

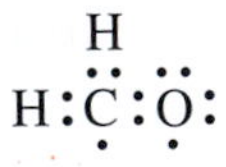

H
$H \! : \! \ddot{C} \! : \! \ddot{O} \! :$

Both the C and the O need one more electron to satisfy the octet rule, so form a second bond between them. The octet rule is satisfied and all the valence electrons have been used, so this is an acceptable structure.

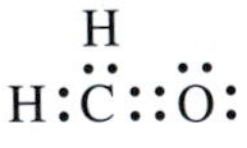

H
$H \! : \! \ddot{C} \! : \! : \! \ddot{O} \! :$

PROBLEM 1.6

Write Lewis structures for these compounds:

a) C_3H_8 b) C_2H_2 c) CH_3N d) NH_3O

1.6   COVALENT IONS

We will encounter many covalent compounds that have a charge. This results in ionic compounds that have covalent cations, covalent anions, or both. For example, consider the following acid–base reaction between ammonia and hydrogen chloride to produce ammonium chloride. How do we determine that NH_4 is a cation with a charge of $+1$ while Cl is an anion with a charge of -1? Several different methods can be used to determine the charge on a Lewis structure. Because charge is determined by the number of protons and electrons, we can simply subtract the total electrons from the protons. In the case of the ammonium cation, there are 7 protons from the nitrogen plus 4 from the 4 hydrogens minus 10 electrons (8 shown in the Lewis structure and the 2 electrons from the $1s$ shell of the nitrogen). This totals to a charge of $+1$ ($7 + 4 - 10 = +1$).

$$\text{H}-\overset{\displaystyle \text{H}}{\underset{\displaystyle \text{H}}{\overset{..}{\text{N}}}}-\text{H} + \text{H}-\overset{..}{\underset{..}{\text{C}}}\text{l}: \longrightarrow \text{H}-\overset{\displaystyle \text{H}}{\underset{\displaystyle \text{H}}{\overset{+}{\text{N}}}}-\text{H} + :\overset{..}{\underset{..}{\text{C}}}\text{l}:^{-}$$

Another way to determine charges is to remember that charge must balance in a chemical reaction. In the previous reaction, the total charge on the left side of the equation is zero. Therefore, the total charge on the right side of the equation must also sum to zero. Because we have already calculated that the charge on the ammonium cation is $+1$, the charge on the chloride anion must be -1. Learn to visualize reactions in terms of the charges involved. In the preceding reaction, a hydrogen, without any electrons, is transferred from Cl to N. Because a hydrogen without electrons is a proton, the reaction has transferred one unit of positive charge from Cl to N.

PROBLEM 1.7
Using Lewis structures, show a balanced equation for the reaction of H_2O with HCl.

The final method of determining charge, and the one most useful to us, is the method of formal charges.

1.7   FORMAL CHARGES

Formal charge is a method used to assign an approximate charge distribution among the atoms of a covalent species, regardless of whether it is ionic or neutral overall. The **formal charge** on an atom is defined as the number of valence electrons in the neutral atom before any bonding (this is the same as the group number of the atom) minus the number of unshared electrons on the atom in the structure of interest and also minus one-half the number of shared electrons on that atom:

$$\begin{pmatrix}\text{formal} \\ \text{charge}\end{pmatrix} = \begin{pmatrix}\text{valence electrons} \\ \text{in the atom}\end{pmatrix} - \begin{pmatrix}\text{number of} \\ \text{unshared electrons}\end{pmatrix} - \frac{1}{2}\begin{pmatrix}\text{number of} \\ \text{shared electrons}\end{pmatrix}$$

Calculation of the formal charges on the atoms of the ammonium ion is shown in Figure 1.7. Note that the sum of the formal charges on all the atoms of a covalent molecule gives the total charge on that molecule.

Figure 1.7

CALCULATING THE FORMAL
CHARGES FOR AMMONIUM
ION, NH₄⁺.

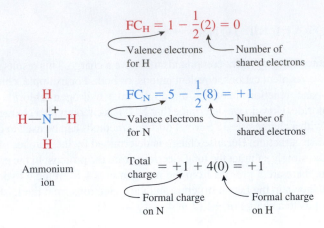

It will help us to understand formal charges if we analyze the process of calculating these charges. Basically, we are determining charge by counting protons and electrons as we did in the first method described earlier. If all the valence electrons were somehow stripped from the atom, the remaining ion would have a total positive charge equal to the group number. We subtract from this positive charge the "average" number of electrons (negative charges) that are around the atom in the Lewis structure. The average number of electrons around the atom is the sum of the unshared electrons (these spend all of their time around the atom) plus one-half of the shared electrons (these spend only one-half of their time around each of the atoms sharing them).

Another way to view this process is to imagine symmetrically breaking all the bonds to an atom—that is, keeping one electron from each bond with each atom. The formal charge can then be calculated by determining the charge on the resulting species. This process is shown for hydroxide ion in Figure 1.8.

Figure 1.8

CALCULATING THE FORMAL
CHARGES FOR HYDROXIDE
ION, OH⁻.

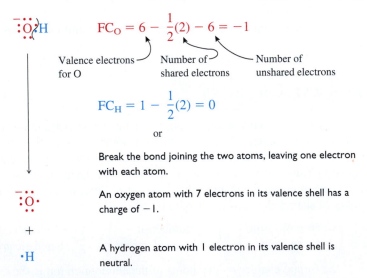

PRACTICE PROBLEM 1.3

Calculate the formal charges on each atom of this species:

$$
\begin{array}{c}
\ddot{\text{:O:}} \\
\parallel \\
\ddot{\text{:O}}\!-\!\overset{\displaystyle N}{}\!-\!\ddot{\text{O:}}
\end{array}
$$

Solution

The formal charges can be calculated from the equation:

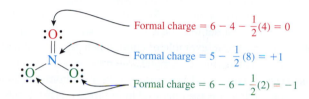

$$\text{Formal charge} = 6 - 4 - \frac{1}{2}(4) = 0$$

$$\text{Formal charge} = 5 - \frac{1}{2}(8) = +1$$

$$\text{Formal charge} = 6 - 6 - \frac{1}{2}(2) = -1$$

ORGANIC
Chemistry Now™
Click *Coached Tutorial Problems*
to practice **Calculating**
Formal Charges.

The formal charge can also be calculated from the species produced by breaking each bond so that one electron of the bond remains with each atom:

$$\text{Charge} = 0$$

$$\text{Charge} = +1$$

$$\text{Charge} = -1$$

The total charge is $0 + 1 + (-1) + (-1) = -1$.

PROBLEM 1.8

Calculate the formal charges on each of the atoms, except hydrogens, of these molecules and determine the total charge of the species:

a)
$$
\begin{array}{c}
\text{H} \\
\mid \\
\text{H}\!-\!\text{C}\!-\!\ddot{\text{O}}\!-\!\text{H} \\
\mid \\
\text{H} \quad \text{H}
\end{array}
$$

b)
$$
\begin{array}{c}
\text{H} \\
\mid \\
\text{H}\!-\!\text{C}\!-\!\ddot{\text{O}}\!: \\
\mid \\
\text{H}
\end{array}
$$

c)
$$
\begin{array}{c}
\text{:}\ddot{\text{F}}\text{:} \\
\mid \\
\text{:}\ddot{\text{F}}\!-\!\text{B}\!-\!\ddot{\text{F}}\text{:} \\
\mid \\
\text{:}\ddot{\text{F}}\text{:}
\end{array}
$$

d)
$$
\begin{array}{c}
\text{H} \\
\diagdown \\
\text{C}\!=\!\text{N}\!=\!\ddot{\text{N}}\text{:} \\
\diagup \\
\text{H}
\end{array}
$$

e)
$$
\begin{array}{c}
\qquad \qquad \ddot{\text{O}}\text{:} \\
\qquad \qquad \diagup \\
\text{H}\!-\!\ddot{\text{O}}\!-\!\text{N} \\
\qquad \qquad \diagdown \\
\qquad \qquad \ddot{\text{O}}\text{:}
\end{array}
$$

f)
$$
\begin{array}{c}
\text{H} \quad \text{H} \quad \ddot{\text{O}}\text{:} \\
\mid \quad \mid \quad \parallel \\
\text{H}\!-\!\text{N}\!-\!\text{C}\!-\!\text{C}\!-\!\ddot{\text{O}}\text{:} \\
\mid \quad \mid \\
\text{H} \quad \text{H}
\end{array}
$$

Glycine (an amino acid)

BioLink

As you continue to study organic chemistry, you will begin to recognize the formal charges on many atoms without doing any calculations. Most atoms in covalent compounds have formal charges of zero. For example, a carbon with four bonds is neutral, as is an oxygen with two bonds and two unshared pairs of electrons. An oxygen with only one bond and three unshared pairs of electrons has a formal charge of -1.

The most important criterion for estimating the stability of a compound represented by a particular Lewis structure is whether or not the structure satisfies the octet rule. Formal charges can be used to refine estimates of stability. Because unlike charges attract, energy is required to separate positive and negative charges. (Similarly, like charges repel, so energy is required to bring them together.) Therefore, the presence of formal charges will destabilize a molecule. This does not mean that any structure with formal charges represents an unstable molecule. It does mean that structures with fewer formal charges will be more stable, other things being equal.

As an example, consider the molecule formed from one atom each of carbon, hydrogen, and nitrogen. We can assemble these atoms in two different ways and still satisfy the octet rule. (This is another example of compounds that have the same molecular formula but a different arrangement of bonded atoms. Such compounds are called **constitutional isomers** or **structural isomers** and are very common in organic chemistry.) The structure on the left was the one used in Figure 1.6. The structure on the right is obtained if the hydrogen is bonded to the N rather than the C.

$$H\!-\!C\!\equiv\!N\!: \qquad\qquad H\!-\!\overset{+}{N}\!\equiv\!\overset{-}{C}\!:$$

Because HCN has formal charges of zero on all the atoms, whereas HNC has a formal charge on nitrogen of $+1$ and a formal charge on carbon of -1, we would expect HCN to be more stable than HNC. Indeed, this is the case. We will encounter many cases of constitutional isomers in which the difference in stability is much less than in this example. In such cases, much more subtle arguments must be used to predict which is more stable, if a prediction can be made at all without experimental measurements.

PROBLEM 1.9

Predict which of the following constitutional isomers for the compound that is formed from one atom each of hydrogen, oxygen, and chlorine is more stable:

$$H\!-\!\ddot{\underset{..}{C}}l\!-\!\ddot{\underset{..}{O}}\!: \quad \text{or} \quad H\!-\!\ddot{\underset{..}{O}}\!-\!\ddot{\underset{..}{C}}l\!:$$

Figure 1.9 illustrates the construction of a Lewis structure for a more complex example, nitromethane. In this case there are a number of reasonable (according to what we know now) constitutional isomers. Therefore, we must be told the **connectivity,** or which atoms are bonded to which. Once the connectivity shown in Figure 1.9 is given, the best structure that satisfies the octet rule has some formal charges. This still represents a quite stable molecule.

The structure shown for nitromethane in Figure 1.9 is not quite consistent with experimental data. According to this structure, the two oxygens are different. One has a single bond to the nitrogen and a formal charge of -1; the other has a double bond to nitrogen and a formal charge of zero. However, various experiments have shown that these two oxygens are, in fact, identical. Their bonds to nitrogen have identical lengths and strengths, between those of single and double bonds, and they have the same

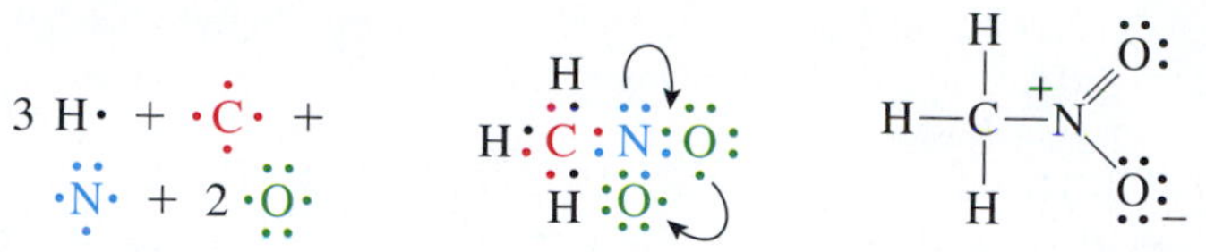

Figure 1.9

WRITING A LEWIS STRUCTURE FOR NITROMETHANE, CH$_3$NO$_2$.

❶ Start with the uncombined atoms. There are a number of plausible structural isomers, so we must be told the connectivity. Nitromethane has the H's bonded to the C, the C bonded to the N, and both O's bonded to the N.

❷ This first attempt results in a structure that needs two more electrons to satisfy the octet rule at all of the atoms. Therefore, a double bond must be present. The arrows show how we mentally picture moving the electrons to accomplish this.

❸ This structure satisfies the octet rule. Although it has a formal charge on N of +1 and a formal charge on one of the O's of −1, it is the best that can be done with the given connectivity. Nitromethane is a stable molecule that can be isolated.

charge. To explain situations like this, in which the Lewis structure does not give an accurate picture of the molecule, the concept of resonance is employed.

1.8 RESONANCE

As shown in Figure 1.10, two Lewis structures can be written for nitromethane. These structures are different because one has a double bond to one oxygen, whereas the other has a double bond to the other oxygen. Obviously, these structures are equally good representations for nitromethane, because they are of equal stability.

In cases such as this, neither structure is an accurate representation of the actual structure. The actual structure is a **resonance hybrid** of these two structures. It is a blend of the extremes represented by these two structures. For example, one of the structures has a double bond between the nitrogen and one of the oxygens, but these same two atoms are connected by a single bond in the other structure. The actual bond between the nitrogen and this oxygen is more than a single bond but less than a double bond. The same analysis applies to the other oxygen, so the bonds to both of these oxygens are identical. Furthermore, because each oxygen is neutral in one resonance structure and has a negative charge in the other structure, the actual charge on each in the resonance hybrid is one-half of a full negative charge.

The last structure in Figure 1.10 is an attempt to show a structure for the resonance hybrid. Because there are partial bonds (represented by the dashed lines) between the nitrogen and the oxygens and there is a pair of electrons that is shared by the two oxygens, even though they are not directly bonded, this structure looks strange. It is difficult to determine whether the octet rule has been satisfied for the oxygens and the nitrogen. Rather than trying to represent such molecules by structures that might be confusing, normal Lewis structures and resonance arrows are used.

Just as the first structure in Figure 1.10 is not an accurate representation of the actual structure of nitromethane, it also does not provide a basis for an accurate estimate of the stability of nitromethane. The compound is more stable than either of the resonance structures would indicate. This extra stability is called the **resonance stabilization energy.**

Figure 1.10

RESONANCE STRUCTURES FOR NITROMETHANE.

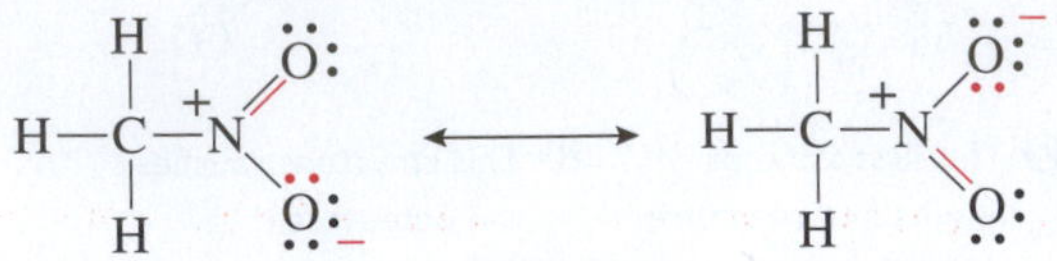

Nitromethane

Two equivalent structures can be written for nitromethane. These are different if it is specified that no atoms have moved.

The double-headed arrow is used to indicate that the actual structure is a resonance hybrid of these two structures.

Because both structures are equivalent, they make equal contributions to the hybrid.

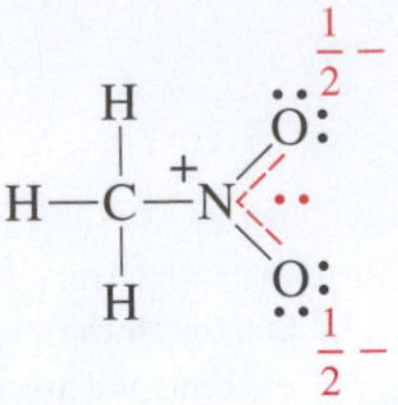

The actual structure is a blend of the two resonance structures. This structure is an attempt to represent the hybrid structure using dashed lines to indicate partial bonds. It has a partial double bond to each oxygen and an unshared pair of electrons that is located partly on each oxygen. Rather than using confusing structures like this, we use regular Lewis structures and the resonance arrow.

We will use resonance often, so it is important to understand exactly what it means. Resonance does not mean that the structure of the molecule is changing back and forth between the structures in the case of nitromethane. It does not mean that the bond to one of those oxygens is a double bond part of the time and a single bond the rest of the time. Instead, the structure is static, an average of the individual structures. *The double-headed **resonance arrow** is used only for resonance and should not be confused with the **equilibrium arrows,** which are used to represent an equilibrium reaction in which the molecules are indeed changing structures.*

Important Convention ▶

Resonance arrow   Equilibrium arrows

Resonance must be used whenever more than one reasonable Lewis structure can be written for a molecule, provided that the Lewis structures have *identical positions of all atoms.* Only the positions of unshared electrons and multiple bonds are changed in writing different resonance structures. When a better picture of bonding is developed in Chapter 3, we will get a better understanding of what resonance means and when it must be used.

PROBLEM 1.10
Experimental evidence indicates that the two oxygens of acetate ion are identical. Use resonance to explain this observation.

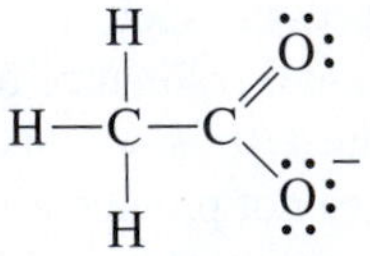

Acetate ion

1.9 POLAR BONDS

In calculations of formal charge, we assume that a shared pair of electrons spends equal time around both of the atoms that form the bond. This is true only if the atoms are the same. More commonly, the bond is between two different atoms, and the electrons are not necessarily shared equally. The atom that has a greater share of the electrons can be determined from the **electronegativities** (electron-attracting abilities) of the atoms involved in the bonds. The electron density represented by a pair of shared electrons is greater around the atom with the greater electronegativity.

Table 1.2 lists the electronegativities of the atoms that are of most interest to organic chemists. Remember that electronegativities increase from left to right (excluding the noble gases) in a period of the periodic table and also from bottom to top in a column. For our purposes we can consider carbon and hydrogen to have roughly the same electronegativity. The order of electronegativities of the atoms of most interest to us is as follows:

$$H \approx C < Br < Cl = N < O < F$$

increasing electronegativity

Because the electrons in a covalent bond spend more time around the more electronegative atom, this atom has more negative charge around it than the formal charge indicates. Consider the hydrogen chloride molecule. The formal charges on both atoms in the Lewis structure are zero. However, chlorine is more electronegative than hydrogen. Therefore, the shared electrons spend more time around the chlorine than around the hydrogen. The chlorine has more electron density around it than the Lewis structure indicates and thus has some negative charge associated with it. Similarly, the hydrogen has less electron density and some positive charge. Because the electrons of the covalent bond are still shared, although unequally, the amount of charge on each atom is less than 1. There is a partial negative charge on the chlorine and a partial positive charge on the hydrogen. The lowercase Greek letter delta (δ) followed by the sign of the charge is used to represent such partial charges:

$$\overset{\delta+}{H}-\overset{\delta-}{\ddot{\underset{..}{Cl}}}: \quad \text{or} \quad \overset{\longrightarrow}{H-Cl}$$

Table 1.2 Pauling Scale of Electronegativity Values for Some Elements

Group						
1A	2A	3A	4A	5A	6A	7A
H 2.1						
Li 1.0	Be 1.5	B 2.0	C 2.5	N 3.0	O 3.5	F 4.0
Na 0.9	Mg 1.2		Si 1.8	P 2.1	S 2.5	Cl 3.0
						Br 2.8
						I 2.5

When unequal electronegativities of two atoms involved in a bond result in charge separation as just described, we say that the bond is **polar.** Hydrogen chloride has a polar bond. The charge separation results in a **dipole,** that is, a positive and a negative "pole" in the molecule. The product of the amount of charge separation (e) times the distance of the charge separation (d) is called the **dipole moment** (μ).

$$\mu = (e)(d)$$

The dipole moment is a vector; that is, it has direction and magnitude. The dipole moment is usually represented as an arrow pointing from the positive end to the negative end of the dipole, as shown earlier for hydrogen chloride.

The amount of the charge, e, is on the order of 10^{-10} electrostatic units (esu), and the distance, d, is on the order of 10^{-10} m, so μ is on the order of 10^{-20} esu m. The debye unit (D) is defined as 1×10^{-20} esu m. The dipole moment of HCl is 1.1 D.

The direction and relative magnitude of the dipole of any bond can be predicted from the electronegativities of the atoms involved in the bond. For example, the electrons of a bond between oxygen and carbon will spend more time around the oxygen because it is more electronegative; therefore, the oxygen will be the negative end of the bond dipole. In the case of a bond between carbon and hydrogen, the two elements are of similar electronegativities. Therefore, the electrons are shared nearly equally, resulting in a slightly polar bond that is usually considered to be a **nonpolar bond**.

PROBLEM 1.11

Show the direction of the dipoles, if any, of these bonds:

a) C—N   **b)** O—N   **c)** O—Cl   **d)** C—Cl
e) B—O   **f)** C—Mg   **g)** C—C   **h)** C—H

The concept of bond polarities is very important, because much of chemistry, both the physical properties of compounds and their chemical reactions, depends on the interaction of charges. For example, a reagent that is seeking positive charge will likely be attracted to the carbon of a carbon–oxygen bond.

So far, we are able to predict the dipoles of individual bonds. The overall dipole moment of a molecule is the vector sum of these individual bond dipoles. Before the bond dipoles can be used to predict the overall dipole moment of a molecule, however, the three-dimensional orientation of the bonds must be known. That is, we need to know the shapes of molecules.

1.10   SHAPES OF MOLECULES

The shapes of molecules are determined by actual experiments, not by theoretical considerations. But we do not want to have to memorize the shape of each molecule. Instead, we would like to be able to look at a Lewis structure and predict the shape of the molecule. Several models enable us to do this. One of the easiest to use is **valence shell electron pair repulsion theory,** which is often referred to by its acronym *VSEPR* (pronounced "vesper"). As the name implies, the theory states that pairs of electrons in the valence shell repel each other and try to stay as far apart as possible. You probably remember this theory from your general chemistry class. The parts of VSEPR theory that are most important to organic chemistry will be reviewed here. The rules that are needed are as follows:

▶ **RULE 1**

Pairs of electrons in the valence shell repel each other and therefore stay as far apart as possible. For example, if four pairs of electrons are arranged around a central atom so that they are as far apart as possible, they will be located at the corners of a tetrahedron. The geometries resulting from other numbers of electron pairs, arranged as far apart as possible, are given in Figure 1.11.

When we attempt to show the shapes of molecules, we are faced with the problem of how to represent such three-dimensional objects on paper. Some bonds extend in

Number of Electron Pairs	Geometry	Example	
2	linear	Cl—Be—Cl 	
3	trigonal planar		
4	tetrahedral		
5	trigonal bipyramid		
6	octahedral		

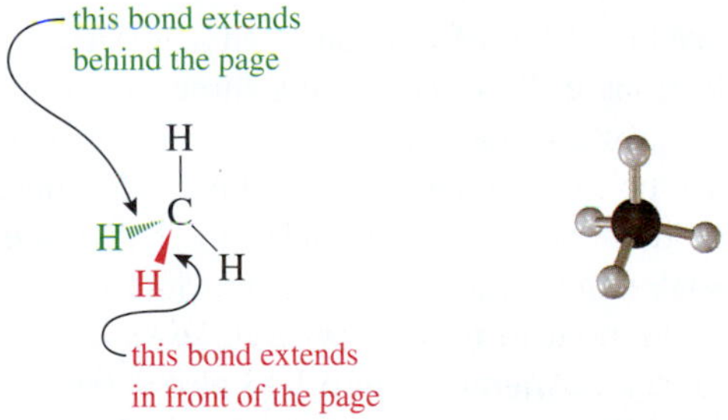

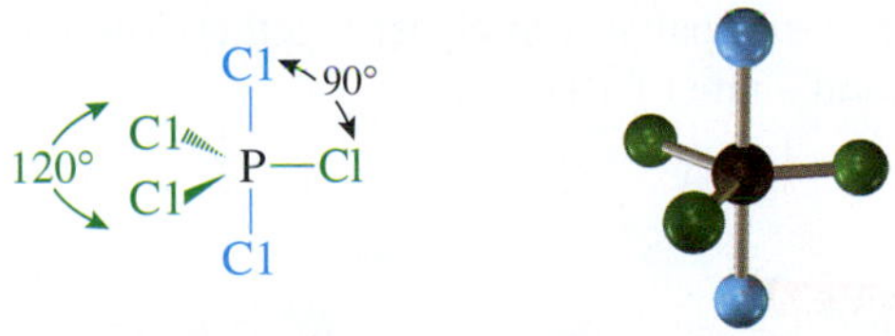

Figure 1.11

SHAPES OF MOLECULES FROM VSEPR THEORY.

ORGANIC
Chemistry ⚛ Now™
Click *Molecular Models* to view the molecules in this book as interactive three-dimensional models.

Figure 1.12

THE SHAPE OF METHANE, CH₄.

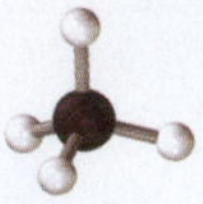

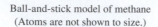

Ball-and-stick model of methane
(Atoms are not shown to size.)

Space-filling model of methane
(Correct relative sizes of atoms are shown.)

Important ▐▐▐▶
Convention

front of the page; others extend behind the page. *Commonly, a heavy wedged line is used to represent a bond that extends in front of the plane of the page and a dashed wedge line is used to represent a bond that extends behind the plane of the page.* These conventions are used to show the shapes of the molecules in Figure 1.11. Figure 1.12 shows some other pictures for the tetrahedral molecule CH₄. It is common for students to have difficulty picturing the actual three-dimensional shapes of the molecules from these two-dimensional drawings. It is a good idea to work with a set of models until the process becomes easier. The computer-generated molecular models on the web page (http://now.brookscole.com/hornback2) are also very useful for picturing shapes.

> **RULE 2**
>
> **An unshared pair of electrons repels other pairs more than a shared pair of electrons does.** This seems reasonable because a shared pair, which also spends time around another atom, should not offer as much charge concentration for repulsion as a pair of electrons that is not shared and therefore spends all of its time around the atom.

Consider the shape of ammonia, NH₃, shown in Figure 1.13. Ammonia has four pairs of electrons in its valence shell: three shared pairs and one unshared pair. These four pairs of electrons have a basic tetrahedral arrangement. However, because the unshared pair repels more than the shared pairs do, the shared pairs are pushed closer together, and the bond angles are 107° rather than the exact tetrahedral bond angle of 109.5°. Usually, the location of the atoms in a molecule, not the position of electrons, is shown, so ammonia is termed a *pyramidal molecule;* that is, the nitrogen and the three hydrogens form a pyramid. Figure 1.13 also shows the shape of water. Again there are four pairs of electrons that have a basic tetrahedral arrangement. The two unshared pairs push the shared pairs even closer together. The result is that water is a bent molecule with a bond angle of 105°.

> **RULE 3**
>
> **Double and triple bonds are treated as one shared pair of electrons.**

Several examples of the use of this rule are shown in Figure 1.14.

By using additional rules from VSEPR theory, we can make more exact predictions about the shapes of molecules. For example, the HCO bond angle of CH₂O can be pre-

107°   Ball-and-stick model   Space-filling model

a The three shared pairs of electrons and the unshared pair of ammonia are arranged in a tetrahedral manner around the nitrogen. The resulting geometry of the nitrogen and the three hydrogens is a pyramid. The HNH bond angle is less than 109.5° (the tetrahedral bond angle) because the unshared electron pair behaves as though it is larger than the electron pairs in the bonds.

105°   Ball-and-stick model   Space-filling model

b The two shared pairs of electrons and the two unshared pairs of water are also arranged in a tetrahedron. The resulting geometry of the three atoms is bent, with the HOH angle even smaller (105°) because of the repulsion of two unshared electron pairs.

Figure 1.13

THE SHAPES OF **a** AMMONIA AND **b** WATER MOLECULES.

Ethyne   Carbon dioxide   Formaldehyde

a The triple bond counts as one pair of electrons for the purposes of VSEPR theory. Therefore, each carbon has two pairs. The geometry is linear around each carbon and linear overall.

b Each double bond counts as one pair of electrons. Therefore, the carbon has two pairs, and the geometry is linear.

c The double bond counts as one pair, so the geometry about the carbon is trigonal planar. The predicted bond angles are 120° Actually, the HCO angle is 122°. Such minor deviations should be expected because the bonds are not the same and therefore are not necessarily the same "size." In fact, a double bond to an oxygen might be expected to be "larger" than a single bond to a hydrogen.

Figure 1.14

THE SHAPES OF **a** ETHYNE, C_2H_2; **b** CARBON DIOXIDE, CO_2; AND **c** FORMALEDHYDE, CH_2O.

dicted to be slightly greater than 120°, as shown in Figure 1.14. However, we will not need to know geometries that accurately. It is enough to know that the geometry of CH_2O is approximately trigonal planar.

PROBLEM 1.12
Predict the geometry at the carbon of these compounds:

a) $H—C≡N:$ b) 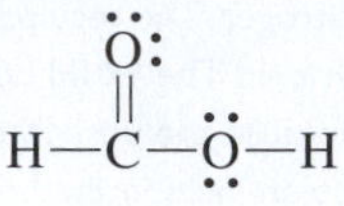   c)

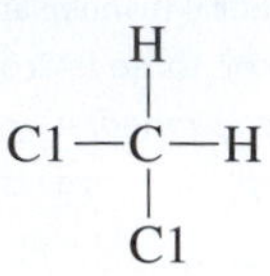

PROBLEM 1.13
Predict the geometry of the following compounds at the indicated atoms:

a) 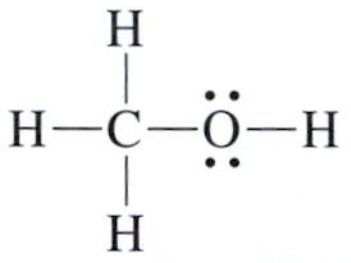   b)   c)

At the carbon   At the carbon   At the carbon
At the oxygen   At the nitrogen   At the oxygen

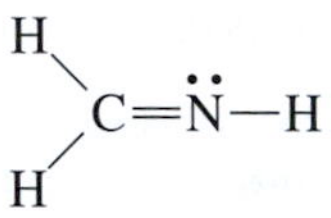

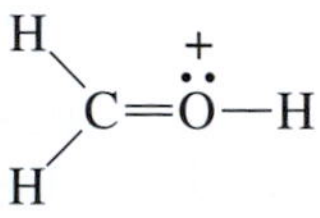

MODEL-BUILDING PROBLEM 1.1
Build models of these compounds with a handheld model kit and note their geometries:
a) CH_4 b) BF_3 c) NH_3 d) H_2O e) PCl_5
f) C_2H_2 g) CH_2O

MODEL-BUILDING PROBLEM 1.2
Build models of the compounds shown in problem 1.13 and note their geometries.

1.11   DIPOLE MOMENTS

The overall dipole moment of a molecule is the vector sum of the individual bond dipoles. If the shape of a molecule is known, vector addition can be used to predict the direction of the dipole moment of that molecule. Several examples are shown in Figure 1.15. The predictions agree with experimental results. For example, CO_2 is found by experiment to have a dipole moment of zero. It is a nonpolar molecule. Because the bonds are polar, this is possible only if the bond dipoles cancel. Therefore, CO_2 must be a linear molecule. On the other hand, because water is polar, with a dipole moment of 1.8 D, it cannot be a linear molecule. A molecule that has only relatively nonpolar carbon–carbon and carbon–hydrogen bonds has only a small dipole moment, if any, and is said to be nonpolar.

These computer-generated electrostatic potential maps use color to show the calculated electron density in a molecule. Electron-rich regions are red and electron-poor regions are blue.

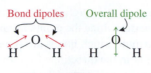

Electrostatic potential map for water (H_2O)

The individual bond dipoles of water are shown in the left structure. The vector sum shown in the right structure indicates the direction of the overall dipole moment of the molecule. Its magnitude is 1.8 D.

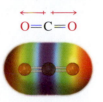

Electrostatic potential map for carbon dioxide (CO_2)

The individual bond dipoles of CO_2 point in opposite directions. These cancel on vector addition, so this compound has a dipole moment of zero.

Electrostatic potential map for carbon tetrachloride (CCl_4)

Although it is more difficult to see, the individual bond dipoles of CCl_4 also cancel on vector addition, and the overall dipole moment is zero.

Electrostatic potential map for trichloromethane ($HCCl_3$)

For $HCCl_3$ the individual bond dipoles, shown in the first structure, add to give an overall dipole moment of 1.9 D with the direction shown in the second structure.

Active Figure 1.15

ORGANIC **Chemistry·Now**™

OBTAINING DIPOLE MOMENTS FROM BOND DIPOLES. Test yourself on the concepts in this figure at **OrganicChemistryNow.**

PRACTICE PROBLEM 1.4

Predict the direction of the dipole moment in this compound:

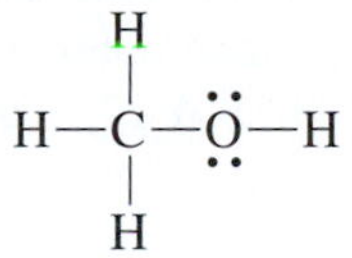

Strategy

Use VSEPR theory to determine the geometry at each atom and the overall shape of the molecule. Then determine which bonds are polar based on the electronegativity differences of the atoms. Put in the directions for individual bond dipoles of these bonds. Estimate the result of vector addition of the bond dipoles to get the approximate direction for the overall dipole for the molecule.

Solution

In problem 1.13 we found that the geometry at the carbon of this molecule, with four bonding pairs of electrons, is tetrahedral. The geometry of the electron pairs at the oxygen (two unshared pairs and two shared pairs) is also tetrahedral, so the geometry of the C—O—H atoms is bent. The C—H bonds are nonpolar, but the O—H and O—C bonds are both polar with the negative end of the bond dipole at the oxygen because oxygen is more electronegative than either carbon or hydrogen. Vector addition gives the overall dipole moment shown here.

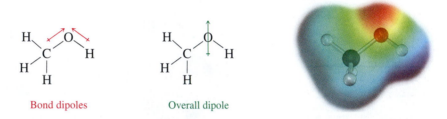

Bond dipoles

Overall dipole

PROBLEM 1.14

Predict the direction of the dipole moments of these compounds:

a)

b)

c)

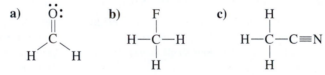

In the next chapter the simple bonding picture that has been developed here will be used to help understand and predict which molecules are stable. We will also begin to learn how the structure of a molecule affects its physical and chemical properties.

Click *Mastery Goal Quiz* to test how well you have met these goals.

Review of Mastery Goals

After completing this chapter, you should be able to:

- Write the best Lewis structure for any molecule or ion. This includes determining how many electrons are available and whether multiple bonds are necessary and satisfying the octet rule if possible. For complex molecules, however, the connectivity must be known. (Problems 1.15, 1.16, 1.17, 1.24, 1.29, 1.30, and 1.31)

■ Calculate the formal charge on any atom in a Lewis structure. (In fact, you should be starting to recognize the formal charges on some atoms in some situations without doing a calculation.) (Problems 1.18, 1.26, and 1.30)

■ Estimate the stability of a Lewis structure by whether it satisfies the octet rule and by the number and the distribution of the formal charges in the structure. (Problems 1.19, 1.20, and 1.36)

■ Recognize some simple cases in which resonance is necessary to describe the actual structure of a molecule. However, a better understanding of resonance will have to wait until Chapter 3. (Problems 1.23, 1.25, 1.34, 1.35, and 1.39)

■ Arrange the atoms that are of most interest to organic chemistry in order of their electronegativities and assign the direction of the dipole of any bond involving these atoms. (Problems 1.21 and 1.37)

■ Determine the shape of a molecule from its Lewis structure by using VSEPR theory. (Problems 1.21, 1.22, 1.24, 1.29, 1.34, 1.35, and 1.36)

■ Determine whether a compound is polar or not and assign the direction of its dipole moment. (Problems 1.21, 1.24, and 1.38)

Additional Problems

1.15 Explain whether the bonds in these compounds would be ionic or covalent and show Lewis structures for them:
a) KCl b) NCl_3 c) NaCN d) KOH

1.16 What is the formula for the simplest neutral compound formed from P and H? Show a Lewis structure for this compound and predict its shape.

1.17 Show Lewis structures for these compounds:
a) CH_5N b) C_2H_5Cl c) N_2 d) CH_2S
e) C_2H_3F f) CH_4S

1.18 Calculate the formal charges on all of the atoms, except hydrogens, in these compounds:

a) $H-\overset{\cdot\cdot}{\underset{\cdot\cdot}{N}}-N\equiv N\colon$

b) $H-\overset{\cdot\cdot}{N}=N=\overset{\cdot\cdot}{N}\colon$

c) $H-\overset{H}{\underset{|}{\overset{\cdot\cdot}{C}}}-N\equiv N\colon$

d)
$$H-\overset{H}{\underset{|}{\overset{|}{C}}}-\overset{\overset{\cdot\cdot}{O}\colon}{\underset{}{\overset{||}{C}}}-\overset{\cdot\cdot}{\underset{\cdot\cdot}{O}}\colon$$

e) $H-\overset{\cdot\cdot}{O}=\overset{H}{\underset{|}{C}}-H$

f)
$$H-\overset{H}{\underset{\underset{H}{|}}{\overset{|}{B}}}-H$$

1.19 Explain which of the two following structures would be more stable. Explain whether they represent isomers or are resonance structures.

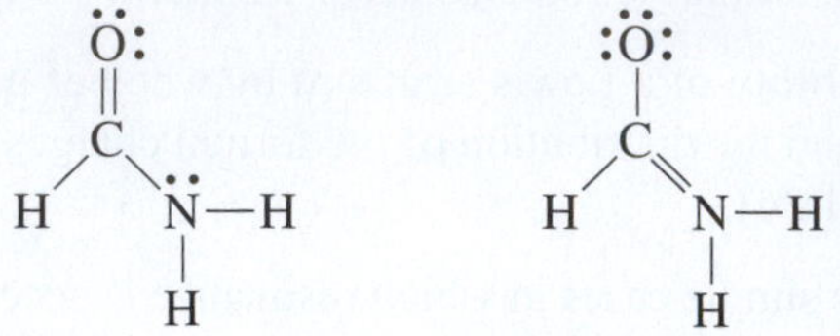

1.20 Draw a Lewis structure for carbon monoxide (CO). Calculate the formal charges on the atoms and comment on the stability of this compound.

1.21 Use heavy and dashed wedged lines to show the shapes of the following molecules. Show the bond dipole of each polar bond and show the overall dipole of each molecule.

a)

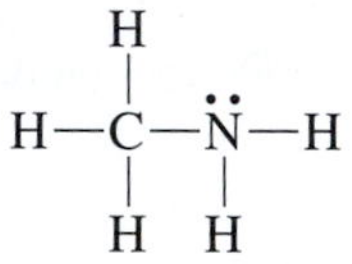

b)

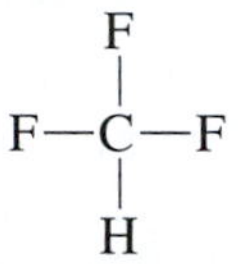

c)

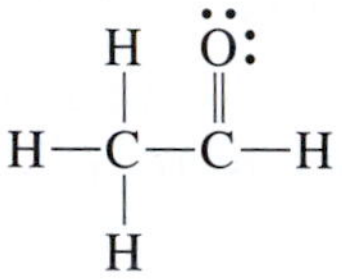

1.22 Predict the geometry at each atom, except hydrogens, in these compounds:

a)

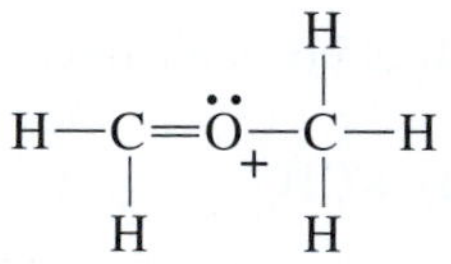

b) 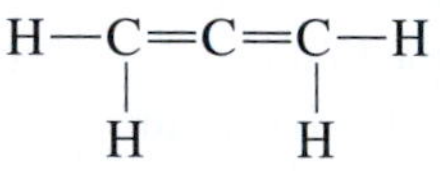 H—C=C=C—H

1.23 a) Show the unshared electron pairs on the following anion. The S has a formal charge of −1, and the formal charges of the other atoms are zero.

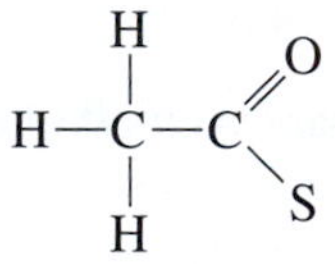

b) Draw a resonance structure for this ion.

1.24 Show a Lewis structure for C_2H_6O in which both carbons are bonded to the oxygen. What is the geometry of this molecule at the oxygen? Show the direction of the dipole for the molecule.

1.25 Show a Lewis structure for NO_2^-. (Both oxygens are bonded to the nitrogen.) Show a resonance structure also.

1.26 Amino acids, from which proteins are formed, exist as "dipolar ions." The structure of the dipolar ion of the amino acid alanine is

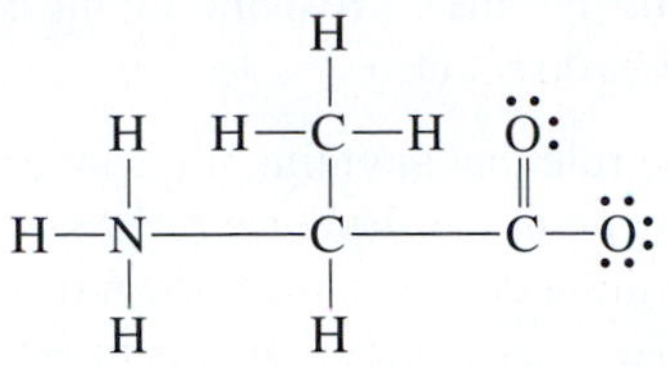

Alanine

a) Calculate the formal charges on all of the atoms, except hydrogens, of alanine.

b) What is the overall charge of alanine?

c) Explain whether or not you expect the two oxygens to be different.

1.27 A covalent ion can also have polar bonds. Consider the ammonium cation. How are its bonds polarized? Do you think that the N of the ammonium cation is more or less "electronegative" than the N of ammonia (NH_3)? Would the hydrogens of NH_4^+ or NH_3 have a larger partial positive charge?

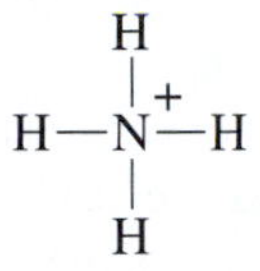

Ammonium cation

1.28 You need to know the melting point for $CaCl_2$ for a lab report you are writing. Your lab partner says that the *Handbook of Chemistry and Physics* lists this as 68°C. Do you think you should trust that your lab partner has looked up the value correctly or should you look it up for yourself?

1.29 Show a Lewis structure for $AlCl_4^-$. What are the formal charges on the atoms of this anion? What is its shape?

1.30 Ammonium cyanate is composed of an ammonium cation (NH_4^+) and a cyanate anion (OCN^-). Show a Lewis structure for the cyanate anion. (Both O and N are bonded to C.) Which atom has the negative formal charge in your structure? What is the shape of the ion? Show a resonance structure for this ion.

1.31 **a)** Show a Lewis structure for urea, CH_4N_2O. Both N's and the O are bonded to the C. The H's are bonded to the N's. None of the atoms has a formal charge.

b) Show a Lewis structure of an isomer of urea that still has both N's and the O bonded to the C and has formal charges of zero at all atoms.

1.32 Phosphorus forms two compounds with chlorine, PCl_3 and PCl_5. The former follows the octet rule, but the latter does not. Show Lewis structures for each of these compounds. For the corresponding nitrogen compounds, explain why NCl_3 exists but NCl_5 does not.

1.33 On the basis of the rule that anything unusual about a structure must be shown explicitly, the nitrogen in the structure NH_3 is seen to have an unshared pair of electrons whether these electrons are shown or not. Suppose you wanted to discuss the unstable species NH_3 that was missing the unshared pair of electrons. How would you draw the structure so that it would be obvious to another person that the unshared pair was absent?

1.34 Ozone, O_3, is a form of oxygen found in the upper atmosphere. It has the connectivity O—O—O and is neutral.
a) Show a Lewis structure for ozone.
b) Calculate the formal charge on each oxygen of ozone.
c) What is the shape of ozone?
d) Experimental observations show that both bonds of ozone are identical. Explain how this is possible.

1.35 **a)** Draw a Lewis structure for the carbonate anion CO_3^{2-}. Each oxygen is bonded only to the carbon.
b) Calculate the formal charge on each atom.
c) What is the shape of this species?
d) Experimental evidence shows that all of the oxygens are identical. Explain.

1.36 Consider the species CH_3, which has three normal carbon–hydrogen bonds and no other electrons on the carbon.
a) What is the charge of this species?
b) What is its geometry?
c) Discuss the stability of this species. Do you think it is more or less stable than the species shown in practice problem 1.1? Explain.
d) Show the Lewis structure of the product of the reaction of this species with hydroxide ion (OH^-).

1.37 Explain how the dipole moments of FCl (0.9 D) and ICl (0.7 D) can be so similar.

1.38 Chlorine is more electronegative than phosphorus. Predict the dipole moment of PCl_5.

1.39 Although carbon–carbon double bonds are shorter than carbon–carbon single bonds, all of the carbon–carbon bonds of benzene are the same length. Explain.

Benzene

1.40 Explain whether or not these molecules are polar (have a dipole moment).

    a) CBr_4 **b)** NH_3 **c)** CH_3OCH_3

    d) CH_2Cl_2 **e)** CO_2

Problems Using Online
Three-Dimensional Molecular Models

ORGANIC
Chemistry Now™
Click *Molecular Model Problems* to view the models needed to work these problems.

1.41 For each model, draw a structure that fits the geometry of the molecule. Then explain the geometry at each atom other than the hydrogens.

1.42 Draw a structure for each model. Explain whether the molecule represented by each model has a dipole moment or not.

1.43 Draw a structure for each model. Show the direction of the dipole moment for the molecule represented by each model.

1.44 Draw a structure for the model. Is the molecule polar or nonpolar? Explain.

 Do you need a live tutor for homework problems? Access vMentor at Organic ChemistryNow at **http://now.brookscole.com/hornback2** for one-on-one tutoring from a chemistry expert.

Organic Compounds: A First Look

Now that you have refreshed your basic understanding of bonding, you are ready to examine organic compounds in more detail. The purpose of this chapter is to provide more experience with simple organic molecules. You will learn about how their atoms are connected (the structure of the molecule) and how their structures affect some of their properties.

The chapter begins with a description of how atoms usually bond in molecules, including a discussion of bond lengths and bond strengths. Next, structural formulas, which show how the atoms of a molecule are connected, are discussed in detail. Structural formulas are very important in organic chemistry because atoms for most molecular formulas can be assembled in many different ways, resulting in the possibility of isomers. Therefore, a structural formula for a molecule is needed to specify a particular compound. A useful aid in drawing structural formulas, called the degree of unsaturation, is presented. Then we learn how examining the structure of a compound enables us to estimate the physical properties of that compound. It is possible to predict whether a compound is likely to be a solid, a liquid, or a gas at room temperature; whether it is likely to be water soluble; and so forth. Finally, an extremely important organizing concept, the functional group, is introduced.

MASTERING ORGANIC CHEMISTRY

▶ Recognizing Stable Bonding Arrangements

▶ Understanding the Trends in Bond Strengths and Bond Lengths

▶ Drawing Constitutional Isomers

▶ Calculating and Using the Degree of Unsaturation

▶ Using Line Structures, Condensed Structures, and Skeletal Structures

▶ Understanding How the Attractive Forces between Molecules Affect Melting Points, Boiling Points, and Solubilities

▶ Recognizing Functional Groups

2.1 COMMON BONDING SITUATIONS

In Chapter 1 we learned that molecules that satisfy the octet rule are likely to be stable. Furthermore, we learned that the presence of formal charges in a molecule is often a destabilizing factor. The common ways in which atoms are bonded can be understood by using these two criteria.

ORGANIC
Chemistry Now™
Look for this logo in the chapter and go to OrganicChemistryNow at
http://now.brookscole.com/hornback2 for tutorials, simulations, problems, and molecular models.

ⓐ Hydrogen almost always has one bond. Exceptions are very rare.

$$H-\overset{..}{\underset{..}{O}}-H$$

Figure 2.1

BONDING ARRANGEMENTS FOR ⓐ HYDROGEN AND ⓑ CARBON.

ⓑ Carbon in its most stable form has four bonds. These may be single, double, or triple bonds. These are by far the most common bonding situations for carbon.

$$H-\overset{\overset{\displaystyle H}{|}}{\underset{\underset{\displaystyle H}{|}}{C}}-\overset{\overset{\displaystyle H}{|}}{\underset{\underset{\displaystyle H}{|}}{C}}-H \qquad \overset{H}{\underset{H}{\diagdown}}C=C\overset{H}{\underset{H}{\diagup}}$$

$$H-C\equiv C-H$$

Carbon is sometimes encountered with only three bonds and a negative charge. Such **carbanions** are less stable than the compounds above, but they are still important.

$$H-\overset{\overset{\displaystyle H}{|}}{\underset{\underset{\displaystyle H}{|}}{C}}\!\!:^{-}$$

A carbanion

Carbon **radicals**, with only seven electrons in the valence shell for carbon, and **carbocations**, with only six electrons and a positive charge on the carbon, do not satisfy the octet rule and are quite unstable. These species are only encountered as highly reactive, transient intermediates in certain chemical reactions.

$$H-\overset{\overset{\displaystyle H}{|}}{\underset{\underset{\displaystyle H}{|}}{C}}\cdot \qquad H-\overset{\overset{\displaystyle H}{|}}{\underset{\underset{\displaystyle H}{|}}{C}}+$$

A radical   A carbocation

Thus, carbon usually forms four bonds. Any other arrangement either does not satisfy the octet rule or has a charge on the carbon. However, carbon is occasionally encountered in one of these less stable bonding arrangements. Both the common and more unusual bonding situations for various atoms are shown in Figures 2.1, 2.2, and 2.3.

PROBLEM 2.1

Discuss the stability of each of these species based on the octet rule and formal charges:

a) $\overset{H}{\underset{H}{\diagdown}}C=\overset{..}{C}\overset{-}{\underset{H}{\diagup}}$

b) $H-\overset{\overset{\displaystyle H}{|}}{\underset{\underset{\displaystyle H}{|}}{C}}-C\equiv C-H$

c) $H-\overset{\overset{\displaystyle H}{|}}{\underset{\underset{\displaystyle H}{|}}{C}}-\overset{\overset{\displaystyle ..}{}}{\underset{\underset{\displaystyle +}{}}{N}}-H$

d) $H-\overset{\overset{\displaystyle H}{|}}{\underset{\underset{\displaystyle H}{|}}{C}}-O-\overset{\overset{\displaystyle H}{|}}{\underset{\underset{\displaystyle H}{|}}{C}}-H$

e) $\overset{H}{\underset{H}{\diagdown}}C=\overset{..}{N}-\overset{\overset{\displaystyle H}{|}}{\underset{\underset{\displaystyle H}{|}}{C}}-H$

Figure 2.2

BONDING ARRANGEMENTS
FOR ⓐ NITROGEN,
ⓑ OXYGEN, AND THE
ⓒ HALOGENS.

ⓐ Nitrogen usually has one unshared pair of electrons and three bonds. The bonds may be single, double, or triple.

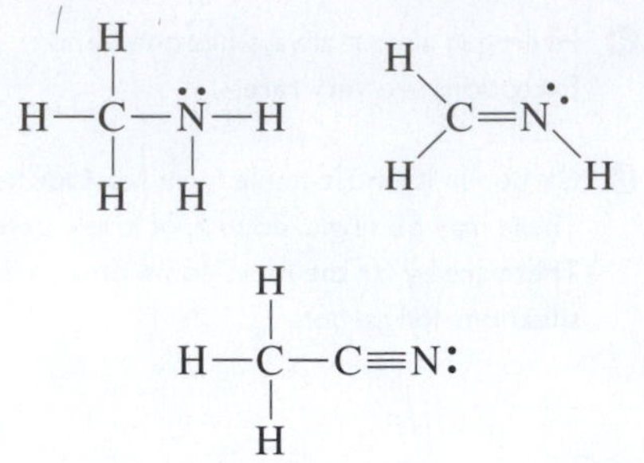

Nitrogen is also encountered with four bonds and a positive charge or with two bonds, two unshared pairs of electrons, and a negative charge. Because both of these species satisfy the octet rule, they are relatively common.

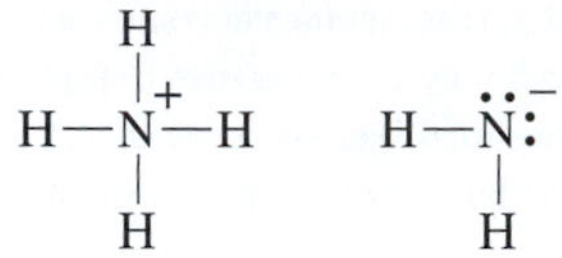

ⓑ Oxygen usually has two bonds and two unshared pairs of electrons. The bonds may be single or double.

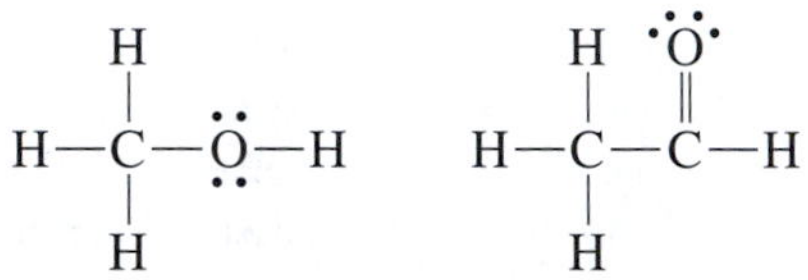

Oxygen with one bond, three pairs of electrons, and a negative charge or with three bonds, one unshared pair of electrons, and a positive charge is also relatively common.

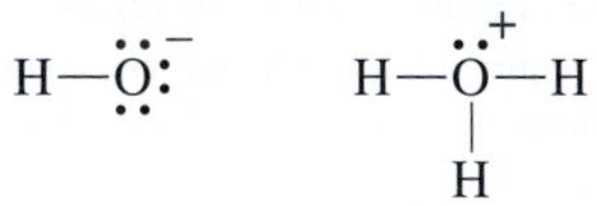

ⓒ The halogens usually have one bond and three unshared pairs of electrons. They are also found as stable negative ions. In addition, Cl, Br, and I can have more than eight electrons in their valence shells.

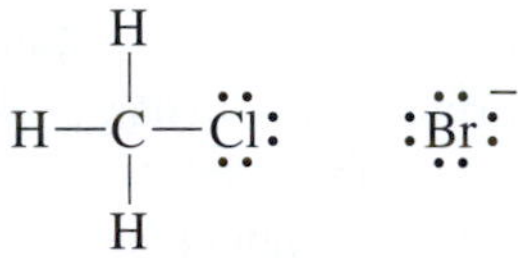

Figure 2.3

BONDING ARRANGEMENTS
FOR ⓐ PHOSPHORUS AND
ⓑ SULFUR.

ⓐ Because phosphorus is beneath nitrogen in the periodic table, it has similar bonding tendencies. Thus, phosphorus often has three bonds and one unshared pair of electrons. However, because phosphorus is in the third row of the periodic table, it can have more than eight electrons in its valence shell.

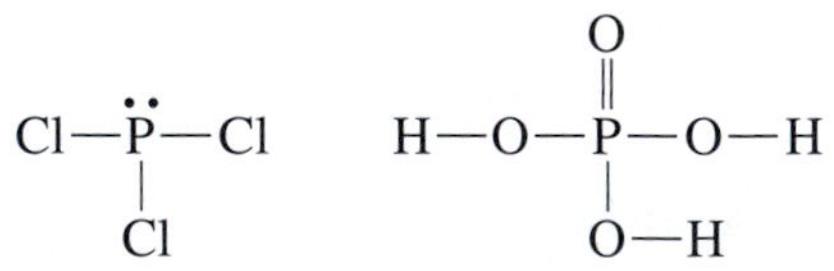

ⓑ Sulfur is beneath oxygen in the periodic table. Therefore, it often has two bonds and two unshared pairs of electrons. Sulfur can also have more than eight electrons in its valence shell.

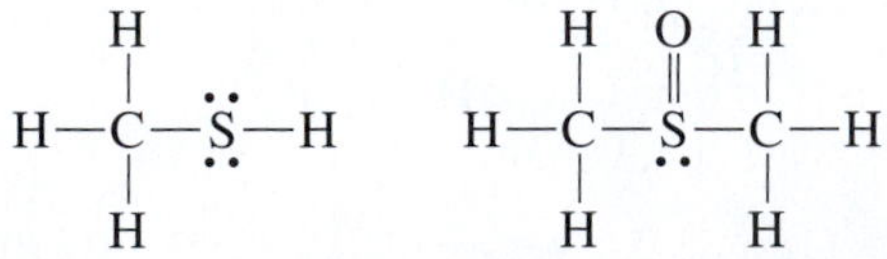

2.2 BOND STRENGTHS AND BOND LENGTHS

In the common bonding situations illustrated in Figures 2.1, 2.2, and 2.3, the bonds are stable. This implies that a considerable amount of energy must be added to a compound to break one of these bonds. One measure of the strength of a bond is the **bond dissociation energy.** This is defined as the amount of energy that must be added in the gas phase to break the bond in a homolytic fashion, that is, breaking the bond in a symmetrical manner, with one electron remaining with each of the atoms. For example, homolytic cleavage of one of the carbon–hydrogen bonds of ethane, as shown next, requires an input of 98 kcal/mol (410 kJ/mol) of energy. Therefore, the bond dissociation energy for one C—H bond in ethane is 98 kcal/mol (410 kJ/mol).

$$+98 \text{ kcal/mol}$$
$$(+410 \text{ kJ/mol})$$

The C—H bond dissociation energy is not the same for all compounds. It depends on the other bonds to the carbon also. For example, the C—H bond dissociation energy for methane, CH_4, is 104 kcal/mol (435 kJ/mol), compared to 98 kcal/mol (410 kJ/mol) for ethane. Of course, other bond dissociation energies also depend on the other bonds to the atoms involved.

Table 2.1 lists typical bond dissociation energies for many of the bonds of interest. Table 2.2 lists approximate bond lengths for these same bonds. Rather than memorizing the numbers, let's see what trends can be discerned from this information.

1. A correlation exists between bond length and bond strength: stronger bonds tend to be shorter.

2. Bonds between H and the second-row elements C, N, and O are all strong—close to 100 kcal/mol (418 kJ/mol). The bond lengths are about 1 Å (1 Å = 10^{-10} m = 100 pm).

3. Bonds between C and the other second-row elements are also reasonably strong—in the vicinity of 70 to 80 kcal/mol (290–335 kJ/mol). The bond lengths are all near 1.5 Å. For these multivalent atoms, note that only carbon forms strong bonds to itself and to the other atoms. For example, even in situations in which the octet rule is satisfied, the O—O bond dissociation energy is only 34 kcal/mol (142 kJ/mol), and that for N—O is only 39 kcal/mol (163 kJ/mol). These bonds are weaker than C—C bonds because of repulsion between the unshared pairs of electrons on the bonded atoms. For this reason, only chains of bonded carbons are commonly encountered.

4. The carbon–halogen bonds become weaker and longer as the atomic number of the halogen increases. This is also true for bonds involving other atoms from the third and subsequent periods of the periodic table. For example, the Si—Si bond is much weaker than the C—C bond.

5. Double and triple bonds are stronger and shorter than single bonds. However, note that a C—C double bond is not twice as strong as a C—C single bond, nor is a triple bond three times as strong. On the other hand, a C—O double bond is very strong, more than twice as strong as a C—O single bond. Therefore, the C—O double bond is a common and important bond.

Chapter 3 discusses the reasons behind these trends.

Table 2.1 Typical Bond Dissociation Energies

Bond	Bond Dissociation Energy
C—H	98 (410)
N—H	92 (385)
O—H	109 (456)
C—C	81 (339)
C—N	66 (276)
C—O	79 (330)
C—F	116 (485)
C—Cl	79 (330)
C—Br	66 (276)
C—I	52 (217)
C=C	145 (606)
C≡C	198 (828)
C=O	173 (723)
C≡N	204 (854)

Values are in units of kcal/mol. Values in parentheses are in units of kJ/mol.

Table 2.2 Approximate Bond Lengths

Bond	Bond Length
C—H	1.10
N—H	1.00
O—H	1.00
C—C	1.54
C—N	1.47
C—O	1.41
C—F	1.38
C—Cl	1.78
C—Br	1.94
C—I	2.14
C=C	1.34
C≡C	1.20
C=O	1.20
C≡N	1.16

Values are in units of angstroms (1 Å = 10^{-10} m).

Most organic compounds are based primarily on covalent bonds to carbon. These carbons are bonded to other carbons in chains or rings. They may also be bonded to hydrogen. In addition, compounds with C bonded to O and N (these may have bonds to H) and to the halogens (especially Cl and Br) are common. Compounds involving S and P or the other elements are stable but less common. Carbon may be doubly bonded to C, N, or O or triply bonded to C or N. Although these are neither the only stable bonds nor the only important bonds, they are the ones that are encountered most often.

2.3   CONSTITUTIONAL ISOMERS

With the experience we have gained so far, it should be fairly easy to draw a structure for any formula. It is also possible to crudely estimate the stability of the compound represented by this structure. As an example, let's show the structure for the compound with the formula C_2H_6O. We quickly discover that there are two ways to assemble these atoms, depending on whether we start with a C—C—O or a C—O—C arrangement of the nonhydrogen atoms.

Ethanol
(ethyl alcohol)

Dimethyl ether

Which of these two structures is correct? Both of them satisfy the octet rule and neither has formal charges, so both are predicted to be of comparable stability. On the basis of what we have discussed so far, we cannot predict which is more stable. In fact, both of these compounds are quite stable and can be "put in a bottle." But they are different compounds. Ethyl alcohol is the "alcohol" found in beverages. It is a liquid at room temperature. In contrast, dimethyl ether is a gas at room temperature and is quite poisonous. As was mentioned in Section 1.7, compounds such as these, with the same molecular formula but different arrangements of bonded atoms (different structures or different connectivities), are called **constitutional isomers** (or structural isomers). Constitutional isomerism is very common in organic compounds. This is another reason why it is necessary to show the structure of the compound under discussion rather than just the molecular formula.

As the number of atoms in a formula increases, the number of possible constitutional isomers increases dramatically. As an illustration, consider the series of **hydrocarbons** (compounds made up of only carbon and hydrogen) shown in Table 2.3. Although there is only one compound with the formula CH_4, there are 75 constitutional isomers with the formula $C_{10}H_{22}$ (all 75 have been prepared in the lab and identified) and more than 4 billion with the formula $C_{30}H_{62}$!

A general method for calculating the number of isomers of a given molecular formula has not yet been developed. The problem is just too complex. However, several

Table 2.3   The Number of Constitutional Isomers for a Series of Hydrocarbons

Formula	Number of Constitutional Isomers
CH_4	1
C_2H_6	1
C_3H_8	1
C_4H_{10}	2
C_5H_{12}	3
C_6H_{14}	5
C_7H_{16}	9
C_8H_{18}	18
C_9H_{20}	35
$C_{10}H_{22}$	75
$C_{20}H_{42}$	366,319
$C_{30}H_{62}$	4,111,846,763

methods have been developed that work for specific cases. One of these was used to determine the number of isomers with the formula $C_{30}H_{62}$. More recently, a type of mathematics called graph theory has been applied with some success to some types of molecular formulas. Nevertheless, a general solution has remained elusive. So we will have to rely on drawing all of the different structures if we want to determine the number of isomers of a particular formula. Obviously, this can be done only for simple cases.

How many of the virtually limitless number of organic compounds are known? The organization Chemical Abstracts Service (CAS) has the task of reviewing every article in every chemical journal that is published and tabulating all of the compounds that have been characterized and reported in these articles. Most of these compounds are organic. The CAS registry listed more than 22 million compounds as of 2003.

Beginning students usually have difficulty drawing all the isomers for a particular formula without omitting or duplicating any. However, we can do this for simpler examples if a systematic approach is used. The process is illustrated in Figure 2.4 for C_6H_{14}. It is important to recognize that these structures show only the connectivity of the isomer; they do not show the shapes of the molecules. (Chapter 6 deals with that.) No matter how an individual structure is twisted or turned, as long as the connectivity remains the same, the structure represents the same isomer.

PROBLEM 2.2

Draw the three constitutional isomers that have the formula C_5H_{12}.

For the first isomer, start with a straight chain of six carbons. Then add the hydrogens to produce isomer ❶. Note that this isomer has a continuous chain of C's. It does not make any difference whether the C's are drawn straight or bent, since we are not trying to show the shape of the molecule. The drawings below are other ways to represent isomer ❶.

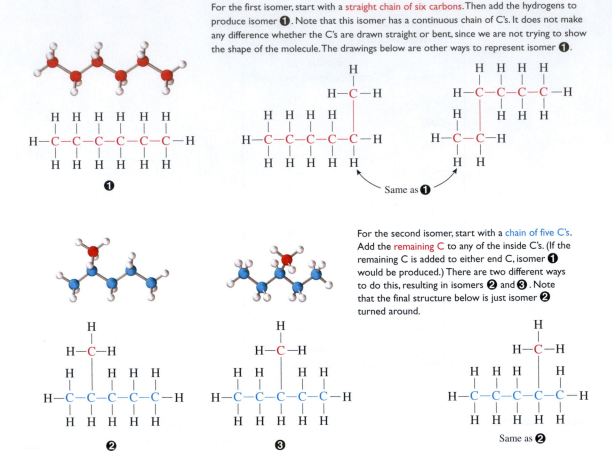

❶

Same as ❶.

For the second isomer, start with a chain of five C's. Add the remaining C to any of the inside C's. (If the remaining C is added to either end C, isomer ❶ would be produced.) There are two different ways to do this, resulting in isomers ❷ and ❸. Note that the final structure below is just isomer ❷ turned around.

❷   ❸   Same as ❷

The total is five isomers. To get the final two, begin with a chain of four C's. Add the two remaining C's to the middle C's, as shown in isomers ❹ and ❺. Any other arrangement will produce one of these five isomers. For example, if both the remaining C's are added as a two-carbon chain to one of the middle C's, a bent version of isomer ❸ is produced.

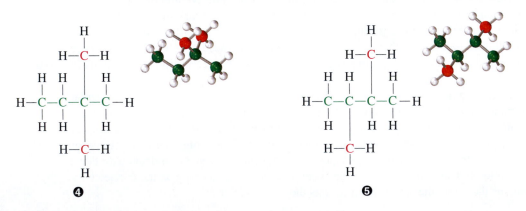

❹   ❺

Figure 2.4

CONSTITUTIONAL ISOMERS FOR C_6H_{14}.

PRACTICE PROBLEM 2.1

There are eight constitutional isomers that have the formula $C_5H_{11}Cl$. Show a structure for each of them.

Strategy

In problems like this one, it is important to be as systematic as possible or you will never get all of the isomers. It is helpful to recognize that Cl has a valence of 1, just like H. Therefore, $C_5H_{11}Cl$ can be viewed as resulting from replacing one H of C_5H_{12} with a Cl. C_5H_{12} has three isomers (see problem 2.2). Each of these will give rise to a different set of $C_5H_{11}Cl$ isomers. So start with each isomer of C_5H_{12} and replace each different H with a Cl.

Solution

Start with the straight-chain isomer of C_5H_{12}. This compound has three different types of hydrogens, so three isomers of $C_5H_{11}Cl$ result by replacing these hydrogens with chlorine.

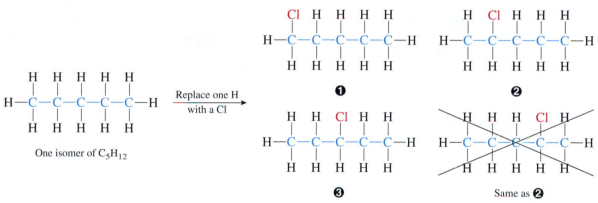

The second isomer of C_5H_{12} gives four isomers of $C_5H_{11}Cl$:

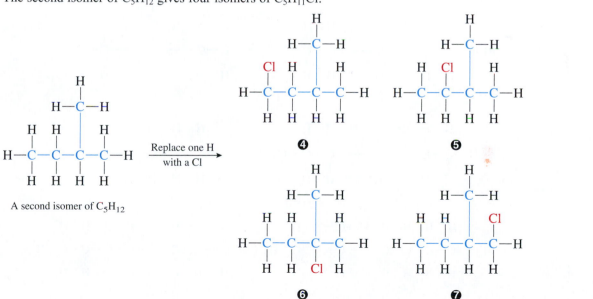

The third isomer of C_5H_{12} is very symmetrical. All of its hydrogens are identical, so it gives only one isomer of $C_5H_{11}Cl$. We get eight isomers in total.

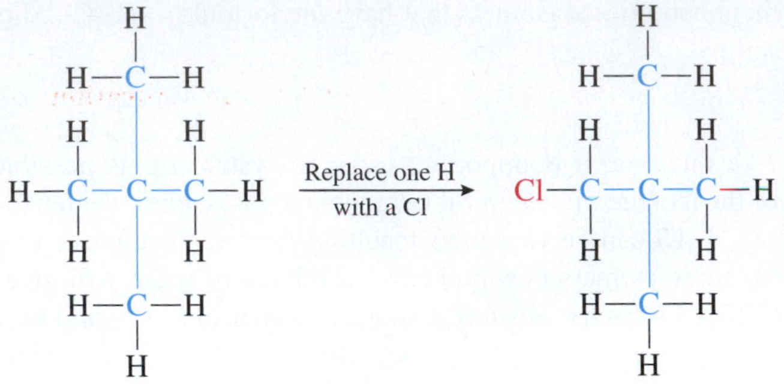

The third isomer of C_5H_{12} ❽

PROBLEM 2.3
The formula $C_4H_{10}O$ results in seven isomers. See how many you can draw.

PROBLEM 2.4
Determine whether these structures represent the same compound or isomers:

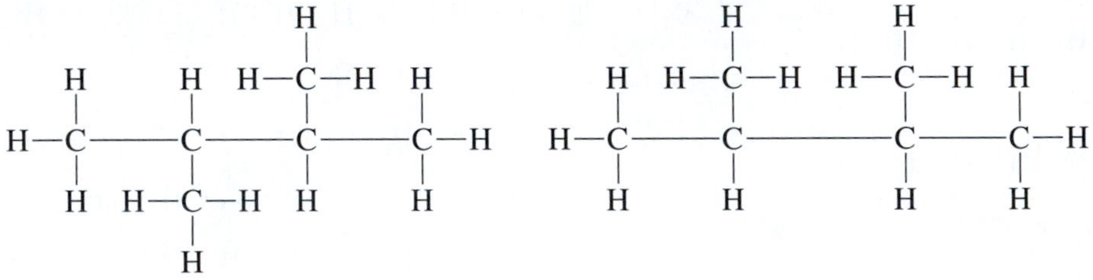

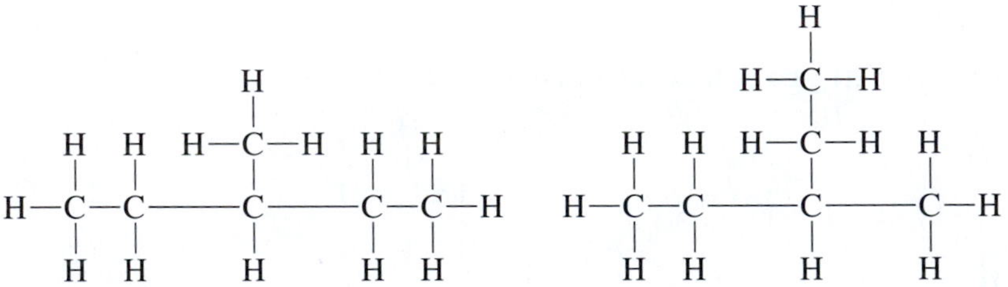

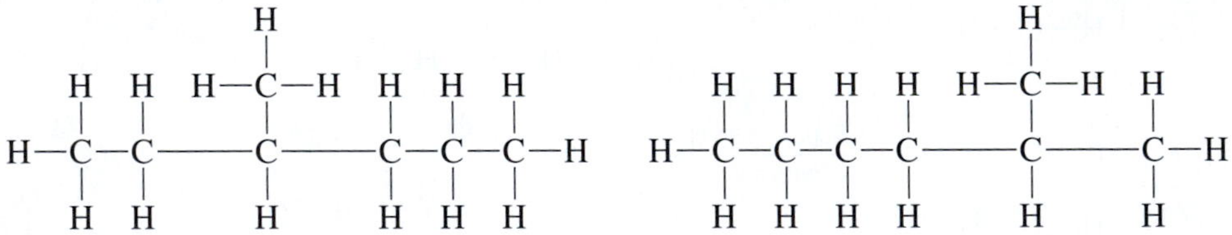

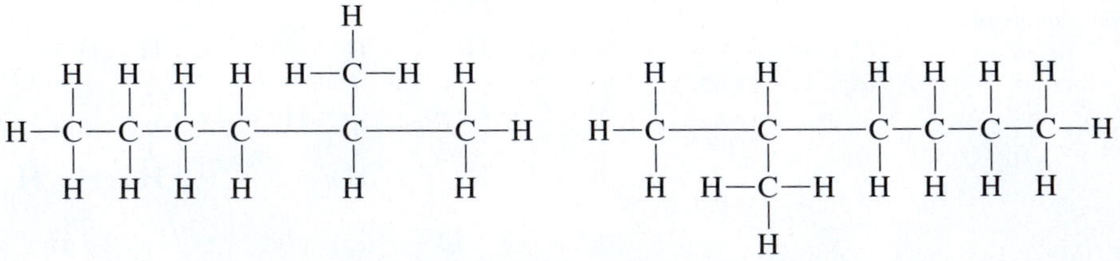

2.4 DEGREE OF UNSATURATION

The task of drawing isomers for a particular formula can be made easier by comparing the number of hydrogens with the number of carbons in the formula. Such a comparison allows the calculation of the **degree of unsaturation,** which furnishes useful information about possible structures that will fit the formula (such as whether double bonds can be present) and provides a starting point for drawing isomers.

The isomers shown in Figure 2.4 have the maximum number of hydrogens possible for a compound with six carbons. The maximum number of H's can easily be calculated from the number of C's present. To see how the formula arises, consider a straight, or linear, chain of some number of C's. Each C has two H's bonded to it, with the exception that the two end C's have one additional H, because they are bonded to only one C. Therefore, for *n* carbons, the maximum number of hydrogens is $2n + 2$. The general formula for a hydrocarbon with the maximum number of hydrogens is C_nH_{2n+2}. As an example, C_5H_{12} has the maximum number of H's for 5 C's [$2(5) + 2 = 12$].

Now let's see what happens to the number of hydrogens in a compound if a double bond is present. To form a double bond, a hydrogen must be removed from each of two adjacent carbons. Therefore, a compound with one double bond has two fewer H's than the maximum.

C_5H_{12} C_5H_{10}

Similarly, to form a ring from a chain of carbons, one H must be removed from both end C's so that they can be bonded together. A compound with a ring also has two fewer H's than the maximum number.

C_6H_{14} C_6H_{12}

Any compound whose formula has *two* hydrogens less than the maximum number possible ($2n + 2$) must contain *one* double bond or *one* ring. The total number of multiple bonds plus rings is called the **degree of unsaturation (DU)** and is equal to 1 for this case. The DU of a compound can be calculated by using the following formula:

$$DU = \frac{(\text{Maximum possible H's}) - (\text{Actual H's})}{2}$$

For a hydrocarbon with the formula C_nH_x the degree of unsaturation is

$$DU = \frac{(2n + 2) - x}{2}$$

where

n = actual C's present

x = actual H's present

The DU is very useful when drawing isomers. Let's look at a simple example. Suppose we are asked to draw the constitutional isomers with the formula C_5H_{10}. The DU for C_5H_{10} is $[(2(5) + 2) - 10]/2 = 1$. Therefore, this compound must have one double bond or one ring. Any compound that has five C's and one double bond or one ring will fit the formula. Although C_5H_{12} has only three isomers, having two fewer H's actually increases the number of isomers, because there is now a double bond or a ring to vary.

As structures become more complex, drawing them as a **line** or **Kekulé structure,** in which each bond is shown as a line, becomes more time consuming. The method of grouping together atoms that are bonded to the same atom to give a **condensed structure,** which was presented in Chapter 1, can be used, but even this becomes tedious. *An even faster method, called a **skeletal structure,** shows the* **Important Convention** *C—C bonds as lines.* Each line is assumed to have a C at each end unless another atom is shown. Hydrogens on the C's are not shown. (However, if a carbon is shown with a C, the hydrogens on it must also be shown as H's.) All other atoms (N, O, Cl, and so on) must be shown, along with any hydrogens bonded to them. Unshared pairs of electrons may or may not be shown. Sometimes the various methods are mixed to emphasize a particular feature. Figure 2.5 shows four of the possible isomers of $C_5H_{10}O$ using these various methods.

PROBLEM 2.5

Calculate the DU for these formulas and draw two constitutional isomers for each:
a) $C_{10}H_{22}$ **b)** C_9H_{16} **c)** C_6H_6

PROBLEM 2.6

Convert these structures to skeletal structures:

a)

b)

c) $CH_3CH_2CCH_2CH(CH_3)_2$ (with O double-bonded to the third carbon)

d) $CH_3CHCH_2NHCH_2CH_2CH_3$ (with CH_3 on the second carbon)

Line structure or Kekulé structure	Condensed structure	Skeletal structure	Molecular model
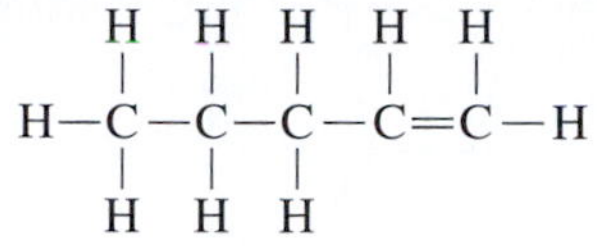 H−C−C−C−C=C−H (with H's)	$CH_3CH_2CH_2CH=CH_2$	 ⓐ ⓑ ⓒ	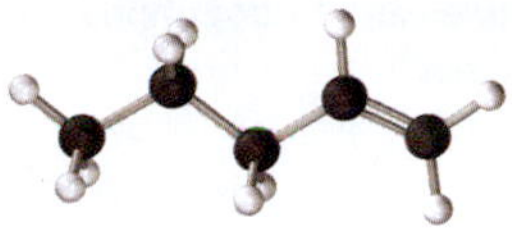
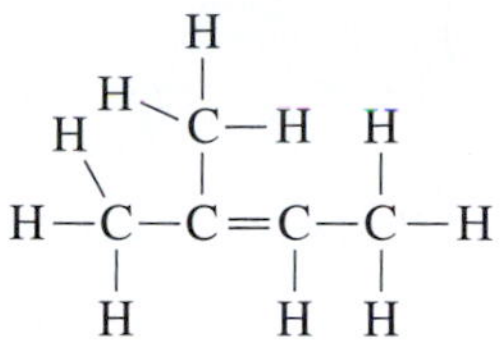 H−C−C=C−C−H (with CH₃ group)	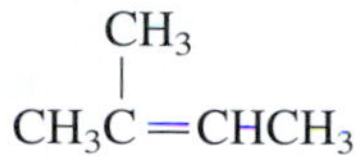 $CH_3C=CHCH_3$ or $(CH_3)_2C=CHCH_3$	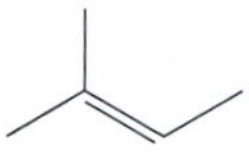	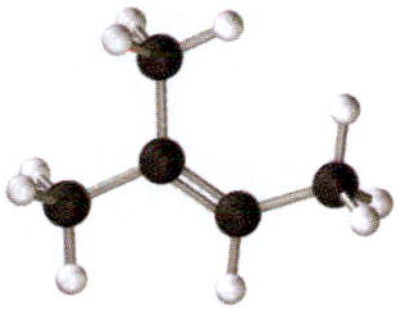
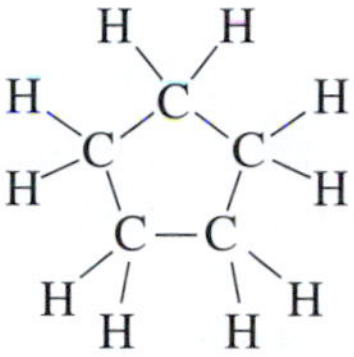 cyclopentane ring structure	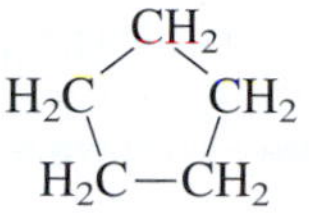 H_2C CH_2 CH_2 $H_2C−CH_2$	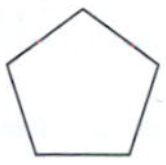 pentagon	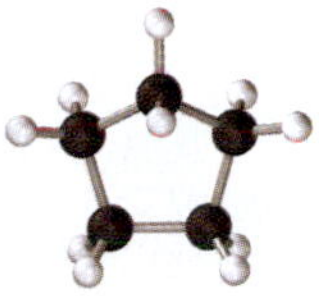
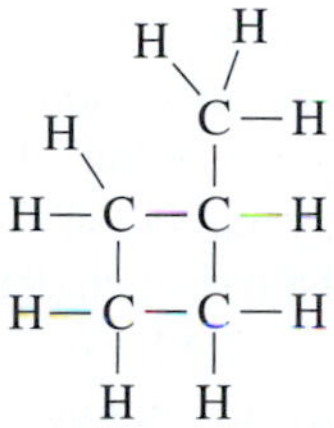 methylcyclobutane structure	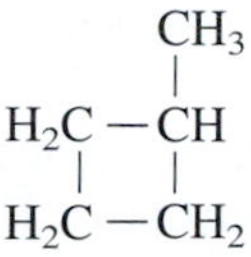 CH_3 $H_2C−CH$ $H_2C−CH_2$	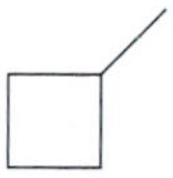 square with methyl	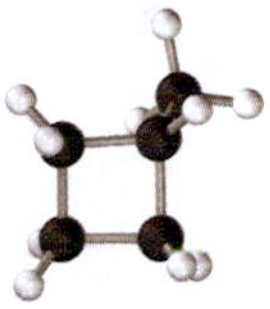

ⓐ This represents a C bonded to one other C (at the other end of the line). It must be bonded to three H's to complete its valence.

ⓑ This represents a C bonded to two other C's. It is also bonded to two H's that are not shown.

ⓒ This represents a C double bonded to another C. It is also bonded to two H's that are not shown.

Figure 2.5

LINE, CONDENSED, AND SKELETAL STRUCTURES AND MOLECULAR MODELS FOR SOME ISOMERS OF C_5H_{10}.

PROBLEM 2.7

Convert these skeletal structures to line structures:

a)

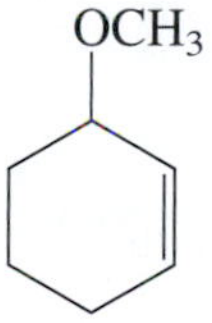

b)

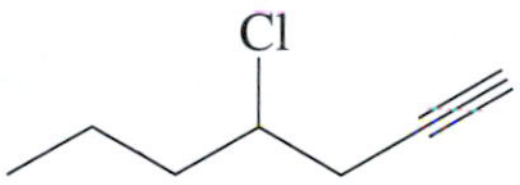

c)

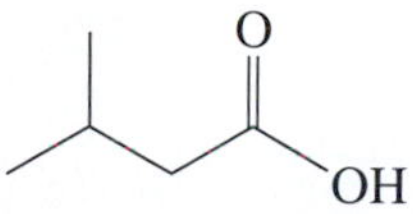

d)

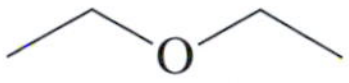

PROBLEM 2.8

Determine whether these structures represent the same compound or isomers:

a) 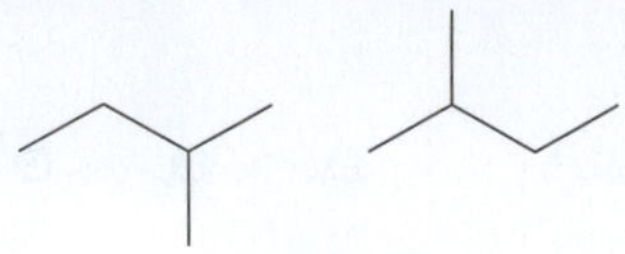   b)

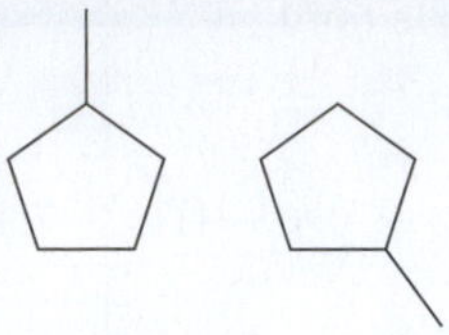

c) 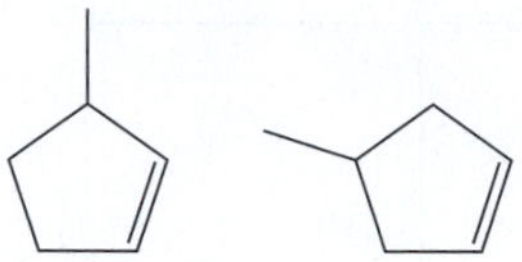   d)

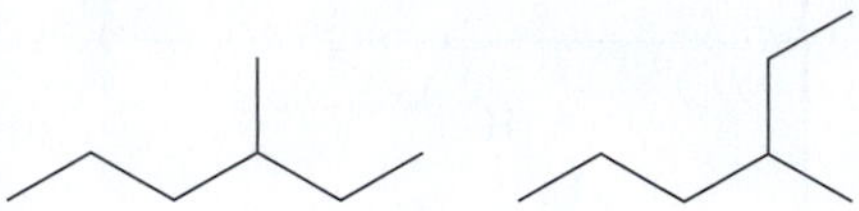

e) CH$_3$CHCH$_2$CH$_2$CH$_3$ CH$_3$CH$_2$CH$_2$CHCH$_3$
    with CH$_3$ on the second carbon and CH$_3$ on the fourth carbon

f) CH$_3$CH$_2$CHCH$_2$CH$_2$CH$_3$ CH$_3$CH$_2$CH$_2$CH$_2$CHCH$_3$
    with CH$_3$ substituents

The DU can also be calculated for formulas that have atoms other than C and H. Because halogens are monovalent, they are counted as hydrogens in the DU calculation. For example, C$_5$H$_{11}$Cl is counted as C$_5$H$_{12}$ and has DU = 0—it is saturated.

Oxygen is divalent. If an O is added to a structure, it can, for example, be inserted between a C and H or between two C's without changing the number of hydrogens. Therefore, we can ignore oxygens when performing a DU calculation. The DU for C$_6$H$_{10}$O is 2, the same as that for C$_6$H$_{10}$. This compound must have two double bonds, one triple bond (a triple bond contributes 2 to the DU), one double bond and one ring, or two rings. Although the presence of oxygen is ignored in calculating the DU, oxygen can, of course, be involved in the features, double bonds or rings, that contribute to the DU. Figure 2.6 shows several isomers for C$_6$H$_{10}$O.

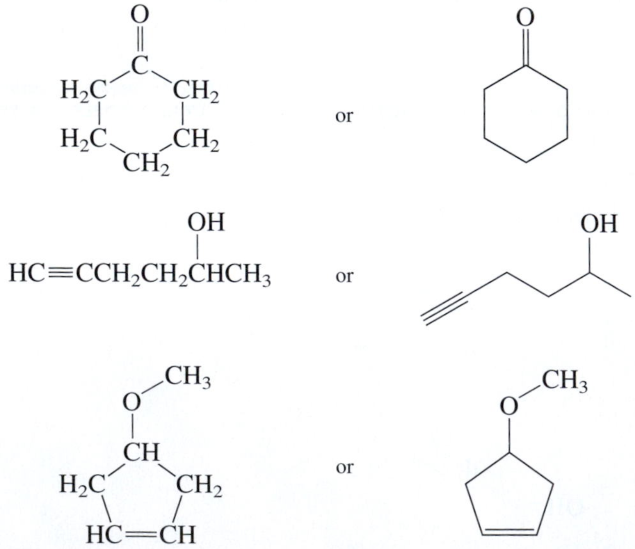

Figure 2.6

THREE (OF MANY)
CONSTITUTIONAL ISOMERS
OF **C$_6$H$_{10}$O.**

Finally, let's consider the effect of nitrogen on a DU calculation. Nitrogen is trivalent. If an N is added to a structure by inserting it between two atoms, one H must also be added to satisfy the third valence of the N. Therefore, each nitrogen that is present in a compound increases the maximum number of H's by one. For example, the maximum number of H's for $C_{10}H_{15}N$ is $2(10) + 2 + 1 = 23$. The DU is $(23 - 15)/2 = 4$.

You must be very careful when using a shorthand method to show structures. Beginning organic chemistry students commonly forget about hydrogens or do not recognize other features of such structures. *It is often a good idea to redraw the structures more completely, showing each carbon and all the hydrogens, until you are very comfortable with them and can automatically picture all the features of the molecule.*

ORGANIC
Chemistry Now™
Click *Coached Tutorial Problems*
to practice **Determining
Degrees of Unsaturation.**

PRACTICE PROBLEM 2.2

Calculate the DU for $C_8H_{13}BrO$.

Solution

Br counts as an H and O can be ignored, so calculate as though the formula were C_8H_{14}. DU $= \frac{1}{2}[(2)(8) + 2 - 14] = 2$.

PROBLEM 2.9

Calculate the DU for these formulas and draw two constitutional isomers for each:

a) $C_{10}H_{20}O$ **b)** C_6H_9N **c)** $C_7H_{14}F_2$
d) C_6H_8ClN **e)** $C_9H_{15}NO$

PROBLEM 2.10

Determine the DU for each of these structures:

a)

b)

c)

d)

Tryptophan (an amino acid)

e)

Acetylsalicylic acid
(aspirin)

BioLink

2.5   PHYSICAL PROPERTIES AND MOLECULAR STRUCTURE

The **physical properties** of a compound are properties such as its melting point, boiling point, or solubility in a particular solvent. These properties all depend on the strength of the attractions between the molecules. The intermolecular attractive forces result from a charge or partial charge on one molecule interacting with a charge or partial charge on another molecule. The attraction between a positive and a negative center becomes stronger as the magnitude of the charges increases and as the distance between them decreases. The various types of charge–charge interactions that are important are shown in Figure 2.7.

Ion–ion interactions are the strongest of these because they involve the most charge. For example, the strength of the forces holding sodium and chloride ions to-

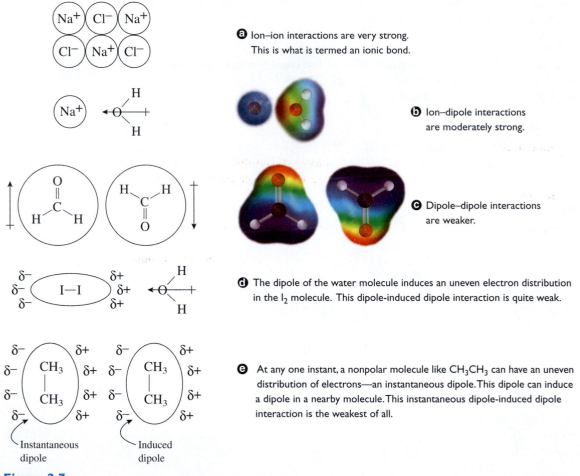

a Ion–ion interactions are very strong. This is what is termed an ionic bond.

b Ion–dipole interactions are moderately strong.

c Dipole–dipole interactions are weaker.

d The dipole of the water molecule induces an uneven electron distribution in the I_2 molecule. This dipole-induced dipole interaction is quite weak.

e At any one instant, a nonpolar molecule like CH_3CH_3 can have an uneven distribution of electrons—an instantaneous dipole. This dipole can induce a dipole in a nearby molecule. This instantaneous dipole-induced dipole interaction is the weakest of all.

Figure 2.7

TYPES OF CHARGE–CHARGE INTERACTIONS.

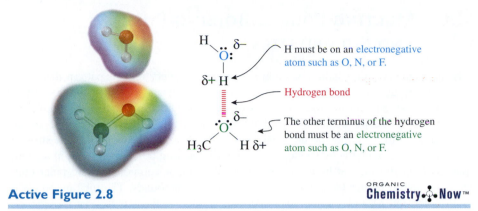

H must be on an electronegative
atom such as O, N, or F.

Hydrogen bond

The other terminus of the hydrogen
bond must be an electronegative
atom such as O, N, or F.

Active Figure 2.8

ORGANIC
Chemistry Now™

AN EXAMPLE OF HYDROGEN BONDING. Test yourself on the concepts in this figure at
OrganicChemistryNow.

gether in the crystal lattice is 188 kcal/mol (788 kJ/mol). (Compare this to a typical covalent bond strength of about 100 kcal/mol [418 kJ/mol].) Because a large amount of
heat energy must be added to disrupt these forces, ionic compounds typically have high
melting and boiling points. For example, NaCl melts at 801°C, whereas CH_3Cl melts at
−98°C.

Because of the lesser amount of charge involved when an ion interacts with a polar
molecule, the magnitude of the resulting **ion–dipole** attraction is much less than that of
an ion–ion attraction. The strength of the attraction continues to decrease as the amount
of charge decreases. Therefore, the attraction of polar molecules for each other
(**dipole–dipole**) is weaker than that of an ion for a polar molecule. The force attracting
a polar molecule to a nonpolar molecule (**dipole-induced dipole**) is still weaker. The
attraction between nonpolar molecules (**instantaneous dipole-induced dipole** or **London force**) is the weakest of all, but the total London force in a large molecule can be
reasonably large and can have an important effect on the physical properties of the
compound. These last three interactions are often called **van der Waals forces** and
range from approximately 5 to 0.5 kcal/mol (20–2 kJ/mol).

Another very important type of charge–charge interaction is **hydrogen bonding.**
This interaction, illustrated in Figure 2.8, is a special type of dipole–dipole attraction.
In hydrogen bonding, a hydrogen on an electronegative atom (O, N, or F) is the positive
part of a dipole and is attracted to the negative end of a dipole (O, N, or F) in another
molecule. Because of the small size of hydrogen, the partial charges are able to approach more closely, and a stronger than usual attraction—a hydrogen bond—results.
The strength of a hydrogen bond is commonly in the range of 3 to 8 kcal/mol
(12–34 kJ/mol). Because C—H bonds are not very polar, these H's do not usually hydrogen bond to any appreciable extent.

PROBLEM 2.11
What kinds of attractive intermolecular forces are found in each of these compounds?
a) $CH_3CH_2CH_3$ **b)** CH_3OCH_3 **c)** $CaCl_2$ **d)** CH_3OH

PROBLEM 2.12
Show the hydrogen bond that is present in liquid ammonia, $NH_3(l)$.

2.6   MELTING POINTS, BOILING POINTS, AND SOLUBILITIES

The molecules of a crystalline solid are arranged in a very regular pattern that maximizes the attractive forces among the molecules. To cause a solid to melt, heat energy must be supplied, causing the vibrations of the molecules to increase. The attractive forces among the molecules are partially overcome, allowing them to move around more freely, although they do not separate very far from their neighbors. The stronger the forces are among the molecules, the higher the melting point. Therefore, ionic compounds typically melt at higher temperatures than polar compounds, and polar compounds melt at higher temperatures than nonpolar compounds. The melting points of both polar and nonpolar compounds increase as their molecular masses increase because the larger molecules have more London forces between them. Finally, the shape of a compound is very important and has a dramatic effect on its melting point. More symmetrical molecules pack into the crystal lattice better, allowing closer approach and larger attractive forces, resulting in higher melting points. Some examples of these effects are shown in Table 2.4.

The compound known as cubane, one of the isomers of the formula C_8H_8, has its eight carbons arranged at the vertices of a cube. It is the ultimate example of the effect of shape on melting point. Octane, a straight-chain saturated hydrocarbon with the formula C_8H_{18}, has a melting point of $-57°C$. Because of its symmetrical shape, cubane packs much better into the crystal lattice. The melting point of cubane is 131°C.

Table 2.4   Examples of the Dependence of Melting Points on Structure

Compound	Melting Point (°C)	Comment
NaCl	801°	Ionic compounds have high melting points.
LiF	842°	Smaller ions are closer together.
$CH_3CH_2CH_3$	−190°	Small (low molecular mass) nonpolar compounds have low melting points.
$CH_3CH_2CH_2CH_2CH_3$ Pentane	−130°	Melting points increase with increasing molecular mass.
$CH_3\overset{\displaystyle CH_3}{\underset{\displaystyle CH_3}{C}}CH_3$ 2,2-Dimethylpropane	−71°	This isomer of pentane has a more compact spherical shape, which allows it to pack better into the crystal lattice, resulting in a significantly higher melting point.
$CH_3CH_2CH_2\overset{\displaystyle O}{\overset{\|}{C}}H$ Butanal	−99°	Butanal has the same molecular mass and same general shape as pentane. It is higher melting than pentane because the dipole–dipole forces of the polar C=O group hold the molecules together more strongly.

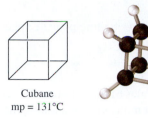

Cubane
mp = 131°C

$CH_3CH_2CH_2CH_2CH_2CH_2CH_2CH_3$

Octane
mp = −57°C

PROBLEM 2.13
Explain whether you would expect KBr or CH_3Br to have the higher melting point.

In the liquid phase, molecules are free to move about, although they are always close to other molecules and interacting with them. To cause a molecule to enter the vapor phase, enough energy must be added to entirely overcome the forces attracting it to other molecules. Therefore, as is the case with melting points, boiling points increase as the forces between the molecules become stronger. However, the effect of the shape of the molecule on its boiling point is quite different from the effect on its melting point. Because liquids have little order, symmetrical compounds do not have higher boiling points. In fact, longer, rod-shaped molecules have more surface area than do spherical molecules of similar molecular mass, so such rod-shaped molecules have slightly higher boiling points owing to increased London forces. In addition, the presence of hydrogen bonding increases boiling points much more than melting points. Some examples of these effects are shown in Table 2.5.

Table 2.5   Examples of the Dependence of Boiling Points on Structure

Compound	Normal Boiling Point (°C)	Comment
NaCl	1413°	Ionic compounds have high boiling points.
LiF	1676°	Smaller ions are closer together.
$CH_3CH_2CH_3$	−42°	Small nonpolar compounds have low boiling points.
$CH_3CH_2CH_2CH_2CH_3$ Pentane	36°	Boiling points increase with increasing molecular mass.
CH_3CCH_3 with CH_3 above and CH_3 below 2,2-Dimethylpropane	10°	Although this isomer is more symmetrical and has a higher melting point, the rod shape of pentane has more surface area, resulting in a higher boiling point due to increased London forces.
$CH_3CH_2CH_2CH$ with O double-bonded Butanal	76°	Butanal has the same molecular mass as pentane. It is higher boiling than pentane because of the polar C=O group.
$CH_3CH_2CH_2CH_2OH$ 1-Butanol	117°	1-Butanol boils significantly higher than the compound above because its molecules can hydrogen bond to each other.

PROBLEM 2.14

Which of these isomers would you expect to have the higher boiling point? Explain.

$$CH_3CH_2CH_2OH \quad \text{or} \quad CH_3CH_2OCH_3$$

When a compound dissolves in a solvent, the molecules of that compound become separated. These molecules then mix among the solvent molecules, separating the latter molecules also. Therefore, energy must be supplied to overcome both the forces holding the solute molecules together and the forces holding the solvent molecules together. Some energy is returned from the interactions between solute and solvent molecules. Usually, the total amount of energy that must be added to overcome the attractive forces among the solute molecules and the attractive forces among the solvent molecules is somewhat greater than the amount of energy that is released by the interactions of the solute with the solvent molecules. Overall, energy must be added to the system—that is, the process is **endothermic:** the enthalpy change is positive ($\Delta H > 0$).

Why, then, does a solution ever form? You may recall from general chemistry that whether a reaction is spontaneous or not depends on both the enthalpy and the entropy of that process. **Entropy** (S) is a measure of disorder. Processes that increase the disorder in a system ($\Delta S > 0$) are favored. Although most solution processes are endothermic (disfavored by enthalpy, $\Delta H > 0$), the solution is more disordered than the separate solute and solvent (favored by entropy, $\Delta S > 0$). Therefore, as long as the process is not too endothermic, the favorable entropy change will cause the solute to dissolve. Whether the process is likely to be too endothermic can be estimated by examining the

---

Focus On

Boiling Points of Fuels

Charles D. Winters

Some camping stoves use butane as a fuel.

The hydrocarbons methane, propane, and butane are all used as fuels in various applications. The boiling points of these compounds increase with increasing molecular mass and help determine the use for each fuel. Methane has a boiling point of $-162°C$ and is the major component of natural gas. Because of its low boiling point, it is difficult to liquefy. Propane boils at $-42°C$ and is more easily liquefied than natural gas; that is, it is a liquid at lower pressures than natural gas. Pressurized tanks containing liquid propane are used as fuel sources for home heating in areas that are not served by natural gas pipelines and for such uses as portable barbecue grills. Butane boils at $-0.5°C$ and is a liquid at pressures only slightly above 1 atmosphere. Plastic containers of liquid butane are used in cigarette lighters. These lighters do not work well if they are allowed to become very cold because then the butane will not vaporize.

CH_4	$CH_3CH_2CH_3$	$CH_3CH_2CH_2CH_3$
Methane	Propane	Butane
bp = $-162°C$	bp = $-42°C$	bp = $-0.5°C$

solute and the solvent. If either the attractive force between solute molecules or that be-tween solvent molecules is considerably larger than the attractive force between a solute and a solvent molecule, the process will be too endothermic and the solution will not form. In other words, the new interactions in the solution must be comparable in strength to the old interactions in the solute and in the solvent. This means that the solute and the solvent must be similar so that the forces between them are similar in magnitude to the forces just in the solute or in the solvent. A common way of stating this is "like dissolves like," which simply means that the polarities of the solute and the solvent must be similar.

Sodium chloride dissolves in water, a very polar solvent. Although the interactions between the water molecules and the sodium and chloride ions are not as strong as the interactions between the ions themselves, they are potent enough to make the process only somewhat endothermic. Entropy then makes the overall process favorable. Sodium chloride does not dissolve in a nonpolar solvent such as pentane ($CH_3CH_2CH_2CH_2CH_3$) because the interactions between the pentane molecules and the ions are too weak in comparison to the forces between the ions, making the overall process too endothermic for entropy to overcome.

If pentane and water are mixed, they form separate layers—they do not dissolve—because the attractive forces between pentane and water molecules are much smaller than the forces between water molecules. Pentane is termed a **hydrophobic** (water-hating) compound. Ethanol, CH_3CH_2OH, has a nonpolar, hydrophobic part (the CH_3CH_2) and a polar, **hydrophilic** (water-loving) part (the OH). The effects of the two parts are in competition, but in this case the hydrophilic part wins, and ethanol is **miscible** (mixes in all proportions) with water. As we might expect, as the hydrophobic part of the molecule becomes larger, the compound becomes less water soluble. Thus, butanol, $CH_3CH_2CH_2CH_2OH$, is only slightly soluble in water (7.4 g/100 mL), and hexanol, $CH_3CH_2CH_2CH_2CH_2CH_2OH$, is even less soluble (0.6 g/100 mL). Many organic compounds have a large hydrophobic part and are essentially insoluble in water. However, they are often soluble in less polar solvents such as ethanol, diethyl ether ($CH_3CH_2OCH_2CH_3$), and dichloromethane (CH_2Cl_2).

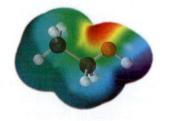

Ethanol, CH_3CH_2OH

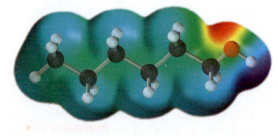

Hexanol, $CH_3CH_2CH_2CH_2CH_2CH_2OH$

PROBLEM 2.15

Which of these compounds would you expect to be more soluble in water? Explain.

$$CH_3CH_2CH_2CH_2\overset{\displaystyle O}{\overset{\displaystyle \|}{C}}OH \quad \text{or} \quad CH_3CH_2CH_2CH_2CH_2\overset{\displaystyle O}{\overset{\displaystyle \|}{C}}OH$$

2.7   INTRODUCTION TO FUNCTIONAL GROUPS

As we have seen, the number of possible organic compounds is virtually limitless. How can anyone learn the chemistry of all of them? Fortunately, we do not need to learn an entire new set of chemical reactions for each new compound encountered. A particular arrangement or group of atoms has very similar chemistry no matter what the remainder of the molecule looks like.

Let's consider the two compounds $CH_3CH_2CH_2OH$ and $CH_3CH_2CH_2CH_2OH$. They are quite similar, and on the basis of the material presented in Section 2.6, they should have similar physical properties. The polarity and hydrogen bonding of the OH group cause both of them to have higher boiling points than similar compounds containing only C and H. Because the second one has a greater molecular mass, it is expected to have a somewhat higher boiling point. The OH group should also confer a degree of water solubility on each. Because the second has a larger nonpolar part, it should be less water soluble than the first. Indeed, these expectations are borne out. It is probably not surprising to discover that these two compounds also undergo very similar chemical reactions. So if we learn the reactions of one, we will know much about the reactions of the other.

As we learn more about organic chemistry, we will find that the attraction between positive and negative charges, or partial charges, is instrumental in many chemical reactions. Often a reaction begins by the reagent being attracted to the polar part of a molecule. Therefore, the reactions of the molecules on the previous page occur at the polar sites: the O—H and C—O bonds. The C—C single bonds and the C—H bonds are not very reactive. In fact, in terms of chemical reactions, it makes little difference how many C—C or C—H bonds the molecule has or exactly how they are arranged. The reactions occur at the OH group. Thus, all of the following compounds exhibit very similar chemical behavior!

$$CH_3CH_2CH_2{-}OH \qquad CH_3CH_2\overset{\displaystyle OH}{\underset{|}{C}}HCH_3$$

It is useful to view a molecule as being composed of two parts. One part only has C's that are singly bonded to other C's (and their associated H's). This group of atoms is called the **alkyl group,** the **carbon skeleton,** or the **carbon framework** and, as we saw earlier, has little effect on the chemical reactions. The other part of the molecule, where the action is, is called the **functional group.** All of the preceding compounds have the same functional group, the OH group, but different alkyl groups. The class of compounds with the OH functional group is called alcohols. *A general alkyl group is represented by R, so all alcohols can be depicted as ROH.* All alcohols have similar chemical reactions.

Important ‖⟫
Convention

The concept of functional groups is a very powerful organizing feature for organic chemistry. All compounds with the same functional group have very similar reactions, regardless of their carbon skeleton. Only a limited number of functional groups are of major importance. These are shown in Table 2.6, along with the class name of each group. *It is of utmost importance to learn these functional groups. Whenever you see an organic compound, you need to focus on and identify the functional group because that*

Table 2.6 Common Functional Groups

General Formula	Example	Name
RH	$CH_3CH_2CH_3$	**Alkane** (Alkanes have no functional group.)
$RCH{=}CH_2$	$CH_3CH_2CH{=}CH_2$	**Alkene** (The functional group is the carbon–carbon double bond.)
$RC{\equiv}CH$	$CH_3C{\equiv}CH$	**Alkyne** (The functional group is the carbon–carbon triple bond.)
ArH		**Arene** (A six-membered ring with three double bonds has different reactions than an alkene so it is given a different name. Arenes or **aromatic rings** can have alkyl or other groups attached to the ring.)
RX	$CH_3CH_2CH_2Cl$	**Alkyl halide** (The functional group is the carbon–halogen bond.)
ROH	CH_3CH_2OH	**Alcohol** (The functional group is the C—O—H.)
ROR	$CH_3CH_2OCH_3$	**Ether** (The functional group is the C—O—C. The alkyl groups on the O can be the same or different.)
RNH_2	$CH_3CH_2CH_2NH_2$	**Amine** (The C—N is the functional group. The other H's on the N can be replaced with alkyl groups.)
$\overset{\displaystyle O}{\overset{\|}{R}}CH$	$\overset{\displaystyle O}{\overset{\|}{CH_3CH_2}}CH$	**Aldehyde** (The functional group is the C=O with at least one H on the C.)
$\overset{\displaystyle O}{\overset{\|}{R}}CR$	$\overset{\displaystyle O}{\overset{\|}{CH_3}}CCH_3$	**Ketone** (The functional group is the C=O with two alkyl groups on the C. The alkyl groups need not be the same.)
$\overset{\displaystyle O}{\overset{\|}{R}}COH$	$\overset{\displaystyle O}{\overset{\|}{CH_3CH_2}}COH$	**Carboxylic acid** (The functional group is the $\overset{\displaystyle O}{\overset{\|}{}}COH$.)
$\overset{\displaystyle O}{\overset{\|}{R}}CCl$	$\overset{\displaystyle O}{\overset{\|}{CH_3}}CCl$	**Acyl chloride** (The functional group is the $\overset{\displaystyle O}{\overset{\|}{}}CCl$.)
$\overset{\displaystyle O\ \ O}{\overset{\|\ \ \ \|}{R}}COCR$	$\overset{\displaystyle O\ \ O}{\overset{\|\ \ \ \|}{CH_3}}COCCH_3$	**Acid anhydride** (The functional group is the $\overset{\displaystyle O\ \ O}{\overset{\|\ \ \ \|}{}}COC$. The alkyl groups may be different.)
$\overset{\displaystyle O}{\overset{\|}{R}}COR$	$\overset{\displaystyle O}{\overset{\|}{CH_3}}COCH_2CH_3$	**Ester** (The functional group is the $\overset{\displaystyle O}{\overset{\|}{}}COC$.)
$\overset{\displaystyle O}{\overset{\|}{R}}CNH_2$	$\overset{\displaystyle O}{\overset{\|}{CH_3CH_2}}CNHCH_3$	**Amide** (The functional group is the $\overset{\displaystyle O}{\overset{\|}{}}CN$. The other groups on the N may be H's or alkyl groups.)
$RC{\equiv}N$	$CH_3CH_2CH_2C{\equiv}N$	**Nitrile** (The functional group is the carbon–nitrogen triple bond.)

is where the reactions will occur—where the action is. And if you know the reactions of that functional group, you will know much about the reactions of that compound.

More complex compounds can have more than one functional group. For example, alanine, an important amino acid in nature, has both an amine and a carboxylic acid functional group.

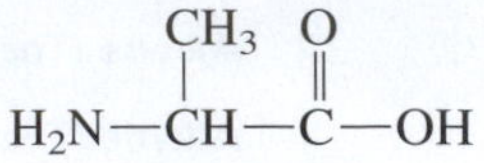

$$
\begin{array}{cc}
CH_3 & O \\
| & \| \\
H_2N-CH-C-OH
\end{array}
$$

Alanine
(an amino acid)

In the next chapter we will reexamine bonding, using orbitals. This will provide a better picture of why some bonds are stable and others are not.

ORGANIC
Chemistry⚛Now™
Click *Coached Tutorial Problems* for more practice in
Identifying Functional Groups.

PROBLEM 2.16

Which functional group is present in each of these compounds?

a) $CH_3CH_2CH_2CH_2OCH_3$

b)

c) $CH_3CH_2CH_2CO_2H$

Careful: how are the O's bonded to the C?

d)

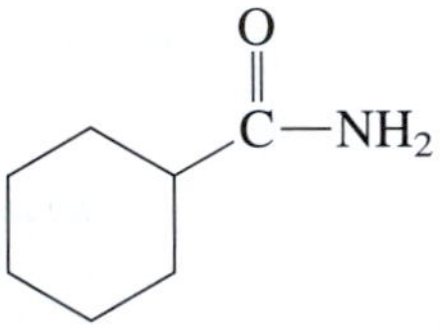

e) 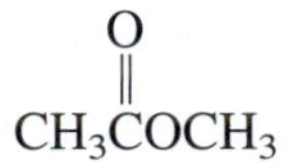 CH_3COCH_3

f)

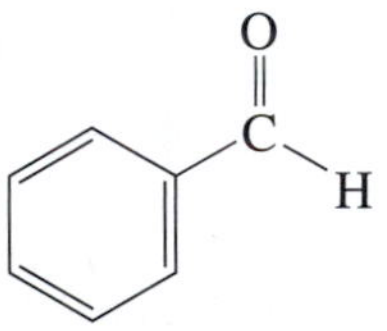

ORGANIC
Chemistry⚛Now™
Click *Mastery Goal Quiz* to test how well you have met these goals.

Review of Mastery Goals

After completing this chapter, you should be able to:

■ Quickly recognize the common ways in which atoms are bonded in organic compounds. You should also recognize unusual bonding situations and be able to estimate the stability of molecules with such bonds.

■ Know the trends in bond strengths and bond lengths for the common bonds. (Problem 2.37)

■ Recognize when compounds are constitutional isomers and be able to draw constitutional isomers for any formula. (Problems 2.21, 2.22, 2.23, 2.24, 2.39, and 2.40)

■ Calculate the degree of unsaturation for a formula and use it to help draw structures for that formula. (Problems 2.25, 2.26, 2.40, 2.45, and 2.46)

■ Draw structures using any of the methods we have seen. You should also be able to examine a shorthand representation for a molecule and recognize all of its features. (Problems 2.17 and 2.18)

■ Examine the structure of a compound and determine the various types of intermolecular forces that are operating. You should be able to crudely estimate the physical properties of the compound. (Problems 2.27, 2.28, 2.29, 2.30, 2.31, 2.32, 2.33, 2.34, 2.35, 2.36, 2.43, and 2.44)

■ Recognize and name all of the important functional groups. (Problems 2.19, 2.20, 2.42, 2.43, and 2.46)

Additional Problems

ORGANIC
Chemistry Now™
Assess your understanding of this chapter's topics with additional quizzing and conceptual-based problems at
http://now.brookscole.com/hornback2

2.17 Convert the following structures to skeletal structures:

a) 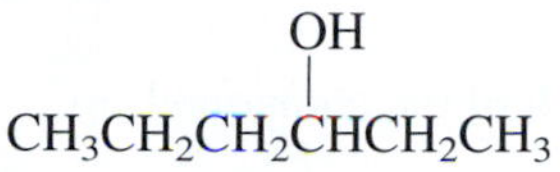 CH₃CH₂CH₂CHCH₂CH₃ with OH on the CH

b) 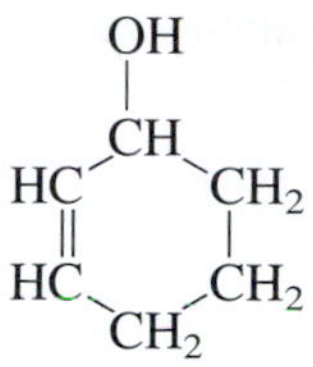
```
      OH
      |
      CH
HC        CH₂
‖         |
HC        CH₂
   CH₂
```

c) 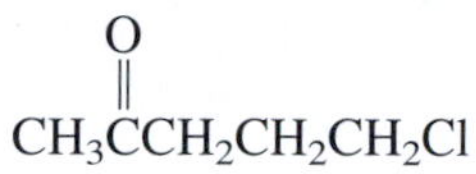
```
        O
        ‖
CH₃CCH₂CH₂CH₂Cl
```

d) 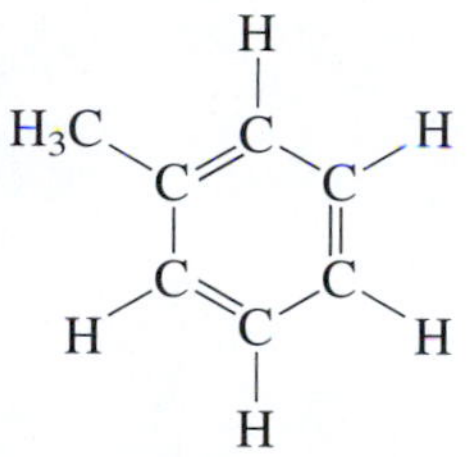
```
        H
        |
H₃C     C      H
   C         C
   ‖         ‖
   C         C
   |    C    |
   H    |    H
        H
```

e) 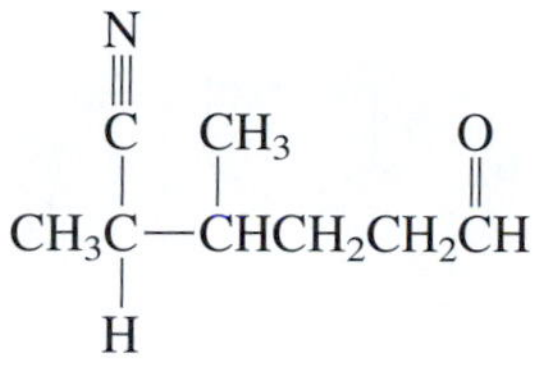
```
        N
        ‖‖‖
        C    CH₃      O
        |    |        ‖
CH₃C—CHCH₂CH₂CH
        |
        H
```

f) 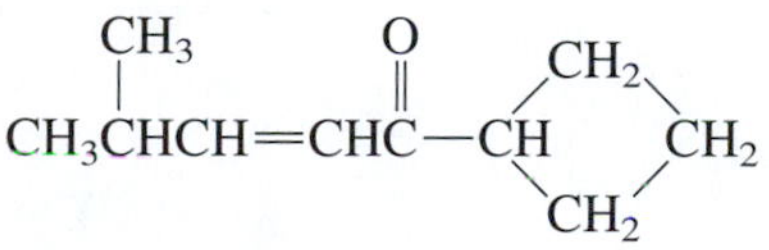
```
      CH₃        O
      |          ‖    CH₂
CH₃CHCH=CHC—CH       CH₂
                  CH₂
```

g) 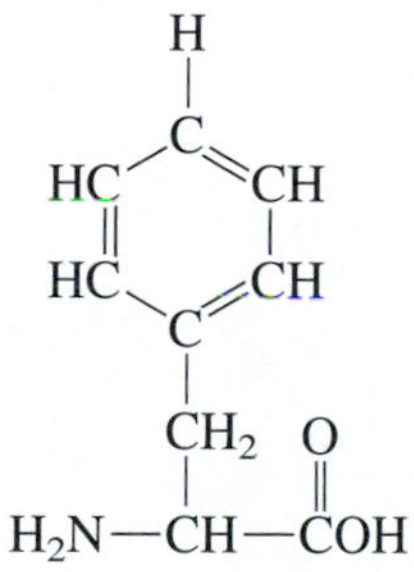
```
            H
            |
            C
HC              CH
HC              CH
            C
            |
            CH₂   O
            |     ‖
   H₂N—CH—COH
```
Phenylalanine
(an amino acid)

BioLink

2.18 Convert the following shorthand representations to structures showing all of the atoms, bonds, and unshared electron pairs:

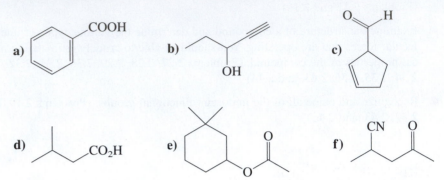

2.19 Name the functional group(s) present in each of the compounds in problem 2.17.

2.20 Name the functional group(s) present in each of the compounds in problem 2.18.

2.21 Determine whether these structures represent the same compound or isomers:

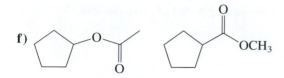

f)

2.22 Draw all the isomers for each of these formulas. The total number for each is given in parentheses.
  a) C_3H_8O (3)   **b)** C_4H_9Cl (4)   **c)** C_4H_8 (5)
  d) C_7H_{16} (9)

2.23 The formula C_4H_8O has many isomers.
  a) Draw three isomers that have a carbon–oxygen double bond. What functional group is present in each of them?
  b) Draw three alcohols with this formula.
  c) Draw an ether with this formula that does not have a carbon–carbon double bond.

2.24 Four of the ten isomers of C_5H_{10} are shown in Figure 2.5. Draw four other isomers with this formula.

2.25 Calculate the DU for each of these formulas and draw a structure that meets the listed restriction:
  a) C_5H_9NO (not an amine)
  b) $C_7H_{12}O$ (a ketone)
  c) $C_6H_{14}O$ (does not hydrogen bond)
  d) C_7H_8 (an aromatic compound)

2.26 Can $C_8H_{17}N$ have a nitrile as its functional group? Explain.

2.27 Show all the different hydrogen bonds that would occur in a mixture of
  a) CH_3CH_2OH and H_2O
  b) CH_3OH and

2.28 One of these isomeric alcohols has mp = 26°C and bp = 82°C; the other has mp = −90°C and bp = 117°C. Explain which isomer has the higher melting point and which has the higher boiling point.

2.29 Explain the differences in the boiling points between the members of each of these pairs of compounds:
  a) $CH_3(CH_2)_6CH_3$ bp: 126°C
      $CH_3(CH_2)_8CH_3$ bp: 174°C
  b) $CH_3CH_2CH_2OH$ bp: 97°C
      $CH_3CH_2OCH_3$ bp: 11°C
  c) $CH_3CH_2CH_3$ bp: −42°C
      CH_3OCH_3 bp: −23°C

2.30 Explain the difference in the melting points of these isomers:

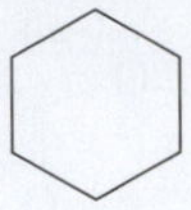

mp = −140°C   mp = 7°C

2.31 Explain which compound you expect to have the higher melting point.

a)

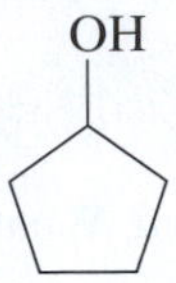

or

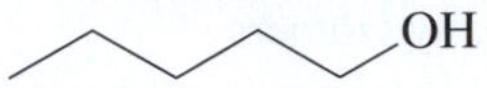

b) $CH_3(CH_2)_{18}CH_3$ or $CH_3(CH_2)_{29}CH_3$

Beeswax

c)

or

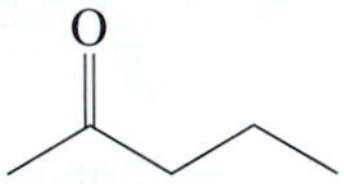

2.32 Explain which compound you expect to have the higher boiling point.

a)

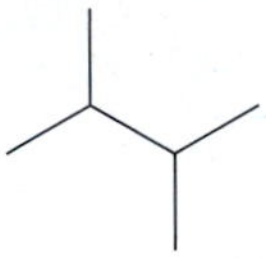

or

b) $CH_3CH_2\overset{\overset{\textstyle O}{\|}}{C}OH$ or $CH_3CH_2CH_2\overset{\overset{\textstyle O}{\|}}{C}H$

c) $CH_3CH_2OCH_2CH_3$ or $CH_3CH_2CH_2OCH_2CH_2CH_3$

d)

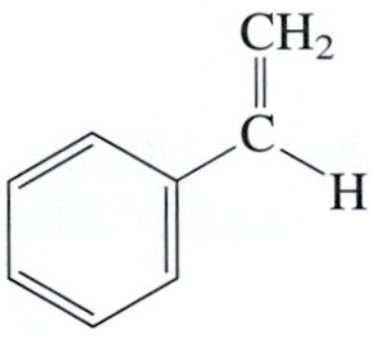

or

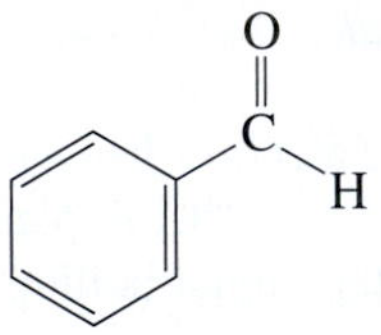

2.33 Which of these two salts would you expect to be more soluble in hexane (C_6H_{14})?

$\overset{+}{NH_4}\ \overset{-}{Cl}$ or $\overset{+}{N}(CH_2CH_2CH_2CH_3)_4\ \overset{-}{Cl}$

2.34 Benzene and hexane are both liquids at room temperature. Do you expect benzene and hexane to be miscible? Do you expect benzene and water to be miscible? Explain.

Hexane    Benzene

2.35 One of these isomers is miscible with water, and the other is nearly insoluble. Explain.

$$CH_3CH_2CH_2\overset{\overset{\displaystyle O}{\|}}{C}OH \qquad CH_3\overset{\overset{\displaystyle O}{\|}}{C}OCH_2CH_3$$

2.36 Because of two hydrogen bonds, carboxylic acids show a very strong attractive force between two molecules that persists even in the gas phase. Show this hydrogen bonding between two carboxylic acid molecules.

2.37 Bond strengths can be used to estimate the relative stability of isomers that have different bonds. The isomer that has the larger total bond energy is more stable. One of the following isomers is more stable than the other. The less stable one is rapidly converted to the more stable one, so it cannot be isolated. On the basis of bond dissociation energies, which of these two isomers is more stable?

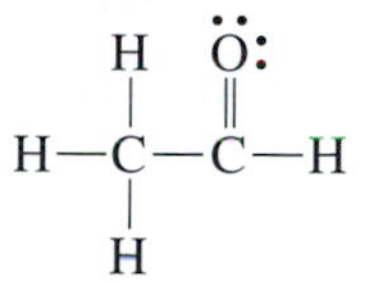

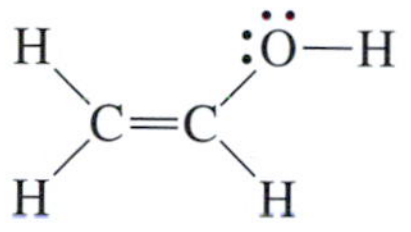

2.38 Bond strengths can also be used to estimate whether a reaction is energetically favorable or not—that is, whether the reactants or the products are more stable. Use bond dissociation energies to determine whether this reaction is energetically favorable. The bond dissociation energy for H_2 is 104 kcal/mol (435 kJ/mol).

$$CH_2{=}CH_2 + H{-}H \longrightarrow CH_3{-}CH_3$$

2.39 One of the isomers of C_5H_{12} reacts with Cl_2 in the presence of light to produce three isomers of $C_5H_{11}Cl$:

$$C_5H_{12} + Cl_2 \xrightarrow{\text{light}} C_5H_{11}Cl + HCl$$

One isomer   Three isomers

This reaction replaces *any one* of the hydrogens of C_5H_{12} with a Cl. What are the structures of the C_5H_{12} isomer and the three $C_5H_{11}Cl$ isomers produced from it?

2.40 On reaction with Cl_2 in the presence of light, an unknown compound with the formula C_5H_{10} gives only one isomer of C_5H_9Cl (see problem 2.39). What is the DU of the unknown compound? Show the structure of the unknown compound and the product of its reaction with Cl_2.

$$C_5H_{10} \; + \; Cl_2 \quad \xrightarrow{\text{light}} \quad C_5H_9Cl \; + \; H\!-\!Cl$$

One isomer

2.41 Explain how the dipole moment for CH_3Cl ($\mu = 1.9$ D) can be larger than the dipole moment for CH_3F ($\mu = 1.8$ D).

2.42 Peptides are smaller versions of proteins and have similar functional groups. What functional groups are present in this peptide?

OH
|
CH₂   O   CH₃   O   O
       ‖   ‖   ‖
H₂N—CH—C—NH—CH—C—NH—CH₂—C—OH

Serine-alanine-glycine

2.43 Glucose is a typical carbohydrate. What functional groups are present in glucose? What would you predict about the water solubility of glucose? Offer a reason why nature uses carbohydrates rather than alkanes as an energy source.

OH   OH   OH   OH   OH   O
| | | | | ‖
CH₂—CH—CH—CH—CH—CH

Glucose

2.44 Amino acids such as alanine actually exist as species called zwitterions, with a positive charge on the nitrogen and a negative charge on the oxygen. Explain what effect you expect this to have on the melting point of alanine.

H   CH₃   O
| | ‖
H—N⁺—CH—C—O⁻
|
H

Alanine zwitterion

2.45 What is the DU of estrone, a female sex hormone? Use the number of carbons and oxygens in estrone to calculate the number of hydrogens it has.

Estrone

2.46 What functional groups are present in these molecules? What is the DU of each?

 **BioLink**

a)

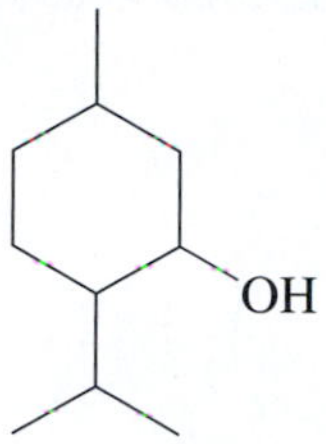

Menthol
(an ingredient of peppermint oil)

b)

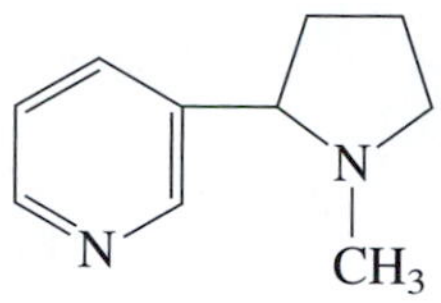

Nicotine
(the addictive substance in tobacco)

c)

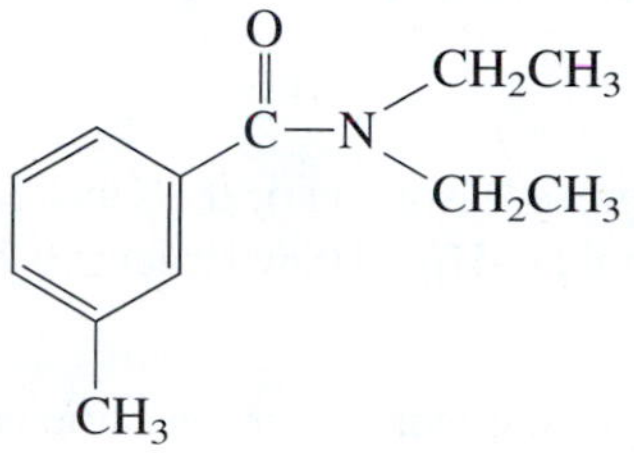

N,N-Diethyl-*m*-toluamide
(DEET, a mosquito repellant)

d)

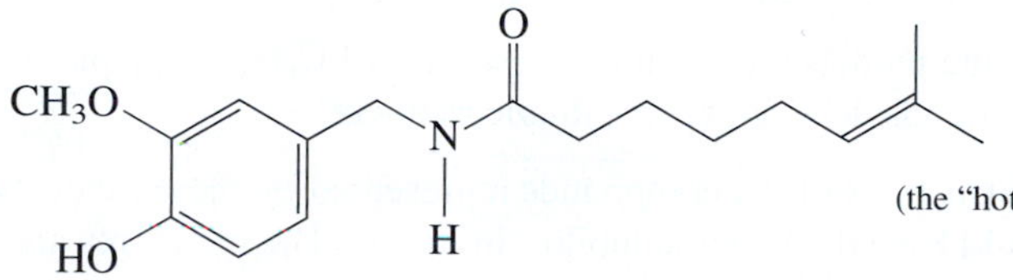

Capsaicin
(the "hot" ingredient in chili peppers)

e)

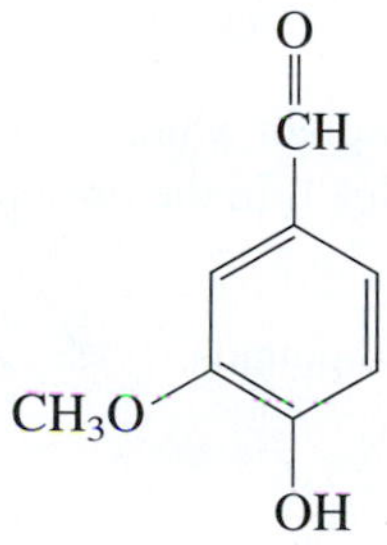

Vanillin
(major component of vanilla flavoring)

f)

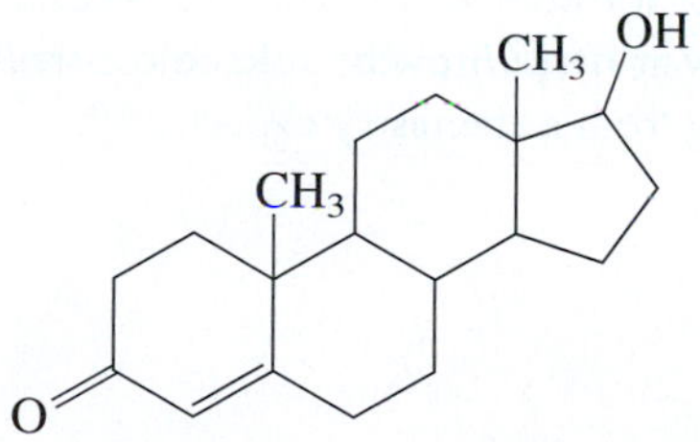

Testosterone
(a male sex hormone)

g) CH$_3$CHCH$_2$— ⟨aromatic ring⟩ —CHCO$_2$H

Ibuprofen
(active ingredient in Advil and Motrin)

h)

Acetaminophen
(active ingredient in Tylenol)

Problems Using Online Three-Dimensional Molecular Models

2.47 Explain whether each pair of models represent isomers or the same compound. (All represent compounds with the formula C_7H_{16}.) Draw structures for each compound represented by the models.

2.48 Explain whether each pair of models represent isomers or the same compound. Draw structures for each compound represented by the models.

2.49 The following models represent three isomers of $C_6H_4Cl_2$. Explain which of these compounds does not have a dipole moment.

2.50 Draw the structures of the compounds represented by these models and explain which would have the larger solubility in water. (The red atoms are oxygens.)

2.51 Draw the structures of the compounds represented by these models and explain which would have the higher melting point.

2.52 These models show the three isomers of C_5H_{12}. Explain which isomer would produce only one $C_5H_{11}Cl$ isomer on reaction with Cl_2 in the presence of light. (See problem 2.39.)

2.53 Determine the degree of unsaturation for these compounds.

Orbitals and Bonding

MASTERING ORGANIC CHEMISTRY

▶ Assigning Electrons to Atomic Orbitals

▶ Recognizing Sigma and Pi Bonds

▶ Using Simple MO Theory to Construct Bonding and Antibonding Molecular Orbitals

▶ Identifying the Hybridization of Atoms in Molecules

▶ Identifying the Atomic Orbitals that Overlap to Form Each Molecular Orbital

▶ Understanding and Using Resonance

▶ Understanding Molecular Orbital Energies

CHAPTER 1 DESCRIBED a covalent bond as a pair of electrons that is shared between two atoms. The purpose of this chapter is to reexamine bonding in more depth using a model that employs electron orbitals. This model will provide us with a better understanding of bonds and reactivity. The chapter begins with a review of atomic orbitals. Then a model where bonding results from atomic orbitals interacting to form molecular orbitals is discussed. Because resonance is so important in organic chemistry, considerable attention is devoted to this topic. The idea of orbitals can help us understand resonance better. Finally, a number of examples of how to use resonance and when it is important are presented.

Most of the subsequent chapters in this book use simple line structures or skeletal structures to represent molecules, as was done in Chapters 1 and 2. However, you should always have in mind the model that is presented in this chapter so that you can call on it whenever a better picture for bonding is needed to explain an observation.

Much of what is discussed in this chapter should be a review of material that you learned in general chemistry. However, the presentation here concentrates more on the aspects of bonding that are important in organic chemistry.

3.1 ATOMIC ORBITALS

You should recall from your general chemistry course that electrons have some of the properties of waves. Chemists use the equations of wave mechanics to describe these electron waves. Solving these wave equations for an electron moving around the nucleus of an atom gives solutions that lead to a series of **atomic**

ORGANIC
Chemistry⚛Now™
Look for this logo in the chapter and go to OrganicChemistryNow at http://now.brookscole.com/hornback2 for tutorials, simulations, problems, and molecular models.

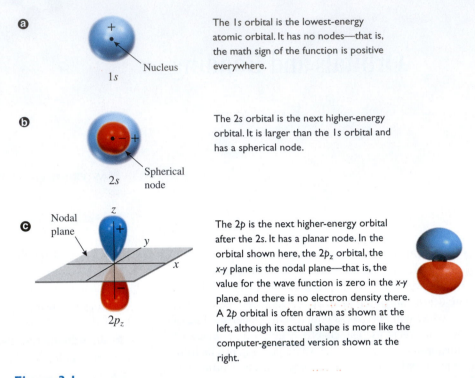

a

The 1s orbital is the lowest-energy atomic orbital. It has no nodes—that is, the math sign of the function is positive everywhere.

1s   Nucleus

b

The 2s orbital is the next higher-energy orbital. It is larger than the 1s orbital and has a spherical node.

2s   Spherical node

c

Nodal plane

The 2p is the next higher-energy orbital after the 2s. It has a planar node. In the orbital shown here, the $2p_z$ orbital, the x-y plane is the nodal plane—that is, the value for the wave function is zero in the x-y plane, and there is no electron density there. A 2p orbital is often drawn as shown at the left, although its actual shape is more like the computer-generated version shown at the right.

$2p_z$

Figure 3.1

SHAPES OF ATOMIC ORBITALS: **a** 1s, **b** 2s, **c** 2p.

orbitals. These orbitals describe the location of the electron charge density when an electron occupies that orbital. Another way of stating this is that the shape of the orbital defines a region about the nucleus where, if the orbital contains an electron, the probability of finding that electron is very high. It is simply the space where the electron spends most of its time.

Figure 3.1 shows the shapes of the orbitals that will be of most concern here. The first orbital pictured is the lowest-energy orbital, the **1s orbital.** It has a spherical shape, as do all s orbitals. Also shown in the figure is the mathematical sign (plus) of the orbital. *This sign does not relate to charge.* It simply indicates that the wave function (the solution to the wave equation) has a positive mathematical value throughout the entire orbital. These math signs are needed when atomic orbitals are combined to make molecular orbitals.

In order of increasing energy the next orbital is the **2s orbital.** Like the 1s orbital, it has a spherical shape. It is larger than the 1s orbital—that is, it extends farther out from the nucleus. Interestingly, the 2s orbital has a region where the mathematical sign of the wave function is positive and another region where the sign is negative. It also has a region where the value of the wave function equals zero. Such a region (in this case a spherical surface) is called a **node.** The probability of finding an electron at a particular point is proportional to the square of the value for the wave function at that point. Whether the wave function has a positive or a negative value at that point does not matter, because the square will always be positive. However, the value of the square of the wave function at a node is zero. This means that the electron density at a node is zero. In general, *the more nodes an orbital contains, the higher energy the orbital is*. Nodes are also very important in molecular orbitals.

The next higher-energy orbital after the 2s orbital is the **2p orbital,** also shown in Figure 3.1. This orbital is not spherically symmetrical like the s orbitals. Its overall shape is something like a dumbbell, with regions of high probability of finding the electron (regions of high electron density) on opposite sides of the nucleus. One lobe of the orbital has a positive math sign and the other has a negative math sign, with a planar node in between. If, as shown in Figure 3.1, the lobes of the orbital are directed along the z-axis, then the orbital is called the $2p_z$ orbital and the plane formed by the x-axis and the y-axis is the plane of the node. There are three 2p orbitals, all of the same energy. Orbitals with the same energy are termed *degenerate*. The three 2p orbitals are mutually perpendicular. If one is directed along the z-axis ($2p_z$), as shown in Figure 3.1, the other two are directed along the x-axis and the y-axis ($2p_x$ and $2p_y$).

The orbitals pictured in Figure 3.1 are the ones of most interest to organic chemistry, because the atoms that are most commonly encountered in organic compounds are H, C, and other second-period atoms. However, atoms belonging to the third or higher periods of the periodic table are sometimes encountered. Orbitals in the third shell have two nodes. The **3s and 3p orbitals** look similar to the 2s and 2p orbitals except that they are larger and have an additional spherical node. The spherical nodes do not affect the picture for bonding, so 3s and 3p orbitals (or any other s and p orbitals) can be treated similarly to 2s and 2p orbitals when they form molecular orbitals. In addition, the atoms of the third period have **3d orbitals** available. We will deal with bonding involving d orbitals later, as necessary.

PROBLEM 3.1
Draw a 3s atomic orbital and compare it to a 2s orbital.

The electron configuration of an atom shows how the electrons are distributed among its orbitals. The electrons are arranged in the orbitals so that the overall energy of the system is minimized. The following set of rules can be used to quickly derive the electron configuration for an atom:

1. Each electron is placed in the lowest-energy orbital available.

2. Each orbital can contain a maximum of two electrons, and these must have opposite spins. This is a result of the **Pauli exclusion principle,** which states that no two electrons can have all four quantum numbers the same. Because two electrons in the same orbital must have three of the quantum numbers the same, the fourth quantum number (the spin quantum number) must be different.

3. When degenerate orbitals are available, the electrons first occupy them singly, with the same (parallel) spins. After all of the degenerate orbitals contain one electron, additional electrons with opposite spins are added to each. This is known as **Hund's rule.**

Figure 3.2 shows the electron configurations for the atoms in the first two rows of the periodic table. This energy level diagram shows the atomic orbitals in order of increasing energy (the lowest-energy orbital is at the bottom) and shows the electrons occupying these orbitals as arrows (the direction of the arrow indicates the spin). This

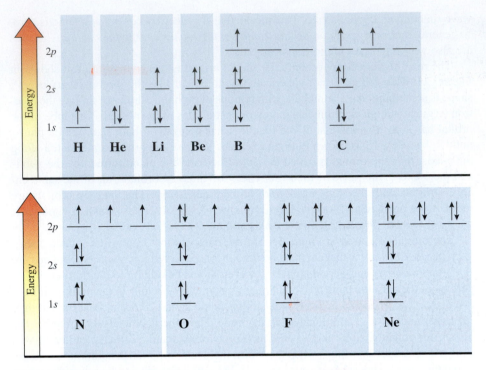

Figure 3.2

ATOMIC ORBITAL ENERGY LEVEL DIAGRAMS. To simplify these diagrams, the orbitals are shown at the same energies for different atoms. Actually, the energy of an orbital decreases as the number of protons in the atom increases. Thus the 2p orbitals of fluorine are lower in energy than the 2p orbitals of oxygen.

figure can easily be constructed from the preceeding rules. For example, carbon has six electrons. The first two go into the lowest available orbital, the 1s orbital (rule 1), with opposite spins (rule 2). Likewise, the next two electrons go into the 2s orbital. The last two electrons must be placed in the 2p orbitals. Because there are three degenerate 2p orbitals, these two electrons are placed in different orbitals, with the same spin (rule 3).

What is the reason behind Hund's rule? Let's consider carbon again. Recall that electrons, having the same charge, repel each other. This electron–electron repulsion is minimized if the electrons are in different regions of space, resulting in a lower-energy situation. If the two electrons in the 2p orbitals of carbon have the same spin, they must be in different 2p orbitals. (If the electrons were together in the same 2p orbital, the Pauli exclusion principle would be violated.) Being in different 2p orbitals keeps the electrons in different regions of space.

Figure 3.2 shows the electron configurations for the **ground state** or the **lowest-energy state** for the atoms. Any other electron arrangement is higher in energy and is termed an **excited state.** For example, a carbon atom with two electrons in the 1s orbital, one electron in the 2s orbital, and three electrons in the 2p orbitals is in an excited state. Such an excited carbon atom will exist for only a very short period of time—only until it can find a way to get rid of that extra energy and return the third 2p electron to the 2s orbital.

PROBLEM 3.2
Show an atomic orbital energy level diagram for these atoms:
a) Si   **b)** Al   **c)** Cl

PROBLEM 3.3
Explain whether the electron arrangement for these atoms is the ground state or an excited state:

a)   **b)**

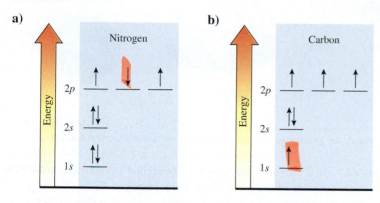

3.2   MOLECULAR ORBITALS

In isolated atoms the electrons are in the atomic orbitals (AOs) of that atom. What happens to the electrons when atoms come together to form bonds? In the simple Lewis model, some of the electrons are pictured as being shared between atoms. In the orbital model, these shared electrons are pictured as being in orbitals that extend around more than one atom. Such orbitals are called **molecular orbitals** (MOs).

The shapes of the MOs indicate the areas of high electron density in the molecule. MOs are useful in understanding reactions because orbitals must overlap to form any new bond. Thus, MOs help us see how molecules must approach each other for their orbitals to overlap in forming new bonds. Molecular orbitals are especially useful in the discussion of spectroscopy (Chapter 15), aromatic compounds (Chapter 16), and pericyclic reactions (Chapter 22).

Most of what was presented previously about AOs also applies to MOs. The shape of the MO, which describes a region around the nuclei of the bonding atoms where the probability of finding an electron is high, is important, as is the energy of the MO. Let's examine H_2, a very simple molecule, to see what happens.

Let's consider the shape of the MO first. The simplest picture considers molecular orbitals as resulting from the overlap of atomic orbitals. When atoms are separated by their usual bonding distance, their AOs overlap. Where this overlap occurs, either the electron waves reinforce and the electron density increases, or the electron waves cancel and the electron density decreases. The left-hand side of Figure 3.3 shows the overlap of the 1s atomic orbitals on two different hydrogens (H_a and H_b) when these hydrogens are separated by their normal bonding distance. The *two* atomic orbitals interact to produce *two* molecular orbitals. The MOs result from a linear combination of the AOs (called the **LCAO approximation**). Simply, this means that the AOs are either added ($1s_a + 1s_b$) or subtracted ($1s_a - 1s_b$) to get the MOs.

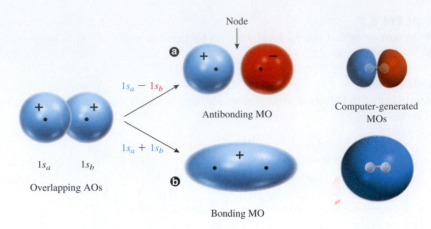

Figure 3.3

COMBINATION OF HYDROGEN 1S ATOMIC ORBITALS TO FORM MOLECULAR ORBITALS.

In Figure 3.3 the AOs have the same math sign in the region of overlap. In the $1s_a$ + $1s_b$ combination the magnitude of the wave function increases in the overlap region, and the electron density also increases here. This increase of electron density between the nuclei results in a more stable orbital—the MO is lower in energy than the AOs. Such MOs are called **bonding MOs.** In the $1s_a$ − $1s_b$ combination the magnitude of the wave function decreases in the region between the nuclei and actually cancels along a plane perpendicular to a line connecting the two nuclei. This results in a node and a decrease in electron density between the nuclei. The resulting MO is less stable than the AOs and is called an **antibonding MO.** In general, the more nodes that are present in a MO, the higher the energy of that MO.

Figure 3.4 shows the energies of these MOs. The energies of the $1s$ AOs of the separated hydrogen atoms are shown on the left and right sides of the diagram. The energies of the MOs are shown in the center of the diagram. The bonding MO is lower in energy than the $1s$ AOs by an amount of energy shown as ΔE in the diagram. In the simple picture we are using, the antibonding MO is higher in energy than the AOs by ΔE also. (A more sophisticated treatment shows that the antibonding MO is actually higher in energy than the $1s$ orbital by more than ΔE.)

Figure 3.4

ENERGY LEVEL DIAGRAM FOR H₂ MOLECULAR ORBITALS.

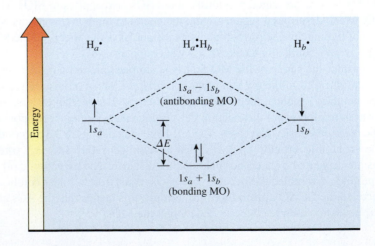

The molecular orbitals are filled with electrons according to the same rules that were used to put electrons in the atomic orbitals of atoms. In this case there are two electrons. These fill the bonding MO. The stability of the molecule is determined by the total energy of the electrons. In the case of H_2 the molecule is more stable than the separated atoms by $2(\Delta E)$. In other words, it would be necessary to add $2(\Delta E)$ of energy to the H_2 molecule to break the covalent bond. As is the case here, the antibonding MOs usually do not have any electrons in them, and they do not affect the energy of the molecule. But they are real and can be occupied by electrons in some situations, such as certain types of spectroscopy and some chemical reactions.

The bond formed between two hydrogen atoms is one example of a **sigma (σ) bond.** In general, the MO of a sigma bond is shaped so that it is symmetrical about the internuclear axis (a line connecting the two nuclei). In other words, if a plane cuts through the MO, perpendicular to a line connecting the nuclei, the intersection of the plane and the MO is a circle. This is shown in Figure 3.5. In a sigma bond, rotation of the AO of one atom of the bond around the internuclear axis does not affect the overlap of the AOs, so it does not affect the energy of the bond.

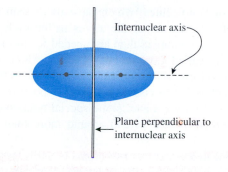

This view of a sigma MO is perpendicular to a line through the two nuclei—that is, perpendicular to the internuclear axis. A plane that cuts the MO perpendicular to the internuclear axis appears as a line in this perspective.

Figure 3.5

SEVERAL VIEWS OF A SIGMA BOND.

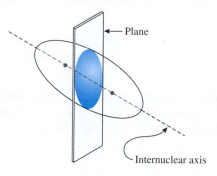

In this view, the MO is slightly tilted. The plane that cuts the MO perpendicular to the internuclear axis can be seen more clearly.

This is a view down the internuclear axis, perpendicular to the plane that cuts through the MO. In this view it can easily be seen that the intersection of the plane with the MO is symmetrical. Therefore, this is classified as a sigma bond.

The MO picture for H_2 agrees completely with the Lewis structure. It shows a pair of electrons shared between the two atoms (in an MO that extends over both atoms), resulting in a molecule that is more stable than the separated atoms. It also explains why the valence for hydrogen is 1. There is no room for additional electrons unless they are placed in the antibonding MO, a destabilizing situation. MO theory can also help us to understand why a molecule such as He_2 does not exist. Like hydrogen, both heliums would use $1s$ AOs to form a bonding and an antibonding MO. However, there would be four electrons to place in these MOs, two from each helium atom. Therefore, both the bonding MO and the antibonding MO would be filled with electrons. The stabilization of the electrons in the bonding MO would be more than offset by the destabilization of the electrons in the antibonding MO. The electrons are actually more stable in the separated atoms.

PROBLEM 3.4
Show an energy level diagram for the MOs for He_2 and show how the electrons would be arranged in these MOs.

What happens with more complicated molecules, which have more than two atoms? To be most accurate, MOs that extend over all the atoms in the molecule should be used. However, such MOs are more complex than are needed for most situations. Instead, we use an approximation called **valence bond theory,** which uses overlap of AOs on the two bonded atoms to form MOs that are localized around these two atoms. These **localized MOs** are much easier to visualize and correspond well to the picture of a bond as a shared pair of electrons. Later, some special situations in which it is necessary to use **delocalized MOs** that extend around more than two atoms will be examined.

If the approximation that MOs are localized between two atoms is used, then the picture is not as complicated as it might seem. The same thing happens every time. The following generalizations can be made:

1. Two AOs on the bonding atoms overlap to produce two MOs. In more complicated situations the number of MOs produced equals the number of AOs initially involved.

2. The overlapping AOs can be used to approximate the shape of the MOs. One combination of the two AOs results in a lower-energy, bonding MO; the other combination results in a higher-energy, antibonding MO with a node between the nuclei. The math signs are a convenient way to keep track of the nodes.

3. The same rules are used to assign electrons to MOs as are used to assign electrons to AOs. Usually, there are just enough electrons to fill the bonding MOs, and the antibonding MOs remain empty.

4. The bond energy is approximately equal to the total amount of energy by which the electrons are lowered in energy in comparison to the electrons in the AOs—that is, the number of electrons times ΔE.

5. The magnitude of ΔE increases with increasing overlap of the AOs. However, if the atoms get too close together, repulsion between the nuclei starts to dominate and the overall energy increases very rapidly.

By using these rules, we can also analyze more complex situations.

3.3 SINGLE BONDS AND sp^3 HYBRIDIZATION

Let's consider methane (CH_4), a simple organic molecule. First, the Lewis structure, which shows four CH bonds, should be examined.

$$H-\underset{\underset{H}{|}}{\overset{\overset{H}{|}}{C}}-H$$

Methane

In the MO picture, there will be a bonding MO (and an antibonding MO) for each bond in the Lewis structure. Furthermore, the MO model must be in accord with experimental observations. Experiments have shown that the bonds in methane are all identical, with tetrahedral geometry. Therefore, methane must have four equivalent bonding MOs, with a tetrahedral arrangement.

Each MO results from an AO on the carbon overlapping with an AO on one of the hydrogens. Only AOs from the valence shell of an atom are used. Each hydrogen uses a $1s$ AO. The four orbitals that are available from the carbon are its $2s$ and three $2p$ AOs. However, problems result if these carbon orbitals are used directly. First, although four equivalent MOs are needed, a MO resulting from a carbon $2s$ AO overlapping with a hydrogen $1s$ AO would obviously be different from a MO resulting from a carbon $2p$ AO overlapping with a hydrogen $1s$ AO. Second, the carbon $2p$ orbitals do not have the correct geometry to point directly at hydrogens that are tetrahedrally arranged around the carbon. This would result in poor overlap between the AOs, a small value for ΔE, and weak bonds.

You probably remember the solution to these problems from general chemistry. The $2s$ and the three $2p$ AOs are mathematically combined in a kind of averaging process to produce four equivalent **hybridized AOs.** Because they result from combining one s orbital and three p orbitals, the new AOs are said to be sp^3 **hybridized.** The hybrid AOs are ideal for bonding. They are all equivalent, and they have tetrahedral geometry. In addition, each has a large lobe of the orbital pointed in the direction where the other atom of the bond will be and only a small lobe on the other side of the carbon. This directionality of the hybrid AO allows for maximum overlap when it interacts with another AO to form a MO. The hybridization process and the sp^3 hybrid AOs are shown in Figure 3.6.

2s and one 2p AO One sp^3 hybrid AO

Part of the 2s and part of the 2p orbitals are added. Where both AOs are plus, the resulting orbital has a larger lobe. Where the p orbital is minus and the s orbital is plus, the size of the lobe decreases.

The four sp^3 hybrid AOs point to the corners of a tetrahedron. The small back lobes of the AOs have been omitted for clarity.

Figure 3.6

sp^3 HYBRID ATOMIC ORBITALS.

We are now ready to see how the MOs for methane are formed. The process is just what we have seen before. As shown in Figure 3.7, an sp^3 hybrid AO on carbon overlaps with a $1s$ AO on hydrogen to produce a lower-energy, bonding MO and a higher-energy, antibonding MO. These are both sigma MOs. *(An asterisk is usually used to designate the antibonding MO, as in σ^*.)* The two electrons of the bond are found in the bonding MO. An identical process occurs for the other three bonds. The four bonding MOs are pictured in Figure 3.7.

Let's consider ethane, C_2H_6, a slightly more complex example, shown in Figure 3.8. The geometry of the molecule tells the hybridization of the atoms involved. In simple organic compounds, any atom that has tetrahedral geometry will be **sp^3 hybridized.** Thus, both carbons are sp^3 hybridized in ethane. The CH bonds are formed in the same way as they were in methane. The CC bond is formed by the overlap of sp^3 AOs on each carbon, as shown here.

Overlapping sp^3 AOs   Sigma bonding MO

A similar picture applies to all sigma bonds, even when the atoms have hybridization other than sp^3.

MO pictures for NH_3 and CH_3OH are also shown in Figure 3.8. For ammonia there are four pairs of electrons around the nitrogen—three bonded pairs and one unshared pair—so on the basis of VSEPR theory the geometry of the electron pairs is tetrahedral. (We will not worry about the small deviations from ideal geometries that commonly occur.) Therefore, the hybridization of the nitrogen is sp^3. The MOs for the NH bonds are formed like those for the CH bonds shown earlier. The only difference is the unshared pair of electrons on N. These electrons are in an sp^3-hybridized, nonbonding AO on the nitrogen. In the picture for methanol, note that there are two pairs of electrons in sp^3, nonbonding AOs on oxygen.

Important Convention ▶

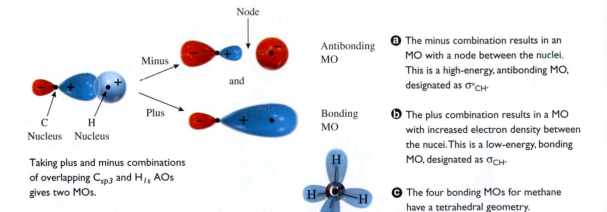

Node

Minus →

and

Plus →

C   H
Nucleus Nucleus

Taking plus and minus combinations of overlapping C_{sp3} and H_{1s} AOs gives two MOs.

Antibonding MO

Bonding MO

ⓐ The minus combination results in an MO with a node between the nuclei. This is a high-energy, antibonding MO, designated as σ^*_{CH}.

ⓑ The plus combination results in a MO with increased electron density between the nucei. This is a low-energy, bonding MO, designated as σ_{CH}.

ⓒ The four bonding MOs for methane have a tetrahedral geometry.

Figure 3.7

Orbital pictures for the sigma bonds of methane. **ⓐ** Antibonding MO, **ⓑ** bonding MO, and **ⓒ** four bonding MOs.

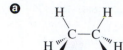

a

Ethane

The CH sigma bond results from a C sp^3 AO overlapping with an H 1s AO ($\sigma_{Csp3+H1s}$). The other CH bonds are the same. The CC sigma bond results from the overlap of C sp^3 AOs on both carbons ($\sigma_{Csp3+Csp3}$).

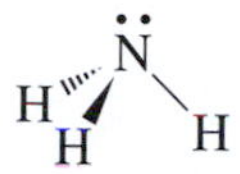

Because each C of ethane has four bonds, the geometry at each is tetrahedral and the hybridization is sp^3.

Figure 3.8

ORBITAL PICTURES FOR THE BONDS OF **a** ETHANE, **b** AMMONIA AND **c** METHANOL. (The small lobes on the back of the various orbitals have been omitted for clarity.)

b

Ammonia

The geometry of the four electron pairs (the three bonding pairs and the unshared pair) of ammonia is tetrahedral, so the hybridization is sp^3.

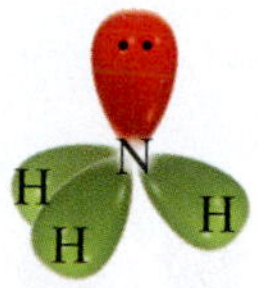

The unshared pair of electrons is in an sp^3-hybridized, nonbonding AO on the nitrogen. The three NH bonds are $\sigma_{Nsp3+H1s}$.

c

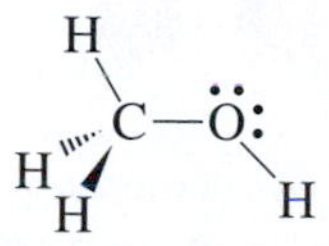

Methanol

The carbon of methanol has four bonding pairs of electrons, so it is tetrahedral and sp^3 hybridized. The geometry of the four electron pairs of oxygen is also tetrahedral, so it is sp^3 hybridized too.

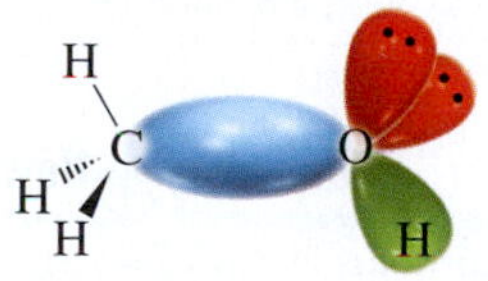

To simplify the drawing, the orbitals for the CH bonds are not shown. They are the same as the CH bonds in ethane. The unshared electrons on the O are in sp^3, nonbonding AOs. The OH bond results from an O sp^3 AO overlapping with an H 1s AO ($\sigma_{Osp3+H1s}$). The CO bond results from the overlap of sp^3 AOs on each atom ($\sigma_{Csp3+Osp3}$).

PROBLEM 3.5

Indicate the type of atomic orbitals that are overlapping to form each of the different kinds of bonds in CH_3OCH_3 (For example, a carbon sp^3 AO and a hydrogen 1s AO). What kinds of orbitals are occupied by the unshared electrons on the oxygen?

3.4 DOUBLE BONDS AND *sp²* HYBRIDIZATION

Next, let's see what happens with a compound that has a double bond. A simple organic compound with a double bond is ethene (ethylene), C_2H_4. Again, we should start by looking at the Lewis structure, which is shown in Figure 3.9a. All the atoms of ethene

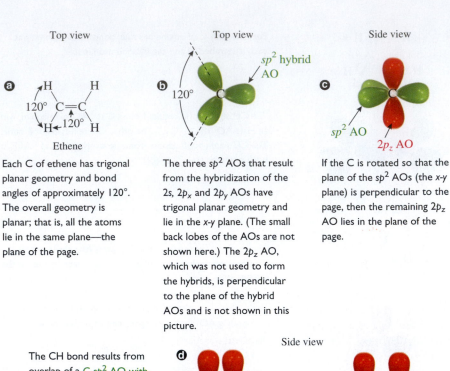

Top view
Top view
Side view
a
H
H
120°
C=C
120°
H
H
Ethene
Each C of ethene has trigonal planar geometry and bond angles of approximately 120°. The overall geometry is planar; that is, all the atoms lie in the same plane—the plane of the page.
b
120°
sp2 hybrid AO
C
The three sp2 AOs that result from the hybridization of the 2s, 2px and 2py AOs have trigonal planar geometry and lie in the x-y plane. (The small back lobes of the AOs are not shown here.) The 2pz AO, which was not used to form the hybrids, is perpendicular to the plane of the hybrid AOs and is not shown in this picture.
c
sp2 AO
2pz AO
If the C is rotated so that the plane of the sp2 AOs (the x-y plane) is perpendicular to the page, then the remaining 2pz AO lies in the plane of the page.
Side view
d
The CH bond results from overlap of a C sp2 AO with an H 1s AO (σCsp2+H1s).
H
H
C
C
H
H
This bond results from overlap of an sp2 AO on one C with an sp2 AO on the other C (σCsp2+Csp2).
This new type of overlap of a 2p AO on each C results in a new type of bond, a pi bond (πC2p+C2p).

Active Figure 3.9 ORGANIC Chemistry•Now™

Bonding and orbital pictures for ethene. Test yourself on the concepts in this figure at **OrganicChemistryNow.**

lie in the same plane. The geometry at each carbon is trigonal planar and the bond angles are approximately 120°. Remember that the MO picture must agree with these experimental observations.

When a compound has a double bond, one of the two bonds is always a sigma bond, formed similarly to the sigma bonds we have already seen. But the second bond is formed in a different fashion. Let's consider the sigma bonds first.

In ethene, each carbon forms three sigma bonds, one to each of the two hydrogens and one to the other carbon. Therefore, three AOs with trigonal planar geometry are needed to form these sigma bonds. This time, the hybrid AOs are formed from the $2s$ and two $2p$ AOs ($2p_x$ and $2p_y$). This results in the formation of three **sp^2-hybridized AOs.** Each of these AOs has the same general shape as the sp^3 AOs we saw previously; that is, each has a larger lobe of the orbital pointed in one direction and a smaller lobe of the orbital pointed in the opposite direction. The three are equivalent and have trigonal planar geometry, as shown in Figure 3.9b.

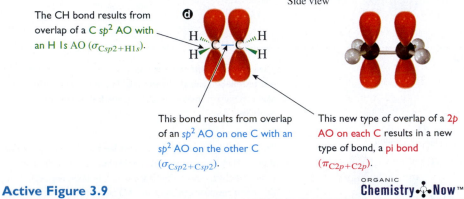

Because formation of the sp^2 hybrid AOs used only two of the three $2p$ orbitals, one $2p$ orbital remains: the $2p_z$ AO. The $2p_z$ AO is perpendicular to the sp^2 hybrid AOs. (The sp^2 hybrid AOs lie in the x-y plane because they are formed from the $2p_x$ and $2p_y$ AOs.) All four of the AOs (three sp^2 AOs and one $2p$ AO) are shown in Figure 3.9c.

The MOs for the sigma bonds of ethene are formed by overlap of the AOs in a manner similar to what we have seen before. Each CH bond results from the overlap of an sp^2 hybrid AO on C with a $1s$ AO on H. One of the two CC bonds is a sigma bond and results from the overlap of sp^2 hybrid AOs on each carbon. The second CC bond results from a very different type of overlap involving the unhybridized $2p$ AOs remaining on each carbon. As shown in Figure 3.9d, these orbitals are parallel to each other. They overlap above and below the plane of the atoms, and they overlap less than orbitals that are pointed directly at each other. This new type of overlap, of orbitals that are parallel rather than orbitals that point directly toward each other, results in a new type of bond, called a **pi bond** (π bond).

Figure 3.10 shows this overlap in more detail. The process of forming the MOs is still the same as we saw before. The two AOs overlap. In the plus combination ($2p_a + 2p_b$) the orbitals reinforce in the region above and below nuclei, resulting in a lower-energy, bonding MO. In the minus combination ($2p_a - 2p_b$), the orbitals cancel in the region between the nuclei, a new node is formed, and a higher-energy, antibonding MO results. Because the orbitals are not symmetrical about the internuclear axis, these are not sigma orbitals. Instead, orbitals with this shape are called pi orbitals.

The energies of the pi MOs are shown in Figure 3.11. Note that there are two electrons (the second pair of electrons of the double bond) to put in these MOs. As usual, the bonding MO is filled with electrons, and the antibonding MO is empty. Because the overlap of the p AOs forming the pi MOs is poorer than in the case of AOs forming sigma MOs, ΔE is less for pi MOs than for sigma MOs. Thus, the poorer overlap results in a pi bond being weaker than a sigma bond. From Table 2.1 the bond dissociation energy for a CC single bond is 81 kcal/mol (339 kJ/mol), whereas that for both bonds of a CC double bond is 145 kcal/mol (607 kJ/mol). If it is assumed that the sigma bond of a double bond has approximately the same bond strength as the sigma bond of a single bond, then the approximate strength of a pi bond can be calculated as $145 - 81 = 64$ kcal/mol (268 kJ/mol), considerably less than the strength of a sigma bond.

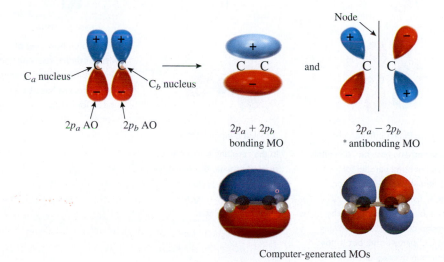

Figure 3.10

FORMATION OF PI BONDING AND PI ANTIBONDING MOLECULAR ORBITALS.

Figure 3.11

ENERGIES OF THE PI
BONDING AND ANTIBONDING
MOLECULAR ORBITALS.

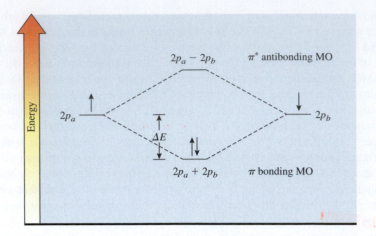

This MO picture helps explain why all the atoms of ethene lie in the same plane. On each carbon the p orbital that is used to form the pi bond is perpendicular to the plane defined by the C and the two attached H's. Unless the plane of one C and its two attached H's is the same as the plane of the other C and its two attached H's, the p orbitals will not be parallel and overlap will be decreased. As we have seen, as overlap decreases, ΔE decreases and the energy of the electrons increases—the molecule is less stable.

Figure 3.12 illustrates what happens as the plane of one HCH group is rotated relative to the plane of the other HCH group. Figure 3.12a shows planar ethene, with maximum overlap of the p orbitals that form the pi bond. If one carbon is rotated 90° about the CC bond, the geometry shown in Figure 3.12b results. This rotation does not change the amount of overlap of the orbitals forming the sigma bond, so its energy is not affected.

Figure 3.12

ROTATION ABOUT THE
CARBON–CARBON BOND
OF ETHENE.

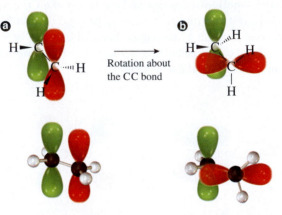

Back p orbital

Front p orbital

This is an end-on view of twisted ethene, showing that the p orbitals do not overlap. The hydrogens have been omited for clarity.

Planar ethene

Planar ethene has the $2p$ orbitals of the pi bond parallel for maximum overlap. Rotation of one C by 90° about the CC bond axis does not affect the overlap of the orbitals forming the CC sigma bond, but it does affect the overlap of the p orbitals forming the pi bond.

Twisted nonplanar ethene

In the resulting nonplanar or twisted ethene, the p orbitals on the two carbons are perpendicular. They no longer overlap at all.

(For this reason a similar rotation about the CC sigma bond of ethane occurs readily.) However, the two p orbitals of the pi bond of ethene are now perpendicular. They no longer overlap at all, so ΔE is zero and the pi bond has been broken. Therefore, to cause such a rotation to occur, enough energy must be added to break the pi bond, approximately 64 kcal/mol (268 kJ/mol). At room temperature, only about 20 kcal/mol (84 kJ/mol) of thermal energy is available. Any process that requires more energy than this will not occur at room temperature. So the rotation pictured in Figure 3.12 does not occur.

Simple organic compounds use **sp^2 hybridization** for any atom that has trigonal planar geometry. Consider CH_2NH, shown in Figure 3.13. The geometry at the C is trigonal planar, so it is sp^2 hybridized. It looks just like one of the carbons of ethene. VSEPR theory treats the N as though it has three pairs of electrons in its valence shell (the unshared pair, the pair of the NH bond, and both pairs of the double bond counting as one pair), so the N also has trigonal planar geometry. It looks just like the C but with an electron pair occupying the position of one of the H's. Overall, CH_2NH looks very similar to ethene. The CN double bond is composed of one sigma bond and one pi bond. The geometry is planar, and the CNH bond angle is approximately 120°. The unshared electron pair on N is in an sp^2-hybridized, nonbonding AO.

Figure 3.13 also shows the structure for CH_2O. Because VSEPR theory predicts a trigonal planar geometry for the double bond and the two electron pairs on the oxygen, we will treat it as sp^2 hybridized. Again, the picture is similar to ethene, with one CO sigma bond and one CO pi bond, but with both unshared electron pairs in

(a)

Methanimine

pi bonding MO

In CH_2NH, all the bond angles are approximately 120°, and the molecule has an overall planar geometry. Both the C and the N are sp^2 hybridized. The CH bonds are the same as those in ethene. The NH bond results from an N sp^2 AO overlapping with a H $1s$ AO ($\sigma_{Nsp2+H1s}$). The unshared electrons are in a nonbonding sp^2 AO on N. The CN sigma bond results from a C sp^2 AO overlapping with a N sp^2 AO ($\sigma_{Csp2+Nsp2}$). The other CN bond is a pi bond resulting from a C $2p$ AO overlapping with a N $2p$ AO ($\pi_{C2p+N2p}$).

(b)

Formaldehyde

In CH_2O, the bond angles are approximately 120°, and the geometry is planar. The C is sp^2 hybridized. The O is also treated as sp^2 hybridized. The two pairs of unshared electrons are each in a nonbonding sp^2 AO on O. The CO sigma bond results from a C sp^2 AO overlapping with an O sp^2 AO ($\sigma_{Csp2+Osp2}$). The other CO bond is a pi bond resulting from a C $2p$ AO overlapping with an O $2p$ AO ($\pi_{C2p+O2p}$).

Figure 3.13

ORBITAL PICTURES FOR (a) METHANIMINE, CH_2NH, AND (b) FORMALDEHYDE, CH_2O.

sp^2-hybridized, nonbonding AOs on the oxygen. (Because there are no bond angles involving the doubly bonded oxygen that can be measured, this O could also be treated as unhybridized. In this case the CO σ bond would be formed from an sp^2 hybrid AO on C and a p AO on O, the π bond would still be formed from p orbitals on C and O, and the unshared pairs of electrons would be in $2s$ and $2p$ nonbonding AOs on O. For us there is no real difference between this picture and the previous one, so we will use the hybridized picture in such cases.)

PROBLEM 3.6

What is the hybridization at the N and each C in this molecule? Indicate the type of bond and the orbitals that are overlapping to form it for each of the designated bonds (for example, $\sigma_{Csp3 + H1s}$).

$$
\begin{array}{ccccccc}
 & H & H & & H & & \\
 & | & |_{1} & & | & & \\
H- & C & -C & = \ddot{N} & -C & -H \\
 & |_{2} & |_{3} & {}_{4} & {}_{5}| & {}_{6} & \\
 & H & & (\text{both}) & H & &
\end{array}
$$

3.5 TRIPLE BONDS AND sp HYBRIDIZATION

A simple compound with a triple bond is ethyne (acetylene), HC≡CH. The Lewis structure for ethyne is shown in Figure 3.14a. It is a linear molecule. One of the CC bonds is a sigma bond. The other two are pi bonds.

Each carbon forms one sigma bond to a hydrogen and one to the other carbon. Therefore, two hybrid AOs with linear geometry are needed. These hybrid AOs are

Figure 3.14

BONDING AND ORBITAL PICTURES FOR ETHYNE.
ⓐ Lewis structure, **ⓑ** two sp AOs, **ⓒ** one sigma and two pi bonds of the triple bond.

ⓐ

H—C≡C—H

Ethyne is a linear molecule with bond angles of 180°.

ⓑ

sp hybrid AOs

Hybridization of the 2s and the $2p_x$ AOs results in the formation of two sp hybrid AOs. (The small back lobes of the AOs are not shown here.) The sp AOs have linear geometry and have a shape similar to the other hybrid AOs we have encountered. The two unused p orbitals (p_y and p_z) are perpendicular to the hybrid AOs.

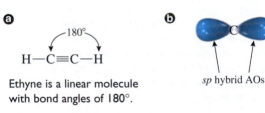

ⓒ

Two pi bonding MOs

Each C—H bond results from overlap of a C sp hybrid AO with a H 1s AO ($\sigma_{Csp+H1s}$). The C—C sigma bond results from overlap of C sp hybrid AOs on each carbon ($\sigma_{Csp+Csp}$). There are two pi bonds (green and red) resulting from a p orbital on one C overlapping with a p orbital on the other C ($\pi_{C2p+C2p}$). The p orbitals of one pi bond are perpendicular to the p orbitals of the other. The three bonds of the triple bond are composed of one sigma bond and two pi bonds.

formed from the 2*s* AO and one 2*p* AO. As shown in Figure 3.14b, the two resulting *sp*-hybridized AOs are equivalent and have linear geometry like the 2*p* AO from which they are formed. They have the same general shape as the other hybrid AOs we have seen, with a large lobe pointed in one direction and a small lobe pointed in the opposite direction.

This time there are two 2*p* AOs, perpendicular to the *sp* hybrid AOs, remaining on each carbon. Each of these overlaps with a 2*p* AO on the other carbon to form two pi bonds. These pi bonds are shown in Figure 3.14c. Because the pi bonds are identical, the MOs have the same energies—that is, they are degenerate.

Rotation of one of the carbons about the CC bond axis has no effect on the geometry of the molecule. None of the bond angles is changed. Therefore, we do not have to worry about whether such rotation will occur.

An atom that has **linear geometry** uses *sp*-hybridized AOs to form the MOs. Other examples are considered in problems 3.7 and 3.8.

PROBLEM 3.7

Consider hydrogen cyanide, H—C≡N.
a) What is the hybridization at the N? at the C?
b) What are the types of the three CN bonds? What orbitals are overlapping to form them?
c) In what type of orbital are the unshared electrons on the N?
d) Draw the molecule showing how the orbitals overlap to form the pi bonds.

PROBLEM 3.8

What is the hybridization at each C in this molecule? Indicate the type of bond and the orbitals that are overlapping to form it for each of the designated bonds.

$$
\begin{array}{cccccc}
& H & H & H & & \\
& | & | & |1 & & \\
H- & C & -C & =C & -C≡C-H \\
& 2\ | & 3 & 4 & 5\quad 6\quad 7 \\
& H & & \text{(both)} & \text{(all} \\
& & & & \text{three)}
\end{array}
$$

PROBLEM 3.9

Show the hybridization at each of the atoms, except H, in these molecules. Indicate the type of each designated bond and the orbitals that are overlapping to form it.

a)
$$
\begin{array}{ccc}
H & \ddot{O}: & \\
| & \| & 4\ \text{(both)} \\
H-C-C-H & & \\
\ \ 1\ |\ 2\ 3 & & \\
H & &
\end{array}
$$

b)
$$
\begin{array}{cccc}
H & H & H & \\
| & | & |5 & \\
H-C-C=C-C≡N: \\
\ \ 1\ |\ 2\ 3\ 4 \\
H & \text{(both)} & \text{(all} \\
& & \text{three)}
\end{array}
$$

c)
$$
\begin{array}{ccccc}
& & \ddot{O}: & H & \\
& \text{(both) 1} & \| & | & \\
H-C≡C-C-C-H \\
\ \ 2\ \ \ 3\ \ 4\ 5\ |\ 6 \\
\text{(all} & & & H \\
\text{three)}
\end{array}
$$

d)
$$
\begin{array}{cccc}
H & H & & \\
| & |1 & \ddot{} \\
H-C-C-\ddot{N}-H \\
\ \ 1\ |\ 2\ |\ 3\ |\ 4 \\
H & H & H
\end{array}
$$

ORGANIC
Chemistry·Now™
Click *Coached Tutorial Problems*
to practice **Identifying Hybridization** of atoms.

3.6   RESONANCE AND MO THEORY

In all the cases presented so far, an adequate picture of bonding was obtained by making the approximation that each MO is localized on only two atoms. Sometimes, however, these localized MOs are not a very good model for certain bonds. The situations in which localized MOs fail are the same ones in which normal Lewis structures are inadequate—situations in which resonance is needed to describe the bonding. In such molecules it is convenient to use **delocalized MOs** (MOs that include AOs from more than two atoms) to describe the bonding. In fact, resonance and delocalized MOs are just different ways to describe the same type of bonds. It is not necessary to delve deeply into delocalized MOs, but a brief discussion of them will provide a much better understanding of resonance and when it should be used. As we continue, we will find the resonance picture adequate for most of our discussions.

Let's consider the formate anion, shown in Figure 3.15. The first Lewis structure (a) does not accurately represent the structure of this covalent ion, because a second Lewis structure can be drawn (b) that is equivalent to the first. The actual structure is a resonance hybrid of these two structures. Experiments confirm that the two CO bonds are identical, with a bond length between that of a single and a double bond, and that the charge on each oxygen is the same, approximately $-\frac{1}{2}$.

Figure 3.15c is an attempt to show how the AOs might overlap to form localized MOs in the formate anion. In this localized MO picture, a p orbital on the carbon overlaps with a p orbital on the upper oxygen to form a pi bond, corresponding to the Lewis structure of Figure 3.15a. In this structure, the lower oxygen has three unshared pairs of electrons. Whenever an atom with an unshared pair of electrons is adjacent to a pi bond, as occurs here, that atom usually assumes a hybridization that places an unshared pair in a p orbital because the overlap of this p orbital with the p orbital of the pi bond on the adjacent atom is stabilizing. It is this overlap that allows resonance to occur. In this case the p orbital with the unshared pair on the lower oxygen overlaps equally well with the p orbital on the carbon so that the pi bond could also be shown using these two orbitals with an unshared pair of electrons in the p orbital on the upper oxygen. This corresponds to the second Lewis structure (b).

In situations like this one, the overlap of the carbon p orbital with one of the oxygen p orbitals cannot be ignored as Figure 3.15c attempts to do. Instead, all three p orbitals must be used to form MOs that involve the carbon and both oxygens. The three AOs interact to form three delocalized pi MOs. Two of these three delocalized MOs contain the four electrons: the pi electrons and an "unshared pair" of electrons from the localized picture. Part d of Figure 3.15 shows how the orbitals overlap in the delocalized picture.

A detailed discussion of the shapes and energies of delocalized MOs will be postponed until Chapters 16 and 22. However, because delocalization lowers the energies of some of the MOs, the total energy of the electrons in the delocalized MOs is lower than what the total energy of these same electrons would be in the hypothetical localized MOs. The electrons are more stable in the delocalized MOs. This stabilization, or energy lowering, is termed *resonance stabilization*.

In general, it is necessary to use resonance (or delocalized MOs) for any molecule that is **conjugated,** that is, any molecule that has a series of three or more overlapping

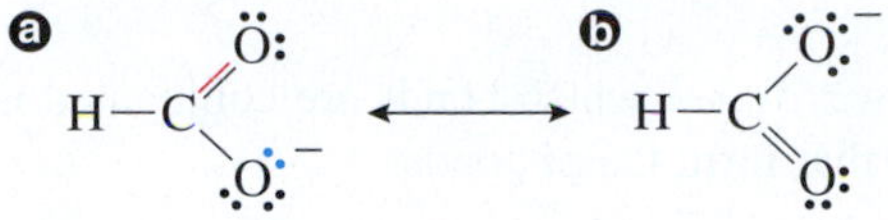

Formate anion

Formate anion is an example of an ion for which two equivalent Lewis structures (ⓐ and ⓑ) can be drawn. The actual structure is a resonance hybrid of these two structures. Remember to use the double-headed arrow only between resonance structures. Never use equilibrium arrows (⇌) between resonance structures.

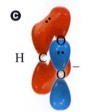

This is an attempt to show an orbital picture for the formate anion. It corresponds to the Lewis structure in part ⓐ. (The two unshared pairs of electrons on each oxygen that are not involved in resonance have been omitted for clarity.) The two red *p* orbitals overlap to form the pi bond. One unshared pair of electrons is in the blue *p* orbital on the other oxygen. This blue *p* orbital overlaps the red *p* orbital on the carbon just like the red *p* orbital on the other oxygen does.

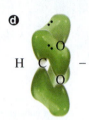

To get an accurate picture for the formate anion, it is necessary to use delocalized MOs that involve all three of the overlapping *p* orbitals (one from each O and one from C). These three AOs overlap to form three pi MOs. There are four electrons (shown in the Lewis structures in parts ⓐ and ⓑ as the pi electrons and one unshared pair on oxygen) in these three MOs. This drawing does not attempt to show the shapes of these three MOs, only the orbitals that overlap to form them. More on their energies and shapes will be presented in later chapters.

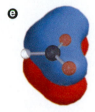

This is a computer-generated picture of the lowest energy pi bonding MO. The three *p* orbitals overlap without any nodes to produce this MO. It looks much like a pi bonding MO, with electron density above and below the plane of the atoms, except that it extends over three atoms, rather than two. Pictures like this, showing the lowest-energy pi MO, are provided throughout the book because they help to visualize how the *p* orbitals overlap to form delocalized MOs.

Figure 3.15

RESONANCE AND ORBITAL PICTURES FOR THE FORMATE ANION.

parallel *p* orbitals on adjacent atoms. The formate anion shown in Figure 3.15 is conjugated because of the three parallel *p* orbitals on the carbon and the two oxygens. However, if the orbitals are not parallel or if there is an extra atom separating them, the overlap is not continuous and the system is not conjugated. Several examples are presented in practice problem 3.1 and problem 3.10.

PRACTICE PROBLEM 3.1

Explain which pi bonds are conjugated in this compound:

$$CH_2=CH-CH=CH-CH_2-CH=CH_2$$

Solution

The best way to see which bonds are conjugated is to draw the molecule showing the *p* orbitals that form the pi bonds:

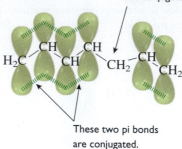

This *sp³*-hybridized C interrupts
the string of parallel *p* orbitals,
so the double bond on the right is
not conjugated with the others.

These two pi bonds
are conjugated.

PROBLEM 3.10

Circle the conjugated pi bonds, if any, in the following compounds.

a) 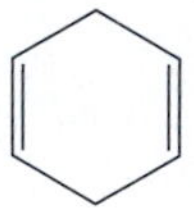 [hexene ring structure]

b) $CH_3—CH=CH—\overset{\overset{\displaystyle O}{\|}}{C}—CH_3$

c) $CH_3—CH=CH—C\equiv N$

d) $H—C\equiv C—CH=CH—\overset{\overset{\displaystyle O}{\|}}{C}—H$

e) $CH_3—CH=CH—CH_2—\overset{\overset{\displaystyle O}{\|}}{C}—CH_3$

PRACTICE PROBLEM 3.2

What is the hybridization at the indicated atoms in this compound?

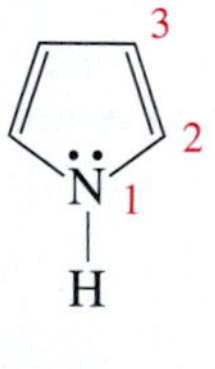

[pyrrole ring structure with positions 3, 2 labeled and N 1, H below]

Solution

Carbons 2 and 3 are *sp²* hybridized. Nitrogen 1 has an unshared pair of electrons and is adjacent to the pi bonds. Therefore, it is *sp²* hybridized, and the "unshared pair" of electrons is in a *p* orbital. This *p* orbital is conjugated with the *p* orbitals of the pi bonds, resulting in additional resonance stabilization.

PROBLEM 3.11

What is the hybridization at the indicated atoms in these compounds?

a) $CH_3CH{=}CH{-}NH{-}CH_3$
      1   2   3   4   5

b)
$$\overset{\displaystyle \overset{..}{\underset{..}{O}}:}{\underset{1}{CH_3}}{-}\overset{..}{\underset{2}{C}}{-}\overset{..}{\underset{..}{\underset{3}{O}}}{-}\underset{4}{CH_3}$$

c) $CH_2{=}CH{-}\overset{..}{\underset{..}{O}}{-}CH_3$
      1   2   3   4

d)
$$\underset{2}{}\ \ \overset{\displaystyle \overset{..}{N}H_2}{\underset{1}{\bigcirc}}$$

3.7 RULES FOR RESONANCE STRUCTURES

Resonance is a very important concept in organic chemistry. It will help us determine how reactive a compound is likely to be and where the most reactive sites in the molecule are located. The following rules will help us to draw resonance structures and determine their importance.

> ▶ **RULE 1**
>
> In drawing resonance structures, the nuclei of atoms may not be moved; only pi and nonbonding electrons in conjugated *p* orbitals may be moved.

As we saw for the formate ion in Figure 3.15, resonance structures are just different ways of arranging pi bonds and electrons among a series of adjacent *p* orbitals that form a conjugated system.

PRACTICE PROBLEM 3.3

Explain whether or not these structures represent resonance structures:

$$CH_3{-}\overset{\displaystyle \overset{..}{\underset{..}{O}}:}{\underset{\displaystyle \underset{..}{\overset{..}{O}}{-}H}{C}} \qquad \text{and} \qquad CH_3{-}\overset{\displaystyle \overset{..}{O}{-}H}{\underset{\displaystyle \underset{..}{\overset{..}{O}}:}{C}}$$

Solution

These are not resonance structures. In addition to changing the positions of some electrons, the H has been moved from the lower O to the upper O.

PROBLEM 3.12

Explain whether or not these structures represent resonance structures:

a) 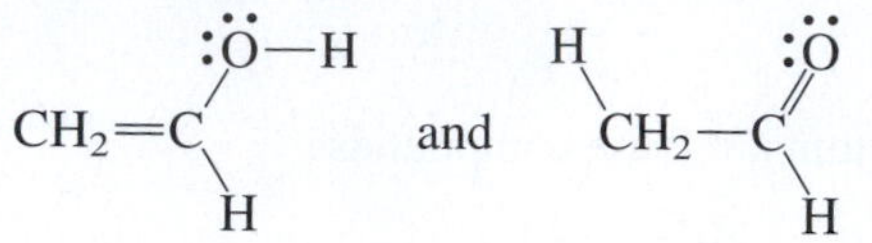 $CH_2{=}C$ with $:\ddot{O}{-}H$ and H groups   and $CH_2{-}C$ with $:\ddot{O}$ and H groups

b) 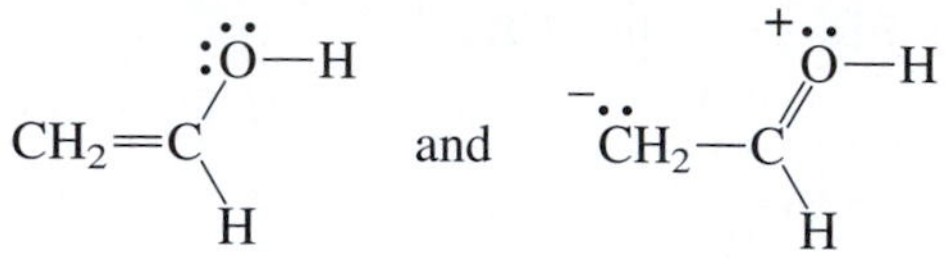 $CH_2{=}C$ with $:\ddot{O}{-}H$ and H groups   and $\bar{\ddot{C}}H_2{-}C$ with $\overset{+}{\ddot{O}}{-}H$ and H groups

c) $CH_3CH{=}CH{-}CH_3$ and $CH_3CH_2{-}CH{=}CH_2$

> ▶ **RULE 2**
>
> Each resonance structure must have the same number of electrons and the same total charge.

A common mistake that students make is to lose track of electrons or charges. For example, if one resonance structure is neutral, then any other resonance structure must be neutral overall. It may have formal charges, even if the original structure does not, but each positive formal charge must be balanced by a negative formal charge. To keep the number of electrons the same in each structure, the sum of the pi and the unshared electrons must be the same. Often, but not always, resonance structures have the same number of pi bonds and the same number of unshared electron pairs.

PRACTICE PROBLEM 3.4

Explain why the structure on the right is not a valid resonance structure for the structure on the left.

$$CH_2{=}\overset{+}{N}{=}\overset{..}{\underset{}{N}}:^{-} \quad \text{and} \quad CH_2{-}\overset{+}{N}{\equiv}N:$$

Solution

The structures do not have the same number of electrons, nor do they have the same total charge. A pair of electrons is missing in the structure on the right, and the formal charge on the C is not correct. (As drawn, the C would have a formal charge of $+1$.) The structure on the left has four pi electrons and four unshared electrons; the one on the right has four pi electrons and two unshared electrons. Correct resonance structures are

$$CH_2=N=N: \longleftrightarrow CH_2-N\equiv N:$$

Note how the *curved arrows* are used to help keep track of electrons. *By convention, the arrows always point from where the electrons are in a structure to where they are going in the next structure.* Here the arrows show the electrons of the CN pi bond of the left structure becoming the pair on the C in the right structure and the pair of electrons on the N in the left structure becoming the electrons of the new NN pi bond in the right structure.

◄ Important Convention

PROBLEM 3.13

Explain why the structure on the right is not a valid resonance structure for the structure on the left.

a) 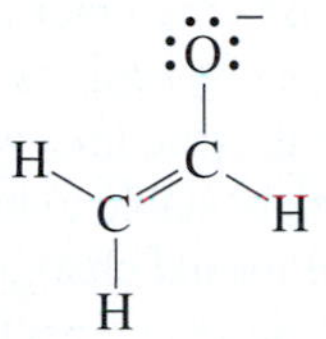 and

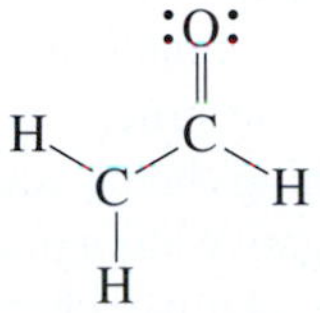

b) $CH_3-CH=CH-\overset{+}{CH_2}$ and $CH_3=CH-CH=CH_2$

c) $H-\overset{+}{N}=N=\overset{-}{N}:$ and $H-N=N-N:$

d) 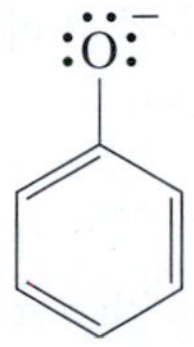 and

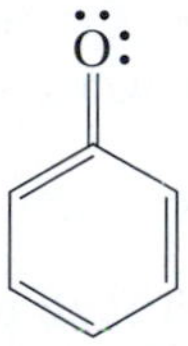

▶ RULE 3

The relative stability of resonance structures can be judged by the same rules that were previously introduced to judge the stability of Lewis structures: the octet rule, the number and location of formal charges, and the interactions between charges in the structure.

Because they come closer to satisfying the octet rule, structures with more bonds are usually more stable than structures with fewer bonds. Structures with fewer formal charges are usually more stable because the separation of positive and negative charge requires energy. The location of the formal charges is also important. Structures with formal charges are more stable when unlike charges are closer together and when like charges are farther apart. Furthermore, it is better to have a negative formal charge on a more electronegative atom and a positive formal charge on a less electronegative atom.

Rank the stability of these resonance structures. Describe the actual structure and the amount of resonance stabilization for this compound.

$$CH_2=\overset{+}{N}=\overset{\cdot\cdot}{\underset{\cdot\cdot}{N}}\colon^- \longleftrightarrow \overset{\cdot\cdot}{\underset{\cdot\cdot}{C}}H_2-\overset{+}{N}\equiv N\colon \longleftrightarrow CH_2=\overset{+}{N}=\overset{\cdot}{N}\colon \longleftrightarrow \overset{+}{C}H_2-\overset{\cdot\cdot}{N}=\overset{\cdot\cdot}{\underset{\cdot\cdot}{N}}\colon^-$$

Strategy

First examine the structures to determine whether they satisfy the octet rule. In general, structures that satisfy the octet rule are considerably more stable than structures that do not. In general, structures with more pi bonds are more likely to satisfy the octet rule and are more stable than structures with fewer pi bonds (as long as the octet rule is not exceeded). Next, look at the formal charges in the structures. In general, structures with fewer formal charges are more stable. Other things being equal, it is more stable to have a negative formal charge on a more electronegative atom or a positive formal charge on a less electronegative atom. Finally, remember that like charges repel, so structures that have like charges close together are destabilized. Likewise, opposite charges attract, so structures that have opposite charges farther apart are destabilized.

Solution

$$CH_2=\overset{+}{N}=\overset{\cdot\cdot}{\underset{\cdot\cdot}{N}}\colon^-$$

Although it has formal charges, this structure has the octet rule satisfied at all of the atoms. It is the most stable of all of these resonance structures.

$$\overset{\cdot\cdot}{\underset{\cdot\cdot}{C}}H_2-\overset{+}{N}\equiv N\colon$$

This structure also has the octet rule satisfied at all of the atoms. (Note that it has the same number of pi bonds as the previous structure.) It is slightly less stable than the previous one because it has the negative formal charge on the less electronegative C rather than on the more electronegative N.

$$CH_2\!\!\!\!\!\diagdown\!\!\!\!\!\!\!\diagup N\equiv N\colon$$

This structure is extremely unstable because the nitrogen in the center has 10 electrons. As you learned in Chapter 1, structures that have more than 8 electrons around a second-row atom should not be written.

$$\overset{+}{C}H_2-\overset{\cdot\cdot}{N}=\overset{\cdot\cdot}{\underset{\cdot\cdot}{N}}\colon^-$$

This structure is considerably less stable than the first two because the octet rule is not satisfied at the C. (Note that it has fewer pi bonds than the first two structures.)

The first two structures make important contributions to the resonance hybrid, whereas the contributions of the last two are unimportant, and they are usually not shown. There is more negative charge on the N than on the C because the first structure is somewhat more stable and contributes more to the resonance hybrid. Because it has two important resonance structures, the compound has considerable resonance stabilization and is significantly more stable than is suggested by either of these structures.

PROBLEM 3.14

Rank the stability of these resonance structures.

a) 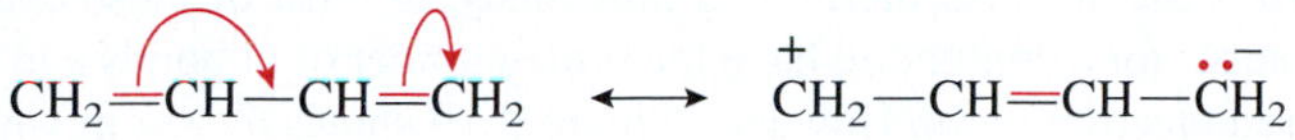 CH_2=CH—CH=CH_2 ⟷ $\overset{+}{CH_2}$—CH=CH—$\overset{..}{CH_2}^{-}$

b)

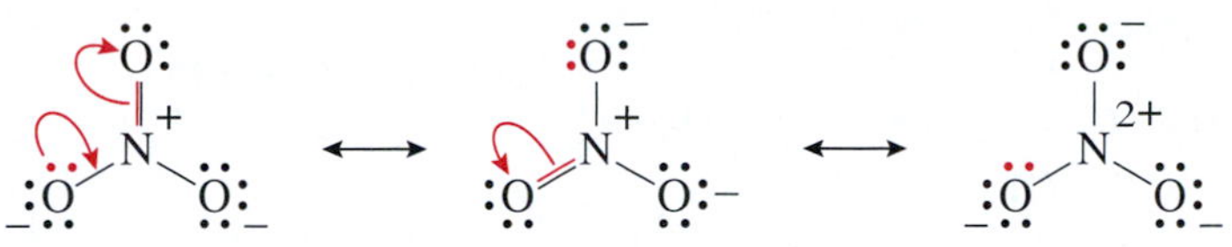

▶ **RULE 4**

The actual structure most resembles the most stable resonance structure.

In other words, a more stable resonance structure contributes more to the resonance hybrid and is said to be more important. Structures of equal stability contribute equally.

▶ **RULE 5**

The resonance stabilization energy increases as the number of important resonance structures increases.

For example, if a compound has two equivalent resonance structures, it is considerably more stable than either. We say that it has a large resonance stabilization energy. In contrast, if a compound has two resonance structures but one is considerably more stable than the other, then the structure more closely resembles the more stable (more important) resonance structure. The energy is also close to that of the more stable structure. The compound is only a little more stable than is indicated by the more stable structure, and it has only a small resonance stabilization.

The use of these rules is illustrated by the examples in the following section.

PROBLEM 3.15

Discuss the actual structure and the amount of resonance stabilization for the examples shown in problem 3.14.

3.8 TYPES OF RESONANCE INTERACTIONS

Figures 3.16 through 3.20 illustrate the common types of bonding situations in which the resonance concept is applied. In each case there is a conjugated series of p orbitals. Some of the p orbitals are part of pi bonds. Other p orbitals may contain a pair of electrons (Figure 3.16), one electron (Figure 3.17), or even no electrons (Figure 3.18). Resonance is also important for pi bonds between atoms of different electronegativities (Figure 3.19) and for cycles of pi bonds (Figure 3.20). The resonance structures are just the various ways the electrons can be arranged among the p orbitals to arrive at different but reasonable Lewis structures. As mentioned previously, in most of these examples the resonance structures for a compound have the same number of pi bonds and the same number of unshared electron pairs. However, Figure 3.19 shows a case in which

Figure 3.16

RESONANCE INVOLVING AN UNSHARED PAIR OF ELECTRONS NEXT TO A PI BOND: ❶ ACETATE ANION, ❷ ACETIC ACID, ❸ ACETONE ANION.

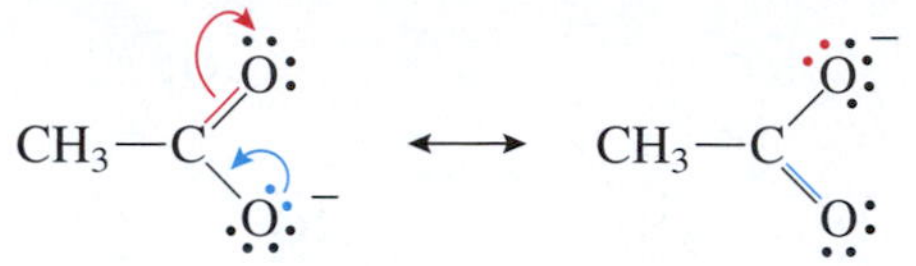

Acetate anion

This is the acetate anion. The curved arrows are used to help keep track of how electrons are moved to get from the first resonance structure to the second. An unshared pair of electrons on the lower oxygen is moved in to become the pi electrons in the second structure. The pi electrons are moved to become an unshared pair on the upper oxygen. Resonance structures must always have the same total charge — in this case −1. These structures happen to be equivalent in other respects also, so they contribute equally to the resonance hybrid. With two important resonance structures, the acetate anion has a large resonance stabilization. It is significantly more stable than would be predicted on the basis of examination of only one of the structures.

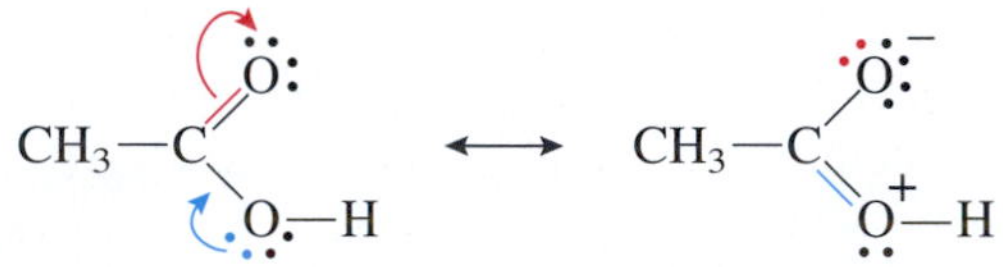

Acetic acid

This is acetic acid, a neutral molecule. Similar resonance structures can be written for acetic acid as are shown in part ❶ for the acetate anion. In this case the two structures are not the same. The second structure is still neutral overall, but it has two formal charges. Therefore, the first structure is more stable and contributes much more to the resonance hybrid than the second does. Acetic acid has a smaller resonance stabilization than that of acetate anion — it is only a little more stable than the first structure would indicate.

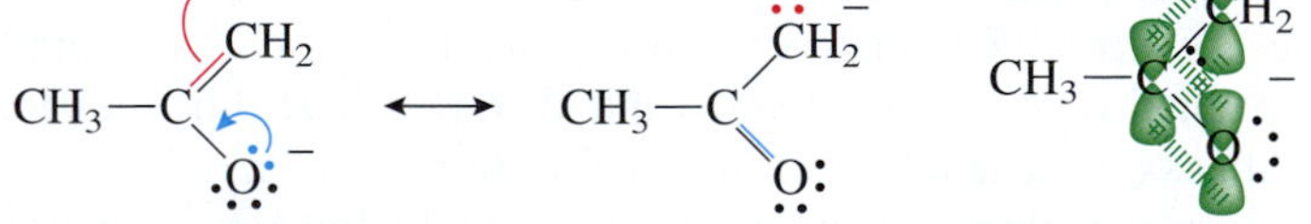

Anion from acetone

This anion resembles acetate ion, but with a CH_2 in place of one O. Again the same type of resonance structures can be drawn. Both are important, but the first one contributes more because the negative charge is located on the more electronegative oxygen atom. This anion has significant resonance stabilization, though less than the acetate anion. The rightmost diagram shows the conjugated p orbitals of this anion.

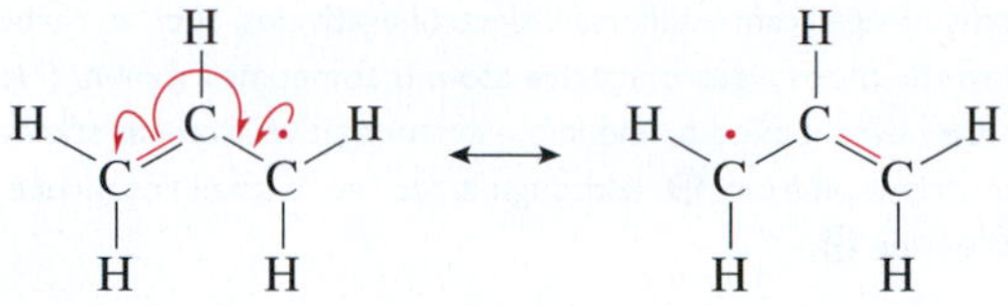

Allyl radical

The allyl radical has an odd number of electrons. The odd electron is in a p orbital, so the species is conjugated. It has two equally important resonance structures. The octet rule is not satisfied, so this radical is an unstable, reactive species. However, because of its large resonance stabilization, it is not as unstable as would be predicted on the basis of examination of a single structure without delocalization. Single-headed arrows are used to show movement of one electron, rather than electron pairs. Radicals are discussed in more detail in Chapter 21.

Figure 3.17

RESONANCE INVOLVING ONE ELECTRON NEXT TO A PI BOND.

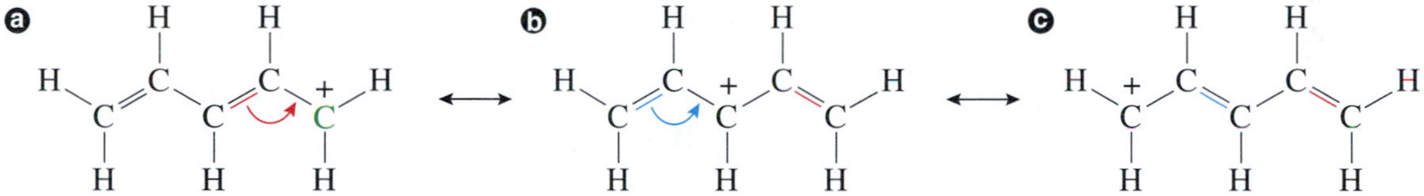

In Lewis structure **ⓐ**, the C with the positive charge is sp^2 hybridized with an empty p orbital. Therefore, the molecule has a series of five conjugated p orbitals, as shown below in **ⓓ**. Moving one pair of pi electrons provides **ⓑ**, with the positive charge located on the center C. Moving the other pair of pi electrons produces **ⓒ**, with the positive charge on the left C. **ⓐ** and **ⓒ** are completely equivalent, and **ⓑ** is quite similar. All three structures are important contributors to the resonance hybrid, and the ion has a large resonance energy. Because the octet rule is not satisfied at one C in any of the structures, this is a reactive species. However, this ion is much less reactive than a similar ion with a positive charge on C but without resonance stabilization. Finally, note that both end carbons and the center carbon have some positive charge and are electron deficient. Reaction with an electron-rich species may occur at any of these three carbons.

These diagrams show the overlap of a series of five conjugated p orbitals in this cation. The hydrogens have been omitted for clarity in the drawing at the left.

Figure 3.18

RESONANCE INVOLVING AN EMPTY p ORBITAL NEXT TO PI BONDS.

When a molecule has a pi bond between atoms of significantly different electronegativities, such as carbon and oxygen in this example, a resonance structure where the pi electrons are moved to the more electronegative atom is sometimes shown. (Note that there are only two parallel *p* orbitals involved in this case.) The electrons are never moved to the less electronegative atom, as shown in **c**. This compound looks most like **a**, with a minor contribution from **b** and no contribution from **c**. Although it has only a small resonance stabilization, the carbon has more positive charge than would be expected without considering **b**.

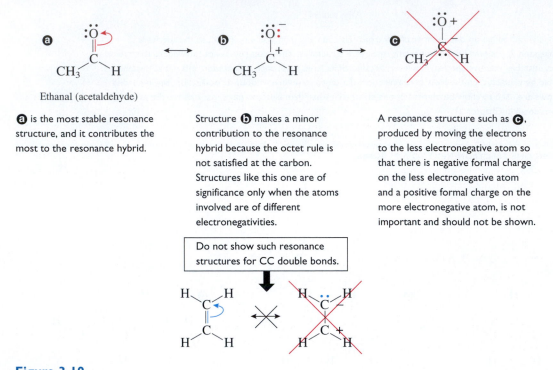

Ethanal (acetaldehyde)

a is the most stable resonance structure, and it contributes the most to the resonance hybrid.

Structure **b** makes a minor contribution to the resonance hybrid because the octet rule is not satisfied at the carbon. Structures like this one are of significance only when the atoms involved are of different electronegativities.

A resonance structure such as **c**, produced by moving the electrons to the less electronegative atom so that there is negative formal charge on the less electronegative atom and a positive formal charge on the more electronegative atom, is not important and should not be shown.

Do not show such resonance structures for CC double bonds.

Figure 3.19

RESONANCE INVOLVING PI BONDS BETWEEN ATOMS OF DIFFERENT ELECTRONEGATIVITIES.

these numbers are not the same. Note that the structures in Figure 3.19 are of very different stability.

Examine these figures carefully so that you gain experience in how and when the resonance model is used. Again note the use of curved arrows to show how the electrons are rearranged in converting one structure to another. *Remember, these arrows point away from where the electrons currently are in a structure to where they will be located in the new structure.* The arrows are a useful device to help keep track of electrons. As mentioned earlier, a common mistake that students make is not being careful enough with electrons so that resonance structures do not have the same number of electrons. The use of the curved arrows and careful attention to details will help you avoid such mistakes. (These arrows will also be used later to show how electrons move in chemical reactions.) Finally, carefully compare the resonance structures in each set to see how their relative importance in contributing to the overall resonance hybrid is determined.

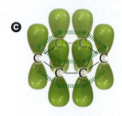

Benzene

A simple structure for benzene shows a cycle of double bonds. As can be seen in part **c**, the *p* orbitals of the double bonds form a conjugated series extending completely around the ring. By moving the electrons in **a** as shown, an equivalent resonance structure is produced, **b**. Structures **a** and **b** make equal contributions to the resonance hybrid. In accord with this, experiments show that all the CC bonds of benzene are identical, with lengths of 1.40 Å, longer than typical double bond (1.34 Å) and shorter than a typical single bond (1.54 Å). Benzene has a very large resonance energy.

Molecules like this, with a ring of parallel *p* orbitals forming a conjugated cycle, are special and often have more stabilization than expected. Chapter 16 treats such molecules in more detail.

Part **c** shows how the *p* orbitals of the pi bonds of benzene overlap in a complete cycle around the ring. The hydrogens have been omitted for clarity. Another version is shown in part **d**.

This computer-generated picture, of the lowest-energy pi MO of benzene shows how the *p* orbitals overlap to form a continous cloud of electron density above and below the plane of the atoms in the ring. This is the lowest-energy pi MO because there is no new node introduced when the *p* orbitals overlap.

Figure 3.20

RESONANCE INVOLVING A CYCLE OF DOUBLE BONDS.

PROBLEM 3.16

Show the important resonance structures for these species. Use the curved arrow convention to show how the electrons are moved to create each new resonance structure.

Many compounds involve combinations of the preceding resonance types. Figure 3.21 shows resonance structures for the anion that results from removing a proton (H^+) from the oxygen of phenol.

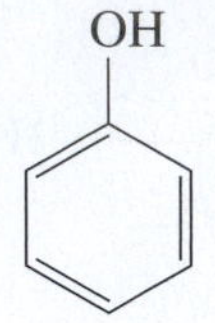

OH

Phenol

This ion has a combination of an unshared pair of electrons (on the O) next to a cycle of double bonds in a benzene ring. The resonance structures are a combination of the types shown in Figures 3.16 and 3.20. The ion has five important resonance structures and thus a large resonance stabilization. Furthermore, the oxygen has decreased electron density in comparison to what might be expected without considering resonance. Certain ring atoms have increased electron density. Later, we will see how these resonance structures accurately predict the chemical behavior of this anion.

The anion derived from phenol by loss of a proton has five important resonance structures that contribute significantly to the resonance hybrid.

Resonance structures **ⓐ** and **ⓑ** are equivalent. Both of these contribute equally to the resonance hybrid.

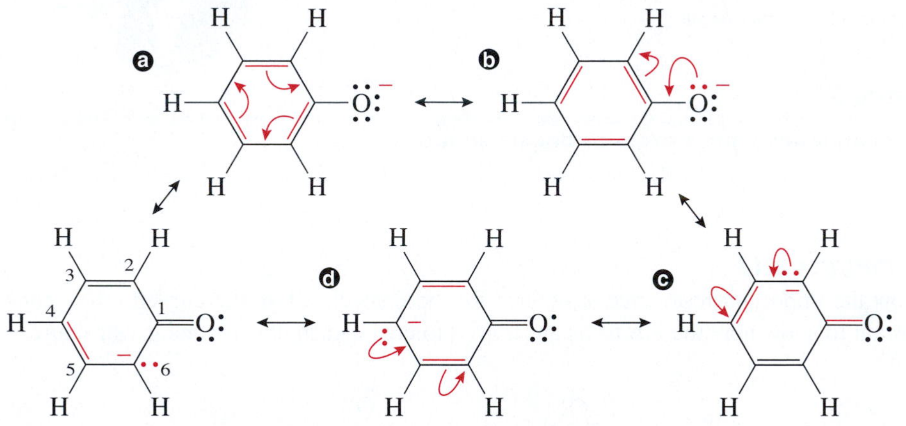

Resonance structures **ⓒ**, **ⓓ**, and **ⓔ** have the negative charge on C rather than the more electronegative O. For this reason they contribute less to the resonance hybrid than do **ⓐ** and **ⓑ**. Two of these are equivalent, and the other is quite similar in stability, so **ⓒ**, **ⓓ**, and **ⓔ** contribute approximately the same to the resonance hybrid.

All of these resonance structures have three pi bonds and three unshared electron pairs. Overall, because of the large number of important resonance structures, this anion has a large resonance stabilization. It is significantly more stable than examination of any single structure would indicate. It is also interesting to note the position of the negative charge in the various structures. Resonance shows that the ring has increased electron density at carbons 2, 4, and 6 but not at carbons 3 or 5. Later we will see how this helps explain some experimental results.

Figure 3.21

RESONANCE STRUCTURES FOR THE ANION DERIVED FROM PHENOL.

Figure 3.17 shows two equivalent resonance structures for the allyl radical. According to rule 5, this radical should have considerable resonance stabilization. Although it is still an unstable species because the octet rule is not satisfied at one of the carbons, it is considerably more stable than a radical that has no resonance stabilization.

This extra stabilization can be demonstrated by examination of bond dissociation energies. Section 2.2 lists the bond dissociation energy of a CH bond of ethane as 98 kcal/mol (410 kJ/mol). This is the energy that must be added to ethane to homolytically break one of its CH bonds.

+98 kcal/mol
(+410 kJ/mol)

Ethane

At first glance, it might seem that a similar amount of energy should be required to break one of the bonds between a hydrogen and the sp^3-hybridized carbon of propene. (Of course, a different bond dissociation energy is expected for the bond between a hydrogen and the sp^2-hybridized carbon.) However, breaking this CH bond requires the addition of only 85 kcal/mol (356 kJ/mol) of energy. It is weaker than the CH bond of ethane by 13 kcal/mol (54 kJ/mol).

+85 kcal/mol
(+365 kJ/mol)

Propene

Allyl radical

The CH bond in propene is weaker than the CH bond of ethane because the allyl radical is stabilized by resonance. The ethyl radical has no such resonance stabilization. The difference between these bond dissociation energies provides an estimate of the resonance stabilization of the allyl radical: 13 kcal/mol (54 kJ/mol).

Figure 3.22 shows resonance structures for a compound that has a CO double bond in conjugation with a CC double bond. The resonance structures are a combination of the types in Figures 3.18 and 3.19. Structure (a) has the octet rule satisfied at all atoms and has no formal charges, so it is more stable than the others and contributes the most to the resonance hybrid. Therefore, the actual structure most resembles this resonance structure (rule 4). In addition, the actual energy of the compound is also closer to the energy of the most important resonance structure. In other words, this compound has only a small resonance stabilization. However, even though structures (b) and (c) make

Figure 3.22

RESONANCE STRUCTURES
FOR 3-BUTEN-2-ONE.

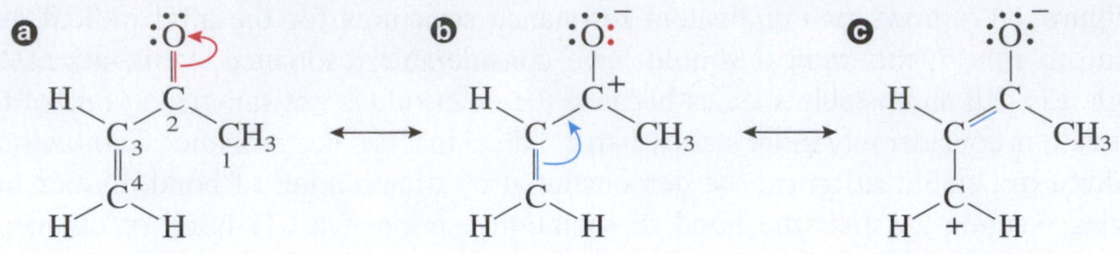

3-Buten-2-one

ⓐ has a CC double bond conjugated with a CO double bond. This resonance structure has the octet rule satisfied at all the atoms, and it has no formal charges. It is by far the major contributor to the resonance hybrid.

In **ⓑ** the electrons of the CO double bond have been moved onto the more electronegative oxygen in the same manner as shown in Figure 3.19. It is a minor contributor to the resonance hybrid because it has fewer pi bonds and the C has only six electrons.

ⓒ is also a minor contributor to the resonance hybrid.

3-Buten-2-one has a small resonance stabilization. It looks most like structure **ⓐ**. Interestingly, the resonance structures show that both carbon 2 and carbon 4 are electron deficient. Later, a confirmation of this in the chemical behavior of this compound will be presented.

only minor contributions to the resonance hybrid, they are important because it is these structures that show that carbons 2 and 4 are electron deficient. The electron-deficient nature of these carbons controls the chemical reactions of this compound.

Remember that the individual resonance structures have no discrete existence. The real structure, the resonance hybrid, is a static structure that is an average of the various resonance structures. The two-headed arrow (⟷) should be used only for resonance.

PRACTICE PROBLEM 3.6

Draw the important resonance structures for aniline. Use the curved arrow convention to show how the electrons are moved to create each new resonance structure. Discuss the relative contribution of each to the resonance hybrid and the overall resonance stabilization of the compound.

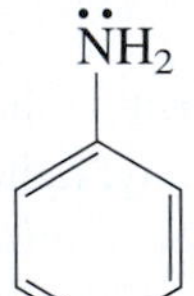

Aniline

Strategy

Examine the structure and determine the conjugated series of orbitals that is involved in resonance. Next, determine the types of resonance interactions that are present (see Figures 3.16–3.20). Draw the resonance structures and evaluate their relative stabilities using the octet rule and the number and location of any formal charges. More stable resonance structures contribute more to the resonance hybrid, and structures of equal stability make equal contributions. If the molecule has several important resonance structures, then it has a large resonance stabilization. If, on the other hand, one of the resonance structures is considerably more stable than the others, then the molecule looks very much like this structure and its energy is close to that of this structure—it has only a small resonance stabilization.

Solution

The nitrogen is sp^2 hybridized with its "unshared" pair of electrons in a p orbital. This is a combination of the resonance types involving a cycle of double bonds and a pair of electrons next to a double bond. In fact, the situation is very similar to that shown for the anion derived from phenol in Figure 3.21. The important resonance structures are as follows:

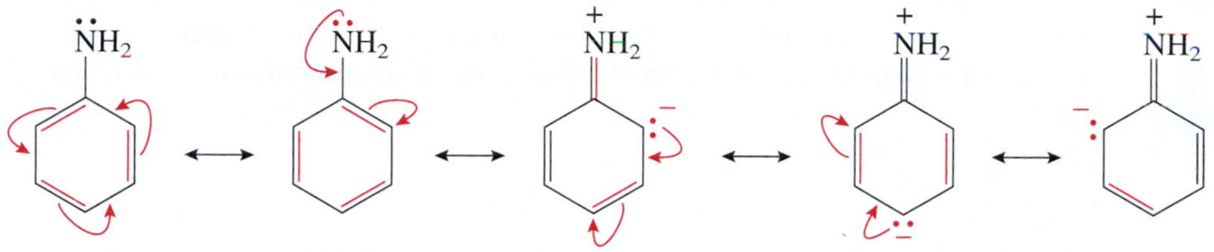

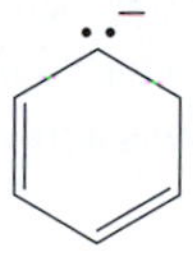

Structures **ⓐ** and **ⓑ** are equivalent and make identical contributions to the resonance hybrid. Because of their formal charges, structures **ⓒ**, **ⓓ**, and **ⓔ** make lesser contributions. Overall, this compound has considerable resonance stabilization.

PROBLEM 3.17

Draw the important resonance structures for these species. Use the curved arrow convention to show how the electrons are moved to create each new resonance structure. Discuss the relative contribution of each to the resonance hybrid and the overall resonance stabilization of the species.

ORGANIC
Chemistry Now™
Click *Coached Tutorial Problems*
to practice **Drawing
Resonance Structures.**

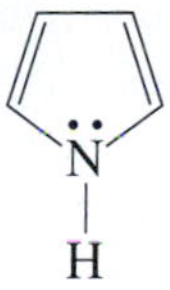

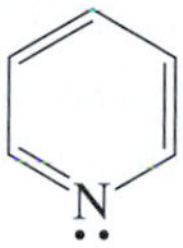

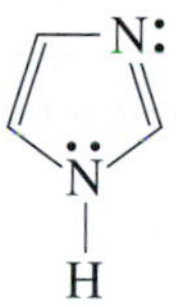

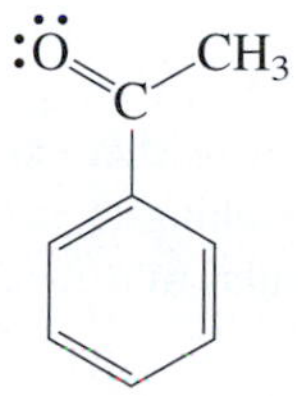

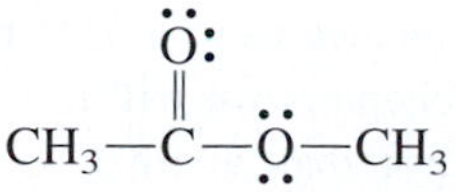

Focus On

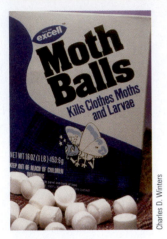

Moth balls are made from napthalene.

Charles D. Winters

Resonance and the Bond Lengths of Naphthalene

Figure 3.20 shows that there are two equivalent resonance structures for benzene. Each has alternating double and single bonds around the ring, so a particular carbon–carbon bond is a single bond in one structure and a double bond in the other. Because both structures contribute equally to the resonance hybrid, all the carbon–carbon bonds of benzene are identical and are intermediate in length between single and double bonds. A typical carbon–carbon single bond has a length of 1.54 Å, and a double bond has a length of 1.34 Å. The carbon–carbon bond lengths of benzene are 1.40 Å, intermediate between those of a single bond and a double bond, as predicted.

Naphthalene, which has two benzene-type rings fused together, provides a more interesting test for the predictive powers of resonance theory. There are three resonance structures for naphthalene, each with alternating single and double bonds around the rings:

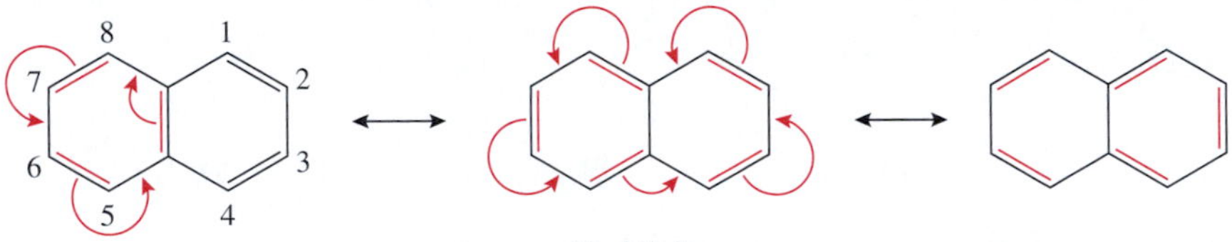

Naphthalene

Although the structure on the left differs slightly from the other two, which are equivalent, all three are quite similar and are expected to make nearly equal contributions to the resonance hybrid.

Now let's examine each bond in the different resonance structures. The bond between C-1 and C-2 is double in two of the resonance structures and single in only one. The same is true for the other bonds that are equivalent to this one, those between C-3 and C-4, between C-5 and C-6, and between C-7 and C-8. All of the other bonds are double in one resonance structure and single in two. Therefore, we would predict that the bond between C-1 and C-2, and the other equivalent bonds, should be shorter than the bonds in benzene but still longer than a normal double bond. The other bonds should be longer than the bonds in benzene but still shorter than a normal single bond. These predictions agree with the experimental bond lengths, which are shown in the following structure:

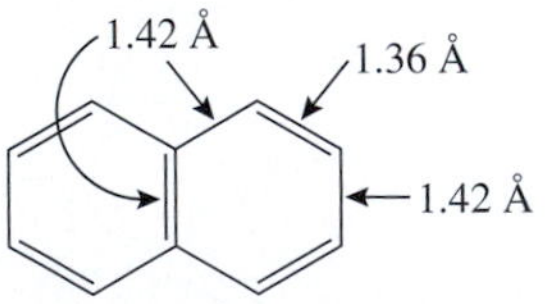

1.42 Å   1.36 Å

1.42 Å

This is just one example of how resonance can be used to explain or predict experimental observations. In subsequent chapters we will use similar reasoning to explain why one compound is more reactive than another, why a particular site in a molecule is more reactive than another, and so on.

3.9 MOLECULAR ORBITAL ENERGIES

Section 3.2 explained that two molecular orbitals are formed when two atomic orbitals overlap. One of these MOs is lower in energy than the interacting AOs (a bonding MO), and the other is higher in energy than the AOs (an antibonding MO). Although this model, which uses localized MOs, is only an approximation for molecules with more than two atoms, it is adequate for most situations. Using this model, let's now see how the energies of the MOs are arranged for some larger molecules.

In general, the lowest-energy MOs are sigma bonding MOs. They are low in energy because good overlap between the AOs causes ΔE to be large. For this same reason, sigma antibonding MOs are usually the highest-energy orbitals. Pi bonding MOs are not as low in energy as sigma bonding MOs because the overlap of the p atomic orbitals is not as good in the pi bonds, resulting in a smaller ΔE. For similar reasons, pi antibonding MOs are not as high in energy as sigma antibonding MOs. Finally, nonbonding orbitals, being neither bonding nor antibonding, are between the bonding and antibonding MOs in energy.

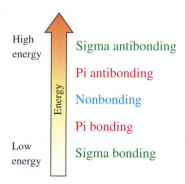

Figure 3.23 is a MO energy level diagram for ethane. This molecule has six CH bonds and one CC bond, so there are seven σ bonding MOs and the corresponding antibonding MOs. Although different sigma bonding MOs (or antibonding MOs) have different energies, these differences are usually not important to us. Therefore, the MO energy level diagrams in this book show all of these MOs at the same energy.

The Lewis structure for ethane shows that there are 14 electrons in this molecule. When these 14 electrons are placed in the MO diagram according to the same rules that

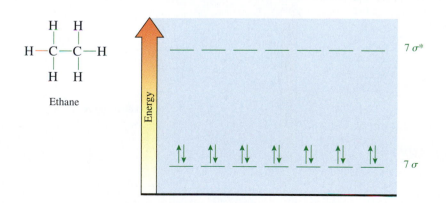

Figure 3.23

MOLECULAR ORBITAL ENERGY LEVEL DIAGRAM FOR ETHANE, C_2H_6.

Figure 3.24

MOLECULAR ORBITAL
ENERGY LEVEL DIAGRAM
FOR ETHENE, C_2H_4.

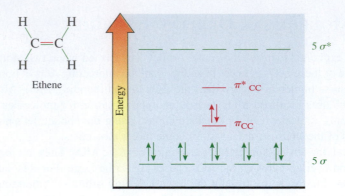

were used previously, the bonding MOs are all filled with electrons and the antibonding MOs are empty. This is typically what happens when an MO energy level diagram is constructed: the bonding MOs are filled, and the antibonding MOs are empty. This is the lowest-energy arrangement of the electrons and is termed the *ground electronic state* for the molecule. Any other arrangement of electrons is less stable and is termed an *excited electronic state.*

Figure 3.24 shows a MO energy level diagram for ethene. This molecule has four CH sigma bonds, one CC sigma bond, and one CC pi bond. There is a bonding MO and an antibonding MO for each of these bonds. The diagram shows the sigma MOs at the same energy. However, the pi bonding MO is significantly higher in energy than the sigma bonding MOs, and the pi antibonding MO is significantly lower in energy than the sigma antibonding MOs. The Lewis structure shows that there are 12 electrons to place in these MOs. Again, the bonding MOs are filled, and the antibonding MOs are empty.

Figure 3.25 shows a similar diagram for formaldehyde. The difference here is the presence of two unshared pairs of electrons on the oxygen. These electrons are in oxygen nonbonding AOs. The nonbonding orbitals are approximately in the middle of the energy level diagram. This time there are 12 electrons in the orbitals. The bonding and nonbonding orbitals are filled, and the antibonding orbitals are empty.

Diagrams similar to Figures 3.23, 3.24, and 3.25 will apply to most of the molecules that are encountered in this text. When delocalized MOs are discussed in Chapters 16 and 22, some modifications to the energy levels of the pi MOs will be necessary. However, these pi MOs will still remain between the sigma bonding and sigma antibonding MOs in energy.

Figure 3.25

MOLECULAR ORBITAL
ENERGY LEVELS FOR
FORMALDEHYDE, CH_2O.

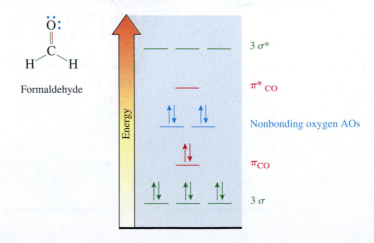

PROBLEM 3.18

Show energy level diagrams for the MOs of these compounds.

a) $H—C\equiv N:$

b) 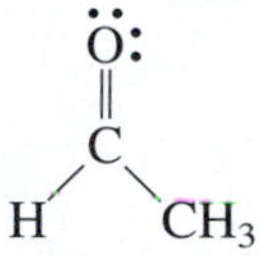 structure: O double bonded to C, with H and CH_3 attached to C

c) $CH_3—\ddot{N}H_2$

Review of Mastery Goals

After completing this chapter, you should be able to:

■ Assign the ground-state electron configuration for simple atoms.

■ Identify any bond as sigma or pi. (Problem 3.34)

■ Draw pictures for various sigma and pi bonding and antibonding MOs.

■ Identify the hybridization of all atoms of a molecule. (Problems 3.20, 3.21, 3.22, 3.40, and 3.44)

■ Identify the type of molecular orbital occupied by each electron pair in a molecule and designate the atomic orbitals that overlap to form that MO. (Problems 3.21 and 3.40)

■ Draw the important resonance structures for any molecule. Assign the relative importance of these structures and estimate the resonance stabilization energy for the molecule. (Problems 3.24, 3.25, 3.26, 3.27, 3.28, 3.29, 3.35, 3.36, 3.37, 3.38, and 3.41)

■ Show a MO energy level diagram for all the orbitals for any molecule. (Problems 3.30, 3.31, 3.32, 3.33, and 3.39)

ORGANIC
Chemistry·Now™
Click *Mastery Goal Quiz* to test how well you have met these goals.

Additional Problems

3.19 Show the location of the two planar nodes in this $3d$ atomic orbital:

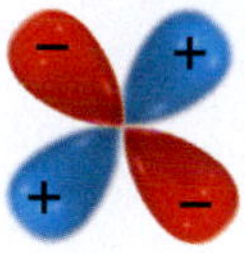

ORGANIC
Chemistry·Now™
Assess your understanding of this chapter's topics with additional quizzing and conceptual-based problems at
http://now.brookscole.com/hornback2

3.20 What is the hybridization at all atoms, except hydrogens, in these compounds?

a) $CH_3—NH_2$ b) $CH_2\!=\!CHCH_2C\equiv N$ c)  structure: cyclopentene ring

d) 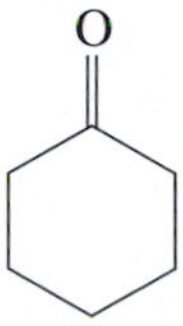 cyclohexanone (ring with O double bond at top)

e) 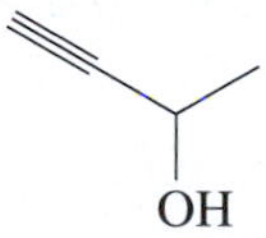 structure: $HC\equiv C$ chain with CH bearing OH

f) 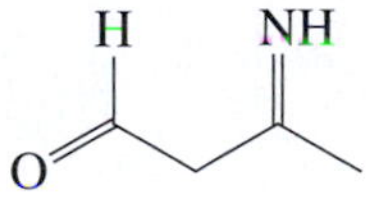 structure: O double bond, chain with H, and $=NH$ imine

3.21   Show the hybridization at each of the atoms, except H, in these molecules. Indicate the type of each designated bond and the orbitals that are overlapping to form it.

a) 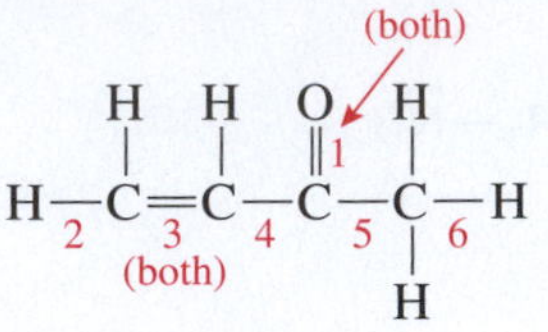
$$H-\underset{2}{C}=\underset{3}{C}-\underset{4}{\overset{O}{\underset{\parallel}{C}}}-\underset{5}{C}-H$$
(both)   (both)

b)
$$H-\underset{1}{C}-\underset{2}{C}-\underset{3}{O}-\underset{4}{H}$$

c)
$$H-C-\underset{1}{C}-\underset{2}{C}\equiv N$$
(all three)

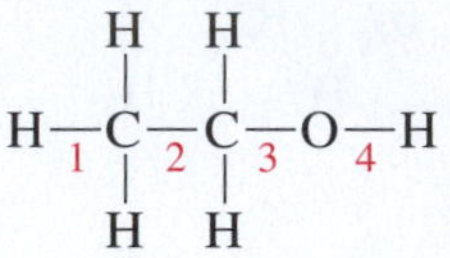

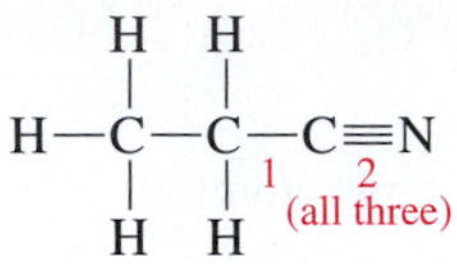

d) 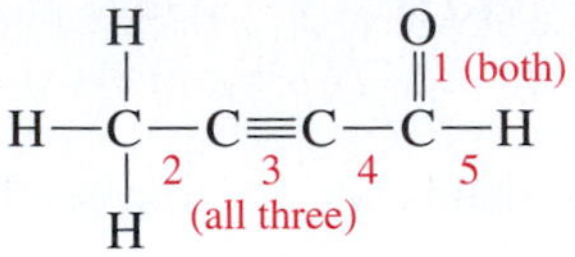
$$H-\underset{}{C}-\underset{2}{C}\equiv\underset{3}{C}-\underset{4}{\overset{O}{\underset{\parallel}{C}}}-\underset{5}{H}$$
1 (both)
(all three)

3.22   What is the hybridization at all atoms, except hydrogens, in these compounds?

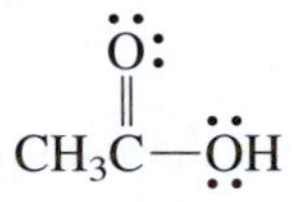

a) 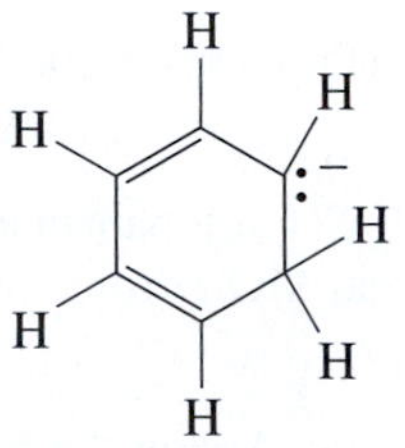   b) 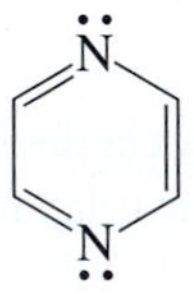   c) 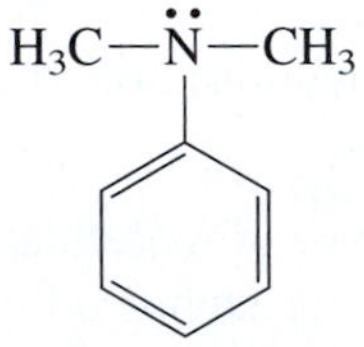 $H_3C-\ddot{N}-CH_3$ d)

e) 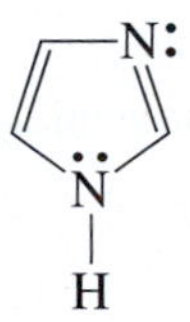   f) 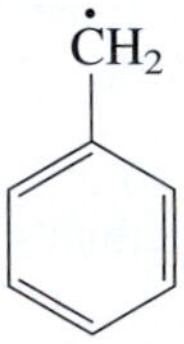 $\dot{C}H_2$ g) $CH_3\overset{\ddot{O}:}{\underset{\parallel}{C}}-\ddot{O}H$ h) $CH_3\overset{\ddot{O}:}{\underset{\parallel}{C}}-\ddot{N}HCH_3$

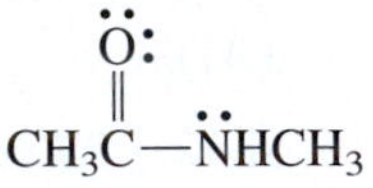

3.23   Draw the *p* orbitals that compose the conjugated part of these molecules:

a) 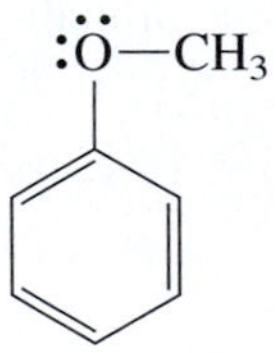 $:\!\ddot{O}-CH_3$ b) $CH_2=CH-\ddot{N}H_2$ c) $H-C\equiv C-CH=CH_2$

3.24   Show the important resonance structures for these compounds. Use the curved arrow convention to show how the electrons are moved to create each new resonance structure.

a) 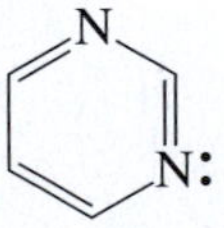   b) 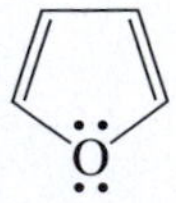   c) 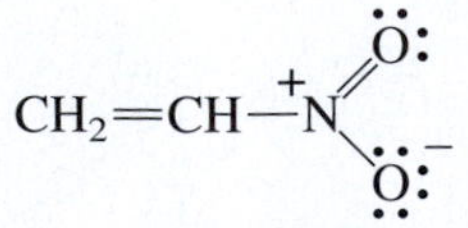 $CH_2=CH-\overset{+}{N}\!\!\begin{array}{c}\ddot{O}:\\\ddot{O}:^-\end{array}$

d) CH$_3$—C $\overset{\overset{\displaystyle :\overset{..}{O}{}^{+}—H}{\big|}}{\underset{\displaystyle \overset{..}{\underset{..}{O}}—H}{}}$ **e)** H—C≡C—$\overset{..}{\underset{..}{C}}H_2$

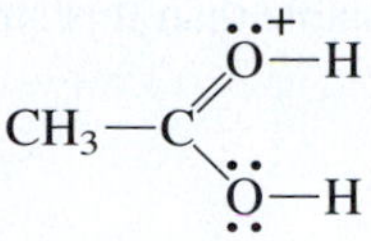

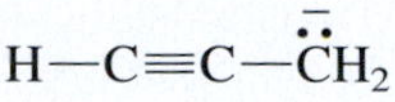

3.25 In these examples the additional structure or structures are not important contributors to the resonance hybrid for the compound represented by the first structure. Explain.

a) ⟷ ⟷

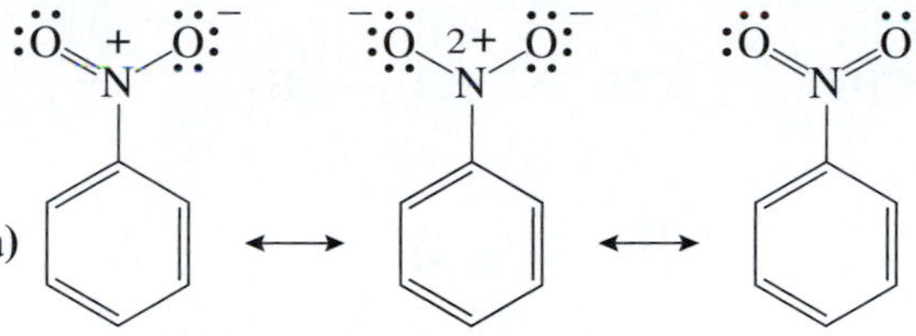

b) ⟷

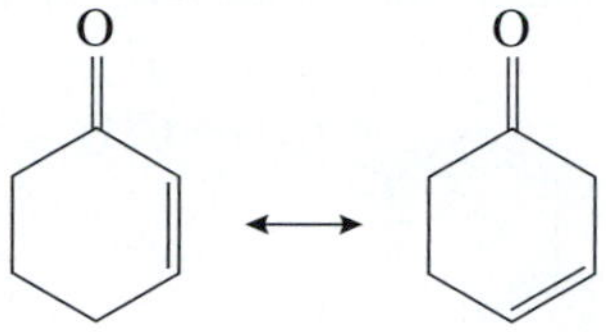

c) $\overset{-}{\underset{..}{C}}H_2$—C≡N: ⟷ CH$_2$=C=N:

d) H—$\overset{\overset{\displaystyle :\overset{..}{\underset{..}{O}}}{\|}}{C}$—CH=CH$_2$ ⟷ H—$\overset{\overset{\displaystyle :\overset{..}{O}{}^{+}}{\|}}{C}$=CH—$\overset{-}{\underset{..}{C}}H_2$

3.26 Draw the important resonance structures for these species and discuss the contribution of each to the resonance hybrid. Explain whether the species has a large or a small amount of resonance stabilization.

a)

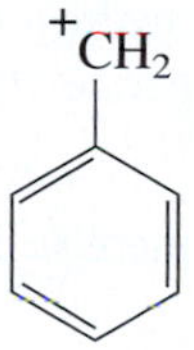

b) $\overset{..}{C}$H$_2$—CH=CH—$\overset{\overset{\displaystyle \overset{..}{\underset{..}{O}}:}{\|}}{C}$—CH$_3$

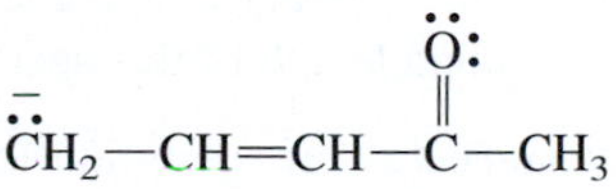

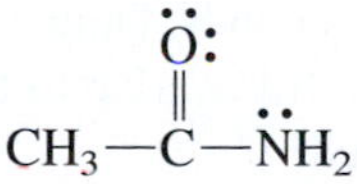

c) CH$_3$—$\overset{\overset{\displaystyle \overset{..}{\underset{..}{O}}:}{\|}}{C}$—$\overset{..}{N}H_2$

d) CH$_3$—$\overset{\overset{\displaystyle \overset{..}{\underset{..}{O}}:}{\|}}{C}$—$\overset{-}{\underset{..}{N}}$H

e) $\overset{-}{\underset{..}{C}}H_2$—CH=CH—$\overset{+}{N}$$\overset{\nearrow \overset{..}{\underset{..}{O}}:}{\searrow \underset{..}{O}:_{-}}$

f) CH$_3$—$\overset{\overset{\displaystyle :\overset{..}{O}}{\|}}{C}$—$\overset{..}{\underset{..}{C}}$H—$\overset{\overset{\displaystyle \overset{..}{O}:}{\|}}{C}$—CH$_3$

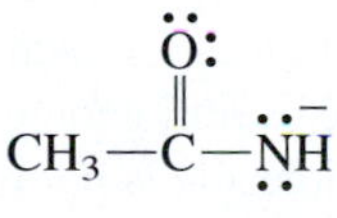

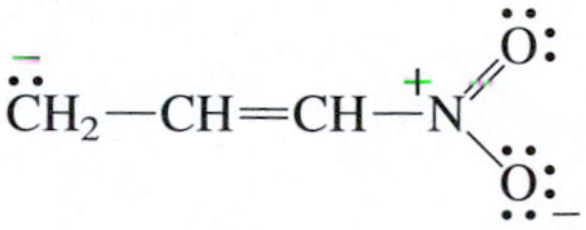

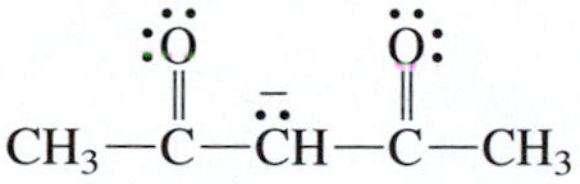

3.27 Explain why this carbocation is considerably more stable than this structure would suggest:

$$\overset{H}{\underset{H}{+C}}-\ddot{\underset{..}{O}}-CH_3$$

3.28 Explain why one of these anions is much more stable than the other:

a) $CH_3-\overset{\overset{..}{O}:}{\overset{\|}{C}}-\overset{..}{C}H-CH_3 \qquad CH_3-\overset{\overset{..}{O}:}{\overset{\|}{C}}-CH_2-\overset{..}{C}H_2$

b) $^-\overset{..}{C}H_2-CH_3 \qquad \overset{..}{C}H_2-C\equiv N:$

3.29 Explain why one of these carbocations is much more stable than the other:

$\overset{+}{C}H-CH_3 \qquad CH_2-\overset{+}{C}H_2$

3.30 Show energy level diagrams for the MOs of these compounds:

a) $H-C\equiv C-CH_3$

b) $CH_3-\overset{..}{\underset{..}{O}}-H$

c) $CH_3-\overset{\overset{..}{N}\diagdown^H}{\overset{\|}{C}}-H$

3.31 The energy level diagram for the MOs of $CH_2{=}CH_2$ is shown in Figure 3.24. Show a similar diagram for the lowest-energy excited state of this molecule.

3.32 Consider the species formed by the addition of an extra electron to H_2 so that there are three electrons and a negative charge. Show an energy level diagram for the MOs of this species. Is there still a bond between the hydrogens—that is, is it still necessary to add energy to cause the atoms to separate? Predict how the bond length of this species compares to that of H_2.

3.33 Draw an energy level diagram for the excited state of H_2. Is there still a bond between the hydrogens?

3.34 Consider $H_2C{=}C{=}CH_2$. What is the hybridization and geometry at each C? Indicate the bond types for each of the carbon–carbon bonds. Draw the molecule, showing the overall geometry and the p orbitals that overlap to form the pi bonds. (The molecule is not planar.) Is it conjugated?

3.35 Draw resonance structures for this anion. Remember, sulfur can have 10 or even 12 electrons in its valence shell.

$$CH_3-\overset{\overset{:\ddot{O}:^-}{|}}{\underset{\underset{:O:^-}{|}}{S}}{}^{2+}\ddot{O}:^-$$

3.36 Phenanthrene has five total resonance structures. One is shown here. Draw the other four. Which carbon–carbon bond of phenanthrene would you predict to be the shortest?

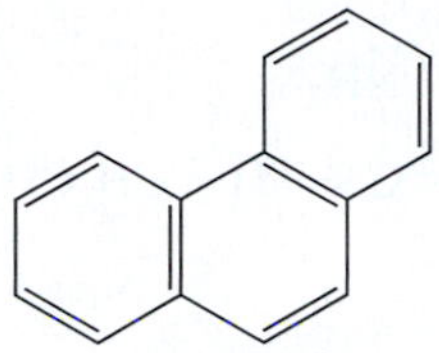

Phenanthrene

3.37 Show the three additional resonance structures for anthracene. Discuss whether the experimental bond lengths shown in the following structure are in accord with predictions based on these resonance structures:

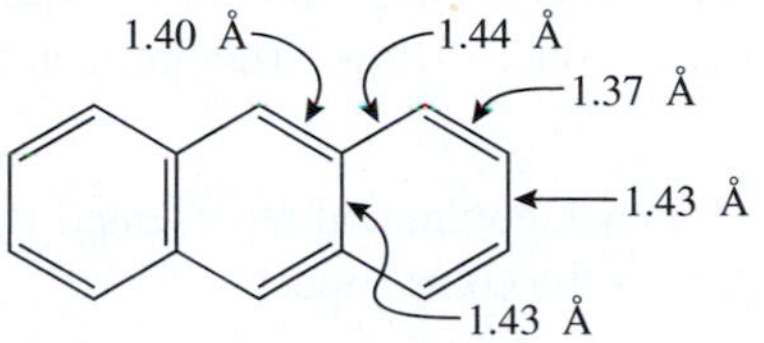

1.40 Å   1.44 Å
1.37 Å
1.43 Å
1.43 Å

Anthracene

3.38 One general reaction of radicals is the coupling of one with another to form a bond, as shown in the following equation:

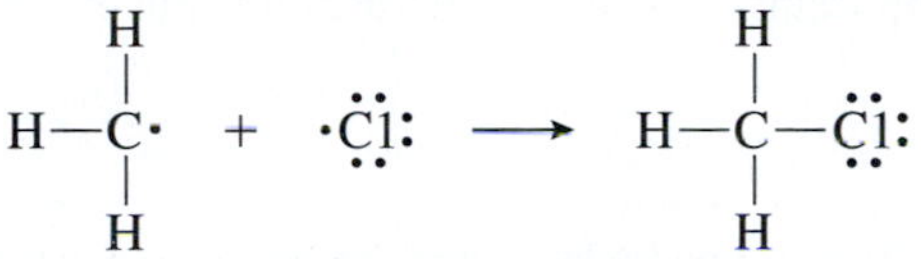

$$\underset{\underset{\text{H}}{|}}{\overset{\overset{\text{H}}{|}}{\text{H}-\text{C}\cdot}} \;+\; \cdot\ddot{\text{C}}\text{l}\text{:} \;\longrightarrow\; \underset{\underset{\text{H}}{|}}{\overset{\overset{\text{H}}{|}}{\text{H}-\text{C}}}-\ddot{\text{C}}\text{l}\text{:}$$

The following coupling reaction gives two products. Show the structures of these products and explain why both are formed.

$$\text{CH}_3-\text{CH}=\text{CH}-\overset{\cdot}{\text{C}}\text{H}_2 \;+\; \cdot\ddot{\text{C}}\text{l}\text{:} \;\longrightarrow\; \text{two products}$$

3.39 Show a MO energy level diagram for the neutral molecule HeH. Use this diagram to explain whether HeH is expected to be stable or not.

3.40 What is the hybridization at each nitrogen of the amino acid histidine? What kind of orbital is occupied by the unshared pair of electrons on each nitrogen? Explain.

$$\text{:N} \cdots \overset{+}{\underset{\overset{|}{\text{N}}}{}} \cdots -\text{CH}_2-\underset{\underset{\text{H}}{}}{\overset{\overset{\text{NH}_3}{|}}{\text{CH}}}-\text{CO}_2^-$$

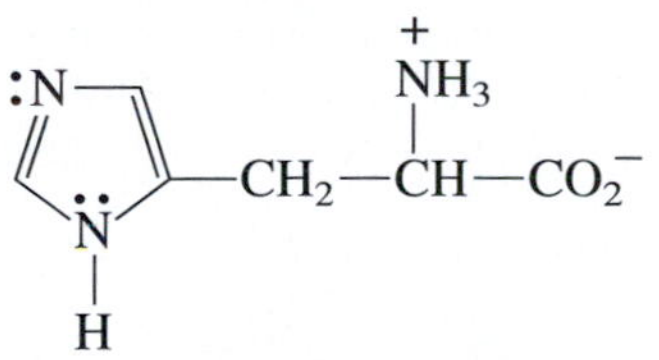

3.41 At a pH of 10.8, the amino acid arginine exists primarily as the following dipolar ion. Show the resonance structures for the cationic part of arginine and discuss their relative contributions to the resonance hybrid.

Arginine

ORGANIC
Chemistry⚛Now™
Click *Molecular Model Problems* to view the models needed to work these problems.

Problems Using Online Three-Dimensional Molecular Models

3.42 Indicate the hybridization at each atom, other than hydrogen, in the compounds represented by these models. Draw structures for each compound represented by the models.

3.43 Draw structures for each compound represented by these models and indicate the conjugated part of the compounds.

3.44 Draw all the important resonance structures for the compounds represented by these models.

3.45 Draw a structure for the neutral molecule represented by the following model. Explain whether the octet rule is satisfied at each atom of the compound. Draw all of the important resonance structures for this molecule.

vMentor™   Do you need a live tutor for homework problems? Access vMentor at Organic ChemistryNow at **http://now.brookscole.com/hornback2** for one-on-one tutoring from a chemistry expert.

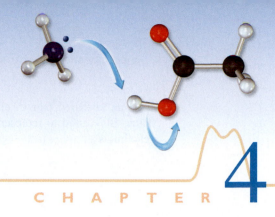

The Acid–Base Reaction

THIS CHAPTER provides our first detailed examination of a chemical reaction, the acid–base reaction or transfer of a proton. Although acid–base reactions are simple, they are very important in organic chemistry because more complicated reactions often involve one or more proton transfer steps. In addition, an important purpose of this chapter is to introduce many concepts about reactions in general. Much of what we learn about the acid–base reaction is applied to other reactions in later chapters.

Acid–base reactions are equilibria, so the concept of equilibrium is reviewed. The rates of these reactions are also discussed. The largest part of the chapter is a discussion of how the structure of the acid or base affects its acid or base strength. We will learn to predict what happens to the strength of an acid when its structure is changed. Not only does this help us to remember how strong an acid (or base) is and how to estimate the strength of an acid or base we have not seen before, but it is of additional importance because exactly the same reasoning is used in later chapters to predict the effects of structural changes on the rates and equilibria of other reactions.

4.1 DEFINITIONS

There are several definitions of acids and bases. According to the **Bronsted-Lowry definition,** an **acid** is a proton donor and a **base** is a proton acceptor. Any compound that has a hydrogen can potentially act as a Bronsted-Lowry acid (although the strength of the acid can vary enormously). Therefore, H—A is used as a general representation for an acid. To accept a proton, most bases

ORGANIC
Chemistry·Now™

Look for this logo in the chapter and go to OrganicChemistryNow at
http://now.brookscole.com/hornback2 for tutorials, simulations, problems, and molecular models.

have an unshared pair of electrons that can be used to form a bond to the proton. Thus, B: is used to represent a general base. The general equation for an acid–base reaction is

$$B: + H-A \rightleftharpoons \overset{+}{B}-H + :\bar{A}$$

Base   Acid   Conjugate   Conjugate
                         acid   base

Equilibrium
arrows

In this reaction, a proton is transferred from the acid to the base. The unshared pair of electrons on the base is used to form the new bond to the proton while the electrons of the H—A bond remain with A as an unshared pair. *Previously, curved arrows have been used to show electron reorganization in resonance structures. Organic chemists also use these arrows to show electron movement in reactions. The arrows are a kind of bookkeeping device that helps us keep track of electrons as the Lewis structures of the reactants are converted to the Lewis structures of the products. Remember, an arrow always points from where the electrons are to where they are going. It does not point from where the hydrogen (or other atom) is to where it is going.*

Important Convention

PRACTICE PROBLEM 4.1

Explain whether ammonia, NH_3, can act as an acid, a base, or both.

Solution

Because NH_3 contains hydrogens, it can act as an acid, and because it has an unshared pair of electrons, it can act as a base. We tend to think of it as a base because that is how it reacts with water. However, in the presence of a strong base, NH_3 can react as an acid. Because many organic reactions involve strongly basic reagents, we need to be aware of the potential of any hydrogen-containing species to donate a proton.

PROBLEM 4.1

Indicate whether each of these species can act as an acid, a base, or both:

a) $H-\overset{\overset{+}{\underset{|}{N}}-H}{\underset{H}{|}}$ (with top H)

b) $H-\ddot{O}-H$

c) $H-\overset{\overset{H}{|}}{\underset{H}{\overset{|}{C}}}-H$

d) $:\ddot{\overset{..}{Cl}}:^-$

e) $H-\overset{\overset{H}{|}}{\underset{H}{\overset{|}{C}}}-\ddot{O}-H$

f) $H-\overset{}{\underset{H}{\overset{|}{C}}}-\overset{\ddot{O}:}{\overset{||}{C}}-\overset{}{\underset{H}{\overset{|}{C}}}-H$

g) $H-\ddot{O}-\overset{\ddot{O}:}{C}-\ddot{O}:^-$

Acid–base reactions are reversible or equilibrium processes. In the reverse reaction, BH^+ acts as the acid and A^- is the base. Therefore, BH^+ is called the **conjugate acid** of the base B, and A⁻ is the **conjugate base** of the acid HA. The charges in a specific acid–base reaction may be different from those shown in the preceding general equation. The proton is positive, so one unit of positive charge is transferred from the acid to the base in the reaction. The initial charge of HA and B can vary, but B is always one unit more negative than BH and HA is always one unit more positive than A.

PROBLEM 4.2

Show the conjugate acids of these species:

a) $CH_3—\ddot{\underset{\cdot\cdot}{O}}—H$ b) $H—\ddot{\underset{\cdot\cdot}{O}}{:}^{-}$ c) $CH_3—\ddot{N}H_2$

PROBLEM 4.3

Show the conjugate bases of these species:

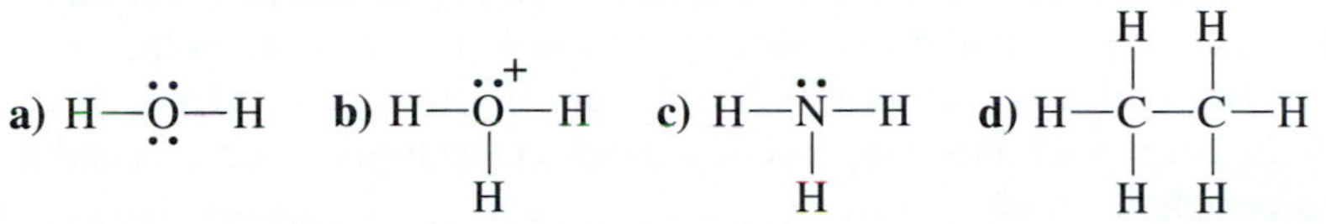

a) $H—\ddot{\underset{\cdot\cdot}{O}}—H$ b) $H—\overset{+}{\underset{\underset{H}{|}}{\ddot{O}}}—H$ c) $H—\underset{\underset{H}{|}}{\ddot{N}}—H$ d) $H—\overset{\overset{H}{|}}{\underset{\underset{H}{|}}{C}}—\overset{\overset{H}{|}}{\underset{\underset{H}{|}}{C}}—H$

Note that a compound that has both a hydrogen and an unshared pair of electrons can potentially react as either an acid or a base, depending on the reaction conditions. Water, ammonia, and alcohols are examples of compounds that react as acids in the presence of strong bases and as bases in the presence of stronger acids. Some specific examples of acid–base reactions are shown in Figure 4.1. Water is the base in the first equation and the acid in the second equation.

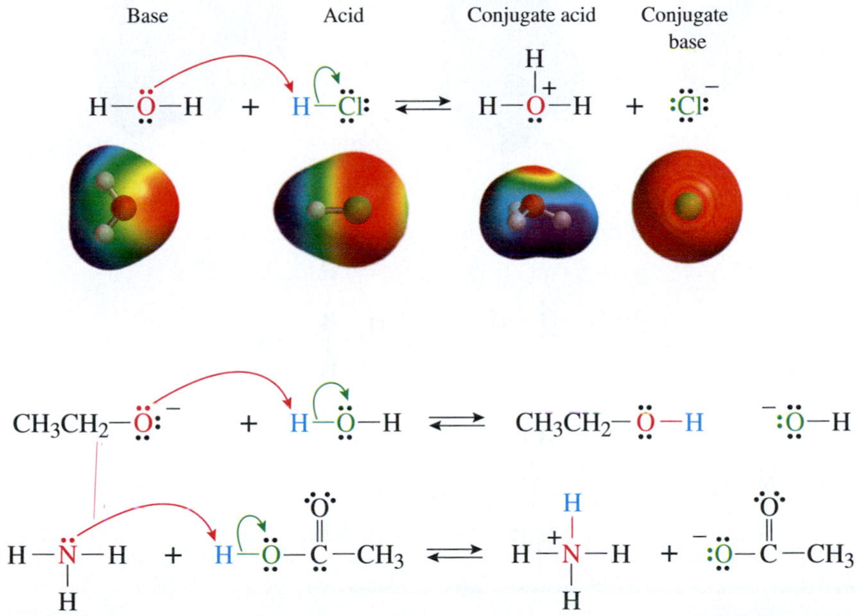

Active Figure 4.1

ORGANIC
Chemistry⚛Now™

SOME ACID–BASE REACTIONS. Test yourself on the concepts in this figure at **OrganicChemistryNow.**

PROBLEM 4.4

Complete these acid–base equations. Use the curved arrow method to show the electron movement in the reactions.

Base	Acid	Conjugate acid	Conjugate base

a) $:\overset{..}{\text{N}}\text{H}_2^-$ + $\text{H}_2\overset{..}{\text{O}}:$ $\rightleftharpoons$

b) $\text{CH}_3\overset{..}{\underset{..}{\text{O}}}:^-$ + $\text{H}_3\overset{+}{\text{O}}:$ $\rightleftharpoons$

According to the **Lewis definition,** an acid is an electron pair acceptor and a base is an electron pair donor. All Bronsted-Lowry bases are also Lewis bases. However, Lewis acids include many species that are not proton acids; instead of H^+, they have some other electron-deficient species that acts as the electron pair acceptor. An example of a Lewis acid–base reaction is provided by the following equation. In this reaction the boron of BF_3 is electron/deficient (it has only six electrons in its valence shell). The oxygen of the ether is a Lewis base and uses a pair of electrons to form a bond to the boron, thus completing boron's octet.

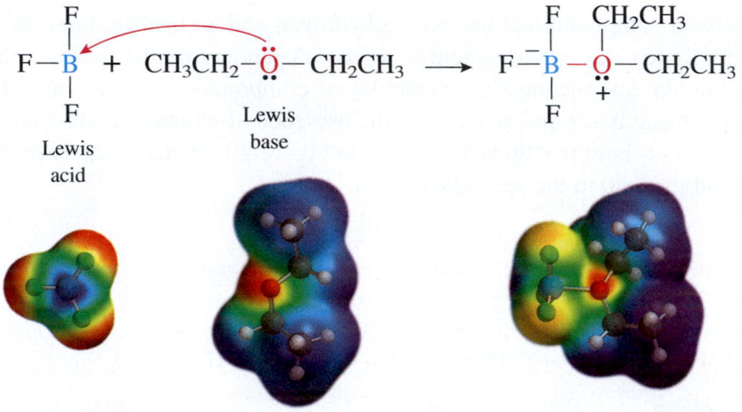

PROBLEM 4.5

Indicate whether each of these species is a Lewis acid, a Lewis base, or both:

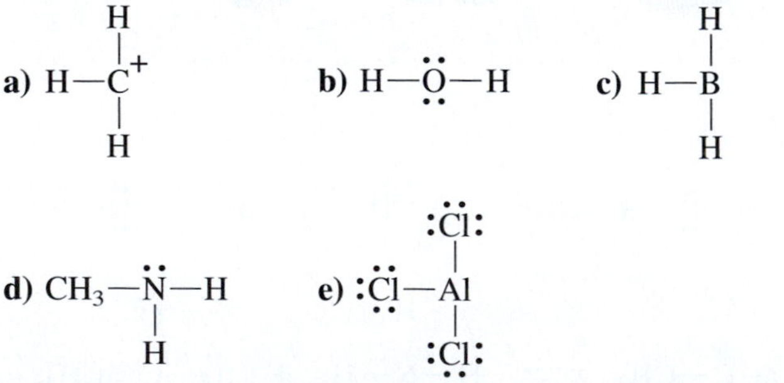

a) $\text{H}-\overset{\text{H}}{\underset{\text{H}}{\text{C}}}^+$

b) $\text{H}-\overset{..}{\underset{..}{\text{O}}}-\text{H}$

c) $\text{H}-\overset{\text{H}}{\underset{\text{H}}{\text{B}}}$

d) $\text{CH}_3-\overset{..}{\underset{\text{H}}{\text{N}}}-\text{H}$

e) $:\overset{:\overset{..}{\text{Cl}}:}{\underset{:\overset{..}{\text{Cl}}:}{\overset{..}{\text{Cl}}}}-\text{Al}$

To avoid confusion, when the term *acid* or *base* is used in this text, it refers to a proton acid or base—that is, a Bronsted-Lowry acid or base. The term *Lewis acid* or *Lewis base* will be used when the discussion specifically concerns this type of acid or base.

4.2 THE ACID–BASE EQUILIBRIUM

The reaction of an acid with a base is in equilibrium with the conjugate base and conjugate acid products. The equilibrium constants, termed *acid dissociation* or *acidity constants,* for the reactions of many acids with water as the base (and solvent) have been determined. They can be found in various reference books. Some selected acidity constants are listed in Table 4.2 in Section 4.9. Let's see how acidity constants are defined.

Consider the following acid–base equilibrium, the ionization (dissociation) of acetic acid in water:

The equilibrium constant for this reaction is

$$K = \frac{[CH_3CO_2^-][H_3O^+]}{[CH_3CO_2H][H_2O]}$$

Because water is also the solvent, it is present in large excess, and its concentration is approximately constant during the reaction. Therefore, a new equilibrium constant, the **acidity constant** (K_a), is used. For the preceding reaction the equation for K_a is

$$K_a = K[H_2O] = \frac{[CH_3CO_2^-][H_3O^+]}{[CH_3CO_2H]} = 1.8 \times 10^{-5}$$

The acidity constant is a measure of the strength of an acid. If the acidity constant for a particular acid is near 1, about equal amounts of the acid and its conjugate base are present at equilibrium. A strong acid, which dissociates nearly completely in water, has an acidity constant significantly greater than 1. A weak acid, which is only slightly dissociated in water, has an equilibrium constant significantly less than 1. The acidity constant for acetic acid is 1.8×10^{-5}—only a small amount of acetic acid actually ionizes in water. It is a weak acid.

The acidity constants that are encountered in organic chemistry vary widely, from greater than 10^{10} to less than 10^{-50}. Because of this wide range, it is convenient to use a logarithmic scale to express these values, as is done with pH. Therefore, pK_a is defined as

$$pK_a = -\log K_a$$

PRACTICE PROBLEM 4.2

If the pK_a of a compound is 10, what is its K_a?

Solution

The calculation is easy when pK_a is an integer. If pK_a = integer, then $K_a = 1 \times 10^{-integer}$. So if $pK_a = 10$, $K_a = 1 \times 10^{-10}$. It is also easy to go in the reverse direction when $K_a = 1 \times 10^x$ because then $pK_a = -x$.

PROBLEM 4.6
Provide the values for the missing K_a or pK_a in the following examples:
a) $pK_a = -4$; $K_a = ?$
b) $K_a = 1 \times 10^{-16}$; $pK_a = ?$
c) $pK_a = 38$; $K_a = ?$
d) $K_a = 1 \times 10^6$; $pK_a = ?$

It is important to be able to quickly recognize the strength of an acid from its K_a or pK_a value. As the strength of the acid increases, the K_a increases and the pK_a decreases (becomes more negative). Base strengths can be determined from the K_a (or pK_a) of the conjugate acid. A strong acid has a weak conjugate base, and a weak acid has a strong conjugate base. These relationships can be summarized as follows:

Strong acid	$K_a > 1$	Negative pK_a	Weak conjugate base
Weak acid	$K_a < 1$	Positive pK_a	Strong conjugate base

PROBLEM 4.7
Indicate whether these compounds are weaker or stronger acids than water (the K_a for water is 1.8×10^{-16}; the pK_a is 15.74):

a) $HClO_4$ $(K_a = 10^{10})$ **b)** $HC{\equiv}CH$ $(pK_a = 25)$

c)
$$\overset{\displaystyle O}{\overset{\|}{HOCOH}}$$
(pK_a = 6.35)   **d)** CH_3CH_3 $(K_a = 10^{-50})$

PROBLEM 4.8
Indicate whether these species are weaker or stronger bases than hydroxide ion. The K_a or pK_a values are for the conjugate acids.

a) $\overset{..}{\underset{..}{:}}\overline{N}H_2$ $(K_a = 10^{-38})$ **b)** $CH_3CH_2\overset{..}{\overline{C}}H_2$ $(pK_a = 50)$

c) $\overset{..}{N}H_3$ $(pK_a = 9.24)$ **d)** $:\overset{..}{\underset{..}{C}l}:^{\overline{}}$ $(pK_a = -7)$

Consider the general acid–base reaction (charges omitted)

$$H{-}A + B: \;\rightleftharpoons\; A: + H{-}B$$

The equilibrium constant (K) for this reaction is

$$K = \frac{K_a \text{ (for HA)}}{K_a \text{ (for HB)}}$$

If HA is a stronger acid than HB, then K is greater than 1 and the right-hand side of the equation is favored at equilibrium; that is, the concentrations of the products (A and

HB) are greater than the concentrations of the reactants (HA and B). If HB is a stronger acid than HA, then K is less than 1 and the equilibrium lies to the left. In general, *the equilibrium favors the formation of the weaker acid and the weaker base.* (Because the stronger acid has the weaker conjugate base, the weaker base is always on the same side of the equation as the weaker acid.)

As an example, consider the following reaction:

$$H{-}Br + H_2O \rightleftharpoons Br^- + H_3O^+$$

HBr is a stronger acid ($K_a = 10^9$ or $pK_a = -9$) than H_3O^+ ($K_a = 55$ or $pK_a = -1.74$), so this equilibrium lies to the right. From the K_a's the equilibrium constant for the reaction can be calculated to be 1.8×10^7. This equilibrium constant is so large that the amount of HBr remaining cannot be measured by most experimental techniques. For all practical purposes the equilibrium lies completely to the right. Strong acids, such as HBr, are said to be completely dissociated in water. Differences in the strengths of very strong acids cannot be detected in water. Equilibrium constants as large as the one for the reaction of HBr with water cannot be measured directly and must be determined by some indirect means. Therefore, the pK_a values of the very strong acids (and also the very weak acids) listed later in Table 4.2 are only approximate. However, they are usually accurate enough for predictions to be made about the position of the equilibrium in an acid–base reaction.

Figure 4.2 shows a scale of the strengths of some of the acids commonly encountered in organic chemistry. At the left side of the scale are *strong acids* such as perchloric ($HClO_4$), hydrobromic (HBr), and sulfuric (H_2SO_4) acids, all of which are significantly stronger than H_3O^+ and are completely ionized in water. Acids that are stronger than H_2O but weaker than H_3O^+ are partially dissociated in water and are

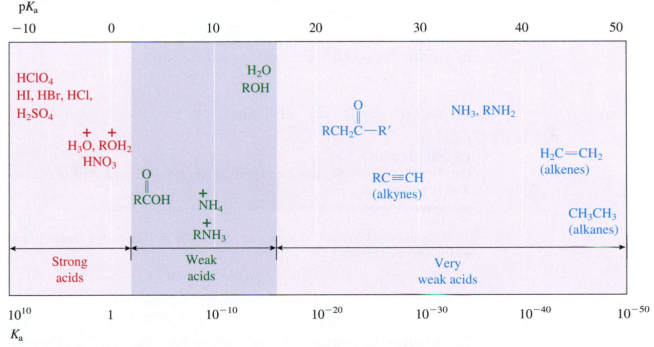

Figure 4.2

SCALE SHOWING SOME ACID STRENGTHS.

termed *weak acids* in this text. Included in this group are carboxylic acids and the conjugate acids of ammonia and the amines. Acids that are comparable to or weaker than H_2O are essentially undissociated in water and are termed *very weak acids*. However, they can react as acids in the presence of very strong bases in solvents that are less acidic than water. The majority of organic compounds fall into this group, ranging from alcohols (pK_a's about 16), which are similar in acid strength to water, to aldehydes and ketones (pK_a's about 20), which are a little less acidic than water, to alkanes (pK_a's about 50), which are extremely weak acids.

PRACTICE PROBLEM 4.3

Use Figure 4.2 to predict whether the reactants or the products of this reaction are favored at equilibrium:

$$CH_3\overset{..}{\underset{..}{C}}H_2^- + HC\equiv CH \;\rightleftharpoons\; CH_3CH_3 + HC\equiv C\overset{..}{\underset{..}{:}}^-$$

Solution

From Figure 4.2, $HC\equiv CH$ is a stronger acid than CH_3CH_3. Because the equilibrium favors the weaker acid, more of the products are present when this reaction reaches equilibrium.

PROBLEM 4.9

Using the information available in Figure 4.2, predict the position of the equilibrium in these reactions; that is, predict whether there is a higher concentration of reactants or products present at equilibrium:

a)  $CH_3\overset{..}{N}H_2 + CH_3\overset{\overset{\textstyle O}{\|}}{C}\overset{..}{\underset{..}{O}}:^- \;\rightleftharpoons\; CH_3\overset{..}{N}H^- + CH_3\overset{\overset{\textstyle O}{\|}}{C}\overset{..}{\underset{..}{O}}H$

b) 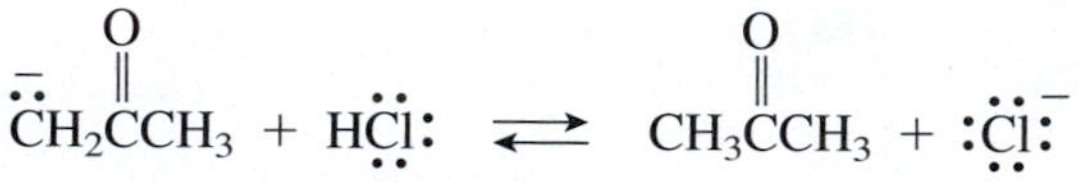 $\overset{..}{\underset{..}{C}}H_2\overset{\overset{\textstyle O}{\|}}{C}CH_3^- + H\overset{..}{\underset{..}{C}}l: \;\rightleftharpoons\; CH_3\overset{\overset{\textstyle O}{\|}}{C}CH_3 + :\overset{..}{\underset{..}{C}}l:^-$

PROBLEM 4.10

Use the information in Figure 4.2 to predict the positions of the equilibria in the reactions in problem 4.4.

An equilibrium constant for a reaction is related to the free-energy change in that reaction by the equation

$$\Delta G^\circ = -RT \ln K$$

Standard free energy change   Gas constant   Absolute temperature

For a reaction to be spontaneous, K must be greater than 1; that is, ΔG° must be negative. Organic chemists find diagrams that show the free energies of the reactants and

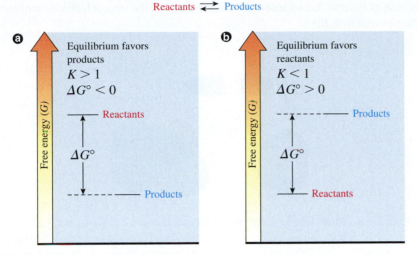

Reactants ⇌ Products

Figure 4.3

FREE ENERGIES OF REACTANTS AND PRODUCTS. **a** The products are more stable than the reactants and are favored at equilibrium. **b** The reactants are more stable and are favored.

products on the same scale to be very useful. Figure 4.3 shows such diagrams for two general reactions. In the diagram on the left-hand side of this figure, the reactants are higher in energy than the products. Therefore, $\Delta G°$ ($G_{\text{products}} - G_{\text{reactants}}$) is negative and K is greater than 1. The reaction proceeds spontaneously to the products; that is, it proceeds spontaneously to the lower energy or more stable compound(s). The situation is reversed in part (b) of the figure. The reactants are more stable, $\Delta G°$ is positive, K is less than 1, and the reactants are favored at equilibrium.

PROBLEM 4.11
Draw diagrams like that in Figure 4.3 for the reactions in problem 4.9.

4.3   RATE OF THE ACID–BASE REACTION

As you should recall from general chemistry, a favorable equilibrium constant is not sufficient to ensure that a reaction will occur. In addition, the rate of the reaction must be fast enough that the reaction occurs in a reasonable period of time. The reaction rate depends on a number of factors. First, the reactants, in this case the acid and the base, must collide. In this collision the molecules must be oriented properly so that the orbitals that will form the new bond can begin to overlap. The orientation required for the orbitals of the reactants is called the **stereoelectronic requirement** of the reaction. (*Stereo* means dealing with the three dimensions of space.) In the acid–base reaction, the collision must occur so that the atomic orbital of the base that is occupied by the unshared pair of electrons can begin to overlap with the 1s orbital of the acidic hydrogen.

In the case of the reaction of ammonia with acetic acid, the stereoelectronic requirement can be pictured as follows:

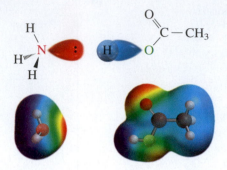

If, instead, the collision were to occur so that the orbital on the nitrogen bumped into some other part of the acetic acid molecule, that collision would not lead to an acid–base reaction.

You might also recall that an energy barrier separates the reactants and the products in most reactions. Therefore, a final requirement for a collision to lead to reaction is that the collision must provide enough energy to surmount this energy barrier. This extra energy is called the **activation energy** for the reaction. Activation energy will be considered in more detail when more complicated reactions are discussed. For now, it is enough to note that the activation energy for acid–base reactions is usually small.

A diagram that shows how the free energy of the system changes as a reaction proceeds is often very informative. Such a diagram has the free energy, G, on the y-axis, just as in Figure 4.3, but has a measure of the progress of the reaction along the x-axis, so the reactants are shown on the left-hand side of the diagram and the products on the right-hand side. Figure 4.4 shows such a diagram for the ionization of HBr in water. Because HBr is a strong acid, ΔG° for this reaction is negative and $K_a > 1$. The products are favored at equilibrium because they are more stable than the reactants. The line connecting the energy of the reactants to that of the products shows how the energy changes during the reaction. As the reaction proceeds, there is initially a slight increase

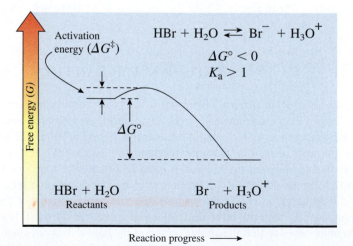

Figure 4.4

FREE ENERGY VERSUS REACTION PROGRESS DIAGRAM FOR THE REACTION OF **HBr** WITH **H₂O**.

in energy, followed by a decrease as the reaction proceeds toward the products. The size of the energy barrier that the reaction must pass over is called the activation energy, $\Delta G^{\ddagger}$. In this case, $\Delta G^{\ddagger}$ is very small, so the energy barrier is easily overcome—the reaction is very fast. These energy versus reaction progress diagrams are very useful and will be encountered often in subsequent chapters.

A **mechanism** for a reaction shows the individual steps in the reaction. It shows how the nuclei and electrons move as the reaction proceeds, how the bonds change, and the order in which the bonds are made and broken. The acid–base reaction has a very simple mechanism. It consists of only one step, which involves breaking the bond between the acid and the hydrogen and forming the new bond between the base and the hydrogen. The free energy versus reaction progress diagram has only a single energy barrier between the reactants and the products. Mechanisms for more complicated organic reactions may involve several different steps in which different bonds are made or broken. Each step has an energy barrier that separates the reactant and product of that step. Reaction mechanisms are very important. Understanding the mechanism for a more complicated reaction will enable us to predict and remember many features of the reaction, such as how to increase the rate, how to maximize the amount of product that is formed, and so forth.

Because the activation energies are small and the stereoelectronic requirements are not difficult to meet, most acid–base reactions are very fast in comparison to other types of organic reactions. Therefore, it is usually not necessary to be concerned with the rates of acid–base reactions. In organic reactions that have mechanisms involving several steps, including an acid–base step, one of the other steps in the mechanism usually controls the rate.

PRACTICE PROBLEM 4.4

Show a free energy versus reaction progress diagram for this reaction.

$$\underset{\text{CH}_3\overset{\displaystyle O}{\overset{\|}{\text{C}}}\text{OH}}{} + \text{H}_2\text{O} \;\rightleftharpoons\; \underset{\text{CH}_3\overset{\displaystyle O}{\overset{\|}{\text{C}}}\text{O}^-}{} + \text{H}_3\text{O}^+$$

Solution

First we must determine whether the reactants or the products are more stable. Acetic acid is a weak acid (see Figure 4.2), so the equilibrium favors the reactants. In other words, $K_a < 1$ and $\Delta G^{\circ} > 0$. The diagram is just the reverse of that shown in Figure 4.4.

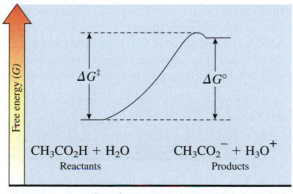

Note that the magnitude of the activation energy, $\Delta G^{\ddagger}$, is slightly larger than that of $\Delta G°$ in this case. This means that there is only a small activation barrier for the reverse reaction.

PROBLEM 4.12
Show a free energy versus reaction progress diagram for the following reaction:

$$HCl + NH_3 \rightleftharpoons Cl^- + {}^+NH_4$$

4.4   EFFECT OF THE ATOM BONDED TO THE HYDROGEN ON ACIDITY

It is important to understand the various factors that determine the strength of an acid— that is, why one compound is a stronger acid than another. Such an understanding makes it much easier to remember whether a particular compound is a strong acid or a weak acid. In addition, it then becomes possible to make qualitative predictions about what a change in the structure of a compound will do to its K_a value. Therefore, most of the remainder of this chapter is a discussion of how the structure of the acid affects its strength. The effects to be discussed are the atom to which the hydrogen is bonded, nearby charges or polar bonds, hydrogen bonding, the hybridization of the atom to which the hydrogen is bonded, and resonance. First, let's consider the effect of the atom bonded to the acidic hydrogen.

Consider the following acid–base reactions:

The hydrogen that is bonded to carbon is 10^{12} times less acidic than the hydrogen that is bonded to nitrogen. This can be explained by considering the electronegativities of the carbon and the nitrogen. Nitrogen is more electronegative than carbon. The unshared pair of electrons on the nitrogen of NH_2^- is in a lower-energy orbital, and is more stable, than the unshared pair of electrons on the carbon of CH_3^-. Therefore, the NH_2^- has less inclination to use its more stable pair of electrons to form a bond to an H^+, so it is a weaker base than CH_3^-, and NH_3 is a stronger acid than the related carbon compounds. Because of this increasing stabilization of the conjugate base as the electronegativity of the atom increases, acidity increases from left to right in a row of the periodic table.

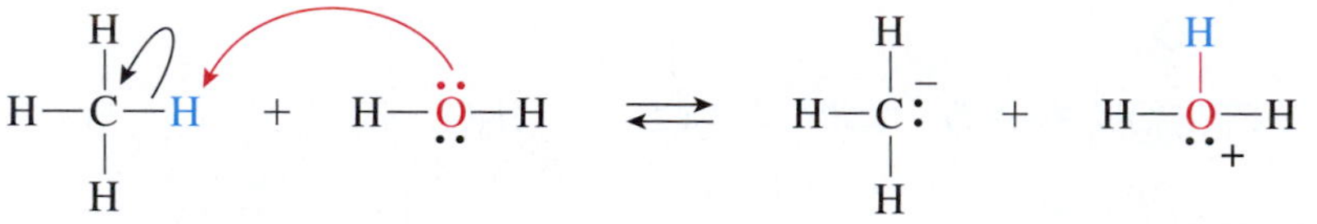

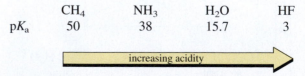

	CH_4	NH_3	H_2O	HF
pK_a	50	38	15.7	3

increasing acidity →

Many useful predictions can be made on the basis of this order. For example, suppose methanol, CH_3OH, reacts with a strong base. Which hydrogen is removed? Which hydrogen is more acidic? Because oxygen is more electronegative than carbon, the hydrogen on the oxygen is considerably more acidic than the hydrogens on the carbon. The reaction that occurs is

$$CH_3 - \overset{..}{\underset{..}{O}} - H + :\bar{B} \rightleftharpoons CH_3 - \overset{..}{\underset{..}{O}}: + H - B$$

In general, oxygen acids (H bonded to O) will be stronger than carbon acids (H bonded to C), other things being equal.

When comparisons are made within a column of the periodic table, it is found that electronegativity is no longer the controlling factor. Acidity increases from top to bottom in a column of the periodic table, whereas electronegativity decreases.

	HF	HCl	HBr	HI
pK_a	3	−7	−9	−10

	H$_2$O	H$_2$S
pK_a	15.7	7

increasing acidity →

Part of the explanation for this trend is that atoms that are lower in a column form weaker bonds to hydrogen because their larger atomic orbitals do not overlap as well with the small hydrogen $1s$ orbital. These weaker bonds make removing the proton easier. The ability of larger anions to better accommodate a negative charge also contributes to this trend.

PROBLEM 4.13

Which species is a stronger acid?

a) H$_2$S or HCl   b) $\overset{+}{P}H_4$ or $\overset{+}{N}H_4$ c) CH$_3$CH$_3$ or H$_2$S

PROBLEM 4.14

Which anion is the stronger base?

a) HO$^-$ or HS$^-$ b) CH$_3\bar{N}$H or CH$_3\bar{O}$

PRACTICE PROBLEM 4.5

Which is the most acidic hydrogen in this compound?

$$\overset{\displaystyle H}{\underset{\displaystyle CH_3NCH_2CH_3}{|}}$$

Solution

If none of the other effects described in subsequent sections are operating, a hydrogen on a nitrogen is more acidic than a hydrogen on a carbon because the nitrogen is more electronegative than the carbon.

This is the most acidic
H in this compound.

$$\overset{\displaystyle H}{\underset{\displaystyle CH_3NCH_2CH_3}{|}}$$

PROBLEM 4.15
Which is the most acidic hydrogen in each of these compounds?

a) $H_2NCH_2CH_2OH$ **b)** CH_3CH_2OH **c)** CH_3SH

4.5 INDUCTIVE EFFECTS

The effect of a nearby dipole in a molecule on a reaction elsewhere in that molecule is termed an **inductive effect.** Consider two carboxylic acids, acetic acid and chloroacetic acid:

$$CH_3COOH \qquad ClCH_2COOH$$
Acetic acid   Chloroacetic acid
$$pK_a = 4.76 \qquad pK_a = 2.86$$

Replacing one of the hydrogens of acetic acid with a chlorine results in an increase in acid strength by almost a factor of 100. Remember that the acidic hydrogen in these molecules is the one bonded to the oxygen, so the chlorine is actually exerting its effect from several bonds away from the reacting bond. Figure 4.5 shows how the presence of the dipole of the C—Cl bond increases the energy of the acid (a), while it lowers the energy of the conjugate base (b). The inductive effect of the chlorine destabilizes the acid and stabilizes the conjugate base. Figure 4.6 shows how this inductive effect changes $\Delta G°$ for the acid–base reaction of chloroacetic acid as compared to acetic acid. Chloroacetic acid is destabilized relative to acetic acid by the inductive effect of the chlorine. In contrast, the conjugate base of chloroacetic acid is stabilized relative to the conjugate base of acetic acid. Therefore, $\Delta G°_2$ for the acid–base reaction of chloroacetic acid is less than $\Delta G°_1$ for that of acetic acid. The value of K_a is larger for chloroacetic acid—it is a stronger acid than acetic acid.

Organic chemists often talk about the inductive effect of a group. Chlorine is an electron-withdrawing group relative to hydrogen because of its inductive effect; that is,

Chloroacetic acid   Conjugate base

In chloroacetic acid the interaction of the dipole of the Cl—C bond with the dipole of the carboxylic acid group is destabilizing because the positive end of one dipole is closer to the positive end of the other. The interaction of these like charges is repulsive and increases the energy of the molecule.

In the conjugate base of chloroacetic acid the interaction of the positive end of the Cl—C dipole with the negative charge of the ionized carboxylate group is a stabilizing interaction. The energy of the anion is lowered by this effect.

Figure 4.5

CHARGE INTERACTIONS IN ⓐ CHLOROACETIC ACID AND ⓑ ITS CONJUGATE BASE.

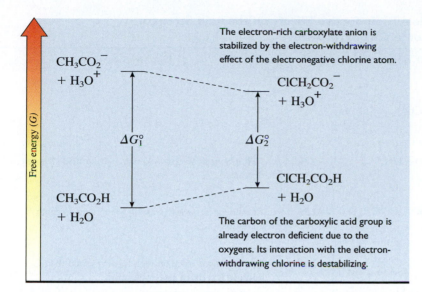

The electron-rich carboxylate anion is stabilized by the electron-withdrawing effect of the electronegative chlorine atom.

The carbon of the carboxylic acid group is already electron deficient due to the oxygens. Its interaction with the electron-withdrawing chlorine is destabilizing.

Figure 4.6

INDUCTIVE EFFECT ON THE FREE-ENERGY CHANGES IN AN ACID–BASE REACTION.

it pulls more electron density away from its bond partner than does hydrogen. Because most functional groups contain atoms that are more electronegative than hydrogen, they are also inductive electron-withdrawing groups. The following is a partial list of inductive electron-withdrawing groups:

Inductive electron-withdrawing groups

In contrast, only a few groups are electron-donating relative to hydrogen because of their inductive effects. Two of these are electron rich because of their negatively charged oxygen atoms. In addition, alkyl groups, such as CH_3 and CH_2CH_3, behave as weak electron-donating groups in many situations:

Inductive electron-donating groups

Replacing a hydrogen with an electron-withdrawing group will destabilize an electron-poor site or stabilize an electron-rich site. Because the conjugate base is always more electron rich than the acid, replacing a hydrogen with an electron-withdrawing group re-

Table 4.1   Inductive Effects on the Acid Strength of Some Carboxylic Acids

Compound	pK_a	Comments
$\overset{\displaystyle O}{\overset{\|}{HCOH}}$	3.75	
$\overset{\displaystyle O}{\overset{\|}{CH_3COH}}$	4.76	CH$_3$ is a weak electron-donating group; acid strength decreases.
$\overset{\displaystyle O}{\overset{\|}{ClCH_2COH}}$	2.86	Cl is an electron-withdrawing group; acid strength increases.
$\overset{\displaystyle O}{\overset{\|}{FCH_2COH}}$	2.66	F is a stronger electron-withdrawing group than Cl because it is more electronegative.
$\overset{\displaystyle O}{\overset{\|}{ICH_2COH}}$	3.12	I is less electronegative and a weaker electron-withdrawing group than Cl.
$\overset{\displaystyle O}{\overset{\|}{ClCH_2CH_2COH}}$	3.98	The Cl is farther from the reaction site; the inductive effect decreases rapidly with increasing distance.
$\overset{\displaystyle O}{\overset{\|}{Cl_3CCOH}}$	0.65	Three Cl's have a stronger inductive effect than one.
$\overset{\displaystyle O\quad O}{\overset{\|\quad\|}{HOC-COH}}$	2.83	The CO$_2$H group is electron withdrawing.
$\overset{\displaystyle O\quad O}{\overset{\|\quad\|}{^-OC-COH}}$	5.69	The CO$_2^-$ group is electron donating; this is a weaker acid than acetic acid.

sults in a stronger acid. Replacing a hydrogen with an electron-donating group has the opposite effect. Table 4.1 provides some examples of inductive effects on the acidities of carboxylic acids.

PROBLEM 4.16
Explain which compound is the stronger acid:

a) CHF$_2$CO$_2$H   or   CH$_2$FCO$_2$H

b) CHF$_2$CO$_2$H   or   CHBr$_2$CO$_2$H

c) CH$_3$OCH$_2$CO$_2$H   or   CH$_3$CO$_2$H

4.6   HYDROGEN BONDING

If the acidic hydrogen forms a hydrogen bond with another atom in the same molecule, the strength of the acid is decreased. It is more difficult for a base to remove the proton because the hydrogen bond must be broken in addition to the regular sigma bond to the hydrogen. However, this effect is complicated by the inductive effect of the group involved in the hydrogen bond (Figure 4.7).

The inductive effect and the hydrogen-bonding effect in *o*-acetylbenzoic acid (see Figure 4.7) are operating in opposite directions. The inductive effect increases the acid strength while the hydrogen bond decreases it. Although the two effects nearly cancel out in this particular case, this cannot usually be predicted in advance. In general, the direction of an effect—that is, whether it is acid strengthening or acid weakening—can readily be determined. However, it is much more difficult to estimate the magnitude of the effect. In a case such as *o*-acetylbenzoic acid, in which the two effects are opposed, it is difficult to predict which one is larger, so it is not possible to predict whether the acid is stronger or weaker than the model compound, in which neither effect is present. Therefore, most of our predictions will be qualitative rather than quantitative—we will be able to determine that one compound is a stronger acid (or reacts faster) than another, but we will not be able to predict exactly how much stronger (or faster).

Benzoic acid
$pK_a = 4.19$

The $CH_3C{=}O$ group (the acetyl group) is electron withdrawing. When it is substituted on the ring position opposite the carboxylic acid group, its inductive effect increases the strength of this acid as compared to benzoic acid.

p-Acetylbenzoic acid
$pK_a = 3.70$

Intramolecular hydrogen bond

o-Acetylbenzoic acid
$pK_a = 4.13$

If the acetyl group is substituted on the position adjacent to the carboxylic acid group, an even stronger acid should result because the inductive effect increases as the groups are brought closer together. However, this acid is weaker than the previous example; its pK_a is similar to that of benzoic acid. The acid-strengthening inductive effect is canceled by the acid-weakening effect of the hydrogen bond formed between the hydrogen of the carboxylic acid group and the oxygen of the acetyl group.

Figure 4.7

INDUCTIVE EFFECTS AND HYDROGEN BONDING EFFECTS.

4.7 HYBRIDIZATION

As the following examples show, the hybridization of the atom bonded to the hydrogen has a large effect on the acidity of that hydrogen:

<table>
<tr><td>Ethane
pK_a = 50</td><td>Ethene
pK_a = 44</td><td>Ethyne
pK_a = 25</td></tr>
</table>

In this series, as the hybridization changes from sp^3 in ethane to sp^2 in ethene and to sp in ethyne, the acidity increases and the pK_a decreases. This is because of the relative stability of the unshared electrons in the conjugate bases of each of these compounds.

Figure 4.8 shows the energies of the hybrid orbitals relative to the s and p orbitals from which they are formed. An sp hybrid orbital is composed of 50% p orbital and 50% s orbital. Therefore, its energy is halfway between the energies of the s orbital and the p orbital. Similarly, the energy of an sp^2 orbital is higher than that of the s orbital by 67% of the difference between the energies of the p orbital and the s orbital, and the energy of an sp^3 orbital is higher than the energy of the s orbital by 75% of this difference.

The unshared pair of electrons of the conjugate base of ethane occupies an sp^3 atomic orbital on the carbon. The unshared pair of electrons of the conjugate base of ethene occupies an sp^2 atomic orbital. Because the sp^2 orbital is lower in energy, the unshared electrons in this orbital in the conjugate base of ethene are more stable (and less basic) than the electrons in the sp^3 orbital of the conjugate base of ethane. Thus, ethene is a stronger acid than ethane. Because the electrons of the conjugate base of ethyne are even lower in energy in an sp orbital, ethyne is an even stronger acid.

This explanation using orbital energies can also be couched in terms of electronegativities. Electronegativity and orbital energy are directly related. For example, the reason why fluorine is more electronegative than oxygen is that the 2p orbital of fluorine is lower in energy than the 2p orbital of oxygen. Because an added electron would prefer to occupy the lower-energy orbital, fluorine is more electronegative. An sp orbital is lower in energy than an sp^2 orbital, so it might be said that an sp-hybridized carbon is more electronegative than an sp^2-hybridized carbon. Therefore, using the same reasoning as in Section 4.4, a hydrogen that is bonded to an sp-hybridized carbon should be more acidic than a hydrogen that is bonded to an sp^2-hybridized carbon.

PROBLEM 4.17
Which is the most acidic hydrogen in $CH_3CH_2C{\equiv}CH$?

4.8 RESONANCE

Stabilization by resonance is used to explain many observations in organic chemistry. Resonance stabilization of a product can shift an equilibrium dramatically to the right, that is, to the product side of the reaction. Resonance can also lower the activation energy for a reaction, resulting in a considerable increase in reaction rate. What we learn

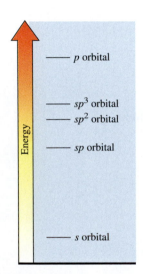

Figure 4.8

ENERGIES OF HYBRID ATOMIC ORBITALS RELATIVE TO s AND p ATOMIC ORBITALS.

Focus On

Calcium Carbide

Because of its *sp* hybridization, ethyne is a strong enough acid that its conjugate base can be readily generated by using one of the strong bases that are available in the laboratory. Calcium carbide, CaC_2, a relatively stable solid, can be viewed as a dianion of ethyne:

$$Ca^{2+} \quad :\overset{-}{C} \equiv \overset{-}{C}:$$

Calcium carbide

Calcium carbide is prepared by the reaction of calcium oxide with carbon, in the form of coke, at very high temperatures:

$$CaO + 3\,C \xrightleftharpoons{2000-3000°C} CaC_2 + CO$$

As expected, the carbide dianion is a very strong base and readily removes protons from water to produce ethyne (acetylene) gas:

$$Ca^{2+} \quad :\overset{-}{C} \equiv \overset{-}{C}: + 2\,H{-}OH \longrightarrow H{-}C \equiv C{-}H + Ca(OH)_2$$

Calcium carbide provides a fairly safe and easy-to-handle source of ethyne. Before the advent of battery-operated lights, portable lamps such as those used on bicycles, carriages, and miner's helmets were fueled by this material. Water was slowly dropped onto solid calcium carbide, and the ethyne that was generated was burned. This reaction is still used as a source of acetylene for welding torches.

Ethyne can be used as a starting material for the preparation of many industrially important organic compounds. Currently, it is more cost-effective to prepare these compounds from petroleum. However, as petroleum supplies dwindle in the future, ethyne prepared from coal via calcium carbide will become more economically attractive as a source of these compounds.

here about how resonance affects the acid–base equilibrium will be very useful in discussions of both equilibria and rates of other reactions in subsequent chapters.

Consider the acid–base reactions of ethanol and acetic acid in water:

$$CH_3CH_2OH + H_2O \rightleftharpoons CH_3CH_2O^- + H_3O^+ \qquad pK_a = 16$$

Ethoxide ion

$$\overset{\displaystyle O}{\overset{\displaystyle \|}{CH_3C}}OH + H_2O \rightleftharpoons \overset{\displaystyle O}{\overset{\displaystyle \|}{CH_3C}}O^- + H_3O^+ \qquad pK_a = 4.76$$

Acetic acid is a stronger acid than ethanol by a factor of about 10^{11}. In both compounds the acidic hydrogen is bonded to an oxygen. Replacing the CH_2 of ethanol with a $C=O$ results in an enormous increase in acidity. Part of this increase is due to the inductive effect of the oxygen of the carbonyl group, but the effect is much too large to be due only to this.

The other factor that is contributing to the dramatic increase in the acidity of acetic acid is resonance stabilization. Neither ethanol nor its conjugate base, which is called ethoxide ion, is stabilized by resonance. The following resonance structures can be written for acetic acid and its conjugate base, acetate anion:

Acetic acid   Acetate anion

As noted in Figure 3.16, acetic acid has only a small amount of resonance stabilization because the lower structure is only a minor contributor to the resonance hybrid. Acetate ion has a large amount of resonance stabilization because it has two equivalent contributors to the hybrid.

Figure 4.9 diagrams these energy changes. Because acetic acid has only a small amount of resonance stabilization, its energy is lowered only a little as compared to

Figure 4.9

ENERGY DIAGRAM FOR THE REACTION OF ETHANOL AND ACETIC ACID AS ACIDS IN WATER.

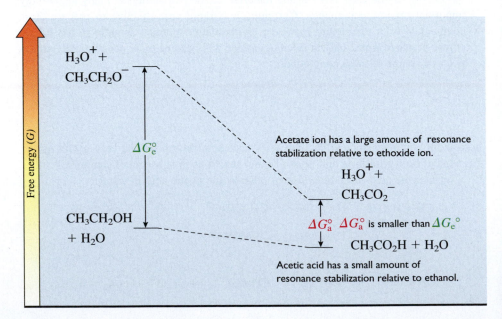

Acetate ion has a large amount of resonance stabilization relative to ethoxide ion.

ΔG_a° ΔG_a° is smaller than ΔG_e°

Acetic acid has a small amount of resonance stabilization relative to ethanol.

ethanol. However, acetate ion has a much larger amount of resonance stabilization so its energy is lowered more relative to ethoxide ion. Therefore, $\Delta G°_a$ is smaller than $\Delta G°_e$, resulting in acetic acid being a stronger acid than ethanol.

In other words, the unshared electron pair of the base, acetate ion, is delocalized (spread over both oxygens) by resonance. This electron pair is stabilized and less available for bonding to the proton, which localizes this electron pair in the sigma bond and costs resonance energy. The most common effect of resonance on an acid–base reaction is to delocalize and stabilize the unshared electron pair of the conjugate base, resulting in a stronger acid.

To make this more clear, consider the following examples. Replacing one H of water with a CH_3CH_2 group to give ethanol should have only a small effect on acidity, so ethanol (and other alcohols) are expected to have pK_a's close to that of water. The pK_a of ethanol is about 16; the pK_a of water is 15.74.

Phenol resembles an alcohol, but with the OH attached to an aromatic ring rather than an alkyl group. However, its pK_a is 10. It is a stronger acid than ethanol by a factor of 10^6. The aromatic ring has caused a large increase in acidity. Five important resonance structures can be written for the conjugate base of phenol. One of them is the benzene resonance structure that is present in all aromatic compounds and does not affect the basicity because it stabilizes both the acid and the conjugate base equally. The others delocalize one unshared electron pair of the basic oxygen, stabilizing the base and making it weaker. (Although similar resonance structures can be written for phenol itself, they are less important because they have formal charges.) Furthermore, note that the delocalized pair of electrons occupies alternating positions around the ring in the various resonance structures. Both of these observations will be useful in subsequent discussions. The last three resonance structures are not as important as the first two because the negative charge is on a carbon rather than the more electronegative oxygen. Therefore, the overall amount of resonance stabilization is not as large as in the case of acetate, which has two equivalent resonance contributors. Phenol is a weaker acid than acetic acid.

CH_3CH_2OH

Ethanol
$pK_a = 16$

OH

Phenol
$pK_a = 10$

At first glance, benzyl alcohol resembles phenol and it might be expected to have a pK_a near 10. However, its pK_a is actually close to 16; it resembles ethanol more than phenol in acid strength. The CH_2 group between the OH and the aromatic ring prevents any resonance interaction between the electrons on the oxygen and the ring—the CH_2 acts something like an insulator. In other words, the electrons on the oxygen are not conjugated with the double bonds of the ring because of the CH_2 group. Remember that resonance does not usually affect acid strength unless it involves the unshared electrons of the conjugate base. Resonance that stabilizes both the acid and the conjugate base equally, as in this case, has no effect on acidity.

The pK_a of *p*-nitrophenol is 7.15; it is a stronger acid than phenol by a factor of 700. This is due in part to the electron-withdrawing inductive effect of the nitro

OH
CH_2

Benzyl alcohol
$pK_a = 16$

p-Nitrophenol
$pK_a = 7.15$

group. However, the nitro group also provides additional delocalization of the un-shared electrons of the conjugate base. When the electrons are placed on the carbon bonded to the nitro group, they can be delocalized onto the oxygen of the nitro group. This extra resonance stabilization is the principal reason for the increase in acidity for this compound.

The pK_a of *m*-nitrophenol is 8.36. The nitro group is not nearly as effective in increasing acidity when it is substituted on this position of the ring as in the case of *p*-nitrophenol, even though its inductive effect should be stronger because it is closer to the OH. The reason for this is the absence of a resonance structure where the electrons are delocalized onto the nitro group in this isomer. Examination of the resonance struc-tures of the conjugate base shows that the electron pair is never on the carbon to which the nitro group is attached, so it cannot be delocalized onto the oxygen of the nitro goup. (Remember that the electron pair is delocalized onto alternating positions.) The nitro group exerts only an inductive effect.

o-Nitrophenol has the OH and NO$_2$ groups bonded to adjacent carbons. Its pK_a is 7.22. This isomer has the same type of resonance stabilization as *p*-nitrophenol. The in-ductive effect of the nitro group is stronger because it is closer to the OH (acid strength-ening) and the acidic hydrogen is hydrogen bonded to the oxygen of the nitro group (acid weakening). These latter two effects nearly cancel in this case.

Picric acid is a strong acid with a pK_a of 0.42. The acid-strengthening effect of the three nitro groups, which all can delocalize the electron pair of the conjugate base, is quite large.

m-Nitrophenol
$pK_a = 8.36$

o-Nitrophenol
$pK_a = 7.22$

2,4,6-Trinitrophenol
(picric acid)
$pK_a = 0.42$

Ammonium ion is a weak acid. Its pK_a is 9.24. On the basis of arguments similar to those presented earlier for phenol, the conjugate acid of aniline is expected to be a stronger acid than ammonium ion because of resonance stabilization of its conjugate base. It is a stronger acid by a factor of more than 10^4; its pKa is 4.63.

NH$_4^+$

pK$_a$ = 9.24

Conjugate acid of aniline
pK$_a$ = 4.63

Which is the stronger base, aniline or ammonia? The same reasoning applies. Aniline is stabilized by resonance. The basic pair of electrons is delocalized into the ring and is less available to form a covalent bond, so it is a weaker base. (Of course, if the conjugate acid of aniline is stronger than ammonium ion, then aniline must be a weaker base than ammonia.)

The cyano group of p-cyanoaniline exerts an electron-withdrawing inductive effect and can delocalize the electron pair of the base by resonance. On the basis of reasoning identical to that presented for p-nitrophenol, the conjugate acid of p-cyanoaniline should be a stronger acid than the conjugate acid of aniline. Indeed, this is the case; the pK$_a$ is 1.74.

Aniline

pK$_a$ = 1.74 p-Cyanoaniline

Alkanes like ethane, with pK$_a$'s around 50, are the weakest acids commonly encountered in organic chemistry. Their conjugate bases are much too strong to be generated by the methods that are used to make other, less strong bases. The pK$_a$ for acetone is 20. The C=O has increased the acidity of the hydrogen on the adjacent C by a factor of 10^{30} compared to ethane. In this case the electrons of the conjugate base are delocalized onto an electronegative oxygen atom, resulting in an enormous increase in acidity. (Note the similarity to the case of acetic acid.) Anions similar to this one are very important in organic chemistry and can be readily generated by treating the conjugate acid with one of the very strong bases that are available in the laboratory. They are encountered often as important reagents or as intermediates in reactions.

CH_3CH_3

Ethane
pK$_a$ = 50

Acetone
pK$_a$ = 20

2,4-Pentanedione
$pK_a = 9$

The most acidic hydrogen of 2,4-pentanedione is on the carbon between the two $C=O$ groups. If this hydrogen is removed, the new unshared electron pair of the conjugate base is delocalized onto both of the oxygens. The additional resonance stabilization causes this compound to be an even stronger acid. Its pKa is 9; it is a stronger acid than water.

These examples illustrate the importance of looking for resonance stabilization in predicting the strength of an acid. Remember that resonance usually stabilizes the conjugate base more than the acid. To do this and to be effective in increasing the acidity, resonance must involve the newly generated unshared pair of electrons of the conjugate base, delocalizing them so that they are less available for bonding to the proton. This makes the conjugate base weaker, and hence the acid is stronger.

PROBLEM 4.18
Show the resonance structures for the conjugate base of the meta isomer of nitrophenol and confirm that the nitro group is less effective at stabilizing this anion than it is in the case of the para isomer.

PRACTICE PROBLEM 4.6

Explain which compound is a stronger acid:

$$CH_3CH_3 \text{ or } CH_3NO_2$$

Strategy

First identify the acidic hydrogen in each compound. Then draw the conjugate base that is formed when this hydrogen is removed. Remember the five factors that can affect the strength of the acid: (1) the atom to which the hydrogen is bonded, (2) an inductive effect, (3) hydrogen bonding involving the acidic hydrogen, (4) the hybridization of the atom to which the hydrogen is bonded, and (5) a resonance effect. Determine which effect is operative in the case being considered. Is the atom to which the hydrogen is bonded different in the compounds being compared (factor 1)? Is there a nearby polar bond that can affect the acidity by its inductive effect (factor 2)? Is the hydrogen near an electronegative atom to which it can hydrogen bond (factor 3)? Is the hybridization of the atom to which the hydrogen is bonded different (factor 4)? Or is the electron pair generated in forming the conjugate base part of a conjugated pi system (factor 5)? Then determine which acid or which conjugate base is stabilized by the effect.

Solution

Each compound has only one type of hydrogen, so the hydrogen that is removed to form the conjugate base is easy to identify. The electron pair of the conjugate base from CH_3NO_2 is adjacent to the NO pi bond, so it can be stabilized by resonance. Because the conjugate base of CH_3CH_3 has no such resonance stabilization, CH_3NO_2 is a stronger acid.

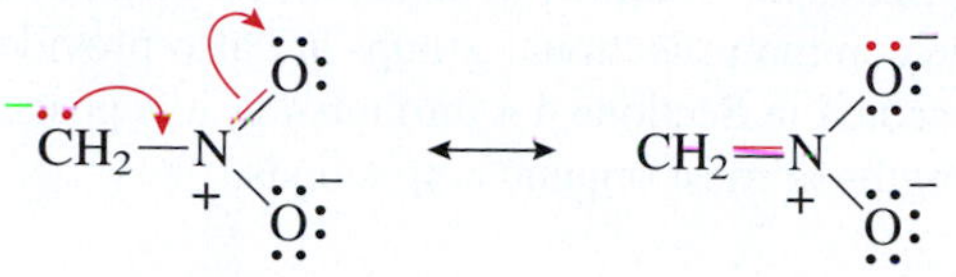

PROBLEM 4.19

Explain which compound is a stronger acid:

a) $CH_3\overset{O}{\overset{\|}{C}}CH_3$ or $CH_3\overset{O}{\overset{\|}{C}}CH_2C\equiv N$

b) 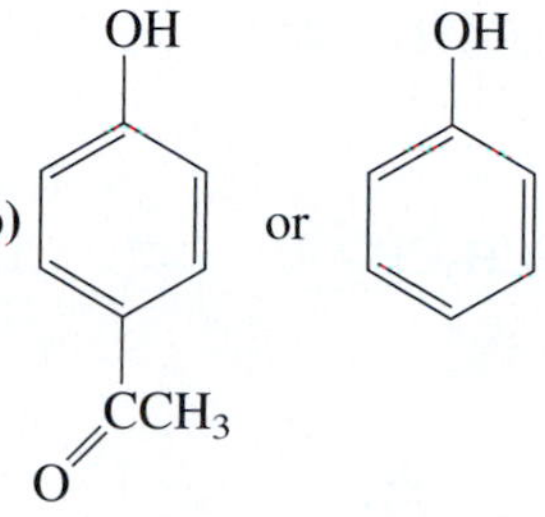 or

c) 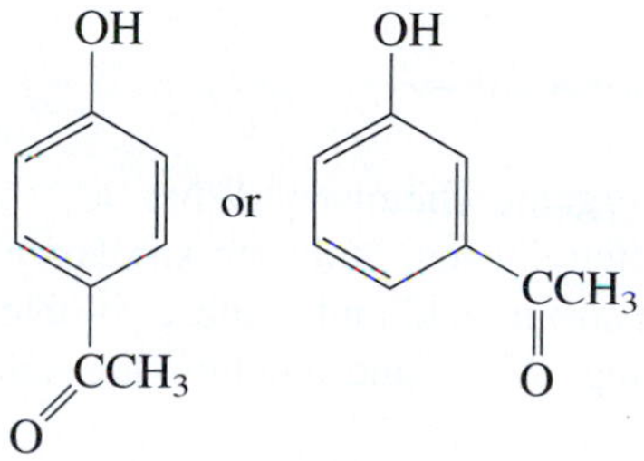 or

d) CH_3CH_3 or $CH_3\overset{O}{\overset{\|}{C}}OCH_3$

e) 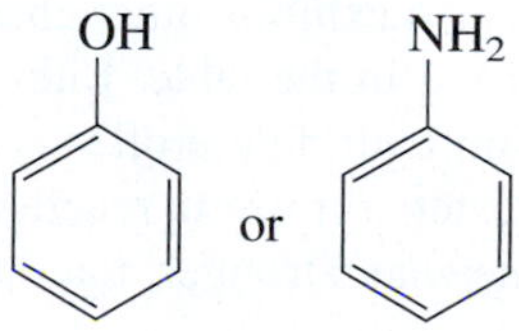 or

PROBLEM 4.20

Explain which compound is the weaker base.

a) 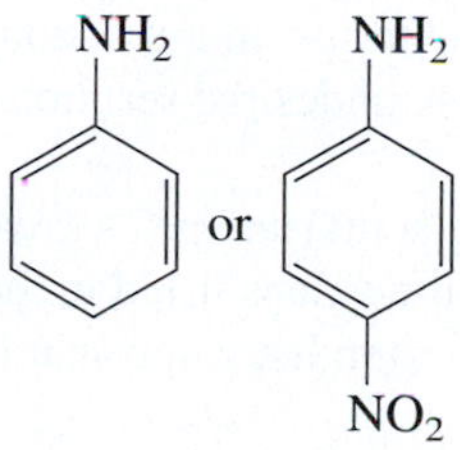 or

b)  or

4.9   TABLES OF ACIDS AND BASES

Table 4.2 lists the pK_a values for a number of acids, some inorganic and some organic. Examples of many of the common functional groups are also provided. By using Table 4.2 and the reasoning presented in Sections 4.4 through 4.8, it is possible to make good estimates of the acid strengths of most organic compounds.

PROBLEM 4.21
Use Table 4.2 to predict whether the equilibrium for these reactions favors the reactants or the products.

a)  :NH$_2$⁻ + CH$_3$CCH$_3$ ⇌ NH$_3$ + CH$_2$CCH$_3$⁻

b)  CH$_3$SCH$_2$⁻ + CH$_3$CH$_2$OH ⇌ CH$_3$SCH$_3$ + CH$_3$CH$_2$O:⁻

c) 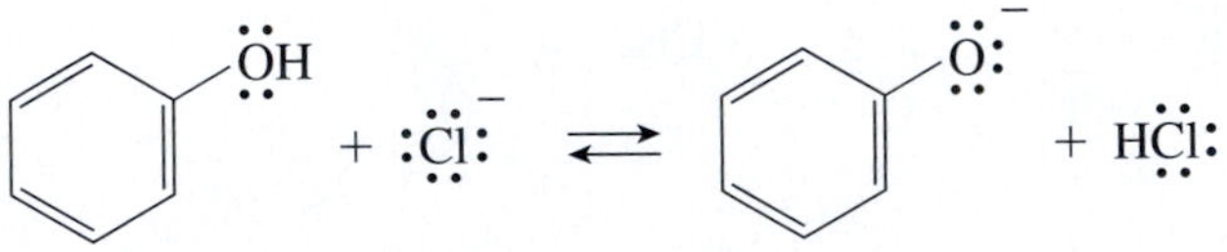 C$_6$H$_5$OH + :Cl:⁻ ⇌ C$_6$H$_5$O:⁻ + HCl:

Table 4.3 lists some common acids used in organic chemistry. When a strong acid is needed, sulfuric or hydrochloric acid is often chosen. Note the similarity of *p*-toluenesulfonic acid to sulfuric acid. It is also a strong acid and is more soluble in organic solvents because of its large organic group. Water and alcohols are acidic enough to react with many strong organic bases.

Table 4.4 lists some common bases used in organic chemistry. Although butyllithium behaves as a very strong base in many reactions, it also exhibits other chemistry, so it is usually used to prepare other strong bases listed in the table. Lithium diisopropylamide, sodium amide, dimsyl anion, and sodium hydride are often used to prepare the conjugate bases of aldehydes, ketones, and esters for use in reactions. Potassium *tert*-butoxide is employed when a base somewhat stronger than the conjugate bases of most alcohols is needed.

Sodium hydroxide and sodium ethoxide are used in many reactions where a moderate base will suffice. Carbonate and bicarbonate are often employed to remove acids in the workup of organic reactions. For many of these bases it does not usually matter whether the cation is sodium or potassium.

Some care must be given to the choice of an acid or a base to use in a reaction. It must be strong enough to do the job but not so strong as to cause undesired reactions at other, less reactive functional groups.

Finally, remember that pK_a's more negative than -2 (strong acids) and pK_a's greater than 16 (very weak acids) can be measured only indirectly, so the values listed in these tables are only approximate. For this reason, different sources often list somewhat different values for these acidity constants.

Table 4.2   Approximate Acidity Constants for Some Selected Compounds

Acid	Conjugate Base	pK_a	Name
FSO_3H	FSO_3^-	<-12	Fluorosulfonic acid (a "super acid")
$HClO_4$	ClO_4^-	-10	Perchloric acid
HI	I^-	-10	Hydroiodic acid
H_2SO_4	HSO_4^-	-10	Sulfuric acid
HBr	Br^-	-9	Hydrobromic acid
HCl	Cl^-	-7	Hydrochloric acid
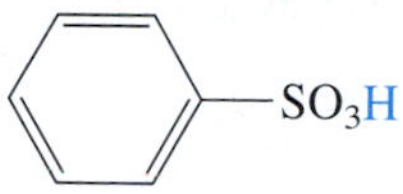 —SO_3H	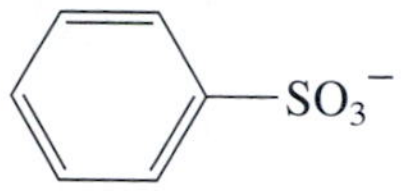 —SO_3^-	-6.5	Benzenesulfonic acid (note similarity to sulfuric acid)
ROH_2^+	ROH	-2	Conjugate acid of an alcohol (note similarity to H_3O^+)
H_3O^+	H_2O	-1.74	Conjugate acid of water (hydronium ion)
HNO_3	NO_3^-	-1.4	Nitric acid
HF	F^-	3.17	Hydrofluoric acid
$\overset{O}{\overset{\|}{R C O H}}$	$\overset{O}{\overset{\|}{R C O^-}}$	$4\text{--}5$	Carboxylic acids
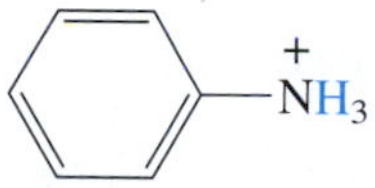 —$\overset{+}{N}H_3$	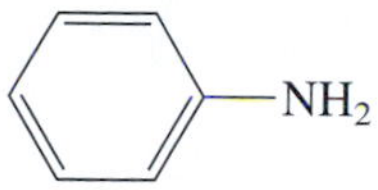 —NH_2	4.63	Conjugate acid of aniline
$\overset{O}{\overset{\|}{H O C O H}}$	$\overset{O}{\overset{\|}{H O C O^-}}$	6.35	Carbonic acid
$\overset{O\quad O}{\overset{\|\quad\|}{CH_3CCH_2CCH_3}}$	$\overset{O\quad O}{\overset{\|\;\;\|}{CH_3C\overset{..}{C}HCCH_3}}$	9	2,4-Pentanedione
NH_4^+	NH_3	9.24	Ammonium ion
$HC\equiv N$	$^-C\equiv N$	9.31	Hydrogen cyanide
$R\overset{+}{N}H_3$	RNH_2	$10\text{--}11$	Conjugate acid of amines
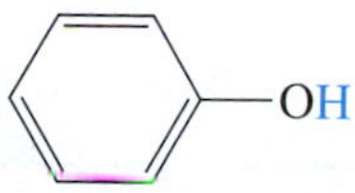 —OH	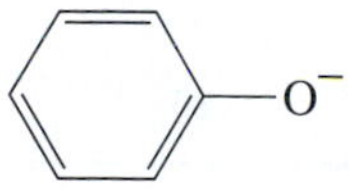 —O^-	10	Phenol
RCH_2NO_2	$R\overset{..}{C}HNO_2$	10	Nitroalkanes

Continued

Table 4.2　Approximate Acidity Constants for Some Selected Compounds—cont'd

Acid	Conjugate Base	pKa	Name
$CH_3CCH_2COCH_2CH_3$ (with two C=O)	$CH_3CCHCOCH_2CH_3$ (with two C=O, carbanion)	11	Ethyl acetoacetate
$CH_3CH_2OCCH_2COCH_2CH_3$ (with two C=O)	$CH_3CH_2OCCHCOCH_2CH_3$ (with two C=O, carbanion)	11	Diethyl malonate
H_2O	HO^-	15.74	Water
RCH_2OH	RCH_2O^-	16	Alcohols (note similarity to water)
$RCNH_2$ (with C=O)	$RCNH^-$ (with C=O)	17	Amides
$RCCH_2R'$ (with C=O)	$RCCHR'^-$ (with C=O)	20	Ketones
$ROCCH_2R'$ (with C=O)	$ROCCHR'^-$ (with C=O)	25	Esters
$RCH_2{-}CN$	$RCH^-{-}CN$	25	Nitriles
$RC{\equiv}CH$	$RC{\equiv}C:^-$	25	Alkynes
H_2	$H:^-$	35	Hydrogen (conjugate base is hydride)
CH_3SCH_3 (with S=O)	$CH_3SCH_2^-$ (with S=O)	38	Dimethylsulfoxide (conjugate base is dimsyl ion)
NH_3	$:NH_2^-$	38	Ammonia (conjugate base is amide ion)
(benzene ring)$-CH_3$	(benzene ring)$-CH_2^-$	41	Toluene
$CH_2{=}CH{-}CH_3$	$CH_2{=}CH{-}CH_2^-$	43	Propene
$CH_2{=}CH_2$	$CH_2{=}CH^-$	44	Ethene
CH_3CH_3	$CH_3CH_2^-$	50	Ethane

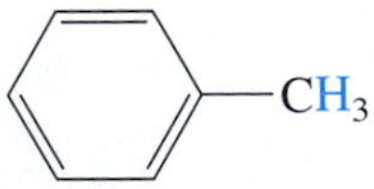

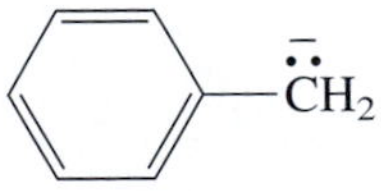

Table 4.3   Common Acids Used in Organic Chemistry

Acid	pK_a	Name
H_2SO_4	-10	Sulfuric acid
HCl	-7	Hydrochloric acid
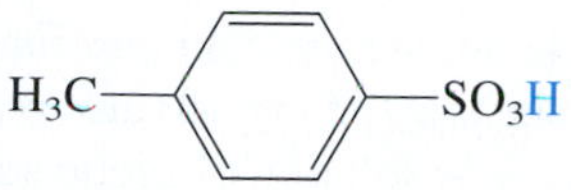 H₃C—⟨benzene ring⟩—SO₃H	-7	p-Toluenesulfonic acid
CH_3CO_2H	4.8	Acetic acid
H_2O	15.7	Water
CH_3OH or CH_3CH_2OH	16	Methanol or ethanol

Table 4.4   Common Bases Used in Organic Chemistry

Base	pK_a of Conjugate Acid	Name
$CH_3CH_2CH_2CH_2Li$	50	Butyllithium
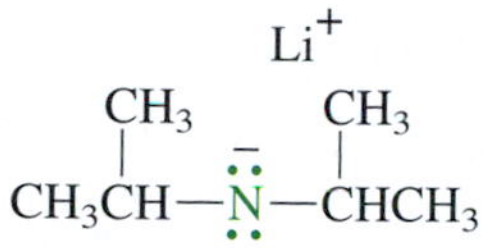 Li^+ CH_3CH—$\overset{..}{\underset{..}{N}}$—$CHCH_3$ (with CH_3 groups)	38	Lithium diisopropylamide (LDA)
$:\overset{-}{N}H_2$ Na^+	38	Sodium amide
$CH_3\overset{O}{\overset{\|}{S}}CH_2^{..-}$ Na^+	38	Dimsyl anion (conjugate base of dimethyl sulfoxide)
$H:^-$ Na^+	35	Sodium hydride
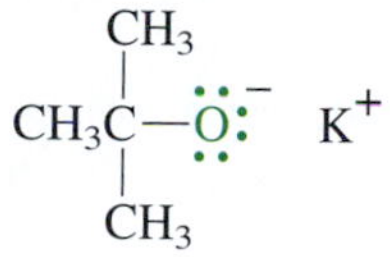 $CH_3\overset{CH_3}{\underset{CH_3}{C}}$—$\overset{..}{\underset{..}{O}}:^-$ K^+	19	Potassium tert-butoxide
$CH_3CH_2\overset{..}{\underset{..}{O}}:^-$ Na^+	16	Sodium ethoxide
$H\overset{..}{\underset{..}{O}}:^-$ Na^+	15.7	Sodium hydroxide
CO_3^{2-} $2 Na^+$	10.3	Sodium carbonate
$HCO3^-$ Na^+	6.4	Sodium bicarbonate
$CH_3\overset{\overset{..}{O}\cdot}{\overset{\|}{C}}\overset{..}{\underset{..}{O}}:^-$ Na^+	4.8	Sodium acetate

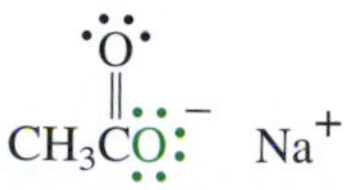

4.10   ACIDITY AND BASICITY OF FUNCTIONAL GROUPS AND SOLVENTS

As can be seen from Table 4.2, most functional groups are only weak acids and/or weak bases. They are too weak to behave as acids or bases in aqueous solution and therefore are said to be neutral. There are three exceptions to this rule among the common functional groups. Carboxylic acids ($pK_a \approx 5$) and substituted phenols ($pK_a \approx 10$) are weak acids. Amines, like ammonia, are weak bases. The pH of a 0.1 M solution of acetic acid (CH_3CO_2H) is about 3. Phenol is a weaker acid and the pH of a 0.1 M solution is about 5.5. And the pH of a 0.1 M solution of CH_3NH_2 is about 11.

These acidic and basic properties can be used to help identify these functional groups. If an unknown compound is water soluble (has a small R group), then the pH of an aqueous solution of the compound provides an important clue to the functional group that it contains. If the unknown is not soluble in water, its solubility behavior in aqueous acid and base provides the same information. Consider, for example, a carboxylic acid that is not soluble in water. If this acid is added to an aqueous solution of sodium hydroxide, the following reaction occurs:

$$\underset{\substack{pK_a = 5 \\ \text{Water insoluble}}}{RC\overset{O}{\overset{\|}{}}OH} + Na^+OH^- \longrightarrow \underset{\substack{pK_a = 16 \\ \text{Water soluble}}}{RC\overset{O}{\overset{\|}{}}O^-\,Na^+} + H_2O$$

Hydroxide ion is a strong enough base that the equilibrium for this reaction lies entirely to the right. The insoluble carboxylic acid is converted to its conjugate base, which, as an ionic compound, is quite soluble in water. So the carboxylic acid dissolves (more accurately, it reacts and the product dissolves) in basic solution. Substituted phenols also dissolve in aqueous sodium hydroxide. They can be distinguished from carboxylic acids by using a weaker base. Bicarbonate anion, HCO_3^-, is a strong enough base to react with carboxylic acids but not with phenols. (The pK_a for carbonic acid, H_2CO_3, is 6.35.) So a compound that dissolves in aqueous sodium hydroxide and in aqueous sodium bicarbonate is a carboxylic acid, whereas one that dissolves in aqueous sodium hydroxide but not in aqueous sodium bicarbonate is a phenol.

An amine that is insoluble in water reacts with acid to form a salt that dissolves in water, as shown in the following equation. Hydrochloric acid is strong enough that this equilibrium lies entirely toward the product side of the equation.

$$RNH_2 + HCl \longrightarrow RNH_3^+\,Cl^-$$

The acidic or basic properties of these functional groups can also be used to advantage to separate them from neutral compounds. For example, suppose we desire to separate a mixture of naphthalene and benzoic acid:

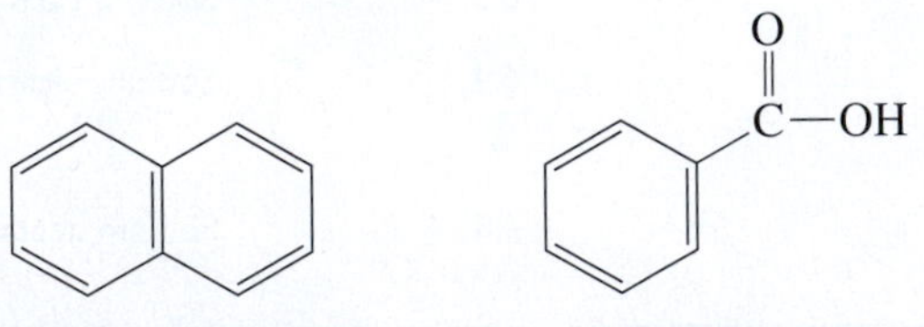

Naphthalene   Benzoic acid

Neither of these compounds is very soluble in water, but both dissolve in a less polar solvent such as dichloromethane (CH_2Cl_2). If this solution is extracted with aqueous sodium hydroxide, the benzoic acid reacts to form a salt that is soluble in the aqueous phase. Therefore, when the two liquids are separated, the carboxylic acid salt is in the aqueous phase and the naphthalene remains in the dichloromethane. The compounds have now been separated. To recover the carboxylic acid, it is necessary to protonate its conjugate base. This can be accomplished by treating the aqueous solution with a strong acid, such as hydrochloric acid. The equilibrium shown in the following equation favors the weaker acid, the carboxylic acid, which then precipitates from the solution:

Precipitates

Phenols can be separated from neutral compounds by a similar process. For an amine an acid extraction step followed by making the solution basic is used to accomplish separation from neutral compounds.

When an organic reaction is being designed, the nature of the solvent, especially whether it contains any acidic hydrogens, must be considered. Many of the reagents that are employed in organic reactions are very strong bases. In using such reagents, even weakly acidic solvents must be avoided. For example, suppose we wanted to use sodium amide, $NaNH_2$, in a reaction. This strong base cannot be used in solvents such as ethanol or water because both of these compounds have a weakly acidic hydrogen on the oxygen. An acid–base reaction, such as the one shown in the following equation, would occur, resulting in the destruction of the basic amide anion:

$$Na^+ \; :\!\overset{..}{N}H_2^- \; + \; H\overset{..}{\underset{..}{O}}CH_2CH_3 \longrightarrow Na^+ \; ^-\!\overset{..}{\underset{..}{O}}CH_2CH_3 \; + \; :NH_3$$

The strongest base that can be used in any particular solvent is the conjugate base of that solvent. Any stronger base will simply react to produce the conjugate base of the solvent. For this reason a base is often used in its conjugate acid as the solvent. For example, NH_2^- is often used in liquid NH_3 as the solvent (NH_3 boils at $-33°C$, so the reaction must be done at low temperature), OH^- is often used in water as the solvent, ethoxide ion ($CH_3CH_2O^-$) is often used in ethanol (CH_3CH_2OH) as the solvent, and the dimsyl anion is often used in DMSO as the solvent:

Dimsyl anion

*Dim*ethyl*sulf*oxide
DMSO

Ethers are also commonly used as solvents for reactions involving strongly basic reagents because they do not have acidic hydrogens but are still polar enough to dissolve the reagents. Ethers that are used include diethyl ether, *tetra*hydro*fur*an (THF), and *di*methoxyethane (DME). These compounds are weak enough acids that they can be used with almost any strong base.

$$CH_3CH_2OCH_2CH_3$$

$$CH_3OCH_2CH_2OCH_3$$

Diethyl ether
"ether"

Tetrahydrofuran
THF

Dimethoxyethane
DME

PROBLEM 4.22

Explain whether each of the following solvents would be acceptable for reactions involving this anion:

$$CH_3C\equiv C:^- \xrightarrow{\text{solvent}}$$

a) Liquid NH_3 **b)** CH_3CH_2OH **c)** $CH_3CH_2OCH_2CH_3$

Just as strong bases react with weakly acidic solvents, strong acids react with even weakly basic solvents. For example, both hydrogen chloride and hydrogen bromide are virtually completely ionized in aqueous solution:

$$HCl + H_2O \longrightarrow H_3O^+ + Cl^- \qquad pK_a = -7$$

$$HBr + H_2O \longrightarrow H_3O^+ + Br^- \qquad pK_a = -10$$

Even though HCl and HBr have different pK_a's, their acid strengths in water are the same because the actual acid that is present in each case is the hydronium ion, H_3O^+. The strongest acid that can be generated in a particular solvent is the conjugate acid of that solvent. Any stronger acid reacts to generate the conjugate acid of the solvent. This is termed the *leveling effect*.

Pure sulfuric acid is a stronger acid than sulfuric acid in water because the acid in aqueous sulfuric acid is actually H_3O^+. Superacids are defined as compounds that are even stronger acids than 100% H_2SO_4. One example of a superacid is fluorosulfonic acid. The inductive effect of the electronegative fluorine makes this a stronger acid than sulfuric acid.

Sulfuric acid

Fluorosulfonic acid

Liquid fluorosulfonic acid can be used to protonate extremely weak bases. It can be made even stronger by adding a Lewis acid, such as antimony pentafluoride, SbF_5, which complexes with the conjugate base of fluorosulfonic acid, decreasing its basicity. This mixture, known as magic acid, is so strong that it can protonate extremely weak bases such as the electron pair of a carbon–carbon pi bond as shown in the following equation. (Note that all of the bases we have seen up to this point have had an unshared pair of electrons that are employed to form a bond to the proton of the acid. It takes an extremely strong acid to protonate the lower-energy electrons in a bonding MO.) The carbocation that is produced is stable enough in a solution of magic acid that its properties can be studied.

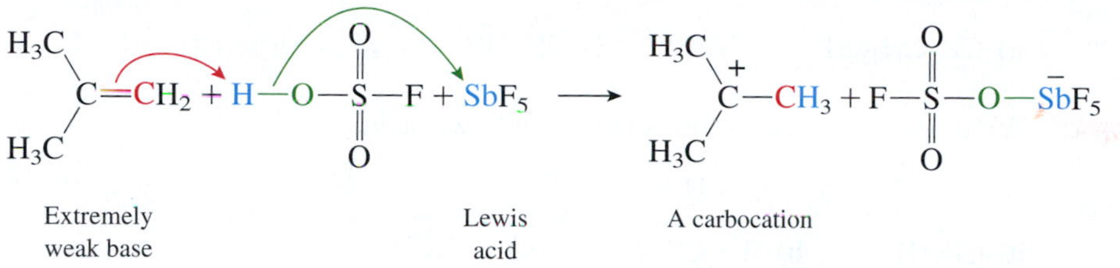

Extremely weak base	Lewis acid	A carbocation

George Olah was awarded the 1994 Nobel Prize in chemistry for his work with carbocations and superacids.

Review of Mastery Goals

After completing this chapter, you should be able to:

ORGANIC
Chemistry Now™
Click *Mastery Goal Quiz* to test how well you have met these goals.

■ Write an acid–base reaction for any acid and base. (Problems 4.35 and 4.36)

■ Recognize Lewis acids, Lewis bases, and the reactions between them. (Problems 4.25, 4.26, and 4.38)

■ Recognize acid or base strengths from K_a or pK_a values and use these to predict the position of an acid–base equilibrium. (Problems 4.23, 4.34, 4.35, 4.36, 4.37, and 4.41)

■ Predict and explain the effect of the structure of the compound, such as the atom bonded to the hydrogen, the presence of an electron-donating or electron-withdrawing group, hydrogen bonding, the hybridization of the atom attached to the hydrogen, or resonance, on the strength of an acid or base. (Problems 4.29, 4.30, 4.35, 4.39, 4.40, 4.41, 4.42, 4.43, and 4.47)

■ Using the same reasoning, arrange a series of compounds in order of increasing or decreasing acid or base strength. (Problems 4.31, 4.32, and 4.33)

■ Identify the most acidic proton in a compound. (Problems 4.24, 4.36, 4.42, and 4.43)

Additional Problems

ORGANIC
Chemistry Now™
Assess your understanding of this chapter's topics with additional quizzing and conceptual-based problems at **http://now.brookscole.com/hornback2**

4.23 Show the conjugate acid of each of these species:

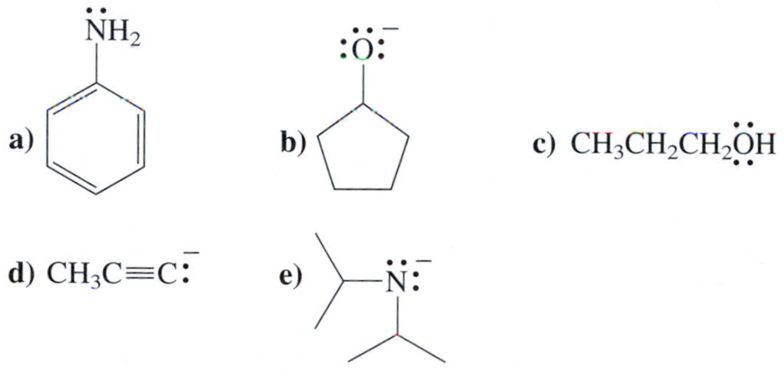

4.24 Show the most stable conjugate base of these compounds:

a) CH_3CH_2OH b) $HOCCH_2CH_2OH$ (with O double bonded above the first C)   c) $H_2NCH_2CH_2OH$

4.25 Which of these species can behave as a Lewis acid?

a) 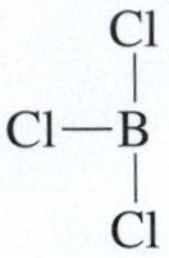 Cl—B (with Cl above and Cl below)   b) H—C—H (with H above and H below)   c) $CH_3CH_2^+$

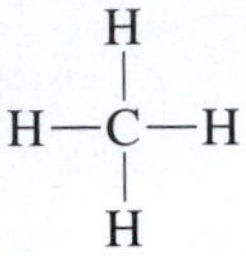

4.26 Which of these species can behave as a Lewis base?

a) $CH_3—\ddot{O}—CH_2CH_3$ b) $CH_3CH_2CH_3$ c) $CH_3\ddot{N}H_2$

d) 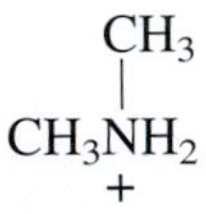 CH_3NH_2 (with CH_3 above and + below)   e) 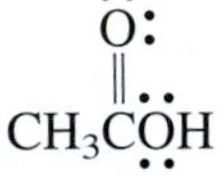 $CH_3C\ddot{O}H$ (with $\ddot{O}$ double bonded above)

4.27 Calculate the pK_a for these compounds.

a) $HCOH$ (with O double bonded above C) $(K_a = 1.75 \times 10^{-4})$ b) CH_3CH_3 $(K_a = 10^{-50})$

4.28 Calculate the K_a for these compounds:

a) $HC\equiv CH$ $(pK_a = 25)$ b) $HC\equiv N$ $(pK_a = 9.31)$

4.29 Explain which compound is the stronger acid:

a) CH_3CNH_2 (with O double bonded above C) or CH_3COH (with O double bonded above C)

b) 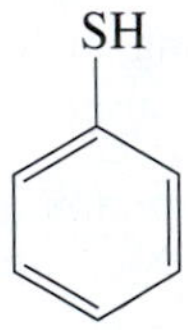 [benzene ring with SH] or [benzene ring with OH]

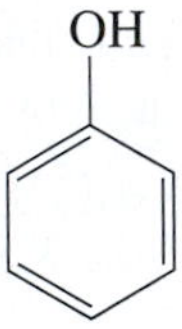

c) $CH_3CH_2CH_2CH_3$ or [benzene ring with CH_3]

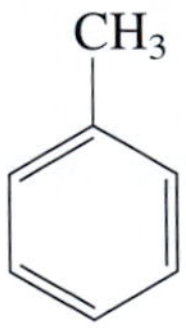

d) 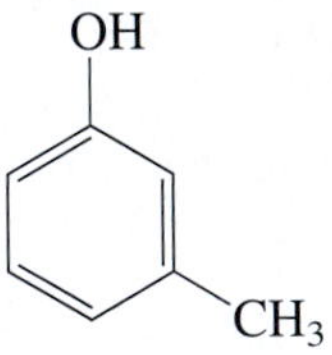 [benzene ring with OH and CH_3] or [benzene ring with OH and CF_3]

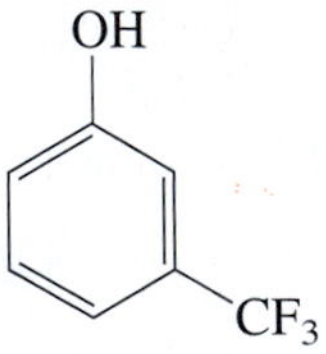

e) CH_3NH_2 or $CH_2=NH$

4.30 Explain which species is the stronger base:

a)

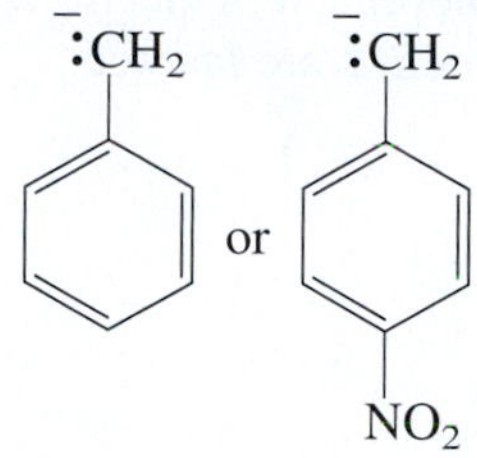

$:CH_2$ (benzyl anion) or $:CH_2$ (p-nitrobenzyl anion, with NO_2)

b) 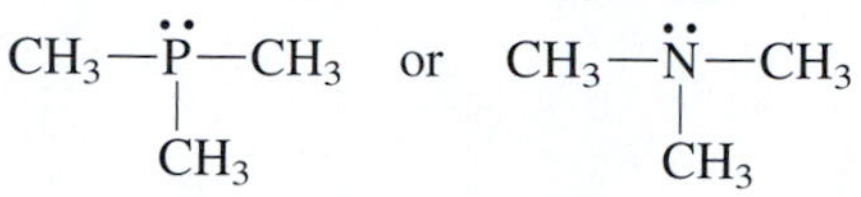 $CH_3-\overset{..}{P}-CH_3$ or $CH_3-\overset{..}{N}-CH_3$
with CH_3 below each

c) $BrCH_2\overset{\overset{..}{O}:}{\underset{}{C}}\overset{..}{O}:^-$ or $ClCH_2\overset{\overset{..}{O}:}{\underset{}{C}}\overset{..}{O}:^-$ **d)** $CH_3\overset{..}{O}:^-$ or $CH_3\overset{-}{\underset{..}{N}}H$

4.31 Arrange these compounds in order of increasing acid strength:

phenol (OH), CH_3OH, p-cyanophenol (OH, CN), m-cyanophenol (OH, CN)

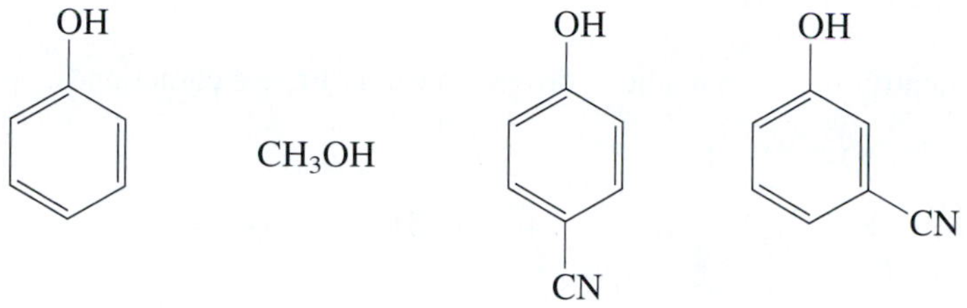

4.32 Arrange these compounds in order of increasing base strength:

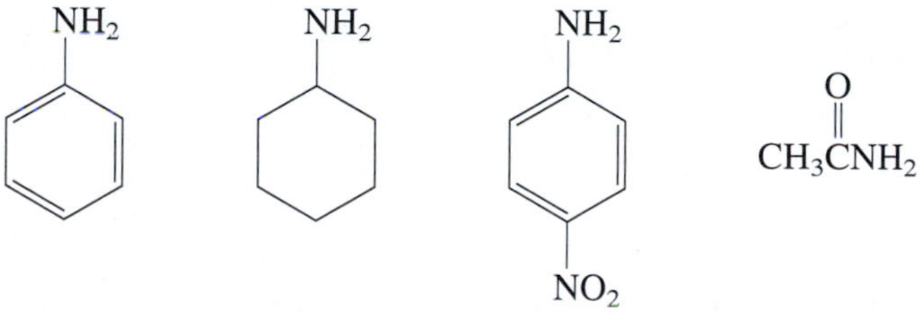

aniline (NH_2), cyclohexylamine (NH_2), p-nitroaniline (NH_2, NO_2), $CH_3\overset{\overset{O}{||}}{C}NH_2$

4.33 Arrange these compounds in order of increasing acid strength:

$CH_3\overset{\overset{O}{||}}{C}OH$ HO$\overset{\overset{O}{||}}{C}-\overset{\overset{O}{||}}{C}OH$ HO$\overset{\overset{O}{||}}{C}CH_2CH_2\overset{\overset{O}{||}}{C}OH$

4.34 Use the tables in this chapter to predict whether these equilibria favor the reactants or the products:

a) 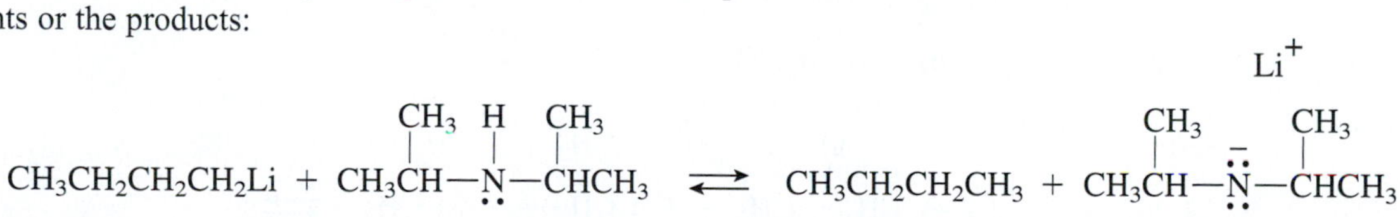 $CH_3CH_2CH_2CH_2Li + CH_3\overset{\overset{CH_3}{|}}{CH}-\overset{\overset{H}{|}}{\underset{..}{N}}-\overset{\overset{CH_3}{|}}{CH}CH_3 \rightleftharpoons CH_3CH_2CH_2CH_3 + CH_3\overset{\overset{CH_3}{|}}{CH}-\overset{Li^+}{\underset{..}{\overset{-}{N}}}-\overset{\overset{CH_3}{|}}{CH}CH_3$

b) $CH_3C\equiv C-H + {}^-\overset{..}{N}H_2 \rightleftharpoons CH_3C\equiv C:^- + :NH_3$

c) $CH_3C\equiv C-H + (CH_3)_3C-\overset{-}{\underset{..}{O}}: \rightleftharpoons CH_3C\equiv C:^- + (CH_3)_3C-\overset{..}{\underset{..}{O}}H$

4.35 Complete these equilibrium reactions in the most reasonable manner possible using the curved arrow convention to show the movement of electrons in the reactions. Predict whether the reactants or the products are favored.

a) $CH_3CH_3 + {}^-\ddot{C}H_2C\equiv N\colon \rightleftharpoons$

b) $CF_3-\underset{\overset{\|}{\ddot{O}\colon}}{C}-\ddot{O}\colon^- + CH_3-\underset{\overset{\|}{\ddot{O}\colon}}{C}-\ddot{O}-H \rightleftharpoons$

c) $CH_3\underset{\overset{\|}{\ddot{O}\colon}}{C}-\ddot{N}H_2 + CH_3\underset{\overset{\|}{\ddot{O}\colon}}{C}-\ddot{C}H_2{}^- \rightleftharpoons$

d) $CH_3CH_2CH_2\underset{\overset{\|}{O}}{C}OH + {}^-OH \rightleftharpoons$

4.36 Identify the most acidic hydrogen in each of these compounds:

a) $HO\underset{\overset{\|}{O}}{C}CH_2CH_2\underset{\overset{\|}{O}}{\overset{\|}{S}}OH$

b) $CH_3CH_2CH_2C\equiv N$

c)

d)

e) $CH_3\underset{\overset{\|}{O}}{C}CH_2\underset{\overset{\|}{O}}{C}OCH_2CH_3$

f) $H_3\overset{+}{N}CH_2\underset{\overset{\|}{O}}{C}OH$

4.37 Show the products of these acid–base reactions and predict whether the equilibria favor the reactants or the products:

a) $CH_3\underset{\overset{\|}{O}}{C}CH_2\underset{\overset{\|}{O}}{C}CH_2CH_3 + {}^-\colon\!\ddot{O}CH_2CH_3 \rightleftharpoons$

b) $CH_3CH_2NO_2 + CH_3\ddot{O}\colon^- \rightleftharpoons$

c) $CH_3\underset{\overset{\|}{O}}{C}OCH_3 + CH_3\underset{\underset{CH_3}{|}}{C}H-\ddot{N}-\underset{\underset{CH_3}{|}}{C}HCH_3 \rightleftharpoons$

d) $FCH_2CH_2-\underset{\overset{\|}{O}}{C}OH + CH_3\underset{\underset{F}{|}}{C}H-\underset{\overset{\|}{O}}{C}O^- \rightleftharpoons$

$$\text{e) } CH_3\overset{\overset{OH}{|}}{CH}-\overset{\overset{O}{\|}}{C}OH + NaOH \rightleftharpoons$$

$$\text{f) } CH_3-\overset{\overset{CH_3}{|}}{\underset{\underset{CH_3}{|}}{C}}-\ddot{\underset{\cdot\cdot}{O}}:^- + H_2O \rightleftharpoons$$

4.38 Which compound is behaving as the Lewis acid and which as the Lewis base in this reaction?

$$CH_3CH_2\overset{\cdot\cdot}{\underset{\cdot\cdot}{O}}CH_2CH_3 + AlCl_3 \rightleftharpoons CH_3CH_2\overset{\overset{^-AlCl_3}{|}}{\underset{+}{O}}CH_2CH_3$$

4.39 Explain which of these compounds is the weaker base:

4.40 Explain why the protonation of an amide occurs at the O rather than the N.

$$CH_3-\overset{\overset{\ddot{O}:}{\|}}{C}-\overset{\cdot\cdot}{N}H_2 + H-Cl \rightleftharpoons CH_3-\overset{\overset{+\overset{\cdot\cdot}{O}\diagup^H}{\|}}{C}-\overset{\cdot\cdot}{N}H_2 + Cl^-$$

4.41 Amino acids contain both a basic functional group, the amine, and an acidic functional group, the carboxylic acid. Thus, they can undergo an internal acid–base reaction as shown in the following equation for the amino acid phenylalanine:

 BioLink

$$H_2N-\overset{\overset{CH_2}{|}}{CH}-\overset{\overset{O}{\|}}{C}OH \rightleftharpoons \overset{+}{H_3}N-\overset{\overset{CH_2}{|}}{CH}-\overset{\overset{O}{\|}}{C}O^-$$

Phenylalanine

Using Table 4.2 and neglecting the effect of one group on the acidity of the other, predict the position of this equilibrium in the case of phenylalanine. Explain whether your prediction is in accord with the experimental observations that phenylalanine melts at 273–276°C and is very soluble in water.

BioLink **4.42** Dipeptides result from the reaction of two amino acids to form an amide. Explain which nitrogen of the following dipeptide is the stronger base:

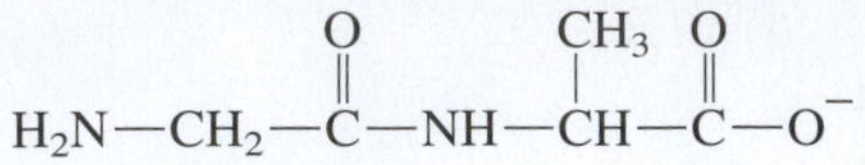

$$\underset{\text{H}_2\text{N}-\text{CH}_2-\overset{\displaystyle\overset{\text{O}}{\|}}{\text{C}}-\text{NH}-\underset{\displaystyle\underset{}{}}{\overset{\displaystyle\overset{\text{CH}_3}{|}}{\text{CH}}}-\overset{\displaystyle\overset{\text{O}}{\|}}{\text{C}}-\text{O}^-}{}$$

BioLink **4.43** Explain which nitrogen in the ring of the amino acid histidine is the stronger base:

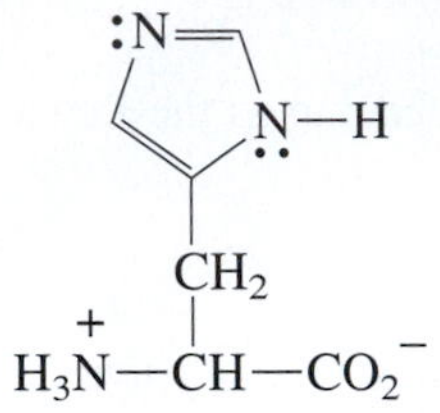

Histidine

BioLink **4.44** The pK_a of the carboxylic acid group of acetic acid is 4.7. The pK_a of the carboxylic acid group of the conjugate acid of the amino acid alanine is 2.3. Explain the difference in these pK_a values.

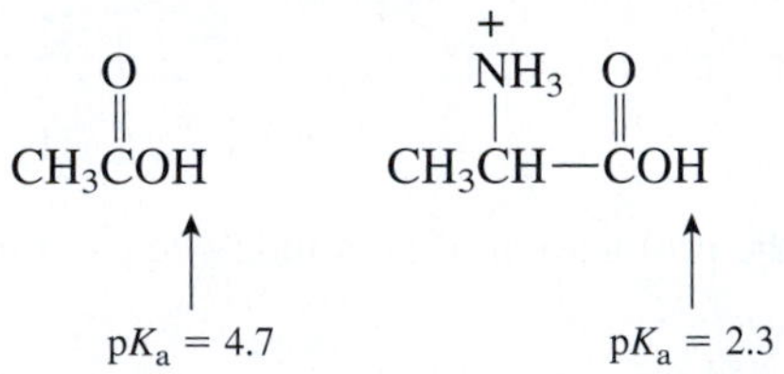

$$\overset{\displaystyle\overset{\text{O}}{\|}}{\text{CH}_3\text{COH}} \qquad\qquad \text{CH}_3\overset{\displaystyle\overset{\overset{+}{\text{NH}_3}}{|}}{\text{CH}}-\overset{\displaystyle\overset{\text{O}}{\|}}{\text{COH}}$$

$$\text{p}K_a = 4.7 \qquad\qquad\qquad \text{p}K_a = 2.3$$

4.45 There are two isomeric conjugate acids that produce the following base. Show the structure of each and explain how they produce the same base upon loss of a proton.

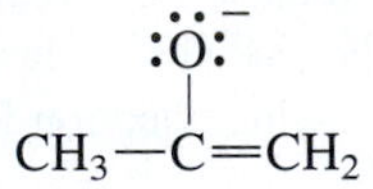

$$\text{CH}_3-\overset{\displaystyle\overset{:\ddot{\text{O}}:^-}{|}}{\text{C}}=\text{CH}_2$$

4.46 When ⓐ is reacted with hydroxide ion, isomer ⓑ is formed.
a) Explain why ⓑ is favored over ⓐ at equilibrium.
b) Show the structure of the conjugate base of ⓐ and explain how this isomerization reaction occurs.

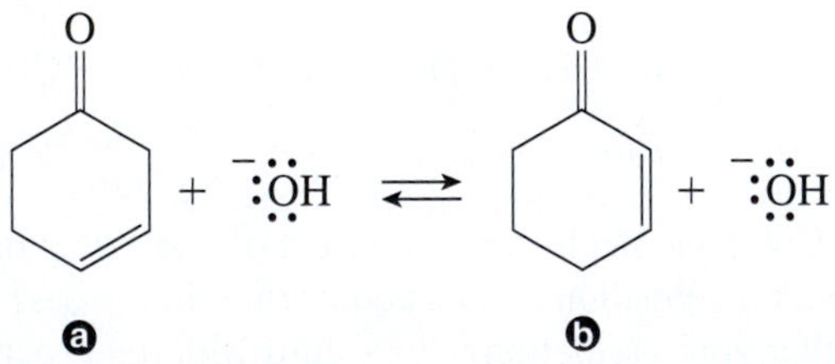

4.47 Compound **c** is a slightly stronger acid than compound **d**. The CH_3O group has both an inductive effect and a resonance effect on the acidity of **d**.

OH

($pK_a = 10.00$)

c

OH

($pK_a = 10.22$)

$:\underset{..}{\overset{..}{O}}-CH_3$

d

a) Explain how the inductive effect of the CH_3O group should affect the acidity of **d**.

b) Show the resonance structures for **d** that involve the CH_3O group. Would you expect the resonance effect of the CH_3O group to cause an increase or decrease in acidity?

c) In situations like this, the resonance effect is usually larger than the inductive effect. Is this consistent with the experimental acidities of these two compounds?

4.48 Explain why the pK_a's of compounds near the middle of Table 4.2 are often listed with two figures to the right of the decimal place (that is, for NH_4^+ the $pK_a = 9.24$), whereas those at the beginning and end of the table are listed without any figures to the right of the decimal point (that is, for H_2SO_4 the $pK_a = -10$).

Problems Using Online Three-Dimensional Molecular Models

ORGANIC

Chemistry·Now™

Click *Molecular Model Problems* to view the models needed to work these problems.

4.49 For each pair of compounds, explain which is the stronger acid.

4.50 Explain why the compound on the left is a stronger acid than the compound on the right.

Do you need a live tutor for homework problems? Access vMentor at Organic ChemistryNow at **http://now.brookscole.com/hornback2** for one-on-one tutoring from a chemistry expert.

Functional Groups and Nomenclature I

C HAPTER 2 showed that the number of organic compounds is virtually limitless. Functional groups were introduced as a useful method for organizing this vast number of compounds because chemical reactions occur at the functional group and compounds with the same functional group undergo similar reactions. Because functional groups are so important, we need to learn more about them. This chapter discusses approximately half of the functional groups in more detail. (A similar treatment of the remainder is provided in Chapter 12.) The physical properties (melting points, boiling points, and solubilities), natural occurrence, and uses of each of these functional groups are presented.

One of the major topics to be discussed in this chapter is organic nomenclature. Every compound needs a name that can be used in talking or writing about it. Once a compound is identified by a unique name, it is then possible to look it up in reference books (such as the *Handbook of Chemistry and Physics* or *Chemical Abstracts*) to find its reported physical and chemical properties. Therefore, a systematic method for naming simple organic compounds is presented. As you might expect, this method is organized on the basis of functional groups.

MASTERING ORGANIC CHEMISTRY

▶ Naming Alkanes Using Systematic Nomenclature

▶ Drawing the Structure of an Alkane from the Name

▶ Naming Complex Alkyl Groups

▶ Naming Cycloalkanes, Alkenes, Alkynes, Alkyl Halides, Alcohols, Ethers, and Amines

▶ Drawing Structures from the Names of These Compounds

▶ Predicting Approximate Physical Properties of Compounds with These Functional Groups

5.1 ALKANES

Organic compounds containing only carbon and hydrogen are called **hydrocarbons.** (In addition to alkanes, the class of hydrocarbons includes alkenes, alkynes, and aromatic compounds that have no additional functional group.) The bonds between the carbons of **alkanes,** the simplest type of hydrocarbons, are all single bonds; alkanes do not have

ORGANIC
Chemistry Now™

Look for this logo in the chapter and go to OrganicChemistryNow at
http://now.brookscole.com/hornback2 for tutorials, simulations, problems, and molecular models.

any double or triple bonds. Alkanes fit the formula C_nH_{2n+2}, so they have a DU of zero. They have no functional group. **Cycloalkanes** are quite similar to alkanes with the exception that they contain rings of carbon atoms. Rings of five, six, and seven carbons are most common. Small rings (three and four carbons) are rarer. Larger rings (those with more than seven carbons) become increasingly less common as their size increases.

$$CH_3CH_2CH_2CH_2CH_3$$

Pentane,
an alkane

Cyclohexane,
a cycloalkane

As you should recall from Chapter 2, the electronegativities of carbon and hydrogen are similar enough that carbon–carbon and carbon–hydrogen bonds have insignificant bond dipoles. Alkanes and cycloalkanes are nonpolar compounds. The only intermolecular interactions are London forces (instantaneous dipole-induced dipole) and are weak. For this reason, alkanes have low melting points and low boiling points. As shown in Table 5.1, methane boils at $-162°C$ and therefore is a gas at room temperature. Boiling points and melting points increase with increasing molecular weight because London forces increase with increasing surface area. Pentane is the smallest alkane that is liquid at room temperature. Very large alkanes, such as eicosane ($C_{20}H_{42}$), are waxy solids. (Beeswax contains $C_{31}H_{64}$.) In general, cycloalkanes melt at significantly higher temperatures than their noncyclic counterparts because they are more symmetrical (compare hexane and cyclohexane in Table 5.1). Cycloalkanes also boil at somewhat higher temperatures.

Because of their nonpolar nature, alkanes are insoluble in water. They are **hydrophobic** (water-hating) compounds. They are composed of atoms of lower mass (C and H) than water (O and H), so liquid alkanes are less dense than water. A mixture of water and a liquid alkane forms two layers, with the alkane as the upper layer. Many of the solvents that are used in the organic laboratory are compounds composed mainly of carbon and hydrogen, and so they tend to be less dense than water. Solvents whose molecules contain a significant fraction of more massive elements, such as chlorine, are more dense than water.

Alkanes are rather unreactive compounds and are sometimes called **paraffins** (from the Latin for "little affinity"). They have strong C—C and C—H bonds, which are hard to break. All their electron pairs are in relatively stable sigma bonding MOs. They have no polar sites or unshared electrons to attract Lewis acid or base reagents. They are extremely weak proton acids. Therefore, only a few of the reactions covered in this text involve alkanes.

Section 2.7 showed that it is often convenient to view a molecule as being composed of two parts: the functional group, where reactions occur, and the alkyl group, sometimes called the backbone of the compound. The alkyl group is just the "alkane" part of the molecule and, like alkanes, does not tend to enter into reactions. In most cases it does not matter what the exact structure of the alkyl group is, because the reactions occur at the functional group.

Alkanes occur in nature in deposits of natural gas and oil. Natural gas is primarily methane, containing progressively smaller amounts of ethane, propane, and butanes. Petroleum is a complex mixture of alkanes and other hydrocarbons. Petroleum is refined by distillation. The various distillation cuts, based on boiling point, are shown in Table 5.2.

Table 5.1   Physical Properties of Some Alkanes

Compound	Name	Melting Point (°C)	Boiling Point (°C)
CH_4	Methane	−183	−162
CH_3CH_3	Ethane	−183	−89
$CH_3CH_2CH_3$	Propane	−187	−42
$CH_3CH_2CH_2CH_3$	Butane	−138	−0.5
$CH_3CH_2CH_2CH_2CH_3$	Pentane	−130	36
$CH_3CH_2CH_2CH_2CH_2CH_3$	Hexane	−94	69
(hexagon)	Cyclohexane	6.5	80
$CH_3(CH_2)_5CH_3$*	Heptane	−91	98
$CH_3(CH_2)_6CH_3$	Octane	−57	126
$CH_3(CH_2)_7CH_3$	Nonane	−51	151
$CH_3(CH_2)_8CH_3$	Decane	−30	174
$CH_3(CH_2)_{18}CH_3$	Eicosane	36	345

*The shorthand used to represent the longer chain alkanes: $CH_3(CH_2)_5CH_3$ means that there are five CH_2's between the CH_3's; the carbon chain has a total of seven carbons.

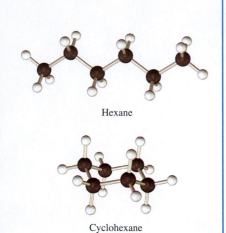

Hexane

Cyclohexane

Table 5.2   Boiling Points for the Distillation Cuts of Petroleum

Distillation Cut	Boiling Point	Comment
Natural gas	Below 20°C	C_1–C_4 alkanes; used primarily in industry and as a fuel for heating homes
Light petroleum and ligroin	20–120°C	C_5–C_7 compounds; large amounts are heated to crack them to small alkenes, a major feedstock for the chemical industry
Gasoline	100–200°C	C_7–C_{10} compounds; contain a large fraction of straight-chain alkanes, which tend to detonate or "knock" when burned; heating over a catalyst breaks bonds, which re-form to produce branched alkanes, which burn more smoothly
Kerosene	200–300°C	C_{12}–C_{18} compounds; used as jet fuel and diesel fuel
Gas oil and lubricating oil	Above 300°C	Larger alkanes; used as heating and lubricating oils; large amounts are used to produce gasoline by heating over a catalyst in a process called *cracking*
Residue	Nonvolatile	Asphalt and bitumen

The most important use of alkanes is as fuels for combustion processes to produce heat and power:

$$C_nH_{2n+2} + (3n + 1)/2 \ O_2 \longrightarrow n \ CO_2 + (n + 1) \ H_2O + \text{heat}$$

Enormous amounts of natural gas and petroleum products are burned daily. Of course, in the combustion process, it is the heat produced that is more important, not the chemical products of the reaction, carbon dioxide and water. However, many scientists are concerned that the production in this reaction of enormous quantities of carbon dioxide, a greenhouse gas, might be initiating a global warming trend.

In the chemical laboratory the major use of alkanes is as solvents. Their property of being chemically unreactive makes them attractive solvents for many reactions because they will not interfere with the desired chemistry. However, their usefulness is limited because they will not dissolve highly polar or ionic compounds because of their completely nonpolar nature.

5.2   COMMON NOMENCLATURE OF ALKANES

The number of organic compounds is virtually limitless. Each needs a name that can be used in discussing the compound or writing about it. Furthermore, it is often necessary to look up the properties of a compound. (You should find out the physical properties of all the compounds that you use in the laboratory.) Very large tables of compounds and indexes use alphabetical listings of compound names.

In the early days of organic chemistry a newly discovered compound was often named according to the source from which it was isolated. The four-carbon carboxylic acid that was isolated from rancid butter was called butyric acid, after *butyrum,* the Latin word for butter. Even the early organic chemists realized the need to be systematic in naming compounds, so the four-carbon alkane that can be prepared from butyric acid was called butane. When an isomeric four-carbon alkane was discovered, it was called isobutane. Butane is a **straight-chain** or **unbranched** alkane. Such alkanes are termed **normal** and in the past were written with a prefix *n-*. Another name, then, that you may encounter in older sources for butane is *n*-butane. However, the *n*- is redundant today because the absence of any prefix implies an unbranched alkane, and it should not be used. Isobutane is a **branched** alkane. There are three isomeric alkanes with five carbons. The unbranched one is called pentane. The one with a single branch is called isopentane. The remaining isomer requires a different prefix. It is called neopentane.

$$
\begin{array}{ccc}
\overset{\displaystyle O}{\underset{\displaystyle \parallel}{}} & & \overset{\displaystyle CH_3}{\underset{\displaystyle |}{}} \\
CH_3CH_2CH_2COH & CH_3CH_2CH_2CH_3 & CH_3CHCH_3 \\
\text{Butyric acid} & \text{Butane} & \text{Isobutane}
\end{array}
$$

$$
\begin{array}{ccc}
 & & \overset{\displaystyle CH_3}{\underset{\displaystyle |}{}} \\
 & \overset{\displaystyle CH_3}{\underset{\displaystyle |}{}} & CH_3CCH_3 \\
CH_3CH_2CH_2CH_2CH_3 & CH_3CHCH_2CH_3 & \underset{\displaystyle CH_3}{\overset{\displaystyle |}{}} \\
\text{Pentane} & \text{Isopentane} & \text{Neopentane}
\end{array}
$$

These names are called **common** or **trivial** names. As the size of the alkane increases, the number of isomers increases dramatically. It would be very cumbersome to continue this process of providing different prefixes for each isomer of a larger alkane.

Furthermore, the task of learning all these prefixes would be daunting indeed. Decane has 75 isomers! Obviously, a systematic nomenclature is needed.

5.3 SYSTEMATIC NOMENCLATURE OF ALKANES

A systematic method for naming alkanes (and other organic compounds) that is simple to use and minimizes memorization was developed by the International Union of Pure and Applied Chemistry and is called the IUPAC nomenclature. To make it easier for the chemists of that time to learn, it incorporated common nomenclature wherever possible.

Before we look at the steps for naming alkanes, let's see what an IUPAC name looks like. Basically, each alkane is considered a straight-chain carbon backbone on which

Focus On

The Energy Content of Fuels

The amount of heat produced by burning (heat of combustion) varies considerably with the type of fuel that is burned. Values for some fuels of interest, in kilocalories (or kilojoules) per gram, are shown in the table that follows.

Some interesting observations can be made from these data. Hydrogen contains considerably more energy per gram than the other fuels. This is because the mass of a hydrogen atom is considerably less than that of a carbon or an oxygen atom, so 1 g of hydrogen contains many more molecules than 1 g of the other fuels. The energy contents of methane, butane, and cyclohexane, typical alkanes, are all similar. (There is a slight decrease in energy content as the percentage of hydrogen decreases from methane to butane to cyclohexane.) Part of the reason why benzene produces less energy than cyclo-

Compound	Heat Evolved kcal/g (kJ/g)	Compound	Heat Evolved kcal/g (kJ/g)
H_2	34.2 (143)	Coal	7.6 (32)
CH_4 Methane	13.4 (56)	CH_3CH_2OH Ethanol	7.2 (30)
$CH_3CH_2CH_2CH_3$ Butane	12.0 (50)	CH_3OH Methanol	5.5 (23)
Cyclohexane	11.2 (47)	Wood	4.5 (19)
Benzene	10.0 (42)	Glucose	3.8 (16)

various branches or groups are attached. The IUPAC name for isopentane is 2-methylbutane. This name consists of a root that designates the number of carbons in the backbone, a number and a prefix that designate the position of the branch on the backbone and the number of carbons in the branching group, and a suffix that designates the functional group.

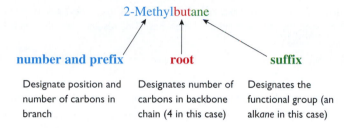

number and prefix	**root**	**suffix**
Designate position and number of carbons in branch	Designates number of carbons in backbone chain (4 in this case)	Designates the functional group (an alk*ane* in this case)

hexane is that benzene has a large resonance stabilization energy. Because benzene is more stable, it produces less heat when burned (more on this in Chapter 16).

One definition of **oxidation** often used in organic chemistry is an increase in the oxygen content of a compound. Combustion is, then, an oxidation process. As can be seen from the heats of combustion of the alkanes as compared to those of ethanol, methanol, and glucose, increasing the initial oxygen content of a compound—that is, increasing its oxidation state—results in a lower energy content. Methanol, for example, can be viewed as resulting from partial oxidation of methane. In methanol there are only three carbon–hydrogen bonds to "burn," compared to four in methane, so the energy content per mole is less for methanol. (The energy content is even less on a per gram basis.)

Gasoline is composed primarily of alkanes, so its energy content is in the 11 to 12 kcal/g (46–50 kJ/g) region. This relatively high energy content makes gasoline an attractive fuel for automobiles. Of course, hydrogen is even better in terms of energy content and burns with less pollution. However, it is much more difficult to handle. Methanol has been proposed as an alternative to gasoline. One drawback is that its energy density is much less than that of gasoline, which means that the miles per gallon would drop substantially if methanol were used. The use of oxygenated gasoline, usually containing 10% or more of ethanol or *m*ethyl *t*ert-*b*utyl *e*ther (MTBE), is mandated in many cities during the winter months because it purportedly reduces pollution due to carbon monoxide. If you live in such an area, you may notice a small decrease in your gas mileage during the winter.

Glucose is a common sugar. Its structure is similar to that of other sugars. Carbohydrates, such as starch, are composed of a large number of glucose units. Therefore, the energy content of glucose is representative of the energy content of a major part of our food. The energy content of glucose is low because of its highly oxygenated structure. It supplies 3.8 kcal/g (16 kJ/g) of energy. Note that the "Calorie" (written with an uppercase C) that is used in nutrition is actually a kilocalorie, so sugars and carbohydrates have about 4 Calories per gram. You may have heard that alcoholic beverages are quite fattening. Ethanol, the alcohol contained in beverages, has nearly twice the energy density of sugar because it contains significantly less oxygen. It supplies 7 Calories per gram.

Table 5.3 Names for the Roots That Designate the Number of Carbons in the Backbone Chain

Number of Carbons	Root
1	meth-
2	eth-
3	prop-
4	but-
5	pent-
6	hex-
7	hept-
8	oct-
9	non-
10	dec-
11	undec-
12	dodec-
13	tridec-
14	tetradec-
15	pentadec-
20	eicos-
21	heneicos-
22	docos-
23	tricos-
30	triacont-

The roots for backbones containing one through four carbons come from the common names for these alkanes. You probably recognize most of these already. For backbones of five or more carbons, systematic roots (derived from Greek) are employed.

Now let's discuss how to name an alkane by the IUPAC method.

STEP 1

Find the longest continuous carbon chain in the compound. The number of carbons in this backbone determines the root (see Table 5.3). If there are two (or more) chains of equal length, choose the one with the greater number of branches. (This results in simpler branching groups that are easier to name.)

It is important to realize that the name of a compound must not depend on how that compound is drawn. When a drawing for a compound is encountered, the longest continuous chain will not necessarily be shown in a straight, horizontal line. Each structure must be carefully examined to identify the longest chain, no matter how it bends and curls around. Some examples, with the longest chain in red, follow:

$$CH_3CH_2$$
$$CH_3CHCH_2CH_3$$

The longest chain has five carbons. The root is pent-.

$$CH_2CH_3$$
$$CH_3CH_2CHCHCH_3$$
$$CH_3CH_2CH_2$$

The longest chain has seven carbons. The root is hept-.

$$CH_3$$
$$CHCH_3$$
$$CH_3CH_2CH_2CHCH_2CH_3$$

Correct

$$CH_3$$
$$CHCH_3$$
$$CH_3CH_2CH_2CHCH_2CH_3$$

Incorrect

There are two different six-carbon chains for this compound. Choose the chain as shown in the left structure because it has two branches rather than in the right one, which has only one branch. The root is hex-.

STEP 2

Attach the suffix. For alkanes, it is -ane.

Steps 1 and 2 are all that are needed to name unbranched alkanes. For example, the straight-chain alkane with seven carbons is heptane.

$$CH_3CH_2CH_2CH_2CH_2CH_2CH_3$$

Heptane

To name branched alkanes, additional steps to name the branching group and locate it on the root chain are needed.

STEP 3

Number the carbons in the root chain. Start from the end that gives the lower number to the carbon where the first branch occurs. If both ends have the first branch at equal distances, choose the end that gives the lower number to the carbon where the next branch occurs.

Examples include the following:

$$
\begin{array}{c}
\quad\;\; CH_3 \;\; CH_3 \\
\quad\;\;\; | \quad\;\; | \\
CH_3CH_2CH_2CH-CHCH_3 \\
6 \quad 5 \quad 4 \quad 3 \quad\;\; 2 \quad 1
\end{array}
$$

For this structure, the correct numbering starts with the right carbon. Then the first branching group is attached to C-2. Incorrect numbering, starting with the left carbon, would result in the first branch being attached to C-4.

$$
\begin{array}{c}
\;\; CH_3 \quad\; CH_2CH_3 \quad\; CH_3 \\
\;\; | \quad\quad\;\; | \quad\quad\quad\; | \\
CH_3CHCH_2CHCH_2CH_2CHCH_3 \\
1 \quad 2 \quad 3 \quad 4 \quad 5 \quad 6 \quad 7 \quad 8
\end{array}
$$

For this structure, the correct numbering starts with the left carbon. The first branch is attached to C-2 and the second branch to C-4. If numbering were to begin with the right carbon, the first branch would be attached to C-2, but the second would be attached to C-5.

$$
\begin{array}{c}
\;\; CH_3 \quad\quad\; CH_3 \\
\;\; | \quad\quad\quad\; | \\
CH_3CHCH_2CH_2CCH_3 \\
6 \quad 5 \quad 4 \quad 3 \quad 2 |\, 1 \\
\quad\quad\quad\quad\; CH_3
\end{array}
$$

For this structure, the correct numbering starts with the right carbon. The first branch is at C-2 regardless of whether the starting point is the right carbon or the left carbon. However, the second branch is also at C-2 when the starting point is the right carbon. It occurs at C-5 when the starting point is the left carbon.

In other words, proceed inward, one carbon at a time, from each end of the root chain. At the first point of difference, where a group is attached to one carbon under consideration but not the other, choose the end nearer that carbon with the attached group to begin numbering.

STEP 4

Name the groups attached to the root. For straight-chain groups, use the same roots as before to designate the number of carbons and add the suffix -yl.

$$
CH_3- \quad\quad\quad CH_3CH_2- \quad\quad\quad CH_3CH_2CH_2CH_2CH_2-
$$

Methyl   Ethyl   Pentyl

A group can be pictured as arising from an alkane by the removal of one hydrogen. It is not a complete compound by itself but requires something to be attached to the remaining valence.

STEP 5

Assemble the name as a single word in the following order: number, group, root, and suffix. Note that a hyphen is used to separate numbers from the name.

$$CH_3$$
$$|$$
$$CH_3CH_2CH_2CHCH_2CH_3$$
$$6 \quad 5 \quad 4 \quad 3 \quad 2 \quad 1$$

3-Methylhexane

If several different groups are present, list them alphabetically. If there are several identical groups, use the prefixes di-, tri-, tetra-, and so on to indicate how many are present. A number is required to indicate the position of each group. For example, 2,2,3-trimethyl indicates the presence of three methyl groups, two attached to C-2 of the backbone and one attached to C-3. Do not use the prefixes that denote the number of groups for alphabetizing purposes. For example, triethyl is listed under *e* and so comes before methyl in a name.

Let's use these steps to name the following compound:

$$CH_3 \quad\quad CH_2CH_3 \quad\quad CH_3$$
$$| \quad\quad\quad | \quad\quad\quad |$$
$$CH_3CHCH_2CHCH_2CH_2CHCH_3$$
$$1 \quad 2 \quad 3 \quad 4 \quad 5 \quad 6 \quad 7 \quad 8$$

4-Ethyl-2,7-dimethyloctane

Previously, it was determined that the longest chain has eight carbons (so the compound is a substituted octane) and that numbering should begin with the leftmost carbon. Three goups are attached to the root chain: two methyl groups and an ethyl group. Therefore the systematic or IUPAC name is 4-ethyl-2,7-dimethyloctane. Note the use of hyphens to separate numbers from letters and the use of commas to separate a series of numbers. Also note that dimethyl is alphabetized under *m* (not *d*) and so is listed after ethyl in the name.

PROBLEM 5.1

Provide IUPAC names for these alkanes:

$$CH_3$$
$$|$$
a) $CH_3CHCH_2CH_2CH_3$

$$CH_3$$
$$|$$
b) $CH_3CH_2CH_2CHCH_3$

$$CH_3$$
$$|$$
c) $CH_3CHCH_2CHCH_3$
$$|$$
$$CH_2CH_3$$

$$CH_2CH_3$$
$$|$$
d) $CH_3CH_2CHCH_2CCH_2CH_2CH_2CH_3$
$$| \quad\quad |$$
$$CH_3 \quad CH_2CH_2CH_3$$

e)

f)

PRACTICE PROBLEM 5.1

Draw the structure of 2,2,5-trimethylheptane.

Strategy

Drawing the structure of a compound when the name is provided is usually a straightforward process. First draw a chain with the number of carbons indicated by the root. Number the chain starting at either end and add the appropriate groups at the appropriate positions.

Solution

The root is hept-, so draw a chain of seven carbons. Number the chain and add three methyl groups: two on C-2 and one on C-5.

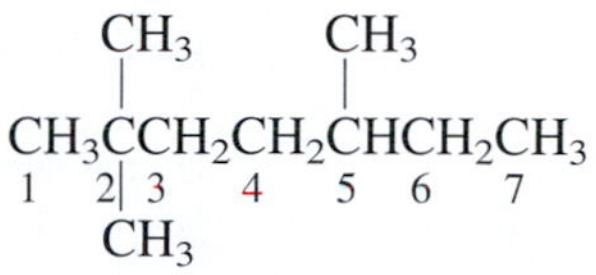

$$
\begin{array}{cc}
CH_3 & CH_3 \\
| & | \\
CH_3CCH_2CH_2CHCH_2CH_3 \\
\overset{1}{}\ \overset{2}{}|\ \overset{3}{}\ \ \overset{4}{}\ \ \ \overset{5}{}\ \ \overset{6}{}\ \ \ \overset{7}{} \\
CH_3
\end{array}
$$

PROBLEM 5.2

Draw the structures of these compounds:
a) 4-Methyloctane
b) 2,4-Dimethyl-5-propyldecane

PRACTICE PROBLEM 5.2

What is wrong with the name 2-ethylpentane? Provide the correct name for this compound.

Solution

First, let's draw the structure suggested by the name:

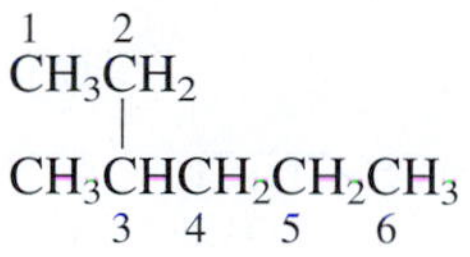

$$
\begin{array}{c}
\overset{1}{CH_3}\overset{2}{CH_2} \\
| \\
CH_3CHCH_2CH_2CH_3 \\
\ \ \ \ \overset{3}{}\ \ \overset{4}{}\ \ \ \overset{5}{}\ \ \ \overset{6}{}
\end{array}
$$

Examination of this structure shows that the longest chain has six carbons rather than five. Therefore, the correct name is 3-methylhexane.

PROBLEM 5.3

What is wrong with these names? Provide the correct name for each.
a) 5,5-Dimethyl-3-ethylhexane
b) 2-Dimethylpentane

Often groups attached to the root chain are branched rather than the straight-chain ones encountered so far. Such complex groups are named in a fashion similar to that used to name compounds, but with a -yl ending to indicate that they are not complete

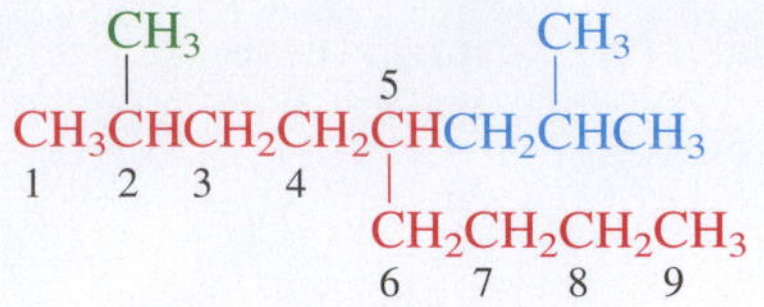

$$CH_3$$
$$|$$
$$\overset{5}{}$$
$$\underset{1\ \ 2\ \ 3\ \ 4}{CH_3CHCH_2CH_2CHCH_2CHCH_3}$$
$$|$$
$$\underset{6\ \ 7\ \ 8\ \ 9}{CH_2CH_2CH_2CH_3}$$

$$CH_3$$
$$|$$
$$\underset{1\ \ \ 2\ \ \ 3}{-CH_2CHCH_3}$$

The longest chain in this compound (in red) has nine carbons. Numbering begins at the left carbon because the first group is then encountered at position 2. The group (in blue) bonded to position 5 has a branch. The compound is named as a substituted nonane, but how is the complex group at position 5 named?

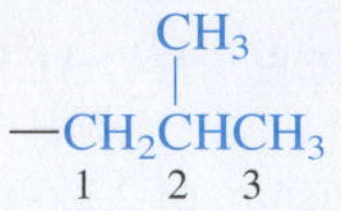

Here is the group. It is named in a manner similar to that used to name alkanes. The longest chain that begins with the carbon attached to the main chain is chosen. As before, the root is determined by the number of carbons in this longest chain (three in this case). The suffix -yl is used to indicate that this is a group rather than a complete compound. So this group is named as a substituted propyl group. Numbering the root chain of a complex group is easy. The carbon attached to the main chain is number 1. The name of the complex group is 2-methylpropyl-.

The name for the complex group is placed in parentheses to avoid confusion when the entire compound is named. The number designating the position of the complex group on the main chain is placed outside the parentheses. Therefore, the name of the compound is 2-methyl-5-(2-methylpropyl)nonane.

Figure 5.1

NAMING A COMPOUND WITH A COMPLEX GROUP.

compounds and need to be attached to something to complete their bonding. Naming complex groups is described in Figure 5.1.

PRACTICE PROBLEM 5.3

What is the name of this complex group?

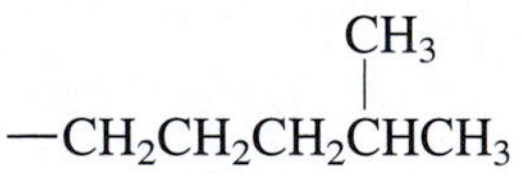

$$CH_3$$
$$|$$
$$-CH_2CH_2CH_2CHCH_3$$

Strategy

Choose the longest chain that begins at the carbon attached to the main chain. Number this chain with the carbon attached to the main chain receiving number 1. Add the groups attached to this chain with the numbers indicating their positions. When the full name of the compound is written, do not forget to place the entire complex group name in parentheses preceded by a number indicating the position of the group on the main chain.

Solution

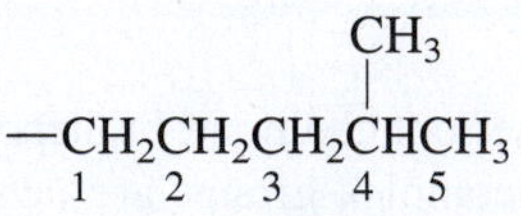

$$CH_3$$
$$|$$
$$\underset{1\ \ \ 2\ \ \ 3\ \ \ 4\ \ 5}{-CH_2CH_2CH_2CHCH_3}$$

In this case the longest chain has five carbons. When the numbering starts at the carbon attached to the main chain, the methyl group is on C-4, so the name of the group is (4-methylpentyl).

PROBLEM 5.4
Provide names for these complex groups:

a) 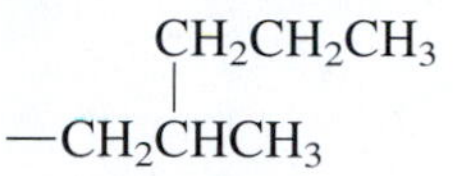 $-CH_2\overset{\overset{\displaystyle CH_2CH_2CH_3}{\vert}}{C}HCH_3$

b) $-\overset{\overset{\displaystyle CH_3}{\vert}}{C}HCH_2CH_3$

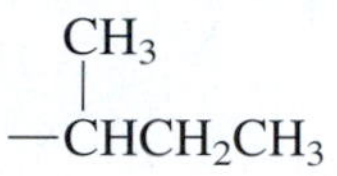

c) $-CH_2\overset{\overset{\displaystyle CH_3}{\vert}}{\underset{\underset{\displaystyle CH_3}{\vert}}{C}}CH_3$

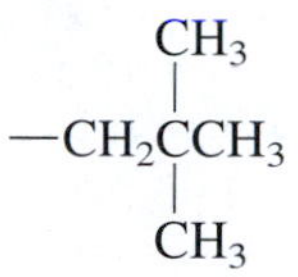

PROBLEM 5.5
Name these compounds:

a) 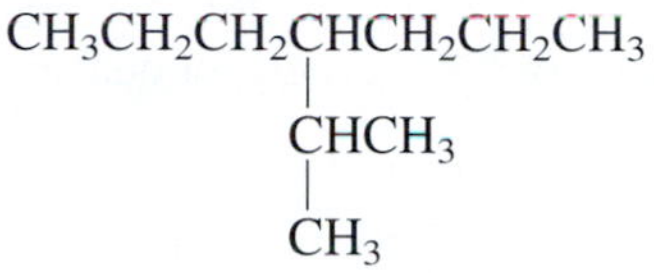 $CH_3CH_2CH_2\overset{\overset{\displaystyle }{}}{C}HCH_2CH_2CH_3$
$\underset{\underset{\displaystyle CH_3}{\vert}}{\overset{\vert}{C}HCH_3}$

b)

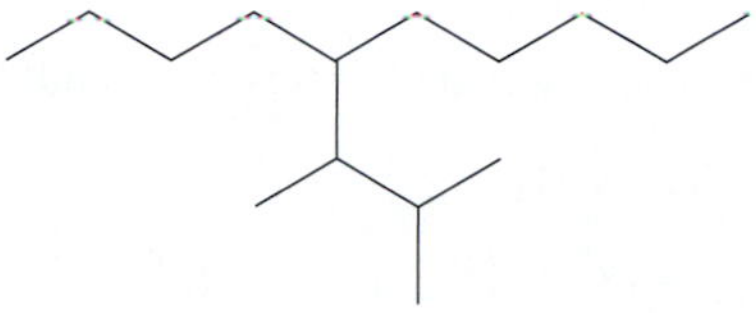

PRACTICE PROBLEM 5.4

Draw a structure for 4-(1,1-dimethylethyl)decane.

Solution

The compound is a decane, so there are 10 carbons in the main chain. On C-4 there is an ethyl group that has two methyl groups attached to the carbon that is attached to the decane chain:

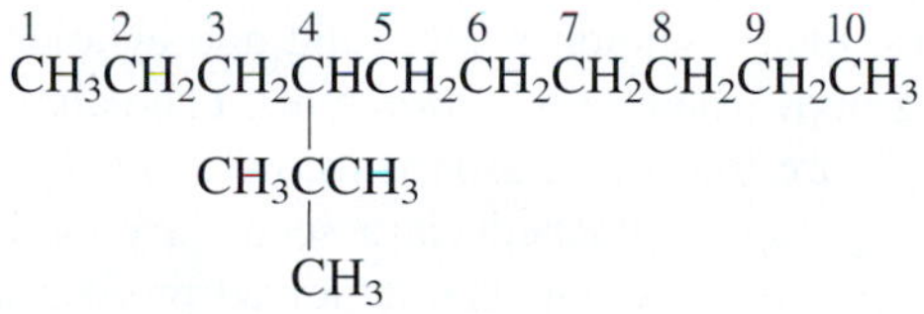

$$\overset{1\quad\ 2\quad\ \ 3\quad\ 4\quad\ 5\quad\ \ 6\quad\ \ 7\quad\ \ 8\quad\ \ 9\quad\ 10}{CH_3CH_2CH_2\underset{\underset{\displaystyle CH_3\underset{\underset{\displaystyle CH_3}{\vert}}{C}CH_3}{\vert}}{C}HCH_2CH_2CH_2CH_2CH_2CH_3}$$

PROBLEM 5.6
Draw structures for these compounds.
a)   4-(1-Methylethyl)heptane
b)   3-Ethyl-7-methyl-5-(1-methylpropyl)undecane

Some additional terminology permeates organic chemistry and must be part of every organic chemist's vocabulary. The chemical reactivity of a carbon or of a functional group attached to that carbon often varies according to the number of other carbons bonded to it. In discussions of chemical reactivity, therefore, it is often useful to distinguish carbons according to how many other carbons are bonded to them. A **primary carbon** is bonded to one other carbon, a **secondary carbon** is bonded to two carbons, a **tertiary carbon** is bonded to three carbons, and a **quaternary carbon** is bonded to four carbons. The following compound contains each of these types of carbons:

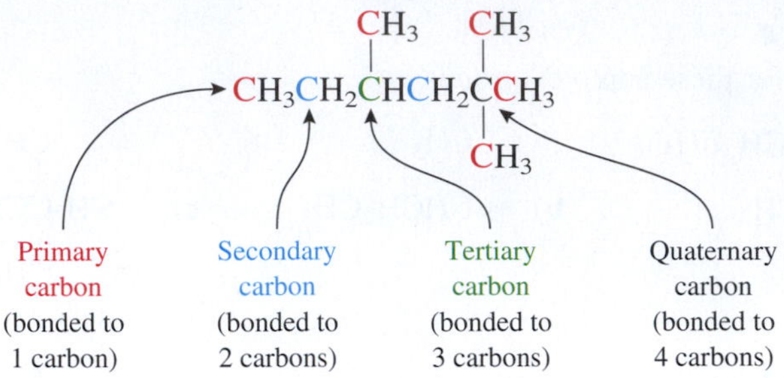

Primary carbon	Secondary carbon	Tertiary carbon	Quaternary carbon
(bonded to 1 carbon)	(bonded to 2 carbons)	(bonded to 3 carbons)	(bonded to 4 carbons)

PROBLEM 5.7

Designate each carbon of these compounds as primary, secondary, tertiary, or quaternary:

a) 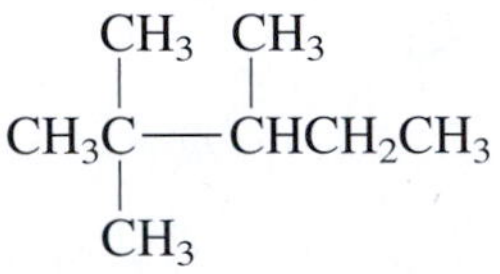 b)

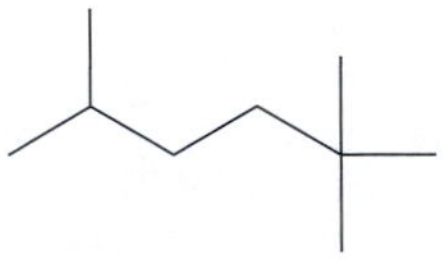

c) 2,3-Dimethylpentane

Finally, in addition to the systematic method for naming groups that we have just seen, we will still encounter some common group names. IUPAC nomenclature allows the use of these common group names as part of the systematic name. (IUPAC nomenclature does allow different names for the same compound; however, no two compounds may have the same name.) Table 5.4 shows a number of groups, along with their systematic names, their common names, and abbreviations that are sometimes used to represent the groups when writing structures. (The abbreviations are not used in nomenclature.) There are four butyl groups in the table. To help remember them, note that the *sec*-butyl group is attached via a secondary carbon and the *tert*-butyl group is attached via a tertiary carbon. The italicized prefixes *sec-* and *tert-* are not used in alphabetizing; *sec*-butyl is alphabetized under *b*. However, isobutyl is alphabetized under *i*.

PROBLEM 5.8

Name this compound using the common name for the group:

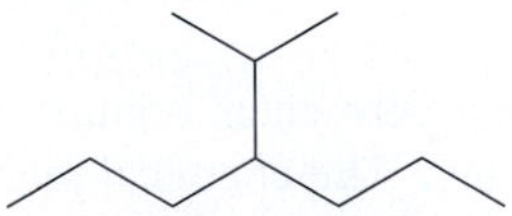

PROBLEM 5.9

Draw the structure of 4-*tert*-butyl-2,3-dimethyloctane.

Table 5.4   Systematic Names, Common Names, and Abbreviations for Some Groups

Group	Systematic Name	Common Name	Abbreviation
CH_3-	Methyl		Me
CH_3CH_2-	Ethyl		Et
$CH_3CH_2CH_2-$	Propyl		Pr
$CH_3\overset{\displaystyle CH_3}{\underset{\vert}{CH}}-$	1-Methylethyl	Isopropyl	*i*-Pr
$CH_3CH_2CH_2CH_2-$	Butyl		Bu
$CH_3\overset{\displaystyle CH_3}{\underset{\vert}{CH}}CH_2-$	2-Methylpropyl	Isobutyl	*i*-Bu
$CH_3CH_2\overset{\displaystyle CH_3}{\underset{\vert}{CH}}-$	1-Methylpropyl	*sec*-Butyl	*sec*-Bu (or *s*-Bu)
$CH_3\overset{\displaystyle CH_3}{\underset{\displaystyle \underset{\vert}{CH_3}}{\underset{\vert}{C}}}-$	1,1-Dimethylethyl	*tert*-Butyl	*tert*-Bu (or *t*-Bu)
$-CH_2-$		Methylene	
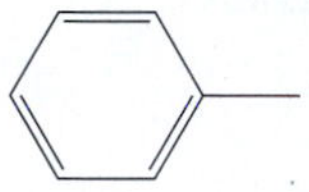 (phenyl ring)	Phenyl		Ph

The common names may be used in IUPAC nomenclature.

5.4   SYSTEMATIC NOMENCLATURE OF CYCLOALKANES

The procedure used to name a cycloalkane by the IUPAC method is very similar to that used for alkanes. Here, the root designates the number of carbons in the ring and the prefix cyclo- is attached to indicate that the compound contains a ring. The rules for numbering the ring carbons are as follows:

No number is needed if only one group is attached to the ring.

For rings with multiple substituents, begin numbering at one substituent and proceed in the direction that gives the lowest numbers to the remaining substituents.

Some examples follow:

Cycloheptane

(There are 7 carbons in the ring.)

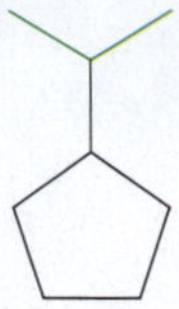

Isopropylcyclopentane or (1-methylethyl)cyclopentane

(No number is needed to locate the isopropyl group because all positions of the ring are identical.)

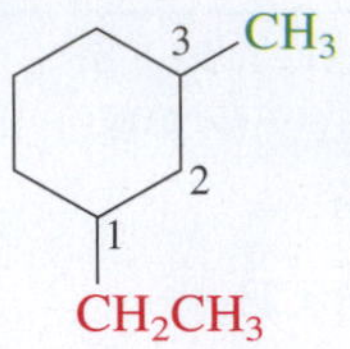

1-Ethyl-3-methylcyclohexane

(To keep the numbers as low as possible, begin at the ethyl group [because it comes first alphabetically] and proceed by the shortest possible path to the methyl group.)

In cases in which the alkyl chain has more carbons than the ring, the compound is named as an alkane with the ring as a substituent group with a -yl suffix.

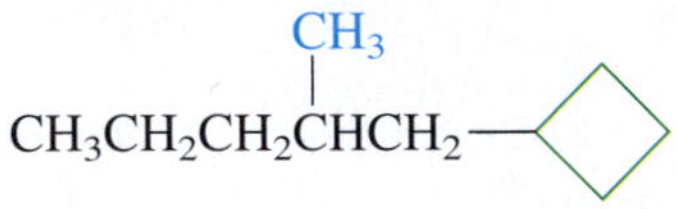

$$CH_3$$
$$|$$
$$CH_3CH_2CH_2CHCH_2{-}\diamond$$

1-Cyclobutyl-2-methylpentane

(There are five carbons in the longest alkyl chain, whereas the ring has only four carbons. Therefore, the ring is named as a substituent group on the alkane chain.)

PROBLEM 5.10
Name the following compounds:

a)

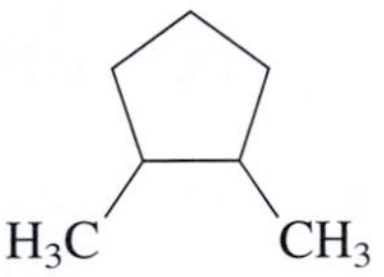

H_3C CH_3

b)

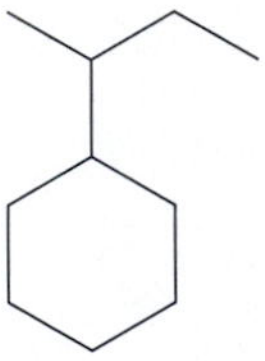

c)

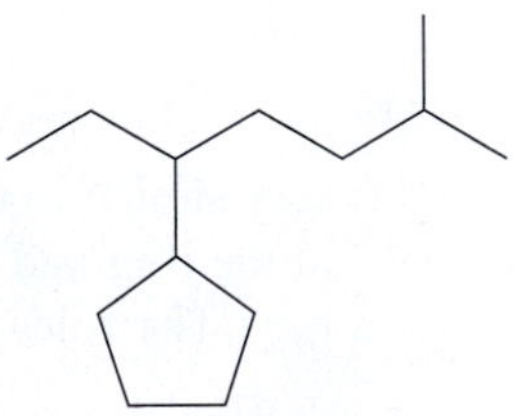

d)

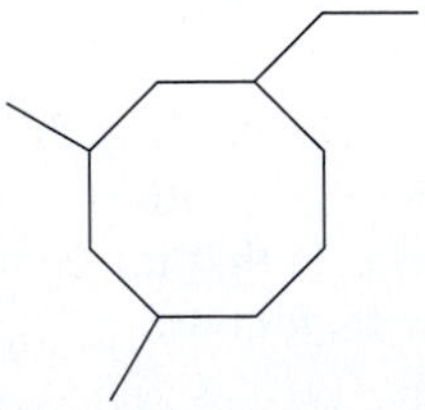

PROBLEM 5.11
Draw structures for these compounds:
a) 1,1-Dimethylcyclohexane   b) Ethylcyclopropane

5.5 ALKENES

Alkenes have one or more carbon–carbon double bonds. In the discussion in Chapter 2 about calculating the degree of unsaturation of a compound, it was shown that each double bond present in an alkene results in a decrease of two hydrogens in the formula when compared to an alkane with the same number of carbons. The term **unsaturated** is used to describe alkenes and actually has a chemical derivation. It is possible to cause compounds containing double or triple bonds to react with hydrogen gas to form alkanes. When the compound will no longer react with hydrogen, it is said to be **saturated.** A compound that will react with hydrogen, such as an alkene or an alkyne, is, then, unsaturated. You have probably heard these terms used a lot in association with cooking oils or margarines. A polyunsaturated oil is composed of compounds that contain several carbon–carbon double bonds. A saturated fat, by contrast, is composed of similar compounds that have no double bonds.

Alkenes are named similarly to alkanes, with the following modifications:

1. The longest continuous chain that includes both carbons of the double bond provides the root.

2. The suffix used for an alkene is -ene. Names for compounds with more than one double bond use the suffixes -diene, -triene, and so on.

3. The root is numbered from the end that gives the lower number to the first carbon of the double bond. The number of this first carbon is used in the name to designate the position of the double bond. This number may be placed before the root or between the root and the suffix.

Some examples include the following:

4-Methyl-2-pentene or 4-methylpent-2-ene

(The double bond, not the methyl group, determines the numbering.)

3-Ethyl-4-methylcyclohexene

(Numbering begins with the carbons of the double bond having numbers 1 and 2 and proceeds in the direction that will give the lowest numbers to the remaining substituents. Because the double bond of a cycloalkene is always located at position 1, the 1 is usually not used. However, some sources do specifically designate the position of the double bond when other groups are present on the ring. Thus, this compound is also properly named 3-ethyl-4-methylcyclohex-1-ene. We will omit the 1 in such situations in this book.)

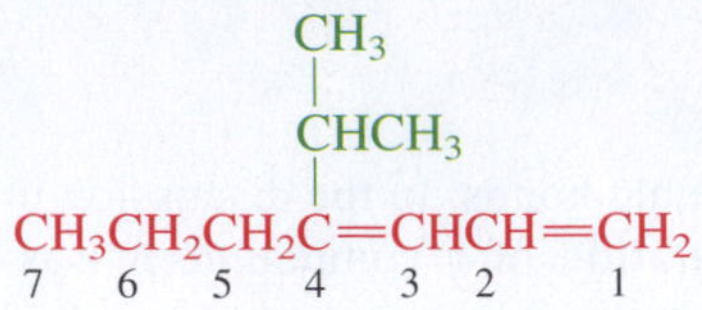

$$CH_3$$
$$|$$
$$CHCH_3$$
$$|$$
$$CH_3CH_2CH_2C{=}CHCH{=}CH_2$$
$$\quad 7 \quad 6 \quad 5 \quad 4 \quad 3 \quad 2 \quad 1$$

$$1\,CH_2$$
$$\|$$
$$CH_3CH_2CCH_2CH_2CH_3$$
$$\quad\quad 2\,3 \quad 4 \quad 5$$

4-Isopropyl-1,3-heptadiene or 4-isopropylhepta-1,3-diene

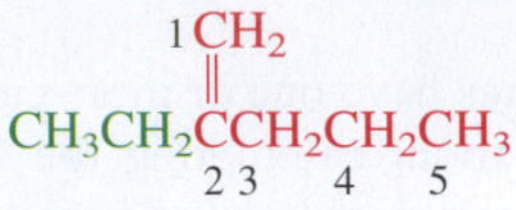

2-Ethyl-1-pentene or 2-ethylpent-1-ene

(Note the *a* in heptadiene. This is added to the root whenever the first letter of the suffix is a consonant to make the name easier to pronounce.)

(Both carbons of the double bond must be part of the root chain.)

PRACTICE PROBLEM 5.5

Name this alkene:

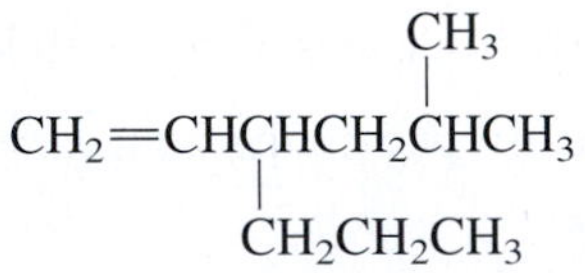

$$CH_3$$
$$|$$
$$CH_2{=}CHCHCH_2CHCH_3$$
$$|$$
$$CH_2CH_2CH_3$$

Solution

Choose the longest chain that contains both of the carbons of the double bond (first priority) and has the most branches (second priority):

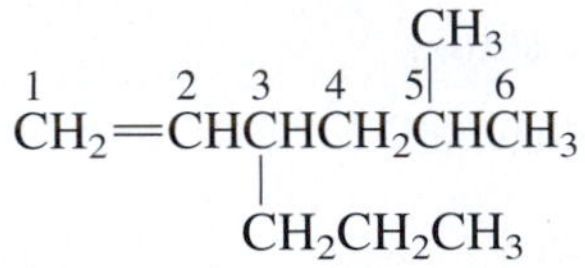

$$CH_3$$
$$\quad\quad\quad\quad\quad\quad\quad\quad 5| \quad$$
$$1 \quad\quad 2 \quad 3 \quad 4 \quad\quad 6$$
$$CH_2{=}CHCHCH_2CHCH_3$$
$$|$$
$$CH_2CH_2CH_3$$

Number it beginning at the left end so that the carbons of the double bond have lower numbers. It has a propyl group on C-3 and a methyl group on C-5, so the name is 5-methyl-3-propyl-1-hexene.

PROBLEM 5.12
Name these compounds:

a) 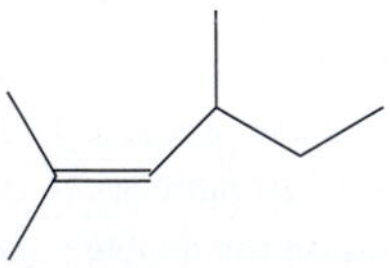   b) 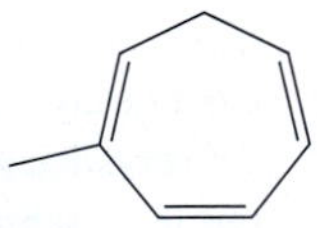   c)

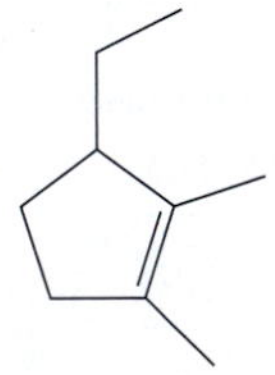

PROBLEM 5.13
Draw structures for these compounds:
a) 3-Ethyl-3-hexene   b) Cyclobutene
c) 3-Propyl-cyclohexa-1,4-diene

Table 5.5 Common Names for Some Compounds and Groups Containing Double Bonds

Compound or Group	Common Name
$CH_2{=}CH_2$	Ethylene
$CH_2{=}CHCH_3$	Propylene
$CH_2{=}CHCH_2CH_3$	Butylene
$CH_2{=}\overset{\displaystyle CH_3}{\underset{\vert}{C}}CH_3$	Isobutylene
$CH_2{=}C{=}CH_2$	Allene
$CH_2{=}CH-$	Vinyl group
$CH_2{=}CHCH_2-$	Allyl group

A few of the common names that are often encountered for alkenes and groups containing double bonds are listed in Table 5.5.

The polarity of an alkene is not much different from the polarity of an alkane. Therefore, the physical properties of an alkene are similar to those of the corresponding alkane. For example, 1-pentene melts at $-138°C$ and boils at $30°C$, values that are comparable to those of pentane, which melts at $-130°C$ and boils at $36°C$.

However, the chemical properties of an alkene are dramatically affected by the presence of the double bond. Recall that a carbon–carbon pi bond is considerably weaker than a carbon–carbon or carbon–hydrogen sigma bond. It is possible to selectively cause a reaction to occur at a pi bond under conditions that do not affect the sigma bonds. The pi bond is the weak spot of an alkene, and it is there that most chemical reactions occur. This is why unsaturated fats spoil more readily than saturated fats. Their pi bonds provide a place for reaction with oxygen to occur, which leads to spoilage.

Alkenes occur in nature, especially among a group of natural products called terpenes that occur in plants. Examples are limonene, which is found in citrus fruits and caraway seeds, and β-carotene, a highly conjugated molecule that is the orange pigment found in carrots and is an important precursor of vitamin A. Interestingly, even the simplest alkene, ethene, has an important role in nature. Ethene has been found to be a plant hormone that causes ripening in fruits.

Limonene

β-Carotene

The two simplest alkenes, ethylene (ethene) and propylene (propene), are the organic compounds that are produced in the largest amounts by the U.S. chemical industry. More than 48 billion pounds of ethylene and more than 28 billion pounds of propylene are produced annually by cracking of hydrocarbons ranging from ethane to heavy gas oil. These alkenes then serve as feedstocks for the production of plastics, such as polyethylene and polypropylene, as well as starting materials for the preparation of numerous other commercial chemicals, such as ethylene glycol (antifreeze).

5.6 ALKYNES

Alkynes are compounds that have a carbon–carbon triple bond. As was discussed in Chapter 3, a triple bond is composed of one sigma bond and two pi bonds. Each triple bond in an alkyne causes it to have four fewer hydrogens than the corresponding alkane. Alkynes are unsaturated compounds.

Alkynes are named in a manner nearly identical to the naming of alkenes except that the suffix is -yne. The same rules for numbering apply. Compounds with several triple bonds use the suffixes -diyne, -triyne, and so on. For compounds that contain both a double bond and a triple bond, both suffixes are used, as in -enyne. Some examples follow:

$$HC \equiv CH$$

Ethyne or acetylene
(Acetylene is the common name. Alkynes as a class are sometimes called acetylenes.)

$$\underset{1\quad2\quad3\quad4\,5\quad6\quad7\quad8}{CH_3CH_2C \equiv CCH_2CH_2\overset{\displaystyle CH_3}{\overset{\displaystyle |}{CH}}CH_3}$$

7-Methyl-3-octyne
or
7-Methyloct-3-yne

(Number so that the first C of the triple bond gets the lowest number.)

1-Cyclodecen-4-yne
or cyclodec-1-en-4-yne

(When there is an equal choice in numbering, double bonds are given lower numbers than triple bonds.)

$$\underset{1\quad2\quad3\quad4\quad5\quad6\;7}{HC \equiv C - CH = CH - CH = CHCH_3}$$

3,5-Heptadien-1-yne
or hepta-3,5-dien-1-yne

(Numbers as low as possible are given to the double and triple bonds even though this results in the triple bond having the lower number. Note the locations in the name of 3 and 5, which refer to diene, and 1, which refers to -yne.)

PROBLEM 5.14

Name these compounds:

a) $HC \equiv CC\overset{\displaystyle CH_3}{\overset{\displaystyle |}{H}}CH_2CH_2CH_2CH_3$

b) $CH_3C \equiv C - \overset{\displaystyle CH_3}{\overset{\displaystyle |}{C}} = CH_2$

c)

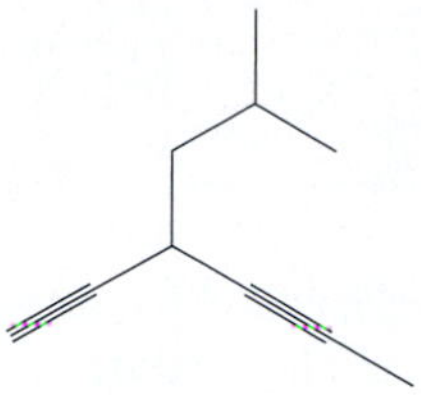

PROBLEM 5.15

Draw structures for these compounds:

a) 1-Pentyne   b) 2,3,4-Trimethyl-5-undecyne

As might be expected, the physical properties of alkynes are very similar to those of alkenes and alkanes with the same number of carbons. For example, the boiling points of hexane, 1-hexene, and 1-hexyne are 69°C, 63°C, and 71°C, respectively.

Because of their pi bonds, the chemical properties of alkynes are very similar to those of alkenes. They undergo many of the same reactions as alkenes and often react at both pi bonds.

Alkynes are less common in nature than are alkenes. Alkynes are also less important in industry. The largest use of acetylene is as a fuel for the oxyacetylene welding torch, which burns at a very high temperature.

5.7 ALKYL HALIDES

The halogens have the same valence as hydrogen. Organic compounds with one or more halogens in place of hydrogens are called **alkyl halides.** These compounds are named as alkanes with the halogen as a substituent. The group names for the halogens are fluoro-, chloro-, bromo-, and iodo-. We may also encounter common names for simple alkyl halides. These use the name of the alkyl group followed by the name of the halogen.

2-Iodopropane	1-Bromobutane	1-Chloro-3-methylcyclohexane
or	or	
isopropyl iodide	butyl bromide	(Other things being equal, chloro gets the lower number because it comes first alphabetically.)

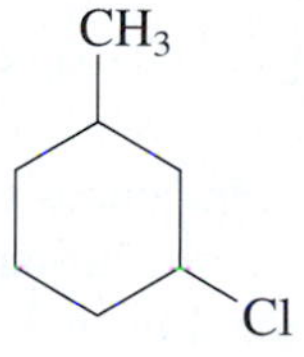

PROBLEM 5.16

Name these compounds:

a) CH_3CHCH_2C——$CHCH_2CH_3$

b)

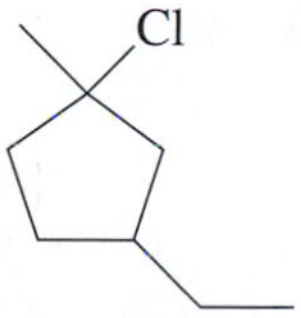

ORGANIC
Chemistry ⚛ Now™
Click *Coached Tutorial Problems* to practice **Naming Alkanes, Alkenes, and Cycloalkanes** or to practice **Drawing Structures** of these compounds.

PROBLEM 5.17

Draw the structure of 3-bromo-4-butylcyclohexene.

The carbon–halogen bond is slightly polar. Overall, however, an alkyl halide is not much more polar than an alkane, so the physical properties of an alkyl halide are not very different from those of an alkane of similar molecular weight. For example, the boiling point of 1-chlorobutane (MW = 92.5 g/mol) is 78°C, whereas that of hexane (MW = 86 g/mol) is 69°C. In general, alkyl halides are insoluble in water. Because of the presence of the more massive halogen atom, the alkyl halide may be more dense than water. For example, when dichloromethane, a common laboratory solvent, and water are mixed, two layers are formed, with dichloromethane as the lower layer.

Halogenated organic compounds are of great industrial importance. Those containing a number of halogens (polyhalogenated) are often chemically inert and thermally stable compounds that have found uses as refrigerants, industrial solvents, degreasers, aerosol propellants, and so on. Some examples include the following:

CF_2Cl_2	Chlorofluorocarbon 12, a freon, was used in refrigeration systems and air conditioners and as an aerosol propellant because of its high stability and low toxicity. Its stability results in environmental problems. It does not decompose until it reaches the upper atmosphere, where ultraviolet light causes the carbon–chlorine bonds to break. The resulting chlorine atoms catalyze the destruction of ozone, resulting in the infamous ozone hole. The use of this and other stable freons has recently been phased out.
CF_3Br	This compound is used in Halon fire extinguishers. Many polychloro and polybromo organics are used as flame retardants in various applications.
CCl_3CH_3	1,1,1-Trichloroethane is the most common solvent used in the dry cleaning of clothes. Carbon tetrachloride (CCl_4) was used in the past, but it has been found to cause liver damage and is carcinogenic.
$CF_3CHBrCl$	Halothane is a popular anesthetic. Chloroform, $CHCl_3$, was one of the first anesthetics, but, like carbon tetrachloride, it is toxic and causes liver damage.

5.8 ALCOHOLS

Alcohols are compounds that contain a hydroxy group (—OH). Common names for simple alcohols use the name of the alkyl group followed by alcohol, such as ethyl alcohol or isopropyl alcohol. In the IUPAC system, alcohols are given the name of the hydrocarbon from which they are derived, with the suffix -ol replacing the final *e* of the name. The longest chain that contains the carbon bonded to the hydroxy group is chosen as the root and numbered so that this carbon has the lowest possible number.

$$\underset{\text{OH}}{|}$$
$$CH_3CHCH_3$$

2-Propanol or isopropyl alcohol

$$\overset{CH_3}{\underset{6\quad 5\quad 4}{CH_3CHCH_2}}\overset{3\quad \text{OH}}{\underset{|}{CHCHCH_3}}$$
$$\underset{CH_2CH_2CH_3}{}$$

5-Methyl-3-propyl-2-hexanol

(The parent chain must contain the carbon bonded to the OH. The chain is numbered so that this carbon has the lower number.)

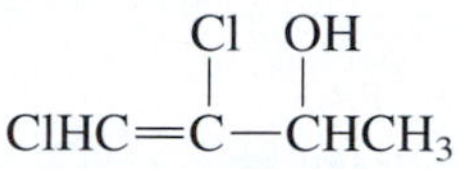

$$\underset{\text{3,4-Dichloro-3-buten-2-ol}}{\text{ClHC}=\overset{\overset{\text{Cl}}{|}}{\text{C}}-\overset{\overset{\text{OH}}{|}}{\text{C}}\text{HCH}_3}$$

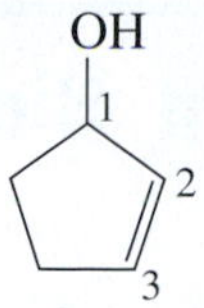

2-Cyclopentenol

(Note the position of the second 3, which refers to the carbon–carbon double bond, and the 2, which refers to the hydroxy group. Numbering is done to give the lower number to the OH, not the double bond. We say that the OH group has a higher priority than the double bond. The alternative style name, 3,4-dichlorobut-3-en-2-ol, makes the location of the double bond and the OH less likely to be confused.)

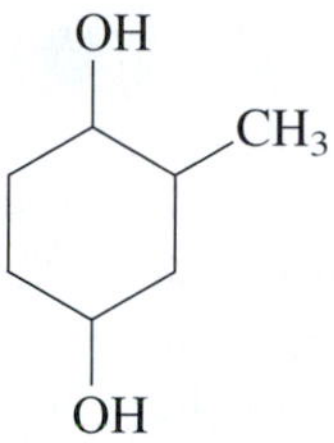

2-Methyl-1,4-cyclohexanediol or 2-methylcyclohexan-1,4-diol

(The suffix -diol is used to show the presence of two hydroxy groups; -triol is used for three, and so on.)

PROBLEM 5.18

Name these compounds:

a) 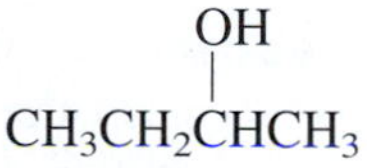 $\underset{}{\text{CH}_3\text{CH}_2\overset{\overset{\text{OH}}{|}}{\text{C}}\text{HCH}_3}$

b) $\text{CH}_3\text{CH}_2\overset{\overset{\text{OH}}{|}}{\underset{\underset{\text{CH}_3}{|}}{\text{C}}}\text{CH}_2\text{CH}_2\text{CH}_3$

c) $-\text{CH}_2\text{CH}_2\overset{\overset{\text{OH}}{|}}{\text{C}}\text{H}_2$

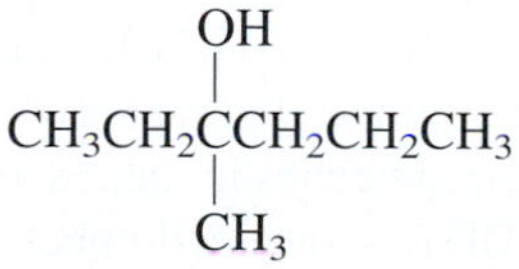

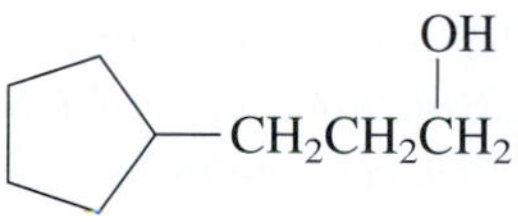

d)

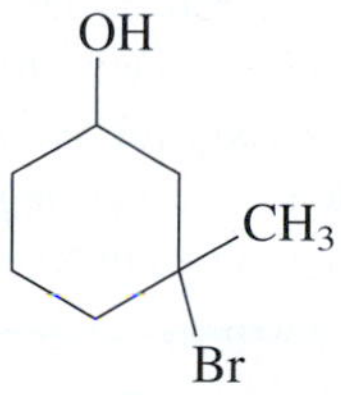

PROBLEM 5.19

Draw structures for these compounds:
a) 2-Methylcyclohexanol   b) 4-Methyl-4-penten-2-ol

Focus On Biological Chemistry

Chlorinated Organic Compounds

The use of chlorinated organic compounds in agriculture and industry has caused a number of environmental problems. A good illustration is provided by the case of 1,1,1-trichloro-2,2-bis(*p*-chlorophenyl)ethane, better known as dichlorodiphenyl-trichloroethane, or DDT.

DDT

DDT was the first synthetic organic pesticide. Although it has received considerable negative publicity, it was probably the most useful insecticide ever developed. The story begins in 1939, when Dr. Paul Müller, a Swiss entomologist, discovered that DDT was extremely effective in controlling flies and mosquitoes. During World War II, DDT was used with great success in Italy against body lice that carry typhus and in the Pacific against mosquitoes that carry malaria. In 1948, Müller received the Nobel Prize in medicine for this discovery.

After the war the agricultural community in the United States used DDT enthusiastically. It had virtually ideal properties as an insecticide; it was effective against a wide variety of insects, it was not very toxic to mammals, it was persistent, and—very important—it was cheap, costing less than 22 cents per pound! In 1961, 160 million pounds of DDT were used in the United States.

However, two of the very properties of DDT that made it valuable as a pesticide led to it also becoming an environmental hazard. DDT is very stable—its biodegradation is very slow. Therefore, it accumulates in the environment. In addition, as examination of its structure suggests, DDT is very hydrophobic. It is very insoluble in water and quite soluble in nonpolar compounds. When an organism ingests DDT, the water insolubility of DDT greatly slows its excretion. Therefore, it accumulates in the organism, specifically in nonpolar regions such as fats or lipids. Thus, while the concentration of DDT in larvae in a lake, for example, may be low, the concentration in trout that feed on these larvae will be much higher because the trout will accumulate all the DDT in all the larvae that they consume. And the concentration in an eagle that feeds on the trout will be higher still. This increase in concentration as one proceeds up the food chain is called biomagnification.

Rachel Carson called attention to the abuse and overuse of pesticides in her 1963 book *Silent Spring*. Decreases in the populations of some wildlife species, especially birds, were attributed to the relatively large concentrations of DDT that were found in them. In 1973 the Environmental Protection Agency banned the use of DDT in the United States. The appropriateness of this action has been hotly debated.

Supporters of the use of pesticides claim that there is not one reported instance of an adverse effect in a human caused by DDT and that it is one of the least toxic and safest of all pesticides to humans and animals. Because of its low cost and effectiveness, DDT is still widely used in countries where malaria is endemic. Some of these poorer countries regard DDT as a very important life-saver.

Regardless of one's position in this debate, it is apparent that the use of compounds that show persistence in the environment must be carefully examined and monitored. Any compound that is persistent and hydrophobic (lipophilic or fat soluble) will be subject to the process of biomagnification and may present special problems. In addition to DDT and other chlorinated pesticides, another example is provided by the polychlorinated biphenyls (PCBs). These compounds have two benzene rings bonded together, with varying numbers of chlorines substituted on the rings. One example is provided by the following structure:

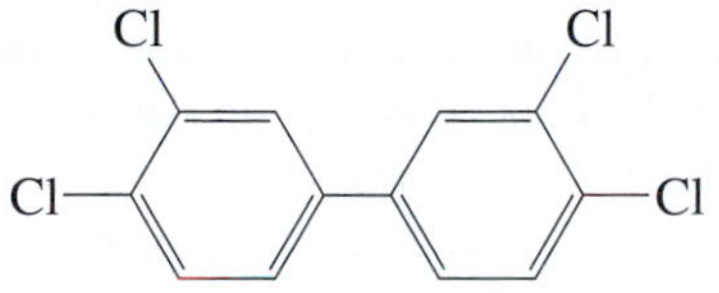

A polychlorinated biphenyl
(PCB)

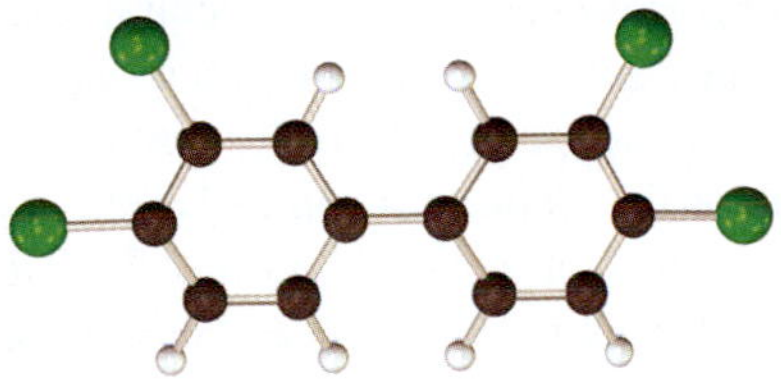

PCBs are so chemically inert that they were extensively used as heat transfer fluids in large electrical transformers, along with other uses. Of course, when they escaped into the environment, they exhibited the two troublesome qualities of persistence and biomagnification. Because PCBs are now suspected carcinogens, they are no longer manufactured in the United States. However, they are now widely distributed in the environment. Furthermore, there is the problem of disposing of the large number of electrical devices that contain these compounds.

The chemical reactions of alcohols differ somewhat depending on how many carbons are attached to the carbon bearing the hydroxy group. Therefore, it is sometimes useful to classify alcohols as primary, secondary, or tertiary. A **primary alcohol** has the hydroxy group on a primary carbon, a **secondary alcohol** has the hydroxy group on a secondary carbon, and a **tertiary alcohol** has the hydroxy group on a tertiary carbon.

$$CH_3CH_2CH_2CH_2OH$$

1-Butanol
A primary alcohol

$$CH_3CH_2CHCH_3 \quad (OH)$$

2-Butanol
A secondary alcohol

$$CH_3CCH_3 \quad (OH, CH_3)$$

2-Methyl-2-propanol
A tertiary alcohol

The physical properties of alcohols are dramatically affected by the presence of the hydroxy group. The polarity of the hydroxy group causes an alcohol to melt at a somewhat higher temperature than an alkane of similar molecular weight; compare the melting points of pentane and 1-butanol in the following table. The effect of the hydroxy group on the boiling point is substantially larger because of the ability of the hydroxy group to form hydrogen bonds, which must be broken in the vaporization process.

$CH_3CH_2CH_2CH_2CH_3$	$CH_3CH_2CH_2CH_2OH$
Pentane	1-Butanol
72 g/mol	74 g/mol
mp −130°C	mp −90°C
bp 36°C	bp 117°C

The hydroxy group is **hydrophilic** (water loving) because of its polarity and ability to form hydrogen bonds, so alcohols are much more soluble in water than alkanes. The smaller alcohols, containing up to three carbons, are miscible with water—that is, they mix with water in all proportions. As expected, as the alkyl group (the hydrophobic part) of the alcohol becomes larger, water solubility decreases. Thus, 1-pentanol dissolves in water to the extent of 2.7 g/100 mL and 1-heptanol to the extent of 0.1 g/100 mL, whereas 1-decanol is essentially insoluble in water.

Alcohols occur widely in nature. Methanol is also known as wood alcohol because it can be obtained by distilling wood in the absence of air. It is very poisonous and can cause blindness or death if ingested. Ethanol is consumed in alcoholic beverages. Other simple alcohols, such as 2-phenylethanol from roses and menthol from peppermint, are constituents of natural flavors and fragrances. Alcohols are important intermediates in chemical synthesis. They are also commonly used as solvents for various chemical processes. Ethylene glycol is used in antifreeze and in the preparation of polymers such as Dacron.

2-Phenylethanol

Menthol

Ethylene glycol
1,2-Ethanediol

5.9 ETHERS

Ethers are compounds with two hydrocarbon groups bonded to an oxygen. Common names are often used for simple ethers. In these, each alkyl group is named, followed by the word *ether*. Thus, $CH_3OCH_2CH_2CH_3$ is methyl propyl ether. In systematic nomenclature the smaller alkyl group and the oxygen are designated as an **alkoxy** substituent on the larger group, which is named as an alkane. Complex ethers must be named by using the IUPAC system. However, IUPAC names for simple ethers are seldom used; diethyl ether (or ethyl ether) is rarely called ethoxyethane, although this name is certainly proper.

$$CH_3CH_2OCH_2CH_3$$

$$CH_3OCCH_3 \quad (with \ CH_3 \ above \ and \ CH_3 \ below)$$

$$OCH_2CH_3$$

Common	Diethyl ether	Methyl *t*-butyl ether	4-Ethoxy-
IUPAC	Ethoxyethane	2-Methoxy-2-methylpropane	cyclohexene

PROBLEM 5.20

Name these compounds:

a) $CH_3OCH_2CH_3$

b)

$$OCH_3$$

$$Cl$$

An ether is somewhat polar because of its carbon–oxygen bonds. However, its melting and boiling points are closer to those of a hydrocarbon of similar molecular weight than to those of an isomeric alcohol because of the inability of the ether to form hydrogen bonds. For example, the boiling point of diethyl ether (35°C) is close to that of pentane (36°C) and considerably lower than that of its isomer 1-butanol (117°C). However, the oxygen of an ether can participate as the Lewis base partner of a hydrogen bond. This results in the solubility of an ether in water being comparable to that of a similar alcohol. For example, the solubility of diethyl ether in water is 8.4 g/100 mL, whereas that of 1-butanol is 7.4 g/100 mL. As was the case with alcohols, the solubility of an ether in water decreases as the sizes of the alkyl groups increase.

You are probably aware that diethyl ether is the "ether" that has been used as an anesthetic. A major drawback is that ether is very flammable and volatile; mixtures of ether and air can be explosive. For this reason, ether has been replaced as an anesthetic by less hazardous compounds such as methoxyflurane:

$$\underset{\underset{\overset{|}{Cl}}{\overset{\overset{Cl}{|}}{H-C}}-\underset{\underset{\overset{|}{F}}{\overset{\overset{F}{|}}{C}}}{}-O-CH_3}{}$$

Methoxyflurane

A major use of ethers in the organic laboratory is as solvents for reactions. Ethers are nonpolar enough to dissolve many organic compounds, and the electrons on the oxygen can interact with alkali metal cations to help solubilize salts. In addition, ethers are nonacidic and are not very reactive. For these reasons they are especially useful in reactions involving strongly basic reagents. In addition to diethyl ether, other ethers that are commonly used as solvents are 1,2-dimethoxyethane (DME) and the cyclic ethers tetrahydrofuran (THF) and 1,4-dioxane:

$CH_3OCH_2CH_2OCH_3$

1,2-Dimethoxyethane
DME

Tetrahydrofuran
THF

1,4-Dioxane

5.10   AMINES

Amines can be considered as derivatives of ammonia in which one or more hydrogens have been replaced by alkyl or aryl groups. **Primary amines** have one alkyl or aryl group bonded to the nitrogen. **Secondary amines** have two groups on the nitrogen, **tertiary amines** have three, and **quaternary ammonium salts** have four. Note that the terms *primary, secondary,* and *tertiary* have different meanings here than they have with other functional groups. In the case of amines they refer to the number of carbon groups bonded to the *nitrogen.* In the case of alcohols and alkyl halides, however, they refer to the number of carbon groups bonded to the *carbon* that is bonded to the hydroxy or halogen substituent.

$CH_3CH_2CH_2CH_2NH_2$

Butylamine
A primary amine

$CH_3CH_2NHCH_2CH_3$

Diethylamine
A secondary amine

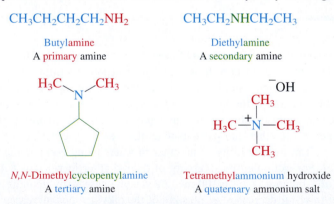

N,N-Dimethylcyclopentylamine
A tertiary amine

Tetramethylammonium hydroxide
A quaternary ammonium salt

Common names are usually employed for simple amines. In these names, the suffix -amine is appended to the name of the alkyl group. The prefixes di-, tri- and tetra- are used when several identical groups are attached to the nitrogen. For secondary and tertiary amines with different groups attached to the nitrogen, the largest group is used with the -amine suffix. An *N*-, rather than a number, is used to indicate other groups that are also attached to the nitrogen. Ionic compounds that are formed by the reaction of amines with acids are named as ammonium salts. Common names for some amines are shown in the preceding examples and in those that follow:

NHCH₂CH₂CH₂CH₃

CH₃
|
CH₃CH₂CH₂NCH₃

CH₃CH₂CH₂NH₃⁺ Cl⁻

N,N-Dimethylpropylamine 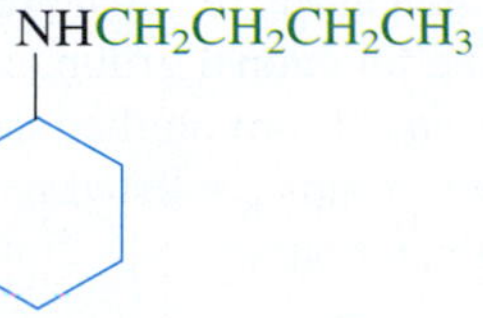   *N*-Butylcyclohexylamine   Propylammonium chloride

PROBLEM 5.21
Name these compounds:

NHCH₂CH₃

a) CH₃CH₂CH₂NH₂   **b)**

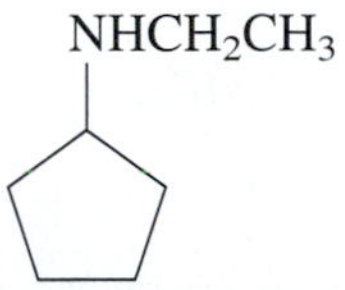

There are many trivial names for amines, especially those involving aromatic rings or where the nitrogen is part of a ring. Several important examples are the following:

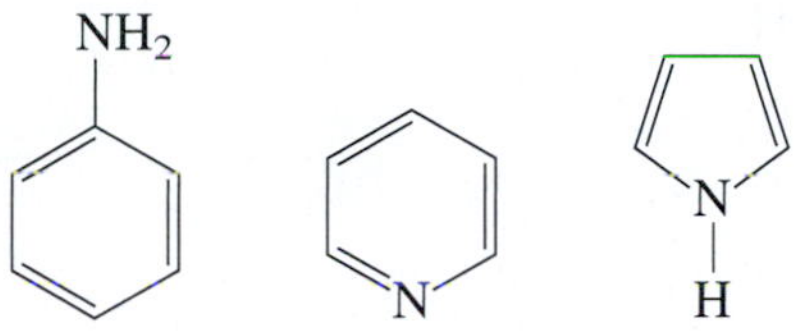

NH₂

Aniline   Pyridine   Pyrrole

For more complex amines, systematic nomenclature is employed. Such names are constructed in a manner very similar to that employed to name alcohols. The largest chain attached to the nitrogen is chosen as the root, numbered so that the carbon attached to the nitrogen has the lower number, and the suffix -amine is attached. Other groups that are attached to the nitrogen are given the prefix *N*-.

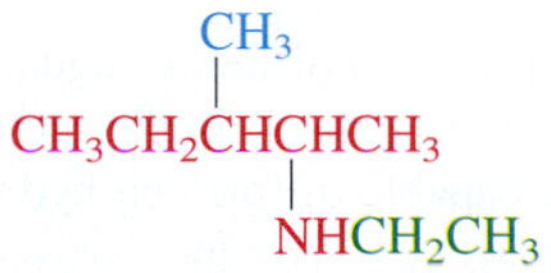

NH₂

CH₃
|
CH₃CH₂CHCHCH₃
|
NHCH₂CH₃

CH₂CH₂CH₃

N-Ethyl-3-methyl-2-pentanamine   3-Propylcyclopentanamine

(The longest chain containing a carbon attached to the nitrogen is chosen as the root and numbered from the end closer to the nitrogen. To show that the ethyl group is also attached to the nitrogen, it is given the prefix *N*-, rather than a number.)

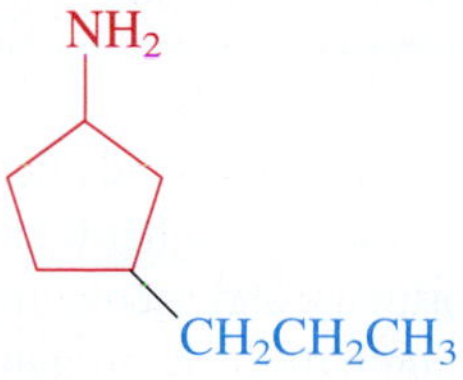

(Numbering begins at the carbon attached to the nitrogen.)

In addition to carbon–carbon double and triple bonds, only one other functional group can be designated as a suffix in the name. For example, if a compound has both an alcohol and an amine functional group, only one of them can be designated with the suffix. The other must be named as a group, using a prefix. The alcohol functional group has higher priority than the amine functional group, so the nitrogen group is named as an amino- group (or an alkylamino- group) on the main chain of the alcohol. (The priorities and group names for other groups are listed in Table 12.3 on p. 492.) An example is provided by the following compound:

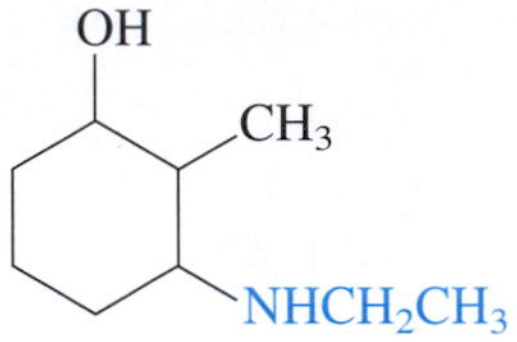

3-Ethylamino-2-methylcyclohexanol

(The hydroxy group has higher priority than the amino group
and is used to determine both the suffix and the numbering.)

PROBLEM 5.22

Name these compounds:

a) $CH_3CHCH_2CH_2CHCH_3$ (with CH_3 and $NHCH_3$ substituents)

b)

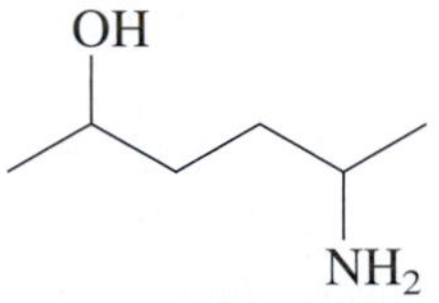

PROBLEM 5.23

Draw structures for these compounds:
a) Diethylammonium bromide
b) *N*-Methyl-3-(1-methylpropyl)-2-octanamine

Amines are polar compounds because of the presence of the nitrogen. Their melting and boiling points are higher than those of hydrocarbons of similar molecular mass. Like alcohols, primary and secondary amines are capable of forming hydrogen bonds, although the strength of the hydrogen bond is somewhat weaker in the case of an amine because nitrogen is less electronegative than oxygen. As a result, the boiling points of primary and secondary amines are somewhat lower than those of a similar alcohol. For example, butylamine boils at 78°C, and 1-butanol boils at 117°C. Because they have no hydrogens bonded to the nitrogen, the molecules of a tertiary amine do not form hydrogen bonds to each other, so the physical properties of tertiary amines resemble those of ethers. Amines of low molecular mass often have ammonia-like or fishy odors, and some have quite unpleasant odors. Perhaps you can imagine the odors of cadaverine and putrescine:

$$NH_2CH_2CH_2CH_2CH_2NH_2 \qquad NH_2CH_2CH_2CH_2CH_2CH_2NH_2$$

Putrescine   Cadaverine

Amines occur widely in nature, both in plants and animals. Natural amines, such as epinephrine (adrenaline), are often physiologically active in animals, as are some synthetic amines, such as amphetamine. Those that occur in plants, such as nicotine and morphine, are called **alkaloids** because they are basic and can be isolated by extraction with acid. When plant matter is extracted with aqueous acid, the amines are protonated according to the following equation:

$$RNH_2 + HCl \ \rightleftharpoons \ RNH_3^+ \ Cl^-$$

The resulting salts dissolve in the aqueous solution and can easily be separated from the rest of the plant material.

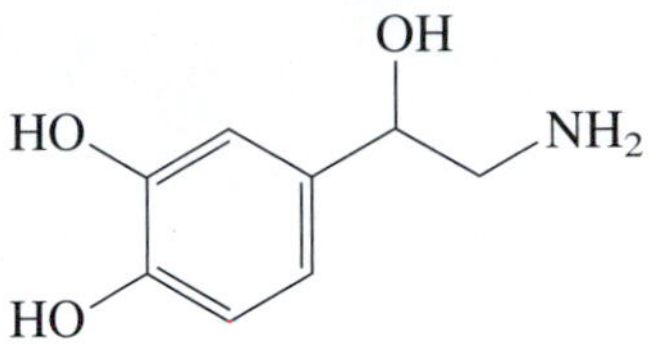

Epinephrine

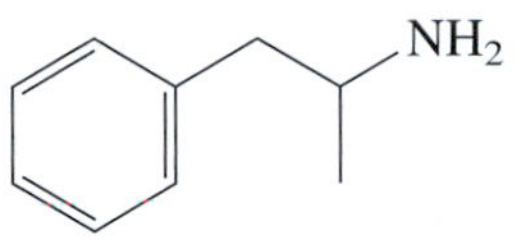

Amphetamine

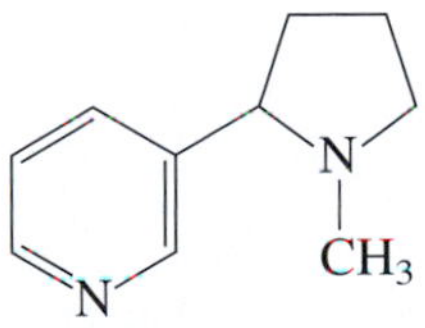

Nicotine

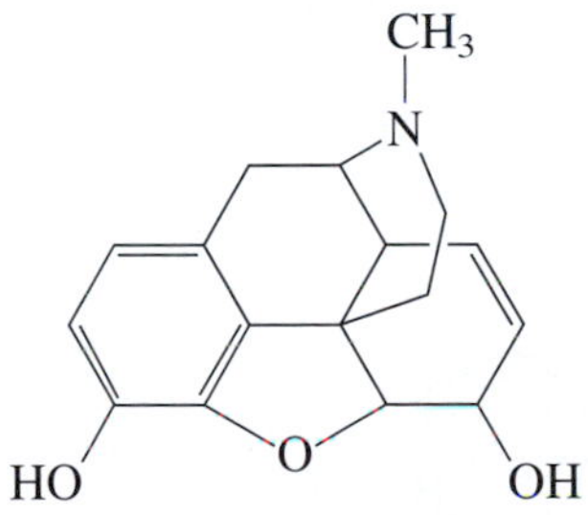

Morphine

Review of Mastery Goals

After completing this chapter, you should be able to:

- Provide the systematic (IUPAC) name for an alkane. (Problems 5.24 and 5.26)

- Draw the structure of an alkane whose name is provided. (Problem 5.25)

- Name a complex group. (Problem 5.30)

- Name a cycloalkane, an alkene, an alkyne, an alkyl halide, an alcohol, an ether, or an amine. (Problems 5.24, 5.29, 5.30, 5.33, 5.34, and 5.43)

ORGANIC
Chemistry Now™
Click *Mastery Goal Quiz* to test how well you have met these goals.

■ Draw the structure of a compound containing one of these functional groups when the name is provided. (Problems 5.25, 5.31, and 5.35)

■ Predict the approximate physical properties of a compound containing one of the functional groups discussed. (Problems 5.36, 5.37, 5.38, 5.39, 5.44, 5.45, and 5.46)

ORGANIC
Chemistry Now™
Assess your understanding of this chapter's topics with additional quizzing and conceptual-based problems at **http://now.brookscole.com/ hornback2**

Additional Problems

5.24 Name these compounds:

$$\underset{a)}{} \ \ CH_3CH_2CH_2\underset{\underset{CH_3}{|}}{CH}-\underset{\underset{CH_3}{|}}{CH}CH_2CH_3$$

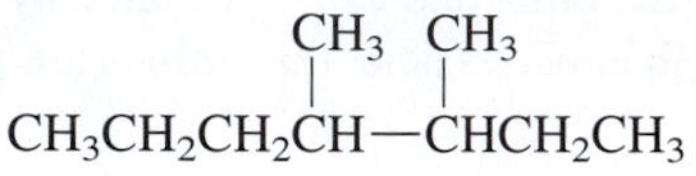

b) 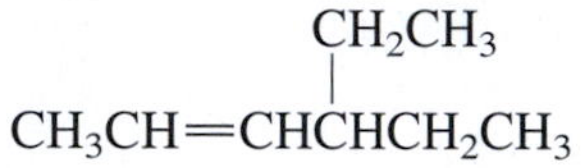 $CH_3CH=CH\underset{\underset{CH_2CH_3}{|}}{CH}CH_2CH_3$

c) 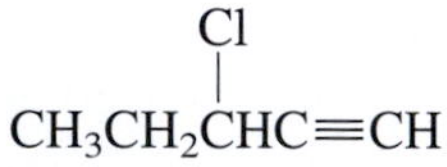 $CH_3CH_2\underset{\underset{Cl}{|}}{CH}C\equiv CH$

d)

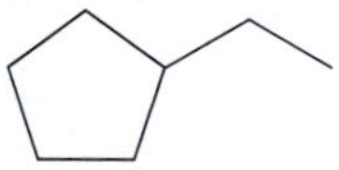

e)

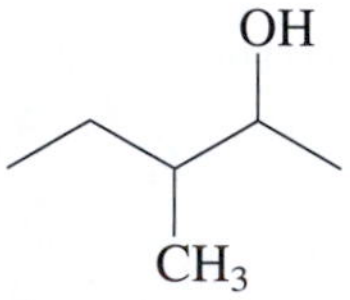

f)

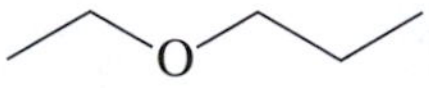

g)

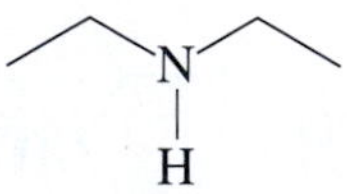

5.25 Draw structures for these compounds:
   a) 5-Ethyl-4-methylnonane
   b) 2-Methyl-1,3-hexadiene
   c) 3-Methylcyclopentanol
   d) 3-Octyne
   e) *sec*-Butylcyclohexane
   f) *tert*-Butyl alcohol

5.26 Name the five isomers of C_6H_{14}.

5.27 Indicate whether each of the indicated carbons is primary, secondary, tertiary, or quaternary:

a) 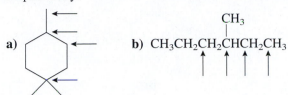 b) CH$_3$CH$_2$CH$_2$CHCH$_2$CH$_3$
with CH$_3$ above the CH carbon and upward arrows pointing to carbons

5.28 Draw compounds that meet these requirements:
a) A primary alcohol   **b)** A tertiary alcohol
c) A secondary alkyl chloride   **d)** A secondary amine

5.29 What is wrong with the names given for these compounds? Provide the correct name for each.

a)

2-Ethyl-2-pentene

b)

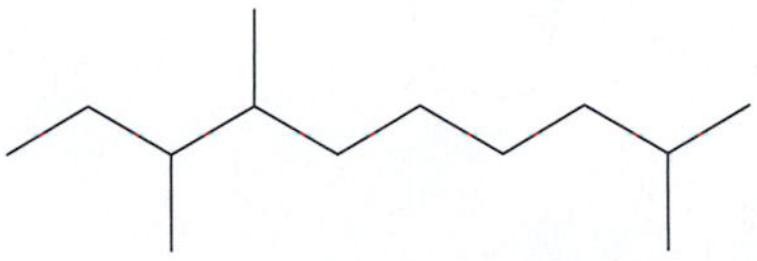

3,4,9-Trimethyldecane

c) 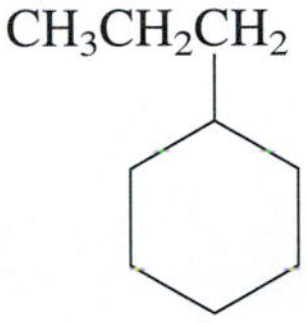 CH$_3$CH$_2$CH$_2$

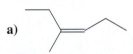

3-Cyclohexylpropane

d)

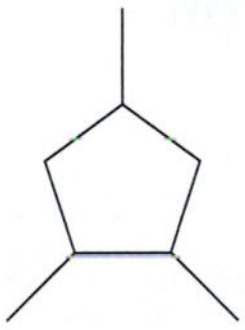

1,3,4-Trimethylcyclopentane

e) 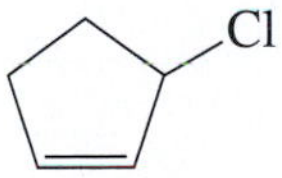 Cl

1-Chloro-2-cyclopentene

f) HC≡CCH$_2$CHCH$_3$ with OH on the CH

1-Pentyn-4-ol

g) 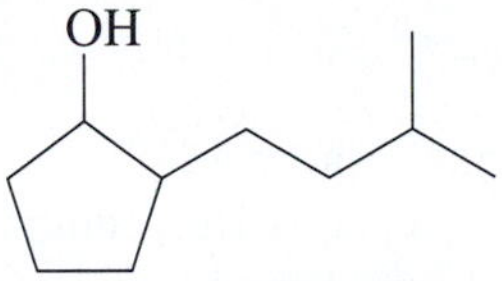 OH

2-2-Methylbutylcyclopentanol

h)  NHCH$_3$

1-Methyl-1-pentenamine

i) 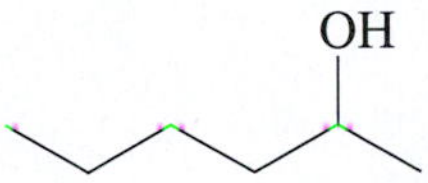 OH

1-Methylpentanol

5.30 Name these compounds:

a)

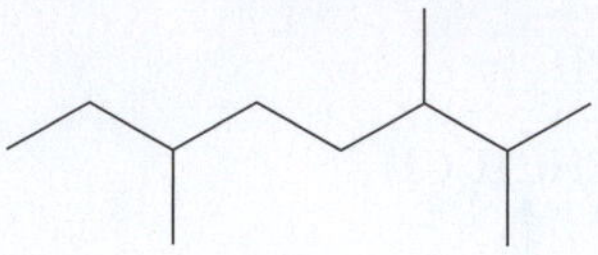

b)

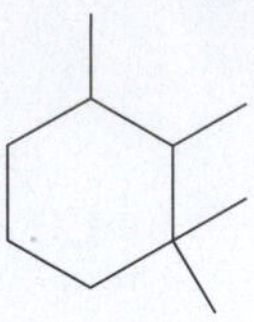

c)

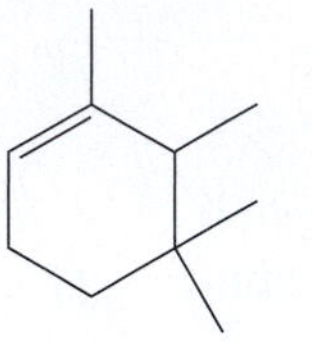

d)

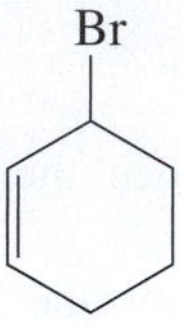

e) 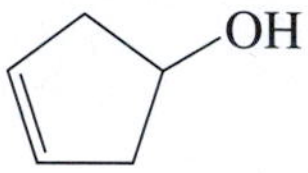 OH

f) $CH_2{=}CHCH_2C{\equiv}CH$

g) 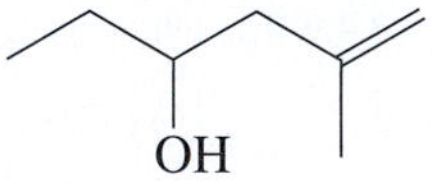 OH

h) 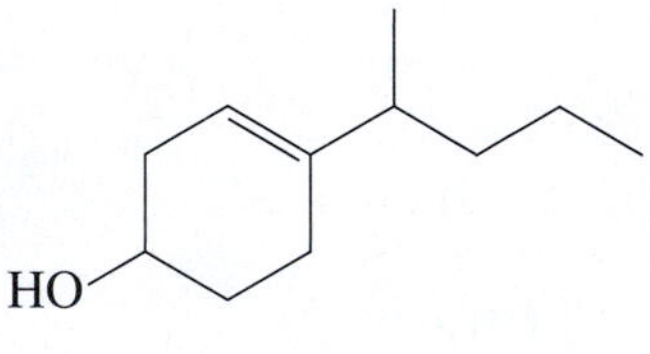 HO

i) 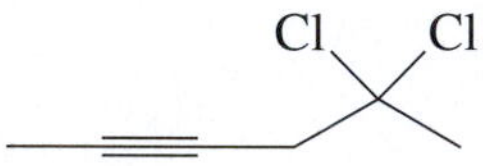 Cl Cl

j) 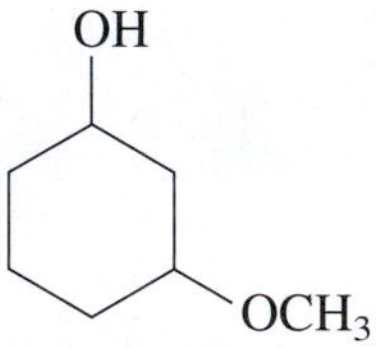 OH OCH$_3$

k) 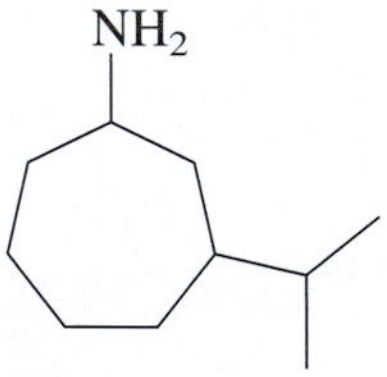 NH$_2$

l)

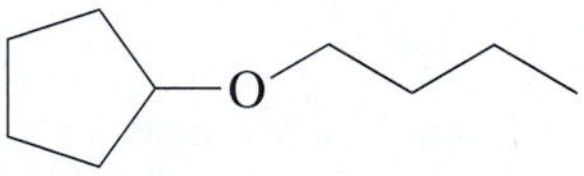

5.31 Draw structures for these compounds:
  a) 1,5-Dibromo-2,2-dichloro-4-ethyl-4-methyl-3-hexanol
  b) 2,2,5,5-Tetramethylcyclohex-3-enol   c) *tert*-Butylamine
  d) 1,2-Cyclopentanediol   e) Dibutyl ether
  f) *N,N*-Diethylbutylamine   g) 3-Isobutylcyclopentanol
  h) 5-(1,2,2-Trimethylpropyl)nonane

5.32 Menthol is a component of oil of peppermint. Label each carbon of menthol as primary, secondary, tertiary, or quaternary. Should menthol be classified as a primary, secondary, or tertiary alcohol?

OH

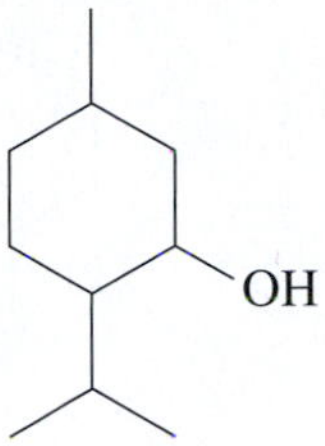

Menthol

5.33 Provide systematic names for these naturally occurring compounds:

a) 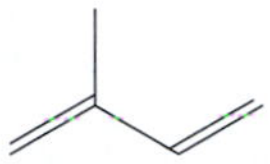 Isoprene

b)
OH   OH   OH
| | |
CH₂—CH—CH₂   Glycerol (obtained from fat)

c) 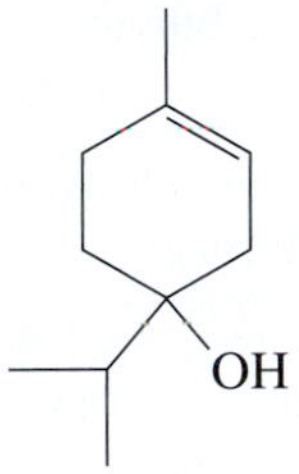 Terpinen-4-ol (a terpene)

OH

d) 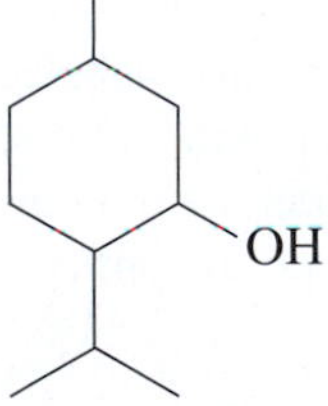 Menthol (a component of peppermint oil)

OH

e) 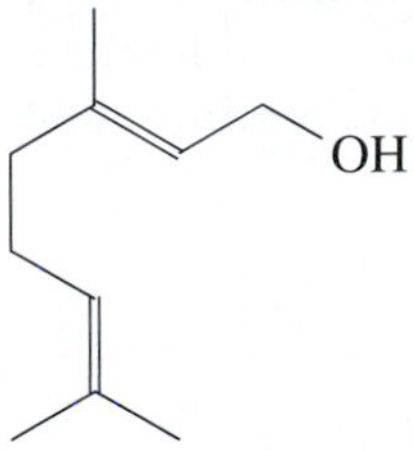 Geraniol (obtained from roses)

OH

f) $H_2NCH_2CH_2CH_2CH_2CH_2NH_2$ Cadaverine

g) $CH_3(CH_2)_2CH{=}CHCH{=}CH(CH_2)_8CH_2OH$ Bombykol
(sex attractant of the female silkworm moth)

BioLink

5.34 The systematic name for the —CH=CH₂ group is ethenyl. Provide a systematic name for limonene, which is found in lemons and other citrus fruits.

Limonene

BioLink

5.35 Vitamin A alcohol is 3,7-dimethyl-9-(2,6,6-trimethyl-1-cyclohexenyl)-2,4,6,8-nonatetraen-1-ol. Draw the structure of vitamin A alcohol.

5.36 Explain which compound has the higher melting point:
a) Cyclopentane or pentane
b) 1-Pentanol or pentane

5.37 Explain which compound has the higher boiling point:
a) Octane or nonane
b) Nonane or 3-nonene
c) 1-Nonyne or 1-nonanol
d) Trimethylamine or propylamine
e) Cyclopentanol or diethyl ether
f) 1-Butene or 1-butyne
g) 1-Chlorobutane or 1-pentanol
h) Cyclopentylamine or cyclopentanol

5.38 Chloroform, $CHCl_3$, is a common solvent in the organic laboratory. It is not miscible with water, so a mixture of these two solvents forms two layers. Which solvent do you expect to form the lower layer?

5.39 Predict which of these compounds has the higher solubility in water:
a) 1-Butanol or 1-chlorobutane
b) 1-Butanol or 1-hexanol
c) Pentane or diethylamine

5.40 While working in the chemical stockroom, you discover an unlabeled bottle containing a liquid compound. You carefully smell the liquid and discover that it has a fishy odor. What functional group do you suspect the unknown compound contains? What simple chemical test could you do to confirm the presence of the suspected functional group?

BioLink

5.41 Turpentine, obtained from pine trees, is composed primarily of α-pinene and β-pinene. Explain whether you expect turpentine to mix with water. If a paint dissolves in turpentine, what does this suggest about the structure of the paint?

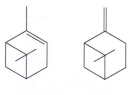

α-Pinene   β-Pinene

5.42 The structure of a typical fat is shown here. Estimate the energy content of fat compared to the other compounds discussed in the Focus On box on p. 146 and explain your reasoning.

$$
\begin{array}{c}
\overset{\displaystyle O}{\overset{\displaystyle \|}{}} \\
CH_2{-}OC(CH_2)_{16}CH_3 \\
\overset{\displaystyle O}{\overset{\displaystyle \|}{}} \\
CH{-}OC(CH_2)_{14}CH_3 \\
\overset{\displaystyle O}{\overset{\displaystyle \|}{}} \\
CH_2{-}OC(CH_2)_7CH{=}CH(CH_2)_7CH_3
\end{array}
$$

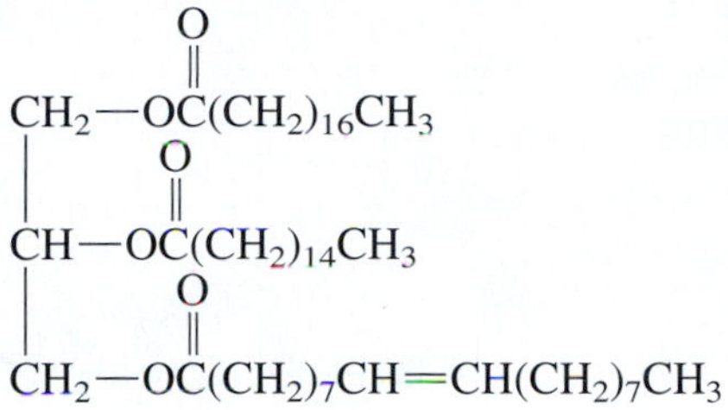

Problems Using Online Three-Dimensional Molecular Models

ORGANIC
Chemistry Now™
Click *Molecular Model Problems* to view the models needed to work these problems.

5.43 Name these compounds.

5.44 Explain which compound has the higher boiling point.

5.45 Explain which compound has the higher melting point.

5.46 Explain which compound has the higher solubility in water.

Do you need a live tutor for homework problems? Access vMentor at Organic ChemistryNow at **http://now.brookscole.com/hornback2** for one-on-one tutoring from a chemistry expert.

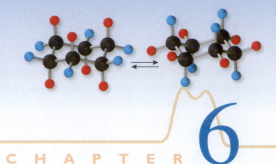

Stereochemistry I

CIS–TRANS ISOMERS AND CONFORMATIONS

I N CHAPTER 1 you learned the geometry of the bonds around an atom. For example, the four bonds of an sp^3-hybridized carbon have a tetrahedral geometry. But what happens when several such carbons are bonded together? What is the geometrical relationship between the bonds on different carbons? What is the overall shape of the molecule? Is more than one shape possible? If so, are they different in energy? Can they interconvert? If so, how fast? These and other questions will be answered in this chapter and the next, which discuss the **stereochemistry,** or three-dimensional structures, of organic molecules. In these chapters you will encounter a new type of isomer: stereoisomers. Unlike the constitutional isomers that you have already seen, **stereoisomers** have the same bonds or connectivity, but the bonds are in a different three-dimensional orientation.

This chapter begins with a discussion of a type of stereoisomer that arises because of the presence of a carbon–carbon double bond in a compound. Then a method to distinguish these stereoisomers when naming them is presented. This is followed by a discussion of the various shapes that molecules can assume by rotating about their single bonds.

An understanding of stereoisomers is important because these compounds often have different physical properties and different chemical reactions. There are reactions that occur with only one stereoisomer and not another. There are other reactions that produce only one stereoisomer and not another. This is especially true for biochemical reactions, both in the laboratory and in living organisms. Therefore, an understanding of stereochemistry is essential to the study of organic chemistry.

It is difficult to appreciate the three-dimensional shapes of organic molecules by examination of only the diagrams or pictures of their structures that are shown in

MASTERING ORGANIC CHEMISTRY

▶ Recognizing Cis–Trans Isomers and Estimating Their Relative Stabilities

▶ Designating the Configuration of Cis–Trans Isomers

▶ Determining Conformations about Single Bonds and Estimating Their Relative Energies

▶ Understanding the Types and Relative Amounts of Strain in Cyclic Molecules

▶ Understanding the Chair Conformations of Cyclohexane Derivatives

▶ Determining the Relative Stabilities of Conformations of Cyclohexane Derivatives

ORGANIC
Chemistry ⚛ Now™

Look for this logo in the chapter and go to OrganicChemistryNow at
http://now.brookscole.com/hornback2 for tutorials, simulations, problems, and molecular models.

two dimensions on the pages of a book. The use of models is an invaluable aid in understanding the material presented in this chapter. You are strongly encouraged to build models of the molecules discussed in this chapter, at least until you become more comfortable with their three-dimensional structures. Be sure to take advantage of the online computer models for the molecules that are discussed in this chapter.

6.1   CIS–TRANS ISOMERS

Suppose you were asked to draw 2-butene. You would quickly write the structure $CH_3CH=CHCH_3$. Now suppose you were asked to carefully show the shape of the molecule, specifically the geometry of the bonds of the carbons involved in the double bond. As discussed in Chapter 3, each carbon of the double bond is sp^2 hybridized and so has a trigonal planar geometry. Furthermore, to have maximum overlap of the p orbitals of the pi bond, the two carbons of the double bond and all four atoms attached to them (the two carbons of the methyl groups and the two hydrogens) must lie in the same plane. You would soon discover that there are two ways to draw 2-butene based on these constraints. One has the two methyl groups on the same side of a line that runs through the doubly bonded carbons (on the same side of the double bond), and the other has the two methyl groups on opposite sides of this line (on opposite sides of the double bond):

cis-2-Butene		*trans*-2-Butene
0.33	Dipole moment (Debyes)	0
4°C	Boiling point	1°C
−139°C	Melting point	−106°C

MODEL BUILDING PROBLEM 6.1
Build and compare handheld models of the two possible 2-butenes.

These structures have the same bonds, so they are not constitutional isomers. However, they have a different arrangement of these bonds in three dimensions—they are stereoisomers. Stereoisomers like these, that differ in the placement of groups on one side or the other of the double bond, are called **cis–trans isomers.** The term *cis*- is used to designate the stereoisomer that has like groups on the same side of the double bond and the term *trans*- is used to designate the stereoisomer that has like groups on opposite sides. The *cis*- and *trans*-isomers of 2-butene are different compounds and have different, although similar, physical and chemical properties. For example, although both are nonpolar compounds, *cis*-2-butene has a small dipole moment. The vectors of any small bond dipoles in *trans*-2-butene must cancel because of its shape, so its dipole moment must be zero. The boiling points of the two isomers are quite similar, but the *trans*-isomer melts at a significantly higher temperature because its more linear shape allows it to pack better into a crystal lattice.

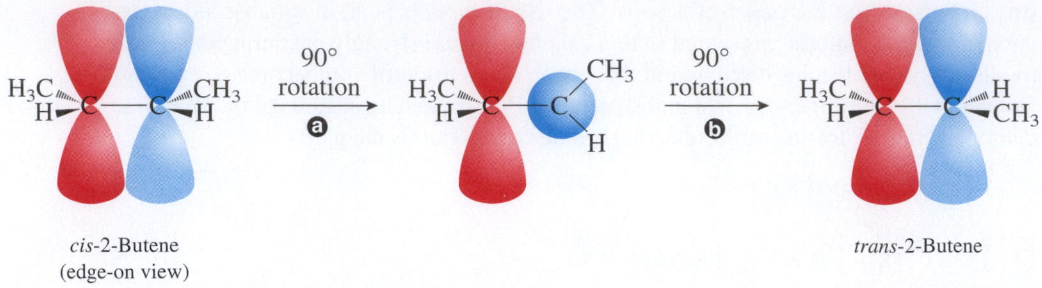

cis-2-Butene
(edge-on view)

trans-2-Butene

The *p* orbitals of the pi bond of *cis*-2-butene are in the plane of the page, and the plane defined by the atoms attached to the carbons of the double bond is perpendicular to the page. To convert to the *trans*-isomer, one of the carbons of the double bond must be rotated about the axis of the double bond.

ⓐ Rotation of the right carbon by 90° produces the middle structure. The plane defined by the left CH_3—C—H is now perpendicular to the plane defined by the CH_3—C—H on the right. The red *p* orbital on the left C is in the plane of the page, and the blue *p* orbital on the right C is pointed directly at you, so the two *p* orbitals are also perpendicular to one another. Therefore, there is no stabilizing overlap of these *p* orbitals—the pi bond has been broken.

ⓑ An additional rotation of the right carbon by 90° produces *trans*-2-butene.

Figure 6.1

INTERCONVERSION OF *CIS*- AND *TRANS*-2-BUTENE BY ROTATION ABOUT THE DOUBLE BOND.

For one cis–trans isomer to convert to the other, rotation about the carbon–carbon double bond must occur as shown in Figure 6.1. As discussed in Chapter 3, such rotation causes the overlap of the *p* orbitals of the pi bond to decrease until, at a rotation of 90°, the *p* orbitals are perpendicular and there is no net overlap. At this point, the pi bond is broken. Continued rotation reverses this process until the pi bond is completely reformed in the other stereoisomer. In Chapter 3 the amount of energy that this process requires was estimated from the difference in the strength of a carbon–carbon double bond and that of a single bond to be approximately 60 kcal/mol (250 kJ/mol). Recall that about 20 kcal/mol (84 kJ/mol) of thermal energy is available at room temperature. Reactions with activation energies less than this amount occur fairly rapidly at room temperature, whereas reactions that require larger amounts of energy are slow. Therefore, interconversion of cis–trans isomers is slow at room temperature (so slow that we say that it does not occur). It is possible to separate cis–trans isomers and study their individual properties.

For an alkene to exhibit cis–trans isomerism, the two groups on one end of the double bond must be different and the two groups on the other end of the double bond must be different. That is, in terms of the following structure, A must be different from B, and D must be different from E. When this is the case, both of the carbons of the double bond are said to be stereocenters. A **stereocenter** or **stereogenic atom** is defined as an atom at which the interchange of two groups produces a stereoisomer.

For cis–trans isomers to exist:

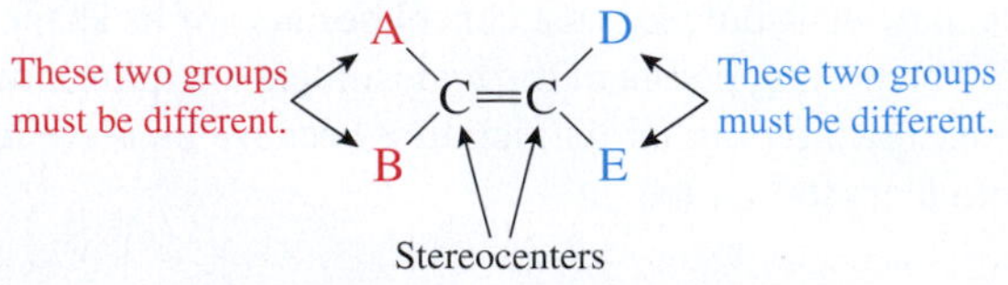

These two groups
must be different.

These two groups
must be different.

Stereocenters

PROBLEM 6.1

Which of these compounds exhibit cis–trans isomerism? Draw both cis–trans isomers when they exist.

a) $CH_3CH_2CH{=}CHCH_3$

b) $CH_3CH_2CH{=}CH_2$

c)
$$\overset{\displaystyle CH_3}{\underset{}{CH_3C}}{=}CHCH_3$$

d)
$$\overset{\displaystyle Cl}{\underset{}{CH_3C}}{=}CHCH_2CH_3$$

As illustrated earlier for the 2-butene isomers, cis–trans isomers usually have different physical properties. It should not be surprising, then, that they also often have somewhat different stabilities. The relative stabilities of *cis*-2-butene and *trans*-2-butene can be determined by comparing the heat that is evolved when each of these compounds reacts with hydrogen. (This reaction is called *catalytic hydrogenation* because a catalyst is required.) In the case of the 2-butenes, both of the cis–trans isomers produce the same product, butane, upon catalytic hydrogenation:

$$\overset{H_3C}{\underset{H}{>}}C{=}C\overset{CH_3}{\underset{H}{<}} + H_2 \xrightarrow{\text{catalyst}} CH_3CH_2CH_2CH_3 \xleftarrow{\text{catalyst}} \overset{H_3C}{\underset{H}{>}}C{=}C\overset{H}{\underset{CH_3}{<}} + H_2$$

When the *cis*-isomer reacts with hydrogen, 28.6 kcal/mol (120 kJ/mol) of heat is produced. When the *trans*-isomer is subjected to the same reaction, 27.6 kcal/mol (115 kJ/mol) of heat is produced. Because the product has the same energy in both reactions, any difference in the heat produced must be due to differences in the energies of the starting alkenes. Because the *cis*-isomer produces more heat, it must be higher in energy (less stable) than the *trans*-isomer. This is best seen by examining a diagram of the energies as shown in Figure 6.2. As can be seen in the diagram, *trans*-2-butene and hydrogen are 27.6 kcal/mol higher in energy than butane, and *cis*-2-butene and hydrogen are 28.6 kcal/mol higher. This means that the *cis*-isomer is 1.0 kcal/mol (4.2 kJ/mol) higher in energy (less stable) than the *trans*-isomer.

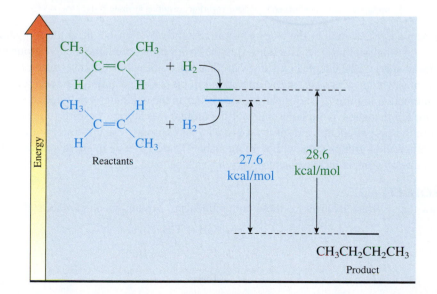

Figure 6.2

ENERGY DIAGRAM FOR THE CATALYTIC HYDROGENATION OF THE ISOMERIC 2-BUTENES.

Figure 6.3

COMPUTER-DRAWN MODELS
OF *CIS*- AND *TRANS*-2-BUTENE.

cis-2-Butene   *trans*-2-Butene

Why is *cis*-2-butene slightly less stable than *trans*-2-butene? Constitutional isomers have different bonds, so it is to be expected that their total bond energies will usually be different. However, in the case of cis–trans isomers the difference must be more subtle than this because the bonds are the same. In this case we must evaluate the increase in energy due to strain in the molecules. **Strain** is any factor that destabilizes a molecule by forcing it to deviate from its optimum bonding geometry.

Different atoms cannot occupy the same region of space. If atoms that are not bonded approach each other too closely (inside their van der Waals radii), they repel each other. A molecule whose structure forces nonbonded atoms too close together will distort its bond lengths and/or bond angles to partially relieve this repulsive force. This causes poorer orbital overlap, resulting in weaker bonds and a higher-energy, less stable molecule. This type of destabilizing strain is called **steric crowding** or **steric strain.** It is this type of strain that causes *cis*-2-butene to be less stable than *trans*-2-butene. In the *cis*-isomer the hydrogens on the methyl groups are forced somewhat too close together because of the rigid geometry of the double bond. This causes a small increase in the energy of the *cis*-isomer due to steric crowding.

$$\begin{array}{ccccccc}
\text{H} & & \text{H} & \text{H} & & \text{H} \\
& \text{C} & & & \text{C} & \\
\text{H} & & \text{C}=\text{C} & & & \text{H} \\
& \text{H} & & \text{H} & &
\end{array}$$

The planar geometry of the double bond causes a hydrogen on each of the methyl groups to be forced too close together, resulting in some steric strain.

The van der Waals radii of the hydrogens only slightly overlap, so the amount of strain energy is small in this case. The *trans*-isomer, however, has the large methyl groups on opposite sides of the double bond and therefore has no destabilization because of steric crowding. The presence of a steric interaction in *cis*-2-butene and the absence of such an interaction in *trans*-2-butene can be seen better in the computer-generated models in Figure 6.3 that show the relative sizes of the atoms. It is straightforward to generalize that in the case of alkenes substituted only with alkyl groups, the cis–trans isomer with the larger alkyl groups trans will be more stable. The difference in energy between the geometrical isomers increases as the alkyl groups become larger.

PROBLEM 6.2

Draw the cis–trans isomers for these compounds and explain which is more stable:

a) $CH_3CH_2CH=CHCH_3$ b) $\begin{array}{cc} CH_3 & CH_3 \\ | & | \\ CH_3C\!\!-\!\!-\!\!-\!\!C=CHCH_3 \\ | \\ CH_3 \end{array}$

MODEL BUILDING PROBLEM 6.2
Build handheld models of both cis–trans isomers of this compound and compare the amount of steric crowding in each.

$$H_3C\diagdown\atop{H_3C\diagdown\atop{}}C=CHCH_3$$

$$\underset{H_3C}{\overset{H_3C\diagup}{}}C\diagdown CH_3$$

6.2 DESIGNATING THE CONFIGURATION OF CIS–TRANS ISOMERS

The three-dimensional arrangement of groups about a stereocenter in a molecule is termed its **configuration.** One way to designate the configurations of cis–trans isomers is to use the terms *cis* and *trans*. However, these terms are ambiguous in many instances. For example, is the following isomer of 1-chloro-1-fluoro-1-propene the *cis*- or the *trans*-stereoisomer?

$$\underset{H}{\overset{H_3C}{}}C=C\underset{F}{\overset{Cl}{}}$$

The problem is that *cis* means that the two groups used as references are on the same side of the double bond and *trans* means that they are on opposite sides. But which are the reference groups? Often, two like groups are used as references, as was the case with the 2-butenes. However, the preceding example does not have two like groups. To designate the configuration of such compounds, a set of rules is needed to determine which of the two groups on each end of the double bond has higher priority and will therefore be used as references. To avoid confusion with the older cis–trans method, the newer method uses different terms to indicate whether the high-priority groups are located on the same or opposite sides of the double bond. If the high-priority groups are on the *same side* of the double bond, the configuration is designated **Z** (from the first letter of the German word *zusammen,* which means "together"), and if the high-priority groups are on *opposite sides,* the configuration is designated as **E** (from the German word *entgegen,* which means "opposite").

High priority⟍   ⟋High priority   High priority⟍   ⟋Low priority
          C=C   C=C
Low priority⟋   ⟍Low priority   Low priority⟋   ⟍High priority

Z   *E*

Priorities are assigned to groups by a series of rules known as the **Cahn-Ingold-Prelog sequence rules,** named after the three chemists who developed them. These rules use the atomic numbers of the atoms attached to the carbons of the double bond.

▶ **RULE 1**

Of the two atoms attached to one carbon of the double bond, the one with the higher atomic number has the higher priority.

This is the only rule needed to assign the configuration of the stereoisomer of chlorofluoropropene shown earlier. Of the two atoms attached to the left carbon of the double bond, C has a higher atomic number than H. Therefore, the methyl group has a higher priority than the hydrogen. On the right carbon, Cl has a higher priority than F. Because the high-priority groups, the CH_3 and the Cl, are on the same side of the double bond, the configuration of this stereoisomer is *Z*. The compound is named (*Z*)-1-chloro-1-fluoro-1-propene.

▶ **RULE 2**

If the two atoms attached to the carbon are the same, compare the atoms attached to them in order of decreasing priority. The decision is made at the first point of difference.

It may be necessary to continue farther out a chain of atoms until the first point of difference is reached. An application of this rule is shown in the following example:

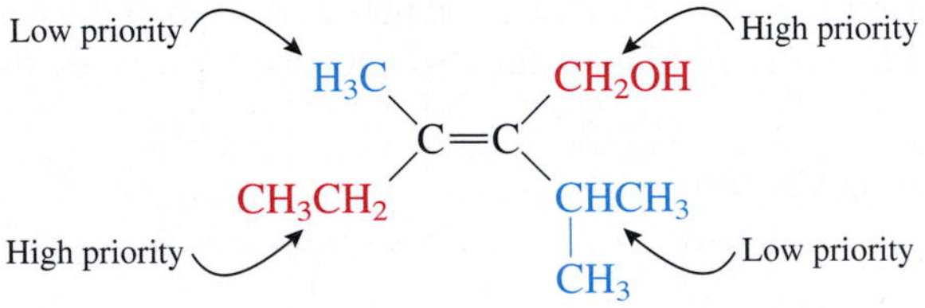

(*E*)-2-Isopropyl-3-methyl-2-penten-1-ol

The atoms attached to the carbon on this end of the double bond are both carbons. Therefore, we must compare the three atoms attached to each of these carbons. The atoms bonded to the upper C (the C of the methyl group) are H, H, and H. The atoms bonded to the lower C (the C of the CH_2 of the ethyl group) are C, H, and H. The highest-priority atom on the upper carbon is H, whereas that on the lower carbon is C, so the lower group, the ethyl group, has the higher priority.

The atoms attached to the carbons on this end of the double bond are also both carbons. The upper carbon is bonded to O, H, and H. The lower carbon is bonded to C, C, and H. In order of decreasing priority the first comparison is O versus C. Because O has a higher atomic number, the top group has the higher priority. Note that the decision was made at the first point of difference. The fact that the second group bonded to the lower carbon (a carbon) has a higher atomic number than the second group attached to the upper carbon (a hydrogen) has no bearing on the outcome.

▶ **RULE 3**

Double and triple bonds that are part of the groups attached to the double bond are treated as though they are constructed from two or three single bonds, respectively.

An easy way to visualize this is to replace the second bond of a double bond with an additional single bond to the same kind of atom. This is done for both partners of the double bond, as illustrated here for the vinyl group:

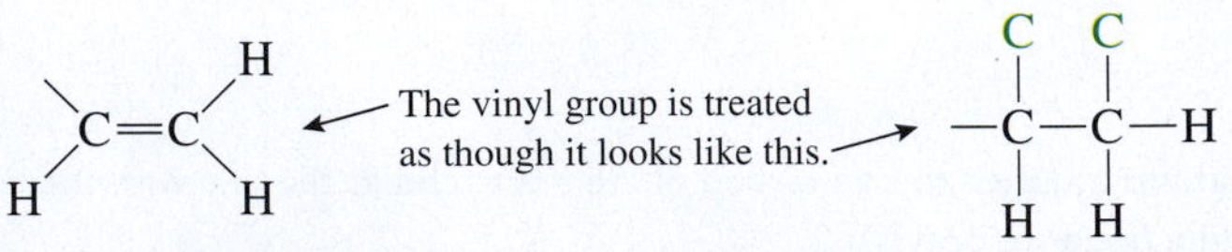

The vinyl group is treated as though it looks like this.

Note that the newly added atoms do not have their valences completed. Similarly, the second and third bonds of a triple bond are replaced with single bonds to the same kind of atom, as shown for the cyano group in the following example:

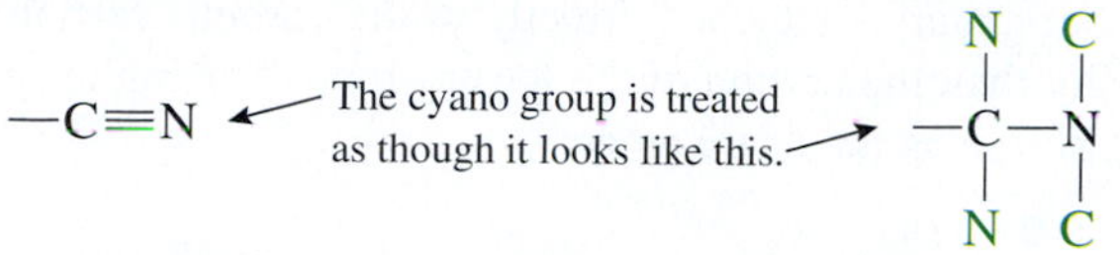

The following example will help illustrate the use of Rule 3:

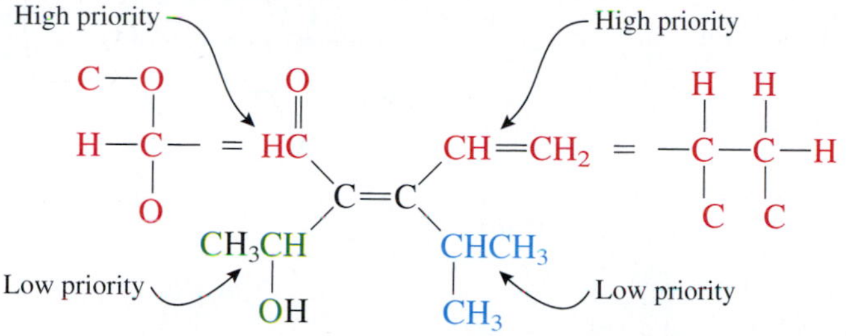

On this side of the double bond, the upper (aldehyde) group has higher priority. Its carbon is bonded to O, O, and H, whereas the carbon of the bottom group is bonded to O, C, and H.

On this side of the double bond, the upper (vinyl) group has higher priority. Its carbon is bonded to C, C, and H, but so is the bottom carbon. The next carbon out on the vinyl group is bonded to C, H, and H, whereas the next carbon on the lower group is bonded to H, H, and H.

Because the high-priority groups are on the same side, this is the (Z)-isomer.

PROBLEM 6.3

Which of these groups has the higher priority?

a) $-CH_3$ or $-CH_2CH_3$ **b)** $-\overset{\overset{\textstyle O}{\|}}{C}OH$ or $-\overset{\overset{\textstyle O}{\|}}{C}NH_2$

c) $-CH_2CH_2CH_3$
         or
    $-CH_2CH_2CH_2CH_3$

d) $-CH_2CH_2CH_3$ or $-C\equiv N$

e) 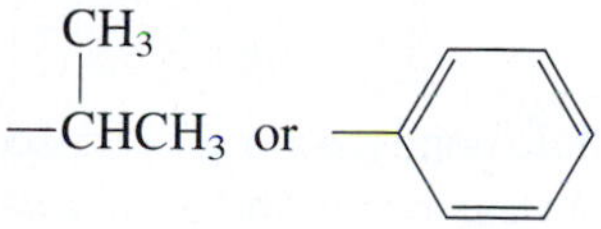 $-\overset{\overset{\textstyle CH_3}{|}}{C}HCH_3$ or

PRACTICE PROBLEM 6.1

Does this alkene have the Z or E configuration?

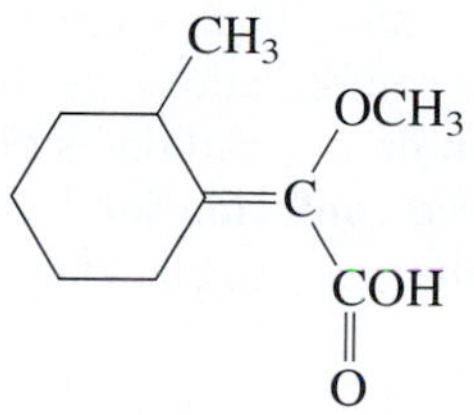

Solution

On the left side the upper carbon has higher priority because it is bonded to two C's and one H, whereas the lower carbon is bonded to one C and two H's. On the right side the oxygen of the upper group is attached directly to the carbon–carbon double bond and has a higher priority than the carbon of the lower group. The higher-priority groups are on the same side, so this is the (Z)-isomer.

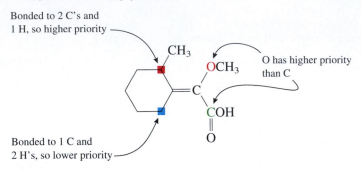

ORGANIC
Chemistry·⚛·Now™
Click *Coached Tutorial Problems* to practice using the **Cahn-Ingold-Prelog Sequence Rules**.

PROBLEM 6.4

Assign these compounds as the Z or E isomers:

a) $\begin{array}{c} H_3C \\ CH_3CH_2 \end{array} C=C \begin{array}{c} CH_2Cl \\ CH_3 \end{array}$

b) $\begin{array}{c} H \\ F \end{array} C=C \begin{array}{c} C\equiv CH \\ CH_2CH_3 \end{array}$

c) (structure with N, vinyl, tert-butyl, CH₂—OH groups)

6.3   CONFORMATIONS

Now let's consider the shapes of molecules containing single bonds. As was discussed in Chapter 3, atoms involved in single bonds can rotate about the axis of the bond without affecting the overlap of the orbitals that form that bond. Because the bond energy is not affected, atoms are free to rotate about single bonds. The various shapes that a molecule can assume by rotation about single bonds are called **conformations.** Derek Barton shared the 1969 Nobel Prize in chemistry for his work in the area of conformations.

As a simple example, let's consider the case of ethane. Rotation about the carbon–carbon sigma bond results in a number of different conformations for this molecule. There are two extremes for these conformations, as shown in Figure 6.4. The **eclipsed conformation** has each hydrogen on one carbon as close as possible to one hydrogen on the other carbon. The **staggered conformation** has the hydrogens on one carbon as far from the hydrogens on the other as possible. Other conformations are intermediate between these two extremes.

ⓐ Eclipsed Conformation

Ethane

Each carbon–hydrogen bond on one carbon of the eclipsed conformation of ethane is directly in line with a carbon–hydrogen bond on the other carbon. This is easier to see if we view the molecule end-on—that is, down the carbon–carbon bond. A drawing of such a view, called a Newman projection, is as follows:

Newman
projection

◀░▌**Important Convention**

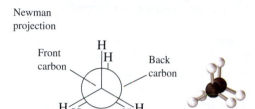

In a Newman projection the molecule is viewed end-on so that one carbon is directly behind the other. By convention, the front carbon is represented by the intersection of its three carbon–hydrogen bonds and the back carbon is represented by a circle. Here it is easy to see that the back carbon–hydrogen bonds lie directly behind the front carbon–hydrogen bonds. The front hydrogens eclipse the back hydrogens.

ⓑ Staggered Conformation

Ethane

In the staggered conformation of ethane the carbon–hydrogen bond on one carbon bisects the angle between two of the carbon–hydrogen bonds on the other carbon. Again, this can better be seen in the Newman projection:

Newman
projection

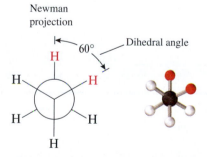

Note how each carbon–hydrogen bond bisects the angle between two carbon–hydrogen bonds on the other carbon. Each hydrogen on the front carbon is perfectly staggered between two hydrogens on the back carbon. The hydrogens on the back carbon are similarly staggered. The **dihedral angle** is defined as the angle between a marker group on the front carbon and one on the back carbon in this flat projection. In this case, if the red hydrogens are the markers, the dihedral angle is 60°.

Figure 6.4

ⓐ **ECLIPSED AND** ⓑ **STAGGERED CONFORMATIONS OF ETHANE AND THEIR NEWMAN PROJECTIONS.**

Experiments have shown that the eclipsed conformation is slightly less stable (by 2.9 kcal/mol [12.1 kJ/mol]) than the staggered conformation. Over the years, a variety of explanations have been proposed for this observation. The current hypothesis is that this difference in stability results from more favorable interactions among the molecular orbitals in the staggered conformation than in the eclipsed conformation. We need not be concerned with the details of this explanation at this point. The important fact is that the staggered conformation is more stable than the eclipsed conformation. The destabilization caused by eclipsed bonds is called **torsional strain.** Because there are three eclipsed pairs of carbon–hydrogen bonds and the total energy increase caused by these interactions is 2.9 kcal/mol (12.1 kJ/mol), each of these eclipsing interactions destabilizes the eclipsed conformation by about 1 kcal/mol (4 kJ/mol).

Now consider what happens to the energy of the conformations as rotation occurs about the carbon–carbon bond of ethane. Beginning with the eclipsed conformation, the energy is at its highest point. As rotation occurs, the energy decreases until a minimum is reached at the staggered conformation after a rotation of 60°. As rotation continues, the energy increases until a new eclipsed conformation is reached at 120°. This process is repeated twice more as rotation proceeds through 360°. A plot of energy versus dihedral angle for this rotation is shown in Figure 6.5. The staggered conformations are located at minima on this plot, separated by 2.9 kcal/mol (12.1 kJ/mol) energy hills at the eclipsed conformations. Because 20 kcal/mol (83.7 kJ/mol) of thermal energy is available at room temperature, there is plenty of energy to allow an ethane molecule in

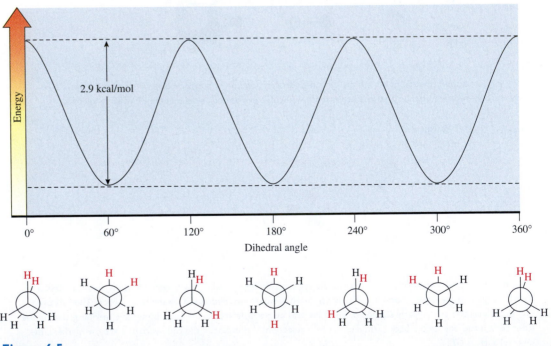

Figure 6.5

PLOT OF ENERGY VERSUS DIHEDRAL ANGLE FOR CONFORMATIONS OF ETHANE.

a staggered conformation to climb over the energy barrier posed by the eclipsed conformation and rotate to another staggered conformation. This rotation occurs very rapidly at room temperature. The barriers for rotations about most single bonds are in the 3 to 5 kcal/mol (12–21 kJ/mol) range, so rotations about most single bonds are fast at room temperature. Therefore, different conformations cannot be separated or isolated and are not usually considered to be isomers. We say that there is *free rotation* about single bonds.

   Next, let's consider the case of propane. Conformational analysis can be done about either of the two identical carbon–carbon bonds. Again there are two limiting conformations: staggered and eclipsed. The only difference between this example and the analysis of ethane done previously is that here there are two hydrogen–hydrogen eclipsing interactions and one hydrogen–methyl eclipsing interaction.

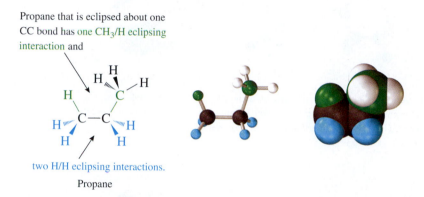

Propane that is eclipsed about one CC bond has one CH₃/H eclipsing interaction and

two H/H eclipsing interactions.

Propane

   It is to be expected that the repulsion between the eclipsed methyl and hydrogen is slightly larger than that between two hydrogens. In fact, the eclipsed conformation of propane is 3.3 kcal/mol (13.8 kJ/mol) higher in energy than the staggered conformation. If each of the H with H interactions contributes about 1 kcal/mol (4 kJ/mol) to this value, as was the case for ethane, then the eclipsing interaction between the methyl and the hydrogen must contribute about 1.3 kcal/mol (5.4 kJ/mol) in repulsion energy. This is due primarily to torsional strain between the C—H bond and the C—C bond, along with a smaller contribution due to steric strain between the hydrogen on the carbon and the hydrogens on the methyl group. The energy versus dihedral angle plot for propane looks just like the one for ethane except that the energy difference between the staggered and eclipsed conformations is 3.3 kcal/mol.

PROBLEM 6.5

Draw a plot of energy versus dihedral angle for the conformations of propane about one of the C—C bonds.

   Butane provides a more complex example. Here there are two different types of carbon–carbon bonds. Analysis of the conformations available by rotation about the bond between carbon 1 and carbon 2 (or carbon 3 and carbon 4) is very similar to the analysis of propane, with the difference that there is an ethyl group on one carbon rather

ⓐ Anti Conformation

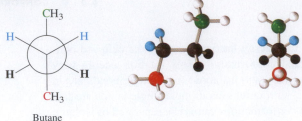

Butane

Two groups, the methyl groups in this case, are said to be anti if the dihedral angle between them is 180°. The anti conformation is the most stable conformation of butane because it is staggered and has the two large methyl groups as far apart as possible.

ⓑ Gauche Conformation

Steric crowding between these methyl groups

Two groups are said to be gauche when the dihedral angle between them is 60°. The gauche conformation is 0.8 kcal/mol (3.3 kJ/mol) higher in energy than the anti conformation. Because it is staggered, it has no torsional strain. Its higher energy is due to a small amount of steric strain caused by interaction between the bulky methyl groups, which are a little too close together.

ⓒ Eclipsed Conformation

This eclipsed conformation has two H with H eclipsing interactions and one CH_3 with CH_3 eclipsing interaction. The eclipsing interaction between the methyl groups is more destabilizing than that between the hydrogens. This conformation is about 4.5 kcal/mol (19 kJ/mol) less stable than the anti conformation. If the two H with H interactions contribute 1 kcal/mol each to this value, as was the case for ethane, then the CH_3 with CH_3 interaction must contribute the remaining 2.5 kcal/mol (10.5 kJ/mol). This value is due to both torsional and steric strain.

ⓓ Eclipsed Conformation

This eclipsed conformation has two CH_3 with H interactions and one H with H interaction. If a CH_3 with H interaction here costs the same amount of energy as it did in the case of propane (1.3 kcal/mol), then this conformation would be expected to be destabilized by 2(1.3) + 1 = 3.6 kcal/mol (15 kJ/mol). This value is in reasonable agreement with the experimental value of 3.7 kcal/mol (15.5 kJ/mol).

Active Figure 6.6

ORGANIC
Chemistry·⚛·Now™

CONFORMATIONS OF BUTANE. Test yourself on the concepts in this figure at **OrganicChemistryNow.**

than a methyl group. However, conformational analysis about the bond between carbon 2 and carbon 3 provides a more interesting situation. In this case, each carbon has one methyl group and two hydrogens. The various conformations lead to a number of different interactions, shown in Figure 6.6.

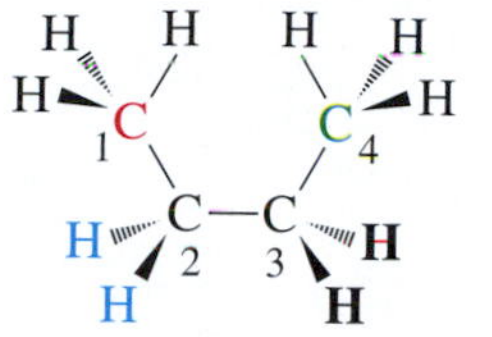

Conformational analysis about the C-2—C-3 bond of butane is more complex than previous examples because each carbon has a methyl and two hydrogens bonded to it. Several different interactions occur in the conformations.

Butane

A plot of the energies of these conformations, with the methyl groups used as the markers for the dihedral angles, is shown in Figure 6.7. The lowest-energy minimum on this plot is located at the anti conformation. There are also minima at the two gauche conformations that are 0.8 kcal/mol (3.3 kJ/mol) less stable than the anti conformation. The highest-energy barrier on the plot, at the conformation where the methyl groups are eclipsed, is 4.5 kcal/mol (19 kJ/mol) higher in energy than the anti conformation. Therefore, rotation about this carbon–carbon bond is fast at room temperature. The gauche and anti conformations are in equilibrium, about 70% of the molecules having the more stable anti conformation at any given instant.

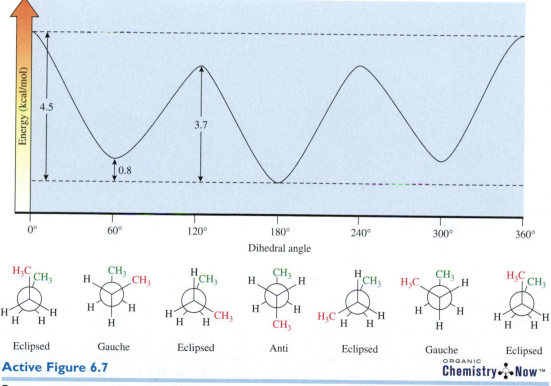

Active Figure 6.7

ORGANIC
Chemistry ⚛ Now™

PLOT OF ENERGY VERSUS DIHEDRAL ANGLE FOR CONFORMATIONS OF BUTANE. Test yourself on the concepts in this figure at **OrganicChemistryNow.**

PRACTICE PROBLEM 6.2

Draw a plot of energy versus dihedral angle for the conformations of 2-methylbutane about the C2—C3 bond.

$$
\begin{array}{c}
\underset{}{CH_3} \\
| \\
\underset{1}{CH_3}-\underset{2}{CH}-\underset{3}{CH_2}-\underset{4}{CH_3}
\end{array}
$$

Strategy

Start by drawing the conformations about the C2—C3 bond using Newman projections. Begin with an eclipsed conformation and rotate the front (or back) carbon by 60°. Continue to rotate the carbon in the same direction, drawing the conformation after each 60° rotation until the original conformation is reached after a total rotation of 360°. Now assign the energies of the conformations. In general, eclipsed conformations are higher in energy than staggered conformations, and conformations that have larger groups eclipsed are higher in energy than those that have smaller groups eclipsed. Next evaluate the energies of the staggered conformations; the energy of the conformation increases as the number of gauche interactions increases. Be careful to place conformations with the same interactions at the same energy.

Solution

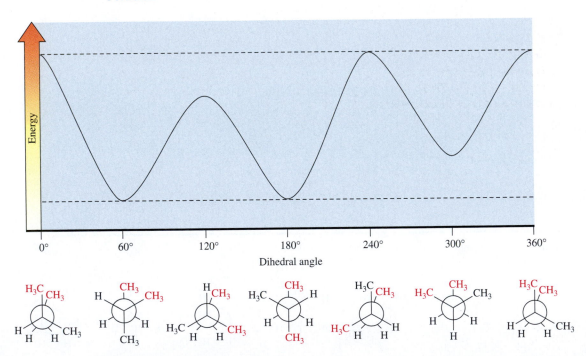

First, let's examine the three eclipsed conformations. The 0° and 240° conformations both have one CH$_3$ with CH$_3$, one CH$_3$ with H, and one H with H eclipsing interaction, so they have the same amount of strain energy. The 120° conformation has three CH_3 with H eclipsing interactions, so it has less strain energy than the others—it is at lower energy. The 60° and 180° staggered conformations have one gauche interaction and therefore are lower in energy than the 300° staggered conformation, which has two gauche interactions.

PROBLEM 6.6
Draw a Newman projection of the highest-energy conformation of 2,3-dimethylbutane about the C2—C3 bond.

Analysis of the conformations of other alkanes can be done in a similar manner. For any linear alkane the most stable conformation is the so-called zigzag conformation, which is anti about all of the carbon–carbon bonds. The zigzag conformation for hexane follows:

Hexane

Free rotation about the carbon–carbon bonds generates a large number of other conformations that are gauche about one or more of these bonds. Although the zigzag conformation is the most stable one, many of the others are only slightly higher in energy and are readily attainable at room temperature. The shape of an individual molecule changes rapidly, twisting and turning among these various possibilities. Finally, the presence of polar substituents can dramatically affect and complicate conformational preferences because of interactions among their dipoles and hydrogen bonding. Factors such as these help control not only the shape but also the function of complex biological molecules, such as enzymes and other proteins.

6.4 CONFORMATIONS OF CYCLIC MOLECULES

Because their carbon chains are confined in rings, cycloalkanes are much less flexible than noncyclic (or acyclic) alkanes. The number of conformations available is dramatically reduced. Furthermore, cycloalkanes are often held in shapes that cause them to have considerable strain energy. One way to measure this strain energy in the laboratory is to burn the alkane in a calorimeter and measure the amount of heat that is produced (the heat of combustion). The heat of combustion must first be corrected for the number of carbons and hydrogens in the ring. For example, cyclohexane (C_6H_{12}) has twice as many atoms as cyclopropane (C_3H_6) and would be expected to produce twice as much heat on a per mole basis, other factors being equal. The easiest way to correct for this is to divide the heat of combustion per mole by the number of CH_2 groups in the ring. This gives a heat of combustion "per mole of CH_2 group." This value can then be

Table 6.1   Heats of Combustion and Strain Energies of Some Cycloalkanes

Ring Size	Heat of Combustion per CH_2	Strain Energy per CH_2	Total Ring Strain
3	166.3 (695.8)	8.9 (37.2)	26.7 (111.6)
4	163.9 (685.8)	6.5 (27.2)	26.0 (108.8)
5	158.7 (664.0)	1.3 (5.4)	6.5 (27.0)
6	157.4 (658.6)	0	0
7	158.3 (662.3)	0.9 (3.7)	6.3 (25.9)
8	158.6 (663.6)	1.2 (5.0)	9.6 (40.0)
9	158.8 (664.4)	1.4 (5.8)	12.6 (52.2)

Units are kcal/mol. Units for values in parentheses are kJ/mol. To obtain the values listed in the second column, the total heats of combustion were divided by the number of carbons in the ring. Values in the third column were obtained by subtracting 157.4 (the heat of combustion per CH_2 for cyclohexane) from the values in the second column. Values in the fourth column were calculated by multiplying values in the third column by the number of carbons in the ring.

compared to that of a standard compound that has no strain, such as an unbranched, long-chain alkane.

Such heat of combustion values for the cycloalkanes ranging in size from 3 to 9 carbons are provided in Table 6.1. The lowest heat of combustion per CH_2 group is found for cyclohexane, which is usually considered to be free of strain. The extra heat produced by the other cycloalkanes can then be attributed to their strain energies. For example, the heat of combustion per CH_2 for cyclopropane is 166.3 kcal/mol (696 kJ/mol), and that for cyclohexane is 157.4 kcal/mol (659 kJ/mol). The difference, 8.9 kcal/mol (37 kJ/mol), is the amount of strain energy per CH_2 group for cyclopropane. The total strain energy for cyclopropane is three times this value, 26.7 kcal/mol (111 kJ/mol), because cyclopropane has three CH_2 groups. Values for the strain energy per CH_2 group and total ring strain for the other cycloalkanes are also provided in Table 6.1. Note that cyclopropane and cyclobutane have large amounts of strain compared to cyclohexane, whereas the other cycloalkanes have much smaller amounts.

By examining the conformations of the cycloalkanes, we are able to determine the origin of these strain energies. Let's begin with the smallest one, cyclopropane, and see what causes the large amount of strain energy that it has (Figure 6.8).

From heat of combustion data, cyclopropane has 26.7 kcal/mol (111.6 kJ/mol) of strain energy. Most of this strain is due to angle strain, but the contribution due to torsional strain is also significant. As we will see later, this strain energy causes cyclopropane to be more reactive than a normal alkane or cycloalkane. However, even though cyclopropane rings are reactive, they are fairly common in organic chemistry.

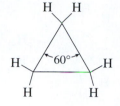

ⓐ The three carbons of cyclopropane form an equilateral triangle. The C—C—C angle is 60°. However, if the hybridization at each carbon is sp^3, as would be expected from the structure, then the angles between the sp^3 hybrid AOs are 109.5°.

The orbitals forming the C—C bonds of cyclopropane

ⓑ Because the angle between the sp^3 hybrid AOs on one carbon is wider than the angle between the carbons to which these orbitals are bonding, the orbitals cannot point directly toward the carbons. Instead, as shown here, they point slightly outside of a line connecting the nuclei. Because the orbitals of a bond do not point directly at each other, the amount of overlap is decreased. This causes the C—C bonds of cyclopropane to be weaker than normal C—C bonds. This type of destabilization is called **angle strain.**

ⓒ In addition to angle strain, cyclopropane has a significant amount of torsional strain. This can best be seen by looking at a Newman projection down any of the C—C bonds. As can be seen in this diagram, each C—C bond is held in an eclipsed conformation by the rigidity of the molecule.

Figure 6.8

CYCLOPROPANE. ⓐ BOND ANGLES, ⓑ ANGLE STRAIN, AND ⓒ TORSIONAL STRAIN.

MODEL BUILDING PROBLEM 6.3

Build a model of cyclopropane and examine its strain.

If the four carbons of cyclobutane lie in a plane, then its carbons form a square. The angles of a square are 90°, so it is expected that cyclobutane also has some angle strain, although not as much as cyclopropane. Planar cyclobutane would also be eclipsed about each C—C bond and would have considerable torsional strain. As the carbons of cyclobutane are distorted from planarity, torsional strain decreases while angle strain increases. For small distortions the increase in angle strain is less than the decrease in torsional strain. Therefore, the lowest-energy conformation of cyclobutane is slightly nonplanar, with an angle between the planes of the carbons of about 35° (Figure 6.9). Cyclobutane has some angle strain and some torsional strain contributing to its total strain energy of 26.0 kcal/mol (108.8 kJ/mol). A Newman projection shows how twisting the carbons out of planarity results in less torsional strain. The hydrogens are no longer exactly eclipsed. Although cyclobutane has less strain than cyclopropane, cyclobutane rings are less common than cyclopropane rings because, as we shall see later, they are more difficult to prepare.

Figure 6.9

CYCLOBUTANE.

Cyclobutane

MODEL BUILDING PROBLEM 6.4

Build a model of cyclobutane. Examine the various types of strain present in the planar and nonplanar geometries.

The angles of a regular pentagon are 108°. Therefore, planar cyclopentane would have little or no angle strain. However, like planar cyclobutane, it would have considerable torsional strain because each C—C bond would be held in an eclipsed conformation. It is to be expected, then, that cyclopentane will distort from planarity to relieve this torsional strain. In one low-energy conformation, one carbon folds out of the plane so that the overall shape is somewhat like an envelope (Figure 6.10). This relieves most of the torsional strain without increasing the angle strain significantly. Overall, cyclopentane has very little strain, 6.5 kcal/mol (27 kJ/mol). It is a very common ring system and is widely distributed among naturally occurring compounds.

MODEL BUILDING PROBLEM 6.5

Build a model of cyclopentane. Examine the various types of strain present in the planar and nonplanar geometries.

6.5   CONFORMATIONS OF CYCLOHEXANE

The cyclohexane ring is very important because it is virtually strain free. This is one of the reasons why compounds containing six-membered rings are very common. If cyclohexane were planar, its C—C—C angles would be 120°—too large for the 109.5° angle of sp^3 hybrid AOs. However, the angles of the ring decrease as it becomes nonplanar. There are two nonplanar conformations, called the **chair conformation** and the **boat conformation,** that are completely free of angle strain. These conformations are shown in Figures 6.11 and 6.12, respectively. The chair conformer of cyclohexane is

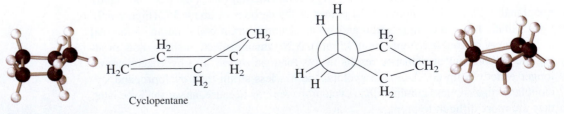

Cyclopentane

Figure 6.10

CYCLOPENTANE.

ⓐ Chair Conformation

All of the C—C—C bond angles are 109.5°, so this comformation has no angle strain. In addition, it has no torsional strain because all of the C—H bonds are perfectly staggered. This can best be seen by examining a Newman projection down C—C bonds on opposite sides of the ring:

ⓑ Newman Projection

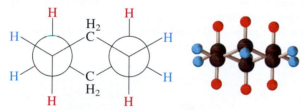

The staggered arrangement of all the bonds can be seen clearly in the Newman projection. This same picture is seen when the projection is viewed down any C—C bond. All the C—C bonds in the molecule are in conformations in which the hydrogens are perfectly staggered.

ⓒ Axial Hydrogens

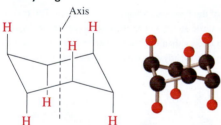

In the chair conformation cyclohexane has two different types of hydrogens. The bonds to one type are parallel to the axis of the ring. These are called axial hydrogens. The axial bonds alternate up and down around the ring.

ⓓ Equatorial Hydrogens

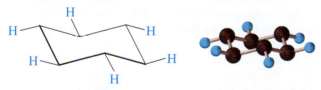

The other hydrogens are directed outward from the ring. They are called equatorial hydrogens because they lie around the "equator" of the ring. Now go back to structure ⓐ, in which both types of hydrogens are shown, and identify the axial hydrogens (red) and the equatorial hydrogens (blue). Also examine the view of the axial and equatorial hydrogens provided by the Newman projection.

Figure 6.11

THE CHAIR CONFORMATION OF CYCLOHEXANE.
ⓐ CHAIR CONFORMATION,
ⓑ NEWMAN PROJECTION,
ⓒ AXIAL HYDROGENS, AND
ⓓ EQUATORIAL HYDROGENS.

Figure 6.12

THE BOAT CONFORMATION OF
CYCLOHEXANE. ⓐ BOAT
CONFORMATION, ⓑ NEWMAN
PROJECTION, AND ⓒ TWIST
BOAT CONFORMATION.

ⓐ **Boat conformation**

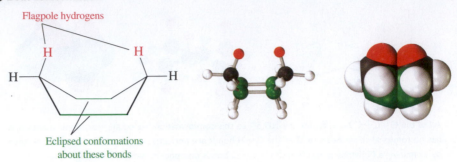

Like the chair conformation, all of the C—C—C bond angles of the boat conformation are 109.5°, so it has no angle strain. However, it does have other types of strain. The two red hydrogens, called flagpole hydrogens, approach each other too closely and cause some steric strain. In addition, the conformations about the green bonds are eclipsed. This can be seen more easily in the Newman projection down these bonds:

ⓑ **Newman Projection**

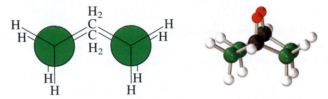

The Newman projection shows that two bonds of the boat conformation are eclipsed. The torsional strain due to these eclipsing interactions and the steric strain due to the interaction of the flagpole hydrogens make the boat conformation higher in energy than the chair conformation. The boat conformation is flexible enough to twist somewhat to slightly decrease its overall strain energy.

ⓒ **Twist Boat Conformation**

In the twist boat conformation the "bow" and the "stern" of the boat have been twisted slightly. Although this decreases the flagpole interaction and relieves some of the torsional strain, angle strain is introduced. Overall, the twist boat conformation is a little more stable than the boat conformation but not nearly as stable as the chair conformation.

perfectly staggered about all of the C—C bonds and therefore has no torsional strain either—it is strain free. The boat conformer, on the other hand, has both steric strain, due to interactions of the flagpole hydrogens, and torsional strain. It is about 6 kcal/mol (25 kJ/mol) less stable than the chair conformer. Some of the steric and torsional strain

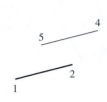

❶ To draw a good cyclohexane chair conformation, first draw two parallel lines, sloping upward and slightly offset. These are the C—C bonds between carbons 1 and 2 and carbons 4 and 5.

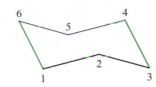

❷ Add carbon 6 to the left and above these lines and carbon 3 to the right and slightly below and draw the C—C bonds. Note how bonds on opposite sides of the ring are drawn parallel (the C-1—C-6 bond is parallel to the C-3—C-4 bond, and so on).

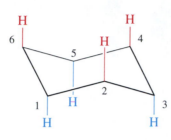

❸ Add the axial C—H bonds. These are drawn up from carbons 2, 4, and 6 and down from carbons 1, 3, and 5.

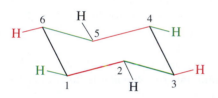

❹ The equatorial C—H bonds should point outward from the ring. They are drawn parallel to the C—C bond once removed, that is, the equatorial C—H bond on C-1 is drawn parallel to the C-2—C-3 (or C-5—C-6) bond, and the C—H bond on C-3 is drawn parallel to the C-1—C-2 (or C-4—C-5) bond, and so on.

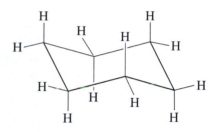

❺ The completed structure should look like this.

Figure 6.13

FIVE STEPS FOR DRAWING CHAIR CYCLOHEXANE.

---

of the boat can be relieved by twisting. The twist boat conformation is about 5 kcal/mol (21 kJ/mol) less stable than the chair conformation.

Figure 6.11 also shows that there are two different types of hydrogens, called **axial** hydrogens and **equatorial** hydrogens, in the chair conformer of cyclohexane. The axial C—H bonds are parallel to the axis of the ring; the equatorial C—H bonds project outward from the ring around its "equator." Steps to help you learn to draw the chair conformation of cyclohexane, including the axial and equatorial hydrogens, are provided in Figure 6.13.

The chair conformation of cyclohexane is not rigid. It can convert to a twist boat conformation and then to a new chair conformation in a process termed ring-flipping, as shown Figure 6.14 (not all the hydrogens are shown for clarity).

In the ring-flipping process, C-1 flips up to give a twist boat. Then C-4 can flip down to produce another chair conformation. When opposite carbons flip like this, all axial and equatorial bonds interconvert; that is, all hydrogens that were axial are converted to equa-

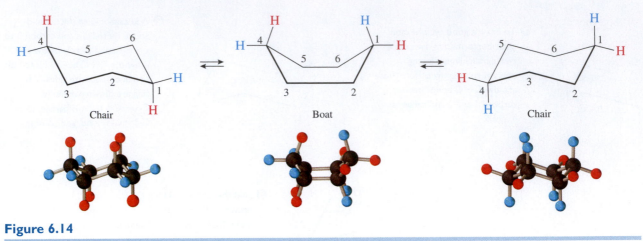

Figure 6.14

RING-FLIPPING. Chair cyclohexane converts to a boat and then to a new chair.

torial, and all hydrogens that were equatorial are converted to axial. This can be seen in Figure 6.14, in which the red hydrogens, which are axial in the left chair, are converted to equatorial hydrogens in the right chair. The energy required for this ring-flipping process is shown in Figure 6.15. The highest barrier, called the half-chair conformation, is about 10.5 kcal/mol (44 kJ/mol) higher in energy than the chair conformation. Again, the 20 kcal/mol (83.7 kJ/mol) of energy that is available at room temperature provides plenty of energy to surmount this barrier; therefore, this ring-flipping is fast. It occurs about 100,000 times per second at room temperature.

Figure 6.15

ENERGY DIAGRAM FOR THE CYCLOHEXANE RING-FLIPPING PROCESS.

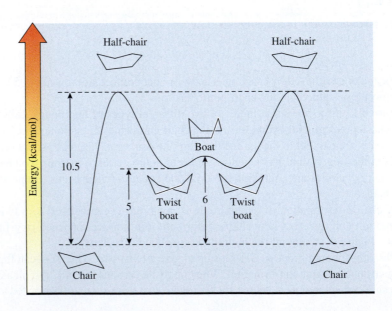

MODEL BUILDING PROBLEM 6.6
Build a model of cyclohexane.

a) Examine the strain present when the geometry is planar.
b) Examine the strain present in the boat conformation.
c) What strain is introduced in the twist boat conformation? What strain is relieved?
d) Examine the strain present in the chair conformation. Examine the conformations about one of the CC bonds. Identify the axial and equatorial hydrogens.
e) Try a ring flip with your model. Label an axial hydrogen and determine what happens to it when the ring flips.

MODEL BUILDING PROBLEM 6.7
Draw a chair cyclohexane. Show the axial and equatorial hydrogens.

6.6 CONFORMATIONS OF OTHER RINGS

Conformational analysis of rings larger than cyclohexane is more complicated. These rings are also less common than cyclohexane, so we discuss their conformations only briefly. As can be seen from Table 6.1, the seven-membered ring compound cycloheptane has only a small amount of strain. Obviously, it is nonplanar to avoid angle strain. It does have some torsional strain, but the overall strain is comparable to that of cyclopentane. It is a fairly common ring system.

Larger rings, having from 8 to 11 carbons, have somewhat more strain than cycloheptane. They are nonplanar, but even so, they seem to have some strain due to bond angles that are too large for the tetrahedral bond angle of 109.5°. They also have some torsional strain. In addition, these rings have a new type of strain, called **transannular** or **cross-ring strain.** This occurs because some C—H bonds are forced to point toward the center of the ring. The hydrogens of these bonds experience steric crowding from their interactions with atoms on the other side of the ring. Rings of 12 or more carbons no longer have transannular strain and are essentially strain-free.

Compounds that contain a benzene ring are also quite common and important. The carbons of benzene are sp^2 hybridized, with bond angles of 120°, which match exactly the angles of a regular hexagon. Thus, benzene is a planar molecule with no angle strain. It is rigid because any deviation from planarity would increase angle strain and decrease the overlap of the p orbitals of the conjugated pi system.

Benzene

In summary, small (3- and 4-membered) rings have a large amount of strain, due primarily to angle strain with some contribution from torsional strain. Rings that have

5, 6, and 7 members have very little strain. Rings that have 8, 9, 10, and 11 members have somewhat more strain, due to a variety of interactions. Large rings (12-membered and larger) have very little strain. However, it is not just the stability of a ring that determines whether it is commonly found but also the probability of its formation. The rings that are most commonly encountered in organic compounds are those that have 3, 5, 6, and 7 members. A discussion of the reasons for this is presented in Section 8.13.

6.7   Conformations of Cyclohexanes with One Substituent

What happens when there is a substituent on the cyclohexane ring? Let's consider methylcyclohexane as a simple example. As before, there are two chair conformations, which interconvert by the ring-flipping process. In this case, however, the two conformations are not identical. As shown in Figure 6.16, the methyl group is equatorial in one conformation and axial in the other. The conformation with the axial methyl is less stable than the conformation with the equatorial methyl by 1.7 kcal/mol (7.1 kJ/mol) because of steric interactions between the methyl and the axial hydrogens on C-3 and C-5. (These are often called **1,3-diaxial interactions.**)

Actually, in the case of a methyl substituent, each of the 1,3-diaxial interactions is identical to the interaction between the two methyl groups in the gauche conformation of butane. The last part of Figure 6.16 shows a Newman projection down the C-1—C-2 bond of the ring. The dihedral angle between the methyl group on carbon 1 and carbon 3 is 60°, just like the dihedral angle between the methyl groups in gauche butane. The axial methyl group is also gauche to C-5, as can be seen by viewing a Newman projection down the C-1—C-6 bond. Because the gauche interaction in butane destabilizes that conformation by 0.8 kcal/mol (3.3 kJ/mol), a logical estimate for the destabilization caused by the axial methyl with its two gauche interactions is twice this value, 1.6 kcal/mol (6.6 kJ/mol). This value is in reasonable agreement with the experimental value of 1.7 kcal/mol (7.1 kJ/mol) for the **axial strain energy** for a methyl group.

PROBLEM 6.7
Draw the two chair conformations for ethylcyclohexane. Which is more stable?

The two conformations of methylcyclohexane are rapidly interconverting—they are in equilibrium. The conformer with the methyl equatorial is more stable than the conformer with the methyl axial, so the equatorial conformer is present in a larger amount in the equilibrium mixture. The axial strain energy is actually the free energy difference between the conformations and can be used to calculate the equilibrium constant for the process by using the equation $\Delta G° = -RT \ln K$. Using the value of -1.7 kcal/mol (-7.1 kJ/mol) for $\Delta G°$, the equilibrium constant is calculated to be 18 at room temperature. Therefore, at any instant, 95% of methylcyclohexane molecules have the methyl group equatorial, and only 5% have the methyl axial.

ⓐ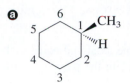

In diagrams like this one, which do not show conformations, the ring is pictured as lying in the plane of the page. The methyl group is above the plane in this case. Of course, the geometry of the molecule is actually a chair conformation.

Methylcyclohexane

ⓑ **Chair Conformations**

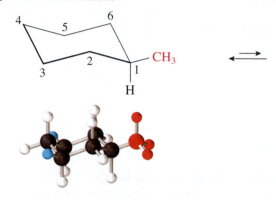

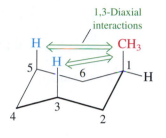

1,3-Diaxial interactions

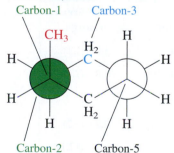

One possible chair conformation of methylcyclohexane.

The chair produced by the ring-flipping process.

It is important to be consistent in drawing chair conformations. In this book, groups that are above the plane of the page in the flat drawing are drawn closer to the top of the page in the chair conformation. The methyl group is above the page in the flat drawing, so it is closer to the top of the page than the hydrogen. In this particular drawing, the methyl group is equatorial.

Recall that ring flips interconvert all equatorial and axial groups. Therefore, the methyl group is now axial. (But note that it is still closer to the top of the page than is the hydrogen.) The steric crowding between the axial methyl and the axial hydrogens on C-3 and C-5 destabilizes this conformer. Actually, these interactions are similar to the gauche interactions in butane. This can better be seen in the Newman projection.

ⓒ **Newman Projection**

The methyl group on C-1 is gauche to C-3 of the ring. If a Newman projection down the C-1—C-6 bond were viewed, a similar gauche interaction between the methyl and C-5 of the ring would be found.

A Newman projection down the C-1—C-2 bond.

Figure 6.16

CONFORMATIONS OF METHYLCYCLOHEXANE. ⓐ FLAT RING PERSPECTIVE, ⓑ CHAIR CONFORMATIONS, AND ⓒ NEWMAN PROJECTION.

PRACTICE PROBLEM 6.3

Given that the equilibrium constant for the interconversion of the axial and equatorial conformations of methylcyclohexane is 18, show how to calculate the percentage of each that is present at equilibrium.

Solution

$$\text{axial} \rightleftharpoons \text{equatorial} \qquad K = \frac{[\text{equatorial}]}{[\text{axial}]} = 18$$

Let [equatorial] $= X\%$, then [axial] $= (100 - X)\%$.

$$\frac{X}{100 - X} = 18$$

$$X = 1800 - 18X$$
$$19X = 1800$$
$$X = 95\% = \text{percent equatorial}$$

In general, substituents larger than hydrogen prefer to be equatorial on a cyclohexane ring to avoid 1,3-diaxial interactions. Axial strain energies for a number of groups are listed in Table 6.2. Note that the values for the ethyl group (1.8 kcal/mol [7.5 kJ/mol]) and isopropyl group (2.2 kcal/mol [9.2 kJ/mol]) are only slightly larger than that for the methyl group (1.7 kcal/mol [7.1 kJ/mol]), while that for the *tert*-butyl group (4.9 kcal/mol [20.5 kJ/mol]) is much larger. The ethyl and propyl groups can be rotated so that a hydrogen is pointed back over the ring to interact with the axial hydrogens, so their effective steric bulk is not much different from that of a methyl group. In contrast, the *tert*-butyl group is forced to have one of its methyl groups pointed over the ring, causing much more severe 1,3-diaxial interactions. Because of

Table 6.2   Axial Strain Energies

Group	Axial Strain Energy	Group	Axial Strain Energy
$-C\equiv N$	0.2 (0.8)	$-CH_3$	1.7 (7.1)
$-F$	0.25 (1.0)	$-CH_2CH_3$	1.8 (7.5)
$-C\equiv CH$	0.4 (1.7)	$-\overset{\displaystyle CH_3}{\underset{\displaystyle}{CHCH_3}}$	2.2 (9.2)
$-Br$	0.5 (2.1)		
$-Cl$	0.5 (2.1)	$-Ph$	2.9 (12.1)
$-OH$	0.9 (3.8)		
$-NH_2$	1.4 (5.9)	$-\overset{\displaystyle CH_3}{\underset{\displaystyle CH_3}{CCH_3}}$	4.9 (20.5)
$-\overset{\displaystyle O}{\overset{\displaystyle \|}{COH}}$	1.4 (5.9)		

Units are kcal/mol. Values in parentheses are in units of kJ/mol.

the resulting large axial strain energy, *tert*-butylcyclohexane exists predominantly (more than 99.9%) in the conformation with the bulky *tert*-butyl group in the equatorial position.

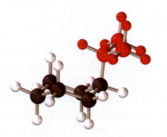

Conformation of isopropylcyclohexane
with the isopropyl group axial

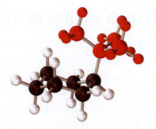

Conformation of *t*-butylcyclohexane
with the *t*-butyl group axial

PROBLEM 6.8

Which of these compounds will have more of the conformation with the substituent on the cyclohexane ring axial present at equilibrium?

a) [structure: cyclohexane with C≡N group] or [structure: cyclohexane with CH₃ group]

b) [structure: cyclohexane with CH₂CH₃ group] or [structure: cyclohexane with Ph group]

c) [structure: cyclohexane with Cl group] or [structure: cyclohexane with CH₂CH₃ group]

PROBLEM 6.9

Bromine is larger than chlorine, yet the two atoms have identical axial destabilization energies. Explain.

6.8 CONFORMATIONS OF CYCLOHEXANES WITH TWO OR MORE SUBSTITUENTS

The presence of two or more substituents on a ring—any size ring—introduces the possibility of stereoisomers. The existence of stereoisomers is independent of conformations and should be analyzed first because different stereoisomers will have different conformations. It is easiest to examine the stereoisomers of cyclic compounds by considering the rings to be flat, even though they may actually exist in chair or other conformations. Once all the stereoisomers have been identified, the conformations of each can be scrutinized.

Let's begin by considering a simple disubstituted cyclohexane, 1,2-dimethylcyclohexane. If the ring is drawn flat, in the plane of the page, then one substituent on each carbon projects above the page and the other projects below the page. The situation is somewhat similar to the cis–trans isomerism that occurs with alkenes. Both

Focus On

How Much Strain Is Too Much?

Molecules that would have large amounts of strain energy can easily be drawn on paper, but can they be prepared—and isolated—in the laboratory? How much strain can a molecule tolerate and still exist? Questions such as these have always fascinated organic chemists and have led them to design preparations of molecules that have very large strain energies. Although some of these syntheses have not been successful, others have led to the preparation of some very interesting, and highly strained, compounds.

For example, we know that cyclobutane has considerable strain because of its four-membered ring. How many four-membered rings can be fused together in the same molecule? A fascinating test case for this question is the molecule that has eight carbon atoms arranged in a cube, known as cubane. (This is a case in which the common name, cubane, rapidly conveys the structure of the compound and, for this reason, is more useful to even experienced chemists than is the systematic name, pentacyclo[4.2.0.0^{2,5}.0^{3,8}.0^{4,7}]octane.) Cubane is also of interest because a cube is a regular polyhedron, one of Plato's perfect solids. Cubane was first prepared in 1964. Although it has considerable strain (its total strain energy has been calculated to be 166 kcal/mol [695 kJ/mol]), it is a relatively stable compound that can easily be isolated and studied. Its melting point, 130–131°C, is quite high for a molecule of this size (the melting point of octane is −57°) owing to its symmetrical shape, which allows it to pack easily into the crystal lattice.

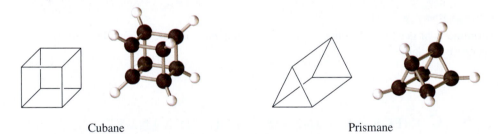

Cubane   Prismane

Prismane is an example of another interesting strained compound. Because it contains three-membered rings fused with four-membered rings, it should be even more strained than cubane. Prismane was prepared in 1973. It is a liquid that is stable at room temperature but explosive under some conditions. In toluene at 90°C its half-life (the time it takes for one-half of the compound to decompose) is 11 h. Note that prismane is isomeric with benzene. In fact, it was one of the structures proposed for benzene in the early days of organic chemistry.

The tetrahedron is another of Plato's perfect solids. The hydrocarbon having this shape is known as tetrahedrane. Because of its three-membered rings, it has considerably more strain than cubane and has, so far, resisted many attempts to prepare it. However, tetrahedrane substituted with *tert*-butyl groups at its vertices was prepared in 1981. It is a stable solid at room temperature.

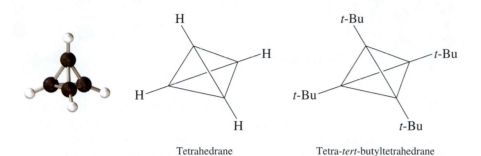

Tetrahedrane   Tetra-*tert*-butyltetrahedrane

Cyclopropane has considerable strain energy. A major part of this strain energy is due to angle strain because the 60° angles of the ring are much smaller than the tetrahedral bond angle of 109.5°. Cyclopropene should have even more angle strain because the ideal angles for a double bond are 120°. Cyclopropene was first prepared in 1960. It is quite reactive and reacts with itself at room temperature. It cannot be stored for any significant period, even at temperatures as low as −78°C. However, 1,2-dimethylcyclopropene is considerably less reactive and is stable at 0°C. The methyl groups help protect the reactive double bond by hindering the approach of other reagents. Even though the cyclopropene ring has so much strain, it does occur naturally in the fatty acid malvalic acid, a component of cottonseed oil. The reactive cyclopropene ring is thought to be one of the causes of abnormalities that develop in animals that ingest cottonseed oil.

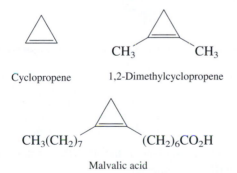

Cyclopropene   1,2-Dimethylcyclopropene

Malvalic acid

Rings of five or more carbons can accommodate a double bond without angle strain as long as that double bond is cis. However, trans double bonds are highly strained in normal-sized rings. In rings that are larger than 11 membered, the *trans*-isomer is more stable than the *cis*-isomer, as is also the case for noncyclic compounds. In rings that are 11 membered and smaller, the *cis*-isomer is more stable than the *trans*-isomer. The smallest simple *trans*-cycloalkene that has been isolated is *trans*-cyclooctene. However, even smaller examples can be generated, although their

Continued

lifetimes are very short. For example, *trans*-1-phenylcyclohexene has a lifetime of 9 × 10⁻⁶ s.

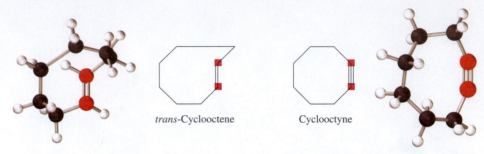

trans-Cyclooctene   Cyclooctyne

The normal bond angles for a triple bond are 180°. Incorporating four atoms (the two carbons of the triple bond and the two carbons attached to them) into a ring in a linear manner can be accomplished only in large rings. In smaller rings the triple bond will cause considerable angle strain. The smallest cycloalkyne that has been isolated is cyclooctyne, which was prepared in 1953. Again, even smaller cycloalkynes have been generated and have a transient existence. For example, 3,3,6,6-tetramethylcyclohexyne can be prepared and studied in a frozen argon matrix at 20 K, and the presence of benzyne as a transient species in a number of reactions is well accepted (see Chapter 17).

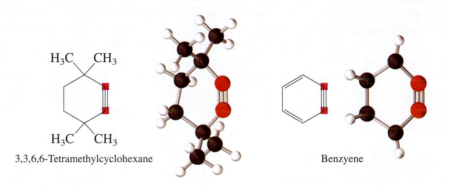

3,3,6,6-Tetramethylcyclohexane   Benzyene

methyl groups can be above the plane of the ring (the *cis*-isomer), or one can be above and one can be below (the *trans*-isomer):

cis-1,2-Dimethylcyclohexane   *trans*-1,2-Dimethylcyclohexane

These two compounds have the same bonds but a different arrangement of these bonds in space. They are stereoisomers. Like cis–trans isomers, they cannot interconvert without breaking a bond, a process that does not occur at room temperature. Note that the same type of isomers can exist with any size ring, not just six-membered ones.

PROBLEM 6.10

Draw the stereoisomers of these compounds:
a) 1,3-Dimethylcyclohexane
b) 1,2-Diethylcyclopropane
c) 1-Chloro-3-methylcyclopentane

Let's examine the conformations of each of these stereoisomers of 1,2-dimethylcyclohexane. Those of the *cis*-isomer are shown in Figure 6.17, and those of the *trans*-isomer are shown in Figure 6.18. As can be seen in these figures, *cis*-1,2-dimethylcyclohexane exists as a mixture of two conformers, both with one axial methyl group and one equatorial methyl group and with identical strain energies of 2.5 kcal/mol (10.4 kJ/mol). The *trans*-isomer exists almost entirely in the conformation with both methyls equatorial that has a strain energy of only 0.8 kcal/mol (3.3 kJ/mol). Thus, the *trans*-isomer is more stable than the *cis*-isomer by about 1.7 kcal/mol (7.1 kJ/mol), the strain energy caused by one axial methyl group.

We must be a little cautious in generalizing these results to 1,3- and 1,4-dimethylcyclohexane. First, let's examine how the axial and equatorial positions vary as we proceed around the ring.

In this representation of the cyclohexane chair conformation, if the hydrogen that is closer to the top of the page is axial on one carbon, it is equatorial on the adjacent carbon. This alternation continues around the entire ring.

In this flat representation, then, the bonds above the plane of the page also alternate between being axial and being equatorial as one proceeds around the ring. Depending on which of the two chair conformations is examined, an "up" bond on a particular carbon may be either axial or equatorial. However, if the "up" bond at a particular carbon is axial, then the "up" bonds at both adjacent carbons are equatorial and vice versa.

On the basis of this understanding, we can now analyze the situation for the other dimethylcyclohexanes. First, consider the case of *cis*-1,3-dimethylcyclohexane. The two methyl groups will both be axial in one conformation (the less stable one), and they will both be equatorial in the other, more stable conformation. The total strain energy of the diaxial conformation cannot be calculated directly from the axial destabilization energies in Table 6.2. Recall that the strain energies listed in that table are for 1,3-diaxial interactions between the groups and the two axial hydrogens on the same side of the ring. However, the destabilization caused by a 1,3-diaxial interaction between two bulky methyl groups is considerably larger. Therefore, the diaxial conformation is expected to have a total strain energy that is larger than twice the destabilization energy due to one

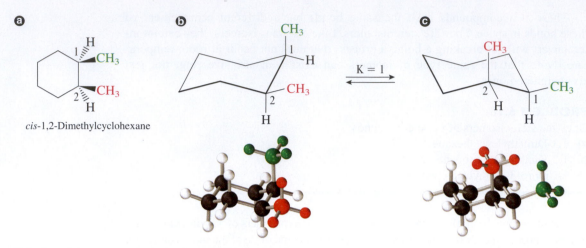

cis-1,2-Dimethylcyclohexane

ⓐ Both methyl groups are above the plane of the page.

ⓑ Note that both methyl groups are drawn closer to the top of the page than the hydrogens, because the methyls are both above the plane of the page in **ⓐ**. The methyl group on C-1 is axial, and the methyl group on C-2 is equatorial. The strain energy of this conformer is 1.7 kcal/mol (7.1 kJ/mol) due to the axial methyl group plus 0.8 kcal/mol (3.3 kJ/mol) due to a gauche interaction between the methyl groups, for a total of 2.5 kcal/mol (10.4 kJ/mol).

ⓒ The chair conformation produced by the ring-flipping process still has both methyl groups closer to the top of the page. The methyl group on C-1 is now equatorial, and the methyl group on C-2 is now axial. Both conformations have one axial methyl group and one equatorial methyl group, so they have identical strain energies. The equilibrium constant for the ring-flipping process is 1.0.

Figure 6.17

CONFORMATIONS OF *CIS*-1,2-DIMETHYLCYCLOHEXANE.

axial methyl group (greater than $2 \times 1.7 = 3.4$ kcal/mol [14.2 kJ/mol]). The other conformation, with both methyl groups equatorial, has no strain energy (there is no gauche interaction between the two methyl groups because they are not on adjacent carbons) and is much more stable than the diaxial conformation. The two conformations are interconverting rapidly, but at any one instant a vast majority (more than 99.9%) of the molecules have the conformation with both methyls equatorial because this conformation is so much more stable.

For *trans*-1,3-dimethylcyclohexane, one methyl is axial and one methyl is equatorial in either conformation. Both conformations have 1.7 kcal/mol (7.1 kJ/mol) of strain energy, and the equilibrium constant for their interconversion is 1.0. The *trans*-isomer is less stable than the *cis*-isomer by 1.7 kcal/mol (7.1 kJ/mol) because of this axial methyl group.

PRACTICE PROBLEM 6.4

Draw both chair conformations of *cis*-1,3-dimethylcyclohexane. Indicate whether each methyl group is axial or equatorial.

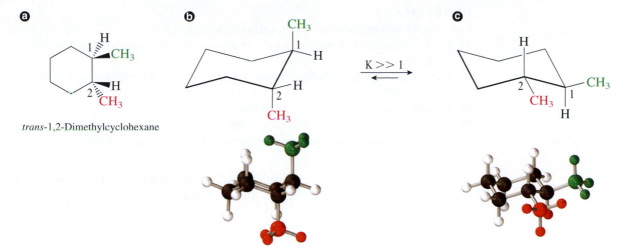

trans-1,2-Dimethylcyclohexane

ⓐ The methyl groups are on opposite sides of the plane of the ring.

ⓑ The methyl on C-1 is closer to the top of the page, and the methyl on C-2 is closer to the bottom of the page. Both methyl groups are axial in this conformation. The strain energy due to an axial methyl is 1.7 kcal/mol (7.1 kJ/mol), so the total strain energy of this conformer is twice this value, or 3.4 kcal/mol (14.2 kJ/mol).

ⓒ The conformation produced by ring-flipping has both methyl groups equatorial and is much more stable than the other. The only strain energy present here is 0.8 kcal/mol (3.3 kJ/mol) due to a gauche interaction between the two methyl groups. Therefore, this conformer is about 2.6 kcal/mol (10.9 kJ/mol) more stable than ⓑ. The equilibrium is greatly in favor of ⓒ (>99%).

Figure 6.18

CONFORMATIONS OF *TRANS*-1,2-DIMETHYLCYCLOHEXANE.

Solution

Very large 1,3-diaxial
destabilization energy

CH₃
CH₃
H

H

Both methyl groups
are axial.

H₃C ⟷ CH₃

H   H

Both methyl groups
are equatorial.

PROBLEM 6.11

Draw both chair conformations of *trans*-1,3-dimethylcyclohexane. Indicate whether each methyl group is axial or equatorial.

PROBLEM 6.12

Consider the two stereoisomers of 1,4-dimethylcyclohexane.

a) Explain whether each methyl is axial or equatorial in the conformations of the *cis*-isomer.

b) Explain whether each methyl is axial or equatorial in the conformations of the *trans*-isomer.

c) Explain which stereoisomer is more stable.

d) Draw the more stable conformation of the more stable stereoisomer.

What happens if there are two nonidentical groups on the ring? Let's consider the case of 1-methyl-4-phenylcyclohexane. The *cis*-isomer will have one group axial and one group equatorial:

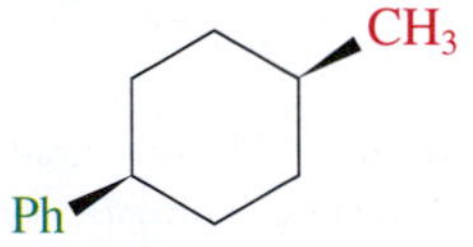

cis-1-Methyl-4-phenylcyclohexane

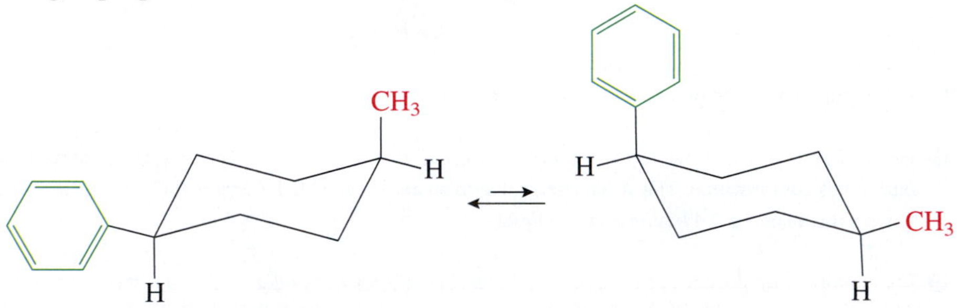

This conformation has an axial methyl and an equatorial phenyl. It has 1.7 kcal/mol (7.1 kJ/mol) of strain energy.

This conformation has an axial phenyl and an equatorial methyl. It has 2.9 kcal/mol (12.1 kJ/mol) of strain energy.

The axial destabilization energy (see Table 6.2) for the phenyl group (2.9 kcal/mol [12.1 kJ/mol]) is larger than that for the methyl group (1.7 kcal/mol [7.1 kJ/mol]), so the conformation with the phenyl equatorial will be more stable by $2.9 - 1.7 = 1.2$ kcal/mol (5.0 kJ/mol), and it will predominate at equilibrium.

The conformations of *trans*-1-methyl-4-phenylcyclohexane have both groups equatorial or both groups axial:

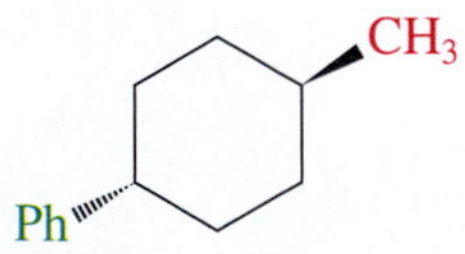

trans-1-Methyl-4-phenylcyclohexane

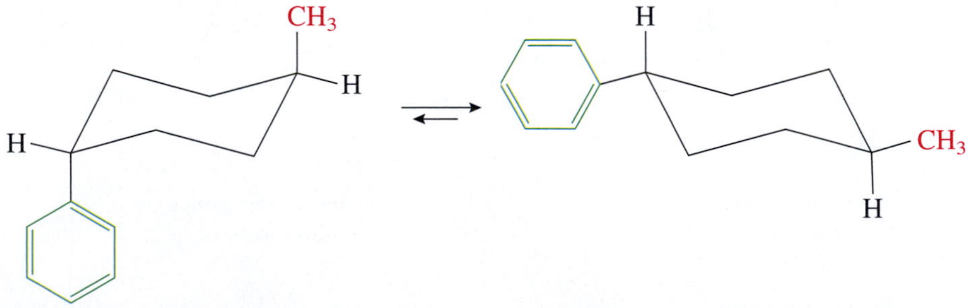

This conformation has both groups axial. It has a total strain energy of $1.7 + 2.9 = 4.6$ kcal/mol (19.2 kJ/mol).

This conformation has both groups equatorial. It has no strain energy.

The conformation with both groups equatorial is obviously much more stable than the other. Overall, the *trans*-stereoisomer is more stable than the *cis*-stereoisomer by the amount of strain due to the axial methyl group in the *cis*-isomer—that is, by 1.7 kcal/mol (7.1 kJ/mol).

Determination of the relative stability of many other substituted cyclohexane stereoisomers can be done in a similar manner. However, examples in which there are complications due to 1,3-diaxial interactions between groups or examples in which the rings are substituted with polar groups, whose dipoles interact, are much more complicated. Recently, computer programs have been developed that enable the most stable conformation of many molecules, cyclic and noncyclic, to be determined. These "molecular mechanics" calculations can provide the most stable shape of even quite complex molecules.

PROBLEM 6.13

Consider the two stereoisomers of 3-isopropylcyclohexanol.
a) Which is the more stable conformation of each stereoisomer?
b) Which is the more stable stereoisomer? By how much?

PRACTICE PROBLEM 6.5

Are the substituents cis or trans in the following conformation of one of the stereoisomers of 1-ethyl-3-methylcyclohexane? Is the ethyl group axial or equatorial? Is the methyl group axial or equatorial? Which is more stable, the conformation shown or the conformation resulting from a ring-flip? Which is more stable, the compound shown or its stereoisomer?

Solution

The methyl is closer to the top of the page than the H on C-3, and the H is closer to the top of the page than the ethyl on C-1, so the ethyl is trans to the methyl. Thus, this is one chair conformation of *trans*-1-ethyl-3-methylcyclohexane. In the conformation shown, the methyl is equatorial and the ethyl is axial. The ring-flipped conformation, with the methyl axial and the ethyl equatorial, is slightly more stable (by only 0.1 kcal/mol, from Table 6.2). *cis*-1-Ethyl-3-methylcyclohexane, the stereoisomer of the compound shown, is more stable because it has a conformation in which both the ethyl and the methyl groups are equatorial.

PROBLEM 6.14

For these compounds, indicate whether the substituents are cis or trans, whether they are axial or equatorial, whether the conformation shown or the other chair conformation is more stable, and whether the compound shown or one of its stereoisomers is more stable.

a)

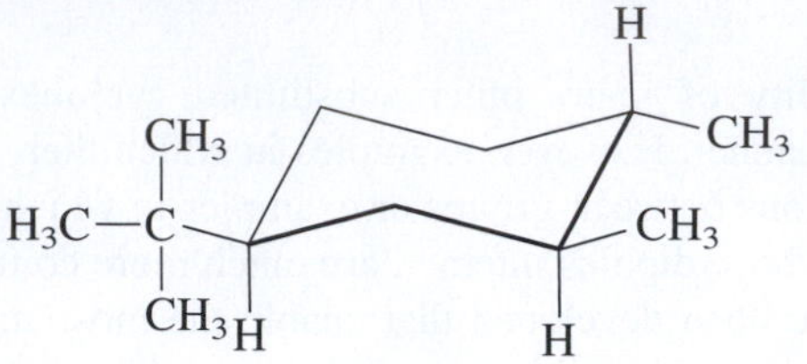

b)

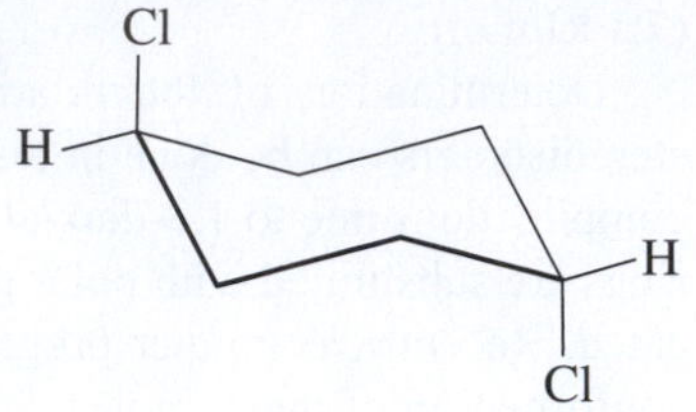

c)

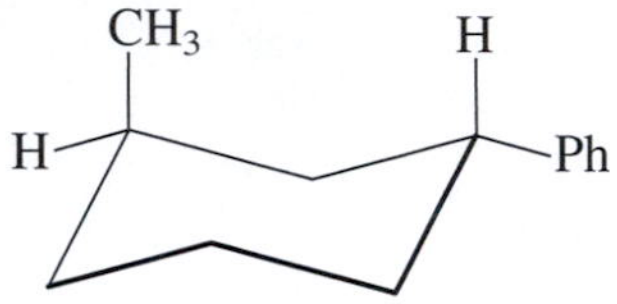

ORGANIC **Chemistry Now**™
Click *Coached Tutorial Problems* for practice with **Conformations of Cyclohexanes.**

Review of Mastery Goals

After completing this chapter, you should be able to:

ORGANIC **Chemistry Now**™
Click *Mastery Goal Quiz* to test how well you have met these goals.

- Recognize compounds that exist as cis–trans isomers and estimate the relative stabilities of these isomers. (Problems 6.15 and 6.31)
- Use the *Z* and *E* descriptors to designate the configurations of cis–trans isomers. (Problems 6.16, 6.17, and 6.27)
- Determine the conformations about a C—C single bond and estimate their relative energies. (Problems 6.18, 6.19, and 6.29)
- Determine the types and relative amounts of strain present in cyclic molecules. (Problems 6.20, 6.30, 6.32, and 6.33)
- Draw the two chair conformations of cyclohexane derivatives and determine which is more stable. (Problems 6.21, 6.22, and 6.25)
- Use analysis of conformations to determine the relative stabilities of stereoisomeric cyclohexane derivatives. (Problems 6.23, 6.24, and 6.26)

Additional Problems

ORGANIC **Chemistry Now**™
Assess your understanding of this chapter's topics with additional quizzing and conceptual-based problems at **http://now.brookscole.com/hornback2**

6.15 Draw all the cis–trans isomers for these compounds:

a) $CH_3CH{=}CHCH{=}CHCH_2CH_3$ **b)** $CH_3CH{=}CHCH{=}CHCH_3$

c) $CH_3CH{=}CHCH{=}CH_2$

6.16 Which of these groups has the higher priority?

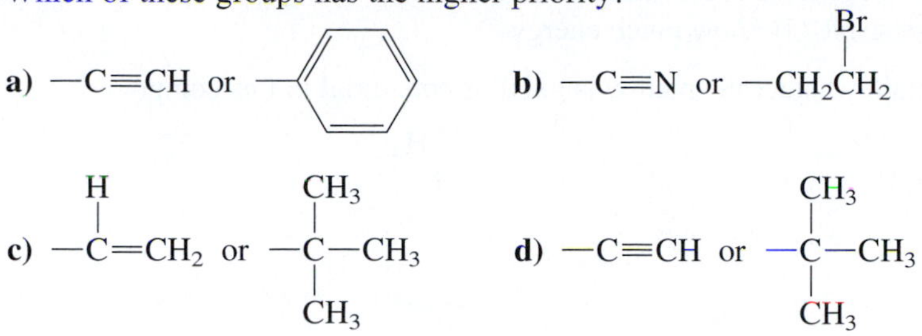

a) —C≡CH or [benzene ring]   **b)** —C≡N or —CH$_2$CH$_2$Br

c) —C=CH$_2$ (with H) or —C—CH$_3$ (with CH$_3$ above, CH$_3$ below)

d) —C≡CH or —C—CH$_3$ (with CH$_3$ above, CH$_3$ below)

6.17 Assign the configurations of these compounds as *Z* or *E*:

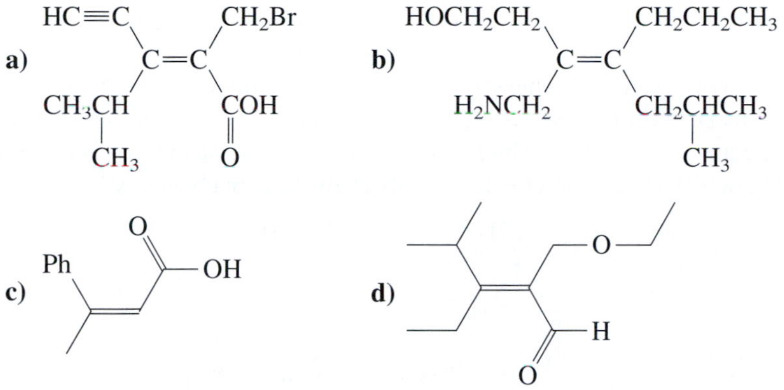

a) HC≡C and CH$_3$CH(CH$_3$) on one carbon; CH$_2$Br and COH(=O) on the other carbon of C=C

b) HOCH$_2$CH$_2$ and H$_2$NCH$_2$ on one carbon; CH$_2$CH$_2$CH$_3$ and CH$_2$CHCH$_3$(CH$_3$) on the other carbon of C=C

c) [Ph-substituted structure with OH and O]

d) [structure with isopropyl, ethyl, O—, and H, O]

6.18 How would the energy versus dihedral angle plot for 2-methylpropane (isobutane) differ from that for propane?

6.19 Draw an energy versus dihedral angle plot for the conformations of 2,3-dimethylbutane about the C-2—C-3 bond.

6.20 Discuss the geometry and the types of strain present in these compounds:
a) Cyclopropane   **b)** Cyclobutane
c) Cyclopentane   **d)** Cyclohexane
e) Cyclodecane

6.21 Explain why the axial strain energies for the —C≡N group and the —C≡CH group are much smaller than that for the CH$_3$ group.

6.22 Draw the chair conformations of 1,1,3-trimethylcyclohexane. Which conformation is more stable? Why is it not possible, on the basis of the material in this chapter, to determine the exact energy difference between these conformations?

6.23 Draw the cis and trans stereoisomers of this compound and explain their relative stabilities.

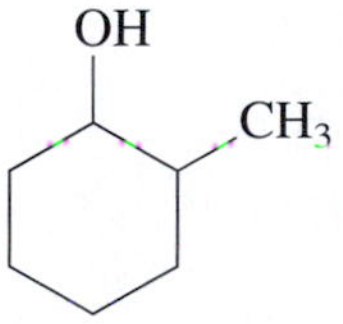

[cyclohexane structure with OH and CH$_3$ substituents]

6.24 Draw both chair conformations of 1-methyl-1-phenylcyclohexane. Which is more stable? By how much energy?

6.25 Explain whether the methyl is axial or equatorial in this compound:

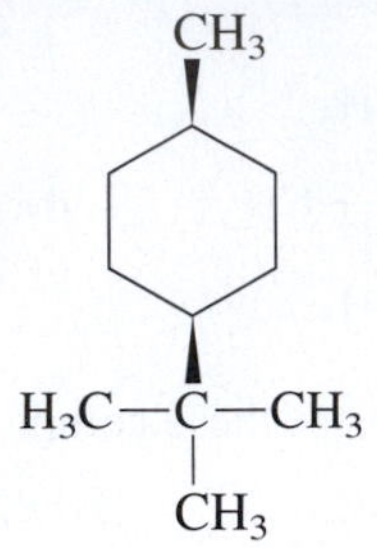

$$CH_3$$

$$H_3C-\underset{\underset{CH_3}{|}}{C}-CH_3$$

 BioLink

6.26 Draw both chair conformations for menthol (a component of peppermint oil) and its stereoisomer, neomenthol. Which groups are axial and which groups are equatorial? Explain which conformation is more stable for each stereoisomer. Which stereoisomer is more stable? By how much energy?

$$CH_3 \qquad\qquad CH_3$$

OH   OH

CHCH₃   CHCH₃

CH₃

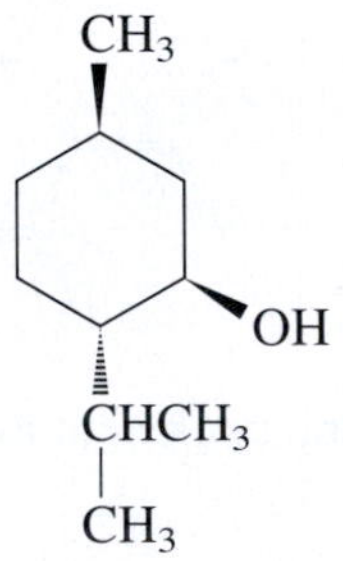

Menthol 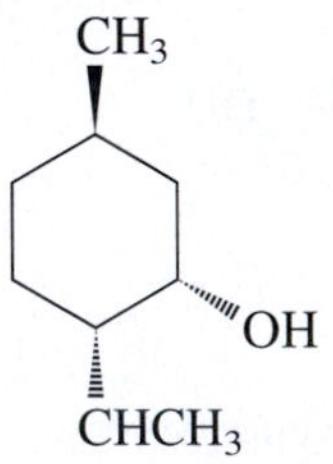   Neomenthol

6.27 Draw the structures of these compounds.
a) (*E*)-3-Hexene
b) (*Z*)-2-Chloro-3-isopropyl-2-heptene

6.28 Which of these compounds can form an intramolecular (within the same molecule) hydrogen bond between the hydrogen of the carboxylic acid and the oxygen of the ether group? Explain. (*Hint:* Construct models.)

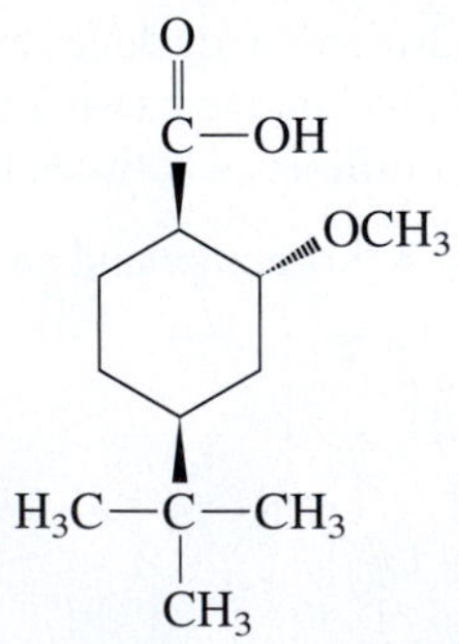

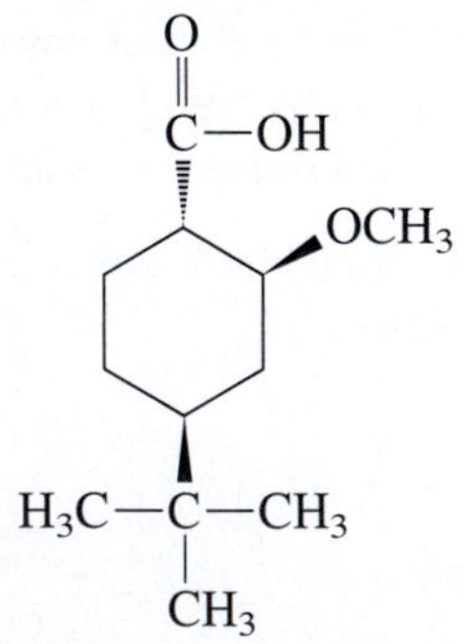

$$\underset{}{\overset{O}{\|}}$$
C—OH   C—OH

OCH₃   OCH₃

$$H_3C-\underset{\underset{CH_3}{|}}{C}-CH_3 \qquad\qquad H_3C-\underset{\underset{CH_3}{|}}{C}-CH_3$$

6.29 Draw Newman projections for the anti and gauche conformations about the C—C bond of these compounds. What other factors, besides steric and torsional strain, influence the stability of these conformations?
a) 1,2-Dichloroethane   **b)** 1,2-Ethanediol

6.30 Explain which is more stable, *cis*-1,2-dimethylcyclopropane or *trans*-1,2-dimethylcyclopropane.

6.31 Linoleic acid and α-linolenic acid are two naturally occurring unsaturated fatty acids that are components of fats. Explain how many cis–trans isomers exist for each of these fatty acids. Interestingly, only the all-cis isomers of each occur naturally. Draw the structure of the naturally occurring stereoisomer of linoleic and α-linolenic acid using skeletal structures.

$$HO\overset{\displaystyle O}{\overset{\|}{C}}(CH_2)_7CH=CHCH_2CH=CH(CH_2)_4CH_3$$

Linoleic acid

$$HO\overset{\displaystyle O}{\overset{\|}{C}}(CH_2)_7CH=CHCH_2CH=CHCH_2CH=CHCH_2CH_3$$

α-Linolenic acid

6.32 Penicillins are relatively unstable compounds and are usually stored in a refrigerator for this reason. Suggest a factor that contributes to the reactivity of penicillin G.

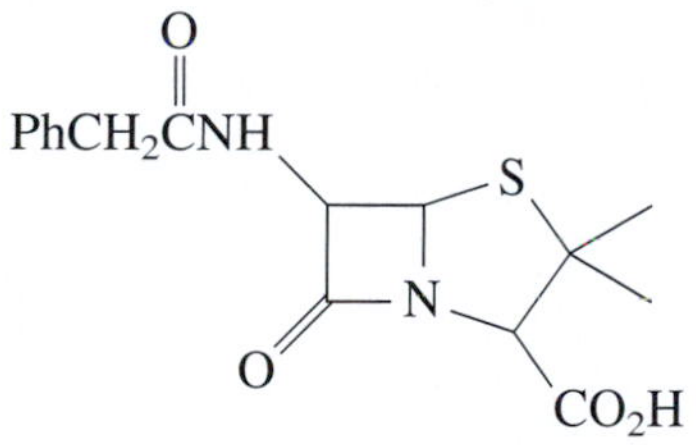

Penicillin G

6.33 Build a handheld model of bicyclo[2.2.1]heptane and discuss the types of strain that are present in this compound.

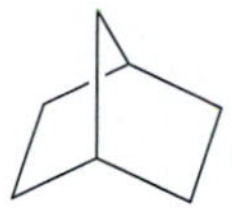

Bicyclo[2.2.1]heptane

Problems Using Online Three-Dimensional Molecular Models

6.34 Explain whether the compound shown is the *Z* or the *E* diastereomer.

6.35 Explain whether the conformation shown is the most stable conformation of each of these molecules.

6.36 Explain which stereoisomer is more stable.

6.37 Explain which isomer has more strain energy in the conformation shown.

6.38 For these compounds, indicate whether the substituents are cis or trans, whether they are axial or equatorial, whether the conformation shown or the other chair conformation is more stable, and whether the compound shown or one of its stereoisomers is more stable.

Do you need a live tutor for homework problems? Access vMentor at Organic ChemistryNow at **http://now.brookscole.com/hornback2** for one-on-one tutoring from a chemistry expert.

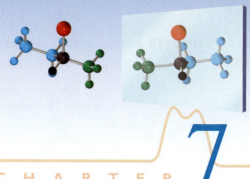

Stereochemistry II

CHIRAL MOLECULES   C H A P T E R   **7**

THIS CHAPTER introduces a new type of stereoisomer, the most subtle that we will encounter. This type of stereoisomer arises because of the tetrahedral geometry of singly bonded carbon. After this stereoisomerism is described, a discussion of how to recognize when these stereoisomers occur is presented. Next, a method to designate the configuration of these stereoisomers is described. After a discussion of when their properties differ, more complex examples are described. Finally, how they are prepared and how they are separated are considered.

Although you are probably getting better at seeing the three-dimensional shapes of molecules when viewing two-dimensional representations, it is still worthwhile to construct models to help you understand the material in this chapter. And remember to take advantage of the online computer models that are available for the molecules discussed in this chapter.

7.1   CHIRAL MOLECULES

Even some fairly simple molecules, such as 2-chlorobutane, exhibit this new type of isomerism just mentioned. The special feature of 2-chlorobutane that causes it to exhibit this type of stereoisomerism is that one of its carbons has four different groups attached to it.

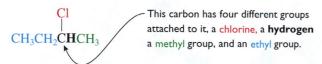

This carbon has four different groups attached to it, a chlorine, a **hydrogen** a methyl group, and an ethyl group.

To see this new type of isomer, we must carefully examine the arrangement of the four groups around this carbon. The following structure shows the arrangement of the groups about this carbon, which is, of course, sp^3 hybridized and has tetrahedral geometry. The **mirror image** of the original structure is also shown.

Mirror

Careful examination of these two structures shows that they are not identical. This can more easily be seen if the right structure is rotated 180° about the axis of the C—Cl bond:

This is the left structure.

This is the right structure, rotated 180° about the C—Cl bond.

The chlorines and the methyl groups occupy the same positions in these two structures, but the hydrogens and the ethyl groups do not. If one structure is rotated so that the ethyl groups and the hydrogens occupy the same positions, the chlorines and the methyl groups will not occupy the same positions. The two structures cannot be superimposed on one another. No amount of rotating of these structures will ever make them the same—they are **nonsuperimposable mirror images.** The only way to make these two structures identical is to interchange two groups (any two) in one of the structures. Of course, this requires bonds to be broken and does not occur at room temperature. Figure 7.1 shows three-dimensional models of these compounds. Many students have some difficulty seeing that these structures are different and cannot be superimposed by examining drawings. The use of handheld models can be invaluable in helping with this visualization.

MODEL BUILDING PROBLEM 7.1
Build a handheld model of a tetrahedral carbon with four differently colored bonds. Build the mirror image of this model. Show that the models cannot be superimposed; that is, show that you cannot line up all the bonds of the same color. Interchange two bonds on one of the models and determine whether they can be superimposed.

ORGANIC
Chemistry✦Now™
Click *Molecular Models* to view computer models of most of the molecules discussed in this chapter.

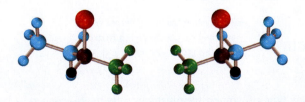

Figure 7.1

STEREOISOMERS OF 2-CHLOROBUTANE.

Molecules such as 2-chlorobutane are termed **chiral.** Chiral molecules exist as either of two stereoisomeric structures. These stereoisomers, a pair of nonsuperimposable mirror images, are called **enantiomers.**

It is easier to understand this kind of stereoisomer if we think about some everyday objects that are chiral. One example of a chiral object is a hand. (In fact, the term *chiral* is derived from the Greek word for hand, *kheir.*) A left hand and a right hand are enantiomers—they are nonsuperimposable mirror images. Other examples of everyday objects that are chiral are feet, gloves, golf clubs, screws, and handwriting. We say that all of these objects exhibit *handedness*. Objects that are **achiral** (not chiral) include socks, some types of mittens, baseball bats, screwdrivers, blank paper, and pencils.

PROBLEM 7.1

Indicate whether each of these objects is chiral or achiral:

a) Golf ball   b) Baseball glove
c) Clock   d) T-shirt
e) Dress shirt   f) Automobile

7.2 RECOGNIZING CHIRAL MOLECULES

How do we recognize when a molecule is chiral? The most certain method is to build a model of the molecule and then build a model of its mirror image. If the two models are superimposable in any conformation, the molecule is achiral; if they are not, it is chiral. However, this method is tedious and does not provide any understanding of what is the cause of a molecule being chiral. A faster, more instructive method is needed.

The feature of 2-chlorobutane that makes it chiral is the presence of a carbon attached to four different groups. Such carbons are another type of stereocenter. The currently accepted term to describe such a carbon, or any other tetrahedral atom attached to four different groups, is **chirality center.** (Some older terms that you may encounter are *chiral carbon atom* or *asymmetric carbon atom*.) Any molecule with one chirality center as its only stereocenter is chiral. (As we shall see shortly, many, but not all, molecules with multiple chirality centers are also chiral.) So, another way to identify a chiral molecule is to look for a single chirality center, which requires some practice. It helps to remember that any carbon that is attached to two identical groups (this includes all doubly and triply bonded carbons) is not a chirality center. Consider these examples:

$$\underset{1 \quad 2 \quad 3 \quad 4 \quad 5}{CH_3CH_2\overset{\overset{\displaystyle OH}{|}}{C}HCH_2CH_3}$$

This compound has no chirality center and is not chiral. Carbons 1, 2, 4, and 5 are all bonded to at least two hydrogens. Carbon 3 has one hydrogen, one hydroxy, and two ethyl groups attached to it.

$$\underset{1 \quad 2 \quad 3 \quad 4 \quad 5 \quad 6}{\underset{\underset{\displaystyle CH_3}{|}}{CH_3\overset{\overset{\displaystyle CH_3}{|}}{C}H\overset{*}{C}HCH_2CH_2CH_3}}$$

Carbons 1, 4, 5, and 6 are all bonded to two or more hydrogens, so they are not chirality centers. Carbon 2 is bonded to two methyl groups, so it is not a chirality center either. However, carbon 3 is bonded to four different groups (hydrogen, methyl, propyl, isopropyl) and is a chirality center. Because the molecule has one chirality center, it is chiral. Chirality centers are sometimes marked with an asterisk (∗).

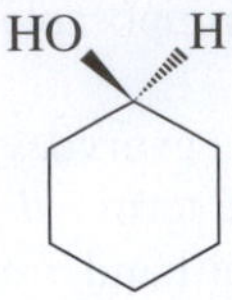

Ring compounds are a little trickier. In this case, all of the carbons except one are bonded to two hydrogens and are not chirality centers. What about the carbon bonded to the hydrogen and the hydroxy group? The other two "groups" that are bonded to this carbon are the carbons of the ring. These groups are identical proceeding around the ring in either direction, so the carbon is not a chirality center and the molecule is not chiral. Note that it is not necessary to consider conformations in this analysis of chirality.

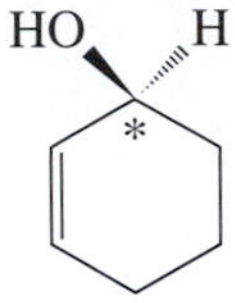

In contrast to the preceding example, the carbon bonded to the hydrogen and the hydroxy group is a chirality center in this molecule. In proceeding around the ring in one direction, the first carbon encountered is part of a double bond; in the other direction, it is not. In ring systems, one must step around the ring in both directions, comparing one carbon at a time. If a difference is found at any point, then the "groups" are different.

PROBLEM 7.2

Determine whether each of these molecules is chiral. For those that are chiral, put an asterisk at the chirality center.

a) 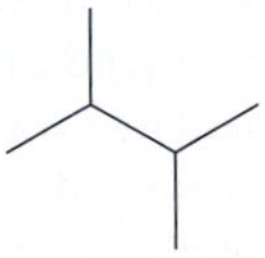   b) 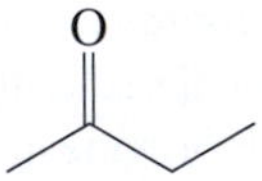   c)

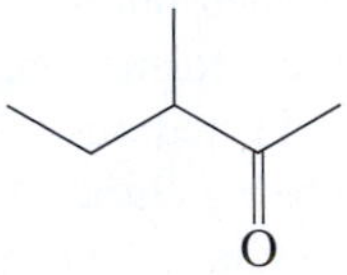

d)

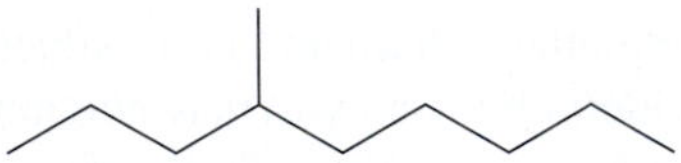

e) 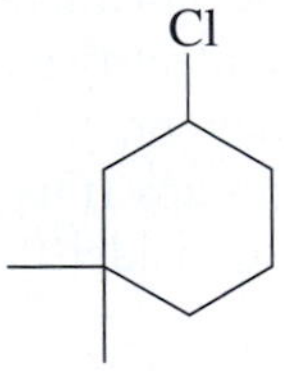   f)

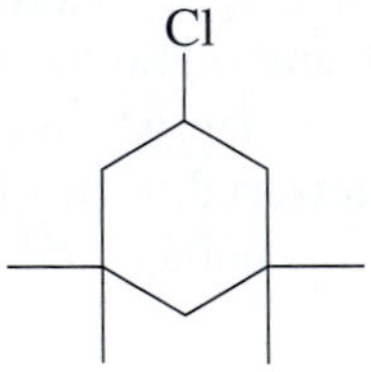

A final way to determine whether a molecule is chiral is to examine its symmetry. Although an object can have several different kinds of symmetry, the only kind that will be of concern to us is called a **plane of symmetry** or a **mirror plane.** An object has a plane of symmetry if an imaginary plane that passes through the center of the object divides the object so that one half is the mirror image of the other half. Any molecule that has a plane of symmetry is not chiral. In most cases (but not all), if no plane of symmetry is present, the molecule is chiral. Some examples are shown in Figure 7.2.

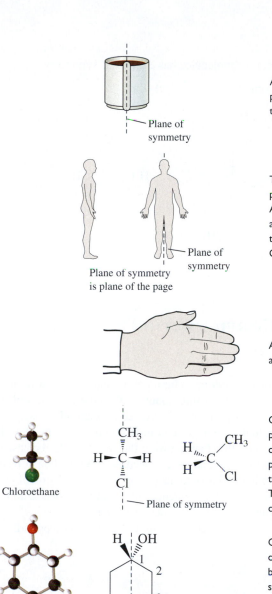

Plane of
symmetry

A cup has a plane of symmetry that
passes vertically through it and bisects
the handle. It is not chiral.

Plane of symmetry
is plane of the page

Plane of
symmetry

Two views of the plane of symmetry
present in an idealized human figure.
Although the figure has chiral parts (such
as hands), they are arranged in pairs so
that one chiral part mirrors the other.
Overall, a human is not chiral.

A hand has no plane of symmetry
and is chiral.

Chloroethane

CH_3

$H \blacktriangleright C \blacktriangleleft H$

Cl

Plane of symmetry

$H_{\text{////}} \overset{CH_3}{\underset{Cl}{C}}$
H

Chloroethane has a plane of symmetry that
passes through the two carbons and the
chlorine. The plane is perpendicular to the
page in the model and left drawing and is
the plane of the page in the right drawing.
The hydrogens mirror each other. This
compound is not chiral.

Cyclohexanol

H | OH

1
2
3
4

Plane of
symmetry

Cyclohexanol is achiral because it has no
chirality center. The same conclusion can
be reached by noting that it has a plane of
symmetry that is perpendicular to the page
and that passes through carbon 1, the
hydrogen and hydroxy group attached to it,
and carbon 4. One side of the ring mirrors
the other side.

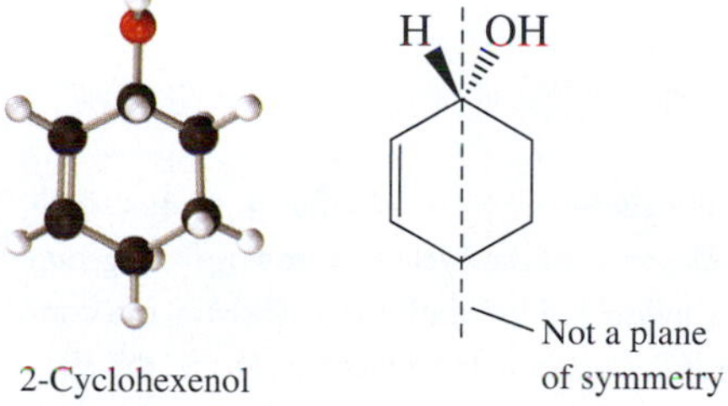

2-Cyclohexenol

H | OH

Not a plane
of symmetry

In 2-cyclohexenol, one side of the ring does
not mirror the other, so it has no plane of
symmetry. It is a chiral molecule.

PROBLEM 7.3

Indicate whether each of these objects or molecules has a plane of symmetry:

a) Idealized human face   **b)** Pencil   **c)** Ear

d) 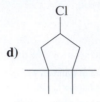   **e)** 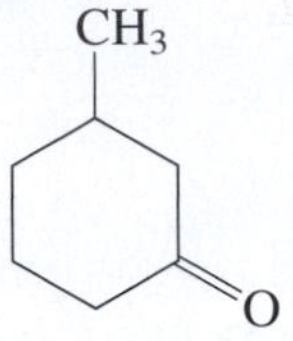   **f)**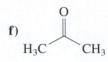

g) 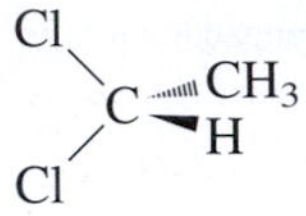 $\underset{Cl}{\overset{Cl}{\diagdown}} C \underset{H}{\overset{CH_3}{\diagup}}$ **h)** $\underset{Cl}{\overset{Br}{\diagdown}} C \underset{H}{\overset{CH_3}{\diagup}}$

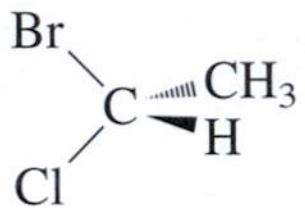

7.3   DESIGNATING CONFIGURATION
OF ENANTIOMERS

Suppose we are working with one enantiomer of a chiral compound. How can we indicate which enantiomer is being used when writing about it? Or how can we look up the properties of this enantiomer in a reference book? We need a method to designate the configuration of the enantiomer, to denote the three-dimensional arrangement of the four groups around the chirality center, other than drawing the structure. In the case of many everyday chiral objects, the terms *right* and *left* are used, as in a left shoe or right-handed golf clubs. In Section 6.2 we learned how to designate the configuration of geometrical isomers using the Cahn-Ingold-Prelog sequence rules and the labels *Z* and *E*. The method to designate the configuration of enantiomers uses these same rules.

The following steps are used to assign the configuration of a chiral compound:

STEP 1

Assign priorities from 1 through 4 to the four groups bonded to the chirality center using the Cahn-Ingold-Prelog sequence rules presented in Section 6.2. The group with the highest priority receives number 1, and the lowest-priority group receives number 4.

STEP 2

View the molecule so that the bond from the chirality center to group number 4 is pointed directly away from you. Now determine whether the direction of the cycle proceeding from group 1 to 2 to 3 and back to 1 is clockwise or counterclockwise. If this rotation is clockwise, the configuration is *R* (from the Latin word for right, *rectus*). If the rotation is counterclockwise, the configuration is *S* (from the Latin word for left, *sinister*).

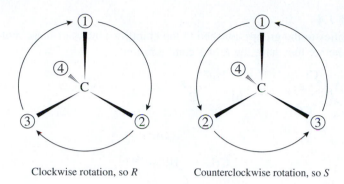

Clockwise rotation, so *R*   Counterclockwise rotation, so *S*

Figure 7.3 gives some examples of how to designate configuration using these rules.

MODEL BUILDING PROBLEM 7.2

Build a model of 2-cyclohexenol with the stereochemistry shown in Figure 7.3 and confirm that it has the *R* configuration.

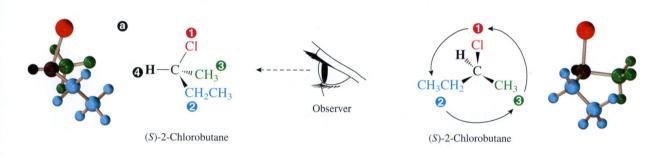

(*S*)-2-Chlorobutane   Observer   (*S*)-2-Chlorobutane

The group priorities are chlorine = 1, ethyl = 2, methyl = 3, and **hydrogen = 4**. Viewing from the right side, so that the bond from the carbon to the hydrogen is pointed away from the observer, 1 → 2 → 3 → 1 is counterclockwise, so the configuration is S. This is easier to see if the structure is rotated.

In this view, the structure has been rotated so that the bond from the carbon to the hydrogen is pointing directly away from you as you view the page. Now it is easier to see that the direction of 1 → 2 → 3 → 1 is counterclockwise.

(*R*)-2-Cyclohexenol

The chirality center is the ring carbon bonded to the OH. The highest-priority group is the OH, and the lowest-priority group is the H. Because the ring carbon on the right of the chirality center is doubly bonded to another carbon, it has higher priority than the other carbon. The lowest-priority group, the hydrogen, is almost pointed directly away from you so the clockwise direction of rotation of 1 → 2 → 3 → 1 can be easily seen.

Active Figure 7.3

ORGANIC
Chemistry ⚛ Now™

DESIGNATING CONFIGURATIONS. ⓐ (*S*)-2-CHLOROBUTANE AND ⓑ (*R*)-2-CYCLOHEXENOL. Test yourself on the concepts in this figure at **OrganicChemistryNow.**

PROBLEM 7.4

Assign priorities to the groups attached to the chirality centers of these molecules and determine whether they have the *R* or *S* configuration:

a)

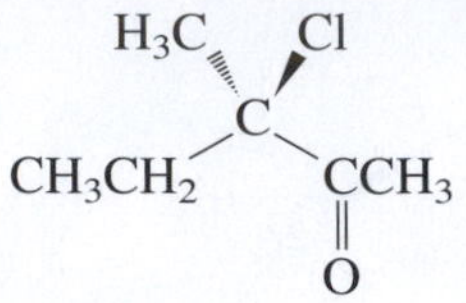

b)

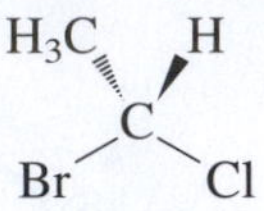

c)

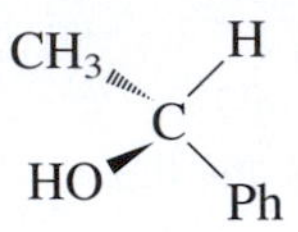

d)

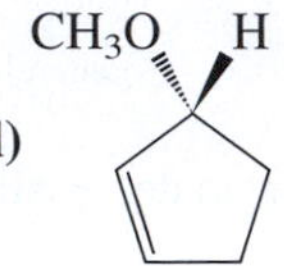

e)

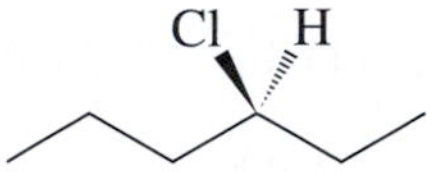

f)

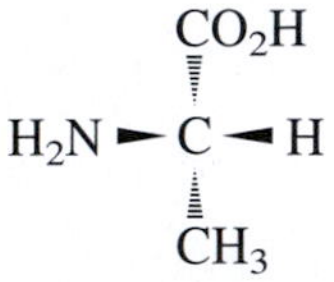

Alanine
(an amino acid)

PRACTICE PROBLEM 7.1

Draw the structure of (*S*)-2-butanol.

Strategy

Draw the structure without stereochemistry, identify the chirality center, and assign priorities to the four groups attached to the chirality center. Then draw a tetrahedral carbon and put the group with the lowest priority on a bond pointed away from you. Put the highest-priority group on any bond. Then put group 2 in the position clockwise from group 1 if it is the (*R*)-enantiomer or in the position counterclockwise from group 1 if it is the (*S*)-enantiomer. Add group 3 to the remaining position.

Solution

The priorities are OH = 1; CH_2CH_3 = 2; CH_3 = 3 ; and H = 4. Put the H on the bond pointed away from you. Put the OH on any bond. Because the (*S*)-enantiomer is desired, put the ethyl group on the bond counterclockwise from the OH group.

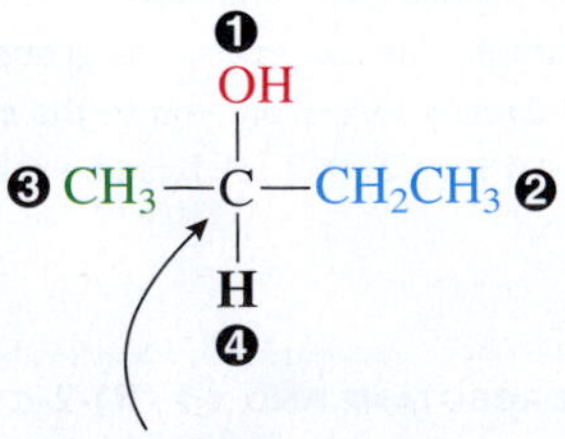

Chirality center

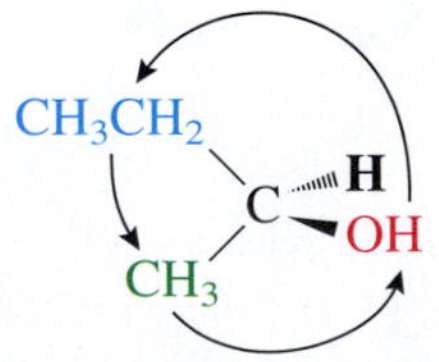

Counterclockwise, so *S*

PROBLEM 7.5

Draw structures for these compounds:

a) (*R*)-3-Ethylcyclohexene

b) (*R*)-2-Bromoheptane

ORGANIC **Chemistry Now**™ Click *Coached Tutorial Problems* to practice using the **Cahn-Ingold-Prelog Sequence Rules.**

The actual three-dimensional arrangement of groups around a chirality center is called the **absolute configuration.** Until a special X-ray technique was developed in 1951, it was impossible to determine the absolute configuration of any compound. Although samples of one enantiomer (or both) of a multitude of compounds were available, no experimental method existed to determine whether that enantiomer had the *R* or the *S* absolute configuration. This was not a major problem for organic chemists, though, because they were able to convert one chiral molecule to another, using reactions whose stereochemical effects were well known. Thus, it was possible to relate the configuration of one compound to that of another. The **relative configurations** of the compounds were known. For example, if one enantiomer of 2-butanol is converted to 2-chlorobutane using a reaction that is known to put the chlorine exactly where the hydroxy group was, then the two compounds have the same relative configuration. If, as shown here, the starting material is (*R*)-2-butanol, then the product is (*R*)-2-chlorobutane:

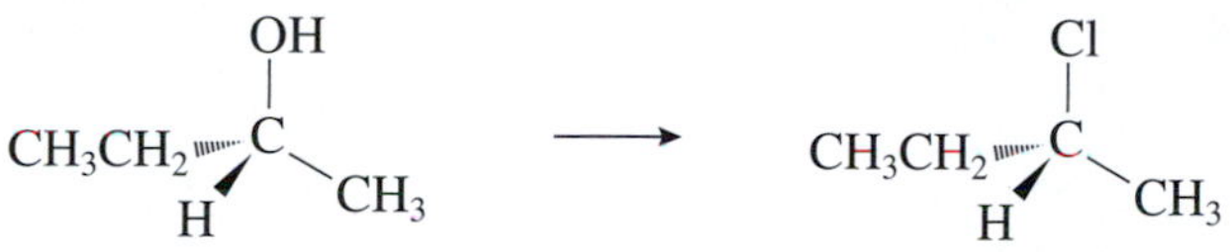

(*R*)-2-Butanol (*R*)-2-Chlorobutane

If the absolute configuration of the starting 2-butanol enantiomer is not known, then the absolute configuration of the product 2-chlorobutane is not known either. However, because the reaction is known to put the chlorine exactly where the hydroxy group was, the two compounds must have the same relative configuration. Often, knowing the relative configurations of the compounds is enough to answer the chemical question under consideration. Of course, once the absolute configuration of one compound has been determined, the absolute configuration of any other compound whose configuration has been related to the first is also known.

7.4 Properties of Enantiomers

When do enantiomers have different properties? Again, it is helpful to draw analogies with everyday objects that are chiral. When do your hands have different properties? They are different when you put on a glove; they are different when you write; they are different when you shake hands. What do these objects or activities have in common? A glove, writing, and shaking hands are all chiral! Hands are different when they interact with one enantiomer of a chiral object or activity. Likewise, enantiomeric molecules are different when they are in a chiral environment. Most commonly, the chiral environment is the presence of one enantiomer of another chiral compound. Otherwise, their properties are identical. For example, the naturally occurring ketones (*R*)- and (*S*)-

carvone have identical melting points, boiling points, solubilities in ethanol, and heats of combustion. However, they have different solubilities in one enantiomer of a chiral solvent, and they have different odors (the odor receptors in the nose are chiral). (*R*)-Carvone smells like caraway and (*S*)-carvone smells like spearmint. An important difference is that *enantiomers have different rates of reaction with one enantiomer of a chiral reagent.*

 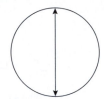

(*S*)-Carvone   (*R*)-Carvone

The most common method used to detect the presence of chiral molecules in a sample employs the interaction of plane-polarized light with the sample.

Regular light waves consist of electromagnetic fields that oscillate in all directions perpendicular to the direction of travel of the wave. If we could see these fields while viewing the light beam coming directly at us, the oscillations would occur along the arrows.

When a regular light beam is passed through a polarizer (polarized sunglasses will work), all of the light waves, except those whose electromagnetic fields oscillate in a single direction, are filtered out. The result is a beam of **plane-polarized light.** The oscillations all occur in a single plane as shown.

When plane-polarized light is passed through a sample containing one enantiomer of a chiral compound, the plane of polarization of the light is rotated. (Samples that rotate plane-polarized light are said to be **optically active.**) A schematic diagram of a simple instrument, called a polarimeter, that can detect this rotation is shown in Figure 7.4. In this instrument the organic compound to be analyzed is placed in the sample tube, either as a pure liquid or as solution in an achiral solvent. When the plane-polarized light passes through the sample, the plane of polarization is rotated. The magnitude of the observed rotation, α, in degrees, is measured by the analyzing polarizer. If the beam has been rotated in a clockwise direction, α is assigned a positive value; if the beam has been rotated in a counterclockwise direction, α is assigned a negative value.

For a particular compound the observed rotation depends on the concentration of the compound, the path length of the sample tube, and the wavelength of the light that is used. Often the yellow light produced by a sodium lamp, called the sodium D line (wavelength = 589 nm), is used. The specific rotation, a constant characteristic of each

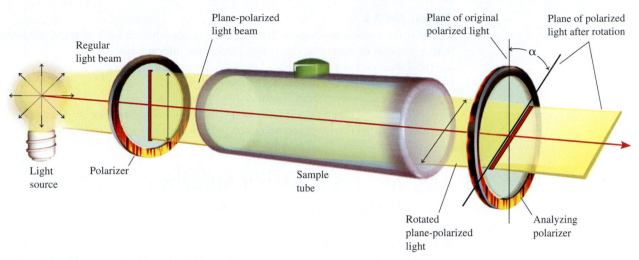

Plane-polarized light beam

Regular light beam

Plane of original polarized light

Plane of polarized light after rotation

α

Light source

Polarizer

Sample tube

Rotated plane-polarized light

Analyzing polarizer

Figure 7.4

SCHEMATIC DIAGRAM OF A POLARIMETER.

chiral compound, can be calculated from the observed rotation obtained in the laboratory by the following equation:

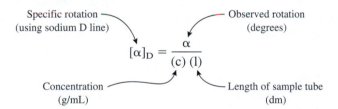

Specific rotation (using sodium D line)

Observed rotation (degrees)

$$[\alpha]_D = \frac{\alpha}{(c)\,(l)}$$

Concentration (g/mL)

Length of sample tube (dm)

Enantiomers rotate plane-polarized light by identical magnitudes but in opposite directions. The enantiomer that rotates the light clockwise is called **dextrorotatory,** (*d*) or (+), while the one that rotates light counterclockwise is called **levorotatory,** (*l*) or (−). A common way to encounter a chiral compound is as an equal mixture of enantiomers, called a **racemic mixture** or a **racemate.** A racemic mixture, often designated as *d,l* or (±), does not rotate plane-polarized light because the rotation due to one enantiomer is canceled by that of the other. Of course, mixtures that have one enantiomer in excess of the other and are neither enantiomerically pure nor completely racemic, may also be encountered.

Methods are available that enable the absolute configuration of some compounds to be predicted from the direction of their rotations, but the process is quite complex. Therefore, the observation that a compound rotates plane-polarized light will indicate to us only that it is chiral and that one enantiomer is present in excess of the other. If the sample contains only one enantiomer, then the specific rotation can be determined. The specific rotation is a constant that can be used to help identify the compound in the same manner as its melting point or boiling point. For example, the specific rotation of sucrose (table sugar) is $[\alpha]_D = +66.37$, and that of (−)-2-butanol is $[\alpha]_D = -13.9$. There is no way for us to tell that the (*l*)-enantiomer of 2-butanol actually has the *R* absolute configuration. In general, there is no relationship between the absolute configuration (*R* or *S*) and the direction of rotation of plane-polarized light (+ or −).

PROBLEM 7.6

Consider the two enantiomers of 2-pentanol. Explain whether each of these statements is true, is false, or cannot be determined from this information.

a) (R)-2-Pentanol is a stronger acid than (S)-2-pentanol.
b) The two enantiomers have different boiling points.
c) The two enantiomers have identical solubilities in water.
d) (S)-2-Pentanol rotates plane-polarized light in the counterclockwise direction.
e) (d)-2-Pentanol rotates plane-polarized light in the clockwise direction.

7.5   MOLECULES WITH MULTIPLE CHIRALITY CENTERS

When a molecule has more than one chirality center, things become somewhat more complicated. Let's consider the simplest possibility: a molecule with two chirality centers. Each chirality center can have either the R or the S absolute configuration. There are four possible combinations: RR, SS, RS, and SR. The mirror image of a chiral molecule, that is, its enantiomer, has the opposite configuration (inverted configuration) at all chirality centers. Therefore, the RR and the SS stereoisomers are enantiomers, as are the RS and SR stereoisomers. The RR and the RS stereoisomers are not mirror images. Such non–mirror-image stereoisomers are called **diastereomers.** They have the opposite configuration at some, but not all, chirality centers. (Note that the cis–trans isomers we saw in Chapter 6 are another type of diastereomers.)

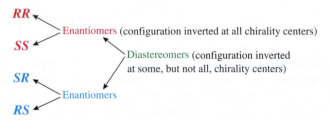

As we saw previously, enantiomers have identical properties unless they are placed in a chiral environment. Diastereomers, on the other hand, are not mirror images and have different properties in all environments. They have different physical properties and different chemical properties.

A similar analysis holds for molecules that have more than two chirality centers. Each chirality center may have either the R or the S configuration. The number of possible stereoisomers can be calculated by using simple probability theory. A molecule with a number of chirality centers equal to n has a maximum of 2^n stereoisomers:

$$\text{maximum number of stereoisomers} = 2^n \quad (n = \text{number of chirality centers})$$

As an example, consider the case of 2-(methylamino)-1-phenyl-1-propanol:

$$\begin{array}{ccc} \text{OH} & \text{NHCH}_3 & \\ | & | & \\ \text{Ph—CH—CH—CH}_3 \\ 1\,* & 2\,* & 3 \end{array}$$

Carbon 1 and carbon 2 are both chirality centers, so there are $2^2 = 4$ stereoisomers. The $(1R,2S)$-stereoisomer has a specific rotation of -6.3 and is known as $(-)$-ephedrine. It is a bronchodilator and is the decongestant used in many cold remedies. It is also the major active ingredient in the herbal supplement ephedra or *ma-huang*. The U.S. Food and Drug Administration has banned the sale of dietary supplements containing ephedra in 2004 because of adverse health effects, including elevated blood pressure and strokes. The enantiomer of $(-)$-ephedrine, the $(1S,2R)$-stereoisomer, does not occur naturally. It has the same melting point as $(-)$-ephedrine but different physiological properties, and it rotates plane-polarized light in the positive direction. The racemic mixture, (d,l)-ephedrine, packs better into a crystal lattice and has a higher melting point (76°C) than either enantiomer.

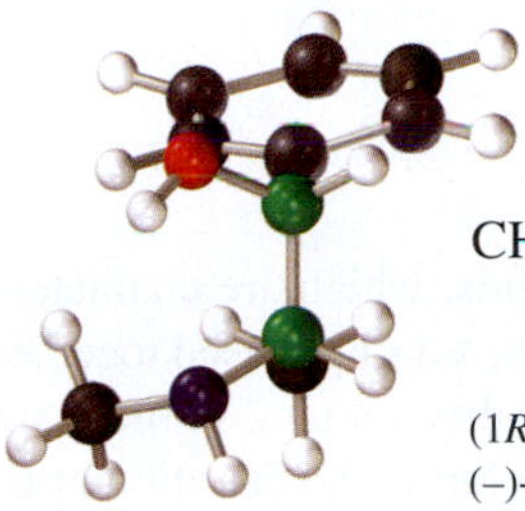

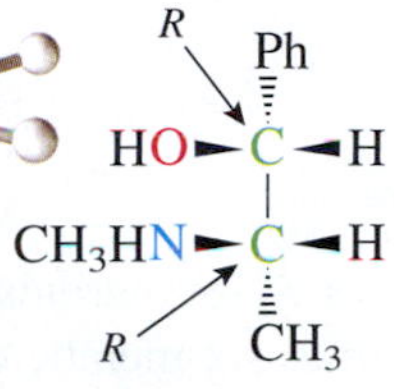

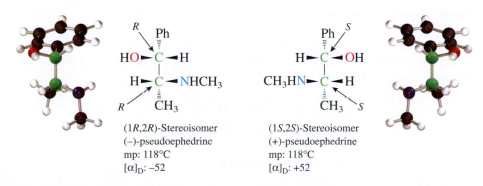

$(1R,2S)$-Stereoisomer
$(-)$-ephedrine
mp: 40°C
$[\alpha]_D$: -6.3

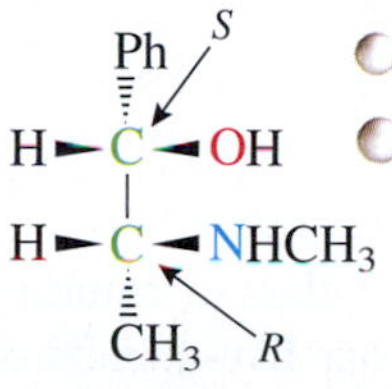

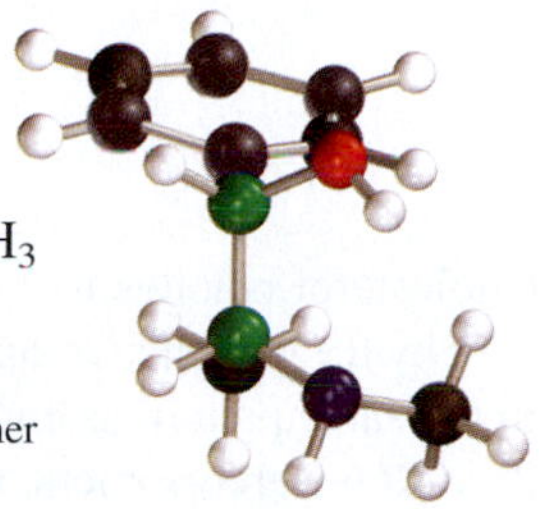

$(1S,2R)$-Stereoisomer
$(+)$-ephedrine
mp: 40°C
$[\alpha]_D$: $+6.3$

One way to produce a drawing of the enantiomer of $(-)$-ephedrine is to imagine a mirror placed just to the right of the $(-)$-ephedrine structure and perpendicular to the page. When the mirror image of the original structure is drawn, the structure on the right, $(+)$-ephedrine, is produced. Another way to construct the enantiomer is to interchange two groups on each chirality center. (Recall that interchanging any two groups at a chirality center inverts the configuration at that chirality center.) Thus, if the OH and H on carbon 1 of $(-)$-ephedrine are interchanged and the CH_3NH and H on carbon 2 are interchanged, then the diagram of the enantiomeric $(+)$-ephedrine on the right is again produced.

To draw a diastereomer of ephedrine, two groups on one chirality center are interchanged, but the configuration at the other chirality center is not changed. For example, if the H and $NHCH_3$ of the preceding diagram of $(-)$-ephedrine are interchanged, the resulting stereoisomer is the $(1R,2R)$-stereoisomer. It is known as $(-)$-pseudoephedrine, which is used as a nasal decongestant. The two enantiomers of pseudoephedrine have very different physical constants from ephedrine. Their chemical and physiological properties are different also.

$(1R,2R)$-Stereoisomer
$(-)$-pseudoephedrine
mp: 118°C
$[\alpha]_D$: -52

$(1S,2S)$-Stereoisomer
$(+)$-pseudoephedrine
mp: 118°C
$[\alpha]_D$: $+52$

PROBLEM 7.7

Draw all the stereoisomers of 2-bromo-3-chlorobutane and indicate whether they are enantiomers or diastereomers.

As another example, consider the compound cholesterol:

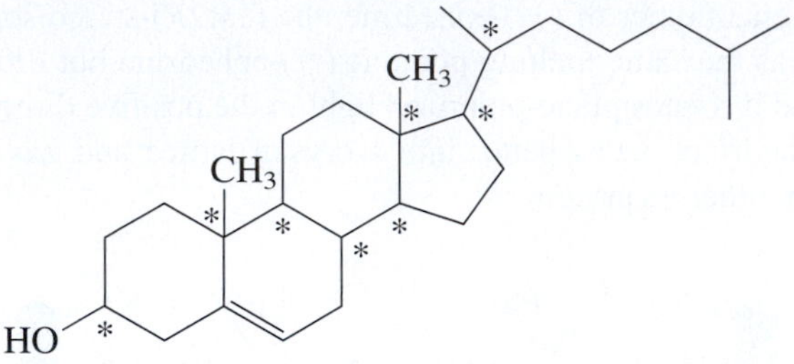

Cholesterol

Cholesterol belongs to a class of natural products called steroids, which are characterized by the presence of one five-membered and three six-membered rings fused together in the same pattern as cholesterol. This structure has eight chirality centers, so there are $2^8 = 256$ stereoisomers. Only one of these is cholesterol. Another is the enantiomer of cholesterol, and the other 254 are diastereomers of cholesterol. Of the 256 possibilities, nature produces only one: cholesterol.

PROBLEM 7.8

Label each chirality center in these compounds with an asterisk and calculate the maximum number of stereoisomers for each:

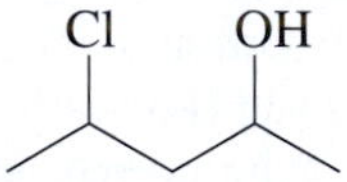

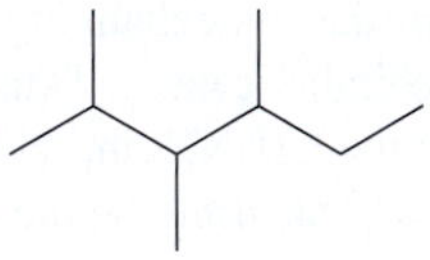

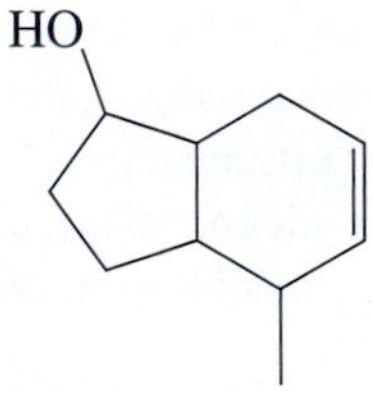

Sometimes there are fewer stereoisomers than predicted by the preceding rule. This occurs when identical chirality centers are symmetrically placed in a compound. As an example, consider the case of tartaric acid.

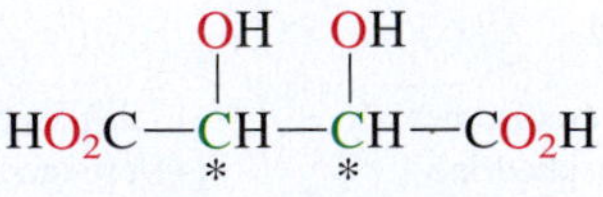

Tartaric acid

There are two chirality centers, so the formula predicts a total of four stereoisomers. However, each of the chirality centers has identical groups attached to the carbon, so fewer than four stereoisomers actually exist. The analysis can be conducted in the same manner as was done previously for ephedrine. We start by drawing one of the stereoisomers, the (2*R*,3*R*)-isomer for example. Then the mirror image of this, the (2*S*,3*S*)-stereoisomer is drawn. These two compounds are nonsuperimposable mirror images—enantiomers.

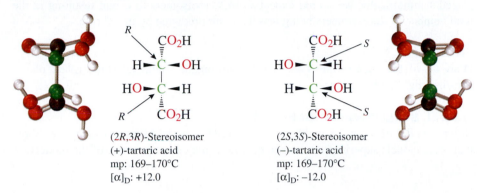

(2*R*,3*R*)-Stereoisomer
(+)-tartaric acid
mp: 169–170°C
[α]$_D$: +12.0

(2*S*,3*S*)-Stereoisomer
(−)-tartaric acid
mp: 169–170°C
[α]$_D$: −12.0

(+)-Tartaric acid occurs naturally in fruits and plants. Its monopotassium salt is called cream of tartar and is a component of baking powder. (−)-Tartaric acid is much less common in nature and has been found only in the fruit of a single West African tree.

A diastereomer of these compounds is constructed by interchanging the H and OH on one of the chirality centers, as shown. The mirror image of this compound is also shown.

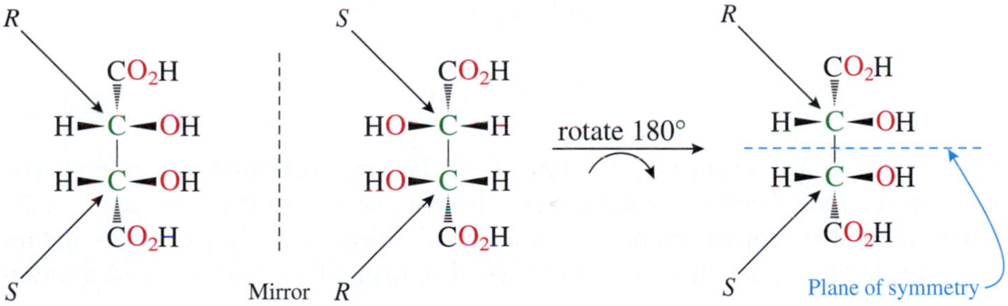

(2*R*,3*S*)-Stereoisomer
meso-tartaric acid
mp: 159–160°C

(2*S*,3*R*)-Stereoisomer
meso-tartaric acid
mp: 159–160°C

Careful examination of these mirror images shows that they are, in fact, identical. If the (2*S*,3*R*)-stereoisomer is rotated 180° in the plane of the paper, the structure on the right is produced and it is identical to the (2*R*,3*S*)-stereoisomer on the left. Because the compound is superimposable on its mirror image, it is not chiral and does not rotate plane-polarized light. Another way to determine that this compound is not chiral is to note that it has a plane of symmetry that bisects the C-2—C-3 bond. Compounds such as this one, which contain chirality centers but are not chiral are called ***meso*-stereoisomers.** *meso*-Tartaric acid is human-made and does not occur in nature. Overall, then, tartaric acid has only three stereoisomers: the two enantiomers of the chiral diastereomer (often called the *d,l*-diastereomer) and the *meso*-diastereomer.

Because of its symmetry, numbering can begin at either end of tartaric acid. It is not surprising, then, that (2R,3S)-tartaric acid is the same as (2S,3R)-tartaric acid. One of the chirality centers is the mirror image of the other. That is why *meso*-tartaric acid has an internal plane of symmetry. As an everyday example, consider an idealized human figure. Although the figure has chiral parts such as hands and feet, they are present in pairs of left and right enantiomers arranged so that the figure has an internal plane of symmetry. Whenever we encounter a compound that has identical chirality centers, placed symmetrically, we should expect *meso*-stereoisomers to occur, resulting in the total number of stereoisomers being less than that predicted by the 2^n rule.

PROBLEM 7.9

Draw all of the stereoisomers for 2,3-dichlorobutane. Indicate which rotate plane-polarized light and which are meso.

MODEL BUILDING PROBLEM 7.3

Build a model of *meso*-2,3-dichlorobutane. Build a model of its mirror image. Show that these models superimpose. Locate the plane of symmetry in one of the models.

7.6   STEREOISOMERS AND CYCLIC COMPOUNDS

The *cis*- and *trans*-stereoisomers of cyclic compounds that were presented previously are actually just special cases of the type of stereoisomers that we have just discussed. For example, consider the case of 1,2-dimethylcyclohexane:

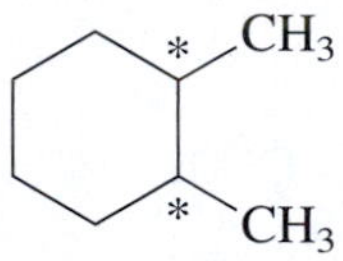

There are two chirality centers, so the formula predicts that there are four stereoisomers. However, because the chirality centers are identical, we expect that there are actually fewer than four stereoisomers. In fact, the analysis is identical to the one we just did for tartaric acid. There are three stereoisomers: a *d,l*-pair of enantiomers and a *meso*-diastereomer.

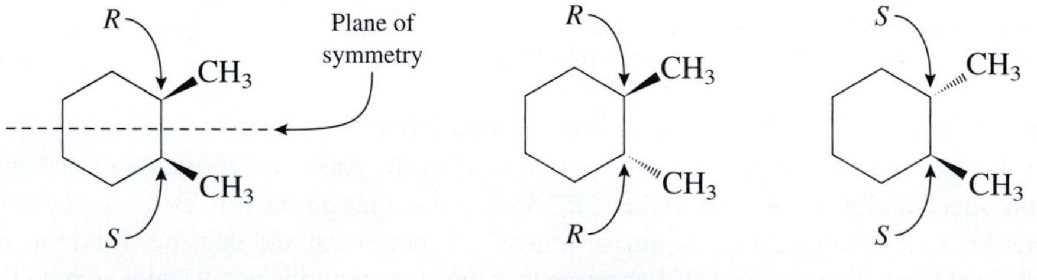

meso-Stereoisomer
(1R,2S) or (1S,2R)
cis-diastereomer

(1R,2R)-Stereoisomer   (1S,2S)-Stereoisomer

Enantiomers of the *trans*-diastereomer

The *cis*-diastereomer is meso. It has a plane of symmetry bisecting the ring bond between the two methyl groups. It is not chiral and does not rotate plane-polarized light.

The *trans*-diastereomer exists as a pair of enantiomers. These stereochemical differences do not depend on the presence of the ring. In fact, suppose the ring is cleaved at the C—C bond opposite the one connecting the chirality centers. The resulting compound, 3,4-dimethylhexane, also has three stereoisomers: a *meso*-diastereomer and two enantiomers of a *d,l*-diastereomer. It is just easier to see that the *cis*- and *trans*-diastereomers of 1,2-dimethylcyclohexane are different than it is to see that the *meso*- and *d,l*-diastereomers of 3,4-dimethylhexane are different.

meso-3,4-Dimethylhexane   (3R,4R)-Dimethylhexane   (3S,4S)-Dimethylhexane

Although cyclohexane rings have chair shapes, rather than being flat, stereochemical analyses such as the preceding one can be done with drawings using planar rings. (This applies to most other rings also.) This is true because the chair conformations are interconverting rapidly and the "average" shape can be considered planar. Careful analysis of one chair conformation of *cis*-1,2-dimethylcyclohexane shows that it is chiral—it is not superimposable on its mirror image. However, the ring-flipped conformation is the enantiomer of the original conformation. Because the conformers interconvert rapidly at room temperature, the compound is not chiral. Again, this result is not unique to ring systems. Some of the conformations of *meso*-3,4-dimethylhexane are also chiral, but there is always an enantiomeric conformation, and interconversion between them is rapid. In looking for internal planes of symmetry in such molecules, it is necessary to use the most symmetrical conformation.

PROBLEM 7.10
Draw all of the stereoisomers of 1,2-dimethylcyclopropane. Explain which rotate plane-polarized light.

PRACTICE PROBLEM 7.2

Explain whether this compound would rotate plane-polarized light:

Solution
Because both chirality centers have the same groups attached, this compound might be meso. To determine this, the conformation must be changed to a more symmetrical one

in which at least one pair of like groups is eclipsed. Rotation of the right carbon by 180° gives this conformation.

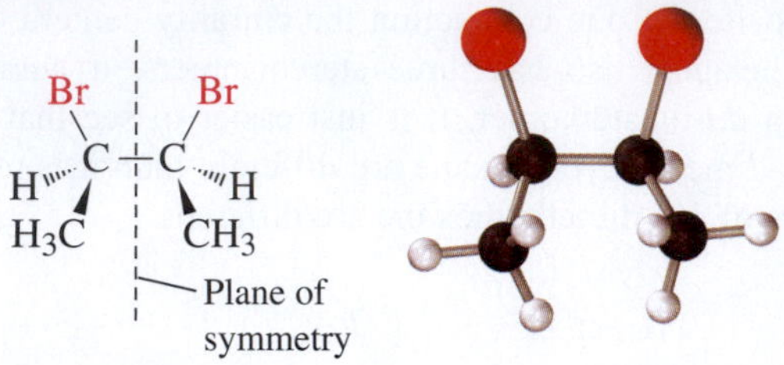

2,3-Dibromobutane

It is readily apparent that there is a plane of symmetry that bisects the carbon–carbon bond, and therefore, this is the *meso*-stereoisomer. *Meso*-stereoisomers are not chiral and do not rotate plane-polarized light.

PROBLEM 7.11

Explain whether these compounds rotate plane-polarized light:

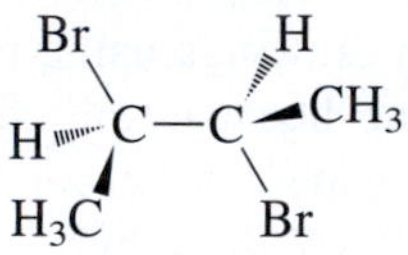

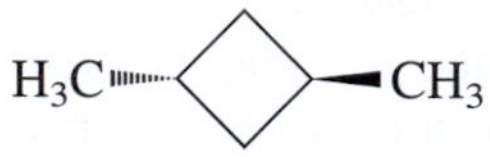

7.7 RESOLUTION: SEPARATING ENANTIOMERS

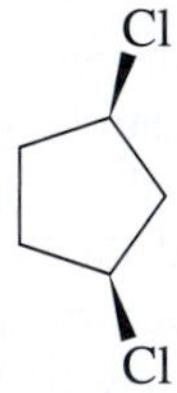

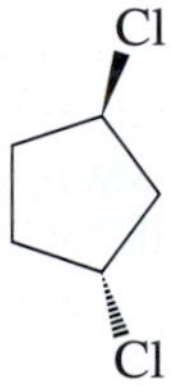

The process of separating the enantiomers of a racemic mixture is called a **resolution.** To accomplish this task, the environment must be made chiral so that the enantiomers have different properties. These different properties can then be employed in the separation process.

The classical method to resolve a racemate is to react the mixture of enantiomers with one enantiomer of some other chiral compound. The products are diastereomers and can be separated by using the usual methods, such as recrystallization or chromatography. Then the separated diastereomers are individually converted back to the enantiomers of the original compound. Figure 7.5 shows how a racemic carboxylic acid can be resolved.

Another resolution method that is sometimes employed involves the selective reaction of one enantiomer of a racemic mixture with one enantiomer of a chiral reagent,

Figure 7.5

RESOLUTION OF A RACEMIC
CARBOXYLIC ACID.

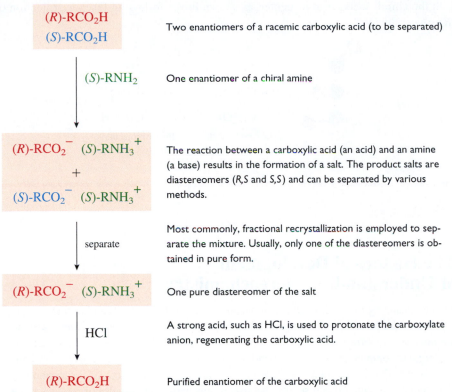

(R)-RCO$_2$H
(S)-RCO$_2$H

Two enantiomers of a racemic carboxylic acid (to be separated)

(S)-RNH$_2$

One enantiomer of a chiral amine

(R)-RCO$_2{}^-$ (S)-RNH$_3{}^+$
+
(S)-RCO$_2{}^-$ (S)-RNH$_3{}^+$

The reaction between a carboxylic acid (an acid) and an amine (a base) results in the formation of a salt. The product salts are diastereomers (R,S and S,S) and can be separated by various methods.

separate

Most commonly, fractional recrystallization is employed to separate the mixture. Usually, only one of the diastereomers is obtained in pure form.

(R)-RCO$_2{}^-$ (S)-RNH$_3{}^+$

One pure diastereomer of the salt

HCl

A strong acid, such as HCl, is used to protonate the carboxylate anion, regenerating the carboxylic acid.

(R)-RCO$_2$H

Purified enantiomer of the carboxylic acid

often an enzyme. Either the unreacted enantiomer of the starting material can be isolated or the single enantiomer of the product can be obtained. For example, Louis Pasteur found that fermentation of ($\pm$)-tartaric acid with *penicillium glaucum* resulted in the metabolism of the ($+$)-enantiomer. The unreacted ($-$)-tartaric acid could be recovered from the fermentation mixture.

Another method that is becoming very important is chromatography using a chiral phase. Often, a chiral stationary phase, prepared by covalently bonding a chiral compound to the surface of silica beads, is used.

7.8 FISCHER PROJECTIONS

Representing these chiral molecules, especially those with more than one chirality center, using only the two dimensions of a piece of paper, requires some special conventions. We have become accustomed to using wedged and dashed bonds for this purpose. Another method was developed by one of the pioneers in the area of organic stereochemistry, Emil Fischer. To construct a Fischer projection, the molecule is first arranged with the horizontal bonds to its chirality center projecting above the plane of the page and the vertical bonds projecting behind the page. In the Fischer projection the bonds are "projected" into the plane of the page, resulting in a cross with the chirality center at its center, as shown in the following Fischer projection of (*R*)-glyceraldehyde:

Important Convention

$$\underset{\text{(R)-Glyceraldehyde}}{\overset{\displaystyle \overset{O}{\underset{\parallel}{CH}}}{H \blacktriangleright C \blacktriangleleft OH}}$$
CH₂OH

$$\underset{\substack{\text{Fisher projection of}\\\text{(R)-glyceraldehyde}}}{\overset{\displaystyle \overset{O}{\underset{\parallel}{CH}}}{H —\!\!\!\!|\!\!\!\!— OH}}$$
CH₂OH

(*R*)-Glyceraldehyde

Fisher projection of
(*R*)-glyceraldehyde

Focus On

The Historical Development of Understanding Stereochemistry

An understanding of the three-dimensional structures of molecules has played an important part in the development of organic chemistry. The first experiments of importance to this area were reported in 1815 by the French physicist J. B. Biot, who discovered that certain organic compounds, such as turpentine, sugar, camphor, and tartaric acid, were optically active: that is, solutions of these compounds rotated the plane of polarization of plane-polarized light. Of course, the chemists of this period had no idea of what caused a compound to be optically active because atomic theory was just being developed and the concepts of valence and stereochemistry would not be discovered until far in the future.

The next major contribution was made in 1848 by the great scientist Louis Pasteur. During the fermentation of wine, large quantities of (+)-tartaric acid precipitate in the barrels. Pasteur was studying a salt of this acid when he discovered that it had a very interesting property. The crystals of this salt had a chiral shape—that is, an individual crystal had a shape that was not superimposable on its mirror image—and all of the crystals had the same handedness. Another tartaric acid, which today is known as racemic tartaric acid, is also produced during the production of wine. It was known that this acid had the same formula as (+)-tartaric acid but was optically inactive, that is, it did not rotate plane-polarized light. Upon careful observation of the salt of this acid, Pasteur found that the individual crystals were chiral, as was the case for (+)-tartaric acid, but in this case the left-handed and the right-handed versions of the crystals were present in equal amounts. Using a tweezers and a magnifying glass (and considerable patience), Pasteur was able to separate these crystals. He found one to be completely identical to the salt of (+)-tartaric acid that he had studied previously. The other had identical physical and chemical properties except that it rotated plane-polarized light in the opposite direction. Pasteur had accomplished the first resolution of a racemic organic compound! Because the salts that gave mirror-image crystals also gave opposite rotations, Pasteur associated optical rotation with chirality. And because solutions of these salts were optically active, he proposed that chirality was not just a macroscopic property of the crystals, but the arrangement of the atoms in a molecule of tartaric acid must also be chiral. He was postulating a chiral shape for the arrangement of these atoms at about the same time that Kekulé was proposing the concept of valence!

PROBLEM 7.12

Draw Fischer projections for these compounds:

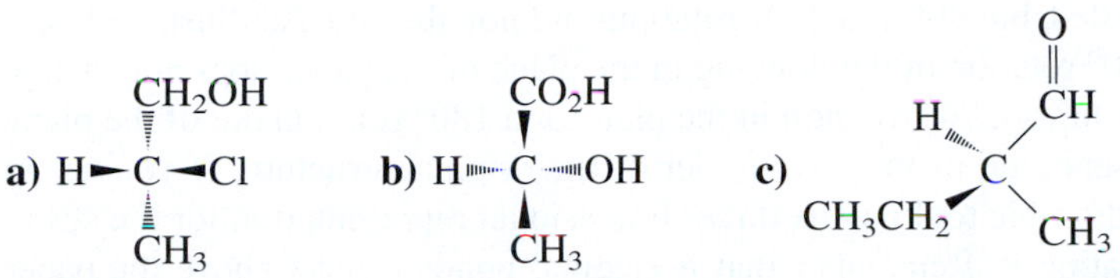

a)
$$\begin{array}{c} CH_2OH \\ | \\ H-C-Cl \\ | \\ CH_3 \end{array}$$

b)
$$\begin{array}{c} CO_2H \\ | \\ H-C-OH \\ | \\ CH_3 \end{array}$$

c)
$$\begin{array}{c} O \\ \| \\ H_{\cdots}\quad CH \\ C \\ CH_3CH_2 \quad CH_3 \end{array}$$

It took about another 20 years for the explanation of chirality to be completed. In 1874, two young chemists, Jacobus van't Hoff from Holland and Joseph Le Bel from France, independently proposed that the four bonds to a carbon were arranged in a tetrahedral manner. Their arguments were based on the number of isomers that exist for various formulas. Although we will not go into all of the details here, the following discussion presents some of the reasoning they used. At the time, it was well accepted that all four of the bonds to a carbon were identical. This was based on the fact that, for a multitude of compounds with the formula CH_3X, only one isomer had ever been found. There is only one CH_3Cl, one CH_3OH, one CH_3CH_3, and so on. Many geometries with low symmetries can be eliminated on the basis of this observation. Two arrangements of the four bonds around a carbon that meet the criterion of having all of the bonds identical are the one with a square planar geometry and the one with a tetrahedral geometry. The square planar geometry can be eliminated on the basis of the observation that, for a multitude of compounds with the formula CH_2X_2, only one isomer has ever been found. There is only one CH_2Cl_2. If carbon had a square planar geometry, then two isomeric compounds with the formula CH_2Cl_2 would be expected, as shown here. However, a tetrahedral geometry predicts only one CH_2Cl_2.

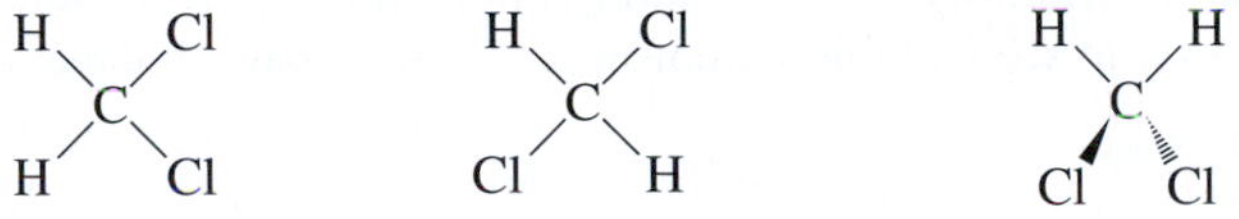

$$\begin{array}{c} H \quad Cl \\ C \\ H \quad Cl \end{array} \qquad \begin{array}{c} H \quad Cl \\ C \\ Cl \quad H \end{array} \qquad \begin{array}{c} H \quad H \\ C \\ Cl \quad Cl \end{array}$$

Two isomers with square planar geometry   One isomer with tetrahedral geometry

At the time of van't Hoff's and Le Bel's work, there were only a few optically active compounds whose structures had been determined. All of these compounds had a carbon bonded to four different groups, a carbon that we today call a chirality center. van't Hoff and Le Bel pointed out that a tetrahedral arrangement of four different groups around a carbon produced a structure that is not superimposable on its mirror image, a chiral structure. Thus, their postulate of a tetrahedral carbon explained the existence of enantiomeric compounds.

Because they are two-dimensional representations of three-dimensional objects, extreme care must be used in manipulating Fischer projections to avoid changing the configuration. Structures may not be "lifted" out of the plane of the paper. A 180° rotation *in the plane* is permitted, but 90° and 270° rotations are not allowed. As illustrated next for (S)-alanine, a 180° rotation of the drawing in the plane of the paper does not change the configuration. A 90° or 270° rotation in the plane or a 180° rotation out of the plane all result in a representation of the enantiomer of the original structure. If you are in doubt, it is always advisable to draw the three-dimensional representation for the structure before manipulating it. Remember that horizontal bonds project above the paper and vertical bonds project behind the paper.

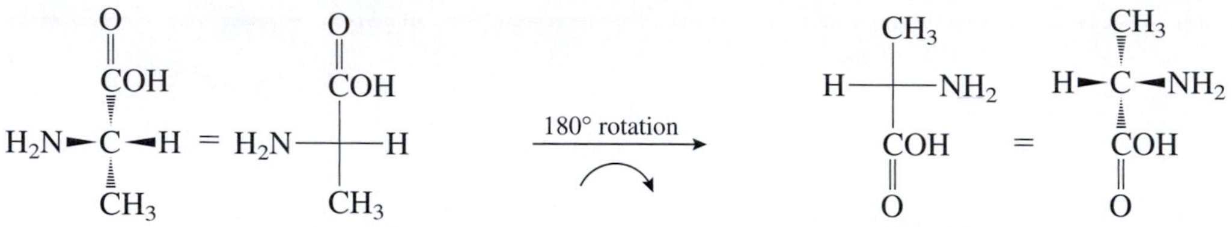

(S)-Alanine   ◄——— Identical ———►   (S)-Alanine

MODEL BUILDING PROBLEM 7.4

Build models of the compounds represented by these Fischer projections. Determine whether the models superimpose. (Note that these Fischer projections are related by a 90° rotation in the plane of the page.)

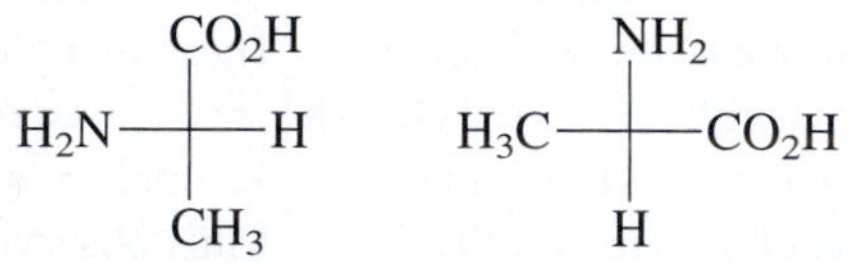

Fischer projections are especially useful in the case of compounds with more than one chirality center. For example, it is easy to see the plane of symmetry in *meso*-tartaric acid. As was the case with regular structures, interchanging any two groups in a Fischer projection results in inversion of configuration at the chirality center. Thus, interchanging the H and OH on the lower chirality center of *meso*-tartaric acid inverts the configuration at that chirality center, resulting in the (2R,3R)-stereoisomer, (+)-tartaric acid. It is also easy to see that this stereoisomer does not have a plane of symmetry.

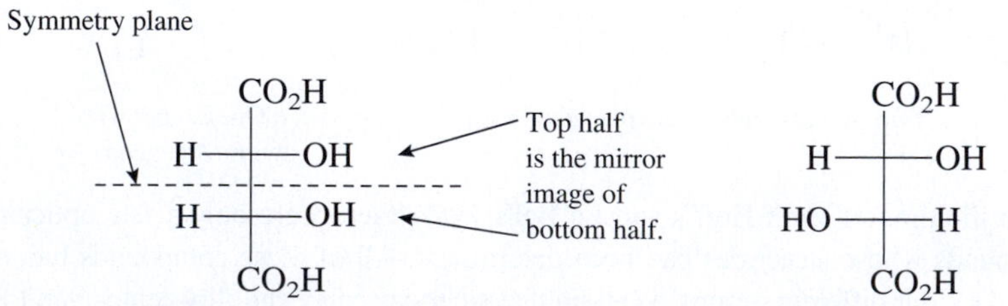

meso-Tartaric acid   (+)-Tartaric acid
                              (2R,3R)-stereoisomer

PRACTICE PROBLEM 7.3

Assign the configuration of this compound as *R* or *S*:

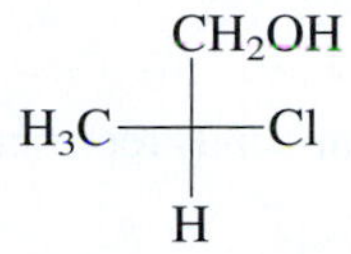

$$CH_2OH$$
$$H_3C-\!\!\!\!-\!\!\!\!-Cl$$
$$H$$

Solution

One way to work this type of problem is to draw the structure, showing its stereochemistry, and then proceed as in previous examples.

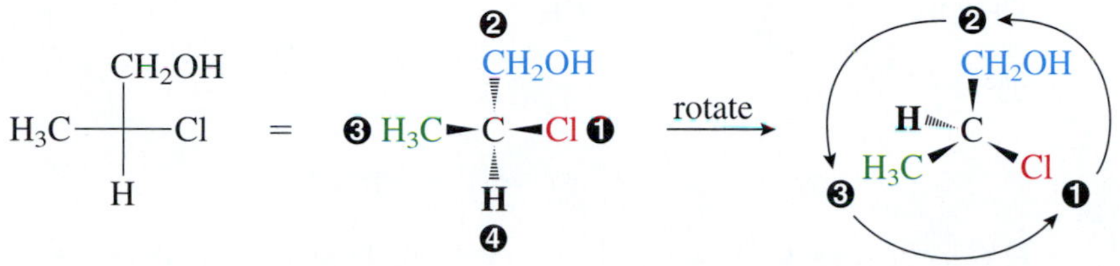

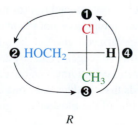

The configuration can also be assigned directly from the Fischer projection. First assign priorities to the groups:

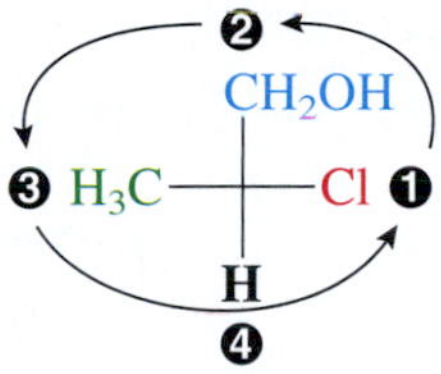

S

If group 4 is attached to a vertical bond, as in this case, it is already pointed away from you. Therefore, the direction of rotation given by proceeding from group 1 to 2 to 3 gives the configuration directly. In this case the rotation is counterclockwise, so the configuration is *S*.

If group 4 is attached to a horizontal bond, it is pointed toward you and you are viewing the molecule from the wrong side.

R

This simply means that the configuration is opposite that given by the direction of rotation proceeding from group 1 to 2 to 3. In this example the direction of rotation is counterclockwise but the H is pointed toward you, so the configuration is *R*. (Try drawing the stereochemistry to confirm this.)

ORGANIC
Chemistry•🔬•Now™
Click *Coached Tutorial Problems*
for more practice using
Fischer Projections.

BioLink 🌀

PROBLEM 7.13

Assign the configurations of the compounds represented by these Fischer projections as *R* or *S*.

a) H_2N─┼─H , top: CO_2H, bottom: CH_2OH

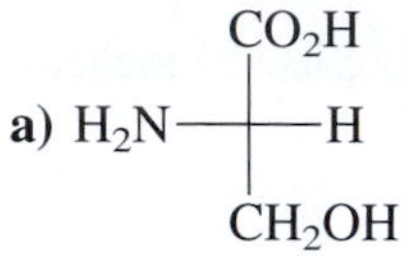

Serine

b) CH_2=CH─┼─CH_2CH_3 , top: $\overset{\text{O}}{\overset{\|}{\text{CH}}}$, bottom: CH_3

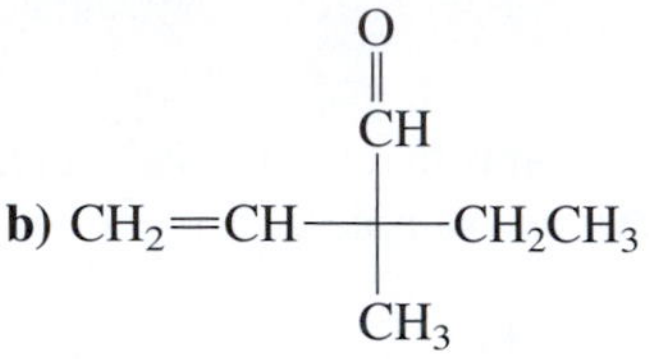

7.9 REACTIONS THAT PRODUCE ENANTIOMERS

A reaction of an achiral molecule may introduce a chirality center, producing a chiral product. For example, reaction of the following ketone with hydrogen in the presence of a catalyst results in addition of the hydrogen to the carbon–oxygen double bond, producing 2-butanol:

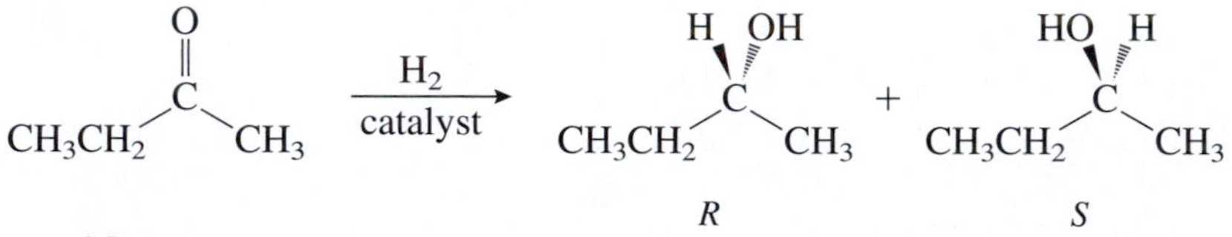

2-Butanone   *R*   *S*
                                                                      2-Butanol

The starting ketone is not chiral, but the product alcohol is chiral. Approach of the hydrogen from above the plane of the ketone (as drawn) produces (*R*)-2-butanol, whereas approach from behind the plane produces (*S*)-2-butanol. There is no apparent reason why the hydrogen should prefer one approach over the other; in fact, the two enantiomers are produced in exactly equal amounts—the product is racemic. As long as there is nothing else that is chiral in the reaction, the enantiomeric products (and the enantiomeric reaction pathways leading to them) have identical energies and must be produced in equal amounts. If all the reagents in a reaction are achiral (or racemic), then the product must be racemic. If the initial reaction mixture does not rotate plane-polarized light, the product mixture cannot rotate plane-polarized light. On the other hand, if one of the components of the initial reaction mixture (a reagent, a catalyst, even the solvent) is chiral and only one enantiomer of it is present, enantiomers of the product may be produced in unequal amounts. Devising methods that produce only one enantiomer of a chiral product, called asymmetric synthesis, is one of the most challenging areas of research facing organic chemists today.

Focus On

Pharmaceuticals and Chirality

Many of the drugs that are so important in medicine today contain chirality centers. If the drug is isolated from a natural source, it is usually obtained as a single enantiomer. But synthetic drugs have most commonly been used as a racemic mixture because obtaining them as single enantiomers is often a time-consuming and expensive process. Because biological processes involve chiral molecules, the effect of drug enantiomers is often different. For example, dextromethorphan is a cough suppressant used in medications such as Robitussin. Its enantiomer, levomethorphan, is a powerful narcotic.

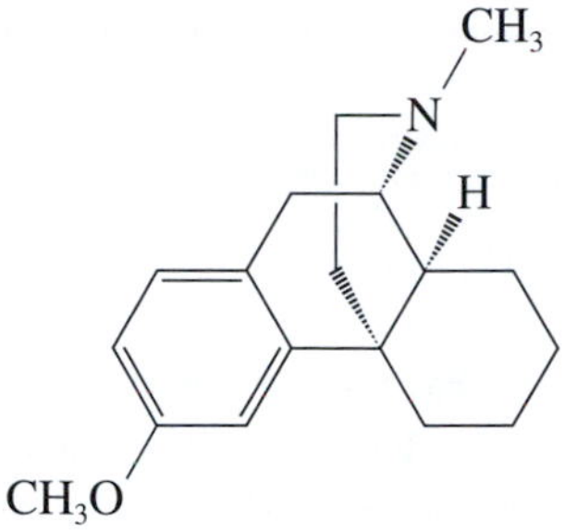

Dextromethorphan 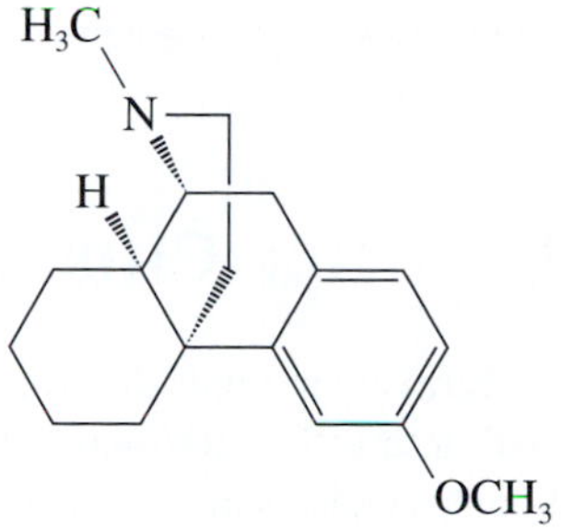   Levomethorphan

If a drug is used as a racemic mixture, often only one of the enantiomers is responsible for the desired pharmacological effect. The other enantiomer may have a lesser effect, or no effect, or may even be responsible for undesired side effects. One example is perhexiline, a racemic drug that was used to treat abnormal heart rhythms. This drug was responsible for a number of deaths in the 1980s because one enantiomer was metabolized much more slowly than the other and accumulated at toxic levels. If perhexiline had been marketed only as the more rapidly metabolized enantiomer, it *might* have been a safer drug.

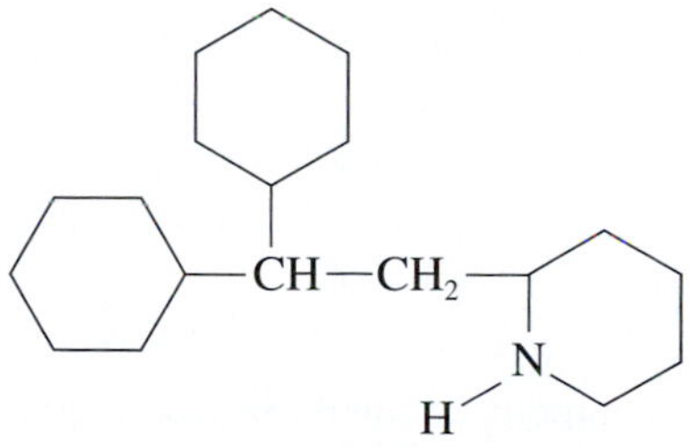

Perhexiline 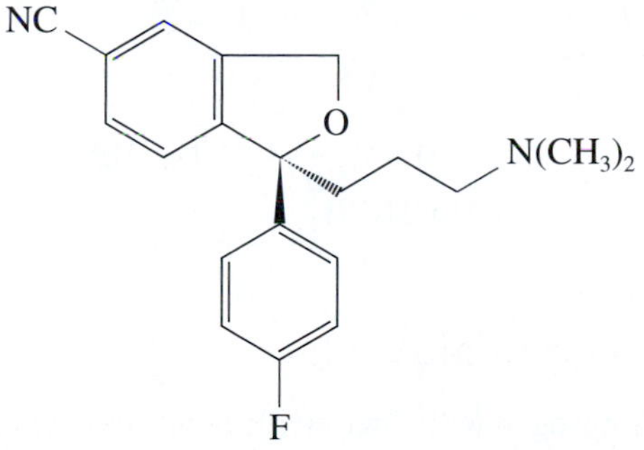   (S)-Citalopram

Continued

An example of improved efficacy of an enantiomerically pure version of a drug over the racemic version is provided by citalopram. The racemic version, known as Celexa, is marketed as an antidepressant. Studies on the resolved enantiomers have shown that the S-enantiomer is the active one and that it has a more rapid onset of action and a more favorable benefit-to-risk ratio than the racemate. As a result, (S)-citalopram (escitalopram or Lexapro) is now being marketed. Not only is this a better and safer drug, but the pharmaceutical company that developed citalopram was able to extend its market exclusivity for an additional 3 years.

Sales of single enantiomer drugs exceeded $159 billion in 2002. Some of these come from biological sources, but the majority are synthetic. For this reason, the development of synthetic methods that produce only a single enantiomer of chiral compounds is a very active research area in both academic and pharmaceutical research labs. Chiral or asymmetric syntheses, which produce only the desired enantiomer, are much preferred over resolution processes, in which at least half of the initial compound is discarded.

7.10   OTHER CHIRAL COMPOUNDS

All of the chiral compounds that we have seen so far have had one or more carbons substituted with four different groups. However, there are other situations that also give rise to chiral compounds. Some of the other possibilities are described here.

Other Tetrahedral Atoms

Of course, any tetrahedral atom, not just carbon, that has four different groups bonded to it is a chirality center, and compounds containing such atoms will exist as a pair of enantiomers. Many such compounds have been prepared and resolved, including the following quaternary ammonium salt and the silicon compound:

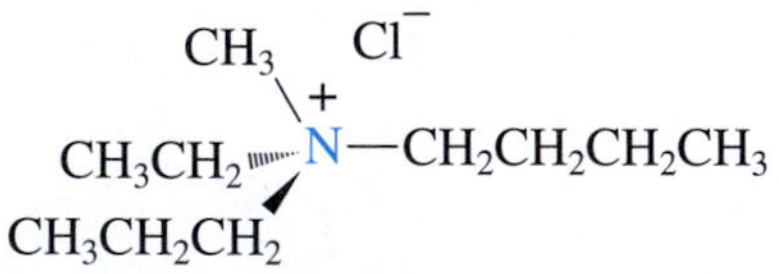

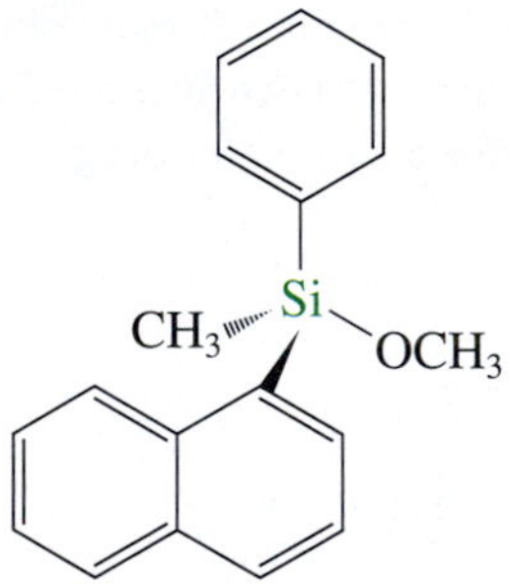

Pyramidal Atoms

A nitrogen with three single bonds has pyramidal geometry. If the three groups attached to the nitrogen are different, then the nitrogen is a stereocenter and the compound exists as a pair of enantiomers. The situation is very similar to a tetrahedral carbon, but with the unshared pair of electrons replacing one of the bonds. However, because there are only three bonds, the groups can move to the other side of the nitrogen in a process reminiscent of an umbrella turning inside out. This results in inversion of configuration. The activation barrier is quite small, only about 5 kcal/mol (21 kJ/mol). Therefore, inversion is quite fast at ambient temperatures, occurring about 10^{11} times per second. Because the two enantiomers interconvert so rapidly, they cannot be separated. The compound behaves as a racemate, and the existence of the rapidly inverting enantiomeric forms can be ignored.

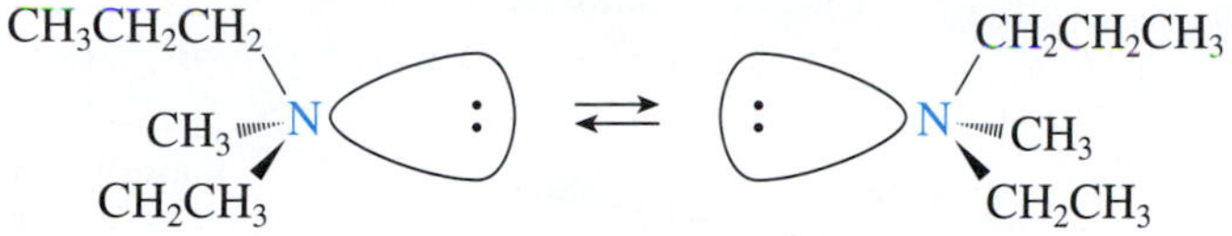

However, compounds containing other pyramidal atoms, with larger barriers to inversion, can be resolved. Examples include the following phosphorus compound (inversion barrier of about 30 kcal/mol [126 kJ/mol]) and the sulfur compound (inversion barrier of about 35 kcal/mol [146 kJ/mol]):

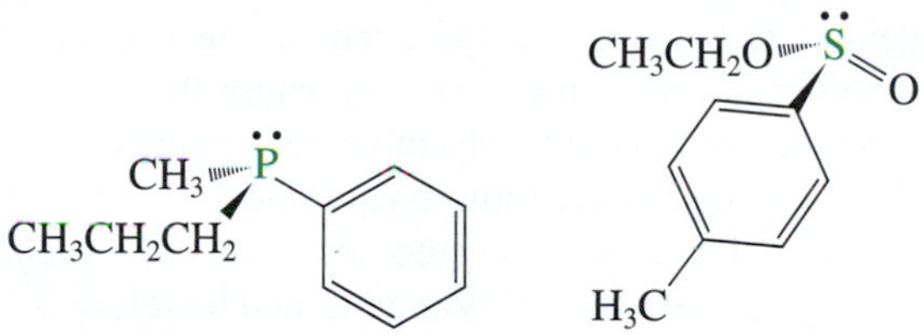

Substituted Allenes

Allene has two adjacent carbon–carbon double bonds. Its geometry is not planar like a normal alkene. The central carbon is *sp* hybridized and has linear geometry. The two *p* orbitals that it uses for the two double bonds are perpendicular, so the planes of the two double bonds are perpendicular. The two hydrogens on one end of allene lie in a plane perpendicular to the two hydrogens on the other end.

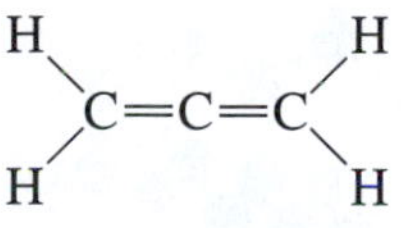

Allene

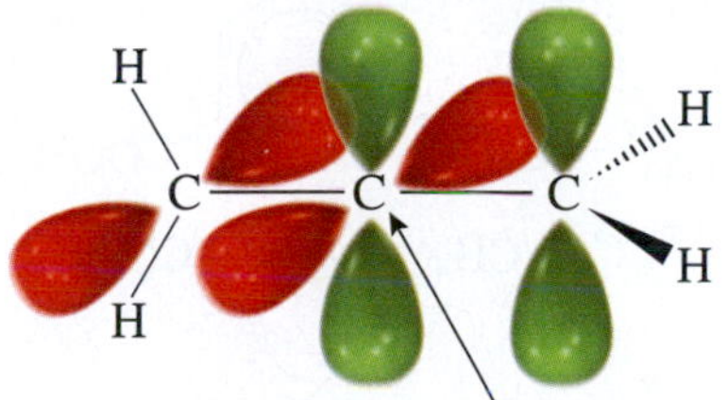

sp Hybridized

As early as 1875, van't Hoff pointed out that properly substituted allenes would be chiral. When the two groups on one end of the allene are different and the two groups on the other end of the allene are also different, the compound is chiral and exists as a pair of enantiomers, rather than as cis–trans isomers as is the case with simple alkenes.

A number of chiral allenes, such as the following dicarboxylic acid, have been prepared and resolved.

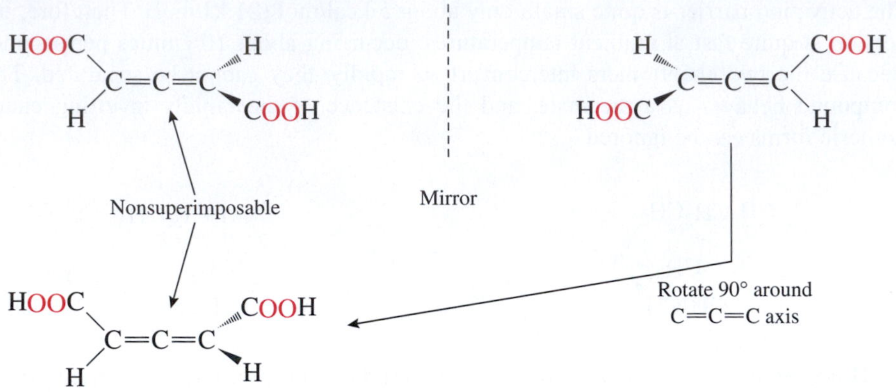

Nonsuperimposable

Mirror

Rotate 90° around
C=C=C axis

Biphenyls

The compound formed by connecting two benzene rings by a single bond is called biphenyl. Steric interactions between the hydrogens on the ortho positions of one ring with those on the ortho positions of the other ring cause the planes of the rings to be perpendicular. If both rings are appropriately substituted at the ortho positions, then the compound is chiral and can be resolved into enantiomers. The groups in the ortho positions serve two purposes. First, the two ortho substituents on each ring must be different so that the biphenyl has no plane of symmetry and therefore is chiral. In addition, because rotation about the single bond connecting the rings (a conformational change) interconverts the enantiomers, the ortho groups must be large enough that their steric interaction raises the energy barrier for this conformational change. If the groups are large enough, rotation is slow at ambient temperatures, and the compound can be resolved. For example, the enantiomers of the following substituted biphenyl do not interconvert at room temperature, and the compound has been resolved. Rotation about the connecting single bond does occur at higher temperatures. The half-life for racemization is 78 min at 118°C.

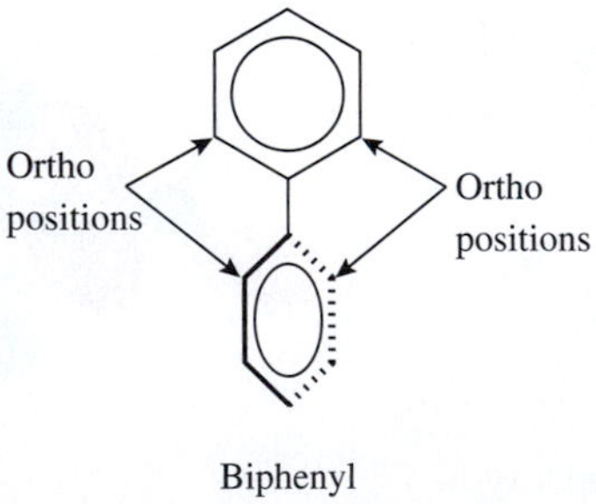

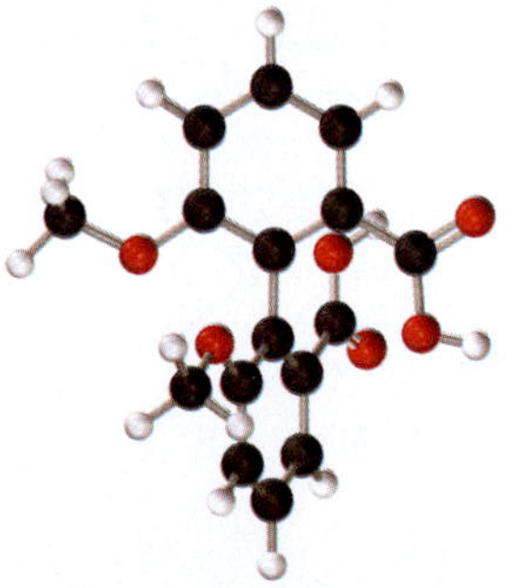

Ortho positions

Ortho positions

Biphenyl

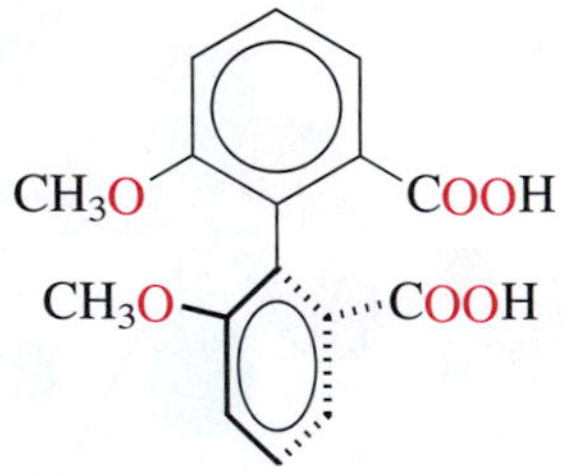

Helical Molecules

Another interesting group of chiral compounds results when molecules are forced to adopt a helical geometry. Like the turn of a screw, the turn of the helix can be either right-handed or left-handed. One example is the compound known as hexahelicene, which has six aromatic rings fused together. The molecule is forced to adopt a helical shape to avoid a severe steric interaction between the end rings. Hexahelicene has been resolved and has an enormous specific rotation of +3640.

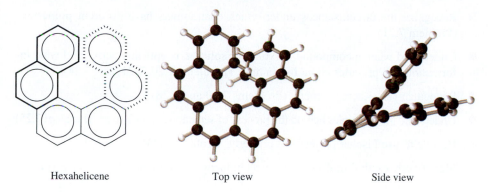

Hexahelicene

Top view

Side view

PROBLEM 7.14

Explain whether each of these compounds is chiral or not:

a)
O
H3C
CCH3
C=C=C
H
H

b)
Br−
CH3CH2
Ph
N+—CH2CH3
H3C

c)
O
H
CCH3
C=C=C
H
CH3

d)
CH3CH2
Ph
Si—OCH2CH3
H3C

e)
H3C
CH3
CH3O
CO2H

f)
CO2H
H3C
CH3
CH3O
CO2H

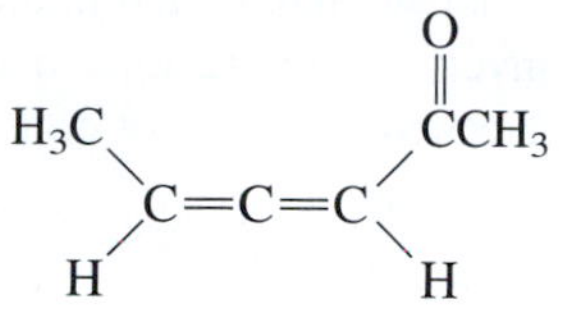

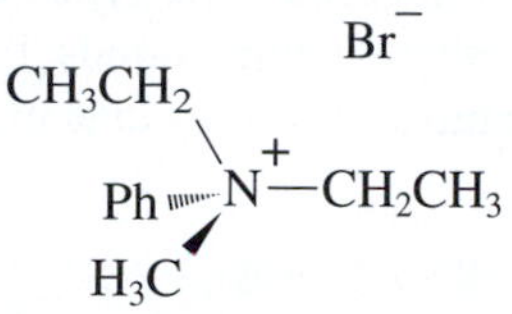

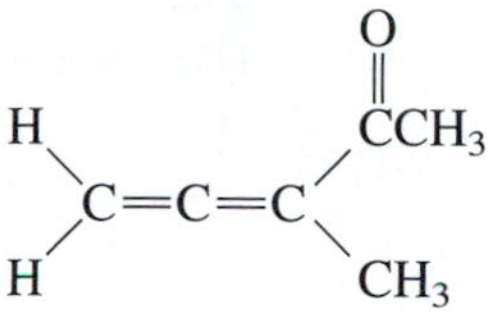

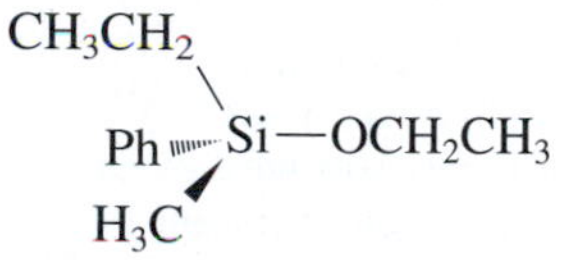

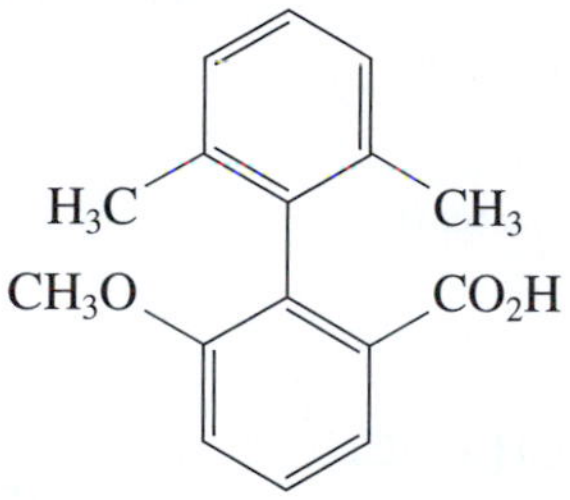

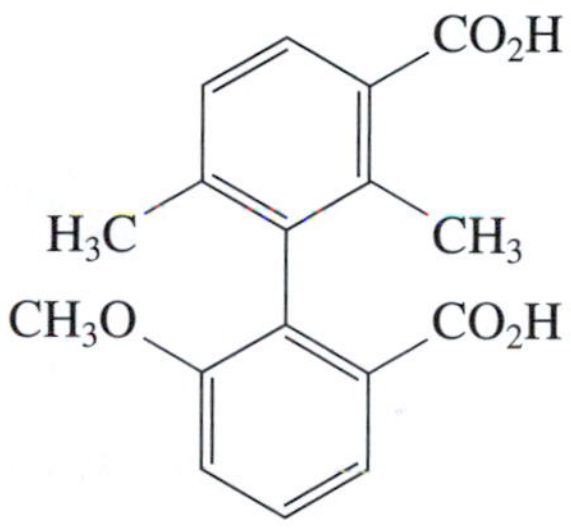

PROBLEM 7.15

Although this biphenyl is chiral, it cannot be resolved. Explain.

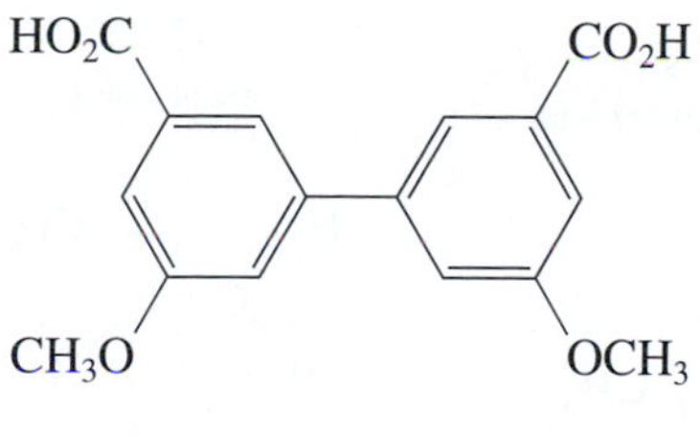

HO2C
CO2H
CH3O
OCH3

Review of Mastery Goals

After completing this chapter, you should be able to:

- Identify chiral compounds, locate chirality centers, and determine how many stereoisomers exist for a particular compound. (Problems 7.18, 7.19, 7.22, 7.27, and 7.29)
- Locate any symmetry planes that are present in a molecule. (Problem 7.20)

ORGANIC Chemistry•Now™
Click *Mastery Goal Quiz* to test how well you have met these goals.

■ Designate the configuration of chirality centers as *R* or *S*. (Problems 7.16, 7.17, and 7.30)

■ Recognize the circumstances under which enantiomers have different properties. (Problem 7.21)

■ Understand when a compound, mixture, or solution is optically active and what information this provides about the sample. (Problems 7.20, 7.21, and 7.24)

■ Recognize *meso*-stereoisomers. (Problems 7.20 and 7.28)

■ Understand the principles behind the process of separating enantiomers. (Problem 7.25)

■ Be able to use Fischer projections properly. (Problem 7.23)

■ Identify other chiral molecules, such as biphenyls and allenes. (Problem 7.31)

Visual Summary of Isomers

The following scheme summarizes the types of isomers that we have encountered. (Because the rotations about sigma bonds that interconvert conformations occur rapidly at room temperature, conformations cannot be separated and are not considered to be isomers.)

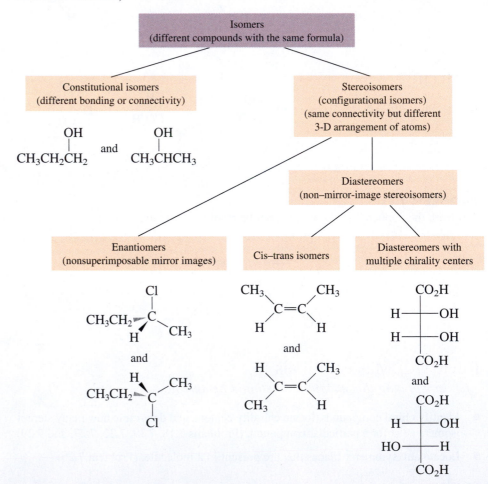

Additional Problems

7.16 Assign the configuration of these compounds as R or S:

a)

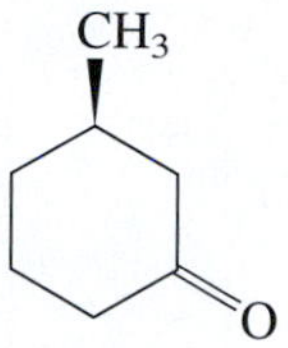

b)

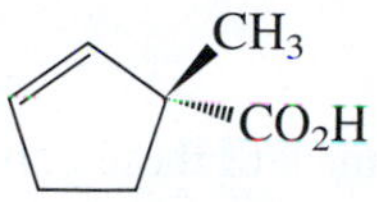

c)

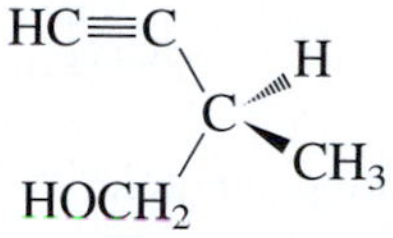

d)

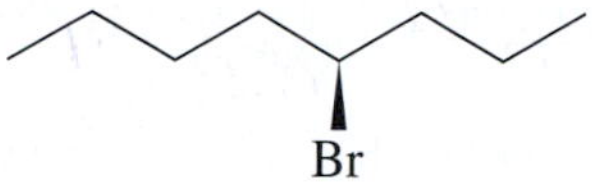

e)

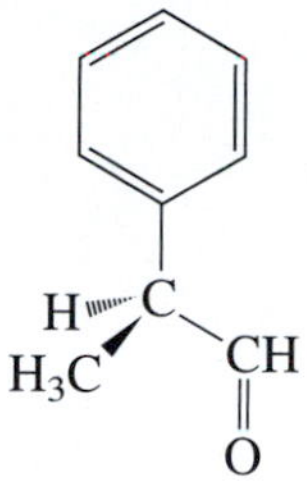

f)

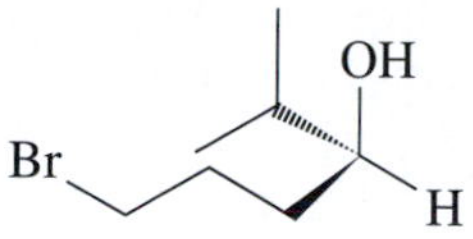

g) 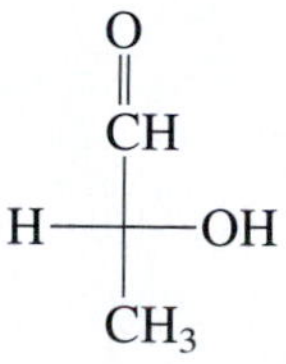 H———OH with CH₃

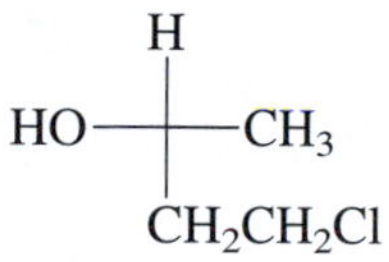

h) HO———CH₃

i) 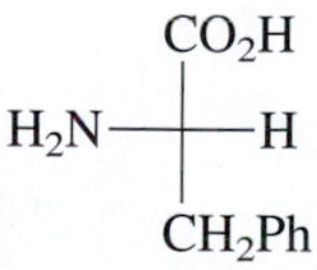 H_2N———H

Phenylalanine

7.17 Draw the structures of these compounds:
a) (R)-3-Chloro-1-pentene
b) (S)-1-Methyl-2-cyclohexenol

7.18 Determine the number of stereoisomers for these compounds:

a)

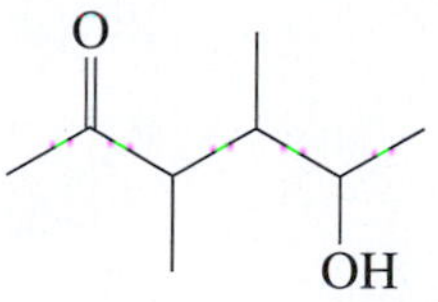

b)

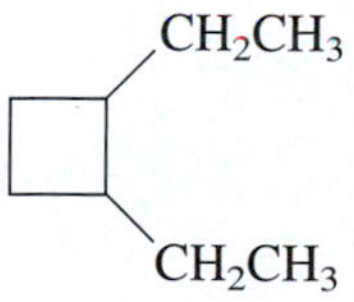

BioLink

c)

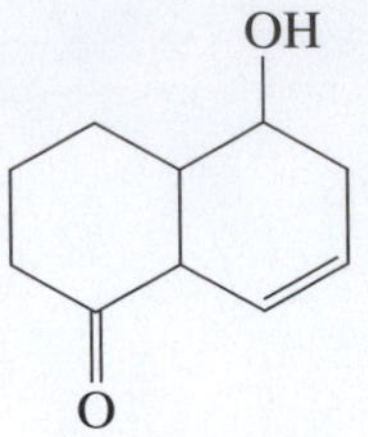

d)

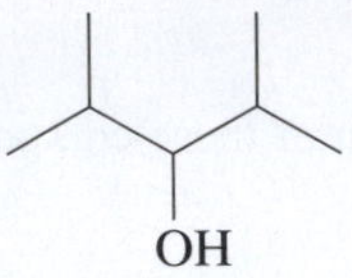

7.19 Draw all of the stereoisomers of these compounds:

a) $CH_3CH{=}CHCHCH_3$ with OH

b) $CH_3CH{=}CHCHCH{=}CHCH_3$ with OH

7.20 Explain whether or not these compounds would rotate plane-polarized light:

a)

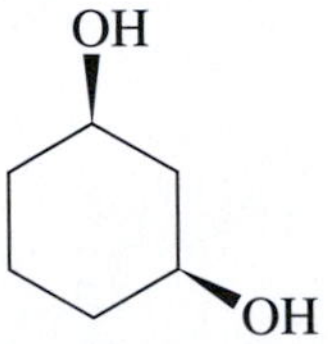

b)

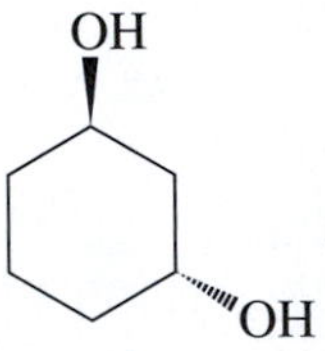

c)

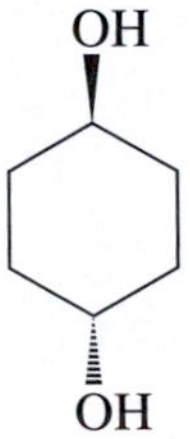

d)

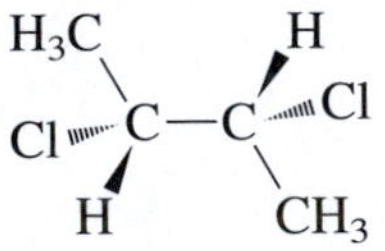

e)

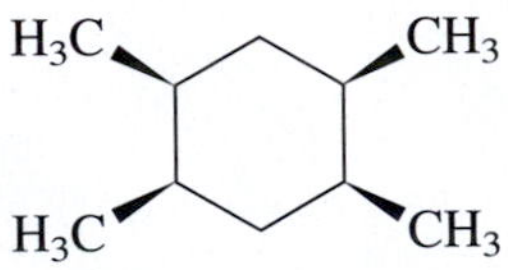

f)

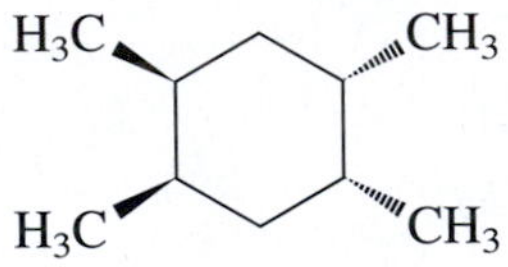

g)

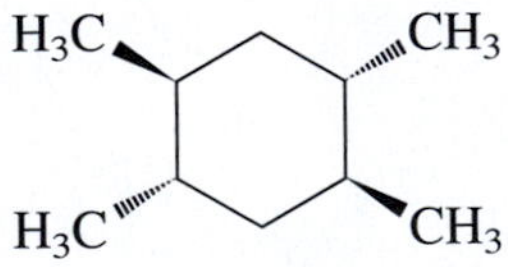

h)

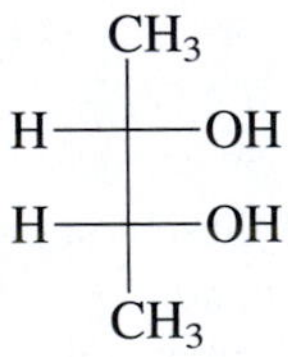

i)

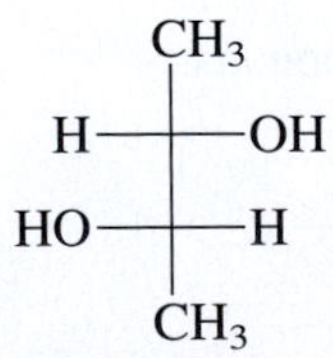

7.21 Consider the two enantiomers of this carboxylic acid:

Explain whether each of the following statements is true, is false, or cannot be determined from this information:

a) The enantiomers have the same melting point.
b) The enantiomers have the same boiling point.
c) The enantiomers have the same solubility in water.
d) The enantiomers have the same magnitude of rotation of plane-polarized light.
e) The enantiomers have the same direction of rotation of plane-polarized light.
f) The enantiomers have the same pK_a.
g) The enantiomers have the same rate of reaction with methanol.
h) The enantiomers have the same pH of an aqueous solution.
i) The enantiomers have the same rate of reaction with (S)-2-butanol.
j) The (R)-enantiomer rotates plane-polarized light in a clockwise direction.
k) This reaction produces more of the (R)-enantiomer than the (S)-enantiomer.

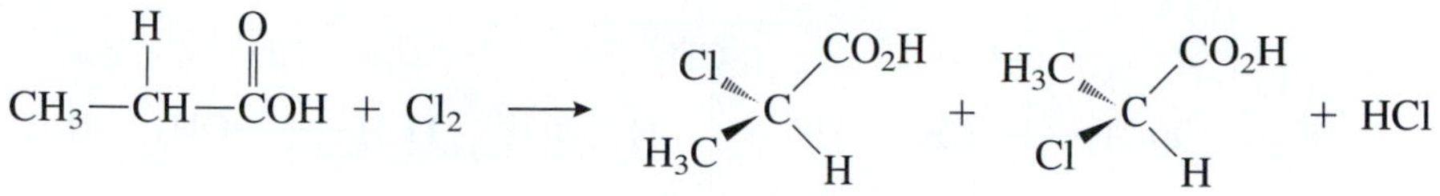

7.22 Identify these pairs of compounds as identical, structural isomers, enantiomers, or diastereomers:

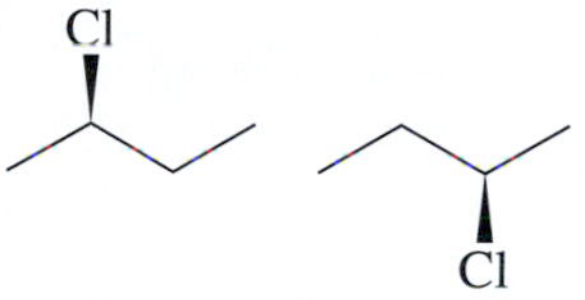

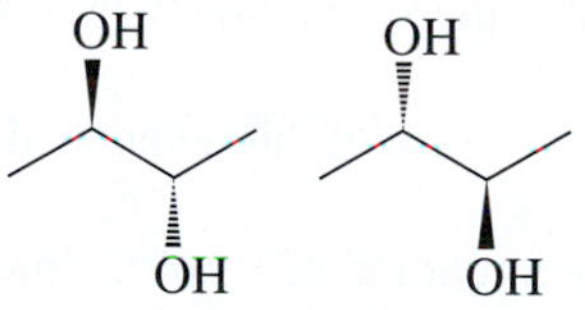

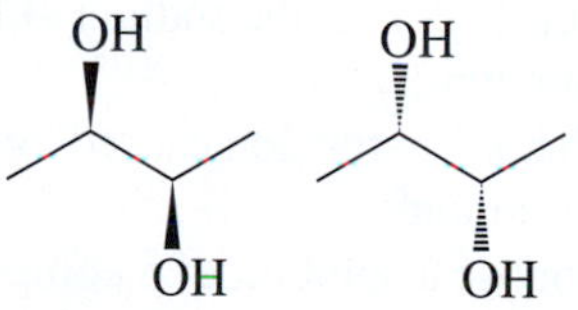

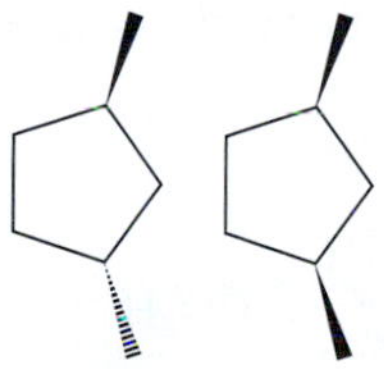

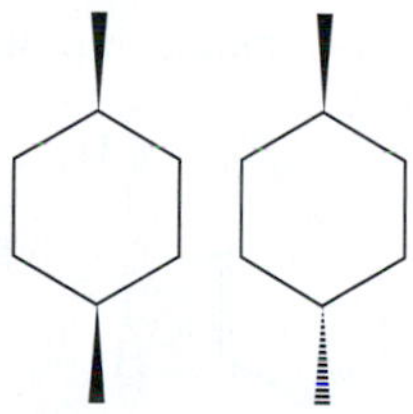

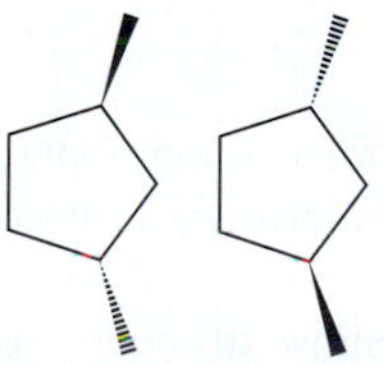

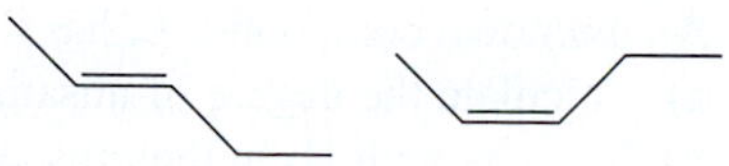

7.23 Identify these pairs of compounds as identical, structural isomers, enantiomers, or diastereomers:

a)

$$
\begin{array}{c}
CH_3 \\
H\!-\!\!-\!OH \\
H\!-\!\!-\!OH \\
CH_3
\end{array}
\qquad
\begin{array}{c}
CH_3 \\
HO\!-\!\!-\!H \\
H\!-\!\!-\!OH \\
CH_3
\end{array}
$$

b)

$$
\begin{array}{c}
CH_3 \\
H\!-\!\!-\!Cl \\
Br\!-\!\!-\!H \\
CH_3
\end{array}
\qquad
\begin{array}{c}
CH_3 \\
H\!-\!\!-\!Br \\
H\!-\!\!-\!Cl \\
CH_3
\end{array}
$$

c)

$$
\begin{array}{c}
CH_3 \\
H\!-\!\!-\!Cl \\
Br\!-\!\!-\!H \\
CH_3
\end{array}
\qquad
\begin{array}{c}
CH_3 \\
Br\!-\!\!-\!H \\
H\!-\!\!-\!Cl \\
CH_3
\end{array}
$$

d)

$$
\begin{array}{c}
O \\
\parallel \\
CH \\
H\!-\!\!-\!OH \\
CH_3
\end{array}
\qquad
\begin{array}{c}
O\quad OH \\
\parallel\ | \\
HC\!-\!\!-\!CH_3 \\
H
\end{array}
$$

e)

$$
\begin{array}{c}
O \\
\parallel \\
CH \\
H\!-\!\!-\!OH \\
CH_3
\end{array}
\qquad
\begin{array}{c}
CH_3 \\
HO\!-\!\!-\!H \\
CH \\
\parallel \\
O
\end{array}
$$

f)

$$
\begin{array}{c}
O \\
\parallel \\
CH \\
H\!-\!\!-\!OH \\
CH_3
\end{array}
\qquad
\begin{array}{c}
O \\
\parallel \\
CH \\
H_3C\!-\!\!-\!OH \\
H
\end{array}
$$

g)

$$
\begin{array}{c}
CO_2H \\
H_2N\!-\!\!-\!H \\
CH_3
\end{array}
\qquad
\begin{array}{c}
CO_2H \\
H\!-\!C\!-\!NH_2 \\
CH_3
\end{array}
$$

7.24 **a)** A solution of 0.2 g/mL of a compound in a 1 dm cell rotates plane-polarized light $+13.3°$ at the sodium D line. What is the specific rotation of this compound?

   b) What is the rotation caused by a solution of 0.1 g of this compound in 10 mL of solution?

   c) Suppose a solution of a compound gave a rotation of $+160°$. How could this rotation be distinguished from one of $-200°$? from one of $+520°$?

7.25 Describe how this amine could be resolved by using this carboxylic acid:

$$
\begin{array}{c}
CH_3 \\
| \\
C\!-\!NH_2 \\
| \\
H
\end{array}
\qquad
\begin{array}{c}
Cl \\
| \\
H_3C\cdots\!C\!-\!CO_2H \\
H
\end{array}
$$

7.26 An unknown compound, **X,** has the formula C_6H_{12}.

   a) Calculate the degree of unsaturation of **X.**

   b) **X** reacts with H_2 in the presence of a catalyst to form a compound, **Y,** with the formula C_6H_{14}. What information does this experiment provide about the structure of **X?**

   c) **X** rotates plane-polarized light, but **Y** does not. Show structures for **X** and **Y.**

7.27 How many stereoisomers exist for this compound? Assign the relative stabilities of each. Is the methyl group axial or equatorial in the more stable conformer of the least stable stereoisomer?

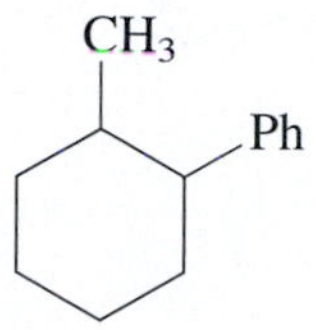

CH₃

Ph

7.28 Draw a stereoisomer of this compound that is chiral, and draw two that are not chiral:

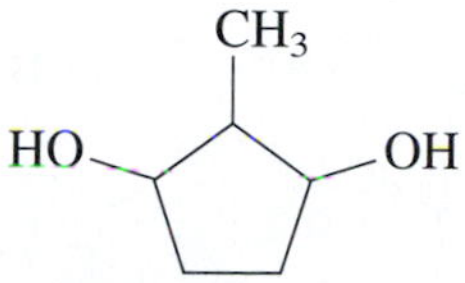

CH₃

HO   OH

7.29 Many compounds are found as a single stereoisomer in nature even though they have numerous chirality centers. Determine how many chirality centers are present in each of the following naturally occurring compounds and how many stereoisomers are possible for each:

a)

```
                Ph              OH
                |               |
    CH₃   O    CH₂   O         CHCH₃
    |     ||   |     ||         |
 NH₂CH—CNHCH—CNHCHCO₂H
```

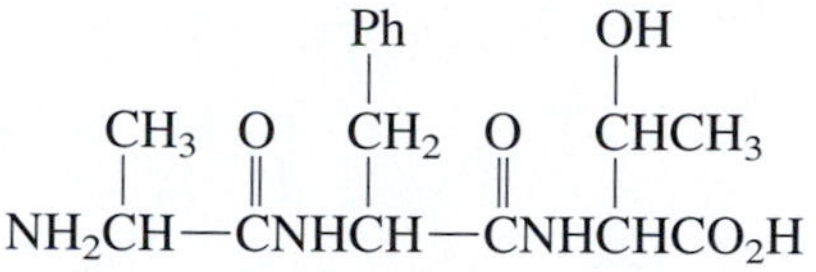

| Alanine residue | Phenylalanine residue | Threonine residue |

(a tripeptide)

b)

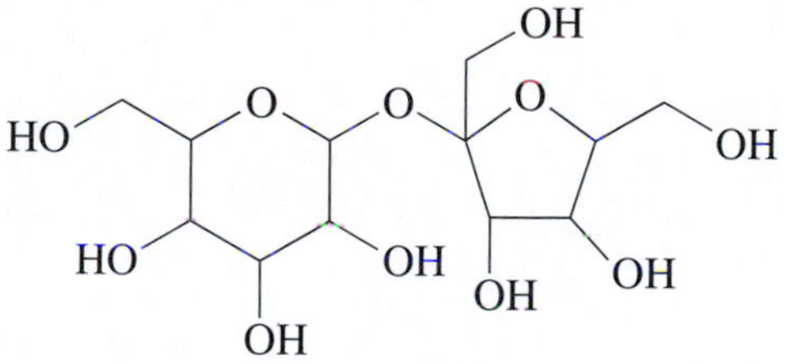

OH

HO

O   O   O.

OH

HO   OH   OH   OH

OH

Sucrose
(table sugar)

c)

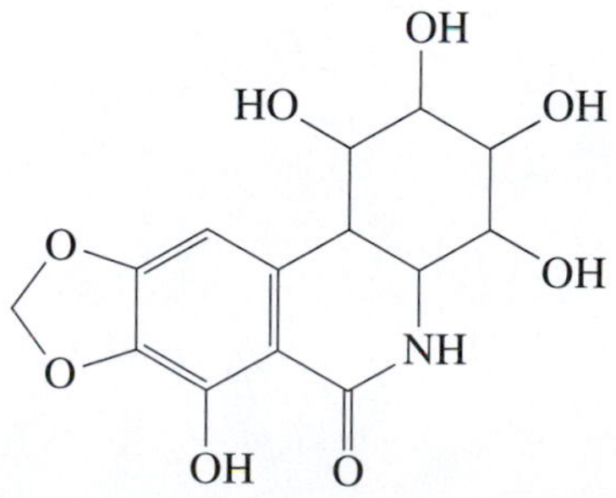

OH

HO   OH

O.

OH

O   OH

O   NH

OH   O

Pancratistatin
(an anticancer drug)

d)

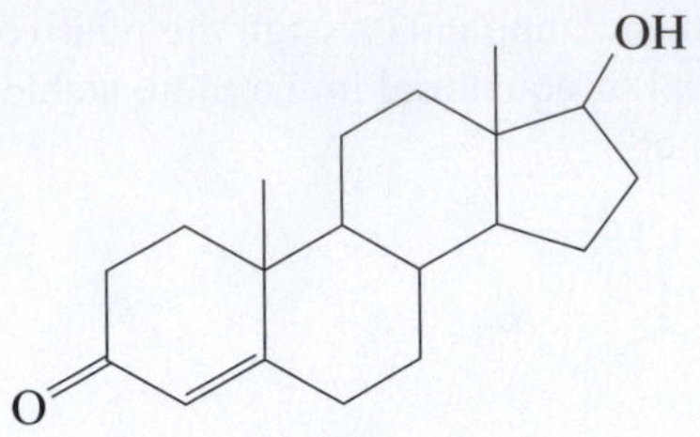

Testosterone
(a steroidal hormone)

e)

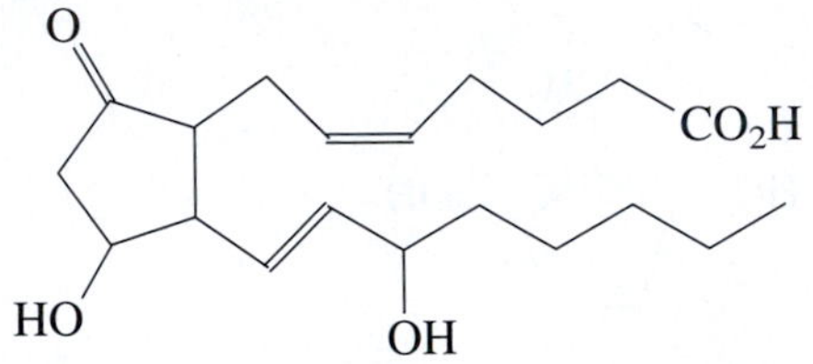

PGE$_2$
(a prostaglandin)

f)

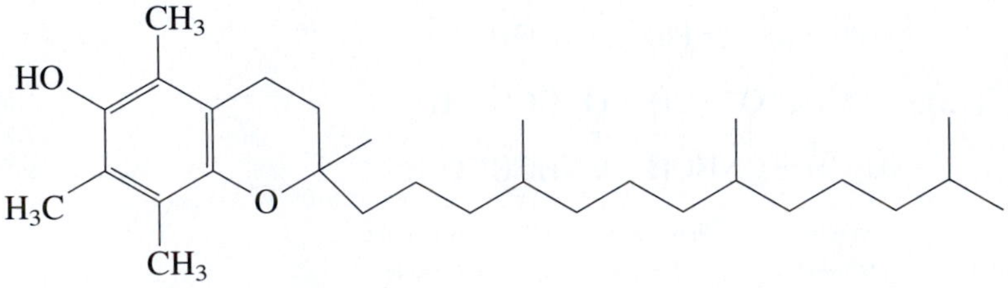

Vitamin E

g)

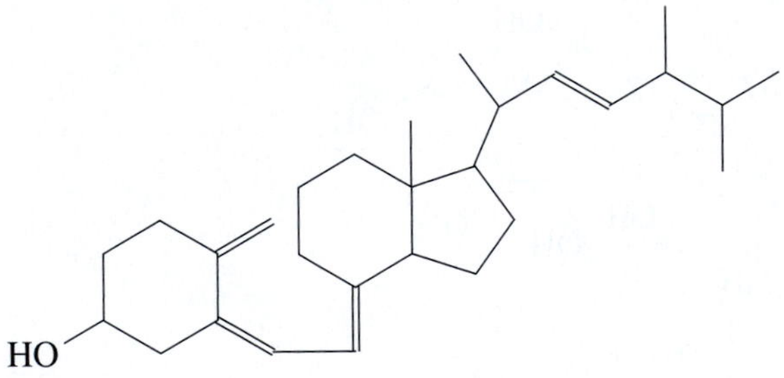

Vitamin D$_2$

h)

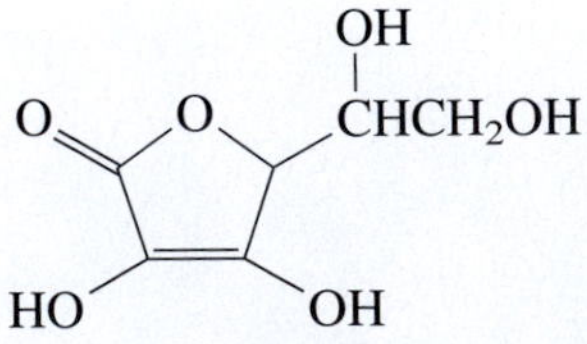

Ascorbic acid
(vitamin C)

i)

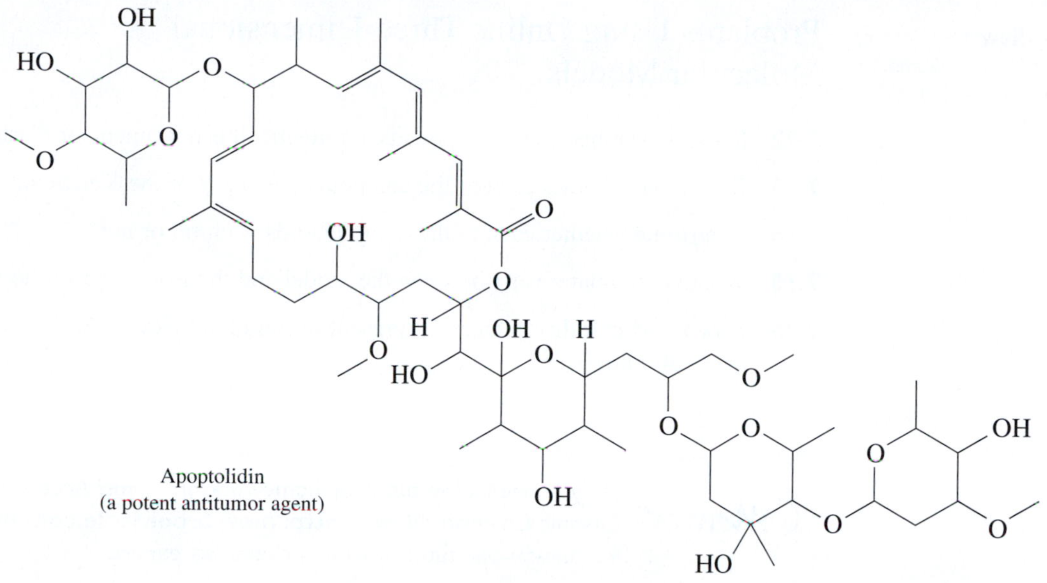

Apoptolidin
(a potent antitumor agent)

7.30 All naturally occurring amino acids have the same relative configuration. All have the S absolute configuration, except for cysteine, which has the R configuration. Explain.

BioLink

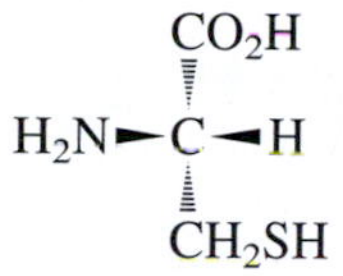

$$H_2N\!-\!\overset{\overset{\displaystyle CO_2H}{|}}{\underset{\underset{\displaystyle CH_2SH}{|}}{C}}\!\!\longrightarrow\!H$$

Cysteine

7.31 Explain whether each of these compounds is chiral or not:

a)

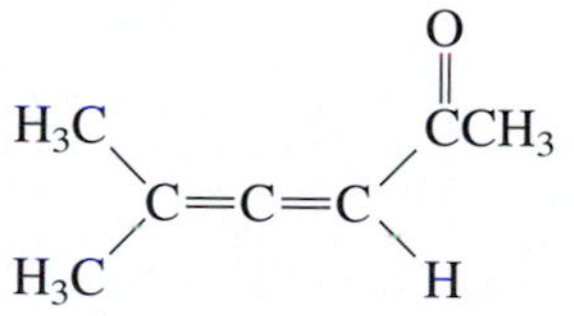

b)

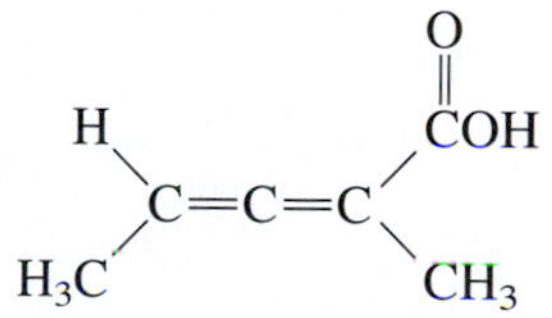

c)

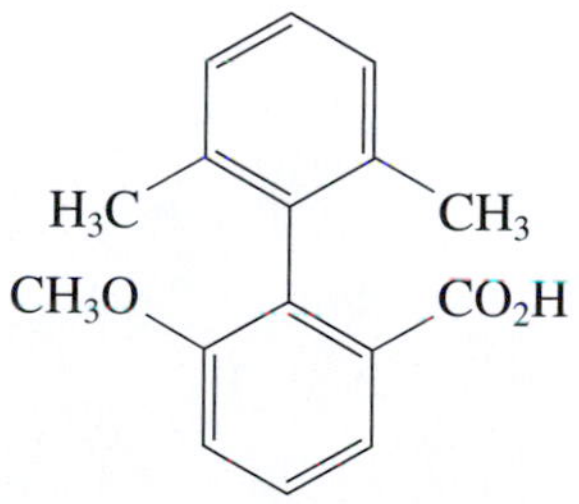

d)

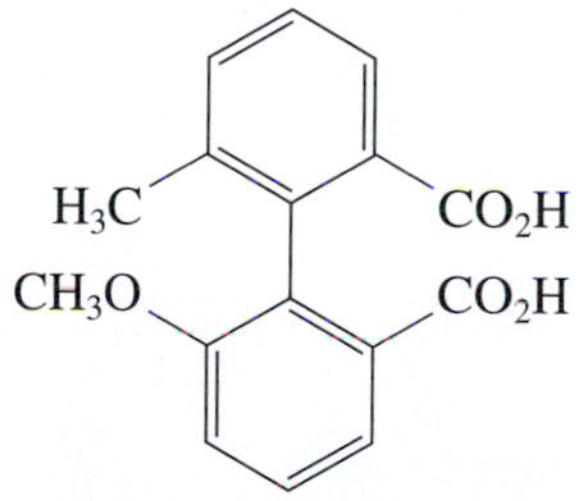

e)

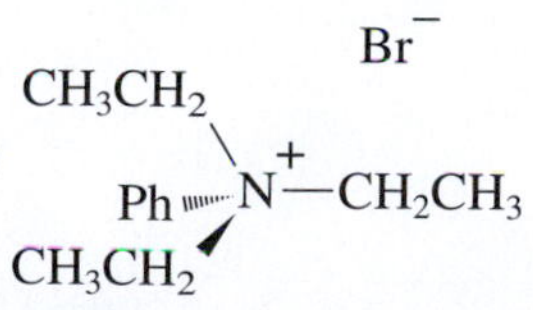

f)

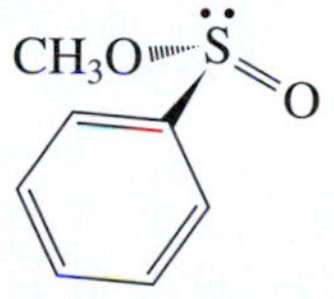

Problems Using Online Three-Dimensional Molecular Models

7.32 Indicate whether these compounds are identical, enantiomers, or diastereomers.

7.33 Determine whether each of the compounds is the *R* or the *S* enantiomer.

7.34 Determine whether each of these compounds is chiral or not.

7.35 What is the relationship between the model and the Fischer projection?

7.36 How many chirality centers are present in estradiol? How many stereoisomers does estradiol have?

Ⓥ**Mentor**™ Do you need a live tutor for homework problems? Access vMentor at Organic ChemistryNow at **http://now.brookscole.com/hornback2** for one-on-one tutoring from a chemistry expert.

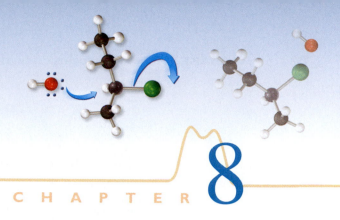

Nucleophilic Substitution Reactions

REACTIONS OF ALKYL HALIDES, ALCOHOLS, AND RELATED COMPOUNDS

THIS CHAPTER provides our first detailed discussion of an organic reaction. In this reaction, a group attached to a carbon is replaced by another group. A simple example of this reaction is

$$HO^- + Cl—CH_3 \longrightarrow HO—CH_3 + Cl^-$$

This reaction is very important and useful because it enables us to exchange one group bonded to carbon (the Cl in this example) with another (the OH). The reaction is examined in considerable detail, using many of the same concepts that were introduced in Chapter 4 to help understand acid–base reactions.

The following features of this reaction are discussed:

- The two basic pathways, or mechanisms, by which the reaction occurs
- The stereochemistry of the reaction, that is, what happens when the carbon is a chirality center
- How other groups on the carbon affect the reaction
- Other groups that can be used in place of chloride ion or hydroxide ion in the reaction
- Competing reactions

In Chapter 9 the competing reactions are examined in more detail. Then, in Chapter 10, applications of these reactions to the preparation or synthesis of other compounds are presented.

MASTERING ORGANIC CHEMISTRY

▶ Learning the Two Mechanisms by Which Substitution Reactions Occur

▶ Recognizing Nucleophiles and Leaving Groups and Understanding the Factors That Control Their Reactivities

▶ Understanding the Factors That Control the Rates of Substitution Reactions

▶ Predicting Which Mechanism Will Occur for a Particular Reaction

▶ Predicting the Products of Substitution Reactions

▶ Predicting the Stereochemistry of the Products

▶ Recognizing When a Rearrangement Reaction Will Occur

▶ Recognizing the Products Formed from Competing Reactions

8.1 THE GENERAL REACTION

Let's examine the reaction of chloromethane with hydroxide ion in detail. We say that hydroxide ion has substituted for chlorine on the methyl group, so this type of reaction is termed a substitution reaction.

Electrophilic carbon

$$H-\overset{..}{\underset{..}{O}}:^{-} \;+\; H-\underset{\underset{H}{|}}{\overset{\overset{H}{|}}{C}}\!\!-\!\!\overset{\delta^{+}}{}\;\overset{\delta^{-}}{\underset{..}{\overset{..}{Cl}}}: \longrightarrow H-\overset{..}{\underset{..}{O}}-\underset{\underset{H}{|}}{\overset{\overset{H}{|}}{C}}\!\!-\!\!H \;+\; :\overset{..}{\underset{..}{Cl}}:^{-}$$

Nucleophile   Leaving group

In this substitution reaction, the hydroxide ion is acting as a Lewis base, using one of its unshared pairs of electrons to form a bond to the carbon, which is reacting as the Lewis acid part of a Lewis acid–base complex. Therefore, these substitution reactions can be viewed as Lewis acid–base reactions. In reactions where carbon is involved, organic chemists have special terms for the Lewis acid and base. The Lewis base is called a **nucleophile,** which is derived from Greek words meaning "nucleus loving." A nucleophile is an electron-rich species that seeks an electron-poor site. The nucleophile can use a pair of its electrons to form a bond to this electron-deficient site. The Lewis acid is called an **electrophile,** which is derived from Greek words meaning "electron loving." An electrophile is an electron-poor species that seeks an electron-rich site, a nucleophile. The electrophile can accept a pair of electrons from the nucleophile and form a bond to it. In the preceding reaction, hydroxide ion is the nucleophile and CH_3Cl is the electrophile. To be more specific, the oxygen is the nucleophilic atom of hydroxide ion. The carbon of CH_3Cl is electron deficient due to the polarization of the carbon–chlorine bond, and it is the electrophilic atom. In the reaction, the nucleophile is attracted to the electrophile, and ultimately a bond forms between the nucleophilic atom and the electrophilic atom. To avoid exceeding the valence of the carbon, the chlorine must depart, taking the electrons of the carbon–chlorine bond with it. The chlorine is called the **leaving group.** The electrostatic potential maps of the reactants in the preceding reaction show electron-rich regions (nucleophilic sites) in red and electron-poor regions (electrophilic sites) in blue.

The concept of nucleophiles and electrophiles is one of the most important in organic chemistry. The reactions in this chapter, as well as most of the reactions that we will study in later chapters, involve a nucleophile bonding to an electrophile. If you can examine a molecule and identify whether it has a nucleophilic or an electrophilic site, not only will you know where that molecule is likely to react but you will also be able to identify what kind of partner is needed for a reaction because nucleophiles react with electrophiles.

Because the nucleophile replaces the leaving group, this type of reaction is termed a **nucleophilic substitution reaction.** A more general equation for a nucleophilic substitution reaction is

$$Nu:^{-} \;+\; R-L \longrightarrow Nu-R \;+\; :L^{-}$$

where Nu^{-} represents a general nucleophile and $R-L$ represents a general leaving group (L) bonded to a carbon group (R).

8.2 REACTION MECHANISMS

As introduced in Chapter 4, a reaction mechanism shows how the nuclei and the electrons move and how the bonds change as the reaction proceeds. It shows the individual steps in a reaction—that is, the order in which the bonds are made and broken. For example, the nucleophilic substitution reaction involves breaking one bond, the bond between the carbon and the leaving group, and forming one bond, the bond between the nucleophile and the carbon. There are three possible timings for these events—three possible mechanisms: (1) the bond to the leaving group may be broken first, followed by formation of the bond to the nucleophile; (2) the bond to the nucleophile may be formed first, followed by breaking the bond to the leaving group; or (3) bond breaking and bond formation may occur simultaneously. Pathways (1) and (3) both occur. Mechanism (2) does not occur, because the intermediate that would form when the nucleophile bonds first has five bonds to the carbon and cannot exist because the valence of the carbon would be exceeded.

Although certain mechanisms for a reaction can be eliminated on the basis of experimental evidence, it is never possible to prove that the reaction follows a particular mechanism. It can only be demonstrated that all the experimental facts are consistent with that mechanism. One piece of experimental information that is of primary importance is the rate law that the reaction follows. The rate law predicted by a possible mechanism must be consistent with the rate law determined in the laboratory. If the two are not consistent, that mechanism can be ruled out. In the case of these nucleophilic substitution reactions, experimental studies have shown that two different rate laws are followed, depending on the substrate (R—L), the nucleophile, and the reaction conditions. This means that there must be two different mechanisms for the reaction. Let's look at each.

8.3 BIMOLECULAR NUCLEOPHILIC SUBSTITUTION

Consider the reaction of hydroxide ion with chloroethane:

$$HO^- + CH_3CH_2—Cl \longrightarrow CH_3CH_2—OH + Cl^-$$

Investigation of this reaction in the laboratory has shown that the reaction rate depends on the concentration of hydroxide ion and on the concentration of chloroethane (EtCl), that is, the reaction follows the second-order rate law:

$$\text{rate} = k[\text{EtCl}][\text{OH}^-]$$

From general chemistry you might recall that the dependence of the rate law on the concentration of a particular species requires that species to be involved in the slowest step of the reaction or a step before the slowest step. Therefore, in this case, both hydroxide ion and chloroethane must be present in the slowest step of the reaction.

This rate law is consistent with mechanism (3), in which the bond to the leaving group (chloride) is broken and the bond to the nucleophile (hydroxide) is formed simultaneously, in the same step. A reaction that occurs in one step is termed a **concerted reaction.** Because two species (hydroxide ion and chloroethane) are involved in this step, the step is said to be bimolecular. This reaction is therefore described as a **bimolecular nucleophilic substitution** reaction, or an S_N2 reaction.

Diagrams that show how the free energy changes as a reaction proceeds were introduced in Chapter 4 and are very useful. Figure 8.1 shows such a diagram for the S_N2 reaction. Recall that the free energy, G, of the system is shown on the y-axis in these diagrams, and the progress of the reaction, which is just a measure of how much the reaction has proceeded from reactants toward products, is shown along the x-axis. The reactants are shown on the left-hand side of the diagram and the products on the right-hand side.

In this case the products are lower in energy than the reactants, so $\Delta G°$ is negative (the reaction is **exergonic**). Such reactions proceed spontaneously to the right. (Reactions in which the free-energy change is positive are **endergonic** and are not spontaneous. The terms **exothermic** and **endothermic** refer to reactions in which the enthalpy change [$\Delta H°$] is negative and positive, respectively.)

Of more importance in terms of the mechanism is what happens to the energy of the system as the reaction progresses from the reactants to the products. As the reaction starts, the bond to the chlorine begins to break and the bond to the hydroxide ion begins to form. Initially, breaking the bond costs more energy than is returned by forming the other bond, so the energy of the system increases. The energy of the system continues to increase until it reaches a maximum where both bonds are approximately half formed (or broken). The structure of the complex at this energy maximum is called the **transition state.** Then the energy begins to drop as the energy decrease from forming the new bond outweighs the energy increase from breaking the bond to the leaving chlorine. The energy difference between the transition state and the reactants is the **free energy of activation, $\Delta G^{\ddagger}$.**

The free energy versus reaction progress diagram shown in Figure 8.1 is typical for a concerted reaction. There is only one energy maximum; there is one transition state, between reactants and products; and there is no minimum between them.

What does the transition state for the S_N2 reaction look like? Because it is a maximum on the free energy versus reaction progress diagram and any change, either forward to products or backward to reactants, is downhill in energy, the transition state has no appreciable lifetime. Because of this, it cannot be observed directly and any information

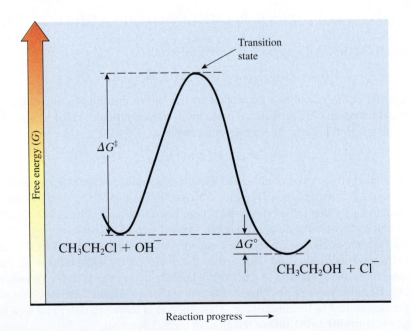

Figure 8.1

FREE ENERGY VERSUS REACTION PROGRESS DIAGRAM FOR THE S_N2 REACTION OF CHLOROETHANE AND HYDROXIDE ION.

about its structure must be obtained by indirect means. However, we do know that the transition state must have a five-coordinate carbon—that is, a carbon with five bonds: three normal bonds and two partial bonds. The geometry of these bonds in the transition state has been determined by investigation of the stereochemistry of the reaction.

8.4 STEREOCHEMISTRY OF THE S$_N$2 REACTION

What happens in the S$_N$2 reaction if the leaving group is attached to a carbon that is a chirality center, that is, one that is bonded to four different groups (the leaving group and three other, different groups)? *Possible* stereochemical outcomes are illustrated in Figure 8.2 for the reaction of hydroxide ion with (*S*)-2-chlorobutane. In possibility 1 the product has the same relative configuration as the reactant. In such a case we say that the reaction has occurred with **retention of configuration.** In possibility 2 the product has the opposite relative configuration to the reactant. In this case the reaction has occurred with **inversion of configuration.** In possibility 3, complete randomization of stereochemistry has occurred in the product. The reaction has occurred with **racemization,** or 50% inversion and 50% retention. (Of course, partial racemization, resulting in different ratios of inversion to retention, is also possible.)

When the reaction of 2-chlorobutane with hydroxide ion is run in the laboratory, the rate is found to depend on the concentration of both species. This indicates that the re-

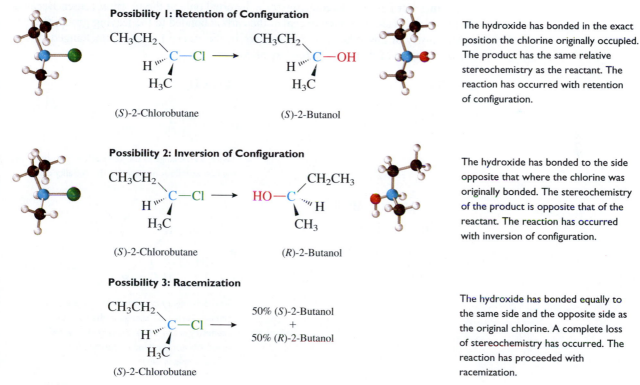

Figure 8.2

POSSIBLE STEREOCHEMICAL OUTCOMES FOR THE REACTION OF (*S*)-2-CHLOROBUTANE WITH HYDROXIDE ION. Only possibility 2 actually occurs.

action is following the S_N2 mechanism. Investigation of the stereochemistry of the reaction shows that the product is formed with inversion of configuration; that is, (S)-2-chlorobutane produces (R)-2-butanol, corresponding to possibility 2 in Figure 8.2. The same result has been found for all S_N2 reactions. S_N2 reactions occur with inversion of configuration at the reaction center.

The fact that S_N2 reactions *always* occur with inversion of configuration enables us to form a better picture of the transition state. The nucleophile must approach the carbon from the side opposite the leaving group **(back-side attack).** The structure of the transition state, with partial bonds to the entering hydroxide and the leaving chloride, is shown in the following structure. Figure 8.3 uses orbitals to show how this process occurs.

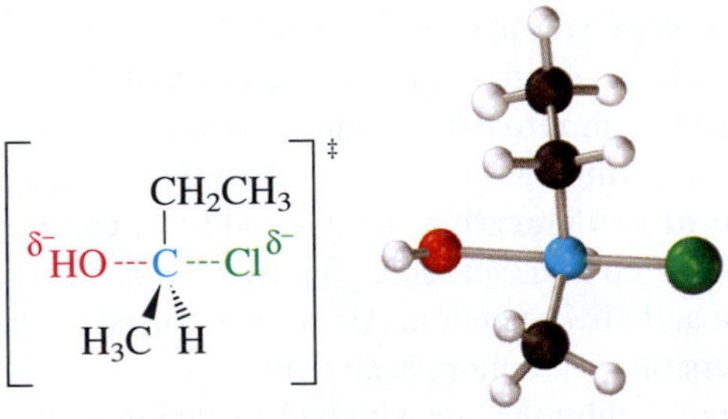

S_N2 transition state

Remember that the stereochemistry is inverted only at the reaction center. Because no bonds are made or broken except at the carbon bonded to the leaving group, the stereochemistry at any other chirality centers in the reactant remains unchanged. Some more examples of the S_N2 reaction are as follows:

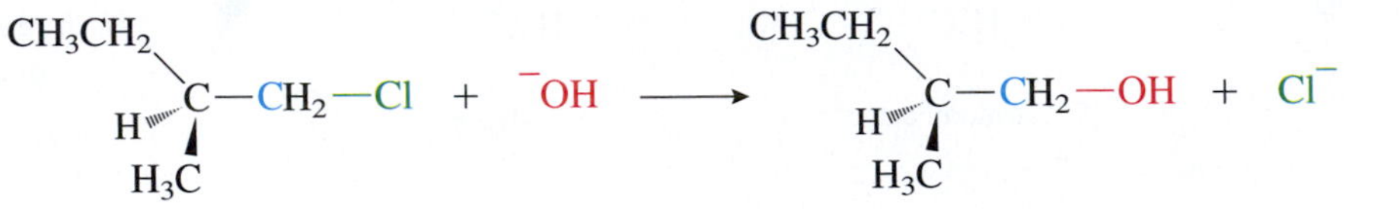

There is no change in the stereochemistry because the reaction does not involve the chirality center.

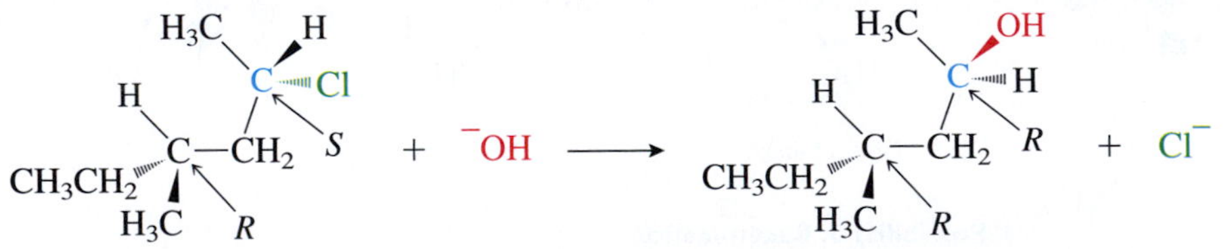

Stereochemistry is inverted at the reaction center; there is no change in the stereochemistry at the other chirality center because it is not involved in the reaction. The next example is similar.

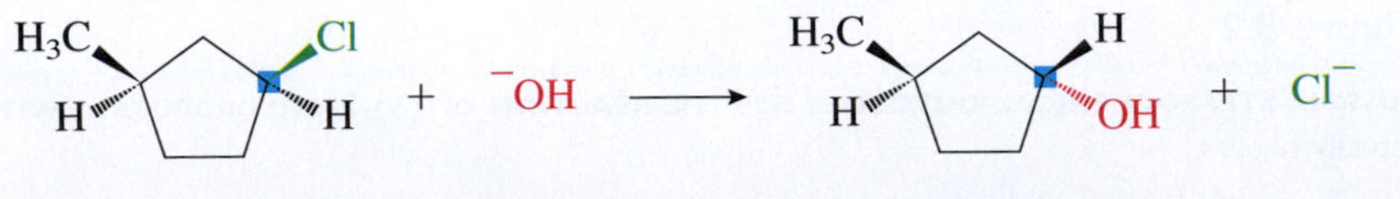

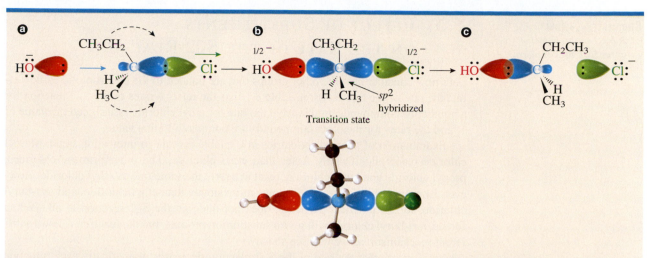

ⓐ The nucleophile approaches the carbon from the side opposite the leaving group (back-side attack). An atomic orbital on the nucleophile, containing an unshared pair of electrons, begins to interact with the back lobe of the sp^3 orbital used by the carbon to form the sigma bond to the leaving group. As the nucleophile continues to approach, the leaving group begins to move away from the carbon, and the three other groups begin to swing to the right as the hybridization of the carbon changes.

ⓑ At the transition state, the carbon has trigonal planar geometry and is sp^2 hybridized. The AO of the nucleophile overlaps with one lobe of the p orbital on the carbon, and the AO of the leaving group overlaps with the other lobe of this orbital. The overall geometry is trigonal bipyramidal, with partial bonds to both Nu and L and approximate charges of negative 1/2 on each.

ⓒ As the nucleophile continues to approach, the distance between the carbon and the leaving group continues to increase, and the other three groups continue to swing to the right. The hybridization of the carbon changes back to sp^3, and its geometry returns to tetrahedral but with the nucleophile bonded to the side opposite to the one to which the leaving group was originally bonded.

Active Figure 8.3

ORGANIC
Chemistry ⬥ Now™

MECHANISM OF THE S$_N$2 REACTION OF (s)-2-CHLOROBUTANE AND HYDROXIDE ION SHOWING ORBITALS. Test yourself on the concepts in this figure at **OrganicChemistryNow.**

PROBLEM 8.1

Show the products, including stereochemistry, of these S$_N$2 reactions:

a) 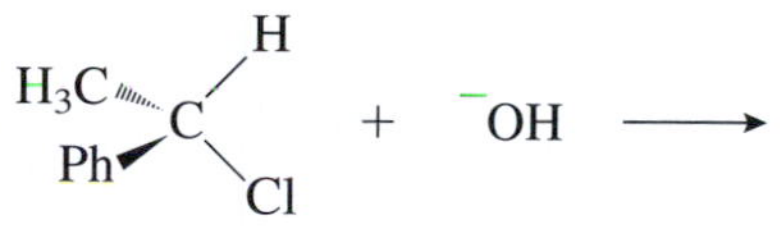 $\underset{Ph}{\overset{H_3C}{\diagdown}} \underset{}{C} \overset{H}{\underset{Cl}{\diagup}}$ + $^-$OH $\longrightarrow$

b) 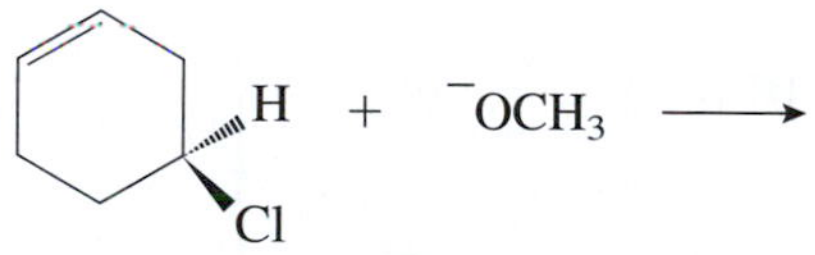 (cyclohexene with H and Cl substituents) + $^-$OCH$_3$ $\longrightarrow$

c) (cyclopentane with CH$_3$, CH$_2$CH$_3$, and CH$_2$CH$_2$Cl substituents) + $^-$OH $\longrightarrow$

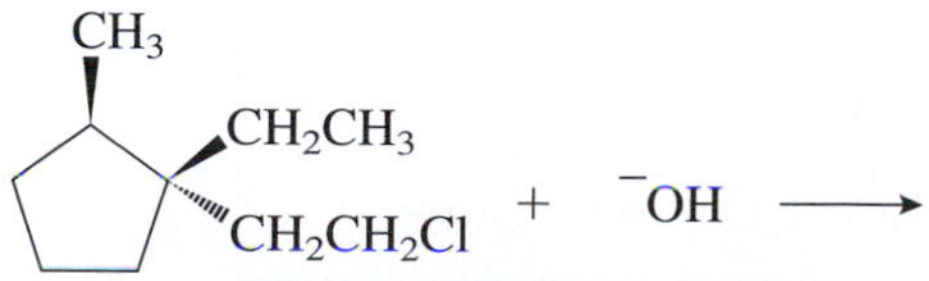

ORGANIC
Chemistry·⚛·Now™
Click *Mechanisms in Motion* to
view this **S$_N$2 Mechanism.**

8.5   EFFECT OF SUBSTITUENTS
ON THE RATE OF THE S$_N$2 REACTION

Let's now consider how the other groups that are bonded to the electrophilic carbon affect the rate of the S$_N$2 reaction. Table 8.1 lists the relative rates of the S$_N$2 reaction for a number of compounds. In this table the rate for ethyl chloride is assigned the value of 1, and the rates for the other compounds are compared to this value.

Examination of the first six entries in the table reveals an interesting trend. Methyl chloride reacts significantly faster than ethyl chloride. The other primary chlorides, propyl chloride and butyl chloride, react at nearly the same rate as ethyl chloride. However, isopropyl chloride reacts 40 times more slowly than ethyl chloride, and *tert*-butyl chloride, for all practical purposes, does not undergo the S$_N$2 reaction at all. (When forced, *tert*-butyl chloride will give a substitution product, but the reaction follows a different mechanism, as we will see shortly.)

You can see that the rate of the S$_N$2 reaction decreases dramatically each time one of the hydrogens on the electrophilic carbon of methyl chloride is replaced with a

Table 8.1   Relative Rates of S$_N$2 Reactions for Selected Compounds

Name	Structure	Relative Rate
Methyl chloride	CH_3Cl	30
Ethyl chloride	CH_3CH_2Cl	1
Propyl chloride	$CH_3CH_2CH_2Cl$	0.4
Butyl chloride	$CH_3CH_2CH_2CH_2Cl$	0.4
Isopropyl chloride	$CH_3\overset{\displaystyle CH_3}{\underset{\vert}{C}}HCl$	0.025
tert-Butyl chloride	$CH_3\overset{\displaystyle CH_3}{\underset{\underset{\displaystyle CH_3}{\vert}}{\overset{\vert}{C}}}Cl$	0
Neopentyl chloride	$CH_3\overset{\displaystyle CH_3}{\underset{\underset{\displaystyle CH_3}{\vert}}{\overset{\vert}{C}}}CH_2Cl$	10^{-5}
Allyl chloride	$CH_2{=}CHCH_2Cl$	40
Benzyl chloride	⬡—CH_2Cl	120
Chloroacetone	$CH_3\overset{\displaystyle O}{\overset{\|}{C}}CH_2Cl$	10^5

methyl group. Thus, replacing one hydrogen with a methyl group, to give ethyl chloride, causes the rate to decrease by a factor of 30. Replacing a second hydrogen with a methyl, to give isopropyl chloride, results in a further rate decrease by a factor of 40. By the time three methyl groups have been added, to give *tert*-butyl chloride, the compound is unreactive in the S_N2 reaction. This effect is a result of the larger size of the methyl group (or other carbon groups) as compared to the size of hydrogen—a **steric effect.** The steric effect is a result of increasing strain energy in the transition state as the size of the groups on the electrophilic carbon increases.

Figure 8.4 illustrates this effect. It shows free energy versus reaction progress diagrams for the S_N2 reactions of methyl chloride, ethyl chloride, and isopropyl chloride with hydroxide ion. In the case of methyl chloride, very little steric strain is introduced into the transition state by the interaction of the hydroxide ion nucleophile with the hydrogens. In the case of ethyl chloride, the interaction between the nucleophile and the methyl group on the electrophilic carbon generates more steric strain in the transition state, causing it to increase in energy. This results in an increase in $\Delta G^{\ddagger}$ and slows the reaction. In the case of isopropyl chloride, the interaction of the nucleophile with two methyl groups causes even more steric strain in the transition state and slows the reaction even more. When three methyl groups are present, as in the case of *tert*-butyl chloride, so much strain is present in the transition state that the rate of the S_N2 reaction is extremely slow.

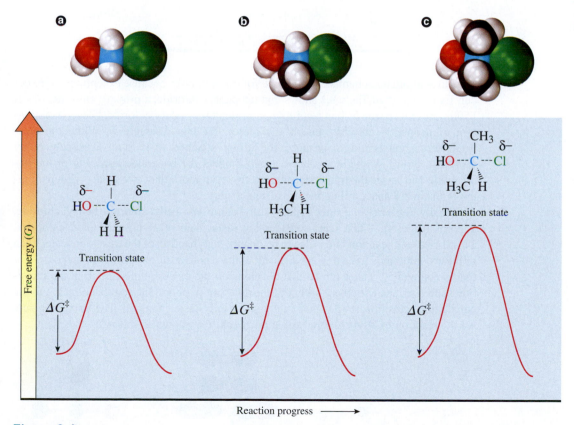

Figure 8.4

FREE ENERGY VERSUS REACTION PROGRESS DIAGRAMS FOR THE S_N2 REACTIONS OF ⓐ METHYL CHLORIDE, ⓑ ETHYL CHLORIDE, AND ⓒ ISOPROPYL CHLORIDE WITH HYDROXIDE ION.

Figure 8.5 provides another view of this effect. It shows computer-generated space-filling models of some of the molecules from Table 8.1 and illustrates the increasing difficulty the nucleophile experiences when approaching the back side of the electrophilic carbon and reaching the transition state as the number of methyl groups bonded to that carbon increases.

Other primary alkyl groups have effects similar to that of the methyl group. Replacing a hydrogen on the electrophilic carbon of methyl chloride with an ethyl group rather than a methyl group causes only a slightly larger rate decrease (compare the relative rates of ethyl chloride and propyl chloride in Table 8.1). This indicates that, as far as this mechanism is concerned, an ethyl group is only slightly "larger" than a methyl group, a result that is consistent with the axial destabilization energies of these groups discussed in Chapter 6.

To summarize, the rate of the S_N2 reaction is controlled by steric factors at the electrophilic carbon. Steric hindrance slows the reaction. Based on the number of carbon groups attached to that carbon, the reactivity order is

Rules in organic chemistry cannot be followed blindly, because exceptions often occur. Examination of Table 8.1 shows that neopentyl chloride, a primary chloride, reacts 2500 times more *slowly* than isopropyl chloride, a secondary chloride. Closer examination of neopentyl chloride reveals the reason for this discrepancy. While the electrophilic carbon is indeed primary, the group attached to it is an extremely bulky *tert*-butyl group. A single *tert*-butyl group hinders the back-side approach of the nucleophile and raises the transition state energy even more than two methyl groups. Figure 8.5 shows a space-filling picture of this group.

Finally, Table 8.1 lists three primary alkyl chlorides—allyl chloride, benzyl chloride, and chloroacetone—that react considerably faster than other primary alkyl chlorides. This increase in reaction rate is due to resonance stabilization of the transition state. Each of these compounds has a pi bond adjacent to the reactive site and forms a transition state that is conjugated. The *p* orbital that develops on the electrophilic carbon in the transition state overlaps with the *p* orbital of the adjacent pi bond. The stabilization due to the conjugated transition state results in a significantly faster reaction. The transition state for the reaction of allyl chloride with a nucleophile is shown as follows:

3-Chloro-1-propene   Transition state
(allyl chloride)

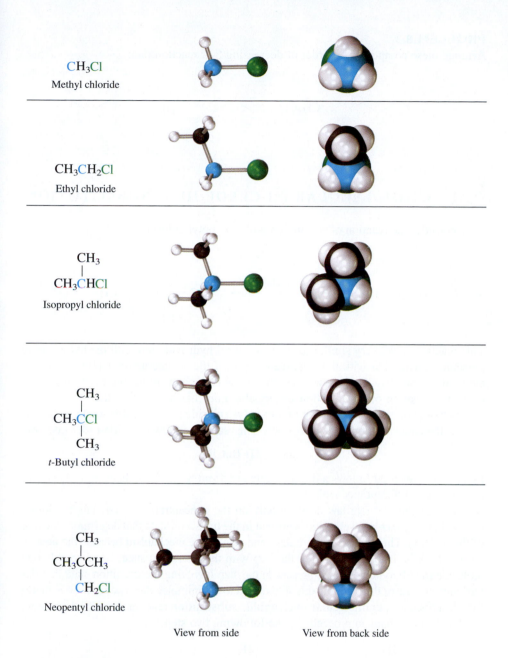

CH_3Cl
Methyl chloride

CH_3CH_2Cl
Ethyl chloride

CH_3
|
CH_3CHCl
Isopropyl chloride

CH_3
|
CH_3CCl
|
CH_3
t-Butyl chloride

CH_3
|
CH_3CCH_3
|
CH_2Cl
Neopentyl chloride

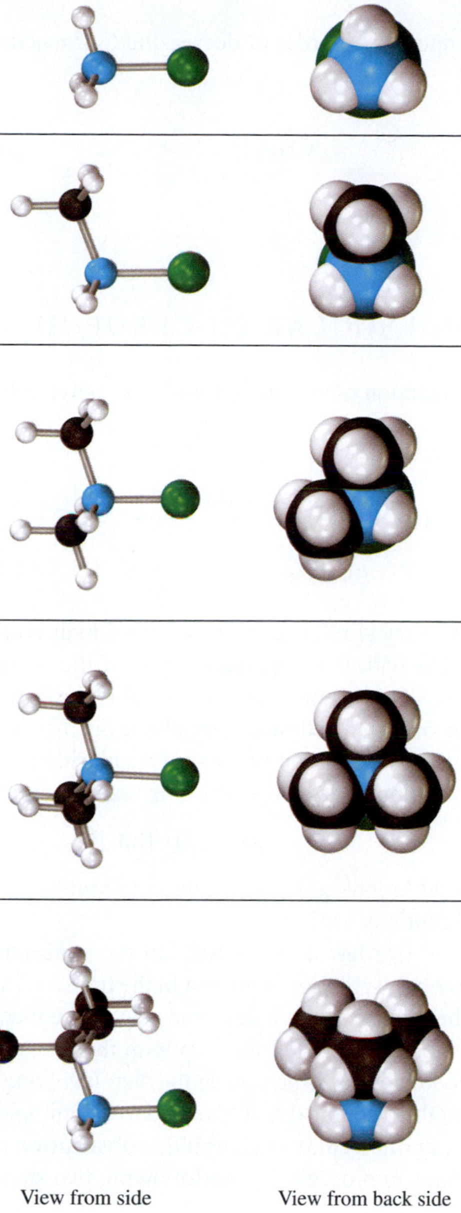
View from side
View from back side

Figure 8.5

SOME ALKYL CHLORIDES. Chlorine is green, and the electrophilic carbon is blue.

PROBLEM 8.2

Explain which compound has a faster rate of S_N2 reaction:

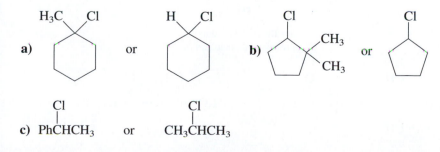

a) 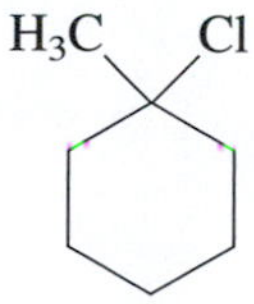 or 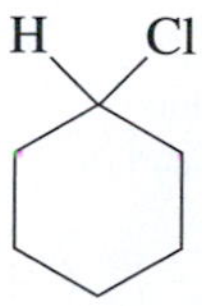

b) 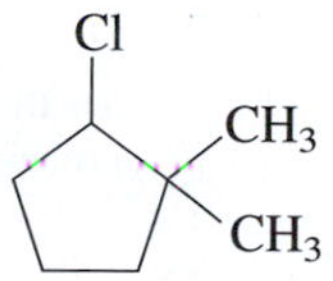 or 

c) PhCH(Cl)CH$_3$ or CH$_3$CH(Cl)CH$_3$

PROBLEM 8.3

Arrange these compounds in order of decreasing S_N2 reaction rate:

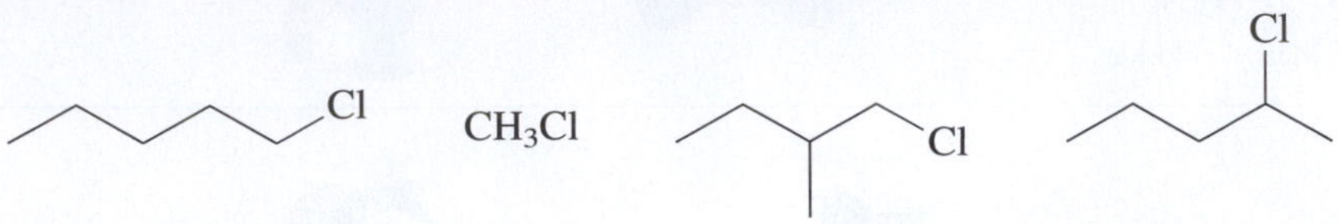

8.6   UNIMOLECULAR NUCLEOPHILIC SUBSTITUTION

Now consider the reaction of acetate ion with *tert*-butyl chloride:

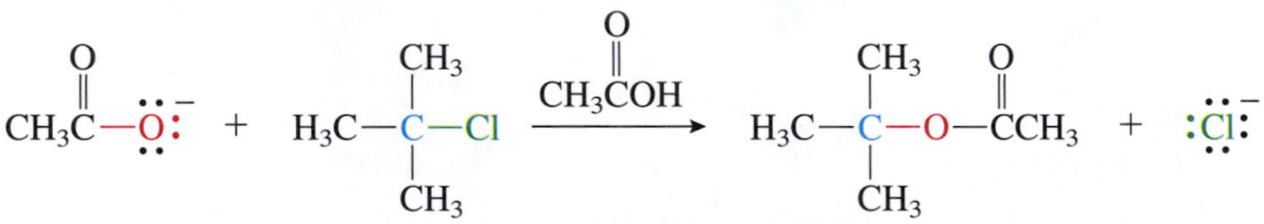

This reaction looks very similar to the reaction of hydroxide ion with methyl chloride presented earlier, but with the negative oxygen of the acetate anion acting as the nucleophile. (The CH_3CO_2H shown over the arrow is the solvent for the reaction.) However, investigation of this reaction in the laboratory has shown that the reaction rate depends only on the concentration of *tert*-butyl chloride (*t*-BuCl). It is totally independent of the concentration of acetate anion. The reaction follows the first-order rate law:

$$\text{rate} = k[\text{t-BuCl}]$$

Because the reaction follows a different rate law from the S_N2 mechanism, it must also proceed by a different mechanism.

The fact that the rate law depends only on the concentration of *tert*-butyl chloride means that only *tert*-butyl chloride is present in the transition state that determines the rate of the reaction. There must be more than one step in the mechanism because the acetate ion must not be involved until after the step with this transition state. Because only one molecule (*tert*-butyl chloride) is present in the step involving the transition state that determines the rate of the reaction, this step is said to be unimolecular. The reaction is therefore described as a **unimolecular nucleophilic substitution** reaction, or an **S_N1** reaction.

The S_N1 mechanism proceeds by the following two steps:

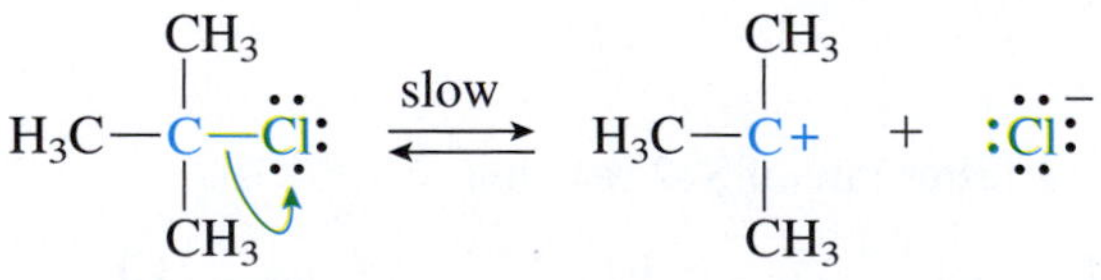

tert-Butyl carbocation
(a reactive intermediate)

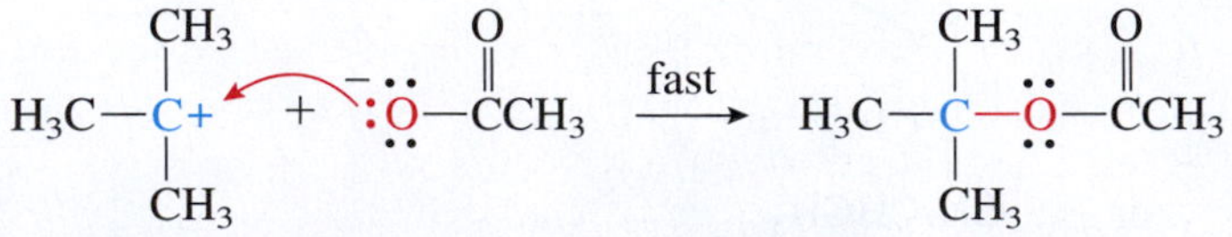

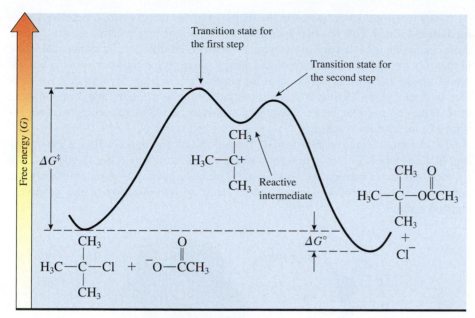

Figure 8.6

**FREE ENERGY VERSUS
REACTION PROGRESS
DIAGRAM FOR THE S$_N$1
REACTION OF *TERT*-BUTYL
CHLORIDE (2-CHLORO-2-
METHYLPROPANE) AND
ACETATE ANION.**

In this mechanism, the bond to the chloride is broken in the first step and the bond to the acetate is formed in the second step. A free energy versus reaction progress diagram for the S$_N$1 mechanism is shown in Figure 8.6. In this reaction, each of the two steps has an energy maximum or transition state separating its reactant and product, which are both at energy minima. The transition state for the first step is at higher energy in this case. Once a molecule makes it over the higher-energy barrier of the first step, it has enough energy to proceed rapidly over the lower-energy barrier of the second step. The first step is called the **rate-limiting** or **rate-determining step** because it determines the rate of the reaction. It acts as a kind of bottleneck for the reaction. The rate of this reaction should depend only on the concentration of *tert*-butyl chloride because it is the only molecule involved in the rate-determining step. Therefore, this mechanism is consistent with the experimentally determined rate law.

In general, for a nonconcerted reaction, that is, a reaction that proceeds in several steps, the free energy versus reaction progress diagram has a separate transition state for each step. One or more intermediates are present along the reaction pathway, each of these located at an energy minimum. These intermediates may be located at relatively high energy and have only a transient existence, such as the carbocation formed in the S$_N$1 reaction, or they may be located at lower energy and have a longer lifetime. If one of the transition states is located at significantly higher energy than the others, then that step is the rate-determining step for the reaction. Molecules that become involved in the mechanism after the rate-determining step do not appear in the rate law for the reaction.

PROBLEM 8.4

Draw a free energy versus reaction progress diagram for a reaction that occurs in two steps with a relatively stable intermediate and in which the transition state for the second step is the highest-energy transition state.

When the chloride ion leaves in the first step of the mechanism, a **reactive intermediate** is formed. This reactive intermediate is a high-energy, reactive species. Under most conditions it has a very short lifetime. However, it differs from a transition state in that it is located at a minimum on the energy curve. It has an activation barrier, although small, that must be surmounted for reaction in either the forward or reverse direction. Although its lifetime is short, it is significantly longer than that of a transition state. It may be possible, under certain circumstances, to obtain experimental observations of a reactive intermediate.

The particular reactive intermediate formed in this reaction is called a **carbocation.** It has a carbon with only three bonds and a positive charge. This carbon has only six electrons in its valence shell and is quite unstable because it does not satisfy the octet rule. It has trigonal planar geometry and sp^2 hybridization at the positively charged carbon.

Empty *p* orbital

$$H_3C \overset{+}{-} C \overset{\text{\tiny\ldots}CH_3}{\underset{CH_3}{\diagdown}}$$

sp^2 hybridized

tert-Butyl carbocation

Carbocations are one of the most important types of reactive intermediates in organic chemistry. They are encountered in many reactions in addition to the S_N1 reaction.

Let's now turn our attention to the transition state for this reaction. What is the structure of the transition state? This is an important question because a better understanding of its structure will help in predicting how various factors affect its stability and therefore will aid in predicting how these same factors affect the rate of the reaction. The transition state has a structure that is intermediate between that of the reactant, *tert*-butyl chloride, and that of the product, the *tert*-butyl carbocation. It has the bond between the carbon and the chlorine partially broken and can be represented as shown in the following structures:

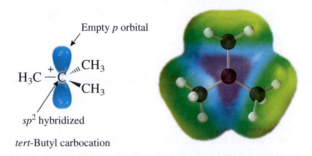

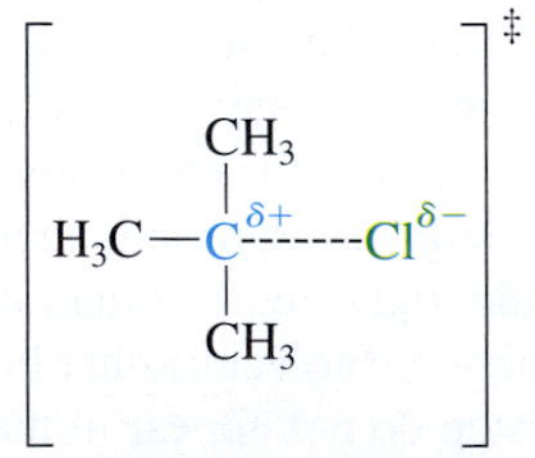

This orbital is between an sp^3 hybrid AO and a *p* orbital.

$$\left[H_3C - \overset{CH_3}{\underset{CH_3}{\overset{|}{\underset{|}{C}}}} \overset{\delta+}{} \text{------} Cl^{\delta-} \right]^{\ddagger}$$

S_N1 transition state

or

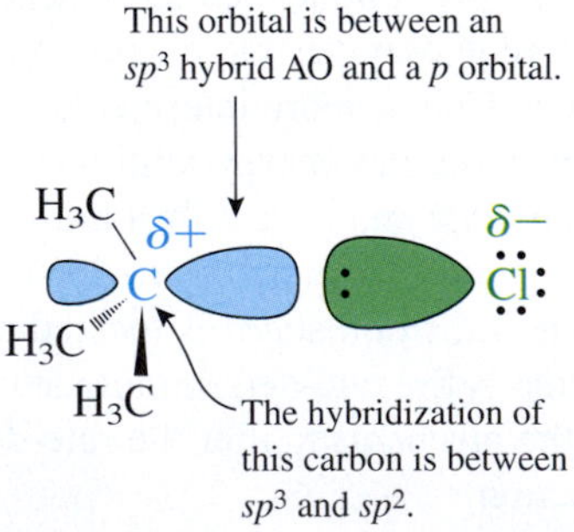

The hybridization of this carbon is between sp^3 and sp^2.

It has a partial positive charge on the carbon and a partial negative charge on the chlorine. The hybridization of the carbon is between that of the reactant, sp^3, and that of the carbocation, sp^2.

Is the bond in the transition state more or less than half broken? The **Hammond postulate** enables questions such as this to be answered. It states that the structure of the

Exergonic Reaction

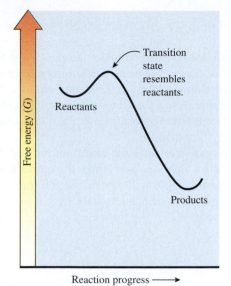

Endergonic Reaction

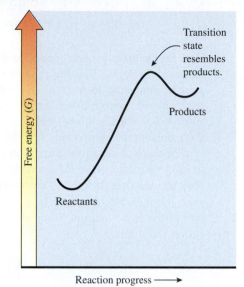

ⓐ For an exergonic reaction, in which the reactants are at higher energy than the products, the energy of the transition state is closer to that of the reactants than that of the products. Therefore, the structure of the transition state resembles that of the reactants more than that of the products. If a bond is forming in the reaction, that bond is less than half formed in the transition state, and if a bond is breaking, it is less than half broken.

ⓑ For an endergonic reaction, in which the products are at higher energy than the reactants, the energy of the transition state is closer to that of the products. Therefore, its structure is also closer to that of the products. Any bonds that are forming in the reaction are more than half formed, and any bonds that are breaking are more than half broken.

Figure 8.7

USING THE HAMMOND POSTULATE TO PREDICT THE STRUCTURE OF A TRANSITION STATE ⓐ EXERGONIC REACTION AND ⓑ ENDERGONIC REACTION.

transition state for a reaction step is closer to that of the species (reactant or product of that step) to which it is closer in energy. If the product of the step is higher in energy than the reactant, the structure of the transition state is more similar to that of the product than it is to that of the reactant. In contrast, if the reactant is higher in energy than the product, the structure of the transition state is more similar to that of the reactant than the product (see Figure 8.7). Because the carbocation is much higher in energy than the starting alkyl halide (the slow step of the mechanism in Figure 8.6 corresponds to the case on the right in Figure 8.7), the structure of the transition state for the S_N1 reaction is closer to that of the carbocation; the bond is more than half broken.

PROBLEM 8.5

Consider the free energy versus reaction progress diagram for the S_N2 reaction shown in Figure 8.1. Does the transition state for this reaction have the C—Cl bond less than half broken, approximately half broken, or more than half broken?

8.7 EFFECT OF SUBSTITUENTS ON THE RATE OF THE S_N1 REACTION

How do the other groups bonded to the electrophilic carbon affect the rate of the S_N1 reaction? Table 8.2 lists the relative rates of the S_N1 reaction for a number of compounds, compared to the rate for isopropyl chloride taken as 1. Methyl chloride and ethyl chloride are not listed in the table because *methyl and simple primary alkyl chlorides do not react by the S_N1 mechanism.* Even under the most favorable S_N1 conditions, these unhindered compounds react by the S_N2 mechanism.

For the S_N1 reaction, formation of the carbocation is the rate-limiting step. We have already seen that the transition state for this step resembles the carbocation. Any change that makes the carbocation more stable will also make the transition state more stable, resulting in a faster reaction. Carbocation stability controls the rate of the S_N1 reaction. Many studies have provided the following order of carbocation stabilities:

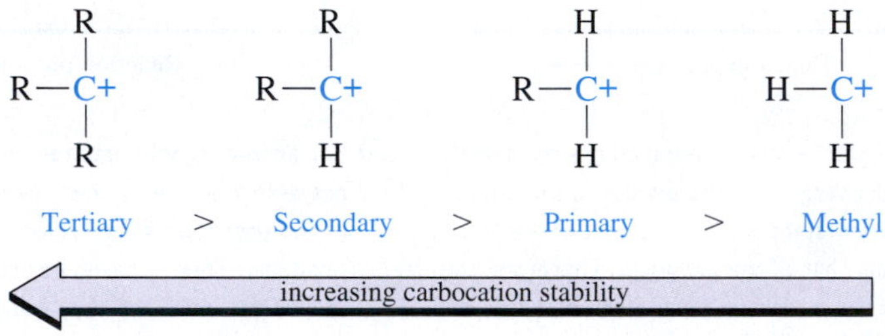

This stability order is important to remember because carbocations occur as intermediates in several other reactions.

Table 8.2   Relative Rates of S_N1 Reactions for Selected Compounds

Name	Structure	Relative Rate
Isopropyl chloride	$CH_3CH{-}Cl$ (with CH_3)	1
tert-Butyl chloride	$CH_3C{-}Cl$ (with CH_3 above and CH_3 below)	1×10^5
Allyl chloride	$CH_2{=}CHCH_2{-}Cl$	3
Benzyl chloride	$PhCH_2{-}Cl$	30
Diphenylmethyl chloride	$Ph_2CH{-}Cl$	1×10^4
Triphenylmethyl chloride	$Ph_3C{-}Cl$	1×10^9

This order shows that the substitution of a methyl group for a hydrogen on a carbocation results in considerable stabilization. (The substitution of other alkyl groups provides a similar stabilization.) This is due to the overlap of a sigma bonding MO from the adjacent carbon with the empty *p* orbital of the carbocation. This overlap forms a conjugated system and allows electron density to flow from the sigma bond to the electron-deficient carbon. This overlap can be illustrated for the ethyl cation as follows:

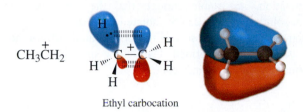

$$CH_3\overset{+}{C}H_2$$

Ethyl carbocation

The sigma MO and the empty *p* AO are coplanar, so they overlap in a manner similar to a pi bond, even though they are not parallel. This overlap provides a path for the electrons of the sigma bond to be delocalized into the empty *p* orbital, thus helping to stabilize the carbocation. Other kinds of sigma bonds can interact with an empty *p* orbital in a similar fashion, as long as the sigma MO and the *p* AO are on adjacent carbons. Such a stabilizing interaction is termed **hyperconjugation.**

Further examination of Table 8.2 shows that allyl chloride and benzyl chloride have much faster rates for S$_N$1 reactions than would be expected for primary systems. Examination of the carbocations reveals that the reason for this enhanced reactivity is the significant resonance stabilization provided by the adjacent double bond or benzene ring. Resonance stabilization increases with the substitution of additional phenyl groups, as illustrated by the reaction rates of diphenylmethyl and triphenylmethyl chloride (Table 8.2).

$$CH_2{=}CH{-}\overset{+}{C}H_2 \longleftrightarrow \overset{+}{C}H_2{-}CH{=}CH_2$$

Allyl carbocation

Benzyl carbocation

Focus On

The Triphenylmethyl Carbocation

Most carbocations are quite unstable and have only a fleeting existence as intermediates in reactions such as the S_N1 substitution. However, some, such as the triphenylmethyl carbocation, are stable enough that they can exist in significant concentrations in solution or even can be isolated as salts.

When triphenylmethanol is dissolved in concentrated sulfuric acid, a solution with an intense yellow color is formed. The yellow species is the triphenylmethyl carbocation, formed by the following reaction:

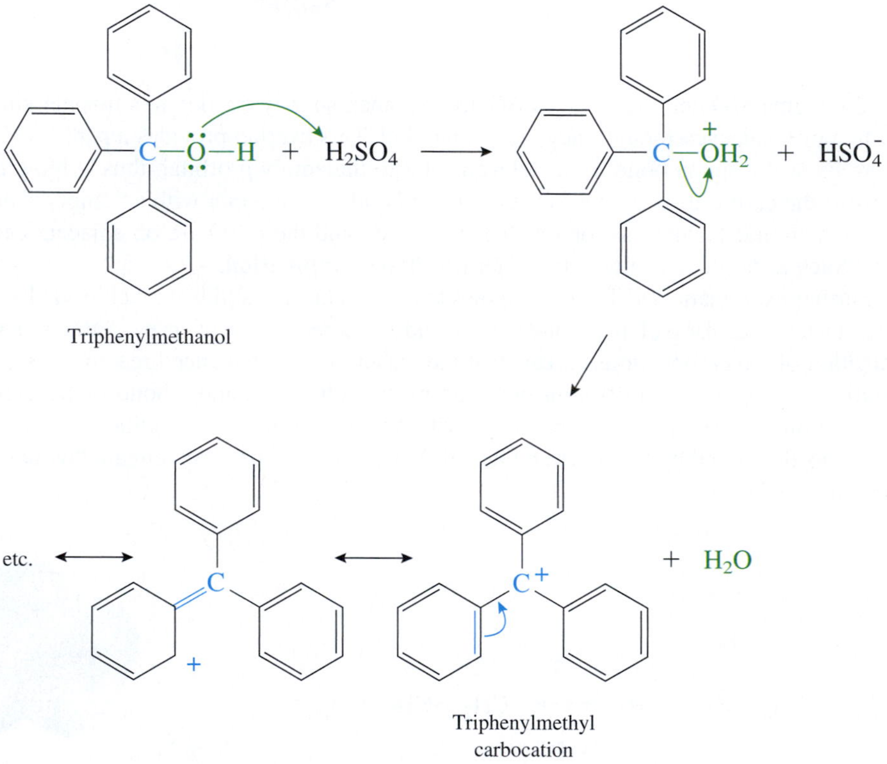

Triphenylmethanol

Triphenylmethyl
carbocation

The extensive resonance stabilization of this carbocation allows it to exist in solution as long as there is no good nucleophile around to react with it.

Other examples of the stability of this carbocation abound. Triphenylmethyl chloride forms conducting solutions in liquid sulfur dioxide because of cleavage of the carbon–chlorine bond (the first step of an S_N1 reaction):

$$Ph_3C-Cl \xrightleftharpoons{SO_2} Ph_3C^+ \ Cl^-$$

If the anion is not very nucleophilic, solid salts containing the triphenylmethyl carbocation can actually be isolated. Thus, the tetrafluoroborate salt, $Ph_3C^+ \ BF_4^-$, can be isolated and stored for years as a stable ionic solid and is even commercially available. The geometry of perchlorate salt, Ph_3C^+ ClO_4^-, has been determined by X-ray crystallography. The central carbon has planar geometry as expected for an sp^2-hybridized carbocation but the rings are twisted out of the plane because of the severe steric crowding that would occur if they were all planar, so the cation has a shape that resembles a propeller. Note that this causes a decrease in resonance sta-bilization (called steric hindrance to resonance) because the p orbitals on the ben-zene rings are not exactly parallel to the p orbital of the central carbon. Still, there is enough resonance stabilization to make this carbocation much more stable than most others.

PROBLEM 8.6

Explain which compound has a faster rate of S_N1 reaction.

a) H₃C, Cl [cyclopentane]   or   Cl [cyclopentane]   **b)** Cl [cyclopentene]   or   Cl [cyclopentene]

c) [CH₂=CH–CH(Cl)–CH₃]   or   [CH₃–CH₂–CH(Cl)–CH₃]   **d)** CH₂Cl [benzene]   or   CH₂Cl [benzene with OCH₃]

PROBLEM 8.7

Arrange these compounds in order of decreasing S_N1 reaction rate.

Ph, Cl | Cl | Cl | Ph, Ph, Cl

8.8   STEREOCHEMISTRY OF THE S_N1 REACTION

What happens in the S_N1 reaction if the leaving group is attached to a carbon that is a chirality center? A common result for the S_N1 reaction is racemization; that is, the product is formed with 50% inversion and 50% retention of configuration. An example, the reaction of (S)-1-chloro-1-phenylethane with water to give racemic 1-phenyl-1-ethanol, is illustrated in Figure 8.8. In this reaction the stereochemical integrity of the reactant is randomized on the pathway to the product. This usually means that there is some intermediate along the reaction pathway that is not chiral. In the case of the S_N1 reaction the carbocation intermediate is sp^2 hybridized and has trigonal planar geometry. Because planar carbons are not chirality centers, this explains why the reaction results in racemization.

Although many S_N1 reactions proceed with racemization, many others result in more inversion of configuration in the product than retention. This is a result of the extremely short lifetime of the carbocation. When the carbocation is first formed, the leav-

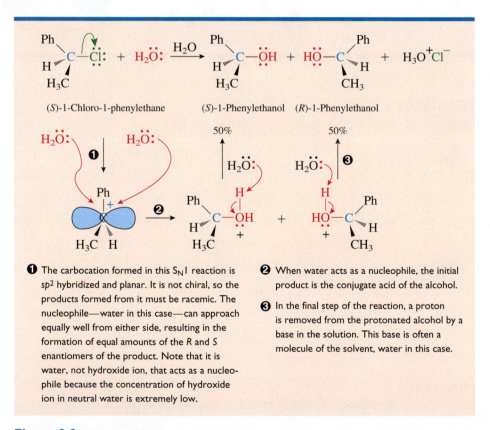

① The carbocation formed in this S_N1 reaction is sp^2 hybridized and planar. It is not chiral, so the products formed from it must be racemic. The nucleophile—water in this case—can approach equally well from either side, resulting in the formation of equal amounts of the R and S enantiomers of the product. Note that it is water, not hydroxide ion, that acts as a nucleophile because the concentration of hydroxide ion in neutral water is extremely low.

② When water acts as a nucleophile, the initial product is the conjugate acid of the alcohol.

③ In the final step of the reaction, a proton is removed from the protonated alcohol by a base in the solution. This base is often a molecule of the solvent, water in this case.

Figure 8.8

MECHANISM AND STEREOCHEMISTRY OF THE S_N1 REACTION OF (S)-1-CHLORO-1-PHENYLETHANE IN AQUEOUS SOLUTION.

ing group is still present on the side of the carbocation where it was originally attached, as shown in Figure 8.9. This species is called an **ion pair.** If the nucleophile attacks the ion pair, the leaving group is still blocking the front side of the carbocation and inversion is favored. After the leaving group has had time to diffuse away, generating a "free" carbocation, the nucleophile can attack equally well from either side, and equal amounts of inversion and retention result. As the lifetime of the carbocation increases, it will more likely reach the free stage, resulting in more complete racemization. The lifetime of the carbocation increases as its stability increases and also depends on the nucleophile and the solvent that are used in a particular reaction. The change of nucleophile and solvent is why the reaction of 1-chloro-1-phenylethane in water (Figure 8.8) gives a different stereochemical result than the reaction of the same compound in

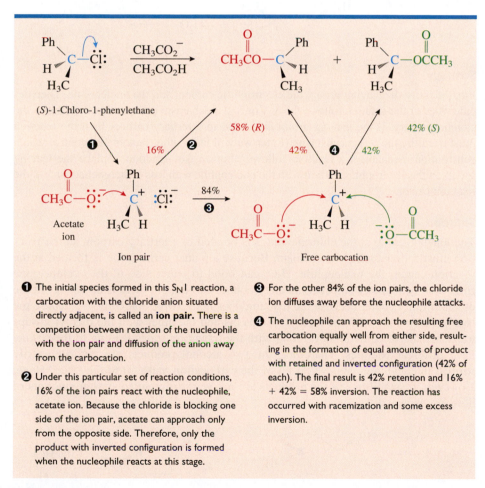

❶ The initial species formed in this S$_N$1 reaction, a carbocation with the chloride anion situated directly adjacent, is called an **ion pair.** There is a competition between reaction of the nucleophile with the ion pair and diffusion of the anion away from the carbocation.

❷ Under this particular set of reaction conditions, 16% of the ion pairs react with the nucleophile, acetate ion. Because the chloride is blocking one side of the ion pair, acetate can approach only from the opposite side. Therefore, only the product with inverted configuration is formed when the nucleophile reacts at this stage.

❸ For the other 84% of the ion pairs, the chloride ion diffuses away before the nucleophile attacks.

❹ The nucleophile can approach the resulting free carbocation equally well from either side, resulting in the formation of equal amounts of product with retained and inverted configuration (42% of each). The final result is 42% retention and 16% + 42% = 58% inversion. The reaction has occurred with racemization and some excess inversion.

Figure 8.9

STEREOCHEMISTRY OF THE S$_N$1 REACTION OF (S)-1-CHLORO-1-PHENYLETHANE IN ACETIC ACID CONTAINING POTASSIUM ACETATE.

acetic acid containing potassium acetate (Figure 8.9). The carbocation has a longer life-time under the reaction conditions of Figure 8.8 than under those of Figure 8.9 (see problem 8.14), allowing the chloride ion time to diffuse away before the nucleophile at-tacks, resulting in the formation of a racemic product. The shorter-lived carbocation of Figure 8.9 reacts partly at the ion-pair stage, resulting in more inversion than retention. In summary, S_N1 reactions occur with racemization, often accompanied by some excess inversion. We will not attempt to predict the exact amount of each enantiomer that is produced.

PRACTICE PROBLEM 8.1

Show the product, including stereochemistry, of this reaction:

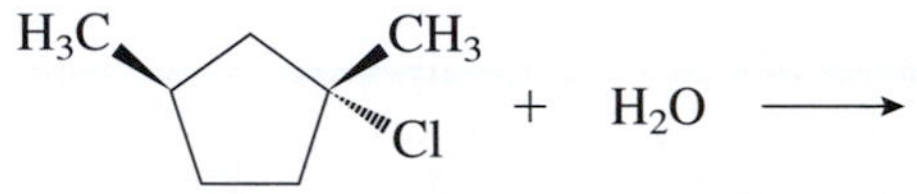

Strategy

First, identify the leaving group, the electrophilic carbon, and the nucleophile. Then de-cide whether the reaction follows the S_N1 or S_N2 mechanism because this determines the stereochemistry. If the leaving group is bonded to a tertiary carbon, then the reaction must occur by the S_N1 mechanism. (Later we will learn other factors that control which substitution mechanism a reaction follows.) For an S_N1 reaction, replace the leaving group on the electrophilic carbon with the nucleophile with loss of stereochemistry at the reaction center.

Solution

The leaving group is the chlorine and it is bonded to a tertiary carbon, so the reac-tion must follow an S_N1 mechanism. Because a planar carbocation is formed at the reaction center, the nucleophile, H_2O, can bond to either side of the cyclopentane ring. Stereochemistry is lost at the electrophilic carbon, but there is no change at the other chirality center because the reaction does not involve that carbon. Because the two products, one with the methyl groups cis and the other with the methyl groups trans, are diastereomers, they, along with the transition states leading to them, may have different energies. Therefore, the two alcohol products are not necessarily formed in equal amounts even if all of the carbocation makes it to the "free" stage. However, the stabilities of the products are not very different so similar amounts of both are expected.

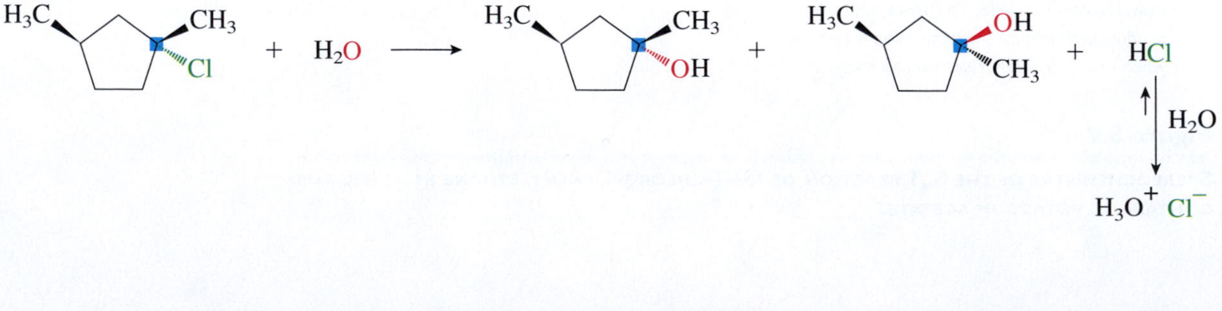

PROBLEM 8.8

Show the products, including stereochemistry, of these S_N1 reactions:

a) $\underset{\overset{|}{Ph}}{\overset{CH_3}{\underset{|}{CH_3CH_2\cdots C}}}\overset{CH_3}{\underset{Cl}{}}$ + CH_3OH ⟶

b) $CH_3CH_2\cdots\overset{H_3C}{\underset{H}{C}}-CH_2-\overset{CH_3}{\underset{CH_3}{C}}-Cl$ + CH_3COH ⟶

8.9 LEAVING GROUPS

The bond to the leaving group is broken during the rate-determining step in both the S_N1 and S_N2 reactions. Therefore, the structure of the leaving group affects the rates of both of these reactions. Although the only leaving group we have seen so far is chloride, there are others that can be used. In general, the more stable the leaving group is as a free species—that is, after it has left—the faster it will leave. This stability is also reflected in the basicity of the species: the more stable it is, the weaker base it is. In general, the leaving groups that are used in the S_N1 and the S_N2 reactions are weak bases. Table 8.3 lists the most important leaving groups and provides their relative reaction rates in an S_N1 reaction. Similar rate effects are found for S_N2 reactions.

As can be seen from Table 8.3, the leaving ability of the halides increases as one goes down the column of the periodic table; that is, Cl^- is the slowest, Br^- is faster, and

Table 8.3 Approximate Reactivities of Some Important Leaving Groups

Structure	Leaving Group	Name	Relative Reactivity
$R-Cl$	Cl^-	Chloride	1
$R-Br$	Br^-	Bromide	10
$R-\overset{+}{\underset{H}{O}}\overset{H}{}$	OH_2	Water	10
$R-I$	I^-	Iodide	10^2
$R-O-\overset{O}{\underset{O}{\overset{\|}{\underset{\|}{S}}}}-CH_3$	$^-O-\overset{O}{\underset{O}{\overset{\|}{\underset{\|}{S}}}}-CH_3$	Mesylate (methanesulfonate)	10^4
$R-O-\overset{O}{\underset{O}{\overset{\|}{\underset{\|}{S}}}}-\bigcirc-CH_3$	$^-O-\overset{O}{\underset{O}{\overset{\|}{\underset{\|}{S}}}}-\bigcirc-CH_3$	Tosylate (p-toluenesulfonate)	10^4

I^- is the fastest. This order parallels the decrease in basicity that occurs as one proceeds down a column of the periodic table. Fluoride ion (F^-) is so slow that it is not commonly used as a leaving group.

PROBLEM 8.9

Explain whether these reactions would follow the S_N1 or the S_N2 mechanism and then explain which reaction is faster:

SN2

a) $CH_3\overset{\text{Cl}}{\underset{|}{CH_2}}$ + ^-OH $\xrightarrow[\text{CH}_3\text{OH}]{\text{H}_2\text{O}}$ or $CH_3\overset{\text{O—SO}_2\text{CH}_3}{\underset{|}{CH_2}}$ + ^-OH $\xrightarrow[\text{CH}_3\text{OH}]{\text{H}_2\text{O}}$

SN1

b) $CH_3\overset{CH_3}{\underset{CH_3}{\overset{|}{\underset{|}{C}}}}\!\!-Br$ + $CH_3\overset{O}{\overset{\|}{C}}O^-$ $\xrightarrow{\text{CH}_3\text{CO}_2\text{H}}$ or $CH_3\overset{CH_3}{\underset{CH_3}{\overset{|}{\underset{|}{C}}}}\!\!-I$ + $CH_3\overset{O}{\overset{\|}{C}}O^-$ $\xrightarrow{\text{CH}_3\text{CO}_2\text{H}}$

Because alcohols are very common and easily prepared by a variety of methods, it would be useful to be able to use OH as a leaving group in nucleophilic substitution reactions. However, OH^- is much too basic to act as a leaving group in S_N1 and S_N2 reactions. It is necessary to modify the OH, converting it to a better leaving group, to use alcohols as substrates for these reactions. Several methods have been developed to accomplish this goal. If the reaction of the alcohol is conducted in acidic solution, the oxygen becomes protonated, producing ROH_2^+. The leaving group is now water, which is comparable to bromide in reactivity. (Of course, to use this leaving group, the nucleophile must be stable in acidic solution.) An example is provided in the following equation:

$$CH_3\!-\!\overset{CH_3}{\underset{CH_3}{\overset{|}{\underset{|}{C}}}}\!-\!\ddot{O}H + H\!-\!\ddot{\underset{\cdot\cdot}{Cl}}\!: \longrightarrow CH_3\!-\!\overset{CH_3}{\underset{CH_3}{\overset{|}{\underset{|}{C}}}}\!-\!\overset{+}{\underset{\cdot\cdot}{O}}H_2 \longrightarrow CH_3\!-\!\overset{CH_3}{\underset{CH_3}{\overset{|}{\underset{|}{C}}}}^+ \;\; :\ddot{\underset{\cdot\cdot}{Cl}}\!:^- \longrightarrow CH_3\!-\!\overset{CH_3}{\underset{CH_3}{\overset{|}{\underset{|}{C}}}}\!-\!\ddot{\underset{\cdot\cdot}{Cl}}\!: + H_2\ddot{\underset{\cdot\cdot}{O}}\!:$$

Another method that can be used to transform the hydroxy group of an alcohol into a leaving group is to replace the hydrogen with some other group that significantly decreases the basicity of the oxygen. A group that is commonly used for this purpose is the SO_2R group. Replacing the hydrogen of the alcohol with this group produces a sulfonate ester, such as the mesylate or tosylate ester shown in Table 8.3. As can be seen by their resemblance to the bisulfate anion (the conjugate base of sulfuric acid), sulfonate anions are weak bases and excellent leaving groups.

$$^-O\!-\!\overset{O}{\underset{O}{\overset{\|}{\underset{\|}{S}}}}\!-\!R \qquad\qquad ^-O\!-\!\overset{O}{\underset{O}{\overset{\|}{\underset{\|}{S}}}}\!-\!OH$$

A sulfonate   Bisulfate
anion   anion

PROBLEM 8.10

a) Show all the steps in the mechanism for this reaction. Don't forget to use curved arrows to show the movement of electrons in each step of the mechanism.

$$\underset{\underset{CH_3}{|}}{\overset{\overset{CH_3}{|}}{Ph-C-OH}} \;+\; HI \;\longrightarrow\; \underset{\underset{CH_3}{|}}{\overset{\overset{CH_3}{|}}{Ph-C-I}} \;+\; H_2O$$

b) Show a free energy versus reaction progress diagram for the reaction of part a.

Primary and secondary alcohols are readily converted to mesylate or tosylate esters by reaction with the corresponding sulfonyl chloride. The mesylate and tosylate esters derived from tertiary alcohols are too reactive and cannot be isolated. (Although we will not go into the mechanism of these reactions in detail at this point, the reactions involve the attack of the oxygen [the nucleophile] of the alcohol at the sulfur [the electrophile], ultimately displacing chloride [the leaving group].) Pyridine is often used as a solvent for these preparations in order to react with the HCl that is produced as a by-product. An example of the preparation of a methanesulfonate (mesylate) ester is shown in the following equation:

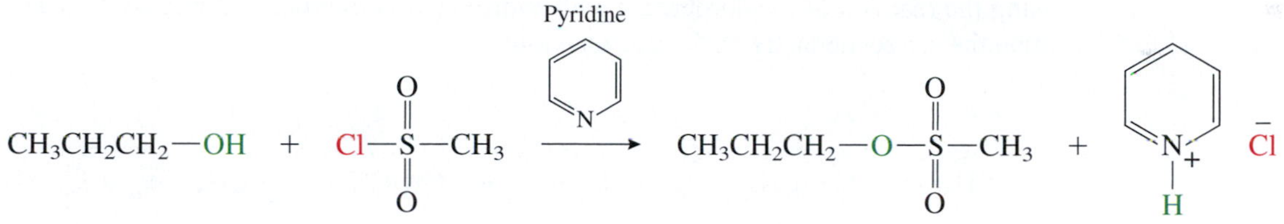

1-Propanol   Methanesulfonyl chloride   Propyl mesylate

or, in more general form:

$$ROH \;+\; MsCl \;\xrightarrow{\text{pyridine}}\; ROMs \qquad Ms \;=\; \overset{\overset{O}{\|}}{\underset{\underset{O}{\|}}{-S-CH_3}}$$

An alcohol   A mesylate ester

An example of the preparation of a tosylate ester is shown in the following equation:

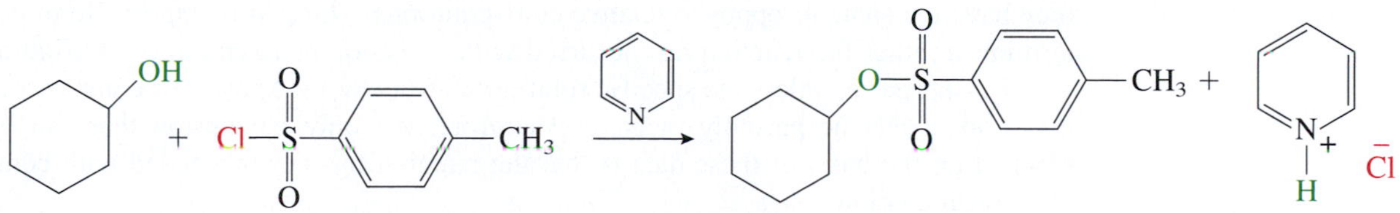

Cyclohexanol   p-Toluenesulfonyl chloride   Cyclohexyl tosylate or, in more general form:

$$\text{ROH} \quad + \quad \text{TsCl} \quad \xrightarrow{\text{pyridine}} \quad \text{ROTs} \qquad \text{Ts} \ = \ \overset{\displaystyle O}{\underset{\displaystyle O}{-\overset{\|}{\underset{\|}{S}}-}}\!\!\!\!\!\!\!\!\!\bigcirc\!\!-\text{CH}_3$$

An alcohol   A tosylate
                                      ester

In summary, the halide ions, Cl^-, Br^-, and I^-, are common leaving groups for these nucleophilic substitution reactions. The OH of an alcohol can be converted to a leaving group by protonation in acid solution or by conversion to a mesylate or tosylate ester. The OR^- of an ether, like the OH^- of an alcohol, is too basic to leave. Likewise, NH_2^- (from an amine), H^- (from breaking C—H bonds of alkanes), and R_3C^- (carbanions from breaking C—C bonds of alkanes) are much too basic and *do not act as leaving groups in these reactions*.

The tosylate leaving group was used in one of the classic experiments that was used to determine the stereochemistry of the S_N2 reaction. Now that we know about this leaving group, let's look at the experiment and see how it helped establish that S_N2 reactions occur with inversion of configuration.

Designing an experiment to demonstrate that an S_N2 reaction occurs with inversion of configuration is not as simple as it might appear at first glance. For example, consider using the reaction of 2-chlorobutane with hydroxide ion to produce 2-butanol to determine the stereochemistry of the S_N2 reaction:

$$\underset{\substack{| \\ H}}{\overset{\substack{Cl \\ |}}{CH_3CH_2-\underset{*}{C}-CH_3}} \quad + \quad {}^-OH \quad \longrightarrow \quad \underset{\substack{| \\ H}}{\overset{\substack{OH \\ |}}{CH_3CH_2-\underset{*}{C}-CH_3}} \quad + \quad Cl^-$$

2-Chlorobutane   2-Butanol

First, it must be established that this reaction proceeds by an S_N2 mechanism. To do this, the experimental rate law for the reaction is determined. Because the reaction is found to follow the second-order rate law, rate = k[2-chlorobutane][hydroxide ion], it is proceeding by the S_N2 mechanism. Then the stereochemistry of the product is investigated. Suppose that the starting 2-chlorobutane has $[\alpha] = +8.5$ and the 2-butanol formed in the reaction has $[\alpha] = -13.9$. What can be deduced from this information? Remember, there is no relationship between the sign of the rotation and the configuration of a compound. So, even though the starting 2-chlorobutane and the product 2-butanol have opposite signs for their rotations, we do not know whether they have the same or opposite relative configurations. Thus, it is impossible to determine whether the reaction has occurred with inversion or retention of configuration. Furthermore, unless the specific rotation of 2-butanol is known for comparison, the product can be partially racemic. Therefore, the only conclusion that can be reached on the basis of these data is that the reaction has not proceeded with complete racemization.

To determine the stereochemistry of the reaction, a method must be found to relate the configuration of the reactant to that of the product. The following experiment shows one method that has been used to accomplish this. A sample of 2-octanol with $[\alpha] = +1.12$ was converted to its tosylate ester, which had $[\alpha] = +0.90$. In this reaction the O of the alcohol displaces the Cl of *p*-toluenesulfonyl chloride. Because the carbon–oxygen bond of the alcohol is not broken in this reaction, the tosylate ester must have the same relative configuration as the alcohol.

The S_N2 reaction of the tosylate ester with acetate ion in acetone as solvent produced an ester with $[\alpha] = -0.83$. (Again, the configuration of the product cannot be determined solely on the basis of this experiment.)

To determine its stereochemistry, the ester was prepared by an alternate pathway. Reaction of the starting alcohol with acetyl chloride (CH_3COCl) produced the ester with $[\alpha] = +0.84$. This reaction, like the formation of the tosylate ester, does not involve breaking the carbon–oxygen bond of the alcohol, so the ester obtained in this reaction must have the same relative configuration as both the alcohol and the *p*-toluenesulfonate ester. Comparison of the rotation of this ester, ($[\alpha] = +0.84$) with that of the ester obtained as the product of the S_N2 reaction ($[\alpha] = -0.83$) demonstrated that they have opposite relative configurations. Therefore, the S_N2 reaction has occurred with complete inversion of configuration, within experimental error. It is interesting to note that it not necessary to know the absolute configuration of any of the compounds in this reaction cycle to determine that the S_N2 reaction has proceeded with inversion of configuration. Nor is it necessary that the original alcohol be enantiomerically pure—that is, that it be composed of a single enantiomer. In fact, in the experiment described here, the enantiomeric purity of the original 2-octanol was only 11%.

A large number of other experiments have been performed, and in each case the results have been the same. S_N2 reactions have always been found to proceed with complete inversion of configuration.

8.10 NUCLEOPHILES

In the case of an S_N1 reaction, the nucleophile does not become involved until after the rate-determining step. Therefore, changing the nucleophile has no effect on the *rate* of an S_N1 reaction, although it may change the *product* of the reaction. In contrast, changing the nucleophile in an S_N2 reaction has a dramatic effect on the reaction rate because the nucleophile attacks in the rate-determining step.

What makes a good nucleophile? Nucleophilic substitutions are somewhat similar to acid–base reactions. In an acid–base reaction, a hydrogen is transferred from one base to another. In a nucleophilic substitution reaction, a carbon group is transferred from one nucleophile (Lewis base) to another. It is not surprising, then, that strength as a nucleophile often parallels strength as a base. However, because nucleophiles react with carbon and bases react with hydrogen, it is also not surprising that there are some differences between nucleophilicity and basicity. The following rules apply.

▶ RULE 1

If the nucleophilic atoms are from the same period of the periodic table, strength as a nucleophile parallels strength as a base. For example,

$$H_2O \quad < \quad NH_3$$

$$CH_3OH \quad \approx \quad H_2O \quad < \quad CH_3CO_2^- \quad < \quad CH_3O^- \quad \approx \quad HO^-$$

increasing base strength/increasing nucleophile strength ⟶

▶ RULE 2

Nucleophile strength increases down a column of the periodic table (in solvents that can hydrogen bond, such as water and alcohols). For example,

$$RO^- \quad < \quad RS^-$$

$$R_3N \quad < \quad R_3P$$

$$F^- \quad < \quad Cl^- \quad < \quad Br^- \quad < \quad I^-$$

increasing nucleophile strength/decreasing base strength ⟶

In these cases the change in nucleophile strength is opposite to the change in base strength. This is due at least in part to the stronger hydrogen bonding that occurs between the smaller ions and solvent molecules of water or alcohols. The resulting tighter arrangement of solvent molecules around the nucleophile makes it more difficult for the electrophilic carbon to approach.

▶ **RULE 3**

Steric bulk decreases nucleophilicity. For example,

$$
\begin{array}{c}
\text{CH}_3 \\
| \\
\text{H}_3\text{C}-\text{C}-\text{O}^- \\
| \\
\text{CH}_3
\end{array}
\quad < \quad \text{HO}^-
$$

| *tert*-Butoxide ion | Stronger nucleophile |
| Weaker nucleophile, stronger base | Weaker base |

Because the presence of bulky groups at the electrophilic carbon slows the rate of S_N2 reactions, it is reasonable that the presence of bulky groups on the nucleophile will also slow the reaction. The steric bulk of the *tert*-butoxide ion causes it to be a much weaker nucleophile than hydroxide ion even though it is a stronger base. Therefore, *tert*-butoxide ion is often used when a strong base that is not very nucleophilic is needed.

PRACTICE PROBLEM 8.2

Explain which reaction proceeds at a faster rate:

$$
\begin{array}{c}
\text{I} \\
| \\
\text{CH}_3\text{CH}_2\text{CH}_2
\end{array}
+ \ ^-\text{OH} \ \xrightarrow[\text{EtOH}]{\text{H}_2\text{O}}
\quad \text{or} \quad
\begin{array}{c}
\text{I} \\
| \\
\text{CH}_3\text{CH}_2\text{CH}_2
\end{array}
+ \ \text{H}_2\text{O} \ \xrightarrow[\text{EtOH}]{\text{H}_2\text{O}}
$$

Strategy

First identify the leaving group, the electrophilic carbon, and the nucleophile. If the leaving group is on a primary carbon, then the mechanism that the reaction follows is S_N2. If the leaving group is on a tertiary carbon, then the mechanism is S_N1. (We will learn how to determine the mechanism if the leaving group is on a secondary carbon in Section 8.12.) Then identify the difference between the two reactions and analyze how this difference will affect the rate.

Solution

Because the leaving group is on a primary carbon, the reaction proceeds by an S_N2 mechanism. The left reaction will be faster because hydroxide ion is a stronger base and, thus, a better nucleophile than water.

PROBLEM 8.11

Explain which of these reactions proceeds at a faster rate:

$$
\textbf{a)}\
\begin{array}{c}
\text{CH}_3 \\
| \\
\text{CH}_3\text{C}-\text{Cl} \\
| \\
\text{CH}_3
\end{array}
+ \ \text{CH}_3\overset{\text{O}}{\overset{\|}{\text{C}}}\text{OH} \ \xrightarrow{\text{CH}_3\text{CO}_2\text{H}}
\quad \text{or} \quad
\begin{array}{c}
\text{CH}_3 \\
| \\
\text{CH}_3\text{C}-\text{Cl} \\
| \\
\text{CH}_3
\end{array}
+ \ \text{CH}_3\overset{\text{O}}{\overset{\|}{\text{C}}}\text{O}^- \ \xrightarrow{\text{CH}_3\text{CO}_2\text{H}}
$$

SN1
same rate

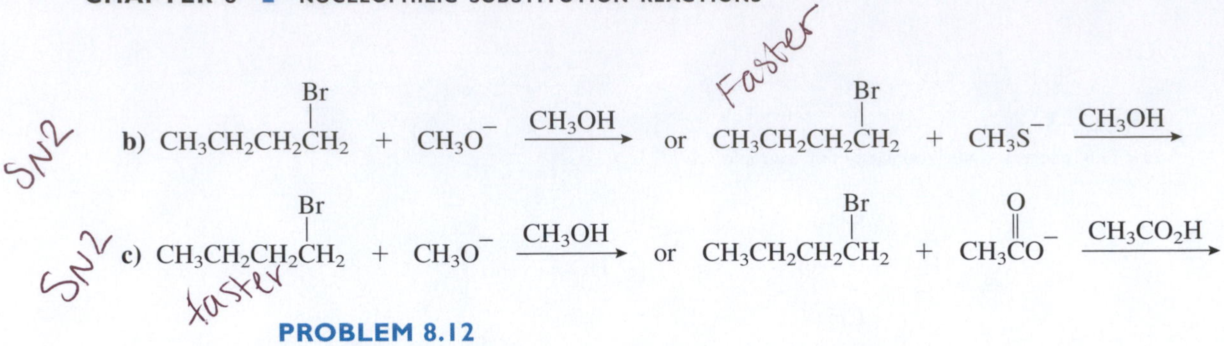

SN2

b) CH₃CH₂CH₂CH₂ (Br) + CH₃O⁻ $\xrightarrow{\text{CH}_3\text{OH}}$ *Faster* or CH₃CH₂CH₂CH₂ (Br) + CH₃S⁻ $\xrightarrow{\text{CH}_3\text{OH}}$

SN2 *faster*

c) CH₃CH₂CH₂CH₂ (Br) + CH₃O⁻ $\xrightarrow{\text{CH}_3\text{OH}}$ or CH₃CH₂CH₂CH₂ (Br) + CH₃CO⁻ (O) $\xrightarrow{\text{CH}_3\text{CO}_2\text{H}}$

PROBLEM 8.12

Show the products and the mechanisms of the following reactions. Don't forget to use curved arrows to show the movement of electrons in each step of the mechanism.

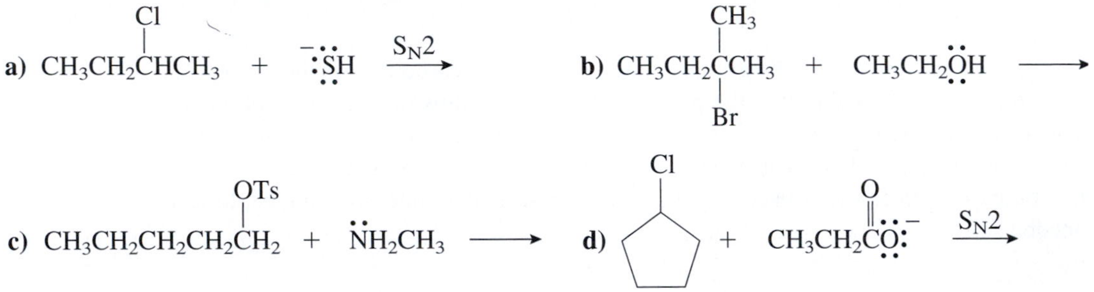

a) CH₃CH₂CHCH₃ (Cl) + ⁻:SH $\xrightarrow{\text{S}_\text{N}2}$

b) CH₃CH₂CCH₃ (CH₃)(Br) + CH₃CH₂ÖH $\longrightarrow$

c) CH₃CH₂CH₂CH₂CH₂ (OTs) + :NH₂CH₃ $\longrightarrow$

d) (cyclopentane ring with Cl) + CH₃CH₂CO:⁻ (O) $\xrightarrow{\text{S}_\text{N}2}$

PROBLEM 8.13

Show all of the steps in the mechanism for this reaction:

$$\text{CH}_3-\underset{\underset{\text{CH}_3}{|}}{\overset{\overset{\text{CH}_3}{|}}{\text{C}}}-\text{Cl} \; + \; \text{CH}_3\text{OH} \longrightarrow \text{CH}_3-\underset{\underset{\text{CH}_3}{|}}{\overset{\overset{\text{CH}_3}{|}}{\text{C}}}-\text{OCH}_3 \; + \; \text{HCl}$$

8.11   EFFECT OF SOLVENT

The solvent has several roles to play in an organic reaction. It must dissolve the reagents so that they can come in contact with one another. It must not react with or decompose any of the reagents. In addition, for reactions that involve ionic or polar molecules (as reactants, intermediates, or products), the polarity of the solvent often dramatically affects the reaction rate.

Polar solvents help to stabilize ions and polar molecules. To understand the effect of the solvent polarity on reaction rates, the polarity of the reactant must be compared with the polarity of the transition state. The one (reactant or transition state) that is more polar (has more charge separation) will be stabilized more by an increase in the polarity of the solvent. If the transition state is more polar than the reactants, increasing the solvent polarity will stabilize the transition state more than the reactants. This will decrease $\Delta G^{\ddagger}$, resulting in a faster reaction. In contrast, if the reactants are more polar than the transition state, increasing the solvent polarity will stabilize the reactants more, resulting in a larger $\Delta G^{\ddagger}$ and a slower reaction.

Figure 8.10 illustrates the results of increasing the solvent polarity on the energy versus reaction progress curve for the S_N1 reaction. Because the transition state, which resembles the carbocation, is more polar than the reactant, the rate of an S_N1 reaction is much faster in a more polar solvent.

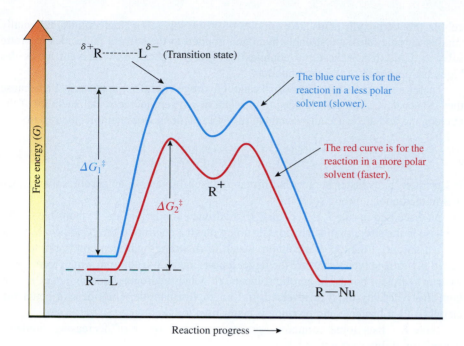

Free energy (G)

$\delta^+ R\text{------}L^{\delta-}$ (Transition state)

The blue curve is for the reaction in a less polar solvent (slower).

The red curve is for the reaction in a more polar solvent (faster).

$\Delta G_1^{\ddagger}$

$\Delta G_2^{\ddagger}$

R^+

R—L

R—Nu

Reaction progress →

Figure 8.10

EFFECT OF CHANGING SOLVENT POLARITY ON THE S_N1 REACTION. Because the transition state is more polar (has more charge separation) than the reactant, the change to a more polar solvent stabilizes the transition state more than it stabilizes the reactant. This results in $\Delta G_1^{\ddagger}$, the activation energy in the less polar solvent, being larger than $\Delta G_2^{\ddagger}$, the activation energy in the more polar solvent. Therefore, the reaction is faster in the more polar solvent. This diagram applies for all S_N1 reactions.

The effect of the solvent polarity on the rate of the S_N2 reaction depends on the charge that is initially present on the nucleophile. If the nucleophile has a negative charge, the reaction can be represented as

$$\overset{-}{Nu:} + \quad C{-}L \quad \longrightarrow \quad \overset{\delta-}{Nu}\text{-----}C\text{-----}L^{\delta-}$$

Reactants
(more polar)

Transition state
(less polar)

Although both the reactants and the transition state have a total charge of -1, this charge is more dispersed in the transition state. Increasing the solvent polarity stabilizes the concentrated charge of the reactant nucleophile more than the transition state, resulting in a slower reaction rate. Therefore, the rate of an S_N2 reaction involving a negative nucleophile is slower in a more polar solvent.

If the nucleophile in an S_N2 reaction is neutral, the reaction can be represented as

$$Nu: + \quad C{-}L \quad \longrightarrow \quad \overset{\delta+}{Nu}\text{-----}C\text{-----}L^{\delta-}$$

Reactants
(less polar)

Transition state
(more polar)

The reactants are neutral, whereas charges have partially formed in the transition state. In this situation the transition state is stabilized more than the reactants as the solvent is changed to a more polar one. Therefore, the rate of an S_N2 reaction involving a neutral nucleophile is faster in a more polar solvent.

In addition to these polarity effects, the ability of certain solvents to form hydrogen bonds to the nucleophile also affects the rate of the S_N2 reaction. Such solvents are termed **protic solvents** and have a hydrogen bonded to nitrogen or oxygen. (Water, alcohols, and carboxylic acids are examples of protic solvents.) Nucleophiles, especially smaller anionic ones, are strongly hydrogen bonded to the solvent molecules in protic solvents. This makes these nucleophiles less reactive because the solvent molecules block the approach of the electrophile.

Therefore, **aprotic solvents,** which cannot hydrogen bond to the nucleophile because they do not have hydrogens bonded to nitrogen or oxygen, increase the reactivity of the nucleophile and are especially favorable for S_N2 reactions. For example, the reaction

$$Cl^- \ + \ CH_3 \!-\! I \ \longrightarrow \ CH_3 \!-\! Cl \ + \ I^-$$

is 1 million times faster in dimethylformamide (see Table 8.4), an aprotic solvent, than it is in methanol, a protic solvent.

As discussed in rule 2 of Section 8.10, hydrogen bonding is also responsible for the increase in the strength of a nucleophile going down a column of the periodic table. The smaller anions at the top of a column, which form strong hydrogen bonds, are less reactive in protic solvents than are the larger anions lower in a column, which form much weaker hydrogen bonds. In aprotic solvents, in which hydrogen bonding is not important, this reactivity order is reversed. In dimethylformamide, chloride is a better nucleophile than bromide and bromide is better than iodide.

Table 8.4 lists some common organic solvents in order of decreasing "ionizing power" (or ability to stabilize ions).

Table 8.4   Some Common Solvents for Substitution Reactions

Name	Structure	
(increasing ionizing power ↑)		
Trifluoroacetic acid	$\overset{\displaystyle O}{\overset{\|}{CF_3COH}}$	These are polar, protic solvents. They are quite good at stabilizing ions and are especially favorable for S_N1 reactions, although they can also be used for S_N2 reactions with favorable substrates.
Water	H_2O	
Methanol	CH_3OH	
Acetic acid	$\overset{\displaystyle O}{\overset{\|}{CH_3COH}}$	
Ethanol	CH_3CH_2OH	
Dimethylsulfoxide (DMSO)	$\overset{\displaystyle O}{\overset{\|}{CH_3SCH_3}}$	These aprotic solvents are still relatively polar, so they can dissolve both the nucleophile and the substrate. They are especially favorable for S_N2 reactions because nucleophiles are more reactive in these solvents than in protic ones.
Dimethylformamide (DMF)	$\overset{\displaystyle O}{\overset{\|}{HCN(CH_3)_2}}$	
Acetone	$\overset{\displaystyle O}{\overset{\|}{CH_3CCH_3}}$	

PRACTICE PROBLEM 8.3

Explain whether the reaction of 1-chloropropane with ammonia would be faster in 20% CH_3OH/80% H_2O or in 40% CH_3OH/60% H_2O as the solvent.

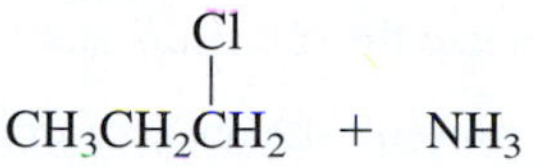

$$\overset{\displaystyle Cl}{\underset{\displaystyle}{CH_3CH_2\overset{|}{C}H_2}} \quad + \quad NH_3$$

Solution

This is an S_N2 reaction with a neutral nucleophile.

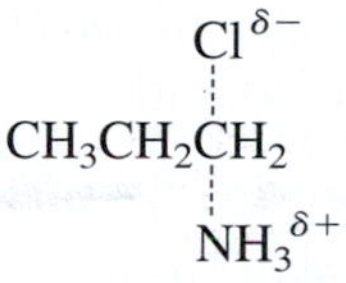

$$\begin{array}{c} Cl^{\delta-} \\ \vdots \\ CH_3CH_2CH_2 \\ \vdots \\ NH_3{}^{\delta+} \end{array}$$

Transition state

The transition state for the reaction is more polar than the reactants so the reaction is faster in a more polar solvent. Water is more polar than methanol, so the reaction is faster in 20% CH_3OH/80% H_2O than in 40% CH_3OH/60% H_2O.

PROBLEM 8.14

Explain why the carbocation shown in Figure 8.8 has a longer lifetime than it does under the conditions shown in Figure 8.9.

PROBLEM 8.15

Explain in which solvent these reactions are faster:

a) 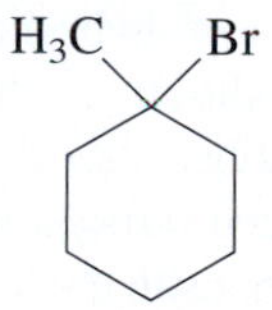 $\underset{\displaystyle}{\overset{\displaystyle H_3C \quad Br}{\bighexagon}}$ in CH_3OH or CH_3CH_2OH

b) 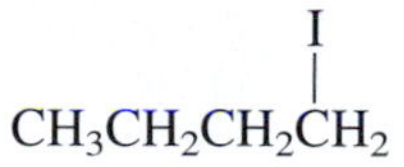 $CH_3CH_2CH_2\overset{I}{\underset{|}{C}}H_2$ + ^-OH in CH_3OH or $\begin{array}{c} 50\% \ CH_3OH \\ 50\% \ H_2O \end{array}$

c) 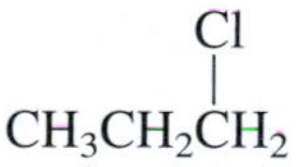 $CH_3CH_2\overset{Cl}{\underset{|}{C}}H_2$ + $:\overset{-}{C}\equiv N:$ in CH_3CH_2OH or DMSO

8.12   COMPETITION BETWEEN S_N1 AND S_N2 REACTIONS

It is possible to examine a particular reaction and determine whether it is expected to proceed by an S_N1 or an S_N2 mechanism.

The S_N1 pathway is favored in the following circumstances:

1. The carbocation is stabilized (tertiary or resonance stabilized carbocations are best; secondary carbocations are acceptable if other factors are favorable; primary carbocations are not formed).

2. The solvent is polar (to stabilize the transition state).

3. Only poor nucleophiles are present (the absence of a good nucleophile slows the rate of a competing S_N2 reaction).

The S_N2 pathway is favored in the following circumstances:

1. The electrophilic carbon is not sterically hindered (reactions at methyl and primary carbons are excellent; reactions at secondary carbons are acceptable; S_N2 reactions do not occur at tertiary carbons).

2. Strong nucleophiles are present.

3. The solvent is aprotic (to make the nucleophile more reactive).

On the basis of these principles, Table 8.5 lists the preferred mechanism followed by different types of electrophilic carbon groups. Compounds with the leaving group on a methyl or a primary carbon react by the S_N2 mechanism. Compounds with the leaving group on a secondary, an allylic, or a benzylic carbon can react by either mechanism, depending on the solvent and the nucleophile. And compounds with the leaving group on a tertiary carbon react by the S_N1 mechanism. Compounds with the leaving group on a neopentyl type carbon do not react by an S_N1 mechanism (because the carbocation formed is primary) and also react very slowly by an S_N2 mechanism (because of the large amount of steric hindrance provided by the *tert*-butyl group). However, acceptable yields in S_N2 reactions can be obtained if an aprotic solvent is used.

The last two examples in Table 8.5 have the leaving group bonded to an sp^2-hybridized carbon, either a vinylic carbon or an aromatic carbon. Under normal conditions, both of these types of compounds are inert to nucleophilic substitution reactions because of the stronger C—L bond, the difficulty in forming carbocations at sp^2-hybridized carbons, and the extra steric hindrance to approach of the nucleophile from the side opposite the leaving group. (Under particularly favorable circumstances, S_N1 reactions of these compounds can be forced to occur.)

PRACTICE PROBLEM 8.4

Explain whether these reactions follow an S_N1 or an S_N2 mechanism:

a) $$CH_3CH_2CH_2\overset{\overset{\displaystyle OTs}{|}}{C}H_2 \ + \ Cl^- \ \xrightarrow{CH_3OH}$$

b) $$CH_3CH_2\overset{\overset{\displaystyle Cl}{|}}{C}HCH_3 \ + \ H_2O \ \xrightarrow[EtOH]{H_2O}$$

**Table 8.5   Preferred Substitution Mechanisms
for Various Carbon Substrates**

Type	Representative Structure	Preferred Mechanism
Methyl	CH_3-L	S_N2
Primary	RCH_2-L	S_N2
Secondary	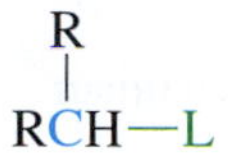 $\overset{\displaystyle R}{\underset{\displaystyle \vert}{R}}CH-L$	S_N2 (with good nucleophiles; solvent can be protic or, better, aprotic) or S_N1 (with poor nucleophiles and polar solvents)
Tertiary	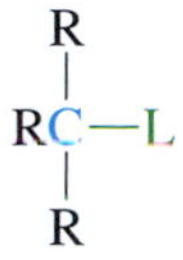 $R\overset{\displaystyle R}{\underset{\displaystyle R}{C}}-L$	S_N1
Allylic	$CH_2{=}CHCH_2-L$	S_N2 (with good nucleophiles; solvent can be protic or, better, aprotic) or S_N1 (with poor nucleophiles and polar solvents)
Benzylic	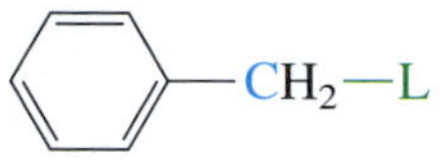 ⟨benzene ring⟩$-CH_2-L$	
Neopentyl	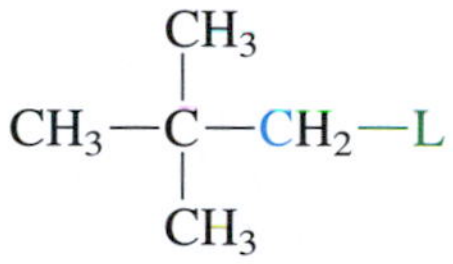 $CH_3-\overset{\displaystyle CH_3}{\underset{\displaystyle CH_3}{C}}-CH_2-L$	Very slow S_N1 and S_N2 (S_N2 in aprotic solvents gives acceptable yields)
Vinylic	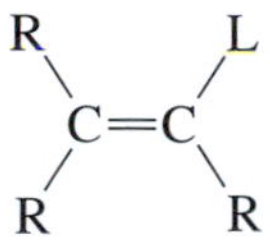 $\overset{R}{\underset{R}{}}C{=}C\overset{L}{\underset{R}{}}$	Inert to both S_N1 and S_N2 under normal reaction conditions
Aromatic	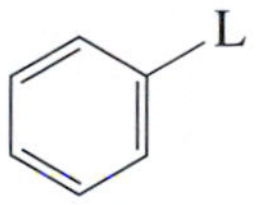 ⟨benzene ring⟩$-L$	

Strategy

Identify the leaving group, the electrophilic carbon (the one bonded to the leaving group), the nucleophile, and the solvent (usually over the arrow). If the electrophilic carbon is methyl or a simple primary carbon, the mechanism is S_N2. If the electrophilic carbon is tertiary, the mechanism is S_N1. If the electrophilic carbon is secondary, allylic, or benzylic, you must examine the nucleophile and the solvent. With good nucleophiles, the mechanism is S_N2. (Aprotic solvents make the nucleophile even stronger.) With poor nucleophiles and polar solvents, the mechanism is S_N1.

Solution

a) The leaving group (OTs) is on a primary carbon, so the reaction follows an S_N2 mechanism. It does not matter what the nucleophile or solvent is.

b) The leaving group (Cl) is on a secondary carbon, so the mechanism depends on the nucleophile and the solvent. Water is both the nucleophile and one component of the solvent. Because water is not a strong nucleophile and the solvent (H_2O/CH_3CH_2OH) is polar, the reaction follows an S_N1 mechanism.

PROBLEM 8.16

Explain whether these reactions follow an S_N1 or an S_N2 mechanism.

a) 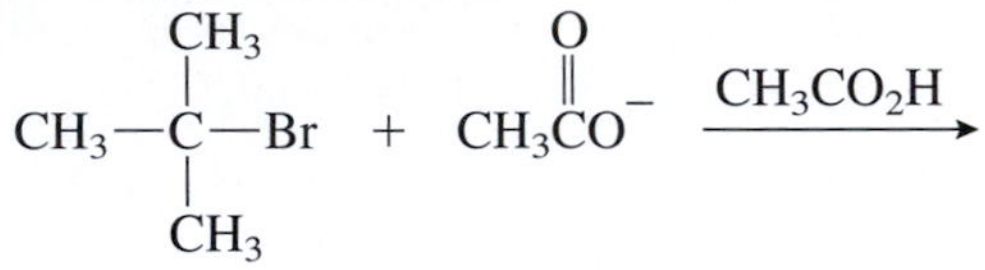
$$CH_3-\overset{\overset{\displaystyle CH_3}{|}}{\underset{\underset{\displaystyle CH_3}{|}}{C}}-Br \ + \ CH_3\overset{\overset{\displaystyle O}{\|}}{C}O^- \ \xrightarrow{CH_3CO_2H}$$

b) 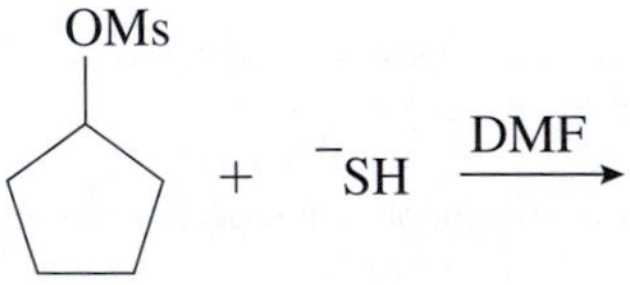
$$\text{(cyclopentyl OMs)} \ + \ {}^-SH \ \xrightarrow{DMF}$$

c)
$$CH_3I \ + \ CH_3-\overset{\overset{\displaystyle CH_3}{|}}{\underset{\underset{\displaystyle CH_3}{|}}{C}}-O^- \ \xrightarrow{t\text{-BuOH}}$$

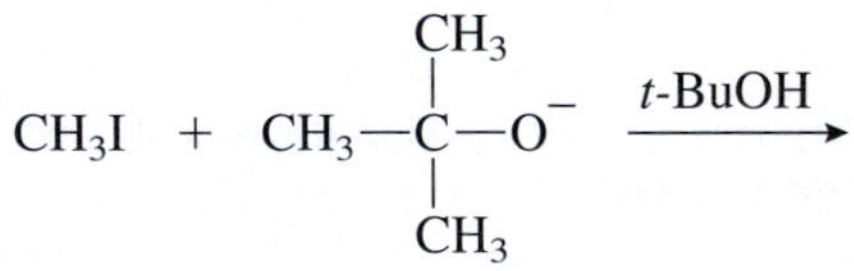

d) 
$$\text{(allyl Cl)} \ + \ CH_3OH \ \xrightarrow{CH_3OH}$$

e) 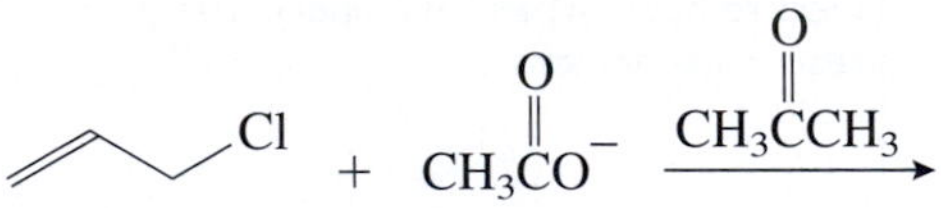
$$\text{(allyl Cl)} \ + \ CH_3\overset{\overset{\displaystyle O}{\|}}{C}O^- \ \xrightarrow{CH_3\overset{\overset{\displaystyle O}{\|}}{C}CH_3}$$

f) 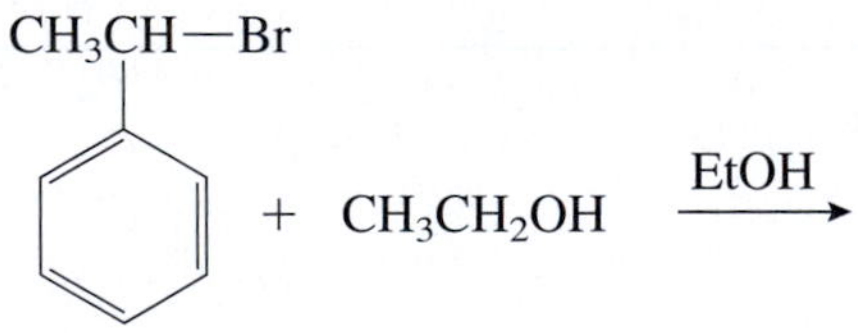
$$\text{CH}_3\text{CH}-\text{Br (benzylic)} \ + \ CH_3CH_2OH \ \xrightarrow{EtOH}$$

8.13 Intramolecular Reactions

The reactions discussed so far have been **intermolecular reactions;** that is, they involve two separate molecules: the nucleophile and the compound with the leaving group. It is also possible for the nucleophile and the leaving group to be part of the same molecule. In such a case the reaction is **intramolecular**—that is, within the same molecule. Examples of an intermolecular nucleophilic substitution reaction and a comparable intramolecular reaction are as follows:

Intermolecular reaction   Relative rate

Intramolecular reaction   Relative rate

Because the nucleophile and the electrophilic carbon are attached by a series of atoms, intramolecular reactions result in the formation of rings. If the ring formed is three, five, or six membered, the intramolecular reaction is much more favorable than its intermolecular counterpart. For example, the preceding intramolecular reaction occurs more than 5000 times faster than its intermolecular counterpart.

Intramolecular reactions are favored by entropy. Recall that entropy is a measure of the disorder of a system. It costs energy to put order into a system—to decrease the entropy of that system. In the case of an intermolecular reaction, the nucleophile and the electrophile must first come together from their initial random positions. This requires an increase in the order of the system, an entropically unfavorable process. In the case of an intramolecular reaction, the nucleophile is held in proximity to the electrophile by the connecting carbon chain. It takes a much smaller increase in the order of the system to position the nucleophile for reaction. In other words, the nucleophile is much closer to the electrophile at all times, and attaining the proper orientation required for the reaction is much more probable.

As mentioned earlier, three-, five-, and six-membered rings are preferred in intramolecular reactions. In the case of a three-membered ring, the increase in strain energy disfavors its formation; that is, it makes the enthalpy change in the reaction more positive (less favorable). However, the reaction centers are held so close that the entropy change for the reaction is quite favorable. In this case, entropy wins over enthalpy; three-membered rings are easily prepared in the laboratory, although they are often quite reactive. In contrast, the strain energy for four-membered rings is still large, whereas the entropy change is not as favorable, because the reaction centers are further apart. The enthalpy effect is larger than the entropy effect. As a result, four-membered rings are more difficult to prepare. For reactions that form five- and six-membered rings, the electrophile and nucleophile are even farther apart. However, because these rings are virtually strain free, there is no unfavorable enthalpy contribution. The entropy effect, although smaller, still enables five- and six-membered rings to be prepared readily in the laboratory. These rings are also the most common ones in naturally occurring compounds. Larger rings are less easily prepared because the entropy effect is no longer of much assistance.

PROBLEM 8.17

Show the products of these reactions. (Remember that acid–base reactions are usually much faster than nucleophilic substitution reactions.)

a) Cl—CH$_2$CH$_2$CH$_2$COH (with O double bond)   +   ⁻OH   ⟶

b) Br⌇⌇⌇⌇⌇⁺NH$_3$ +   ⁻OH   ⟶

8.14   COMPETING REACTIONS

We have already seen that the S$_N$1 and S$_N$2 reactions may be in competition under certain circumstances. In addition, other reactions also compete with these two. If we are trying to prepare a specific compound, these competing reactions often result in a lower yield of the desired compound and may also cause purification problems.

An **elimination reaction,** in which the leaving group and a hydrogen are lost from adjacent carbons, resulting in the formation of a double bond between these two carbons, competes with both the S$_N$1 and S$_N$2 reactions. Figure 8.11 shows a reaction that produces both elimination and substitution products under S$_N$2 conditions. The competition occurs because the nucleophile is also a base. When it reacts as a base, it removes a proton from the carbon adjacent to the leaving group, resulting in the formation of the elimination product.

In the S$_N$1 mechansim, a competition between elimination and substitution also results from the ability of the nucleophile to act as a base. However, in this case the competition occurs at the carbocation stage of the reaction. Figure 8.12 shows an example. Elimination reactions are discussed in more detail in Chapter 9. Chapter 10 presents methods to minimize elimination when the substitution product is desired and methods to maximize elimination when the alkene is the desired product. For now it is important only to recognize that eliminations may decrease the yields in substitution reactions.

Figure 8.11

COMPETITION BETWEEN S$_N$2 AND ELIMINATION REACTIONS.

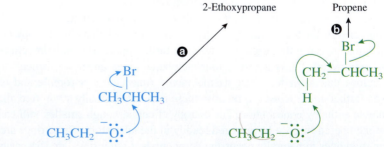

ⓐ Ethoxide ion acts as a nucleophile, attacking the carbon, resulting in the formation of the S$_N$2 product.

ⓑ Ethoxide ion acts as a base, removing a proton to give the elimination product. The by-product is ethanol.

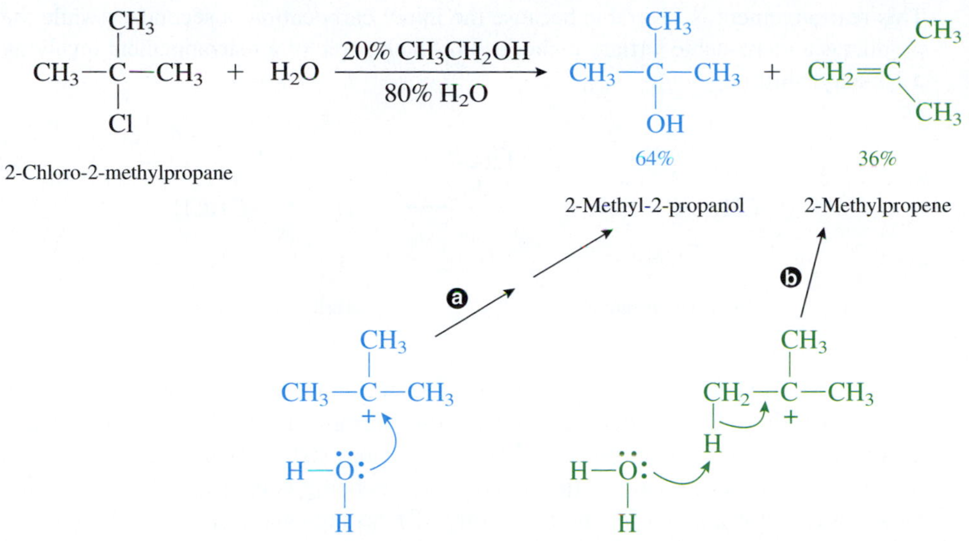

Figure 8.12

COMPETITION BETWEEN S$_N$1 AND ELIMINATION REACTIONS.

2-Chloro-2-methylpropane → 2-Methyl-2-propanol (64%) + 2-Methylpropene (36%)

ⓐ Water acts as a nucleophile, bonding to the cationic carbon to give, after the loss of a proton, the substitution product.

ⓑ Water acts as a base, removing a proton from the carbon adjacent to the cationic carbon to give the elimination product.

PROBLEM 8.18

Show both the substitution and elimination products that are formed in these reactions:

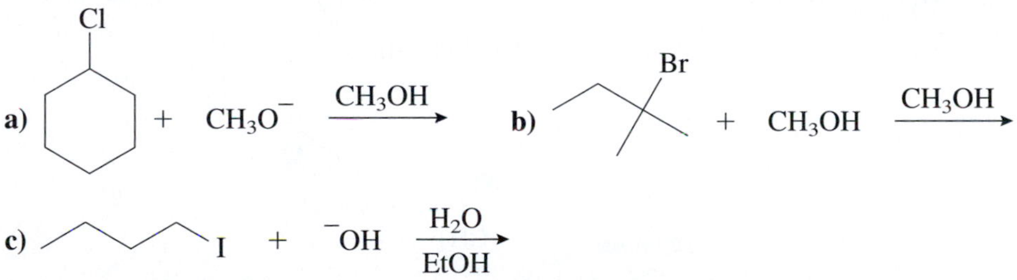

a) (cyclohexyl chloride) + CH_3O^- $\xrightarrow{CH_3OH}$

b) (2-bromo-2-methylbutane) + CH_3OH $\xrightarrow{CH_3OH}$

c) (1-iodobutane) + ^-OH $\xrightarrow[\text{EtOH}]{H_2O}$

An additional complication often occurs in S$_N$1 reactions. The initially formed carbocation may rearrange to a more stable carbocation. Such rearrangements occur by migration of a hydrogen, or other group, with its bonding pair of electrons from an *adjacent* carbon to the positively charged carbon. Because the hydrogen moves as $H:^-$ (which is named **hydride**) between atoms that are adjacent, the process is termed a 1,2-hydride shift. A similar migration of an alkyl group (a 1,2-alkyl shift) can also occur. An example of a rearrangement involving a 1,2-hydride shift is

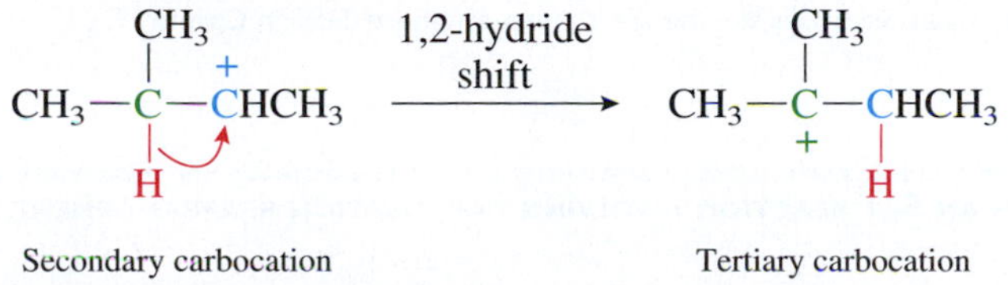

$$CH_3{-}\overset{+}{\underset{|}{C}}{-}\underset{|}{C}HCH_3 \xrightarrow[\text{shift}]{1,2\text{-hydride}} CH_3{-}\underset{+}{\underset{|}{C}}{-}\underset{|}{C}HCH_3$$

Secondary carbocation   Tertiary carbocation

This rearrangement is favorable because the initial carbocation is secondary while the product is a more stable tertiary carbocation. An example of a rearrangement involving a 1,2-alkyl shift is

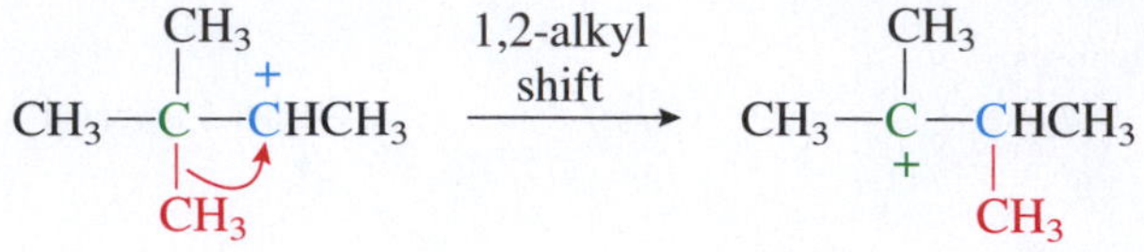

Secondary carbocation   Tertiary carbocation

The change from the secondary carbocation to the more stable tertiary carbocation again makes this rearrangement favorable. These rearrangements are very common and invariably occur if the rearranged carbocation is more stable. Therefore, if a reaction that proceeds through a carbocation intermediate is being considered, the carbocation must always be examined for the possibility of rearrangement. Figure 8.13 shows an S_N1 reaction in which both the substitution and the elimination products are produced from a rearranged carbocation.

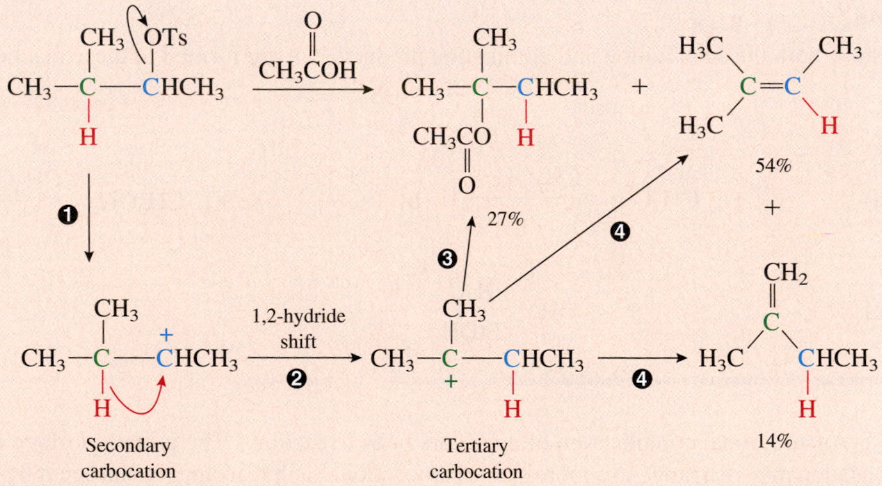

❶ The carbocation formed initially in this reaction is secondary. Only a very small amount of product is formed from this cation.

❷ Instead, it rearranges to a more stable tertiary carbocation by a 1,2-hydride shift.

❸ The tertiary carbocation produces 27% of the substitution product (S_N1) and ❹ 68% of the elimination products. In this case there are two possible elimination products, depending on which adjacent carbon loses the hydrogen. Situations like this are discussed in more detail in Chapter 9.

Figure 8.13

MECHANISM OF AN S_N1 REACTION INVOLVING CARBOCATION REARRANGEMENT.

PROBLEM 8.19

Show the rearranged carbocations that are expected from these carbocations:

a) (cyclopentane ring with $^+CH_2$ substituent)

b) $CH_3\overset{CH_3}{\underset{+}{CH}}CHCH_2CH_3$

c) (cyclohexane ring with CH_3 and $+$)

Finally, it is important to recognize that an S_N1 reaction that forms an allylic carbocation often provides more than one site at which the nucleophile can bond. The nucleophile may bond to either of the carbons that bear the positive charges in the resonance structures. If the allylic cation is not symmetrical, this will result in the formation of two products: one "normal" and one "rearranged." An example of such an "allylic rearrangement" is

$$CH_3CH-CH=CH_2 \xrightarrow{CH_3CH_2OH} CH_3\overset{OCH_2CH_3}{CH}-CH=CH_2 + CH_3CH=CH-\overset{OCH_2CH_3}{CH_2}$$

3-Chloro-1-butene   18%   82%

$$CH_3\overset{+}{CH}-CH=CH_2 \longleftrightarrow CH_3CH=CH-\overset{+}{CH_2}$$

Allylic carbocation

Note that this "rearrangement" is fundamentally different from the previously described carbocation rearrangements. The allyl carbocation has partial positive charge located on two carbons, both of which are expected to react with the nucleophile. One resonance structure does not "rearrange" to the other. It would be quite surprising if two products were not produced from such a resonance stabilized carbocation.

PRACTICE PROBLEM 8.5

Show the substitution products for this reaction:

$$CH_3-\overset{CH_3}{\underset{CH_3}{C}}-\overset{Br}{CH}CH_3 + CH_3OH \xrightarrow{CH_3OH}$$

Strategy

Whenever an S_N1 reaction is encountered, it is important to examine the carbocation for the possibility of rearrangement. Rearrangement will occur if the carbon adjacent to the electrophilic carbon is bonded to more carbon groups than the electrophilic carbon. In such cases, at least part of the product results from the carbocation formed by migration of a hydrogen or an alkyl group from the adjacent C to the electrophilic C.

Solution

The leaving group (Br) is on a secondary carbon. The reaction involves a poor nucleophile (CH_3OH) and a polar solvent (CH_3OH), so it follows an S_N1 mechanism. The initial carbocation is secondary and can rearrange to a tertiary carbocation by migration of a methyl group.

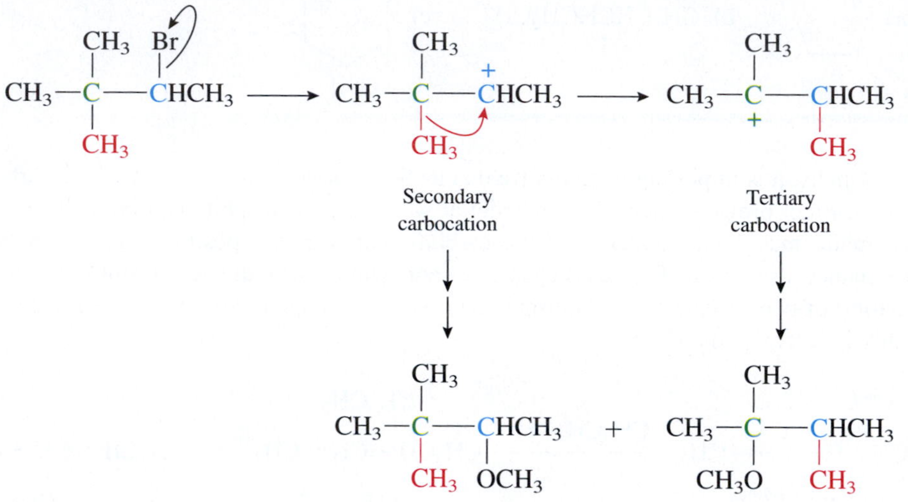

Secondary carbocation   Tertiary carbocation

PROBLEM 8.20

Show the substitution products for these reactions:

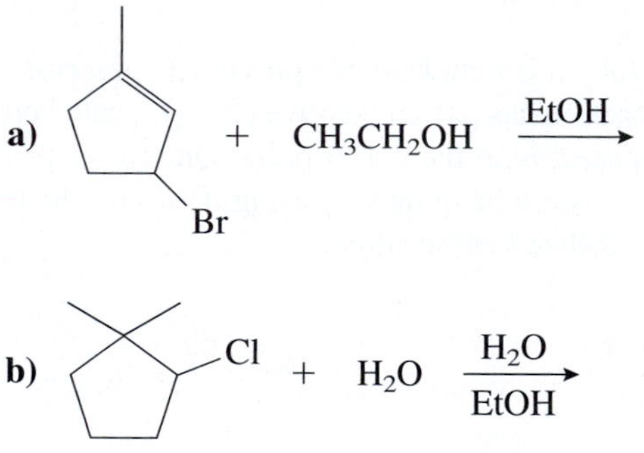

a)

+ CH_3CH_2OH $\xrightarrow{\text{EtOH}}$

b)

+ H_2O $\xrightarrow[\text{EtOH}]{\text{H}_2\text{O}}$

PROBLEM 8.21

a) Show all of the steps in the mechanism for this reaction. Don't forget to use curved arrows to show the movement of electrons in each step of the mechanism.

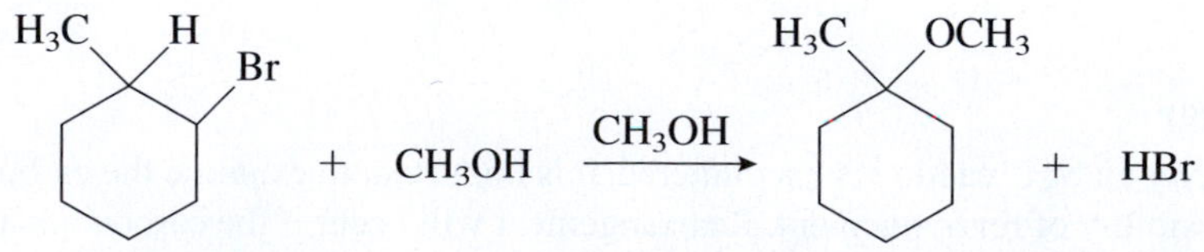

+ CH_3OH $\xrightarrow{\text{CH}_3\text{OH}}$ +   HBr

b) Show a free energy versus reaction progress diagram for this reaction.

Focus On

Carbocation Rearrangements in Superacids

Normally, carbocations are encountered as transient intermediates along the pathway of reactions such as the S_N1 substitution. However, under conditions in which no nucleophiles or bases are available to react with them, carbocations can have significant lifetimes. Because superacids are very weak nucleophiles (see Section 4.10) and are quite polar, they provide an environment in which carbocations have lifetimes long enough to allow them to be studied by a variety of instrumental techniques.

As an example, the *tert*-butyl cation can be generated by treating 2-methyl-2-propanol (*tert*-butyl alcohol) with the superacid FSO_3H/SbF_5 in liquid sulfur dioxide as the solvent. The reaction is shown in the following equation:

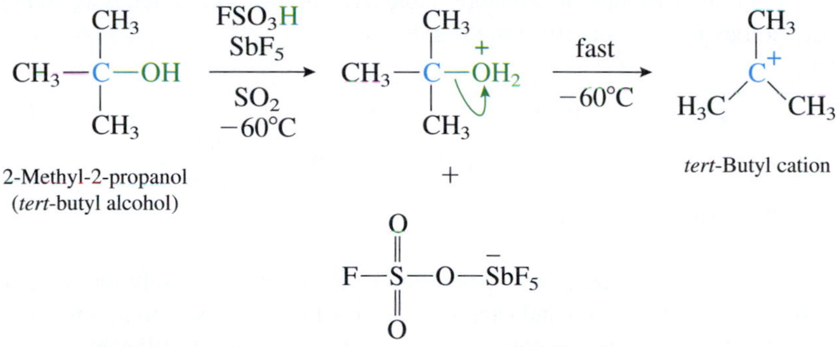

First, the oxygen is protonated to make it a better leaving group. Then water leaves to produce the *tert*-butyl cation. This step is very fast, even at $-60°C$, so the carbocation is the only product that can be detected as soon as the alcohol is added to the superacid medium. Because there is no nucleophile for the carbocation to react with (the H_2O generated in the reaction is protonated by the strong acid to form H_3O^+), its lifetime under these conditions is quite long, and it can be studied by a variety of techniques.

As we have seen, carbocations rearrange even under the conditions of the S_N1 reaction, in which their lifetimes are extremely short and rearrangement must compete with the fast reaction with a nucleophile. Therefore, in superacid solution, in which their lifetimes are much longer, it is not surprising that carbocations undergo extensive rearrangements. Usually, such rearrangements occur until a tertiary carbocation is formed. For example, consider the case in which 1-butanol is dissolved in superacid solution:

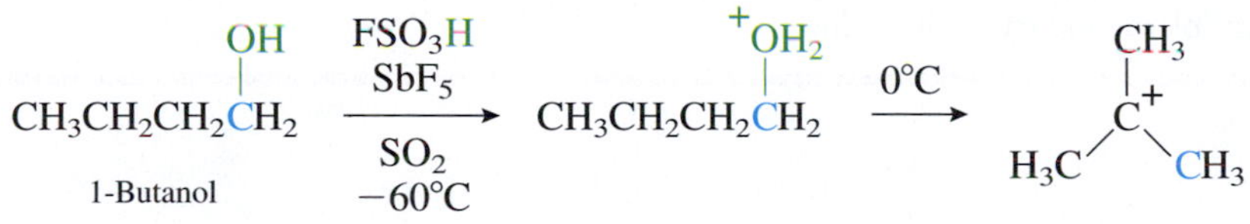

At $-60°C$, the protonated alcohol is formed but water does not leave because the primary carbocation that would be formed is too unstable. When the temperature is

Continued

raised to 0°C, water leaves but the carbocation rearranges rapidly to the more stable *tert*-butyl cation, which is the only carbocation that can be observed. The other two isomeric 4-carbon alcohols behave similarly. At −60°C, 2-methyl-1-propanol is protonated but water does not leave because the carbocation that would be formed is primary. When the temperature of the solution is raised to −30°C, water leaves and the *tert*-butyl cation is again produced by rapid rearrangements.

$$\underset{\text{2-Methyl-1-propanol}}{\underset{\overset{\displaystyle |\quad|}{\text{CH}_3\text{CHCH}_2}}{\overset{\displaystyle \text{H}_3\text{C}\ \ \text{OH}}{}}} \quad \xrightarrow[\substack{\text{SO}_2 \\ -60°\text{C}}]{\substack{\text{FSO}_3\text{H} \\ \text{SbF}_5}} \quad \underset{\overset{\displaystyle |\quad|}{\text{CH}_3\text{CHCH}_2}}{\overset{\displaystyle \overset{+}{\text{H}_3\text{C}\ \ \text{OH}_2}}{}} \quad \xrightarrow{-30°\text{C}} \quad \underset{\text{H}_3\text{C}\quad\text{CH}_3}{\overset{\text{CH}_3}{\overset{|}{\overset{+}{\text{C}}}}}$$

When 2-butanol is treated under the same conditions, the protonated alcohol is again the first species that is observed. However, because a more stable secondary carbocation is produced initially, water slowly leaves even at −60°C. The secondary carbocation rapidly rearranges to the more stable *tert*-butyl cation, which is again the only carbocation that can be observed in the solution.

$$\underset{\text{2-Butanol}}{\underset{\overset{\displaystyle |}{\text{CH}_3\text{CH}_2\text{CHCH}_3}}{\overset{\displaystyle \text{OH}}{}}} \quad \xrightarrow[\substack{\text{SO}_2 \\ -60°\text{C}}]{\substack{\text{FSO}_3\text{H} \\ \text{SbF}_5}} \quad \underset{\overset{\displaystyle |}{\text{CH}_3\text{CH}_2\text{CHCH}_3}}{\overset{\displaystyle \overset{+}{\text{OH}_2}}{}} \quad \xrightarrow[-60°\text{C}]{\text{slow}} \quad \underset{\text{H}_3\text{C}\quad\text{CH}_3}{\overset{\text{CH}_3}{\overset{|}{\overset{+}{\text{C}}}}}$$

In summary, all of the other isomeric butyl carbocations rapidly rearrange to the most stable *tert*-butyl cation and cannot be detected under these conditions.

The 2-butyl cation can be observed at lower temperature. As illustrated in the following equation, it is formed by the reaction of 2-fluorobutane with SbF$_5$ at −110°C. When the temperature of the solution is raised to −40°C, rearrangement to the *tert*-butyl cation occurs.

$$\underset{\text{2-Fluorobutane}}{\underset{\overset{\displaystyle |}{\text{CH}_3\text{CH}_2\text{CHCH}_3}}{\overset{\displaystyle \text{F}}{}}} + \text{SbF}_5 \quad \xrightarrow{-110°\text{C}} \quad \underset{\substack{+ \\ \text{SbF}_6^-}}{\overset{+}{\text{CH}_3\text{CH}_2\text{CHCH}_3}} \quad \xrightarrow{-40°\text{C}} \quad \underset{\text{H}_3\text{C}\quad\text{CH}_3}{\overset{\text{CH}_3}{\overset{|}{\overset{+}{\text{C}}}}}$$

Overall, these and related experiments help confirm the existence of carbocations and show that the activation energies for their rearrangements must be small in most cases because the rearrangements occur rapidly at rather low temperatures. In addition, it is readily apparent that a tertiary carbocation is considerably more stable than are primary or secondary carbocations.

Review of Mastery Goals

After completing this chapter, you should be able to:

ORGANIC
Chemistry•⚛•Now™
Click Mastery Goal Quiz to test how well you have met these goals.

- Write mechanisms for the S_N1 and S_N2 reactions. (Problems 8.29, 8.35, 8.43, 8.44, and 8.45)

- Recognize the various nucleophiles and leaving groups and understand the factors that control their reactivities. (Problems 8.31, 8.32, 8.33, 8.41, 8.42, and 8.52)

- Understand the factors that control the rates of these reactions, such as steric effects, carbocation stabilities, the nucleophile, the leaving group, and solvent effects. (Problems 8.22, 8.23, 8.24, 8.25, 8.34, 8.39, 8.55, 8.59, and 8.60)

- Be able to use these factors to predict whether a particular reaction will proceed by an S_N1 or an S_N2 mechanism and to predict what effect a change in reaction conditions will have on the reaction rate. (Problems 8.28, 8.36, 8.37, and 8.38)

- Show the products of any substitution reaction. (Problems 8.26, 8.27, 8.30, 8.40, 8.50, 8.51, and 8.56)

- Show the stereochemistry of the products. (Problems 8.27, 8.56, 8.57, and 8.58)

- Recognize when a carbocation rearrangement is likely to occur and show the products expected from the rearrangement. (Problems 8.47 and 8.48)

- Show the structures of the products that result from the elimination reactions that compete with the substitution reactions. (Problem 8.46)

Visual Summary of Key Reactions

The importance of identifying the electrophile and the nucleophile in organic reactions cannot be overemphasized. The electrophile is usually a carbon bonded to an electronegative element so that the carbon is electron deficient. The nucleophile most often contains an atom with an unshared pair of electrons, a Lewis base. In this chapter the electrophilic carbon is the one bonded to a leaving group. A few nucleophiles have been presented, but many more will be discussed in Chapter 10. Later chapters will introduce new electrophiles or new nucleophiles, but the new reactions always involve the nucleophile forming a bond to the electrophile.

Table 8.6 provides a summary of the most important features of the S_N1 and S_N2 reactions.

Table 8.6   Summary of the S_N1 and S_N2 Reactions

	S_N1	S_N2
Mechanism	Two step R—L ⟶ R⁺ ⟶ R—Nu	One step R—L + Nu⁻ ⟶ R—Nu + L⁻
Kinetics	First-order rate = k[R—L]	Second-order rate = k[R—L][Nu⁻]
Effects of Nucleophile	No effect on rate (favored by weaker nucleophiles because S_N2 is slower)	Stronger nucleophiles cause faster rate
Effect of Carbon Structure	Tertiary > secondary Resonance stabilization of R⁺ important	Methyl > primary > secondary
Stereochemistry	Racemization (possibly excess inversion)	Inversion
Effect of Solvent	Favored by polar solvents	Favored by aprotic solvents
Competing Reactions	Elimination, rearrangement	Elimination

ORGANIC
Chemistry ⚛ Now™
Assess your understanding of this chapter's topics with additional quizzing and conceptual-based problems at
http://now.brookscole.com/ hornback2

Additional Problems

8.22   Which of these compounds would have a faster rate of S_N2 reaction?

a)

b)

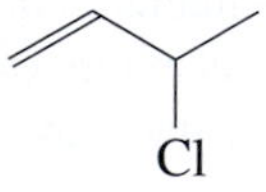

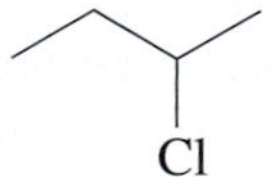

c)

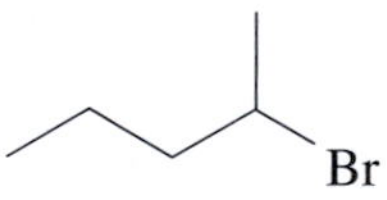

d)

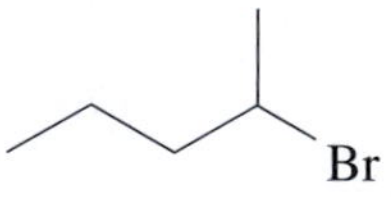

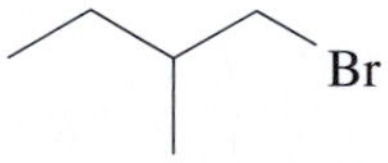

e)

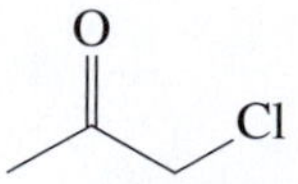

8.23 Which of these compounds would have a faster rate of S_N1 reaction?

a) 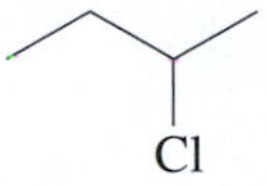 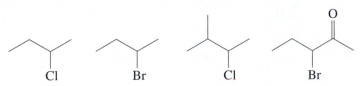 or

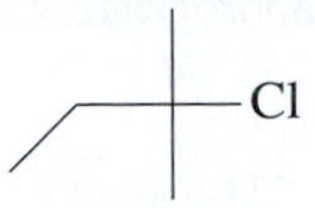

b) 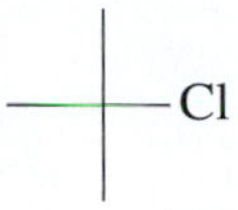 ⎯⎯Cl   or   ⎯⎯I

c) 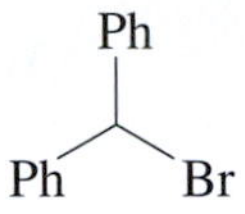
Ph or CH₃
Ph Br Ph Br

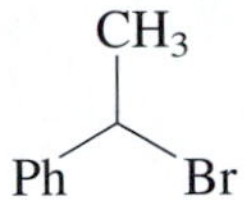

d)

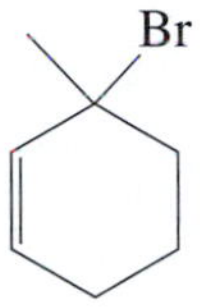

Br or Br

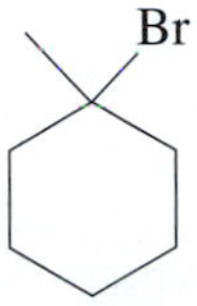

e)

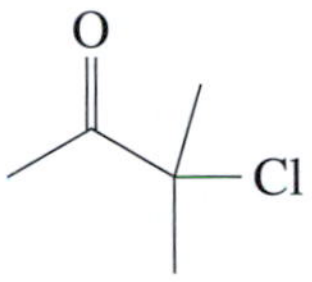

Cl or Cl

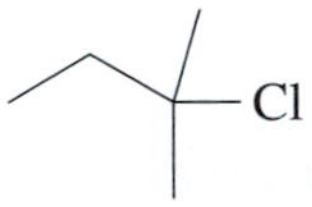

8.24 Arrange these compounds in order of increasing S_N2 reaction rate:

Cl   Br   Cl   Br

8.25 Arrange these compounds in order of increasing S_N1 reaction rate:

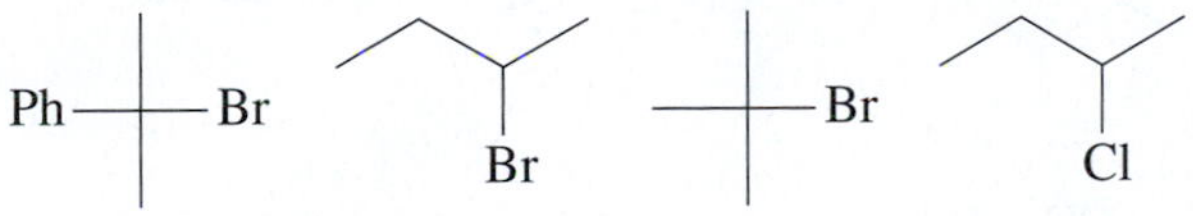

Ph⎯⎯Br   Br   ⎯⎯Br   Cl

8.26 Show the products of these reactions and explain whether each would follow an S_N1 or an S_N2 mechanism:

a) 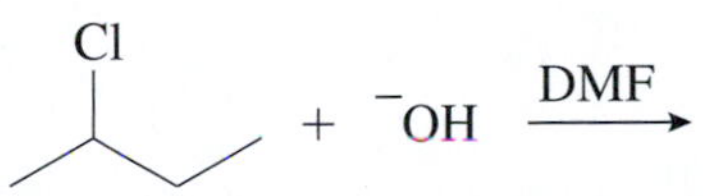
Cl
+ ⁻OH $\xrightarrow{\text{DMF}}$

b) 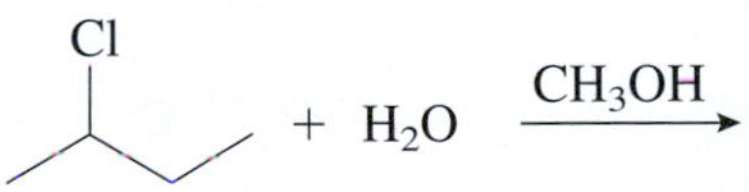
Cl
+ H_2O $\xrightarrow{\text{CH}_3\text{OH}}$

c) 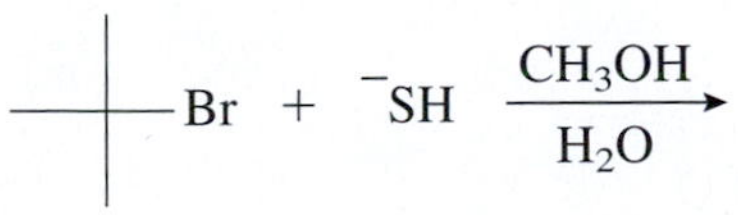
⎯⎯Br + ⁻SH $\xrightarrow[\text{H}_2\text{O}]{\text{CH}_3\text{OH}}$

d) 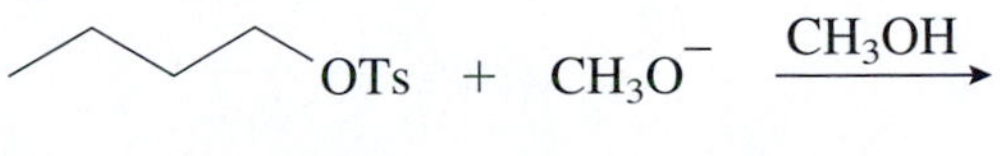
OTs + CH_3O^- $\xrightarrow{\text{CH}_3\text{OH}}$

e) 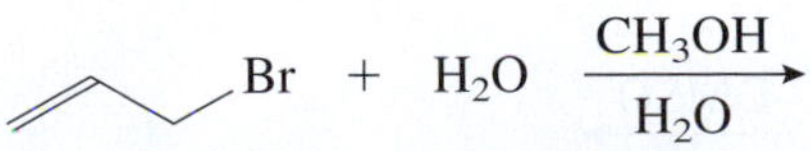
Br + H_2O $\xrightarrow[\text{H}_2\text{O}]{\text{CH}_3\text{OH}}$

f) 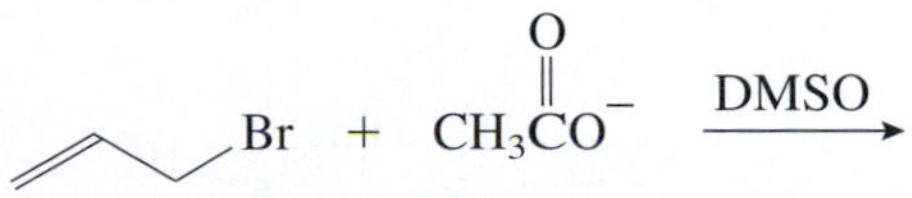
Br + $CH_3\overset{O}{\overset{\|}{C}}O^-$ $\xrightarrow{\text{DMSO}}$

8.27 Show the substitution products for these reactions. Don't forget to show the stereochemistry of the product, where appropriate.

a) 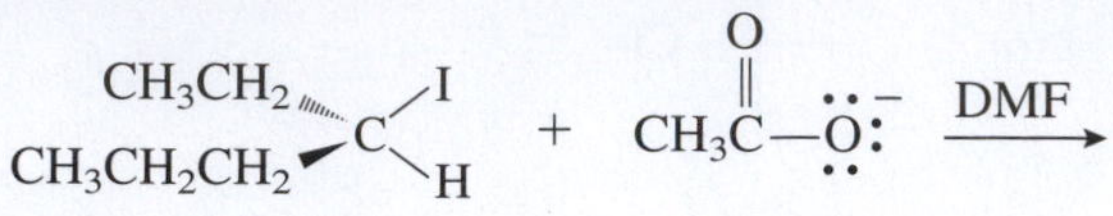

b) 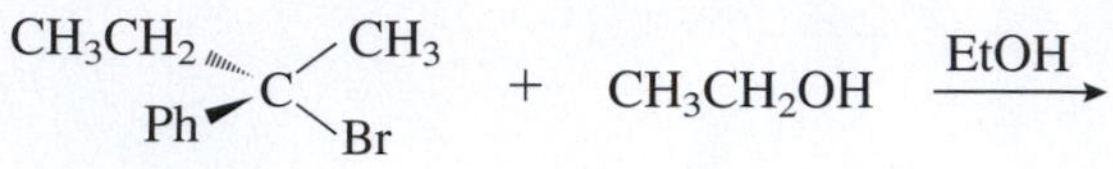

c) 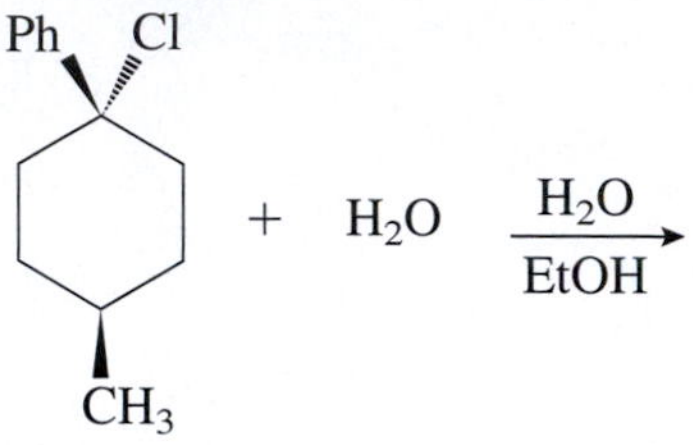

d) 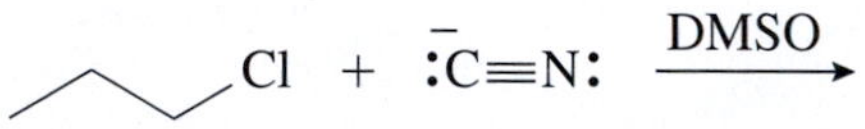

e) 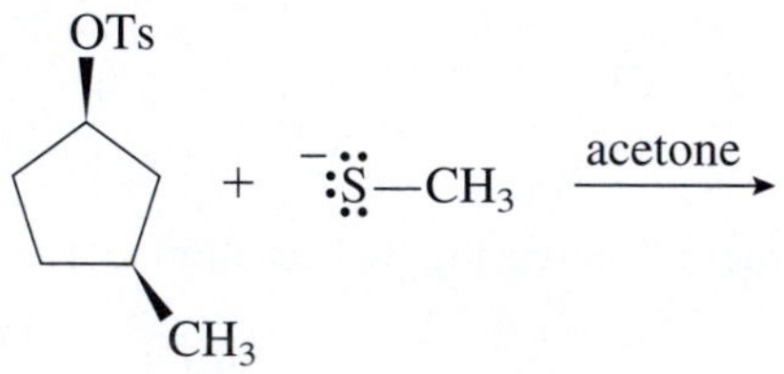

f) 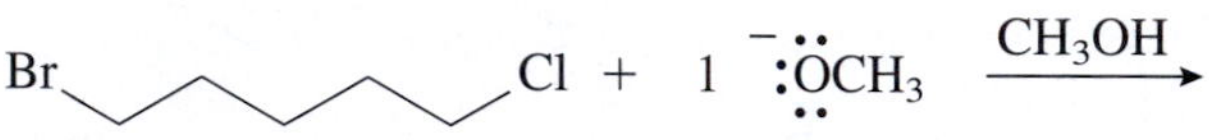

g) 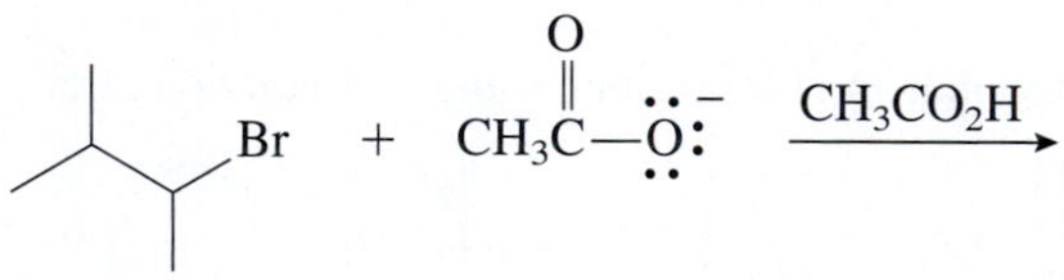

h) 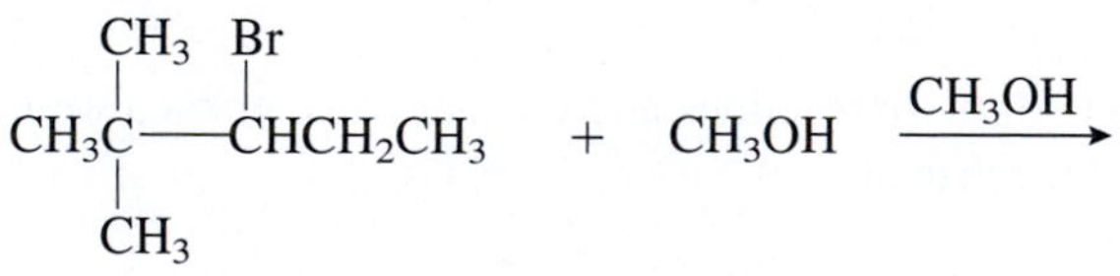

i) 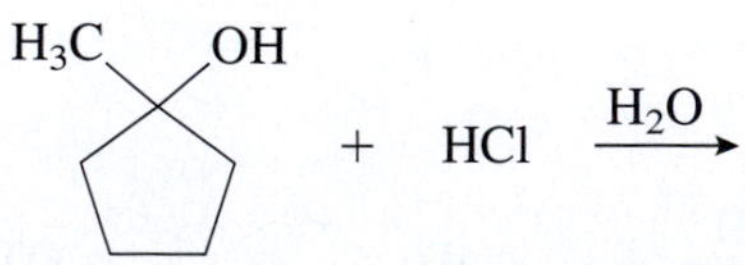

j) 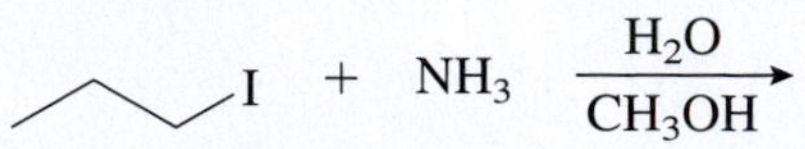

k) 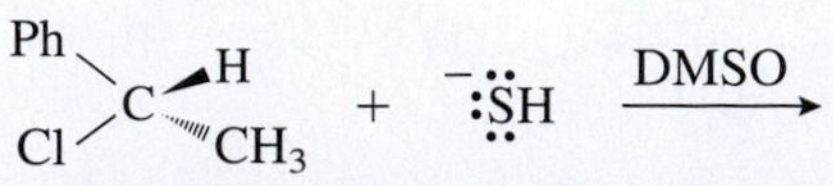

l)
OH
O
Cl—S—CH3
O
pyridine
NaCl
DMF

m) $CH_3CH_2CH_2CH_2Br$ + NaI $\xrightarrow{\text{acetone}}$

8.28 Explain whether each pair of reactions should follow an S_N1 or an S_N2 mechanism. Then explain which member of the pair should proceed at a faster rate.

a)
Cl + CH3OH
CH3OH
or
Br + CH3OH
CH3OH

b)
Br + ⁻CN
acetone
or
Br + ⁻CN
methanol

c)
OTs
CHCH3
+ CH3CH2OH
EtOH
or
OTs
CHCH3
CH3
+ CH3CH2OH
EtOH

d)
I + CH3S⁻
CH3OH
or
I + CH3S⁻
CH3OH

e)
Br + CH3OH
CH3OH
or
Br + CH3OH
CH3OH

f)
OMs + ⁻OH
DMF
or
OMs + CH3CO⁻
O
DMF

g)
Br
+ CH3OH
CH3OH
or
Br
+ CH3CH2OH
CH3CH2OH

h)
:O:⁻
Cl
CH3OH
or
Cl + CH3O⁻
CH3OH

i)
Br +
N
EtOH
or
Br +
N
CH3
EtOH

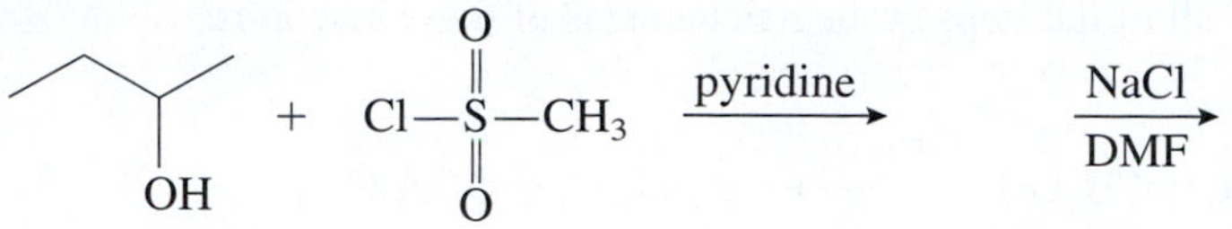

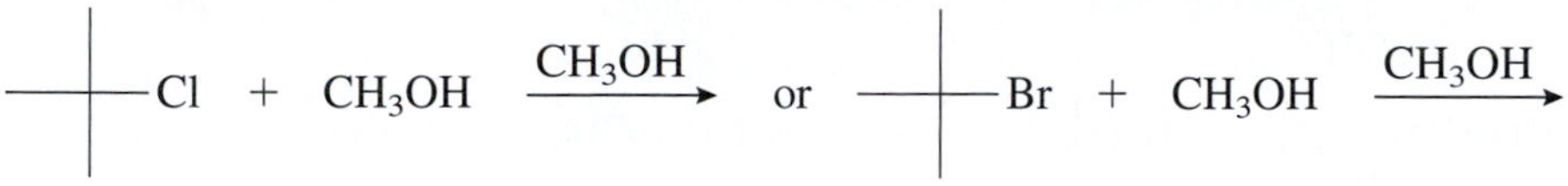

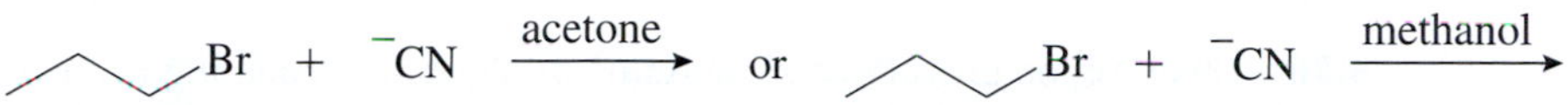

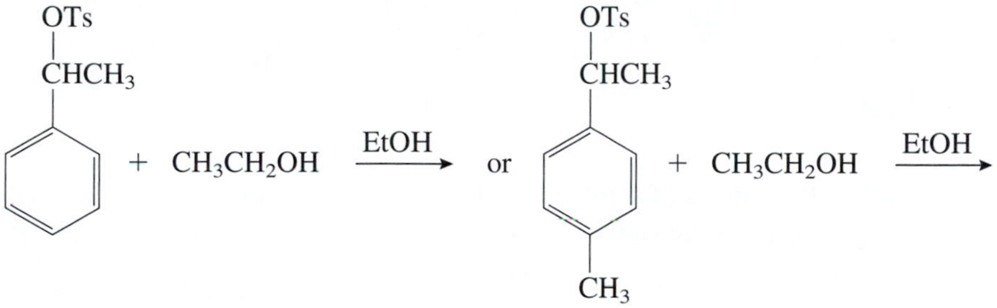

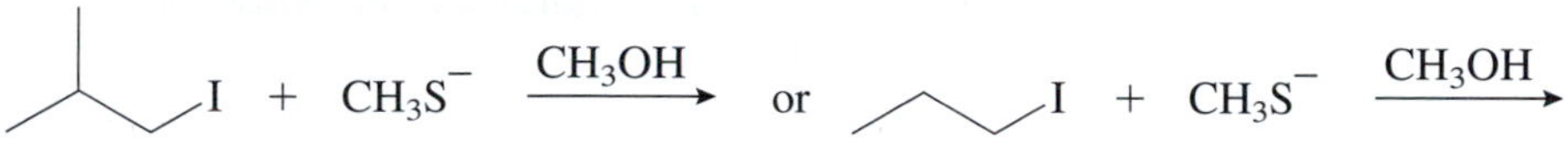

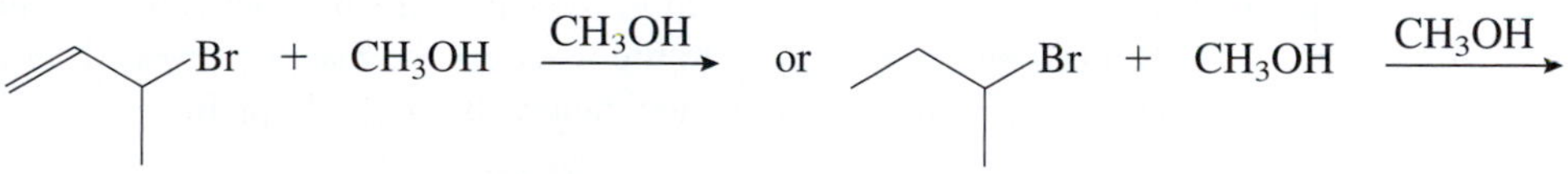

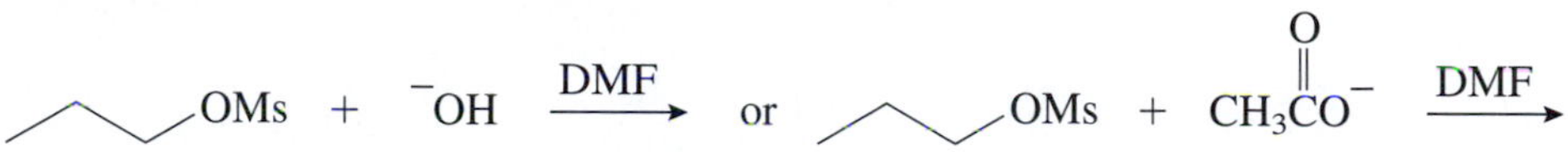

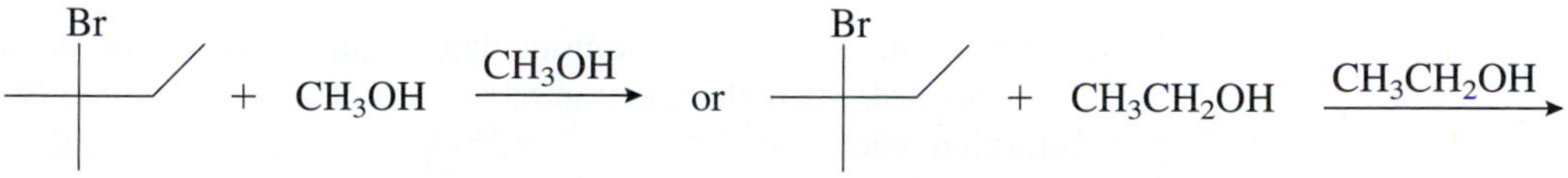

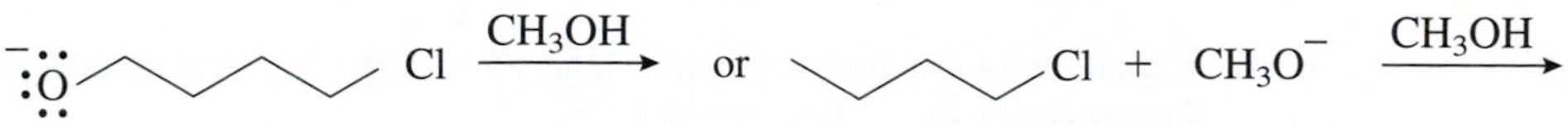

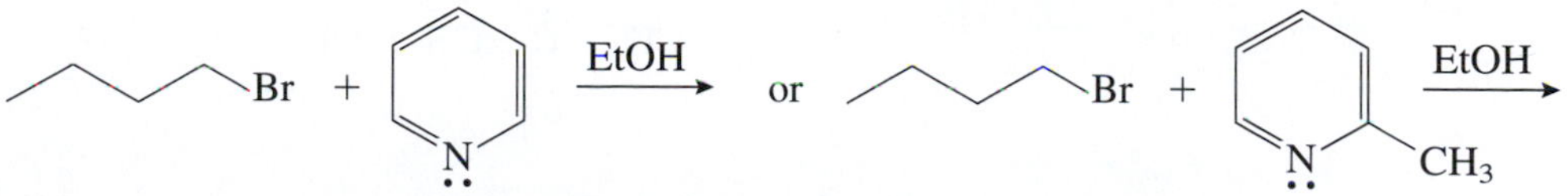

8.29 Show all of the steps in the mechanisms of these reactions:

a)
$$CH_3-\overset{\overset{\displaystyle CH_3}{|}}{\underset{\underset{\displaystyle CH_3}{|}}{C}}-Br \ + \ CH_3CH_2OH \ \xrightarrow{EtOH} \ CH_3-\overset{\overset{\displaystyle CH_3}{|}}{\underset{\underset{\displaystyle CH_3}{|}}{C}}-OCH_2CH_3 \ + \ CH_3CH_2\overset{+}{O}H_2 \ \ \overset{-}{Br}$$

b)
$$CH_3CH_2CH_2\overset{\overset{\displaystyle Cl}{|}}{CH_2} \ + \ CH_3O^- \ \xrightarrow{DMSO} \ CH_3CH_2CH_2\overset{\overset{\displaystyle OCH_3}{|}}{CH_2} \ + \ Cl^-$$

c)
$$CH_3CH_2\overset{\overset{\displaystyle OH}{|}}{\underset{\underset{\displaystyle CH_3}{|}}{C}}CH_3 \ + \ \underset{\underset{\displaystyle \overset{-}{Br}}{}}{H_3O^+} \ \xrightarrow{H_2O} \ CH_3CH_2\overset{\overset{\displaystyle Br}{|}}{\underset{\underset{\displaystyle CH_3}{|}}{C}}CH_3 \ + \ 2\,H_2O$$

8.30 This reaction gives three substitution products (not counting *E/Z* isomers). Show the structures of these products and show the mechanism for their formation:

8.31 Explain why the trifluoromethanesulfonate anion is a better leaving group than the mesylate anion.

$\overset{-}{O}-\overset{\overset{\displaystyle O}{\|}}{\underset{\underset{\displaystyle O}{\|}}{S}}-CF_3$ Trifluoromethanesulfonate anion

8.32 When benzyl tosylate is heated in methanol, the product is benzyl methyl ether. When bromide ion is added to the reaction, the reaction proceeds at exactly the same rate, but the product is now benzyl bromide. Explain.

$$PhCH_2-OTs \ \xrightarrow{CH_3OH} \ PhCH_2-OCH_3$$
$$\xrightarrow[CH_3OH]{Br^-} \ PhCH_2-Br$$

8.33 The substitution reaction of bromomethane with hydroxide ion proceeds about 5000 times faster than the reaction of bromomethane with water. However, the substitution reaction of 2-bromo-2-methylpropane proceeds at the same rate with both of these nucleophiles. Explain.

8.34 Explain why this primary halide reacts very rapidly under conditions that are favorable for the S_N1 mechanism:

$$CH_3-O-CH_2-Cl$$

8.35 Ethers can be cleaved by treatment with strong acids. Show all of the steps in the mechanism for this reaction and explain why these products are formed rather than iodomethane and 2-methyl-2-butanol:

$$CH_3-\underset{\underset{CH_3}{|}}{\overset{\overset{CH_3}{|}}{C}}-O-CH_3 \ + \ H-I \ \xrightarrow{H_2O} \ CH_3-\underset{\underset{CH_3}{|}}{\overset{\overset{CH_3}{|}}{C}}-I \ + \ H-O-CH_3$$

8.36 The reaction of a compound with silver nitrate in ethanol is used as a chemical test to determine if the compound is an alkyl halide. The formation of a precipitate of the silver halide constitutes a positive test.

$$R-X \ + \ Ag^+ \ NO_3^- \ \xrightarrow{CH_3CH_2OH} \ AgX \ (s) \ + \ R^+ \ \longrightarrow \ R-OCH_2CH_3$$

a) Explain why these conditions favor the S_N1 mechanism.
b) Which of these halides would give a precipitate more rapidly when reacted with $AgNO_3$ in ethanol?

i) 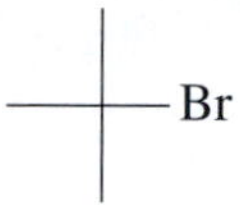 ⟩—Br or ⟩—Cl   **ii)** PhCHClCH$_3$ or CH$_3$CHClCH$_3$

iii)  (cyclopentane-Cl) or (dimethyl-Cl cyclobutane)   **iv)** 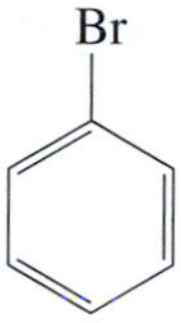 (benzene-Br) or (benzene-CH$_2$Br)

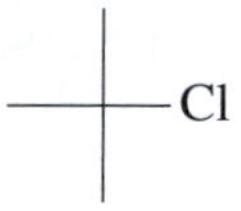

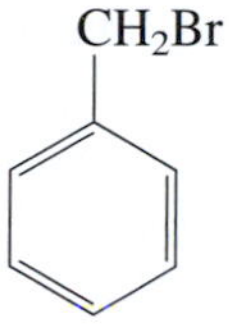

8.37 The reaction of an alkyl chloride (or bromide) with sodium iodide in acetone proceeds according to the following equation:

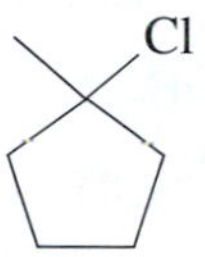

$$R-Cl \ + \ Na^+ \ I^- \ \xrightarrow{acetone} \ R-I \ + \ NaCl \ (s)$$

Sodium iodide is soluble in acetone, whereas both sodium chloride and sodium bromide are insoluble, so the appearance of a precipitate is a positive test for the presence of an alkyl chloride or bromide.

a) Explain why these conditions favor the S_N2 mechanism.
b) Which of these halides would give a precipitate more rapidly when reacted with NaI in acetone?

i) 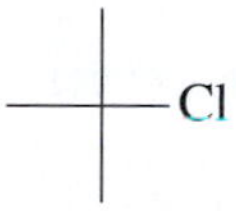 ⟩—Cl or ⟩—Cl   **ii)**  (propyl-Cl) or (propyl-Br)

iii)  (cyclopentane-Cl) or CH$_3$CH$_2$Cl   **iv)**  (cyclopentane-Br) or (dimethyl-Br cyclopentane)

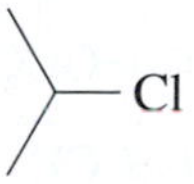

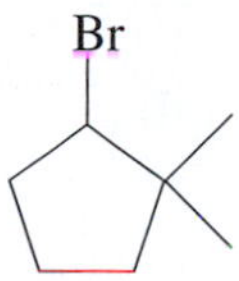

8.38 The Lucas test is used to check for the presence of an alcohol functional group in an unknown compound. The test reaction is shown in the following equation:

$$R-OH \ + \ HCl \ \xrightarrow[\text{H}_2\text{O}]{\text{ZnCl}_2} \ R-Cl \ + \ H_2O$$

Smaller alcohols are soluble in the strongly acidic solution, but the corresponding chlorides are not. A positive test is indicated by the formation of an insoluble layer of the alkyl chloride.

a) Explain why the reaction conditions favor an S_N1 mechanism.

b) Which of these alcohols reacts more rapidly with HCl and $ZnCl_2$ in H_2O?

i) 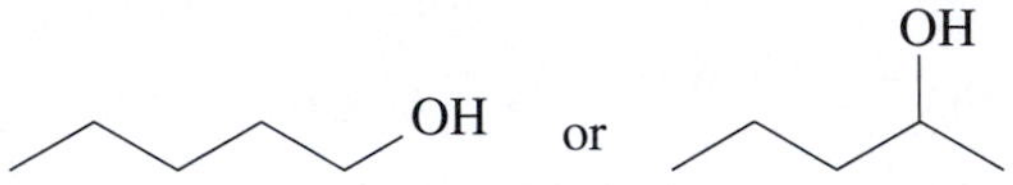   or

ii) 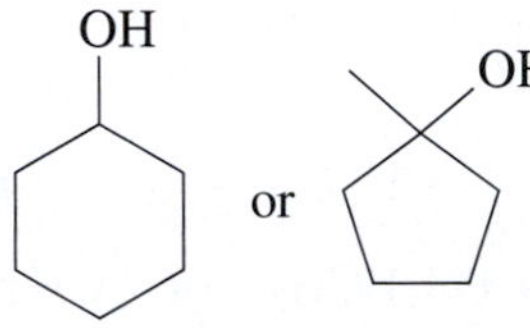   or   iii) 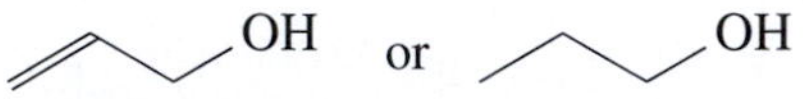   or

8.39 Explain why this secondary alcohol reacts with HCl and $ZnCl_2$ in H_2O at about the same rate as a primary alcohol (see problem 8.38):

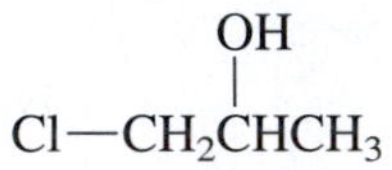

$$Cl-CH_2\overset{\overset{\displaystyle OH}{|}}{C}HCH_3$$

8.40 What reagent and solvent would you use to carry out the following transformations?

a)

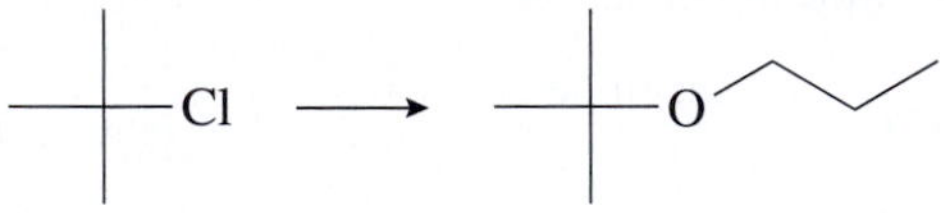

b)

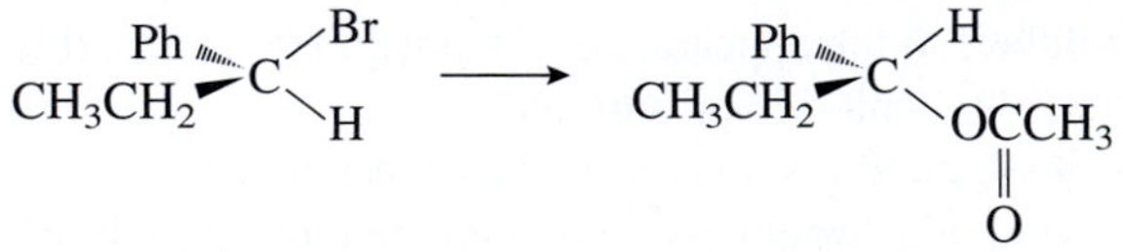

c)

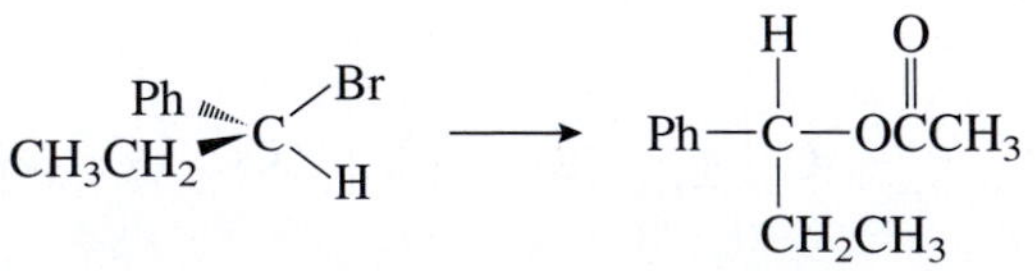

d)

e)

8.41 In most cases, RO$^-$ cannot act as a leaving group in nucleophilic substitution reactions. Explain why the following reaction does occur:

$$R—O—\langle \bigcirc \rangle—NO_2 \ + \ :Nu^- \ \longrightarrow \ R—Nu \ + \ :\ddot{O}^-—\langle \bigcirc \rangle—NO_2$$

8.42 Explain which of these reactions would have the faster rate:

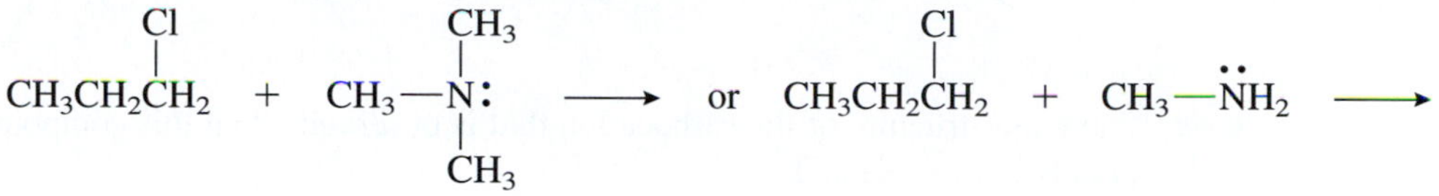

$$CH_3CH_2\overset{\underset{\displaystyle |}{Cl}}{CH_2} \ + \ CH_3—\overset{\underset{\displaystyle |}{CH_3}}{\underset{\underset{\displaystyle |}{CH_3}}{N}}: \ \longrightarrow \quad or \quad CH_3CH_2\overset{\underset{\displaystyle |}{Cl}}{CH_2} \ + \ CH_3—\ddot{N}H_2 \ \longrightarrow$$

8.43 Heating ethanol with sulfuric acid is one method used for the preparation of diethyl ether. Show all of the steps in the mechanism for this reaction:

$$2 \ CH_3CH_2OH \ \xrightarrow{H_2SO_4} \ CH_3CH_2—O—CH_2CH_3 \ + \ H_2O$$

8.44 When an aqueous solution of (R)-2-butanol is treated with a catalytic amount of sulfuric acid, slow racemization of the alcohol occurs. Show all of the steps in the mechanism for this process.

8.45 Show all of the steps in the mechanism and explain the stereochemistry for this reaction:

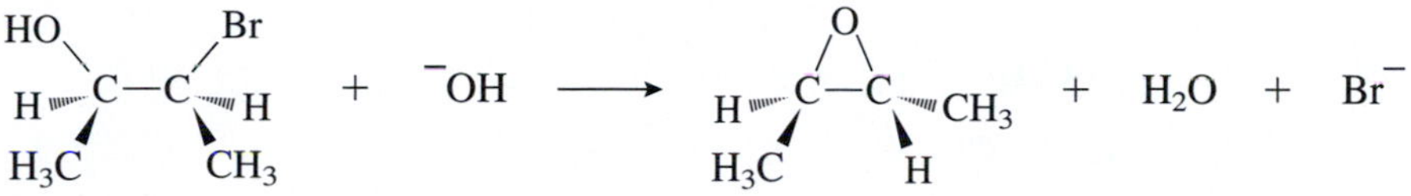

$$\overset{HO}{\underset{H_3C}{\overset{|}{\underset{|}{C}}}}{}_{H}—\overset{Br}{\underset{CH_3}{\overset{|}{\underset{|}{C}}}}{}_{H} \ + \ ^-OH \ \longrightarrow \ \overset{O}{H \cdots \overset{}{C}—\overset{}{C} \cdots CH_3} \ + \ H_2O \ + \ Br^-$$

8.46 Show both the substitution and elimination products that would be formed in these reactions:

a) 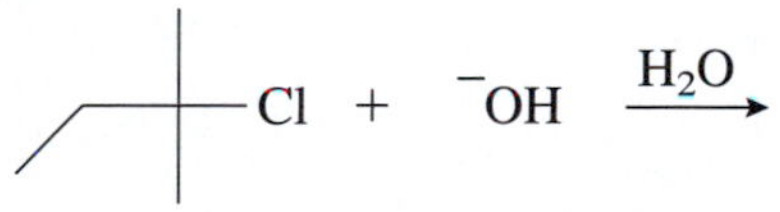 $\quad$ (CH₃)₃C—Cl $\ + \ ^-OH \ \xrightarrow{H_2O}$

b)

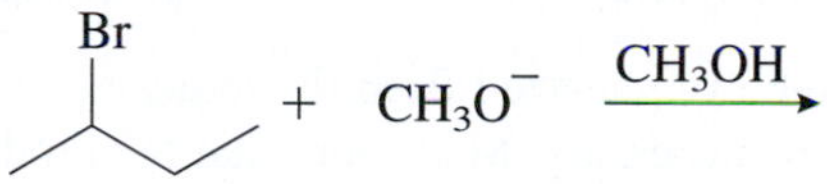

$$\overset{Br}{\underset{}{\Large\vee\!\!\!\diagup}} \ + \ CH_3O^- \ \xrightarrow{CH_3OH}$$

c)

$$\text{(1-chloro-1-methylcyclohexane)} \ + \ CH_3CH_2OH \ \xrightarrow{EtOH}$$

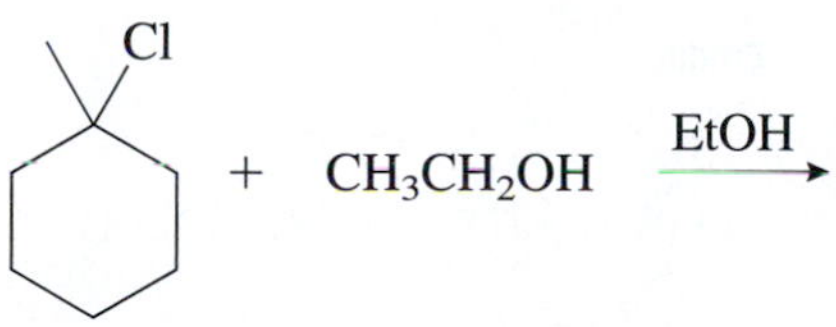

8.47 The reaction of 3-iodo-2,2-dimethylbutane with ethanol gives three elimination products in addition to two substitution products as shown in the following equation. Show all the steps in the mechanism for the formation of the elimination products. Show the structures of the substitution products.

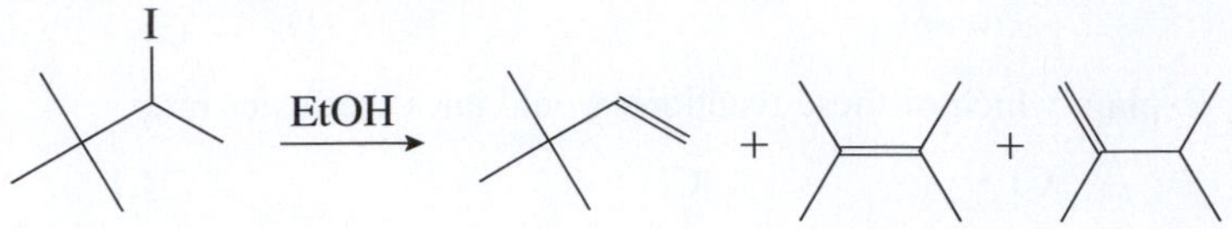
I
EtOH
+
+

8.48 Show the structure of the carbocation that is observed when this compound is dissolved in superacid:

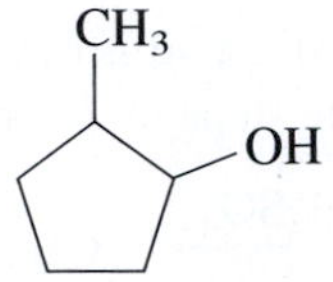
CH3
OH

8.49 How much is the reaction rate for these reactions increased or decreased if the concentration of hydroxide ion is doubled? if the concentrations of both the alkyl chloride and hydroxide ion are halved?

a) 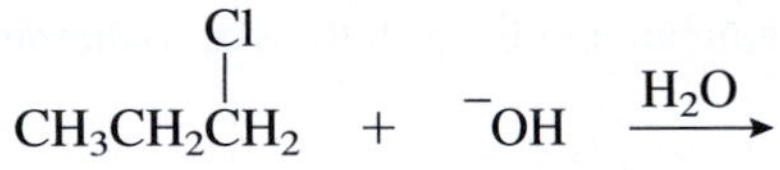
Cl
CH3CH2CH2
+
⁻OH
H2O

b) 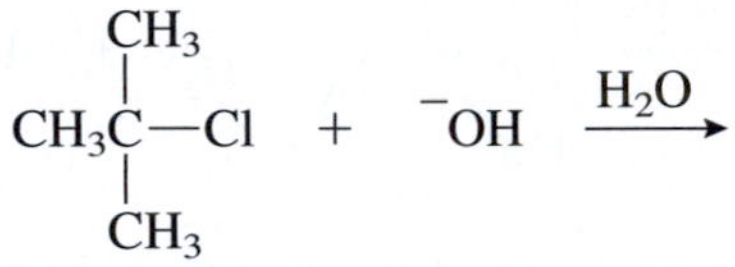
CH3
CH3C—Cl
CH3
+
⁻OH
H2O

8.50 Show how these compounds could be prepared from alkyl halides:

a) 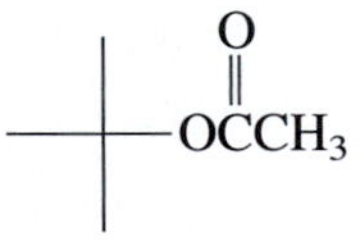
O
OCCH3

b) 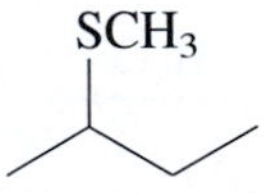
SCH3

Two methods

c) 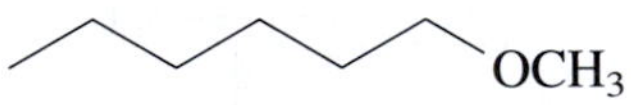
OCH3

Two methods

8.51 Show how these products could be synthesized from the indicated starting material. More than one step may be necessary. Make sure that the product has the stereochemistry shown.

Starting material / Product

a) 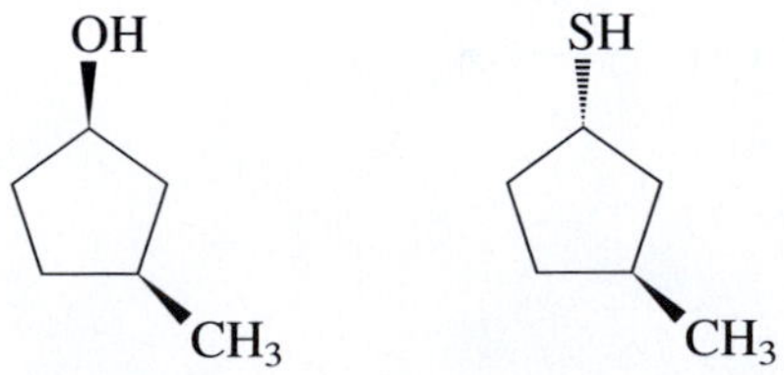
OH
CH3
SH
CH3

b) 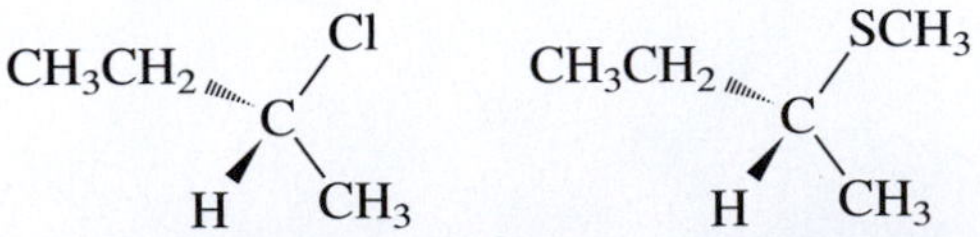
CH3CH2
Cl
C
H
CH3
CH3CH2
SCH3
C
H
CH3

8.52 Cyanide anion has two potential nucleophilic sites: the carbon and the nitrogen. Explain which site is expected to be the stronger nucleophile.

$$:C\equiv N:^-$$

8.53 This S_N1 reaction gives the mixture of products shown. Show the structure of the carbocation formed in this reaction and explain what factor favors the reaction of the nucleophile at the primary carbon and what factor favors reaction at the tertiary carbon.

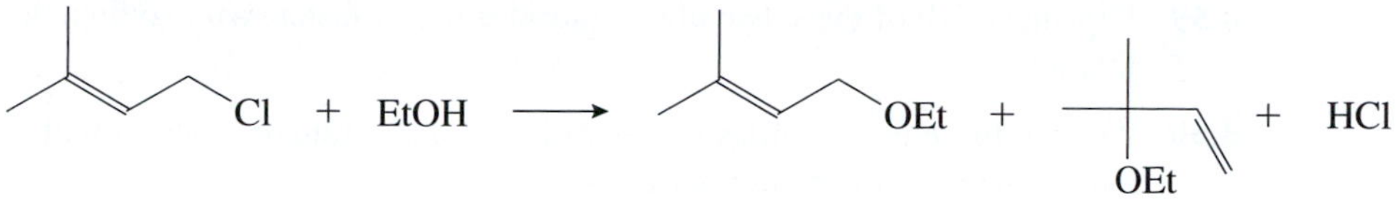

(structure) Cl + EtOH $\longrightarrow$ (structure) OEt + (structure) OEt + HCl

8.54 The rearrangements of both the 1-butyl and 2-butyl carbocations to the *tert*-butyl carbocation occur rapidly in superacid solution. Both of these rearrangements proceed through several steps and must involve an unfavorable secondary carbocation to primary carbocation rearrangement. Show the steps in the rearrangement of the 1-butyl carbocation to the *tert*-butyl carbocation.

8.55 The two nitrogens of the following dipeptide have very different reactivities as nucleophiles. Explain which nitrogen is the better nucleophile.

$$\overset{\ddot{O}:}{\underset{}{\underset{\ddot{N}H_2CH_2\overset{\|}{C}NHCH_2\overset{\|}{C}OCH_3}{}}}$$

BioLink

8.56 Amino acids can be prepared (as their conjugate bases) by the reaction of a bromine substituted carboxylic acid with excess ammonia.

a) The conjugate base of the amino acid alanine is formed in the following reaction. Show the structure of the product and explain why an excess of ammonia is required. Explain which mechanism the reaction follows.

$$\overset{Br\ O}{\underset{}{CH_3\overset{|}{C}H\overset{\|}{C}OH}} + NH_3 \longrightarrow$$

BioLink

b) Show how a similar reaction could be used to prepare the conjugate base of the amino acid phenylalanine.

$$\overset{NH_2\ O}{\underset{}{PhCH_2\overset{|}{C}H-\overset{\|}{C}O^-}}$$

Problems Using Online Three-Dimensional Molecular Models

8.57 Explain which of the two acetate esters, product 1 or product 2, is formed when the alkyl chloride is reacted with sodium acetate in DMSO.

8.58 Explain which of the two methyl ethers, product 1 or product 2, is formed when the alkyl chloride is heated in methanol.

8.59 Explain which of these two alkyl chlorides reacts faster with sodium acetate in DMSO.

8.60 Arrange these nucleophiles in order of decreasing rate of reaction with iodomethane and explain your answer.

ⱴMentor™   Do you need a live tutor for homework problems? Access vMentor at Organic ChemistryNow at **http://now.brookscole.com/hornback2** for one-on-one tutoring from a chemistry expert.

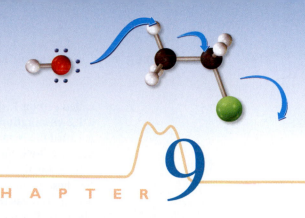

Elimination Reactions

REACTIONS OF ALKYL HALIDES, ALCOHOLS, AND RELATED COMPOUNDS

CHAPTER 9

A S WE SAW in Chapter 8, elimination reactions often compete with nucleophilic substitution reactions. Both reactions can be useful in synthesis if this competition can be controlled. This chapter discusses the two common mechanisms by which elimination reactions occur, the stereochemistry of the reactions, the direction of the elimination, and the factors that control the competition between elimination and substitution. Based on these factors, procedures are presented that can be used to minimize elimination if the substitution product is the desired one or to maximize elimination if the alkene is the desired product.

9.1   THE GENERAL REACTION

As described in Section 8.14, an **elimination reaction** occurs when a proton and the leaving group are lost from adjacent carbons, resulting in the formation of a double bond. In general, the reaction can be represented as

(Note that this general equation is not intended to indicate the order in which these bonds are made or broken; that is, it is not intended to show the mechanism for the reaction.) If the carbon bonded to the leaving group is

ORGANIC
Chemistry ⚛ Now™
Look for this logo in the chapter and go to OrganicChemistryNow at
http://now.brookscole.com/hornback2 for tutorials, simulations, problems, and molecular models.

called carbon 1, then the proton is lost from carbon 2. The reaction is therefore termed a **1,2-elimination.** (Alternatively, because the carbon attached to the functional group is called the α-carbon and the adjacent carbon is the β-carbon, the reaction can also be called a **β-elimination.**)

As was the case with nucleophilic substitution reactions, there are two mechanisms for these elimination reactions. One mechanism is concerted and parallels the S_N2 reaction, whereas the other involves the formation of a carbocation intermediate and parallels the S_N1 reaction. The concerted mechanism is discussed first.

9.2   BIMOLECULAR ELIMINATION

The reaction of ethoxide ion with *tert*-butyl bromide in ethanol as solvent results in the formation of the elimination product 2-methylpropene.

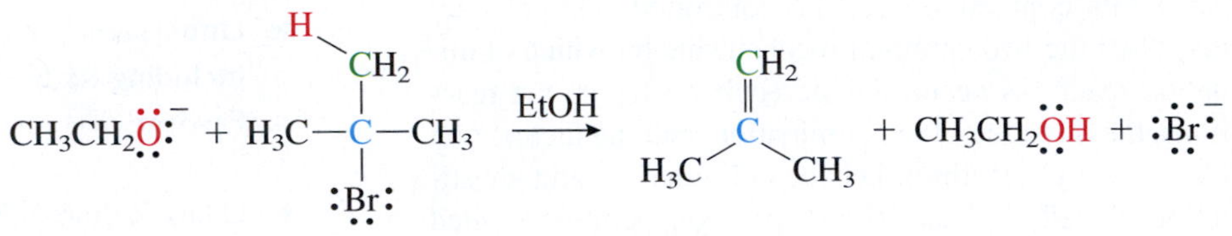

This reaction follows the second-order rate law:

$$\text{rate} = k[\text{EtO}^-][t\text{-BuBr}]$$

The rate law is consistent with a concerted or one-step mechanism. Because two species, ethoxide ion and *tert*-butyl bromide, react in this step, the reaction is described as a **bimolecular elimination** or an **E2 reaction.** This reaction is quite common when alkyl halides are treated with strong bases. Because many nucleophiles are also quite basic, the E2 reaction often competes with the S_N2 reaction.

The mechanism of this concerted reaction involves breaking and forming several bonds simultaneously. The base begins to form a bond to the hydrogen, the carbon–hydrogen bond begins to break, the carbon–bromine bond also begins to break, and the pi bond begins to form. The overall process and the transition state can be represented as follows:

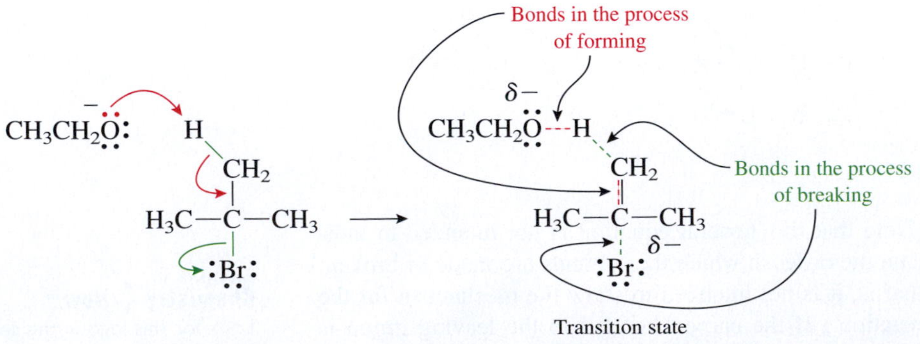

Transition state

PROBLEM 9.1

Show the elimination products of these reactions:

a) CH₃CH₂CHCH₂CH₃ + CH₃CH₂O⁻ $\xrightarrow{\text{EtOH}}$

(Br on the CH)

b) (cyclopentane with Cl) + ⁻OH $\xrightarrow[\text{EtOH}]{\text{H}_2\text{O}}$

c) (cyclohexane with H₃C and Cl) + CH₃CH₂O⁻ $\xrightarrow{\text{EtOH}}$

What evidence is available to support the mechanism shown for the E2 reaction? The experimental rate law tells us that both the base and the alkyl halide are present in the transition state or in some step prior to the transition state. Many other experimental techniques can be used to test whether a mechanism that has been proposed for a reaction is the one that is most plausible. Several of these employ the substitution of a less common isotope for one or more of the atoms of the compound. For example, a normal hydrogen atom ($^1_1$H) can be replaced with a deuterium atom ($^2_1$H or D) or a tritium ($^3_1$H or T) atom. Or a normal carbon ($^{12}_6$C) atom can be replaced with a $^{13}_6$C or $^{14}_6$C atom. Because isotopic substitution has only a very small effect on the chemical behavior of a compound, the isotopically modified compound undergoes the same reactions and follows the same mechanisms as its unmodified counterpart. In one type of experiment, the isotope is used to trace the fate of the labeled atom as the reactant is converted to the product.

In another type of experiment, the effect of the isotope on the reaction rate is studied. Because of their different molecular masses, isotopes can have a small effect on the rate of a reaction. Because deuterium is twice as massive as a normal hydrogen, the bond dissociation energy for a C—D bond is about 1.2 kcal/mol (5.0 kJ/mol) larger than that for a C—H bond. If this bond is broken during the rate-determining step of the reaction, then replacing the hydrogen with a deuterium can result in a significant decrease in the reaction rate because more energy must be supplied to break the stronger C—D bond. This is called a kinetic isotope effect.

As an example, comparison of the experimental rate of this elimination reaction

PhCH—CH₂ + ⁻:ÖCH₂CH₃ $\xrightarrow{\text{EtOH}}$ PhCH=CH₂ + CH₃CH₂OH + Br⁻
(Br, H labeled)

to the rate of the reaction of the deuterated analog

PhCD—CH₂ + ⁻:ÖCH₂CH₃ $\xrightarrow{\text{EtOH}}$ PhCD=CH₂ + CH₃CH₂OD + Br⁻
(Br, D labeled)

showed that the deuterated compound reacted more slowly by a factor of 7.1. This is a large deuterium isotope effect and indicates that the C—D bond is being broken during the rate-determining step of this reaction. Although this experimental result does not

Focus On Biological Chemistry

DDT-Resistant Insects

Some time after the introduction of DDT as an insecticide, it was found that some insects were becoming resistant to the chemical. This was later traced to the presence of an enzyme in the resistant insects that catalyzes the dehydrochlorination of DDT to form dichlorodiphenyldichloroethylene (DDE), which is not toxic to the insects. (As expected on the basis of its structure, DDT also gives this elimination reaction when treated with base.)

DDT   enzyme or ⁻OH   DDE

The presence of this enzyme is a recessive trait that was present in a small part of the insect population before the introduction of DDT. When the insecticide killed the majority of the population that did not possess the enzyme, the insects that had the enzyme became dominant, a classic example of natural selection. (Or is it "unnatural selection" in this case?)

prove that the mechanism is E2 (an experiment cannot prove a mechanism but can disprove it), it is consistent with this mechanism.

As a contrasting example, these reactions occur at nearly identical rates:

$$Ph{-}\underset{\underset{Cl}{|}}{CH}{-}CH_3 \xrightarrow[H_2O]{EtOH} Ph{-}CH{=}CH_2 + \text{Substitution products}$$

$$Ph{-}\underset{\underset{Cl}{|}}{CH}{-}CD_3 \xrightarrow[H_2O]{EtOH} Ph{-}CH{=}CD_2 + \text{Substitution products}$$

The fact that the reaction of the undeuterated compound is only 1.2 times faster than that of the deuterated analog indicates that the C—D bond is not being broken in the rate-determining step. Therefore, this reaction cannot be proceeding by an E2 mechanism.

9.3 STEREOCHEMISTRY OF THE E2 REACTION

Reactions such as the E2 elimination, in which several bonds are made and broken simultaneously, usually have strict requirements for the stereochemical relationship of these bonds as the reaction proceeds. These **stereoelectronic requirements** occur because the orbitals that are going to form the new bonds must begin to overlap at the very start of the reaction. As the reaction proceeds, this overlap increases and provides significant stabilization in the transition state to help offset the energy cost of breaking the other bonds.

In the case of the E2 reaction the carbon sp^3 orbital of the carbon–hydrogen sigma bond and the carbon sp^3 orbital of the carbon-leaving group sigma bond must begin to overlap to form the pi bond. This requires that these two sigma bonds lie in the same plane; they must be coplanar. The two bonds may be on the same side of the C—C bond (**syn-periplanar conformation**) or on opposite sides of the C—C bond (**anti-periplanar conformation**). These conformations are shown in Figure 9.1. As the reaction proceeds, the bonds to the hydrogen and the leaving group begin to lengthen, and the two carbons begin to change hybridization from sp^3 to sp^2. The orbitals that are initially bonded to the hydrogen and the leaving group change from sp^3 orbitals to the p orbitals of the pi bond.

ORGANIC
Chemistry ✦ Now ™
Click *Mechanisms in Motion* to
view the **E2 Mechanism.**

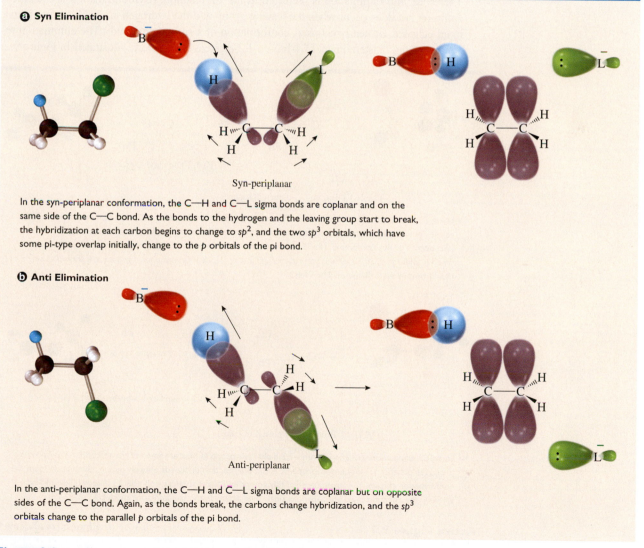

ⓐ **Syn Elimination**

Syn-periplanar

In the syn-periplanar conformation, the C—H and C—L sigma bonds are coplanar and on the same side of the C—C bond. As the bonds to the hydrogen and the leaving group start to break, the hybridization at each carbon begins to change to sp^2, and the two sp^3 orbitals, which have some pi-type overlap initially, change to the p orbitals of the pi bond.

ⓑ **Anti Elimination**

Anti-periplanar

In the anti-periplanar conformation, the C—H and C—L sigma bonds are coplanar but on opposite sides of the C—C bond. Again, as the bonds break, the carbons change hybridization, and the sp^3 orbitals change to the parallel p orbitals of the pi bond.

Figure 9.1

MECHANISM OF **E2** ELIMINATION FROM ⓐ SYN-PERIPLANAR AND ⓑ ANTI-PERIPLANAR CONFORMATIONS.

Closer examination reveals that the syn-periplanar conformation has all the bonds eclipsed, whereas the anti-periplanar conformation has these bonds staggered, as shown in the following Newman projections:

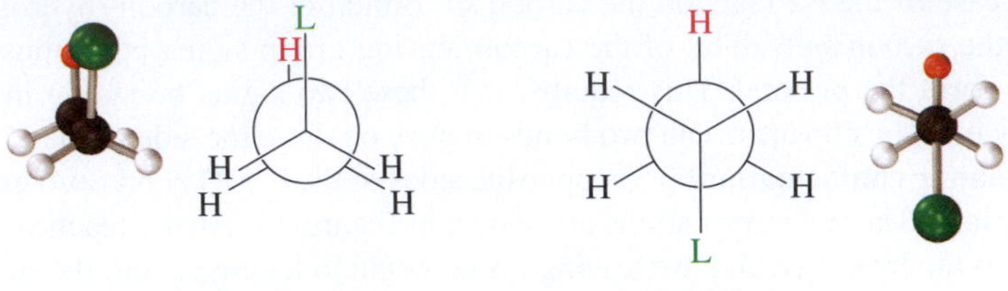

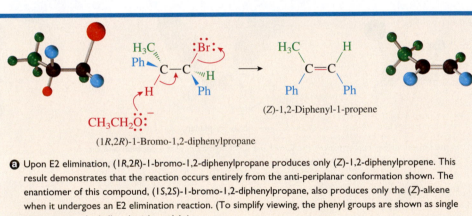

Syn-periplanar
conformation

Anti-periplanar
conformation

The anti-periplanar conformation is more stable because it is staggered rather than eclipsed. Therefore, anti elimination is preferred in the E2 reaction. (Syn elimination is much less common but does occur when the leaving group and the hydrogen are held syn-periplanar in an eclipsed, or nearly eclipsed, conformation of a rigid compound.) The elimination reactions of the diastereomers of 1-bromo-1,2-diphenylpropane are illustrated in Figure 9.2.

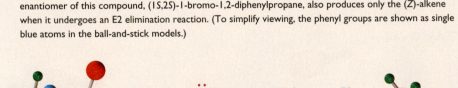

(1R,2R)-1-Bromo-1,2-diphenylpropane

(Z)-1,2-Diphenyl-1-propene

a Upon E2 elimination, (1R,2R)-1-bromo-1,2-diphenylpropane produces only (Z)-1,2-diphenylpropene. This result demonstrates that the reaction occurs entirely from the anti-periplanar conformation shown. The enantiomer of this compound, (1S,2S)-1-bromo-1,2-diphenylpropane, also produces only the (Z)-alkene when it undergoes an E2 elimination reaction. (To simplify viewing, the phenyl groups are shown as single blue atoms in the ball-and-stick models.)

(1S,2R)-1-Bromo-1,2-diphenylpropane

(E)-1,2-Diphenyl-1-propene

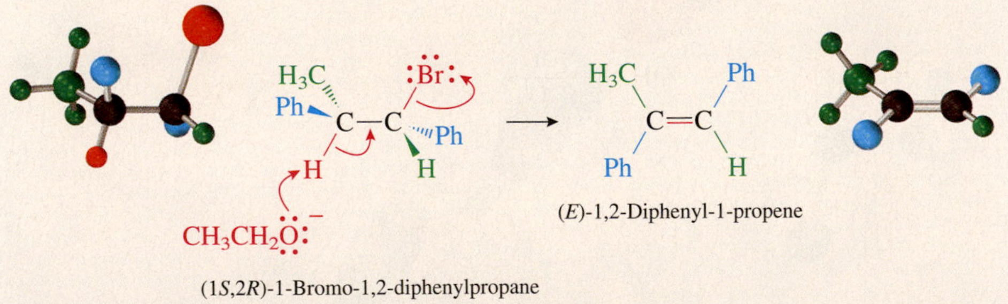

b Upon E2 elimination, (1S,2R)-1-bromo-1,2-diphenylpropane (a diastereomer of the (1R,2R)-stereoisomer) produces only (E)-1,2-diphenyl-1-propene. Again the reaction proceeds entirely by anti elimination from the conformation shown. The enantiomer of this compound also produces only the (E)-alkene. (To simplify viewing, the phenyl groups are shown as single blue atoms in the ball-and-stick models.)

Active Figure 9.2

ORGANIC
Chemistry⚛Now™

MECHANISM AND STEREOCHEMISTRY OF THE E2 ELIMINATION REACTIONS OF THE DIASTEREOMERS OF 1-BROMO-1,2-DIPHENYL-PROPANE TO PRODUCE a THE (Z) STEREOISOMER AND b THE (E) STEREOISOMER OF 1,2-DIPHENYL-1-PROPENE. Test yourself on the concepts in this figure at **OrganicChemistryNow.**

The preference for anti elimination results in the (1R,2R)-diastereomer of the bromide producing only the (Z)-isomer of the alkene, and the (1S,2R)-bromide producing only the (E)-alkene.

PRACTICE PROBLEM 9.1

Show the stereochemistry of the product of this elimination reaction:

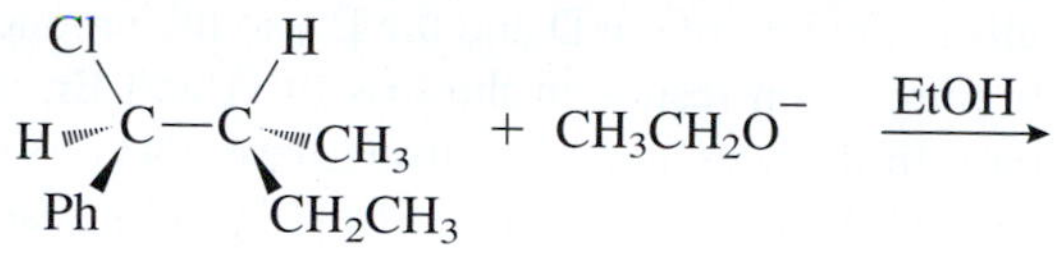

Strategy

First identify the H and the leaving group (L) that are eliminated in the reaction. Remember that E2 elimination requires a conformation that has the H and the L in an anti-periplanar geometry. If they are not in such a conformation as drawn, redraw the molecule so that they are. When the elimination occurs, groups that are on the same side of the plane defined by H-C-C-L in the reactant become cis in the product alkene.

Solution

As originally shown, the H and the Cl that are eliminated are syn, so a rotation of 180° about the C—C bond is needed. Anti elimination then gives the (E)-isomer of the product.

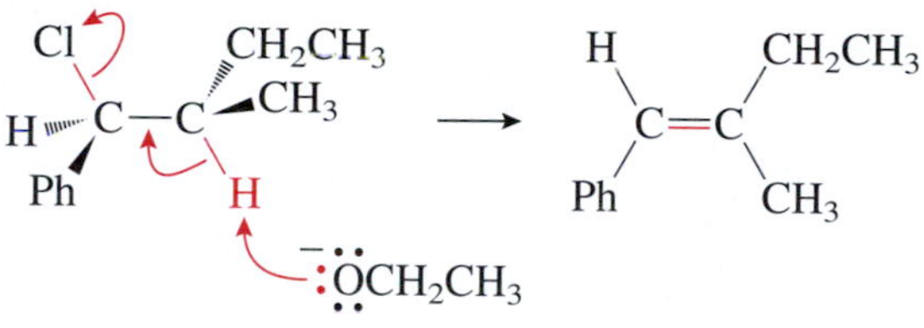

PROBLEM 9.2

Show the products, including stereochemistry, of these elimination reactions.

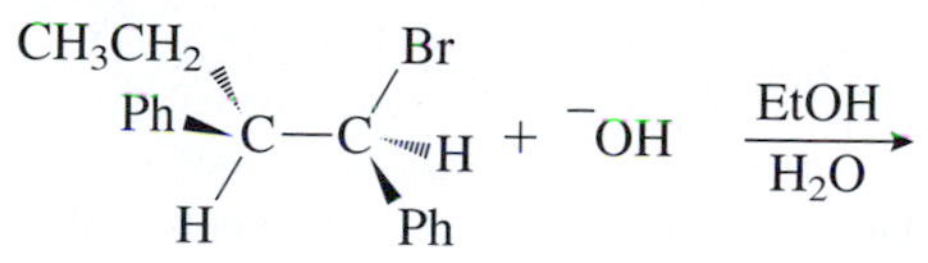

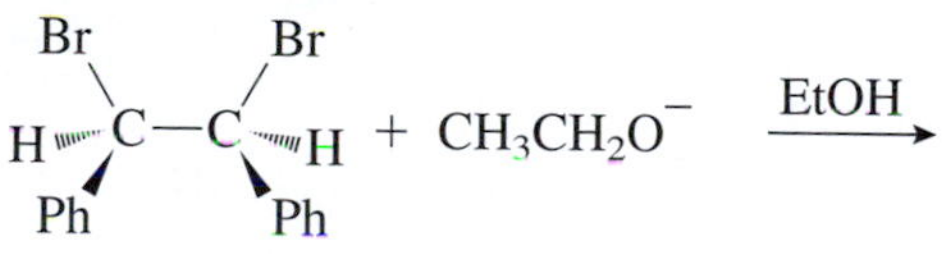

PROBLEM 9.3

What product would be expected from the elimination reaction of (1R,2S)-1-bromo-1,2-diphenylpropane using sodium ethoxide in ethanol as the solvent?

PROBLEM 9.4

Both cis and trans alkenes can be formed from this compound by anti elimination. Draw a Newman projection of the conformation required to form each of these products and, on the basis of these projections, predict which of these products would be formed in larger amounts.

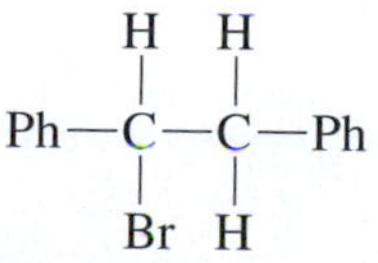

Syn elimination is less favorable than anti elimination because the molecule must be in a higher-energy eclipsed conformation for syn elimination to occur. However, syn elimination does occur in rigid molecules where the leaving group and the hydrogen are held in an eclipsed conformation. For example, the bicyclic compound shown in Figure 9.3 is very rigid. As can be seen in the Newman projection down the C-2—C-3 bond, the Br and the D are eclipsed. (Examination of a model will help you see how rigid this molecule is and that the Br and D groups are eclipsed.) Deuterium is used as a label to enable syn elimination to be distinguished from anti elimination. Syn elimination results in the loss of the D and the Br, so the product contains no deuterium, whereas anti elimination results in the loss of H and Br, so the product still contains the deuterium. In the presence of a strong base, 94% of the product alkene contains no deuterium and therefore has resulted from syn elimination.

Not much of the "anti elimination" product is formed because the hydrogen on C-3 and the bromine are not in an anti relationship (the dihedral angle between them is 120°). Thus, elimination of this hydrogen and the bromine occurs with poor orbital overlap at the early stages of the reaction.

For anti elimination to occur in a cyclohexane ring, the leaving group and the hydrogen must be trans. Furthermore, they will be anti-periplanar only in the conformation where both are axial. This ***trans*-diaxial elimination** is illustrated in Figure 9.4 for menthyl chloride and in Figure 9.5 for neomenthyl chloride. These substituted cyclohexyl chlorides are diastereomers. They differ only in the configuration of the chlorine substituent. However, as shown in Figures 9.4 and 9.5, this difference leads to dramatically different behavior when these two compounds are treated with the strong base ethoxide ion.

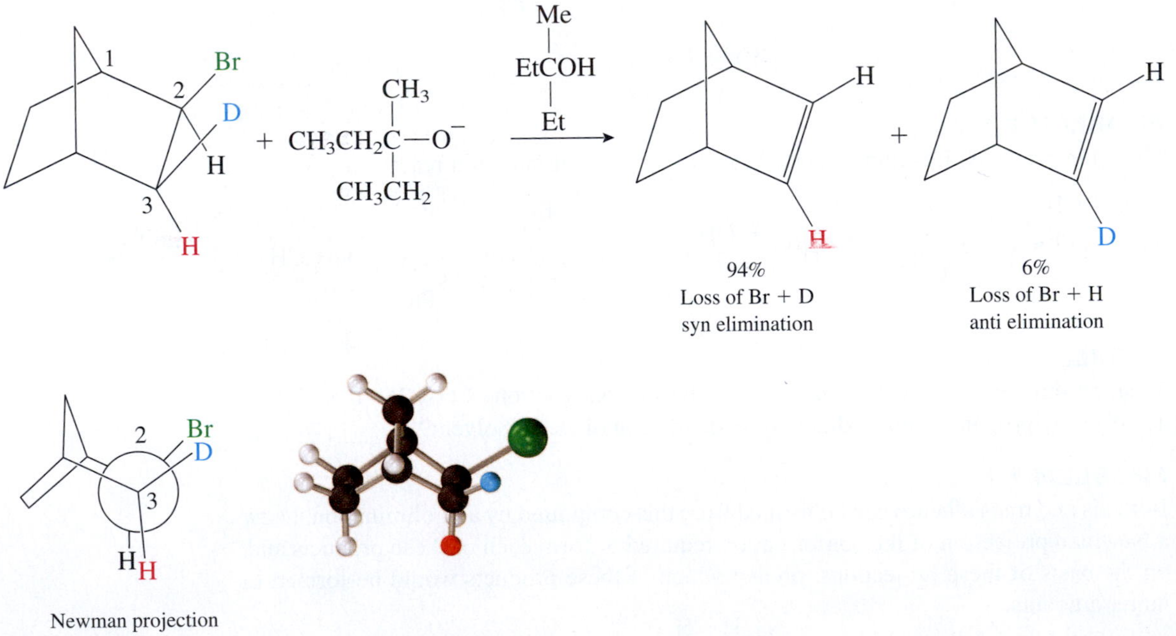

94%
Loss of Br + D
syn elimination

6%
Loss of Br + H
anti elimination

Newman projection

Figure 9.3

SYN ELIMINATION IN A RIGID MOLECULE.

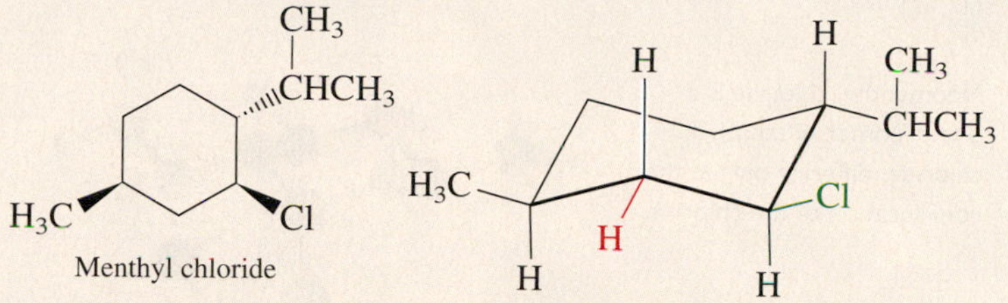

a The most stable chair conformation of menthyl chloride has all three substituents equatorial. However, elimination cannot occur from this conformation because the chlorine is not axial.

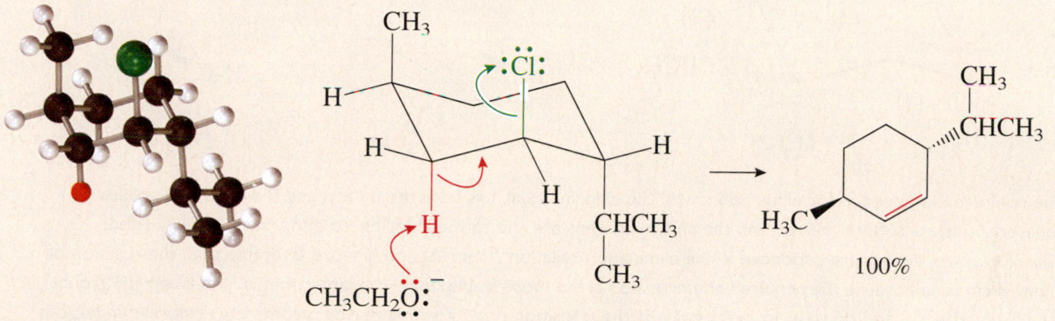

b The other chair conformation is much less stable because all three substituents are axial. It constitutes only a small percentage of the conformational equilibrium (about 0.06%). Anti elimination can occur from this conformation because it has the chlorine and a hydrogen in a *trans*-diaxial arrangement. Only the red hydrogen is anti to the chlorine, so it is the only one that can be involved in the elimination reaction. This results in the production of a single alkene. The reaction is relatively slow because the activation energy includes at least part of the energy needed to ring-flip to this less stable conformation.

Figure 9.4

MECHANISM AND STEROCHEMISTRY OF THE E2 ELIMINATION REACTION OF MENTHYL CHLORIDE.

Analysis of the elimination reactions of these cyclohexane derivatives must be done with care. First, the chair conformation with the chlorine axial must be examined. In the case of menthyl chloride (see Figure 9.4) this conformation has all three groups axial and is much less stable than the ring-flipped conformation. However, elimination must occur from this conformation because only here are a hydrogen and the chlorine anti-periplanar. The reaction has a higher activation energy and is relatively slow because some of the strain energy of the higher-energy conformation is still present in the transition state. In addition, the reactive conformation has a single hydrogen in the required anti-periplanar relationship to the chlorine. Only this hydrogen is lost in the elimination reaction, and therefore, a single alkene is produced.

In contrast, neomenthyl chloride, with the opposite configuration at the carbon bearing the chlorine, has both the isopropyl and the methyl groups equatorial in the reactive conformation, with the chlorine axial (see Figure 9.5). Because both of these groups have larger axial strain energies than chlorine (see Table 6.2), the reactive conformation is the more stable one. There is less steric strain in the transition state for elimination,

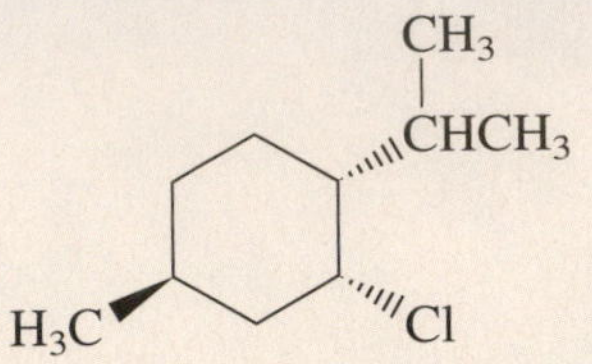

Neomenthyl chloride is a diastereomer of menthyl chloride, differing only in the configuration of the chlorine.

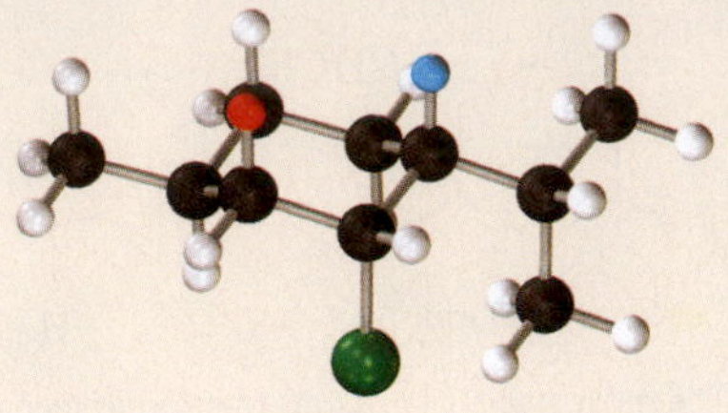

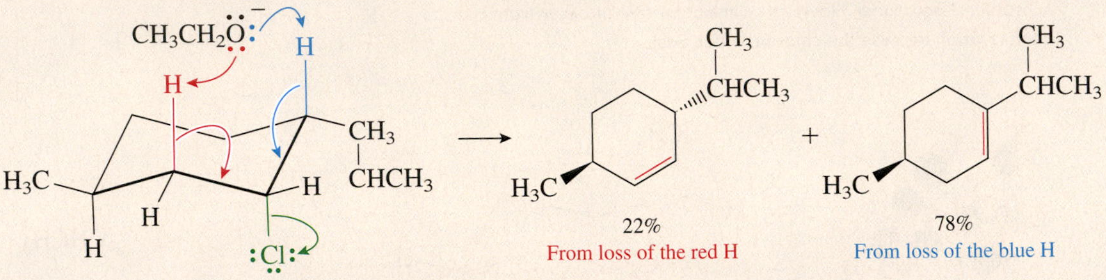

22%
From loss of the red H

78%
From loss of the blue H

The conformation needed for elimination, with the chlorine axial, has both the methyl and the isopropyl groups in equatorial positions. Both the red and the blue hydrogens are anti to the chlorine, so either may be eliminated. Therefore, two alkenes are produced in the elimination reaction. This reaction is more favorable than the reaction of menthyl chloride because the required conformation is the more stable one—the larger methyl and isopropyl groups are both equatorial and there is no extra strain in the transition state. For this reason, neomenthyl chloride undergoes elimination about 40 times faster than menthyl chloride.

Figure 9.5

MECHANISM AND STEROCHEMISTRY OF THE E2 ELIMINATION REACTION OF NEOMENTHYL CHLORIDE.

so the activation energy is smaller. This results in neomenthyl chloride reacting about 40 times faster than menthyl chloride.

As can be seen in Figure 9.5, neomenthyl chloride has two hydrogens in an anti-periplanar geometry with the chlorine. Either of these hydrogens can be lost in the elimination reaction, resulting in the formation of two alkenes. (All of the examples presented up to this point were chosen so that only one alkene could be formed.) Let's now address this issue of the direction of the elimination and learn how to predict which will be the major product when more than one alkene can be formed.

PROBLEM 9.5
Show the products of these elimination reactions:

a) + NaOCH$_2$CH$_3$ $\xrightarrow{\text{EtOH}}$

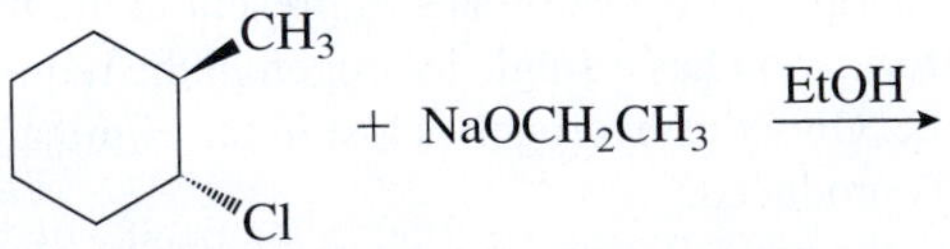

b) + NaOCH$_2$CH$_3$ $\xrightarrow{\text{EtOH}}$

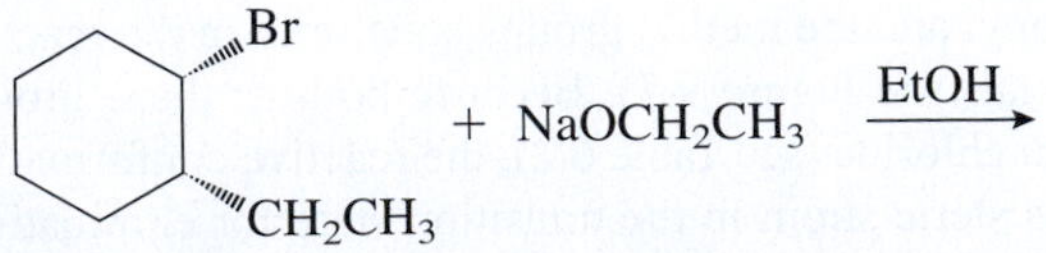

PROBLEM 9.6

Explain why one of these compounds reacts readily by an E2 mechanism when treated with sodium ethoxide in ethanol but the other does not:

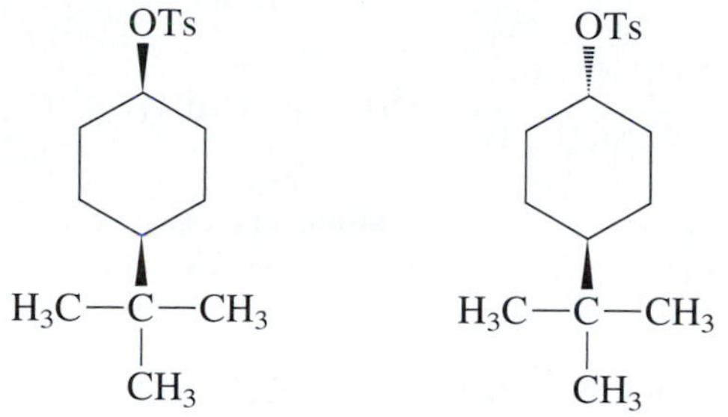

9.4 DIRECTION OF ELIMINATION

Chapter 8 discussed the stereochemistry of substitution reactions—that is, what happened to the stereochemistry when the reaction occurred at a carbon chirality center. This section discusses the **regiochemistry** of the elimination reaction—that is, what happens when a reaction can produce two or more structural isomers. The structural isomers that can often be produced in elimination reactions have the double bond in different positions. As shown in Figure 9.5, elimination of hydrogen chloride from neomenthyl chloride produces two structural isomers but in unequal amounts.

It is possible to predict which product will be formed in larger amounts in these reactions. Most E2 elimination reactions follow **Zaitsev's rule:**

▶ **ZAITSEV'S RULE**

The major alkene product is the one with more alkyl groups on the carbons of the double bond (the more highly substituted product).

In the case of neomenthyl chloride the major product has three carbon groups and one hydrogen as the four groups bonded to the carbons of the double bond, whereas the minor product has two carbon groups and two hydrogens so bonded.

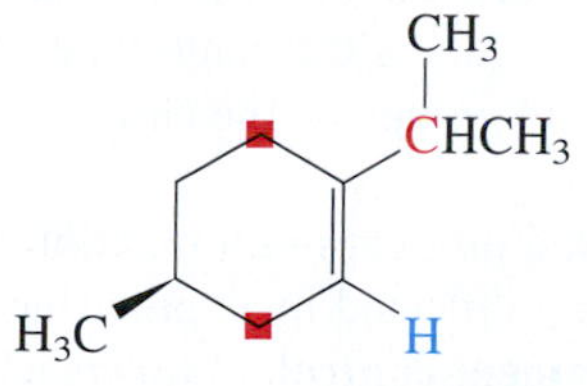

Major product (78%)

This product has three carbons and one hydrogen bonded to the carbons of the double bond.

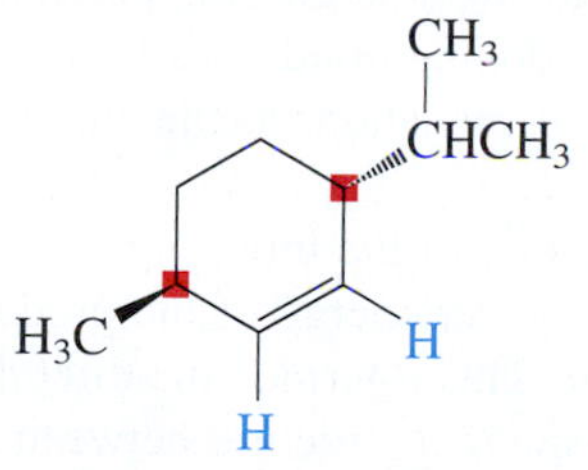

Minor product (22%)

This product has two carbons and two hydrogens bonded to the carbons of the double bond.

Another example is provided by the elimination reaction of 2-bromobutane using ethoxide ion as the base. The major product is the more highly substituted alkene.

$$\underset{\substack{| \\ \text{Br}}}{\text{CH}_3\text{CHCH}_2\text{CH}_3} \quad \xrightarrow[\text{CH}_3\text{CH}_2\text{OH}]{\text{CH}_3\text{CH}_2\text{O}^-} \quad \text{CH}_3\text{CH}\!=\!\text{CHCH}_3 \; + \; \text{CH}_2\!=\!\text{CHCH}_2\text{CH}_3$$

$$\qquad\qquad\qquad\qquad\qquad\qquad\qquad\qquad 81\% \qquad\qquad\qquad 19\%$$

Mixture of cis and
trans isomers

It is important to note, however, that reactions that follow Zaitsev's rule still produce significant amounts of the less highly substituted product.

Zaitsev's rule is an **empirical rule;** that is, it is based on experimental observations, not theory. It was first proposed in 1875, when there was little theoretical basis for any part of organic chemistry. Now that we have learned about mechanisms, we know that the transition state leading to the more highly substituted product must be of lower energy than the one leading to the less highly substituted product. Let's examine why this is the case.

A number of studies using heats of reaction have shown that a more highly substituted alkene is more stable than a less highly substituted alkene. Therefore, the general order of stability for compounds containing carbon–carbon double bonds is

Monosubstituted   Disubstituted   Trisubstituted   Tetrasubstituted

increasing stability

Although there is still some debate, the increasing stabilization is postulated to result from the increasing amount of hyperconjugation that can occur as the number of alkyl groups on the doubly bonded carbons increases. This is the same type of interaction that stabilizes carbocations as the number of alkyl groups on the carbon increases (see Section 8.7).

The formation of the more highly substituted product in elimination reactions results from the lower energy of the product being reflected in a somewhat more stable transition state. This is termed **product development control.** In this reaction the transition state, having a structure between the reactant and the product, is stabilized by the same factors that stabilize the product, although the amount of stabilization is less. This is illustrated in Figure 9.6, an energy versus reaction progress diagram for the elimination reaction of 2-bromobutane. The more stable alkene, 2-butene, is the major product (81%), but because the energy difference between the transition states for the

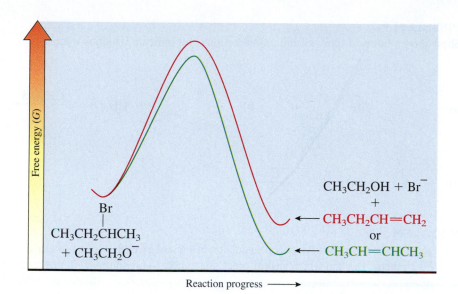

Figure 9.6

FREE ENERGY VERSUS
REACTION PROGRESS
DIAGRAM FOR THE
ELIMINATION REACTION
OF 2-BROMOBUTANE.

two products is not very large, a significant amount (19%) of the minor product, 1-butene, is also formed.

PRACTICE PROBLEM 9.2

Show the products of this elimination reaction and explain which is the major product:

$$\underset{\underset{Br}{|}}{\overset{\overset{CH_3}{|}}{CH_3CCH_2CH_3}} + CH_3CH_2O^- \xrightarrow{\text{EtOH}}$$

Solution

Two alkenes are expected to be formed in this reaction. According to Zaitsev's rule, the more highly substituted alkene is the major product.

$$\underset{\underset{Br}{|}}{\overset{\overset{CH_3}{|}}{CH_3CCH_2CH_3}} + CH_3CH_2O^- \xrightarrow{\text{EtOH}} \quad \underset{H_3C}{\overset{H_3C}{>}}C=C\underset{H}{\overset{CH_3}{<}} \quad + \quad \underset{H}{\overset{H}{>}}C=C\underset{CH_2CH_3}{\overset{CH_3}{<}}$$

Major   Minor

The major product is a trisubstituted alkene with three alkyl
groups on the carbons of the
double bond.

The minor product is a disubstituted alkene with two alkyl
groups on the carbons of the
double bond.

PROBLEM 9.7

Show the products of these elimination reactions and indicate which is major:

a) [structure: 1-chloro-1-methylcyclohexane] + ⁻OH $\xrightarrow{\text{H}_2\text{O} \atop \text{EtOH}}$

b) [structure with OTs group] + CH₃O⁻ $\xrightarrow{\text{CH}_3\text{OH}}$

c) [structure with Cl] + CH₃CH₂O⁻ $\xrightarrow{\text{EtOH}}$

PROBLEM 9.8

The reaction of 2-bromobutane with ethoxide ion in ethanol gives 81% of a mixture of (Z)- and (E)-2-butene. Explain which stereoisomer you expect to predominate in this mixture.

Although most E2 reactions follow Zaitsev's rule, there are some exceptions. One major exception is a reaction known as the **Hofmann elimination.** Compounds that can undergo this reaction have a quaternary nitrogen atom—that is, a nitrogen that is positively charged because it is bonded to four alkyl groups. The nitrogen, with three of the alkyl groups, acts as a leaving group when the ion is heated in the presence of a base such as hydroxide ion. An example of the Hofmann elimination reaction is provided in the following equation:

$$\underset{\underset{\underset{\displaystyle\text{HO:}^-}{}}{}}{\underset{\text{H}_3\text{C}-\overset{\overset{\displaystyle\text{CH}_3}{|}}{\overset{+}{\text{N}}}-\text{CH}_3}{\text{CH}_3\text{CH}_2\text{CH}_2\text{CH}-\text{CH}_2 \atop |\atop\text{H}}} \longrightarrow \text{CH}_3\text{CH}_2\text{CH}_2\text{CH}=\text{CH}_2 + \text{H}_3\text{C}-\overset{\overset{\displaystyle\text{CH}_3}{|}}{\text{N}}-\text{CH}_3 + \text{HO}-\text{H}$$

1-Pentene

In this example, 1-pentene and 2-pentene are possible products. The reaction produces almost exclusively 1-pentene, the less highly substituted product. This elimination reaction follows **Hofmann's rule:**

▶ **HOFMANN'S RULE**

The major alkene product has fewer alkyl groups bonded to the carbons of the double bond (the less highly substituted product).

As exemplified by the high yield of 1-pentene in this reaction, the preference for the less highly substituted product is often quite strong.

In the previous example the elimination could occur only in the pentyl group because the other three substituents on the nitrogen are methyl groups, which do not have β-carbons. However, if more than one of the alkyl groups bonded to the nitrogen is larger than methyl, then elimination can, in principle, involve any of these groups.

Again, Hofmann's rule predicts that the major pathway will be the one that produces the less highly substituted alkene. An example is shown in the following equation:

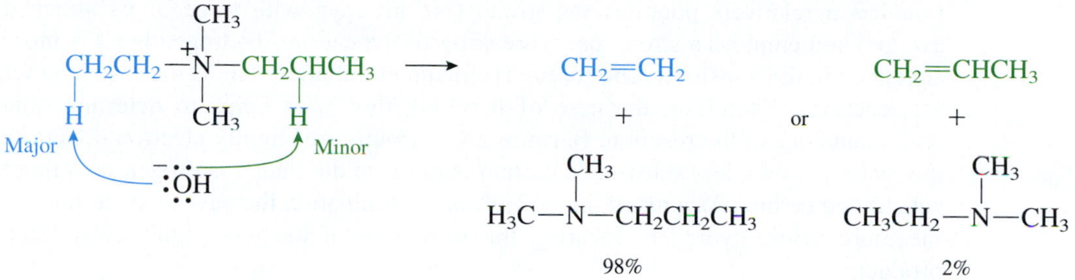

In this case a β-hydrogen (a hydrogen on the β-carbon) could be lost from either the ethyl or the propyl group. In accord with Hofmann's rule the less highly substituted alkene, ethene, is found to be the major product.

PRACTICE PROBLEM 9.3

Show the products of this reaction and explain which is the major product:

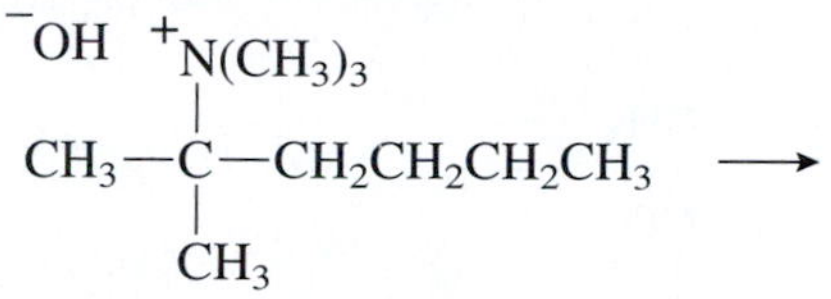

Solution

This elimination reaction follows Hofmann's rule, so the less substituted alkene should be the major product.

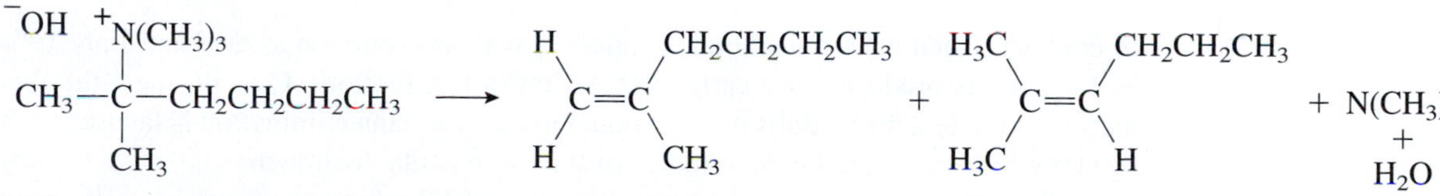

PROBLEM 9.9

Show the products of these reactions and indicate which is major:

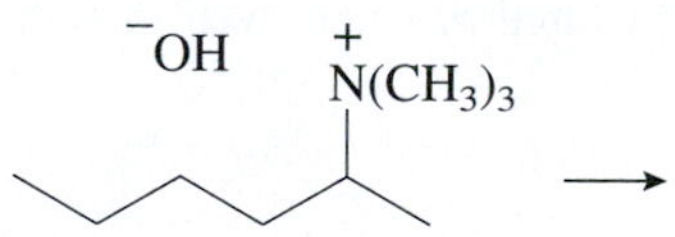

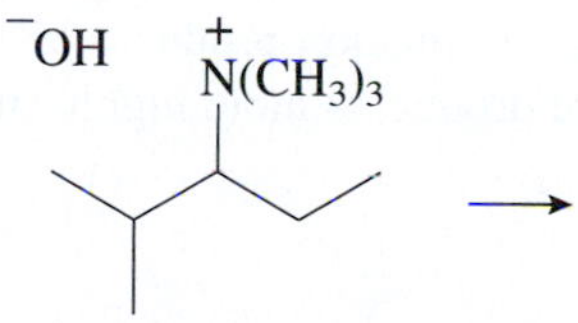

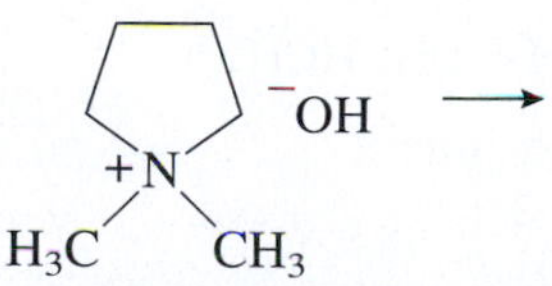

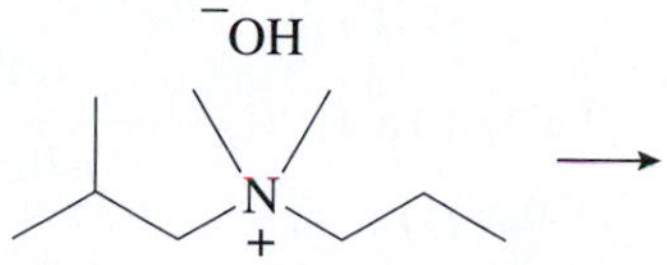

The reasons why the Hofmann elimination produces more of the less highly sub-stituted alkene (which is also the less stable alkene) are complex. Because the reaction has a relatively poor leaving group (the nitrogen with three of its attached groups) and employs a strong base, breaking of the carbon–hydrogen bond is more advanced in the transition state for the Hofmann elimination than is the case in other E2 reactions. Therefore, the ease of breaking this bond helps to determine the regiochemistry of the reaction. Because alkyl groups are slightly electron donating, the hydrogen on a less substituted carbon is more acidic than a hydrogen on a more substituted carbon. As a result, in the Hofmann elimination the base tends to remove the more acidic hydrogen, favoring the formation of the less highly substituted product.

Steric effects are also postulated to be important in determining the regiochemistry of this reaction. It is proposed that the large size of the leaving group in the Hofmann elimination (a tertiary amine) causes the base to attack the less sterically hindered hydrogen—one on the carbon with fewer alkyl groups. Steric effects can be important in reactions other than the Hofmann elimination also. For example, the elimination re-action of 2-bromobutane using *tert*-butoxide ion as the base, an E2 reaction employing a strong, sterically hindered base, gives more 1-butene than 2-butene, as shown in the following equation:

$$
\underset{\substack{|\\ \text{Br}}}{CH_3\overset{\displaystyle Br}{\underset{\displaystyle }{C}HCH_2CH_3}} \quad \xrightarrow[\;t\text{-BuOH}\;]{\overset{\displaystyle CH_3}{\underset{\displaystyle CH_3}{H_3C\overset{|}{\underset{|}{C}}O^-}}} \quad CH_3CH{=}CHCH_3 \;+\; CH_2{=}CHCH_2CH_3
$$

$$47\% \qquad\qquad\qquad 53\%$$

In contrast, when the elimination is conducted with ethoxide ion as the base, only 19% of 1-butene is produced (see earlier discussion in this section). Overall, then, the for-mation of the less highly substituted product in the Hofmann elimination is favored both by steric hindrance and by the removal of the more acidic hydrogen.

A final factor that affects the regiochemistry of E2 reactions occurs when the new double bond can be conjugated with another double bond or a benzene ring. Because of resonance stabilization, a conjugated product is considerably more stable than a non-conjugated one. The partial development of conjugation in the transition state results in enough stabilization that the conjugated product is always the major one. For example, the following elimination produces 98% of the conjugated product, even though the other possible product is more highly substituted:

$$
\underset{\substack{|\\ \text{OTs}}}{PhCH_2\overset{\displaystyle CH_3}{\underset{\displaystyle }{C}HCHCH_3}} \quad \xrightarrow[\;t\text{-BuOH}\;]{\overset{\displaystyle CH_3}{\underset{\displaystyle CH_3}{H_3C\overset{|}{\underset{|}{C}}O^-}}} \quad PhCH{=}\overset{\displaystyle CH_3}{\underset{\displaystyle }{C}HCHCH_3}
$$

$$98\%$$

In summary, the regiochemistry of E2 reactions can be predicted by using the following rules:

1. Most E2 reactions follow Zaitsev's rule; the more highly substituted alkene is the major product.

2. Hofmann eliminations follow Hofmann's rule; the less highly substituted product is the major product.

3. A conjugated alkene is always preferred to an unconjugated alkene.

PROBLEM 9.10
Show the major products of these elimination reactions:

a) 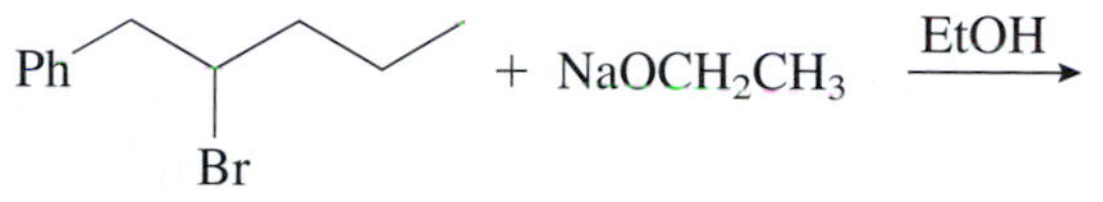 Ph—CH₂—CH(Br)—CH₂—CH₂—CH₃ + NaOCH₂CH₃ —EtOH→

b) 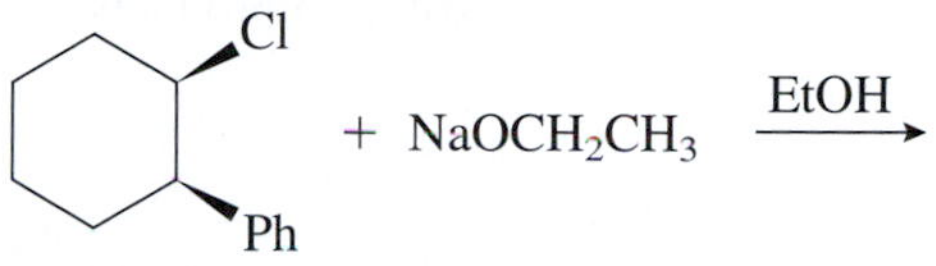 (cyclohexane with Cl and Ph substituents) + NaOCH₂CH₃ —EtOH→

c) 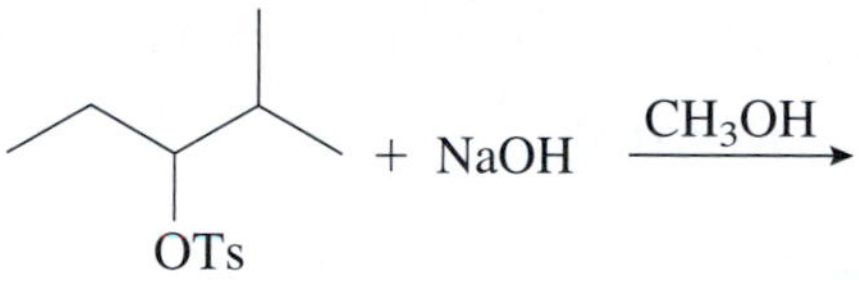 (CH₃CH₂—CH(OTs)—CH(CH₃)₂) + NaOH —CH₃OH→

d) 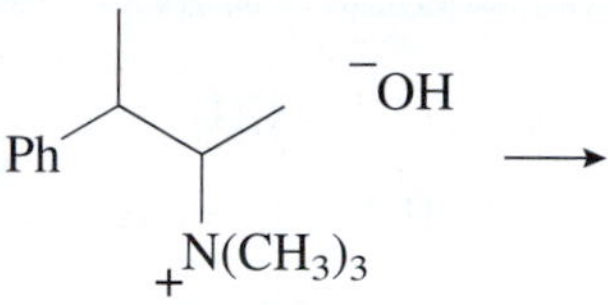 Ph—CH₂—C(CH₃)(⁺N(CH₃)₃)—CH₃ + ⁻OH →

9.5 UNIMOLECULAR ELIMINATION

The reaction of *tert*-butyl bromide with ethoxide ion, shown at the beginning of Section 9.2, resulted in the formation of 2-methylpropene in an E2 reaction. If ethoxide ion is not present in the reaction, E2 elimination is greatly slowed, and both substitution and elimination products are formed.

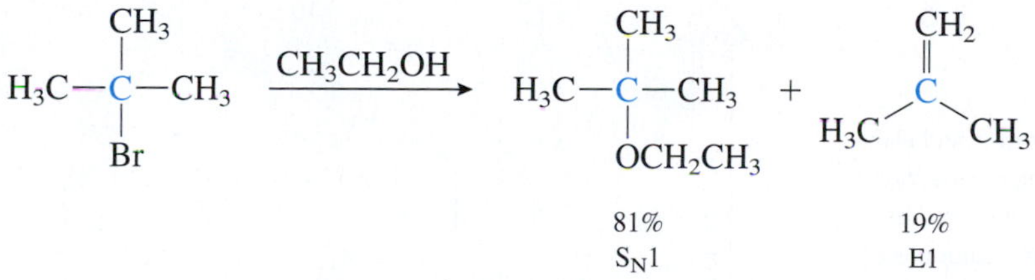

$$H_3C-\underset{\underset{Br}{|}}{\overset{\overset{CH_3}{|}}{C}}-CH_3 \xrightarrow{CH_3CH_2OH} H_3C-\underset{\underset{OCH_2CH_3}{|}}{\overset{\overset{CH_3}{|}}{C}}-CH_3 + \underset{H_3C}{\overset{\overset{CH_2}{\|}}{C}}\diagup CH_3$$

81%   19%
S_N1   E1

The rate of appearance of both products is found to follow the first-order rate law:

$$\text{rate} = k\,[t\text{-BuBr}]$$

It is easy to recognize that the substitution product is formed by an S_N1 mechanism, because the starting alkyl bromide is tertiary. Because the elimination reaction follows the same rate law, its rate-determining step, like that of the S_N1 reaction, must also involve only a molecule of *tert*-butyl bromide. Therefore, the elimination reaction is described as a **unimolecular elimination** or an **E1 reaction.**

The E1 reaction follows the same rate law as the S_N1 reaction because both mechanisms have the same rate-determining step: the formation of the carbocation, as shown in Figure 9.7. In the second step of the mechanism the ethanol can react with the carbocation as a nucleophile to give the S_N1 product or as a base to give the E1 product. Under these particular conditions the transition state (for the second step) leading to substitution is slightly lower in energy than the transition state leading to elimination, so the substitution product is the major one. If an elimination is possible (if there is a β-hydrogen), then some elimination product always accompanies the substitution product in S_N1 reactions. Because the amount of elimination is difficult to control, this detracts from the usefulness of these reactions to prepare other compounds. In addition, carbocation rearrangements (see Section 8.14) may lead to rearranged elimination products as well as rearranged substitution products.

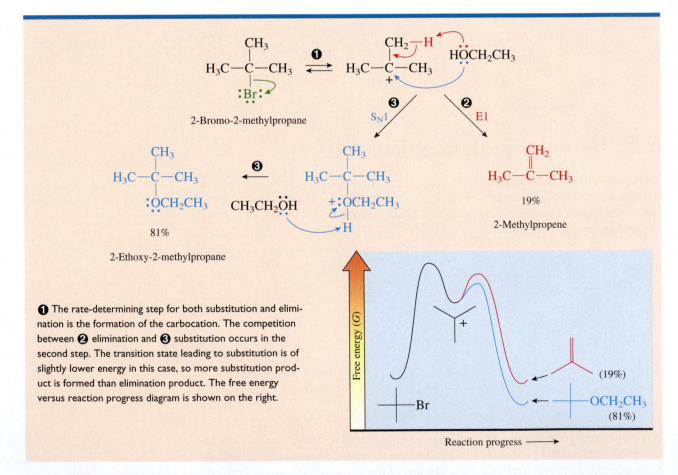

❶ The rate-determining step for both substitution and elimination is the formation of the carbocation. The competition between ❷ elimination and ❸ substitution occurs in the second step. The transition state leading to substitution is of slightly lower energy in this case, so more substitution product is formed than elimination product. The free energy versus reaction progress diagram is shown on the right.

Figure 9.7

MECHANISM OF THE S_N1 SUBSTITUTION AND E1 ELIMINATION REACTIONS OF *TERT*-BUTYL BROMIDE (2-BROMO-2-METHYLPROPANE).

PROBLEM 9.11

When 2-methyl-2-propanol is treated with sulfuric acid, 2-methylpropene is formed. Show all of the steps in the mechanism for this reaction. Don't forget to use curved arrows to show the movement of electrons in each step of the mechanism.

$$
\underset{\underset{CH_3}{|}}{\overset{\overset{CH_3}{|}}{H_3CCOH}} \quad \xrightarrow{H_2SO_4} \quad \underset{\underset{H}{}}{\overset{\overset{H}{}}{}}C=C\underset{\underset{CH_3}{}}{\overset{\overset{CH_3}{}}{}} \quad + \; H_2O
$$

PROBLEM 9.12

Show all of the steps in the mechanism for this reaction. What other products would you expect to be formed?

$$
\underset{\underset{CH_3}{|}}{\overset{\overset{CH_3 \quad Br}{|\quad\;|}}{CH_3C\!-\!\!-\!CHCH_2CH_3}} \quad \xrightarrow{CH_3OH} \quad \underset{H_3C}{\overset{H_3C}{}}C=C\underset{CH_2CH_3}{\overset{CH_3}{}}
$$

9.6 REGIOCHEMISTRY AND STEREOCHEMISTRY OF THE E1 REACTION

When regioisomers are possible in an E1 reaction, the product distribution is found to follow Zaitsev's rule. The reaction of 2-bromo-2-methylbutane under $S_N1/E1$ conditions (in a polar solvent mixture of ethanol and water with no good base or nucleophile present) gives 64% of the substitution products (water acts as the nucleophile to give an alcohol, or ethanol acts as a nucleophile to give an ether), 30% of the more highly substituted alkene, and 6% of the less highly substituted alkene.

$$
\underset{\underset{CH_3}{|}}{\overset{\overset{CH_3}{|}}{CH_3CH_2CBr}} \quad \xrightarrow[H_2O]{CH_3CH_2OH} \quad \underset{\underset{CH_3}{|}}{\overset{\overset{CH_3}{|}}{CH_3CH_2COR}} \; + \; CH_3CH=C\underset{CH_3}{\overset{CH_3}{}} \; + \; \underset{CH_3CH_2}{\overset{\overset{CH_2}{||}}{C}}CH_3
$$

(R = H or CH₂CH₃)

64%   30%   6%

As in the E2 reaction, the relative yields of the alkenes are under product development control—the more stable alkene predominates.

Unlike the E2 reaction, the relative stereochemistry of the leaving group and the hydrogen is not important in the E1 elimination reaction. In the first step of the reaction, the leaving group departs, producing a planar carbocation. Only at this point must the C—H bond be aligned parallel to the empty p orbital of the carbocation so that pi overlap can occur as the C—H bond begins to break. Therefore, *trans*-diaxial elimination is not required for E1 reactions involving cyclohexane rings. For example, the reaction of menthyl chloride under $S_N1/E1$ conditions gives the product distribution shown in the following equation:

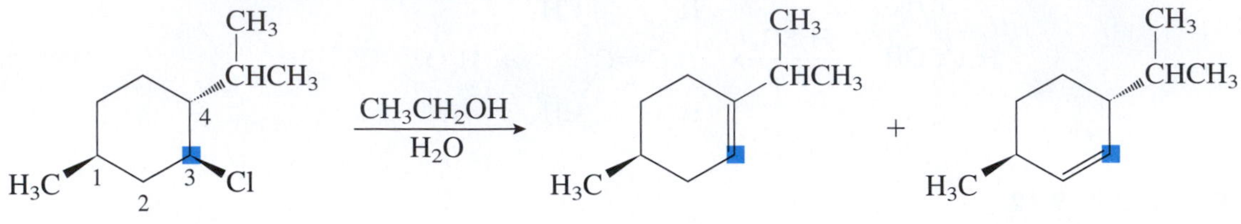

Menthyl chloride		3-Menthene	2-Menthene
		47%	22%

+ 31% Substitution products

Recall that the E2 reaction of menthyl chloride (Figure 9.4) produced 100% of 2-menthene because it was the only alkene that could be formed by *trans*-diaxial elimination. In contrast, the E1 reaction produces both alkenes because the stereochemistry is unimportant. The major product is 3-menthene (the more stable product), in accord with Zaitsev's rule.

PRACTICE PROBLEM 9.4

Show the products of this reaction:

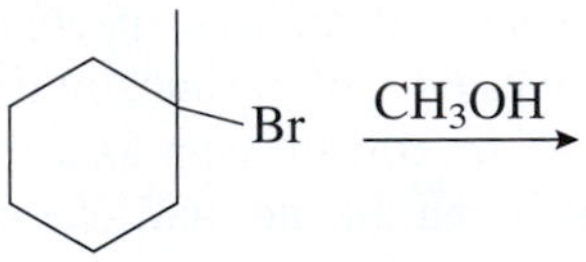

Solution

The bromine is bonded to a tertiary carbon and there is not a strong base present, so the reaction will proceed by an $S_N1/E1$ mechanism. The substitution product should predominate. The E1 reaction follows Zaitsev's rule, so more 1-methylcyclohexene should be formed than methylenecyclohexane.

	S_N1 major	1-Methylcyclohexene	Methylenecyclohexane
			E1 minor

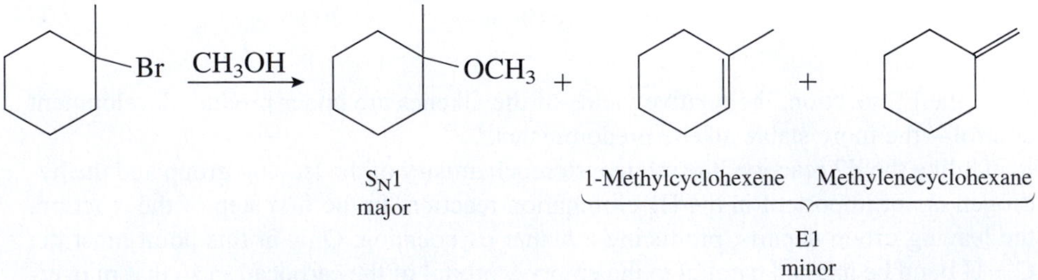

PROBLEM 9.13

Show the products of these reactions:

a) Ph─│─Br $\xrightarrow{CH_3CH_2OH}$

b) [cyclohexane with CH₃, Br, Ph] $\xrightarrow{CH_3OH}$

c) [cyclohexane with CH₃, Br, Ph] $\xrightarrow{CH_3OH}$

d) [cyclopentane with CH₃, Br] $\xrightarrow[EtOH]{H_2O}$

Focus On

The E1cb Mechanism

In the E1 mechanism the bond to the leaving group breaks during the first step and the bond to the hydrogen breaks in a second step. In the E2 mechanism, both of these bonds break in a single step. There is a third possible mechanism in which the bond to the hydrogen breaks during the first step and the bond to the leaving group breaks in a second step, as illustrated in this equation:

$$\bar{B}: \quad \underset{|}{\overset{H \quad L}{-C-C-}} \quad \rightleftharpoons \quad -\overset{..}{C}-\underset{|}{\overset{L}{C}}- \quad \longrightarrow \quad \overset{}{C}=\overset{}{C} \; + \; :\bar{L}$$

$$+$$

$$B-H$$

When this mechanism does occur, the second step often determines the rate. Because this step involves a unimolecular reaction of the conjugate base of the initial reactant, the mechanism is designated as elimination, unimolecular, conjugate base—or E1cb. The mechanism can be made more favorable by the presence of substituents that stabilize the intermediate carbanion. In addition, a poorer leaving group, which makes the second step less likely to be concerted with the first, also favors this mechanism.

The presence of a good anion-stabilizing group, such as a carbonyl group, attached to the carbon from which the proton is lost makes the E1cb mechanism quite favorable. In such situations, even a poor leaving group, such as hydroxide ion, can be eliminated as shown in the following equation:

$$\underset{H-C-CH-CHCH_3}{\overset{O \quad H \quad OH}{\|}} \quad :\overset{..}{O}H \quad \rightleftharpoons \quad \underset{H-C-CH-CHCH_3}{\overset{O \qquad OH}{\|}} \quad \longrightarrow \quad \underset{H-C-CH=CHCH_3}{\overset{O}{\|}}$$

Continued

It is not always easy to distinguish an elimination reaction that is following the E1cb mechanism from one that follows the E2 pathway because the E1cb reaction usually exhibits second-order kinetics also. However, because the E1cb reaction is not concerted, there are no strict requirements concerning the stereochemistry of the reaction. In contrast to the preferred anti elimination that occurs in the E2 mechanism, E1cb reactions often produce a mixture of stereoisomers, as illustrated in the following equation:

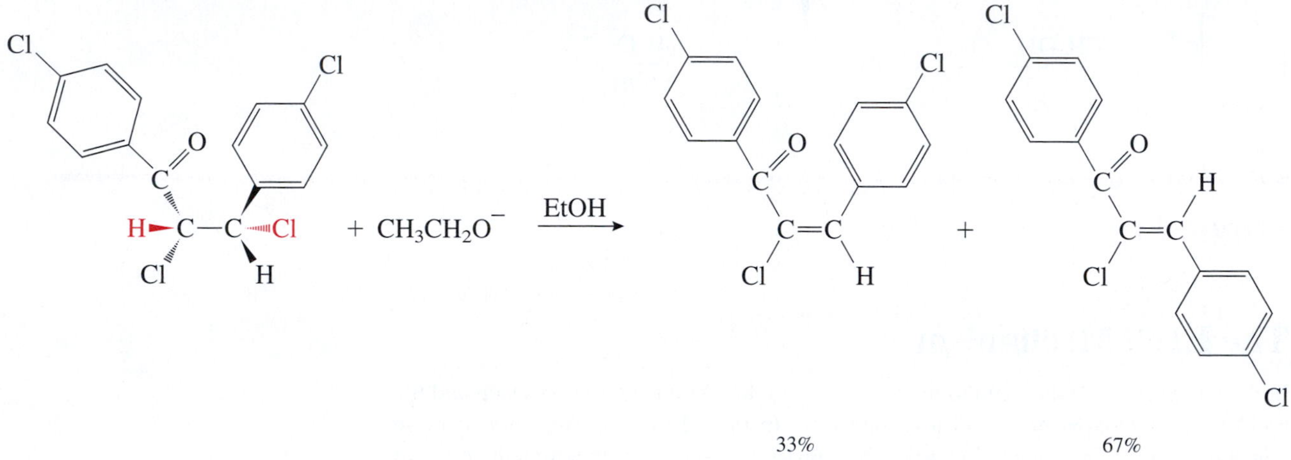

33%   67%

PROBLEM 9.14

Show the products of these reactions:

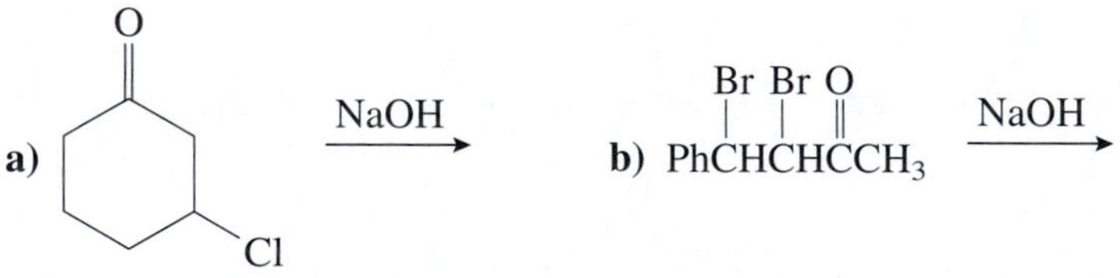

a)   b) $PhCHCHCCH_3$ with Br Br O

9.7   THE COMPETITION BETWEEN ELIMINATION AND SUBSTITUTION

As the previous discussion has shown, there is a competition among four different mechanisms in these reactions: S_N1, S_N2, E1, and E2. The amount of each that occurs depends on multiple factors, such as the substrate, the identity of the base or nucleophile, and the solvent. When these factors favor a particular one of these mechanisms, then it is possible to predict which one will predominate. However, when these factors favor different mechanisms, then predictions are very difficult. In fact, it is quite possible for several of the mechanisms to occur at the same time in a particular reaction. However, when these reactions are used in the laboratory, we are usually attempting to prepare a particular substitution product or a particular elimination product. It is usually possible to choose conditions carefully so that one mechanism is faster than the others and the yield of the desired substitution or elimination product is maximized.

The following generalizations provide a useful summary of the factors that control the competition among these mechanisms.

S_N2 and E2

These two pathways require the reaction of the carbon substrate with a nucleophile or a base in the rate-determining step of the reaction. So the presence of a good base or nucleophile favors these mechanisms over the S_N1 and E1 pair. Steric hindrance is an important factor in the competition between the S_N2 and E2 pathways. The fact that substitution is slowed by steric hindrance, whereas elimination is not, is reflected in the following examples. In these examples the presence of ethoxide ion, which is a strong base and a strong nucleophile, favors the S_N2/E2 mechanisms over the S_N1/E1 pathways. As the electrophilic carbon is changed from primary to secondary to tertiary, the mechanism changes from nearly complete S_N2 to complete E2 as the increasing steric hindrance at the electrophilic carbon slows the rate of the S_N2 reaction. Remember, E2 is not slowed by steric hindrance, so substrates with the leaving group on a tertiary carbon give E2 elimination rather than S_N1 substitution in the presence of a strong base.

$$CH_3CH_2Br + {}^-OCH_2CH_3 \xrightarrow{EtOH} CH_3CH_2OCH_2CH_3 + CH_2{=}CH_2$$
$$\qquad\qquad\qquad\qquad\qquad\qquad\qquad 99\% \qquad\qquad 1\%$$

$$\underset{\underset{\displaystyle CH_3CHCH_3}{|}}{\overset{\displaystyle Br}{}} + {}^-OCH_2CH_3 \xrightarrow{EtOH} \underset{\underset{\displaystyle CH_3CHOCH_2CH_3}{|}}{\overset{\displaystyle CH_3}{}} + CH_3CH{=}CH_2$$
$$\qquad\qquad\qquad\qquad\qquad\qquad\qquad 20\% \qquad\qquad 80\%$$

$$\underset{\underset{\displaystyle CH_3}{|}}{\overset{\overset{\displaystyle Br}{|}}{CH_3CCH_3}} + {}^-OCH_2CH_3 \xrightarrow{EtOH} \underset{\underset{\displaystyle 97\%}{}}{\overset{\overset{\displaystyle CH_3}{|}}{CH_3C{=}CH_2}}$$

Chapter 8 discussed the observation that nucleophile strength increases as base strength increases (as long as the basic/nucleophilic atoms are from the same period of the periodic table). Therefore, the rates of both S_N2 and E2 reactions increase as the strength of the base and nucleophile increases. However, many experimental observations have shown that a stronger base (which is also a stronger nucleophile) tends to give a higher ratio of elimination to substitution than does a weaker base (which is also a weaker nucleophile). For example, the reaction of 2-bromopropane with ethoxide ion, a strong base and a strong nucleophile, results in 20% substitution and 80% elimination. In contrast, the reaction of 2-bromobutane with acetate ion, a weaker base and a weaker nucleophile, results in a high yield of the substitution product.

$$\underset{\begin{array}{c}|\\[-2pt]Br\end{array}}{CH_3\overset{|}{C}HCH_3} \;+\; CH_3CH_2O^- \xrightarrow{\;CH_3CH_2OH\;} \underset{\begin{array}{c}|\\[-2pt]OCH_2CH_3\end{array}}{CH_3\overset{|}{C}HCH_3} \;+\; CH_3CH\!=\!CH_2$$

$$\qquad\qquad\qquad\qquad\qquad\qquad\qquad\qquad\qquad\qquad 20\% \qquad\qquad\qquad 80\%$$

$$\underset{\begin{array}{c}|\\[-2pt]Br\end{array}}{CH_3CH_2\overset{|}{C}HCH_3} \;+\; CH_3\overset{O}{\overset{||}{C}}O^- \xrightarrow{\;DMF\;} CH_3CH_2\overset{\begin{array}{c}O\\||\\OCCH_3\end{array}}{\underset{}{C}HCH_3}$$

$$\qquad\qquad\qquad\qquad\qquad\qquad\qquad\qquad\qquad\qquad 96\%$$

The temperature of the reaction also affects the relative amounts of substitution and elimination. Usually, the percentage of elimination increases at higher temperatures. This results from the fact that elimination breaks the molecule into fragments and therefore is favored more by entropy than is substitution. As the temperature is increased, the entropy term in the equation $\Delta G = \Delta H - T\Delta S$ becomes more important and the amount of elimination increases. This effect is illustrated in the following equations:

$$\underset{\begin{array}{c}|\\[-2pt]Br\end{array}}{CH_3\overset{|}{C}HCH_3} \;+\; NaOH \xrightarrow[\begin{array}{c}40\%\ H_2O\\45°C\end{array}]{60\%\ EtOH} \underset{\begin{array}{c}|\\[-2pt]OH\end{array}}{CH_3\overset{|}{C}HCH_3} \;+\; CH_3CH\!=\!CH_2$$

$$\qquad\qquad\qquad\qquad\qquad\qquad\qquad\qquad\qquad\qquad 47\% \qquad\qquad\qquad 53\%$$

$$\qquad\qquad\qquad\qquad\qquad\qquad\qquad \xrightarrow{\;100°C\;} \qquad 29\% \qquad\qquad\qquad 71\%$$

In summary, in the competition between S_N2 and E2, nucleophiles that are weak bases, minimum steric hindrance, and lower temperatures are used to maximize substitution; strong bases, maximum steric hindrance, and higher temperatures are used to maximize elimination.

S_N1 and E1

Both of these mechanisms involve rate-determining formation of a carbocation, so they most commonly occur with tertiary (best) or secondary substrates in polar solvents. The reaction conditions are often neutral or acidic to avoid the presence of any strong base or strong nucleophile that might favor the S_N2 or E2 pathways. Because the step that controls which product is formed occurs after the rate-determining step, it is much more difficult to influence the ratio of substitution to elimination here. In general, some elimination always accompanies an S_N1 reaction and must be tolerated. An example is provided in the equation in Figure 9.7.

When substitution or elimination reactions are used in organic synthesis—that is, to prepare organic compounds—it is usually possible to control whether the substitution product or the elimination product is the major compound that is formed. From the viewpoint of the carbon substrate the use of these reactions in synthesis can be summarized in the following manner.

Methyl Substrates: CH_3L

Methyl substrates are excellent for S_N2 reactions. There is no β-carbon, so elimination cannot occur. Also, the methyl carbocation is very unstable, so the S_N1 mechanism does not occur either.

Primary Substrates: RCH_2L

Because of their low amount of steric hindrance, these compounds are excellent for S_N2 reactions with almost any nucleophile. However, they can be forced to follow a pre-dominantly E2 pathway by the use of a strong base that is not a very good nucleophile because of its steric bulk. Most commonly, potassium *tert*-butoxide is used in this role, as shown in the following equation:

$$CH_3(CH_2)_5CH_2\overset{\overset{\displaystyle Br}{|}}{C}H_2 \;+\; CH_3\overset{\overset{\displaystyle CH_3}{|}}{\underset{\underset{\displaystyle CH_3}{|}}{C}}O^- K^+ \xrightarrow{\;t\text{-BuOH}\;} \underset{85\%}{CH_3(CH_2)_5CH=CH_2}$$

Primary substrates do not follow the S_N1 or E1 mechanisms unless they are allylic or benzylic because primary carbocations are too high in energy.

Secondary Substrates: R_2CHL

Secondary substrates can potentially react by any of the four mechanisms. They give good yields of substitution products by the S_N2 mechanism when treated with nucle-ophiles that are not too basic ($CH_3CO_2^-$, RCO_2^-, CN^-, RS^-, and others to be dis-cussed in Chapter 10). They give predominantly elimination products by the E2 mechanism when treated with strong bases such as OH^- or OR^-. (Many examples were provided in the preceding discussion.) They give predominantly substitution by the S_N1 mechanism, accompanied by some elimination by the E1 mechanism, when reacted in a polar solvent in the absence of a good nucleophile or base (acidic or neu-tral conditions). As an example, the following equation illustrates the use of an alco-hol as both nucleophile and solvent (called a **solvolysis reaction**) in the preparation of an ether by an S_N1 process. Note that the strongly basic EtO^- nucleophile cannot be used because it is a strong base and would cause elimination by the E2 mechanism to be the major process.

$$CH_3\overset{\overset{\displaystyle Br}{|}}{C}HCH_3 \;+\; CH_3CH_2OH \xrightarrow{\;EtOH\;} \underset{97\%}{CH_3\overset{\overset{\displaystyle OCH_2CH_3}{|}}{C}HCH_3} \;+\; \underset{3\%}{CH_3CH=CH_2}$$

Tertiary Substrates: R_3CL

Tertiary substrates do not give S_N2 reactions because they are too hindered. Therefore, if a substitution reaction is desired, it must be done under S_N1 conditions (polar solvent, absence of base). Acceptable yields are usually obtained, but some elimination by the E1 pathway usually occurs also. Excellent yields of elimination product can be obtained by the E2 mechanism if the substrate is treated with a strong base (usually OH⁻ or OR⁻). Examples are provided in the following equations:

$$\underset{\underset{CH_3}{|}}{\overset{\overset{Br}{|}}{CH_3CCH_3}} + CH_3CH_2OH \xrightarrow{\text{EtOH}} \underset{\underset{CH_3}{|}}{\overset{\overset{OCH_2CH_3}{|}}{CH_3CCH_3}} + \underset{}{\overset{\overset{CH_2}{\|}}{CH_3CCH_3}}$$

$$\qquad\qquad\qquad\qquad\qquad\qquad\qquad\qquad 81\% \qquad\qquad 19\%$$

$$\underset{\underset{CH_3}{|}}{\overset{\overset{Br}{|}}{CH_3CCH_3}} + CH_3CH_2O^- \xrightarrow{\text{EtOH}} \overset{\overset{CH_2}{\|}}{CH_3CCH_3}$$

$$\qquad\qquad\qquad\qquad\qquad\qquad\qquad\qquad\qquad 97\%$$

Chapter 10 discusses the synthetic uses of all of these reactions in considerably more detail.

PRACTICE PROBLEM 9.5

Show the substitution and/or elimination products for these reactions. Explain which mechanisms are occurring and which product you expect to be the major one.

a) $CH_3CH_2CH_2CH_2Cl + CH_3O^- \xrightarrow{\text{CH}_3\text{OH}}$

b) $\underset{\underset{Br}{|}}{\overset{\overset{CH_3}{|}}{CH_3CCH_2CH_3}} + {}^-OH \xrightarrow{\text{CH}_3\text{CH}_2\text{OH}}$

c) $\underset{}{\overset{}{\diagup\!\!\!\diagdown}} Cl + CH_3CH_2OH \xrightarrow{\text{EtOH}}$

Strategy

First identify the leaving group, the electrophilic carbon, and the nucleophile or base. Then, examine the nature of the electrophilic carbon because this often limits the possible mechanisms. If this carbon is methyl, the mechanism is S_N2. If it is primary, the mechanism is S_N2 unless a very sterically hindered base, such as potassium *tert*-butoxide, is used to promote E2 elimination. If the electrophilic carbon is tertiary, the mechanism is S_N1 and E1 under acidic or neutral conditions (absence of strong base) and E2 in the presence of a strong base (usually HO^-, CH_3O^-, or $CH_3CH_2O^-$). If the electrophilic carbon is secondary, all of the mechanisms are available, so the nucleophile and the solvent must be considered. The presence of a good nucleophile that is only a moderately strong base ($CH_3CO_2^-$, RCO_2^-, CN^-, RS^-) suggests that the mechanism will be S_N2, especially if the solvent is aprotic. In the presence of a strong base (usually HO^-, CH_3O^-, or $CH_3CH_2O^-$), the mechanism will be E2. The mechanism will be S_N1 in the absence of good nucleophiles and strong bases (usually acidic or neutral conditions) in a polar solvent. Do not forget about possible resonance stabilization of the carbocation, which makes an S_N1 reaction more favorable. Under favorable conditions (absence of good nucleophiles and strong bases), a compound that can form a resonance-stabilized carbocation can react by an S_N1 mechansim even when the electrophilic carbon is primary. Once you have decided on the mechanism that will be followed, consider the stereochemistry and regiochemistry of the reaction if they make a difference in the reaction (inversion for S_N2; racemization for S_N1; anti elimination and Zaitsev's rule for E2; and Zaitsev's rule for E1).

Solutions

a) The leaving group, Cl, is bonded to a primary carbon. Methoxide ion (CH_3O^-) is a strong nucleophile and a strong base but not sterically hindered, so the reaction will follow the S_N2 mechanism. Not much elimination should occur. There is no stereochemistry visible at the primary carbon.

$$CH_3CH_2CH_2CH_2Cl + CH_3O^- \xrightarrow{CH_3OH} CH_3CH_2CH_2CH_2OCH_3$$

b) The leaving group (Br) is bonded to a tertiary carbon. Hydroxide ion is a strong base, so the E2 mechanism should be followed. According to Zaitsev's rule, the more highly substituted alkene should be the major product.

Major   Minor

c) The leaving group is on a tertiary carbon and there is no strong base present, so the reaction follows the S_N1 and E1 mechanisms. Substitution is usually the major product, but a significant amount of elimination product is also formed. According to Zaitsev's rule, the more highly substituted product should predominate among the alkenes.

Major (S_N1)   Z + E   Trace (E1)
                   Minor (E1)

PROBLEM 9.15

Show the substitution and/or elimination products for these reactions. Explain which mechanisms are occurring and which product you expect to be the major one.

a) $CH_3CH_2\overset{\overset{\displaystyle OTs}{|}}{C}HCH_3 + CH_3CH_2\overset{\overset{\displaystyle O}{\|}}{C}O^- \xrightarrow{CH_3CH_2CO_2H}$

b) $CH_3CH_2\overset{\overset{\displaystyle OTs}{|}}{C}HCH_3 + CH_3CH_2O^- \xrightarrow{EtOH}$ c) $Ph\overset{\overset{\displaystyle CH_3}{|}}{\underset{\underset{\displaystyle CH_3}{|}}{C}}Cl + CH_3OH \xrightarrow{CH_3OH}$

PROBLEM 9.16

Explain which mechanism is preferred in these reactions and show the major products:

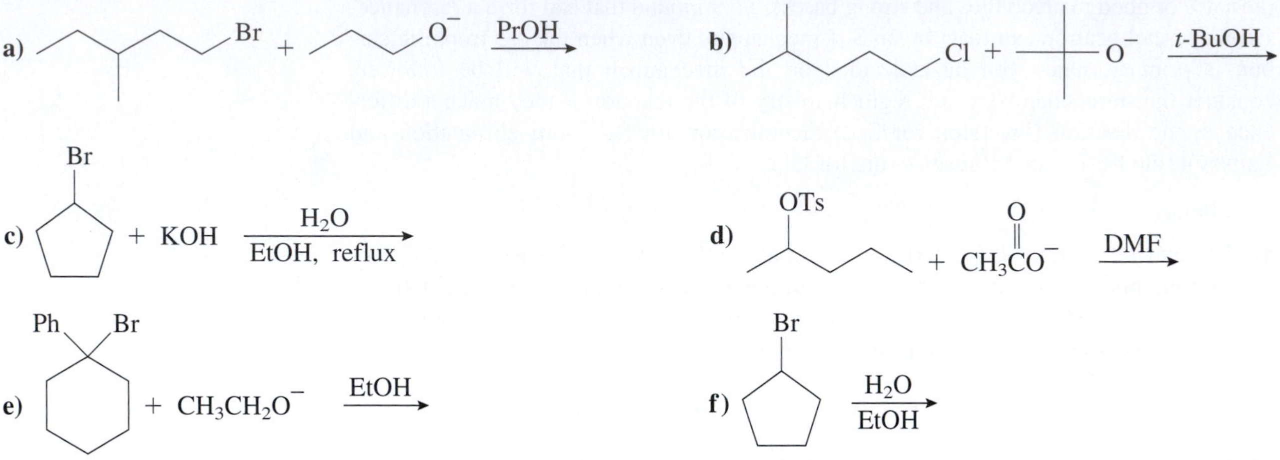

a) + $\xrightarrow{\text{PrOH}}$

b) Cl + $\xrightarrow{t\text{-BuOH}}$

c) + KOH $\xrightarrow[\text{EtOH, reflux}]{\text{H}_2\text{O}}$

d) + $CH_3\overset{\text{O}}{\overset{\|}{C}}O^-$ $\xrightarrow{\text{DMF}}$

e) + $CH_3CH_2O^-$ $\xrightarrow{\text{EtOH}}$

f) $\xrightarrow[\text{EtOH}]{\text{H}_2\text{O}}$

Focus On Biological Chemistry

Biological Elimination Reactions

Elimination reactions occur in living organisms also. One important example is the conversion of 2-phosphoglycerate to phosphoenolpyruvate during the metabolism of glucose:

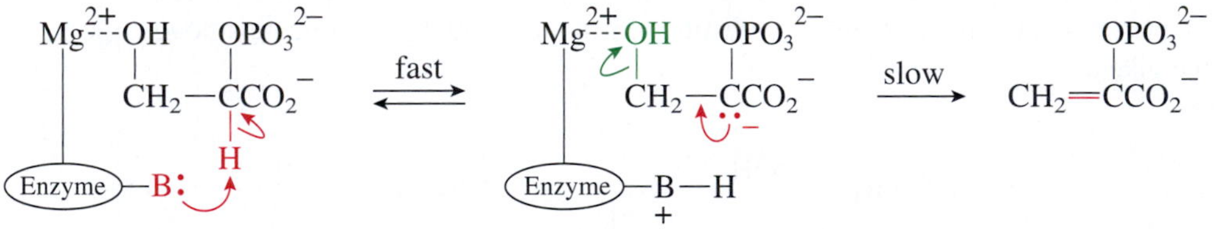

2-Phosphoglycerate   Phosphoenolpyruvate

This elimination is catalyzed by the enzyme enolase and follows an E1cb mechanism. The enzyme supplies a base to remove the acidic proton and generate a carbanion in the first step. In addition, a Mg^{2+} cation in the enzyme acts as a Lewis acid and bonds to the hydroxy group, making it a better leaving group.

Another example occurs in the citric acid cycle, where the enzyme aconitase catalyzes the elimination of water from citrate to produce aconitate:

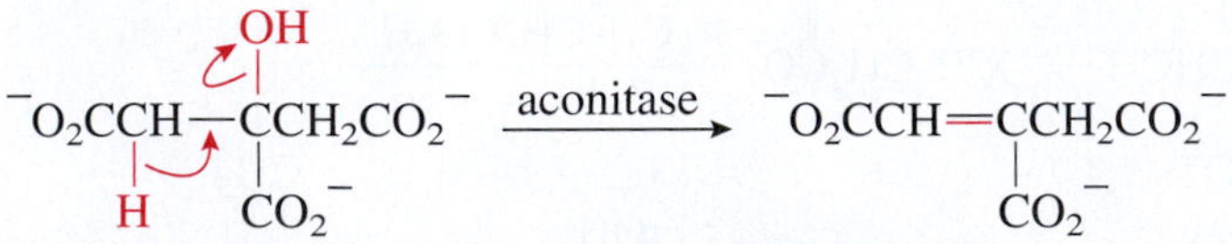

Citrate   Aconitate

Review of Mastery Goals

After completing this chapter, you should be able to:

ORGANIC
Chemistry Now™
*Click Mastery Goal Quiz to test
how well you have met these
goals.*

■ Provide a detailed description, including stereochemistry, for the E2 mechanism and summarize the conditions that favor its occurrence. (Problems 9.20, 9.21, 9.23, and 9.24)

■ Provide similar information about the E1 mechanism. (Problems 9.19, 9.28, and 9.33)

■ Understand Zaitsev's rule and Hofmann's rule and when each applies. (Problem 9.17)

■ Show the expected products, including stereochemistry and regiochemistry, of any elimination reaction. (Problems 9.17, 9.22, 9.27, 9.34, 9.36, and 9.37)

■ Predict whether the major pathway that will be followed under a particular set of reaction conditions will be S_N1, S_N2, E1, or E2. (Problems 9.18, 9.25, and 9.26)

Visual Summary of Key Reactions

The two mechanisms for the elimination reactions that compete with substitutions are summarized in Table 9.1.

Table 9.1   Summary of Elimination Mechanisms

	E1	E2
Mechanism	Two steps R—L $\longrightarrow$ R$^+$ $\longrightarrow$ alkene	One step R—L + B$^-$ $\longrightarrow$ alkene + BH + L$^-$
Kinetics	First order rate = k[RL]	Second order rate = k[RL][B$^-$]
Competes with	S_N1	S_N2
Stereochemistry	No preferred conformation	Anti-periplanar elimination (*trans*-diaxial in cyclohexanes)
Regiochemistry	Zaitsev's rule: favors more highly substituted alkene	Zaitev's rule: favors more highly substituted alkene (Hofman elimination follows Hofman's rule: favors less highly substituted alkene)
	Conjugated always preferred	Conjugated always preferred
Competition with Substitution	Some E1 always accompanies S_N1	E2 favored by strong bases, steric hindrance, and higher temperatures

Additional Problems

ORGANIC
Chemistry Now™
Assess your understanding of
this chapter's topics with
additional quizzing and
conceptual-based problems at
**http://now.brookscole.com/
hornback2**

9.17 Show the elimination products of these reactions. When more than one product is possible, indicate which product is major.

a) —Br + EtO$^-$ $\xrightarrow{\text{EtOH}}$

b) $\overset{\underset{|}{CH_3}}{CH_3CH_2\overset{+}{N}CH_2CH_2CH_2CH_3}$ ^-OH $\longrightarrow$
$\underset{CH_3}{}$

c) 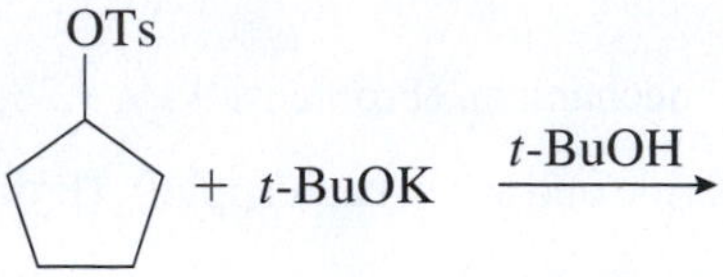 [cyclopentane with OTs] + t-BuOK $\xrightarrow{t\text{-BuOH}}$

d) 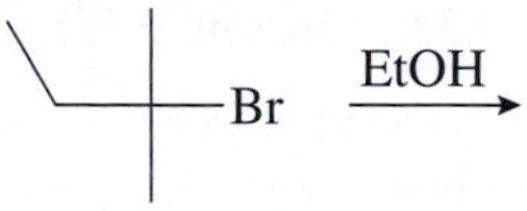 [structure]—Br $\xrightarrow{EtOH}$

e) Ph_2CHCH_2OTs + t-BuOK $\xrightarrow{t\text{-BuOH}}$

f) 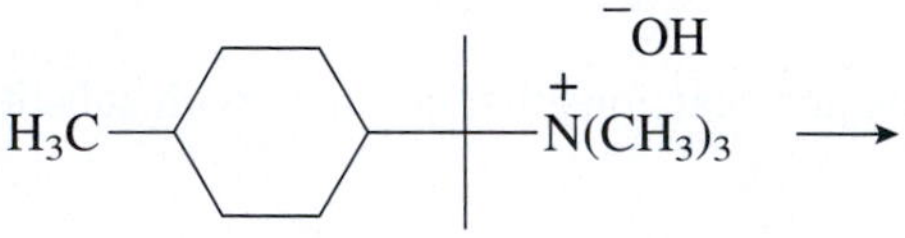 H_3C—[cyclohexane]—[C]—$\overset{+}{N}(CH_3)_3$ ^-OH $\longrightarrow$

g) 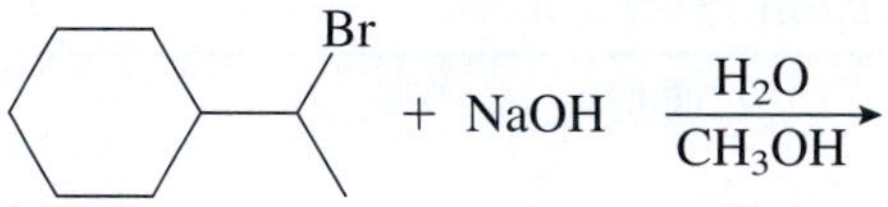 [cyclohexyl]—CH(Br)CH₃ + NaOH $\xrightarrow[CH_3OH]{H_2O}$

h) [structure]—Cl + NaOCH₃ $\xrightarrow{CH_3OH}$

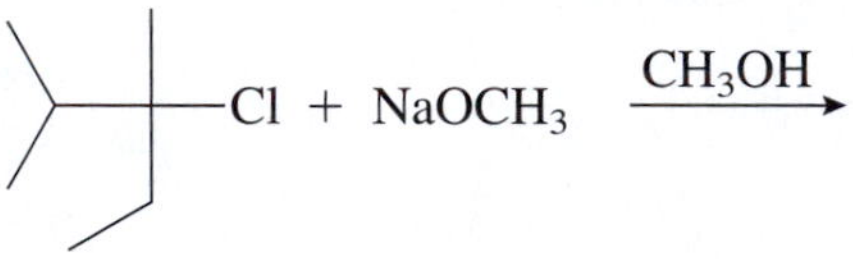

i) 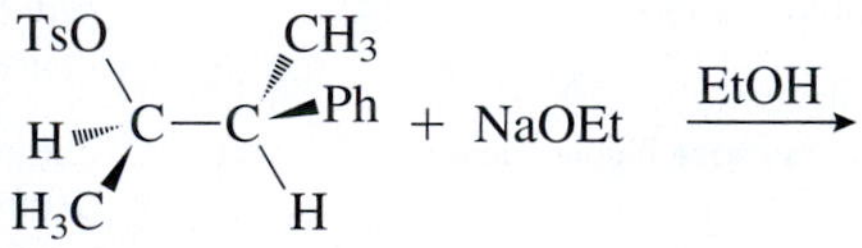 $\underset{H_3C}{\overset{TsO}{H\cdots C}}-\overset{CH_3}{\underset{H}{C\cdots Ph}}$ + NaOEt $\xrightarrow{EtOH}$

j) 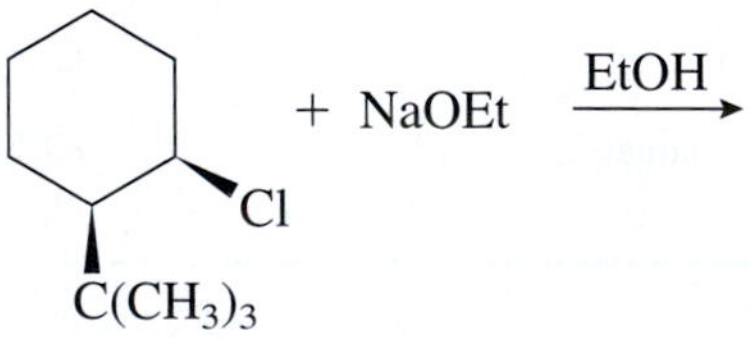 [cyclohexane with Cl and C(CH₃)₃] + NaOEt $\xrightarrow{EtOH}$

k) 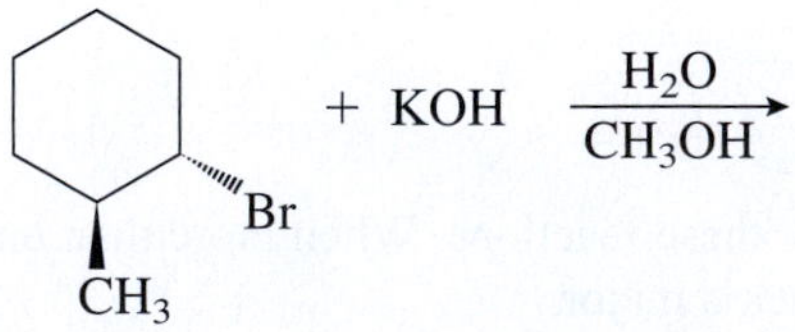 [cyclohexane with Br and CH₃] + KOH $\xrightarrow[CH_3OH]{H_2O}$

l)

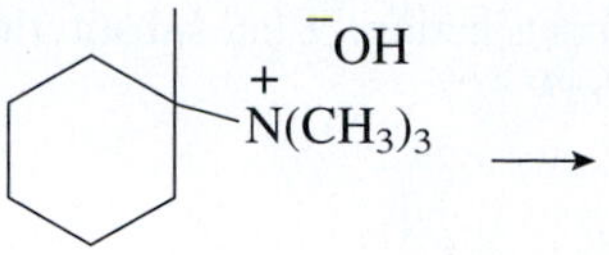

m)

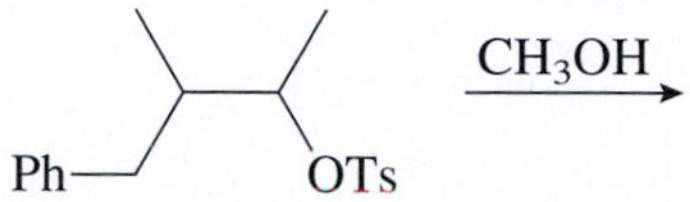

9.18 Explain which reaction mechanism (E1, S_N1, E2, S_N2) these reactions follow, and show the major products:

a)

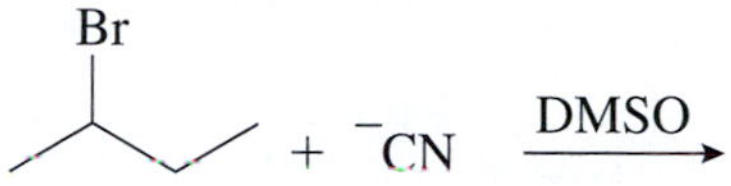

b)

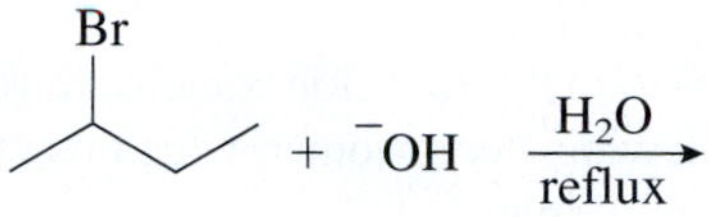

c)

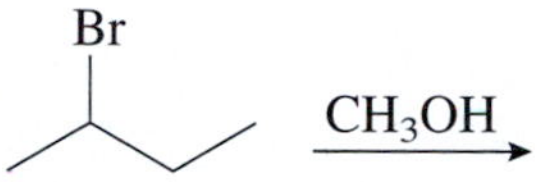

d)

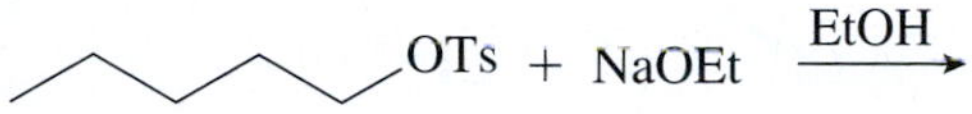

e)

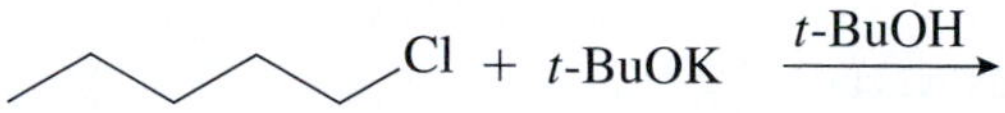

f)

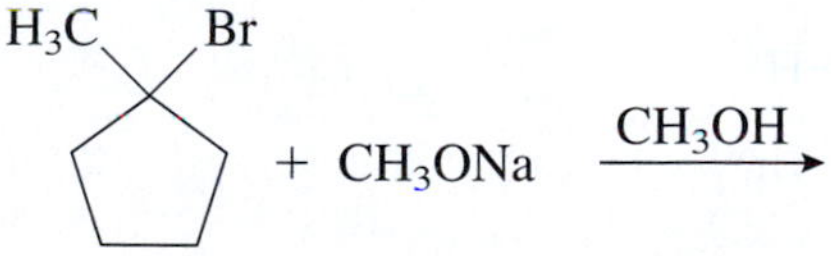

g)

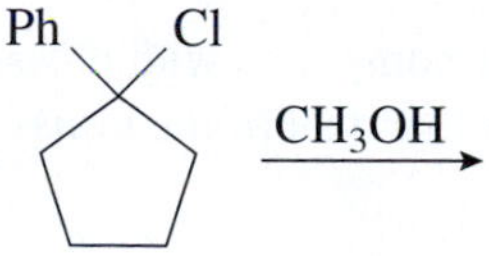

9.19   Show all the steps in the mechanism for this reaction. What substitution product(s) would also be formed in this reaction?

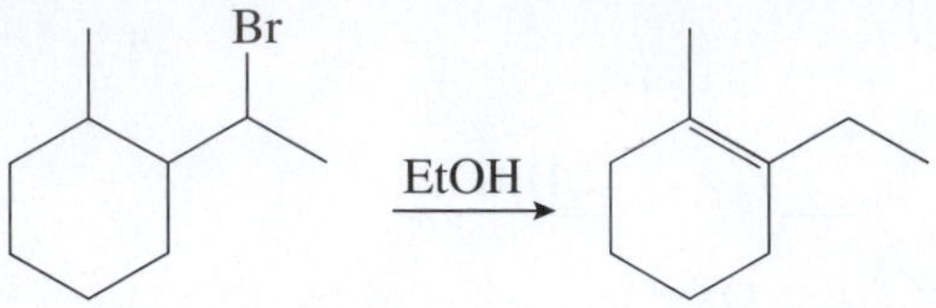

9.20   Show the products of this reaction. How would the composition of the products change if *t*-BuO⁻ in *t*-BuOH were used in place of ethoxide ion in ethanol?

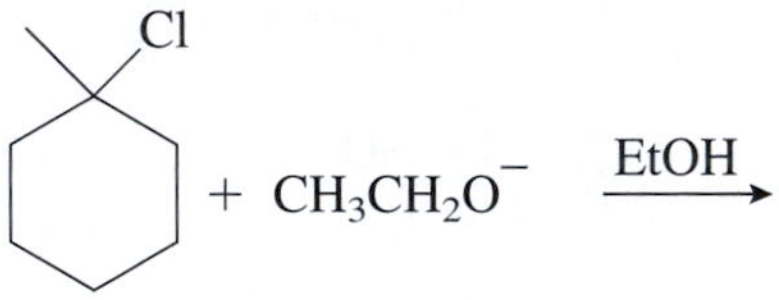

9.21   All of the stereoisomers of 1,2,3,4,5,6-hexachlorocyclohexane have very similar rates of E2 reaction except the following stereoisomer, which reacts about 7000 times more slowly than the others. Explain.

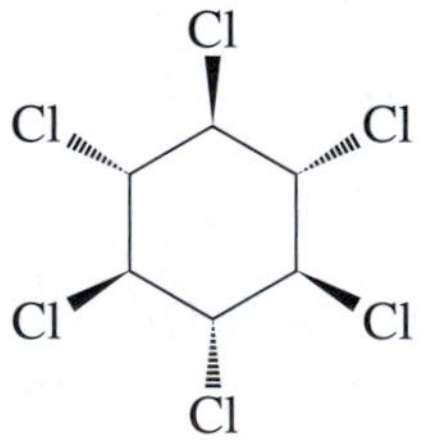

9.22   Show the product of this reaction:

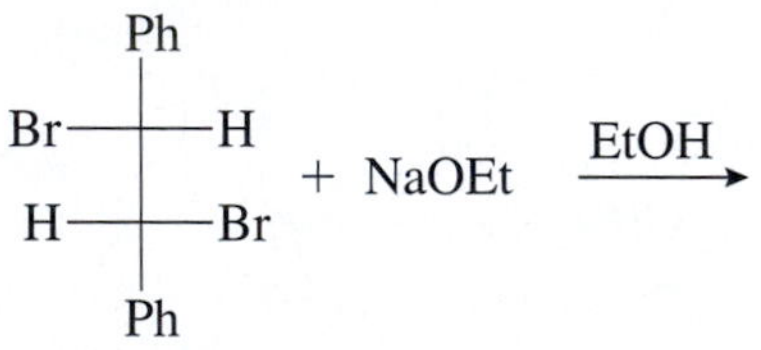

9.23   Explain why the reaction of the *cis*-isomer of this compound with potassium *tert*-butoxide in *tert*-butanol is about 500 times faster than that of the *trans*-isomer.

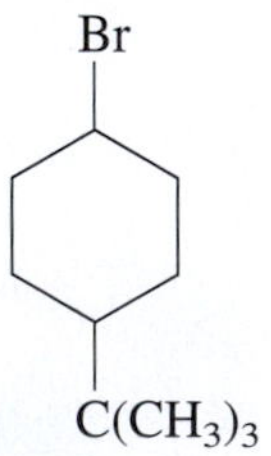

9.24 Explain which of these compounds has a faster rate of E2 elimination:

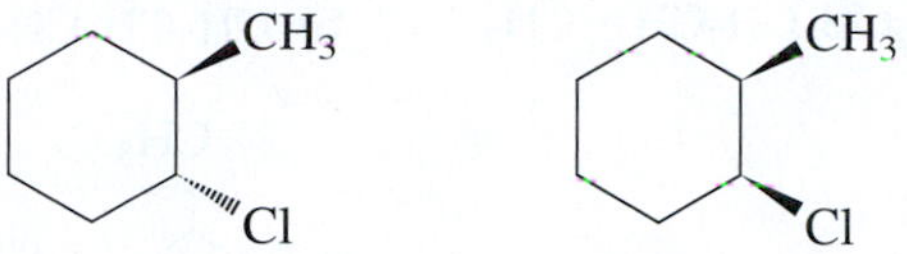

9.25 Frequently, several different routes can potentially be used to synthesize a desired compound. For example, the following two routes can be envisioned for the preparation of cyclopentyl methyl ether. Explain which of these two routes you expect to give the higher yield of the desired ether.

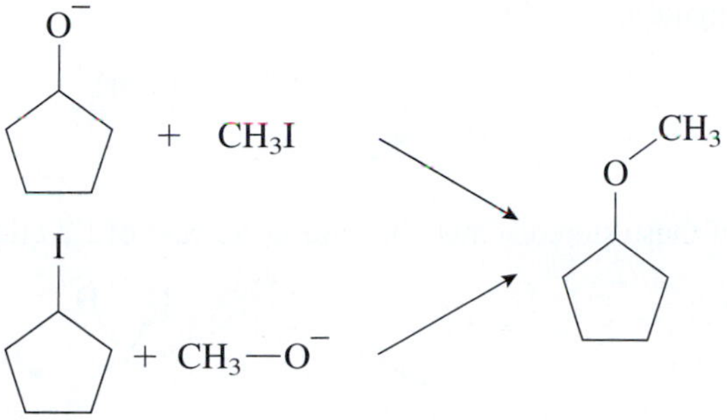

9.26 What reagents and reaction conditions could be used to carry out the following transformations?

a) 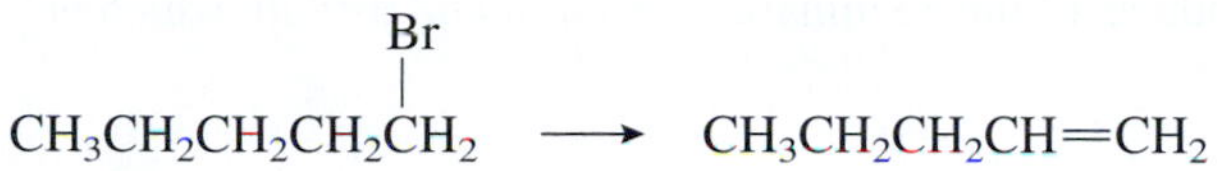
$$\underset{\underset{Br}{|}}{CH_3CH_2CH_2CH_2CH_2} \longrightarrow CH_3CH_2CH_2CH=CH_2$$

b) 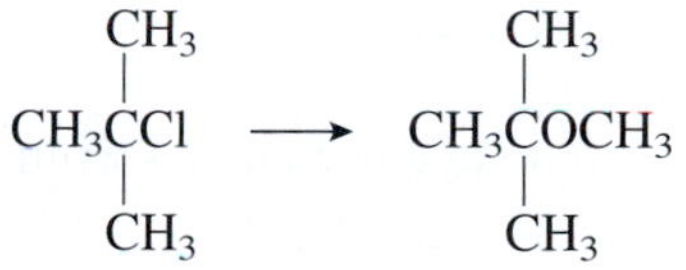
$$\underset{\underset{CH_3}{|}}{\overset{\overset{CH_3}{|}}{CH_3CCl}} \longrightarrow \underset{\underset{CH_3}{|}}{\overset{\overset{CH_3}{|}}{CH_3COCH_3}}$$

c) 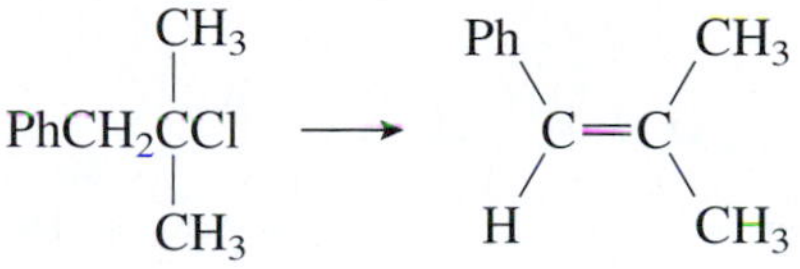
$$\underset{\underset{CH_3}{|}}{\overset{\overset{CH_3}{|}}{PhCH_2CCl}} \longrightarrow \underset{H}{\overset{Ph}{}}C=C\underset{CH_3}{\overset{CH_3}{}}$$

d) 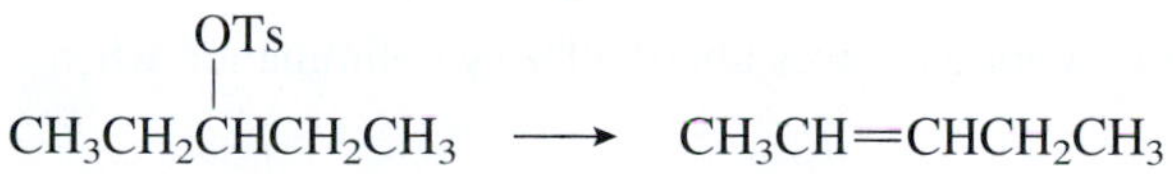
$$\underset{\underset{OTs}{|}}{CH_3CH_2CHCH_2CH_3} \longrightarrow CH_3CH=CHCH_2CH_3$$

e)

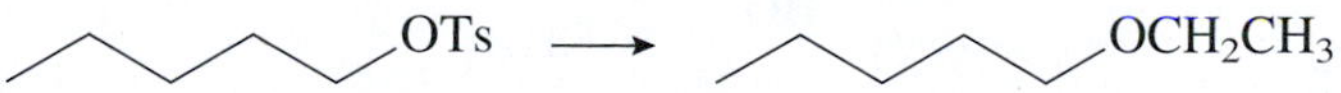

f)

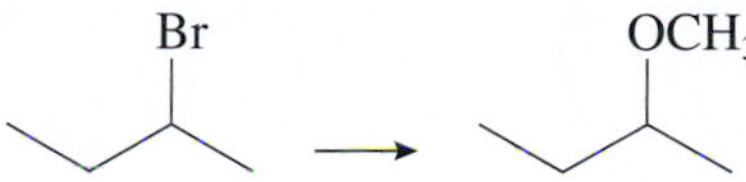

9.27 Suggest the best way to prepare these alkenes from alkyl halides.

a) $CH_3CH_2CH_2CH_2CH=CH_2$ b) $CH_3CH_2CH_2CH=CHCH_2CH_3$

c)

d)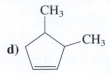

9.28 When heated in ethanol, this alkyl halide gives two substitution and two elimination products. Show the structures of these products and the mechanism for their formation.

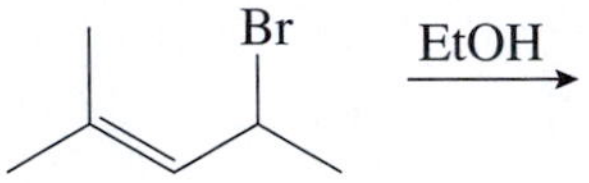

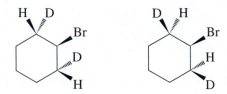

9.29 Which of these stereoisomers has the faster rate of E2 elimination?

9.30 Show the products of this elimination reaction and explain which is major:

$$\overset{\overset{\displaystyle D}{\displaystyle |}}{PhCHCH_2Cl} + t\text{-BuO}^- \xrightarrow{\;t\text{-BuOH}\;}$$

9.31 Explain why deuterium is lost in preference to hydrogen in this Hofmann elimination reaction:

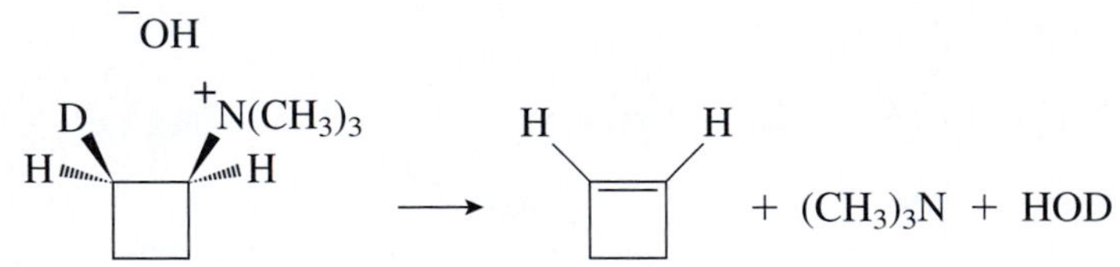

$+ (CH_3)_3N + HOD$

9.32 Explain why this compound gives about 50% syn elimination when heated:

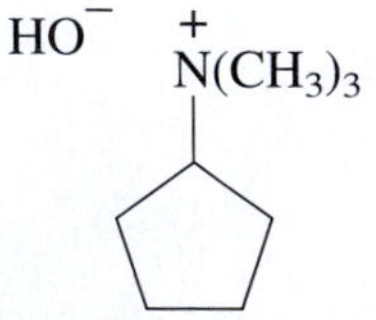

HO^- $\overset{+}{N}(CH_3)_3$

9.33 When 2-bromo-2-methylbutane is heated in a mixture of ethanol and water, it gives a 64% yield of substitution products and a 36% yield of elimination products.

    a) What mechanisms does this reaction follow?

    b) Show the structures of the substitution products.

    c) Show the structures of the elimination products.

    d) Under the same reaction conditions, how would 2-iodo-2-methylbutane differ from the bromide in its rate of reaction? in the products of its reaction?

9.34 This elimination reaction gives a single product. Show its structure and explain why it is the only product formed.

$$\underset{\underset{\displaystyle CH_2CH_3}{|}}{CH_3\overset{\overset{\displaystyle O}{\|}}{C}CH_2\overset{\overset{\displaystyle Cl}{|}}{C}CH_2CH_3} \; + \; KOH \; \xrightarrow[\text{EtOH}]{\text{H}_2\text{O}}$$

9.35 How might the structure of DDT be modified to make it again effective against resistant insects?

BioLink

Problems Using Online Three-Dimensional Molecular Models

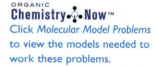
ORGANIC
Chemistry Now™
Click *Molecular Model Problems* to view the models needed to work these problems.

9.36 Explain which product is formed when each of these alkyl bromides reacts with sodium ethoxide in ethanol.

9.37 Explain which of these alkyl chlorides reacts faster with sodium ethoxide in ethanol.

9.38 Explain which product is formed when each of these alkyl chlorides reacts with sodium ethoxide in ethanol.

9.39 Explain which elimination product is formed when this alkyl chloride is heated in ethanol.

Ⓥ**Mentor**™   Do you need a live tutor for homework problems? Access vMentor at Organic ChemistryNow at **http://now.brookscole.com/hornback2** for one-on-one tutoring from a chemistry expert.

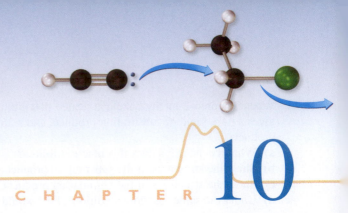

Synthetic Uses of Substitution and Elimination Reactions

INTERCONVERTING FUNCTIONAL GROUPS

THE PREVIOUS two chapters examined the mechanisms of substitution and elimination reactions in considerable detail. The purpose of this chapter is to see how to use these reactions to make other organic compounds, our first exploration into the area of organic synthesis. In this chapter the reactions are organized according to the type of compound that is produced. First, substitution reactions are covered, reactions that can be used to convert alkyl halides and alcohols into a host of other compounds, including ethers, esters, amines, and alkanes, by using different nucleophiles. In addition, two carbon nucleophiles are introduced. The use of carbon nucleophiles in substitution reactions results in the formation of carbon–carbon bonds, a very important part of the synthesis of organic compounds. For each of these substitution reactions, the factors that affect the yield are discussed, along with any limitations the reaction might have. Then the use of elimination reactions to prepare compounds containing double and triple bonds is presented, along with a discussion of the limitations of these reactions.

The number of reactions in this chapter may seem overwhelming at first. The key to success is to remember that nucleophiles react with electrophiles. If you can identify the nucleophile or base and the electrophilic carbon (the one bonded to the leaving group) in each reaction and recall the factors that affect the competition between the two substitution mechanisms and the two elimination mechanisms, the material you have to learn will be much more manageable.

MASTERING ORGANIC CHEMISTRY

▶ Predicting the Major Products of Substitution and Elimination Reactions

▶ Predicting the Stereochemistry of Substitution and Elimination Reactions

▶ Writing the Mechanisms for Substitution and Elimination Reactions

▶ Using Substitution and Elimination Reactions to Synthesize Compounds

10.1   SUBSTITUTION REACTIONS

In general, reactions proceeding by the S_N2 mechanism are more commonly employed for syntheses than are reactions proceeding by the S_N1 mechanism because of potential

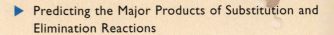

ORGANIC
Chemistry⚛Now™

Look for this logo in the chapter and go to OrganicChemistryNow at
http://now.brookscole.com/hornback2 for tutorials, simulations, problems, and molecular models.

complications, such as loss of stereochemistry or rearrangement, that often accompany reactions involving carbocations. The major factor that decreases the yields of S_N2 reactions is competing elimination reactions. Elimination can often be decreased by minimizing steric hindrance at the reaction site and decreasing the basicity of the nucleophile.

Compounds in which the leaving group is attached to a methyl or primary carbon usually give excellent yields in S_N2 substitutions. If the leaving group is attached to a secondary carbon, yields are still acceptable for nucleophiles that are only weakly basic. However, nucleophiles that are stronger bases often cause unacceptable amounts of elimination with secondary substrates. In such cases it is necessary to prepare the desired compound by an S_N1 pathway, if possible, or by an indirect route involving more than one step. When the leaving group is attached to a tertiary carbon, the S_N2 mechanism cannot occur. In such situations the substitution must be S_N1, and some elimination must be tolerated.

With these caveats in mind, let's see how to use these substitution reactions to prepare a variety of functional groups.

10.2 PREPARATION OF ALCOHOLS

Alcohols are widely available from a number of reactions that are described in subsequent chapters. For this reason they are often the starting materials for the preparation of other functional groups using substitution reactions. However, they can be prepared from alkyl halides, when necessary, by using either water or hydroxide ion as the nucleophile. A general equation for the reaction using hydroxide ion as the nucleophile is

$$H-\ddot{\underset{..}{O}}:^- \; + \; R-L \; \longrightarrow \; R-\ddot{\underset{..}{O}}H \; + \; :L^-$$

Hydroxide ion, the conjugate base of water, is a strong base and a strong nucleophile and reacts by the S_N2 mechanism. As illustrated in the following example, hydroxide ion gives good yields of alcohols with primary alkyl halides:

$$N\equiv C-\langle\!\!\langle\;\;\rangle\!\!\rangle-CH_2-Cl \; + \; :\ddot{\underset{..}{O}}-H \; \xrightarrow{H_2O} \; N\equiv C-\langle\!\!\langle\;\;\rangle\!\!\rangle-CH_2-OH \; + \; :\ddot{\underset{..}{Cl}}:^- \quad (85\%)$$

Yields are also acceptable for reactions of hydroxide ion with secondary alkyl halides if the compound is especially favorable for S_N2 reactions (halides that are allylic, benzylic, or adjacent to a carbonyl group), as shown in the following example:

$$\xrightarrow[H_2O]{NaOH} \quad (88\%)$$

Hydroxide ion is seldom used as a nucleophile with unactivated secondary halides and never with tertiary halides because of competing E2 elimination reactions. For such compounds, replacement of the halide with OH can be accomplished by using water as a nucleophile and S_N1 conditions:

$$H_2\ddot{O}: \; + \; R-L \; \longrightarrow \; R-OH \; + \; H-L$$

As shown in the following equation, the alkyl halide is heated in water as the solvent. (A Δ over or under a reaction arrow is used to indicate the application of heat.)

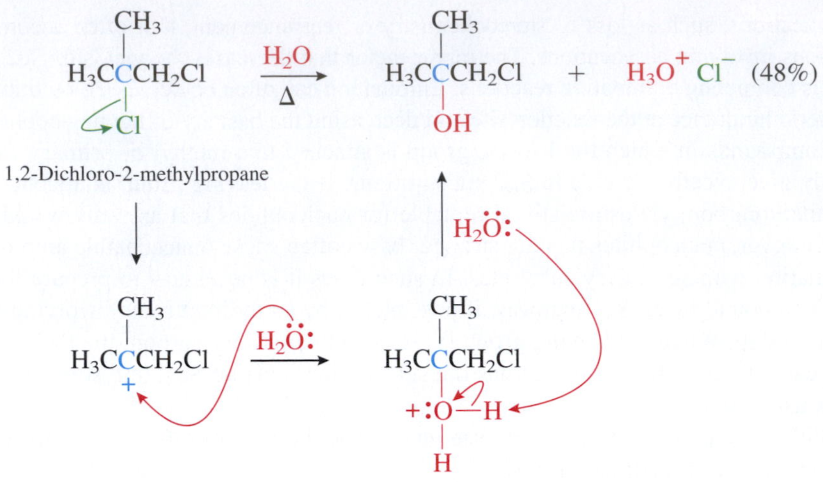

1,2-Dichloro-2-methylpropane

The mechanism (S_N1) for this reaction is also shown. Note that the nucleophile is water, not hydroxide ion. The second proton on the oxygen is not lost until after the oxygen has bonded to the carbon. A reaction such as this one, in which the nucleophile is also the solvent, is called a **solvolysis reaction.** In this specific case, where water is both the nucleophile and solvent, it is a **hydrolysis reaction.**

PROBLEM 10.1

Show the products of these reactions:

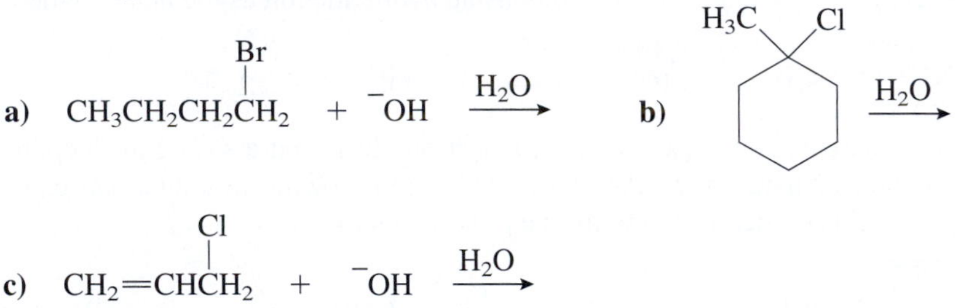

PROBLEM 10.2

Explain why only one of the two chlorines of 1,2-dichloro-2-methylpropane is replaced by a hydroxy group when the compound is heated in water (see the preceding hydrolysis reaction).

The reaction of hydroxide ion with secondary alkyl halides gives poor yields of the substitution product because hydroxide ion, a strong base, causes too much elimination. It is not unusual in organic chemistry for a reagent to give a poor yield of the desired product in a reaction because it is too reactive. The organic chemist's solution to such a problem is to modify the reactivity of the reagent—that is, to make it less reactive so that the yield in the desired reaction is higher. This is often accomplished by attaching a group to the reagent that decreases its reactivity. This modified reagent is then used in the desired reaction. Finally, the modifying group is removed from the reagent.

This technique can be used to prepare alcohols in better yields using substitution reactions. The hydrogen of hydroxide ion is replaced with a group that makes the oxygen less basic. Decreasing the basicity of the nucleophile slows the elimination reaction more than it slows the substitution reaction, resulting in a higher proportion of substitution in the product mixture (see Section 9.7). After the substitution is accomplished, the group that was used to replace the hydrogen must be removed so that the overall transformation is the replacement of the halide with OH.

A group that can be used to replace the hydrogen and decrease the basicity of the oxygen is the acetyl group. Thus, acetate anion is used as the nucleophile rather than hydroxide ion.

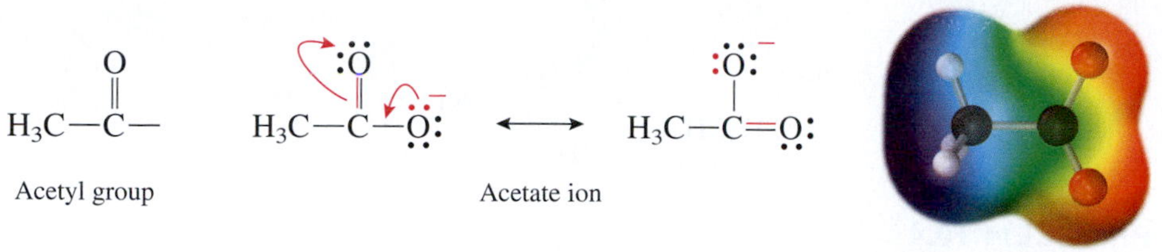

Acetyl group   Acetate ion

Acetate ion is a weaker base than hydroxide ion because of its resonance stabilization. For this reason, acetate ion gives higher yields of substitution products when used as a nucleophile in S_N2 reactions with secondary substrates. The resulting acetate ester can be converted to the desired alcohol in good yield by reaction with base and water. This step is not an S_N2 substitution. Instead, it begins with hydroxide ion, a nucleophile, attacking at the carbon of the carbonyl group, which is an electrophile. Ultimately, the single bond between the carbonyl carbon and the oxygen is broken. (The details of the mechanism for this reaction, called ester hydrolysis, are covered in Chapter 19.) Because the ester hydrolysis does not involve the carbon–oxygen bond that is formed in the initial S_N2 step, elimination is not a problem and the stereochemistry of that carbon–oxygen bond remains unchanged. An example of this process is shown in Figure 10.1.

In the process illustrated in Figure 10.1, acetate ion is termed a **synthetic equivalent** of hydroxide ion because the final product is the same as if hydroxide ion were used directly. But the two-step process results in a higher yield of 2-butanol than could be obtained by a direct substitution reaction of 2-bromobutane with hydroxide ion. The use of a carbonyl group to decrease the basicity or nucleophilicity of a reagent in order to

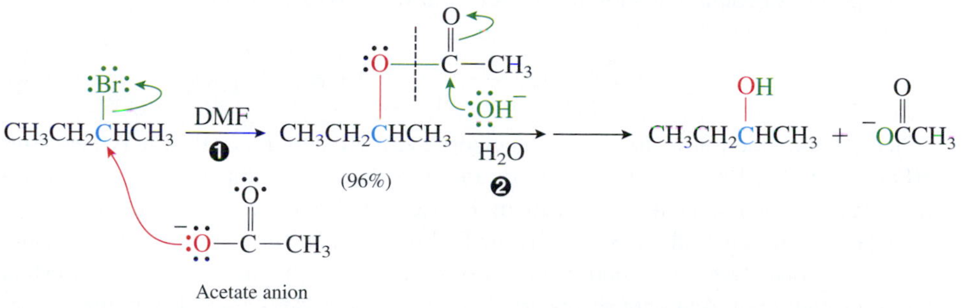

Acetate anion

Figure 10.1

PREPARATION OF AN ALCOHOL BY THE USE OF ACETATE ION AS THE NUCLEOPHILE.

❶ Acetate ion gives good yields in S_N2 reactions at secondary carbons, especially when an aprotic solvent, such as dimethylformamide, is used.

❷ The second step is not an S_N2 reaction. Instead, the green bond between the oxygen and the carbonyl carbon is cleaved by a different mechanism. (The details of this mechanism are discussed in Chapter 19 and need not concern us here. However, to help you remember what happens, note that this process begins with the hydroxide ion nucleophile attacking the carbonyl carbon, which is an electrophile.) The red CO bond is not cleaved, so its stereochemistry does not change in this step, nor does elimination occur under these conditions. The net effect of this two-step procedure is to substitute an OH group for the Br.

provide better control of the reaction and a higher yield is a common strategy in organic synthesis.

PROBLEM 10.3

On the basis of the bond cleavage shown for this reaction in Figure 10.1, predict the stereochemistry of the product. Explain.

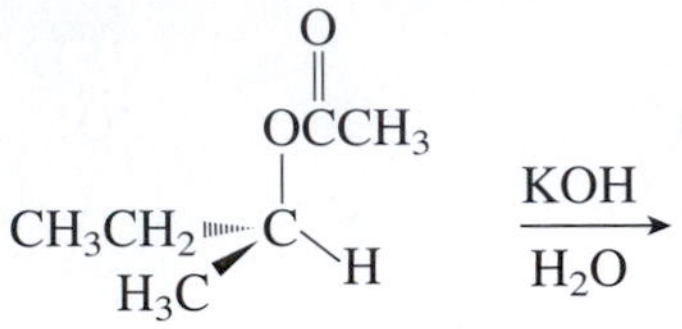

PROBLEM 10.4

Show the products of these reactions:

a)

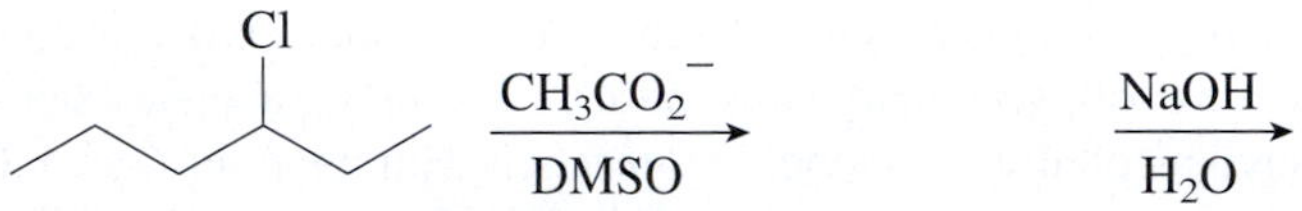

b)

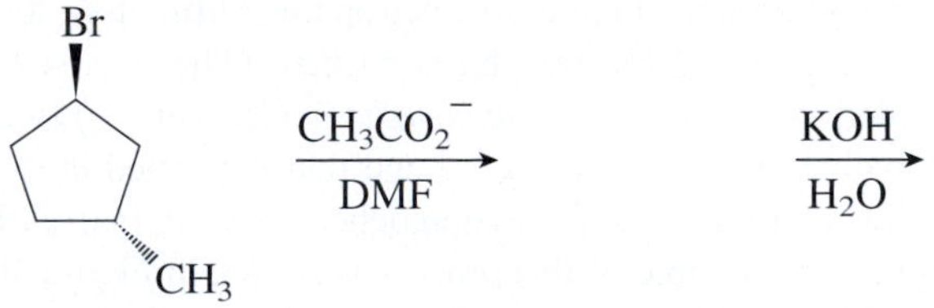

10.3   PREPARATION OF ETHERS

Ethers can be prepared by using an alcohol or its conjugate base, an alkoxide ion, as the nucleophile. A general equation for the reaction with alkoxide ion is

$$R-\ddot{\underset{..}{O}}:^{-} + R'-L \longrightarrow R-O-R' + :L^{-}$$

When an alkoxide ion is used as the nucleophile, the reaction is called a **Williamson ether synthesis.** Because the basicity of an alkoxide ion is comparable to that of hydroxide ion, much of the discussion about the use of hydroxide as a nucleophile also applies here. Thus, alkoxide ions react by the S_N2 mechanism and are subject to the usual S_N2 limitations. They give good yields with primary alkyl halides and sulfonate esters but are usually not used with secondary and tertiary substrates because elimination reactions predominate.

The alkoxide ion nucleophile is often prepared from the alcohol by reaction with sodium metal, as shown in the following equation for the formation of ethoxide ion from ethanol:

$$2\ CH_3CH_2\ddot{\underset{..}{O}}-H\ +\ 2\ Na\cdot \longrightarrow 2\ CH_3CH_2\ddot{\underset{..}{O}}:^{-}\ Na^{+}\ +\ H-H$$

Because phenols are stronger acids than alcohols, nucleophilic phenoxide ions can be prepared by reacting the phenol with bases such as hydroxide ion or carbonate ion.

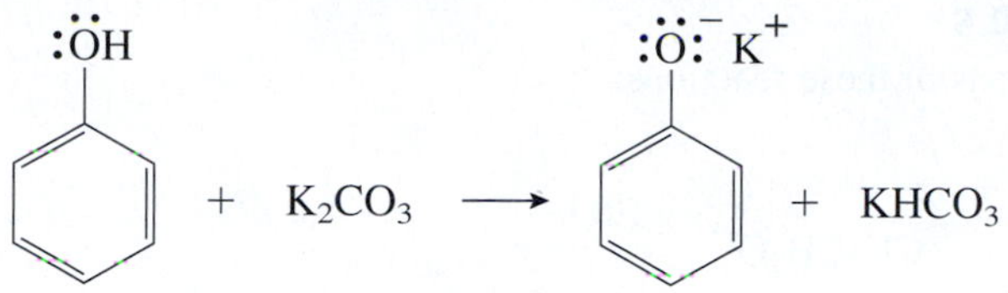

Several examples of the Williamson ether synthesis are given in the following equations:

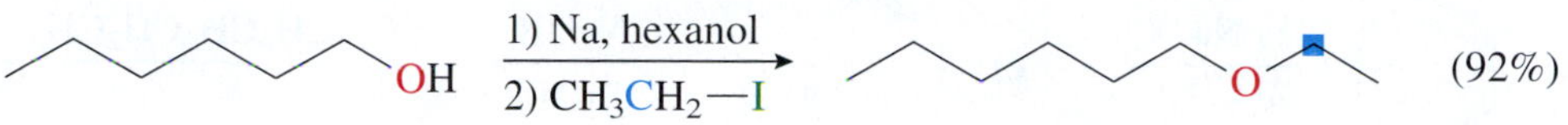

In equations like this, the reagents over the arrow are added in a sequence of separate steps, not all at once. Thus, in step 1, sodium metal is added to excess hexanol, which is both a reactant and the solvent for the reaction. Only after the reaction of the sodium and the alcohol is complete and the conjugate base of the alcohol has formed is the reagent shown in step 2 added. In the second step, the alkoxide ion acts as a nucleophile, replacing the leaving group of iodoethane to form the ether.

Important Convention

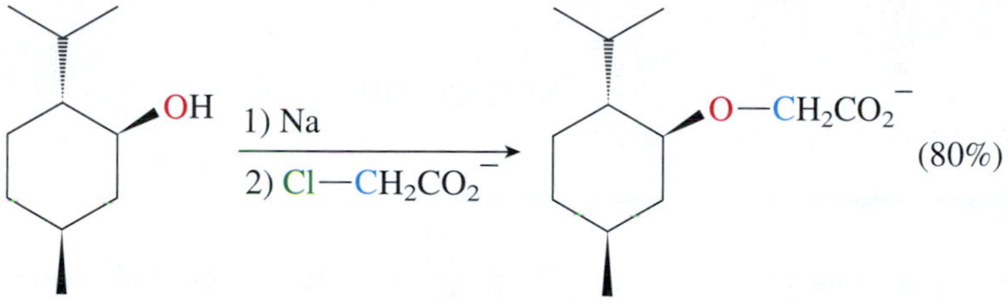

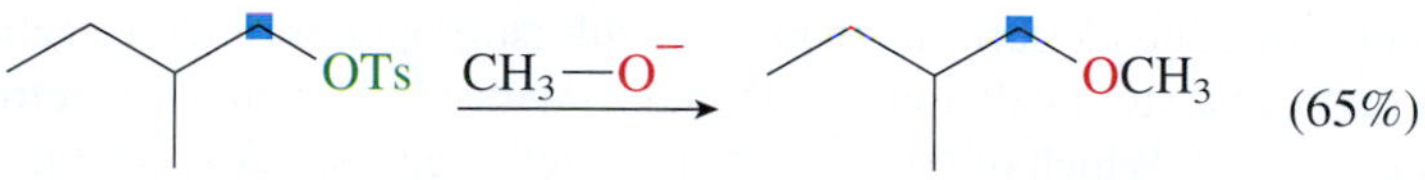

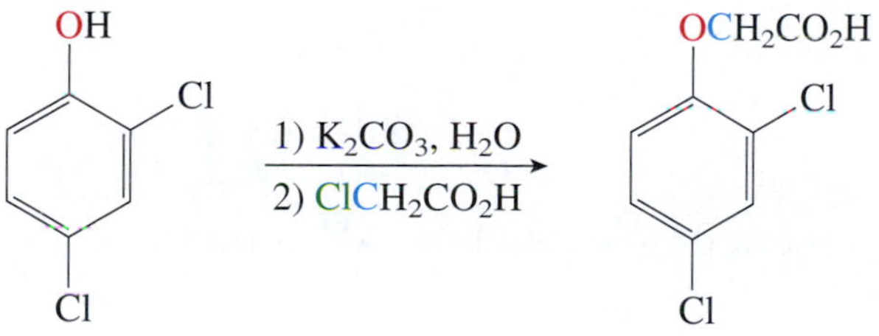

2,4-Dichlorophenoxyacetic acid
(2,4-D, an important herbicide)

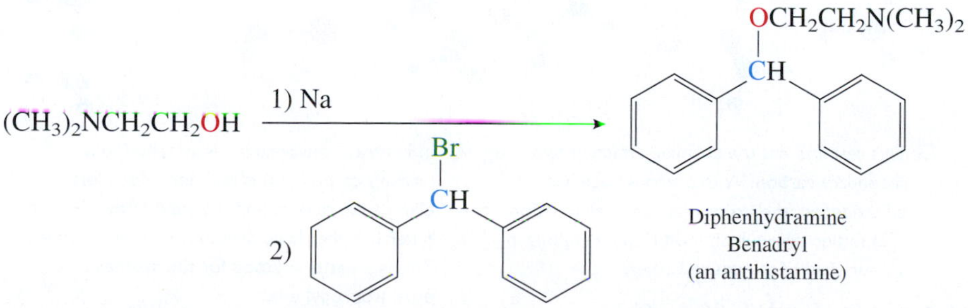

Diphenhydramine
Benadryl
(an antihistamine)

PROBLEM 10.5
Show the products of these reactions:

a) $\overset{\displaystyle Br}{\underset{\displaystyle |}{CH_3CH_2CH_2}}$ + $CH_3CH_2O^-$ $\xrightarrow{\ EtOH\ }$

b)   [cyclohexanol structure with OH] $\xrightarrow[\text{2) CH}_3\text{I}]{\text{1) Na}}$

c)   [phenol with OH and CH₃ substituent] $\xrightarrow[\text{EtOH}]{\text{NaOH}}$ $\overset{\displaystyle Cl}{\underset{\displaystyle |}{CH_3CH_2CH_2CH_2}}$ $\xrightarrow{\ \ \ \ }$

PROBLEM 10.6
Diphenhydramine can also be synthesized by heating bromodiphenylmethane with the amino alcohol shown here. Offer a reason why the oxygen, rather than the nitrogen, of this compound acts as the nucleophile. What factor favors the N? What factor favors the O? Which factor is winning in this case?

$$\begin{array}{c} H_3C \\[-2pt] \\[-2pt] H_3C \end{array}\!\!\!\!> N CH_2CH_2OH$$

An unsymmetrical ether can usually be prepared by two different Williamson ether syntheses. For example, the preparation of ethyl isopropyl ether could be accomplished by the reaction of ethoxide ion (nucleophile) with isopropyl bromide (electrophile) or by the reaction of isopropoxide ion (nucleophile) with ethyl bromide (electrophile), as shown in Figure 10.2. Which of these routes is better? Because alkoxide ions are strong

Figure 10.2

TWO POSSIBLE SYNTHESES OF ETHYL ISOPROPYL ETHER.

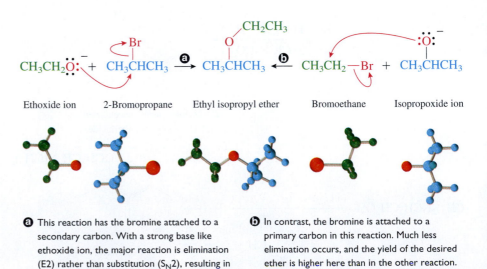

ⓐ This reaction has the bromine attached to a secondary carbon. With a strong base like ethoxide ion, the major reaction is elimination (E2) rather than substitution (S_N2), resulting in a poor yield of the desired ether.

ⓑ In contrast, the bromine is attached to a primary carbon in this reaction. Much less elimination occurs, and the yield of the desired ether is higher here than in the other reaction. This is a better method for the synthesis of ethyl isopropyl ether.

bases, an unacceptable amount of elimination occurs if the leaving group is attached to a secondary carbon. Therefore, the route using the primary halide (ethyl bromide) will give a higher yield of the substitution product.

PROBLEM 10.7

Explain which route would provide a better synthesis of these ethers:

a) 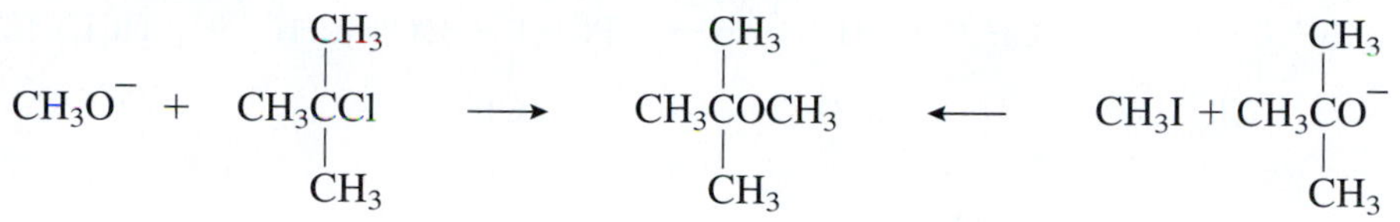 CH_3O^- + $CH_3\overset{\overset{\displaystyle CH_3}{|}}{\underset{\underset{\displaystyle CH_3}{|}}{C}}Cl$ $\longrightarrow$ $CH_3\overset{\overset{\displaystyle CH_3}{|}}{\underset{\underset{\displaystyle CH_3}{|}}{C}}OCH_3$ $\longleftarrow$ $CH_3I + CH_3\overset{\overset{\displaystyle CH_3}{|}}{\underset{\underset{\displaystyle CH_3}{|}}{C}}O^-$

b)

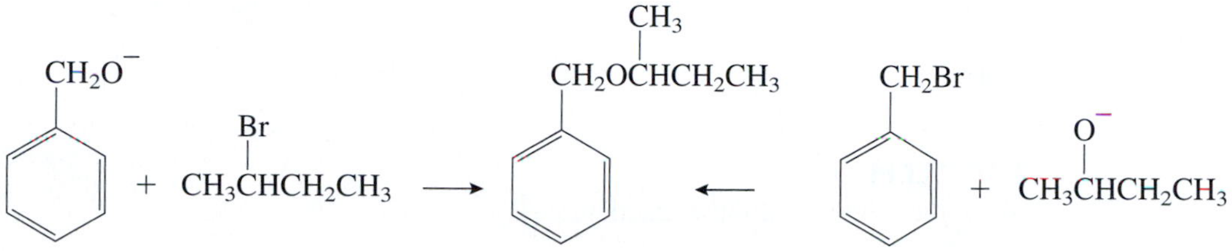

PRACTICE PROBLEM 10.1

Show a method for synthesizing this ether from an alcohol and an alkyl halide:

OCH_2CH_3

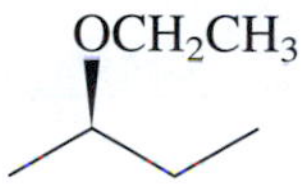

Solution

To minimize competing elimination by the E2 mechanism, treat the conjugate base of the secondary alcohol with the primary alkyl halide:

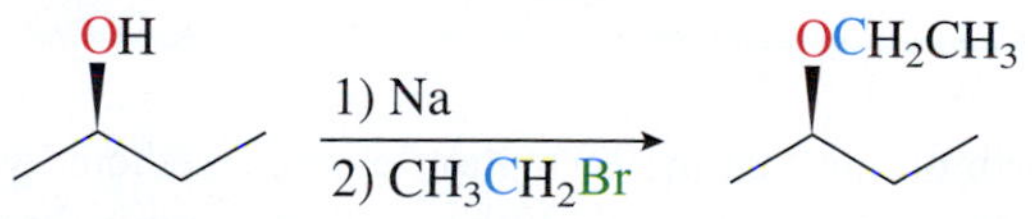

OH $\xrightarrow[\text{2) } CH_3CH_2Br]{\text{1) Na}}$ OCH_2CH_3

PROBLEM 10.8

Suggest a synthesis of these ethers starting with an alcohol and an alkyl halide:

$OCH_2CH_2CH_3$

a) $CH_3OCH_2CH_2CH_2CH_3$ b) 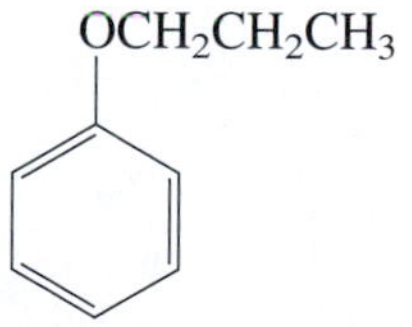   c)

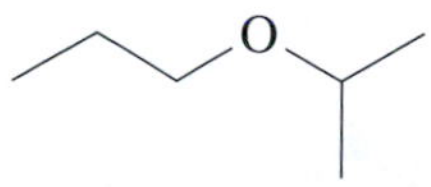

Ethers can also be prepared by using alcohols as the nucleophiles:

$$R—O—H + R'—L \longrightarrow R—O—R' + HL$$

If the leaving group is bonded to a secondary or tertiary carbon, the reaction usually follows the S_N1 mechanism and is the preferred method in order to avoid problems with elimination. An alcohol must also be used as the nucleophile when the reaction is run under acidic conditions because alkoxide ions cannot exist in acid. Examples are provided by the following equations. In the first example, in which ethanol is the solvent, the reaction is an **ethanolysis.**

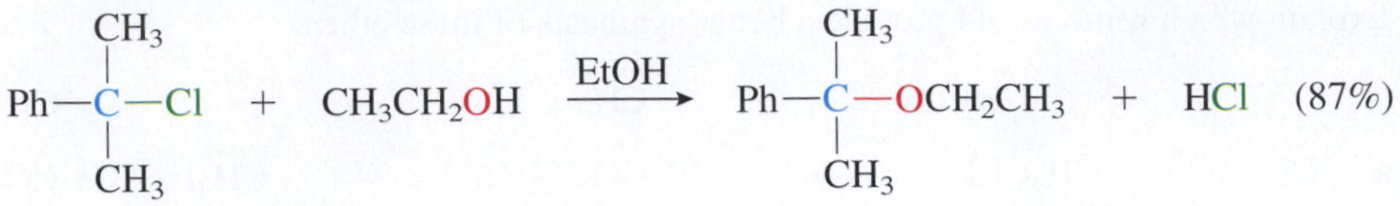

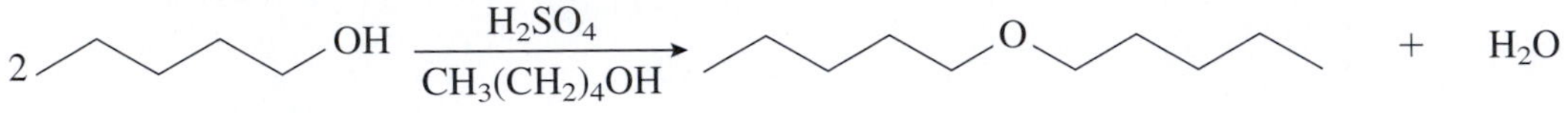

$$\underset{\overset{|}{\underset{CH_3}{\overset{CH_3}{|}}}}{Ph-C-Cl} \ + \ CH_3CH_2OH \ \xrightarrow{EtOH} \ \underset{\overset{|}{\underset{CH_3}{\overset{CH_3}{|}}}}{Ph-C-OCH_2CH_3} \ + \ HCl \quad (87\%)$$

$$2 \underset{\text{1-Pentanol}}{\diagdown\diagup\diagdown\diagup\diagdown OH} \ \xrightarrow[CH_3(CH_2)_4OH]{H_2SO_4} \ \underset{\text{Dipentyl ether}}{\diagup\diagdown\diagup\diagdown O\diagup\diagdown\diagup} \ + \ H_2O$$

PROBLEM 10.9
Show the products of these reactions:

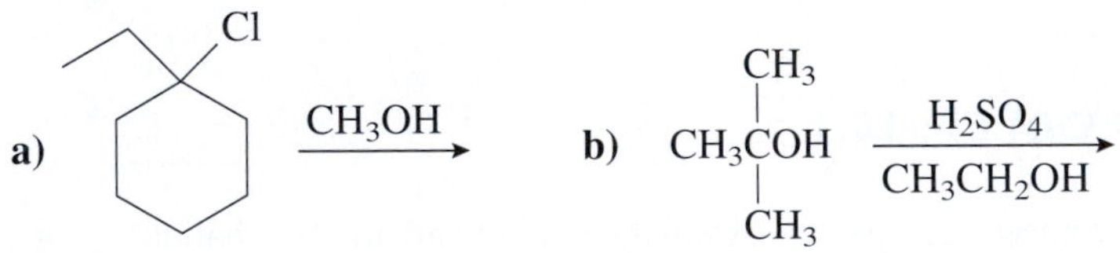

a) $\xrightarrow{CH_3OH}$

b) $\underset{\overset{|}{\underset{CH_3}{}}}{\overset{\overset{CH_3}{|}}{CH_3COH}} \ \xrightarrow[CH_3CH_2OH]{H_2SO_4}$

c) $Ph_2CHOH \ + \ HOCH_2CH_2Cl \ \xrightarrow{H_2SO_4}$

PROBLEM 10.10
Show all the steps in the mechanism for the reaction of 1-pentanol with sulfuric acid to form dipentyl ether.

Finally, it is worth noting that the formation of cyclic ethers by intramolecular nucleophilic substitutions is quite favorable if the resulting ring is three, five, or six membered, as shown in the following reactions:

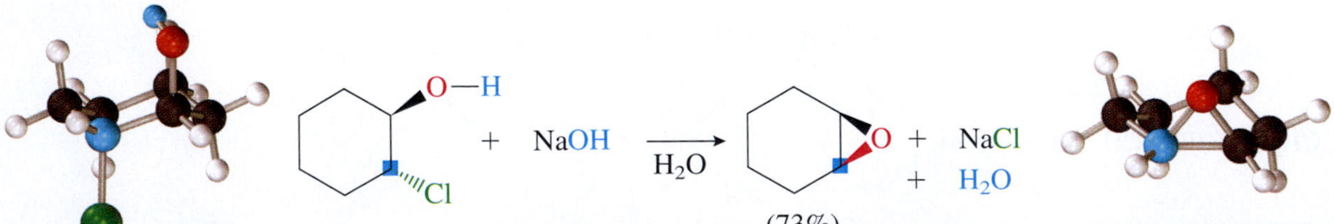

$\underset{H_2O}{\overset{+ \ NaOH}{\longrightarrow}}$ $+ \ NaCl$
$+ \ H_2O$
(73%)

$\underset{OH \qquad\qquad OH}{\diagup\diagdown\diagup\diagdown\diagup\diagdown\diagup} \ \xrightarrow{H_3PO_4} \ \diagdown\diagup\diagdown\diagup\diagdown \ + \ H_2O \quad (97\%)$

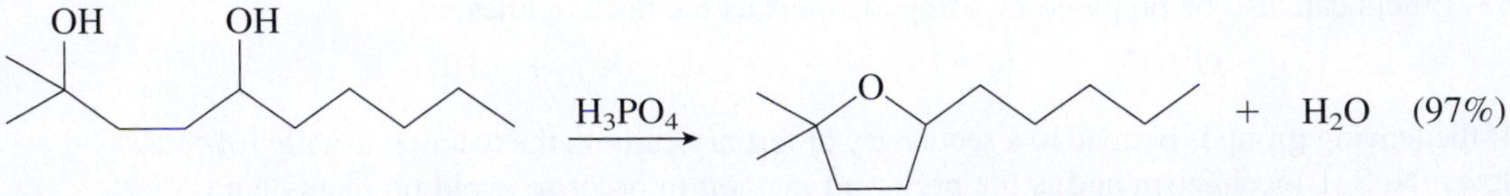

PROBLEM 10.11
Show the steps in the mechanism for the reaction of *trans*-2-chlorocyclohexanol with sodium hydroxide shown in the previous equation. Explain why *cis*-2-chlorocyclohexanol does not give a similar reaction.

PROBLEM 10.12
Show the product, including stereochemistry, for this reaction:

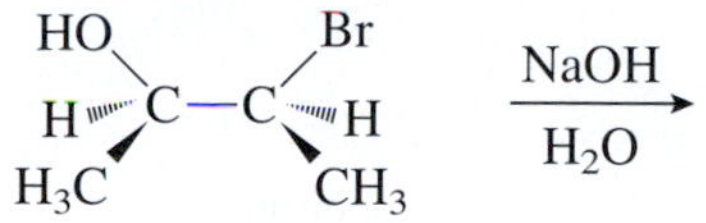

PROBLEM 10.13
Because of the acidic conditions, this reaction proceeds by an S_N1 mechanism. Which hydroxy group acts as the leaving group in the reaction? Show all the steps in the mechanism for this reaction:

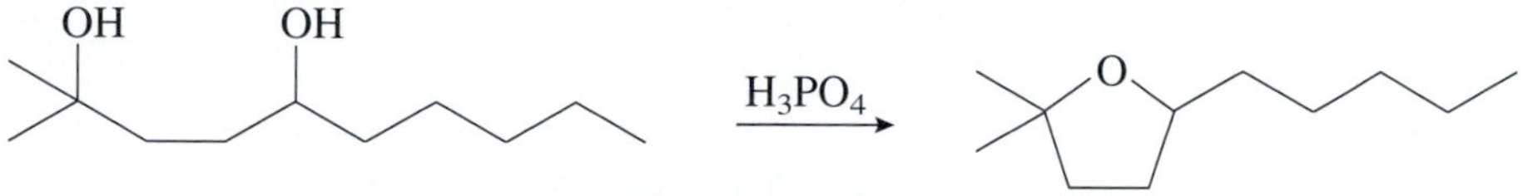

10.4 PREPARATION OF ESTERS

Esters can be prepared by employing carboxylate salts as nucleophiles, as shown in the following equation:

Because carboxylate salts are only weakly basic, elimination is not a problem when the leaving group is attached to a primary or secondary carbon. Several examples are provided in the following equations:

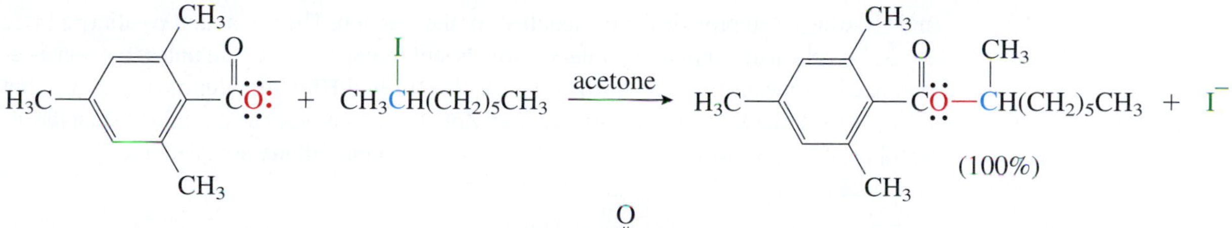

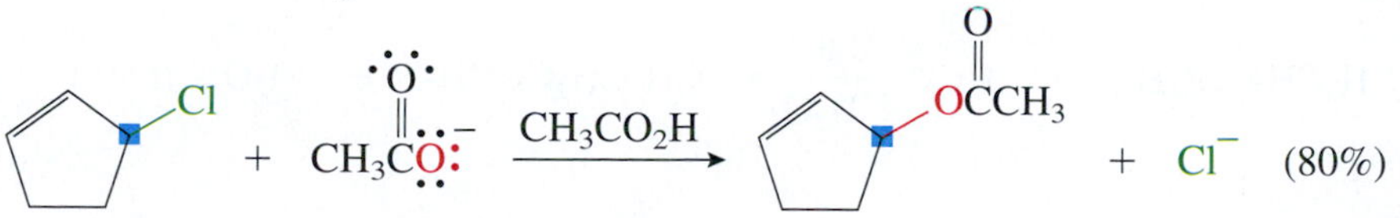

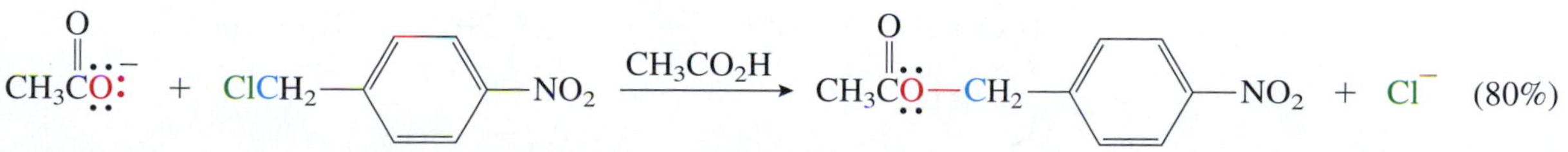

PROBLEM 10.14

Show the products of these reactions:

a)

$$\text{cyclopentane with Br and CH}_3 \quad + \quad CH_3CO_2^- \quad \xrightarrow{\text{DMSO}}$$

b)

$$\text{(structure with Cl)} \quad + \quad CH_3CH_2CO_2^- \quad \xrightarrow{\text{acetone}}$$

c)

$$CH_3CH_2CH_2CH_2Br \quad + \quad \text{(benzene ring with } CO_2^-) \quad \xrightarrow{\text{DMF}}$$

10.5 PREPARATION OF ALKYL HALIDES

The preparation of alkyl halides by substitution reactions usually starts from alcohols because alcohols are widely available. Hydroxide ion is a poor leaving group, so the OH must first be converted into a better leaving group, either by protonation in acid or by conversion to a sulfonate or similar ester (see Section 8.9), as illustrated in the following equations:

$$R-\overset{..}{\underset{..}{O}}-H \quad \xrightarrow{H-A} \quad R-\overset{H}{\underset{+}{\overset{|}{O}}}-H \quad \xrightarrow{:\overset{..}{X}\overset{..}{:}^-} \quad R-X \quad + \quad H_2O$$

$$R-\overset{..}{\underset{..}{O}}-H \quad \xrightarrow{R'SO_2Cl} \quad R-O-SO_2R' \quad \xrightarrow{:\overset{..}{X}\overset{..}{:}^-} \quad R-X \quad + \quad R'SO_3^-$$

Protonation of the alcohol can be accomplished by using the halogen acids, HCl, HBr, and HI, which also provide the nucleophile for the reaction. These reaction conditions favor the S$_N$1 mechanism, although primary alcohols still follow the S$_N$2 path unless a resonance-stabilized carbocation can be formed. The acids HBr and HI work with most alcohols, but HCl, a weaker acid, requires the presence of ZnCl$_2$ (a Lewis acid) as a catalyst when the alcohol is primary or secondary. Examples are shown in the following equations:

$$CH_3CH_2\overset{OH}{\overset{|}{C}}HCH_3 \quad + \quad HCl \quad \xrightarrow[H_2O]{ZnCl_2} \quad CH_3CH_2\overset{Cl}{\overset{|}{C}}HCH_3 \quad + \quad H_2O \quad (65\%)$$

$$H_3C\overset{OH}{\underset{\overset{|}{CH_3}}{\overset{|}{C}}}CH_3 \quad + \quad HCl \quad \xrightarrow{H_2O} \quad H_3C\overset{Cl}{\underset{\overset{|}{CH_3}}{\overset{|}{C}}}CH_3 \quad + \quad H_2O \quad (88\%)$$

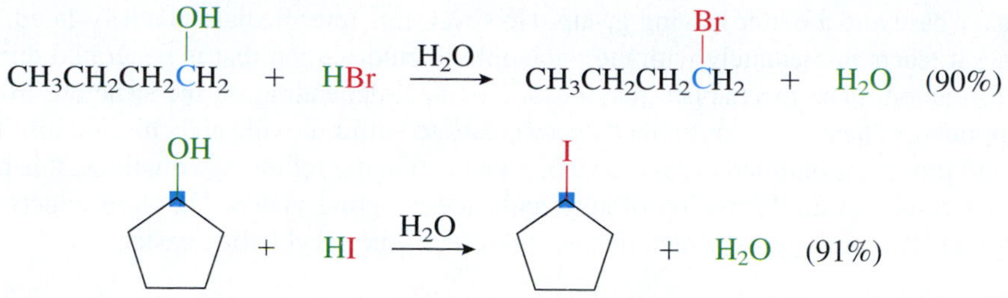

OH

CH$_3$CH$_2$CH$_2$CH$_2$ + HBr $\xrightarrow{\text{H}_2\text{O}}$ CH$_3$CH$_2$CH$_2$CH$_2$ + H$_2$O (90%)

Br

OH

(cyclopentanol) + HI $\xrightarrow{\text{H}_2\text{O}}$ (iodocyclopentane) + H$_2$O (91%)

PRACTICE PROBLEM 10.2

Show all the steps in the mechanism for the reaction of 1-butanol with HBr in water.

Solution

The reactant is a primary alcohol, so the mechanism must be S$_N$2. First the hydroxy group is protonated. Then bromide ion acts as a nucleophile.

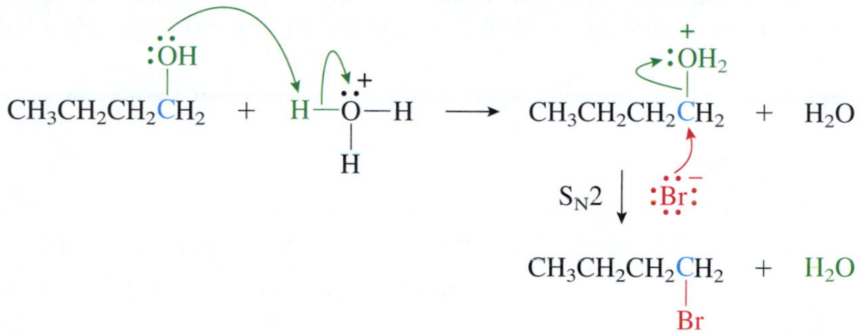

$$\ddot{\text{O}}\text{H}$$

CH$_3$CH$_2$CH$_2$CH$_2$ + H—$\overset{+}{\underset{\text{H}}{\ddot{\text{O}}}}$—H ⟶ CH$_3CH_2CH_2CH_2$ + H$_2$O

$$\overset{+}{:}\text{OH}_2$$

S$_N$2 ↓ :Br:⁻

CH$_3$CH$_2$CH$_2$CH$_2$ + H$_2$O

Br

PROBLEM 10.15

Show all the steps in the mechanism for the reaction of 2-methyl-2-butanol with HCl in water.

Conversion of the alcohol into a sulfonate ester followed by an S$_N$2 substitution using a halide nucleophile is another method that is commonly employed. Examples are provided in the following equations:

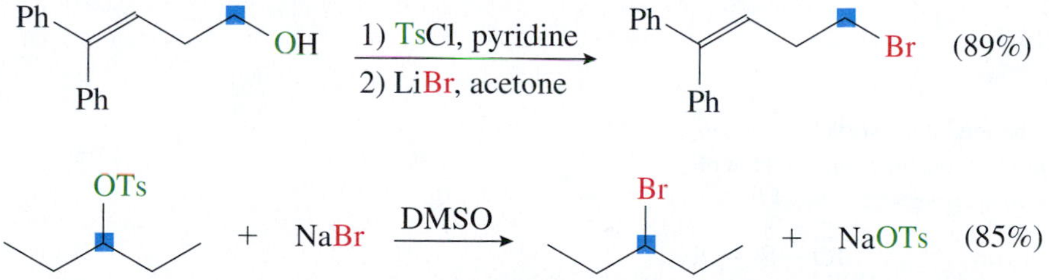

Ph⎯ ⎯OH $\xrightarrow[\text{2) LiBr, acetone}]{\text{1) TsCl, pyridine}}$ Ph⎯ ⎯Br   (89%)
Ph   Ph

OTs   Br

⎯⎯ + NaBr $\xrightarrow{\text{DMSO}}$ ⎯⎯ + NaOTs (85%)

The sulfonate ester method requires two steps for the conversion of an alcohol into an alkyl chloride. A reagent that can accomplish this transformation in one step is thionyl chloride, SOCl$_2$. In a reaction very similar to the formation of sulfonate esters, this reagent replaces the hydrogen of the alcohol with a group that makes the oxygen a weaker base and a better leaving group. However, this intermediate is not isolated. Instead, it reacts immediately with the nucleophilic chloride ion that is generated during its formation. (The mechanism may be S_N1 or S_N2, depending on the structure of the compound.) The leaving group then decomposes to sulfur dioxide and chloride ion. The overall process is outlined in Figure 10.3. As shown in the following equations, this procedure results in the formation of alkyl chlorides in good yields. The by-products are SO_2 and HCl, both gases, which makes isolation of the alkyl halide easier.

$$CH_3(CH_2)_4CH_2\!\!-\!\!OH \ + \ SOCl_2 \ \longrightarrow \ CH_3(CH_2)_4CH_2\!\!-\!\!Cl \ + \ SO_2\,(g) \ + \ HCl\,(g)$$

$$(61\%)$$

$$\text{[cyclic structure]} + SOCl_2 \longrightarrow \text{[cyclic structure]} + SO_2\,(g) + HCl\,(g)$$

$$(75\%)$$

The reagents PBr_3 and PI_3 can be used to convert alcohols to alkyl bromides and alkyl iodides in one step. The reactions are very similar to those described for thionyl chloride. First, the oxygen of the alcohol attacks the phosphorus, replacing a halogen

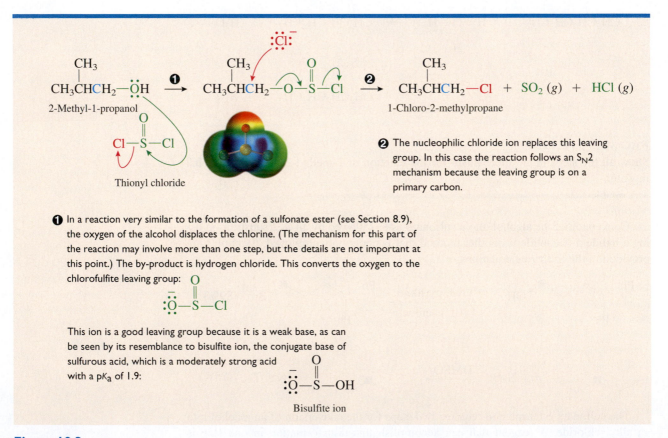

① In a reaction very similar to the formation of a sulfonate ester (see Section 8.9), the oxygen of the alcohol displaces the chlorine. (The mechanism for this part of the reaction may involve more than one step, but the details are not important at this point.) The by-product is hydrogen chloride. This converts the oxygen to the chlorofulfite leaving group:

This ion is a good leaving group because it is a weak base, as can be seen by its resemblance to bisulfite ion, the conjugate base of sulfurous acid, which is a moderately strong acid with a pK_a of 1.9:

Bisulfite ion

② The nucleophilic chloride ion replaces this leaving group. In this case the reaction follows an S_N2 mechanism because the leaving group is on a primary carbon.

Figure 10.3

MECHANISM OF THE REACTION OF AN ALCOHOL WITH THIONYL CHLORIDE TO PRODUCE AN ALKYL CHLORIDE.

and making the oxygen a better leaving group. Then the halide ion replaces this leaving group to produce the alkyl halide product. Several examples are provided in the following equations:

$$CH_3CH_2CH_2CH_2-OH \xrightarrow{PBr_3} CH_3CH_2CH_2CH_2-Br \quad (93\%)$$

$$CH_3CH_2\overset{\overset{\displaystyle OH}{|}}{C}HCH_2CH_3 \xrightarrow{PBr_3} CH_3CH_2\overset{\overset{\displaystyle Br}{|}}{C}HCH_2CH_3 \quad (90\%)$$

$$CH_3CH_2CH_2-OH \xrightarrow{PI_3} CH_3CH_2CH_2-I \quad (90\%)$$

A final method sometimes employed to prepare alkyl halides uses an S_N2 reaction with one halogen as the leaving group and a different halide ion as the nucleophile, as shown in the following general equation:

$$:\!X:^- + R-\!X: \rightleftharpoons R-\!X: + :\!X:^-$$

However, this is an equilibrium reaction; the product can react with the displaced halide ion and reform the starting material. If the reaction is to be useful in synthesis, some method must be found to favor the product at equilibrium. If acetone is used as the solvent, the reaction of sodium iodide with alkyl chlorides or bromides can be used to prepare alkyl iodides. In this case the equilibrium favors the alkyl iodide because sodium chloride and sodium bromide (but not sodium iodide) are insoluble in acetone and precipitate, thus driving the equilibrium to the right according to Le Chatelier's principle.

$$CH_3\overset{\overset{\displaystyle Br}{|}}{C}HCH_2C\equiv N + NaI \xrightarrow{acetone} CH_3\overset{\overset{\displaystyle I}{|}}{C}HCH_2C\equiv N + NaBr\,(s) \quad (96\%)$$

PROBLEM 10.16
Show the products of these reactions:

a) 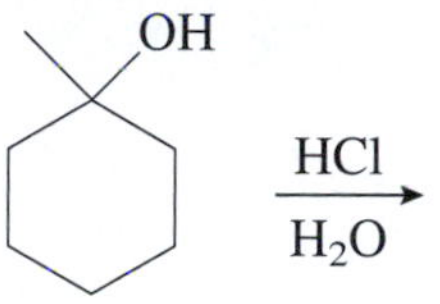 HCl / H₂O

b) 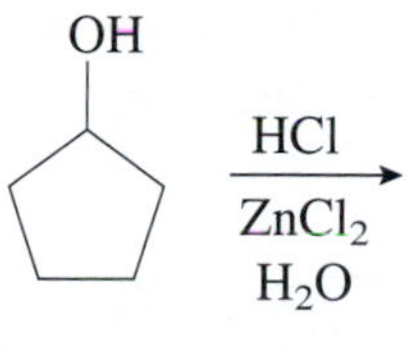 HCl / ZnCl₂ / H₂O

c) 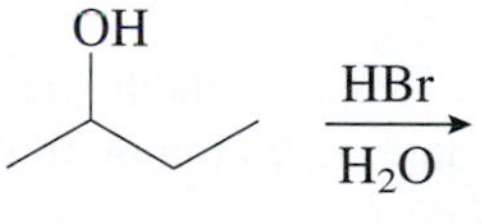 HBr / H₂O

d) $CH_3CH_2CH_2OH \xrightarrow{HI, H_2O}$

e) 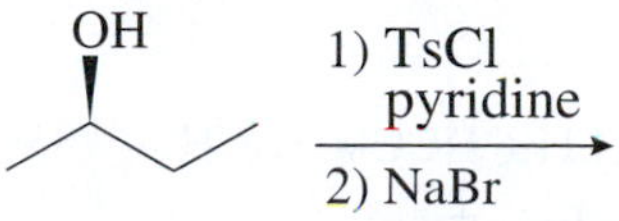 1) TsCl pyridine 2) NaBr

f) 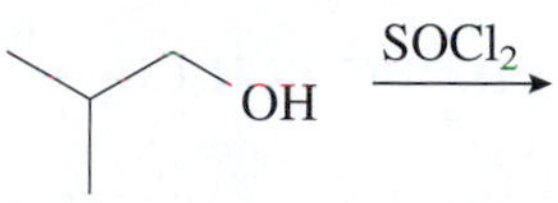 SOCl₂

g)

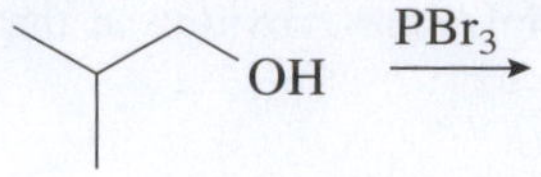

$$\xrightarrow{\text{PBr}_3}$$

h)

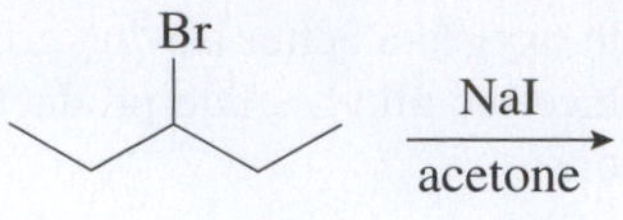

$$\xrightarrow[\text{acetone}]{\text{NaI}}$$

i)

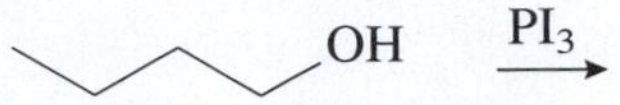

$$\xrightarrow{\text{PI}_3}$$

j)

$$\xrightarrow{\text{SOCl}_2}$$

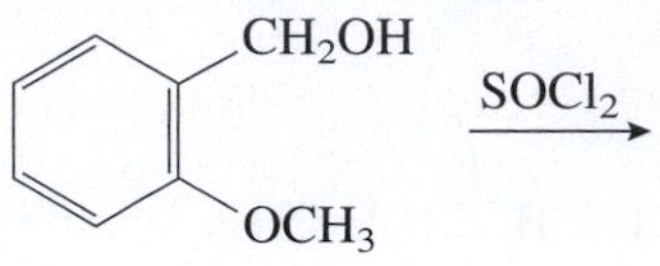

PROBLEM 10.17

Suggest reagents that could be used to prepare these alkyl halides from alcohols:

a)

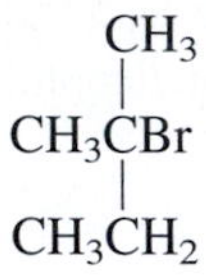

$$\begin{array}{c} \text{CH}_3 \\ | \\ \text{CH}_3\text{CBr} \\ | \\ \text{CH}_3\text{CH}_2 \end{array}$$

b)

$$\begin{array}{c} \text{CH}_3 \\ | \\ \text{CH}_3\text{CCl} \\ | \\ \text{Ph} \end{array}$$

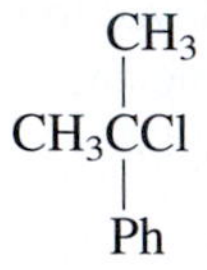

c)

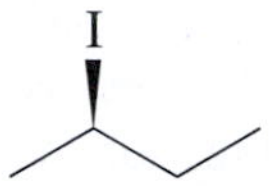

d)

e)

f)

10.6   PREPARATION OF AMINES

Ammonia and unhindered amines are good nucleophiles. Therefore, it would appear that amines should be readily prepared by reacting these nucleophiles with the appropriate alkyl halide or sulfonate ester in an S_N2 reaction, according to the following general equations:

$$\ddot{\text{N}}\text{H}_3 \;+\; \text{R}-\text{L} \;\longrightarrow\; \text{R}-\overset{+}{\text{N}}\text{H}_3 \;+\; :\!\bar{\text{L}}$$

$$\text{R'}\ddot{\text{N}}\text{H}_2 \;+\; \text{R}-\text{L} \;\longrightarrow\; \text{R}-\overset{+}{\text{N}}\text{H}_2\text{R'} \;+\; :\!\bar{\text{L}}$$

$$\text{R'}_2\ddot{\text{N}}\text{H} \;+\; \text{R}-\text{L} \;\longrightarrow\; \text{R}-\overset{+}{\text{N}}\text{HR'}_2 \;+\; :\!\bar{\text{L}}$$

$$\text{R'}_3\ddot{\text{N}} \;+\; \text{R}-\text{L} \;\longrightarrow\; \text{R}-\overset{+}{\text{N}}\text{R'}_3 \;+\; :\!\bar{\text{L}}$$

As illustrated in the following reaction, this method provides acceptable yields of tertiary amines, using secondary amines as nucleophiles. Quaternary ammonium salts can also be prepared from tertiary amines as nucleophiles.

$$+ \;\; \text{CH}_3\text{CHCO}_2\text{CH}_2\text{CH}_3 \;\xrightarrow{\text{benzene}}\; \text{CH}_3\text{CHCO}_2\text{CH}_2\text{CH}_3 \quad (85\%)$$

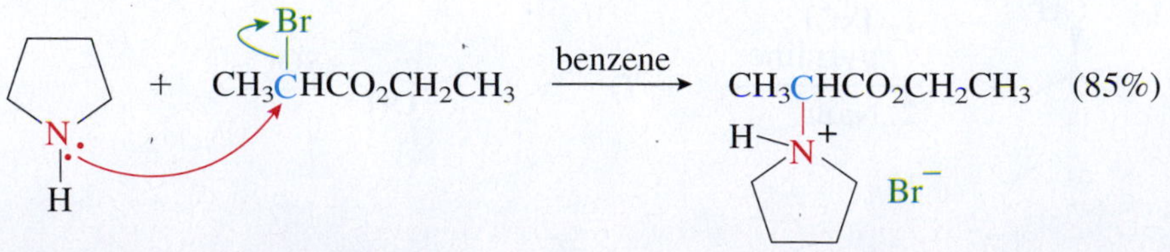

However, this reaction is much less useful when ammonia is the nucleophile because the initial product, a primary amine, is a stronger base and a stronger nucleophile than is ammonia. Therefore, the primary amine preferentially reacts as the nucleophile, producing a secondary amine as a by-product (see Figure 10.4). This problem is termed **multiple alkylation** because more than one alkyl group becomes attached to the nucleophile. Even when a large excess of ammonia is used to favor its reaction as the nucleophile, a significant amount of secondary amine is often formed. For similar reasons the use of a primary amine as the nucleophile results in the formation of a tertiary amine in addition to the desired secondary amine. (However, because of steric effects, a tertiary amine is not a stronger nucleophile than a secondary amine, so multiple alkylation is not a problem when a secondary amine is used as the nucleophile.)

PROBLEM 10.18

Show the products of these reactions:

a) $(CH_3CH_2)_2NH$ + CH_3CH_2Br $\xrightarrow{CH_3OH}$

b) $CH_3CH_2\overset{\displaystyle |}{\underset{\displaystyle CH_3}{N}}CH_3$ + CH_3I $\xrightarrow{\text{ether}}$

---

❷ In an equilibrium process, the resulting ammonium salt can lose a proton to a base such as ammonia to produce a primary amine. The primary amine is a stronger base and, therefore, a better nucleophile than ammonia.

❸ Even when a large excess of ammonia is present, some of the primary amine reacts to produce a secondary amine.

❶ Ammonia acts as the nucleophile in an S_N2 reaction, replacing the bromine.

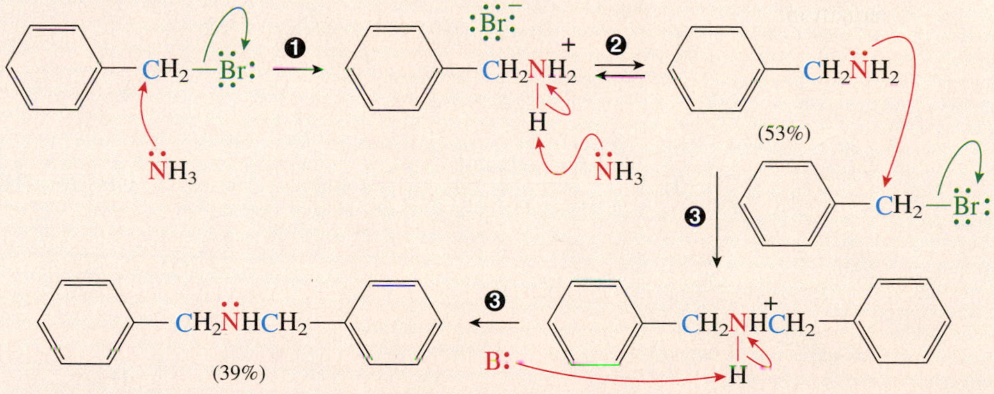

In this particular case, in which an eightfold excess of ammonia is used, the product mixture consists of 53% of the primary amine and 39% of the secondary amine.

Figure 10.4

MECHANISM OF MULTIPLE ALKYLATION USING AMMONIA AS A NUCLEOPHILE.

Because of the problem of multiple alkylation when ammonia reacts with alkyl halides, a multistep method, called the Gabriel synthesis, has been developed to prepare primary amines. This procedure resembles the acetate method for preparing alcohols (Section 10.2) in that carbonyl groups are attached to the nitrogen to decrease its reactivity. After the substitution has been accomplished, the carbonyl groups are removed to provide the desired primary amine. In the Gabriel synthesis the synthetic equivalent for ammonia is phthalimide. (An imide has two carbonyl groups bonded to the nitrogen.) The electrons on the nitrogen of phthalimide are not very basic or nucleophilic because of resonance involving both carbonyl groups:

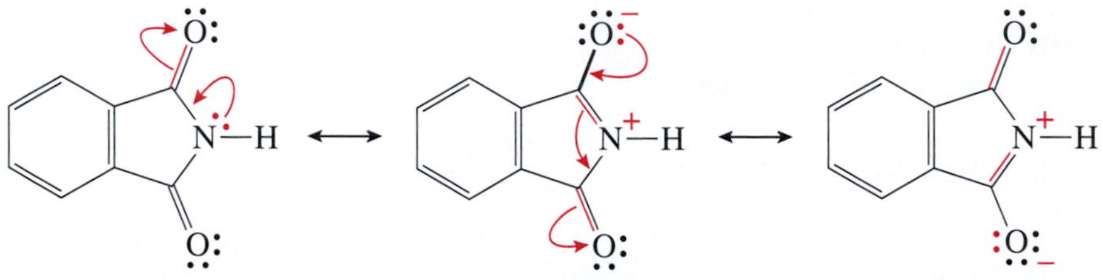

Phthalimide

Therefore, the proton on the nitrogen must be removed to use this nitrogen as a nucleophile. This hydrogen is relatively acidic (pK_a = 9.9) because of resonance stabilization of the conjugate base, similar to that shown for phthalimide. Hydroxide ion is a strong enough base to remove this proton and generate the conjugate base of phthalimide. The reaction of this nucleophile with an alkyl halide or an alkyl sulfonate ester, by an S_N2 mechanism, produces a substituted phthalimide with an alkyl group bonded to the nitrogen. The electrons on the nitrogen of this alkylated phthalimide are not nucleophilic, so there is no danger of multiple alkylation. The carbonyl groups are then removed to give the desired primary amine in a reaction that is very similar to the ester hydrolysis described in Figure 10.1. The process is outlined in Figure 10.5 and an additional example is provided by the following equation:

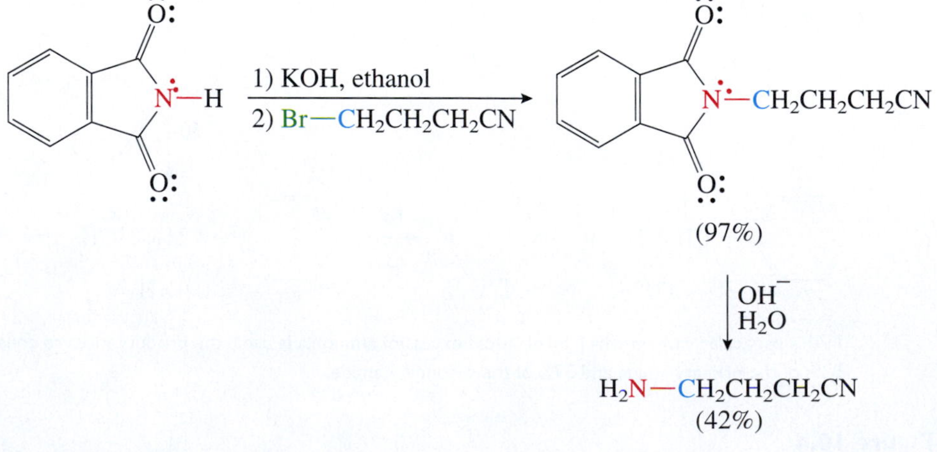

❶ Because of resonance, the electron pair on the nitrogen of phthalimide is not very basic or nucleophilic. The hydrogen on the nitrogen is much more acidic than a hydrogen of a normal amine because the conjugate base is stabilized by resonance. It is acidic enough to be removed completely by a base such as hydroxide ion.

❷ The resulting phthalimide anion is a good nucleophile in the S_N2 reaction.

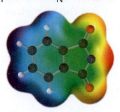

Figure 10.5

THE GABRIEL SYNTHESIS OF A PRIMARY AMINE.

Phthalimide

1-Bromopropane

Propylamine

❸ The desired amine is generated by reaction of the phthalimide with an aqueous base. The conditions are very similar to those of the ester cleavage shown in Figure 10.1. Again, the mechanism begins by attack of a hydroxide ion nucleophile at the electrophilic carbonyl carbon, ultimately breaking the bond between the carbonyl carbon and the nitrogen. (The mechanism for this reaction is covered in detail in Chapter 19.) A similar reaction occurs at the other carbonyl carbon. Overall, phthalimide is a synthetic equivalent for ammonia.

Like phthalimide itself, the alkylated phthalimide is not nucleophilic, so there is no problem with multiple alkylation occurring. The next step of the process is to replace the carbonyl groups on the nitrogen with hydrogens.

Focus On Biological Chemistry

Biological Alkylations and Poisons

Many of the reagents that are routinely used as substrates for S_N2 reactions in the laboratory are poisonous and must be used with caution. These compounds have a leaving group bonded to an unhindered carbon, so they are very reactive toward nucleophiles. They are called *alkylating agents* because they attach an alkyl group to the nucleophile.

Iodomethane is a prime example of a reactive alkylating agent. Because of its lack of steric hindrance and excellent leaving group, it is very reactive toward nucleophiles.

Continued

It is a common reagent and often is the first choice when a chemist desires to attach a methyl group to a nucleophile. (Bromomethane and chloromethane might serve as well except that they are gases at room temperature and are therefore much more difficult to handle than liquid iodomethane.) Like the other reactive alkylating agents, iodomethane is poisonous because it reacts with nucleophiles, such as NH_2 and SH groups, in the organism, attaching a methyl group to them. Iodomethane can deactivate an enzyme and interfere with its biological function by alkylating a nucleophile at the active site and changing its nucleophilicity:

$$\text{(Enzyme)}-\ddot{N}H_2 \;+\; CH_3-I \longrightarrow \text{(Enzyme)}-\overset{+}{N}H_2-CH_3 \quad I^-$$

In addition, iodomethane and similar reagents can act as carcinogens by alkylating the nitrogens in the bases of DNA. This can change how the base hydrogen bonds, resulting in a mutation.

Benzyl chloride is a powerful lachrymator (tear gas) that is intensely irritating to the skin, eyes, and mucous membranes. (Recall from Section 8.5 that the phenyl group increases the reactivity of this compound toward the S_N2 mechanism by resonance stabilization of the transition state.) Chloroacetophenone is the active ingredient in mace and is used in tear gas. (Like the phenyl group of benzyl chloride, the carbonyl group of chloroacetophenone greatly increases its reactivity toward nucleophiles.)

Benzyl chloride   Chloroacetophenone

One of the most infamous reactive alkylating agents is mustard gas, which was used as a chemical warfare agent during World War I. The sulfur of mustard gas acts as an intramolecular nucleophile to generate a cyclic sulfonium ion that is even more reactive as an alkylating agent. Note that it has two reactive electrophilic sites in each molecule. In heavy doses it can cause blindness and death, but its delayed effects, including cough; respiratory damage; and reddening, itching, and blistering of the skin, are more insidious.

When we desire to attach a methyl group to a nucleophile in the laboratory, we often choose a simple reagent such as iodomethane. Living organisms cannot use this reagent because it is too reactive and too indiscriminate. Iodomethane will react with almost any nucleophile. Nature's iodomethane is a much more complex molecule called *S*-adenosylmethionine, or SAM. The leaving group in SAM is a disubstituted sulfur atom and confers just the right reactivity on the compound. SAM is used to methylate the nitrogen of norepinephrine in the biosynthesis of epinephrine (adrenaline) and also serves as the methylating agent in the biosynthesis of the important lipid phosphatidylcholine (lecithin) from phosphatidylethanolamine.

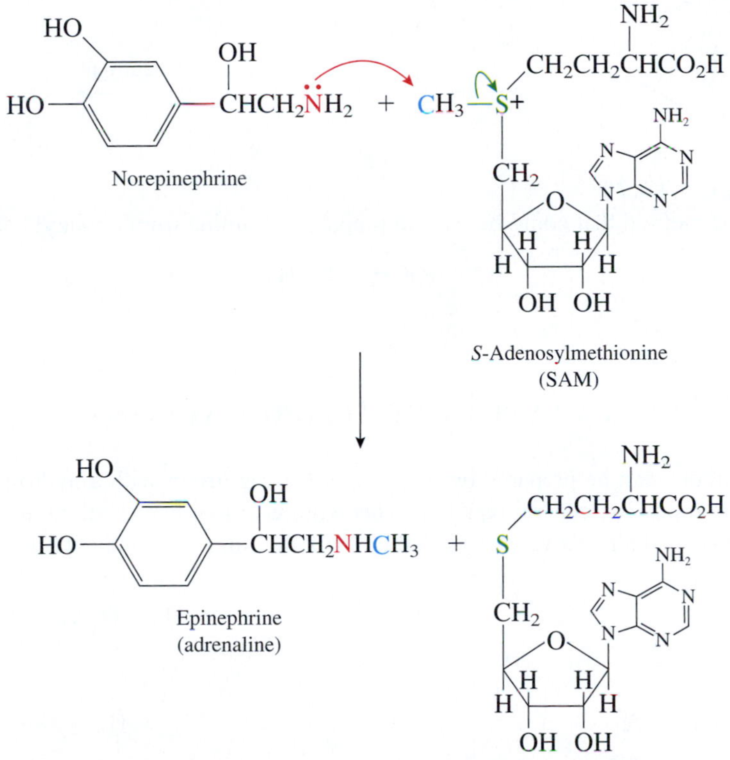

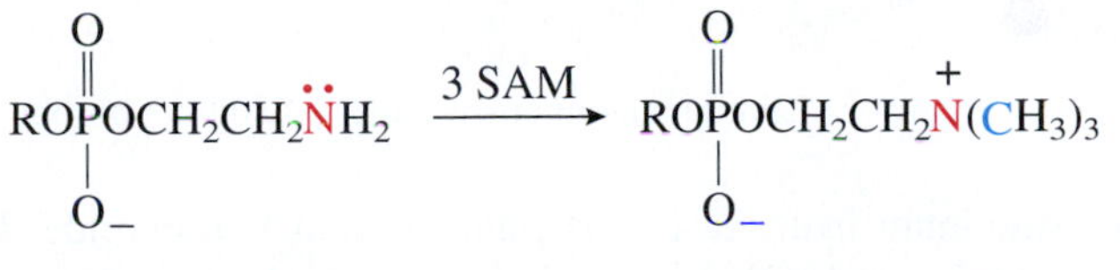

Norepinephrine

S-Adenosylmethionine
(SAM)

Epinephrine
(adrenaline)

$$ROPOCH_2CH_2\ddot{N}H_2 \xrightarrow{\text{3 SAM}} ROPOCH_2CH_2\overset{+}{N}(CH_3)_3$$

Phosphatidylethanolamine

Phosphatidylcholine

PROBLEM 10.19

Show the products of these reactions:

a)
$$\begin{array}{c} \text{phthalimide} \end{array} \xrightarrow[\text{2) CH}_3(\text{CH}_2)_4\text{CH}_2\text{Br}]{\text{1) KOH}} \quad \xrightarrow[\text{H}_2\text{O}]{\text{NaOH}}$$

b)
$$\begin{array}{c} \text{(S)-2-bromohexane} \end{array} + \begin{array}{c} \text{potassium phthalimide} \end{array} \longrightarrow \quad \xrightarrow[\text{H}_2\text{O}]{\text{NaOH}}$$

PROBLEM 10.20

Suggest a method that could be used to prepare this amine from an alkyl halide:

$$\text{PhCH}_2\text{CH}_2\text{NH}_2$$

10.7   PREPARATION OF HYDROCARBONS

Hydrocarbons can be prepared by replacing a leaving group with a hydrogen, according to the following general equation. This requires a hydrogen with an unshared pair of electrons and a negative charge, that is, **hydride ion,** as the nucleophile.

$$\text{H:}^- + \text{R—L} \longrightarrow \text{R—H} + \text{:L}^-$$
Hydride ion

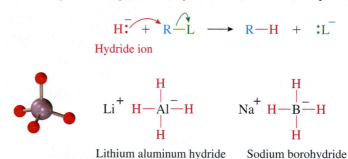

Li$^+$ H—Al—H   (Lithium aluminum hydride)

Na$^+$ H—B—H   (Sodium borohydride)

Both lithium aluminum hydride, LiAlH$_4$, and sodium borohydride, NaBH$_4$, react as though they contain a nucleophilic hydride ion, although the hydrogens are covalently bonded to the metal atoms, either aluminum or boron. However, hydrogen is more electronegative than either of these metals, resulting in each hydrogen having a partial negative charge. Because of this polarization, the compounds react as sources of hydride ion. Lithium aluminum hydride is a very reactive compound and reacts vigorously (often explosively) with even weakly acidic compounds such as water and alcohols. It must be used in inert solvents such as ethers. Sodium borohydride is much less reactive and is often used in alcohols or alkaline water as solvent. With either reagent the reactions have the usual S$_N$2 limitations; that is, they work well only when the leaving group is bonded to a primary or secondary carbon. The following equations provide examples of the use of these reagents to replace a leaving group with hydrogen:

$$CH_3(CH_2)_4\overset{Br}{\underset{|}{C}}HCH_3 + LiAlH_4 \xrightarrow{Et_2O} CH_3(CH_2)_4\overset{H}{\underset{|}{C}}HCH_3 \quad (92\%)$$

$$PhCH_2{-}Br + NaBH_4 \xrightarrow{(CH_3OCH_2CH_2)_2O} PhCH_2{-}H \quad (86\%)$$

PROBLEM 10.21
Show the products of these reactions:

a)

$\xrightarrow[\text{ether}]{LiAlH_4}$

b)

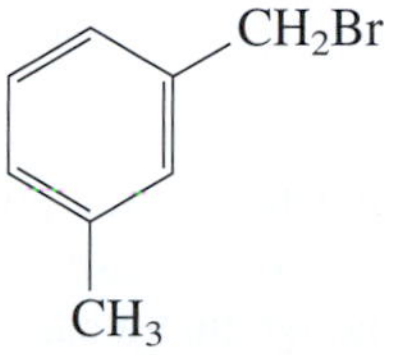

$\xrightarrow[\text{CH}_3\text{OH}]{NaBH_4}$

10.8 FORMATION OF CARBON–CARBON BONDS

Reactions that form carbon–carbon bonds are of great importance in organic synthesis because they enable smaller compounds to be converted to larger compounds. Forming these bonds by nucleophilic substitution reactions requires a carbon nucleophile—a **carbanion** (carbon anion), as shown in the following general equation:

$$R'{:}^- + R{-}L \longrightarrow R'{-}R + {:}L^-$$

Two useful carbon nucleophiles are introduced in this section. Other important carbon nucleophiles are discussed in later chapters, especially Chapter 20.

The first of these carbon nucleophiles, **cyanide ion,** is a moderate base and a good nucleophile:

$${:}\overset{-}{C}{\equiv}N{:}$$
Cyanide ion

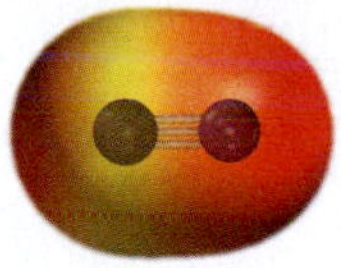

Cyanide ion reacts by the S_N2 mechanism and aprotic solvents are often employed to increase its reactivity. Yields of substitution products are excellent when the leaving group is attached to a primary carbon. Because of competing elimination reactions, yields are lower, but still acceptable, for secondary substrates. As expected for an S_N2 process, the reaction does not work with tertiary substrates. Substitution with cyanide ion adds one carbon to the compound while also providing a new functional group for additional synthetic manipulation. Some examples are given in the following equations:

$$CH_3CH_2CH_2CH_2{-}Cl + Na^+ {:}\overset{-}{C}{\equiv}N{:} \xrightarrow{DMSO} CH_3CH_2CH_2CH_2{-}C{\equiv}N{:} + NaCl \quad (92\%)$$

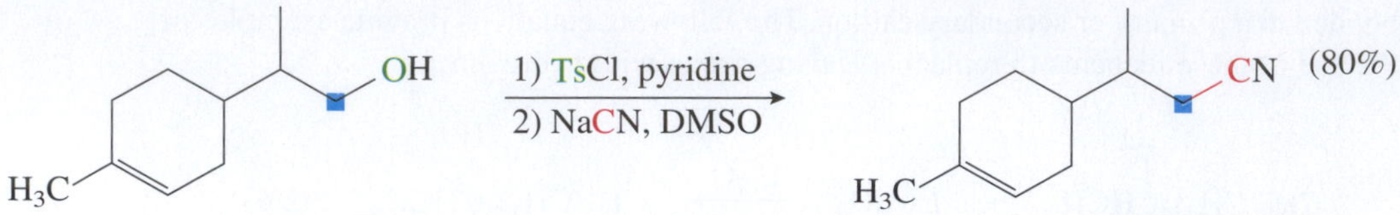

A second group of important carbon nucleophiles are the **acetylide anions**. These nucleophiles are generated by treating 1-alkynes with a very strong base, such as amide ion:

Amide ion

An acetylide anion

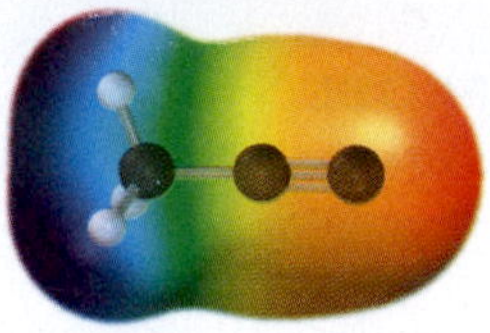

As discussed in Chapter 4, a proton on a carbon–carbon triple bond is relatively acidic ($pK_a = 25$) because of the sp hybridization of the carbon to which it is bonded. The proton is acidic enough that it can be removed with some of the strong bases that are available to the organic chemist. Usually, sodium amide ($NaNH_2$), the conjugate base of ammonia ($pK_a = 38$), often in liquid ammonia as the solvent, is used to remove the proton. Amide ion is a strong enough base so that the equilibrium in the above equation lies entirely to the right. (Note that carbanions generated by removing protons from sp^3- and sp^2-hybridized carbons are not generally available for use as nucleophiles in S_N2 reactions because the protons attached to them are not acidic enough to be removed in this manner.)

Because acetylide anions are strong nucleophiles, they react by the S_N2 mechanism. Good yields of substitution products are obtained only when the leaving group is attached to a primary carbon; secondary substrates give mainly elimination because the anion is also a strong base. Several examples are provided in the following equations. The last example shows how ethyne can be alkylated twice—both hydrogens can be replaced with alkyl groups in sequential steps!

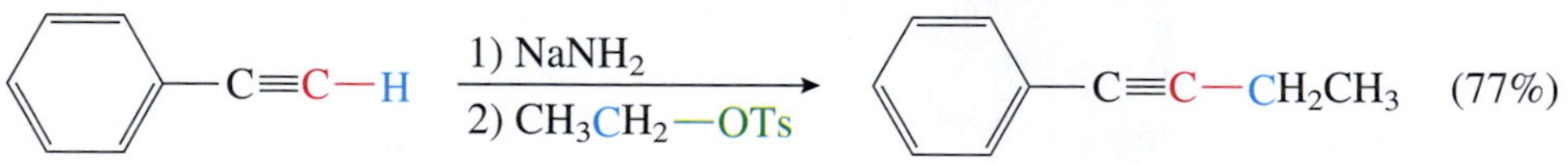

PROBLEM 10.22

Show the products of these reactions:

a) 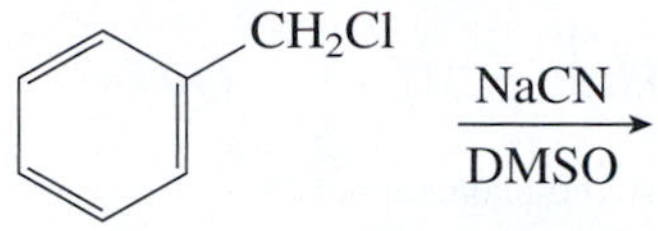 [benzyl chloride, CH₂Cl] $\xrightarrow[\text{DMSO}]{\text{NaCN}}$

b) $CH_3C\equiv C-H$ $\xrightarrow[\text{2) CH}_3\text{CH}_2\text{CH}_2\text{Br}]{\text{1) NaNH}_2, \text{NH}_3 \ (l)}$

c) 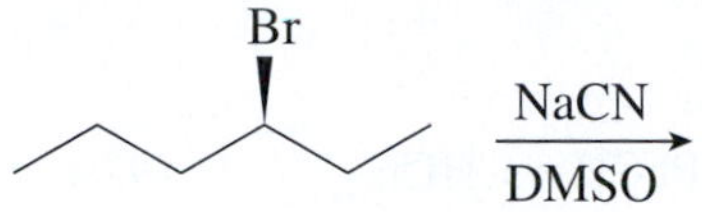 [2-bromohexane, Br] $\xrightarrow[\text{DMSO}]{\text{NaCN}}$

d) $HC\equiv CH$ $\xrightarrow[\text{2) CH}_3\text{CH}_2\text{Br}]{\text{1) NaNH}_2, \text{NH}_3 \ (l)}$ $\xrightarrow[\text{2) CH}_3\text{I}]{\text{1) NaNH}_2}$

e)  $Cl\diagup\diagdown\diagup Br$ + 1 NaCN $\xrightarrow{\text{DMF}}$

f) 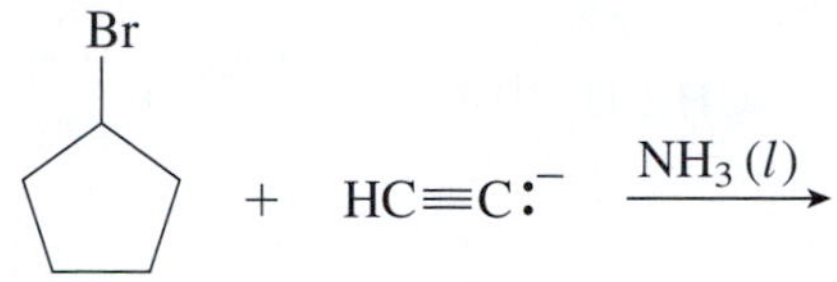 [bromocyclopentane, Br] + $HC\equiv C:^-$ $\xrightarrow{\text{NH}_3 \ (l)}$

PROBLEM 10.23

Suggest methods for preparing these compounds from alkyl halides:

a) 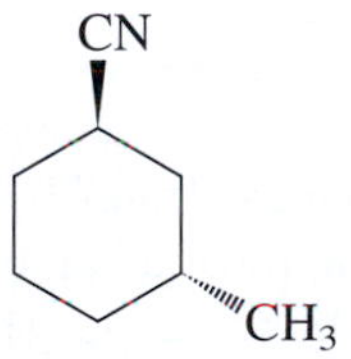 [cyclohexane with CN and CH₃ substituents]

b) $HC\equiv CCH_2CH_2\overset{\overset{\displaystyle CH_3}{|}}{C}HCH_3$

c) $CH_3C\equiv CCH_2Ph$

10.9 PHOSPHORUS AND SULFUR NUCLEOPHILES

Sulfur occurs directly beneath oxygen in the periodic table. Therefore, sulfur compounds are weaker bases but better nucleophiles than the corresponding oxygen compounds. Sulfur compounds are excellent nucleophiles in S_N2 reactions, and because they are relatively weak bases, elimination reactions are not usually a problem. Yields are good with primary and secondary substrates. For similar reasons, phosphorus compounds also give good yields when treated with primary and secondary substrates in S_N2 reactions. The following equations provide examples of the use of these nucleophiles:

$$CH_3-\ddot{\underset{\displaystyle ..}{S}}:^- \ + \ Cl-CH_2CH_2OH \xrightarrow{\text{ethanol}} CH_3-\ddot{\underset{\displaystyle ..}{S}}-CH_2CH_2OH \ + \ Cl^- \quad (80\%)$$

$$:\ddot{S}:^{2-} \quad + \quad 2\ CH_3CH_2CH_2\text{—}Br \quad \xrightarrow{\text{ethanol}} \quad CH_3CH_2CH_2\text{—}S\text{—}CH_2CH_2CH_3 \quad + \quad 2\ Br^- \quad (85\%)$$

$$Ph_3P: \quad + \quad CH_3\text{—}Br \quad \xrightarrow{\text{benzene}} \quad Ph_3\overset{+}{P}\text{—}CH_3 \ \ Br^- \quad (99\%)$$

Triphenylphosphine   A phosphonium salt

$$Ph_3P: \quad + \quad PhCH{=}CH\overset{\overset{\displaystyle Cl}{|}}{C}H_2 \quad \xrightarrow{\text{xylene}} \quad PhCH{=}CH\overset{\overset{\displaystyle \overset{+}{P}Ph_3}{|}}{C}H_2 \ \ Cl^- \quad (92\%)$$

Triphenylphosphine is probably the most important phosphorus nucleophile for organic chemists because it produces phosphonium salts (see the preceding two equations). These phosphonium salts are starting materials for an important preparation of alkenes that will be discussed in Chapter 18.

PROBLEM 10.24
Show the products of these reactions:

a) (structure) + PhS⁻ Na⁺ $\xrightarrow{CH_3OH}$ b) Ph_3P + $CH_3CH_2CH_2Br$ $\xrightarrow{\text{benzene}}$

c) $CH_3CH_2CH_2CH_2S^-$ + CH_3I $\xrightarrow{EtOH}$ d) $NaSCH_2CH_2SNa$ + $BrCH_2CH_2Br$ $\xrightarrow{CH_3OH}$

10.10   RING OPENING OF EPOXIDES

Section 8.9 discussed the generation of a leaving group, water, from an alcohol by protonation of the oxygen of the hydroxy group. In a similar fashion, protonation of the oxygen of an ether also generates a leaving group—an alcohol in this case—as shown in the following equation:

$$R\text{—}\ddot{O}\text{—}R' \quad \xrightarrow{H\text{—}A} \quad R\text{—}\overset{\overset{\displaystyle H}{|}}{\underset{+}{\ddot{O}}}\text{—}R' \quad \xrightarrow{:\bar{Nu}} \quad R\text{—}\ddot{O}\text{—}H \quad + \quad R'\text{—}Nu$$

This reaction requires more vigorous conditions than the reaction of alcohols, resulting in low yields in many cases. For this reason the reaction is not commonly used in synthesis.

An **epoxide** (also known as an oxirane) is a three-membered cyclic ether:

$$\underset{\text{An epoxide}}{H_2C\overset{\displaystyle \overset{O}{\diagup\;\diagdown}}{\text{—}}CH_2}$$

Like cyclopropane, epoxides have a large amount of ring strain and are much more reactive than normal ethers. Because of this ring strain, one carbon–oxygen bond of an epoxide can be broken in a nucleophilic substitution reaction. The following equation shows an example:

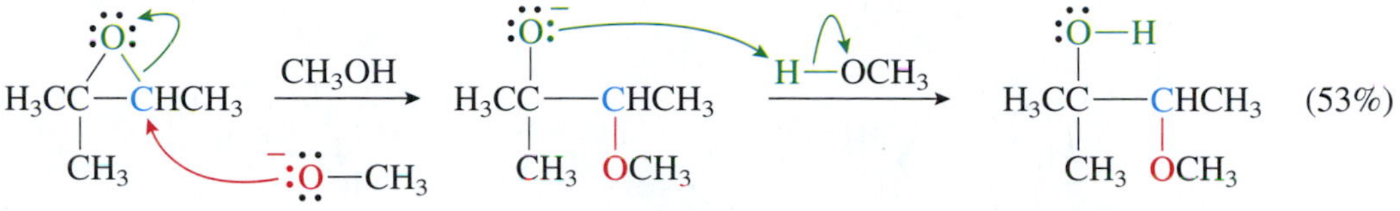

Both of the carbons of the epoxide ring are electrophilic, so at first glance, either might be expected to react with the nucleophile, methoxide ion. However, reactions of epoxides under basic or neutral conditions, as in this case, usually follow an S_N2 mechanism. Therefore, the nucleophile reacts at the less hindered secondary carbon, with inversion of configuration.

In the preceding reaction the leaving group (RO^-) is a very strong base. As discussed in Chapter 8, HO^- and RO^- are much too basic to act as leaving groups in normal nucleophilic substitution reactions. In the special case of epoxides, however, even RO^- can act as a leaving group because of the large amount of strain that is relieved when the carbon–oxygen bond is broken and the ring is opened.

Nucleophilic ring opening of epoxides can also be accomplished in acid solution. The oxygen is first protonated, making it a much better leaving group. Although these are typical S_N1 conditions, the actual mechanism is somewhere between S_N1 and S_N2—the reaction has characteristics of both mechanisms. The stereochemistry is that predicted for an S_N2 mechanism; the nucleophile approaches from the side opposite the leaving oxygen. The regiochemistry is that predicted for an S_N1 mechanism; the substitution occurs at the carbon that would be more stable as a carbocation. This often results in the carbon–oxygen bond that is broken under acidic conditions being different from the one that is broken under basic conditions, as can be seen by comparing the product in the following reaction with the one from the preceding reaction:

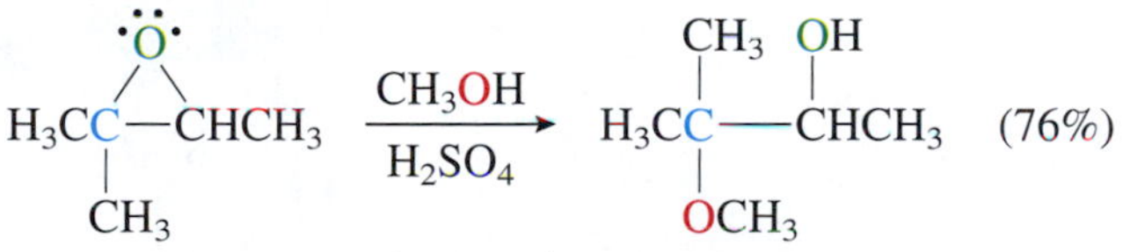

Another example, in which it can be seen that the reaction proceeds with inversion of configuration, is provided in the following reaction:

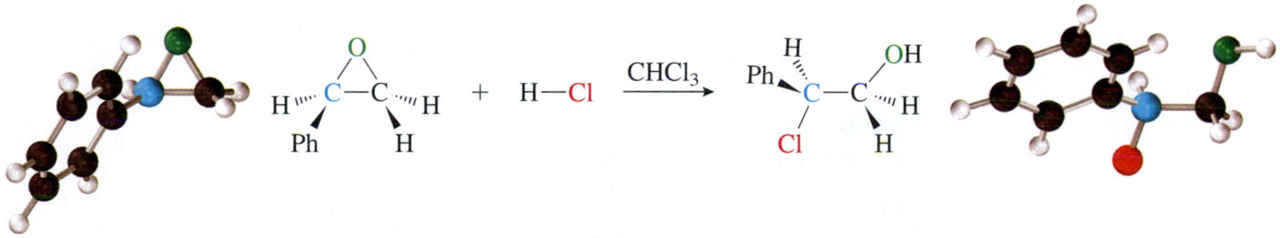

Because such reactions have features of both the S_N2 mechanism (stereochemistry) and the S_N1 mechanism (regiochemistry), they are said to follow a borderline S_N2 mechanism. The transition state geometry resembles that for an S_N2 reaction, but the bond to the leaving group is broken to a greater extent than the bond to the nucleophile is formed, resulting in considerable positive charge buildup on the carbon. Therefore, the transition state that has this positive charge buildup on the carbon that would be the more stable carbocation is favored. The two possible transition states for the preceding reaction are as follows:

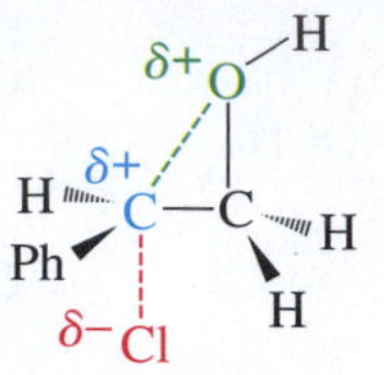

This transition state has a buildup of positive charge on the carbon attached to the phenyl group. The phenyl group helps stabilize the positive charge, making this transition state more stable. The reaction pathway resulting in the observed product proceeds through this transition state.

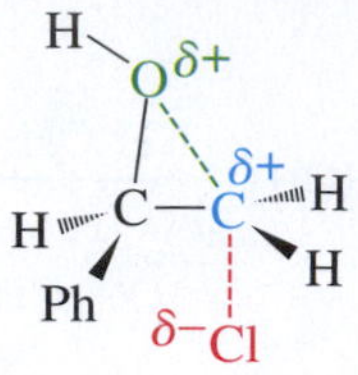

This transition state has a buildup of positive charge on the primary carbon, where it is less stable. As a result, no product is observed from this transition state.

Other examples of nucleophilic substitutions on epoxides are given in the following equations:

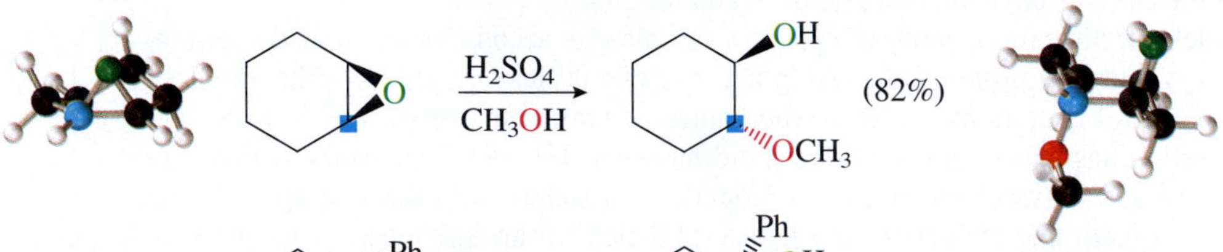

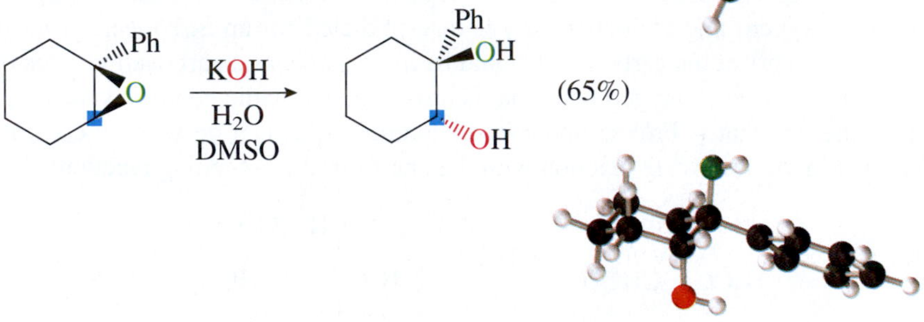

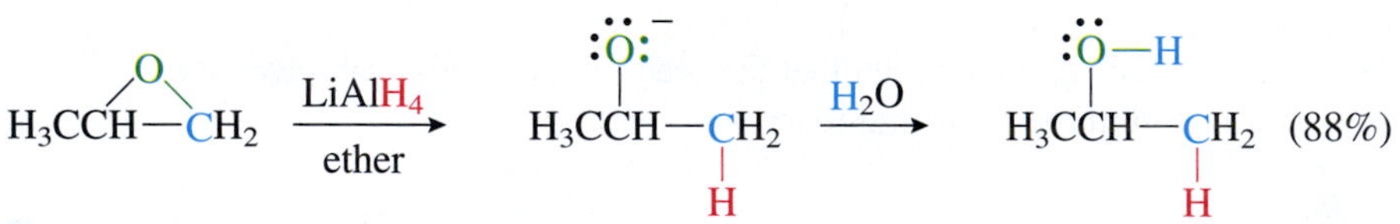

$$H_3CCH\overset{O}{\frown}CH_2 \xrightarrow[\text{ether}]{LiAlH_4} H_3CCH\overset{:\ddot{O}:^-}{-}CH_2 \xrightarrow{H_2O} H_3CCH\overset{:\ddot{O}-H}{-}CH_2 \ (88\%)$$

PROBLEM 10.25

Show the products of these reactions:

a) 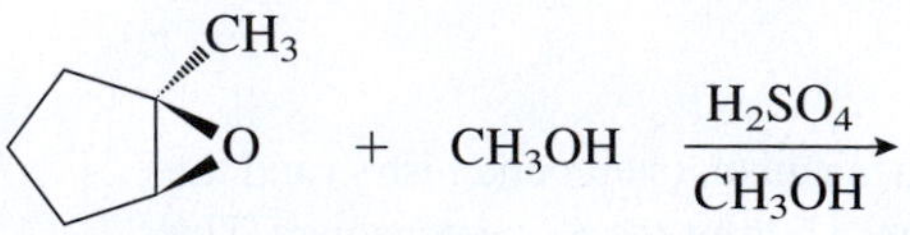 [cyclopentane epoxide with CH₃] + CH_3OH $\xrightarrow[CH_3OH]{H_2SO_4}$

b) 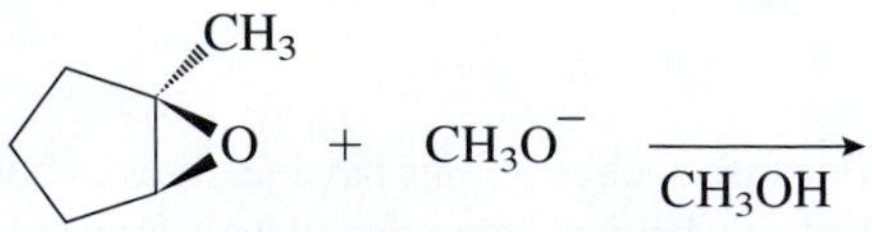 [cyclopentane epoxide with CH₃] + CH_3O^- $\xrightarrow{CH_3OH}$

c) 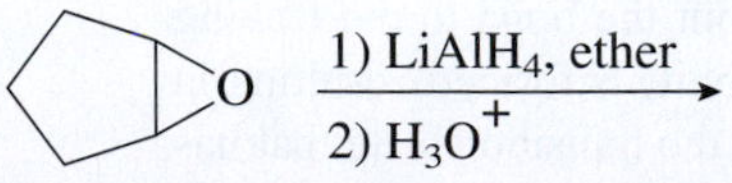 [cyclopentane epoxide] $\xrightarrow[\text{2) } H_3O^+]{\text{1) LiAlH}_4, \text{ ether}}$

Focus On

Uses of Epoxides in Industry

Epoxides are important intermediates in many industrial processes. For example, the reaction of the simplest epoxide, ethylene oxide, with water is employed to produce ethylene glycol, which is used in antifreeze and to prepare polymers such as Dacron. One method for the preparation of ethylene oxide employs an intramolecular nucleophilic substitution reaction of ethylene chlorohydrin:

| Ethylene chlorohydrin | Ethylene oxide | Ethylene glycol |

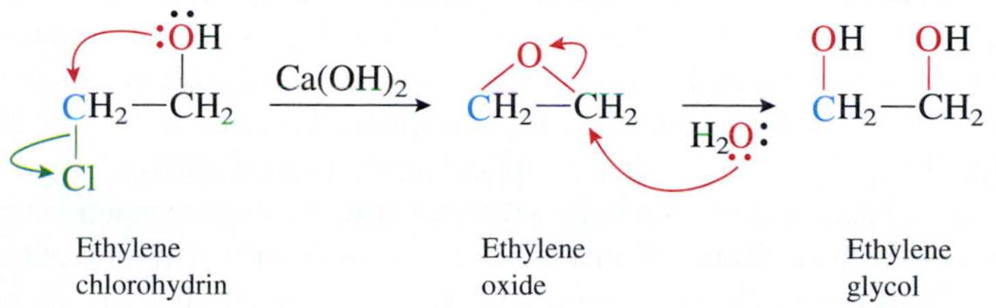

Nucleophilic cyanide ion can also be used to open the epoxide ring. This reaction was employed in a now obsolete pathway for the preparation of acrylonitrile, which is used to make Orlon:

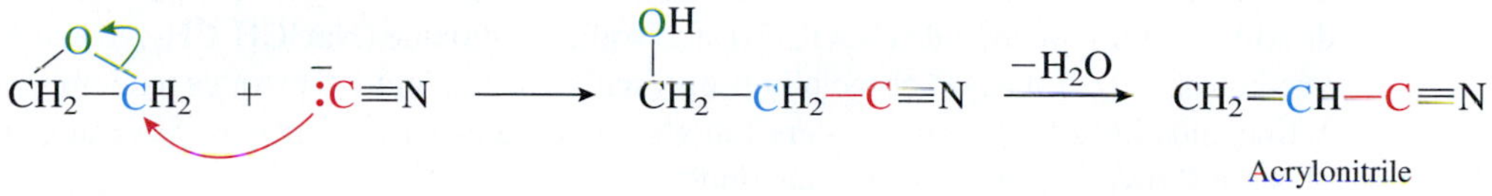

Acrylonitrile

Propranolol, a drug that is used to lower blood pressure, is prepared from the epoxide epichlorohydrin. First, the oxygen of 1-naphthol displaces the chlorine of epichlorohydrin in an S_N2 reaction. Then the epoxide ring is opened by the nucleophilic nitrogen of isopropylamine in another S_N2 reaction to form propranolol.

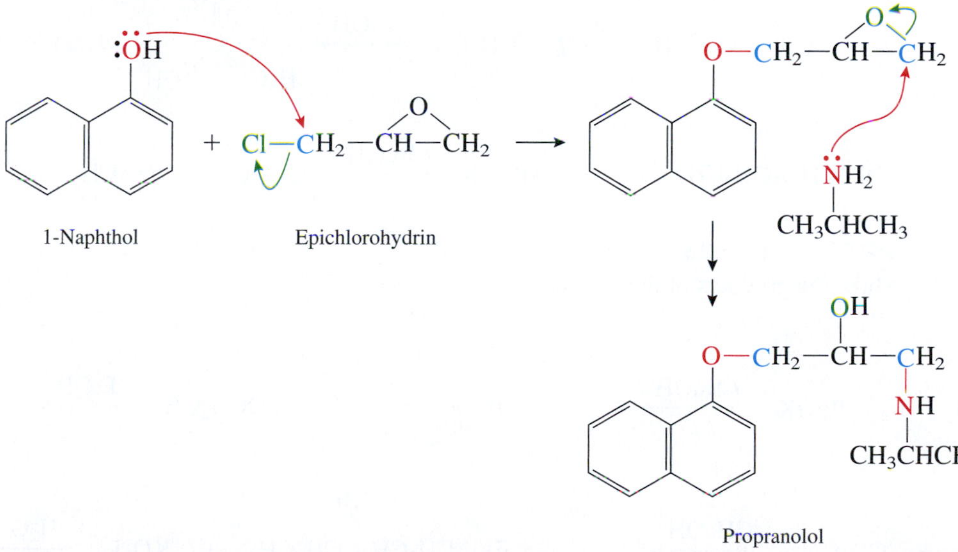

1-Naphthol   Epichlorohydrin

Propranolol

10.11   ELIMINATION OF HYDROGEN HALIDE (DEHYDRAHALOGENATION)

Elimination reactions are a useful method for the preparation of alkenes, provided that certain limitations are recognized. One problem is the competition between substitution and elimination. The majority of eliminations are done under conditions that favor the E2 mechanism. In these cases, steric hindrance can be used to slow the competing S_N2 pathway. Tertiary substrates and most secondary substrates give good yields of the elimination product when treated with strong bases. Sterically hindered bases can be employed with primary substrates to minimize substitution.

Another problem that occurs with eliminations is the regiochemistry of the reaction. As we saw in Chapter 9, most eliminations follow Zaitsev's rule and produce the more highly substituted alkene as the major product. However, a significant amount of the less highly substituted product is also formed. In addition, mixtures of cis and trans isomers are produced when possible, further complicating the product mixture. Because separating a mixture of such isomers is usually a difficult task, elimination reactions are often not the best way to prepare alkenes. (Other methods will be described in subsequent chapters.) However, if only one product can be formed, or if one is expected to greatly predominate in the reaction mixture, then these elimination reactions can be quite useful.

Thus, reaction of an alkyl halide with a strong base can be employed for the preparation of an alkene, provided that a mixture of isomers is not produced. The strong bases that are commonly used for these eliminations are sodium hydroxide, potassium hydroxide, sodium methoxide ($NaOCH_3$), and sodium ethoxide ($NaOCH_2CH_3$). Potassium *tert*-butoxide (*t*-BuOK) is especially useful with less hindered substrates to avoid competing substitution. Sulfonate esters can also be used as leaving groups. Several examples are shown in the following reactions:

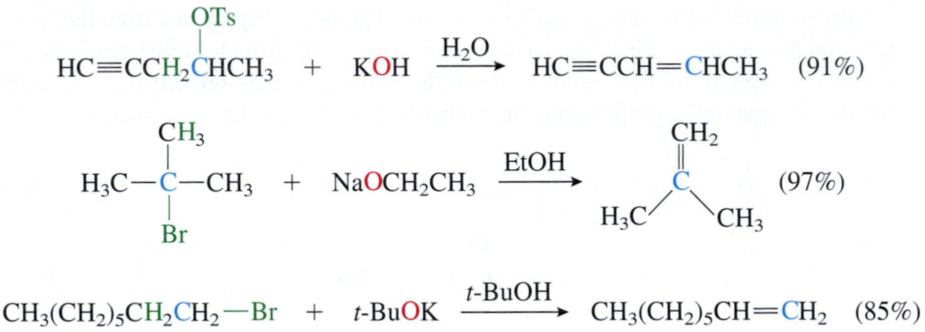

$$HC\equiv CCH_2\overset{\displaystyle OTs}{\underset{\displaystyle |}{C}}HCH_3 \ + \ KOH \ \xrightarrow{H_2O} \ HC\equiv CCH=CHCH_3 \quad (91\%)$$

$$H_3C-\overset{\displaystyle CH_3}{\underset{\displaystyle |}{\underset{\displaystyle |}{\overset{\displaystyle |}{C}}}}-CH_3 \ + \ NaOCH_2CH_3 \ \xrightarrow{EtOH} \ \overset{\displaystyle CH_2}{\underset{\displaystyle H_3C \diagup \ \diagdown CH_3}{\overset{\displaystyle \|}{C}}} \quad (97\%)$$

(Br below central carbon)

$$CH_3(CH_2)_5CH_2CH_2-Br \ + \ t\text{-BuOK} \ \xrightarrow{t\text{-BuOH}} \ CH_3(CH_2)_5CH=CH_2 \quad (85\%)$$

PROBLEM 10.26

Show the products of these reactions:

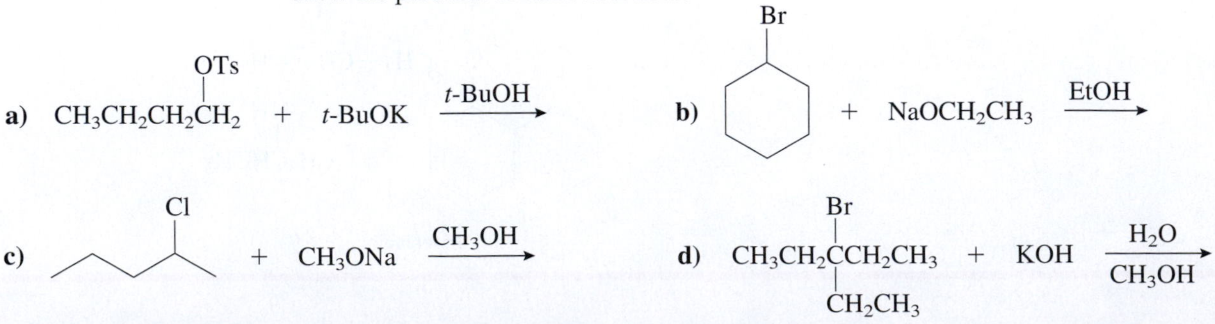

a) $CH_3CH_2CH_2\overset{\displaystyle OTs}{\underset{\displaystyle |}{C}}H_2$ + *t*-BuOK $\xrightarrow{t\text{-BuOH}}$

b) (cyclohexane ring with Br) + $NaOCH_2CH_3$ $\xrightarrow{EtOH}$

c) (structure with Cl) + CH_3ONa $\xrightarrow{CH_3OH}$

d) $CH_3CH_2\overset{\displaystyle Br}{\underset{\displaystyle |}{\underset{\displaystyle CH_2CH_3}{\overset{\displaystyle |}{C}}}}CH_2CH_3$ + KOH $\xrightarrow[CH_3OH]{H_2O}$

PROBLEM 10.27
Explain whether these elimination reactions would be a good way to prepare these alkenes:

a)

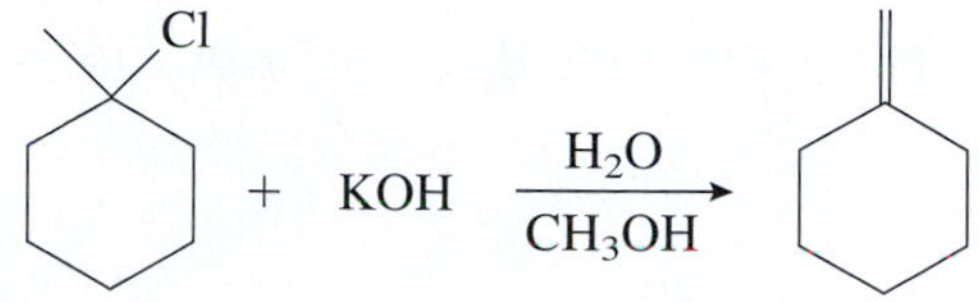

+ KOH $\xrightarrow[\text{CH}_3\text{OH}]{\text{H}_2\text{O}}$

b) PhCH$_2$CHCH$_3$ + NaOEt $\xrightarrow{\text{EtOH}}$ PhCH=CHCH$_3$
    with Cl on the CH

PROBLEM 10.28
Explain which of these reactions would provide a better synthesis of 2-pentene:

CH$_3$CH$_2$CHCH$_2$CH$_3$ + CH$_3$O$^-$ $\xrightarrow{\text{CH}_3\text{OH}}$ CH$_3$CH=CHCH$_2$CH$_3$
with Br on middle carbon

CH$_3$CHCH$_2$CH$_2$CH$_3$ + CH$_3$O$^-$ $\xrightarrow{\text{CH}_3\text{OH}}$ CH$_3$CH=CHCH$_2$CH$_3$
with Br on 2nd carbon

10.12   PREPARATION OF ALKYNES

Alkynes can be prepared from dihaloalkanes by elimination of two molecules of HX. This reaction requires very strongly basic conditions so potassium hydroxide at elevated temperatures or the stronger base sodium amide (NaNH$_2$) is commonly employed. Examples are provided by the following equations:

PhCH—CHPh + 2 KOH $\xrightarrow[\Delta]{\text{EtOH}}$ PhC≡CPh + 2 H—O—H + 2 KBr (68%)
with Br Br

KOH ↘   ↗ KOH

[PhCH=CPh]
     with Br

A vinyl halide

PhCH—CH$_2$ + 3 NaNH$_2$ $\xrightarrow{\text{NH}_3\,(l)}$ PhC≡C: $^-$ Na$^+$ + 3 H—NH$_2$ + 2 NaBr
with Br Br

$\downarrow$ H$_2$O

PhC≡CH   (50%)

In these reactions, elimination of the first molecule of HX results in the formation of a vinyl halide—an alkene with a halogen bonded to one of the carbons of the double bond. A second, more difficult elimination (this is why the strong base is necessary) produces the triple bond. Therefore, it is not surprising that vinyl halides can also be used to prepare alkynes, as shown in the following reactions:

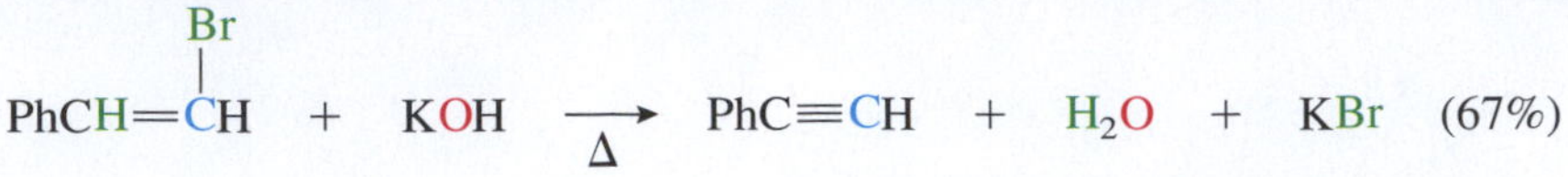

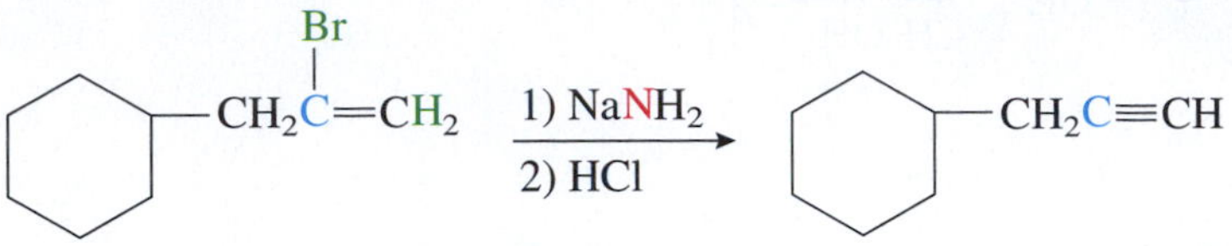

$$PhCH=\overset{\overset{\displaystyle Br}{|}}{C}H \ + \ KOH \ \xrightarrow{\Delta} \ PhC\equiv CH \ + \ H_2O \ + \ KBr \quad (67\%)$$

PROBLEM 10.29

Show the products of these reactions:

a)

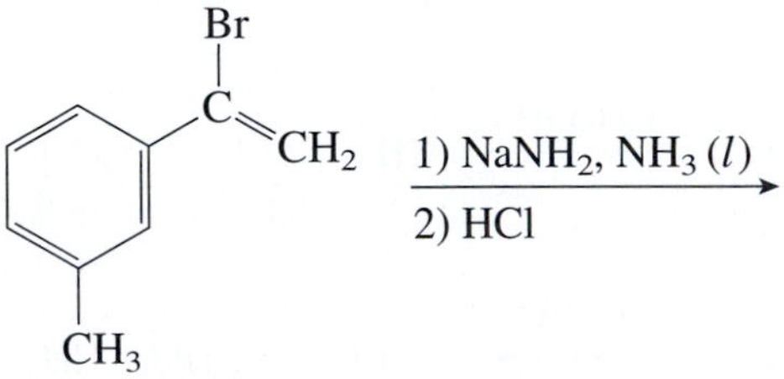

b)

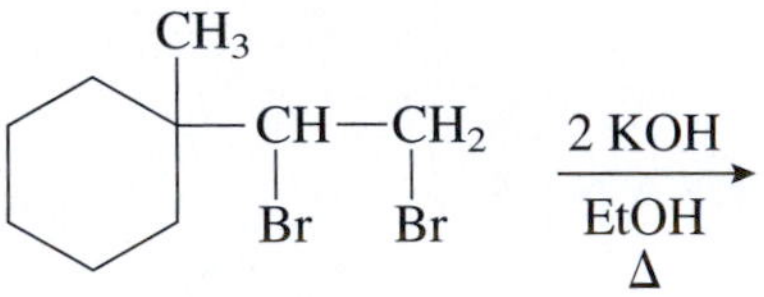

10.13 DEHYDRATION

Section 10.5 described the reaction of alcohols with the halogen acids, HX, to produce alkyl halides. If, instead of a halogen acid, a catalytic amount of sulfuric or phosphoric acid is used, the reaction takes a different pathway and an elimination product is formed. Because water is eliminated, the reaction is termed *dehydration*.

The mechanism for the dehydration of cyclohexanol to produce cyclohexene is shown in Figure 10.6. In general, these reactions follow the E1 mechanism, so tertiary alcohols are more reactive than secondary alcohols. (Note that this is one of the few cases in which the E1 mechanism is favored over the S_N1 mechanism.) At the carbocation stage, there is a competition between substitution and elimination. Under the conditions used for the dehydration reaction, elimination is favored, because there are no good nucleophiles present to cause substitution. The conjugate bases of sulfuric and phosphoric acids (HSO_4^- and $H_2PO_4^-$) are not very nucleophilic. Only a small amount of acid is needed because the reaction is acid catalyzed; that is, the acid is regenerated in the final step of the mechanism.

The dehydration reaction has some limitations. Because the mechanism is E1 and involves a carbocation, rearrangements are possible. Figure 10.7 shows an example of a dehydration involving a carbocation rearrangement. In addition, the reaction is not

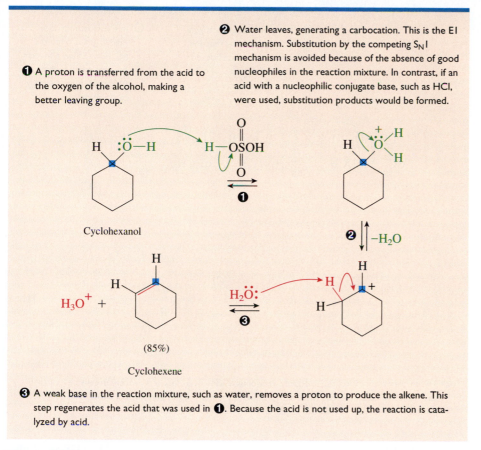

① A proton is transferred from the acid to the oxygen of the alcohol, making a better leaving group.

② Water leaves, generating a carbocation. This is the E1 mechanism. Substitution by the competing S_N1 mechanism is avoided because of the absence of good nucleophiles in the reaction mixture. In contrast, if an acid with a nucleophilic conjugate base, such as HCl, were used, substitution products would be formed.

Cyclohexanol

Cyclohexene

(85%)

③ A weak base in the reaction mixture, such as water, removes a proton to produce the alkene. This step regenerates the acid that was used in ①. Because the acid is not used up, the reaction is catalyzed by acid.

Figure 10.6

MECHANISM OF THE E1 DEHYDRATION OF CYCLOHEXANOL.

practical when isomeric products can be formed unless there is some factor that causes one product to greatly predominate. As long as these limitations are recognized, the reaction can be useful, as illustrated in the following examples:

(81%)

(88%)   (4%)

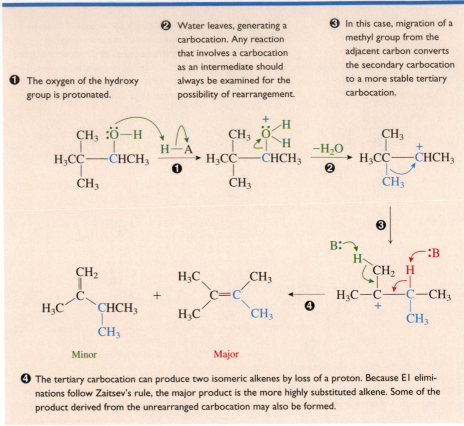

❷ Water leaves, generating a carbocation. Any reaction that involves a carbocation as an intermediate should always be examined for the possibility of rearrangement.

❸ In this case, migration of a methyl group from the adjacent carbon converts the secondary carbocation to a more stable tertiary carbocation.

❶ The oxygen of the hydroxy group is protonated.

Minor

Major

❹ The tertiary carbocation can produce two isomeric alkenes by loss of a proton. Because E1 eliminations follow Zaitsev's rule, the major product is the more highly substituted alkene. Some of the product derived from the unrearranged carbocation may also be formed.

Active Figure 10.7

ORGANIC
Chemistry ·⚛· Now™

MECHANISM OF AN E1 DEHYDRATION INVOLVING REARRANGEMENT. Test yourself on the concepts in this figure at **OrganicChemistryNow.**

PROBLEM 10.30
Show the products of these reactions:

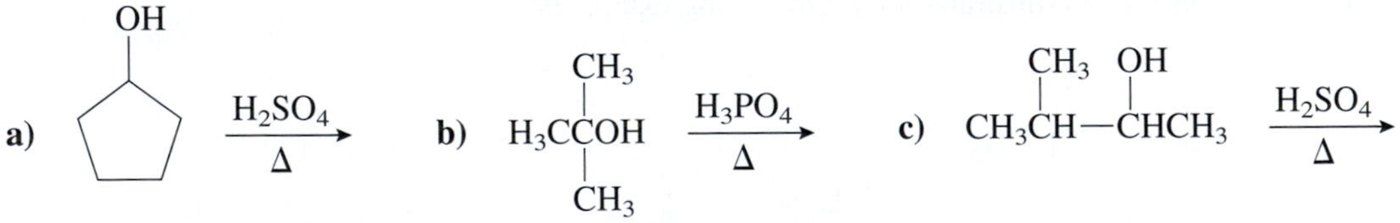

a) [cyclopentanol] $\xrightarrow[\Delta]{H_2SO_4}$

b) $H_3C\overset{\overset{\displaystyle CH_3}{|}}{\underset{\underset{\displaystyle CH_3}{|}}{C}}OH \xrightarrow[\Delta]{H_3PO_4}$

c) $CH_3\overset{\overset{\displaystyle CH_3}{|}}{CH}-\overset{\overset{\displaystyle OH}{|}}{CH}CH_3 \xrightarrow[\Delta]{H_2SO_4}$

10.14   ELIMINATIONS TO FORM CARBON–OXYGEN DOUBLE BONDS; OXIDATION REACTIONS

In beginning chemistry courses, oxidation is defined as a loss of electrons and reduction as a gain in electrons. To use these definitions with covalent compounds, oxidation states must be assigned to all the atoms. Although this can be done for organic compounds, we will use a simpler definition. In an organic chemist's vocabulary an **oxidation** is a reaction that results in an increase in oxygen content of the compound and/or a decrease in hydrogen content. Similarly, a **reduction** is a reaction that results in a de-

crease in oxygen content of the compound and/or an increase in hydrogen content. According to these definitions, the conversion of an alcohol to a carbonyl group

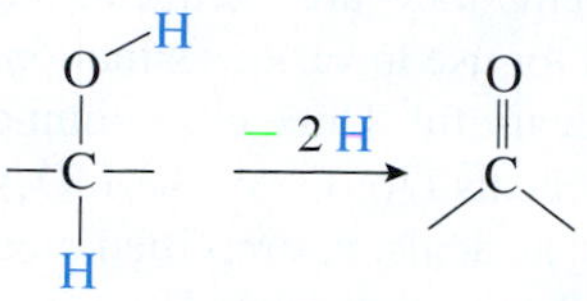

is an example of an oxidation reaction because two hydrogens are lost during the reaction. The reverse of this reaction, the addition of two hydrogens to the carbon–oxygen
double bond, is a reduction.

Oxidation of a primary alcohol produces an aldehyde. Further oxidation of an aldehyde to produce a carboxylic acid occurs readily. Therefore, if it is desired to stop the
reaction at the aldehyde stage, special reagents must be employed. Secondary alcohols
are oxidized to ketones. Tertiary alcohols are inert to most oxidizing reagents.

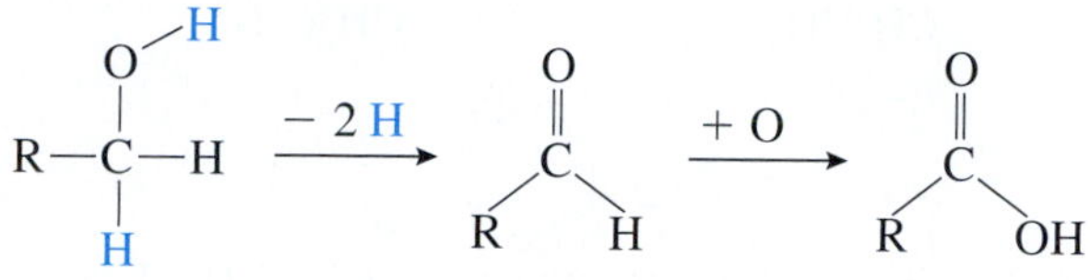

Primary   Aldehyde   Carboxylic
alcohol   acid

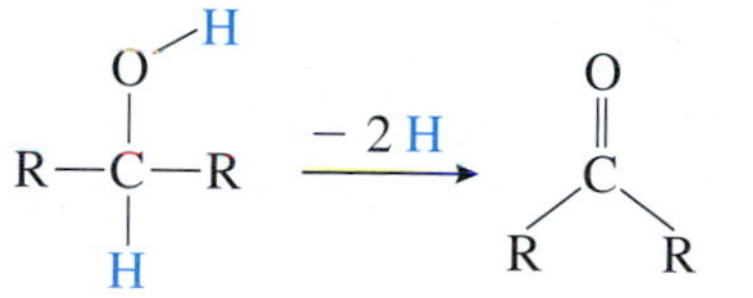

Secondary   Ketone
alcohol

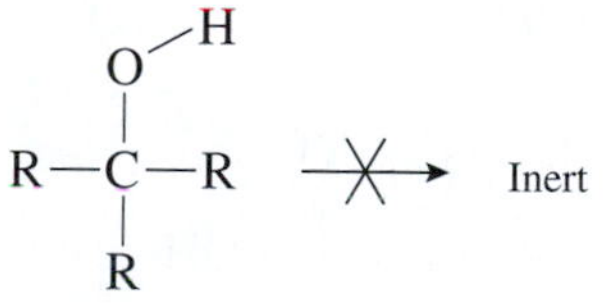

Tertiary
alcohol

To accomplish the preceding reactions, it is necessary to replace the hydrogen on
the oxygen with some group that can act as a leaving group—that is, a group that can
leave with the bonding pair of electrons. The following equation shows the similarity of
this process to the other eliminations presented in this chapter. The difference here is
that the "leaving group" is on an oxygen rather than a carbon.

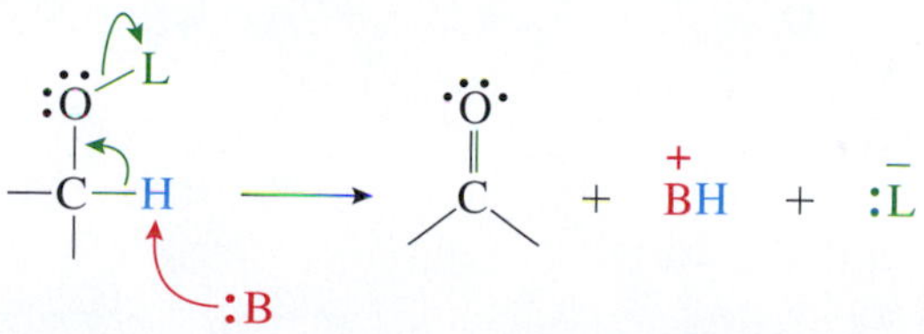

The species that are used as leaving groups in this reaction are most commonly metals in high oxidation states. When the metals leave, taking the electron pair of the metal–oxygen bond with them, they are "reduced." A large number of oxidation reagents have been developed for use in various situations. Ones based on chromium in the +6 oxidation state are very useful. Three chromium oxidation reagents are listed in Table 10.1 along with two reagents (Ag_2O and $KMnO_4$) that are effective for the oxidation of aldehydes to carboxylic acids. A simplified mechanism for the oxidation of 2-propanol to 2-propanone with chromic acid, H_2CrO_4, is illustrated in Figure 10.8. Examples of oxidations using these reagents are shown in the following equations:

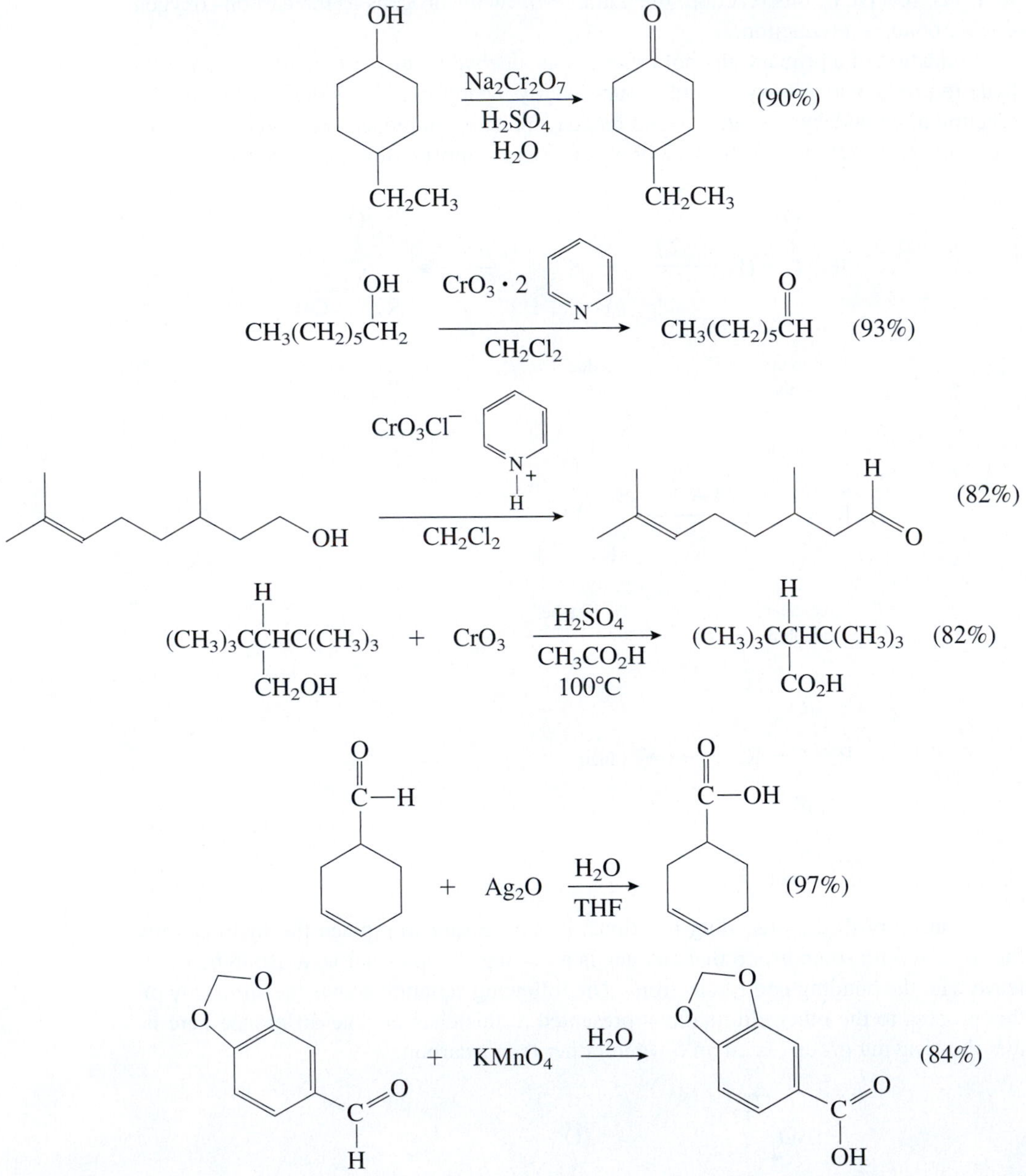

Table 10.1   Some Useful Oxidizing Reagents

Reagent	Comments
$Na_2Cr_2O_7$, $K_2Cr_2O_7$, or CrO_3 in H_2SO_4 and H_2O	Used for simple alcohols that can tolerate acidic conditions; good for oxidizing secondary alcohols to ketones; can be used to oxidize primary alcohols to carboxylic acids.
$CrO_3 \cdot 2$ [pyridine]	Chromium trioxide–pyridine complex is used when nonacidic conditions are needed; good for converting secondary alcohols to ketones or primary alcohols to aldehydes without overoxidation.
CrO_3Cl^- [pyridinium]	Pyridinium chlorochromate (PCC) is good for sensitive compounds; good for converting secondary alcohols to ketones or primary alcohols to aldehydes without overoxidation.
$KMnO_4$ or Ag_2O	Potassium permanganate and silver oxide are used for oxidation of aldehydes to carboxylic acids.

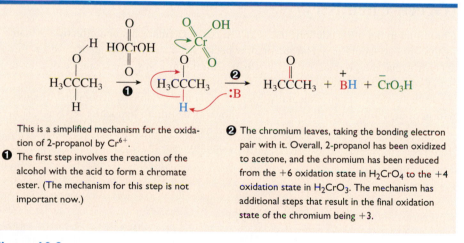

This is a simplified mechanism for the oxidation of 2-propanol by Cr^{6+}.

❶ The first step involves the reaction of the alcohol with the acid to form a chromate ester. (The mechanism for this step is not important now.)

❷ The chromium leaves, taking the bonding electron pair with it. Overall, 2-propanol has been oxidized to acetone, and the chromium has been reduced from the +6 oxidation state in H_2CrO_4 to the +4 oxidation state in H_2CrO_3. The mechanism has additional steps that result in the final oxidation state of the chromium being +3.

Figure 10.8

MECHANISM OF THE OXIDATION OF 2-PROPANOL TO 2-PROPANONE.

PROBLEM 10.31
Show the products of these reactions.

a) [structure: cyclooctanol] $\xrightarrow[\substack{H_2O \\ acetone}]{\substack{CrO_3 \\ H_2SO_4}}$

b) [structure: 2,4-dimethyl-3-pentenol] $\xrightarrow[CH_2Cl_2]{PCC}$

c)

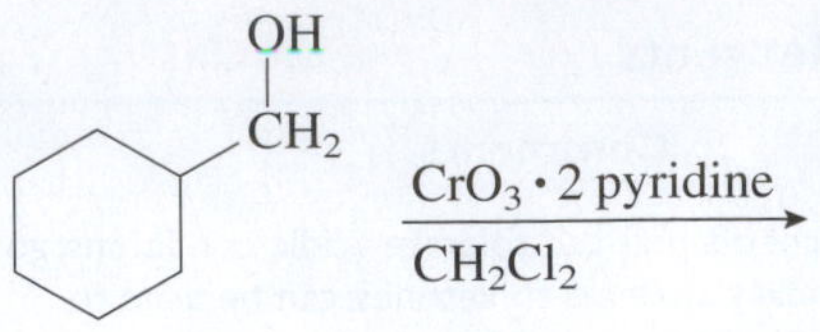

$$\underset{\text{CH}_2\text{Cl}_2}{\overset{\text{CrO}_3 \cdot 2 \text{ pyridine}}{\longrightarrow}}$$

d)

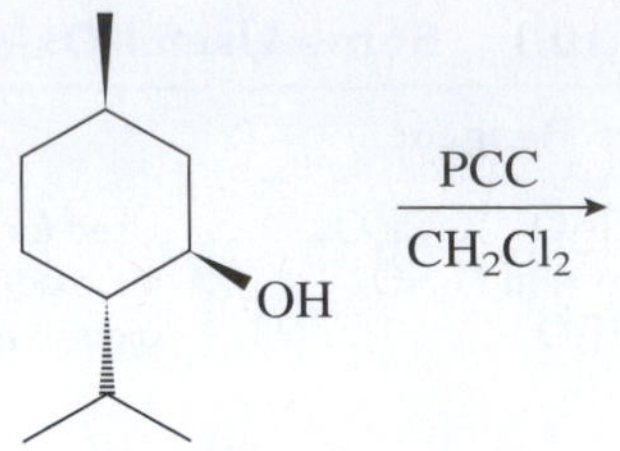

$$\underset{\text{CH}_2\text{Cl}_2}{\overset{\text{PCC}}{\longrightarrow}}$$

e)

$$\underset{\text{THF}}{\overset{\text{Ag}_2\text{O}}{\underset{\text{H}_2\text{O}}{\longrightarrow}}}$$

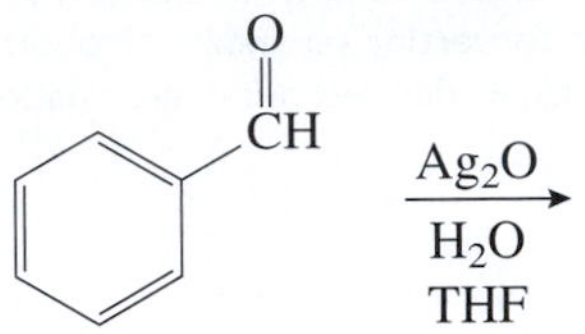

Focus On

Environmentally Friendly Chemistry (Green Chemistry)

Chromium in the +6 oxidation state, Cr(VI), is a very important and effective oxidant in the organic laboratory. The major drawback to the use of reagents based on this species is that the product, Cr(III), is toxic. Chromium is just one example of a toxic heavy metal that requires quite expensive disposal procedures.

Is there a more environmentally friendly reagent available to accomplish the oxidation of alcohols? Recently, it has been shown that sodium hypochlorite (NaOCl) in acidic solution is an excellent reagent for the oxidation of secondary alcohols to ketones. Examples are shown in the following equations:

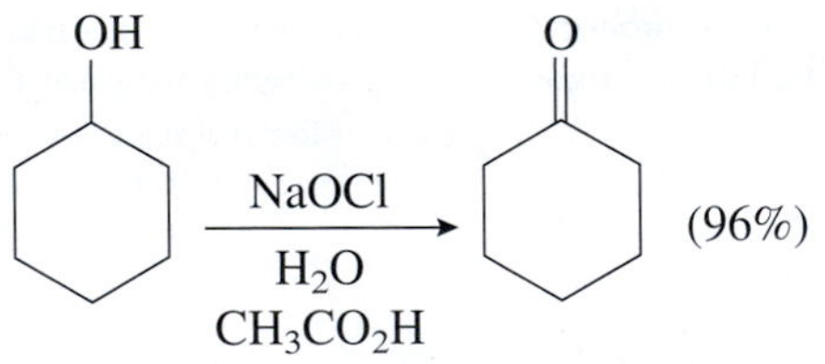

$$\underset{\underset{\text{CH}_3\text{CO}_2\text{H}}{\text{H}_2\text{O}}}{\overset{\text{NaOCl}}{\longrightarrow}} \quad (96\%)$$

Cyclohexanol   Cyclohexanone

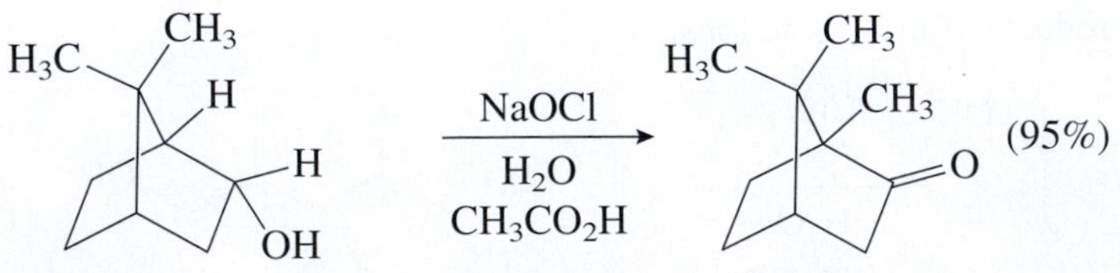

$$\underset{\underset{\text{CH}_3\text{CO}_2\text{H}}{\text{H}_2\text{O}}}{\overset{\text{NaOCl}}{\longrightarrow}} \quad (95\%)$$

Borneol   Camphor

Sodium hypochlorite is an inexpensive, environmentally benign reagent that is available in the form of "swimming pool chlorine," the material that is used to disinfect swimming pools. An even more convenient source is laundry bleach, a 5.25% solution of sodium hypochlorite that is available in most grocery stores.

The mechanism for this reaction is thought to involve initial chlorination of the hydroxy group to form an alkyl hypochlorite intermediate. An E2 elimination of HCl from this intermediate produces the ketone:

Cyclohexanol   An alkyl hypochlorite   Cyclohexanone

The development of environmentally safe reagents that can be used to replace more toxic materials in organic reactions is an area that deserves and is receiving considerable research attention, especially in industrial laboratories. All chemists need to be conscious of the effect on the environment of each reaction that they run.

10.15 THE STRATEGY OF ORGANIC SYNTHESIS

A common problem that an organic chemist faces in the laboratory is the lack of availability of a compound. Perhaps the compound is needed to test as a new pharmaceutical or to test a postulated reaction mechanism. If the compound is not available from a chemical supply house, the chemist is faced with the task of synthesizing it. The first step is to check the chemical literature to determine whether anyone else has ever prepared that compound. If the compound has never been prepared or if the reported preparation is difficult or of low yield, the chemist must design a new synthesis of the compound. How does an organic chemist approach such a problem?

The chemist does not begin by considering how to convert some compound that is available into the desired compound, the **target.** Instead, the question that is asked is "What reaction could I use to make the target compound from a simpler compound?" The simpler compound then becomes the new target compound, and the process is repeated until a commercially available compound is reached. Overall, many steps may be involved in the proposed synthesis. This process of working backward from the target compound is called **retrosynthetic analysis.** Often, several routes to the target compound can be envisioned. The best route depends on a number of factors, such as the number of steps, the yield of each step, and the overall cost. In fact, which route is "best" often depends on why the compound is needed. When a small amount of the compound is needed in a research laboratory, time is often the most important consideration. In contrast, cost is of utmost importance for a compound that is to be prepared on a large scale for commercial purposes.

Some examples will help clarify this process. Suppose the target is benzyl cyclopentyl ether:

Benzyl cyclopentyl ether

and we are asked to make it starting from alkyl halides. (Because you do not know which compounds can be purchased or are readily available, this book will specify the starting compound for the synthesis in some other fashion, such as here where the type of functional group is designated.)

First, we recall that ethers can be prepared by substitution reactions of alkoxide anion nucleophiles with alkyl halide electrophiles—the Williamson ether synthesis. The two ways to prepare the target ether are as follows:

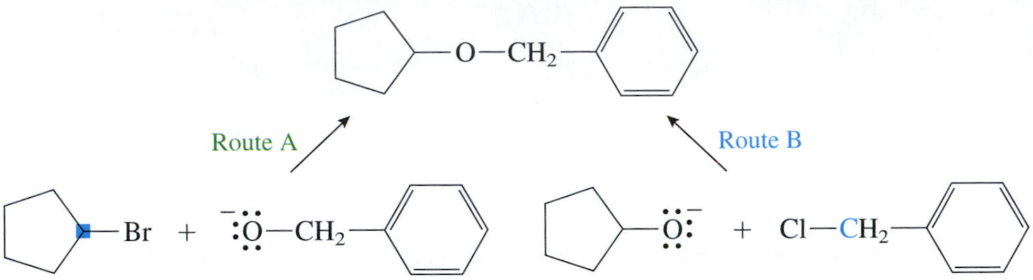

Next, we examine the two reactions to determine whether both are expected to give a good yield of the target compound. Because route A combines a strongly basic nucleophile and a secondary alkyl halide, we expect the major product to result from elimination by the E2 mechanism. Route B, on the other hand, employs a primary alkyl halide that cannot give elimination (it has no hydrogen on the β-carbon) and that is an excellent substrate for an S$_N$2 substitution because it is benzylic. Route B is the obvious choice.

Benzyl chloride is an acceptable starting material, because the problem has specified that we must start with alkyl halides. However, we must still prepare the alkoxide anion. This is the conjugate base of cyclopentanol and can be made by the reaction of the alcohol with sodium metal:

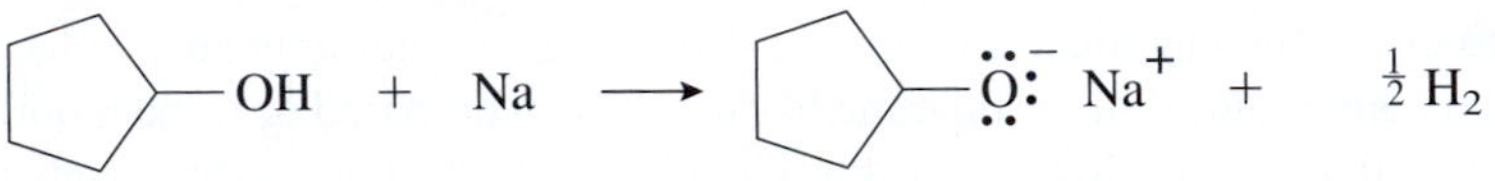

Cyclopentanol

The target is now cyclopentanol. Alcohols can be prepared from alkyl halides by reaction with hydroxide ion as the nucleophile. Again, however, the combination of a strongly basic nucleophile and a secondary alkyl halide will result in an unacceptable amount of elimination. A better plan is to treat bromocyclopentane with acetate ion in an aprotic solvent such as DMSO, followed by cleavage of the ester to cyclopentanol:

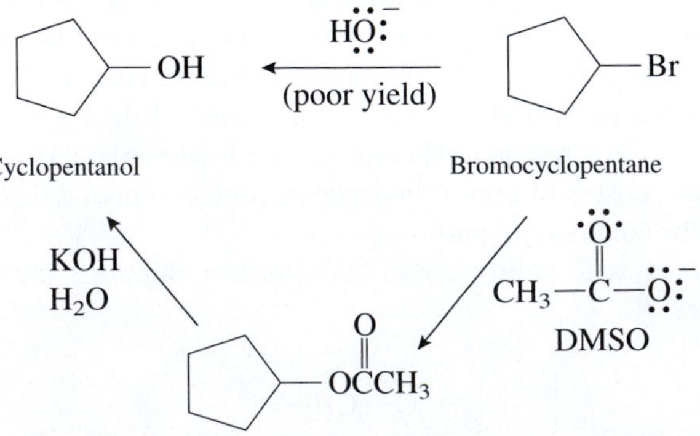

Written in the forward direction, our proposed synthesis of benzyl cyclopentyl ether is as follows:

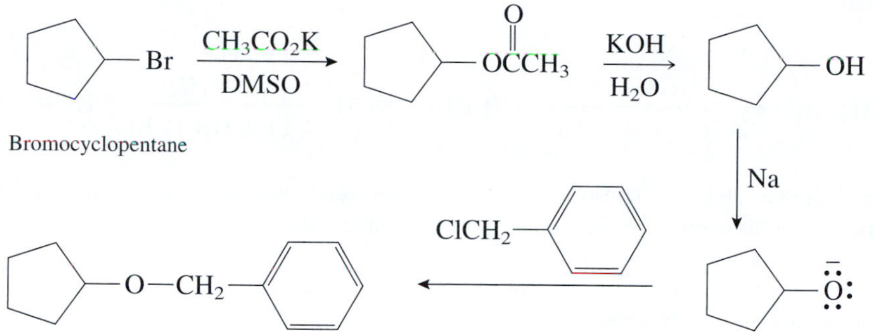

Bromocyclopentane

Benzyl cyclopentyl ether

Only a few reactions have been presented so far, so these synthesis problems are fairly easy. But as you learn additional reactions, the syntheses will become longer and more complex, and the value of using retrosynthetic analysis will be more apparent. Because synthesis problems such as this one require a somewhat different thought process than you have been using, they are an excellent way to determine whether you have a good command of the reactions.

PRACTICE PROBLEM 10.3

Show syntheses of these compounds from the indicated starting materials:

a) $CH_3CH_2C \equiv CCH_2CH_2CH_3$ from $HC \equiv CH$

b) 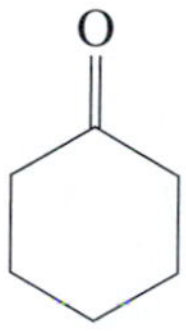 [structure] from an alkyl halide   c) 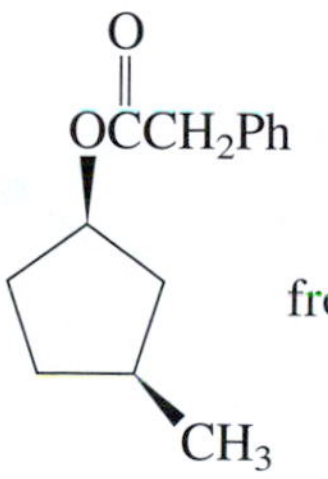 [structure] from [structure]

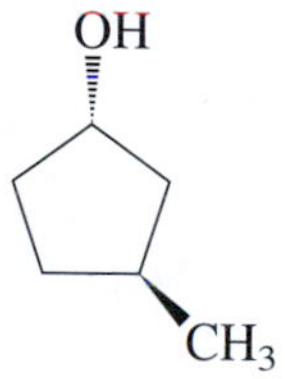

Strategy

Remember that it helps to use retrosynthetic analysis in synthesis problems. This means working backward to simpler and simpler compounds until an available compound is reached. These problems offer an additional clue in that the starting material is specified. In such cases it is often useful to identify which carbons in the target come from the carbons of the starting material. It is usually advisable to change these carbons as little as possible. It is also useful to identify which carbon–carbon bonds must be formed in the synthesis and how any functional groups need to be modified. In some cases the entire path will be apparent after this examination. In others it will be necessary to proceed backward one step at a time and repeat the examination at each step. Let's try some examples.

Solutions

a) $CH_3CH_2C\equiv CCH_2CH_2CH_3$ $\xleftarrow[\text{2) } CH_3CH_2CH_2I]{\begin{array}{c}\text{1) } NaNH_2\\ NH_3\,(l)\end{array}}$ $CH_3CH_2C\equiv CH$ $\xleftarrow[\text{2) } CH_3CH_2Br]{\begin{array}{c}\text{1) } NaNH_2\\ NH_3\,(l)\end{array}}$ $HC\equiv CH$
Ethyne

The initial target is a disubstituted alkyne. It is probably best that the two carbons of the triple bond of the starting material, ethyne, become the two carbons of the triple bond of the target. Thus, we need to form new carbon–carbon bonds at both the carbons of the triple bond. This suggests that we use an acetylide nucleophile and the appropriate alkyl group attached to a halogen leaving group in an S_N2 reaction. The order in which we add the alkyl groups does not make much difference. The reaction is accomplished by treating the 1-alkyne with sodium amide followed by the alkyl halide.

This process is repeated to put the other alkyl group on ethyne.

b)

Cyclohexanol

First, note that we do not have to change the carbon skeleton. We merely have to change the functional group from a bromide to a ketone. We do not know how to do this directly, but we do know how to make a ketone by the oxidation of a secondary alcohol.

The target is now cyclohexanol, a secondary alcohol. The reaction of a secondary alkyl halide with hydroxide ion gives an unacceptably high amount of elimination by the E2 mechanism. A better choice is to use acetate ion as the synthetic equivalent of hydroxide ion.

c)

Comparison of the target to the starting material shows that we need to substitute the ester group for the hydroxy group with inversion of configuration. So we need to convert the OH to a leaving group and do an S_N2 reaction. We can convert the OH to a tosylate or mesylate ester. Then do an S_N2 substitution using the carboxylate anion as the nucleophile in an aprotic solvent such as DMSO.

PROBLEM 10.32

Show syntheses of these compounds from the indicated starting materials.

a) $CH_3CH_2CH_2CH_2\overset{+}{N}(CH_3)_3 \ \ I^-$ from compounds with none of the CN bonds of the final product

b) $CH_3CH_2O\overset{\underset{\displaystyle |}{CH_3}}{C}HCH_2CH_3$ from alkyl halides

c) 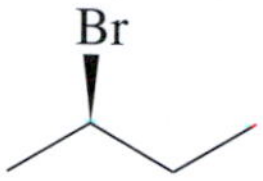 from

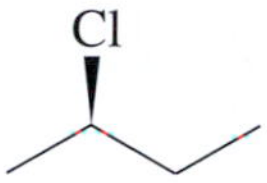

d) 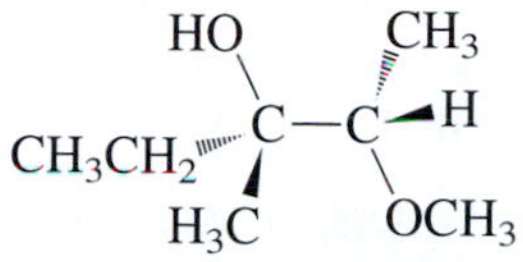   from an epoxide

e) $PhCH_2C{\equiv}CCH_3$ from $HC{\equiv}CH$

Review of Mastery Goals

After completing this chapter, you should be able to:

- Show the major product(s) of any of the reactions discussed in this chapter. (Problems 10.33, 10.36, and 10.40)

- Show the stereochemistry of the product(s). (Problems 10.33, 10.43, 10.55, 10.56, 10.57, 10.58, and 10.59)

- Write the mechanisms of these reactions. (Problems 10.42, 10.44, 10.46, 10.50, 10.51, 10.52, and 10.53)

- Synthesize compounds using these reactions. (Problems 10.34, 10.35, 10.37, 10.38, 10.39, 10.41, 10.47, and 10.54)

ORGANIC
Chemistry Now™
Click *Mastery Goal Quiz* to test how well you have met these goals.

Visual Summary of Key Reactions

This chapter has presented a large number of reactions. Yet most of them fall into one of two classes. A nucleophile or base reacts with a substrate containing a leaving group in either a substitution reaction, in which the nucleophile replaces the leaving group on the electrophilic carbon, or an elimination reaction, in which a double bond is formed between the electrophilic carbon and an adjacent carbon:

$$\underset{\overset{|}{\underset{|}{C}}-\overset{L}{\underset{|}{C}}}{\overset{H}{\underset{|}{C}}}\quad+\quad :Nu\quad\longrightarrow\quad \underset{\overset{|}{\underset{|}{C}}-\overset{Nu}{\underset{|}{C}}}{\overset{H}{\underset{|}{C}}}\quad\text{or}\quad \underset{}{C}=\underset{}{C}$$

It is very important to be able to recognize the nucleophile, the leaving group, and the electrophilic carbon. So, rather than just attempting to memorize each reaction on its own, you will have more success in learning these reactions if you take this approach:

1. Identify the leaving group (halides, sulfonate and related esters, or hydroxy groups under acidic conditions).

2. Identify the electrophilic carbon (the one bonded to the leaving group).

3. Identify the base or nucleophile (has an atom with an unshared pair of electrons).

4. Identify the mechanism by which the reaction proceeds (S_N1, S_N2, E1, or E2).

5. For substitutions, replace the leaving group with the nucleophile with inversion for S_N2 and racemization for S_N1.

6. For eliminations, form the carbon–carbon double bond according to Zaitsev's rule (except the Hofmann elimination) and use anti elimination to determine the stereochemistry of E2 reactions.

7. Consider the possibility of carbocation rearrangements for all S_N1 and E1 reactions.

The substitution reactions covered in this chapter are summarized in Table 10.2, and the elimination reactions are summarized in Table 10.3.

Table 10.2   Summary of Substitution Reactions

Reaction	Comments
Section 10.2 Preparation of Alcohols	
R—X + :Ö—H ⟶ R—OH	S_N2 conditions; strong base, so secondary substrates give significant elimination
R—X + H_2O ⟶ R—OH	S_N1 conditions (unless primary); hydrolysis (solvolysis)
R—X $\xrightarrow[\text{2) KOH, }H_2O]{\text{1) }CH_3CO_2^-}$ R—OH	S_N2 conditions; weak base, so fewer elimination problems with secondary substrates; synthetic equivalent of hydroxide ion
Section 10.3 Preparation of Ethers	
R—L + :Ö—R' ⟶ R—O—R'	S_N2 conditions; Williamson ether synthesis; strong base, so secondary substrates give mostly elimination; often two routes to same product
ROH + Na ⟶ R—Ö:⁻ Na⁺	Preparation of alkoxide nucleophile and base
PhOH + K_2CO_3 ⟶ Ph—Ö:⁻ K⁺	Preparation of phenoxide nucleophile
R—L + HOR' ⟶ R—O—R'	S_N1 conditions (unless primary); solvolysis
Section 10.4 Preparation of Esters	
R—L + :Ö—CR'(=O) ⟶ R—Ö—CR'(=O)	S_N2 conditions; weak base, so good yields with secondary substrates
Section 10.5 Preparation of Alkyl Halides	
R—OH + HCl ⟶ R—Cl or HBr ⟶ R—Br or HI ⟶ R—I	S_N1 conditions (unless primary); watch for rearrangements; HCl requires $ZnCl_2$ for primary and secondary alcohols
R—OH $\xrightarrow[\text{2) :X:}^-]{\text{1) TsCl}}$ R—X	S_N2 conditions
R—OH + $SOCl_2$ ⟶ R—Cl or PBr_3 ⟶ R—Br or PI_3 ⟶ R—I	S_N1 or S_N2; thionyl chloride, phosphorus tribromide, and phosphorus triiodide provide a good one-step procedure for conversion of an alcohol to an alkyl chloride, bromide, or iodide
R—X: + :X:⁻ ⟶ R—X: + :X:⁻	S_N2 conditions; must drive equilibrium

Continued

Table 10.2   Summary of Substitution Reactions—cont'd

Reaction	Comments
Section 10.6 Preparation of Amines	
$R-L + :NH_3 \longrightarrow R-\overset{+}{N}H_3$	S_N2 conditions; problem with multiple alkylation; acceptable yields for tertiary amines and quaternary ammonium salts
$R-L \xrightarrow[\text{2) KOH, H}_2\text{O}]{\text{1) phthalimide anion}} R-NH_2$	S_N2 conditions; Gabriel synthesis; synthetic equivalent of ammonia; yields acceptable with secondary substrates
Section 10.7 Preparation of Hydrocarbons	
$R-L \xrightarrow{\text{LiAlH}_4 \text{ or NaBH}_4} R-H$	S_N2 conditions; lithium aluminum hydride and sodium borohydride give acceptable yields with primary and secondary substrates
Section 10.8 Formation of Carbon–Carbon Bonds	
$R-L + :C\equiv N: \longrightarrow R-CN$	S_N2 conditions; cyanide ion is a weak base, so yields are acceptable with secondary substrates
$R-L + :C\equiv C-R' \longrightarrow R-C\equiv C-R'$	S_N2 conditions; acetylide anion is a strong base, so yields are satisfactory with primary substrates only
$H-C\equiv C-R' + :NH_2^- \longrightarrow :C\equiv C-R'$	Amide anion is commonly used as the base to prepare acetylide anions
Section 10.9 Phosphorus and Sulfur Nucleophiles	
$R-L + :\overset{-}{S}-H \longrightarrow R-S-H$ or $:\overset{-}{S}-R'$ $R-S-R'$	S_N2 conditions; sulfides are weak bases but good nucleophiles; good yields with primary and secondary substrates
$R-L + :PPh_3 \longrightarrow R-\overset{+}{P}Ph_3$	S_N2 conditions; phosphines are weak bases but good nucleophiles; good yields with primary and secondary substrates
Section 10.10 Ring Opening of Epoxides	
epoxide $+ :Nu \longrightarrow$ $\overset{OH}{\underset{Nu}{\vert\quad\vert}}$	In base, nucleophile bonds to less hindered carbon with inversion (S_N2); in acid, nucleophile bonds to more substituted carbon with inversion (borderline mechanism)

Table 10.3   Summary of Elimination Reactions

Reaction	Comments
Section 10.11 Elimination of Hydrogen Halide (Dehydrohalogenation)	
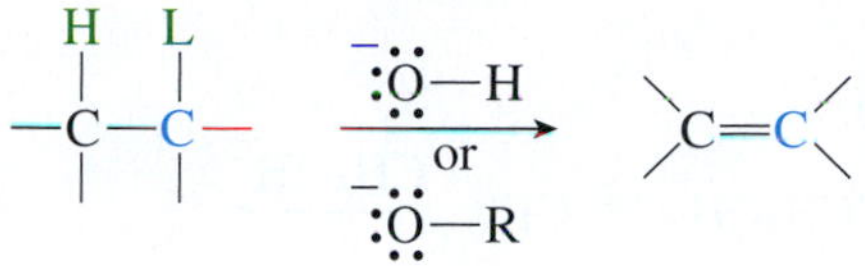	E2 conditions; anti elimination and Zaitsev's rule; steric hindrance in either base or substrate slows competing S_N2 reaction
Section 10.12 Preparation of Alkynes	
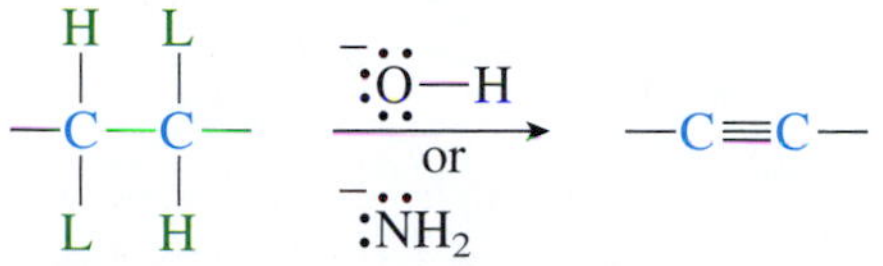	E2 conditions
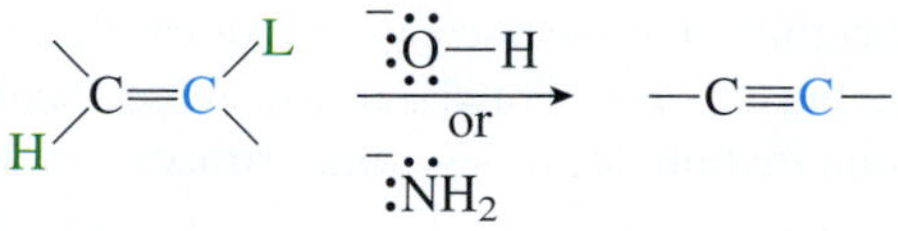	E2 conditions; vinyl halides are intermediates in the preceding reaction
Section 10.13 Dehydration	
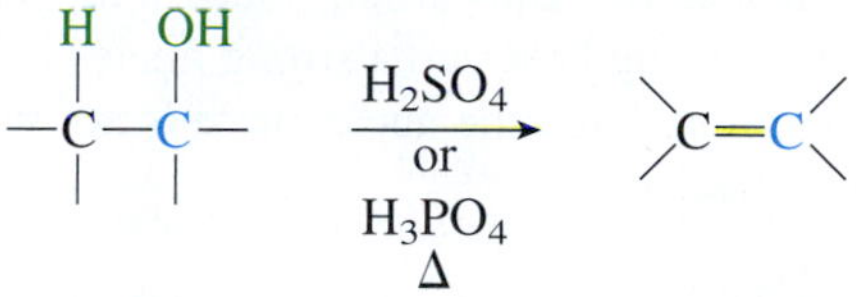	E1 conditions; dehydration; Zaitsev's rule; rate depends on carbocation stability; watch for rearrangement
Section 10.14 Eliminations to Form Carbon–Oxygen Double Bonds; Oxidation Reactions	
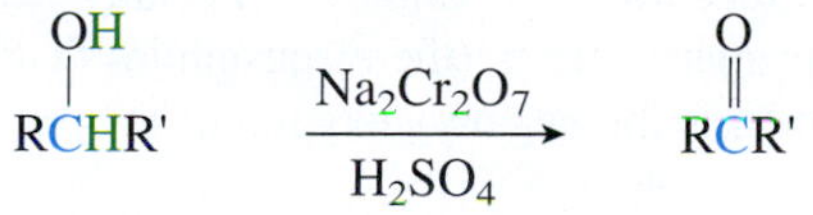	Good for converting secondary alcohols to ketones and primary alcohols to carboxylic acids; can use $K_2Cr_2O_7$ or CrO_3 also
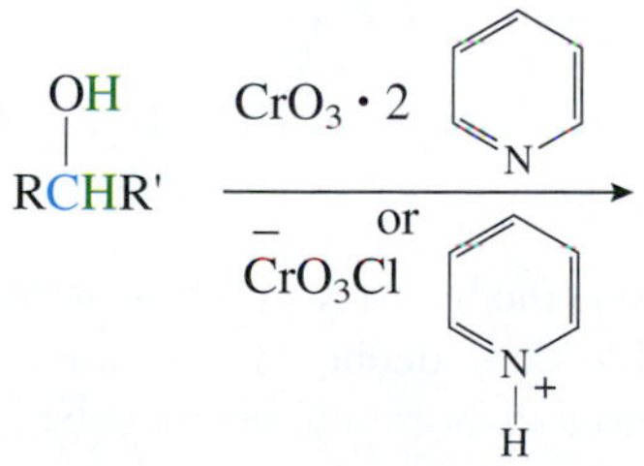	Both of these reagents work well for preparation of aldehydes from primary alcohols also
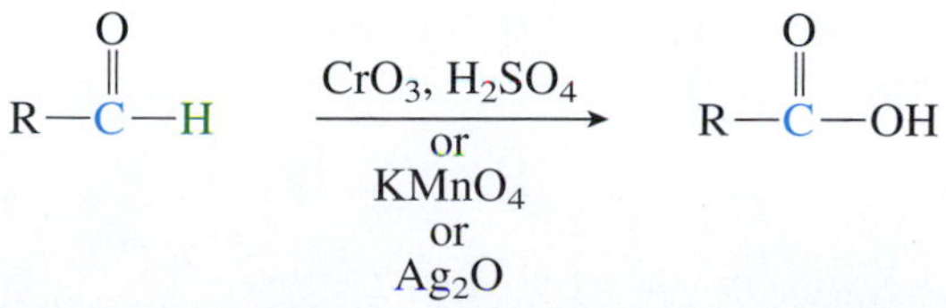	Chromium (VI), permanganate ion, or silver oxide can all be used to oxidize an aldehyde to a carboxylic acid; Cr^{6+} also oxidizes primary alcohols to carboxylic acids

Integrated Practice Problem

ORGANIC **Chemistry Now™** Click *Coached Tutorial Problems* for more practice with **Substitution and Elimination Reactions.**

Show the products of these reactions.

a) 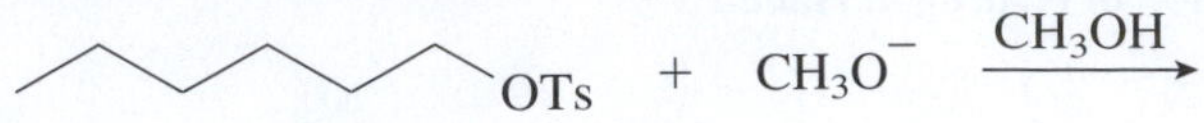

b) 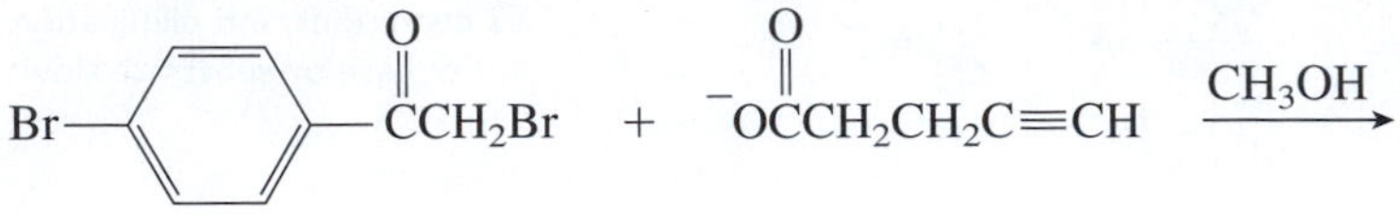

c) 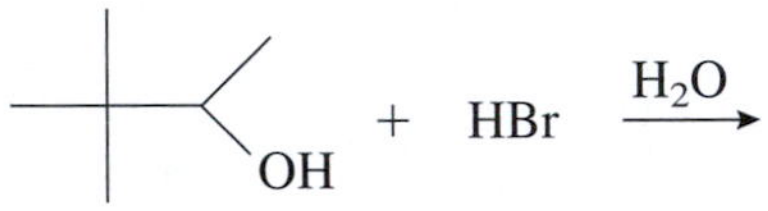

Strategy

Follow the steps listed in the preceding Visual Summary of Key Reactions section. Identify the leaving group, the electrophilic carbon, and the nucleophile (or base). Then determine which mechanism is favored (see Section 9.7). Watch out for stereochemistry where important, regiochemistry in elimination reactions, and carbocation rearrangements when the mechanism is S_N1 or E1.

Solutions

a) The leaving group is the tosylate group. The electrophilic carbon (blue) is the one bonded to the OTs group. Methoxide ion is a strong base and a strong nucleophile. Because the electrophilic carbon is primary and the nucleophile is not sterically hindered, the reaction follows an S_N2 mechanism:

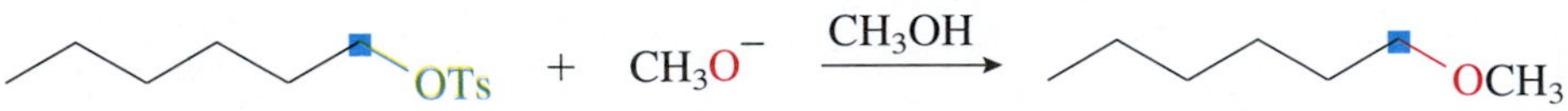

b) There are two bromines in the reactant. However, the one bonded to the benzene ring is inert to substitution and elimination reactions. The other is especially reactive in the S_N2 reaction because it is on a primary carbon (the electrophile) that is adjacent to a carbonyl group. The nucleophile is the negative oxygen of the carboxylate anion:

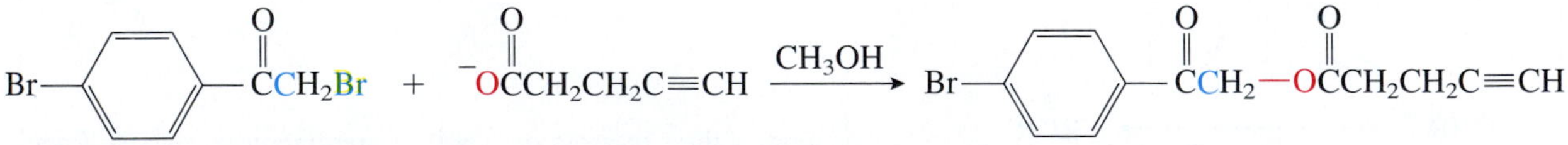

c) In the reaction of an alcohol under acidic conditions the hydroxy group is protonated and acts as a leaving group. Secondary and tertiary alcohols follow the S_N1/E1 mechanisms. Because the reaction involves a carbocation, we must be aware of the possibility of a rearrangement. In this case, rearrangement does oc-

cur, converting the original secondary carbocation to a more stable tertiary carbocation. Because bromide ion is a good nucleophile, the major product results from
substitution. Some elimination product is also formed.

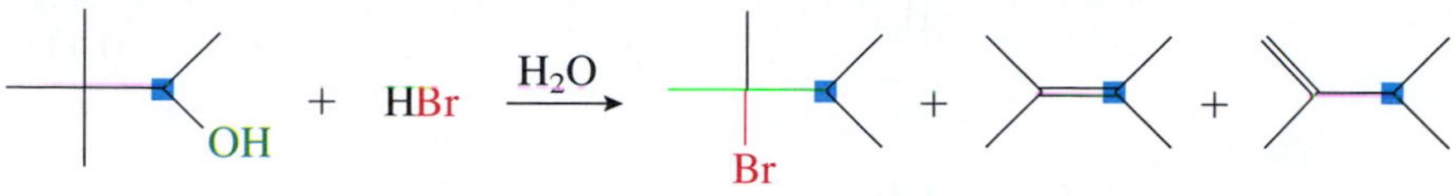

Additional Problems

ORGANIC
Chemistry ✦ Now ™
Assess your understanding of
this chapter's topics with
additional quizzing and
conceptual-based problems at
**http://now.brookscole.com/
hornback2**

10.33 Show the products of these reactions:

a) 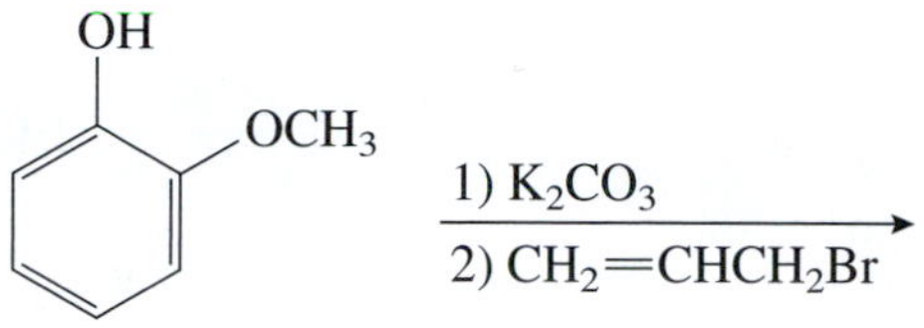   1) K_2CO_3 2) CH_2=$CHCH_2Br$

b) 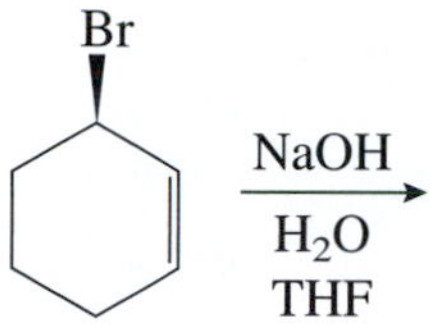 $\dfrac{NaOH}{H_2O}$ THF

c) 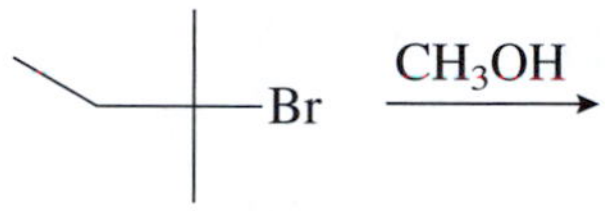 —Br $\xrightarrow{CH_3OH}$

d) 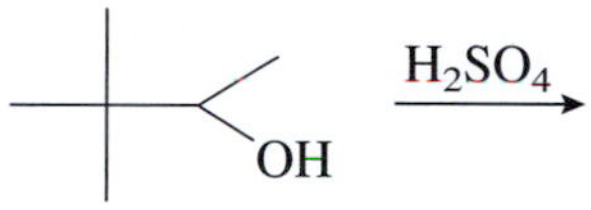 $\xrightarrow{H_2SO_4}$

e) $CH_3\overset{\overset{\displaystyle OH}{|}}{C}HCH_3$ $\xrightarrow{SOCl_2}$

f) 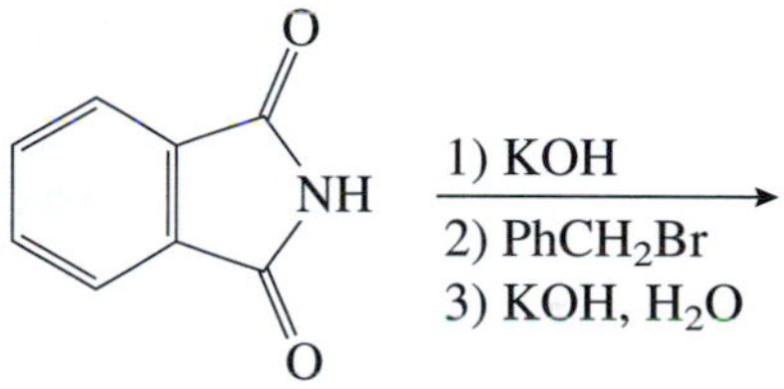   1) KOH   2) $PhCH_2Br$ 3) KOH, H_2O

g) 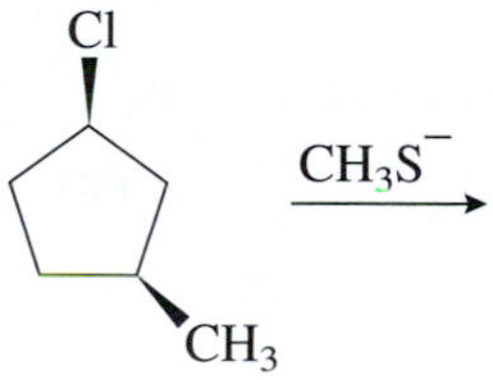 $\xrightarrow{CH_3S^-}$

h) $\dfrac{LiAlH_4}{ether}$

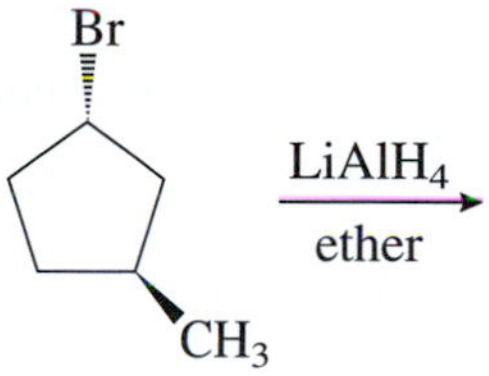

i) $CH_3CH_2CH_2CH_2Br$ $\dfrac{NaCN}{DMSO}$

j) $CH_3CH_2C{\equiv}CH$ 1) $NaNH_2$, NH_3 (l)   2) $PhCH_2CH_2CH_2Br$

k) 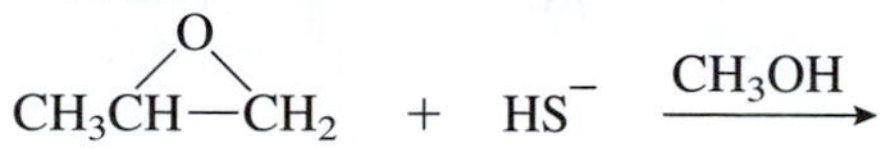 $CH_3\overset{\displaystyle O}{\overset{\diagup\!\!\!\diagdown}{CH-CH_2}}$ + HS^- $\xrightarrow{CH_3OH}$

l) 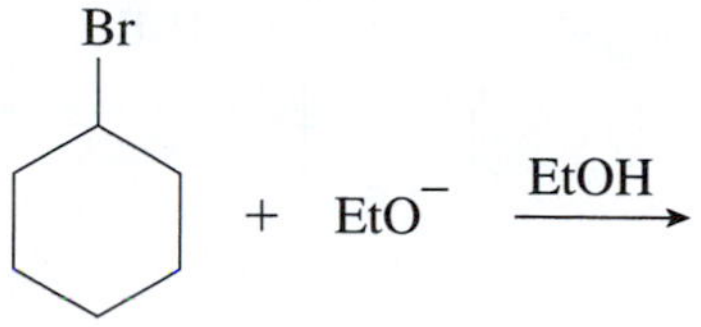 + EtO^- $\xrightarrow{EtOH}$

m) 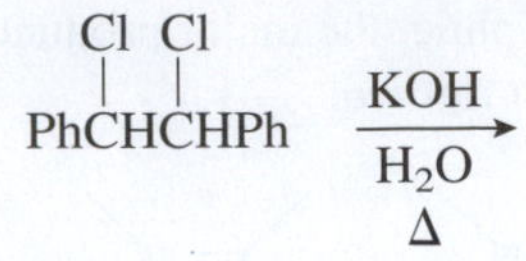 PhCHCHPh $\xrightarrow[\Delta]{\text{KOH} \atop \text{H}_2\text{O}}$

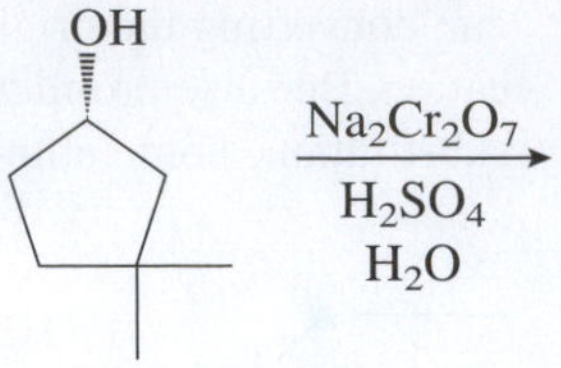

(with Cl Cl above the central carbons)

n) $\xrightarrow{\text{Na}_2\text{Cr}_2\text{O}_7 \atop \text{H}_2\text{SO}_4 \atop \text{H}_2\text{O}}$

o) CH₃CH₂CHCH₂CH₂ $\xrightarrow{\text{PCC} \atop \text{CH}_2\text{Cl}_2}$

(with CH₃ and OH substituents)

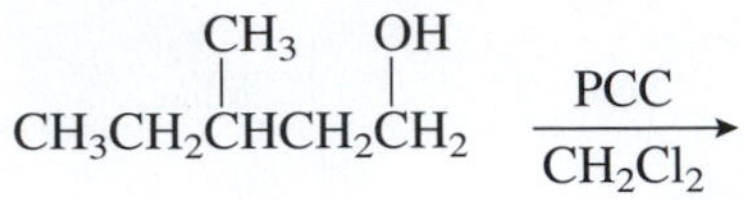

10.34 Show reactions that could be used to convert 1-butanol to these compounds:

a)    b)

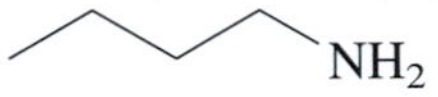

c) 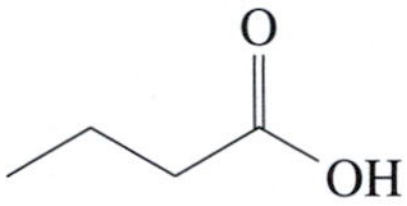   d)

e)    f)

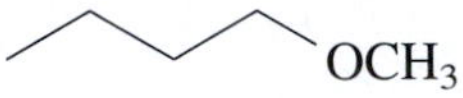

g)    h)

i) 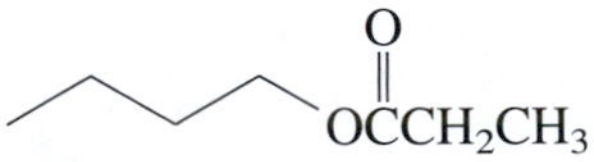   j)

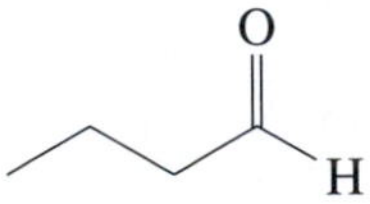

10.35 Show reactions that could be used to convert the epoxide

to these compounds. More than one step may be necessary.

a) 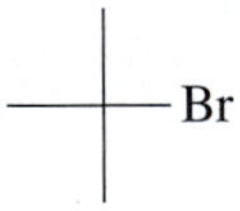   b) 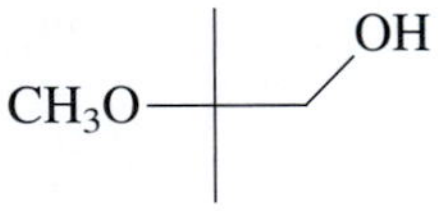 CH₃O   c)

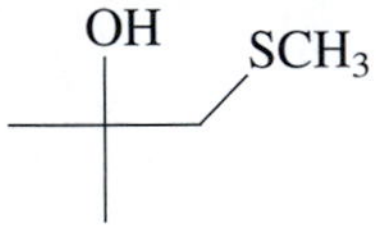

10.36 Show the products of these reactions:

a) 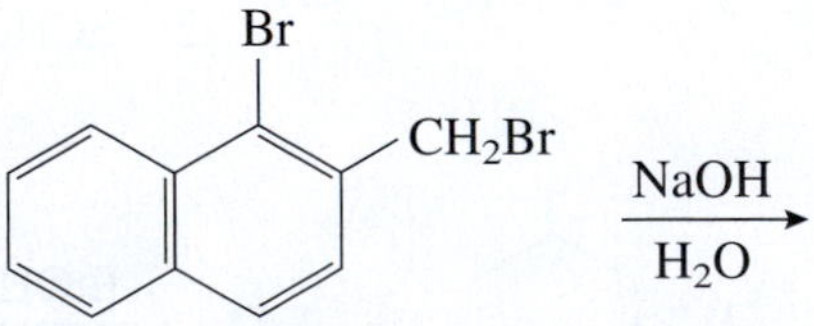 $\xrightarrow{\text{NaOH} \atop \text{H}_2\text{O}}$

b) 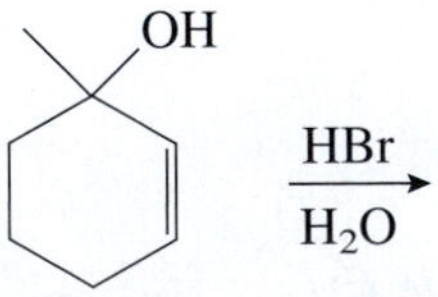 $\xrightarrow{\text{HBr} \atop \text{H}_2\text{O}}$ (2 products)

c) 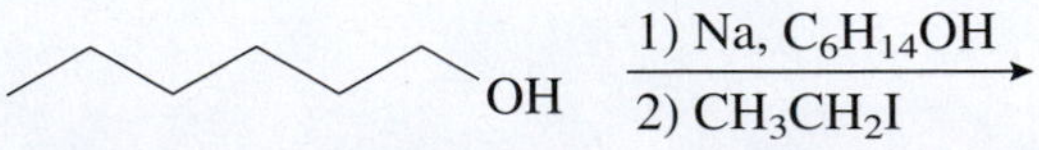 $\xrightarrow{\text{1) Na, C}_6\text{H}_{14}\text{OH} \atop \text{2) CH}_3\text{CH}_2\text{I}}$

d) 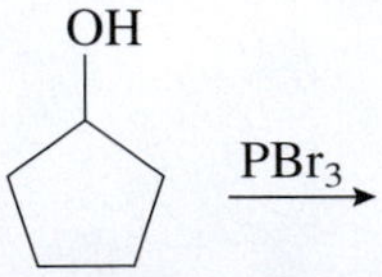 $\xrightarrow{\text{PBr}_3}$

e) 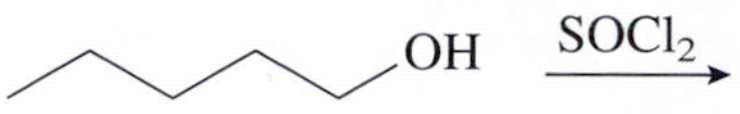 CH₃CH₂CH₂CH₂CH₂OH $\xrightarrow{\text{SOCl}_2}$

f) 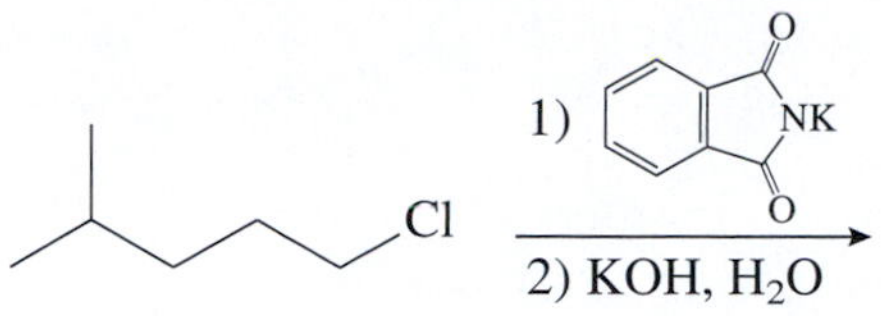 (CH₃)₂CHCH₂CH₂CH₂Cl $\xrightarrow[\text{2) KOH, H}_2\text{O}]{\text{1)} \ \text{(phthalimide NK)}}$

g) Ph₃P + CH₃CH₂Br $\xrightarrow{\text{benzene}}$

h) 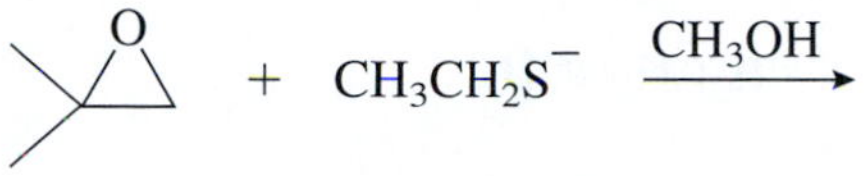 (2-methyloxirane) + CH₃CH₂S⁻ $\xrightarrow{\text{CH}_3\text{OH}}$

i) PhĊHOH (CH₃) $\xrightarrow[\Delta]{\text{H}_2\text{SO}_4}$

j) 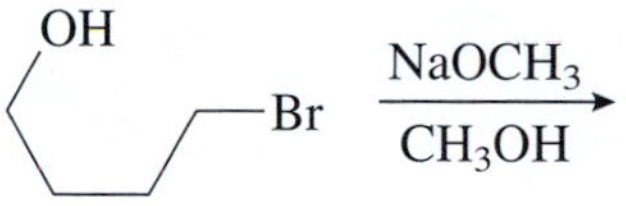 (HO–CH₂CH₂CH₂CH₂–Br) $\xrightarrow[\text{CH}_3\text{OH}]{\text{NaOCH}_3}$ (product has a ring)

k) 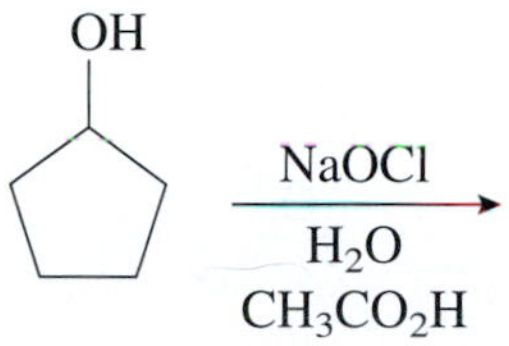 (cyclopentanol, OH) $\xrightarrow[\begin{array}{c}\text{H}_2\text{O}\\ \text{CH}_3\text{CO}_2\text{H}\end{array}]{\text{NaOCl}}$

l) PhCH₂Cl + CH₃CH₂CO⁻ (C=O) $\xrightarrow{\text{CH}_3\text{CH}_2\text{CO}_2\text{H}}$

m) (cyclohexanol, OH) $\xrightarrow[\text{H}_2\text{SO}_4]{\text{CrO}_3}$

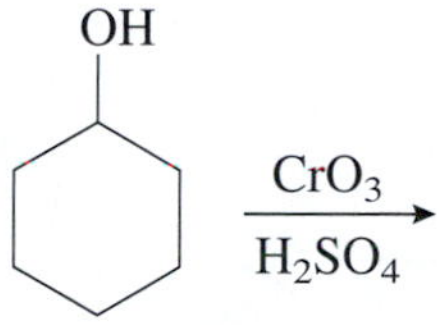

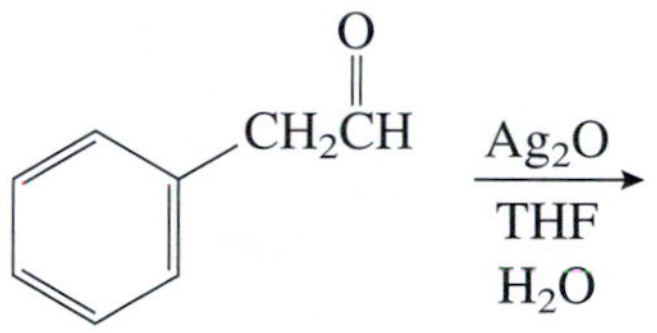

n) (PhCH₂CH with C=O, phenylacetaldehyde) $\xrightarrow[\begin{array}{c}\text{THF}\\ \text{H}_2\text{O}\end{array}]{\text{Ag}_2\text{O}}$

o) (2-chloropentane, Cl) $\xrightarrow[\text{DMSO}]{\text{NaCN}}$

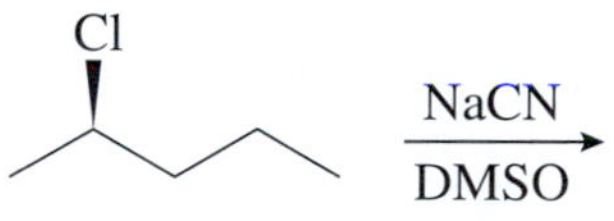

p) CH₃CH₂CH₂C≡CH $\xrightarrow[\text{2) CH}_3\text{CH}_2\text{CH}_2\text{Br}]{\text{1) NaNH}_2,\ \text{NH}_3\,(l)}$

q) 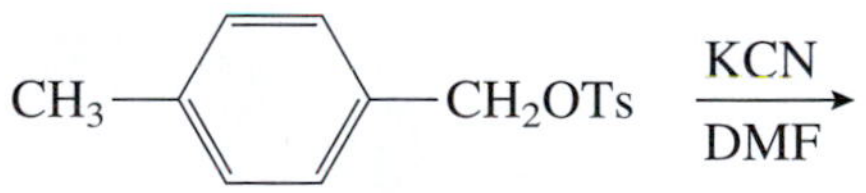 CH₃—(C₆H₄)—CH₂OTs $\xrightarrow[\text{DMF}]{\text{KCN}}$

r) PhCH₂CH₂CH₂ (with OH) $\xrightarrow[\text{CH}_2\text{Cl}_2]{\text{PCC}}$

10.37 Suggest reagents that could be used to accomplish the following transformations.

a)  (CH₃CH₂CH₂CH₂OH) ⟶ (CH₃CH₂CH₂CH₂Br)

b) 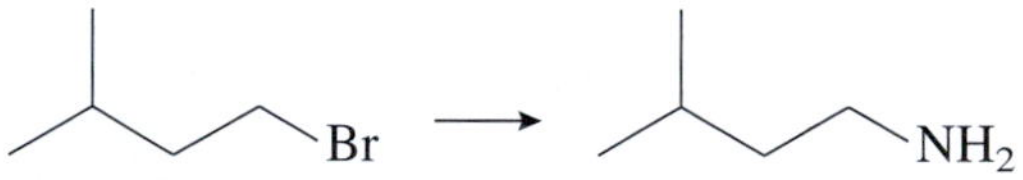 ((CH₃)₂CHCH₂CH₂Br) ⟶ ((CH₃)₂CHCH₂CH₂NH₂)

c)  (OMs on isopropyl) ⟶ (CN on isopropyl)

d) 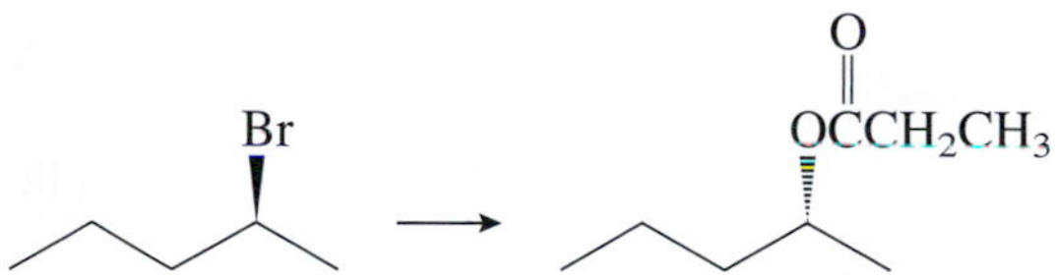 (Br stereocenter) ⟶ (OCCH₂CH₃ ester, C=O)

e)  (CH₃CH₂CH₂CH₂Cl) ⟶ (CH₃CH₂CH₂CH₂OH)

f) 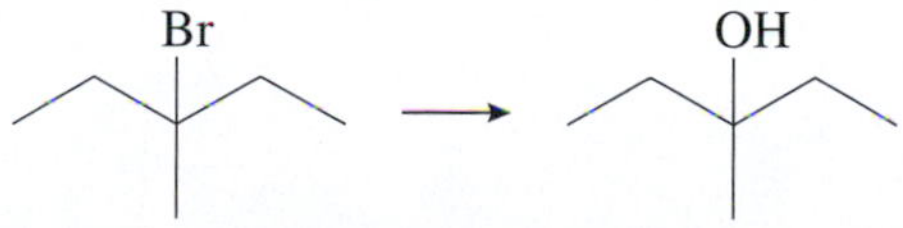 (Br on tertiary) ⟶ (OH)

g) 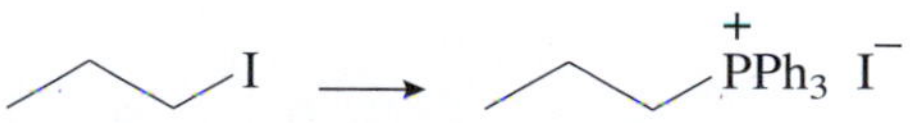 (CH₃CH₂CH₂I) ⟶ (CH₃CH₂CH₂⁺PPh₃ I⁻)

h)

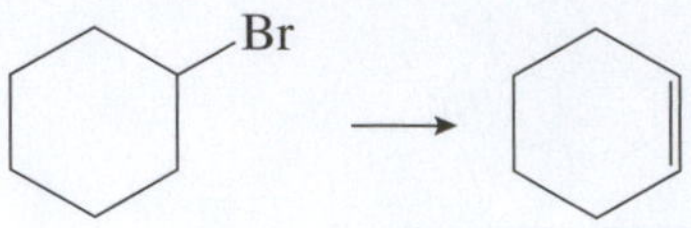

i) 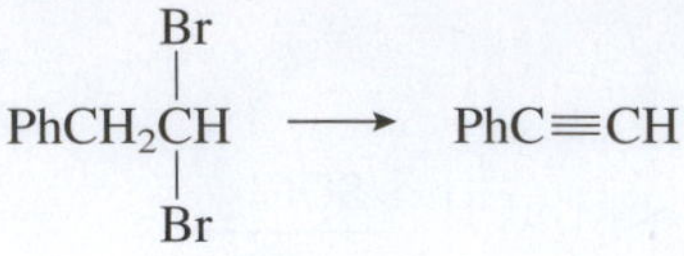 $PhCH_2CH(Br)_2 \longrightarrow PhC\equiv CH$

j) $CH_3CH_2Cl \longrightarrow CH_3CH_2OCH_3$

k)

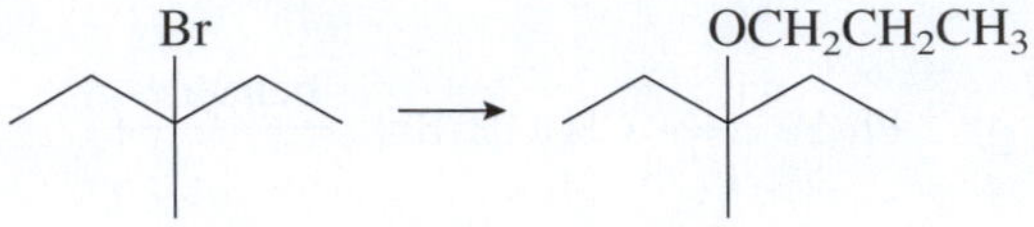

l)

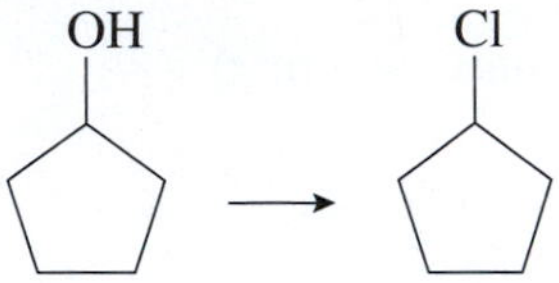

m)

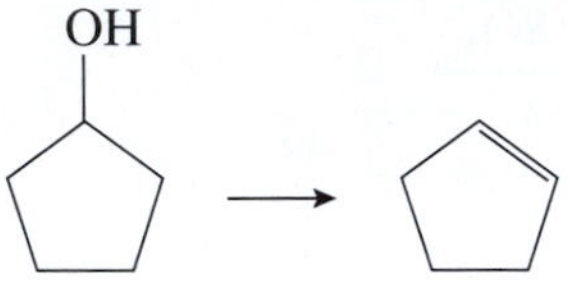

n)

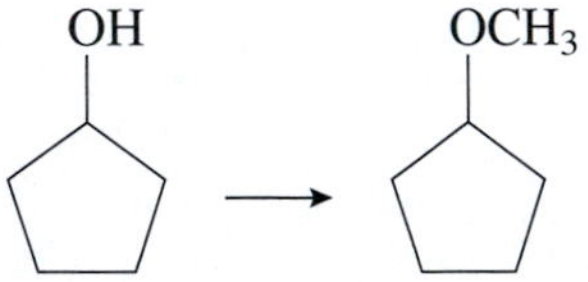

o)

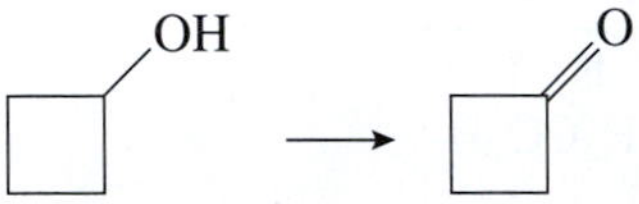

p)

q)

r)

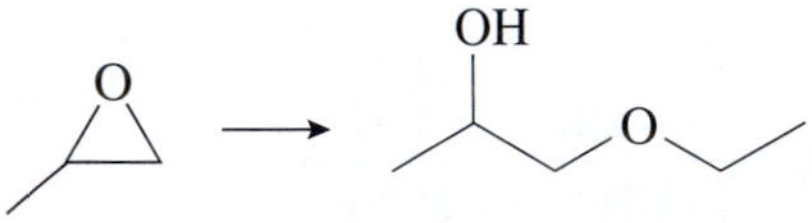

s)

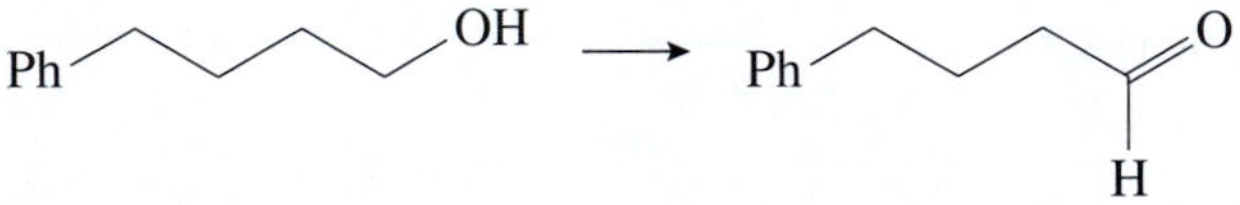

t)

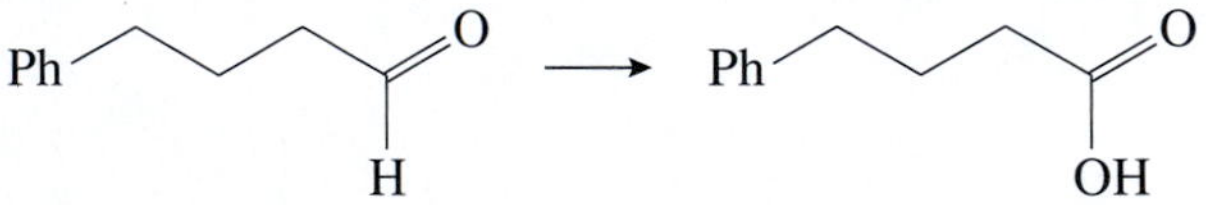

10.38 Show how these compounds could be synthesized from alkyl halides:

a) Heptane

b) $PhCH_2CH_2CH_2NH_2$

c) $PhCH_2C\equiv CH$

d)

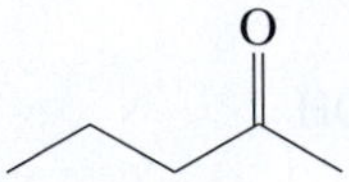

e)

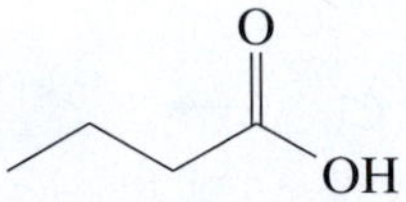

f)

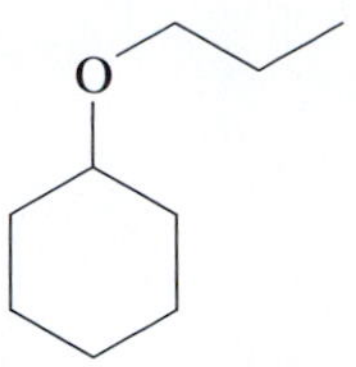

g)

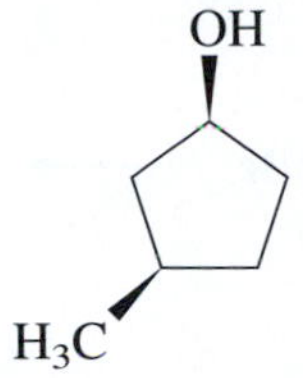

10.39 Show how this synthesis might be accomplished:

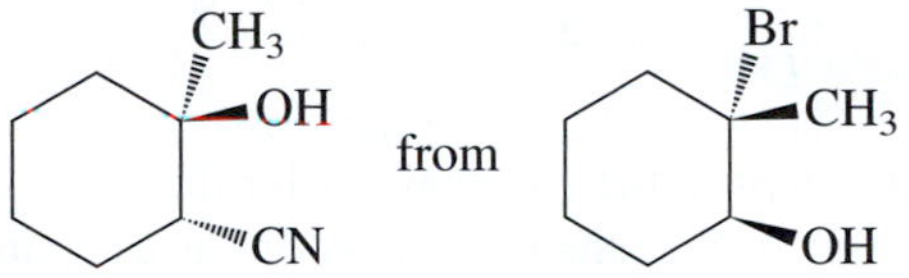

from

10.40 What is wrong with these reactions? Explain.

a)  —C—Cl + NaOCH₃ ⟶ —C—OCH₃ + NaCl

b) (phenol)—OH + HBr ⟶ (bromobenzene)—Br + H₂O

c) (epoxide) + CH₃O⁻ →[CH₃OH] —C—OH with OCH₃

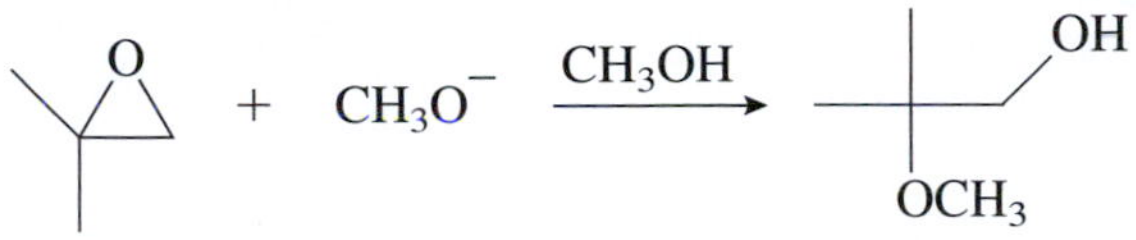

d)  ∿—OH →[HCl] ∿—Cl

10.41 What is wrong with these syntheses? Explain.

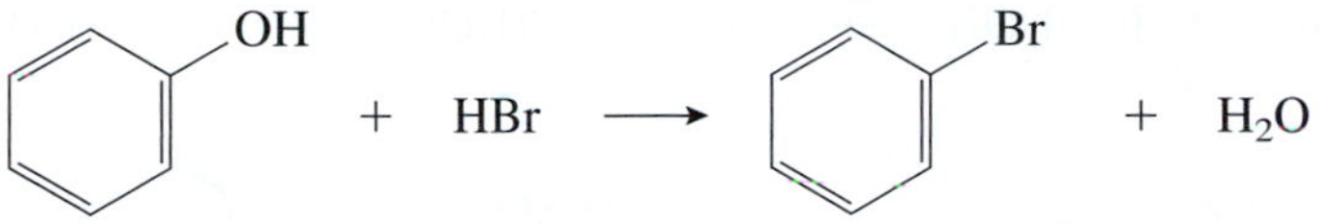

a) 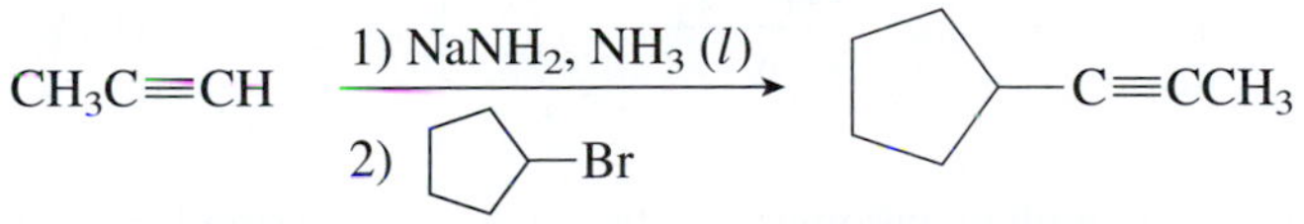 CH₃C≡CH →[1) NaNH₂, NH₃ (l)][2) cyclopentyl—Br] cyclopentyl—C≡CCH₃

b) CH₃CH₂I + NH₃ →[H₂O] CH₃CH₂—N⁺H₃ I⁻

c) 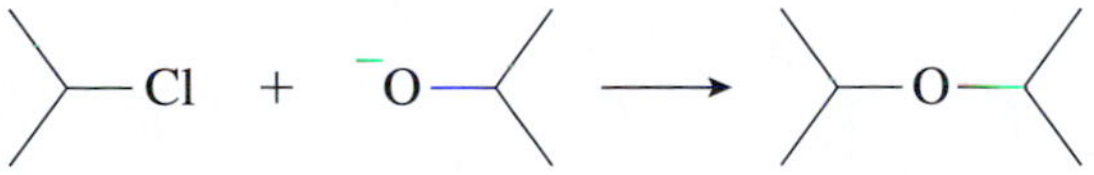 ⟩—Cl + ⁻O—⟨ ⟶ ⟩—O—⟨

d) 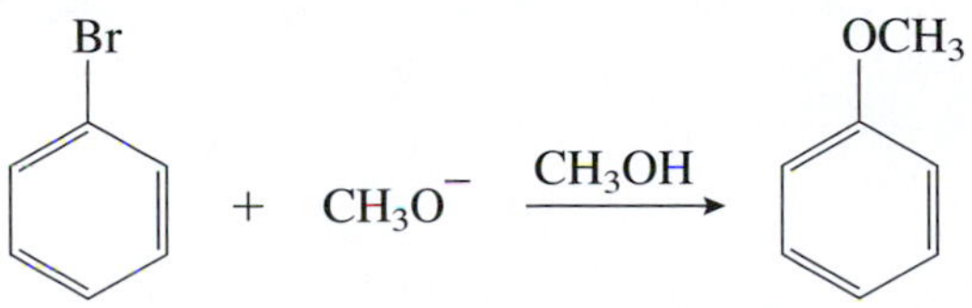 (bromobenzene)—Br + CH₃O⁻ →[CH₃OH] (anisole)—OCH₃

e) 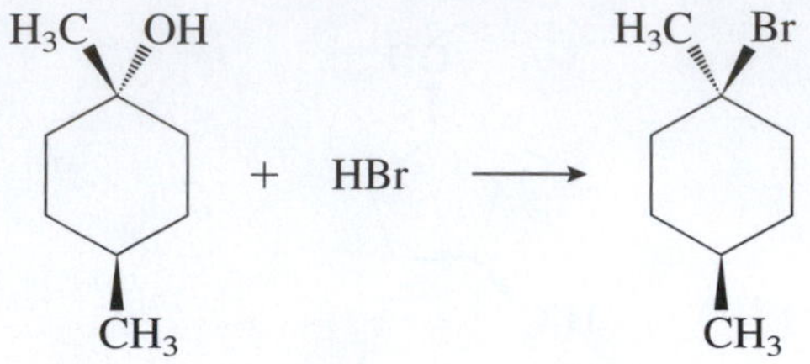 + HBr ⟶

f)

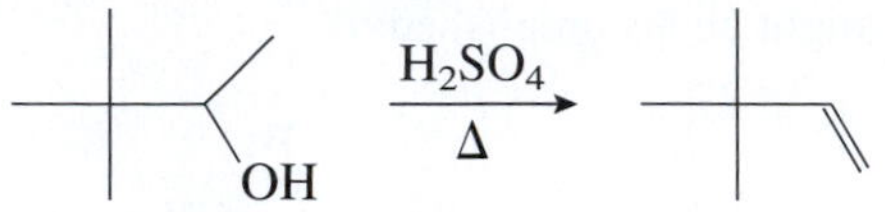

10.42 Show all the steps in the mechanisms for these reactions. Don't forget to use curved arrows to show the movement of electrons in each step.

a) 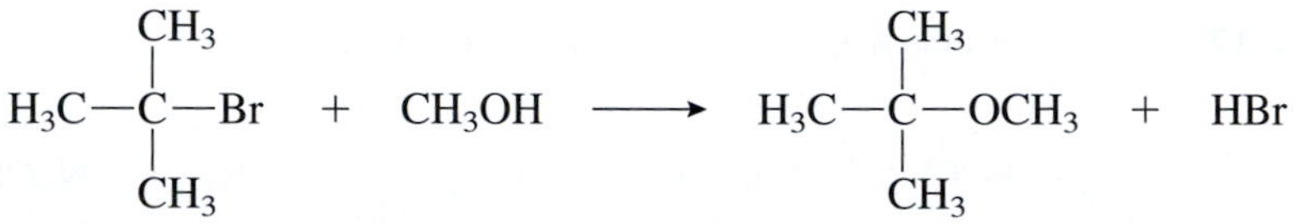 H_3C—$\underset{\underset{CH_3}{|}}{\overset{\overset{CH_3}{|}}{C}}$—Br + CH_3OH ⟶ H_3C—$\underset{\underset{CH_3}{|}}{\overset{\overset{CH_3}{|}}{C}}$—$OCH_3$ + HBr

b) $2\ CH_3CH_2OH$ $\xrightarrow{H_2SO_4}$ $CH_3CH_2OCH_2CH_3$ + H_2O

c) 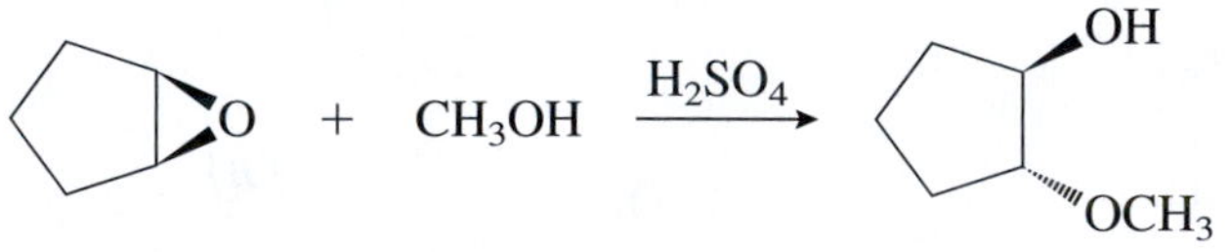 + CH_3OH $\xrightarrow{H_2SO_4}$

d) 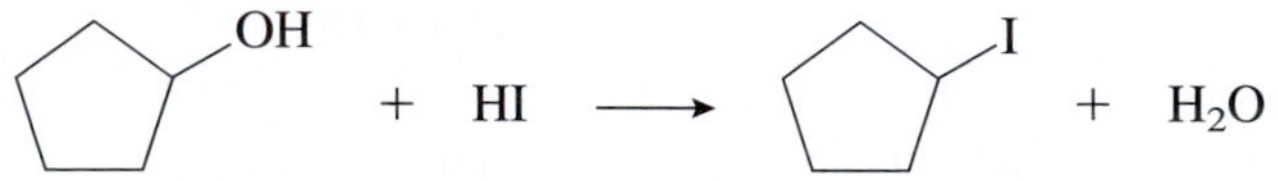 + HI ⟶ + H_2O

e) 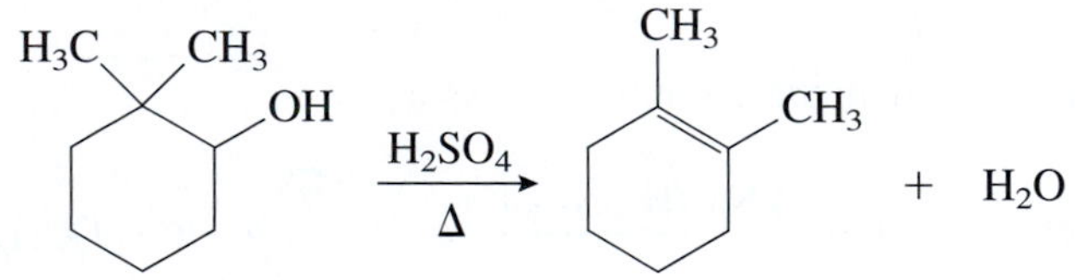 $\xrightarrow[\Delta]{H_2SO_4}$ + H_2O

10.43 Explain how both enantiomers of the product are formed in the reaction shown in problem 10.42c.

10.44 Show all the steps in the mechanism for this reaction:

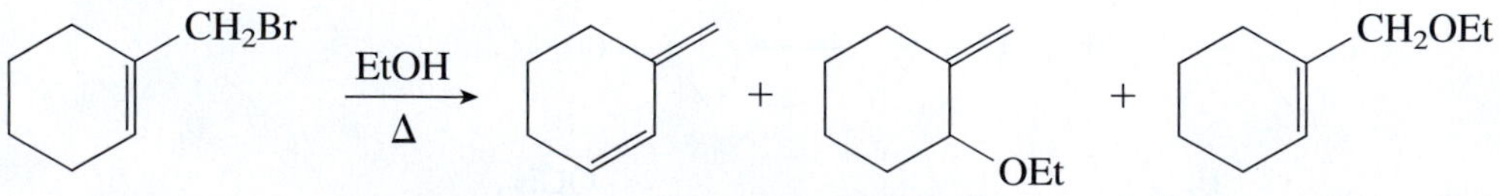

10.45 Classify these transformations as oxidations or reductions:

a)

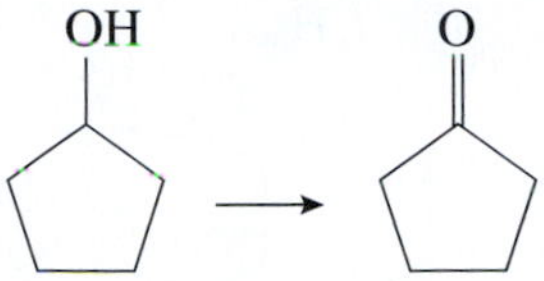

b)

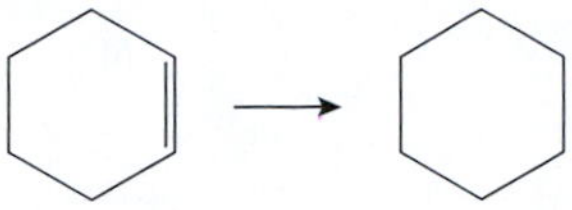

c)

d) $CH_3CH_2\overset{\displaystyle O}{\overset{\|}{C}}OH \longrightarrow CH_3CH_2CH_2OH$

10.46 Show a mechanism for this reaction:

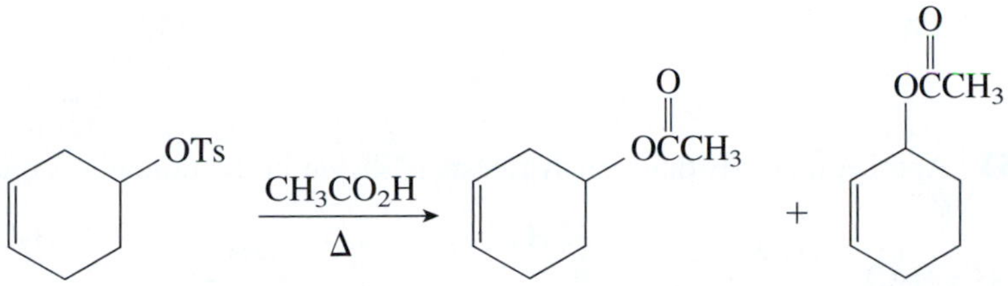

10.47 Show how mustard gas could be prepared from ethylene oxide and sodium sulfide (Na_2S).

Ethylene oxide

10.48 Explain why one of the oxygens preferentially acts as the nucleophile in this reaction:

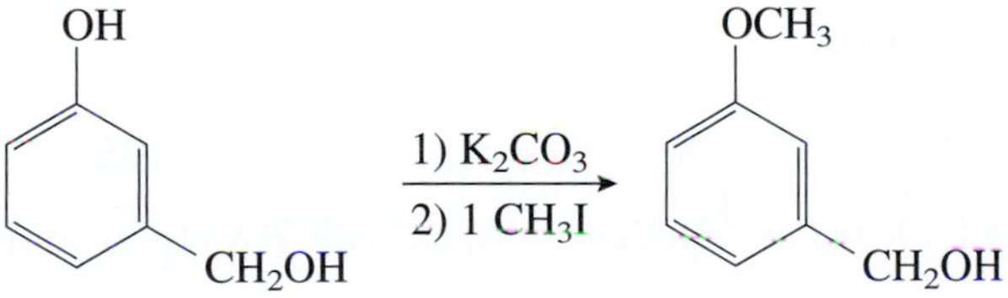

$$\begin{array}{c} 1)\ K_2CO_3 \\ \hline 2)\ 1\ CH_3I \end{array}$$

10.49 Only one of the chlorines acts as a leaving group in this reaction. Explain.

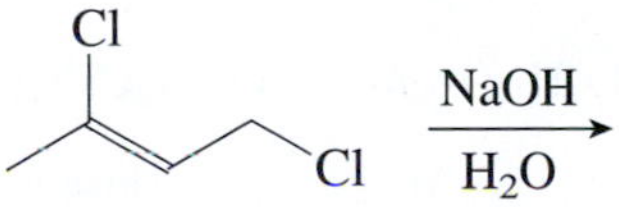

$$\begin{array}{c} NaOH \\ \hline H_2O \end{array}$$

10.50 This reaction gives two substitution products. Show the structures for these products and provide a mechanism for their formation.

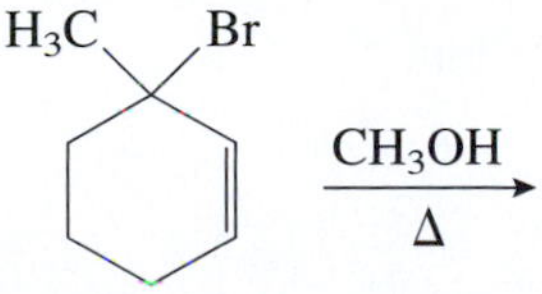

$$\begin{array}{c} CH_3OH \\ \hline \Delta \end{array}$$

10.51 Suggest a mechanism for this reaction:

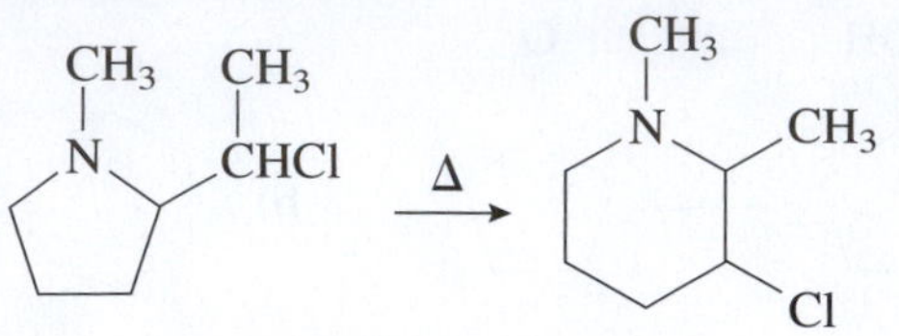

BioLink

10.52 In addition to the reaction shown on p. 353, diphenhydramine can also be prepared by heating bromodiphenylmethane and 2-(dimethylamino)-1-ethanol in a polar solvent. Show a mechanism for this reaction:

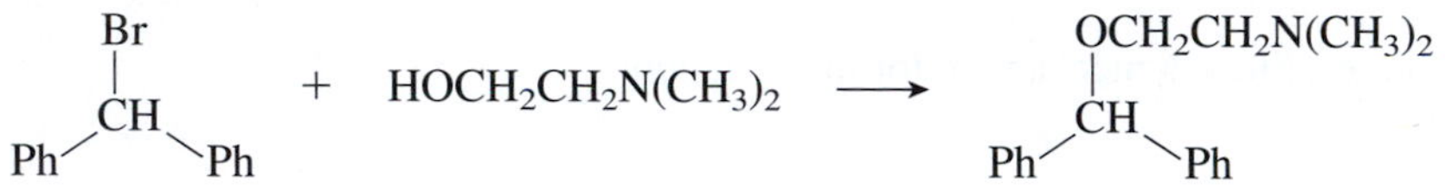

Diphenhydramine

BioLink

10.53 Another diphenhydramine synthesis is shown in the following equation:

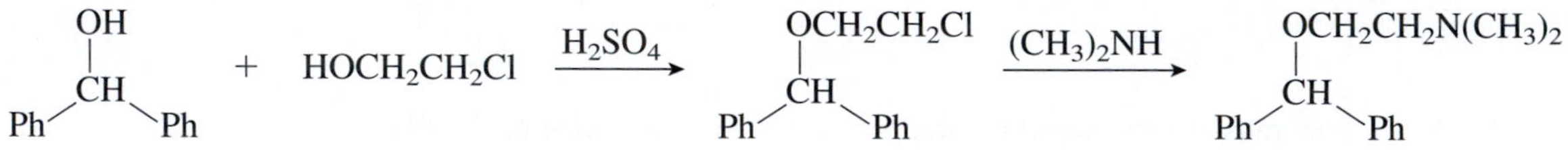

Diphenhydramine

a) Show a mechanism for the first step in this synthesis.
b) Explain which mechanism is occurring in the second step.

BioLink

10.54 Suggest syntheses of these amino acids from the indicated starting materials:

a) PhCH$_2$CHCO$_2$H (NH$_2$)   from   PhCH$_2$CHCO$_2$H (Br)

Phenylalanine

b) CH$_3$SCH$_2$CH$_2$CHCO$_2$H (NH$_2$)   from   HSCH$_2$CH$_2$CHCO$_2$H (NH$_2$)

Methionine   Homocysteine

c) HSCH$_2$CHCO$_2$H (NH$_2$)   from   HOCH$_2$CHCO$_2$H (NH$_2$)

Cysteine   Serine

Problems Using Online Three-Dimensional Molecular Models

ORGANIC
Chemistry•⚛•**Now**™
Click *Molecular Model Problems*
to view the models needed to
work these problems.

10.55 a) Explain which of the following two products is formed when the reactant alkyl chloride reacts with sodium acetate in DMSO.

b) Explain which of the following two product alcohols is formed when the product from part a reacts with sodium hydroxide in water.

10.56 Explain which of the following two ether products is formed when the reactant alcohol reacts with sodium followed by reaction with iodomethane.

10.57 Explain which of the following chlorohydrins forms an epoxide more readily upon treatment with base.

10.58 Explain which of the following epoxide products is formed when the chlorohydrin reactant is treated with base.

10.59 a) Explain which of the following products is formed when the epoxide reactant is treated with methoxide ion in methanol.

b) Explain which of the products from part a is formed when the epoxide reactant from part a is treated with acid in methanol.

Do you need a live tutor for homework problems? Access vMentor at Organic ChemistryNow at **http://now.brookscole.com/hornback2** for one-on-one tutoring from a chemistry expert.

Additions to Carbon–Carbon Double and Triple Bonds

REACTIONS OF ALKENES AND ALKYNES   C H A P T E R 11

THE REACTION that is presented in this chapter results in the addition of two groups—an electrophile and a nucleophile—to the carbons of a carbon–carbon double or triple bond. In the following example of this addition reaction, a proton (the electrophile) and a chloride ion (the nucleophile) add to the carbons of ethene to produce chloroethane:

The features of this reaction that will be examined include the following:

The mechanism of the reaction

The effect of substituents on the double bond on the rate of the reaction

The regiochemistry of the reaction

The stereochemistry of the reaction

The various combinations of electrophiles and nucleophiles that undergo this reaction

The variations in mechanism, regiochemistry, and stereochemistry that occur with these different reagents

Similar reactions that occur with carbon–carbon triple bonds

A large number of reactions are presented in this chapter. As was the case with the previous reactions we have seen, identifying the electrophile and the nucleophile will make the task of remembering all of these reactions much easier. Keep the general mechanism in mind and note what changes are caused in the mechanism by the different reagents as each is discussed.

11.1 THE GENERAL MECHANISM

The simplest version of the mechanism for this addition reaction occurs in two steps. First, the electrophile adds to the double bond, producing a carbocation intermediate. In the second step the nucleophile adds to the carbocation. This step is identical to the second step of the S_N1 reaction. Because the initial species that reacts with the double bond is an electrophile, the reaction is called an **electrophilic addition reaction.**

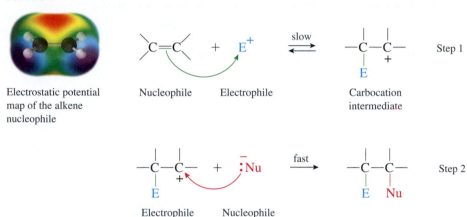

Electrostatic potential map of the alkene nucleophile

Nucleophile   Electrophile   Carbocation intermediate   Step 1

Electrophile   Nucleophile   Step 2

In the first step the nucleophile is the alkene, or, more specifically, the highest-energy pair of electrons of the alkene: the pi electrons of the double bond. Because these electrons, are in a bonding MO rather than in a higher-energy nonbonding MO, they are only weakly nucleophilic, so a relatively strong electrophile, such as H^+ from a strong acid, is needed to react with them. Most of these addition reactions are run under acidic or neutral conditions to avoid destroying the electrophiles, which are all fairly strong Lewis acids.

The product of the first step is a high-energy carbocation intermediate. Forming this reactive intermediate is the slow step of the mechanism. As was the case in the S_N1 reaction, the transition state for this step resembles the carbocation (recall the Hammond postulate; see Section 8.6). Therefore, structural features that stabilize the carbocation also stabilize the transition state leading to it, thus lowering the activation barrier and resulting in a faster rate of reaction. We know from Chapter 8 the various features that stabilize carbocations. The presence of electron-donating alkyl groups on the positive carbon stabilizes the carbocation; tertiary carbocations are more stable than secondary carbocations, and secondary carbocations are more stable than primary carbocations. Therefore, alkyl substituents on the double bond accelerate the reaction. Resonance stabilization of the carbocation also speeds up the reaction. The presence of groups that withdraw electrons destabilizes the carbocation and results in a slower reaction rate. We also know that carbocations are prone to re-arrange if a more stable cation can result. Rearrangements do occur in these electrophilic addition reactions, so we must always examine the carbocation to see whether rearrangement is likely.

Because the second step is the same as the second step in the S_N1 mechanism, similar nucleophiles, such as H_2O and the halide ions, are found here. In addition, there are electrophiles and nucleophiles that we have not yet encountered that undergo this reaction. Some of these cause variations on the mechanism presented earlier. However, the general theme of the electrophile adding first and ultimate formation of a product with the electrophile bonded to one carbon of the initial double bond and the nucleophile bonded to the other remains unchanged. Let's look at the various combinations of electrophiles and nucleophiles that are commonly employed and see how the details of the reaction are affected in each case.

PROBLEM 11.1

Arrange these alkenes in order of increasing rate of reaction with HCl:

a) $CH_3CH_2CH{=}CH_2$ $CH_2{=}CH_2$ $\overset{\overset{\displaystyle CH_3}{\displaystyle |}}{CH_3CH_2C}{=}CH_2$

b) [benzene ring]—$CH{=}CH_2$ $CH_3CH{=}CH_2$ CH_3O—[benzene ring]—$CH{=}CH_2$

11.2   ADDITION OF HYDROGEN HALIDES

All of the halogen acids, HF, HCl, HBr, and HI, add to alkenes to give alkyl halides, as shown in the following example in which hydrogen chloride adds to 2-butene:

$$CH_3CH{=}CHCH_3 + HCl \longrightarrow CH_3\overset{\overset{\displaystyle H}{\displaystyle |}}{C}H{-}\overset{\overset{\displaystyle :\overset{..}{C}l:}{\displaystyle |}}{C}HCH_3$$

2-Butene   2-Chlorobutane

$$\overset{\overset{\displaystyle H}{\displaystyle |}}{:\overset{..}{\underset{..}{C}}l:} \longrightarrow CH_3\overset{\overset{\displaystyle H}{\displaystyle |}}{C}H{-}\overset{+}{C}HCH_3 \quad :\overset{..}{\underset{..}{C}}l:^{-}$$

In this example the electrophile is a proton and the nucleophile is a chloride anion. The mechanism is just as described in the previous section; first the electrophilic proton adds to produce a carbocation intermediate, and then the chloride nucleophile bonds to the carbocation.

Because 2-butene is a symmetrical alkene, it does not matter which carbon initially bonds to the proton. Only one product, 2-chlorobutane, is possible. In the case of an unsymmetrical alkene—that is, one with different substituents on the two carbons of the double bond—two products could be formed, depending on which carbon bonds to the electrophile and which bonds to the nucleophile. For example, the reaction of hydrogen chloride with propene could produce 1-chloropropane or 2-chloropropane:

$$CH_2{=}CHCH_3 + HCl \longrightarrow \underset{\text{2-Chloropropane}}{\overset{\overset{\displaystyle H \quad Cl}{|\quad\;|}}{CH_2-CHCH_3}} + \underset{\text{1-Chloropropane}}{\overset{\overset{\displaystyle Cl \quad H}{|\quad\;|}}{CH_2-CHCH_3}}$$

Propene

Only product   None of this is formed.

When this reaction is run in the laboratory, the only product formed is 2-chloropropane. No 1-chloropropane is observed. A reaction such as this one that produces only one of two possible orientations of addition is termed a **regiospecific reaction.** (A reaction that produces predominantly one possible orientation but does form some of the product with the other orientation is termed a **regioselective reaction.**)

In 1869 Vladimir Markovnikov studied the regiochemistry of a large number of these addition reactions. On the basis of his observations, he postulated an empirical rule that can be used to predict the orientation of additions to alkenes:

▶ **MARKOVNIKOV'S RULE**

In addition reactions of HX to alkenes, the H bonds to the carbon with more hydrogens (fewer alkyl substituents) and the X bonds to the carbon with fewer hydrogens (more alkyl substituents).

As an example, Markovnikov's rule predicts that the addition of hydrogen bromide to 2-methylpropene should produce 2-bromo-2-methylpropane:

$$\underset{H}{\overset{H}{}}C{=}C\overset{CH_3}{\underset{CH_3}{}} + HBr \longrightarrow \overset{\overset{\displaystyle H \quad CH_3}{|\quad\;|}}{H-C-C-CH_3}\underset{\underset{\displaystyle H \quad Br}{}}{}$$

2-Bromo-2-methylpropane

C-1 is bonded to more hydrogens (two), so the H should bond here.

C-2 is bonded to fewer hydrogens (none), so the Br should bond here.

When the reaction is run in the laboratory, only the product predicted by Markovnikov's rule is observed.

Markovnikov's rule was empirical; that is, it was based on observation only. At the time it was proposed, the concept of organic reaction mechanisms had not yet been imagined; so it was impossible to provide a theoretical basis for why the H added to one carbon and the X to the other. Now that we know the mechanism for the reaction, it is easy to understand why these reactions are regiospecific. In fact, if we write the mechanism for the additions to give both possible products, we can predict the preferred product simply on the basis of what we already know about carbocations. It is much better to make predictions based on the mechanism of the reaction because cases that are exceptions to the empirical rule will be readily apparent.

The mechanisms for the two possible orientations of addition of HCl to propene are as follows:

Propene   A secondary carbocation   2-Chloropropane

A primary carbocation   1-Chloropropane

The first mechanism, which leads to the Markovnikov product 2-chloropropane, proceeds via a secondary carbocation. The second mechanism, which would lead to the unobserved product 1-chloropropane, proceeds via a primary carbocation. Because a secondary carbocation is lower in energy than a primary carbocation, the first mechanism should be preferred over the second mechanism. In fact, formation of the secondary carbocation is enough faster than formation of the primary carbocation that the first mechanism occurs to the exclusion of the second.

We can now see the reason behind Markovnikov's rule. If the proton adds to the carbon with more hydrogens, the positive carbon, where the nucleophile will ultimately bond, is bonded to more alkyl groups, resulting in a more stable carbocation. A more modern version of Markovnikov's rule, based on this mechanistic reasoning is as follows:

▶ **MECHANISM-BASED RULE**
The electrophile adds so as to form the more stable carbocation.

This means that the product has the electrophile attached to the carbon that would be less stable as a carbocation and the nucleophile attached to the carbon that would be more stable as a carbocation. Later, we will encounter exceptions to Markovnikov's rule. However, these exceptions are still in accord with this mechanistically based rule.

Some examples of additions of hydrogen halides are provided by the following equations. As you look at each example, try to predict the regiochemistry of the product before looking at the actual product that is formed.

$$CH_3CH{=}CH_2 + HF \xrightarrow{\text{no solvent}} CH_3\overset{\overset{\displaystyle F}{|}}{C}H\overset{\overset{\displaystyle H}{|}}{C}H_2 \quad (61\%)$$

$$Ph{-}\overset{\overset{\displaystyle CH_3}{|}}{C}{=}CH_2 + HCl \xrightarrow{\text{no solvent}} Ph{-}\overset{\overset{\displaystyle CH_3}{|}}{\underset{\underset{\displaystyle Cl}{|}}{C}}{-}\overset{}{\underset{\underset{\displaystyle H}{|}}{C}}H_2 \quad (98\%)$$

$$CH_3CH_2CH_2CH_2CH{=}CH_2 + HBr \xrightarrow{H_2O} CH_3CH_2CH_2CH_2\overset{\overset{\displaystyle Br}{|}}{CH}{-}\overset{\overset{\displaystyle H}{|}}{CH_2} \quad (88\%)$$

$$\text{[cyclohexene]} + HI \xrightarrow{H_2O} \text{[iodocyclohexane]} \quad (90\%)$$

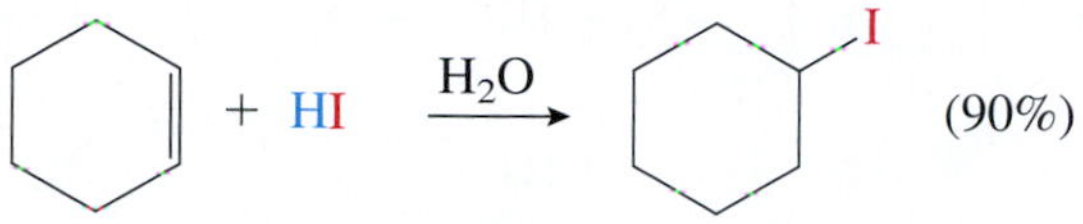

$$CH_3CH_2CH_2CH{=}CHCH_3 + HBr \xrightarrow{H_2O} CH_3CH_2CH_2\overset{\overset{\displaystyle Br}{|}}{CH}{-}CH_2CH_3 \quad (29\%)$$

$$+$$

$$CH_3CH_2CH_2CH_2{-}\overset{\overset{\displaystyle Br}{|}}{CH}CH_3 \quad (57\%)$$

PROBLEM 11.2

Show the structures of the carbocations that are formed in the reaction of HBr with 2-hexene and explain why two products are formed.

PRACTICE PROBLEM 11.1

Show the product of this reaction:

$$PhCH{=}CHCH_3 + HBr \longrightarrow$$

Strategy

Identify the electrophile and the nucleophile. In this problem the electrophile is H^+ and the nucleophile is bromide anion. Add the H^+ to one carbon so that the more stable carbocation is formed at the other carbon. Add the bromide nucleophile to the other carbon.

Solution

The two possible carbocations that can be formed by the addition of H^+ are shown in the following equations:

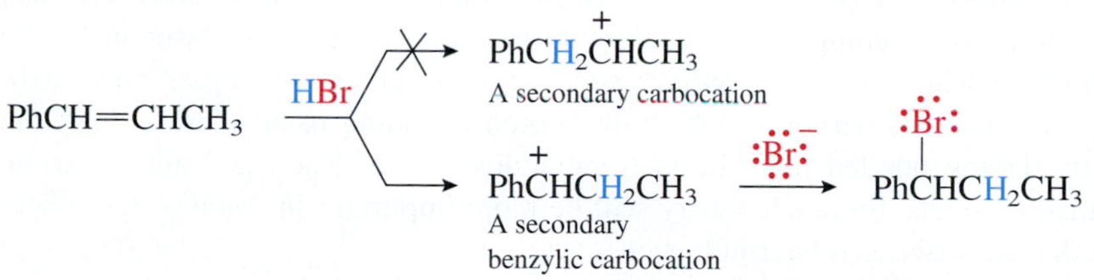

$$PhCH{=}CHCH_3 \xrightarrow{HBr} \begin{cases} \times\ \overset{+}{PhCH_2}CHCH_3 \quad \text{A secondary carbocation} \\[2ex] \overset{+}{PhCHCH_2CH_3} \xrightarrow{\;:\!\overset{..}{\underset{..}{Br}}\!:^-\;} Ph\overset{\overset{\displaystyle :\overset{..}{\underset{..}{Br}}:}{|}}{CH}CH_2CH_3 \end{cases}$$

A secondary benzylic carbocation

Because of its resonance stabilization, the secondary benzylic carbocation is more stable than the secondary carbocation, so the product has the bromine (nucleophile) bonded to the carbon attached to the phenyl group.

PROBLEM 11.3

Show the products of these reactions:

a) 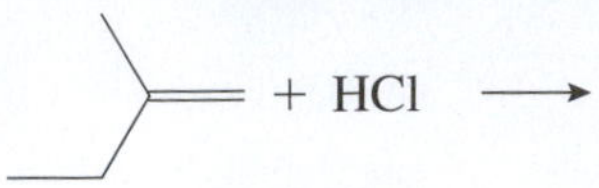 + HCl ⟶   b) 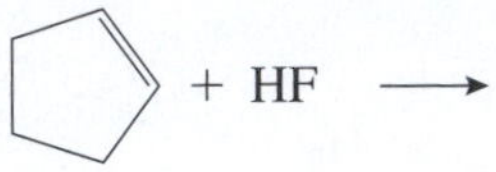 + HF ⟶   c) 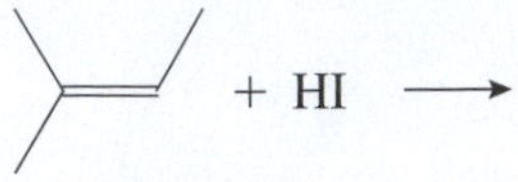 + HI ⟶

d) 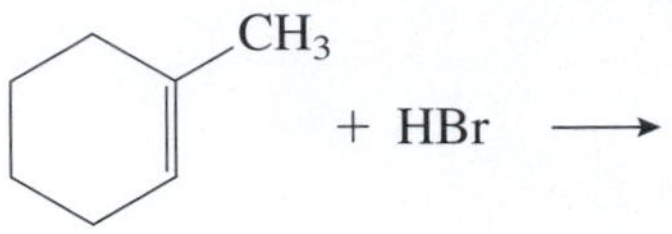 + HBr ⟶   e) 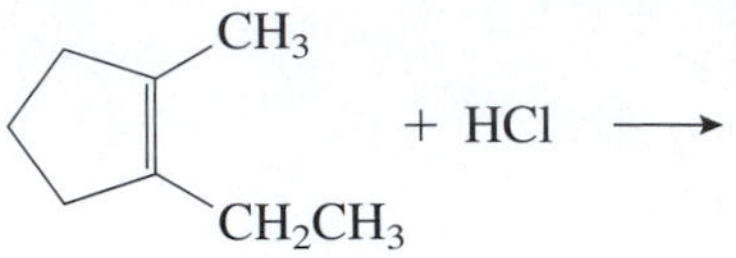 + HCl ⟶

Next, let's consider the stereochemistry of these reactions. Overall, addition reactions are the reverse of the elimination reactions we saw in Chapter 9. As was the case with the eliminations, there are two possible stereochemistries for the addition—syn and anti:

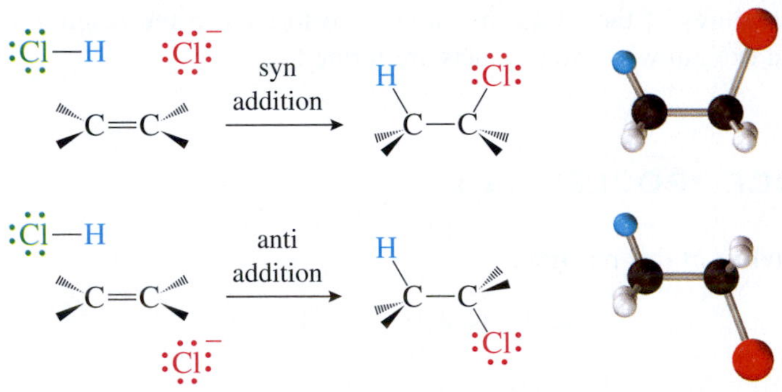

In a syn addition the electrophile and the nucleophile both add from the same side of the plane of the double bond; in an anti addition they add from opposite sides of this plane.

The mechanism for the addition of the hydrogen halides to alkenes proceeds through a carbocation intermediate. As was the case in the S_N1 reaction, the nucleophile can approach the planar carbocation equally well from either side, so we expect that the products should result from a mixture of syn and anti addition. Indeed, this is often the case. Under some conditions, however, the stereochemisty results from predominant syn addition, whereas anti addition is the favored pathway under other conditions. This occurs because these reactions are often conducted in nonpolar solvents in which ion pair formation is favored. The details of how this may affect the stereochemistry of these reactions are complex. Fortunately, stereochemistry is not an issue in most of the reactions in which hydrogen halides add, including all the examples previously presented, because the carbon to which the proton is adding usually has at least one hydrogen already bonded to it. In such situations, syn addition and anti addition give identical products. Stereochemistry will be more important in some of the other reactions that are discussed later in this chapter.

Because carbocations are intermediates in these reactions, rearrangements can occur. The carbocation formed initially in the following example is secondary. Part of the time (17%), the chloride nucleophile intercepts the carbocation before it has a chance to rearrange. But a majority of the time (83%), the carbocation rearranges to a more stable tertiary carbocation, which, on reaction with chloride ion, produces the rearranged product.

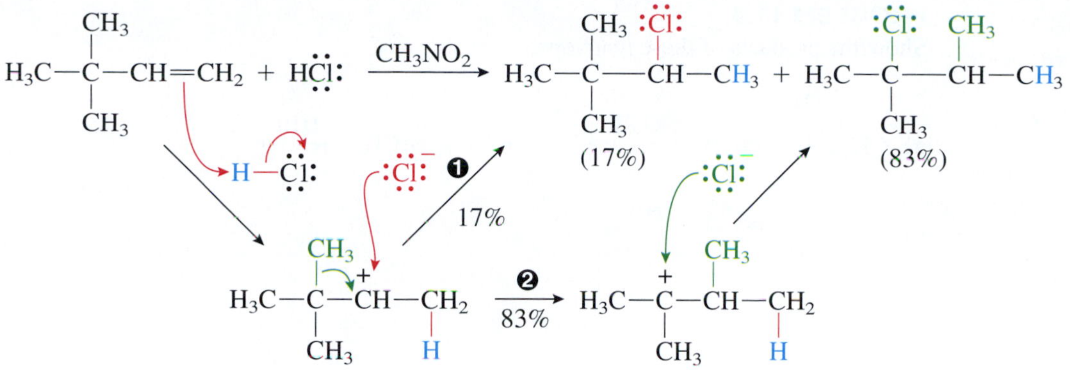

The reaction scheme at the top shows:

$$CH_3\text{—}C(CH_3)_2\text{—}CH=CH_2 + H\ddot{C}l: \xrightarrow{CH_3NO_2}$$

Products:

$H_3C\text{—}C(CH_3)\text{—}CH\ddot{C}l\text{—}CH_3$ (17%) + $H_3C\text{—}C\ddot{C}l(CH_3)\text{—}CH\text{—}CH_3$ (83%)

❶ This secondary carbocation reacts with chloride (17%)   A tertiary carbocation
❷ and rearranges to the more stable tertiary carbocation (83%).

The hydrogen halides also add to the triple bond of alkynes. The regiochemistry of the reactions follows Markovnikov's rule. It is usually possible to add to just one of the pi bonds, producing a vinyl halide, or to both of the pi bonds, producing a dihaloalkane. Some examples are provided in the following equations:

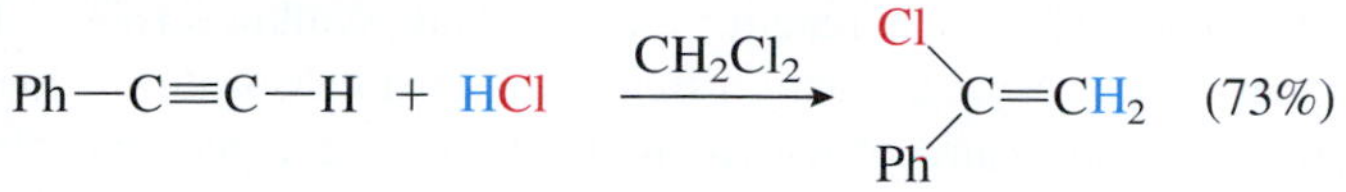

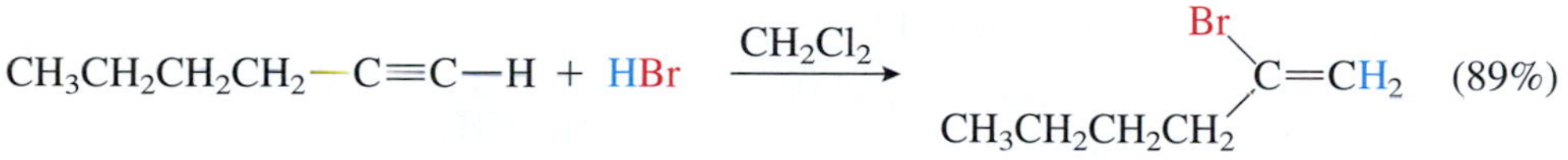

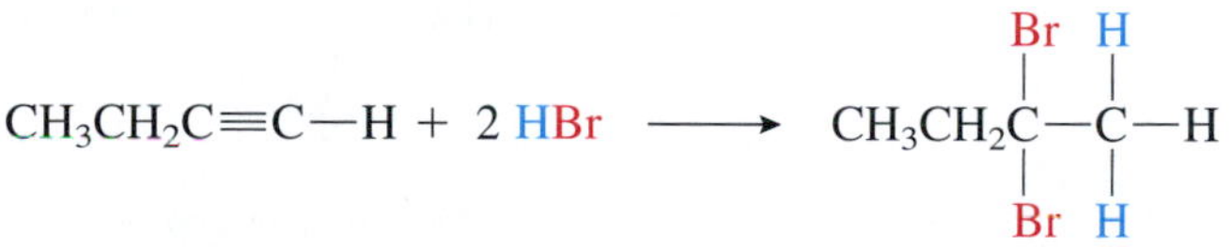

$$Ph\text{—}C\equiv C\text{—}H + HCl \xrightarrow{CH_2Cl_2} \underset{Ph}{\overset{Cl}{\diagdown}}C=CH_2 \quad (73\%)$$

$$CH_3CH_2CH_2CH_2\text{—}C\equiv C\text{—}H + HBr \xrightarrow{CH_2Cl_2} \underset{CH_3CH_2CH_2CH_2}{\overset{Br}{\diagdown}}C=CH_2 \quad (89\%)$$

$$CH_3CH_2C\equiv C\text{—}H + 2\,HBr \longrightarrow CH_3CH_2\underset{\underset{Br}{|}}{\overset{\overset{Br}{|}}{C}}\text{—}\underset{\underset{H}{|}}{\overset{\overset{H}{|}}{C}}\text{—}H$$

PRACTICE PROBLEM 11.2

Show the products of this reaction:

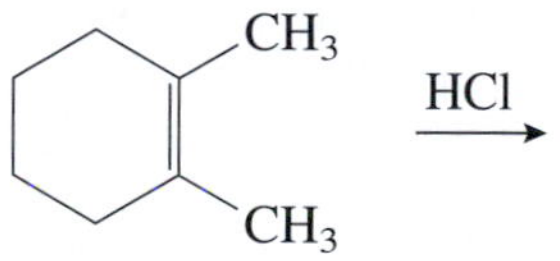

[Structure: cyclohexene ring with two CH_3 groups on the double-bonded carbons] $\xrightarrow{HCl}$

Solution

Because the addition of HCl to an alkene proceeds through a carbocation intermediate, products from both syn and anti addition are usually formed. This means that products are expected with the H and the Cl both cis and trans.

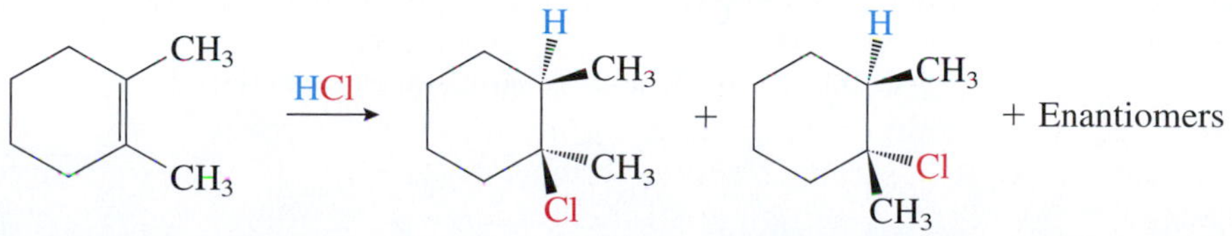

[Structures showing the dimethylcyclohexene reacting with HCl to give two stereoisomeric products with H, CH_3, and Cl substituents] + Enantiomers

PROBLEM 11.4

Show the products of these reactions:

a) $CH_3\underset{\underset{CH_3}{|}}{CH}CH{=}CH_2 \xrightarrow{\text{HBr}}$ b) $PhC{\equiv}CH \xrightarrow{\text{HBr}}$

c) $\xrightarrow{\text{2 HBr}}$

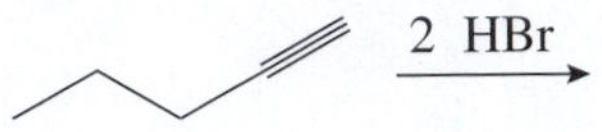

11.3 ADDITION OF WATER (HYDRATION)

Treatment of an alkene with a strong acid, such as sulfuric acid, that has a relatively nonnucleophilic conjugate base results in the addition of the elements of water (H and OH) to the double bond. This reaction has many similarities to the addition of the halogen acids described in Section 11.2. First H⁺ adds to produce a carbocation and then water acts as the nucleophile. The reaction follows Markovnikov's rule and the stereochemistry is that expected for a reaction that involves a carbocation—loss of stereochemistry. Some examples are provided in the following equations. Note that the mechanism is the exact reverse of the E1 mechanism for acid-catalyzed dehydration of alcohols described in Section 10.13.

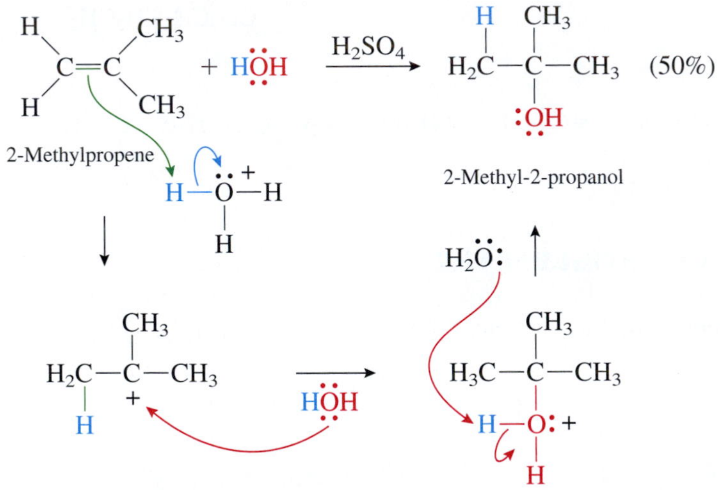

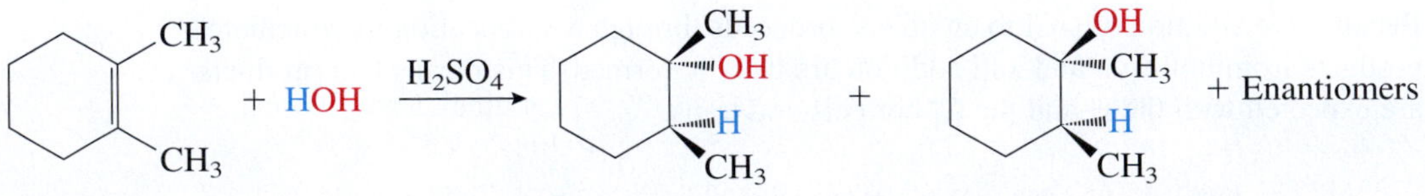

2-Methylpropene

2-Methyl-2-propanol

1,2-Dimethylcyclohexene   *cis*-1,2-Dimethylcyclohexanol   *trans*-1,2-Dimethylcyclohexanol

The yields in this reaction are often relatively low because of the strongly acidic conditions. In addition, the reaction conditions are very favorable for carbocation rearrangements as shown in the following example:

3-Methyl-1-butene

2-Methyl-2-butanol

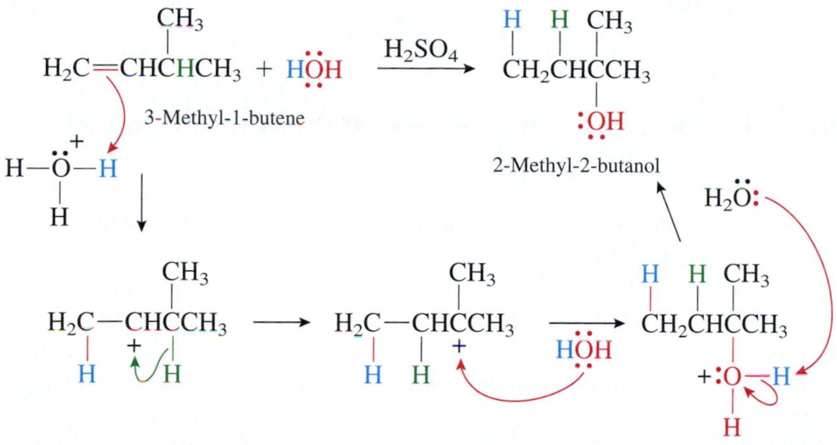

For these reasons, acid-catalyzed hydration is often not the method of choice for preparing alcohols from alkenes. Another reaction that accomplishes this same transformation, often in higher yield, is described in Section 11.6.

PROBLEM 11.5
Show the products of these reactions:

a)  [structure] + H$_2$O $\xrightarrow{\text{H}_2\text{SO}_4}$

b) 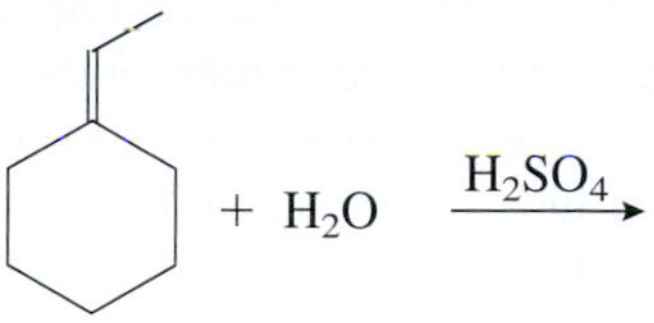 [structure] + H$_2$O $\xrightarrow{\text{H}_2\text{SO}_4}$

c) 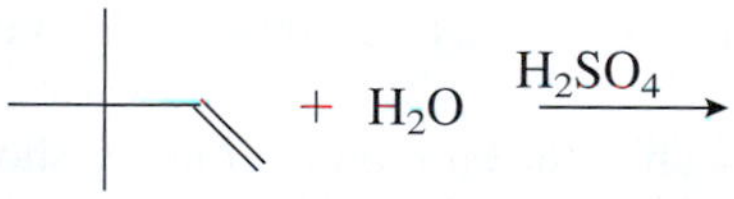 [structure] + H$_2$O $\xrightarrow{\text{H}_2\text{SO}_4}$

PROBLEM 11.6
Show all of the steps in the mechanism for the addition of water to propene catalyzed by sulfuric acid. Explain whether propene or phenylethene (PhCH=CH$_2$) has a faster rate in this reaction.

11.4 ADDITION OF HALOGENS

Both Cl$_2$ and Br$_2$ add to carbon–carbon double bonds to produce dihalides as illustrated in the following examples. The other halogens are not commonly used—F$_2$ because it is too reactive and I$_2$ because it is not reactive enough. These reactions are usually run in an inert solvent such as CCl$_4$, CHCl$_3$, or CH$_2$Cl$_2$.

A solution of Br$_2$ in carbon tetrachloride is red. When a few drops of an alkene are added, the color disappears.

$$CH_3CH=CHCH_2CH_3 + Cl_2 \xrightarrow{CHCl_3} CH_3CH-CHCH_2CH_3 \quad (81\%)$$

2-Pentene

2,3-Dichloropentane

$$PhCH=CH-\overset{\overset{\displaystyle O}{\|}}{C}OCH_2CH_3 + Br_2 \xrightarrow{CCl_4} PhCH-CH-\overset{\overset{\displaystyle O}{\|}}{C}OCH_2CH_3 \quad (85\%)$$

The reaction with bromine is a classical test for the presence of double (or triple) bonds in an unknown compound. In the test, the unknown is added dropwise to a solution of bromine in a solvent such as CCl$_4$ or CH$_2$Cl$_2$. The bromine solution has a red-brown color. If the unknown contains carbon–carbon double or triple bonds, the addition reaction is nearly instantaneous. Because the addition products are colorless, the rapid disappearance of the bromine color constitutes a positive test for the presence of unsaturation.

The mechanism of these additions has an interesting variation from the one for the addition of the hydrogen halides. It begins with electrophilic attack of the halogen at the pi electrons of the double bond, as shown in Figure 11.1. As the bromine–bromine bond breaks in a heterolytic manner, the bromine that is becoming electron deficient acts as the electrophile and adds to the double bond. However, a carbocation is not produced in this addition. Instead, an unshared pair of electrons on the attacking bromine forms a bond to the other carbon. These three events happen simultaneously: the bromine–bromine bond breaks, the pi electrons form a bond between one carbon and the electrophilic bromine, and an unshared pair of electrons on the adding bromine forms a bond to the other carbon. The result is a three-membered ring containing a positively charged bromine—a **bromonium ion.** The bromonium ion is formed instead of a carbocation because the bromonium ion is more stable. It has an octet of electrons around both carbons and also the bromine. Having the octet rule satisfied provides more stabilization even though the positive charge is on the more electronegative bromine. Chlorine reacts by a similar mechanism with simple alkenes.

When we first encounter a new intermediate, such as the bromonium ion, we should compare it to other similar species that we have learned about to predict how it is likely to behave. What other three-membered rings containing a positively charged heteroatom have we seen? The answer is the protonated epoxide that was discussed in Section 10.10. The chemistry of the bromonium ion is very similar to that of the protonated epoxide. It reacts with nucleophiles—bromide ion, in this case—resulting in opening of the ring. As was the case with the protonated epoxide, this nucleophilic attack has characteristics of both the S$_N$2 and S$_N$1 reactions. The stereochemistry is that of an S$_N$2 reaction: the nucleophilic bromide approaches from the side opposite the leaving bromine. This results in an overall anti addition of the two bromines. The regiochemistry is that of an S$_N$1 reaction: the nucleophile bonds to the carbon that would be more stable as a carbocation because there is more positive charge located there. Of course, we cannot tell which bromine came in as the electrophile and which came in as the nucleophile in these reactions. However, we will encounter similar reactions later in which the electrophile and the nucleophile can be distinguished.

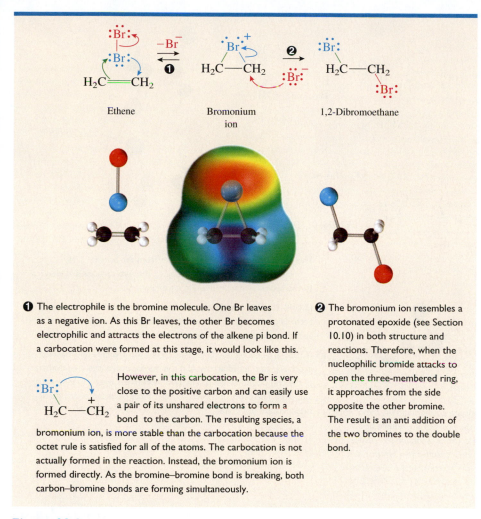

❶ The electrophile is the bromine molecule. One Br leaves as a negative ion. As this Br leaves, the other Br becomes electrophilic and attracts the electrons of the alkene pi bond. If a carbocation were formed at this stage, it would look like this.

However, in this carbocation, the Br is very close to the positive carbon and can easily use a pair of its unshared electrons to form a bond to the carbon. The resulting species, a bromonium ion, is more stable than the carbocation because the octet rule is satisfied for all of the atoms. The carbocation is not actually formed in the reaction. Instead, the bromonium ion is formed directly. As the bromine–bromine bond is breaking, both carbon–bromine bonds are forming simultaneously.

❷ The bromonium ion resembles a protonated epoxide (see Section 10.10) in both structure and reactions. Therefore, when the nucleophilic bromide attacks to open the three-membered ring, it approaches from the side opposite the other bromine. The result is an anti addition of the two bromines to the double bond.

Figure 11.1

MECHANISM OF THE ADDITION OF BROMINE TO ETHENE.

Let's examine the stereochemistry of the reaction of bromine with 2-butene:

$$CH_3CH=CHCH_3 + Br_2 \xrightarrow{CCl_4} \overset{\overset{\displaystyle Br \quad\;\; Br}{\displaystyle |\qquad |}}{CH_3CH-CHCH_3}$$

2-Butene   2,3-Dibromobutane

Considering stereochemistry, there are two stereoisomers of the starting alkene (cis and trans) and three stereoisomers of the product dibromide (the *d*-, *l*-, and *meso*-isomers). The reaction of bromine with (*Z*)-2-butene is shown in Figure 11.2, and the reaction of

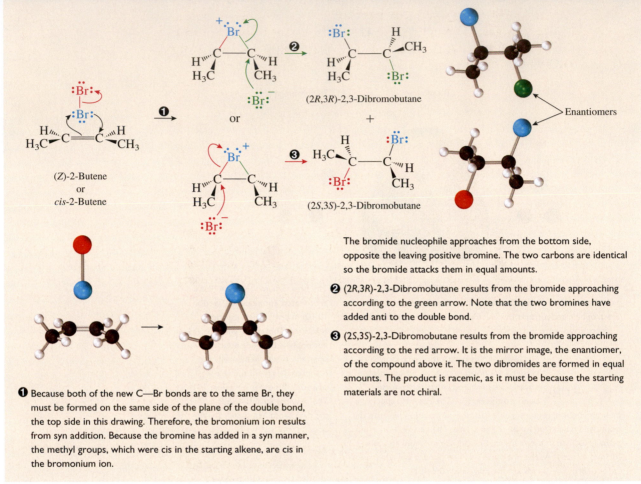

The bromide nucleophile approaches from the bottom side, opposite the leaving positive bromine. The two carbons are identical so the bromide attacks them in equal amounts.

❷ (2R,3R)-2,3-Dibromobutane results from the bromide approaching according to the green arrow. Note that the two bromines have added anti to the double bond.

❸ (2S,3S)-2,3-Dibromobutane results from the bromide approaching according to the red arrow. It is the mirror image, the enantiomer, of the compound above it. The two dibromides are formed in equal amounts. The product is racemic, as it must be because the starting materials are not chiral.

❶ Because both of the new C—Br bonds are to the same Br, they must be formed on the same side of the plane of the double bond, the top side in this drawing. Therefore, the bromonium ion results from syn addition. Because the bromine has added in a syn manner, the methyl groups, which were cis in the starting alkene, are cis in the bromonium ion.

Active Figure 11.2

ORGANIC
Chemistry ⚛ Now™

MECHANISM OF THE ADDITION OF BROMINE TO (Z)-2-BUTENE (CIS-2-BUTENE). Test yourself on the concepts in this figure at **OrganicChemistryNow.**

bromine with (E)-2-butene is shown in Figure 11.3. As can be seen from these figures, anti addition to (Z)-2-butene produces only (d,l)-2,3-dibromobutane as a racemic mixture. (Remember, if the reactants are not chiral, the products, if chiral, must be produced as a racemic mixture.) Anti addition to (E)-2-butene produces only the *meso*-diastereomer. The reactions are **stereospecific;** that is, one diastereomer of the reactant produces only one diastereomer of the product.

As expected for an electrophilic addition, the reaction rate increases as alkyl groups are substituted on the double bond. The electron-donating alkyl groups make the alkene more nucleophilic. Table 11.1 lists the relative rates of bromination of a series of alkenes. As can be seen from this table, replacing all four of the hydrogens of ethene with methyl groups results in an increase in the rate of the reaction by a factor of 2 million.

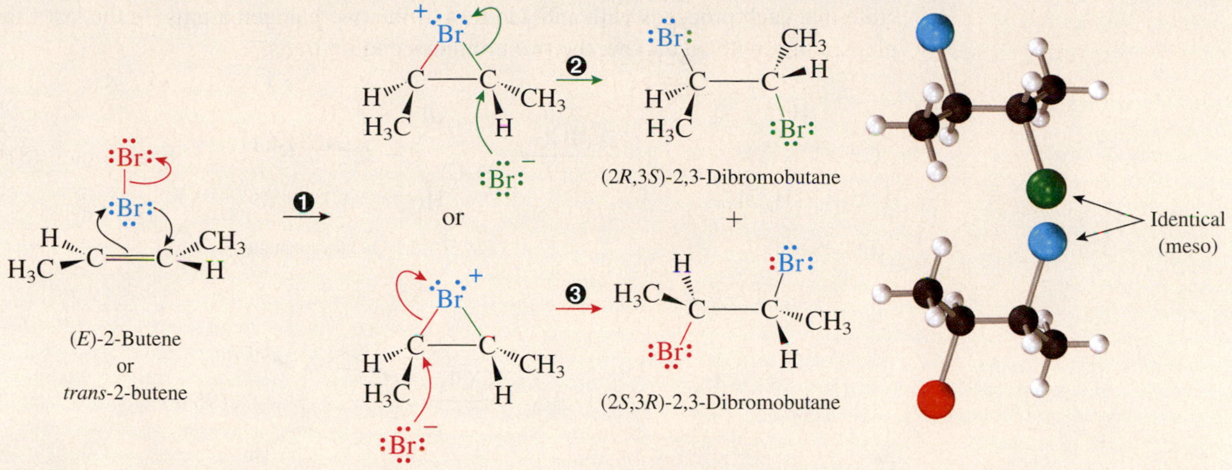

Active Figure 11.3

MECHANISM OF THE ADDITION OF BROMINE TO (*E*)-2-BUTENE (*TRANS*-2-BUTENE). Test yourself on the concepts in this figure at **OrganicChemistryNow.**

❶ The bromine has added in a syn manner, the methyl groups, which were trans in the starting alkene, are trans in the bromonium ion. The bromide nucleophile then approaches from the bottom side, opposite the leaving positive bromine. The two carbons are identical so the bromide attacks them in equal amounts.

❷ (2*R*,3*S*)-2,3-Dibromobutane results from the bromide approaching according to the green arrow. Note that the two bromines have added anti to the double bond.

❸ (2*S*,3*R*)-2,3-Dibromobutane results from the bromide approaching according to the red arrow. It is a superimposable mirror image of the compound above it. The two dibromides are identical—the product is *meso*-2,3-dibromobutane. If the compound is shown in its most symmetrical conformation, the internal plane of symmetry is apparent.

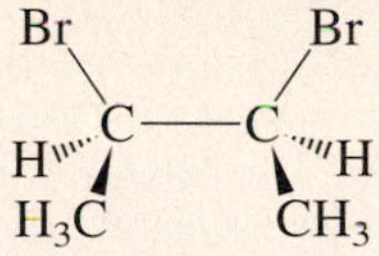

Table 11.1   Relative Rates of Reaction of Alkenes with Bromine

Alkene	Relative Rate	Alkene	Relative Rate
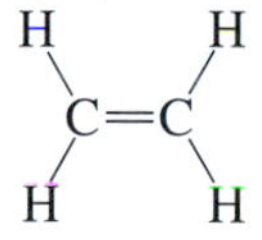 H₂C=CH₂	1	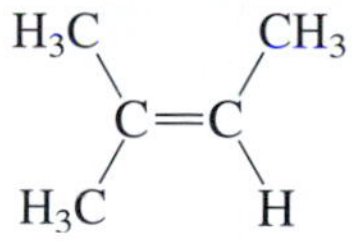 (CH₃)₂C=CHCH₃ (H₃C, CH₃ / C=C / H₃C, H)	1×10^5
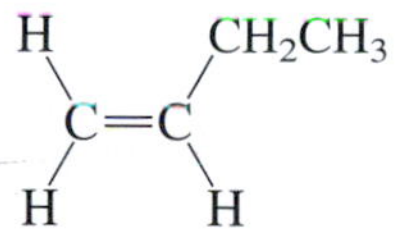 H,CH₂CH₃ / C=C / H,H	1×10^2	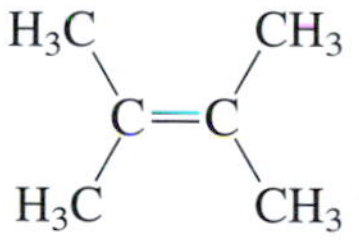 H₃C, CH₃ / C=C / H₃C, CH₃	2×10^6
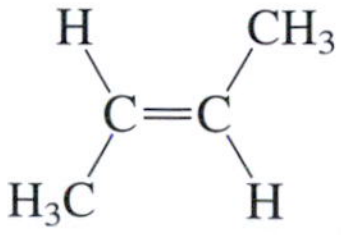 H,CH₃ / C=C / H₃C,H	2×10^3		

The reaction is the alkene plus Br₂ in methanol as solvent.

Some examples of these addition reactions are provided in the following equations. Note that each proceeds with anti addition of the two halogen atoms. In the last example, starting with an alkyne, the two bromines end up trans.

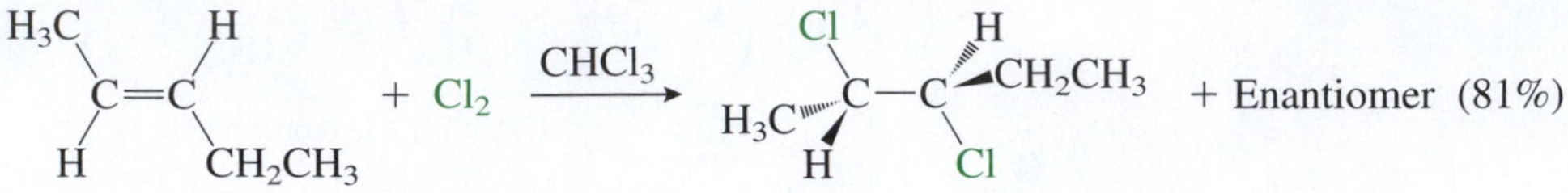

(E)-2-Pentene   (2S,3R)-2,3-Dichloropentane

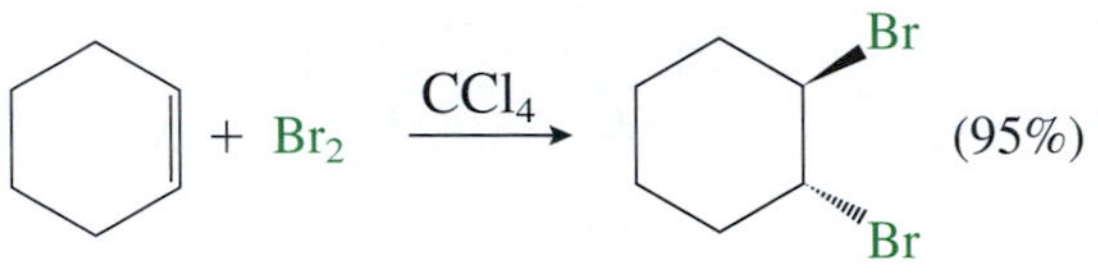

Cyclohexene   trans-1,2-Dibromocyclohexane

Racemic

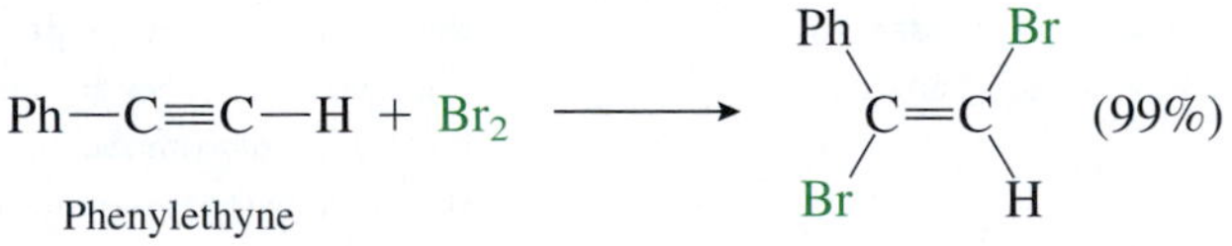

Phenylethyne   (E)-1,2-Dibromo-1-phenylethene

PROBLEM 11.7

Show the products of these reactions:

a) 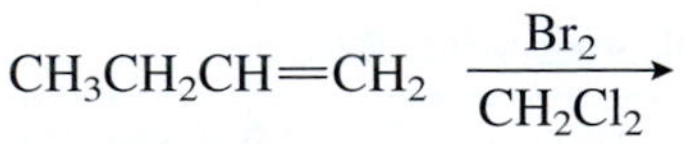 $CH_3CH_2CH{=}CH_2$ $\xrightarrow[CH_2Cl_2]{Br_2}$

b)  $\xrightarrow[CH_2Cl_2]{Br_2}$

c)  $\xrightarrow[CH_2Cl_2]{Cl_2}$

d) 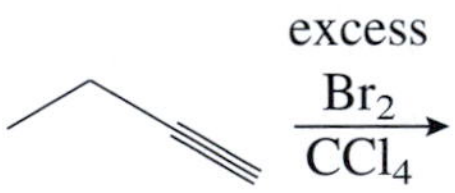 $\xrightarrow[CCl_4]{\text{excess} \atop Br_2}$

PROBLEM 11.8

Show all of the steps, including stereochemistry, in the mechanism for this reaction:

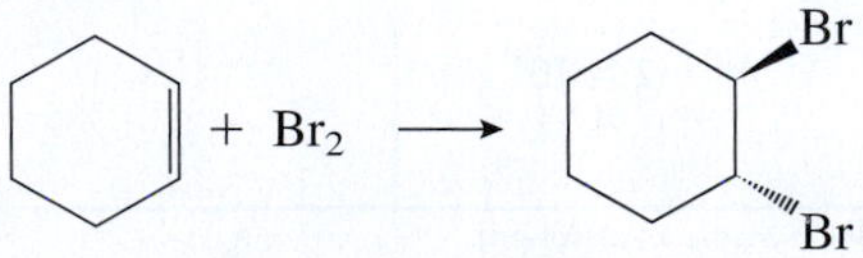

PROBLEM 11.9

Explain which of these compounds has the faster rate of reaction with Br_2:

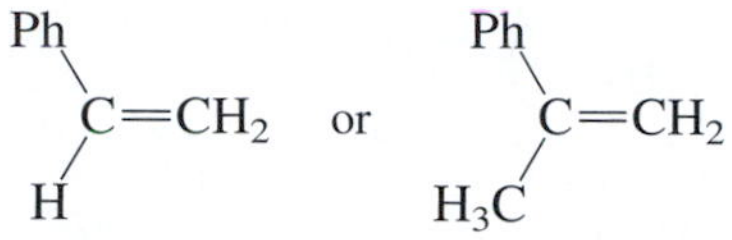

11.5 Halohydrin Formation

The reactions of chlorine and bromine with alkenes described in the previous section are conducted in inert solvents such as CCl_4 and CH_2Cl_2. In these reactions the only nucleophile that is present to react with the halonium ion is the halide anion. However, if the reaction is performed in a nucleophilic solvent, such as water, then a water molecule can act as the nucleophile, resulting in the addition of a halogen and a hydroxy group to the double bond.

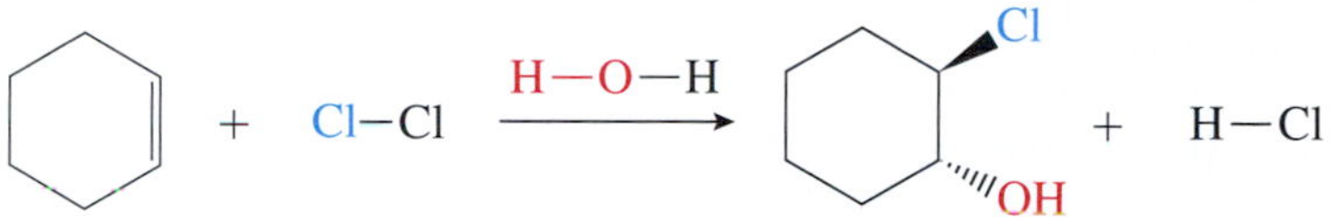

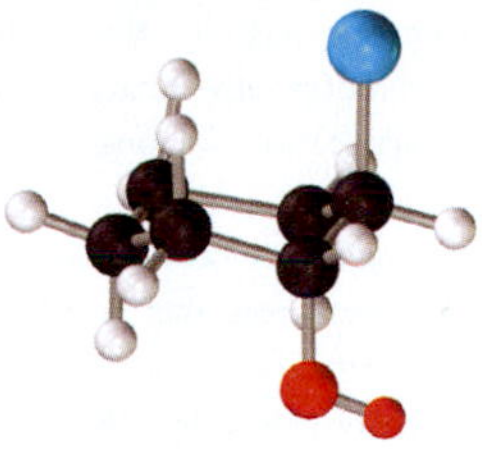

Cyclohexene

trans-2-Chlorocyclohexanol
+ Enantiomer
(a halohydrin)

The product is called a halohydrin. Because the concentration of the water molecules is so much higher than the concentration of the halide anions (water is the solvent), water wins the competition to act as the nucleophile, and only the halohydrin is formed in significant amounts.

The mechanism for this reaction is very similar to that for the addition of halogens as described in Section 11.4. First the chloronium ion is formed in exactly the same manner as before. Water then acts as a nucleophile, approaching the carbon from the side opposite the chlorine that it is displacing. This results in an anti addition of the chlorine and hydroxy group. As was the case for both the opening of the halonium ion ring (Section 11.4) and the opening of a protonated epoxide ring (Section 10.10), this opening of the halonium ion ring by water has characteristics of both an S_N1 and an S_N2 reaction. It proceeds with S_N2 stereochemistry (inversion) and S_N1 regiochemistry (the nucleophile attaches to the carbon that is more stable as a carbocation). The electrophile and the nucleophile are different here, so it is possible to see to which carbon each has added. The mechanism for this reaction is given in Figure 11.4 for an example in which both stereochemistry and regiochemistry of the reaction can be observed in the product.

One of the major uses of these halohydrins is for the preparation of epoxides. Treatment of the halohydrin with base, such as NaOH or KOH, results in deprotonation of the alcohol followed by an intramolecular nucleophilic substitution (see Section 10.3), as shown in the following example. Remember that the nucleophilic oxygen must displace the chlorine from the opposite side, resulting in inversion of configuration at that carbon.

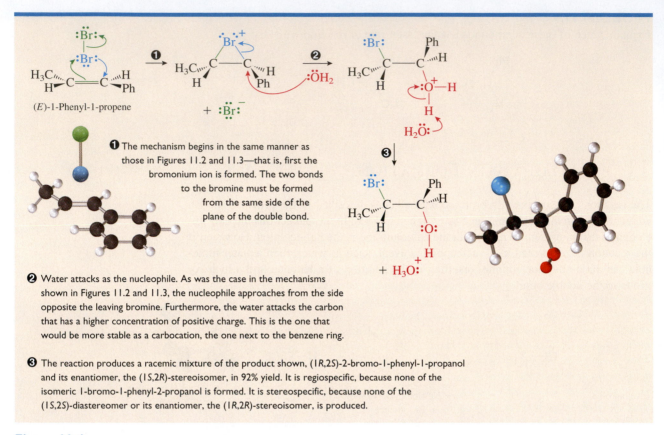

❶ The mechanism begins in the same manner as those in Figures 11.2 and 11.3—that is, first the bromonium ion is formed. The two bonds to the bromine must be formed from the same side of the plane of the double bond.

❷ Water attacks as the nucleophile. As was the case in the mechanisms shown in Figures 11.2 and 11.3, the nucleophile approaches from the side opposite the leaving bromine. Furthermore, the water attacks the carbon that has a higher concentration of positive charge. This is the one that would be more stable as a carbocation, the one next to the benzene ring.

❸ The reaction produces a racemic mixture of the product shown, (1R,2S)-2-bromo-1-phenyl-1-propanol and its enantiomer, the (1S,2R)-stereoisomer, in 92% yield. It is regiospecific, because none of the isomeric 1-bromo-1-phenyl-2-propanol is formed. It is stereospecific, because none of the (1S,2S)-diastereomer or its enantiomer, the (1R,2R)-stereoisomer, is produced.

Figure 11.4

MECHANISM OF THE ADDITION OF BROMINE TO AN ALKENE IN WATER.

PRACTICE PROBLEM 11.3

Show the product of this reaction:

$$CH_3CH_2CH{=}CH_2 \xrightarrow[H_2O]{Cl_2}$$

Solution

First a chloronium ion is formed. Then water acts as a nucleophile and attacks at the carbon of the chloronium ion that has more positive charge. In this case it is the secondary carbon rather than the primary carbon that is attacked by water:

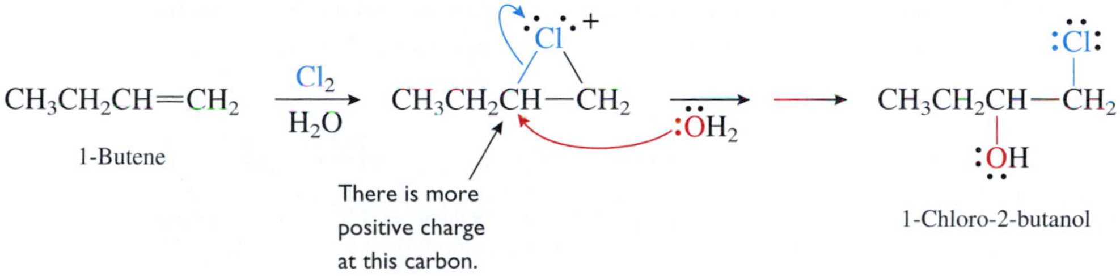

1-Butene

There is more
positive charge
at this carbon.

1-Chloro-2-butanol

PROBLEM 11.10

Show the products of these reactions.

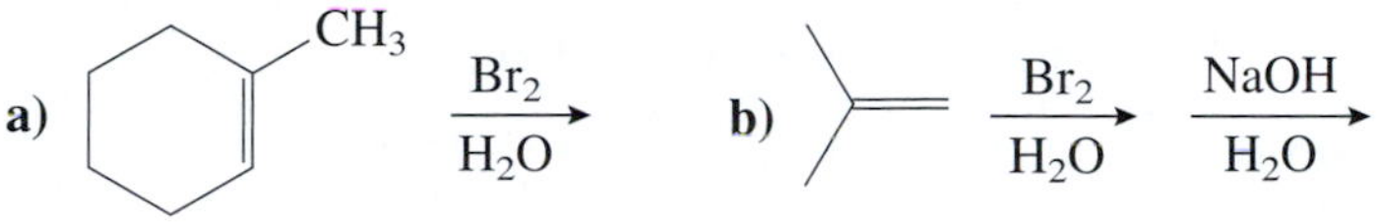

a)

$$\xrightarrow[H_2O]{Br_2}$$

b)

$$\xrightarrow[H_2O]{Br_2} \xrightarrow[H_2O]{NaOH}$$

PROBLEM 11.11

The reaction of an alkene with bromine in an alcohol as solvent produces an ether as the product. Show a mechanism for the following reaction and explain the stereochemistry of the product:

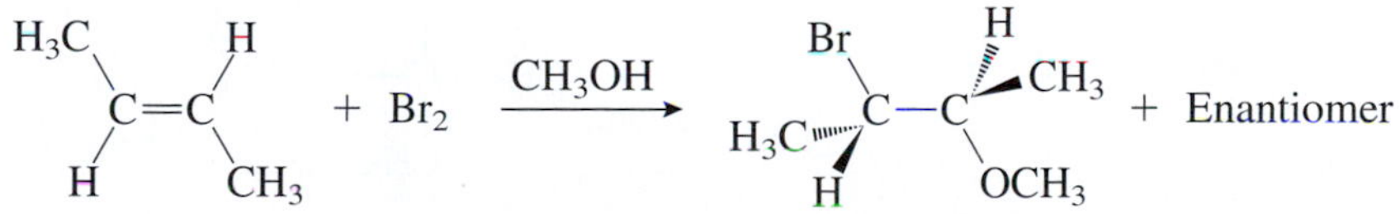

Focus On

Industrial Addition Reactions

Several addition reactions have been or are currently used on a large scale in industrial chemical plants. For example, an older method for the preparation of ethylene oxide employed the addition of chlorine to ethylene in water to form ethylene chlorohydrin or 2-chloroethanol. (In industry, ethene is almost always called ethylene.) Treatment of the chlorohydrin with calcium hydroxide results in the formation of ethylene oxide, which is an important intermediate in the manufacture of ethylene glycol and other products (see the Focus On box on page 375). However, this method is wasteful of

Continued

chlorine (Cl$_2$ is added, but there is no chlorine in the final product), so it has been re-placed with a more efficient process.

$$CH_2{=}CH_2 + Cl_2 \xrightarrow{\text{H}_2\text{O}} \underset{\underset{\text{Cl}}{|}}{CH_2}{-}\overset{\overset{\text{OH}}{|}}{CH_2} \xrightarrow{\text{Ca(OH)}_2} CH_2{-}CH_2$$

2-Chloroethanol (ethylene chlorohydrin)	Oxirane (ethylene oxide)

More than 18 billion pounds of ethylene dichloride (1,2-dichloroethane) are cur-rently manufactured each year by the addition of chlorine to ethylene. Elimination of HCl from ethylene dichloride produces vinyl chloride, which is the starting material for the production of poly(vinyl chloride), an important polymer:

$$CH_2{=}CH_2 + Cl_2 \longrightarrow \underset{\underset{\text{Cl}}{|}}{CH_2}{-}\underset{\underset{\text{Cl}}{|}}{CH_2} \xrightarrow{-\text{HCl}} CH_2{=}\overset{\overset{\text{Cl}}{|}}{CH}$$

1,2-Dichloroethane (ethylene dichloride)	Chloroethene (vinyl chloride)

An obsolete method for the production of vinyl chloride employed the addition of HCl to acetylene:

$$HC{\equiv}CH + H{-}Cl \longrightarrow CH_2{=}\overset{\overset{\text{Cl}}{|}}{CH}$$

Ethyne
(acetylene)

One of the fastest-growing petrochemicals of recent times was methyl *tert*-butyl ether (MTBE). More than 13 billion pounds per year of this chemical were manufac-tured by the acid-catalyzed addition of methanol to isobutylene:

$$\underset{}{\overset{\overset{\text{CH}_3}{|}}{CH_3C}}{=}CH_2 + CH_3OH \xrightarrow{\text{[HA]}} CH_3{-}\underset{\underset{\text{CH}_3}{|}}{\overset{\overset{\text{CH}_3}{|}}{C}}{-}O{-}CH_3$$

2-Methylpropene
(isobutylene)

MTBE

MTBE was initially added to unleaded gasoline to increase its octane rating when leaded gasoline was being phased out. Later it was added for environmental reasons. Gasoline that contains a small percentage of oxygen in its molecules burns cleaner and produces less car-bon monoxide than gasoline without oxygen. Therefore, certain large urban areas that suf-fer from carbon monoxide pollution require the addition of oxygenated compounds, either ethanol or MTBE, to gasoline during the winter months when this pollution is at its worst. Recently, however, MTBE has been found in drinking water supplies due to leakage from gasoline storage tanks. Because MTBE makes water smell and taste foul, even at very low concentrations—in addition to having health concerns—its use is being phased out.

PROBLEM 11.12
Show all of the steps in the mechanism for the formation of MTBE from methanol and isobutylene.

11.6 OXYMERCURATION–REDUCTION

In addition to the hydration reaction described in Section 11.3, the oxymercuration–reduction reaction can be used to add the elements of water to a carbon–carbon double bond in a two-step process. First the alkene is reacted with mercuric acetate, $Hg(O_2CCH_3)_2$, in water, followed by treatment with sodium borohydride in sodium hydroxide solution:

Cyclopentene   Cyclopentanol

Although this reaction involves two steps, they can be run sequentially in the same flask. This procedure is usually the preferred method for the hydration of an alkene because the yields are higher than the acid-catalyzed addition described in Section 11.3, and rearrangements do not occur.

Figure 11.5 shows a mechanism that has been postulated for this reaction. First, an electrophilic mercury species adds to the double bond to form a cyclic mercurinium ion. Note how similar this mechanism is, including its stereochemistry and regiochemistry, to that shown in Figure 11.4 for the formation of a halohydrin. The initial product results from anti addition of Hg and OH to the double bond. In the second step, sodium borohydride replaces the mercury with a hydrogen with random stereochemistry. (The mechanism for this step is complex and not important to us at this time.) The overall result is the addition of H and OH with Markovnikov orientation.

Some additional examples are provided by the following equations. Note the excellent yields in all of the examples. Also note that the last example proceeds without rearrangement because the intermediate is a mercurinium ion rather than a carbocation. Attempts to prepare this alcohol by acid-catalyzed addition of water result in completely rearranged product.

1-Hexene   2-Hexanol

Methylenecyclohexane   1-Methylcyclohexanol

3,3-Dimethyl-1-butene   3,3-Dimethyl-2-butanol

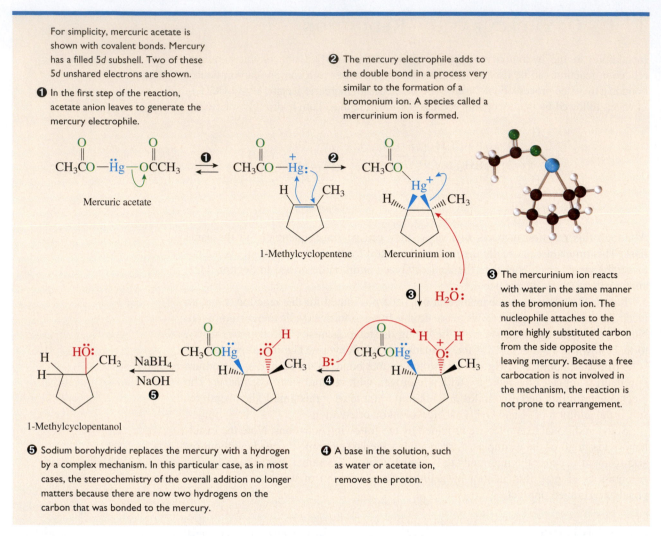

For simplicity, mercuric acetate is shown with covalent bonds. Mercury has a filled 5d subshell. Two of these 5d unshared electrons are shown.

❶ In the first step of the reaction, acetate anion leaves to generate the mercury electrophile.

❷ The mercury electrophile adds to the double bond in a process very similar to the formation of a bromonium ion. A species called a mercurinium ion is formed.

1-Methylcyclopentene

Mercurinium ion

❸ The mercurinium ion reacts with water in the same manner as the bromonium ion. The nucleophile attaches to the more highly substituted carbon from the side opposite the leaving mercury. Because a free carbocation is not involved in the mechanism, the reaction is not prone to rearrangement.

1-Methylcyclopentanol

❺ Sodium borohydride replaces the mercury with a hydrogen by a complex mechanism. In this particular case, as in most cases, the stereochemistry of the overall addition no longer matters because there are now two hydrogens on the carbon that was bonded to the mercury.

❹ A base in the solution, such as water or acetate ion, removes the proton.

Figure 11.5

MECHANISM OF THE HYDRATION OF AN ALKENE USING MERCURIC ACETATE.

The addition of water to alkynes is also aided by the presence of mercury (II) salts. The reaction is usually conducted in water, with the presence of a strong acid, such as sulfuric acid, and a mercury salt, such as $HgSO_4$ or HgO. In this case the mercury is spontaneously replaced by hydrogen under the reaction conditions, so a second step is not necessary. The addition occurs with a Markovnikov orientation; stereochemistry is not an issue.

$$CH_3CH_2CH_2CH_2C\equiv CH \xrightarrow[\substack{H_2SO_4 \\ Hg^{2+}}]{H_2O} \left[CH_3CH_2CH_2CH_2\overset{HO}{C}=\overset{H}{CH} \right] \longrightarrow CH_3CH_2CH_2CH_2\overset{O}{\overset{\|}{C}}CH_3$$

1-Hexyne   An enol   (90%)

2-Hexanone

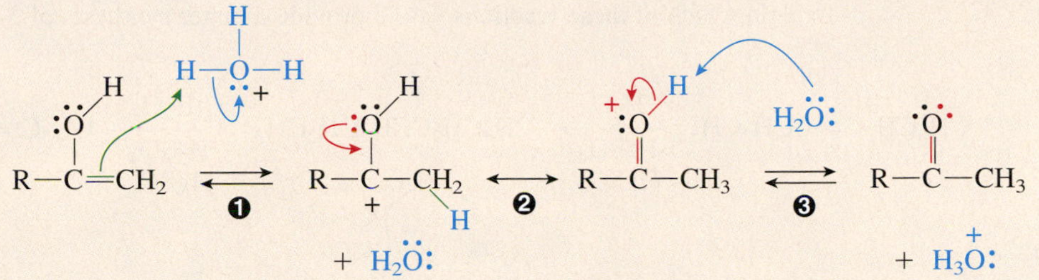

① In the presence of acid the double bond of the enol gets protonated. This step of the mechanism is identical to the first step of the mechanism for addition of the hydrogen halides and the acid-catalyzed addition of water to alkenes. The addition occurs so that the positive charge is located on the carbon that is bonded to the hydroxy group because this carbocation is stabilized by resonance.

② The resonance structure on the right is much more stable than the one on the left because the octet rule is satisfied for all of the atoms. The cation is actually the conjugate acid of a ketone. Because this cation is so much lower in energy than the usual carbocation, the transition state leading to it is also lower in energy (Hammond postulate). Thus, it is formed readily and the initial addition of the proton is very fast.

③ To complete the tautomerization, it is necessary only for a base, such as water, to remove the proton on the oxygen. The tautomerization is fast in both directions, but the equilibrium greatly favors the ketone tautomer.

Figure 11.6

MECHANISM OF THE TAUTOMERIZATION OF AN ENOL TO A KETONE.

The initial product has a hydroxy group attached to a carbon–carbon double bond. Compounds such as this are called **enols** (ene + ol) and are very labile—they cannot usually be isolated. Enols such as this spontaneously rearrange to the more stable ketone isomer. The ketone and the enol are termed **tautomers.** This reaction, which simply involves the movement of a proton and a double bond, is called a **keto–enol tautomerization** and is usually very fast. In most cases the ketone is much more stable, and the amount of enol present at equilibrium is not detectable by most methods. The mechanism for this tautomerization in acid is shown in Figure 11.6. The mercury-catalyzed hydration of alkynes is a good method for the preparation of ketones, as shown in the following example:

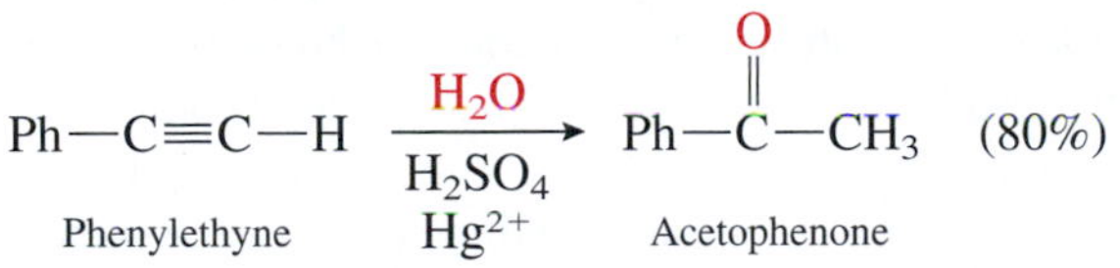

PROBLEM 11.13

Show the products of these reactions:

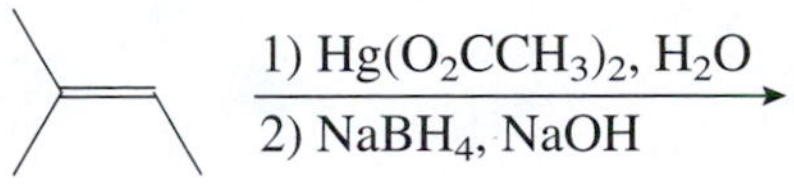

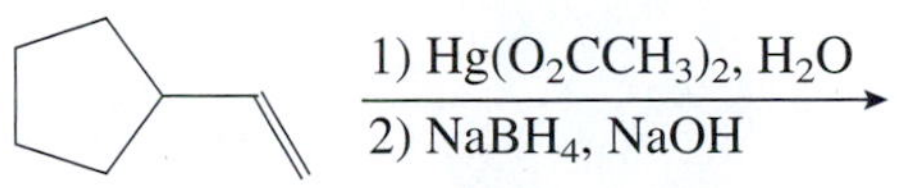

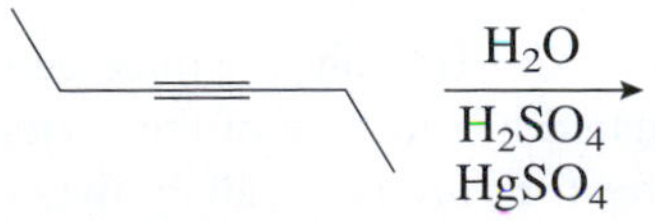

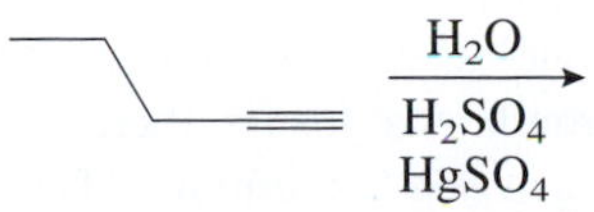

PROBLEM 11.14

Explain which of these reactions would provide a better synthesis of 3-hexanone.

$$CH_3CH_2C{\equiv}CCH_2CH_3 \xrightarrow[\substack{H_2SO_4 \\ HgSO_4}]{H_2O} \underset{\text{3-Hexanone}}{CH_3CH_2\overset{\overset{\displaystyle O}{\|}}{C}CH_2CH_2CH_3} \xleftarrow[\substack{H_2SO_4 \\ HgSO_4}]{H_2O} CH_3C{\equiv}CCH_2CH_2CH_3$$

11.7 HYDROBORATION–OXIDATION

Another method for the addition of the elements of water, H and OH, to an alkene was developed by H. C. Brown, who shared the 1979 Nobel Prize in chemistry for this work. In this reaction the alkene is first allowed to react with a complex of borane (BH_3) in tetrahydrofuran (THF). The initial product is then allowed to react with a basic solution of hydrogen peroxide. An example is shown in the following equation:

$$\underset{\text{1-Pentene}}{CH_3CH_2CH_2CH{=}CH_2} \xrightarrow[\text{2) } H_2O_2\text{, NaOH}]{\text{1) } BH_3\text{, THF}} \underset{\text{1-Pentanol}}{CH_3CH_2CH_2\overset{\overset{\displaystyle H}{|}}{C}H{-}\overset{\overset{\displaystyle OH}{|}}{C}H_2} \quad (95\%)$$

The most interesting feature of this reaction is its regiochemistry. The hydrogen has added to the carbon with more alkyl groups, and the hydroxy group has added to the carbon with more hydrogens, in opposition to Markovnikov's rule. Thus, the hydroboration–oxidation reaction results in **anti-Markovnikov addition** of water to the carbon–carbon double bond. It is just this feature, the opposite regiochemistry as compared to the hydration reaction or the oxymercuration–reduction process, that makes this reaction so valuable. Now we have a choice: we can add H and OH to an alkene with either regiochemistry!

Although the hydroboration–oxidation reaction gives a product with a regiochemistry opposite to that predicted by Markovnikov's rule, the regiochemistry is in accord with the mechanistic version of this rule—that is, the electrophile adds to the less substituted carbon. Let's look at the mechanism of this reaction.

Borane has the same structure as a carbocation. The boron is sp^2 hybridized, with trigonal planar geometry, and has an empty p orbital. Although neutral, it is electron deficient because there are only six electrons around the boron. It is a strong Lewis acid. An electron-deficient compound often employs unusual bonding to alleviate somewhat its instability. In the case of borane, two molecules combine to form one molecule of diborane:

Borane   Diborane

Diborane has some very unusual bonds. The two B—H—B bonds are three-center, two-electron bonds; that is, there are two electrons shared by all three of these atoms, one hydrogen and two borons. That is the only way for the borons to satisfy the octet rule in this electron-deficient species. In tetrahydrofuran solution, borane is present as a complex with the oxygen of the tetrahydrofuran, which acts as a Lewis base.

Borane–tetrahydrofuran complex

However, it is easiest to understand the hydroboration reaction by considering the reactive species to be BH_3 itself.

In the first step of the reaction, as shown in Figure 11.7, borane adds to the alkene. The reaction is initiated by the electrophilic boron reacting with the pi electrons of the carbon–carbon double bond. However, a carbocation is not formed. Instead, as positive charge begins to build up on the carbon, the hydrogen is delivered from the boron to this carbon as a nucleophile. The overall process is concerted—two bonds (C—B and C—H) are made and two bonds (B—H and C—C pi) are broken simultaneously in a cyclic, four-membered transition state (see Figure 11.8). Each of the boron–hydrogen bonds is reactive, so the process is repeated twice more to produce a trialkylborane. The second step of the reaction employs hydrogen peroxide to convert the carbon–boron bonds to carbon–oxygen bonds.

The regiochemistry of the reaction is in accord with the rule that the electrophile—the boron, in this case—adds to the carbon that is bonded to more hydrogens. The reason given previously for this orientation was that the electrophile adds so as to produce the more stable carbocation. As shown in Figure 11.7, no carbocation is involved in the

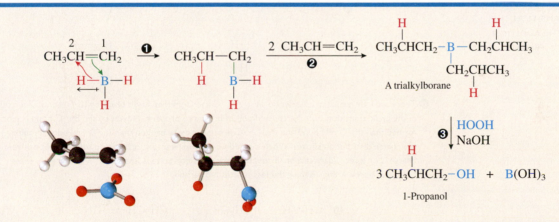

❶ Boron, with its empty *p* orbital, that acts as the electrophile. As positive charge begins to build up on C-2, the hydrogen, which has a partial negative charge because it is more electronegative than the boron, acts as a nucleophile. The C—H and C—B bonds are formed and the B—H and C—C pi bonds are broken in a single step. In accord with the mechanistic version of Markovnikov's rule, the electrophile, the boron, adds to the carbon bonded to more hydrogens.

❷ Both of the other two B—H bonds also add to alkene molecules. The overall result is that one borane molecule reacts with three alkene molecules to produce a boron attached to three alkyl groups, a trialkylborane.

❸ In the next step of the reaction, hydrogen peroxide breaks each C—B bond, replacing them with C—OH and B—OH bonds. (An understanding of the mechanism of this step is not important.)

Figure 11.7

MECHANISM FOR THE HYDROBORATION OF PROPENE.

mechanism of the hydroboration reaction. Why, then, is the rule still followed? Although a fully charged carbocation is not formed in the reaction, there is some charge buildup in the transition state. The boron–carbon bond is somewhat more formed in the transition state than is the hydrogen–carbon bond. In addition, the pi bond is somewhat more broken than is the boron–hydrogen bond. As shown in Figure 11.8, this results in some negative charge buildup on the boron and some positive charge buildup on the carbon. The more stable transition state has the positive charge buildup on the carbon that can best stabilize it, the one with more alkyl substituents. To form this transition state, the electrophile must add to the carbon attached to more hydrogens. Because the charge on the carbon is less than a full unit of positive charge, the difference in the energies of the two transition states is smaller than was the case for the addition reactions that proceed through carbocation intermediates. A small amount of the reaction does proceed through the less stable transition state, and the reaction is regioselective but not regiospecific. However, the amount of the minor product is usually small enough that it can be ignored. Steric factors are also important in these reactions. Thus, the boron, being larger than the hydrogen, prefers to add to the less hindered carbon of the double bond. This steric effect favors the same transition state as the electronic factors—the less hindered carbon is the one bonded to more hydrogens.

What about the stereochemistry of the reaction? Because the boron and the hydrogen are bonded to each other, in a concerted reaction they must add in a syn manner. There is no way that the boron could add to one side of the double bond and the hydrogen bonded to it could add to the other side. However, of more concern to us is the stereochemistry of the hydroxy group. The hydroxy group replaces the boron with complete retention of con-

Figure 11.8

POSSIBLE TRANSITION STATES FOR THE ADDITION OF BORANE TO PROPENE.

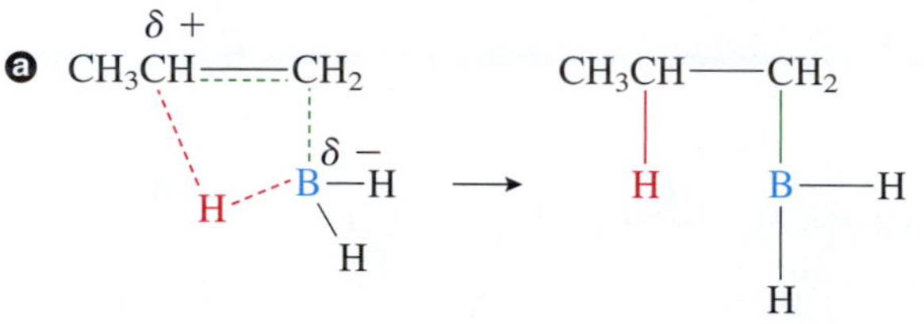

ⓐ There are four partial bonds, two forming (B—C and C—H) and two breaking (B—H and C—C pi) in the cyclic transition state for the addition of borane to propene. However, the B—C bond is more formed than the C—H bond and the C—C pi bond is more broken than the B—H bond, resulting in the charge distribution shown. When the addition occurs with this orientation, the partial positive charge is on a secondary carbon. In addition, this transition state is less sterically hindered because the larger boron is approaching the less substituted carbon.

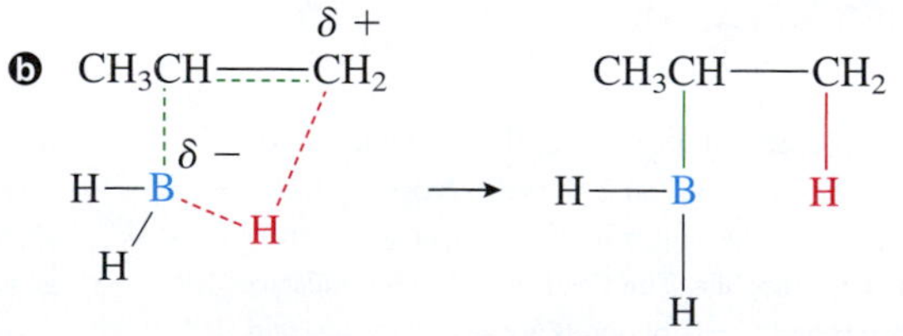

ⓑ When the addition occurs with this orientation, the partial positive charge is on a primary carbon. The charge is less stabilized here than in case ⓐ, so this transition state is less favorable than that of case ⓐ. This transition state is also less favorable for steric reasons.

The difference in transition state energies is enough that most of the reaction follows the orientation of case ⓐ where the electrophile has added to the carbon attached to more hydrogens. Although some of the product is formed by the orientation of case ⓑ, the amount is small enough that it can be ignored for most purposes.

figuration. Therefore, the reaction occurs in a regioselective and stereospecific manner: the H and the OH are added with anti-Markovnikov orientation and syn stereochemistry. Some illustrative examples are provided in the following equations. Pay particular attention to both the regiochemistry and the stereochemistry in the last example.

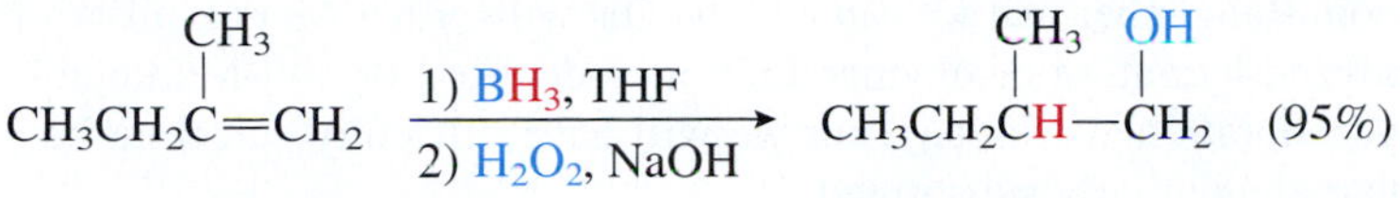

$$
\underset{\text{2-Methyl-1-butene}}{\text{CH}_3\text{CH}_2\overset{\overset{\displaystyle \text{CH}_3}{|}}{\text{C}}=\text{CH}_2}
\quad
\xrightarrow[\text{2) H}_2\text{O}_2,\ \text{NaOH}]{\text{1) BH}_3,\ \text{THF}}
\quad
\underset{\text{2-Methyl-1-butanol}}{\text{CH}_3\text{CH}_2\overset{\overset{\displaystyle \text{CH}_3}{|}}{\text{CH}}-\overset{\overset{\displaystyle \text{OH}}{|}}{\text{CH}_2}}
\quad (95\%)
$$

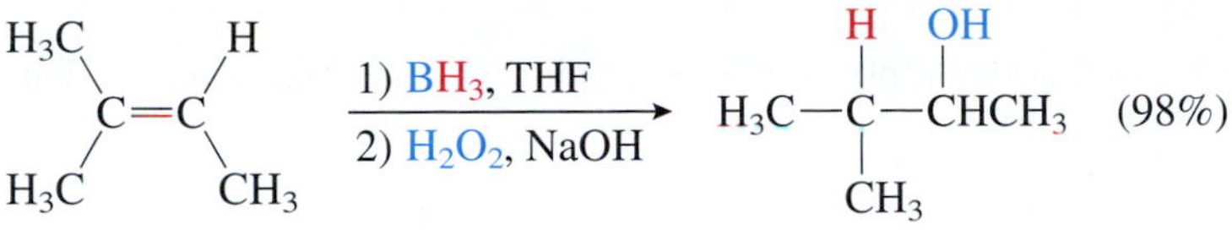

$$
\underset{\text{2-Methyl-2-butene}}{\underset{\text{H}_3\text{C}}{\overset{\text{H}_3\text{C}}{>}}\text{C}=\text{C}\underset{\text{CH}_3}{\overset{\text{H}}{<}}}
\quad
\xrightarrow[\text{2) H}_2\text{O}_2,\ \text{NaOH}]{\text{1) BH}_3,\ \text{THF}}
\quad
\underset{\text{3-Methyl-2-butanol}}{\text{H}_3\text{C}-\overset{\overset{\displaystyle \text{H}}{|}}{\underset{\underset{\displaystyle \text{CH}_3}{|}}{\text{C}}}-\overset{\overset{\displaystyle \text{OH}}{|}}{\text{CH}}\text{CH}_3}
\quad (98\%)
$$

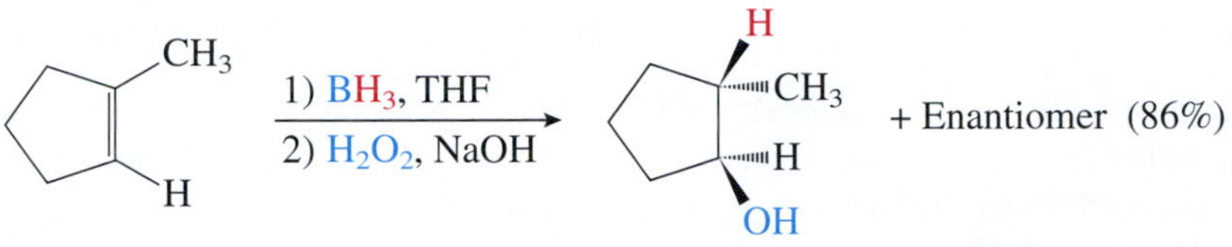

1-Methylcyclopentene → trans-2-Methylcyclopentanol + Enantiomer (86%)

PROBLEM 11.15

Show the products of these reactions:

a) 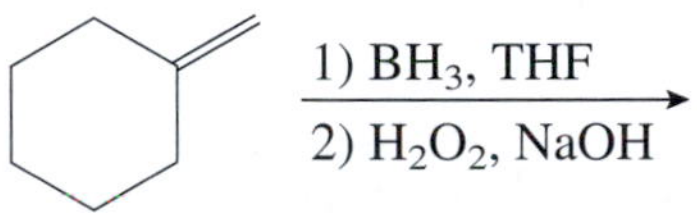 [structure] 1) BH$_3$, THF   2) H$_2$O$_2$, NaOH

b) 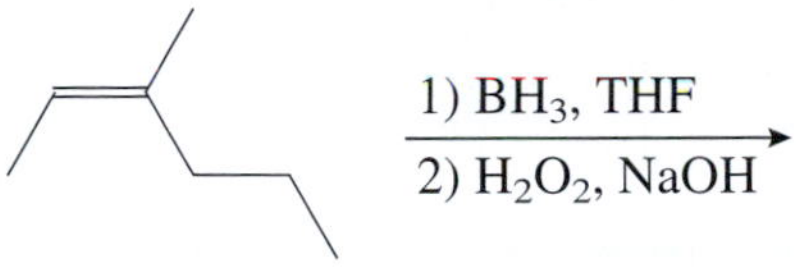 [structure] 1) BH$_3$, THF   2) H$_2$O$_2$, NaOH

c) 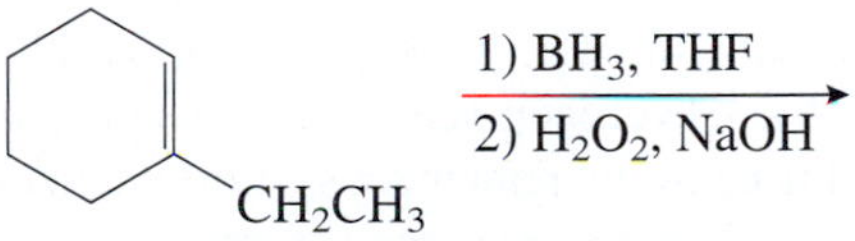 [structure] CH$_2$CH$_3$ 1) BH$_3$, THF   2) H$_2$O$_2$, NaOH

PROBLEM 11.16

This hydroboration reaction forms two products. Show these products and explain which one you expect to be major.

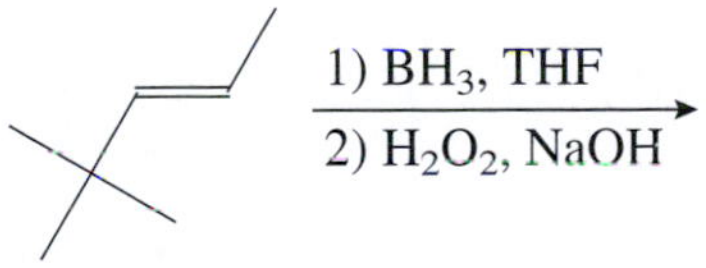

[structure] 1) BH$_3$, THF   2) H$_2$O$_2$, NaOH

PRACTICE PROBLEM 11.4

Show how to prepare this alcohol from an alkene:

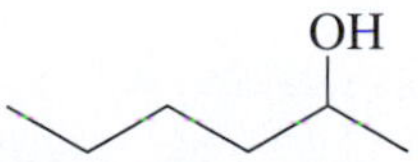

[structure with OH]

Strategy

The hydroxy group must be located on one of the doubly bonded carbons of the original alkene, so first draw all of the alkenes that meet this criterion. Examine the alkenes to determine whether it is possible to selectively add the OH group to the desired carbon. Remember that we can add the OH with either Markovnikov orientation (acid-catalyzed hydration or oxymercuration–reduction) or anti-Markovnikov orientation (hydroboration–oxidation), but we will have difficulty selecting between two carbons that are similarly substituted.

Solution

The two alkenes that could potentially be used to prepare 2-hexanol are 1-hexene and 2-hexene:

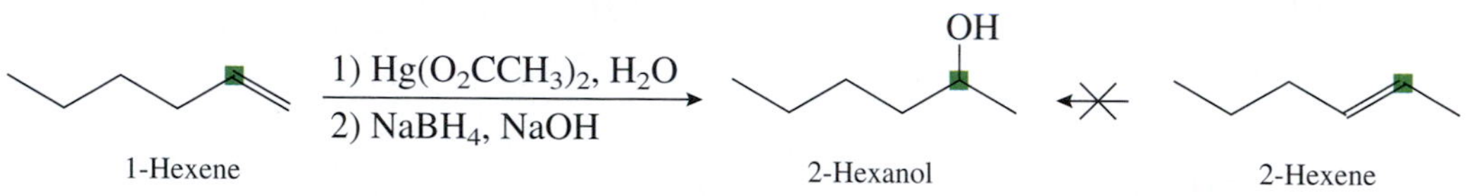
1) Hg(O2CCH3)2, H2O
2) NaBH4, NaOH
OH
1-Hexene
2-Hexanol
2-Hexene

It is possible to regiospecifically add the hydroxy group to carbon 2 of 1-hexene by using either the oxymercuration reaction or acid-catalyzed hydration. However, it is not possible to selectively add a hydroxy group only to carbon 2 of 2-hexene because both carbons are monosubstituted. Therefore, the path starting from 1-hexene should be used.

PROBLEM 11.17

Show preparations of these alcohols from alkenes.

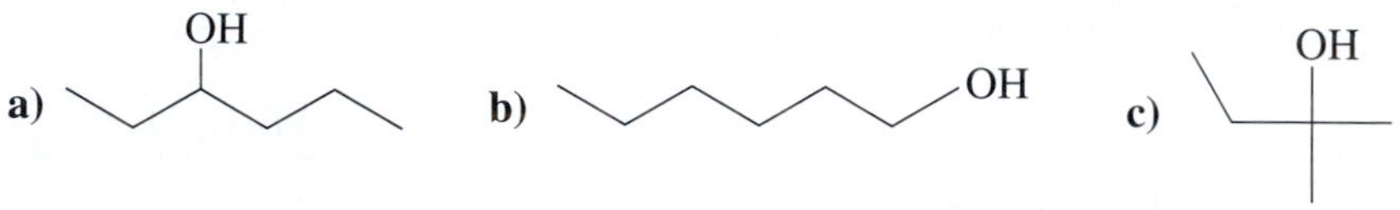
a)
OH
b)
OH
c)
OH

The hydroboration reaction is also useful with alkynes. As is shown in the following example, the product after treatment with basic hydrogen peroxide is an enol. As we have seen before, the enol cannot be isolated because it spontaneously tautomerizes to a ketone. This provides another way to hydrate alkynes to produce ketones.

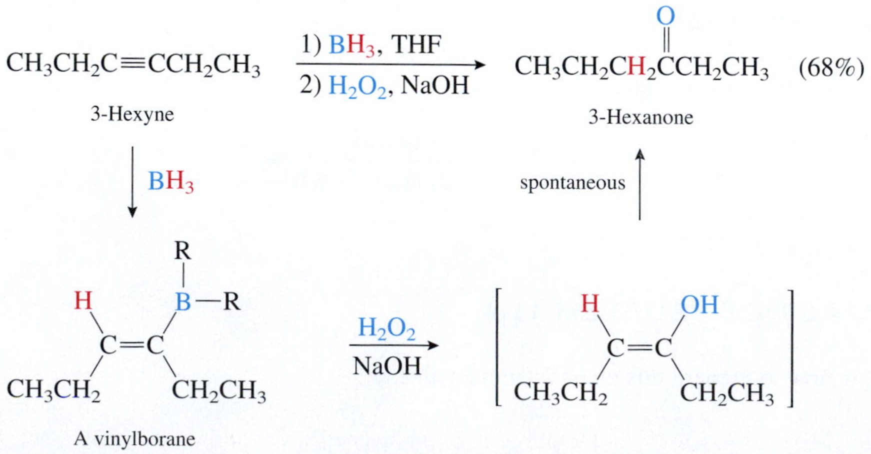
CH3CH2C≡CCH2CH3
1) BH3, THF
2) H2O2, NaOH
O
CH3CH2CH2CCH2CH3
(68%)
3-Hexyne
3-Hexanone
BH3
spontaneous
R
H
B—R
C=C
CH3CH2
CH2CH3
H2O2
NaOH
H
OH
C=C
CH3CH2
CH2CH3
A vinylborane

As mentioned previously, the addition of borane, or an alkyl-substituted borane, to a carbon–carbon double bond is very sensitive to steric hindrance. The preceding vinylborane does not add a second boron because of the steric bulk of the ethyl group and the boron group on the end of the double bond. If the boron is considered to be about the same size as a carbon, then this vinylborane corresponds to a trisubstituted alkene.

3-Hexyne has the triple bond in the middle of a carbon chain and is termed an internal alkyne. If, instead, an alkyne with the triple bond at the end of the carbon chain, a 1-alkyne or a terminal alkyne, were used in this reaction, then the reaction might be useful for the synthesis of aldehydes. The boron is expected to add to the terminal carbon of a 1-alkyne. Reaction with basic hydrogen peroxide would produce the enol resulting from anti-Markovnikov addition of water to the alkyne. Tautomerization of this enol would produce an aldehyde. Unfortunately, the vinylborane produced from a 1-alkyne reacts with a second equivalent of boron as shown in the following reaction. The product, with two borons bonded to the end carbon, does not produce an aldehyde when treated with basic hydrogen peroxide.

$$CH_3CH_2CH_2CH_2C\equiv C-H \xrightarrow{BH_3} \quad \text{(vinylborane)} \longrightarrow \quad \text{(diboron product)}$$

1-Hexyne

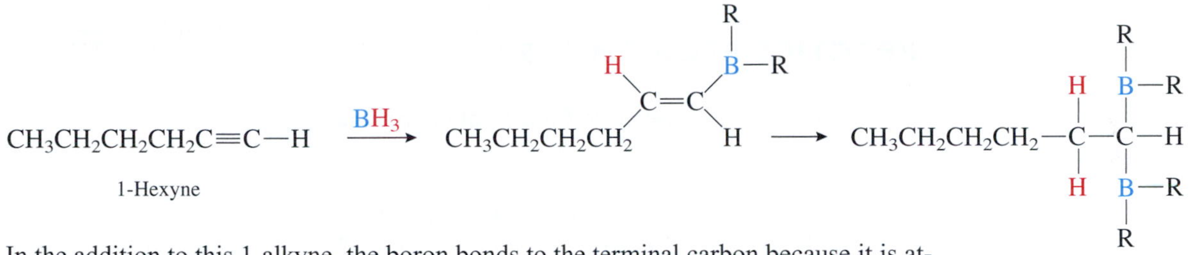

In the addition to this 1-alkyne, the boron bonds to the terminal carbon because it is attached to more hydrogens. The resulting vinylborane can be viewed as a disubstituted alkene and is less hindered than the vinylborane produced from an internal alkyne (a trisubstituted alkene). Because the vinylborane is less hindered, it adds a second boron to produce an alkane substituted on the end carbon with two boron groups.

If the hydroboration reaction is to be used to convert 1-alkynes into aldehydes, some way to stop the addition at the vinylborane stage is needed. The problem is that there is not enough steric hindrance at the end carbon of the vinylborane. The solution is to build extra steric hindrance into the other alkyl groups attached to the boron of the vinylborane. A borane, R_2BH, with two bulky R groups already attached to the boron is used as the hydroboration reagent. One such reagent is prepared by the reaction of two equivalents of 2-methyl-2-butene (also known by the common name of isoamylene) with borane to produce a dialkylborane called disiamylborane (a shortened version of diisoamylborane):

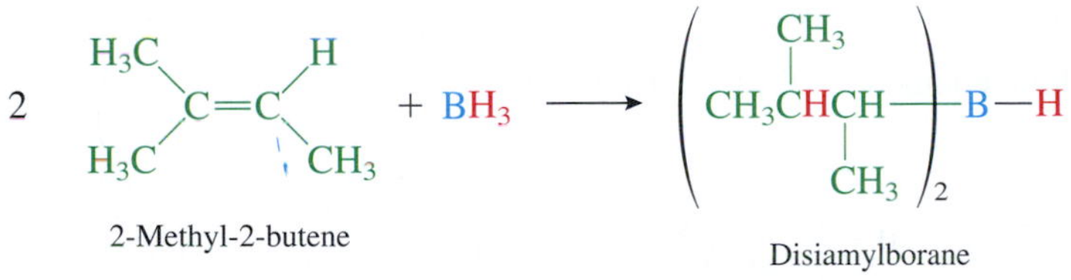

$$2 \quad \begin{array}{c} H_3C \\ H_3C \end{array}\!\!\!C\!\!=\!\!C\!\!\!\begin{array}{c} H \\ CH_3 \end{array} + BH_3 \longrightarrow \left(CH_3CHCH\!\!-\!\!\!\begin{array}{c} CH_3 \\ | \\ | \\ CH_3 \end{array} \right)_2 \!\!\!\!B\!-\!H$$

2-Methyl-2-butene   Disiamylborane

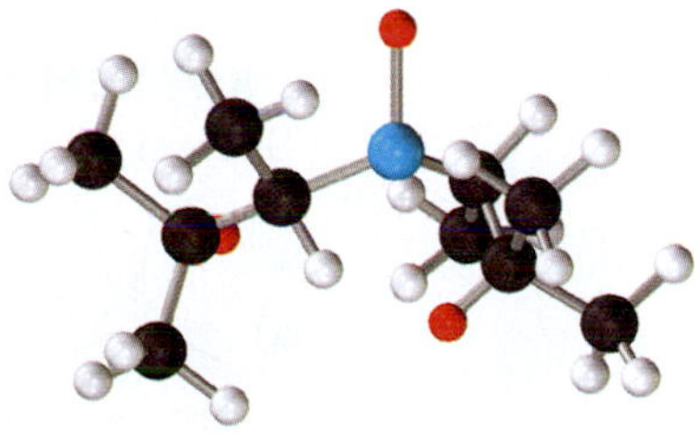

Once there are two of the bulky alkyl groups attached to the boron, it is difficult to add a third group. Disiamylborane is not reactive enough to add to isoamylene, a trisubstituted alkene, because of these steric effects. However, it is reactive enough to add to a 1-alkyne to produce a vinylborane, but it is too sterically hindered to react with the vinylborane product. Oxidation of the vinylborane with basic hydrogen peroxide produces an aldehyde in excellent overall yield, as shown in the following example:

$$CH_3CH_2CH_2CH_2C\equiv CH \xrightarrow[\text{2) } H_2O_2,\text{ NaOH}]{\text{1) disiamylborane}} CH_3CH_2CH_2CH_2CH_2\overset{\overset{\displaystyle O}{\|}}{C}H \quad (88\%)$$

1-Hexyne   Hexanal

PROBLEM 11.18

Show the products of these reactions:

a)  $\xrightarrow[\text{2) } H_2O_2,\text{ NaOH}]{\text{1) } BH_3,\text{ THF}}$

b) 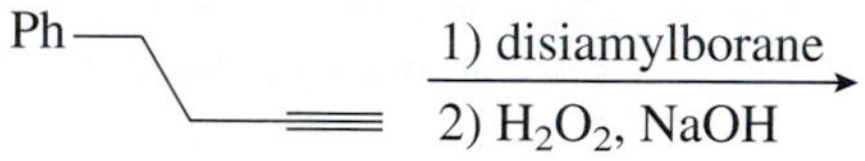   Ph $\xrightarrow[\text{2) } H_2O_2,\text{ NaOH}]{\text{1) disiamylborane}}$

PRACTICE PROBLEM 11.5

Show a synthesis of 2-pentanone from 1-chloropropane:

$$CH_3CH_2CH_2\overset{\overset{\displaystyle O}{\|}}{C}CH_3$$

Solution

Remember to work backward. The target ketone has five carbons, whereas the designated starting material has only three, so it is necessary to form a carbon–carbon bond. A nucleophilic substitution reaction can be done at C-1 of 1-chloropropane, so a two-carbon nucleophile that can be ultimately converted to a ketone is required. A carbon–carbon bond-forming reaction that meets these requirements is the alkylation of an acetylide anion (see Section 10.8). Once the carbon–carbon bond has been formed, hydration of the alkyne can be used to convert the triple bond to a ketone:

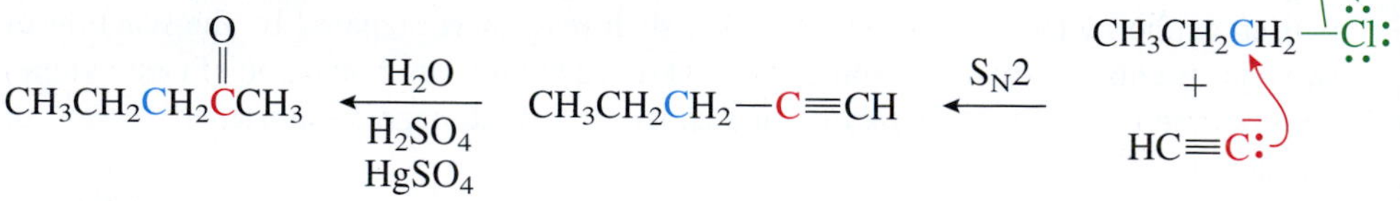

$$CH_3CH_2CH_2\overset{\overset{\displaystyle O}{\|}}{C}CH_3 \xleftarrow[\substack{H_2SO_4\\HgSO_4}]{H_2O} CH_3CH_2CH_2-C\equiv CH \xleftarrow{S_N2} \begin{array}{c} CH_3CH_2CH_2-\ddot{C}l\!: \\ + \\ HC\equiv C\!:\end{array}$$

PROBLEM 11.19

Show syntheses of these compounds from 1-bromobutane:

a) 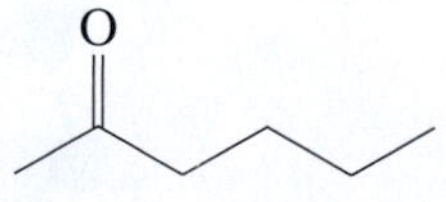   b)

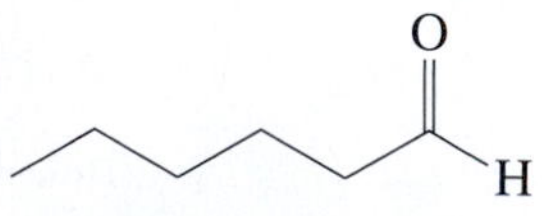

c)

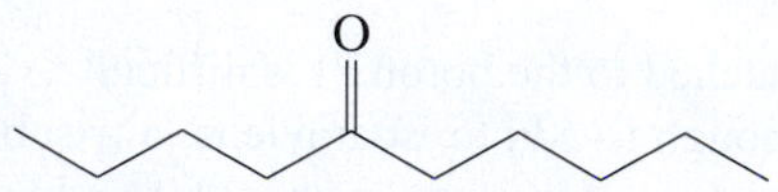

Focus On

Chiral Boranes in Organic Synthesis

As we saw in Chapter 7, one of the goals of synthetic organic chemistry is to develop methods that produce only a single enantiomer of a desired chiral compound rather than the usual racemic mixture that is the result of most reactions. Recently, a method has been developed that employs one enantiomer of a chiral borane to prepare a single enantiomer of a chiral alcohol from an achiral alkene. The chiral borane that is used is *trans*-2,5-dimethylborolane.

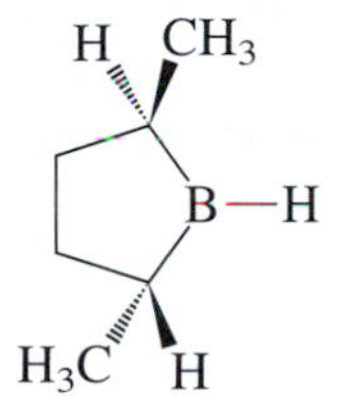

trans-2,5-Dimethylborolane

The stereoisomer of *trans*-2,5-dimethylborolane shown here has the *R* configuration at both stereocenters. Reaction of this chiral borane with *trans*-2-butene produces 2-butanol in 71% yield. The 2-butanol is almost enantiomerically pure. It consists of 99.5% of the (*S*)-enantiomer and only 0.5% of the (*R*)-enantiomer. This occurs because the transition state leading to the (*S*)-enantiomer is much more favorable on steric grounds than is the transition state leading to the (*R*)-enantiomer.

The approach of *trans*-2-butene to the chiral borane is shown in the following diagram.

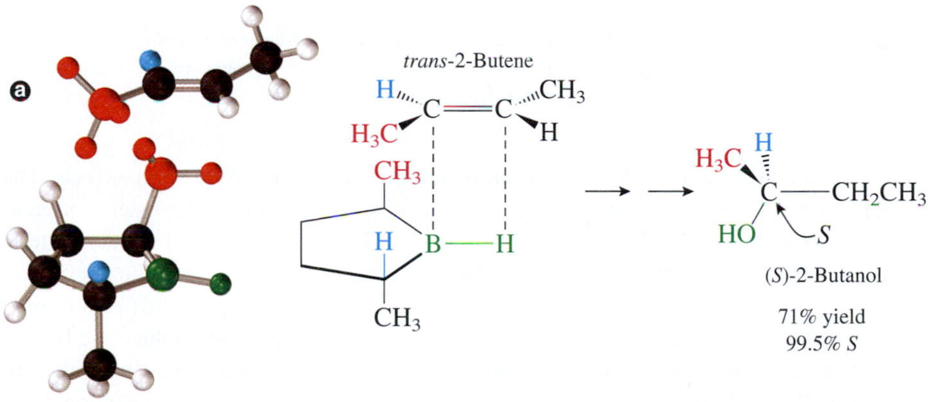

ⓐ The blue hydrogen of the alkene interacts with the red methyl group of the borane while the red methyl group of the alkene interacts with the blue hydrogen of the borane. This geometry minimizes the steric repulsions as the molecules approach to form the transition state. After oxidation of the carbon–boron bond to a carbon–oxygen bond, which occurs with retention of configuration, 2-butanol is produced in 71% yield. The reaction is remarkably selective. The product is almost entirely (*S*)-2-butanol (99.5%). It is contaminated with only a very small amount (0.5%) of the enantiomeric (*R*)-2-butanol.

Continued

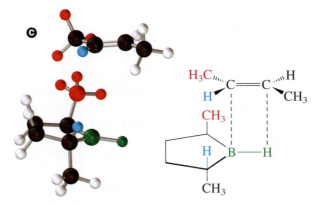

ⓑ This is another view of the geometry of the reaction, with the alkene approaching from above the borane. Again it can be seen that the steric interactions between the borane and the alkene are minimized in this approach; the smaller group on the alkene (the blue hydrogen) interacts with the larger group on the borane (the red methyl) and the larger group on the alkene (the red methyl) interacts with the smaller group (the blue hydrogen) on the borane.

ⓒ This approach would lead to the product that is not formed, (*R*)-2-butanol. It is much less sterically favorable than **ⓐ** because the larger groups (the red methyl groups) on the alkene and the borane are interacting and hinder the formation of the transition state.

As we have seen, the hydroboration reaction is very sensitive to steric effects. The chiral borane approaches the alkene in such a manner as to minimize steric interactions. As shown in **ⓐ**, the transition state that leads to (*S*)-2-butanol has fewer, steric interactions than the transition state in **ⓒ** that leads to (*R*)-2-butanol. The overall selectivity of the reaction is very high; the product is almost entirely the (*S*)-enantiomer. Note that, to be successful in producing a single enantiomer of the product, the borane must be enantiomerically pure. Any of the (*S,S*)-enantiomer that is present in the borane will result in the formation of an equal amount of the other enantiomer of 2-butanol, the (*R*)-enantiomer in this case.

This reaction meets one of the criteria of a successful chiral synthesis; that is, it produces the product in high enantiomeric purity. Its drawback is that the chiral borane must be prepared and resolved, a somewhat laborious and expensive process. A more ideal process would employ the chiral reagent as a catalyst in a reaction that produces the desired product in high enantiomeric purity but requires the investment of only a small amount of the expensive chiral reagent (see the Focus On box on asymmetric hydrogenation later in this chapter).

PROBLEM 11.20

Show the product of this reaction:

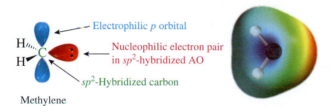

PROBLEM 11.21

Explain how a similar hydroboration reaction could be used to prepare (*R*)-2-butanol in good enantiomeric excess.

11.8 ADDITION OF CARBENES

A **carbene** is a reactive species having a carbon with only two bonds and an unshared pair of electrons. It is quite unstable and, like a carbocation, exists only as a transient intermediate in certain reactions. The simplest carbene, CH_2, is called methylene. It is sp^2 hybridized. The unshared pair of electrons occupies one of the sp^2-hybridized AOs and the other two are used to form the bonds to the hydrogens. The remaining AO on the carbon is an unoccupied *p* orbital.

Electrophilic *p* orbital

Nucleophilic electron pair in sp^2-hybridized AO

sp^2-Hybridized carbon

Methylene

The empty *p* orbital is electrophilic, and the unshared pair of electrons is nucleophilic. A carbene resembles both the electrophilic bromine that we encountered in the addition of Br_2 to alkenes and the electrophilic mercury species that we encountered in the oxymercuration reaction in that it has both an electrophilic site (an empty orbital) and a nucleophilic site (an unshared electron pair) on the same atom. Recall that both of these species reacted as both electrophile and nucleophile, forming three-membered rings—a bromonium ion and a mercurinium ion, respectively. It is not surprising, then, that a carbene also forms a three-membered ring, a cyclopropane. However, because the ring is not charged in this case, the reaction stops at this stage, rather than continuing by reaction with a nucleophile:

Cyclopropane

Of course, because the electrophile and the nucleophile are the same atom, the addition must proceed in a syn manner.

How are these reactive carbene species generated? Two groups must be removed from a tetravalent carbon, leaving behind one pair of electrons. We will examine three of the more important methods that accomplish this. The first is the elimination of N_2 from diazo compounds. The simplest diazo compound is diazomethane:

Diazomethane

Diazo compounds are relatively unstable and readily eliminate nitrogen to form carbenes. The loss of the stable N_2 molecule provides the driving force for the formation of the high-energy carbene. The elimination can be induced by heating, by irradiation with light, or by the presence of certain metal cations, such as Cu^{2+}. If the carbene is produced in the presence of an alkene, a cyclopropane ring is produced as shown in the following example. Note the stereochemistry of the product.

(E)-1-Phenylpropene

trans-1-Methyl-2-phenylcyclopropane

A second method to generate a carbene is to eliminate a proton and a leaving group, such as chloride or bromide ion, *from the same carbon:*

Trichloromethane
(chloroform)

Dichlorocarbene

Such a reaction is termed a **1,1-elimination** (or an ***α*-elimination**) because both groups are removed from the same carbon. (The eliminations to produce alkenes, discussed in Chapter 9, are called **1,2-eliminations** or ***β*-eliminations.**) Because they give carbenes as products, these 1,1-eliminations are inherently less favorable than 1,2-eliminations and occur only when the latter are precluded. In the presence of a strong base, chloroform ($CHCl_3$) and bromoform ($CHBr_3$) undergo 1,1-elimination to produce dichlorocarbene and dibromocarbene, respectively. (No 1,2-elimination is possible with these compounds.) The mechanism is somewhat different from those for the 1,2-eliminations that were presented in Chapter 9 in that first the proton is removed and then the halide leaves. The electron-withdrawing effect of the halogens makes the hydrogen acidic enough to be removed by a strong base such as hydroxide or alkoxide ion. A halide ion then acts as a leaving group from the conjugate base, producing the carbene. If the carbene is generated in the presence of an alkene, a cyclopropane derivative is produced, as shown in the following examples:

(83%)

(65%)

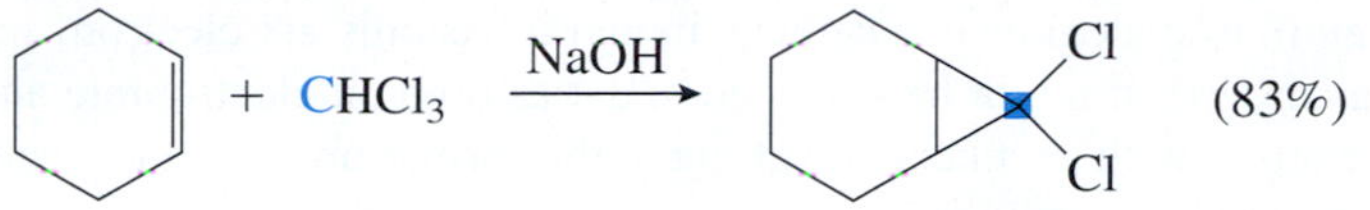

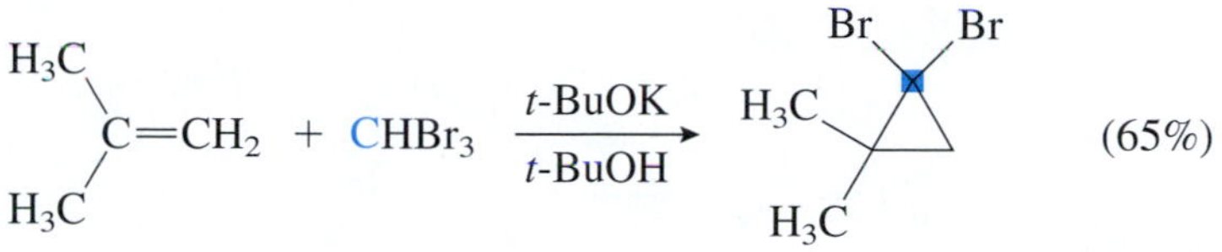

Another reaction that builds a cyclopropane ring onto a carbon–carbon double bond is called the **Simmons-Smith reaction.** This reaction does not actually involve a carbene, but rather a **carbenoid,** an organometallic species that reacts like a carbene. This species is generated by reaction of diiodomethane with zinc metal from a special alloy of zinc and copper:

$$CH_2I_2 \ + \ :Zn(Cu) \ \longrightarrow \ I-CH_2-Zn-I$$

Diiodomethane   A carbenoid
Reacts like $\overset{..}{C}H_2$

The mechanism for the formation of this carbenoid and for its reaction with alkenes need not concern us here. Just remember that it reacts as though it is methylene. The Simmons-Smith reaction is an excellent way to prepare cyclopropane derivatives from alkenes, as shown in the following examples. Note the stereochemistry in the second equation.

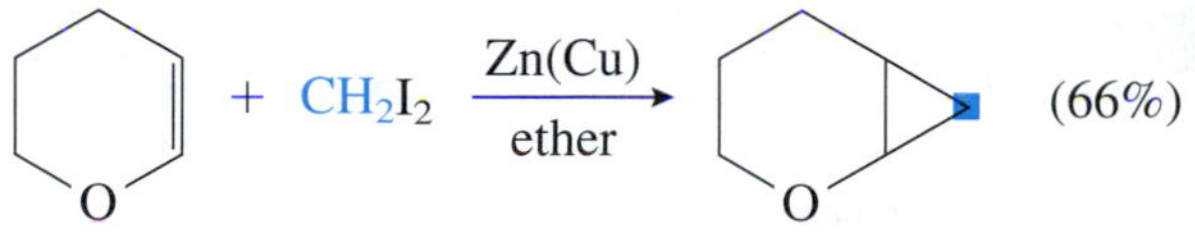

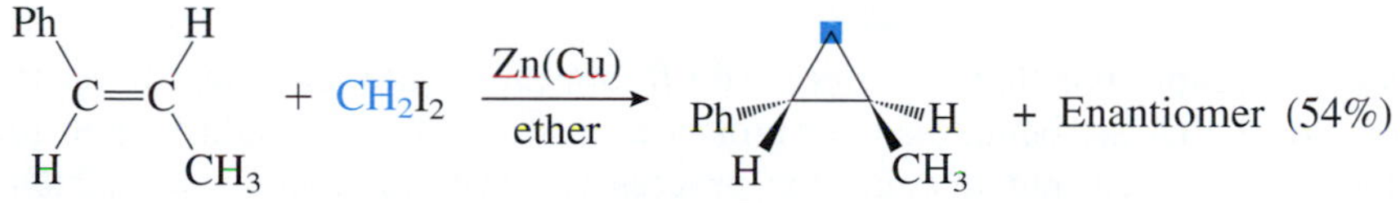

(66%)

+ Enantiomer (54%)

The reactions of carbenes with alkenes, as illustrated in this section, are the best methods for the synthesis of cyclopropanes.

PROBLEM 11.22

Show the products of these reactions:

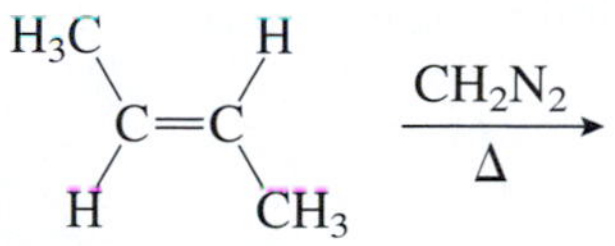

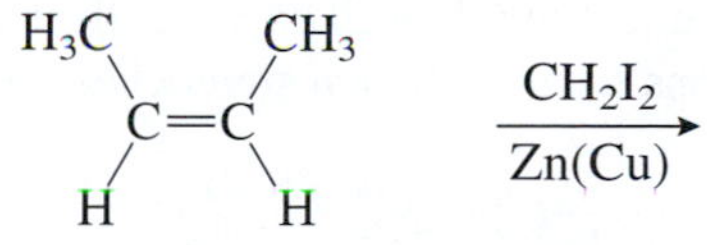

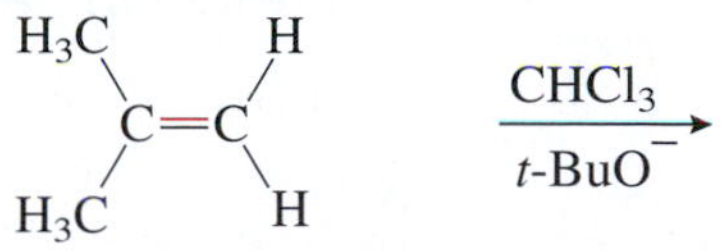

11.9 Epoxidation

An oxygen atom is analogous to a carbene in that it has only six electrons in its valence shell. Like a carbene, it might be expected to act as both an electrophile and a nucleophile in its reaction with an alkene, resulting in the formation of a three-membered ring containing an oxygen—an epoxide.

An epoxide

Although bottles of oxygen atoms are not available in the laboratory, **percarboxylic acids** are reagents that are able to accomplish this transformation. Recall that hydrogen peroxide, H—O—O—H, has an oxygen–oxygen bond. A percarboxylic acid also has an oxygen–oxygen bond and can act as a source of electrophilic oxygen when reacting with an alkene. The product of this reaction is an epoxide.

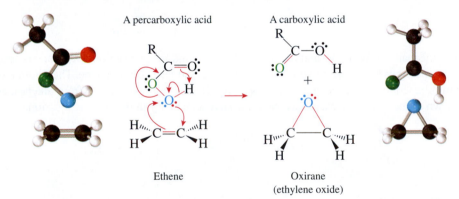

A percarboxylic acid

A carboxylic acid

Ethene

Oxirane
(ethylene oxide)

This epoxidation reaction is concerted. It appears quite complex because of all the bonds that are being made or broken in this one step. It might help to focus on the blue oxygen and imagine the process to occur in a stepwise fashion. A 1,1-elimination of the proton and the green oxygen would generate a normal carboxylic acid and an oxygen atom. Addition of the oxygen atom electrophile to the alkene, as illustrated previously, would then produce the epoxide. All of this simply happens in a single step.

Examination of this mechanism suggests that the nature of the R group should not make much difference in the reaction. In fact, a number of different percarboxylic acids can be used to epoxidize alkenes, as illustrated in the following examples. As expected, the additions occur with syn stereochemistry.

$$
\text{(cyclopentene)} + \underset{\substack{\text{Perbenzoic acid}}}{\text{PhCOOH}} \longrightarrow \text{(epoxide)} \quad (90\%)
$$

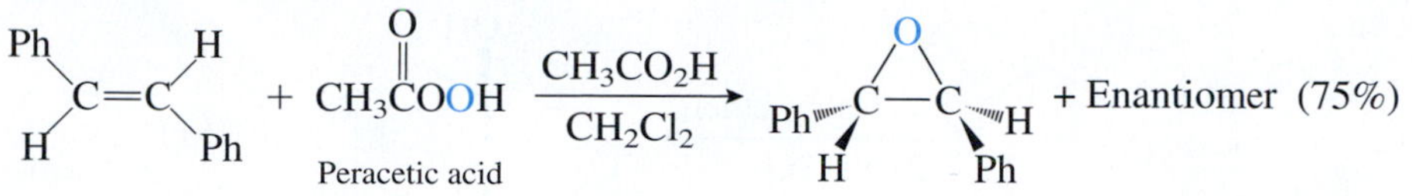

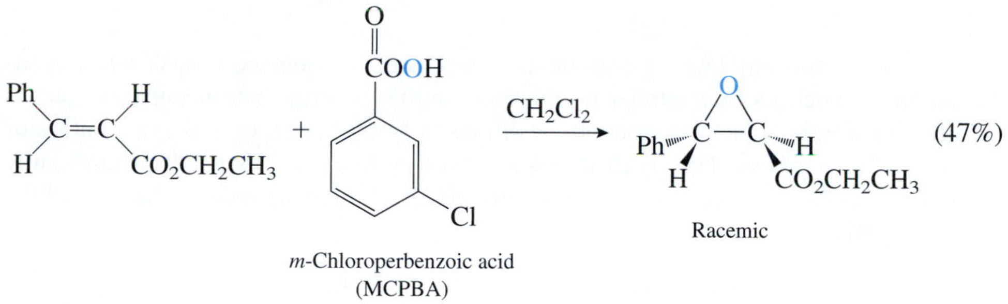

Peracetic acid

+ Enantiomer (75%)

m-Chloroperbenzoic acid
(MCPBA)

(47%)

Racemic

PROBLEM 11.23
Show the products of these reactions:

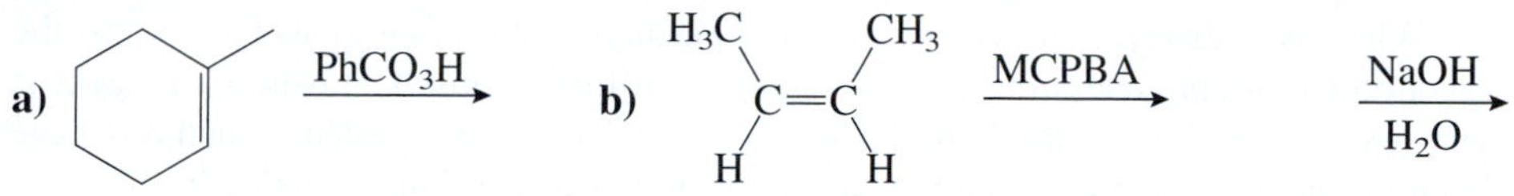

a)

$$\xrightarrow{\text{PhCO}_3\text{H}}$$

b)

$$\xrightarrow{\text{MCPBA}} \xrightarrow[\text{H}_2\text{O}]{\text{NaOH}}$$

11.10 HYDROXYLATION

There are two reagents that add hydroxy groups to both carbons of a carbon–carbon double bond: osmium tetroxide, OsO_4, and potassium permanganate, $KMnO_4$. These operate through similar mechanisms in which one of the oxygens that is bonded to the metal acts as an electrophile and another acts as a nucleophile, as illustrated in the following equation for the reaction of osmium tetroxide:

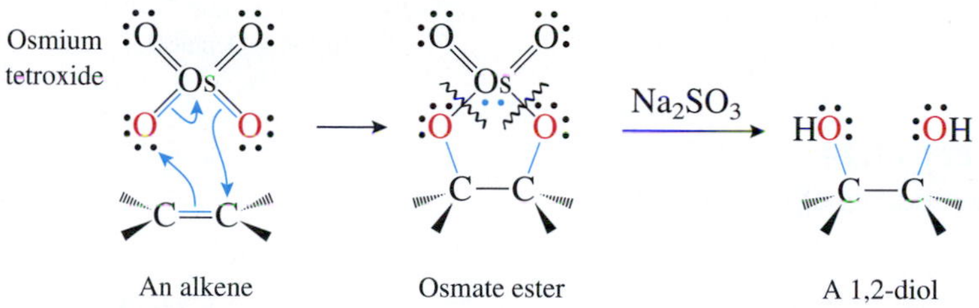

Osmium
tetroxide

$$\xrightarrow{\text{Na}_2\text{SO}_3}$$

An alkene   Osmate ester   A 1,2-diol

The cyclic intermediate, called an osmate ester, is not isolated; instead, the osmium–oxygen bonds are cleaved by using a reagent such as sodium sulfite, Na_2SO_3, resulting in the formation of a 1,2-diol. (The mechanistic details of the cleavage step need not concern us.) Because both the electrophilic and nucleophilic oxygens are attached to the same metal atom, both are delivered from the same side of the plane of the double bond—the reaction is a syn addition.

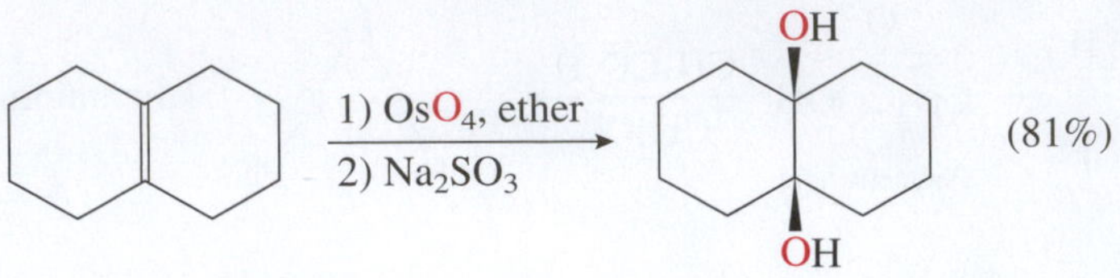

(81%)

The reaction employing potassium permanganate is conducted in basic aqueous solution. It proceeds by a similar mechanism, and the intermediate manganate ester is cleaved directly under the reaction conditions, resulting in an overall syn addition of two hydroxy groups. The yields in this reaction are often less than 50%, significantly lower than the reaction using osmium tetroxide. The following reaction has one of the better yields:

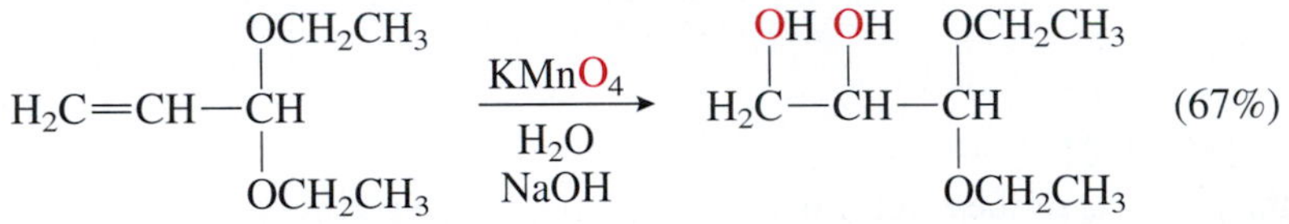

(67%)

Whereas hydroxylation using potassium permanganate often gives low yields, the reaction employing osmium tetroxide has its own limitations. The osmium reagent is very expensive and extremely toxic. To help alleviate these problems, methods have been developed that require only a catalytic amount of osmium tetroxide in the presence of some additional oxidizing agent, such as *tert*-butyl peroxide. The reaction proceeds to the cyclic osmate ester as before. The peroxide serves to cleave the ester to the diol and, importantly, oxidizes the osmium back to the tetroxide so that the cycle can be repeated. Therefore, only a small amount of osmium tetroxide is needed. The yields are respectable, as shown in the following example. Note the stereochemistry of the product.

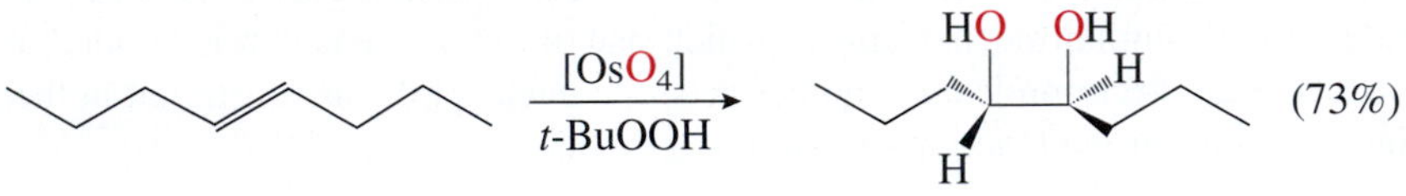

(73%)

(*E*)-4-Octene

Racemic
(*d*,*l*)-4,5-Octanediol

PROBLEM 11.24

Show the products of these reactions:

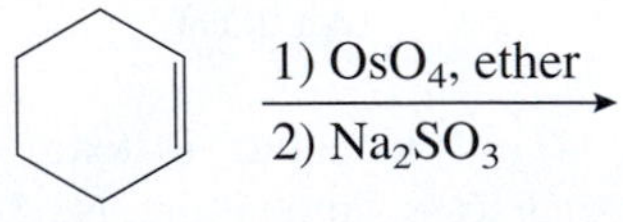

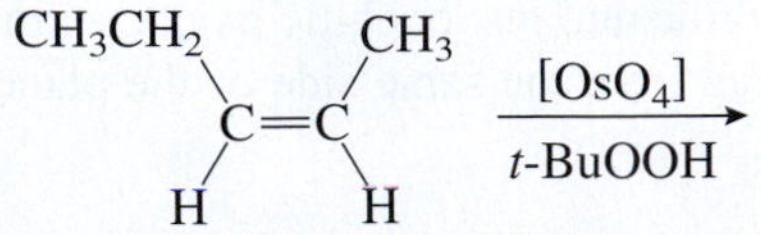

PROBLEM 11.25

Show syntheses of these compounds from (Z)-2-butene:

a)

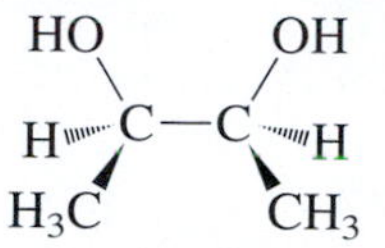

HO, OH
H····C—C····H
H$_3$C, CH$_3$

b)

HO, OH
H····C—C····CH$_3$
H$_3$C, H

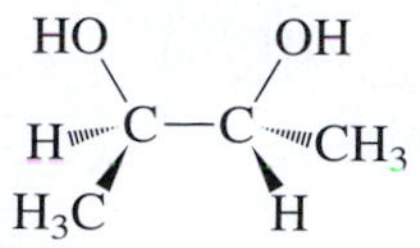

11.11 OZONOLYSIS

Ozone, O_3, is a form of oxygen found in the upper atmosphere. Its presence there is beneficial because it absorbs some harmful ultraviolet radiation, preventing it from reaching the surface of the earth. Ozone is also found at the surface of the earth as a component of smog. Here, its presence is detrimental because of its high reactivity. One of the important reactions of ozone occurs with alkenes and results in the cleavage of the carbon–carbon double bond as illustrated in the following example. The ozone first adds to the double bond. This addition can be viewed as though one oxygen atom acts as an electrophile and the other acts as a nucleophile, although the mechanism is actually an example of a pericyclic reaction (see Chapter 22). The initial product, called a molozonide, spontaneously rearranges to an ozonide. The ozonide is not usually isolated. Instead, it is treated with a reducing agent (dimethyl sulfide works nicely) that cleaves the O—O bond and results in the formation of two carbonyl groups. This sequence results in the cleavage of the carbon–carbon double bond and the formation of two carbon–oxygen double bonds.

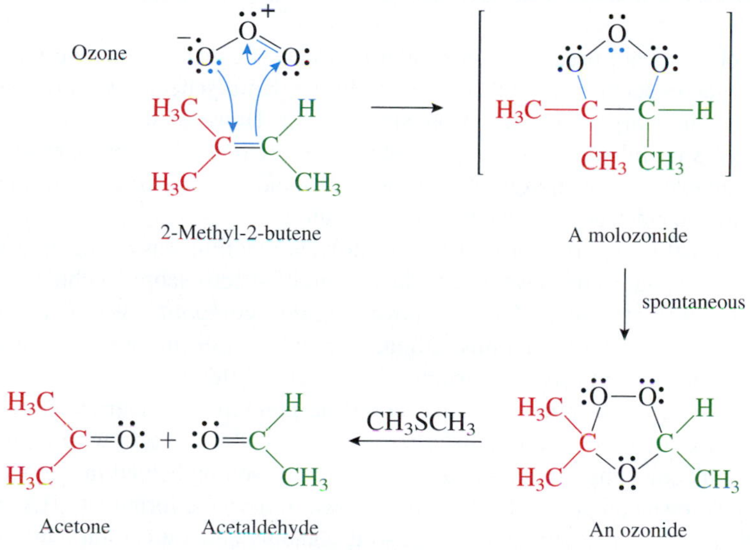

2-Methyl-2-butene   A molozonide

spontaneous

Acetone   Acetaldehyde   An ozonide

Some examples are shown in the following equations:

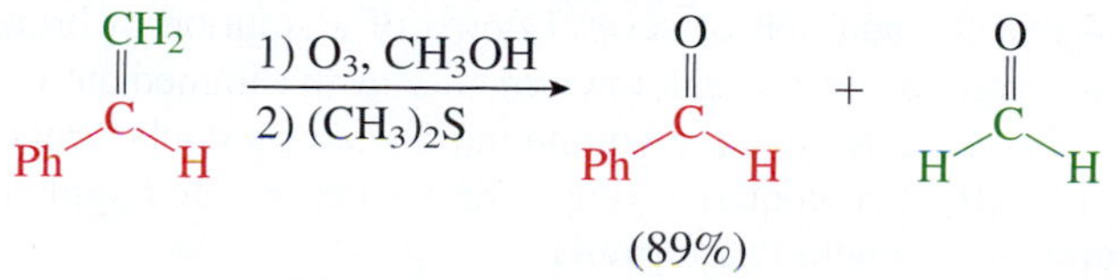

(89%)

PROBLEM 11.26
Show the products of these reactions:

a) $CH_3CH_2CH{=}CH_2$ $\xrightarrow[\text{2) }(CH_3)_2S]{\text{1) }O_3,\ CH_3OH}$

b) $\xrightarrow[\text{2) }(CH_3)_2S]{\text{1) }O_3,\ CH_3OH}$

c) $\xrightarrow[\text{2) }(CH_3)_2S]{\text{1) }O_3,\ CH_3OH}$

d) $\xrightarrow[\text{2) }(CH_3)_2S]{\text{1) }O_3,\ CH_3OH}$

e) $\xrightarrow[\text{2) }(CH_3)_2S]{\text{1) }O_3,\ CH_3OH}$

Ozonolysis is not used often in synthesis because it is a degradative reaction—it breaks larger molecules into smaller ones. In synthetic schemes we are usually attempting to build larger molecules from smaller ones. However, the ozonolysis reaction can provide a useful way to prepare an aldehyde or ketone if the appropriate alkene is readily available. One such example is provided by the cleavage of cyclohexene to produce the dialdehyde shown in the previous equation.

A historically important use of the ozonolysis reaction was in the area of structure determination. In the days before the advent of spectroscopic techniques (Chapters 13–15), the structure of an unknown organic compound was determined by submitting it to a host of reactions. Often, a complex molecule was broken into several fragments to simplify the structural problem. After the individual fragments were identified, the original molecule could be mentally reconstructed from them. Alkenes were often cleaved to aldehydes and ketones by reaction with ozone.

Let's look at a simple example to illustrate the reasoning behind this process. Consider an unknown compound that has been shown to have the formula C_5H_{10}. Note that it has DU (degree of unsaturation) = 1, so it contains either a carbon–carbon double bond or a ring. How could we determine which of these two structural features is present? One way to do this is to use the test described in Section 11.4. When the test is conducted, we find that the addition of several drops of a solution of bromine in carbon tetrachloride to a solution of the unknown results in the immediate discharge of the bromine color. This indicates that the compound contains a double bond. Although all of the isomers of C_5H_{10} that contain a ring have been eliminated by this test, there are still five isomeric alkenes with this formula.

The next step is to react the unknown alkene with ozone, followed by treatment with dimethyl sulfide. The products of this reaction are isolated and identified (by other chemical tests and their physical properties) as the two aldehydes shown in the following equation:

$$\text{Unknown alkene } C_5H_{10} \xrightarrow[\text{2) } (CH_3)_2S]{\text{1) } O_3} \underset{\text{Acetaldehyde}}{CH_3\overset{O}{\overset{\|}{C}}H} + \underset{\text{Propanal}}{H\overset{O}{\overset{\|}{C}}CH_2CH_3}$$

Now we must apply some reverse reasoning. What alkene will produce these two aldehydes upon cleavage with ozone? The two carbons of the carbonyl groups of these aldehydes (the blue carbons) must have been doubly bonded together in the original alkene. Therefore, the unknown alkene must be 2-pentene:

$$CH_3CH{=}CHCH_2CH_3$$

2-Pentene

The problem is not quite completed, because we still do not know whether the original alkene was the *cis*- or *trans*-isomer of 2-pentene. More information is needed to answer this question.

PRACTICE PROBLEM 11.6

What alkene would produce 2 equivalents of propanal upon ozonolysis?

$$H\overset{O}{\overset{\|}{C}}CH_2CH_3 \quad \text{(2 equivalents)}$$

Propanal

Solution

The ozonolysis reaction cleaves the carbon–carbon double bond into two carbonyl groups. In working backward from the carbonyl compounds to the alkene that produces them, just connect the two carbons of the carbonyl groups by a double bond. Because 2 equivalents of the aldehyde are formed in this example, the two carbonyl compounds are the same. The original alkene was the symmetrical 3-hexene. Of course, we have no way of determining whether it was the *E* or the *Z* stereoisomer on the basis of this information.

$$CH_3CH_2\overset{O}{\overset{\|}{C}}H + H\overset{O}{\overset{\|}{C}}CH_2CH_3 \xleftarrow[\text{2) } (CH_3)_2S]{\text{1) } O_3,\ CH_3OH} CH_3CH_2CH{=}CHCH_2CH_3$$

PROBLEM 11.27

Show the alkenes that produce these compounds on ozonolysis:

a) $CH_3\overset{O}{\overset{\|}{C}}CH_2CH_3 + CH_3\overset{O}{\overset{\|}{C}}H$ **b)**

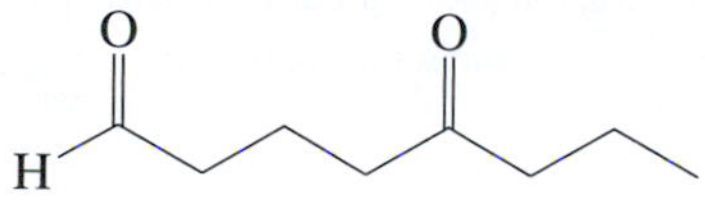

c) $CH_3\overset{O}{\overset{\|}{C}}CH_2CH_2\overset{O}{\overset{\|}{C}}H + H\overset{O}{\overset{\|}{C}}H + CH_3\overset{O}{\overset{\|}{C}}H$

11.12   CATALYTIC HYDROGENATION

The reaction of an alkene with hydrogen in the presence of a metal catalyst results in the addition of hydrogen to the carbon–carbon double bond.

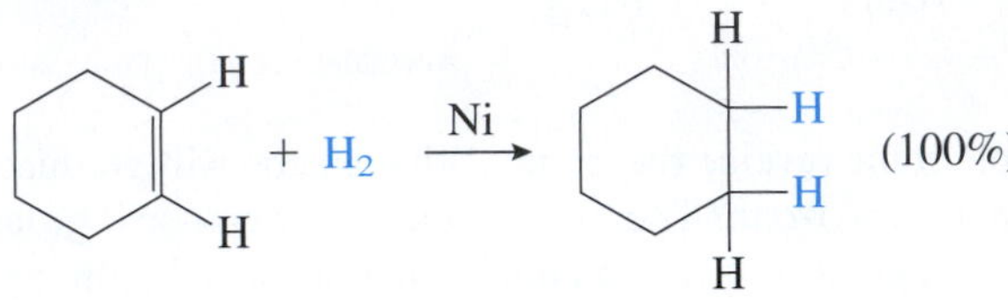

Dihydropyran   Tetrahydropyran

The catalysts that are commonly used for this reaction are the members of the nickel group: nickel, palladium, and platinum. Although nickel is sometimes used, as in the previous example, palladium and platinum usually work better.

This reaction does not involve an electrophilic addition to the alkene and, technically, does not belong in this chapter. However, it is convenient to include it here because it does result in addition to the alkene. The reaction occurs on the surface of the metal, so the catalyst must be present in a very finely divided or powdered state to maximize its surface area. A simplified version of the reaction mechanism on a platinum catalyst is presented in Figure 11.9. The unused valences of the atoms at the surface of the metal are used to break the hydrogen–hydrogen bonds, forming metal–hydrogen

❶ The atoms at the surface of any metal have unused bonding capability (or valence) that can be used to form bonds to other species that approach the surface. For example, when H_2 molecules adsorb to the surface of platinum, the H—H bond is broken and the hydrogen atoms form bonds to the platinum atoms.

❷ Likewise, an alkene can form a complex with the atoms at the metal surface, using the *p* orbitals of its pi bond.

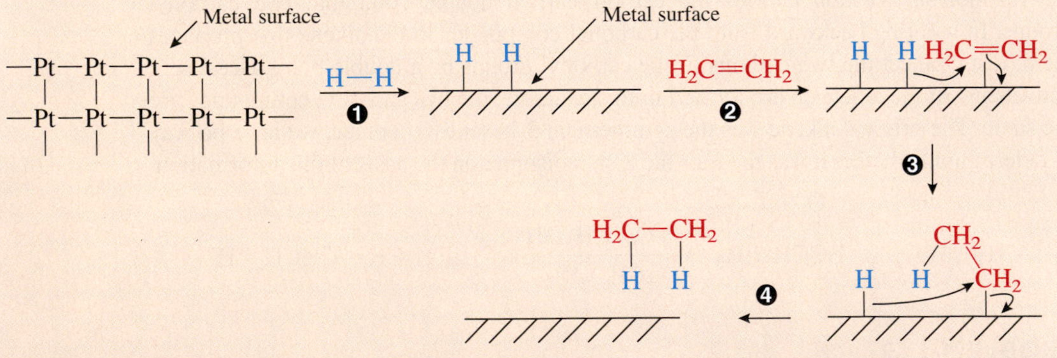

❹ A second hydrogen is transferred to the other carbon. Because no bond exists between them anymore, the alkane product diffuses away from the metal surface so that the surface is available to repeat the cycle.

❸ A hydrogen atom is then transferred to one carbon of the alkene. Now there is an alkyl group bonded to the metal surface.

Figure 11.9

MECHANISM OF THE CATALYTIC HYDROGENATION OF AN ALKENE.

bonds, and to complex with the carbon–carbon pi bonds of the alkene. The hydrogens are then transferred to the carbons, and the newly produced alkane detaches from the surface so that the metal atoms are free to repeat the process. The stereochemistry of the addition to alkenes only makes a difference when the alkene is substituted with four alkyl groups—that is, with tetrasubstituted alkenes. In such cases the major product is found to be the one resulting from syn addition, although the reaction is seldom completely stereospecific. This is consistent with a mechanism by which both hydrogens attach to the side of the double bond that initially bonds to the metal surface.

When the catalytic hydrogenation reaction is run under relatively mild conditions (room temperature and a pressure of hydrogen gas of several atmospheres or less), the reaction is very selective. Carbon–carbon double bonds of alkenes and carbon–carbon triple bonds of alkynes react readily, whereas carbon–carbon double bonds of aromatic rings and carbon–oxygen double bonds are usually inert under these reaction conditions. Some examples are provided in the following equations. Note that the stereochemistry of the addition reaction makes no difference in the first two examples. In the last example the major product results from syn addition.

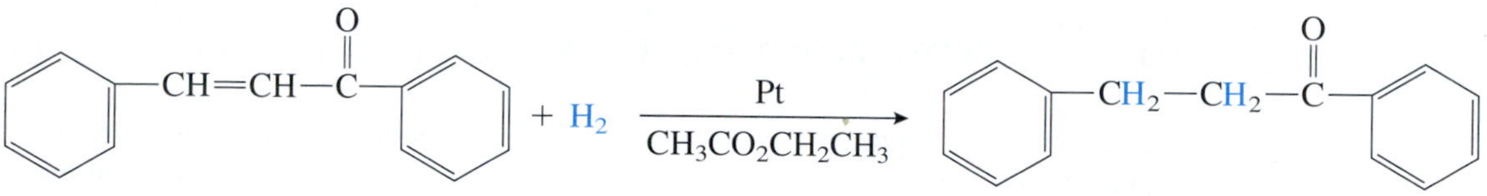

(95%)

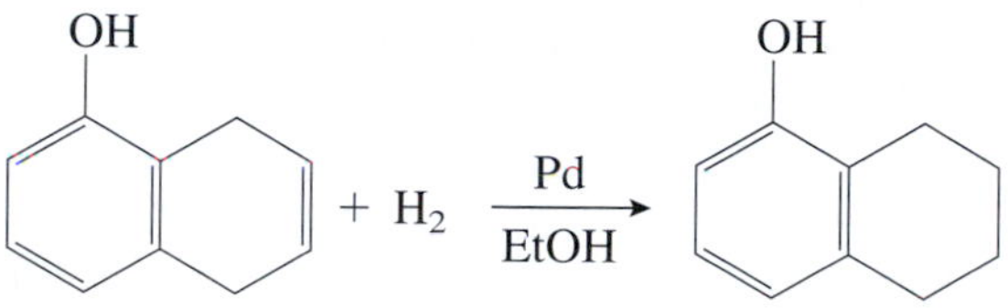

(88%)

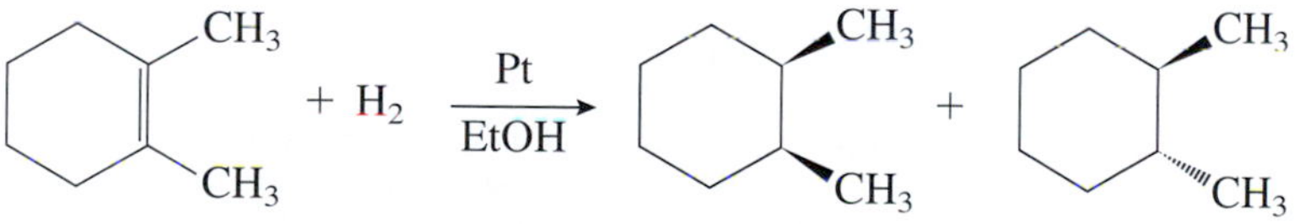

Major   Minor

Alkynes normally react with two equivalents of hydrogen to produce alkanes. However, it is also possible to react an alkyne with only one equivalent of hydrogen and stop the reaction at the alkene stage if a special catalyst, called Lindlar catalyst, is used. The Lindlar catalyst is a deactivated form of palladium that is less reactive than normal catalysts. This method provides good yields of the *cis*-alkene, resulting from syn addition of hydrogen:

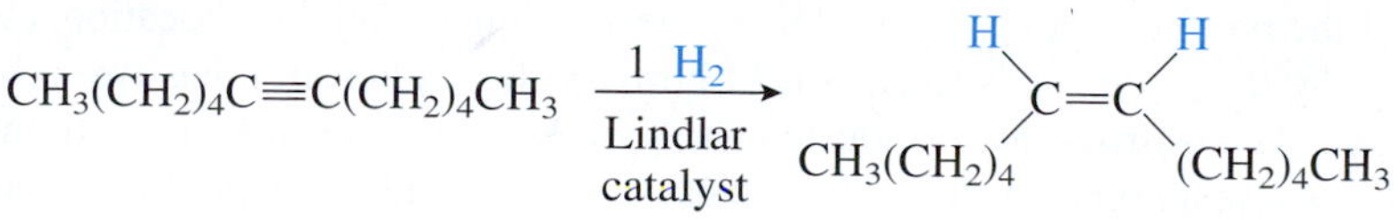

(87%)

A *cis*-alkene

PROBLEM 11.28
Show the products of these reactions:

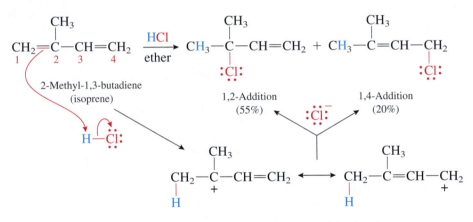

a) [structure: 1-phenyl-1-butene] $\xrightarrow[\text{Pt}]{\text{H}_2}$

b) [structure: cyclohexenone] $\xrightarrow[\text{Pd}]{\text{H}_2}$

c) $CH_3C{\equiv}CCH_2CH_3$ $\xrightarrow[\text{Pd}]{2\ H_2}$

d) $CH_3C{\equiv}CCH_2CH_3$ $\xrightarrow[\text{Lindlar catalyst}]{1\ H_2}$

e) [structure: 1,3-cyclohexadiene] $\xrightarrow[\text{Pt}]{2\ H_2}$

11.13   ADDITIONS TO CONJUGATED DIENES

The addition of one equivalent of a hydrogen halide to a conjugated diene results in a rapid reaction. Often, a mixture of products is formed as illustrated by the following example:

Electrostatic potential map
of the 1,1-dimethylallyl carbocation

[reaction scheme]

$$\underset{\substack{1\quad 2\quad 3\quad 4\\ \text{2-Methyl-1,3-butadiene}\\ \text{(isoprene)}}}{CH_2{=}\overset{\overset{\textstyle CH_3}{|}}{C}{-}CH{=}CH_2}\quad\xrightarrow[\text{ether}]{\text{HCl}}\quad \underset{\substack{\text{1,2-Addition}\\(55\%)}}{CH_3{-}\overset{\overset{\textstyle CH_3}{|}}{\underset{\underset{\textstyle :\ddot{C}l:}{|}}{C}}{-}CH{=}CH_2}\ +\ \underset{\substack{\text{1,4-Addition}\\(20\%)}}{CH_3{-}\overset{\overset{\textstyle CH_3}{|}}{C}{=}CH{-}\underset{\underset{\textstyle :\ddot{C}l:}{|}}{CH_2}}$$

$$\underset{\text{The 1,1-dimethylallyl carbocation}}{CH_2{-}\overset{\overset{\textstyle CH_3}{|}}{\underset{\underset{\textstyle H}{|}}{C}}{-}CH{=}\overset{+}{C}H_2 \quad\longleftrightarrow\quad \overset{+}{C}H_2{-}\overset{\overset{\textstyle CH_3}{|}}{\underset{\underset{\textstyle H}{|}}{C}}{=}CH{-}CH_2}$$

This reaction of hydrogen chloride with 2-methyl-1,3-butadiene results in the formation of 55% of the product in which the proton has added to carbon 1 and the chloride has added to carbon 2 (called 1,2-addition) and 20% of the product in which the proton has added to carbon 1 and the chloride has added to carbon 4 (called 1,4-addition). Using our knowledge of the mechanism of this reaction and what we have learned about carbocation behavior, we can easily explain these results. The reaction begins with the addition of the proton to carbon 1, producing the 1,1-dimethylallyl carbocation. (We will address why the proton adds to carbon 1 shortly.) This carbocation has two resonance structures. The positive charge is located partly on carbon 2 and partly on carbon 4. The chloride nucleophile can attack either of these carbons, resulting in the formation of the two products. In fact, given the distribution of positive charge in the carbocation, it would be very unusual if a mixture of products were not formed. The resonance structure on the left should contribute more to the resonance hybrid because it has the positive charge on a tertiary carbon, while the other structure has it on a primary carbon. In this particular case the major product results from the chloride nucleophile attacking the carbon with the larger amount of positive charge. Do not confuse this reaction with one that involves a carbocation rearrangement. This carbocation is not rearranging; it is a resonance hybrid that provides two electrophilic sites.

Now let's address the issue of why carbon 1 is the one that is initially protonated. According to the mechanistic rule, the electrophile—the proton—should add so as to produce the most stable carbocation. We have already seen that addition of the proton to carbon 1 produces a resonance stabilized carbocation. Addition to the other carbons produces the following carbocations:

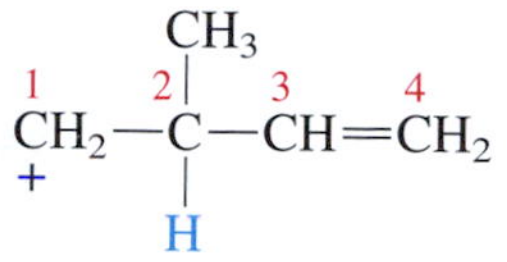

The carbocation that results from addition of the proton to carbon 2 has the positive charge localized on a primary carbon, and there is no resonance stabilization. This carbocation is quite unstable and is not formed in the reaction.

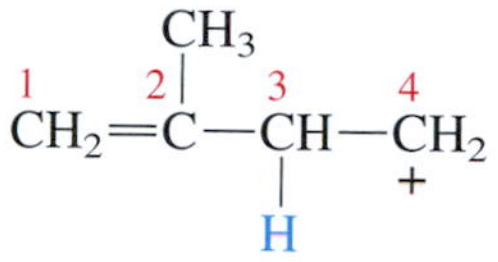

The carbocation that results from the addition of the proton to carbon 3 is similar to the previous cation in that the positive charge is localized on a primary carbon, and it has no resonance stabilization. It, also, is too unstable to be formed in the reaction.

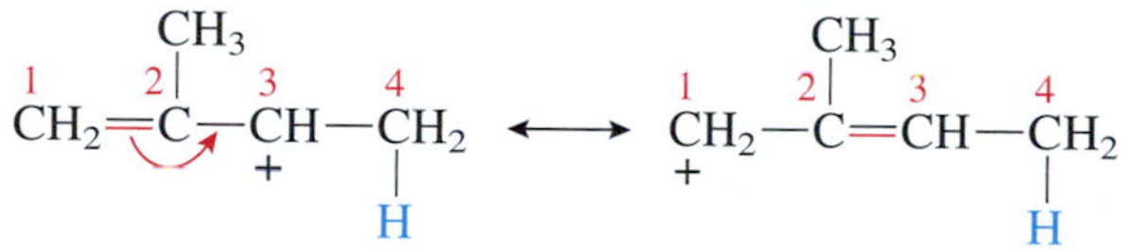

The carbocation that results from the addition of the proton to carbon 4 has two resonance structures, so it is much more stable than the previous two carbocations. In the resonance structures for this carbocation, the positive charge is located on a secondary carbon and a primary carbon. Therefore, it is somewhat less stable than the carbocation produced by addition of the proton to carbon 1, which has the positive charge located on a tertiary carbon and a primary carbon. In accord with the mechanistic rule, the proton adds so as to produce the most stable carbocation—it adds to carbon 1.

PROBLEM 11.29

Show all of the steps in the mechanism for this reaction and explain the regiochemistry of the addition:

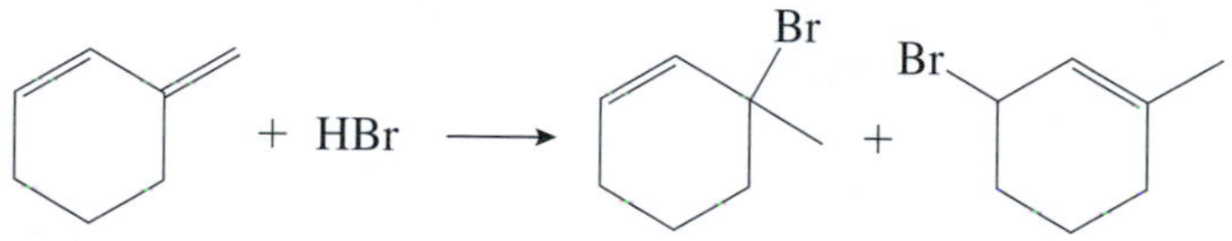

PROBLEM 11.30

Show the products of these reactions:

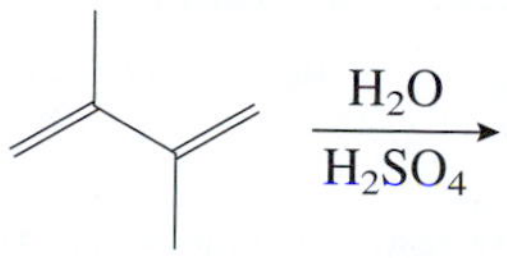

The composition of the product mixture formed upon electrophilic addition to conjugated dienes often changes with the temperature at which the reaction is conducted. For example, when HBr is added to 1,3-butadiene at $-80°C$, the major product is formed by 1,2-addition. In contrast, when the reaction is run at $45°C$, the major product is the one resulting from 1,4-addition:

	1,2-Addition	1,4-Addition
$CH_2=CH-CH=CH_2$ $\xrightarrow{H-Br}$	$CH_2-CH-CH=CH_2$ (with H and Br)	$CH_2-CH=CH-CH_2$ (with H and Br)
1,3-Butadiene	3-Bromo-1-butene	1-Bromo-2-butene
At $-80°C$	80%	20%
At $45°C$	15%	85%

Figure 11.10 shows an energy versus reaction progress diagram for this reaction. At the lower temperature, the reaction is under **kinetic control.** The rate of formation of the product determines which product is the major one. In this case, the activation energy ($\Delta G^{\ddagger}$) for the formation of the 1,2-addition product (called the kinetic product) is lower than that for formation of the 1,4-addition product, so the 1,2-addition product is formed faster, even though the 1,4-addition product is more stable.

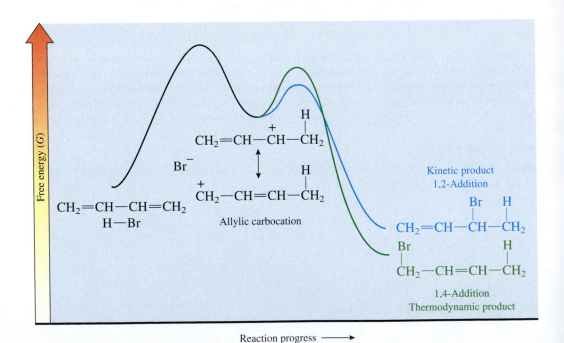

Figure 11.10

FREE ENERGY VERSUS REACTION PROGRESS DIAGRAM FOR THE ADDITION OF **HBr** TO **1,3-**BUTADIENE.

At the higher temperature, the reaction becomes reversible and is under **thermodynamic control.** This means there is enough energy available for either product to reform the allylic carbocation by an S_N1 type ionization and then form the other product. As a result the products are in equilibrium. At equilibrium, the relative amount of the products is controlled only by the difference in energy between them (ΔG). In this case, the 1,4-addition product (called the thermodynamic product) is more stable than the 1,2-addition product, so more of it is present in the equilibrium mixture. This same equilibrium mixture of products (15% of the 1,2-addition product and 85% of the 1,4-addition product) is produced when the low-temperature reaction product mixture (80% of the 1,2-addition product and 20% of the 1,4-addition product) is heated to 45°C.

In later chapters, we will encounter several other reactions where competition occurs between the formation of the kinetic product and the thermodynamic product. However, most reactions are conducted under conditions where only one of these factors controls the product distribution. In addition, the kinetic product and the thermodynamic product are often the same, so no competition occurs in these reactions either.

Focus On

Asymmetric Hydrogenation

The Focus On box on page 433 described a hydroboration reaction that produces a single enantiomer of a chiral alcohol as the product. The chirality of one enantiomer of the boron hydride reagent is used to control the formation of a single enantiomer of the product. As discussed in that Focus On box, the drawback to this reaction is that it requires one mole of the chiral borane for each mole of chiral alcohol that is produced. The chiral reagent is rather expensive because it must be resolved or prepared from another enantiomerically pure compound. A more desirable process would use the expensive chiral reagent as a catalyst so that a much smaller amount could be employed to produce a larger amount of the chiral product.

This goal has been successfully achieved in catalytic hydrogenation reactions. However, rather than use a solid metal catalyst, these reactions employ a soluble metal-containing species called a homogeneous catalyst. A homogeneous catalyst has a metal atom bonded (coordinated) to several other groups, called ligands. Because this coordinated metal atom is soluble, the reaction occurs in a single, homogeneous phase. The mechanism for the reaction is similar to that shown in Figure 11.9 for a heterogeneous catalytic hydrogenation. The metal atom still serves as a place to bond the hydrogen atoms and the alkene, helping to break bonds and stabilize intermediates so that the reaction can occur by a lower-energy pathway.

If some of the ligands bonded to the metal atom in a homogeneous catalyst are chiral, then the hydrogenation can, in theory, produce an excess of one enantiomer of the reduction product. One catalyst that has been found to be effective in such an asymmetric hydrogenation reaction is this chiral rhodium complex:

Continued

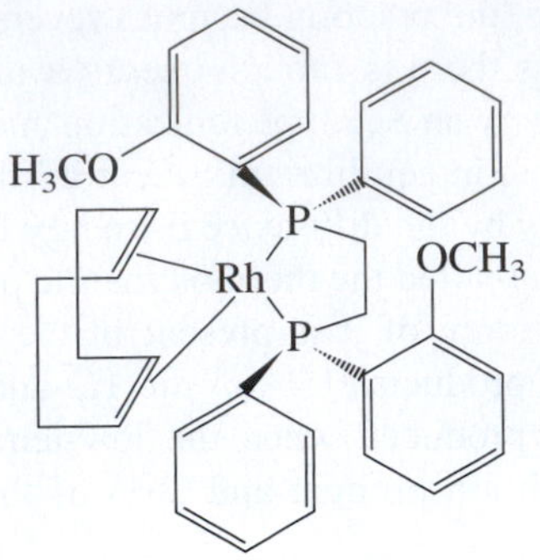

In the hydrogenation reaction the 1,5-cyclooctadiene ligand is replaced by hydrogen and the alkene and the chirality of the phosphorus ligand causes a single enantiomer of the reduced product to be formed. This process is now used industrially by Monsanto to produce L-dopa, which is used as a treatment for Parkinson's disease, from an achiral starting material:

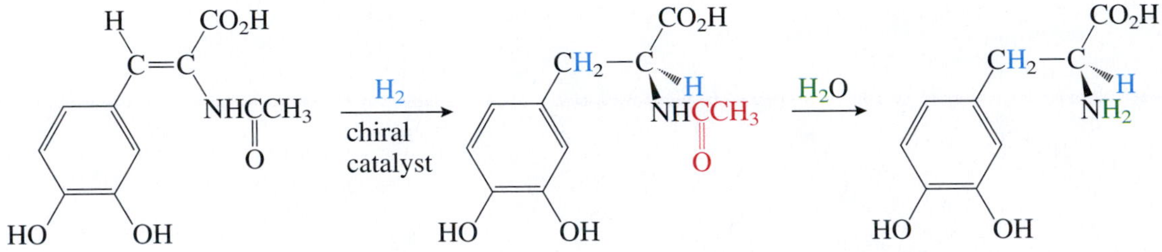

Achiral — *S* only — (L)-3,4-Dihydroxyphenylalanine (L-dopa)

Monsanto also uses a similar process to produce a single enantiomer of the arthritis drug Naproxen, a nonsteroidal anti-inflammatory drug (NSAID). Note that this asymmetrical hydrogenation produces only the (*S*)-enantiomer of the drug in a yield of 98.5% from an achiral precursor:

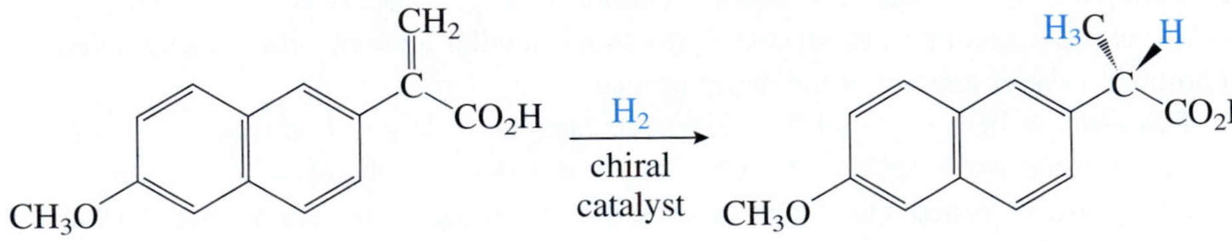

Naproxen
(*S*)-Enantiomer only
(98.5%)

The 2001 Nobel Prize in chemistry was shared by W. S. Knowles, who developed these chiral hydrogenations at Monsanto; R. Noyori, who also worked in the area of chiral hydrogenations; and K. B. Sharpless, who developed methods for chiral epoxidation reactions.

11.14 SYNTHESIS

As described in Section 10.15, a common task confronted by an organic chemist is the synthesis of a desired compound that is not available from commercial sources. Now that we have the addition reactions of this chapter available, let's practice designing some more complicated syntheses.

Recall that the best way to approach planning a synthesis is to use retrosynthetic analysis, that is, to work backward from the desired compound, the target, to simpler compounds, until an available compound is reached. *A special type of reaction arrow, called the* **retrosynthetic arrow,** *is useful in such a synthetic analysis. As shown in the following example, the retrosynthetic arrow points from the target to the reactant.* Usually, the reagents that cause the transformation are not specified, although it is necessary to have some method in mind that will accomplish the transformation. After retrosynthetic analysis has led to a reasonable starting material, the synthesis is written in the forward direction and all the necessary reagents are included.

◀▓▓ **Important Convention**

Target
compound   Starting
material

Retrosynthetic
arrow

Reactions that form carbon–carbon bonds are extremely important in synthesis because they enable larger compounds, containing more carbons, to be constructed from smaller compounds. This requires the reaction of a carbon nucleophile with a carbon electrophile. The most important carbon nucleophiles that we have encountered so far are cyanide ion and acetylide anions (see Section 10.8). If we remember that acetylide anions can be reduced to *cis*-alkenes (see Section 11.12), then all of the addition products of this chapter are accessible from simple alkynes.

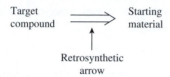

$$\begin{array}{ccc} \text{H}\diagdown\text{C}=\text{C}\diagup\text{H} & & \\ \text{R}\diagup\quad\diagdown\text{R}' & \Longrightarrow & \text{R}-\text{C}\equiv\text{C}-\text{R}' & \Longrightarrow & \text{H}-\text{C}\equiv\text{C}-\text{H} \end{array}$$

Section 11.12   Section 10.8

Let's try a synthesis. Suppose we are asked to synthesize 2-heptanone from ethyne.

$$\underset{\text{2-Heptanone}}{CH_3CH_2CH_2CH_2CH_2\overset{\displaystyle O}{\overset{\|}{C}}CH_3} \quad \text{from} \quad \underset{\text{Ethyne}}{HC\equiv CH}$$

First we note that it is necessary to form a carbon–carbon bond because the starting material has only two carbons and the target has seven. Because the starting material is an alkyne, we can probably use an acetylide anion as the nucleophile to form the carbon–carbon bond (see Section 10.8). How can a ketone functional group be introduced? Section 11.6 described the hydration of an alkyne to produce a ketone. Our retrosynthetic analysis then becomes:

$$CH_3CH_2CH_2CH_2CH_2\overset{\displaystyle O}{\overset{\|}{C}}CH_3 \Longrightarrow CH_3CH_2CH_2CH_2CH_2C\equiv CH \Longrightarrow HC\equiv CH$$

Written in the forward direction, the synthesis is as follows:

$$HC\equiv CH \xrightarrow[\text{2) CH}_3\text{(CH}_2\text{)}_3\text{CH}_2\text{Cl}]{\text{1) NaNH}_2} CH_3CH_2CH_2CH_2CH_2C\equiv CH \xrightarrow[\substack{\text{H}_2\text{SO}_4 \\ \text{HgSO}_4}]{\text{H}_2\text{O}} CH_3CH_2CH_2CH_2CH_2\overset{\overset{\displaystyle O}{\|}}{C}CH_3$$

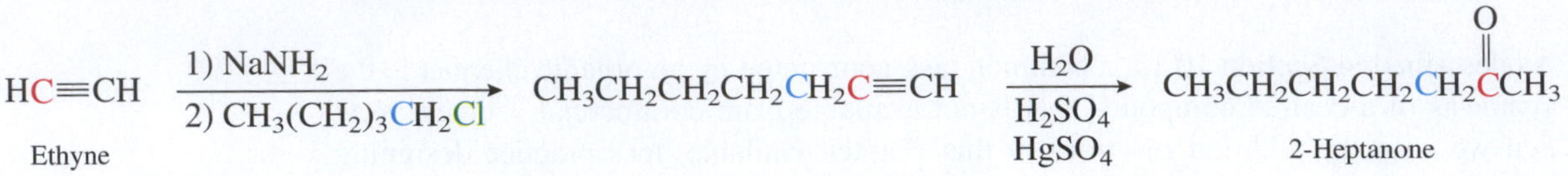

Ethyne   2-Heptanone

Let's try another example. This time our assignment is to prepare *meso*-2,3-butanediol from propyne:

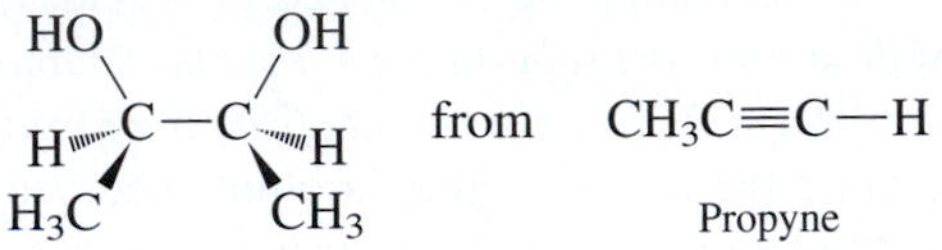

meso-2,3-Butanediol   from   CH₃C≡C—H   Propyne

The diol can be prepared from syn hydroxylation of (*Z*)-2-butene. The *cis*-alkene can be prepared by hydrogenation of 2-butyne, and 2-butyne can be prepared by alkylation of propyne. The retrosynthetic analysis is:

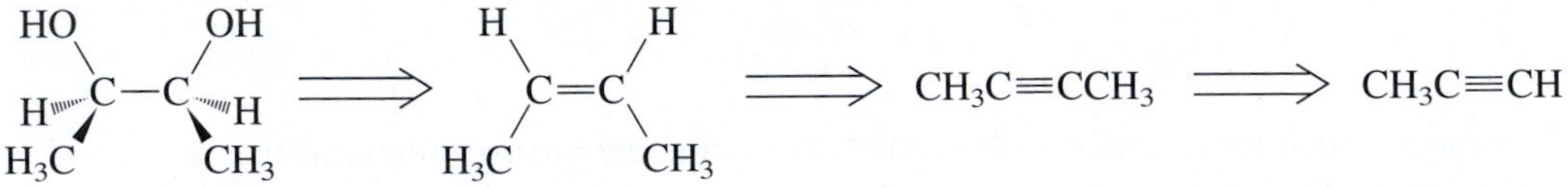

Written in the forward direction, the synthesis is as follows:

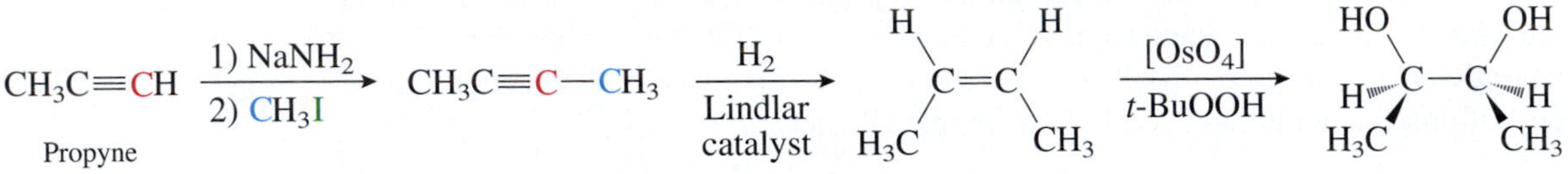

PROBLEM 11.31

Show syntheses of these compounds from the indicated starting materials. Your syntheses may produce both enantiomers of any target that is chiral.

a) $CH_3CH_2CH_2CH_2CH_2OCH_3$ from $CH_3CH_2CH_2CH{=}CH_2$

b) $CH_3\overset{\overset{\displaystyle CH_3}{|}}{C}HCH_2CH_2\overset{\overset{\displaystyle O}{\|}}{C}CH_3$ from $HC{\equiv}CH$

c) $PhCH_2CH_2CH_2\overset{\overset{\displaystyle O}{\|}}{C}H$ from $HC{\equiv}CH$

d) $CH_3CH_2\overset{\overset{\displaystyle O}{\|}}{C}CH_2CH_2CH_3$ from $HC{\equiv}CH$

e) 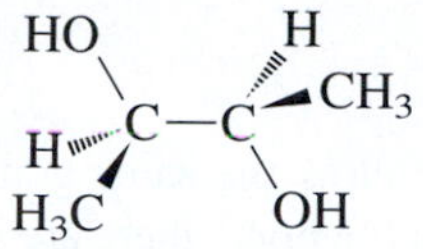 $\underset{H_3C}{\overset{HO}{\underset{}{\bigg|}}} C - C \overset{H}{\underset{OH}{\bigg|}} CH_3$ from $CH_3C{\equiv}CCH_3$

f) 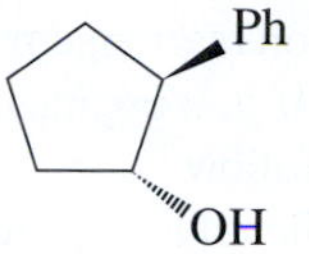 (cyclopentane with Ph and OH substituents)   from   (cyclopentane with OH and Ph substituents)

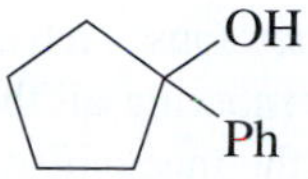

g) 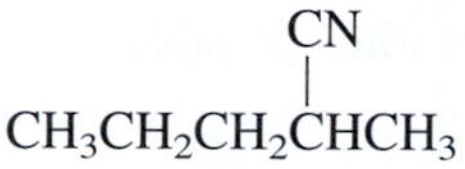 $\underset{}{CH_3CH_2CH_2\overset{\overset{\displaystyle CN}{|}}{C}HCH_3}$ from $CH_3CH_2CH_2CH{=}CH_2$

h) $CH_3CH_2CH_2CH_2 \overset{H}{\underset{Cl}{\bigg|}} C - C \overset{Cl}{\underset{CH_3}{\bigg|}} H$ 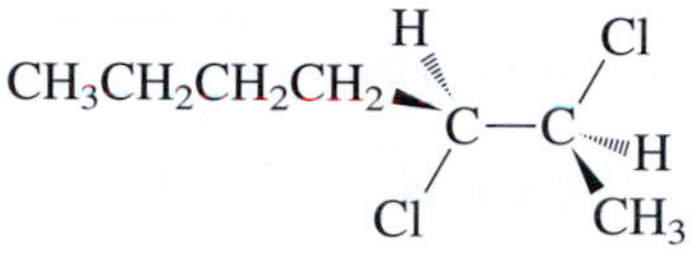   from $CH_3CH_2CH{=}CH_2$

Review of Mastery Goals

After completing this chapter, you should be able to:

ORGANIC
Chemistry·❖·Now™
Click *Mastery Goal Quiz* to test
how well you have met these
goals.

■ Show the products, including regiochemistry and stereochemistry, resulting from the addition to alkenes of all of the reagents listed in this chapter. (Problems 11.32, 11.33, 11.34, 11.35, 11.43, 11.55, 11.56, 11.57, 11.58, 11.59, 11.60, 11.61, and 11.62)

■ Show the products, including regiochemistry and stereochemistry, resulting from the addition to alkynes of all of the reagents listed in this chapter. These reagents include HX, X_2, $Hg^{2+}/H^+/H_2O$, disiamylborane, and H_2. (Problems 11.32 and 11.33)

■ Show the mechanisms for any of these reactions. (Problems 11.36, 11.39, 11.41, 11.42, and 11.44)

■ Show rearranged products when they are likely to occur. (Problems 11.32f and 11.33c)

■ Predict how the rate of addition varies with the structure of the alkene. (Problem 11.40)

■ Predict the products from additions to conjugated dienes. (Problem 11.33o)

■ Use these reactions in synthesis. (Problems 11.37 and 11.38)

Visual Summary of Key Reactions

This chapter presented many reactions. Although most of them follow the same general mechanism in which an electrophile adds to the carbon–carbon double bond, there are several variations on this mechanism. In addition, the regiochemistry and/or the stereochemistry of the reaction may be important. Complications, such as carbocation rearrangements, may occur. You will have an easier time remembering all these details if you organize the reactions according to the three variations of the mechanism that they follow.

Table 11.2 summarizes the reactions in which a proton electrophile adds to produce a carbocation intermediate (hydrohalogenation and hydration). These reactions follow Markovnikov's rule and tend to give mixtures of stereoisomers when possible. They are very prone to carbocation rearrangements.

Table 11.2   Addition Reactions Following a Carbocation Mechanism

These reactions proceed by initial addition of a proton to the alkene to give a carbocation intermediate. The nucleophile adds in the second step.

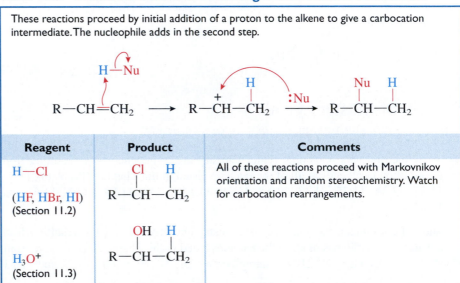

Reagent	Product	Comments
H—Cl (HF, HBr, HI) (Section 11.2)	Cl H | | R—CH—CH₂	All of these reactions proceed with Markovnikov orientation and random stereochemistry. Watch for carbocation rearrangements.
H₃O⁺ (Section 11.3)	OH H | | R—CH—CH₂	

Table 11.3 summarizes the reactions that proceed through a three-membered cyclic intermediate. In each of these reactions the electrophile has an unshared pair of electrons, so it can also act as a nucleophile. The electrophile adds to form the three-membered ring in a syn manner. If the electrophilic atom is uncharged, then the reaction stops at this cyclic stage (carbenes, epoxidation). If the electrophile has a positive charge in the ring, then a nucleophile adds in a second step (halogenation, halohydrin formation, oxymercuration, mercury-catalyzed hydration of alkynes). The regiochemistry of the addition is such that the nucleophile attaches to the carbon that would be more stable as a carbocation and the addition occurs with anti stereochemistry.

Table 11.4 summarizes the reactions in which the electrophile and the nucleophile are linked in the same molecule (hydroboration, hydroxylation, ozonolysis). These additions occur in a concerted manner. The regiochemistry of the addition is such that the nucleophile attaches to the carbon that would be more stable as a carbocation and the addition occurs with syn stereochemistry.

Table 11.5 summarizes the catalytic hydrogenation reactions that can be used to add hydrogens to double or triple bonds, converting alkenes and alkynes to alkanes or alkynes to *cis*-alkenes.

Table 11.3 Addition Reactions Proceeding through a Three-Membered Cyclic Intermediate

When the electrophile also has an unshared pair of electrons, addition initially produces a three-membered ring. If the ring is uncharged, the reaction stops here. If the ring has a positive charge, a nucleophile attacks and opens the ring.

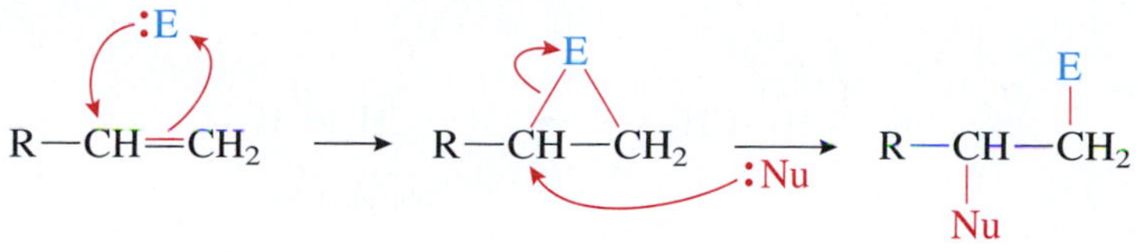

Syn addition   Anti addition

Stops here if neutral;
adds Nu if E has +.

Reagent	Product	Comments
CH_2N_2 (CHX_3/OH^-) $(CH_2I_2, Zn/Cu)$ (Section 11.8)	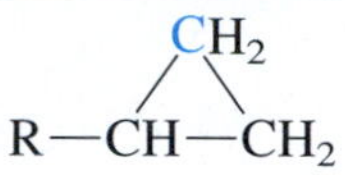 $R-CH-CH_2$ with CH_2 bridge	Carbenes add to give cyclopropane derivatives. These are syn additions. The Simmons-Smith reaction generates a carbenoid that reacts like methylene.
$R'CO_3H$ (Section 11.9)	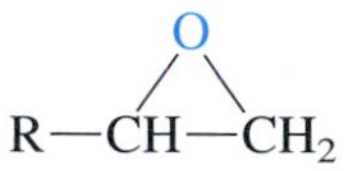 $R-CH-CH_2$ with O bridge	Percarboxylic acids add to alkenes to give epoxides in a syn addition.
Cl_2 (Br_2)	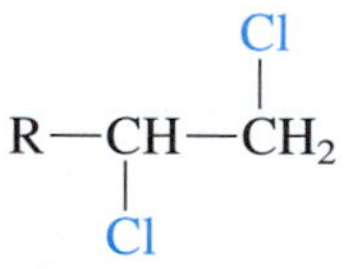 $R-CH-CH_2$ with Cl on each carbon	Chlorine and bromine add with anti stereochemistry.
Cl_2/H_2O (Br_2/H_2O) (Section 11.4)	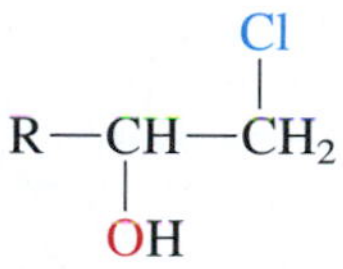 $R-CH-CH_2$ with Cl on CH_2 and OH on CH	In water as solvent, Cl_2 and Br_2 add to give halohydrins, with the OH on the more substituted carbon and anti stereochemistry.
1) $Hg(O_2CCH_3)_2, H_2O$ 2) $NaBH_4, NaOH$ (Section 11.6)	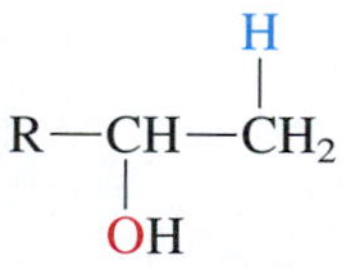 $R-CH-CH_2$ with H on CH_2 and OH on CH	The oxymercuration reaction is a method for the Markovnikov addition of water without rearrangement.
Hg^{2+}, H_2O, H_2SO_4 (Section 11.6)	$R-\overset{O}{\overset{\|}{C}}-CH_3$ from $R-C\equiv CH$	The initial product, an enol, tautomerizes to a ketone.

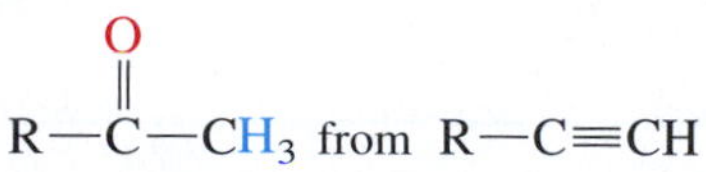

Table 11.4   Addition Reactions Where the Nucleophile and Electrophile Are Linked

When the electrophile and nucleophile are part of the same molecule, concerted additions occur with syn stereochemistry.

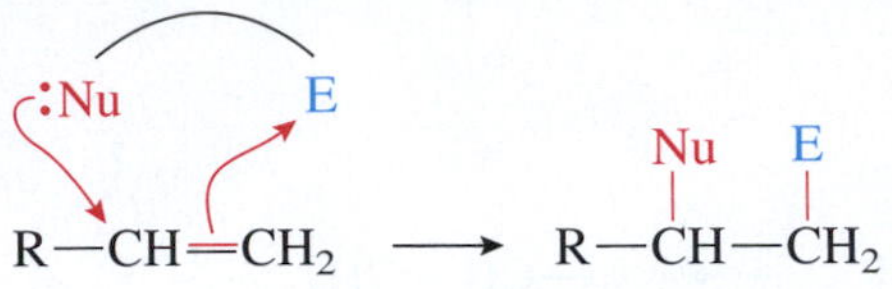

Syn addition

Reagent	Product	Comments
1) BH$_3$, THF 2) H$_2$O$_2$, NaOH (Section 11.7)	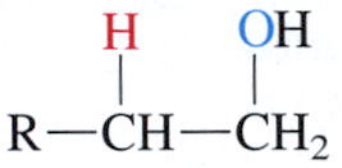 H OH R—CH—CH$_2$	The hydroboration reaction results in anti-Markovnikov addition of water with syn stereochemistry.
1) Disiamylborane 2) H$_2$O$_2$, NaOH (Section 11.7)	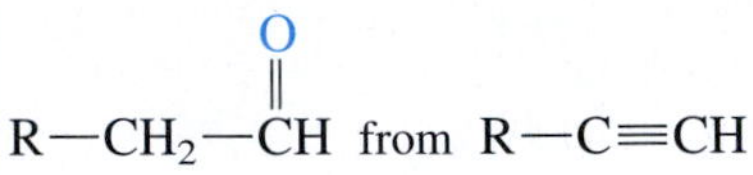 O ‖ R—CH$_2$—CH from R—C≡CH	The reaction of 1-alkynes with the hindered borane produces aldehydes.
OsO$_4$, t-BuOOH or KMnO$_4$, H$_2$O, NaOH (Section 11.10)	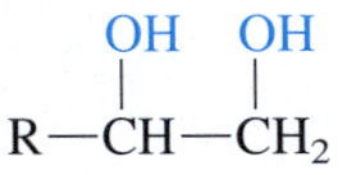 OH OH R—CH—CH$_2$	Osmium tetroxide and permanganate result in the syn addition of hydroxy groups to the alkene.
1) O$_3$ 2) (CH$_3$)$_2$S (Section 11.11)	 O O ‖ ‖ R—CH + HCH	Ozone can be used to cleave the alkene to two carbonyl compounds.

Table 11.5   Catalytic Hydrogenation Reactions

Reagent	Product	Comments
H$_2$, cat. (Ni, Pd, Pt) (Section 11.12)	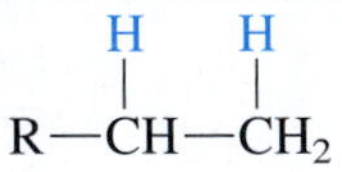 H H R—CH—CH$_2$	Catalytic hydrogenation results in the addition of hydrogen to the alkene.
H$_2$ Lindlar catalyst (Section 11.12)	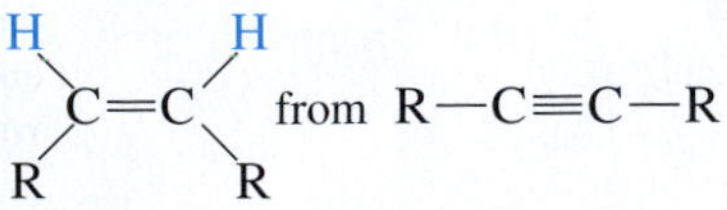 H H C=C from R—C≡C—R R R	Addition of hydrogen to an alkyne using Lindlar catalyst produces a cis-alkene.

Integrated Practice Problem

Show the products of these reactions:

a) 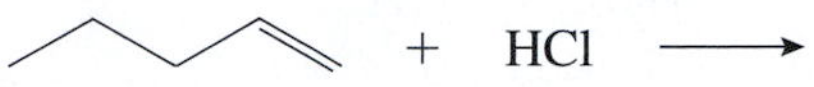 + HCl ⟶

b) 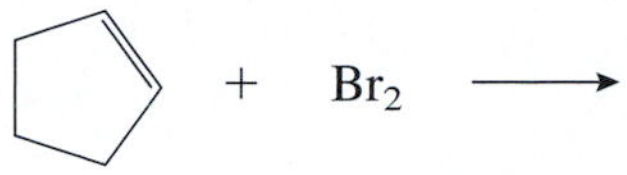 + Br$_2$ ⟶

c) 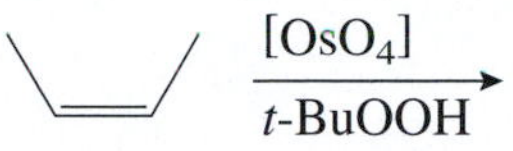 $\dfrac{[OsO_4]}{t\text{-BuOOH}}$

Strategy

As usual, the key is to identify the electrophile and the nucleophile. Add these to the pi bond with the nucleophile bonded to the carbon that would be more stable as a carbocation. Identification of the mechanism type that is being followed will help you remember the details of the reaction, such as the stereochemistry.

- If the electrophile is H$^+$, then the reaction is one from Table 11.2 and proceeds through a carbocation intermediate. Watch for rearrangements.
- If the electrophile has an unshared pair of electrons, the reaction is one from Table 11.3 and proceeds through a three-membered cyclic intermediate, which is formed by syn addition. If the cyclic intermediate is neutral, the reaction stops here. If the intermediate is charged, the nucleophile adds with inversion (borderline S$_N$2 mechanism), resulting in overall anti addition.
- If the electrophile and nucleophile are part of the same molecule, the reaction is one from Table 11.4 and the addition is syn.

Solutions

a) The electrophile is H$^+$ and the nucleophile is Cl$^-$. The reaction is one from Table 11.2. It proceeds through a carbocation, but rearrangement will not occur because a more stable carbocation cannot be readily generated. The nucleophile is bonded to carbon 2 because this carbon (secondary) would be more stable as a carbocation than carbon 1 (primary).

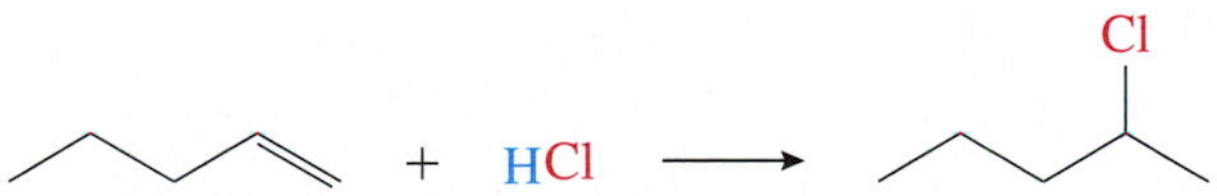

b) The electrophile is Br$^+$ and the nucleophile is Br$^-$. Because the electrophile has an unshared electron pair, the reaction is one from Table 11.3. The cyclic bromonium ion is charged, so the bromide nucleophile adds in the second step resulting in overall anti addition.

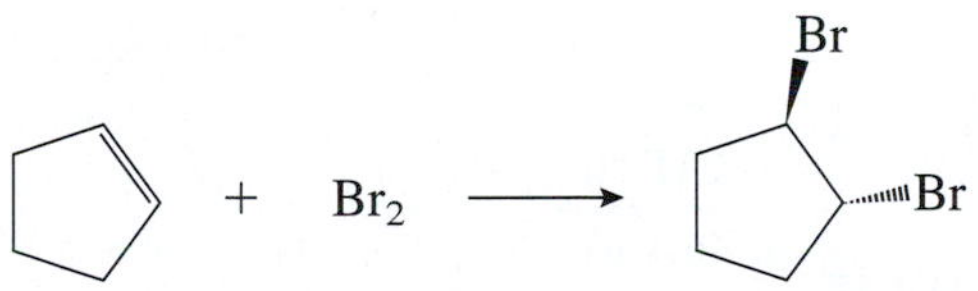

c) Both the electrophile and the nucleophile are oxygens bonded to osmium. The reaction is one from Table 11.4. The reaction results in syn addition of two OH groups to the alkene.

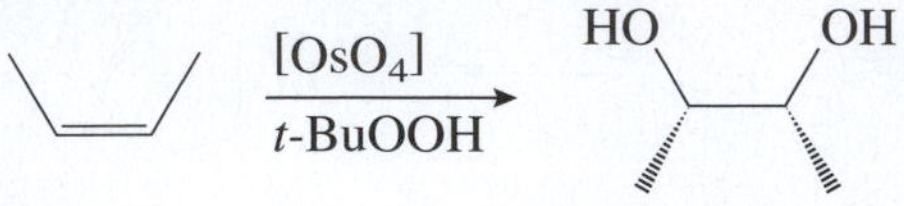

Additional Problems

11.32 Show the products of these reactions:

a) 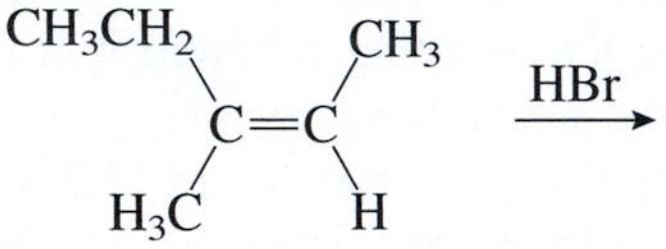 HBr →

b)  $\xrightarrow{H_2}{Pt}$

c) 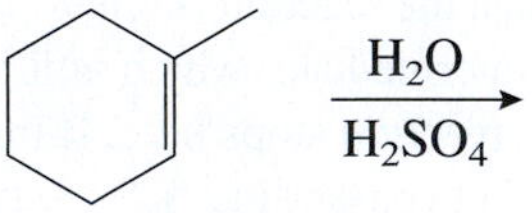 $\xrightarrow{H_2O}{H_2SO_4}$

d) 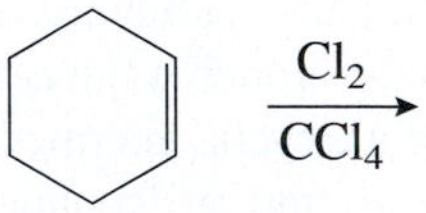 $\xrightarrow{Cl_2}{CCl_4}$

e) $CH_3CH_2CH_2C\equiv CH$ $\xrightarrow{H_2O}{H_2SO_4 \\ HgSO_4}$

f) $\xrightarrow{H_2O}{H_2SO_4}$

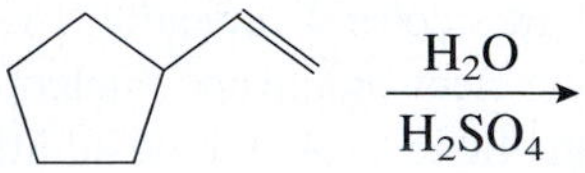

g) 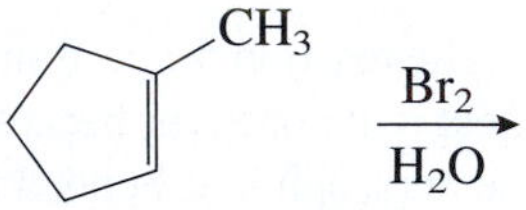 $\xrightarrow{Br_2}{H_2O}$

h) 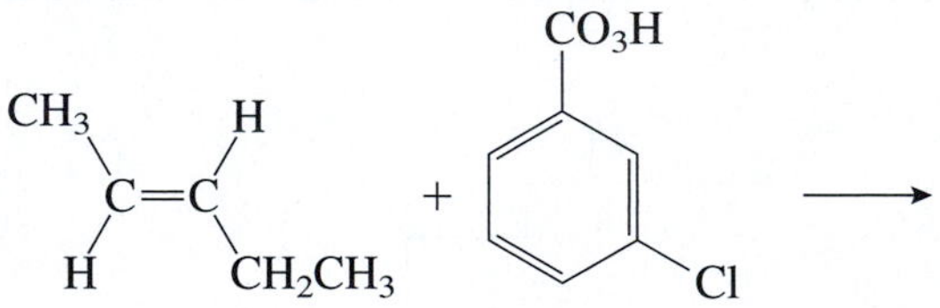 $+$ →

i) 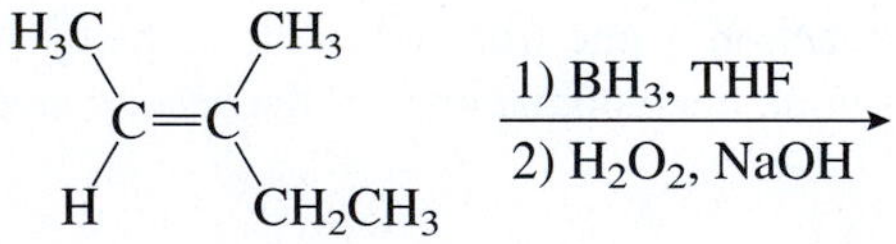 $\xrightarrow{1) BH_3, THF}{2) H_2O_2, NaOH}$

j) 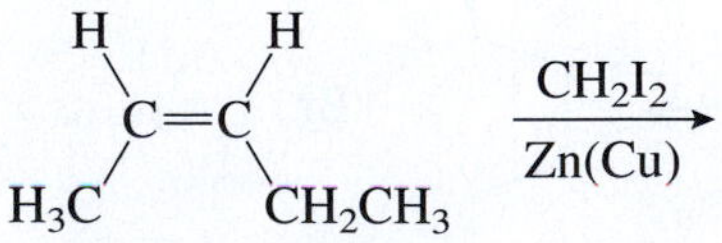 $\xrightarrow{CH_2I_2}{Zn(Cu)}$

k) 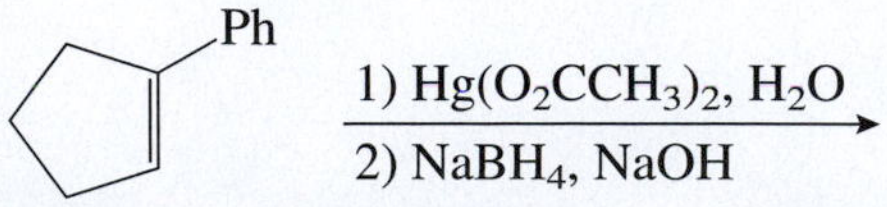 $\xrightarrow{1) Hg(O_2CCH_3)_2, H_2O}{2) NaBH_4, NaOH}$

l) $PhC\equiv CH$ $\xrightarrow{2\ HCl}$

m) $CH_3CH_2C\equiv CCH_2CH_3$ $\xrightarrow[\substack{Lindlar \\ catalyst}]{1\ H_2}$ $\xrightarrow[t\text{-BuOOH}]{[OsO_4]}$

n) $PhCH_2C\equiv CH$ $\xrightarrow[\text{2) } H_2O_2,\ NaOH]{\text{1) disiamylborane}}$

11.33 Show the products of these reactions.

a) 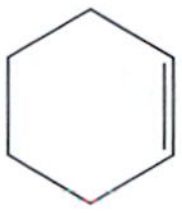 $\xrightarrow[\text{KOH}]{\text{CHBr}_3}$

b) 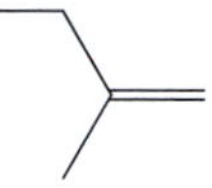 $\xrightarrow[H_2SO_4]{H_2O}$

c) 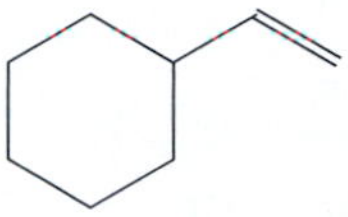 $\xrightarrow{\text{HCl}}$

d) $PhC\equiv CPh$ $\xrightarrow[\substack{H_2SO_4 \\ HgSO_4}]{H_2O}$

e) 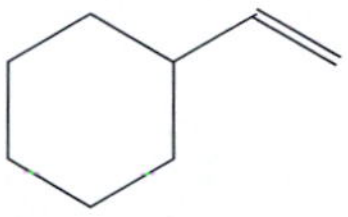 $\xrightarrow[\text{2) NaBH}_4,\ NaOH]{\text{1) Hg(O}_2\text{CCH}_3)_2,\ H_2O}$

f) $CH_3\overset{\overset{\displaystyle CH_3}{|}}{C}HCH_2CH=CH_2$ $\xrightarrow[\text{2) } H_2O_2,\ NaOH]{\text{1) BH}_3,\ THF}$

g) $CH_3CH_2CH_2C\equiv CH$ $\xrightarrow[\text{2) } H_2O_2,\ NaOH]{\text{1) } \left(\substack{H_3C\ \ CH_3 \\ |\ \ \ \ | \\ CH_3CHCH}\right)_2 -BH}$

h) 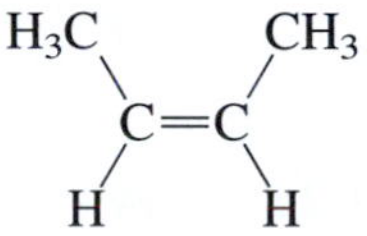 $\xrightarrow{CH_3CO_3H}$

i) 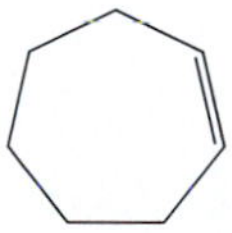 $\xrightarrow[CH_2Cl_2]{Br_2}$

j) 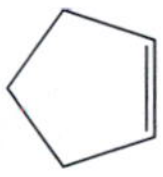 $\xrightarrow[t\text{-BuOOH}]{[OsO_4]}$

k) 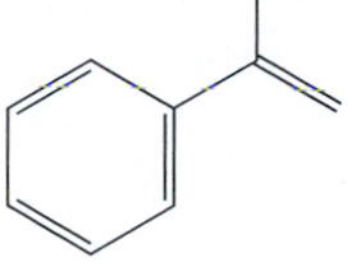 $\xrightarrow[Pt]{H_2}$

l) 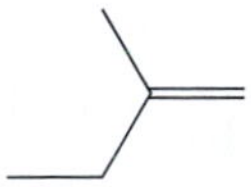 $\xrightarrow[H_2O]{Br_2}$

m) 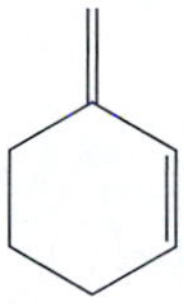 $\xrightarrow[\text{2) (CH}_3)_2S]{\text{1) O}_3}$

n) $CH_3C\equiv CCH_3$ $\xrightarrow[CCl_4]{1\ Br_2}$

o) 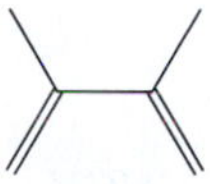 $\xrightarrow{\text{HBr}}$

11.34 Show the products of the reactions of 1-propylcyclopentene with these reagents:

a) Br_2, CCl_4

b) Br_2, H_2O

c) 1) BH_3, THF; 2) H_2O_2, NaOH

d) HBr

e) H_2O, H_2SO_4

f) [OsO_4], t-BuOOH

g) 1) O_3; 2) $(CH_3)_2S$

11.35 Show the alkenes that would give these products. More than one answer may be possible in some cases.

a) 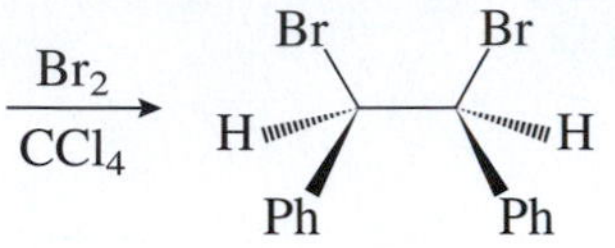 $\xrightarrow[\text{CCl}_4]{\text{Br}_2}$

b) 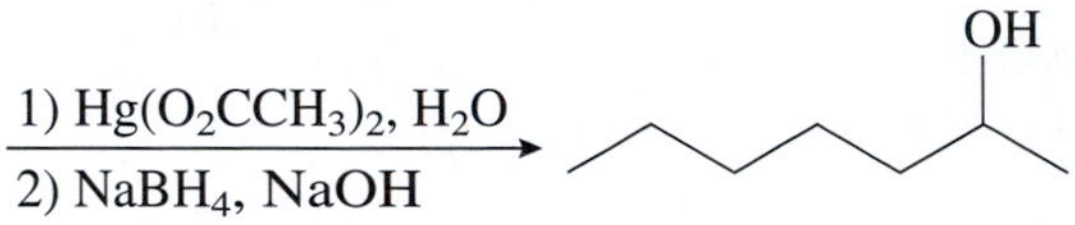 $\xrightarrow[\text{2) NaBH}_4,\ \text{NaOH}]{\text{1) Hg(O}_2\text{CCH}_3)_2,\ \text{H}_2\text{O}}$

c) 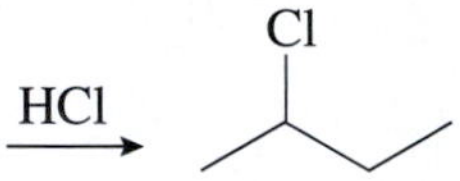 $\xrightarrow{\text{HCl}}$

d) 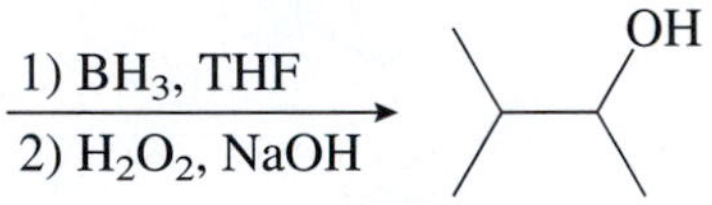 $\xrightarrow[\text{2) H}_2\text{O}_2,\ \text{NaOH}]{\text{1) BH}_3,\ \text{THF}}$

e) 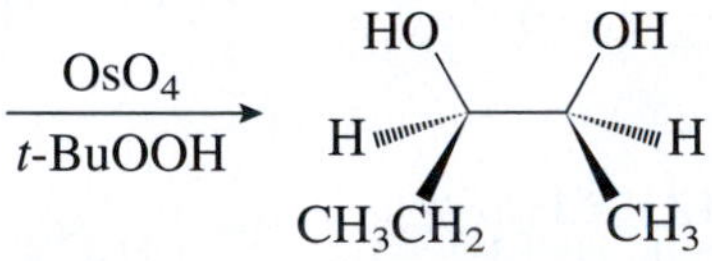 $\xrightarrow[t\text{-BuOOH}]{\text{OsO}_4}$

Racemic

f) 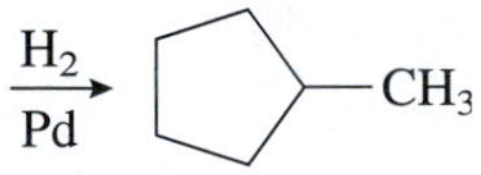 $\xrightarrow[\text{Pd}]{\text{H}_2}$

11.36 Show all the steps in the mechanisms for these reactions. Include stereochemistry where it is important.

a) 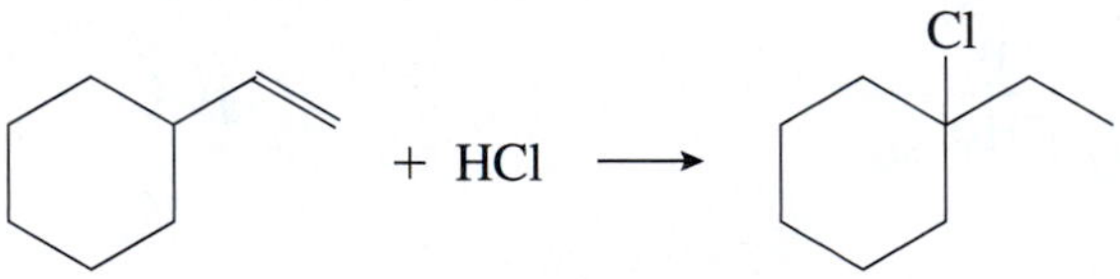 + HCl ⟶

b) 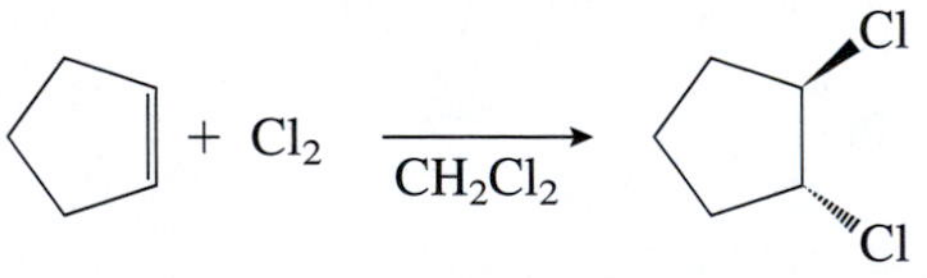 + Cl_2 $\xrightarrow{\text{CH}_2\text{Cl}_2}$

c) 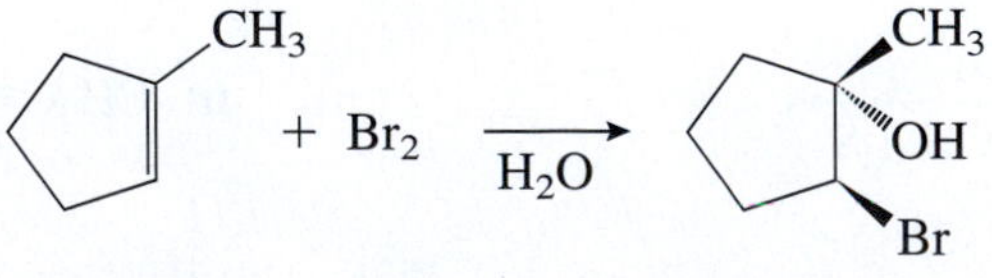 + Br_2 $\xrightarrow{\text{H}_2\text{O}}$

d) 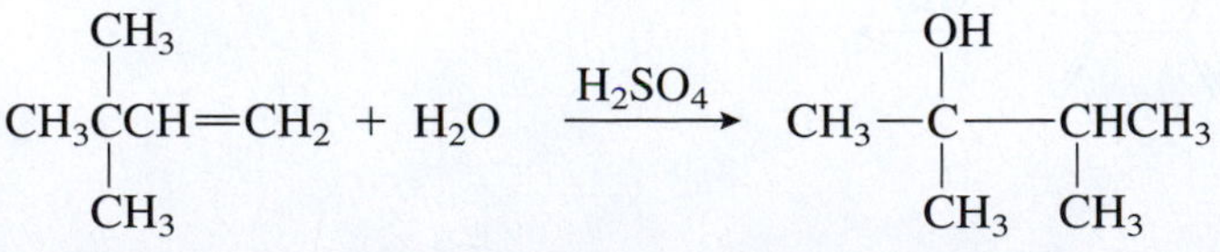 $CH_3\overset{\underset{\textstyle|}{CH_3}}{\underset{\underset{\textstyle|}{CH_3}}{C}}CH{=}CH_2$ + H_2O $\xrightarrow{\text{H}_2\text{SO}_4}$ $CH_3\overset{\underset{\textstyle|}{OH}}{\underset{\underset{\textstyle|}{CH_3}}{C}}{-}\overset{}{\underset{\underset{\textstyle|}{CH_3}}{C}}HCH_3$

e) 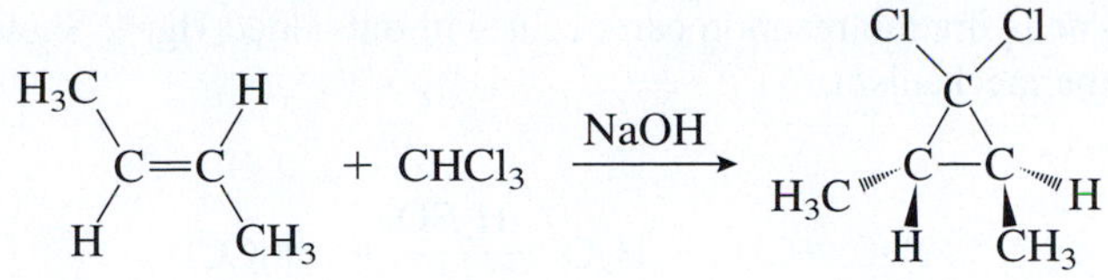 + CHCl₃ $\xrightarrow{\text{NaOH}}$

11.37 Show reactions that could be used to convert 1-pentene to these compounds. More than one step may be necessary.

a)

b)

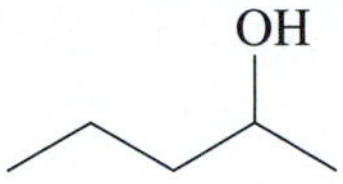

c)

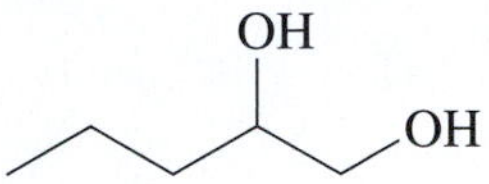

d)

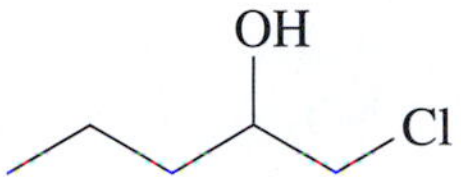

e)

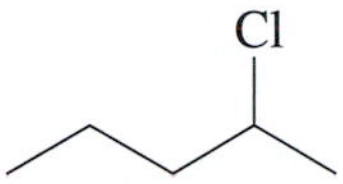

f)

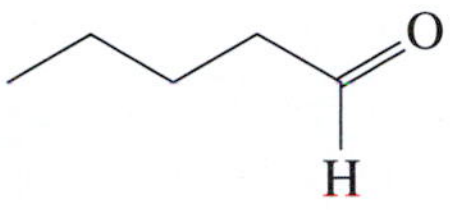

g)

h)

11.38 Show syntheses of these compounds from the indicated starting materials. More than one step may be necessary. Your syntheses may produce both enantiomers of any target that is chiral.

a) $CH_3\overset{\text{O}}{\overset{\|}{C}}CH_2CH_3$ from $CH_3C{\equiv}CH$

b) 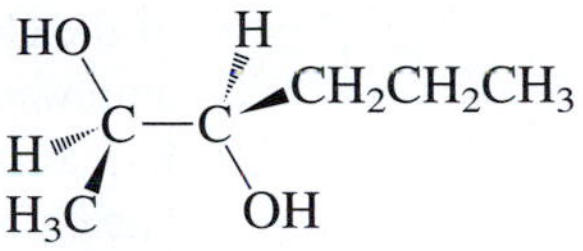 from $CH_3C{\equiv}CH$

c) 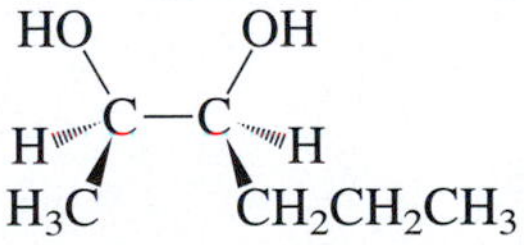 from $CH_3C{\equiv}CH$

d) 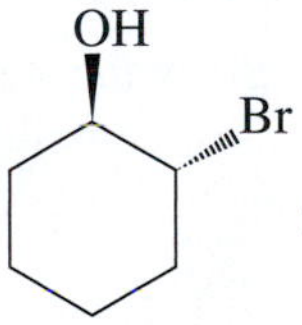 from

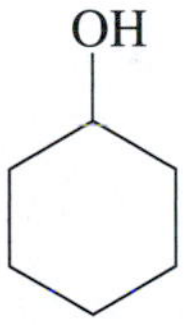

e) $CH_3CH_2CH_2CH_2\overset{\text{OH}}{\overset{|}{C}}H_2$ from $HC{\equiv}CH$

f) 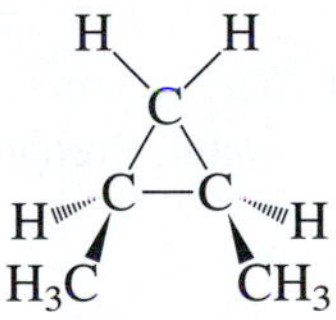 from $CH_3C{\equiv}CH$

g) 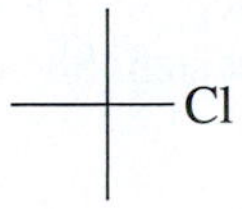 from

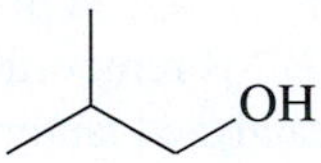

h) 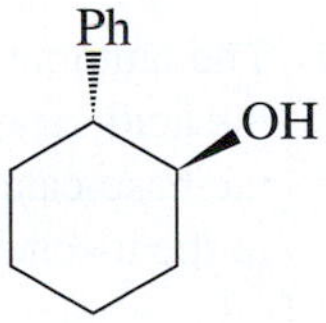 from

11.39 This alkyne hydration reaction can occur without added Hg^{2+}. Show all the steps in the mechanism.

$$PhC{\equiv}CH + H_2O \xrightarrow{H_2SO_4} Ph\overset{\overset{\displaystyle O}{\|}}{C}CH_3$$

11.40 Explain which compound has a faster rate of reaction with HCl:

a) 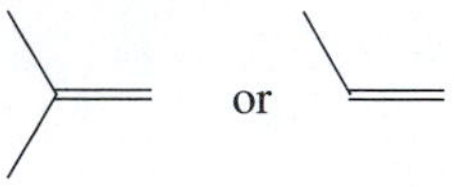 or

b) 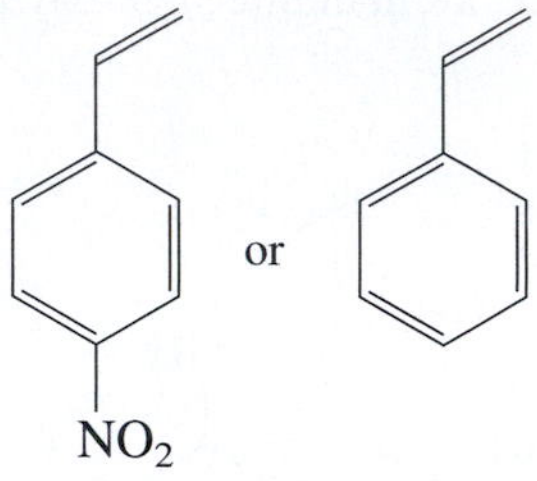 or

c) 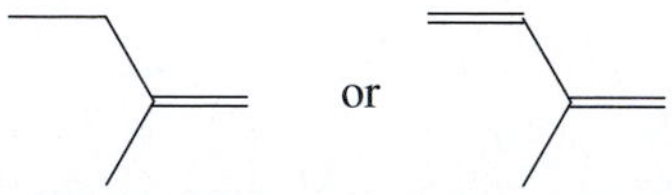 or

11.41 The addition of Cl_2 to (E)-2-pentene produces a racemic mixture of (2R,3S)-2,3-dichloropentane and its enantiomer, (2S,3R)-2,3-dichloropentane.

a) Show the structures of the two chloronium ions that are formed in this reaction. What is the relationship between them?

b) What is the relationship of the products that are formed by attack of the chloride nucleophile at each carbon of the two chloronium ions?

c) Explain why the percentages of nucleophile attack at the two carbons of one chloronium ion are not necessarily identical.

d) Explain why the product that is formed must be racemic.

11.42 Show the steps in the mechanism and predict the product that would be formed in this reaction:

$\xrightarrow[CH_3OH]{Br_2}$

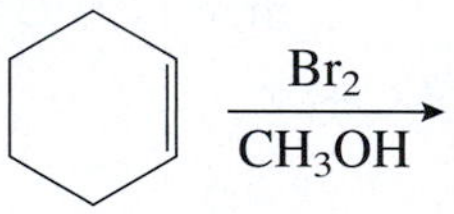

11.43 The oxymercuration reaction can be run in methanol as the solvent rather than water. Predict the product of this reaction:

$$CH_3CH_2CH_2CH{=}CH_2 \xrightarrow[\text{2) NaBH}_4, \text{ NaOH}]{\text{1) Hg(O}_2\text{CCH}_3)_2, \text{ CH}_3\text{OH}}$$

11.44 The tautomerization of an enol to a ketone is catalyzed by either acid or base. In the acid-catalyzed mechanism, H^+ is added in the first step (see Figure 11.6). In the base-catalyzed mechanism, H^+ is removed in the first step. Show the steps in the mechanism for the base-catalyzed tautomerization.

$$CH_2{=}\overset{\overset{\displaystyle OH}{|}}{C}{-}CH_3 \underset{}{\overset{^-OH}{\rightleftharpoons}} CH_3\overset{\overset{\displaystyle O}{\|}}{C}CH_3$$

11.45 An unknown compound has the formula C_6H_{10}.
   a) What is the DU for this compound?
   b) When a solution of Br_2 in CCl_4 is added to the unknown, the bromine color disappears. What information does this provide about the structure of the unknown?
   c) The unknown reacts with excess H_2 in the presence of Pt to give C_6H_{12}. What information does this provide about the structure of the unknown?
   d) This ketone is one of the products isolated from the ozonolysis of the unknown. What is the structure of the unknown?

11.46 An unknown compound has the formula C_7H_{12}.
   a) What is the DU for this compound?
   b) The unknown reacts with H_2 in the presence of Pd to give C_7H_{16}. What information does this provide about the structure of the unknown?
   c) The unknown reacts with H_2 in the presence of Lindlar catalyst to give C_7H_{14}. What information does this provide about the structure of the unknown?
   d) The product from part c, C_7H_{14}, gives these two aldehydes upon ozonolysis. Show the structure of the original unknown.

$$\underset{CH_3CH-CH}{\overset{\overset{CH_3}{|}\ \ \overset{O}{\|}}{}} + \underset{HCCH_2CH_3}{\overset{\overset{O}{\|}}{}}$$

11.47 Explain the difference in the percentages of the products in these two hydroboration reactions:

$$\underset{CH_3CHCH=CHCH_3}{\overset{\overset{CH_3}{|}}{}} \xrightarrow[\text{2) H}_2\text{O}_2,\ \text{NaOH}]{\text{1) BH}_3,\ \text{THF}} \underset{CH_3CH-CHCH_2CH_3}{\overset{\overset{CH_3}{|}\ \ \overset{OH}{|}}{}} + \underset{CH_3CHCH_2CHCH_3}{\overset{\overset{CH_3}{|}\ \ \overset{OH}{|}}{}}$$

(43%)   (57%)

$$\xrightarrow[\text{2) H}_2\text{O}_2,\ \text{NaOH}]{\text{1) disiamylborane}}$$

(5%)   (95%)

11.48 Explain why this reaction occurs with anti-Markovnikov regiochemistry:

$$CF_3CH=CH_2 + HCl \longrightarrow \underset{CF_3CH_2CH_2}{\overset{\overset{Cl}{|}}{}}$$

11.49 Explain why the hydration of this alkene occurs 10^{15} times faster than the hydration of ethene:

$$CH_3CH_2OCH=CH_2 \xrightarrow[\text{H}_2\text{SO}_4]{\text{H}_2\text{O}} \underset{CH_3CH_2OCHCH_3}{\overset{\overset{OH}{|}}{}}$$

11.50 The addition of HCl to alkynes proceeds through a vinyl cation intermediate. Explain which of the two possible vinyl cations that could be formed from the addition of HCl to propyne is more stable.

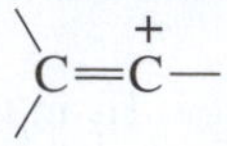

$$\underset{\text{A vinyl cation}}{\overset{+}{C}=\overset{}{C}-}$$

11.51 Suggest a mechanism for this reaction:

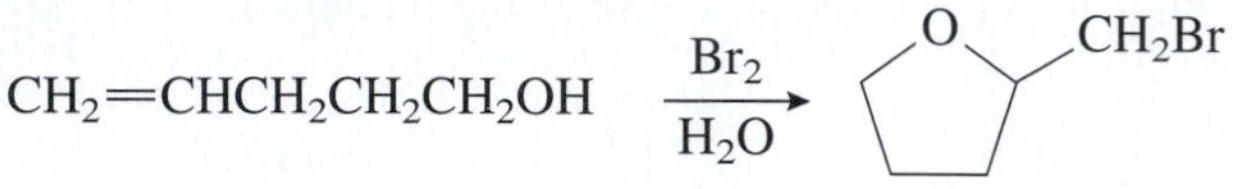

$$CH_2{=}CHCH_2CH_2CH_2OH \xrightarrow[H_2O]{Br_2}$$

11.52 Limonene, a major component of lemon oil, has the formula $C_{10}H_{16}$.
  a) On reaction with excess H_2 in the presence of Pt, limonene produces $C_{10}H_{20}$. What information does this provide about the structure of limonene?
  b) On ozonolysis, limonene produces these compounds. Suggest possible structures for limonene.

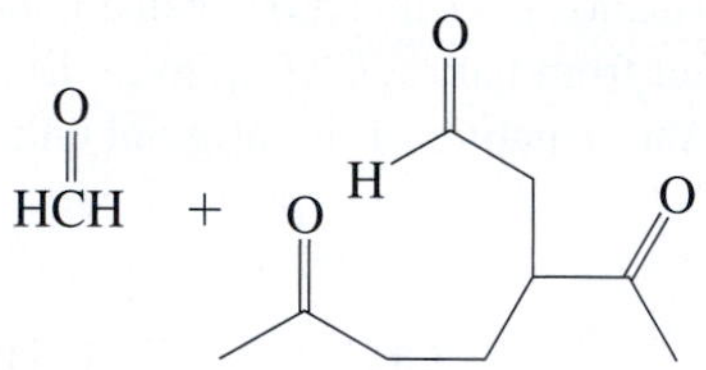

11.53 Show the structures of **A, B, C,** and **D** in the following reaction scheme:

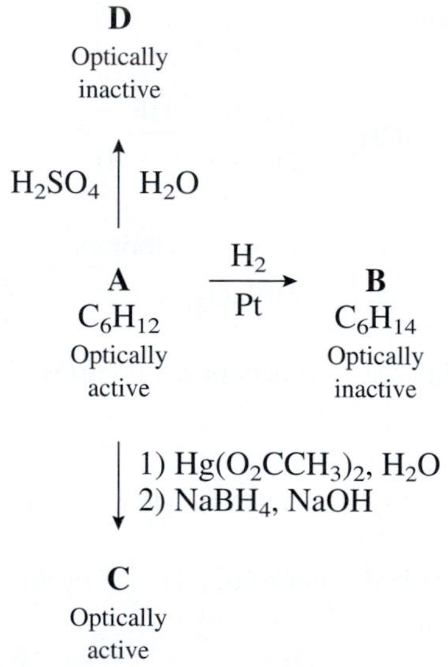

D
Optically
inactive

H_2SO_4 | H_2O

A
C_6H_{12}
Optically
active

$\xrightarrow[\text{Pt}]{H_2}$

B
C_6H_{14}
Optically
inactive

1) $Hg(O_2CCH_3)_2$, H_2O
2) $NaBH_4$, NaOH

C
Optically
active

11.54 In Figure 11.3, suppose Br_2 adds to the alkene from the bottom, rather than from the top as shown. Analyze the stereochemistry of the reaction in this case and explain which products are formed.

BioLink

Problems Using Online Three-Dimensional Molecular Models

11.55 Explain which of the three products shown is formed when 1-butene reacts with HCl.

11.56 Explain which of the four products shown is formed when *cis*-2-pentene reacts with Cl_2.

11.57 Explain which of the four products shown is formed when cyclopentene reacts with Cl_2 and water.

11.58 Explain which of the five products shown is formed when 1-ethylcyclopentene reacts with BH_3 in THF, followed by treatment with NaOH and H_2O_2.

11.59 Explain which of the three products shown is formed when *trans*-2-butene reacts with CH_2I_2 and Zn(Cu).

11.60 Explain which of the three products shown is formed when *cis*-2-butene reacts with OsO_4 and *t*-BuOOH.

11.61 The hydroboration–oxidation of α-pinene gives the product shown. Carefully explain the regiochemistry and the stereochemistry of this reaction.

11.62 The catalytic hydrogenation of the alkene shown gives the product shown. The hydrogens that added to the double bond are blue in the product. Explain the stereochemistry of this reaction.

Functional Groups and Nomenclature II

THIS CHAPTER BEGINS where Chapter 5 left off and considers the nomenclature, physical properties, natural occurrences, and uses of the principal functional groups that were not discussed in Chapter 5. The nomenclature rules are extensions of those in Chapter 5, with different suffixes for the new functional groups. We will also see how compounds that have more than one functional group are named.

12.1 AROMATIC HYDROCARBONS

A complete definition of **aromatic compounds** must wait until Chapter 16. For the present we will define them as benzene and its substituted derivatives. They are also called **arenes.** These aromatic compounds have a six-membered ring with three conjugated double bonds. It is this cycle of conjugated double bonds that makes arenes special. Examples include the following:

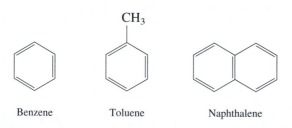

Benzene   Toluene   Naphthalene

If you have spent much time in an organic laboratory, you are well aware that many organic compounds have rather strong (and sometimes disagreeable) odors. Aromatic compounds, however, tend to have more fragrant odors than other compounds. If you have a chance, compare the odor of toluene, for example, with that of cyclohexene. In fact, the term *aromatic* was originally given to these compounds, even before their structures were known, because of their fragrant odors. Today, however, *aromatic* has evolved to mean something entirely different. To an organic chemist, an **aromatic compound** now means a compound that is especially stable because of its cycle of conjugated *p* orbitals. (The explanation for this special stabilization is provided in Chapter 16.)

Aromatic compounds are named by using benzene as the parent or root and designating the substituents attached to the ring in the same manner that is used to name groups attached to an alkane chain. (Cyclohexatriene is *never* used to name benzene, nor is it used in the names of any aromatic compounds.) Because of their stability, aromatic compounds are very common, and many were isolated in the very early days of organic chemistry. Many of the common names that were originally given to these compounds remain entrenched in the vocabulary of organic chemistry. For example, you will seldom hear methylbenzene referred to by this name; it will invariably be called toluene. Some of the more important of these common names will be introduced as the compounds are encountered.

When several substituents are present on a benzene ring, the ring is numbered in the same manner as the rings of cycloalkanes—that is, so that the numbers for the substituents are as low as possible. In addition, some special terms are used with *disubstituted benzenes only*. Two substituents on adjacent carbons (positions 1 and 2) are said to be *ortho,* or *o-*. Two substituents on positions 1 and 3 are *meta,* or *m-*. And two substituents on positions 1 and 4 are *para,* or *p-*. Finally, if an alkyl group with six or more carbons is attached to a benzene ring, the compound is named as an alkane with a **phenyl** substituent. Some examples are as follows:

m-Propyltoluene
or 3-propyltoluene

p-Dimethylbenzene
or 1,4-dimethylbenzene

(The common name for the
dimethylbenzenes is xylene,
so this is also called *p*-xylene.)

4-Butyl-1-ethyl-2-isopropylbenzene

(The numbering starts with the ethyl,
so the numbers are 1, 2, and 4. If
numbering started with the butyl, the
numbers would be higher—1, 3, and 4.)

The phenyl group

(Recall that this group can be
abbreviated as Ph in drawing
structures.)

The benzyl group

3-Methyl-2-phenylhexane

(The larger or more highly substituted
group, the alkyl group in this case, is
chosen as the root.)

PROBLEM 12.1

Provide names for these compounds:

a) $CH_3CHCH_2CH_3$

b) CH_3 Cl

c) Br OCH_3

d)   Cl

e)

f)   OH

PROBLEM 12.2

Draw structures for these compounds:

a) *p*-Ethyltoluene
b) *m*-Dichlorobenzene
c) 2-Phenyl-3-heptyne
d) 2-Bromo-1-chloro-3-pentylbenzene
e) *o*-Xylene
f) Benzyl methyl ether

Aromatic hydrocarbons are nonpolar, and their physical properties resemble those of alkanes of similar molecular mass. However, as was the case with cycloalkanes, the symmetrical shapes of many aromatic hydrocarbons often result in higher melting points. For example, the melting and boiling points of benzene are nearly identical to those of cyclohexane. (Recall that cyclohexane melts at considerably higher temperatures than does hexane.) As expected, a mixture of benzene and water forms two layers, with benzene as the upper layer.

On initial inspection, aromatic compounds might be expected to resemble alkenes in their chemical reactions. However, their reactions are dramatically different. Aromatic compounds are much less reactive than alkenes. The reason for this decrease in reactivity is the large amount of stabilization present in aromatic compounds as a result of their special cycle of conjugated *p* orbitals. This great difference in reactivity is the reason why aromatic compounds are classified separately from the alkenes. As an example of this reactivity difference, the treatment of benzene with Cl_2 under conditions in which the Cl_2 would rapidly add to an alkene does not result in any reaction. Under more drastic conditions, in the presence of a Lewis acid catalyst such as $AlCl_3$, a reaction does occur, but the product retains the conjugated system of the aromatic ring. Rather than the addition reaction that would have resulted with an alkene, a chlorine replaces one of the hydrogens on the ring. This reaction, in which one group replaces another, is another type of substitution reaction and is the most important reaction of aromatic compounds.

Substitution Reaction:

Benzene   Chlorobenzene

Because of their stability, aromatic compounds are found in a variety of natural sources. Often the aromatic ring occurs in combinations with other functional groups. Benzene itself was first isolated, by Michael Faraday in 1825, from the oily residue that condensed from the gas that was used to light the street lamps of London. A few examples of other naturally occurring compounds that contain aromatic rings are cumene (isopropylbenzene), which occurs in petroleum; estrone, a female sex hormone that has an aromatic ring as part of a complex ring system; and benzo[a]pyrene, which consists of a series of fused aromatic rings. Benzo[a]pyrene is a carcinogenic (cancer-causing) substance produced upon combustion of many materials. It is found in soot and is one of the major carcinogens found in tobacco smoke.

Cumene   Estrone   Benzo[a]pyrene   Styrene

Benzene is one of the major chemicals produced by the petroleum industry. More than 1.6 billion gallons are produced each year by cracking and reforming various petroleum fractions. Most of this is used in the production of styrene, which is then polymerized to polystyrene. Other arenes that are made in large amounts include toluene (830 million gallons), cumene, *o*-xylene, and *p*-xylene. At one time, benzene was an important solvent in the organic laboratory. Recently, however, its use has been phased out because of its potential adverse health effects. Long exposure to benzene has been shown to lead to bone marrow depression and leukemia.

Focus On

Structure Proof by the Number of Isomers

How did organic chemists identify the structures of organic compounds before the advent of spectroscopy? Basically, the structure had to be consistent with all the facts known about a compound. Often, these facts included the results of a number of chemical reactions. Let's examine the case of the substitution products that occur on reaction of benzene with bromine.

Continued

In 1866, August Kekulé, one of the true pioneers of organic chemistry, proposed a structure for benzene that is remarkably similar to the structure used today. From experiments, he knew that substitution reactions of benzene always gave a single mono-substitution product. For example, the reaction of benzene (C_6H_6) with bromine in the presence of aluminum tribromide as a catalyst gave only one monobromide (C_6H_5Br). This indicates that all of the hydrogens of benzene must be identical. One of the structures that satisfies the valence rules and this observation is a six-membered ring with three double bonds:

Benzene   Bromobenzene

In this structure for benzene, all the carbons are identical, so it does not matter which one bonds to the bromine in the substitution reaction; only one monobromo substitution product is possible.

Kekulé next considered what would happen if bromobenzene were further substituted with a second bromine. From experiment, it was known that reaction of C_6H_5Br in this same reaction resulted in the formation of three isomers of $C_6H_4Br_2$. Today, this reaction would be written as shown in the following equation:

ortho   meta   para

However, to make the results of this reaction consistent with his theory, Kekulé had to make some modifications. To see the problem, it is necessary only to recall that the concept of resonance had not yet been proposed at this point in the development of organic chemistry. Kekulé realized that, according to his structural theory, there should be two products related to the ortho product, one with a single bond between the carbons attached to the bromines and one with a double bond between these carbons.

To make his theoretical prediction (four products) consistent with experimental observation (three products), Kekulé proposed that the two "*ortho*-isomers" were actually in rapid equilibrium and therefore behaved as a single compound. Of course, today we know that these are two resonance structures of *ortho*-dibromobenzene rather than distinct isomers, but Kekulé's proposal was not too far removed from our current ideas.

Now let's carry our considerations one step further. As we know, bromination of bromobenzene produces three isomers of dibromobenzene. One of these is the *ortho*-isomer, one is the *meta*, and one is the *para*, but which is which? This question was answered by Wilhelm Korner in 1874, who reacted each of the $C_6H_4Br_2$ isomers under substitution conditions that produced the tribrominated products, $C_6H_3Br_3$. By determining the number of isomeric tribromides produced from each dibromide, he was able to assign the structures of all of the compounds. He found that one $C_6H_4Br_2$ isomer, call it **A**, produced a single isomer of $C_6H_3Br_3$, call it **D**. Another isomer of $C_6H_4Br_2$, **B**, produced two isomers of $C_6H_3Br_3$: **D** (obtained previously) and a new isomer, **E**. The final isomer of $C_6H_4Br_2$, **C**, produced three isomers of $C_6H_3Br_3$: **D**, **E**, and a new isomer, **F**. These results are summarized in the following equations:

Dibromide		Tribromide	
A	⟶	**D**	(one product)
B	⟶	**D + E**	(two products)
C	⟶	**D + E + F**	(three products)

The products predicted to result from the bromination of the *ortho*-, *meta*-, and *para*-isomers of dibromo benzene are as follows:

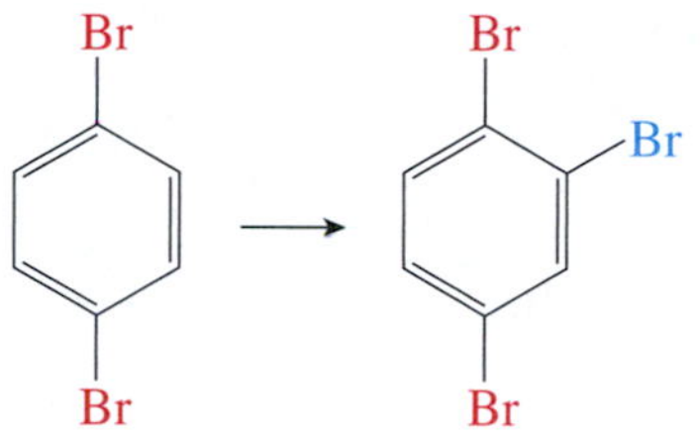

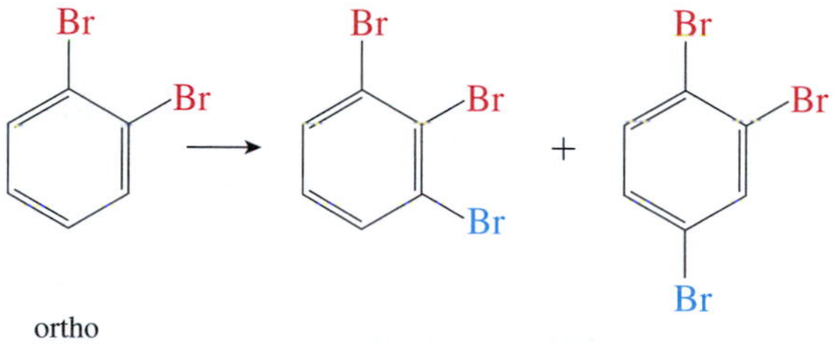

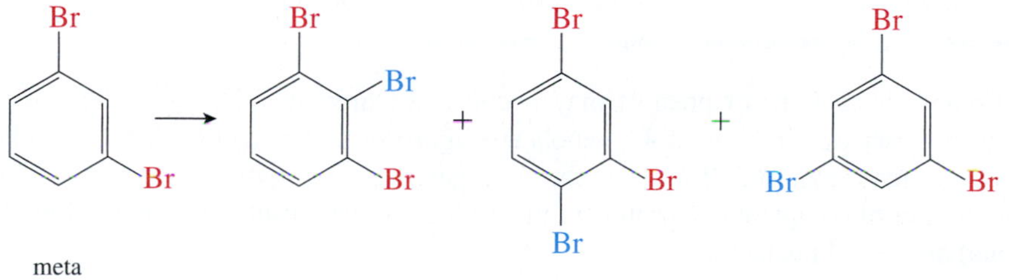

The *para*-isomer is predicted to give a single tribromo product. Therefore, the *para*-isomer must be **A**, and its product, 1,2,4-tribromobenzene, must be **D**. The *ortho*-isomer gives two products, so it must be **B**. Its two products are **D** and 1,2,3-tribromobenzene, **E**. Finally, the *meta*-isomer gives three products, so it must be **C**. Its three products are **D**, **E**, and the new product, **F**, which must be 1,3,5-tribromobenzene. Using this logic, Korner was able to assign the structures of not only all of the dibromobenzenes, but also all of the tribromobenzenes!

12.2   PHENOLS

Phenols are compounds that have a hydroxy group bonded directly to an aromatic ring. The delocalization of the electrons on the oxygen onto the aromatic ring changes the reactivity of the hydroxy group enough that phenols are given their own functional group rather than being classified with alcohols.

The simplest member of this class of compounds is named phenol. Others can be named as substituted phenols, although numerous common names may be encountered.

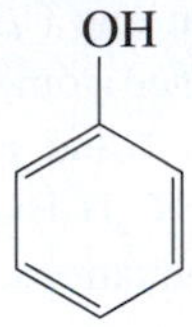

Phenol 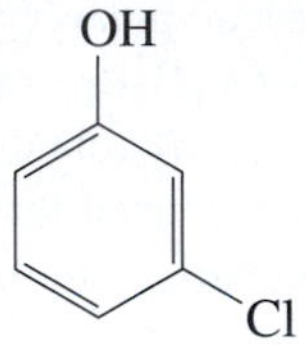   *m*-Chlorophenol 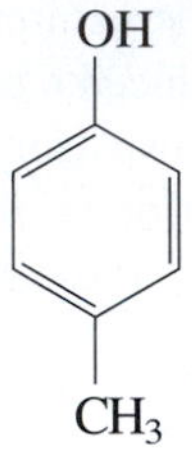   *p*-Methylphenol
                                                 (*p*-cresol)

PROBLEM 12.3
Provide names for these compounds:

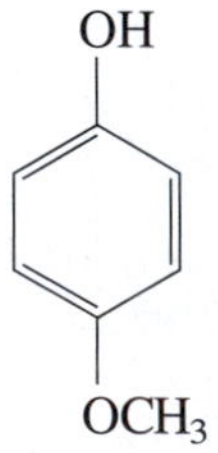

PROBLEM 12.4
Draw structures for these compounds:
a) *o*-Ethylphenol   **b)** *m*-Cresol   **c)** 2,6-Dinitrophenol

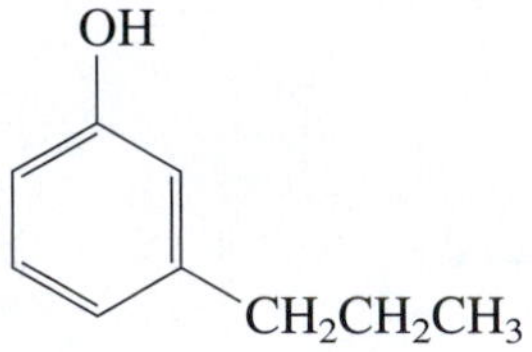

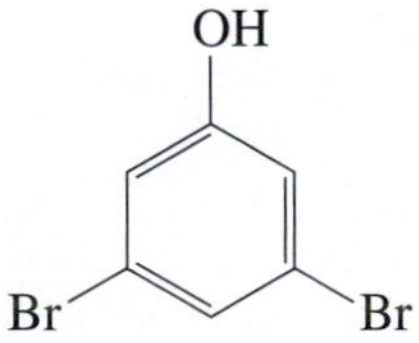

Chemical reactions of phenols may occur at the aromatic ring or at the hydroxy group. As discussed in Chapter 4, phenols are significantly stronger acids than alcohols because of resonance stabilization of the conjugate base. Sodium hydroxide is a strong enough base to completely deprotonate most phenols. The resulting anions (phenolate anions) are useful nucleophiles.

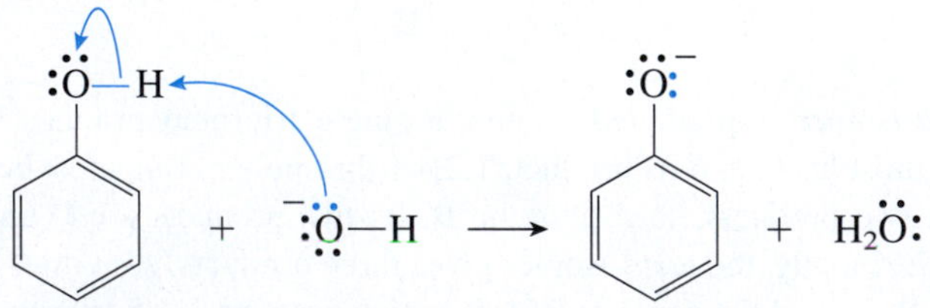

Phenol   Phenolate
                                            anion

PROBLEM 12.5

Show the resonance structures for the conjugate base of phenol.

PROBLEM 12.6

The pK_a for phenol is 10, the pK_a for ethanol is 16, and the pK_a for carbonic acid (H_2CO_3) is 6.35. Complete these equations and predict whether the reactants or the products are favored at equilibrium.

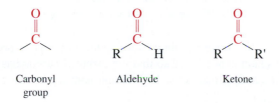

Phenol is an important industrial chemical. More than 3 billion pounds are produced each year. The major uses of phenol are as a disinfectant and in the production of polymers. Complex phenols, with multiple substituents and functional groups, are common in nature, although the simple phenols are seldom encountered.

12.3 ALDEHYDES AND KETONES

A carbon–oxygen double bond is called a **carbonyl group.** Compounds that contain a carbonyl group are among the most important in organic chemistry because of the varied reactions that they undergo. **Aldehydes** have at least one hydrogen atom bonded to the carbonyl group (the other atom can be hydrogen or carbon). In the case of **ketones,** both of the atoms bonded to the carbonyl group must be carbons.

| Carbonyl group | Aldehyde | Ketone |

As with the other functional groups encountered so far, the first step in naming an aldehyde is finding the longest chain that contains the functional group—the carbonyl group in this case. Because the carbonyl group of an aldehyde must be at the end of the chain, numbering *always* begins with the carbonyl carbon. The suffix used to designate an aldehyde is -al.

Common names for simple aldehydes are frequently encountered. These common names are derived from the common names for the related carboxylic acid (see Section 12.4) by replacing the suffix -ic acid with the suffix -aldehyde. Thus, the aldehyde related to acetic acid is acetaldehyde. If the carbonyl group of an aldehyde is attached to a ring system, the compound can be named as a hydrocarbon with the suffix -carbaldehyde. (Some sources use -carboxaldehyde.)

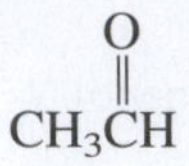

$$\underset{\text{CH}_3\text{CH}}{\overset{\overset{\displaystyle O}{\|}}{}}$$

Ethanal
(The common name is acetaldehyde, derived from acetic acid.)

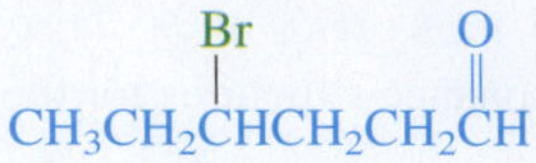

$$\underset{\text{CH}_3\text{CH}_2\text{CHCH}_2\text{CH}_2\text{CH}}{\overset{\overset{\displaystyle Br\qquad\qquad O}{|\qquad\qquad\|}}{}}$$

4-Bromohexanal
(No number is needed to designate the position of the carbonyl carbon of an aldehyde because it is always located at position 1.)

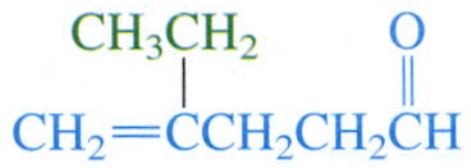

$$\underset{\text{CH}_2=\text{CCH}_2\text{CH}_2\text{CH}}{\overset{\overset{\displaystyle \text{CH}_3\text{CH}_2\qquad O}{|\qquad\quad\|}}{}}$$

4-Ethyl-4-pentenal
(The longest chain containing the carbonyl carbon *and* the carbon–carbon double bond is chosen as the root. Again, numbering begins with the carbonyl carbon.)

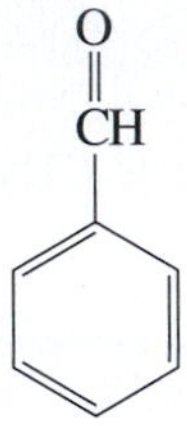

Benzaldehyde
(This name, derived from benzoic acid, is also the systematic name for this aldehyde. It is used to name substituted derivatives.)

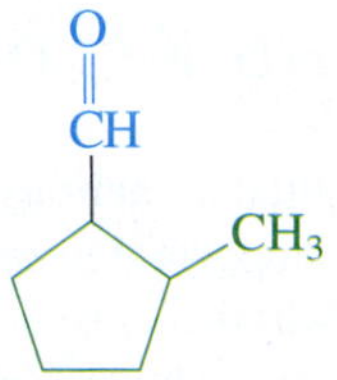

2-Methylcyclopentanecarbaldehyde
(The suffix -carbaldehyde indicates that the —CHO group is attached to the parent ring.)

To name ketones, the longest chain containing the carbonyl carbon is again chosen as the parent. This chain is numbered so that the carbonyl carbon has the lowest possible number, and the suffix -one is used. Common names are also encountered occasionally.

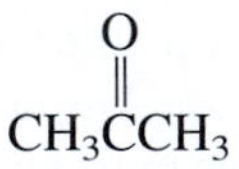

$$\underset{\text{CH}_3\text{CCH}_3}{\overset{\overset{\displaystyle O}{\|}}{}}$$

2-Propanone
(This compound is usually called by its common name, acetone.)

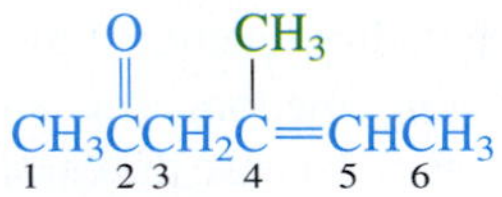

$$\underset{\underset{1\quad\ \ 2\ 3\quad\ 4\quad\ \ 5\ \ 6}{\text{CH}_3\text{CCH}_2\text{C}=\text{CHCH}_3}}{\overset{\overset{\displaystyle O\qquad\ \text{CH}_3}{\|\qquad\ \ |}}{}}$$

4-Methyl-4-hexen-2-one
or
4-methylhex-4-en-2-one
(Choose the parent so that it includes both the carbonyl group and the carbon–carbon double bond. Number so that the carbonyl carbon gets the lower number.)

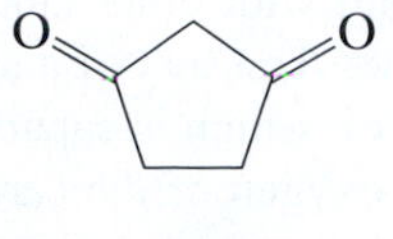

1,3-Cyclopentanedione
or
cyclopentane-1,3-dione
(Note the "e" that is added between
"cyclopentan" and "dione" to aid in
pronunciation.)

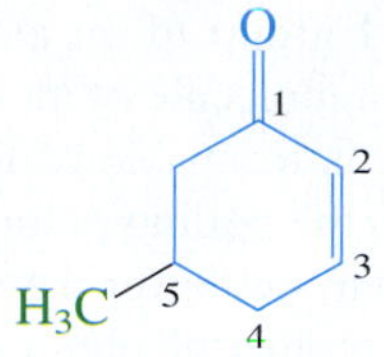

5-Methyl-2-cyclohexenone
or
5-methylcyclohex-2-enone
(Number so that first the carbonyl group
and then the carbon–carbon double bond
get the lowest possible numbers.)

PROBLEM 12.7
Provide names for these compounds:

a) $CH_3CH_2CH_2CH_2CH_2CH$

b)

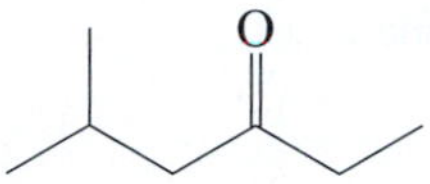

c)

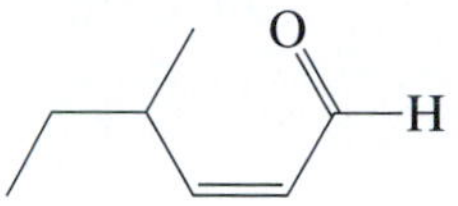

d)

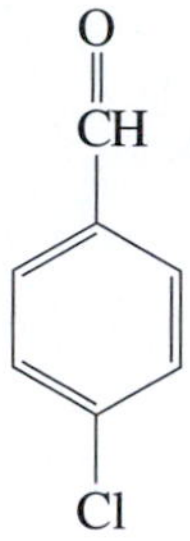

e)

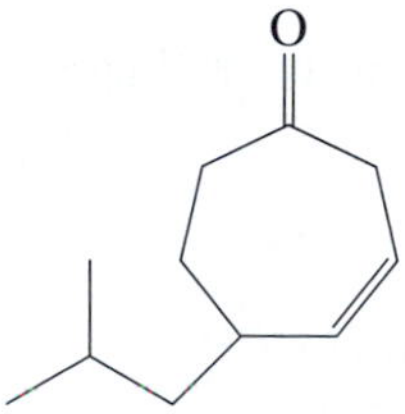

f) $CH_3CCH_2CCH_3$

g)

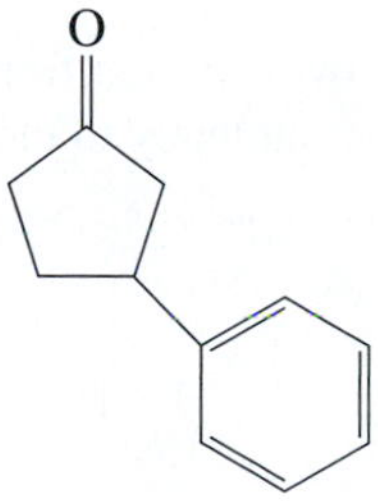

h)

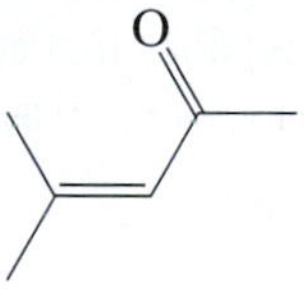

PROBLEM 12.8
Draw structures for these compounds:
a) (Z)-Oct-3-en-2-one
b) 3-Ethylheptanal
c) 2,4-Pentadienal
d) 3,4-Dimethylbenzaldehyde
e) 1-Phenyl-1-propanone
f) 2,2,6,6-Tetramethylcyclohexanone

ORGANIC
Chemistry⚛Now™
Click *Coached Tutorial Problems*
for more practice **Drawing
Structures of Aldehydes
and Ketones from IUPAC
Names.**

The carbonyl group of an aldehyde is more polar than a carbon–oxygen single bond. However, a molecule of an aldehyde does not have a hydrogen on an electronegative atom, so it is incapable of forming a hydrogen bond with other aldehyde molecules. Therefore, the boiling point of an aldehyde is higher than that of a similar ether, which is less polar, but lower than that of a similar alcohol, which is capable of hydrogen bonding with other alcohol molecules. Because the oxygen of the carbonyl group can act as the Lewis base partner in a hydrogen bond, lower molecular weight aldehydes are, like alcohols, relatively water soluble. Similar arguments apply to ketones.

Aldehydes and ketones share many chemical reactions because both have a carbonyl group. Often their reactions begin with an acid–base step. As discussed in Chapter 4, the protons on the carbon adjacent to a carbonyl group, called the **α-carbon,** are weakly acidic, with a pK_a of about 20. A strong base can remove a proton from the α-carbon, as illustrated in the following equation. One feature that distinguishes aldehydes from ketones is the fact that aldehydes can be oxidized to carboxylic acids (see Section 10.14), whereas ketones are inert to most oxidizing agents. Aldehydes and ketones are among the most useful functional groups in the organic laboratory because of the wide variety of reactions that they undergo.

$$CH_3CCH_2{-}H \ + \ :B^- \ \longrightarrow \ CH_3CCH_2^- \ + \ H{-}B$$

α-Carbon

PROBLEM 12.9

Explain which are the most acidic hydrogens in these compounds:

a) [cyclopentanone structure]

b) $PhCH_2CCH_3$

c) $CH_3CCH_2CCH_3$

PROBLEM 12.10

a) Is sodium hydroxide a strong enough base to completely remove a proton from the α-carbon of acetone; that is, does this equilibrium lie nearly completely to the right when sodium hydroxide is the base?

$$CH_3{-}\overset{O}{\underset{}{C}}{-}\overset{H}{\underset{}{C}}H_2 \ + \ :B^- \ \rightleftharpoons \ CH_3{-}\overset{O}{\underset{}{C}}{-}CH_2^- \ + \ H{-}B$$

b) Which common bases can be used to completely remove a proton from acetone? (See Table 4.4 on page 131.)

Aldehydes and ketones often have pleasant odors. They are found as components of many perfumes and flavorings, both natural and artificial. For example, citral has a strong lemon odor and is found in lemon and orange oils, cinnamaldehyde has a strong cinnamon odor and is found in cinnamon oil, and vanillin is a major component of vanilla flavoring. Camphor, isolated from the camphor tree, is used in liniments and inhalants, and muscone, which has a "musky" aroma, is used in many perfumes.

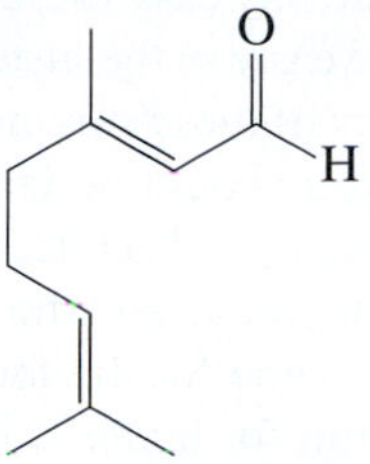

Citral 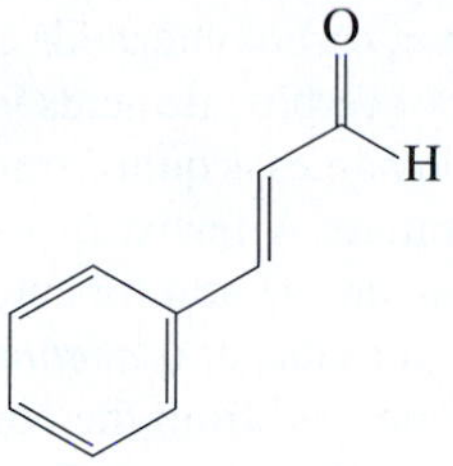   Cinnamaldehyde 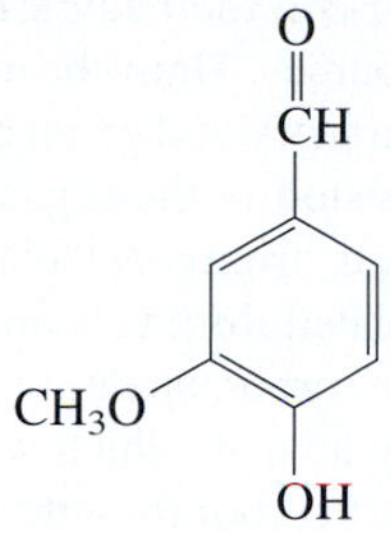   Vanillin

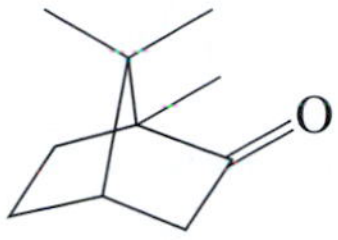

Camphor 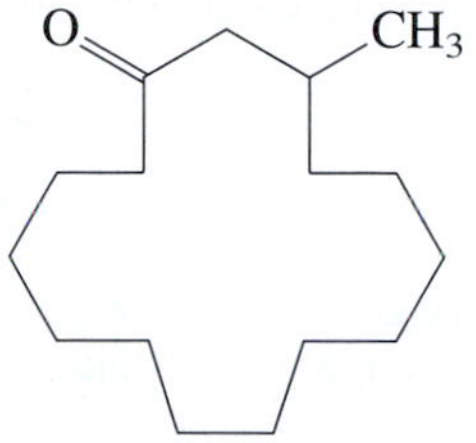   Muscone
(3-methylcyclopentadecanone)

12.4 CARBOXYLIC ACIDS

Carboxylic acids are compounds containing a carboxy group, that is, a carbonyl group with an attached hydroxy group. The systematic name for a carboxylic acid uses as the root the longest chain with the carboxy group at one end with the suffix -oic acid added. As with aldehydes, the chain is always numbered beginning with the carbon of the carboxy group.

Carboxy group   Carboxylic acid

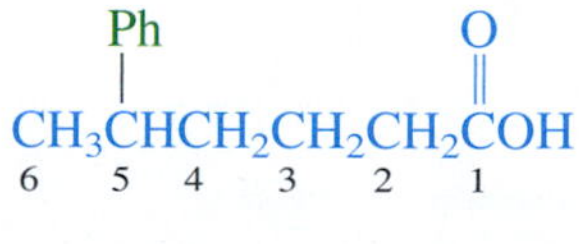

5-Phenylhexanoic acid
(Numbering always begins with the carboxy carbon.)

3-Butynoic acid

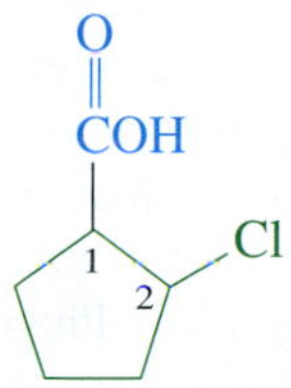

2-Chlorocyclopentanecarboxylic acid
(In similar fashion to the use of the -carbaldehyde suffix in naming aldehydes, cyclic compounds with the carboxy group attached to the ring use the name of the ring with the suffix -carboxylic acid.)

Because of their acidic character, carboxylic acids are relatively easy to isolate from natural sources. Thus, the number of carboxylic acids isolated early in the history of organic chemistry and given common names is quite large. Many of these common names have persisted in the organic literature. A few with which you should be familiar are formic acid, originally isolated from ants (Latin for ant is *formica*) and acetic acid, originally isolated from vinegar (Latin for vinegar is *acetum*). As mentioned earlier, the root for a four-carbon chain, but-, was derived from the common name for the four-carbon acid, butyric acid, which was isolated from rancid butter (Latin for butter is *butyrum*). Similarly, the root for a three-carbon chain, prop-, was derived from the common name for the three-carbon acid, propionic acid. This acid was considered to be the smallest one derived from fats and its name is derived from the Greek words *pro-* (first) and *pion* (fat). Aromatic carboxylic acids are named as derivatives of benzoic acid.

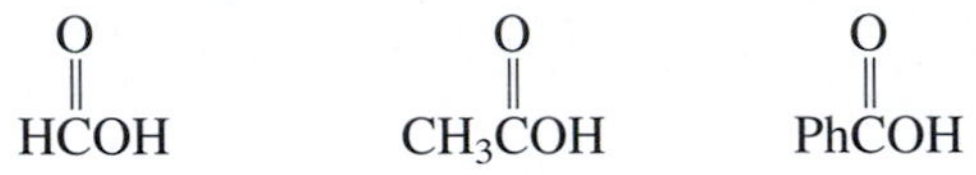

$$\underset{\substack{\text{Formic acid}\\\text{(methanoic acid)}}}{\overset{\overset{\displaystyle O}{\|}}{HCOH}} \qquad \underset{\substack{\text{Acetic acid}\\\text{(ethanoic acid)}}}{\overset{\overset{\displaystyle O}{\|}}{CH_3COH}} \qquad \underset{\substack{\text{Benzoic acid}}}{\overset{\overset{\displaystyle O}{\|}}{PhCOH}}$$

PROBLEM 12.11

Provide names for these compounds:

a)

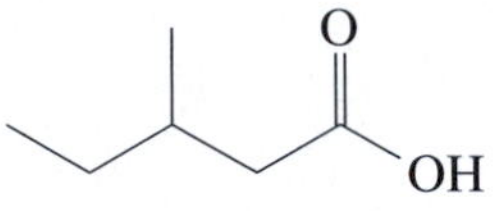

b)

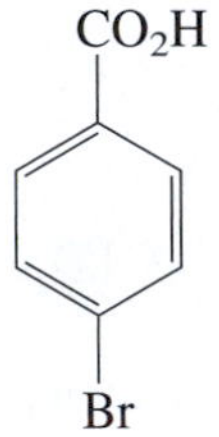

c)

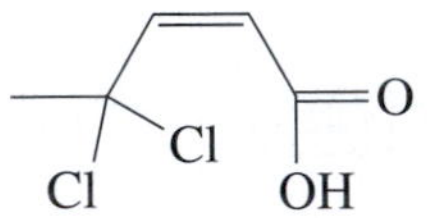

d)

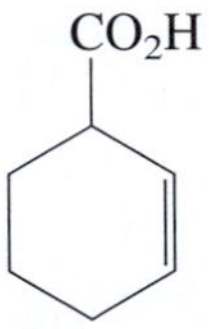

e)

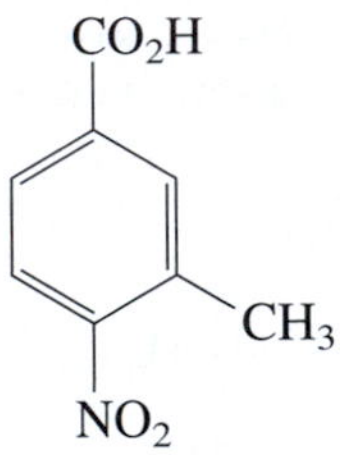

f)

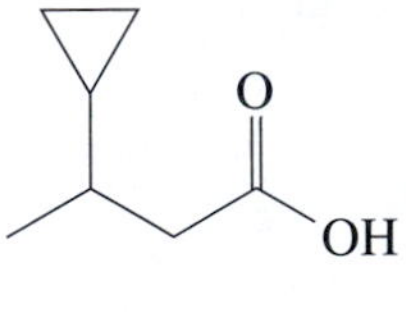

PROBLEM 12.12

Draw structures for these compounds:

a) 6-Bromo-3,5-dichlorohexanoic acid
b) Cyclobutanecarboxylic acid
c) *m*-Chlorobenzoic acid
d) (*E*)-3-Phenyl-2-propenoic acid

The carboxy group is more polar than a carbonyl group or a hydroxy group, and it can form hydrogen bonds to other carboxy groups. Therefore, carboxylic acids melt and boil at somewhat higher temperatures than alcohols of similar molecular weight. As was the case with alcohols, the smaller carboxylic acids are miscible with water. As expected, their solubility decreases as the size of the hydrocarbon group increases.

PROBLEM 12.13

Hydrogen bonding is quite strong in the case of acetic acid and persists even in the gas phase where two molecules form a dimer held together by two hydrogen bonds. Suggest a structure for the hydrogen-bonded dimer of acetic acid.

Carboxylic acids are weak acids ($pK_a \approx 5$) and react by donating the proton attached to the oxygen of the carboxy group as illustrated in the following equation:

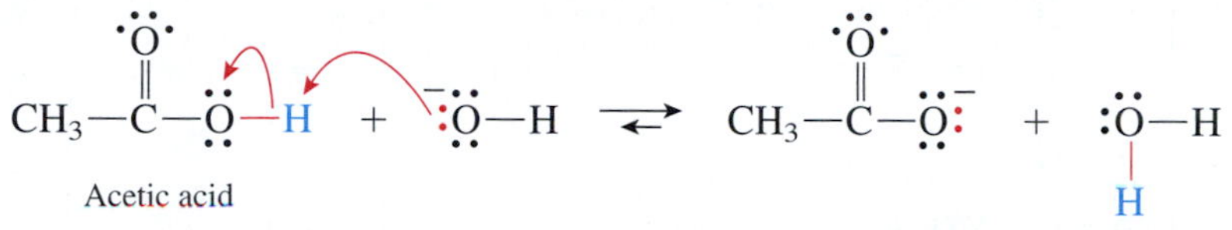

Acetic acid

Although carboxylic acids do occur naturally, they most often are found as their ester or amide derivatives. Acetic acid is a major industrial chemical, produced in excess of 3 billion pounds annually. Some acids, such as acetylsalicylic acid (aspirin) and ibuprofen, have found considerable use in the medical field.

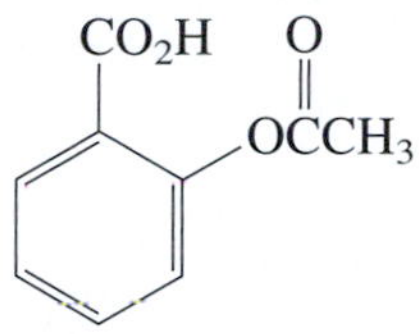

Acetylsalicylic acid
(aspirin)

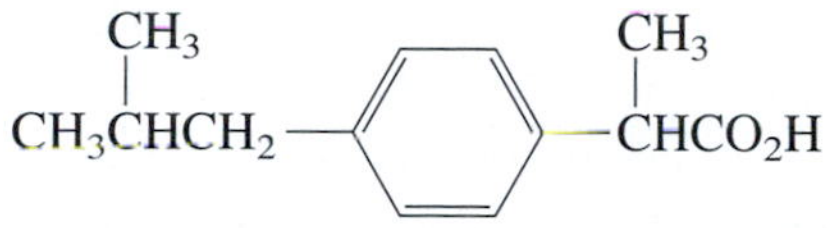

Ibuprofen

12.5 DERIVATIVES OF CARBOXYLIC ACIDS

Removal of the hydroxy group of a carboxylic acid leaves a carbonyl group with an attached alkyl group, which is called an **acyl group.** If the acyl group is bonded to a hydrogen, an aldehyde results. If it is bonded to a carbon, a ketone results. And if it is bonded to a hydroxy group, a carboxylic acid is produced. However, if the acyl group is bonded to another **heteroatom** (not C or H) group, such as Cl or NH_2, a series of compounds called carboxylic acid derivatives is produced. The most important of these, along with their names and the suffixes used in their nomenclature, are listed in Table 12.1.

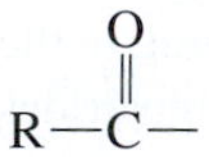

Acyl group

Table 12.1   Carboxylic Acid Derivatives

Structure	Name of Functional Group	Suffix
$R-\overset{\displaystyle O}{\overset{\|}{C}}-Cl$	Acid chloride or acyl chloride	-yl chloride
$R-\overset{\displaystyle O}{\overset{\|}{C}}-O-\overset{\displaystyle O}{\overset{\|}{C}}-R'$	Acid anhydride	-ic anhydride
$R-\overset{\displaystyle O}{\overset{\|}{C}}-O-R'$	Ester	-ate
$R-\overset{\displaystyle O}{\overset{\|}{C}}-NH_2$	Amide	-amide
$R-C\equiv N$	Nitrile	-nitrile
$R-\overset{\displaystyle O}{\overset{\|}{C}}-O^-$	Carboxylate salt	-ate

Carboxylic acid derivatives are all named by using the same root as the carboxylic acid from which they are derived. Note that the root always begins with the carbon of the carbonyl group and this carbon is always given the number 1.

Acid chlorides (or **acyl chlorides**) result from substituting a chlorine for the hydroxy group of a carboxylic acid. Although other acid halides also exist, they are seldom encountered. Acid chlorides are named by replacing the -ic acid of the carboxylic acid name (either common or systematic) with -yl chloride.

$$CH_3\overset{\displaystyle O}{\overset{\|}{C}}-Cl$$

$$\underset{4\quad 3\qquad 2\quad 1}{CH_3CH=CHC}\overset{\displaystyle O}{\overset{\|}{}}-Cl$$

Acetyl chloride
(ethanoyl chloride)

2-Butenoyl chloride

(The common and most used name for the related carboxylic acid is acetic acid. The -ic acid is replaced by -yl chloride.)

(The corresponding carboxylic acid is 2-butenoic acid.)

Acid anhydrides result from substituting the acyl group of one acid for the hydroxy hydrogen of another. They are called anhydrides because they can be viewed as resulting from the loss of water from two carboxylic acid molecules (removing H from one and OH from the other). Symmetrical anhydrides derived from two molecules of the same carboxylic acid are most often encountered. These are simply named by replacing *acid* in the name of the carboxylic acid with *anhydride*.

$$CH_3\overset{\displaystyle O}{\overset{\|}{C}}\!-\!O\!-\!\overset{\displaystyle O}{\overset{\|}{C}}CH_3$$

Acetic anhydride
(ethanoic anhydride)

Benzoic anhydride

Esters can be viewed as resulting from the combination of a carboxylic acid with an alcohol, using the alkoxy (OR) group of the alcohol to replace the hydroxy (OH) group of the acid. The name must therefore designate both the alcohol part and the acid part of the ester. The name uses two separate words. First the R group of the alcohol is named just like other *groups* we have encountered, using a -yl suffix. This is the first word in the name. Then the acid part is named as usual, and the -ic acid suffix is replaced with -ate. This is the second word in the name.

Propyl benzoate
(The red part of this ester is derived from the alcohol [propanol], and the blue part from the carboxylic acid [benzoic acid]. Be careful to correctly identify which part of each ester comes from the acid and which from the alcohol. Note that *propyl* and *benzoate* are separate words.)

(2-Methylbutyl) 3-methylcyclohexanecarboxylate
(The complex group from the alcohol portion of the ester is named in the same manner as the complex groups of alkanes.)

Amides can be viewed as resulting from the replacement of the OH of a carboxylic acid with an NH_2 (primary amide), NHR (secondary amide), or NR_2 (tertiary amide) group. To name an amide, the longest carbon chain having the carbonyl group at one terminus is chosen as the root, as usual. Systematic names are formed by replacing the final -e of the *hydrocarbon name* for this root with -amide. Amide names can also be derived from common names of acids by replacing -ic acid or -oic acid with -amide. Other groups attached to the nitrogen are designated with the prefix *N-*, as was done in the case of amines.

$$O$$
$$CH_3CH_2CH_2CH_2C{-}NH_2$$

Pentanamide
a primary amide

(This name is derived from the five
carbon hydrocarbon, pentane.)

$$CH_3$$ $$O$$
$$CH_2{=}CHCHCH{=}CHCNHCH_2CH_3$$
6   5   4   3   2   1

N-Ethyl-4-methyl-2,5-hexadienamide
or *N*-ethyl-4-methylhexa-2,5-dienamide
a secondary amide

(Note the *N*- that is used to show that
the ethyl group is bonded to the nitrogen.)

$$O\quad CH_3$$
$$H{-}C{-}N{-}CH_3$$

N,N-Dimethylformamide
(DMF), a tertiary amide

(The common name for the
one-carbon carboxylic acid is
formic acid.)

On first consideration, **nitriles** do not appear to be related to the other carboxylic acid derivatives because they have no acyl group. However, they can be viewed as resulting from the removal of H_2O from a primary amide (loss of both H's from the N and the O from the carbonyl group), and their chemical reactions are related to other carboxylic acid derivatives. Therefore, it is convenient to include them with the other carboxylic acid derivatives. They are named in a similar manner to amides; that is, -nitrile is appended to the hydrocarbon name. (Do not forget to count the carbon of the —CN group and to give this carbon the number 1.) Common names are obtained from the common name of the carboxylic acid by replacing the -ic acid or -oic acid with -onitrile. In complex compounds the —CN group can be named as a cyano group.

$$CH_3C{\equiv}CCH_2C{\equiv}N$$
5   4   3 2   1

3-Pentynenitrile
(This is a systematic name. Because
the C of the CN must be given the
number 1, the related hydrocarbon is
3-pentyne.)

Benzonitrile
(The related carboxylic acid is benzoic
acid.)

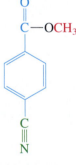

Methyl 4-cyanobenzoate
(In a complex compound like this, with
more than one functional group, the
—CN is named as a substituent group, a
cyano group, attached to the root in the
same manner as a halogen.)

Carboxylate salts consist of a carboxylate anion (the anion formed by removal of the proton from the OH of a carboxylic acid) and a cation. They are named in a manner similar to esters, using two words. The first word designates the cation. The second word designates the carboxylate anion, using the -ate suffix, just as is done for esters.

$$O$$
$$CH_3C{-}O^-\ Na^+$$

Sodium acetate

$$O$$
$$C{-}O^-\ NH_4^+$$

$$CH_3$$

Ammonium 3-methylbenzoate

PROBLEM 12.14

Provide names for these compounds:

a)

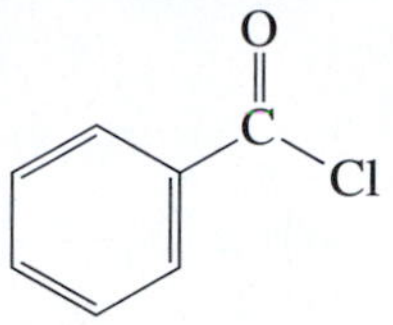

b)  $CH_3CH_2COCCH_2CH_3$ (with two O above)

c) 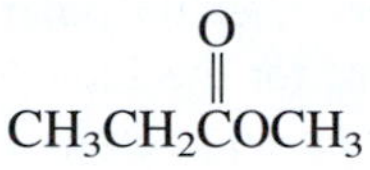 $CH_3CH_2COCH_3$ (with O above)

d)

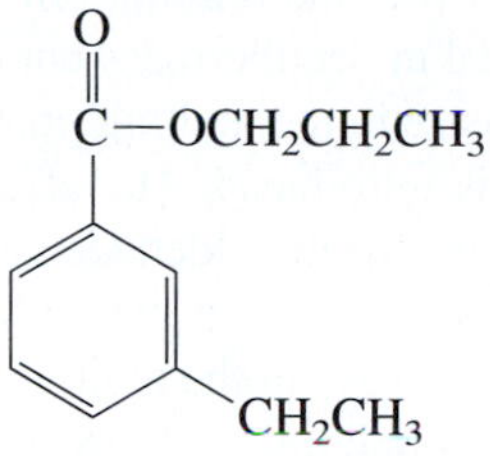

e)

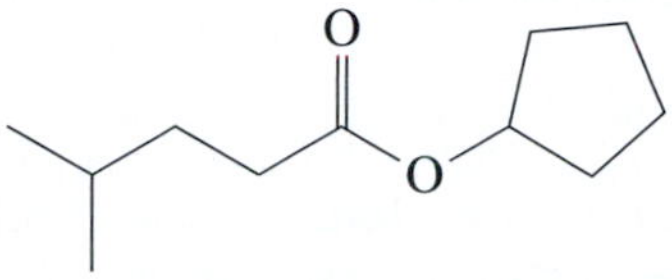

f)

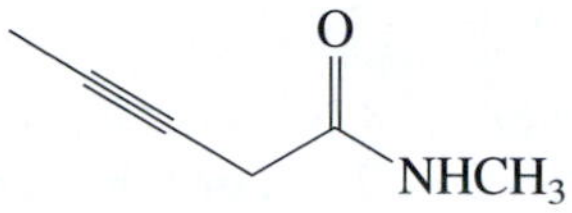

g)

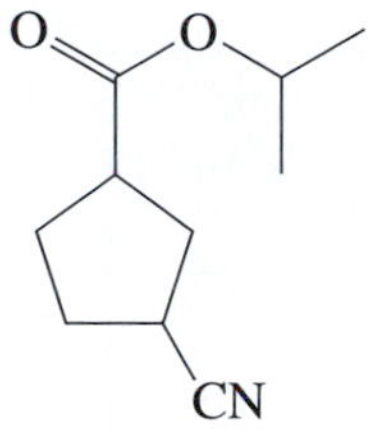

h)

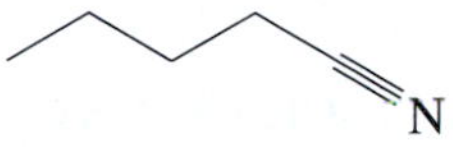

i)

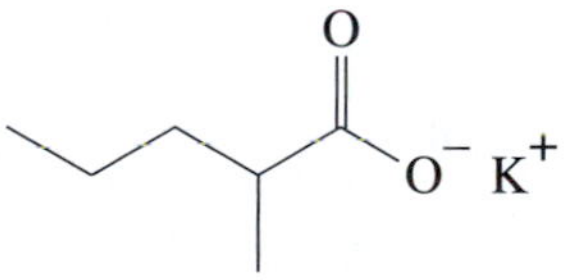

PROBLEM 12.15

Draw structures for these compounds:

a) Propanoyl chloride
b) *N,N*-Dimethylacetamide
c) Pentanoic anhydride
d) Sodium *p*-nitrobenzoate
e) Hexanamide
f) Isopropyl acetate
g) Benzyl benzoate
h) Ethyl cyclopentanecarboxylate
i) 3-Chlorobenzonitrile
j) 3-Methylheptanenitrile

ORGANIC
Chemistry⋅⚛⋅Now™
Click *Coached Tutorial Problems*
for more practice **Drawing**
Structures of Carboxylic
Acids and Derivatives
from **IUPAC Names.**

Acid chlorides, acid anhydrides, and esters all contain the carbonyl group but the presence of this polar group has only a small effect on their melting and boiling points. Amides, however, are considerably more polar because of the significant contribution of a charged resonance structure to the resonance hybrid. In addition, primary and secondary amides, with hydrogens bonded to the nitrogen, can also form hydrogen bonds among themselves. For these reasons, amides melt and boil at even higher temperatures than do carboxylic acids of similar molecular mass.

$$R-\underset{\displaystyle \overset{\displaystyle \overset{..}{\overset{..}{O}}}{\|}}{C}-\ddot{N}H_2 \quad \longleftrightarrow \quad R-\underset{\displaystyle \overset{..}{\underset{..}{O}}\text{:}^-}{\overset{|}{C}}=\overset{+}{N}H_2$$

An amide

Table 12.2 lists the melting points and boiling points of a series of compounds of nearly identical molecular mass but containing a variety of different functional groups. Note the increase in the melting point and boiling point that occurs when the polar carbonyl group is introduced. The effect is small (or may be absent) for the ester but is more significant for the aldehyde and ketone, especially on their boiling points. There is an additional increase in the boiling point for the alcohol due to hydrogen bonding and a further increase in the melting and boiling points of the carboxylic acid due to its polar, hydrogen-bonding carboxy group. Finally, the amide has the highest melting and boiling points because of its highly polar nature in combination with its ability to form hydrogen bonds.

Table 12.2   Comparison of the Effect of Functional Groups on Melting and Boiling Points of Compounds of Comparable Molecular Mass

Structure	Functional Group	mp (°C)	bp (°C)
$CH_3CH_2CH_2CH_2CH_3$	Alkane	−130	36
$CH_3CH_2OCH_2CH_3$	Ether	−116	35
$CH_3\overset{\displaystyle O}{\overset{\|}{C}}OCH_3$	Ester	−98	57
$CH_3CH_2CH_2\overset{\displaystyle O}{\overset{\|}{C}}H$	Aldehyde	−99	76
$CH_3CH_2\overset{\displaystyle O}{\overset{\|}{C}}CH_3$	Ketone	−86	80
$CH_3CH_2CH_2CH_2OH$	Alcohol	−90	117
$CH_3CH_2\overset{\displaystyle O}{\overset{\|}{C}}OH$	Carboxylic acid	−21	141
$CH_3CH_2\overset{\displaystyle O}{\overset{\|}{C}}NH_2$	Amide	81	213

PROBLEM 12.16

Explain which compound has the higher melting point or boiling point:

a) Higher mp $\quad CH_3CH_2CH_2\overset{O}{\underset{\|}{C}}NH_2 \quad$ or $\quad CH_3\overset{O}{\underset{\|}{C}}N(CH_3)_2$

b) Higher bp $\quad CH_3CH_2CH_2\overset{O}{\underset{\|}{C}}OH \quad$ or $\quad CH_3CH_2\overset{O}{\underset{\|}{C}}OCH_3$

c) Higher bp

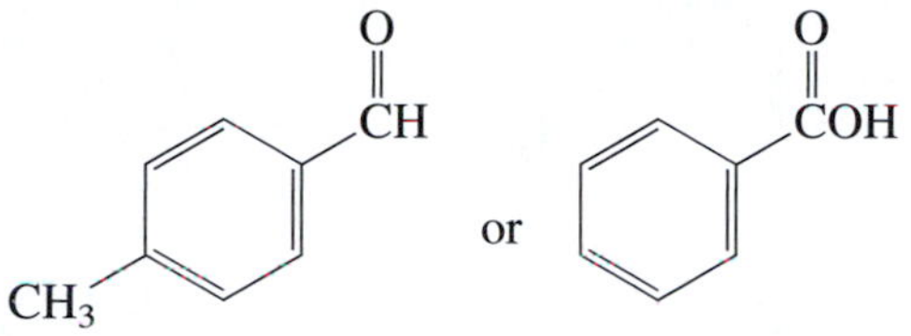

or

d) Higher mp $\quad CH_3CH_2CH_2\overset{O}{\underset{\|}{C}}OH \quad$ or $\quad CH_3CH_2CH_2\overset{O}{\underset{\|}{C}}NH_2$

In general, acid chlorides and acid anhydrides are too reactive to occur naturally, and nitriles are rare in nature. Esters and amides, on the other hand, are very common. Many esters have pleasant odors, often sweet or fruity, and are responsible for the fragrant odors of fruits and flowers. They are components of many flavorings, both natural and artificial. For example, isopentyl acetate has a strong banana odor, and methyl butanoate is used as an artificial rum flavoring. Typical fats are triesters formed from long-chain "fatty" acids and the triol glycerol.

$$CH_3\overset{O}{\underset{\|}{C}}-OCH_2CH_2\overset{CH_3}{\underset{|}{C}HCH_3} \qquad CH_3CH_2CH_2\overset{O}{\underset{\|}{C}}-OCH_3$$

Isopentyl acetate Methyl butanoate

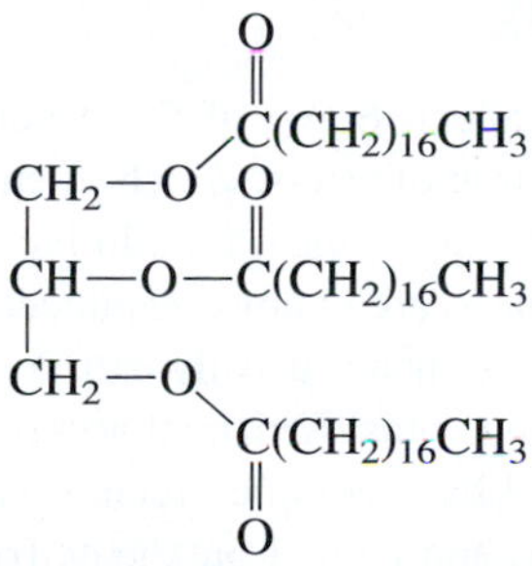

A fat 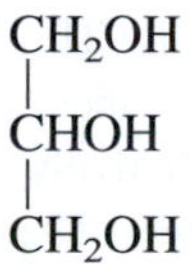 Glycerol

Amides often have pronounced physiological activity; acetaminophen, a common pain reliever, and the diethyl amide of lysergic acid (LSD), a hallucinogen, are two examples. The important peptide bond of proteins is actually an amide bond. Dimethylformamide (DMF) and dimethylacetamide (DMA) are important solvents in the organic laboratory. They are highly polar and are capable of dissolving many ionic reagents in addition to less polar organic compounds. They are fairly unreactive because the bond between the carbonyl group and the nitrogen of an amide is reasonably strong, and they do not have hydrogens on the nitrogens that might react as acids. They are especially good solvents for S_N2 reactions because they are aprotic.

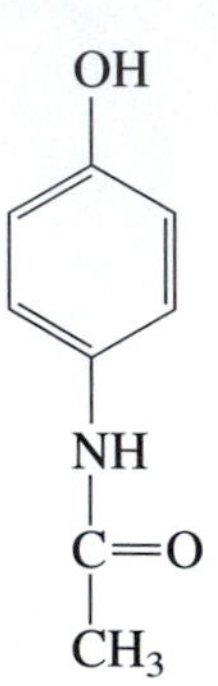

Acetaminophen

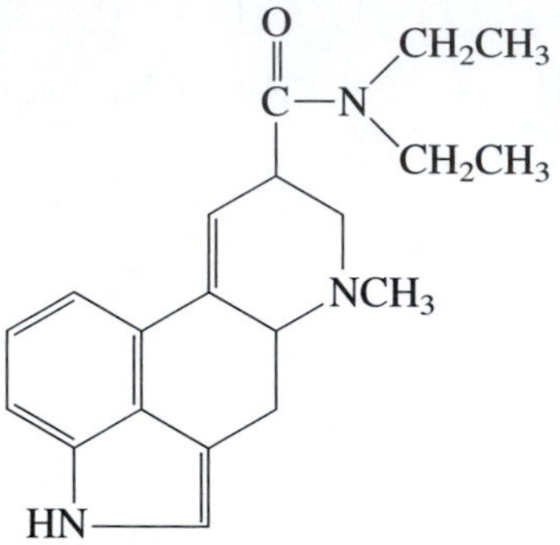

Lysergic acid diethylamide
(LSD)

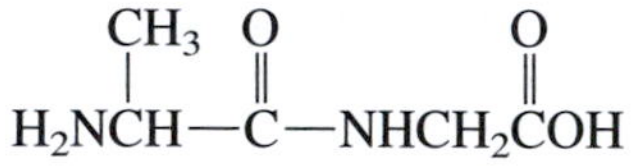

A dipeptide

Dimethylformamide
(DMF)

Dimethylacetamide
(DMA)

Focus On

Fragrant Organic Compounds

Many organic compounds have very powerful odors. Some of these odors are disagreeable; others are pleasant. This is especially true of esters, which often have fruity odors, and aldehydes and ketones, many of which have floral odors. In fact, aldehydes, ketones, and esters, with relatively simple structures, are major components of numerous natural scents and flavors. These natural materials are usually extremely complex mixtures, sometimes containing hundreds of compounds. Artificial scents and flavors usually contain fewer components, but many still have complex recipes consisting of dozens of ingredients. Isopentyl acetate (banana) and methyl butanoate (rum) are examples of fragrant esters that were mentioned earlier. Other examples include isopentenyl acetate, the flavoring used in Juicy Fruit gum; ethyl phenylacetate, which has a honey odor; methyl salicylate, a major component of oil of wintergreen; and benzyl acetate, which composes more than 60% of jasmine oil. Coumarin has an odor that is often described as "new-mown hay" or "woody" and is used in men's toiletries. Along with vanillin (see page 477), it is a component of natural vanilla. A combination of these two compounds, prepared synthetically in the laboratory, is used in artificial vanilla, and the flavoring in cream soda consists of coumarin and vanillin.

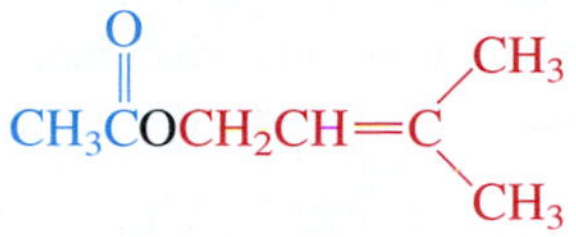

Isopentenyl acetate
(Juicy Fruit gum)

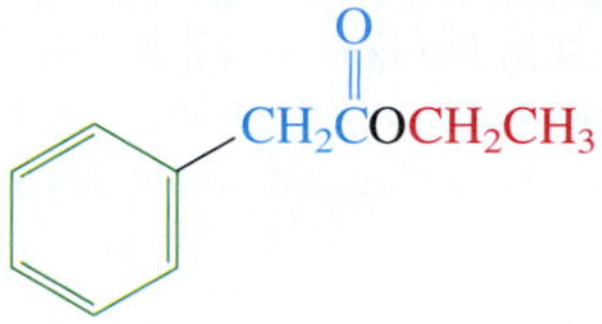

Ethyl phenylacetate
(honey)

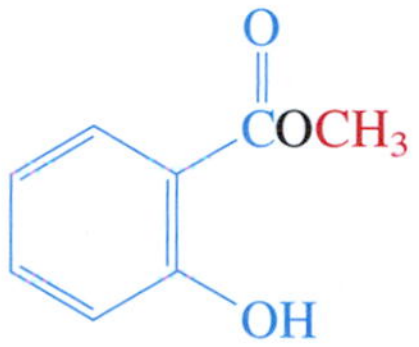

Methyl salicylate
(wintergreen)

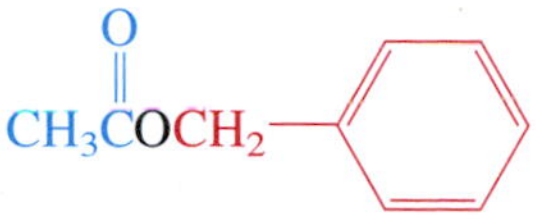

Benzyl acetate
(jasmine)

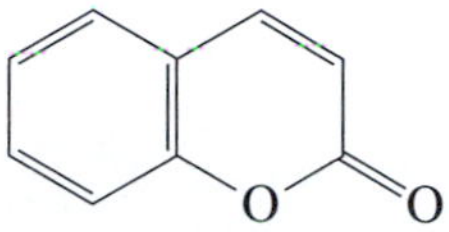

Coumarin
(new-mown hay)

Although many esters are used in artificial flavorings, they are less often employed as ingredients of perfumes because they are slowly hydrolyzed to a carboxylic acid and an alcohol on the skin, as shown in the following equation:

$$R{-}\overset{O}{\overset{\|}{C}}{-}O{-}R' + H_2O \longrightarrow R{-}\overset{O}{\overset{\|}{C}}{-}O{-}H + H{-}O{-}R'$$

This is quite undesirable in a perfume because many carboxylic acids have objectionable odors. For example, the sharp, penetrating odor of vinegar is due to acetic acid, butanoic acid (butyric acid) smells like rancid butter, and 2-methylpropanoic acid (isobutyric acid) is a component of sweat. A common name for hexanoic acid is caproic acid, derived from the Latin word *caper,* which means goat. If you have ever been around goats, perhaps you can imagine the odor of hexanoic acid.

$$CH_3CH_2CH_2\overset{O}{\overset{\|}{C}}OH$$

Butanoic acid
(butyric acid)
(rancid butter)

$$CH_3\overset{CH_3}{\overset{|}{C}}H{-}\overset{O}{\overset{\|}{C}}OH$$

2-Methylpropanoic acid
(isobutyric acid)
(sweat)

$$CH_3CH_2CH_2CH_2CH_2\overset{O}{\overset{\|}{C}}OH$$

Hexanoic acid
(caproic acid)
(goat)

Aldehydes and ketones are also important components of many fragrances and flavors. For example, butanal (butyraldehyde) is used to impart a buttery flavor to margarine and other foods. Because aldehydes are slowly oxidized to carboxylic acids by the oxygen of air, it is readily apparent how the odor of rancid butter arises. Although α-pentylcinnamaldehyde does not occur naturally, it has been found to have a powerful jasmine odor and is used in many perfumes and soaps. α-Ionone is a naturally oc-

Continued

curring ketone with an odor resembling violets. Many large-ring ketones have a musky odor and are prized ingredients in perfumes. Muscone (see page 477), which was first isolated from the musk deer, has a 15-membered ring, and civetone, from the civet cat, has a 17-membered ring. These compounds were extremely expensive when they could only be obtained from natural sources. Once their structures were determined, they were prepared in the laboratory. Although the preparation of such large rings is difficult, these synthetic materials are still considerably less expensive than their natural counterparts. Other fragrant aldehydes and ketones are mentioned on pages 476–477.

Butanal
butyraldehyde
(buttery)

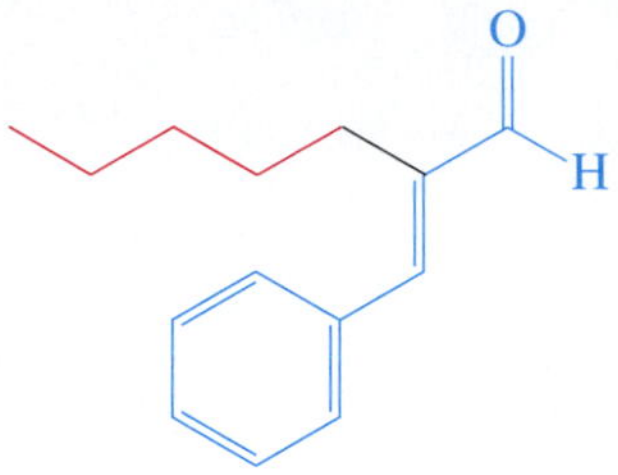

α-Pentylcinnamaldehyde
(jasmine)

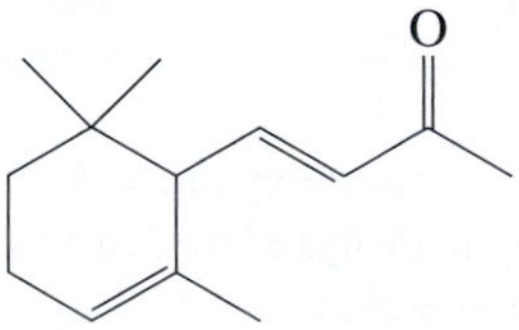

α-Ionone
(violets)

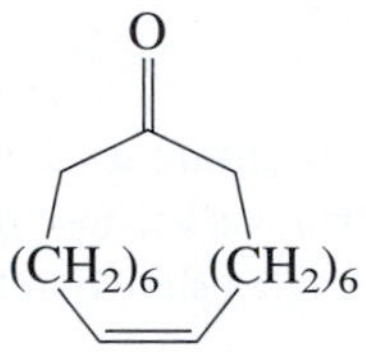

Civetone
(musky)

12.6   SULFUR AND PHOSPHORUS COMPOUNDS

Sulfur occurs directly beneath oxygen in the periodic table, and, like oxygen, it often exhibits a valence of two. Therefore, sulfur analogs of alcohols and ethers are often encountered. However, because sulfur is in the third period of the periodic table, it can also have a higher valence. Structures with four or six bonds to a sulfur are common. In organic chemistry the most important of these "expanded valence" compounds have the sulfur bonded to one or two extra oxygens.

Similarly, phosphorus occurs directly beneath nitrogen in the periodic table and therefore often exhibits a valence of three. Again, structures with an expanded valence, having five bonds to the phosphorus, are common, especially when the extra bonds are to oxygen. This book is not concerned with all the possible sulfur and phosphorus compounds, nor does it spend much time on their nomenclature. Instead, it concentrates on those of most importance in organic chemistry and biochemistry. Let's begin with a discussion of some common sulfur compounds.

Sulfur analogs of alcohols are called **thiols** or **mercaptans.** They are named in the same general manner as alcohols but with the suffix -thiol added to the name of the hydrocarbon. (In chemistry, thi- or thio-, from the Greek word for sulfur, *theion,* is used to indicate a compound that contains sulfur.) Thus, the sulfur analog of ethanol, CH_3CH_2SH, is named ethanethiol. Note that the "e" at the end of the hydrocarbon name is retained to aid in pronunciation. Sulfur analogs of ethers are called **sulfides.** They are named in the same general manner as ethers but with sulfide replacing ether in the name. Examples of thiols and sulfides include the following:

$$\underset{\text{3-Methyl-1-butanethiol}}{\overset{\displaystyle CH_3}{\underset{|}{CH_3CHCH_2CH_2}}-SH} \qquad \underset{\text{2-Butene-1-thiol}}{CH_3CH=CHCH_2-SH} \qquad \underset{\text{Dimethyl sulfide}}{CH_3-S-CH_3}$$

A characteristic of organic sulfur compounds, especially volatile (low molecular mass) thiols, is their disagreeable odors. For example, 3-methyl-1-butanethiol and 2-butene-1-thiol are ingredients of a skunk's "perfume," and methanethiol or ethanethiol is usually added, in small amounts, to natural gas, which is odorless by itself, so that leaks can be readily detected. The chemical properties of thiols and sulfides differ from those of alcohols and ethers in that thiols are somewhat stronger acids than alcohols and the sulfur atoms of these compounds are considerably more nucleophilic than the oxygen of their analogs. They are excellent nucleophiles in substitution reactions.

Other sulfur compounds of importance in organic chemistry have additional oxygens bonded to the sulfur, as in the **sulfoxide** and **sulfone** functional groups. (You may encounter sulfoxides written as either of their resonance structures.) The chemistry of these compounds is not covered in detail in this text. It is worth noting, however, that dimethylsulfoxide (DMSO) is an important solvent that is used in many organic reactions. It is quite polar, so it will dissolve both organic compounds and inorganic reagents, and it is fairly unreactive, so it will not interfere with many reactions. It is also aprotic.

$$\underset{\text{A sulfoxide}}{\overset{\displaystyle :\overset{..}{O}}{\underset{\displaystyle \underset{..}{S}}{R-\underset{|}{S}-R}}} \longleftrightarrow \underset{+}{\overset{\displaystyle :\overset{..}{O}:^{-}}{R-\overset{|}{\underset{..}{S}}-R}} \qquad \underset{\text{A sulfone}}{\overset{\displaystyle O}{\underset{\displaystyle O}{R-\overset{||}{\underset{||}{S}}-R}}} \qquad \underset{\substack{\text{Dimethylsulfoxide} \\ \text{(DMSO)}}}{\overset{\displaystyle O}{CH_3-\overset{||}{S}-CH_3}}$$

Another important group of sulfur-containing compounds can be viewed as being derived from sulfuric acid. Replacing one of the OH groups of sulfuric acid with a carbon group results in a new functional group called a **sulfonic acid.** Sulfonic acids are strong acids, comparable to sulfuric acid in strength. *p*-Toluenesulfonic acid is often used when a strong acid that is soluble in organic solvents is needed. The chemistry of sulfonic acids has some similarities to that of carboxylic acids. Thus, derivatives of sulfonic acid, like derivatives of carboxylic acids, include sulfonyl chlorides, sulfonate esters, and sulfonamides. As shown in the following examples, nomenclature of these compounds attaches -sulfon- to the hydrocarbon name and adds the same final suffixes as are used for the carboxylic acid derivatives.

$$HO-\overset{\overset{\displaystyle O}{\|}}{\underset{\underset{\displaystyle O}{\|}}{S}}-OH$$

Sulfuric acid

$$R-\overset{\overset{\displaystyle O}{\|}}{\underset{\underset{\displaystyle O}{\|}}{S}}-OH$$

A sulfonic acid

$$CH_3-\bigcirc-\overset{\overset{\displaystyle O}{\|}}{\underset{\underset{\displaystyle O}{\|}}{S}}-OH$$

p-Toluenesulfonic acid
(The suffix -sulfonic acid is added
to the hydrocarbon name.)

$$CH_3\overset{\overset{\displaystyle O}{\|}}{\underset{\underset{\displaystyle O}{\|}}{S}}-Cl$$

Methanesulfonyl chloride
(a sulfonyl chloride)

$$CH_3\overset{\overset{\displaystyle O}{\|}}{\underset{\underset{\displaystyle O}{\|}}{S}}-O-\bigcirc$$

Cyclopentyl methanesulfonate
(a sulfonate ester)

$$NH_2-\bigcirc-\overset{\overset{\displaystyle O}{\|}}{\underset{\underset{\displaystyle O}{\|}}{S}}-NH_2$$

p-Aminobenzenesulfonamide
(a sulfonamide)
(This is better known as sulfanilamide,
a sulfa drug.)

The phosphorus analogs of amines are called **phosphines.** The parent, PH_3, is called phosphine and is a toxic gas with an unpleasant odor. Triphenylphosphine is a good nucleophile that is employed in certain organic reactions. Esters derived from phosphoric acid play an important role in living organisms. One, two, or all three of the OH groups of phosphoric acid can be replaced with OR groups to form various phosphate esters. The anions produced by the ionization of dialkyl hydrogen phosphates in water by an acid–base reaction are especially important in biological systems. This group provides the backbone for DNA and RNA. As another example, phosphatidylcholine is a phospholipid that is an important constituent of cell membranes.

Triphenylphosphine

$$HO-\overset{\overset{\displaystyle O}{\|}}{\underset{\underset{\displaystyle OH}{\|}}{P}}-OH$$

Phosphoric acid

Phosphatidylcholine

$$RO-\overset{\overset{\displaystyle O}{\|}}{\underset{\underset{\displaystyle OH}{\|}}{P}}-OR \; + \; H_2O \longrightarrow RO-\overset{\overset{\displaystyle O}{\|}}{\underset{\underset{\displaystyle O_-}{\|}}{P}}-OR \; + \; H_3O^+$$

A dialkyl hydrogen
phosphate

PROBLEM 12.17

Provide names for these compounds:

a)

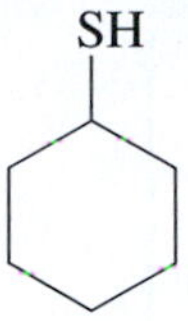

SH

b)

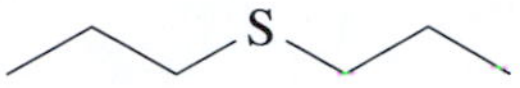

S

c)

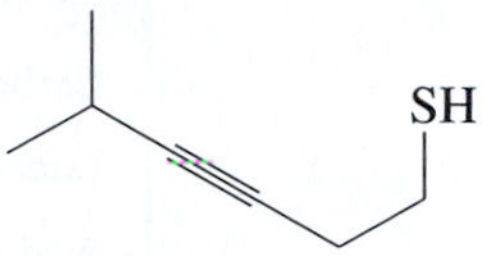

SH

d)

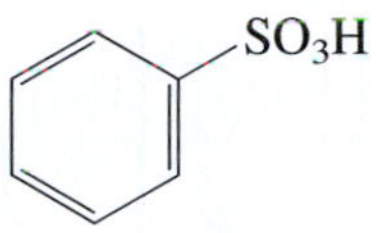

SO₃H

e) 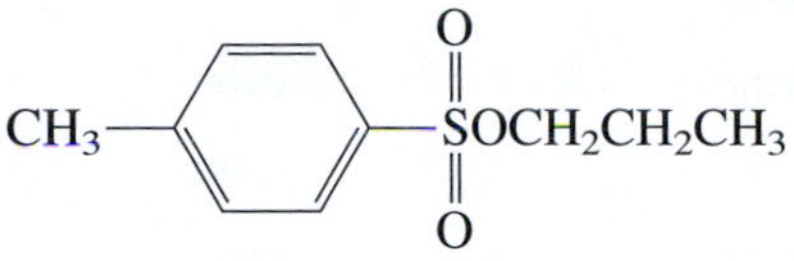   CH₃—

$$O$$
$$\parallel$$
—SOCH₂CH₂CH₃
$$\parallel$$
$$O$$

PROBLEM 12.18

Draw structures for these compounds:

a) 2-Butanethiol
b) Benzenethiol
c) Isopropyl methanesulfonate
d) *p*-Bromobenzenesulfonic acid
e) Phenyl trichloromethyl sulfide

12.7 NOMENCLATURE OF COMPOUNDS WITH SEVERAL FUNCTIONAL GROUPS

Most of the compounds that you have encountered in this text so far have been fairly simple. In addition to any carbon–carbon double or triple bonds, they have contained only one functional group and could be named by using suffixes such as -ynol or -dienone. However, for a compound that contains more than one heteroatom functional group, only one of these functional groups can be designated in the suffix. For example, consider the following compound, which has both alcohol and ketone functional groups:

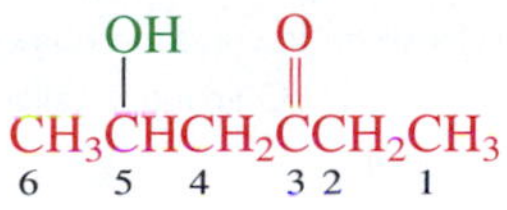

$$\underset{6\quad\;5\quad\;4\quad\;\;3\;2\quad\;1}{CH_3\overset{OH}{\underset{|}{C}}HCH_2\overset{O}{\underset{\parallel}{C}}CH_2CH_3}$$

5-Hydroxy-3-hexanone
or
5-hydroxyhexan-3-one

Because both functional groups cannot be denoted in the suffix, one must be chosen as higher priority and used to control both the numbering and the suffix. A prefix is then used to indicate the presence of the lower-priority functional group on the main chain. In the case of the preceding compound, the ketone group has a higher priority than the alcohol group, so an -one suffix is used. Numbering begins with the right carbon so that the carbon of the carbonyl group gets the lower number. The group name for OH is hydroxy. Therefore, the compound is named as 5-hydroxy-3-hexanone. Table 12.3 lists the

Table 12.3 **Order of Priority for Selected Functional Groups**

Functional Group	Group Prefix	
Carboxylic acid		Highest priority
Ester		
Acid chloride		
Amide		
Aldehyde	oxo-	
Nitrile	cyano-	
Ketone	oxo-	
Alcohol	hydroxy-	
Amine	amino-	Lowest priority
Ether	alkoxy-	
Halogen	fluoro-, chloro-	
	bromo-, iodo-	
$-NO_2$	nitro-	

Note that the ether, halogen, and $-NO_2$ groups are always denoted by prefixes in systematic nomenclature.

priorities and group names for selected other functional groups. Some examples of naming other complex molecules follow:

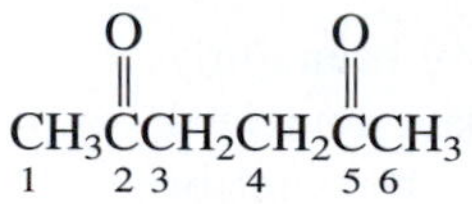

$$\underset{1 \quad 2\,3 \quad 4 \quad 5\,6}{CH_3\overset{O}{\overset{\|}{C}}CH_2CH_2\overset{O}{\overset{\|}{C}}CH_3}$$

2,5-Hexanedione
or
hexan-2,5-dione
(The presence of several identical functional groups can be denoted with di-, tri-, etc.)

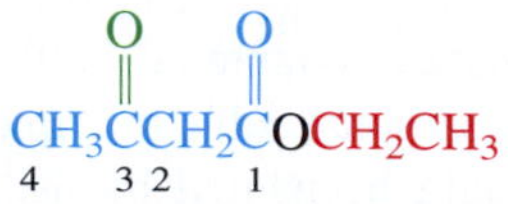

$$\underset{4 \quad 3\,2 \quad 1}{CH_3\overset{O}{\overset{\|}{C}}CH_2\overset{O}{\overset{\|}{C}}OCH_2CH_3}$$

Ethyl 3-oxobutanoate
(ethyl acetoacetate)
(The ester group has higher priority, so it determines the numbering and the suffix. The ketone group is designated by the prefix oxo-. In common nomenclature the $CH_3C{=}O$ group is called acetyl or aceto.)

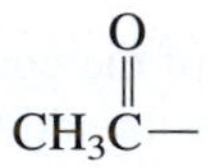

$$CH_3\overset{O}{\overset{\|}{C}}-$$

The acetyl or aceto group

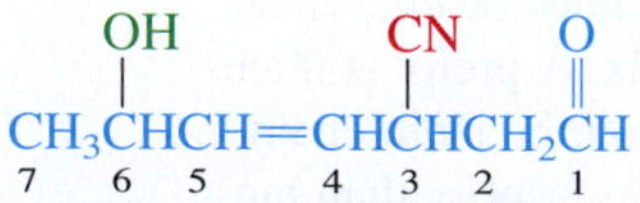

$$\underset{7 \quad 6\,5 \quad 4 \quad 3 \quad 2 \quad 1}{CH_3\overset{OH}{\underset{|}{C}}H\,CH{=}CH\,\overset{CN}{\underset{|}{C}}H\,CH_2\overset{O}{\overset{\|}{C}}H}$$

3-Cyano-6-hydroxy-4-heptenal
(The aldehyde is the highest-priority functional group, so the OH and the CN must be denoted by their group prefixes.)

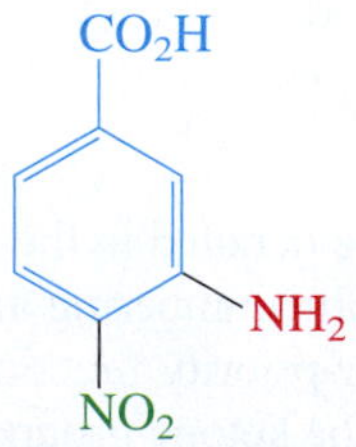

3-Amino-4-nitrobenzoic acid
(The acid group has the highest priority.)

PROBLEM 12.19

Provide systematic names for these compounds:

a)

b)

Leucine
(an amino acid)

c) $HOCCH_2CH_2COH$

d)

e)

f)

g) $CH_3CCH_2C{\equiv}N$

h)

Ph

PROBLEM 12.20

Draw structures for these compounds:
a) 1,6-Hexanedioic acid
b) Ethyl 2-ethyl-2-hydroxybutanoate
c) 2-Amino-3-cyclohexyl-1-propanol
d) *tert*-Butyl 2-hydroxy-5-octenoate
e) *N,N*,3-Trimethyl-2-oxobutanamide
f) 2-Amino-3-phenylpropanoic acid (an amino acid commonly called phenylalanine)

As you can see, organic nomenclature is a complex subject. Many of the rules and examples are beyond the scope of this book. You should not expect to be able to name every compound that you might encounter. However, based on the rules and examples given here and in Chapter 5, your knowledge of nomenclature should be sufficient to allow you to read and discuss organic chemistry.

Finally, you should realize that organic nomenclature is not an exact science. In complicated molecules there will often be ambiguities, and more than one correct systematic name may result. In addition, many complex molecules are best identified by a short common name rather than by a purely systematic name that is unmanageably long or complicated. As an example, the systematic name for codeine would be so complex that it would convey little immediate information about the compound to most chemists. In practice, therefore, common names continued to be coined for many molecules. For example, the systematic name pentacyclo[4.2.0.0^{2,5}.0^{3,8}.0^{4,7}]octane must be carefully analyzed by the average organic chemist, probably using pencil and paper, before its structure can be recognized. However, the common name given to this compound when it was first synthesized is very descriptive and immediately summons a mental picture for the compound. It is best known as cubane!

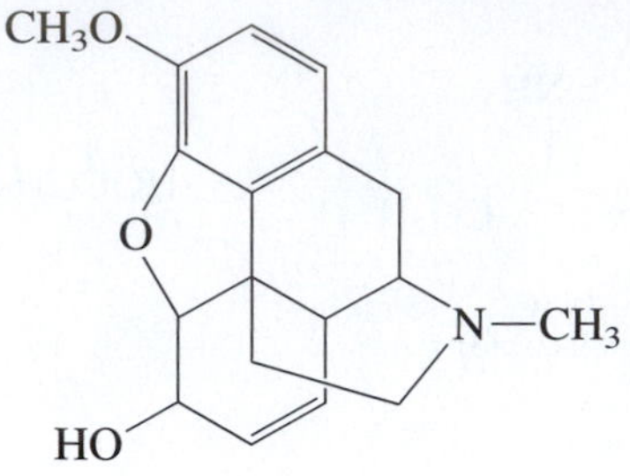

Codeine

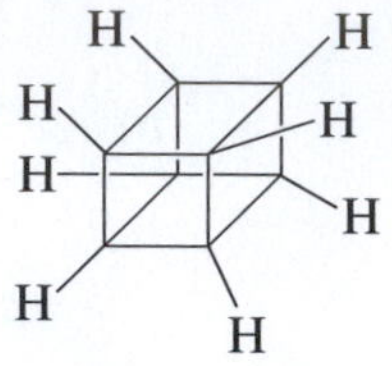

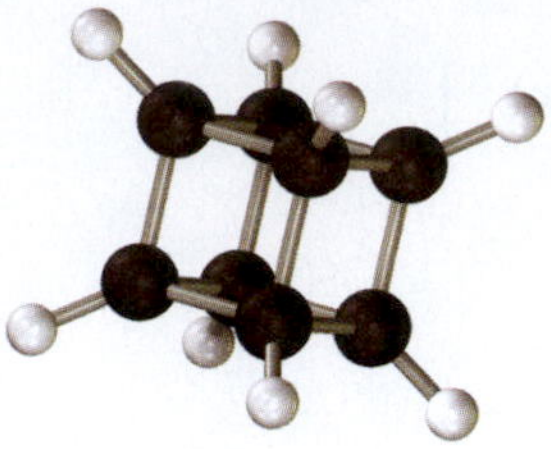

Cubane

Review of Mastery Goals

After completing this chapter, you should be able to:

■ Name an aromatic compound, a phenol, an aldehyde, a ketone, a carboxylic acid, an acid chloride, an anhydride, an ester, an amide, a nitrile, and a carboxylic acid salt. (Problems 12.21, 12.31, and 12.33)

■ Draw the structure of a compound containing one of these functional groups when the name is provided. (Problems 12.22 and 12.32)

■ Recognize the common functional groups that contain sulfur or phosphorus. (Problem 12.25)

■ Name a compound containing more than one functional group or draw the structure of such a compound when the name is provided. (Problem 12.23)

■ Understand how the physical properties of these compounds depend on the functional group that is present. (Problems 12.24, 12.34, 12.35, and 12.36)

Additional Problems

12.21   Provide names for these compounds:

a)

b) 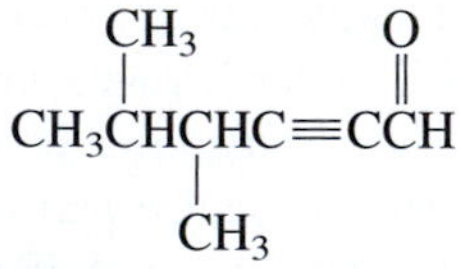 $CH_3CHCHC{\equiv}CCH$ with CH_3 and CH_3 groups, and O

c)

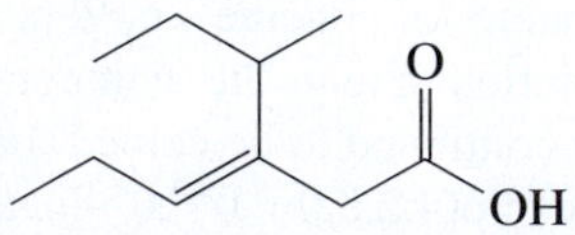

d)

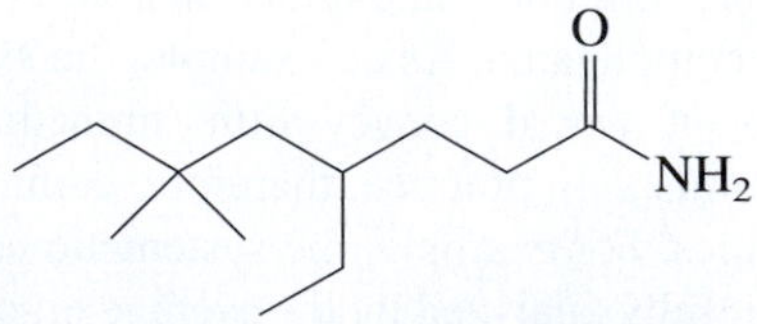

e)

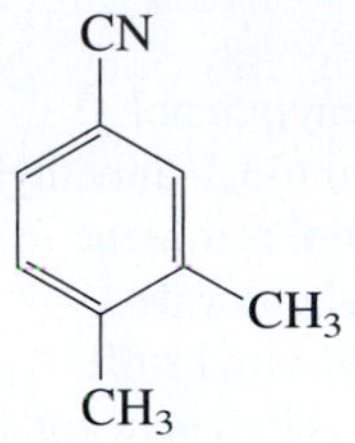

f) 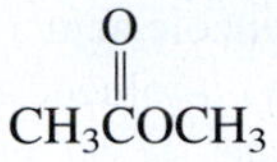 $CH_3\overset{O}{\overset{\|}{C}}OCH_3$

g)

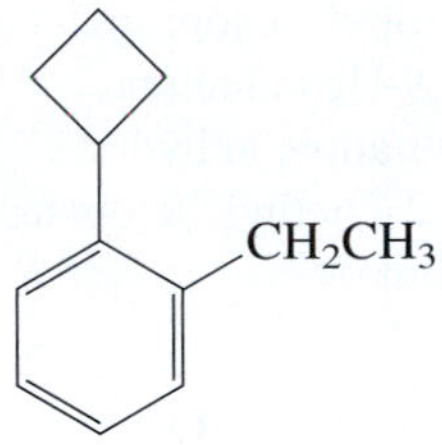

h)

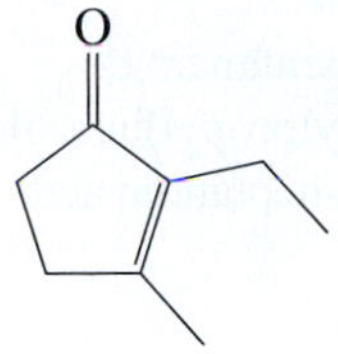

i)

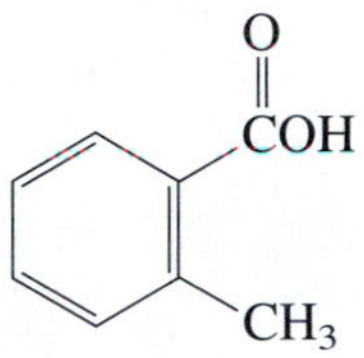

j)

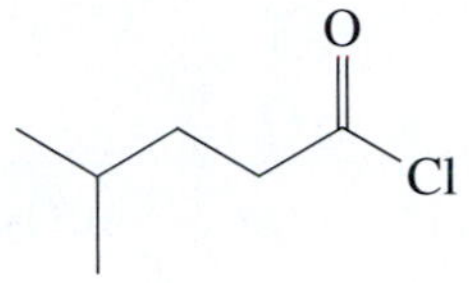

k)

l)

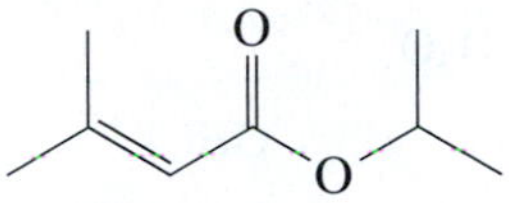

m)

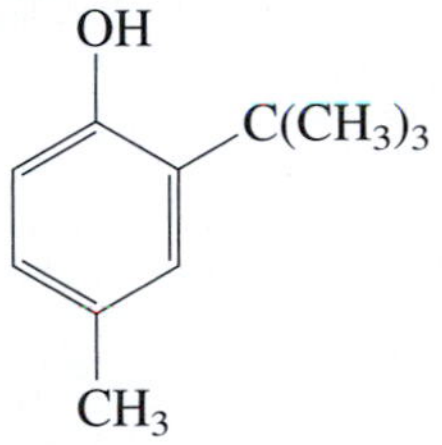

n)

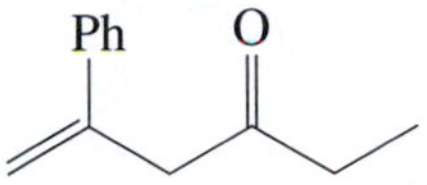

o)

p)

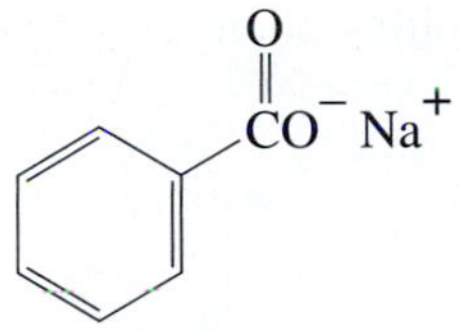

q)

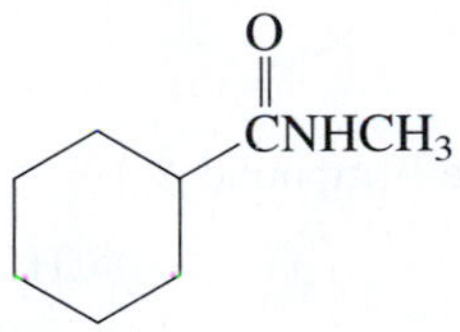

r)

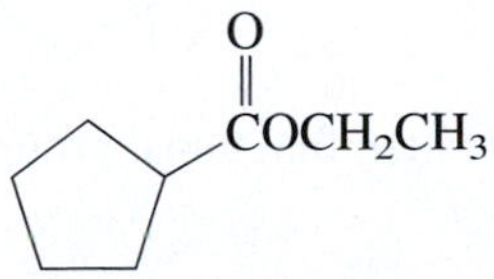

12.22 Draw structures for these compounds:

a) 3-Methoxybenzoic acid

b) *p*-Phenylphenol

c) 2-Methylbutyl *p*-toluenesulfonate

d) 4-Cyano-3,5-dimethyloctanal

e) Sodium acetate

f) 2,4-Dinitrotoluene

g) Ethyl 3-hydroxy-4-oxohexanoate

h) Acetyl chloride

i) Ethyl 2-chlorobenzoate

j) Phenyl vinyl sulfide

k) 1,2,4,5-Tetramethylbenzene

l) 2,4-Cyclooctadienone

m) Acetonitrile

n) Propyl 3-isopropyl-4-heptenoate

o) *N*-Methylpentanamide

p) 2,5-Hexanedione

q) 4-(1-Methylpropyl)benzaldehyde

r) Butanoic anhydride

s) 4-Hydroxyheptanoic acid

t) 2,2-Diethylcyclobutanecarboxylic acid

12.23 Provide names for these compounds:

a)

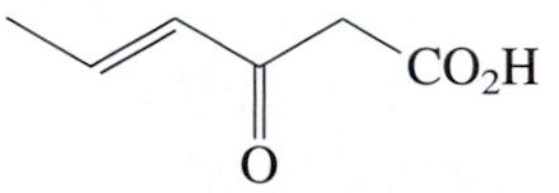

b)

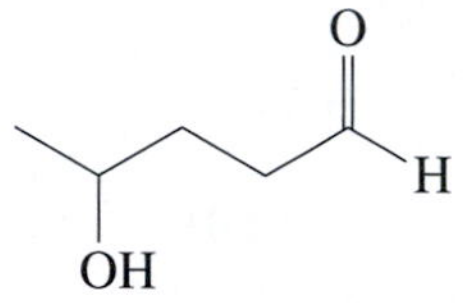

c)

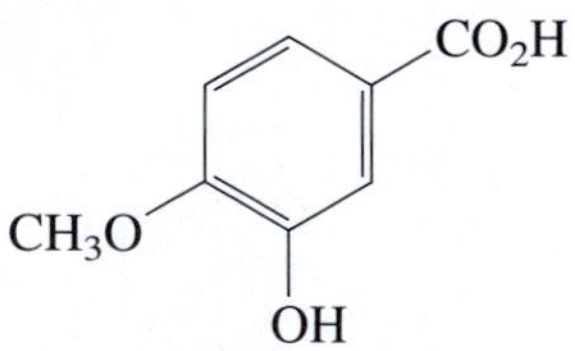

d)

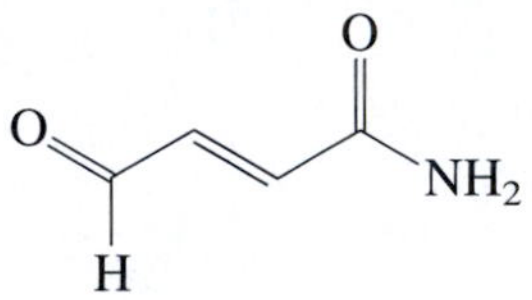

12.24 Explain which compound has the higher melting point or boiling point:

a)   Melting point 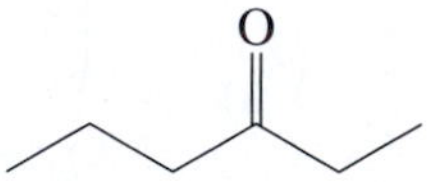   or

b)   Boiling point 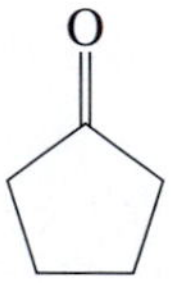   or

c)   Boiling point 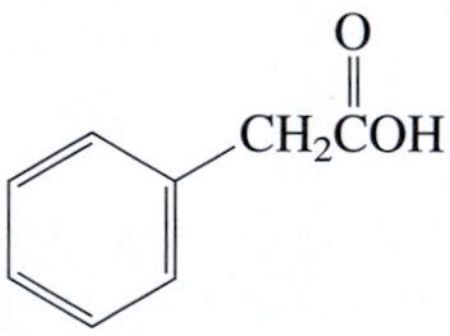   or

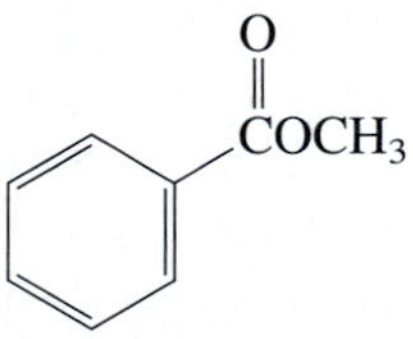

12.25 What is the functional group present in these compounds?

a) CH_3CH_2—P—Ph
                    |
                    CH_3

b)

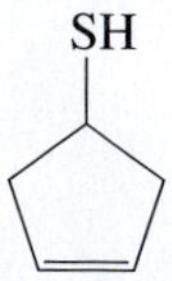

c)

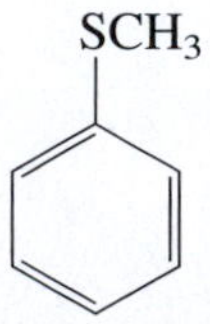

d) CH_3CH_2SPh **e)** $CH_3\overset{\displaystyle O}{\underset{\displaystyle O}{\overset{\|}{\underset{\|}{S}}}}CH_3$ **f)** $CH_3CH_2\overset{\displaystyle O}{\underset{\displaystyle O}{\overset{\|}{\underset{\|}{S}}}}OH$

12.26 Identify the most acidic site in these compounds:

a) (benzamide structure) $\overset{\displaystyle O}{\overset{\|}{C}}\!-\!NH_2$

b) (3-methylphenol structure with OH and CH_3)

c) $CH_3\overset{\displaystyle O}{\overset{\|}{C}}OCH_2CH_2CH_3$

d) $CH_3CH_2\overset{\displaystyle O}{\overset{\|}{C}}CH_2CH_3$ **e)** $CH_3CH_2CH_2\overset{\displaystyle O}{\overset{\|}{C}}OH$

12.27 Suggest explanations for the origins of "ibu," "pro," and "fen" in the name ibuprofen. Provide a systematic name for this compound (see page 479).

12.28 The pK_a for picric acid is 0.42. Explain why it is such a strong acid.

(structure of picric acid: phenol with OH at top, O_2N and NO_2 ortho, NO_2 para)

Picric acid

12.29 Use Table 4.2 to find a base that is strong enough to deprotonate benzoic acid but not *p*-methylphenol. Then explain how this base might be used to separate these two compounds in the laboratory.

12.30 The hydrogens of a hydrocarbon can be randomly replaced by chlorine by reaction of the hydrocarbon with Cl_2 in the presence of light. An unknown compound, **A,** with the formula C_4H_{10}, produces two isomeric monochlorides, **B** and **C,** with the formula C_4H_9Cl, when submitted to these reaction conditions. Monochloride **B** produces three isomeric dichlorides, **D, E,** and **F,** with the formula $C_4H_8Cl_2$, when submitted to these reaction conditions. Monochloride **C** gives a single dichloride, **D,** when submitted to these reaction conditions. Show the structures of **A, B, C, D, E,** and **F.** Are there any ambiguities in these structure assignments?

$$
\begin{array}{ccccccc}
\mathbf{A} & + & Cl_2 & \xrightarrow{\text{light}} & \mathbf{B} & + & \mathbf{C} \\
C_4H_{10} & & & & C_4H_9Cl & & C_4H_9Cl \\
\mathbf{B} & + & Cl_2 & \xrightarrow{\text{light}} & \mathbf{D} & + \; \mathbf{E} \; + \; \mathbf{F} \\
 & & & & C_4H_8Cl_2 & \;\; C_4H_8Cl_2 \;\; C_4H_8Cl_2 \\
\mathbf{C} & + & Cl_2 & \xrightarrow{\text{light}} & \mathbf{D} & \\
\end{array}
$$

BioLink   12.31 Provide systematic names for these naturally occurring compounds:

a) 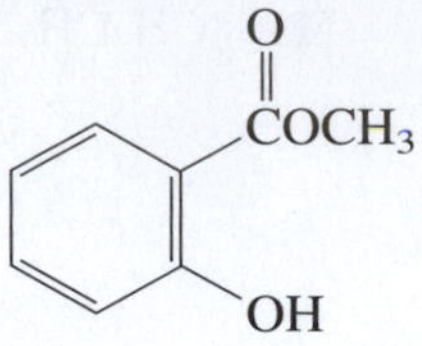
Methyl salicylate (from
oil of wintergreen)

b) 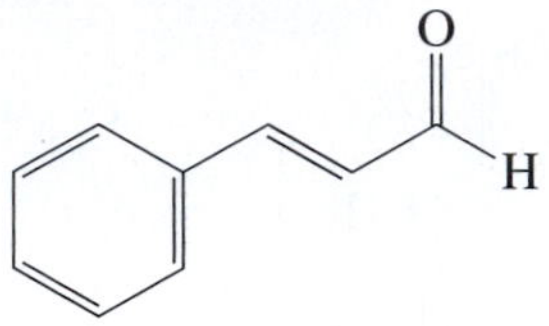
Cinnamaldehyde
(from cinnamon)

c) 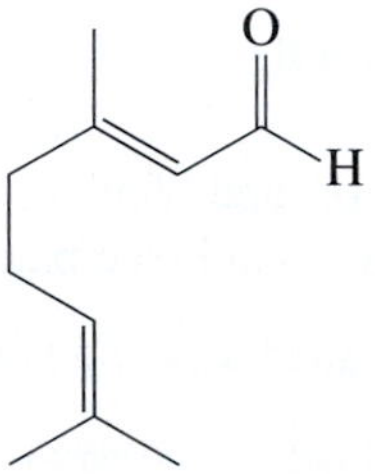
Citral (from lemons)

d) 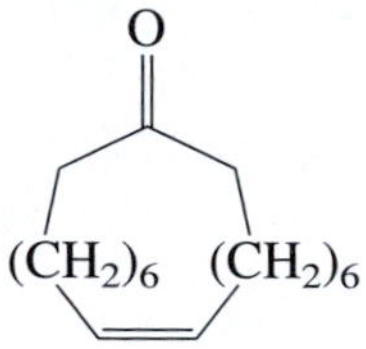
Civetone (a perfume ingredient from
the civet cat) (heptadec = 17)

e) 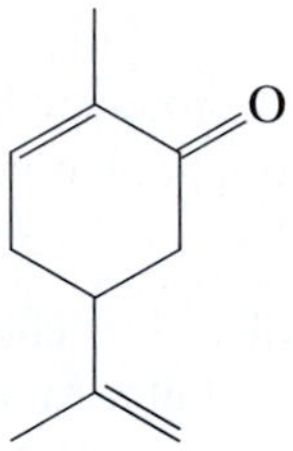
(From caraway seeds)
(see problem 5.34)

f) 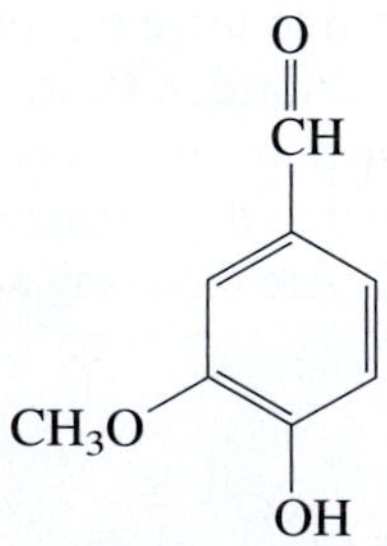
Vanillin
(from vanilla beans)

g) 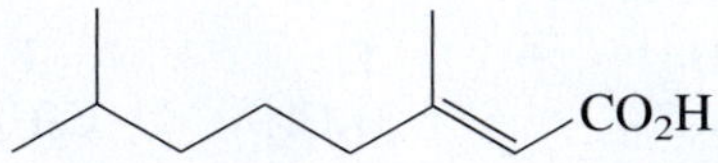
(Sex attractant of the
bean weevil)

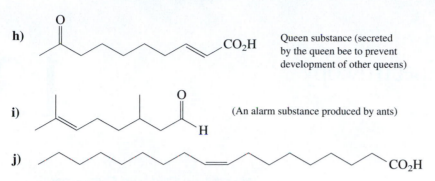

h) Queen substance (secreted by the queen bee to prevent development of other queens)

i) (An alarm substance produced by ants)

j)

Oleic acid (an unsaturated fatty acid) (octadec = 18)

12.32 Provide structures for these naturally occurring compounds:

 BioLink

a) 2,6-Diaminohexanoic acid (lysine, an amino acid)

b) Hex-2-en-1-yl acetate (sex attractant of Indian water bug)

c) (*Z*)-7-Dodecen-1-yl acetate (sex attractant of elephants and moths)

d) (*Z,Z*)-9,12-Octadecadienoic acid (linoleic acid, a polyunsaturated fatty acid) (octadec = 18)

e) (*all E*)-3,7-Dimethyl-9-(2,6,6-trimethylcyclohex-1-en-1-yl)nona-2,4,6,8-tetraenal (all-*trans*-retinal, or vitamin A aldehyde)

f) (*E*)-*N*-(4-Hydroxy-3-methoxybenzyl)-8-methyl-6-nonenamide (capsaicin, the hot ingredient in chili peppers) (benzyl is $PhCH_2$)

Problems Using Online Three-Dimensional Molecular Models

ORGANIC
Chemistry Now™
Click *Molecular Model Problems* to view the models needed to work these problems.

12.33 Name these compounds.

12.34 Explain which compound has the higher melting point.

12.35 Explain which compound has the higher boiling point.

12.36 Explain whether each compound is soluble in aqueous NaOH, aqueous $NaHCO_3$, both, or neither.

 Do you need a live tutor for homework problems? Access vMentor at Organic ChemistryNow at **http://now.brookscole.com/hornback2** for one-on-one tutoring from a chemistry expert.

Infrared Spectroscopy

MASTERING ORGANIC CHEMISTRY

▶ Predicting the Important Absorption Bands in an Infrared Spectrum

▶ Determining the Functional Group of a Compound from Its IR Spectrum

O NE OF THE MAJOR TASKS that an organic chemist faces in the laboratory is the determination of the structure of an unknown compound. This could be a compound isolated from a reaction mixture or a sample of a suspected illicit drug. In these cases the chemist usually has some idea of what the structure might be. Or it could be a compound isolated from a plant or animal source, in which case the information concerning its structure may be quite limited. In the not too distant past, the determination of the structure of such a compound was an enormous task, sometimes requiring years of work. The general procedure was to submit the unknown to numerous different chemical reactions. Each of these reactions supplied a little more information about the structure. Eventually, like a jigsaw puzzle, the structure was assembled. A simple example of this process, using the ozonolysis reaction, was described in Section 11.11.

Today, organic chemists rely on an array of very powerful instruments that enable them to identify compounds in much less time. With use of these instruments, it is often possible to determine the structure of an unknown compound in less than an hour. Three of the most powerful techniques are presented in this and the following chapters. They are infrared spectroscopy and two related techniques: proton and carbon-13 nuclear magnetic resonance spectroscopy. **Spectroscopy** is the study of the interaction of electromagnetic radiation (light) with molecules.

This chapter begins with a discussion of electromagnetic radiation and spectroscopy in general. Then infrared spectroscopy is presented. We will learn how the functional groups that are present in a compound can be identified by examination of its infrared spectrum. In the next chapter, we will see how nuclear magnetic resonance spectroscopy complements infrared spectroscopy by pro-

ORGANIC
Chemistry Now™

Look for this logo in the chapter and go to OrganicChemistryNow at **http://now.brookscole.com/hornback2** for tutorials, simulations, problems, and molecular models.

viding information about the hydrocarbon part of the compound. A combination of these techniques often provides enough information for the identification of an unknown.

13.1 ELECTROMAGNETIC RADIATION

Electromagnetic radiation is a general term for light, not just the visible light that our eyes detect but many other types also, including infrared, ultraviolet, and microwave. You are probably aware that light has properties of both particles and waves. Light is characterized by its **wavelength**, λ, which is the distance of one complete cycle of the light wave, that is, the distance between successive crests (or troughs) of the wave. Wavelength has units of length, such as centimeter and nanometer.

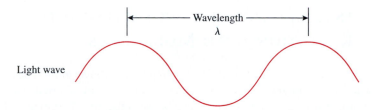

Light is also characterized by its **frequency**, ν, which is the number of wave cycles that pass a point in a second. The unit for frequency is seconds^{-1} (s^{-1}), also called cycles per second or hertz (Hz). The product of the wavelength times the frequency equals the speed of light (c):

$$\lambda\nu = c = 3.00 \times 10^8 \text{ m s}^{-1}$$

The energy of one light photon (ϵ) is equal to the frequency times Planck's constant or Planck's constant times the speed of light divided by the wavelength:

$$\epsilon = h\nu = hc/\lambda$$
$$h = \text{Planck's constant} = 1.58 \times 10^{-37} \text{ kcal s } (6.63 \times 10^{-37} \text{ kJ s})$$

You have probably begun to remember some energy values in units of kcal/mol (or kJ/mol). For example, typical covalent single-bond strengths are in the range of 50 to 100 kcal/mol (210–420 kJ/mol), and the amount of thermal energy available at room temperature is about 20 kcal/mol (84 kJ/mol). For comparison purposes, therefore, it is more convenient to use the energy of a mole of photons (E) in units of kcal/mol (or kJ/mol). This requires that ϵ be multiplied by Avogadro's number ($N = 6.02 \times 10^{23}$ mol^{-1}). The appropriate equations then become in units of kcal/mol:

$$E = (9.53 \times 10^{-14} \text{ s kcal/mol})(\nu)$$
$$\text{or} \quad E = (2.86 \times 10^{-5} \text{ m kcal/mol})(1/\lambda)$$

and in units of kJ/mol:

$$E = (3.99 \times 10^{-13} \text{ s kJ/mol})(\nu)$$
$$\text{or} \quad E = (1.20 \times 10^{-4} \text{ m kJ/mol})(1/\lambda)$$

where the frequency, ν, is in units of s^{-1} or the wavelength, λ, is in units of meters (m). It is important to remember that the energy of light is directly proportional to its frequency and inversely proportional to its wavelength.

Low-Energy Light	High-Energy Light
Low frequency	High frequency
Long wavelength	Short wavelength

PROBLEM 13.1

Calculate these quantities:

a) The wavelength of light (in centimeters) with a frequency of 9.00×10^{12} Hz.

b) The frequency of light with a wavelength of 310 nm.

c) The energy of light (in kcal/mol or kJ/mol) with a frequency of 9.00×10^{12} Hz.

d) The energy of light (in kcal/mol or kJ/mol) with a wavelength of 310 nm.

13.2 INTERACTION OF ELECTROMAGNETIC RADIATION WITH MOLECULES

Chapter 3 discussed the orbitals that an electron may occupy in an atom. The energies of these orbitals are **quantized**—that is, only certain energies are allowed. For example, there is no orbital with an energy intermediate between that of a $1s$ orbital and that of a $2s$ orbital. Not only are the energies of the electron orbitals quantized, but all of the energy states of a molecule are quantized. It is this fact that makes spectroscopy possible.

To illustrate, let's consider a hypothetical molecule that has only two energy states (see Figure 13.1). Initially, the molecule has the lower amount of energy—it is in the **ground state.** When the molecule interacts with light that has an energy exactly equal

Figure 13.1

TWO ENERGY STATES OF A HYPOTHETICAL MOLECULE.

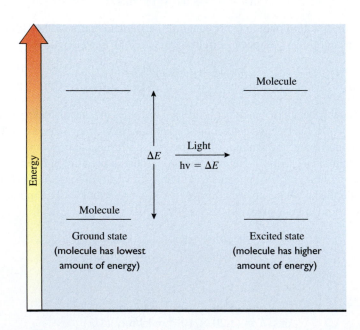

to the energy difference between the two states—that is, light with a frequency such that $h\nu = \Delta E$—the molecule absorbs the light and is raised to the higher-energy state, known as an **excited state.** If the energy of the light does not match the energy difference between the two states, then the light is not absorbed. A **spectrum** is just a plot of the amount of light that is absorbed versus the frequency (or wavelength) of the light. The spectrum provides information about the spacing of the energy levels of the molecule. Because these energy levels depend on the structure of the molecule, this information can, with practice, be used to determine the structure of the compound.

The following is a simplified schematic representation of an absorption spectrometer:

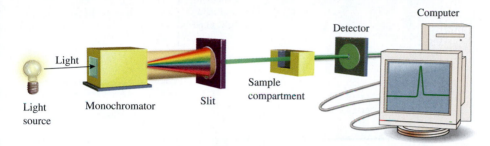

Light from an appropriate source is passed through a monochromator. A monochromator is a device, such as a prism, that spreads or disperses the light into its various wavelengths. The light from the monochrometer is then passed through the sample of the compound under investigation to see which wavelengths of light are absorbed by the sample. A detector is used to determine the intensity at each wavelength, and the results are sent to a recorder that plots the spectrum. Today, this entire process of gathering and storing the data and plotting the resulting spectrum is usually computerized.

In some types of spectroscopy there are only a few, well-separated energy levels. In such cases, only a very narrow range of wavelengths is absorbed each time the molecule is excited from its lowest-energy state to some higher-energy state, resulting in an absorption line for each of these transitions. The spectrum consists of a number of these absorption lines. More commonly, however, there are a number of energy sublevels in each energy state. In such cases a number of closely spaced wavelengths are absorbed. The lines are often so close together that they cannot be resolved and the absorption appears as a broad peak or band. Figure 13.2 illustrates both of these situations.

13.3   THE ELECTROMAGNETIC SPECTRUM

The spectrum of electromagnetic radiation, ranging from very energetic cosmic rays to low-energy radio waves, is shown in Figure 13.3. The energy of cosmic and gamma radiation is so large that this radiation passes right through most matter because there are no states separated by enough energy that these high-energy photons can be absorbed. However, most other types of radiation are absorbed, resulting in the formation of a variety of different excited states.

The types of spectroscopy that are of most use to organic chemists employ infrared light or radiation in the radio region. Light in the infrared region has energy that

ⓐ When there are only a few energy levels and the spacing between them is large enough, a line spectrum is produced.

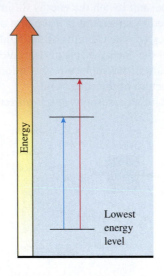

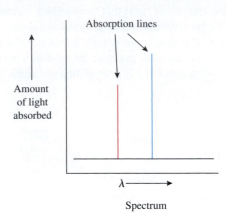

ⓑ When there are many, closely spaced sublevels in each energy level, the absorptions occur in broad bands because the individual lines are not resolved.

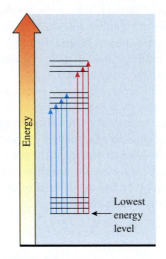

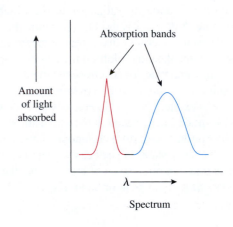

Figure 13.2

TYPES OF ABSORPTION SPECTRA: ⓐ LINE SPECTRUM AND ⓑ BROADBAND SPECTRUM.

matches the energy separation of the vibrational energy states of covalent bonds. An infrared spectrum provides information about the types of bonds and therefore the functional groups that are present in the molecule. Nuclear magnetic resonance spectroscopy employs very low energy radiation in the radio region. It involves transitions between nuclear spin states and provides information about the hydrogens and carbons in the molecule.

Cosmic and gamma rays are very high energy and do not interact much with matter because no energy states with ΔE this large are available. The energy of X-rays corresponds to ΔE between electron shells and can excite and eject inner-shell electrons.

Light from the ultraviolet and visible regions has the correct energy to excite electrons from bonding or nonbonding MOs to antibonding MOs in some molecules. An example of such an electronic transition is the excitation of an electron in a bonding pi MO to an antibonding pi MO. Ultraviolet and visible spectroscopy is useful in organic chemistry and is discussed in Chapter 15.

Energy

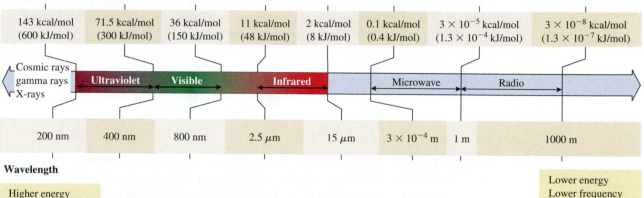

Higher energy
Higher frequency
Shorter wavelength

The energy of infrared light matches the ΔE between the vibrational energy levels of covalent bonds. When a molecule absorbs light, its bonds vibrate more rapidly. IR spectroscopy is discussed in this chapter.

Radar employs microwave radiation. Microwave ovens use this radiation to excite rotational energy states of water and other molecules in food.

Lower energy
Lower frequency
Longer wavelength

Radio and TV broadcasts use radio frequency radiation. In addition, nuclear magnetic resonance spectroscopy, which causes transitions between nuclear spin states, uses radiation from this region. One typical NMR spectrometer operates at 2×10^8 Hz or 200 MHz (1.9×10^{-5} kcal/mol or 8×10^{-5} kJ/mol). NMR spectroscopy is discussed in Chapter 14.

Figure 13.3

THE ELECTROMAGNETIC SPECTRUM.

PROBLEM 13.2

What kind of light has a frequency of 9.00×10^{13} Hz?

PROBLEM 13.3

Some NMR spectrometers operate at 4×10^8 Hz (400 MHz). What is the energy of this radiation?

13.4 INFRARED SPECTROSCOPY

The atoms of a covalent bond never sit still, even at absolute zero. They are constantly vibrating, somewhat like balls connected by springs. The energies of the vibrations, like all energies on a molecular scale, are quantized. At room temperature, most molecules are in the lowest vibrational energy level. The separation between the energy levels ranges from 1 to 14 kcal/mol (4–60 kJ/mol). This energy corresponds to the energy of infrared (IR) radiation. Therefore, when IR radiation is passed through a sample of a compound, the light is absorbed and the molecule is excited to a higher vibrational state when the energy of the light matches the energy separation between two vibrational energy levels. Because each vibrational level has a number of closely spaced rotational energy levels, IR absorptions appear as bands rather than lines.

Two types of vibrations occur in molecules: stretches, where the distance between bonded atoms oscillates, and bends, where the bond angles oscillate. These oscillations are often coupled among several atoms. As examples, one stretching vibration and one bending vibration of the CH_2 group are shown here:

Stretch   Bend

A typical organic molecule has a large number of possible stretches, bends, and combinations, so there are many absorption bands. However, it is not necessary to know the particular vibration that is responsible for each of these bands. One way in which IR is used to identify an unknown compound is to compare the spectrum of the unknown compound with those of known compounds. Like humans and fingerprints, no two compounds have identical IR spectra. Therefore, if the spectrum of the unknown is identical to that of a known compound, the identity of the unknown is established.

However, the feature that makes IR spectroscopy really useful in identifying an unknown compound is that a particular type of covalent bond always produces some absorption bands in consistent positions in the spectrum. It is possible, with practice, to identify the functional group or groups that are present in a compound by examination of its IR spectrum. Therefore, it is possible to identify an unknown sample as a ketone, or ester, or alcohol, for example, just by examination of its IR spectrum.

Figure 13.4 shows a typical IR spectrum, that of cyclohexanone. The percentage of light transmitted (percent transmittance, $\%T$) is plotted on the y-axis, with 100% transmittance (no light absorbed) at the top of the spectrum. Absorption bands therefore come downward from the top of the spectrum. The intensity of a band is often described qualitatively as w (weak, high $\%T$), m (medium), or s (strong, low $\%T$). The positions of the absorption bands are usually expressed as wavenumbers ($\overline{\nu}$), rather than wavelengths or frequencies. The **wavenumber**, whose unit is cm^{-1}, is the reciprocal of wavelength:

$$\overline{\nu} = 1/\lambda$$

Note that wavenumbers are directly proportional to energy; that is, light of higher wavenumber has higher energy. The wavenumber range covered in most of the IR spectra in this book is from 4000 cm^{-1} (higher energy) to 500 cm^{-1} (lower energy).

To give a brief preview of how IR spectroscopy is useful, note the strong absorption band at 1714 cm^{-1} in the spectrum of cyclohexanone (Figure 13.4). This band is due to the presence of the carbonyl group. As we will see shortly, all compounds that have a carbonyl group show a similar absorption in the general region of 1700 cm^{-1}. Therefore, the presence of such an absorption band in the spectrum of an unknown compound is strong evidence that the unknown has a carbonyl group.

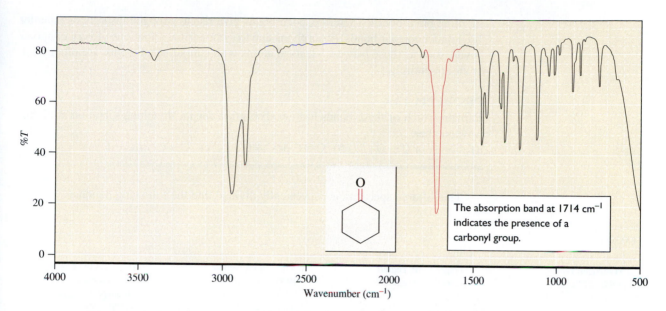

The absorption band at 1714 cm^{-1} indicates the presence of a carbonyl group.

Percent transmittance (%T) is plotted along the y-axis with 100% T at the top. Absorption bands come down from the top of the spectrum.

Wavenumber, in units of cm^{-1}, is plotted along the x-axis. The IR spectrometer that was used to obtain this spectrum covers the region from 4000 to 500 cm^{-1}.

Figure 13.4

THE INFRARED SPECTRUM OF CYCLOHEXANONE.

13.5   GENERALIZATIONS

An approximate value for the wavenumber for a particular IR transition can be calculated from the following equation:

$$\text{wavenumber} = \bar{\nu} = \frac{1}{2\pi c}\sqrt{\frac{k}{M}}$$

where k is the force constant for the bond and M is the reduced mass of the bonded atoms. The force constant, k, is related to the bond strength and is larger for stronger bonds. It is also larger for bond-stretching vibrations than for bending vibrations. The reduced mass, M, is larger for atoms of larger atomic mass. On the basis of this information the following generalizations about the position of an absorption can be made:

1. Because of their larger force constants, stronger bonds absorb at higher wavenumbers than weaker bonds.

2. For similar reasons, bond stretches absorb at higher wavenumbers than bends.

3. Bonds involving a light hydrogen atom have smaller reduced masses and therefore absorb at higher wavenumbers than bonds involving only heavier atoms.

For IR energy to be absorbed and result in a transition between vibrational energy levels, the dipole moment of the molecule must change as the vibration occurs. The intensity of the absorption is proportional to the magnitude of this change and is larger for more polar bonds. Therefore, polar bonds give rise to stronger absorptions (smaller %T) than nonpolar bonds. Absorptions due to symmetrical vibrations of very nonpolar bonds, such as those between identical atoms with similar substituents, are often weak or absent entirely.

PROBLEM 13.4
Explain which of these bonds has the absorption for its stretching vibration at higher wavenumber:

a) C—H or C—D   **b)** C=C or C≡C   **c)** C—Cl or C—I

The infrared spectrum is conveniently divided into the following regions:

Higher energy $\bar{\nu} = 4000$ cm^{-1}			Lower energy $\bar{\nu} = 500$ cm^{-1}
Hydrogen region 3800–2700 cm^{-1}	**Triple-bond region** 2300–2000 cm^{-1}	**Double-bond region** 1900–1500 cm^{-1}	**Single-bond region** **Bend region** **Fingerprint region** 1500–500 cm^{-1}

The highest-energy region, from 3800 to 2700 cm^{-1}, is called the **hydrogen region.** Absorptions due to O—H, N—H, and C—H bond stretches occur here. These absorption bands are very high in energy because hydrogen is such a light atom. All of the bonds involving only heavier atoms, such as C—C and C—O bonds, absorb at lower energies. Next comes the **triple-bond region,** from 2300 to 2000 cm^{-1}. Absorptions due to CN and CC triple-bond stretches occur here. These are the strongest bonds, so they occur at higher wavenumber than double and single bonds. The **double-bond region,** from 1900 to 1500 cm^{-1}, is lower in energy because double bonds are weaker than triple bonds. Absorptions due to CO, CN, and CC double-bond stretches occur in this region. Finally, at lowest energy is the **single-bond region,** from 1500 to 500 cm^{-1}. In addition to the absorptions due to stretches of single bonds, absorptions due to bends also occur in this region. There are usually so many bands here that it is difficult to assign each to a specific vibration. However, the large number of bands make this region especially useful in comparing spectra, so it is also known as the **fingerprint region.** Let's consider each of these regions in more detail.

13.6   THE HYDROGEN REGION

Bonds between a light hydrogen atom and oxygen, nitrogen, and carbon atoms have absorptions in this region. The position of the band depends on the bond strength, which increases in the order C—H < N—H < O—H. The absorptions for C—H bonds appear at lower energy, in the region of 3330 to 2700 cm^{-1}. The C—H bond strength varies slightly because of the hybridization of the carbon atom, increasing in the order H—C$_{sp3}$ < H—C$_{sp2}$ < H—C$_{sp}$. The absorptions for a bond between a hydrogen and an sp^3-hybridized carbon appears in the region of 3000 to 2850 cm^{-1}. Most

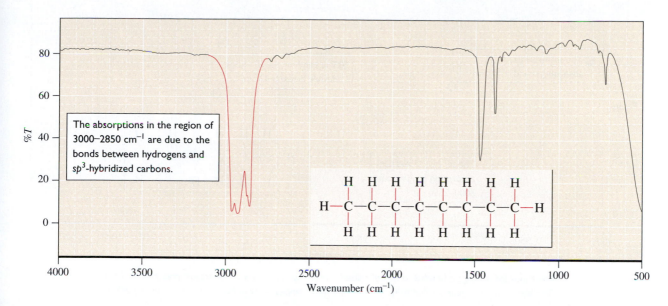

The absorptions in the region of 3000–2850 cm^{-1} are due to the bonds between hydrogens and sp^3-hybridized carbons.

Alkanes: Most compounds have an alkyl part and therefore have absorption bands in the 3000–2850 cm^{-1} region due to their CH bonds. The feature that distinguishes alkanes is the absence of bands for hydroxy groups, carbonyl groups, and so on. The spectrum of an alkane usually has many fewer absorptions than those of compounds with other functional groups.

Figure 13.5

THE INFRARED SPECTRUM OF OCTANE.

organic compounds have several bands in this region, owing to the numerous C—H bonds of this type. Although these bonds are not very polar, the absorptions are relatively strong because of the number of bonds. An example is provided in the spectrum of octane, shown in Figure 13.5. For now, do not attempt to identify all of the absorption bands when you examine the IR spectra in these figures. Instead, focus on the particular part of the spectrum that is currently being discussed—that is, the absorption due to the C—H bonds in this particular spectrum. Most of the spectra have additional bands identified. Subsequent sections will refer back to these bands. In addition, each spectrum has a summary of the absorptions due to the particular functional group illustrated by that spectrum. Later, we will return to examine all of the important bands in each of these spectra.

The bond between a hydrogen and an sp^2-hybridized carbon is somewhat stronger than that between a hydrogen and an sp^3-hybridized carbon and thus appears at higher wavenumbers, in the region of 3100 to 3000 cm^{-1}. Note that 3000 cm^{-1} is a convenient dividing line for alkene and alkane C—H absorptions. The spectrum of 1-hexene, shown in Figure 13.6, shows bands for both types of hydrogens.

The strongest C—H bond is that of a 1-alkyne, where the hydrogen is bonded to an sp-hybridized carbon. Therefore, the C—H stretch of a 1-alkyne occurs at the highest wavenumber of all C—H bonds, near 3300 cm^{-1}. The spectrum of 1-hexyne is shown in Figure 13.7.

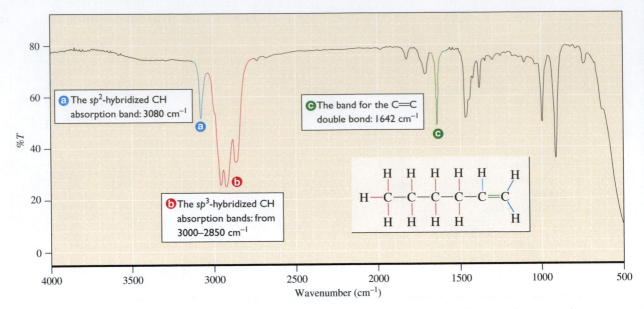

The bands for hydrogens bonded to sp^2-hybridized carbons appear at wavenumbers just greater than 3000 cm⁻¹. The bands for hydrogens bonded to sp^3-hybridized carbons appear at wavenumbers just less than 3000 cm⁻¹.

Alkenes: Most alkenes have absorption bands in the 3100–3000 cm⁻¹ region, due to the CH bonds where the carbon is sp^2 hybridized, and an absorption in the 1660–1640 cm⁻¹ region due to the CC double bond. However, the CH absorption is absent if the double bond is tetrasubstituted. Furthermore, the CC double-bond absorption is often weak because the bond is not very polar and may be difficult to discern.

Figure 13.6

THE INFRARED SPECTRUM OF 1-HEXENE.

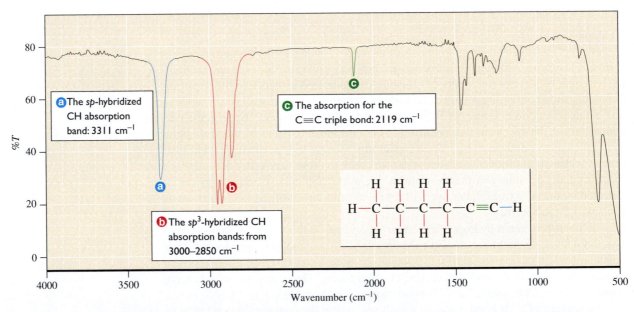

Alkynes: Terminal alkynes, where the triple bond occurs at the end of a carbon chain, are readily identified by a band near 3300 cm⁻¹ due to the bond between the hydrogen and the sp-hybridized carbon and a band in the region of 2150–2100 cm⁻¹ due to the CC triple bond. Alkynes in which the triple bond is in the middle of a carbon chain are more difficult to recognize because the CH bond is absent and the band due to the triple bond may be quite weak.

Figure 13.7

THE INFRARED SPECTRUM OF 1-HEXYNE.

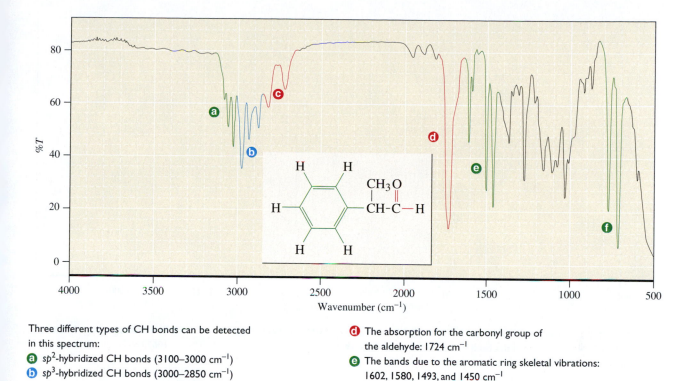

Three different types of CH bonds can be detected in this spectrum:

ⓐ sp^2-hybridized CH bonds (3100–3000 cm^{-1})
ⓑ sp^3-hybridized CH bonds (3000–2850 cm^{-1})
ⓒ Aldehyde CH bond (two bands at 2820 and 2720 cm^{-1})

ⓓ The absorption for the carbonyl group of the aldehyde: 1724 cm^{-1}
ⓔ The bands due to the aromatic ring skeletal vibrations: 1602, 1580, 1493, and 1450 cm^{-1}
ⓕ The bands due to the CH bending vibrations of the aromatic ring: 760 and 700 cm^{-1}

Aldehydes: The appearance of two bands in the region of 2830–2700 cm^{-1} (hydrogen to carbonyl carbon bond) and the presence of an absorption due to a carbonyl group provide evidence for the presence of an aldehyde. The band for the carbonyl of an aldehyde occurs near 1730 cm^{-1}, although it is shifted to lower wavenumbers when conjugated.

Figure 13.8

THE INFRARED SPECTRUM OF 2-PHENYLPROPANAL.

Finally, the C—H bond between the carbonyl carbon and hydrogen of an aldehyde normally results in two absorptions in the region of 2830 to 2700 cm^{-1} and can usually be distinguished from the alkane C—H absorptions. Three different types of absorptions due to C—H bonds can be seen in the spectrum of 2-phenylpropanal shown in Figure 13.8.

An O—H bond is stronger than a C—H bond, so its absorption band appears at higher energy, in the region of 3600 cm^{-1}. However, a band at this position for a hydroxy group is observed only in the vapor phase or in very dilute solution, where intermolecular hydrogen bonding between O—H groups is negligible. Most IR spectra are not run under such conditions. Instead, the samples are considerably more concentrated, often even pure (also termed neat) liquids. Under these conditions, hydrogen bonding is very important and weakens the O—H bond, causing the absorption to shift to lower wavenumbers, from 3550 to 3200 cm^{-1}. The band is very broad because the strength of each hydrogen bond differs slightly, depending on the exact separation and geometry of

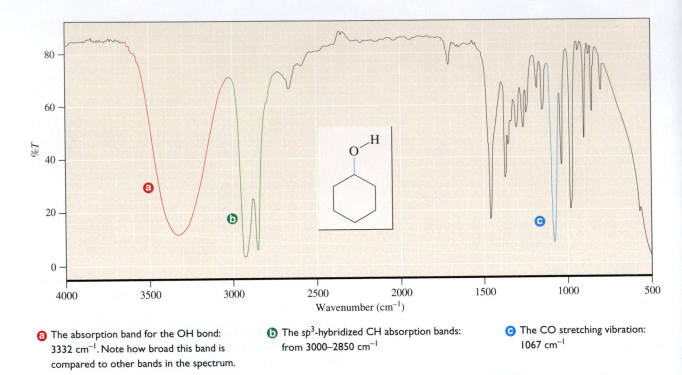

a The absorption band for the OH bond: 3332 cm^{-1}. Note how broad this band is compared to other bands in the spectrum.

b The sp^3-hybridized CH absorption bands: from 3000–2850 cm^{-1}

c The CO stretching vibration: 1067 cm^{-1}

Alcohols: Alcohols are characterized by a strong and very broad absorption in the 3550–3200 cm^{-1} region due to the hydrogen-bonded OH group. They also show a strong band for the CO bond in the 1300–1000 cm^{-1} region, although this absorption can be difficult to identify because it occurs in the fingerprint region with many other bands.

Figure 13.9

THE INFRARED SPECTRUM OF CYCLOHEXANOL.

the two partners. Because the O—H bond is quite polar, this absorption is very strong. Figure 13.9 shows the IR spectrum of cyclohexanol. The band for the O—H occurs at 3332 cm^{-1}. Note that this band is considerably broader than the other bands in the spectrum.

Because a carboxylic acid forms very strong hydrogen bonds, the band for its O—H group is extremely broad and occurs at even lower wavenumbers, usually centered near 3000 cm^{-1}. This band overlaps the C—H absorptions but is readily distinguished from them because it is so broad. As illustrated in Figure 13.10, the C—H absorptions can often be seen superimposed on the intense O—H absorption.

The N—H bond strength is between those of O—H and C—H bonds, so it absorbs between these two groups, in the region of 3400 to 3250 cm^{-1}. Like the O—H absorption, it is broadened because of hydrogen bonding, but to a lesser extent because its hydrogen bonds are weaker. Although the region for N—H bands overlaps with the region for O—H bands, with some experience it is usually possible to distinguish the two types because the N—H bands are narrower and less intense than the O—H bands. Primary amines show two bands in this region, whereas secondary amines show only one. Examples are provided in Figures 13.11 and 13.12.

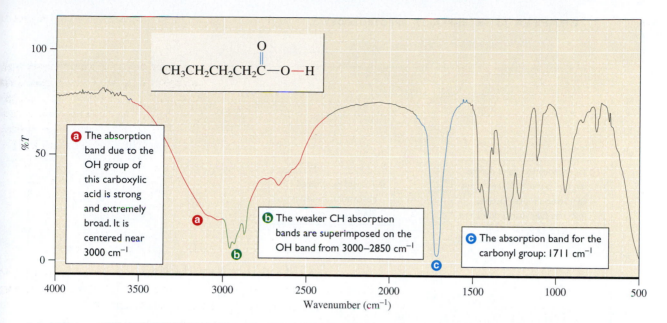

Carboxylic acids: A carboxylic acid is readily recognized from its carbonyl absorption, near 1710 cm⁻¹, and its extremely broad OH band, often centered near 3000 cm⁻¹. This band is much broader and appears at lower wavenumbers than the OH band of an alcohol. Often, the CH bands are superimposed on the OH absorption of a carboxylic acid.

Figure 13.10

THE INFRARED SPECTRUM OF PENTANOIC ACID.

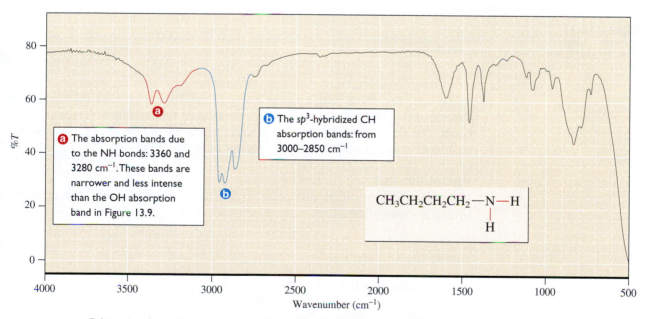

Primary amines: Primary amines are characterized by two absorption bands that appear in the 3400–3250 cm⁻¹ region due to the NH bonds. These absorptions appear in the same region as those due to an OH group but are narrower and less intense. The absorption due to the CN bond stretch is very difficult to use because it occurs in the fingerprint region with only weak to moderate intensity.

Figure 13.11

THE INFRARED SPECTRUM OF BUTYLAMINE.

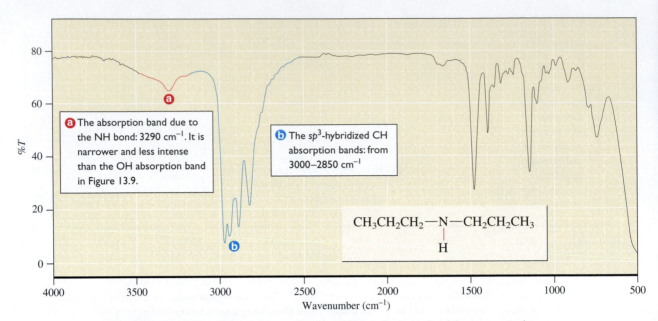

a The absorption band due to the NH bond: 3290 cm⁻¹. It is narrower and less intense than the OH absorption band in Figure 13.9.

b The sp^3-hybridized CH absorption bands: from 3000–2850 cm⁻¹

$$CH_3CH_2CH_2-N-CH_2CH_2CH_3$$
$$\quad\quad\quad\quad\quad\quad | $$
$$\quad\quad\quad\quad\quad\quad H$$

Secondary amines: Secondary amines are characterized by an absorption band that appears in the 3400–3250 cm⁻¹ region due to the NH bond. In contrast to a primary amine, a secondary amine shows only one absorption in this region. Although it appears in the same region as the absorption due to an OH group, it is narrower and less intense so that it can be distinguished with some practice. The absorption due to the CN bond is very difficult to use because it occurs in the fingerprint region with only weak to moderate intensity.

Figure 13.12

THE INFRARED SPECTRUM OF DIPROPYLAMINE.

Overall, then, the hydrogen region is very useful. We can establish the presence or absence of O—H and N—H groups, and we can identify four different types of C—H bonds. The exact functional group or groups that are present in the compound can be confirmed by examination of the other regions of the spectrum.

PROBLEM 13.5

Indicate the positions of the absorption bands and any other noteworthy features in the hydrogen region of the IR spectra of these compounds:

a) $CH_3CH=CHCOOH$

b) [structure: phenyl ring with C(=O)CH₃]

c) [structure: phenyl ring with NH₂]

d) $CH_2=CHCH_2CH_2OH$

e) $CH_3CH_2NHCH_2CH_3$ f) $CH_3CH_2CH_2\overset{\displaystyle O}{\overset{\|}{C}}H$

13.7   THE TRIPLE-BOND REGION

The large force constant of strong triple bonds results in absorptions in the 2300 to 2000 cm^{-1} region. The CN triple-bond stretch of a cyano group or nitrile occurs in the region of 2260 to 2220 cm^{-1}. An example of the absorption band for a CN triple bond is shown in Figure 13.13. The CC triple-bond stretch of an alkyne usually occurs in the region of 2150 to 2100 cm^{-1}. This band is most visible for 1-alkynes (see Figure 13.7). If the triple bond has similar substituents on the two carbons, the absorption may be very weak or absent altogether because of the lack of change of the dipole moment of the bond during the vibration.

PROBLEM 13.6

Explain why the presence of a triple bond is much easier to detect in the IR spectrum of 1-hexyne than it is in the spectrum of 3-hexyne.

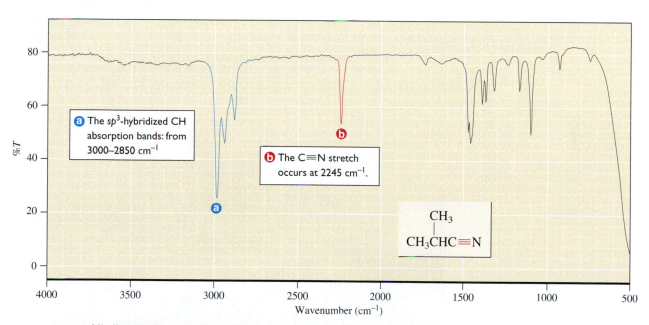

Nitriles: Nitriles are readily identified by the absorption due to the CN triple bond that appears in the region of 2260–2220 cm^{-1}. This band is relatively strong because the bond is polar.

Figure 13.13

THE INFRARED SPECTRUM OF 2-METHYLPROPANENITRILE.

Focus On

Remote Sensing of Automobile Pollutants

Most types of spectroscopy can be used for quantitative analysis because the intensity of an absorption band—that is, the amount of light absorbed at a particular wavelength—is proportional to the amount of compound in the sample. A group of chemists at the University of Denver has developed a device that can measure the amount of carbon monoxide and hydrocarbons in the exhaust of an automobile by remote sensing—that is, in the street as the automobile passes by.

Poorly maintained automobiles pollute by emitting carbon monoxide and hydrocarbons, products of incomplete combustion. Because carbon monoxide is very poisonous and hydrocarbons contribute to the formation of smog, it is desirable to minimize the emission of both of these pollutants.

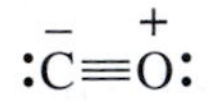

Carbon monoxide

As shown in the following figure, the remote sensing device consists of a source of IR light on one side of the road and a detector on the other, with the beam at tailpipe height. As the auto passes through the beam, the amount of CO in the exhaust is measured by monitoring the intensity of the absorption at 2170 cm^{-1}, where carbon monoxide has a strong absorption due to its triple-bond stretch. The amount of hydrocarbons is measured by monitoring the intensity of the absorption in the region of 3000 to 2900 cm^{-1}, where the C—H stretching absorptions of hydrocarbons occur. More advanced systems can also monitor nitric oxide (NO) emissions using the absorption of ultraviolet light.

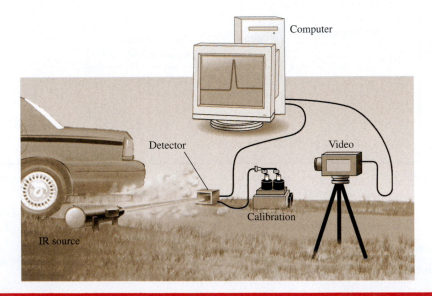

Prof. Donald Stedman, University of Denver

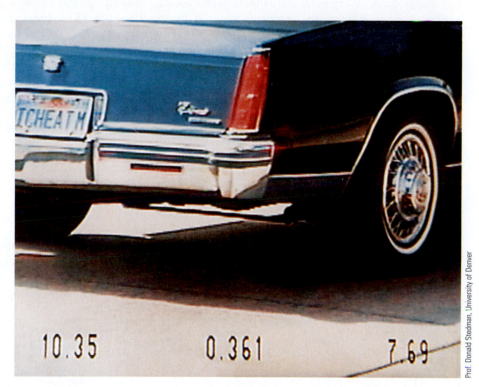

The amounts of carbon monoxide, hydrocarbons, and carbon dioxide are recorded by a computer, and a video camera records a picture of the license plate of the car. The accompanying picture shows the video screen with the percent CO in the exhaust listed at the lower left, the percent hydrocarbon in the middle, and the percent CO_2 at the right. This particular vehicle was a high polluter, emitting 10.35% CO, 0.36% hydrocarbons, and 7.69% CO_2. The emission of a well-maintained, low-polluting vehicle contains about 0% CO, 0% hydrocarbons, and 15% CO_2. Because about 10% of the vehicles emit about 65% of the pollution, the identification and repair of these high polluters could substantially reduce the air pollution caused by automobiles.

PROBLEM 13.7

The exhaust from a poorly maintained automobile may contain a wide variety of different hydrocarbon pollutants. Why is the 3000 to 2900 cm^{-1} region a good place to monitor the amount of these compounds?

13.8   THE DOUBLE-BOND REGION

The double-bond region extends from 1900 to 1500 cm^{-1}. Carbonyl groups have an absorption band in the general region of 1700 cm^{-1}. This band is very strong because the CO double bond is very polar and is quite useful because carbonyl groups occur in a variety of common functional groups. In addition, the exact position of the band provides information about the other groups bonded to the carbonyl carbon. Table 13.1 lists the approximate position for the carbonyl absorption in common functional groups. If the carbonyl group is conjugated with a CC double bond or with an aromatic ring, the position of the absorption band is decreased by 20 to 40 cm^{-1}. This is due to the contribution to the resonance hybrid of a structure that has a single bond between the carbon and the oxygen:

$$CH_3-\overset{\overset{\displaystyle :\ddot{O}:}{\|}}{C}-CH=CH_2 \longleftrightarrow CH_3-\overset{\overset{\displaystyle :\ddot{O}:^-}{|}}{C}=CH-\overset{+}{C}H_2$$

If the carbonyl carbon is part of a five-membered ring, the position of the absorption band is increased by about 30 cm^{-1}.

Table 13.1   Infrared Absorption Band Positions of Various Carbonyl Groups

Functional Group		Base Position of Carbonyl Absorption (cm^{-1})
$R-\overset{\overset{\displaystyle O}{\|}}{C}-R'$	Ketone	1715
$R-\overset{\overset{\displaystyle O}{\|}}{C}-H$	Aldehyde	1730
$R-\overset{\overset{\displaystyle O}{\|}}{C}-OH$	Carboxylic acid	1710
$R-\overset{\overset{\displaystyle O}{\|}}{C}-OR'$	Ester	1740
$R-\overset{\overset{\displaystyle O}{\|}}{C}-O-\overset{\overset{\displaystyle O}{\|}}{C}-R$	Anhydride	1820 and 1750 (two bands)
$R-\overset{\overset{\displaystyle O}{\|}}{C}-Cl$	Acyl chloride	1800
$R-\overset{\overset{\displaystyle O}{\|}}{C}-NH_2$	Amide	1690–1630 (often has additional bands at slightly lower wavenumbers)

PRACTICE PROBLEM 13.1

Predict the position of the absorption band for the carbonyl group of this compound in its IR spectrum:

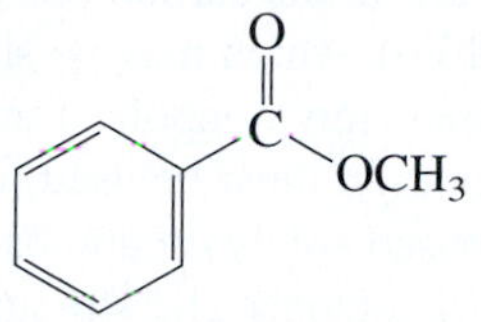

Solution

This compound is an ester, and the carbonyl group is conjugated with the aromatic ring. The predicted position is $1740 - (20 \text{ to } 40) = 1720$ to 1700 cm^{-1}.

PROBLEM 13.8

Predict the positions of the absorption bands in the IR spectra for the carbonyl groups of these compounds.

a) 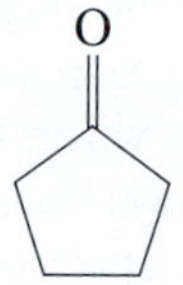 b) $CH_3CH=CHCH$ c) 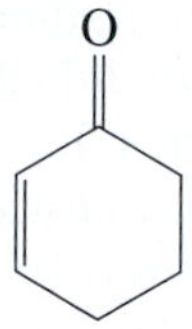 d) 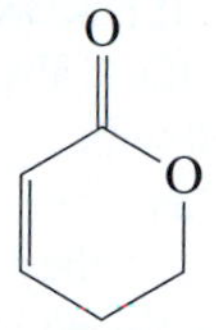 e)

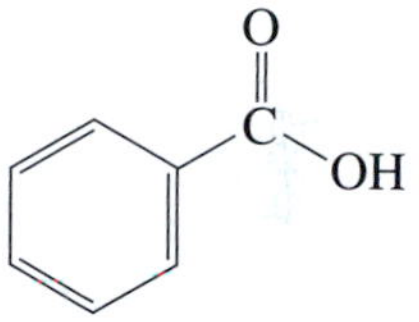

The presence of a carbonyl group in a compound is readily apparent from its IR spectrum. In addition, it is often possible to determine which carbonyl-containing functional group is present from the exact position of the peak *and* an examination of the rest of the spectrum for other absorptions associated with a particular functional group. For example, in addition to a band for the carbonyl group, carboxylic acids have a very broad band in the O—H region, and aldehydes have bands due to the C—H bond in the 2830 to 2700 cm^{-1} region (see Figures 13.8 and 13.10). Procedures to distinguish among all of these carbonyl-containing functional groups are described in Section 13.10. Remember to check that all the absorptions for a particular functional group appear in the spectrum before deciding that that functional group is actually present in the compound.

PROBLEM 13.9

Explain how IR spectroscopy could be used to distinguish between these compounds:

a) $CH_3CH_2CH_2CCH_3$ and $CH_3CH_2CH_2CH_2CH$ b) 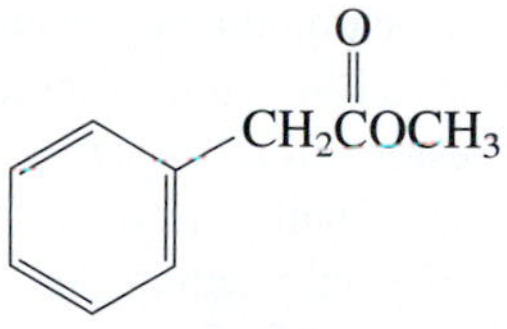   and

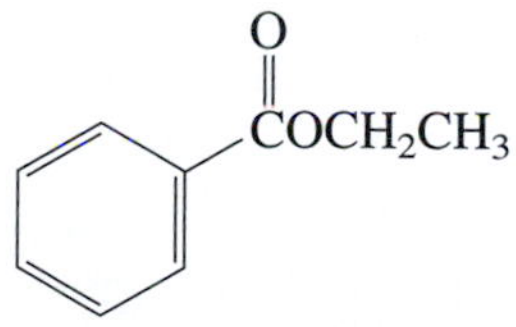

c) $CH_3CH=CHCOCH_3$ and $CH_3CH=CHCH_2COH$

Absorption bands for carbon–nitrogen double bonds occur at somewhat lower wavenumbers than those for carbonyl groups. However, compounds containing this functional group are relatively uncommon, so their IR spectra are not discussed in detail in this book.

Alkenes show an absorption due to the carbon–carbon double bond in the region of 1660 to 1640 cm^{-1} (see Figure 13.6), which may be shifted to lower wavenumbers because of conjugation. This is a relatively nonpolar functional group, so the absorption band is often weak. Therefore, caution must be used in deciding whether a CC double bond is present in a compound based solely on an absorption in this region. (Look for confirming evidence in the C—H region.) The strength of the absorption increases if the double bond is substituted with a polar group or is conjugated with a carbonyl group.

Aromatic compounds usually show four absorptions due to skeletal vibrations of the benzene ring. These occur near 1600, 1580, 1500, and 1450 cm^{-1} but vary in intensity (see Figure 13.8). Again caution must be used in assigning these bands, and confirming evidence must be found in other regions of the spectrum before the presence of an aromatic ring can be established.

A final functional group that shows absorptions in this region is the nitro group (NO$_2$). It has two strong absorption bands near 1550 and 1380 cm^{-1}. Like the carbonyl bands, these absorptions are shifted to lower wavenumbers when the nitro group is conjugated with an aromatic ring. Even though these bands appear in a region where there are many others, it is usually possible to recognize them because they always occur together and they are very strong because the nitro group is very polar. The spectrum of 2-nitrotoluene is shown in Figure 13.14.

13.9   THE FINGERPRINT REGION

The region below 1500 cm^{-1} contains numerous absorptions due to single-bond stretches and a variety of bending vibrations. Because of the large number of bands, it is difficult to assign these absorptions to a particular functional group. In this book we will make only limited use of this region. However, as its name implies, this is the region where comparison of the spectrum of an unknown to that of a known compound can establish the identity of the unknown. Although the entire spectra must match, the fingerprint region, with its numerous absorptions, provides the crucial test of identity.

One band that is often recognizable in the fingerprint region is the C—O stretch that occurs in the region of 1300 to 1000 cm^{-1}. This is a strong absorption because the bond is quite polar. Especially useful is the absorption due to the C—O single bond between the carbonyl carbon and oxygen of an ester. This band is often broader and stronger than the absorption of the carbonyl group and provides a method to distinguish between an ester and a ketone.

Finally, most aromatic compounds have one or more strong bands in the 900 to 675 cm^{-1} region. As an example, note the strong bands at 760 and 700 cm^{-1} in the spectrum of 2-phenylpropanal (Figure 13.8). The presence of these absorptions together with those in the 1600 to 1400 cm^{-1} and 3100 to 3000 cm^{-1} regions is usually sufficient to establish the presence of a benzene ring in a compound.

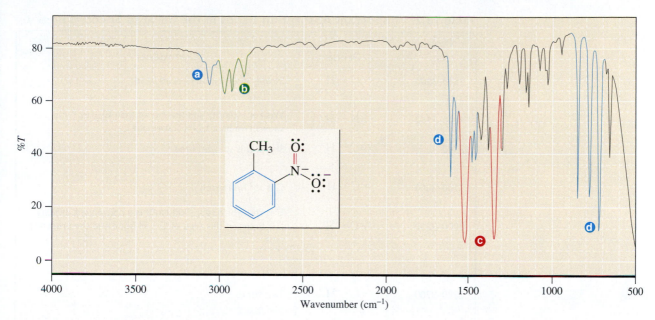

Two types of CH bonds can be detected in this spectrum:

ⓐ sp^2-hybridized CH bonds (3100–3000 cm^{-1})

ⓑ sp^3-hybridized CH bonds (3000–2850 cm^{-1})

ⓒ The absorption bands due to the nitro group: 1523 and 1347 cm^{-1}. They are at lower wavenumbers than usual because the nitro group is conjugated with the benzene ring.

ⓓ The aromatic ring is responsible for the absorptions due to the sp^2 CH stretching vibrations (3100 – 3000 cm^{-1}); the ring skeletal vibrations at 1612, 1577, 1500, and 1461 cm^{-1}; and the CH bending vibrations at 859, 788, and 728 cm^{-1}.

Nitro compounds: Compounds containing nitro groups are identified by the appearance of two strong bands near 1550 and 1380 cm^{-1}. These absorptions appear at lower wavenumbers if the nitro group is conjugated with a benzene ring.

Figure 13.14

THE INFRARED SPECTRUM OF 2-NITROTOLUENE.

13.10 INTERPRETATION OF IR SPECTRA

Interpretation of the IR spectrum of an unknown compound is an art that requires experience and practice. The more spectra you examine, the easier it will become to recognize the absorption due to an O—H group and to differentiate between that band and one that results from an N—H group.

Table 13.2 summarizes the positions of the various absorption bands that have been discussed so far. On the basis of these absorptions, it is usually possible to determine the nature of the functional group that is present in the compound whose spectrum is being considered. Many functional groups require the presence of several characteristic absorptions, whereas the absence of a band in a particular region of the spectrum can often be used to eliminate the presence of a particular group.

Infrared spectra of compounds belonging to each of the major functional group classes are provided in the figures in this chapter. Each figure has a summary of the im-

Table 13.2 **Important Absorption Bands in the Infrared Spectral Region**

Position (cm^{-1})	Group	Comments
3550–3200	—O—H	Strong intensity, very broad band
3400–3250	—N—H	Weaker intensity and less broad than O—H; NH$_2$ shows two bands, NH shows one
3300	≡C—H	Sharp, C is *sp* hybridized
3100–3000	=C—H	C is *sp*^{2} hybridized
3000–2850	—C—H	C is *sp*^{3} hybridized; 3000 cm^{-1} is a convenient dividing line between this type of C—H bond and the preceding type
2830–2700	O‖—C—H	Two bands
2260–2200	—C≡N	Medium intensity
2150–2100	—C≡C—	Weak intensity
1820–1650	O‖—C—	Strong intensity, exact position depends on substituents; see Table 13.1
1660–1640	C=C	Often weak intensity
1600–1450	 ⬡	Four bands of variable intensity
1550 and 1380	—NO$_2$	Two strong intensity bands
1300–1000	—C—O—	Strong intensity
900–675	 ⬡	Strong intensity

portant absorption bands for that functional group. Some of these figures are found on previous pages; others appear on later pages. Table 13.3 provides a list of the important functional groups and the figure(s) that show IR spectra of typical compounds containing that functional group. Now that all of the important absorption bands have been discussed, this is a good time for you to examine all of these spectra to become more familiar with the combination of bands caused by each functional group.

Table 13.3   IR Spectra of Functional Groups

Functional Group	Figure	Page	Functional Group	Figure	Page
Alkane	13.5	509	Alkene	13.6	510
Alkyne	13.7	510	Arene	13.8	511
				13.14	521
Alcohol	13.9	512		13.15	523
				13.18	526
Ether	13.16	524		13.22	529
Primary amine	13.11	513			
			Nitro compound	13.14	521
Secondary amine	13.12	514	Nitrile	13.13	515
Tertiary amine	13.17	525			
			Aldehyde	13.8	511
Ketone	13.4	507	Carboxylic acid	13.10	513
	13.18	526			
			Anhydride	13.20	528
Ester	13.19	527			
Amide	13.21	528	Acyl chloride	13.22	529

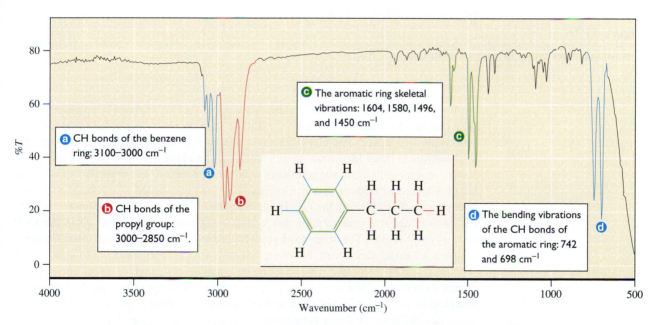

Arenes: Arenes that have hydrogens on the aromatic ring show absorptions in the 3100 and 3000 cm⁻¹ region. They also show four absorptions of variable intensity near 1600, 1580, 1500, and 1450 cm⁻¹ due to skeletal vibrations of the benzene ring. In addition, most aromatic compounds have at least one strong absorption in the 900 and 675 cm⁻¹ region due to bending vibrations of the CH bonds of the aromatic ring. Although care must be used in assigning some of these bands because they occur in the fingerprint region, the presence of all of them provides evidence that the compound contains an aromatic ring.

Figure 13.15

THE INFRARED SPECTRUM OF PROPYLBENZENE.

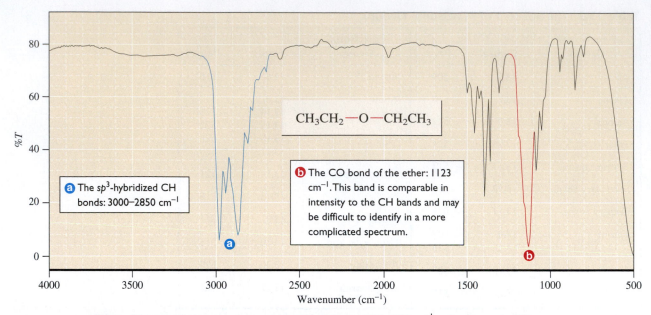

Ethers: Ethers usually have one or more strong bands in the 1300–1000 cm^{-1} region due to the CO bond. As was the case with alcohols, this band is often difficult to identify because it occurs in the fingerprint region with many other absorptions. If oxygen is known to be present from the formula of a compound, the presence of an ether can be inferred by the absence of absorptions due to OH, carbonyl, or other possible oxygen-containing functional groups.

Figure 13.16

THE INFRARED SPECTRUM OF DIETHYL ETHER.

It is usually possible to distinguish among the various carbonyl-containing functional groups by careful examination of the IR spectrum. Of course, all have carbonyl absorptions. However, acid chlorides show this band at unusually high wavenumbers; acid anhydrides have two bands in this region; and amides have this band at unusually low wavenumbers, often have additional bands in this region, and may also show N—H bands. Carboxylic acids are easily recognized by their very broad O—H band, and aldehydes are distinguished by the C—H bands in the 2830 to 2700 cm^{-1} region. To identify an ester, the strong absorption due to the C—O bands must be located in the 1300 to 1000 cm^{-1} region. The absence of any of these other features suggests that the compound is a ketone. Recognize, however, that assignment of an unknown to a particular functional group class must sometimes be only tentative. In such cases, confirming evidence must be obtained from other sources, such as other spectroscopic techniques or chemical tests. Be careful not to become so sure of a functional group assignment that other, contradictory evidence is ignored.

In attempting to identify the functional group that is present in an unknown compound from its IR spectrum, it is usually best to begin by examining the region from 4000 to 2700 cm^{-1} to determine what type of bonds involving hydrogen are present. Then the region from 2300 to 2100 cm^{-1} should be examined to look for indications of the presence of triple-bond groups. Next, you should look for absorptions due to carbonyl groups,

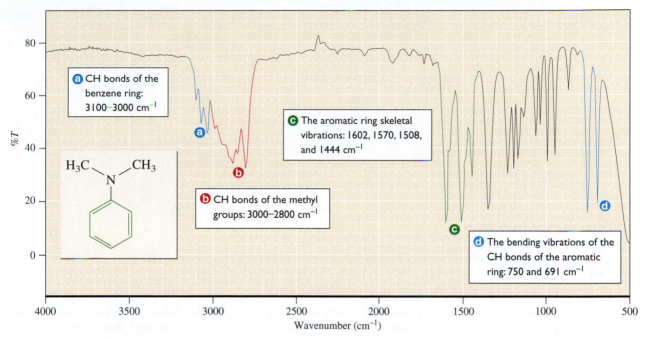

a CH bonds of the benzene ring: 3100–3000 cm^{-1}

c The aromatic ring skeletal vibrations: 1602, 1570, 1508, and 1444 cm^{-1}

b CH bonds of the methyl groups: 3000–2800 cm^{-1}

d The bending vibrations of the CH bonds of the aromatic ring: 750 and 691 cm^{-1}

%T

Wavenumber (cm^{-1})

Only the alkyl groups and the aromatic ring can be detected in the spectrum of this tertiary amine. The absorptions due to the aromatic ring skeletal vibrations are stronger than usual, suggesting that the ring is substituted with a polar substituent.

Tertiary amines: Tertiary amines do not have a NH bond, so there is no evidence for the amine group in the 3400–3250 cm^{-1} region. Because the CN bond-stretching vibration is difficult to assign in the fingerprint region, tertiary amines are not readily identified from their IR spectra. Chemical tests are helpful in such cases.

Figure 13.17

THE INFRARED SPECTRUM OF *N,N*-DIETHYLANILINE.

CC double bonds, aromatic rings, and nitro groups in the 1800 to 1350 cm^{-1} region. Finally, you should look for C—O absorptions in the 1300 to 1000 cm^{-1} region and for aromatic ring bands in the 900 to 675 cm^{-1} region.

Two cautions must be given. First, do not overinterpret the spectrum. Be very careful in assigning the presence of a functional group when only weak bands occur in the appropriate region. It is helpful to compare the spectrum of the unknown to that of a known compound that has that same functional group. Second, make sure that your conclusions are consistent with all of the data. A spectrum that has two bands in the 2830 to 2700 cm^{-1} region cannot be that of an aldehyde unless it also shows an absorption due to a carbonyl group.

Let's try a problem. The IR spectrum of an unknown compound is shown in Figure 13.23. First, let's examine the hydrogen region. The absence of absorptions in the 3600 to 3100 cm^{-1} region indicates that the compound does not have any O—H or N—H groups. The bands in the 3100 to 3000 cm^{-1} region indicate the presence of

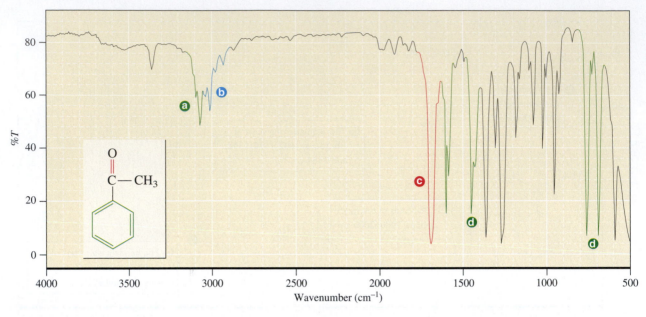

ⓐ CH bonds of the benzene ring:
3100–3000 cm⁻¹

ⓑ CH bonds of the methyl group:
3000–2850 cm⁻¹

ⓒ The carbonyl group: 1685 cm⁻¹. This is
the position predicted for a carbonyl group
of a ketone that is conjugated to an aromatic
ring, 1715 − (20 to 40) = 1695 to 1675 cm⁻¹.

ⓓ The aromatic ring vibrations: 1599, 1579, 1449,
760, and 691 cm⁻¹. Note that the band around
1500 cm⁻¹ is very weak in this spectrum.

Ketones: The carbonyl of a ketone has an absorption band near 1715 cm⁻¹. This band is shifted to
lower wavenumbers if the carbonyl group is conjugated, and it is shifted to higher wavenumbers if the
carbonyl group is part of a five-membered ring. Ketones have no other characteristic bands and often
can only be distinguished from the other carbonyl-containing functional groups by the absence of the
bands required for those other groups. Again, chemical tests can be very useful in confirming the
presence of a ketone.

Figure 13.18

THE INFRARED SPECTRUM OF ACETOPHENONE.

hydrogens bonded to sp^2-hybridized carbons, so the compound must have one or more
CC double bonds. The absorptions in the 3000 to 2850 cm⁻¹ region indicate that there
are also hydrogens bonded to sp^3-hybridized carbons in the compound. Next, examination of the triple-bond region shows no indications of the presence of any triple-
bonded functional group. Continuing to the double-bond region, the strong
absorption at 1722 cm⁻¹ indicates the presence of a carbonyl group. This is not part
of a carboxylic acid (no O—H) or an aldehyde (absence of absorptions in the
2830–2700 cm⁻¹ region). Nor does the unknown appear to be an amide (no N—H,
carbonyl absorption too high), an anhydride (absence of a second carbonyl band), or
an acyl chloride (carbonyl position too low). This leaves a ketone or an ester as possibilities. The strong absorption at 1282 cm⁻¹ suggests that the unknown is an ester.
The bands at 1607, 1591, 1489, and 1437 cm⁻¹ along with the absorptions at 3100 to
3000 and 746 cm⁻¹ suggest the presence of an aromatic ring. The carbonyl of the ester

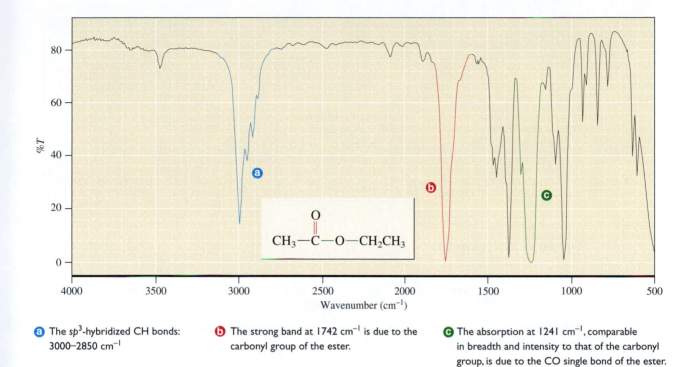

a The sp^3-hybridized CH bonds: 3000–2850 cm^{-1}

b The strong band at 1742 cm^{-1} is due to the carbonyl group of the ester.

c The absorption at 1241 cm^{-1}, comparable in breadth and intensity to that of the carbonyl group, is due to the CO single bond of the ester.

Esters: Esters show a carbonyl band near 1740 cm^{-1}. They also show a strong absorption in the CO single bond region, from 1300–1000 cm^{-1}, that can be used to differentiate them from ketones. This band is usually of comparable breadth and intensity to the carbonyl band. (However, a compound containing both a ketone and an ether group also has both of these absorptions.) Chemical tests can be very useful in distinguishing a ketone from an ester.

Figure 13.19

THE INFRARED SPECTRUM OF ETHYL ACETATE.

occurs at slightly lower wavenumbers (1722 cm^{-1}) than the usual position (1740 cm^{-1}), indicating that it might be conjugated. Therefore, we conclude that the unknown is probably an ester, that it may have an aromatic ring, and that the ester may be conjugated (with the aromatic ring?). However, these conclusions must be considered tentative until confirming evidence is obtained from other sources. The unknown is actually methyl 3-methylbenzoate:

Methyl 3-methylbenzoate

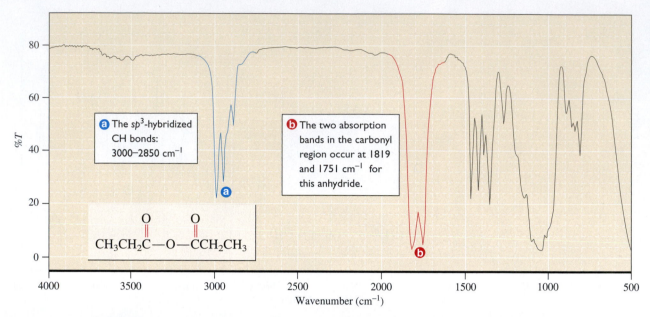

Anhydrides: Anhydrides are characterized by the presence of two bands in the carbonyl region, one near 1820 and one near 1750 cm^{-1}.

Figure 13.20

THE INFRARED SPECTRUM OF PROPANOIC ANHYDRIDE.

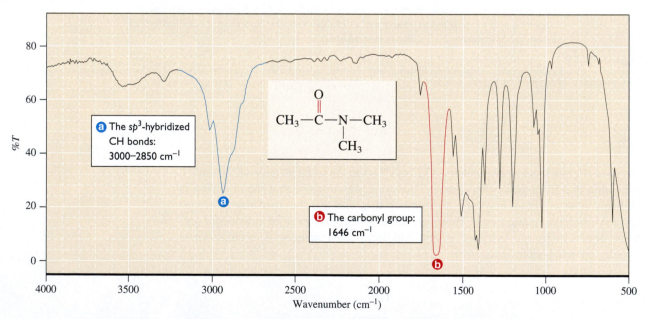

Amides: The absorption for the carbonyl group of an amide appears in the region of 1690–1630 cm^{-1}, lower wavenumbers than most other carbonyl bands. In addition, other relatively strong bands often appear at slightly lower wavenumbers. Amides derived from ammonia or primary amines have bands in the hydrogen region due to their NH bonds.

Figure 13.21

THE INFRARED SPECTRUM OF *N,N*-DIMETHYLACETAMIDE.

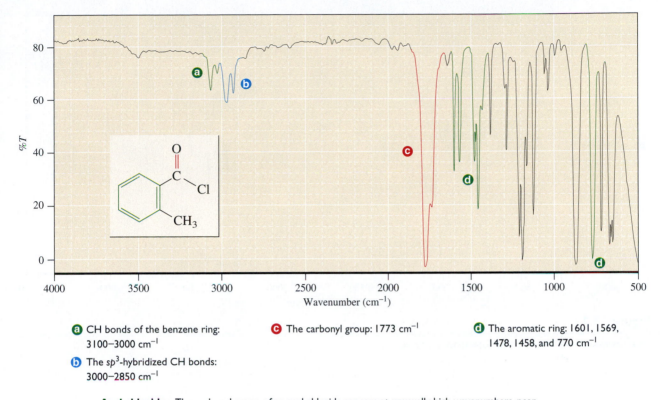

a CH bonds of the benzene ring: 3100–3000 cm⁻¹

c The carbonyl group: 1773 cm⁻¹

d The aromatic ring: 1601, 1569, 1478, 1458, and 770 cm⁻¹

b The sp^3-hybridized CH bonds: 3000–2850 cm⁻¹

Acyl chlorides: The carbonyl group of an acyl chloride appears at unusually high wavenumbers, near 1800 cm⁻¹. Because the carbonyl group is conjugated with the benzene ring in this compound, it is shifted to lower wavenumbers by about 30 cm⁻¹.

Figure 13.22

THE INFRARED SPECTRUM OF O-TOLUYL CHLORIDE.

The exact identity of an unknown cannot be established only on the basis of its IR spectrum (unless, of course, the spectrum of a known exactly matches the spectrum of the unknown). However, information about the functional group that is present in a compound is available. The presence or absence of the following groups can be determined:

O—H	N—H	Type of C—H
C≡C	C≡N	
Type of C=O	C=C	Aromatic ring
C—O	NO$_2$	

The IR spectrum provides little information about the hydrocarbon part of the compound. However, this is exactly the information provided by nuclear magnetic resonance spectroscopy, discussed in Chapter 14. The combination of these two types of spectroscopy is of enormous value in organic chemistry.

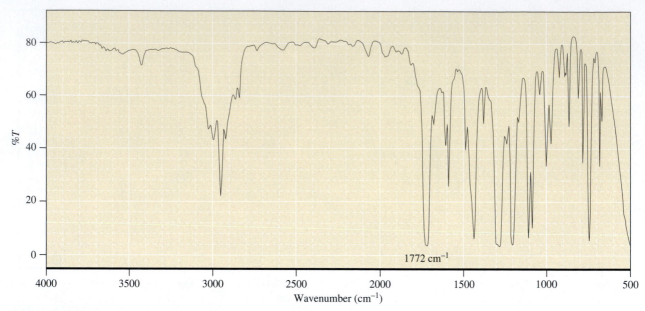

Figure 13.23

THE INFRARED SPECTRUM OF AN UNKNOWN COMPOUND.

PROBLEM 13.10

Predict the positions of the major absorption bands in the IR spectra of these compounds:

a) $CH_3CH{=}CHCCH_3$ (with $C{=}O$)

b) benzene with CH_2NH_2 substituent

c) cyclohexene with CH_3O substituent

d) $CH_3CH_2CH_2OH$

e) $CH_2CH_2C{\equiv}C{-}H$ with NO_2

f) benzaldehyde with CH_3 substituent ($C{-}H$, O)

PROBLEM 13.11

Explain how IR spectroscopy could be used to distinguish between these compounds:

a) $CH_3CH_2CH{=}CH_2$ and $CH_3CH_2C{\equiv}CH$

b) $CH_3{-}\underset{\underset{CH_3}{|}}{\overset{\overset{CH_3}{|}}{C}}{-}OH$ and   benzene with CH_2OH substituent

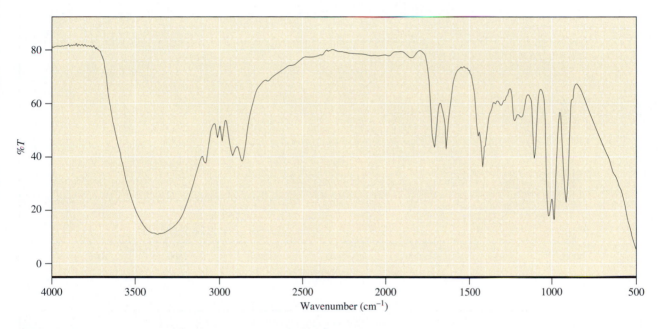

c) 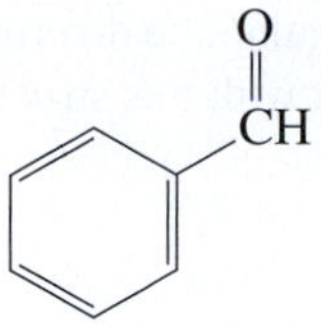 and

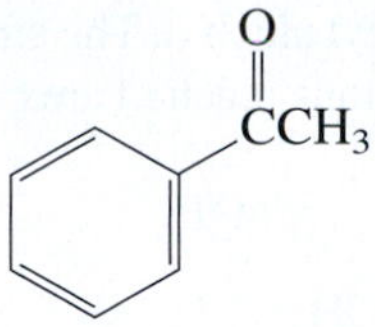

d) $CH_3CH_2CH_2CH_2NH_2$ and $CH_3CH_2NHCH_2CH_3$

PRACTICE PROBLEM 13.2

Explain which functional groups are present in this compound on the basis of its IR spectrum:

Solution

The broad absorption centered near 3300 cm⁻¹ indicates the presence of a hydroxy group. The absorption at 3005 cm⁻¹ suggests the presence of H's bonded to sp^2-hybridized C's. (Note that you are not expected to read peak positions this exactly from any of these spectra.) This is supported by the absorption for a CC double bond at 1646 cm⁻¹. The absorptions in the region of 3000 to 2850 cm⁻¹ indicate the presence of H's bonded to sp^3-hybridized C's. Although the compound has a CC double bond, there is no indication of the presence of an aromatic ring due to the absence of the four bands in the 1600 to 1450 cm⁻¹ region and the absence of bands in the 900 to 675 cm⁻¹ region.

   In summary, the structural features that can be identified from the IR spectrum are as follows:

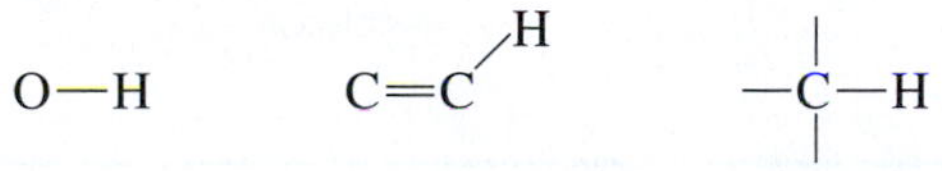

$$O-H \qquad C=C^{\diagup H} \qquad -\overset{|}{\underset{|}{C}}-H$$

(This is the spectrum of 2-propen-1-ol or allyl alcohol. The structure cannot be determined only from this IR spectrum, but the conclusions reached are consistent with this structure.)

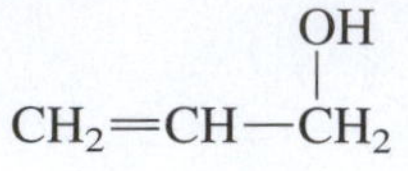

$$CH_2{=}CH{-}CH_2\!\!-\!\!\overset{\displaystyle OH}{|}$$

2-Propen-1-ol (allyl alcohol)

PROBLEM 13.12

Explain which functional groups are present in these compounds on the basis of their IR spectra:

a)

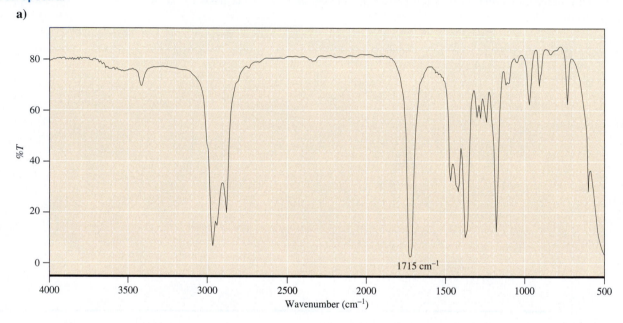

b)

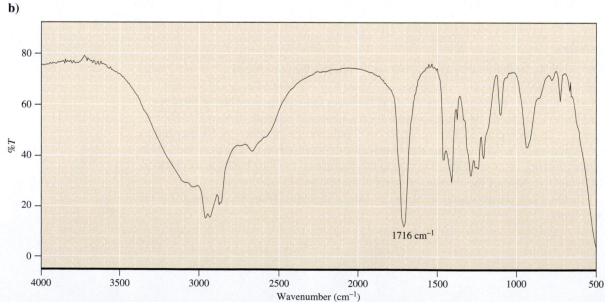

c)

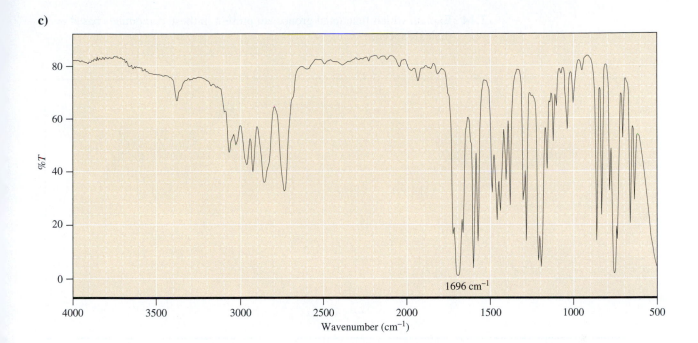

Review of Mastery Goals

After completing this chapter, you should be able to:

■ Predict the important absorption bands in the IR spectrum of a compound. (Problems 13.13, 13.15, 13.26, 13.27, 13.28, and 13.29)

■ Determine the functional group that is present in a compound by examination of its infrared spectrum. (Problems 13.14, 13.16, 13.17, 13.18, 13.19, 13.20, 13.21, 13.22, 13.23, 13.24, and 13.25)

Additional Problems

13.13 List the positions of the important absorption bands in the IR spectra of these compounds:

a) HC≡CCH$_2$CH$_2$NH$_2$

b) HOCCH$_2$CH$_2$CH$_2$C≡N
    (with O double-bonded above the C)

c) CH$_3$CH$_2$COCH$_2$CH$_2$CH$_3$
    (with O double-bonded above the second C)

d) CCH$_2$CH$_3$ (phenyl group attached, with O double-bonded above the C)

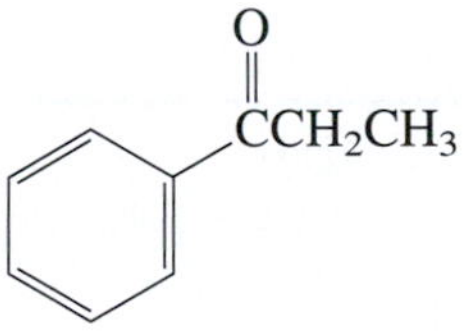

13.14 Explain which functional groups are present in these compounds based on their IR spectra:

a)

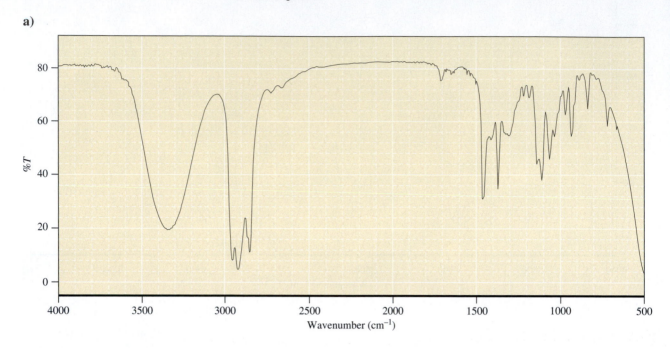

b)

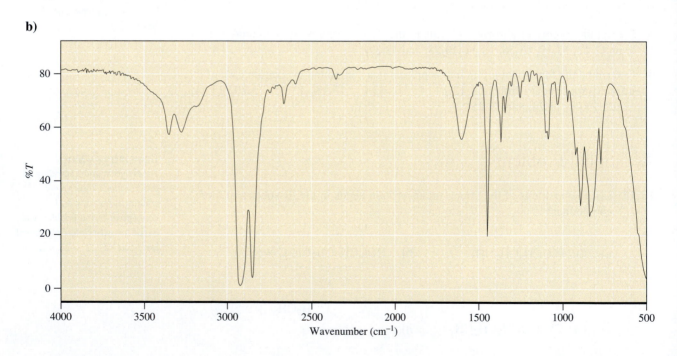

c)

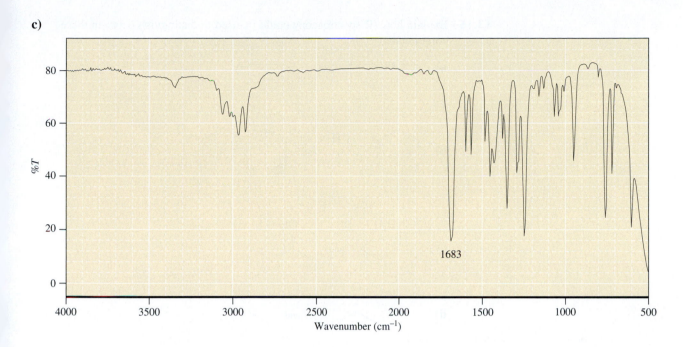

d)

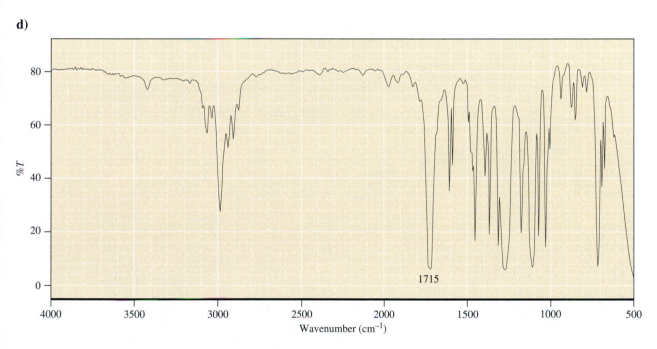

13.15 Explain how IR spectroscopy could be used to distinguish between these compounds:

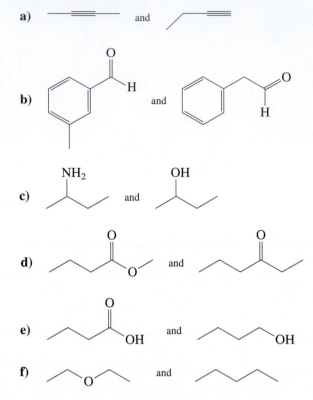

a) [structure] and [structure]

b) [structure] and [structure]

c) [structure] NH_2 and [structure] OH

d) [structure] and [structure]

e) [structure] OH and [structure] OH

f) [structure] and [structure]

13.16 Explain which functional group(s) is present in the compound that has this IR spectrum:

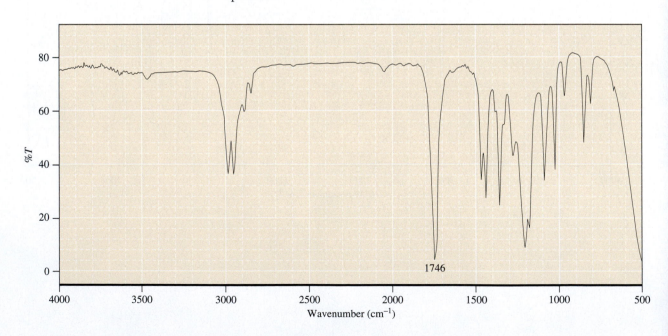

13.17 Explain which functional group(s) is present in the compound that has this IR spectrum:

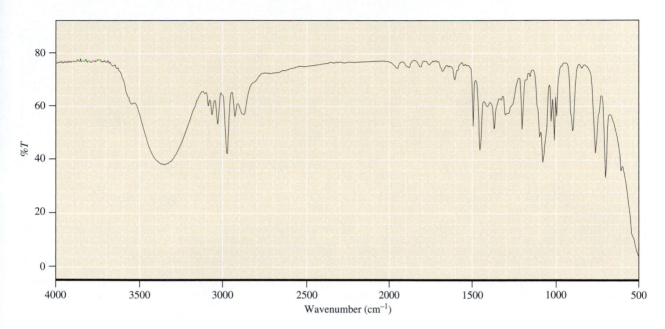

13.18 Explain which functional group(s) is present in the compound that has this IR spectrum:

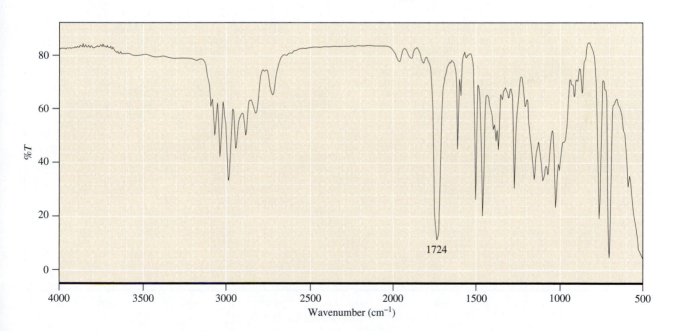

13.19 Explain which functional group(s) is present in the compound that has this IR spectrum:

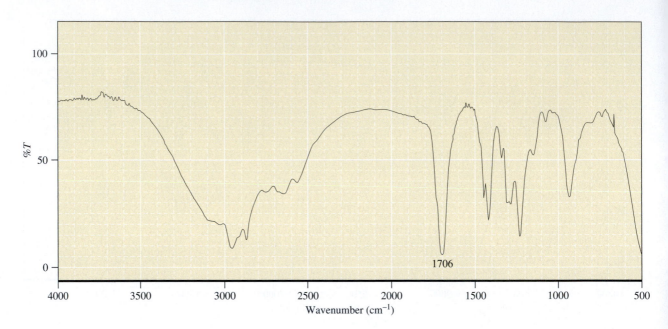

13.20 Explain which functional group(s) is present in the compound that has this IR spectrum:

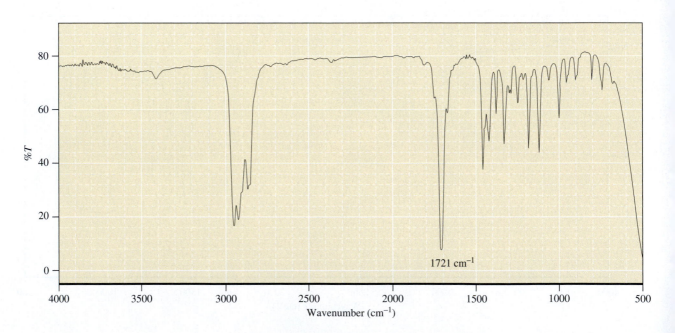

13.21 Suggest a possible structure for a compound with the formula $C_5H_{12}O$ that has the following IR spectrum and explain your reasoning:

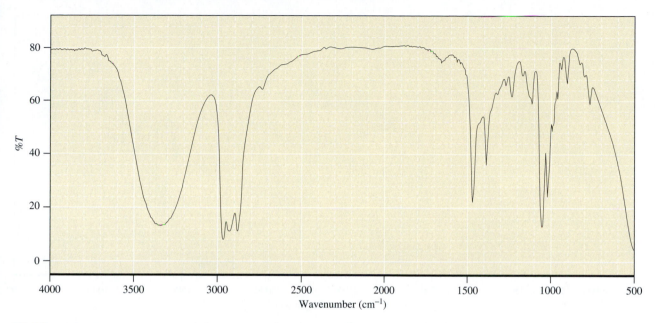

13.22 Suggest a possible structure for a compound with the formula $C_6H_{12}O_2$ that has the following IR spectrum and explain your reasoning:

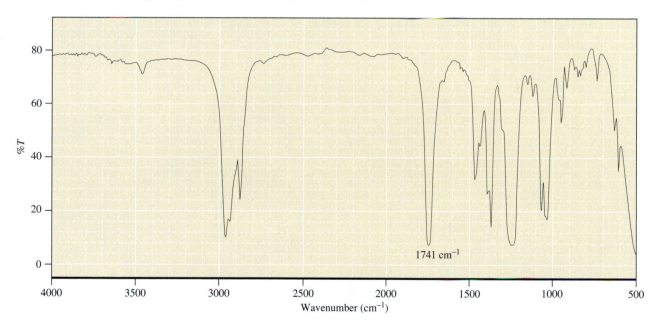

1741 cm^{-1}

13.23 Suggest a possible structure for a compound with the formula $C_5H_7NO_2$ that has the following IR spectrum and explain your reasoning:

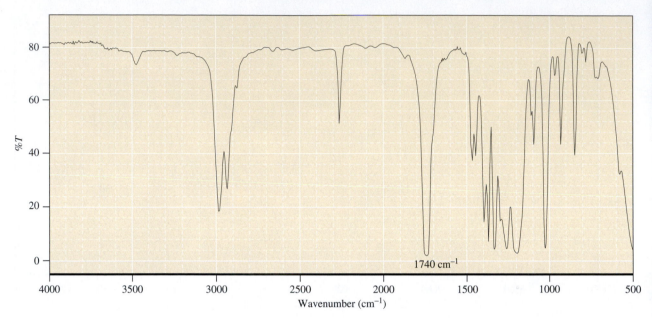

13.24 Suggest a possible structure for a compound with the formula $C_7H_{12}O$ that has the following IR spectrum and explain your reasoning:

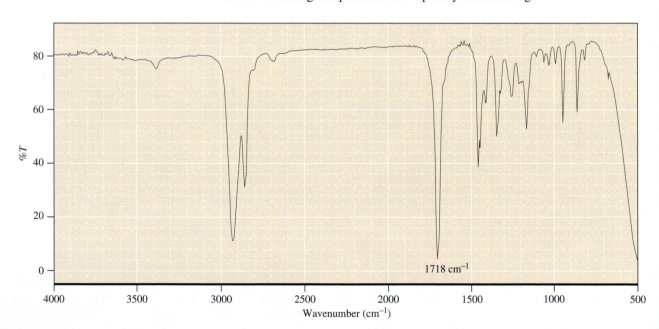

13.25 Suggest a possible structure for a compound with the formula $C_9H_{10}O$ that has the following IR spectrum and explain your reasoning:

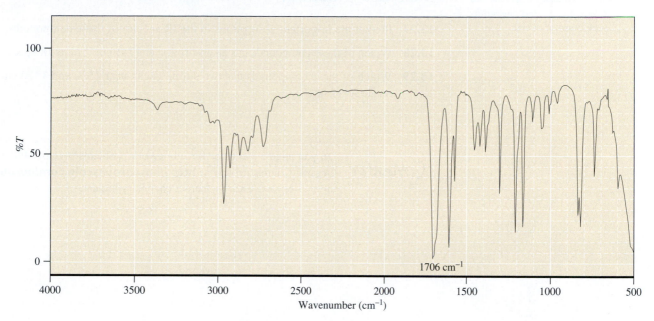

1706 cm⁻¹

%T

Wavenumber (cm⁻¹)

13.26 Forensic laboratories often have to identify various illicit drug samples. Explain how IR spectroscopy could be used to help distinguish between morphine and heroin.

BioLink

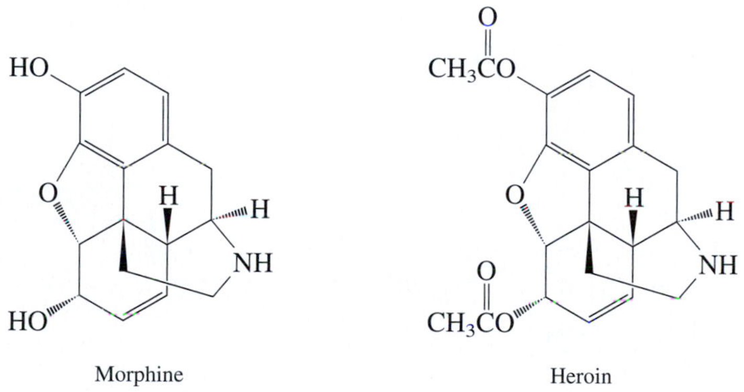

Morphine

Heroin

Problems Using Online Three-Dimensional Molecular Models

13.27 Predict the important absorptions in the IR spectra of these compounds.

13.28 Which of these compounds has an absorption at 1741 cm⁻¹ in its IR spectrum?

13.29 Which of these compounds has two strong peaks near 1520 and 1350 cm⁻¹ in its IR spectrum?

Do you need a live tutor for homework problems? Access vMentor at Organic ChemistryNow at **http://now.brookscole.com/hornback2** for one-on-one tutoring from a chemistry expert.

Nuclear Magnetic Resonance Spectroscopy

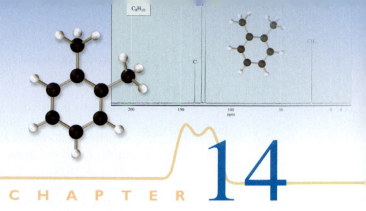

A S WE SAW IN CHAPTER 13, infrared (IR) spectroscopy can be used to determine which functional group is present in a compound. This chapter discusses proton and carbon-13 nuclear magnetic resonance (NMR) spectroscopy. These techniques complement IR spectroscopy because they provide information about the hydrocarbon part of the molecule. The combination of IR and NMR spectroscopy often provides enough data to determine the structure of an unknown compound.

Proton magnetic spectroscopy is discussed first. After a brief discussion of the theory behind this technique, the use of ^{1}H-NMR spectroscopy to determine the structure of the hydrocarbon part of the compound is described. Next the use of ^{13}C-NMR spectroscopy to gain information about the carbons in the compounds is presented. Finally, examples of the use of various combinations of these techniques to identify unknown organic compounds are discussed.

14.1 PROTON MAGNETIC RESONANCE SPECTROSCOPY

The nuclei of certain isotopes of some elements have two (or more) energy states available when they are in a magnetic field. The transitions between these energy states can be investigated by using the technique of NMR spectroscopy. Although many nuclei exhibit this phenomenon, the two that are of most use to organic chemists are the hydrogen nucleus (a proton, ^{1}H) and the nucleus of the isotope of carbon with an atomic mass number of 13 (^{13}C).

Proton magnetic resonance spectroscopy provides information about the relative numbers of different kinds of hydrogens in the compound, the nature of the carbons bonded to them, and which hydrogens are nearby. From this information it is possible to get a good idea about the structure of the hydrocarbon part of the compound. In combination with the knowledge of the functional group obtained from the IR spectrum, the NMR spectrum often enables the structure of a compound to be assigned with certainty.

Before we look at the theory behind the NMR technique, let's look at a sample spectrum to find out what kind of information it provides. Figure 14.1 shows the ¹H-NMR spectrum of 3-pentanone. Note that the absorption peaks extend up from the baseline at the bottom of the spectrum, in contrast to IR spectra.

Three types of information are present in a NMR spectrum:

1. The position on the *x*-axis. Called the **chemical shift,** this provides information about the carbon (or other atom) to which the hydrogen is attached.

2. The number of peaks in each group. Called the **multiplicity,** this provides information about the other hydrogens that are near the hydrogen or hydrogens that produce the peaks.

3. The **integral.** The area under a group of peaks is proportional to the number of hydrogens that produce that group of peaks.

Many NMR spectrometers provide the integral in the form of a line drawn on the spectrum that increases in height in proportion to the total area of the peaks. The height of each "step" is proportional to the number of hydrogens that produce the peaks under

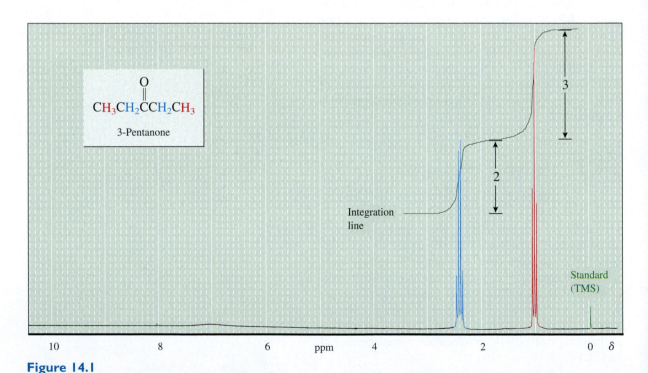

Figure 14.1

THE ¹H-NMR SPECTRUM OF 3-PENTANONE.

that step. The heights must be put into small whole number ratios to provide the relative numbers of each type of hydrogen. To simplify interpretation, the spectra in this book have already been integrated, and the relative numbers of hydrogens for each group of peaks are provided. However, remember that the actual numbers of hydrogens can be multiples of the numbers provided by the integrals.

14.2 THEORY OF ¹H-NMR

The following discussion of the theory behind ¹H-NMR is adequate for our needs. The charge of some nuclei, including ¹H, "spins" on the nuclear axis. This spinning charge generates a small magnetic field. For the purposes of further discussion, the nucleus can be considered as a small bar magnet.

The hydrogen nucleus has two possible spin states that have identical energies under normal circumstances. However, if the atoms are placed in an external magnetic field—that is, between the poles of a large magnet in the laboratory—then the spin states have different energies. As shown in Figure 14.2, one state has its magnetic field oriented in the same direction as the external magnetic field, B_0, and is lower in energy than the other state that has its field oriented in opposition to the external field. With two states of different energy, spectroscopy can be done.

The difference in energy between the two states is related to the strength of the external magnetic field by the equation

$$\Delta E = h\gamma B_0/2\pi$$

where B_0 is the strength of the external magnetic field and γ is the magnetogyric ratio, which differs for each kind of atomic nucleus.

The magnetogyric ratio is extremely small. Therefore, ΔE is very small even when very large magnets are used. For example, early NMR spectrometers employed a 14,000-gauss (or 1.4-tesla) magnet. (The magnetic field of the Earth is about 0.5 gauss, whereas small magnets, such as those used to hold notes on a refrigerator, have fields

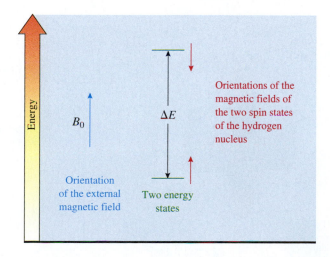

Figure 14.2

ENERGY LEVEL DIAGRAM FOR THE TWO SPIN STATES OF A HYDROGEN NUCLEUS IN A MAGNETIC FIELD.

of hundreds of gauss.) For these instruments, ΔE is about 10^{-6} kcal/mol, requiring radiation with a frequency of 60×10^6 s^{-1} or 60 MHz. Because the two energy states differ by such a small amount of energy, the number of nuclei with the lower energy is only slightly larger than the number with the higher energy.

Better spectra are obtained with larger ΔE's, which require stronger magnetic fields. Therefore, many current NMR instruments use superconducting magnets, cooled with liquid helium, with fields that are substantially greater than 1.4 tesla. Instruments that operate at 200 MHz (4.67-tesla magnet) and 400 MHz (9.33-tesla magnet) are relatively common, and some with even larger magnets, although expensive, are also available. The spectra in this book were obtained on instruments operating at 200 or 400 MHz.

Other nuclei, such as ^{13}C, ^{19}F, ^{2}H, and ^{31}P, also have nuclear spins and can be studied with NMR techniques. However, because γ is different for these nuclei, they appear in a very different region of the spectrum from hydrogen and are not seen in a ^{1}H-NMR spectrum. Both ^{12}C and ^{16}O, which are very common in organic compounds, do not have nuclear spin and therefore have no NMR absorptions.

Figure 14.3 shows a schematic diagram of a NMR spectrometer. The sample is dissolved in a suitable solvent and placed in a thin glass tube. The tube is placed between the poles of a powerful magnet. In the original type of instrument, the frequency of the electromagnetic radiation in the radio region is held constant and the strength of the magnetic field is slowly varied. When the magnetic field strength is such that the difference between the energy states of the hydrogen nucleus matches the energy of the radiation, the hydrogen absorbs the energy of the radiation and is said to be in resonance. It takes several minutes to scan the entire spectrum. Because of the time it takes to obtain a spectrum, this type of instrument is seldom used today.

In a modern NMR instrument, the magnetic field is held constant and the sample is irradiated with a brief pulse of radio-frequency irradiation. All the nuclei are excited simultaneously. As the nuclei return to their equilibrium population, a complex signal is generated. A computer converts this signal to a normal spectrum using a mathematical treatment called a Fourier transformation (FT). The advantage of an FT-NMR is that a spectrum can be obtained within a few seconds. This allows signal averaging to be used to increase the quality of the spectrum. In signal averaging, many spectra are summed by the computer. Because noise in each spectrum is random, it tends to cancel, whereas

Figure 14.3

A SCHEMATIC DIAGRAM OF A **NMR** SPECTROMETER.

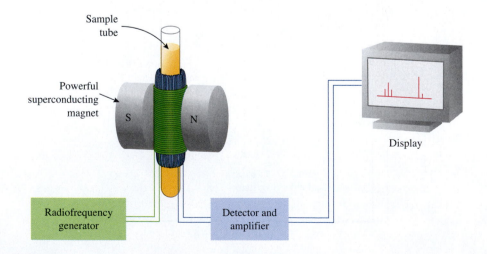

Sample tube

Powerful superconducting magnet

S   N

Radiofrequency generator

Detector and amplifier

Display

the signals of the nuclei get stronger as the spectra are added. In this way it is possible to get a good spectrum with a very small amount of compound in a relatively short period of time.

If all of the hydrogen nuclei in a compound absorbed at identical magnetic field strengths, the NMR technique would not be very useful. However, the exact field required for resonance depends on the local environment around the hydrogen. The electrons in the molecule circulate and create magnetic fields that oppose the external magnetic field. Therefore, the magnetic fields due to the electrons partially screen the hydrogen from the external magnetic field. Because the electron density varies throughout a molecule, different hydrogens require different field strengths to absorb the fixed frequency radiation. The variation is quite small, about 2500 Hz on an instrument that operates at 200 MHz, or about 0.001%, so a NMR spectrum must be obtained at very high resolution to be able to detect these differences.

14.3 THE CHEMICAL SHIFT

As discussed earlier, the chemical shift of a hydrogen signal—that is, the field required for the hydrogen to be in resonance—varies slightly with the chemical environment of the hydrogen. To measure chemical shifts, a small amount of a reference compound, usually tetramethylsilane (TMS), is added to the sample. The separation, in hertz, between the peak of interest and the peak due to TMS is measured. TMS is chosen as the reference because it has only one NMR peak and this peak occurs in a region of the spectrum where it does not usually overlap with other absorptions. Figure 14.4 illustrates the use of TMS as a reference compound in the spectrum of acetone.

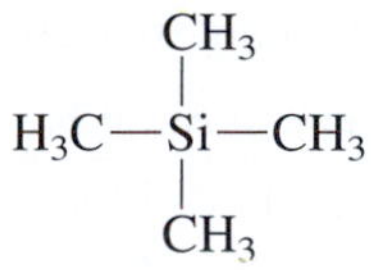

$$\begin{array}{c} CH_3 \\ | \\ H_3C-Si-CH_3 \\ | \\ CH_3 \end{array}$$

Tetramethylsilane (TMS)

In the plot of a typical NMR spectrum, the field strength increases from left to right. The protons of TMS absorb at higher field than most other protons, so the TMS signal occurs at the right edge of the spectrum, as can be seen in Figure 14.4. The signals for most other types of hydrogens appear to the left of the TMS peak. Left on a NMR spectrum is termed the **downfield** direction; right is termed the **upfield** direction. The absorption for acetone occurs 436 Hz downfield from TMS in this spectrum.

Recall that the energy separation of the two nuclear spin states of the hydrogen is directly proportional to the magnetic field strength, B_0, of the NMR instrument. This means that the chemical shift, in hertz, also is directly proportional to the magnetic field strength. On an instrument with a 4.67-T magnet, which operates at a frequency of 200 MHz, the peak for acetone occurs 436 Hz downfield from TMS. On a spectrometer with a magnet that is twice as strong, which operates at a frequency of 400 MHz, the peak for acetone occurs 872 Hz downfield from TMS, farther downfield by a factor of 2. Chemical shifts that do not depend on the particular instrument that is used to acquire the spectrum are obtained by dividing the chemical shift, in hertz, by the operating frequency of the instrument. The result is multiplied by 10^6 to get a number of a more con-

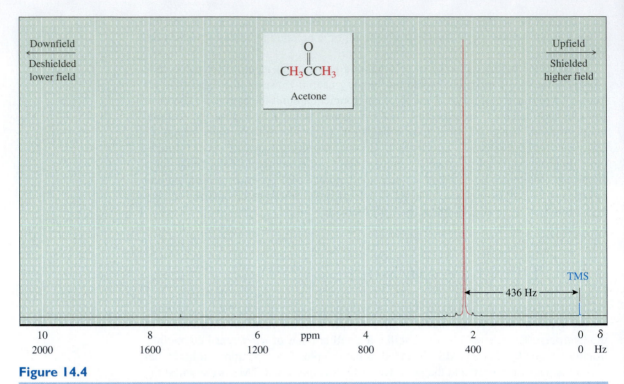

Figure 14.4

THE ¹H-NMR SPECTRUM OF ACETONE AT 200 MHz.

venient magnitude. The resulting values for chemical shifts, called **parts per million** (ppm) or δ, do not depend on the operating frequency of the instrument.

$$\delta = \frac{\text{Observed position of peak (Hz)}}{\text{Operating frequency of instrument (Hz)}} \times 10^6$$

The peak for acetone on a 200-MHz instrument occurs at 436 Hz, so the chemical shift is 2.18 δ:

$$\frac{(436 \text{ Hz})(10^6)}{(200 \times 10^6 \text{ Hz})} = 2.18 \ \delta$$

This peak also occurs at 2.18 δ on any other instrument, regardless of operating frequency. Figure 14.4 also shows δ values.

PROBLEM 14.1

The absorption for the hydrogens of benzene appears 444 Hz downfield from TMS on an instrument that operates at 60 MHz.

a) Calculate the position of this absorption in δ units.

b) Calculate the position of this absorption, in hertz, on 200- and 400-MHz instruments.

c) What is the position of this absorption, in δ units, on a 400-MHz instrument?

As discussed previously, the chemical shift of a hydrogen in a molecule is affected by the electrons surrounding it. The moving electrons generate their own small magnetic field that usually opposes the external magnetic field. The electrons shield the hydrogen from the external magnetic field, so the strength of the field experienced by the hydrogen is decreased. A hydrogen surrounded by a large amount of electron density requires a stronger external magnetic field to reach the resonance condition. Such a hydrogen is said to be *shielded*, and its absorption appears at a more *upfield* chemical shift (lower δ). Similarly, a hydrogen surrounded by a small amount of electron density is *deshielded*, and its absorption appears at a more *downfield* chemical shift (higher δ).

Inductive Effects

Consider a hydrogen bonded to a carbon. Its chemical shift depends on the other atoms bonded to that carbon. If these atoms are electronegative, they pull electron density away from the hydrogen and deshield it, resulting in absorption at higher δ. Approximate chemical shifts of the hydrogens of methyl groups bonded to atoms of varying electronegativities are as follows:

$$F-CH_3 \qquad -O-CH_3 \qquad -\overset{|}{\underset{|}{N}}-CH_3 \qquad -\overset{|}{\underset{|}{C}}-CH_3 \qquad -\overset{|}{\underset{|}{Si}}-CH_3$$

$$4.3\ \delta \qquad\qquad 3.3\ \delta \qquad\qquad 2.2\ \delta \qquad\qquad 0.9\ \delta \qquad\qquad 0\ \delta$$

⟵ increasing electronegativity of group bonded to methyl group

Electronegative atoms deshield nearby hydrogens, resulting in a downfield shift.

As the atom that is attached to the methyl group becomes more electronegative, the absorption for the hydrogen is shifted farther downfield. The peak for TMS occurs upfield from the peaks of most common organic compounds because silicon is less electronegative than most of the other elements encountered in organic compounds.

Carbon is slightly more electronegative than hydrogen. Thus, the absorption for hydrogens on a secondary carbon appears slightly downfield (approximately 0.3 δ) from that for hydrogens on a primary carbon, and the peak for a hydrogen on a tertiary carbon appears even farther downfield (approximately 0.7 δ from the absorption for a primary hydrogen).

$$C-\overset{\overset{\displaystyle C}{|}}{\underset{\underset{\displaystyle H}{|}}{C}}-C \qquad\qquad C-\overset{\overset{\displaystyle H}{|}}{\underset{\underset{\displaystyle H}{|}}{C}}-C \qquad\qquad C-\overset{\overset{\displaystyle H}{|}}{\underset{\underset{\displaystyle H}{|}}{C}}-H$$

Tertiary   Secondary   Primary

1.6 δ 1.2 δ 0.9 δ

Pi Electron Effects

Electrons in pi MOs circulate more readily in the external magnetic field than do those in sigma MOs because they are less strongly held by the nuclei. Depending on the exact geometry of the molecule, the magnetic field of these circulating electrons at the hydrogen of interest may be aligned with the external magnetic field, causing a downfield shift, or opposed to it, causing an upfield shift. This effect is especially pronounced in benzene derivatives, in which the magnetic field generated by the circulation of the electrons in the ring of conjugated orbitals opposes the external magnetic field in the center of the ring but is aligned with it around the periphery of the ring, where the hydrogen is located.

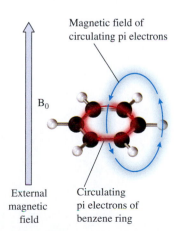

Magnetic field of circulating pi electrons

B_0

External magnetic field

Circulating pi electrons of benzene ring

This "ring current effect" results in a large downfield shift for hydrogens attached to aromatic rings. The absorption for the hydrogens of benzene appears at 7.4 δ. A similar but smaller effect causes the hydrogens on the double bond of an alkene to appear downfield also. The absorptions for the vinyl hydrogens of cyclohexene appear at 5.7 δ. A hydrogen attached to the carbonyl carbon of an aldehyde appears even farther downfield, near 10 δ, owing to the additional inductive effect of the oxygen. Surprisingly, a hydrogen on carbon 1 of a terminal alkyne appears upfield, near 1.8 δ. In this case the field due to the circulating electrons opposes the external magnetic field at the hydrogen, causing an upfield shift.

Approximate chemical shifts for hydrogens on pi bonded carbons

Hydrogens Bonded to Heteroatoms

The chemical shift of hydrogens bonded to oxygen and nitrogen depends on concentration and temperature because the extent of hydrogen bonding varies with these factors. The peak for an alcohol usually occurs in the region of 2 to 5 δ, and the peak for an amine appears in the region of 1 to 3 δ. The peak for the hydrogen of a carboxylic acid, which forms dimers with strong hydrogen bonds, appears in the region of 10 to 13 δ and is quite characteristic.

So far, the issue of how many signals occur in a ¹H-NMR spectrum has not been directly addressed. Hydrogens have different chemical shifts (although coincidental overlaps do sometimes occur) unless they are chemically equivalent. *Chemically equivalent* means that the hydrogens would have identical chemical reactions. For example, it is readily apparent that all three hydrogens of a methyl group are chemically equivalent. However, for many compounds it may be less obvious which hydrogens are chemically equivalent. One way to make this determination is to picture replacing the hydrogens in question, one at a time, with an imaginary group, A. If the resulting imaginary compounds are identical, the hydrogens are chemically equivalent and have identical chemical shifts; if the resulting imaginary compounds are isomers, then the hydrogens are different and *may* result in different NMR signals. As an example, consider 1,2-dimethylbenzene. How many NMR signals are expected for this compound? Let's begin by determining whether the hydrogens on the two methyl groups are chemically equivalent. The two structures that result from replacing a hydrogen on one or the other methyl group with imaginary group A are identical, so the hydrogens of the methyl groups are chemically equivalent.

Similar transformations show that the two hydrogens on C-3 and C-6 are chemically equivalent, as are the two on C-4 and C-5. Therefore, three absorptions are expected in the NMR spectrum: one for all six methyl hydrogens, one for the two hydrogens on C-3 and C-6, and one for the two hydrogens on C-4 and C-5. The latter two types of hydrogens are very similar (all are bonded to the aromatic ring) and are expected to have very similar chemical shifts. Although such hydrogens might absorb at the same position in some cases, they *may* have different chemical shifts in other cases because they are not chemically equivalent.

Some cases are a little more subtle. Consider the two hydrogens of the CH_2 of bromoethane. Are they chemically equivalent? The two imaginary compounds that result from replacing these hydrogens are enantiomers. The hydrogens are termed **enantiotopic** and, like enantiomers, are different only in the presence of something else that is chiral. Thus, they have identical chemical shifts. Two NMR absorptions are expected for this compound: one for the CH_3 group and one for the CH_2 group.

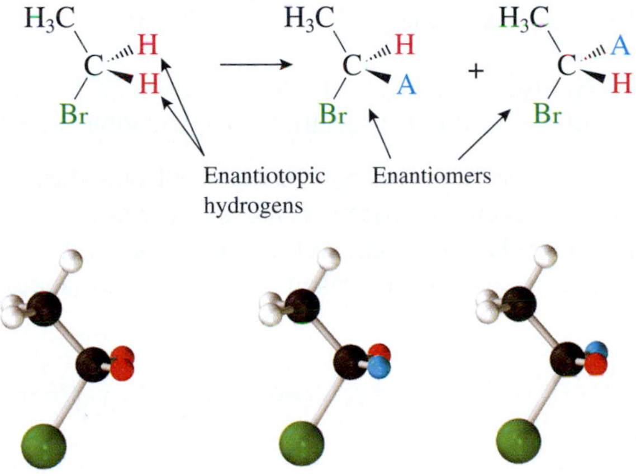

How many signals are expected for 2-bromobutane? The methyl groups on each end are obviously different, as is the single H on C-2. At first glance, the two H's on C-3 appear to be identical, but in fact replacement with an imaginary group produces diastereomers. Such hydrogens are termed **diastereotopic** and have different chemical shifts.

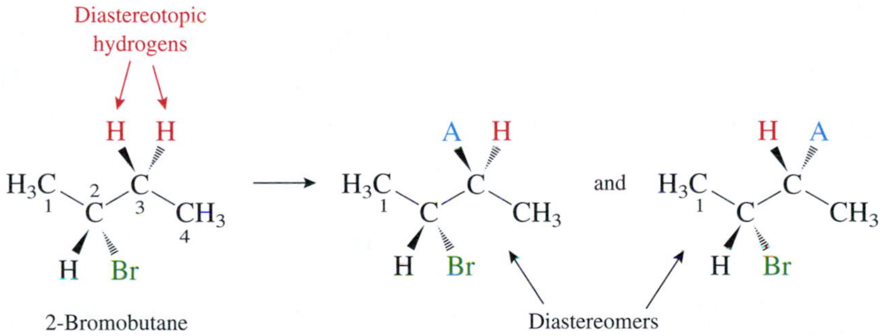

Therefore, there are five different types of hydrogens in 2-bromobutane. However, the chemical shifts of the diastereotopic hydrogens on C-3 will be very similar. In general, the two hydrogens of a CH_2 group are diastereotopic when a chirality center is present.

PRACTICE PROBLEM 14.1

How many absorptions are expected in the ^{1}H-NMR spectra of these compounds?

a) 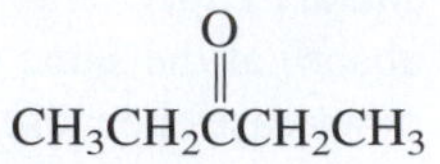 $CH_3CH_2CCH_2CH_3$ (with $=O$ on central C)

b)

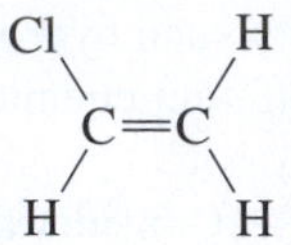

Solutions

a) The CH$_2$ groups are chemically equivalent because these imaginary compounds are identical.

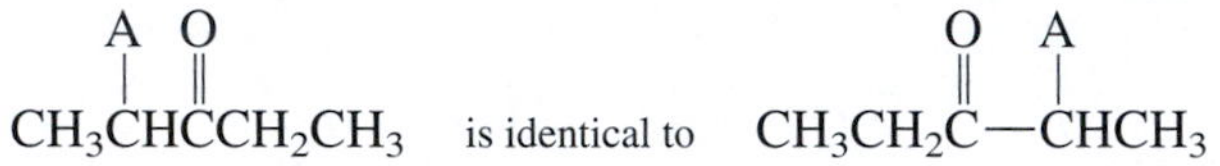

$$CH_3\overset{A}{C}H\overset{O}{C}CH_2CH_3 \quad \text{is identical to} \quad CH_3CH_2\overset{O}{C}-\overset{A}{C}HCH_3$$

Likewise, the CH$_3$ groups are identical. Therefore, there are only two absorptions (ignoring the multiplicities) in the spectrum of 3-pentanone (see Figure 14.1).

b) The hydrogens on C-2 are diastereotopic because replacing them produces diastereomers. (Remember, *cis*- and *trans*-isomers are one type of diastereomers.) Diastereotopic hydrogens have different chemical shifts, so all three hydrogens of chloroethene are chemically nonequivalent and three signals appear in the spectrum.

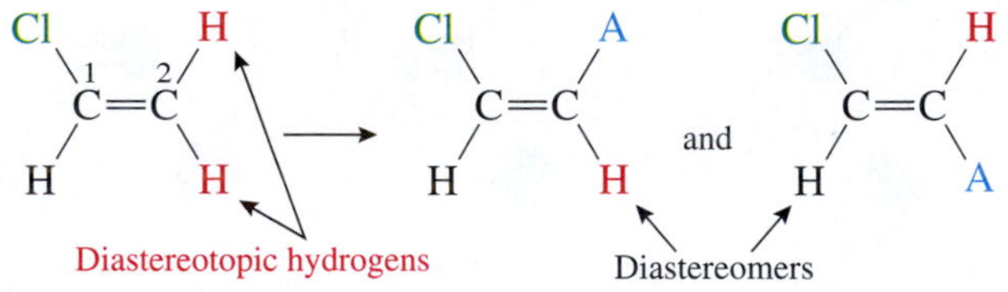

Diastereotopic hydrogens   Diastereomers

PROBLEM 14.2

How many absorptions are expected in the ^{1}H-NMR spectra of these compounds?

a) 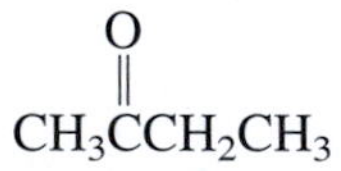 $CH_3CCH_2CH_3$ (with $=O$ on second C)

b)

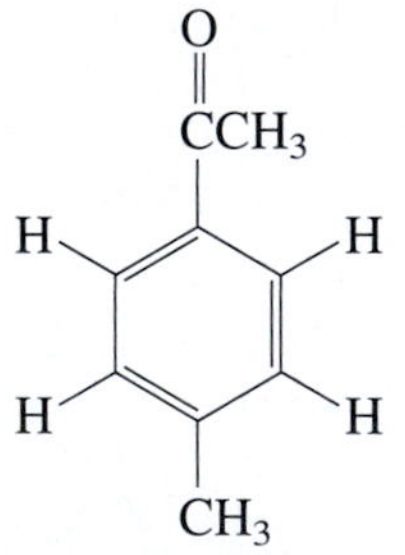

c) $CH_3CH_2CHCH_3$ (with OH on third C)

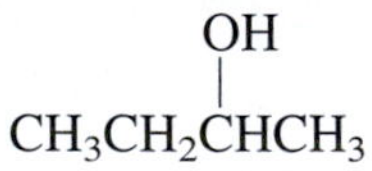

d)

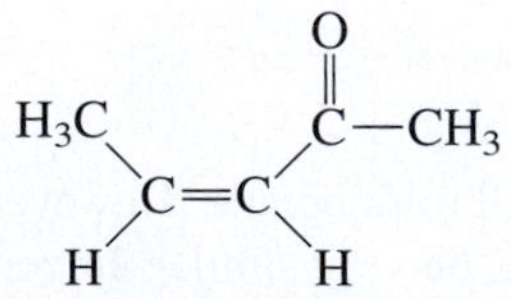

e)

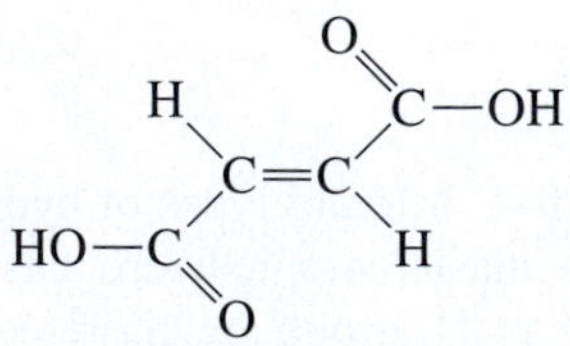

A useful summary of the chemical shift regions where various types of hydrogens appear in NMR spectra is provided by the following diagram:

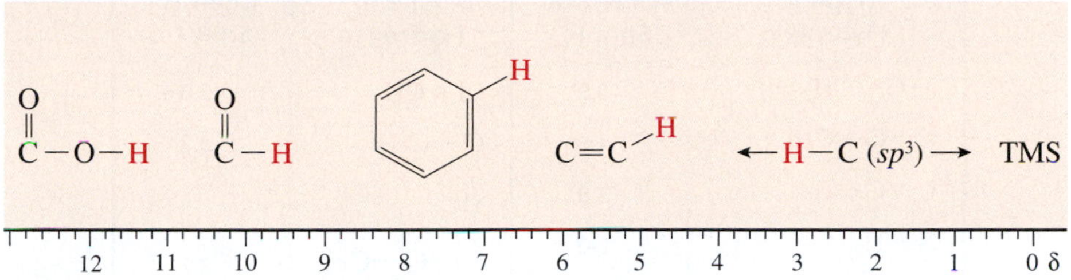

Although the regions overlap somewhat, peak positions often suggest the presence of certain types of hydrogens. For example, a peak in the region of 10 δ is usually due to the hydrogen of an aldehyde, whereas absorptions in the region of 7 to 8 δ suggest the presence of hydrogens on an aromatic ring. These assignments can often be confirmed by examination of the IR spectrum.

More advanced texts on NMR spectroscopy contain detailed tables listing chemical shift values for numerous different kinds of hydrogens. With use of these tables, a good prediction of the chemical shift of almost any proton can be made. Recently, computer programs that make reasonably accurate predictions of chemical shifts have become available. However, we do not need to know chemical shifts that accurately in order to use NMR effectively. Table 14.1 is a much abbreviated version of more complete tables and is sufficient for our needs. It lists the approximate chemical shifts for the most common types of hydrogens. The values in Table 14.1 are for CH_3 hydrogens (methyl groups); CH_2 hydrogens (methylene groups) appear about 0.3 δ downfield and CH hydrogens (methine groups) appear about 0.7 δ downfield from the CH_3 values. Also note that the presence of additional functional groups near the hydrogen causes additional downfield shifts. If the additional functional group is attached to the same carbon as the hydrogen under consideration, then the downfield shift is relatively large. The effect decreases rapidly with distance, so if the additional functional group is attached to the carbon adjacent to the carbon to which the hydrogen is attached, the downfield shift is small. If the distance is larger than this, then the shift can be neglected.

Let's see how Table 14.1 can be used to assign chemical shifts in a spectrum to the hydrogens in a compound. Consider 1-propanol. The chemical shift for the hydroxy hydrogen depends on the concentration of the sample and the temperature, so it cannot be predicted exactly, but it is expected to be in the region of 2 to 5 δ. (In this particular sample it actually appeared at 2.3 δ.) The hydrogens on C-1 should appear furthest downfield because they are closest to the electronegative oxygen. From Table 14.1, the predicted position for the hydrogens of $O—CH_3$ is 3.3 δ. The hydrogens on C-1 should appear slightly downfield from this because they are part of a CH_2 group rather than a CH_3 group. (Their actual position is 3.5 δ.) The hydrogens on C-2 should appear further upfield because they are not as close to the electronegative oxygen. They are expected to appear somewhat downfield from 0.9 δ, the position for $C—CH_3$ in Table 14.1, because they are part of a CH_2 group and are also slightly shifted by the oxygen. (Their actual position is 1.5 δ.) The hydrogens on C-3 should appear furthest upfield because they are too far from the hydroxy group to be affected by it. The position predicted from Table 14.1 for $C—CH_3$ is 0.9 δ. (The actual position is 0.9 δ.)

Table 14.1 **Approximate Chemical Shifts of Hydrogens in ^{1}H-NMR Spectra**

Type of Hydrogen	Chemical Shift (δ)	Type of Hydrogen	Chemical Shift (δ)
$-C-CH_3$	0.9	$Cl-CH_3$	3.0
$C=C-CH_3$	1.6	$O-CH_3$	3.3
$C\equiv C-H$	1.8		
$N-H$	1–3	$\overset{\overset{O}{\|\|}}{C}-O-CH_3$	3.7
$O-H$	2–5		
		O_2N-CH_3	4.1
		$F-CH_3$	4.2
$R-O-\overset{\overset{O}{\|\|}}{C}-CH_3$	2.0	$C=C\overset{H}{<}$	5.5–6.5
$\overset{\overset{O}{\|\|}}{C}-CH_3$	2.2	⬡$-H$	7–8
$N-CH_3$	2.2		
$I-CH_3$	2.2	$\overset{\overset{O}{\|\|}}{C}-H$	10
$N\equiv C-CH_3$	2.2		
$Ph-CH_3$	2.3	$\overset{\overset{O}{\|\|}}{C}-O-H$	12
$Br-CH_3$	2.7		

Note that these positions are only approximate. Furthermore, most of these positions are given for CH_3 groups.
CH_2 groups appear farther downfield by about 0.3 ppm and CH groups by about 0.7 ppm.

OH ⟵ 2.3 δ (Predict 2–5 δ)

^{1}CH$_2$ ⟵ 3.5 δ (Predict slightly downfield from 3.3 δ)

^{2}CH$_2$ ⟵ 1.5 δ (Predict somewhat downfield from 0.9 δ)

^{3}CH$_3$ ⟵ 0.9 δ (Predict 0.9 δ)

As another example, consider 1-ethyl-4-methylbenzene (*p*-ethyltoluene). This compound has five different types of hydrogens: the CH_3 attached directly to the ring, the CH_3 of the ethyl group, the CH_2 of the ethyl group, the two hydrogens ortho to the methyl group, and the two hydrogens meta to the methyl group. Both types of hydrogens on the benzene ring are predicted to appear in the 7 to 8 δ region. They should have very similar chemical shifts because the ethyl and methyl groups are both nonpolar. (They actually appear at 7.1 δ.) The methyl group on the aromatic ring is predicted to appear at 2.3 δ, the value from Table 14.1 for $Ph-CH_3$. (The actual position is 2.3 δ.) The CH_2 of the ethyl group is predicted to appear slightly downfield from

2.3 δ because it is a CH$_2$ group rather than a CH$_3$ group. (The actual position is 2.6 δ.) The CH$_3$ of the ethyl group should appear most upfield. It is expected to be shifted slightly downfield from 0.9 δ (C—CH$_3$) due to the nearby aromatic ring. (The actual position is 1.2 δ.)

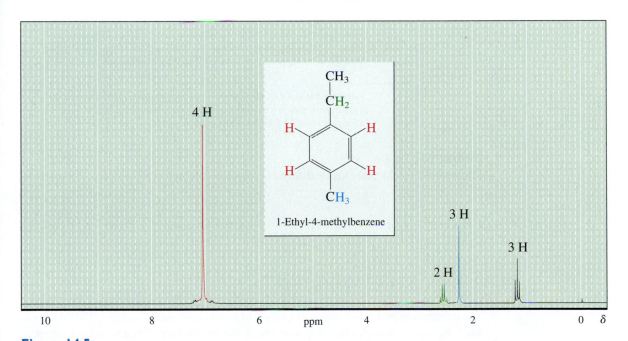

The actual NMR spectrum of 1-ethyl-4-methylbenzene is shown in Figure 14.5. Note that the two types of aromatic hydrogens are so similar that they appear as a single peak, at least at the resolution of this spectrometer. Also note that the CH$_2$ of the ethyl group appears as a set of four lines, termed a *quartet,* and the CH$_3$ of the ethyl group appears as a set of three lines, termed a *triplet.* These multiple peaks result from a process termed *spin coupling.* Analysis of this coupling pattern can be used to tell us that there are three hydrogens near the CH$_2$ group and two hydrogens near the CH$_3$ group and enables us to determine that the compound contains an ethyl group. Let's see how spin coupling works.

Figure 14.5

THE ^{1}H-NMR SPECTRUM OF 1-ETHYL-4-METHYLBENZENE. The numbers above each group of peaks are the integrals and provide the relative numbers for hydrogens responsible for that group of peaks.

PRACTICE PROBLEM 14.2

Predict the approximate chemical shifts for the different hydrogens in these compounds.

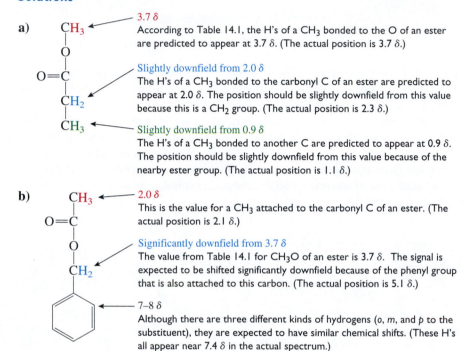

a) CH₃CH₂COCH₃   **b)**

Solutions

a)   CH₃ ◄——— 3.7 δ

According to Table 14.1, the H's of a CH₃ bonded to the O of an ester are predicted to appear at 3.7 δ. (The actual position is 3.7 δ.)

Slightly downfield from 2.0 δ

The H's of a CH₃ bonded to the carbonyl C of an ester are predicted to appear at 2.0 δ. The position should be slightly downfield from this value because this is a CH₂ group. (The actual position is 2.3 δ.)

Slightly downfield from 0.9 δ

The H's of a CH₃ bonded to another C are predicted to appear at 0.9 δ. The position should be slightly downfield from this value because of the nearby ester group. (The actual position is 1.1 δ.)

b)   CH₃ ◄——— 2.0 δ

This is the value for a CH₃ attached to the carbonyl C of an ester. (The actual position is 2.1 δ.)

Significantly downfield from 3.7 δ

The value from Table 14.1 for CH₃O of an ester is 3.7 δ. The signal is expected to be shifted significantly downfield because of the phenyl group that is also attached to this carbon. (The actual position is 5.1 δ.)

7–8 δ

Although there are three different kinds of hydrogens (o, m, and p to the substituent), they are expected to have similar chemical shifts. (These H's all appear near 7.4 δ in the actual spectrum.)

PROBLEM 14.3

Predict the approximate chemical shifts for the different hydrogens in these compounds:

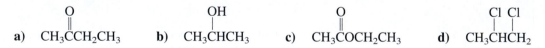

a) CH₃CCH₂CH₃   **b)** CH₃CHCH₃   **c)** CH₃COCH₂CH₃   **d)** CH₃CHCH₂

14.4   SPIN COUPLING

The information provided by spin coupling is often the most useful part of a ¹H-NMR spectrum. Recall that each hydrogen in a molecule can be considered as a little bar magnet. A particular hydrogen experiences the small magnetic fields of other nearby hydrogens in addition to the large magnetic field of the external magnet. This results in the splitting of the absorption into multiple peaks. From the multiplicity, or number of peaks, of a signal, it is possible to determine which hydrogens are near the hydrogen responsible for the signal.

To spin couple, the hydrogens must have different chemical shifts, and they must be relatively close together because the effect of the small magnetic fields of the hydrogens decreases rapidly with distance. Hydrogens bonded to the same carbon (called **geminal** hydrogens) couple if they are diastereotopic and have different chemical shifts. Hydrogens bonded to adjacent atoms (called **vicinal** hydrogens) also couple. Coupling is not commonly observed between hydrogens that are farther apart than this unless a pi bond is involved, in which case a small coupling between hydrogens separated by three carbons is sometimes observed. Coupling between vicinal hydrogens is by far the most common type.

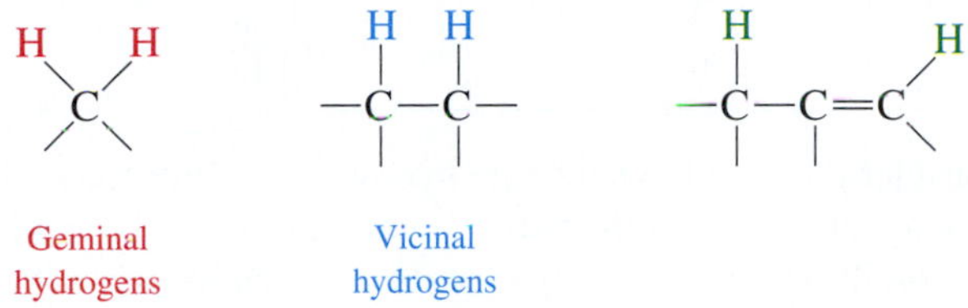

Geminal hydrogens

Vicinal hydrogens

Let's begin by considering the simplest possible case, in which one hydrogen, call it H_a, is interacting with one other hydrogen, call it H_x. In addition to the large external magnetic field, H_a also experiences the small magnetic field of H_x. Because the energy difference between the two spin states of a proton is extremely small, the magnetic field of H_x is oriented in the same direction as the external magnetic field in approximately 50% of the molecules, while the fields are oriented in opposite directions in the other 50% of the molecules. (Basically, the hydrogens in the two spin states are in equilibrium. Because the difference in energy between the two states is so small, the "equilibrium constant" is very close to 1.) The total magnetic field experienced by H_a is the sum of the field of the external magnet and the field due to the nearby hydrogen, H_x. In 50% of the molecules the field of H_x is oriented in the same direction as the external field and increases the field strength at H_a. In this situation a weaker external magnetic field is needed for the total field strength at H_a to have the correct value for resonance to occur. This causes the absorption for H_a to appear at a slightly downfield position. In the other 50% of the molecules the field of H_x is oriented in opposition to the external magnetic field and subtracts from the field at H_a. A larger external magnetic field is necessary for resonance and the absorption appears at a slightly upfield position. As a result, the absorption for H_a appears as two closely spaced peaks, a **doublet.** Both peaks of the doublet have equal areas because the magnetic field orientations that produce them are equally probable. This process is visually represented in the following illustration, called a splitting diagram or a tree diagram:

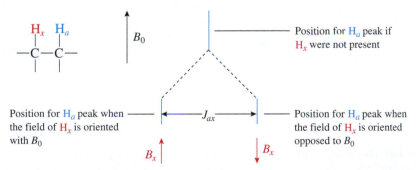

The separation between the two peaks is called the **coupling constant,** J_{ax}. The coupling constant has units of hertz and does not depend on the operating frequency of the instrument. It ranges from 0 to 20 Hz, with 6 Hz being a typical value for vicinal hydrogens.

The analysis for H_x in the preceding example is identical. It is split into a doublet by H_a, and it is important to note that the coupling constant is exactly the same. The NMR spectrum of this system has two doublets: one for H_a and one for H_x. The separation between the two peaks of each doublet, J_{ax}, is identical. The chemical shift of either hydrogen is the center of its doublet.

Next let's consider the case in which one hydrogen, H_a, interacts with two identical protons, H_x. The magnetic fields of the two H_x's can be oriented both with, one with and one against (two possibilities), or both against the external magnetic field. Following is the splitting diagram:

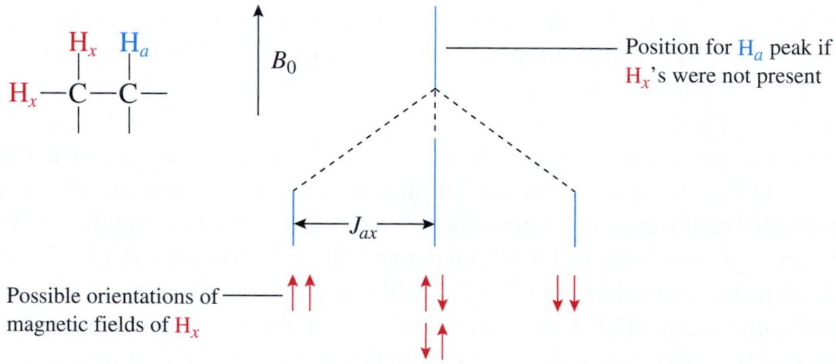

$$H_x—C—C—$$

B_0 — Position for H_a peak if H_x's were not present

J_{ax}

Possible orientations of magnetic fields of H_x — ↑↑ ↑↓ ↓↑ ↓↓

The absorption for H_a is split into a triplet by the two H_x's, with the relative heights of the peaks in a 1:2:1 ratio. The absorption for the H_x's is split into a doublet by the single H_a. The coupling constant, J_{ax}, is the separation between two adjacent peaks of the triplet or between the peaks of the doublet. An example of such a system is shown in Figure 14.6, the NMR spectrum of 1,1,2-trichloroethane.

As a more complicated example, let's consider the case in which three hydrogens (H_x) on one carbon are coupled to two hydrogens (H_a) on an adjacent carbon. The following splitting diagram shows that the signal for H_a is split into four peaks, a quartet, with relative heights of 1:3:3:1:

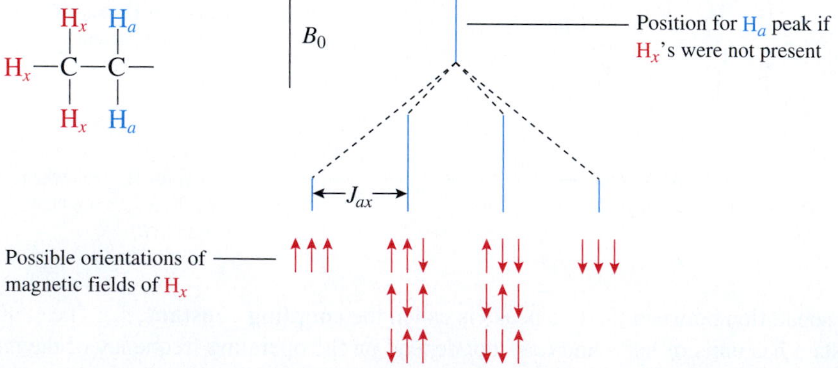

$$H_x—C—C—$$

B_0 — Position for H_a peak if H_x's were not present

J_{ax}

Possible orientations of magnetic fields of H_x — ↑↑↑ ↑↑↓ ↑↓↓ ↓↓↓
↑↓↑ ↓↑↓
↓↑↑ ↓↓↑

The signal for H_x is split into a triplet by the two H_a's. The NMR spectrum of bromoethane, shown in Figure 14.7, provides an example of such a system. This pattern of a downfield quartet, due to two hydrogens, and an upfield triplet, due to three hydro-

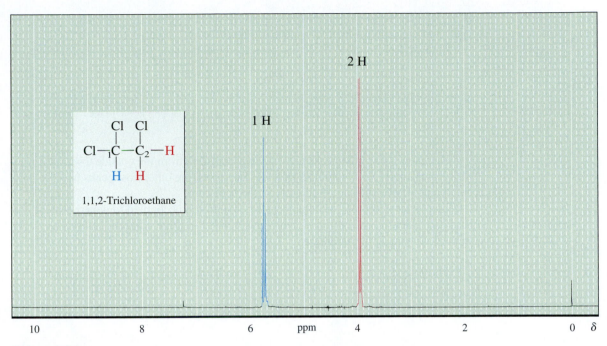

Figure 14.6

THE ¹H-NMR SPECTRUM OF 1,1,2-TRICHLOROETHANE. The hydrogen on C-1 appears farther downfield because there are two electronegative Cl's on this carbon. Its signal is split into a triplet by the two H's on C-2. The signal for the two H's on C-2 is upfield and is split into a doublet by the single H on C-1.

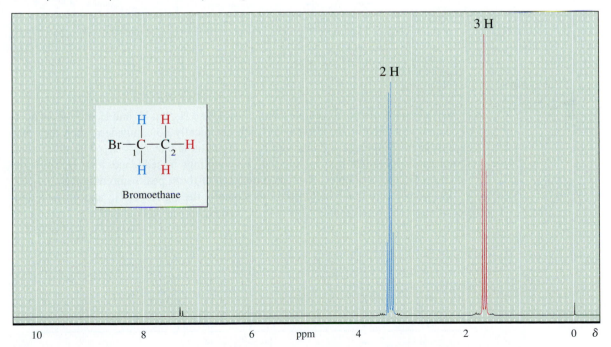

Figure 14.7

THE ¹H-NMR SPECTRUM OF BROMOETHANE. The absorption for the two H's on C-1 is downfield because of the electronegative Br attached to this carbon. The signal is split into a quartet by the three nearby H's on C-2. The signal for the three H's on C-2 is split into a triplet by the two nearby H's on C-1.

gens, is characteristic of an ethyl group and is quite common in NMR spectra (see the spectrum of 1-ethyl-4-methylbenzene in Figure 14.5).

An analysis similar to those just presented can be done for other possible coupling combinations. In general,

A hydrogen(s) that is coupled to n equivalent hydrogens is split into $n + 1$ peaks.

The intensities of the peaks increase in proceeding from the outside toward the middle of the pattern. As the number of peaks increases, the outermost peaks become quite small and may be difficult to identify in the spectrum.

Finally, let's consider what happens when a proton is coupled to nonequivalent protons. Consider the case of three hydrogens bonded to three adjacent carbons as shown in the following splitting diagram:

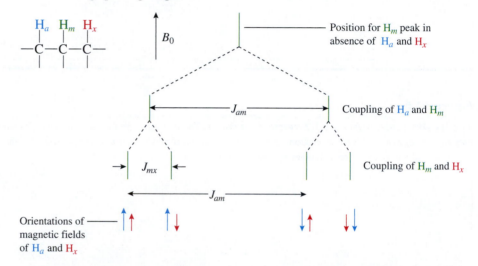

In such an arrangement, H_m couples with both H_a and H_x, but H_a and H_x are too far apart to couple with each other. If H_a and H_x are chemically nonequivalent, then they may have different coupling constants with H_m. Let's assume that $J_{am} > J_{mx}$. The splitting diagram can be obtained by applying the two couplings in sequence. First H_a splits the signal for H_m into a doublet. Then H_x splits each of those peaks into doublets.

In the spectrum that results from this situation the signal for H_m appears as a doublet of doublets, whereas those for both H_a and H_x appear as doublets.

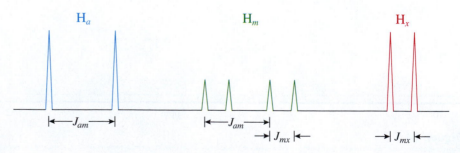

PROBLEM 14.4

Predict the multiplicity of the absorption for H_m if $J_{am} = J_{mx}$. Explain.

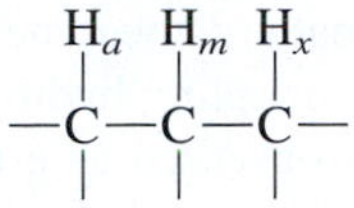

$$H_a \quad H_m \quad H_x$$
$$-\overset{|}{\underset{|}{C}}-\overset{|}{\underset{|}{C}}-\overset{|}{\underset{|}{C}}-$$

PROBLEM 14.5

Construct a tree diagram for the absorption of H_m. Assume that $J_{am} < J_{mx}$.

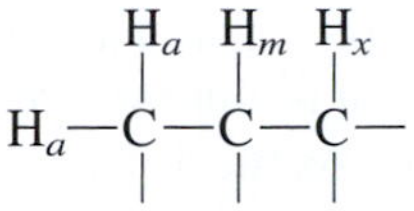

$$\quad\quad H_a \quad H_m \quad H_x$$
$$H_a-\overset{|}{\underset{|}{C}}-\overset{|}{\underset{|}{C}}-\overset{|}{\underset{|}{C}}-$$

PRACTICE PROBLEM 14.3

Predict the multiplicities of the absorptions for the hydrogens in this group:

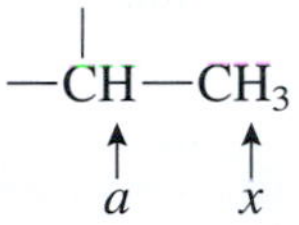

$$-\overset{|}{\underset{\underset{a}{\uparrow}}{C}}H-\underset{\underset{x}{\uparrow}}{C}H_3$$

Solution

H_a is split by three H_x's, so its absorption should appear as a quartet. The H_x's are split by one H_a, so their absorption should appear as a doublet.

PROBLEM 14.6

Predict the multiplicities of the absorptions for the hydrogens of these groups. Assume that hydrogens labeled a are different from those labeled x but that all of those labeled a are identical and all of those labeled x are identical.

a) 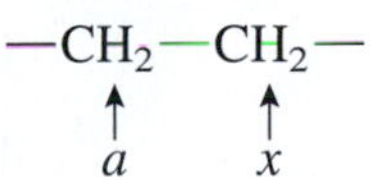 $-\underset{\underset{a}{\uparrow}}{C}H_2-\underset{\underset{x}{\uparrow}}{C}H_2-$

b) $\underset{\underset{x}{\uparrow}}{C}H_3-\overset{|}{\underset{\underset{a}{\uparrow}}{C}}H-\underset{\underset{x}{\uparrow}}{C}H_3$

c) $-\overset{|}{\underset{\underset{a}{\uparrow}}{C}}H-\overset{|}{\underset{\underset{x}{\uparrow}}{C}}H_2$

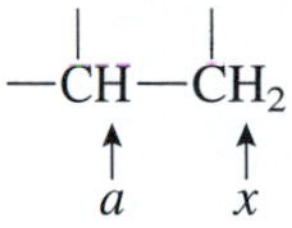

d) 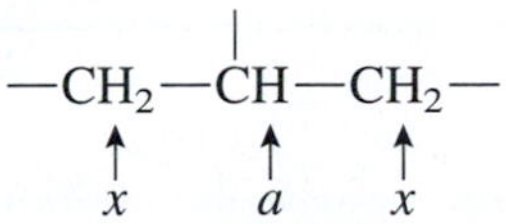 $-\underset{\underset{x}{\uparrow}}{C}H_2-\overset{|}{\underset{\underset{a}{\uparrow}}{C}}H-\underset{\underset{x}{\uparrow}}{C}H_2-$

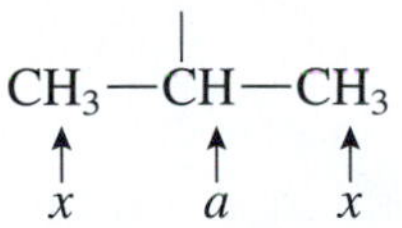

e) 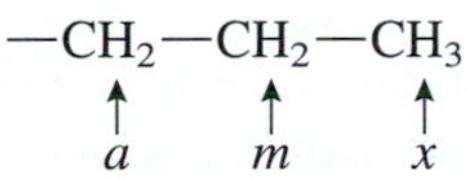 $-\underset{\underset{a}{\uparrow}}{C}H_2-\underset{\underset{m}{\uparrow}}{C}H_2-\underset{\underset{x}{\uparrow}}{C}H_3$

Assume $J_{am} = J_{mx}$

ORGANIC
Chemistry❖Now™
Click *Coached Tutorial Problems*
for more practice interpreting
Spin Coupling in ^{1}H-NMR Spectroscopy.

14.5   COMPLEX COUPLING

The splitting patterns that have been discussed so far result from what is called first-order coupling. However, splitting patterns become much more complex as the difference in chemical shifts between the coupling hydrogens becomes smaller. The first indication of this, termed **leaning**, is observed as a distortion of the peak heights of a multiplet. As an example, note how the downfield peak of the triplet in the spectrum of bromoethane (Figure 14.7) is slightly larger than the upfield peak. The triplet "leans" toward the quartet of the other hydrogens. Similarly, the quartet "leans" toward the triplet. However, the deviations from first-order coupling are quite small in this case and the pattern of a triplet and a quartet is easily recognized.

Figure 14.8 shows a spectrum in which the chemical shifts of the coupling hydrogens are closer. The two hydrogens on the aromatic ring spin couple and should appear as doublets. In this case the distortion is more severe, and the outside peaks are considerably smaller than the inside peaks. As the difference in chemical shifts becomes even smaller, the outside peaks become quite small relative to the inside peaks. If more than two hydrogens are involved in such complex coupling, extra peaks appear and the pattern becomes quite complex. Although such patterns, termed **multiplets**, are often encountered, their interpretation is difficult. However, the presence of a multiplet in the spectrum does supply information about the structure in that structural features resulting in complex coupling must be present.

Coupling patterns are more likely to be first order on a NMR instrument with a large magnetic field strength and a high operating frequency because the chemical shift dif-

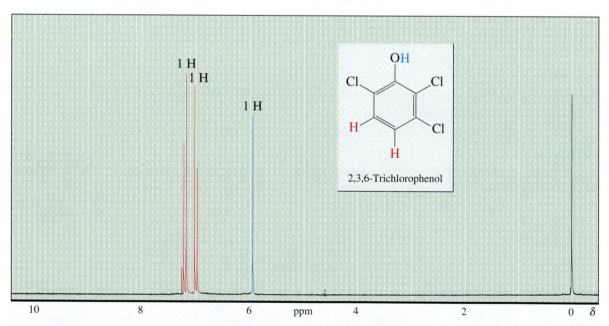

Figure 14.8

THE ^{1}H-NMR SPECTRUM OF 2,3,6-TRICHLOROPHENOL. The H of the hydroxy group appears at 5.9 δ and is not split. The two H's on the aromatic ring do couple, and each should appear as a doublet. Because their chemical shifts are similar, the doublets are distorted, the outside peaks being significantly smaller than the inside peaks.

ference, measured in hertz, increases as the magnetic field strength increases. (A chemical shift difference of 0.1 δ is 20 Hz on an instrument operating at 200 MHz and 40 Hz on an instrument operating at 400 MHz.) As a result, given two different instruments, most spectra will be easier to interpret on the instrument operating at the higher frequency. This is one reason why NMR instruments with larger and larger magnets are constantly being built.

14.6   CHEMICAL EXCHANGE

NMR has been compared to a slow camera—that is, one in which the shutter is open for a relatively long period. An object that moves while the camera shutter is open produces a blurred picture. Similarly, a hydrogen that is rapidly switching between two environments appears at a chemical shift that is the average of the chemical shifts of the two environments. For example, the axial and equatorial hydrogens of cyclohexane have different chemical shifts. However, at room temperature they are interchanging so rapidly by the ring-flipping process that the NMR spectrum of cyclohexane shows only one peak at the average of the two positions. If the spectrum is run at very low temperature, the ring-flips can be slowed enough that separate peaks appear for the two types of hydrogens.

The spectrum of methanol, CH_3OH, provides another example of this effect. The hydroxy hydrogen is expected to appear as a quartet and the methyl hydrogens as a doublet if coupling occurs. However, this coupling is not observed unless the methanol is extremely pure. A trace of acid (or base) that is present in normal samples of this alcohol causes a rapid exchange of the hydroxy hydrogens by the following acid–base reaction:

$$CH_3\overset{H}{\underset{+}{\overset{|}{O}}}{-}H \; + \; CH_3{-}\overset{..}{\underset{..}{O}}{-}H \; \rightleftharpoons \; CH_3{-}\overset{..}{\underset{..}{O}}{-}H \; + \; CH_3\overset{H}{\underset{+}{\overset{|}{O}}}{-}H$$

This reaction is so rapid that during the time that the NMR is examining the hydrogens of the methyl group, a large number of different hydrogens have been bonded to the oxygen. Because 50% of these hydrogens have their magnetic fields oriented in one direction and 50% have them oriented in the opposite direction, the average field is zero. Therefore, the peak due to the methyl group is not split, nor is the peak due to the hydroxy hydrogen. The spectrum shows two singlets.

In general, hydrogens on oxygen or nitrogen are subject to this rapid exchange process and do not couple to nearby hydrogens. However, caution must be exercised because coupling does occur in some samples.

14.7   DEUTERIUM

Although the nucleus of deuterium (the isotope of hydrogen with an atomic mass of 2) has spin quantum states, its magnetogyric ratio is different from that of hydrogen, so it does not appear in a ^{1}H-NMR spectrum. In addition, spin coupling between deuterium and hydrogen is very small, so splitting is usually negligible. Therefore, deuterium is essentially invisible in ^{1}H-NMR spectra, a useful feature. For example, the solvents used to dissolve samples for NMR spectra should not have any signals. Carbon tetrachloride can be used, but it is not very polar and many organic compounds are not very soluble in it. The deuterium analog of chloroform, $CDCl_3$, is a more versatile solvent and is commonly used to obtain NMR spectra. More expensive solvents, such as the deuterated analogs of acetone, benzene, or DMSO, are available for special applications. Deuterium is also useful as a label in the study of organic reaction mechanisms. A particular hydrogen in the reactant is replaced with deuterium. Then NMR spectroscopy is used to determine the position of this deuterium in the product by noting which signal is missing.

PRACTICE PROBLEM 14.4

Predict the ^{1}H-NMR spectrum of 2-butanone. Include the approximate chemical shift, multiplicity, and integral for each type of hydrogen.

$$\overset{\displaystyle O}{\underset{\displaystyle \parallel}{CH_3CCH_2CH_3}}$$

2-Butanone

Solution

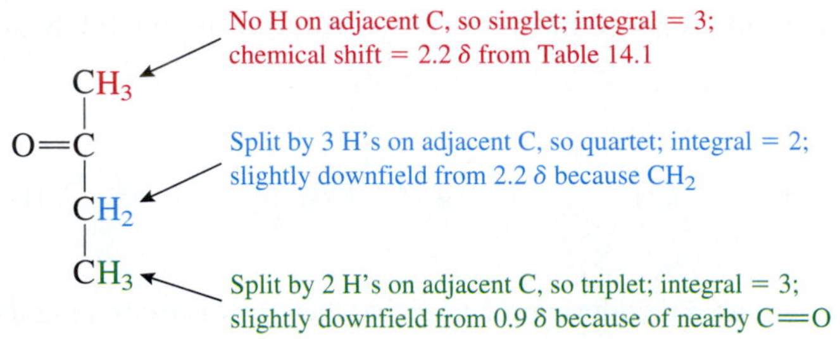

No H on adjacent C, so singlet; integral = 3; chemical shift = 2.2 δ from Table 14.1

Split by 3 H's on adjacent C, so quartet; integral = 2; slightly downfield from 2.2 δ because CH$_2$

Split by 2 H's on adjacent C, so triplet; integral = 3; slightly downfield from 0.9 δ because of nearby C=O

PROBLEM 14.7

Predict the ^{1}H-NMR spectra of these compounds. Include the approximate chemical shift, multiplicity, and integral for each type of hydrogen.

a) CH_3CH_2OH

b) $CH_3\overset{\displaystyle Cl}{\underset{\displaystyle |}{CH}}CH_3$

c) $CH_3CH_2OCH_2CH_3$

d) 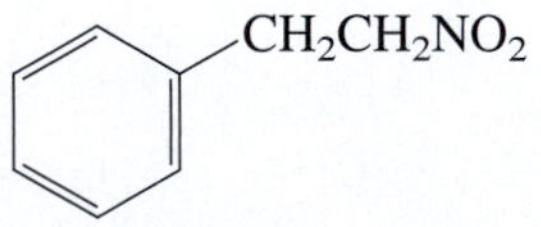 $CH_2CH_2NO_2$

e) $CH_3\overset{\displaystyle O}{\underset{\displaystyle \parallel}{C}}OCH_2CH_3$

f) $CH_3CH_2CH_2Cl$

g) CH_3CHCl_2

Focus On

NMR Spectroscopy of Carbocations in Superacid

The Focus On box in Chapter 8 on page 298 showed that when carbocations are generated in superacid solution, they undergo extensive rearrangements, usually forming a relatively stable tertiary carbocation. As an example, when 1-butanol is dissolved in superacid at $-60°C$, the protonated alcohol is formed. Water does not leave at this temperature because the carbocation that would be formed is primary. When the temperature is raised to $0°C$, water leaves but the carbocation rearranges rapidly to the more stable *tert*-butyl carbocation:

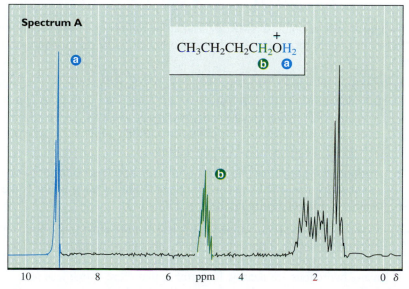

^1H-NMR spectroscopy was used to conduct this investigation. Spectrum A shows the spectrum of 1-butanol dissolved in superacid at $-60°C$. **a** The two hydrogens on the oxygen of the protonated alcohol appear as a triplet near 9.5 δ. **b** The two hydrogens on C-1 are shifted downfield by the electron-withdrawing positive oxygen and appear near 5 δ. They are split by the two hydrogens on the oxygen and the two hydrogens on C-2 and so should appear as a triplet of triplets. These overlap so that only seven lines are observed.

Continued

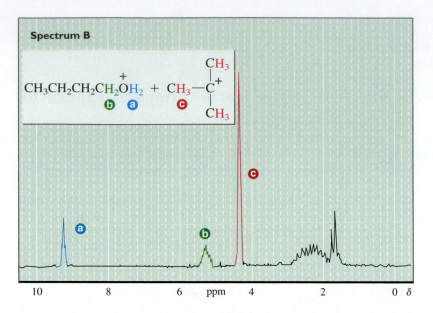

When the temperature is raised above 0°C, the absorptions for the protonated alcohol begin to decrease and a new signal **c** begins to appear near 4 δ (see Spectrum B). This is the absorption due to the *tert*-butyl carbocation, in which all of the hydrogens are identical. No absorptions that could be attributed to other carbocations are observed, indicating that the rearrangement to the *tert*-butyl carbocation is very fast.

14.8   INTERPRETATION OF ¹H-NMR SPECTRA

There is a lot of information in a NMR spectrum. The following series of steps provides an effective procedure for using this information to determine the structure of a compound:

Step 1.   *Examine the general positions of the peaks.* The purpose of this examination is to determine the general features of the compound, such as the presence of an aldehyde hydrogen, aromatic hydrogens, and alkyl groups. Do not attempt to be too specific at this point, such as determining that a peak is due to a methyl group next to a carbonyl group.

Step 2.   *Examine the integral for the ratios of the different kinds of hydrogens.* Remember that the actual numbers can be multiples of these ratios. If the formula is available, determine the actual numbers.

Step 3.   *Examine the coupling patterns.* This is often the most important step because it enables fragments of the compound to be identified. The number of peaks in the signal for a particular hydrogen(s) tells how many hydrogens are nearby. Remember that there is one less nearby hydrogen than the number of peaks; that is, if a signal is split into four lines (a quartet), then there are three nearby hydrogens. Keep in mind that coupling is reciprocal; that is, if there is a signal for two hydrogens split into a doublet because of coupling to one hydrogen, then somewhere else in the spectrum there must appear a signal for this one hydrogen that is split into a triplet by the two hydrogens. (Actually, this signal could contain more than three lines if the one hydrogen is also coupled to other hydrogens.)

Step 4. *Construct a tentative structure.* At this point, various fragments have been identified from the coupling patterns, the chemical shifts, the IR spectrum (if available), the formula, and so on. Assemble these fragments into a tentative structure.

Step 5. *Determine whether all of the information is consistent with this structure.* See whether the chemical shifts are consistent for each type of hydrogen. Check the multiplicity and integral of each type. Look for any different way to assemble the fragments that would be consistent with all of the data.

Let's try an example. Figure 14.9 shows the NMR spectrum of an unknown compound with the formula C_9H_{12}. When the formula is known, it is useful to first calculate the degree of unsaturation. In this case the DU is 4, so the compound contains a total of four rings and pi bonds. Examination of the general positions shows a multiplet in the aromatic region (7.2 δ) and two multiplets upfield in the alkyl region. The integral indicates that these hydrogens are in a 5:1:6 ratio. Because there are 12 hydrogens, the

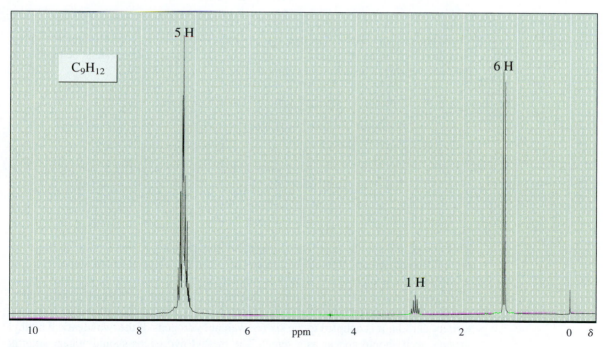

Figure 14.9

THE ¹H-NMR SPECTRUM OF AN UNKNOWN COMPOUND WITH THE FORMULA C_9H_{12}.

actual numbers must be 5, 1, and 6. The five hydrogens in the aromatic region suggest the presence of a monosubstituted benzene ring:

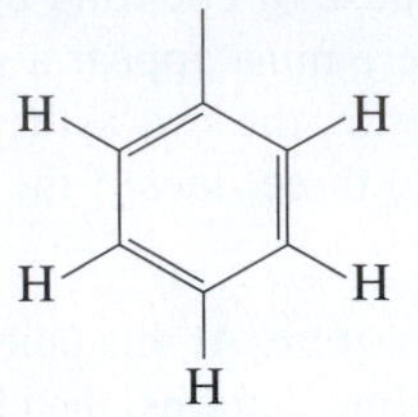

(Note that a benzene ring has three pi bonds and one ring, accounting for the DU of 4.) The substituent on the ring must be nonpolar so that all five of the hydrogens have similar chemical shifts. The signal at 2.9 δ, resulting from a single hydrogen, contains seven lines (a septet). It must be coupled to six nearby hydrogens. (When a signal has a large number of lines, it is often difficult to count the exact number because the outside lines are so small. It is difficult to be sure that the small outermost lines are real and not part of the noise in the baseline. However, internal consistency requires the presence of seven lines here, as will become apparent shortly.) The signal at 1.2 δ, due to six identical hydrogens, is a doublet, so these hydrogens must be coupled to a single hydrogen. Note the internal consistency: one hydrogen at 2.9 δ coupled to six and six hydrogens at 1.2 δ coupled to one. From this information an isopropyl fragment must be present:

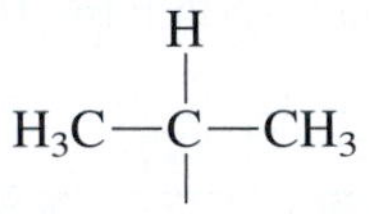

$$H_3C-\overset{\overset{\displaystyle H}{|}}{\underset{|}{C}}-CH_3$$

There are no other signals, so we are now ready to assemble these fragments into a structure. In this case this is particularly easy—the isopropyl group must be bonded to the benzene ring. The unknown compound is

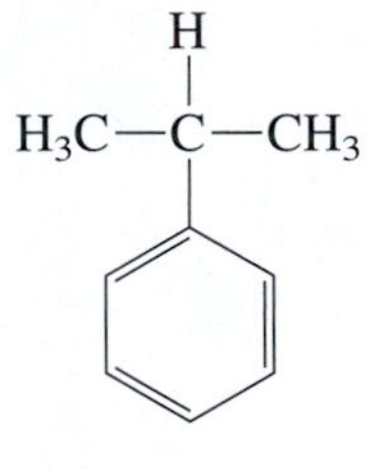

Isopropylbenzene

This compound should have five hydrogens in the aromatic region. Although there are three different types of aromatic hydrogens (those ortho, meta, and para to the isopropyl group), their chemical shifts should be similar, so it is not surprising that they appear close together as a complex multiplet. The single hydrogen on the carbon attached to the ring should appear somewhat downfield from 2.3 δ (Ph—CH$_3$) because it is on a tertiary carbon. It is coupled to the six equivalent hydrogens of the two identical methyl groups, so it should appear as a septet. The methyl hydrogens should appear slightly downfield from 0.9 δ (C—CH$_3$), due to the effect of the nearby phenyl group. They are coupled to a single proton, so they should appear as a doublet. All of the features of the spectrum are consistent with this structure.

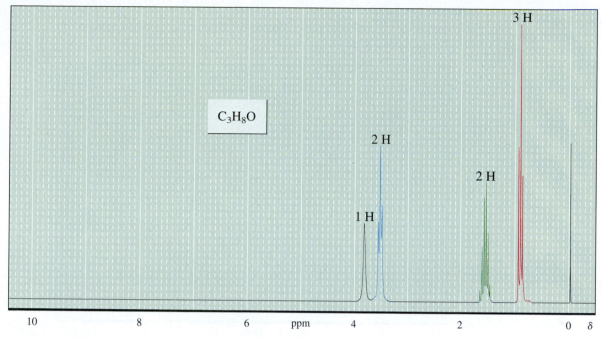

Figure 14.10

THE ¹H-NMR SPECTRUM OF AN UNKNOWN COMPOUND, C_3H_8O. This compound shows an OH in its IR spectrum. Its DU equals 0. Examination of its NMR spectrum indicates only alkyl-type H's. The integral provides the actual number of H's in this case, since they total eight. The broad singlet due to one H at 3.8 δ is probably due to the hydroxy H, which is not coupled due to rapid chemical exchange. The two H's at 3.5 δ appear as a triplet and must be coupled to two H's—the two H's at 1.5 δ. The three H's at 0.9 δ appear as triplet and must be coupled to the two H's at 1.5 δ also. If this is the case, then the two H's at 1.5 δ are coupled to 3 + 2 = 5 H's and should appear as six lines, a sextet. This information allows the fragment $CH_2CH_2CH_3$ to be written. Combining this with the HO indicates that the unknown is 1-propanol. The predicted chemical shifts (see page 554) agree well with those in the spectrum.

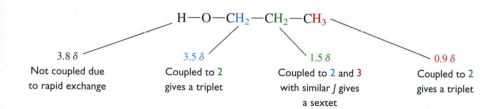

H—O—CH₂—CH₂—CH₃			
3.8 δ	3.5 δ	1.5 δ	0.9 δ
Not coupled due to rapid exchange	Coupled to 2 gives a triplet	Coupled to 2 and 3 with similar J gives a sextet	Coupled to 2 gives a triplet

Figure 14.10 provides another example. An interesting feature of this case is the multiplicity of the signal for the center CH_2 group. These hydrogens are coupled to two hydrogens on one side and three different hydrogens on the other side and could appear as 3 × 4 = 12 lines. However, because the coupling constants are very similar, the signal behaves as though there were five identical hydrogens doing the splitting, and it appears as six lines. This is typical for coupling involving hydrogens on alkyl chains with minimum conformational restrictions.

Problems that use both IR and NMR spectra to determine the identity of unknowns are provided in Section 14.10.

PROBLEM 14.8

Determine the structures of these compounds from their ¹H-NMR spectra:

a) The formula is C_3H_6O.

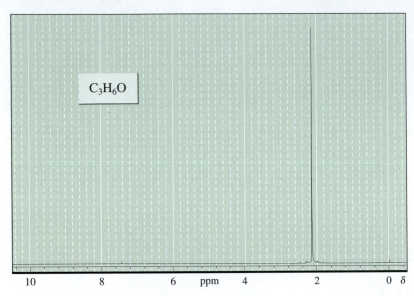

b) The formula is C_8H_{10}.

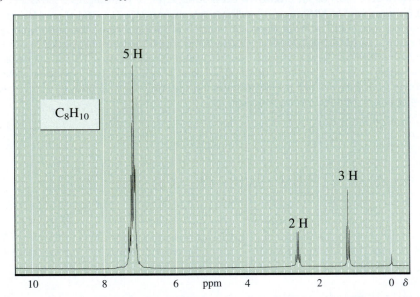

c) The formula is C_2H_4O.

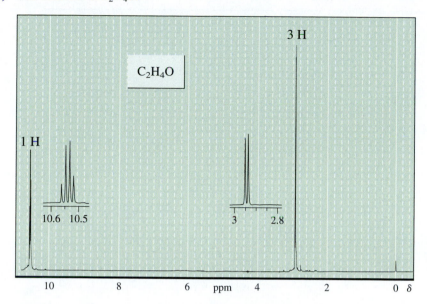

d) The formula is C_5H_{10}.

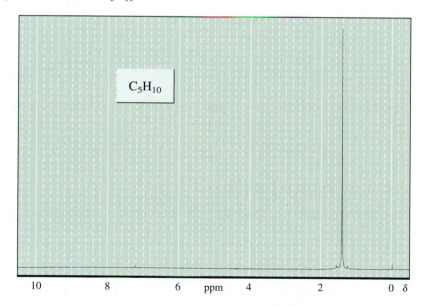

e) The formula is $C_5H_{10}O_2$.

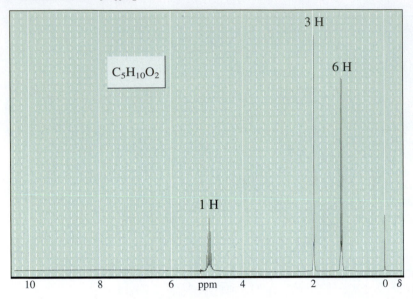

Focus On

Magnetic Resonance Imaging

Physicians are always looking for methods for viewing the internal organs of the human body without invasive techniques such as surgery. One method that has found considerable use is computed tomography (CT), also known as computed axial tomography (CAT). In a CAT scan, X-rays are used to generate the images that are collected and processed by computer. X-rays interact more strongly with atoms of larger atomic mass, so imaging agents must often be administered to the patient to enhance the pictures of soft tissues, which are composed primarily of C, H, N, and O. Another potential disadvantage of the CAT technique is the high energy of the radiation that is used. X-rays are called ionizing radiation because they have enough energy to eject electrons from the orbitals of atoms. Chemical reactions caused by the resulting ions can cause damage to living tissue. Although ionizing radiation techniques pose little hazard if done properly, techniques using less energetic radiation are desirable.

Magnetic resonance imaging (MRI) is a newer technique based on the same principles as ^{1}H-NMR. The 2003 Nobel Prize in physiology or medicine was awarded to Paul C.

f) The formula is C_4H_9Cl.

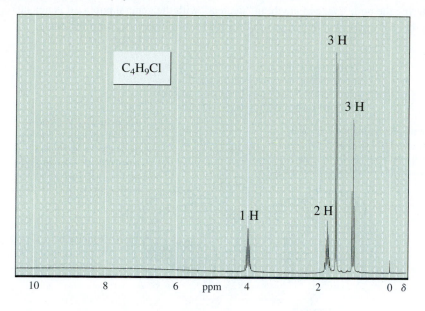

Lauterbur and Sir Peter Mansfield for their discoveries concerning this technique. The patient is placed within the field of a huge magnet and radio-frequency radiation is used to excite hydrogen nuclei to their higher-energy spin state. (The low energy of this radiation poses no danger to the patient.) The magnetic field of an MRI instrument is not nearly as uniform as that of a NMR spectrometer, so the signal for the protons is a very broad peak rather than the individual multiplets that we have seen in a NMR spectrum. The instrument detects differences in the intensity of the proton signal. The intensity depends on the concentration of hydrogens in the small area being sampled and on the "relaxation times," that is, the time that it takes a hydrogen in the higher-energy spin state to return to the ground state. Both of these factors cause different environments, such as fluids, tissues, and even diseased tissues, to produce signals of different intensities. The data are gathered by a computer, which produces a map or picture of the intensities. Because MRI is looking at hydrogens, it gives a particularly good image of soft tissue and therefore complements CAT. The accompanying figure shows a MRI image of a human skull.

Continued

Actually, this technique should be called nuclear magnetic resonance imaging. Considering the poor image of anything "nuclear" with the general public, it is not surprising that the medical community has decided to drop this term from the name.

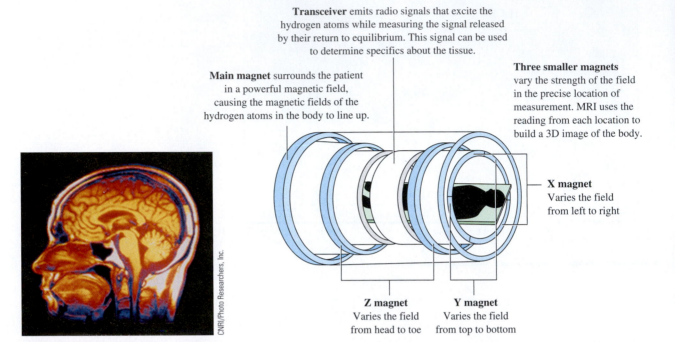

Transceiver emits radio signals that excite the hydrogen atoms while measuring the signal released by their return to equilibrium. This signal can be used to determine specifics about the tissue.

Main magnet surrounds the patient in a powerful magnetic field, causing the magnetic fields of the hydrogen atoms in the body to line up.

Three smaller magnets vary the strength of the field in the precise location of measurement. MRI uses the reading from each location to build a 3D image of the body.

X magnet Varies the field from left to right

Z magnet Varies the field from head to toe

Y magnet Varies the field from top to bottom

CNRI/Photo Researchers, Inc.

14.9 CARBON-13 MAGNETIC RESONANCE SPECTROSCOPY

The isotope of carbon with seven neutrons, ^{13}C, composes about 1.1% of carbon atoms. It is similar to hydrogen in that it has two nuclear spin states of different energy when it is in an external magnetic field. The spectroscopy that is done using this nucleus, ^{13}C-NMR, provides direct information about the carbon chains in the compound, information that is often complementary to that obtained from ^{1}H-NMR spectroscopy.

Figure 14.11 shows the ^{13}C-NMR spectrum of 3-buten-2-one. In contrast to ^{1}H-NMR, the peak for each carbon appears as a sharp singlet. Chemical shifts are measured by using the carbons of TMS as a standard. The chemical shift range is much larger in ^{13}C-NMR than in ^{1}H-NMR—peaks appear as far as 240 ppm downfield from TMS. Therefore, overlap of peaks resulting from different carbons occurs much less often than overlap of hydrogen peaks. It is usually possible to count all of the different types of carbons in a compound by examination of its ^{13}C-NMR spectrum.

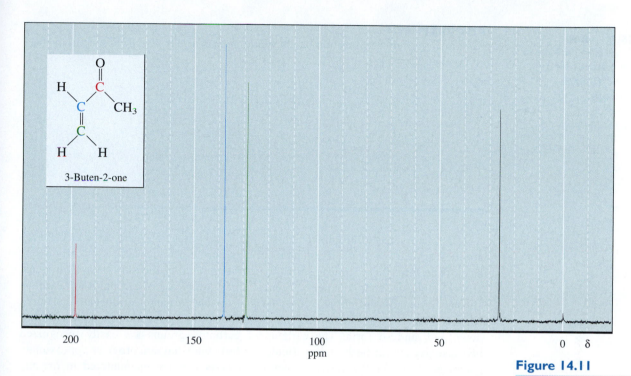

Figure 14.11

THE ¹³C-NMR SPECTRUM
OF 3-BUTEN-2-ONE.

PRACTICE PROBLEM 14.5

How many different absorption bands would appear in the ¹³C-NMR spectrum of benzoic acid?

Benzoic acid

Solution

The compound has a plane of symmetry, so some carbons are chemically equivalent to others. There are five different types of carbons, C-7, C-1, C-2 = C-6, C-3 = C-5, and C-4, so there are five absorptions in the ¹³C-NMR spectrum.

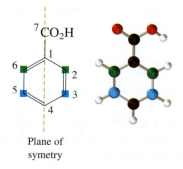

Plane of
symetry

ORGANIC **ChemistryNow™**
Click *Coached Tutorial Problems* for more practice counting carbons in **^{13}C-NMR Spectroscopy.**

PROBLEM 14.9

How many different absorption bands would appear in the ^{13}C-NMR spectra of these compounds?

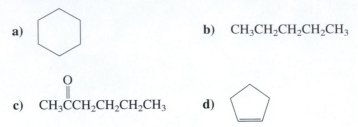

a)

b) $CH_3CH_2CH_2CH_2CH_3$

c) $CH_3\overset{\overset{O}{\|}}{C}CH_2CH_2CH_2CH_3$

d)

The factors that control the chemical shifts of hydrogens, such as the electronegativities of nearby atoms, have similar effects on the chemical shifts of carbon signals. Therefore, the chemical shifts for carbons parallel the shifts for hydrogens, although they are approximately 20 times larger in the case of carbon. Alkyl carbons appear in the most upfield positions, carbons attached to an electronegative element such as oxygen are shifted downfield, the carbons of an aromatic ring appear farther downfield, and so forth. The carbons of carbonyl groups are easily recognized because they appear farthest downfield. The following diagram provides approximate chemical shifts for the various types of carbons that are encountered in organic compounds.

Note the similarity of this diagram to that for proton chemical shifts on page 553, with the difference that the carbon chemical shifts are about 20 times larger than the hydrogen chemical shifts. Table 14.2 provides a somewhat more detailed summary of carbon chemical shifts.

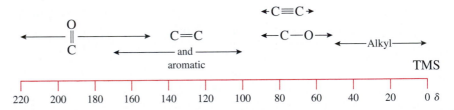

Numerous tables, empirical equations, and even computer programs are available that enable the chemical shifts of the carbons of most compounds to be predicted rather accurately. However, it is possible to assign the carbons responsible for the peaks in many spectra based only on the limited information presented here. For example, the peak at 198.0 δ in Figure 14.11 is assigned to the carbonyl carbon, and the peak at 26.1 δ is due to the methyl carbon. The two alkene carbons appear at 137.5 and 128.5 δ.

PROBLEM 14.10

Assign the absorptions in the ^{13}C-NMR spectra of these compounds to the appropriate carbons:

a) 1-Butanol; absorptions at 61.4, 35.0, 19.1, and 13.6 δ

b) Cyclohexanone; absorptions at 209.7, 41.9, 26.6, and 24.6 δ

Table 14.2   Approximate Chemical Shifts of Carbons in ^{13}C-NMR Spectra

Type of Carbon	Chemical Shift (δ)
1° Alkyl, RCH_3	0–40
2° Alkyl, RCH_2R	10–50
3° Alkyl, $RCHR_2$	15–50
Alkyl halide or amine, $-\overset{\mid}{\underset{\mid}{C}}-X$ (X = Cl, Br, or N—)	10–65
Alcohol or ether, $-\overset{\mid}{\underset{\mid}{C}}-O$	50–90
Alkyne, $-C\equiv$	60–90
Alkene, $\diagdown C=$	100–170
Aryl, $\bigcirc\!\!-C-$	100–170
Nitriles, $-C\equiv N$	120–130
Amides, $-\overset{O}{\overset{\|}{C}}-\overset{\mid}{N}-$	150–180
Carboxylic acids, esters, $-\overset{O}{\overset{\|}{C}}-O$	160–185
Aldehydes, ketones, $-\overset{O}{\overset{\|}{C}}-$	180–215

Let's now deal with the issue of spin coupling in ^{13}C-NMR spectra. No ^{13}C—^{13}C coupling is observed because of the low natural abundance of this isotope. A particular ^{13}C has another ^{13}C adjacent to it in only 1% of the situations, so the split peaks are very small in comparison to the unsplit peaks. However, coupling between carbon and hydrogen is strong and occurs even when they are separated by several intervening bonds. Because the resulting spectra are usually complex and difficult to interpret, C—H coupling is removed by a technique called **broadband decoupling.** In this technique, as the carbon spectrum is being obtained, the sample is simultaneously irradiated with a band of radio-frequency radiation that excites all of the hydrogens. This causes each of the hydrogens to flip rapidly between its two spin states, so its two magnetic field orientations average to zero. No coupling occurs with the carbon and each peak appears as a singlet.

Several techniques have been developed that enable the number of hydrogens attached to the carbon to be determined. An older technique, called **off-resonance decoupling,** allows hydrogens and carbons that are directly bonded to couple but removes any longer-range coupling. In an off-resonance decoupled spectrum, a CH_3 appears as a quartet, a CH_2 appears as a triplet, a CH appears as a doublet, and a C that has no hydrogens bonded to it appears as a singlet. A newer and more convenient technique, called **DEPT-NMR** (<u>d</u>istortionless <u>e</u>nhancement by <u>p</u>olarization <u>t</u>ransfer), also allows the determination of the number of hydrogens attached to each carbon. In a DEPT experiment, three spectra are obtained. One is a normal broadband decoupled spectrum. Another spectrum (DEPT 90° spectrum) is obtained under special conditions in which only carbons bonded to a single hydrogen (CH's) appear. A third spectrum (DEPT 135° spectrum) is obtained under conditions in which CH's and CH_3's appear as positive absorptions and CH_2's appear as negative absorptions. By combining the information in these spectra, each peak can be assigned as resulting from a CH_3, CH_2, CH, or C group. Thus, signals that appear only in the broadband decoupled spectrum are due to C's with no attached H's. Signals that appear in the DEPT 90° spectrum are due to CH groups. Signals that appear as negative peaks in the DEPT 135° spectrum are due to CH_2 groups, and signals that appear as positive peaks in the DEPT 135° spectrum but are absent from the DEPT 90° spectrum are due to CH_3 groups. DEPT and broadband decoupled spectra of ethyl 2-propenoate are shown in Figure 14.12. With the use of computer addition and subtraction of spectra, many modern NMR spectrometers automatically report the results of a DEPT experiment as three spectra: one that shows only peaks due to CH_3 groups, one that shows only peaks due to CH_2 groups, and one that shows only peaks due to CH groups. When combined with the broadband decoupled spectrum, the number of hydrogens on each carbon is readily identified. The remaining ^{13}C-NMR spectra in this book are broadband decoupled spectra with the information obtained from the DEPT spectra indicated above each peak as C, CH, CH_2, or CH_3.

As you examine more ^{13}C-NMR spectra, you will find that the heights (which are proportional to the areas) of the peaks often do not directly correspond to the number of carbons responsible for those peaks. Factors other than the number of carbons also affect the areas of the peaks. Although it is possible to obtain an accurate integral experimentally, the process is time consuming and is not usually done.

In summary, considerable information is available from a ^{13}C-NMR spectrum. First, the number of different carbons can be counted, providing information about the symmetry of the molecule. Second, the chemical environment of the carbons can be deduced from their chemical shifts. Third, information from the DEPT spectra tells how many hydrogens are bonded to each carbon. Some of this same information is provided by the ^1H-NMR spectrum. However, the two types of spectra often provide complementary information and help solidify deductions made with one alone. ^{13}C-NMR is especially useful when the ^1H-NMR spectrum is too complex for ready interpretation.

Let's look at some examples of structure determination using ^{13}C-NMR spectroscopy. Figure 14.13 shows the spectrum of C_8H_{10}. First, calculation indicates that the DU is 4. The compound has some symmetry, because the spectrum shows the presence of only four different types of carbons. When there is symmetry, it is often useful to also count the number of hydrogens indicated by the spectrum, because this may tell us which signals are the result of more than one identical carbon in the compound. In this case the formula obtained by adding the fragments in the spectrum is C_4H_5. This accounts for half of the carbons and half of the hydrogens, suggesting that there are two carbons of each type. Examination of the chemical shifts shows one alkyl type and three alkene/aromatic types. A benzene ring is consistent with the DU of 4. The signal at 19.5 δ is due to a methyl group. The signals at 125.9 and 129.7 δ are due to carbons bonded to one hydrogen, and the signal at 136.2 δ is due to a carbon that is not bonded to any hydrogens. If

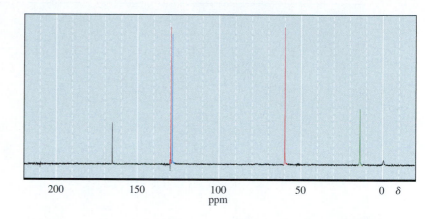

Ethyl 2-propenoate

Broadband decoupled spectrum

DEPT 90° spectrum
(Only the carbons of CH groups appear.)

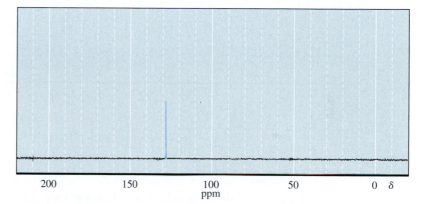

DEPT 135° spectrum
(The carbons of CH₂ groups appear as negative absorptions, whereas the carbons of CH and CH₃ groups appear as positive absorptions.)

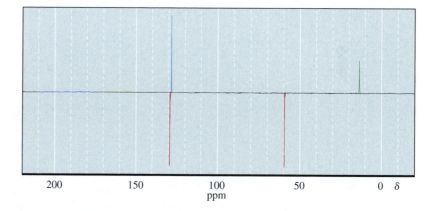

Figure 14.12

BROADBAND DECOUPLED AND DEPT ¹³C-NMR SPECTRA OF ETHYL 2-PROPENOATE. The peaks are assigned as follows: the peak at 166 δ is due to a C with no attached H's (C-1) because it appears only in the broadband decoupled spectrum; the peak at 129.7 δ is due to an alkene CH₂ (C-3) because it appears as a negative peak in the DEPT 135° spectrum; the peak at 128.7 δ is due to an alkene CH (C-2) because it appears in the DEPT 90° spectrum; the peak at 60 δ is due to a CH₂ (C-4) because it appears as a negative peak in the DEPT 135° spectrum; and the peak at 14 δ is due to a CH₃ (C-5) because it appears as a positive peak in the DEPT 135° spectrum but does not appear in the DEPT 90° spectrum.

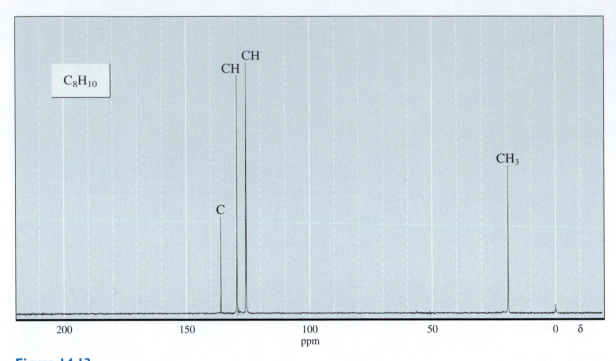

Figure 14.13

THE ^{13}C-NMR SPECTRUM OF C_8H_{10}.

there are indeed two of each of these types of carbons, then the fragments 2 CH_3, 2 CH, 2 CH, and 2 C can readily be assembled to form dimethylbenzene (xylene). However, there are three isomers of xylene: ortho, meta, and para.

o-Xylene	*m*-Xylene	*p*-Xylene
4 ^{13}C-NMR signals	5 ^{13}C-NMR signals	3 ^{13}C-NMR signals

These isomers can be distinguished on the basis of their ^{13}C-NMR spectra because they have different symmetries. The *ortho*-isomer has four different carbons, the *meta*-isomer has five, and the *para*-isomer has only three. The unknown must be *ortho*-xylene. Note that this compound could be identified as one of the xylene isomers on the basis of its ^{1}H-NMR spectrum, but it would be difficult to establish which isomer it is from just that information.

As another example, consider the spectrum of $C_6H_{12}O$, shown in Figure 14.14. The DU is 1. Examination of the ^{13}C-NMR spectrum shows five different types of carbons.

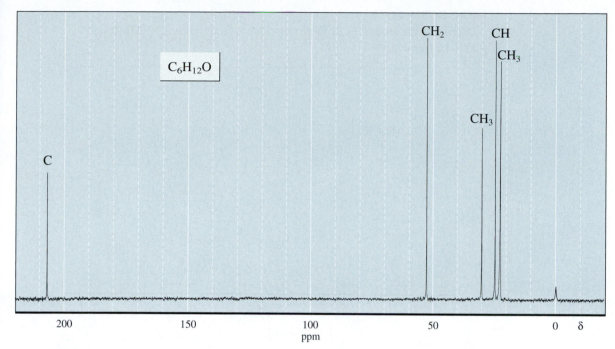

Figure 14.14

THE ^{13}C-NMR SPECTRUM OF $C_6H_{12}O$.

Therefore, there must be two carbons that appear at identical chemical shifts because they are chemically equivalent. The spectrum accounts for only nine hydrogens, so the extra carbon is probably a CH_3 group that is identical to one of the other CH_3 groups. (However, remember that OH and NH hydrogens will not appear in the ^{13}C-NMR spectrum.) The peak at 207.3 δ is due to a carbonyl carbon. (This accounts for the one degree of unsaturation.) It has no hydrogens bonded to it, so it is bonded to two carbons—the unknown is a ketone. From the remaining peaks in the spectrum the following fragments can be deduced:

$$
\underset{207.3\,\delta}{\overset{\displaystyle O \atop \displaystyle \|}{-C-}} \qquad \underset{53.0\,\delta}{CH_2} \qquad \underset{30.4\,\delta}{CH_3} \qquad \underset{25.0\,\delta}{CH} \qquad \underset{22.8\,\delta}{CH_3}
$$

These fragments total to C_5H_9O, so there must indeed be an additional CH_3 group to account for the actual formula of $C_6H_{12}O$. The two alkyl carbons that appear farthest downfield, the CH_2 at 53.0 δ and the CH_3 at 30.4 δ, are probably bonded to the electronegative carbonyl group. By putting the remaining fragments together, we see that the compound is 4-methyl-2-pentanone. The two methyl groups of the isobutyl group are chemically equivalent, and both appear at 22.8 δ.

$$
\underset{\text{4-Methyl-2-pentanone}}{CH_3-\overset{\displaystyle O \atop \displaystyle \|}{C}-CH_2-\overset{\displaystyle CH_3 \atop \displaystyle |}{CH}-CH_3}
$$

PROBLEM 14.11

Determine the structures of these compounds from their ^{13}C-NMR spectra:

a) The formula is C_7H_5ClO.

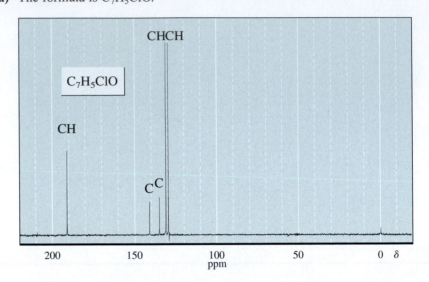

b) The formula is C_5H_{10}.

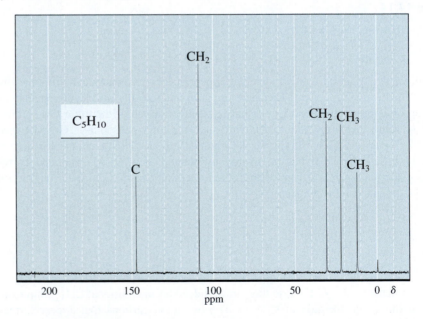

c) The formula is C_6H_{10}.

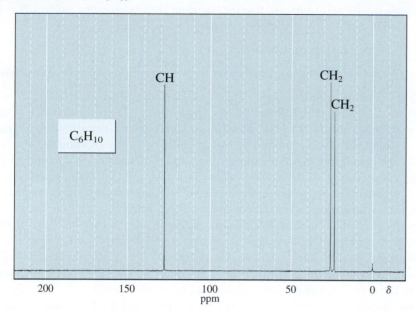

14.10   SOLVED PROBLEMS EMPLOYING
IR AND NMR SPECTRA

Most structure determination problems that are encountered in real-life situations in the laboratory rely on a combination of these spectral methods for solution. A general strategy that is often successful is as follows:

Step 1.   *Calculate the degree of unsaturation if the formula is available.*

Step 2.   *Examine the IR spectrum to determine the functional group.* Do not forget the information provided by the formula. For example, a compound with only one oxygen cannot be a carboxylic acid, nor can a compound with a DU of 3 contain a benzene ring. Some structural features suggested by the IR spectrum can be rapidly confirmed by examination of the ^{1}H-NMR spectrum. For example, it is often difficult to be confident of the presence of an aromatic ring based solely on examination of the IR spectrum. However, hydrogens on a benzene ring appear in a characteristic region of the NMR spectrum. Likewise, the H of an aldehyde group and the H of a carboxylic acid can be readily identified in the NMR spectrum.

Step 3.   *Examine the ^{1}H-NMR spectrum as described on pages 566–567.* Examine the general positions of the peaks. Examine the integral for the ratios of the different kinds of hydrogens. Examine the coupling patterns. Remember that the number of nearby hydrogens coupled to the hydrogen(s) being examined is one less than the number of peaks.

Step 4.   *Examine the ^{13}C-NMR spectrum.* Count the number of peaks to see if there are any identical carbons. Use the chemical shift information to identify carbonyl, aromatic, and alkene carbons.

Step 5.   *Construct a tentative structure.*

Step 6.   *Check to see if all of the information is consistent with the proposed structure.*

Let's try determining the structure of an unknown using IR and ^{1}H-NMR. Figure 14.15 shows the IR and ^{1}H-NMR spectra of $C_4H_8O_2$. The DU is 1. Examination of the IR spectrum shows only hydrogens bonded to sp^3-hybridized carbons in the hydrogen region. A strong carbonyl absorption occurs at 1746 cm^{-1}. The CO double bond accounts for the one degree of unsaturation. The absence of bands in the appropriate regions for carboxylic acids, aldehydes, and so on, indicates that the compound is a ketone or an ester. The strong absorption at 1204 cm^{-1} suggests that the compound is an ester. This is supported by the carbonyl position (see Table 13.1).

The only absorptions in the NMR spectrum occur in the alkyl region. The 3 H singlet at 3.66 δ suggests a methyl group. The quartet at 2.32 δ is due to a CH$_2$ group that is split by three H's (four lines minus one equals three H's), probably the three H's of a CH$_3$ group. This CH$_3$ group appears as the triplet (split by the two H's of the CH$_2$ group) at 1.12 δ. These two signals indicate the presence of an ethyl group.

On the basis of the IR and the NMR spectra the following fragments are known to be present:

$$\underset{\overset{\|}{O}}{-\text{C}}-\text{O}- \qquad\qquad \text{CH}_3- \qquad\qquad \text{CH}_3-\text{CH}_2-$$

Two possible esters can be assembled from these fragments:

$$\text{CH}_3-\underset{\overset{\|}{O}}{\text{C}}-\text{O}-\text{CH}_2\text{CH}_3 \qquad\qquad \text{CH}_3\text{CH}_2-\underset{\overset{\|}{O}}{\text{C}}-\text{O}-\text{CH}_3$$

$$\text{Ethyl acetate} \qquad\qquad\qquad\qquad \text{Methyl propanoate}$$

These esters can be readily distinguished on the basis of chemical shifts. On the basis of Table 14.1, the methyl group attached to the carbonyl of ethyl acetate is predicted to appear near 2.0 δ, and the CH$_2$ of the ethyl group should appear slightly downfield from 3.7 δ. For methyl propanoate the predictions are 3.7 δ for the methyl group attached directly to the oxygen of the ester and slightly downfield from 2.0 δ for the CH$_2$ bonded to the carbonyl group. In other words, because the signal for the CH$_3$ is further downfield than the signal for the CH$_2$ of the ethyl group, the CH$_3$ must be bonded to the oxygen. The compound is methyl propanoate.

Let's now try an example using IR and both types of NMR. Figure 14.16 shows the IR, ^{1}H-NMR, and ^{13}C-NMR spectra for $C_8H_{19}N$. The DU for this compound is zero, so it has no pi bonds or rings. On the basis of the presence of nitrogen and DU = 0, the unknown must be an amine. The absorption bands from 3000 to 2800 cm^{-1} in the IR spectrum show hydrogens bonded to sp^3-hybridized carbons, as expected. The small absorption at 3280 cm^{-1} suggests that the compound is a secondary amine, although caution must be exercised when assigning a weak band such as this one.

In the ^{1}H-NMR spectrum the peak at 2.61 δ can be recognized as a distorted triplet. This suggests that the hydrogens that are responsible for this signal are near two hydrogens, but the coupling is beginning to deviate from first order. The complex multi-

Figure 14.15

THE ⓐ IR AND ⓑ ¹H-NMR
SPECTRA OF C₄H₈O₂.

ⓐ

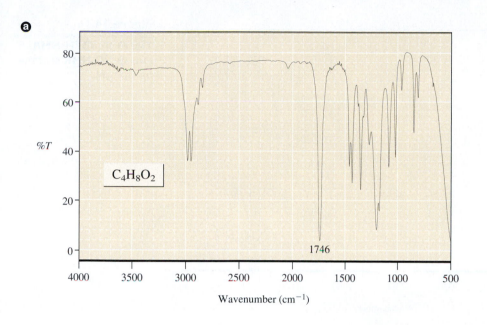

ⓑ

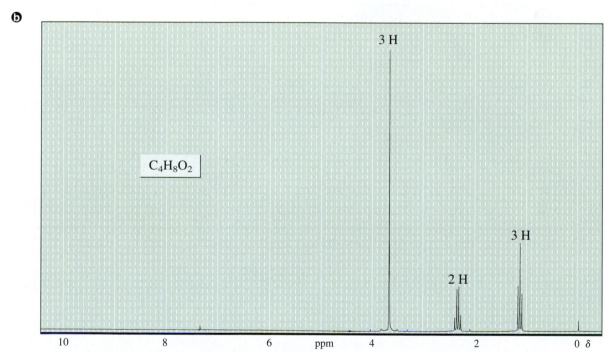

plet that appears from 1.2 to 1.6 δ indicates eight hydrogens with complex coupling. The peak near 0.9 δ resembles a distorted triplet, with, perhaps, an overlapping peak on the downfield side.

Although the ¹H-NMR spectrum has not provided much helpful information in this case, the ¹³C-NMR spectrum is quite useful. The spectrum has only four peaks, indicating that the compound has symmetry. There are probably two carbons of each type.

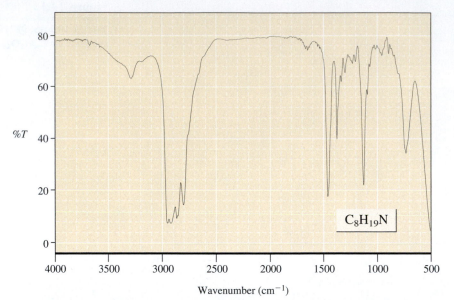

Figure 14.16

The ⓐ IR, ⓑ ^{1}H-NMR, and ⓒ ^{13}C-NMR spectra of $C_8H_{19}N$.

Wavenumber (cm^{-1})

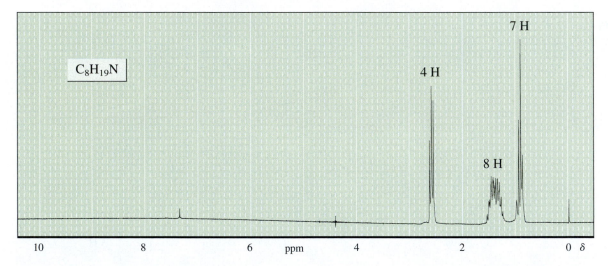

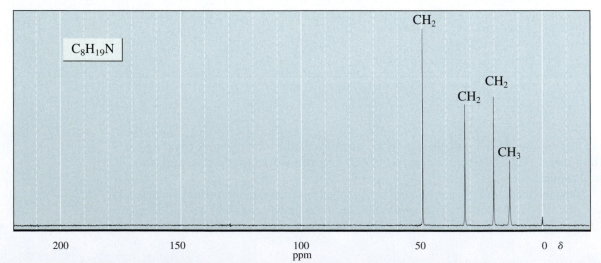

There are three CH_2 groups and one CH_3 group. These fragments can be assembled only into a butyl group. This group is attached to a nitrogen. The carbons closer to the nitrogen are shifted farther downfield:

$$-CH_2-$$

$$-CH_2- \qquad CH_3- \quad \longrightarrow \quad N-CH_2-CH_2-CH_2-CH_3$$

$$-CH_2- \qquad\qquad\qquad\qquad \uparrow \quad\ \uparrow \quad\ \uparrow \quad\ \uparrow$$

$$\qquad\qquad\qquad\qquad\qquad\qquad 50.0 \quad 32.7 \quad 20.7 \quad 14.1\ \delta$$

On the basis of the symmetry of the molecule, it must have two chemically equivalent butyl groups. This accounts for 18 hydrogens. The 19th hydrogen must be bonded to the nitrogen (recall the NH absorption in the IR spectrum). These fragments can now be assembled to produce dibutylamine:

$$CH_3CH_2CH_2CH_2- \\ \qquad\qquad\qquad\qquad N-H \quad \longrightarrow \qquad CH_3CH_2CH_2CH_2 \\ CH_3CH_2CH_2CH_2- \qquad\qquad\qquad\qquad\qquad\qquad CH_3CH_2CH_2CH_2 \quad NH$$

Dibutylamine

At this point we should check to make certain that this structure is consistent with the chemical shifts and coupling in the ^{1}H-NMR spectrum. Note that the H on the N is responsible for the peak near 1 δ on the downfield side of the distorted triplet.

This chapter has provided some examples and general guidelines for the use of IR and NMR spectra to solve structure problems. However, each problem is unique, and the exact procedure will vary from problem to problem. The best way to become adept at using these techniques is to work as many problems as possible.

PROBLEM 14.12

Determine the structures of these compounds from their IR and ^{1}H-NMR spectra:

a) The formula is $C_9H_{10}O$.

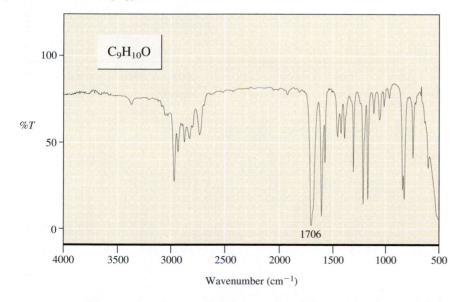

C₉H₁₀O

1706

Wavenumber (cm^{-1})

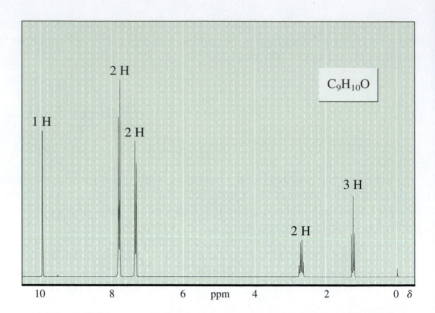

b) The formula is $C_5H_9BrO_2$.

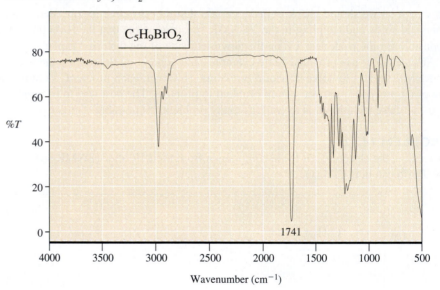

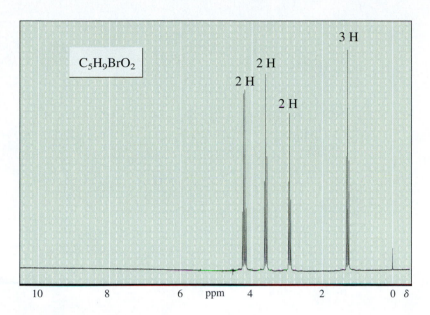

c) The formula is $C_8H_{10}O$.

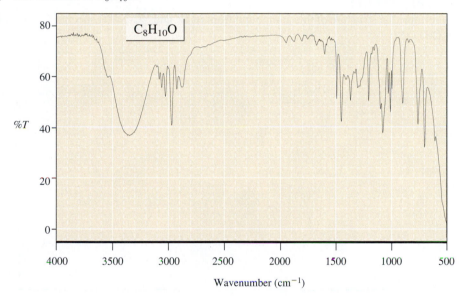

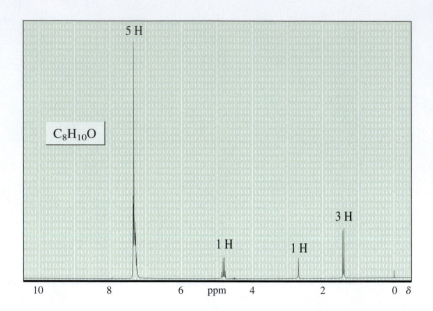

PROBLEM 14.13

Determine the structure of this compound from its IR and ^{13}C-NMR spectra. Its formula is $C_7H_{16}O_2$:

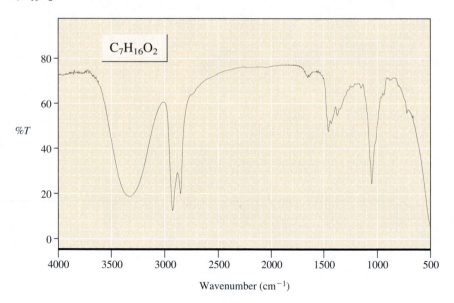

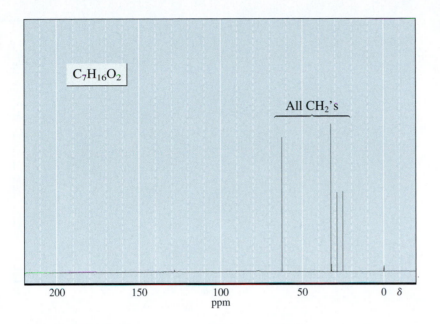

PROBLEM 14.14

Determine the structures of these compounds from their IR, ^{1}H-NMR, and ^{13}C-NMR spectra:

a) The formula is $C_5H_{12}O$.

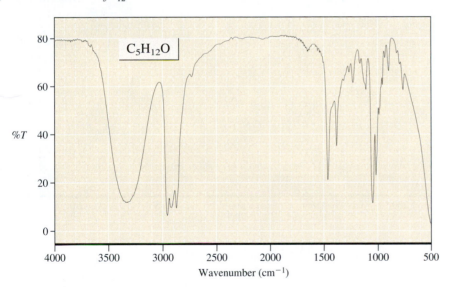

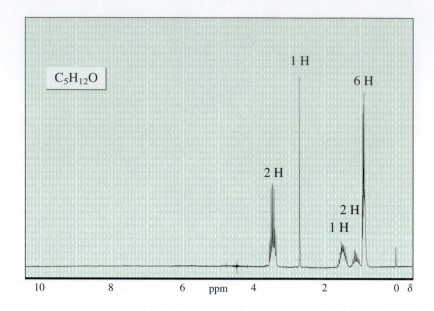

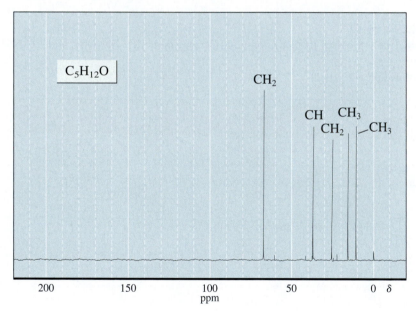

b) The formula is $C_7H_{12}O$.

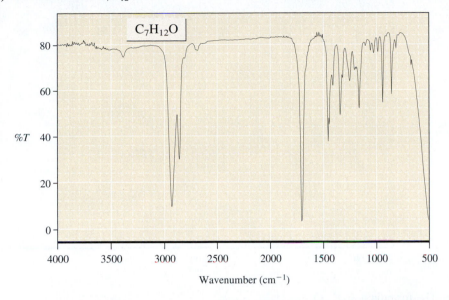

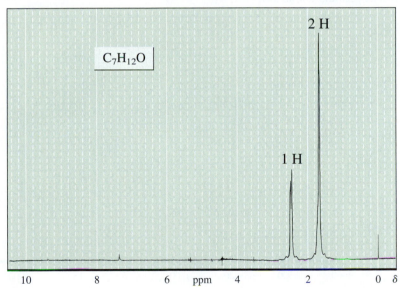

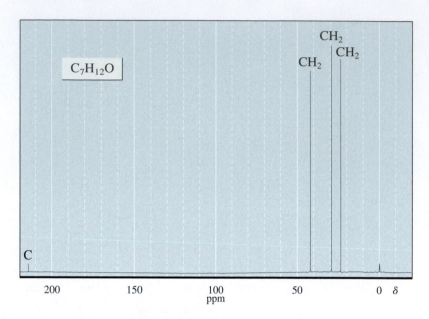

c) The formula is $C_9H_{10}O$.

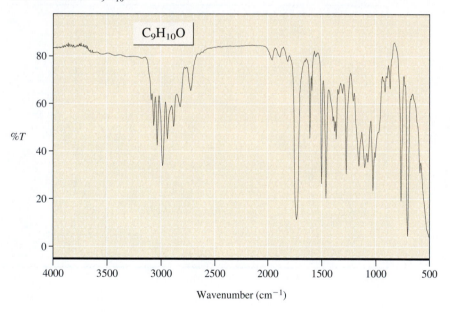

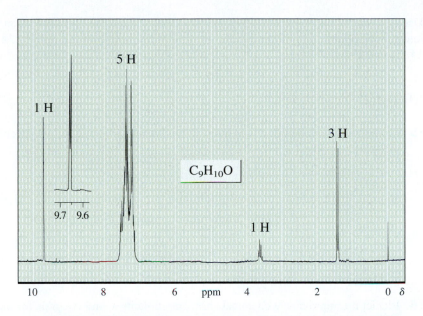

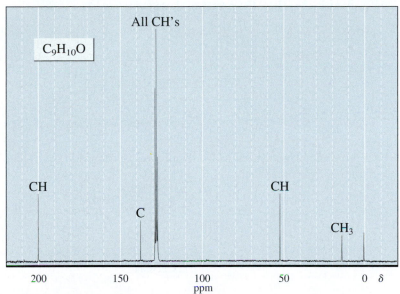

Review of Mastery Goals

After completing this chapter, you should be able to:

- Predict the approximate chemical shifts, multiplicity, and integrals of peaks in the ^{1}H-NMR spectrum of a compound. (Problems 14.15, 14.16, 14.30, and 14.32)

- Predict the number and approximate chemical shifts of peaks in the ^{13}C-NMR spectrum of a compound. (Problems 14.17 and 14.33)

- Determine the hydrocarbon skeleton of a compound by examination of its ^{1}H and/or ^{13}C-NMR spectrum. (Problems 14.18, 14.19, 14.20, and 14.28)

ORGANIC
Chemistry ⚛ Now™
Click *Mastery Goal Quiz* to test
how well you have met these
goals.

■ Use a combination of IR and NMR spectra to determine the structure of an unknown compound. (Problems 14.21, 14.22, 14.23, 14.24, 14.25, 14.26, 14.27,
14.29, and 14.31)

Additional Problems

14.15 Predict the multiplicities of the indicated hydrogens in the ^{1}H-NMR spectra of
these compounds:

a)

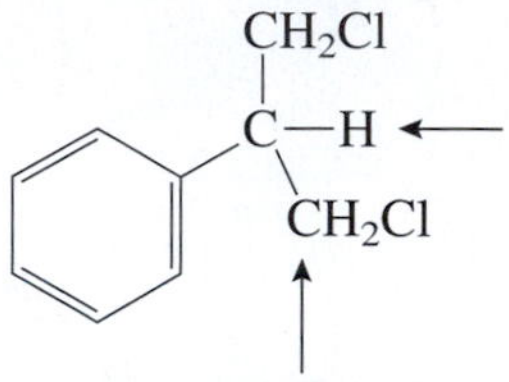

CH$_2$Cl

C—H ←

CH$_2$Cl

b) H—C—C—C—H

Cl H Cl

Cl H Cl

c)

Cl   H ←

C=C

H   H ←

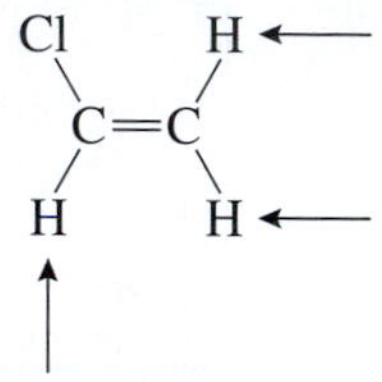

14.16 Predict the approximate chemical shifts, multiplicities, and integrals for the
absorptions in the ^{1}H-NMR spectra of these compounds:

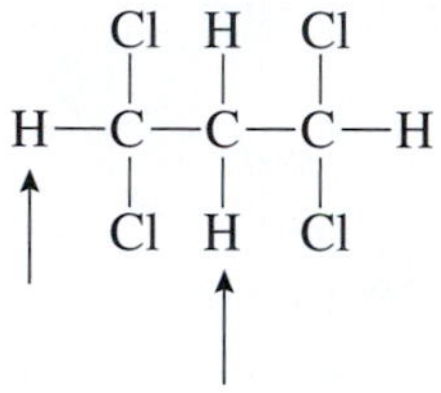

a) 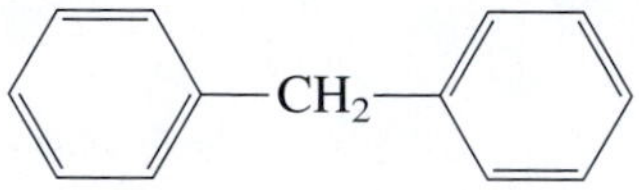 —CH$_2$—

b) 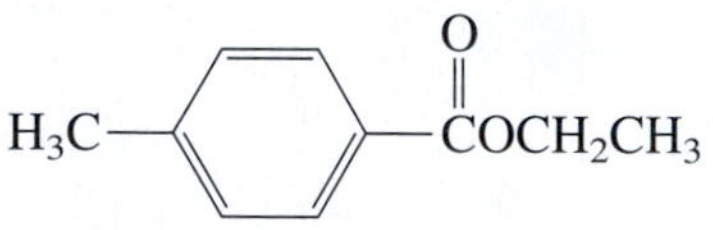 H$_3$C—

O
‖
COCH$_2$CH$_3$

c) CH$_3$CH$_2$NHCH$_2$CH$_3$

d) CH$_3$CHCH$_2$CH$_3$

OH

14.17 Predict the approximate chemical shifts of the absorptions in the ^{13}C-NMR
spectrum of this compound:

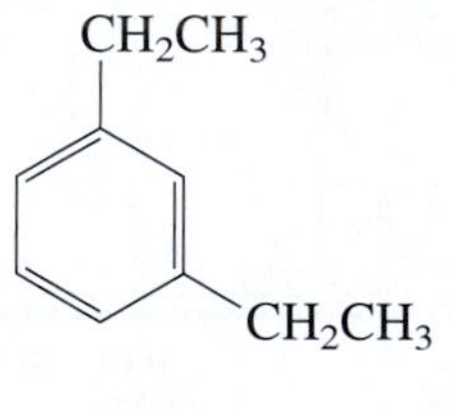

CH$_2$CH$_3$

CH$_2$CH$_3$

14.18 Suggest how these compounds could be distinguished by using NMR spectroscopy:

a)

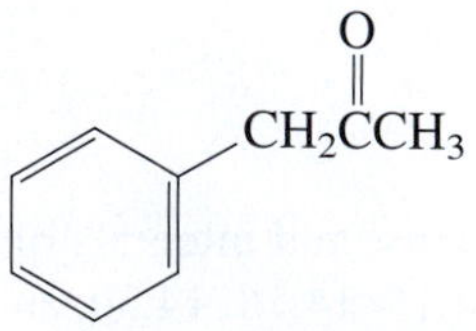

O
‖
CH$_2$CCH$_3$

and

O
‖
CCH$_2$CH$_3$

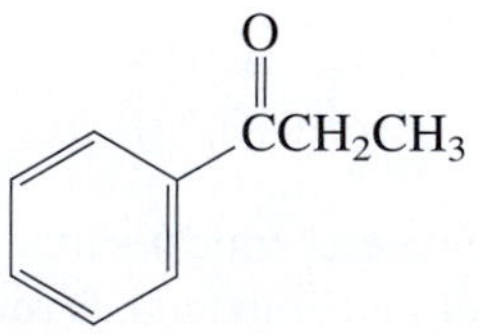

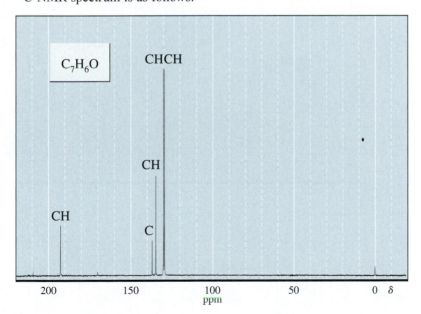

b) 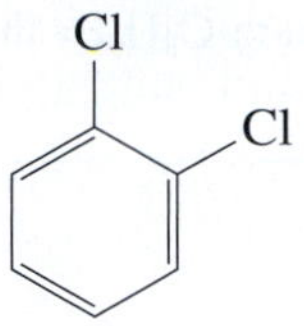   and

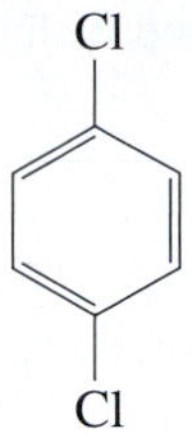

c) 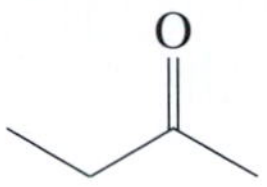   and

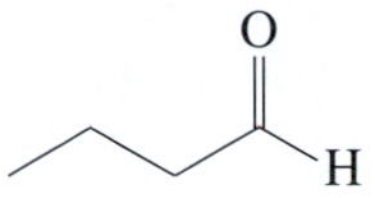

d) 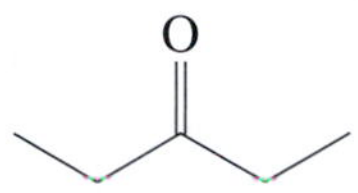   and

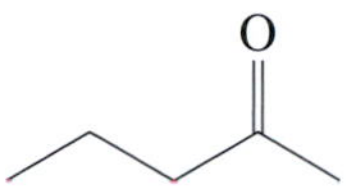

14.19 Suggest structures for these compounds. First the formula is given. Then the absorptions in the ^{1}H-NMR are listed as chemical shift (multiplicity, integral). Some compounds also have an important IR peak given.

a) C_4H_9Cl; 3.35 δ (doublet, 2), 2.0 δ (multiplet, 1), 1.0 δ (doublet, 6).

b) $C_3H_6Br_2$; 3.6 δ (triplet, 2), 2.4 δ (quintuplet, 1).

c) $C_4H_{10}O$; 3.3 δ (doublet, 2), 2.4 δ (singlet, 1), 1.7 δ (multiplet, 1), 0.9 δ (doublet, 6); 3330 cm^{-1}, broad.

d) $C_5H_{10}O$; 2.6 δ (septet, 1), 2.1 δ (singlet, 3), 1.1 (doublet, 6); 1716 cm^{-1}.

14.20 Suggest a structure for the compound with the formula C_7H_6O whose ^{13}C-NMR spectrum is as follows:

14.21 Suggest a structure for the compound with the formula C_8H_9Br that has the following IR and ^{1}H-NMR spectra:

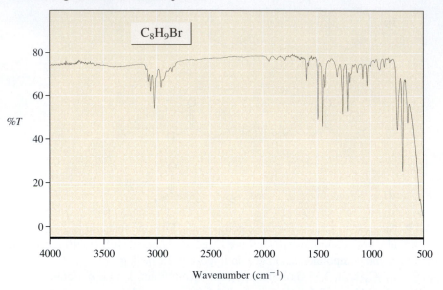

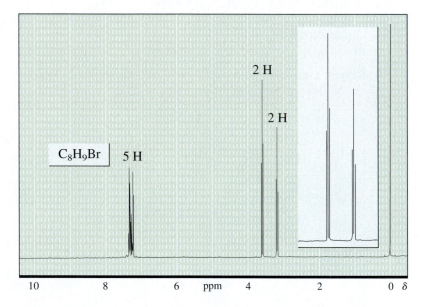

14.22 Suggest a structure for the compound with the formula $C_5H_7NO_2$ that has the following IR and ^{1}H-NMR spectra:

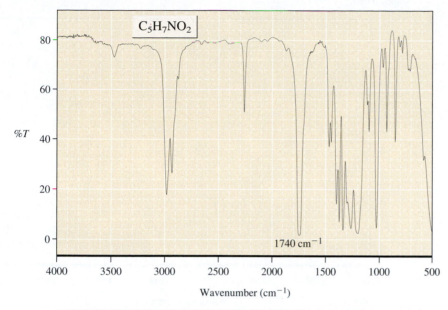

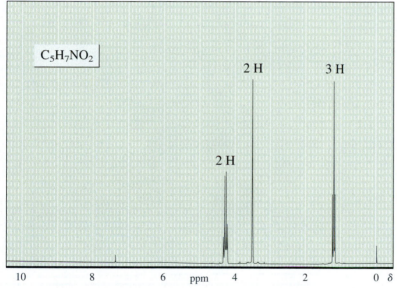

14.23 Suggest a structure for the compound with the formula $C_5H_{10}O_2$ that has the following IR and ^{1}H-NMR spectra. (Some absorptions overlap in the NMR spectrum.)

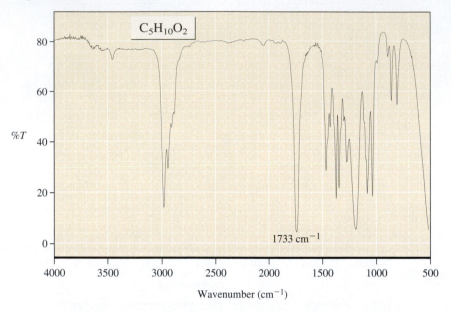

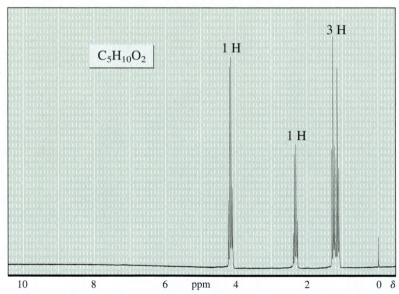

14.24 Suggest a structure for the compound with the formula $C_6H_{10}O_2$ that has the following IR, ^{1}H-NMR, and ^{13}C-NMR spectra:

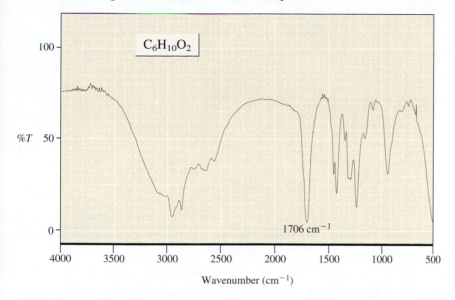

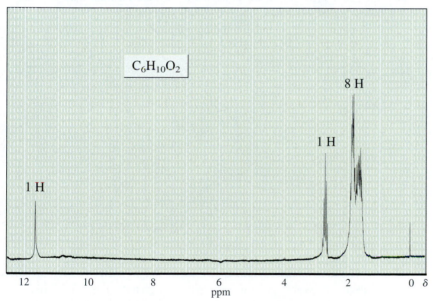

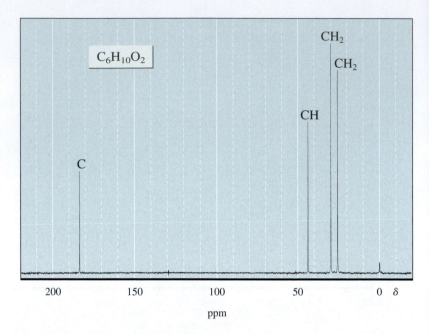

14.25 Suggest a structure for the compound with the formula $C_6H_{12}O_2$ that has the following IR, ^{1}H-NMR, and ^{13}C-NMR spectra:

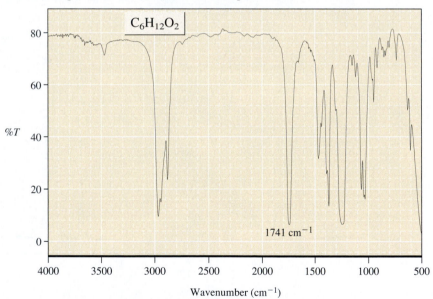

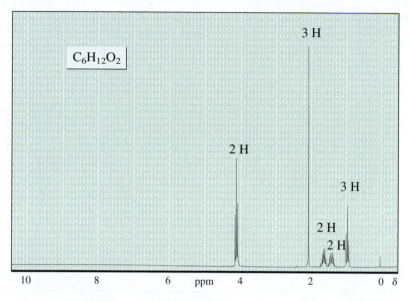

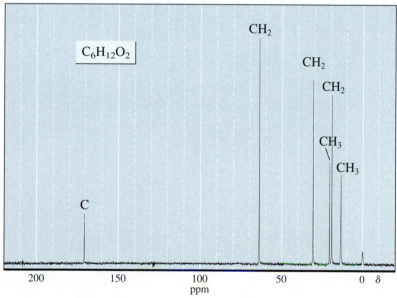

14.26 Suggest a structure for the compound with the formula $C_8H_{18}O$ that has the following IR, ^1H-NMR, and ^{13}C-NMR spectra:

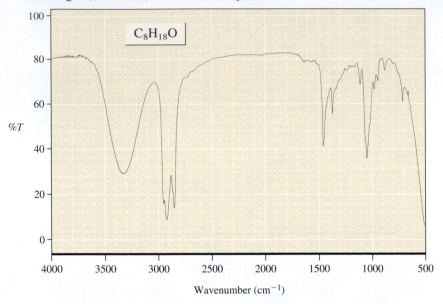

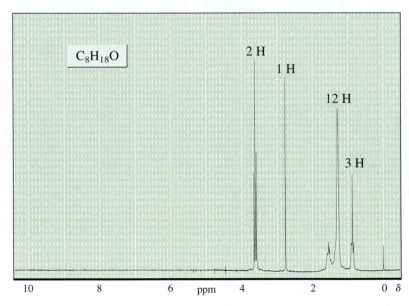

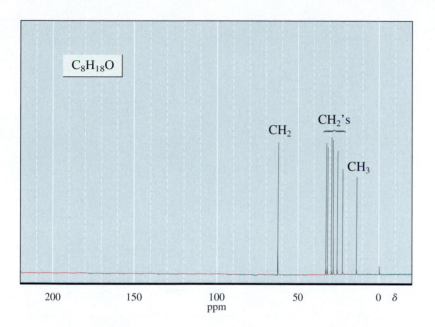

14.27 Suggest a structure for the compound with the formula $C_7H_{12}O$ that has the following IR, ^1H-NMR, and ^{13}C-NMR spectra:

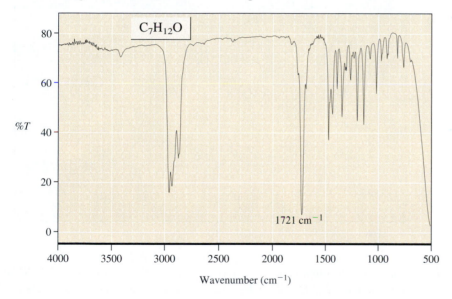

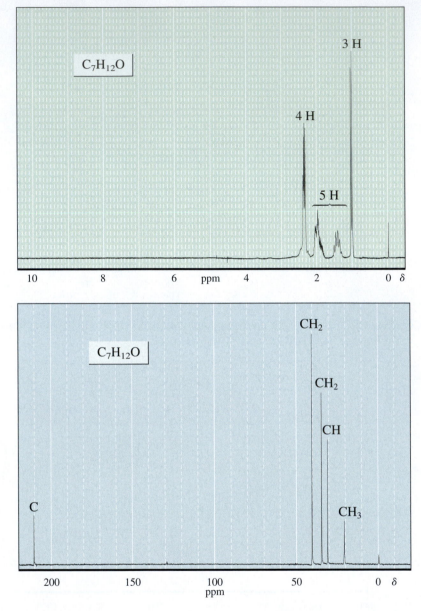

14.28 An unknown compound, **A** (C_7H_{10}), shows four absorptions in its ^{13}C-NMR spectrum, at 22 (CH_2), 24 (CH_2), 124 (CH), and 126 (CH) δ. On reaction with excess H_2 and a Pt catalyst, **A** produces **B** (C_7H_{14}). **B** shows a single peak at 28.4 δ in its ^{13}C-NMR spectrum. Ozonolysis of **A** gives $C_2H_2O_2$ and $C_5H_8O_2$. Suggest structures for **A** and **B**.

14.29 The product of the following reaction has a broad absorption at 3330 cm^{-1} in its IR spectrum. Its ^{13}C-NMR spectrum shows absorptions at 70 (C), 34 (CH$_2$), 30 (CH$_3$), and 15 (CH$_3$). Suggest a structure for this compound.

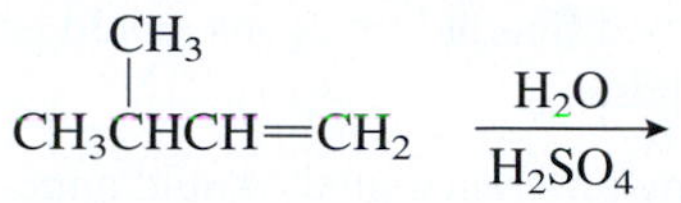

$$\underset{\displaystyle CH_3CHCH=CH_2}{\overset{\displaystyle CH_3}{|}} \quad \xrightarrow[\;H_2SO_4\;]{\;H_2O\;}$$

14.30 Explain how many absorptions appear in the ^{1}H-NMR spectrum of this compound:

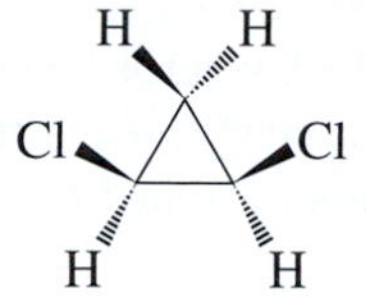

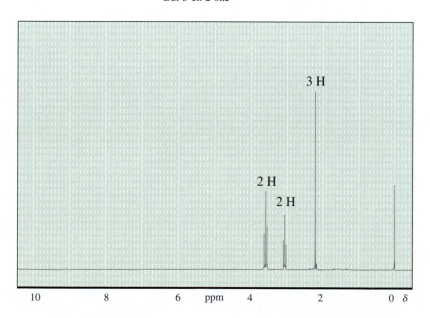

14.31 The addition of HCl to but-3-en-2-one gives a product with the following ^{1}H-NMR spectrum. Show the structure of this product, show a mechanism for its formation, and explain the regiochemistry of the reaction (see Section 11.2).

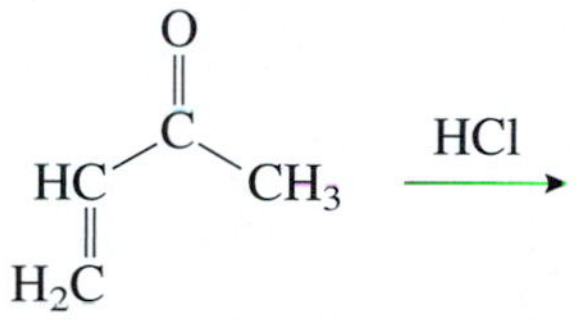

But-3-en-2-one

Problems Using Online Three-Dimensional Molecular Models

14.32 Explain how many different hydrogens would appear in the ^{1}H-NMR spectra of these compounds.

14.33 Explain how many different signals would appear in the ^{13}C-NMR spectra of these compounds.

Do you need a live tutor for homework problems? Access vMentor at Organic ChemistryNow at **http://now.brookscole.com/hornback2** for one-on-one tutoring from a chemistry expert.

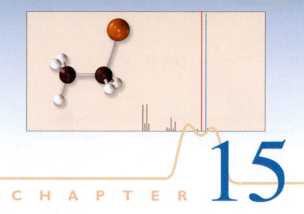

Ultraviolet-Visible Spectroscopy and Mass Spectrometry

CHAPTER 15

Mastering Organic Chemistry

- Determining Whether a Compound Will Absorb Ultraviolet or Visible Light
- Identifying the Chromophore and Type of Transition Responsible for Absorption of UV Radiation
- Determining the Presence of Sulfur, Chlorine, Bromine, and Nitrogen from a Mass Spectrum
- Understanding the Major Fragmentation Pathways for Compounds Containing Some Simple Functional Groups

In addition to IR and NMR spectroscopy there are many other instrumental techniques that are useful to the organic chemist. Two of these, ultraviolet-visible spectroscopy and mass spectrometry, are discussed in this chapter. Ultraviolet-visible spectroscopy is presented first. The use of this technique to obtain information about the conjugated part of a molecule is described. Then mass spectrometry is discussed. This technique provides the molecular mass and formula for a compound. In addition, the use of the mass spectrum to provide structural information about the compound under investigation is presented.

15.1 Ultraviolet-Visible Spectroscopy

As discussed in Chapter 13, the ultraviolet (UV) region of the spectrum covers the range from 200 to 400 nm, and the visible region covers the range from 400 to 800 nm. The amount of energy available in this radiation, ranging from 143 kcal/mol (600 kJ/mol) to 36 kcal/mol (150 kJ/mol), is enough to cause an **electronic transition** in a molecule, that is, to excite an electron from an occupied MO to an antibonding MO.

Figure 15.1 shows a general diagram for such an electronic transition. The lowest-energy electron arrangement, the ground electronic state, is illustrated on the left side of the diagram. Only the highest-energy orbital that is occupied by electrons (the **highest occupied MO,** or **HOMO**) and the lowest-energy empty MO (**lowest unoccupied MO,** or **LUMO**) are shown in this diagram. In general, there are other occupied MOs at lower energies than the HOMO and other unoccupied MOs at higher energies than the LUMO (see Section 3.9). The HOMO

ORGANIC Chemistry•Now™
Look for this logo in the chapter and go to OrganicChemistryNow at **http://now.brookscole.com/hornback2** for tutorials, simulations, problems, and molecular models.

Figure 15.1

GENERAL DIAGRAM FOR AN
ELECTRONIC TRANSITION.

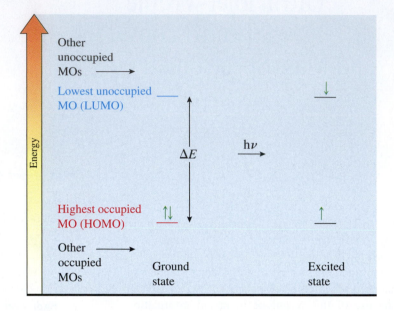

may be a bonding or a nonbonding MO; the LUMO is usually an antibonding MO. When the energy of the light matches the energy difference between the HOMO and the LUMO, the light is absorbed and an electron is promoted to the LUMO, producing the excited state. The electronic transition shown in Figure 15.1 is the lowest-energy one. Other transitions, resulting from the excitation of an electron from the HOMO to higher-energy unoccupied MOs or from the excitation of an electron from lower-energy occupied MOs to any of the unoccupied MOs, are also possible. Of course, all of these higher-energy transitions result in the absorption of shorter wavelengths of light.

Figure 15.2 shows the UV spectrum of 2,5-dimethyl-2,4-hexadiene. The wavelength of the light, in nanometers, is plotted along the x-axis, and the absorption band comes up from this axis, as was the case for NMR spectra. The wavelength range for most UV spectra begins at 200 nm because the O_2 of air and quartz glass absorb light with wavelengths shorter than this. Most absorption bands due to electronic transitions are broad and rather featureless like this one because each electronic energy level has numerous vibrational and rotational sublevels (see Figure 13.2).

The amount of light absorbed by the sample is plotted along the y-axis as the **absorbance** (A), which is defined as

$$A = \log\left(\frac{I_0}{I}\right)$$

where I_0 is the intensity of the light striking the sample and I is the intensity of the light emerging from the sample. Absorbance is directly proportional to the concentration of the sample and the path length through the sample. The equation expressing this proportionality, known as the Lambert-Beer law, is

$$A = \epsilon c l$$

where ϵ is the **molar absorptivity** or the **molar extinction coefficient,** in units of M^{-1} cm^{-1}, c is the concentration of the compound in moles per liter, and l is the path length

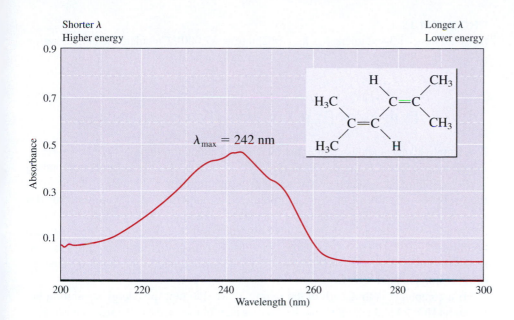

Figure 15.2

THE ULTRAVIOLET SPECTRUM
OF 2,5-DIMETHYL-
2,4-HEXADIENE.

in centimeters. (The Lambert-Beer law is very useful for quantitative analysis. You probably used this equation in the general chemistry laboratory to measure the concentration of a dissolved solute that absorbed UV or visible light.)

For a particular compound, the wavelength at the maximum of the absorption band (λ_{max}) and the extinction coefficient at the maximum (ϵ_{max}) are characteristic constants for that compound and are often listed in reference books along with the melting point, boiling point, and other physical constants of the compound. The solvent in which the sample is dissolved is also reported because λ_{max} and ϵ_{max} vary slightly with the solvent. 2,5-Dimethyl-2,4-hexadiene, whose UV spectrum is shown in Figure 15.2, has λ_{max} = 242 nm and ϵ_{max} = 13,100 in methanol as solvent.

Although the λ_{max} and ϵ_{max} can be used like a melting point to aid in the identification of a compound, they also provide information about the energy separation between the MOs of the compound. Let's see how to use this to obtain information about the structure of the compound.

PROBLEM 15.1

Anthracene has $\epsilon = 1.80 \times 10^5 \ M^{-1} \ cm^{-1}$ at λ_{max} = 256 nm. Calculate the absorbance of a $1.94 \times 10^{-6} \ M$ solution of anthracene in a 1 cm cell.

PROBLEM 15.2

A solution of 0.0014 g of benzophenone in 1 L of ethanol has A = 0.153 (1 cm cell) at λ_{max} = 252 nm. Calculate the molar absorptivity of benzophenone.

Benzophenone

PROBLEM 15.3

trans-1-Phenyl-1,3-butadiene has λ_{max} = 280 (ϵ = 27,000). Calculate the concentration of a solution that has A = 0.643 at 280 nm in a 1 cm cell.

15.2 Types of Electronic Transitions

In a simple alkane such as ethane, the HOMO is a sigma bonding MO (it is not really important whether it is the σ_{CC} or a σ_{CH} MO) and the LUMO is a sigma antibonding MO (see Figure 3.23). The lowest-energy electronic transition that can occur involves the excitation of an electron from a bonding to an antibonding MO and is termed a $\sigma \longrightarrow \sigma^*$ transition. This transition occurs at high energy in ethane (λ_{max} = 135 nm) and other alkanes because sigma bonds are strong and the energy separation between bonding and antibonding MOs is large. Therefore, alkanes are transparent in the accessible UV region above 200 nm and are often used as solvents for obtaining UV spectra of other compounds.

If a compound is to absorb in the region above 200 nm, the energy separation between its HOMO and LUMO must be smaller than in the case of alkanes. The bonding and antibonding MOs for pi bonds, which are weaker than sigma bonds, are closer together in energy (see Figure 3.24). Although the $\pi \longrightarrow \pi^*$ transition for ethene (λ_{max} = 165 nm, ϵ_{max} = 10,000) still does not occur in the accessible UV region, it does occur at lower energy and longer wavelength than the $\sigma \longrightarrow \sigma^*$ transition of ethane.

Delocalized pi MOs must be used to describe compounds with conjugated pi bonds. This topic is covered in more detail in Chapters 16 and 22. For now it is enough to note that the energy of the highest bonding pi MO increases and the energy of the lowest antibonding pi MO decreases as the number of conjugated pi bonds in the compound increases. Figure 15.3 illustrates this for the series ethene, 1,3-butadiene, and 1,3,5-hexatriene. As would be expected based on this figure, λ_{max} for the $\pi \longrightarrow \pi^*$ transition increases as the number of conjugated pi bonds increases and moves into the accessible UV region for dienes and trienes. As the **chromophore** (the part of the molecule that is responsible for the absorption of ultraviolet or visible light) becomes more conjugated, the absorption maximum moves to even longer wavelengths. Note also that the molar absorptivities for these $\pi \longrightarrow \pi^*$ transitions are all quite large.

In the case of an aldehyde or a ketone, the chromophore is the carbonyl group. The $\pi \longrightarrow \pi^*$ transition for acetone, like that for ethene, occurs below 200 nm (λ_{max} = 188 nm) because there is no conjugation. However, acetone has electrons in nonbonding orbitals that are higher in energy than the pi electrons. The longest wavelength absorption for acetone is due to an $n \longrightarrow \pi^*$ transition (λ_{max} = 279 nm, ϵ_{max} = 15). The molar absorptivity for the $n \longrightarrow \pi^*$ transition is much smaller than that for a typical $\pi \longrightarrow \pi^*$ transition because the n orbital and the π^* occupy different regions of space, resulting in a less probable transition.

Another chromophore that is commonly encountered in organic compounds is the benzene ring. Benzene itself has $\pi \longrightarrow \pi^*$ transitions with λ_{max} = 204 nm (ϵ_{max} = 7,900) and 256 nm (ϵ_{max} = 200).

PROBLEM 15.4

Nitromethane has λ_{max} = 275 nm (ϵ = 15). What kind of transition is responsible for this absorption?

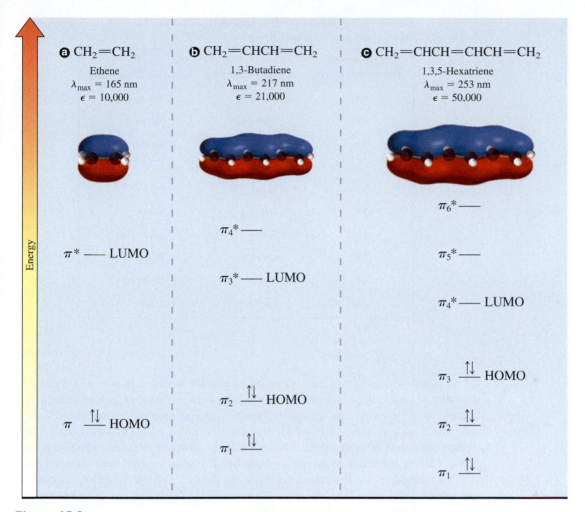

Figure 15.3

ENERGIES OF THE PI MOs OF ⓐ ETHENE, ⓑ 1,3-BUTADIENE, AND ⓒ 1,3,5-HEXATRIENE.

PROBLEM 15.5

3-Buten-2-one has λ_{max} = 213 nm (ϵ = 7080) and λ_{max} = 320 nm (ϵ = 21). What kind of transition is responsible for each of these absorptions?

For a compound to be colored, it must absorb light in the visible region. The color of the compound is complementary to the color of the light that is absorbed (see Table 15.1). Most simple organic compounds are colorless because the amount of energy separating their HOMOs and LUMOs is too large for the absorption of visible light to occur. In the case of some compounds with λ_{max} in the ultraviolet region, the tail of the absorption band extends into the visible region. Such compounds absorb some of the violet light and appear yellow. 1,2-Benzanthracene (λ_{max} = 386 nm) is an example of such a yellow compound. Compounds with even more extensive conjugation absorb visible light and appear colored. β-Carotene, which absorbs at 452 nm, is an example

Table 15.1 Relationship between Color Absorbed and Color Observed

Wavelength Absorbed (nm)	Color Absorbed	Color Observed
400	Violet	Yellow-green
425	Blue-violet	Yellow
450	Blue	Orange
490	Blue-green	Red
510	Green	Purple
530	Yellow-green	Violet
550	Yellow	Blue-violet
590	Orange	Blue
640	Red	Blue-green
730	Purple	Green

of a compound in which a long series of conjugated double bonds results in the absorption of visible light. Because it absorbs blue light, β-carotene appears orange. It is the pigment responsible for the color of carrots. Indigo absorbs yellow-orange light at 600 nm. It appears blue and is used to dye denim for blue jeans. Azo dyes, which constitute more than 60% of the manufactured dyes, contain a nitrogen–nitrogen double bond. Usually they have an aromatic group bonded to each nitrogen of the azo group. The wavelength of the absorption maximum of the dye, and thus its color, depends on the substituents on the aromatic rings. An example is FD&C Red No. 40, a red azo dye (λ_{max} = 508 nm) that is used in food coloring.

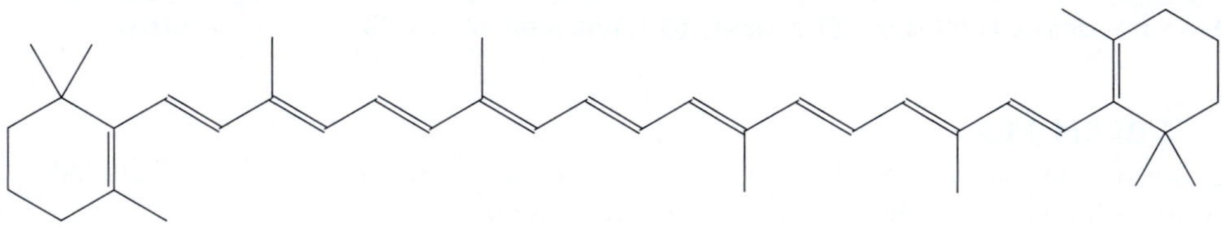

β-Carotene λ_{max} = 452 nm

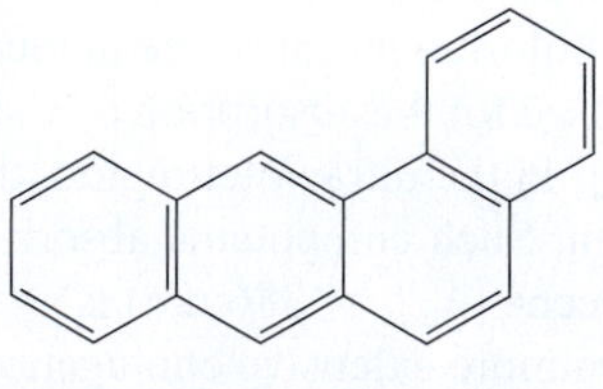

1,2-Benzanthracene λ_{max} = 386 nm

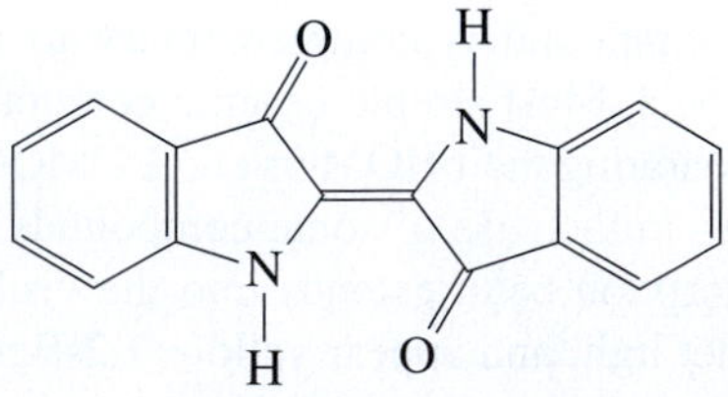

Indigo λ_{max} = 600 nm

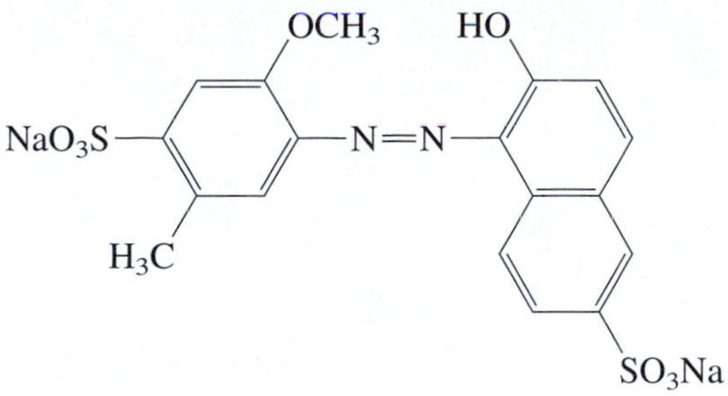

FD&C Red No. 40 (an azo dye)
λ_{max} = 508 nm

15.3 UV-Visible Spectroscopy in Structure Determination

Although UV-visible spectroscopy is not as useful as IR or NMR, it does have value in structure determination because it provides information about the conjugated part of the compound. If a compound has no absorptions in the 200 to 800 nm region, then features such as conjugated carbon–carbon double bonds, carbonyl groups, and benzene rings must be absent. The presence of an absorption band usually does not allow, by itself, the identification of the responsible chromophore. However, the proposed chromophore must be consistent with the spectrum. The absorption maxima of a variety of compounds can be calculated with good accuracy by using several sets of empirical rules, so the UV spectrum can often be used to distinguish among isomeric chromophores.

In addition, structural features that are not directly attached to the chromophore have little effect on the spectrum. Therefore, smaller compounds that contain the same chromophore can be used as models to predict the spectra of larger compounds. As an example, the following two compounds have very similar UV spectra:

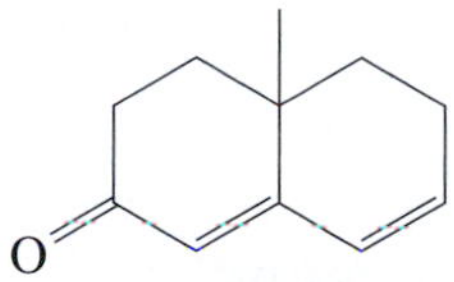

λ_{max} = 283 nm 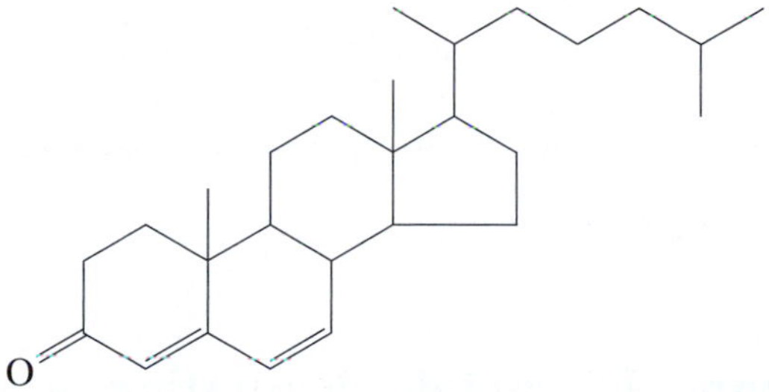 λ_{max} = 284 nm

Studies using model compounds can be useful in structure determinations. For example, the presence of a 1,4-naphthoquinone chromophore in vitamin K_1 was suggested by the similarity of its UV spectrum to those of simpler compounds such as 2-*tert*-butyl-1,4-naphthoquinone:

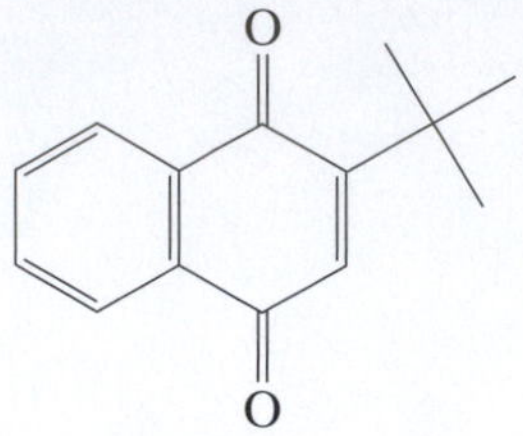

2-*tert*-Butyl-1,4-naphthoquinone

λ_{max} (nm)	ϵ
249	19,600
260	18,000
325	2,400

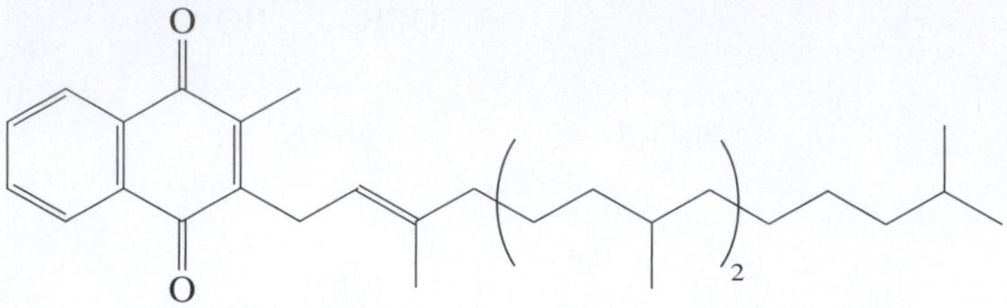

Vitamin K$_1$

λ_{max} (nm)	ϵ
248	18,600
264	14,200
331	2,730

PROBLEM 15.6

Explain how UV spectroscopy could be used to distinguish between these compounds.

a) 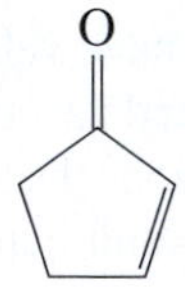   and

b) 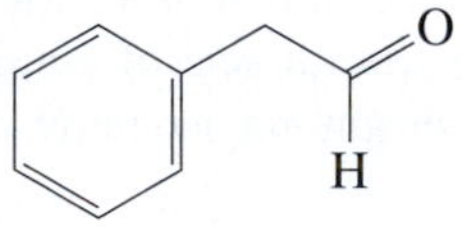   and

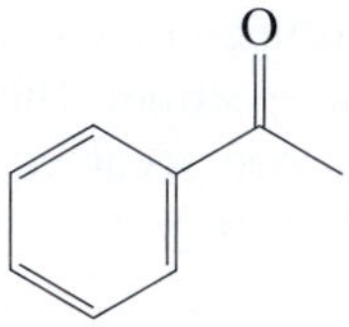

c) 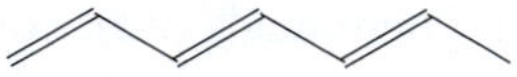   and

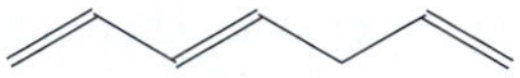

d) 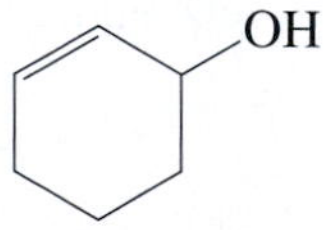   and

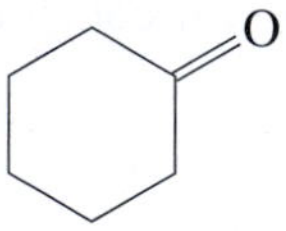

Focus On

Ozone, Ultraviolet Radiation, and Sunscreens

Ozone is an extremely reactive chemical. Its reaction with alkenes was discussed in Section 11.11. Although it is considered a pollutant in the lower atmosphere, its presence in the upper atmosphere has beneficial health effects due to its absorption of ultraviolet light.

The UV spectrum of ozone is shown in the following figure. There is enough ozone in the upper atmosphere that essentially all of the ultraviolet radiation of wavelengths less than 295 nm is absorbed. The tail of the ozone band absorbs part of the light in the region from 295 to 320 nm. This radiation, known as UV-B, is responsible for sunburn and also causes damage to DNA that may result in skin cancer. Any decrease in the ozone concentration in the upper atmosphere allows more of this damaging UV-B radiation to reach the surface of the earth. This is why there is so much concern about the release of ozone-destroying chemicals, such as chlorofluorocarbons.

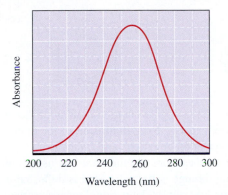

One way to protect against the damaging effects of UV-B radiation is to use a sunscreen. The active ingredient in a sunscreen is a compound that absorbs harmful ultraviolet radiation. Early sunscreens contained *p*-aminobenzoic acid, PABA, which has λ_{max} = 289 nm and absorbs strongly in the UV-B region. Newer formulations contain ingredients such as oxybenzone, which has λ_{max} = 284 and 324 nm. These current sunscreens are designed to absorb UV-A (wavelengths of 320–400 nm) in addition to UV-B radiation. Although it does not cause much burning, UV-A is thought to cause wrinkling and premature aging of the skin over long periods of exposure.

H_2N—⬡—CO_2H

p-Aminobenzoic acid
PABA
λ_{max} = 289 nm

Oxybenzone
λ_{max} = 284 and 324 nm

15.4 MASS SPECTROMETRY

Mass spectrometry is used to measure the molecular mass of a compound and provides a method to obtain the molecular formula. It differs from the other instrumental techniques presented thus far because it does not involve the interaction of electromagnetic radiation with the compound. Instead, molecules of the compound being studied are bombarded with a high-energy beam of electrons in the vapor phase. When an electron from the beam impacts on a molecule of the sample, it knocks an electron out of the molecule. The product, called the **molecular ion** (*represented as* M^{+}), has the same mass as the original molecule but has one less electron. It has both an odd number of

Important Convention

electrons and a positive charge. Species that have an odd number of electrons are called **radicals** (see Chapter 21). Therefore, the molecular ion is a **radical cation.** This process is illustrated in the following equation for the formation of the radical cation from benzene:

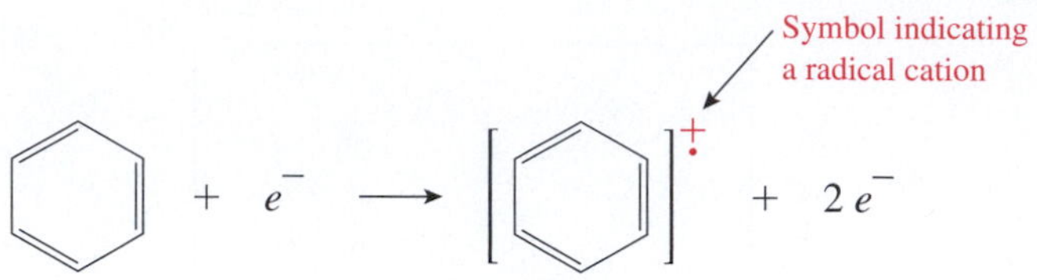

Symbol indicating
a radical cation

Because the electron beam is highly energetic, many of the molecular ions are formed with considerable excess energy. **Fragmentation** of the molecular ions with excess energy produces a number of other cations and radicals.

The masses of all of these ions are measured by using a magnetic field. A charged particle that is moving through a magnetic field is deflected in a direction perpendicular to the field. The amount of this deflection depends on the ratio of the charge to the mass of the ion. Heavier ions are deflected less than lighter ions. In a mass spectrometer, the **mass to charge ratio** (m/z) is measured. The charge on almost all of the ions is $+1$, so their masses can be determined. The mass of the molecular ion provides the molecular mass of the compound, whereas the masses of the fragment ions provide information about its structure. A schematic diagram of a mass spectrometer is shown in Figure 15.4.

The mass spectrum is a report of the relative numbers or abundances of ions of each m/z that are detected, and it is often presented in the form of a bar graph. The most abundant ion, the **base ion,** is assigned a value of 100, and the amounts of the other ions are expressed as percentages of this. A mass spectrum for benzene is shown in

Figure 15.4

SCHEMATIC DIAGRAM OF A MASS SPECTROMETER. In the ionization chamber, electron bombardment is used to eject electrons from sample molecules, creating radical cations. These are accelerated away from the positively charged repeller plate. A beam of these ions is passed into the magnetic field, which is perpendicular to the plane of the page. As the magnetic field strength is varied, ions of different mass to charge ratio pass through the slit and reach the detector. The data are sent to a computer, which records the relative numbers of ions of each mass.

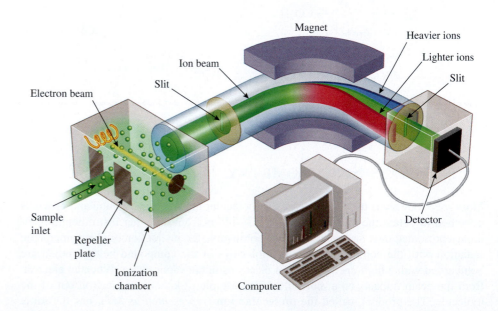

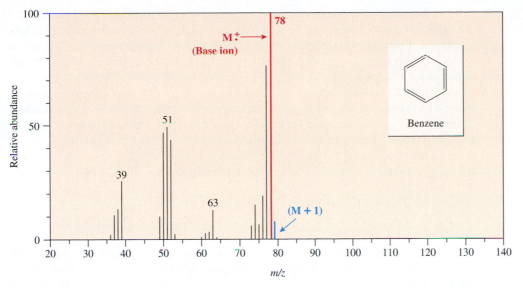

Figure 15.5

THE MASS SPECTRUM OF BENZENE.

Figure 15.5. In this particular case the molecular ion, at m/z 78, is also the base ion. More commonly, the base ion and the molecular ion are not the same.

15.5   DETERMINING THE MOLECULAR FORMULA

Two general types of mass spectrometers are available. The most common type, a low-resolution mass spectrometer, provides the masses of the ions to the nearest whole number. A more expensive instrument, a high-resolution mass spectrometer, provides the masses to several decimal places. The more accurate mass provided by a high-resolution instrument is especially useful in determining the molecular formula of the compound because the atomic masses of the various atoms are not exact integers. Consider the compounds CO_2, C_2H_4O, and C_3H_8. On a low-resolution mass spectrometer, these compounds all have a molecular ion at m/z 44. However, the exact masses of these compounds are all slightly different. Table 15.2 provides the exact masses of the major isotopes of a few elements that commonly occur in organic compounds. The exact mass of a CO_2 molecule is 43.9898, that of C_2H_4O is 44.0262, and that of C_3H_8 is 44.0626. These masses can be readily distinguished with a high-resolution instrument.

PROBLEM 15.7

Butane and acetone both have a molecular mass of 58. Calculate the exact masses of these compounds and explain whether they can be distinguished by high-resolution mass spectrometry.

Table 15.2   Exact Masses of Some Common Isotopes

Isotope	Atomic Mass (amu)
^1H	1.00783
^{12}C	12.00000
^{14}N	14.0031
^{16}O	15.9949

These values are not the same as the atomic masses in the periodic table because these are the exact masses of individual isotopes. The masses in the periodic table are average masses of the element based on the masses and natural abundances of the isotopes of which it is composed.

If a high-resolution mass spectrum is not available, it is still possible to obtain information about the molecular formula from the low-resolution spectrum. In the mass spectrum of benzene, shown in Figure 15.5, the molecular ion appears at m/z 78. In addition, there is a smaller peak at m/z 79, called the M + 1 peak, that is 6.8% of the intensity of the M‡ peak. The M + 2 peak, at m/z 80, is 0.2% of the M‡ peak. The M + 1 and M + 2 peaks are caused by the presence of isotopic atoms of heavier mass in some of the molecules. Their intensities relative to the M‡ peak can be used to deduce information about the formula. Let's look at how these peaks arise in more detail.

Most elements are composed of more than one isotope. In the case of carbon the relative abundances of the two major isotopes, ^{12}C and ^{13}C, are 100 to 1.1. In a sample of methane, then, for every 100 molecules composed of $^{12}CH_4$ there are 1.1 molecules of $^{13}CH_4$. Therefore, in the mass spectrum of methane the m/z 17 peak, which is due to $^{13}CH_4$, has 1.1% of the intensity of the m/z 16 peak, which is due to $^{12}CH_4$. Benzene has six carbons, each with a 1.1% chance of being ^{13}C. Therefore, the M + 1 peak at m/z 79 should be approximately $6 \times 1.1 = 6.6\%$ of the M‡ peak at 78 m/z, in good agreement with the observed value of 6.8%. The M + 2 peak, with 0.2% of the intensity of the M‡ peak, is due primarily to molecules that contain two atoms of ^{13}C.

PROBLEM 15.8

What are the predicted intensities of the M + 1 peaks in the mass spectra of butane and acetone, relative to the intensity of the M‡ peak? Do you think that these intensities could be used to distinguish between these compounds?

Table 15.3 lists the relative abundances of the major isotopes of the elements that are commonly encountered in organic compounds. In principle, it is possible to use the

Table 15.3   Natural Abundances of Isotopes of Some Common Elements

Element	Major Isotope	RA	M + 1 Isotope	RA	M + 2 Isotope	RA
Hydrogen	^1H	100				
Carbon	^{12}C	100	^{13}C	1.1		
Nitrogen	^{14}N	100	^{15}N	0.4		
Oxygen	^{16}O	100			^{18}O	0.2
Fluorine	^{19}F	100				
Sulfur	^{32}S	100	^{33}S	0.8	^{34}S	4.4
Chlorine	^{35}Cl	100			^{37}Cl	32.5
Bromine	^{79}Br	100			^{81}Br	98.0
Iodine	^{127}I	100				

The relative abundance (RA) of the most abundant isotope is listed as 100, and the abundances of the other isotopes are listed relative to that number. The M + 1 isotope is the one that is responsible for the peak at m/z one unit higher than the peak for M‡.

information in this table to determine the molecular formula of a compound on the basis of the intensities of its M + 1 and M + 2 peaks. In practice, this is difficult because of various inaccuracies that arise in measuring these intensities. Nevertheless, the presence of certain elements in a compound can readily be inferred from the size of the M + 2 peak:

The presence of one sulfur in the formula is indicated by an M + 2 peak that is about 4% of the $M^{\ddagger}$ peak.

The presence of one chlorine in the formula is indicated by an M + 2 peak that is about one-third as large as the $M^{\ddagger}$ peak (see Figure 15.6).

The presence of one bromine in the formula is indicated by an M + 2 peak that is about the same intensity as the $M^{\ddagger}$ peak (see Figure 15.7).

The presence of two of these atoms in the molecule results in the appearance of a characteristic pattern in the $M^{\ddagger}$, M + 2, and new M + 4 peak, whereas the presence of three results in a different pattern, including an M + 6 peak.

The presence of one nitrogen in the formula is readily apparent because the molecular ion has an odd mass. To help understand why this is so, recall the degree of unsaturation calculation discussed in Section 2.4. Compounds consisting of carbon (even mass) and oxygen (even mass) have an even number of hydrogens and thus have $M^{\ddagger}$ at an even m/z. If a hydrogen (odd mass) is replaced by a halogen (odd mass), $M^{\ddagger}$ is still even. The presence of a single nitrogen (even mass) in a compound requires one additional hydrogen be present also. Because the total number of hydrogens is now odd, the molecular ion occurs at an odd m/z. In general, the appearance of the

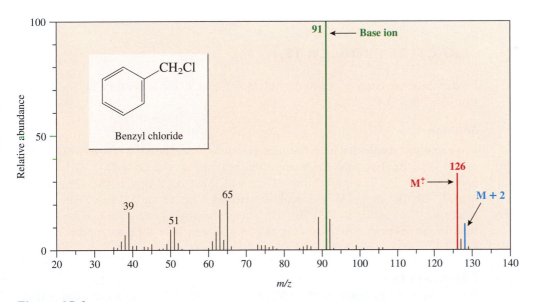

Figure 15.6

THE MASS SPECTRUM OF BENZYL CHLORIDE.

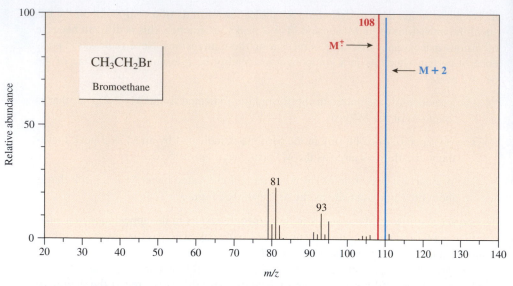

Figure 15.7

THE MASS SPECTRUM OF BROMOETHANE.

molecular ion at an odd m/z indicates the presence of an odd number of nitrogens in the formula.

PRACTICE PROBLEM 15.1

Predict the relative intensities of the M⁺, M + 2, and M + 4 peaks for CH_2Br_2. Assume that $^{79}Br/^{81}Br = 1/1$.

Solution

It is equally probable that each bromine is ^{79}Br or ^{81}Br. The M⁺ peak (m/z 172) is due to $CH_2{}^{79}Br{}^{79}Br$. Two molecules cause the peak at M + 2 (m/z 174): $CH_2{}^{79}Br{}^{81}Br$ and $CH_2{}^{81}Br{}^{79}Br$. The probability of each of these occurring is the same as that of $CH_2{}^{79}Br{}^{79}Br$, so the M + 2 peak is twice as large as the M⁺ peak. The M + 4 peak (m/z 176) is due to $CH_2{}^{81}Br{}^{81}Br$. The probability for this peak is the same as that for the M⁺ peak. Therefore, the M⁺, M + 2, and M + 4 peaks appear in a ratio of 1:2:1.

PROBLEM 15.9

Predict the relative intensities of the M⁺, M + 2, and M + 4 peaks for these compounds. Assume that $^{79}Br/^{81}Br = 1/1$ and $^{35}Cl/^{37}Cl = 3/1$.

a) CH_2Cl_2 **b)** CH_2BrCl

PROBLEM 15.10

What conclusions can be drawn about these compounds from their mass spectra?

a)

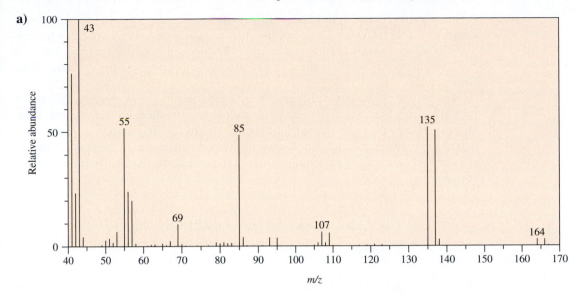

b)

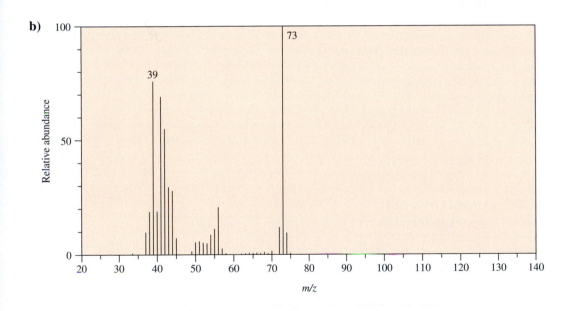

15.6 FRAGMENTATION OF THE MOLECULAR ION

As mentioned earlier, fragmentation of the molecular ion results in the formation of a number of other radicals and cations. In some cases, fragmentation is so facile that essentially all the molecular ions react and no $M^{\ddot{+}}$ peak appears in the spectrum. Let's see how the masses of these fragments can be used to help determine the structure of the compound.

First, we need to discuss the structure of the molecular ion in more detail. It is a radical cation, resulting from the loss of one electron from the original molecule. What electron is lost? It must be the highest-energy electron in the compound. If the compound has only sigma bonds, the location of the odd electron and the positive charge is uncertain and the radical cation is usually represented as shown in the following equation:

$$CH_3CH_2CH_3 \;+\; e^- \;\longrightarrow\; \left[CH_3CH_2CH_3\right]^{\ddot{+}} \;+\; 2\,e^-$$

Propane

The radical cation from compounds that have nonbonding or pi electrons can also be represented in this manner. However, because the highest-energy electron is known in these cases, the structure of the radical cation can be drawn more specifically as illustrated in the following equations:

$$CH_3-\overset{\displaystyle ..}{\underset{\displaystyle ..}{O}}-H \;+\; e^- \;\longrightarrow\; CH_3-\overset{\displaystyle .+}{\underset{\displaystyle ..}{O}}-H \;+\; 2\,e^-$$

Methanol

Ethene

$$\underset{H}{\overset{H}{>}}C=C\underset{H}{\overset{H}{<}} \;+\; e^- \;\longrightarrow\; \underset{H}{\overset{H}{>}}\overset{\displaystyle .}{C}-\overset{\displaystyle +}{C}\underset{H}{\overset{H}{<}} \;+\; 2\,e^-$$

Ethene

PROBLEM 15.11

Show the molecular ions formed from these compounds:

a) $CH_3\overset{..}{N}HCH_3$ b) ⬡ (cyclohexene)

There are two general types of fragmentation reactions. In one type the radical cation fragments to a neutral molecule and a new radical cation. This process is especially favorable when the neutral product is a small, stable molecule. For example, the loss of water from the molecular ion of alcohols is very facile. For this reason the $M^{\ddot{+}}$ peak is very small for primary and secondary alcohols, and it is usually undetectable for tertiary alcohols. An example of this fragmentation process is provided in the following equation. *Note the use of arrows with only half of the arrowhead to indicate movement of one electron rather than two.*

Important Convention

$$m/z\ 60 \qquad\qquad m/z\ 42 \qquad\quad \text{mass } 18$$

$$\text{CH}_3\text{CH}{-}\text{CH}_2 \longrightarrow \text{CH}_3\overset{\cdot}{\text{CH}}{-}\overset{+}{\text{CH}}_2 \;+\; \text{H}_2\ddot{\text{O}}{:}$$

or

$$\left[\text{CH}_3\text{CH}{=}\text{CH}_2\right]^{\overset{+}{\cdot}}$$

In the other type of fragmentation, a bond is cleaved so that the positive charge remains with one fragment and the odd electron goes with the other. Only the positive fragment is detected and appears in the mass spectrum. The stability of the product cation and radical determine the favorableness of this type of cleavage. You are already quite familiar with the factors that affect the stability of cations, especially carbocations. Although radicals are inherently more stable than carbocations because they are less electron deficient, they are stabilized by the same factors that stabilize carbocations. Thus, tertiary radicals are more stable than secondary radicals, and secondary radicals are more stable than primary radicals. Resonance stabilization is also important.

Let's see how these stabilities help to explain the fragmentations in some simple examples.

Alkanes

The mass spectrum of octane is shown in Figure 15.8. The molecular ion appears at m/z 114. Cleavage of the sigma bond between C-1 and C-2 would lead to the fragments shown in the following equations:

$$m/z\ 114 \qquad\qquad\qquad m/z\ 99 \qquad\qquad \text{mass } 15$$

$$\left[\text{CH}_3(\text{CH}_2)_5\text{CH}_2{\text{\large\{}}\text{CH}_3\right]^{\overset{+}{\cdot}} \longrightarrow \text{CH}_3(\text{CH}_2)_5\overset{+}{\text{CH}}_2 \;+\; {\cdot}\text{CH}_3$$

Octane radical cation

A primary carbocation   A methyl radical

$$\text{mass } 99 \qquad\qquad m/z\ 15$$

$$\text{CH}_3(\text{CH}_2)_5\overset{\cdot}{\text{CH}}_2 \;+\; \overset{+}{\text{CH}}_3$$

A primary radical   A methyl carbocation

If the charge stays with the larger fragment, a primary carbocation, m/z 99 (M − 15), and a methyl radical are formed. This fragmentation is not favorable because the methyl radical is not very stable, so no peak at m/z 99 is observed in the spectrum. The process

Figure 15.8

THE MASS SPECTRUM OF OCTANE.

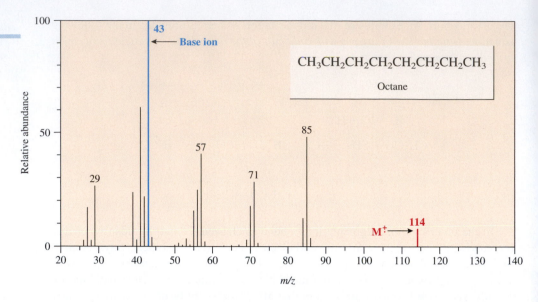

in which the charge is on the smaller fragment produces a highly unstable methyl carbocation and is even less favorable.

Cleavage of the sigma bond between C-2 and C-3 leads to the fragments shown in the following equations:

$$m/z\ 114 \qquad m/z\ 85 \qquad \text{mass } 29$$

$$\left[CH_3(CH_2)_4CH_2 \overset{\xi}{\underset{\xi}{\vdash}} CH_2CH_3 \right]^{+\cdot} \longrightarrow \underset{\text{A primary carbocation}}{CH_3(CH_2)_4\overset{+}{C}H_2} + \underset{\text{A primary radical}}{\cdot CH_2CH_3}$$

$$\text{mass } 85 \qquad m/z\ 29$$

$$\longrightarrow \underset{\text{A primary radical}}{CH_3(CH_2)_4\overset{\cdot}{C}H_2} + \underset{\text{A primary carbocation}}{\overset{+}{C}H_2CH_3}$$

Regardless of whether the charge is on the larger or the smaller fragment, a primary carbocation and a primary radical are produced. Both of these fragmentations occur as indicated by peaks at m/z 85 and m/z 29. (Remember that the radical products do not appear in the spectrum; only the cations are detected.) A similar cleavage between C-3 and C-4 results in peaks at m/z 71 and 43, whereas one between C-4 and C-5 gives a peak at m/z 57. The series of peaks separated by 14 units, the mass of the CH_2 fragment, is characteristic of straight-chain alkanes.

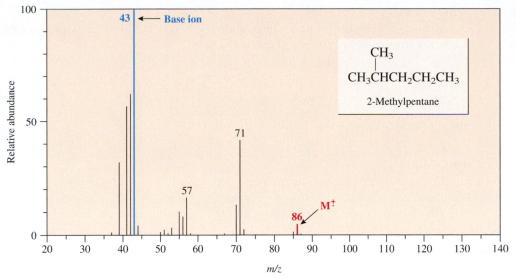

Figure 15.9

THE MASS SPECTRUM OF
2-METHYLPENTANE.

Figure 15.9 shows the mass spectrum of 2-methylpentane. Branched alkanes tend to cleave on either side of the branch because more stable carbocations result. The two primary cleavage pathways for 2-methylpentane are shown in the following equations:

Although the fragmentation required to produce a methyl radical and a primary carbocation did not occur in the case of octane, the loss of a methyl radical does occur here because the resulting carbocation, with m/z 71, is secondary. Cleavage on the other side of the branch results in the formation of a primary radical and a secondary carbocation, with m/z 43. This is the most favorable fragmentation, so m/z 43 is the base ion.

PROBLEM 15.12

Show equations for the major fragmentations you would expect from the molecular ions of these compounds. List the *m/z* of the product ions.

a) $CH_3CH_2CH_2CH_2CH_3$

b)
$$\begin{array}{c} CH_3 \\ | \\ CH_2 \\ | \\ CH_3CH_2CHCH_2CH_3 \end{array}$$

PROBLEM 15.13

a) The base ion in the mass spectrum of 3-ethyl-2-methylpentane occurs at *m/z* 43. Show the fragmentation that produces this ion.

b) What other fragment would you predict to provide a major peak in the spectrum?

Alkenes

The fragmentation of an alkene to produce an allylic carbocation is usually a major pathway because of resonance stabilization:

$$\left[R \wr CH_2{-}CH{=}CH_2\right]^{+\cdot} \longrightarrow R\cdot + \overset{+}{C}H_2{-}CH{=}CH_2 \longleftrightarrow CH_2{=}CH{-}\overset{+}{C}H_2$$

An allylic carbocation

The base ion in the spectrum of 1-hexene, Figure 15.10, is the allylic carbocation with *m/z* 41, which is produced according to the following equation:

$$\underset{\text{1-Hexene radical cation}}{\overset{m/z\ 84}{\left[CH_3CH_2CH_2 \wr CH_2CH{=}CH_2\right]^{+\cdot}}} \longrightarrow \overset{\text{mass } 43}{CH_3CH_2\dot{C}H_2} + \overset{m/z\ 41}{\overset{+}{C}H_2CH{=}CH_2}$$

Benzylic Cleavage

The benzyl carbocation, like the allyl carbocation, has substantial resonance stabilization. The base peak at *m/z* 91 in the spectrum of butylbenzene (see Figure 15.11) results from a fragmentation that produces a benzylic carbocation as illustrated in the following equation. (We will see in Chapter 16 that the benzylic carbocation rearranges to an isomeric carbocation, which is even more stable.)

m/z 134 *m/z* 91 mass 43

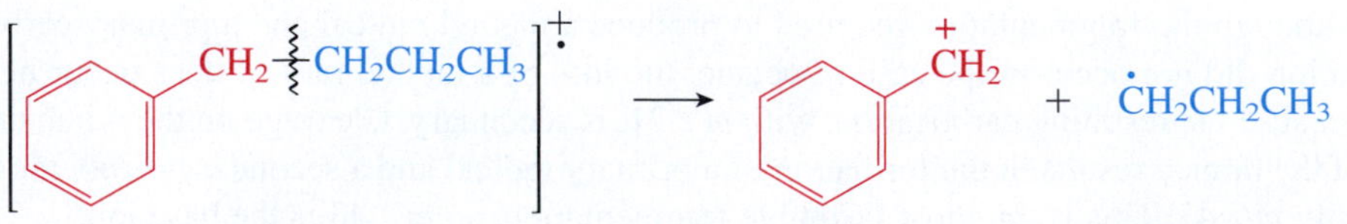

Butylbenzene radical cation

A benzylic carbocation

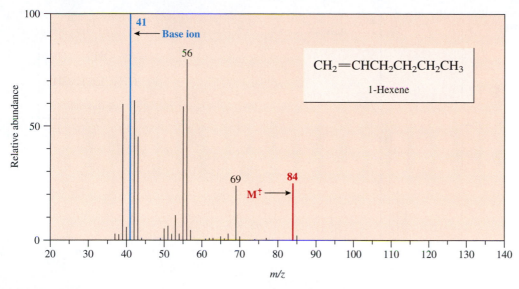

Figure 15.10

THE MASS SPECTRUM OF 1-HEXENE. Note that many of the peaks in this spectrum and others are difficult to explain. This is because of rearrangements and other fragmentations that we have not discussed.

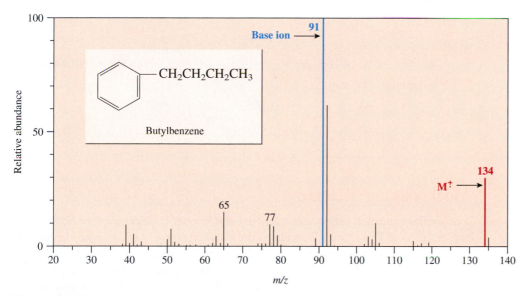

Figure 15.11

THE MASS SPECTRUM OF BUTYLBENZENE.

Alcohols

In addition to the elimination of water, alcohols tend to fragment by cleaving one of the bonds between the hydroxy carbon and an adjacent carbon. The resulting cations are the conjugate acids of aldehydes and ketones and are stabilized because they satisfy the octet rule at all of the atoms. Figure 15.12 shows the mass spectrum of 2-butanol.

Figure 15.12

THE MASS SPECTRUM OF 2-BUTANOL.

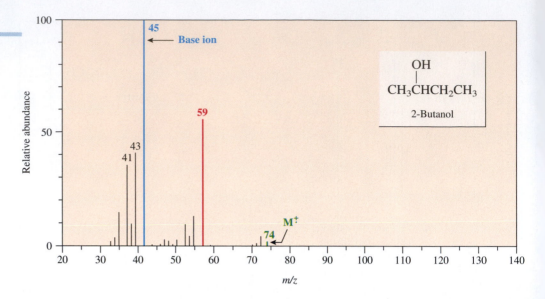

The major fragments for this alcohol, at m/z 59 and m/z 45, are formed by this type of fragmentation, as illustrated in the following equations:

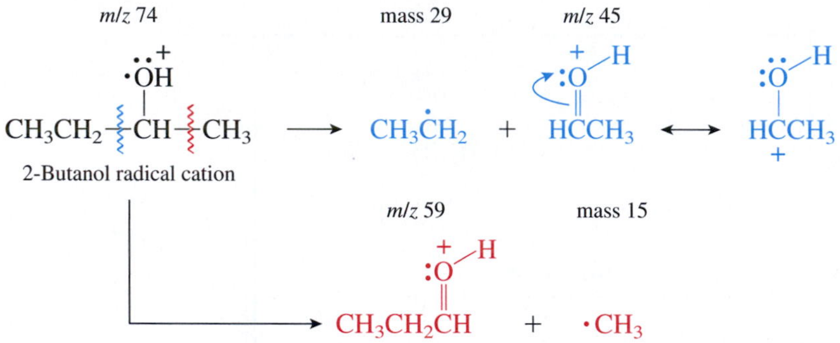

Aldehydes and Ketones

A major fragmentation pathway in aldehydes and ketones is cleavage of the bonds to the carbonyl carbon. This process is similar to the fragmentation that occurs with alcohols and produces cations that are relatively stable because the octet rule is satisfied at all of the atoms. The major peaks at m/z 85 and m/z 43 in the spectrum of 4-methyl-2-pentanone (see Figure 15.13) result from this cleavage, as illustrated in the following equations:

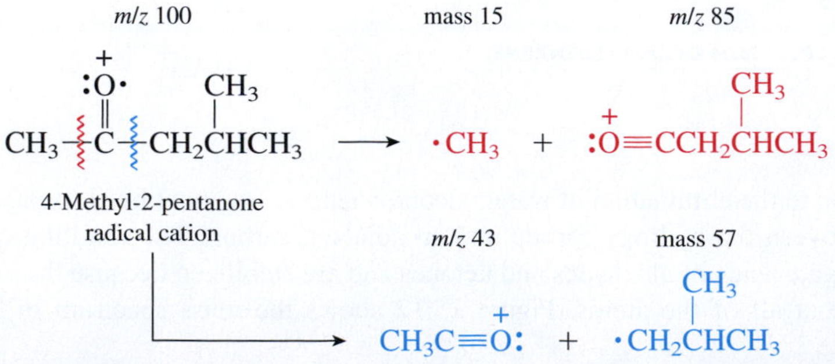

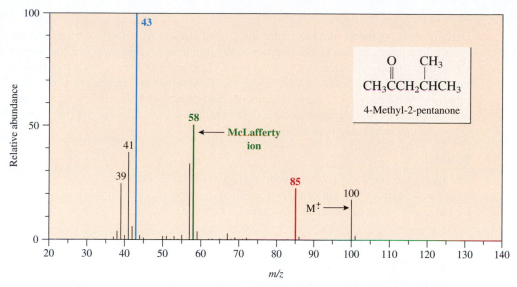

Figure 15.13

THE MASS SPECTRUM OF
4-METHYL-2-PENTANONE.

Aldehydes and ketones that have a hydrogen separated from the carbonyl carbon by three intervening carbons undergo a fragmentation called the McLafferty rearrangement. This process is illustrated in the following equation for 4-methyl-2-pentanone. It is responsible for the peak at m/z 58 in the mass spectrum shown in Figure 15.13. The McLafferty rearrangement is especially favorable because the cyclic transition state contains six atoms, a very favorable ring size.

These fragmentations serve to illustrate many of the major types. The driving force behind all of them is the formation of stable cations and radicals. Fragmentations of functional groups that have not been covered here are often similar to those described earlier. Although this has been only a very brief introduction to mass spectrometry, the power and utility of this technique should be apparent.

PRACTICE PROBLEM 15.2

Show equations to account for the major fragment ions that occur at m/z 97 and m/z 57 in the mass spectrum of 2,4,4-trimethyl-1-pentene.

$$CH_3-\underset{\underset{CH_3}{|}}{\overset{\overset{CH_3}{|}}{C}}-CH_2-\underset{\overset{|}{CH_3}}{\overset{\overset{CH_3}{|}}{C}}=CH_2$$

2,4,4-Trimethyl-1-pentene

Strategy

Determine the mass for the molecular ion. It sometimes helps to then determine the mass of the fragment that is lost to produce the fragment ion. For example, the loss of a fragment of mass 15 indicates the loss of a CH_3 group and the loss of a fragment of mass 29 indicates the loss of a CH_3CH_2 group. Remember that fragmentation to form relatively stable carbocations and radicals is greatly favored. This results in fragmentations at branches in a chain, at carbons next to double bonds and benzene rings, and at carbons singly or doubly bonded to oxygens.

Solution

The molecular ion for this compound (C_8H_{16}) has m/z 112. The peak at m/z 97 results from the loss of a fragment of mass 15, which must be a CH_3 group. This fragmentation is important because it produces a tertiary carbocation.

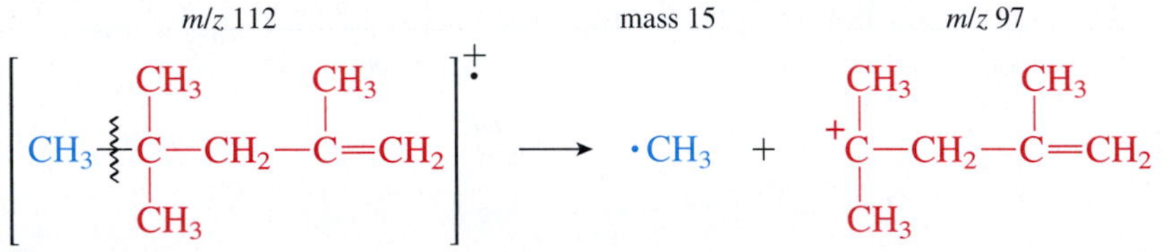

The peak at m/z 57 results from the fragmentation to produce a tertiary carbocation and an allylic radical according to the following equation. (As might be expected, the peak at m/z 57 is the base ion in the mass spectrum of this compound.)

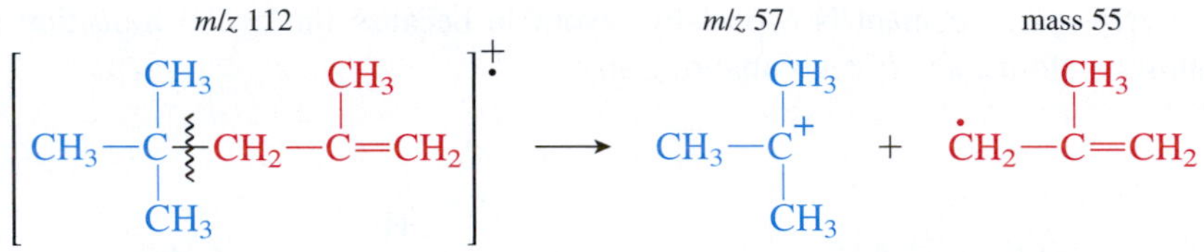

PROBLEM 15.14

Show equations to account for the major fragment ions that occur at the indicated m/z for these compounds:

a)

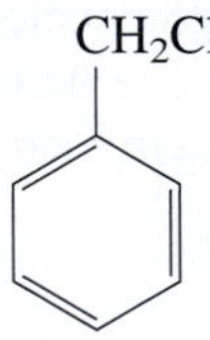

CH_2Cl

m/z 91

b) $CH_3CH_2CH_2CH_2OH$ m/z 31

c) $CH_3CH_2\overset{\displaystyle O}{\overset{\|}{C}}CH_2CH_2CH_2CH_3$ m/z 85, 72, 57

Focus On

Gas Chromatography and Mass Spectrometry

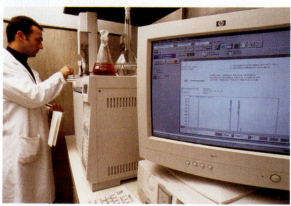

The techniques of gas chromatography (GC) and mass spectrometry (MS) have been combined into a powerful instrument, a GC/MS, for the identification of organic compounds. In this technique a small sample, containing a mixture of compounds, is injected onto the gas chromatograph, which separates the compounds. As the individual components elute from the column, they are passed directly into a mass spectrometer. The mass spectrum can be obtained without ever isolating the pure component.

Gas chromatography is an extremely powerful separation method. It requires only a very small amount of material and is able to separate even a complex mixture into its numerous components. The time that it takes for a particular compound to elute from the column under a given set of conditions, called the retention time, is characteristic for that compound. However, the identity of a component cannot be established solely on the basis of its retention time because another compound may coincidentally have an identical retention time. Whenever an analysis has a potential for legal or disciplinary consequences, the identity of the compound must be established by some other technique. Mass spectrometry is an ideal partner because it can establish the identity of a compound with certainty by using only the small amount of sample obtained from the GC.

Extensive drug screening is done at many athletic events, such as the Olympic Games. Usually, separate analyses, using different extraction procedures, are done for stimulants, narcotics, anabolic steroids, diuretics, and peptide hormones. In the analysis for stimulants, which are amines such as amphetamine and cocaine, a 5 mL urine sample is first made basic with KOH to ensure that the amines are present as the neutral molecules rather than as salts. The free amines are then extracted from the sample with diethyl ether. To save time and expense, the sample is first analyzed by gas chromatography only. If a peak appears with the retention time of one of the proscribed stimulants, then the sample is reanalyzed by GC/MS to confirm the identity of the suspected compound.

Amphetamine

Cocaine

Review of Mastery Goals

After completing this chapter, you should be able to:

■ Determine whether a compound will absorb light in the ultraviolet or visible region. (Problems 15.17, 15.28, 15.29, and 15.30)

■ Identify the chromophore and type of transition responsible for absorption of UV-visible radiation. (Problems 15.16 and 15.26)

■ Determine whether sulfur, chlorine, bromine, or nitrogen is present in a compound by examination of the M, M + 1, and M + 2 peaks in its mass spectrum. (Problems 15.23 and 15.27)

■ Explain the major fragmentation pathways for compounds containing some of the simple functional groups. (Problems 15.18, 15.19, 15.20, 15.21, 15.22, 15.23, 15.24, and 15.25, and 15.31)

Additional Problems

15.15 A student wishes to record the UV spectrum of *trans*-stilbene, which has λ_{max} = 308 nm (ϵ = 25,000). What concentration should be prepared if the desired absorbance is 0.5 at the maximum?

15.16 Indicate the types of transitions responsible for the absorptions of these compounds:

a)

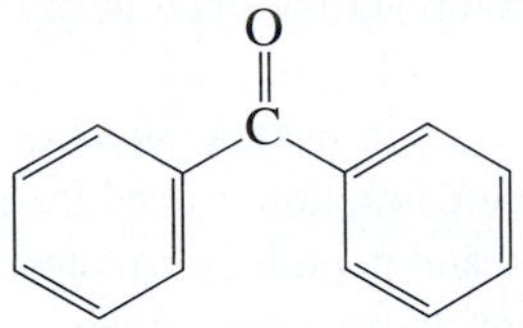

λ_{max} = 252 nm (ϵ = 20,000)
λ_{max} = 325 nm (ϵ = 180)

b)

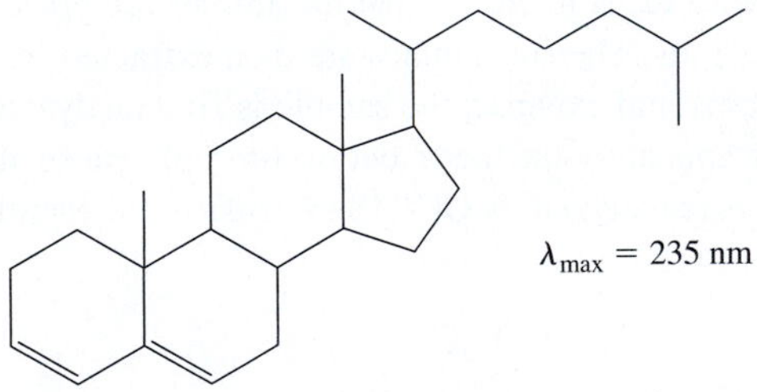

λ_{max} = 235 nm (ϵ = 19,000)

c)

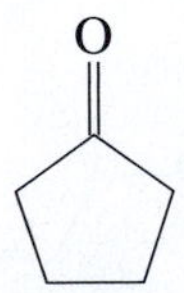

λ_{max} = 299 nm (ϵ = 20)

d) $CH_3C{\equiv}C{-}C{\equiv}CCH_3$ λ_{max} = 227 nm (ϵ = 360)

15.17 Which of these compounds are expected to have an absorption maximum in the region of 200 to 400 nm in their UV spectra?

a) $CH_3\overset{\overset{\displaystyle O}{\|}}{C}CH_2CH_3$

b)

c)

d) 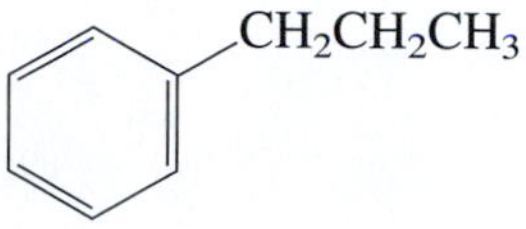 $CH_2CH_2CH_3$

e)

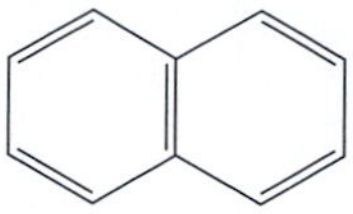

f) $CH_3CH_2OCH_2CH_3$

g)    OH

h)

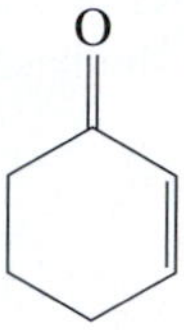

15.18 Predict the major fragments and their m/z that would appear in the mass spectra of these compounds:

a)

b)

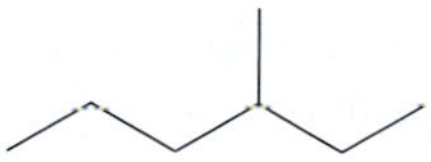

c) 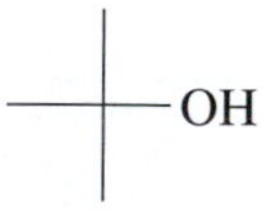   —OH

d)

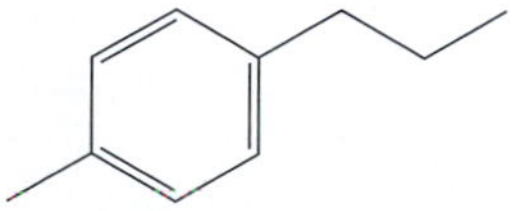

e)

f) 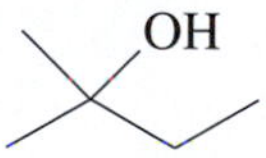   OH

g)

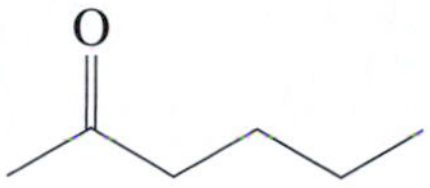

h) 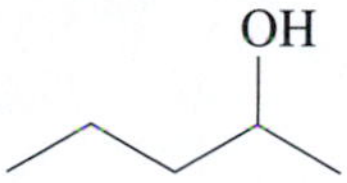   OH

i)

15.19 Explain how mass spectrometry could be used to distinguish between these compounds:

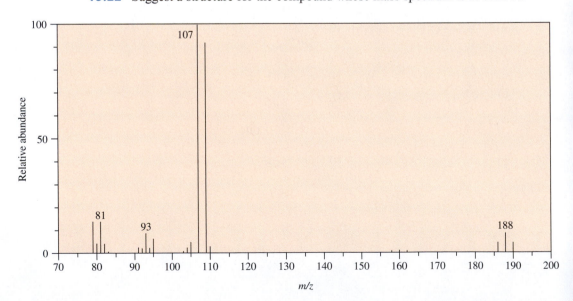

a) [structure] and [structure]

b) [structure] and [structure]

c) [structure] and [structure]

15.20 Explain why neopentane shows no molecular ion in its mass spectrum. Predict the structure and m/z for the base ion in its mass spectrum.

$$H_3C-\underset{\underset{CH_3}{|}}{\overset{\overset{CH_3}{|}}{C}}-CH_3$$

Neopentane

15.21 Explain how the peaks at m/z 115, 101, and 73 arise in the mass spectrum of 3-methyl-3-heptanol.

15.22 Suggest a structure for the compound whose mass spectrum is as follows:

15.23 The mass spectra of 3-methyl-2-pentanone and 4-methyl-2-pentanone are as follows. Explain which spectrum goes with which compound. What is the structure of the ion responsible for the peak at *m/z* 43 in each spectrum?

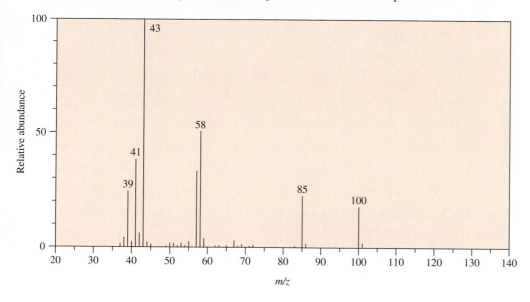

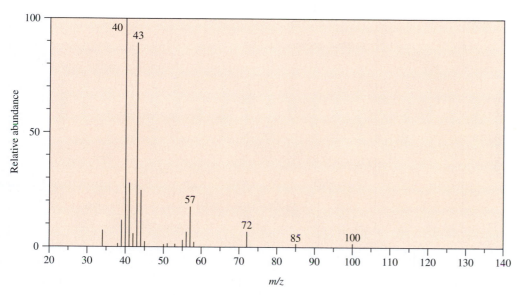

15.24 Compounds **A** and **B** are isomers with the formula C_3H_6O. **A** has a peak at 1730 cm^{-1} in its IR spectrum and **B** has a peak at 1715 cm^{-1}. The mass spectra of **A** and **B** are as follows. Show the structures of **A** and **B**.

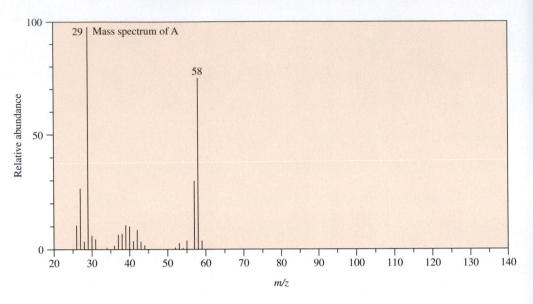

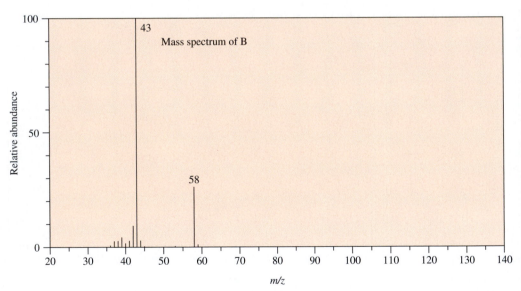

15.25 Compounds **C** and **D** are isomers with the formula C_9H_{12}. In addition to other absorption peaks, both compounds show a peak near 7.25 δ (area 5) in their ^1H-NMR spectra. Their mass spectra are as follows. Show the structures of **C** and **D**.

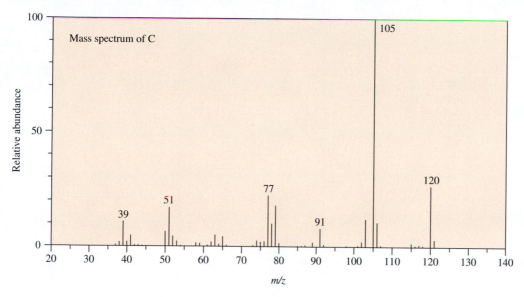

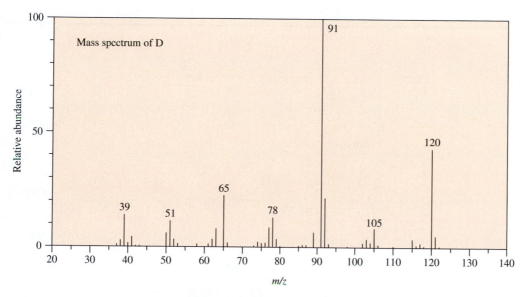

15.26 Compounds **E** and **F** are isomers with the formula C_6H_8. Both react with H_2 in the presence of Pt to give **G** (C_6H_{12}). **G** shows a single peak in its ^{13}C-NMR spectrum. **E** has no absorption maximum above 200 nm in its UV spectrum, whereas **F** has $\lambda_{max} = 259$ nm ($\epsilon = 10,000$). Show the structures of **E**, **F**, and **G**.

15.27 The ^{1}H-NMR spectrum for compound **H** is as follows. The peaks at highest m/z in the mass spectrum of **H** appear at 122 and 124 with intensities in a ratio of about 1 to 1. Show the structure of **H**.

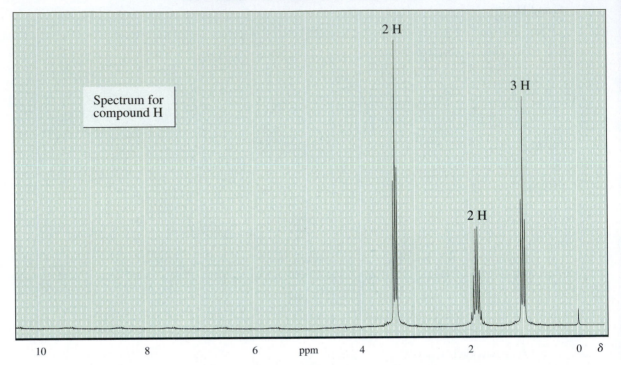

Spectrum for compound H

2 H   3 H   2 H

10   8   6   ppm   4   2   0 δ

15.28 Ultraviolet spectroscopy is often used to monitor the amount of a protein in a sample. The amount of protein is correlated with the absorbance at 280 nm. Explain which of the following amino acids you expect to have an absorption at 280 nm. (The carbonyl groups of the peptide [amide] bonds of the protein do not absorb at this wavelength.)

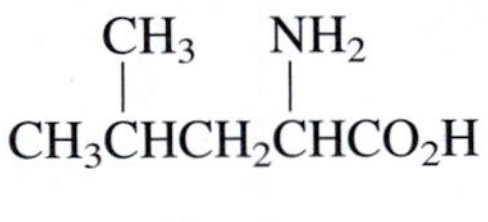

CH$_3$ NH$_2$
CH$_3$CHCH$_2$CHCO$_2$H

Leucine

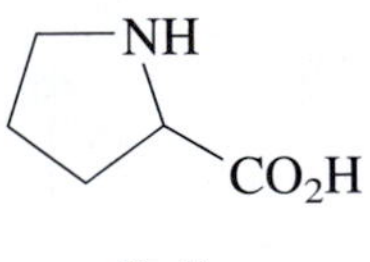

NH
CO$_2$H

Proline

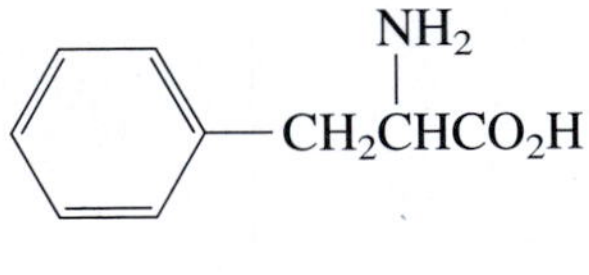

NH$_2$
CH$_2$CHCO$_2$H

Phenylalanine

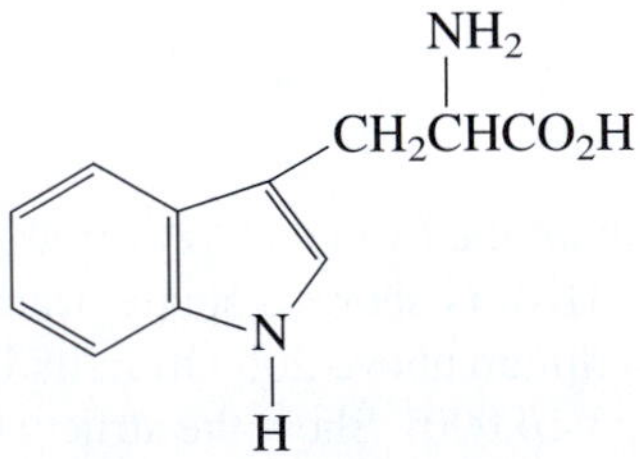

NH$_2$
CH$_2$CHCO$_2$H

NH$_2$
HOCH$_2$CHCO$_2$H

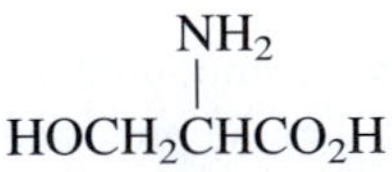

Serine

Tryptophan

Problems Using Online Three-Dimensional Molecular Models

ORGANIC
Chemistry-Now™
Click *Molecular Model Problems* to view the models needed to work these problems.

15.29 Explain why the ultraviolet spectrum of one of these dienes has its maximum absorption at a longer wavelength than that of the other.

15.30 Explain why the ultraviolet spectrum of one of these compounds has its maximum absorption at a longer wavelength than that of the other.

15.31 Show the structures of the fragment ions that occur at *m/z* 57, 86, and 99 in the mass spectrum of this compound.

Do you need a live tutor for homework problems? Access vMentor at Organic ChemistryNow at **http://now.brookscole.com/hornback2** for one-on-one tutoring from a chemistry expert.

Benzene and Aromatic Compounds

CHAPTER 16

MASTERING ORGANIC CHEMISTRY

▶ Understanding the MO Energy Levels for Planar, Cyclic, Conjugated Compounds

▶ Using Hückel's Rule to Predict Whether a Compound Is Aromatic, Antiaromatic, or Neither

▶ Understanding How Aromaticity and Antiaromaticity Affect the Chemical Behavior of Compounds

BENZENE OCCUPIES a special place in the field of organic chemistry. It is an especially stable compound, and because of this stability, substituted benzenes are widely distributed among natural products and industrial chemicals. Efforts by organic chemists to understand this stability have contributed significantly to our current models for the electronic structure of organic compounds and have led to the development of theories that not only explain the special properties of benzene but also help to explain and predict which other compounds have this special stability that has come to be called aromaticity.

This chapter begins with a discussion of some experimental observations that support the conclusion that benzene is especially stable. Then a model based on molecular orbital theory is presented to explain this stability. This model is generalized so that it can be applied to other compounds that are especially stable and also to some that are especially unstable. Several different classes of such compounds are discussed, along with examples of a variety of experimental observations that can be rationalized based on this theory.

16.1 BENZENE

Benzene was discovered in 1825 by Michael Faraday, who isolated it from the liquid that condensed from the gas that was burned in the street lamps of London. Although Faraday was able to deduce that the formula of benzene is C_6H_6, it was not until 1866 that the correct structure was proposed by Kekulé (see the Focus On box in Chapter 12 on page 469).

From the beginning, it has been apparent that benzene does not behave like an alkene. For example, the addition

ORGANIC
Chemistry Now™
Look for this logo in the chapter and go to OrganicChemistryNow at
http://now.brookscole.com/hornback2 for tutorials, simulations, problems, and molecular models.

of bromine to alkenes and the use of this reaction as a test for the presence of a carbon–carbon double bond was described in Section 11.4. Thus, when bromine is added to a solution of an alkene, such as cyclohexene, in dichloromethane, a rapid reaction causes the red-brown color of the bromine to quickly disappear. The product results from addition of the bromine to the double bond.

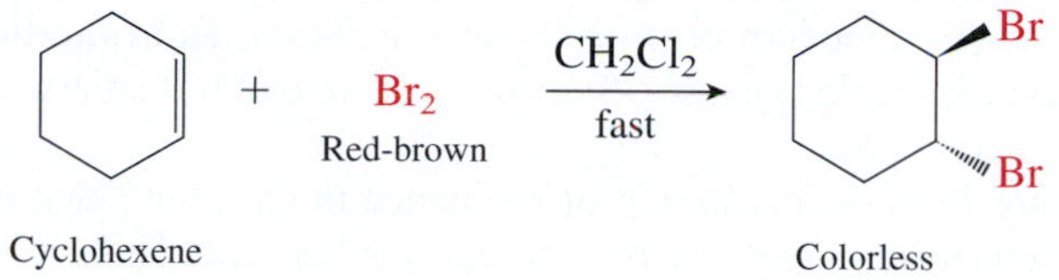

Cyclohexene

Red-brown

Colorless

In sharp contrast, benzene is unreactive with bromine under these same conditions. Under more vigorous reaction conditions (the presence of a Lewis acid catalyst, such as aluminum tribromide, and higher temperature), benzene does react with bromine. However, the product results from a substitution of a Br for a H, rather than addition of Br_2 to one of the double bonds. The product retains the benzene structure of three conjugated double bonds in a six-membered ring:

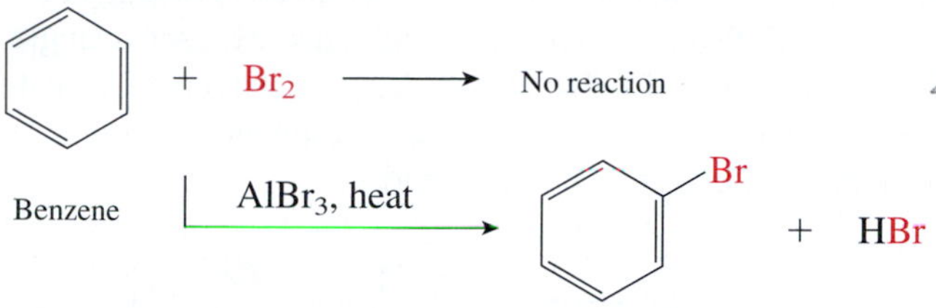

Benzene

AlBr$_3$, heat

+ Br_2 ⟶ No reaction

+ HBr

Another example is provided by the catalytic hydrogenation reaction described in Section 11.12. When a compound such as styrene is reacted with hydrogen and a catalyst at room temperature and relatively low pressure, only the alkene double bond reacts. The aromatic ring is inert under these conditions. Styrene reacts with hydrogen to form a cyclohexane ring only under much more vigorous conditions employing higher temperature and pressure:

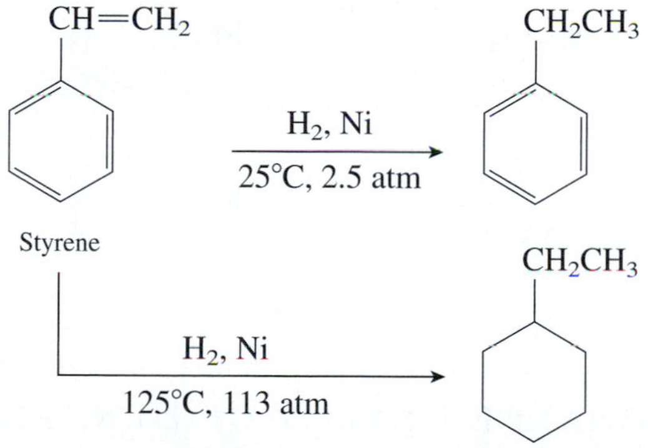

Styrene

H$_2$, Ni
25°C, 2.5 atm

H$_2$, Ni
125°C, 113 atm

The slowness of the reaction of benzene with bromine and the vigorous conditions required for its catalytic hydrogenation show that it is much less reactive than an alkene. In fact, until recent concerns about its carcinogenicity developed, benzene was used as a solvent for many organic reactions because of its low reactivity. When reaction with bromine is forced to occur, the formation of a substitution product, rather than an addition product, illustrates the stability of the six-membered aromatic ring. Such substitution reactions are typical of aromatic compounds and are discussed extensively in Chapter 17.

As mentioned in Section 12.1, the term *aromatic* was originally applied to substituted benzene derivatives because they have more pleasant odors than do many other organic compounds. To a modern organic chemist, however, an **aromatic compound** is one that is especially stable because of resonance, one that has an especially large resonance energy.

You are aware from the discussion of resonance in Chapter 3 that no single Lewis structure satisfactorily represents the true structure of benzene. There is no alternation in bond length as one proceeds around the ring, as would be implied by a single structure. All of the carbon–carbon bonds of benzene are the same length (1.4 Å), intermediate between the length of a single bond (1.5 Å) and the length of a double bond (1.3 Å).

Benzene has two major resonance structures that contribute equally to the resonance hybrid. These are sometimes called **Kekulé structures** because they were originally postulated by Kekulé in 1866. You may also encounter benzene written with a circle inside the six-membered ring rather than the three double bonds. This representation is meant to show that the bonds in benzene are neither double nor single. However, the circle structure makes it difficult to count electrons. This text uses a single Kekulé structure to represent benzene or its derivatives. You must recognize that this does not represent the true structure and picture the other resonance structure or call upon the MO model presented in Section 16.3 when needed.

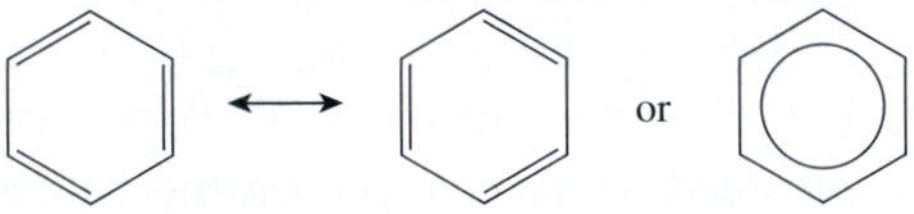

Kekulé structures

PROBLEM 16.1
Show the products of these reactions:

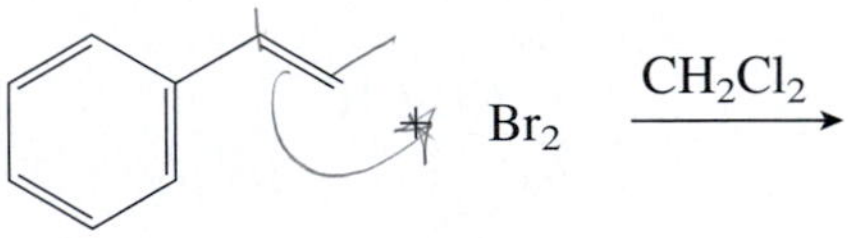

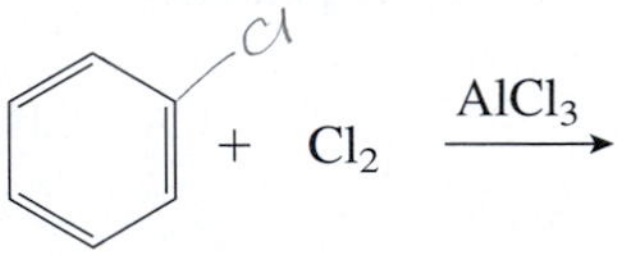

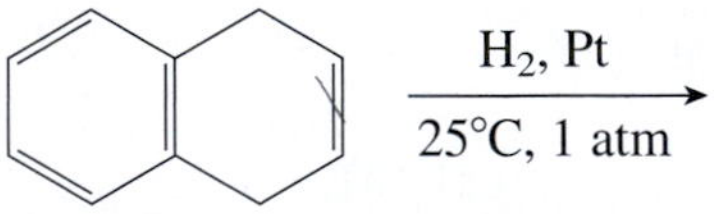

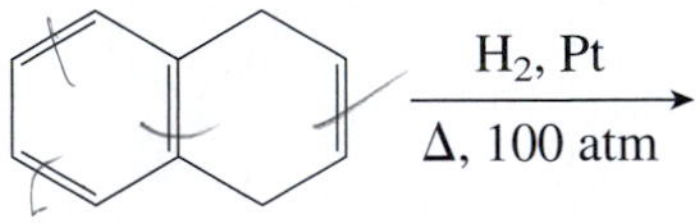

16.2   RESONANCE ENERGY OF BENZENE

The bromination and hydrogenation reactions just discussed, along with much other evidence, indicate that benzene has an especially large resonance stabilization. Can the magnitude of this resonance stabilization be determined? One way to measure such a stabilizing effect in a compound is to measure the amount of heat evolved in some reaction that destroys the effect and compare it to that evolved in the same reaction of some model compound where the stabilizing effect is absent. In the case of benzene the resonance stabilization is lost when it is saturated by catalytic hydrogenation. Reaction of one mole of benzene with three moles of hydrogen is an exothermic reaction and produces cyclohexane and 49.8 kcal/mol (208 kJ/mol) of heat.

Benzene + 3 H$_2$ → + 49.8 kcal/mol (208 kJ/mol)

What compound should be used in the hydrogenation as the model for the hypothetical compound "cyclohexatriene"—that is, benzene without any resonance stabilization? The model should be as similar to benzene as possible but without any possible resonance stabilization. The best that can be done is to use the double bond of cyclohexene as the model for one double bond of benzene. Hydrogenation of cyclohexene produces cyclohexane and 28.6 kcal/mol (120 kJ/mol) of heat:

Cyclohexene + H$_2$ → + 28.6 kcal/mol (120 kJ/mol)

According to this model, the hydrogenation of the three "double bonds" of benzene should produce $3 \times 28.6 = 85.8$ kcal/mol (359 kJ/mol) of heat if there were no resonance stabilization. The difference between the amount calculated on the basis of three cyclohexene double bonds and the amount that is actually produced is the resonance stabilization for benzene. By using these reactions, the resonance energy or resonance stabilization of benzene is calculated to be 36.0 kcal/mol (151 kJ/mol):

Heat evolved from the reaction of 3 C=C without resonance stabilization (3 × 28.6 kcal/mol)	85.8 kcal/mol (359 kJ/mol)
Minus heat evolved from the reaction of 3 C=C of benzene	−49.8 kcal/mol (208 kJ/mol)
Equals the resonance stabilization of benzene	36.0 kcal/mol (151 kJ/mol)

This is a significantly larger resonance stabilization than is found in other types of compounds that are not termed aromatic. Figure 16.1 shows a diagram of the energies of these hydrogenation reactions.

PROBLEM 16.2

The catalytic hydrogenation of naphthalene produces 80.0 kcal/mol (335 kJ/mol) of heat. Calculate the resonance stabilization of naphthalene. Do you think naphthalene should be termed aromatic?

Naphthalene + 5 H$_2$ $\xrightarrow{\text{catalyst}}$ + 80.0 kcal/mol (335 kJ/mol)

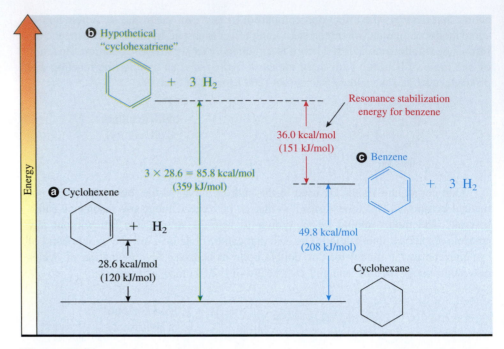

Figure 16.1

**DIAGRAM OF THE ENERGIES OF HYDROGENATION OF ⓐ CYCLOHEXENE, ⓑ HYPOTHETICAL
"CYCLOHEXATRIENE," AND ⓒ BENZENE.**

16.3 MOLECULAR ORBITAL MODEL FOR CYCLIC CONJUGATED MOLECULES

Although resonance is one of the most useful concepts in organic chemistry, one of the areas in which it is inadequate is in the explanation of aromaticity. For example, resonance structures similar to those of benzene can be written for cyclobutadiene:

 Cyclobutadiene

On the basis of this resonance picture only, organic chemists initially expected that cyclobutadiene, like benzene, would have a large resonance stabilization and would be especially stable. Yet cyclobutadiene proved to be an extraordinarily elusive compound. Many unsuccessful attempts were made to prepare this compound before it was finally synthesized at very low temperature in 1965. The compound is quite unstable and reacts rapidly at temperatures above 35 K. As we shall see, cyclobutadiene is a member of an unusual group of compounds that are actually destabilized by resonance. To understand why benzene is so stable while cyclobutadiene is so unstable, we must examine a molecular orbital picture for these compounds.

Benzene is a planar, hexagonal molecule. The 120° angles of a regular hexagon match exactly the trigonal planar bond angles of the sp^2-hybridized carbons of benzene, so it has no angle strain. There is a p orbital on each carbon that is perpendicular to the plane of the ring carbons. This results in an overlapping cycle of conjugated p orbitals that extends completely around the ring.

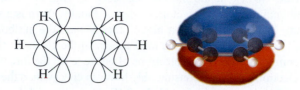

Recall from Section 3.6 that molecules such as benzene that have a series of conjugated p orbitals cannot be described very well by localized molecular orbitals that extend over only two atoms at a time. Instead, delocalized pi MOs that extend over the entire group of conjugated p orbitals must be employed. For benzene, all six of the p atomic orbitals combine to form six delocalized MOs. The number of MOs equals the number of AOs that overlap to form them.

Let's compare the molecular orbital picture for benzene to that for the imaginary "cyclohexatriene" in which the double bonds are not conjugated. As shown in Figure 16.2, each of the three "nonconjugated" double bonds of "cyclohexatriene" would have a pi bonding MO and a pi antibonding MO at the same energies as the pi MOs of ethene. This results in a total of six MOs: three degenerate bonding MOs and three degenerate antibonding MOs. (Recall that MOs that have the same energy are said to be degenerate.) The energies of the bonding and antibonding MOs are symmetrically placed about the energy of a nonbonding p orbital, which is, by convention, assigned a value of zero.

Calculations using a simple version of molecular orbital theory show that the energies of the six delocalized pi MOs of benzene are arranged somewhat differently. The lowest-energy MO, designated π_1, is significantly lower in energy than the bonding

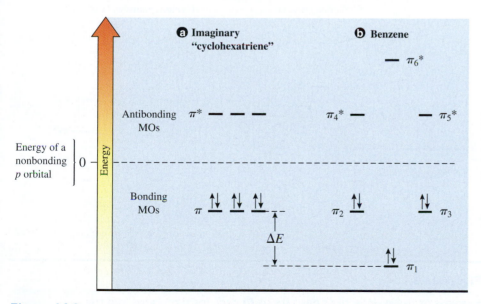

Figure 16.2

ENERGIES OF THE PI MOLECULAR ORBITALS OF ⓐ IMAGINARY "CYCLOHEXATRIENE" COMPARED TO THOSE OF ⓑ BENZENE.

MOs of "cyclohexatriene" (see Figure 16.2). Then there are two degenerate bonding MOs, designated π_2 and π_3, that have the same energy as the bonding MOs of "cyclohexatriene." Next are two degenerate antibonding MOs, π_4^* and π_5^*, that have the same energy as the antibonding MOs of "cyclohexatriene." Finally, there is a single highest-energy antibonding MO, π_6^*, that has a higher energy than the antibonding MOs of "cyclohexatriene." Remember that it is the total energy of the electrons that determines the stability of a compound. The six pi electrons of benzene fill the three bonding MOs. Because the two electrons in π_1 are considerably lower in energy than their counterparts in "cyclohexatriene," benzene is more stable by $2\Delta E$. This extra stabilization is what we have previously termed the resonance energy of benzene.

Calculations for other cyclic conjugated molecules show that the energies of their pi molecular orbitals are arranged in patterns similar to that of benzene. All such compounds have one lowest-energy MO, followed by pairs of degenerate MOs at increasingly higher energies. If the total number of MOs is even, then there is one highest-energy MO and the MOs are arranged symmetrically about zero energy, as is the case with benzene. This arrangement is illustrated for the eight MOs of cyclooctatetraene in Figure 16.3. (For the MOs of cyclooctatetraene to be symmetrically

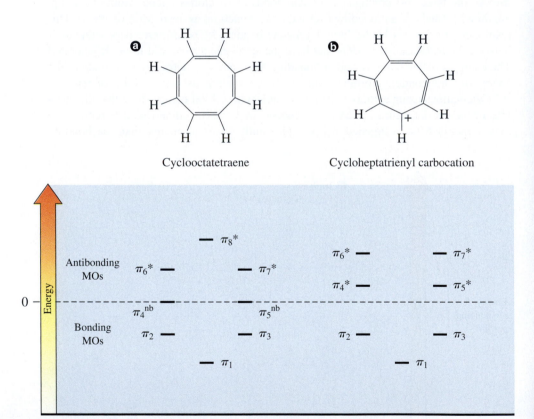

Figure 16.3

ENERGIES OF THE PI MOLECULAR ORBITALS OF ⓐ CYCLOOCTATETRAENE AND THE ⓑ CYCLOHEPTATRIENYL CARBOCATION. Note that the electrons are not yet shown in the MOs in these diagrams.

arranged about zero energy, it is necessary for one degenerate pair, π_4^{nb} and π_5^{nb}, to be located at zero energy. MOs at zero energy are termed nonbonding.) If the total number of MOs is odd, the pattern is the same with the exception that the highest-energy MO is absent. In this case the MOs can no longer be symmetrically arranged about zero energy. This arrangement is also illustrated in Figure 16.3 for the seven MOs of the cycloheptatrienyl carbocation. The exact energy of each MO is not as important as is the pattern of one lowest-energy MO and degenerate pairs with increasing energy.

PROBLEM 16.3
Show the patterns for the pi MO energy levels for these compounds:

a)   b)   c)   d)

16.4 CYCLOBUTADIENE

Let's now examine the MO picture for cyclobutadiene and see whether we can discern why it is so much less stable than benzene. There are four p orbitals in cyclobutadiene, so there are four pi MOs. These must be arranged with one MO at lowest energy, two degenerate MOs at zero energy, and one MO at highest energy. There are four electrons in these MOs. Two of these electrons occupy the lowest-energy MO. According to Hund's rule, the two remaining electrons have the same spin and each occupies a different member of the degenerate pair of nonbonding MOs.

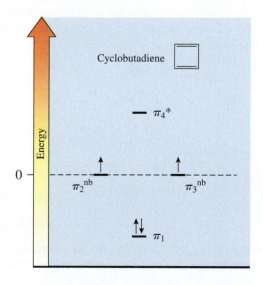

The difference between the electron arrangement in benzene and that in cyclobutadiene is the key to whether a cyclic, conjugated compound will be especially stable, like benzene, or especially unstable, like cyclobutadiene. Benzene has its *highest occupied molecular orbitals* (the highest-energy MOs that contain electrons, also known as **HOMOs**) completely filled with electrons. Cyclic compounds, completely conjugated around the ring, with filled HOMOs, such as benzene, are especially stable and are aromatic. In contrast, cyclobutadiene has only enough electrons that its HOMOs are half filled. Compounds with half-filled HOMOs such as cyclobutadiene are found to be especially unstable and are termed **antiaromatic.**

Cyclobutadiene is a highly reactive compound. As mentioned previously, numerous attempts to prepare it failed because of this high reactivity. However, it can be prepared and studied at very low temperatures, below 35 K. Such studies indicate that it does not have a square geometry, as would be suggested by the resonance structures shown on page 646. Rather, it has a rectangular geometry, with shorter double bonds alternating with longer single bonds. Such bond alternation relieves some of the antiaromatic destabilization by decreasing the overlap between the *p* orbitals where the longer bonds occur and is characteristic of compounds that are not aromatic. When a sample of cyclobutadiene is allowed to warm above 35 K, the molecules react rapidly to form dimers that are no longer conjugated and therefore are no longer antiaromatic.

PROBLEM 16.4

Add the appropriate number of electrons to the MO energy level diagram for cyclooctatetraene in Figure 16.3. Is this compound aromatic or antiaromatic?

PROBLEM 16.5

Add the appropriate number of electrons to the MO energy level diagram for the cycloheptatrienyl cation in Figure 16.3. Is this ion aromatic or antiaromatic? The cycloheptatrienyl anion has two more electrons than the cation. Do you expect this anion to be a stable species? Explain.

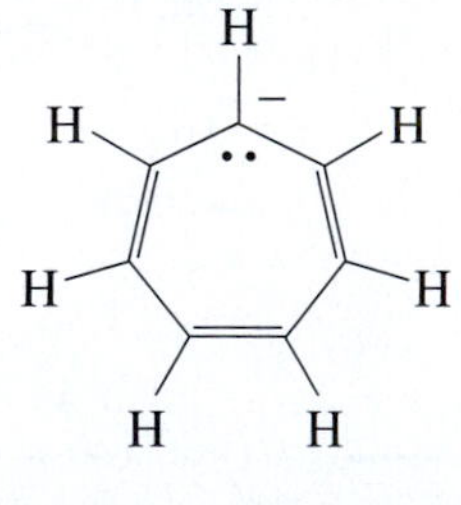

Cycloheptatrienyl anion

PROBLEM 16.6

Add electrons to the MO energy level diagrams for the compounds in parts b, c, and d of Problem 16.3, and predict whether each is aromatic or antiaromatic.

16.5 HÜCKEL'S RULE

All cyclic, conjugated molecules have one lowest-energy pi MO and pairs of degenerate MOs at higher energies. Whether a compound is aromatic or antiaromatic depends on the number of electrons occupying these MOs. An aromatic compound has its HOMOs completely filled with electrons. Therefore, it must have two electrons in the lowest-energy MO plus some multiple of four electrons so that the HOMOs are filled. (Another way of stating this is that it must have an odd number of pairs of electrons, one pair for the lowest-energy MO and an even number of pairs to fill the occupied degenerate MOs.)

The criteria for a compound to be aromatic were developed by Erich Hückel.

> ▶ **HÜCKEL'S RULE**
>
> Cyclic, fully conjugated, planar molecules with $4n + 2$ pi electrons (n = any integer including zero) are aromatic.

Let's analyze each aspect of this rule. First, the molecule must have a ring with a series of conjugated p orbitals that extends completely around the cycle, like benzene. If the cycle of conjugated orbitals is interrupted, as in the case of cyclopentadiene, then the compound is neither aromatic nor antiaromatic. It is just an alkene.

This sp^3-hybridized carbon interrupts the cycle of p orbitals, so cyclopentadiene is neither aromatic nor antiaromatic.

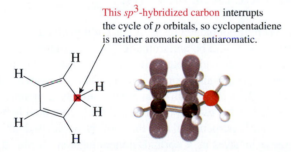

Second, the ring must be planar so that the p orbitals overlap in pi fashion completely around the cycle. If the ring is not planar, the p orbitals are twisted so that they are not parallel, resulting in a decrease in overlap. This decreases or even eliminates the aromatic or antiaromatic effect of the conjugation.

Finally, the number of pi electrons must equal 2 plus a multiple of 4 (or $4n + 2$ pi electrons). Some of the possible numbers are as follows:

n	$4n + 2$	Number of pairs
0	2	1
1	6	3
2	10	5
3	14	7

As we have seen, benzene, with its six pi electrons (sometimes called an aromatic sextet), is the prototypical aromatic compound.

The criteria for an antiaromatic compound can be generalized in a similar manner. The requisite number of electrons to have the HOMOs half filled is a multiple of four (an even number of pairs). Therefore, cyclic, fully conjugated, planar molecules with $4n$ pi electrons are antiaromatic. Some of the possible numbers of electrons are as follows:

n	$4n$	**Number of pairs**
1	4	2
2	8	4
3	12	6

Some additional examples will help clarify this concept and illustrate its usefulness in predicting and explaining experimental observations.

PROBLEM 16.7

Use Hückel's rule to predict whether each of the compounds in problem 16.3 is aromatic or not.

16.6 CYCLOOCTATETRAENE

Cyclooctatetraene has four double bonds, so it has eight pi electrons. This is a multiple of 4, so cyclooctatetraene would be antiaromatic if it were planar. However, planar cyclooctatetraene would have considerable angle strain because its bond angles would be 135° rather than the trigonal planar bond angle of 120°. To relieve both antiaromatic destabilization and angle strain, cyclooctatetraene adopts a nonplanar, tub-shaped geometry. Each double bond is twisted relative to the adjacent double bonds so that the *p* orbitals of one are nearly perpendicular to those of the adjacent double bonds. Therefore, cyclooctatetraene behaves as a normal, nonconjugated alkene. It can be readily prepared and isolated, and it does not show any of the special instability associated with antiaromatic compounds. Nor does it have any features of the special stability associated with aromatic compounds. It exhibits bond alternation—that is, shorter double bonds alternating with longer single bonds—and reacts with bromine to give addition products. It is said to be nonaromatic.

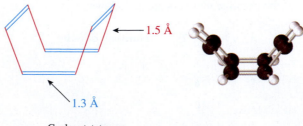

1.5 Å

1.3 Å

Cyclooctatetraene

16.7 HETEROCYCLIC AROMATIC COMPOUNDS

A heterocyclic compound is one that has an atom other than carbon as one of the ring atoms. The compound with a five-membered ring that has two double bonds and a nitrogen atom is called pyrrole.

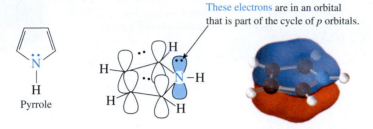

These electrons are in an orbital that is part of the cycle of *p* orbitals.

It is a planar molecule, and the pair of electrons on the nitrogen is in a *p* orbital that is parallel to the *p* orbitals of the ring double bonds. Therefore, pyrrole has a series of conjugated *p* orbitals that extend completely around the ring and that contain a total of six electrons: two from the nitrogen and four from the double bonds. It is an aromatic compound. On the basis of its heat of combustion, its aromatic resonance energy is calculated to be 21 kcal/mol (88 kJ/mol). It undergoes substitution reactions like benzene, rather than addition reactions like alkenes. Furthermore, it is much less basic than other amines. As can be seen in the following equation, the nitrogen of its conjugate acid does not have an electron pair to contribute to the cycle, so it is no longer aromatic. In other words, the electrons on the nitrogen of pyrrole are part of an aromatic sextet. If they are used to form a bond to a proton in an acid–base reaction, the aromatic stabilization of pyrrole is lost.

$$\text{Pyrrole} + H-O-H \rightleftharpoons \text{(conjugate acid)} + {}^-OH$$

Aromatic   Nonaromatic

Pyrrole is about 10^{14} times weaker as a base than is pyrrolidine, the five-membered nitrogen heterocycle that has no double bonds, because the basic pair of electrons on the nitrogen of pyrrolidine is not part of an aromatic cycle.

$$\text{Pyrrolidine} + H-O-H \rightleftharpoons \text{(conjugate acid)} + {}^-OH$$

The oxygen analog of pyrrole is furan. In this case, one pair of electrons on the oxygen is part of an aromatic sextet while the other is in an sp^2 hybrid AO that lies in the plane of the ring and is not part of the aromatic cycle. The sulfur analog, thiophene, has a similar structure. Both furan and thiophene are aromatic compounds that exhibit substitution reactions. From their heats of combustion the resonance energies of furan and thiophene are calculated to be 16 kcal/mol (67 kJ/mol) and 29 kcal/mol (121 kJ/mol), respectively.

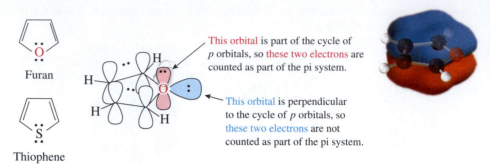

The differing amounts of aromatic stabilization for benzene, pyrrole, furan, and thiophene demonstrate that aromatic stabilization occurs in varying degrees, depending on the structure of the compound. Some compounds have a large aromatic stabilization that dramatically affects their stabilities and chemical reactions. Others may have only a small stabilization and have stabilities and reactions that are more comparable to a normal alkene.

Pyridine has a six-membered ring containing a nitrogen atom and is the nitrogen analog of benzene. The electrons on the nitrogen are in an sp^2 orbital in the plane of the ring like the electrons of the carbon–hydrogen bonds and are not part of the cycle of pi electrons. Pyridine has six electrons in its pi MOs and is aromatic.

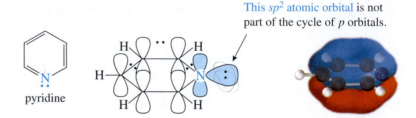

Because the unshared electron pair on the nitrogen is not part of the aromatic pi electron system, pyridine is a much stronger base than pyrrole. Recall that pyridine is used as a basic solvent in a number of reactions.

PRACTICE PROBLEM 16.1

Determine which electrons of imidazole are in orbitals that are part of the conjugated cycle of p orbitals and which are not. Explain whether imidazole is aromatic or not.

Imidazole

Solution

Nitrogen 1 (red) is like the nitrogen of pyrrole. Its electron pair is in a *p* orbital that is part of the conjugated cycle. In contrast, nitrogen 3 (blue) is like the nitrogen of pyridine. Its electron pair is in an sp^2 hybrid AO that is perpendicular to the conjugated cycle of *p* orbitals, so these electrons are not counted. (Only one orbital on an atom can be part of the conjugated cycle. Because the *p* orbital of the double bond is part of the cycle, the other orbital on N-3 cannot be part of the cycle.) Overall imidazole has six electrons in the cycle, the four electrons of the two double bonds, and the two electrons on N-1, so it is aromatic.

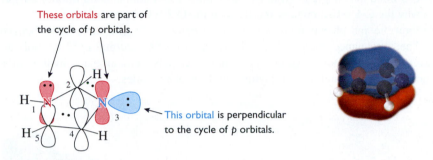

These orbitals are part of the cycle of *p* orbitals.

This orbital is perpendicular to the cycle of *p* orbitals.

PROBLEM 16.8

Explain whether each of these compounds is aromatic, antiaromatic, or nonaromatic:

a)

b)

c)

d)

e)

f) $5e^-$

16.8 POLYCYCLIC AROMATIC HYDROCARBONS

Polycyclic aromatic hydrocarbons have two or more benzene rings fused together. The simplest example is naphthalene. Hückel's rule does not apply to such fused ring systems. However, if the individual rings that are fused to form the polycyclic compound are aromatic, then the fused compound is also considered to be aromatic. Naphthalene, formed by fusing two benzene rings, is aromatic, although it is not expected to have as much resonance stabilization as two benzenes. This is confirmed experimentally. On the basis of its heat of combustion, the resonance energy for naphthalene has been calculated to be 61 kcal/mol (255 kJ/mol), a value that is larger than that of benzene (36 kcal/mol [151 kJ/mol]), although not twice as large.

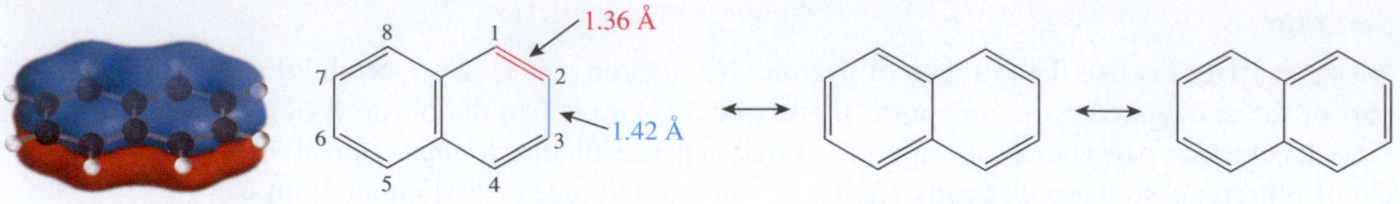

Naphthalene

Three resonance structures can be written for naphthalene. Note that the C-1—C-2 bond is a double bond in two of these structures and a single bond in one, while the C-2—C-3 bond is a single bond in two structures and a double bond in one. This explains why the C-1—C-2 bond is shorter than the C-2—C-3 bond.

Anthracene and phenanthrene are isomeric compounds with three fused benzene rings. Their resonance energies are calculated to be 84 kcal/mol (352 kJ/mol) and 92 kcal/mol (385 kJ/mol), respectively. Many other polycyclic aromatic hydrocarbons are known. Chrysene and benzo[a]pyrene are typical examples.

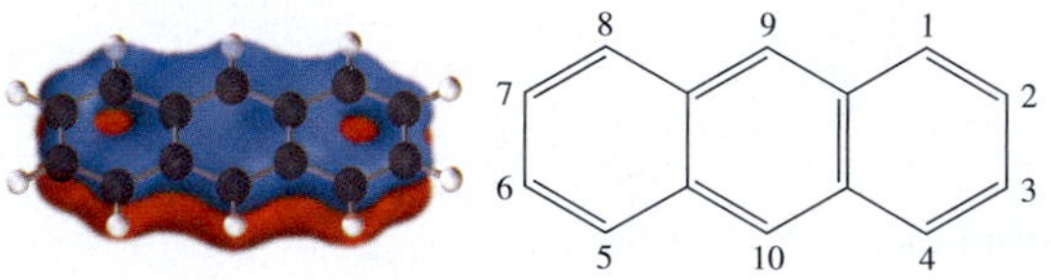

Anthracene

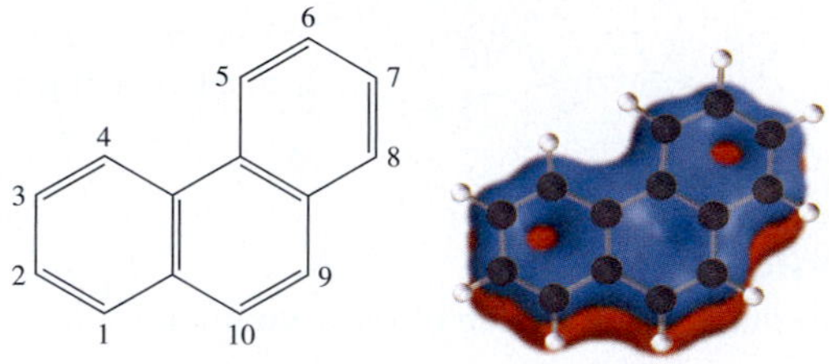

Phenanthrene

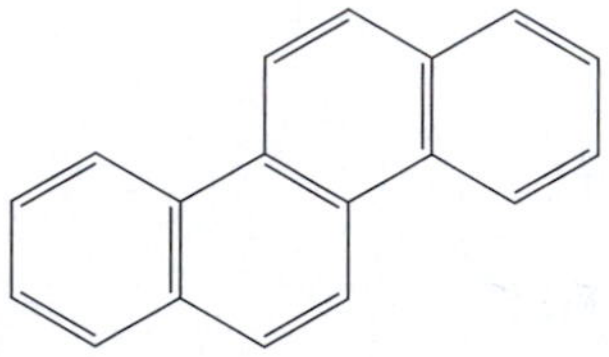

Chrysene

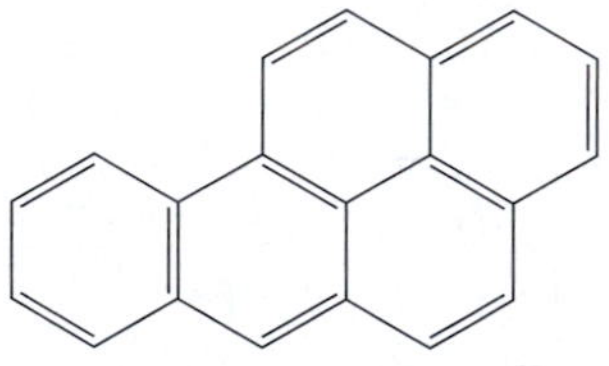

Benzo[a]pyrene

These compounds all possess considerable aromatic stabilization. They undergo substitution reactions like benzene, although some of them have pi bonds that are more reactive than those of benzene.

PROBLEM 16.9

Draw the five resonance structures for phenanthrene. Based on examination of these structures, which carbon–carbon bond of phenanthrene should be the shortest?

PROBLEM 16.10

Reaction of phenanthrene with Br_2 produces $C_{14}H_{10}Br_2$. This reaction occurs at the bond with the most double bond character. Show the structure of this product. Qualitatively compare the amount of resonance energy lost on formation of this product to the amount that would be lost if the addition were to occur at a different bond.

Focus On

Carcinogenic Polycyclic Aromatic Hydrocarbons

The first case of an environmental carcinogen was identified by an English surgeon, Percivall Pott, in 1775. He recognized a high incidence of scrotal cancer among chimney sweeps and correctly identified the causative agent as the coal soot to which they were continuously exposed. In the 1930s, some of the polycyclic aromatic hydrocarbons (PAH) found in coal soot were proved to be carcinogenic.

Incomplete combustion of carbonaceous material produces a wide variety of polycyclic aromatic hydrocarbons because these compounds are relatively stable and have a high ratio of carbon to hydrogen. Some of the highly carcinogenic compounds that are produced are benzo[a]pyrene, dibenz[a,h]anthracene, and dibenz[a,h]pyrene. These compounds are produced by the combustion of fossil fuels and are also found in tobacco smoke and automobile exhaust.

Dibenz[a,h]anthracene   Dibenz[a,h]pyrene

Although the mode of action of these compounds is not completely known, one idea is that they bind to DNA by sliding between its aromatic bases. Then oxidation of a reactive double bond in the PAH produces an epoxide intermediate that reacts with the DNA to initiate the carcinogenic process.

16.9 NMR AND AROMATICITY

It can be difficult to determine whether some compounds have any aromatic character or not based on their chemical reactions, especially if the amount of aromatic stabilization is small. In such situations, NMR spectroscopy provides another useful criterion for aromaticity. As discussed in Chapter 14, the hydrogens on a benzene ring usually appear in the region of 7 to 8 δ in the ^{1}H-NMR spectrum, significantly downfield from the position for hydrogens on alkene double bonds. This downfield shift is a result of a "ring current" that results from circulation of the pi electrons when the molecule is placed in the external magnetic field of the NMR instrument. The circulating electrons generate

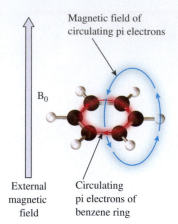

Magnetic field of circulating pi electrons

B_0

External magnetic field

Circulating pi electrons of benzene ring

a magnetic field that is opposed to the external magnetic field in the center of the ring but is parallel to the external magnetic field outside the ring in the region where the hydrogens are located.

Because the induced field is parallel to the external field where the hydrogens are located, less external field is needed to reach the total field required for the absorption of the electromagnetic radiation and the hydrogens appear at a downfield position. Of course, if a hydrogen is held near the center of the ring, an upfield shift is observed. As an example, the bridges in the following compound force the red proton to sit directly above the benzene ring. This hydrogen appears upfield from TMS at the extremely high field position of $-4.03\ \delta$!

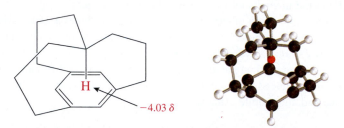

H
$-4.03\ \delta$

This ring current in benzene is termed **diamagnetic** and is characteristic of aromatic compounds in general. The presence of a diamagnetic ring current provides a useful experimental criterion for the presence of aromaticity in a compound. Other examples of the use of this method are provided in Section 16.10.

Antiaromatic compounds exhibit a different ring current, termed **paramagnetic,** that induces a magnetic field that is parallel to the external magnetic field in the center of the ring and opposed to it outside the ring. This causes hydrogens on the outside of the ring to appear upfield from the position of normal alkene hydrogen, a result that is exactly the opposite of the effect found with aromatic compounds.

An example of this effect is provided by tri-*tert*-butylcyclobutadiene. This compound is stable at room temperature for a brief time because the bulky *tert*-butyl groups retard the dimerization reaction that destroys less hindered cyclobutadienes. The ring hydrogen of this compound appears at $5.38\ \delta$, a position somewhat upfield from that of the hydrogens of a nonaromatic model compound such as cyclobutene ($5.95\ \delta$).

t-Bu   *t*-Bu   H ← $5.95\ \delta$

t-Bu   H ← $5.38\ \delta$

Tri-*tert*-butylcyclobutadiene

Because of its nonplanar geometry, cyclooctatetraene is not antiaromatic and its hydrogens appear at $5.75\ \delta$, a value typical for alkenes. However, the triple bonds of the compound called benzo-1,5-cyclooctadiene-3,7-diyne force this molecule to assume a nearly planar geometry. The pi system of its eight-membered ring contains eight electrons. (Only two of the electrons of each triple bond are part of the conjugated system.)

The hydrogens on the double bond of the eight-membered ring appear upfield at 4.93 δ, indicating that this compound has some antiaromatic character.

Cycloactatetraene — 5.75 δ

Benzo-1,5-cyclooctadiene-3,7-diyne — 4.93 δ

16.10 ANNULENES

The general name **annulene** is sometimes given to rings that contain alternating single and double bonds in a single Lewis structure. Thus, benzene can be called [6]annulene, and cyclooctatetraene can be called [8]annulene. A number of larger annulenes have been prepared to determine whether they follow Hückel's rule and are aromatic when they have $4n + 2$ electrons in the cycle.

The larger members of this series would have considerable angle strain if they were planar and had only cis double bonds. The incorporation of trans double bonds provides a way to relieve this angle strain, although this often introduces steric strain resulting from atoms on opposite sides of the ring being forced into the same region of space. Consider, for example, [10]annulene. With two trans double bonds it has no angle strain, but the two hydrogens that point into the interior of the ring cause so much steric strain that attempts to prepare this compound have not yet been successful. However, the compound with a CH_2 bridge in place of the offending hydrogens has been prepared. Although the bridge causes the ring to be somewhat distorted from planarity, the compound does show the presence of a diamagnetic ring current typical of an aromatic compound. The hydrogens on the periphery of the ring appear at 6.9 to 7.3 δ, and the hydrogens on the bridge, which are held over the face of the ring, appear at the substantially upfield position of −0.5 δ. (The hydrogens of a typical CH_2 group attached to a carbon–carbon double bond appear near 2 δ.)

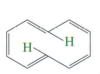

[10]Annulene (not yet prepared)

A bridged [10]annulene — H H ← −0.5 δ — 6.9−7.3 δ

The hydrogens inside the ring of the [14]annulene with four trans double bonds appear at 0.0 δ and the hydrogens on the outside of the ring appear at 7.6 δ, indicating the presence of a diamagnetic ring current. However, the steric strain caused by the hydrogens inside the ring makes this compound quite reactive. The bridged [14]annulene, where these steric interactions are absent, is stable and has many characteristics of an aromatic compound. The bond distances are all near 1.4 Å, it undergoes substitution reactions rather than addition reactions, the outer hydrogens appear at 8.1 to 8.7 δ, and the hydrogens of the methyl groups appear at -4.25 δ. As a final example, the hydrogens on the inside of the ring of [18]annulene appear at -3 δ and the outside hydrogens appear at 9 δ.

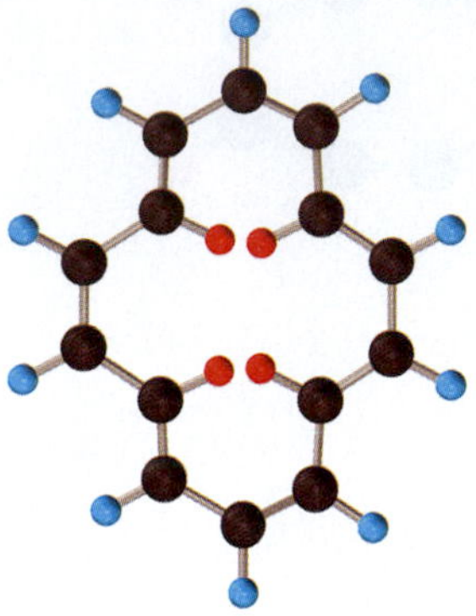

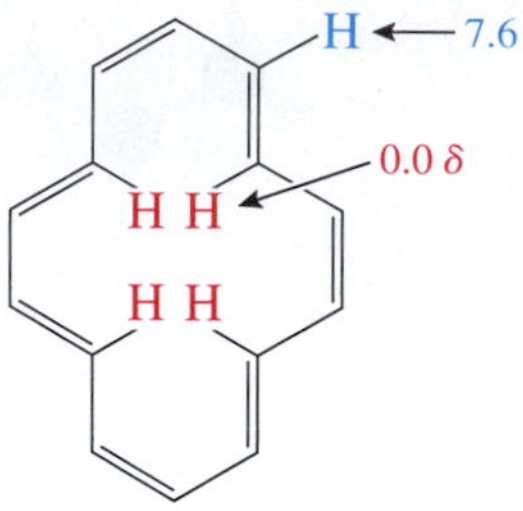

[14]Annulene

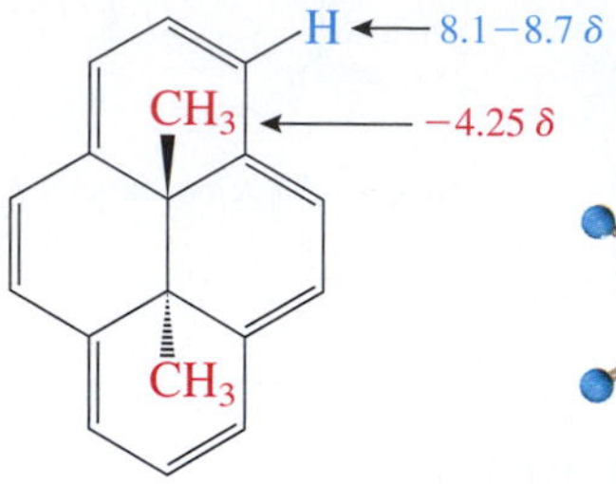

A bridged [14]annulene

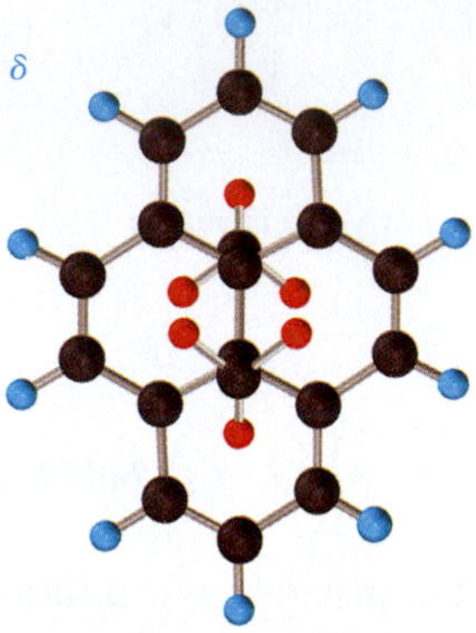

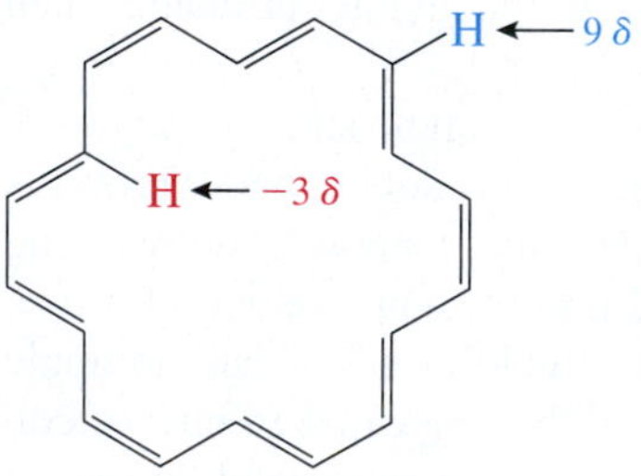

[18]Annulene

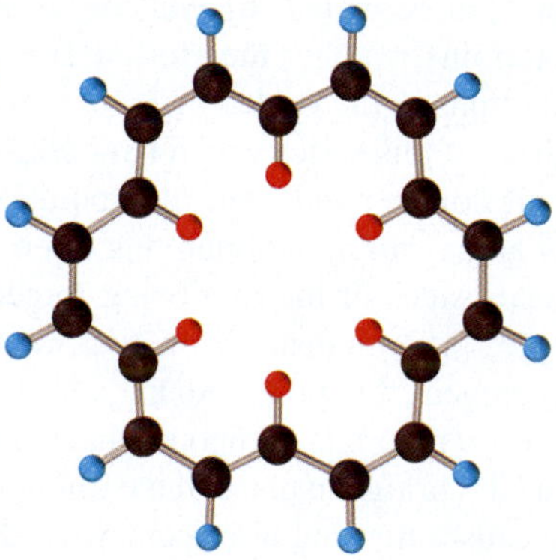

PROBLEM 16.11

The ^{1}H-NMR spectrum of this compound shows absorptions in the region of 9.5 δ and other absorptions in the region of -7 δ. Explain which hydrogens are responsible for each of these absorptions.

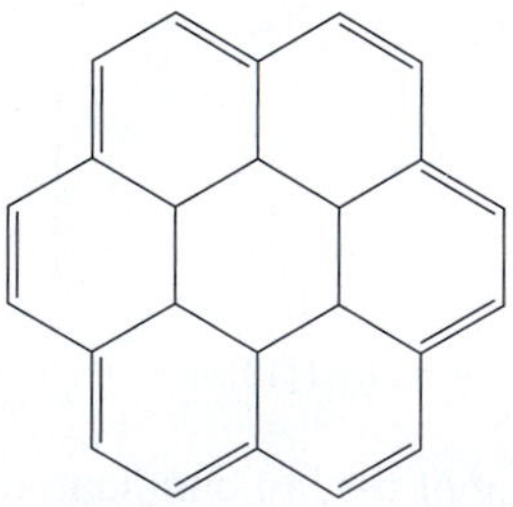

MODEL BUILDING PROBLEM 16.1

Build models of [10]annulene and the bridged [10]annulene discussed on the previous page and examine the strain and planarity of each.

16.11   AROMATIC AND ANTIAROMATIC IONS

Rings containing an odd number of carbon atoms can be aromatic or antiaromatic, if they are planar and have a conjugated p orbital on each ring atom. To have an even number of electrons in their odd number of p orbitals, these species must be ionic. They must be carbocations or carbanions.

The simplest example of such an ion is the cyclopropenyl carbocation:

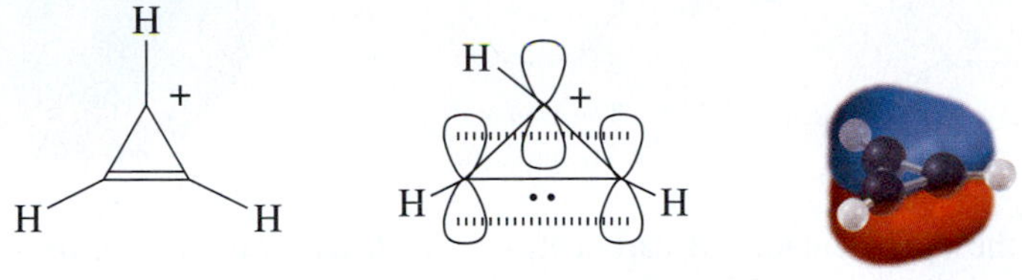

The cyclopropenyl carbocation

Because a carbocation is sp^2 hybridized, with trigonal planar geometry and an empty p orbital, this ion has a cycle of three p orbitals. (Remember that it is not the number of orbitals that determines whether a compound is aromatic or not, but rather the number of electrons in the pi MOs.) The cyclopropenyl carbocation has two electrons in its three pi MOs, so it fits Hückel's rule and should be aromatic. In fact, cyclopropenyl carbocations are significantly more stable than other carbocations, even though they have considerable angle strain. For example, most carbocations react rapidly with water, a weak nucleophile. In contrast, tri-*tert*-butyl-cyclopropenyl perchlorate, a carbocation salt, is stable enough to be recrystallized from water.

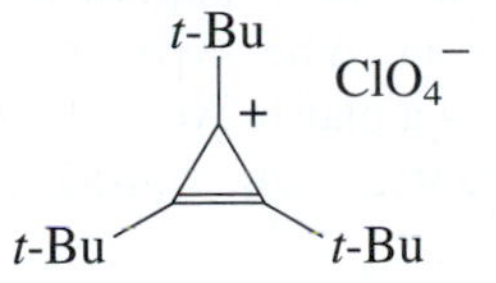

Tri-*tert*-butylcyclopropenyl perchlorate

Another example is provided by the acidity of cyclopentadiene. This compound is approximately as strong an acid as water and is many orders of magnitude more acidic than other hydrocarbons because its conjugate base, with six pi electrons, is aromatic.

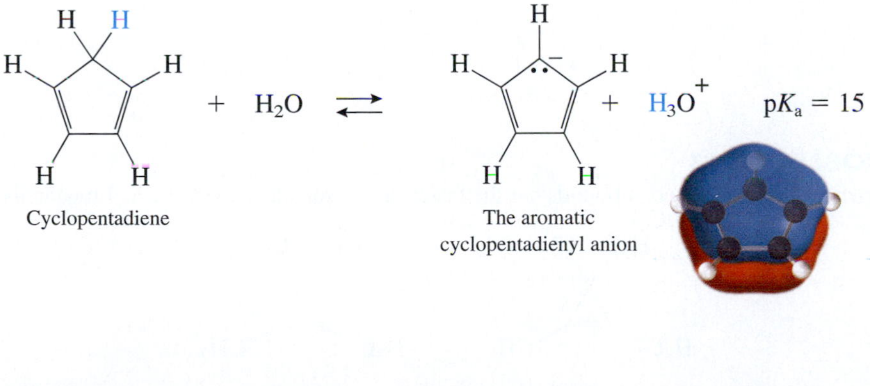

Cyclopentadiene   The aromatic
                     cyclopentadienyl anion

The cycloheptatrienyl carbocation, also known as the tropylium cation, has six pi electrons. It is also aromatic and is quite stable. In fact, 7-bromo-1,3,5-cycloheptatriene actually exists as an ionic compound.

H
Br
H
+
Br⁻

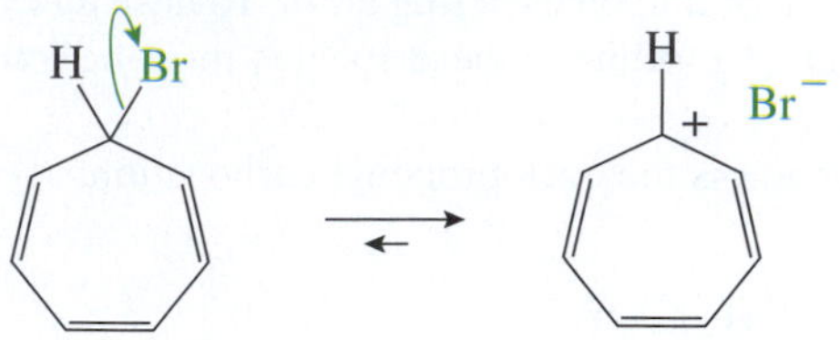

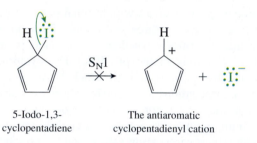

7-Bromo-1,3,5-cycloheptatriene

The aromatic tropylium cation

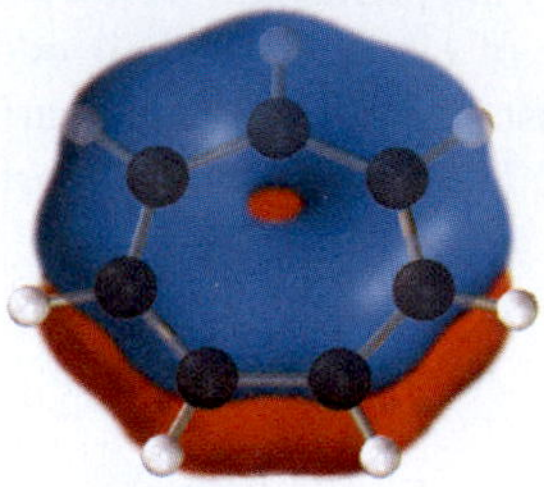

In contrast, the cyclopentadienyl carbocation, which has four pi electrons and is antiaromatic, is quite unstable. Thus, 5-iodo-1,3-cyclopentadiene is unreactive under conditions in which iodocyclopentane reacts rapidly by an S_N1 mechanism.

H
:I:
S_N1
H
+
+
:I:⁻

5-Iodo-1,3-cyclopentadiene

The antiaromatic cyclopentadienyl cation

ORGANIC **Chemistry Now**™ Click *Coached Tutorial Problems* for additional practice identifying **Aromatic Compounds.**

The concept of aromaticity is very important in understanding the chemical behavior of cyclic, conjugated compounds. It is most important with benzene and its derivatives, but it also has applications to many other types of compounds. Whenever a reactant, product, or intermediate contains a planar cycle of *p* orbitals, the effect of aromaticity (or antiaromaticity) on the reaction must be considered.

PROBLEM 16.12

Explain which of these compounds is a stronger acid:

a)
or

b)
CN
Ph
Ph
or
CN
Ph
Ph

c)
or

PROBLEM 16.13

Explain which of these compounds has the faster rate of substitution by the S_N1 mechanism:

H₃C Cl
H₃C
CH₃
H₃C Cl
H₃C
CH₃

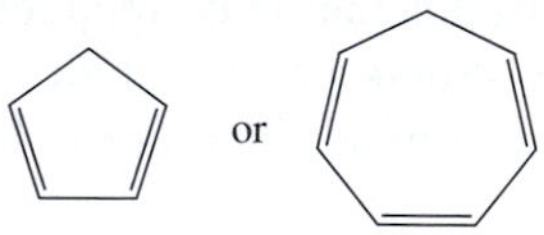

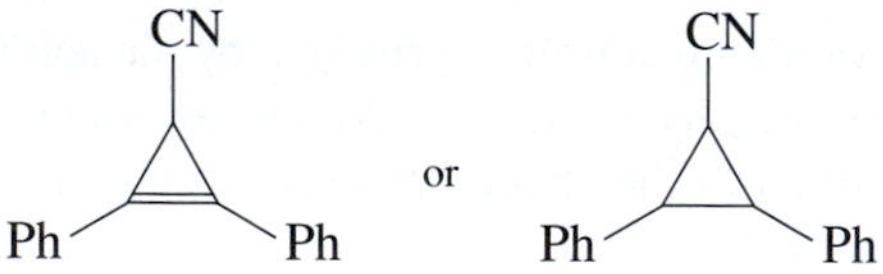

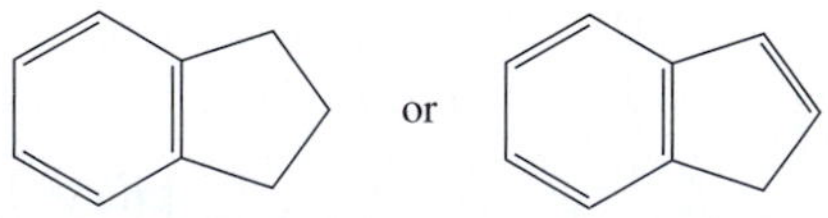

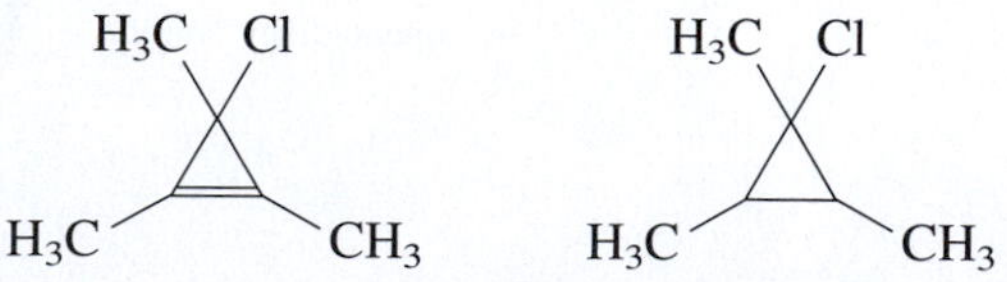

Focus On

Buckminsterfullerene, a New Form of Carbon

Until recently, elemental carbon was thought to occur in only two forms: diamond and graphite. Diamond consists of sp^3-hybridized carbons arranged in a three-dimensional, tetrahedral network. Graphite consists of large arrays of sp^2-hybridized carbons arranged in planar sheets of fused benzenoid rings. It can be viewed as the ultimate polycyclic aromatic hydrocarbon. The ability of one graphite sheet to slide over another gives it its lubricating properties.

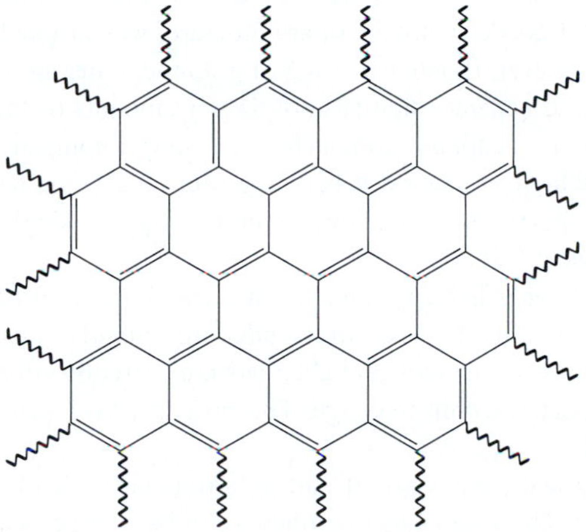

Graphite

Who would have thought that another form of carbon might exist and remain undiscovered until the mid-1980s? In 1985, R. E. Smalley, R. F. Curl, Jr., H. W. Kroto, and their collaborators reported their studies on the products from the vaporization of carbon by a laser. (Smalley, Curl, and Kroto shared the 1996 Nobel Prize in chemistry for this work.) Under certain conditions a relatively large amount of a compound with the formula C_{60} (determined by mass spectrometry) is produced. These scientists proposed that C_{60} is ball shaped and has the geometry of a truncated icosahedron, that is, a polygon with 12 pentagonal faces and 20 hexagonal faces. (The seams of a soccer ball have this geometry.) The structure of this C_{60} molecule is related to that of graphite, with its six-membered benzenoid type rings. However, the presence of the five-membered rings causes the surface of the atoms to curve back on itself so that a spherical shape can be attained. In this manner, all of the carbons can have their valences satisfied without introducing any significant strain into the molecule.

Continued

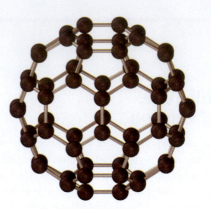

Buckyball

Buckminsterfullerene (or buckyball), named after R. Buckminster Fuller, who was famous for his use of geodesic domes in architecture, was originally produced in minuscule amounts. However, resistive heating of graphite under an inert atmosphere has recently been found to generate significantly larger amounts of this fascinating compound. (In fact, it is now available from a chemical supply company for about $250 for 1 g.) Because of its high symmetry, it has only four absorption bands in its IR spectrum. Its ^{13}C-NMR spectrum is even simpler, consisting of a single peak at 143.2 δ— all 60 carbons are identical!

Research on fullerenes is being reported at a frantic pace. It has been doped with metals to produce materials that are superconducting. Small atoms have been trapped within the hollow cavity of the ball, and chemists are currently attempting to trap other atoms, especially metals, within the cage. The properties of such caged metal atoms may be truly unique.

Larger, nonspherical assemblies of carbon atoms have also been prepared, some with a tubular shape. These so-called nanotubes can be viewed as a rolled-up graphite sheet, perhaps capped with half of a buckyball in some cases. Nanotubes have many potential applications. They may be useful in constructing faster and smaller electronic devices because they can be doped to become semiconducting or metallic and they can be made to carry electrical current at higher densities than metals. They can also be spun into incredibly strong fibers. However, before they can reach their true potential, methods to produce them inexpensively must be developed.

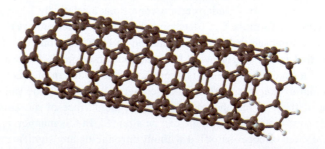

Review of Mastery Goals

After completing this chapter, you should be able to:

ORGANIC
Chemistry-Now™

Click *Mastery Goal Quiz* to test how well you have met these goals.

■ Show the MO energy levels for planar, cyclic, conjugated compounds. (Problem 16.14)

■ Apply Hückel's rule and recognize whether a particular compound is aromatic, antiaromatic, or neither. (Problems 16.15, 16.19, 16.20, 16.23, 16.29, and 16.30)

■ Understand how aromaticity and antiaromaticity affect the chemistry (and NMR spectra) of compounds. (Problems 16.16, 16.17, 16.18, 16.21, 16.22, 16.24, 16.25, 16.26, 16.28, and 16.31)

Additional Problems

ORGANIC
Chemistry-Now™

Assess your understanding of this chapter's topics with additional quizzing and conceptual-based problems at
http://now.brookscole.com/hornback2

16.14 Show the energy levels for the pi MOs of these species. Show the electrons occupying these MOs.

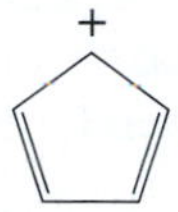

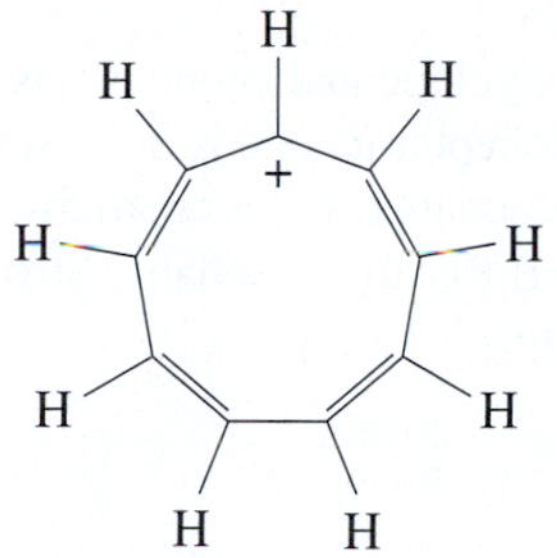

16.15 Explain whether each of these species is aromatic, antiaromatic, or neither:

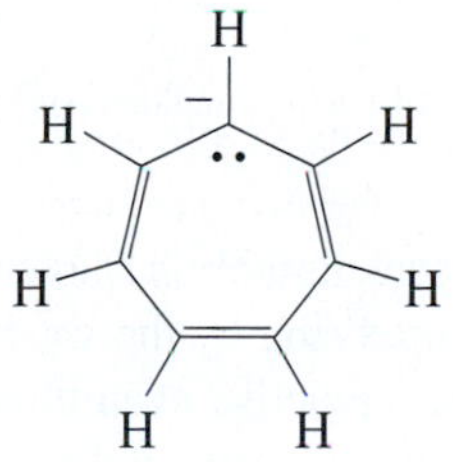

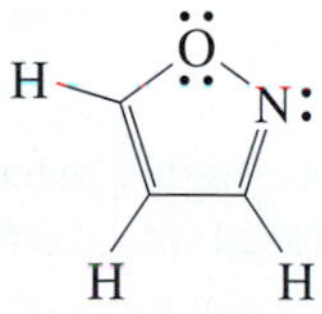

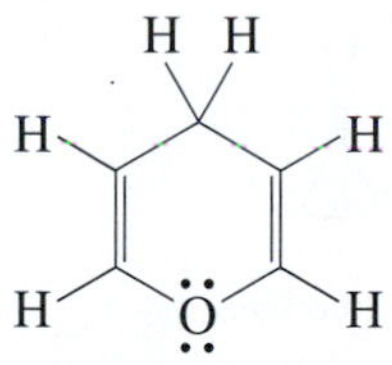

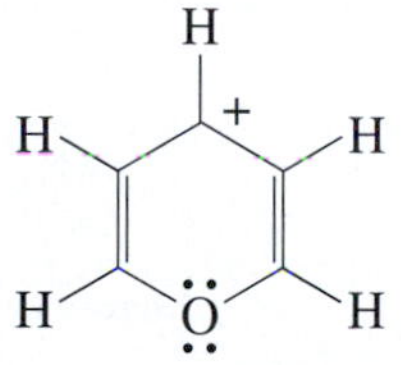

e)

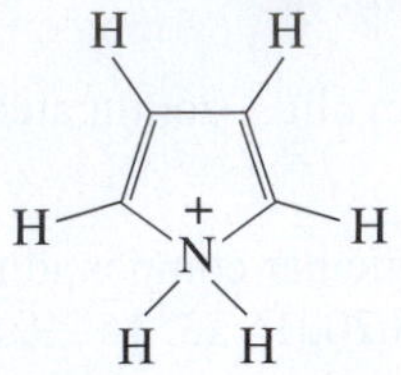

f)

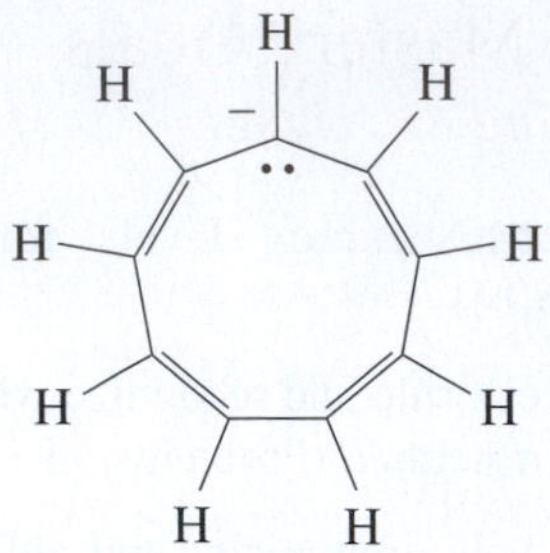

16.16 Cyclooctatetraene readily reacts with potassium metal to form a dianion. Discuss the electronic structure and geometry of this dianion and explain why it is formed so readily.

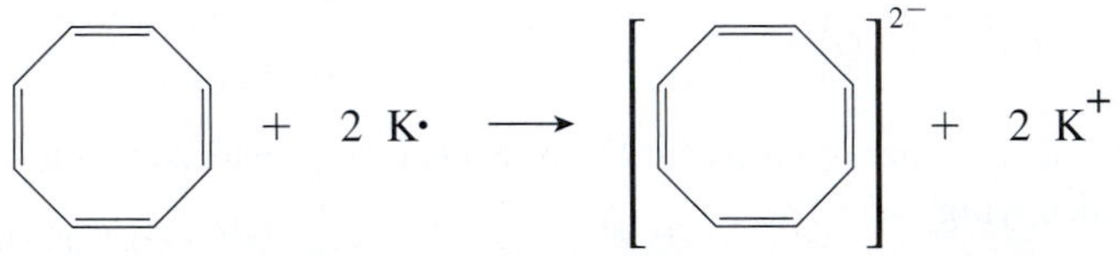

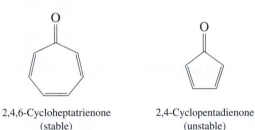

16.17 2,4-Cyclopentadienone is unstable and cannot be isolated, whereas 2,4,6-cycloheptatrienone is quite stable and is readily isolated. Use arguments based on resonance and aromaticity to explain these experimental observations. (*Hint*: Recall a resonance structure that is commonly written for the carbonyl group.)

2,4,6-Cycloheptatrienone
(stable)

2,4-Cyclopentadienone
(unstable)

16.18 Cyclopropanone is a highly reactive ketone, presumably because of the extra angle strain introduced into the three-membered ring by the sp^2-hybridized carbonyl carbon. Cyclopropenone is much less reactive even though it has more angle strain. Offer an explanation for this experimental observation.

More reactive   Less reactive

16.19 Hückel's rule applies only to compounds with a single ring, such as benzene and cyclobutadiene. However, it can be used with multiple-ring compounds if the resonance structure with all of the double bonds on the periphery of the ring is considered. For example, using such a structure for naphthalene shows 10 electrons in the cycle, so it is predicted to be aromatic.

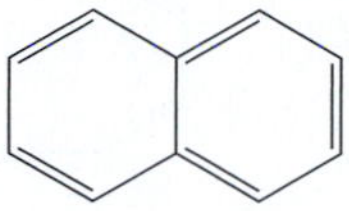

Use this to explain why two of the following compounds are very reactive, whereas one is quite stable:

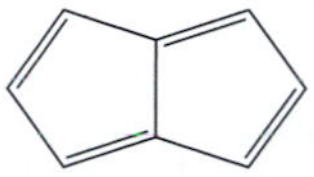

16.20 Coumarin is an important natural flavoring. Explain whether the oxygen-containing ring has any aromatic character.

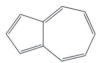

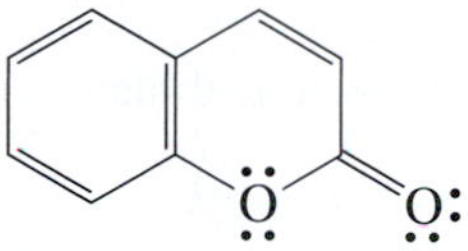

Coumarin

16.21 The conjugate acid of this ketone is quite stable. Explain.

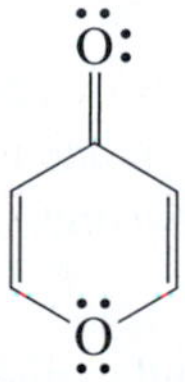

16.22 Benzocyclobutadiene is a very reactive compound. Explain.

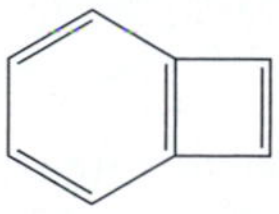

Benzocyclobutadiene

16.23 Explain whether these compounds are aromatic or not:

a) 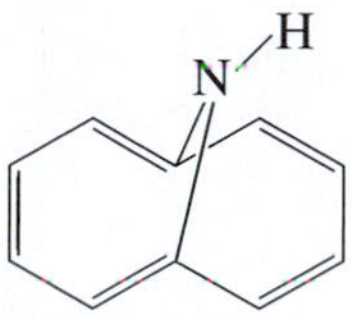   b)

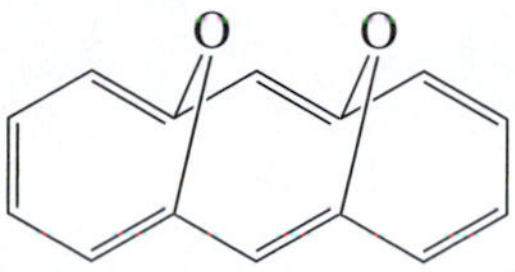

16.24 Predict the approximate positions for the absorptions in the ^{1}H-NMR spectrum of this compound.

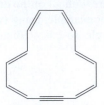

BioLink

16.25 The amino acid histidine contains an imidazole ring. In many enzymes this ring acts as a basic catalyst. Explain which nitrogen of the imidazole ring is more basic and show the structure of the conjugate acid of imidazole.

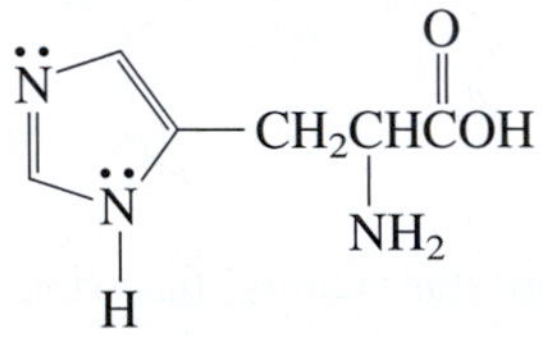

Histidine

16.26 This compound is a stronger acid than nitric acid. Explain.

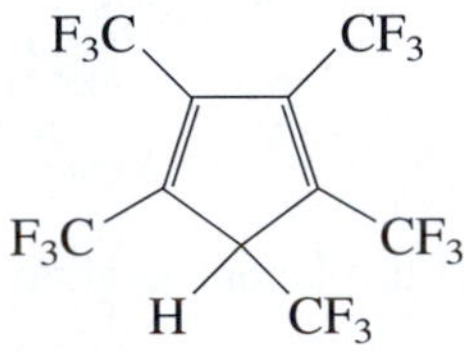

16.27 The heat of hydrogenation of pyrrole is 31.6 kcal/mol (132 kJ/mol). The heat of hydrogenation of cyclopentene is 26.6 kcal/mol (111 kJ/mol). Calculate the resonance stabilization of pyrrole.

16.28 The following compound reacts with butyllithium to form $C_8H_6^{2-}$. Suggest a structure for this dianion and explain its ready formation.

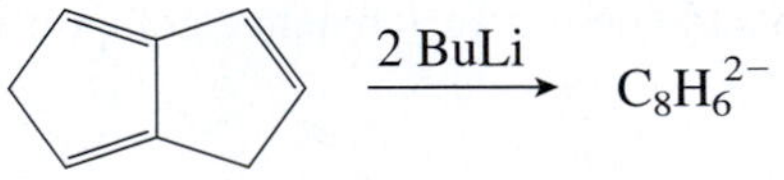

$$\xrightarrow{\text{2 BuLi}} C_8H_6^{2-}$$

16.29 Fused heterocyclic compounds are similar to polycyclic aromatic compounds except that one or more of the fused rings is a heterocycle. Explain whether or not the heterocyclic rings of these compounds are aromatic.

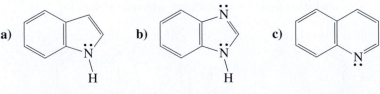

a) Indole   b) Benzimidazole   c) Quinoline

16.30 Adenine is an important base that is found as a component of DNA. Explain whether adenine is aromatic or not.

Adenine

16.31 Derivatives of this compound have been found to have large dipole moments. Use resonance and the theory of aromaticity to explain this observation.

Problems Involving Spectroscopy

16.32 When 7-fluoro-1,3,5-cycloheptatriene is dissolved in SbF_5/SO_2, a species is produced that shows a singlet near 9.2 δ in its ^1H-NMR spectrum. Show the structure of this species and explain why the absorption is so far downfield.

$$\xrightarrow[\quad SO_2 \quad]{\quad SbF_5 \quad}$$

16.33 When this dichloride is dissolved in SbF_5/SO_2, a species is produced that shows only two absorptions in its ^{13}C-NMR spectrum. One of these signals appears at 209 δ. Show the structure of this species and explain why the signal is so far downfield.

$$\xrightarrow[\quad SO_2 \quad]{\quad SbF_5 \quad}$$

16.34 The fragmentation in the mass spectrometer of the molecular ion of aromatic compounds to produce benzylic carbocations was discussed in Section 15.6. Actually, the benzylic carbocation is thought to rearrange to an even more stable carbocation, also with m/z 91, that is the actual species that is detected. Suggest a structure for this carbocation and explain why it is so stable.

Problems Using Online Three-Dimensional Molecular Models

16.35 Explain which of the carbon–carbon bonds in this model of cyclooctatetraene are single bonds and which are double bonds.

16.36 Purine is a heterocyclic compound with four nitrogen atoms. Each N has a pair of electrons on it in the Lewis structure. Explain which of these pairs are part of the pi system and which are not. Explain whether purine is aromatic or not.

16.37 For each of these heterocyclic compounds, explain which electron pairs are part of the pi system and which are not. Explain whether each compound is aromatic or not.

16.38 Explain why fluorene is a much stronger acid than most hydrocarbons. Its pK_a is 23.

Do you need a live tutor for homework problems? Access vMentor at Organic ChemistryNow at **http://now.brookscole.com/hornback2** for one-on-one tutoring from a chemistry expert.

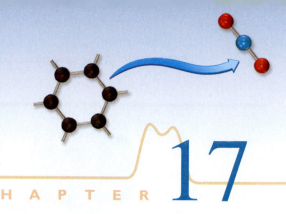

Aromatic Substitution Reactions

MOST OF THE REACTIONS discussed in this chapter involve the attack of an electrophile on an aromatic compound. Although the initial step of the mechanism resembles that of the electrophilic addition reactions of carbon–carbon double bonds discussed in Chapter 11, the final product here results from substitution of the electrophile for a hydrogen on the aromatic ring rather than addition. Therefore, these reactions are called **electrophilic aromatic substitutions.**

First, the general mechanism for these reactions is presented. This is followed by a specific example, the substitution of a nitro group onto a benzene ring. Then the effect of a group that is already present on the ring on the rate of the reaction and its regiochemistry is discussed in detail. Next, reactions that add halogens, sulfonic acid groups, alkyl groups, and acyl groups to the aromatic ring are presented. In each case the required reagents, the mechanism for generating the electrophile, the usefulness, and the limitations of the reactions are discussed. These reactions are very important and constitute the majority of the chapter.

Next, three different mechanisms for nucleophilic substitutions on aromatic rings are presented. These are followed by several other reactions that are useful in synthesis because they interconvert groups attached to aromatic rings. Finally, the use of combinations of all of these reactions to synthesize a variety of substituted aromatic compounds is discussed.

MASTERING ORGANIC CHEMISTRY

▶ Predicting the Products of Aromatic Substitution Reactions

▶ Understanding the Mechanisms of Aromatic Substitution Reactions

▶ Predicting the Effect of a Substituent on the Rate and Regiochemistry of an Electrophilic Aromatic Substitution Reaction

▶ Synthesizing Aromatic Compounds

17.1 MECHANISM FOR ELECTROPHILIC AROMATIC SUBSTITUTION

A general mechanism for the electrophilic aromatic substitution reaction is outlined in Figure 17.1. The process begins by reaction of the electrophile with a pair of pi electrons of the aromatic ring, which acts as the nucle-

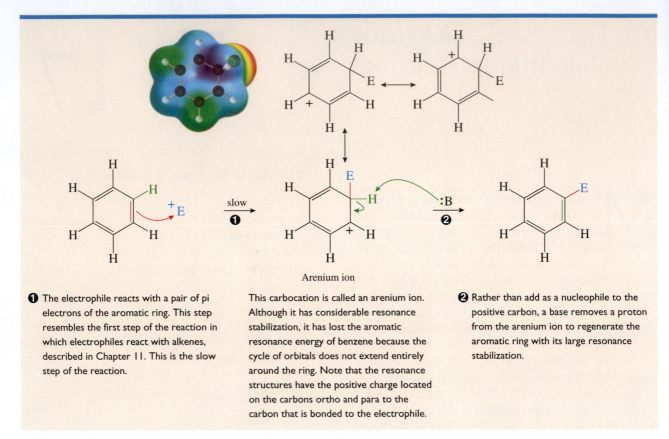

❶ The electrophile reacts with a pair of pi electrons of the aromatic ring. This step resembles the first step of the reaction in which electrophiles react with alkenes, described in Chapter 11. This is the slow step of the reaction.

This carbocation is called an arenium ion. Although it has considerable resonance stabilization, it has lost the aromatic resonance energy of benzene because the cycle of orbitals does not extend entirely around the ring. Note that the resonance structures have the positive charge located on the carbons ortho and para to the carbon that is bonded to the electrophile.

❷ Rather than add as a nucleophile to the positive carbon, a base removes a proton from the arenium ion to regenerate the aromatic ring with its large resonance stabilization.

Figure 17.1

MECHANISM OF A GENERAL ELECTROPHILIC AROMATIC SUBSTITUTION REACTION.

ophile, in a fashion very similar to the addition reactions described in Chapter 11, which begin by reaction of an electrophile with the pi electrons of an alkene. This results in the formation of a carbocation called an **arenium ion.** Removal of a proton from the arenium ion by some weak base that is present restores the aromatic ring and results in the substitution of the electrophile for a hydrogen on the aromatic ring.

It is instructive to examine the energetics of this reaction and compare them to those of the addition reactions of Chapter 11 (see Figure 17.2). Because of its aromatic resonance energy, benzene is considerably more stable than the alkene. Because of resonance, the arenium ion that is produced in the electrophilic aromatic substitution reaction is more stable than the carbocation produced in the addition reaction. However, the arenium ion is no longer aromatic, so its stabilization relative to the carbocation is less than the stabilization of benzene relative to the alkene. Because the transition states for both of these reactions resemble the carbocation intermediates (recall the Hammond postulate), the transition state leading to the arenium ion must have lost most of its aromatic stabilization also. This causes the activation energy for the electrophilic aromatic substitution reaction, $\Delta G_s^{\ddagger}$, to be larger than the activation energy for the addition reaction, $\Delta G_a^{\ddagger}$. In other words, the loss of aromatic resonance energy that occurs on going to the transition state for substitution results in a higher activation barrier. The substitution reaction usually requires much stronger electrophiles than the addition reaction.

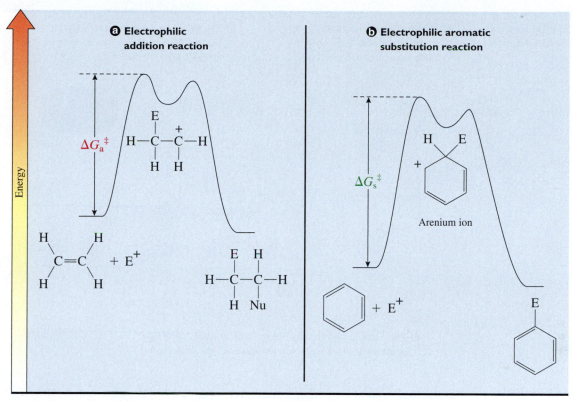

Figure 17.2

REACTION ENERGY DIAGRAMS FOR ⓐ AN ELECTROPHILIC ADDITION TO AN ALKENE AND ⓑ AN ELECTROPHILIC AROMATIC SUBSTITUTION REACTION.

The arenium ion, like any carbocation, has two reaction pathways available. It could react with a nucleophile to give an addition product, or it could lose a proton to some base in the system to give a substitution product. If addition were to occur, the product would be a cyclohexadiene, which is no longer aromatic because it has only two double bonds in the ring. It has lost at least 35 kcal/mol (146 kJ/mol) of stabilization, the difference in resonance energy between an aromatic ring and a cyclohexadiene. Obviously, the formation of the aromatic ring in the substitution product is greatly favored.

All of the electrophilic aromatic substitution reactions follow this same general mechanism. The only difference is the structure of the electrophile and how it is generated. Let's look at a specific example, the nitration of benzene. This reaction is accomplished by reacting benzene with nitric acid in the presence of sulfuric acid:

$$+ \ HNO_3 \ \xrightarrow{\ H_2SO_4\ } \ + \ H_2O$$

Benzene   Nitrobenzene

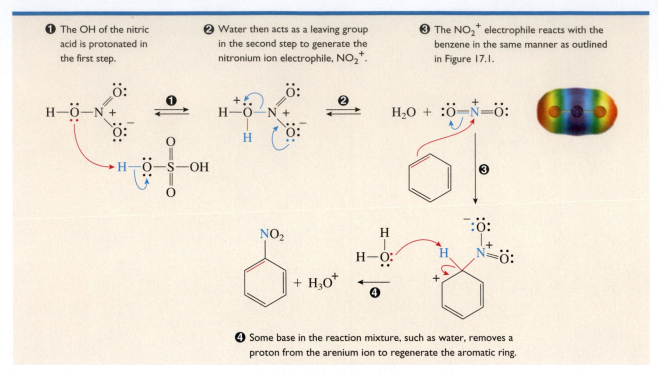

① The OH of the nitric acid is protonated in the first step.

② Water then acts as a leaving group in the second step to generate the nitronium ion electrophile, NO_2^+.

③ The NO_2^+ electrophile reacts with the benzene in the same manner as outlined in Figure 17.1.

④ Some base in the reaction mixture, such as water, removes a proton from the arenium ion to regenerate the aromatic ring.

Figure 17.3

MECHANISM OF THE NITRATION OF BENZENE.

The mechanism for this reaction is presented in Figure 17.3. The electrophile, NO_2^+ (nitronium ion) is generated from the nitric acid by protonation of an OH group. Water then acts as a leaving group to generate the electrophile. The rest of the mechanism is identical to that outlined in Figure 17.1.

17.2 EFFECT OF SUBSTITUENTS

When a substituted benzene is nitrated, the substituent on the ring has an effect on the rate of the reaction. In addition, the NO_2^+ electrophile can attach ortho, meta, or para to the substituent. For example, when toluene is nitrated, it is found to react 17 times faster than benzene. Substitution occurs primarily ortho and para to the methyl group.

Toluene   o-Nitrotoluene   m-Nitrotoluene   p-Nitrotoluene

60%   3%   37%

The methyl group accelerates the reaction compared to benzene and directs the incoming electrophile to the ortho and para positions. Both the rate enhancement and the regiochemistry of the reaction can be understood by examination of the three possible arenium ions produced by attack of the electrophile at the positions ortho, meta, and para to the methyl group. Consider the case of attack at an ortho position first:

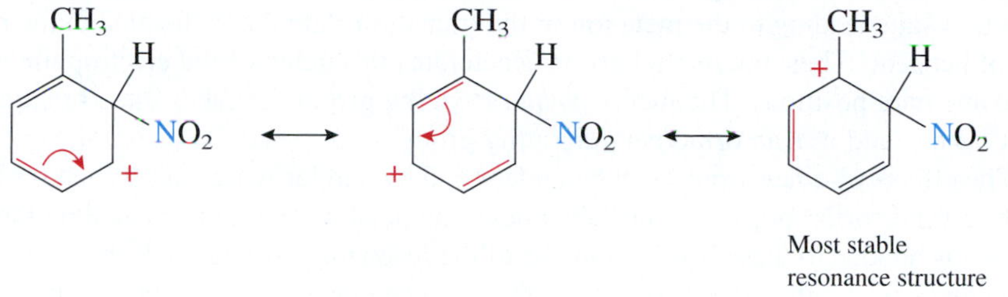

Arenium ion from ortho attack

In one of the resonance structures, the positive charge is located on the carbon bonded to the methyl substituent. As we are well aware, the methyl group will stabilize the carbocation, so this arenium ion is somewhat lower in energy (more stable) than the arenium ion produced in the nitration of benzene itself.

Consider next the arenium ion produced by attack of the electrophile at a position meta to the methyl group:

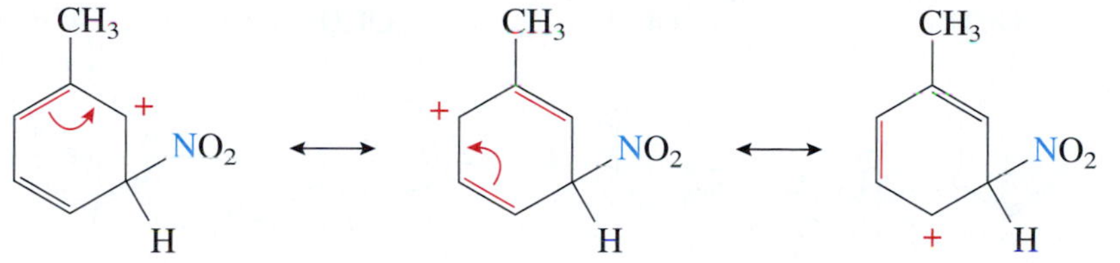

Arenium ion from meta attack

In this case, none of the resonance structures has the positive charge located on the carbon bonded to the methyl group. This ion has no extra stabilization when compared to the arenium ion formed from benzene.

Finally, consider the case of attack of the electrophile at the para position:

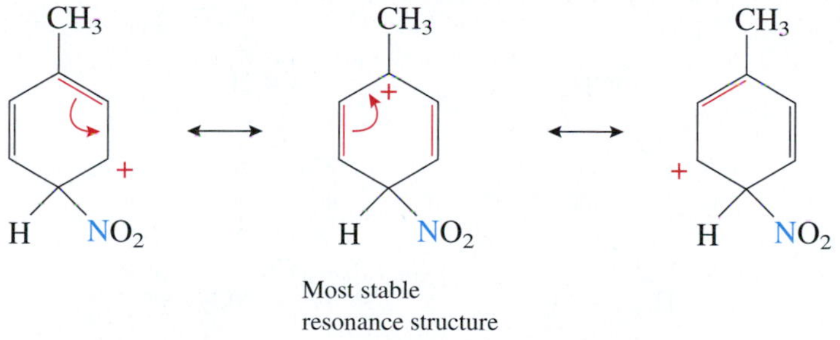

Arenium ion from para attack

This arenium ion is similar to that produced by attack at the ortho position in that the positive charge is located on the carbon bonded to the methyl group in one of the resonance structures. Therefore, it is more stable than the arenium ion formed from benzene.

Although all arenium ions are unstable, reactive intermediates, those resulting from ortho and para attack of the electrophile on toluene, but not that from meta attack, are more stable than the arenium ion produced from benzene itself. Therefore, for toluene the transition states leading to the ortho and para ions are at lower energy than is the transition state leading to the meta ion or the transition state that is formed in the nitration of benzene. Thus, the methyl group accelerates the attack of the electrophile at the ortho and para positions. The methyl is an activating group (it makes the aromatic ring react faster), and it is an ortho/para directing group.

The effects of other groups can be understood by similar reasoning. When the electrophile bonds ortho or para to the substituent, the positive charge is located on the carbon that is bonded to the substituent in one of the resonance structures. If the substituent is one that can stabilize the carbocation, then it accelerates the reaction and directs the incoming electrophile to the ortho and para positions. If, on the other hand, the substituent is one that destabilizes the carbocation, then it slows the reaction and directs the electrophile to the meta position so that the positive charge is never on the carbon directly attached to the substituent. The meta arenium ion is less destabilized than the ortho or para ions in this case. Some other examples will help clarify this reasoning.

The nitration of methoxybenzene (anisole) proceeds 10,000 times faster than does nitration of benzene and produces predominantly the ortho and para isomers of nitroanisole.

Methoxybenzene (anisole) → 31% + 2% + 67%

From these results it can be seen that the methoxy group is also an ortho/para director and is a much stronger activating group than is the methyl group. Examination of the resonance structure for the para arenium ion that has the positive charge located on the carbon bonded to the methoxy group explains these results.

Especially stable resonance structure

Two electronic effects are operating in this case: an inductive effect and a resonance effect. Because of the high electronegativity of the oxygen, the methoxy group withdraws electrons by its inductive effect (see Section 4.5). If this were the only effect operating, then it would be a deactivating group. However, in this case there is also a resonance effect. As shown in the resonance structure on the right, the methoxy group is a resonance electron-donating group. This resonance structure is especially stable because the octet rule is satisfied for all of the atoms. A similar, especially stable resonance structure can be written for the arenium ion that is produced by reaction of the electrophile at the ortho position. However, when the electrophile reacts at the meta position, the positive charge is never located on the carbon bonded to the methoxy group, so this especially stable resonance structure cannot be formed. Overall, the resonance effect dominates the inductive effect in this case so the methoxy group is a strongly activating group and an ortho/para director. With a few exceptions that are discussed shortly, any group that has an unshared pair of electrons on the atom bonded to the ring, represented by the general group Z in the following equation, has a similar resonance effect and acts as an activating group and an ortho/para director:

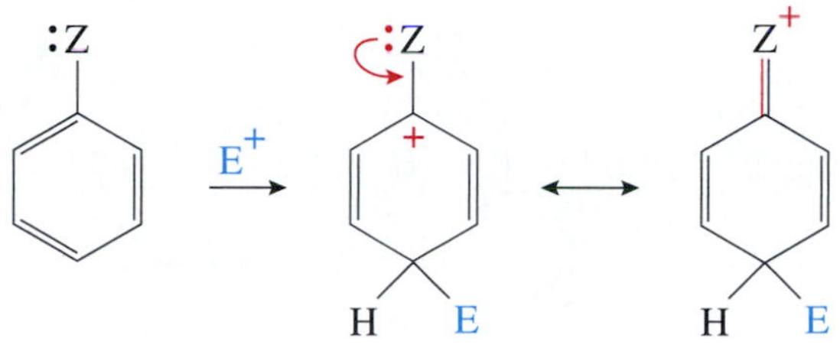

Especially stable
resonance structure

PROBLEM 17.1

Show all of the resonance structures for the arenium ion that is produced by attack of the NO_2^+ electrophile at the ortho position of anisole. Which of these structures is especially stable?

PROBLEM 17.2

Show all of the resonance structures for the arenium ion that is produced by attack of the NO_2^+ electrophile at the meta position of anisole. Is there an especially stable resonance structure in this case?

PROBLEM 17.3

Explain why these compounds react faster than benzene in electrophilic aromatic substitution reactions and give predominantly ortho and para products:

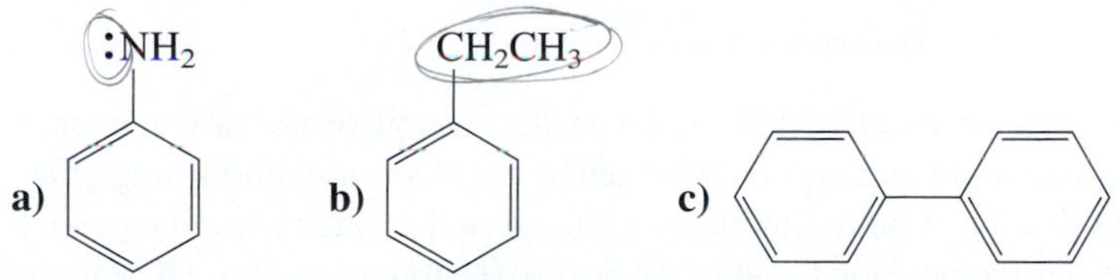

a) :NH_2 b) CH_2CH_3 c)

Now let's consider an example where the substituent slows the reaction. The nitration of nitrobenzene occurs approximately 10^7 times more slowly than the nitration of benzene and gives predominantly the *meta*-isomer of dinitrobenzene.

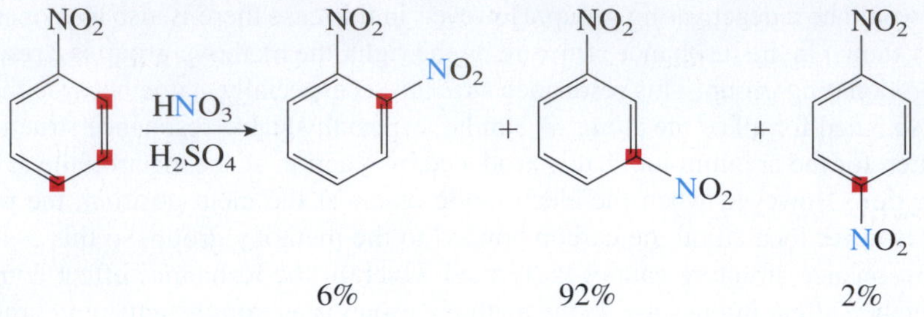

6%   92%   2%

The nitro group is deactivating and directs the incoming electrophile to the meta position. Again, examination of the arenium ions provides an explanation for these results. First, consider the arenium ion produced by attack of the electrophile at an ortho position:

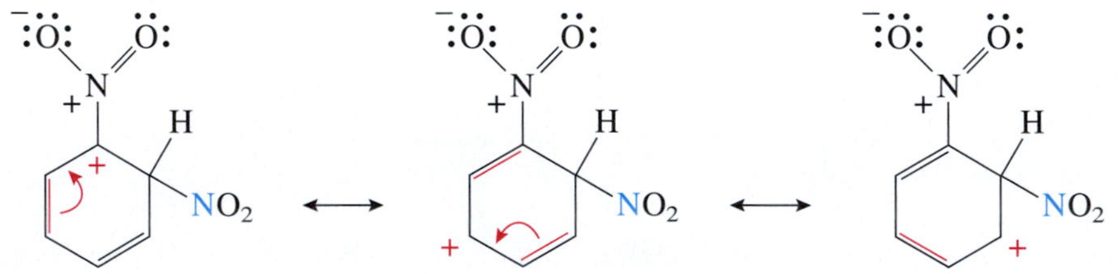

Arenium ion from ortho attack

The nitro group is an electron-withdrawing group both by its inductive effect and by its resonance effect. The first resonance structure is especially destabilized because the positive charge is located directly adjacent to the electron-withdrawing nitro group. Thus, the presence of the nitro group on the ring dramatically slows attack of an electrophile at the ortho position.

Now, consider attack of the electrophile at a meta position:

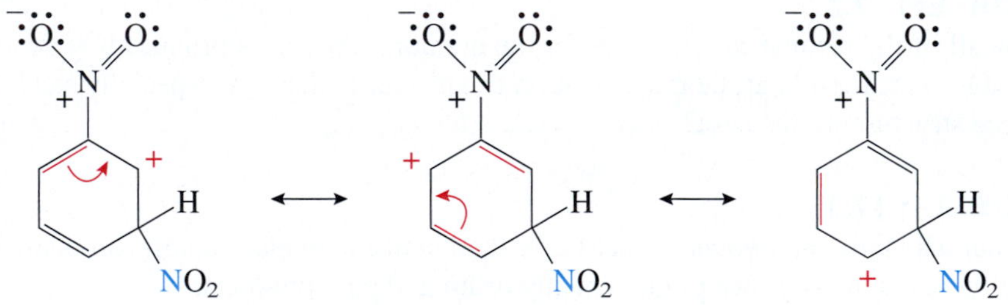

Arenium ion from meta attack

This time there is no resonance structure that has the positive charge on the carbon bonded to the nitro group. The arenium ion is still destabilized by the electron-withdrawing effect of the nitro group, but this ion is not destabilized as much as the ion produced by attack of the electrophile at the ortho position because the positive charge is never as close to the nitro group.

The arenium ion produced by attack of the electrophile at the para position resembles that produced by attack at the ortho position in that it also has a resonance structure that has the positive charge on the carbon that is bonded to the nitro group. The destabilization of this ion is comparable to that of the ortho ion.

Overall, then, the nitro group slows the reaction, but it slows attack at the meta position less than it slows attack at the ortho and para positions. It is a deactivating group and a meta-directing group. In general, any group that withdraws electrons from the ring by an inductive and/or resonance effect behaves similarly.

PROBLEM 17.4

Explain why these compounds react more slowly than benzene in electrophilic aromatic substitution reactions and give predominantly meta products:

deactivating

a) CF₃ EW

b) ⊕N(CH₃)₃ Cl⁻ *deactivating* EW

So far, groups have been either activating and ortho/para directors or deactivating and meta directors. The halogens are exceptions to this generalization. They are slightly deactivating compared to benzene but still direct to the ortho and para positions. For example, chlorobenzene is nitrated 17 times slower than benzene and produces predominantly *ortho*- and *para*-chloronitrobenzene.

| Chlorobenzene | 35% | 1% | 64% |

Why do the halogens have this unusual behavior? Like the methoxy group, the inductive and resonance effects of the halogens are in competition. In the case of the halogens, however, the inductive electron-withdrawing effect is slightly stronger than the resonance electron-donating effect. The high electronegativity of fluorine is responsible for its inductive effect being stronger than its resonance effect. The other halogens are weaker resonance electron-donating groups because their p orbitals do not overlap well with the $2p$ AO of the ring carbon, owing to the longer length of the carbon–halogen bond and the size of the $3p$, $4p$, or $5p$ AO. As a result, the halogens are weakly deactivating groups. But because resonance electron donation is most effective at the ortho and para positions, these positions are deactivated less than the meta position. Therefore, the halogens are slightly deactivating ortho/para directors.

The effect of almost any substituent can be understood on the basis of similar reasoning. Table 17.1 lists the effect on both the reaction rate and the regiochemistry of the substituents most commonly found on benzene rings. Rather than just memorizing this table, try to see the reasons why each group exhibits the behavior that it does. The strongly activating, ortho/para directors all have an unshared pair of electrons on the atom attached to the ring that is readily donated by resonance. Alkyl and aryl groups

Table 17.1   Effect of Substituents on the Rate and Regiochemistry of Electrophilic Aromatic Substitution Reactions

Substituent	Rate Effect	Regiochemistry	Comments
—NR₂ —NHR —NH₂ —OH	Strongly activating	ortho and para	Resonance donating effect is stronger than inductive withdrawing effect
—OR —SR O O ‖ ‖ —NHCCH₃ —OCCH₃	Moderately activating	ortho and para	Resonance donating effect is stronger than inductive withdrawing effect, but not as much so as above
—R Alkyl groups —Ar Aryl groups	Weakly activating	ortho and para	Weak inductive or resonance donors
—X: Halogens	Weakly deactivating	ortho and para	Resonance donating effect controls regio-chemistry but is weaker than inductive withdrawing effect that controls rate
O O O ‖ ‖ ‖ —CH —CR —COH O O ‖ ‖ —COR —CNH₂	Moderately deactivating	meta	Inductive and resonance withdrawers
—CN —SO₃H —CX₃ + + —NR₃ —NH₃ —NO₂	Strongly deactivating	meta	Inductive and/or resonance withdrawers

stabilize a positive charge on an adjacent carbon, so they are activating ortho/para directors, although they are less activating than the preceding groups. The deactivating meta-directing groups all have a positive or partial positive charge on the atom attached to the ring. The halogens are unusual because they are weakly deactivating, ortho/para directors.

PROBLEM 17.5

Predict the effect of these substituents on the rate and regiochemistry of electrophilic aromatic substitution reactions:

a) —S̈—CH₃   b) O‖—S—CH₃‖O   c) O‖—C—Cl

Finally, let's compare the amount of ortho product to the amount of para product produced in a reaction of an aromatic compound that has an ortho/para director on the ring. There are two ortho positions and only one para position, so if statistics were the only important factor, the ratio of ortho to para products should be 2 to 1. However, attack at the ortho position can be disfavored by the steric effect of the group. Obviously, this depends partly on the size of the group and the size of the electrophile. In addition, some reactions are more sensitive to steric effects than others. In general, then, the ratio of ortho to para product ranges from 2 to 1 in favor of the ortho product to predominantly para for reactions involving bulky substituents or reactions that are very sensitive to steric hindrance. An example of this steric effect can be seen by comparing the following reaction to the nitration of toluene presented earlier (page 674). The major product from toluene is the *ortho-* isomer (60% ortho and 37% para). In contrast, the bulky group of *t*-butylbenzene causes the major product from its nitration to be the *para*-isomer.

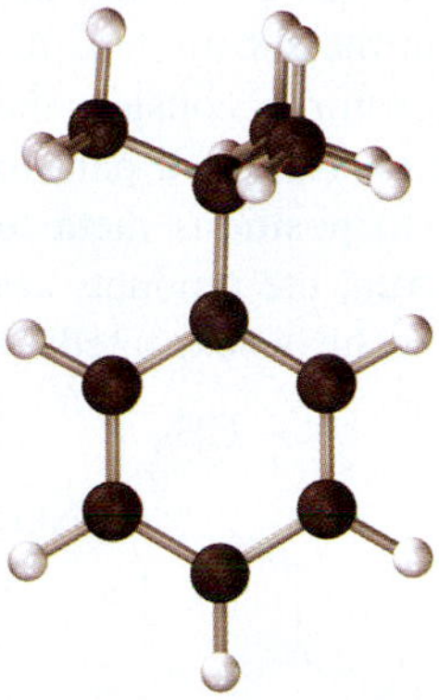

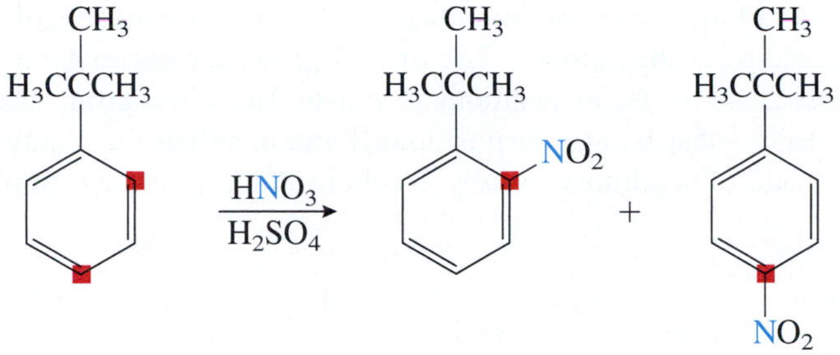

t-Butylbenzene   18%   82%

PRACTICE PROBLEM 17.1

Show the products of the reaction of ethylbenzene with nitric acid and sulfuric acid.

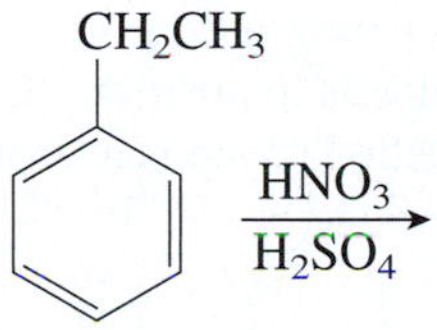

Solution

The ethyl group, like other alkyl groups, is weakly activating and directs to the ortho and para positions. The small amount of meta product that is formed is usually not shown.

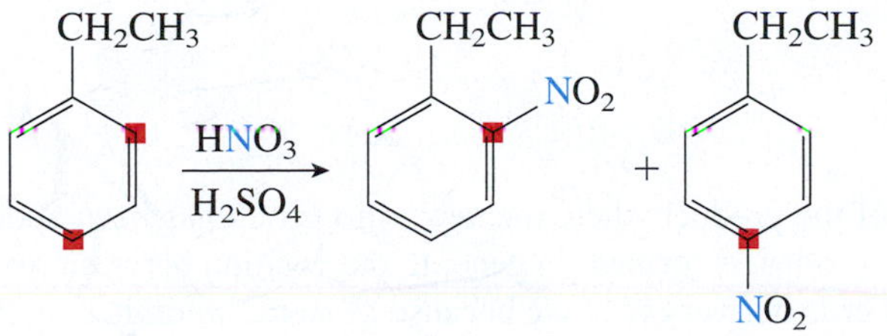

PROBLEM 17.6

Show the products of these reactions:

a) [structure: nitrobenzene with NO$_2$] *deactivating* *meta* *NO$_2$* $\xrightarrow[\text{H}_2\text{SO}_4]{\text{HNO}_3}$ b) [structure: bromobenzene with Br] *deactivating* *ortho para* *NO$_2$* $\xrightarrow[\text{H}_2\text{SO}_4]{\text{HNO}_3}$

17.3   EFFECT OF MULTIPLE SUBSTITUENTS

The situation is more complicated if there is more than one substituent on the benzene ring. However, it is usually possible to predict the major products that are formed in an electrophilic aromatic substitution reaction. When the substituents direct to the same position, the prediction is straightforward. For example, consider the case of 2-nitrotoluene. The methyl group directs to the positions ortho and para to itself—that is, to positions 4 and 6. The nitro group directs to positions meta to itself—that is, also to positions 4 and 6. When the reaction is run, the products are found to be almost entirely 2,4-dinitrotoluene and 2,6-dinitrotoluene, as expected:

2-Nitrotoluene   29%   70%   1%

If the groups direct to different positions on the ring, usually the stronger activating group controls the regiochemistry. Groups that are closer to the top of Table 17.1 control the regiochemistry when competing with groups lower in the table. In the case of 3-nitrotoluene the methyl group directs to positions 2, 4, and 6 while the nitro group directs to position 5. Because the methyl group is a stronger activating group than the nitro group, it controls the regiochemistry:

3-Nitrotoluene   38%   60%   1%

Note that none of the product where the new nitro group has been added to position 2, between the two groups, is formed. In general, the position between two groups that are meta to each other is not very reactive because of steric hindrance by the groups on either side of this position.

PROBLEM 17.7

Explain which positions would be preferentially nitrated in the reaction of these compounds with nitric acid and sulfuric acid:

a) 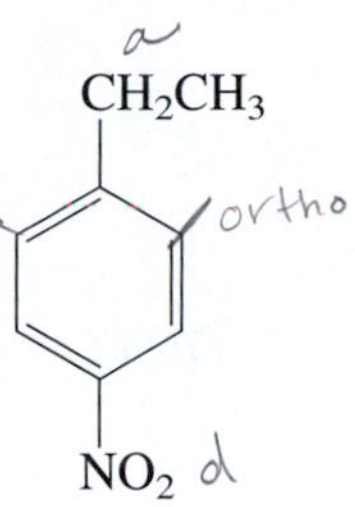 CH$_2$CH$_3$ ... NO$_2$ (ortho)

b) 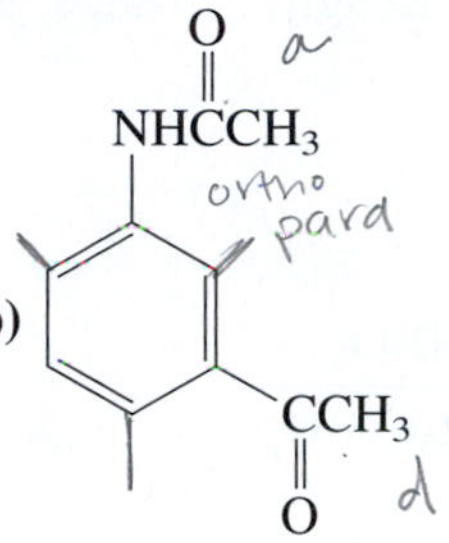 NHCCH$_3$ (O) ... CCH$_3$ (O) (ortho, para)

c) 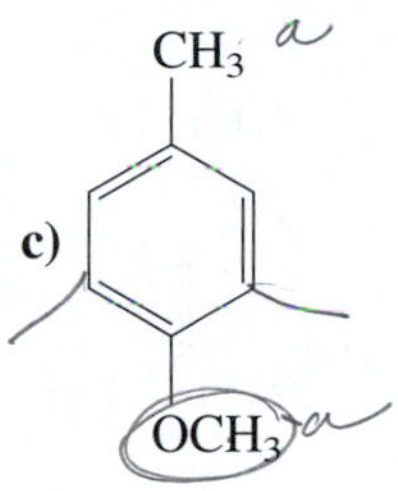 CH$_3$ ... OCH$_3$

d) 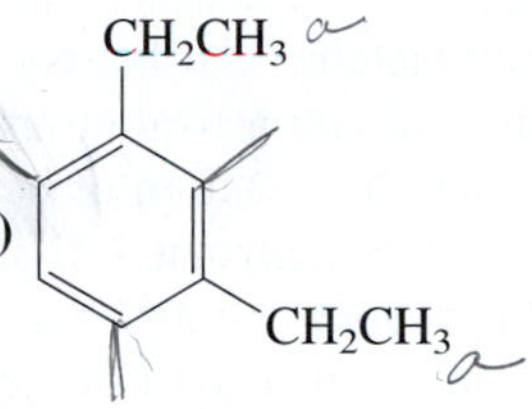 CH$_2$CH$_3$ ... CH$_2$CH$_3$

e) 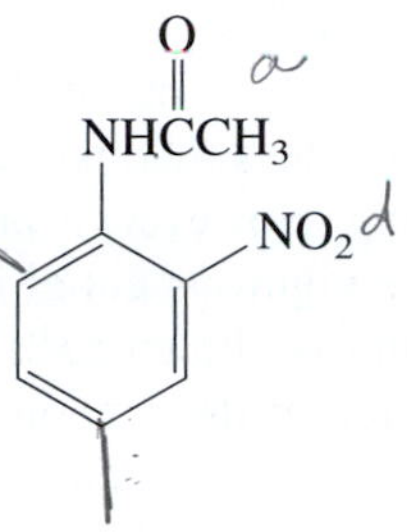 NHCCH$_3$ (O) ... NO$_2$

f) 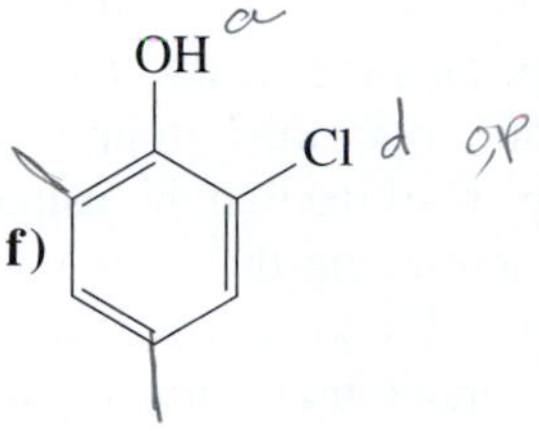 OH ... Cl

PROBLEM 17.8

Show the major products of these reactions:

a) 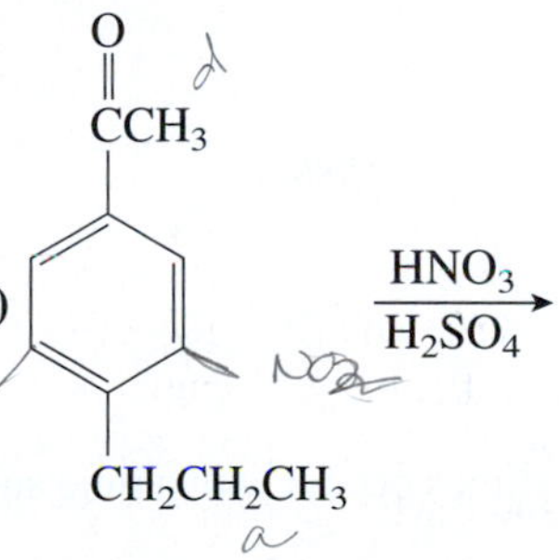 CCH$_3$ (O) ... CH$_2$CH$_2$CH$_3$ $\xrightarrow{\text{HNO}_3}{\text{H}_2\text{SO}_4}$

b) 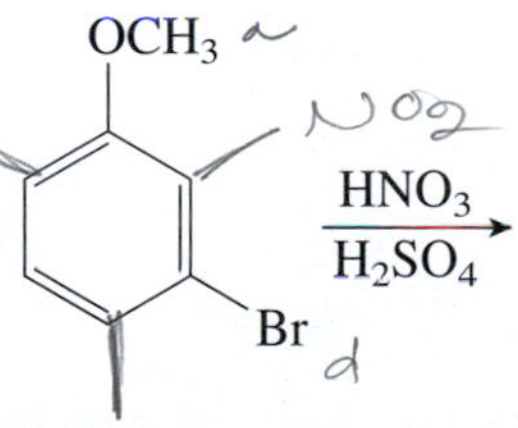 OCH$_3$ ... Br $\xrightarrow{\text{HNO}_3}{\text{H}_2\text{SO}_4}$

17.4 NITRATION

The reagents and the mechanism for the nitration of an aromatic ring have already been discussed. The reaction is very general and works with almost any substituent on the ring, even strongly deactivating substituents.

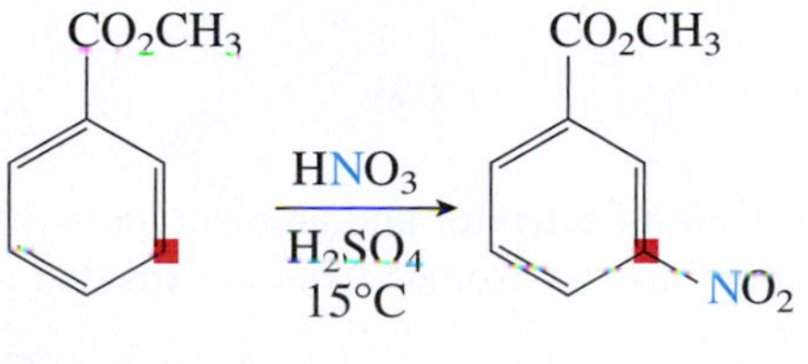

CO$_2$CH$_3$ $\xrightarrow[\substack{\text{H}_2\text{SO}_4 \\ 15°\text{C}}]{\text{HNO}_3}$ CO$_2$CH$_3$ ... NO$_2$ (85%)

Methyl benzoate

The concentration of the NO_2^+ electrophile is controlled by the strength of the acid that is used in conjunction with nitric acid. Milder conditions, such as nitric acid without sulfuric acid, nitric acid and acetic acid, or nitric acid in acetic anhydride, are employed when the ring is strongly activated, as illustrated in the following example:

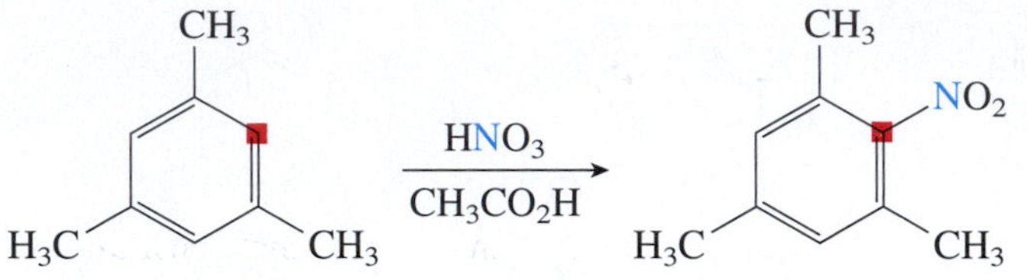

1,3,5-Trimethylbenzene

The nitro group that is added to the ring in these reactions is a deactivating group. This means that the product is less reactive than the reactant, so it is easy to add only one nitro group to the ring. However, it is possible to add a second nitro group, if so desired, by using more vigorous conditions. Thus, the reaction of benzoic acid using the same conditions as shown earlier for methyl benzoate (HNO_3, H_2SO_4, 15°C) results in the formation of the mononitration product, *m*-nitrobenzoic acid. Under more drastic conditions (higher temperature and higher sulfuric acid concentration), two nitro groups can be added, as illustrated in the following equation:

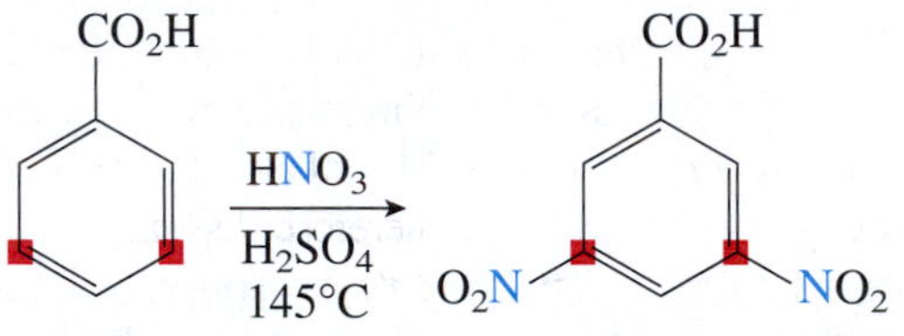

Benzoic acid

The following example illustrates a problem that sometimes occurs with amino substituents:

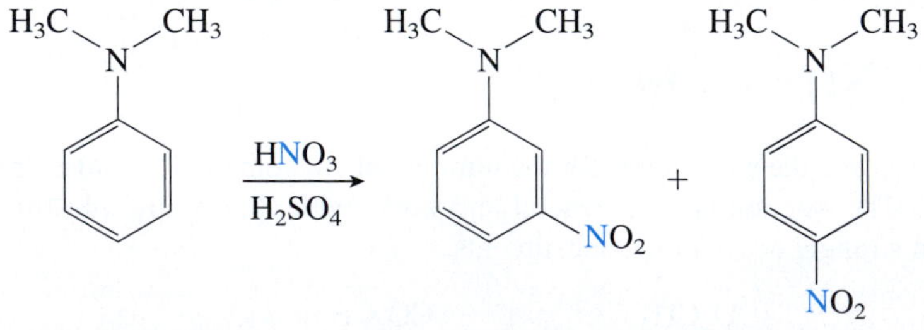

N,N-Dimethylaniline   63%   18%

The dimethylamino group is a strong activator and an ortho/para director, yet the major product from the reaction is the *meta*-isomer. This unexpected result is due to the basicity of the amino group. In the strongly acidic reaction mixture, the nitrogen is protonated:

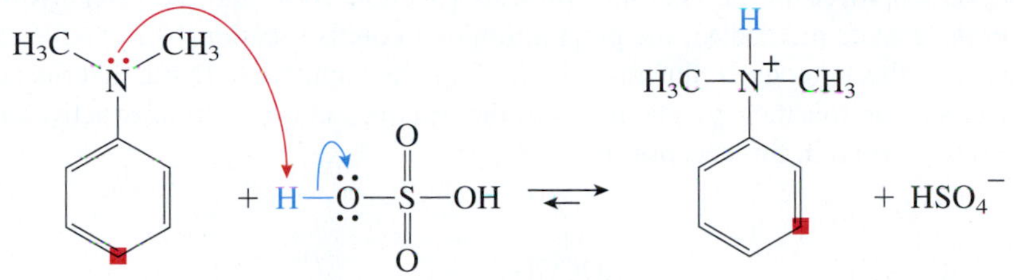

The $NH(CH_3)_2{}^+$ group deactivates the ring and directs to the meta position. The major product, the *meta*-isomer, results from the reaction of the protonated amine. The minor product, the *para*-isomer, results from the reaction of a very small amount of unprotonated amine. Although its concentration in the strongly acidic solution is extremely small, the unprotonated amine is many orders of magnitude more reactive than its conjugate acid, so some of the para product is formed.

It is fairly common for the electron pair on an amino group, which is a good Lewis base, to react with a Lewis acid under the strongly electrophilic conditions of these substitution reactions. This changes the substituent from a strong activating group to a strong deactivating group. As a result, the reaction often has the undesired regiochemistry, and in some cases the desired reaction may not occur at all. Because the exact result is difficult to predict or control, the amino substituent is usually modified to decrease its reactivity. The strategy is similar to that employed in the Gabriel synthesis (see Section 10.6). A "protecting group" that makes the electrons on the nitrogen less basic is bonded to the amino group. After the desired substitution reaction has been accomplished, the protecting group is removed and the amino group is regenerated.

The most common method to decrease the reactivity of an unshared pair of electrons on an atom is to attach a carbonyl group to that atom. Therefore, the amine is first reacted with acetyl chloride to form an amide. (This reaction and its mechanism are described in detail in Section 19.6. To help you remember the reaction for now, note that the nitrogen nucleophile attacks the carbonyl carbon electrophile, displacing the chloride leaving group.)

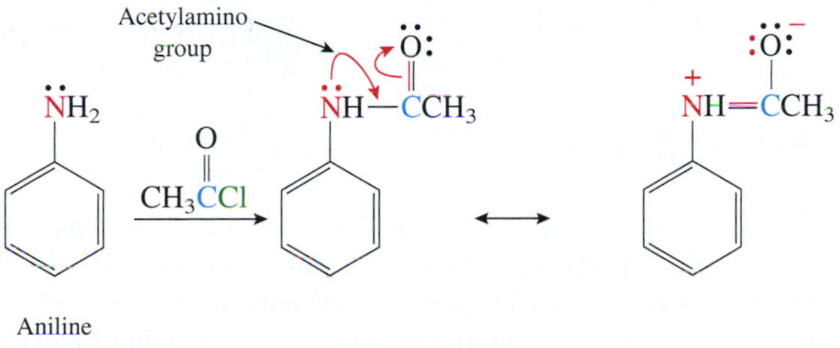

Because of delocalization of the nitrogen's electron pair onto the carbonyl oxygen, the electrons of the acetylamino group are less available for delocalization into the ring by resonance. (This is why the acetylamino group is a weaker activator than the amino group.) In addition, the electron pair on the nitrogen of an amide is much less basic, and reactions with Lewis acids in the substitution reactions are not usually a problem. However, the acetylamino group is still an activator and an ortho/para director, so substitution reactions work well. After the substitution has been completed, the acetyl group can be removed by hydrolysis of the amide bond (This reaction is very similar to the imide hydrolysis employed in the Gabriel synthesis [Section 10.6] and the ester hydrolysis used in the acetate method for the preparation of alcohols [Section 10.2].) An example of the use of this strategy is illustrated in the following synthesis. (Note that the acetyl-amino group controls the regiochemistry of the reaction, so it is a stronger activator than the methoxy group in this reaction.)

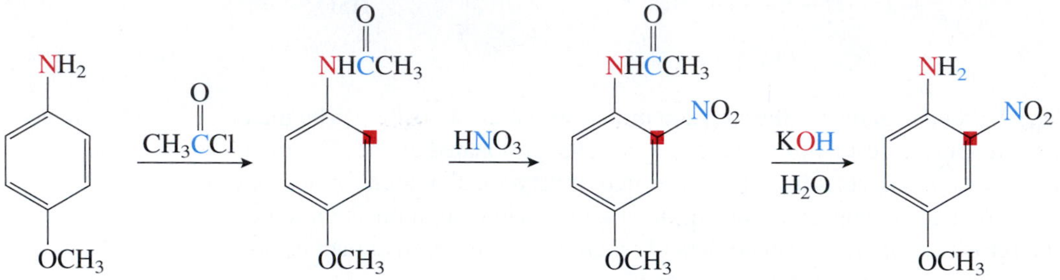

p-Methoxyaniline

PROBLEM 17.9
Show the products of these reactions:

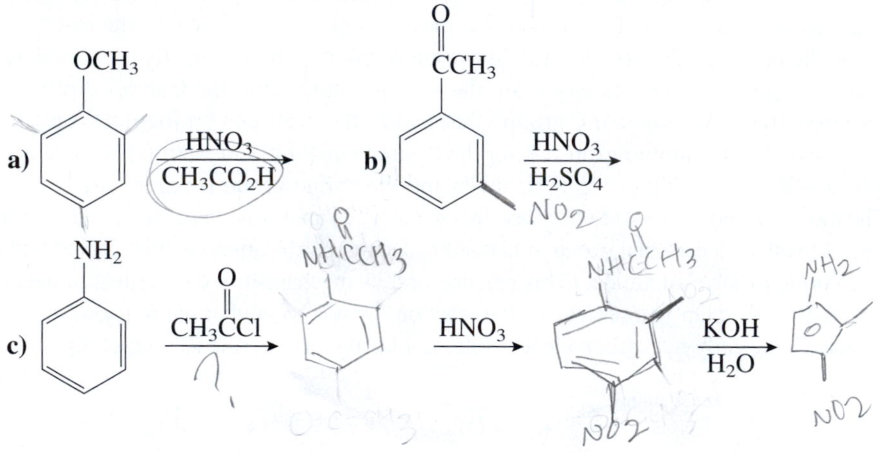

17.5 HALOGENATION

Chlorine and bromine can be substituted onto an aromatic ring by treatment with Cl_2 or Br_2. With all but highly activated aromatic rings (amines, phenols, polyalkylated rings), a Lewis acid catalyst is also required to make the halogen electrophile strong enough to accomplish the reaction. The most common catalysts are the aluminum and iron halides, $AlCl_3$, $AlBr_3$, $FeCl_3$, and $FeBr_3$. An example is provided by the following equation:

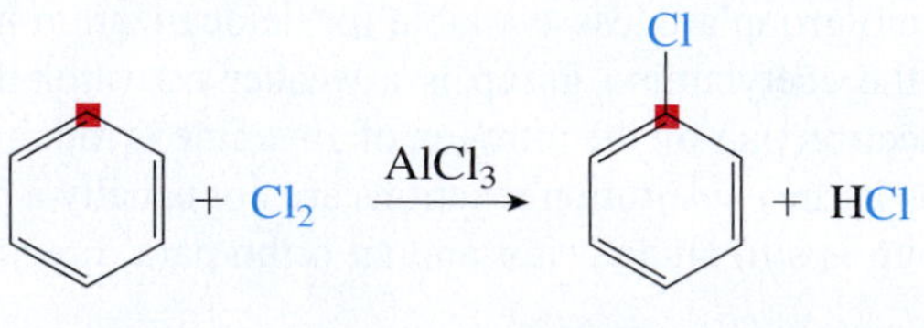

Chlorobenzene

As shown in the following equation, the Lewis acid, $AlCl_3$ in this case, bonds to one of the atoms of Cl_2 to produce the electrophilic species:

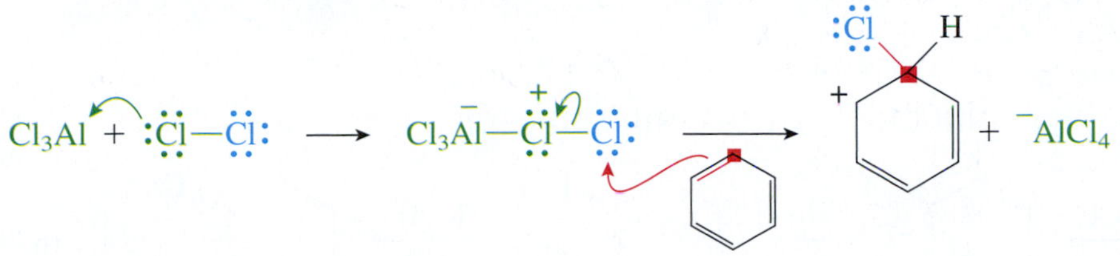

A pair of pi electrons of the aromatic ring then bonds to the electrophilic chlorine as $AlCl_4^-$ leaves. ($AlCl_4^-$ is a weaker base and a better leaving group than Cl^-.) The remainder of the mechanism is the same as that illustrated in Figure 17.1.

This substitution reaction provides a general method for adding chlorine or bromine to an aromatic ring. Because both are deactivating substituents, the product is less reactive than the starting aromatic compound, so it is possible to add a single halogen. The reaction works with deactivated substrates, as illustrated in the following example:

ORGANIC **Chemistry Now™** Click *Mechanisms in Motion* to view **Electrophilic Aromatic Bromination.**

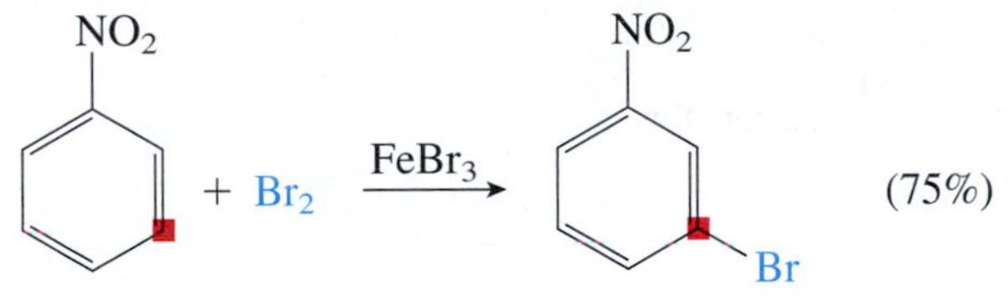

Nitrobenzene

As mentioned previously, halogenation of highly reactive substrates can be accomplished without the use of a Lewis acid catalyst. Thus, the bromination of mesitylene (1,3,5-trimethylbenzene) is readily accomplished by reaction with bromine in carbon tetrachloride:

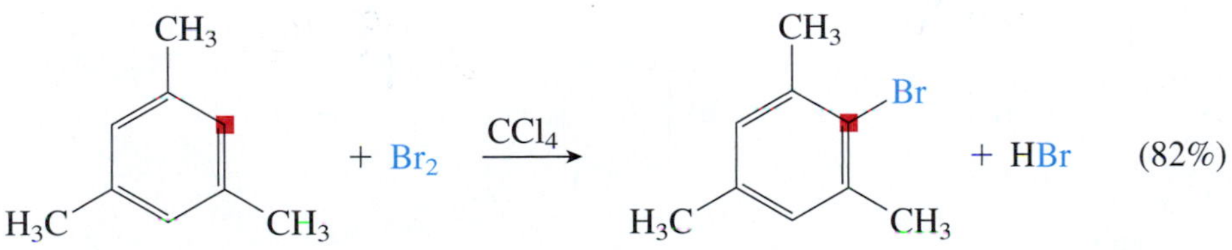

Mesitylene
(1,3,5-trimethylbenzene)

With very reactive compounds, such as anilines and phenols, it is often difficult to stop the reaction after only one halogen has added to the ring. In such cases the product that is isolated usually has reacted at all of the activated positions:

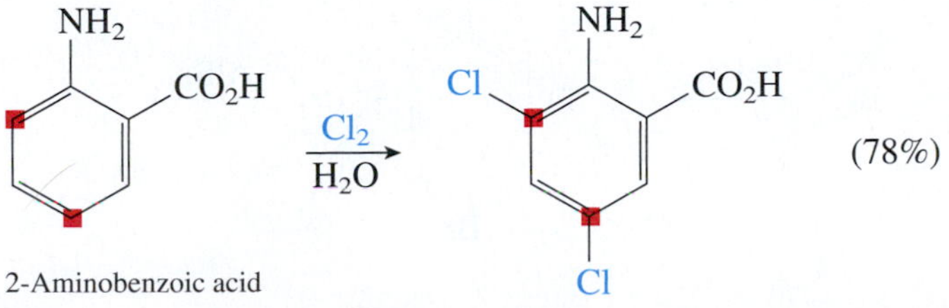

2-Aminobenzoic acid

If this is a problem, the solution again is to decrease the reactivity of the ring by modification of the activating group. The carbonyl protecting group is removed after the halogenation is accomplished.

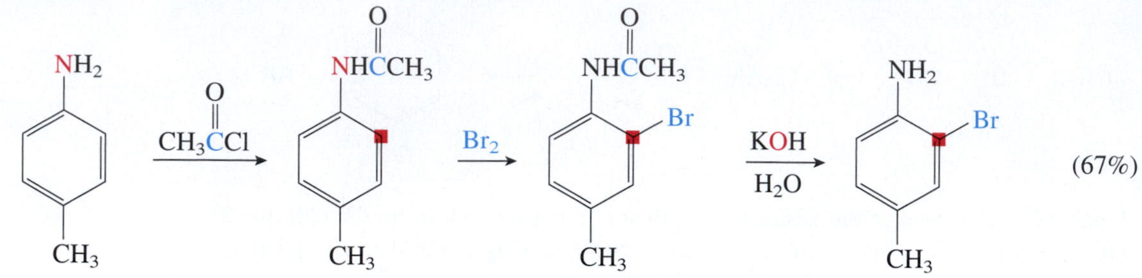

p-Methylaniline

PROBLEM 17.10

Show all of the steps in the mechanism for this reaction:

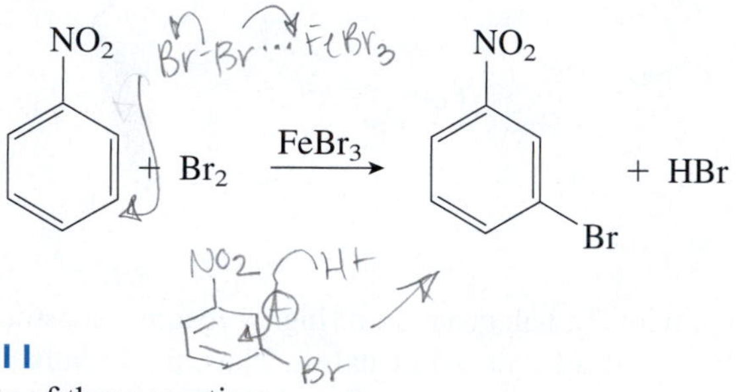

PROBLEM 17.11

Show the products of these reactions:

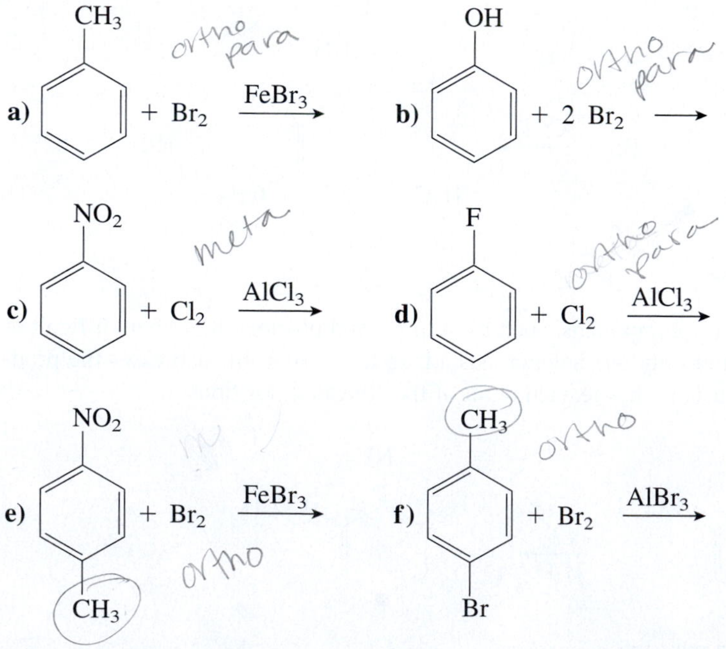

17.6 SULFONATION

A sulfonic acid group can be substituted onto an aromatic ring by reaction with concentrated sulfuric acid as shown in the following example:

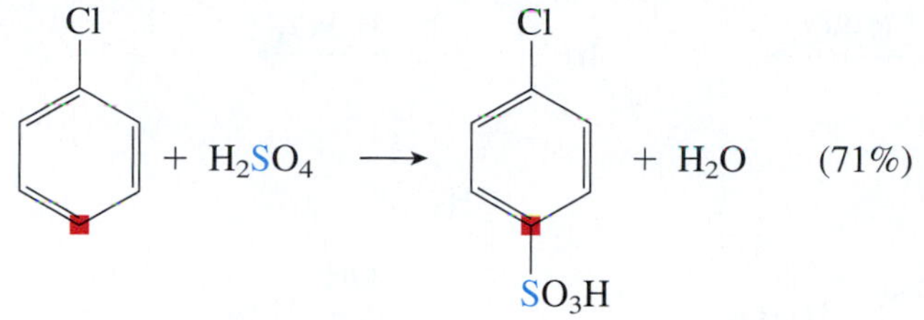

Chlorobenzene   p-Chlorobenzenesulfonic acid

Although the electrophile varies depending on the exact reaction conditions, it is often sulfur trioxide, SO_3, that is formed from sulfuric acid by the loss of water. The mechanism for the addition of this electrophile proceeds according to the following equation:

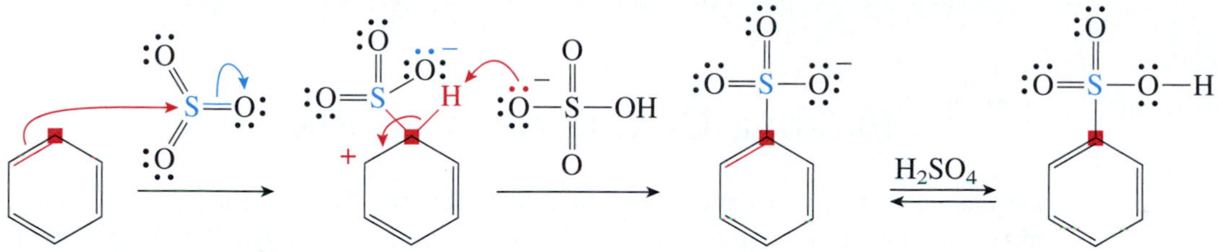

Benzenesulfonic acid

Again, this is a general reaction that works for deactivated as well as activated rings. The SO_3H group that is added is a deactivator, so the reaction can be halted after the substitution of a single group.

In contrast to the other reactions that have been presented so far, this substitution is readily reversible. Reaction of a sulfonic acid in a mixture of water and sulfuric acid results in removal of the sulfonic acid group. In this case a proton is the electrophile. An example is provided by the following equation:

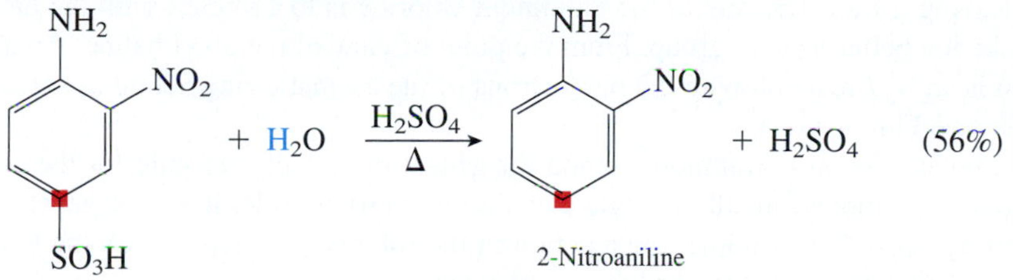

2-Nitroaniline

PROBLEM 17.12
Show all of the steps in the mechanism for this reaction:

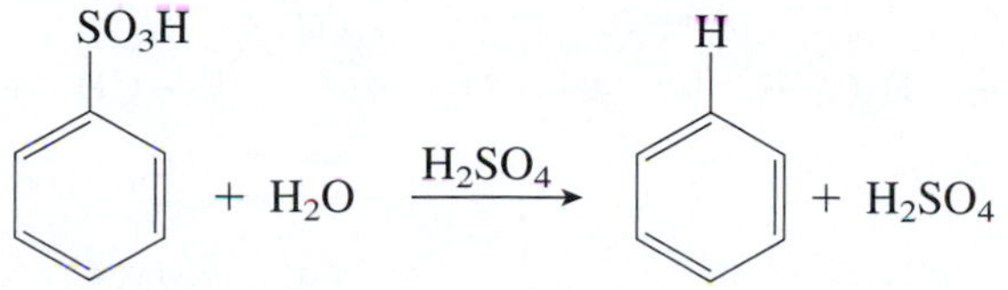

PROBLEM 17.13

Show the products of these reactions:

a)

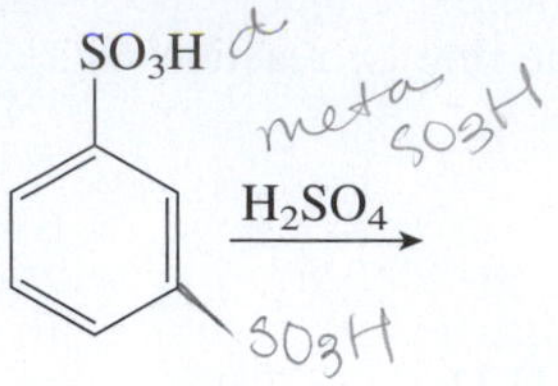

b)

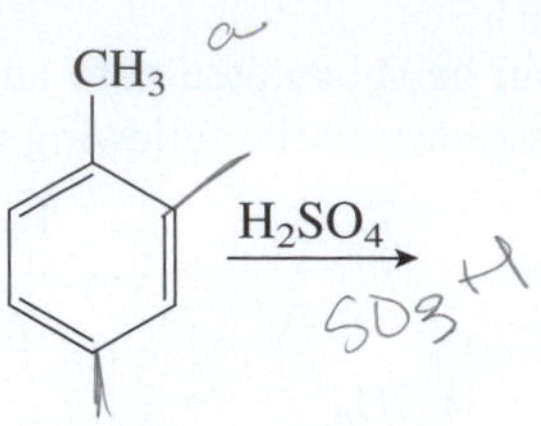

c)

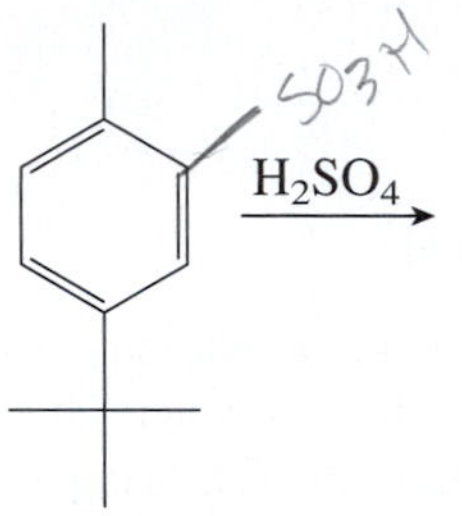

d)

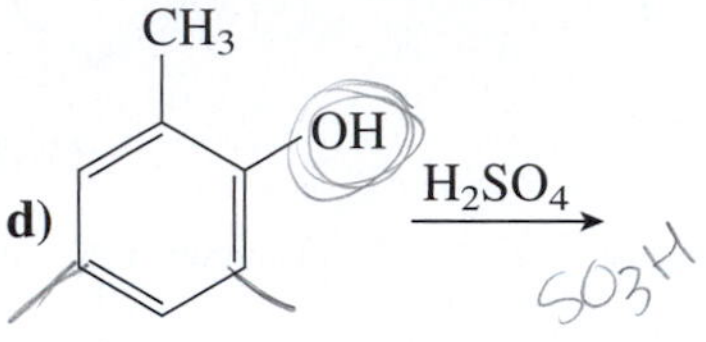

17.7 FRIEDEL-CRAFTS ALKYLATION

Developed by C. Friedel and J. M. Crafts, the reaction of an alkyl halide with an aromatic compound in the presence of a Lewis acid catalyst, usually $AlCl_3$, results in the substitution of the alkyl group onto the aromatic ring:

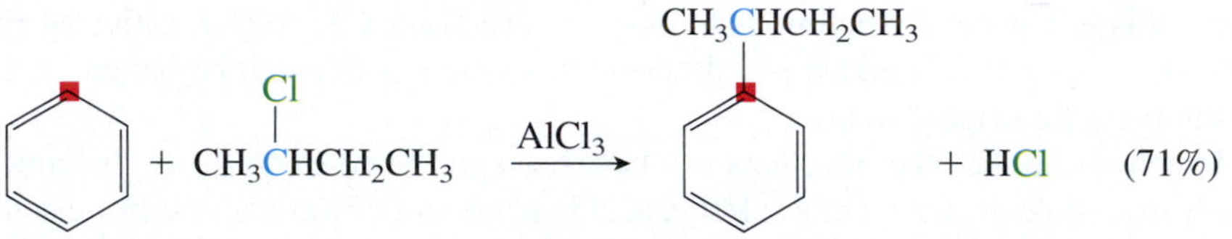

In most cases the electrophile is the carbocation that is generated when the halide acts as a leaving group. The role of the aluminum chloride is to complex with the halogen to make it a better leaving group. From the point of view of the alkyl halide, the mechanism is an S_N1 reaction with the pi electrons of the aromatic ring acting as the nucleophile (see Figure 17.4).

Although the most common method for generating the electrophile for the alkylation reaction employs an alkyl halide and aluminum trichloride, it can be generated in other ways also. For example, the reaction in the following equation uses the reaction of an alcohol and an acid to produce the carbocation:

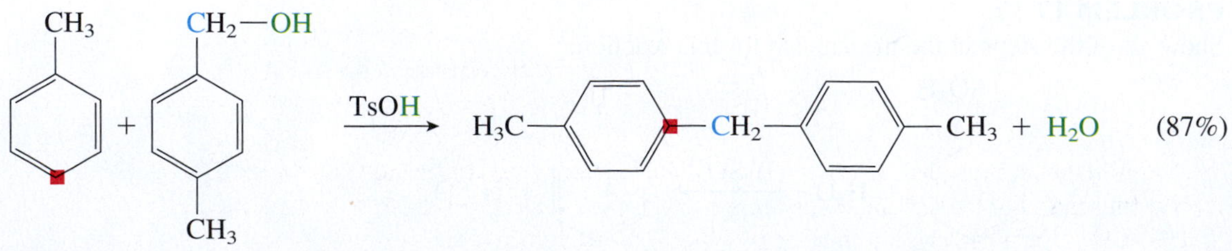

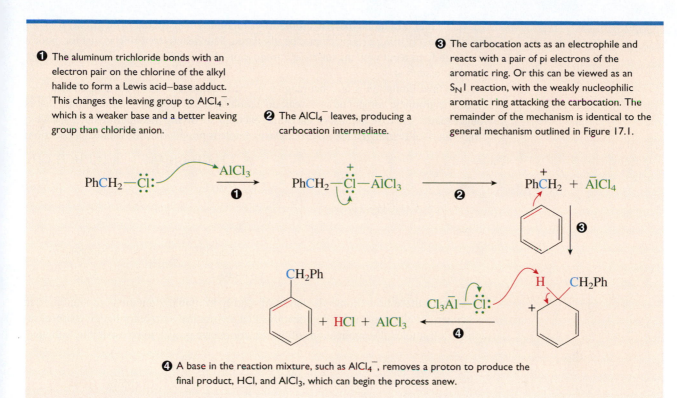

❶ The aluminum trichloride bonds with an electron pair on the chlorine of the alkyl halide to form a Lewis acid–base adduct. This changes the leaving group to $AlCl_4^-$, which is a weaker base and a better leaving group than chloride anion.

❷ The $AlCl_4^-$ leaves, producing a carbocation intermediate.

❸ The carbocation acts as an electrophile and reacts with a pair of pi electrons of the aromatic ring. Or this can be viewed as an S_N1 reaction, with the weakly nucleophilic aromatic ring attacking the carbocation. The remainder of the mechanism is identical to the general mechanism outlined in Figure 17.1.

❹ A base in the reaction mixture, such as $AlCl_4^-$, removes a proton to produce the final product, HCl, and $AlCl_3$, which can begin the process anew.

Active Figure 17.4

ORGANIC
Chemistry ⚛ Now ™

MECHANISM OF THE FRIEDEL-CRAFTS ALKYLATION REACTION. Test yourself on the concepts in this figure at **OrganicChemistryNow.**

Alternatively, the carbocation can be generated by protonation of an alkene. This reaction resembles the additions to alkenes discussed in Chapter 11. An example is provided by the following equation:

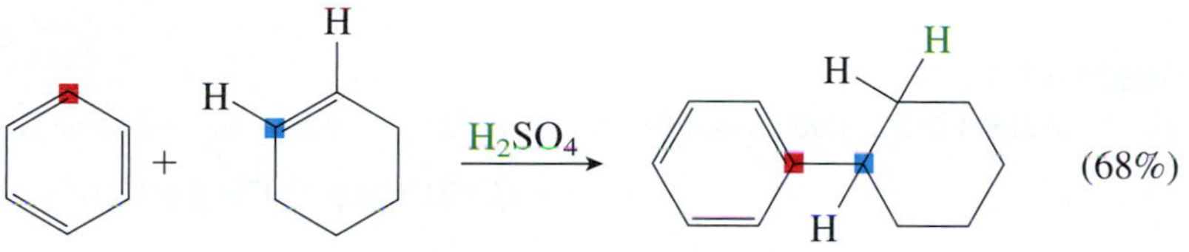

(68%)

Several limitations occur with the Friedel-Crafts alkylation reaction. First, the alkyl group that is added to the ring is an activating group. This causes the alkylated product to be more reactive (by a factor of about 2) than the starting aromatic compound. Therefore, a significant amount of product where two or more alkyl groups have been added is commonly formed. The best solution to this problem is to use a large excess of the aromatic compound that is to be alkylated. This can easily be accomplished for compounds that are readily available, such as benzene or toluene, by using them as the solvent for the reaction. Note that the Friedel-Crafts alkylation is the only one of these electrophilic aromatic substitution reactions in which the product is more reactive than the starting material. All of the other reactions put deactivating groups on the ring, so they do not suffer from the problem of multiple substitution.

A second limitation is that aromatic compounds substituted with moderately or strongly deactivating groups cannot be alkylated. The deactivated ring is just too poor a nucleophile to react with the unstable carbocation electrophile before other reactions occur that destroy it.

The final limitation is one that plagues all carbocation reactions: rearrangements. Because the aromatic compound is a weak nucleophile, the carbocation has a lifetime that is longer than is the case in most of the other reactions involving this intermediate, allowing ample time for rearrangements to occur. An example is provided by the following equation:

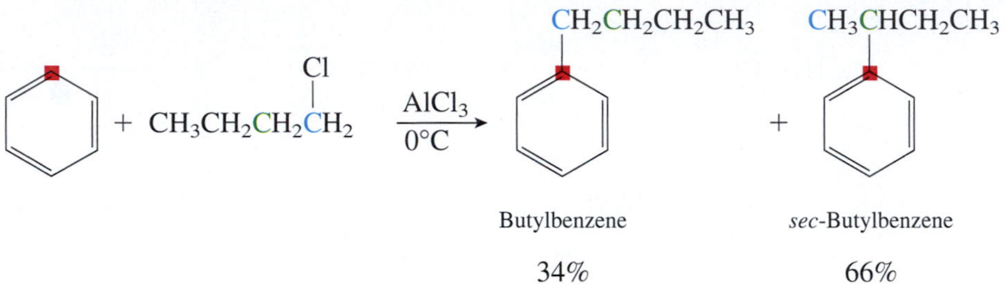

Butylbenzene

34%

sec-Butylbenzene

66%

Despite these limitations, alkylation of readily available aromatic compounds, such as benzene and toluene, using carbocations that are not prone to rearrange, is a useful reaction. Intramolecular applications of this reaction have proven to be especially valuable.

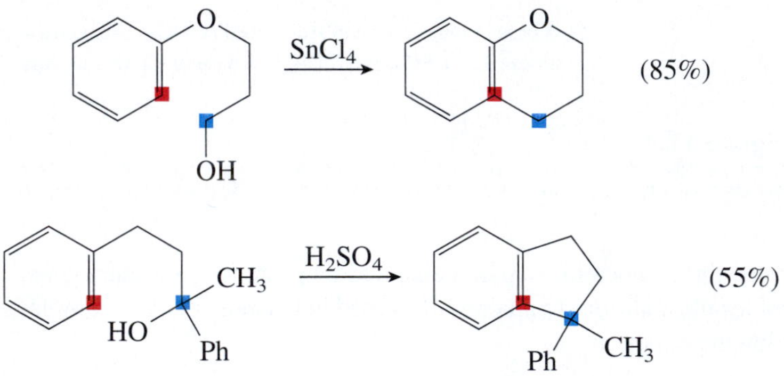

(85%)

(55%)

PROBLEM 17.14

Show all of the steps in the mechanism for the formation of both products in this reaction:

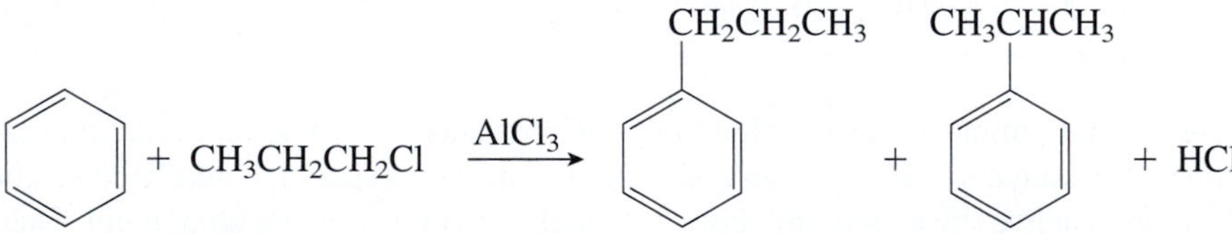

PROBLEM 17.15

Show the products of these reactions:

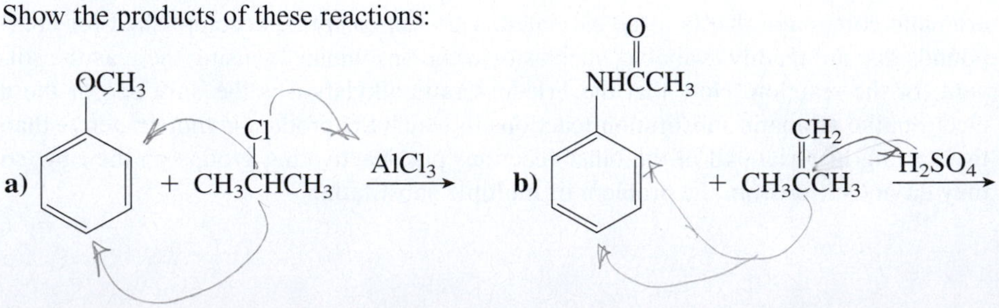

c) 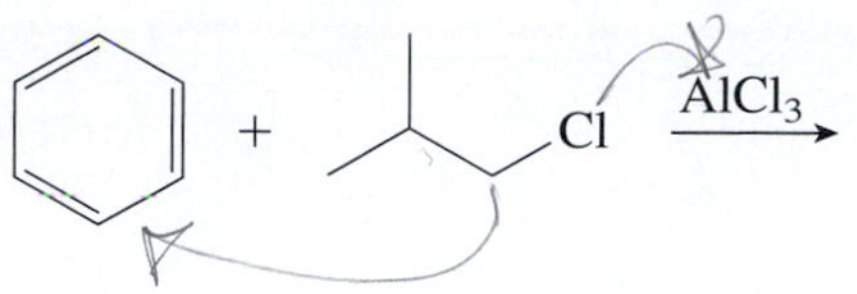   + $\xrightarrow{\text{AlCl}_3}$

d) 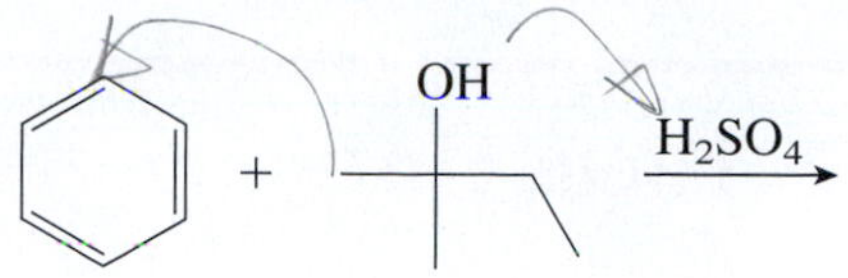   + $\xrightarrow{\text{H}_2\text{SO}_4}$

e) 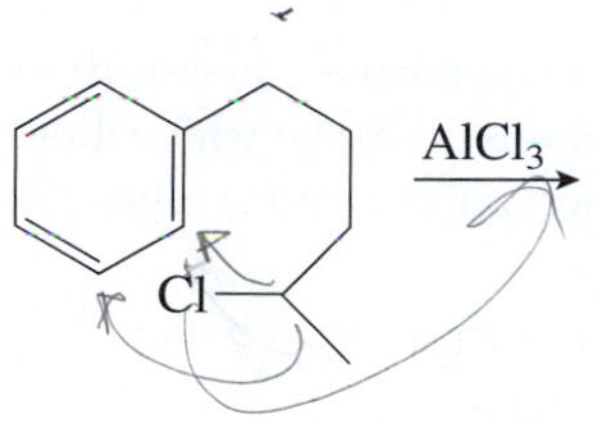 $\xrightarrow{\text{AlCl}_3}$

f) 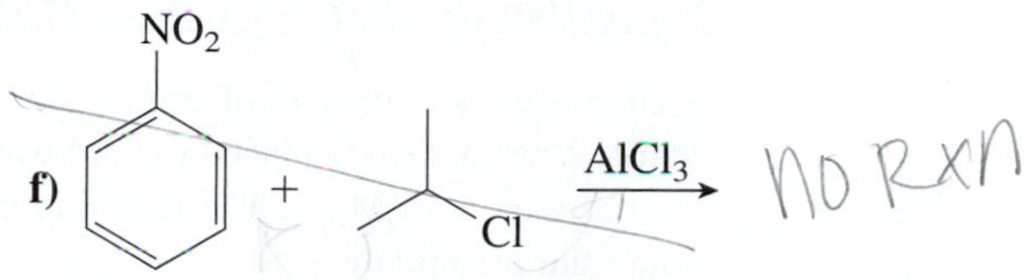   + $\xrightarrow{\text{AlCl}_3}$ no Rxn

PROBLEM 17.16

Show syntheses of these compounds from benzene:

a) 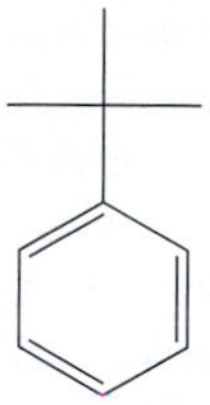   b)

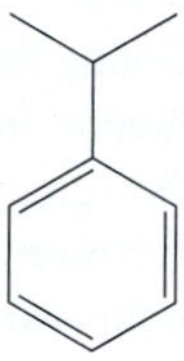

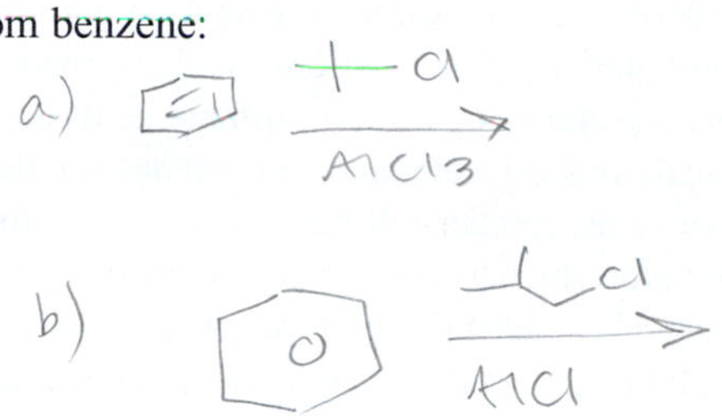

a) [benzene] $\xrightarrow[\text{AlCl}_3]{\text{+—Cl}}$

b) [benzene] $\xrightarrow[\text{AlCl}]{\text{—Cl}}$

PRACTICE PROBLEM 17.2

Explain which of these routes would provide a better method for the preparation of *p*-nitrotoluene:

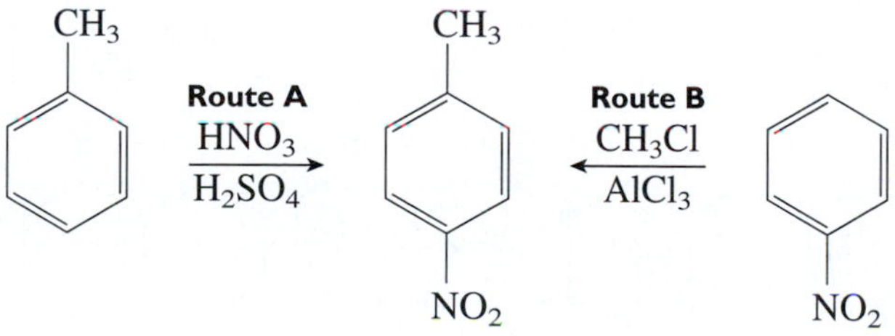

p-Nitrotoluene

Solution

Route A works fine. Toluene is readily nitrated, and the methyl group is an ortho/para director. The only problem is that both the desired compound and its *ortho*-isomer are produced and must be separated. (This is a common problem, and we usually assume that the separation can be accomplished, although it is not always easy in the laboratory.) Route B is unsatisfactory because the Friedel-Crafts alkylation reaction does not work with deactivated compounds such as nitrobenzene. Furthermore, even if the alkylation could be made to go, the nitro group is a meta director, so the desired product would not be formed.

Focus On

Synthetic Detergents, BHT, and BHA

A soap is the sodium salt of carboxylic acid attached to a long, nonpolar hydrocarbon chain. When a soap is placed in hard water, the sodium cations exchange with cations such as Ca^{2+} and Mg^{2+}. The resulting calcium and magnesium salts are insoluble in water and precipitate to form "soap scum."

$$2\ CH_3(CH_2)_{16}CO_2^-\ Na^+ + Ca^{2+} \longrightarrow [CH_3(CH_2)_{16}CO_2^-]_2\ Ca^{2+} + 2\ Na^+$$
$$\text{Precipitates}$$

Synthetic detergents were invented to alleviate this problem. Rather than use the anion derived from a carboxylic acid with a large nonpolar group, detergents employ the anion derived from a sulfonic acid attached to a large nonpolar group. The calcium and magnesium salts of these sulfonic acids are soluble in water, so detergents do not precipitate in hard water and can still accomplish their cleaning function.

Two of the reactions that are used in the industrial preparation of detergents are electrophilic aromatic substitution reactions. First, a large hydrocarbon group is attached to a benzene ring by a Friedel-Crafts alkylation reaction employing tetrapropene as the source of the carbocation electrophile. The resulting alkylbenzene is then sulfonated by reaction with sulfuric acid. Deprotonation of the sulfonic acid with sodium hydroxide produces the detergent.

Tetrapropene

A detergent

The exact structure of the alkyl group on the benzene ring is not important as long as it is large enough to confer the necessary hydrophobic character. Tetrapropene was used in the early versions of detergents because it was readily and cheaply available from the treatment of propene with acid. In this reaction, four propenes combine to form tetrapropene through carbocation intermediates. (In addition to the compound shown in the equation, an isomer with the double bond between carbon 2 and carbon 3 is also formed. If you are interested in the mechanism for this reaction, it is a variation of the cationic polymerization mechanism described later in Section 24.3.)

$$4 \quad \text{(alkene)} \xrightarrow[\substack{205°C \\ 1000 \text{ psi}}]{H_3PO_4} \text{(branched alkene product)}$$

However, the detergent prepared from tetrapropene caused a problem in sewage treatment plants. The microorganisms that degrade such compounds start from the end of the hydrocarbon chain and seem to have trouble proceeding through tertiary carbons. The presence of several tertiary carbons in the tetrapropene chain slows its biodegradation to the point at which a significant amount passes through a treatment plant unchanged. This causes the resulting effluent and the waterways into which it is discharged to become foamy, an environmentally unacceptable result.

To solve this problem, most modern detergents are prepared from straight-chain alkenes. The resulting linear alkylbenzenesulfonate detergents are more easily degraded, and our rivers are no longer foamy. An example of a typical alkylation is shown in the following equation:

$$CH_3(CH_2)_6CH{=}CH(CH_2)_5CH_3 \; + \; \text{(benzene)} \xrightarrow[AlCl_3]{HCl} CH_3(CH_2)_6CH(CH_2)_6CH_3$$

PROBLEM 17.17
What isomeric alkyl benzene should also be formed in this reaction?

Butylated hydroxytoluene (BHT) and butylated hydroxyanisole (BHA) are antioxidants that are added to foods and many other organic materials to inhibit decomposition caused by reactions with oxygen. Perhaps you have seen these compounds listed among the ingredients on your cereal box at breakfast. (The mechanism of operation for these antioxidants is described in Section 21.8.) Both of these compounds are prepared by Friedel-Crafts alkylation reactions. BHT is synthesized by the reaction of p-methylphenol with 2-methylpropene in the presence of an acid catalyst.

$$\text{p-Methylphenol} + 2\; CH_3\underset{CH_3}{\overset{CH_3}{C}}{=}CH_2 \xrightarrow[HCl]{AlCl_3} \text{Butylated hydroxytoluene (BHT)}$$

p-Methylphenol
(p-hydroxytoluene)

2-Methylpropene
(isobutylene)

Butylated hydroxytoluene
(BHT)

Continued

Addition of a proton to 2-methylpropene produces the *t*-butyl carbocation, which then alkylates the ring. Conditions are adjusted so that two *t*-butyl groups are added. BHA is prepared in a similar manner by the reaction of *p*-methoxyphenol with 2-methylpropene and an acid catalyst. In this case conditions are adjusted so that only one *t*-butyl group is added. Because the hydroxy group and the methoxy group are both activating groups, a mixture of products is formed in this case.

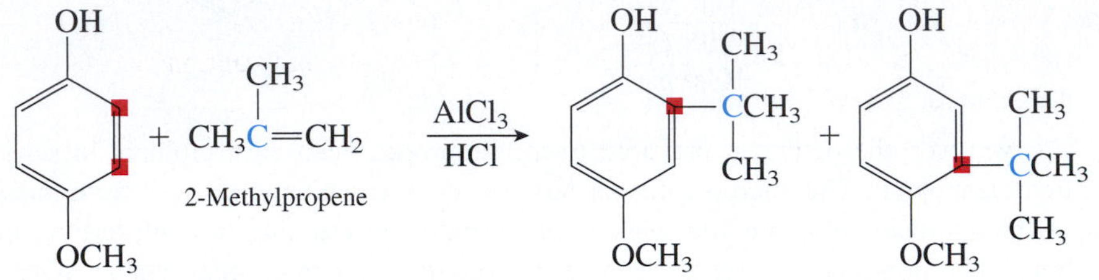

p-Methoxyphenol (*p*-hydroxyanisole) Butylated hydroxyanisole (BHA)

17.8 Friedel-Crafts Acylation

The reaction of an aromatic compound with an acyl chloride in the presence of a Lewis acid (usually $AlCl_3$) results in the substitution of an acyl group onto the aromatic ring. An example of this reaction, known as the Friedel-Crafts acylation, is provided by the following equation:

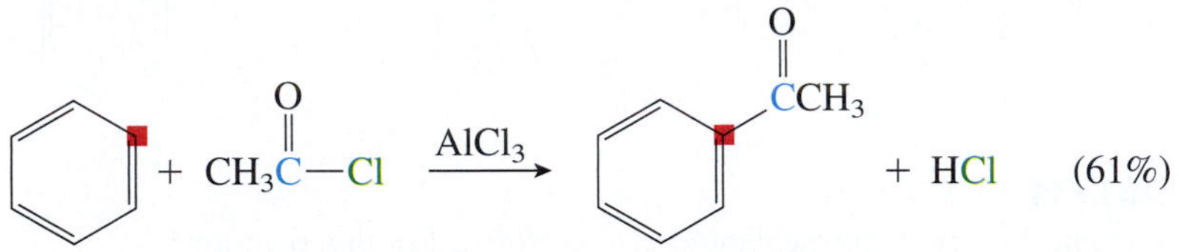

Benzene Acetyl chloride Acetophenone

ORGANIC **Chemistry•Now™** Click *Coached Tutorial Problems* to quiz yourself on **Mechanisms of Electrophilic Aromatic Substitution.**

The electrophile, an acyl cation, is generated in a manner similar to that outlined in Figure 17.4 for the generation of the carbocation electrophile from an alkyl halide. First the Lewis acid, aluminum trichloride, complexes with the chlorine of the acyl chloride. Then $AlCl_4^-$ leaves, generating an acyl cation. The acyl cation is actually more stable than most other carbocations that we have encountered because it has a resonance structure that has the octet rule satisfied for all of the atoms:

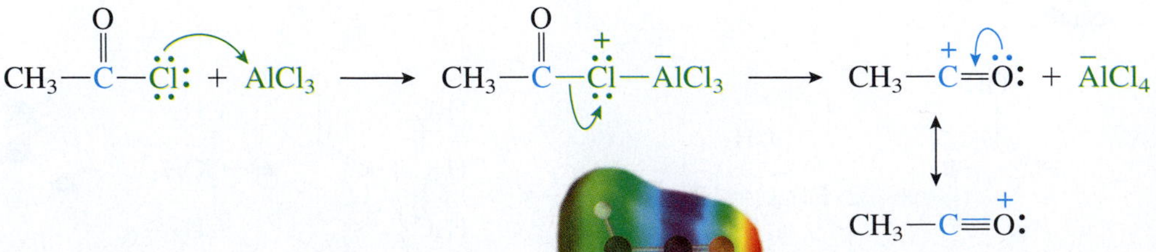

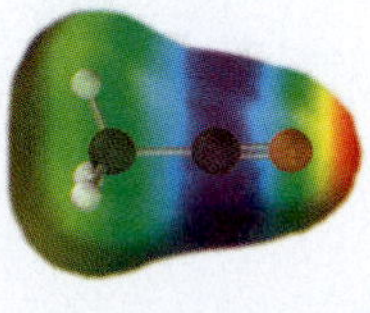

Acyl cation

The Friedel-Crafts acylation reaction does not have most of the limitations of the alkylation reaction. Because of the stability of the acyl cation, rearrangements do not occur in this reaction. In addition, the acyl group that is added to the ring is a deactivator, so the product is less reactive than the starting aromatic compound. Therefore, there is no problem with multiple acyl groups being added to the ring. However, like alkylations, the acylation reaction does not work with moderately or strongly deactivated substrates—that is, with rings that are substituted only with meta directing groups. As a consequence of fewer limitations, the Friedel-Crafts acylation reaction is more useful than the alkylation reaction.

Anhydrides can be used in place of acyl chlorides as the source of the electrophilic acyl cation:

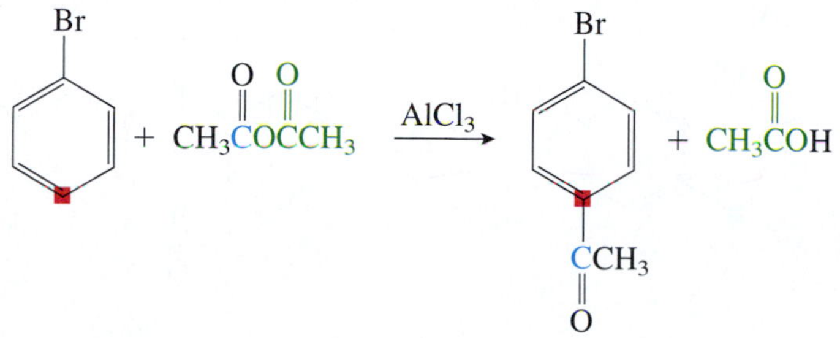

As seen in this example, the acylation reaction is more sensitive to steric effects than the other reactions that have been discussed so far and tends to give predominantly the para product. Some additional examples are provided by the following equations:

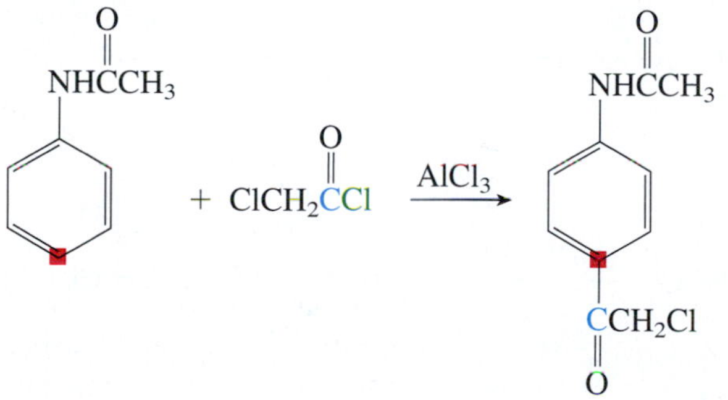

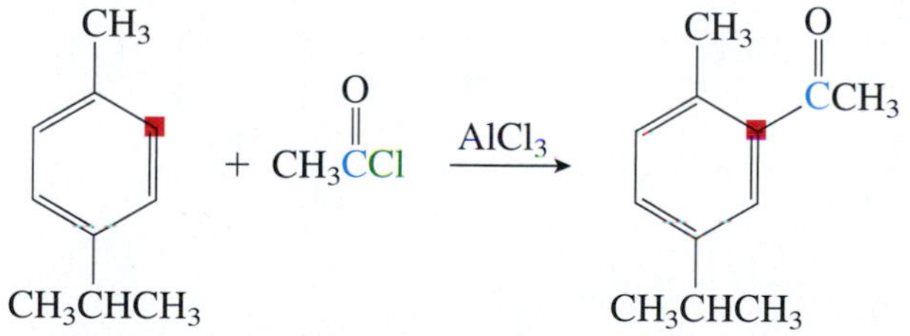

p-Isopropyltoluene

Finally, the intramolecular version of the Friedel-Crafts acylation reaction has proved to be very valuable in the construction of polycyclic compounds, as illustrated in the following equation:

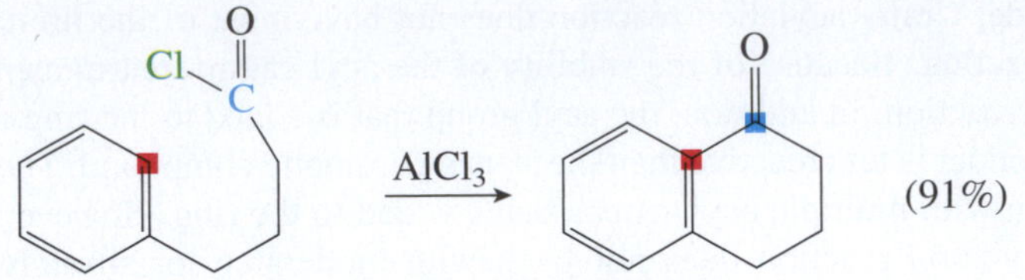

(91%)

For intramolecular reactions, treatment of a carboxylic acid with sulfuric acid or polyphosphoric acid is sometimes used to generate the acyl cation electrophile. This method is usually too mild for intermolecular acylations but works well for intramolecular examples, as shown in the following equation:

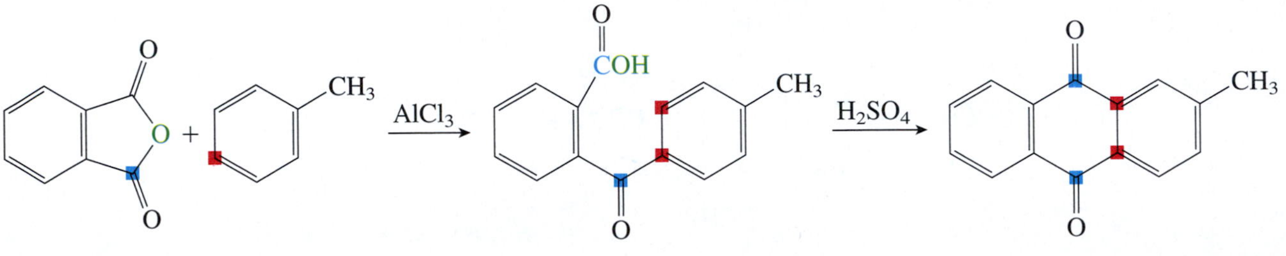

(90%)

PROBLEM 17.18

Explain why the Friedel-Crafts acylation of *p*-isopropyltoluene shown on the previous page results in the substitution of the acyl group at the position ortho to the methyl group.

PROBLEM 17.19

Show the products of these reactions:

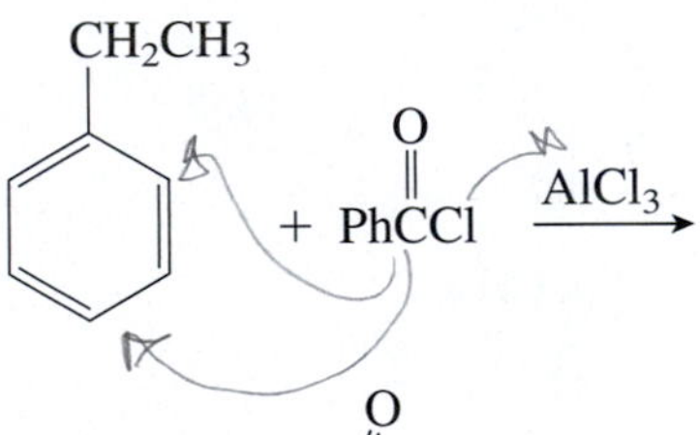

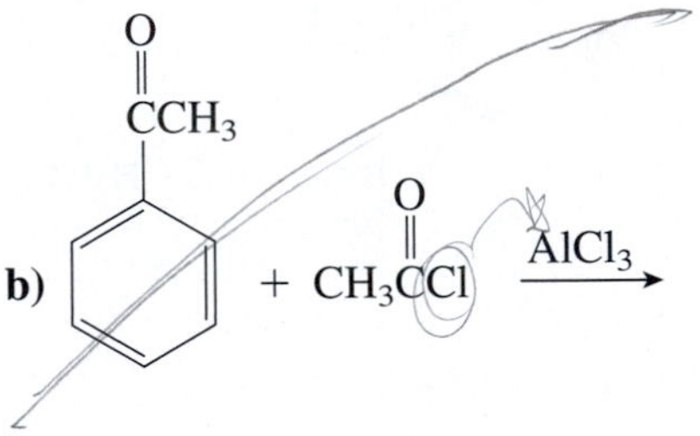

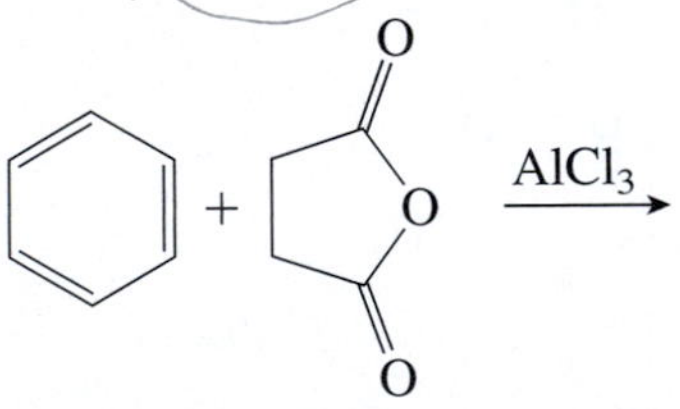

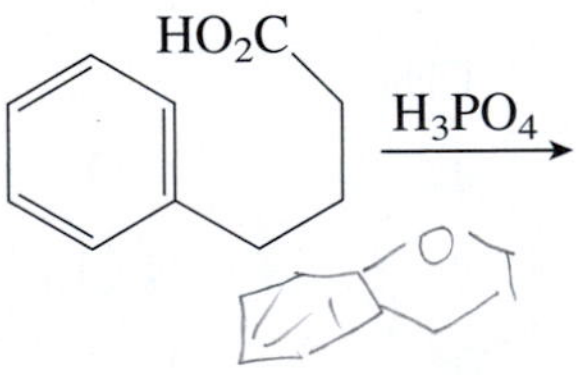

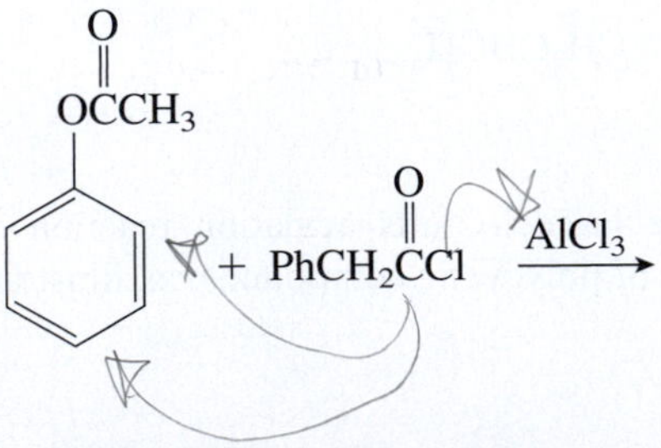

PROBLEM 17.20

Suggest syntheses of these compounds using Friedel-Crafts acylation reactions:

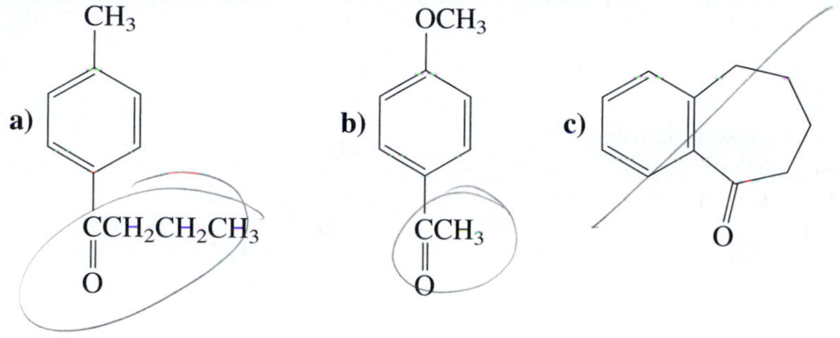

a) b) c)

ORGANIC
Chemistry⚛Now™
Click *Coached Tutorial Problems* for additional practice showing the products of **Electrophilic Aromatic Substitution Reactions.**

17.9 ELECTROPHILIC SUBSTITUTIONS OF POLYCYCLIC AROMATIC COMPOUNDS

Polycyclic aromatic compounds also undergo electrophilic aromatic substitution reactions. Because the aromatic resonance energy that is lost in forming the arenium ion is lower, these compounds tend to be more reactive than benzene. For example, the bromination of naphthalene, like that of other reactive aromatic compounds, does not require a Lewis acid catalyst:

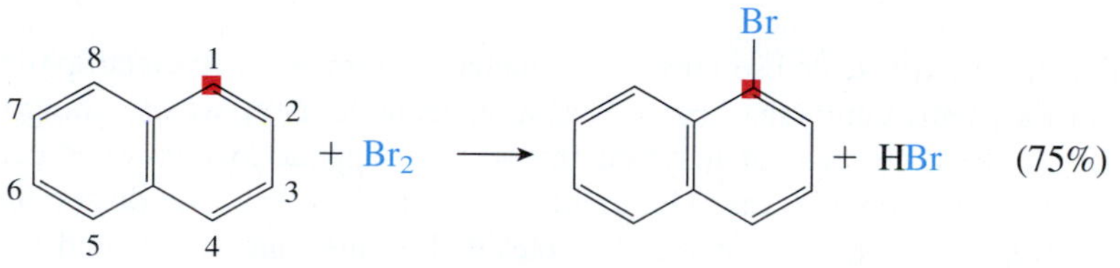

Naphthalene

Naphthalene also undergoes the other substitution reactions described for benzene. For example, it is acylated under standard Friedel-Crafts conditions:

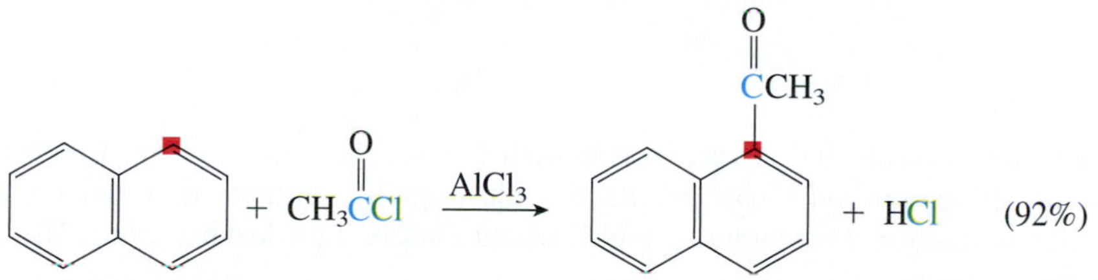

Note that both the bromination and the acylation of naphthalene result in the substitution of the electrophile at the 1 position. None of the isomeric product with the electrophile bonded to the 2 position is isolated in either case. The higher reactivity of the 1 position can be understood by examination of the resonance structures for the arenium ion. When the electrophile adds to the 1 position, the arenium ion has a total of seven resonance structures, whereas only six exist for the arenium ion resulting from addition of the electrophile to the 2 position.

PROBLEM 17.21

Draw the seven resonance structures for the arenium ion formed in the bromination of naphthalene at the 1 position and the six resonance structures formed in bromination at the 2 position.

PROBLEM 17.22

Show the products of these reactions:

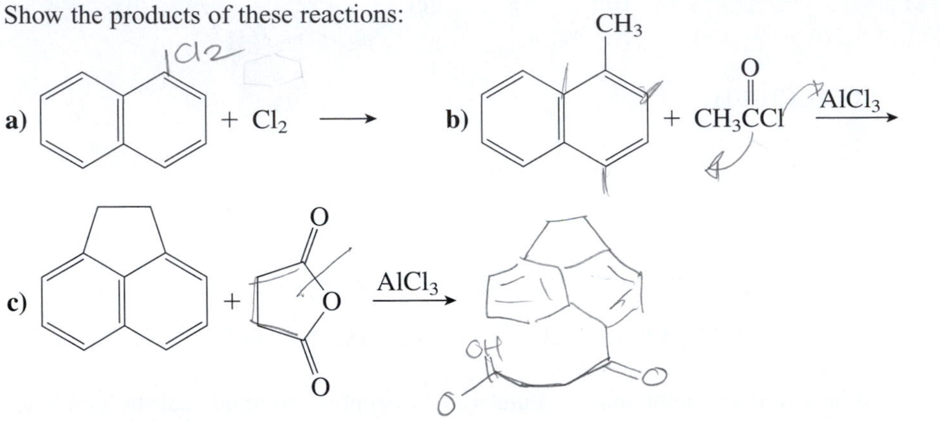

a) [naphthalene] + Cl_2 $\longrightarrow$

b) [methylnaphthalene] + CH_3CCl $\xrightarrow{AlCl_3}$

c) [acenaphthylene] + [maleic anhydride] $\xrightarrow{AlCl_3}$ [product]

17.10   NUCLEOPHILIC AROMATIC SUBSTITUTION: DIAZONIUM IONS

All of the reactions presented so far in this chapter have involved an electrophile reacting with an aromatic compound that acts as a nucleophile. Now we are going to consider several reactions that, at least on the surface, appear to involve attack by a nucleophile on the aromatic ring. The first of these involves aromatic diazonium ions, which are prepared from the reaction of amines with sodium nitrite and acid:

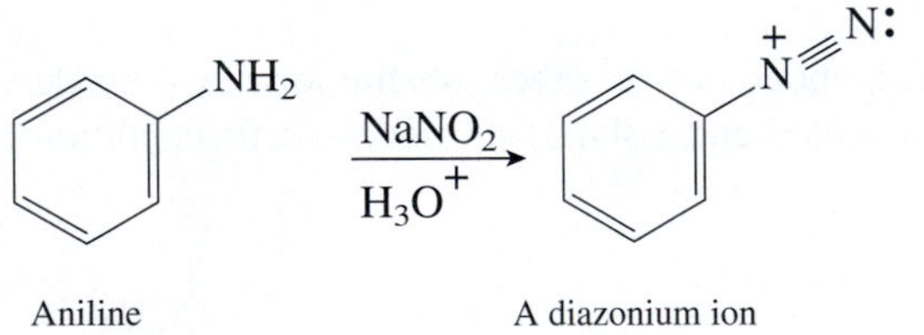

[structure: benzene ring with NH_2] $\xrightarrow[H_3O^+]{NaNO_2}$ [structure: benzene ring with $N \equiv N$ diazonium]

Aniline   A diazonium ion

Although aromatic halides are inert to both S_N1 and S_N2 reactions (see Chapter 8), aromatic diazonium ions can act as the electrophilic partner in a nucleophilic substitution reaction. These ions are highly reactive because the leaving group, N_2, is an extremely weak base:

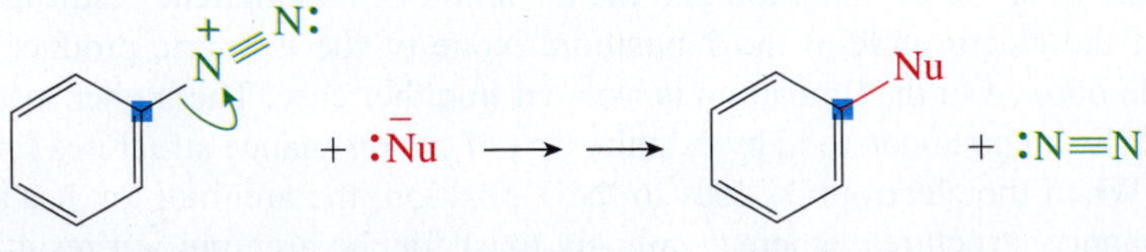

[structure: benzene ring with diazonium] + :Nu $\longrightarrow$ $\longrightarrow$ [structure: benzene ring with Nu] + :N≡N:

The mechanisms for these reactions are not well understood, but there is evidence that some actually follow an S_N1 pathway. Many others are known to involve radicals (odd electron species). We will not be concerned with the details of the various mechanisms here.

The diazo group can be replaced by a number of different nucleophiles. Although several different mechanisms may operate, it is easiest to remember the reactions if you consider them all to be simple nucleophilic substitutions, even though most are not. The following equations provide examples of the various substitutions that can be accomplished with diazonium ions.

The diazo group can be replaced with chlorine or bromine by reaction with cuprous chloride or cuprous bromide. [The Cu(I) aids in the formation of the radicals that are involved in these particular reactions.]

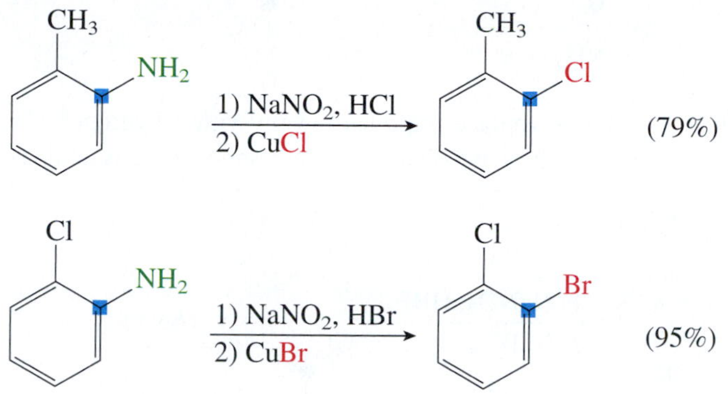

(79%)

(95%)

The diazo group can be replaced with iodine by reaction with potassium iodide:

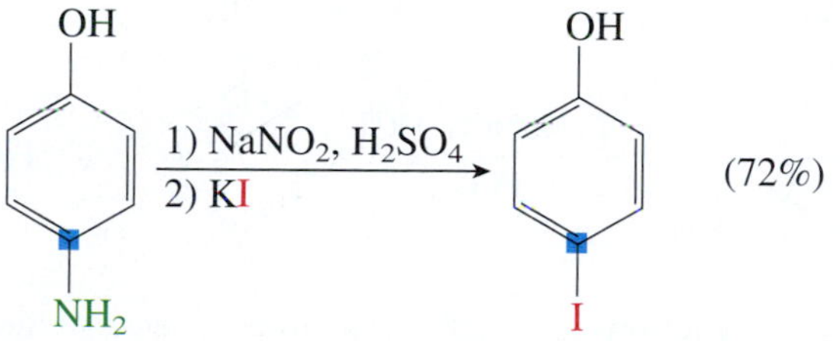

(72%)

If the diazonium ion is heated in the presence of tetrafluoroborate ion, the diazo group is replaced with a fluorine:

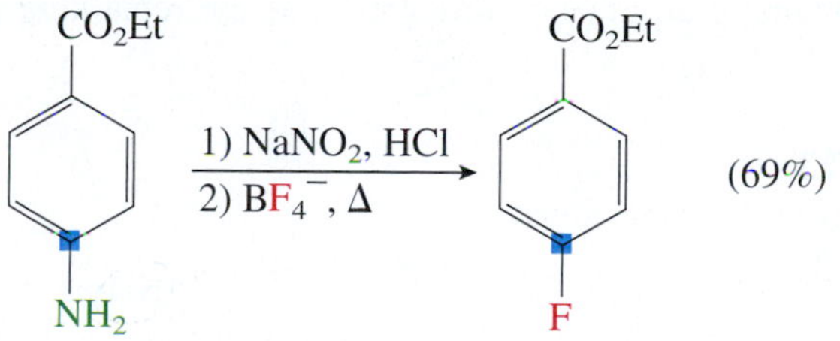

(69%)

If the diazonium ion is heated in aqueous acid, the diazo group is replaced with a hydroxy group:

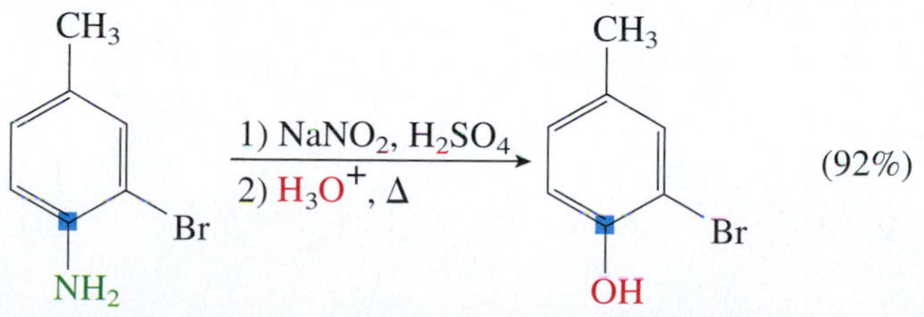

(92%)

The diazo group can be replaced with a cyano group by reaction with cuprous cyanide. This reaction is very similar to the reactions with cuprous chloride and cuprous bromide:

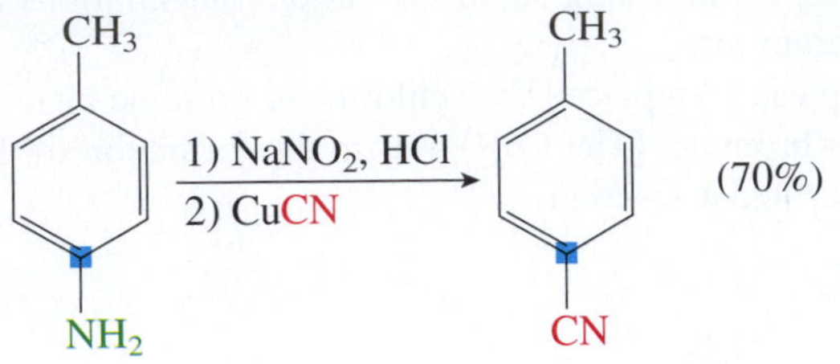

Finally, it is possible to replace the diazo group with a hydrogen. This is accomplished by reaction with sodium borohydride or hypophosphorous acid (H₃PO₂):

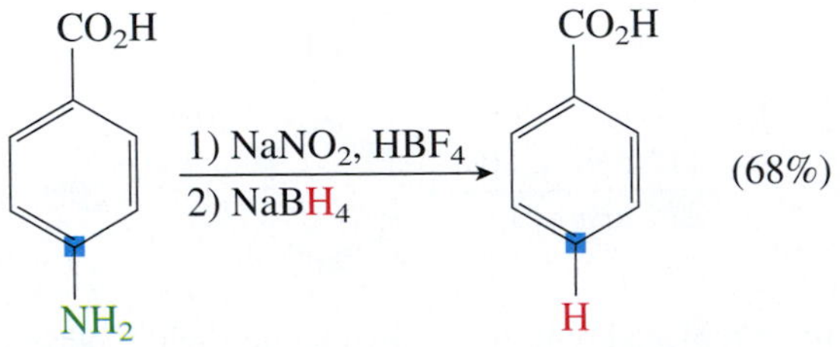

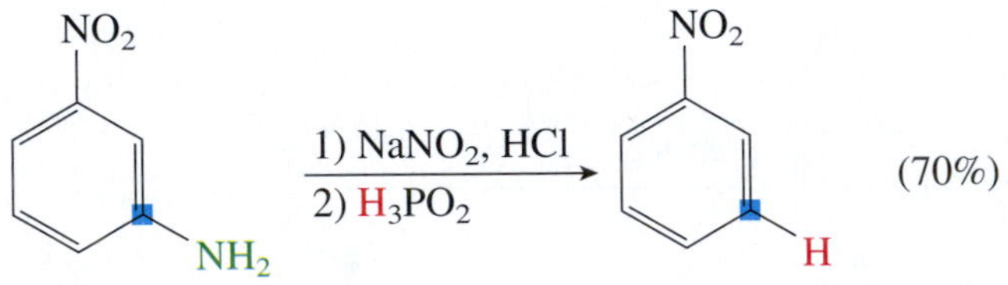

These substitution reactions are useful in synthesis because they are the only direct methods for adding several substituents (I, F, OH, CN) to an aromatic ring. (A method to introduce the precursor amino substituent onto the ring will be described shortly.) In addition, the ability to replace an amino group with a hydrogen can be very useful in obtaining an unusual orientation of the remaining substituents (see Section 17.14).

PROBLEM 17.23
Show the products of these reactions:

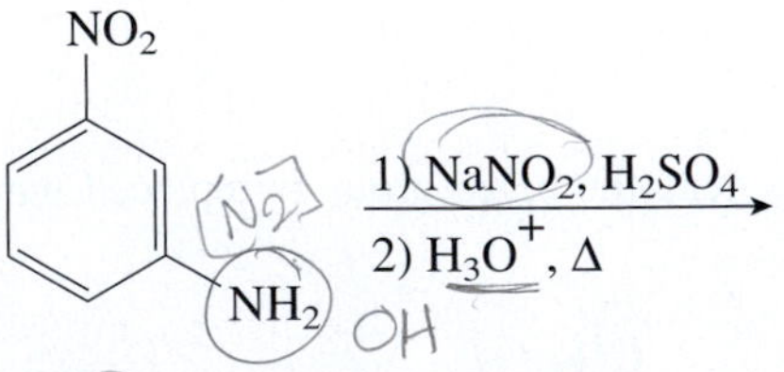

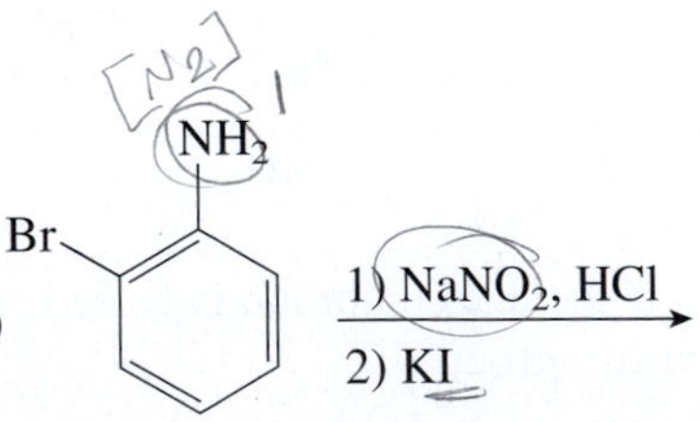

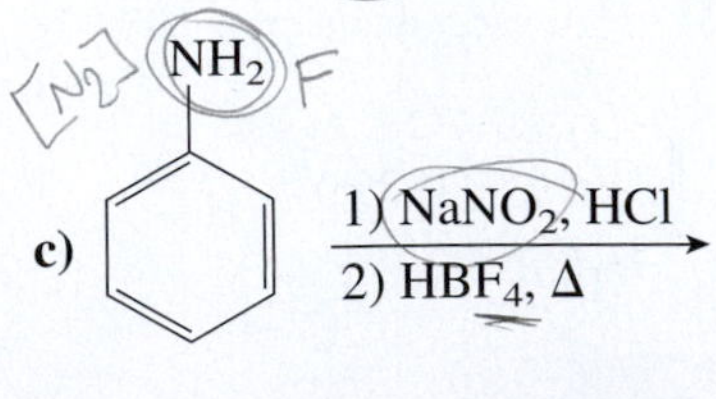

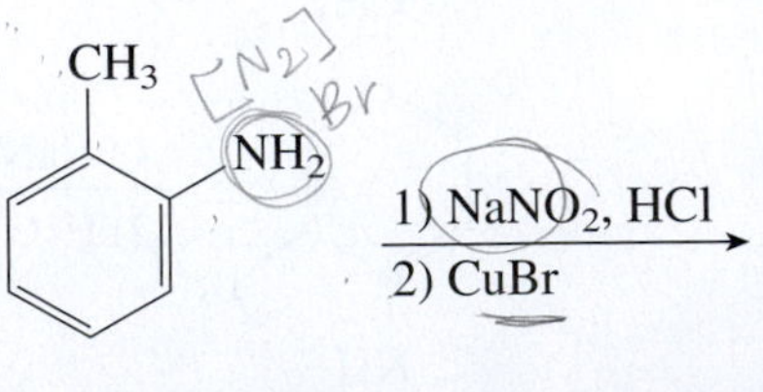

e)
O
CH
NH2
1) NaNO2, HCl
2) CuCl

f)
CO2H
Br
Br
NH2
Br
1) NaNO2, H2SO4
2) H3PO2

g)
CH2CH3
NH2
1) NaNO2, HCl
2) CuCN

h)
NH2
NO2
CH3
1) NaNO2, HCl
2) NaBH4

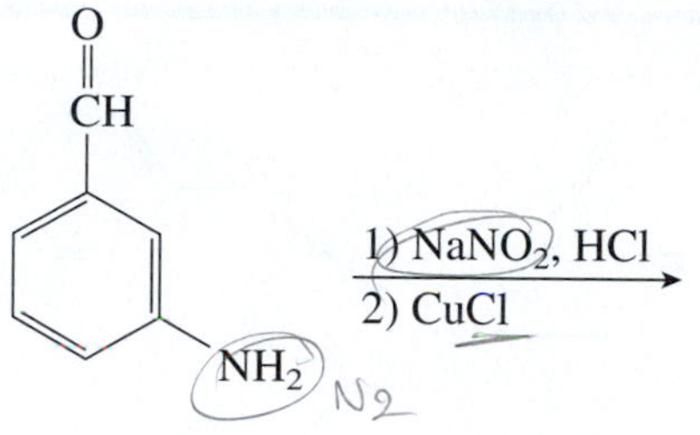

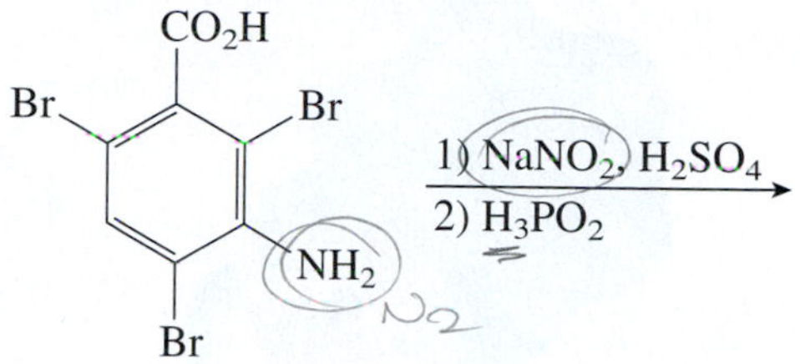

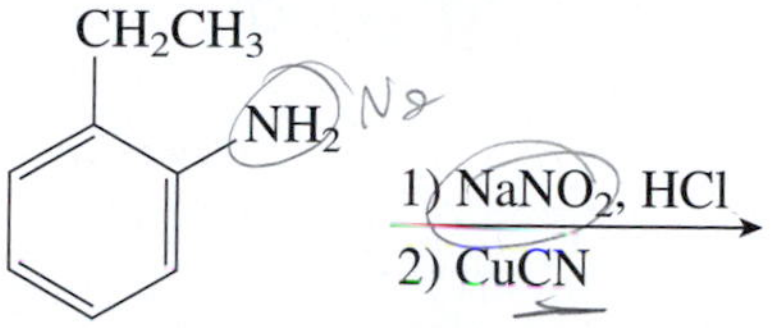

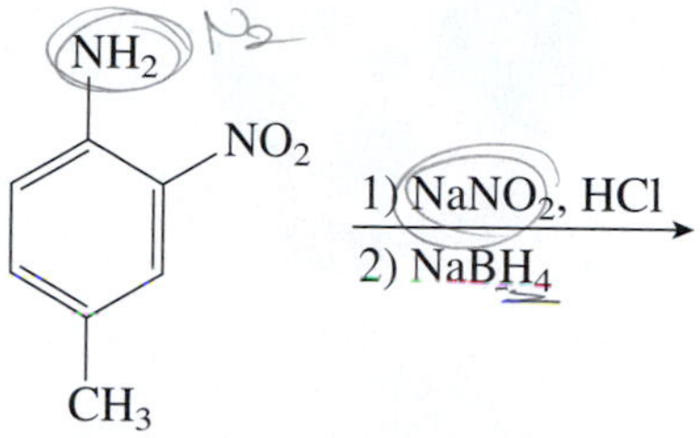

17.11 Nucleophilic Aromatic Substitution: Addition–Elimination

The reaction of a nucleophile with an aromatic halide that has a strong electron withdrawing group ortho and/or para to the halogen results in substitution of the nucleophile for the halogen.

O2N
Cl
+
O
O2N
O
+ Cl⁻
(82%)

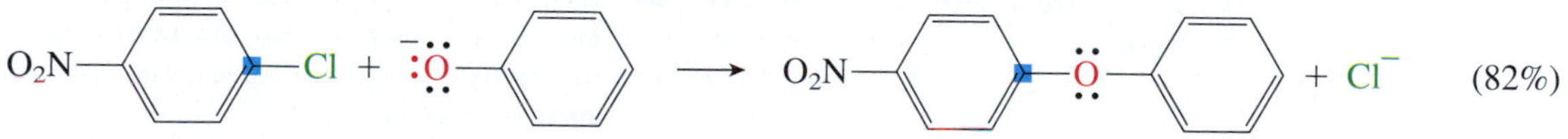

p-Chloronitrobenzene

This reaction at first might appear to be a normal S_N2 reaction because its rate depends on the concentration of both the nucleophile and the aromatic halide. However, experiments have shown that the mechanism consists of two steps. Addition of the nucleophile occurs in the first step, followed by departure of the leaving group in the second step, as shown in Figure 17.5. Therefore, this is called the **addition–elimination mechanism.**

One indication that this is not a normal S_N2 reaction is the requirement that electron-withdrawing groups be attached to the ring. As shown in Figure 17.5, the intermediate that is formed in this substitution reaction is a carbanion. This carbanion must have substituents, such as nitro or carbonyl groups, attached to the positions ortho or para to the leaving group so that they can delocalize the electron pair by resonance, thus stabilizing the anionic intermediate.

Another indication that the mechanism is different from those encountered in Chapter 8 is the effect of changing the leaving group. For this reaction the order of reactivity is

Fastest leaving group						Slowest leaving group
F	>	Cl	>	Br	>	I

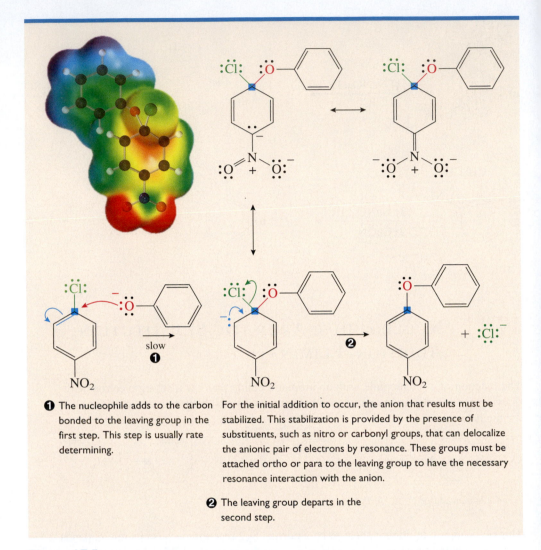

The nucleophile adds to the carbon bonded to the leaving group in the first step. This step is usually rate determining.

For the initial addition to occur, the anion that results must be stabilized. This stabilization is provided by the presence of substituents, such as nitro or carbonyl groups, that can delocalize the anionic pair of electrons by resonance. These groups must be attached ortho or para to the leaving group to have the necessary resonance interaction with the anion.

❷ The leaving group departs in the second step.

Figure 17.5

MECHANISM OF NUCLEOPHILIC SUBSTITUTION ON AN AROMATIC RING BY ADDITION–ELIMINATION.

Note that this is the opposite of the order found for S_N1 and S_N2 reactions. The reason for this reversal is that the leaving group does not depart in the rate-determining step in this reaction. Instead, the leaving group exerts its effect by helping stabilize the carbanion intermediate by its inductive effect. Because fluorine is the most electronegative of these atoms, it helps the most to stabilize the carbanion and the transition state leading to it.

Although the requirement for the presence of certain electron-withdrawing groups ortho and/or para to the leaving group limits the generality of this reaction, it works well with appropriately substituted compounds. Additional examples are shown in the following equations:

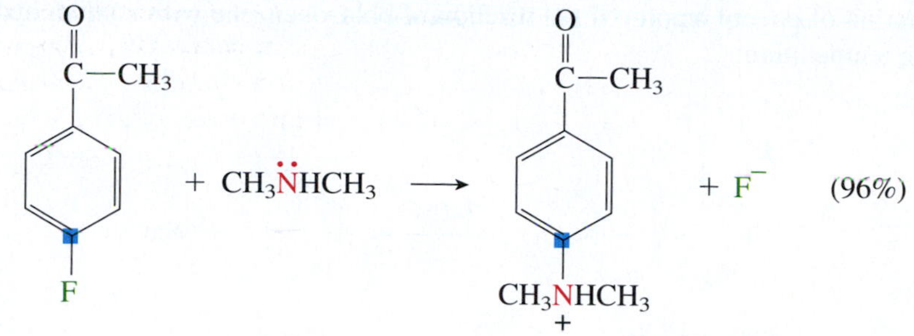

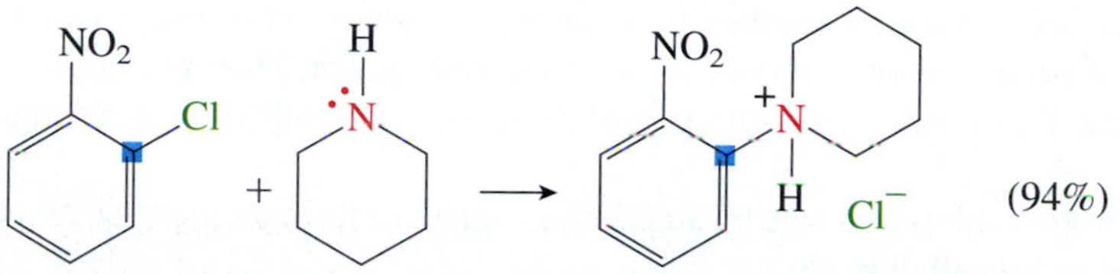

PROBLEM 17.24

Arrange these compounds in order of increasing rate of reaction with sodium hydroxide by the addition–elimination mechanism. Remember that the first step is rate determining.

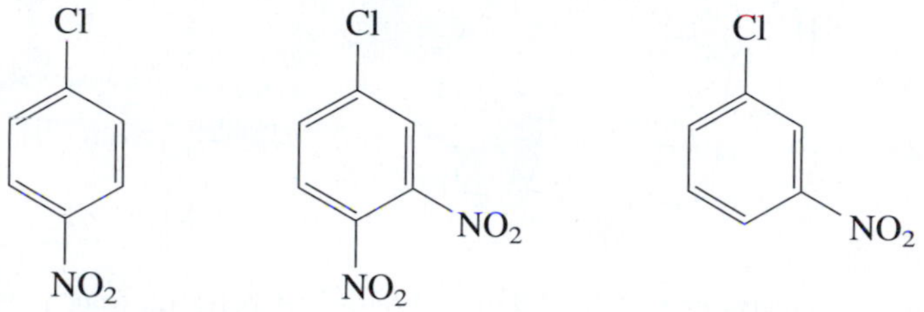

Addition – Elimination

PROBLEM 17.25

Show the products of these reactions:

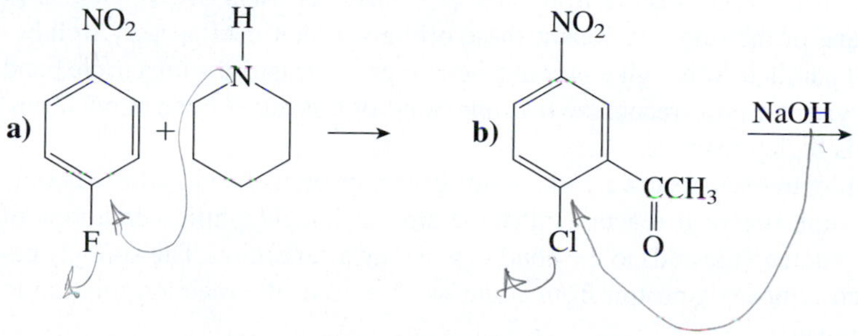

17.12 NUCLEOPHILIC AROMATIC SUBSTITUTION: ELIMINATION–ADDITION

A final mechanism for nucleophilic aromatic substitution occurs when aromatic halides are reacted with very strong bases, such as amide anion, or with weaker bases, such as hydroxide ion, at high temperatures. For example, an older industrial method for the preparation of phenol employed the reaction of chlorobenzene with sodium hydroxide at high temperature:

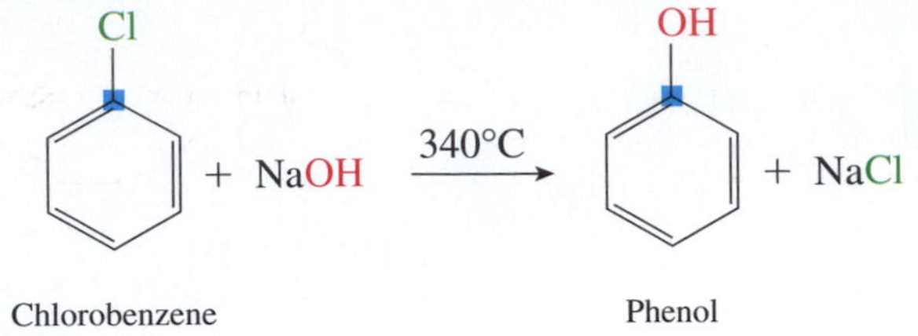

Although this reaction appears as though it might be a simple substitution, experiments indicate that it occurs by an elimination–addition mechanism. This mechanism is outlined in Figure 17.6 for the reaction of chlorobenzene with amide anion in liquid ammonia as solvent.

First, hydrogen chloride is lost by an E2 elimination to form an unusual, and highly reactive, compound called **benzyne:**

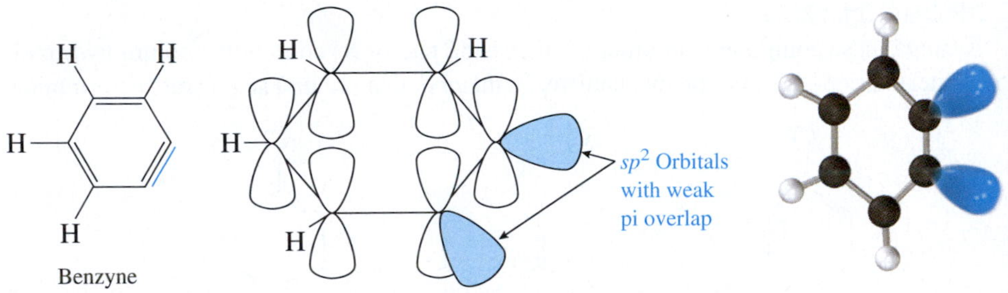

The elimination converts one of the "double bonds" of benzene into a "triple bond." This triple bond is quite different from a normal triple bond because of the angle constraints imposed by the six-membered ring. An orbital picture of benzyne shows the normal benzene arrangement of p orbitals perpendicular to the plane of the ring. The other bond of the triple bond results from two sp^2-hybridized AOs overlapping in pi fashion in the plane of the ring. Obviously, these orbitals do not overlap very well because they are not parallel. Although it is convenient to use a structure with a triple bond to represent benzyne, we must recognize that one bond of this triple bond is not a typical pi bond and is highly reactive.

Benzyne is an extremely reactive compound. It cannot be isolated and exists only for a very short time before it reacts. Under the strongly nucleophilic conditions of these reactions, a nucleophile adds to the bond to generate a carbanion. The strongly basic carbanion then removes a proton from some weak acid in the reaction mixture to form the final product.

One of the characteristics of reactions involving benzyne intermediates is that the nucleophile can bond to the same carbon to which the leaving group was bonded, or it can bond to the carbon adjacent to the one to which the leaving group was bonded. This often results in the formation of isomeric products when substituted aromatic halides are used. For example, the reaction of p-bromotoluene with sodium dimethylamide in dimethylamine as the solvent gives a 50:50 mixture of the meta and para

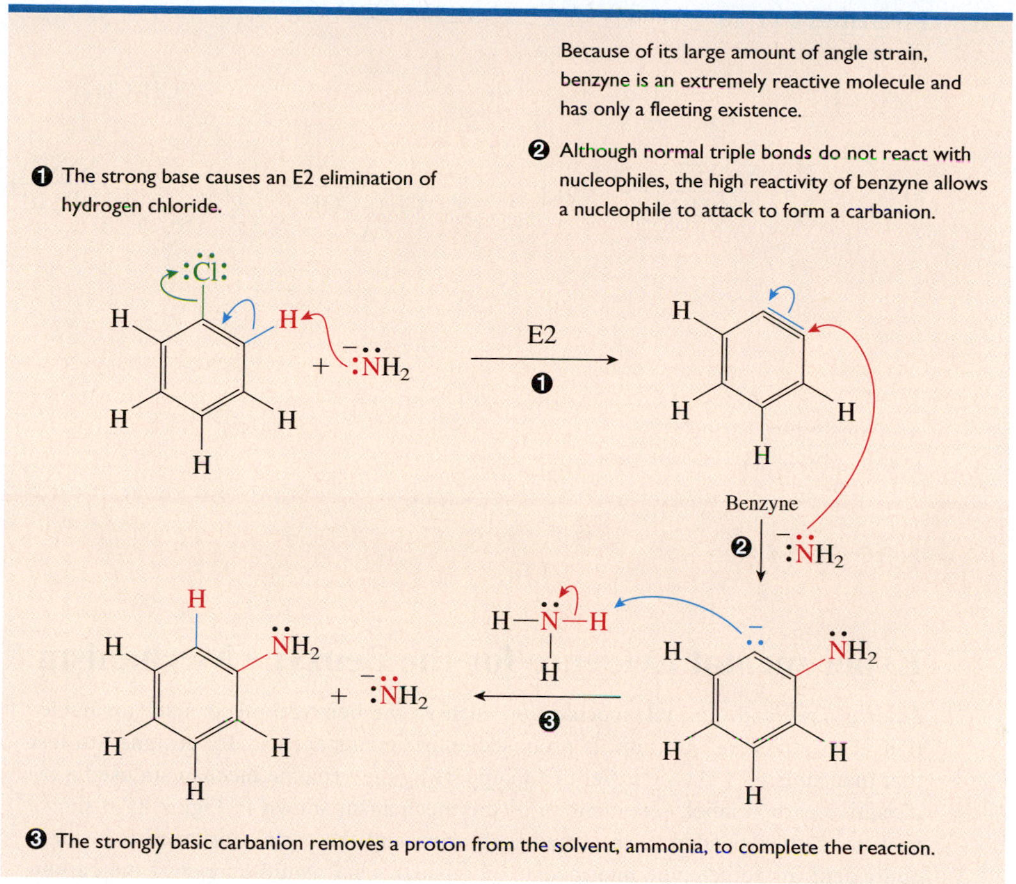

① The strong base causes an E2 elimination of hydrogen chloride.

Because of its large amount of angle strain, benzyne is an extremely reactive molecule and has only a fleeting existence.

❷ Although normal triple bonds do not react with nucleophiles, the high reactivity of benzyne allows a nucleophile to attack to form a carbanion.

❸ The strongly basic carbanion removes a proton from the solvent, ammonia, to complete the reaction.

Figure 17.6

MECHANISM OF NUCLEOPHILIC AROMATIC SUBSTITUTION BY ELIMINATION–ADDITION (THE BENZYNE MECHANISM).

products because the nucleophile can bond to either carbon of the benzyne triple bond. Such a product mixture is common whenever an asymmetrical benzyne intermediate is formed.

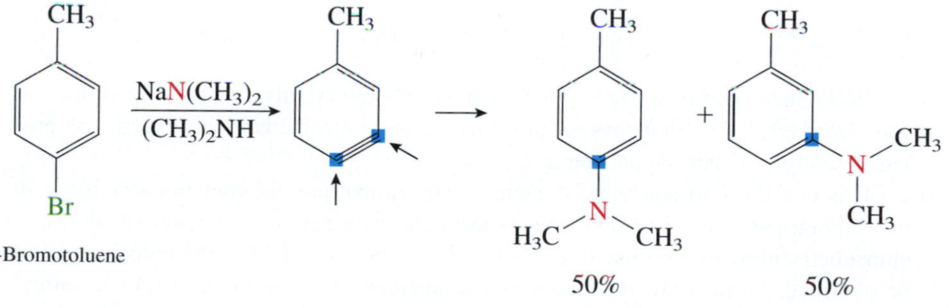

p-Bromotoluene

50%   50%

PROBLEM 17.26

Show the products of these reactions:

a) [structure: 2-methyl-bromobenzene (ortho-bromotoluene) with CH_3 group] with reagents $\dfrac{NaN(CH_3)_2}{HN(CH_3)_2}$

b) [structure: bromobenzene derivative with Br] with reagents $\dfrac{(CH_3)_3CO^-}{(CH_3)_3COH}$, Δ

c) [structure: ethylbenzene with Br, CH_2CH_3 group] with reagents $\dfrac{NaNH_2}{NH_3 \, (l)}$

Focus On

Experimental Evidence for the Benzyne Mechanism

After a new (and unusual) mechanism, such as the benzyne mechanism for nucleophilic aromatic substitution, is proposed, experiments are usually designed to test that mechanism. A classic experiment supporting the benzyne mechanism used a radioactive carbon label. Examination of the mechanism shown in Figure 17.6 shows that the carbon bonded to the leaving chlorine and the carbon ortho to it become equivalent in the benzyne intermediate. Consider what would happen if the carbon bonded to the chlorine were a radioactive isotope of carbon (^{14}C) rather than the normal isotope of carbon (^{12}C). If we follow the position of the radioactive carbon label through the mechanism of Figure 17.6, we find that the label should be equally distributed between the carbon attached to the amino group in the product and the carbon ortho to it.

[reaction scheme: chlorobenzene (with Cl and labeled ¹⁴C carbon) → $\dfrac{NaNH_2}{NH_3 \, (l)}$ → aniline with NH_2 (label on ipso carbon) (50%) + aniline with NH_2 (label on ortho carbon) (50%)]

■ Represents a radioactive ^{14}C atom

When the experiment was conducted in the laboratory, this is exactly the result that was observed. Although it does not prove the benzyne mechanism, this experiment provides strong evidence supporting it.

It is one thing to conceive of such an experiment and another to carry it out in the laboratory. Let's see how the experiment was actually accomplished. First, chlorobenzene with a radioactive label at the carbon attached to the chlorine had to be obtained. Fortunately, this material was available from a commercial laboratory.

(Recognize, however, that the preparation of this compound from a source of radioactive carbon, such as $^{14}CO_2$ or $H^{14}CN$, is not a trivial task.) Once the preceding reaction had been run, the product (aniline) had to be degraded in a controlled manner to determine the position of the label. The degradation was accomplished in the following manner:

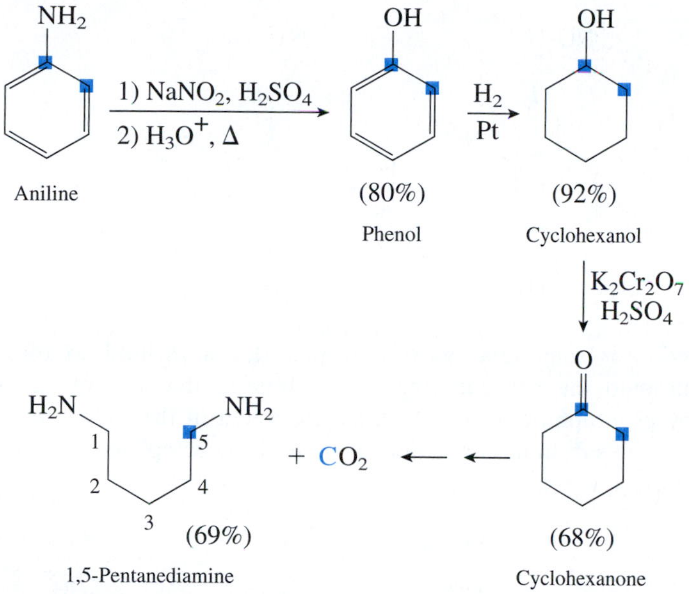

First the amino group was converted to a hydroxy group via a diazonium ion (Section 17.10). The benzene ring was reduced with hydrogen and a catalyst to produce cyclohexanol. Oxidation with potassium dichromate (Section 10.14) gave cyclohexanone. The bonds between the carbonyl carbon and both α-carbons were then cleaved by a series of reactions not covered in this book. The carbon of the carbonyl group was converted to carbon dioxide in this process. One-half of the original radioactivity was found in the carbon dioxide, and the other one-half was found in the other product, 1,5-pentanediamine. Additional experiments showed that the ^{14}C in the diamine product was located at C-1 or C-5.

17.13   SOME ADDITIONAL USEFUL REACTIONS

This section presents several additional reactions that are very useful in the synthesis of aromatic compounds because they provide methods to convert substituents that can be attached by electrophilic substitution reactions to other substituents that cannot be attached directly. The mechanisms of these reactions need not concern us here.

The first of these reactions converts a nitro group to an amino group. This reduction can be accomplished using hydrogen and a catalyst or by using acid and a metal (Fe, Sn, or $SnCl_2$). Examples are provided in the following equations:

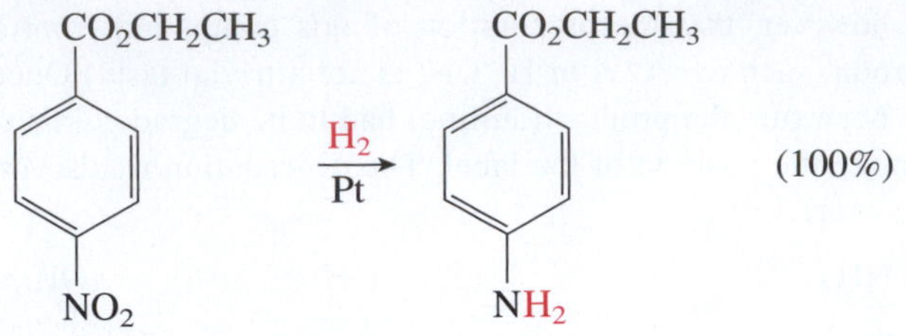

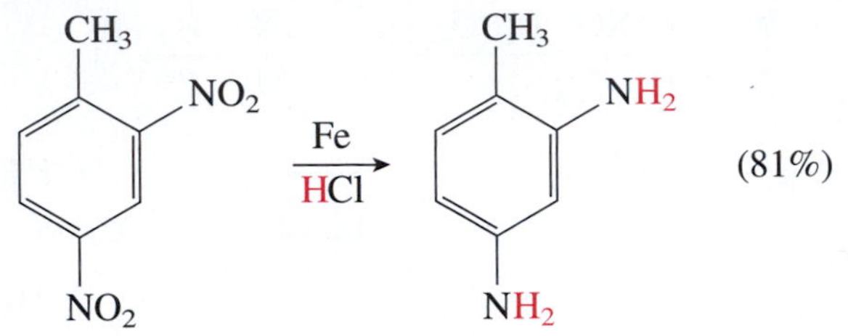

$$(100\%)$$

$$(81\%)$$

This reaction is important because it provides a method to place an amino substituent onto the benzene ring, a substitution that cannot be accomplished directly by electrophilic attack. And, as illustrated in the following example, this opens all of the substitution reactions that can be accomplished through diazonium ion reactions.

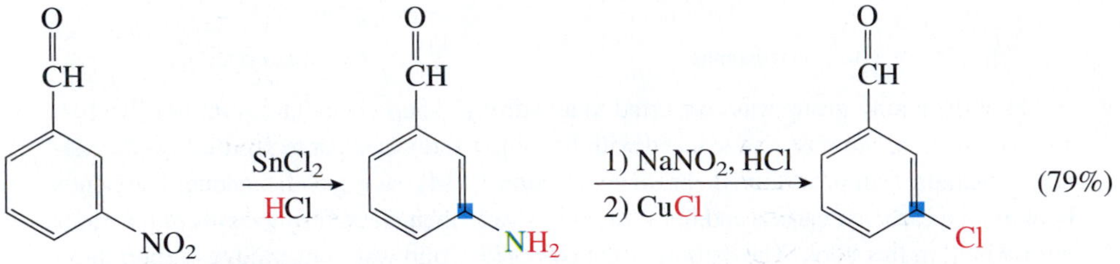

$$(79\%)$$

Several procedures can be used to convert the carbonyl group of an aldehyde or ketone to a methylene group. One reaction, known as the Clemmensen reduction, employs amalgamated zinc (zinc plus mercury) and hydrochloric acid as the reducing agent. An example is provided by the following equation:

$$(88\%)$$

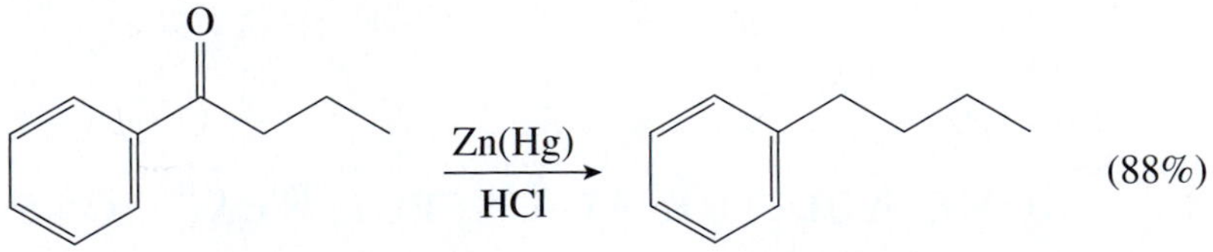

1-Phenyl-1-butanone   Butylbenzene

Another reaction that can be used to accomplish the same transformation is the Wolff-Kishner reduction. In this procedure the aldehyde or ketone is heated with hydrazine and potassium hydroxide in a high boiling solvent. An example is provided in the following equation. (The mechanism for the Wolff-Kishner reduction is presented in Section 18.8.) The Clemmensen reduction and the Wolff-Kishner reduction are complementary because one employs acidic conditions and the other employs basic conditions.

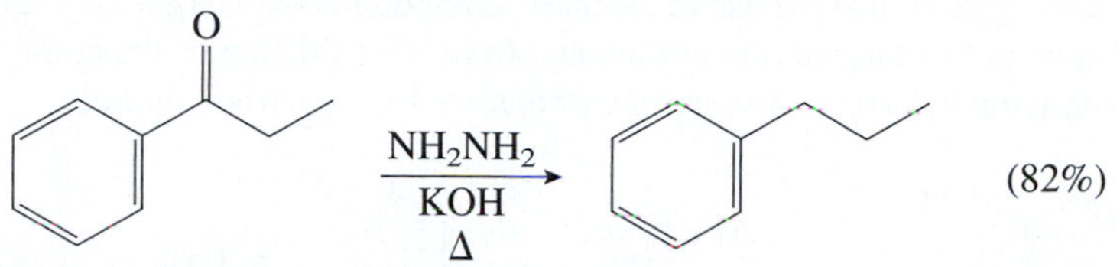

(82%)

The reduction of the carbonyl group of an aromatic ketone to a methylene group can also be accomplished by catalytic hydrogenation. An example of this method is shown in the following equation. Note that the carbonyl group in this reaction must be attached directly to the aromatic ring. The Clemmensen and Wolff-Kishner reductions do not have this restriction.

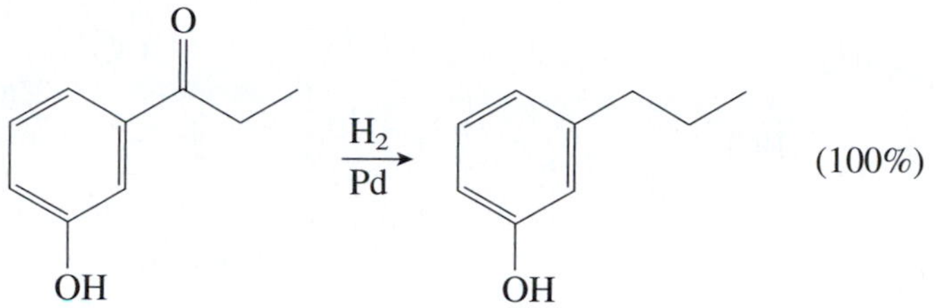

(100%)

These reactions are quite useful in the preparation of aromatic compounds substituted with primary alkyl groups. For example, suppose a synthesis of butylbenzene is required. We might first consider preparing this compound by a Friedel-Crafts alkylation reaction. However, using a primary alkyl halide in this reaction invariably results in carbocation rearrangement. The reaction of benzene with 1-chlorobutane produces a mixture of butylbenzene (34%) and *sec*-butylbenzene (66%) (see page 692). The low yield of the desired primary product and the difficulty in obtaining it pure from the product mixture make this an unacceptable synthetic route. A much better synthesis can be accomplished in two steps by first preparing 1-phenyl-1-butanone by a Friedel-Crafts acylation reaction using benzene and butanoyl chloride, followed by conversion of the carbonyl group to a methylene group by one of these reduction reactions. As shown in the equation on the preceding page, the Clemmensen reduction accomplishes this transformation in 88% yield.

The final reaction in this section provides a method to prepare aromatic rings bonded to a carboxylic acid group. Because we do not have a direct way to attach this group, this procedure is very useful. The reaction is usually accomplished by oxidation of a methyl group to the carboxylic acid employing hot potassium permanganate in basic solution:

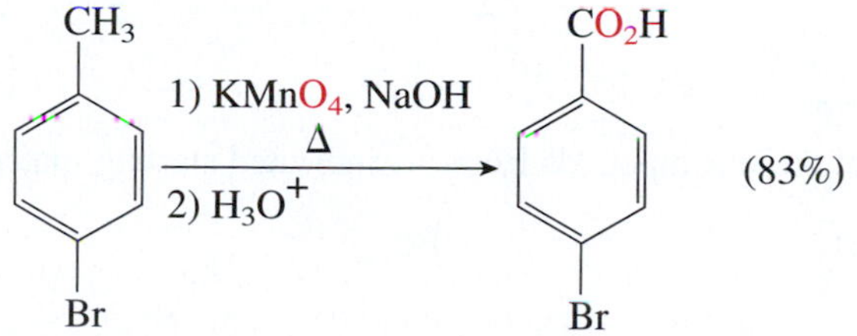

(83%)

Although methyl groups are most commonly oxidized in these reactions, other alkyl groups can also be employed, as long as the carbon that is bonded to the aromatic ring is not quaternary. Note that the use of aromatic compounds with larger alkyl groups still gives the same product as would be produced from the oxidation of the compound substituted with a methyl group. The extra carbons are lost as carbon dioxide:

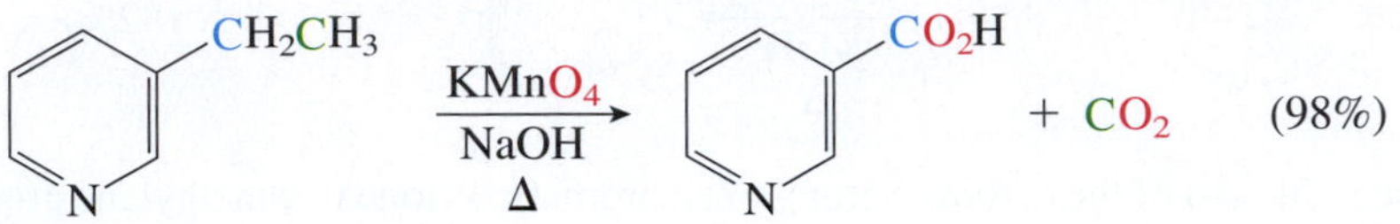

$$\text{(3-ethylpyridine)} \xrightarrow[\substack{\text{NaOH} \\ \Delta}]{\text{KMnO}_4} \text{(nicotinic acid)} + CO_2 \quad (98\%)$$

PROBLEM 17.27
Show the products of these reactions:

a) 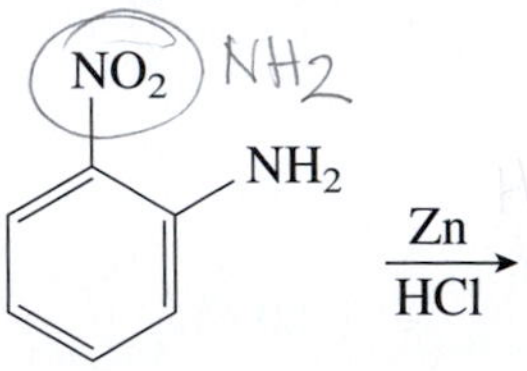 [structure: benzene with NO₂, NH₂, NH₂ substituents] $\xrightarrow[\text{HCl}]{\text{Zn}}$ *H₂O*

b) 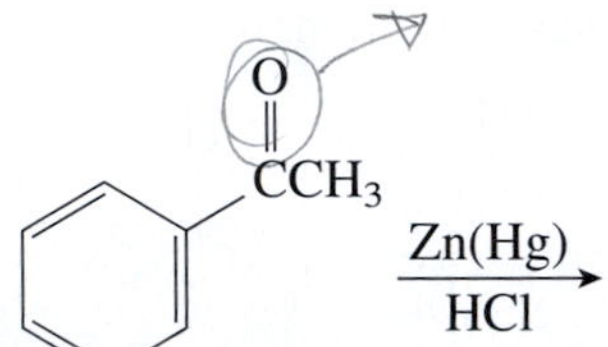 [structure: acetophenone, CCH₃ with O] $\xrightarrow[\text{HCl}]{\text{Zn(Hg)}}$

c) 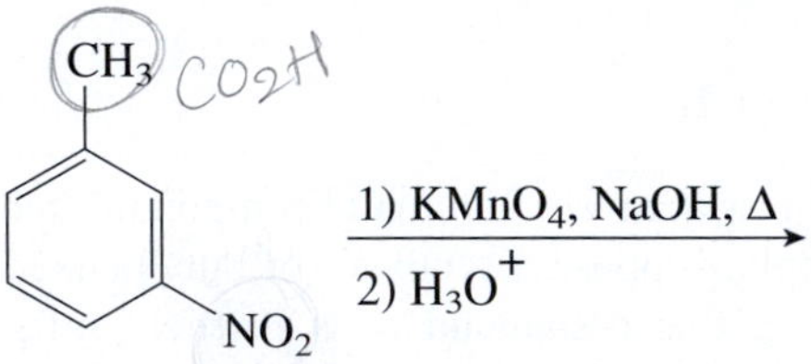 [structure: benzene with CH₃ (CO₂H) and NO₂] $\xrightarrow[\text{2) H}_3\text{O}^+]{\text{1) KMnO}_4, \text{NaOH}, \Delta}$

d) 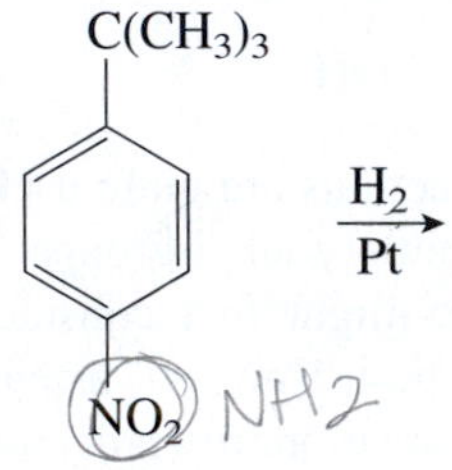 [structure: benzene with C(CH₃)₃ and NO₂ (NH₂)] $\xrightarrow[\text{Pt}]{\text{H}_2}$

e) 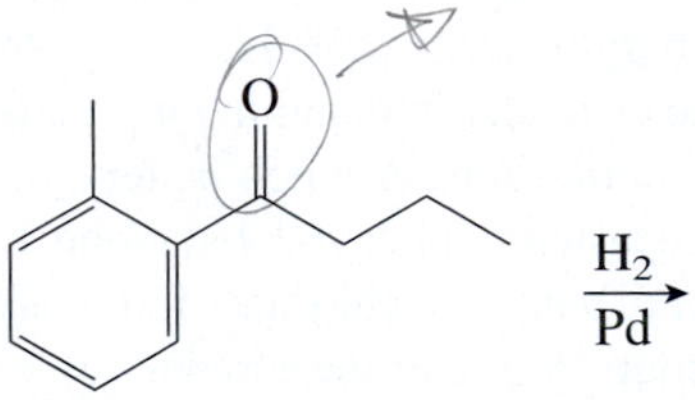 [structure: o-methyl phenyl butyl ketone] $\xrightarrow[\text{Pd}]{\text{H}_2}$

f) 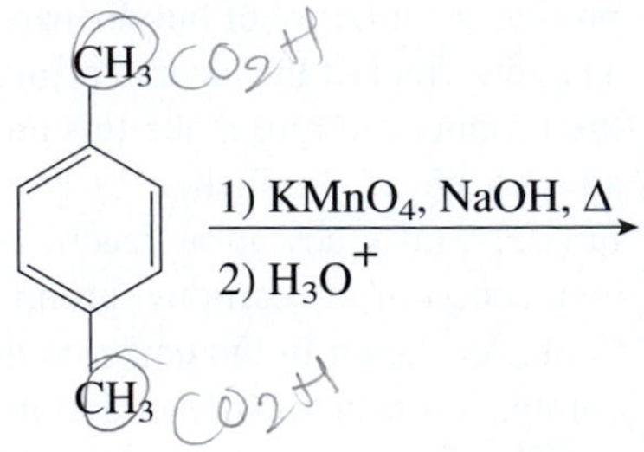 [structure: benzene with CH₃ (CO₂H) and CH₃ (CO₂H)] $\xrightarrow[\text{2) H}_3\text{O}^+]{\text{1) KMnO}_4, \text{NaOH}, \Delta}$

g) 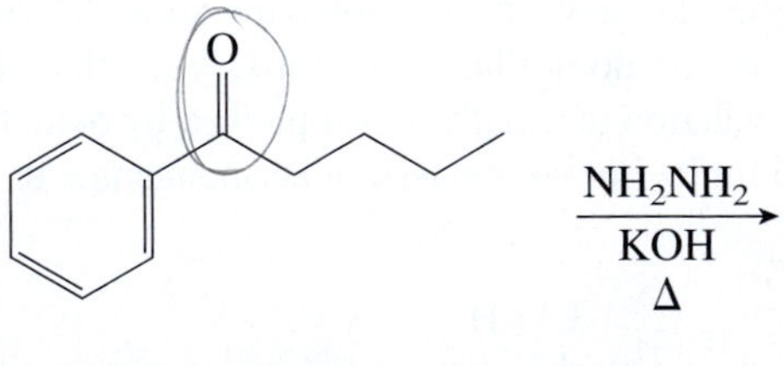 [structure: phenyl butyl ketone] $\xrightarrow[\substack{\text{KOH} \\ \Delta}]{\text{NH}_2\text{NH}_2}$

PROBLEM 17.28
Show syntheses of these compounds from the indicated starting materials:

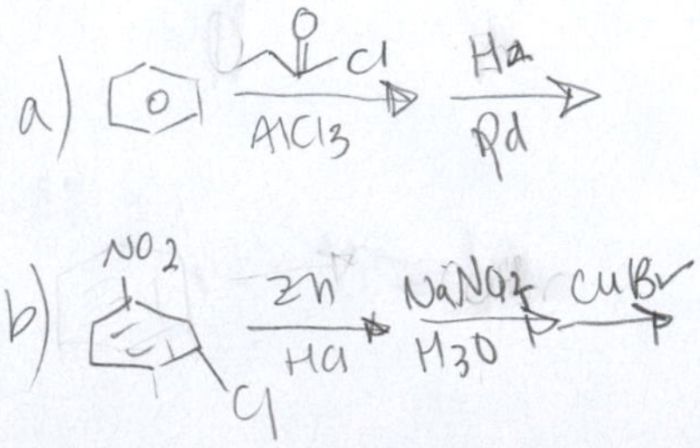

a) 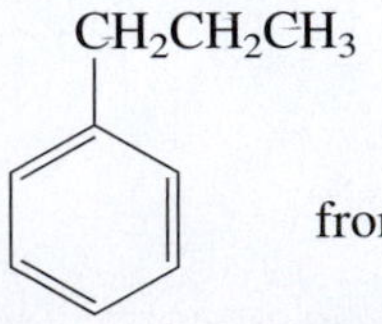 [structure: propylbenzene, CH₂CH₂CH₃]   from benzene

b) 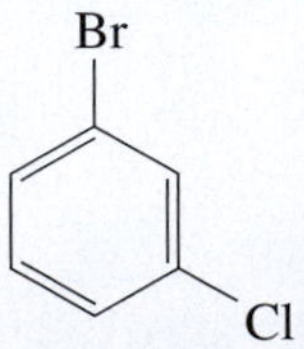 [structure: benzene with Br and Cl, m-bromochlorobenzene]   from *m*-chloronitrobenzene

a) [handwritten] [benzene] $\xrightarrow[\text{AlCl}_3]{\overset{O}{\text{Cl}}}$ $\xrightarrow[\text{Pd}]{\text{H}_2}$

b) [handwritten] [benzene with NO₂ and Cl] $\xrightarrow[\text{HCl}]{\text{Zn}}$ $\xrightarrow[\text{H}_3\text{O}]{\text{NaNO}_2}$ $\xrightarrow{\text{CuBr}}$

17.14 SYNTHESIS OF AROMATIC COMPOUNDS

The previous sections presented a powerful array of reactions that can be used to substitute almost any type of group onto an aromatic ring. Figure 17.7 summarizes the transformations that can be accomplished by using these reactions. Several syntheses of aromatic compounds using these reactions are presented in this section. This provides an excellent opportunity to develop and practice the strategy used in synthesis of relatively complex compounds.

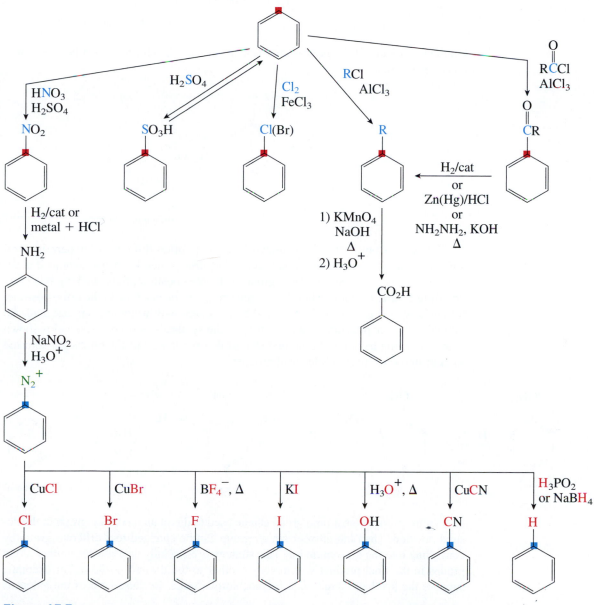

Figure 17.7

A SUMMARY OF REACTIONS AVAILABLE FOR USE IN THE SYNTHESIS OF AROMATIC COMPOUNDS.

The first examples illustrate that the order of addition of the substituents is important in controlling their orientation. For example, suppose we needed to prepare *m*-chloronitrobenzene from benzene. Because the chlorine is an ortho/para director and the nitro group is a meta director, it is apparent that the nitro group must be added first if the meta product is desired:

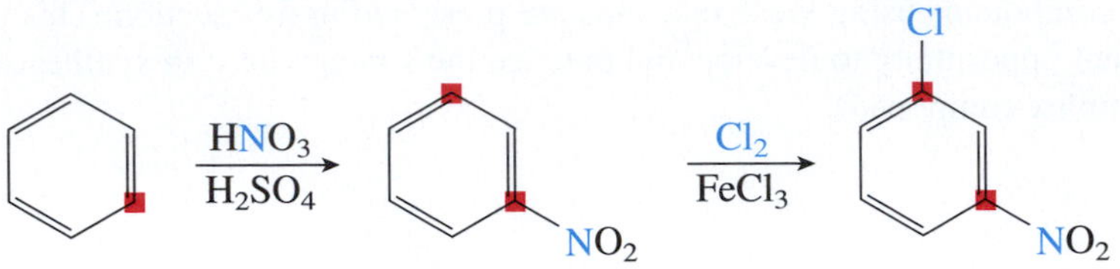

m-Chloronitrobenzene

On the other hand, if the *para*-isomer is desired, then the chlorine must be substituted onto the ring first:

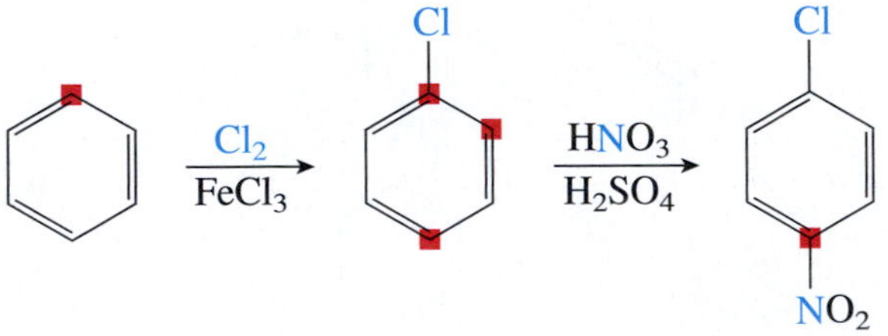

p-Chloronitrobenzene

The *ortho*-isomer of a disubstituted benzene is often difficult to prepare in good yield because of the steric effect that favors the *para*-isomer. In such situations it is advantageous to place a sulfonic acid group at the para position, thus blocking this position from further reaction. After the desired group has been added to the ortho position, the sulfonic acid group can be removed by treatment with water and sulfuric acid. An example of the application of this strategy to the synthesis of *o*-bromophenol is shown in the following equation. In the first step of this synthesis the conditions are adjusted so as to introduce two sulfonic acid groups:

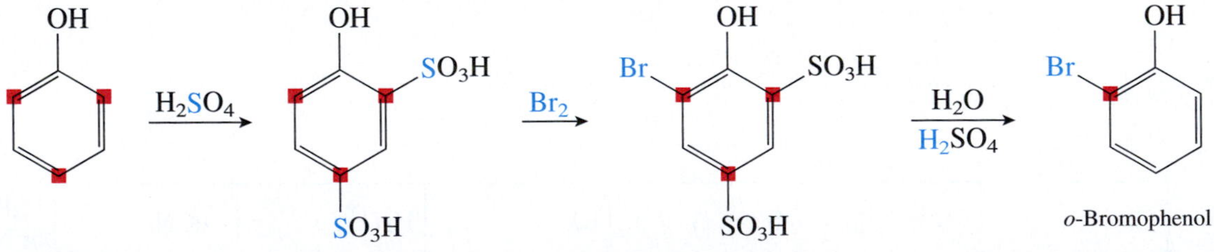

o-Bromophenol

A nitro group or an amino group can be used to direct an incoming group to the desired position. Then the nitro or amino group can be changed to a different group by converting it to the diazonium ion. This strategy is especially useful when neither of the groups in the final product will direct the other to the desired position. For example, suppose the synthetic target is *m*-bromochlorobenzene. Because both of the halogens are ortho/para directors, how can they be placed in a meta orientation? The solution to this problem is to use the nitro group to direct one of the halogens to the meta position and then change the nitro group into the other halogen. This synthesis is outlined in the following equation. A variation of this strategy can be employed to place two meta directing groups para to each other.

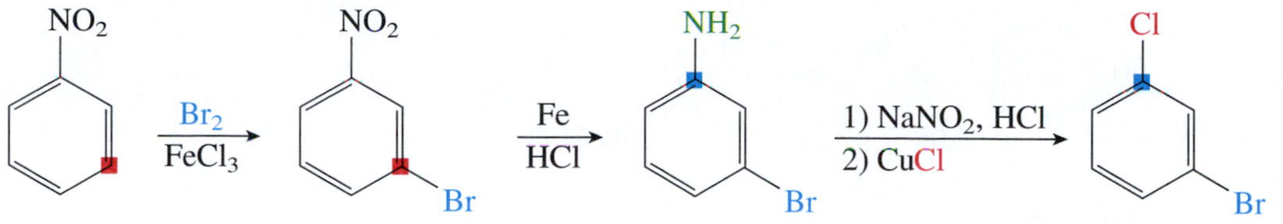

m-Bromochlorobenzene

In another approach, an amino group can be used to obtain the desired regiochemistry for the product. Then it can be removed via the diazonium ion. For example, suppose that we want to prepare *m*-bromotoluene starting from toluene. The difficulty is that the methyl group is an ortho/para director. The solution to this problem is to add an amino group para to the methyl group. This strong activating group can then be used to direct the bromine to the position ortho to itself and meta to the methyl group. The amino group is then removed. An application of this strategy is illustrated in the following example. Note that it is necessary to decrease the reactivity of the amine by converting it to the amide before the bromine is added:

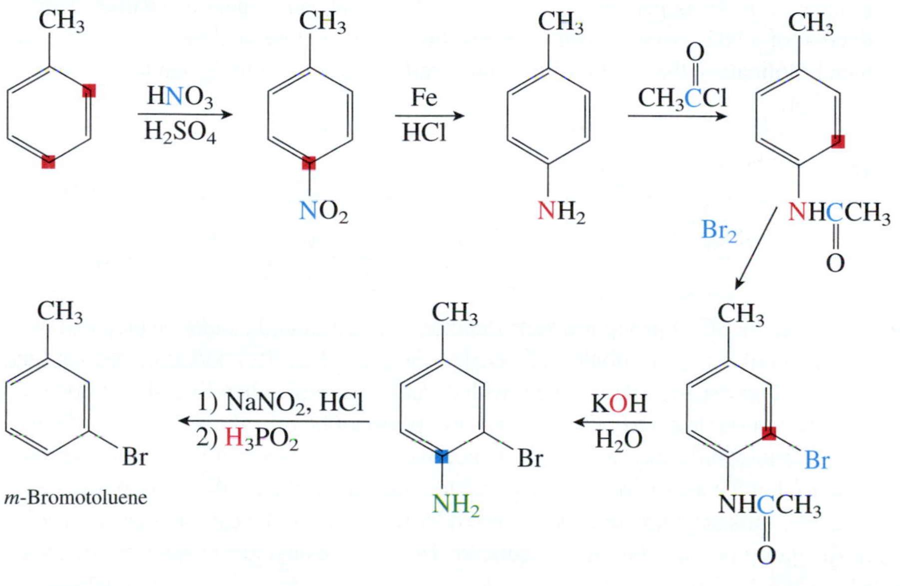

m-Bromotoluene

These examples show that a synthesis must be carefully planned. The regiochemistry of each step must be considered as well as the compatibility of the substituents already on the ring with the reaction conditions. However, when completed, a cleverly crafted synthesis is a thing of beauty! Chemists often describe such a synthesis as elegant.

PRACTICE PROBLEM 17.3

Show syntheses of these compounds from benzene:

a)

b)

Strategy

Examine the target to see what groups have to be added to the aromatic ring. Review Figure 17.7 to determine which groups can be added by various reactions. Remember that NO_2, SO_3H, Cl_2, Br_2, alkyl, and acyl groups can be added by electrophilic aromatic substitution reactions and that a NO_2 group can be reduced to a NH_2 group, which can then be converted to a variety of other substituents via a diazonium ion. Pay particular attention to directive effects and the order in which the groups must be added to obtain the desired isomer. Recall that the Friedel-Crafts reactions do not work on strongly de-activated rings. If it is necessary to attach a primary alkyl group on the ring, it is best to use a Friedel-Crafts acylation, followed by a reduction.

Solutions

a) Both the NH_2 group and the Br are ortho/para directing groups, but they are in a meta orientation in the target compound. However, the NH_2 substituent is obtained by re-duction of a NO_2 group, which is a meta director. Therefore, add the nitro group first, then brominate at the meta position, and finally reduce the nitro group to an amine.

b) Although the SO_3H group is a meta director, this compound cannot be prepared by alkylation of benzenesulfonic acid because the Friedel-Crafts alkylation reaction (and the acylation reaction) do not work with deactivated rings. Therefore, the carbon group must be put on first. The primary butyl group cannot be directly added in good yield by a Friedel-Crafts alkylation reaction because of rearrangement of the intermediate primary butyl carbocation. The best way to attach the primary alkyl group is with a Friedel-Crafts acylation reaction followed by reduction of the carbonyl group. The SO_3H group is added before the reduction because the acyl group is a meta director.

PROBLEM 17.29

Show syntheses of these compounds from benzene:

a) 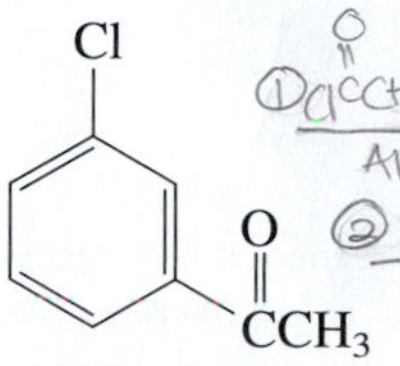 (Cl, CCH₃ structure) ① CH₃CCl / AlCl₃   ② Cl₂ / AlCl₃

b) 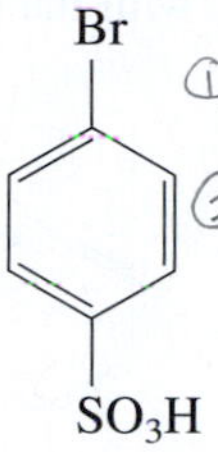 (Br, SO₃H structure) ① Br₂ / AlBr₃   ② H₂SO₄

c) 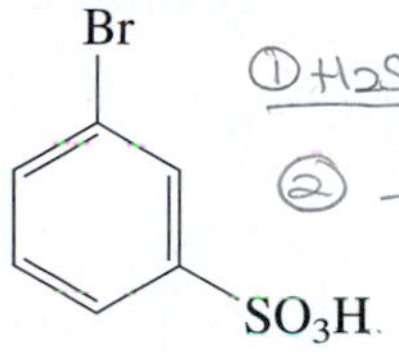 (Br, SO₃H structure) ① H₂SO₄   ② Br₂ / AlBr₃

d) 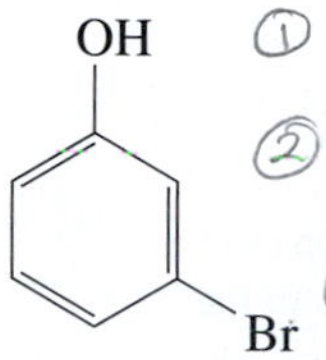 (OH, Br structure) ① HNO₃ / H₂SO₄   ② Br₂ / AlBr₃   ③ NaNO₂ / H₃O   ④ H₃O Δ

e) 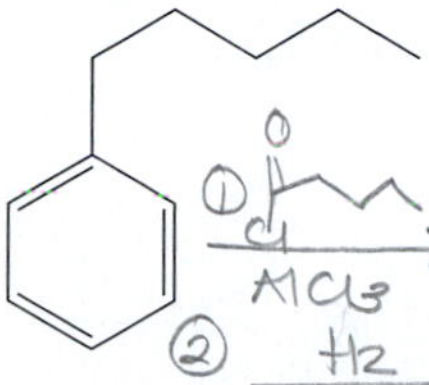 (pentylbenzene structure) ① CH₃(CH₂)₃CCl / AlCl₃   ② H₂ / Pt

f) 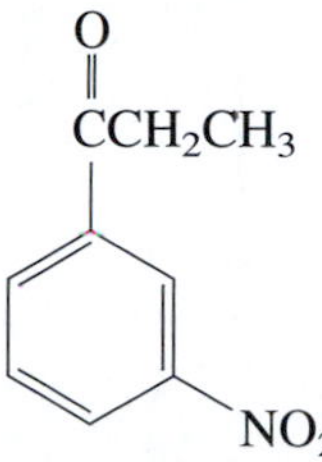 (CCH₂CH₃, NO₂ structure) ClCCH₂CH₃ / AlCl₃   HNO₃ / H₂SO₄

g) 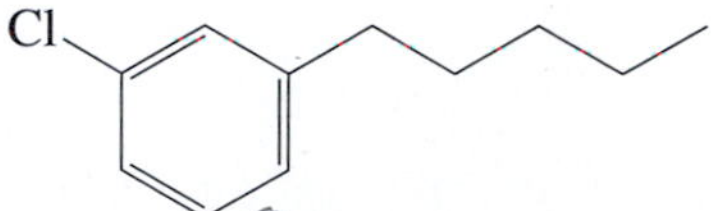 (Cl, pentyl structure) ① CH₃(CH₂)₃CCl / AlCl₃   ② Cl₂ / AlCl₃   ③ H₂ / Pt

h) 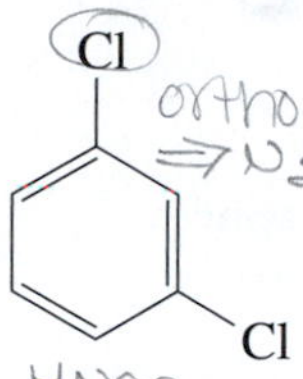 (Cl, Cl structure) ortho, para ⇒ N₂ ⇒ NH₂ ⇒ NO₂ meta
① HNO₃ / H₂SO₄   ② Cl₂ / AlCl₃   ③ H₂ / Pt   ④ NaNO₂ / H₃O   ⑤ CuCl

i) 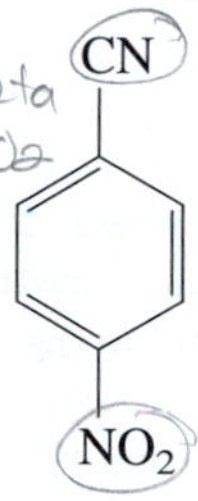 (CN, NO₂ structure) CN Meta ⇒ N₂ ⇒ NH₂ ortho para
NO₂ Meta
① HNO₃ / H₂SO₄   ② H₂ / Pt   ③ HNO₃ / H₂SO₄ keep para   ④ NaNO₂ / H₃O   ⑤ CuCN

PRACTICE PROBLEM 17.4

Show a synthesis of this compound from benzene:

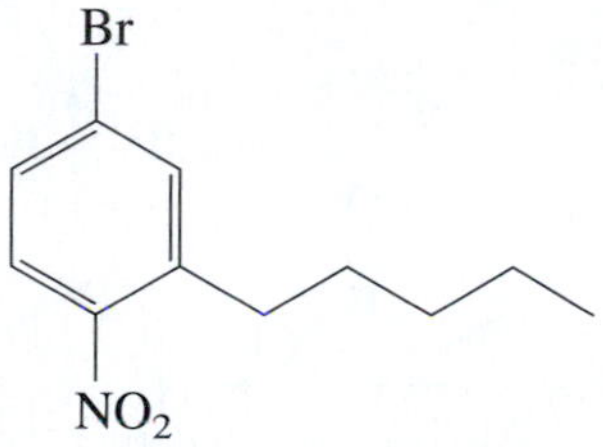

(Br, NO₂, pentyl substituted benzene structure)

Strategy

With more complicated syntheses like this one, it is best to use retrosynthetic analysis to determine the order in which the groups should be added. Start by analyzing the addition of each different group to determine whether the reaction with the desired orientation is feasible. For each reaction that appears feasible, analyze the addition of the other groups in the starting material for that path in the same manner.

Solution

Start by analyzing whether the bromo, nitro, or pentyl groups can be added to the appropriate disubstituted benzene with the required orientation.

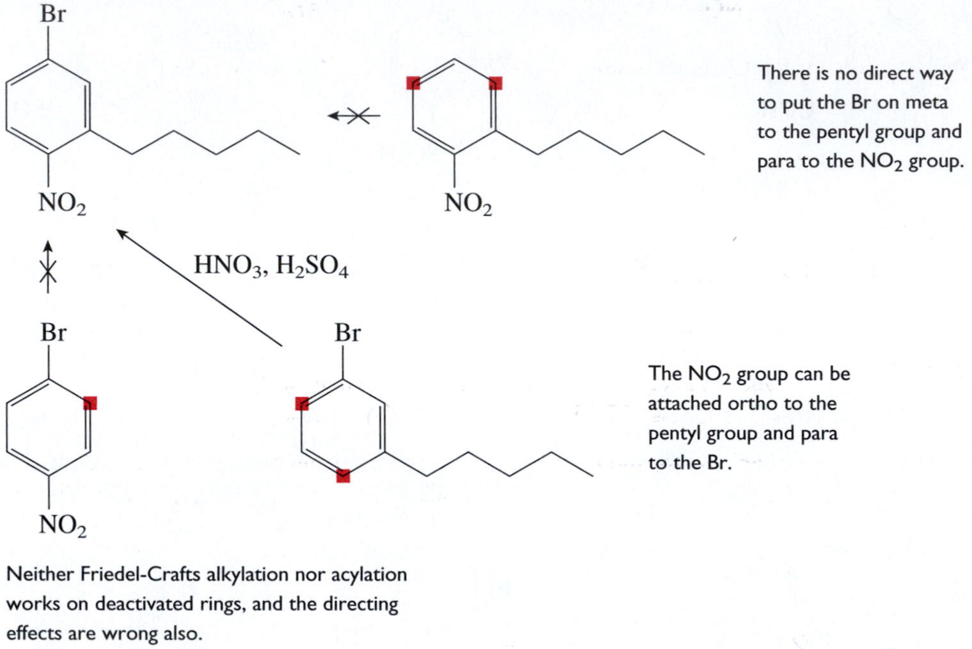

Br
NO_2
NO_2
There is no direct way to put the Br on meta to the pentyl group and para to the NO_2 group.
HNO_3, H_2SO_4
Br
Br
NO_2
The NO_2 group can be attached ortho to the pentyl group and para to the Br.
Neither Friedel-Crafts alkylation nor acylation works on deactivated rings, and the directing effects are wrong also.

The pentyl group must be put on by a Friedel-Crafts acylation reaction followed by a reduction to avoid rearrangement. The bromine must be added at the acyl stage to get meta orientation.

Br
Zn(Hg)
HCl
Br
O
Br_2
$FeBr_3$
O
Cl
$AlCl_3$
O

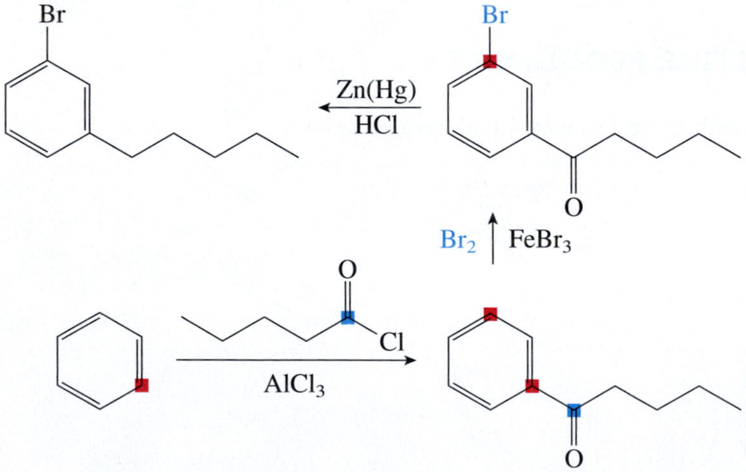

PROBLEM 17.30

Show syntheses of these compounds from benzene:

a) 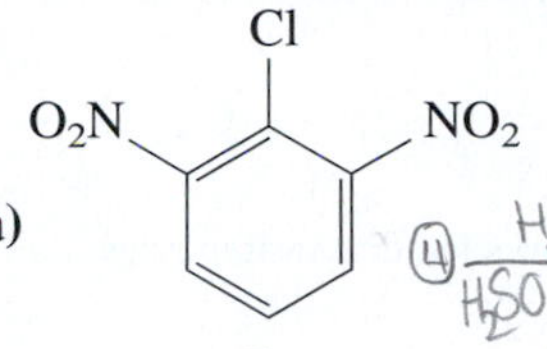

b) 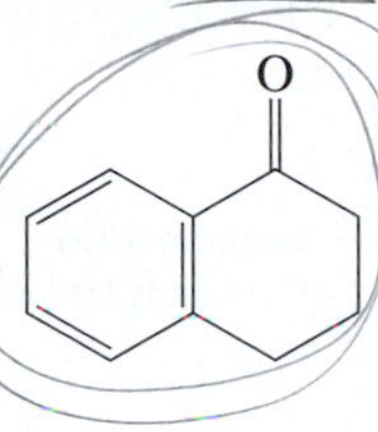

c) 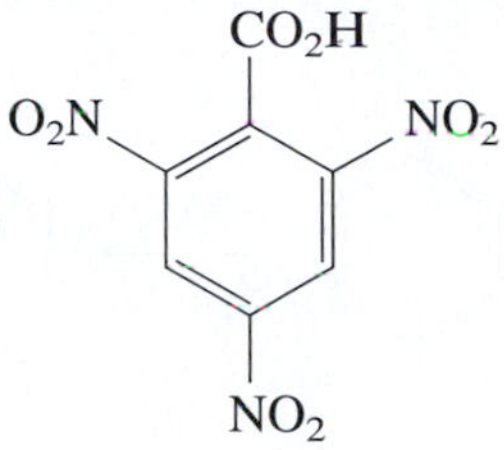

Review of Mastery Goals

After completing this chapter, you should be able to:

- Show the products of any of the reactions discussed in this chapter. (Problems 17.31, 17.32, 17.33, 17.34, 17.35, 17.39, 17.40, 17.41, 17.66, and 17.67)
- Show the mechanisms for the reactions whose mechanisms were discussed. (Problems 17.45, 17.46, 17.47, 17.48, 17.49, 17.50, 17.51, 17.59, 17.60, 17.61, 17.62, 17.63, 17.64, and 17.65)
- Predict the effect of a substituent on the rate and regiochemistry of an electrophilic aromatic substitution reaction. (Problems 17.31, 17.32, 17.33, 17.34, 17.35, 17.36, 17.37, 17.39, 17.40, 17.41, 17.42, 17.52, 17.53, 17.55, 17.56, 17.69, 17.70, and 17.71)
- Use these reactions to synthesize aromatic compounds. (Problems 17.38, 17.43, 17.44, 17.54, 17.57, and 17.58)

ORGANIC Chemistry•Now™
Click *Mastery Goal Quiz* to test how well you have met these goals.

Visual Summary of Key Reactions

The reactions in this chapter can be placed into three groups. First, there are the electrophilic aromatic substitution reactions, in which an electrophile attacks the benzene ring. These reactions are summarized in Table 17.2. Next are reactions that involve nucleophiles reacting with aromatic compounds. The reactions of diazonium ions are summarized in Table 17.3. Nucleophilic aromatic substitution reactions proceeding through the addition–elimination and elimination–addition mechanisms are summarized in Table 17.4. Finally, three other reactions that are useful in interconverting groups on an aromatic ring are summarized in Table 17.5.

Table 17.2   **Electrophilic Aromatic Substitution Reactions**

Reaction	Comments
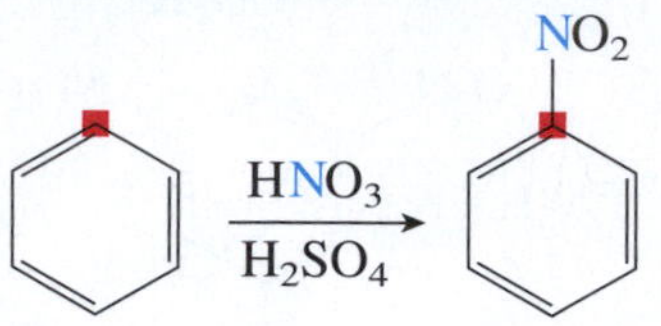 $\xrightarrow[\text{H}_2\text{SO}_4]{\text{HNO}_3}$	**Section 17.4** Product is less reactive; works for deactivated rings
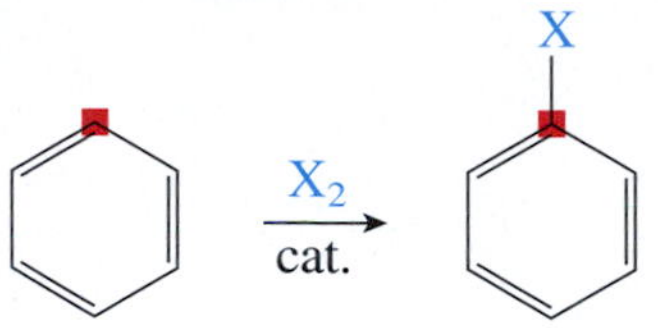 $\xrightarrow[\text{cat.}]{\text{X}_2}$	**Section 17.5** Works for Cl_2 and Br_2; catalysts are $AlCl_3$, $FeCl_3$, $AlBr_3$, $FeBr_3$; product is less reactive; works for deactivated rings
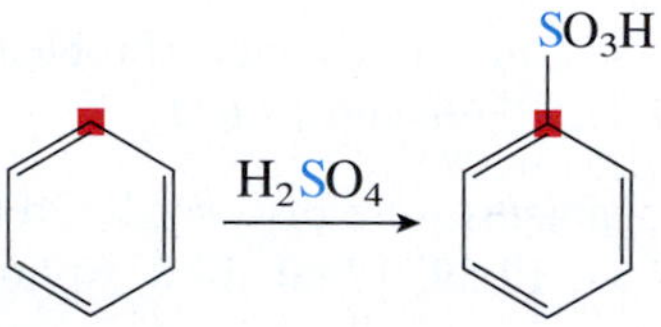 $\xrightarrow{\text{H}_2\text{SO}_4}$	**Section 17.6** Product is less reactive; works for deactivated rings; can be reversed by using H_2SO_4 and H_2O
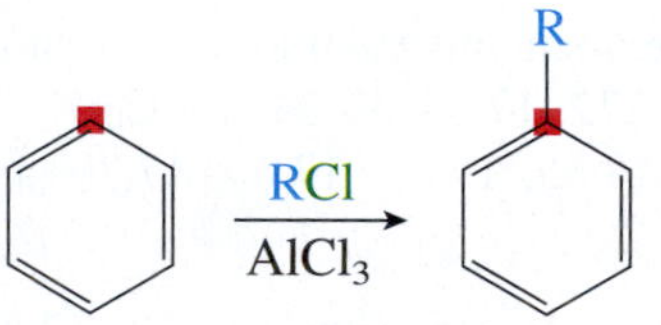 $\xrightarrow[\text{AlCl}_3]{\text{RCl}}$	**Section 17.7** Friedel-Crafts alkylation; product is more reactive; does not work for strongly deactivated rings; rearrangements are common; can also use acid plus alkene or alcohol to generate electrophile
$\xrightarrow[\text{AlCl}_3]{\overset{\text{O}}{\overset{\|}{\text{R}\text{C}}}\text{Cl}}$	**Section 17.8** Friedel-Crafts acylation; product is less reactive; does not work for strongly deactivated rings; can use anhydride to generate electrophile

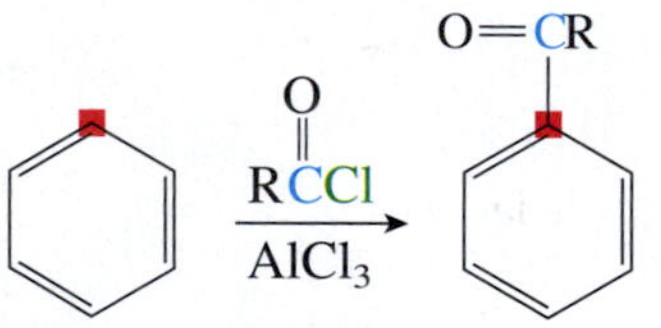

Table 17.3   Substitutions Using Diazonium Ions

Reaction	Comments
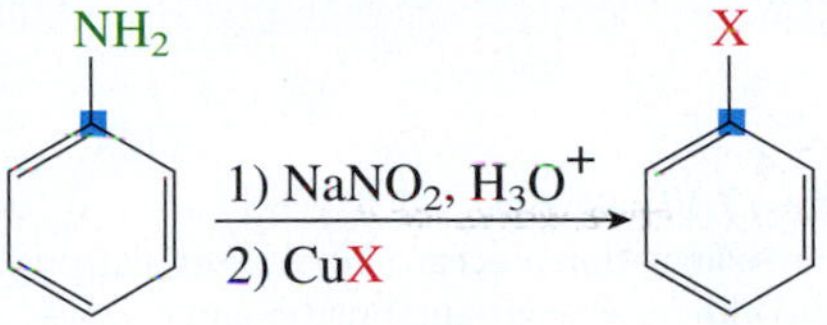	**Section 17.10** Replaces NH_2 with Cl or Br
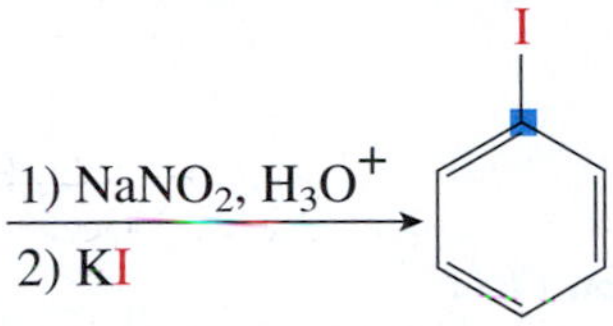	Replaces NH_2 with I
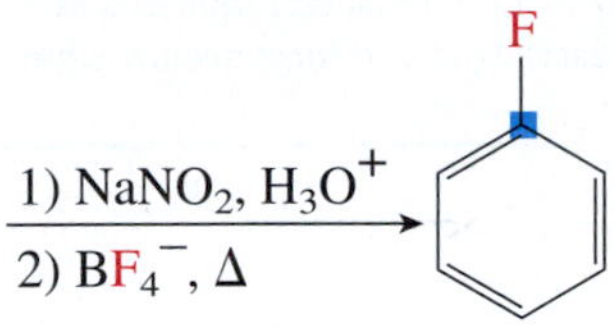	Replaces NH_2 with F
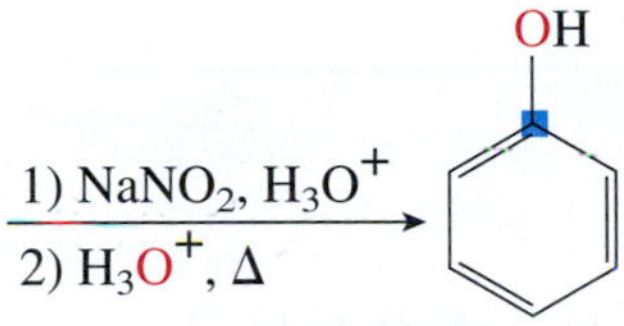	Replaces NH_2 with OH
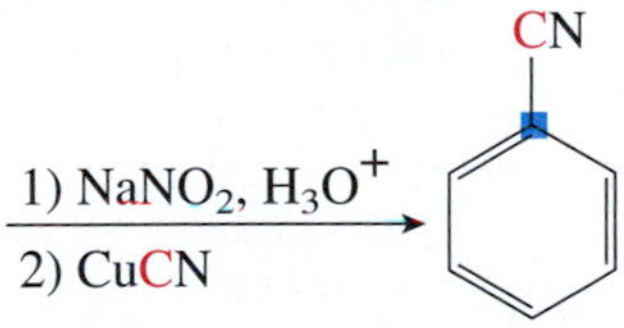	Replaces NH_2 with CN
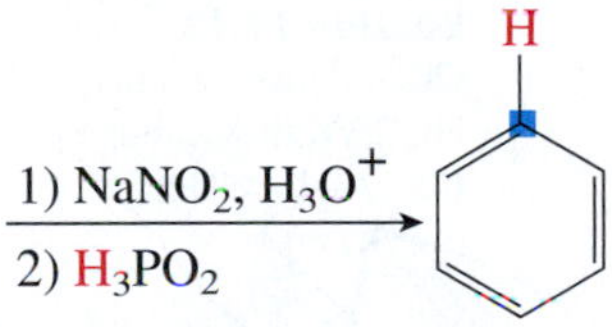	Replaces NH_2 with H; can also be accomplished with $NaBH_4$

Table 17.4   **Nucleophilic Substitution Reactions Proceeding through the Addition–Elimination and Elimination–Addition Mechanisms**

Reaction	Comments
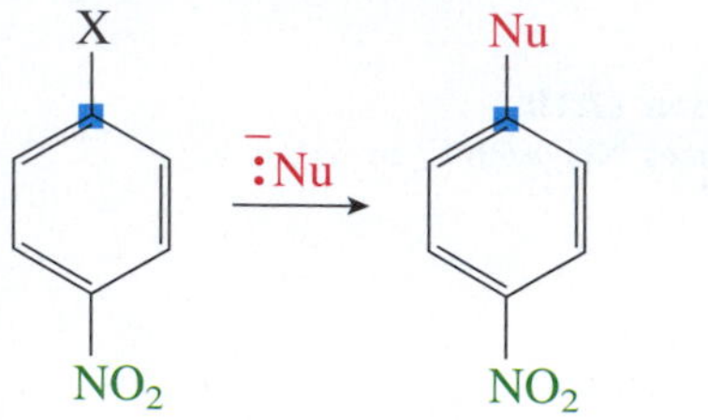	**Section 17.11** Addition–elimination mechanism; requires a strong electron-withdrawing group(s) ortho and/or para to the halogen
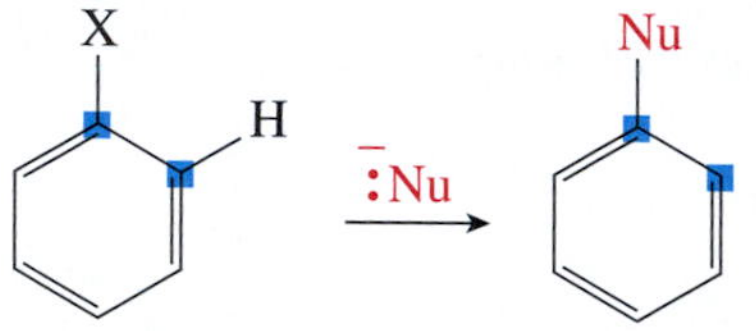	**Section 17.12** Elimination–addition mechanism; proceeds through a benzyne intermediate; requires a very strong base; rearranged products are possible

Table 17.5   **Other Useful Reactions for the Synthesis of Aromatic Compounds**

Reaction	Comments
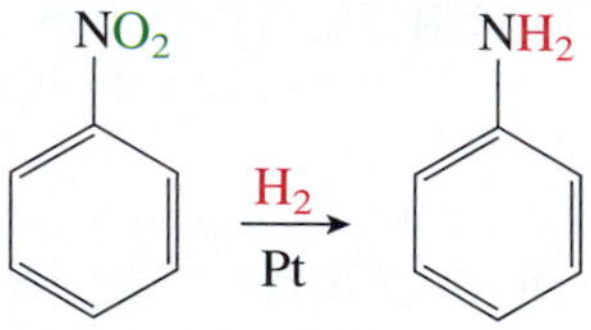	**Section 17.13** Can also be accomplished by using acid and Fe, Sn, or $SnCl_2$
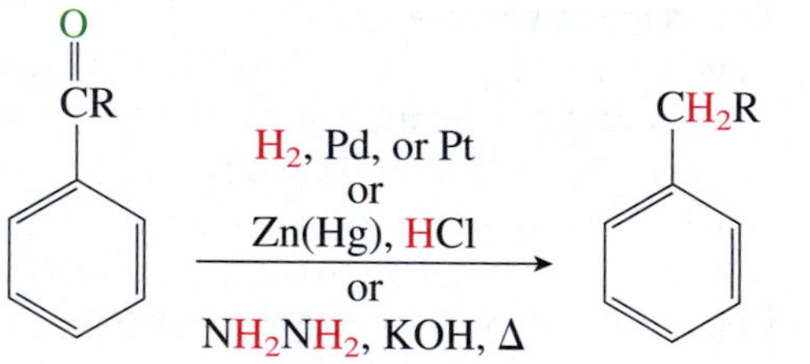	**Section 17.13** Clemmensen reduction [Zn(Hg), HCl]; Wolff-Kishner reduction (NH_2NH_2, KOH)
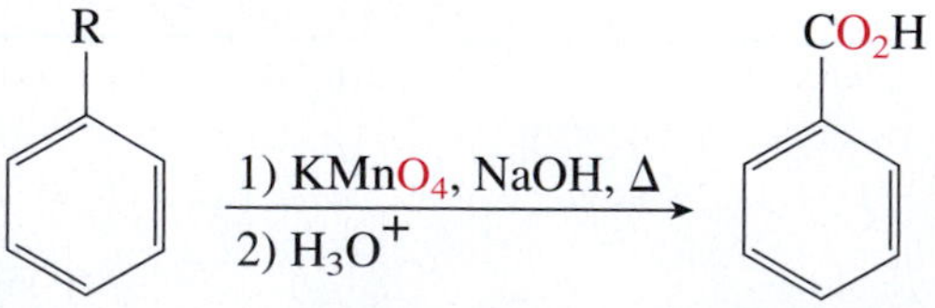	**Section 17.13** OK as long as R is not tertiary

Integrated Practice Problem

Show the products of these reactions:

a)

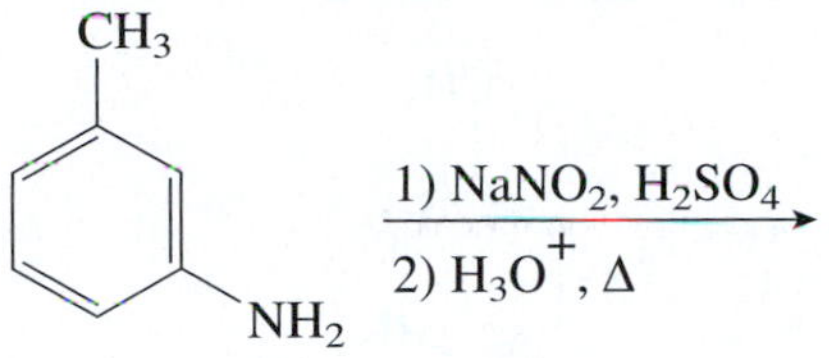

$$\xrightarrow[\text{2) H}_3\text{O}^+, \Delta]{\text{1) NaNO}_2, \text{H}_2\text{SO}_4}$$

b)

$+ \ CH_3CH_2\overset{\overset{\displaystyle O}{\|}}{C}Cl \quad \xrightarrow{\text{AlCl}_3}$

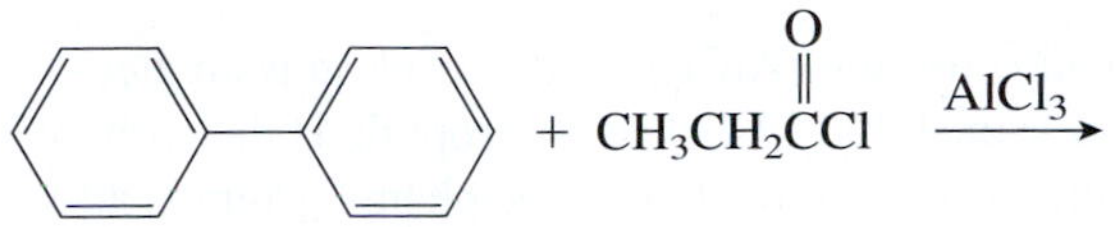

c)

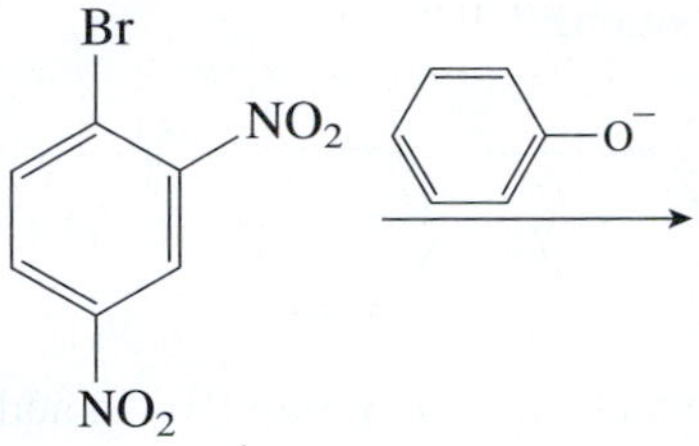

Strategy

As usual, the key is to identify the electrophile and the nucleophile. This is a little more difficult in this chapter because the aromatic compound can be either the nucleophile or the electrophile. Therefore, you need to look at the other reactants.

If the other reactant is an electrophile and a strong Lewis acid or proton acid is present, then the aromatic ring acts as the nucleophile and the reaction is one of the electrophilic aromatic substitution reactions listed in Table 17.2. Do not forget to consider the directive and rate effects of substituents on the aromatic ring.

If there is a diazonium ion leaving group on the ring ($ArNH_2 + NaNO_2 + acid \rightarrow ArN_2^+$), then the reaction is one of the nucleophilic aromatic substitution reactions listed in Table 17.3. The nucleophile replaces the diazonium ion leaving group.

If a strongly basic nucleophile is present and the ring has a halogen leaving group, then the reaction is one of the nucleophilic aromatic substitution reactions listed in Table 17.4. If there is a strong electron-withdrawing group ortho and/or para to the leaving group, the nucleophile replaces the leaving group by an addition–elimination mechanism. If there is no strong electron-withdrawing group on the ring, then the reaction follows the elimination–addition mechanism via a benzyne intermediate. Remember that the nucleophile can bond to either carbon of the benzyne.

Finally, the reaction can be a reduction (H_2 and a catalyst, or a metal and acid) or an oxidation ($KMnO_4$), as listed in Table 17.5.

Solutions

a) The presence of the NH_2 group along with sodium nitrite and acid indicates that the reaction is a nucleophilic substitution proceeding through a diazonium ion (see Table 17.3). The nucleophile is water, so the product is a phenol.

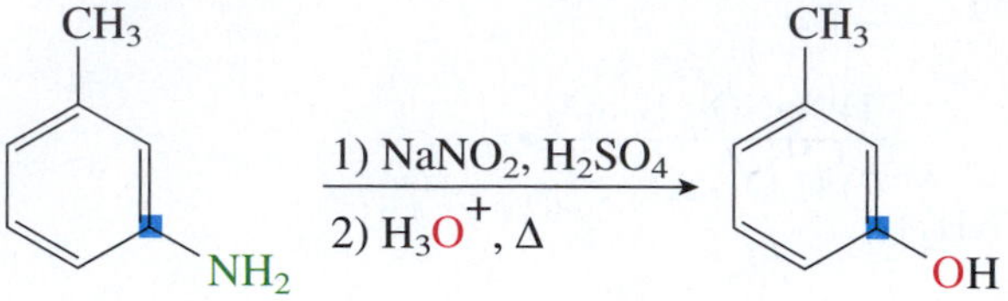

b) In this case a strong Lewis acid is present ($AlCl_3$), so the reaction is an electrophilic aromatic substitution from Table 17.2. The electrophile is the carbonyl carbon and the reaction is a Friedel-Crafts acylation. The phenyl group is an ortho/para director and the reaction is very sensitive to steric effects, so the major product has the acyl group added to the para position.

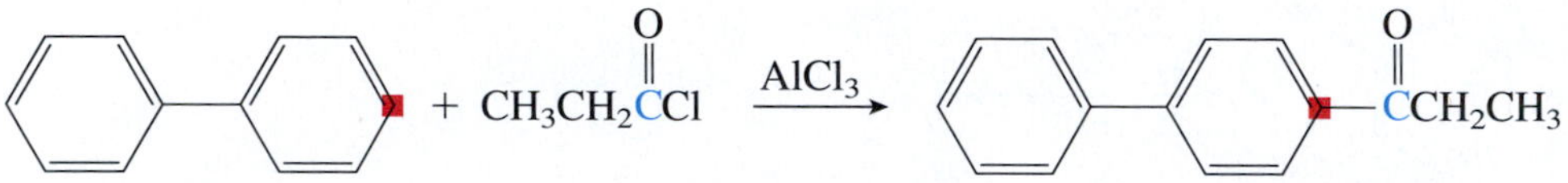

c) In this case there is a strongly basic nucleophile and a leaving group (Br) in addition to strong electron-withdrawing groups (NO_2) on the aromatic ring, so the reaction is an addition–elimination nucleophilic substitution from Table 17.4.

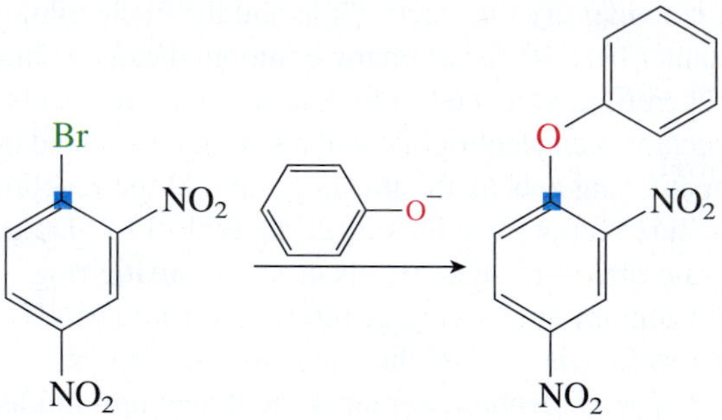

Additional Problems

17.31 Show the products of these reactions:

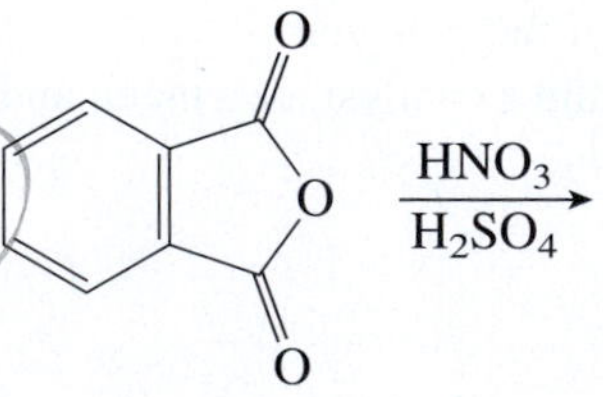

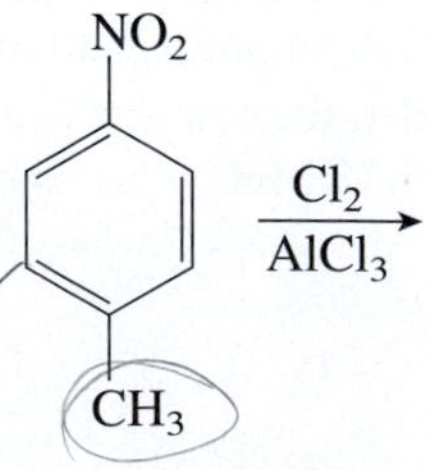

c) [aniline with NH$_2$ (handwritten CN), CH$_3$ meta] $\xrightarrow[\text{2) CuCN}]{\text{1) NaNO}_2\text{, H}_3\text{O}^+}$

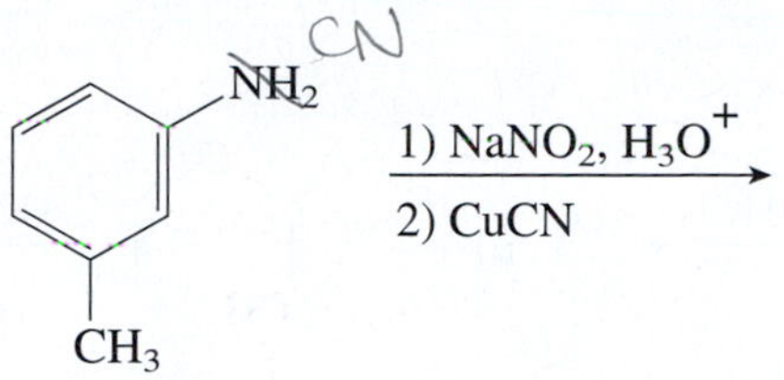

d) O$_2$N— [benzene ring with NO$_2$] —Cl + [pyrrolidine N–H] $\longrightarrow$

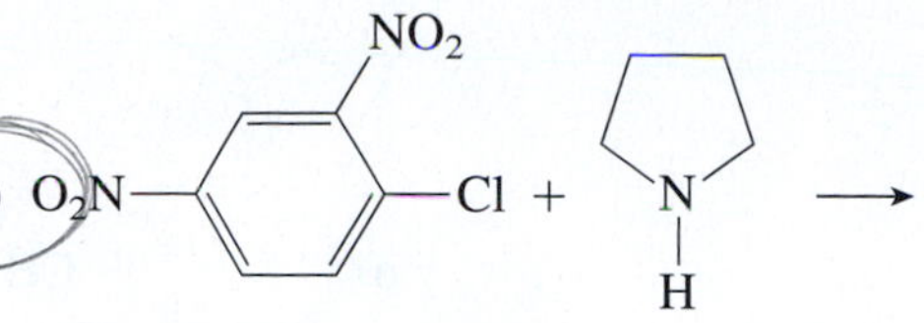

e) [benzene with CH$_2$CO$_2$H (handwritten) and CH$_3$ / CO$_2$H] $\xrightarrow[\text{2) H}_3\text{O}^+]{\text{1) KMnO}_4\text{, NaOH, }\Delta}$

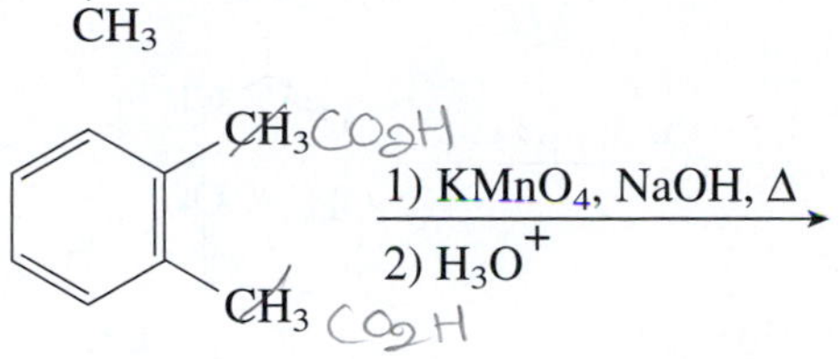

f) [trimethylbenzene ring, CH$_3$ groups, handwritten NO$_2$] $\xrightarrow[\text{H}_2\text{SO}_4]{\text{HNO}_3}$

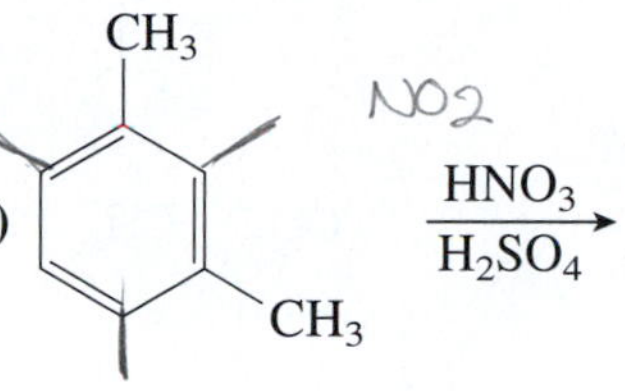

g) [benzene with OCH$_3$, CH$_3$] + PhCH$_2$Cl $\xrightarrow{\text{AlCl}_3}$

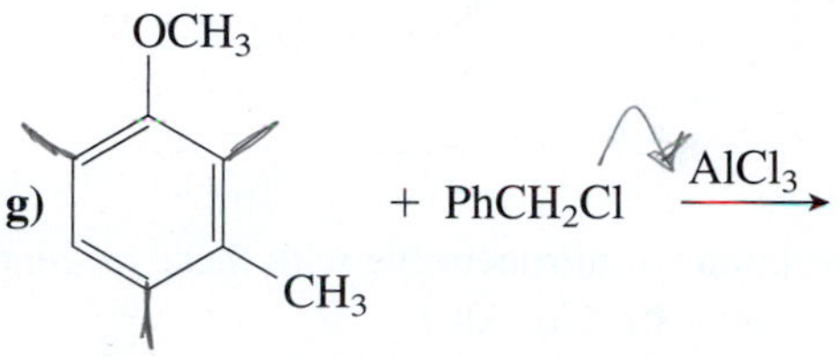

h) [benzene] + PhCH$_2$CH$_2$Cl $\xrightarrow{\text{AlCl}_3}$

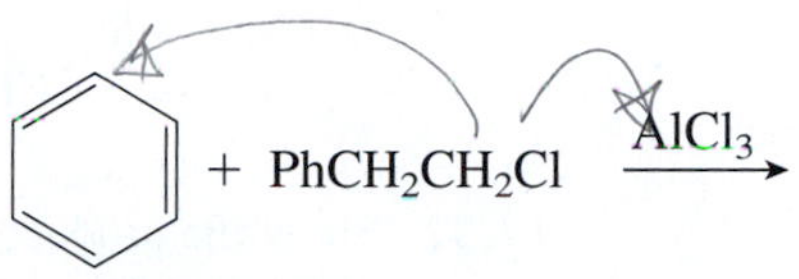

i) [benzene with CHO and NH$_2$ (handwritten H), meta] $\xrightarrow[\text{2) H}_3\text{O}^+\text{, }\Delta]{\text{1) NaNO}_2\text{, H}_3\text{O}^+}$

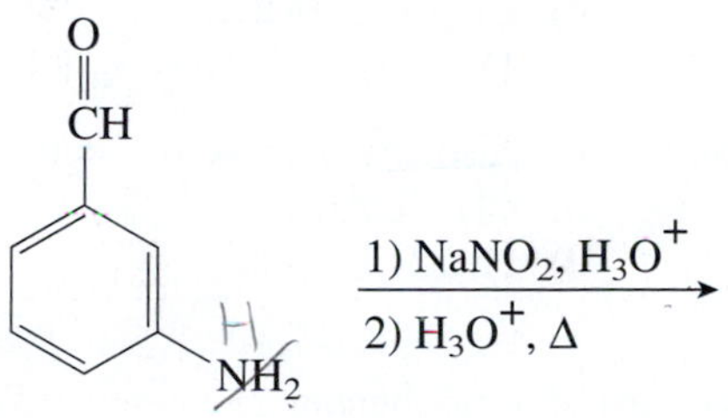

j) [benzene with NHCCH$_3$ (amide, O), handwritten Br, CH$_3$] $\xrightarrow{\text{Br}_2}$

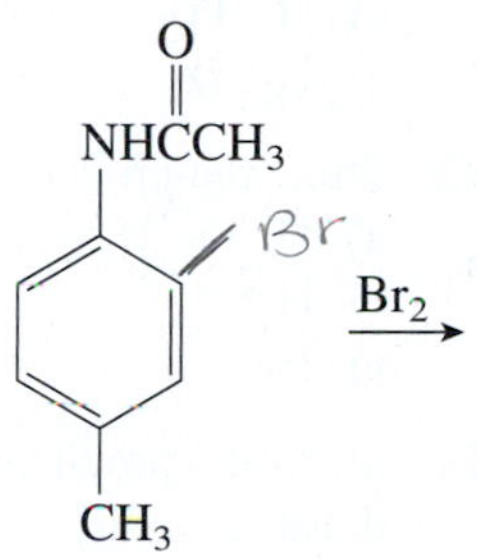

k) [benzene with OCH$_3$ para to Cl (handwritten NH$_2$)] $\xrightarrow[\text{NH}_3\ (l)]{\text{NaNH}_2}$

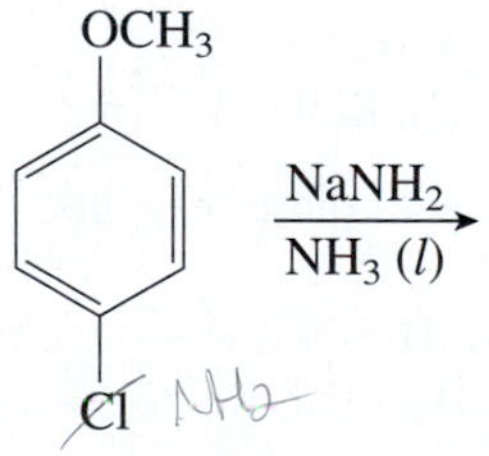

l) [biphenyl]—NO$_2$ $\xrightarrow[\text{HCl}]{\text{Fe}}$

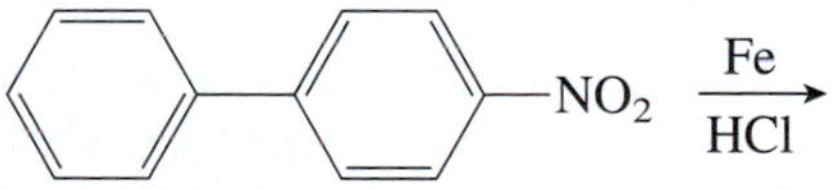

m) [toluene, CH$_3$] + CH$_3$CH=CH$_2$ $\xrightarrow[\text{AlCl}_3]{\text{HCl}}$

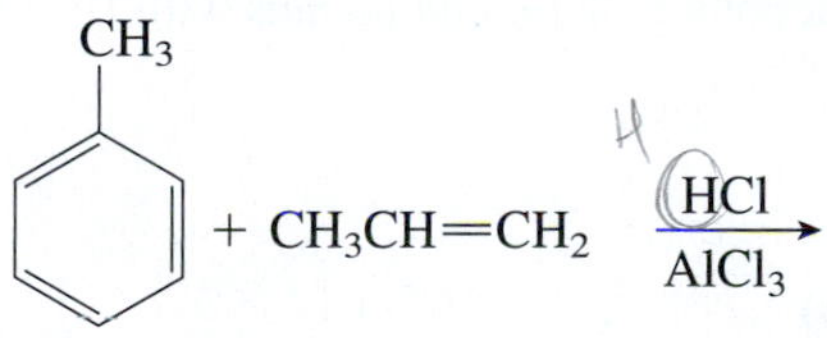

n) [benzene with Br, handwritten SO$_2$H, CH$_3$] $\xrightarrow{\text{H}_2\text{SO}_4}$

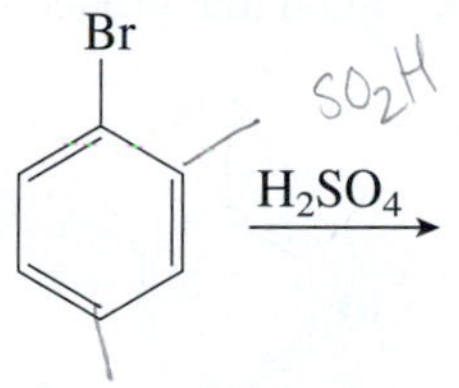

o) 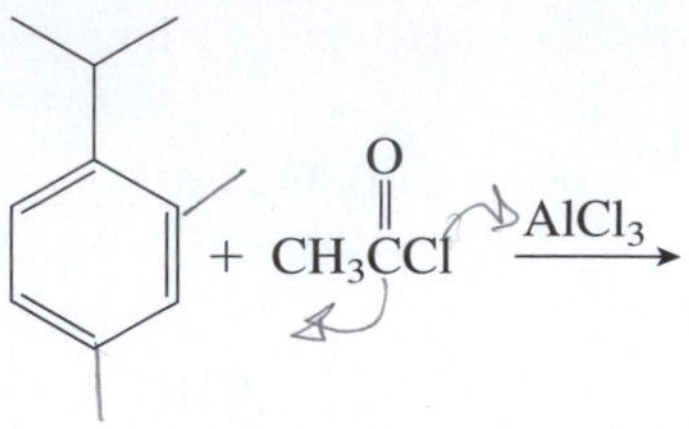 [structure of 1-isopropyl-2,4-dimethylbenzene] + CH₃CCl →(AlCl₃)

p) 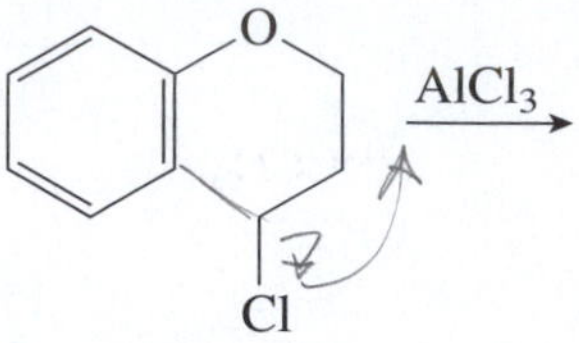 [chroman structure with Cl] →(AlCl₃)

q) 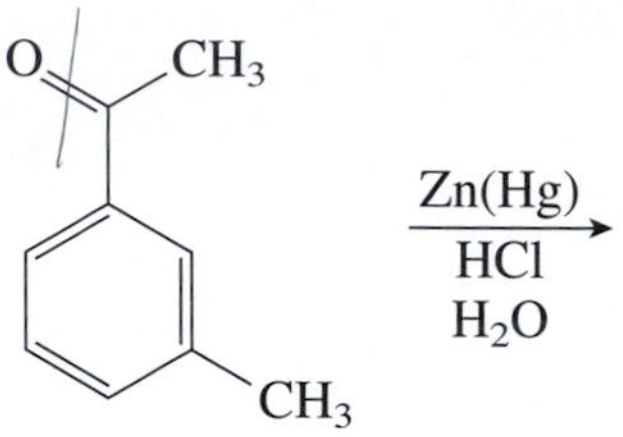 [3-methylacetophenone structure] →(Zn(Hg) / HCl / H₂O)

r) 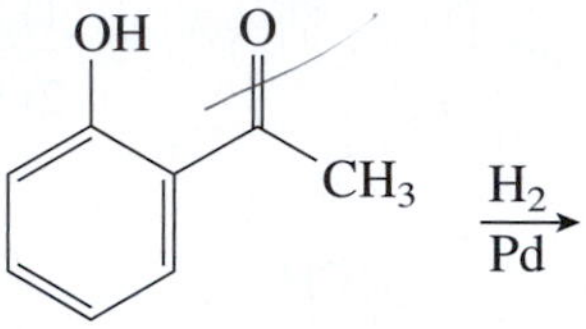 [2-hydroxyacetophenone structure] →(H₂ / Pd)

17.32 Show the products of the reactions of nitrobenzene with these reagents:
a) HNO₃, H₂SO₄   *meta NO₂*   b) Cl₂, AlCl₃   *Cl*
c) H₂, Pt   *H*   d) CH₃COCl, AlCl₃
e) H₂SO₄   *meta SO₃H*   f) 1) Zn, HCl; 2) NaNO₂, HBr; 3) CuBr

17.33 Show the products of the reactions of anisole with these reagents:
a) HNO₃, H₂SO₄   *NO₂*   b) CH₃Cl, AlCl₃   *CH₃*
c) H₂SO₄   *SO₃H*   d) PhCOCl, AlCl₃
e) Br₂   *Br*

17.34 Show the products of the reactions of acetophenone (1-phenylethanone) with these reagents:
a) H₂SO₄   *SO₃H*   b) Br₂, AlBr₃   *Br*
c) HNO₃, H₂SO₄   *NO₂*   d) Zn(Hg), HCl
e) H₂, Pd   f) NH₂NH₂, KOH, Δ

17.35 Show the products of the reaction of toluene with these reagents:
a) KMnO₄, NaOH, Δ   *CO₂H*   b) H₂SO₄   *SO₃H*
c) CH₃CH₂Cl, AlCl₃   *CH₃CH₂*   d) HNO₃, H₂SO₄   *NO₂*
e) Cl₂, AlCl₃   *Cl*   f) CH₃CH₂COCl, AlCl₃

17.36 Show the products of the reactions of these compounds with Br₂ and FeBr₃:

a) 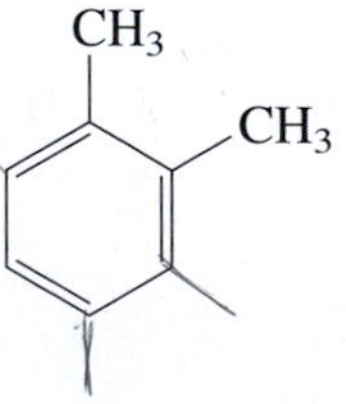 [1,2-dimethyl-4-methylbenzene structure]

b) 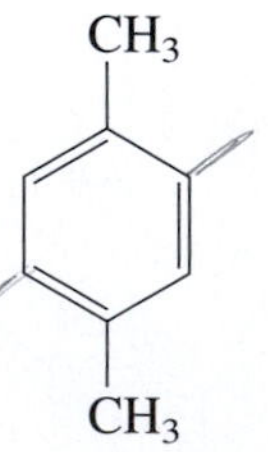 [1,2,4,5-tetramethylbenzene structure with CH₃ groups]

c) 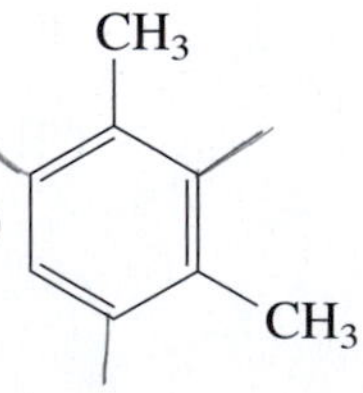 [1,2,3-trimethylbenzene structure with CH₃ groups]

d) 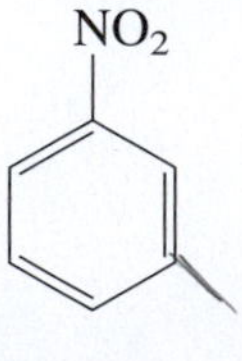 [3-methylnitrobenzene, NO₂]

e) 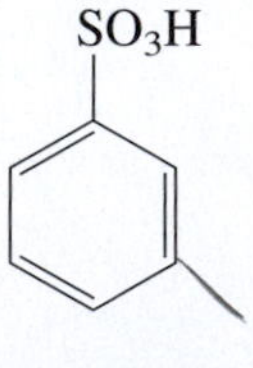 [3-methylbenzenesulfonic acid, SO₃H]

f) 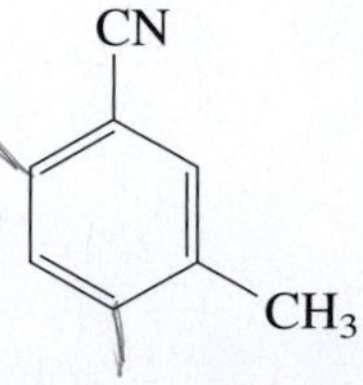 [substituted benzonitrile, CN and CH₃]

17.37 Show the products of the reactions of these compounds with propanoyl chloride and aluminum trichloride:

a) 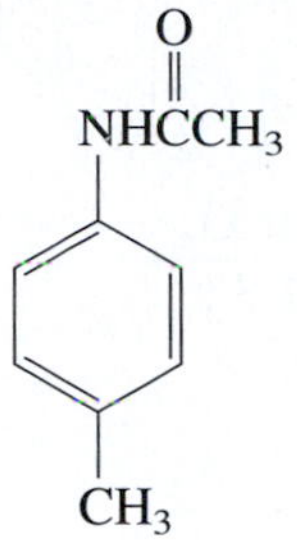
NHCCH₃ (with O double bond), CH₃

b) 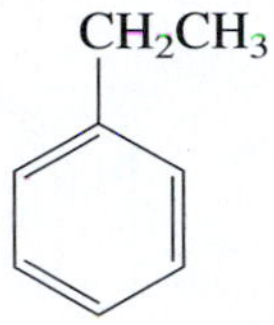
CH₂CH₃

c)

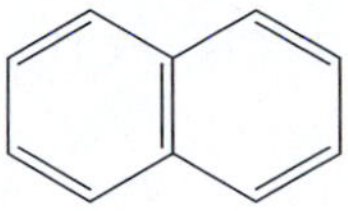

d) 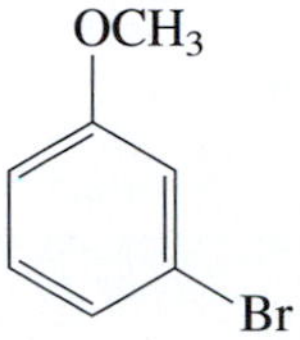
OCH₃, Br

e)

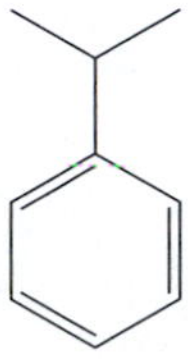

f)

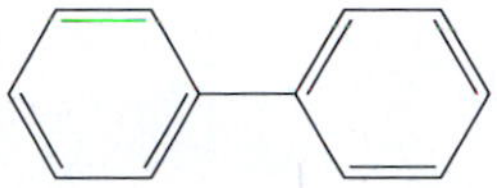

17.38 Show the reagents that could be used to accomplish these transformations. More than one step may be necessary in some cases.

a) 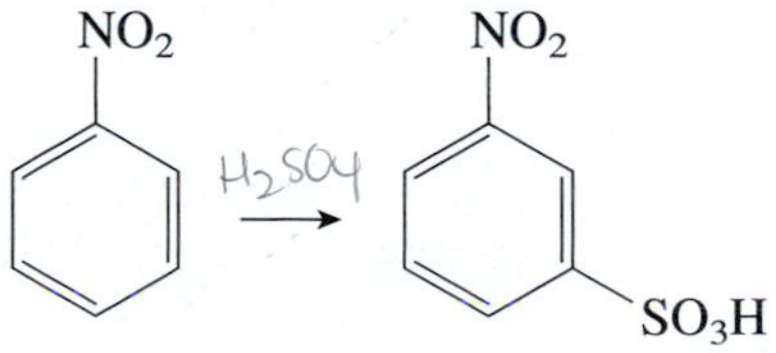 NO₂ →(H₂SO₄)→ NO₂, SO₃H

b) 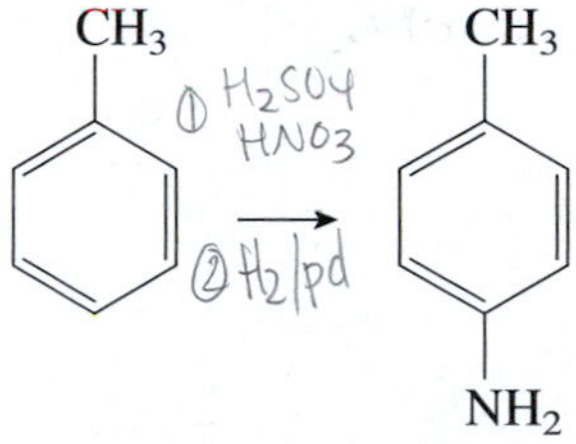 CH₃ →(① H₂SO₄ HNO₃ ② H₂/pd)→ CH₃, NH₂

c) 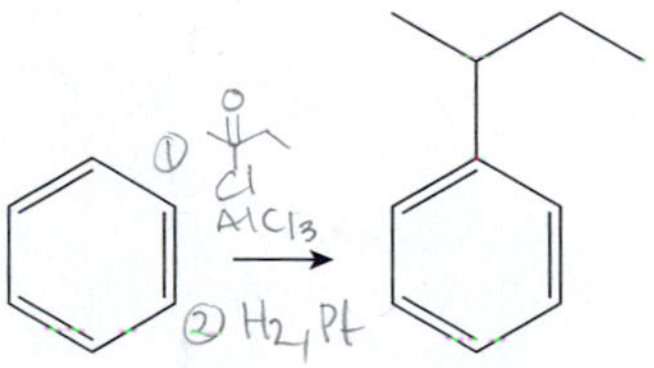 →(① (propanoyl chloride) AlCl₃ ② H₂, Pt)→ (isobutylbenzene)

d) 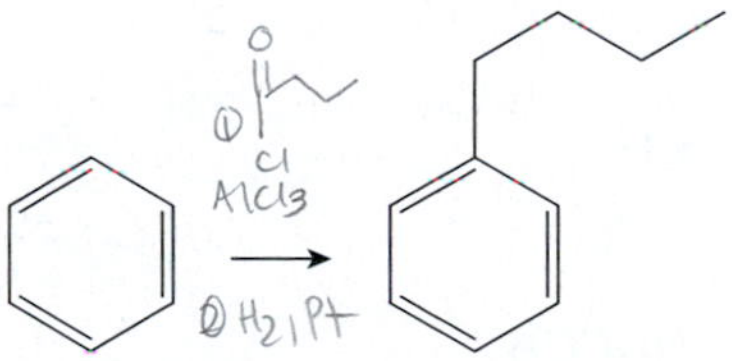 →(① (butanoyl chloride) AlCl₃ ② H₂, Pt)→ (butylbenzene)

e) 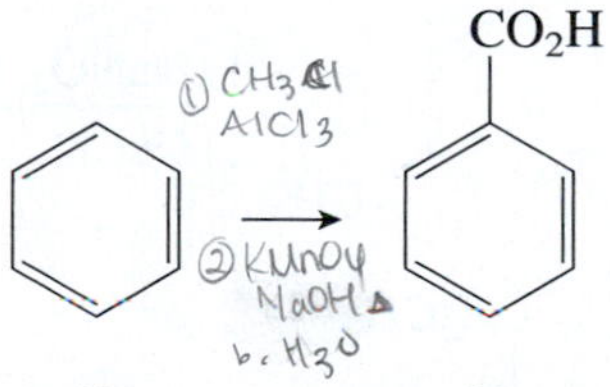 →(① CH₃Cl AlCl₃ ② KMnO₄ NaOH Δ b. H₃O)→ CO₂H

f) 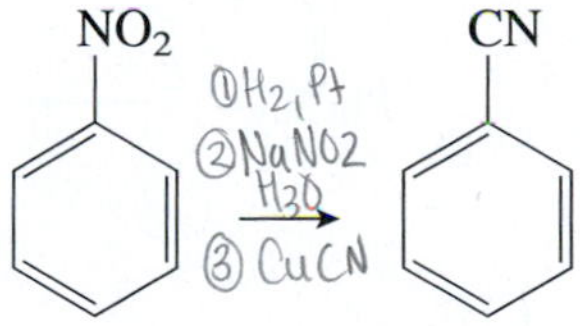 NO₂ →(① H₂, Pt ② NaNO₂ H₂O ③ CuCN)→ CN

g) 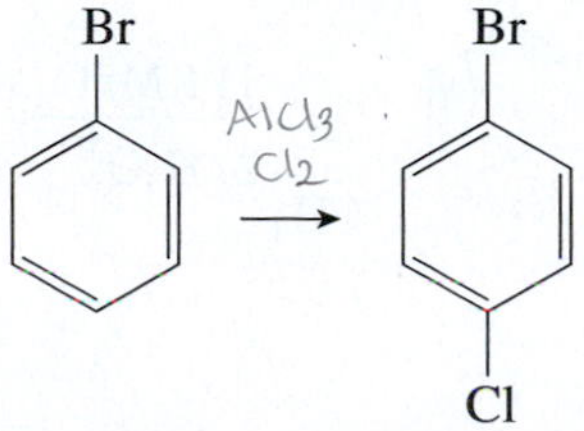 Br →(AlCl₃ Cl₂)→ Br, Cl

h) 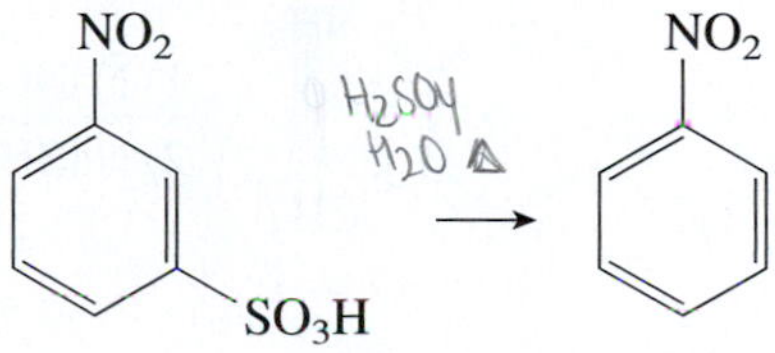 NO₂, SO₃H →(H₂SO₄ H₂O Δ)→ NO₂

17.39 Show the products of these reactions:

a)

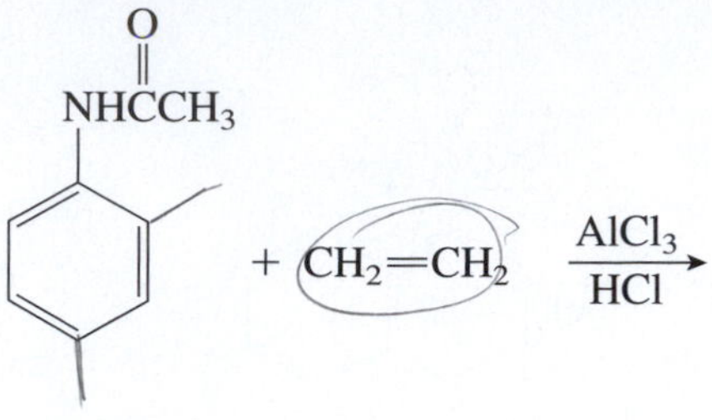

$NHCCH_3$ (with O above)

+ $CH_2{=}CH_2$ $\xrightarrow[\text{HCl}]{\text{AlCl}_3}$

(ring with CH_3 and CH_3 substituents)

b)

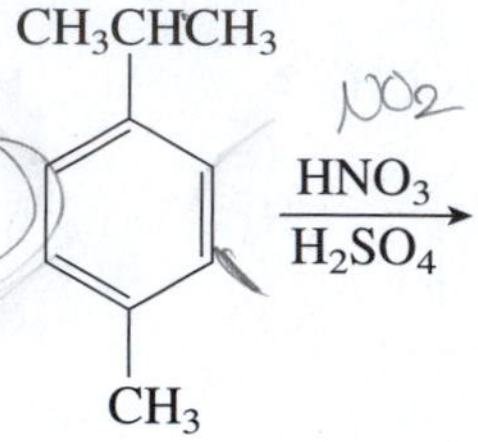

CH_3CHCH_3

NO_2

CH_3

$\xrightarrow[\text{H}_2\text{SO}_4]{\text{HNO}_3}$

c)

CH_3

OH

$\xrightarrow{\text{H}_2\text{SO}_4}$ SO3H

d)

(naphthalene) Br

$\xrightarrow{\text{Br}_2}$

e)

(benzene) + (methylcyclohexene) $\xrightarrow{\text{H}_2\text{SO}_4}$

f)

CO_2H

$\xrightarrow[\text{FeBr}_3]{\text{Br}_2}$ Br

g)

CN

CH_3

$\xrightarrow[\text{AlCl}_3]{\text{Cl}_2}$ Cl

h)

$OCCH_3$ (with O above)

+ (cyclopentanol) OH $\xrightarrow{\text{H}_2\text{SO}_4}$

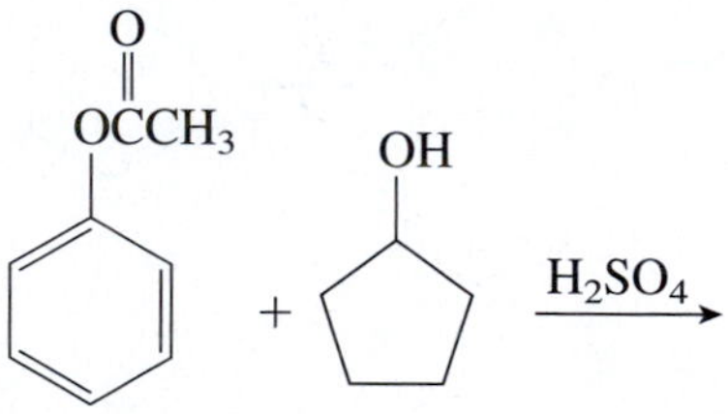

i)

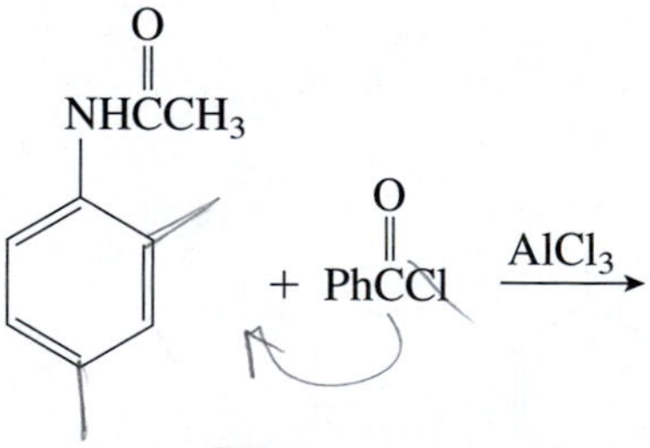

$NHCCH_3$ (with O above)

+ $PhCCl$ (with O above) $\xrightarrow{\text{AlCl}_3}$

j)

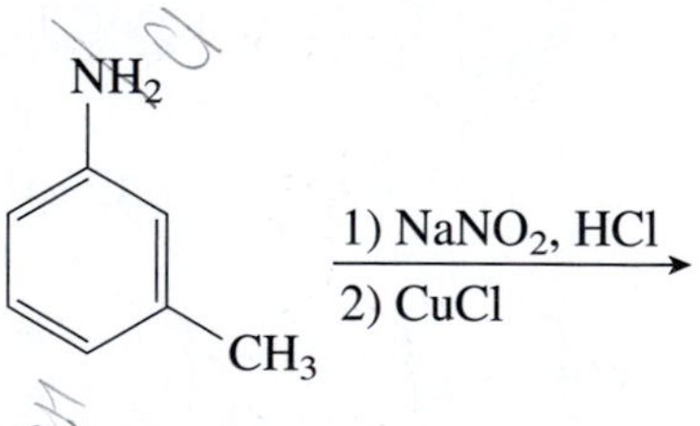

NH_2 Cl

CH_3

$\xrightarrow[\text{2) CuCl}]{\text{1) NaNO}_2\text{, HCl}}$

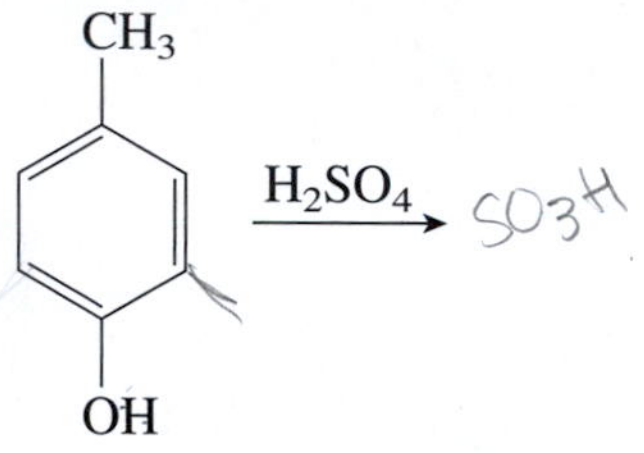

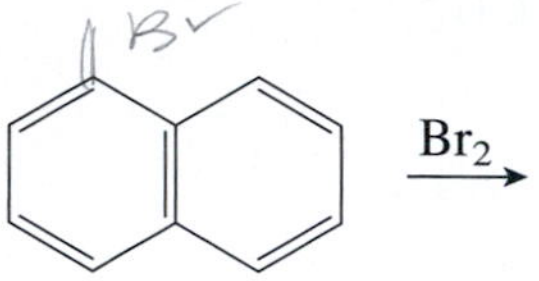

k)

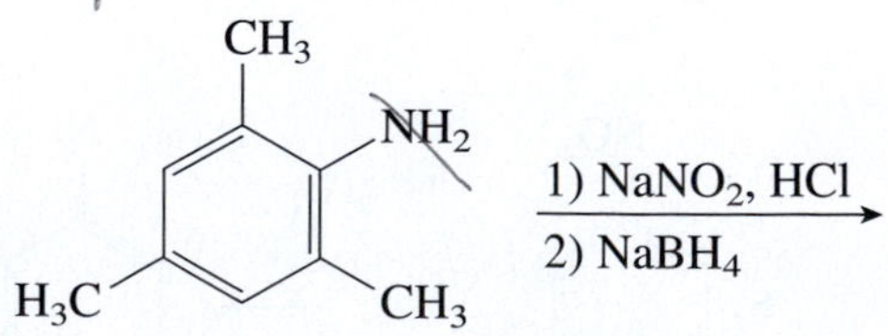

CH_3

NH_2

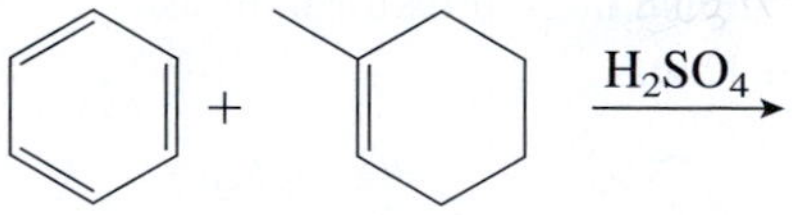

H_3C CH_3

$\xrightarrow[\text{2) NaBH}_4]{\text{1) NaNO}_2\text{, HCl}}$

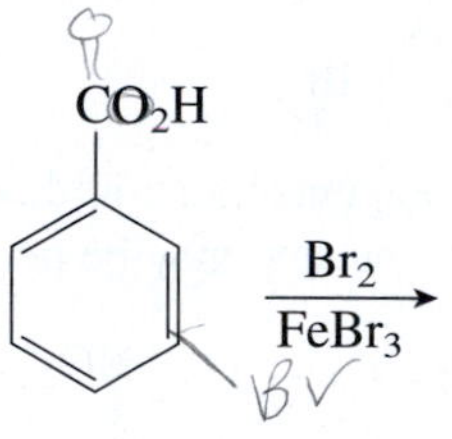

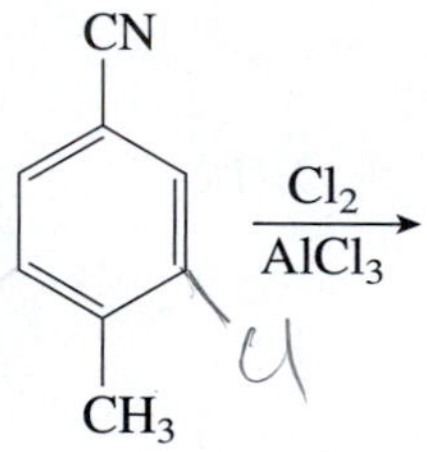

l)

CH_2CH_3 CO2H

CH_3 CO2H

$\xrightarrow[\text{2) H}_3\text{O}^+]{\text{1) KMnO}_4\text{, NaOH, }\Delta}$

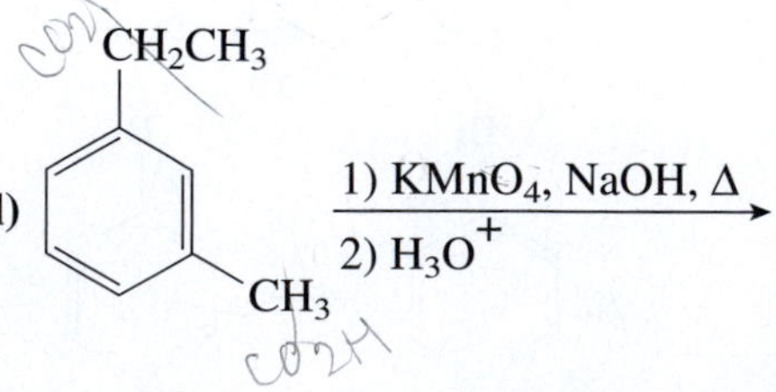

17.40 Show the products of these reactions:

a) 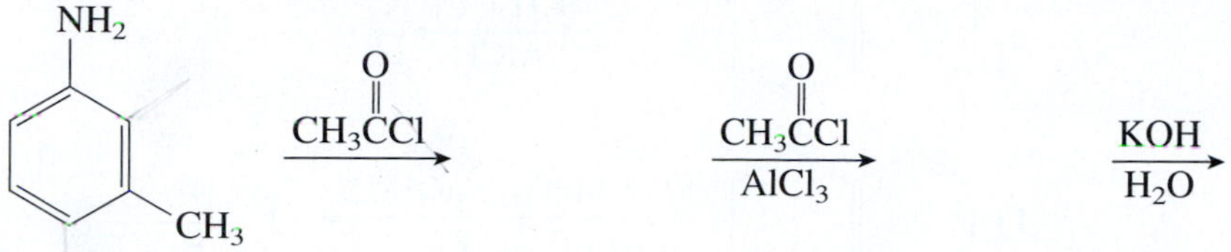
NH2
CH3
O
CH3CCl
O
CH3CCl
AlCl3
KOH
H2O

b) 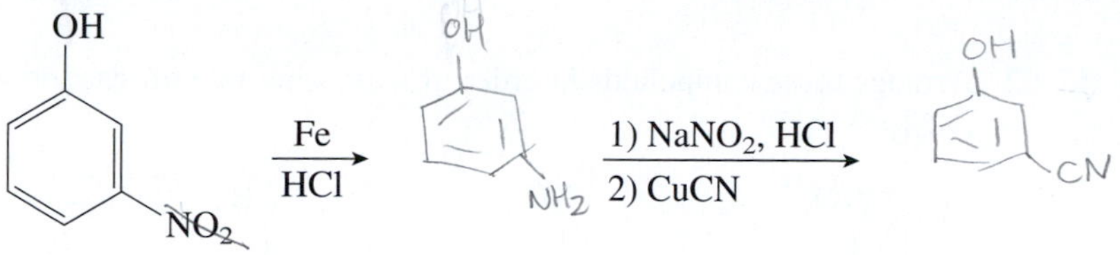
OH
NO2
Fe
HCl
1) NaNO2, HCl
2) CuCN

c) 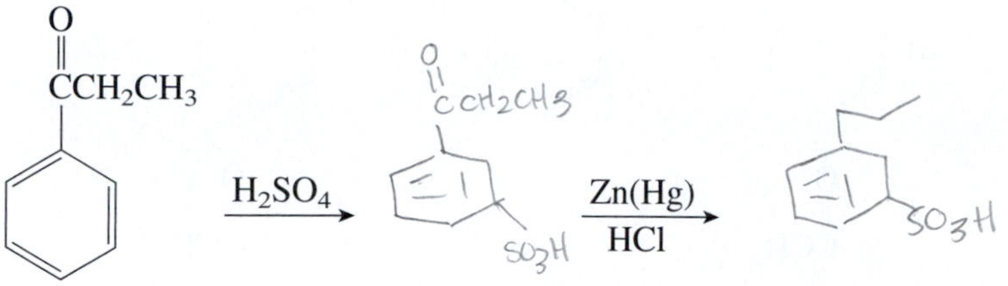
O
CCH2CH3
H2SO4
Zn(Hg)
HCl

17.41 Show the products of these reactions:

a) 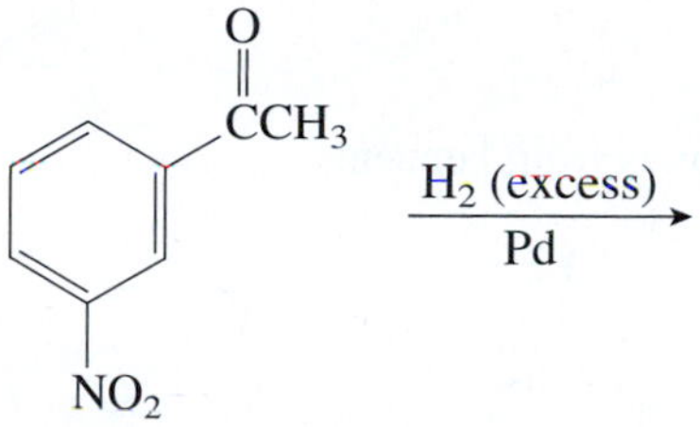
O
CCH3
NO2
H2 (excess)
Pd

b) 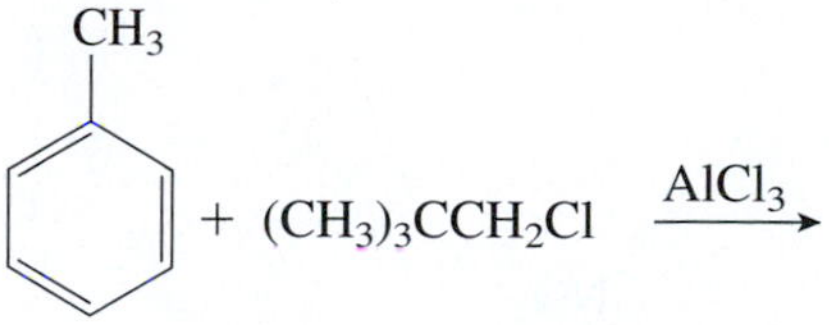
CH3
+ (CH3)3CCH2Cl
AlCl3

c) 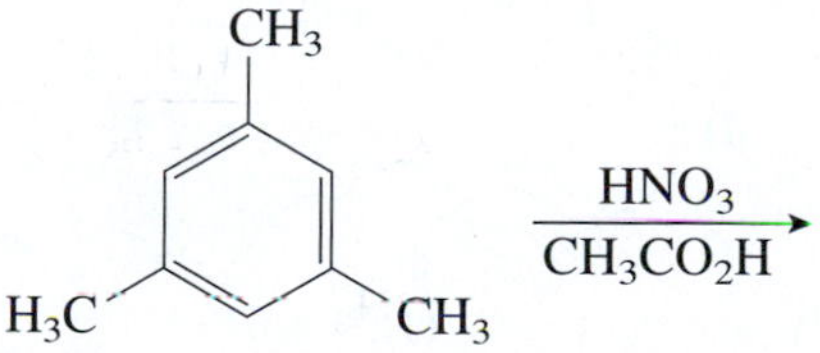
CH3
H3C
CH3
HNO3
CH3CO2H

d) 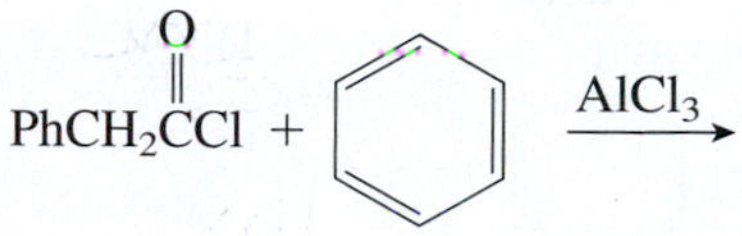
O
PhCH2CCl +
AlCl3

e) 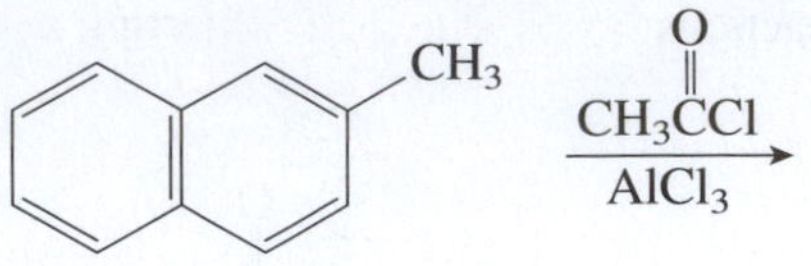 [structure: 2-methylnaphthalene] $\xrightarrow[\text{AlCl}_3]{\overset{\displaystyle O}{\text{CH}_3\text{CCl}}}$

f) 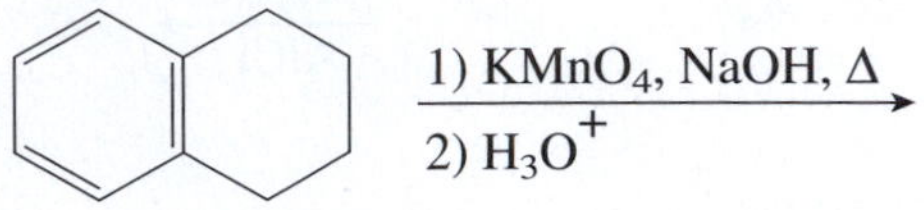 [structure: tetralin] $\xrightarrow[\text{2) H}_3\text{O}^+]{\text{1) KMnO}_4,\ \text{NaOH},\ \Delta}$

17.42 Arrange these compounds in order of increasing rate of reaction with Cl_2 and $AlCl_3$:

a)

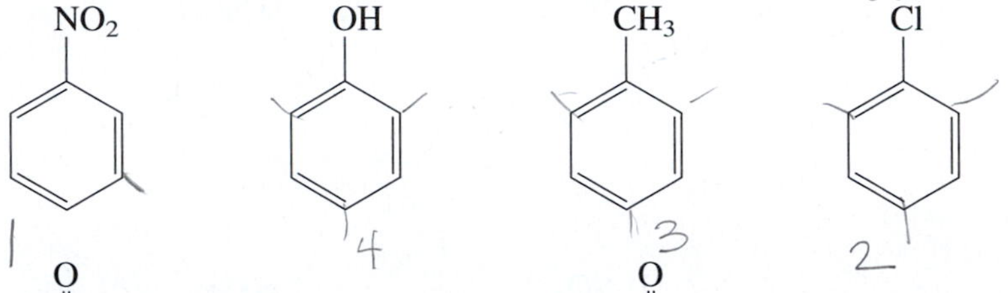

NO$_2$ — I

OH — 4

CH$_3$ — 3 (weak)

Cl — 2 (d.)

b)

$\overset{O}{\text{CCH}_3}$ — 4

CH$_3$ — 3

$\overset{O}{\text{OCCH}_3}$ — 1

OCH$_3$ — 2

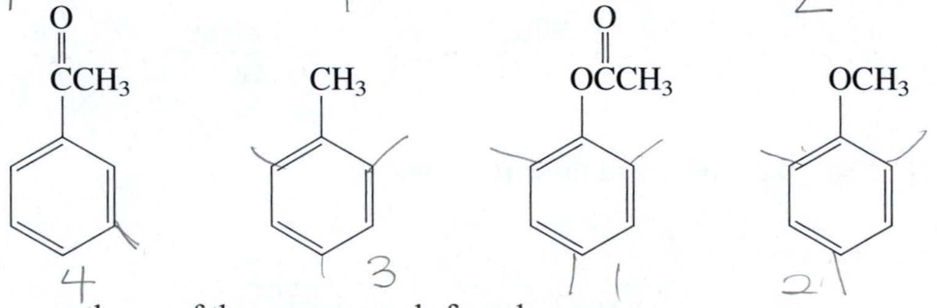

17.43 Show syntheses of these compounds from benzene:

a) 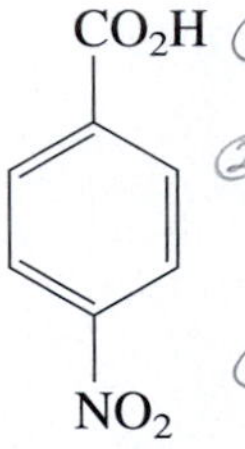 [structure with CO$_2$H top, NO$_2$ bottom]
① CH3Cl / AlCl3
② HNO3 / H2SO4 Keep para
③ a KMnO4 / NaOH, Δ   b H3O

b) [structure with Br and ethyl group]
① [acyl chloride] / AlCl3
② Br2 / AlBr3
③ H2 / Pt

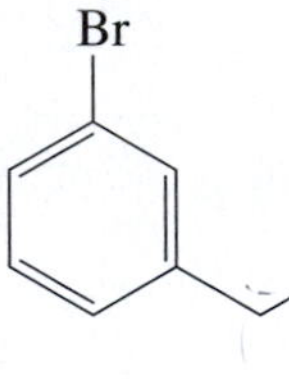

c) 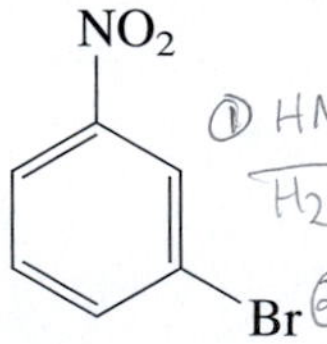 [structure with NO$_2$ top, Br bottom] CH$_3$CHCH$_3$ ...
① HNO3 / H2SO4
② Br2 / AlBr3

d) 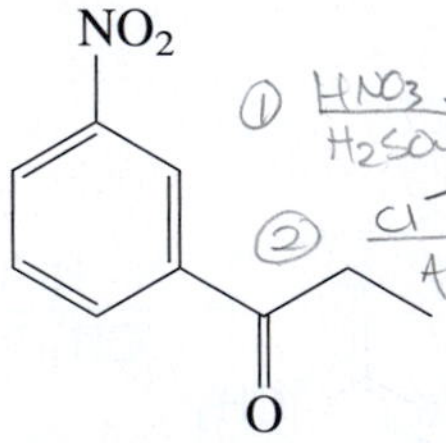 [structure with NO$_2$ top, $\overset{O}{}$ bottom]
① HNO3 / H2SO4
② Cl / AlCl3

e) 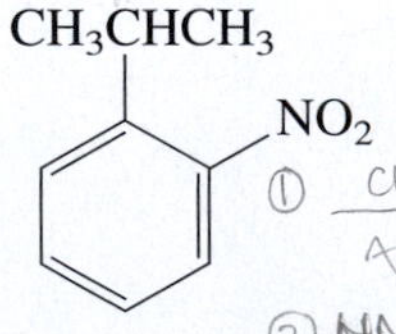 [structure with CH$_3$CHCH$_3$ and NO$_2$]
① CH3CClCH3 / AlCl3
② HNO3 / H2SO4 keep ortho
③ H2 / Pt

f) 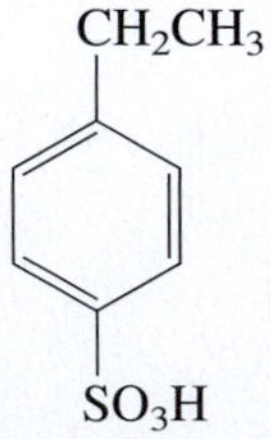 [structure with CH$_2$CH$_3$ top, SO$_3$H bottom]
① ClCH2CH3 / AlCl3
② H2SO4

17.44 Show syntheses of these compounds from the indicated starting materials:

a)

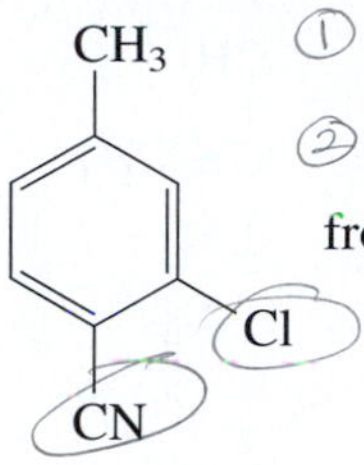

CH$_3$

Cl

CN

from toluene

① $\dfrac{HNO_2}{H_2SO_4}$ para

② $\dfrac{H_2}{Pt}$ → CuCN ⑤

③ $\dfrac{Cl_2}{AlCl_3}$

④ $\dfrac{NaNO_2}{H_3O}$

b)

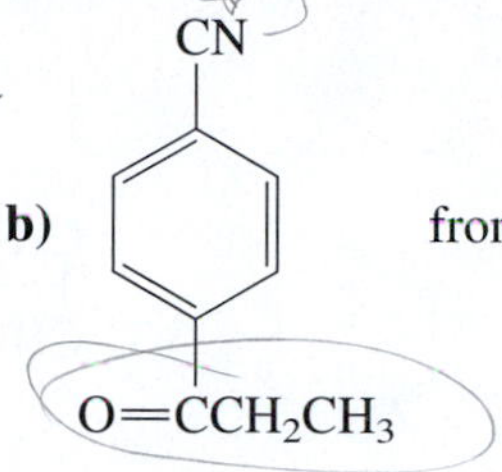

CN

O=CCH$_2$CH$_3$

from nitrobenzene

① $\dfrac{H_2}{Pt}$

② $\dfrac{Cl\overset{O}{C}CH_2CH_3}{AlCl_3}$

③ $\dfrac{NaNO_2}{H_3O}$

④ $\dfrac{}{CuCN}$

c)

OH

Cl

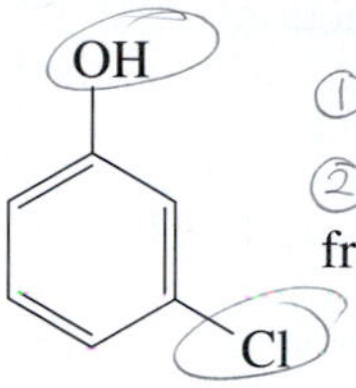

from benzene

① $\dfrac{HNO_2}{H_2SO_4}$

② $\dfrac{Cl_2}{AlCl_3}$

③ $\dfrac{a. H_2, Pt}{b. NaNO_2, H_3O}$

④ $\dfrac{}{H_3O}$

d)

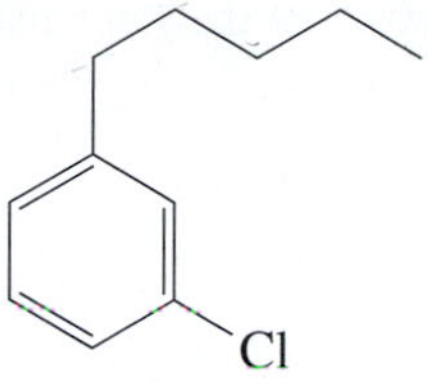

Cl

from pentylbenzene

e)

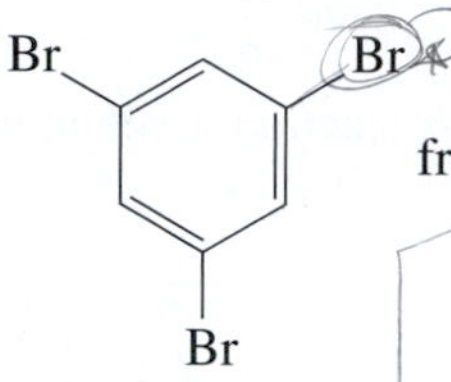

Br

Br

Br

from aniline

① $\dfrac{Br_2}{AlBr_3}$ excess

≡ NH$_2$

f)

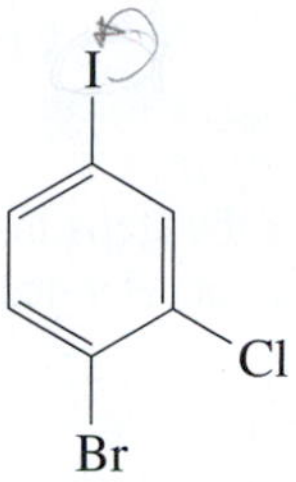

I

Cl

Br

from nitrobenzene

① $\dfrac{Cl_2}{AlCl_3}$

② $\dfrac{H_2, Pt}{}$

③ $\dfrac{Br_2}{AlBr_3}$

④ $\dfrac{a. NaNO_2}{H_3O}$

b. KI

g)

OH

CH$_3$

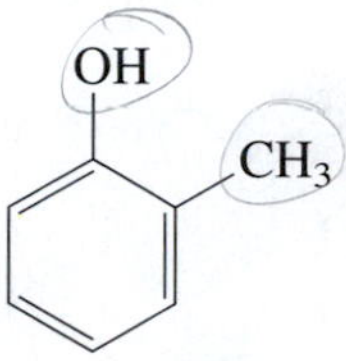

from benzene (no para product)

② $\dfrac{a. H_2, Pt}{b. NaNO_2, H_3O}$ c. CuBr

17.45 Show all of the steps in the mechanism for this reaction:

+ Br$_2$ $\xrightarrow{FeBr_3}$

Br

+ HBr

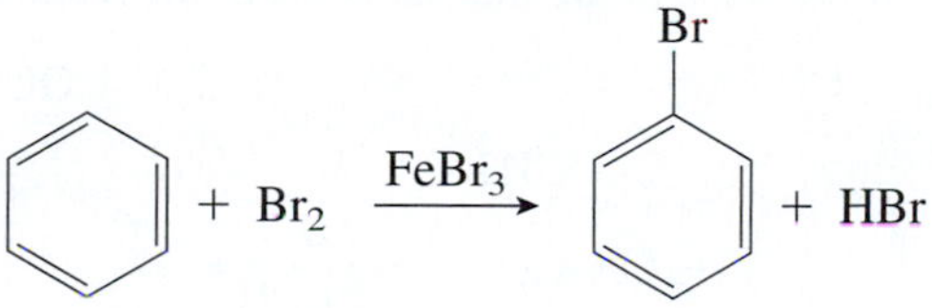

17.46 Show all of the steps in the mechanism for this reaction:

CH$_3$

+ HNO$_3$ $\xrightarrow{H_2SO_4}$

CH$_3$

NO$_2$

+ H$_2$O

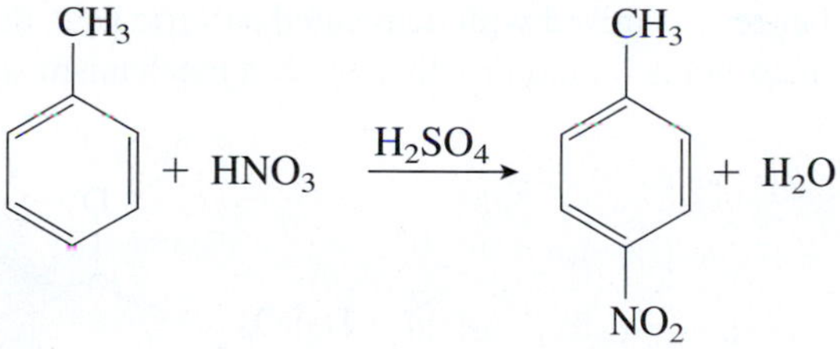

17.47 Show all of the steps in the mechanism for this reaction:

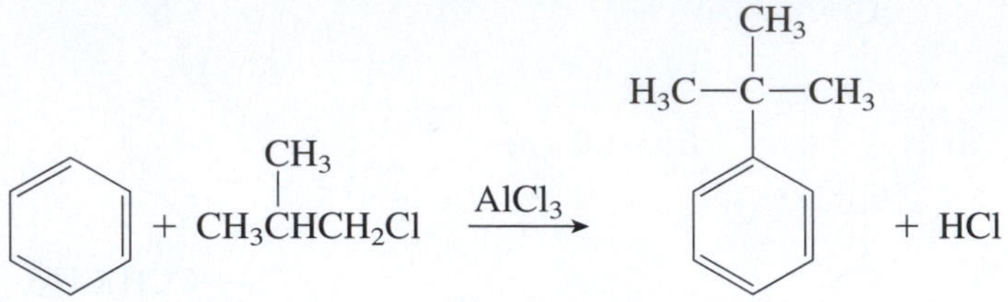

17.48 Show all of the steps in the mechanism for this reaction:

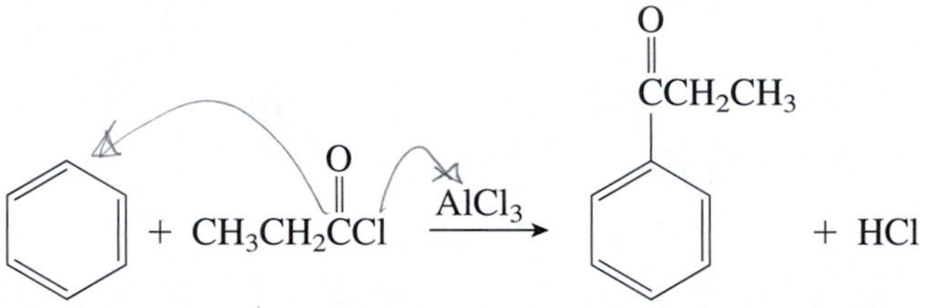

17.49 Show all of the steps in the mechanism for this reaction. Explain why the acetyl group accelerates the reaction.

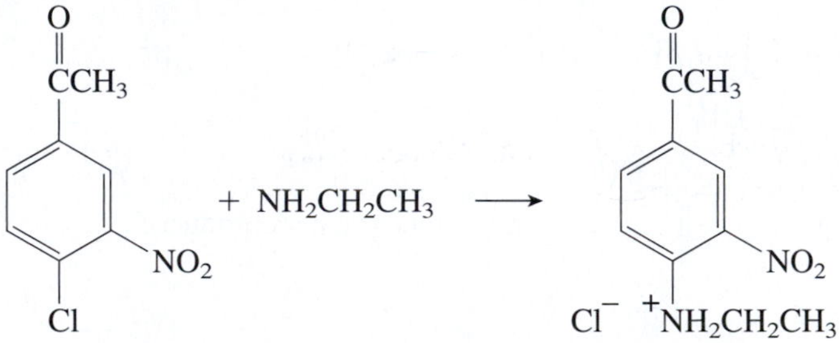

17.50 Show all of the steps in the mechanism for this reaction:

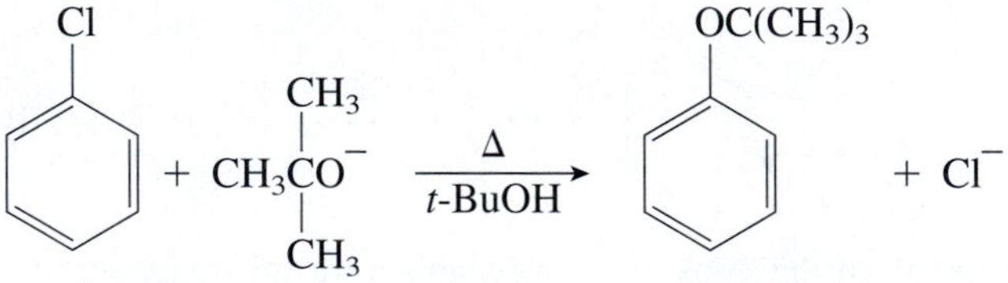

17.51 When benzene is mixed with deuterated sulfuric acid, deuterium is slowly incorporated onto the ring. Show a mechanism for this process:

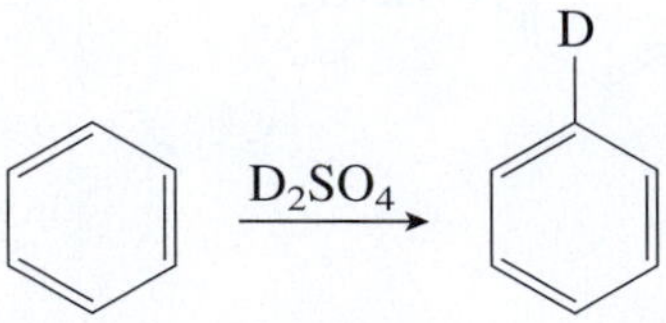

17.52 Explain the changes in the amounts of the products that occur in these bromination reactions:

		ortho product	para product

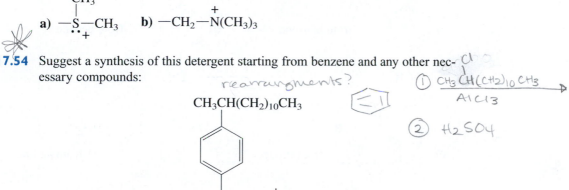

			ortho product	para product
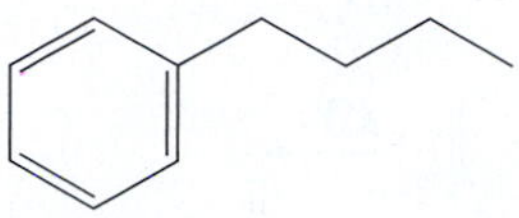 (phenyl-propyl)	$+ Br_2$ $\xrightarrow{FeBr_3}$		37%	63%
(isobutylbenzene)			25%	75%
(sec-butylbenzene)			21%	79%

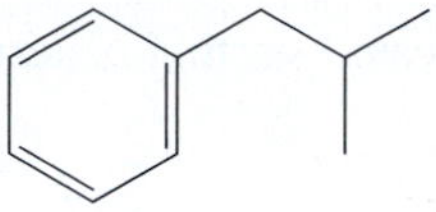

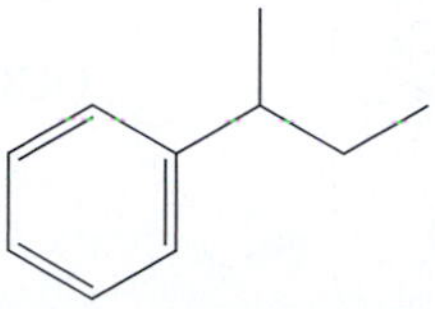

17.53 Predict the effect of these substituents on the rate and regiochemistry of an electrophilic aromatic substitution reaction:

a) $-\overset{\overset{\text{CH}_3}{|}}{\underset{\overset{..}{+}}{\text{S}}}-\text{CH}_3$ b) $-\text{CH}_2-\overset{+}{\text{N}}(\text{CH}_3)_3$

17.54 Suggest a synthesis of this detergent starting from benzene and any other necessary compounds:

$\text{CH}_3\text{CH}(\text{CH}_2)_{10}\text{CH}_3$

rearrangements? $\Longleftarrow$

① $\underset{\text{AlCl}_3}{\underline{\text{CH}_3\text{CH}(\text{CH}_2)_{10}\text{CH}_3}} \rightarrow$

② H_2SO_4

(para-substituted benzene with $\text{SO}_3^- \text{Na}^+$)

17.55 Nitrosobenzene reacts slightly slower than benzene in an electrophilic aromatic substitution reaction and gives predominantly ortho and para products. Explain this behavior.

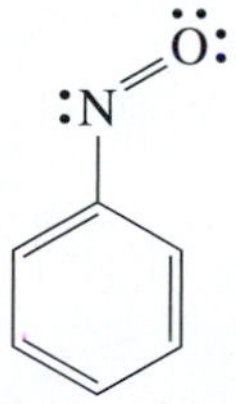

Nitrosobenzene

17.56 Predict the products that would be formed in this reaction and explain your answer:

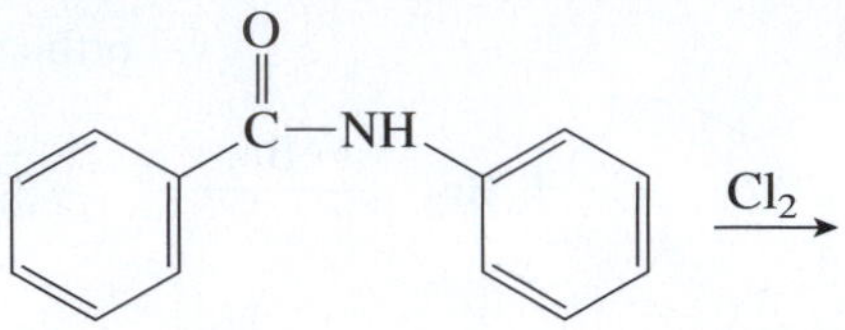

17.57 Show syntheses of these compounds from the indicated starting materials:

a) 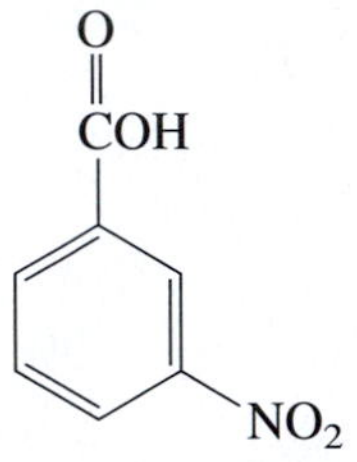 from benzene

b) 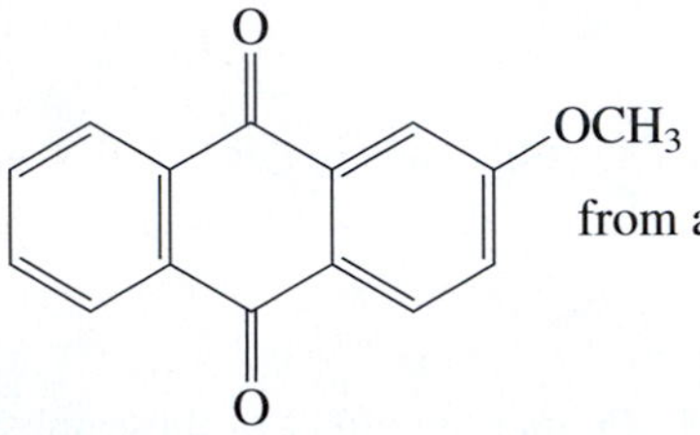 from anisole

c) 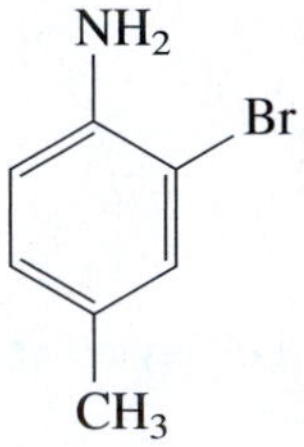 from benzene

d) 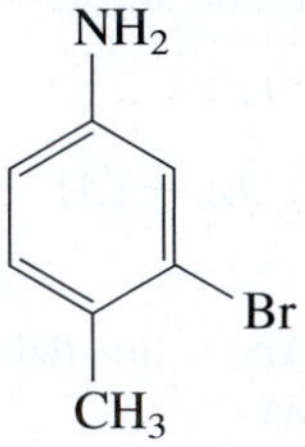 from benzene

e) 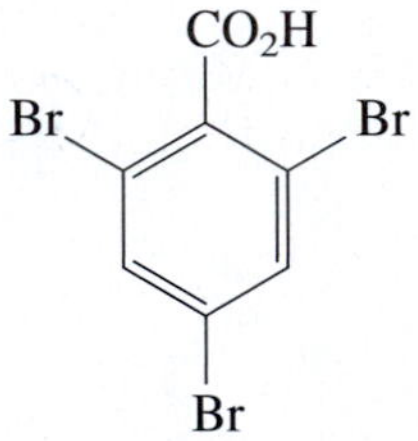 from benzene

f) 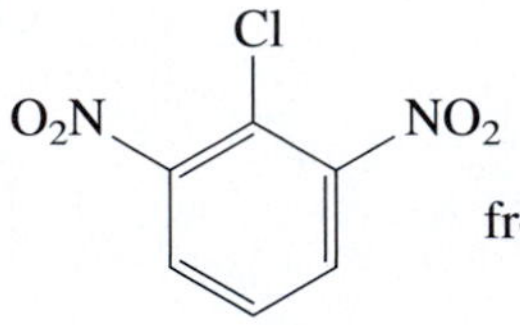 from aniline

17.58 Show syntheses of these compounds from the indicated starting materials. Use of reactions from previous chapters may be necessary.

a)

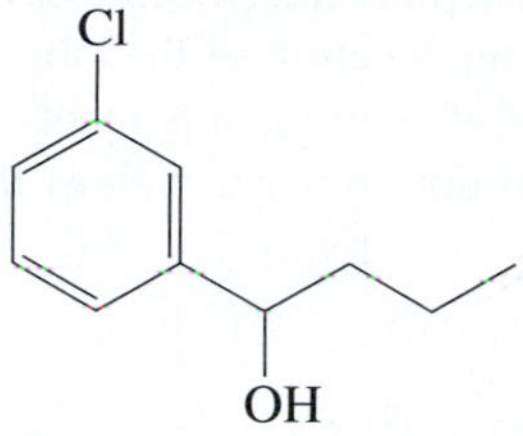

from benzene

b)

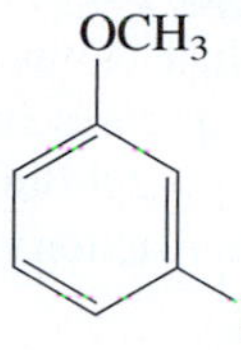

from benzene

c)

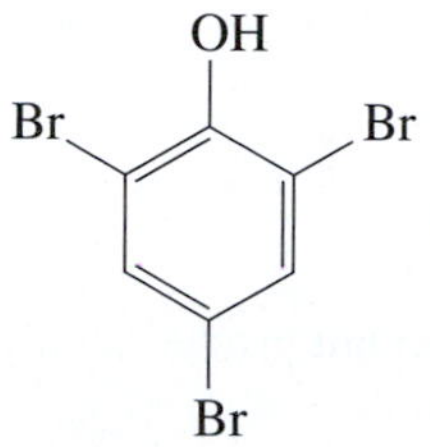

from benzene

d)

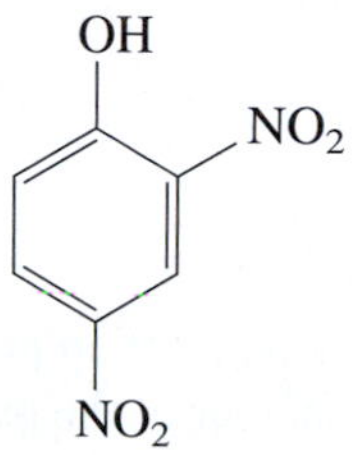

from chlorobenzene

e)

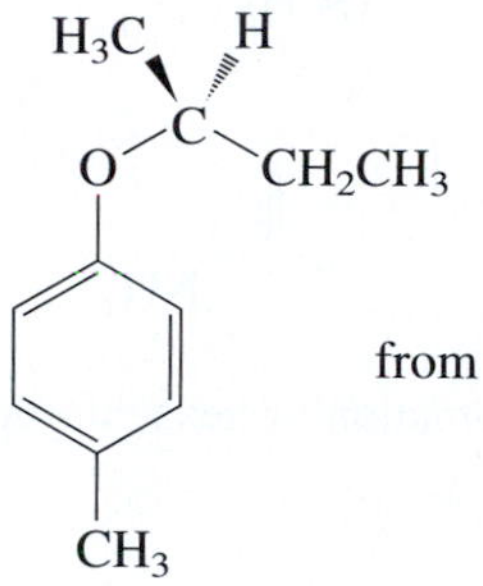

from benzene

f)

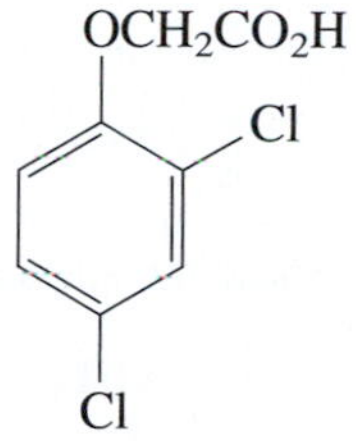

from phenol

2,4-Dichlorophenoxyacetic acid
(2,4-D, a herbicide)

17.59 Show the steps in the mechanism for this reaction:

$$+ \ ClCH_2CH_2CH_2Br \quad \xrightarrow[\text{2) NaOH}]{\text{1) 150°C}}$$

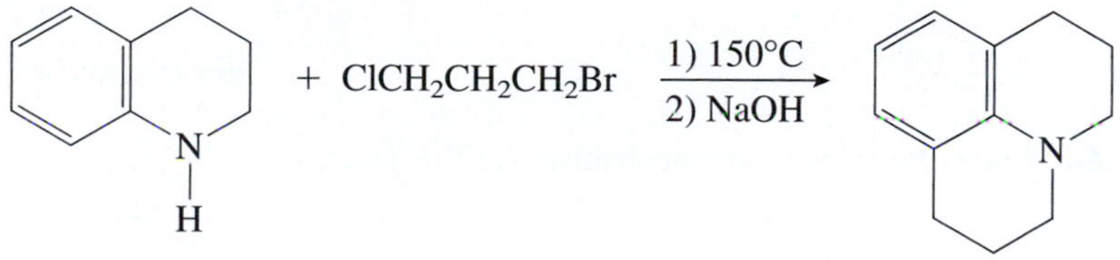

17.60 Show the steps in the mechanism for this reaction:

$$\xrightarrow[\substack{\text{CH}_3\text{CO}_2\text{H} \\ \text{reflux} \\ \text{2 min}}]{\text{HBr}}$$

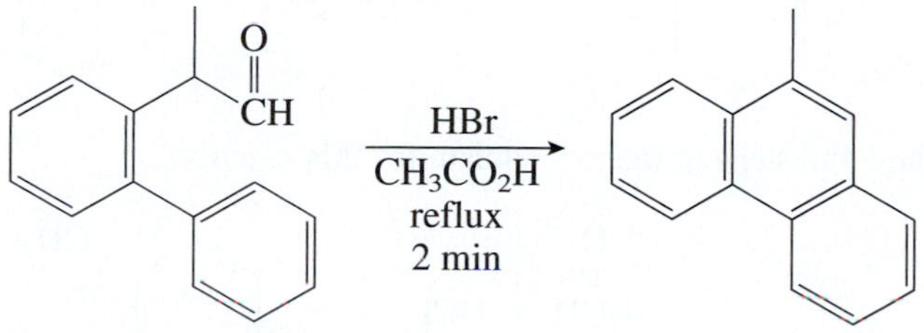

BioLink

17.61 1-Fluoro-2,4-dinitrobenzene is sometimes used to label or tag the amino acid at one end (the end with a free amino group) in a peptide or protein (see Chapter 26). After the amide bonds of the peptide have been cleaved, the "N-terminal" amino acid can be identified by the location of the label. In the following representation of a peptide, explain why nitrogen A reacts with 1-fluoro-2,4-dinitrobenzene and nitrogen B does not. Then show the product of the reaction.

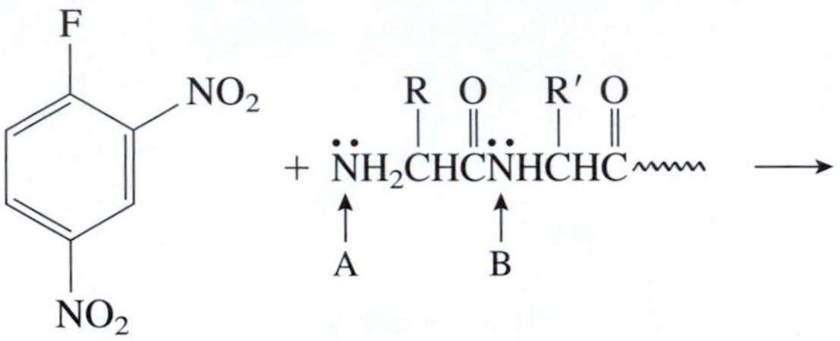

17.62 This reaction proceeds by a benzyne mechanism but produces only the *meta*-isomer of the product. Explain this observation.

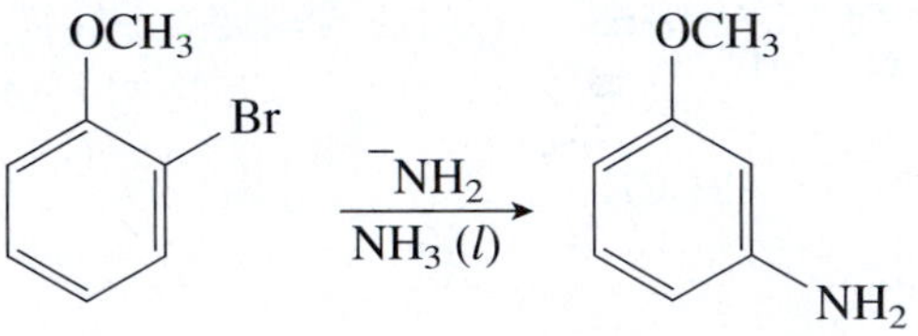

17.63 Show the steps in the mechanism for the formation of hexachlorophene:

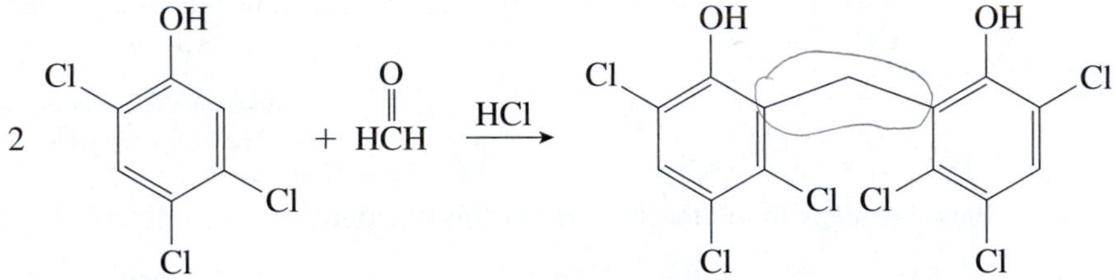

Hexachlorophene

17.64 Show the steps in the mechanism for this reaction:

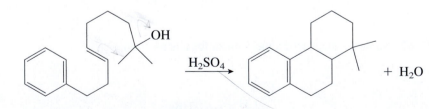

17.65 Show the steps in the mechanism for this reaction:

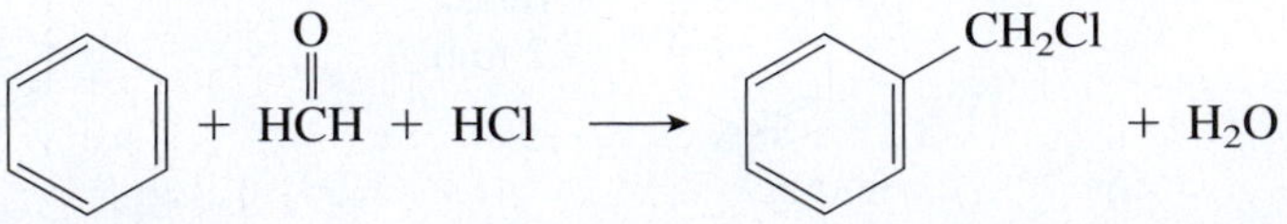

Problems Involving Spectroscopy

17.66 Compound **A**, $C_9H_{11}Cl$, gives compound **B**, C_9H_{10}, when treated with $AlCl_3$. The ^{1}H-NMR spectrum of **B** follows. Show the structures of **B** and **A** and show a mechanism for the reaction.

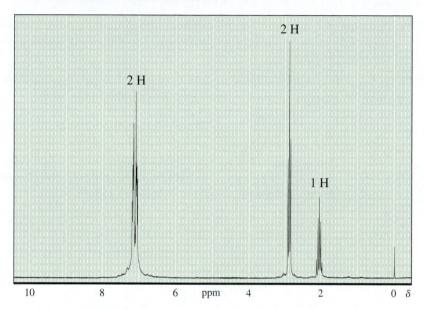

17.67 The product of the reaction of excess benzene with chloroform in the presence of $AlCl_3$ has the formula $C_{19}H_{16}$. This product shows only five absorptions in its ^{13}C-NMR spectrum. Show the structure of this product.

$$\bigcirc + CHCl_3 \xrightarrow{AlCl_3}$$

17.68 The bromination of bromobenzene gives three dibromobenzenes in the yields shown in the following equation. **A** shows three peaks in its ^{13}C-NMR spectrum, **B** shows two, and **C** shows four. Show the structures of **A**, **B**, and **C**.

$$\text{Br}\!\!-\!\!\bigcirc + Br_2 \xrightarrow{FeBr_3} \mathbf{A} + \mathbf{B} + \mathbf{C}$$
$$\qquad\qquad\qquad 14\% \quad 84\% \quad 2\%$$

Problems Using Online Three-Dimensional Molecular Models

17.69 Arrange these compounds in order of decreasing amount of ortho product formed when they are reacted with Br_2 and $FeBr_3$.

17.70 Nitration of 1-isopropyl-4-methylbenzene with nitric acid and sulfuric acid gives 75% of one of the products below and 7% of the other. Explain which is the major product.

17.71 *N,N*-2,6-Tetramethylaniline reacts slower than *N,N*-dimethylaniline in electrophilic aromatic substitution reactions. Explain this observation.

Do you need a live tutor for homework problems? Access vMentor at Organic ChemistryNow at **http://now.brookscole.com/hornback2** for one-on-one tutoring from a chemistry expert.

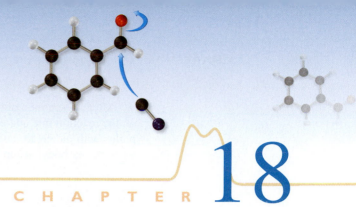

Additions to the Carbonyl Group

REACTIONS OF ALDEHYDES AND KETONES

CHAPTER 8 DISCUSSED nucleophilic substitution reactions that occur at sp^3-hybridized carbons: the S_N1 and S_N2 reactions. In those reactions the nucleophile replaces a leaving group bonded to the electrophilic carbon. The carbon is electrophilic because the leaving group, being more electronegative, pulls the electrons of its bond with the carbon toward itself, making the carbon electron deficient. This chapter introduces a new electrophile, the carbon of a carbonyl group. This carbon is quite electrophilic because of the electronegativity of the oxygen of the carbonyl group. In addition, a charged resonance structure, where the pi electrons have moved onto the oxygen, makes a minor contribution to the structure of carbonyl-containing compounds and causes the carbonyl carbon to be even more electrophilic.

MASTERING ORGANIC CHEMISTRY

▶ Predicting the Products of the Addition of Nucleophiles to Aldehydes and Ketones

▶ Predicting the Products of the Addition of Nucleophiles to α,β-Unsaturated Compounds

▶ Understanding the Mechanisms for These Reactions

▶ Predicting the Effect of the Structure of the Aldehyde or Ketone on the Position of the Equilibrium for These Reactions

▶ Using These Reactions to Synthesize Compounds

▶ Using Acetals as Protecting Groups in Organic Synthesis

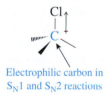

Electrophilic carbon in S_N1 and S_N2 reactions

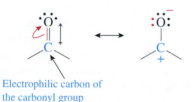

Electrophilic carbon of the carbonyl group

The reactions presented in this chapter involve nucleophiles attacking at the electrophilic carbon of the carbonyl group of aldehydes and ketones. Chapter 19 discusses the reactions in which nucleophiles react at the carbonyl carbon of carboxylic acid derivatives.

ORGANIC
Chemistry⚛Now™

Look for this logo in the chapter and go to OrganicChemistryNow at
http://now.brookscole.com/hornback2 for tutorials, simulations, problems, and molecular models.

After a discussion of the general mechanisms for the reactions, the effect of structure on the reaction rate and equilibrium is presented. Then the reactions of various nucleophiles with aldehydes and ketones are discussed. Finally, a related reaction that occurs when the carbonyl group is conjugated with a CC double bond is considered.

18.1   GENERAL MECHANISMS

The simplest version of these reactions results in the addition of the nucleophile to the carbonyl carbon and a proton to the oxygen, as illustrated in the following general equation:

$$\overset{-}{Nu}{:} \; + \; \underset{R \quad\quad R'}{\overset{:\overset{\cdot\cdot}{O}:}{C}} \; + \; H{-}A \; \longrightarrow \; R{-}\underset{Nu}{\overset{:\overset{\cdot\cdot}{O}\diagup^{H}}{\underset{|}{\overset{|}{C}}}}{-}R' \; + \; {:}\overset{-}{A}$$

The reaction stops at this stage with some nucleophiles, but with others it proceeds further to somewhat different products.

There are two mechanisms for this reaction that differ only in the order of Nu—C and O—H bond formation. Under acidic conditions—that is, in the presence of an acid that is strong enough to protonate the oxygen of the carbonyl group—the O—H bond is formed first. Under basic conditions—that is, in the absence of such an acid—the Nu—C bond is formed first. Each of these mechanisms is outlined in Figure 18.1.

Some of the nucleophiles that are employed in these reactions are quite strong bases, so the presence of even weak acids, such as water or ethanol, must be avoided. In these reactions a solvent that has no acidic hydrogens, such as diethyl ether, is used. The reaction follows the basic conditions mechanism and stops at the anionic stage until an acid is added during the workup. Reactions involving less basic nucleophiles can be conducted with water or alcohols as solvent. These reactions also follow the basic conditions mechanism with the solvent supplying the proton that is added to the negative oxygen in the second step. Reactions of even weaker nucleophiles that are also only weakly basic are conducted in the presence of acid, so the acidic conditions mechanism is followed. Protonation of the oxygen of the carbonyl group occurs first, making the carbon more electrophilic, thus facilitating the addition of the weak, neutral nucleophile in the second step.

Although these additions to CO double bonds have some superficial similarities to the electrophilic additions to CC double bonds that were presented in Chapter 11, there are many differences. The acidic conditions mechanism here resembles the mechanism for addition to carbon–carbon double bonds in that the electrophile (the proton) adds first, followed by addition of the nucleophile. However, in this case the first step is fast because it is a proton transfer involving oxygen, a simple acid–base reaction. The second step, the attack of the nucleophile, is the rate-determining step. (Recall that it is the first step, the addition of the electrophile, that is slow in the additions to CC double bonds.) Furthermore, in the case of additions to simple alkenes there is no mechanism comparable to the one that operates here under basic conditions, in which the nucleophile adds first. Because the nucleophile adds in the slow step, the reactions presented in this chapter are termed *nucleophilic additions,* even if the protonation occurs first. In

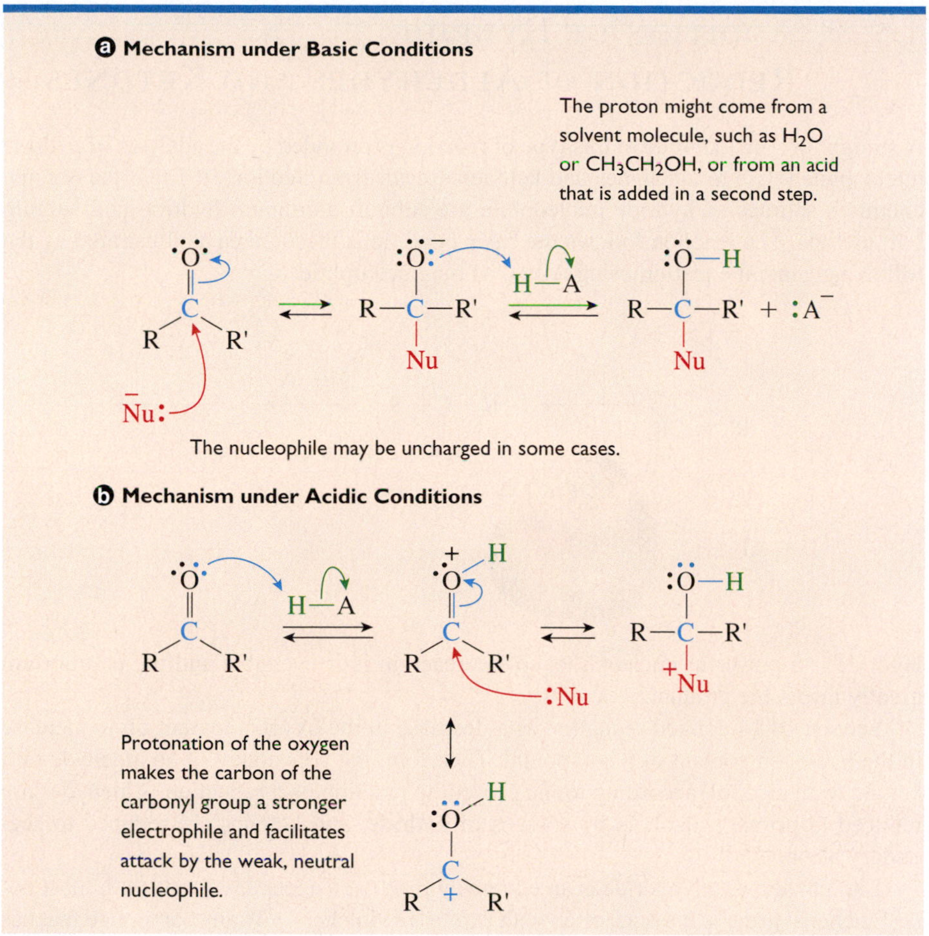

ⓐ Mechanism under Basic Conditions

The proton might come from a solvent molecule, such as H_2O or CH_3CH_2OH, or from an acid that is added in a second step.

The nucleophile may be uncharged in some cases.

ⓑ Mechanism under Acidic Conditions

Protonation of the oxygen makes the carbon of the carbonyl group a stronger electrophile and facilitates attack by the weak, neutral nucleophile.

Figure 18.1

MECHANISMS OF NUCLEOPHILIC ADDITIONS TO A CARBONYL GROUP UNDER ⓐ BASIC AND ⓑ ACIDIC CONDITIONS.

contrast, the reactions presented in Chapter 11 are called *electrophilic additions* because the electrophile adds in the slow step.

In some ways the reactions described in this chapter are simpler than those described in Chapter 11. The stereochemistry of the addition is not a concern here because there is no way to determine whether the addition occurs in a syn or an anti manner. Furthermore, the regiochemistry of these reactions is simple: the nucleophile always adds to the carbon of the carbonyl group.

The equilibrium in these reactions may favor the products or the reactants, depending on the strength of the nucleophile and the structure of the carbonyl compound. Stronger nucleophiles shift the equilibrium toward the products, and very strong nucleophiles give an irreversible reaction, that is, one that proceeds in only one direction, from the reactants to the products. The structure of the aldehyde or ketone exerts its influence through resonance, steric, and inductive effects, as usual. It is easiest to see these effects in examples, so let us proceed to examine some of the nucleophiles that can be used in these addition reactions.

18.2   ADDITION OF HYDRIDE; REDUCTION OF ALDEHYDES AND KETONES

A straightforward example of this type of reaction is provided by the addition of hydride nucleophile ($H:^-$) to aldehydes and ketones. Recall from Section 10.7 that the organic chemist's sources of hydride nucleophile are lithium aluminum hydride and sodium borohydride. The reaction follows the basic conditions mechanism as illustrated by the following general equation using AlH_4^- as the nucleophile:

Hydride is a powerful nucleophile, so the reaction is irreversible and the equilibrium greatly favors the product.

Section 10.14 defined reduction as a decrease in the oxygen content or an increase in the hydrogen content of a compound. Therefore, the conversion of an aldehyde or a ketone to an alcohol, according to the preceding reaction, is a reduction. Aldehydes are reduced to primary alcohols by sources of hydride, and ketones are reduced to secondary alcohols.

Lithium aluminum hydride is an extremely reactive reagent and reacts with most polar functional groups. It reacts, often with explosive violence, with any compound that has a relatively acidic hydrogen, such as an OH or NH, as shown in the following equation:

$$LiAlH_4 \ + \ 4\,ROH \ \longrightarrow \ LiAl(OR)_4 \ + \ 4\,H_2 \qquad \text{Highly exothermic}$$

Therefore, reductions employing $LiAlH_4$ are usually conducted using ether or THF as the solvent. After addition of the hydride is complete, acid is carefully added to decompose any excess reagent and protonate the initially formed alkoxide anion. In contrast, sodium borohydride is much less reactive and reacts selectively with aldehydes and ketones in preference to many other functional groups. Its reaction with alcohols is slow enough that methanol or ethanol are often used as solvents for its reactions. The following equations provide some examples of the use of $LiAlH_4$ and $NaBH_4$ to reduce aldehydes and ketones. One mole of each of these reagents is capable of reducing four moles of carbonyl compound, although an excess of the hydride is usually employed.

Heptanal   1-Heptanol

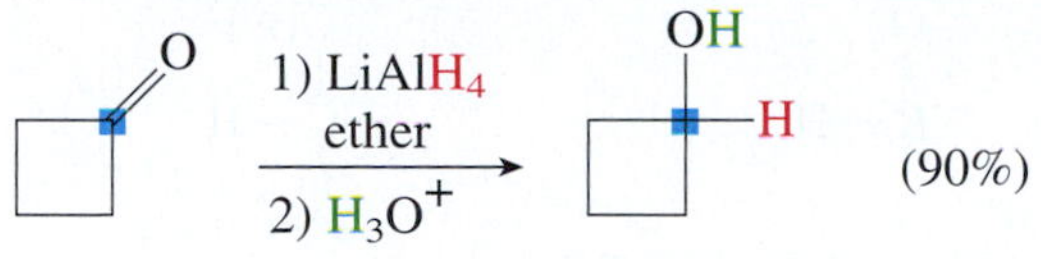

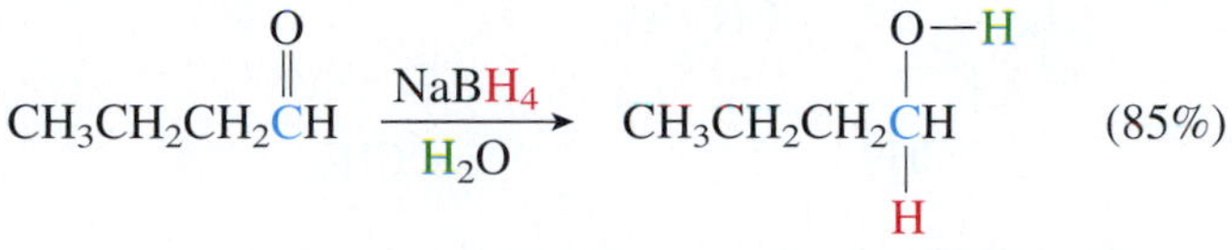

$$CH_3CH_2CH_2\overset{\displaystyle O}{\overset{\|}{C}}H \xrightarrow[\text{H}_2\text{O}]{\text{NaBH}_4} CH_3CH_2CH_2\overset{\displaystyle O-H}{\underset{\displaystyle H}{C}H} \quad (85\%)$$

Butanal   1-Butanol

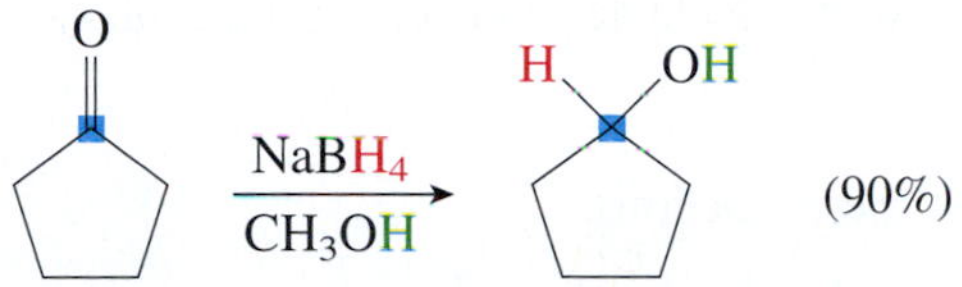

Cyclopentanone   Cyclopentanol

PROBLEM 18.1

Show the steps involved in the mechanism for this reaction:

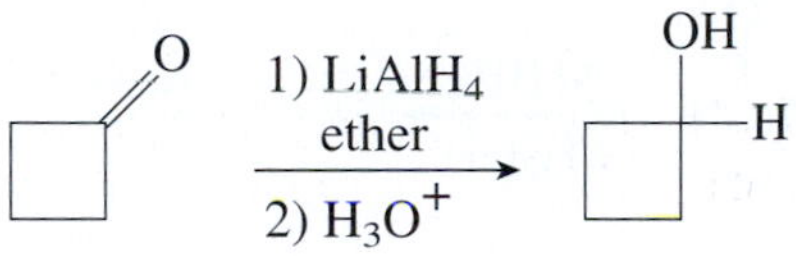

The hydride nucleophile from LiAlH$_4$ and NaBH$_4$ attacks only electrophilic carbons. Isolated carbon–carbon double bonds and the double bonds of aromatic rings do not react with these reagents. This means that carbonyl groups can be selectively reduced in the presence of CC double bonds as illustrated in the following examples. (Recall from Section 11.12 that carbon–carbon double bonds can be selectively reduced in the presence of carbonyl groups using hydrogen and a catalyst.)

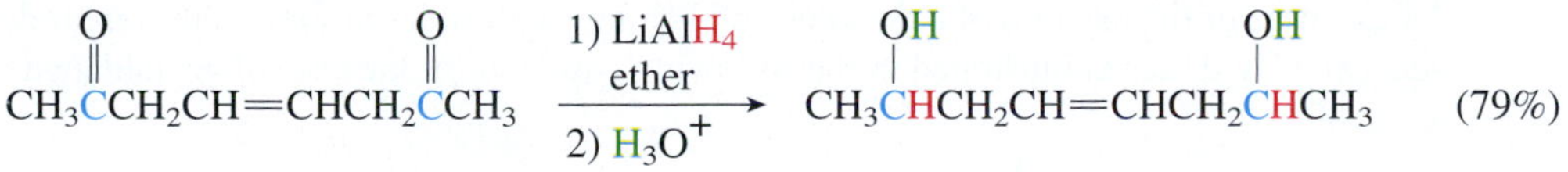

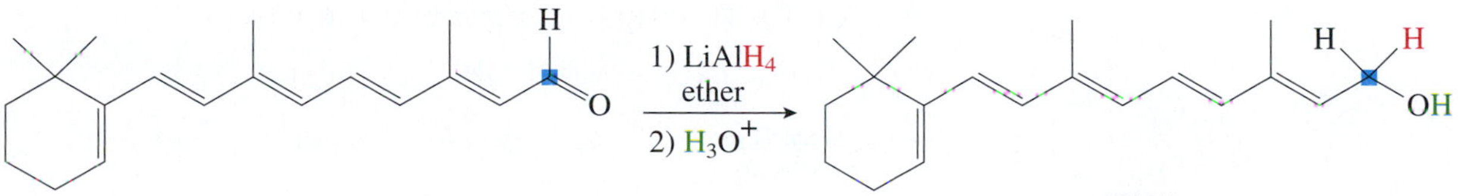

$$CH_3\overset{\displaystyle O}{\overset{\|}{C}}CH_2CH=CHCH_2\overset{\displaystyle O}{\overset{\|}{C}}CH_3 \xrightarrow[\text{2) H}_3\text{O}^+]{\text{1) LiAlH}_4 \;\; \text{ether}} CH_3\overset{\displaystyle OH}{\overset{|}{C}}HCH_2CH=CHCH_2\overset{\displaystyle OH}{\overset{|}{C}}HCH_3 \quad (79\%)$$

Vitamin A aldehyde   (75%)
(retinal)   Vitamin A
                            (retinol)

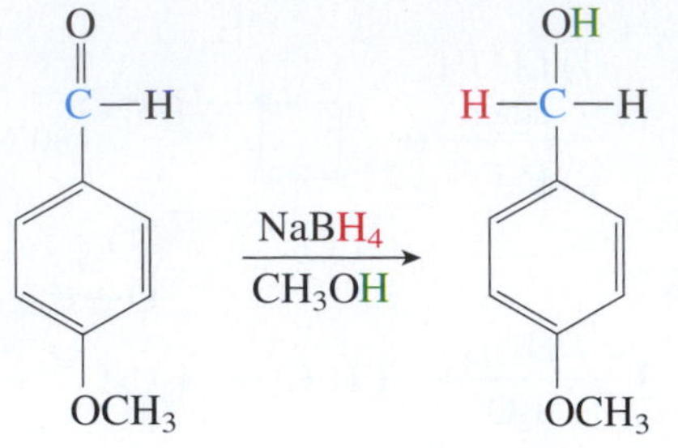

p-Methoxybenzaldehyde   p-Methoxybenzyl alchohol

The use of lithium aluminum hydride or sodium borohydride provides the best method for the reduction of aldehydes and ketones to alcohols.

PROBLEM 18.2
Show the products of these reactions:

a)

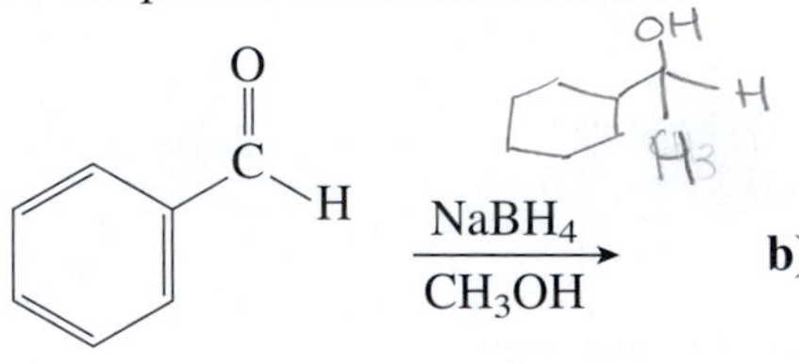

$\xrightarrow[\text{CH}_3\text{OH}]{\text{NaBH}_4}$

b) $\text{CH}_3\text{CH}_2\overset{\text{O}}{\overset{\|}{\text{C}}}\text{CH}_3 \xrightarrow[\text{2) H}_3\text{O}^+]{\text{1) LiAlH}_4}$

$CH_2 CH_2 \overset{OH}{\overset{|}{C}} CH$

c) $\text{CH}_3\text{CH}=\text{CHCH}_2\text{CH}_2\overset{\text{O}}{\overset{\|}{\text{C}}}\text{H} \xrightarrow[\text{CH}_3\text{OH}]{\text{NaBH}_4}$

$CH_3CH=CHCH_2CH_2\overset{OH}{\overset{|}{C}}H_2$

d)

$\xrightarrow[\text{2) H}_3\text{O}^+]{\text{1) LiAlH}_4}$

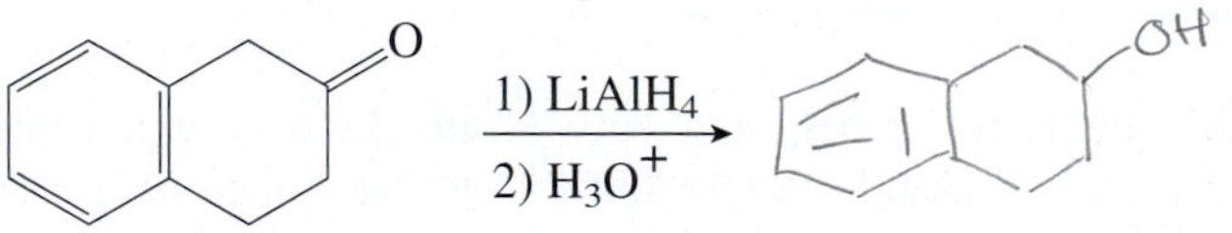

EXAM 2

18.3   ADDITION OF WATER

An example of the reaction of aldehydes and ketones with water to form addition products called hydrates is illustrated in the following equation for the case of acetaldehyde (ethanal):

$$\text{CH}_3\overset{\text{O}}{\overset{\|}{\text{C}}}-\text{H} + \text{H}_2\text{O} \underset{[^-\text{OH}]}{\overset{[\text{H}_3\text{O}^+]}{\rightleftharpoons}} \text{ or } \quad \text{CH}_3\overset{\text{OH}}{\underset{\text{OH}}{\overset{|}{\underset{|}{\text{C}}}}}-\text{H}$$

Acetaldehyde   A hydrate
(ethanal)

The reaction can occur by either the acidic conditions mechanism or the basic conditions mechanism, as shown in the following equations:

Mechanism in base:

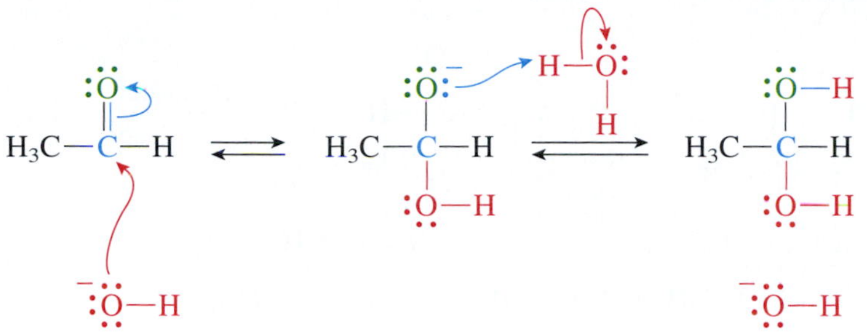

ORGANIC **Chemistry**Now™
Click *Mechanisms in Motion* to view **Hydration under Base Conditions** or **Hydration under Acid Conditions.**

Mechanism in acid:

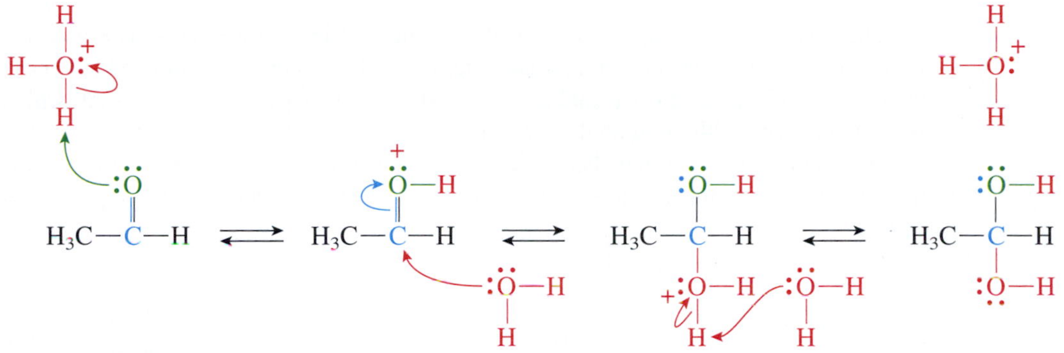

Note that in both cases the base or acid that is consumed in the first step is regenerated in the last step, so the reactions are base catalyzed or acid catalyzed.

The reaction, by either mechanism, is fast at room temperature. However, because water is a relatively weak nucleophile, the equilibrium does not favor the product in most cases. As a result, the hydrates of aldehydes and ketones usually cannot be isolated because removal of the water solvent also drives the equilibrium back to the left. Therefore, the hydration reaction is not useful in synthesis.

Because the equilibrium constants are neither too large nor too small, the hydration reaction provides an excellent opportunity to examine the effect of the structure of the carbonyl compound on the equilibrium constant. Let's consider inductive effects first.

The interaction of the dipole of an **electron-withdrawing group** with the large dipole of the carbonyl group destabilizes an aldehyde or ketone more than it destabilizes the hydrate product. This causes the equilibrium to shift to the right, toward the product, resulting in a larger equilibrium constant. In contrast, an **electron-donating group** stabilizes the aldehyde or ketone more than the product and causes the equilibrium to shift toward the reactants. As an illustration of these effects, let's compare the equilibrium constant for hydrate formation for acetaldehyde ($K = 1.3$) to that for 2-chloroacetaldehyde ($K = 37$). The presence of an electronegative chlorine substituent in 2-chloroacetaldehyde causes the equilibrium constant to increase substantially. (Recall that an equilibrium constant near 1 means that nearly equal amounts of reactant and product are present at equilibrium. In the case of an aqueous solution of acetaldehyde, with $K = 1.3$, the hydrate constitutes 57% of the mixture at equilibrium and the aldehyde constitutes 43%. In the case of

2-chloroacetaldehyde, with $K = 37$, the products are favored at equilibrium; the hydrate constitutes 97% of the mixture.)

$$\underset{\text{Acetaldehyde}}{CH_3\overset{\displaystyle O}{\overset{\|}{C}}{-}H} + H_2O \;\rightleftharpoons\; CH_3\underset{\text{OH}}{\overset{\text{OH}}{\underset{|}{\overset{|}{C}}}}{-}H \qquad K = \frac{[\text{hydrate}]}{[\text{aldehyde}]} = 1.3$$

$$\underset{\text{2-Chloroacetaldehyde}}{H_2C{-}\overset{\displaystyle O}{\overset{\|}{C}}{-}H} + H_2O \;\rightleftharpoons\; \underset{\text{OH}}{H_2C{-}\overset{\text{OH}}{\underset{|}{\overset{|}{C}}}{-}H} \qquad K = 37$$

The interaction between the dipoles destabilizes the reactant.

The position of the equilibrium is also influenced by steric effects. The product, which is sp^3 hybridized and approximately tetrahedral, has the groups bonded to the hydrate carbon closer together (bond angles ~ 109°) than they are in the reactant, which is sp^2 hybridized with bond angles of ~ 120°. There is more steric hindrance between the groups in the product than there is in the reactant. Therefore, larger substituents shift the equilibrium toward the less crowded carbonyl form and decrease the equilibrium constant.

Bond angle = 120°   Bond angle = 109°

An important example of this effect is the decrease in equilibrium constants for ketones as compared to aldehydes. The replacement of the aldehyde hydrogen of acetaldehyde ($K = 1.3$) with a methyl group, to produce acetone ($K = 2 \times 10^{-3}$), results in a decrease in the equilibrium constant for hydration by a factor of approximately 1000. The inductive effect of the electron-donating alkyl group also helps shift the equilibrium for ketones toward the reactant.

$$\underset{\text{Acetone}}{CH_3\overset{\displaystyle O}{\overset{\|}{C}}CH_3} + H_2O \;\rightleftharpoons\; CH_3\underset{\text{OH}}{\overset{\text{OH}}{\underset{|}{\overset{|}{C}}}}CH_3 \qquad K = 2 \times 10^{-3}$$

In general, the equilibrium lies more toward the addition product for an aldehyde than it does for a ketone in all of the reactions in this chapter. These factors affect the rate of the reaction in the same manner, so aldehydes also react more rapidly than ketones because it is easier for the nucleophile to approach the less hindered (and more electrophilic) carbonyl group of the aldehyde. (The transition state for nucleophilic attack on an aldehyde has less steric strain than the transition state for nucleophilic attack on a ketone.) Additional examples of both steric and electronic effects are provided in Table 18.1. These same electronic and steric effects also apply to the other nucleophilic additions discussed subsequently in this chapter.

PROBLEM 18.3

When acetone is treated with $H_2^{18}O$ (water which has some of its ^{16}O atoms replaced with ^{18}O atoms), the ^{18}O atoms are incorporated into the carbonyl group of the ketone. Show the steps in the mechanism for this reaction in the presence of acid:

$$\underset{\text{CH}_3\overset{\overset{\text{O}}{\|}}{\text{C}}\text{CH}_3}{} + H_2^{18}O \; \overset{H_3O^+}{\underset{}{\rightleftharpoons}} \; \underset{\text{CH}_3\overset{\overset{^{18}\text{O}}{\|}}{\text{C}}\text{CH}_3}{} + H_2O$$

Table 18.1   Some Equilibrium Constants for Hydrate Formation

Compound	K	Comments
O‖ HCH Formaldehyde	2×10^3	Formaldehyde has a large equilibrium constant for hydrate formation because it has no bulky, electron-donating alkyl groups. It is more than 99.9% in the hydrated form in aqueous solution. The "formaldehyde" or formalin used to preserve biological samples is actually a concentrated solution of the hydrate in water. Formaldehyde itself is a gas.
O‖ CH₃CH Acetaldehyde	1.3	The more hindered carbonyl carbon of acetaldehyde is less reactive. Acetaldehyde is slightly more than 50% hydrated in aqueous solution.
O‖ CH₃CCH₃ Acetone	2×10^{-3}	Acetone, with an even more hindered carbonyl carbon, forms only a negligible amount of hydrate.
O‖ CH₃CH₂CH	0.71	As can be seen by comparing these two examples to acetaldehyde, an increase in steric hindrance further from the carbonyl carbon results in only a small decrease in the equilibrium constant.
CH₃ O| ‖ CH₃CH—CH	0.44	
O‖ ClCH₂CH	37	As can be seen by comparing these two examples to acetaldehyde, the inductive effect of chlorine shifts the equilibrium toward the hydrate. When three chlorines are present, the product, known as chloral hydrate, can be isolated (mp = 57°C). It is a powerful hypnotic and is the active ingredient of a "Mickey Finn," or knockout drops.
O‖ Cl₃CCH	2.8×10^4	

PROBLEM 18.4

Explain which compound has the larger equilibrium constant for hydrate formation:

a) (cyclohexanone) or (2,2-difluorocyclohexanone, with F, F substituents)

b) (cyclopentyl–C(=O)–CH₃) or (cyclopentyl–C(=O)–H)

$$\text{c) } CH_3CH_2\overset{O}{\overset{\|}{C}}H \text{ or } (CH_3)_3C\overset{O}{\overset{\|}{C}}H$$

$$\text{d) } CH_3\overset{O}{\overset{\|}{C}}CH_3 \text{ or } Cl_3C\overset{O}{\overset{\|}{C}}CH_3$$

$$\text{e) } CH_3CH_2CH_2\overset{O}{\overset{\|}{C}}H \text{ or } CH_3\overset{CH_3}{\overset{|}{C}}HCH_2\overset{O}{\overset{\|}{C}}H$$

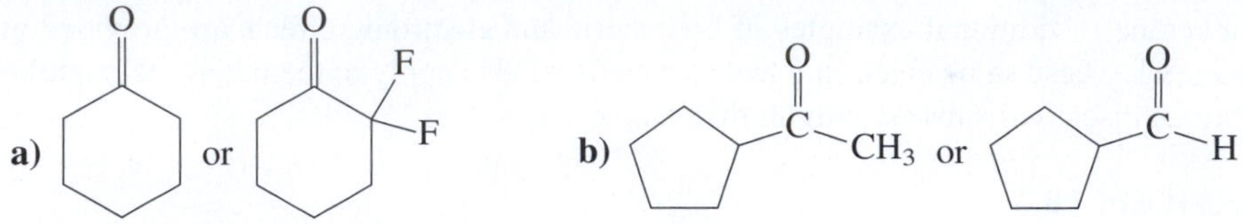

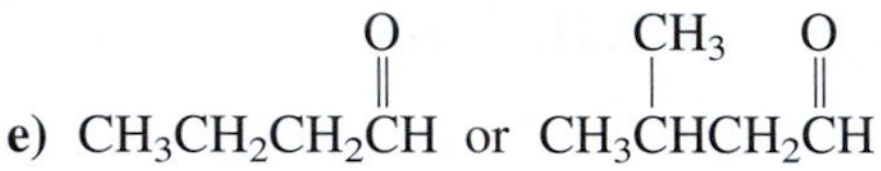

PROBLEM 18.5

Arrange these compounds in order of increasing equilibrium constant for hydrate formation:

$$CH_3\overset{Cl}{\overset{|}{C}}H\overset{O}{\overset{\|}{C}}H \qquad \overset{Cl}{\overset{|}{C}}H_2CH_2\overset{O}{\overset{\|}{C}}H \qquad CH_3CH_2\overset{O}{\overset{\|}{C}}H \qquad CH_3CF_2\overset{O}{\overset{\|}{C}}H$$

18.4 ADDITION OF HYDROGEN CYANIDE

Hydrogen cyanide adds to aldehydes and ketones to form products known as cyanohydrins, as shown in the following examples. The reaction proceeds by the basic conditions mechanism and is catalyzed by cyanide ion.

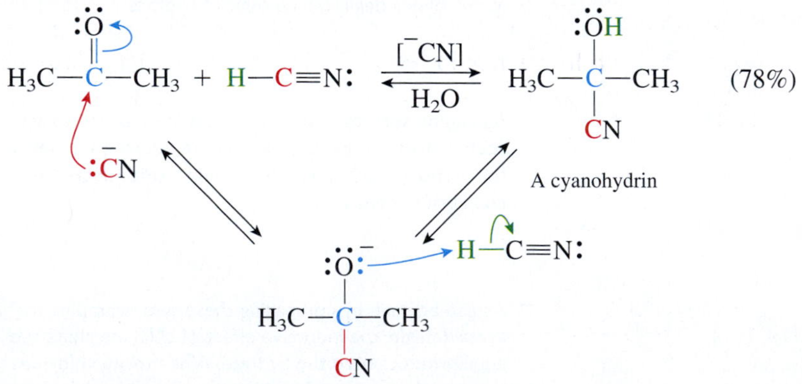

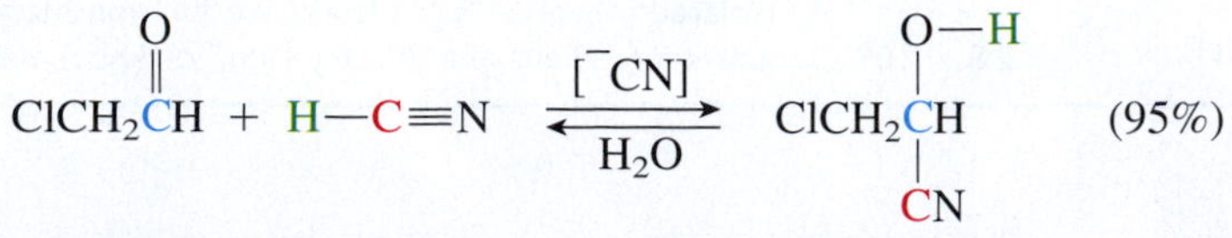

A cyanohydrin

$$ClCH_2\overset{O}{\overset{\|}{C}}H + H-C\equiv N \underset{H_2O}{\overset{[^-CN]}{\rightleftarrows}} ClCH_2\overset{O-H}{\overset{|}{C}}H \quad (95\%) \atop CN$$

The equilibrium constants are somewhat larger for cyanohydrin formation than they were for hydrate formation because cyanide ion is a stronger nucleophile than water. Therefore, aldehydes and many ketones give good yields of the addition product. However, ketones that are conjugated with benzene rings have unfavorable equilibrium constants. In such compounds the reactant has resonance stabilization due to the conjugation of the carbonyl group with the aromatic ring (see the following structures). This stabilization is lost in the product, resulting in the product being less favored at equilibrium.

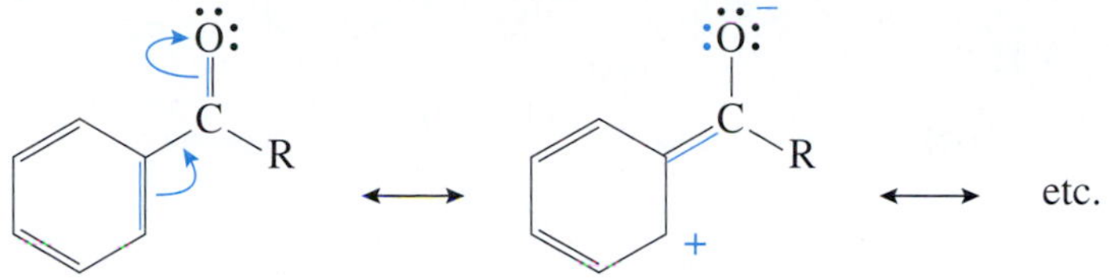

As an example of this effect, the equilibrium constant for the reaction of acetone with hydrogen cyanide is 32, whereas the equilibrium constant for a similar reaction of acetophenone is 0.77:

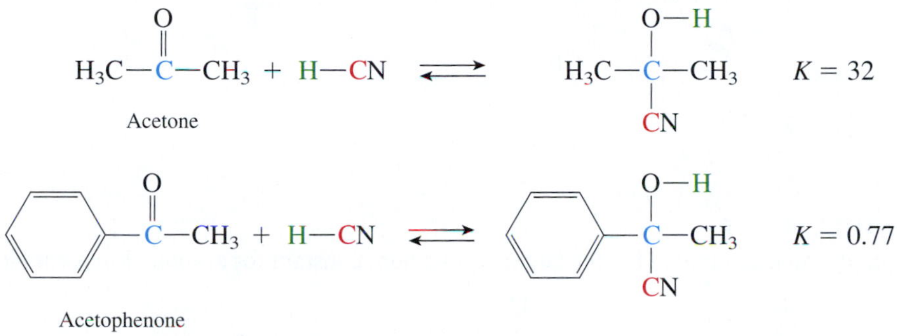

Under typical conditions the reaction of acetone with hydrogen cyanide ($K = 32$) has most of the reactants converted to the product at equilibrium. This allows the cyanohydrin to be obtained in acceptable isolated yield (78%). In contrast, the amount of cyanohydrin product that is present at equilibrium in the reaction of acetophenone ($K = 0.77$) is too low for the reaction to be synthetically useful unless some method is used to drive the equilibrium toward the product.

PROBLEM 18.6
Show the products of these reactions:

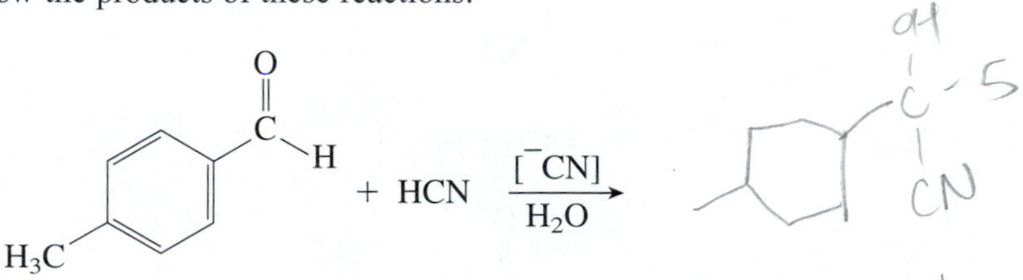

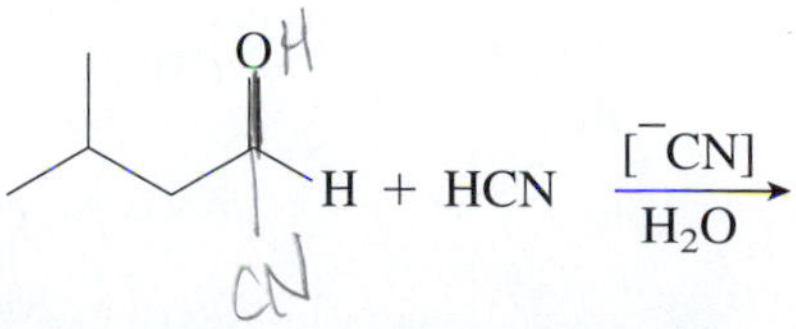

PRACTICE PROBLEM 18.1

Explain why *p*-methoxyacetophenone has a smaller equilibrium constant for cyanohydrin formation than does acetophenone.

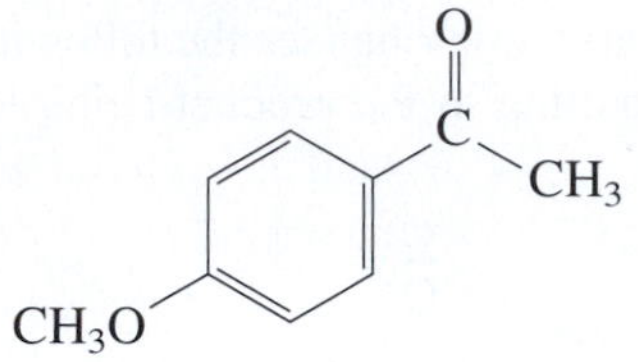

p-Methoxyacetophenone 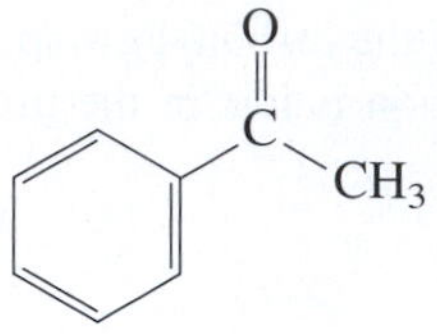   Acetophenone

Solution

The methoxy group donates electrons to the carbonyl carbon by resonance, making it less electrophilic and less reactive. Therefore, the equilibrium constant for cyanohydrin formation is larger for acetophenone than for *p*-methoxyacetophenone.

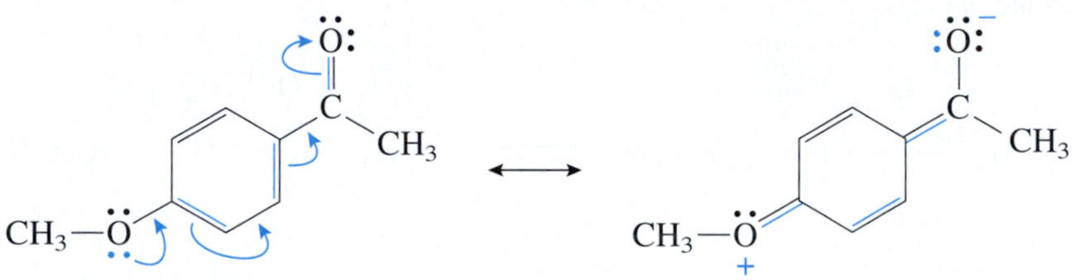

PROBLEM 18.7

Explain which compound has the larger equilibrium constant for cyanohydrin formation:

a) $CH_3CH_2CCH_3$ or $CH_3CH_2CH_2CH$ *Less hindered*

b) $CH_3CH_2CH_2CH$ or   *resonance*

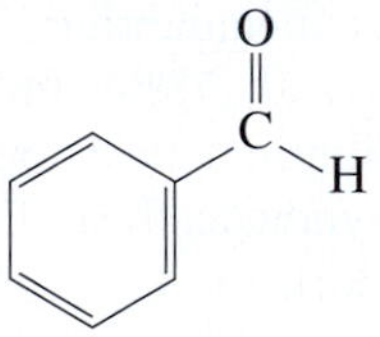

c) 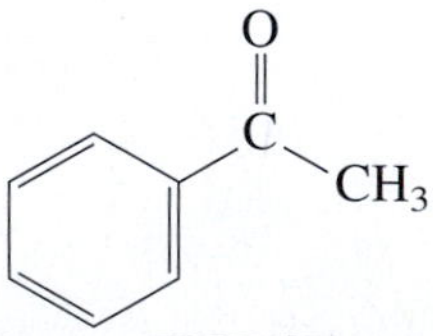 or

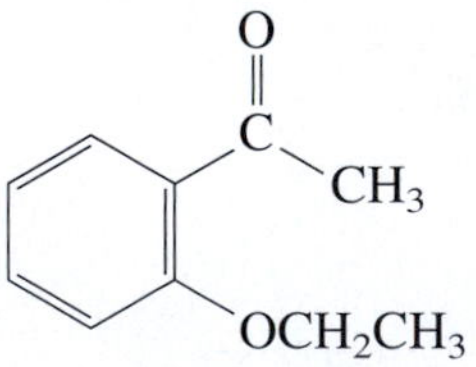

d) 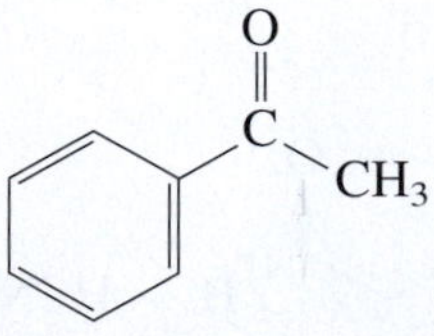 or   *more e-philic*

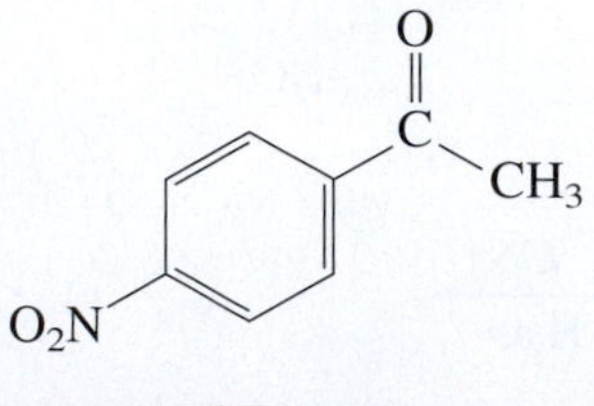

18.5 PREPARATION AND PROPERTIES OF ORGANOMETALLIC NUCLEOPHILES

Carbon nucleophiles are very useful species because their reactions with carbon electrophiles result in the formation of carbon–carbon bonds. Section 10.8 introduced acetylide anions as nucleophiles that could be used in S_N2 reactions. These nucleophiles are prepared by reacting 1-alkynes with a strong base such as sodium amide. The relatively acidic hydrogen on the sp-hybridized carbon is removed in this acid–base reaction:

$$R-C\equiv C-H + {}^{-}\!:NH_2 \longrightarrow R-C\equiv C:^{-} + \ddot{N}H_3$$

Hydrogens bonded to sp^3- and sp^2-hybridized carbons are not acidic enough to be removed by bases that are commonly available in the laboratory. However, the carbon–halogen bond of many organic halides can be converted to a carbon–metal bond, resulting in the formation of an **organometallic compound.** Because the metal is less electronegative than carbon, the bond is polarized in the direction opposite to that found in most organic compounds; that is, the negative end of the dipole is on the carbon and the positive end is on the metal. Although the carbon–metal bond is covalent, many organometallic compounds react as though they are carbanions and are useful as carbon nucleophiles.

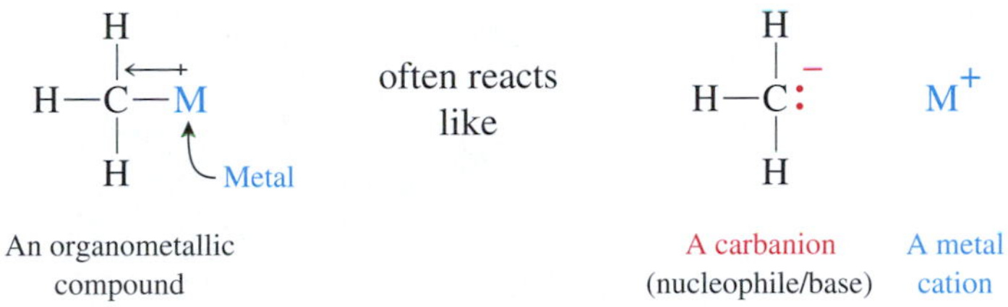

An organometallic compound   often reacts like   A carbanion (nucleophile/base)   A metal cation

Perhaps the most useful of all of the organometallic reagents are the **organomagnesium halides,** known as **Grignard reagents.** These reagents were developed by Victor Grignard, who was awarded the Nobel Prize in chemistry in 1912 for this work. They are readily prepared by reacting organic halides with magnesium metal in a solvent such as diethyl ether or THF. The ether is necessary because it acts as a Lewis base to help stabilize the organomagnesium halide, as shown in the following equation:

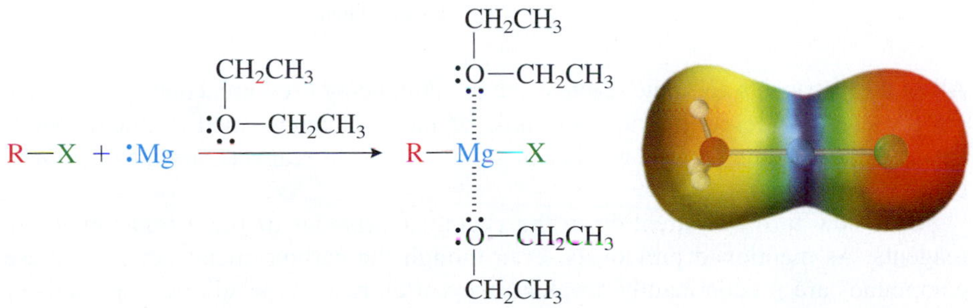

The structure of the R group of RX can vary widely as long as no other reactive functional group is present. The halogen of RX can be iodine (most reactive), bromine, or chlorine (least reactive); fluorine is not commonly used. Several examples are provided

in the following equations. (To simplify writing these reactions, the ether molecules co-ordinated to the magnesium are not shown.) In similar fashion, organolithium reagents can be prepared by reactions of lithium metal with organic halides. Their reactions are nearly identical to those of the Grignard reagents, although the organolithium reagents are slightly more reactive. Note that magnesium, which has two electrons in its valence shell, forms two bonds (not counting its bonds with the ether molecules), whereas lithium, with a single electron in its valence shell, forms only one bond. Therefore, 2 moles of lithium are needed for this reaction.

$$CH_3—I + Mg: \xrightarrow{\text{ether}} CH_3MgI$$

Iodomethane
(methyl iodide)

Methylmagnesium
iodide

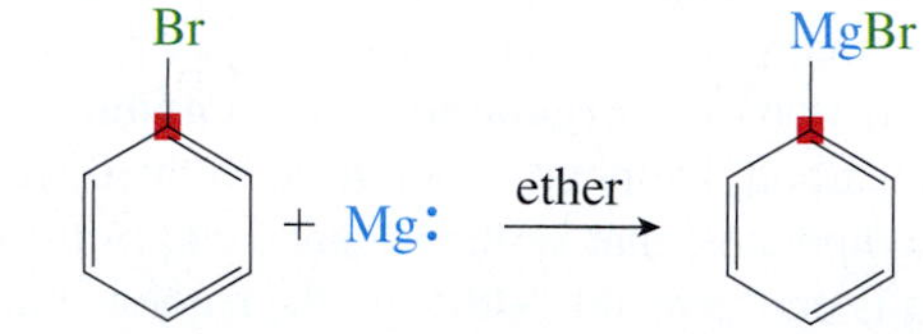

Bromobenzene

Phenylmagnesium
bromide

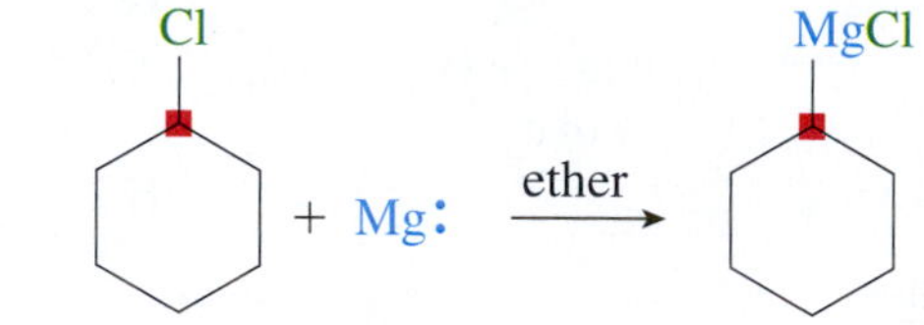

Chlorocyclohexane

Cyclohexylmagnesium
chloride

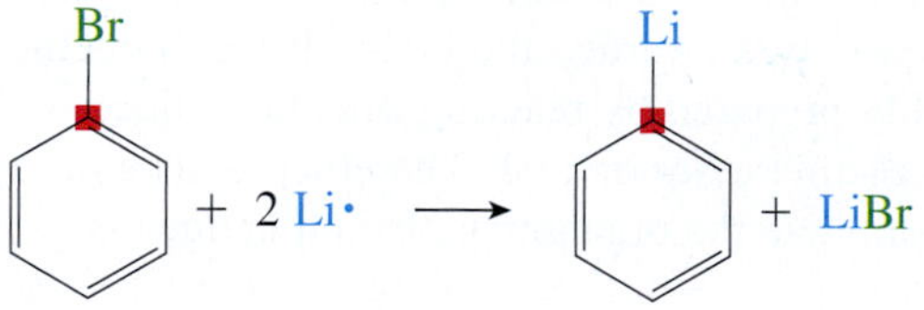

Phenyllithium

All of these reagents are quite reactive and are commonly used immediately after they are prepared. However, solutions of some of them, including methyllithium, butyl-lithium, phenyllithium, and a number of simple Grignard reagents, are now commer-cially available.

Let's now turn our attention to the chemical behavior of these organometallic reagents. As mentioned previously, even though the carbon–metal bonds in these compounds are predominantly covalent, they often react as would be predicted for the corresponding carbanion, although the identity of the metal certainly modifies their behavior somewhat. As expected for carbanions whose pK_a's would be in the range of 45 to 50, all of these organometallic reagents behave as strong bases and react rapidly with even fairly weak acids, such as water and alcohols, as illustrated in the following equation. Therefore, the solvents that are used for their preparation must be scrupulously dried, and compounds containing OH or NH groups must be avoided.

$$CH_3CH_2CH_2CH_2MgBr + H\text{—}OH \longrightarrow CH_3CH_2CH_2\overset{\overset{\displaystyle H}{|}}{C}H_2 + Mg(OH)Br$$

Butylmagnesium bromide   Butane

Even a 1-alkyne is acidic enough to react with a Grignard reagent. This reaction is the most common method of preparing Grignard reagents derived from these alkynes:

$$H\text{—}C\equiv C\text{—}H + CH_3CH_2MgBr \longrightarrow H\text{—}C\equiv C\text{—}MgBr + CH_3\overset{\overset{\displaystyle H}{|}}{C}H_2$$

Ethyne   Ethylmagnesium bromide   Ethane

PROBLEM 18.8

Show the products of these reactions:

a) (Cl on CH₃CHCH₂CH₃) $\xrightarrow[\text{2) } H_2O]{\text{1) Mg, ether}}$

b) (benzene ring with Br) $\xrightarrow[\text{2) } D_2O]{\text{1) Mg, ether}}$

c) CH_3CH_2Br $\xrightarrow[\text{2) } CH_3CH_2C\equiv CH]{\text{1) Mg, ether}}$

18.6 ADDITION OF ORGANOMETALLIC NUCLEOPHILES

The reaction of Grignard reagents and organolithium compounds with aldehydes and ketones is perhaps the most useful method for the preparation of alcohols. The reaction is conducted under basic conditions and proceeds according to the following general mechanism:

$$\overset{\overset{\displaystyle :\ddot{O}:}{\|}}{\underset{\underset{\displaystyle R\text{—}MgX}{}}{R'\text{—}C\text{—}R''}} \longrightarrow \overset{\overset{\displaystyle :\ddot{O}:^-}{|}}{\underset{\underset{\displaystyle R}{|}}{R'\text{—}C\text{—}R''}} \xrightarrow[\text{(workup)}]{H\text{—}A} \overset{\overset{\displaystyle :\ddot{O}\text{—}H}{|}}{\underset{\underset{\displaystyle R}{|}}{R'\text{—}C\text{—}R''}}$$

Because these organometallic reagents are powerful nucleophiles, the reaction is irreversible. After the addition is complete, acid is added in the workup step to protonate the alkoxide ion and produce the alcohol. Reactions employing formaldehyde as the carbonyl component produce primary alcohols, those using other aldehydes produce secondary alcohols, and those using ketones produce tertiary alcohols. Solutions of hydrochloric or sulfuric acid are commonly used in the protonation step. However, to avoid alkene formation, the weaker acid, ammonium chloride ($pK_a = 9$) in aqueous solution, is used when the product is a tertiary or other alcohol that readily undergoes acid-catalyzed E1 elimination (Section 10.13). Of course, if the alkene is the desired product, the elimination can be accomplished during the workup without isolating the alcohol. A number of examples are provided in the following equations.

The reaction of an organometallic reagent with formaldehyde produces a primary alcohol:

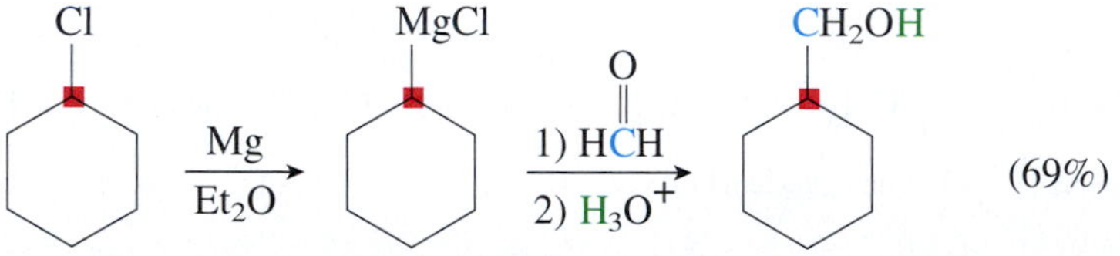

Reactions with other aldehydes produce secondary alcohols. Although the presence of most other functional groups (OH, NH, carbonyl groups, and so on) must be avoided because they react with Grignard or organolithium reagents, ether and alkene groups can be present:

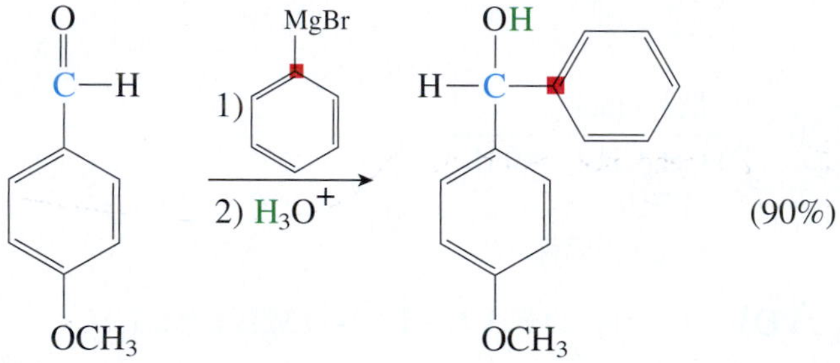

p-Methoxybenzaldehyde

In the next example the bromine, being more reactive than chlorine, selectively reacts to form the Grignard reagent. The product alcohol is especially prone to E1 elimination (the carbocation would be stabilized by resonance), so the weak acid, ammonium chloride, is used in the workup step.

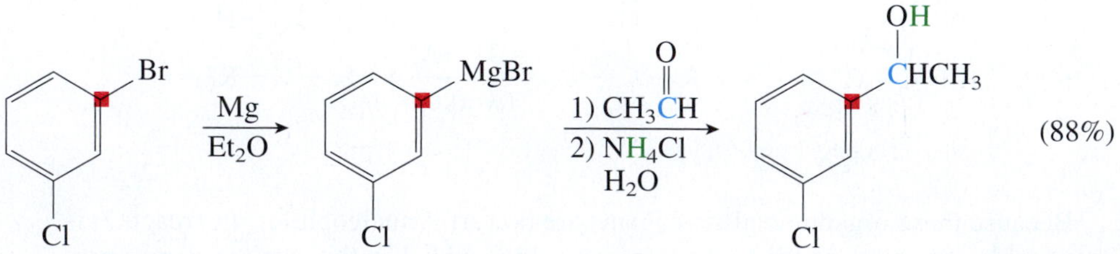

m-Bromochlorobenzene

Reactions with ketones give tertiary alcohols. These are very prone to E1 elimination, so weak acid is used in the workup.

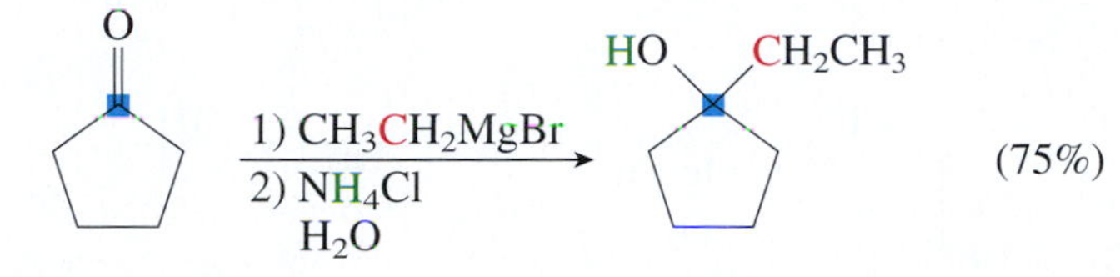

Cyclopentanone

(75%)

Organolithium reagents work well in any of these reactions:

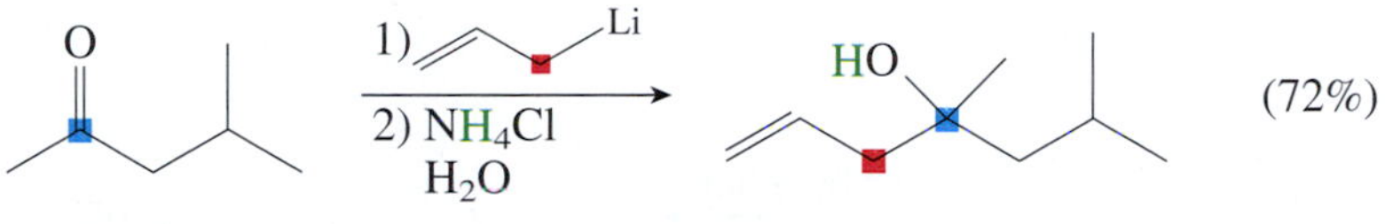

4-Methyl-2-pentanone

(72%)

If the product resulting from elimination of water from the alcohol is desired, then the workup can be conducted by using a stronger acid so that the reaction proceeds directly to the alkene without the isolation of the alcohol.

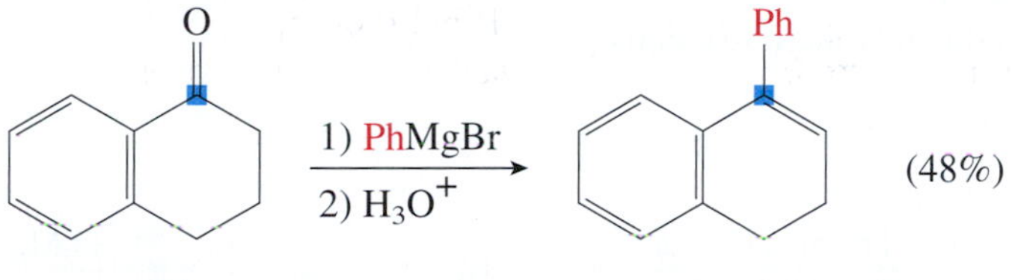

α-Tetralone

(48%)

Acetylenic Grignard reagents are commonly prepared by reaction of the appropriate 1-alkyne with an alkyl Grignard reagent, such as ethylmagnesium bromide.

$H-C \equiv C-H$ + CH_3CH_2MgBr ⟶ $H-C \equiv C-MgBr$

Ethyne   Ethylmagnesium bromide   + CH_3CH_3

(69%)

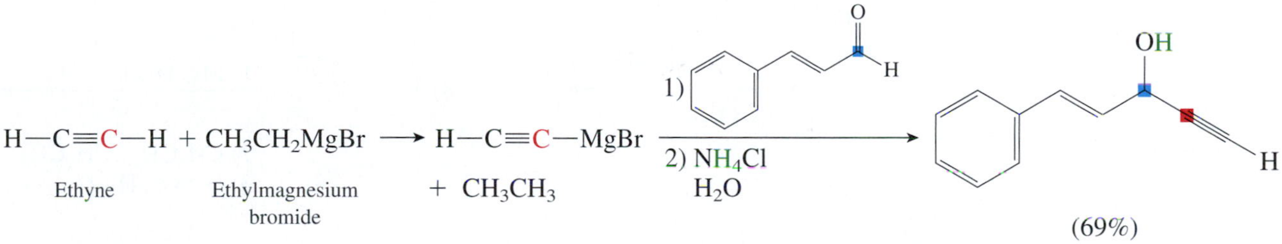

The carbon of carbon dioxide is electrophilic, similar to a carbonyl carbon. Grignard reagents react with carbon dioxide to form salts of carboxylic acids:

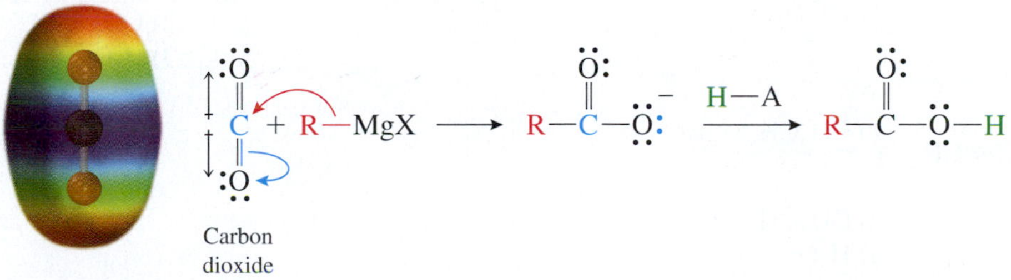

Carbon dioxide

Acidification of the reaction mixture produces a carboxylic acid. Examples are provided in the following equations:

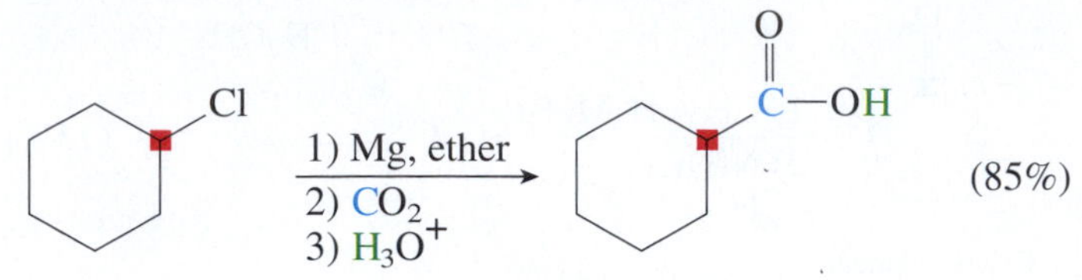
Cl
O
C—OH
1) Mg, ether
2) CO2
3) H3O+
(85%)

Chlorocyclohexane

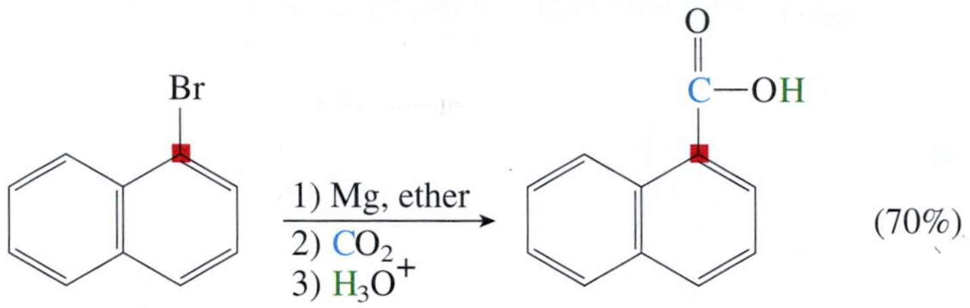
Br
O
C—OH
1) Mg, ether
2) CO2
3) H3O+
(70%)

1-Bromonaphthalene

PROBLEM 18.9

Show the products of these reactions:

a) 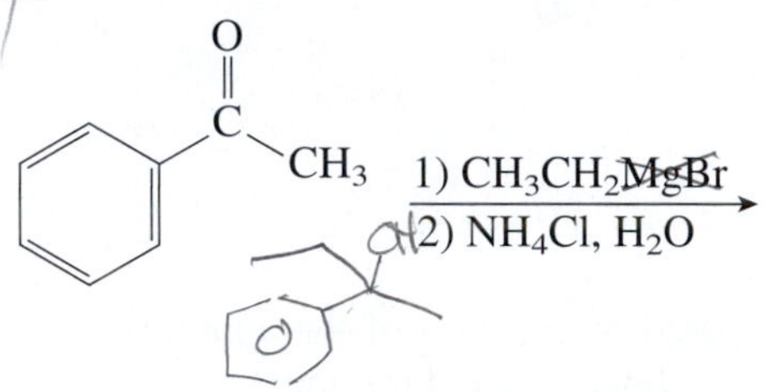
O
C
CH3
1) CH3CH2MgBr
2) NH4Cl, H2O

b) 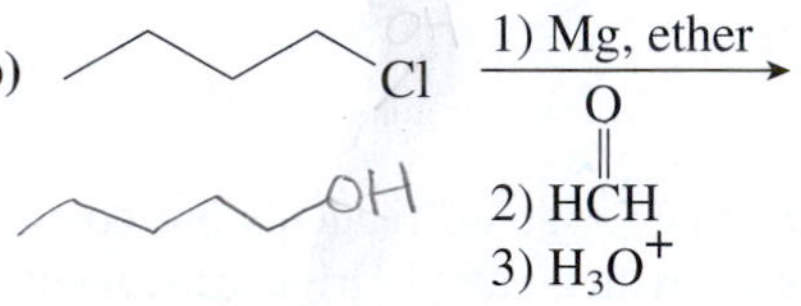
Cl
1) Mg, ether
O
2) HCH
3) H3O+

c) 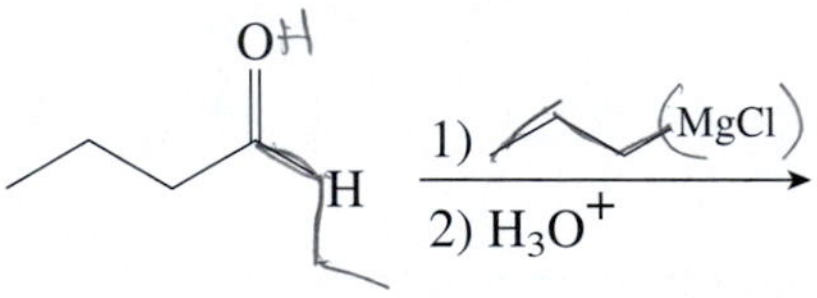
H
1) MgCl
2) H3O+

d) 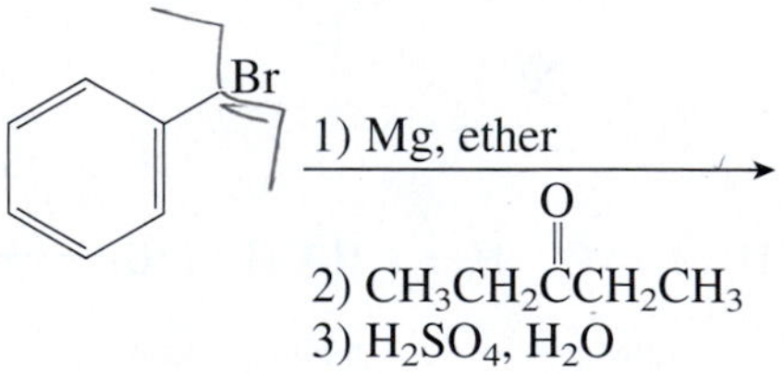
Br
1) Mg, ether
O
2) CH3CH2CCH2CH3
3) H2SO4, H2O

e) 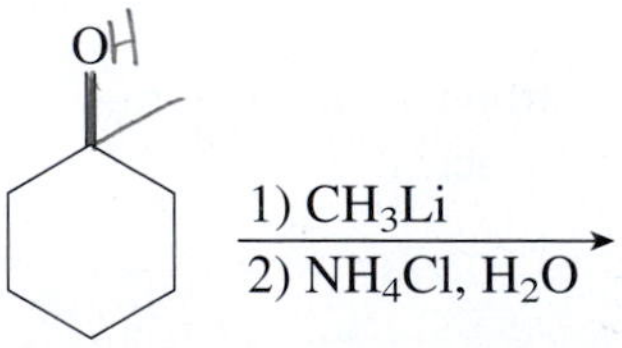
1) CH3Li
2) NH4Cl, H2O

f) 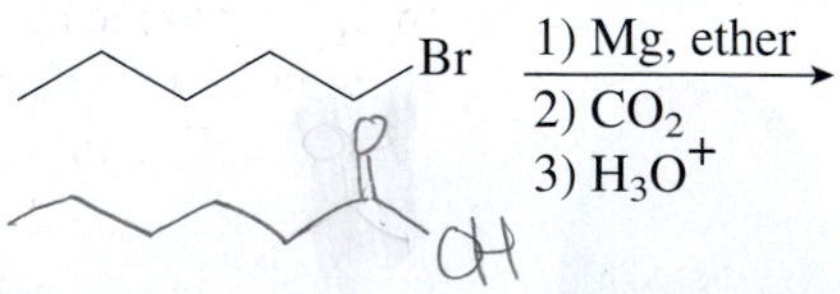
Br
1) Mg, ether
2) CO2
3) H3O+

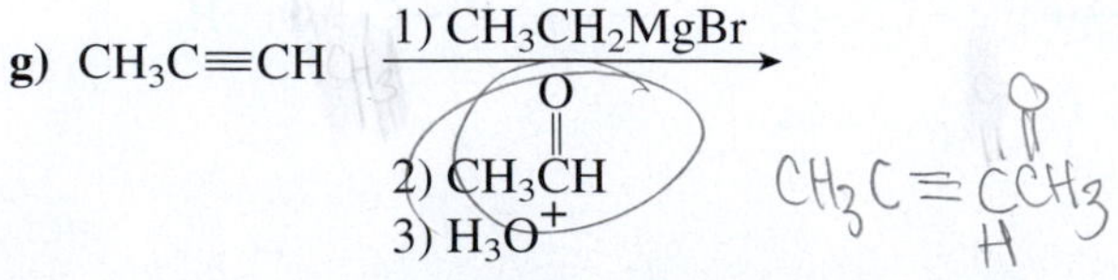
g) CH3C≡CH
1) CH3CH2MgBr
O
2) CH3CH
3) H3O+

The Grignard reaction is an important and versatile way to prepare alcohols. Whenever an alcohol is encountered as a synthetic target, this reaction should be considered because it forms a carbon–carbon bond, building the alcohol from smaller compounds, and it often allows more than one route to the desired product. For example, suppose the target is 2-phenyl-2-butanol. This alcohol can be prepared by three different Grignard reactions:

$$
\begin{array}{c}
\text{OH} \\
| \\
\text{H}_3\text{C}-\text{C}-\text{CH}_2\text{CH}_3 \\
| \\
\text{Ph}
\end{array}
$$

2-Phenyl-2-butanol

$$
\begin{array}{ccc}
\text{O} & \text{O} & \text{O} \\
\parallel & \parallel & \parallel \\
\text{CH}_3\text{CCH}_2\text{CH}_3 & \text{PhCCH}_2\text{CH}_3 & \text{PhCCH}_3 \\
+ \ \text{PhMgBr} & + \ \text{CH}_3\text{MgI} & + \ \text{CH}_3\text{CH}_2\text{MgBr}
\end{array}
$$

Because the product is a tertiary alcohol, each of these reactions must be acidified with a weak acid (NH_4Cl/H_2O) to avoid elimination. Which of these pathways is the best depends on a number of factors, such as the availability of the ketone and the halide needed to prepare the Grignard reagent and the yield of the reaction.

PRACTICE PROBLEM 18.2

Show two ways to prepare 2-butanol using Grignard reagents.

$$
\begin{array}{c}
\text{OH} \\
| \\
\text{CH}_3\text{CH}_2\text{CHCH}_3
\end{array}
$$

2-Butanol

Strategy

When you encounter an alcohol as a synthetic target, consider using a Grignard reaction to synthesize it. The carbon bonded to the hydroxy group was the electrophilic carbon of the carbonyl group of the reactant. One of the alkyl groups bonded to this carbon was the nucleophilic carbon of the Grignard reagent or alkyllithium reagent. There are often several ways to accomplish the synthesis, depending on which alkyl group is added as the nucleophile. Remember to work up the reaction with the weak acid NH_4Cl if the product alcohol is prone to E1 elimination.

Solution

The carbon (blue) bonded to the hydroxy group in the alcohol comes from the carbonyl carbon electrophile of the starting material. This carbon is also bonded to a methyl group and an ethyl group. We can add the ethyl group, using the reaction of ethylmagnesium bromide and ethanal, or we can add the methyl group, using methylmagnesium iodide and propanal.

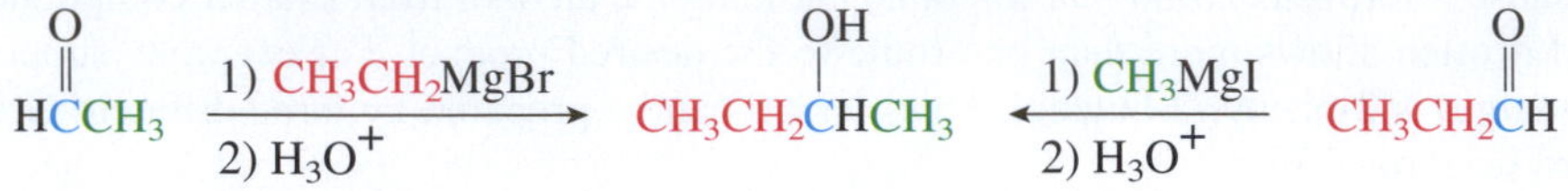

$$\underset{HCCH_3}{\overset{O}{\|}} \quad \overset{1) \ CH_3CH_2MgBr}{\underset{2) \ H_3O^+}{\longrightarrow}} \quad \underset{CH_3CH_2CHCH_3}{\overset{OH}{|}} \quad \overset{1) \ CH_3MgI}{\underset{2) \ H_3O^+}{\longleftarrow}} \quad \underset{CH_3CH_2CH}{\overset{O}{\|}}$$

PROBLEM 18.10

Suggest syntheses of these compounds using Grignard reagents:

a) 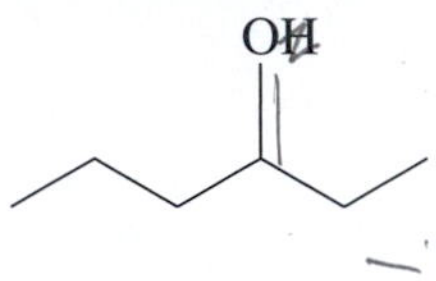 (2 ways)

—MgBr
H3O+

b)

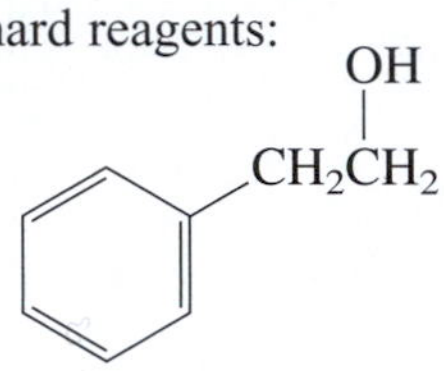

c) 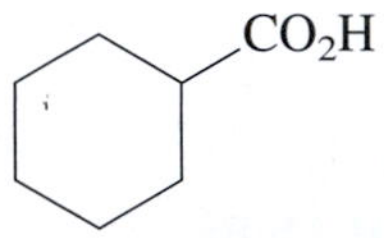 CO₂H

d) Ph₃COH

e) 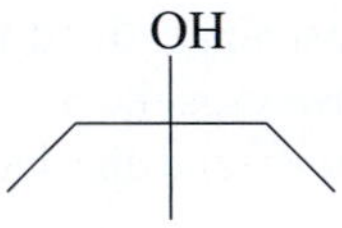 OH (2 ways)

f) 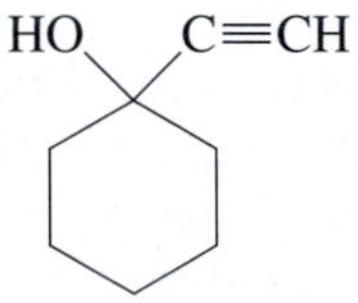 HO C≡CH

g) 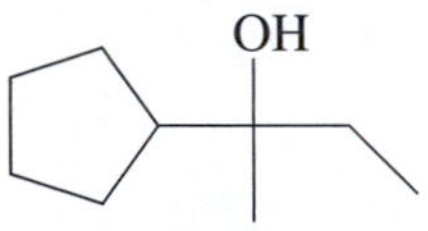 OH (3 ways)

PROBLEM 18.11

When the Grignard reaction shown in this equation is attempted, the products are benzene and the starting hydroxyaldehyde. Explain this result.

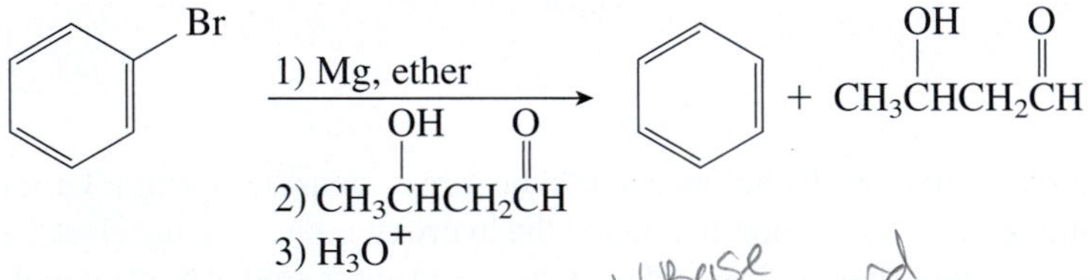

$$\overset{Br}{\bigcirc} \quad \overset{1) \ Mg, \ ether}{\underset{\begin{array}{c}2) \ CH_3CHCH_2CH \\ 3) \ H_3O^+\end{array}}{\longrightarrow}} \quad \bigcirc \ + \ CH_3\overset{OH}{\overset{|}{C}}HCH_2\overset{O}{\overset{\|}{C}}H$$

+H
⟨⟩⁻ mgBr

Acid/Base
no Grignard

18.7 ADDITION OF PHOSPHORUS YLIDES; THE WITTIG REACTION

In Chapters 9 and 10 the use of elimination reactions to prepare alkenes was described. The major problem with that method is that a mixture of alkenes is often produced, resulting in lower yields and separation problems. The Wittig reaction provides an alternative method for the synthesis of alkenes. It is especially useful because it results in carbon–carbon bond formation and the position of the double bond is completely controlled. Georg Wittig shared the 1979 Nobel Prize in chemistry for developing this reaction. (He shared the award with H. C. Brown, who developed the hydroboration reaction; see Section 11.7.)

The nucleophile used in this reaction is called an **ylide.** It is a carbanion that is bonded to a positive phosphorus group that helps to stabilize it:

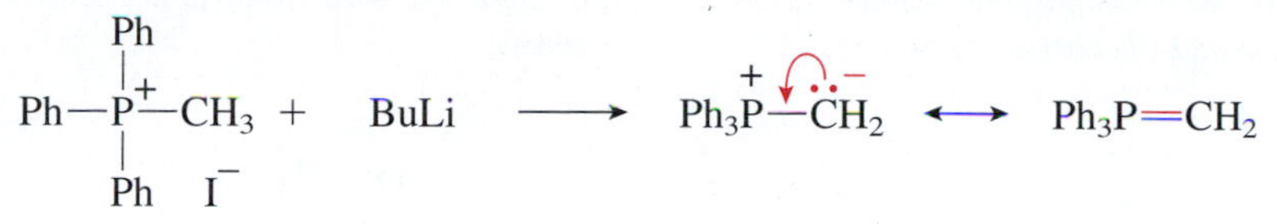

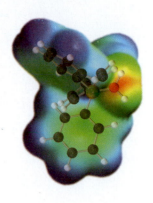

Electrostatic potential
map of $Ph_3\overset{+}{P}{-}\overset{..}{\underset{.}{\overset{-}{C}}}H_2$

| Ph
\|
Ph—$\overset{+}{P}$—CH₃ + | BuLi | $\longrightarrow$ | $Ph_3\overset{+}{P}{-}\overset{..}{\overset{-}{C}}H_2$ | $\longleftrightarrow$ | $Ph_3P{=}CH_2$ |

(with Ph and I⁻ below the phosphorus)

| Phosphonium salt | Butyllithium | | Ylide | Minor contributor | |

The ylide is prepared by deprotonating a triphenylalkylphosphonium salt with a strong base, commonly an organometallic base such as butyllithium or phenyllithium. The hydrogens on the carbon that is bonded to the phosphorus of the salt are somewhat acidic because the carbanion of the conjugate base (the ylide) is stabilized by the inductive effect of the positive phosphorus atom. In addition, a resonance structure with five bonds to phosphorus makes a minor contribution to the structure and provides some additional stabilization. The triphenylalkylphosphonium salt can be prepared by an S_N2 reaction of triphenylphosphine with the appropriate alkyl halide (see Section 10.9).

$$Ph_3P\colon\;+\;H_3C{-}I\;\xrightarrow{S_N2}\;Ph_3\overset{+}{P}{-}CH_3\overset{I^-}{}\quad(99\%)$$

PROBLEM 18.12

Show a preparation of this phosphonium salt from an alkyl halide:

$$Ph_3\overset{+}{P}{-}CH_2Ph\quad Cl^-$$

The reaction of the ylide with an aldehyde or ketone results in the formation of an alkene with the double bond connecting the carbonyl carbon of the reactant to the anionic carbon of the ylide, as shown in the following example:

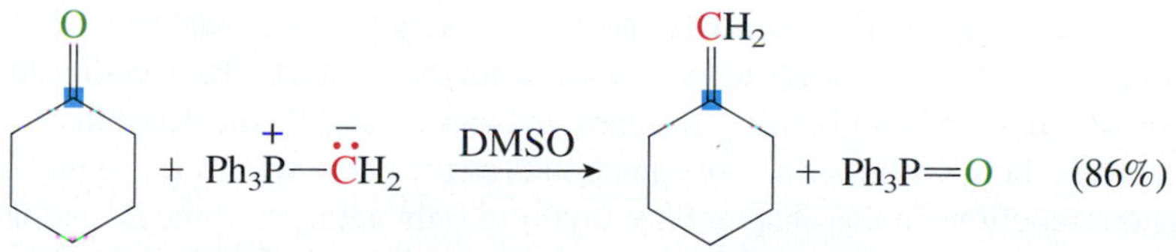

$$\text{(cyclohexanone)} + Ph_3\overset{+}{P}{-}\overset{..}{\overset{-}{C}}H_2 \xrightarrow{DMSO} \text{(methylenecyclohexane)} + Ph_3P{=}O\quad(86\%)$$

The by-product is triphenylphosphine oxide. The ylide is a strong nucleophile, so the equilibrium greatly favors the products and the reaction is irreversible.

The mechanism for this reaction is shown in Figure 18.2. The first part of the mechanism is similar to the others that have been presented so far: the nucleophile (the negative carbon of the ylide) attacks the carbonyl carbon. However, unlike the previous cases, this

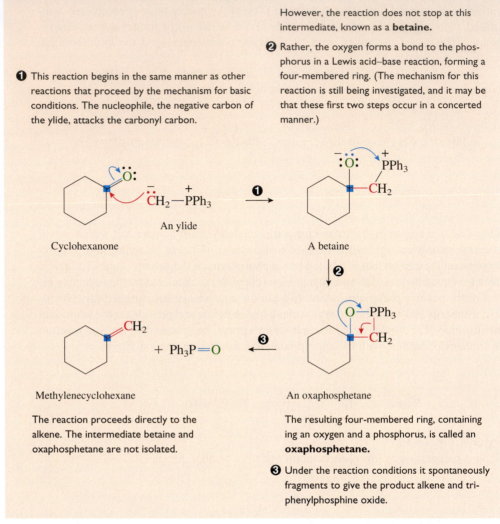

① This reaction begins in the same manner as other reactions that proceed by the mechanism for basic conditions. The nucleophile, the negative carbon of the ylide, attacks the carbonyl carbon.

However, the reaction does not stop at this intermediate, known as a **betaine.**

② Rather, the oxygen forms a bond to the phosphorus in a Lewis acid–base reaction, forming a four-membered ring. (The mechanism for this reaction is still being investigated, and it may be that these first two steps occur in a concerted manner.)

Cyclohexanone

An ylide

A betaine

Methylenecyclohexane

An oxaphosphetane

The reaction proceeds directly to the alkene. The intermediate betaine and oxaphosphetane are not isolated.

The resulting four-membered ring, containing an oxygen and a phosphorus, is called an **oxaphosphetane.**

③ Under the reaction conditions it spontaneously fragments to give the product alkene and triphenylphosphine oxide.

Active Figure 18.2

ORGANIC
Chemistry ⚛ Now™

MECHANISM OF THE WITTIG REACTION. Test yourself on the concepts in this figure at **OrganicChemistryNow.**

reaction proceeds beyond this step. The phosphorus, acting as a Lewis acid, bonds to the basic oxygen and ultimately leads to its removal from the product. The overall result is the formation of a double bond between the carbonyl carbon and the nucleophilic carbon of the ylide, while both of the carbon–oxygen bonds of the carbonyl group are broken.

Previous reactions in this chapter have involved only addition of the nucleophile and a hydrogen to the carbonyl group. In this reaction, addition is followed by elimination of the oxygen to form a double bond between the carbonyl carbon and the nucleophile. Such an **addition–elimination** reaction occurs when the nucleophile has or can generate (by the loss of a proton or a phosphorus group) a second pair of electrons that can be used to form a second bond to the electrophilic carbon. In the case of the Wittig reaction, the phosphorus and the oxygen are eliminated to form the alkene. The formation of the strong phosphorus–oxygen bond helps make this step of the reaction favorable. (Note that the previous reactions in this chapter, employing hydride, cyanide, and organometallic nucleophiles, did not have any way to form a second bond between the nucleophile and the electrophilic carbon, so only addition occurred.)

Because the location of the double bond in the product is well defined, the Wittig reaction provides probably the most useful general method for the preparation of alkenes. Some examples are provided in the following equations:

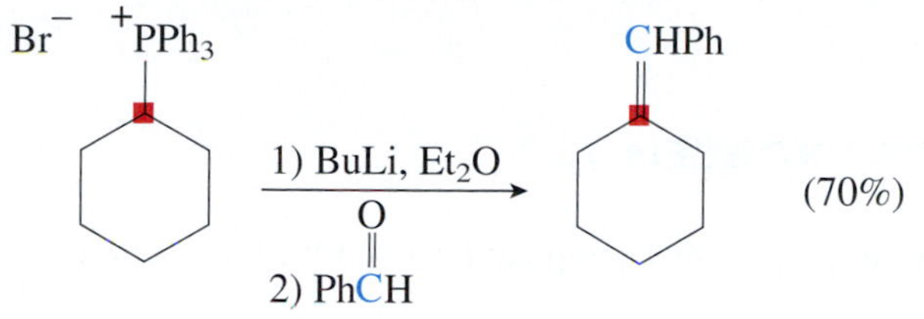

$$PhCH=CHCH + Ph_3P-\overset{..}{\overset{-}{C}}HPh \xrightarrow{EtOH} PhCH=CHCH=CHPh \quad (84\%)$$

Because the electron pair of the carbanion of the ylide in the following example is stabilized by resonance delocalization with the carbonyl group, it is a weaker nucleophile. Such ylides react readily with aldehydes but do not react well with ketones.

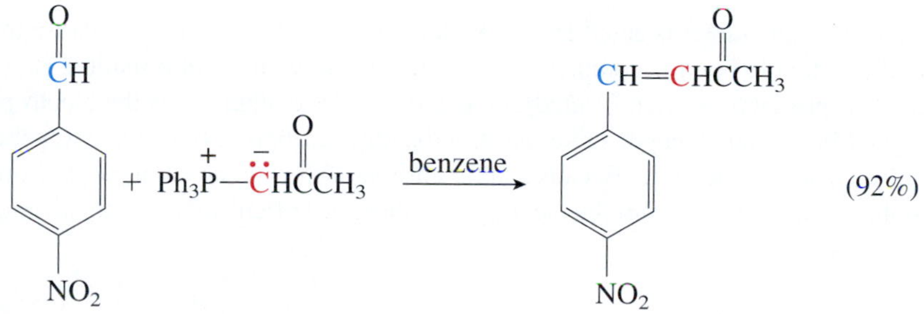

PROBLEM 18.13

Show the products of these reactions:

a) $PhCH + \left(CH_3CH_2\overset{..}{\overset{-}{C}}H-\overset{+}{P}Ph_3 \right) \longrightarrow$

Ph–CH=CH₂CH₃

b) 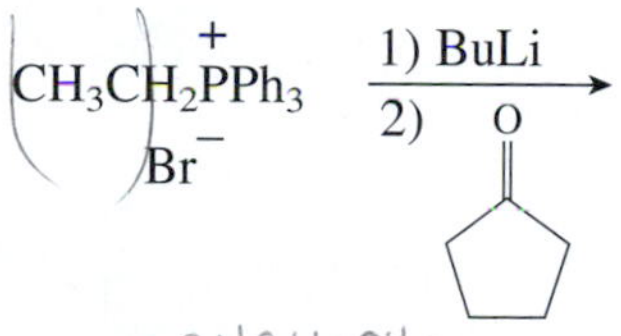 $\left(CH_3CH_2\overset{+}{P}Ph_3 \atop Br^- \right) \xrightarrow{1) BuLi \atop 2)}$

CHCH₂CH₃

c) $HC\equiv CCH + Ph_3P-\overset{..}{\overset{-}{C}}H(CH_2CH_2CH_3) \longrightarrow$

HC≡CCH CHCH₂CH₃

d) 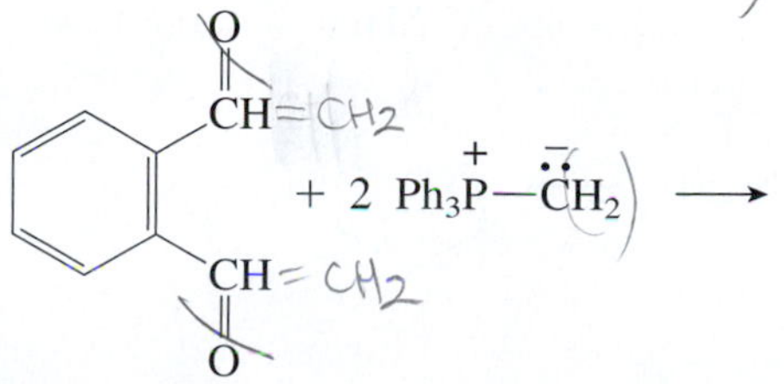 [structure] $+ 2 Ph_3P-\overset{..}{\overset{-}{C}}H_2 \longrightarrow$

CH=CH₂ CH=CH₂

e) $PhCH + Ph_3P-\overset{..}{\overset{-}{C}}HCOEt \longrightarrow$

CHCOEt PhCH

PROBLEM 18.14

Explain why the phosphonium salt shown in the following equation can be deprotonated by using sodium ethoxide, a much weaker base than the butyllithium that is usually needed to deprotonate other phosphonium salts.

$$\overset{+}{Ph_3P}-CH_2Ph \;\; \overset{Cl^-}{} \; + \; NaOEt \;\; \longrightarrow \;\; \overset{+}{Ph_3P}-\overset{\displaystyle ..}{\overset{..}{C}}HPh \; + \; HOEt \; + \; NaCl$$

PRACTICE PROBLEM 18.3

Show two ways to prepare this compound using a Wittig reaction.

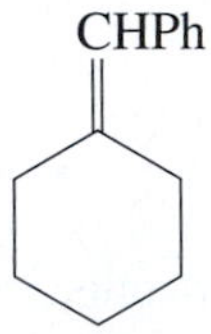

Strategy

When the synthetic target is an alkene, consider using a Wittig reaction, because the location of the double bond is completely controlled (unlike an E2 elimination, in which a mixture of products is often formed). One carbon of the alkene was the electrophilic carbonyl carbon of an aldehyde or ketone and the other carbon of the alkene was the nucleophilic carbon of the ylide. Because either carbon of the alkene could come from the nucleophile and the other from the electrophile, there are often two ways to accomplish the synthesis.

Solution

Break the double bond, making one carbon the nucleophile of an ylide and the other carbon part of a carbonyl group. The two possible ways to do this in this case are as follows:

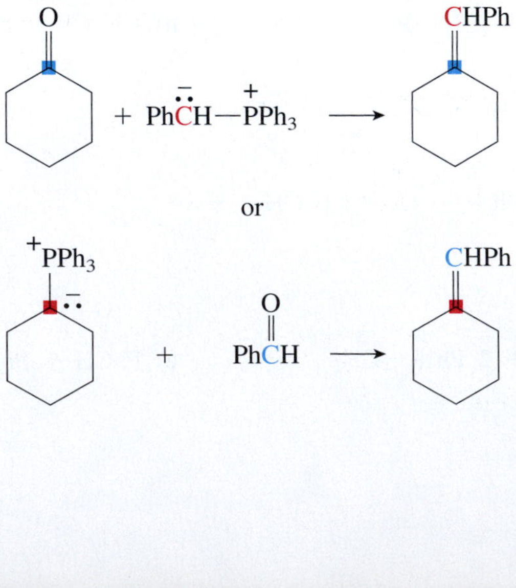

PROBLEM 18.15

Suggest syntheses of these compounds using the Wittig reaction:

a) $PhCH_2CH{=}CHCH_2CH_3$ (2 ways)

b) $CH_3CH_2CH{=}CHCCH_2CH_3$ (with carbonyl O)

c) (aromatic ring with H_3C substituent)$-CH{=}CHCOCH_3$

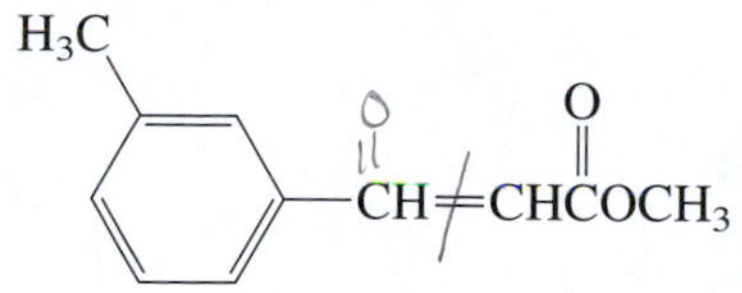

Focus On

Synthesis of Vitamin A

Vitamin A, also known as retinol, is essential for vision in mammals (see the Focus On box on page 773) and is involved in a number of other important biological functions, such as bone growth and embryonic development. A deficiency in vitamin A leads to night blindness, in which the eye cannot see in dim light. Our bodies are capable of converting compounds such as β-carotene, an orange pigment that is present in many vegetables, to vitamin A. As its structure suggests, vitamin A is relatively nonpolar and therefore is not very soluble in water. It accumulates in fat deposits and is not readily excreted. For this reason, too much vitamin A is toxic.

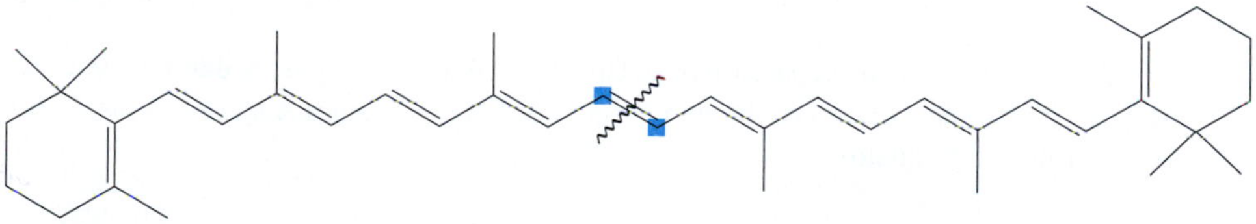

β-Carotene

Oxidative cleavage

2

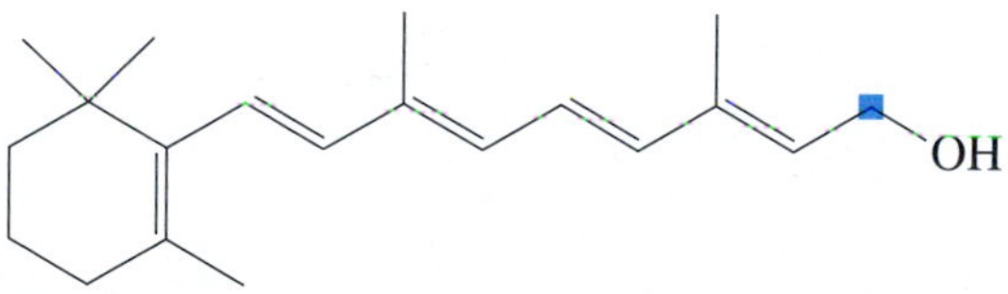

Vitamin A (retinol)

The Wittig reaction has proved to be especially useful in the synthesis of natural products, such as vitamin A, which contain a number of carbon–carbon double bonds. An industrial synthesis of vitamin A is outlined in the following equations:

Continued

❶ β-Ionone can be isolated from natural sources or synthesized in the laboratory. It is reacted with ethynyl magnesium bromide, or some other source of the anion derived from ethyne, to produce an alcohol (see Section 18.6).

❷ The triple bond is reduced to a double bond using hydrogen and the Lindlar catalyst (see Section 11.12).

❸ Then, reaction of the alcohol with triphenylphosphine under acidic conditions produces the phosphonium salt by an S_N1 reaction. A resonance-stabilized carbocation is formed and reacts at the terminal position of the chain.

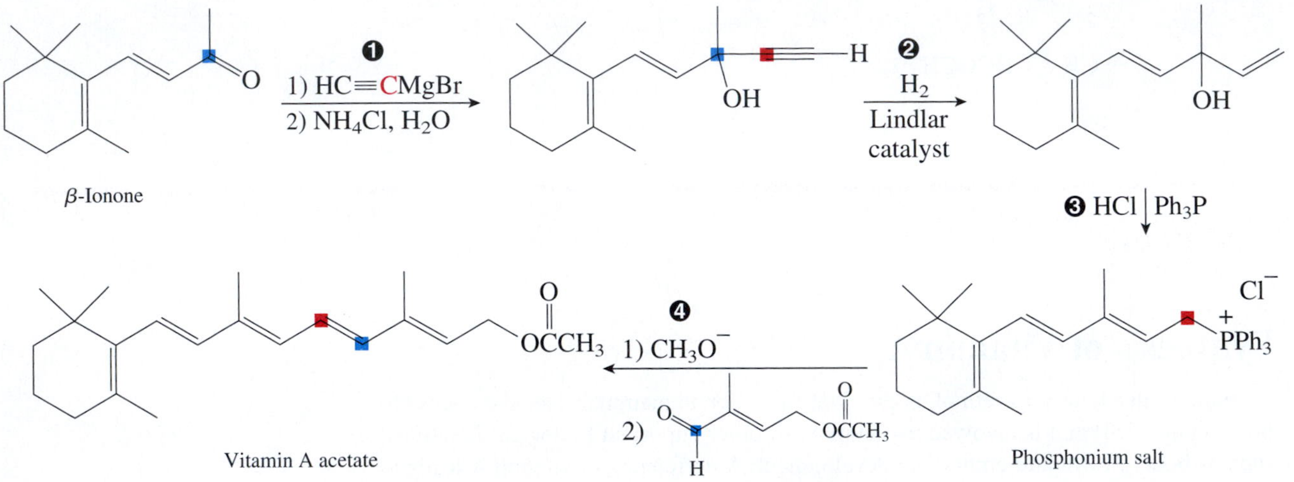

β-Ionone

Vitamin A acetate

Phosphonium salt

❹ The phosphonium salt is more acidic than usual because its conjugate base, the ylide, is stabilized by resonance involving the double bonds. Therefore, methoxide ion, a weaker base than usual, can be used to form the ylide. Reaction of the ylide with the aldehyde that has its hydroxy group protected as an ester produces vitamin A acetate. The acetate group can readily be removed to complete the synthesis of vitamin A (see Section 10.2).

β-Carotene is also prepared industrially by a Wittig reaction. A dialdehyde is reacted with two equivalents of the same ylide used in the vitamin A synthesis, as shown in the following equation:

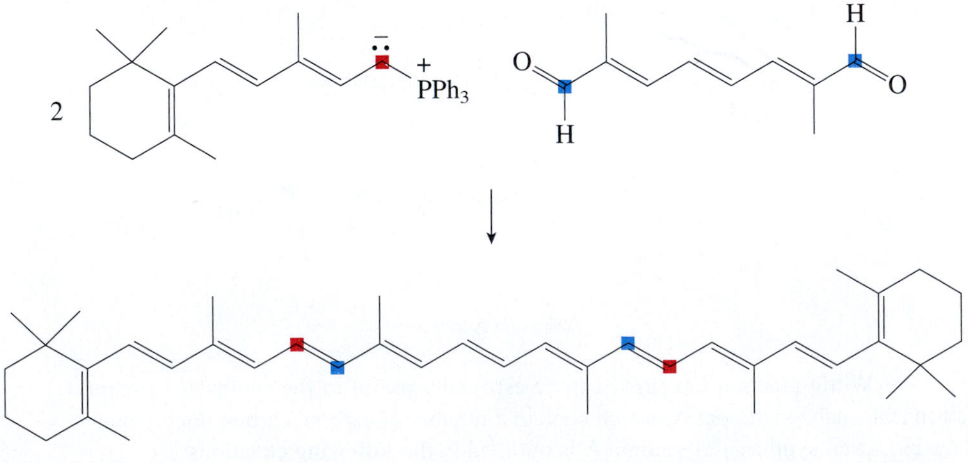

β-Carotene

It is interesting to note that an alkene can be prepared by two different Wittig pathways, depending on which of the doubly bonded carbons was originally the carbon of the carbonyl group and which was the carbon of the ylide. Thus, the synthesis of β-carotene has also been accomplished by using the reaction of a diylide with two equivalents of an aldehyde, as illustrated in the following equations:

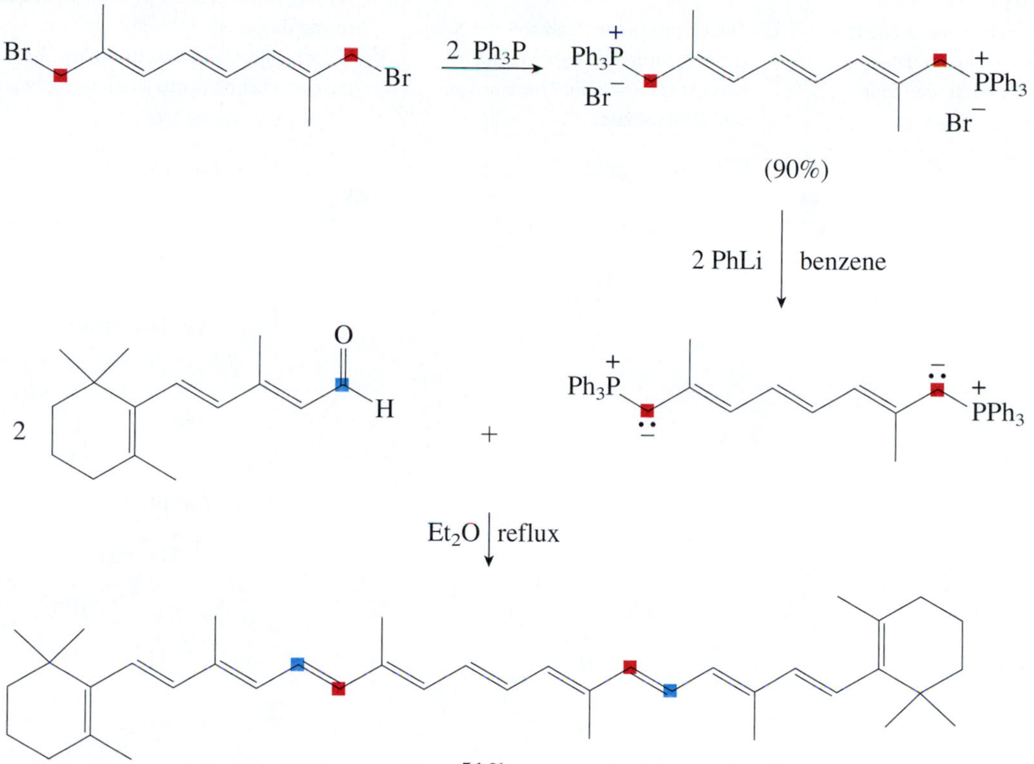

(90%)

2 PhLi | benzene

Et₂O | reflux

51%
β-Carotene

18.8   ADDITION OF NITROGEN NUCLEOPHILES

Amines add to the carbonyl groups of aldehydes or ketones to produce compounds containing CN double bonds and water. These nitrogen analogs of aldehydes and ketones are called **imines**.

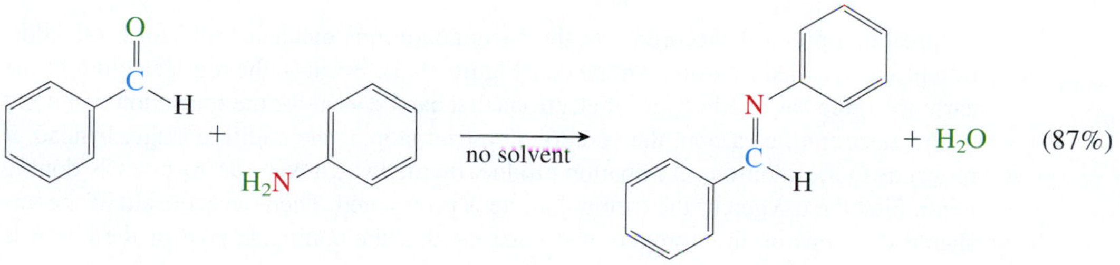

Benzaldehyde   Aniline   An imine

❶ The addition step occurs according to the basic conditions mechanism. If the solution is too acidic, the amine is protonated and is no longer nucleophilic. The rate of this first step then becomes very slow.

❷ The O gets protonated and the N gets deprotonated by acids and bases in the solution. These proton transfers are fast.

The intermediate is called a **carbinolamine.** (Most of the other addition reactions presented so far stop at this stage where the nucleophile has added to the carbonyl carbon and a proton has added to the oxygen.)

❸ To proceed onward to the imine, the oxygen must be protonated so that water can act as a leaving group.

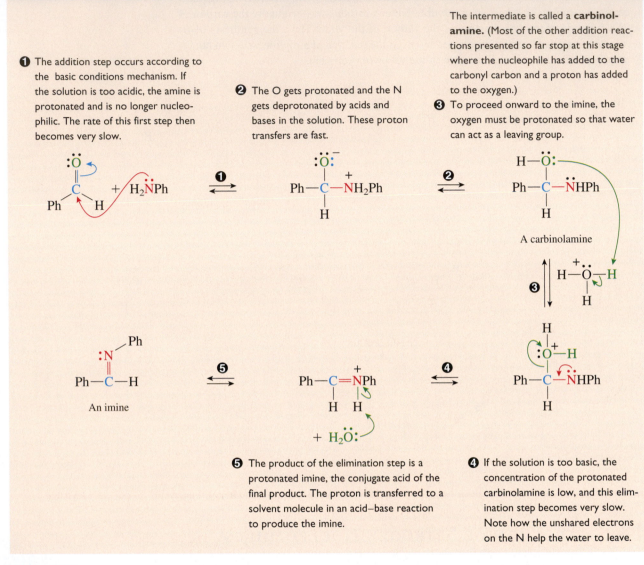

❺ The product of the elimination step is a protonated imine, the conjugate acid of the final product. The proton is transferred to a solvent molecule in an acid–base reaction to produce the imine.

❹ If the solution is too basic, the concentration of the protonated carbinolamine is low, and this elimination step becomes very slow. Note how the unshared electrons on the N help the water to leave.

Figure 18.3

MECHANISM OF THE ADDITION OF AN AMINE TO AN ALDEHYDE TO FORM AN IMINE.

ORGANIC
Chemistry ♦ Now™
Click *Mechanisms in Motion* to view the **Mechanisms of Imine Formation.**

The reaction proceeds according to the basic conditions mechanism to form the addition product, called a carbinolamine (see Figure 18.3). Because the nitrogen atom of the carbinolamine has another pair of electrons that can be used for the formation of a bond to the electrophilic carbon, the reaction does not stop at the addition stage. Instead, it proceeds to the addition–elimination product by the loss of water to form a CN double bond. First the oxygen of the carbinolamine is protonated. Then, with the aid of the unshared electrons on the nitrogen, water leaves, and the conjugate acid of the imine is formed. Loss of a proton gives the imine.

The rate of this reaction has an interesting dependence on the pH of the solution. At low pH the reaction is slow because the amine nucleophile is protonated in the strongly acidic solution. The low concentration of the nucleophile makes the addition step slow. At higher pH the concentration of the unprotonated amine is larger, and therefore the reaction is faster. However, if the solution is too basic, then the reaction is again slow because the concentration of the protonated carbinolamine is low, and therefore the elimination of water is slow. The maximum reaction rate occurs in the pH range of 4 to 6.

Because a CO double bond is considerably stronger than a CN double bond, the equilibrium in these reactions often favors the carbonyl compound rather than the imine. In such cases it is necessary to drive the equilibrium to the product. This is usually accomplished by removing the water as it is formed. Some additional examples of imine formation are provided in the following equations:

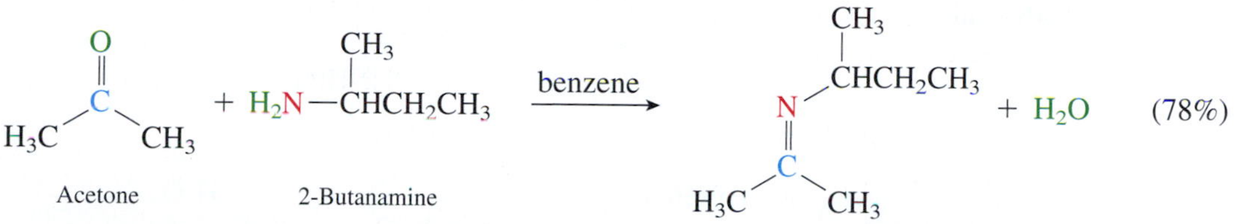

The product in the following reaction results from the formation of two CN double bonds. In this case the equilibrium favors the product because of the additional resonance stabilization provided by the new aromatic ring.

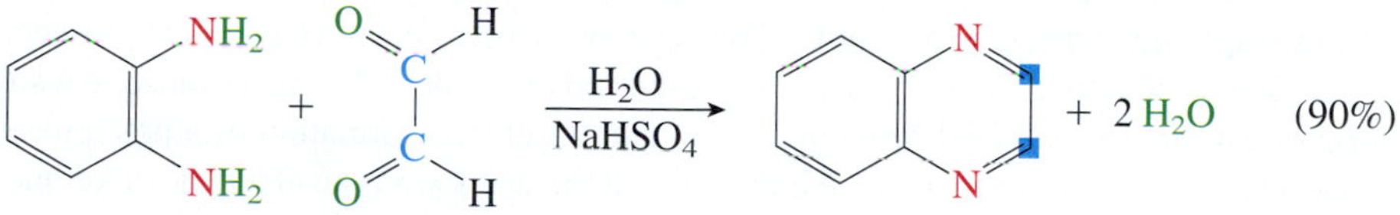

In the days before the advent of spectroscopic techniques, reactions that form CN double bonds were often used in determining the structure of unidentified aldehydes and ketones. After the functional group of an unknown compound was determined and its possible identity was narrowed to a few choices, the final step in the structure determination was often the conversion of the unknown to a solid derivative using a standard chemical reaction of that functional group. The melting point of the derivative was then compared with the melting points that had been reported in the literature for the derivatives of the possible candidates. A match between the melting points was considered strong evidence in establishing the identity of the unknown.

In the case of aldehydes and ketones, several C=N forming reactions were used to make derivatives. The most common of these are shown in the following equations. Note that each reagent has an electronegative group substituted on the NH_2. This helps shift the equilibrium toward the product and makes it more likely to be a solid. Tables of the melting points of these derivatives for common aldehydes and ketones can be found in many reference books.

The reaction of an aldehyde or a ketone with **hydroxylamine** produces an **oxime** derivative:

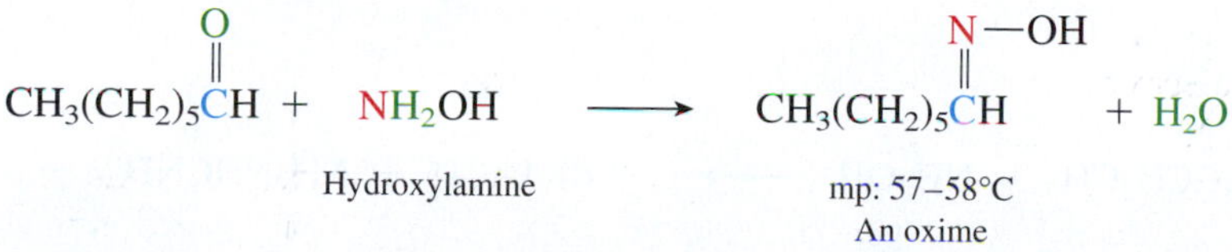

The reaction of an aldehyde or a ketone with **semicarbazide** produces a **semicarbazone** derivative:

$$CH_3\overset{O}{\underset{\|}{C}}CH_3 \ + \ NH_2NH\overset{O}{\underset{\|}{C}}NH_2 \ \longrightarrow \ CH_3\overset{N NHC NH_2}{\underset{\|}{C}}CH_3 \ + \ H_2O$$

Semicarbazide   mp: 187°C
                                       A semicarbazone

The reaction of an aldehyde or a ketone with **phenylhydrazine** produces a **phenylhydrazone** derivative:

(cyclohexanone with O) + PhNH**NH₂** ⟶ (cyclohexylidene with NNHPh) + H₂O

Phenylhydrazine   mp: 77°C
                                       A phenylhydrazone

The reaction of an aldehyde or a ketone with **2,4-dinitrophenylhydrazine** produces a **2,4-dinitrophenylhydrazone** derivative. This reaction is also used as a test for the presence of an aldehyde or ketone. A drop or two of a suspected aldehyde or ketone is added to a solution of 2,4-dinitrophenylhydrazine in ethanol and water. The formation of a precipitate, usually orange or red, of the derivative indicates that the unknown is an aldehyde or ketone:

$$CH_3\overset{O}{\underset{\|}{C}}CH_2CH_3 \ + \ \text{(2,4-Dinitrophenylhydrazine)} \ \longrightarrow \ \text{(2,4-dinitrophenylhydrazone)} \ + \ H_2O$$

2,4-Dinitrophenylhydrazine   mp: 117°C
                                                    A 2,4-dinitrophenylhydrazone

PROBLEM 18.16
Show the products of these reactions:

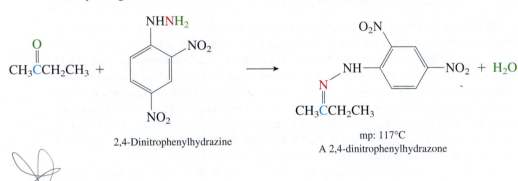

a) (cyclopentanone) + PhNH₂ ⟶

b) (3-methoxy-1,2-benzoquinone) + (1,2-diaminobenzene, H₂N, H₂N) ⟶

c) $CH_3\overset{O}{\underset{\|}{C}}CH_2CH_3$ + NH₂OH ⟶

d) $Ph\overset{O}{\underset{\|}{C}}H$ + $NH_2NH\overset{O}{\underset{\|}{C}}NH_2$ ⟶

e)

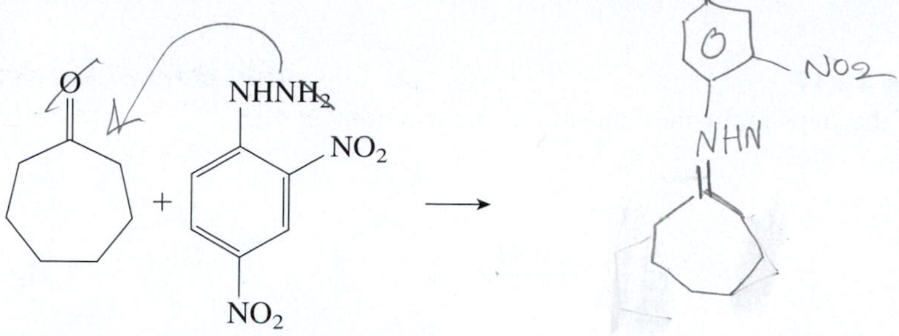

So far, all of the examples have involved primary amines. The reaction of ammonia with aldehydes and ketones also forms imines, but the products are unstable and cannot usually be isolated. If a secondary amine is used, an enamine, rather than an imine, is formed. An **enamine** has an amino group bonded to one of the carbons of a CC double bond. It is related to the imine in the same manner as an enol is related to a ketone (see Section 11.6). The mechanism for its formation can be outlined as follows:

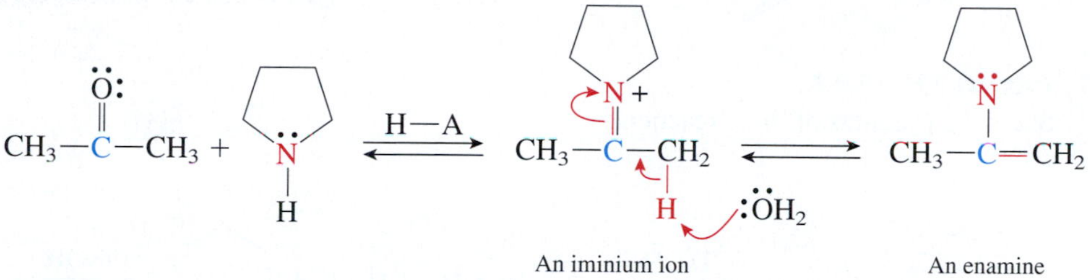

An iminium ion   An enamine

The reaction of an aldehyde or a ketone with a secondary amine follows exactly the same mechanism as the reaction with a primary amine (see Figure 18.3) until the final step. Unlike the case with a primary amine, the nitrogen of the iminium ion does not have a proton that can be removed to produce a stable imine. Therefore, a proton is removed from an adjacent carbon, resulting in the formation of an enamine. Enamine formation is illustrated in the following equations. In each case the equilibrium is driven toward the products by removal of water.

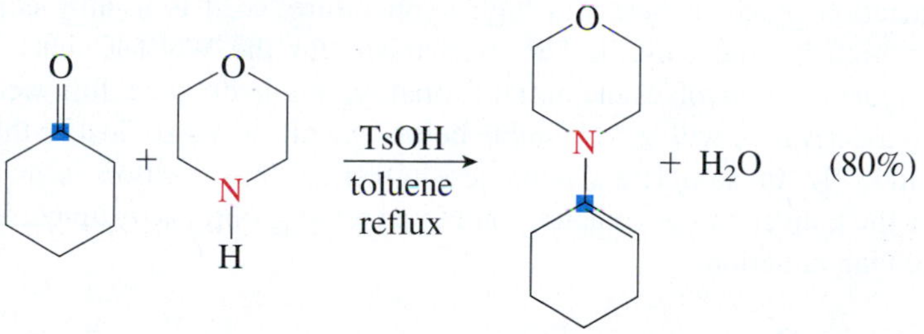

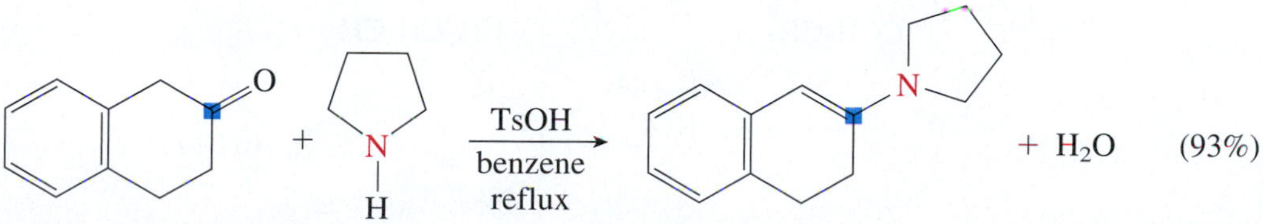

PROBLEM 18.17

Show all of the steps in the mechanism for this reaction:

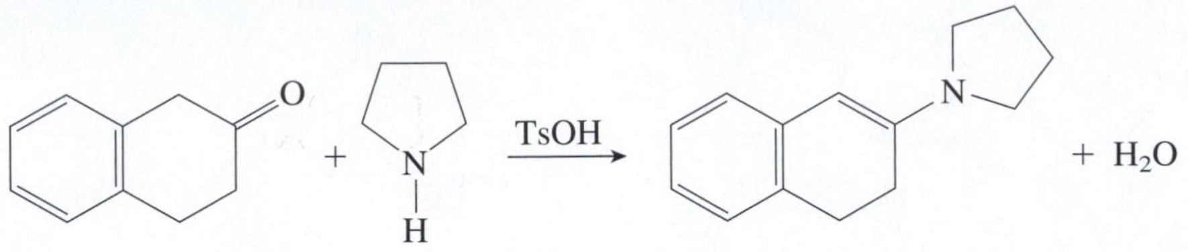

PROBLEM 18.18

Explain why the reaction in problem 18.17 produces the enamine shown in that equation rather than the following enamine:

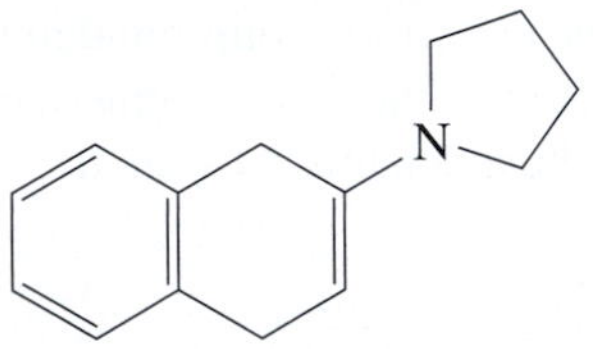

Not formed

PROBLEM 18.19

Show the products of these reactions:

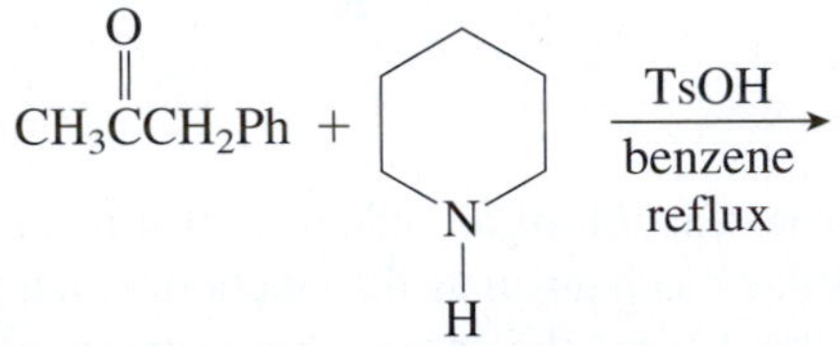

The Wolff-Kishner reduction (see Section 17.13) is a useful reaction that involves an imine as an intermediate. In this procedure the carbonyl group of an aldehyde or ketone is converted to a CH_2 group by treatment with hydrazine and potassium hydroxide. The reaction proceeds best at a high temperature, so it is usually conducted at reflux in a high boiling solvent. The mechanism for the Wolff-Kishner reduction, shown in Figure 18.4, involves the initial formation of a hydrazone, followed by isomerization to a derivative with a NN double bond. The nitrogens are lost as the very stable molecule, N_2, to complete the process. Overall, this reaction is an important method for the conversion of an acyl group to an alkyl group. An example is provided in the following equation:

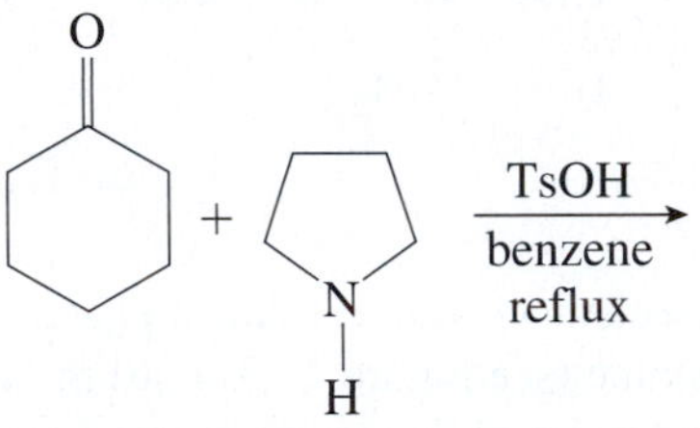

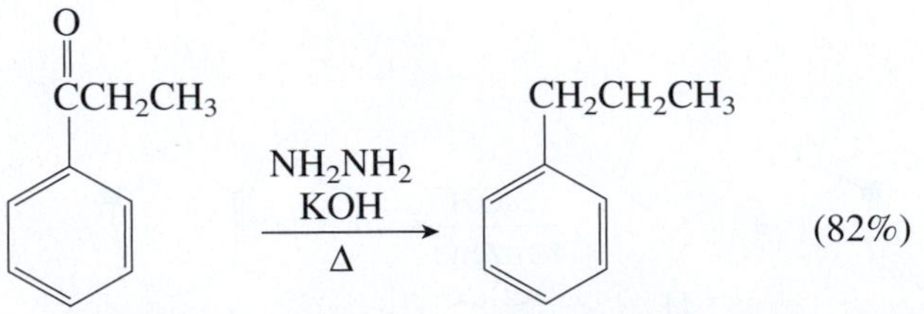

1-Phenyl-1-propanone
(propiophenone)

Propylbenzene

(82%)

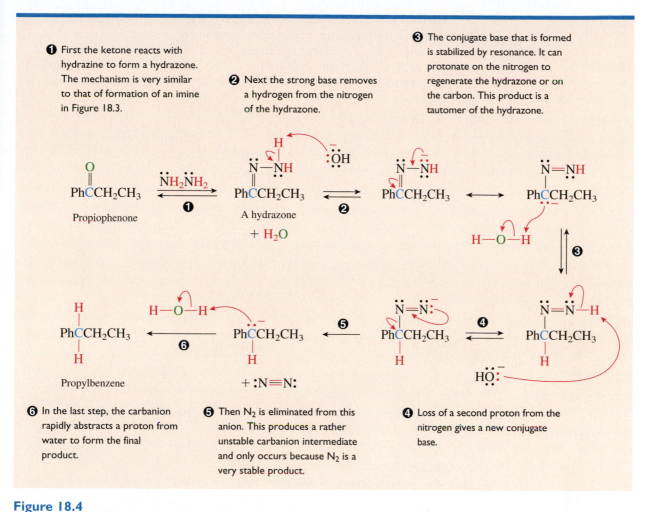

① First the ketone reacts with hydrazine to form a hydrazone. The mechanism is very similar to that of formation of an imine in Figure 18.3.

② Next the strong base removes a hydrogen from the nitrogen of the hydrazone.

③ The conjugate base that is formed is stabilized by resonance. It can protonate on the nitrogen to regenerate the hydrazone or on the carbon. This product is a tautomer of the hydrazone.

⑥ In the last step, the carbanion rapidly abstracts a proton from water to form the final product.

⑤ Then N_2 is eliminated from this anion. This produces a rather unstable carbanion intermediate and only occurs because N_2 is a very stable product.

④ Loss of a second proton from the nitrogen gives a new conjugate base.

Figure 18.4

MECHANISM OF THE WOLFF-KISHNER REDUCTION.

Imines are also intermediates in a useful process called reductive amination. In this reaction an aldehyde or a ketone is reacted with an amine to form an imine. A reducing agent, such as hydrogen and a catalyst, is also present in the reaction mixture. The reducing agent chosen is one that will not reduce the carbonyl compound but will reduce the more reactive imine. The imine is reduced as rapidly as it is formed and is not isolated. The reaction is illustrated in the following equation:

Another reducing agent that can be used in this reaction is sodium cyanoborohydride, a derivative of sodium borohydride with one of the hydrogens replaced by a cyano group. Sodium cyanoborohydride is less nucleophilic than sodium borohydride and does not react with aldehydes or ketones under these conditions. However, it does react with the protonated form of the imine, which is considerably more electrophilic:

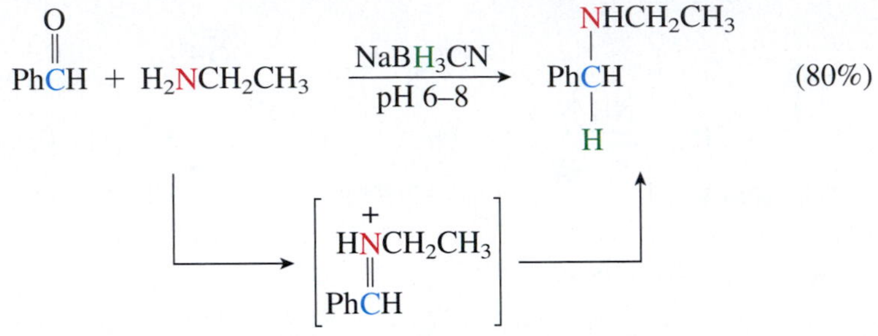

$$\underset{\text{O}}{\text{PhCH}} + H_2NCH_2CH_3 \xrightarrow[\text{pH 6–8}]{\text{NaBH}_3\text{CN}} \underset{\overset{|}{\text{H}}}{\overset{\text{NHCH}_2\text{CH}_3}{\text{PhCH}}} \qquad (80\%)$$

$$\left[\underset{\text{PhCH}}{\overset{+}{\text{HNCH}_2\text{CH}_3}} \right]$$

Protonated imine

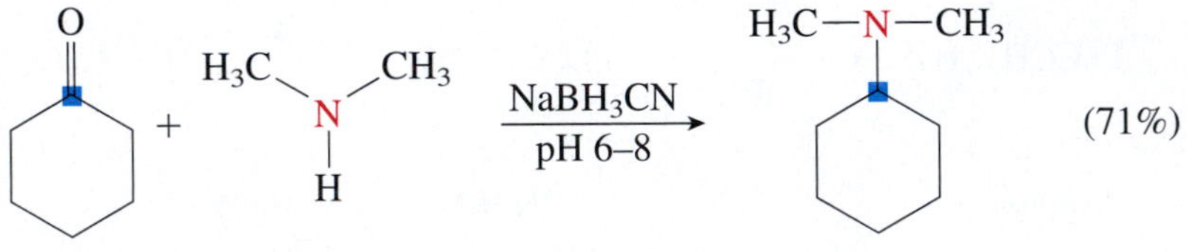

Cyclohexanone   N,N-Dimethylcyclohexanamine

(71%)

PROBLEM 18.20

Show the products of these reactions:

a) 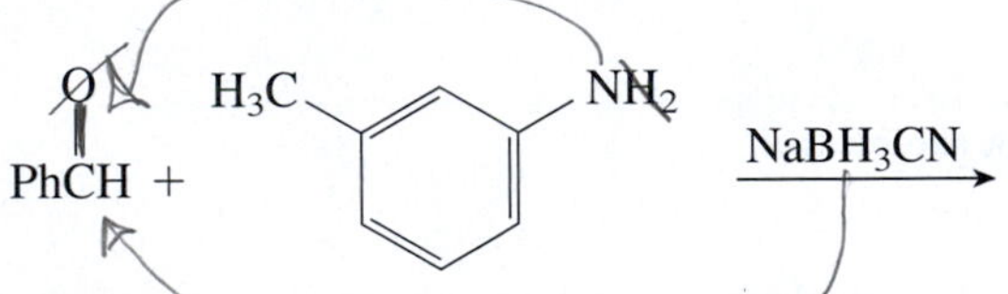 PhCH + [3-methylaniline, H₃C— —NH₂] $\xrightarrow{\text{NaBH}_3\text{CN}}$

b) 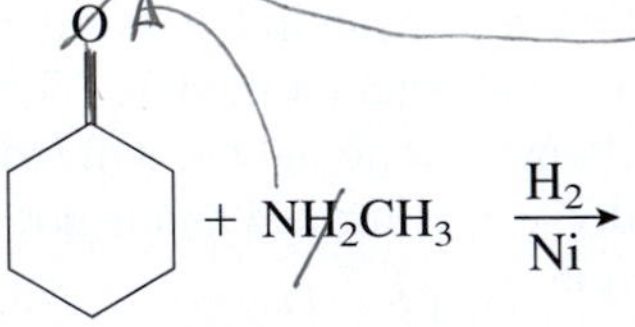 [cyclohexanone] + NH₂CH₃ $\xrightarrow[\text{Ni}]{\text{H}_2}$

c) 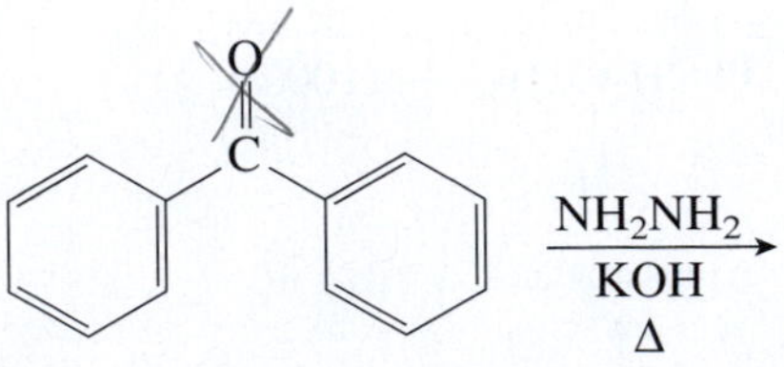 [benzophenone, Ph₂C=O] $\xrightarrow[\substack{\text{KOH} \\ \Delta}]{\text{NH}_2\text{NH}_2}$

Focus On Biological Chemistry

Imines in Living Organisms

Imines are readily formed from amines and aldehydes or ketones under physiological conditions. In addition, they are readily hydrolyzed back to the amine and the carbonyl compound. These properties provide a common way in biochemical reactions to temporarily link an aldehyde or a ketone to a protein by reacting the carbonyl with a free amino group of the protein to form an imine.

One example of this process is found in the chemistry of vision. In the rods and cones of the eye, the aldehyde 11-*cis*-retinal forms an imine by reaction with an amino group of the protein opsin. Studies have shown that the nitrogen of the imine of the product, called rhodopsin, is protonated.

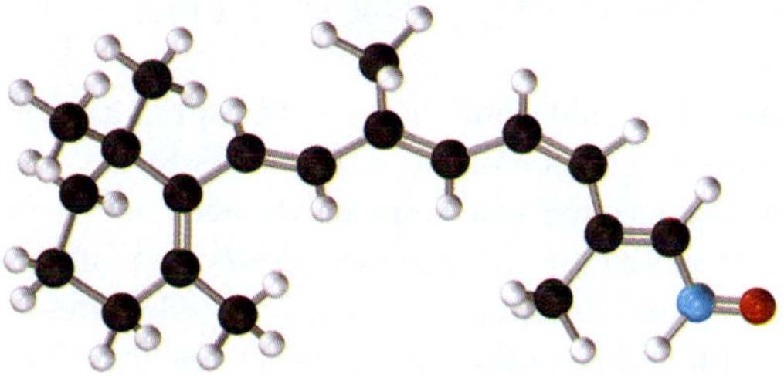

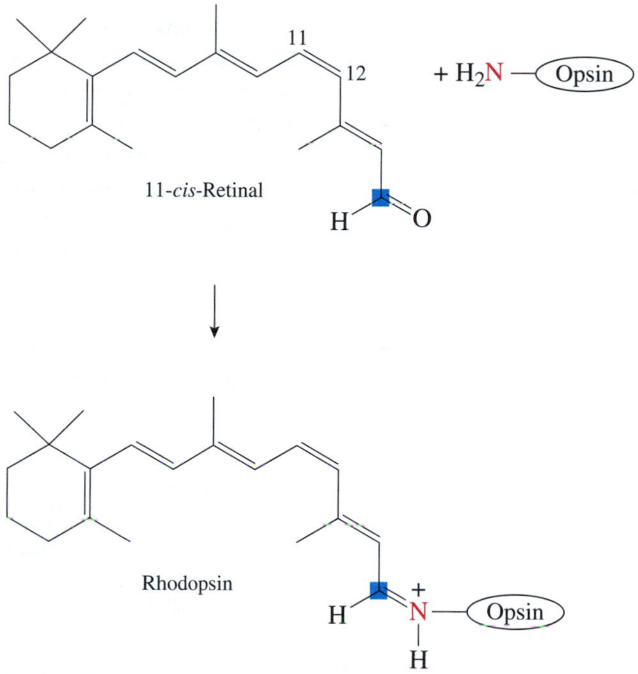

11-*cis*-Retinal

$+ H_2N$ — Opsin

Rhodopsin

When rhodopsin absorbs light in the vision process, the cis double bond between carbons 11 and 12 isomerizes to a trans double bond. This isomerization triggers a nerve impulse telling the brain that light has been absorbed by the eye. The imine of the isomerized product is unstable and is hydrolyzed to opsin and the all-*trans* form of retinal (also known as vitamin A aldehyde). All-*trans* retinal is converted back to 11-*cis*-retinal by enzymes so that it can be used again in rhodopsin formation.

Imine formation is also important in the enzymatic decarboxylation of acetoacetate anion to form acetone, which occurs during the metabolism of glucose. Initial

Continued

formation of a protonated imine facilitates the loss of carbon dioxide to form an enamine. Hydrolysis of the enamine produces acetone and regenerates the enzyme catalyst:

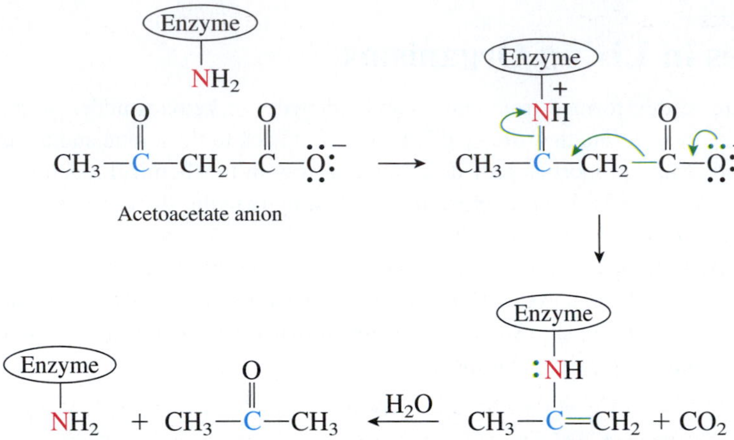

Acetoacetate anion

Many enzymes require the presence of an additional compound, called a coenzyme, to carry out their catalytic functions. The coenzyme often bonds to the substrate, modifying its structure so that the enzyme can more easily accomplish its catalytic reaction. Pyridoxal, also known as vitamin B_6, acts as a coenzyme by forming an imine between its aldehyde group and an amine group of the substrate. In these biological reactions it is important that the equilibrium constant for imine formation not be too large so that the CN double bond can be cleaved by hydrolysis when the reaction is finished, thus completing the catalytic cycle. The compounds that are used to form derivatives of carbonyl compounds—hydroxylamine, hydrazine, phenylhydrazine, and semicarbazide—are poisonous because they form imine derivatives of pyridoxal. The equilibrium constant for the formation of these derivatives is so large that there is not enough free pyridoxal available to carry out its catalytic functions.

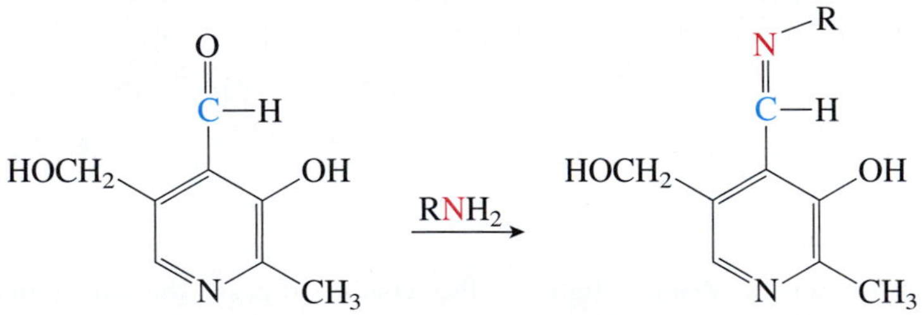

Pyridoxal
(vitamin B_6)

18.9 ADDITION OF ALCOHOLS

Aldehydes and ketones add two equivalents of alcohols to form **acetals.** (The term *ketal,* which was formerly used to describe the product formed from a ketone, may still be encountered.)

m-Nitrobenzaldehyde   An acetal

The mechanism for this reaction, shown in Figure 18.5, is as long as any that you will encounter in this text. You can make the task of learning this mechanism much easier by recognizing its similarities to other mechanisms in this chapter and the similarities among the steps within this mechanism. The initial addition follows the acidic conditions mechanism. Steps 1, 2, and 3 are nearly identical to the mechanism for acid-catalyzed hydration (Section 18.3) with the exception that the nucleophile is the oxygen of methanol rather than the oxygen of water. The addition product, a **hemiacetal,** is similar to the hydrate or the carbinolamine. Steps 4 and 5 are very similar to steps in imine formation (Figure 18.3). First the oxygen is protonated, and then water leaves with the help of the unshared electrons on the other oxygen. However, in contrast to imine formation, the product of step 5 cannot be stabilized by the loss of a proton. Instead, the oxygen of a second molecule of alcohol acts as a nucleophile.

Like hydrates, hemiacetals are not favored at equilibrium and, in general, cannot be isolated. The equilibrium is shifted in their favor by the inductive effects of electron-withdrawing groups, similar to the case of hydrates. In addition, the equilibrium is shifted in their favor if the alcohol nucleophile and carbonyl electrophile are part of the same molecule as in the following example:

11.4%   88.6%
A cyclic hemiacetal

In this example the oxygen of the hydroxy group acts as an intramolecular nucleophile. Recall from Section 8.13 that intramolecular reactions are favored by entropy. Therefore, the formation of a cyclic hemiacetal has a larger equilibrium constant than a comparable intermolecular reaction. This reaction is especially important in the area of carbohydrates (sugars) because sugars contain both carbonyl and hydroxy functional

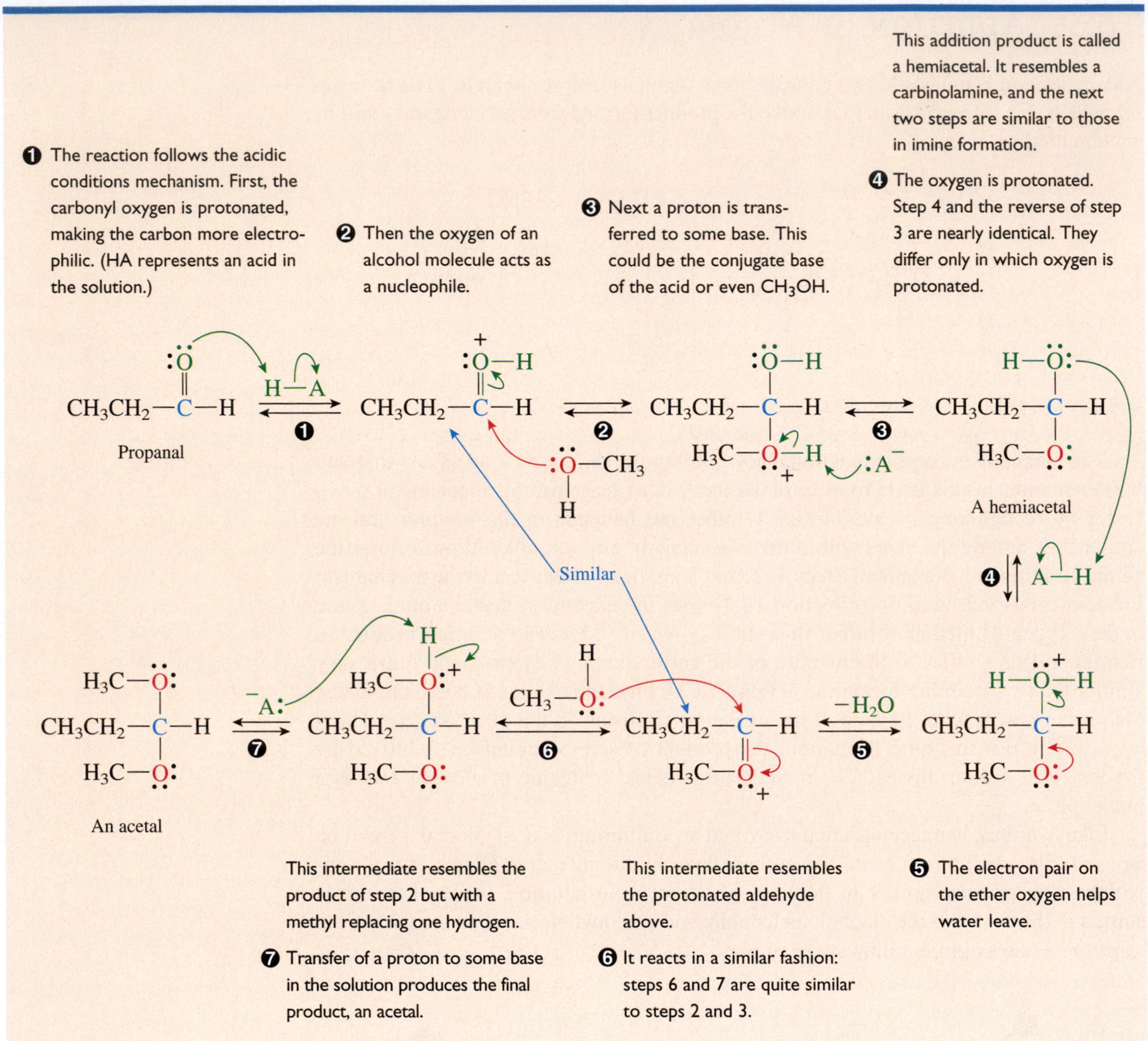

❶ The reaction follows the acidic conditions mechanism. First, the carbonyl oxygen is protonated, making the carbon more electrophilic. (HA represents an acid in the solution.)

❷ Then the oxygen of an alcohol molecule acts as a nucleophile.

❸ Next a proton is transferred to some base. This could be the conjugate base of the acid or even CH₃OH.

This addition product is called a hemiacetal. It resembles a carbinolamine, and the next two steps are similar to those in imine formation.

❹ The oxygen is protonated. Step 4 and the reverse of step 3 are nearly identical. They differ only in which oxygen is protonated.

An acetal

This intermediate resembles the product of step 2 but with a methyl replacing one hydrogen.

❼ Transfer of a proton to some base in the solution produces the final product, an acetal.

This intermediate resembles the protonated aldehyde above.

❻ It reacts in a similar fashion: steps 6 and 7 are quite similar to steps 2 and 3.

❺ The electron pair on the ether oxygen helps water leave.

Figure 18.5

MECHANISM OF THE FORMATION OF AN ACETAL.

ORGANIC
Chemistry••Now™
Click *Mechanisms in Motion* to view the **Mechanism of Acetal Formation.**

groups. For example, in aqueous solution, glucose exists almost entirely (~99.98%) as the cyclic hemiacetal:

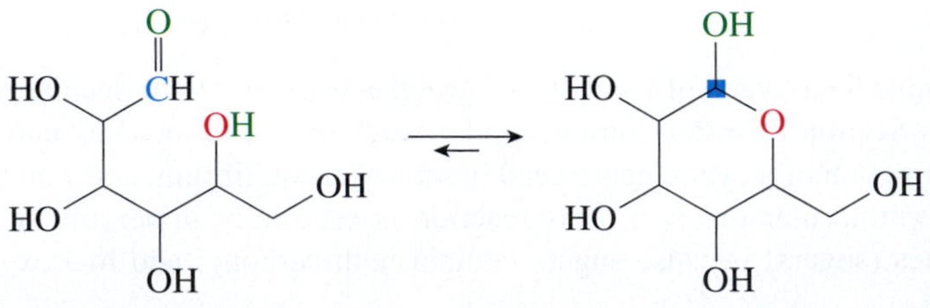

PROBLEM 18.21

Explain which of these compounds has more cyclic hemiacetal present at equilibrium:

The equilibrium between an aldehyde or a ketone and an acetal usually favors the aldehyde or ketone. Therefore, to prepare an acetal, the equilibrium must somehow be driven toward the product. This is usually accomplished by removing the water that is formed as a by-product. Acetals are stable in the absence of acid, because the reverse of step 7 (see Figure 18.5) is a protonation and cannot occur under basic or neutral conditions. Once the equilibrium has been driven to completion, the reaction mixture is neutralized and the products can then be isolated. Often a diol, such as ethylene glycol, is used in the preparation of acetals. In such cases the reaction is more favorable because the conversion of the hemiacetal to the acetal is intramolecular.

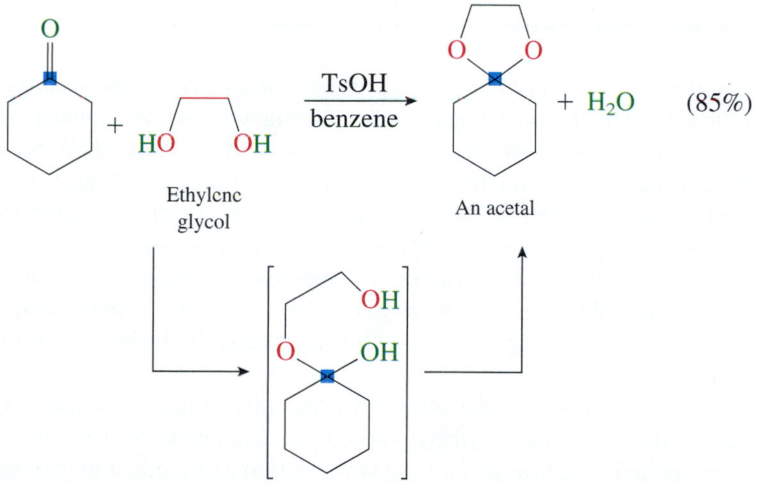

Similar chemistry can be used to prepare thioacetals. In Chapter 20, we will examine an important carbon nucleophile that can be generated from thioacetals.

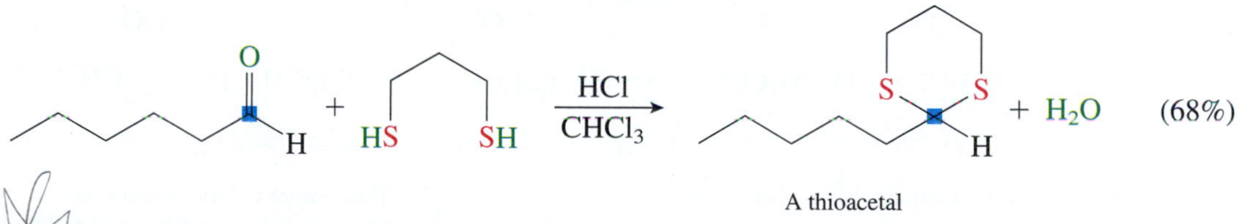

PROBLEM 18.22

Show all of the steps in the mechanism of this reaction:

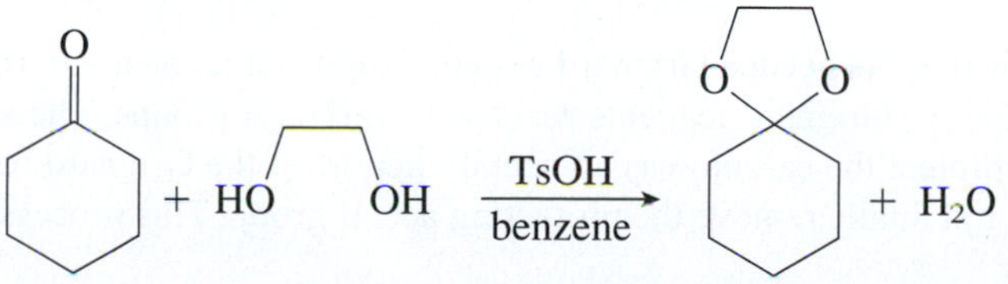

PROBLEM 18.23

Show the products of these reactions:

a) $CH_3CH_2CH_2\overset{O}{\overset{\|}{C}}H$ + CH_3CH_2OH $\xrightarrow{H_2SO_4}$

b) [benzaldehyde structure] + $HO\frown OH$ $\xrightarrow{TsOH}$

c) [cyclopentyl carbaldehyde structure] + $HS\frown SH$ $\xrightarrow{HCl}$

A major use of acetals is as protecting groups for aldehydes and ketones. A protecting group is used to prevent a reagent from reacting with one functional group while it reacts with another functional group somewhere else in the molecule. First, the protecting group is put on one of the functional groups. After the desired reaction has been performed elsewhere in the molecule, the protecting group is removed, regenerating the original functional group. Aldehydes and ketones are readily converted to acetals. Many reagents that would react with the carbonyl compound do not react with acetals, because this group is stable to nucleophiles and bases. After the desired reaction is accomplished, the carbonyl group can readily be regenerated by hydrolysis of the acetal with aqueous acid.

Let's look at an example of the utility of protecting groups in organic synthesis. Suppose the synthetic target is 6-hydroxy-6-methyl-2-heptanone. Retrosynthetic analysis on this compound might suggest a Grignard reaction as a method to prepare the alcohol group while also forming a carbon–carbon bond:

$$\underset{\substack{|\\CH_3}}{\overset{\substack{OH\\|}}{CH_3C}}CH_2CH_2CH_2\overset{O}{\overset{\|}{C}}CH_3 \quad \longleftarrow \quad CH_3\overset{O}{\overset{\|}{C}}CH_3 + BrMgCH_2CH_2CH_2\overset{O}{\overset{\|}{C}}CH_3$$

6-Hydroxy-6-methyl-2-heptanone
(synthetic target)

This reagent cannot be prepared because the carbonyl group and the Grignard reagent are incompatible.

The problem is that the needed Grignard reagent is not stable, owing to the presence of the carbonyl group: Grignard reagents react with carbonyl groups. The solution to the problem is to protect the carbonyl as an acetal, then form the Grignard reagent, react it with acetone, and finally remove the protecting acetal group. This process is outlined in Figure 18.6.

An attempt to form a Grignard reagent from 5-bromo-2-pentanone is doomed to failure because the Grignard will react with the carbonyl group.

① Therefore, the carbonyl group is first protected as an ethylene glycol acetal.

② Because the acetal group does not react with Grignard reagents (or other basic or nucleophilic reagents), the Grignard reagent can be prepared from this compound. The acetal is being used as a protecting group for the carbonyl group.

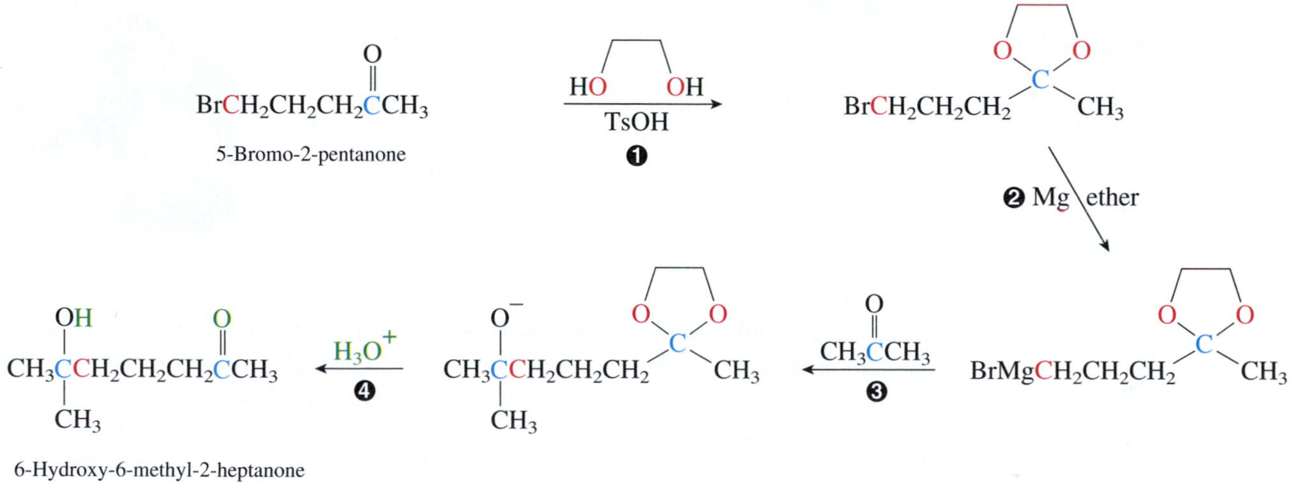

$$BrCH_2CH_2CH_2\overset{\overset{\displaystyle O}{\|}}{C}CH_3$$

5-Bromo-2-pentanone

$$\xrightarrow[\text{TsOH}]{\text{HO}\qquad\text{OH}}$$

①

$$BrCH_2CH_2CH_2\overset{\displaystyle O\qquad O}{\underset{\displaystyle C}{\diagdown\diagup}}CH_3$$

② Mg ether

$$BrMgCH_2CH_2CH_2\overset{\displaystyle O\qquad O}{\underset{\displaystyle C}{\diagdown\diagup}}CH_3$$

③ This Grignard reagent reacts like any other Grignard reagent.

$$\overset{\displaystyle OH}{\underset{\displaystyle CH_3}{CH_3\overset{|}{C}CH_2CH_2CH_2\overset{\overset{\displaystyle O}{\|}}{C}CH_3}}$$

6-Hydroxy-6-methyl-2-heptanone

$$\xleftarrow[\text{④}]{H_3O^+}$$

$$\overset{\displaystyle O^-}{\underset{\displaystyle CH_3}{CH_3\overset{|}{C}CH_2CH_2CH_2}}\overset{\displaystyle O\qquad O}{\underset{\displaystyle C}{\diagdown\diagup}}CH_3$$

$$\xleftarrow[\text{③}]{CH_3\overset{\overset{\displaystyle O}{\|}}{C}CH_3}$$

④ When the reaction is worked up with aqueous acid, not only is the alkoxide group protonated but the acetal is also hydrolyzed back to the ketone and ethylene glycol. Easy removal is an important feature of protecting groups.

Figure 18.6

USE OF AN ACETAL AS A PROTECTING GROUP.

PROBLEM 18.24

Because aldehydes are more reactive than ketones, it is possible to selectively form an acetal group at the aldehyde group without also forming one at the ketone group of a compound that contains both functional groups. Using this information, suggest how the following transformation could be accomplished:

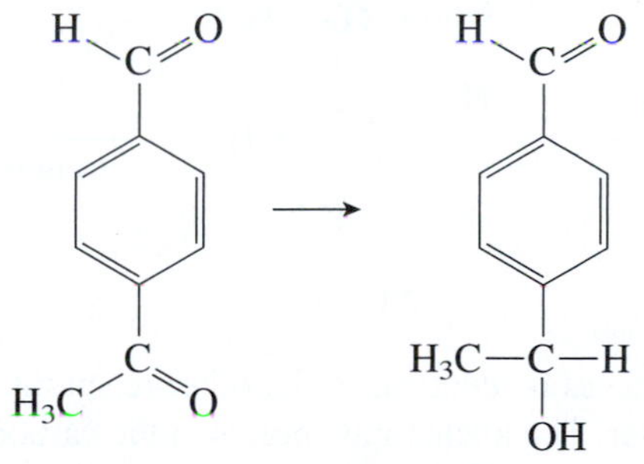

$$\text{(structure with } H\!-\!C\!=\!O \text{ aldehyde and } H_3C\!-\!C\!=\!O \text{ ketone on benzene ring)} \longrightarrow \text{(structure with } H\!-\!C\!=\!O \text{ aldehyde and } H_3C\!-\!\overset{\displaystyle OH}{\underset{}{\overset{|}{C}}}\!-\!H \text{)}$$

18.10   CONJUGATE ADDITIONS

When a CC double bond is conjugated with a carbonyl group, the double bond often has chemical reactions that are similar to those of the carbonyl group. (Recall from Section 12.3 that the carbon adjacent to the carbonyl group is sometimes called the

α-carbon. The next carbon is called the β-carbon, and so on. Therefore, a compound with a double bond conjugated to the carbonyl group is said to be α,β-unsaturated.) The β-carbon of an α,β-unsaturated carbonyl compound is electrophilic, very similar to the carbonyl carbon. This can best be understood by examination of the resonance structures for such a conjugated compound:

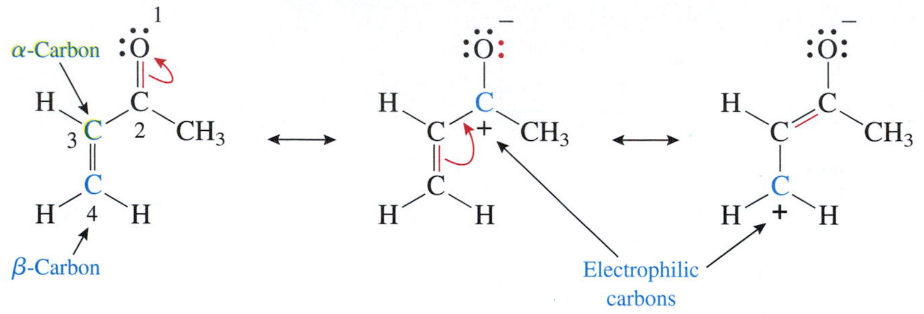

When a nucleophile reacts with an α,β-unsaturated carbonyl compound, it may bond to either of the two electrophilic carbons. If it bonds to the carbonyl carbon, the reaction is termed a normal addition or a 1,2-addition (because the nucleophile and electrophile have added to adjacent positions). If, instead, it bonds to the β-carbon, the reaction is termed a conjugate addition or a 1,4-addition. The following mechanism operates under basic conditions:

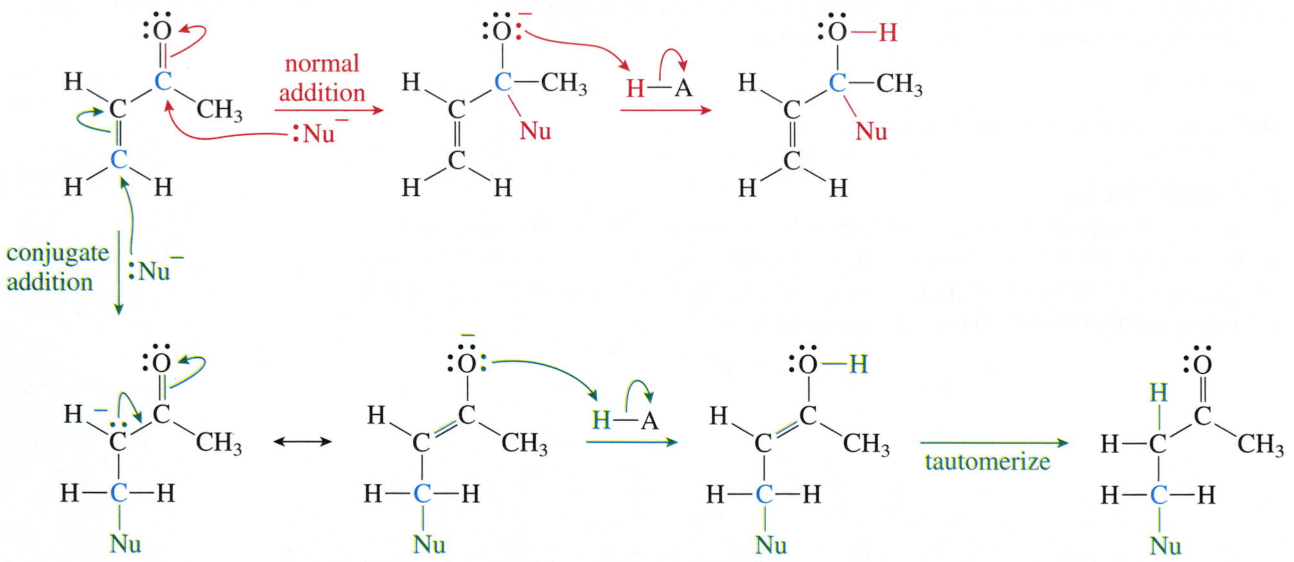

The normal addition process is identical to the other reactions that have been encountered so far in this chapter: The nucleophile bonds to the carbonyl carbon and the electrophile bonds to the oxygen of the carbonyl group. In a conjugate addition the nucleophile bonds to the β-carbon. The electrophile, a proton, can bond to either the α-carbon or the oxygen of the resonance stabilized anion. It actually reacts faster at the oxygen, producing an enol in an overall 1,4-addition. However, as discussed in Section 11.6, enols are less stable than the carbonyl tautomers, so the product that is isolated contains the carbonyl group.

The overall result of a conjugate addition is the addition of a proton and a nucleophile to the CC double bond. However, this reaction differs greatly from the additions discussed in Chapter 11, in which the electrophile adds first. Here, the nucleophile adds in the first step. This reaction does not occur unless there is a group attached to the double bond that can help stabilize, by resonance, the carbanion intermediate. In many cases this is the carbonyl group of an aldehyde or a ketone. However, other groups, such as the carbonyl group of an ester or a cyano group, also enable this reaction to occur.

Whether a normal or a conjugate addition occurs depends on both the structure of the α,β-unsaturated electrophile and the nature of the nucleophile. The less reactive nucleophiles usually give conjugate addition because the reactions are reversible and the conjugate addition product is more stable. Thus, cyanide ion and amines add in a conjugate manner as shown in the following examples:

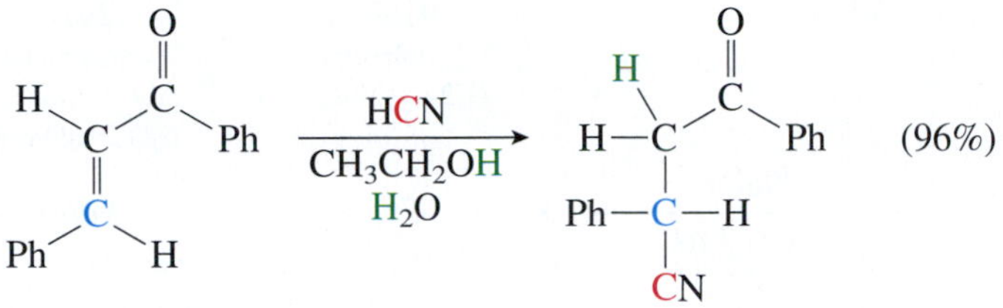

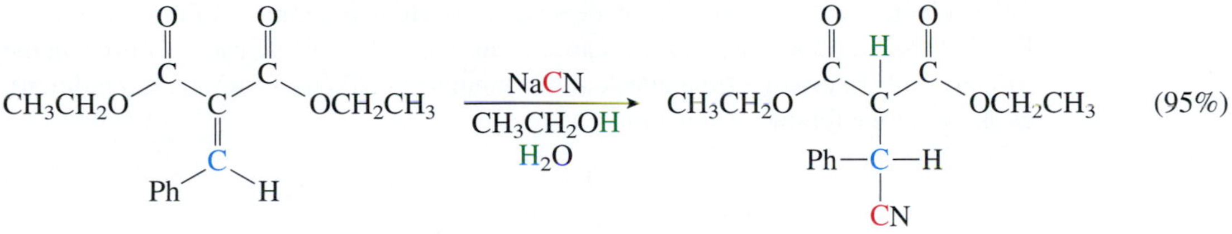

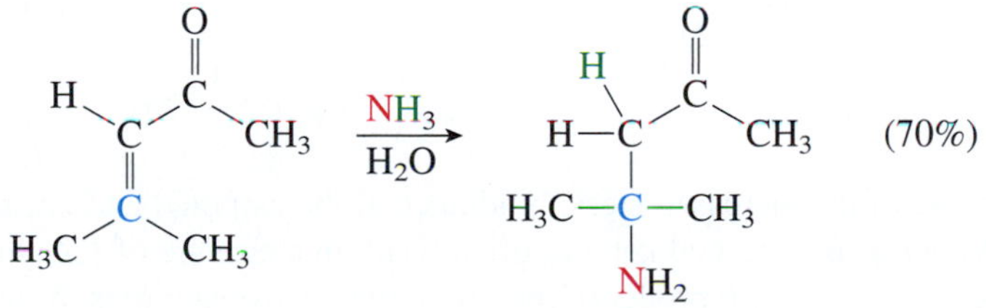

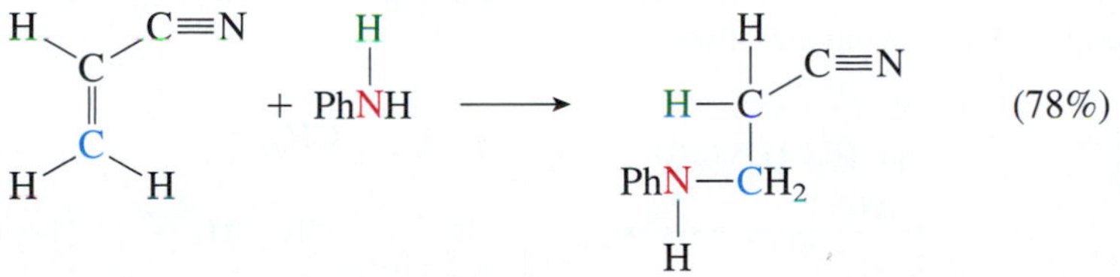

For the more reactive nucleophiles, where addition is essentially irreversible, whether 1,2-addition or 1,4-addition occurs depends on the relative rates of addition to the two electrophilic sites, the carbonyl carbon and the β-carbon. Lithium aluminum hydride usually gives predominantly 1,2-addition and provides a useful way to reduce the carbonyl group of an α,β-unsaturated compound. Sodium borohydride, on the other hand, often gives a mixture of 1,2-addition and the completely reduced product, where 1,4-addition followed by 1,2-addition has occurred. Thus, the reaction of 2-cyclohexenone with lithium aluminum hydride gives a good yield of the 1,2-addition product, 2-cyclohexenol. In contrast, reaction with sodium borohydride is less useful because of the mixture of products that is formed. Part of the reaction occurs by 1,2-addition to produce 2-cyclohexenol and part by 1,4-addition to produce cyclohexanone, which is then reduced further to cyclohexanol.

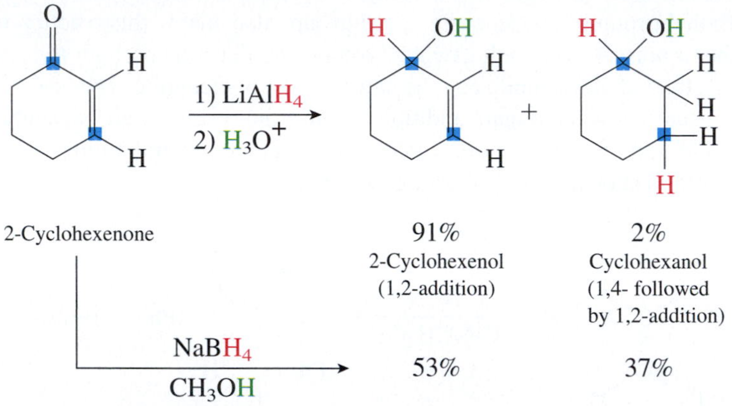

2-Cyclohexenone

91%
2-Cyclohexenol
(1,2-addition)

2%
Cyclohexanol
(1,4- followed
by 1,2-addition)

NaBH₄
CH₃OH
53%
37%

In the case of a Grignard reagent reacting as the nucleophile, the amounts of normal addition and conjugate addition depend on the steric hindrance at the carbonyl carbon and the β-carbon. Reaction with an α,β-unsaturated aldehyde usually results in the formation of the product from attack at the unhindered aldehyde carbon (1,2-addition), as shown in the following equation:

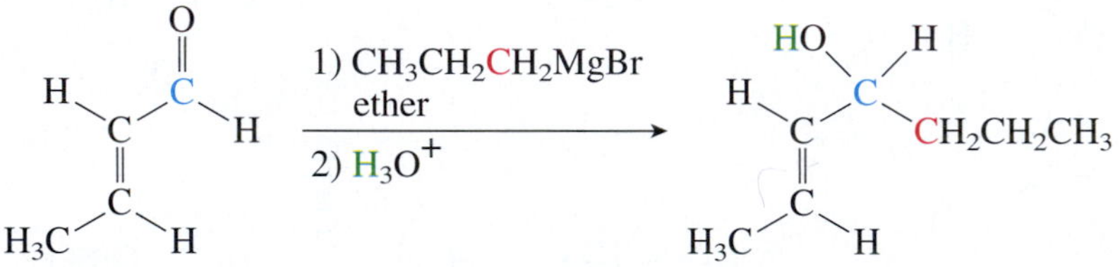

Because of the increased steric hindrance at the carbonyl carbon, similar reactions involving α,β-unsaturated ketones often result in a mixture of 1,2- and 1,4-addition. The exact amount of each product depends on the relative amounts of steric hindrance at the two electrophilic carbons and may be difficult to predict in advance. An example is provided in the following equation:

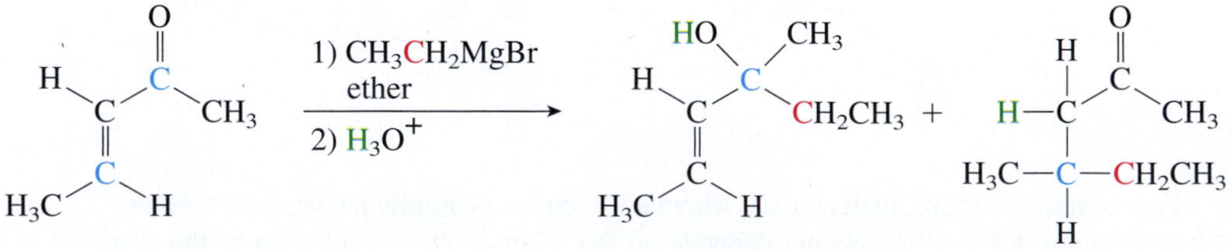

An α,β-unsaturated
ketone

41%
1,2-Addition

39%
1,4-Addition

If conjugate addition of an organometallic nucleophile is desired, this can be accomplished by using a lithium diorganocuprate reagent, which has the organic group attached to a copper atom. These reagents are prepared by the reaction of organolithium compounds with copper(I) salts such as cuprous iodide, as illustrated in the following equation for the preparation of lithium dimethylcuprate:

$$2\ CH_3Li + CuI \longrightarrow (CH_3)_2CuLi + LiI$$

Methyllithium — Lithium dimethylcuprate

The reactions of lithium diorganocuprates with α,β-unsaturated carbonyl compounds give excellent yields of 1,4-addition products:

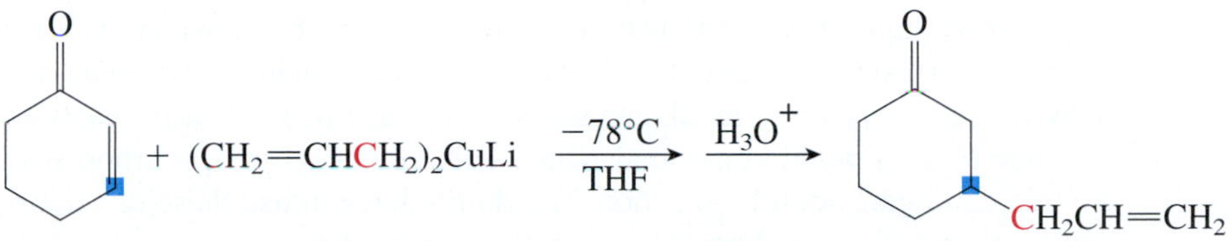

(94%)

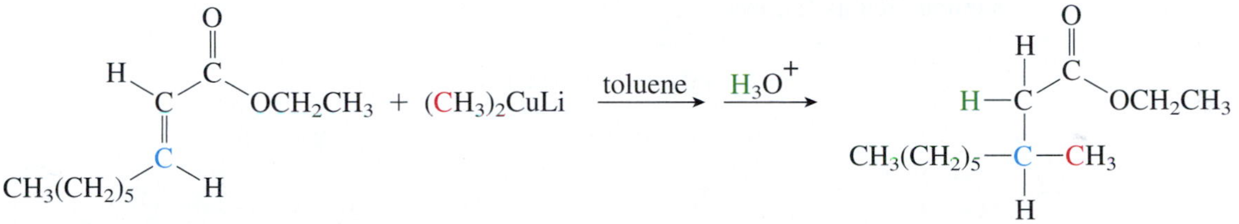

(86%)

In summary, lithium aluminum hydride usually gives good yields of 1,2-addition products, that is, unsaturated alcohols. The reaction of Grignard reagents with α,β-unsaturated aldehydes also gives good yields of 1,2-addition products. Nucleophiles that give good yields of 1,4-addition products are cyanide ion, amines, and lithium diorganocuprate reagents. The cuprate reagents provide an excellent method to add a carbon nucleophile to the β-carbon of α,β-unsaturated carbonyl compounds, a process that is very useful in synthesis.

PROBLEM 18.25

Show the products of these reactions:

a) 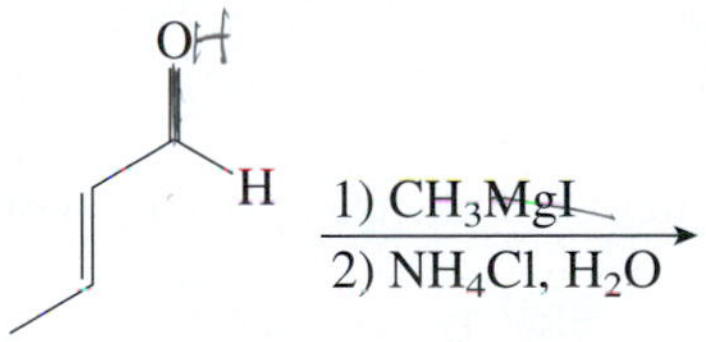

b) 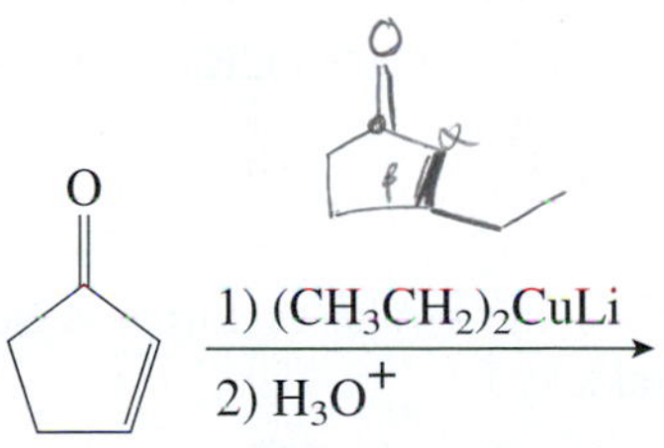

c) $CH_2{=}CH\overset{O}{\overset{\|}{C}}CH_3 \xrightarrow[2)\ H_3O^+]{1)\ LiAlH_4}$

d) $CH_2{=}CH\overset{O}{\overset{\|}{C}}OCH_3 \xrightarrow[CH_3OH]{HCN}$

e) 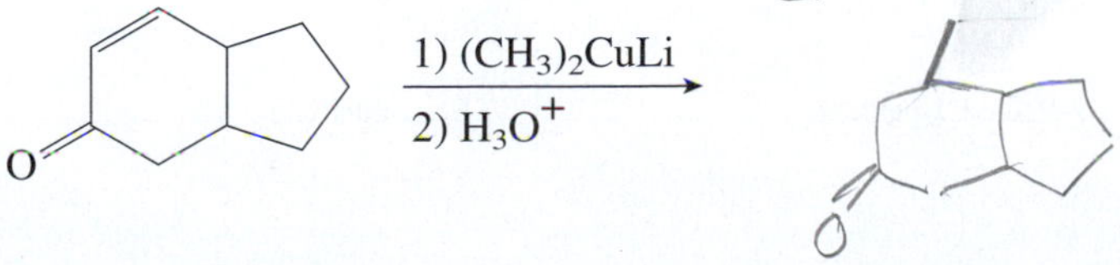

18.11 SYNTHESIS

By now, you should be fairly comfortable using retrosynthetic analysis to design the synthesis of a target molecule. Remember that carbon–carbon bond forming reactions are especially important in synthesis. Several extremely useful reactions that result in the formation of carbon–carbon bonds have been introduced in this chapter. These are the Grignard reaction to prepare alcohols, the Wittig reaction to prepare alkenes, and the conjugate addition of organocuprate reagents to α,β-unsaturated carbonyl compounds.

When you are designing a synthesis, the presence of certain structural features in the target suggests that the use of certain reactions be considered. Whenever an alcohol is the synthetic target, you should consider using a Grignard reaction to make it. The presence of an alkene suggests the use of the Wittig reaction, and the presence of a carbonyl compound with a substituent on the β-carbon suggests the use of a conjugate addition reaction. You do not have to use these reactions, but you should consider using them. Using retrosynthetic notation, these possibilities are summarized as follows:

Let's try some syntheses. Suppose we need to prepare 2-phenyl-l-pentene, starting from benzaldehyde:

2-Phenyl-1-pentene   Benzaldehyde

Applying retrosynthetic analysis, the presence of the alkene group in the target suggests using a Wittig reaction in its preparation.

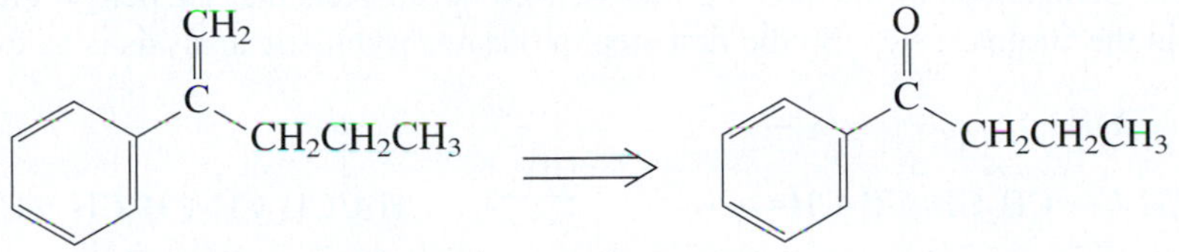

The new target is a ketone. We need to somehow add a propyl group to the carbonyl carbon of benzaldehyde to make this ketone. At this point we do not know of a reaction that will accomplish this transformation directly, but we recognize that a ketone can be prepared by oxidation of an alcohol. We can prepare the alcohol using the Grignard reaction. Our retrosynthetic analysis is as follows:

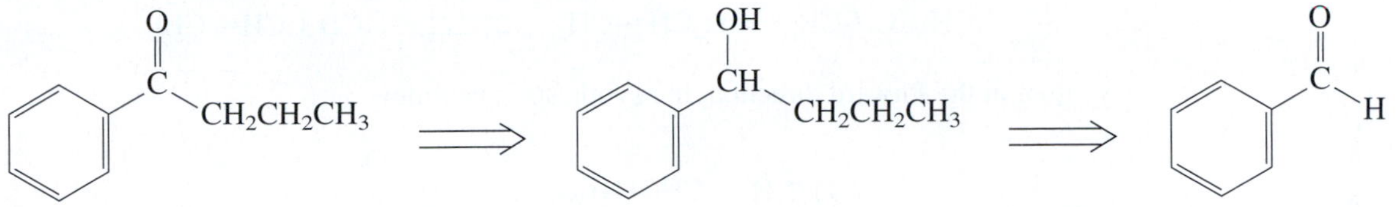

We are now ready to write the synthesis in the forward direction:

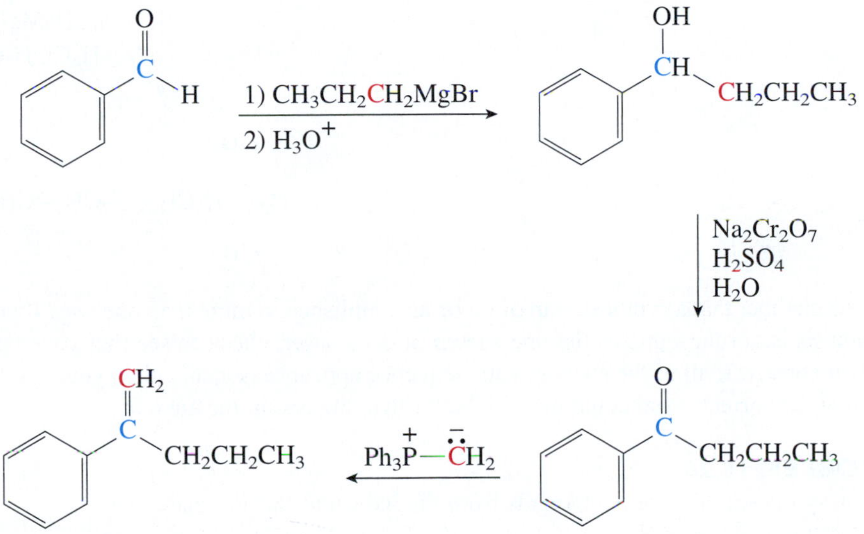

Let's try another example. This time our task is to prepare 2-methyl-1-phenylhept-6-en-2-ol from but-3-en-2-one.

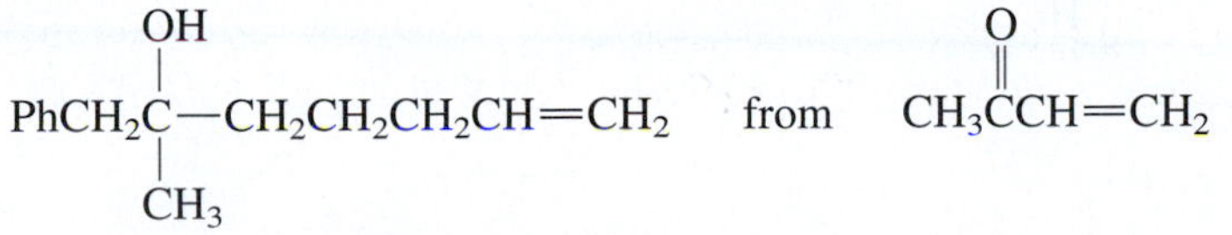

2-Methyl-1-phenylhept-6-en-2-ol   But-3-en-2-one

The target is an alcohol, so we should consider using a Grignard reaction. Any of the three groups attached to the carbon bonded to the hydroxy group could potentially be attached. Comparison of the target to but-3-en-2-one suggests that the benzyl group be added in the Grignard step. So the first step in our retrosynthetic analysis is as follows:

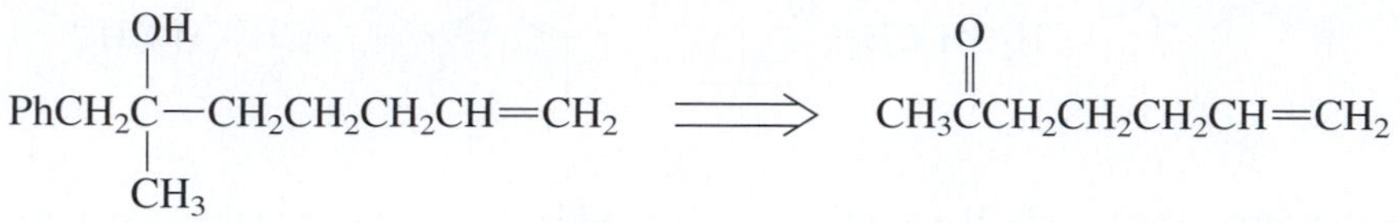

$$\underset{\underset{CH_3}{|}}{\overset{\overset{OH}{|}}{PhCH_2C}}-CH_2CH_2CH_2CH=CH_2 \implies CH_3\overset{\overset{O}{\|}}{C}CH_2CH_2CH_2CH=CH_2$$

Our new target is a ketone that is related to but-3-en-2-one by the presence of an allyl group on the β-carbon. This suggests the use of a conjugate addition reaction:

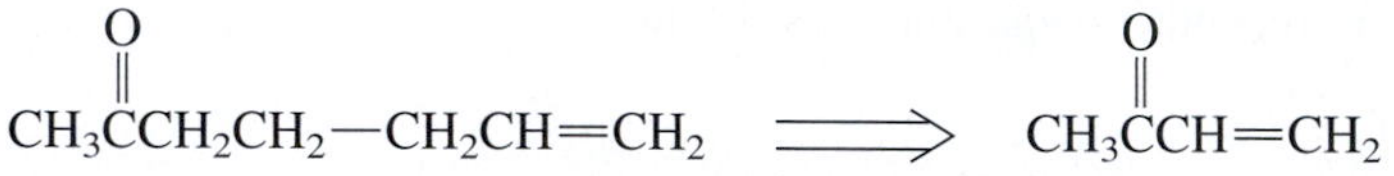

$$CH_3\overset{\overset{O}{\|}}{C}CH_2CH_2-CH_2CH=CH_2 \implies CH_3\overset{\overset{O}{\|}}{C}CH=CH_2$$

Written in the forward direction, the synthesis is as follows:

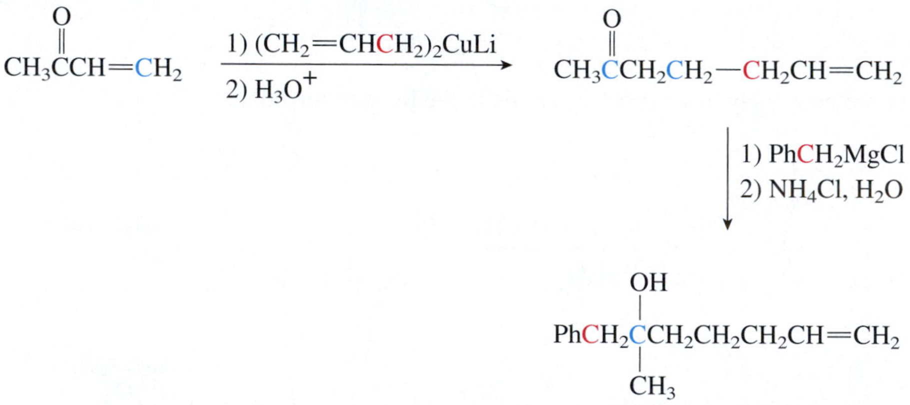

$$CH_3\overset{\overset{O}{\|}}{C}CH=CH_2 \xrightarrow[\text{2) } H_3O^+]{\text{1) } (CH_2=CHCH_2)_2CuLi} CH_3\overset{\overset{O}{\|}}{C}CH_2CH_2-CH_2CH=CH_2$$

1) PhCH$_2$MgCl
2) NH$_4$Cl, H$_2$O

$$\underset{\underset{CH_3}{|}}{\overset{\overset{OH}{|}}{PhCH_2C}}CH_2CH_2CH_2CH=CH_2$$

Remember that a synthesis can often be accomplished in more than one way. If your synthesis is not the same as the one shown in the answer, check to see that your steps are all correct. If all of the steps in your sequence appear reasonable, then your synthesis may be correct—and could even be better than the one in the answer.

PROBLEM 18.26

Show syntheses of these compounds from the indicated starting materials:

a) 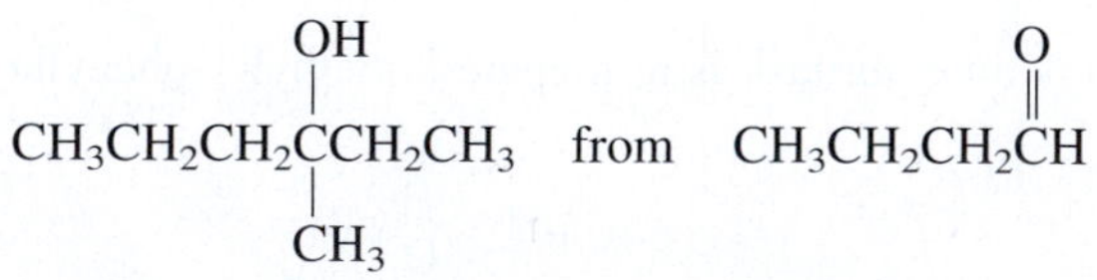 $\underset{\underset{CH_3}{|}}{\overset{\overset{OH}{|}}{CH_3CH_2CH_2C}}CH_2CH_3$ from $CH_3CH_2CH_2\overset{\overset{O}{\|}}{C}H$

b) 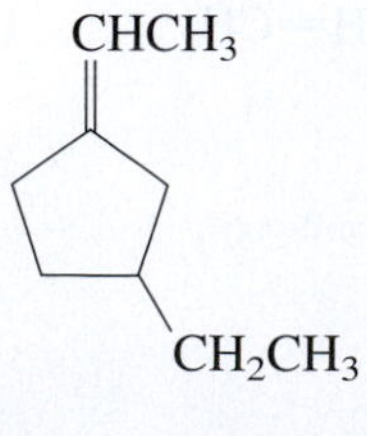 from

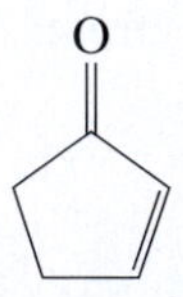

c) $CH_3\overset{\overset{\displaystyle CH_2}{\|}}{C}CH_2CH_2CH_2CH_3$ from $H\overset{\overset{\displaystyle O}{\|}}{C}CH{=}CH_2$

d) [structure of cyclopentyl-CH₂-CO-CH₂CH₂CH₃] from [structure of cyclopentyl-CH₂-OH]

e) [structure of keto-acid with O, O and OH] from [structure of keto-bromide with O and Br]

Review of Mastery Goals

After completing this chapter, you should be able to:

ORGANIC
Chemistry Now™
Click Mastery Goal Quiz to test how well you have met these goals.

■ Show the products resulting from the addition to aldehydes and ketones of all of the reagents discussed in this chapter. (Problems 18.27, 18.28, 18.32, 18.33, and 18.36)

■ Show the products resulting from the addition of certain of these reagents to α,β-unsaturated compounds, noting whether 1,2- or 1,4-addition predominates. (Problems 18.29, 18.32, and 18.36)

■ Show the mechanisms for any of these additions. (Problems 18.39, 18.41, 18.45, 18.47, 18.48, and 18.52)

■ Predict the effect of the structure of the aldehyde or ketone on the position of the equilibrium for these reactions. (Problems 18.30, 18.31, 18.34, 18.35, 18.58, and 18.59)

■ Use these reactions, in combination with the reactions from previous chapters, to synthesize compounds. (Problems 18.38, 18.42, and 18.43)

■ Use acetals as protecting groups in syntheses. (Problem 18.43)

Visual Summary of Key Reactions

The reactions in this chapter begin with the addition of a nucleophile to the carbon of a carbonyl group and an electrophile, usually a proton, to the oxygen. Under basic conditions the nucleophile adds first, whereas the proton adds first under acidic conditions. Depending on the nature of the nucleophile, the reaction may stop at this stage or proceed further. Figure 18.7 summarizes the mechanisms followed by the various nucleophiles. Table 18.2 lists the nucleophiles and the products that result from their reactions with aldehydes and ketones.

Table 18.2 Additions to the Carbonyl Group

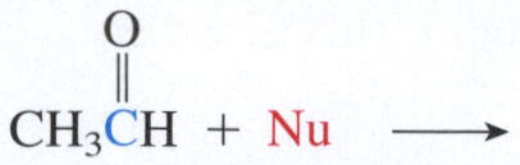

$$\underset{\substack{\text{O} \\ \parallel}}{CH_3CH} + Nu \longrightarrow$$

Nucleophile	Product	Comments
H—Al⁻—H (with H above and H below)	OH | CH₃CH | H	**Section 18.2** Reaction with $NaBH_4$ or $LiAlH_4$ proceeds to stage 1 (see Figure 18.7) and follows the basic conditions mechanism.
H_2O	OH | CH₃CH | OH	**Section 18.3** This reaction proceeds to stage 1. Hydrates usually cannot be isolated because of the unfavorable equilibrium. The reaction follows either the acidic or basic conditions mechanism.
HCN	OH | CH₃CH | CN	**Section 18.4** This reaction proceeds to stage 1 and follows the basic conditions mechanism.
R—MgX	OH | CH₃CH | R	**Section 18.6** Reaction with organometallic nucleophiles (Grignard reagents and organolithium reagents) proceeds to stage 1 and follows the basic conditions mechanism.
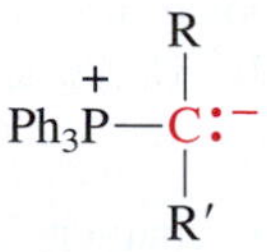 Ph₃P⁺—C:⁻ (with R above and R′ below)	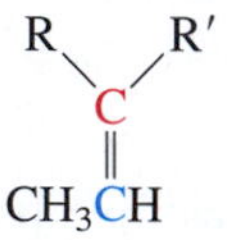 R R′ \C=C/ CH₃CH	**Section 18.7** The Wittig reaction proceeds to stage 2 and follows the basic conditions mechanism.
RNH_2	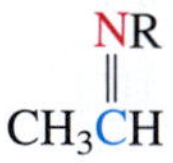 NR ‖ CH₃CH	**Section 18.8** Imine formation proceeds to stage 2 with primary amines. Addition follows the basic conditions mechanism, but acid is needed to remove the oxygen. Secondary amines give enamines.
ROH	OR | CH₃CH | OR	**Section 18.9** Acetals are formed at stage 3. Thiols react in a very similar manner. The unfavorable equilibrium must be driven to products. The reaction follows the acidic conditions mechanism.

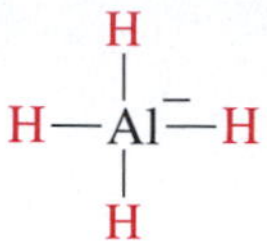

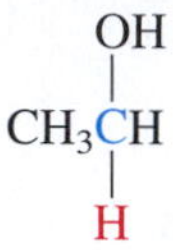

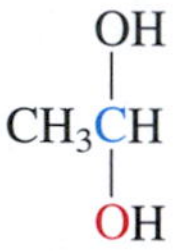

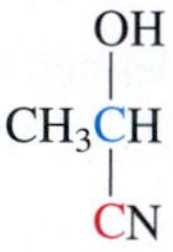

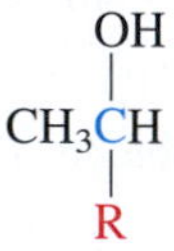

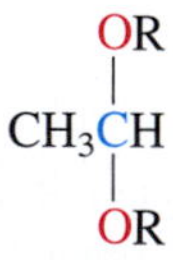

❶ All of these reactions begin this way. The electrophile (E) is usually hydrogen, but in the case of the Wittig reaction, it is phosphorus. Under basic conditions, Nu adds first. Under acidic conditions, E (a proton) adds first.

The reaction stops at stage 1 if the original Nu has only one unshared pair of electrons (CN⁻, hydrides, organometallic nucleophiles).

❷ If the nucleophile has a second pair of electrons (or can generate one), then the oxygen is eliminated. The O must be protonated first, unless E is phosphorus.

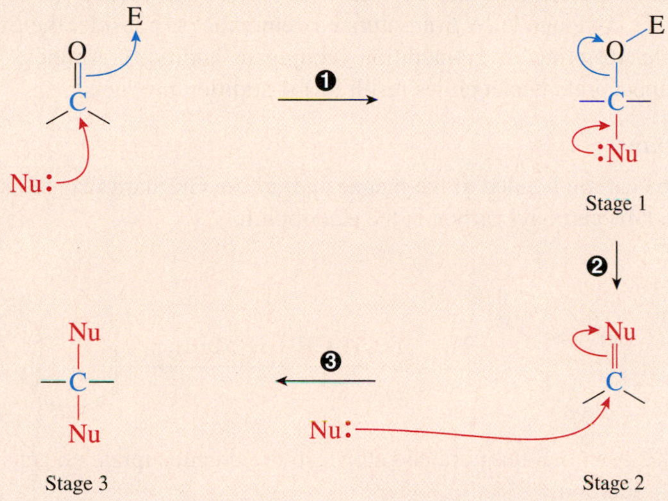

Stage 1

Stage 3

Stage 2

The reaction proceeds to stage 3 for alcohols and thiols as nucleophiles.

The reaction stops at stage 2 if this species is uncharged (Wittig reaction, imines).

❸ When the doubly bonded Nu has a positive charge, the reaction proceeds further. If Nu is a secondary amine, a proton is lost to form an enamine. If Nu is ROH or RSH, a second Nu attacks.

Figure 18.7

MECHANISMS BEGINNING WITH ADDITION OF A NUCLEOPHILE TO A CARBONYL GROUP.

Integrated Practice Problem

Show the products of these reactions:

a) 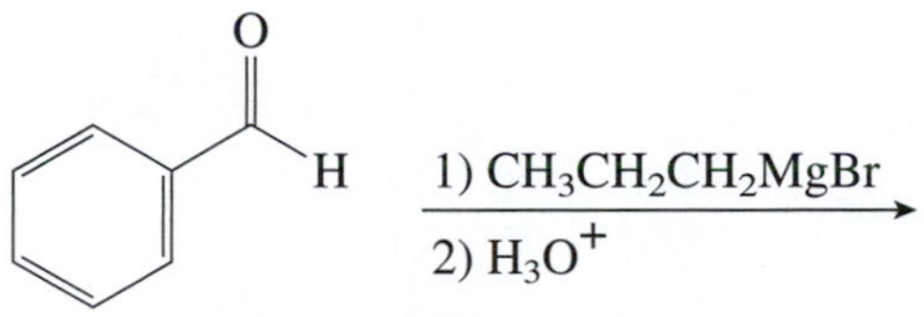 $\xrightarrow[\text{2) H}_3\text{O}^+]{\text{1) CH}_3\text{CH}_2\text{CH}_2\text{MgBr}}$

b) 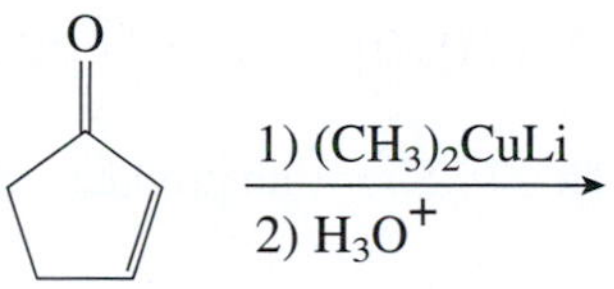 $\xrightarrow[\text{2) H}_3\text{O}^+]{\text{1) (CH}_3)_2\text{CuLi}}$

c) 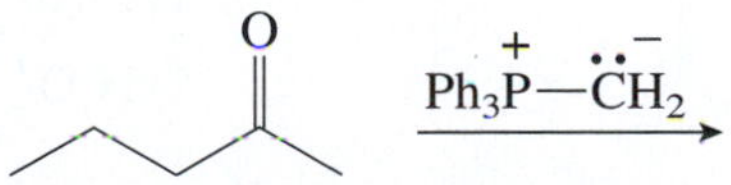 $\xrightarrow{\text{Ph}_3\overset{+}{\text{P}}-\overset{..}{\overset{-}{\text{C}}}\text{H}_2}$

Strategy

As usual, the best strategy is to identify the nucleophile and the electrophile. This chapter introduced a new electrophile, the carbonyl carbon of an aldehyde or ketone. The nucleophiles are listed in Table 18.2. Hydride, water, HCN, and organometallic nucleophiles result in the addition of the nucleophile to the carbon and a hydrogen to the oxygen of the carbonyl group. Ylides and nitrogen nucleophiles result in the formation of a double bond between the carbonyl carbon and the nucleophile. And alcohols and thiols add two nucleophiles to the carbonyl carbon.

If the compound is α,β-unsaturated, remember to consider the β-carbon as the electrophile, resulting in 1,4-addition (conjugate addition). Amines, HCN, and lithium diorganocuprate nucleophiles result in 1,4-addition products.

Solutions

a) The carbon bonded to the magnesium of the Grignard reagent is the nucleophile and the carbonyl carbon is the electrophile.

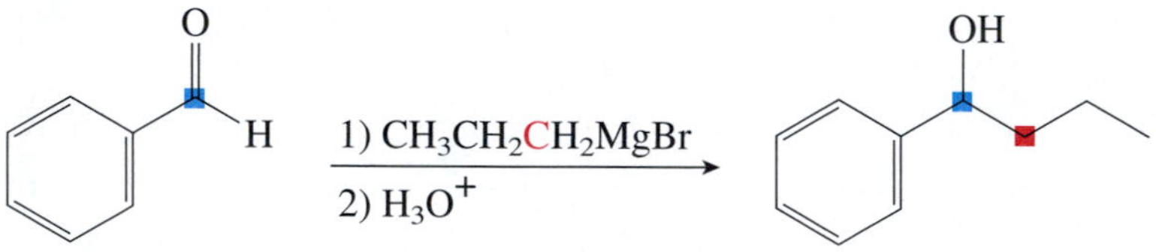

b) This is an α,β-unsaturated ketone, so the organocuprate reagent results in a 1,4-addition.

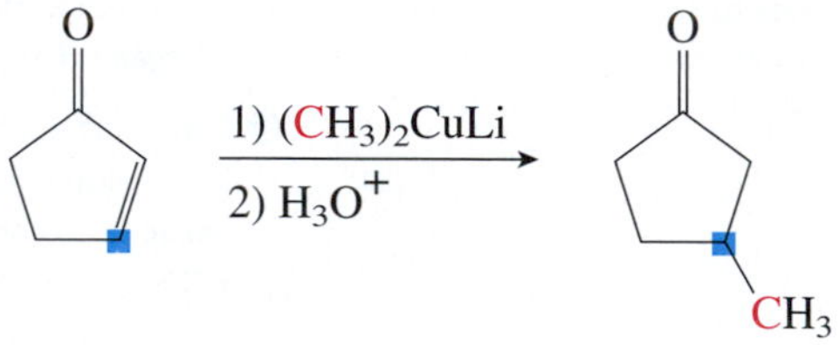

c) The carbonyl carbon is the electrophile and the carbanion of the ylide is the nucleophile. The Wittig reaction results in a double bond between the electrophilic and nucleophilic carbons.

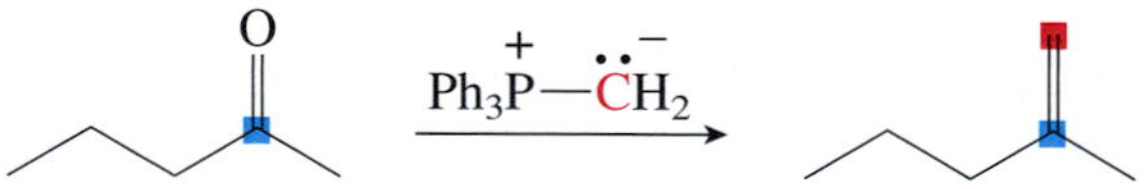

Additional Problems

18.27 Show the products of these reactions.

a) PhCH$_2$MgBr 1) CO$_2$ 2) H$_3$O$^+$

b) 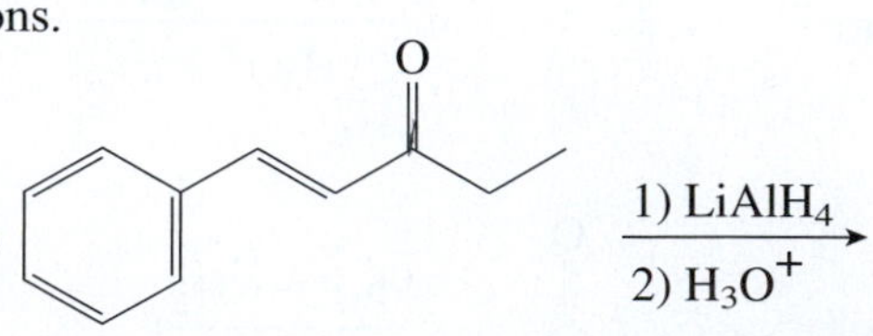 1) LiAlH$_4$ 2) H$_3$O$^+$

c) $CH_3\overset{O}{\overset{\|}{C}}CH_3$ + $NH_2NH\overset{O}{\overset{\|}{C}}NH_2$ $\xrightarrow{H_3O^+}$

d) 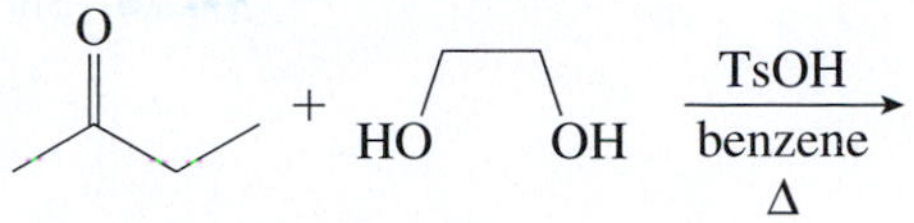 + $\underset{HO}{\qquad}\overset{\qquad}{\underset{OH}{\qquad}}$ $\xrightarrow[\underset{\Delta}{benzene}]{TsOH}$

e) 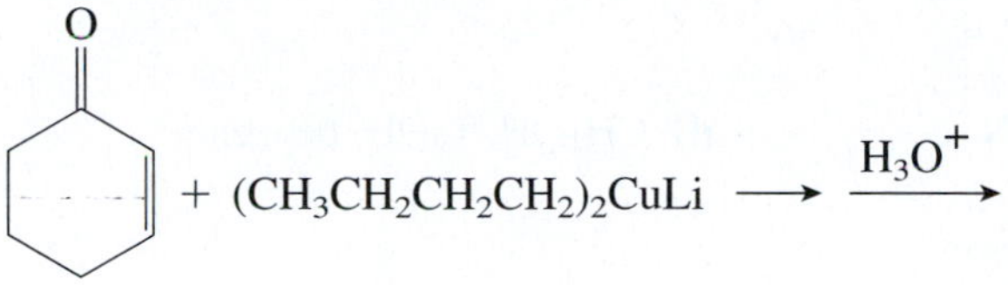 + $(CH_3CH_2CH_2CH_2)_2CuLi$ $\xrightarrow{H_3O^+}$

f) 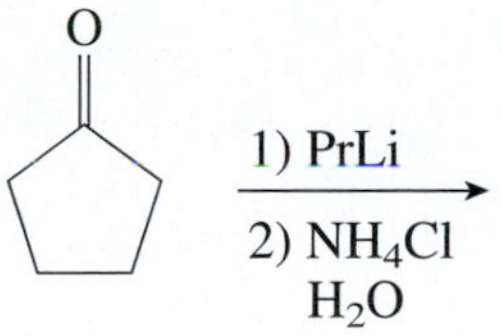 $\xrightarrow[\substack{2)\ NH_4Cl \\ H_2O}]{1)\ PrLi}$

g) 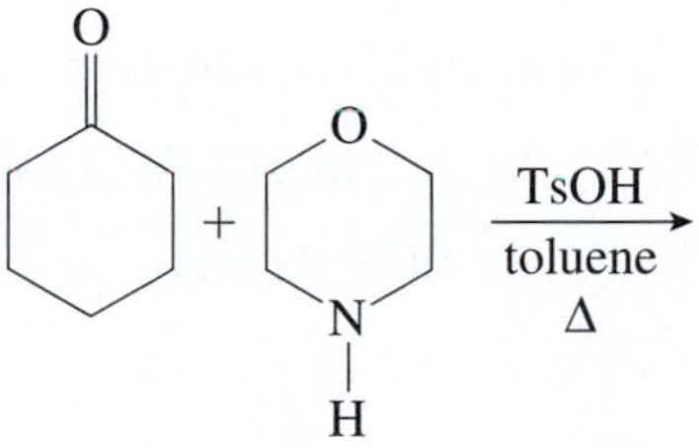 + $\xrightarrow[\underset{\Delta}{toluene}]{TsOH}$

h) $\xrightarrow[HCl]{CH_3OH}$

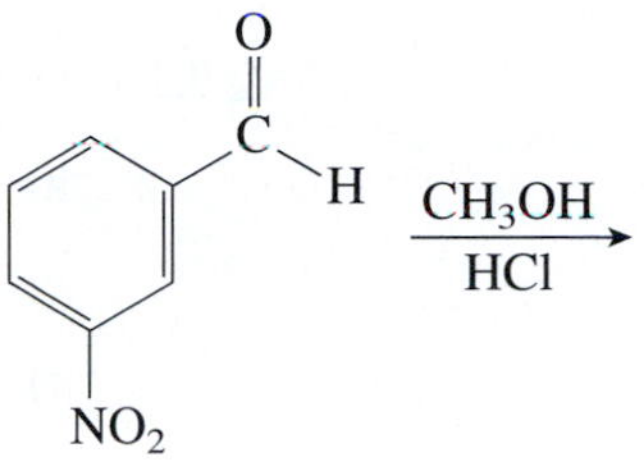

i) $Ph-\overset{O}{\overset{\|}{C}}-\overset{CH_3}{\overset{|}{CH}}CH_3$ $\xrightarrow[NaBH_3CN]{CH_3NH_2}$

j) $PhCH_2Cl$ $\xrightarrow[\substack{2)\ Ph_2CO \\ 3)\ H_3O^+}]{1)\ Mg,\ ether}$

k) $CH_3CH_2CH_2CH_2\overset{O}{\overset{\|}{C}}H$ $\xrightarrow[[CN^-]]{HCN}$

l) $CH_3CH_2CH_2CH_2\overset{O}{\overset{\|}{C}}H$ $\xrightarrow[2)\ H_3O^+]{1)\ LiAlH_4}$

m) 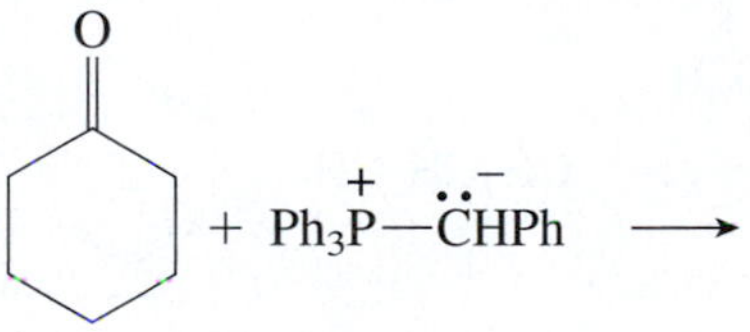 + $Ph_3\overset{+}{P}-\overset{..}{\overset{-}{C}}HPh$ $\longrightarrow$

n) $Ph\overset{O}{\overset{\|}{C}}Ph$ + NH_2OH $\xrightarrow{H_3O^+}$

o) $PhCH=CH\overset{O}{\overset{\|}{C}}Ph$ $\xrightarrow[\substack{H_2O \\ EtOH}]{KCN}$

p) 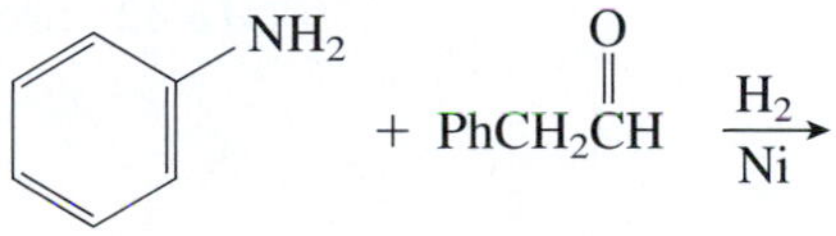 + $PhCH_2\overset{O}{\overset{\|}{C}}H$ $\xrightarrow{H_2\ Ni}$

q) $HC\equiv C-\overset{O}{\overset{\|}{C}}-OCH_3$ $\xrightarrow[2)\ H_3O^+]{1)\ Bu_2CuLi}$

r) 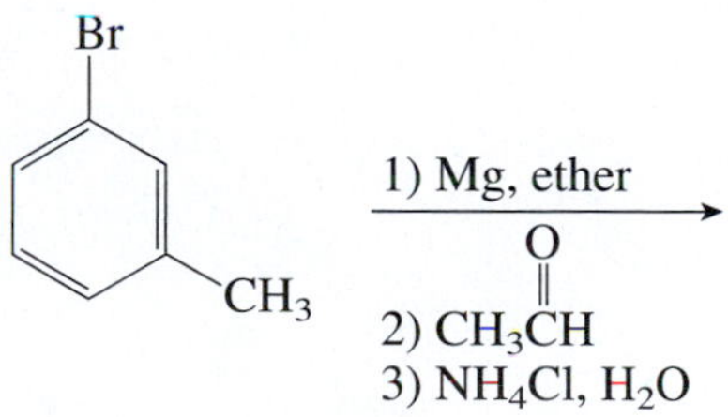 $\xrightarrow[\substack{2)\ CH_3\overset{O}{\overset{\|}{C}}H \\ 3)\ NH_4Cl,\ H_2O}]{1)\ Mg,\ ether}$

s) 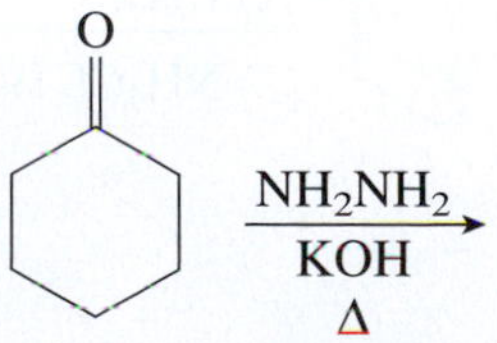 $\xrightarrow[\underset{\Delta}{KOH}]{NH_2NH_2}$

18.28 Show the products of the reactions of benzaldehyde with these reagents:

a) NaBH$_4$, CH$_3$OH

b) 1) 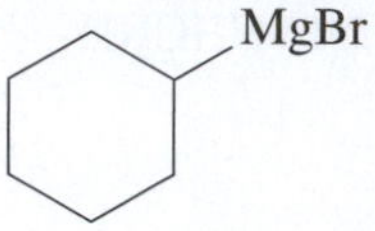 [cyclohexyl-MgBr]   2) H$_3$O$^+$

c) HCN, [CN$^-$]

d) CH$_3$OH, TsOH, benzene

e) NH$_2$NHCNH$_2$ (with O double bond)

f) Ph$_3$P—$\overset{+}{C}$HCH$_2$CH$_2$CH$_3$

18.29 Show the products of the reactions of 3-penten-2-one with these reagents:

a) 1) CH$_3$Li 2) H$_3$O$^+$

b) 1) LiAlH$_4$ 2) H$_3$O$^+$

c) 1) (CH$_3$CH$_2$)$_2$CuLi 2) H$_3$O$^+$

d) HCN, CH$_3$OH

e) PhNH$_2$

18.30 Explain why this reaction gives a poor yield of the cyanohydrin product:

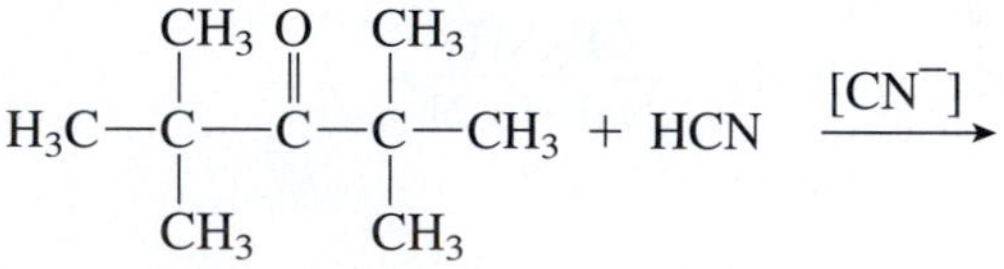

$$H_3C-\overset{\overset{CH_3}{|}}{\underset{\underset{CH_3}{|}}{C}}-\overset{O}{\overset{||}{C}}-\overset{\overset{CH_3}{|}}{\underset{\underset{CH_3}{|}}{C}}-CH_3 + HCN \xrightarrow{[CN^-]}$$

18.31 Arrange these compounds in order of increasing equilibrium constant for cyanohydrin formation and explain your reasoning:

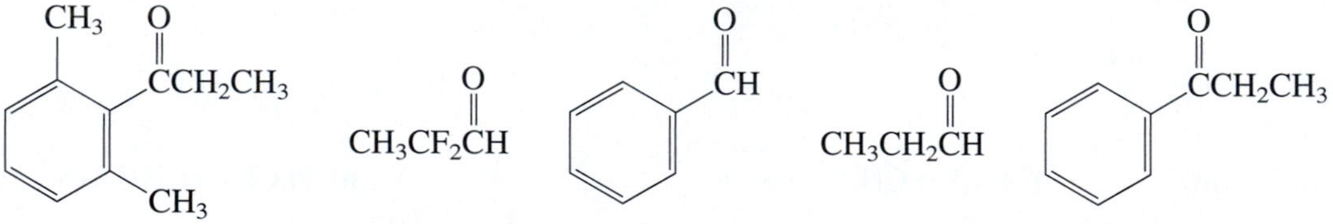

CH$_3$ ... CCH$_2$CH$_3$ (with O), CH$_3$ (disubstituted benzene)

CH$_3$CF$_2$CH (with O)

(benzaldehyde) CH (with O)

CH$_3$CH$_2$CH (with O)

(benzene) CCH$_2$CH$_3$ (with O)

18.32 Show the products of these reactions:

a) 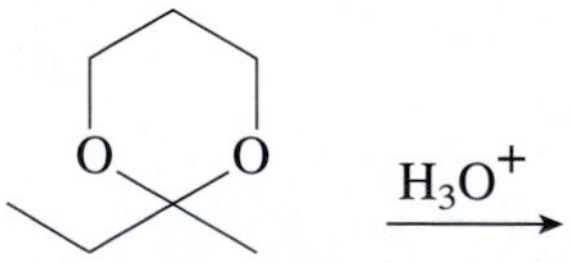 [1,3-dioxane with dimethyl] $\xrightarrow{H_3O^+}$

b) 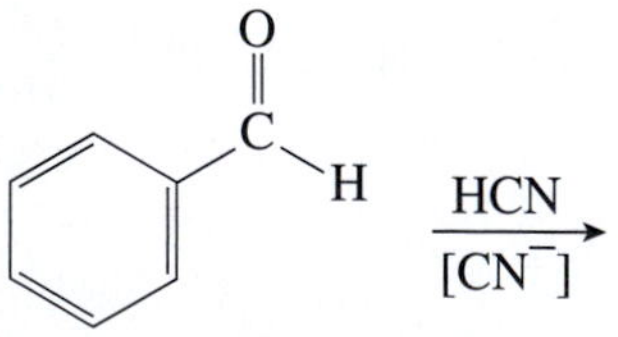 [benzaldehyde] $\xrightarrow[\text{[CN}^-\text{]}]{HCN}$

c) 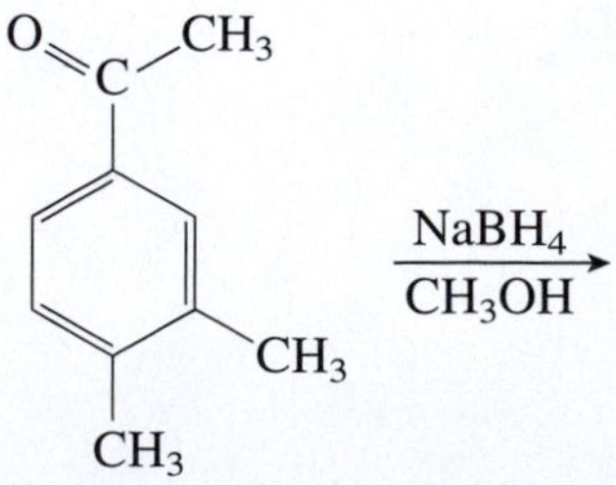 [substituted acetophenone] $\xrightarrow[\text{CH}_3\text{OH}]{NaBH_4}$

d) 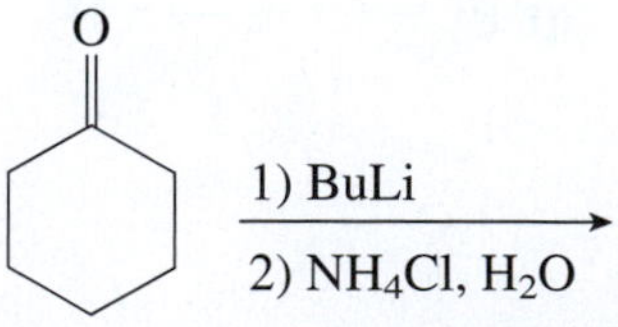 [cyclohexanone] $\xrightarrow[\text{2) NH}_4\text{Cl, H}_2\text{O}]{\text{1) BuLi}}$

e)
$$\underset{\text{O}}{\overset{\text{O}}{\text{PhCH}}} + \text{Ph}_3\overset{+}{\text{P}}-\overset{..}{\text{CHCH}}=\text{CH}_2 \longrightarrow$$

f)
$$\underset{\text{O}}{\overset{\text{O}}{\text{PhCPh}}} + \text{NH}_2\text{OH} \longrightarrow$$

g)
1) $(\text{CH}_3\text{CH}_2)_2\text{CuLi}$
2) H_3O^+

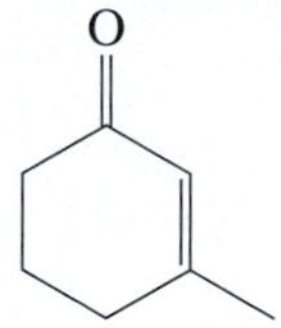

18.33 Show the products of these reactions:

a) 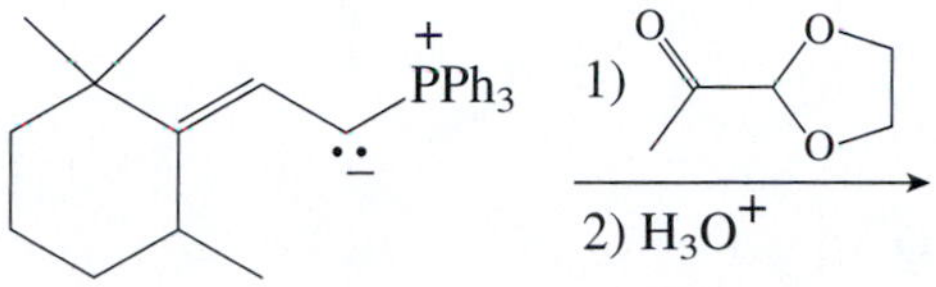
1) [dioxolane structure]
2) H_3O^+

b)
1) LiAlH_4
2) H_3O^+

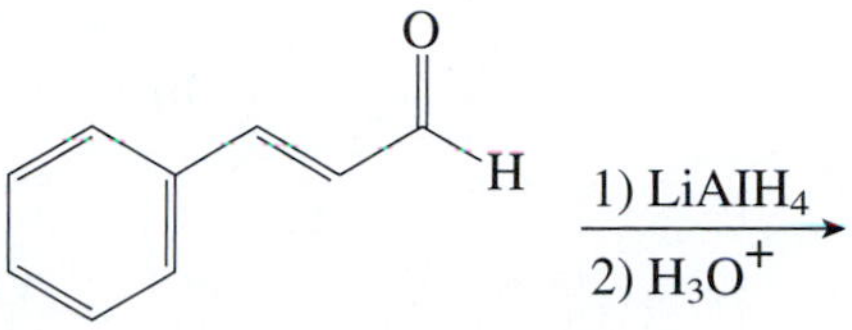

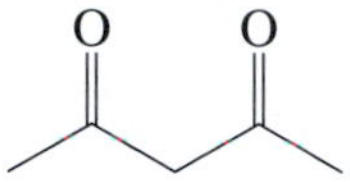

c)
$\dfrac{\text{NaBH}_4}{\text{H}_2\text{O}}$

d)
$$\text{CH}_3\overset{\overset{\displaystyle \text{Br}}{|}}{\text{CH}}\text{CH}_2\text{CH}_3$$
1) Mg, ether
2) CO_2
3) H_3O^+

e) 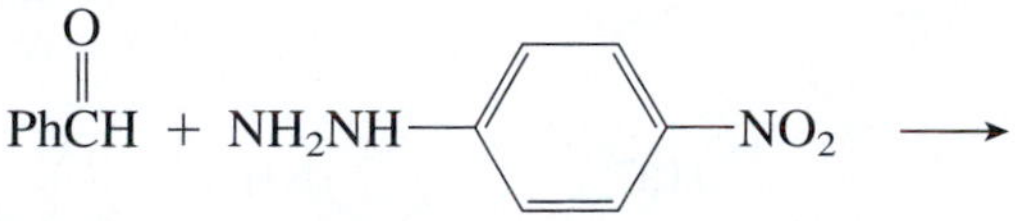
$$\underset{\text{O}}{\overset{\text{O}}{\text{PhCH}}} + \text{NH}_2\text{NH}-\!\!\!\!\!\!\text{—}\!\!\!\!\!\!-\text{NO}_2 \longrightarrow$$

f) 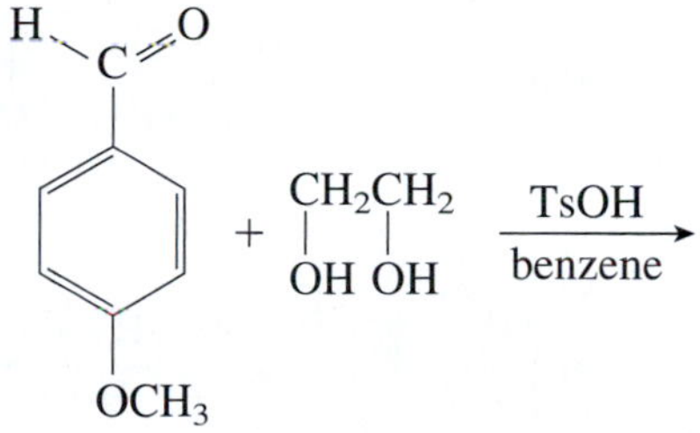
$+ \underset{\text{OH OH}}{\text{CH}_2\text{CH}_2} \quad \dfrac{\text{TsOH}}{\text{benzene}}$

g) $\text{PhNH}_2 + \text{CH}_2=\text{CH}-\text{C}\equiv\text{N} \longrightarrow$

18.34 Explain the difference in the equilibrium constants for cyanohydrin formation for these compounds:

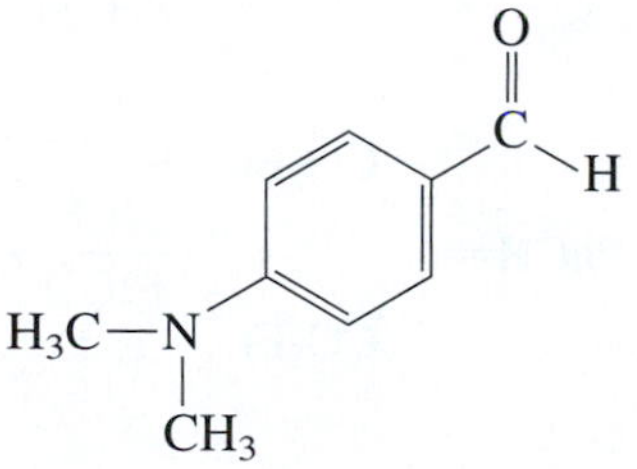

$K = 5$

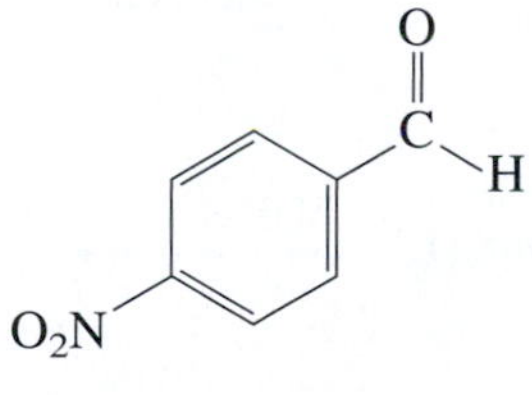

$K = 1820$

18.35 Explain which of the following compounds would have the larger equilibrium constant for cyanohydrin formation:

a) 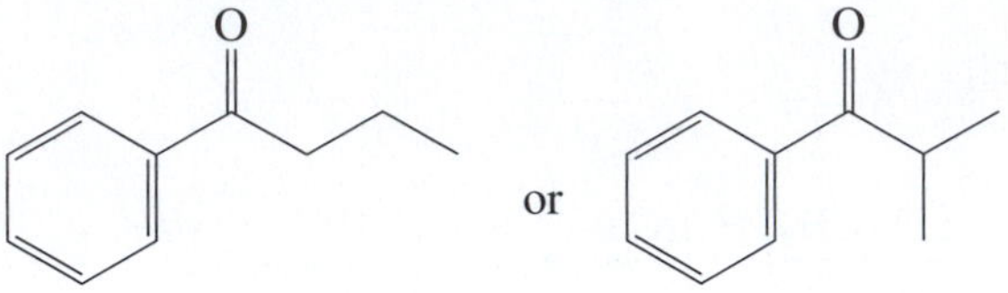   or

b) 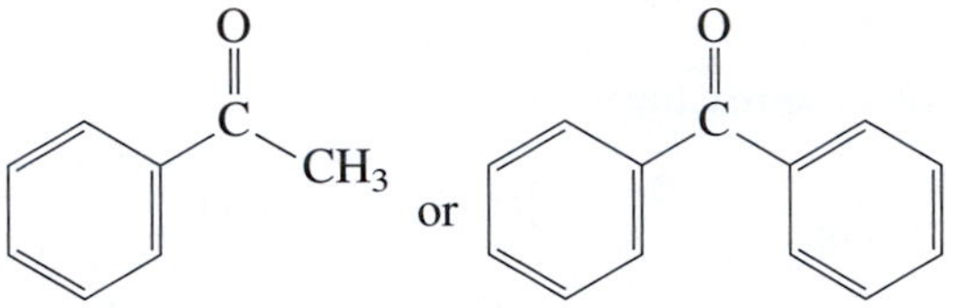   or

c) $CH_3\overset{O}{\underset{||}{C}}CH_3$ or $Cl_3\overset{O}{\underset{||}{C}}CCH_3$

18.36 Show the products of these reactions:

a) 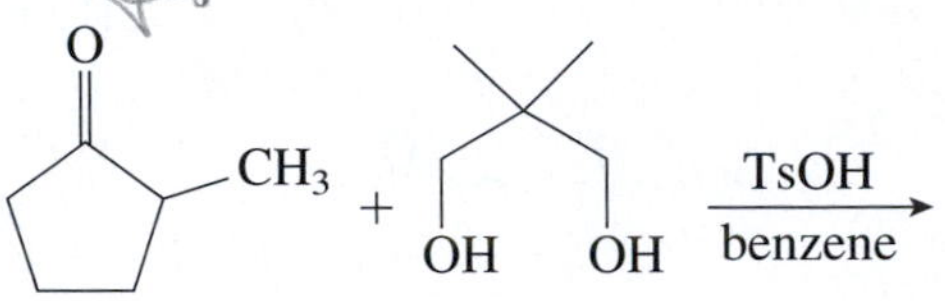 $+$ $\xrightarrow[\text{benzene}]{\text{TsOH}}$

b) 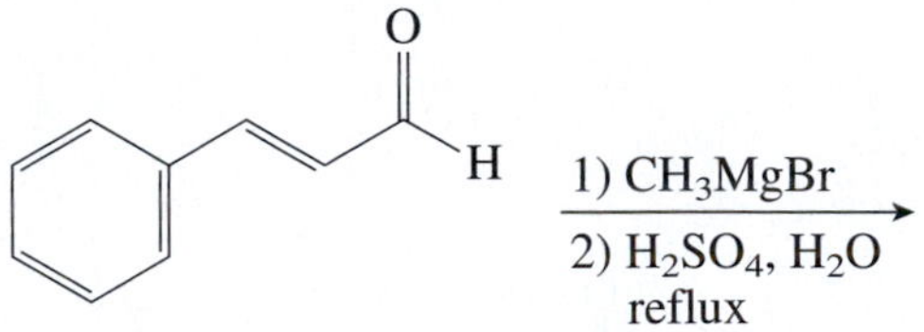 $\xrightarrow[\substack{\text{2) } H_2SO_4,\ H_2O \\ \text{reflux}}]{\text{1) } CH_3MgBr}$

c) 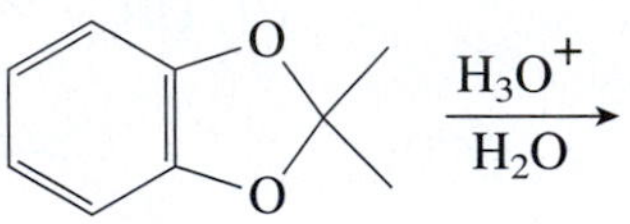 $\xrightarrow[H_2O]{H_3O^+}$

d) 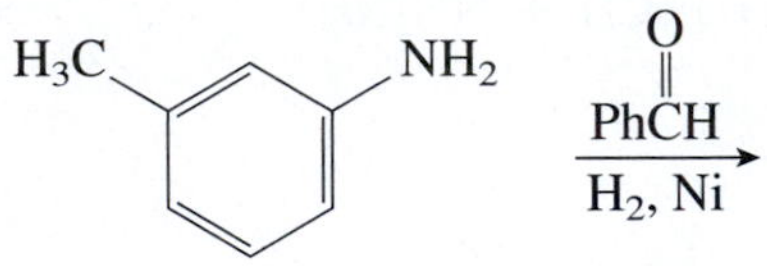 $\xrightarrow[H_2,\ Ni]{\overset{O}{\underset{||}{PhCH}}}$

e) $CH_3CH_2CH_2\overset{O}{\underset{||}{CH}} + NH_2Ph \xrightarrow[\text{benzene}]{\text{TsOH}}$

f) 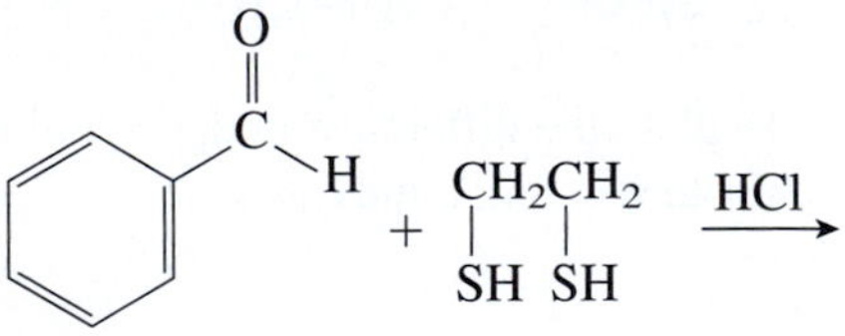 $+ \underset{\text{SH SH}}{\overset{}{CH_2CH_2}} \xrightarrow{HCl}$

g) 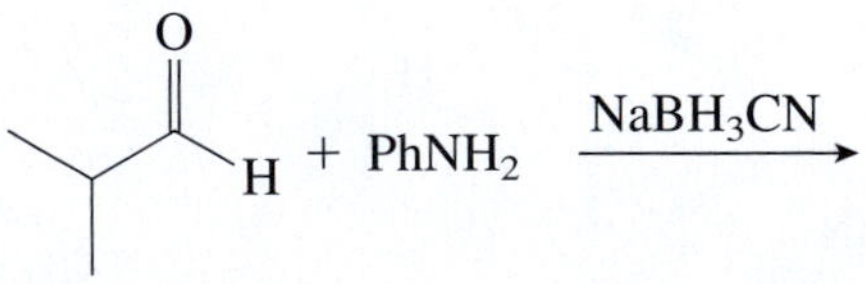 $+ PhNH_2 \xrightarrow{NaBH_3CN}$

h) $PhCH{=}C\overset{CO_2Et}{\underset{CO_2Et}{\diagup}} \xrightarrow[\substack{\text{EtOH} \\ H_2O}]{\text{KCN}}$

18.37 Show the aldehyde or ketone and any other reagents that are necessary to give these products:

a)

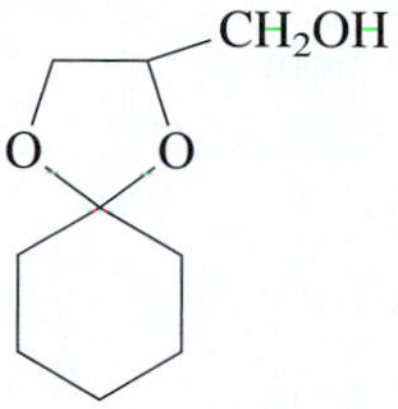

$+ H_2O$

b)

$$CH_3 \quad OH$$
$$CH_3CHCH_2CHCN$$

c)

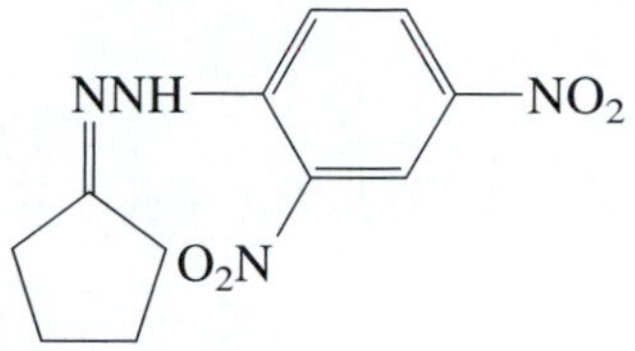

d)

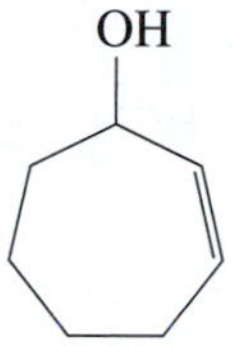

e)

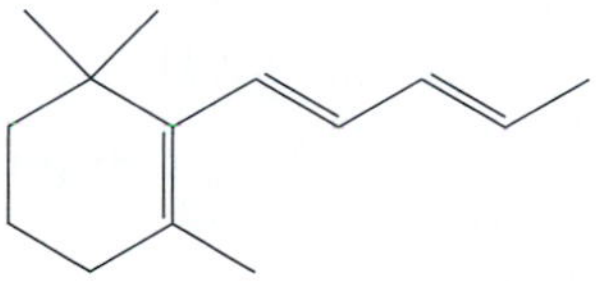

f)

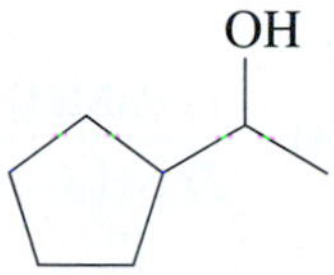

18.38 Show syntheses of these compounds from the indicated starting materials:

a) $CH_3\overset{\text{OH}}{\underset{}{C}}HCH_2CH_3$ from $CH_3\overset{\text{O}}{\underset{\|}{C}}H$

b) 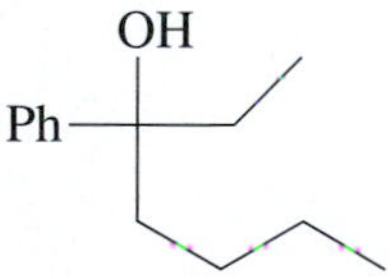 Ph

from PhBr

c) 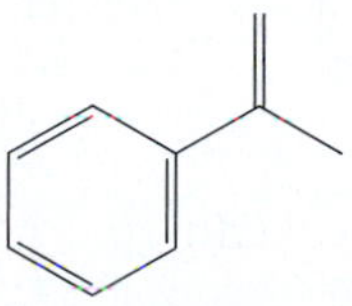 from 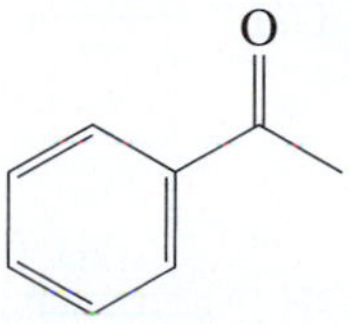 (2 ways)

d) 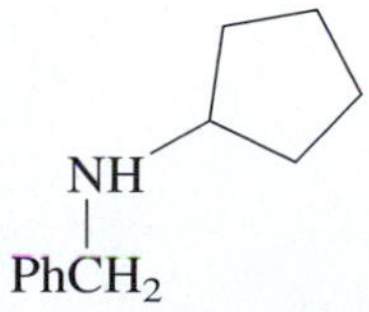 from $Ph\overset{\text{O}}{\underset{\|}{C}}H$

e) [structure: CH₃CH=CH-CH(OH)-CH₂CH₃ with OH] from [structure: CH₃CH=CH-CHO] (an α,β-unsaturated aldehyde, O, H)

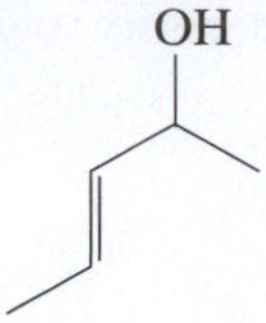

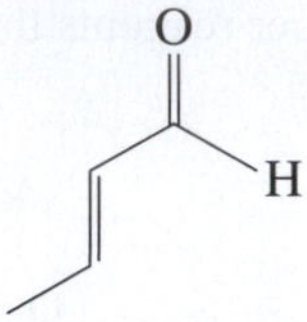

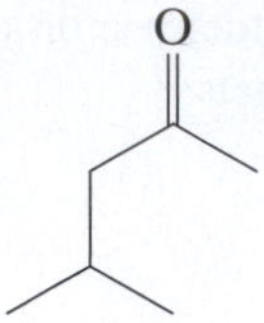

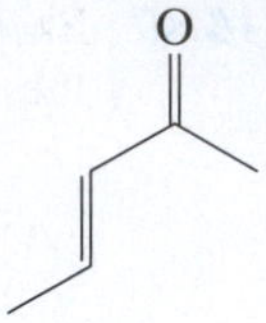

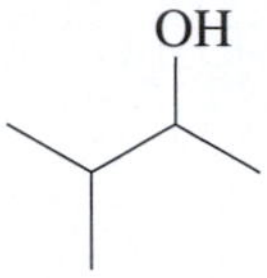

f) [structure: ketone with isobutyl group] from [structure: CH₃CH=CH-C(=O)-CH₃, an enone, O]

g) [structure: (CH₃)₂CH-CH(OH)-CH₃ with OH] from [structure: (CH₃)₂CH-Br, Br]

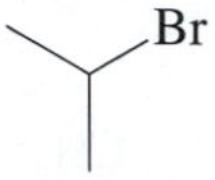

18.39 Show all the steps in the mechanisms for these reactions:

$$
\textbf{a)} \ \underset{\text{CH}_3\text{CH}}{\overset{\text{O}}{\|}} + \text{H}_2\text{O} \underset{}{\overset{\text{H}_3\text{O}^+}{\rightleftarrows}} \ \underset{\text{CH}_3\text{CH}}{\overset{\text{OH}}{\underset{\text{OH}}{|}}}
$$

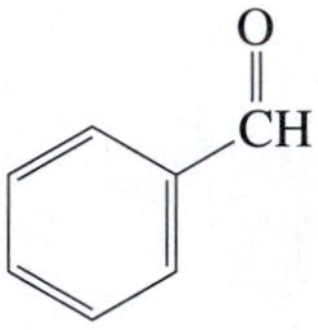

$$
\textbf{b)} \ \underset{\text{CH}_3\text{CH}}{\overset{\text{O}}{\|}} + \text{HCN} \ \underset{\text{H}_2\text{O}}{\overset{[\text{CN}^-]}{\rightleftarrows}} \ \underset{\text{CH}_3\text{CH}}{\overset{\text{CN}}{\underset{\text{OH}}{|}}}
$$

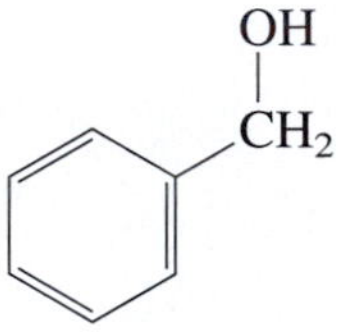

c) [benzaldehyde: C₆H₅-CH=O] + NaBH₄ $\xrightarrow{\text{CH}_3\text{OH}}$ [benzyl alcohol: C₆H₅-CH₂-OH]

$$
\textbf{d)} \ \underset{\text{CH}_3\text{CH}}{\overset{\text{O}}{\|}} \ \xrightarrow[\text{2) NH}_4\text{Cl, H}_2\text{O}]{\text{1) PhMgBr}} \ \underset{\text{CH}_3\text{CH}}{\overset{\text{OH}}{\underset{\text{Ph}}{|}}}
$$

$$
\textbf{e)} \ \underset{\text{CH}_3\text{CH}_2\text{CH}}{\overset{\text{O}}{\|}} + \text{Ph}_3\overset{+}{\text{P}}{-}\overset{..}{\overset{-}{\text{C}}}\text{H}_2 \ \longrightarrow \ \underset{\text{CH}_3\text{CH}_2\text{CH}}{\overset{\text{CH}_2}{\|}} + \text{Ph}_3\text{PO}
$$

f) [benzaldehyde: C₆H₅-CH=O] + NH₂OH $\xrightarrow{\text{H}_3\text{O}^+}$ [oxime: C₆H₅-CH=NOH] + H₂O

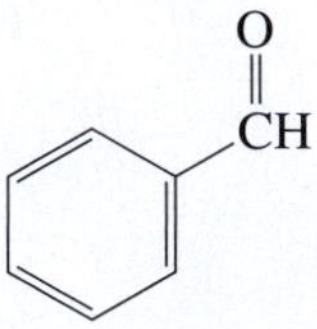

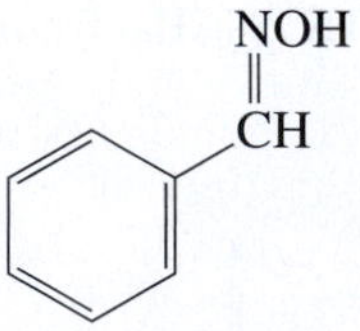

$$
\textbf{g)} \ \underset{\text{H}_3\text{C} \quad \text{CH}_3}{\overset{\text{O}}{\underset{}{\|}}}\text{C} + \text{EtOH} \ \underset{}{\overset{\text{TsOH}}{\rightleftarrows}} \ \underset{\text{H}_3\text{C} \quad \text{CH}_3}{\overset{\text{EtO} \quad \text{OEt}}{\underset{}{}}}\text{C} + \text{H}_2\text{O}
$$

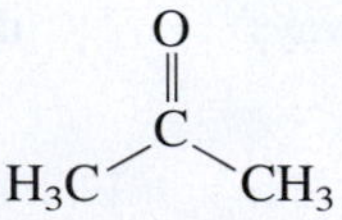

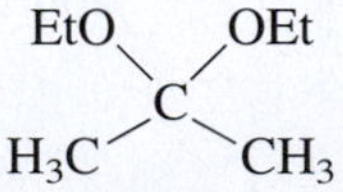

18.40 Explain why this reaction does not occur:

$$PhNH_2 + CH_2{=}CH{-}CH_3 \;\;\xcancel{\longrightarrow}\;\; PhNH{-}CH_2{-}CH_2{-}CH_3$$

but this one does occur:

$$PhNH_2 + CH_2{=}CH{-}NO_2 \;\longrightarrow\; PhNH{-}CH_2{-}CH_2{-}NO_2$$

18.41 Show the steps in the mechanism for this reaction:

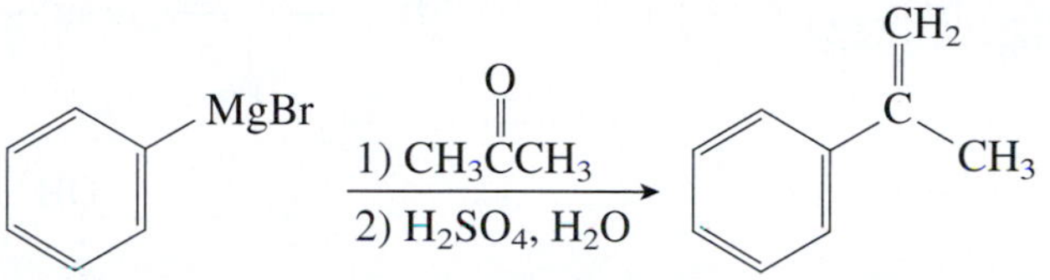

18.42 Show syntheses of these compounds from the indicated starting materials. Several steps are required. You may need to use reactions from previous chapters.

a) 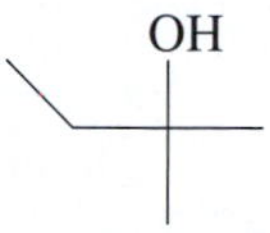 from

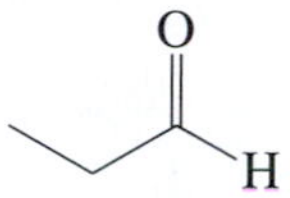

b) 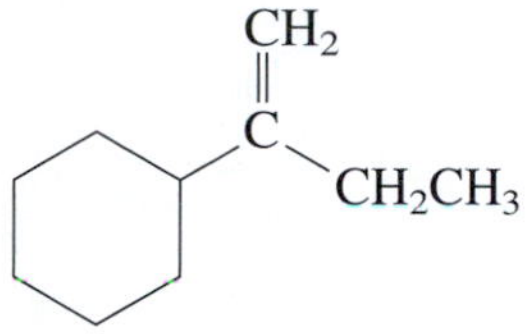 from

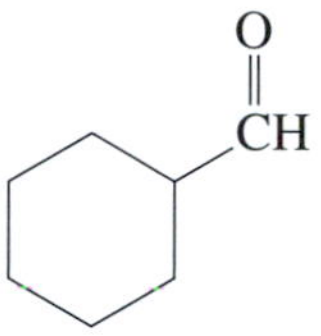

c) 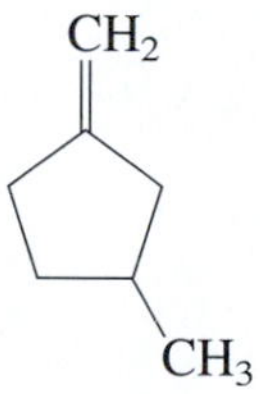 from

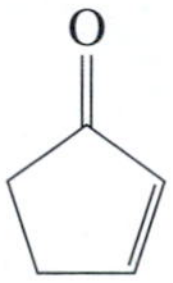

d) 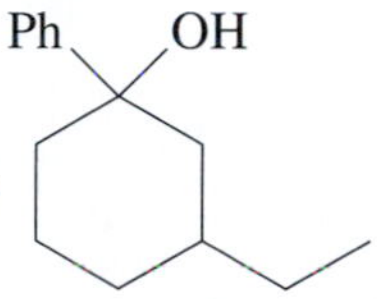 from

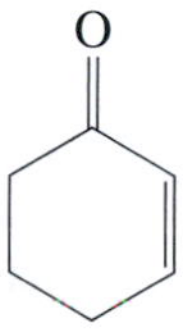

e) 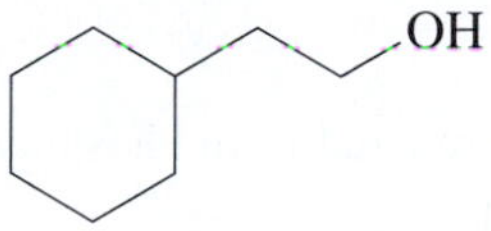 from

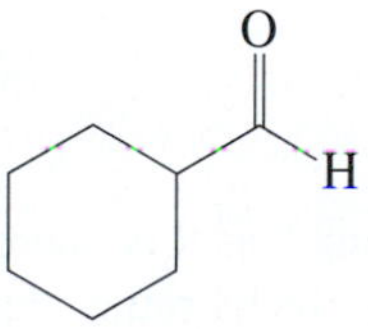

f) 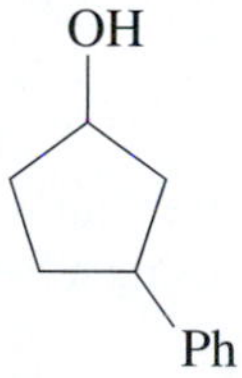 from

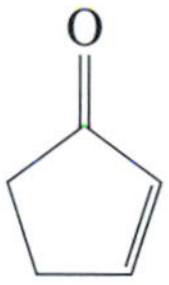

18.43 Show syntheses of these compounds from the indicated starting materials. You may need to use reactions from previous chapters.

a) 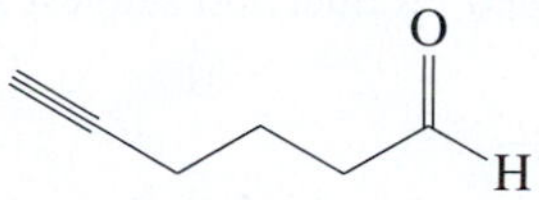 from

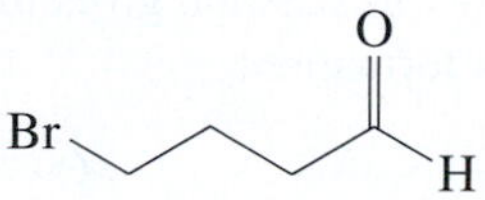

b) 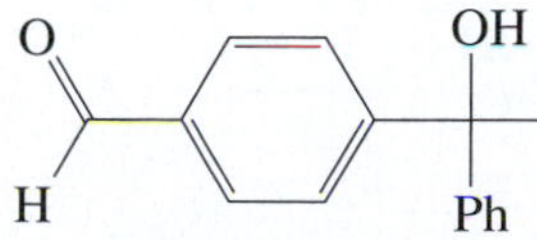 from

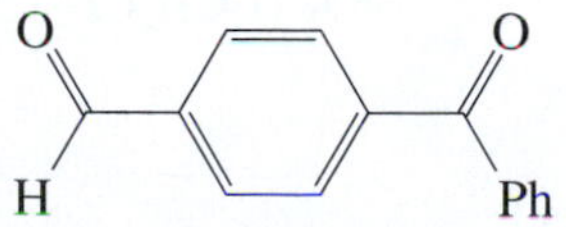

(See problem 18.24)

c) 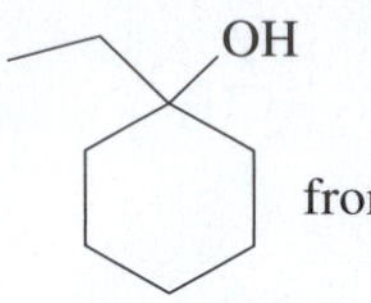 [structure: 1-ethylcyclohexan-1-ol]   from   [structure: cyclohex-2-enone]

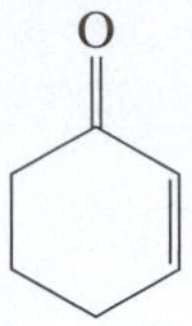

d)  [structure: 2-methylpent-1-ene]   from   [structure: acetaldehyde]

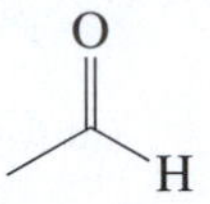

e) 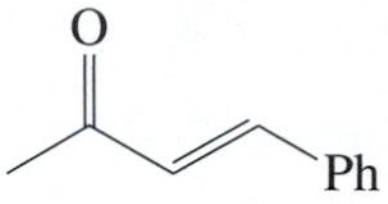 [structure: (E)-4-phenylbut-3-en-2-one]   from   [structure: 1-bromoacetone (CH₃COCH₂Br)]

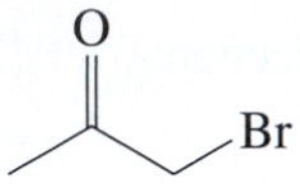

f) 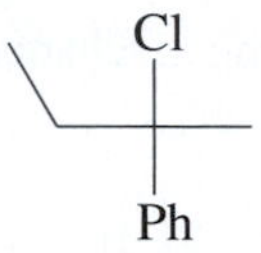 [structure: 2-chloro-2-phenylbutane]   from   [structure: butan-2-one / methyl propyl ketone]

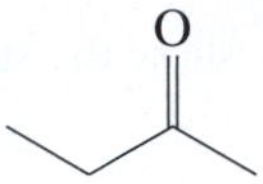

g) 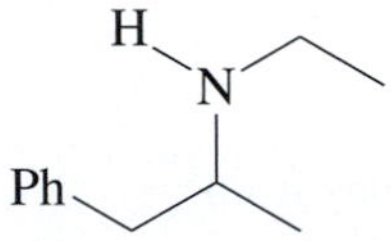 [structure: N-ethyl-N-(1-phenylpropan-2-yl)amine]   from   [structure: PhCH₂CHO (phenylacetaldehyde)]

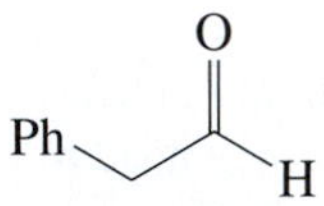

h) 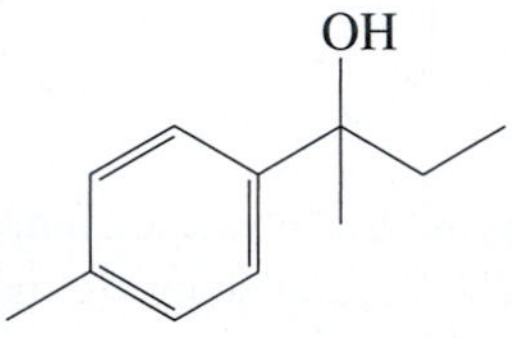 [structure: 2-(4-methylphenyl)butan-2-ol]   from benzene

i) 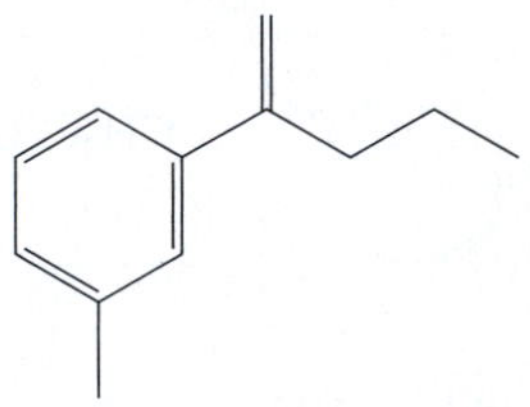 [structure: 1-(but-1-en-2-yl)-3-methylbenzene]   from benzene

j) 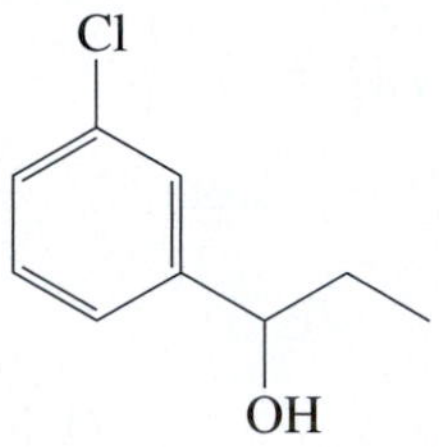 [structure: 1-(3-chlorophenyl)propan-1-ol]   from benzene

k) 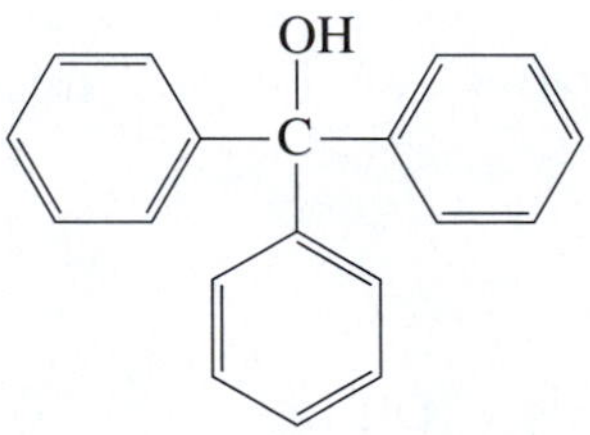 [structure: triphenylmethanol (Ph₃COH)]   from benzene

18.44 Explain why ylides such as this one can be prepared from phosphonium salts using a base such as NaOH rather than BuLi:

$$\overset{+}{Ph_3P}-\overset{..}{\overset{-}{C}H}\overset{O}{\overset{\|}{C}}CH_3$$

18.45 Explain why this reaction gives the indicated product and suggest a mechanism for its formation:

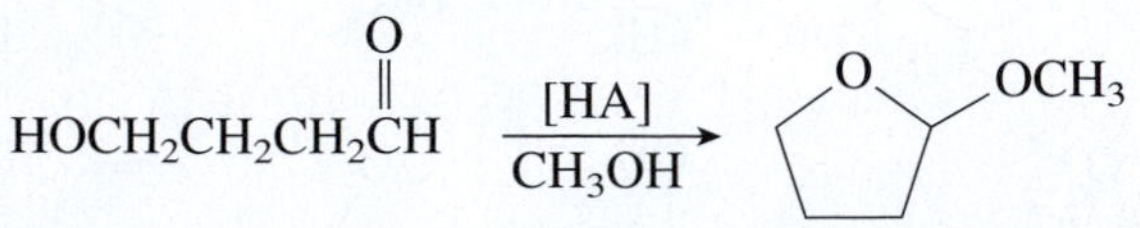

$$HOCH_2CH_2CH_2\overset{O}{\overset{\|}{C}}H \quad \xrightarrow[CH_3OH]{[HA]} \quad \text{[structure: 2-methoxytetrahydrofuran]}$$

18.46 Brevicomin, an aggregation pheromone of the Western pine beetle, is an acetal. Show the structure of the dihydroxyketone that reacts spontaneously to produce brevicomin.

Brevicomin

18.47 The sugar fructose is an isomer of glucose. Like glucose, fructose forms a cyclic hemiacetal, but in this case the ring is five membered rather than six membered. Show the structure for the hemiacetal formed from fructose and show a mechanism for its formation in acidic solution.

H_3O^+

Fructose

18.48 Glycosides are naturally occurring acetals formed from sugars and alcohols. The glycoside salicin, found in willow bark, is formed from glucose and a phenol. Show the structure of the phenol and the steps in the mechanism for the formation of salicin from glucose hemiacetal.

+ ? H_3O^+

Glucose hemiacetal   Salicin

18.49 Suggest a structure for **A** in this reaction and suggest a mechanism for the conversion of **A** to **B**:

$$PhCH(=O) + CH_3NH_2 \xrightarrow[\text{reflux}]{\text{benzene}} \underset{(95\%)}{\textbf{A}} \xrightarrow[\text{2) H}_2\text{O}]{\text{1) PhCH}_2\text{MgCl}} \textbf{B}$$

B structure: benzene ring with CHCH₂Ph bearing NHCH₃ substituent

B
(96%)

18.50 The Strecker synthesis is used to prepare amino acids in the laboratory. As shown in the following equation, an aldehyde is reacted with sodium cyanide and ammonium chloride in water to produce a cyanoamine. Conversion of the cyano group to a carboxylic acid completes the synthesis. Show the structure of the intermediate, **A,** in the following synthesis, and show the steps in the mechanism for the formation of **A** and for the conversion of **A** to the cyanoamine. (*Hint:* Remember that NH_4^+ and H_2O are in equilibrium with NH_3 and H_3O^+.)

$$CH_3CH(=O) \xrightarrow[\substack{NH_4Cl \\ H_2O \quad C_2H_5N}]{NaCN} \textbf{[A]} \longrightarrow \underset{NH_2}{CH_3CHC\equiv N} \longrightarrow \underset{\substack{NH_2 \\ \text{Alanine}}}{CH_3CHCO_2H}$$

18.51 Explain the formation of the product that results when two equivalents of benzaldehyde are reacted with one equivalent of ammonia in the presence of hydrogen and a catalyst.

$$2\ PhCH(=O) + 1\ NH_3 \xrightarrow[Ni]{H_2} PhCH_2NHCH_2Ph \quad (81\%)$$

18.52 Suggest a mechanism for this reaction, which appeared in the Focus On box on page 764.

reaction scheme: allylic alcohol (with OH) + HCl / Ph₃P → phosphonium salt (Cl⁻ +PPh₃)

18.53 The following compound, known as CS, is a component of tear gas. It is believed that it exerts its effect by reacting with nucleophilic SH groups of proteins. Suggest a structure for the product of this reaction.

structure of CS: benzene ring with Cl substituent and CH=C(C≡N)₂ group

$$+ \ HS-\text{Protein} \longrightarrow$$

CS

Problems Involving Spectroscopy

18.54 A student set out to synthesize the following compound in the laboratory. When the compound was isolated, its IR spectrum did not show a strong absorption band in the region of 1730 cm^{-1}. Was the synthesis unsuccessful? Explain.

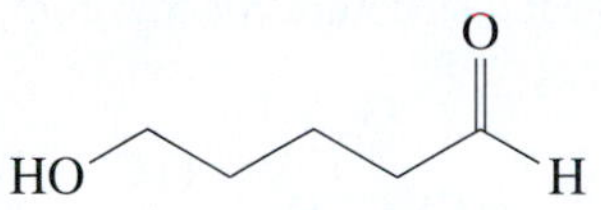

18.55 The reaction of 2 moles of benzaldehyde with 1 mole of hydrazine gives a product with the formula $C_{14}H_{12}N_2$. This compound shows only five absorptions in its ^{13}C-NMR spectrum. Suggest a structure for the product of this reaction.

$$2 \ \text{PhCH} + 1 \ \text{NH}_2\text{NH}_2 \longrightarrow C_{14}H_{12}N_2$$

18.56 A graduate student ran the Grignard reaction shown in the following equation. After adding the acid in the workup step, the student took a break for lunch. After lunch, the workup was completed and the product was isolated. However, the product did not show an absorption in the 3500 to 3200 cm^{-1} region of its IR spectrum. The ^{1}H-NMR spectrum of the product is shown here. Suggest a structure for the product and explain its formation.

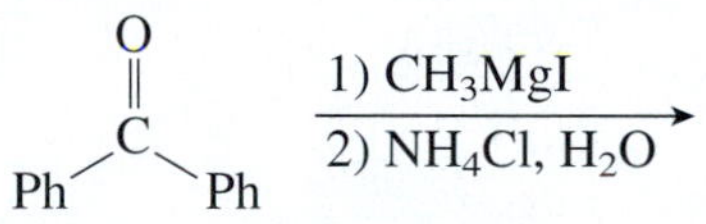

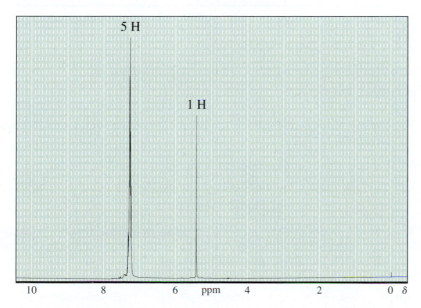

18.57 A graduate student ran the Grignard reaction shown in the following equation. The student was unaware that the flask that was used for the reaction was contaminated with some copper(I) salt. The product that was isolated had no absorption in the 3500 to 3200 cm^{-1} region of its IR spectrum but did have a strong absorption near 1715 cm^{-1}. The ^{1}H-NMR spectrum of the product is shown here. Suggest a structure for this compound.

$$CH_2{=}CH\overset{\overset{\displaystyle O}{\|}}{C}CH_3 \xrightarrow[\text{2) } NH_4Cl,\ H_2O]{\text{1) } CH_3MgI} C_5H_{10}O$$

ORGANIC **Chemistry Now**™
Click Molecular Model Problems to view the models needed to work these problems.

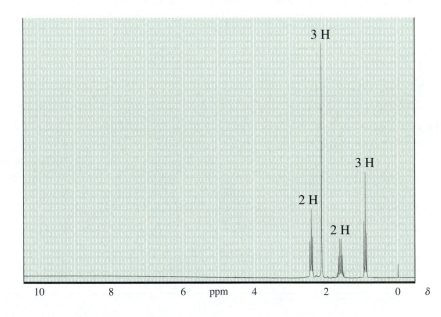

Problems Using Online Three-Dimensional Molecular Models

18.58 Explain which compound has the larger equilibrium constant for cyanohydrin formation.

18.59 Explain which compound reacts faster with sodium borohydride.

18.60 The reduction of camphor with lithium aluminum hydride gives 90% isoborneol and 10% borneol. Explain these results.

18.61 One of these ketones gives 100% conjugate addition (1,4-addition) when reacted with phenylmagnesium bromide, whereas the other gives 100% normal addition (1,2-addition). Explain these results.

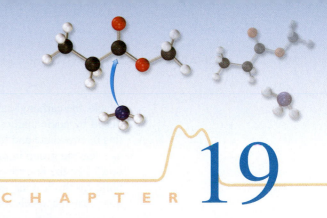

Substitutions at the Carbonyl Group

REACTIONS OF CARBOXYLIC ACIDS
AND DERIVATIVES

C H A P T E R 19

THE PREVIOUS CHAPTER discussed the attack of nucleophiles at the electrophilic carbonyl carbon of aldehydes and ketones. In most of the reactions this resulted in the addition of the nucleophile to the carbon and a proton to the oxygen of the carbonyl group. In this chapter the electrophile is the carbonyl carbon of a carboxylic acid or a related derivative such as an acyl chloride, anhydride, ester, or amide. The reactions in this chapter begin in exactly the same way as those in the last chapter—that is, by attack of a nucleophile at the carbonyl carbon. Here, however, the presence of the other heteroatom (an atom other than carbon or hydrogen, such as oxygen, nitrogen, or chlorine) on the carbonyl carbon causes the reaction to diverge at this point. Rather than the addition reactions of Chapter 18, the heteroatom group leaves, resulting in a substitution.

First, the general mechanisms for these reactions are presented. Then the reactivity of these carboxylic acid derivatives is discussed. As expected, the factors that control the reactivity are very similar to those that affect the addition reactions of Chapter 18. Next, reactions with nucleophiles that interconvert all of the members of the carboxylic acid family are presented. Finally, the reactions of hydride and organometallic nucleophiles with these electrophiles are discussed.

19.1 THE GENERAL MECHANISM

As was the case for nucleophilic additions to aldehydes and ketones, two mechanisms occur for the reactions in this chapter: one under basic conditions and one under acidic conditions. Both of these start in the same manner as the addition reaction mechanisms. The basic conditions mechanism is shown in Figure 19.1. First the nucleophile attacks the carbonyl carbon, generating an intermediate that is very similar to that produced by nucleophilic attack on an aldehyde or ketone. The major difference between this intermediate and that produced from an aldehyde or ketone is the ability of L to act as a leaving group in this case. In the second step, L leaves in a reaction that looks very much like the reverse of the first step. The overall result is a substitution of the nucleophile for the leaving group. Although this reaction is a substitution, it is important to note that the mechanism is quite different from the S_N2 mechanism that occurs at sp^3-hybridized carbons. This mechanism occurs in two steps, with the nucleophile bonding to the carbon first. Such a mechanism is impossible at an sp^3-hybridized carbon because the intermediate would have five bonds to the carbon and would violate the octet rule. At a carbonyl carbon a pair of electrons can be displaced onto the oxygen when the nucleophile bonds, resulting in the formation of an sp^3-hybridized carbon in the intermediate (often called the **tetrahedral intermediate** because of its geometry).

For this mechanism to occur for an aldehyde or ketone, hydride ion or a carbanion would have to act as the leaving group. These species are much too basic to leave under normal circumstances. In the case of carboxylic acid derivatives the leaving group is one of the following less basic species:

ORGANIC
Chemistry﹣Now™
Click *Mechanisms in Motion* to view the **Mechanism of Nucleophilic Substitution at a Carbonyl Group under Basic Conditions.**

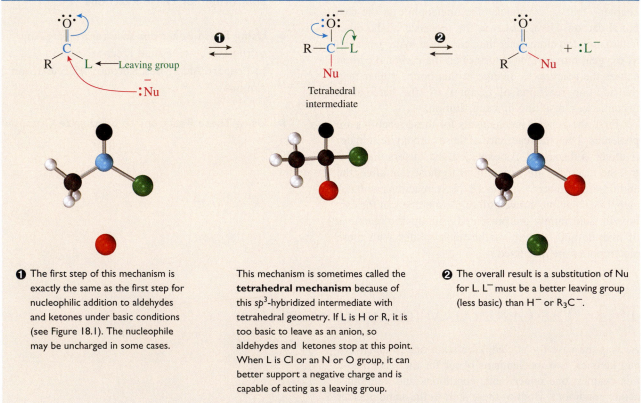

① The first step of this mechanism is exactly the same as the first step for nucleophilic addition to aldehydes and ketones under basic conditions (see Figure 18.1). The nucleophile may be uncharged in some cases.

This mechanism is sometimes called the **tetrahedral mechanism** because of this sp^3-hybridized intermediate with tetrahedral geometry. If L is H or R, it is too basic to leave as an anion, so aldehydes and ketones stop at this point. When L is Cl or an N or O group, it can better support a negative charge and is capable of acting as a leaving group.

② The overall result is a substitution of Nu for L. L⁻ must be a better leaving group (less basic) than H⁻ or R₃C⁻.

Figure 19.1

MECHANISM OF NUCLEOPHILIC SUBSTITUTION AT A CARBONYL GROUP UNDER BASIC CONDITIONS.

Except for chloride anion, these anions are too basic to act as leaving groups in the S_N2 reaction. However, poorer leaving groups can be used in this reaction because the first step, attack of the nucleophile, is usually the rate-determining step. The leaving group is lost in the subsequent step, which is favored by entropy because it is intramolecular.

As was the case for the addition reaction of Chapter 18, another version of the mechanism operates under acidic conditions. In this version the carbonyl oxygen is protonated before the nucleophile attacks. Because this makes the carbonyl carbon more electrophilic, weaker nucleophiles can be used (often the conjugate acids of those used under basic conditions). In addition, the leaving group may be protonated before it leaves. Examples of the acidic conditions mechanism are provided in later sections.

Now let's address the reactivity of these compounds—that is, how the rate and the position of the equilibrium for the reaction are affected by the structure of the compound. The first step, attack of the nucleophile, is usually the rate-determining step. Because this step is the same as the first step in the mechanism for additions to aldehydes and ketones, steric, resonance, and inductive effects control the rate of this reaction in exactly the same manner as was described in Sections 18.3 and 18.4. These effects can be summarized as follows:

Steric effects: Steric hindrance slows the approach of the nucleophile and causes a decrease in the reaction rate. (However, the position of the equilibrium is not affected if the starting material and the substitution product have similar steric interactions.)

Inductive effects: Electron-withdrawing groups make the carbonyl carbon more electrophilic and increase the reaction rate; electron-donating groups make the carbonyl carbon less electrophilic and decrease the reaction rate.

Resonance effects: Resonance electron-withdrawing groups make the carbonyl carbon more electrophilic and increase the rate; resonance electron donors make the carbonyl carbon less electrophilic and decrease the reaction rate.

PRACTICE PROBLEM 19.1

Explain which of these compounds would have the faster rate of nucleophilic substitution at its carbonyl group:

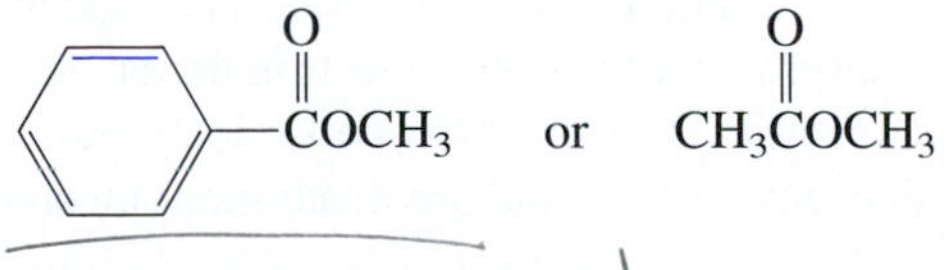

Solution

The benzene ring donates electrons to the carbonyl carbon by resonance, thus making the carbon less electrophilic and slowing the reaction rate:

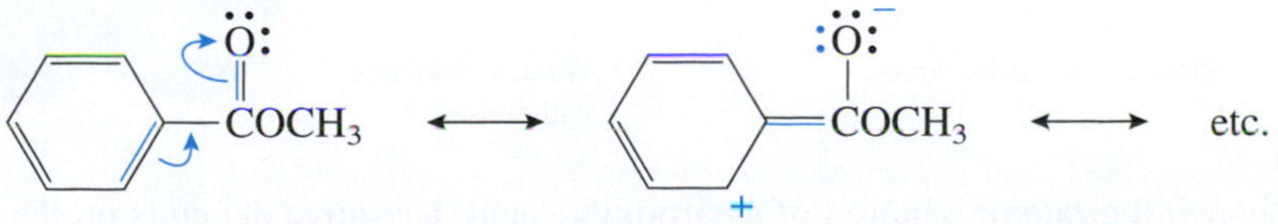

PROBLEM 19.1

Explain which compound would have the faster rate of nucleophilic substitution at its carbonyl group:

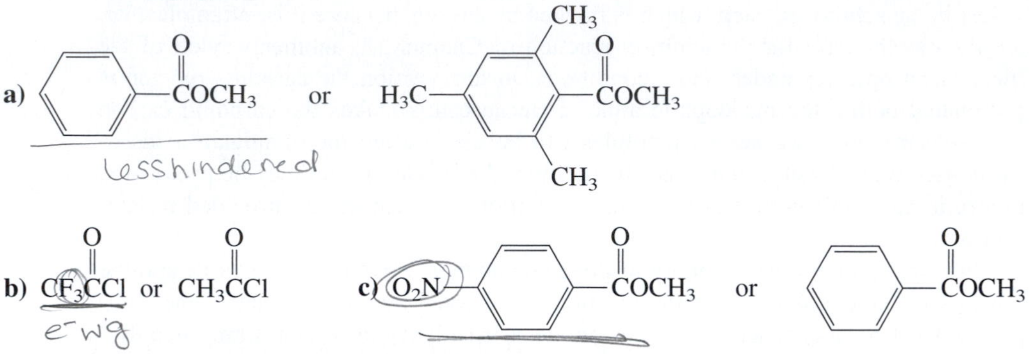

a)

[benzene ring]—COCH$_3$ or   H$_3$C—[benzene ring with CH$_3$ groups]—COCH$_3$

lesshindered

b) CF$_3$CCl or CH$_3$CCl   **c)** O$_2$N—[benzene ring]—COCH$_3$ or   [benzene ring]—COCH$_3$

e⁻wg

Although the leaving group does not come off during the rate-determining step, it does affect the reaction rate by its inductive and resonance effects. Furthermore, its stability after it has left affects the position of the equilibrium. Leaving groups that are weaker bases are more stable and cause the products to be more favored at equilibrium.

Consider the case of an acyl chloride. The chlorine is an inductive electron withdrawer and a resonance electron donor. As we saw in Chapter 17, the inductive effect is stronger. (Recall that chlorine is not a very strong resonance electron donor because the long C—Cl bond and the size difference between the $3p$ AO on the Cl and the $2p$ AO on the C result in poor overlap of these orbitals.) In addition, chloride anion is a very weak base. Overall, acyl chlorides are the most reactive of the carboxylic acid derivatives discussed here and are the least favored at equilibrium.

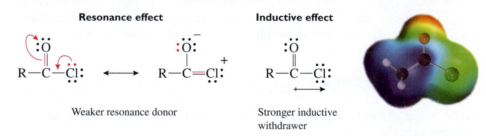

Now consider the case of an amide. The same effects are operating. The nitrogen is an inductive electron withdrawer and a resonance electron donor. As we saw in Chapter 17, the resonance effect is stronger in this case. Also, amide ion is quite basic. Overall, amides are much less reactive than acyl halides and are much more favored at equilibrium.

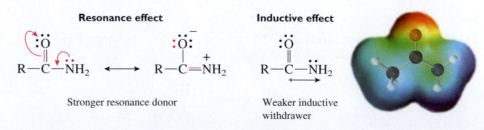

Although the rate of reaction of a carboxylic acid derivative depends on the inductive and resonance effects of the leaving group and the position of the equilibrium depends on the stability of the leaving group, both of these effects are related to the basicity of the leaving group. Acid derivatives with leaving groups that are weaker bases react faster and are less favored at equilibrium. Table 19.1 lists the reactivity order for all of these compounds. Note that aldehydes and ketones have been included in this reactivity scale, but for the rate of nucleophilic attack only. The second step, in which the leaving group comes off, does not occur for these compounds.

The reactivity scale of Table 19.1 is very important. Not only does it tell where a nucleophile will react when faced with a choice between two different carbonyl groups, but it also enables the position of the equilibrium to be predicted for many reactions. For example, a compound that has both an aldehyde and an ester group is more likely to be attacked by a nucleophile at the aldehyde carbonyl carbon. The table also indicates that the equilibrium in the reaction

$$
\underset{\substack{\| \\ \text{O}}}{R-C-Cl} + H-O-R' \rightleftharpoons \underset{\substack{\| \\ \text{O}}}{R-C-O-R'} + HCl
$$

favors the products because the ester is lower on the reactivity scale than is the acyl chloride. Any carboxylic acid derivative can readily be prepared by reaction of the appropriate nucleophile with a derivative that is higher on the scale. For example, an ester is often prepared by reaction of an acyl chloride with an alcohol, as shown in the preceding equation. Let's examine the details of how to prepare acid derivatives.

Table 19.1   Reactivity Scale for Carbonyl Compounds

	Compound	Structure	Leaving Group	Comment	
Most reactive compound	Acyl chloride	R—C(=O)—Cl	Cl⁻		**Less favored at equilibrium**
	Anhydride	R—C(=O)—O—C(=O)—R	⁻O—C(=O)—R		
	Aldehyde	R—C(=O)—H		First step only	
Increasing reaction rate	Ketone	R—C(=O)—R'		First step only	**Increasing equilibrium constant**
	Ester	R—C(=O)—O—R'	⁻O—R'	Esters and acids are very similar in both rate and equilibrium position	
	Acid	R—C(=O)—O—H	⁻O—H		
	Amide	R—C(=O)—NH₂	⁻NH₂		
Least reactive compound	Carboxylate anion	R—C(=O)—O⁻	O²⁻	Poor leaving group; seldom leaves	**More favored at equilibrium**

PROBLEM 19.2

Explain whether these equilibria favor the reactants or the products:

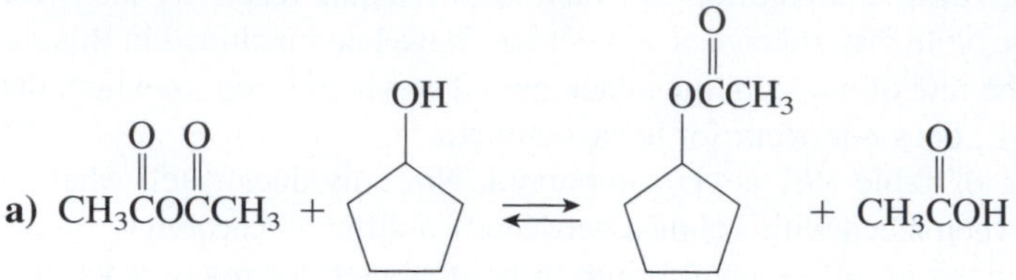

a) CH_3COCCH_3 + [cyclopentanol] ⇌ [cyclopentyl acetate] + CH_3COH

b) $CH_3CH_2CNH_2$ + CH_3OH ⇌ $CH_3CH_2COCH_3$ + NH_3

c) [benzoyl chloride] + $CH_3CH_2NH_2$ ⇌ [N-ethylbenzamide] + HCl

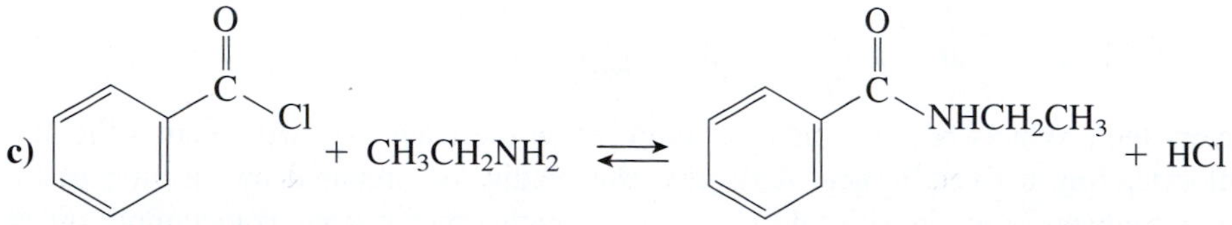

PROBLEM 19.3

Explain which of these reactions is faster:

$$CH_3COCH_3 + {}^-OH \longrightarrow CH_3CO^- + CH_3OH$$

or

$$CH_3CNHCH_3 + {}^-OH \longrightarrow CH_3CO^- + CH_3NH_2$$

PROBLEM 19.4

Suggest a reaction that could be used to prepare this amide:

$$CH_3CH_2CH_2CNHCH_2CH_2CH_3$$

19.2 PREPARATION OF ACYL CHLORIDES

Because they are readily available from a number of synthetic reactions, carboxylic acids are the most common starting materials for the preparation of the other members of this family. Conversion of a carboxylic acid to an acyl chloride provides access to any of the other derivatives because the acyl chloride is at the top of the reactivity scale. But how can the acyl chloride be prepared from the acid when the acid is lower on the reactivity scale? This can be accomplished by using an even more reactive compound to drive the equilibrium in the desired direction. The reagent that is employed in the vast majority of cases is thionyl chloride, $SOCl_2$. Phosphorus trichloride, PCl_3, and phosphorus pentachloride, PCl_5, are also used occasionally. Examples are provided in the following equations:

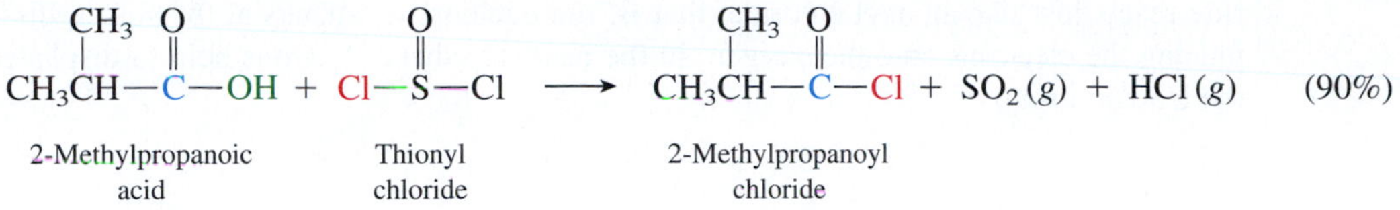

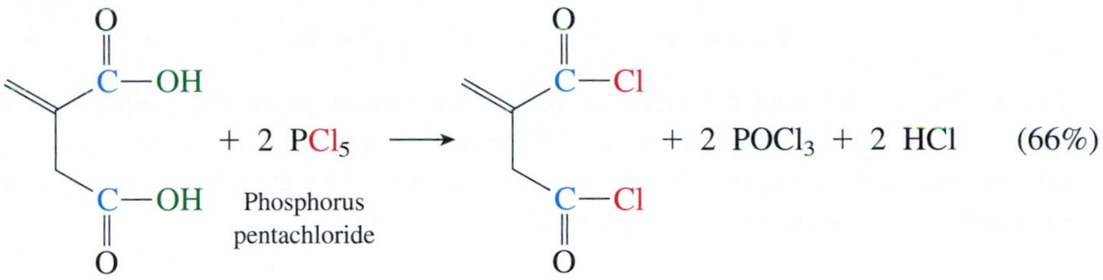

CH₃ O
CH₃CH—C—OH + Cl—S—Cl ⟶ CH₃CH—C—Cl + SO₂(g) + HCl(g)   (90%)

2-Methylpropanoic   Thionyl   2-Methylpropanoyl
     acid   chloride   chloride

O
‖
C—OH
          + 2 PCl₅ ⟶   + 2 POCl₃ + 2 HCl   (66%)
C—OH   Phosphorus
‖   pentachloride
O

Thionyl chloride can be viewed as the di(acid chloride) of sulfurous acid. It is even more reactive than the acyl halide. Its reactivity, along with the formation of gaseous products (SO₂ and HCl), serves to drive the equilibrium to the acyl chloride. The mechanism for this reaction is shown in Figure 19.2. It is interesting to note that thionyl chlo-

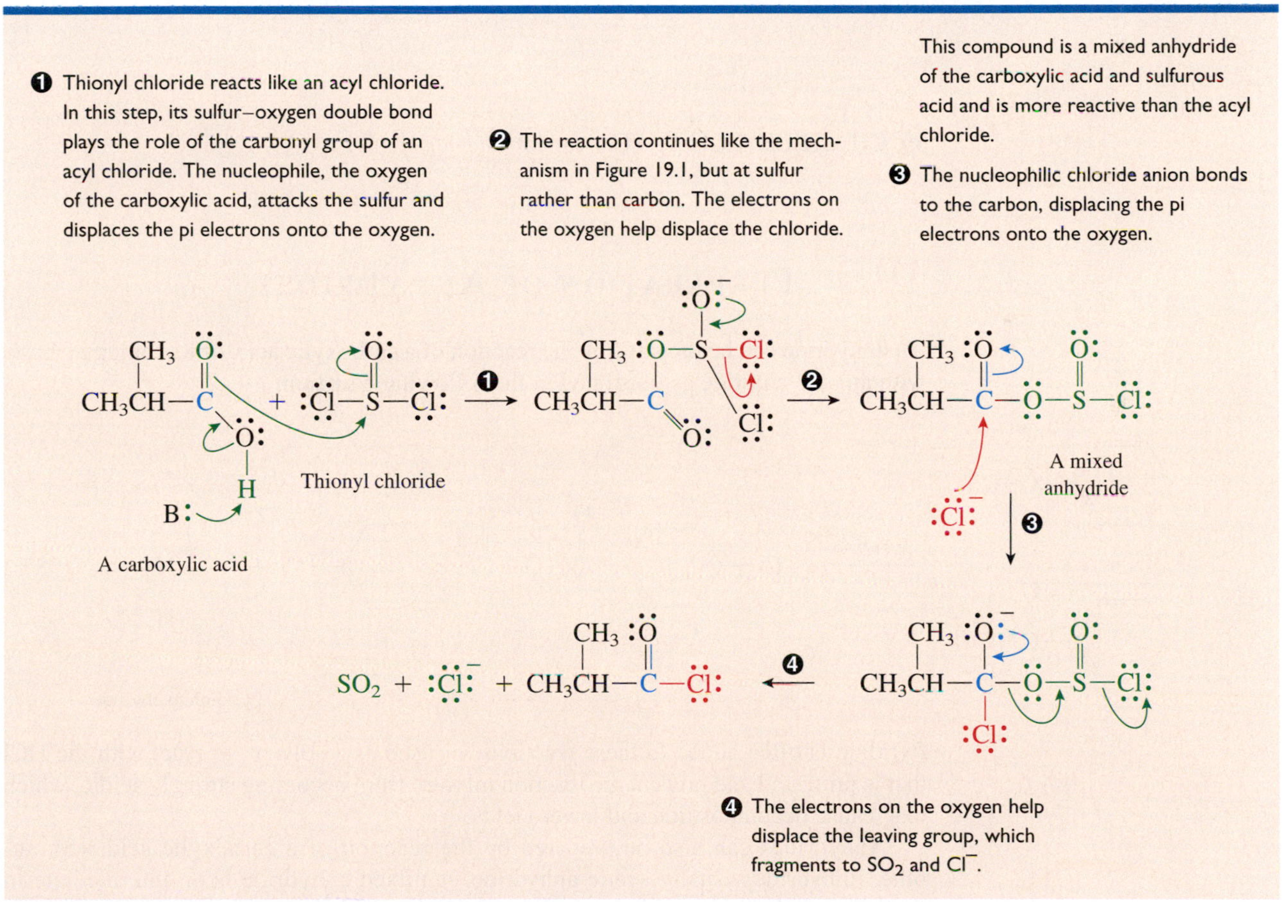

❶ Thionyl chloride reacts like an acyl chloride. In this step, its sulfur–oxygen double bond plays the role of the carbonyl group of an acyl chloride. The nucleophile, the oxygen of the carboxylic acid, attacks the sulfur and displaces the pi electrons onto the oxygen.

❷ The reaction continues like the mechanism in Figure 19.1, but at sulfur rather than carbon. The electrons on the oxygen help displace the chloride.

This compound is a mixed anhydride of the carboxylic acid and sulfurous acid and is more reactive than the acyl chloride.

❸ The nucleophilic chloride anion bonds to the carbon, displacing the pi electrons onto the oxygen.

A carboxylic acid   Thionyl chloride

A mixed anhydride

❹ The electrons on the oxygen help displace the leaving group, which fragments to SO₂ and Cl⁻.

Figure 19.2

MECHANISM OF THE REACTION OF A CARBOXYLIC ACID WITH THIONYL CHLORIDE.

ride reacts just like an acyl chloride; that is, the nucleophile attacks at the sulfur, displacing the electrons onto the oxygen. In the next step these electrons help to displace the leaving group.

$$HO-\overset{\overset{\displaystyle O}{\|}}{S}-OH \qquad Cl-\overset{\overset{\displaystyle O}{\|}}{S}-Cl$$

Sulfurous acid   Thionyl chloride

The major use for acyl chlorides is as starting materials for the preparation of the other carboxylic acid derivatives. Acyl fluorides, bromides, and iodides could potentially be employed to prepare the other derivatives also. However, because they offer no advantages over the acyl chlorides, they are seldom used.

PROBLEM 19.5
Show the products of these reactions:

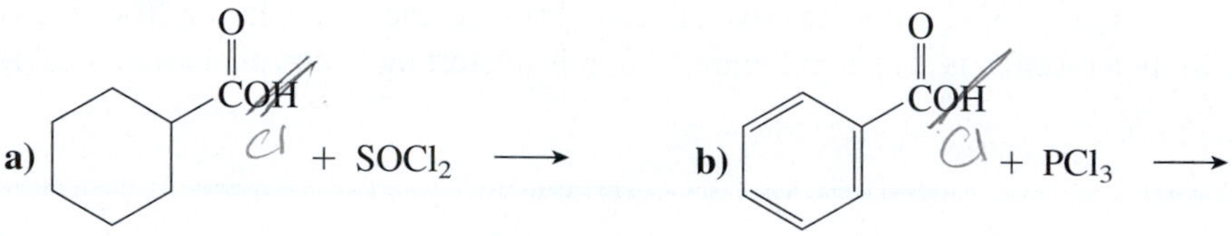

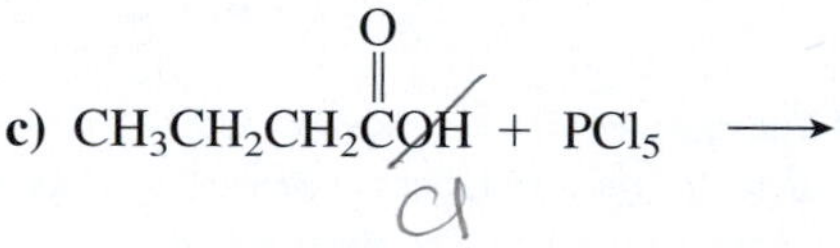

a) cyclohexane-$\overset{\overset{\displaystyle O}{\|}}{C}OH$ + SOCl$_2$ $\longrightarrow$

b) benzene-$\overset{\overset{\displaystyle O}{\|}}{C}OH$ + PCl$_3$ $\longrightarrow$

c) CH$_3$CH$_2$CH$_2$$\overset{\overset{\displaystyle O}{\|}}{C}OH$ + PCl$_5$ $\longrightarrow$

19.3 PREPARATION OF ANHYDRIDES

An anhydride can be prepared by the reaction of a carboxylic acid, or its conjugate base, with an acyl chloride as illustrated in the following equation:

$$Cl-C_6H_4-\overset{\overset{\displaystyle O}{\|}}{C}-Cl \; + \; Cl-C_6H_4-\overset{\overset{\displaystyle O}{\|}}{C}-OH \xrightarrow{\text{pyridine}} Cl-C_6H_4-\overset{\overset{\displaystyle O}{\|}}{C}-O-\overset{\overset{\displaystyle O}{\|}}{C}-C_6H_4-Cl \; + \; \text{pyridinium} \quad (98\%)$$

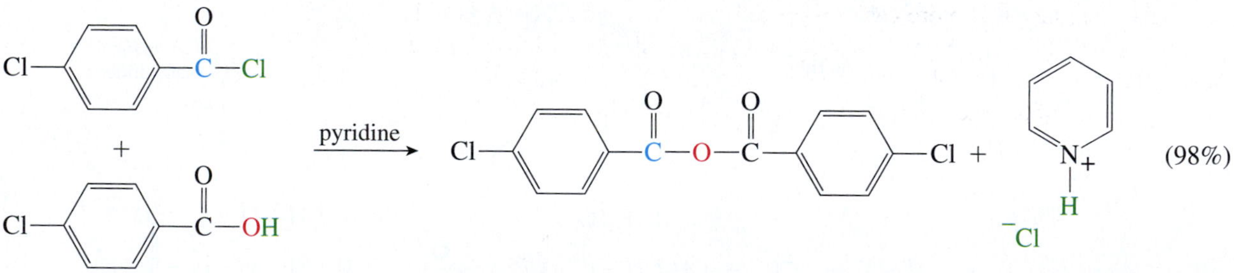

Pyridinium chloride

Pyridine is often added to these reactions, or used as a solvent, to react with the HCl that is produced and prevent the reaction mixture from becoming strongly acidic, which may cause decomposition and lower yields.

Anhydrides can also be prepared by the reaction of a carboxylic acid with another anhydride, usually acetic anhydride. A mixed anhydride is an intermediate in this reaction:

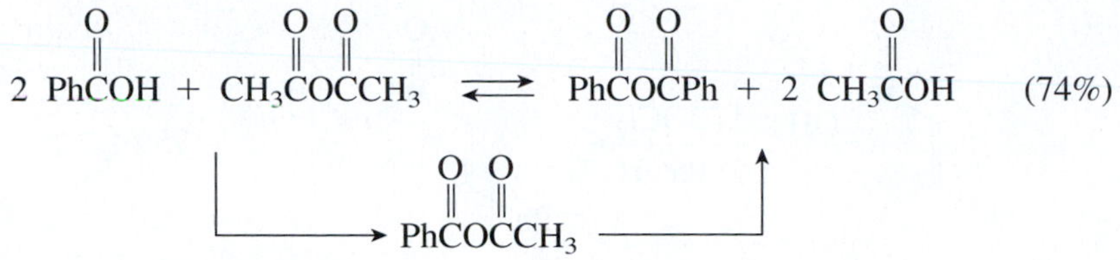

$$2 \; PhCOH \; + \; CH_3COCCH_3 \; \rightleftharpoons \; PhCOCPh \; + \; 2 \; CH_3COH \quad (74\%)$$

$$\longrightarrow \quad PhCOCCH_3$$

To make this preparation useful, the equilibrium is driven to the right by removal of the acetic acid by careful distillation. This is often the method of choice for the preparation of cyclic anhydrides, in which the equilibrium is favored by entropy:

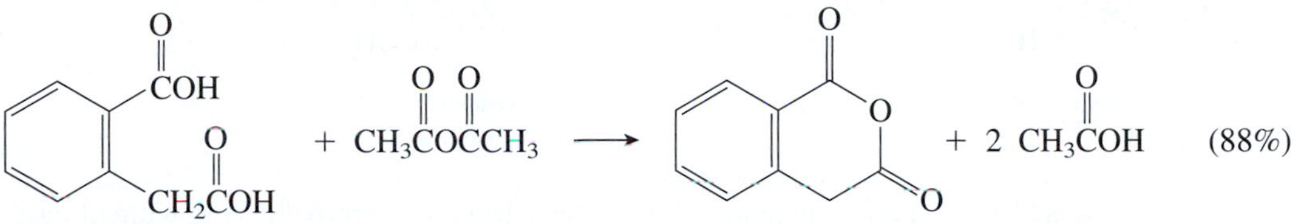

$$\text{(structure)} \; + \; CH_3COCCH_3 \; \longrightarrow \; \text{(structure)} \; + \; 2 \; CH_3COH \quad (88\%)$$

PROBLEM 19.6

Show the products of these reactions:

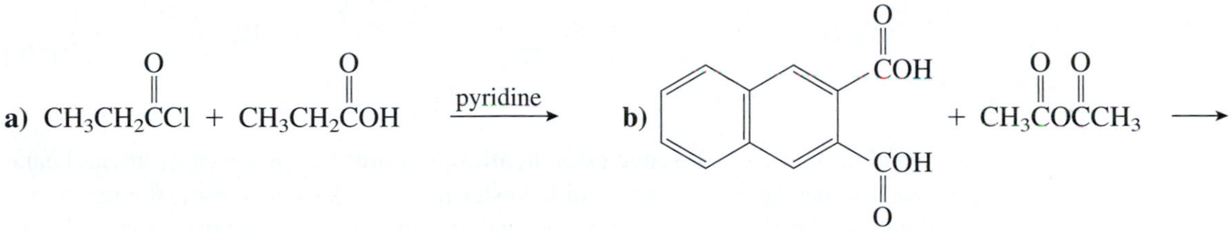

a) $CH_3CH_2CCl \; + \; CH_3CH_2COH \xrightarrow{\text{pyridine}}$

b) $\text{(structure)} \; + \; CH_3COCCH_3 \; \longrightarrow$

19.4 PREPARATION OF ESTERS

Esters are readily prepared by reaction of an alcohol with either an acyl chloride or an anhydride. Because it is more easily prepared from the acid, the acyl chloride is commonly employed. Again, a base, such as pyridine, is often added to react with the HCl that is produced. Acetic anhydride, which is commercially available, is often used for the preparation of acetate esters. Following are several examples.

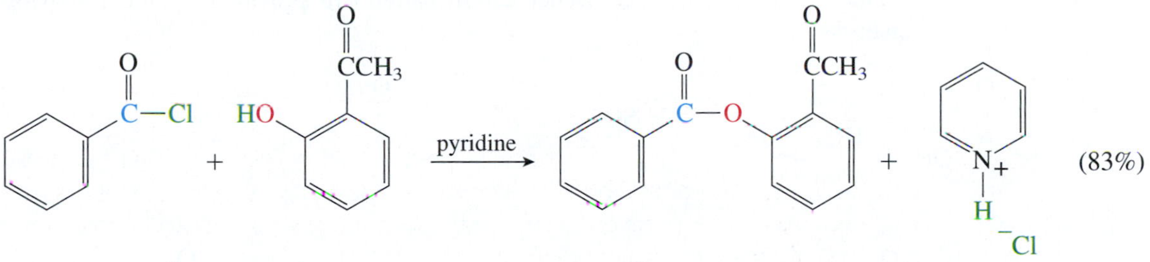

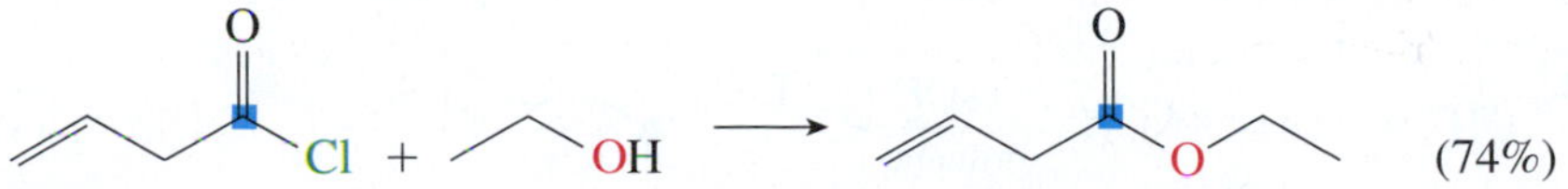

$$\text{(structure)} \; + \; \text{(structure)} \xrightarrow{\text{pyridine}} \; \text{(structure)} \; + \; \text{(pyridinium chloride)} \quad (83\%)$$

$$\text{(structure)} \; + \; \text{(structure)OH} \; \longrightarrow \; \text{(ester structure)} \quad (74\%)$$

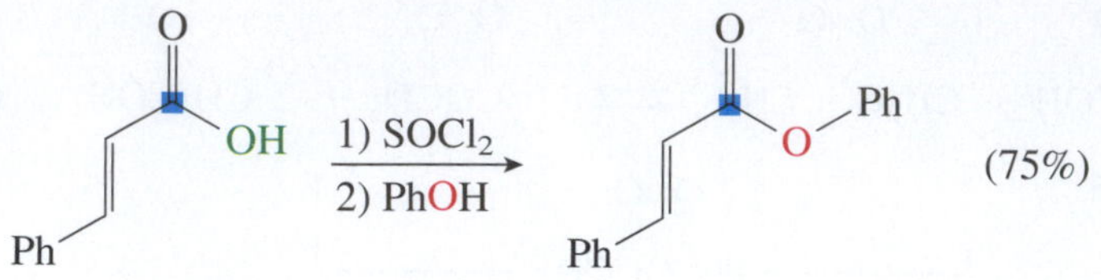

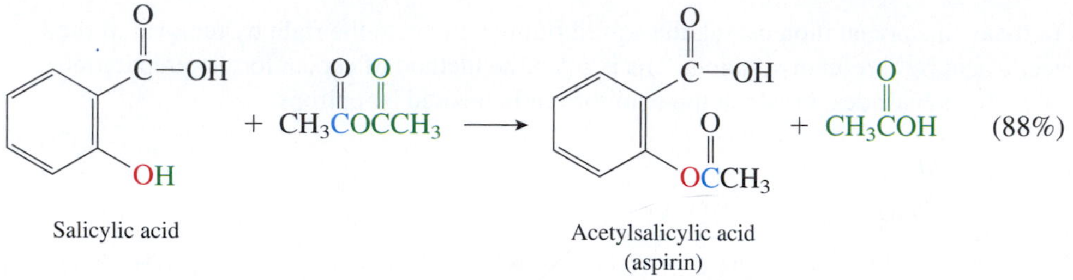

(75%)

Salicylic acid   Acetylsalicylic acid
                                         (aspirin)

(88%)

It is also common to prepare esters directly from the carboxylic acid without passing through the acyl chloride or the anhydride, as follows:

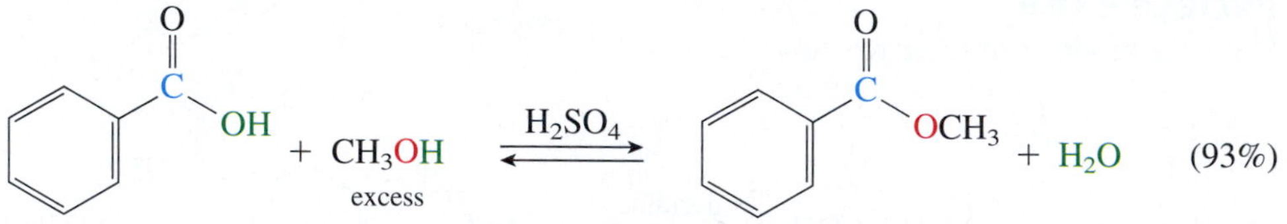

(93%)

This reaction, known as **Fischer esterification,** requires the presence of an acid catalyst. Because the carboxylic acid and the ester have similar reactivities, the reaction is useful only if a method can be found to drive the equilibrium in the direction of the desired product—the ester. In accord with Le Chatelier's principle, this is accomplished by using an excess of one of the reactants or by removing one of the products. An excess of the alcohol is used if it is readily available, as is the case for methanol or ethanol. Or water can be removed by azeotropic distillation with a solvent such as toluene.

The mechanism for the Fischer esterification is shown in Figure 19.3. Sulfuric acid, hydrochloric acid, or *p*-toluenesulfonic acid is most often used as a catalyst. The mechanism will be easier to remember if you note the similarities to other acid-catalyzed mechanisms, such as the one for the formation of acetals in Figure 18.5. Also note that the steps leading from the tetrahedral intermediate to the carboxylic acid and alcohol starting materials and to the ester and water products are very similar.

Other examples of the Fischer esterification are provided by the following equations:

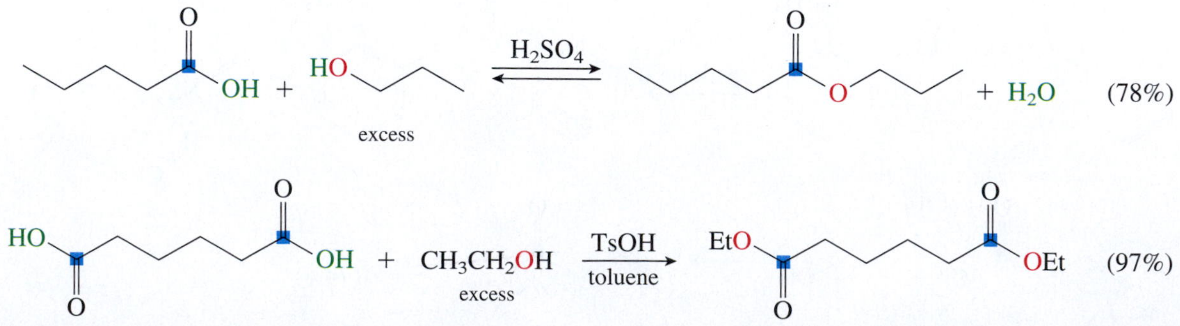

(78%)

(97%)

Lactones are cyclic esters. Their formation is very favorable when a hydroxy group and a carboxylic acid in the same molecule can react to form a five- or six-membered ring:

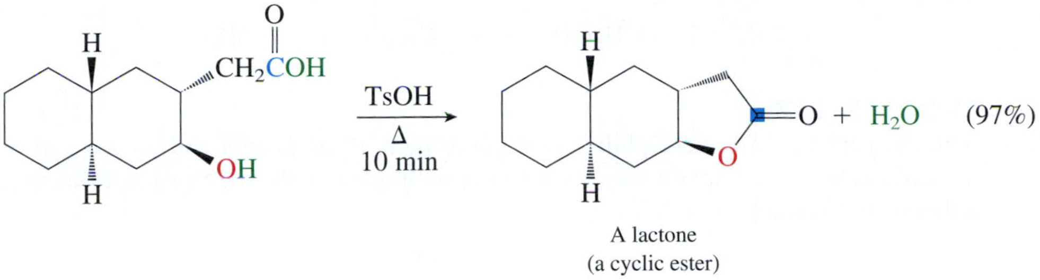

A lactone
(a cyclic ester)

ORGANIC
Chemistry—Now™
Click *Mechanisms in Motion* to view the **Mechanism of Fischer Esterification.**

❶ Acid-catalyzed reactions begin by protonation of the oxygen of the carbonyl group to make the carbon more electrophilic.

❷ Then the nucleophile, the oxygen of the alcohol, bonds to the carbonyl carbon.

❸ Next, a proton is transferred to some base in the solution.

This is the tetrahedral intermediate in the mechanism for acidic conditions. It differs from the one in Figure 19.1 only in that the oxygen is protonated. Note the similarity of the steps leading away from this intermediate in both directions.

Tetrahedral intermediate

❻ Only a proton transfer is needed to complete the reaction. This step resembles the reverse of step 1. The acid, HA, is regenerated here, so the reaction is acid catalyzed.

❺ Then water leaves in a step that resembles the reverse of step 2.

❹ Before the oxygen leaves, it is protonated to make it a better leaving group, water. This resembles the reverse of step 3.

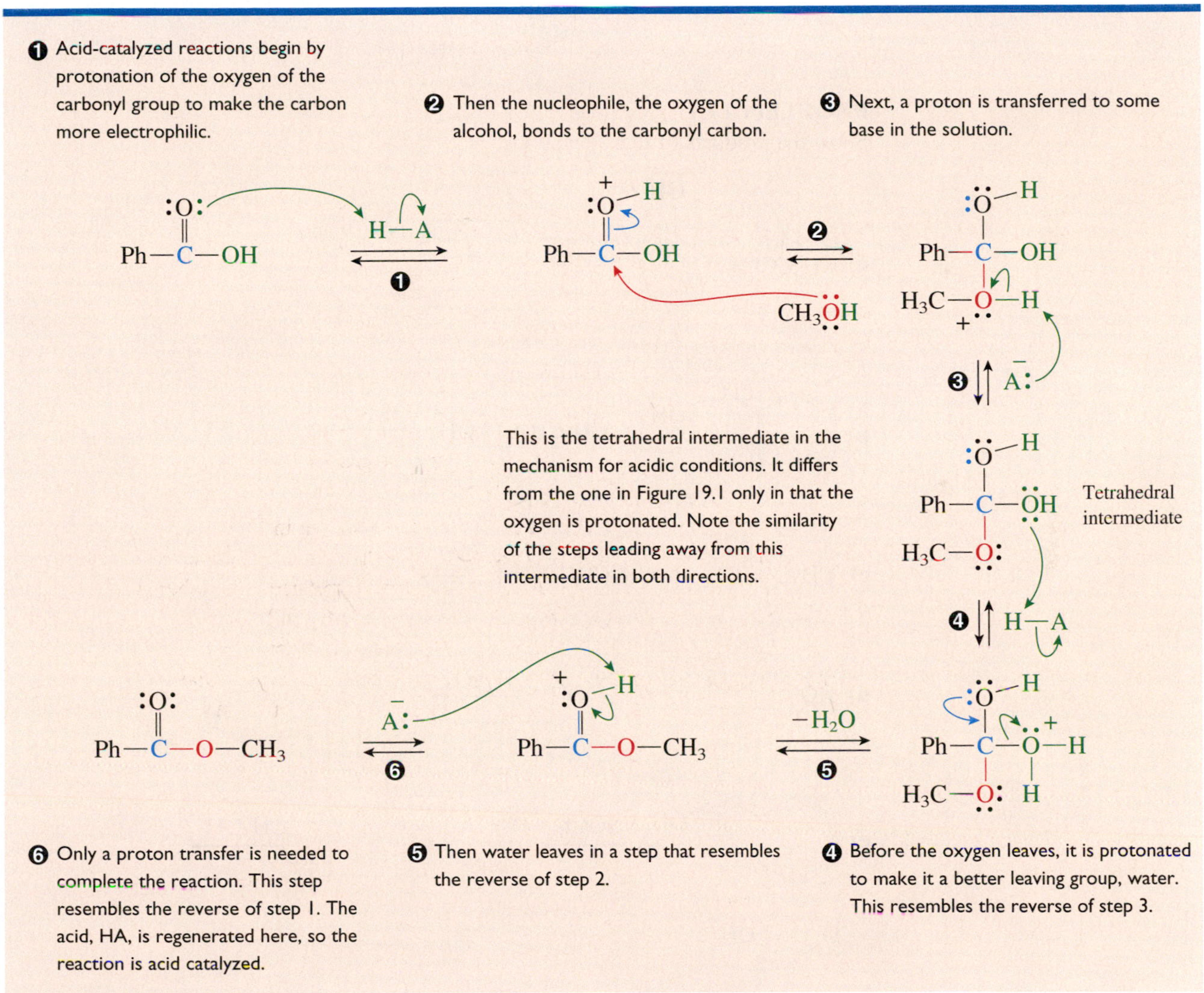

Figure 19.3

MECHANISM OF FISCHER ESTERIFICATION.

PROBLEM 19.7

Show all of the steps in the mechanism for this reaction:

$$CH_3\overset{O}{\overset{\|}{C}}Cl + CH_3OH \longrightarrow CH_3\overset{O}{\overset{\|}{C}}OCH_3 + HCl$$

PROBLEM 19.8

Salicylic acid has two nucleophilic sites: the oxygen of the phenol and the oxygen of the carboxylic acid. Explain why its reaction with acetic anhydride produces aspirin rather than this compound:

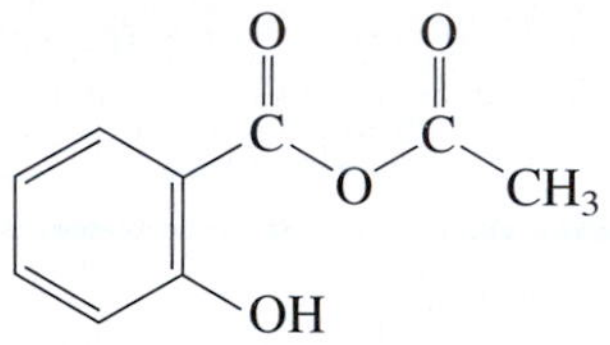

PROBLEM 19.9

Show the products of these reactions:

a)

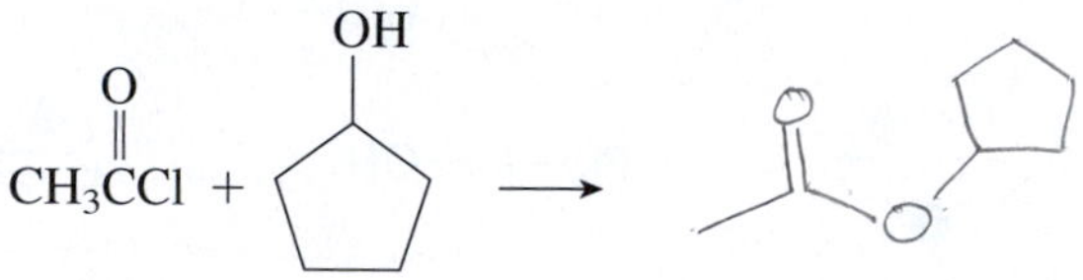

b)

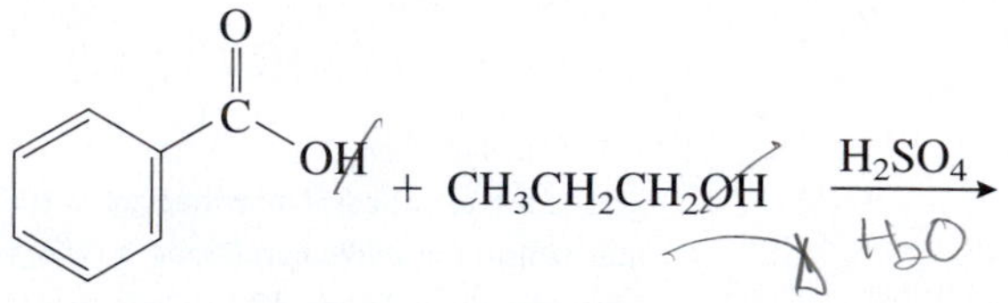

c)

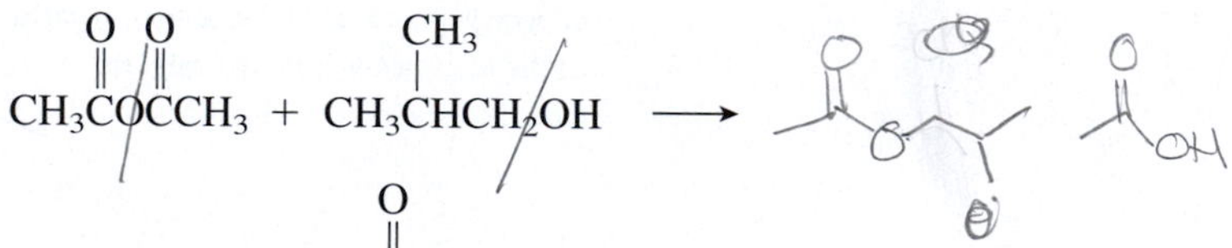

d)

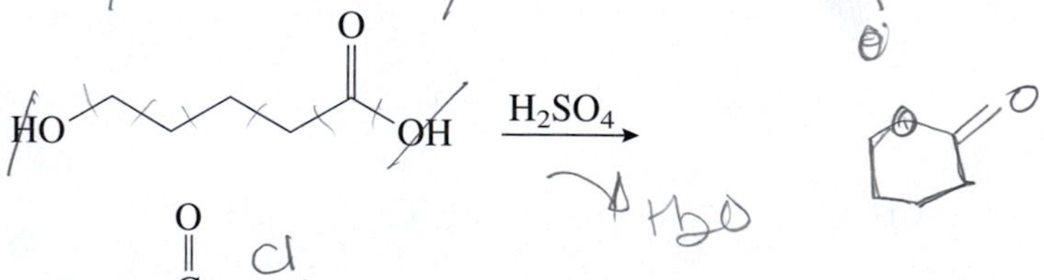

e)

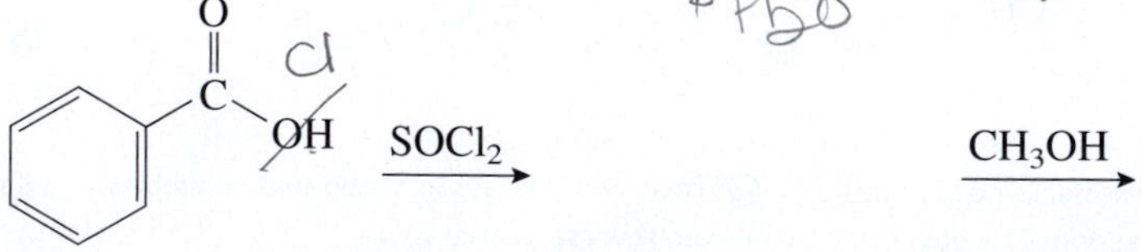

f)

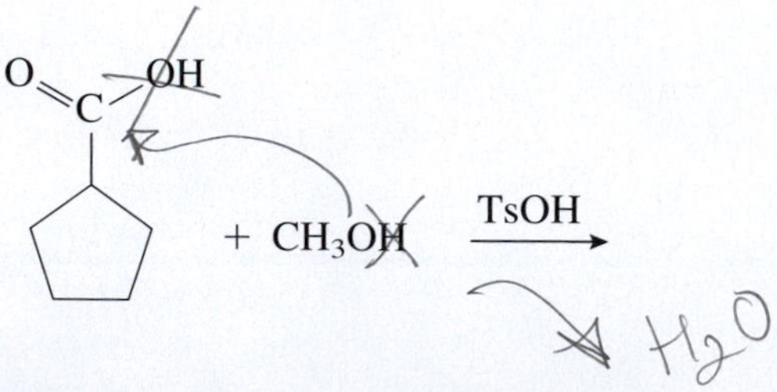

19.5 PREPARATION OF CARBOXYLIC ACIDS

Carboxylic acids commonly are used as the starting materials for the preparation of the other acid derivatives. However, any of the acid derivatives can be hydrolyzed to the carboxylic acid by reaction with water under the appropriate conditions. Acid or base catalysis is necessary for the less reactive derivatives.

Acyl Chloride and Anhydride Hydrolysis

Acyl chlorides and anhydrides must be protected from water because they react readily, often vigorously, with water to produce carboxylic acids. This reaction is not of much synthetic usefulness because the acyl chloride or anhydride is usually prepared from the acid. However, the hydrolysis reaction is occasionally used for the preparation of a carboxylic acid if the acyl chloride or anhydride is available from some other source. The following equation provides an example:

(94%)

Ester Hydrolysis

Esters can be hydrolyzed to carboxylic acids under either acidic or basic conditions. Under acidic conditions the mechanism is the exact reverse of the Fischer esterification mechanism shown in Figure 19.3. Again, because the acid and the ester have comparable reactivities, some method must be used to drive the equilibrium toward the desired product—the acid in this case. This can be accomplished by using water as the solvent, providing a large excess of this reagent that, by Le Chatelier's principle, shifts the equilibrium toward the carboxylic acid.

PROBLEM 19.10

Show all of the steps in the mechanism for this reaction:

$$CH_3\overset{O}{\overset{\|}{C}}OCH_2CH_3 + H_2O \underset{}{\overset{H_2SO_4}{\rightleftharpoons}} CH_3\overset{O}{\overset{\|}{C}}OH + CH_3CH_2OH$$

It is more common to hydrolyze esters under basic conditions because the equilibrium is favorable. The mechanism for this process, called **saponification,** is presented in Figure 19.4. The production of the conjugate base of the carboxylic acid, the carboxylate anion, which is at the bottom of the reactivity scale, drives the equilibrium in the desired direction. To isolate the carboxylic acid, the solution must be acidified after the hydrolysis is complete. Some examples are provided in the following equations. We saw another example of this hydrolysis reaction in Chapter 10, where it was

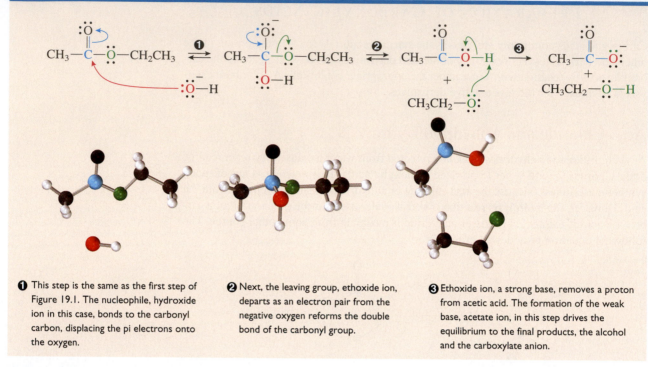

❶ This step is the same as the first step of Figure 19.1. The nucleophile, hydroxide ion in this case, bonds to the carbonyl carbon, displacing the pi electrons onto the oxygen.

❷ Next, the leaving group, ethoxide ion, departs as an electron pair from the negative oxygen reforms the double bond of the carbonyl group.

❸ Ethoxide ion, a strong base, removes a proton from acetic acid. The formation of the weak base, acetate ion, in this step drives the equilibrium to the final products, the alcohol and the carboxylate anion.

Figure 19.4

MECHANISM OF THE BASE-CATALYZED HYDROLYSIS (SAPONIFICATION) OF AN ESTER.

used in the preparation of alcohols by the acetate method (see Section 10.2 and Figure 10.1 on page 351).

Click *Mechanisms in Motion* to view the **Mechanism of Hydrolysis of an Ester by Base.**

ORGANIC
Chemistry⋅Now™

As we have seen before, a variety of experiments are used to support or disprove a mechanism that has been postulated for a particular reaction. The mechanism for the basic hydrolysis of esters, shown in Figure 19.4, involves cleavage of the bond between the ether oxygen of the ester and the carbonyl carbon rather than the cleavage of the bond between this oxygen and the carbon of the alcohol part of the ester.

$$CH_3C\overset{O}{-}O-CH_2CH_2CH_2CH_2CH_3$$

The acyl C—O
bond is cleaved.

The alkyl C—O bond
remains intact.

Proof that it is indeed this bond that is cleaved was provided by conducting the hydrolysis in water with an enriched content of ^{18}O. Pentyl acetate was hydrolyzed in this isotopically enriched water. The 1-pentanol product was shown by mass spectrometry to contain only the normal amount of ^{18}O, demonstrating that the alkyl C—O bond was not broken during the reaction.

$$CH_3C\overset{O}{-}O-CH_2CH_2CH_2CH_2CH_3 \xrightarrow[\overline{HO}^{18}]{H_2O^{18}} CH_3C\overset{O}{-}\overline{O}^{18} + HO-CH_2CH_2CH_2CH_2CH_3$$

Pentyl acetate

1-Pentanol

No O^{18} found here

Amide Hydrolysis

Amides are less reactive than esters, and their hydrolysis often requires vigorous heating in either aqueous acid or base. The mechanism for acidic conditions is quite similar to the reverse of the Fischer esterification mechanism shown in Figure 19.3. The mechanism for basic conditions is related to that depicted in Figure 19.4 for ester saponification and a similar tetrahedral intermediate is formed in the first step. However, NH_2^- is a strong base and a poor leaving group, so the nitrogen must usually be protonated before it can leave. The exact timing of the various proton transfers that occur in this mechanism (and many others) is difficult to establish and depends on both the structure of the amide and the reaction conditions. One possible mechanism is shown in Figure 19.5.

Under acidic conditions the equilibrium for the hydrolysis of an amide is driven toward the products by the protonation of the ammonia or amine that is formed. Under basic conditions the equilibrium is driven toward the products by the formation of the carboxylate anion, which is at the bottom of the reactivity scale. The pH of the final solution may need to be adjusted, depending on which product is to be isolated. If the carboxylic acid is desired, the final solution must be acidic, whereas isolation of the amine requires that the solution be basic. Several examples are shown in the following equations. Also, note that the last step of the Gabriel amine synthesis, the hydrolysis of the phthalimide (see Section 10.6 and Figure 10.5 on page 365), is an amide hydrolysis.

(98%)

$$+ NH_3 \qquad\qquad + NH_4^+ \; Cl^-$$

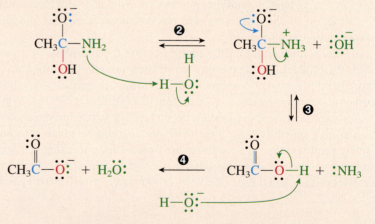

Figure 19.5

POSSIBLE MECHANISM OF THE HYDROLYSIS OF AN AMIDE UNDER BASIC CONDITIONS.

Focus On

The Preparation of Soap

The preparation of a solution of soap by the reaction of fat with water in the presence of base was probably one of the earliest chemical processes discovered by humans. Although the details of this discovery are lost in antiquity, we can imagine early humans finding that water that had been in contact with wood ashes from the campfire could be used to remove grease from hands and other objects and that this water became a more effective cleaning agent as it was used. The water leaches some alkaline compounds from the ashes, and this basic water hydrolyzes the esters of the fat or grease to alcohols and soap. This is why the hydrolysis of esters under basic conditions is called saponification (the Latin word for soap is *sapo*).

Fats are triesters formed from a triol (glycerol) and fatty acids, which are carboxylic acids with long, unbranched alkyl chains. These alkyl chains may be saturated, or they may have one or more double bonds. Usually, the three carboxylic acids of a fat molecule are not the same. The following equation illustrates the hydrolysis of a representative fat molecule to glycerol and the conjugate bases of stearic acid, palmitoleic acid, and linoleic acid. Addition of NaCl causes the fatty acid salts to precipitate. The resulting solid is formed into bars of soap.

The cleaning action of soap is due to the dual nature of the conjugate base of the fatty acid molecule. On one end is the ionic carboxylate anion group; the rest of the molecule consists of a nonpolar hydrocarbon chain. The ionic part, called the head, is attracted to a polar solvent such as water and is hydrophilic, whereas the long hydrocarbon tail is hydrophobic. In water, soap molecules tend to group together in clusters called micelles, with their ionic heads oriented toward the water molecules and their hydrocarbon tails in the interior of the cluster so that unfavorable interactions with water are avoided. Non-

Continued

polar grease and fat molecules dissolve in the interior of the micelle. The ionic heads keep the micelle in solution and allow water to wash it away, along with the grease.

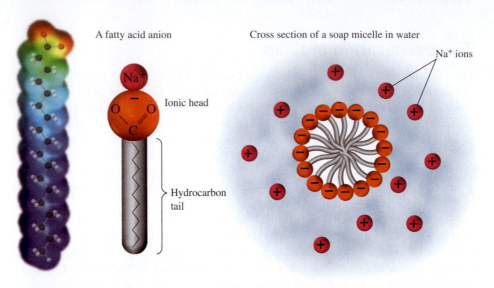

A fatty acid anion

Ionic head

Hydrocarbon tail

Cross section of a soap micelle in water

Na⁺ ions

PROBLEM 19.11

Show the products that are formed when this fat is saponified:

$$
\begin{array}{l}
\overset{\displaystyle O}{\overset{\displaystyle \|}{}} \\
CH_2-OC(CH_2)_7CH=CH(CH_2)_7CH_3 \\
\overset{\displaystyle O}{\overset{\displaystyle \|}{}} \\
CH-OC(CH_2)_{14}CH_3 \\
\overset{\displaystyle O}{\overset{\displaystyle \|}{}} \\
CH_2-OC(CH_2)_{12}CH_3
\end{array}
$$

As mentioned previously, acyl chlorides and anhydrides are hydrolyzed to carboxylic acids by water without the need for acid or base catalysis. If an acyl chloride or anhydride is stored for an extended period, it is often found to be contaminated with the corresponding carboxylic acid because of contact with water from the air or some other source.

Esters and amides, on the other hand, require the presence of an acid or base catalysis to react with water. These reactions are not instantaneous but require rather strongly acidic or basic conditions and heat to proceed at a reasonable rate. For example, a typical ester saponification is usually conducted with 10% NaOH in water, and the solution is refluxed until the ester layer disappears. (Most esters are not soluble in water.) This may require from 15 minutes up to several hours of reflux. Similarly, a typical amide hydrolysis is often conducted by refluxing the amide in concentrated hydrochloric acid for a period ranging from 15 minutes up to several hours. Esters and amides are relatively stable to the near-neutral conditions found in living organisms, which is one reason why they are important functional groups in biochemistry.

PRACTICE PROBLEM 19.2

Show all of the steps in the mechanism for this reaction:

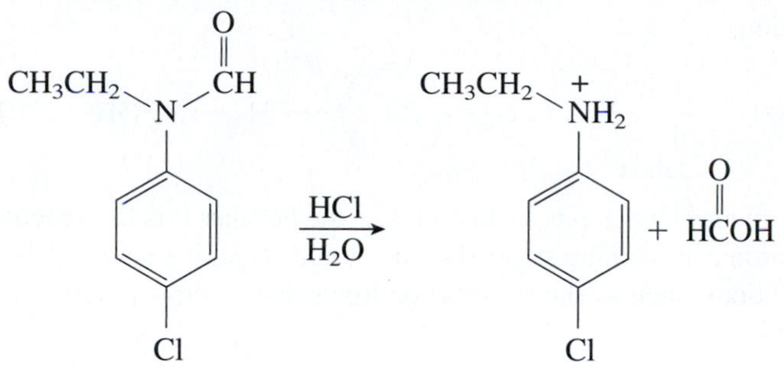

Solution

This acid-catalyzed mechanism resembles the reverse of the mechanism for Fischer esterification:

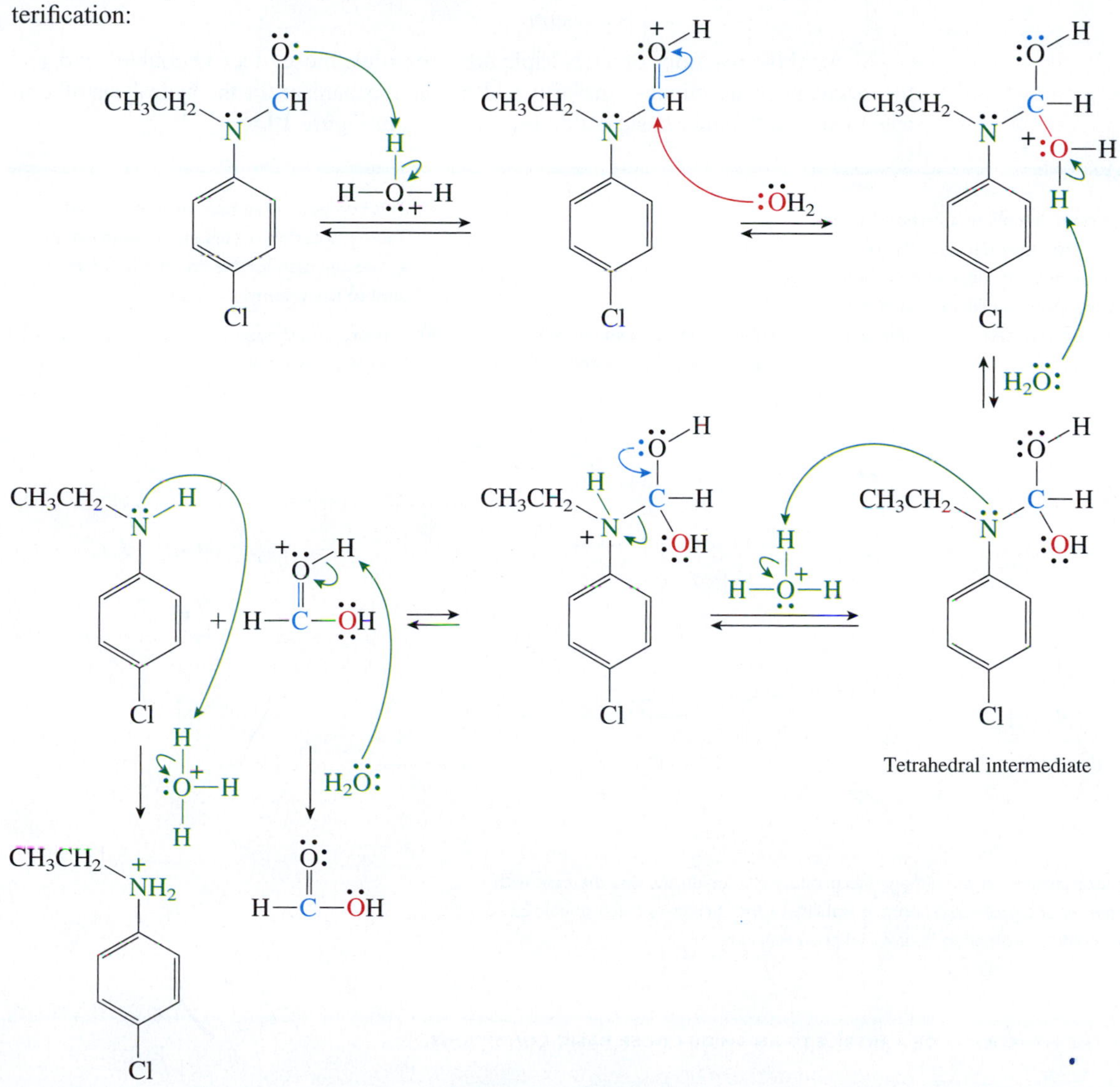

Tetrahedral intermediate

Nitrile Hydrolysis

Nitriles are often considered as derivatives of carboxylic acids because they can be hydrolyzed to the carboxylic acid under acidic or basic conditions, as illustrated in the following equation:

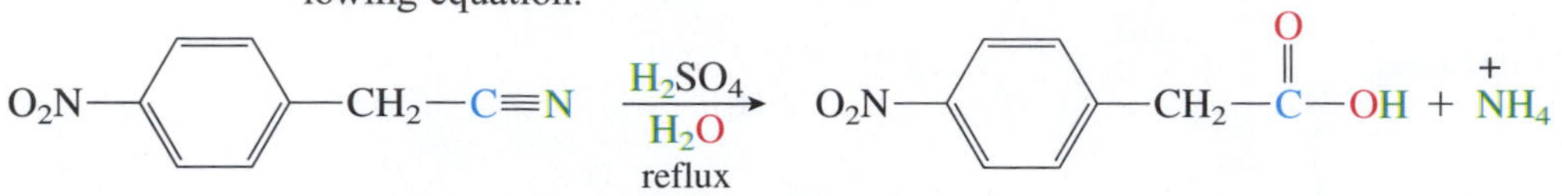

$$O_2N-\!\!\!\!\bigcirc\!\!\!\!-CH_2-C\!\equiv\!N \xrightarrow[\substack{H_2O \\ reflux}]{H_2SO_4} O_2N-\!\!\!\!\bigcirc\!\!\!\!-CH_2-\overset{\displaystyle O}{\overset{\|}{C}}-OH + \overset{+}{N}H_4 \quad (95\%)$$

The amide is an intermediate in the hydrolysis, and because it is less reactive than the nitrile, the reaction can often be stopped at the amide stage, if so desired, by using milder reaction conditions, such as shorter reaction times, lower temperatures, or weaker base:

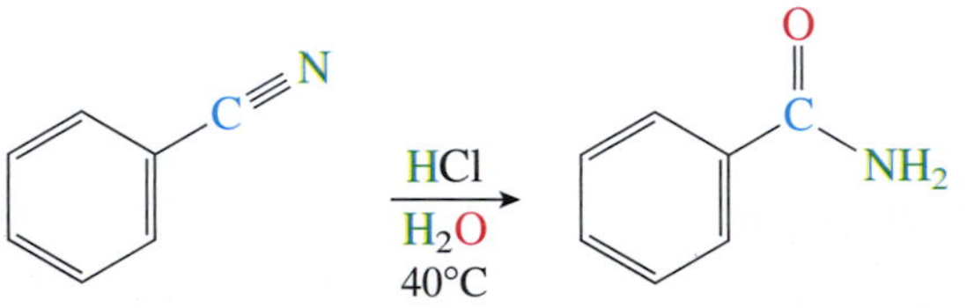

$$\bigcirc\!\!\!\!-C\!\equiv\!N \xrightarrow[\substack{H_2O \\ 40°C}]{HCl} \bigcirc\!\!\!\!-\overset{\displaystyle O}{\overset{\|}{C}}-NH_2 \quad (86\%)$$

Many of the reactions of a CN triple bond resemble those of a CO double bond, and the mechanisms have many similarities also. The mechanism for the hydrolysis of a nitrile to an amide under basic conditions is shown in Figure 19.6.

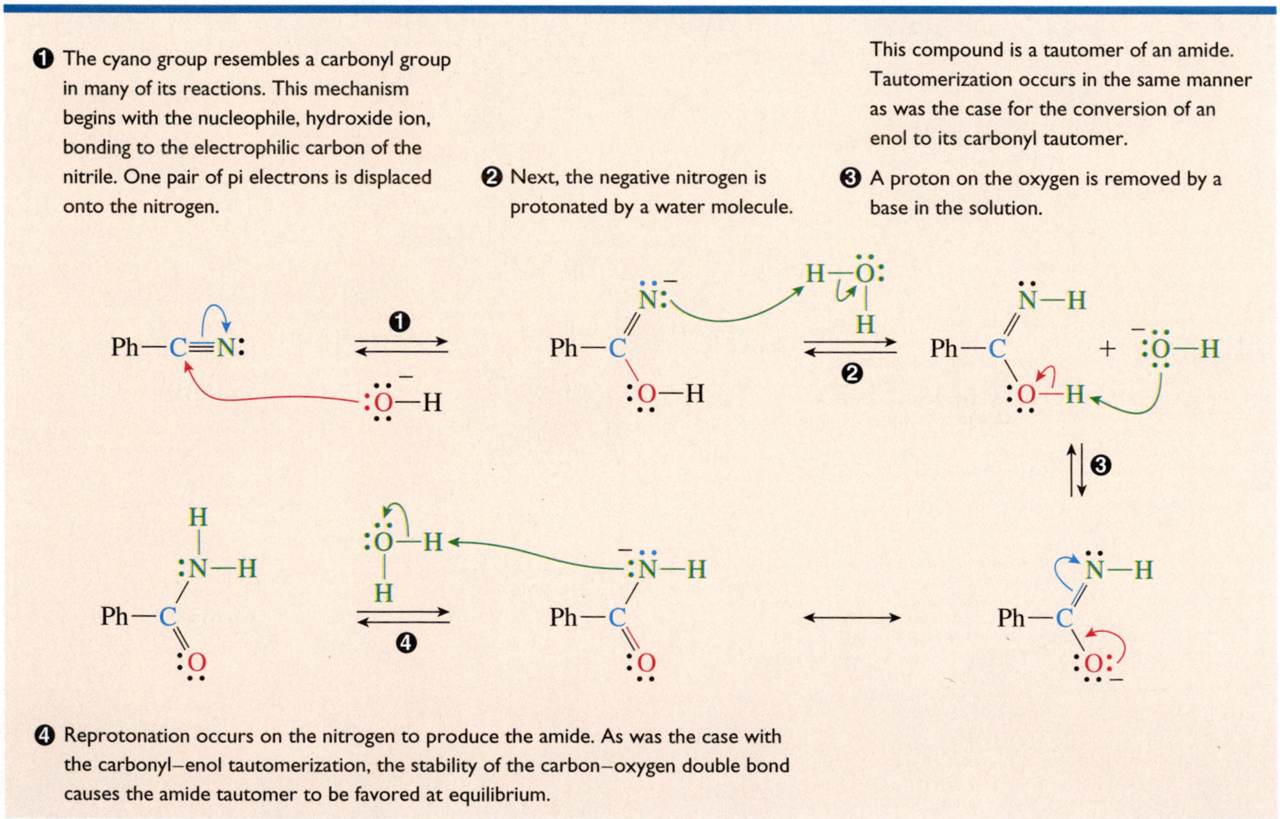

❶ The cyano group resembles a carbonyl group in many of its reactions. This mechanism begins with the nucleophile, hydroxide ion, bonding to the electrophilic carbon of the nitrile. One pair of pi electrons is displaced onto the nitrogen.

❷ Next, the negative nitrogen is protonated by a water molecule.

This compound is a tautomer of an amide. Tautomerization occurs in the same manner as was the case for the conversion of an enol to its carbonyl tautomer.

❸ A proton on the oxygen is removed by a base in the solution.

❹ Reprotonation occurs on the nitrogen to produce the amide. As was the case with the carbonyl–enol tautomerization, the stability of the carbon–oxygen double bond causes the amide tautomer to be favored at equilibrium.

Figure 19.6

MECHANISM OF THE HYDROLYSIS OF A NITRILE TO AN AMIDE UNDER BASIC CONDITIONS.

The use of cyanide ion as a nucleophile in an S_N2 reaction (see Section 10.8), followed by hydrolysis of the product nitrile, provides a useful preparation of carboxylic acids that contain one more carbon than the starting compound:

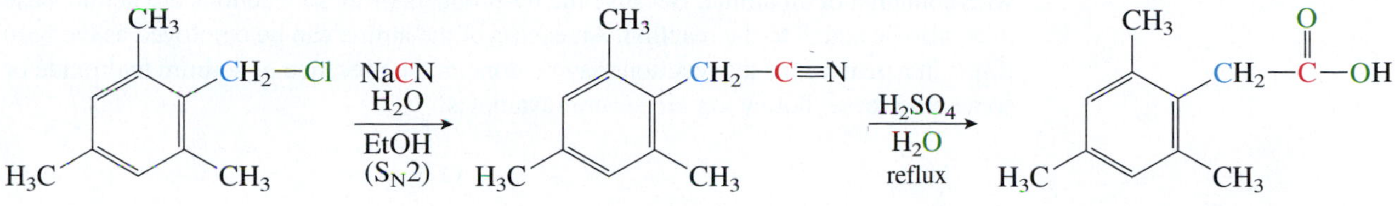

(93%)   (87%)

PROBLEM 19.12

Show the products of these reactions:

a) $CH_3COCCH_3 + H_2O$ →

b) $PhC-OCH_2CH_3 + H_2O$ $\xrightarrow{NaOH}$

c) $CH_3CNHPh + H_2O$ $\xrightarrow{HCl}$

d) 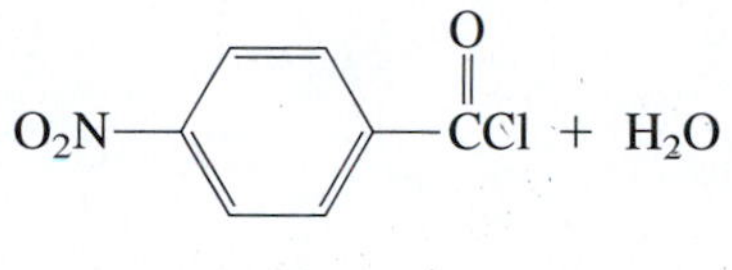 $O_2N-\!\!\!\!\bigcirc\!\!\!\!-CCl + H_2O$ →

e) $CH_3CH_2COCH_2CH_3) + H_2O$ $\xrightarrow{H_2SO_4}$

f) 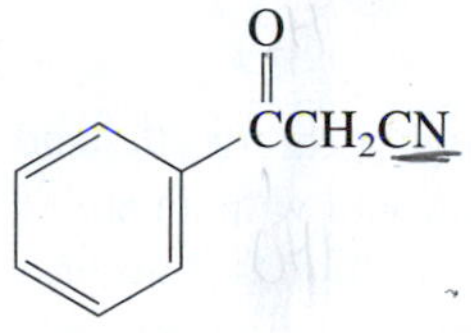 (benzene ring)$-CCH_2CN$ $+ H_2O$ $\xrightarrow[\substack{low\ temp.\\short\ time}]{NaOH}$

g) $CH_3CH_2CHCH_3$ $\xrightarrow[\substack{2)\ NaOH,\ H_2O}]{1)\ CH_3CO^-,\ DMF}$

h) 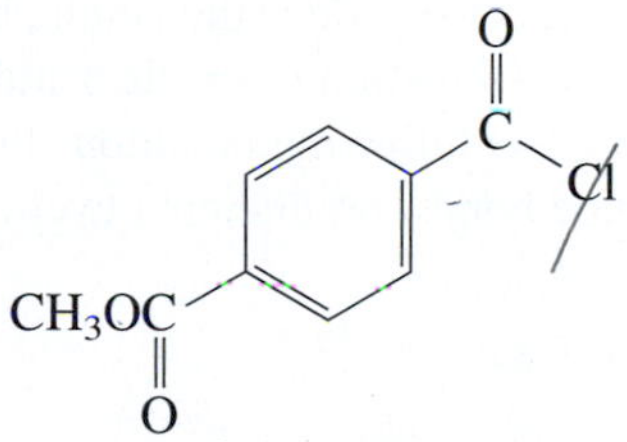 (see structure) $+ H_2O$ →

i) $CH_3CH_2CH_2CN + H_2O$ $\xrightarrow[\substack{high\ temp.\\long\ time}]{HCl}$

PROBLEM 19.13

Suggest a synthesis of 2-phenylacetic acid from benzyl bromide:

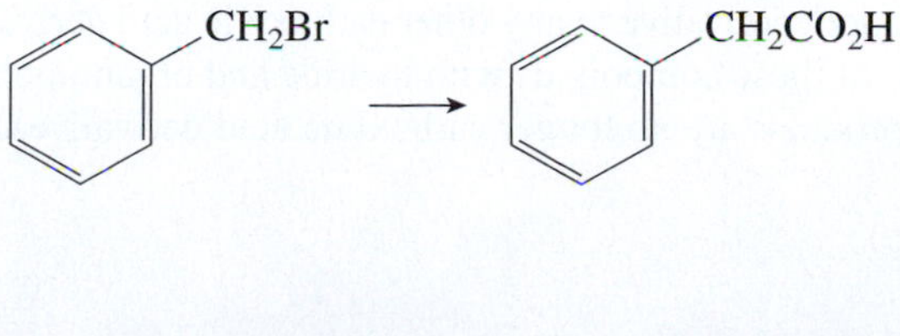

CH_2Br → CH_2CO_2H

19.6 Preparation of Amides

Amides are most commonly prepared by reaction of an acyl chloride or an anhydride with ammonia or an amine. Because the by-products of these reactions are acidic, base must also be added to the reaction. An excess of the amine can be employed as the base if it is inexpensive, or the reaction may be done in the presence of sodium hydroxide or some other base. Following are several examples.

$$\text{PhCH}_2\text{NH}_2 + \text{Cl}-\overset{\overset{\text{O}}{\|}}{\text{C}}-\text{OCH}_2\text{CH}_3 \xrightarrow[\text{H}_2\text{O}]{\text{NaOH}} \text{PhCH}_2\text{NH}-\overset{\overset{\text{O}}{\|}}{\text{C}}-\text{OCH}_2\text{CH}_3 \quad (96\%)$$

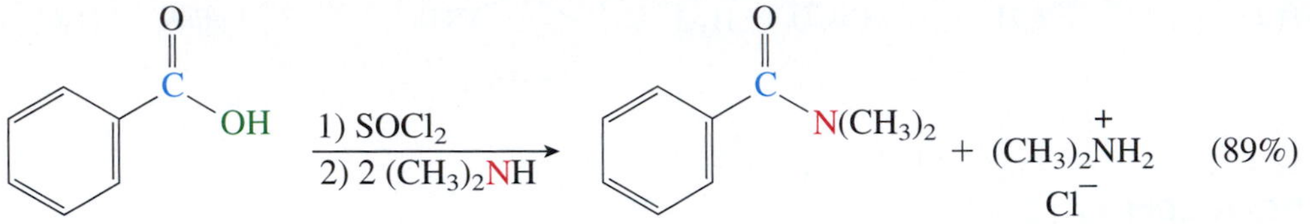

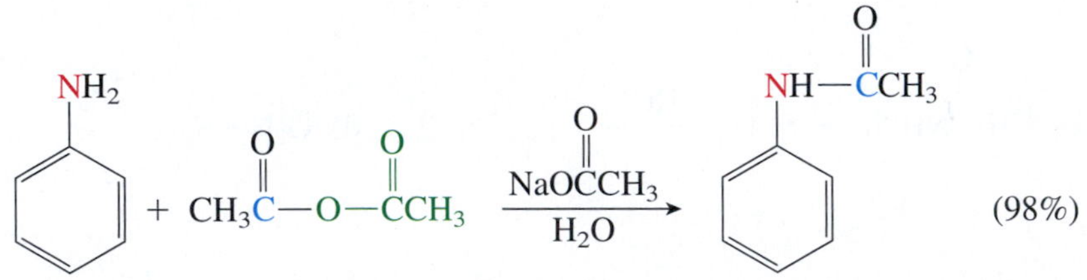

Although a carboxylic acid is higher on the reactivity scale than an amide, the reaction of an acid with an amine does not directly produce the amide because of the fast acid–base reaction that occurs with these compounds to form the carboxylate anion and the ammonium cation derived from the amine. Vigorous heating of this salt does drive off water to produce the amide, but this procedure is seldom of preparative value except for the formation of **cyclic amides** (called **lactams**) containing five or six atoms in the ring. The following equation shows such a reaction of a dicarboxylic acid to form two amide bonds, producing a cyclic **imide:**

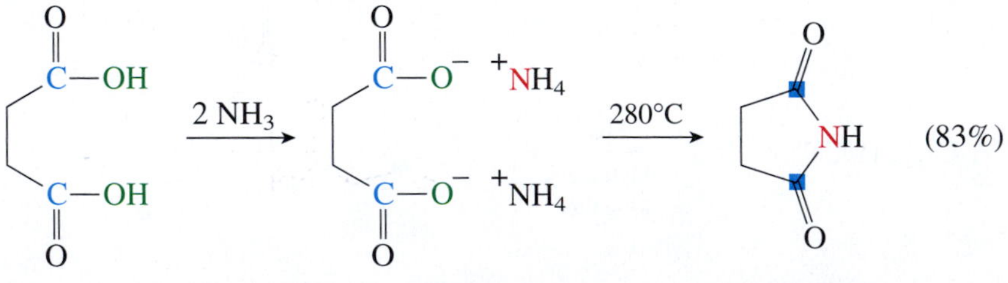

An imide

By using the reactions described in Sections 19.2 through 19.6, it is possible to convert one carboxylic acid derivative to any other carboxylic acid derivative. Now let's examine the reactions of these compounds with hydride and organometallic nucleophiles. In these cases the products are no longer carboxylic acid derivatives.

PROBLEM 19.14

Show the products of these reactions:

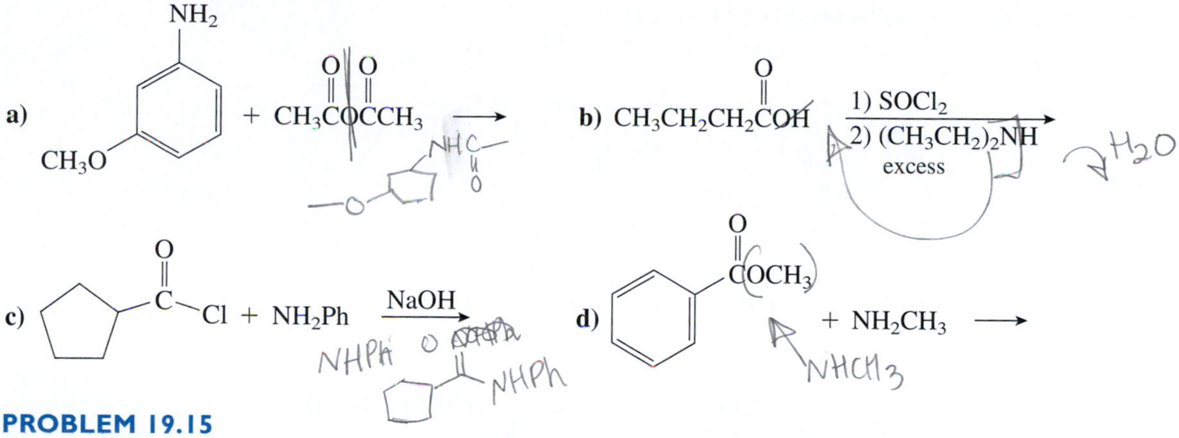

a) [structure: 3-methoxyaniline with NH₂ group] + CH₃COCCH₃ (anhydride) ⟶

[handwritten: NHC with ring and —O— structure]

b) CH₃CH₂CH₂COH → 1) SOCl₂ 2) (CH₃CH₂)₂NH excess [handwritten: H₂O]

c) [cyclopentyl structure]—C(=O)—Cl + NH₂Ph → NaOH

[handwritten: NHPh, cyclopentyl-C(=O)-NHPh]

d) [benzene ring with COCH₃ group] + NH₂CH₃ ⟶

[handwritten: NHCH₃]

PROBLEM 19.15

Show all of the steps in the mechanism for this reaction and explain why the indicated product is formed and the other is not:

PhCH₂NH₂ + ClCOCH₂CH₃ ⟶ PhCH₂NHCOCH₂CH₃   PhCH₂NHCCl

                                    Formed   Not formed

PRACTICE PROBLEM 19.3

Suggest reactions that could be used to convert this amide to the methyl ester:

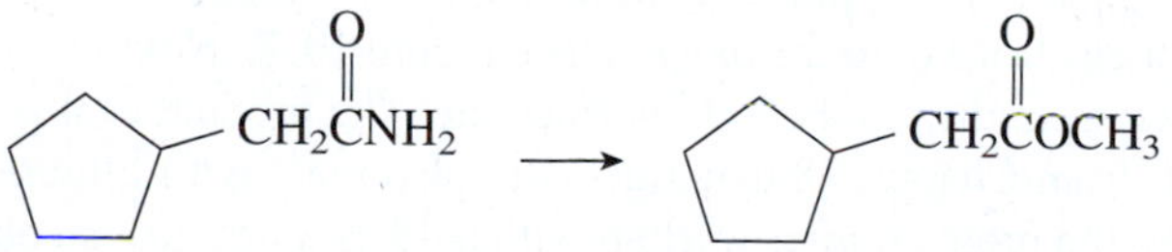

[cyclopentyl]—CH₂CNH₂ (amide) ⟶ [cyclopentyl]—CH₂COCH₃ (ester)

Strategy

Any carboxylic acid derivative can be converted to any other using one or more of the reactions discussed in the previous sections. If the conversion requires going from a less reactive derivative to a more reactive one, then an indirect route may be necessary. Remember that any derivative can be hydrolyzed to a carboxylic acid using water and acid or base. Also remember that the carboxylic acid can be converted to the acyl chloride with thionyl chloride, providing access to the other derivatives.

Solution

There is no reaction that directly converts an amide to an ester because the amide is less reactive. Therefore, the amide is first hydrolyzed to the acid and then the acid is esterified:

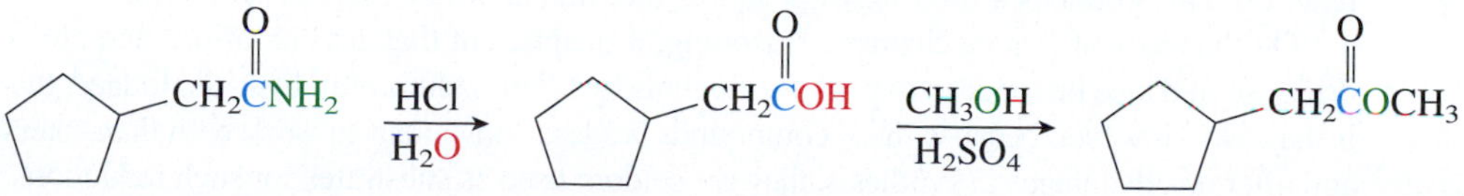

[cyclopentyl]—CH₂CNH₂ →(HCl, H₂O)→ [cyclopentyl]—CH₂COH →(CH₃OH, H₂SO₄)→ [cyclopentyl]—CH₂COCH₃

PROBLEM 19.16

Suggest reactions that would accomplish these transformations. More than one step may be necessary in some cases.

a)

b)

c)

19.7 REACTION WITH HYDRIDE NUCLEOPHILES

On the basis of what we have already learned about the reactions of lithium aluminum hydride with aldehydes and ketones (Chapter 18) and the mechanisms presented so far in this chapter, we can readily predict the product that results when hydride reacts with a carboxylic acid derivative. Consider, for example, the reaction of ethyl benzoate with lithium aluminum hydride. As with all of the reactions in this chapter, this reaction begins with attack of the nucleophile, hydride ion, at the carbon of the carbonyl group, displacing the pi electrons onto the oxygen (see Figure 19.7). Next, these electrons help displace ethoxide from the tetrahedral intermediate. The product of this step is an aldehyde. But recall from Chapter 18 that aldehydes also react with lithium aluminum hydride. Therefore, the product, after workup with acid, is a primary alcohol.

From the mechanism it can be seen that it takes two hydride nucleophiles to accomplish this reduction. Therefore, 1 mole of lithium aluminum hydride will reduce 2 moles of the ester.

$$2 \ PhC\!\!\overset{\displaystyle O}{\overset{\displaystyle \|}{-}}\!\!OCH_2CH_3 \quad \xrightarrow[]{LiAlH_4 \quad H_3O^+} \quad 2 \ PhCH_2OH \ + \ 2 \ CH_3CH_2OH \quad (90\%)$$

What happens if less hydride reagent is used? Even though the aldehyde is produced first in the mechanism, the reaction cannot be stopped at the aldehyde stage because the aldehyde is higher on the reactivity scale than the ester. The aldehyde reacts with the hydride nearly as fast as it is formed. If less than 1 mole of hydride per 2 moles of ester is used, a mixture of the ester and the alcohol is produced, along with some of the aldehyde. A reaction that produces a mixture of products like this is not synthetically useful.

On the basis of this mechanistic reasoning, it is apparent that acyl chlorides and anhydrides should also be reduced to primary alcohols by lithium aluminum hydride. Indeed, this is the case. However, because these compounds are less convenient to work with than esters and offer no advantages in synthesis, they are seldom used as substrates for such reductions.

Carboxylic acids are also reduced to primary alcohols by lithium aluminum hydride. The first step in the mechanism for this reaction is an acid–base reaction between the

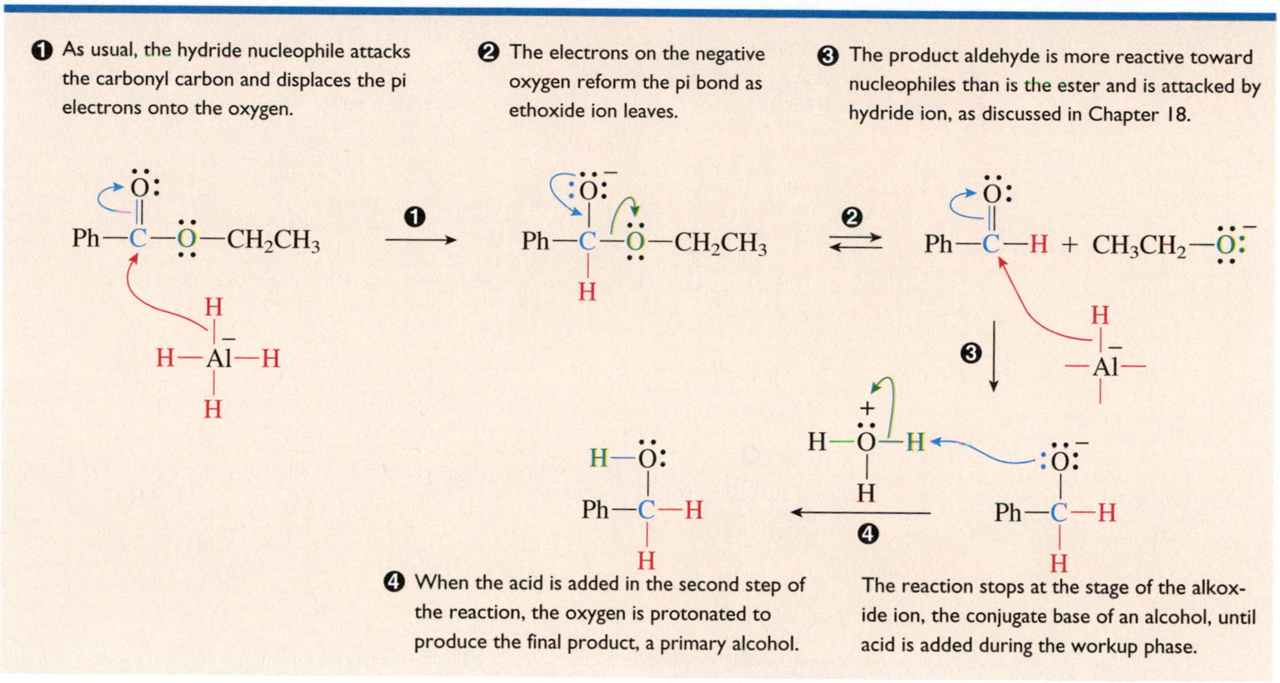
❶ As usual, the hydride nucleophile attacks the carbonyl carbon and displaces the pi electrons onto the oxygen.
❷ The electrons on the negative oxygen reform the pi bond as ethoxide ion leaves.
❸ The product aldehyde is more reactive toward nucleophiles than is the ester and is attacked by hydride ion, as discussed in Chapter 18.
❹ When the acid is added in the second step of the reaction, the oxygen is protonated to produce the final product, a primary alcohol.
The reaction stops at the stage of the alkoxide ion, the conjugate base of an alcohol, until acid is added during the workup phase.

Figure 19.7

MECHANISM OF THE REDUCTION OF AN ESTER WITH LITHIUM ALUMINUM HYDRIDE.

hydride and the acid, as shown in Figure 19.8. Lithium aluminum hydride is reactive enough to attack the relatively unreactive carbonyl group of the carboxylate anion. For the reaction to proceed further, one of the negatively charged oxygens must act as a leaving group. The oxygen is assisted in this process by Lewis acid–base coordination with aluminum. This makes the oxygen less basic and better able to leave. (Similar Lewis acid–base coordination occurs in many other reactions. This is usually ignored in this text unless it makes a particularly important contribution to the reaction.)

In summary, reductions of carboxylic acid derivatives to primary alcohols are usually accomplished by reaction of esters or acids with lithium aluminum hydride. The following equations provide several examples:

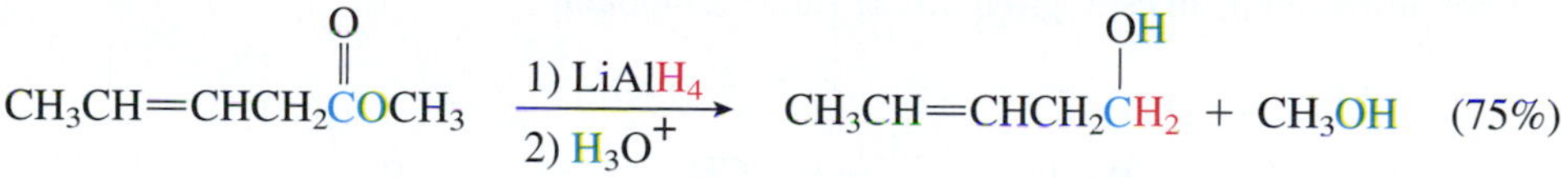
1) LiAlH4
2) H3O+
(75%)

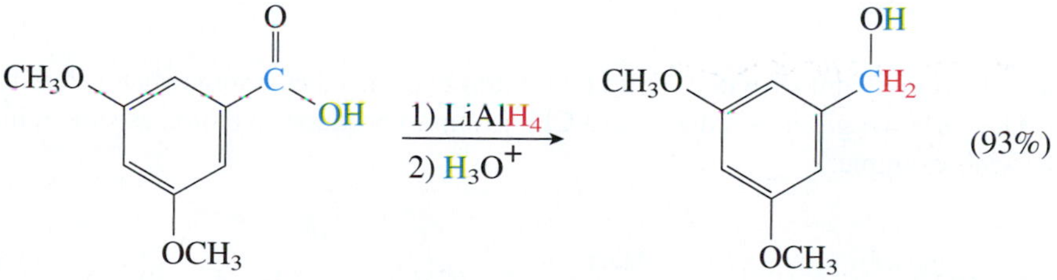
1) LiAlH4
2) H3O+
(93%)

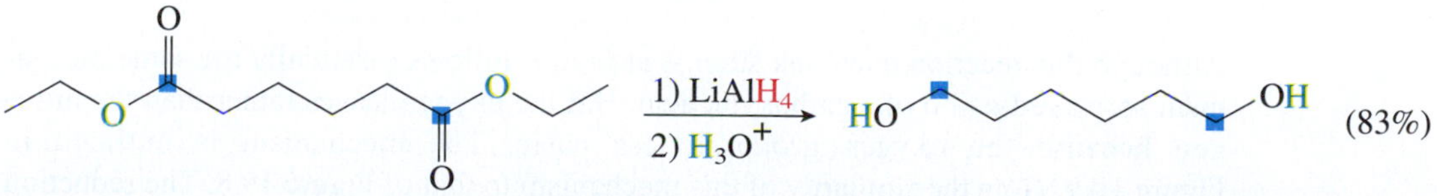
1) LiAlH4
2) H3O+
(83%)

❶ The strongly basic hydride ion removes the acidic proton from the carboxylic acid.

❷ Even though the carboxylate anion is at the bottom of the reactivity scale, lithium aluminum hydride is still powerful enough to react with it.

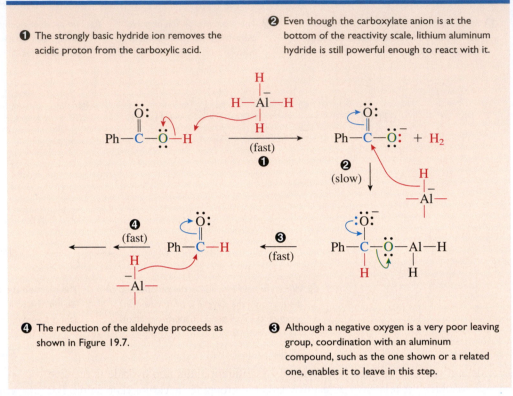

❹ The reduction of the aldehyde proceeds as shown in Figure 19.7.

❸ Although a negative oxygen is a very poor leaving group, coordination with an aluminum compound, such as the one shown or a related one, enables it to leave in this step.

Figure 19.8

Mechanism of the reduction of a carboxylic acid with lithium aluminum hydride—the first steps.

Sodium borohydride is much less reactive than lithium aluminum hydride and reacts only slowly with esters, so it is seldom used for their reduction. However, this property can be used to advantage when we want to reduce an aldehyde or ketone group without reducing an ester group in the same compound.

$$\text{CH}_3\text{OC}-\!\!\!\!\!\bigcirc\!\!\!\!\!=\!\text{O} \quad \xrightarrow[\substack{\text{CH}_3\text{OH} \\ 0°\text{C}}]{\text{NaBH}_4} \quad \text{CH}_3\text{OC}-\!\!\!\!\!\bigcirc\!\!\!\!\!\substack{\text{O}-\text{H} \\ \text{H}} \qquad (87\%)$$

The reduction of amides with lithium aluminum hydride takes a somewhat different course. The carbonyl group is reduced to a CH_2 group to produce an amine as shown in the following example:

$$\text{CH}_3(\text{CH}_2)_{10}\overset{\overset{\text{O}}{\|}}{\text{C}}\text{NHCH}_3 \quad \xrightarrow[\substack{\text{2) H}_2\text{O}}]{\text{1) LiAlH}_4} \quad \text{CH}_3(\text{CH}_2)_{10}\text{CH}_2\text{NHCH}_3 \quad (95\%)$$

Although this reaction may look strange at first, it follows essentially the same mechanism as the reduction of a carboxylic acid. But the oxygen leaves rather than the nitrogen because the oxygen group is less basic. The mechanism is outlined in Figure 19.9. Note the similarity of this mechanism to that of Figure 19.8. The reduction

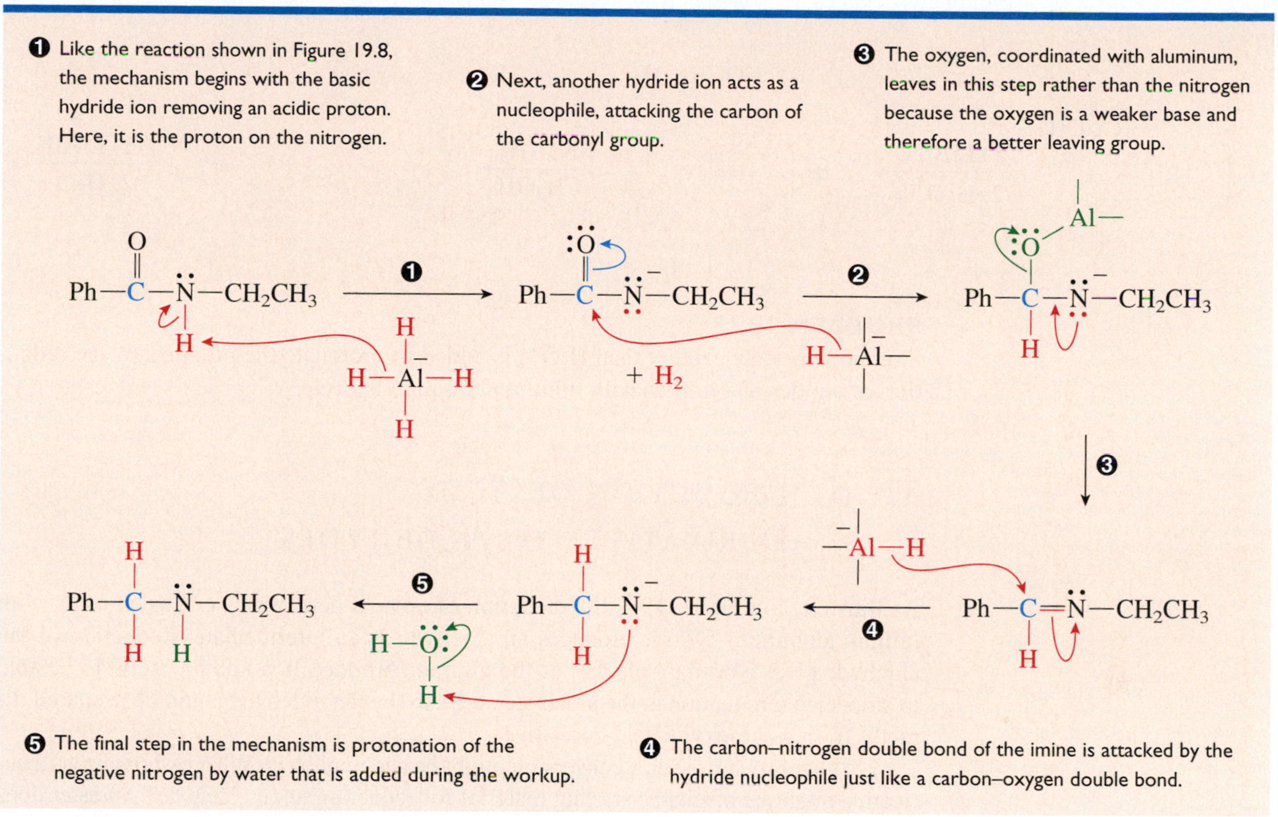
❶ Like the reaction shown in Figure 19.8, the mechanism begins with the basic hydride ion removing an acidic proton. Here, it is the proton on the nitrogen.
❷ Next, another hydride ion acts as a nucleophile, attacking the carbon of the carbonyl group.
❸ The oxygen, coordinated with aluminum, leaves in this step rather than the nitrogen because the oxygen is a weaker base and therefore a better leaving group.
❺ The final step in the mechanism is protonation of the negative nitrogen by water that is added during the workup.
❹ The carbon–nitrogen double bond of the imine is attacked by the hydride nucleophile just like a carbon–oxygen double bond.

Figure 19.9

MECHANISM OF THE REDUCTION OF AN AMIDE TO AN AMINE BY LITHIUM ALUMINUM HYDRIDE.

of amides with lithium aluminum hydride provides an important method for the preparation of amines. Primary, secondary, or tertiary amines can be prepared in this manner.

Another method that can be used to prepare primary amines is the reduction of nitriles with lithium aluminum hydride. The mechanism for this reaction involves sequential addition of two hydride nucleophiles to the electrophilic carbon of the cyano group. The addition of water in the workup step supplies the two protons on the nitrogen in the product. An example follows:

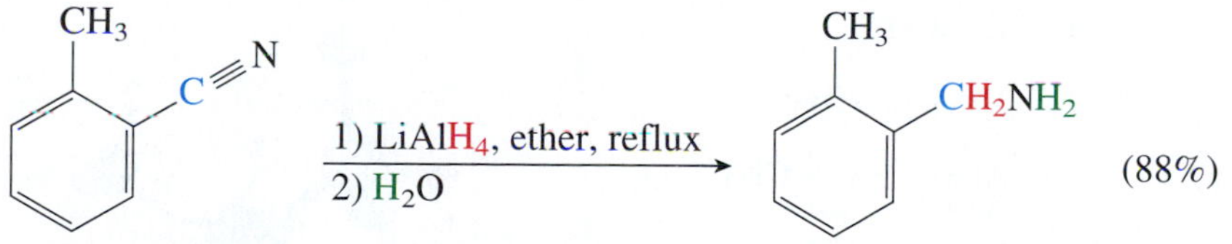
1) $LiAlH_4$, ether, reflux
2) H_2O
(88%)

PROBLEM 19.17

Show the products of these reactions:

a) 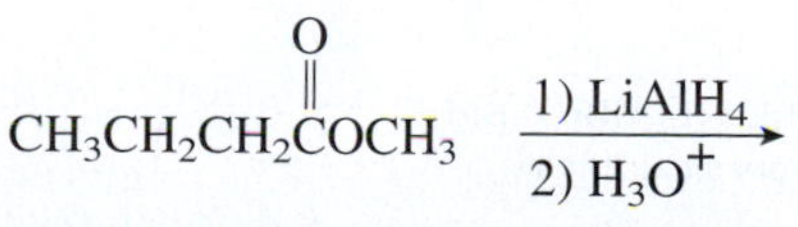
1) $LiAlH_4$
2) H_3O^+

b) 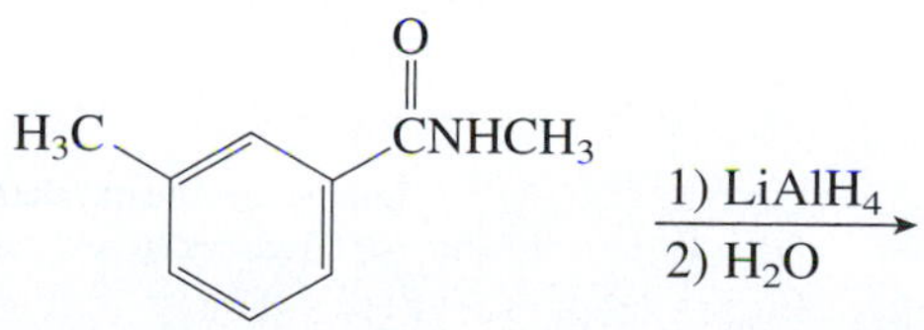
1) $LiAlH_4$
2) H_2O

c)

(structure of cyclohexenyl acetate with OCCH₃ ester group)

$\xrightarrow[\text{2) H}_3\text{O}^+]{\text{1) LiAlH}_4}$

d)

(benzene ring with COCH₂CH₃ and CCH₃ ketone substituents)

$\xrightarrow[\text{CH}_3\text{OH}]{\text{NaBH}_4}$

e)

(pentyl nitrile chain with terminal N)

$\xrightarrow[\text{2) H}_2\text{O}]{\text{1) LiAlH}_4}$

PROBLEM 19.18

Explain why water, rather than H_3O^+, is added to protonate the products of the reduction of amides and nitriles with lithium aluminum hydride.

19.8   REDUCTION OF ACID DERIVATIVES TO ALDEHYDES

As illustrated in Figure 19.7, the reduction of an acid derivative, such as an ester, with lithium aluminum hydride produces an aldehyde as an intermediate. Reduction of the aldehyde gives a primary alcohol as the ultimate product. It would be useful to be able to stop such a reduction at the aldehyde stage so that an aldehyde could be prepared directly from a carboxylic acid derivative.

Let's analyze how such a conversion might be accomplished. What carboxylic acid derivative might be a suitable starting material for reduction to an aldehyde? An ester does not appear to be a good choice because it is less reactive than an aldehyde. Any reagent that is reactive enough to reduce the ester will also be reactive enough to reduce the aldehyde product. However, a more reactive carboxylic acid derivative, such as an acyl chloride, might prove suitable. An acyl chloride is more reactive than an aldehyde, so it is possible, at least in theory, to find a reducing agent that is reactive enough to reduce the acyl chloride but not reactive enough to reduce the aldehyde product. In fact, lithium tri-*t*-butoxyaluminum hydride, LiAlH(O*t*-Bu)₃, has been found to have just such reactivity:

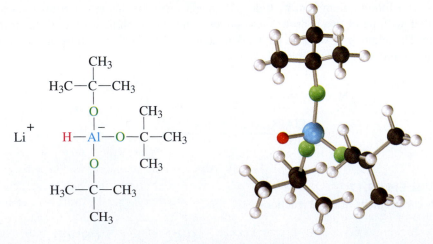

Lithium tri-*t*-butoxyaluminum hydride [LiAlH(Ot-Bu)₃]
(reduces an acyl chloride but not an aldehyde)

The steric hindrance caused by the bulky tertiary butoxy groups makes this reagent much less reactive than lithium aluminum hydride. At low temperature ($-78°C$) it is reactive enough to attack the carbonyl carbon of an acyl chloride, but it reacts only slowly with the aldehyde product. Examples of the use of this reagent to prepare aldehydes are provided in the following equations:

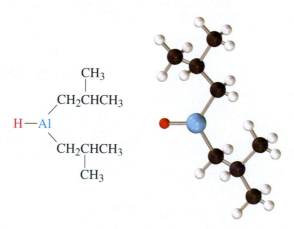

Numerous other metal hydride reagents have been developed to accomplish a variety of specialized reductions. The only other one that will be discussed here is diisobutylaluminum hydride, i-Bu$_2$AlH or DIBALH:

Diisobutylaluminum hydride
(DIBALH or i-Bu$_2$AlH)

This reagent reduces esters to aldehydes at low temperatures. On the basis of the preceding discussion, we might wonder why the aldehyde is not also reduced by this reagent, because it is reactive enough to attack the ester. The reason is that the aldehyde is not produced until acid is added to the reaction mixture during the workup. The initial intermediate, with a diisobutylaluminum group bonded to the oxygen, is stable at low temperature and does not expel ethoxide ion to produce the aldehyde. Therefore, the aldehyde is never in the presence of the reducing agent. The following equation provides an example:

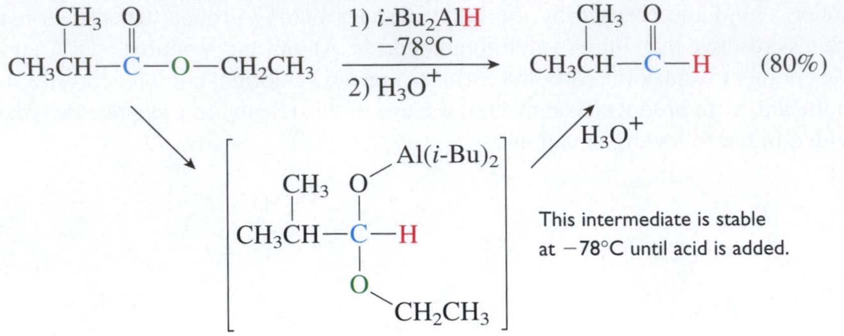

PROBLEM 19.19

Show the products of these reactions:

a) 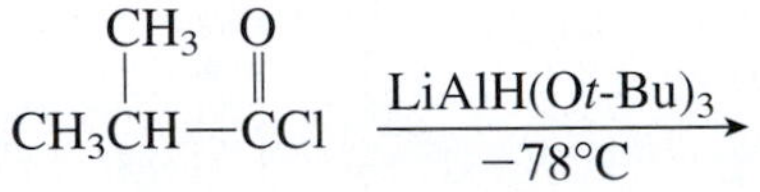

b) 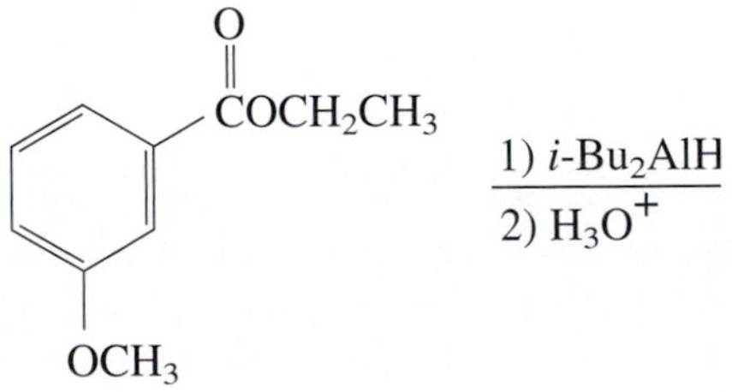

c) 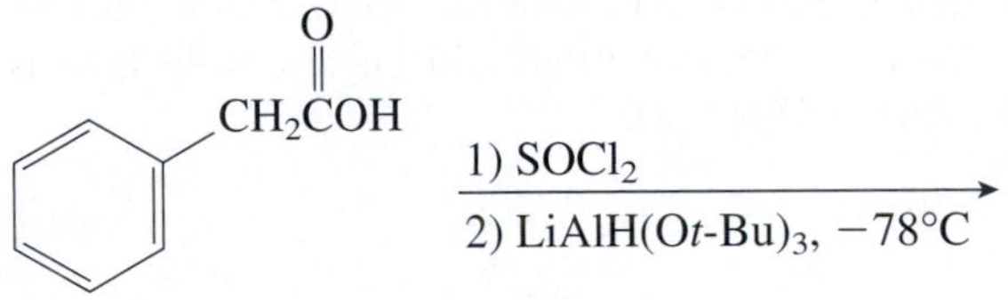

PROBLEM 19.20

Suggest methods to accomplish the following transformations. More than one step may be necessary in some cases.

a) $PhCOOH \longrightarrow PhCHO$

b) $PhCOOH \longrightarrow PhCH_2NHPh$

c) 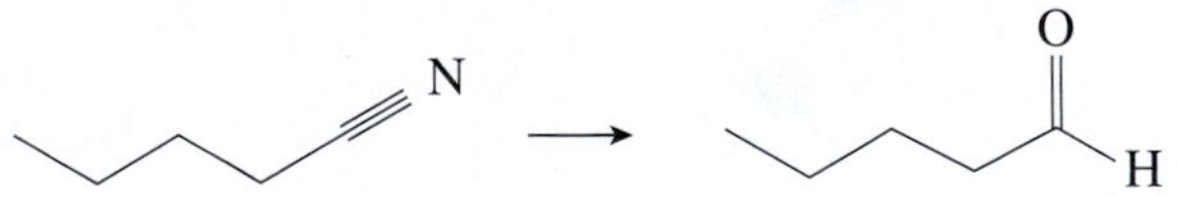

d) 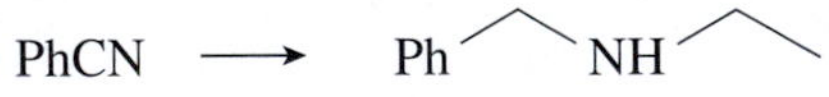

19.9 Reactions with Organometallic Nucleophiles

Similar to the case for its reaction with lithium aluminum hydride, an ester reacts with a Grignard or organolithium reagent to produce a ketone as the initial product. But because the ketone also reacts with the organometallic reagent, an alcohol is the final product. The mechanism for this reaction is shown in Figure 19.10. Note the similarities between this mechanism and that shown in Figure 19.7 for the reduction of an ester with lithium aluminum hydride.

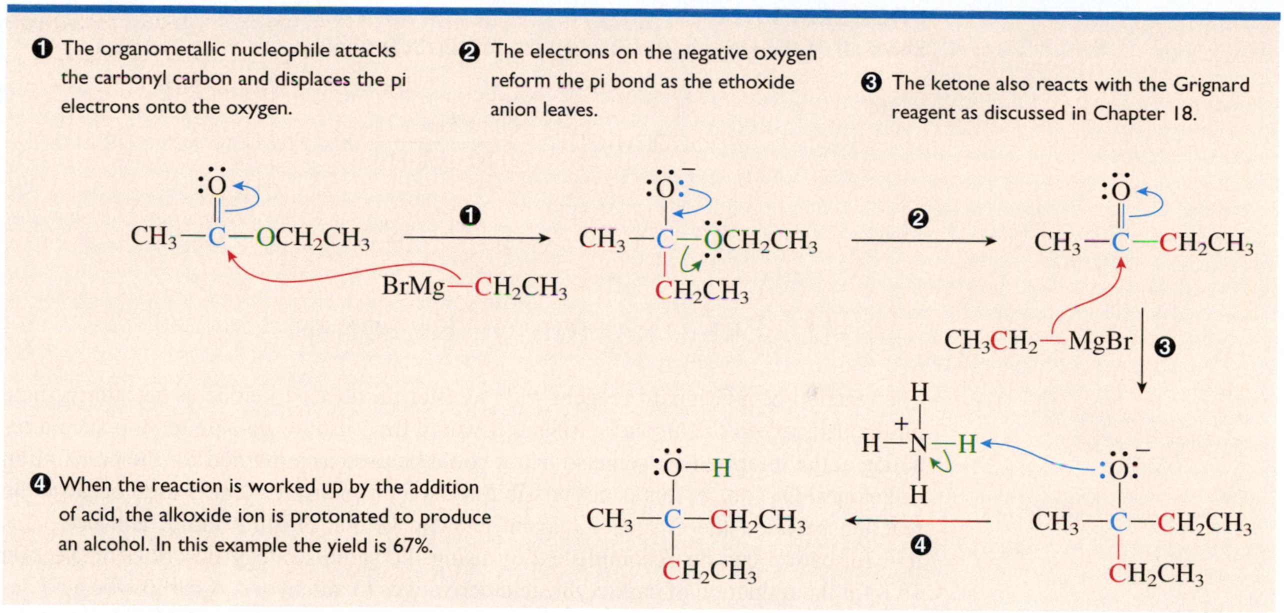

❶ The organometallic nucleophile attacks the carbonyl carbon and displaces the pi electrons onto the oxygen.

❷ The electrons on the negative oxygen reform the pi bond as the ethoxide anion leaves.

❸ The ketone also reacts with the Grignard reagent as discussed in Chapter 18.

❹ When the reaction is worked up by the addition of acid, the alkoxide ion is protonated to produce an alcohol. In this example the yield is 67%.

Figure 19.10

MECHANISM OF THE REACTION OF A GRIGNARD REAGENT WITH AN ESTER.

This reaction is a useful method to prepare alcohols with two identical groups on the carbon bonded to the hydroxy group. Formate esters produce secondary alcohols; other esters produce tertiary alcohols. Examples are provided in the following equations. Again, acyl chlorides and anhydrides also give this reaction, but they are seldom used because they offer no advantages over esters.

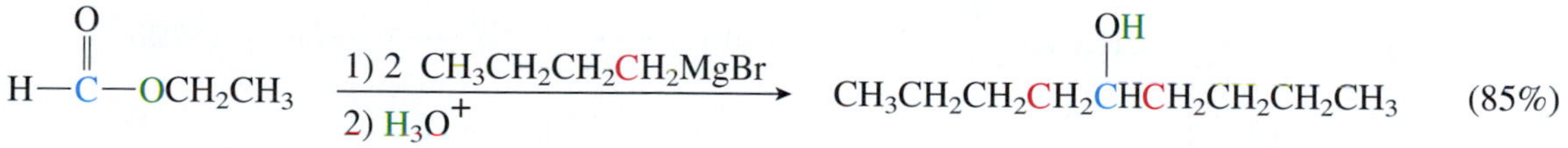

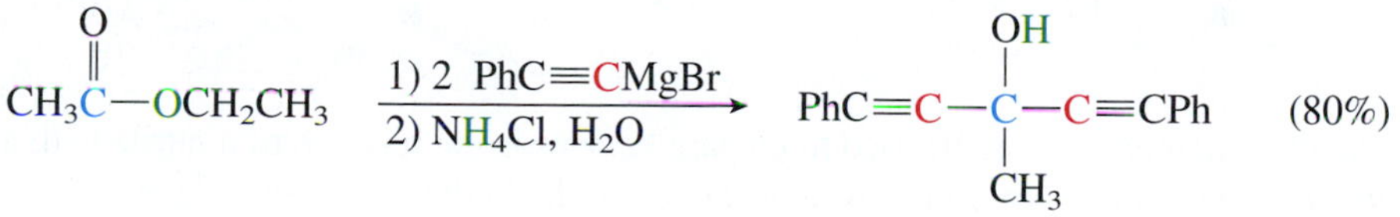

$$\text{H}-\overset{\overset{\displaystyle O}{\|}}{\text{C}}-\text{OCH}_2\text{CH}_3 \quad \xrightarrow[\text{2) H}_3\text{O}^+]{\text{1) 2 CH}_3\text{CH}_2\text{CH}_2\text{CH}_2\text{MgBr}} \quad \text{CH}_3\text{CH}_2\text{CH}_2\text{CH}_2\overset{\overset{\displaystyle OH}{|}}{\text{CH}}\text{CH}_2\text{CH}_2\text{CH}_2\text{CH}_3 \quad (85\%)$$

$$\text{CH}_3\overset{\overset{\displaystyle O}{\|}}{\text{C}}-\text{OCH}_2\text{CH}_3 \quad \xrightarrow[\text{2) NH}_4\text{Cl, H}_2\text{O}]{\text{1) 2 PhC}\equiv\text{CMgBr}} \quad \text{PhC}\equiv\text{C}-\overset{\overset{\displaystyle OH}{|}}{\underset{\underset{\displaystyle CH_3}{|}}{\text{C}}}-\text{C}\equiv\text{CPh} \quad (80\%)$$

PROBLEM 19.21

Show the products of these reactions:

a) 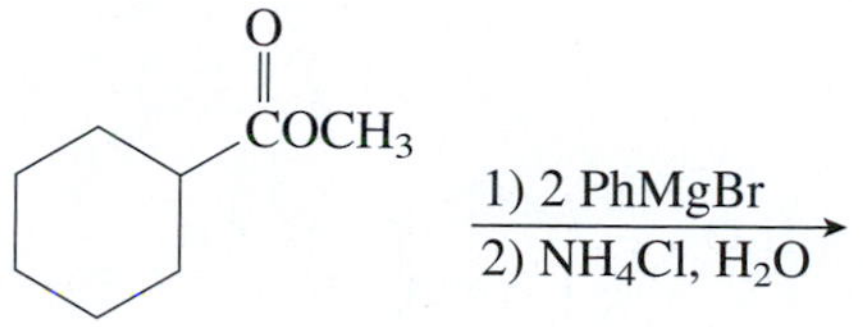
$$\xrightarrow[\text{2) NH}_4\text{Cl, H}_2\text{O}]{\text{1) 2 PhMgBr}}$$

b) 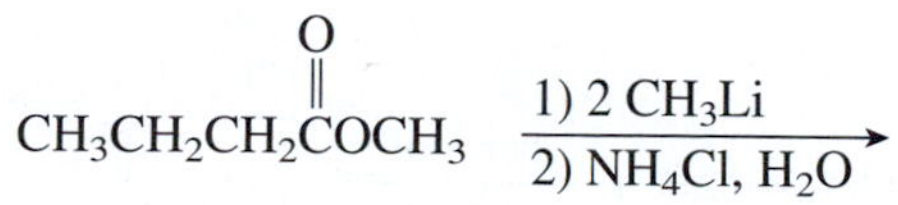 $\text{CH}_3\text{CH}_2\text{CH}_2\overset{\overset{\displaystyle O}{\|}}{\text{C}}\text{OCH}_3 \quad \xrightarrow[\text{2) NH}_4\text{Cl, H}_2\text{O}]{\text{1) 2 CH}_3\text{Li}}$

c) $\text{HC}\overset{\overset{\displaystyle O}{\|}}{}\text{OCH}_3 \quad \xrightarrow[\text{2) H}_3\text{O}^+]{\text{1) 2}} $

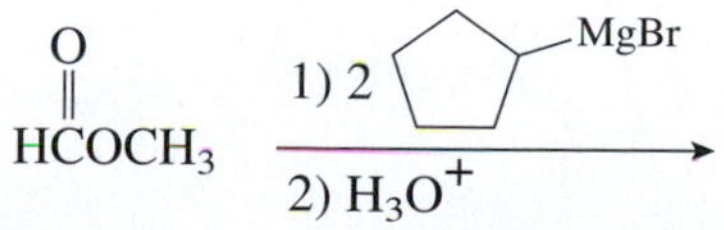

PROBLEM 19.22

Show all of the steps in the mechanism for this reaction:

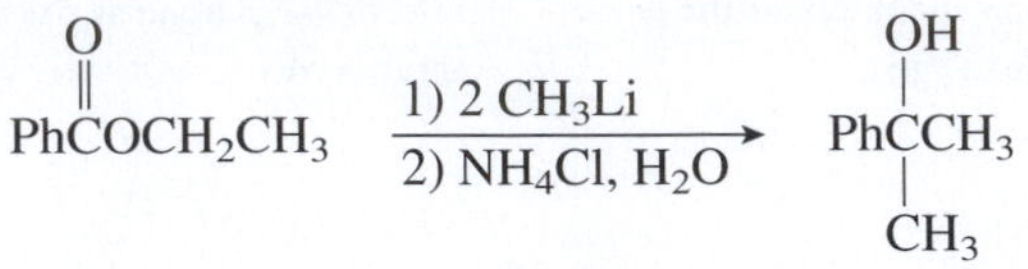

$$
\underset{\substack{O \\ \parallel}}{PhCOCH_2CH_3} \xrightarrow[\substack{2)\ NH_4Cl,\ H_2O}]{1)\ 2\ CH_3Li} \underset{\substack{OH \\ | \\ CH_3}}{PhCCH_3}
$$

19.10 PREPARATION OF KETONES

The reaction of a Grignard reagent with an ester produces a ketone as an intermediate on the pathway to the alcohol product. It would be useful to be able to stop such a reaction at the intermediate stage so that it could be used as a method for the preparation of ketones. Of course, this is not possible when a Grignard reagent is used because the reaction between the Grignard reagent and the ketone is quite rapid. However, this transformation can be accomplished by using the same strategy described in Section 19.8 for the reduction of carboxylic acid derivatives to aldehydes. A carboxylic acid derivative that is more reactive than a ketone, such as an acyl chloride, and an organometallic nucleophile that will react with the acyl chloride but not the ketone are required.

This transformation can be accomplished by the reaction of a lithium diorgano-cuprate reagent with an acyl chloride. Recall that this reagent is prepared from the organolithium reagent by reaction with cuprous iodide, CuI (see Section 18.10). Those containing methyl, primary alkyl, aryl, or vinyl groups add to acyl chlorides to give ketones as illustrated in the following examples:

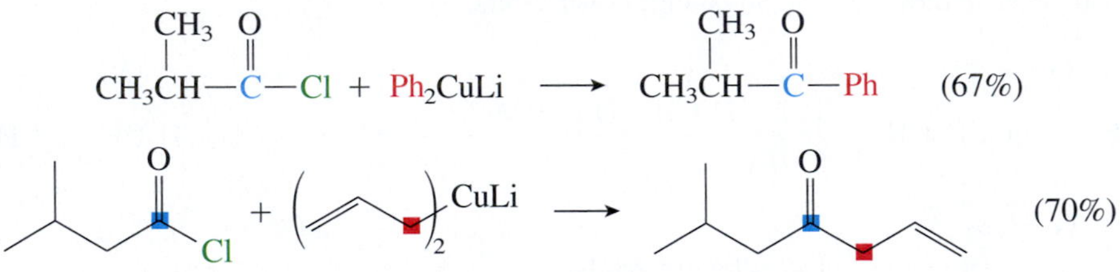

Another method that can be used to prepare ketones is the reaction of a nitrile with a Grignard reagent. In this case the CN double bond of the initial adduct (see Figure 19.11) does not react with the Grignard reagent because the negative charge on the nitrogen makes the carbon too weak as an electrophile. Addition of acid during the workup produces an imine, which is hydrolyzed to a ketone by the reverse of the mechanism shown in Figure 18.3 on page 766. An example is shown in the following equation:

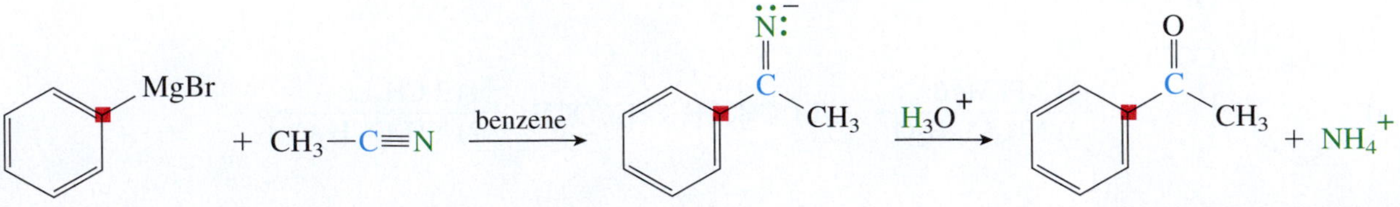

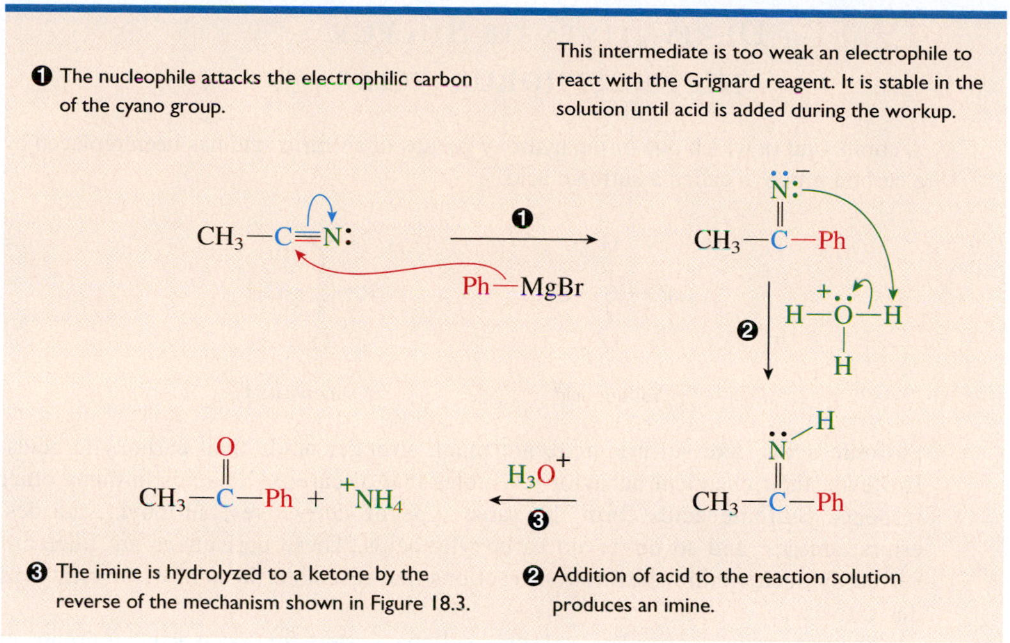

❶ The nucleophile attacks the electrophilic carbon of the cyano group.

This intermediate is too weak an electrophile to react with the Grignard reagent. It is stable in the solution until acid is added during the workup.

$CH_3-C\equiv N:$ $\xrightarrow{\text{❶}}$

Ph—MgBr

$CH_3-\overset{\overset{\displaystyle N:^-}{\|}}{C}-Ph$

❷

$H-\overset{+}{\overset{\displaystyle \cdot\cdot}{O}}-H$

$\overset{\displaystyle H}{|}$

$\overset{O}{\overset{\|}{CH_3-C-Ph}} + {}^+NH_4$ $\xleftarrow[\text{❸}]{H_3O^+}$ $CH_3-\overset{\overset{\displaystyle \cdot\cdot}{N}-H}{\underset{}{\overset{\|}{C}}}-Ph$

❸ The imine is hydrolyzed to a ketone by the reverse of the mechanism shown in Figure 18.3.

❷ Addition of acid to the reaction solution produces an imine.

Figure 19.11

MECHANISM OF THE REACTION OF A GRIGNARD REAGENT WITH A NITRILE TO PRODUCE A KETONE.

PROBLEM 19.23

Show the products of these reactions:

a) 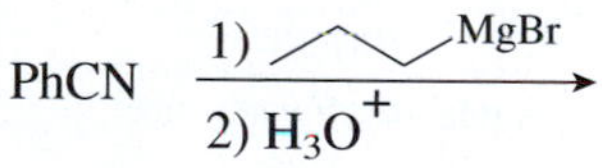 PhCN $\xrightarrow[\text{2) } H_3O^+]{\text{1) } \diagup\diagdown\diagup\text{MgBr}}$

b) 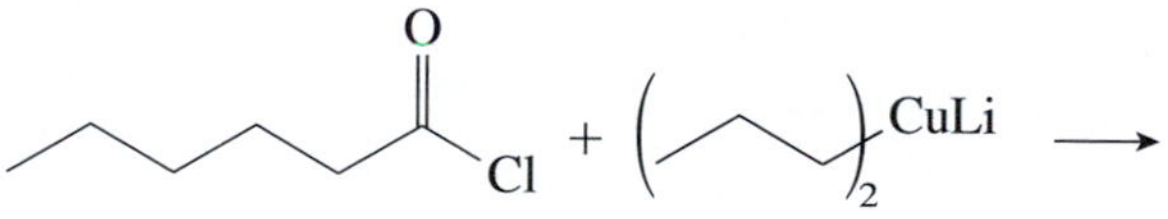 (pentanoyl chloride) $+$ (dibutyl)$_2$CuLi $\longrightarrow$

PROBLEM 19.24

Suggest methods to accomplish the following transformations. More than one step may be necessary in some cases.

a) 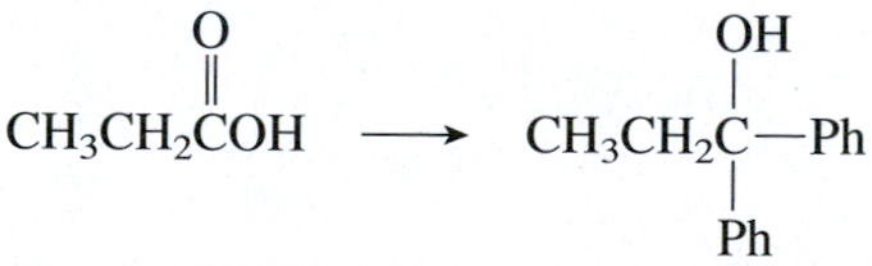 $CH_3CH_2\overset{O}{\overset{\|}{C}}OH \longrightarrow CH_3CH_2\overset{\overset{\displaystyle OH}{|}}{\underset{\underset{\displaystyle Ph}{|}}{C}}-Ph$

b) $CH_3CH_2\overset{O}{\overset{\|}{C}}OH \longrightarrow CH_3CH_2\overset{O}{\overset{\|}{C}}-\!\!\!\!\bigcirc\!\!\!\!-CH_3$

c) (phenyl bromide) $\longrightarrow$ (phenyl cyclohexyl ketone)

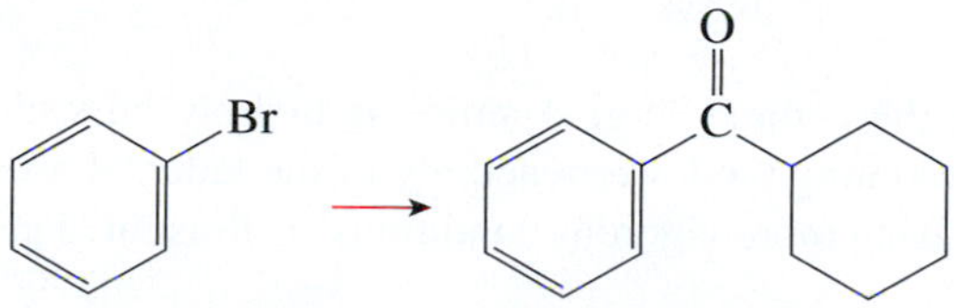

d) 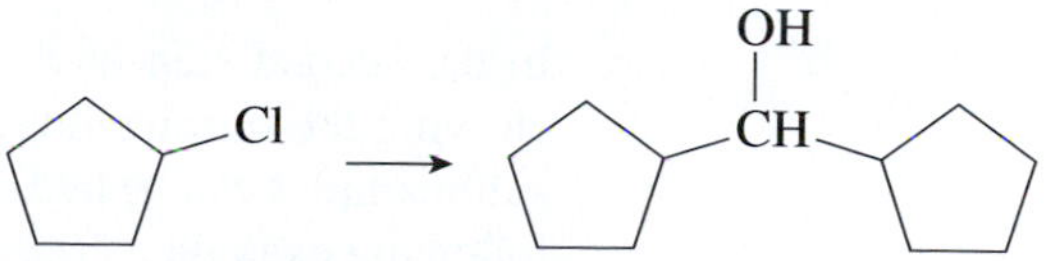 (cyclopentyl chloride) $\longrightarrow$ (dicyclopentyl carbinol)

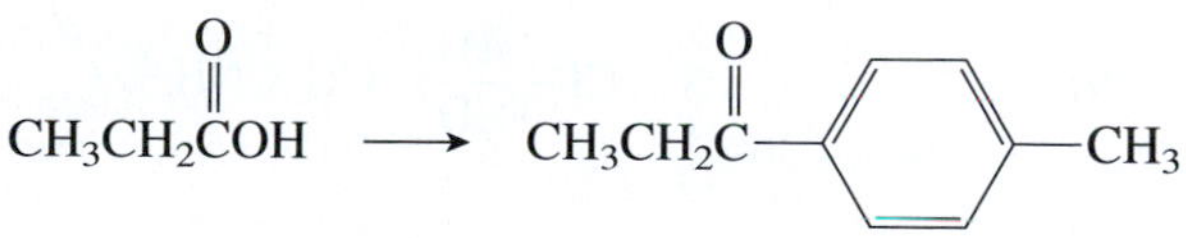

19.11 DERIVATIVES OF SULFUR AND PHOSPHORUS ACIDS

A compound in which one of the hydroxy groups of sulfuric acid has been replaced by a carbon group is called a sulfonic acid.

Sulfuric acid   A sulfonic acid

Sulfonic acids, like sulfuric acid, are much stronger acids than carboxylic acids. However, their chemical behavior resembles that of carboxylic acids in many other respects. Sulfonic acids form the same type of derivatives, sulfonyl chlorides, esters, amides, and so on, as do carboxylic acids. These derivatives are interconverted by nucleophilic substitution reactions that resemble those of carboxylic acid derivatives.

The preparation of tosylate and other sulfonate esters for use as leaving groups in nucleophilic substitution reactions (see Section 8.9) employs the reaction of a sulfonyl chloride (an acid chloride of a sulfonic acid) with an alcohol. Another example is shown in the following equation. Note the similarity of this reaction to the reaction of an acyl chloride with an alcohol to form an ester.

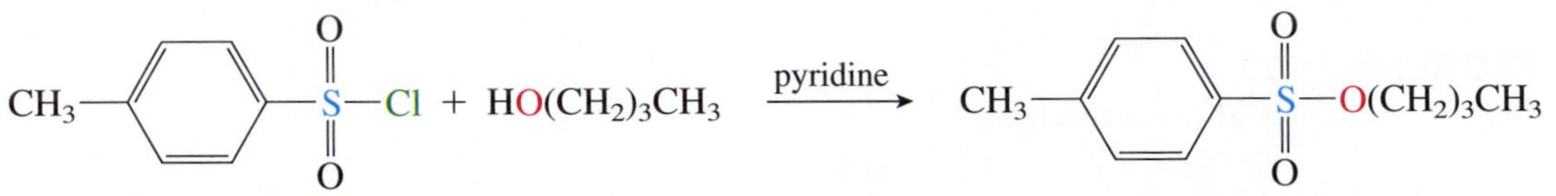

p-Toluenesulfonyl chloride   (90%)
A toluenesulfonate ester

As expected, the reaction of a sulfonyl chloride with ammonia produces a sulfonamide. This reaction is used in the preparation of sulfanilamide, a sulfa drug that was one of the first antibacterial agents:

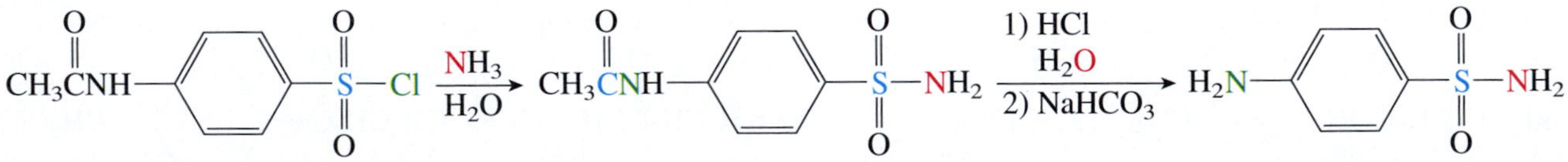

p-Aminobenzenesulfonamide
(sulfanilamide)

In the second step of this process, the carboxylic acid amide is hydrolyzed without cleaving the sulfonamide group, illustrating the lower reactivity of the latter. However, sulfonamides can be hydrolyzed by using more vigorous conditions, as illustrated in the following example:

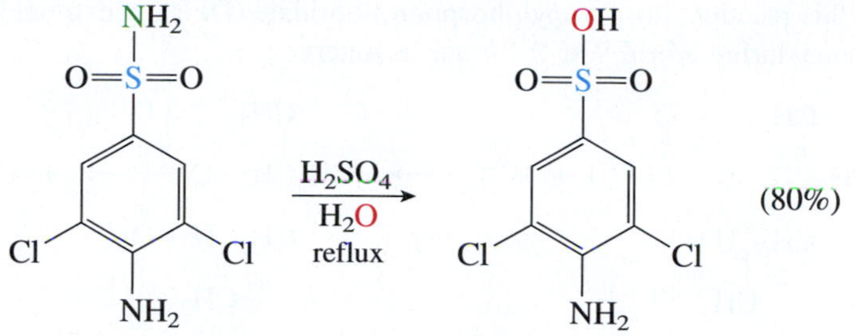

(80%)

Derivatives of phosphoric acid, pyrophosphoric acid, and related compounds are very important in biological systems. Pyrophosphoric acid is an anhydride of phosphoric acid. Adenosine triphosphate, an energy carrier that is universally found in living organisms, has a phosphorus dianhydride connected to an adenosine group by a phosphate ester linkage. Phosphorus ester bonds are used to form the polymeric backbone of DNA (see Chapter 27).

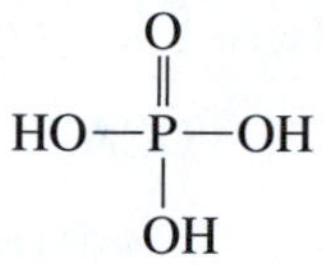

Phosphoric acid 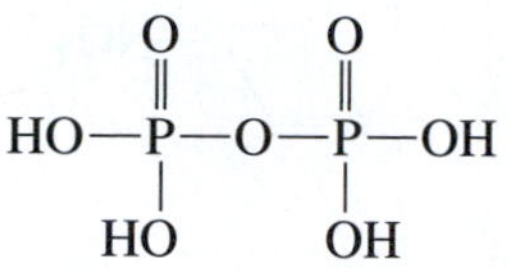   Pyrophosphoric acid

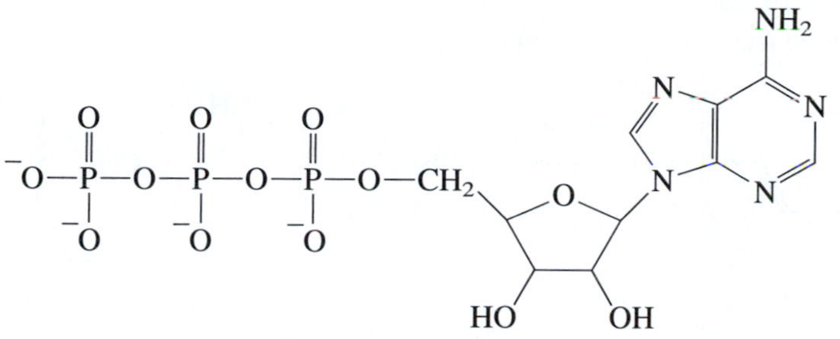

Adenosine triphosphate
(ATP)

Although phosphoric acid has three hydroxy groups that can be used to form derivatives, the various reactions involved are again similar to those of carboxylic acids. For example, the reaction of phosphorus oxychloride, which can be viewed as a tri(acid chloride) of phosphoric acid, with excess dimethyl amine produces a triamide, hexamethylphosphoric triamide:

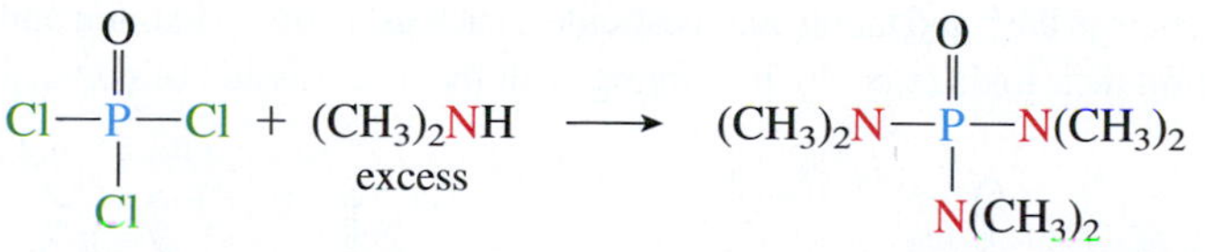

Phosphorus   Hexamethylphosphoric triamide
oxychloride   (HMPA)

In the reaction illustrated in the following equation, a fluoride nucleophile replaces the chlorine of a derivative that can be viewed as both a diester and an acid chloride. The product of this reaction, diisopropylphosphorofluoridate (DFP), is extremely toxic and was developed during World War II for use as a nerve gas.

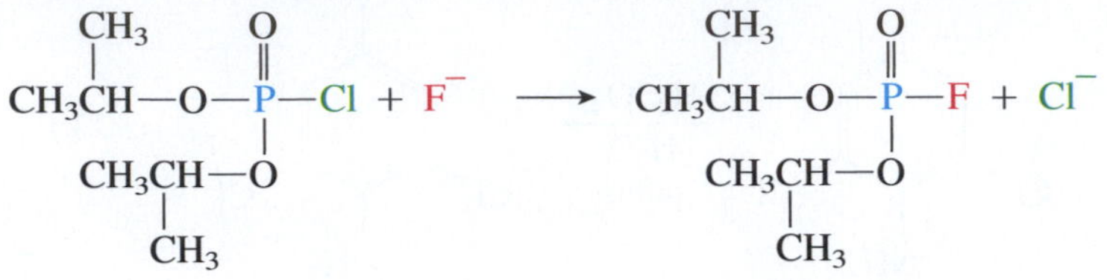

Diisopropylphosphorofluoridate
(DFP)

Methyl parathion, an insecticide, is prepared by a similar reaction. The change to a phosphorus–sulfur double bond, rather than a phosphorus–oxygen double bond, does not have a dramatic effect on the chemistry of the compound.

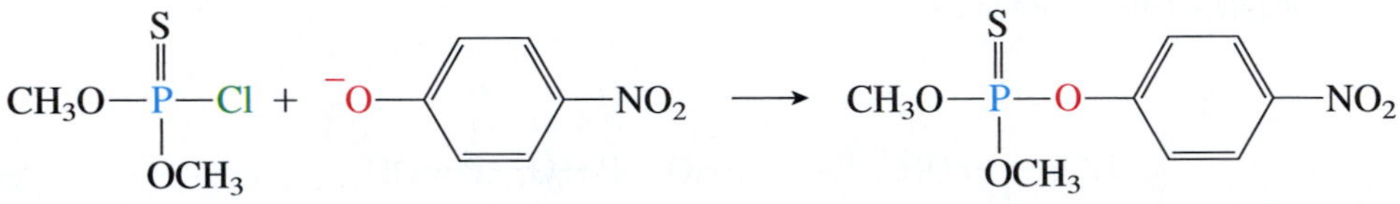

Methyl parathion

PROBLEM 19.25
Show the products of these reactions:

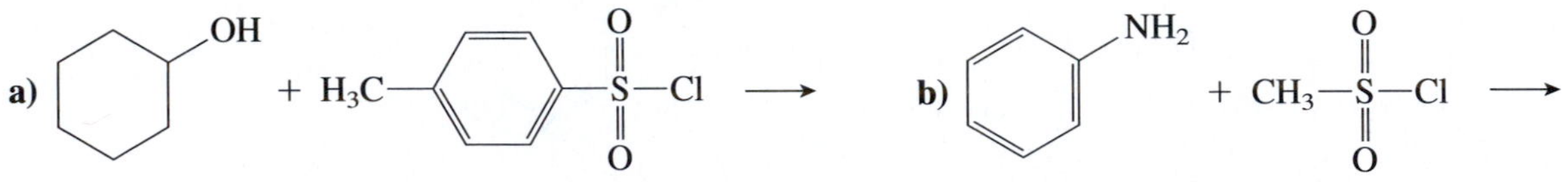

a)

b)

c) POCl$_3$ + 3 CH$_3$CH$_2$OH $\longrightarrow$

Focus On Biological Chemistry

Nerve Gases and Pesticides

Nerve gases, such as DFP and tabun, and pesticides, such as methyl parathion and diazinon, generally exhibit their toxic effect by interfering with the transmission of nerve signals.

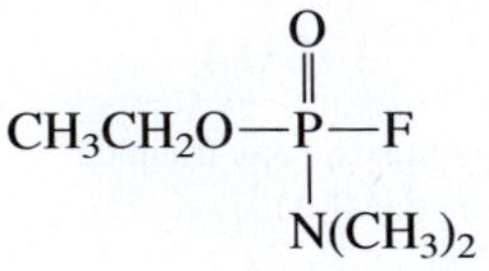

Tabun 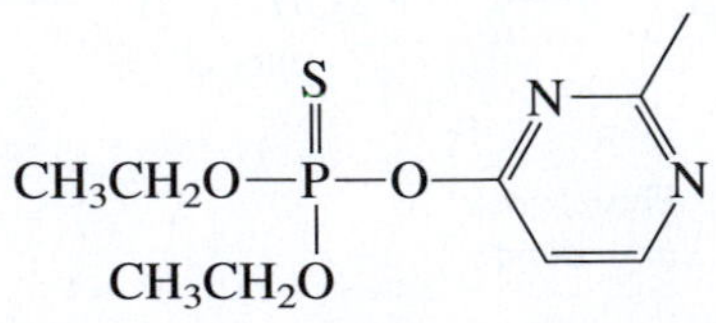   Diazinon

The signals between nerve cells are transmitted by chemicals known as neurotransmitters. Acetylcholine is an important neurotransmitter that is involved in the signal from nerves to muscles. The nerve releases acetylcholine, which binds to receptors on the muscle cell, causing it to contract. An enzyme, acetylcholinesterase, catalyzes the hydrolysis of acetylcholine, decomposing it so that the muscle can relax until another contraction is caused by another release of acetylcholine from the nerve cell. If the acetylcholine is not hydrolyzed, the muscle is paralyzed.

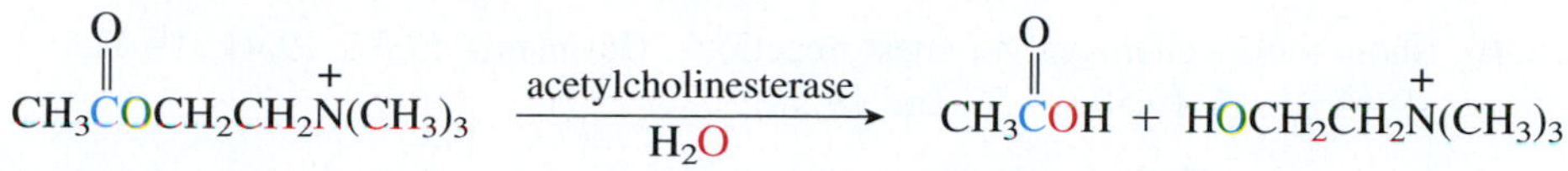

The mechanism for the hydrolysis that is catalyzed by the enzyme involves the hydroxy group of a serine amino acid residue in the protein acting as a nucleophile and attacking the carbonyl carbon of the acetylcholine ester. The ester is cleaved, and the acetyl group becomes bonded to the enzyme. Then the acetyl group is hydrolyzed off the enzyme, enabling it to perform another catalytic cycle. This hydrolysis is very facile, so a single enzyme molecule can catalyze the hydrolysis of many acetylcholine molecules:

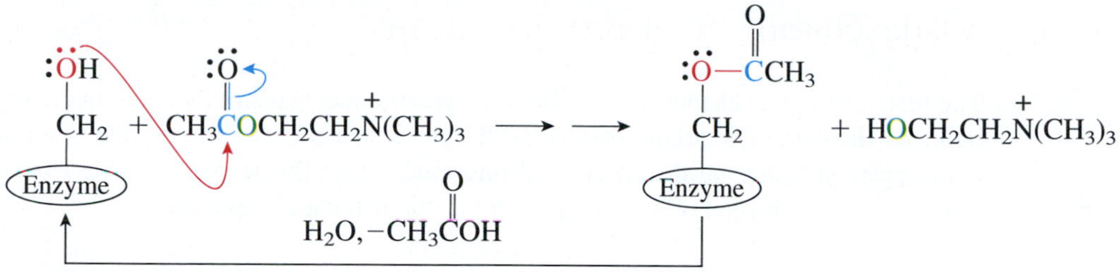

The phosphorus-based nerve gases and insecticides act by deactivating acetylcholinesterase. Note that all of these compounds have a good leaving group on the phosphorus. They react readily with the nucleophilic hydroxy group of the enzyme to form a phosphate triester in a reaction that is very similar, both in its mechanism and its product, to the reaction of an acyl chloride with an alcohol to form an ester:

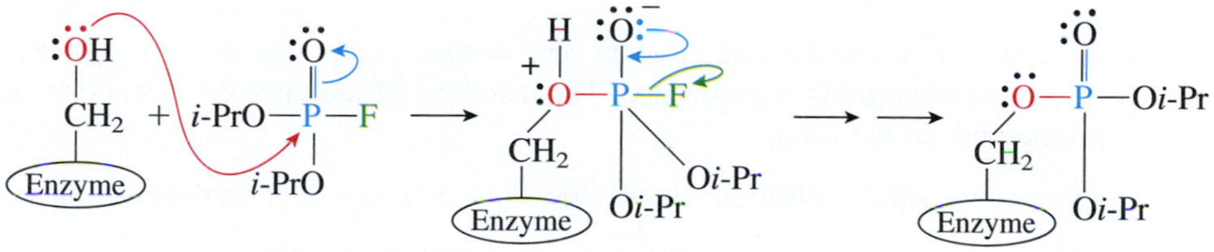

A phosphate triester

Unlike the acetate ester, the phosphate ester is not readily hydrolyzed off the enzyme. Therefore, the enzyme can no longer catalyze the hydrolysis of acetylcholine, and the muscle remains paralyzed. Paralysis of the muscles that are involved in breathing results quickly in asphyxiation and death.

Although there is considerable similarity among the structures of the nerve gases and the pesticides, their toxicity to insects and warm-blooded animals is often quite different. Pesticides that have found commercial applications are much less toxic to humans than are nerve gases. Nevertheless, they are toxic to some extent and must be used with caution. Chemists are continually seeking the perfect pesticide that is very toxic to insects (ideally to a single species of insect) yet completely harmless to higher organisms.

Review of Mastery Goals

After completing this chapter, you should be able to:

■ Show the products of any of the reactions discussed in this chapter. (Problems 19.26, 19.27, 19.28, 19.29, 19.33, 19.34, 19.35, 19.36, and 19.49)

■ Show the mechanisms for these reactions. (Problems 19.37, 19.41, 19.42, 19.43, 19.44, 19.45, 19.50, 19.55, and 19.56)

■ Predict the effect of a change in structure on the rate and equilibrium of a reaction. (Problems 19.30, 19.31, 19.40, 19.47, 19.54, 19.59, and 19.65)

■ Use these reactions to interconvert any of the carboxylic acid derivatives and to prepare aldehydes, ketones, alcohols, and amines. (Problems 19.32 and 19.46)

■ Use these reactions in combination with reactions from previous chapters to synthesize compounds. (Problems 19.38 and 19.39)

Visual Summary of Key Reactions

The reactions in this chapter follow the same general mechanism. First, the nucleophile bonds to the carbonyl carbon, displacing the pi electrons of the CO double bond onto the oxygen and forming the tetrahedral intermediate. In the second step the unshared electrons on the oxygen reform the pi bond as the leaving group leaves:

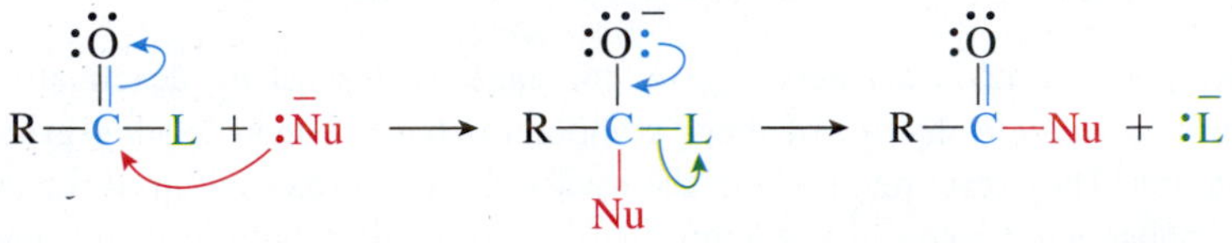

Tetrahedral
intermediate

The mechanism for acidic conditions is very similar, except that the carbonyl oxygen and/or the leaving group is protonated. The reactivity of the carboxylic acid derivative is affected by the following:

Resonance effects: Electron donors slow the reaction; withdrawers accelerate it.

Inductive effects: Electron withdrawers accelerate the reaction.

Steric effects: Steric hindrance slows the approach of the nucleophile.

The overall reactivity order for the carboxylic acid derivatives (see Table 19.1) is as follows:

Most reactive   Least reactive

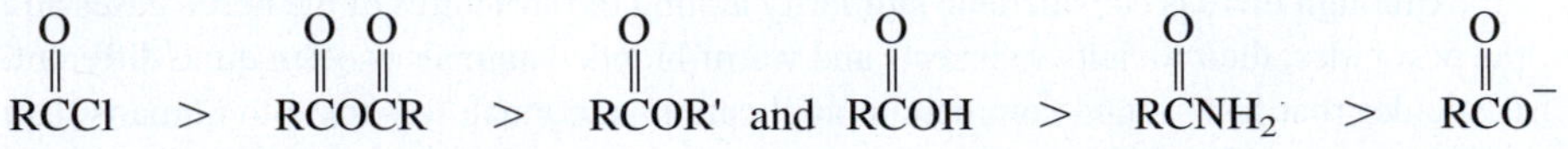

Table 19.2 provides a summary of the reactions presented in this chapter.

Table 19.2   Nucleophilic Substitution Reactions at Carbonyl Carbons

Reaction	Comment
$\underset{\text{O}}{\overset{\text{O}}{\parallel}}$ RCOH + SOCl$_2$ $\longrightarrow$ RCCl (PCl$_3$ or PCl$_5$)	**Section 19.2** Preparation of acyl chlorides. Acyl chlorides are commonly used to prepare other carboxylic acid derivatives.
RCOH + RCCl $\longrightarrow$ RCOCR	**Section 19.3** Preparation of anhydrides.
2 RCOH + CH$_3$COCCH$_3$ $\longrightarrow$ RCOCR + 2 CH$_3$COH	**Section 19.3** Preparation of anhydrides by exchange.
RCCl + R'OH $\longrightarrow$ RCOR' or RCOCR	**Section 19.4** Preparation of esters.
RCOH + R'OH $\xrightarrow{\text{HA}}$ RCOR'	**Section 19.4** Preparation of esters by Fischer esterification. The equilibrium must be driven to favor the ester. Requires an acid catalyst.
RCCl + H$_2$O $\longrightarrow$ RCOH or RCOCR	**Section 19.5** Hydrolysis of acyl chlorides and anhydrides. These derivatives must be protected from water to avoid these reactions.
RCOR' + H$_2$O $\xrightarrow[\substack{\text{or} \\ ^-\text{OH}}]{\text{H}^+}$ RCOH + R'OH	**Section 19.5** Hydrolysis of esters. Base is most commonly used in the process known as saponification.
RCNH$_2$ + H$_2$O $\xrightarrow[\substack{\text{or} \\ ^-\text{OH}}]{\text{H}^+}$ RCOH + NH$_3$	**Section 19.5** Hydrolysis of amides. This reaction can be accomplished by using either acid or base.
RC$\equiv$N + H$_2$O $\xrightarrow[\substack{\text{or} \\ ^-\text{OH}}]{\text{H}^+}$ RCNH$_2$ $\longrightarrow$ RCOH	**Section 19.5** Hydrolysis of nitriles. This reaction can be stopped at the amide or carried to the carboxylic acid.
R—L + $^-$CN $\longrightarrow$ RCN	The preparation of nitriles by S$_N$2 reactions combined with hydrolysis of nitriles provides a carboxylic acid preparation.

Continued

Table 19.2   Nucleophilic Substitution Reactions at Carbonyl Carbons—cont'd

Reaction	Comment
$\underset{\text{O}}{\overset{\text{O}}{\parallel}}$ RCCl + R'NH$_2$ $\longrightarrow$ RCNHR' or RCOCR	**Section 19.6** Preparation of amides.
RCOR' or RCOH $\xrightarrow[\text{2) H}_3\text{O}^+]{\text{1) LiAlH}_4}$ RCH$_2$OH	**Section 19.7** Reduction of esters or acids to alcohols.
RCNHR' $\xrightarrow[\text{2) H}_2\text{O}]{\text{1) LiAlH}_4}$ RCH$_2$NHR'	**Section 19.7** Reduction of amides to amines.
RC≡N $\xrightarrow[\text{2) H}_2\text{O}]{\text{1) LiAlH}_4}$ RCH$_2$NH$_2$	**Section 19.8** Reduction of nitriles to primary amines.
RCCl $\xrightarrow[-78°\text{C}]{\text{LiAlH(O}t\text{-Bu)}_3}$ RCH	**Section 19.8** Reduction of acyl chlorides to aldehydes.
RCOR' $\xrightarrow[\text{2) H}_3\text{O}^+]{\text{1) DIBALH, } -78°\text{C}}$ RCH	**Section 19.8** Reduction of esters to aldehydes.
RCOR' $\xrightarrow[\text{2) H}_3\text{O}^+]{\text{1) 2 R"MgX}}$ R—C—R" with OH and R"	**Section 19.9** Preparation of alcohols from esters.
RCCl $\xrightarrow{\text{R'}_2\text{CuLi}}$ RCR'	**Section 19.10** Preparation of ketones from acyl chlorides.
RC≡N $\xrightarrow[\text{2) H}_3\text{O}^+]{\text{1) R'MgX}}$ RCR'	**Section 19.10** Preparation of ketones from nitriles.

Integrated Practice Problem

Show the products of these reactions:

b) $PhCH_2\overset{O}{\overset{\|}{C}}OEt + 2\ CH_3CH_2MgI \xrightarrow[H_2O]{NH_4Cl}$

Strategy

A lot of reactions were presented in this chapter. Remember to identify the electrophile (usually the carbonyl carbon of a carboxylic acid derivative) and the nucleophile. The nucleophile bonds to the carbonyl carbon to form the tetrahedral intermediate. Then the leaving group departs as the CO double bond reforms. The product may be subject to further nucleophilic attack.

ORGANIC Chemistry•Now™ Click *Coached Tutorial Problems* to practice predicting the products of **Reactions of Carboxylic Acid Derivatives.**

Solutions

a) The electrophile is the carbonyl carbon of the anhydride; the nucleophile is the nitrogen of the aromatic amine. The nitrogen replaces the acetate group, so the ultimate products are an amide and acetic acid.

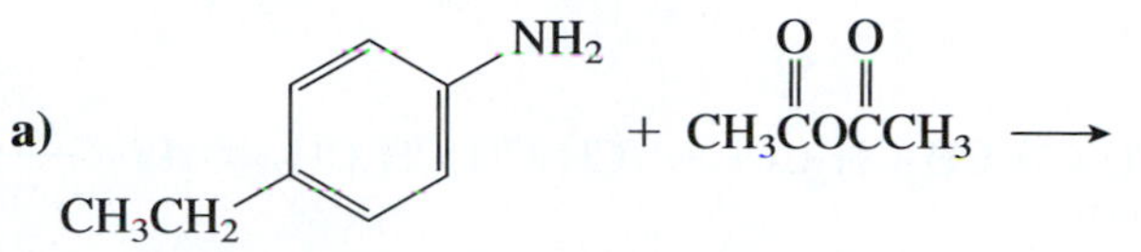

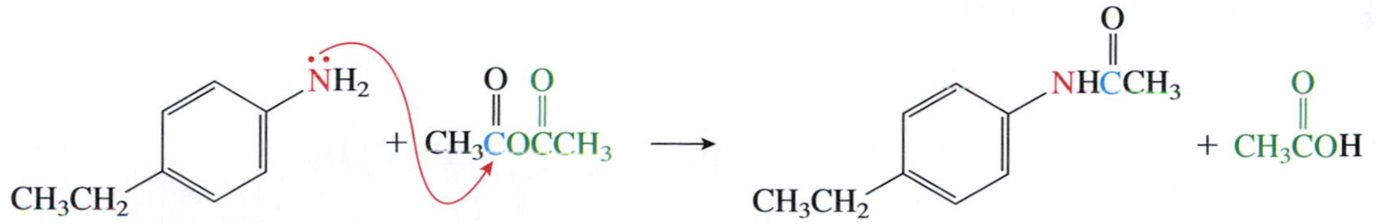

b) The electrophile is the carbonyl carbon of the ester. The nucleophile is the carbon bonded to the magnesium of the Grignard reagent. In the first part of the reaction, the nucleophile bonds to the carbonyl carbon, replacing the EtO^- group. The product of this step is a ketone, which still has an electrophilic carbonyl carbon that reacts with the second Grignard reagent. The final product is a tertiary alcohol with two ethyl groups bonded to the original carbonyl carbon of the ester group.

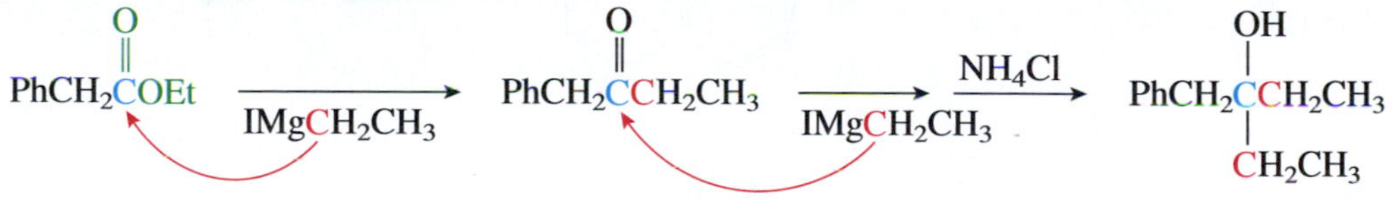

Additional Problems

ORGANIC Chemistry•Now™ Assess your understanding of this chapter's topics with additional quizzing and conceptual-based problems at **http://now.brookscole.com/hornback2**

PROBLEM 19.26

Show the products of these reactions:

a) $2\ CH_3CH_2OH + Cl{-}\overset{O}{\overset{\|}{C}}{-}Cl \longrightarrow$

b) $CH_3CH_2\overset{O}{\overset{\|}{C}}OH + SOCl_2 \longrightarrow$

c) PhCH₂COH $\xrightarrow{SOCl_2}$ PhCH₂CO⁻ $\xrightarrow{}$

d) PhCO₂H + CH₃CH₂CH₂CH₂OH $\xrightarrow{H_2SO_4}$

e)

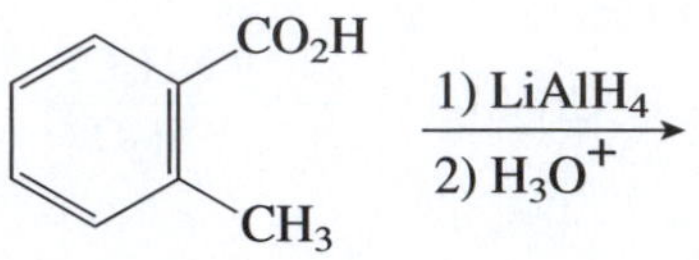

CO₂H / CH₃ (aromatic) $\xrightarrow[\text{2) H}_3\text{O}^+]{\text{1) LiAlH}_4}$

f) CH₃CH₂CH₂CH₂CCl + (CH₃CH₂CH₂CH₂)₂CuLi $\longrightarrow$

g) CH₃CH₂CH₂C≡N $\xrightarrow[\text{2) H}_2\text{O}]{\text{1) LiAlH}_4}$

h) CH₂=CHCCl + CH₃CH₂OH $\longrightarrow$

i) CH₃CHCO₂H (CH₃) $\xrightarrow{SOCl_2}$ $\xrightarrow[\text{H}_2\text{O}]{\text{NH}_3}$

j) CH₃(CH₂)₁₀CH₂OH + ClSO₂—⟨aryl⟩—CH₃ $\xrightarrow{\text{pyridine}}$

k) NCCH₂CH₂CH₂CN $\xrightarrow[\Delta]{\text{HCl} \atop \text{H}_2\text{O}}$

l) H₃C—⟨aryl⟩—CH₂COCH₃ $\xrightarrow[\text{2) H}_3\text{O}^+]{\text{1) LiAlH}_4}$

m) COCH₂CH₃ / OCH₃ (aromatic) $\xrightarrow[\text{2) H}_3\text{O}^+]{\text{1) DIBALH, }-78°\text{C}}$

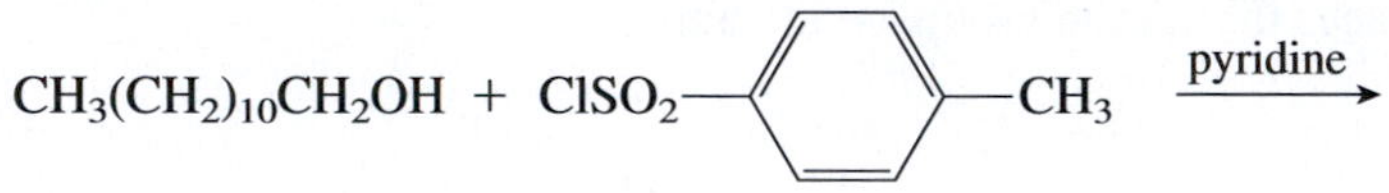

n) naphthalene-C≡N $\xrightarrow[]{\text{CH}_3\text{MgI}}$ $\xrightarrow{\text{H}_3\text{O}^+}$

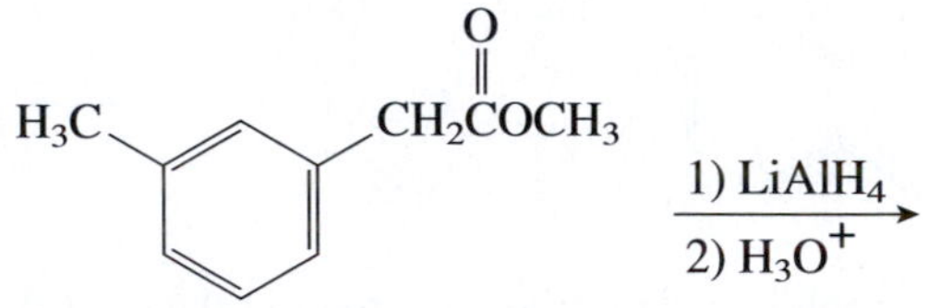

o) PhCH(OH)—CPh + CH₃COCCH₃ $\longrightarrow$

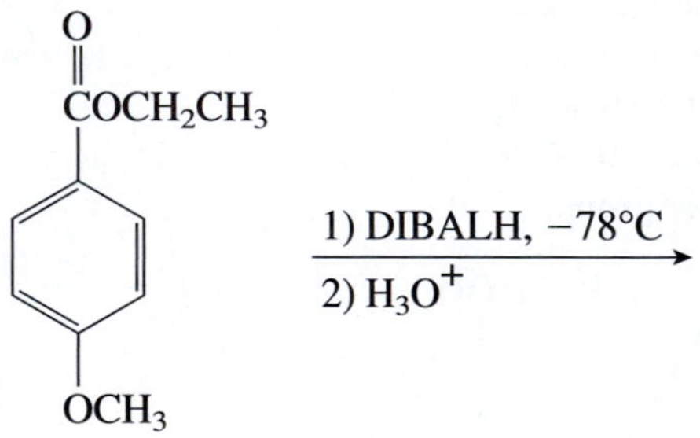

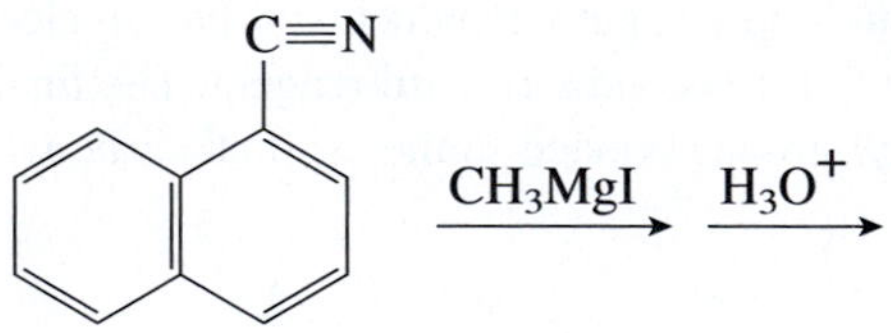

p) ⟨aryl⟩ CH₃ / NHCCH₃(O) $\xrightarrow[\Delta]{\text{HCl} \atop \text{H}_2\text{O}}$ $\xrightarrow{\text{NaOH}}$

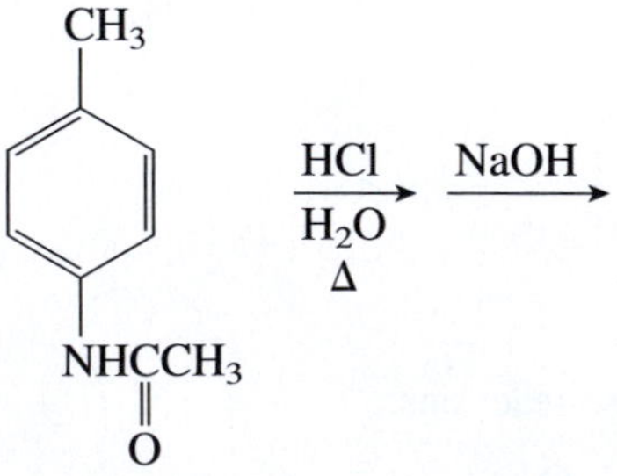

q) 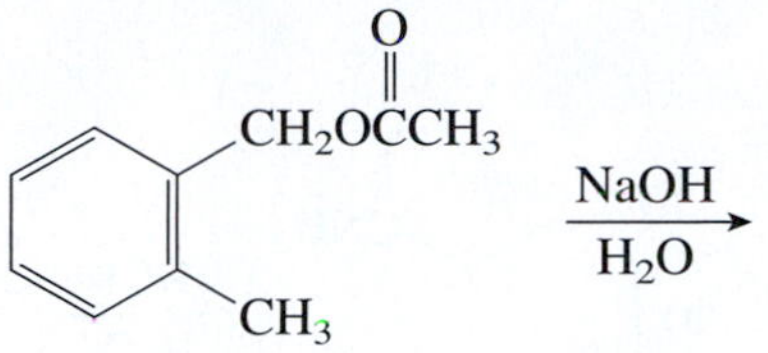 (o-methylbenzyl) CH_2OCCH_3 ester with CH_3 group $\xrightarrow[\text{H}_2\text{O}]{\text{NaOH}}$

r) 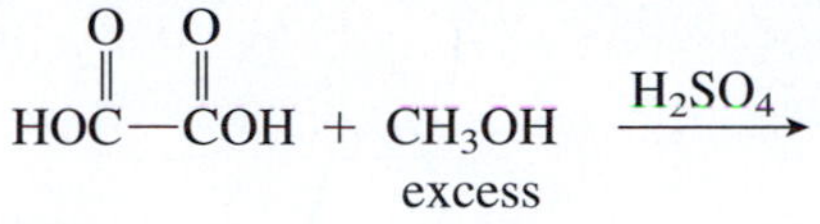 $HOC-COH + CH_3OH$ (excess) $\xrightarrow{\text{H}_2\text{SO}_4}$

s) 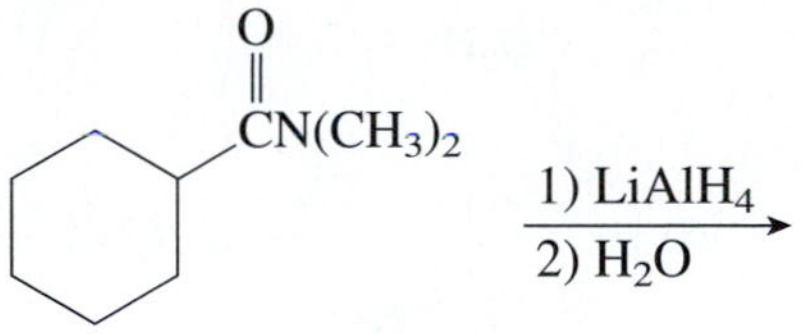 cyclohexyl-$CN(CH_3)_2$ amide $\xrightarrow[\text{2) H}_2\text{O}]{\text{1) LiAlH}_4}$

t) 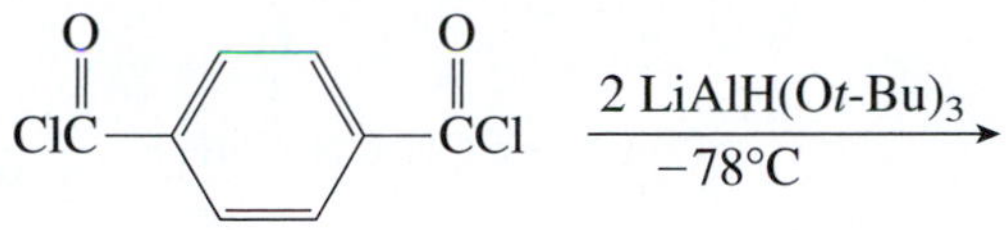 ClC—(benzene)—CCl (terephthaloyl dichloride) $\xrightarrow[-78°C]{\text{2 LiAlH(O}t\text{-Bu)}_3}$

19.27 Show the products of these reactions:

a) 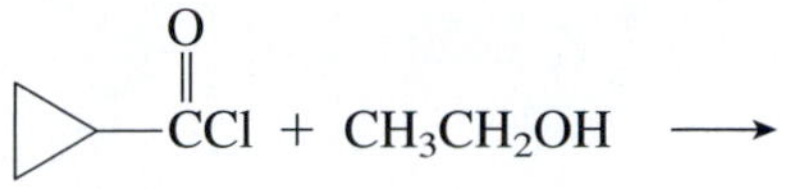 cyclopropyl-CCl + $CH_3CH_2OH \longrightarrow$

b) 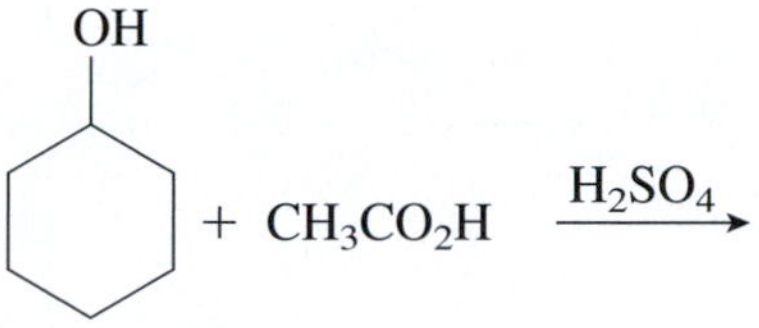 cyclohexanol (OH) + CH_3CO_2H $\xrightarrow{\text{H}_2\text{SO}_4}$

c) 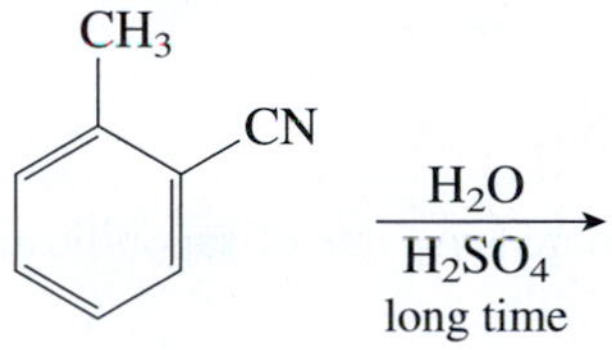 o-methylbenzonitrile (CH_3, CN) $\xrightarrow[\substack{\text{H}_2\text{SO}_4 \\ \text{long time}}]{\text{H}_2\text{O}}$

d) 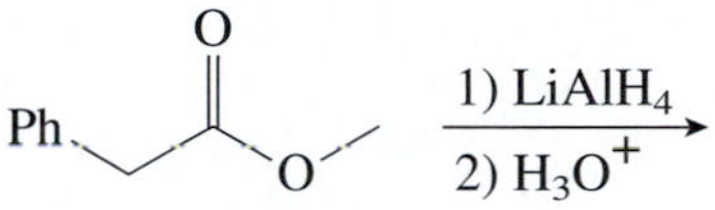 Ph—CH_2—$C(=O)$—O—CH_3 ester $\xrightarrow[\text{2) H}_3\text{O}^+]{\text{1) LiAlH}_4}$

e) 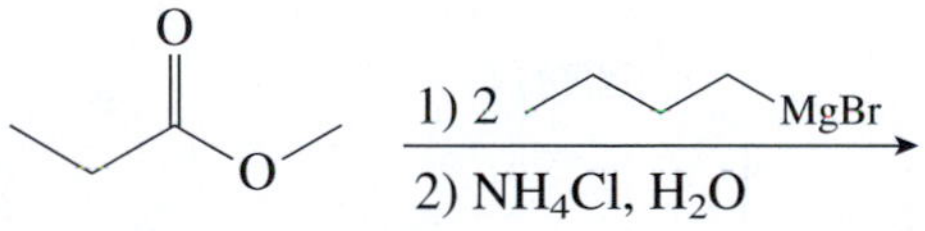 CH_3CH_2—$C(=O)$—O—CH_3 ester $\xrightarrow[\text{2) NH}_4\text{Cl, H}_2\text{O}]{\text{1) 2 } CH_3CH_2CH_2CH_2MgBr}$

19.28 Show the products of these reactions:

a) 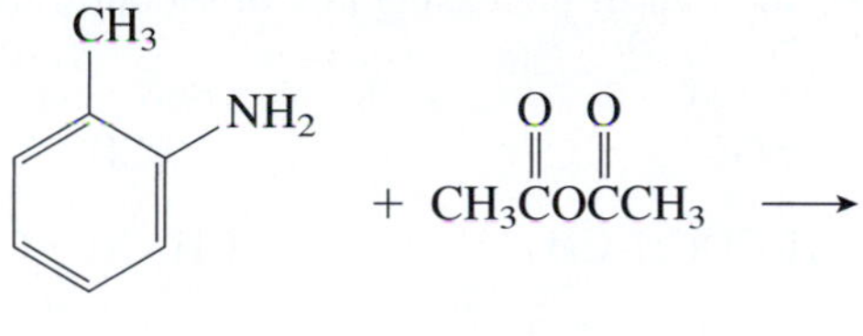 o-methylaniline (CH_3, NH_2) + CH_3COCCH_3 $\longrightarrow$

b) 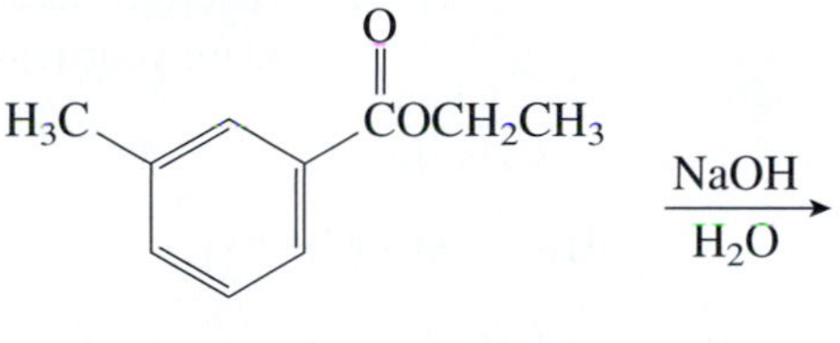 H_3C—(benzene)—$COCH_2CH_3$ (m-methyl) $\xrightarrow[\text{H}_2\text{O}]{\text{NaOH}}$

c) 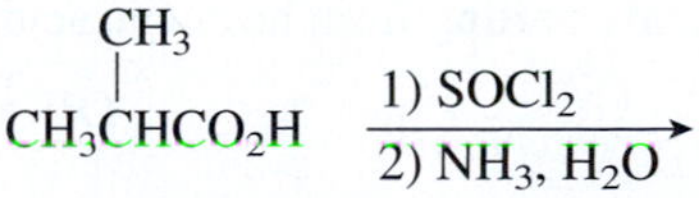 CH_3CHCO_2H (CH_3) $\xrightarrow[\text{2) NH}_3, \text{H}_2\text{O}]{\text{1) SOCl}_2}$

d) 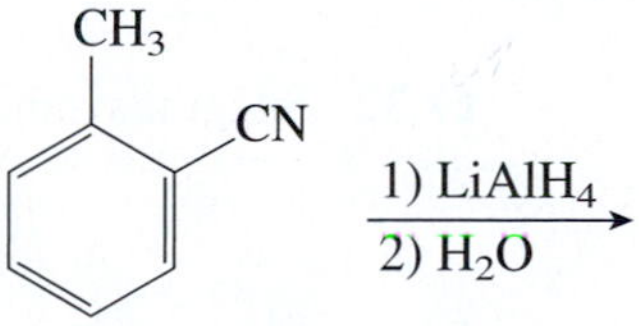 o-methylbenzonitrile (CH_3, CN) $\xrightarrow[\text{2) H}_2\text{O}]{\text{1) LiAlH}_4}$

e) 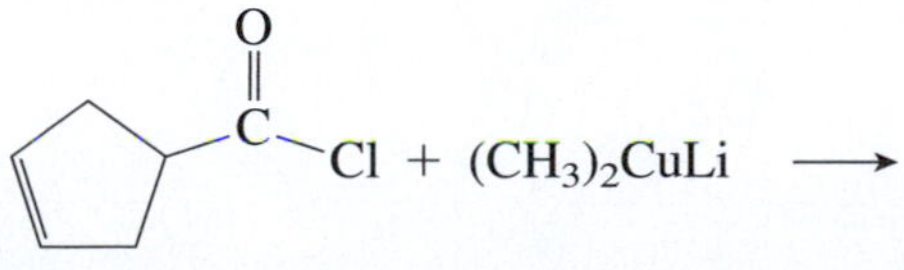 cyclopentenyl-$C(=O)$—Cl + $(CH_3)_2CuLi \longrightarrow$

19.29 Show the products of these reactions:

a) 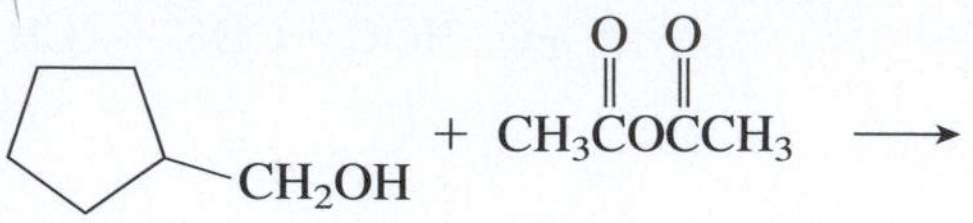 + $CH_3\overset{O}{\underset{\|}{C}}\overset{O}{\underset{\|}{C}}CH_3$ ⟶

b) $\xrightarrow[H_2O]{HCl}$

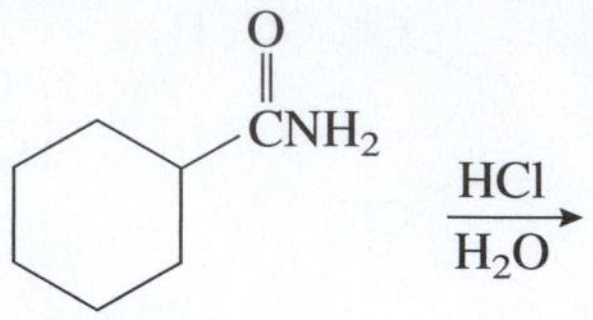

c) $CH_3CH_2\overset{O}{\underset{\|}{C}}NH_2$ $\xrightarrow[\text{2) } H_2O]{\text{1) LiAlH}_4}$

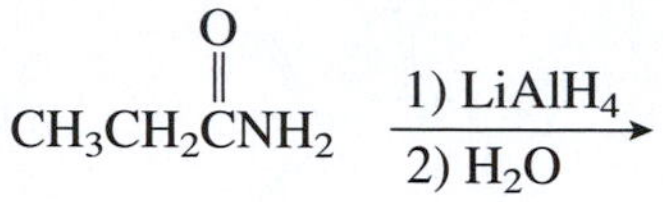

d) 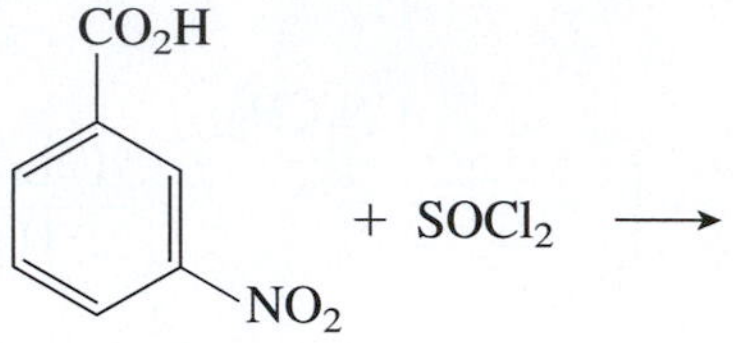 + $SOCl_2$ ⟶

e) 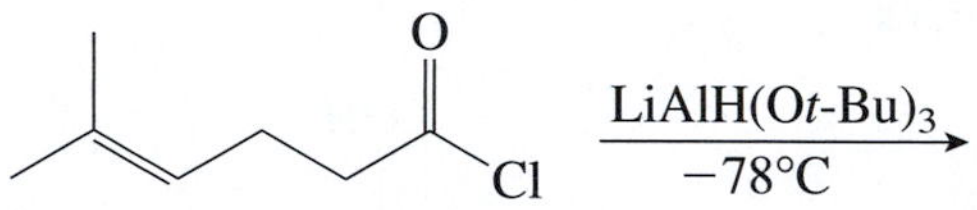 $\xrightarrow[-78°C]{\text{LiAlH(O}t\text{-Bu)}_3}$

f) 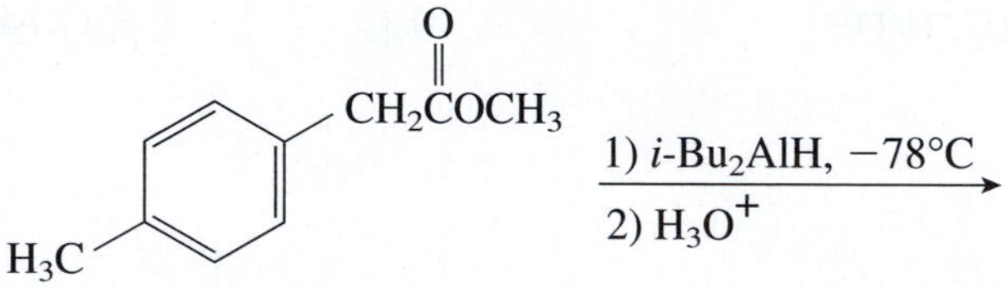 $\xrightarrow[\text{2) } H_3O^+]{\text{1) } i\text{-Bu}_2\text{AlH, } -78°C}$

19.30 Arrange these compounds in order of increasing rate of saponification and explain your reasoning:

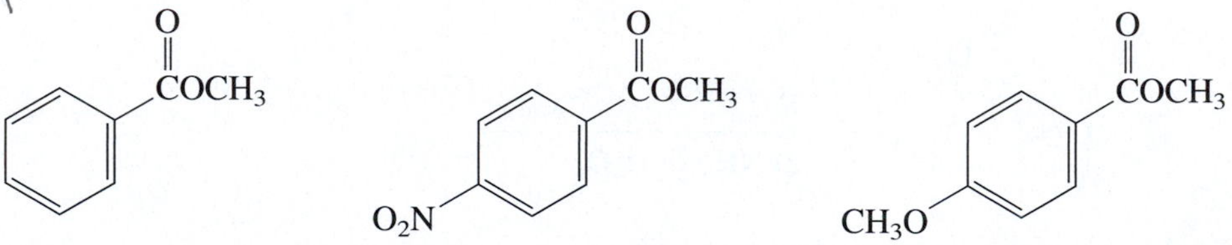

19.31 Arrange these compounds in order of increasing rate of saponification and explain your reasoning:

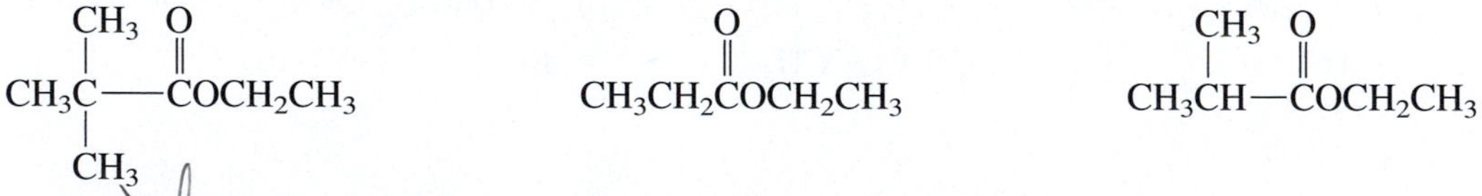

19.32 Suggest syntheses of these compounds starting from hexanoic acid:

a) 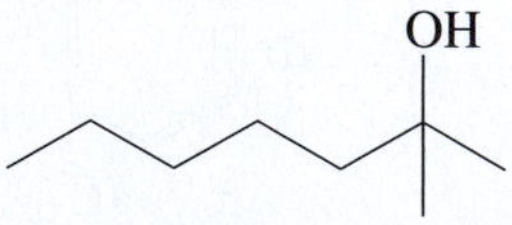   b)

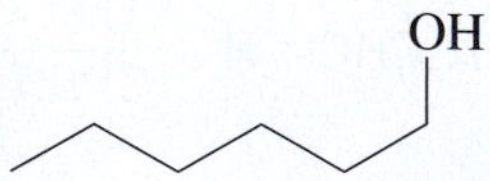

c) [structure: pentanal, O double bond with H]   **d)** [structure: chain with N, methyl]

e) [structure: ketone, hexanone]   **f)** [structure: chain with OH and Ph]

19.33 Show the products of the reaction of benzoic acid with each of these reagents:

a) 1) SOCl₂ 2) CH₃OH   **b)** CH₃$\overset{\overset{\displaystyle CH_3}{|}}{C}$HOH, H₂SO₄

c) 1) LiAlH₄ 2) H₃O⁺   **d)** CH₃NH₂

e) CH₃NH₂, Δ

19.34 Show the products of the reaction of benzoyl chloride with each of these reagents:

a) PhNH₂   **b)** CH₃$\overset{\overset{\displaystyle OH}{|}}{C}$HCH₃

c) PhC̈O⁻   **d)** LiAlH(O*t*-Bu)₃, −78°C

e) H₂O   **f)** (CH₃)₂CuLi

19.35 Show the products of the reaction of ethyl benzoate with each of these reagents:
a) 1) 2 CH₃CH₂MgBr 2) NH₄Cl, H₂O
b) 1) *i*-Bu₂AlH, −78° 2) H₃O⁺
c) NaOH, H₂O
d) 1) LiAlH₄ 2) H₃O⁺
e) H₃O⁺, H₂O

19.36 Show the products of the reaction of *N*-methylbenzamide with each of these reagents:
a) HCl, H₂O
b) 1) LiAlH₄ 2) H₂O

19.37 Show all of the steps in the mechanisms for these reactions:

a) CH₃C≡N + H₂O $\xrightarrow{H_3O^+}$ CH₃$\overset{\overset{\displaystyle O}{||}}{C}$NH₂

b) CH₃CH₂$\overset{\overset{\displaystyle O}{||}}{C}$OCH₃ + H₂O $\xrightarrow{^-OH}$ CH₃CH₂$\overset{\overset{\displaystyle O}{||}}{C}$O⁻ + CH₃OH

c) $CH_3CH_2\overset{O}{\overset{\|}{C}}OCH_3 + H_2O \xrightarrow{H_3O^+} CH_3CH_2\overset{O}{\overset{\|}{C}}OH + CH_3OH$

d) 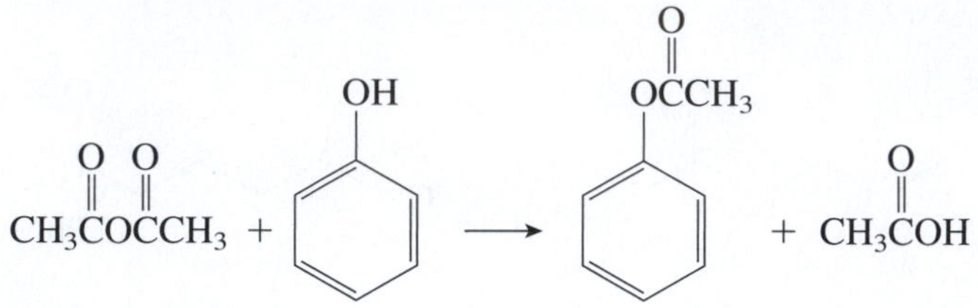 $CH_3\overset{O}{\overset{\|}{C}}O\overset{O}{\overset{\|}{C}}CH_3 +$ [phenol with OH] $\longrightarrow$ [phenyl acetate] $+ CH_3\overset{O}{\overset{\|}{C}}OH$

e) 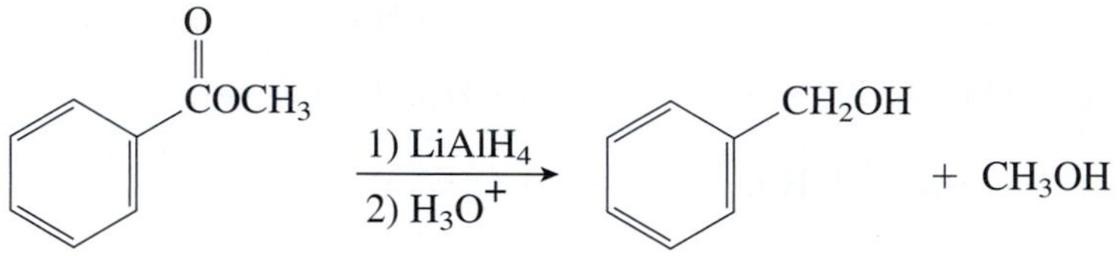 [acetophenone $\overset{O}{\overset{\|}{C}}OCH_3$] $\xrightarrow[\text{2) }H_3O^+]{\text{1) LiAlH}_4}$ [benzyl alcohol CH_2OH] $+ CH_3OH$

f) $CH_3CH_2CH_2C{\equiv}N \xrightarrow[\text{2) }H_2O]{\text{1) LiAlH}_4} CH_3CH_2CH_2CH_2NH_2$

g) $CH_3CH_2\overset{O}{\overset{\|}{C}}OCH_2CH_3 \xrightarrow[\text{2) }H_3O^+]{\text{1) CH}_3\text{Li}} CH_3CH_2\overset{OH}{\underset{CH_3}{\overset{|}{C}}}CH_3 + CH_3CH_2OH$

h) 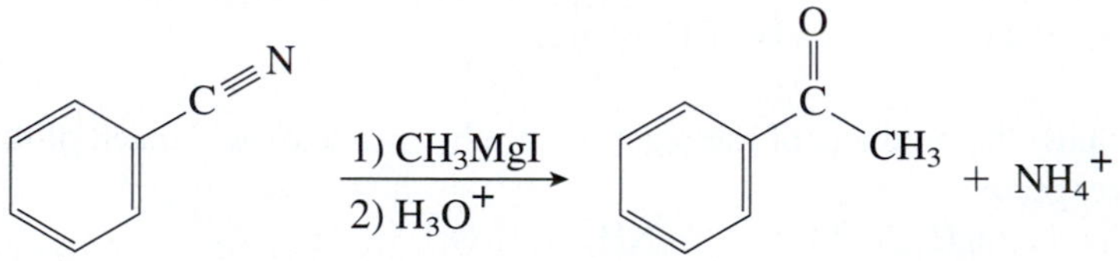 [benzonitrile $C{\equiv}N$] $\xrightarrow[\text{2) }H_3O^+]{\text{1) CH}_3\text{MgI}}$ [acetophenone $\overset{O}{\overset{\|}{C}}CH_3$] $+ NH_4^+$

i) 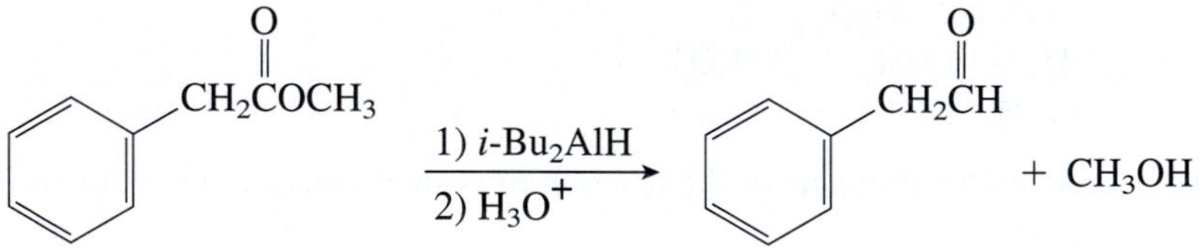 [PhCH₂COCH₃ $CH_2\overset{O}{\overset{\|}{C}}OCH_3$] $\xrightarrow[\text{2) }H_3O^+]{\text{1) }i\text{-Bu}_2\text{AlH}}$ [PhCH₂CHO $CH_2\overset{O}{\overset{\|}{C}}H$] $+ CH_3OH$

j) $CH_3CH_2{-}C{\equiv}N \xrightarrow[\text{H}_2\text{O}]{\text{NaOH}} CH_3CH_2{-}\overset{O}{\overset{\|}{C}}{-}O^- + NH_3$

19.38 Show syntheses of these compounds from the indicated starting materials:

a) 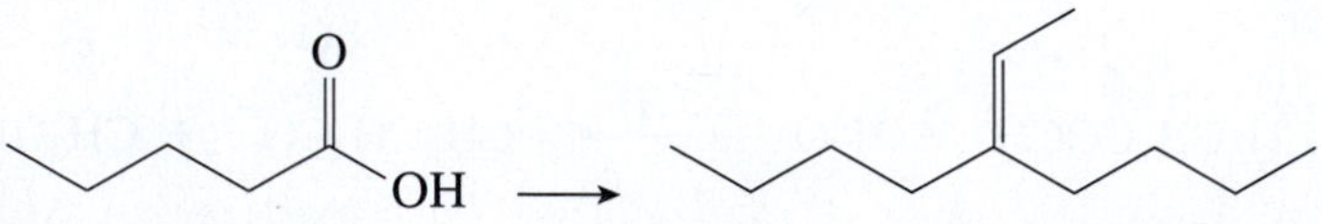 [CH₃CH₂CH₂CH₂COOH] $\longrightarrow$ [branched alkene structure]

b)

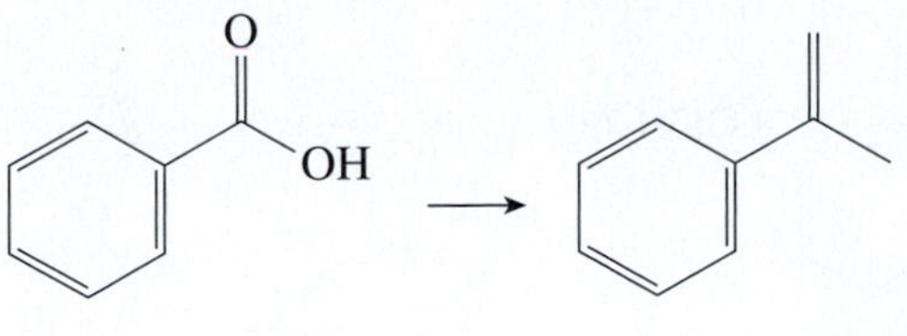

c)

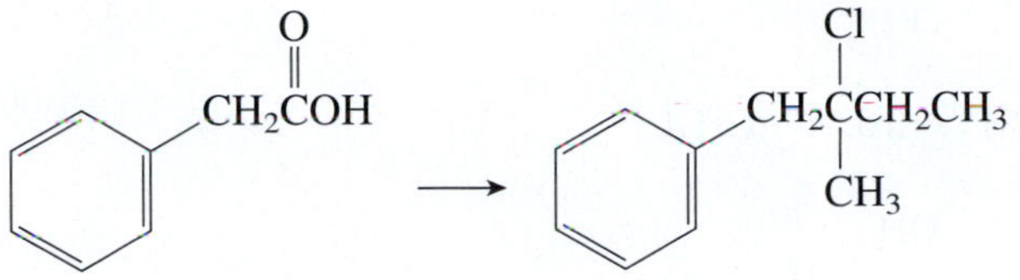

d)

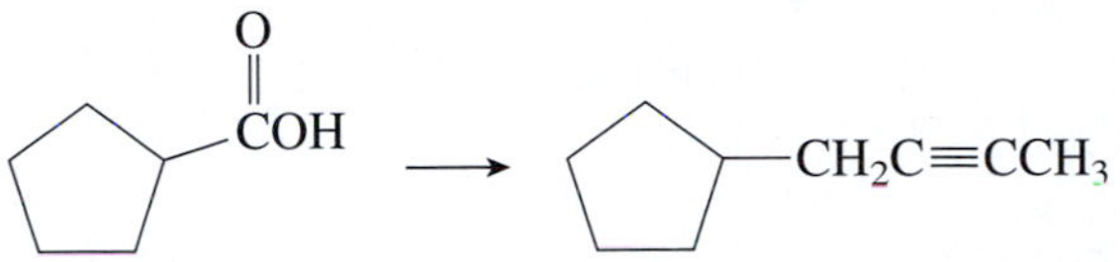

e)

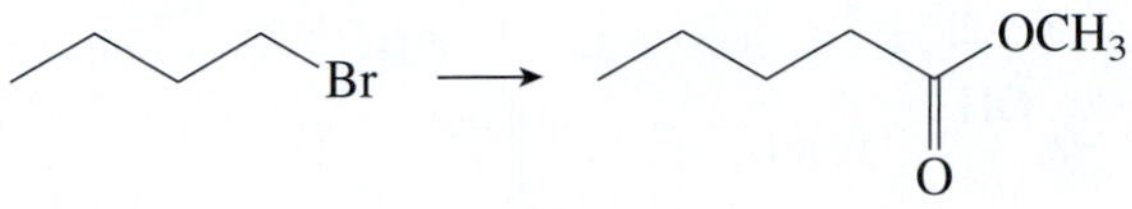

19.39 Show syntheses of these compounds from the indicated starting materials:

a) 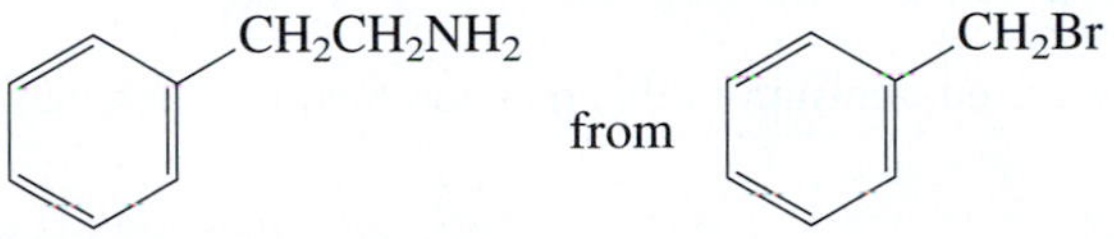 from

b) 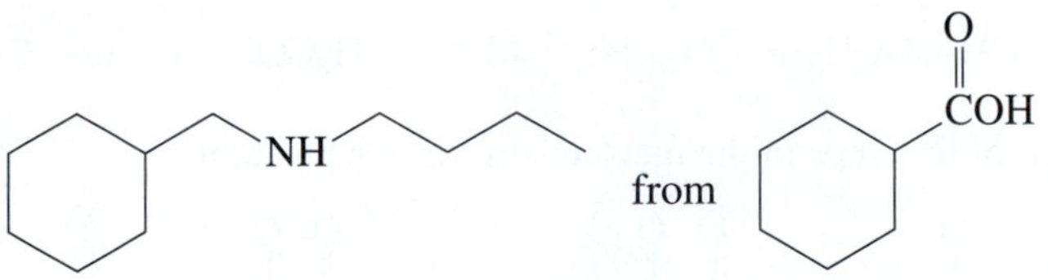 from

c) 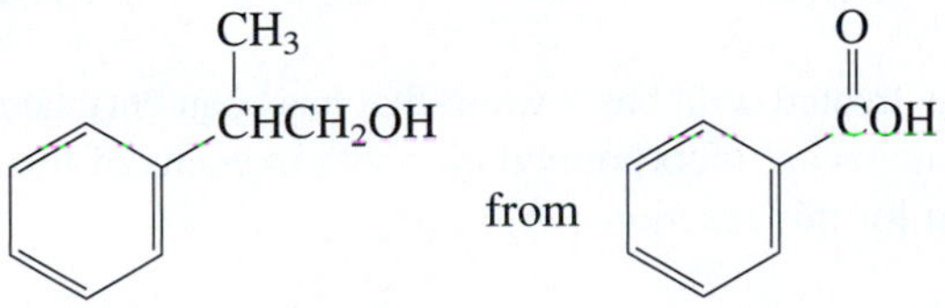 from

d) 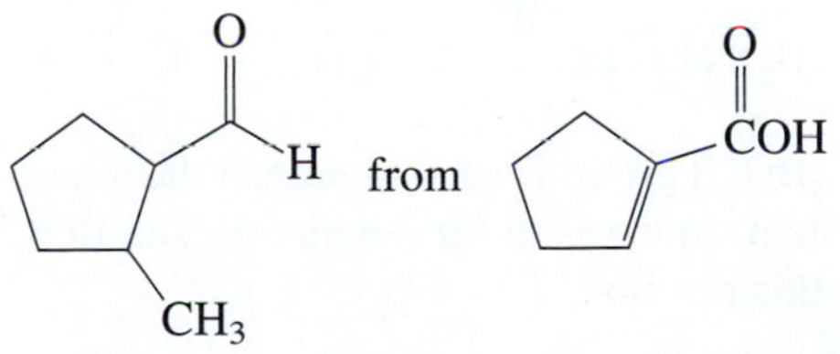 from

e) 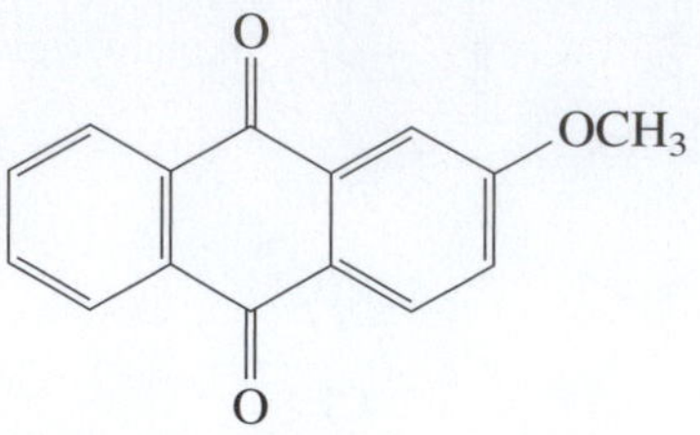 [structure: 2-methoxyanthraquinone with OCH₃] from [structure: phthalic anhydride]

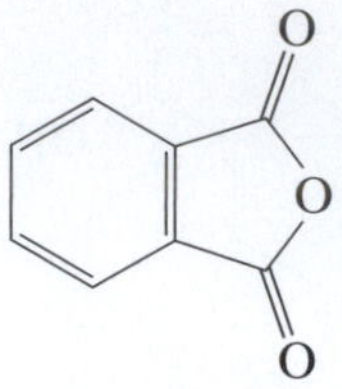

f) Ph₃COH from CH₃I

g) 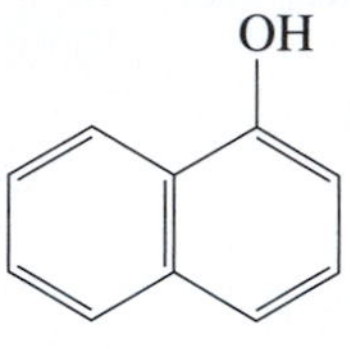 [structure: 1-naphthol with OH] from benzene

h) 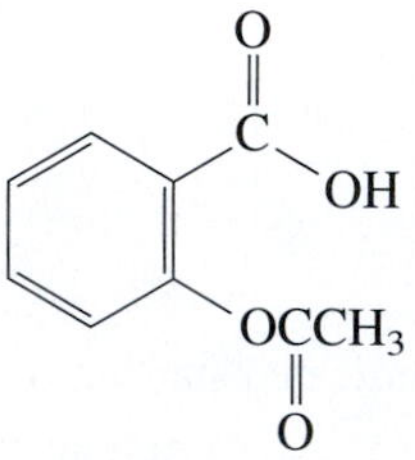 [structure: acetylsalicylic acid, with C(=O)OH and OCCH₃(=O)] from [structure: o-cresol, with CH₃ and OH]

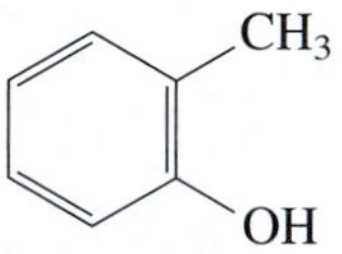

19.40 Explain whether the equilibrium in this reaction favors the reactants or the products:

$$\underset{\text{O}}{\text{CH}_3\overset{\parallel}{\text{C}}\text{OCH}_3} + {}^-\text{O}-\text{Ph} \rightleftharpoons \underset{\text{O}}{\text{CH}_3\overset{\parallel}{\text{C}}\text{OPh}} + {}^-\text{O}-\text{CH}_3$$

19.41 Show all of the steps in the mechanism for this reaction:

$$2\ \text{PhCOH} + \text{CH}_3\text{COCCH}_3 \rightleftharpoons \text{PhCOCPh} + 2\ \text{CH}_3\text{COH}$$

19.42 When acetate ion is treated with basic water that has been enriched with ¹⁸O, the heavy oxygen atoms are incorporated into both oxygens of the acetate ion. Show a mechanism for this reaction.

$$\underset{\text{O}}{\text{CH}_3\overset{\parallel}{\text{C}}-\text{O}^-} + \text{H}_2{}^{18}\text{O} \xrightarrow{{}^{18}\text{OH}^-} \underset{{}^{18}\text{O}}{\text{CH}_3\overset{\parallel}{\text{C}}-{}^{18}\text{O}^-} + \text{H}_2\text{O}$$

 BioLink

19.43 The *tert*-butoxycarbonyl (BOC) group is used to protect the amino group of an amino acid during protein synthesis in the laboratory. The BOC group is attached to the amine by this reaction:

$$t\text{-BuOCOCOt-Bu} + \text{NH}_2-\text{R} \longrightarrow t\text{-BuOCNH}-\text{R} + \left[\text{HOCOt-Bu} \right]$$

Show the steps in the mechanism for this reaction and explain why the following product is not formed:

$$t\text{-BuOCOCNHR} \qquad \text{Not formed}$$

19.44 Show the steps in the mechanism for this reaction:

$$\xrightarrow{200°C} \quad + \quad CH_3OH$$

19.45 Esters are occasionally prepared by an interchange reaction as shown in this equation:

$$\xrightleftharpoons{\text{TsOH}} \quad + \quad CH_3OH$$

a) Show the steps in the mechanism for this reaction.
b) What is the approximate equilibrium constant for this reaction?
c) How could the equilibrium be driven toward the products?

19.46 Amides are occasionally prepared by the reaction of an amine with an ester. Show how this amide could be prepared by this procedure. Is it necessary to drive the equilibrium toward the products in this case?

$$CH_3CH_2CH_2\overset{\displaystyle O}{\overset{\|}{C}}{-}N$$

19.47 Penicillins have a four-membered lactam ring. These antibiotics apparently work by reacting with an NH_2 nucleophile at the active site of an enzyme that is important in the bacterium, as shown in the following equation. What makes the equilibrium favorable for this reaction?

 BioLink

19.48 Grignard reagents react with carbonate esters in a similar manner to their reactions with other esters. Predict the products of this reaction:

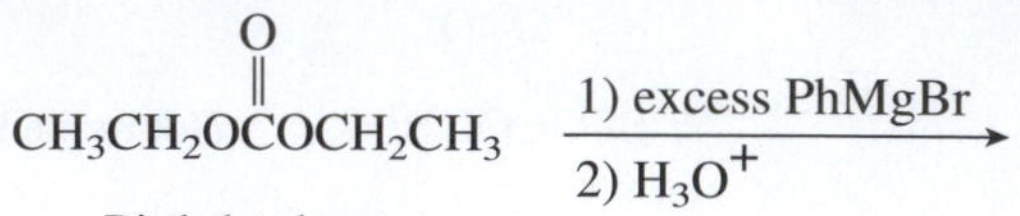

$$CH_3CH_2O\overset{\displaystyle O}{\overset{\|}{C}}OCH_2CH_3 \quad \xrightarrow[\text{2) } H_3O^+]{\text{1) excess PhMgBr}}$$

Diethyl carbonate
(a carbonate ester)

19.49 Show the products of these reactions:

a)

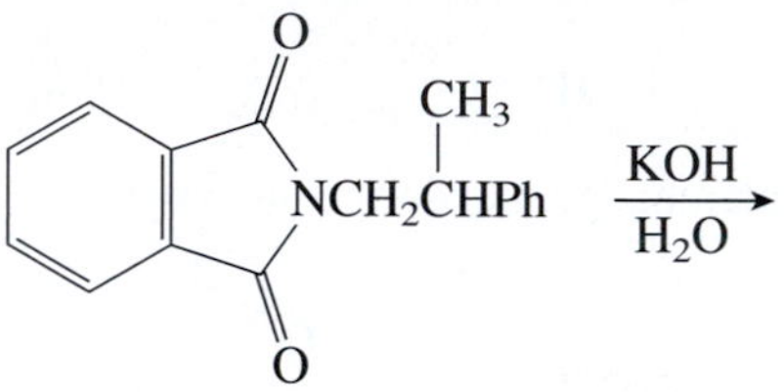

$$\text{(phthalimide)} \quad N CH_2\overset{\displaystyle CH_3}{\underset{}{|}}CHPh \quad \xrightarrow[H_2O]{KOH}$$

b)

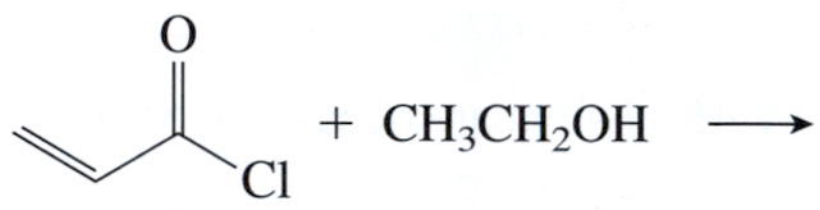

$$\overset{\displaystyle O}{\underset{Cl}{\|}} + CH_3CH_2OH \longrightarrow$$

c) 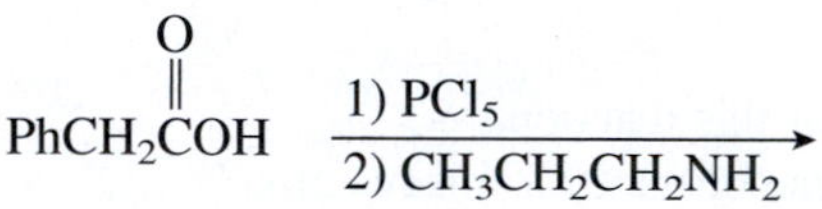 $PhCH_2\overset{\displaystyle O}{\overset{\|}{C}}OH \quad \xrightarrow[\text{2) } CH_3CH_2CH_2NH_2]{\text{1) } PCl_5}$

d)

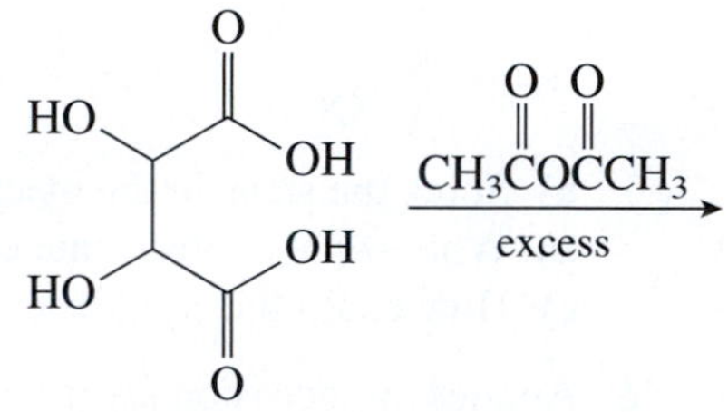

$$\xrightarrow[\text{excess}]{CH_3\overset{O}{\overset{\|}{C}}O\overset{O}{\overset{\|}{C}}CH_3}$$

e) 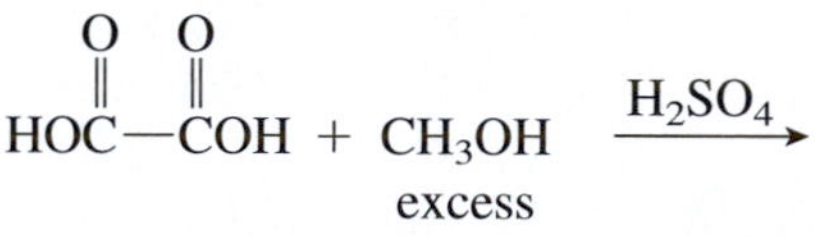 $HO\overset{\displaystyle O}{\overset{\|}{C}}-\overset{\displaystyle O}{\overset{\|}{C}}OH + CH_3OH \quad \xrightarrow{H_2SO_4}$

excess

f)

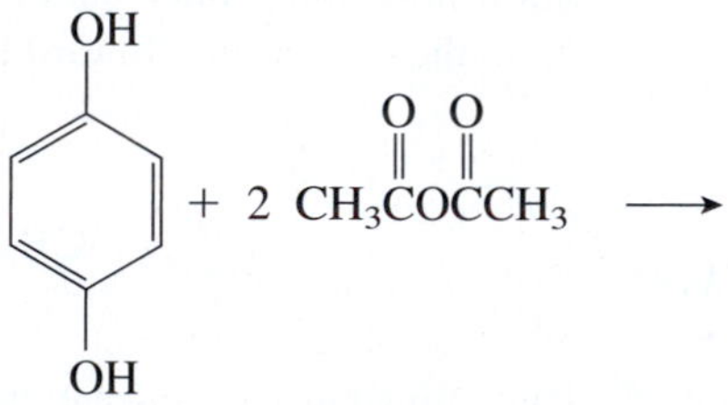

$$+ 2\ CH_3\overset{O}{\overset{\|}{C}}O\overset{O}{\overset{\|}{C}}CH_3 \longrightarrow$$

g)

$$\xrightarrow[H_2O]{HCl}$$

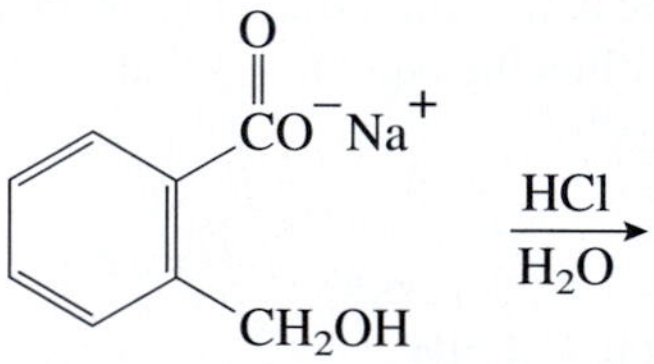

19.50 As indicated by the position of ^{18}O in the product, the acid-catalyzed hydrolysis of *t*-butyl acetate in water enriched in ^{18}O does not follow the mechanism for the reverse of Fischer esterification, shown in Figure 19.3. Suggest a mechanism that explains the position of the ^{18}O in the product and explain why this mechanism is favored in this case.

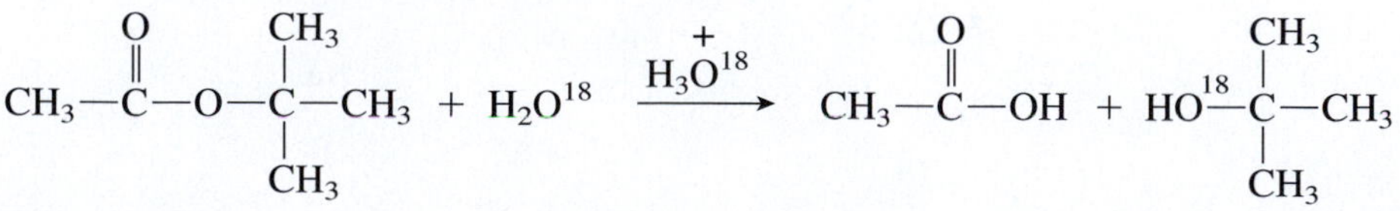

$$CH_3-\overset{\displaystyle O}{\overset{\|}{C}}-O-\overset{\displaystyle CH_3}{\underset{\displaystyle CH_3}{\overset{|}{\underset{|}{C}}}}-CH_3 + H_2O^{18} \xrightarrow{H_3O^{18}} CH_3-\overset{\displaystyle O}{\overset{\|}{C}}-OH + HO^{18}-\overset{\displaystyle CH_3}{\underset{\displaystyle CH_3}{\overset{|}{\underset{|}{C}}}}-CH_3$$

19.51 Explain why one of these compounds forms a lactone when heated and the other does not. Show the structure of the lactone.

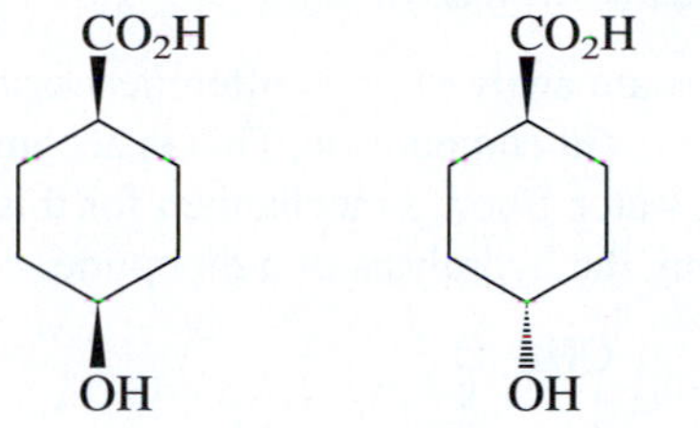

19.52 Even though nitrogen is usually more basic than oxygen, when an amide is protonated the proton becomes bonded to the oxygen. Explain this observation.

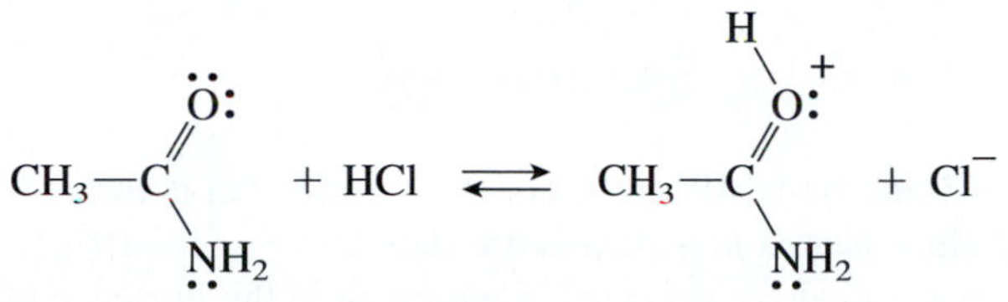

$$CH_3-C \underset{\ddot{N}H_2}{\overset{\ddot{O}:}{|}} + HCl \rightleftharpoons CH_3-C \underset{\ddot{N}H_2}{\overset{\overset{H}{\underset{|}{O}}:^+}{|}} + Cl^-$$

19.53 Suggest a synthesis of sulfathiazole, a sulfa drug, from the indicated starting material:

 BioLink

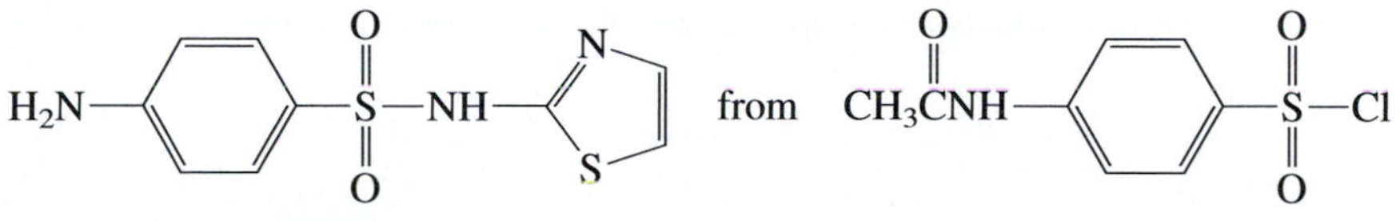

Sulfathiazole

19.54 The hydrolysis shown in the following equation occurs readily in cold water. Explain why this reaction is so much faster than the hydrolysis of an amide, which requires heat and the presence of acid or base.

$$CH_3\overset{O}{\overset{||}{C}}-\overset{+}{N}(CH_3)_3 \; \overset{Cl^-}{} + H_2O \longrightarrow CH_3\overset{O}{\overset{||}{C}}-OH + \overset{+}{H}N(CH_3)_3 \; \overset{Cl^-}{}$$

19.55 Sodium 4-hydroxybutanoate, also known as gamma hydroxybutyrate (GHB) and liquid ecstacy, is an illegal drug that produces a euphoric effect similar to drunkenness. Overdoses can cause vomiting, muscle spasms, and unconsciousness. Because of its knockout capability, it is associated with sexual assaults and date rape. GHB can be prepared from gamma butyrolactone (GBL) by reaction with sodium hydroxide as shown in the following equation:

 BioLink

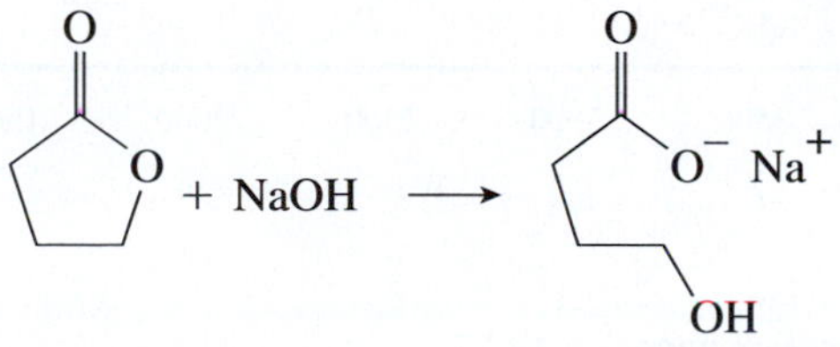

GBL   GHB

Show a mechanism for the conversion of GBL to GHB. Explain why the equilibrium favors GHB in this reaction. Explain why GHB is converted back to GBL when treated with acid.

BioLink

19.56 When proteins are analyzed, it is often necessary to cleave them into their individual amino acid components. This is accomplished by refluxing the protein in acidic water. Show a mechanism for this process using the following model reaction, the hydrolysis of a dipeptide:

$$\underset{\text{H}_2\text{NCH}_2\overset{\overset{\displaystyle\text{O}}{\|}}{\text{C}}-\text{NHCH}-\overset{\overset{\displaystyle\text{O}}{\|}}{\text{C}}\text{OH}}{\overset{\text{CH}_3}{|}} \quad \xrightarrow[\text{H}_2\text{O}]{\text{HCl}} \quad \text{H}_3\overset{+}{\text{N}}\text{CH}_2\overset{\overset{\displaystyle\text{O}}{\|}}{\text{C}}\text{OH} \; + \; \text{H}_3\overset{+}{\text{N}}\overset{\text{CH}_3}{\underset{|}{\text{CH}}}-\overset{\overset{\displaystyle\text{O}}{\|}}{\text{C}}\text{OH}$$

Problems Involving Spectroscopy

19.57 The IR and ^{1}H-NMR spectra of the product that is formed when succinic anhydride is heated in methanol are shown in Figures 19.12a and 19.12b. When succinic anhydride is heated in methanol in the presence of a catalytic amount of sulfuric acid, the product with the IR spectrum shown in Figure 19.12c is formed instead. This compound shows only three absorptions in its ^{13}C-NMR spectrum. Show the structures of both of these products and explain why different products are formed under the different conditions.

Succinic anhydride

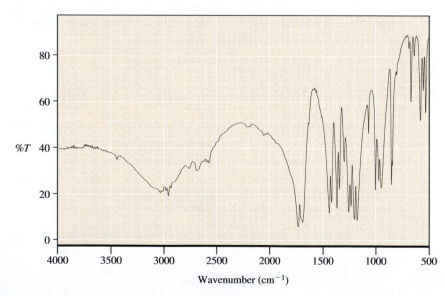

Figure 19.12a

SPECTRA TO ACCOMPANY PROBLEM 19.57.

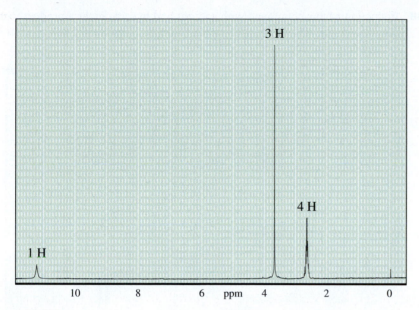

Figure 19.12b

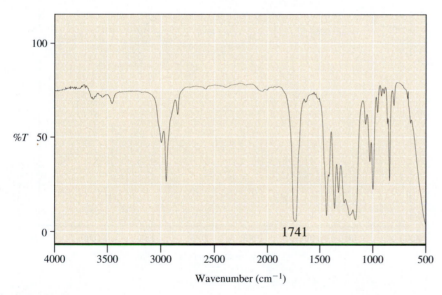

Figure 19.12c

19.58 Reduction of this compound with sodium borohydride gives a product with the following ¹H-NMR and IR spectra. Show the structure of this product and explain its formation.

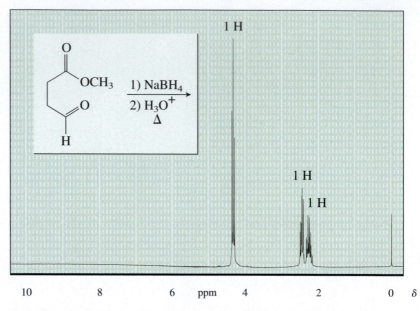

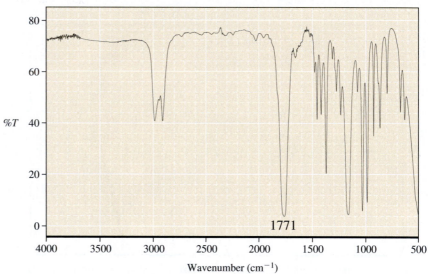

Problems Using Online Three-Dimensional Molecular Models

19.59 Arrange these carboxylic acids in order of increasing rate of reaction with ethanol and acid to form their ethyl esters and explain your reasoning.

19.60 This methyl ester is very resistant to saponification under conditions where most esters react rapidly. Explain this observation.

19.61 These four stereoisomers contain both a carboxylic acid group and a hydroxy group. Explain which isomers can form a lactone and which cannot.

19.62 Fatty acids (the carboxylic acids produced by saponification of fats) usually have an even number of carbons and may also have one or more cis double bonds. The fatty acid that has 18 carbons with a cis double bond at the 9 position, (Z)-9-octadecenoic acid, is called oleic acid. Catalytic hydrogenation of oleic acid produces stearic acid, or octadecanoic acid, another fatty acid. One of these fatty acids melts at 13.4°C and the other melts at 69.6°C. Explain which fatty acid melts at the higher temperature.

19.63 When treated with HCl, one of these isomeric carboxylate anions forms a compound with a pK_a of about 5. The other carboxylate anion forms a compound that has no proton with a pK_a of less than 20. Explain these results.

19.64 One of these stereoisomeric hydroxy acids readily forms a lactone upon heating, whereas the other does not. Explain these results.

19.65 The *trans*-stereoisomer of the ester shown undergoes saponification about 20 times faster than the *cis*-stereoisomer. Explain this observation.

Enolate and Other Carbon Nucleophiles

R EACTIONS THAT form carbon–carbon bonds are very important in organic chemistry because they enable larger, more complex organic molecules to be assembled from smaller, simpler ones. This process requires the reaction of a carbon nucleophile with a carbon electrophile. We have encountered a variety of carbon electrophiles, including alkyl halides in S_N1 and S_N2 reactions and the carbonyl carbon of aldehydes, ketones, and carboxylic acid derivatives in addition and substitution reactions. However, the carbon–carbon bond-forming reactions that we have seen have been limited because only a few carbon nucleophiles have been introduced so far. Cyanide and acetylide nucleophiles for the S_N2 reaction were presented in Chapter 10, and organometallic and ylide nucleophiles for reactions with compounds that contain the carbonyl group were discussed in Chapters 18 and 19.

Grignard reagents and related nucleophiles such as organolithium compounds are not very useful as nucleophiles in S_N2 reactions because they are too reactive. However, if there is a group bonded to the nucleophilic carbon that can help stabilize the pair of electrons, the resulting reagent can be very useful in reactions with alkyl halide electrophiles. One group that is particularly good at stabilizing a pair of electrons on an adjacent carbon is the carbonyl group. A carbanion adjacent to a carbonyl group is called an **enolate ion.** Enolate anions are perhaps the most important type of carbon nucleophile. These nucleophiles also give very useful reactions with carbonyl electrophiles.

MASTERING ORGANIC CHEMISTRY

▶ Predicting the Products of the Reactions of Enolate and Other Carbon Nucleophiles

▶ Understanding the Mechanisms of These Reactions

▶ Using These Reactions to Synthesize Compounds

$$CH_3-\overset{\overset{\ddot{O}}{\|}}{C}-\overset{\cdot\cdot}{C}H_2 \longleftrightarrow CH_3-\overset{\overset{\ddot{O}:^-}{|}}{C}=CH_2$$

Enolate anion derived from acetone

First, this chapter discusses the structure, stability, and methods for generation of enols and enolate anions. The reactions of these species as nucleophiles with halo-

gens as electrophiles are discussed next. Then the use of enolate anions as nucleophiles in S_N2 reactions is presented. Their reactions with aldehyde, ketone, and carboxylic acid derivative electrophiles are discussed next. This is followed by the introduction of several other valuable carbon nucleophiles. A discussion of the use of these nucleophiles in conjugate additions precedes the final section, in which the use of these nucleophiles in synthesis is presented.

20.1 ENOLS AND ENOLATE ANIONS

Section 11.6 discussed the acid-catalyzed addition of water to alkynes. The initial product of this reaction, called an enol, has a hydroxy group attached to one of the carbons of a CC double bond. The enol is unstable and rapidly converts to its tautomer, a carbonyl compound. The carbonyl and enol tautomers of acetone are shown in the following equation:

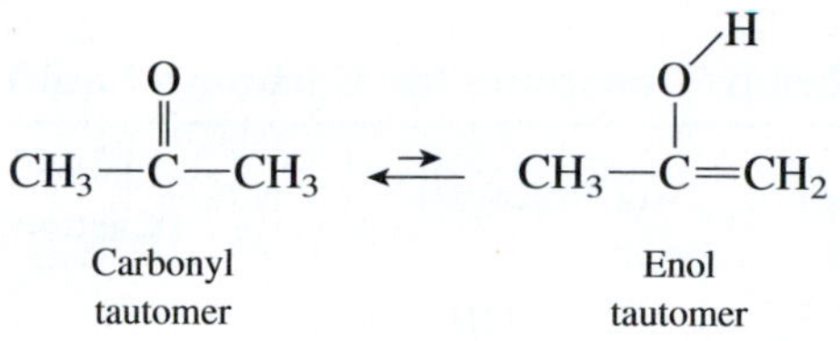

Carbonyl tautomer   Enol tautomer

The interconversion of the carbonyl and enol tautomers is catalyzed by either acid or base and occurs rapidly under most circumstances. The process requires only the addition of a proton to either the carbon or the oxygen atom and the removal of a proton from the other atom. In the acid conditions mechanism the proton is added first, while the base conditions mechanism involves removal of the proton in the first step. These mechanisms are shown in Figures 20.1 and 20.2.

Table 20.1 shows that the amount of enol tautomer that is present at equilibrium in the case of simple aldehydes and ketones is very small. Simple carboxylic acid derivatives, such as esters, have an even smaller enol content. However, 1,3-dicarbonyl compounds

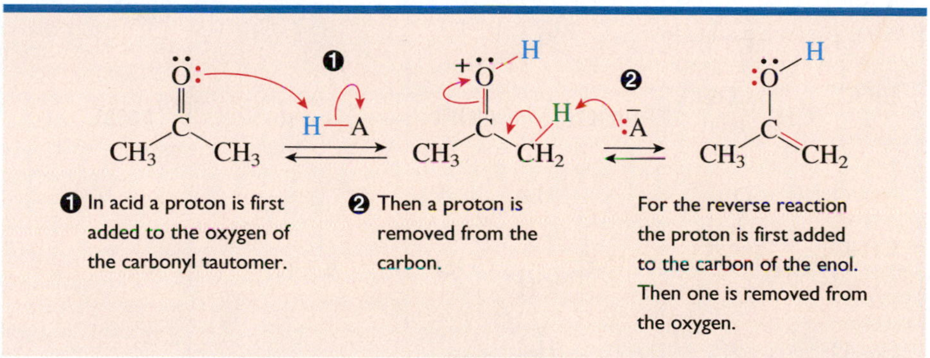

❶ In acid a proton is first added to the oxygen of the carbonyl tautomer.

❷ Then a proton is removed from the carbon.

For the reverse reaction the proton is first added to the carbon of the enol. Then one is removed from the oxygen.

Figure 20.1

MECHANISM OF INTERCONVERSION OF CARBONYL AND ENOL TAUTOMERS UNDER ACID CONDITIONS. This mechanism should be familiar to you. It was shown previously in Figure 11.6.

❶ In base a proton is first removed from the carbon of the carbonyl tautomer.

Resonance stabilization of the enolate anion makes it a weaker base than other carbanions, so it is easier to generate by acid–base reactions.

❷ Addition of a proton to the oxygen of the enolate anion produces the enol tautomer. If, instead, the proton is added to the carbon, the carbonyl tautomer forms in the reverse of the first step.

Figure 20.2

MECHANISM OF INTERCONVERSION OF CARBONYL AND ENOL TAUTOMERS UNDER BASE CONDITIONS.

Table 20.1   Equilibrium Constants for Carbonyl–Enol Tautomerization

Carbonyl Tautomer	Enol Tautomer	$K = \dfrac{[Enol]}{[Carbonyl]}$	Enol Present
O ‖ CH₃CH	OH CH₂=CH	6×10^{-7}	0.00006%
O ‖ CH₃CCH₃	OH CH₂=CCH₃	5×10^{-9}	0.0000005%
(cyclohexanone)	(cyclohexenol, OH)	1×10^{-8}	0.000001%
O O ‖ ‖ EtOC–CH₂–COEt	EtOC=CH–COEt (enol, H···O)	8×10^{-5}	0.008%
O O ‖ ‖ CH₃C–CH₂–COEt	CH₃C=CH–COEt (enol, H···O)	9×10^{-2}	8%
O O ‖ ‖ CH₃C–CH₂–CCH₃	CH₃C=CH–CCH₃ (enol, H···O)	3	76%

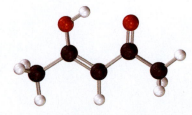

(also called β-dicarbonyl compounds), such as the bottom three entries in Table 20.1, have a significantly larger amount of enol present at equilibrium. In these cases the enol is stabilized by conjugation of the CC double bond with the remaining carbonyl group and by intramolecular hydrogen bonding of the hydrogen of the hydroxy group with the carbonyl oxygen. The progressive increase in the enol content proceeding from the diester (0.008% enol) to the ketoester (8% enol) to the diketone (76% enol) again illustrates that enols involving ester carbonyl groups are less favorable than those involving ketone or aldehyde carbonyl groups.

PROBLEM 20.1

Show all of the enol tautomers of these compounds. If more than one is possible, explain which is more stable.

a) $PhCH_2CCH_3$ b) $CH_3CH_2CCH_2CH_3$ c)

If a simple enol is generated by some reaction, such as the addition of water to an alkyne described in Section 11.6, the enol cannot be isolated because it rapidly converts to the more stable carbonyl tautomer. The lifetime of $CH_2{=}CHOH$, the simplest enol, is about 1 minute in aqueous solution at pH 7, about 1 second in acidic solution, and about 10^{-6} second in basic solution.

A hydrogen on a carbon adjacent to a carbonyl group (the α-carbon) is relatively acidic because the conjugate base is an enolate anion and is stabilized by resonance. A hydrogen on a carbon adjacent to a cyano group is relatively acidic for similar reasons. (It is important to remember that only hydrogens on the α-carbon are acidic because the carbanion is stabilized by resonance only when it is directly attached to the carbonyl or cyano group.) The pK_a's for the α-hydrogens of these compounds are in the range of 20 to 25.

RCH_2CR'	RCH_2COR'	$RCH_2C{\equiv}N$
Ketone	Ester	Nitrile
(or aldehyde)	$pK_a = 25$	$pK_a = 25$
$pK_a = 20$		

Reaction of these compounds with a strong enough base removes a hydrogen to generate an enolate anion. The following bases, whose conjugate acids have pK_a's greater than 30, are all strong enough to completely deprotonate these compounds.

$NaNH_2$	$\overset{\displaystyle Li^+}{CH_3CH{-}\bar{N}{-}CHCH_3}$	NaH
Sodium amide	Lithium diisopropylamide	Sodium hydride
$pK_a = 38$	(LDA)	$pK_a = 35$
	$pK_a = 38$	

A hydrogen that is on a carbon adjacent to two carbonyl groups is even more acidic. The following β-dicarbonyl compounds are acidic enough that they can be completely deprotonated by bases such as ethoxide ion. (The pK_a of its conjugate acid, ethanol, is 16.)

$$
\begin{array}{ccc}
\underset{\substack{\text{2,4-Pentanedione}\\ \text{p}K_a = 9}}{\overset{\overset{O}{\parallel}\qquad\overset{O}{\parallel}}{CH_3\;\;C\;\;CH_2\;\;C\;\;CH_3}}
&
\underset{\substack{\text{Ethyl acetoacetate}\\ \text{p}K_a = 11}}{\overset{\overset{O}{\parallel}\qquad\overset{O}{\parallel}}{CH_3\;\;C\;\;CH_2\;\;C\;\;OEt}}
&
\underset{\substack{\text{Diethyl malonate}\\ \text{p}K_a = 13}}{\overset{\overset{O}{\parallel}\qquad\overset{O}{\parallel}}{EtO\;\;C\;\;CH_2\;\;C\;\;OEt}}
\end{array}
$$

PROBLEM 20.2

Show the enolate or related anion formed in these acid–base reactions:

a) PhCCH$_2$CH$_3$ + LDA $\longrightarrow$

b) CH$_3$CH$_2$CCH$_2$COCH$_2$CH$_3$ + NaOEt $\longrightarrow$

c) PhCH$_2$CCH$_2$CH$_3$ + NaNH$_2$ $\longrightarrow$ *resonance stable*

d) CH$_3$CH$_2$CH$_2$CN + LDA $\longrightarrow$ *resonance stable*

20.2 HALOGENATION OF THE α-CARBON

The reaction of an aldehyde or ketone with Cl$_2$, Br$_2$, or I$_2$, under either acidic or basic conditions, results in the replacement of a hydrogen on the α-carbon with a halogen.

$$
\underset{}{\overset{\overset{O}{\parallel}}{R-C-CH_3}} + X-X \xrightarrow[\substack{\text{or}\\ ^-OH}]{HX} \overset{\overset{O}{\parallel}\;\;\overset{X}{|}}{R-C-CH_2} + HX
$$

Under acidic conditions, the enol, generated according to the mechanism shown in Figure 20.1, acts as the nucleophile and attacks the electrophilic halogen.

$$
\underset{}{\overset{\overset{O}{\parallel}}{CH_3CCH_3}} \underset{}{\overset{HBr}{\rightleftharpoons}} \underset{\text{Enol}}{\overset{:\ddot{O}H}{CH_3C=CH_2}} \xrightarrow{Br-Br} \overset{\overset{+}{O}\;\;Br}{CH_3C-CH_2} \xrightarrow{:\ddot{Br}:} \overset{\ddot{O}:\;\;Br}{CH_3C-CH_2} + HBr
$$

The presence of the halogen retards enolization, so it is possible to stop the reaction after the addition of a single halogen. This reaction can provide a useful way to add a halogen to the α-carbon of a ketone, as long as the ketone is symmetrical or has only one reactive site. Some examples follow:

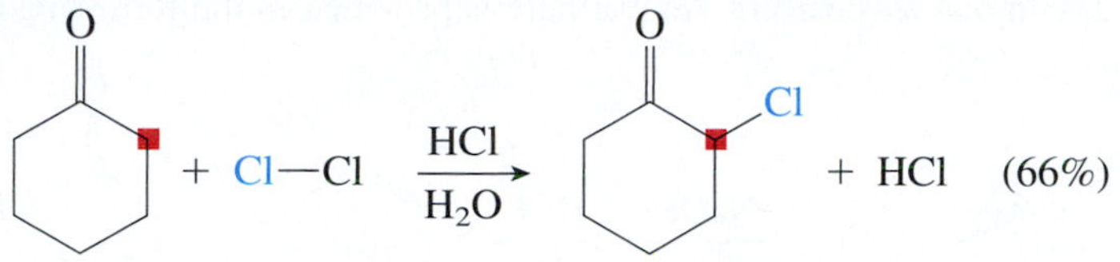

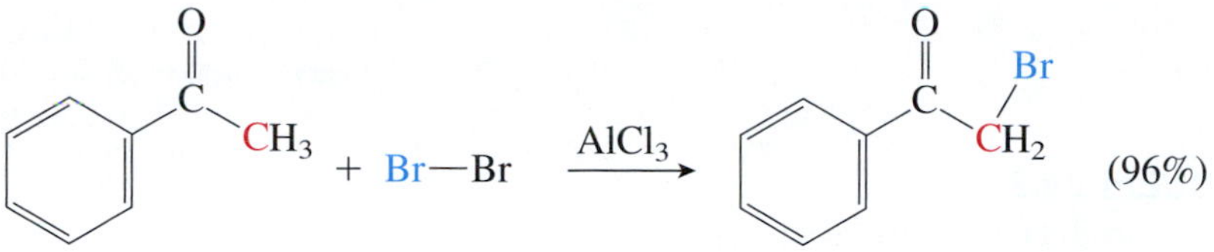

Under basic conditions, it is the enolate anion that acts as the nucleophile.

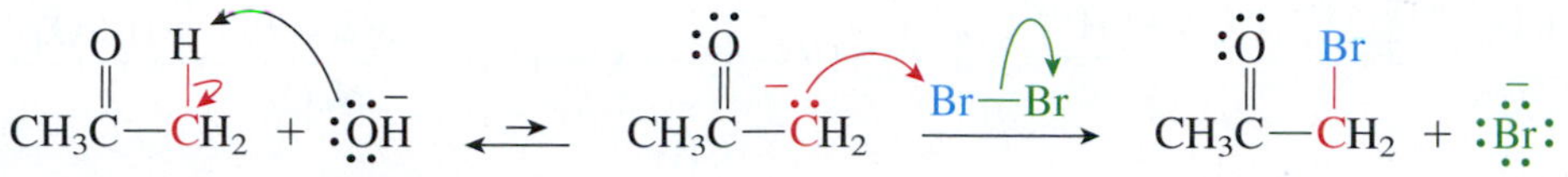

Enolate anion

In the **haloform reaction,** a ketone with a methyl group bonded to the carbonyl carbon is reacted with an excess of halogen in base. The methyl group is removed, and the ketone is converted to a carboxylic acid with one less carbon. This reaction first replaces a hydrogen on the methyl group with a halogen by the base mechanism just described. Because of the inductive electron-withdrawing effect of the halogen, the hydrogens on the α-carbon become more acidic after the first halogen is added, and a second halogen is added more rapidly than the first. This is followed by addition of a third halogen. Then hydroxide ion attacks the carbonyl carbon. The three halogens help the methyl carbon leave by stabilizing the resulting carbanion.

Haloform Reaction
(Iodoform Reaction)

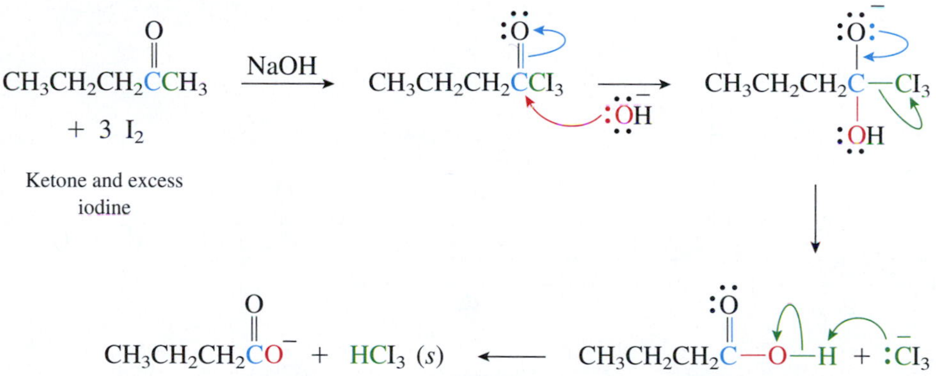

The cleavage reaction does not occur unless there are three halogens on the carbon, so only methyl groups are removed in this manner. The reaction with I_2 (iodoform reaction) has been used as a test for methyl ketones. The formation of iodoform (CHI_3), which precipitates as a yellow solid, provides a positive test for the presence of a methyl ketone. The reaction can also be used in synthesis to convert a methyl ketone to a carboxylic acid with one less carbon. An example is provided in the following equation:

A methyl ketone   A carboxylic acid   (88%)

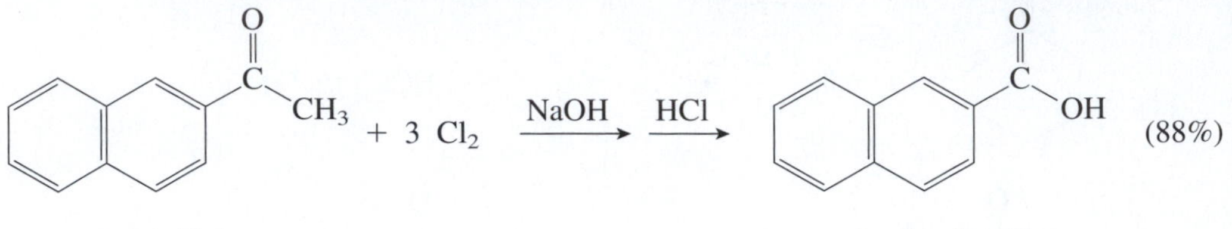

PROBLEM 20.3

Show the products of these reactions:

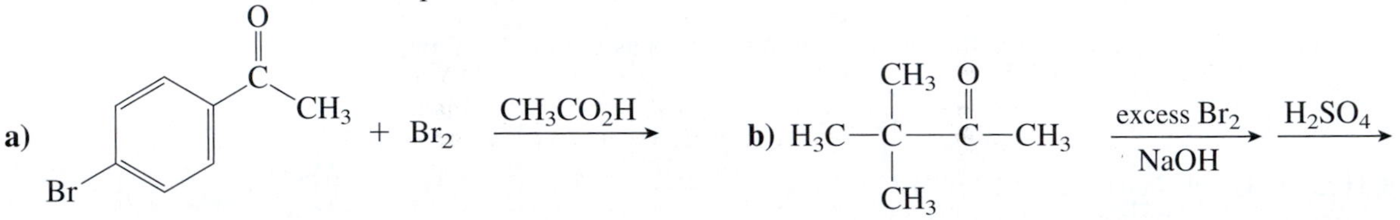

20.3 ALKYLATION OF ENOLATE ANIONS

Enolate anions generated from ketones, esters, and nitriles can be used as nucleophiles in S_N2 reactions. This results in the attachment of an alkyl group to the α-carbon in a process termed *alkylation*. Aldehydes are too reactive and cannot usually be alkylated in this manner. Alkylation of cyclohexanone is illustrated in the following equation:

(62%)

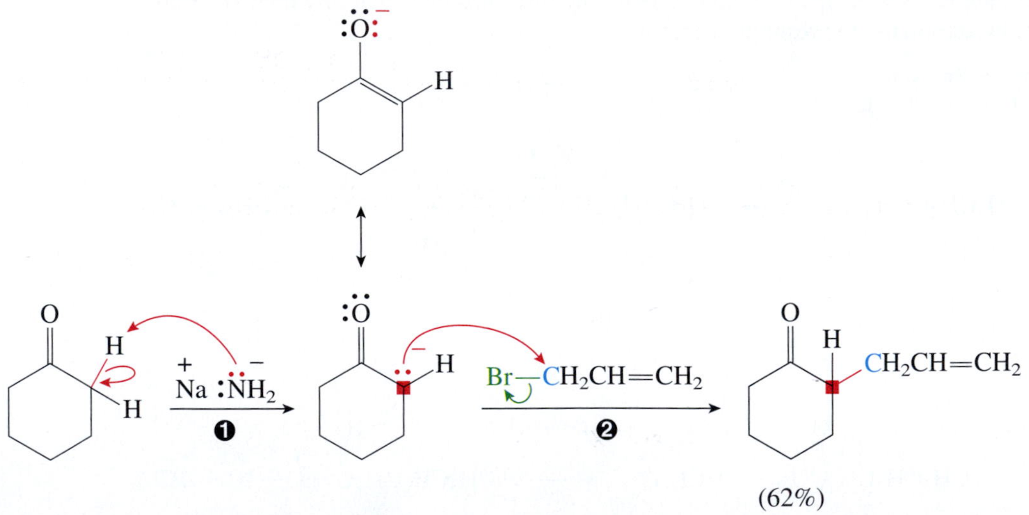

❶ A strong base must be used to ensure complete deprotonation in this step. The solvent must not have any acidic hydrogens. An ether (diethyl ether, DME, THF, dioxane) or DMF is commonly used.

❷ Because this is an S_N2 reaction, it works only when the leaving group is attached to an unhindered carbon (primary or secondary). When the leaving group is attached to a tertiary carbon, E2 elimination occurs rather than substitution.

The base that is used must be strong enough to convert all of the starting ketone (or ester or nitrile) to the enolate anion. If it is not strong enough to do so, unwanted reactions of the enolate nucleophile with the electrophilic carbon of the remaining ketone may occur (see Sections 20.5 and 20.6). Lithium diisopropylamide (LDA) is often the base of choice. It is a very strong base but is not prone to give side reactions in which it acts as a nucleophile because of the steric hindrance provided by the bulky isopropyl groups. The alkyl halide (or tosylate or mesylate ester) is subject to the usual restrictions of the S_N2 mechanism. The leaving group may be bonded to a primary or secondary carbon but not to a tertiary carbon.

PROBLEM 20.4

What reaction would occur if one attempted to use butyllithium to form the enolate anion of cyclohexanone?

To avoid the formation of two products, deprotonation of the ketone must produce a single enolate ion. Therefore, the ketone must be symmetrical, like cyclohexanone in the preceding example, or have a structure that favors the formation of the enolate ion at only one of the α-carbons, as is the case in the following example:

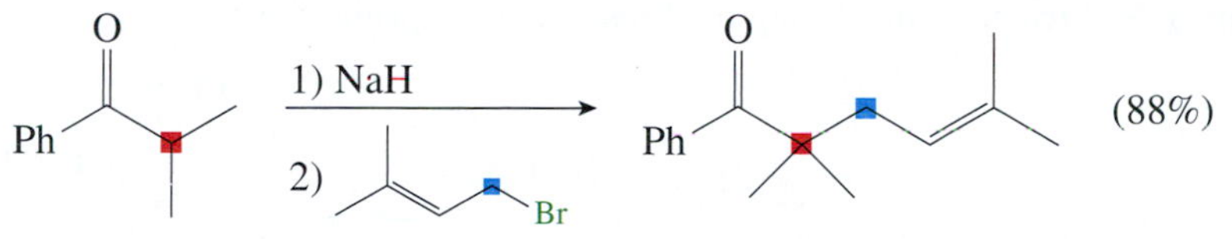

Esters and nitriles can also be alkylated by this procedure:

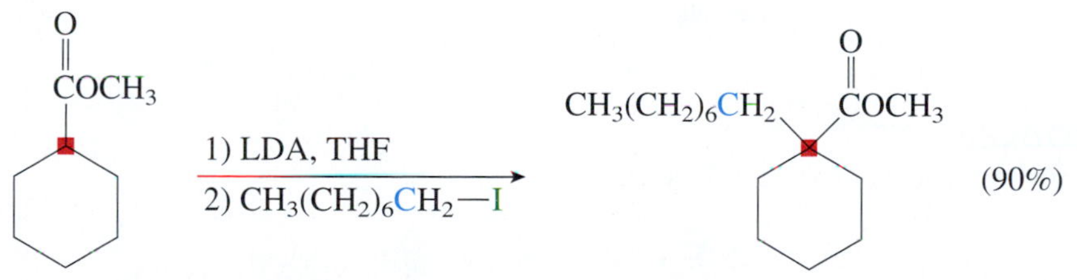

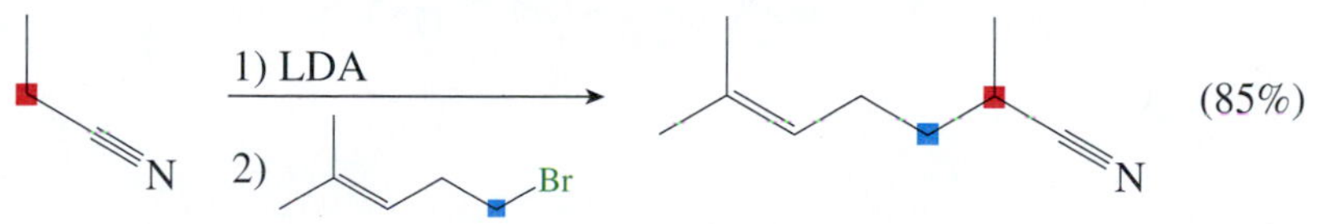

This last example shows how two alkyl groups can be added in sequence:

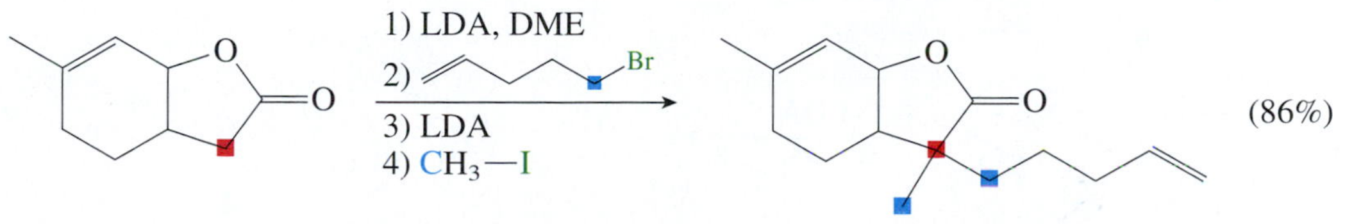

PRACTICE PROBLEM 20.1

Show the product of this reaction:

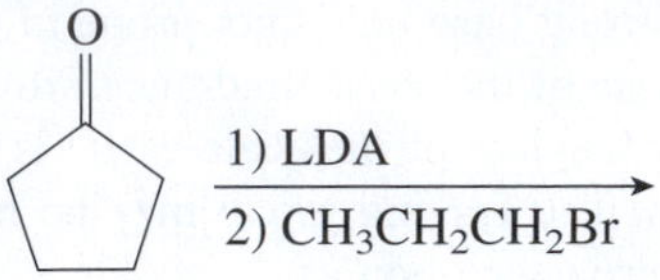

1) LDA
2) $CH_3CH_2CH_2Br$

Strategy

The key to working problems of this type is the same as it has been in previous chapters. First identify the nucleophile and the electrophile. In these particular reactions the nucleophile is generated by removal of the most acidic hydrogen (one on the carbon α to the carbonyl or cyano group) to generate an enolate or related anion. This is the nucleophile in an S_N2 reaction.

Solution

The diisopropylamide anion acts as a base and removes an acidic hydrogen from the carbon adjacent to the carbonyl group. The resulting enolate anion reacts as a nucleophile in an S_N2 reaction, displacing the bromine at the primary carbon.

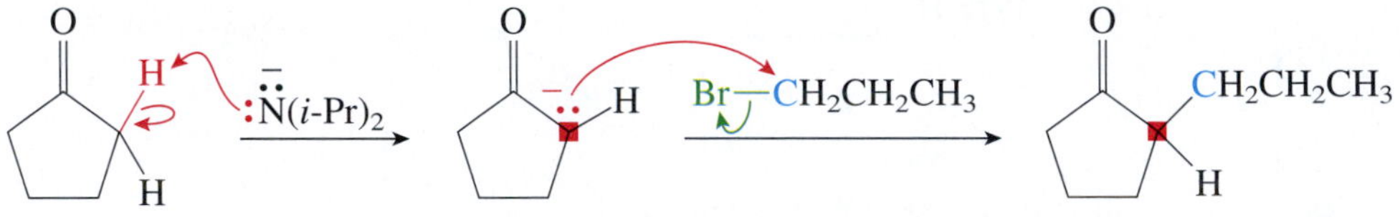

Enolate ion
nucleophile

PROBLEM 20.5
Show the products of these reactions:

a) $PhCH_2CN$
1) $NaNH_2$
2) [image: cyclohexyl bromide] Br
→ [image: PhCHCN-cyclohexyl]

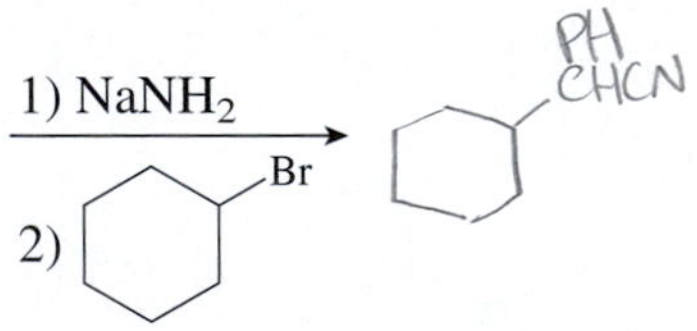

b) [image: ethyl ester structure]
1) LDA
2) [image: bromide chain] Br

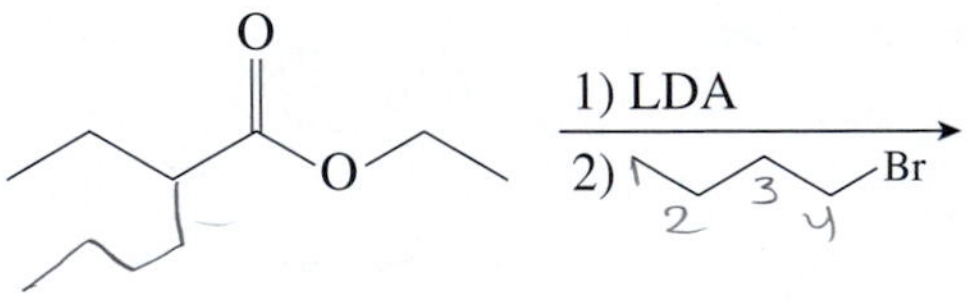

c) $PhCCH_2CH_2CH_3$
1) LDA
2) CH_3CH_2Br

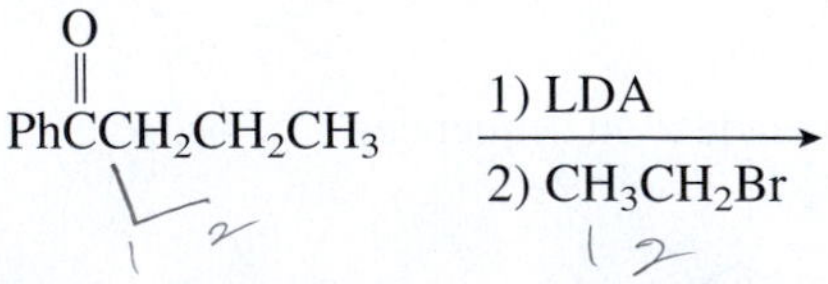

d) $ClCH_2CH_2CH_2CN$
$NaNH_2$

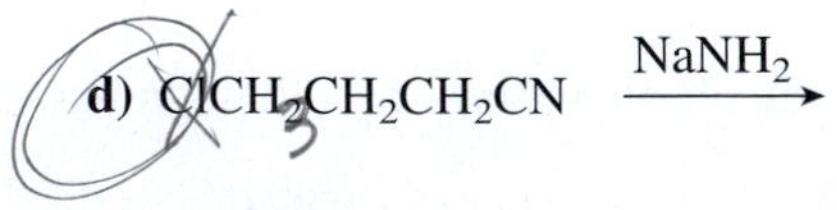

e) [image: ethyl ester structure]
1) LDA
2) [image: allyl bromide] Br

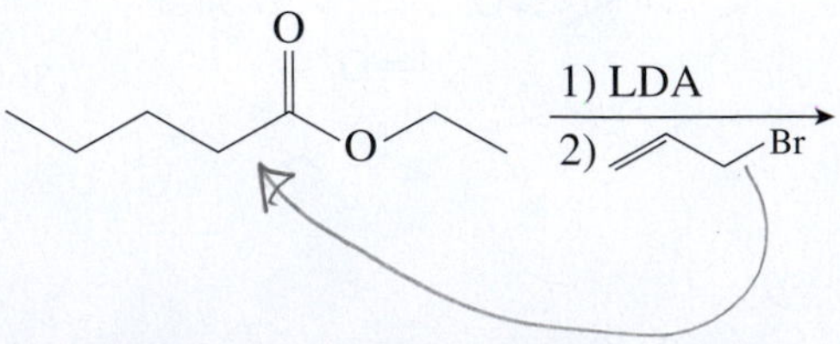

20.4 ALKYLATION OF MORE STABILIZED ANIONS

The nucleophiles described in the preceding section are strong bases and therefore are quite reactive. This high reactivity sometimes causes problems. We have seen before that one way to solve problems caused by a nucleophile that is too reactive is to attach a group to the nucleophilic site that decreases its reactivity. After this new, less reactive reagent, which causes fewer side reactions, is used in the substitution, the extra group is removed. Although it involves more steps, this overall process provides the same product, often in higher yield, as would be obtained by using the original nucleophile.

One example of this strategy is the preparation of alcohols using acetate anion as the synthetic equivalent of hydroxide ion in S_N2 reactions (see Section 10.2 and Figure 10.1).

$$CH_3-\overset{\overset{\displaystyle O}{\|}}{C}-\ddot{\ddot{O}}:^- \quad \text{is the synthetic equivalent of} \quad H-\ddot{\ddot{O}}:^-$$

In a similar manner the conjugate base of phthalimide is used as the synthetic equivalent of amide ion for the preparation of primary amines in the Gabriel synthesis (see Section 10.6 and Figure 10.5).

is the synthetic equivalent of $:\ddot{N}H_2$

In both of these cases a carbonyl group(s) is attached to the nucleophilic atom. Resonance delocalization of the electron pair makes the anion more stable. It is easier to generate, and its reactions are easier to control. After the substitution reaction has been accomplished, the carbonyl group(s) is removed, unmasking the desired substitution product.

Now suppose that we want to use the enolate anion derived from acetone ($pK_a = 20$) as a nucleophile in a substitution reaction. This anion requires the use of a very strong base to generate it, and its high reactivity often causes low yields of the desired product. Instead, we may choose to use its synthetic equivalent, the enolate anion derived from ethyl acetoacetate ($pK_a = 11$):

$$CH_3 \quad \overset{\overset{\displaystyle O}{\|}}{C} \quad \overset{-}{\underset{\displaystyle CH}{\ddot{}}} \quad \overset{\overset{\displaystyle O}{\|}}{C} \quad OCH_2CH_3$$

Enolate anion of
ethyl acetoacetate

is the synthetic equivalent of

$$CH_3 \quad \overset{\overset{\displaystyle O}{\|}}{C} \quad \overset{-}{\underset{\displaystyle CH_2}{\ddot{}}}$$

Enolate anion of
acetone

Alkylation of the enolate anion derived from ethyl acetoacetate followed by removal of the ester group is known as the **acetoacetic ester synthesis** and is an excellent method for the preparation of methyl ketones. The product of an acetoacetic ester synthesis is the same as the product that would be produced by the addition of the same alkyl group to the α-carbon of acetone. First, the enolate ion is generated from the β-ketoester by the use of a moderate base such as sodium ethoxide. Then the enolate ion is alkylated by reaction with an alkyl halide (or alkyl sulfonate ester) in an S_N2 reaction. This step is subject to the usual S_N2 restriction that the leaving group be on a primary or secondary carbon.

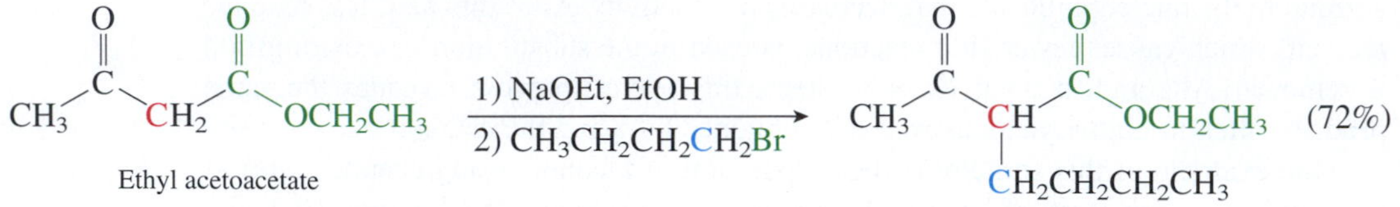

Ethyl acetoacetate

The ester group is removed by treating the alkylated β-ketoester with aqueous base, followed by treatment with acid and heat:

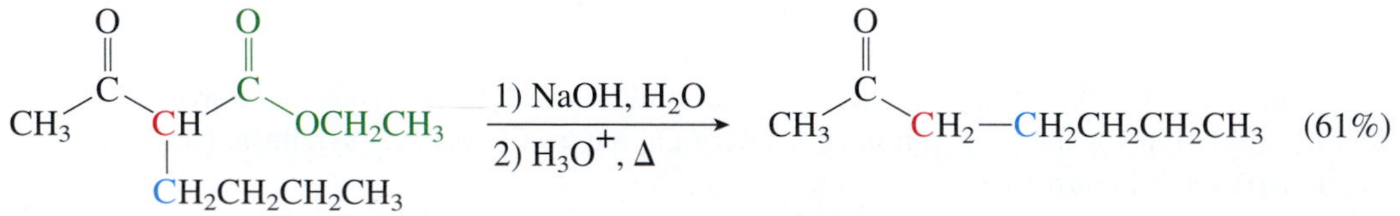

Let's examine the mechanism of the last part of the acetoacetic ester synthesis, which results in the loss of the ester group. Treatment of the β-ketoester with aqueous base results in saponification of the ester to form, after acidification, a β-ketoacid. The mechanism for this step was described in Chapter 19.

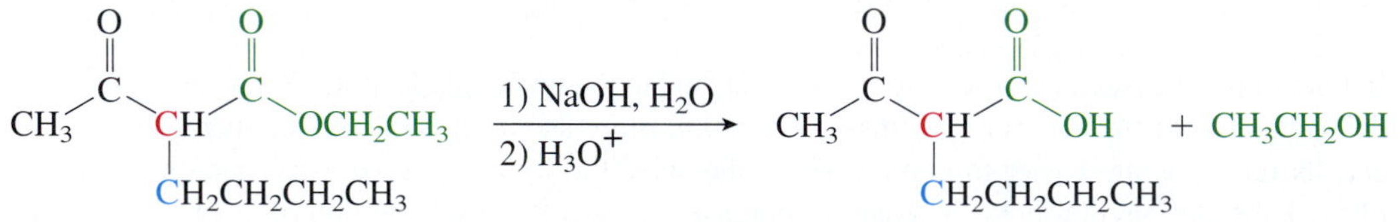

A β-ketoester   A β-ketoacid

When the β-ketoacid is heated, carbon dioxide is lost. This step, a decarboxylation, occurs by a mechanism that is quite different from any other that we have encountered so far. Three bonds are broken and three bonds are formed in a concerted reaction that proceeds through a cyclic, six-membered transition state. The product of this step is an enol, which tautomerizes to the final product, a ketone:

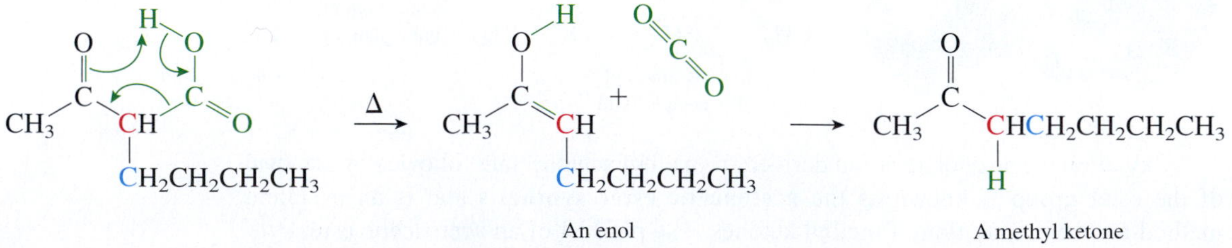

An enol   A methyl ketone

Note that simple carboxylic acids are quite stable and do not lose carbon dioxide when heated. For carbon dioxide to be eliminated, the acid must have a carbonyl group at the β-position so that the cyclic mechanism can occur.

Another example of the acetoacetic ester synthesis is shown in the following equation:

2-Hexanone

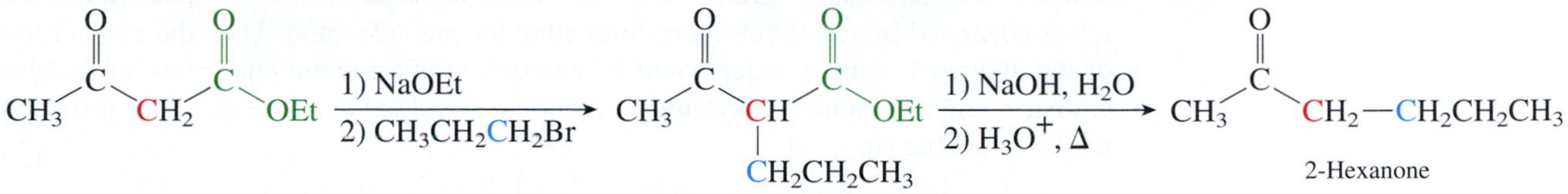

(47%)

Note, again, that the final product, 2-hexanone, is the same product that would result from the reaction of the enolate anion of acetone with 1-bromopropane.

The **malonic ester synthesis** is similar to the acetoacetic ester synthesis. It begins with deprotonation of diethyl malonate ($pK_a = 11$) to produce an enolate anion that is the synthetic equivalent of the enolate anion derived from acetic acid:

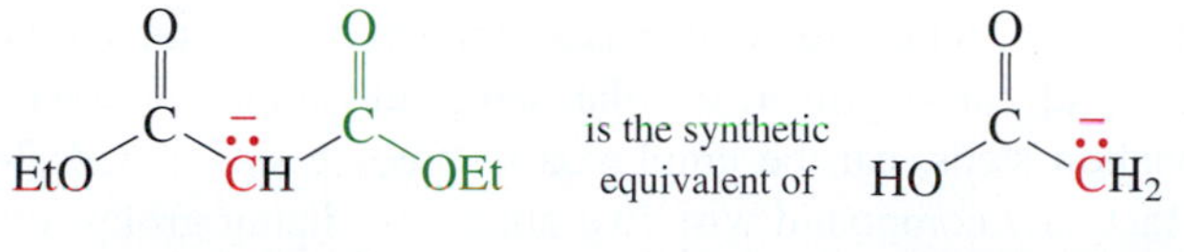

Enolate ion of   Enolate ion of
diethyl malonate   acetic acid

In the malonic ester synthesis this enolate ion is alkylated in the same manner as in the acetoacetic ester synthesis. Saponification of the alkylated diester produces a diacid. The carbonyl group of either of the acid groups is at the β-position relative to the other acid group. Therefore, when the diacid is heated, carbon dioxide is lost in the same manner as in the acetoacetic ester synthesis. The difference is that the product is a carboxylic acid in the malonic ester synthesis rather than the methyl ketone that is produced in the acetoacetic ester synthesis. The loss of carbon dioxide from a substituted malonic acid to produce a monoacid is illustrated in the following equation:

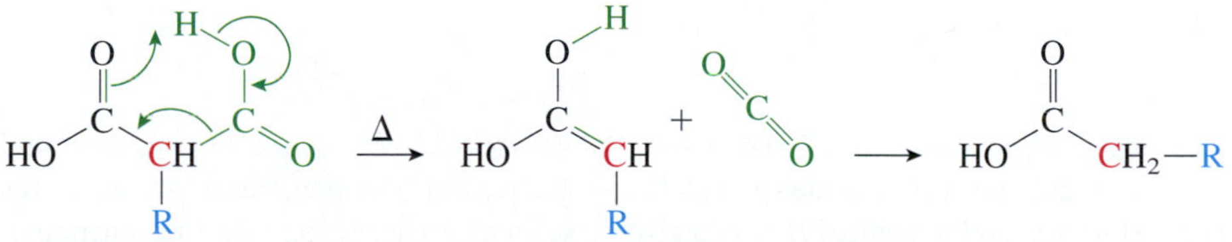

As was the case in the decarboxylation that occurs in the acetoacetic ester synthesis, it is the presence of a carbonyl group at the β-position of the carboxylic acid that allows carbon dioxide to be lost when the compound is heated.

Examples of the malonic ester synthesis are provided in the following equations:

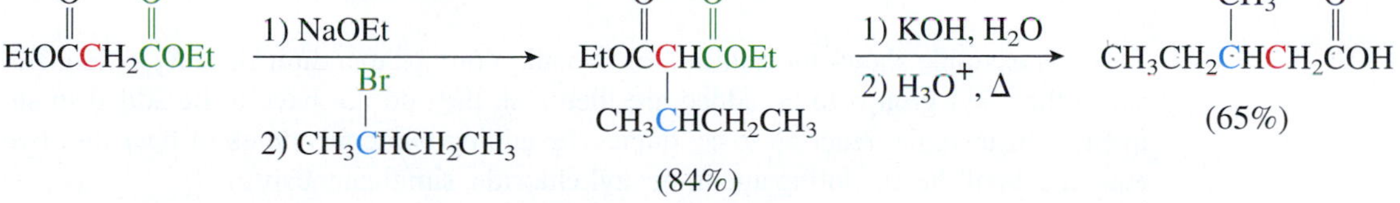

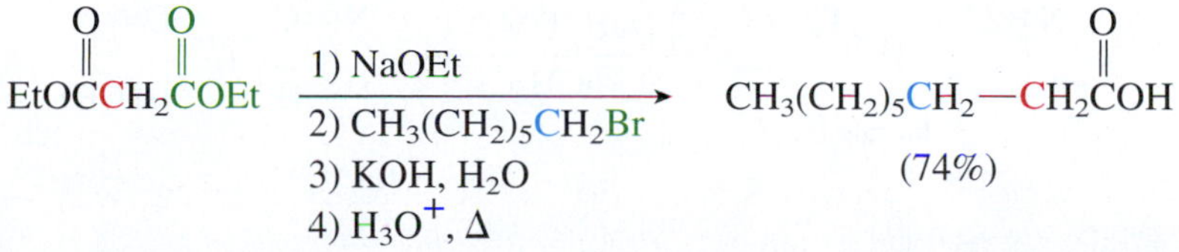

(84%)   (65%)

(74%)

In both the acetoacetic ester synthesis and the malonic ester synthesis, it is possible to add two different alkyl groups to the α-carbon in sequential steps. First the enolate ion is generated by reaction with sodium ethoxide and alkylated. Then the enolate ion of the alkylated product is generated by reaction with a second equivalent of sodium ethoxide, and that anion is alkylated with another alkyl halide. An example is provided by the following equation:

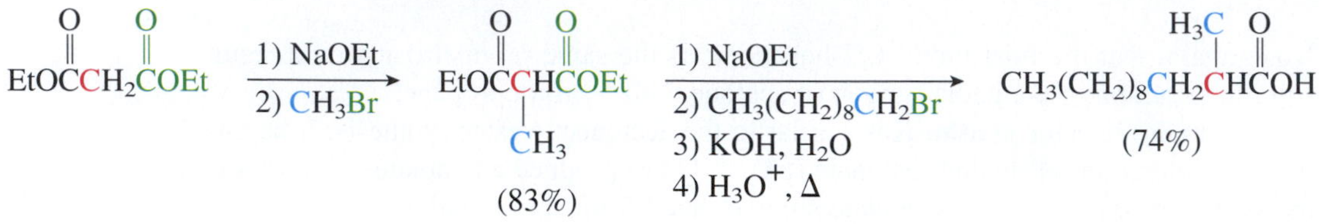

Although the acetoacetic ester synthesis and the malonic ester synthesis are used to prepare ketones and carboxylic acids, the same alkylation, without the hydrolysis and decarboxylation steps, can be employed to prepare substituted β-ketoesters and β-diesters. In fact, any compound with two anion stabilizing groups on the same carbon can be deprotonated and then alkylated by the same general procedure. Several examples are shown in the following equations. The first example shows the alkylation of a β-ketoester. Close examination shows the similarity of the starting material to ethyl acetoacetate. Although sodium hydride is used as a base in this example, sodium ethoxide could also be employed.

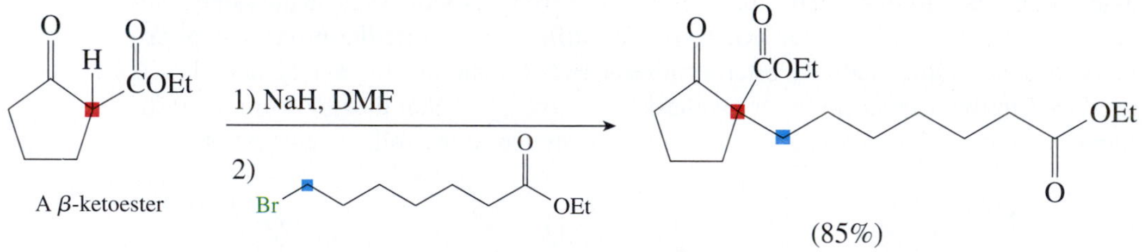

This next example shows the alkylation of a β-diketone ($pK_a = 9$). Because this compound is more acidic than a β-ketoester or a β-diester, the weaker base potassium carbonate was used. However, sodium ethoxide would also be satisfactory as the base for this reaction:

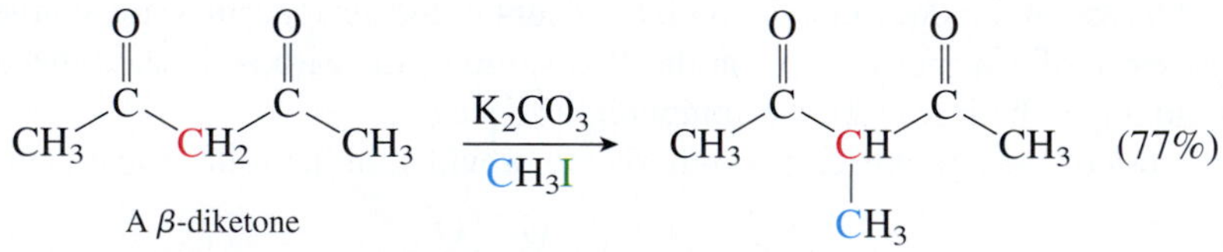

This last example shows the addition of two alkyl groups to a dinitrile ($pK_a = 11$). Because the alkyl groups to be added are identical, they do not have to be added in sequence. Instead, the reaction is conducted by adding two equivalents of base and two equivalents of the alkylating agent, benzyl chloride, simultaneously:

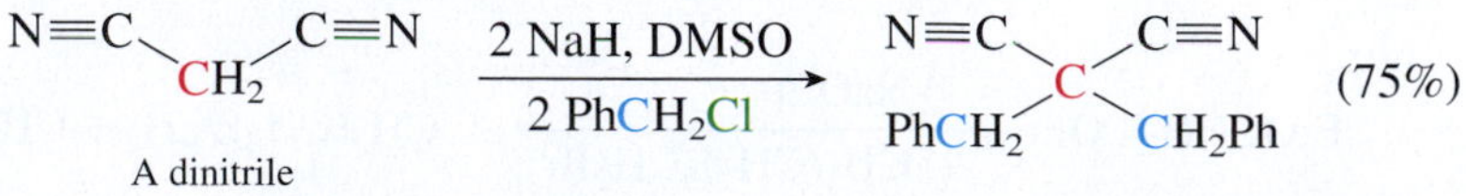

PROBLEM 20.6

Show the products of these reactions:

a) CH₃CCH₂COEt   1) NaOEt, EtOH
                      2) PhCH₂Br
                      1) NaOH, H₂O
                      2) H₃O⁺, Δ

b) EtOCCH₂COEt   1) NaOEt, EtOH
                      2) CH₃CH₂CH₂Br
                      1) NaOH, H₂O
                      2) H₃O⁺, Δ

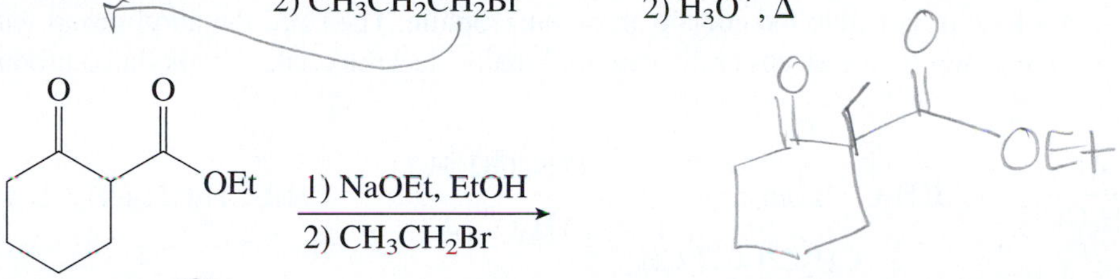

c)   ...OEt   1) NaOEt, EtOH
                  2) CH₃CH₂Br

d) CH₃CCH₂COEt   1) NaOEt, EtOH
                      2) CH₃CH₂CH₂CH₂Br
                      3) NaOEt, EtOH
                      4) CH₃I
                      1) NaOH, H₂O
                      2) H₃O⁺, Δ

e) N≡CCH₂CCH₃   1) NaOEt, EtOH
                     2) CH₃CH₂CH₂Cl

PRACTICE PROBLEM 20.2

Show a synthesis of 2-heptanone using the acetoacetic or malonic ester synthesis:

$$CH_3CH_2CH_2CH_2CH_2CCH_3$$

2-Heptanone

Strategy

Decide which synthesis to use. The acetoacetic ester synthesis is used to prepare methyl ketones, and the malonic ester synthesis is used to prepare carboxylic acids. Both syntheses provide a method to add alkyl groups to the α-carbon. Therefore, next identify the group or groups that must be added to the α-carbon. Remember that the α-carbon is the nucleophile, so the groups to be attached must be the electrophile in the S_N2 reaction; they must have a leaving group bonded to the carbon to which the new bond is to be formed.

Solution

The acetoacetic ester synthesis is used to prepare methyl ketones such as this. In this example, a butyl group must be attached to the enolate nucleophile.

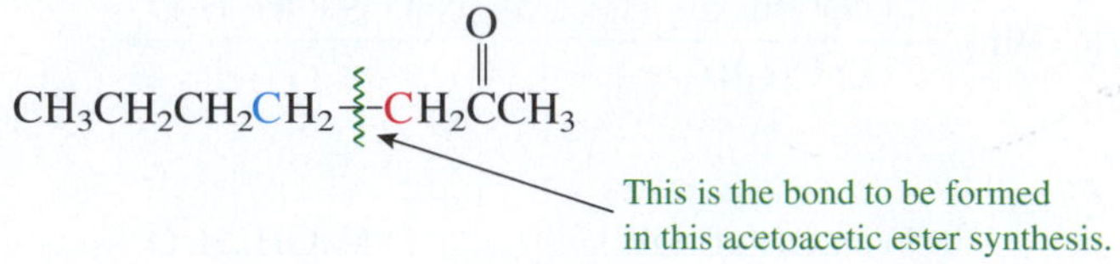

$$CH_3CH_2CH_2CH_2 \overset{\wr}{-} CH_2\overset{O}{\overset{||}{C}}CH_3$$

This is the bond to be formed in this acetoacetic ester synthesis.

Use a base to generate the enolate anion nucleophile. Then add the alkyl group with a leaving group on the electrophilic carbon. Finally, decarboxylate the alkylated product.

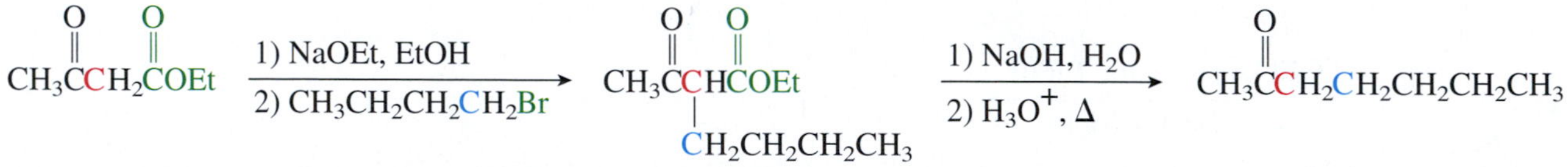

$$CH_3\overset{O}{\overset{||}{C}}CH_2\overset{O}{\overset{||}{C}}OEt \xrightarrow[\text{2) } CH_3CH_2CH_2CH_2Br]{\text{1) NaOEt, EtOH}} CH_3\overset{O}{\overset{||}{C}}\underset{CH_2CH_2CH_2CH_3}{\overset{O}{\overset{||}{C}}HC}OEt \xrightarrow[\text{2) } H_3O^+, \Delta]{\text{1) NaOH, } H_2O} CH_3\overset{O}{\overset{||}{C}}CH_2CH_2CH_2CH_2CH_3$$

PROBLEM 20.7

Show syntheses of these compounds using the acetoacetic or malonic ester syntheses:

a) 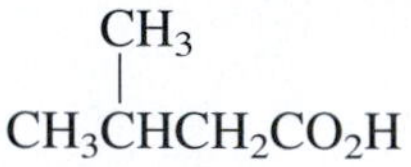 $CH_3\underset{\overset{|}{CH_3}}{CH}CH_2CO_2H$

b)

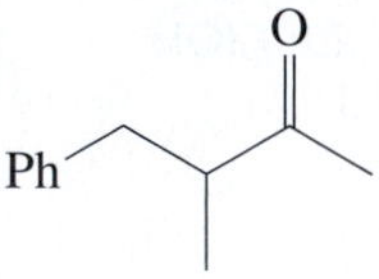

c)

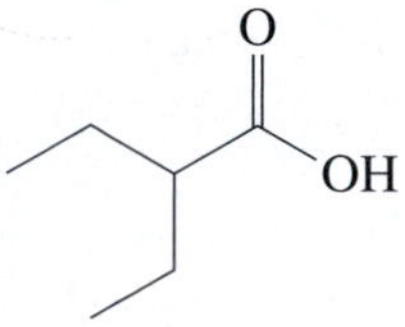

d)

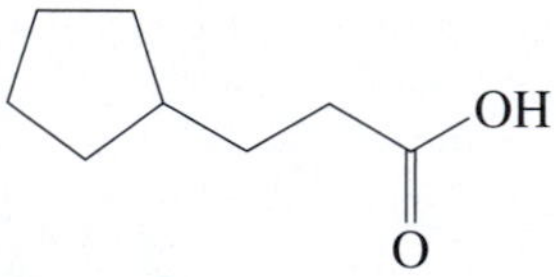

e)

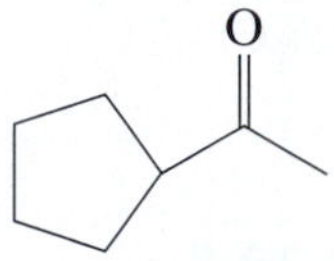

PROBLEM 20.8

Show how these compounds could be synthesized using alkylation reactions:

a)

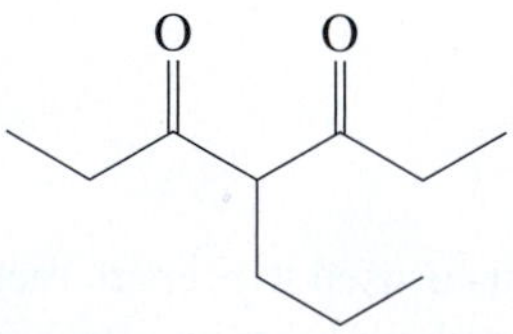

b)

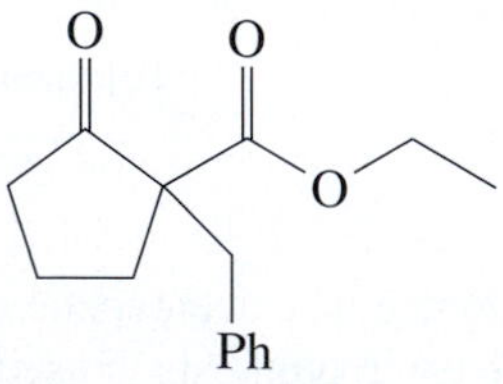

c)

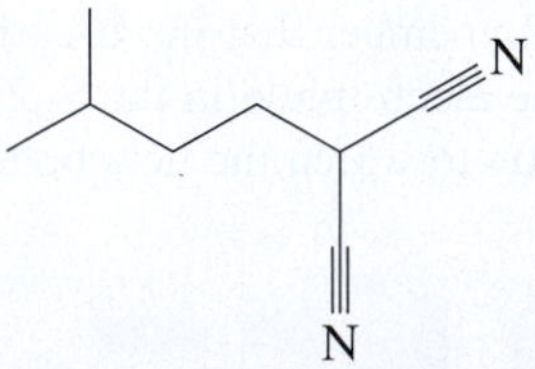

d)

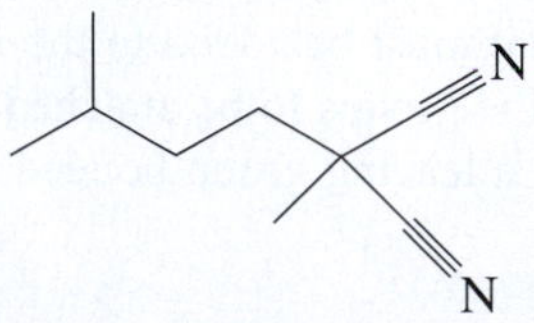

20.5 THE ALDOL CONDENSATION

You may have noticed that aldehydes were conspicuously absent from the examples of alkylation reactions presented in Sections 20.3 and 20.4. This is due to the high reactivity of the carbonyl carbon of an aldehyde as an electrophile. When an enolate anion nucleophile is generated from an aldehyde, under most circumstances it rapidly reacts with the electrophilic carbonyl carbon of an un-ionized aldehyde molecule. Although this reaction, known as the **aldol condensation,** interferes with the alkylation of aldehydes, it is a very useful synthetic reaction in its own right. The aldol condensation of ethanal is shown in the following equation:

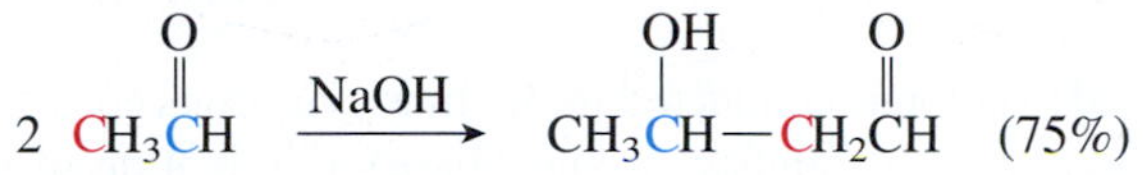

$$2 \ CH_3CH \ \overset{\text{O}}{\underset{}{\parallel}} \quad \xrightarrow{\text{NaOH}} \quad CH_3\underset{\text{OH}}{\overset{}{\underset{\mid}{C}}}H - CH_2\overset{\text{O}}{\underset{}{\overset{\parallel}{C}}}H \quad (75\%)$$

Ethanal

Aldol
(3-hydroxybutanal)

The product, 3-hydroxybutanal, is also known as aldol and gives rise to the name for the whole class of reactions.

The mechanism for this reaction is shown in Figure 20.3. For this mechanism to occur, both the enolate anion derived from the aldehyde and the un-ionized aldehyde must be present. To ensure that this is the case, hydroxide ion is most commonly used as the base. Because hydroxide ion is a weaker base than the aldehyde enolate anion, only a small amount of the enolate anion is produced. Most of the aldehyde remains

❶ The base, hydroxide ion, removes an acidic hydrogen from the α-carbon of the aldehyde. The conjugate base of the aldehyde is a stronger base than hydroxide, so the equilibrium for this first step favors the reactants.

However, enough enolate ion nucleophile is present to react with the electrophilic carbonyl carbon of a second aldehyde molecule.

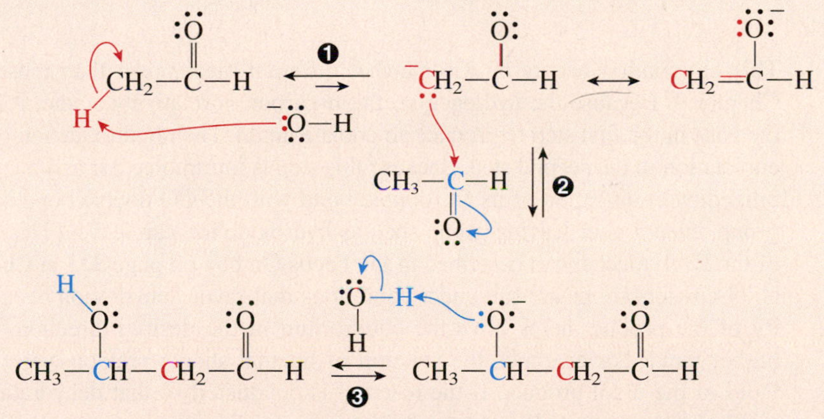

This part of the mechanism is just like the mechanism for the addition reactions of Chapter 18. **❷** The enolate nucleophile adds to the carbonyl carbon of a second aldehyde molecule, and **❸** the negative oxygen removes a proton from water. This step regenerates hydroxide ion, so the reaction is base catalyzed.

Figure 20.3

MECHANISM OF THE ALDOL CONDENSATION.

un-ionized and is available for reaction as the electrophile. Note that the strong bases described in Section 20.3, which would tend to convert most of the aldehyde molecules to enolate ions, are not used in aldol condensations. The addition of the enolate nucleophile to the aldehyde follows the same mechanism as the addition of other nucleophiles that were described in Chapter 18. Remember that the α-carbon of one aldehyde molecule bonds to the carbonyl carbon of a second aldehyde molecule, as illustrated in the following example:

$$2\ CH_3CH_2CH_2\overset{O}{\overset{\|}{CH}} \xrightarrow{\text{KOH}} CH_3CH_2CH_2\overset{OH}{\underset{|}{CH}}-\overset{O}{\underset{|}{\overset{\|}{CH}}}\overset{\|}{CH}\ (75\%)$$
$$\underset{CH_3CH_2}{|}$$

If the aldol condensation is conducted under more vigorous conditions (higher temperature, longer reaction time, and/or stronger base), elimination of water to form an α,β-unsaturated aldehyde usually occurs. This elimination is illustrated in the following example. Note that the α-carbon of one molecule is now doubly bonded to the carbonyl carbon of the other. (This text uses the symbol for heat, Δ, to indicate the vigorous conditions that cause eliminations to occur in these aldol condensations, even though other conditions might have been used.)

This elimination occurs by a somewhat different mechanism than those described in Chapter 9. Because the hydrogen on the α-carbon is relatively acidic, it is removed by the base in the first step to produce an enolate anion. Then hydroxide ion is lost from the enolate ion in the second step. Because this step is intramolecular and the product is stabilized by conjugation of its CC double bond with the CO double bond of the carbonyl group, even a poor leaving group such as hydroxide ion can leave. (This is an example of the E1cb mechanism described in the Focus On box on page 333 in Chapter 9.) Most aldol condensations are run under conditions that favor dehydration because the stability of the product helps drive the equilibrium in the desired direction, resulting in a higher yield. For example, the reaction of butanal shown previously results in a 75% yield of the aldol product. If the reaction is conducted so that dehydration occurs, the yield of the conjugated product is 97%.

$$2\ CH_3CH_2CH_2\overset{O}{\overset{\|}{CH}} \xrightarrow[\Delta]{\text{NaOH}} CH_3CH_2CH_2CH=\underset{|}{\overset{O}{\overset{\|}{C}}}CH + H_2O\ (97\%)$$
$$\underset{CH_3CH_2}{|}$$

Another example is shown in the following equation:

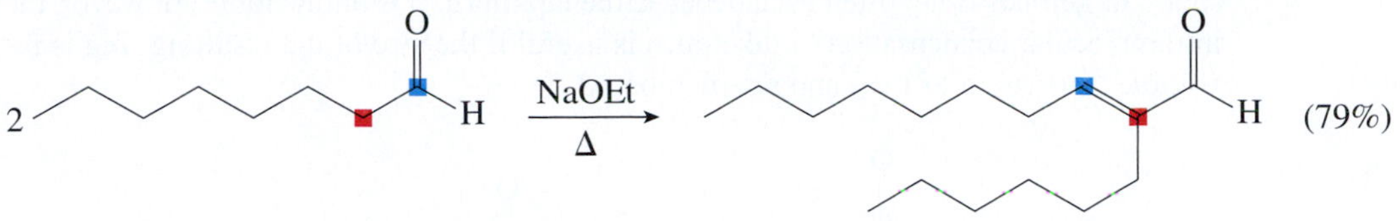

PRACTICE PROBLEM 20.3

Show the product of this reaction:

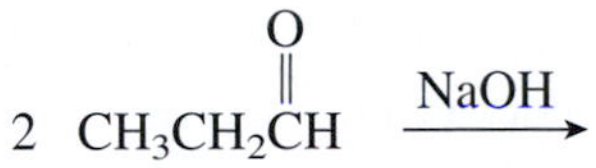

Strategy

The key to determining the products of an aldol condensation is to remember that the nucleophile is an enolate anion, which is formed at the α-carbon of the aldehyde, and the electrophile is the carbonyl carbon of another aldehyde molecule. Therefore the product has the α-carbon of one aldehyde molecule bonded to the carbonyl carbon of another aldehyde molecule. Under milder conditions an OH group remains on the carbonyl carbon of the electrophile, whereas under vigorous conditions the α-carbon and the carbonyl carbon are connected by a double bond.

Solution

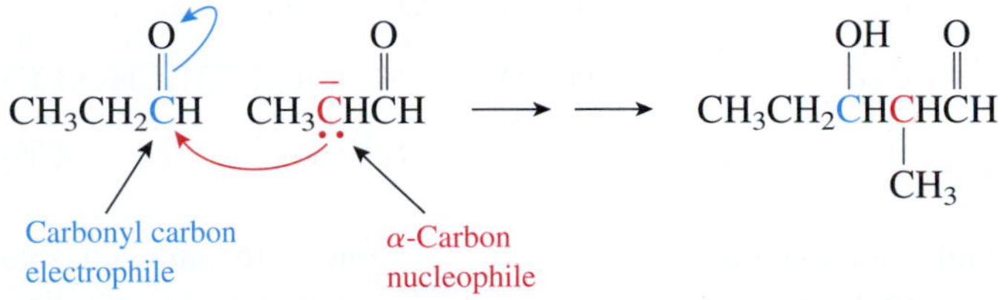

PROBLEM 20.9
Show the products of these reactions:

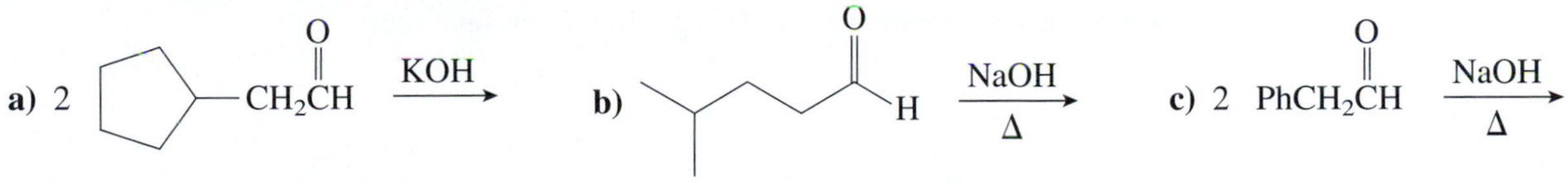

PROBLEM 20.10
Show all of the steps in the mechanism for this reaction:

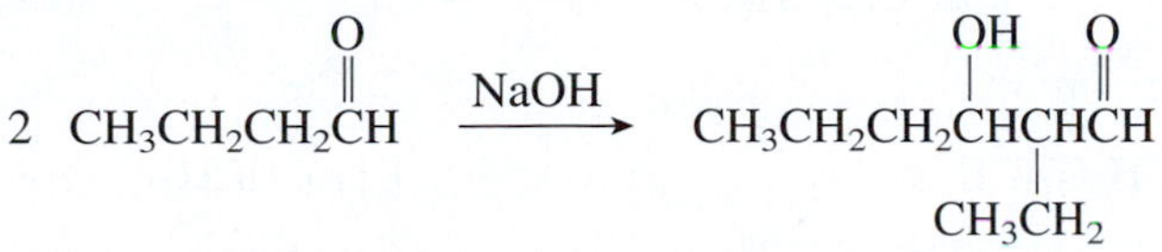

Ketones are less reactive electrophiles than aldehydes. Therefore, the aldol condensation of ketones is not often used because the equilibrium is unfavorable. However, the intramolecular condensation of diketones is useful if the size of the resulting ring is favorable (formation of five- and six-membered rings).

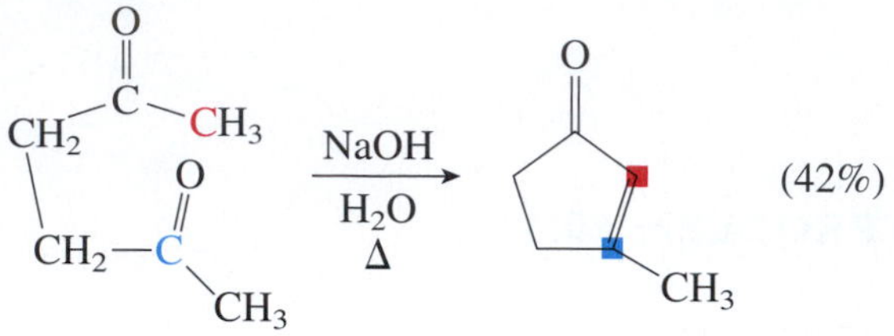

A diketone

Often, it is desirable to conduct an aldol condensation in which the nucleophile and the electrophile are derived from different compounds. In general, such **mixed aldol condensations,** involving two different aldehydes, result in the formation of several products and for this reason are not useful. For example, the reaction of ethanal and propanal results in the formation of four products because there are two possible enolate nucleophiles and two carbonyl electrophiles:

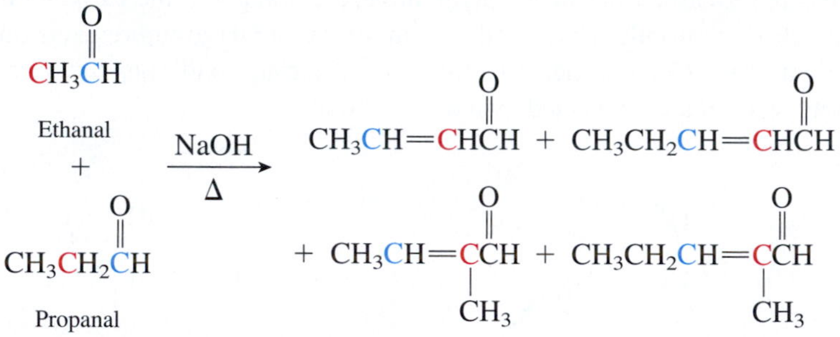

Mixed aldol condensations can be employed if one of the aldehydes has no hydrogens on the α-carbon, so it cannot form an enolate ion and can only act as the electrophilic partner in the reaction. Aromatic aldehydes are especially useful in this role because the dehydration product has additional stabilization from the conjugation of the newly formed CC double bond with the aromatic ring. This stabilization makes the equilibrium for the formation of this product more favorable.

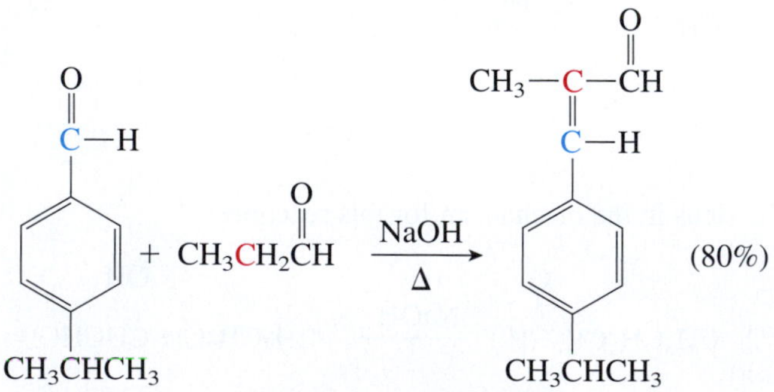

With an aromatic aldehyde as the electrophilic partner, the nucleophilic enolate ion can also be derived from a ketone or a nitrile. As illustrated in the following examples, this enables the aldol condensation to be used to form a wide variety of compounds:

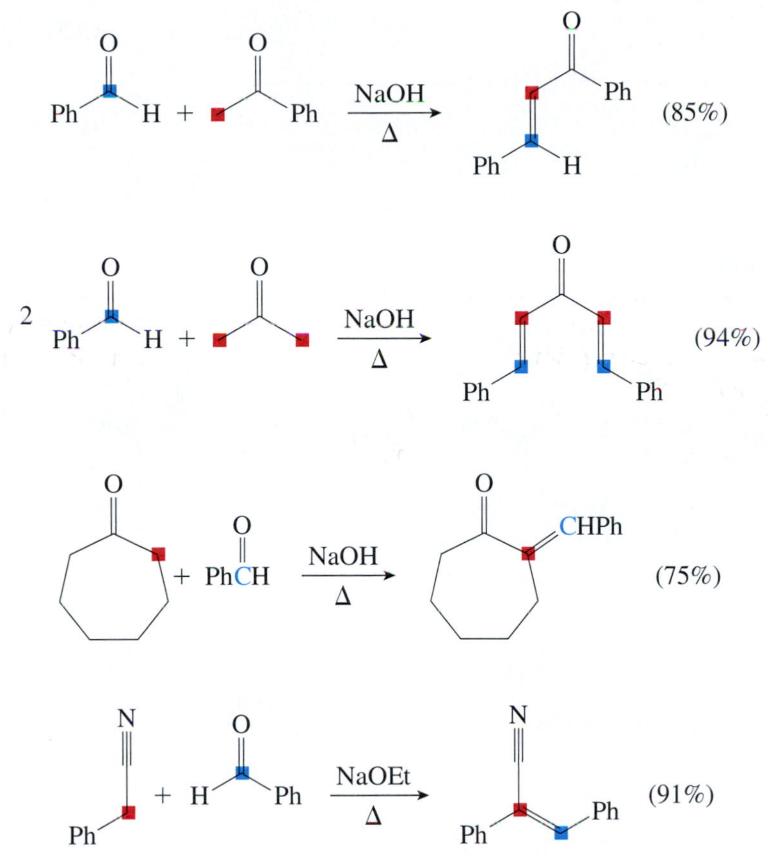

PRACTICE PROBLEM 20.4

Show the product of this reaction:

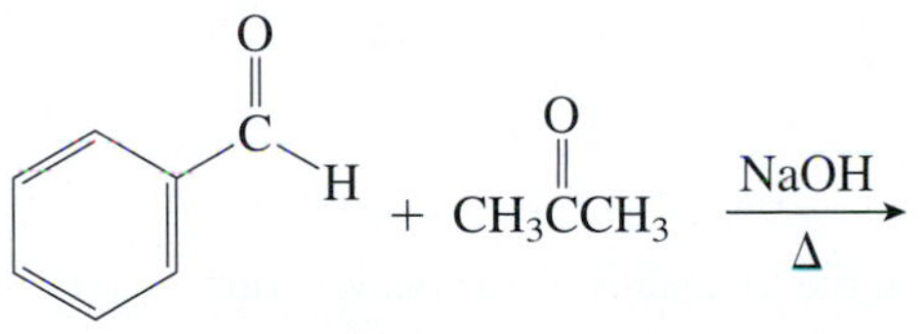

Strategy

Again, the key is to identify the nucleophile (the enolate anion) and the electrophile (the carbonyl carbon).

Solution

In this example the enolate anion can be derived only from acetone. The electrophile is the more reactive carbonyl carbon, that of benzaldehyde.

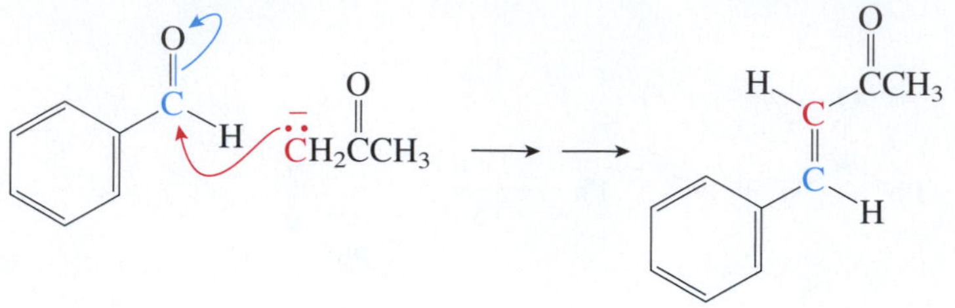

PROBLEM 20.11

Show the products of these reactions:

a)

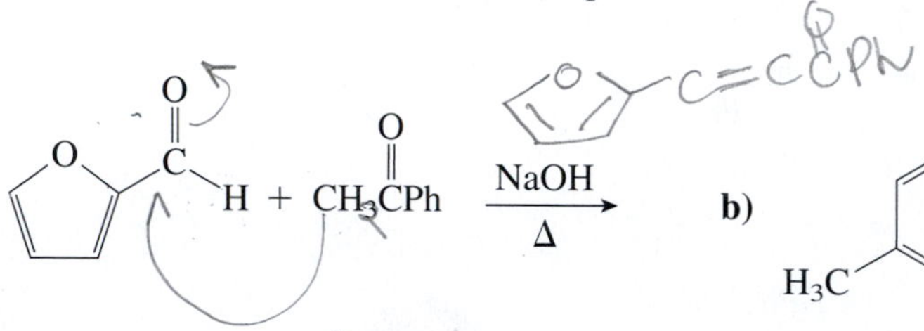

b)

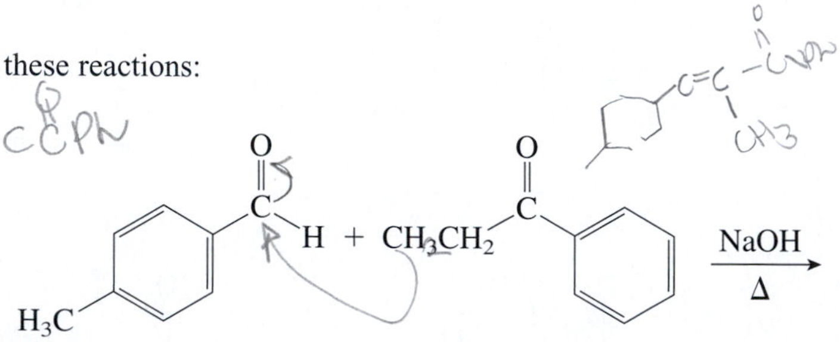

c)

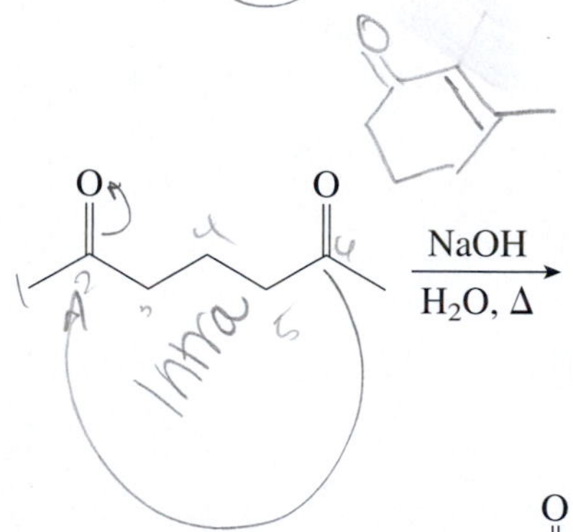

d) 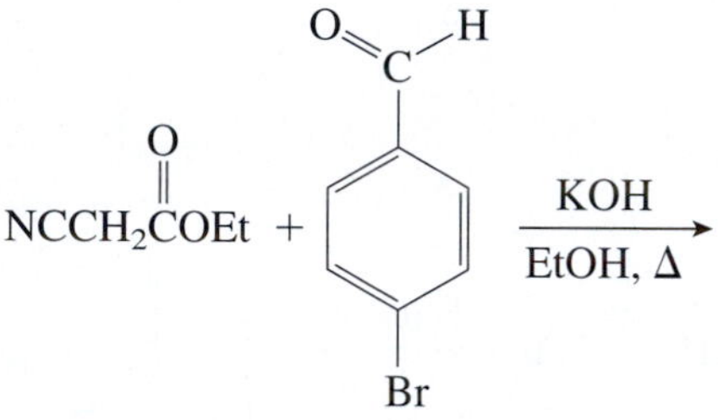 NCCH₂COEt +

e)

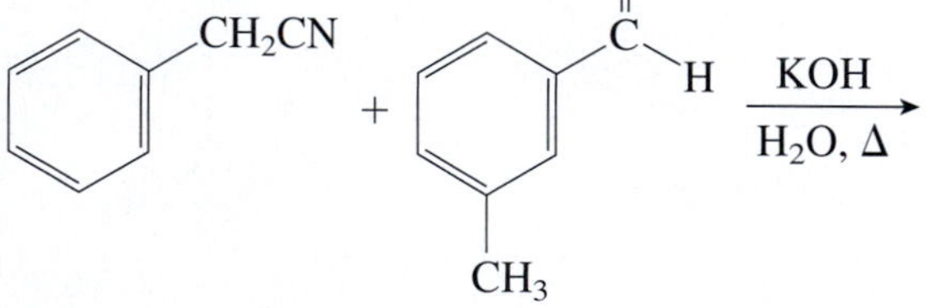

PROBLEM 20.12

Show all of the steps in the mechanism for this reaction:

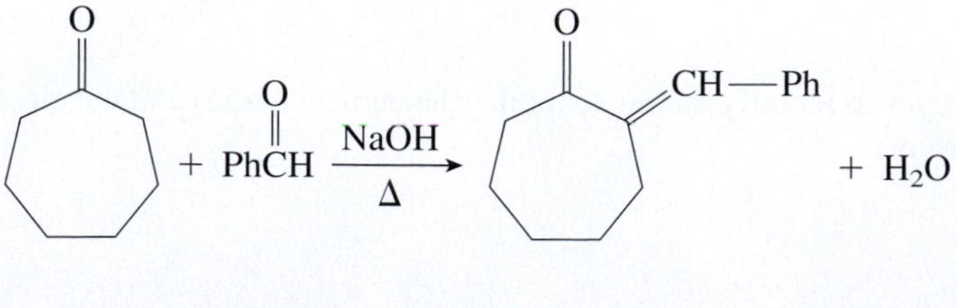

PRACTICE PROBLEM 20.5

Show how the aldol condensation could be used to synthesize this compound:

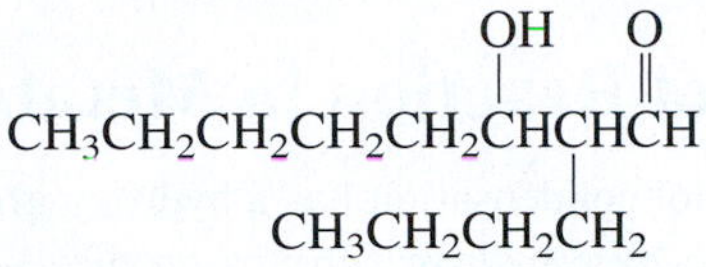

$$
\underset{CH_3CH_2CH_2CH_2}{\underset{|}{\overset{\overset{OH}{|} \quad \overset{O}{\|}}{CH_3CH_2CH_2CH_2CH_2CHCHCH}}}
$$

Strategy

First identify the carbon–carbon bond that could be formed in an aldol condensation. This is the bond between the α-carbon of the carbonyl group of the product and the carbon that is either doubly bonded to it or has a hydroxy substituent. Disconnection of this bond gives the fragments needed for the aldol condensation. Remember, the α-carbon of the product is the nucleophilic carbon of the enolate anion and the carbon to which it is bonded is the electrophilic carbonyl carbon.

Solution

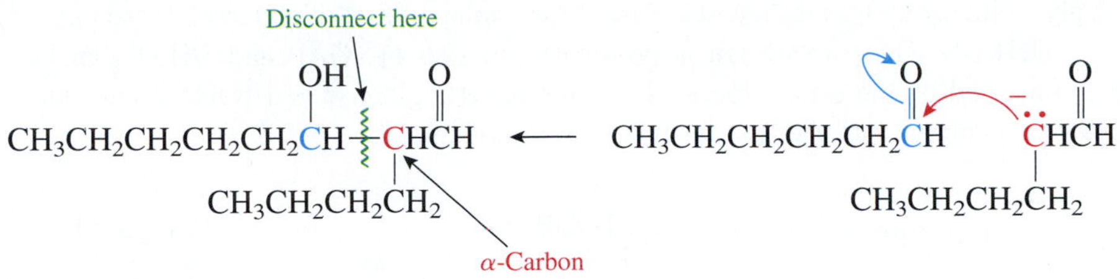

Disconnect here

$$
\underset{CH_3CH_2CH_2CH_2}{CH_3CH_2CH_2CH_2CH_2CH\text{-}CHCH} \longleftarrow CH_3CH_2CH_2CH_2CH_2CH \quad \underset{CH_3CH_2CH_2CH_2}{CHCH}
$$

α-Carbon

So the synthesis is

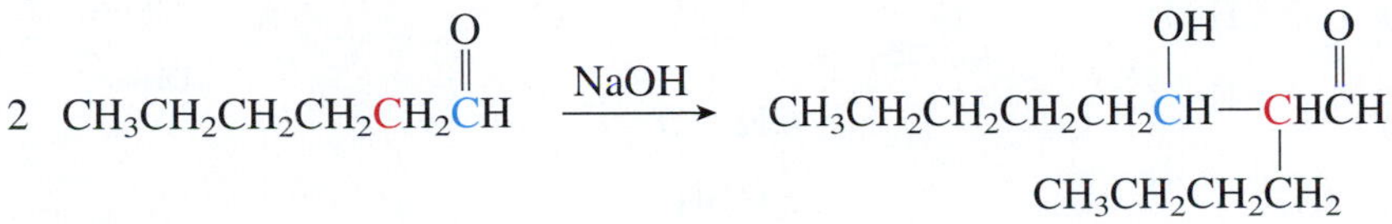

$$
2 \ CH_3CH_2CH_2CH_2CH_2CH \xrightarrow{\ NaOH\ } \underset{CH_3CH_2CH_2CH_2}{CH_3CH_2CH_2CH_2CH_2CH\text{—}CHCH}
$$

PROBLEM 20.13

Show how the aldol condensation could be used to synthesize these compounds.

a) 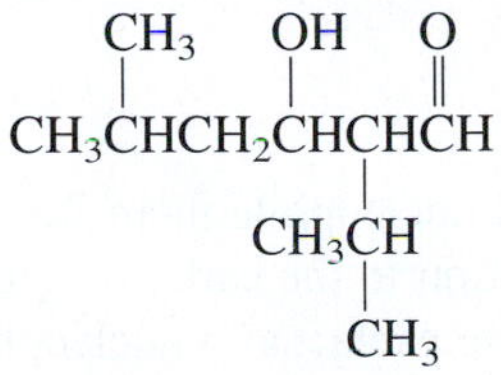 $\underset{\quad CH_3CH}{\underset{|}{\overset{\overset{CH_3}{|} \ \overset{OH}{|} \ \overset{O}{\|}}{CH_3CHCH_2CHCHCH}}}$ ... CH₃CH ... CH₃

b)

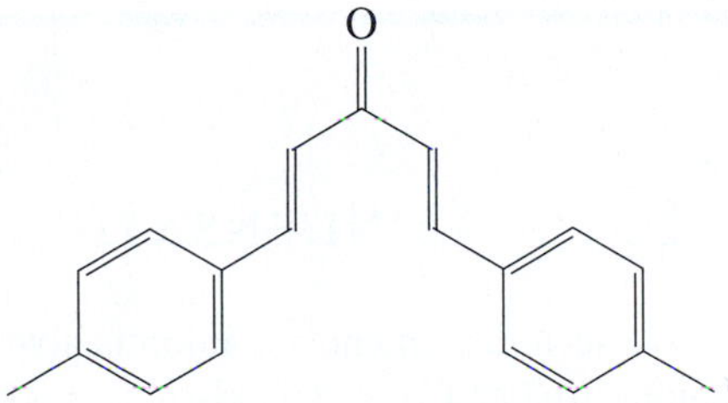

c) 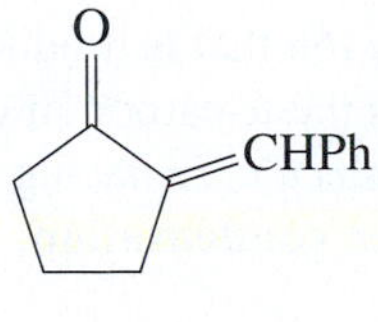 (cyclopentanone with =CHPh)

d)

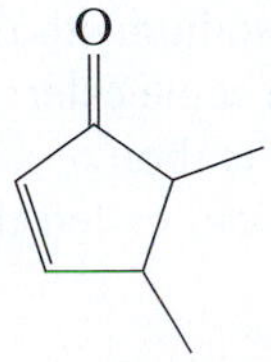

ORGANIC
Chemistry Now™
Click Coached Tutorial Problems
to practice more Aldol
Condensations.

Focus On Biological Chemistry

The Reverse Aldol Reaction in Metabolism

The initial product of an aldol condensation has a hydroxy group on the β-carbon to a carbonyl group. Sugars also have hydroxy groups on the β-carbon to their carbonyl groups, so they can be viewed as products of aldol condensations. In fact, a reverse aldol condensation is used in the metabolism of glucose (glycolysis) to cleave this six-carbon sugar into two three-carbon sugars.

To cleave a six-carbon sugar into two three-carbon fragments by a reverse aldol condensation, there must be a carbonyl group at C-2 and a hydroxy group at C-4. Therefore, glucose is first isomerized to fructose during its metabolism. The substrate for the cleavage reaction is the diphosphate ester, fructose-1,6-bisphosphate. In step ❶, a proton is removed from the hydroxy group on C-4. The bond between C-3 and C-4 is then broken in step ❷, which is the reverse of the aldol condensation, producing glyceraldehyde-3-phosphate (GAP) and the enolate ion of dihydroxyacetone phosphate (DHAP). This enolate ion is protonated in step ❸. GAP and DHAP can be interconverted by the same process that interconverts glucose and fructose and thus provide a common intermediate for further metabolism.

Fructose-1,6-bisphosphate   GAP

20.6   ESTER CONDENSATIONS

So far, we have seen that an enolate anion is able to act as a nucleophile in an S_N2 reaction (Sections 20.3 and 20.4) and also in an addition reaction to the carbonyl group of an aldehyde in the aldol condensation (Section 20.5). It also can act as a nucleophile in a substitution reaction with the carbonyl group of an ester as the electrophile. When an ester is treated with a base such as sodium ethoxide, the enolate ion that is produced can react with another molecule of the same ester. The product has the α-carbon of one ester molecule bonded to the carbonyl carbon of a second ester molecule, replacing the alkoxy group. Examples of this reaction, called the **Claisen ester condensation,** are provided by the following equations:

The reverse aldol reaction is catalyzed by an enzyme called aldolase. One of the roles of the enzyme is to stabilize the enolate anion intermediate because such ions are too basic to be produced under physiological conditions. In animals, aldolase accomplishes this task by forming an imine bond between the carbonyl group of fructose-1,6-bisphosphate and the amino group of a lysine amino acid of the enzyme. As a result, the product of the reverse aldol step is an enamine derived from DHAP rather than its enolate anion. (Section 20.8 shows that enamines are the synthetic equivalents of enolate anions.) The formation of the strongly basic enolate anion is avoided. This process is outlined here:

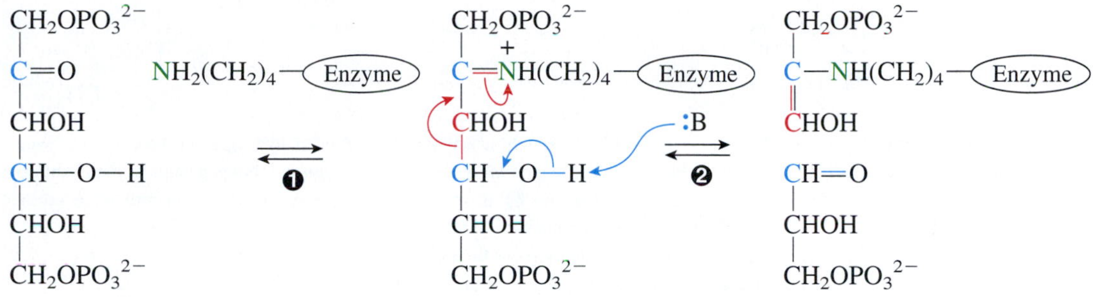

Fructose-1,6-bisphosphate

❶ An amino group of a lysine amino acid of the enzyme reacts with the carbonyl group of fructose-1,6-bisphosphate to form a protonated imine. See Figure 18.3 for the mechanism of this reaction.

❷ When the reverse aldol reaction occurs, an enamine, rather than a strongly basic enolate anion, is produced.

This enamine is hydrolyzed to DHAP, freeing the enzyme to catalyze another reverse aldol cleavage.

$$2\ CH_3-\overset{\overset{\displaystyle O}{\|}}{C}-OEt \xrightarrow[2)\ H_3O^+]{1)\ NaOEt,\ EtOH} CH_3-\overset{\overset{\displaystyle O}{\|}}{C}-CH_2-\overset{\overset{\displaystyle O}{\|}}{C}-OEt + EtOH \quad (80\%)$$

$$2\ CH_3CH_2CH_2\overset{\overset{\displaystyle O}{\|}}{C}OEt \xrightarrow[2)\ H_3O^+]{1)\ NaOEt,\ EtOH} CH_3CH_2CH_2\overset{\overset{\displaystyle O}{\|}}{C}\underset{\underset{\displaystyle CH_3CH_2}{|}}{\overset{\overset{\displaystyle O}{\|}}{C}H}COEt + EtOH \quad (76\%)$$

The mechanism for this reaction, shown in Figure 20.4, has similarities to those of both an aldol condensation (see Figure 20.3) and an ester saponification (see Figure 19.4). As was the case with the aldol condensation, the presence of both the enolate ion and the

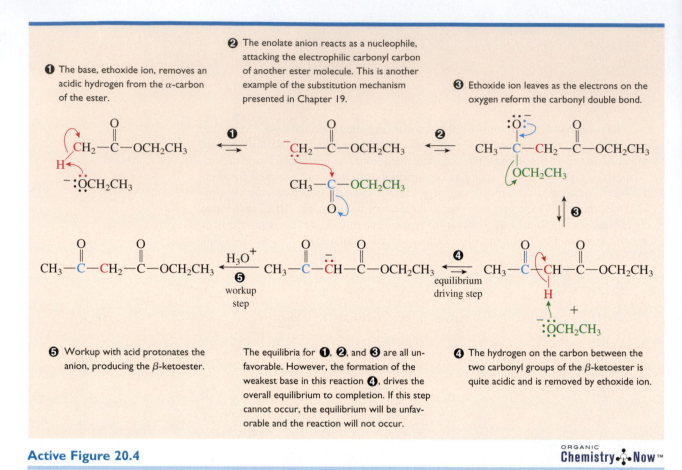

❶ The base, ethoxide ion, removes an acidic hydrogen from the α-carbon of the ester.

❷ The enolate anion reacts as a nucleophile, attacking the electrophilic carbonyl carbon of another ester molecule. This is another example of the substitution mechanism presented in Chapter 19.

❸ Ethoxide ion leaves as the electrons on the oxygen reform the carbonyl double bond.

❺ Workup with acid protonates the anion, producing the β-ketoester.

The equilibria for **❶**, **❷**, and **❸** are all unfavorable. However, the formation of the weakest base in this reaction **❹**, drives the overall equilibrium to completion. If this step cannot occur, the equilibrium will be unfavorable and the reaction will not occur.

❹ The hydrogen on the carbon between the two carbonyl groups of the β-ketoester is quite acidic and is removed by ethoxide ion.

Active Figure 20.4

ORGANIC
Chemistry⚛Now™

MECHANISM OF THE CLAISEN ESTER CONDENSATION. Test yourself on the concepts in this figure at **OrganicChemistryNow**.

neutral ester is necessary for the reaction to occur. Therefore, ethoxide ion is used as the base because it is a weaker base than the enolate ion. Step 4 of the mechanism, in which ethoxide ion removes the acidic hydrogen, is of critical importance. The equilibria for the first three steps of the reaction are all unfavorable. But the equilibrium for step 4 is very favorable because the product of this step, the conjugate base of a β-ketoester, is the weakest base in the reaction. The formation of this weak base drives the equilibria to the product. If step 4 cannot occur, no significant amount of the β-ketoester is present in the reaction mixture and the condensation fails.

When sodium ethoxide is used as the base, the ester condensation fails with esters that have only one hydrogen on the α-carbon. The equilibrium favors the reactants because the equilibrium driving step (step 4 of the mechanism in Figure 20.4) cannot occur.

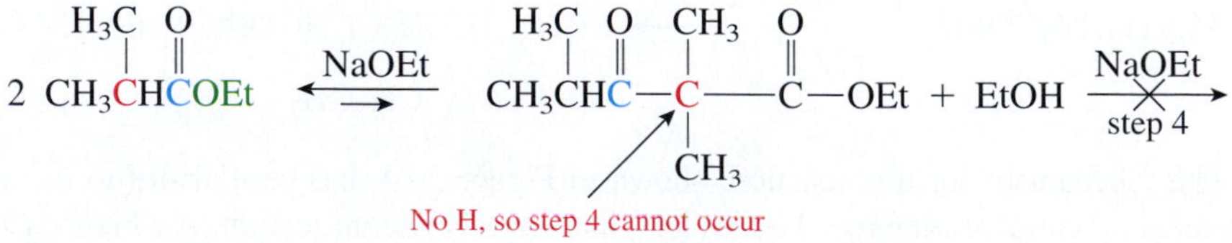

No H, so step 4 cannot occur

However, good yields of the condensation product can be obtained if a very strong base is used. In this case the equilibrium is driven toward the products because the ethoxide ion that is formed on the product side of the equation is weaker than the base on the re-actant side of the equation. Note that only enough base is used to deprotonate one-half of the ester.

$$2 \ \underset{\underset{CH_3}{|}}{CH_3CHCOEt} + KH \longrightarrow CH_3CHC\!-\!C\!-\!-\!C\!-\!OEt + H_2 + K^+ \ ^-OEt$$

In fact, even with an ester that gives an acceptable yield of the condensation product with sodium ethoxide as the base, a better yield is often obtained when a stronger base is employed.

Intramolecular ester condensation reactions are called **Dieckmann condensations** and are very useful ring-forming reactions. Examples are shown in the following equations. In the second equation the yield is only 54% if sodium ethoxide is used as the base.

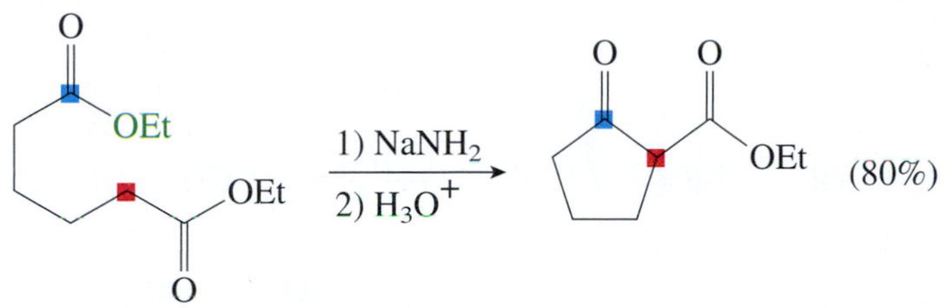

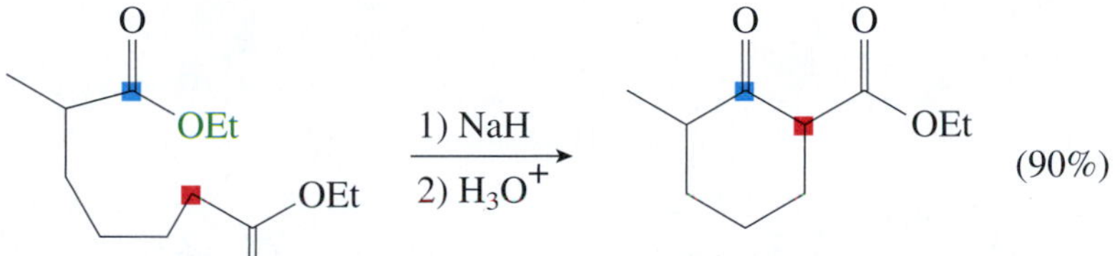

$$\xrightarrow[\text{2) } H_3O^+]{\text{1) NaNH}_2} \quad (80\%)$$

$$\xrightarrow[\text{2) } H_3O^+]{\text{1) NaH}} \quad (90\%)$$

Ethyl 3-methyl-2-oxo-cyclohexanecarboxylate

PROBLEM 20.14
Show the products of these reactions:

a) $2 \ CH_3CH_2COEt \xrightarrow[\text{2) } H_3O^+]{\text{1) NaOEt, EtOH}}$

b) $2 \ \underset{\underset{CH_3}{|}}{CH_3CH_2CHCOEt} \xrightarrow[\text{2) } H_3O^+]{\text{1) LDA}}$ *Adds to self*

c) $2 \ PhCH_2COEt \xrightarrow[\text{2) } H_3O^+]{\text{1) NaOEt, EtOH}}$

d) $EtOC(CH_2)_5COEt \xrightarrow[\text{2) } H_3O^+]{\text{1) NaOEt, EtOH}}$

PROBLEM 20.15

The second example of a Dieckmann condensation shown earlier produces ethyl 3-methyl-2-oxocyclohexanecarboxylate in 90% yield. What other cyclic product might have been formed in this reaction? Explain why the actual product is favored rather than this other product.

As was the case with the aldol condensation, mixed ester condensations can be useful if one of the components can only act as the electrophile—that is, if it cannot form an enolate anion (no hydrogens on the α-carbon). The following esters are most commonly employed in this role:

| Ethyl formate | Diethyl carbonate | Diethyl oxalate | Ethyl benzoate |

The nucleophile, an enolate or related anion, can be obtained by deprotonation of an ester, ketone, or nitrile. Examples are provided by the following equations:

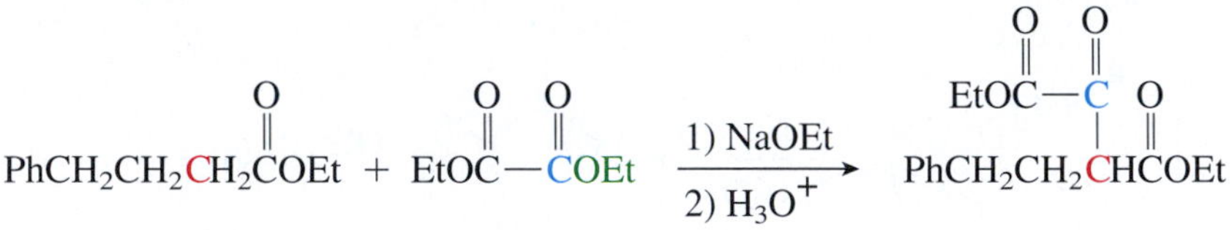

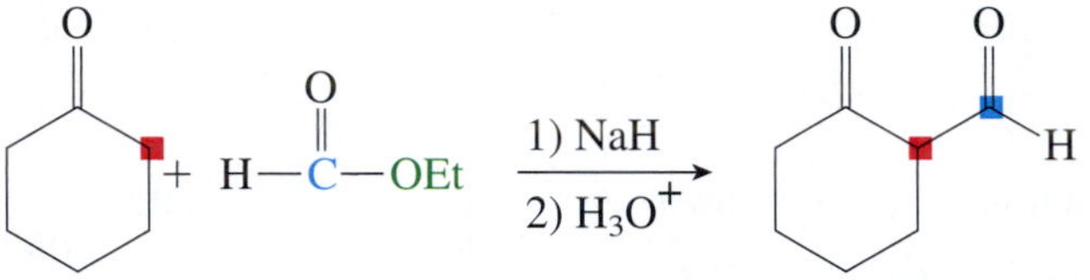

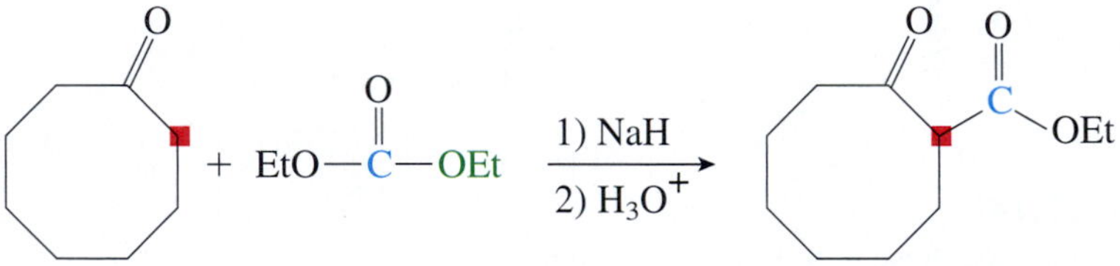

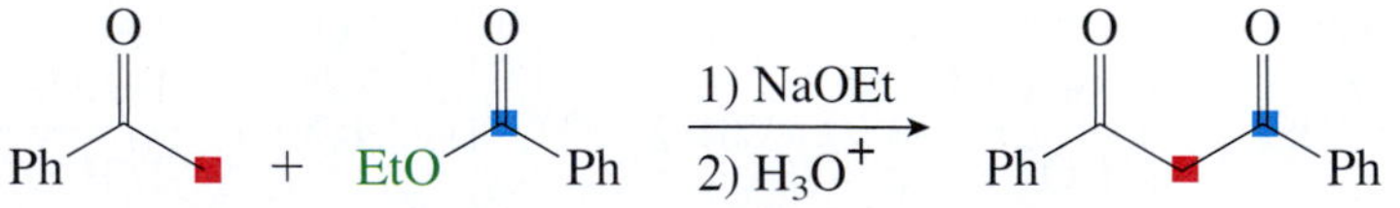

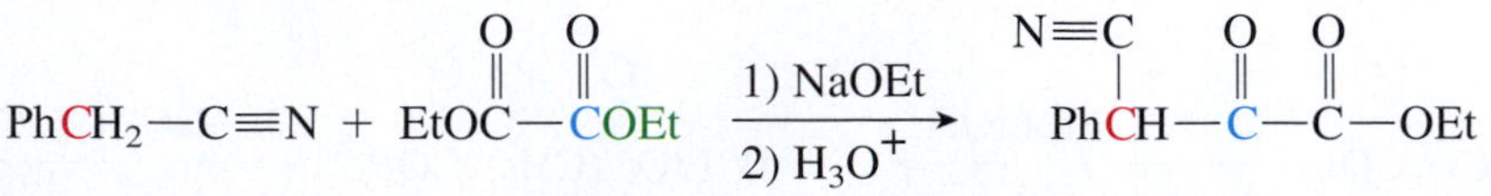

Finally, note that the products of most of these reactions are β-dicarbonyl compounds. They can be alkylated in the same manner as ethyl acetoacetate and diethyl malonate, and they can also be decarboxylated if one of the two carbonyl groups is an ester group. This makes them quite useful in synthesis.

PRACTICE PROBLEM 20.6

Show the product of this reaction:

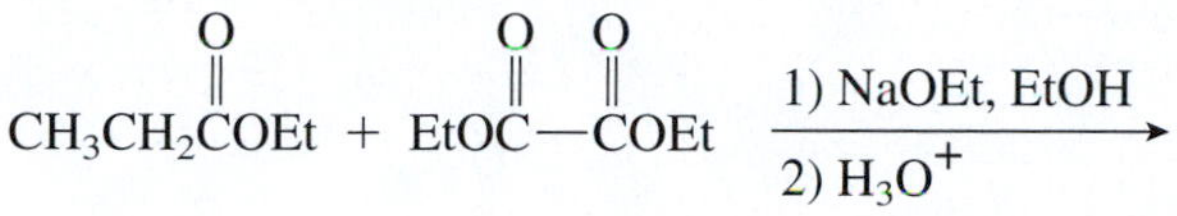

$$CH_3CH_2\overset{\overset{\displaystyle O}{\|}}{C}OEt \ + \ EtO\overset{\overset{\displaystyle O}{\|}}{C}-\overset{\overset{\displaystyle O}{\|}}{C}OEt \quad \xrightarrow[\text{2) H}_3\text{O}^+]{\text{1) NaOEt, EtOH}}$$

Strategy

Again the best approach is to identify the site where the nucleophilic enolate anion forms, the α-carbon with the most acidic hydrogen. This carbon becomes bonded to the carbonyl carbon of the ester electrophile in the final product.

Solution

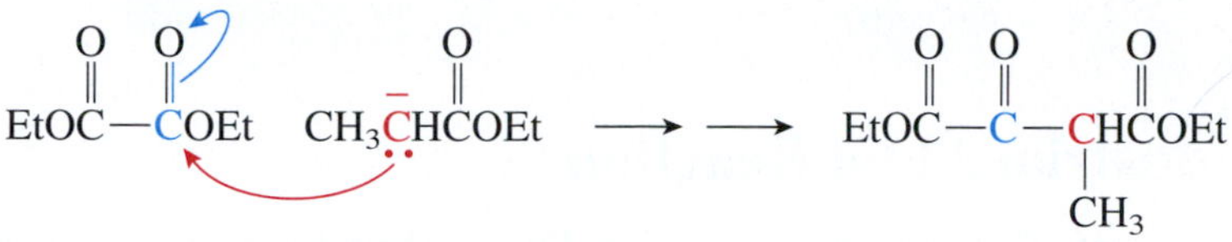

PROBLEM 20.16

Show the products of these reactions.

a) 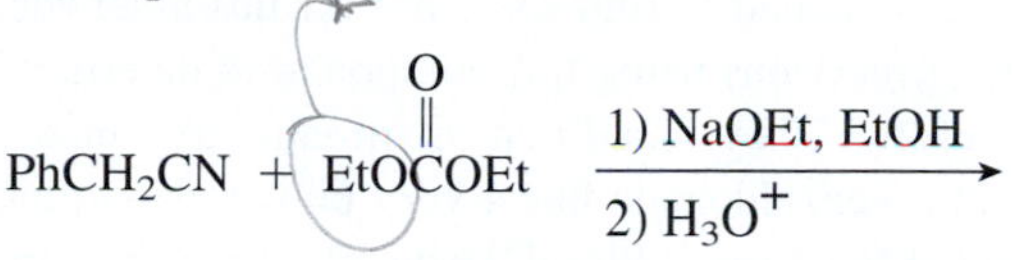 $PhCH_2CN \ + \ EtO\overset{\overset{\displaystyle O}{\|}}{C}OEt \quad \xrightarrow[\text{2) H}_3\text{O}^+]{\text{1) NaOEt, EtOH}}$

b) $PhC\overset{\overset{\displaystyle O}{\|}}{C}H_3 \ + \ H\overset{\overset{\displaystyle O}{\|}}{C}OEt \quad \xrightarrow[\text{2) H}_3\text{O}^+]{\text{1) NaOEt, EtOH}}$

c) $+ \ PhC\overset{\overset{\displaystyle O}{\|}}{}OEt \quad \xrightarrow[\text{2) H}_3\text{O}^+]{\text{1) NaOEt, EtOH}}$

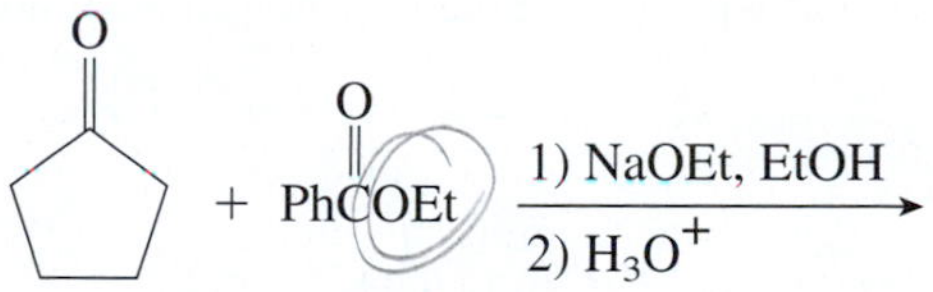

d) 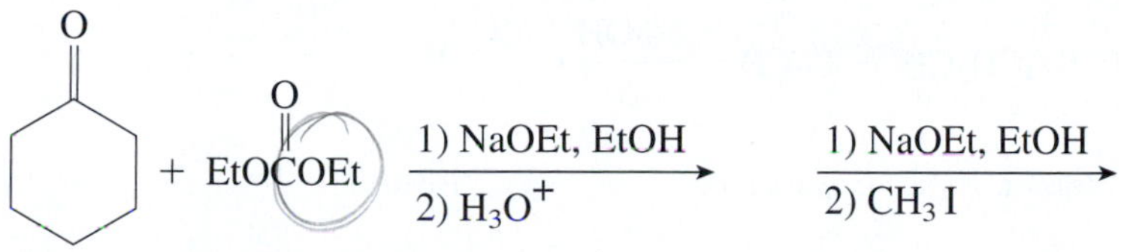 $+ \ EtO\overset{\overset{\displaystyle O}{\|}}{C}OEt \quad \xrightarrow[\text{2) H}_3\text{O}^+]{\text{1) NaOEt, EtOH}} \quad \xrightarrow[\text{2) CH}_3\text{I}]{\text{1) NaOEt, EtOH}}$

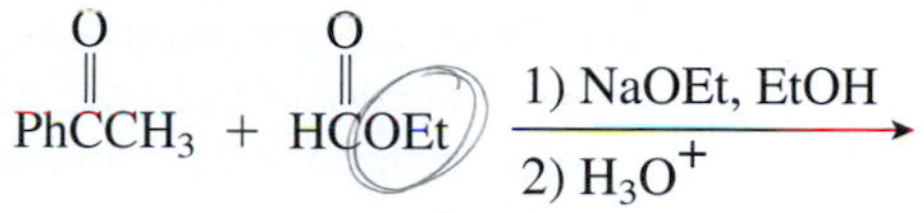

PROBLEM 20.17

Show all of the steps in the mechanism for this reaction:

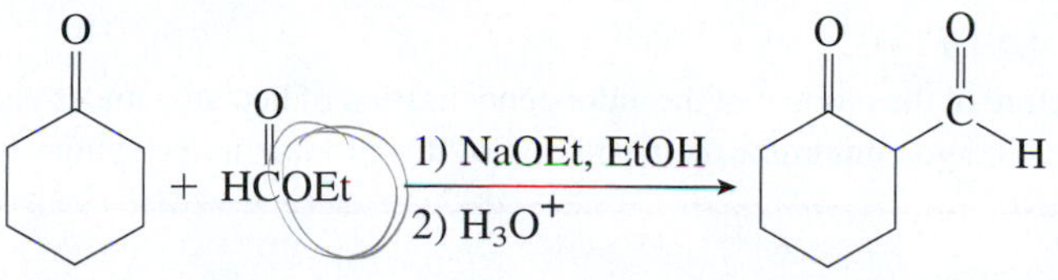

PROBLEM 20.18

Show how ester condensation reactions could be used to synthesize these compounds:

a)

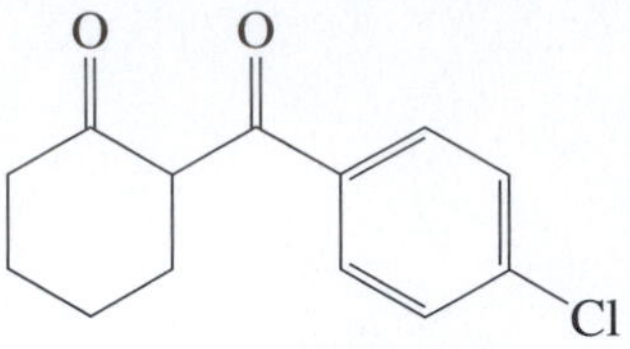

b)

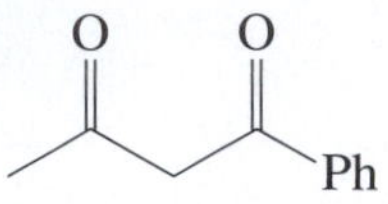

c)

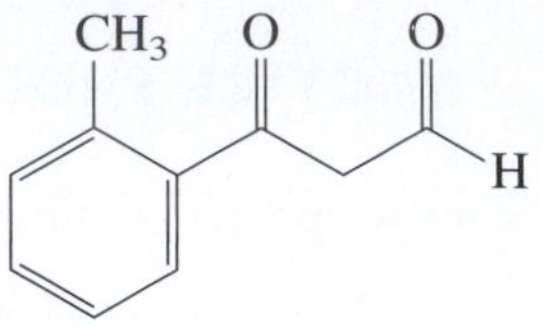

d)

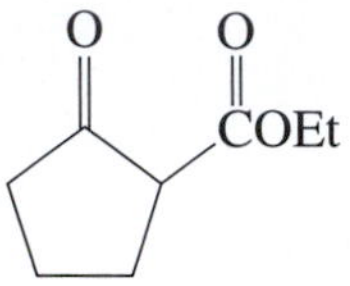

e)

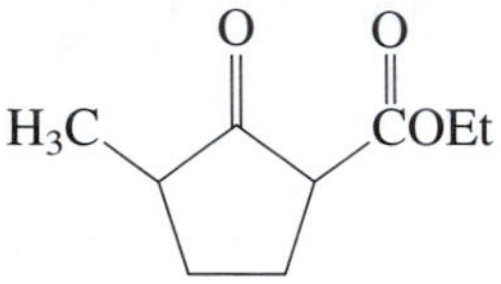

Focus On

An Industrial Aldol Reaction

The development of the perfumery ingredient Flosal, which has a strong jasminelike odor, provides an interesting example of how discoveries are sometimes made in an industrial setting. A chemical company was producing substantial amounts of heptanal as a by-product of one of its processes and needed to find some use for this aldehyde. The company's chemists ran a number of reactions using this compound to determine whether any of the products might be useful. They found that the product of a mixed aldol condensation between heptanal and benzaldehyde has a very powerful jasmine odor. This compound, which was given the trade name Flosal, became an important ingredient in soaps, perfumes, and cosmetics and was at one time prepared in amounts in excess of 100,000 pounds per year.

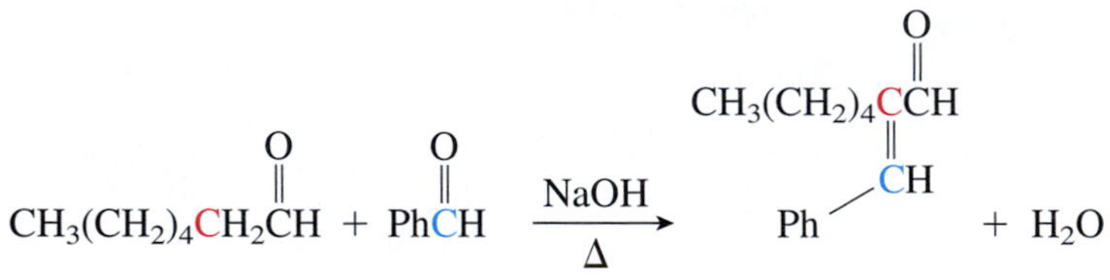

$$CH_3(CH_2)_4CH_2CH + PhCH \xrightarrow[\Delta]{NaOH} \begin{array}{c} CH_3(CH_2)_4CCH \\ Ph \end{array} CH + H_2O$$

Heptanal   Benzaldehyde   Flosal

The synthesis must be carefully controlled to minimize the formation of the aldol product that results from the reaction of two molecules of heptanal (2-pentyl-2-nonenal) because this compound has an unpleasant, rancid odor.

PROBLEM 20.19

Show the structure of the product of the aldol condensation of heptanal under vigorous conditions. How would you minimize the formation of this product in the synthesis of Flosal?

20.7 CARBON AND HYDROGEN LEAVING GROUPS

Previously we learned that the carbonyl group of an aldehyde or a ketone does not undergo the substitution reactions of Chapter 19 because hydride ion and carbanions are strong bases and poor leaving groups. There are, however, some special situations in which these species do leave. Some of these exceptions are described in this section.

A carbanion can act as a leaving group if it is stabilized somehow. We have already seen several examples of this. For example, CX_3^-, a methyl carbanion stabilized by the inductive effect of three electronegative halogen atoms, leaves in the haloform reaction (see Section 20.2).

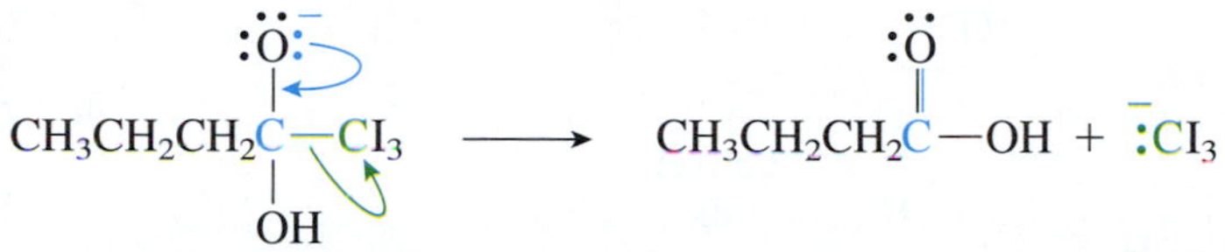

$$CH_3CH_2CH_2\overset{:\overset{..}{O}:^-}{\underset{OH}{C}}{-}CI_3 \longrightarrow CH_3CH_2CH_2\overset{:\overset{..}{O}}{C}{-}OH + :CI_3^-$$

Another example is provided by the equilibrium in the aldol condensation. Examination of the mechanism for this reaction (see Figure 20.3) shows that an enolate anion leaves in the reverse of the second step of this reaction. Again, it is the stabilization of the carbanion, this time by resonance, that enables the enolate anion to leave.

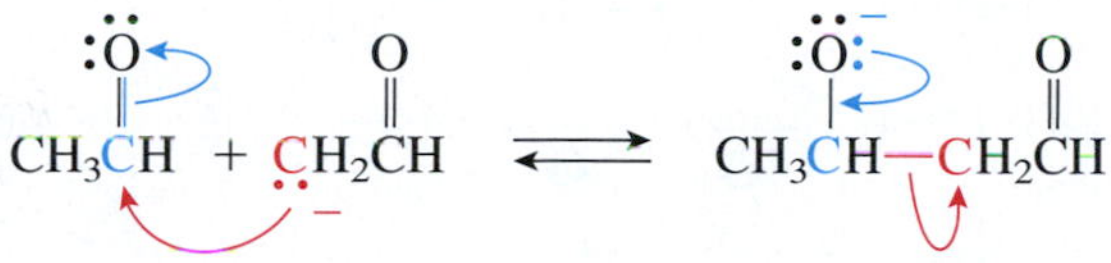

$$CH_3\overset{:\overset{..}{O}}{C}H + \overset{O}{\overset{\|}{C}}H_2CH \rightleftharpoons CH_3\overset{:\overset{..}{O}:^-}{C}H{-}CH_2\overset{O}{\overset{\|}{C}}H$$

A similar process is described in the Focus On box titled "The Reverse Aldol Reaction in Metabolism" on page 880. The Claisen ester condensation also has an equilibrium step in which an enolate anion leaves in the reverse of the step (see Figure 20.4).

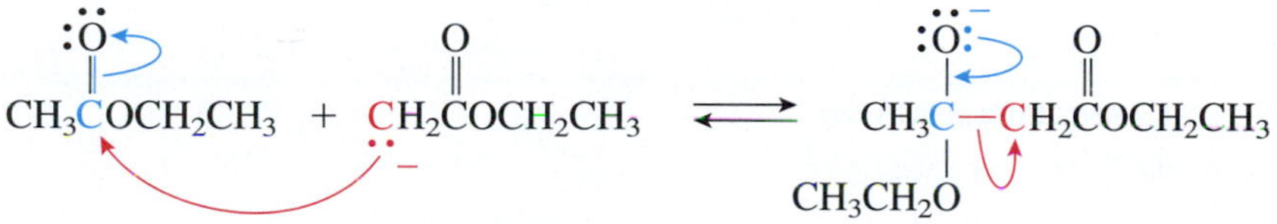

$$CH_3\overset{:\overset{..}{O}}{C}OCH_2CH_3 + \overset{O}{\overset{\|}{C}}H_2COCH_2CH_3 \rightleftharpoons CH_3\overset{:\overset{..}{O}:^-}{\underset{CH_3CH_2O}{C}}{-}CH_2\overset{O}{\overset{\|}{C}}OCH_2CH_3$$

Reactions in which hydride leaves are less common but can occur if other reactions are precluded and the hydride is transferred directly to an electrophile. One example occurs when an aldehyde without any hydrogens on its α-carbon is treated with NaOH or KOH. (If the aldehyde has hydrogens on its α-carbon, the aldol condensation is faster and occurs instead.) In this reaction, called the Cannizzaro reaction, two molecules of aldehyde react. One is oxidized to a carboxylate anion and the other is reduced to a primary alcohol. The mechanism for this reaction is shown in Figure 20.5. The reaction begins in the same manner as the reactions described in Chapter 18; a hydroxide ion nucleophile attacks the carbonyl carbon of the aldehyde to form an anion. The reaction now begins to resemble the reactions in Chapter 19.

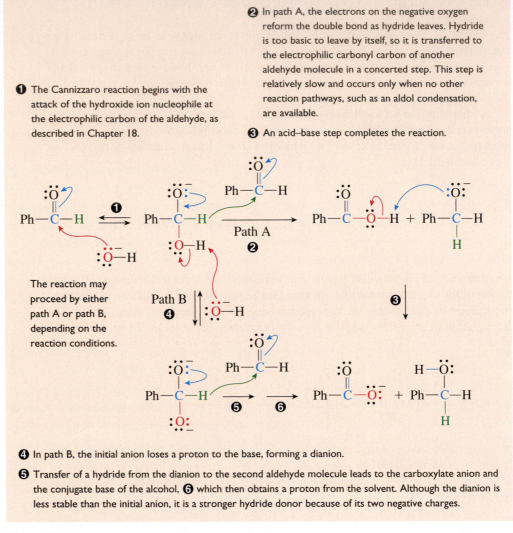

❶ The Cannizzaro reaction begins with the attack of the hydroxide ion nucleophile at the electrophilic carbon of the aldehyde, as described in Chapter 18.

❷ In path A, the electrons on the negative oxygen reform the double bond as hydride leaves. Hydride is too basic to leave by itself, so it is transferred to the electrophilic carbonyl carbon of another aldehyde molecule in a concerted step. This step is relatively slow and occurs only when no other reaction pathways, such as an aldol condensation, are available.

❸ An acid–base step completes the reaction.

The reaction may proceed by either path A or path B, depending on the reaction conditions.

❹ In path B, the initial anion loses a proton to the base, forming a dianion.

❺ Transfer of a hydride from the dianion to the second aldehyde molecule leads to the carboxylate anion and the conjugate base of the alcohol, ❻ which then obtains a proton from the solvent. Although the dianion is less stable than the initial anion, it is a stronger hydride donor because of its two negative charges.

Figure 20.5

MECHANISM OF THE CANNIZZARO REACTION.

As the electrons on the negative oxygen reform the double bond, hydride ion begins to leave. But hydride ion is too poor a leaving group to leave without help, so it is transferred directly to the carbonyl carbon of a second aldehyde molecule, as shown in path A of the mechanism. An acid–base reaction completes the process. Under more strongly basic conditions, the mechanism may change slightly and follow path B. In this case the initially formed anion loses a proton to the base to form a dianion. Although the concentration of the dianion is less than that of the monoanion, it donates hydride more rapidly than the monoanion because of its two negative charges. After protonation of the alkoxide anion by the solvent, the same products are produced from path B as from path A. An example of the Cannizzaro reaction is provided in the following equation:

63%   63%

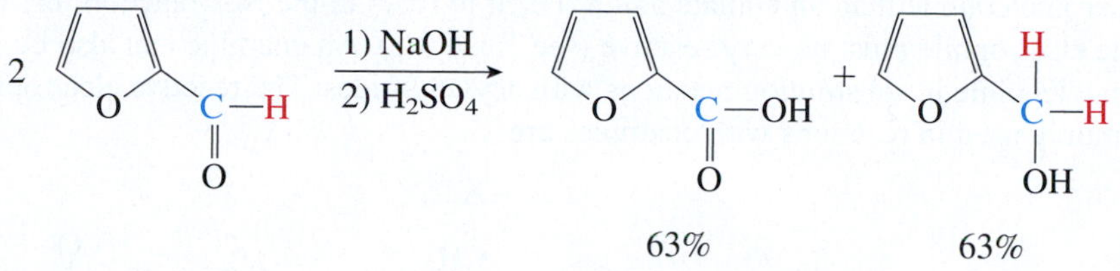

PROBLEM 20.20

Show the product of this reaction:

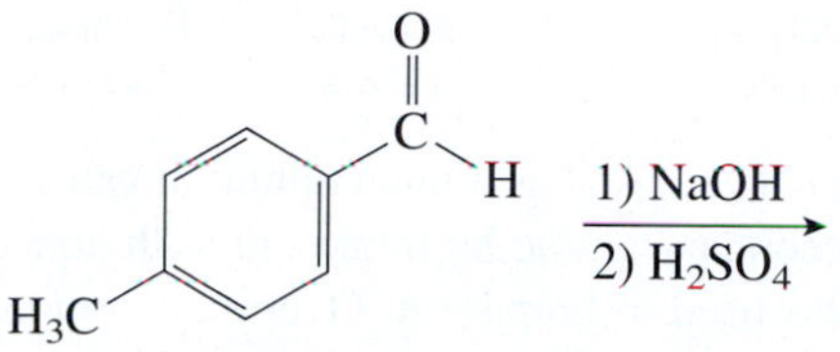

PROBLEM 20.21

β-Ketoesters that have two substituents on the α-carbon undergo fragmentation when treated with ethoxide anion as shown in the following equation. Suggest a mechanism for this reaction.

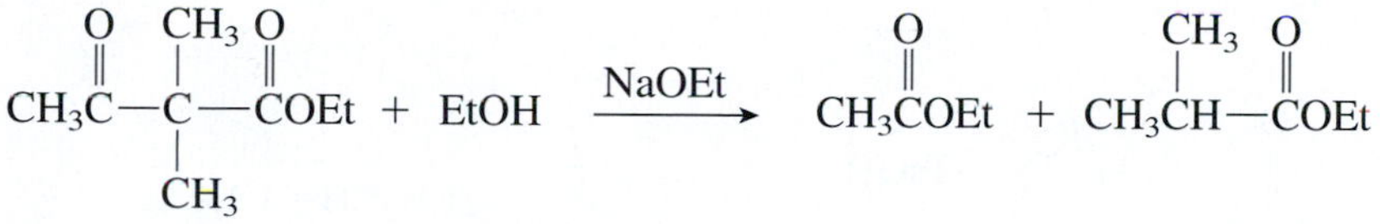

20.8 ENAMINES

Because of the importance of carbon nucleophiles in synthesis, organic chemists have spent considerable effort developing others in addition to the enolate anions that have already been described. Several of these other carbon nucleophiles are presented in this and the following section. This section describes the use of enamines.

As discussed in Section 18.8, enamines are prepared by the reaction of a secondary amine with a ketone in the presence of an acid catalyst. The equilibrium is usually driven toward product formation by removal of water.

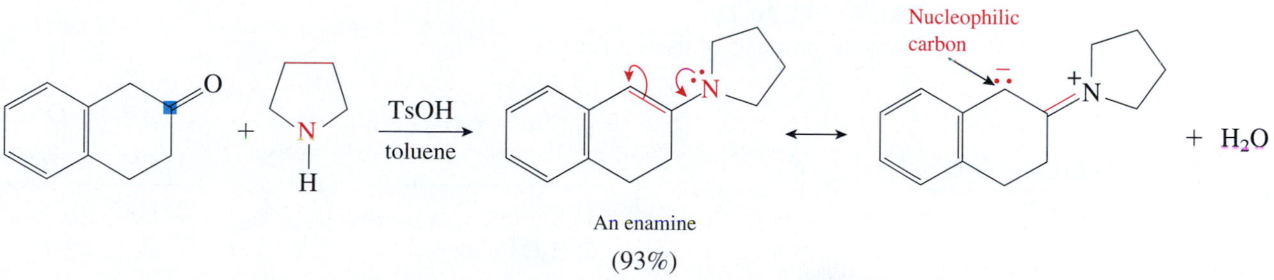

An enamine

(93%)

Because of the contribution of structures such as the one on the right to the resonance hybrid, the α-carbon of an enamine is nucleophilic. However, an enamine is a much weaker nucleophile than an enolate anion. For it to react in the S$_N$2 reaction, the alkyl halide electrophile must be very reactive (see Table 8.1). An enamine can also be used as a nucleophile in substitution reactions with acyl chlorides. The reactive electrophiles commonly used in reactions with enamines are:

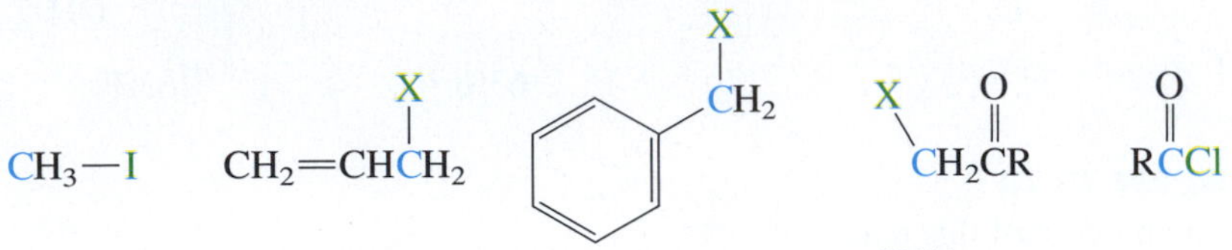

| Methyl iodide | Allylic halides | Benzylic halides | Halides on α-carbons of ketones and esters | Acyl chlorides |

After the enamine has been used as a nucleophile, it can easily be hydrolyzed back to the ketone and the secondary amine by treatment with aqueous acid. This is simply the reverse of the process used to prepare it. Overall, enamines serve as the synthetic equivalent of ketone enolate anions. Examples are provided in the following equations:

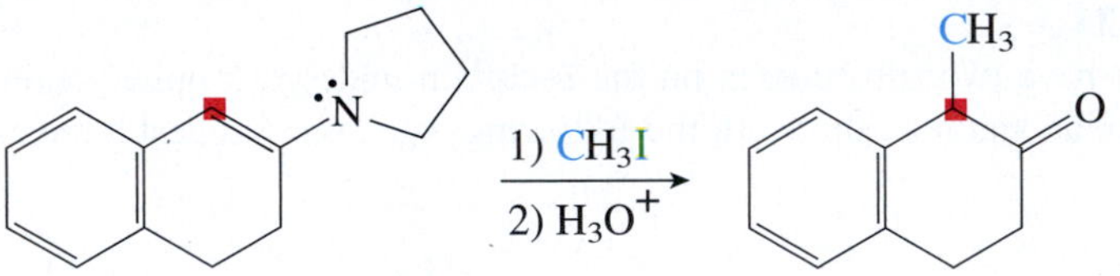

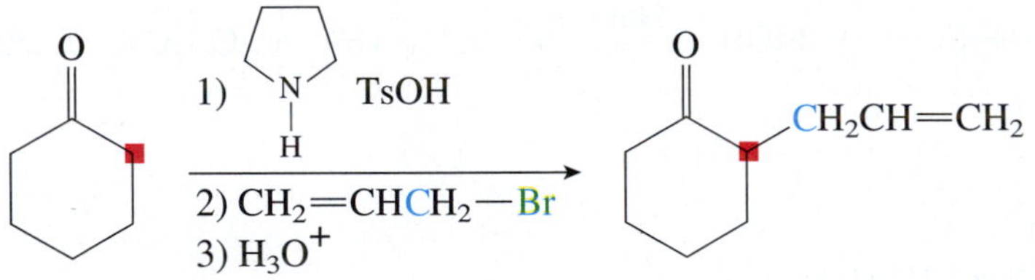

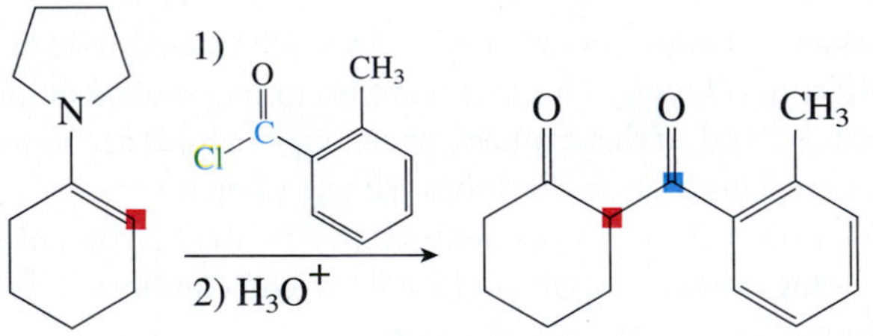

PROBLEM 20.22
Show the products of these reactions:

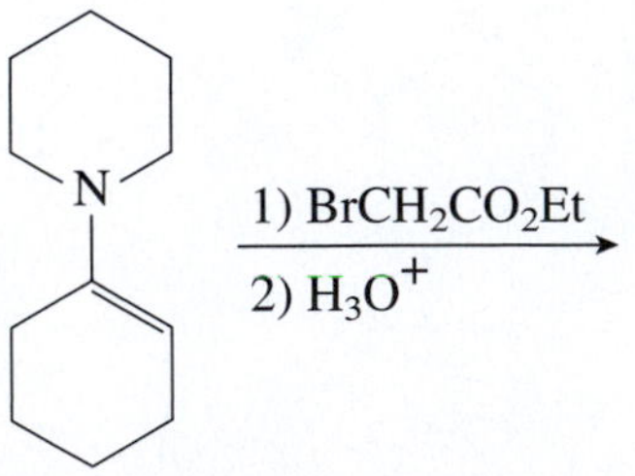

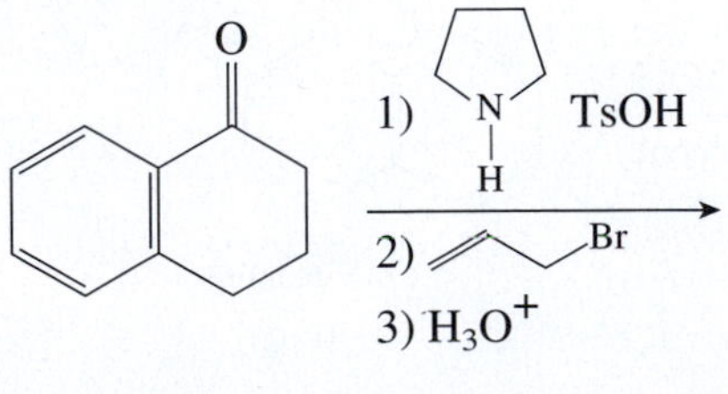

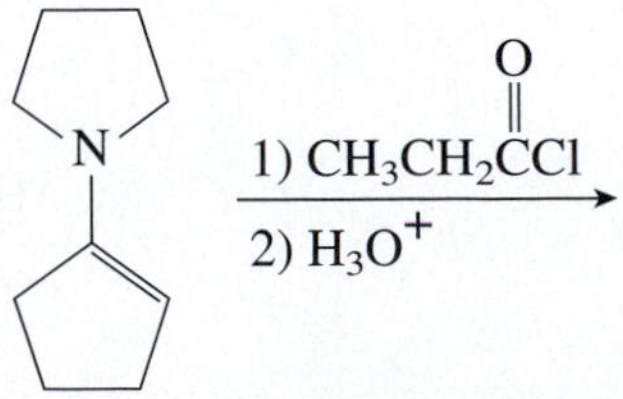

20.9 OTHER CARBON NUCLEOPHILES

Many other carbon nucleophiles have been developed. Only two additional types are introduced here, but both provide interesting variations on the themes that have been presented so far.

The first of these nucleophiles is derived from a dithiane. A dithiane can be prepared by the reaction of an aldehyde with 1,3-propanedithiol. This reaction, described in Section 18.9, is the sulfur analog of acetal formation and requires a proton or Lewis acid catalyst:

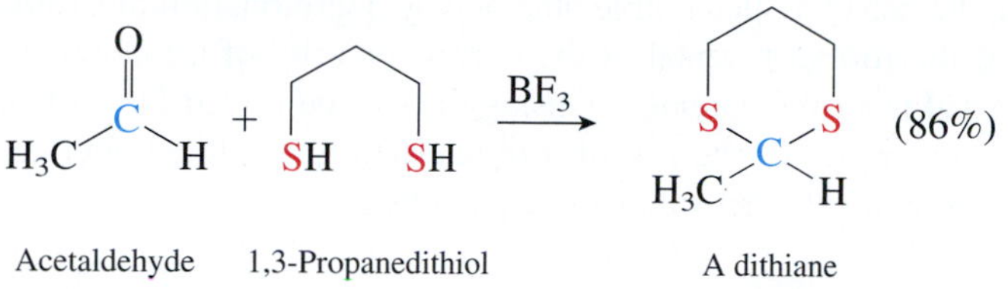

Acetaldehyde   1,3-Propanedithiol   A dithiane

The hydrogen on the carbon attached to the two sulfur atoms is weakly acidic ($pK_a =$ 31) and can be removed by reaction with a strong base, such as butyllithium. (Butyllithium is also a nucleophile, and therefore it is not used to generate enolate anions from carbonyl compounds. However, the dithiane is not electrophilic, so butyllithium can be used as the base in this reaction.)

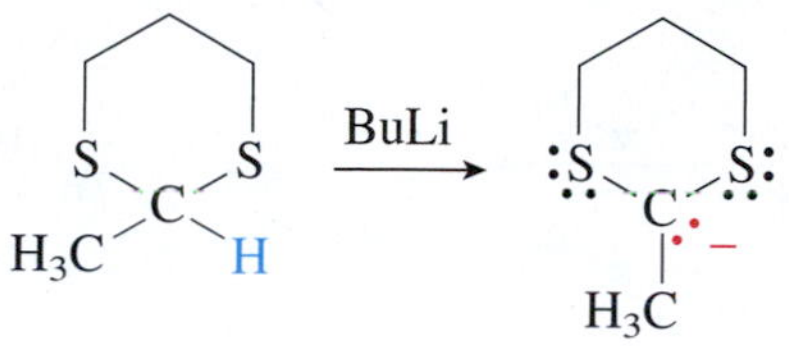

The acidity of the dithiane can be attributed to the stabilization of the conjugate base by the inductive effect of the sulfurs.

The dithiane anion is a good nucleophile in S_N2 reactions. After it has been alkylated, the thioacetal group can be removed by hydrolysis using Hg^{2+} as a Lewis acid catalyst.

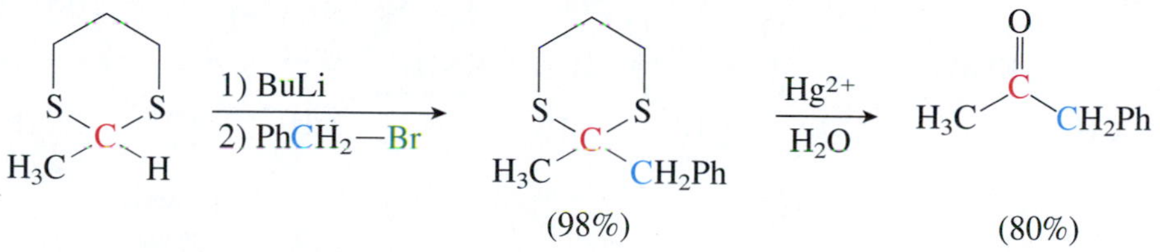

The following equation provides an example of the overall process:

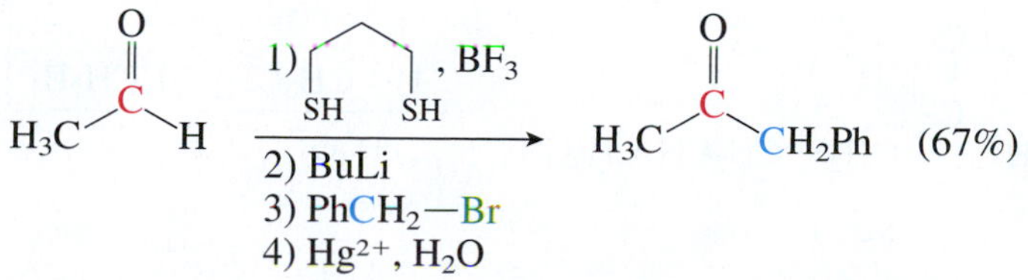

The nucleophile obtained by deprotonation of a dithiane serves as the synthetic equivalent of an acyl anion, a species that is too unstable to be prepared directly.

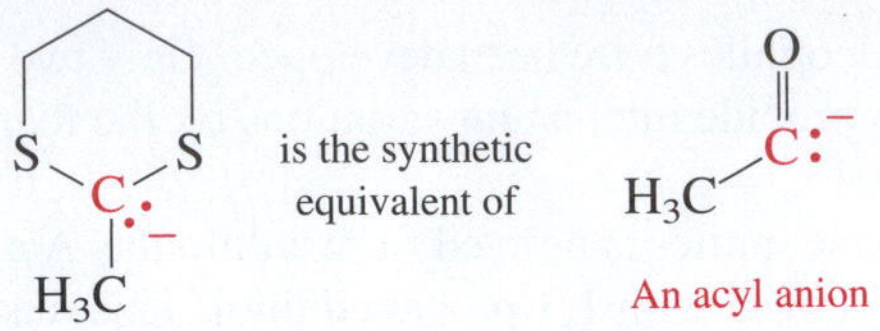

In all of the reactions that have been presented until this one, a carbonyl carbon has always reacted as an electrophile. An acyl anion, however, has a nucleophilic carbonyl carbon. Thus, the use of a nucleophile obtained by deprotonation of a dithiane provides an example of the formal reversal of the normal polarity of a functional group. Such **polarity reversal** is termed *umpolung,* using the German word for reversed polarity.

Additional examples of the use of a dithiane to generate an acyl anion synthetic equivalent are provided by the following equations:

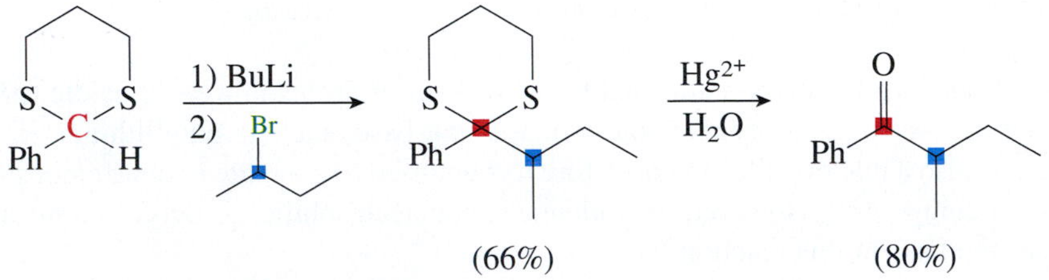

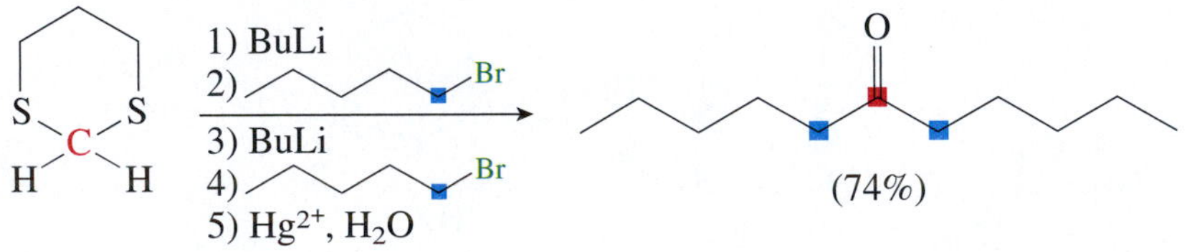

The final type of carbon nucleophile that is discussed in this chapter is a dianion. In some cases, treatment of an anion with a very strong base can remove a second proton to form a dianion. As an example, the reaction of 2,4-pentanedione with one equivalent of base removes a proton from the carbon between the two carbonyl groups. If this anion is treated with a second equivalent of a strong base, such as potassium amide, a second proton can be removed to form a dienolate anion:

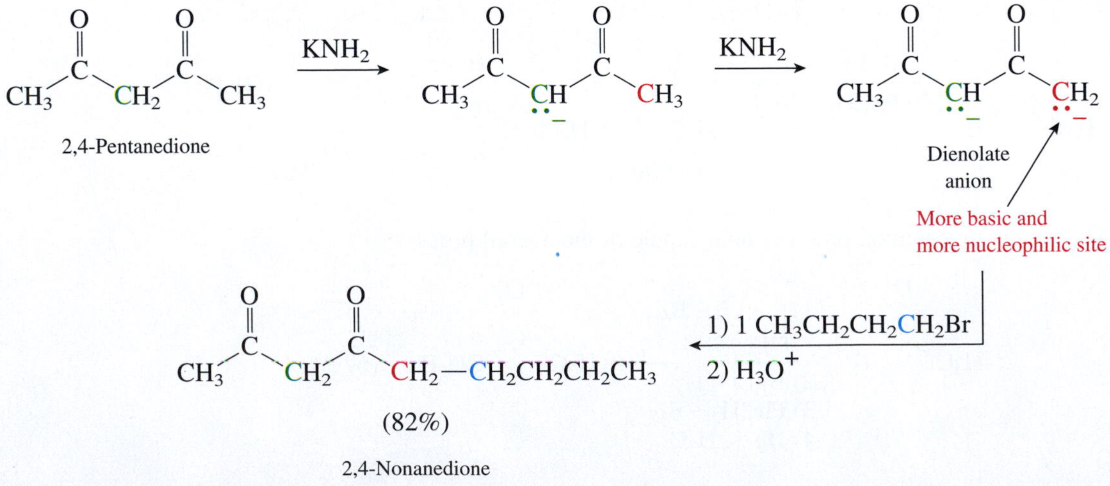

When this dianion is reacted with one equivalent of an alkyl halide, the more basic site acts as a nucleophile. The addition of acid neutralizes the remaining anion. Overall, this process allows the more basic site to be alkylated preferentially. The following equation shows another example:

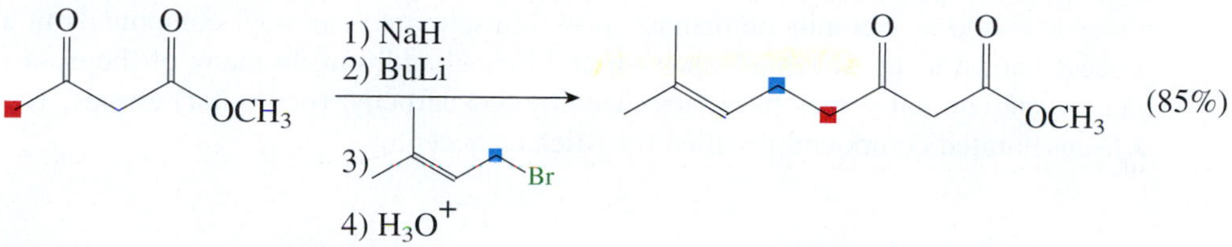

The same strategy can be extended to the alkylation of carboxylic acids at the α-carbon, as illustrated in the following example:

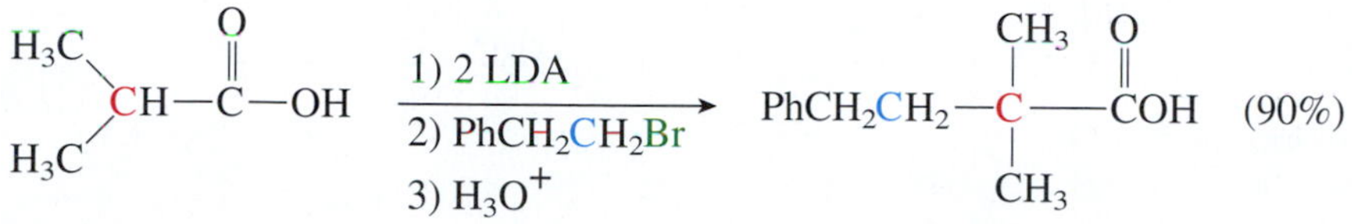

PROBLEM 20.23
Show the products of these reactions:

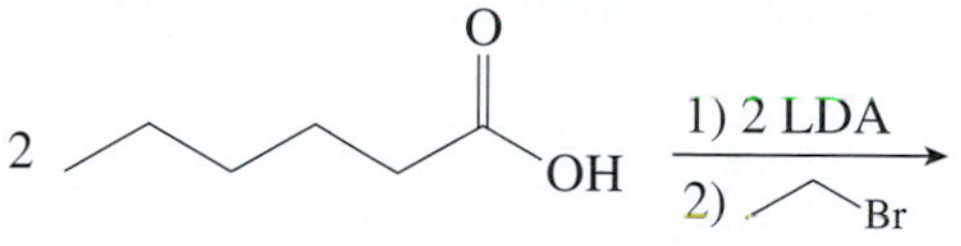

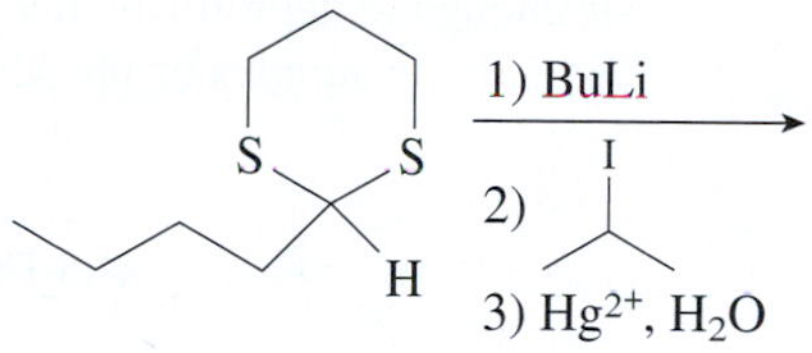

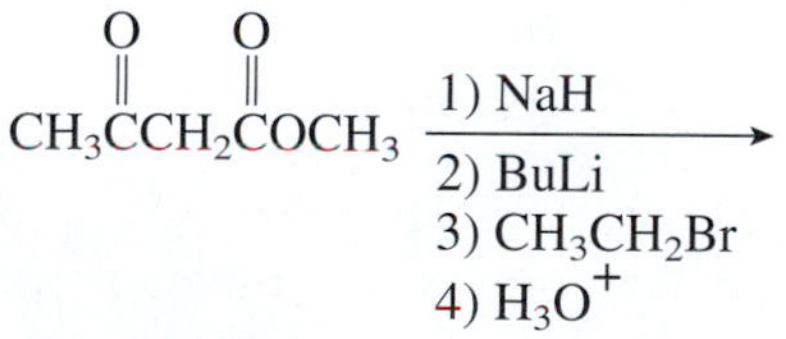

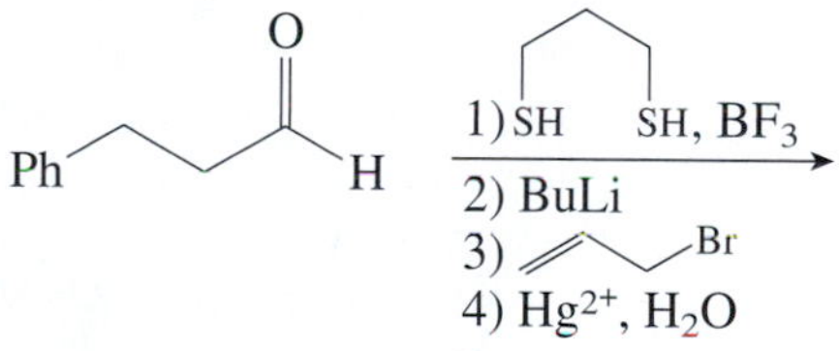

PROBLEM 20.24
Suggest methods to accomplish these transformations. More than one step may be necessary.

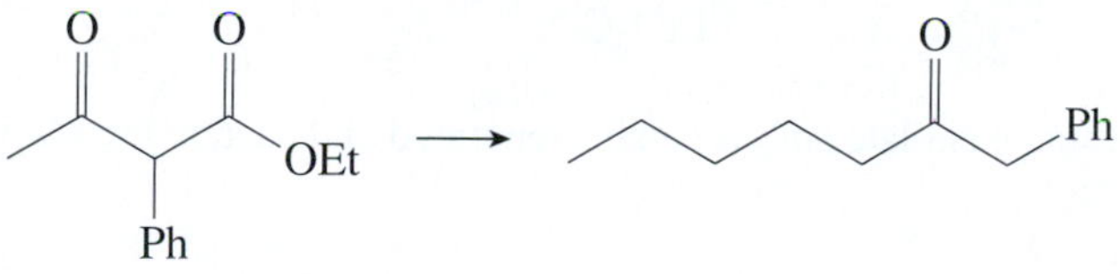

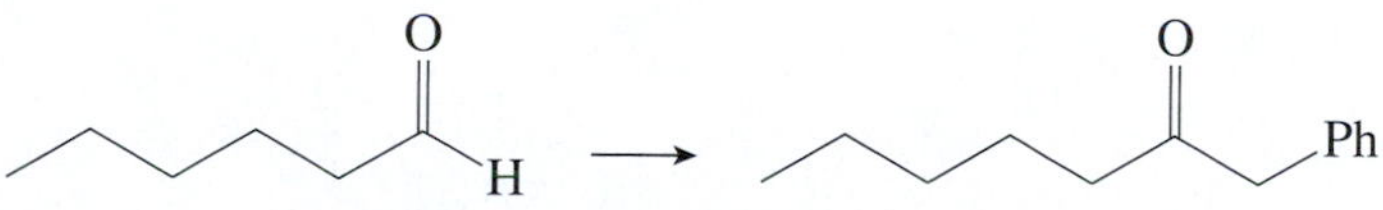

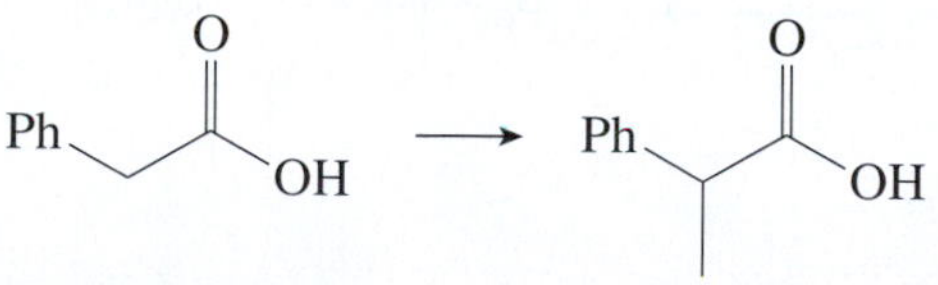

20.10 CONJUGATE ADDITIONS

The conjugate addition of nucleophiles to α,β-unsaturated carbonyl compounds at the β-position was described in Section 18.10. Enolate and related carbanion nucleophiles also add in a conjugate manner to α,β-unsaturated carbonyl compounds in a process known as the **Michael reaction** or Michael addition. In many of the examples the enolate ion is one that is stabilized by two carbonyl (or similar) groups. The α,β-unsaturated compound is called the Michael acceptor.

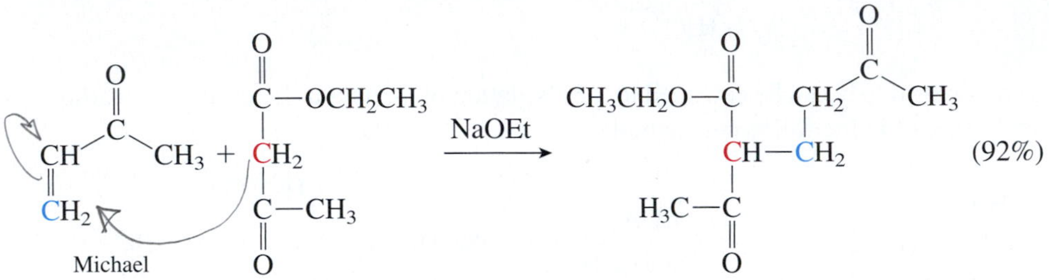

The mechanism for the Michael reaction is shown in Figure 20.6. Only a catalytic amount of base is needed because the initial adduct is itself an enolate anion and is basic enough to deprotonate the dicarbonyl compound, allowing additional reaction to occur. Other examples of the Michael reaction are provided in the following equations:

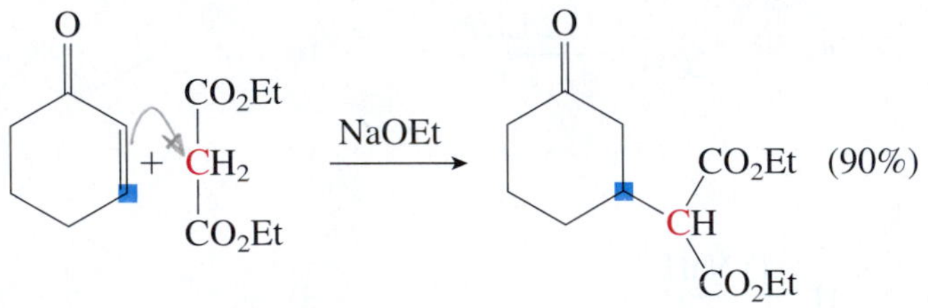

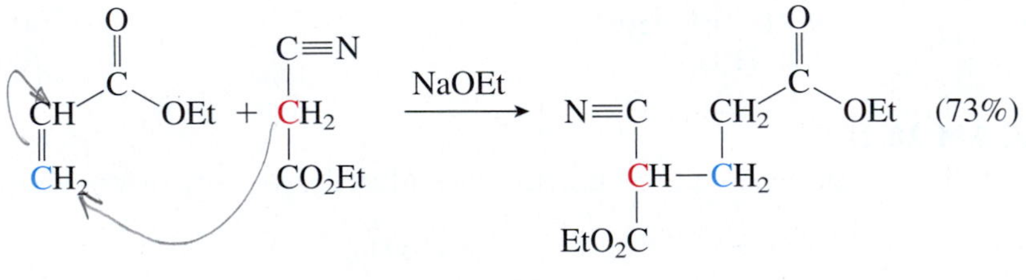

With stronger bases, less stable enolate anions can be generated and used in the Michael reaction:

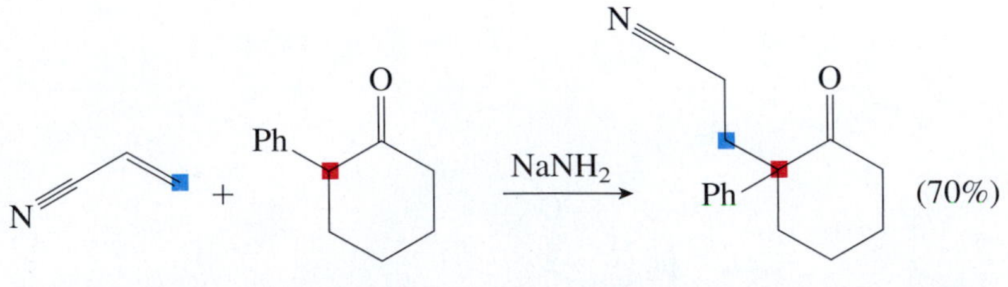

2-Propenenitrile   2-Phenylcyclohexanone

Figure 20.6

MECHANISM OF THE
MICHAEL REACTION.

❶ The enolate anion of the β-dicarbonyl compound is generated in the usual manner.

❷ The enolate nucleophile adds to the β-carbon of an α,β-unsaturated ketone (or ester or nitrile), which is called the Michael acceptor.

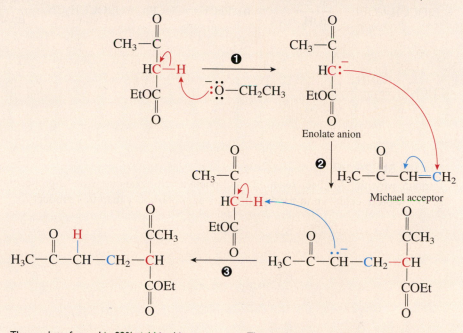

The product, formed in 92% yield in this case, has a bond from the α-carbon of the original enolate ion to the β-carbon of the Michael acceptor.

The product of this addition is an enolate ion.

❸ This ion reacts as a base with ethyl aceto-acetate to regenerate another enolate ion, so only a catalytic amount of base is needed for the reaction.

PRACTICE PROBLEM 20.7

Show the product of this reaction:

$$CH_2{=}CHCOEt \ + \ CH_3CCH_2COEt \ \xrightarrow[\text{EtOH}]{\text{NaOEt}}$$

Solution

Identify the electrophile and the nucleophile. The base (NaOEt) removes the most acidic hydrogen, the one on the carbon between the carbonyl groups of ethyl acetoac-etate, to generate the enolate anion. This nucleophile then attacks at the β-carbon of the Michael acceptor:

$$CH_3C{-}\overset{\_}{C}H{-}COEt \ + \ CH_2{=}CHCOEt \ \longrightarrow \ \longrightarrow \ CH_3CCH{-}CH_2CH_2COEt$$
$$CO_2Et$$

PROBLEM 20.25

Show the products of these reactions:

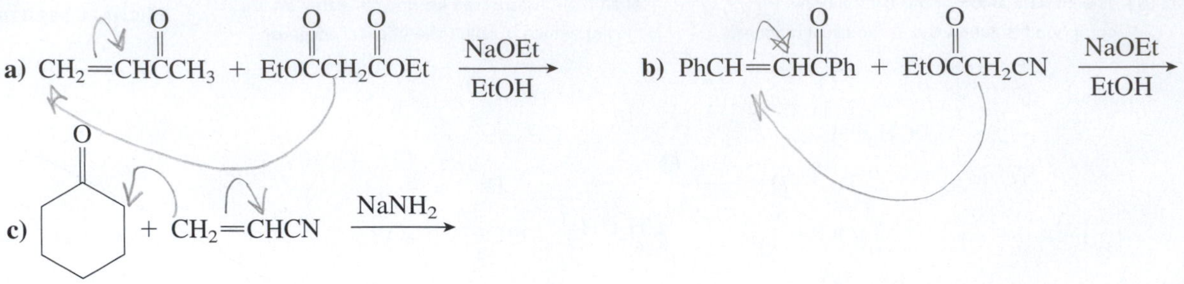

a) $CH_2{=}CHCCH_3$ + $EtOCCH_2COEt$ $\xrightarrow[\text{EtOH}]{\text{NaOEt}}$

b) $PhCH{=}CHCPh$ + $EtOCCH_2CN$ $\xrightarrow[\text{EtOH}]{\text{NaOEt}}$

c) + $CH_2{=}CHCN$ $\xrightarrow{\text{NaNH}_2}$

PROBLEM 20.26

Explain why the Michael reaction of 2-phenylcyclohexanone with 2-propenenitrile gives the product shown in the equation on page 894 rather than this product:

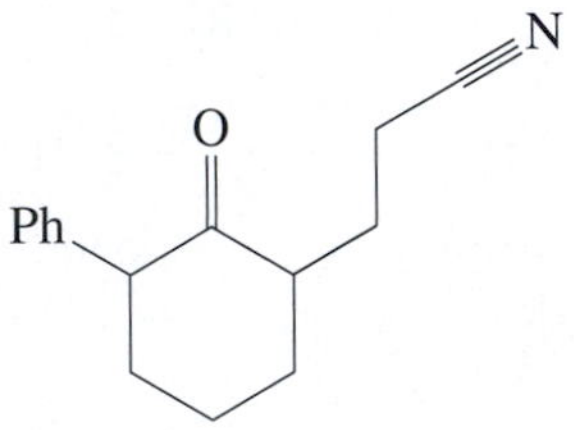

The Michael reaction in combination with an aldol condensation provides a useful method for the construction of six-membered rings in a process termed the **Robinson annulation.** In the following example a tertiary amine is used as the base to catalyze the conjugate addition. Then, treatment with sodium hydroxide causes an intramolecular aldol condensation to occur.

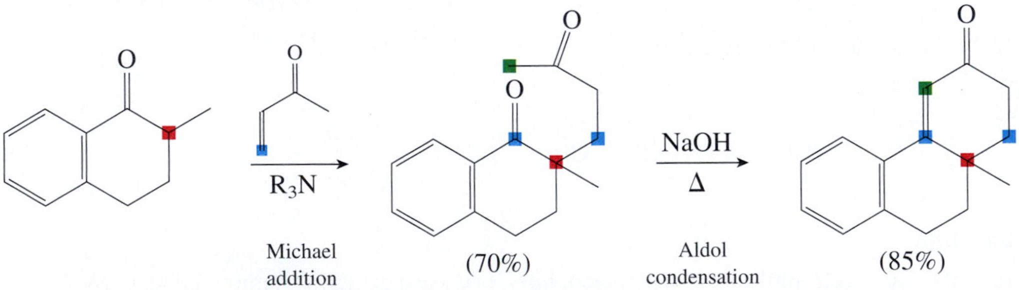

$\xrightarrow{\text{R}_3\text{N}}$ $\xrightarrow[\Delta]{\text{NaOH}}$

Michael   (70%)   Aldol   (85%)
addition   condensation

Often, the Michael addition product is not isolated. Instead, the intramolecular aldol condensation occurs immediately, and the new six-membered ring is formed, as shown in the following equation. (However, when you are attempting to write the product of such a reaction, it is best to first write the product of Michael addition and then write the final product that results from the aldol condensation.)

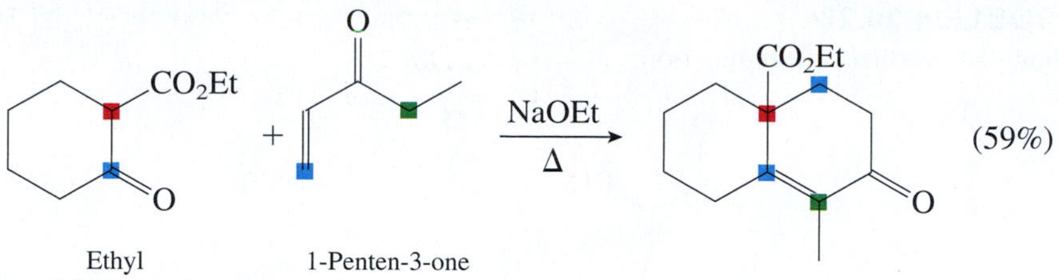

Ethyl
2-oxocyclohexanecarboxylate   1-Penten-3-one

PROBLEM 20.27

Show the intermediate aldol product in the Robinson annulation reaction of ethyl 2-oxocyclohexanecarboxylate with 1-penten-3-one.

PRACTICE PROBLEM 20.8

Show the product of this Robinson annulation:

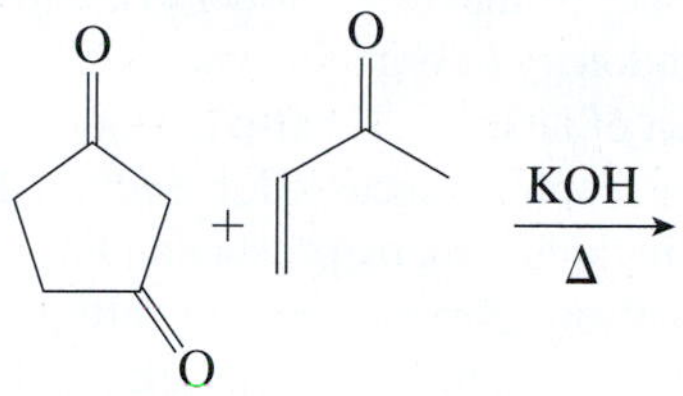

Solution

The base removes the most acidic hydrogen, and the resulting enolate anion undergoes a conjugate addition with the Michael acceptor:

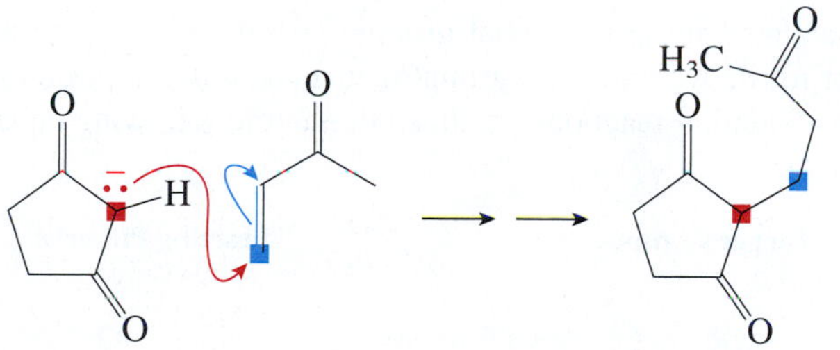

This product then undergoes an aldol condensation to give the final product:

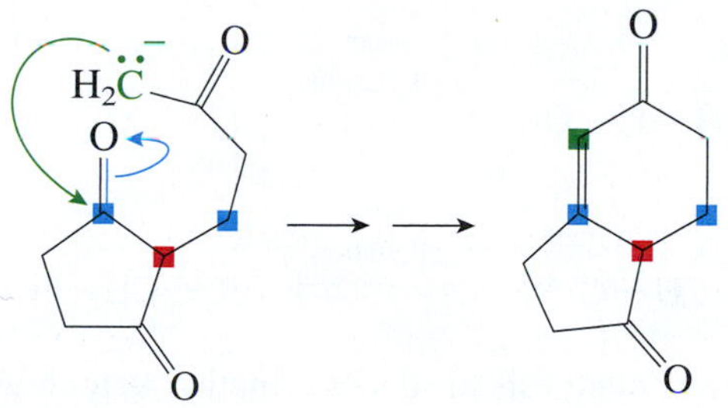

PROBLEM 20.28

Show the product of this reaction:

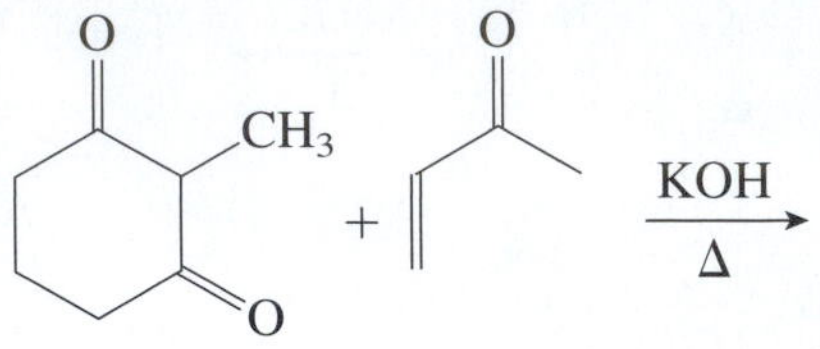

20.11 SYNTHESIS

The reactions presented in this chapter are very important in synthesis because they all result in the formation of carbon–carbon bonds. As we have seen, the best way to approach a synthesis problem is to employ retrosynthetic analysis, that is, to work backward from the target molecule to simpler compounds until a readily available starting material is reached. (Elias J. Corey, winner of the 1990 Nobel Prize in chemistry for "his development of the theory and methodology of organic synthesis," coined the term *retrosynthetic analysis* and formalized much of its logic. He also developed numerous new reagents, including the dithiane anion nucleophile discussed in Section 20.9, and synthesized a large number of natural products, including many prostaglandins [Section 28.9].) Recall that it is helpful in retrosynthetic analysis to recognize that certain structural features in the target suggest certain reactions. For example, we learned in Chapter 18 that an alcohol target compound suggests that a Grignard reaction might be used in its synthesis.

Let's look at the reactions presented in this chapter in terms of their products so that we might more easily recognize the synthetic reactions that are suggested by the presence of certain features in the target compound.

Alkylations of ketones, esters, and nitriles add an alkyl group to the α-carbon of compounds containing these functional groups. Therefore, a target molecule that is a ketone, ester, or nitrile with an alkyl group(s) attached to its α-carbon suggests the use of one of these alkylation reactions, as illustrated in the following equations using retrosynthetic arrows:

Target Compound		**Starting Material**
$R'-CH(R)-C(=O)-R''$	ketone alkylation $\Longrightarrow$	$R'-CH_2-C(=O)-R''$
$R'-CH(R)-C(=O)-OEt$	ester alkylation $\Longrightarrow$	$R'-CH_2-C(=O)-OEt$
$R'-CH(R)-C\equiv N$	nitrile alkylation $\Longrightarrow$	$R'-CH_2-C\equiv N$

The acetoacetic ester synthesis produces a methyl ketone with an alkyl group(s) substituted on the α-carbon, whereas the malonic ester synthesis produces a carboxylic acid with an alkyl group(s) substituted on the α-carbon. Note that these targets can sometimes also be synthesized by the direct alkylation of a ketone or ester:

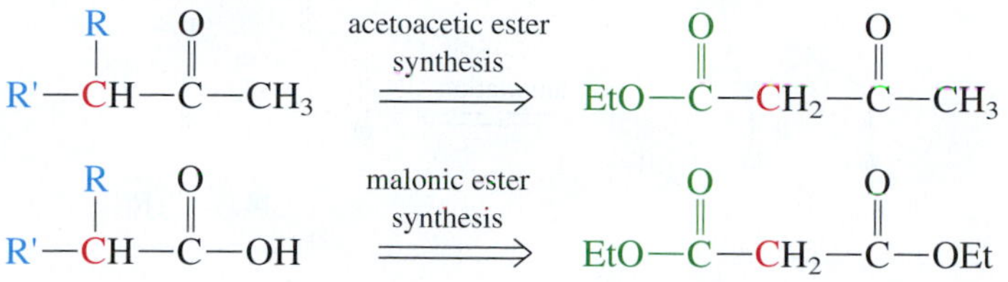

The presence of a carbon–carbon double bond conjugated to a carbonyl group (an α,β-unsaturated aldehyde, ketone, and so on) in the target compound suggests that an aldol condensation be employed:

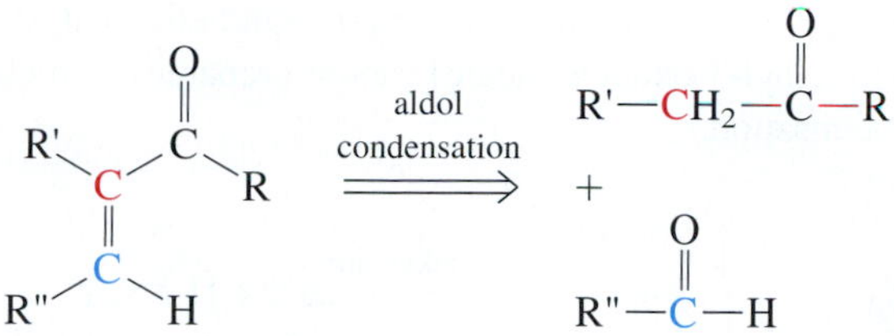

If the target compound has two carbonyl groups attached to the same carbon, an ester condensation is suggested. Note that such a target could also be prepared by alkylation of a β-dicarbonyl compound:

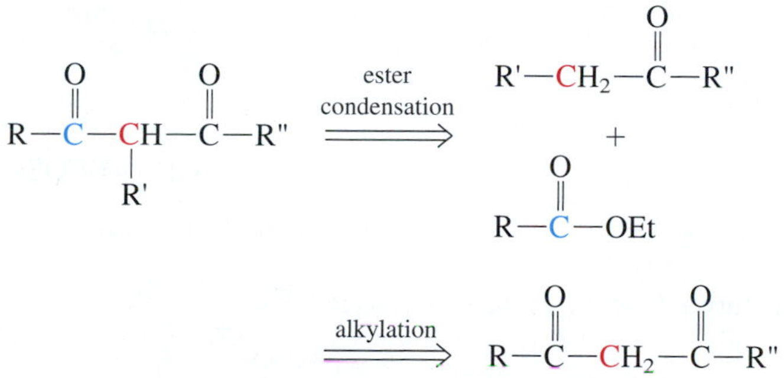

The presence of two carbonyl groups (or other functional groups that are capable of stabilizing a carbanion) in a 1,5-relationship suggests the use of a Michael addition to prepare that target compound:

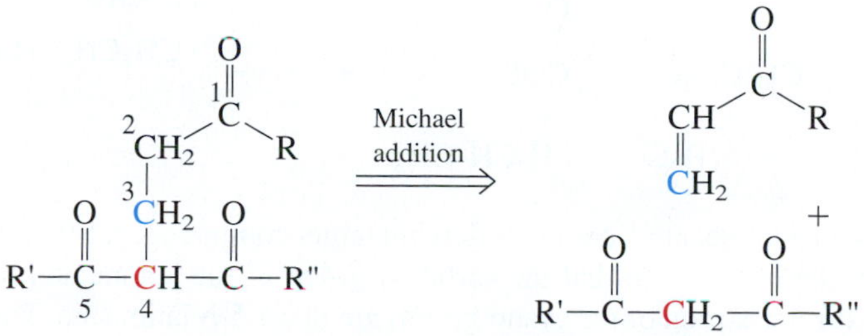

Finally, the presence of a cyclohexenone ring in the target suggests that a Robinson annulation might be employed in its synthesis:

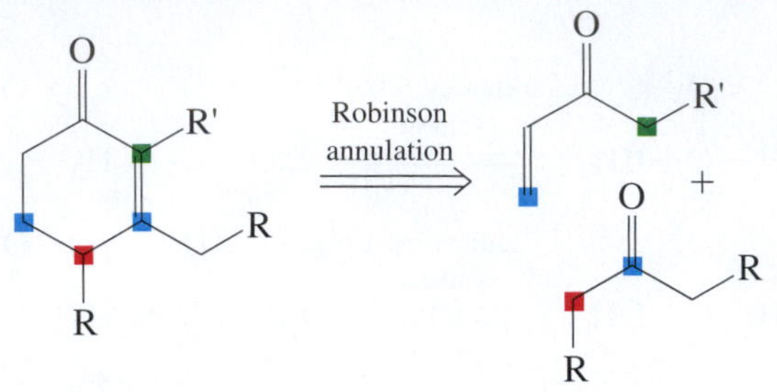

Let's try a synthesis. Suppose the target is ethyl 2-methyl-3-oxo-2-propylpentanoate. The presence of the β-ketoester functionality suggests employing an alkylation reaction and/or an ester condensation. In one potential pathway, the propyl group can be attached by alkylation of a simpler β-ketoester. Further retrosynthetic analysis suggests that the new target (ethyl 2-methyl-3-oxopentanoate) can be prepared from ethyl propanoate by a Claisen ester condensation.

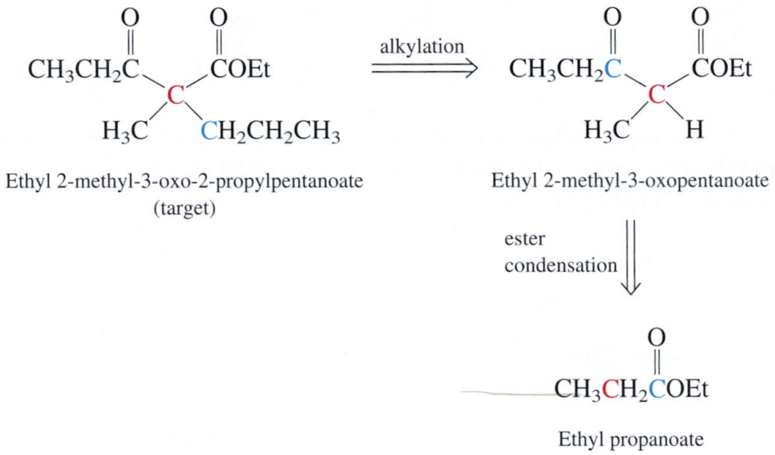

Ethyl 2-methyl-3-oxo-2-propylpentanoate
(target)

Ethyl 2-methyl-3-oxopentanoate

Ethyl propanoate

Written in the forward direction, the synthesis is

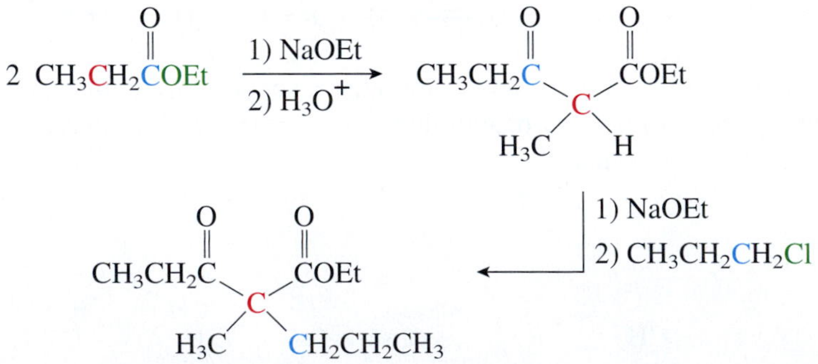

As another example, consider the following target compound. A Michael reaction is suggested by the observation that the carbonyl group of the ketone and the ester carbonyl group (or the carbon of the cyano group) are in a 1,5-relationship. The compound required for the Michael addition is an α,β-unsaturated ketone, suggesting that a mixed aldol condensation be used to prepare it.

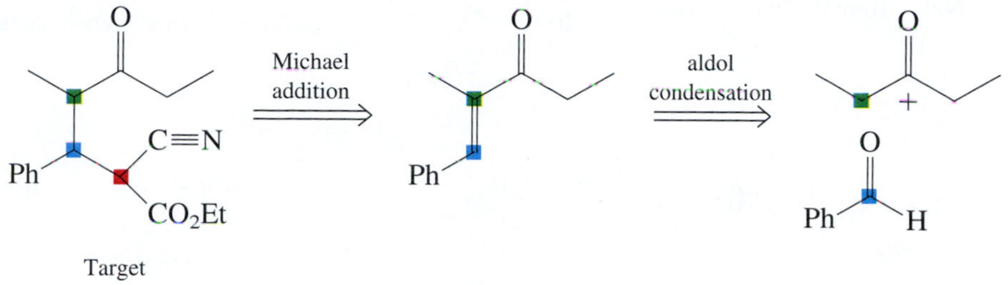

Target

Written in the forward direction, the synthesis is

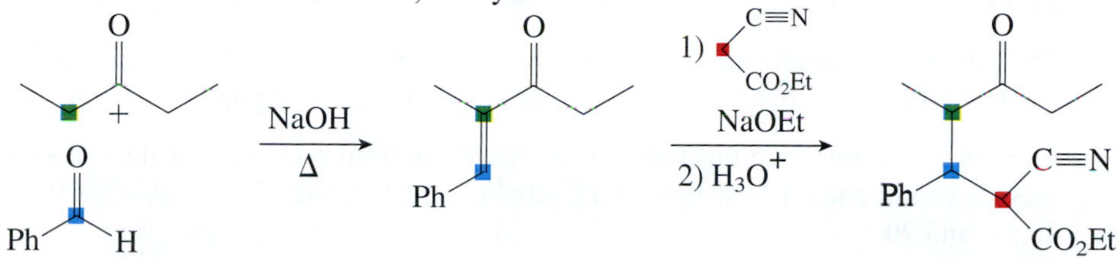

Like any other endeavor, the only way to become proficient in designing syntheses is practice. Work the problems and, when working them, examine each target compound for structural features that suggest certain reactions. Remember that most targets can be reached by numerous pathways, so if your route does not match the one presented in the answers, do not despair. Check all of the reactions that you use to ensure that they are appropriate. If your pathway seems a reasonable route to the target, then it may be as good as, or even better than, the one shown in the answer.

PROBLEM 20.29

Show syntheses of these compounds from the indicated starting materials:

a) 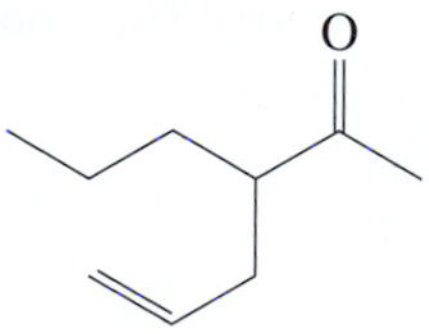 from ethyl acetoacetate

b) 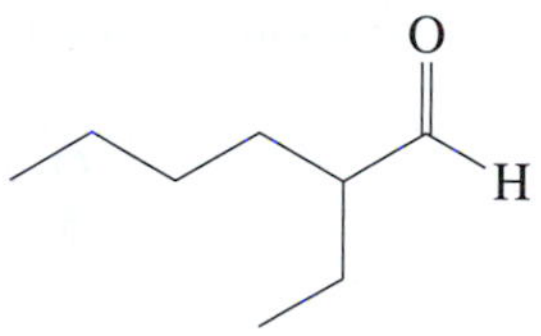 from butanal

c) 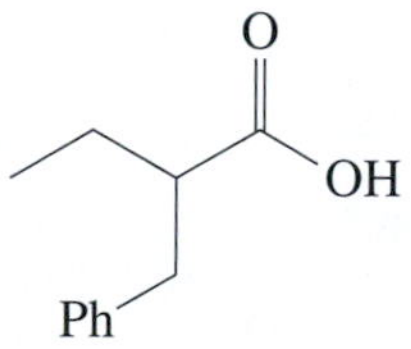 from ethyl butanoate

d) 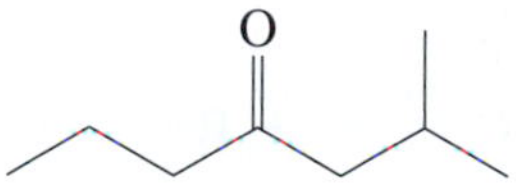 from butanal

e) 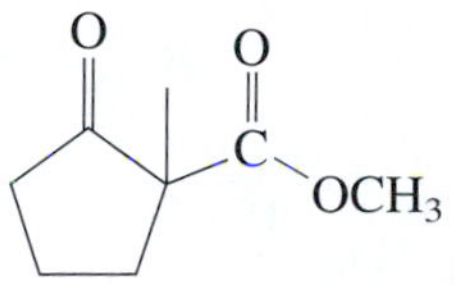 from cyclopentanone

f) 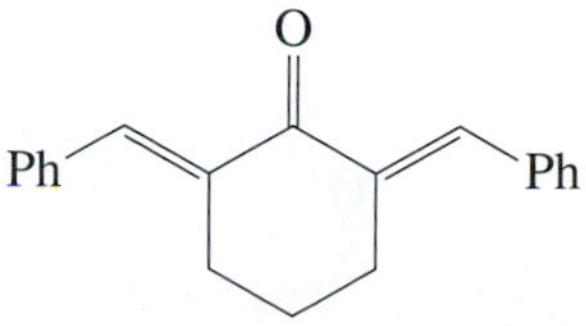 Ph from cyclohexanone

g) 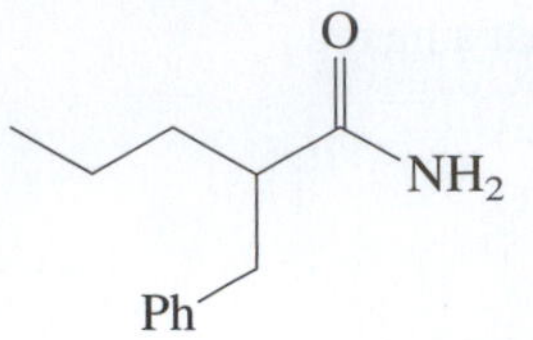 [structure: pentanamide with benzyl substituent, Ph] NH$_2$ from diethyl malonate

h) 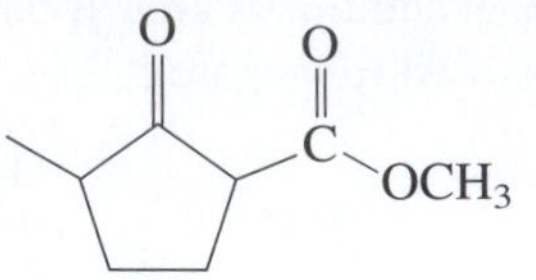 [structure: methylcyclopentanone with C(=O)OCH$_3$ group] from cyclopentanone

Review of Mastery Goals

After completing this chapter, you should be able to:

■ Show the products of any of the reactions discussed in this chapter. (Problems 20.30, 20.31, 20.32, 20.41, 20.47, 20.54, and 20.55)

■ Show the mechanism for any of these reactions. (Problems 20.33, 20.34, 20.35, 20.36, 20.40, 20.44, 20.45, 20.46, 20.51, 20.52, 20.53, 20.56, 20.58, 20.62, and 20.63)

■ Use these reactions in combination with reactions from previous chapters to synthesize compounds. (Problems 20.37, 20.38, 20.39, 20.42, 20.43, 20.48, 20.49, 20.50, and 20.57)

Visual Summary of Key Reactions

A large number of reactions have been presented in this chapter. However, all of these reactions involve an enolate ion (or a related species) acting as a nucleophile (see Table 20.2). This nucleophile reacts with one of the electrophiles discussed in Chapters 8, 18, and 19 (see Table 20.3). The nucleophile can bond to the electrophilic carbon of an alkyl halide (or sulfonate ester) in an S$_N$2 reaction, to the electrophilic carbonyl carbon of an aldehyde or ketone in an addition reaction (an aldol condensation), to the electrophilic carbonyl carbon of an ester in an addition reaction (an ester condensation) or to the electrophilic β-carbon of an α,β-unsaturated compound in a conjugate addition (Michael reaction). These possibilities are summarized in the following equations:

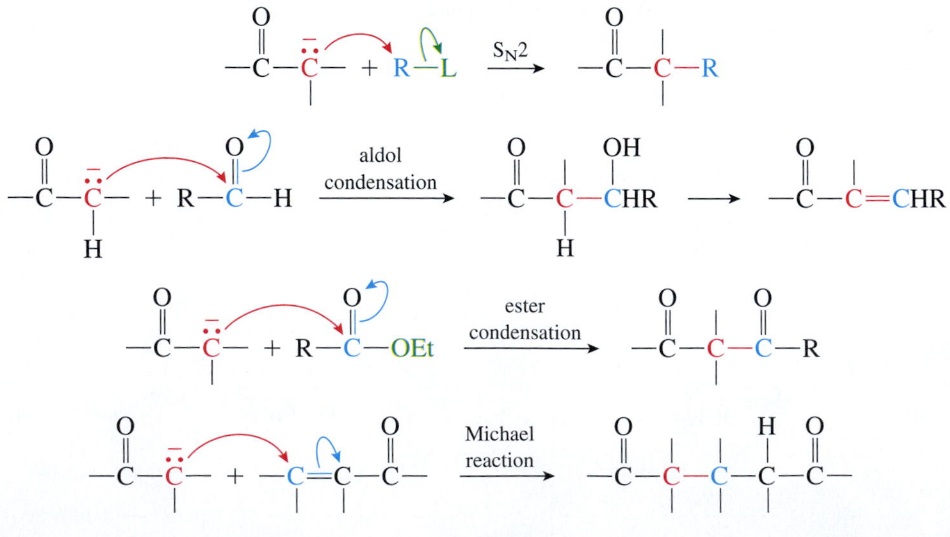

[reaction scheme: enolate + R–L, S$_N$2, giving alkylated product]

[reaction scheme: enolate + R–CH(=O), aldol condensation, giving β-hydroxy carbonyl then α,β-unsaturated carbonyl (=CHR)]

[reaction scheme: enolate + R–C(=O)–OEt, ester condensation, giving β-keto product]

[reaction scheme: enolate + α,β-unsaturated carbonyl, Michael reaction, giving 1,5-dicarbonyl product]

Table 20.2 Carbon Nucleophiles

Nucleophile	Comments
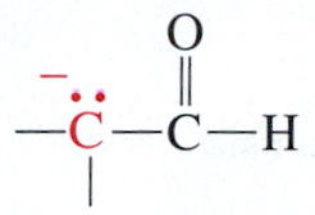	Because aldehydes are so reactive as electrophiles, enolate anions derived from them are primarily restricted to use in the aldol condensation.
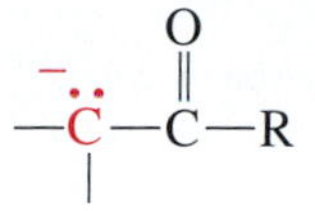	Enolate anions derived from ketones, esters, and nitriles can be alkylated, used in the aldol or ester condensations, or used in the Michael reaction.
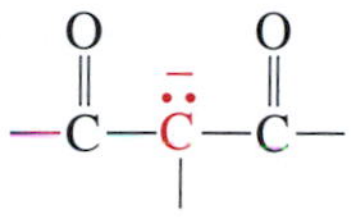	Enolate anions that are stabilized by two carbonyl groups (or cyano groups) can be alkylated and give excellent yields in the Michael reaction.
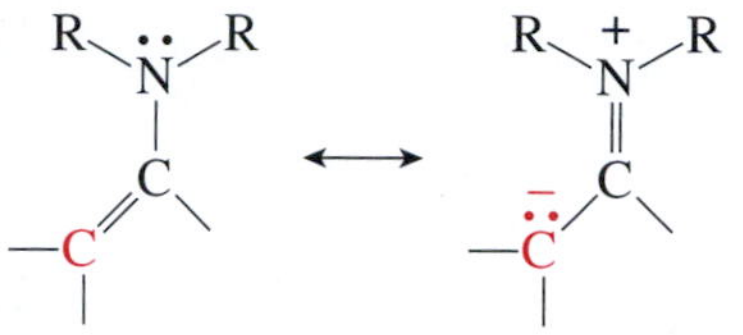	Enamines react only with very reactive electrophiles.
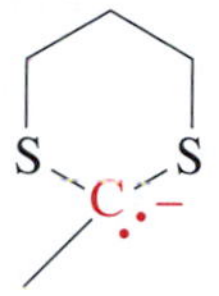	Dithiane anions are acyl anion equivalents and can be readily alkylated.
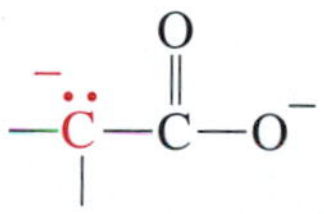 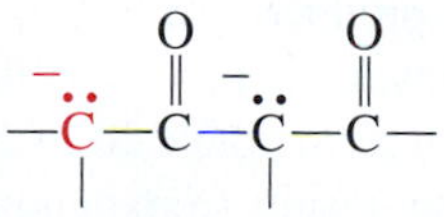	These dianions can be selectively alkylated at the more basic site.

Table 20.3 Electrophiles

Electrophiles	Comments
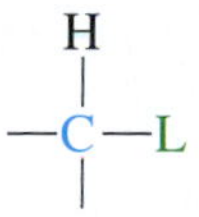	The carbon must be primary or secondary to use these compounds in an S_N2 reaction. The leaving group can be a halide or a sulfonate group.
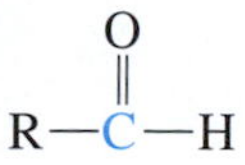	Aldehydes are very reactive electrophiles in the aldol condensation.
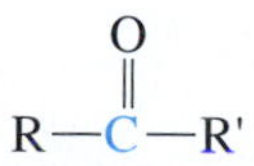	Ketones are less reactive and are most useful when the aldol condensation is intramolecular.

Continued

Table 20.3   Electrophiles—cont'd

Electrophiles	Comments
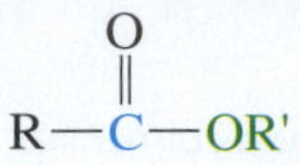 $\underset{\displaystyle R-\overset{\displaystyle O}{\overset{\displaystyle \|}{C}}-OR'}{}$	Esters are useful in the Claisen ester condensation. Most often, R′ = Me or Et.
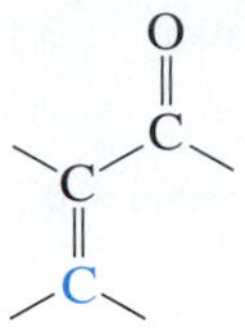 (α,β-unsaturated carbonyl structure)	α,β-Unsaturated compounds are useful in the Michael reaction. A cyano group can be used in place of the carbonyl group.

Integrated Practice Problem

Show the products of these reactions:

a) 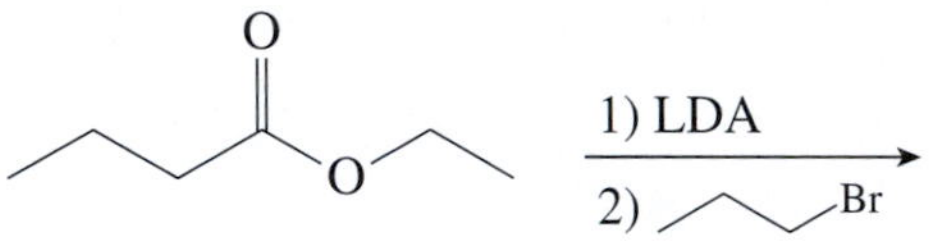 (ethyl butanoate) $\xrightarrow[\text{2)} \diagup\diagdown\text{Br}]{\text{1) LDA}}$

b) 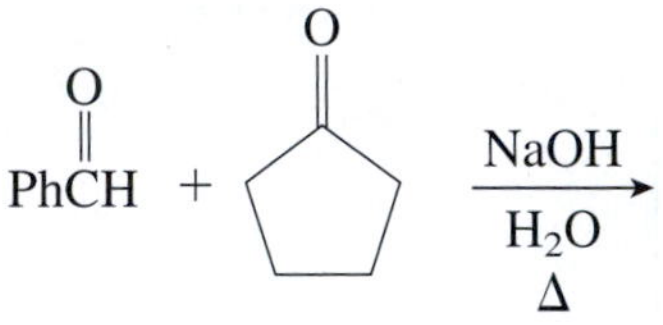 PhCH $+$ (cyclopentanone) $\xrightarrow[\substack{\text{H}_2\text{O} \\ \Delta}]{\text{NaOH}}$

c) 2 (ethyl butanoate) $\xrightarrow[\text{2) H}_3\text{O}^+]{\text{1) NaOEt, EtOH}}$

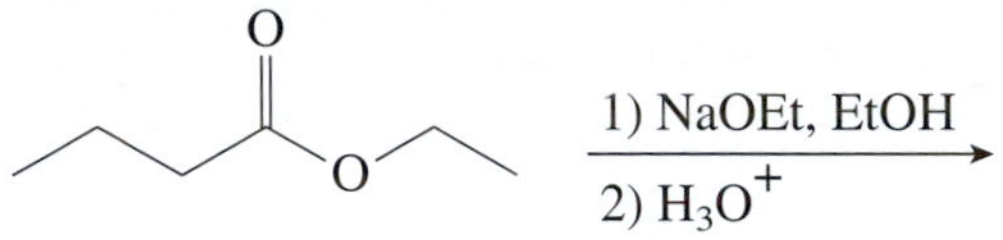

Strategy

Students often have difficulty with these reactions because the products are large and rather complex. You will have a much easier time remembering these reactions if you first identify the site where the enolate anion or related nucleophile will form (Table 20.2) and then identify the electrophilic site (Table 20.3). The product simply results from bonding the nucleophilic carbon to the electrophilic carbon.

Solutions

a) LDA is a strong base and removes a proton from the α-carbon of the ester. The resulting enolate anion acts as a nucleophile in an S_N2 reaction with the alkyl bromide.

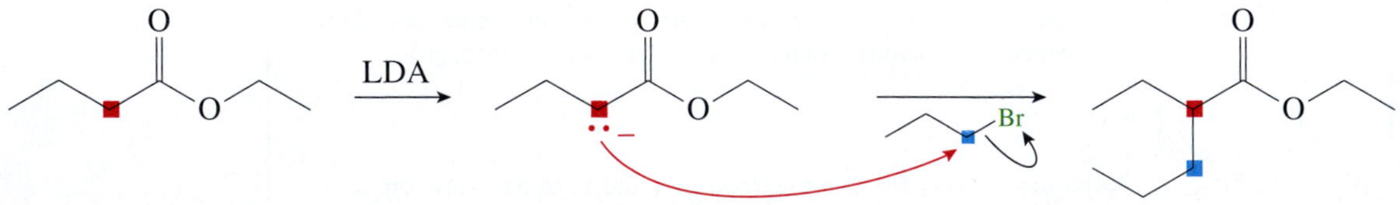

b) The base removes the acidic hydrogen on the α-carbon of cyclopentanone. The resulting enolate anion nucleophile bonds to the electrophilic carbonyl carbon of benzaldehyde in an aldol condensation.

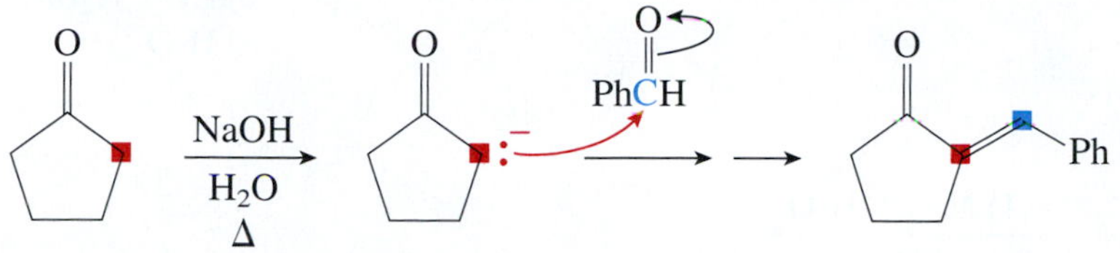
NaOH
H_2O
Δ
PhCH
Ph

c) The base removes the acidic hydrogen on the α-carbon of the ester. The resulting enolate anion nucleophile bonds to the electrophilic carbonyl carbon of another ester molecule in an ester condensation.

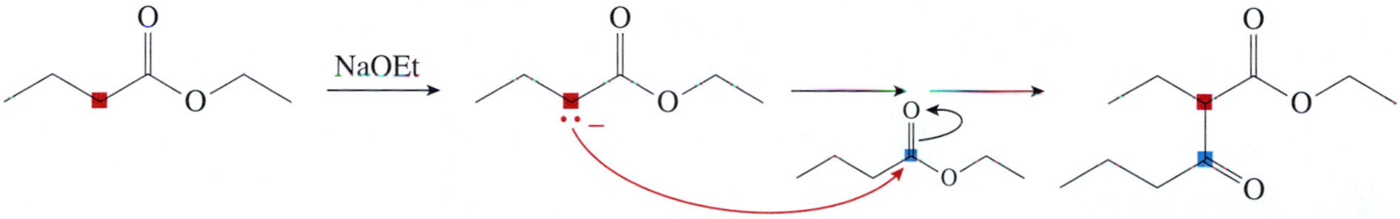
NaOEt

Additional Problems

ORGANIC Chemistry Now™ Assess your understanding of this chapter's topics with additional quizzing and conceptual-based problems at **http://now.brookscole.com/hornback2**

20.30 Show the products of these reactions:

a)
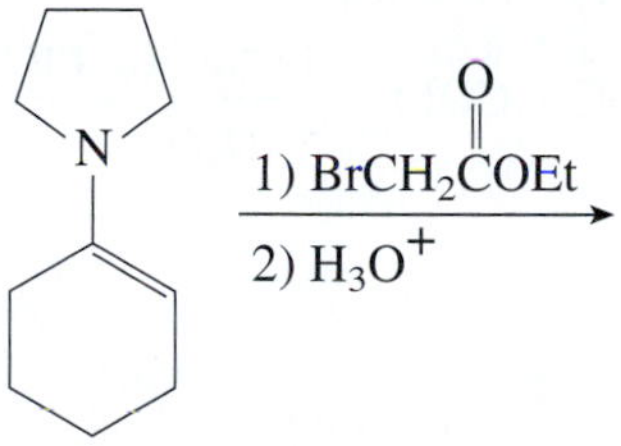
1) $BrCH_2\overset{O}{\overset{\|}{C}}OEt$
2) H_3O^+

b)
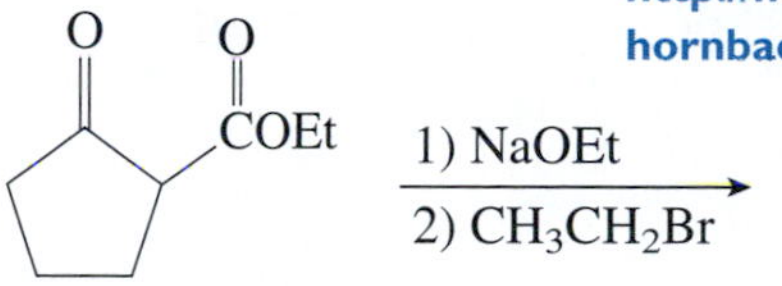
COEt
1) NaOEt
2) CH_3CH_2Br

c)
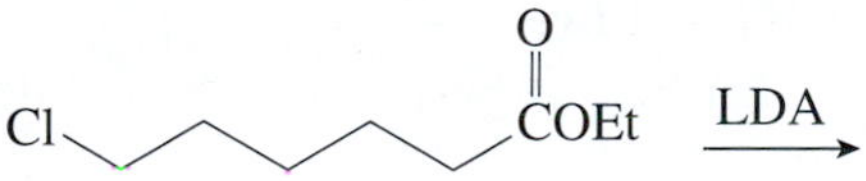
Cl
COEt
LDA

d) $2\ CH_3\overset{CH_3}{\overset{|}{C}}H{-}\overset{O}{\overset{\|}{C}}H \xrightarrow[H_2O]{NaOH}$

e) $EtO\overset{O}{\overset{\|}{C}}CH_2CH_2CH_2\underset{CH_3}{\underset{|}{C}}H\overset{O}{\overset{\|}{C}}OEt \xrightarrow[2)\ H_3O^+]{1)\ NaOEt,\ EtOH}$

f)
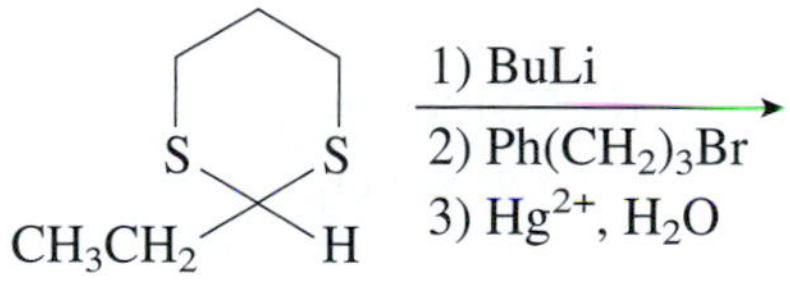
S
S
CH_3CH_2
H
1) BuLi
2) $Ph(CH_2)_3Br$
3) Hg^{2+}, H_2O

g) $PhCH_2\overset{O}{\overset{\|}{C}}OEt \xrightarrow[2)\ PhCH_2CH_2Br]{1)\ NaNH_2}$

h) $CH_3\overset{O}{\overset{\|}{C}}(CH_2)_4\overset{O}{\overset{\|}{C}}H \xrightarrow[H_2O,\ \Delta]{NaOH}$

i)

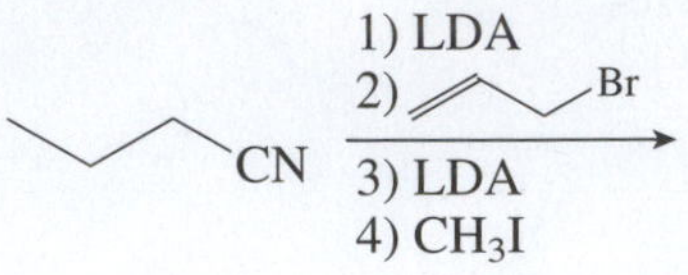

$$\text{CH}_3\text{CH}_2\text{CH}_2\text{CN} \xrightarrow[\substack{3)\ \text{LDA} \\ 4)\ \text{CH}_3\text{I}}]{\substack{1)\ \text{LDA} \\ 2)\ \diagup\diagdown\text{Br}}}$$

j)

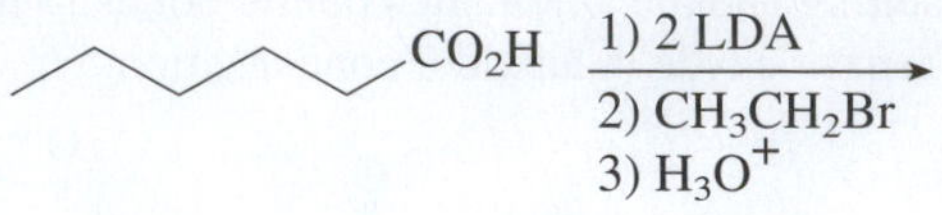

$$\text{CO}_2\text{H} \xrightarrow[\substack{2)\ \text{CH}_3\text{CH}_2\text{Br} \\ 3)\ \text{H}_3\text{O}^+}]{1)\ 2\ \text{LDA}}$$

k)

$$\text{EtO-CO-CH}_2\text{-CO-OEt} \xrightarrow[\substack{2)\ \diagup\diagdown\diagup\text{Br}}]{1)\ \text{NaOEt}} \xrightarrow[\substack{2)\ \text{H}_3\text{O}^+,\ \Delta}]{1)\ \text{NaOH, H}_2\text{O}}$$

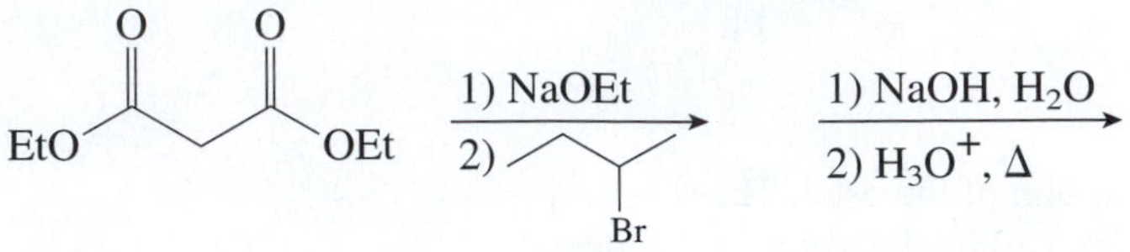

l)

$$\text{CH}_3\text{-CO-CH}_2\text{-CO-OEt} \xrightarrow[\substack{\text{Br}\diagup\diagdown\diagup\diagdown\text{Br}}]{2\ \text{NaOEt}} \xrightarrow[\substack{2)\ \text{H}_3\text{O}^+,\ \Delta}]{1)\ \text{NaOH, H}_2\text{O}}$$

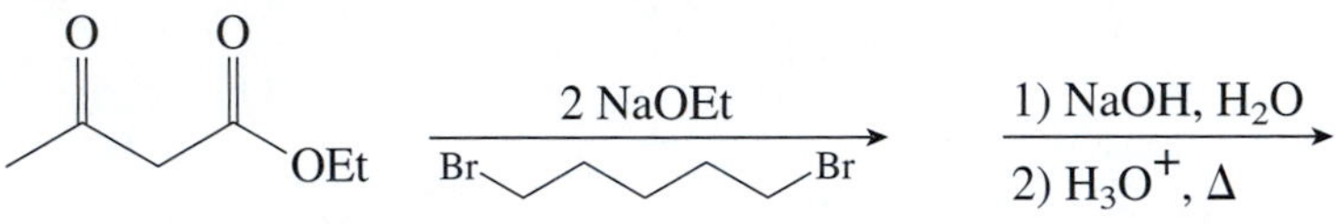

m)

$$\text{CH}_3\text{-CO-CH}_2\text{-CO-CH}_3 \xrightarrow[\substack{2)\ \diagup\diagdown\text{Br} \\ 3)\ \text{H}_3\text{O}^+}]{1)\ 2\ \text{NaNH}_2}$$

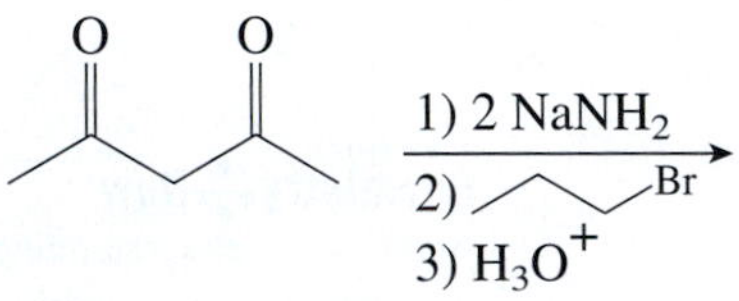

n) furan-CHO $+\ \text{CH}_3\overset{O}{\overset{\|}{\text{C}}}\text{CH}_3 \xrightarrow[\substack{\text{H}_2\text{O} \\ \Delta}]{\text{NaOH}}$

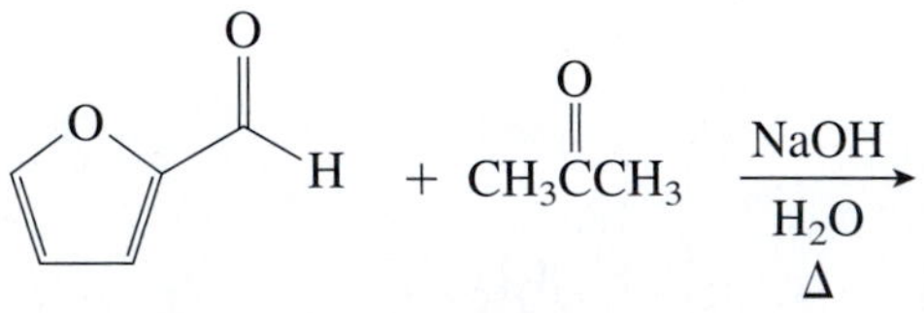

o) $\text{CH}_2(\text{CO}_2\text{Et})_2 + \text{Ph}\diagup\diagdown\text{CO}_2\text{Et} \xrightarrow{\text{NaOEt}}$

(with =CH$_2$)

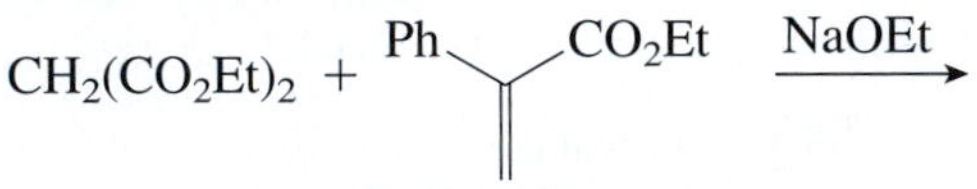

p) 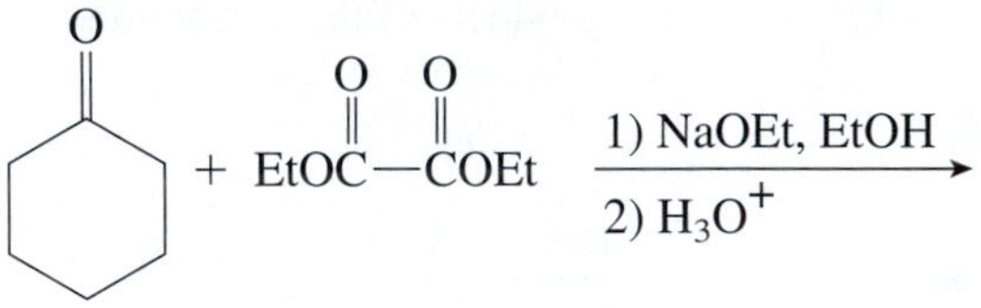 cyclohexanone $+\ \text{EtOC-COEt}\ (\text{with two } O) \xrightarrow[\substack{2)\ \text{H}_3\text{O}^+}]{1)\ \text{NaOEt, EtOH}}$

q)

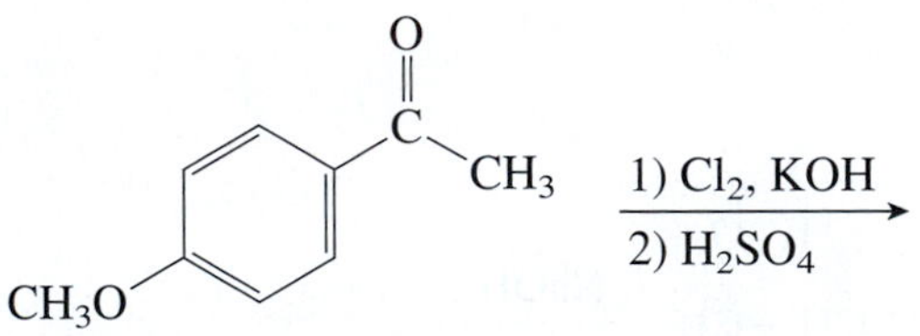

$$\text{CH}_3\text{O-}\diagup\diagdown\text{-}\overset{O}{\overset{\|}{\text{C}}}\text{-CH}_3 \xrightarrow[\substack{2)\ \text{H}_2\text{SO}_4}]{1)\ \text{Cl}_2,\ \text{KOH}}$$

r) 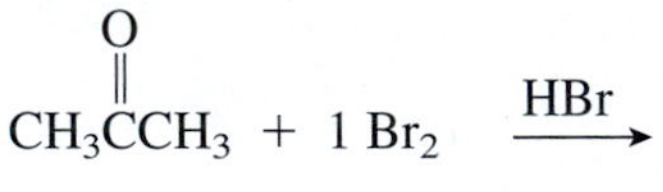 $\text{CH}_3\overset{O}{\overset{\|}{\text{C}}}\text{CH}_3 + 1\ \text{Br}_2 \xrightarrow{\text{HBr}}$

s)

$$\text{H}_3\text{C-}\diagup\diagdown\text{-}\overset{O}{\overset{\|}{\text{C}}}\text{-H} \xrightarrow[\substack{2)\ \text{H}_2\text{SO}_4}]{1)\ \text{NaOH}}$$

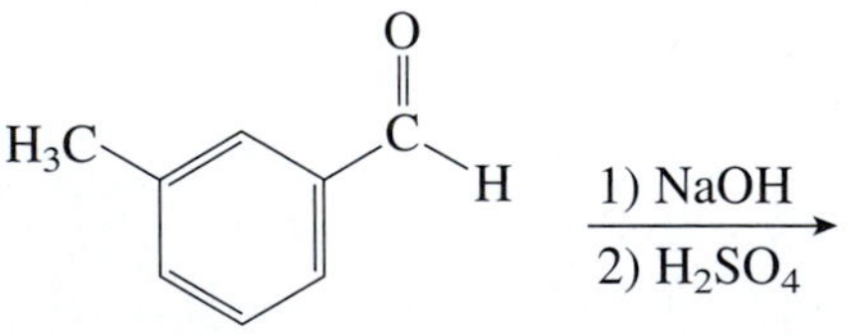

20.31 Show the products of these reactions:

a) 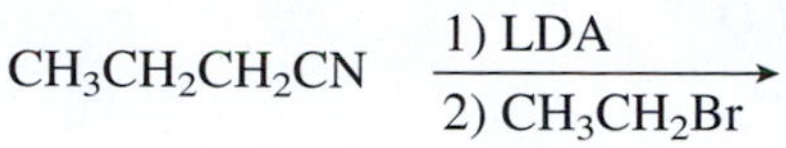 $CH_3CH_2CH_2CN$ $\xrightarrow{\text{1) LDA} \atop \text{2) } CH_3CH_2Br}$

b) 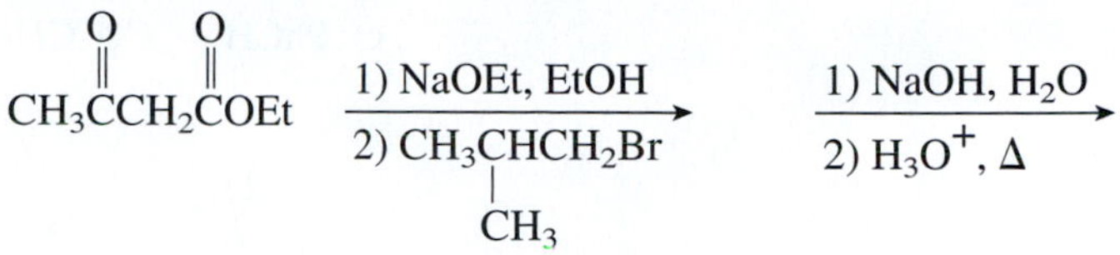 $CH_3\overset{O}{\overset{||}{C}}CH_2\overset{O}{\overset{||}{C}}OEt$ $\xrightarrow{\text{1) NaOEt, EtOH} \atop \text{2) } CH_3CHCH_2Br}$ $\xrightarrow{\text{1) NaOH, H}_2O \atop \text{2) } H_3O^+, \Delta}$
$\underset{CH_3}{|}$

c) 2 $CH_3CH_2CH_2CH_2\overset{O}{\overset{||}{C}}H$ $\xrightarrow[\substack{H_2O \\ \Delta}]{\text{NaOH}}$

d) 2 [cyclopentyl]$-CH_2\overset{O}{\overset{||}{C}}OEt$ $\xrightarrow{\text{1) NaOEt, EtOH} \atop \text{2) } H_3O^+}$

e) [cyclohexanone] $\xrightarrow{\substack{\text{1) } \overset{\displaystyle N}{\underset{H}{\diagup}} \text{ TsOH} \\ \text{2) PhCCl (}\overset{O}{\overset{||}{}}\text{)} \\ \text{3) } H_3O^+}}$

f) $CH_3\overset{O}{\overset{||}{C}}CH_2\overset{O}{\overset{||}{C}}OEt$ $\xrightarrow{\substack{\text{1) NaH} \\ \text{2) BuLi} \\ \text{3) PhCH}_2Br \\ \text{4) } H_3O^+}}$

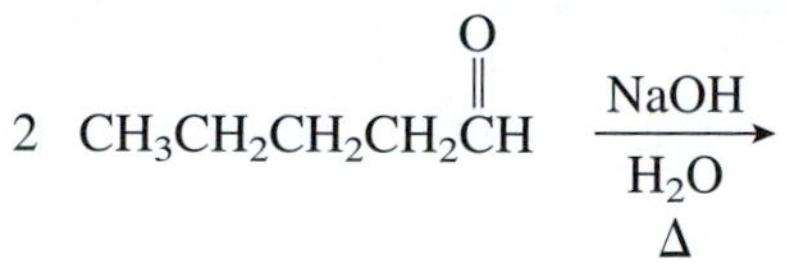

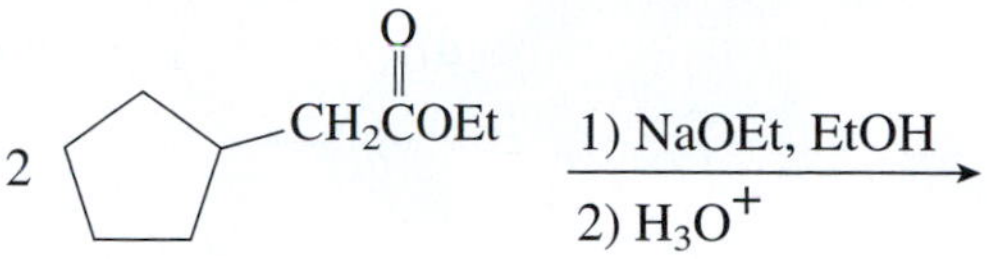

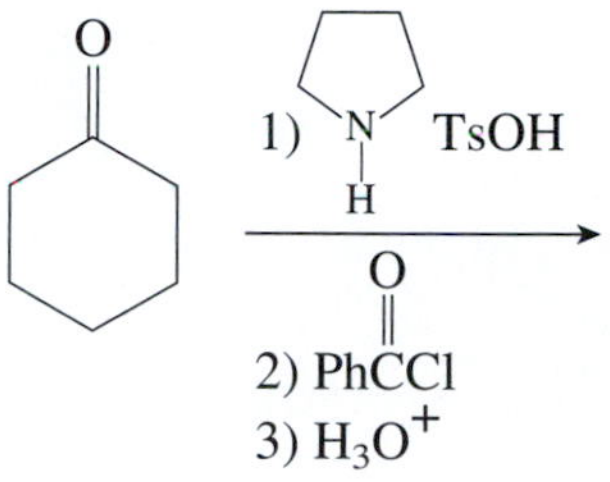

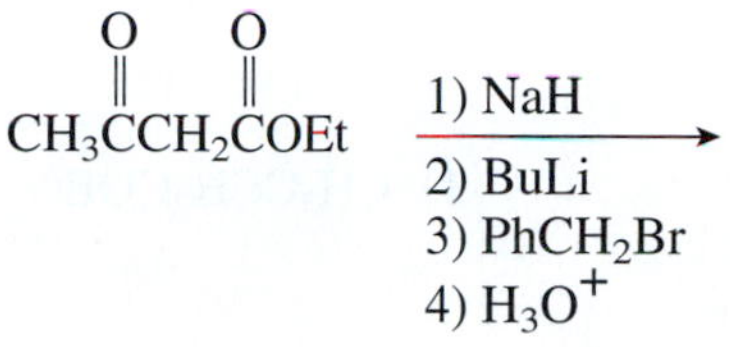

g) [cyclohexanone] + $EtO\overset{O}{\overset{||}{C}}CH_2\overset{O}{\overset{||}{C}}OEt$ $\xrightarrow[\Delta]{\text{NaOEt}}$

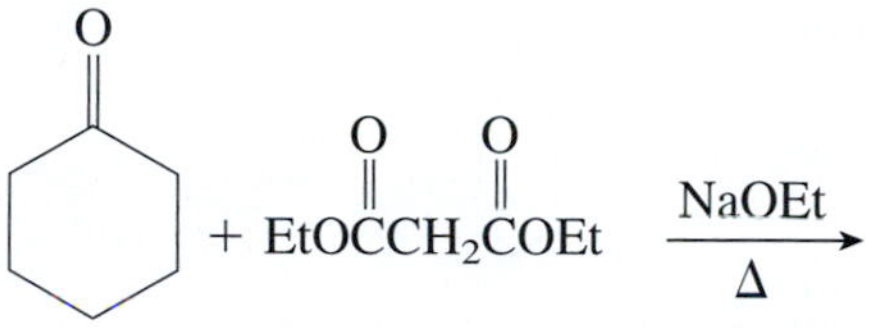

h) $EtO\overset{O}{\overset{||}{C}}CH_2\overset{O}{\overset{||}{C}}OEt$ $\xrightarrow{\substack{\text{1) NaOEt, EtOH} \\ \text{2) } \text{[cyclopentenyl]} -Cl}}$

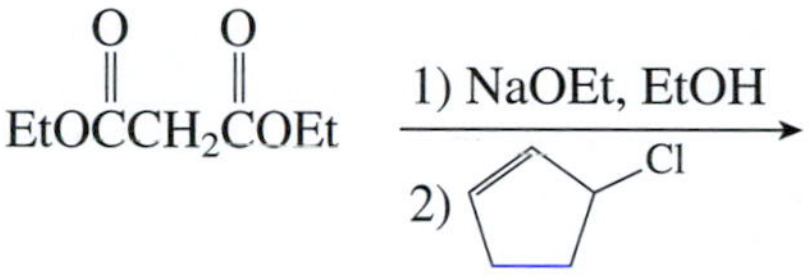

i) 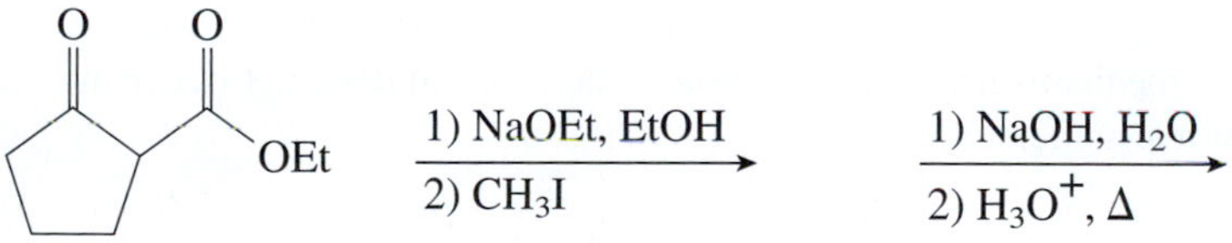 [cyclopentanone with $\overset{O}{\overset{||}{C}}$OEt] $\xrightarrow{\text{1) NaOEt, EtOH} \atop \text{2) } CH_3I}$ $\xrightarrow{\text{1) NaOH, H}_2O \atop \text{2) } H_3O^+, \Delta}$

20.32 Show the products of these reactions:

a) 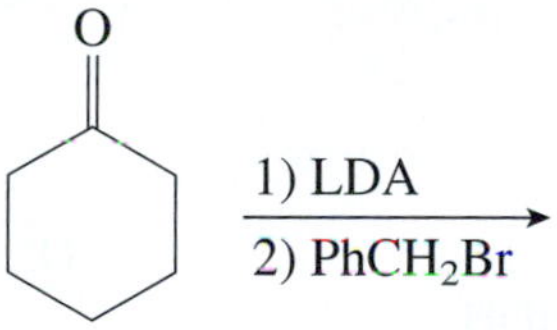 [cyclohexanone] $\xrightarrow{\text{1) LDA} \atop \text{2) PhCH}_2Br}$

b) 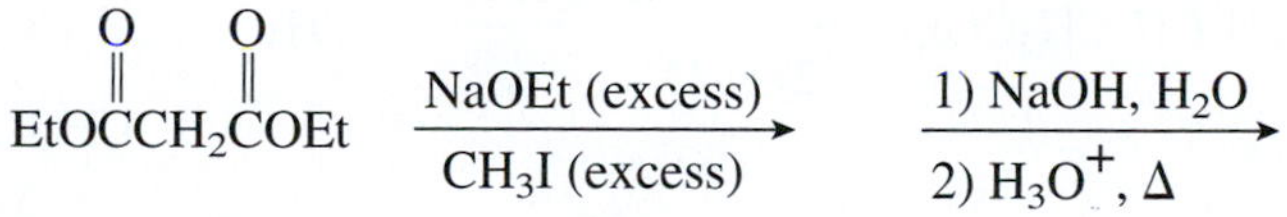 $EtO\overset{O}{\overset{||}{C}}CH_2\overset{O}{\overset{||}{C}}OEt$ $\xrightarrow{\text{NaOEt (excess)} \atop \text{CH}_3I \text{ (excess)}}$ $\xrightarrow{\text{1) NaOH, H}_2O \atop \text{2) } H_3O^+, \Delta}$

c) $PhCH + CH_3CH_2CCH_2CH_3$ $\xrightarrow[\substack{H_2O \\ \Delta}]{NaOH}$

d) 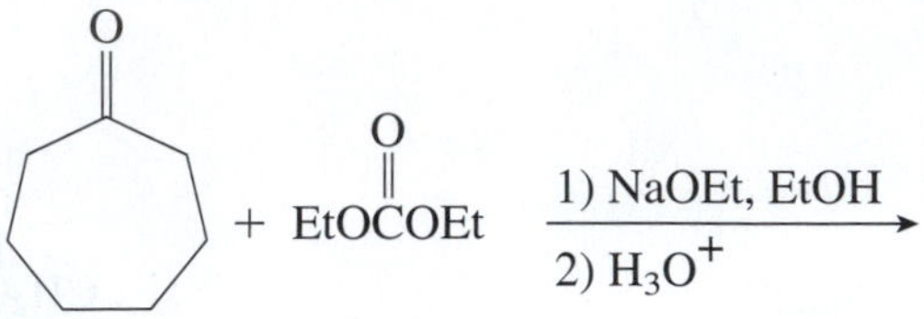 + $EtOCOEt$ $\xrightarrow[2) H_3O^+]{1) NaOEt, EtOH}$

e) $PhCH_2COH$ $\xrightarrow[\substack{2) PhCH_2Br \\ 3) H_3O^+}]{1) 2\ LDA}$

f) CH_3CCH_2COEt $\xrightarrow[\substack{2) BuLi \\ 3) PhCH_2CH_2CH_2Cl \\ 4) H_3O^+}]{1) NaH}$

20.33 Suggest a mechanism for this reaction:

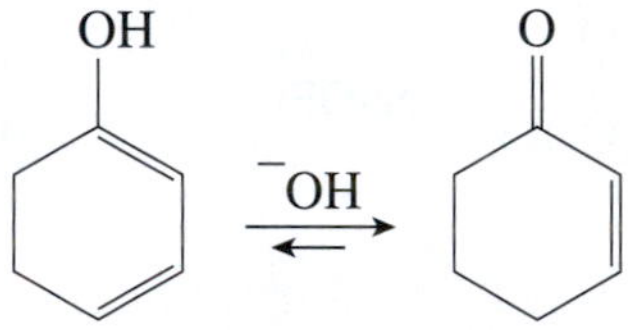

20.34 Suggest a mechanism for this reaction. (The reaction does not occur by a carbocation rearrangement.)

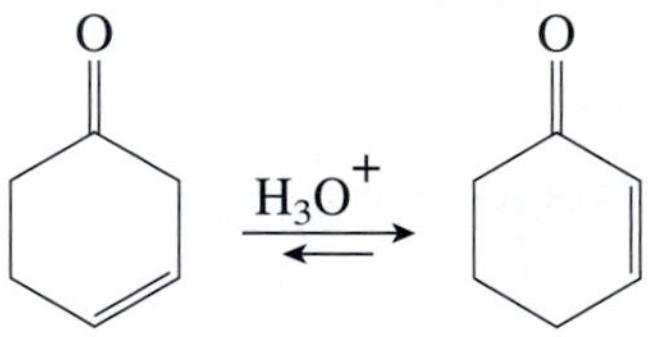

20.35 Show all of the steps in the mechanism for this reaction:

$2\ CH_3CH_2CH_2COEt$ $\xrightarrow[2) H_3O^+]{1) NaOEt, EtOH}$ $CH_3CH_2CH_2CCHCOEt$
  CH_3CH_2

20.36 Show all of the steps in the mechanism for this reaction:

$2\ CH_3CH_2CH$ $\xrightarrow{NaOH}$ $CH_3CH_2CHCHCH$
  CH_3

20.37 Show syntheses of these compounds from propanal:

a)

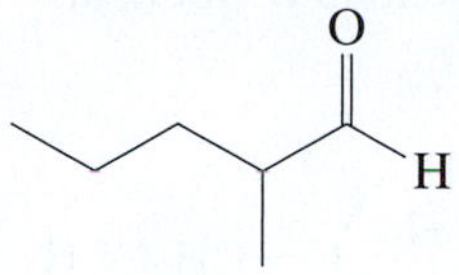

b)

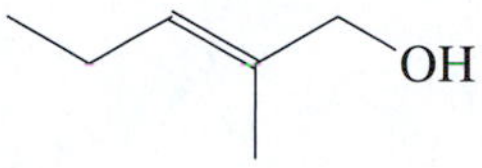

c)

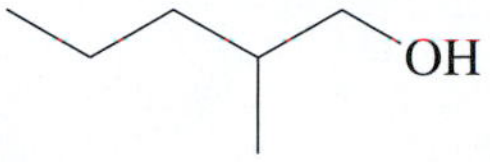

d)

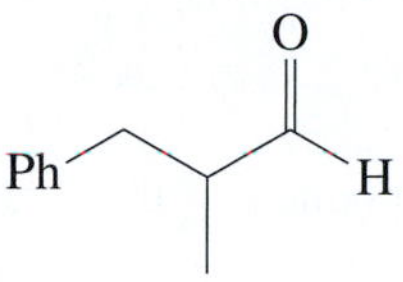

20.38 Show syntheses of these compounds from ethyl propanoate:

a)

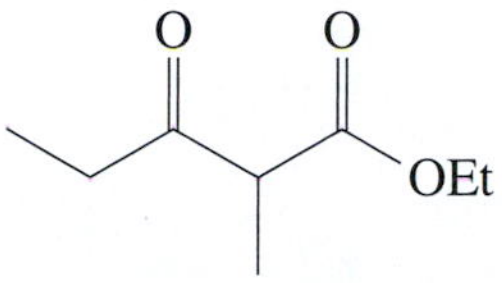

b)

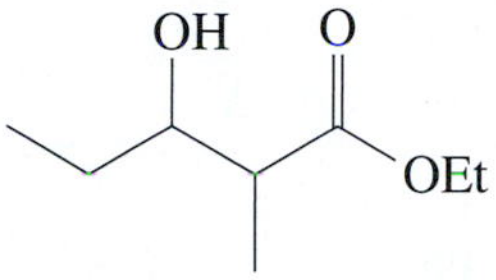

c)

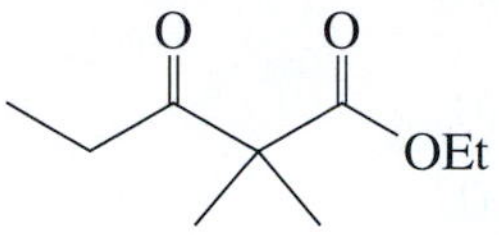

d)

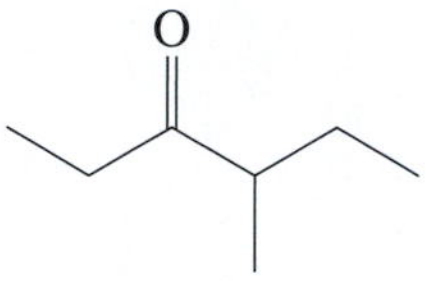

e)

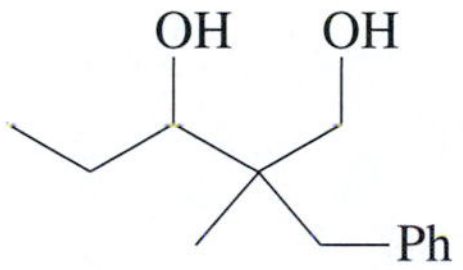

f)

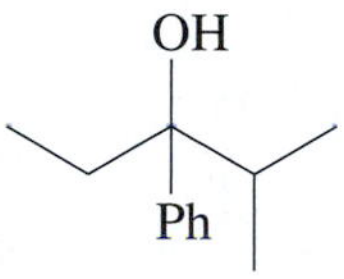

g)

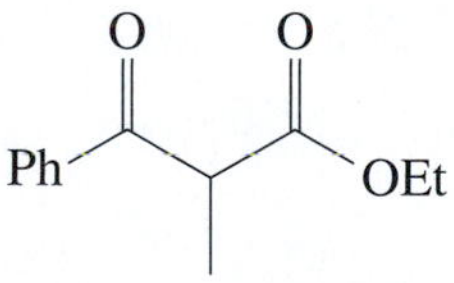

20.39 Show syntheses of these compounds from ethyl acetoacetate:

a)

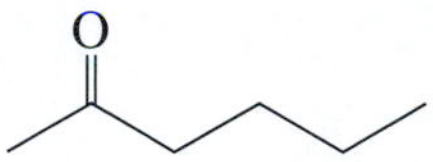

b)

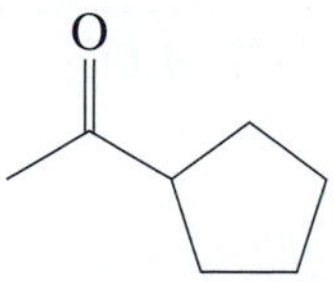

c)

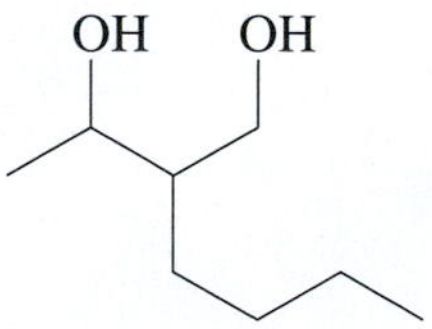

d)

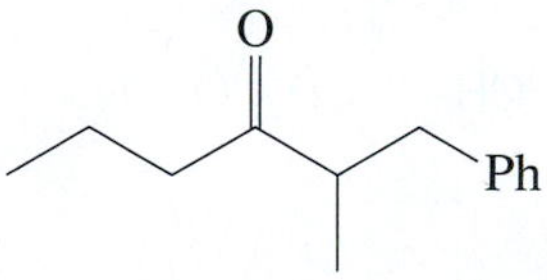

20.40 Optically active ketone **A** undergoes racemization in basic solution. Show a mechanism for this process. Explain whether ketone **B** would also racemize in basic solution.

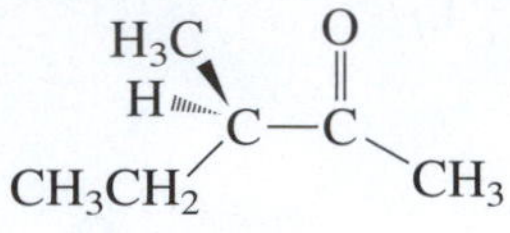

A 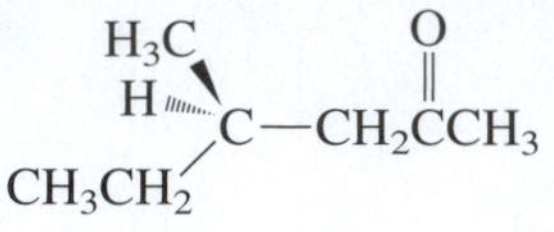 **B**

20.41 Show the products of these reactions:

a)

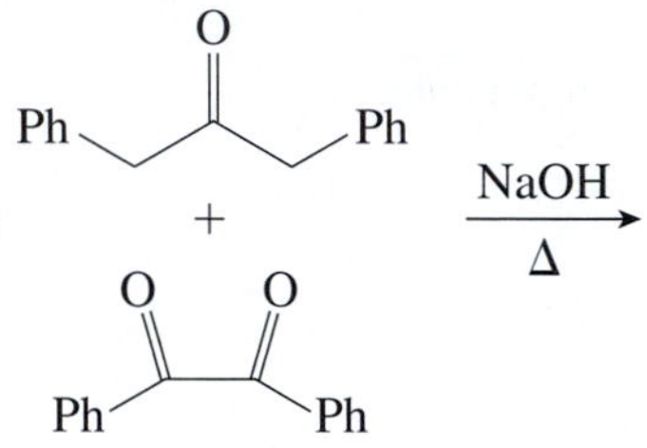

$\xrightarrow[\Delta]{\text{NaOH}}$

b)

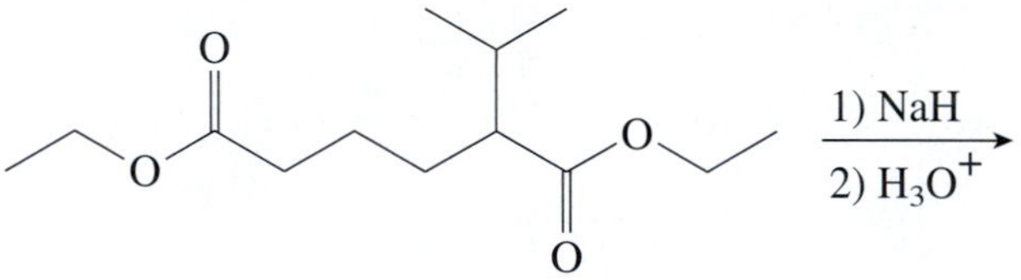

$\xrightarrow{\begin{array}{c}1)\ \text{NaH}\\2)\ \text{H}_3\text{O}^+\end{array}}$

c)

$\xrightarrow[\Delta]{\text{NaOH}}$

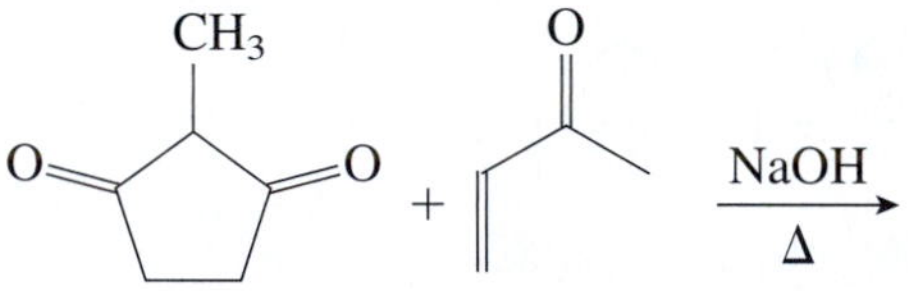

d)

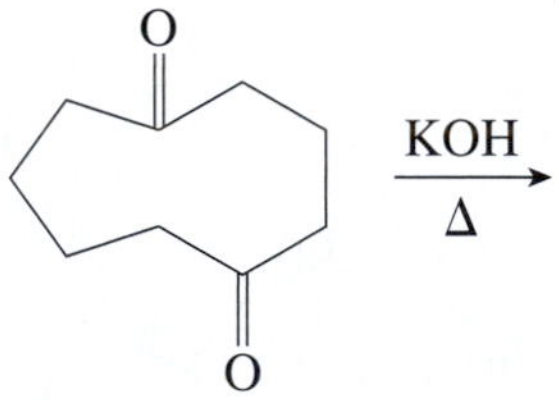

$\xrightarrow[\Delta]{\text{KOH}}$

e)

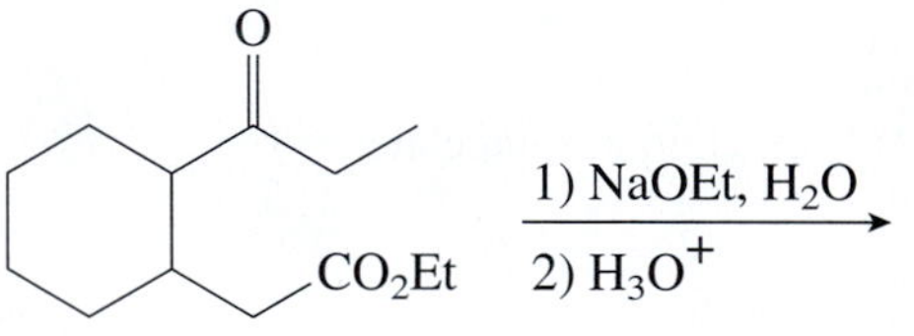

$\xrightarrow{\begin{array}{c}1)\ \text{NaOEt, H}_2\text{O}\\2)\ \text{H}_3\text{O}^+\end{array}}$

f) 2

$\xrightarrow[\Delta]{\text{KOH}}$

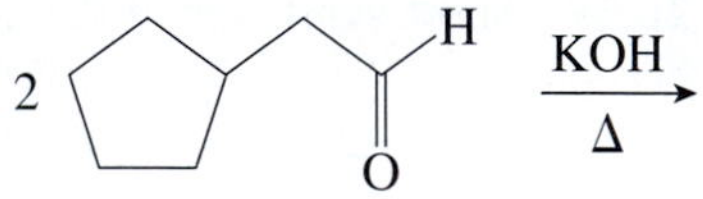

g)

$+ \text{CH}_3\text{OCCH}_2\text{COCH}_3 \xrightarrow{\text{NaOCH}_3}$

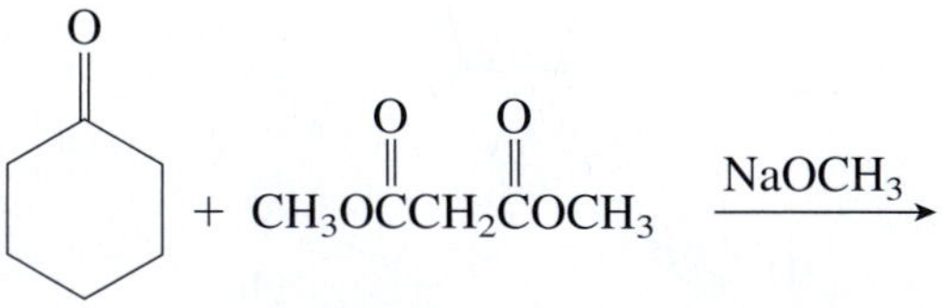

20.42 Show syntheses of these compounds from the indicated starting materials:

a) 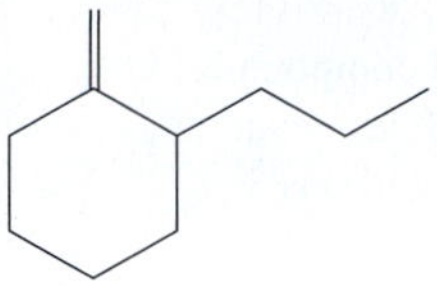 from cyclohexanone

b) 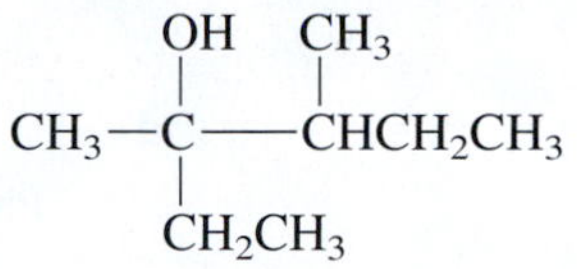
OH
CH3
CH3—C—CHCH2CH3
CH2CH3
from ethyl acetoacetate

c) 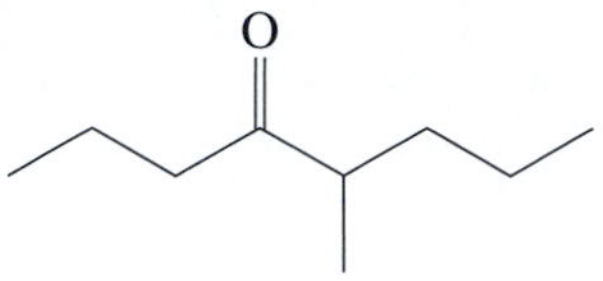
O
from butanoic acid

d) 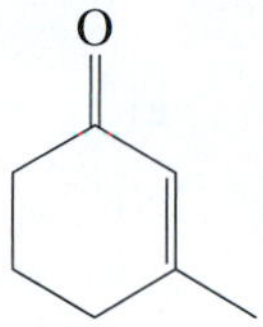
O
from compounds without a ring

e) 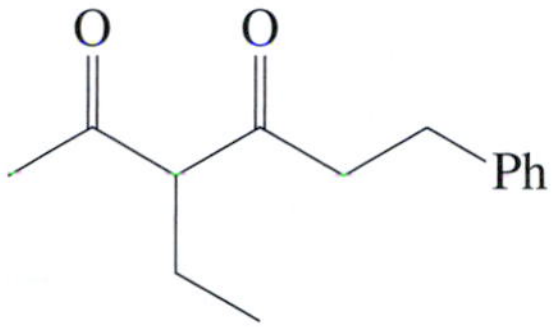
O
O
Ph
from 2,4-pentanedione

f) 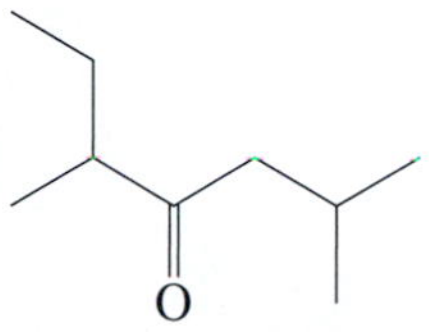
O
from propanoic acid

g) 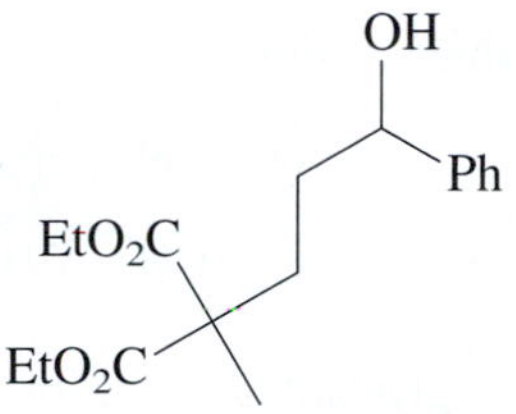
OH
Ph
EtO2C
EtO2C
from diethyl malonate

20.43 Show syntheses of these compounds from the indicated starting materials:

a) 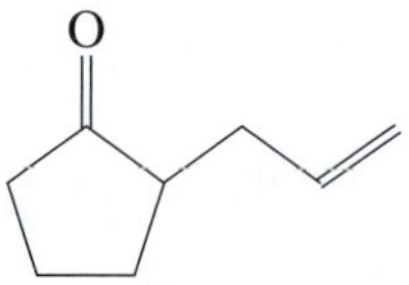
O
from 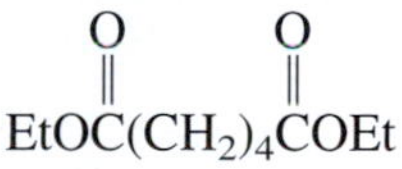
O
O
EtOC(CH2)4COEt

b) 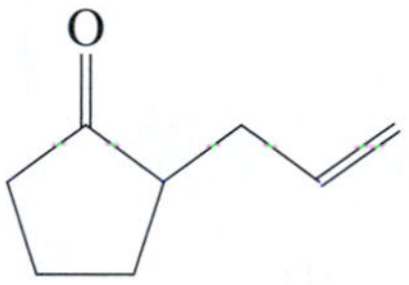
O
from 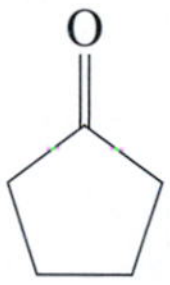
O

c) 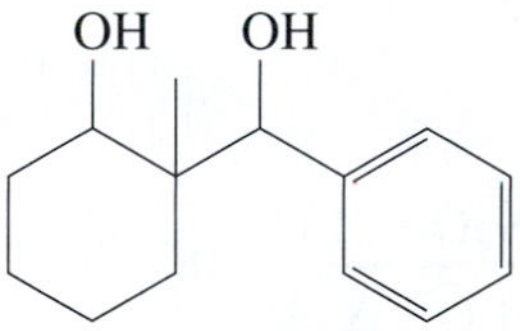
OH
OH
from 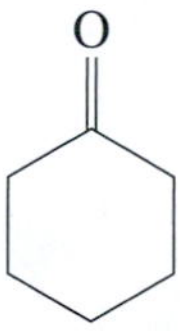
O

d) 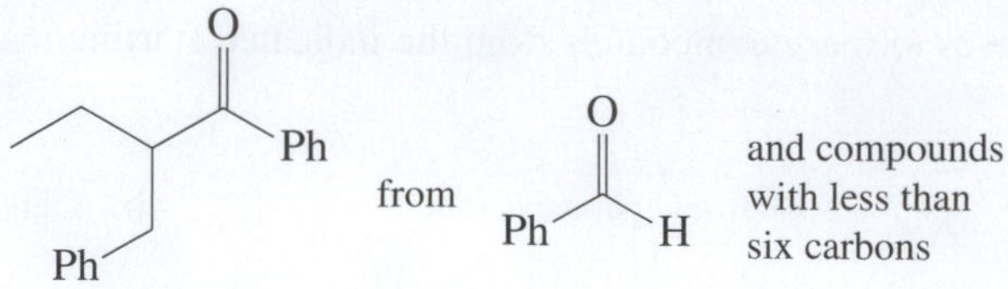 [structure] from [structure: Ph—C(=O)—H] and compounds with less than six carbons

e) 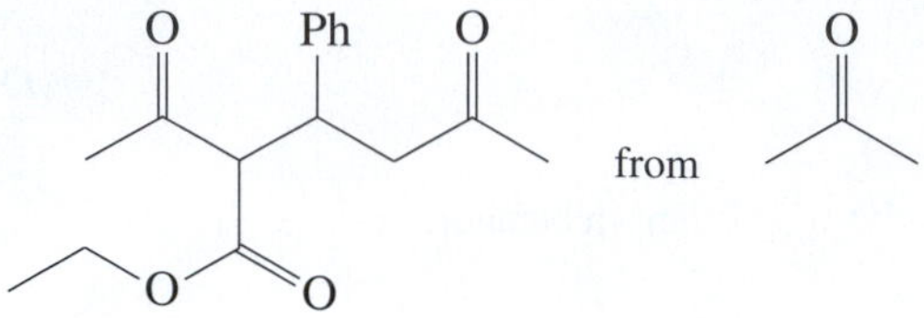 [structure] from [structure: acetone]

20.44 Show all of the steps in the mechanism for this reaction:

[structure] $\xrightarrow[\text{2) H}_3\text{O}^+, \Delta]{\text{1) NaOH, H}_2\text{O}}$ [structure] $+\ CO_2\ +\ EtOH$

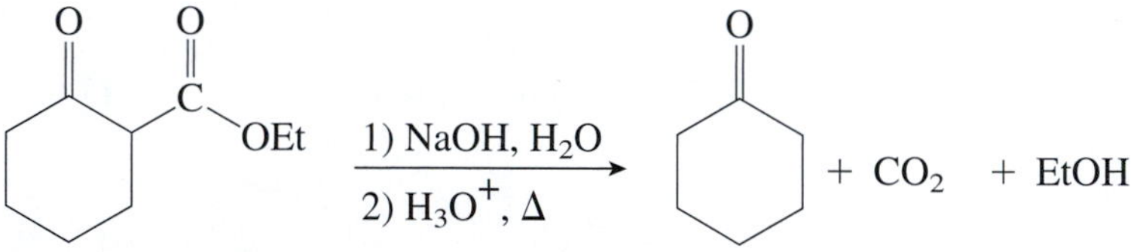

20.45 Show all of the steps in the mechanism for this reaction:

$$CH_2{=}CHCOEt + NCCH_2COEt \xrightarrow{\text{NaOEt}} EtOC{-}CH{-}CH_2CH_2COEt$$

with O, CN, O labels over the structure.

20.46 Show a mechanism for the interconversion of glyceraldehyde-3-phosphate (GAP) and dihydroxyacetone phosphate (DHAP) in basic solution:

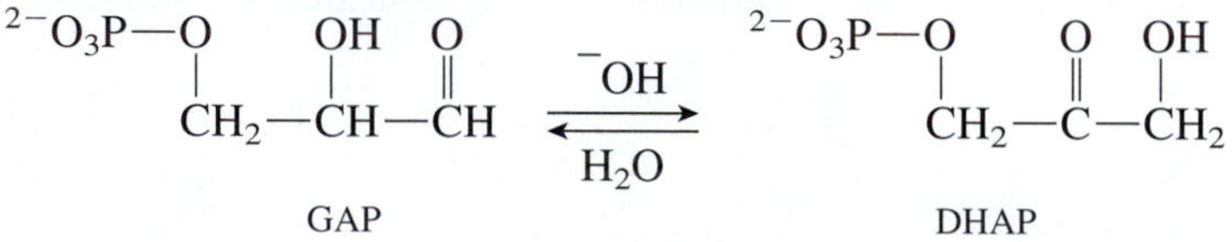

[structure GAP] $\underset{\text{H}_2\text{O}}{\overset{^-\text{OH}}{\rightleftarrows}}$ [structure DHAP]

GAP   DHAP

20.47 Show the missing products, **A** and **B,** in this reaction scheme and explain the regiochemistry of the reactions.

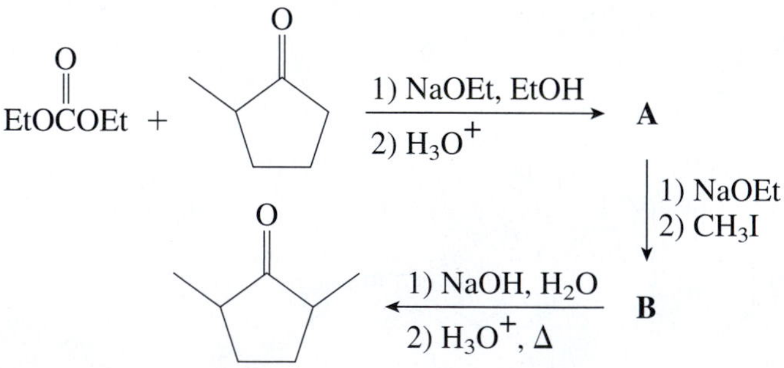

[structure: EtOCOEt] + [structure: 2-methylcyclopentanone] $\xrightarrow[\text{2) H}_3\text{O}^+]{\text{1) NaOEt, EtOH}}$ **A**

$\downarrow$ 1) NaOEt 2) CH$_3$I

[structure: 2,5-dimethylcyclopentanone] $\xleftarrow[\text{2) H}_3\text{O}^+, \Delta]{\text{1) NaOH, H}_2\text{O}}$ **B**

20.48 Show syntheses of these compounds using the Robinson annulation reaction:

a)

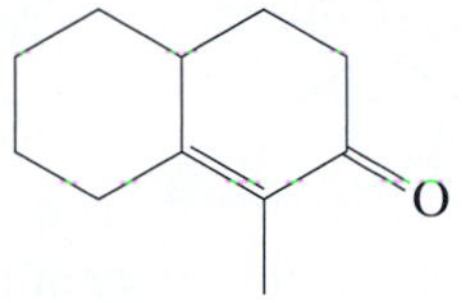

b)

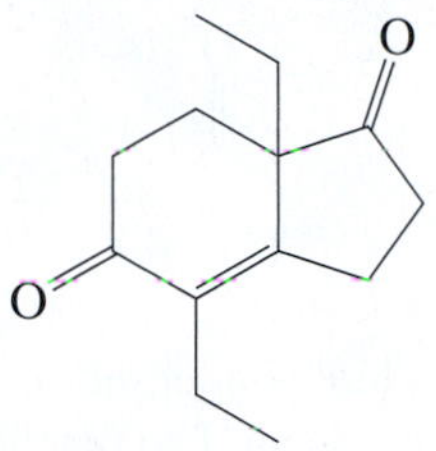

20.49 2-Ethyl-1,3-hexanediol is the active ingredient in the insect repellant "6-12." Suggest a synthesis of this compound from precursors with four or fewer carbons.

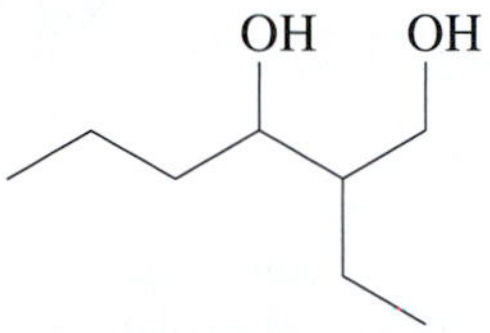

2-Ethyl-1,3-hexanediol

20.50 2-Ethyl-1-hexanol is used industrially as a plasticizer in the manufacture of plastics. Suggest a synthesis of this compound from precursors with four or fewer carbons.

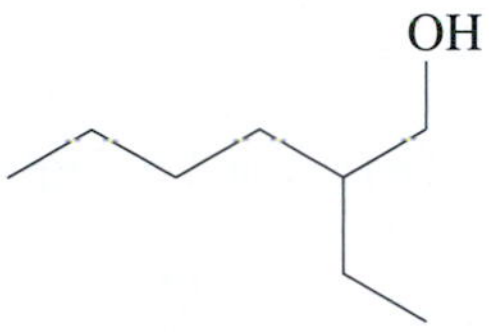

2-Ethyl-1-hexanol

20.51 Suggest a mechanism for this reaction:

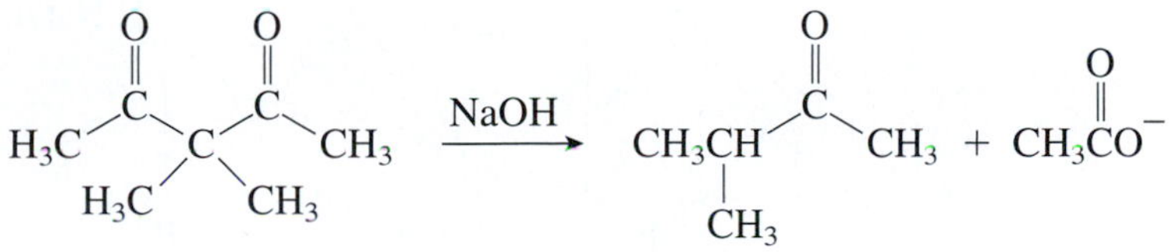

$$\text{(diketone)} \xrightarrow{\text{NaOH}} CH_3CH(CH_3)\text{-CO-}CH_3 + CH_3CO^-$$

20.52 Deuterium can be incorporated at the positions α to a carbonyl group by reaction with D_2O in the presence of acid. Show a mechanism for this process. If the reaction were continued, what is the maximum number of deuterium atoms that would be incorporated into a single molecule?

$$CH_3CH_2CCH_2CH_3 + D_2O \xrightarrow{D_3O^+} CH_3CH_2CCHDCH_3$$

20.53 Suggest a mechanism for this reaction:

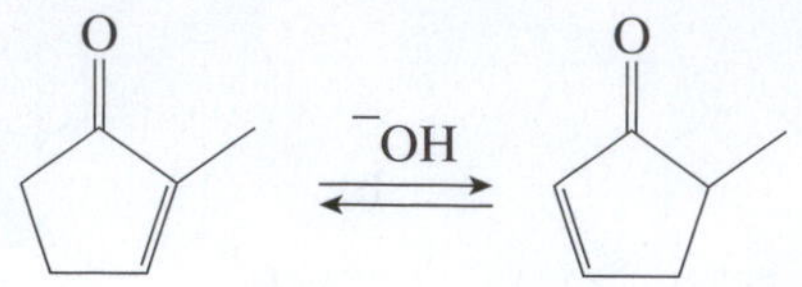

20.54 Intramolecular aldol condensations often present the possibility of the formation of several products. The reaction of 6-oxoheptanal gives the product shown in the following equation:

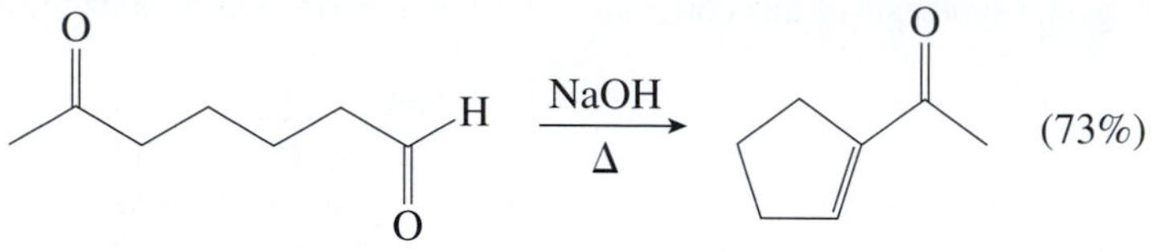

a) Show structures for the other α,β-unsaturated products that could be formed in this reaction and explain why the observed product is formed preferentially.

b) Predict the preferred product in this aldol cyclization:

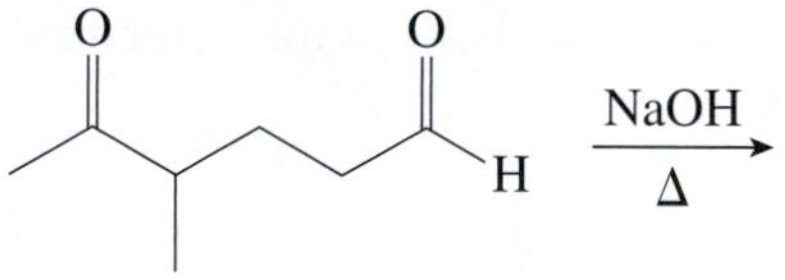

20.55 Show the missing products, **A** and **B,** in this reaction scheme:

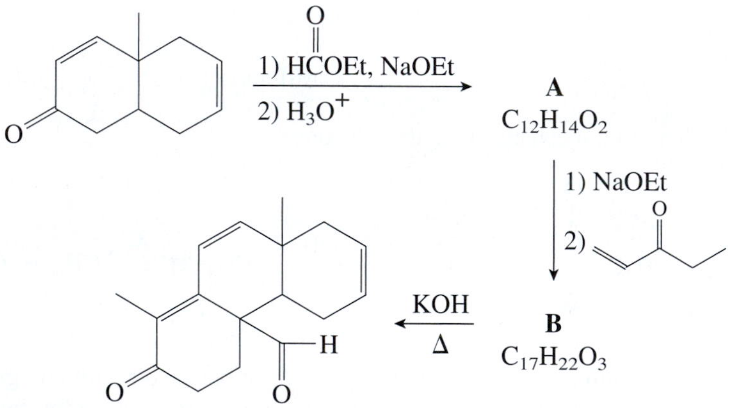

20.56 Suggest a mechanism for this reaction:

$$Ph-\underset{\underset{O}{\parallel}}{C}-\underset{\underset{O}{\parallel}}{C}-H + NaOH \longrightarrow Ph-\underset{\underset{H}{|}}{\overset{\overset{OH}{|}}{C}}-\underset{\underset{O}{\parallel}}{C}-\overset{-}{O} \; Na^{+}$$

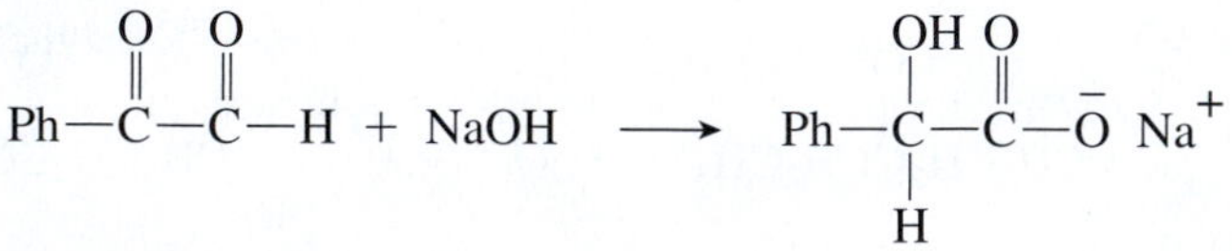

20.57 Suggest a synthesis of CS, a component of tear gas:

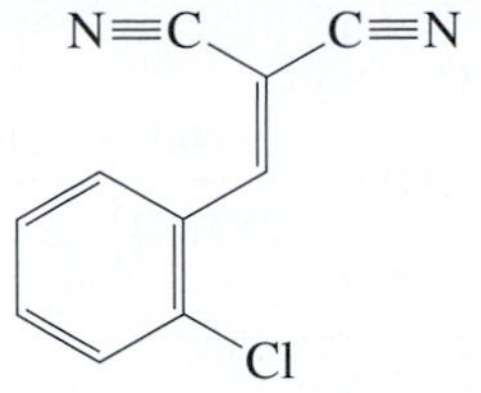

CS

20.58 Thalidomide was used to treat morning sickness in the late 1950s, but was soon discovered to have caused a number of birth defects. For a time it was believed that the birth defects were due to one enantiomer of this drug and that the other enantiomer was harmless. However, it was later found that either enantiomer will racemize under the acidic conditions of the stomach so the harmless enantiomer is converted to the harmful one. Suggest a mechanism for this racemization.

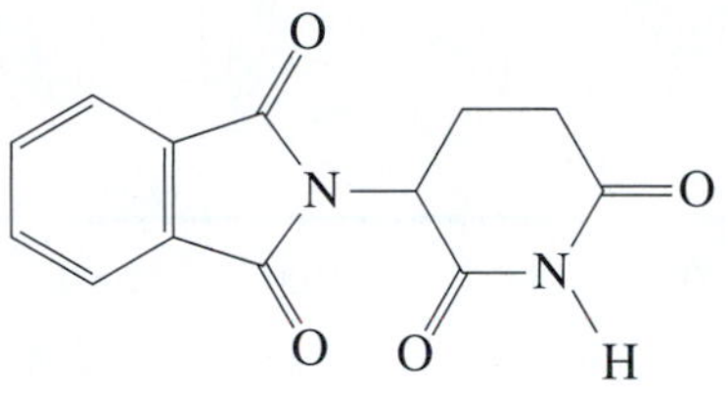

Thalidomide

Problems Involving Spectroscopy

20.59 2,5-Heptanedione forms two products upon reaction with NaOH. Both products show a strong absorption near 1715 cm^{-1} in their IR spectra, and neither shows any absorption bands in the region of 3600 to 3300 cm^{-1}. The major product has singlets at 1.90 and 1.65 δ in its ^{1}H-NMR spectrum. Show structures for these products.

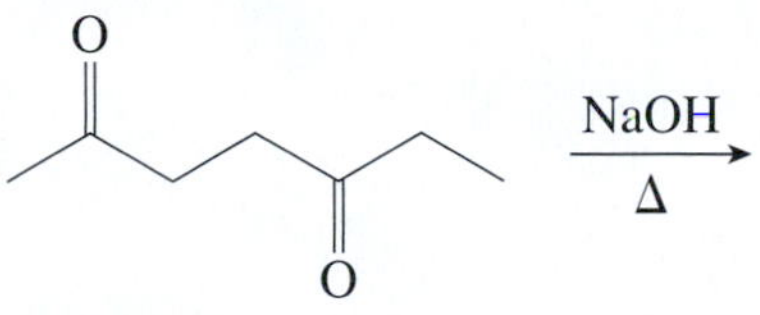

$$\xrightarrow[\Delta]{\text{NaOH}}$$

20.60 The product from the reaction of 1-phenyl-2-butanone with LDA and methyl iodide shows a quartet (1 H), a quartet (2 H), a doublet (3 H), and a triplet (3 H) in the alkyl region of its ^{1}H-NMR spectrum. Show the structure of this product and explain the regiochemistry of the reaction.

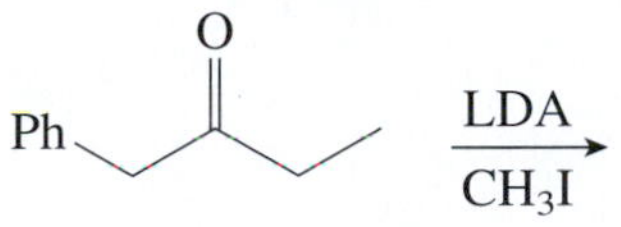

$$\text{Ph} \xrightarrow[\text{CH}_3\text{I}]{\text{LDA}}$$

20.61 The IR and ^{1}H-NMR spectra of the product of this reaction follow. The formula of the product is $C_9H_8O_2$. Show the structure of the product.

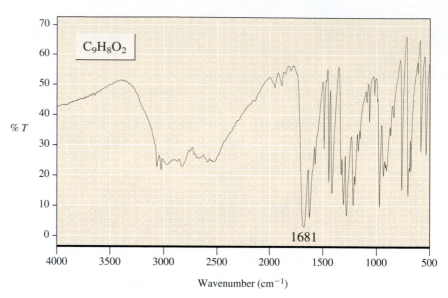

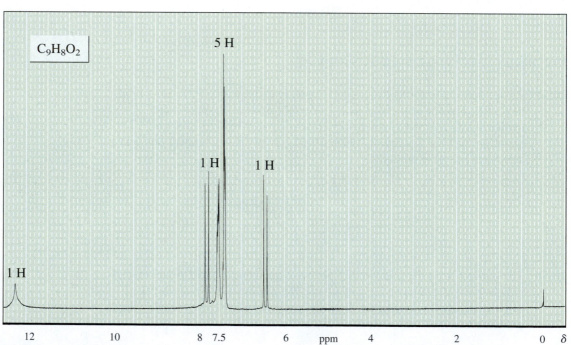

Problems Using Online Three-Dimensional Molecular Models

ORGANIC
Chemistry ⚡ Now™
Click *Molecular Model Problems*
to view the models needed to
work these problems.

20.62 When either of these two stereoisomeric ketones is treated with NaOH, a mixture containing both of them is formed. Suggest a mechanism for this process. Explain which stereoisomer should be the major component of the mixture at equilibrium.

20.63 Reaction of 2-(3-bromopropyl)cyclopentanone with potassium *t*-butoxide gives the three products shown. Suggest a mechanism for the formation of each of the products.

20.64 The aldol condensation of 2-propanone (acetone) with two molecules of benzaldehyde forms the product shown. Explain the stereochemistry of the carbon–carbon double bonds in the product.

Do you need a live tutor for homework problems? Access vMentor at Organic ChemistryNow at **http://now.brookscole.com/hornback2** for one-on-one tutoring from a chemistry expert.

The Chemistry of Radicals

CHAPTER 21

ALL OF THE MECHANISMS that have been presented so far have involved the reaction of electrophiles with nucleophiles. Carbocations and carbanions have been encountered as intermediates. In this chapter the chemistry of a new reactive intermediate, called a *radical* (or *free radical*), is presented. A **radical** is a species with an odd number of electrons. After a discussion of the structure of radicals, including their stability and geometry, various methods of generating them are described. Next, the general reactions that they undergo are presented. Finally, specific reactions involving radical intermediates are discussed.

MASTERING ORGANIC CHEMISTRY

▶ Understanding the Effect of the Structure of a Radical on Its Stability and the Rate and Regiochemistry of Its Reactions

▶ Predicting the Products of Radical Reactions

▶ Understanding the Mechanisms of Radical Reactions

▶ Using Radical Reactions in Synthesis

21.1 RADICALS

The reactions that we have encountered up to this point have involved the movement of pairs of electrons. For example, when a bond was broken, both electrons of that bond remained with one of the atoms. This process is termed **heterolytic bond cleavage:**

$$X \overset{\frown}{\underset{}{-}} Y \longrightarrow \overset{+}{X} \ + \ \overset{-}{Y} :$$

Heterolytic bond cleavage

Radicals are formed by **homolytic bond cleavage,** in which one electron of the bond remains with each of the atoms:

$$X \overset{\wedge\wedge}{\underset{}{-}} Y \longrightarrow X\cdot \ + \ Y\cdot$$

Homolytic bond cleavage

Note that an arrow with only half of an arrowhead is used to show the movement of a single electron that occurs in radical reactions, whereas the normal arrow shows the movement of a pair of electrons.

◀▎▎ **Important Convention**

ORGANIC
Chemistry Now™
Look for this logo in the chapter and go to OrganicChemistryNow at
http://now.brookscole.com/hornback2 for tutorials, simulations, problems, and molecular models.

Some examples of radicals are the following:

$$H\cdot \qquad :\ddot{\underset{..}{C}}l\cdot \qquad H-\underset{\underset{H}{|}}{\overset{\overset{H}{|}}{C}}\cdot \qquad CH_3-\overset{..+}{\underset{..}{O}}-H$$

| Hydrogen atom | Chlorine atom | Methyl radical | Methanol radical cation |

Hydrogen and chlorine atoms have an odd number of electrons and are radicals. The methyl radical is the simplest organic radical. It has one more electron than a carbocation and one fewer than a carbanion. The last example is a radical cation, which results from the loss of one electron from a normal molecule. Radical cations are important in mass spectrometry (see Chapter 15).

PROBLEM 21.1

Both nitrogen oxide (NO) and nitrogen dioxide (NO_2) are radicals. Show Lewis structures for these compounds.

21.2   STABILITY OF RADICALS

Because most radicals have an odd number of electrons on an atom, the octet rule cannot be satisfied at that atom. It is no surprise, then, that most radicals are unstable species and are quite reactive. They are most often encountered, like carbocations, as transient intermediates in reactions. However, alkyl radicals tend to have longer lifetimes than carbocations because they are less electron deficient, and therefore more stable. In fact, the lifetime of a radical can be appreciable in an environment where nothing is available with which to react. For example, hydrogen atoms are the principal type of matter in interstellar space. And the methyl radical has a lifetime of about 10 min when frozen in a methanol matrix at 77 K.

A comparison of the stabilities of different carbon radicals is provided by the bond dissociation energies of the bond between the carbon and a hydrogen. This is the energy that must be added when the reaction shown in the following equation occurs:

$$R\overset{\frown\frown}{-}H \quad \longrightarrow \quad R\cdot \ + \ H\cdot$$

Bond dissociation energies for some carbon–hydrogen bonds are shown in Table 21.1.

The bond dissociation energy is the energy that must be supplied to generate a hydrogen atom and the carbon radical. Because a hydrogen radical is produced in each case, the difference between two bond dissociation energies reflects the difference in stability between the two carbon radicals. For example, because $104 - 98 = 6$ kcal/mol (25 kJ/mol) less energy must be supplied to ethane than to methane to dissociate a hydrogen atom, the ethyl radical is 6 kcal/mol (25 kJ/mol) more stable than the methyl radical. Examination of Table 21.1 shows that the order of radical stabilities roughly parallels the order of carbocation stabilities presented in Section 8.7:

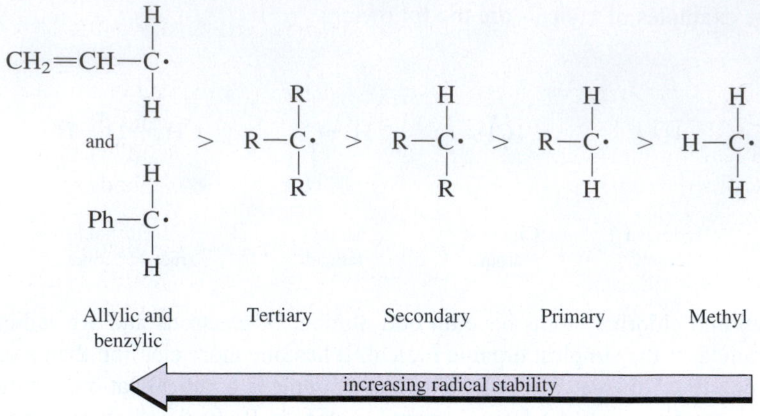

$$CH_2=CH-\underset{\underset{H}{|}}{\overset{\overset{H}{|}}{C}}\cdot$$

and $\quad > \quad R-\underset{\underset{R}{|}}{\overset{\overset{R}{|}}{C}}\cdot \quad > \quad R-\underset{\underset{R}{|}}{\overset{\overset{H}{|}}{C}}\cdot \quad > \quad R-\underset{\underset{H}{|}}{\overset{\overset{H}{|}}{C}}\cdot \quad > \quad H-\underset{\underset{H}{|}}{\overset{\overset{H}{|}}{C}}\cdot$$

$$Ph-\underset{\underset{H}{|}}{\overset{\overset{H}{|}}{C}}\cdot$$

Allylic and   Tertiary   Secondary   Primary   Methyl
benzylic

⟵ increasing radical stability

Table 21.1   Bond Dissociation Energies for Some Carbon–Hydrogen Bonds

Bond	Bond Dissociation Energies (kcal/mol [kJ/mol])
(phenyl)–H	110 (460)
$CH_2=CH$–H	108 (452)
CH_3–H	104 (435)
CH_3CH_2–H	98 (410)
$CH_3CH(CH_3)$–H	95 (397)
$CH_3C(CH_3)(CH_3)$–H	92 (385)
$CH_2=CHCH_2$–H	89 (372)
$PhCH_2$–H	85 (356)

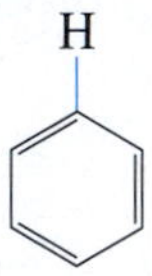

The reasons for this order of stabilities are the same for radicals as for carbocations. A primary radical is more stable than the methyl radical because overlap of a sigma MO on the adjacent carbon of the primary radical with the AO containing the odd electron provides a pathway for the electrons of the sigma bond to be delocalized onto the electron-deficient radical carbon, thus stabilizing it. (This is the same as the stabilization of carbocations by hyperconjugation described on page 273.) Secondary and tertiary radicals have additional stabilizing interactions of this type. Resonance stabilization is important in allylic and benzylic radicals. However, because they are more stable to begin with, the difference in stabilities between radicals is smaller than the difference in stabilities between the corresponding carbocations. In other words, the difference in stability between a secondary and tertiary radical is considerably smaller than that between a secondary and tertiary carbocation. This is one reason why radicals are not as prone to rearrangements as carbocations are.

PROBLEM 21.2

Arrange these radicals in order of increasing stability:

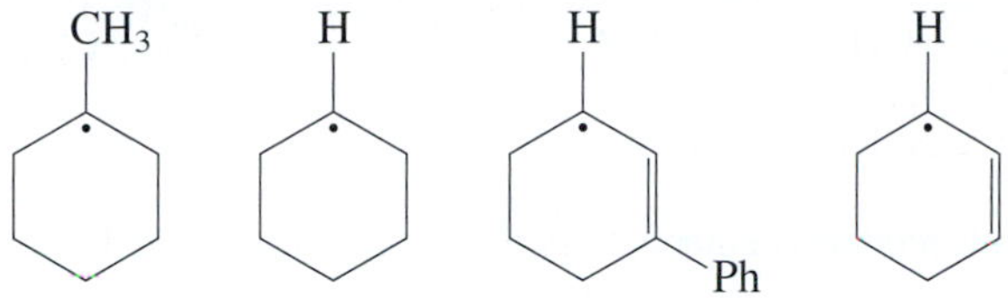

21.3 GEOMETRY OF CARBON RADICALS

If an attempt were made to apply the rules of valence shell electron pair repulsion theory to radicals, it would not be clear how to treat the single electron. Obviously, a single electron should not be as "large" as a pair of electrons, but it is expected to result in some repulsion. Therefore, it is difficult to predict whether a radical carbon should be sp^2 hybridized with trigonal planar geometry (with the odd electron in a p orbital), sp^3 hybridized with tetrahedral geometry (with the odd electron in an sp^3 AO), or somewhere in between. Experimental evidence is also somewhat uncertain. Studies of the geometry of simple alkyl radicals indicate that either they are planar or, if they are pyramidal, inversion is very rapid.

This model shows the electron density for the odd electron of the planar methyl radical. The radical electron is in a p orbital perpendicular to the plane of the atoms.

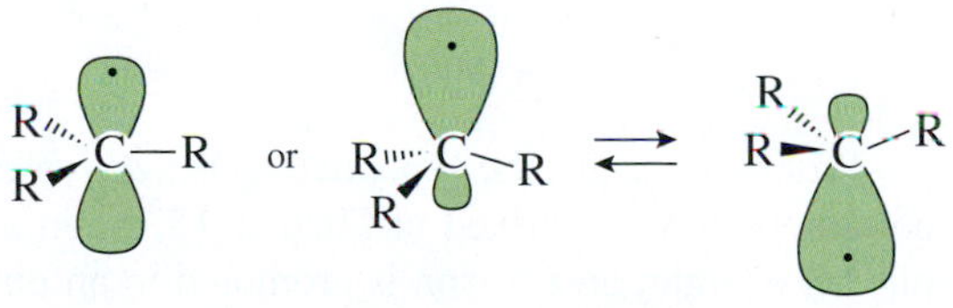

A trigonal planar radical

A rapidly inverting pyramidal radical

The important consequence of this is that reactions that involve radicals, like reactions that involve carbocations, result in the loss of stereochemistry (racemization) at the radical carbon.

21.4 GENERATION OF RADICALS

When a compound that has an especially weak bond is heated, the weak bond is selectively cleaved to produce radicals. Because the bond energy of the oxygen–oxygen bond is small, only about 30 kcal/mol (126 kJ/mol), peroxides readily undergo bond homolysis when they are heated to relatively low temperatures (80°–100°C). Commercially available peroxides, such as benzoyl peroxide and *tert*-butyl peroxide, are commonly used as sources of radicals.

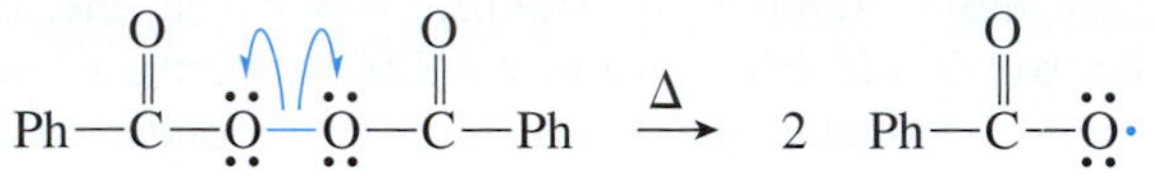

Benzoyl peroxide

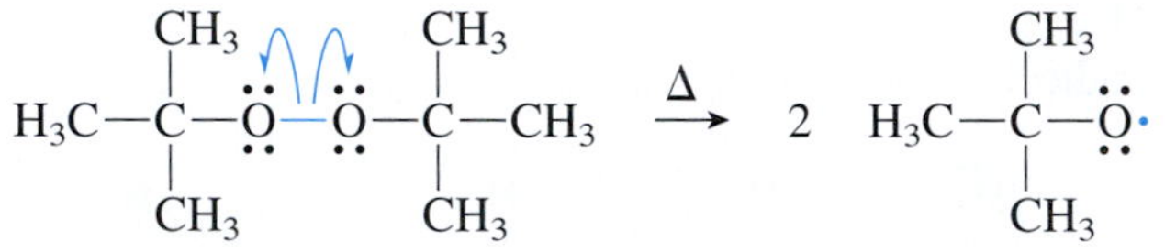

tert-Butyl peroxide

Azo compounds provide another common source for the thermal generation of radicals.

An azo compound

In this case it is not that the carbon–nitrogen bond is so weak; rather, it is the formation of the strong nitrogen–nitrogen triple bond of the N_2 product that enables the reaction to occur at relatively low temperatures. Azobis(isobutyronitrile), also known as AIBN, has been widely used as a radical source because it is commercially available. In addition, it undergoes bond homolysis at lower temperatures than other azo compounds (below 100°C) because the product radicals are tertiary and are stabilized by resonance.

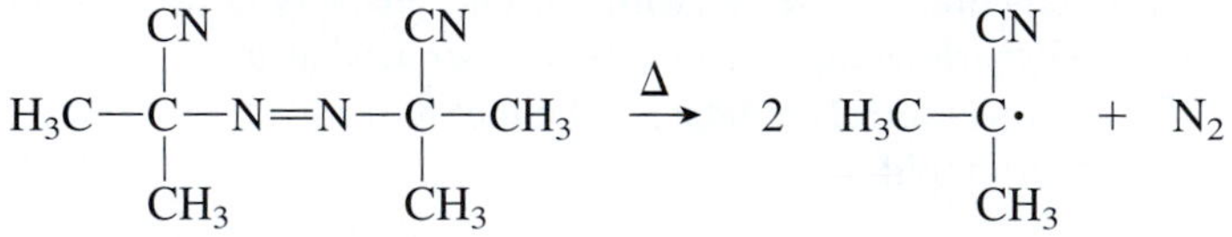

Azobis(isobutyronitrile)
AIBN

Radicals can also be generated by the action of ultraviolet or visible light on certain compounds. As described in Chapter 15, when a compound is excited by absorbing a photon of light, an electron is promoted to an unoccupied orbital. Because this orbital is usually antibonding in character, some bond in the excited molecule is weakened and may cleave in a homolytic fashion. For this reason, many photochemical reactions involve radicals. Examples of photochemically induced homolytic bond cleavages are provided in the following equations. *(Recall that hν is used to indicate the action of light on a compound.)*

> ⚑⦚ **Important Convention**

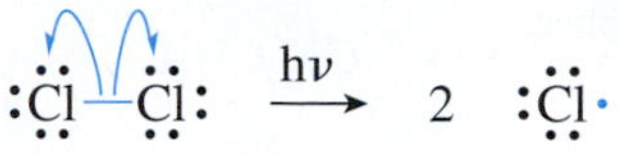

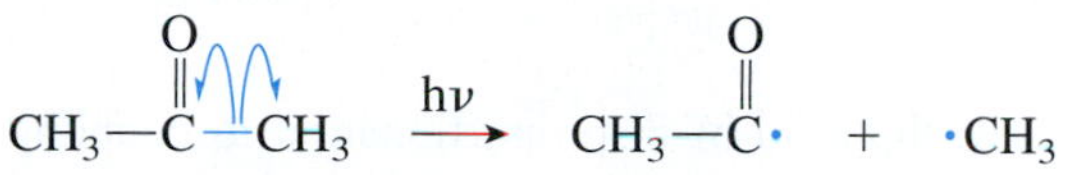

$$:\ddot{C}l\!-\!\ddot{C}l: \xrightarrow{h\nu} 2\ :\ddot{C}l\cdot$$

$$CH_3\!-\!\overset{O}{\overset{\|}{C}}\!-\!CH_3 \xrightarrow{h\nu} CH_3\!-\!\overset{O}{\overset{\|}{C}}\!\cdot \ + \ \cdot CH_3$$

PROBLEM 21.3

Show the radicals produced in these reactions:

a) 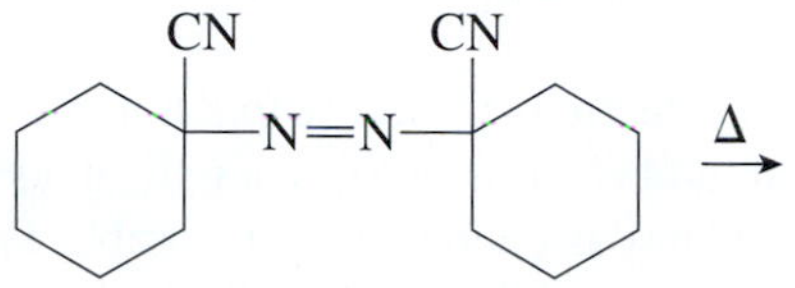 $\xrightarrow{\Delta}$

b) $Br\!-\!Br \xrightarrow{h\nu}$

c) $CH_3\overset{CH_3}{\overset{|}{C}H}\!-\!O\!-\!O\!-\!\overset{CH_3}{\overset{|}{C}H}CH_3 \xrightarrow{\Delta}$

21.5 GENERAL RADICAL REACTIONS

A radical is unstable because of its odd electron. To form a stable molecule, the radical needs to use that odd electron to form a bond to another atom. The reaction of two radicals, each with an odd number of electrons, allows the formation of a molecule (or molecules) that is stable because it has an even number of electrons and has the octet rule satisfied for all of its atoms.

A common way for two radicals to form a stable product is by **coupling** to form a bond between their radical centers.

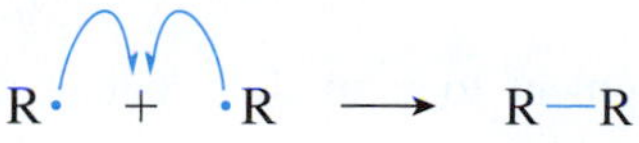

$$R\cdot \ + \ \cdot R \longrightarrow R\!-\!R$$

Coupling

Another way that radicals can react to form stable products is for one radical to abstract a hydrogen atom from the carbon adjacent to the radical center of another radical in a process called **disproportionation.** This results in the formation of a pi bond in one of the radicals:

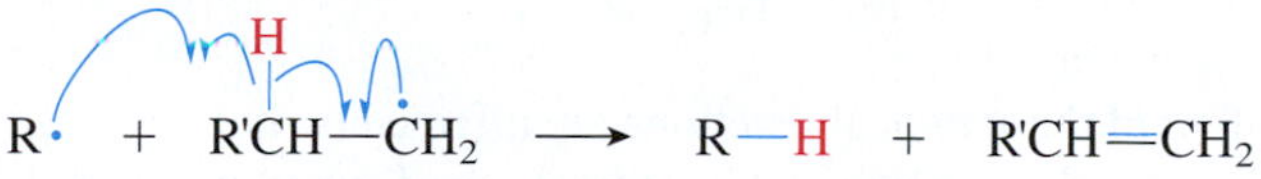

$$R\cdot \ + \ R'CH\!-\!CH_2 \longrightarrow R\!-\!H \ + \ R'CH\!=\!CH_2$$

Disproportionation

Many radicals can react by both of these pathways, as illustrated in the following equation showing the coupling and disproportionation of propyl radicals:

$$CH_3CH_2\overset{\cdot}{C}H_2 \;+\; \overset{\cdot}{C}H_2CH_2CH_3 \begin{cases} \xrightarrow{\text{coupling}} & CH_3CH_2CH_2\!-\!CH_2CH_2CH_3 \\ \xrightarrow[\text{disproportionation}]{} & CH_3CH_2CH_3 \;+\; CH_3CH\!=\!CH_2 \end{cases}$$

As might be expected, the ratio of disproportionation to coupling increases with increasing steric hindrance at the radical centers.

Coupling and disproportionation are energetically very favorable and tend to occur nearly every time two radicals collide. These reactions would dominate radical chemistry but for the fact that the concentration of radicals is usually very small. Therefore, the rate of coupling and disproportionation reactions is often slow because the collision of two radicals is rare.

The product of a reaction of a radical, an odd electron species, with a normal molecule, an even electron species, must produce an odd electron species, a radical, as one of the products. Although this type of reaction is not as energetically favorable as the reaction of two radicals, it is quite common because the collision of a radical with a normal molecule is more probable than the collision of two radicals.

Two different reactions are possible when a radical collides with a normal molecule. The radical may abstract an atom, usually hydrogen or a halogen, from the normal molecule:

$$R\cdot \;+\; H\!-\!\overset{|}{\underset{|}{C}}\!- \;\longrightarrow\; R\!-\!H \;+\; \cdot\overset{|}{\underset{|}{C}}\!-$$

Abstraction

If the normal molecule has a double bond, the radical can add to the pi bond:

$$R\cdot \;+\; {\Large >}C\!=\!C{\Large <} \;\longrightarrow\; R\!-\!\overset{|}{\underset{|}{C}}\!-\!\overset{|}{\underset{|}{C}}\!\cdot$$

Addition

Finally, radicals can fragment to a smaller radical and a normal molecule in a process that is the reverse of addition. As an example, radicals derived from carboxylic acids eliminate carbon dioxide very rapidly.

$$R\!-\!\overset{\overset{\displaystyle O}{\|}}{C}\!-\!\overset{..}{\underset{..}{O}}{:} \;\longrightarrow\; R\cdot \;+\; CO_2$$

Fragmentation

The rates of all of these radical reactions are affected by the same factors that affect the rates of other reactions. More stable radicals are formed more readily, react more slowly, and have longer lifetimes. In addition, steric hindrance can prevent two radicals from approaching close enough to couple, thus increasing the lifetime of the radical. As an example of the importance of steric factors, the 2,4,6-tri-*tert*-butylphenoxy radical does not couple in solution and can actually be isolated as a solid.

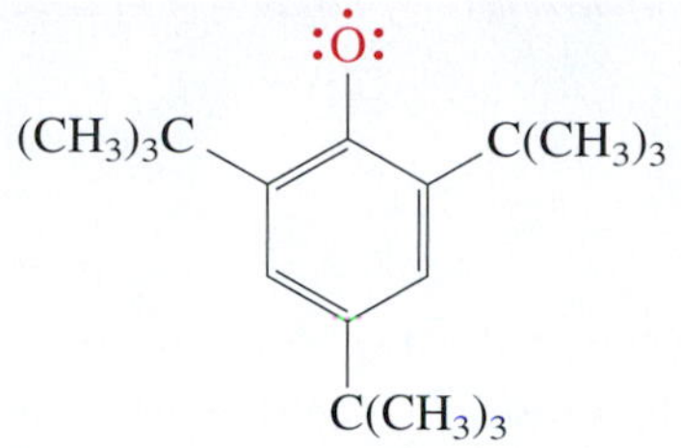

:Ö:

(CH₃)₃C ⟍⟍⟍ C(CH₃)₃

C(CH₃)₃

2,4,6-tri-*tert*-Butylphenoxy radical

The green areas in this model show the locations of the odd electron density in the radical, primarily on the oxygen and the positions ortho and para to it. (This is consistent with the location of the odd electron in the resonance structures that can be written for this radical.) Note how the positions with odd electron density are shielded by the bulky *tert*-butyl groups.

PROBLEM 21.4

Explain which of these reactions would be faster:

a) :Br· + CH₃CH₂CH₃ ⟶ HBr + CH₃ĊHCH₃   2°

or

:Br· + CH₃CHCH₃ ⟶ HBr + CH₃ĊCH₃   ✓ 3°
      | |
     CH₃   CH₃

b)

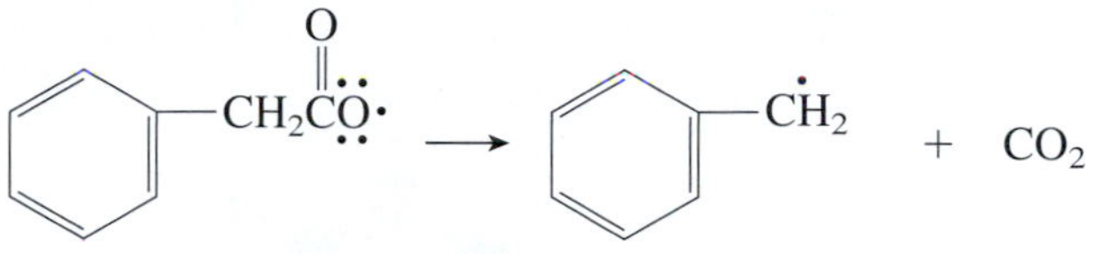

⬡—CH₂CÖ· ⟶ ⬡—ĊH₂ + CO₂   ✓ benzylic
      ‖
      O

or

     O
     ‖
CH₃CH₂CÖ· ⟶ CH₃ĊH₂ + CO₂   1°

PROBLEM 21.5

Suggest structures for products that might be formed in these reactions:

a) :Br· + CH₃CH=CH₂ ⟶

b) CH₃COOCCH₃ —Δ→
       ‖   ‖
       O   O

c) 2 CH₃CH₂CH₂ĊH₂ ⟶

d) 2 NO₂ ⇌

e) CH₃C—N=N—CCH₃ —Δ→
        | |
       CH₃   CH₃
        | |
       CH₃   CH₃

Now let's examine some actual radical reactions that result from combinations of these general reactions.

Focus On

The Triphenylmethyl Radical

The existence of a carbon radical was first proposed by Moses Gomberg in 1900. Gomberg was trying to prepare hexaphenylethane from triphenylmethyl chloride according to the following equation:

$$
2 \quad Ph-\underset{\underset{Ph}{|}}{\overset{\overset{Ph}{|}}{C}}-Cl \quad + \quad 2 \quad Ag \quad \longrightarrow \quad Ph-\underset{\underset{Ph}{|}}{\overset{\overset{Ph}{|}}{C}}-\underset{\underset{Ph}{|}}{\overset{\overset{Ph}{|}}{C}}-Ph \quad + \quad 2 \quad AgCl
$$

Triphenylmethyl chloride   Hexaphenylethane

He obtained a colorless hydrocarbon, which he thought was hexaphenylethane. However, this compound is highly reactive, much more so than would be expected for hexaphenylethane. Furthermore, when this compound is dissolved in solvents such as benzene, it gives a yellow solution. The intensity of the yellow color increases as the solution is heated, indicating that the concentration of the colored species increases as the temperature increases. Furthermore, the color is discharged when oxygen is bubbled through the solution.

Gomberg correctly hypothesized that the hydrocarbon is in equilibrium with triphenylmethyl radicals:

Gomberg's
hydrocarbon $\rightleftarrows$ 2 $Ph-\underset{\underset{Ph}{|}}{\overset{\overset{Ph}{|}}{C}}\cdot$

The equilibrium constant has been determined to be 2×10^{-4} M, which means that about 1% of a 1 M solution of the hydrocarbon is dissociated into radicals. These radicals react with any oxygen in the solution to give colorless products.

Today, we recognize that the ability of triphenylmethyl radicals to exist free in solution is due to two factors. First, the radical has considerable resonance stabilization. Second, and more important, there is considerable steric hindrance to the dimerization of the radical due to the three bulky phenyl groups. In fact, it has recently been shown that Gomberg's hydrocarbon is not hexaphenylethane but actually results from one triphenylmethyl radical bonding to the para position of a phenyl group on the second, even though this results in the loss of the aromatic resonance energy of one ring:

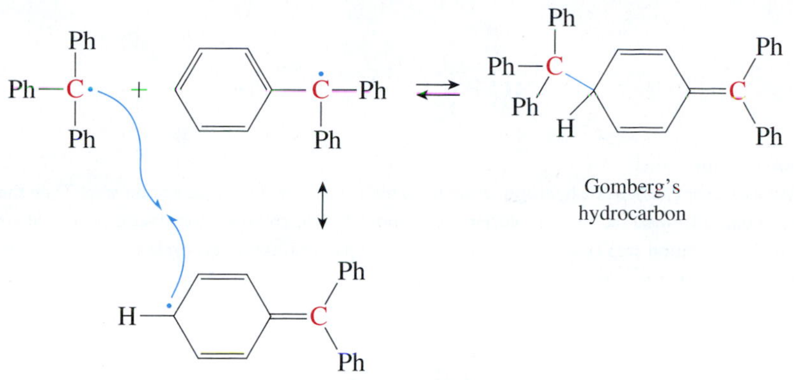

Gomberg's hydrocarbon

Gomberg's proposal was the first suggestion that carbon is not always tetravalent. The acceptance of the triphenylmethyl radical by the scientific community helped facilitate the development of mechanistic organic chemistry with its trivalent carbocations, carbanions, and radicals.

21.6 HALOGENATION

The reaction of methane with chlorine to form chloromethane is induced by light:

This substitution reaction follows a radical chain mechanism, as outlined in Figure 21.1. Radical chain mechanisms involve three types of steps:

Initiation step: In this step, a normal compound undergoes homolytic bond cleavage to generate radicals.

Propagation steps: A radical chain reaction often has several propagation steps. In each of these steps, a radical reacts with a normal compound to produce a new radical. The final propagation step of a chain reaction produces the same radical that reacts in the initial propagation step, so the process can begin anew.

Termination steps: In termination steps, two radicals react to give nonradical products by coupling or disproportionation. These steps destroy radicals.

The radicals produced by one initiation step cause a very large number of propagation steps to occur before they are finally destroyed by a termination step. Termination steps are slower than propagation steps because the concentration of radicals is small. Therefore, the probability of one radical encountering another so that a termination can occur is low.

As shown in Figure 21.1, the initiation step in the chlorination of methane is light-induced cleavage of Cl_2 to produce two chlorine atoms. A chlorine atom then abstracts a hydrogen from methane to produce a methyl radical. The methyl radical abstracts a chlorine

Initiation
The weak chlorine–chlorine bond is broken by light.

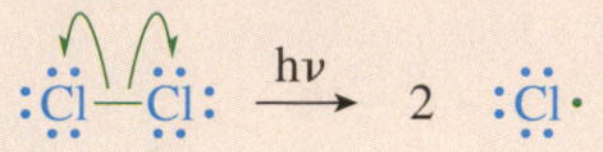

$$:\overset{\cdot\cdot}{\underset{\cdot\cdot}{Cl}}\!-\!\overset{\cdot\cdot}{\underset{\cdot\cdot}{Cl}}: \quad \xrightarrow{h\nu} \quad 2 \quad :\overset{\cdot\cdot}{\underset{\cdot\cdot}{Cl}}\cdot$$

Propagation
A chlorine atom abstracts a hydrogen atom from methane in the first propagation step. Then the methyl radical that is formed abstracts a chlorine atom from Cl_2. The chlorine atom that is produced in the second propagation step reacts again as in the first propagation step. This cycle of two propagation steps is repeated many times in a chain reaction.

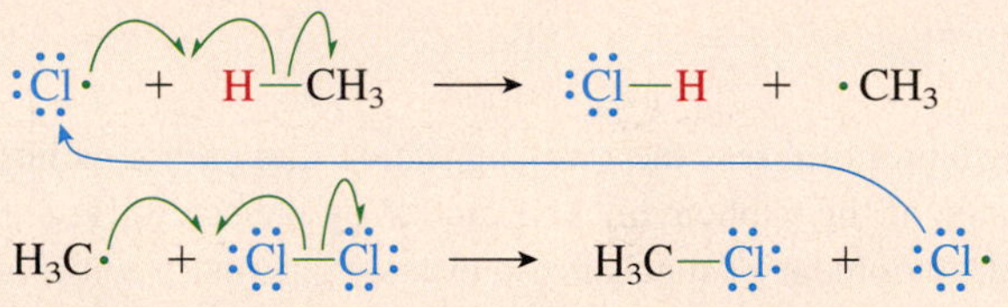

$$:\overset{\cdot\cdot}{\underset{\cdot\cdot}{Cl}}\cdot \;+\; H\!-\!CH_3 \;\longrightarrow\; :\overset{\cdot\cdot}{\underset{\cdot\cdot}{Cl}}\!-\!H \;+\; \cdot CH_3$$

$$H_3C\cdot \;+\; :\overset{\cdot\cdot}{\underset{\cdot\cdot}{Cl}}\!-\!\overset{\cdot\cdot}{\underset{\cdot\cdot}{Cl}}: \;\longrightarrow\; H_3C\!-\!\overset{\cdot\cdot}{\underset{\cdot\cdot}{Cl}}: \;+\; :\overset{\cdot\cdot}{\underset{\cdot\cdot}{Cl}}\cdot$$

Termination
The termination steps include all of the possible radical coupling reactions.

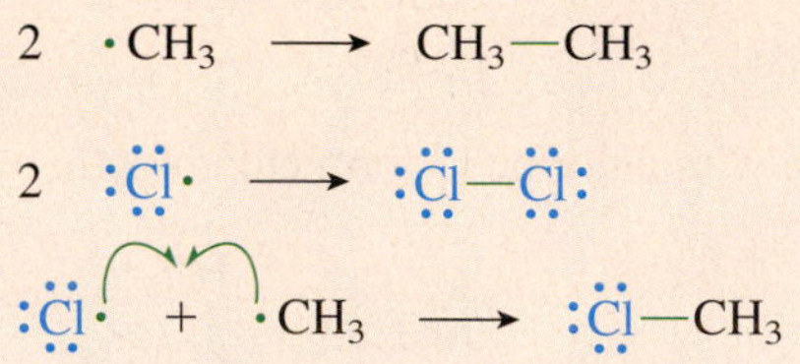

$$2 \;\;\cdot CH_3 \;\longrightarrow\; CH_3\!-\!CH_3$$

$$2 \;\;:\overset{\cdot\cdot}{\underset{\cdot\cdot}{Cl}}\cdot \;\longrightarrow\; :\overset{\cdot\cdot}{\underset{\cdot\cdot}{Cl}}\!-\!\overset{\cdot\cdot}{\underset{\cdot\cdot}{Cl}}:$$

$$:\overset{\cdot\cdot}{\underset{\cdot\cdot}{Cl}}\cdot \;+\; \cdot CH_3 \;\longrightarrow\; :\overset{\cdot\cdot}{\underset{\cdot\cdot}{Cl}}\!-\!CH_3$$

Active Figure 21.1

ORGANIC
Chemistry ⚛ Now™

RADICAL CHAIN MECHANISM OF THE CHLORINATION OF METHANE. Test yourself on the concepts in this figure at **OrganicChemistryNow.**

atom from Cl_2 to produce a molecule of the product, chloromethane, and another chlorine atom. This chlorine atom abstracts a hydrogen from another methane molecule, and the cycle is repeated. A single initiation step causes a very large number (as many as 1000 or more) of propagation cycles to occur. The large number of propagation cycles is characteristic of a chain reaction. Eventually, one of the radicals encounters another radical, and a termination reaction occurs. It is important to recognize that a chain reaction follows the pattern

one initiation reaction,

many propagation cycles,

one termination reaction.

Although the initiation and termination steps are important in the mechanism of a chain reaction, almost all of the products are formed as the result of the propagation steps because there are so many more of them. An equation for the overall reaction can be obtained by summing the propagation steps and canceling species that appear on both sides of the equation:

$$:\overset{..}{\underset{..}{Cl}}\cdot \;+\; CH_4 \;\longrightarrow\; HCl \;+\; \cdot CH_3$$

$$\cdot CH_3 \;+\; Cl_2 \;\longrightarrow\; CH_3Cl \;+\; :\overset{..}{\underset{..}{Cl}}\cdot$$

$$\overline{\quad CH_4 \;+\; Cl_2 \;\longrightarrow\; CH_3Cl \;+\; HCl \quad}$$

The amount of chloromethane that is produced by the termination steps is negligible in comparison to the amount produced by the propagation steps. Likewise, the amount of ethane produced by termination steps is so small that it is hard to detect and can be neglected in the balanced equation for the reaction (but not in the mechanism).

Radical chlorination is a difficult reaction to control. As the reaction proceeds and the initial product, chloromethane, accumulates, it can also undergo hydrogen abstraction by a chlorine atom, resulting in the formation of dichloromethane. Chloroform is formed from dichloromethane and carbon tetrachloride from chloroform in a similar manner. The reaction of a 1:1 ratio of methane and chlorine at 440°C (at this high temperature, homolytic fission of the chlorine–chlorine bond occurs without light) results in the product mixture shown in the following equation:

$$CH_4 \;+\; Cl_2 \xrightarrow{\;440°C\;} CH_4 \;+\; CH_3Cl \;+\; CH_2Cl_2 \;+\; CHCl_3 \;+\; CCl_4 \;+\; HCl$$

| | 47% | 20% | 22% | 10% | 1% | |

Because of the number of products that are produced, this reaction is not very useful in the laboratory. However, the chlorination of methane by this method is used industrially for the production of the various chloromethanes. After the reaction is completed, the hydrogen chloride is removed by treatment with water, and the mixture of products is separated by distillation. The methane is recycled, and the chlorinated products are sold.

PROBLEM 21.6

Show all of the steps in the mechanism for the formation of dichloromethane from chloromethane:

$$CH_3Cl \;+\; Cl_2 \;\longrightarrow\; CH_2Cl_2 \;+\; HCl$$

Another factor that limits the usefulness of the chlorination reaction in the laboratory is the lack of selectivity exhibited by the chlorine atom when more than one type of hydrogen is available to be abstracted. As an example, the reaction of isobutane with chlorine produces 1-chloro-2-methylpropane and 2-chloro-2-methylpropane in a 2:1 ratio, in addition to products containing more than one chlorine.

Isobutane		1-Chloro-2-methylpropane	2-Chloro-2-methylpropane
		2 : 1	

The formation of this mixture of products limits the usefulness of this reaction for laboratory synthesis.

Let's carefully analyze the reasons why these products are formed in this ratio. When a chlorine atom encounters an isobutane molecule, it may abstract a primary hydrogen, leading ultimately to the formation of 1-chloro-2-methylpropane:

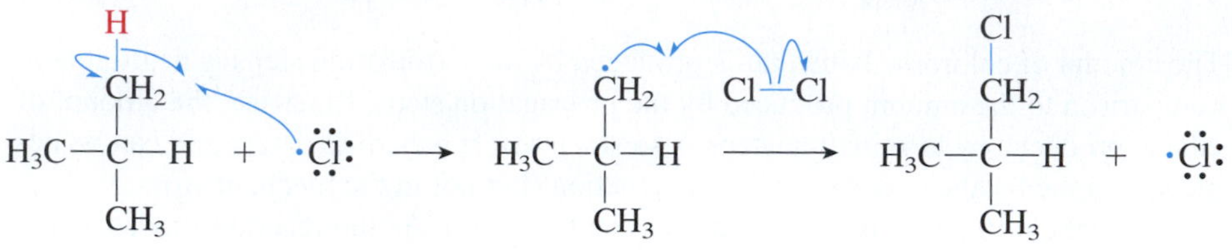

A primary radical

Or it may abstract the tertiary hydrogen, leading to the formation of 2-chloro-2-methylpropane:

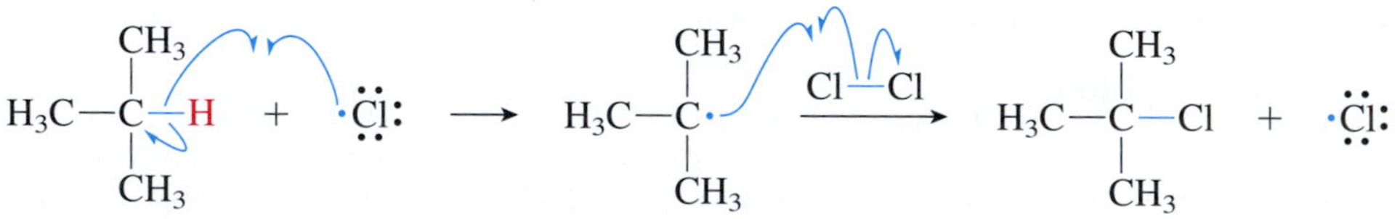

A tertiary radical

There are nine primary hydrogens and only one tertiary hydrogen, so if the hydrogens were all equally reactive, the products should be formed in a 9:1 ratio. Because the observed ratio is 2:1, the tertiary hydrogen must be 4.5 times as reactive as a primary hydrogen under this set of experimental conditions. The tertiary hydrogen is more reactive than a primary hydrogen because the tertiary radical that is produced is more stable than a primary radical. Therefore, the transition state leading to the tertiary radical is somewhat lower in energy than the transition state leading to the primary radical. Although a chlorine atom has a slight preference to abstract a tertiary hydrogen over a secondary hydrogen and a slight preference to abstract a secondary hydrogen over a primary hydrogen, these preferences are small enough, and usually offset somewhat by statistical factors, that a mixture of products is usually formed.

Bromine also reacts with alkanes in the presence of light to give products resulting from the substitution of a bromine for a hydrogen:

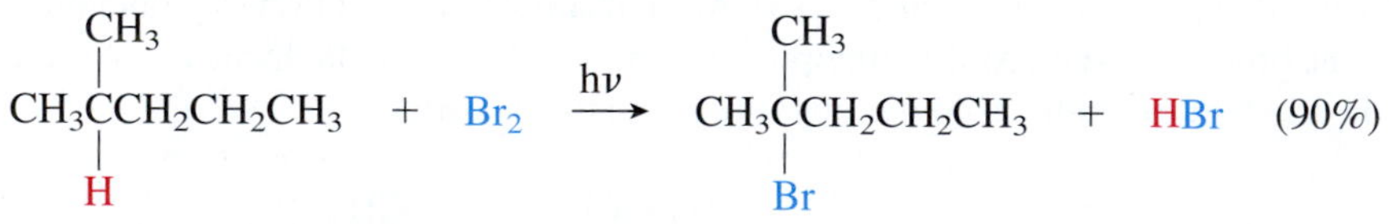

2-Methylpentane

This reaction follows a radical chain mechanism that is analogous to the chlorination mechanism shown in Figure 21.1. However, as illustrated in this example, a bromine radical is much more selective than a chlorine radical. It greatly prefers to abstract a hydrogen from a tertiary carbon rather than from a primary or secondary carbon. This preference is large enough that a single product dominates in favorable cases such as the preceding example. Therefore, the bromination reaction is much more useful in synthesis.

Why is a bromine atom so much more selective than a chlorine atom in these hydrogen abstraction reactions? The answer can be found by examination of the energetics of each reaction. Because the bond strength of the hydrogen–chlorine bond (103 kcal/mol [413 kJ/mol]) is larger than the bond strength of the hydrogen–bromine bond (87 kcal/mol [364 kJ/mol]), the abstraction of a hydrogen by a chlorine atom is exothermic, whereas the abstraction of a hydrogen by a bromine atom is slightly endothermic. For example, the enthalpies for the abstraction of the tertiary hydrogen of isobutane by chlorine and by bromine are provided in the following equations:

$$
\begin{array}{c}
\text{CH}_3 \\
| \\
\text{H}_3\text{C}-\overset{\displaystyle |}{\underset{\displaystyle |}{\text{C}}}-\text{H} \;+\; \cdot\ddot{\text{Cl}}\!: \;\longrightarrow\; \text{H}_3\text{C}-\overset{\displaystyle |}{\underset{\displaystyle |}{\text{C}}}\cdot \;+\; \text{H}-\text{Cl} \quad \Delta H \;=\; -12\,\text{kcal/mol} \\
\text{CH}_3 \qquad\qquad\qquad\qquad\qquad \text{CH}_3 \qquad\qquad\qquad\qquad (-50\,\text{kJ/mol})
\end{array}
$$

$$
\begin{array}{c}
\text{CH}_3 \\
| \\
\text{H}_3\text{C}-\overset{\displaystyle |}{\underset{\displaystyle |}{\text{C}}}-\text{H} \;+\; \cdot\ddot{\text{Br}}\!: \;\longrightarrow\; \text{H}_3\text{C}-\overset{\displaystyle |}{\underset{\displaystyle |}{\text{C}}}\cdot \;+\; \text{H}-\text{Br} \quad \Delta H \;=\; +4\,\text{kcal/mol} \\
\text{CH}_3 \qquad\qquad\qquad\qquad\qquad \text{CH}_3 \qquad\qquad\qquad\qquad (+17\,\text{kJ/mol})
\end{array}
$$

According to the Hammond postulate, the transition state for abstraction by chlorine resembles the reactant because this is an exothermic reaction. In contrast, the transition state for abstraction by bromine resembles the product because it is an endothermic reaction (see Figure 21.2). In the case of abstraction by chlorine the carbon–hydrogen bond is only slightly broken in the transition state, and the stability

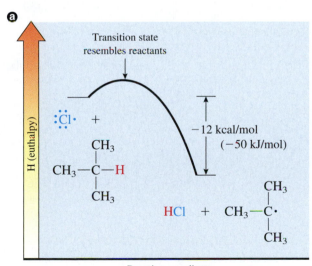

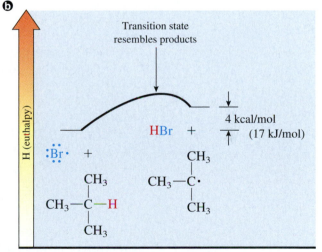

By the Hammond postulate, the transition state for abstraction of a hydrogen by a chlorine atom resembles the reactants and has only a small amount of radical character. Therefore, the transition state leading to a tertiary radical is only slightly more stable than the transition state leading to a primary radical.

In contrast, the transition state for abstraction of a hydrogen atom by a bromine atom resembles the product and has a large amount of radical character. Therefore, the transition state leading to a tertiary radical is considerably more stable than the transition state leading to a primary radical.

Figure 21.2

ENTHALPIES OF HYDROGEN ABSTRACTION BY ⓐ CHLORINE AND ⓑ BROMINE ATOMS.

of the product radical has only a small effect on the transition state energy. When a hydrogen is abstracted by bromine, the carbon–hydrogen bond is more broken in the transition state. The transition state has more radical character, so its energy is more affected by the stability of the radical product. In other words, a chlorine atom is more reactive than a bromine atom. Usually, the more reactive a reagent is, the less selective that reagent is.

PROBLEM 21.7

The bond strengths for the HF and HI bonds are 135 kcal/mol (565 kJ/mol) and 71 kcal/mol (297 kJ/mol), respectively. Explain why F_2 reacts explosively with alkanes whereas I_2 does not react at all.

Because of the greater selectivity of the bromine atom, radical brominations can be useful in synthesis as long as the compound to be brominated has one hydrogen that is considerably more reactive than the others. The reaction of 2-methylpentane with bromine, shown previously, gives predominantly a single product because there is only one tertiary hydrogen. Because allylic and benzylic radicals are stabilized by resonance, bromination at these positions can also be successfully accomplished. An example is provided by the following equation:

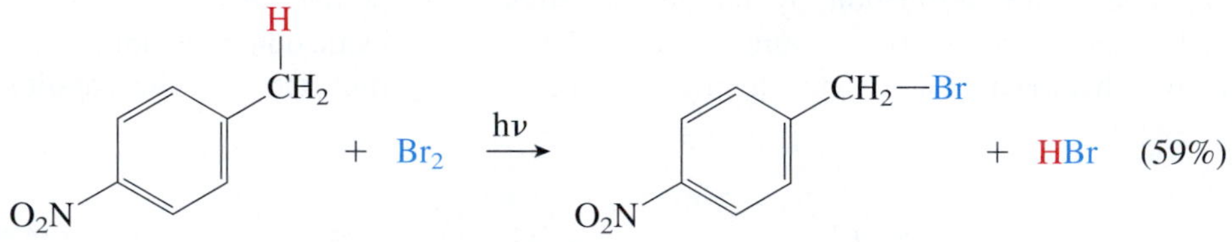

Another reagent that is quite useful in substituting bromine for hydrogen at allylic and benzylic positions is *N*-bromosuccinimide (NBS):

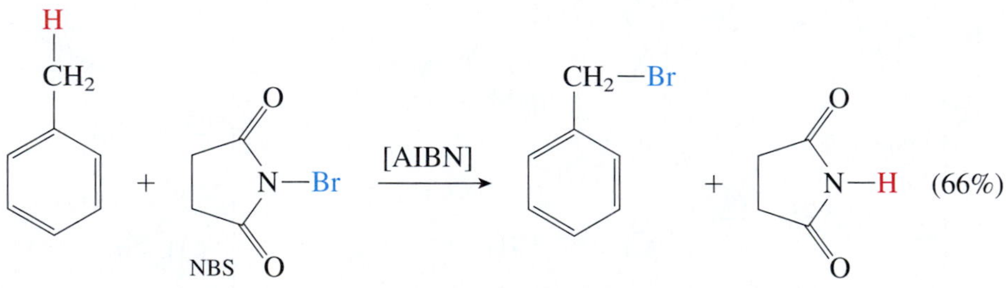

The actual brominating agent in these reactions is thought to be Br_2, which is produced in low concentration by the reaction of NBS with the HBr that is produced in the reaction:

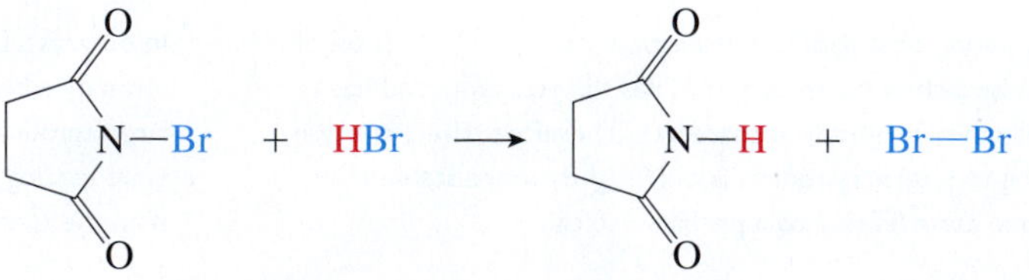

The mechanism then follows the same radical chain mechanism that is followed in other brominations with the exception that a radical initiator, such as AIBN or a peroxide, is usually employed to initiate the reaction. The initiation steps are shown in the following equations:

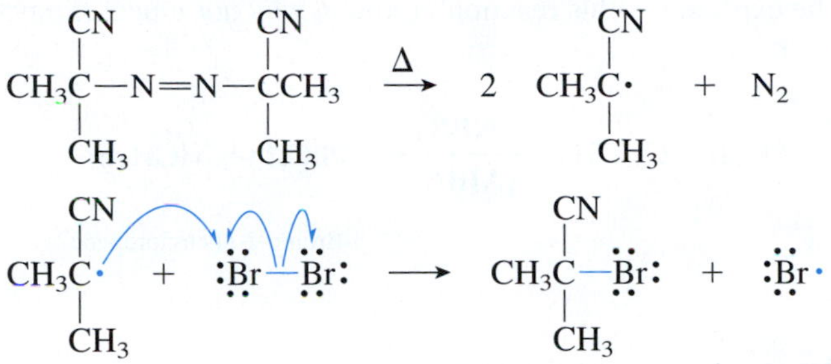

NBS has proved to be especially useful in brominations at allylic positions because competition from the addition of bromine to the double bond is not a problem. Apparently, the fact that only a low concentration of Br_2 is ever present in NBS brominations somehow inhibits the addition reaction. Examples are provided in the following equations. Note that the reaction is best if only a single type of allylic hydrogen is available to be abstracted. In addition, the resonance-stabilized allylic radical provides two sites that can abstract a bromine atom. If these two sites are different, a mixture of products is formed, as shown in the second example:

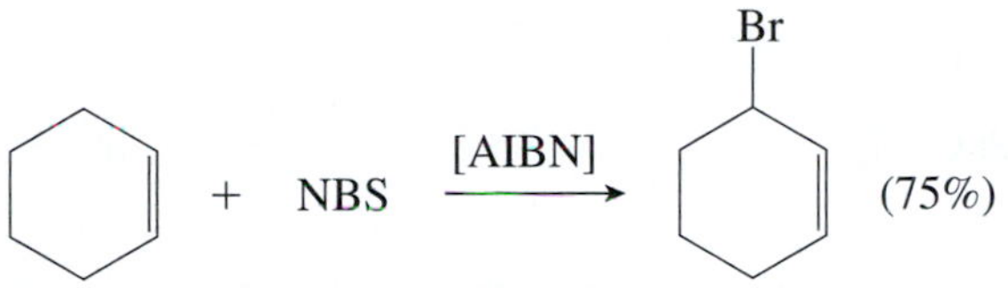

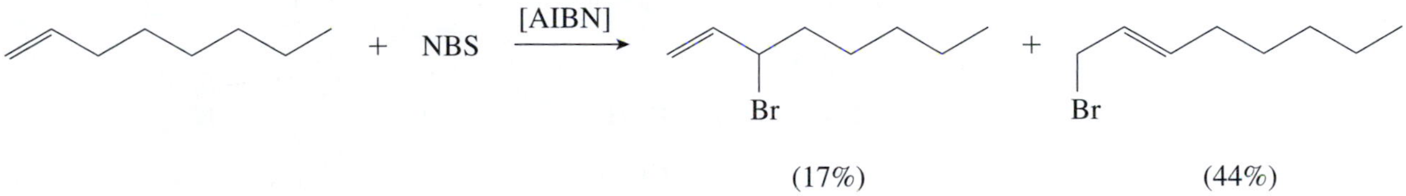

PROBLEM 21.8
Show the products of these reactions:

a)

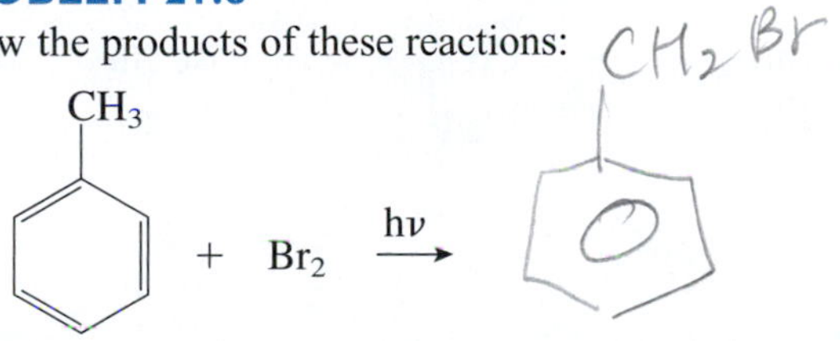

b)

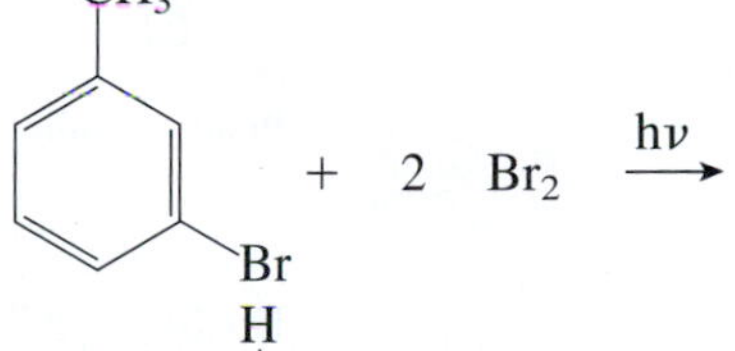

c)

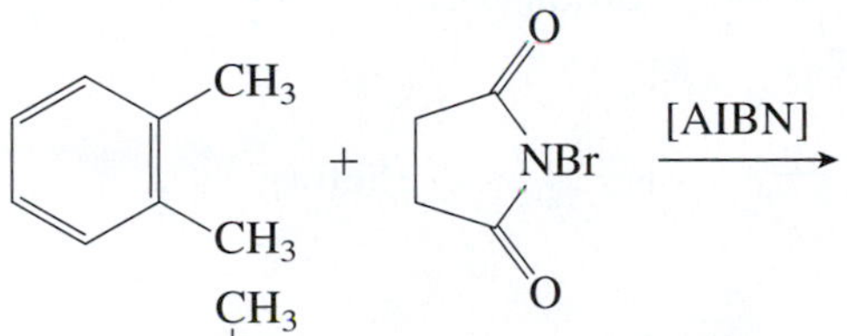

d)

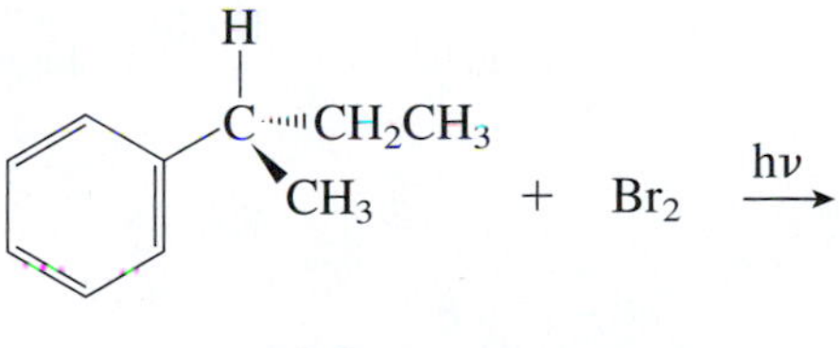

e)

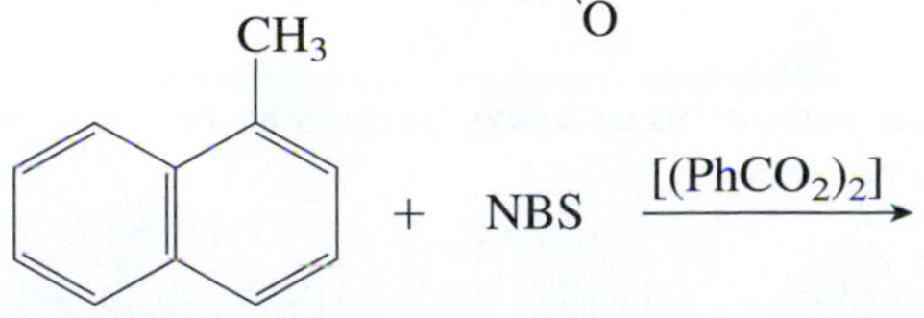

f)

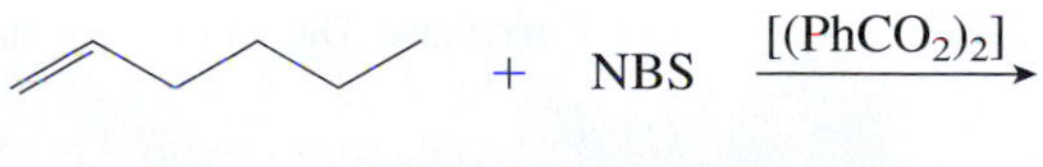

PROBLEM 21.9

The following reaction produces 3-bromo-1-phenylpropene in good yield. What other product might be expected in this reaction? Explain why not much, if any, of this other product is formed.

$$PhCH=CHCH_3 \xrightarrow[\text{[AIBN]}]{\text{NBS}} PhCH=CHCH_2Br$$

3-Bromo-1-phenylpropene

21.7 DEHALOGENATION

The reaction of an alkyl halide with tributyltin hydride, using a radical initiator, results in the replacement of the halogen by hydrogen. The reaction follows a radical chain mechanism as outlined in Figure 21.3. Examples are provided in the following equations:

$$PhCH_2CH_2Br + Bu_3SnH \xrightarrow{\text{[AIBN]}} PhCH_2\overset{H}{\underset{|}{C}}H_2 + Bu_3SnBr \quad (85\%)$$

$$\overset{O}{\underset{\|}{Ph C}}CH_2Br + Bu_3SnH \xrightarrow{\text{[AIBN]}} \overset{O}{\underset{\|}{PhC}}{-}\overset{H}{\underset{|}{C}}H_2 + Bu_3SnBr \quad (84\%)$$

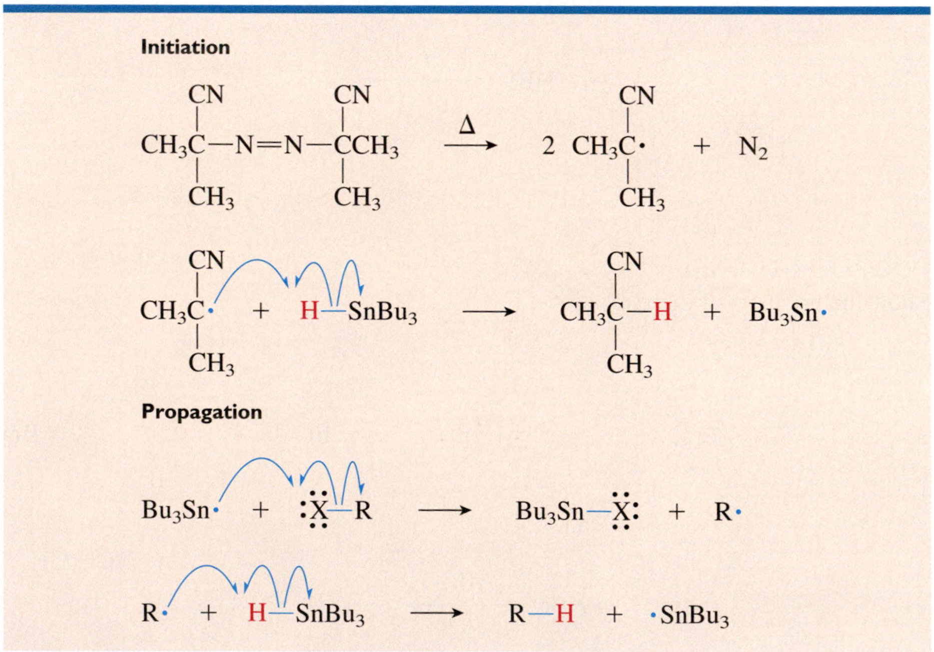

Figure 21.3

MECHANISM OF THE RADICAL CHAIN DEHALOGENATION OF ALKYL HALIDES BY TRIBUTYLTIN HYDRIDE. The various termination steps are not shown.

PROBLEM 21.10

Show the products of these reactions:

a) 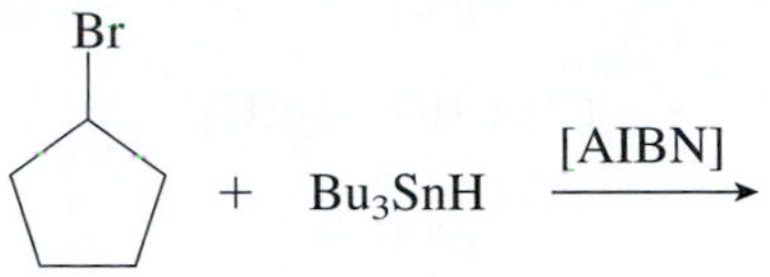 [structure: cyclopentane with Br substituent] + Bu₃SnH —[AIBN]→

b)  [structure: Ph-CH(Cl)-CH₃] + Bu₃SnH —[AIBN]→

21.8 AUTOXIDATION

The slow oxidation of organic materials that are exposed to oxygen in the atmosphere is termed **autoxidation.** A simple example of this reaction is provided by the following equation:

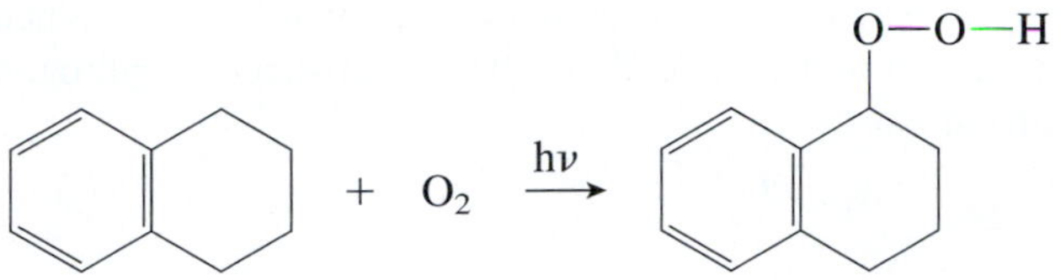

[structure: tetralin] + O_2 —$h\nu$→ [structure: tetralin with O—O—H group]

This process follows a radical chain mechanism and is catalyzed by light. (This is one reason why compounds are often sold and stored in brown glass bottles.) In the first step of the mechanism an initiating radical (In·) is generated by light or some other means. This radical abstracts a hydrogen from the substrate to produce a carbon radical (R·):

Initiation

$$R—H + ·In \longrightarrow R· + H—In$$

In the propagation steps, the carbon radical first adds to the oxygen–oxygen double bond to produce a peroxide radical. This radical abstracts a hydrogen from another molecule of the substrate, generating a hydroperoxide and another carbon radical that can repeat the propagation cycle:

Propagation

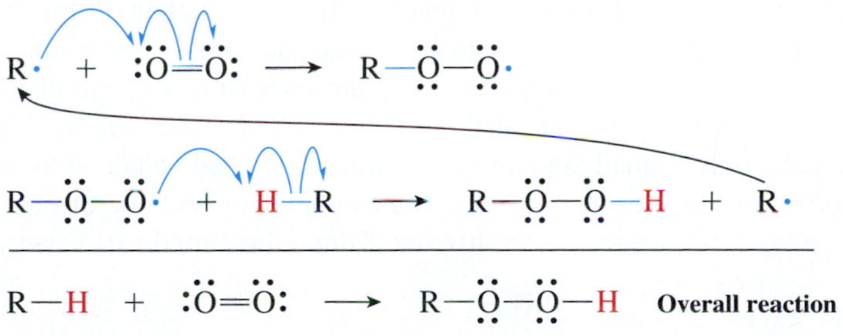

$$R· + :\ddot{O}=\ddot{O}: \longrightarrow R—\ddot{O}—\ddot{O}·$$

$$R—\ddot{O}—\ddot{O}· + H—R \longrightarrow R—\ddot{O}—\ddot{O}—H + R·$$

$$R—H + :\ddot{O}=\ddot{O}: \longrightarrow R—\ddot{O}—\ddot{O}—H \quad \text{Overall reaction}$$

The autoxidation reaction is difficult to control, so it is not often used for synthetic purposes. However, it is a very important natural process. The slow deterioration of organic materials, such as rubber, paint, and oils, and that of many foods, such as butter and fats, is due to autoxidation. As one example, peroxides are formed in solvents such as diethyl ether or THF that are stored for long periods of time in contact with air:

$$CH_3CH_2OCH_2CH_3 \ + \ O_2 \ \longrightarrow \ \overset{\overset{\displaystyle OOH}{\displaystyle |}}{CH_3CHOCH_2CH_3}$$

Diethylether   A peroxide

Because peroxides are explosive, the use of such contaminated solvents leads to a very dangerous situation. Numerous explosions have resulted in the laboratory when peroxides have been concentrated as the solvent is removed by distillation during the workup of a reaction. Therefore, it is important to be certain that ether and THF are free of peroxides before they are employed as solvents. Solutions of ether or THF should never be distilled to dryness.

Antioxidants, such as 2,6-di-*tert*-butyl-4-methylphenol (also known as <u>b</u>utylated <u>h</u>ydroxy<u>t</u>oluene or BHT) and 2-*tert*-butyl-4-methoxyphenol (also known as <u>b</u>utylated <u>h</u>ydroxy<u>a</u>nisole or BHA), are added to many organic materials to prevent autoxidation. They function by interfering with the autoxidation chain reaction. When a radical encounters an antioxidant molecule, such as BHA, it abstracts a hydrogen to produce a resonance-stabilized radical:

BHT

BHA

Because of this resonance stabilization and the steric hindrance provided by the bulky *tert*-butyl group, this radical is not very reactive. It acts as a chain terminator because it is not reactive enough to abstract a hydrogen or add to an oxygen–oxygen double bond and continue the autoxidation chain. The presence of a single molecule of BHA or BHT can prevent the oxidation of thousands of other molecules by terminating a chain. Therefore, only a small amount of an antioxidant need be added to a compound to provide protection against autoxidation. For example, the addition of a small amount of BHA to butter increases its storage lifetime from a few months to a few years.

PROBLEM 21.11

Predict the major product from autoxidation of these compounds:

a)   b)

Focus On Biological Chemistry

Vitamin E and Lipid Autoxidation

Glycerophospholipids are major components of biological membranes. They resemble the fats described in the Focus On box "Preparation of Soaps" on page 819 in that they are composed of esters of the triol glycerol with carboxylic acids (fatty acids) that have long hydrocarbon chains. In contrast to fats, however, glycerophospholipids have only two fatty acid groups attached to the glycerol. The third site is occupied by an ionic or highly polar group attached to the glycerol by an ester linkage. The hydrocarbon chains contribute hydrophobic character to the lipid, whereas the polar groups are hydrophilic. This combination of properties enables glycerophospholipids to form the bilayer membranes that are so important in biological systems.

The hydrocarbon chains of fatty acids often contain one or more cis double bonds. These unsaturated lipids are especially prone to autoxidation. An example is provided by the following equation, which shows an autoxidation reaction of the ester of α-linoleic acid. Note that this is a chain reaction, so one initiator can cause numerous lipid molecules to be oxidized.

❶ Some radical initiates the chain by abstracting a hydrogen atom. The allylic hydrogens of unsaturated lipids are especially susceptible to this process.

❷ The resulting radical adds to an oxygen molecule.

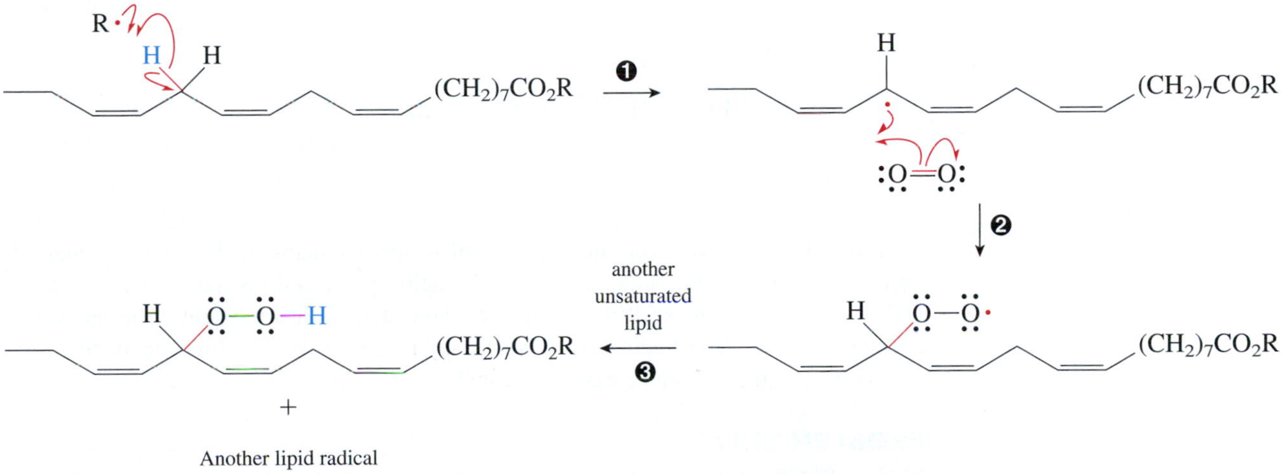

another unsaturated lipid

Another lipid radical

The new lipid radical continues the chain process.

❸ This peroxy radical abstracts a hydrogen from a neighboring lipid molecule.

The peroxidized tails of these lipids are more hydrophilic and try to migrate to the surface of the membrane. This disrupts the structure of the membrane and makes it

Continued

"leaky." Similar peroxidation of unsaturated lipids in low-density lipoproteins is thought to contribute to arteriosclerosis and heart disease.

Nature has many defenses against unwanted oxidation reactions. Vitamin E, also known as α-tocopherol, helps to serve this role in human membranes. It is a hindered phenol, somewhat resembling BHT and BHA, and it is soluble in the membrane because of its nonpolar hydrocarbon tail. The radical that is produced when the hydrogen of its hydroxy group is abstracted is not very reactive, so vitamin E terminates autoxidation chains:

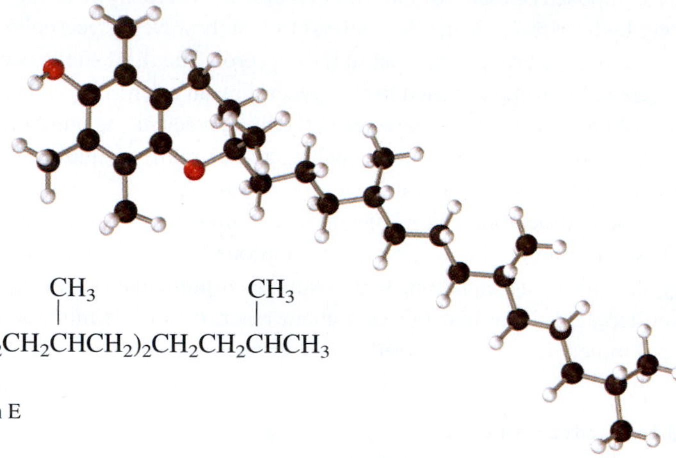

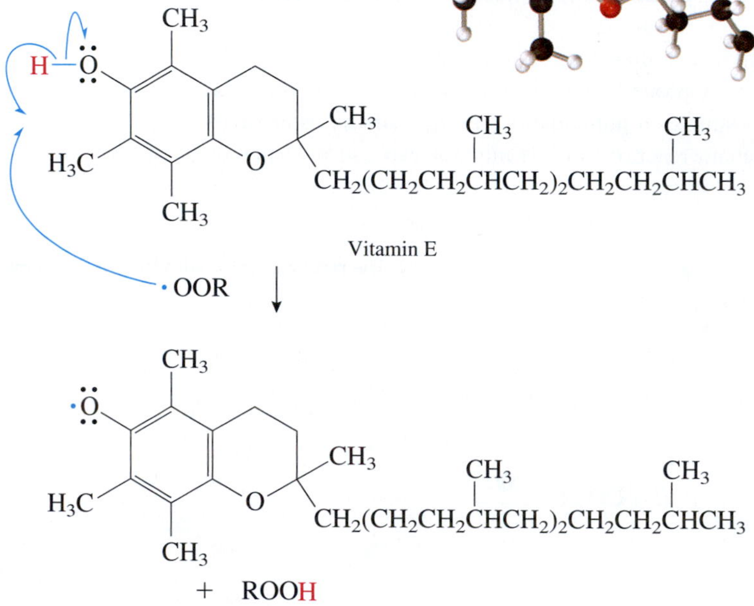

Vitamin E

Radicals may have a role in a number of health problems, including those mentioned previously, along with cancer and the aging process. This has recently led to widespread use of antioxidant nutritional supplements. Unfortunately, little has yet been proved about the effectiveness of such supplements. A healthy lifestyle, including a good diet, will probably be more beneficial.

PROBLEM 21.12

Explain why the radical produced from vitamin E is not very reactive and thus acts as a chain terminator.

21.9 RADICAL ADDITIONS TO ALKENES

In Chapter 11 the electrophilic addition of hydrogen bromide to alkenes, which proceeds by an ionic mechanism, was discussed. Recall that this reaction follows Markovnikov's rule and the bromine adds to the more highly substituted carbon, as illustrated in the following example:

$$CH_3CH_2CH_2CH_2CH{=}CH_2 \ + \ HBr \ \longrightarrow \ CH_3CH_2CH_2CH_2\overset{Br}{\underset{}{C}}\overset{H}{\underset{}{H}}CH_2 \ \ (88\%)$$

When a similar reaction occurs under conditions favoring the formation of radicals—that is, in the presence of light or a peroxide that can initiate the reaction—the addition still occurs, but with the opposite regiochemistry. The bromine adds to the less highly substituted carbon, and the addition is said to occur in an **anti-Markovnikov** manner. Examples are provided by the following equations:

$$CH_3CH_2CH{=}CH_2 \ + \ HBr \ \xrightarrow[\substack{\text{or} \\ h\nu}]{[ROOR]} \ CH_3CH_2\overset{H}{\underset{}{C}}HCH_2Br \ \ (95\%)$$

$$PhCH_2CH{=}CH_2 \ + \ H{-}Br \ \xrightarrow{[ROOR]} \ PhCH_2\overset{H}{\underset{}{C}}HCH_2{-}Br \ \ (80\%)$$

The change in regiochemistry is a result of a change in the mechanism of the reaction, from an ionic mechanism in the Markovnikov reaction to a radical chain mechanism in the anti-Markovnikov reaction. The radical chain mechanism for the addition of hydrogen bromide to 1-butene is outlined in the following equations:

Initiation

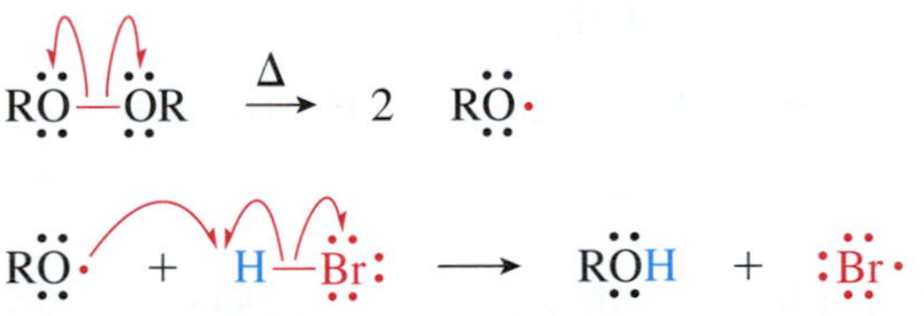

$$RO{-}OR \ \xrightarrow{\Delta} \ 2 \ RO\cdot$$

$$RO\cdot \ + \ H{-}Br\!: \ \longrightarrow \ ROH \ + \ :Br\cdot$$

Propagation

$$:Br\cdot \ + \ CH_2{=}CHCH_2CH_3 \ \longrightarrow \ :Br{-}CH_2{-}\dot{C}HCH_2CH_3$$

$$:Br{-}CH_2{-}\dot{C}HCH_2CH_3 \ + \ H{-}Br\!: \ \longrightarrow \ :Br\cdot \ + \ :Br{-}CH_2{-}\overset{H}{\underset{}{C}}HCH_2CH_3$$

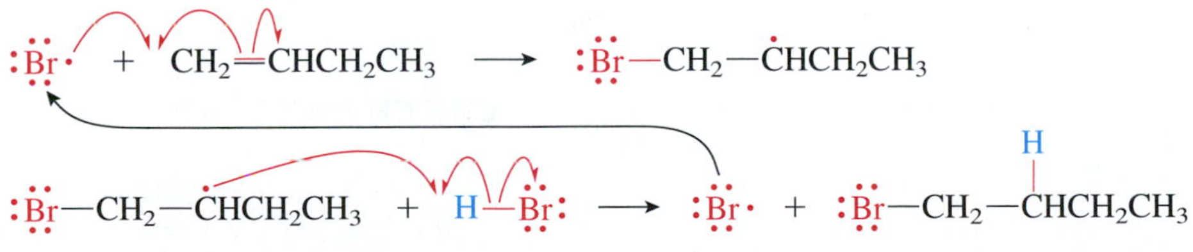

In the propagation steps, a bromine atom first adds to the carbon–carbon double bond. The resulting carbon radical then abstracts a hydrogen from hydrogen bromide, generating a molecule of 1-bromobutane and a bromine atom that can continue the chain.

The regiochemistry of the product is controlled by the energetics of the step in which the bromine atom adds to the double bond. The bromine adds to the less substituted carbon so that a more stable radical, with the odd electron located on the more highly substituted carbon, is formed. Thus, radical stability controls the regiochemistry of this reaction in the same manner that carbocation stability controls the regiochemistry of the ionic reaction. In both reactions the adding species bonds to the less substituted carbon (the carbon that is bonded to more hydrogens), so the resulting carbocation, or radical, is more stable because the positive charge, or odd electron, is located on the more highly substituted carbon (the carbon bonded to more carbons). A proton electrophile adds in the ionic reaction, whereas a bromine radical adds in the radical reaction, so the two reactions have the opposite regiochemistry.

Hydrogen chloride and hydrogen iodide do not readily give this reaction because one of the propagation steps is slow. In the case of HCl the strong hydrogen–chlorine bond (bond dissociation energy = 103 kcal/mol [431 kJ/mol]) causes the abstraction step to be slow:

$$ClCH_2\overset{\cdot}{C}HR \;+\; H\!-\!\overset{..}{\underset{..}{Cl}}: \quad\xrightarrow{\text{slow}}\quad ClCH_2\overset{\overset{H}{|}}{C}HR \;+\; \cdot\overset{..}{\underset{..}{Cl}}:$$

In the case of HI the weak carbon–iodine bond (bond dissociation energy = 52 kcal/mol [217 kJ/mol]) causes the addition step to be slow:

$$:\overset{..}{\underset{..}{I}}\cdot \;+\; CH_2\!=\!CHR \quad\xrightarrow{\text{slow}}\quad :\overset{..}{\underset{..}{I}}\!-\!CH_2\!-\!\overset{\cdot}{C}HR$$

Only in the case of HBr do both propagation steps occur readily.

Other reagents can add to carbon–carbon double bonds in synthetically useful yields by similar radical chain mechanisms. The reaction of an alkene with a thiol under radical conditions results in the addition of the sulfur and a hydrogen to the carbons of the double bond:

$$CH_3\overset{\overset{CH_3}{|}}{C}\!=\!CH_2 \;+\; CH_3CH_2S\!-\!H \quad\xrightarrow{h\nu}\quad CH_3\overset{\overset{CH_3}{|}}{\underset{\underset{H}{|}}{C}}\!-\!CH_2\!-\!SCH_2CH_3 \quad (94\%)$$

Likewise, several tetrahalomethanes add a trihalomethyl group and a halogen to the carbons of the double bond of an alkene under radical conditions. Examples are provided in the following equations:

$$CH_3\overset{\overset{CH_3}{|}}{C}\!=\!CH_2 \;+\; Cl\!-\!CCl_3 \quad\xrightarrow{h\nu}\quad CH_3\overset{\overset{CH_3}{|}}{\underset{\underset{Cl}{|}}{C}}CH_2\!-\!CCl_3 \quad (70\%)$$

$$CH_3(CH_2)_5CH\!=\!CH_2 \;+\; Br\!-\!CBr_3 \quad\xrightarrow{h\nu}\quad CH_3(CH_2)_5\overset{\overset{Br}{|}}{C}HCH_2\!-\!CBr_3 \quad (88\%)$$

$$PhCH\!=\!CH_2 \;+\; CCl_3Br \quad\xrightarrow{h\nu}\quad Ph\overset{\overset{Br}{|}}{C}HCH_2\!-\!CCl_3 \quad (78\%)$$

PROBLEM 21.13

Show all of the steps in the mechanism for this reaction:

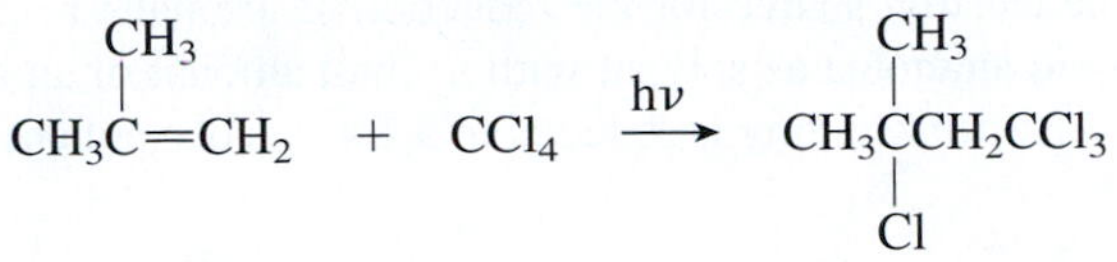

$$CH_3C{=}CH_2 \ + \ CCl_4 \ \xrightarrow{h\nu} \ CH_3\overset{\displaystyle CH_3}{\underset{\displaystyle Cl}{C}}CH_2CCl_3$$

(with CH₃ substituent above the central carbon of the alkene and of the product, and Cl below the product carbon)

PROBLEM 21.14

Show the products of these reactions:

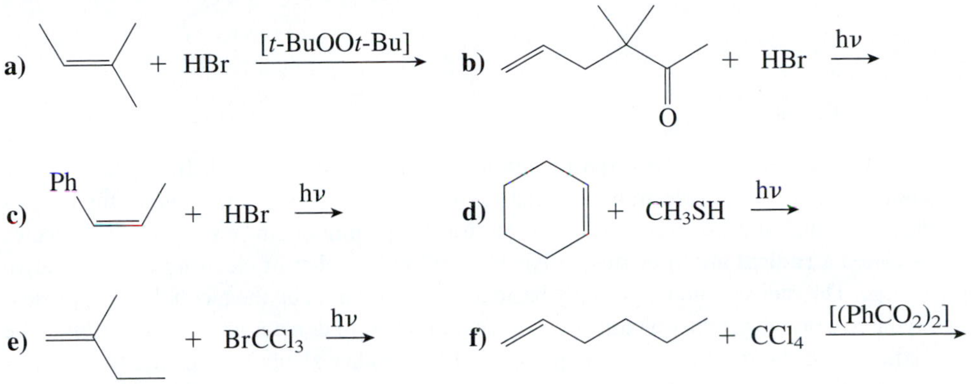

a) [structure] + HBr $\xrightarrow{[t\text{-BuOO}t\text{-Bu}]}$

b) [structure] + HBr $\xrightarrow{h\nu}$

c) Ph [structure] + HBr $\xrightarrow{h\nu}$

d) [cyclohexene structure] + CH₃SH $\xrightarrow{h\nu}$

e) [structure] + BrCCl₃ $\xrightarrow{h\nu}$

f) [structure] + CCl₄ $\xrightarrow{[(PhCO_2)_2]}$

PROBLEM 21.15

Show methods to prepare these compounds from alkenes:

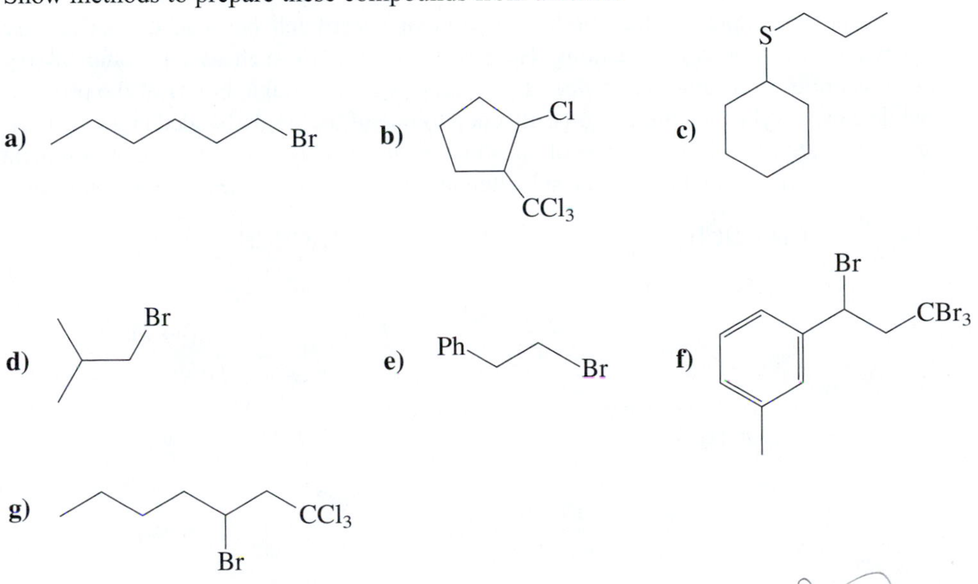

a) [structure] Br

b) [cyclopentane structure with Cl and CCl₃]

c) [cyclohexane structure with S-propyl group]

d) [structure with Br]

e) Ph [structure] Br

f) [structure with Br and CBr₃]

g) [structure with CCl₃ and Br]

21.10 REDUCTIONS AND RADICAL ANIONS

Many organic compounds can be reduced by reaction with a source of electrons and a source of protons. (Previous examples include the Clemmensen reduction and the reduction of nitro groups to amino groups described in Chapter 17.) Often this is accomplished by employing electropositive metals as the electron source in the presence of a weak acid. A useful application of this process, known as the **Birch reduction,** employs sodium or lithium metal as the electron source for the reduction of aromatic rings. The reaction is usually done in liquid ammonia as solvent with a small amount of an alcohol present as the proton source. The benzene ring is reduced to a 1,4-cyclohexadiene derivative:

Benzene   1,4-Cyclohexadiene

The mechanism of the Birch reduction is shown in Figure 21.4. In the first step, a sodium atom gives an electron to a benzene molecule. This electron occupies the lowest-energy orbital in the benzene that has room for it, a pi antibonding (π^*) MO. This species is called a **radical anion** because it has both an odd number of electrons and a negative charge. The radical anion is strongly basic and is protonated by the alcohol that is present in the reaction to produce a radical. Another sodium atom donates a second electron to the radical, and the resulting anion is protonated by another alcohol molecule. In this final protonation step, the delocalized anion is protonated so as to form the nonconjugated product, 1,4-cyclohexadiene, rather than the conjugated product, 1,3-cyclohexadiene. The reasons for this preference are complex and beyond the scope of this book.

Several examples of the Birch reduction of substituted benzene derivatives are shown in the following equations. Note that substituents such as alkyl and alkoxy groups prefer to be attached to one of the carbons of the double bonds of the product, while a carboxyl group prefers to be attached to one of the singly bonded carbons. Benzene derivatives with other types of substituents are usually not employed as reactants in the Birch reduction because the substituents are not stable to the reaction conditions.

Figure 21.4

MECHANISM OF THE
BIRCH REDUCTION OF
BENZENE.

❶ An electron is transferred from a sodium atom to a benzene molecule. This electron is added to the lowest-energy orbital that is available, a pi antibonding MO in this case.

The product of the first step has an odd number of electrons and a negative charge so it is called a radical anion. It has seven electrons in its pi MOs. This Lewis structure is one way to represent it. It has a number of resonance structures.

❸ A second electron is added to the radical, producing a carbanion.

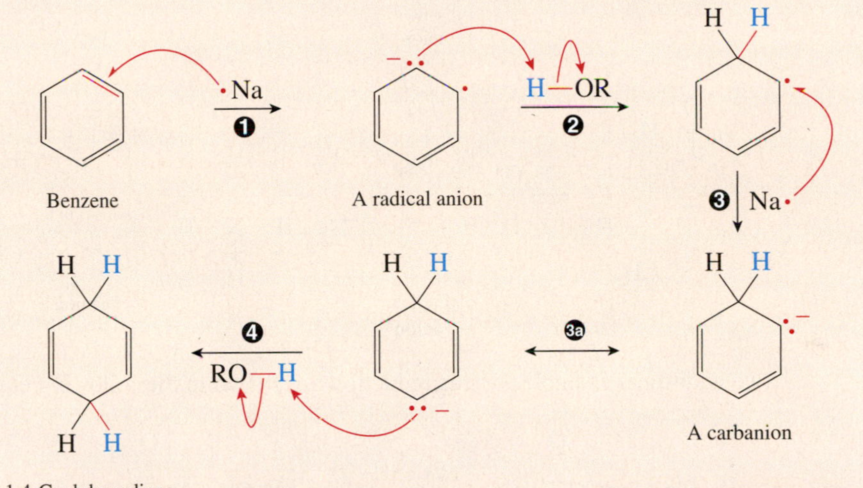

❷ The radical anion is a very strong base and removes a proton from the alcohol. The product of this step is a radical.

Benzene

A radical anion

A carbanion

1,4-Cyclohexadiene

❹ The carbanion is also a strong base and removes a proton from the alcohol. Although it could potentially be protonated either ortho or para to the site of the first protonation, para protonation is faster. The product is a nonconjugated diene, 1,4-cyclohexadiene.

❸ₐ This carbanion is stabilized by resonance. Two resonance structures are shown here. A third resonance structure that resembles the first is not shown.

A normal alkene is not reduced under the conditions of the Birch reduction because its pi antibonding MO is too high in energy for an electron from a sodium atom to be readily added to it. However, as shown in the following example, a carbon–carbon double bond that is conjugated with a carbonyl group is readily reduced by lithium or sodium metal in liquid ammonia solvent, without any added alcohol.

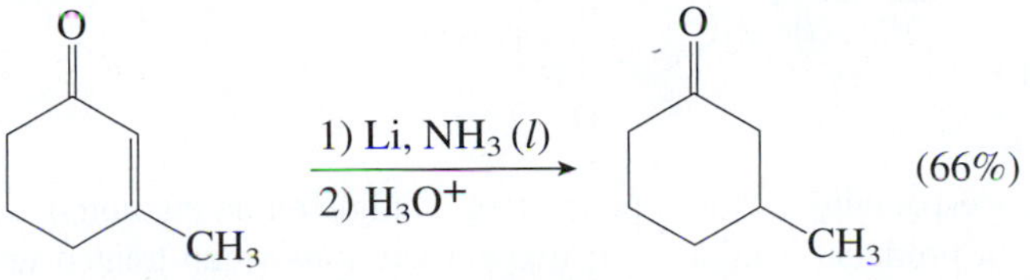

1) Li, NH_3 (l)
2) H_3O^+

(66%)

The mechanism for this reaction is slightly different from that presented previously for the reduction of benzene. First an electron is added to the pi antibonding MO to produce a radical anion. Because there is no alcohol to protonate this anion, a second electron is then added to produce a dianion. The dianion is a strong enough base to remove a proton from the ammonia solvent, producing an enolate anion. The enolate anion is stable to the reaction conditions until acid is added to work up the reaction.

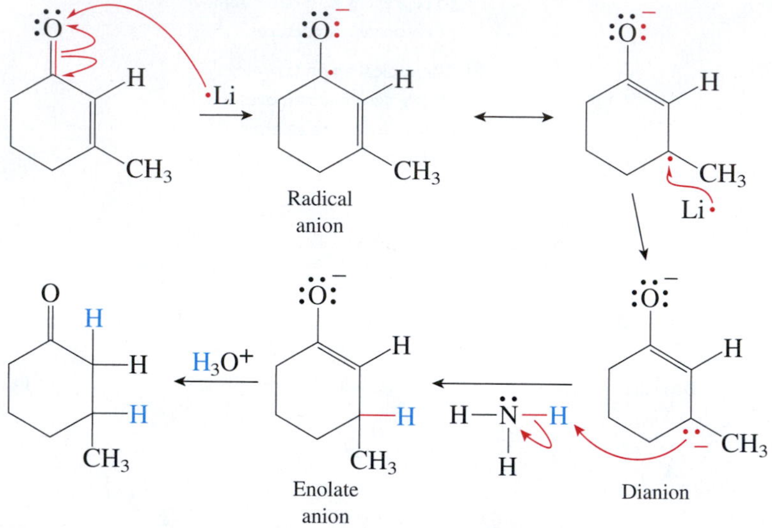

Two additional examples of this reduction are shown in the following equations:

$$PhCH=CHCOH \xrightarrow[\text{2) H}_3\text{O}^+]{\text{1) Li, NH}_3 (l)} PhCH_2CH_2COH \quad (95\%)$$

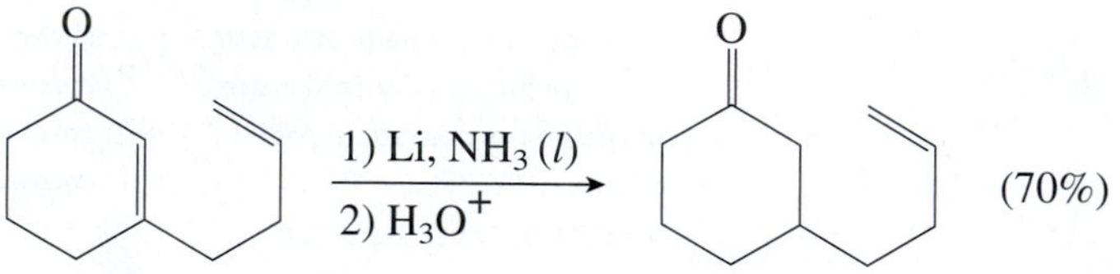

Because an enolate anion is an intermediate in these reductions, it can be used in synthesis in much the same manner as enolate anions that are generated by deprotonation (see Section 20.3). Thus, the addition of an alkyl halide to the reaction mixture after the reduction has been completed results in the alkylation of the enolate anion:

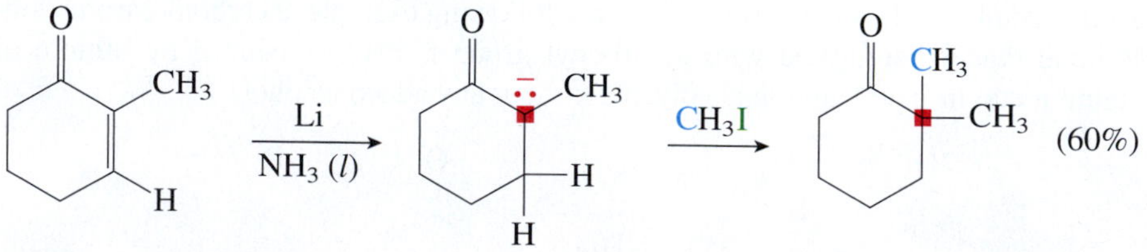

This reaction is especially useful in generating a single enolate anion of an unsymmetrical ketone that produces a mixture of two enolate ions when treated with base. This allows the position of alkylation of unsymmetrical ketones to be controlled. As an example, suppose we want to alkylate 3-methylcyclohexanone at the 2 position. Direct alkylation is synthetically unattractive because treatment of the ketone with base results in the formation of two enolate ions, and two alkylation products are produced:

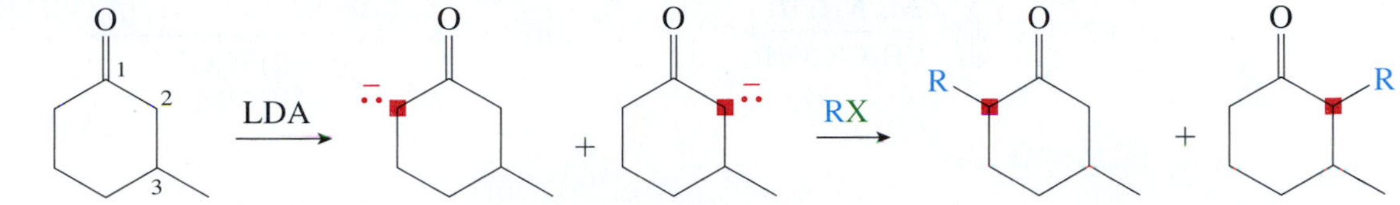

3-Methylcyclohexanone

However, the enolate anion can be generated specifically at the 2 position by reduction of 3-methyl-2-cyclohexenone. Addition of the alkyl halide results in the formation of a single alkylation product as illustrated in the following equation:

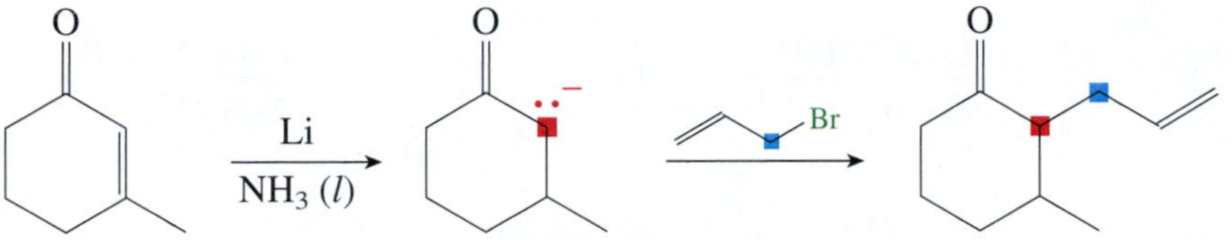

Finally, the reduction of the carbon–carbon triple bond of an alkyne can also be accomplished by using sodium or lithium in liquid ammonia. This reaction is especially useful because it produces the (E)-isomer of the alkene product. (Recall that the (Z)-isomer can be prepared by catalytic hydrogenation of the alkyne; see Section 11.12.)

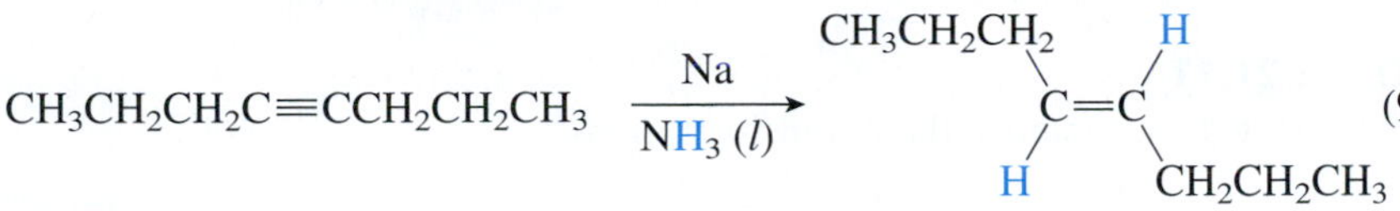

$$CH_3CH_2CH_2C{\equiv}CCH_2CH_2CH_3 \xrightarrow[\text{NH}_3\,(l)]{\text{Na}} \quad \begin{matrix} CH_3CH_2CH_2 \\ \diagdown \\ C{=}C \\ \diagup \quad \diagdown \\ H \quad \quad CH_2CH_2CH_3 \end{matrix} \quad (90\%)$$

Again, the mechanism proceeds by initial formation of a radical anion, which is then protonated by the ammonia solvent. The resulting radical accepts another electron from the sodium to produce an anion that has the bulky alkyl groups in a trans orientation. This anion removes a proton from another ammonia molecule to give the final product.

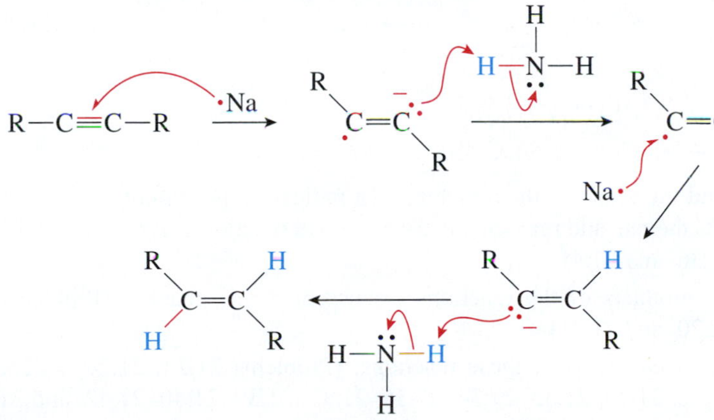

PROBLEM 21.16

Show the products of these reactions:

a) 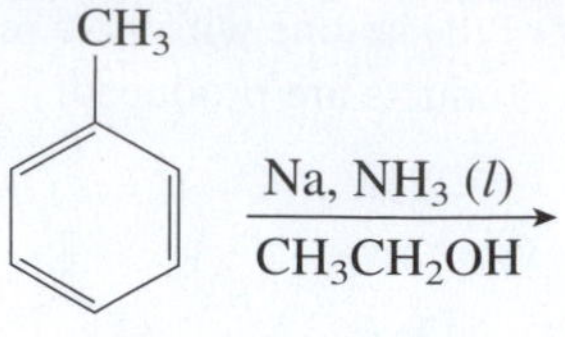
CH3
Na, NH3 (l)
CH3CH2OH

b) 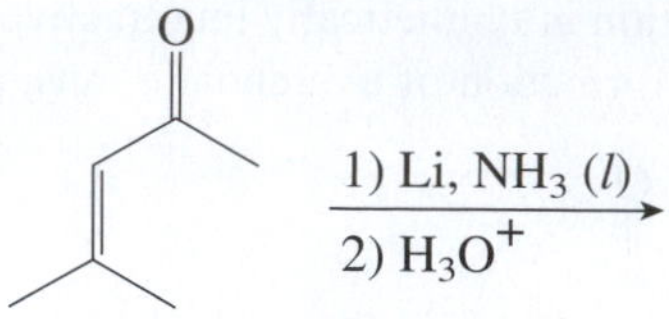
O
1) Li, NH3 (l)
2) H3O+

c) 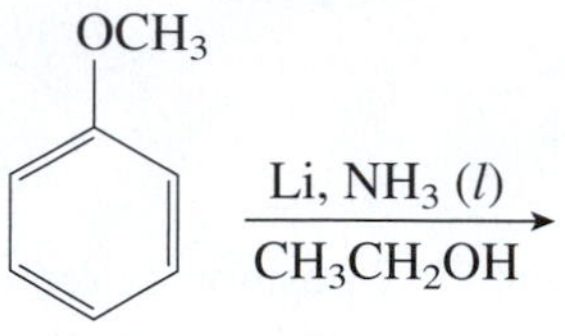
OCH3
Li, NH3 (l)
CH3CH2OH

d) 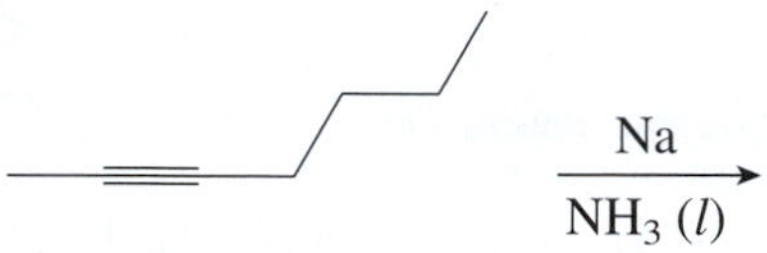
Na
NH3 (l)

e) 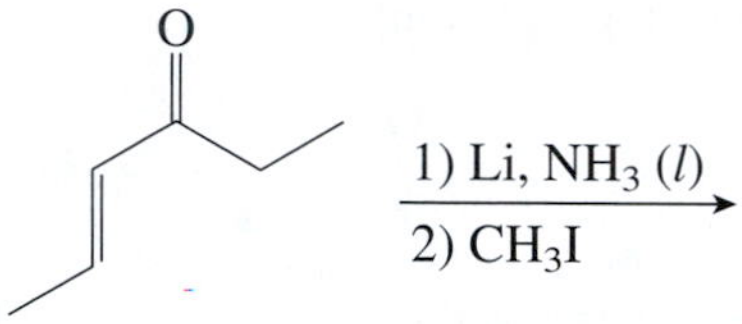
O
1) Li, NH3 (l)
2) CH3I

f) 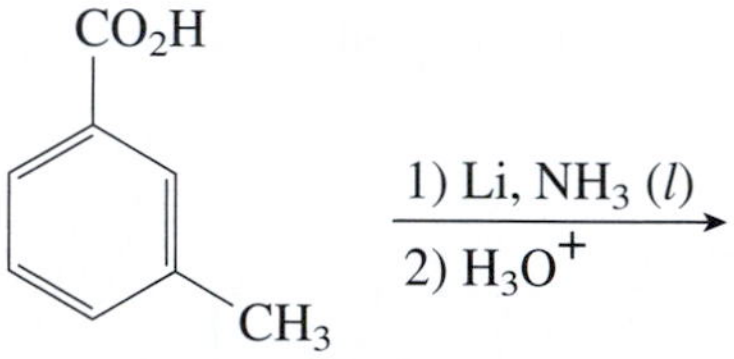
CO2H
CH3
1) Li, NH3 (l)
2) H3O+

g) 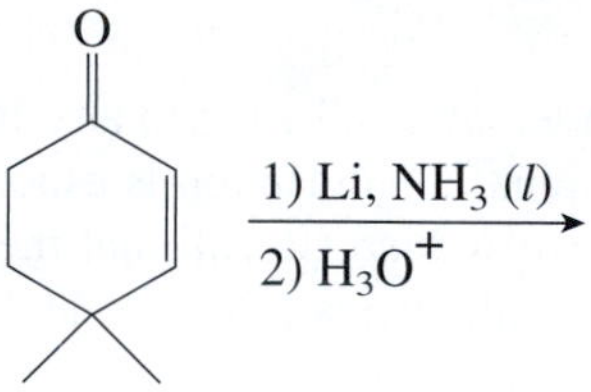
O
1) Li, NH3 (l)
2) H3O+

h) 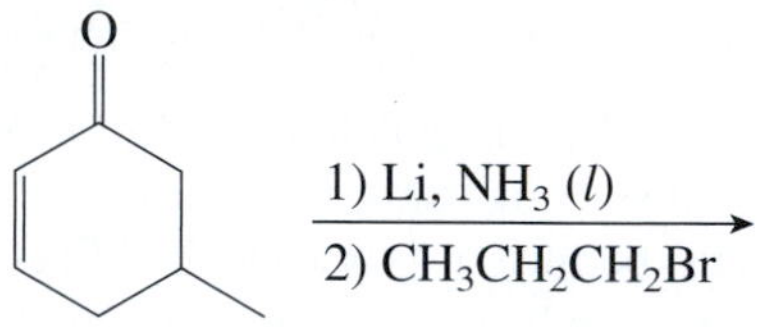
O
1) Li, NH3 (l)
2) CH3CH2CH2Br

PROBLEM 21.17

Suggest reagents to accomplish these transformations:

a) 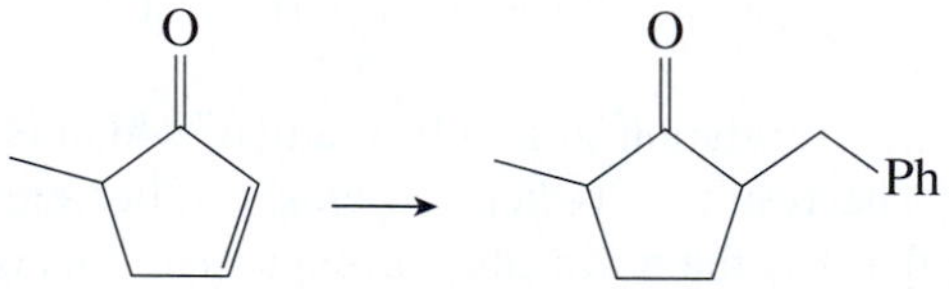
O
O
Ph

b) 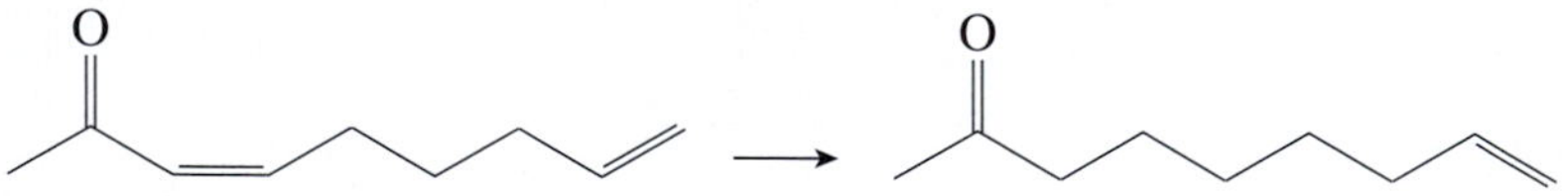
O
O

ORGANIC **Chemistry Now**™ Click *Mastery Goal Quiz* to test how well you have met these goals.

Review of Mastery Goals

After completing this chapter, you should be able to:

- Understand the effect of the structure of a radical on its stability and explain how this affects the rate and regiochemistry of radical reactions. (Problems 21.30, 21.31, 21.33, 21.36, and 21.41)
- Show the products of the reactions discussed in this chapter. (Problems 21.18, 21.19, 21.20, and 21.21)
- Show the mechanisms of these reactions. (Problems 21.23, 21.24, 21.25, 21.26, 21.27, 21.28, 21.29, 21.32, 21.34, 21.35, 21.37, 21.39, 21.40, 21.42, and 21.43)
- Use these reactions in synthesis. (Problem 21.22)

Visual Summary of Key Reactions

The reactions presented in this chapter involve radicals as intermediates. They are summarized in Table 21.2. The use of light or an initiator, such as a peroxide or an azo compound (AIBN), is a key to recognizing radical reactions. The effect of structure on the stability of radical intermediates parallels the effect on carbocation stability.

Table 21.2   Summary of Radical Reactions

Reaction	Comments
$R{-}H \ + \ Cl_2 \xrightarrow{h\nu} R{-}Cl$	**Section 21.6** Chain mechanism, not very selective, multiple chlorinations occur
$R{-}H \ + \ Br_2 \xrightarrow{h\nu} R{-}Br$	**Section 21.6** Chain mechanism, more selective, can use NBS + initiator
$R{-}X \ + \ Bu_3Sn{-}H \xrightarrow{[AIBN]} R{-}H$	**Section 21.7** Chain mechanism
$R{-}H \ + \ O_2 \xrightarrow{h\nu} R{-}OOH$	**Section 21.8** Autoxidation, chain mechanism, not very selective
$R{-}CH{=}CH_2 \ + \ HBr \xrightarrow{[ROOR]} R{-}\overset{\text{H}}{\underset{}{C}}H{-}\overset{\text{Br}}{\underset{}{C}}H_2$	**Section 21.9** Anti-Markovnikov addition, chain mechanism, HCl and HI do not add
$R{-}CH{=}CH_2 \ + \ R'SH \xrightarrow{h\nu} R{-}\overset{\text{H}}{\underset{}{C}}H{-}\overset{\text{SR'}}{\underset{}{C}}H_2$	**Section 21.9** Anti-Markovnikov addition, chain mechanism
$R{-}CH{=}CH_2 \ + \ CX_4 \xrightarrow{h\nu} R{-}\overset{\text{X}}{\underset{}{C}}H{-}\overset{\text{CX}_3}{\underset{}{C}}H_2$	**Section 21.9** Chain mechanism
(benzene ring with R) $\xrightarrow[\text{ROH}]{\text{Na, NH}_3\,(l)}$ (cyclohexadiene with R)	**Section 21.10** Birch reduction, Li can also be used, radical anion mechanism
enone $\xrightarrow[\text{2) H}_3\text{O}^+]{\text{1) Li, NH}_3\,(l)}$ ketone	**Section 21.10** Radical anion mechanism
enone $\xrightarrow[\text{2) R'X}]{\text{1) Li, NH}_3\,(l)}$ alkylated ketone	**Section 21.10** Alkylation of enolate anion
$R{-}C{\equiv}C{-}R \xrightarrow[\text{NH}_3\,(l)]{\text{Na}}$ trans-alkene	**Section 21.10** *Trans*-alkene is formed

Many of the reactions presented in this chapter follow radical chain mechanisms. Such mechanisms consist of three types of steps: initiation, propagation, and termination. One initiation step results in a very large number of propagation steps, followed by a single termination step. The presence of a small amount of a chain inhibitor, such as an antioxidant, can prevent a chain reaction from occurring.

Additional Problems

21.18 Show the products of these reactions.

a) 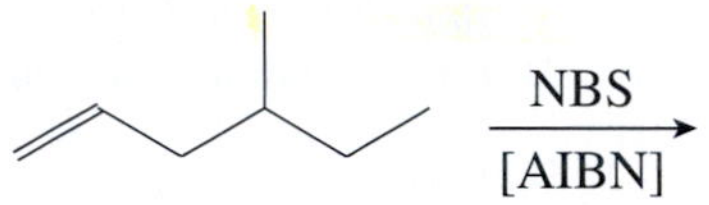 $\xrightarrow[\text{[AIBN]}]{\text{NBS}}$

b) 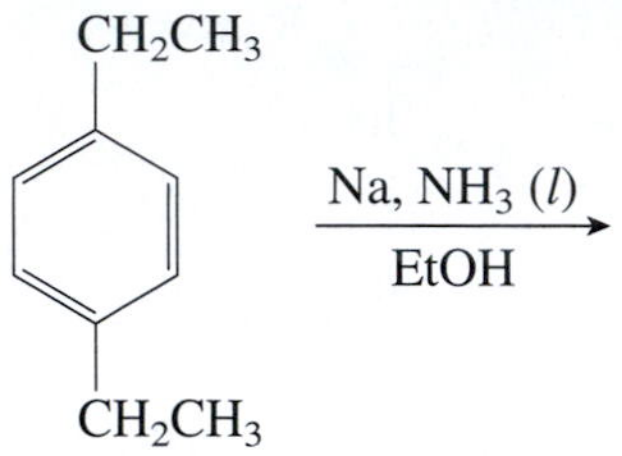 $\xrightarrow[\text{EtOH}]{\text{Na, NH}_3\ (l)}$

c) 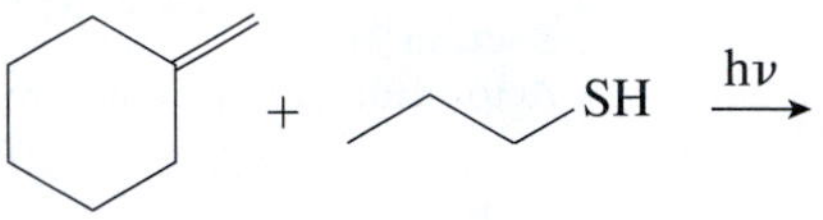 $+$ SH $\xrightarrow{h\nu}$

d) 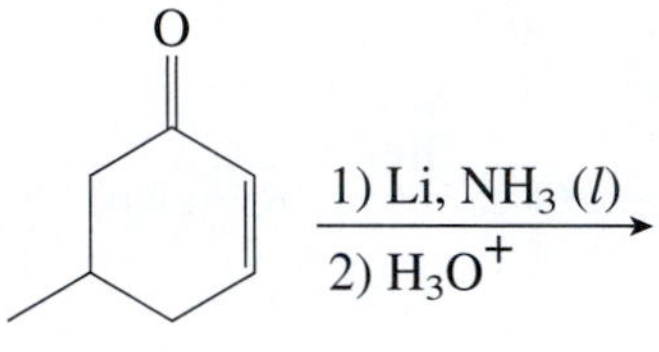 $\xrightarrow[\text{2) H}_3\text{O}^+]{\text{1) Li, NH}_3\ (l)}$

e) 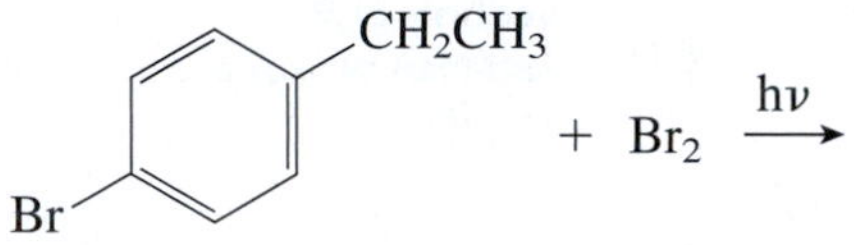 $+\ \text{Br}_2 \xrightarrow{h\nu}$

f) $\text{CH}_3\text{C}\equiv\text{CCH}_2\text{CH}_2\text{CH}_3 \xrightarrow[\text{NH}_3\ (l)]{\text{Na}}$

g) 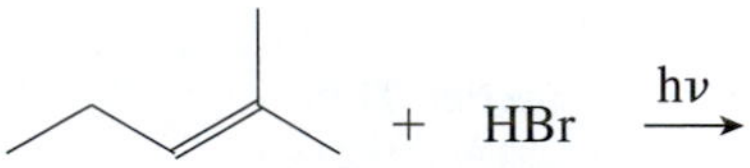 $+\ \text{HBr} \xrightarrow{h\nu}$

h) $\xrightarrow[\text{2) H}_3\text{O}^+]{\text{1) Li, NH}_3\ (l)}$

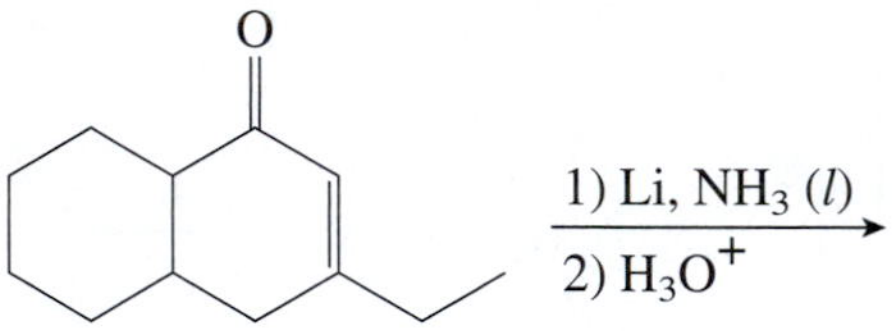

i) 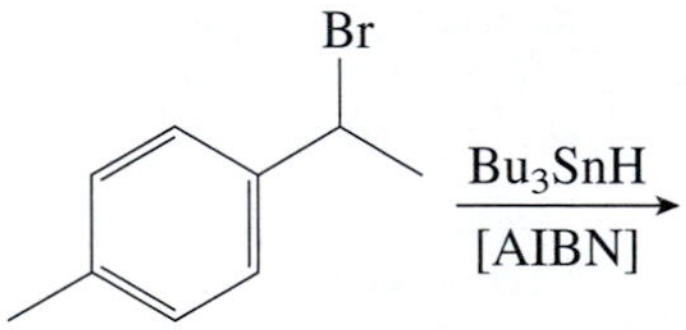 $\xrightarrow[\text{[AIBN]}]{\text{Bu}_3\text{SnH}}$

j) 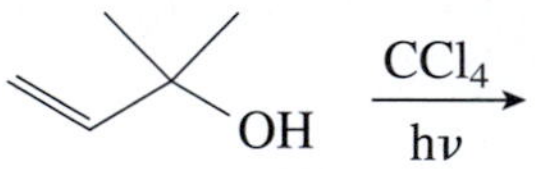 $\xrightarrow[h\nu]{\text{CCl}_4}$

k) 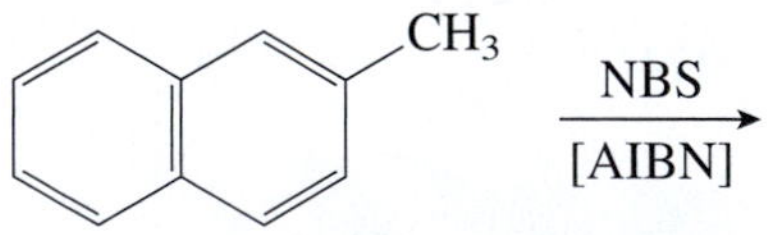 $\xrightarrow[\text{[AIBN]}]{\text{NBS}}$

l) $\xrightarrow[h\nu]{\text{CCl}_3\text{Br}}$

21.19 Show the products of the reactions of 1-hexene with each of these reagents:
    a) NBS, [AIBN]
    b) HBr, [t-BuOOt-Bu]
    c) CH₃CH₂SH, hν
    d) CCl₄, hν
    e) CBrCl₃, hν

21.20 Show the products of these reactions:

a)
+ NBS
[(PhCO2)2]

b)
O
O
+ Br2
hν

c)
+ HBr
[(PhCO2)2]

d)
CH3
Br
CH3
+ Bu3SnH
[AIBN]

e)
+
SH
hν

f)
+ CCl4
hν

g)
O
OH
1) Li, NH3 (l)
2) H3O+

h)
CH3
CH3
Na, NH3 (l)
EtOH

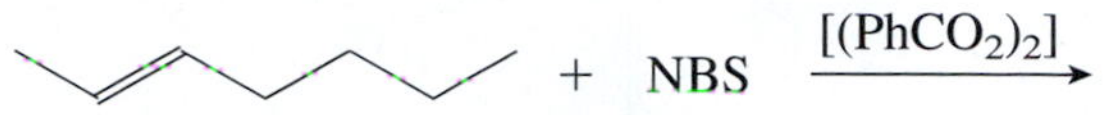

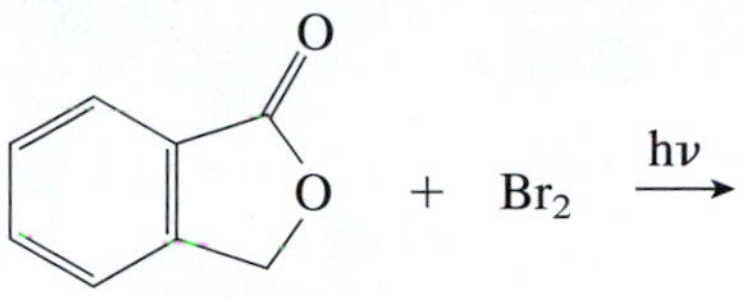

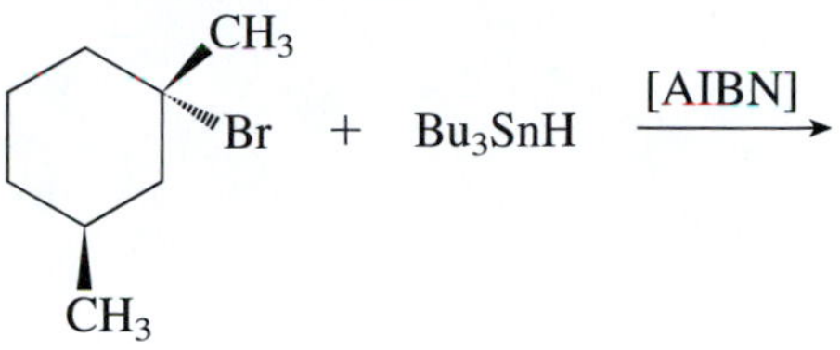

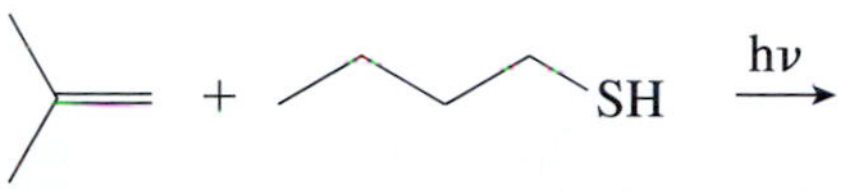

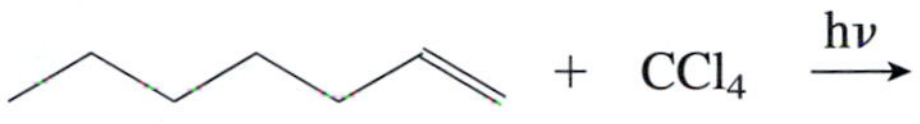

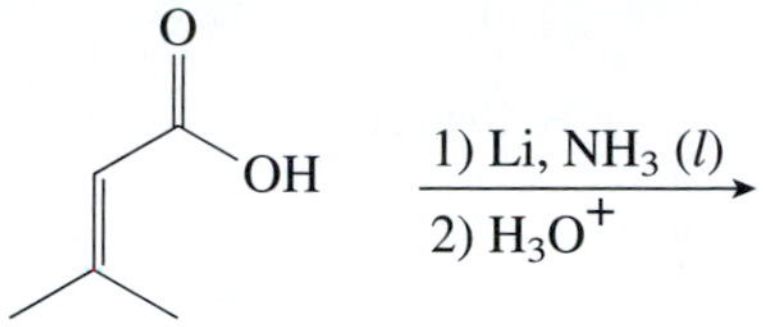

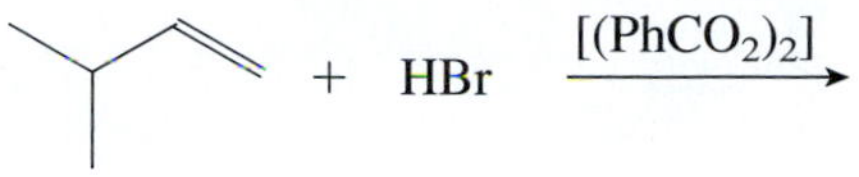

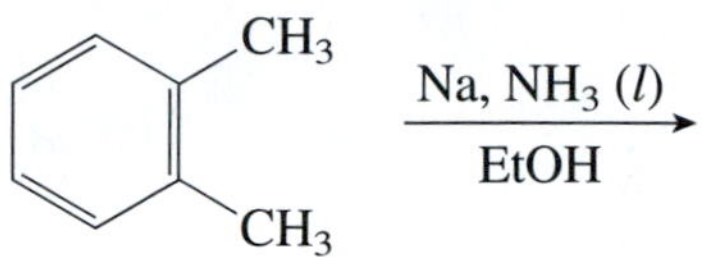

21.21 Show the products of these reactions:

a)
CH3
Cl
+ Cl2
hν

b)
CH2CH3
OCH2CH3
+ NBS
[AIBN]

c)
+
O
CH3CSH
hν

d)
+ CBrCl3
hν

e)
Li, NH3 (l)
EtOH

f)
+ HBr
[AIBN]

g)
Na
NH3 (l)

h)
O
1) Li, NH3 (l)
2)
Br

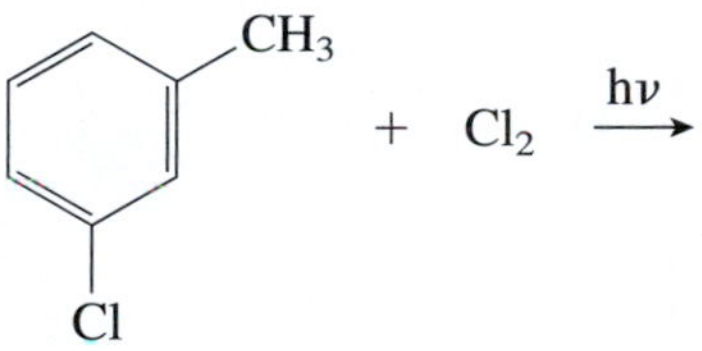

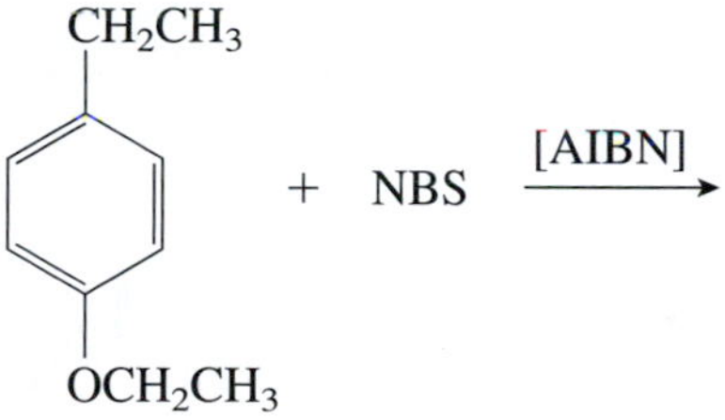

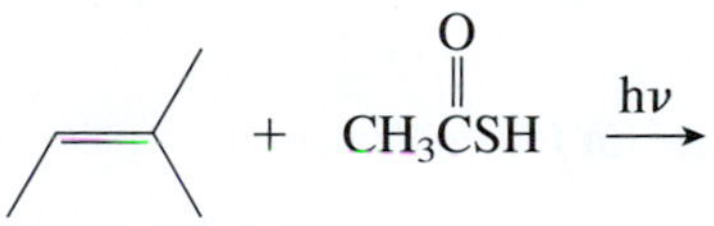

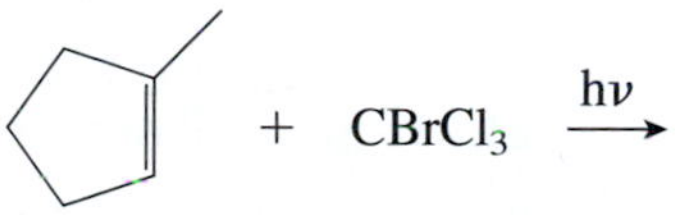

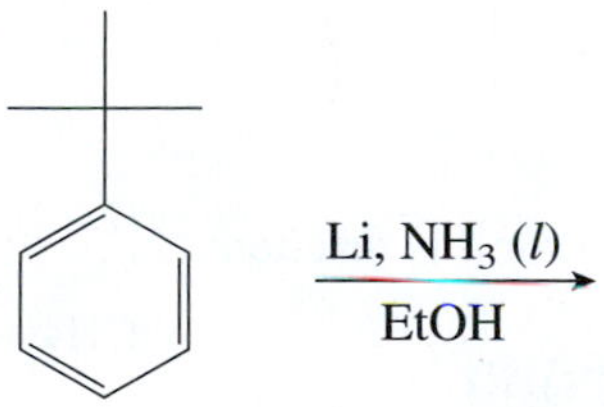

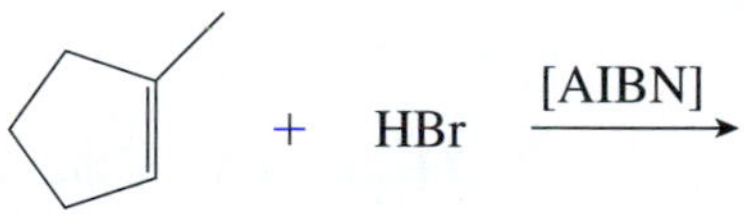

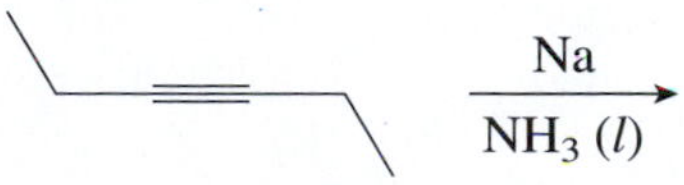

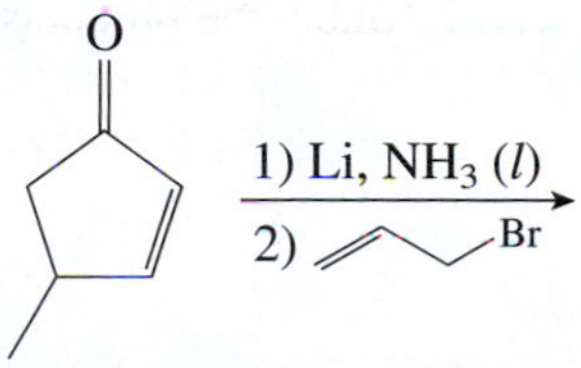

21.22 Show syntheses of these compounds from the indicated starting materials. Reactions from previous chapters may be needed in some cases.

a)

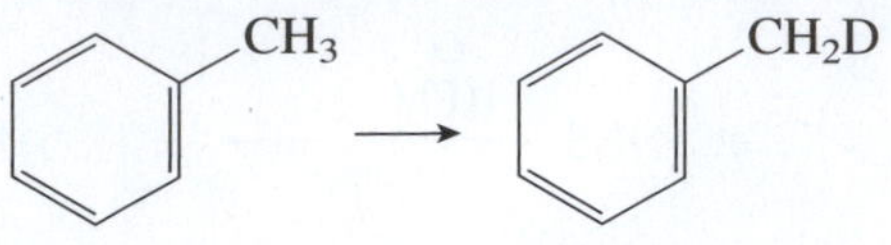

b)

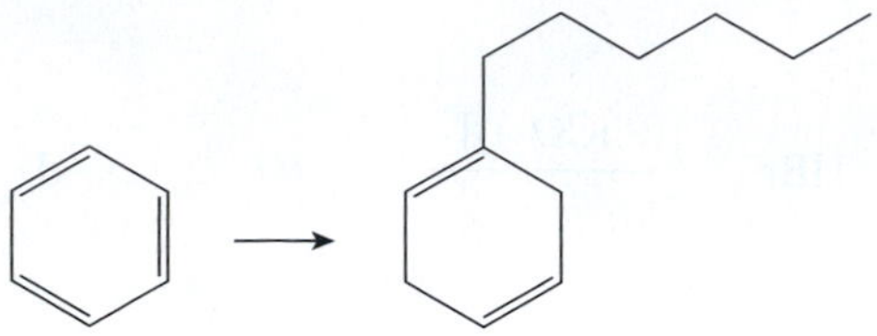

c)

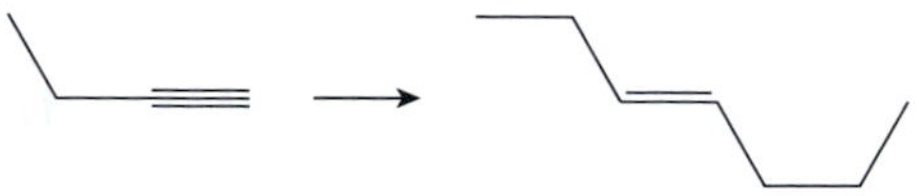

d)

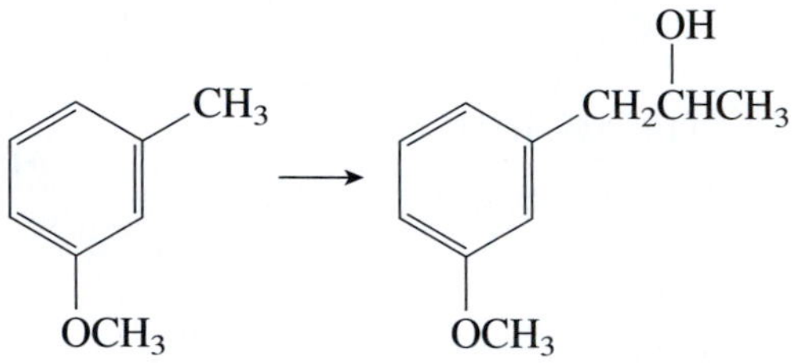

e)

f)

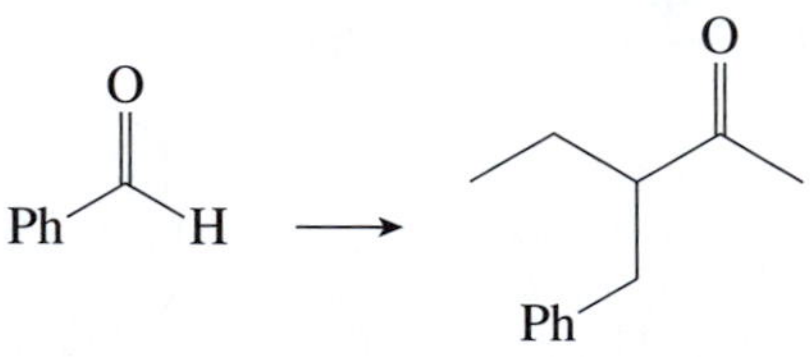

21.23 Show all of the steps in the mechanism for this reaction:

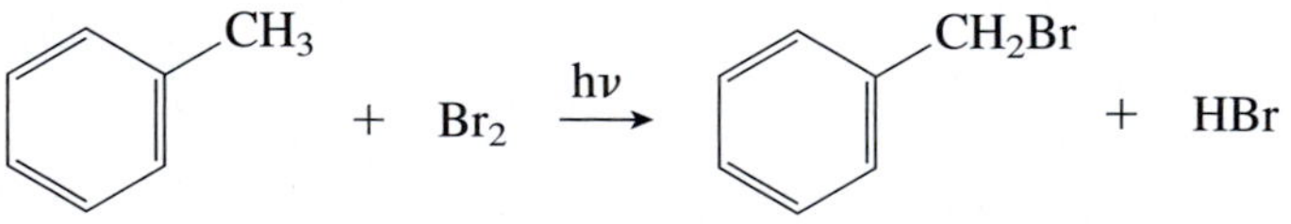

21.24 Show all of the steps in the mechanism for this reaction:

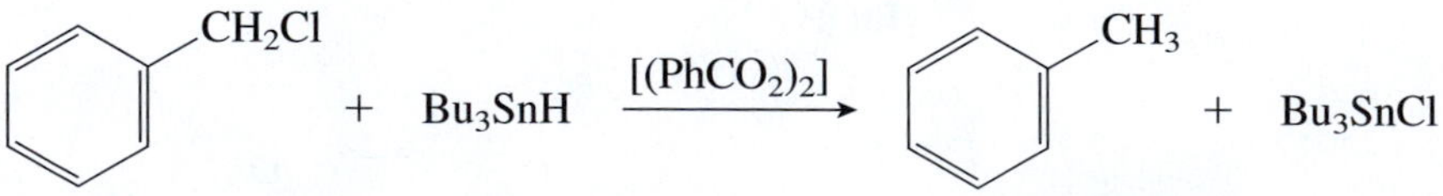

21.25 Show all of the steps in the mechanism for this reaction:

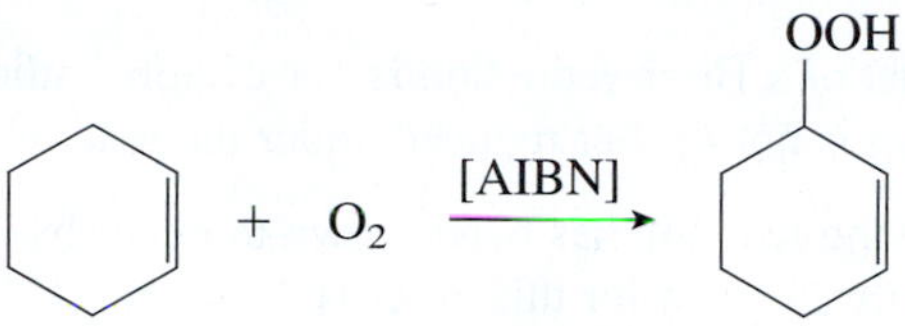

21.26 Show all of the steps in the mechanism for this reaction:

$$CH_3\overset{\underset{|}{CH_3}}{C}=CH_2 \ + \ CH_3CH_2SH \ \xrightarrow{h\nu} \ CH_3\overset{\underset{|}{CH_3}}{CH}CH_2SCH_2CH_3$$

21.27 Show all of the steps in the mechanism for this reaction:

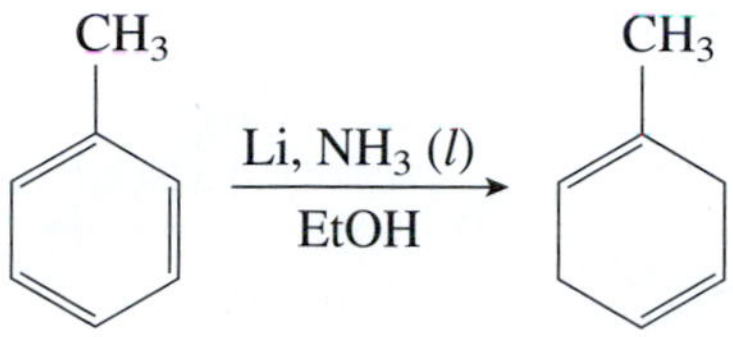

21.28 Show all of the steps in the mechanism for this reaction:

$$H_2C=CH\overset{\overset{O}{\|}}{C}CH_3 \ \xrightarrow[\text{2) } H_3O^+]{\text{1) Li, NH}_3 \ (l)} \ CH_3CH_2\overset{\overset{O}{\|}}{C}CH_3$$

21.29 The following reaction gives three monobromo products. Show the structures of these products and explain their formation.

21.30 Radicals do not usually rearrange because the increase in stability is usually too small to provide sufficient driving force. Use the data in Table 21.1 to estimate the amount of heat that would be given off if this rearrangement were to occur:

$$CH_3\overset{\underset{|}{CH_3}}{CH}\underset{\cdot}{C}HCH_3 \ \longrightarrow \ CH_3\overset{\underset{|}{CH_3}}{\underset{\cdot}{C}}CH_2CH_3$$

21.31 Explain why azobenzene is not very useful as an initiator for radical reactions.

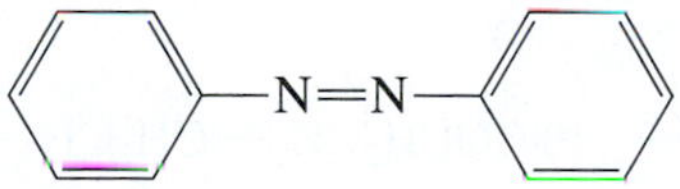

Azobenzene

21.32 What reaction do you suppose occurs when O_2 is bubbled through a solution of the triphenylmethyl radical?

21.33 The product of a Birch reduction is 1,4-cyclohexadiene. Offer a reason why this product is not further reduced under the reaction conditions.

21.34 The following reaction has been shown to occur by a radical chain process. Suggest a mechanism for this process.

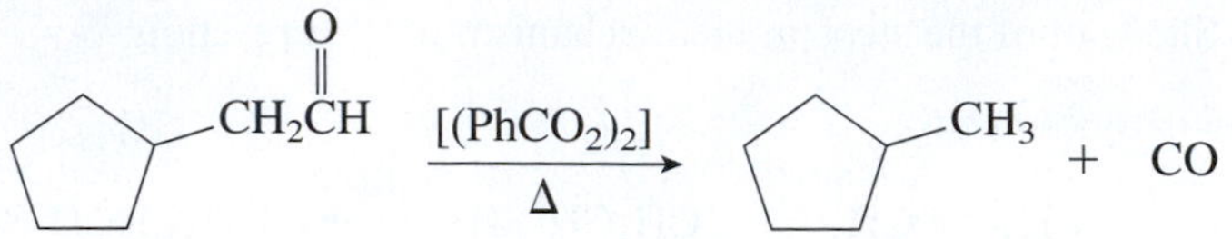

21.35 Suggest a mechanism that explains the formation of both of the products of this reaction:

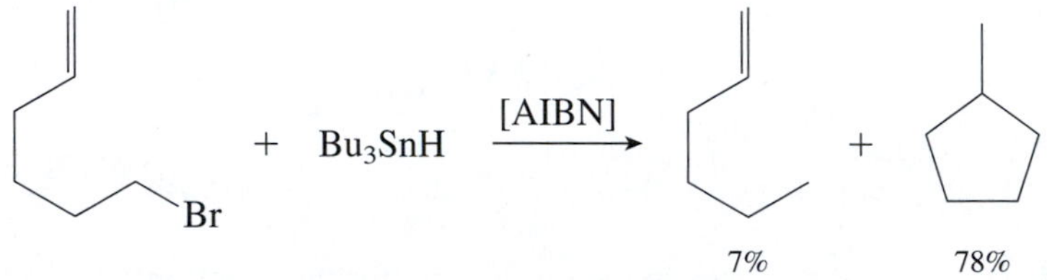

21.36 Explain why the substituent on 1-naphthol causes the reduction to occur in the unsubstituted ring, whereas the substituent on 1-ethoxynaphthalene causes the reduction to occur in the substituted ring.

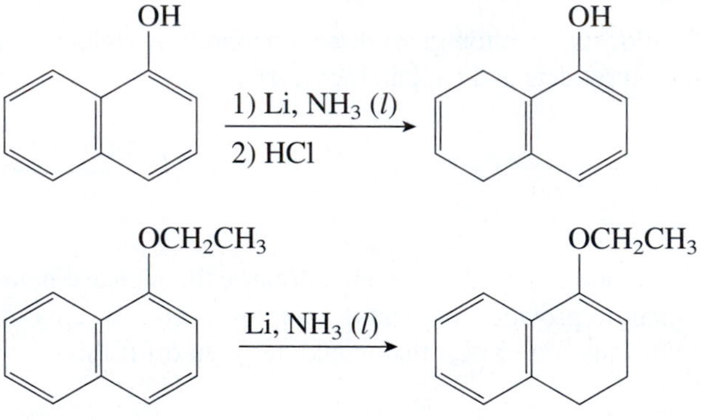

21.37 The reaction of alkenes with a small amount of a radical initiator is a common method used to prepare polymers (see Chapter 24). An intermediate stage in the formation of polyethylene is shown in this equation. Show a mechanism for the formation of this intermediate.

$$CH_2{=}CH_2 \xrightarrow{[(PhCO_2)_2]} Ph\overset{O}{\overset{\|}{C}}OCH_2CH_2{-}CH_2CH_2{-}CH_2\overset{\bullet}{C}H_2 \longrightarrow \text{Polyethylene}$$

21.38 The Birch reduction of anisole, followed by reaction with acid, provides a useful method for the preparation of cyclohexenones. Show a mechanism for the hydrolysis step of this process.

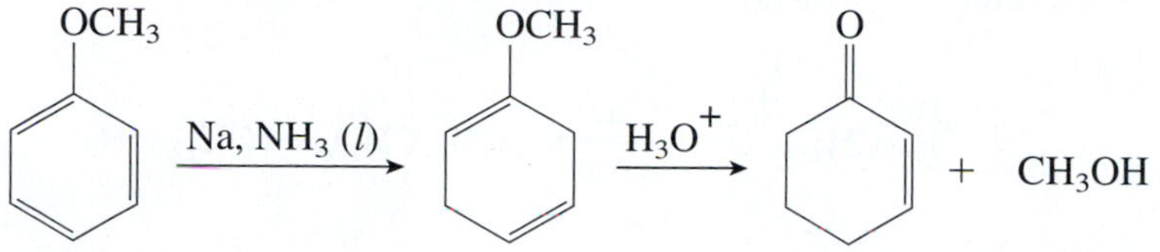

OCH₃ — Na, NH₃ (*l*) → OCH₃ — H₃O⁺ → O + CH₃OH

21.39 The following reaction, known as the Hunsdiecker reaction, proceeds by a radical chain mechanism. Show the steps in this mechanism. (*Hint:* The weakest bond in the compound, the oxygen–bromine bond, is broken in the initiation step.)

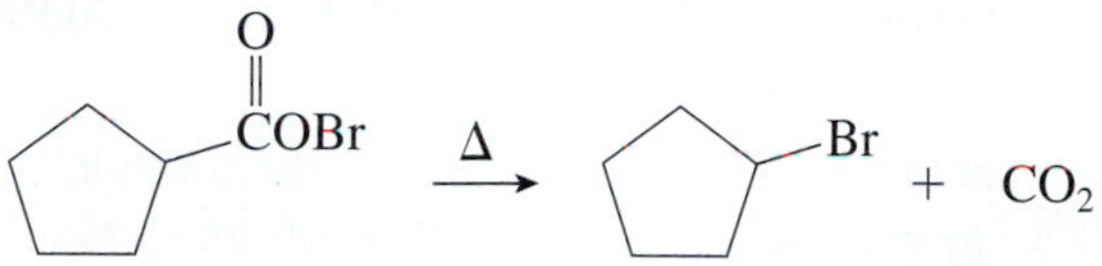

$$\text{cyclopentyl—COBr} \xrightarrow{\Delta} \text{cyclopentyl—Br} + CO_2$$

21.40 Suggest a mechanism for this photochemical reaction:

$$PhCH_2CCH_2Ph \xrightarrow{h\nu} PhCH_2CH_2Ph + CO$$

21.41 Esters of the fatty acid oleic acid are components of membranes that are subject to autoxidation. Explain which hydrogens of an oleate ester you expect to be abstracted most readily by a radical. Show the structures of the major autoxidation products that would be formed from an oleate ester.

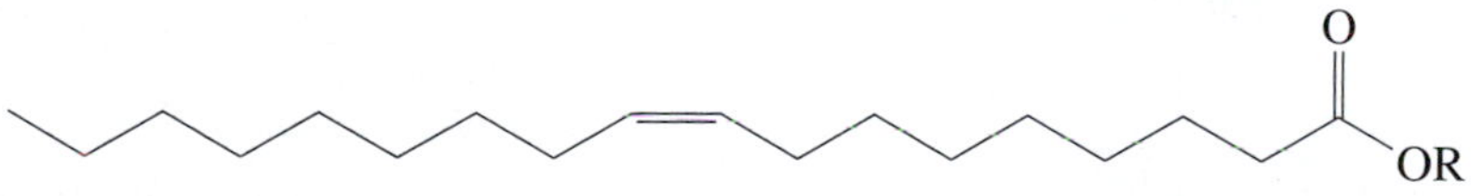

An oleate ester

21.42 Ozone, produced in the stratosphere by the action of ultraviolet light on O_2, helps shield the surface of the earth from harmful ultraviolet radiation. Ozone is slowly decomposed by reaction with oxygen atoms according to the following equation:

$$O_3 + O \xrightarrow{\text{slow}} 2\,O_2$$

Chlorofluorocarbons, such as CF_3Cl, catalyze this reaction and are responsible for the formation of the "ozone hole." The decomposition is a chain reaction involving chlorine atoms as the chain-carrying species. Suggest a mechanism for this reaction.

21.43 Carboxylate anions are oxidized at the anode of an electrochemical cell to produce radicals and ultimately hydrocarbons in a reaction known as the Kolbe electrolysis. Suggest a mechanism for the Kolbe electrolysis shown in the following equation:

$$2\ CH_3(CH_2)_6\overset{O}{\underset{}{\overset{\|}{C}}}O^- \xrightarrow[\text{anode}]{-2e^-} CH_3(CH_2)_6-(CH_2)_6CH_3\ +\ 2\ CO_2$$

Problems Involving Spectroscopy

21.44 A hydrocarbon with the formula C_5H_{12} gives a single monochloro substitution product when reacted with Cl_2 in the presence of light. This product shows two singlets in a 9:2 ratio in its ^{1}H-NMR spectrum. Suggest structures for the hydrocarbon and its chlorination product.

21.45 The product of the following reaction has the formula $C_7H_{12}O_2$. Its IR and ^{1}H-NMR spectra follow. Its ^{13}C-NMR spectrum shows six absorptions. Show the structure of the product and suggest a mechanism for its formation.

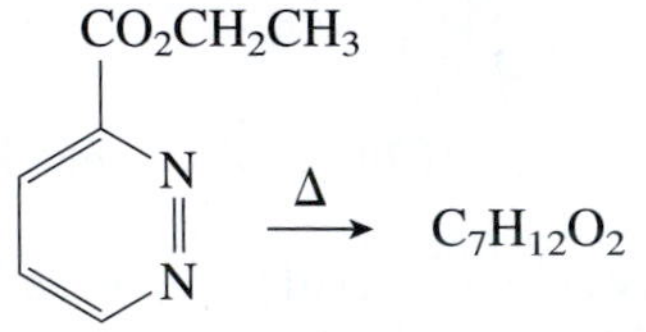

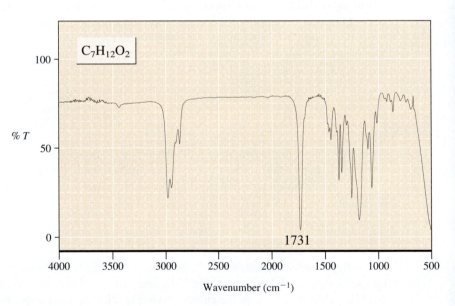

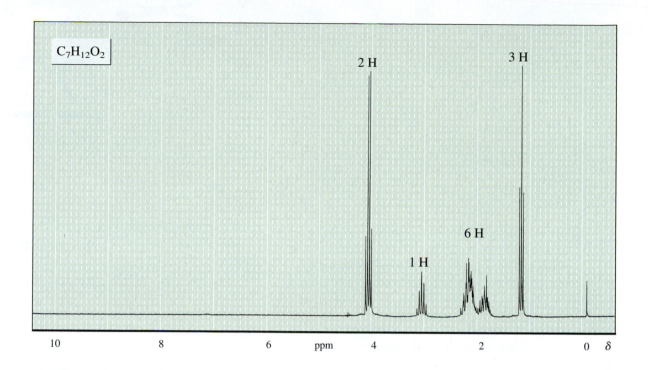

Problems Using Online Three-Dimensional Molecular Models

21.46 Radical A, shown here, is related to the triphenylmethyl radical by having oxygen atoms bridging the ortho positions of the phenyl groups. This radical dimerizes much faster than the triphenylmethyl radical. Explain this observation.

21.47 Galvinoxyl is a relatively unreactive radical. It is stable to oxygen, stable for extended storage as a solid, and decomposes only slowly in solution. Explain which atoms have radical electron density in galvinoxyl and explain why it is so unreactive.

21.48 Free radical chlorination (Cl_2 and light) of (R)-1-chloro-2-methylbutane gives the two products shown, in addition to other products. These two products are always formed in exactly equal amounts. Explain this observation.

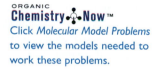

ORGANIC
Chemistry-Now™
Click *Molecular Model Problems* to view the models needed to work these problems.

Do you need a live tutor for homework problems? Access vMentor at Organic ChemistryNow at **http://now.brookscole.com/hornback2** for one-on-one tutoring from a chemistry expert.

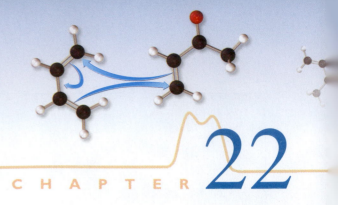

Pericyclic Reactions

Most of the reactions presented in previous chapters involved nucleophiles and electrophiles and occurred in several steps involving cationic, anionic, or, in the last chapter, radical intermediates. In this chapter a group of concerted (one-step) reactions, called pericyclic reactions, that involve none of these intermediates is discussed. The mechanisms of these reactions are exceedingly simple because they consist of a single step. Yet, as we shall see, pericyclic reactions are amazingly selective, both in terms of when they occur and also in their stereochemical requirements.

First, pericyclic reactions are defined, and an example of their unusual stereochemical selectivity is presented. A theoretical treatment of pericyclic reactions requires examination of the MOs for the conjugated molecules that participate in these reactions, so MO theory for these compounds is developed next. Then a theoretical explanation for the selectivity and stereochemistry observed in each of the three classes of pericyclic reactions is presented, along with a number of common examples of reactions of each kind.

MASTERING ORGANIC CHEMISTRY

▶ Understanding the Energies and Nodal Properties of the Pi MOs of a Conjugated System

▶ Recognizing Electrocyclic Reactions, Cycloaddition Reactions, and Sigmatropic Rearrangements

▶ Using Pi MOs to Determine Whether These Reactions Are Allowed or Forbidden

▶ Predicting the Products, Including Stereochemistry, of These Reactions

▶ Understanding the Mechanisms of the Pinacol, Beckmann, Hofmann, and Baeyer-Villiger Rearrangements

▶ Understanding the Mechanism, Selectivity, and Stereochemistry of Pericyclic Reactions

▶ Using These Reactions in Synthesis

22.1 PERICYCLIC REACTIONS

A **pericyclic reaction** is concerted and proceeds through a cyclic transition state in which two or more bonds are made and/or broken. As a concerted reaction, it does not involve any detectable ionic or radical intermediates. The effect of changing the polarity of the solvent is usually small, indicating the lack of significant charge buildup in the transition state. These reactions are highly stereospecific, and the stereochemistry often changes when the energy for the reaction is supplied by heat (**thermolysis** or **pyrolysis**) as compared to when it is supplied by light (**photolysis**).

An example of a reaction, first presented in Section 20.4, that falls under the pericyclic classification is the decarboxylation of β-ketoacids produced in the malonic and acetoacetic ester syntheses:

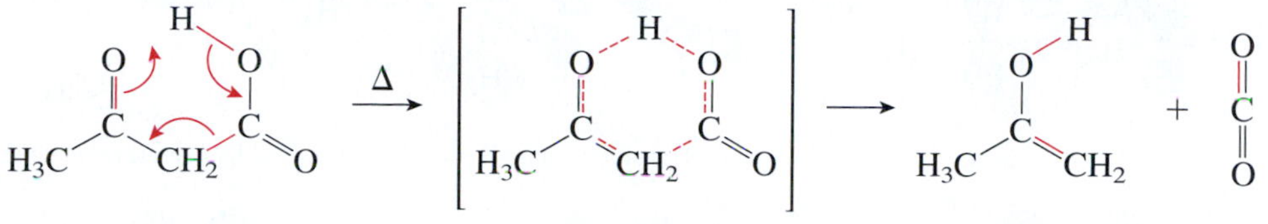

Transition state

This cyclic movement of electrons and a transition state that involves a cycle of breaking and forming bonds are characteristics of a pericyclic reaction.

An example of a pericyclic reaction that illustrates the rather amazing stereoselectivity that these reactions often exhibit is provided by the thermal and photochemical interconversions of dienes and cyclobutenes. When $(2E,4E)$-hexadiene is heated, it cyclizes to form *trans*-3,4-dimethylcyclobutene. None of the *cis*-isomer is produced. In the reverse reaction the cyclobutene opens to produce only the (E,E)-isomer of the hexadiene. The reaction is completely stereospecific in both directions:

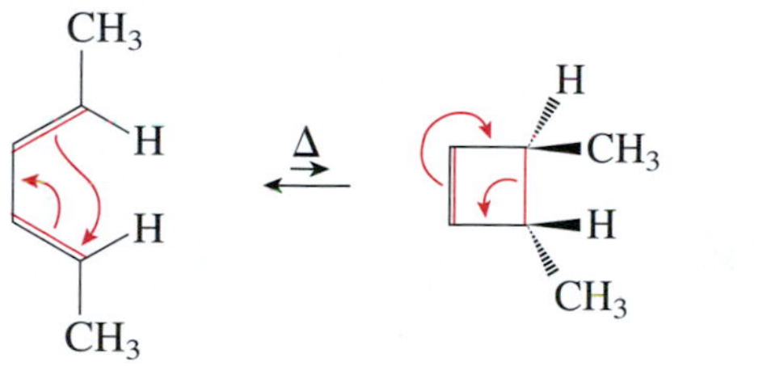

$(2E,4E)$-Hexadiene   *trans*-3,4-Dimethylcyclobutene

Making the stereochemical results even more remarkable, if the same stereoisomer of the hexadiene is photolyzed, rather than heated, *cis*-3,4-dimethylcyclobutene is the only product formed:

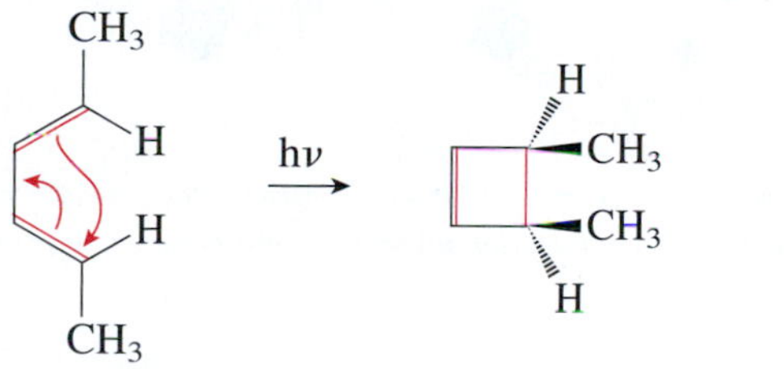

$(2E,4E)$-Hexadiene   *cis*-3,4-Dimethylcyclobutene

The stereochemistry of the photochemical reaction is the opposite of that of the thermal reaction!

Let's carefully compare the movements of the atoms that must occur in these reactions to cause such different stereochemical results. As shown in Figure 22.1, the process is quite simple. The two end carbons of the conjugated pi system need only rotate so that their *p* orbitals begin to overlap to form the new sigma bond of the cyclobutene. In the case of the thermal reaction, both carbons rotate in the same direction, a process that is

Conrotation

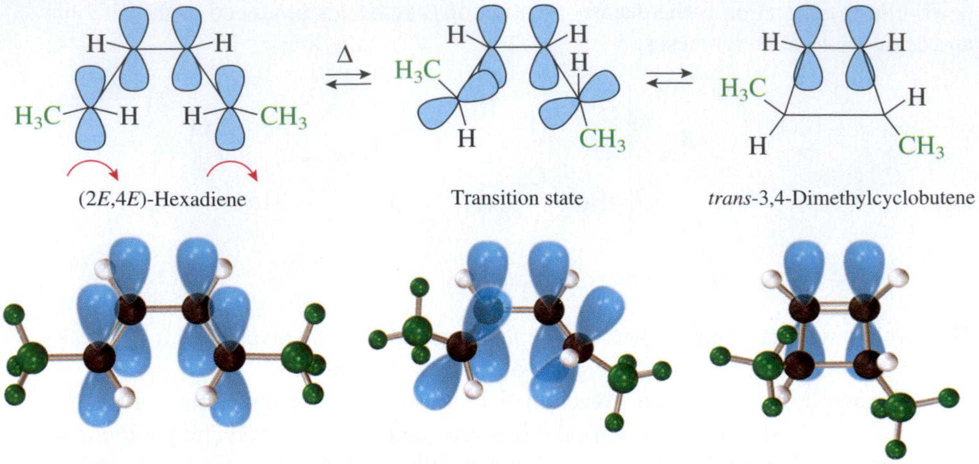

(2E,4E)-Hexadiene   Transition state   *trans*-3,4-Dimethylcyclobutene

When (2E,4E)-hexadiene is heated, only *trans*-3,4-dimethylcyclobutene is produced. The end carbons of the conjugated diene system rotate in the same direction (both clockwise or both counterclockwise). This is called **conrotation.**

Disrotation

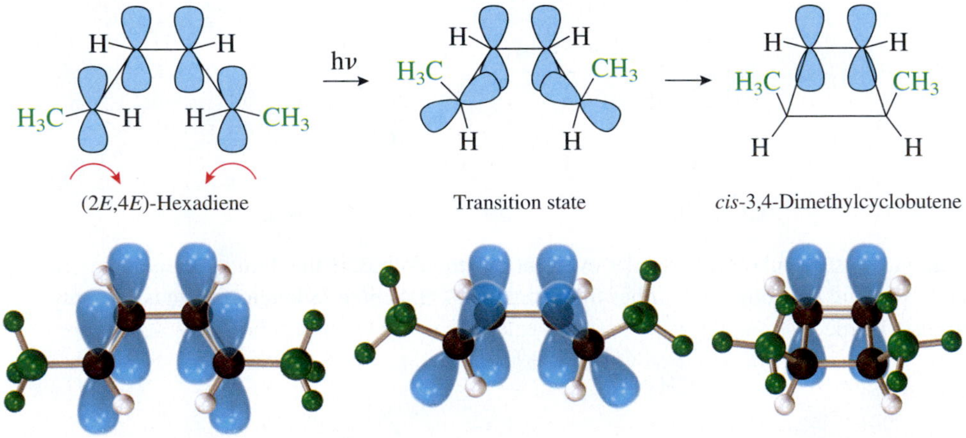

(2E,4E)-Hexadiene   Transition state   *cis*-3,4-Dimethylcyclobutene

In the photochemical reaction, the end carbons of the conjugated diene system rotate in opposite directions (one clockwise and one counterclockwise) so that cis-3,4-dimethylcyclobutene is produced. This is called **disrotation.**

Active Figure 22.1

ORGANIC
Chemistry Now™

BOND ROTATIONS IN THE REACTIONS OF 2,4-HEXADIENE TO PRODUCE 3,4-DIMETHYLCYCLOBUTENE.
Test yourself on the concepts in this figure at **OrganicChemistryNow.**

termed **conrotation,** so the methyl groups have a trans orientation in the cyclobutene product. The hybridization of these carbons changes from sp^2 to sp^3 as the rotation occurs. The two *p* orbitals on the central carbons of the diene system form the new pi bond. The process is very similar for the photochemical reaction, with the exception that the end carbons rotate in opposite directions, a process termed **disrotation,** so that the methyl groups have a cis orientation in the cyclobutene product. There is no obvious reason why conrotation should be preferred over disrotation in the thermal reaction, much less why this preference is reversed in the photochemical reaction.

PROBLEM 22.1

Determine whether these reactions occur by conrotation or disrotation:

a)

b)

MODEL BUILDING PROBLEM 22.1

You will find a set of handheld models invaluable in helping to visualize the stereochemistry of the reactions in this chapter. Build a model of (2E,4E)-hexadiene and examine both conrotation to give *trans*-3,4-dimethylcyclobutene and disrotation to give *cis*-3,4-dimethylcyclobutene.

ORGANIC
Chemistry ⚛ Now™
Click *Molecular Models* to view computer models of the molecules in this chapter.

These apparently simple reactions were poorly understood by organic chemists until 1965, when R. B. Woodward and R. Hoffmann developed a theoretical explanation for the selectivities. A related theory was developed by K. Fukui at about the same time. (Fukui and Hoffmann shared the 1981 Nobel Prize in chemistry for this work. Woodward died in 1979 and so did not share in this award because the Nobel Prize is not given posthumously. However, he had already won a Nobel Prize in 1965 for his work in the area of organic synthesis.)

Both of these theories explain whether the transition state for the reaction under consideration is favorable or not by examining how the reactant MOs are converted to the product MOs. To understand and use these theories, we need to first discuss the MOs for the conjugated molecules that participate in these reactions.

22.2   MO THEORY FOR CONJUGATED MOLECULES

A model for localized MOs—that is, MOs that extend over only two atoms—was developed in Chapter 3. To describe conjugated molecules, MOs that extend over more than two atoms, delocalized MOs, are needed. Some aspects of the theory for delocalized MOs were presented in Chapters 15 and 16; all of the important aspects are reiterated here.

To explain pericyclic reactions, the energies and the location of the nodes of the pi MOs of noncyclic, linear, conjugated molecules are needed. These can be obtained by the application of the following rules:

1. The number of MOs equals the number of AOs combining to form them.

2. The energies of the MOs are symmetrically placed about the energy of an isolated p orbital (arbitrarily taken as zero energy).

3. The energy of an MO increases as the number of nodes increases.

4. Nodes are symmetrically placed in a molecule.

Let's see how to apply these rules to the MOs of conjugated molecules.

First, consider the pi MOs of ethene, shown in Figure 22.2. The two p orbitals combine to form two pi MOs. The energies of these pi MOs are symmetrically arranged about zero energy; one MO is bonding, and one is antibonding. The bonding MO has no node, and the antibonding MO has one node. The bonding MO is occupied by a pair of electrons, while the antibonding MO is empty.

The pi MOs of 1,3-butadiene, also shown in Figure 22.2, are constructed in the same manner. There are two double bonds, so four p orbitals combine to form four pi MOs. The energies of these MOs are symmetrically arranged about zero energy, so

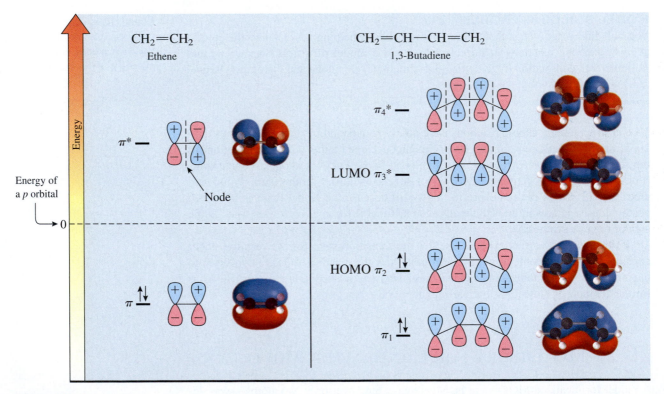

Figure 22.2

ENERGIES AND NODAL PROPERTIES FOR THE PI MOs OF ETHENE AND 1,3-BUTADIENE.

there are two bonding MOs and two antibonding MOs. The lowest-energy MO, π_1, has zero nodes. All of its AOs overlap in a bonding fashion; that is, a plus lobe of one orbital overlaps with the plus lobe of the adjacent orbital (or a minus lobe overlaps with a minus lobe). The next MO, π_2, has one node, placed symmetrically in the molecule, between the two central carbons. In this MO, two of the overlaps of the p AOs are bonding in nature, and one of the overlaps, in the center of the molecule, is antibonding in nature because the plus lobe of one orbital overlaps with the minus lobe of the other. This orbital is overall bonding because it has two favorable (bonding) overlaps and only one unfavorable (antibonding) overlap. The next MO, π_3^*, has two nodes that are again symmetrically placed in the molecule. It is a net antibonding MO because two of the overlaps are antibonding in nature, while only one is bonding. Finally, π_4^* has three nodes. It is the highest-energy MO because the overlaps of all of the p AOs are antibonding in nature. There are four pi electrons, so the two bonding MOs, π_1 and π_2, are both occupied by a pair of electrons. The highest-energy orbital that contains an electron, π_2 in this case, is termed the **highest occupied molecular orbital (HOMO),** and the lowest-energy orbital that contains no electrons, π_3^* in this case, is termed the **lowest unoccupied molecular orbital (LUMO).**

Figure 22.3 shows the pi molecular orbitals for the system of six conjugated p orbitals, 1,3,5-hexatriene. The six p AOs combine to form six pi MOs. These are arranged symmetrically about zero energy, so there are three bonding MOs and three antibonding MOs. The lowest-energy MO has zero nodes, the next has one node, and so on. These pi MOs have six electrons. The lowest-energy arrangement of electrons, called the **ground state,** has these six electrons in the three bonding pi MOs. (Any other arrangement of electrons is higher in energy and is termed an **excited state.)** For the ground state of 1,3,5-hexatriene the HOMO is π_3 and the LUMO is π_4^*.

As described in the discussion of UV-visible spectroscopy in Section 15.1, the absorption of a photon of light by a compound causes an electron to be excited from an occupied MO to an unoccupied MO. The lowest-energy excitation occurs when an electron in the HOMO is excited to the LUMO. In the case of 1,3,5-hexatriene an electron in π_3 is promoted to π_4^* to form the lowest-energy excited state (see Figure 22.3). In the excited state, the HOMO is π_4^* and the LUMO is π_5^*. It is this excited state that reacts in a photochemical reaction.

Conjugated carbocations, carbanions, and radicals that have an odd number of orbitals are also important. The same rules are used to construct the MOs for these odd orbital systems. As a simple example, consider the allyl radical, shown in Figure 22.4. The three p AOs result in the formation of three pi MOs. For these orbitals to be symmetrically placed about zero energy, one must occur at zero energy. Therefore, the allyl radical has a bonding MO, a nonbonding MO at zero energy, and an antibonding MO. The presence of a nonbonding MO is a characteristic of all odd orbital systems.

Another characteristic feature of odd orbital systems is nodes that pass through atoms. For example, π_2^{nb} of the allyl radical has one node. For this node to be symmetrically placed in the radical, it must pass through C-2, as shown in Figure 22.4. This means that the p orbital on C-2 does not contribute to the nonbonding MO at all. The p orbitals on C-1 and C-3 are too far apart to interact, so this MO has the same energy as an isolated p orbital and is nonbonding.

The allyl radical has three electrons, so two occupy π_1 and one occupies the nonbonding MO. The allyl cation and the allyl anion have exactly the same MOs. The

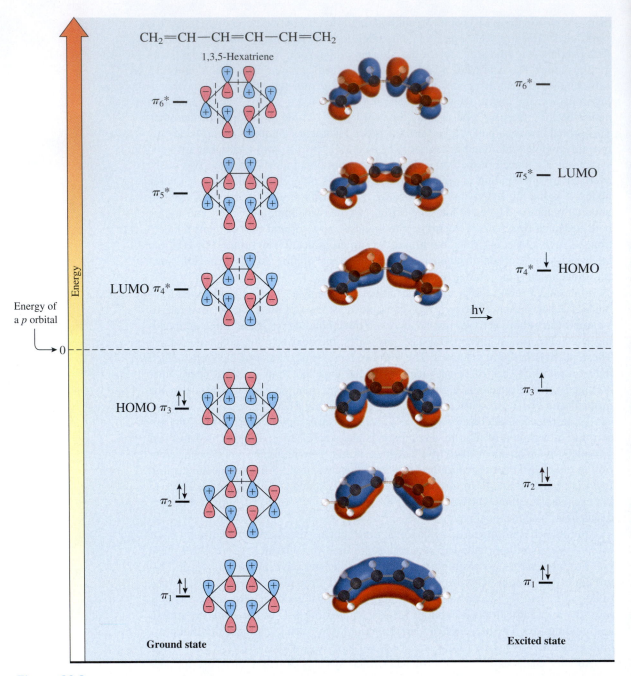

Figure 22.3

ENERGIES AND NODAL PROPERTIES OF THE PI MOS FOR THE GROUND STATE AND THE EXCITED STATE OF 1,3,5-HEXATRIENE.

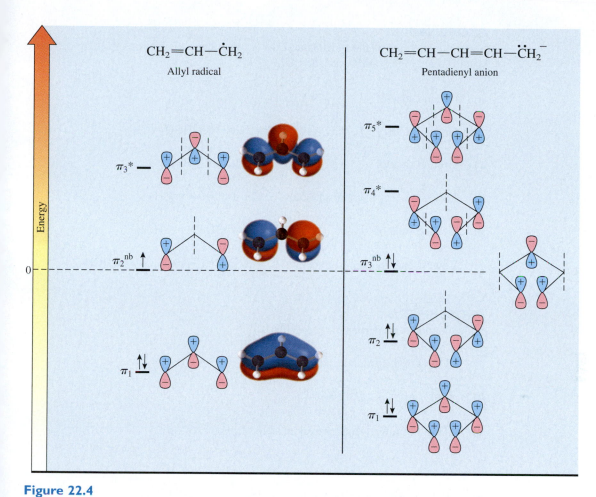

Figure 22.4

ENERGIES AND NODAL PROPERTIES FOR THE PI MOS OF THE ALLYL RADICAL AND THE PENTADIENYL ANION.

cation, however, has only two electrons, which occupy π_1, whereas the anion has four electrons, which fill both the bonding MO and the nonbonding MO.

Figure 22.4 also shows the MOs for the pentadienyl anion. The five p orbitals of this carbanion combine to form five MOs: two bonding MOs, two antibonding MOs, and one nonbonding MO. The lowest-energy MO, π_1, has no nodes. The next MO, π_2, has one node that passes through the center of the carbanion at C-3. The nonbonding MO, π_3^{nb}, has two nodes, at C-2 and C-4. The next MO, π_4^*, has three nodes, one of which passes through the center of the molecule at C-3. Finally, π_5^* has four nodes. There are six electrons in this carbanion, so the nonbonding MO is occupied by a pair of electrons.

In general, these odd orbital systems all have a nonbonding MO that has nodes passing through the even-numbered carbons. A cation has no electrons in the non-bonding MO, the radical has one electron in this MO, and the anion has two electrons occupying this MO.

PROBLEM 22.2

Show the energies of the pi MOs and the arrangement of electrons in them for these species. Label the HOMO and the LUMO.

a)   Ground state of $CH_2{=}CH\overset{+}{C}H_2$

b)   Ground state of $CH_2{=}CHCH{=}CHCH{=}CHCH{=}CH_2$

c)   Ground state of $CH_2{=}CHCH{=}CHCH{=}CH\overset{+}{C}H_2$

d)   Lowest-energy excited state of $CH_2{=}CHCH{=}CH_2$

e)   Ground state of $CH_2{=}CHCH{=}CH\overset{\bullet}{C}H_2$

f)   Lowest-energy excited state of $CH_2{=}CHCH{=}CHCH{=}CH\overset{..\,-}{C}H_2$

PROBLEM 22.3

Show the nodal properties of these MOs.

a)   LUMO of the ground state of $+$

b) π_2 of

c)   LUMO of lowest-energy excited state of

d) $\pi_4{}^{nb}$ of $+$

e)   HOMO of the ground state of

To determine whether a pericyclic reaction is favorable, we need to evaluate how the total energy of the electrons changes during the course of the reaction. As the reaction occurs, the MOs of the reacting molecule are converted into the MOs of the product. If this conversion is energetically favorable—that is, if the electrons in the occupied MOs do not increase in energy—then the reaction is likely to occur and is said to be *allowed*. If the electrons in the occupied MOs increase in energy, the reaction is unfavorable and is said to be *forbidden*. Therefore, we need to evaluate what happens to the energies of the occupied MOs as the reaction proceeds.

The method developed by Woodward and Hoffmann uses the symmetry of the placement of the nodes in the MOs to determine how the reactant MOs are converted to the product MOs and how the total electron energy changes during the course of the reaction. Fukui's method concentrates on the so-called **frontier MOs**, the HOMO and the LUMO, because the energy changes of these are the key to whether the overall energy change is favorable. This method examines how the orbitals of the HOMO, or in some cases the orbitals of the HOMO of one component and the LUMO of the other, overlap to form the new bonds. If the new overlaps are favorable (bonding overlaps), then the reaction is allowed. And if the new overlaps are unfavorable (antibonding overlaps), then the reaction is forbidden. Fukui's method (the **frontier orbital method**) is a little simpler, so it is used in this book. The examples presented in the following sections illustrate how this method is applied.

22.3   ELECTROCYCLIC REACTIONS

Pericyclic reactions are commonly divided into three classes: electrocyclic reactions, cycloaddition reactions, and sigmatropic rearrangements. An **electrocyclic reaction** forms a sigma bond between the end atoms of a series of conjugated pi bonds within a molecule. The 1,3-butadiene to cyclobutene conversion is an example, as is the similar reaction of 1,3,5-hexatriene to form 1,3-cyclohexadiene:

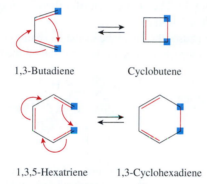

1,3-Butadiene   Cyclobutene

1,3,5-Hexatriene   1,3-Cyclohexadiene

As can be seen from these examples, the product has one more sigma bond and one less pi bond than the reactant.

Let's begin by considering the simplest electrocyclic reaction, the thermally induced interconversion of a diene and a cyclobutene. As illustrated in the following example, the reaction is remarkably stereospecific, occurring only by a conrotatory motion:

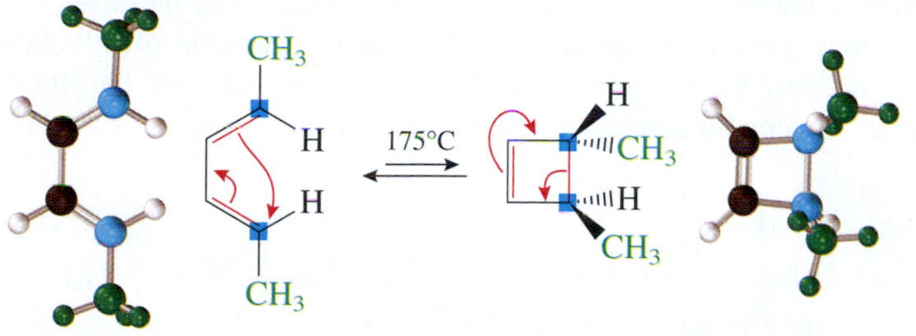

In electrocyclic reactions the end carbons of the conjugated system must rotate for the p orbitals on these carbons to begin to overlap to form the new carbon–carbon sigma bond. The preference for the stereochemistry of the rotation in these reactions can be understood by examination of the new orbital overlap in the HOMO as the rotation occurs. For the formation of the new sigma bond to be favorable, rotation must occur so that the overlap of the orbitals forming this bond is bonding in the HOMO.

For the ground-state reaction of a conjugated diene the HOMO is π_2. Conrotation of this MO causes the plus lobe of the p orbital on one end of the pi system to overlap with the plus lobe of the p orbital on the other end of the pi system:

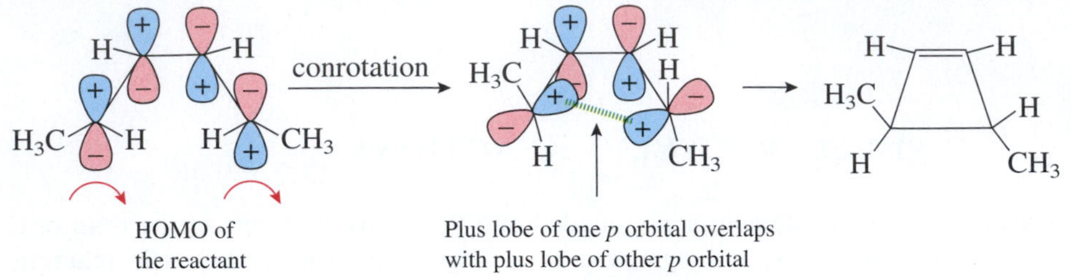

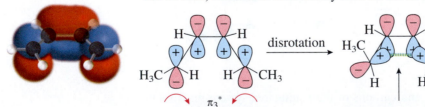

HOMO of   Plus lobe of one p orbital overlaps
the reactant   with plus lobe of other p orbital

Conrotation of a diene is thermally allowed.

This bonding overlap in the HOMO when the rotation occurs makes the formation of the new sigma bond favorable. The conversion of a diene to a cyclobutene by a conrotatory motion is *thermally allowed*.

In contrast, disrotation causes the plus lobe of the p orbital on one end of the pi system of the diene to overlap with the minus lobe of the p orbital on the other end in the HOMO:

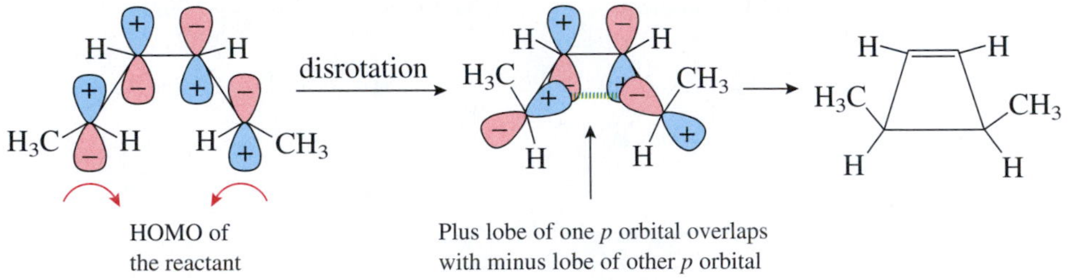

HOMO of   Plus lobe of one p orbital overlaps
the reactant   with minus lobe of other p orbital

Disrotation of a diene is thermally forbidden.

The antibonding overlap in the HOMO when disrotation occurs makes the formation of the new sigma bond unfavorable. The disrotatory closure of a diene to a cyclobutene is *thermally forbidden*.

The requirement for a favorable bonding overlap of the orbitals forming the new sigma bond in the HOMO allows the preferred rotation to be predicted for any electrocyclic reaction. Let's consider the photochemical conversion of a butadiene to a cyclobutene. Because the reaction occurs through the excited state, the HOMO is $\pi_3{}^*$. As can be seen, disrotation is necessary here for the new overlap to be bonding:

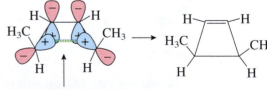

$\pi_3{}^*$
HOMO of the excited   Plus lobe of one p orbital overlaps
state of the reactant   with plus lobe of other p orbital

Disrotation of a diene is photochemically allowed.

In accord with this analysis, numerous experiments have shown that the disrotatory closure of a diene to a cyclobutene is indeed the pathway that occurs when the compound is irradiated with UV light.

PROBLEM 22.4

Use orbital drawings to show that conrotation of a diene is photochemically forbidden.

A similar analysis also correctly predicts the stereochemistry of the formation of a cyclohexadiene from a triene. For the thermal reaction the HOMO is π_3. Examination of this MO shows that disrotation is necessary for the overlap forming the new sigma bond to be bonding:

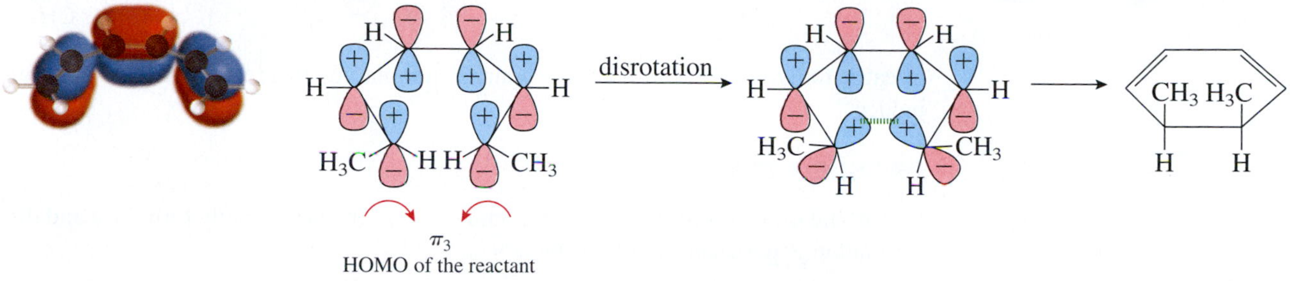

π_3
HOMO of the reactant

Disrotation of a triene is thermally allowed.

For the excited state reaction the HOMO is π_4^*. Here conrotation is the favored pathway:

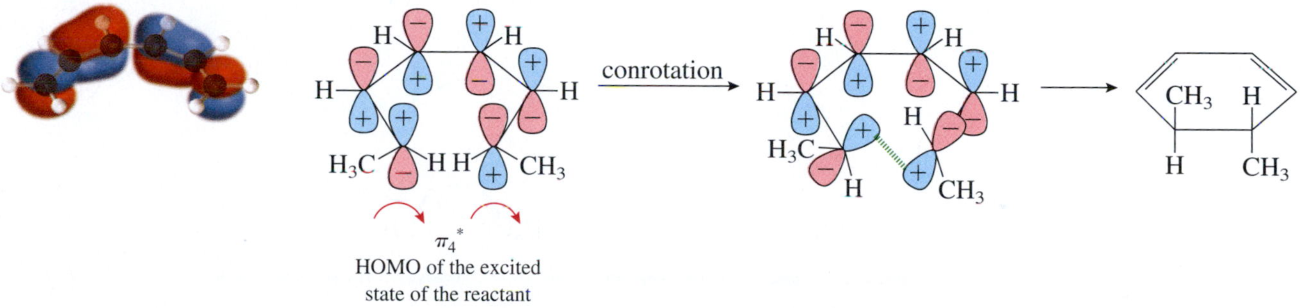

π_4^*
HOMO of the excited
state of the reactant

Conrotation of a triene is photochemically allowed.

Both of these conclusions are in accord with the experimental results presented in the next section.

PRACTICE PROBLEM 22.1

Use orbital drawings to determine whether this reaction is allowed or forbidden:

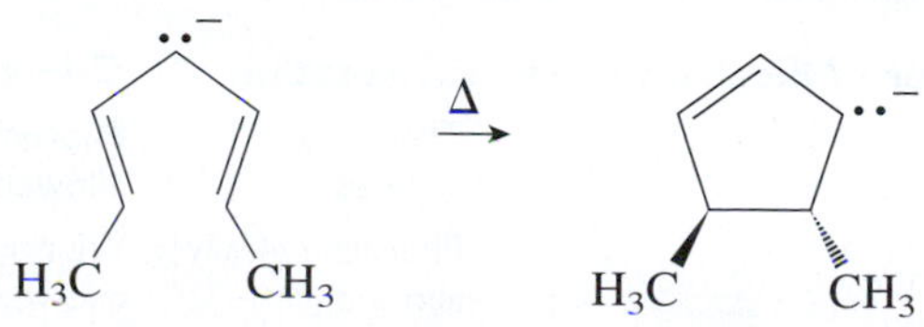

Solution

This anion has five *p* orbitals, so there are five MOs (see Figure 22.4). There are six electrons in these MOs, so the HOMO for the thermal (ground state) reaction is π_3^{nb}. Conrotation is necessary to form the product with the methyl groups trans.

Because the overlap to form the new bond is antibonding, this reaction is thermally forbidden.

PROBLEM 22.5

Use orbital drawings to show that conrotation of a triene is thermally forbidden and that disrotation is photochemically forbidden.

ORGANIC
Chemistry·ᵅ·Now™
Click *Coached Tutorial Problems* for more practice using MOs to predict the stereochemistry of **Electrocyclic Reactions.**

PROBLEM 22.6

Use orbital drawings to determine whether these reactions are allowed or forbidden:

a)

b)

It is possible to generalize the preferences for conrotation or disrotation based on the number of electron pairs in the pi MOs of the reacting molecule. A diene, with two pi electron pairs (two pi bonds), has π_2 as its HOMO in the ground state. Because this MO has one node, conrotation is favored for the thermal reaction. A triene, with three pi electron pairs (three pi bonds), has π_3 as its HOMO. This MO has two nodes (one more than π_2 of a diene), so the opposite rotation, disrotation, is preferred. It is apparent that the favored rotation alternates with the number of pi electron pairs. Therefore, a molecule with four pi bonds (π_4 is the HOMO) prefers conrotation. Furthermore, the preference for the excited state is just reversed from that of the ground state because the excited-state HOMO always has one more node than the ground-state HOMO. These preferences are summarized in the following chart:

Number of Electron Pairs	Disrotation	Conrotation
Odd	Thermally allowed	Photochemically allowed
Even	Photochemically allowed	Thermally allowed

(Note that an odd number of electron pairs equals $4n + 2$ electrons, whereas an even number of electron pairs equals $4n$ electrons, where n is any integer, including zero.)

PROBLEM 22.7

Use the preceding chart to determine whether the reactions of problem 22.6 are allowed or forbidden.

PRACTICE PROBLEM 22.2

Determine the number of electron pairs in this reaction, the type of rotation, and whether the reaction is allowed or forbidden:

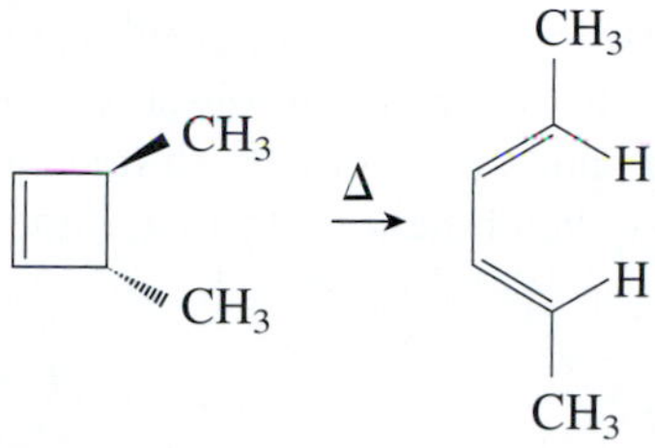

Solution

We have usually analyzed this type of reaction from the other direction—that is, the conversion of a diene to a cyclobutene. However, the same analysis works for either direction of a reaction. In this case an even number (two) of electron pairs are involved; the sigma bond and the pi bond of the reactant are converted to the two pi bonds of the product. The reaction occurs by a conrotation:

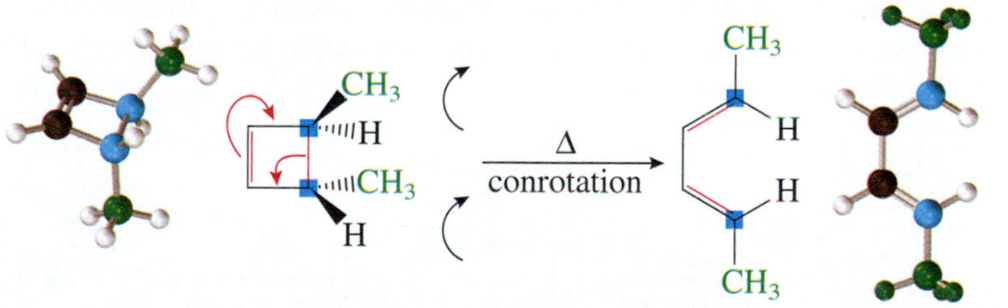

According to the chart, a conrotation involving an even number of electron pairs is thermally allowed. Therefore, the reaction shown is allowed. Note that the same reaction is forbidden under photochemical conditions.

PROBLEM 22.8

Indicate the number of electron pairs and the type of rotation for these reactions and determine whether each is allowed or forbidden:

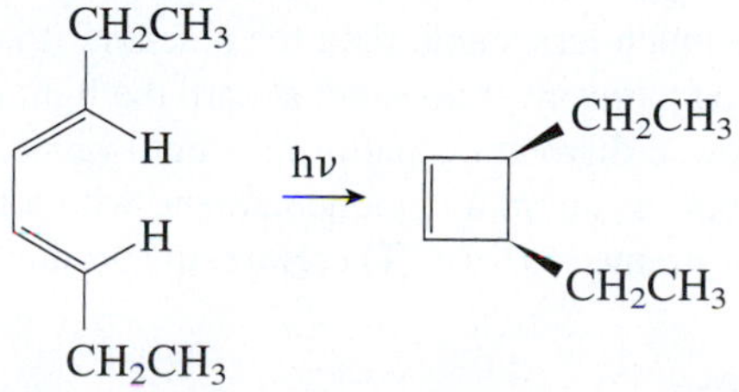

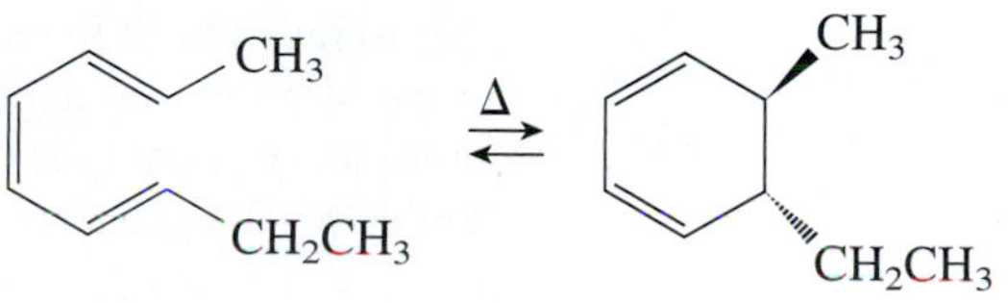

c) 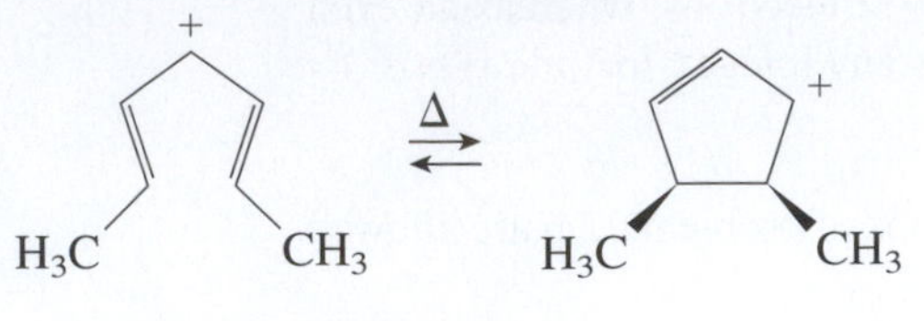   d)

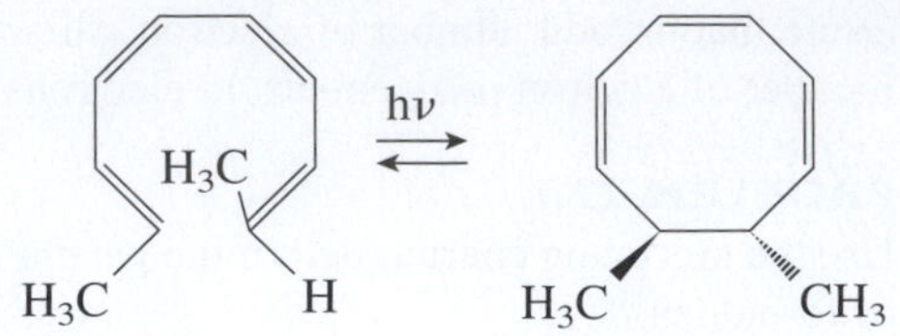

22.4 EXAMPLES OF ELECTROCYCLIC REACTIONS

The analysis in Section 22.3 indicates that the thermal interconversion of a diene with a cyclobutene should occur by conrotation. The reaction is allowed in both directions, as long as a conrotatory motion is followed. However, usually only the conversion of the cyclobutene to the diene is observed because the cyclobutene is destabilized by angle strain and is present only in trace amounts at equilibrium. An example of the opening of a cyclobutene to form a diene is provided by the following equation:

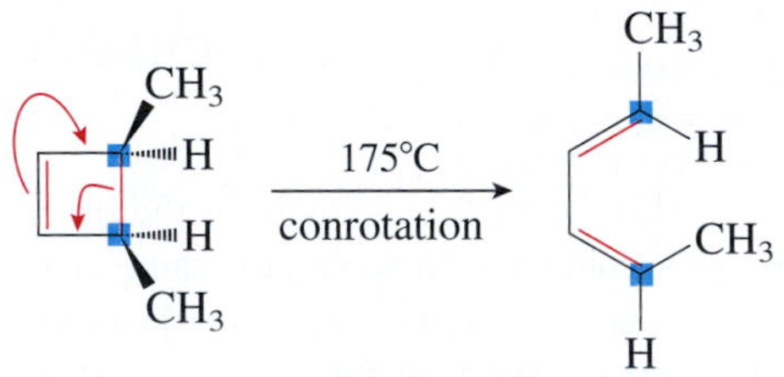

The reverse process, the conversion of a diene to a cyclobutene, can be accomplished photochemically. Although the cyclobutene is less stable, it is possible to selectively excite the diene because it absorbs longer-wavelength light (see Section 15.2). An example is shown in the following equation:

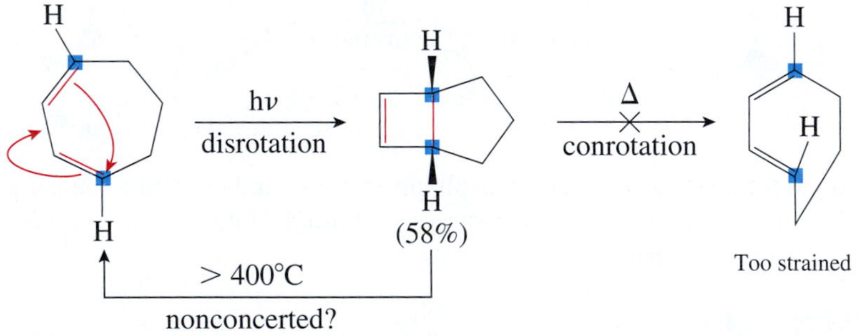

In this reaction, light of appropriate energy is used to selectively excite 1,3-cycloheptadiene. The diene closes to a cyclobutene by a disrotatory motion. Although the product, because of its strained cyclobutene ring, is much less stable than the reactant, it is unable to revert back to the diene by an allowed pathway. It does not absorb the light used in the reaction, so the photochemically allowed disrotatory pathway is not available. A conrotatory opening is thermally allowed but results in a cycloheptadiene with a trans double bond. Such a compound is much too strained to form. Therefore, the product can be readily isolated. If it is heated to a high enough temperature, greater than 400°C, the strained ring does break to produce 1,3-cycloheptadiene. This reaction might be occurring by the forbidden disrotatory pathway, or, more likely, it may involve a nonconcerted mechanism.

MODEL BUILDING PROBLEM 22.2

Build a model of the following compound. Then replace the connector for the bond that is part of both rings with two separate connectors. Do a conrotation and a disrotation to see the strain that is incurred in each process.

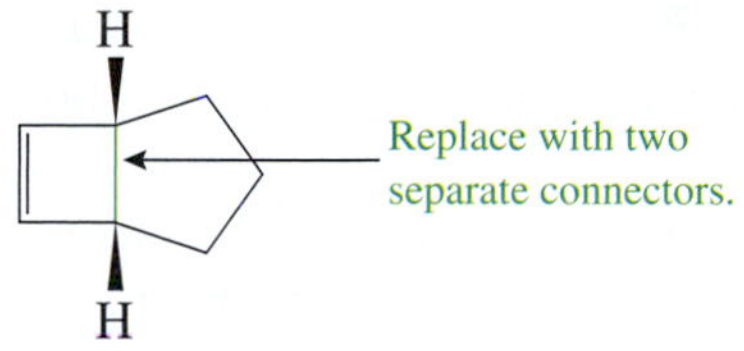

The thermally allowed cyclization of a triene to form a cyclohexadiene occurs by a disrotatory motion, as illustrated in the following equation. In this case the product is favored at equilibrium because it has one more sigma bond and one fewer pi bond than the reactant. (Sigma bonds are stronger than pi bonds.)

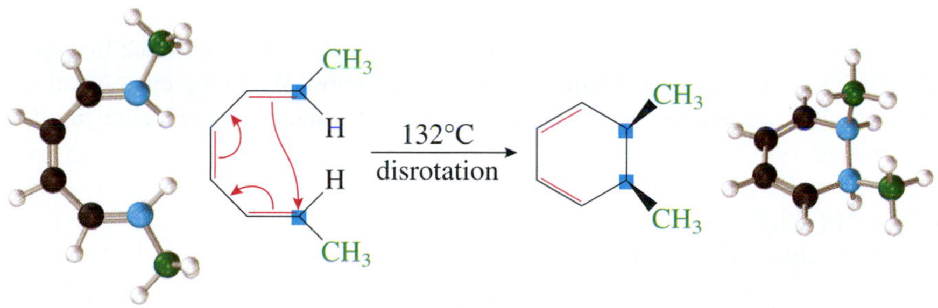

As another example, the electrocyclic interconversion of 1,3,5-cycloheptatriene and norcaradiene also occurs by a disrotatory motion:

1,3,5-Cycloheptatriene   Norcaradiene

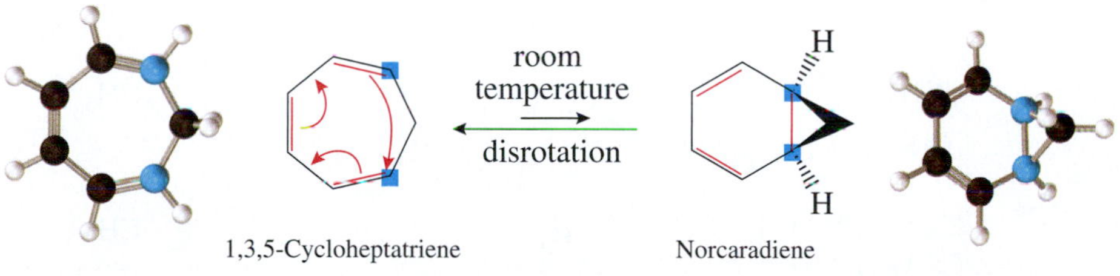

Norcaradiene cannot be isolated because its conversion to cycloheptatriene is fast at room temperature and its concentration in the equilibrium mixture is very low because of ring strain. In a related example, 1,3,5-cyclononatriene cyclizes by a disrotatory motion, as shown in the following equation. Again, the reaction is quite facile, as illustrated by its half-life of about 1 h at 50°C. In this case the product, with a five- and a six-membered ring, is more stable than the starting material with its nine-membered ring:

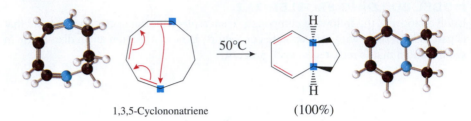

1,3,5-Cyclononatriene   (100%)

In the case of a tetraene, conrotation is again the thermally allowed pathway for cyclization. The following equation provides an example:

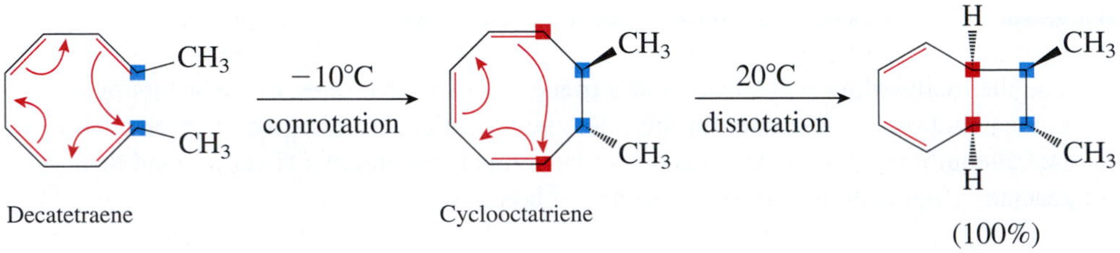

Decatetraene   Cyclooctatriene   (100%)

The decatetraene cyclizes at −10°C, by a conrotatory motion, to produce the cyclooctatriene with the two methyl groups in a trans orientation. When the cyclooctatriene is warmed to 20°C, it undergoes a thermally allowed disrotatory ring closure to give the final product.

PROBLEM 22.9
Show the products of these reactions:

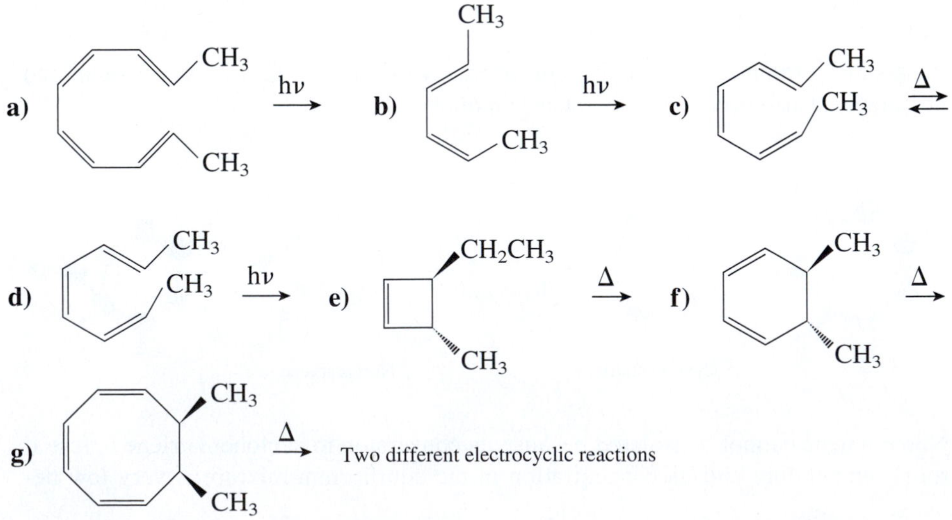

a) $\xrightarrow{h\nu}$ b) $\xrightarrow{h\nu}$ c) $\underset{\rightleftarrows}{\xrightarrow{\Delta}}$

d) $\xrightarrow{h\nu}$ e) $\xrightarrow{\Delta}$ f) $\xrightarrow{\Delta}$

g) $\xrightarrow{\Delta}$ Two different electrocyclic reactions

Focus On

Dewar Benzene

In the mid-1800s, chemists were struggling to determine the structure of benzene. One of the structures that was proposed at that time is known as Dewar benzene, after the chemist who suggested it:

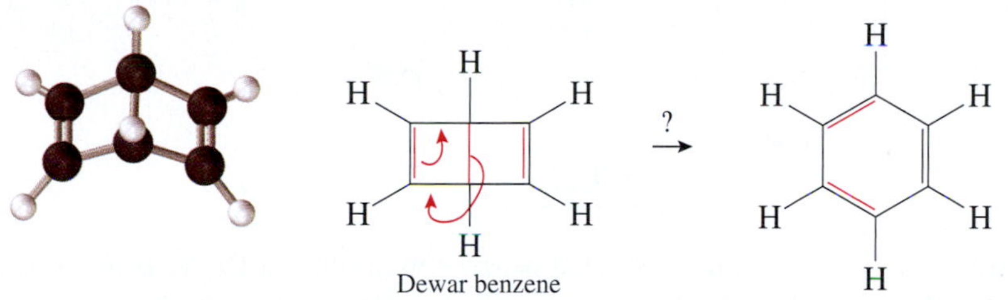

Dewar benzene

As chemists learned more about the effects of structure on the stability of organic compounds, it became apparent that Dewar benzene is much less stable than benzene. Not only does it have a considerable amount of angle strain, but it also has none of the stabilization due to aromaticity that benzene has. Because of these factors, Dewar benzene is 71 kcal/mol (297 kJ/mol) less stable than benzene. Because the conversion of Dewar benzene to benzene is so exothermic and involves an apparently simple electron reorganization, many chemists believed that the isolation of this strained isomer would prove to be impossible. They thought that if it were prepared, it would rapidly convert to benzene. In support of this view, numerous attempts to synthesize Dewar benzene met with failure.

Once the theory of pericyclic reactions was developed, it was recognized that the conversion of Dewar benzene to benzene is an electrocyclic reaction. This conversion involves two pairs of electrons: one pair of pi electrons and one pair of sigma electrons of the Dewar benzene. (The third pair of electrons is located in exactly the same place in both the reactant and the product and so is not involved in the reaction.) An electrocyclic reaction involving two pairs of electrons must occur by a conrotatory motion if it is to be thermally allowed. However, the conrotatory opening of Dewar benzene is geometrically impossible, because it would result in a benzene with a trans double bond, a compound with too much angle strain to exist.

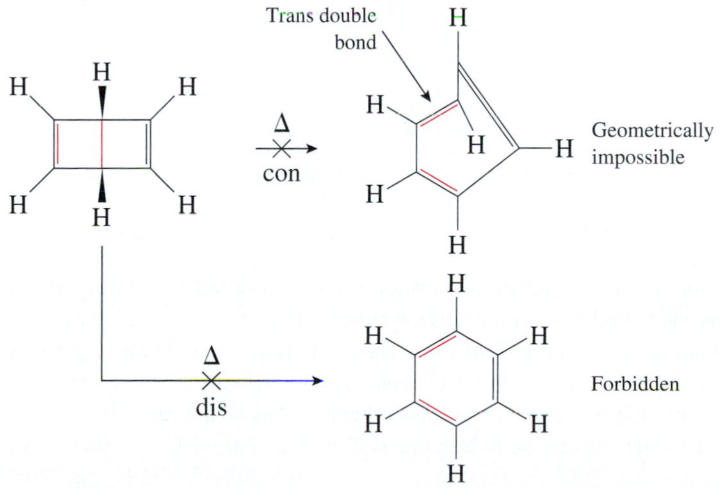

Continued

Disrotatory opening of Dewar benzene, which would produce benzene, is thermally forbidden. Therefore, even though the conversion of Dewar benzene to benzene is quite exothermic, it might indeed prove possible to isolate Dewar benzene because there is no low-energy pathway for its conversion to benzene.

The first derivative of Dewar benzene was prepared in 1962 by irradiation of 1,2,4-tri-*t*-butylbenzene:

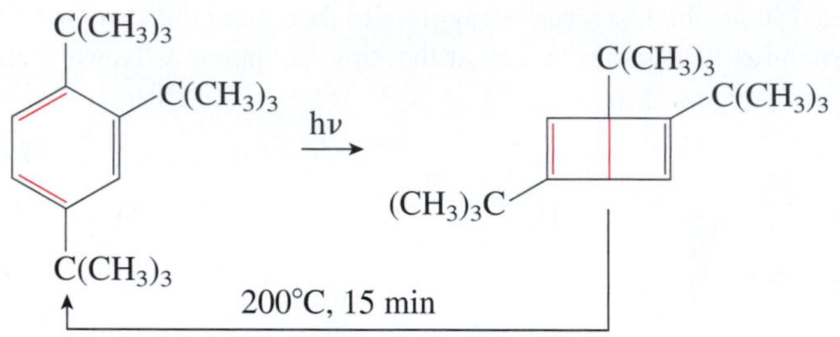

Disrotatory closure of the substituted benzene to produce a Dewar benzene is photochemically allowed, as is, of course, the reverse process. However, because benzene is conjugated, it absorbs UV light at longer wavelengths than the Dewar benzene isomer. Therefore, it is possible to selectively excite the benzene chromophore and produce the less stable Dewar isomer. In this particular case the *tert*-butyl groups favor the reaction because they destabilize the benzene isomer somewhat, owing to steric hindrance. Because the two adjacent *tert*-butyl groups in the Dewar isomer do not lie in the same plane, this steric strain is decreased in the product. Because of this steric effect and the forbidden nature of the conversion back to benzene, the Dewar isomer is relatively stable. However, when it is heated to 200°C, it is rapidly converted to the benzene isomer, probably by a nonconcerted pathway.

Shortly after this, in 1963, the parent Dewar benzene was prepared by the following reaction sequence:

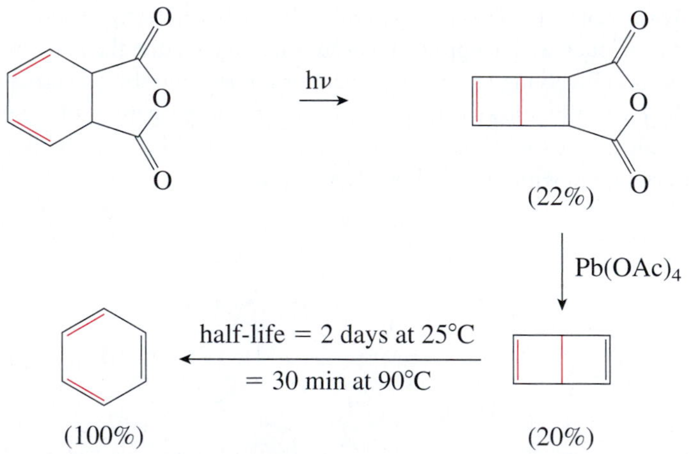

Irradiation of the substituted cyclohexadiene resulted in the formation of the Dewar benzene skeleton by a disrotatory ring closure. Reaction with lead tetraacetate (a reaction that is not covered in this book) was used to remove the anhydride group and introduce the final double bond of Dewar benzene. Again, because of the forbidden nature of the conrotatory opening to benzene, Dewar benzene has an appreciable lifetime. At 25°C the half-life for its conversion to benzene is 2 days, and at 90°C its half-life is 30 min.

MODEL BUILDING PROBLEM 22.3

Build a model of Dewar benzene. Then replace the connector for the bond that is part of both rings with two separate connectors. Do a conrotation and a disrotation to see the strain that is incurred in each process. Also build a model of 1,2,4-tri-*t*-butylbenzene and the Dewar benzene formed from it and examine the steric strain that is present in each compound.

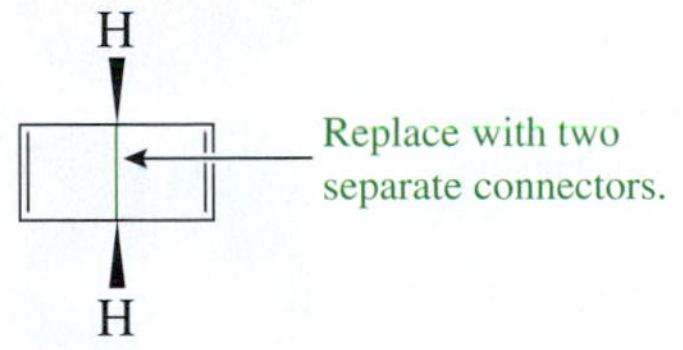

Explain why the following Dewar benzene is not produced in the photochemical reaction of 1,2,4-tri-*t*-butylbenzene:

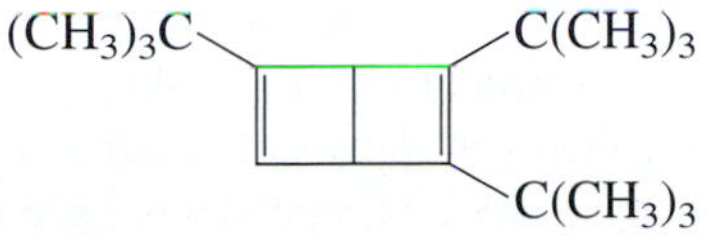

22.5 CYCLOADDITION REACTIONS

A **cycloaddition reaction** most commonly involves two molecules reacting to form two new sigma bonds between the end atoms of their pi systems, resulting in the formation of a ring. The product has two more sigma bonds and two fewer pi bonds than the reactants. The reactions are classified according to the number of pi electrons in each of the reactants. Thus, the reaction of two alkenes to form a cyclobutane derivative is termed a [2 + 2] cycloaddition reaction, and the reaction of a diene with an alkene to form a cyclohexene derivative is termed a [4 + 2] cycloaddition reaction:

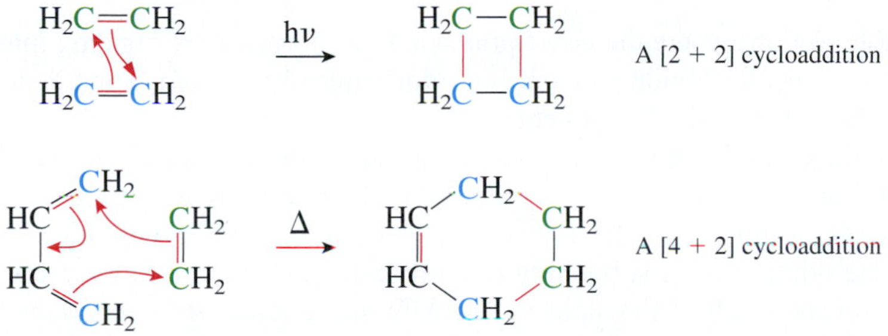

Cycloaddition reactions can be viewed as involving the flow of electrons from the HOMO of one reactant to the LUMO of the other. Therefore, examination of the interaction of these MOs is used to determine whether the reaction is favorable. For a cycloaddition to be allowed, the overlap of the orbitals of the HOMO of one component and the LUMO of the other must be bonding where the new sigma bonds are to be formed. Some examples will help make the application of this rule clear.

Let's consider first the thermal [2 + 2] cycloaddition. The overlaps between the HOMO of one component (π) and the LUMO of the other (π^*) are as follows:

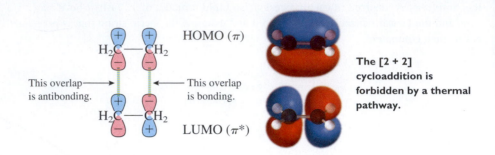

Because one of these overlaps is bonding and the other is antibonding, the reaction is thermally forbidden—that is, it does not occur when the compounds are heated.

Next, let's consider the photochemical version of this reaction. A photochemical reaction between two molecules involves the excited state of one component and the ground state of the other. (Because most excited states have extremely short lifetimes, the chances of two excited molecules colliding are exceedingly small, so reactions where both components are excited are highly improbable.) The HOMO of the excited ethene is π^*, and the LUMO of the ground-state ethene is also π^*. The overlaps are as follows:

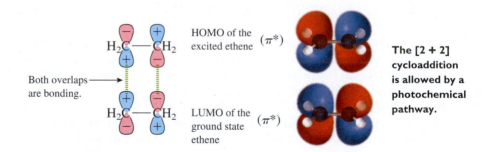

Both of the overlaps where the new sigma bonds are forming are bonding interactions, so the [2 + 2] cycloaddition reaction is photochemically allowed—that is, it does occur when the compounds absorb light.

Two points should be noted in analyzing the orbital overlaps that occur in these cycloadditions. First, always start with a bonding overlap where one of the new sigma bonds is forming. The nodal properties of the MOs will then determine whether the other overlap is bonding or antibonding. (If both new overlaps are antibonding, reversing all of the signs in one MO makes both of the overlaps bonding, so the reaction is allowed.) Second, it does not matter whether the analysis is done by employing the HOMO of component A and the LUMO of component B or the LUMO of component A and the HOMO of component B. Both analyses give the same results.

Let's consider the [4 + 2] cycloaddition. The interaction of the HOMO of 1,3-butadiene (π_2) with the LUMO of ethene (π^*) is as follows:

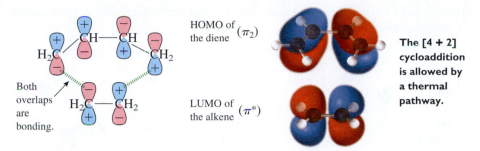

HOMO of the diene (π_2)

LUMO of the alkene (π^*)

Both overlaps are bonding.

The [4 + 2] cycloaddition is allowed by a thermal pathway.

Both overlaps are bonding, so the reaction is thermally allowed.

PROBLEM 22.10

Use orbital drawings to show that the [4 + 2] cycloaddition reaction is photochemically forbidden.

PROBLEM 22.11

Use orbital drawings to determine whether the thermal [4 + 4] cycloaddition reaction is allowed or forbidden.

If another pi bond is added to either component of the previous cycloaddition, an additional node is introduced into either the HOMO or the LUMO, resulting in the reaction being thermally forbidden and photochemically allowed. Recognition of this pattern enables the preferences for cycloadditions to be summarized in the chart below.

Number of Electron Pairs	Allowed Cycloaddition
Odd	Thermal
Even	Photochemical

Examples of cycloaddition reactions are presented in the next two sections.

PROBLEM 22.12

Indicate how many electron pairs are involved in these reactions and determine whether each reaction is allowed or forbidden:

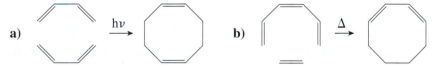

a) $\xrightarrow{h\nu}$

b) $\xrightarrow{\Delta}$

22.6   THE DIELS-ALDER REACTION

The [4 + 2] cycloaddition was discovered long before the theory of pericyclic reactions was developed. It is more commonly known as the *Diels-Alder reaction,* named after O. Diels and K. Alder, who shared the 1950 Nobel Prize in chemistry for developing this reaction. The Diels-Alder reaction occupies a very important place among the tools of the synthetic organic chemist because it provides a method for the construction of six-membered rings from acyclic precursors with excellent control of stereochemistry.

In general, the Diels-Alder reaction involves the combination of a diene with an alkene, termed a dienophile, to form a cyclohexene derivative. The simplest example, the reaction of 1,3-butadiene (the diene) with ethene (the dienophile), is illustrated in the following equation:

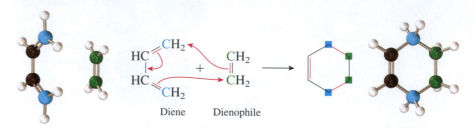

The yield of cyclohexene in this particular reaction is not very high. In general, the yields in the Diels-Alder reaction are much better if the diene and the dienophile are substituted with groups of opposite polarity—that is, with electron-withdrawing groups on one component and electron-donating groups on the other. The vast majority of examples that have been reported have employed electron-withdrawing groups on the dienophile. Some examples of the many alkenes, substituted with electron-withdrawing groups, that are excellent dienophiles in the Diels-Alder reaction are as follows:

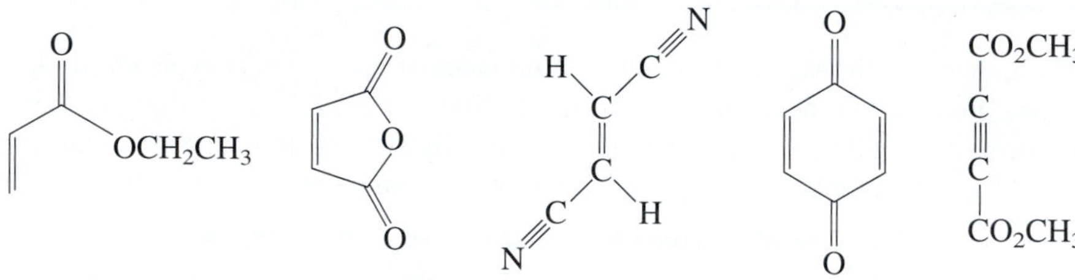

The following equation provides an example of the use of a good dienophile in a Diels-Alder reaction that proceeds in excellent yield:

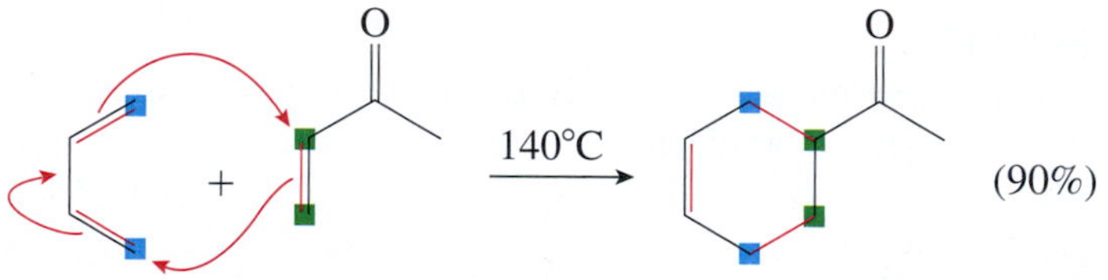

$$140°C \quad (90\%)$$

Next, let's address the stereochemistry of the reaction. If the orbitals overlap as illustrated in Section 22.5, the addition is syn on both the diene and the dienophile. An example showing that the addition is indeed syn on the dienophile is illustrated in the following equation. In this reaction the *cis*-stereochemistry of the carboxylic acid groups in the dienophile is preserved in the adduct:

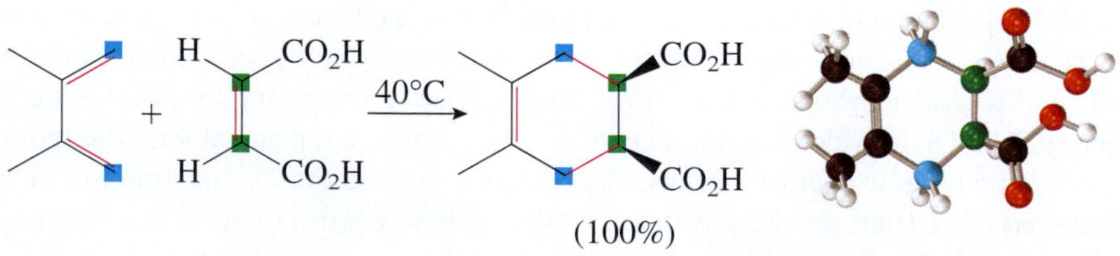

$$40°C$$

(100%)

Now let's consider the three-dimensional shape of the diene. To maximize the stabilization due to conjugation, the *p* orbitals on the central carbons of a diene must be parallel. For an acyclic diene, such as 1,3-butadiene, there are two planar conformations for which these orbitals are parallel:

$$
\begin{array}{ccc}
\underset{1}{H_2C}\!=\!\underset{2}{\overset{C}{\underset{|}{}}}\!\!\!\!\!\!\!\! & & \\
\text{s-trans} & \rightleftharpoons & \text{s-cis} \\
\text{Unreactive} & & \text{Reactive}
\end{array}
$$

The conformation that has C-1 and C-4 of the double bonds on opposite sides of the single bond between C-2 and C-3 is termed the **s-trans conformation,** whereas the conformation that has them on the same side of the single bond is called the **s-cis conformation.** The *s*-trans conformation of 1,3-butadiene is more stable because it has less steric strain—the larger groups are farther apart. Although the interconversion of these two conformations is fast, only the *s*-cis conformation can react in the Diels-Alder cycloaddition. In the *s*-trans conformation, C-1 and C-4 are too far apart to bond simultaneously to the dienophile.

The reactivity of a particular diene depends on the concentration of the *s*-cis conformation in the equilibrium mixture. Factors that increase the concentration of this conformation make the diene more reactive. As an example of this effect, consider 2,3-dimethyl-1,3-butadiene:

$$
\begin{array}{ccc}
& & \\
\text{s-trans} & \rightleftharpoons & \text{s-cis} \\
\text{Unreactive} & & \text{Reactive}
\end{array}
$$

The methyl groups on C-2 and C-3 cause the *s*-trans and *s*-cis conformations to have similar amounts of steric strain. Therefore, more of the *s*-cis conformer is present at equilibrium for 2,3-dimethyl-1,3-butadiene than for 1,3-butadiene itself. For this reason, 2,3-dimethyl-1,3-butadiene reacts about 10 times faster than 1,3-butadiene in the Diels-Alder reaction.

In the case of 3-methylenecyclohexene the double bonds are held in the *s*-trans conformation. This compound cannot react as a diene in the Diels-Alder reaction.

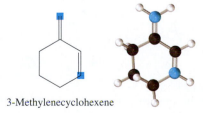

3-Methylenecyclohexene

In contrast, the double bonds of cyclopentadiene are held in the *s*-cis conformation. This makes cyclopentadiene so reactive as a diene in the Diels-Alder reaction that it dimerizes at room temperature. One molecule reacts as the diene and the other as the dienophile to form dicyclopentadiene, as shown in the following equation:

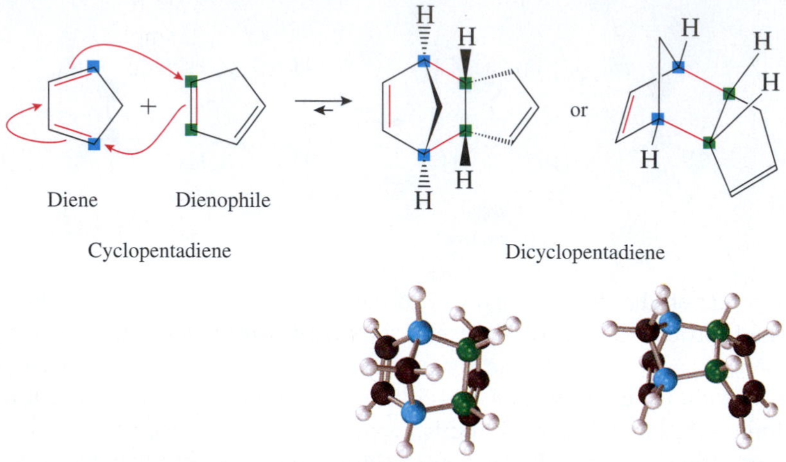

Cyclopentadiene cannot be purchased because it is too reactive and dimerizes to dicyclopentadiene upon storage. If it is needed, it is prepared by heating dicyclopentadiene. Cyclopentadiene is produced in a reverse Diels-Alder reaction. It is distilled from the hot reaction mixture as it is formed and used immediately.

PROBLEM 22.13

Explain which of these isomers of 2,4-hexadiene is more reactive as a diene in the Diels-Alder reaction:

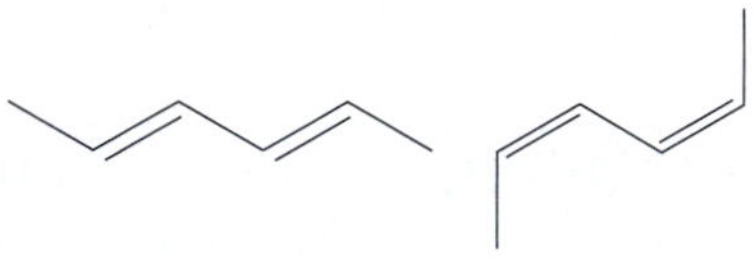

The dimerization of cyclopentadiene introduces a new aspect of the Diels-Alder reaction, the stereochemical relationship between the diene and the dienophile. When cyclopentadiene reacts as a diene, the newly formed six-membered ring of the product has a one-carbon bridge connecting positions 1 and 4:

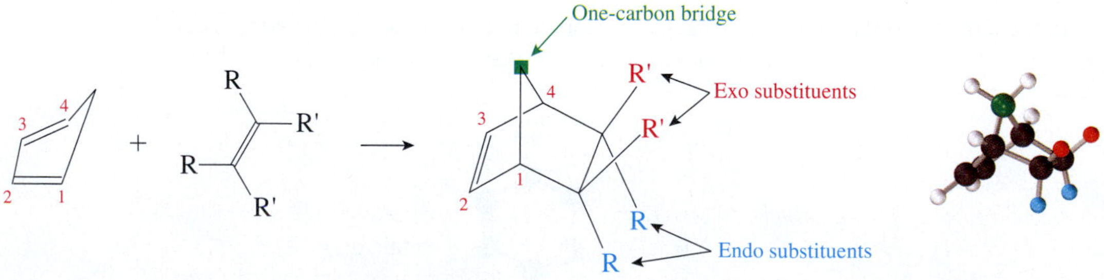

Groups that are cis to this bridge are termed **exo** substituents, and groups that are trans to this bridge are termed **endo** substituents.

The dicyclopentadiene that is formed in the dimerization of cyclopentadiene has the ring of the dienophile in an endo orientation to the cyclopentadiene ring that acts as the diene. Usually, substituents on the dienophile are found to be endo in the adduct if the substituents contain pi bonds. Another example is provided by the reaction of cyclopentadiene and maleic anhydride illustrated in the following equation:

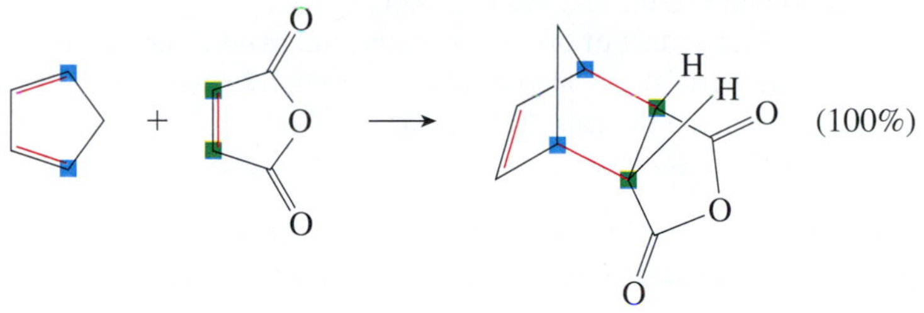

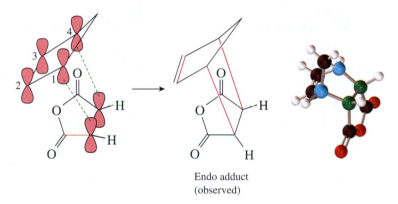

The orientation of the reactants that leads to the endo product is as follows:

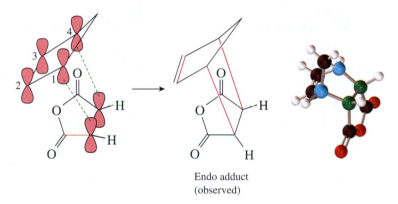

Endo adduct
(observed)

and the orientation that leads to the exo product is as follows:

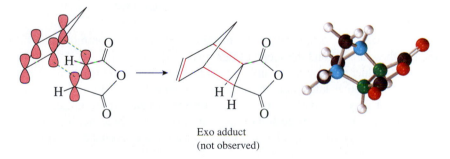

Exo adduct
(not observed)

(Models may be useful to help you visualize these different orientations.) Because the orientation leading to the endo adduct is more sterically congested than that leading to the exo adduct, it is apparent that steric effects are not controlling the orientation. Rather, it appears that a stabilizing interaction between the orbitals on C-2 and C-3 of the diene with pi orbitals on the substituents of the dienophile causes the endo orientation to be preferred. As a result, the group with the p orbitals prefers to be cis to the double bond bridge and trans to the one-carbon bridge.

MODEL BUILDING PROBLEM 22.4

Build these models:

a) The Diels-Alder adduct of ethene and cyclopentadiene. Identify the exo and endo positions.

b) The endo Diels-Alder adduct of cyclopentadiene and maleic anhydride. Examine the model for any steric strain and possible interaction between the pi systems of the CC double bond and the anhydride group.

c) The exo Diels-Alder adduct of cyclopentadiene and maleic anhydride. Examine the model for any steric strain and possible interaction between the pi systems of the CC double bond and the anhydride group.

When both the diene and the dienophile are unsymmetrically substituted, regio-isomeric products are possible. When the diene is substituted on C-1, the "ortho-like" product is preferred:

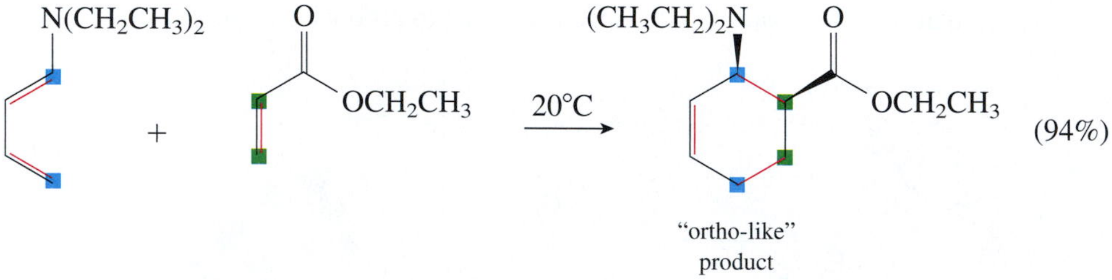

"ortho-like"
product

When the diene is substituted on C-2, the "para-like" product is preferred.

"para-like"
product

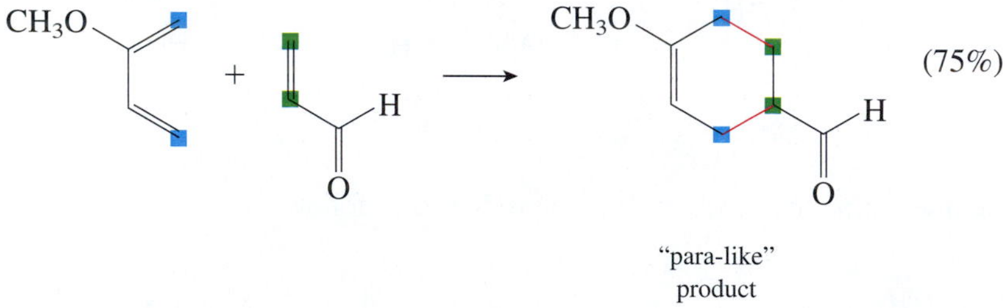

These preferences can be rationalized by using arguments based on molecular orbital theory, but these are beyond the scope of this book.

The Diels-Alder reaction is certainly one of the most important reactions in organic chemistry. A few other interesting examples are provided in the following equations. Benzene is not very reactive as a diene because the product would not be aromatic. However, reactive dienophiles do add to the central ring of anthracene. In this case the product, with two benzene rings, has not lost much aromatic resonance energy.

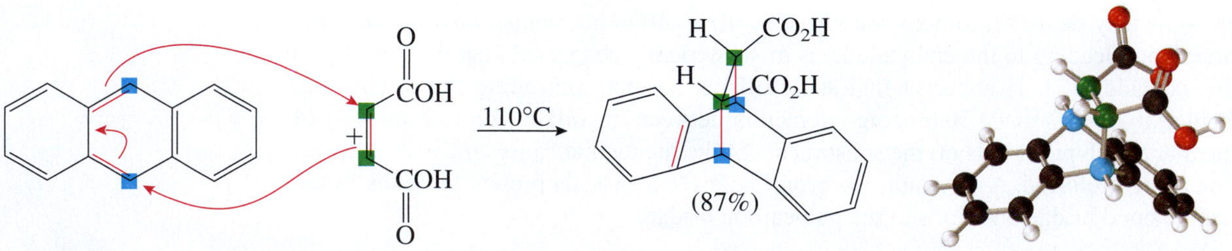

Intramolecular Diels-Alder reactions can be used to construct several rings simultaneously:

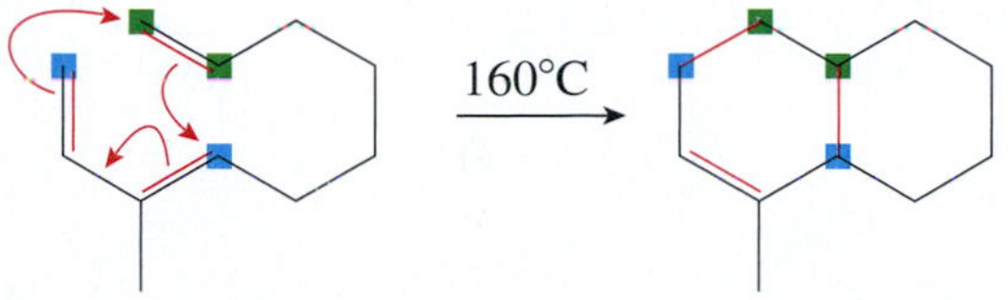

(95%)

The dienophile can be an alkyne:

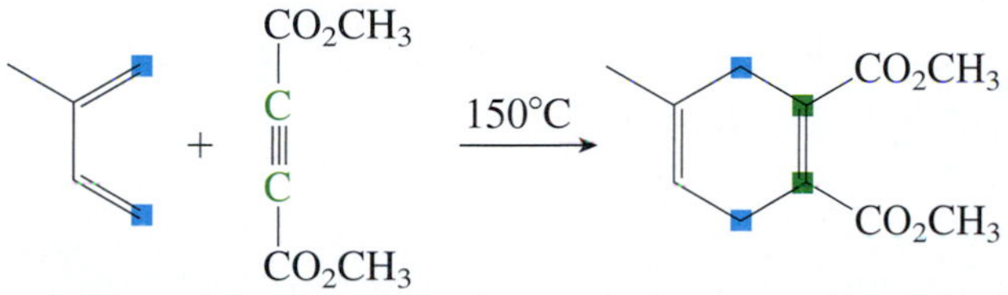

(87%)

Atoms other than carbon can even be part of the diene or dienophile:

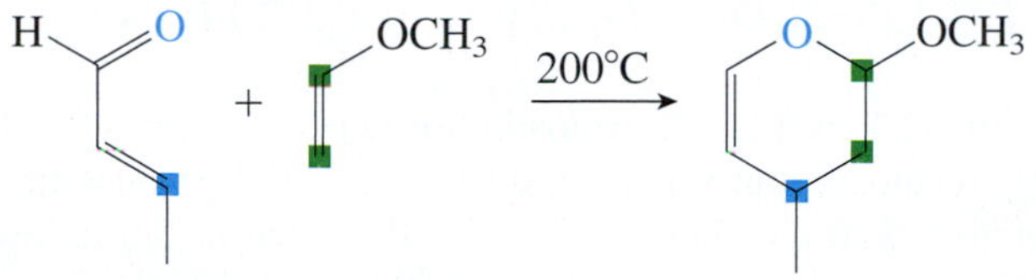

(57%)

ORGANIC
Chemistry ••• Now™
Click Coached Tutorial Problems
to practice additional Diels-
Alder Reactions.

PROBLEM 22.14
Show the products of these reactions.

a)

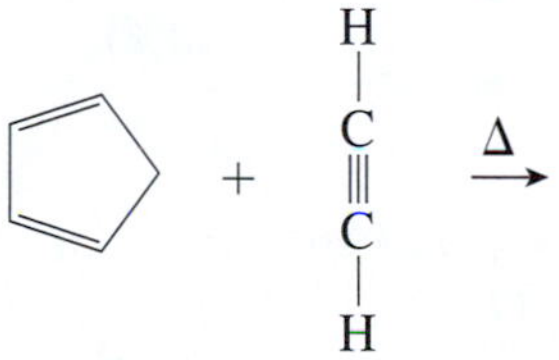

b)

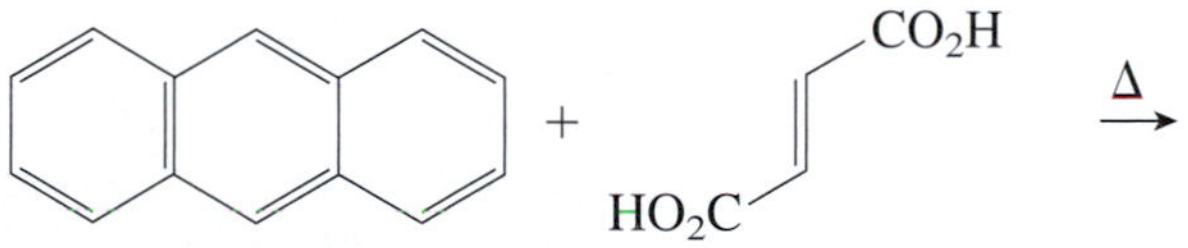

c)

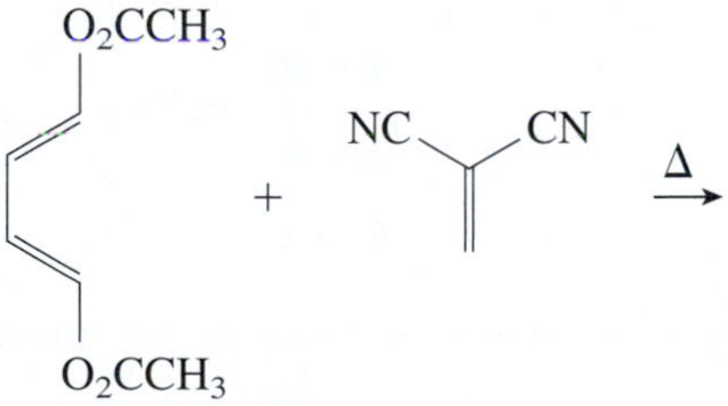

d)

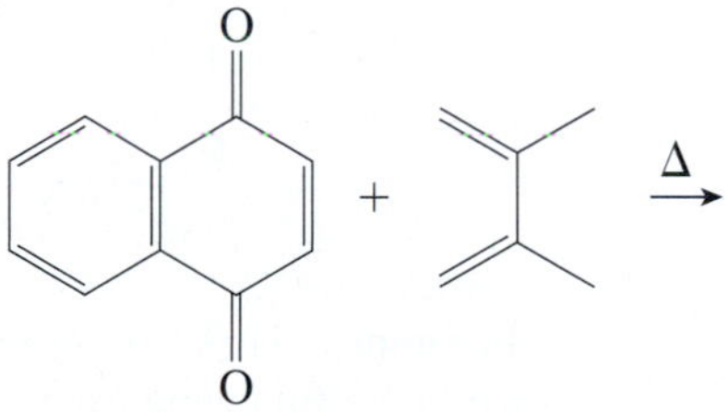

e)

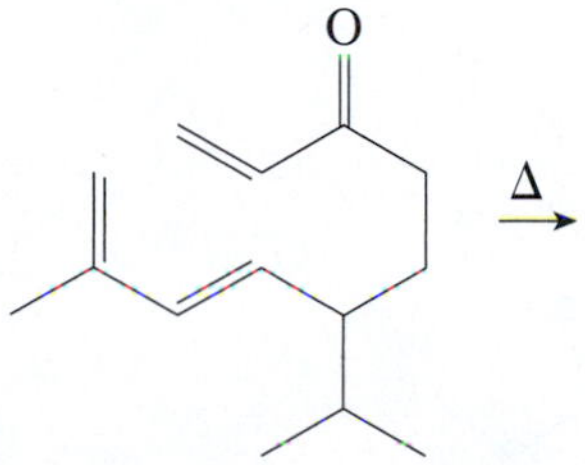

f)

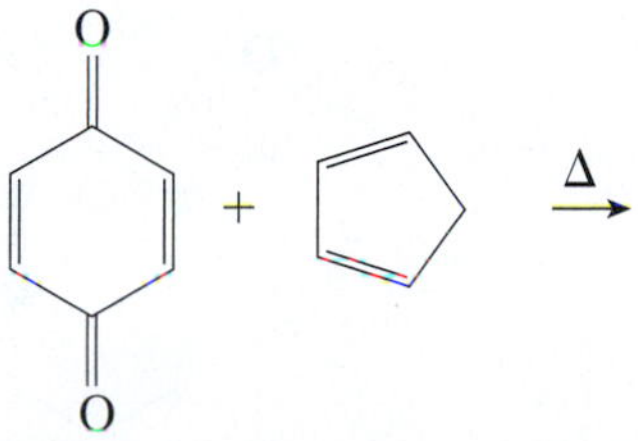

g)

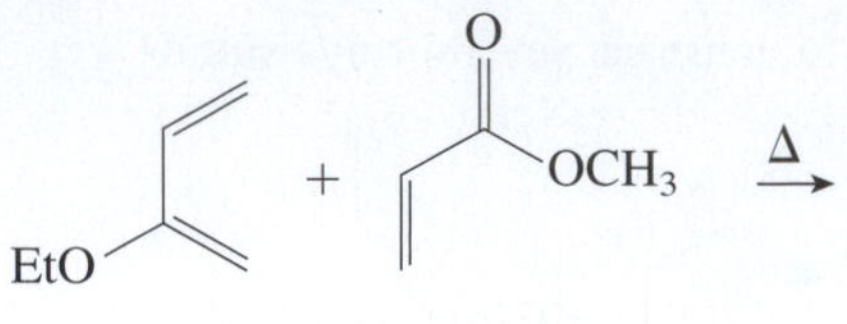

h)

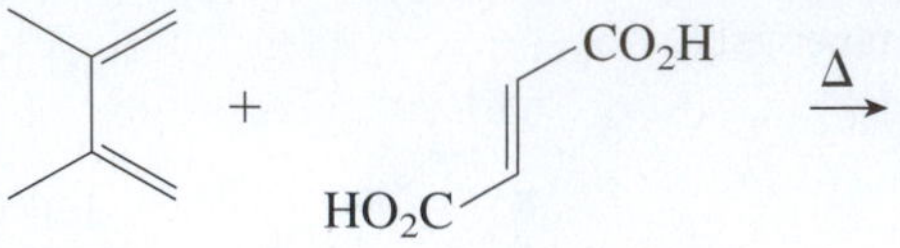

i)

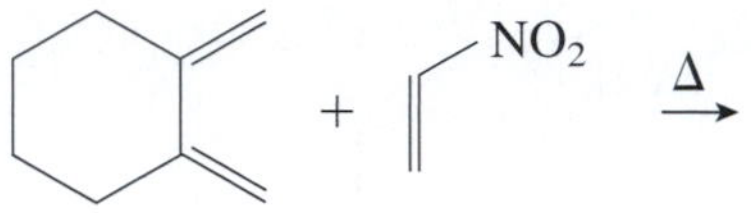

j)

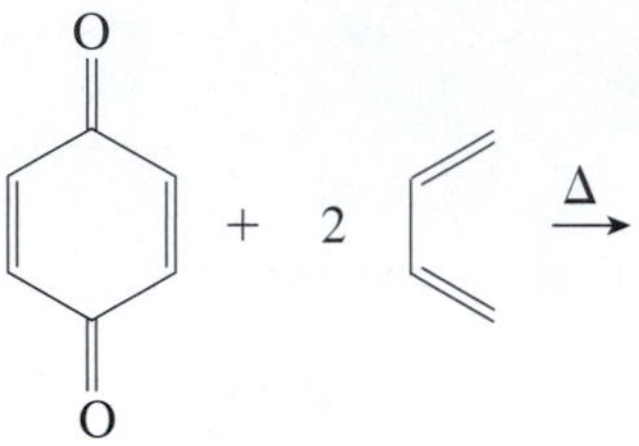

k)

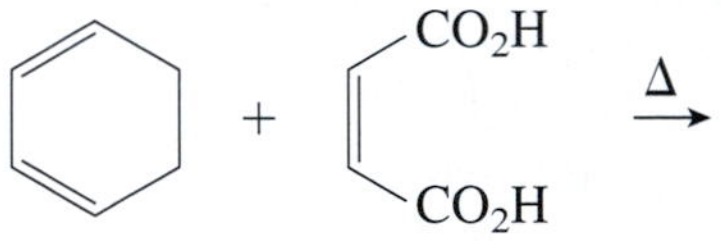

22.7 OTHER CYCLOADDITION REACTIONS

As discussed in Section 22.5, the [2 + 2] cycloaddition is photochemically allowed. The yields are often only mediocre, but this reaction is still useful because there are few good methods to prepare four-membered rings. As illustrated in the following equations, the cycloaddition can be used to dimerize two identical alkenes or to cyclize different alkenes:

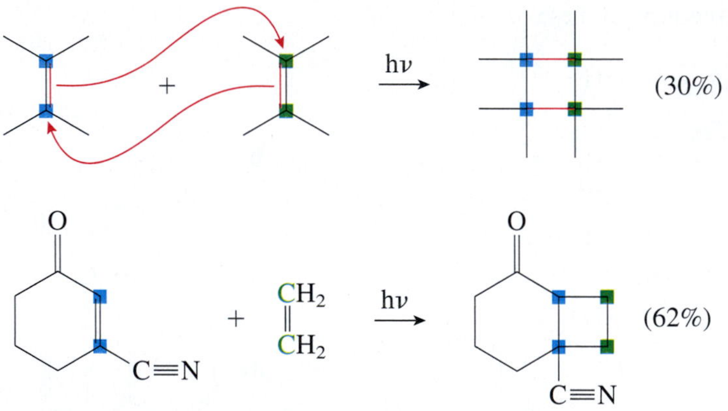

Intramolecular [2 + 2] cycloadditions often give good yields of the adduct, as shown in the following example:

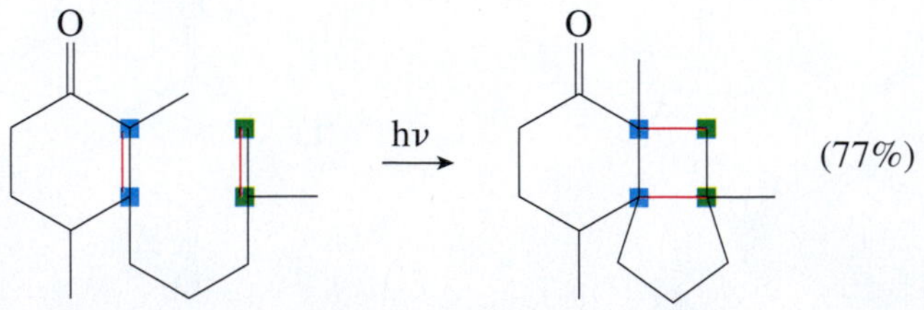

The CO double bond of an aldehyde or a ketone can act as one component in a [2 + 2] cycloaddition with an alkene. The product, a four-membered ring containing an oxygen, is called an oxetane.

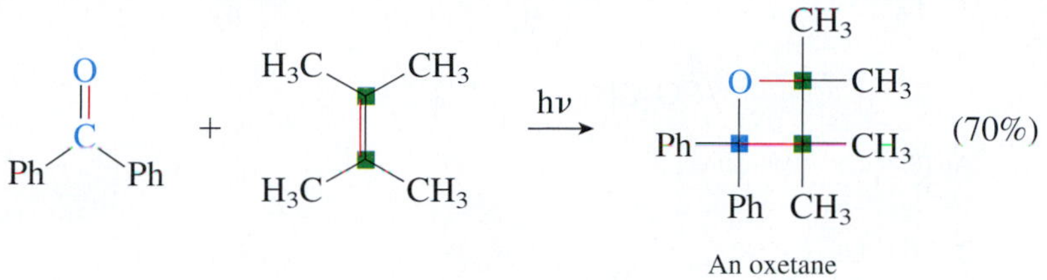

An oxetane

Cycloadditions to form rings larger than six-membered are much less common because of the unfavorable entropy of reactions that form large rings. In simple terms, such reactions are not favorable because collisions with the required geometry—that is, with both ends of the two pi systems colliding simultaneously—are not very probable. However, such reactions are known in cases in which other bonds in the molecule hold the pi system in a reactive geometry. For example, the following cycloaddition has been reported:

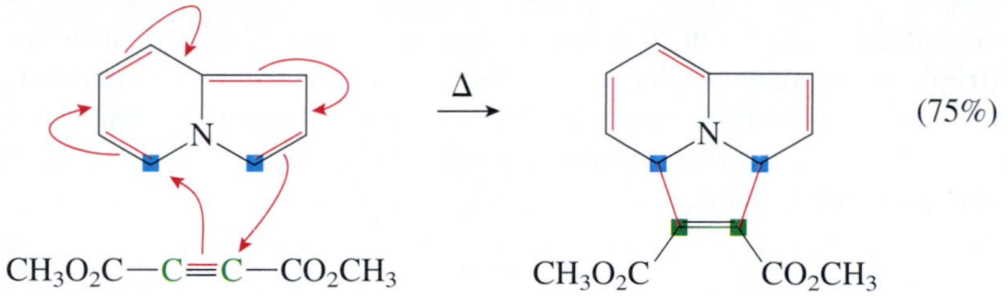

This reaction is an [8 + 2] cycloaddition and is thermally allowed. In terms of the pi systems, a 10-membered ring is formed, an entropically unfavorable process. However, the nitrogen holds the ends of the eight-electron pi system close together, so the two-electron component can easily reach them. Counting via the nitrogen, a 5-membered ring is formed, a process that is much more favorable.

PROBLEMS 22.15
Explain whether you would use heat or light to accomplish this cycloaddition reaction:

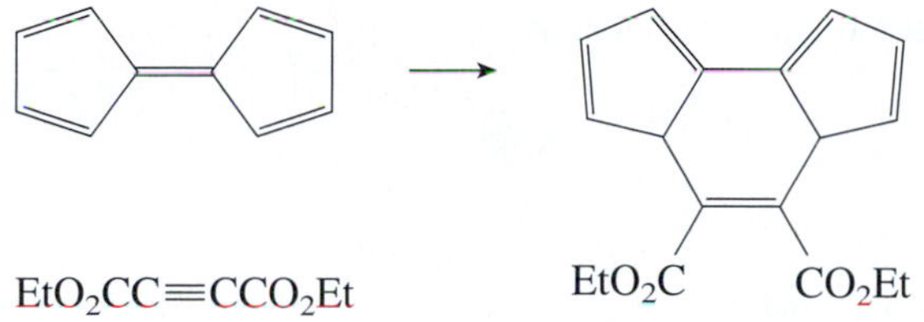

PROBLEMS 22.16
Show the products of these reactions:

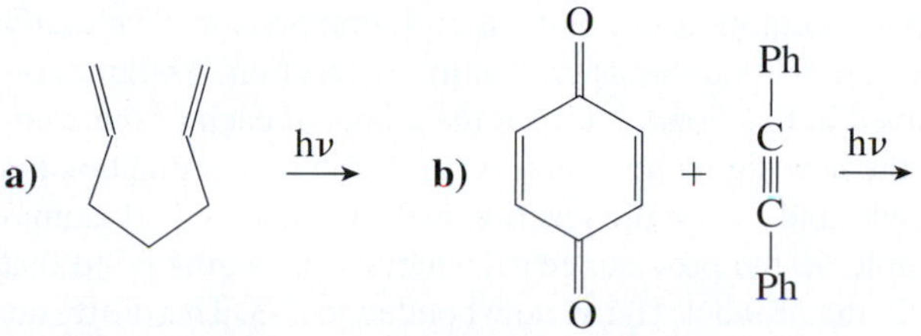

c)

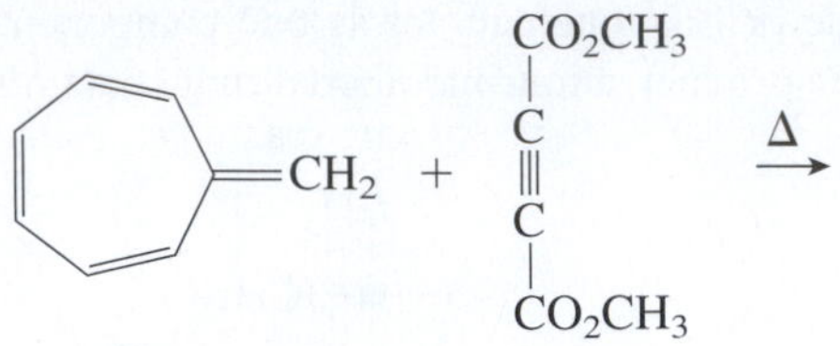

An [8 + 2] cycloaddition

d)

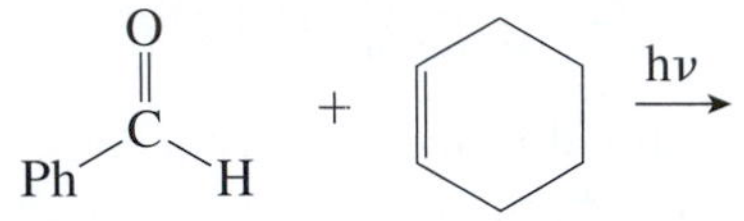

22.8 SIGMATROPIC REARRANGEMENTS

An intramolecular migration of a group along a conjugated pi system is termed a **sigmatropic rearrangement.** Basically, a sigma bond adjacent to a conjugated pi system is broken, the pi bonds are reorganized, and a new sigma bond is formed at the end of the pi system. The reaction shown in the following equation is a simple example of a sigmatropic rearrangement:

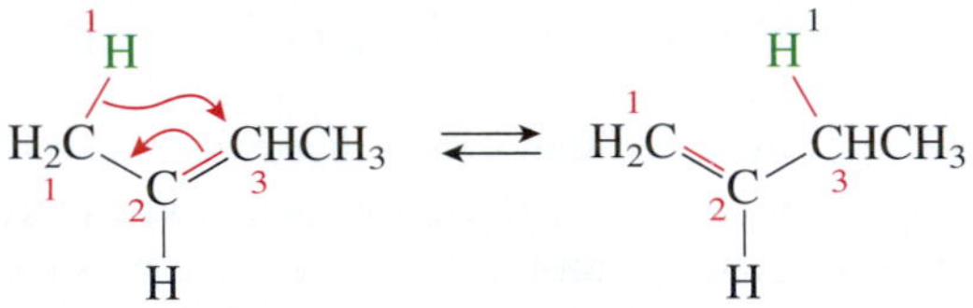

A [1,3] sigmatropic rearrangement

In this reaction a hydrogen that is sigma bonded to the carbon that is adjacent to the pi bond (C-1) migrates to the end carbon of the pi bond (C-3). As this occurs, the pi bond "migrates" from C-2—C-3 to C-1—C-2. A picture of the orbitals involved in this reaction may help make the process clearer:

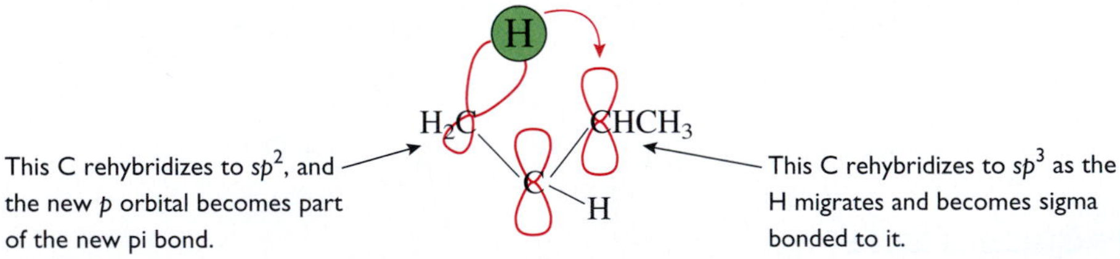

This C rehybridizes to sp^2, and the new p orbital becomes part of the new pi bond.

This C rehybridizes to sp^3 as the H migrates and becomes sigma bonded to it.

Sigmatropic rearrangements are classified according to the number of bonds separating the migration origin and the migration terminus in each component. To classify the rearrangement, first find the sigma bond that is broken in the reaction. Assign number 1 to *both* of the atoms involved in this bond. Number the atoms of each of the components up to the atoms where the new sigma bond is formed. If the new sigma bond is formed between atoms numbered i and j, then the reaction is designated an [i,j] sigmatropic rearrangement. For example, in the previous rearrangement the sigma bond that is broken connects H-1 to C-1. In the product, H-1 is now bonded to C-3. Therefore, the reaction is designated a [1,3] sigmatropic rearrangement. The following reaction is similar, except that there is one more pi bond in one component and the hydrogen has migrated to C-5. This reaction is an example of a [1,5] sigmatropic rearrangement.

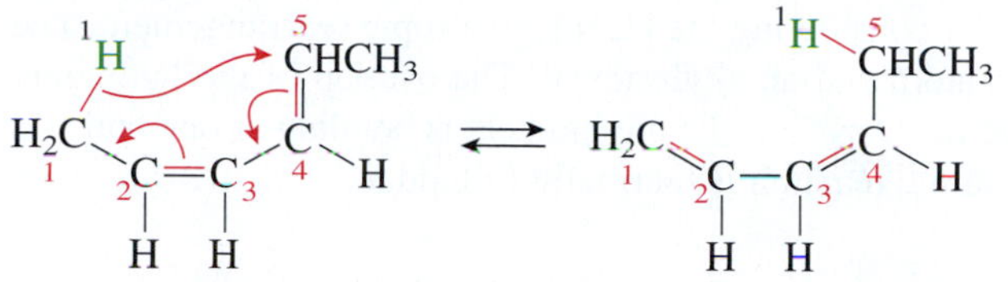

A [1,5] sigmatropic rearrangement

In addition to rearrangements such as the previous two, it is possible for both of the components to contain pi bonds. In the following reaction the carbon–oxygen bond is broken and the new sigma bond is formed between the carbons at the other end of each of the components. This is an example of a [3,3] sigmatropic rearrangement.

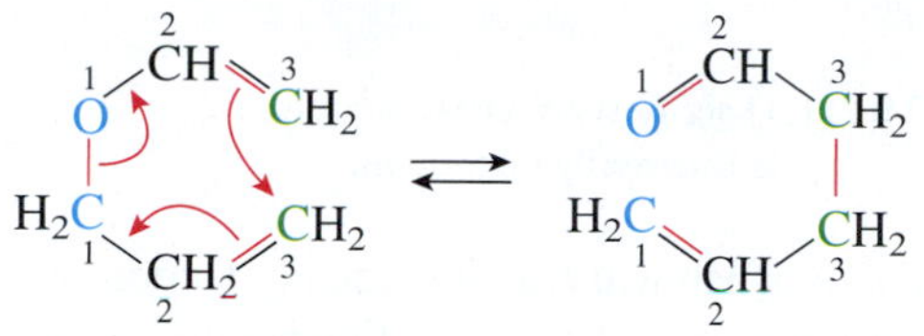

A [3,3] sigmatropic rearrangement

PROBLEM 22.17

Assign the values for *i* and *j* for these [*i,j*] sigmatropic rearrangements:

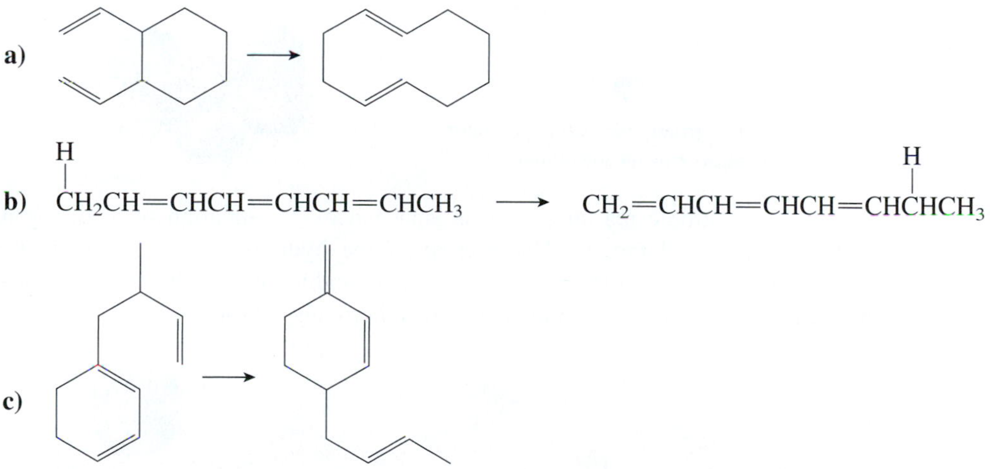

a)

b) $\overset{\text{H}}{\underset{|}{\text{CH}_2}}\text{CH}=\text{CHCH}=\text{CHCH}=\text{CHCH}_3 \longrightarrow \text{CH}_2=\text{CHCH}=\text{CHCH}=\text{CHCH}=\overset{\text{H}}{\underset{|}{\text{CHCHCH}_3}}$

c)

To determine whether a sigmatropic rearrangement is allowed or forbidden, the compound is first imagined to form two radicals by homolytic cleavage of the sigma bond that is broken in the reaction. (Although this analysis pictures the formation of radical intermediates, remember that the reaction is actually concerted.) The interaction of the orbitals of the two radical fragments at the migration origin and the migration terminus are then examined. If the overlap of the orbitals of the HOMOs is bonding at both the migration origin and terminus, the reaction is allowed. If one of these overlaps is antibonding, the reaction is forbidden. The radical fragments that are analyzed usually have an odd number of orbitals. In such a case the HOMOs for a thermal reaction are the nonbonding MOs.

Let's begin by considering the [1,3] sigmatropic rearrangement. The two fragments are a hydrogen atom and an allyl radical. The overlap of the hydrogen $1s$ orbital with the nonbonding MO (π_2^{nb}) of the allyl radical is bonding at one end and antibonding at the other end, so the reaction is thermally forbidden.

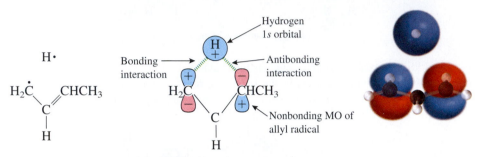

**The [1,3] sigmatropic rearrangement
is thermally forbidden.**

For analysis of the photochemical reaction, the interaction of the hydrogen $1s$ orbital with π_3^* of the allyl system is used. The interaction is bonding at both the migration origin and terminus, so the [1,3] sigmatropic rearrangement is photochemically allowed.

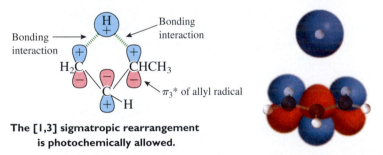

**The [1,3] sigmatropic rearrangement
is photochemically allowed.**

The [1,5] sigmatropic rearrangement is analyzed by examination of a hydrogen atom and a pentadienyl radical. The overlap of the hydrogen $1s$ orbital with the nonbonding MO of the five-orbital system (π_3^{nb}) is bonding at both the migration origin and the migration terminus, so the reaction is thermally allowed.

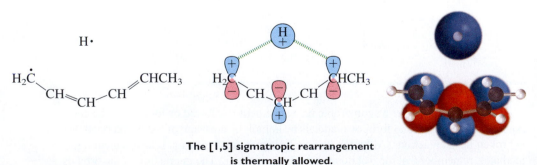

**The [1,5] sigmatropic rearrangement
is thermally allowed.**

A compound rearranging by a [3,3] sigmatropic pathway is split into two allyl radicals. Interaction of the nonbonding MOs of these allyl radicals is bonding at both ends so the reaction is thermally allowed.

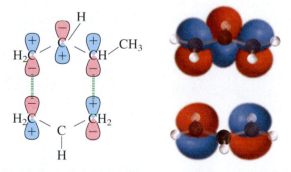

**The [3,3] sigmatropic rearrangement
is thermally allowed.**

The interaction of the nonbonding MO of one allyl radical with π_3^* of the other must have one antibonding interaction, so the photochemical [3,3] sigmatropic rearrangement is forbidden.

**The [3,3] sigmatropic rearrangement
is photochemically forbidden.**

PROBLEM 22.18
Use orbital drawings to show that a [1,5] sigmatropic rearrangement is photochemically forbidden.

PROBLEM 22.19
Use orbital drawings to show that a [1,7] sigmatropic rearrangement is thermally forbidden.

The [1,3] sigmatropic rearrangement involves four electrons (two electron pairs) and is photochemically allowed. Both the [1,5] and [3,3] sigmatropic rearrangements involve six electrons (three electron pairs) and are thermally allowed. (Note that the number of electrons involved in an $[i,j]$ sigmatropic rearrangement equals $i + j$.) These results can be generalized in the form of the chart below that can be used to predict whether a particular sigmatropic rearrangement is allowed or forbidden on the basis of whether the number of electron pairs involved in the rearrangement is even or odd.

Number of Electron Pairs	Allowed Sigmatropic Rearrangement
Odd	Thermal
Even	Photochemical

PROBLEM 22.20
Explain whether the sigmatropic rearrangements of problem 22.17 are allowed thermally or photochemically.

22.9   EXAMPLES OF SIGMATROPIC REARRANGEMENTS

Because they involve two electron pairs, [1,3] sigmatropic rearrangements are photochemically allowed and thermally forbidden. An example that occurs upon irradiation is shown in the following equation:

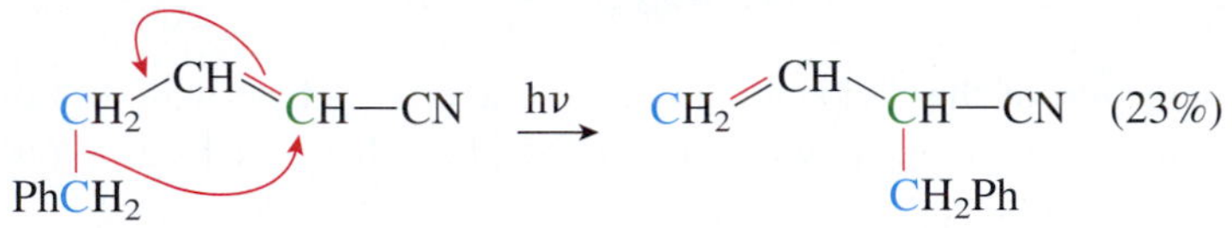

The forbidden nature of the thermal [1,3] sigmatropic rearrangement is illustrated by the following example:

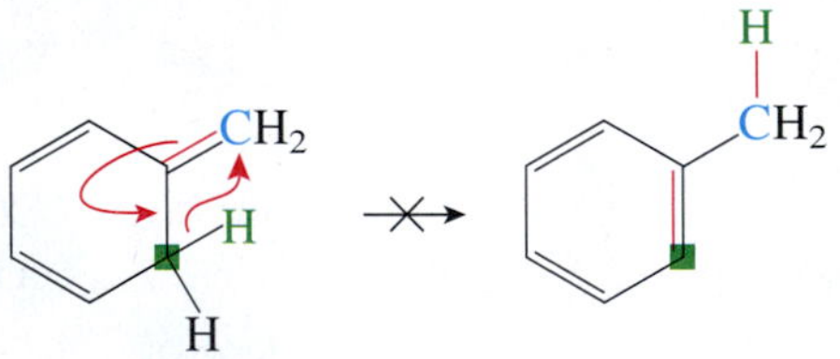

Although the rearrangement of the triene to toluene is highly exothermic, it does not occur at room temperature because the necessary migration is forbidden.

Thermal [1,5] sigmatropic rearrangements are allowed and are quite common. This process is responsible for the interconversion of the methylcyclopentadiene isomers, which proceeds with a half-life of about 1 h at 20°C.

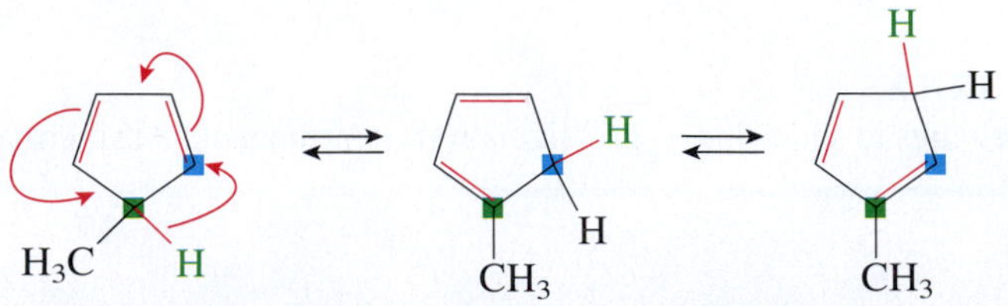

[3,3] Sigmatropic rearrangements are also thermally allowed and are quite important in synthesis. When the six atoms involved are all carbons, the reaction is known as the **Cope rearrangement.** An example is shown in the following equation:

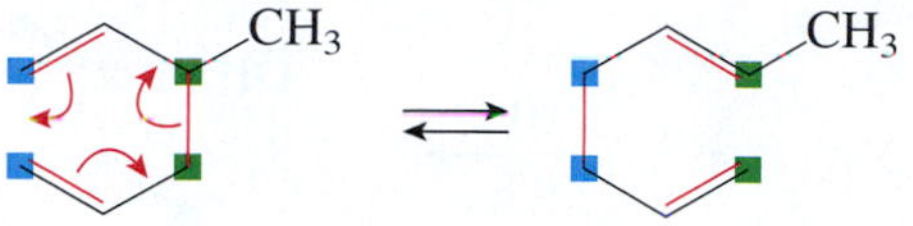

15%   85%

As is the case with all of these reactions, the process is allowed in both directions, so a mixture of the reactant and product results if the equilibrium constant does not dramatically favor one over the other. Many Cope rearrangements produce such a mixture because the reactant and the product are comparable in stability.

However, factors such as ring strain or conjugation can make one product predominate at equilibrium. For example, the cyclopropane derivative shown in the following equation is stable at −20°C but rearranges completely to 1,4-cycloheptadiene at room temperature. The equilibrium is driven toward the product by relief of ring strain.

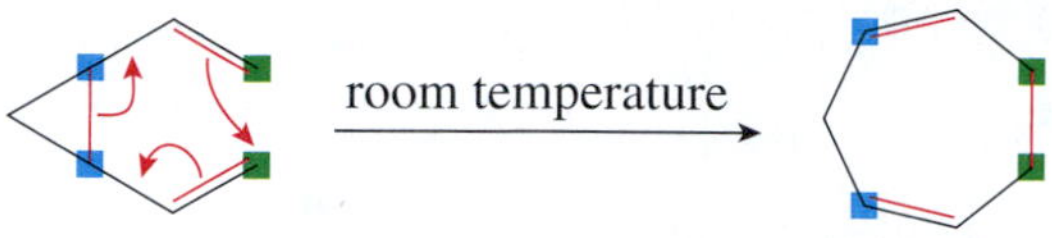

room temperature

1,4-Cycloheptadiene

The following two equations provide examples in which the product of the Cope rearrangement is favored because it is stabilized by conjugation:

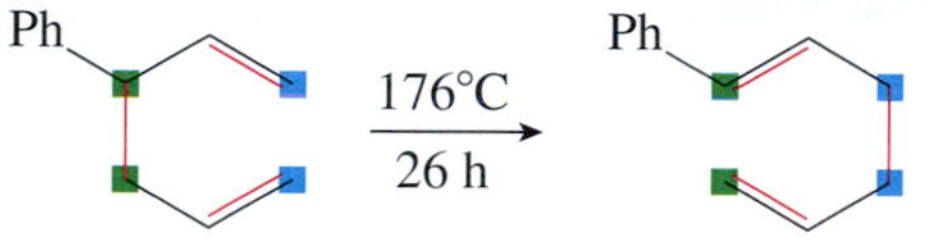

$\dfrac{176°C}{26 \text{ h}}$ (72%)

$\dfrac{175°C}{1.5 \text{ h}}$ (96%)

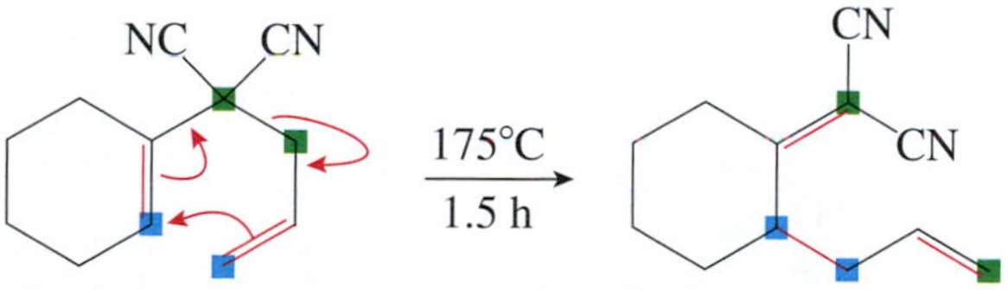

The oxygen analog of the Cope rearrangement is called the **Claisen rearrangement.** Often, one of the pi bonds is part of an aromatic ring, as shown in the following example:

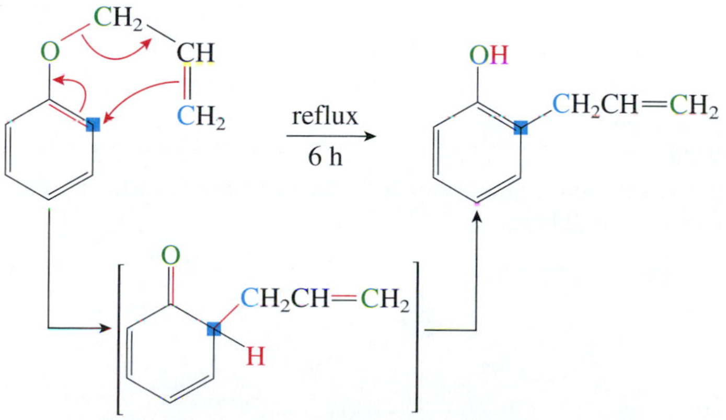

$\dfrac{\text{reflux}}{6 \text{ h}}$ (77%)

In these cases the initially formed product spontaneously tautomerizes to regain its aromatic stabilization. Another example is provided in the following equation:

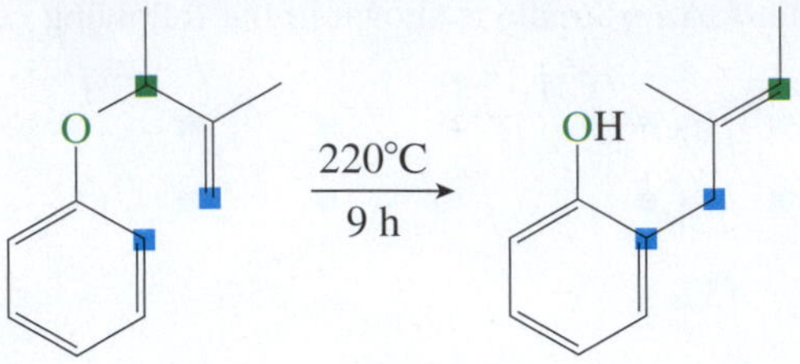

$$\xrightarrow[\text{9 h}]{220°C}$$ (94%)

Claisen rearrangements also occur with nonaromatic substrates. In these reactions the formation of the strong carbon–oxygen double bond drives the equilibrium to completion:

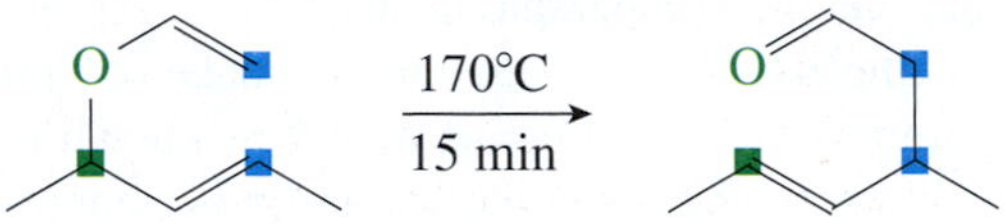

$$\xrightarrow[\text{15 min}]{170°C}$$ (90%)

$$\xrightarrow[\text{15 min}]{190°C}$$ (93%)

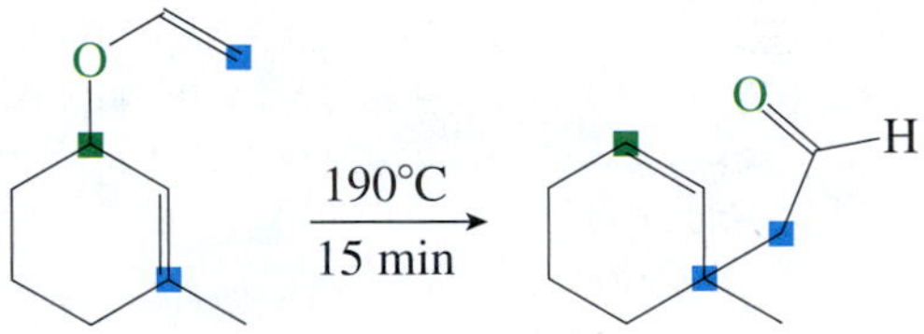

ORGANIC
Chemistry⦿Now™
Click Coached Tutorial Problems
for more practice with
Sigmatropic
Rearrangements.

PROBLEM 22.21
Show the products of these reactions:

a) 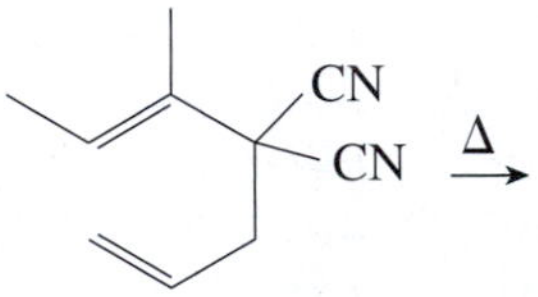 $\xrightarrow{\Delta}$

b) 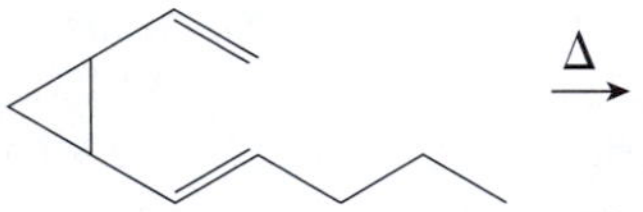 $\xrightarrow{\Delta}$

c) 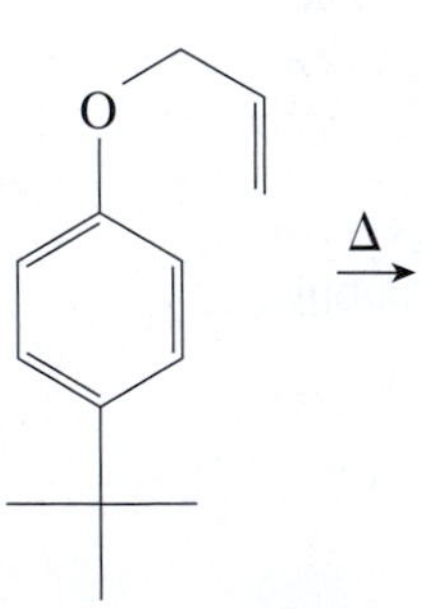 $\xrightarrow{\Delta}$

d) 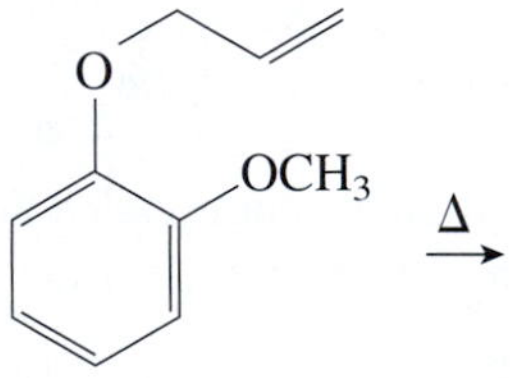 $\xrightarrow{\Delta}$

e) 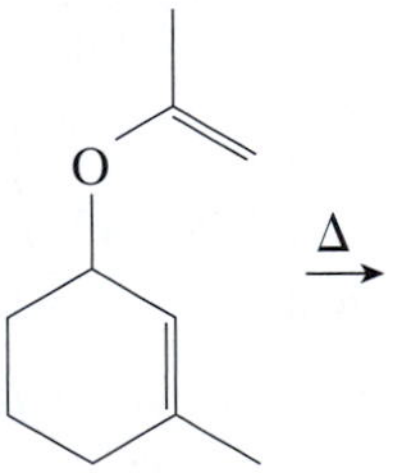 $\xrightarrow{\Delta}$

PROBLEM 22.22
Explain whether this reaction is allowed or forbidden and predict whether the reactant or product is favored at equilibrium.

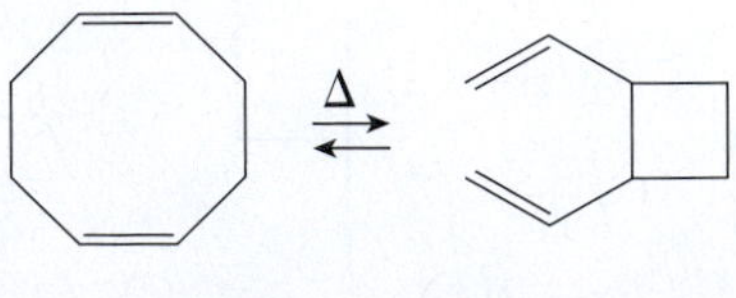

Focus On Biological Chemistry

Pericyclic Reactions and Vitamin D

Vitamin D is necessary for the proper deposition of calcium in growing bones. A deficiency of this vitamin leads to the disease known as rickets. You may have heard that sunlight contains vitamin D. Of course, there are no chemical compounds, or vitamins, in sunlight, but animals require sunlight to make vitamin D.

In the skin of animals, 7-dehydrocholesterol is converted to vitamin D_3 by the reaction sequence that follows. The first step in this process, the conversion of 7-dehydrocholesterol to pre-cholecalciferol, requires light. This is an electrocyclic reaction and must occur by a conrotatory motion to avoid the formation of a highly strained trans double bond in one of the rings. Conrotation involving three pairs of electrons must occur photochemically to be allowed.

❶ In the skin of animals, 7-dehydrocholesterol is converted to pre-cholecalciferol by the action of sunlight. This electrocyclic reaction must occur in a conrotatory fashion to avoid forming a trans double bond in one of the rings. Because three electron pairs are involved, this reaction is photochemically allowed.

7-Dehydrocholesterol

Pre-cholecalciferol

❷ The second step is a [1,7] sigmatropic rearrangement and has been shown to occur thermally. Because it involves four electron pairs, to be thermally allowed it must occur with an unusual geometry, where the hydrogen migrates from the top of the pi system on one end to the bottom of the pi system on the other.

Cholecalciferol
(vitamin D_3)

The second step of this process, the conversion of pre-cholecalciferol to cholecalciferol or vitamin D_3, is also a pericyclic reaction. It is a [1,7] sigmatropic rearrangement that has been demonstrated to occur thermally. To be thermally allowed, a sigmatropic rearrangement involving four pairs of electrons must occur with an unusual geometry where the hydrogen migrates from the top of the pi system on one end

Continued

to the bottom of the pi system on the other, as shown in the following diagram. The required geometry is readily attainable for pre-cholecalciferol.

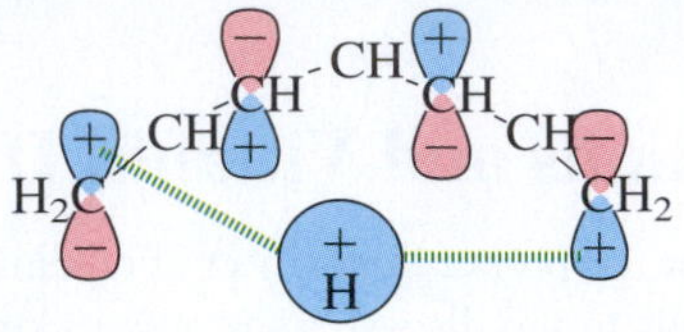

H. Steenbock, at the Wisconsin Agricultural Research Station, discovered that irradiation of rat food was able to cure rickets in rats. He patented the process of using ultraviolet irradiation to enrich the compound that helps prevent rickets in a variety of foods. (During 1925–1945, this patent generated about 14 million dollars, which was used to support research in Wisconsin.) Eventually, it was determined that the vitamin (vitamin D_2) that is present in these foods was produced from ergosterol, a steroid that is very similar to 7-dehydrocholesterol, differing only in the structure of its side chain. On irradiation, ergosterol is converted to calciferol or vitamin D_2 by the same mechanism as 7-dehydrocholesterol. This is the "vitamin D" that is added to milk.

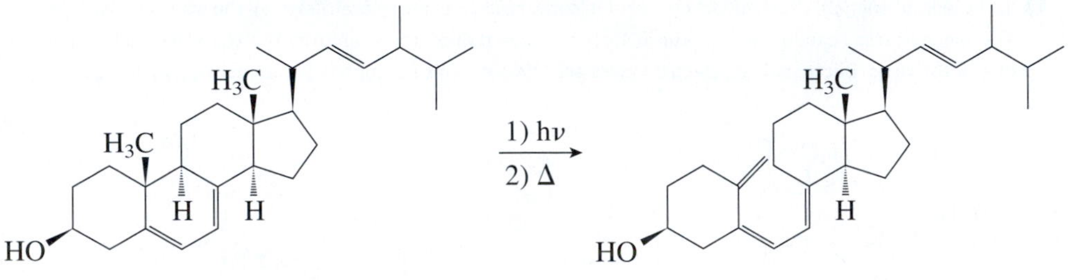

Ergosterol
    Calciferol
    (vitamin D_2)

22.10 REARRANGEMENTS TO ELECTRON-DEFICIENT CENTERS

The rearrangements of carbocations that were first encountered in Chapter 8 can be classified as [1,2] sigmatropic rearrangements:

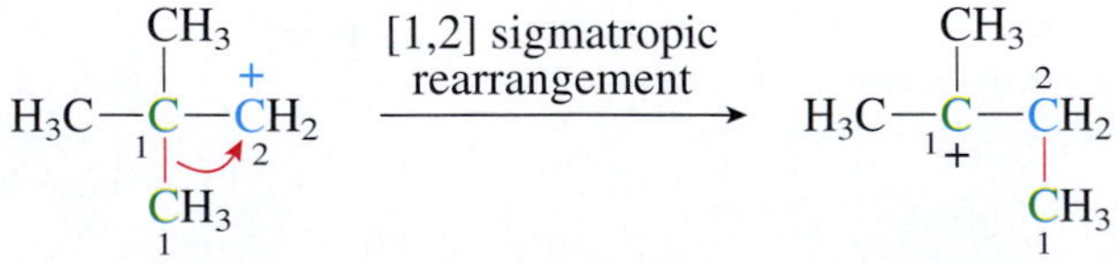

These rearrangements involve one pair of electrons; therefore, they are thermally allowed. We have already seen that they are very common. In contrast, the corresponding rearrangement of carbanions involves two electron pairs and is thermally forbidden. This is why carbanion rearrangements are rare.

The **pinacol rearrangement** is a useful reaction that proceeds via a carbocation rearrangement. Treatment of 2,3-dimethyl-2,3-butanediol, also known as pinacol, with acid results in the formation of a ketone, pinacolone:

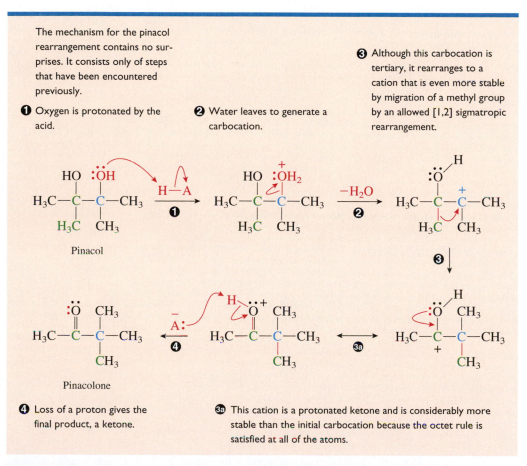

The mechanism for this reaction, shown in Figure 22.5, involves a carbocation rearrangement that occurs by an allowed [1,2] sigmatropic shift. The product of this rearrangement, a protonated ketone, is considerably more stable than the initial carbocation, so the migration is quite favorable. Another example of the pinacol rearrangement is provided in the following equation:

The mechanism for the pinacol rearrangement contains no surprises. It consists only of steps that have been encountered previously.

❸ Although this carbocation is tertiary, it rearranges to a cation that is even more stable by migration of a methyl group by an allowed [1,2] sigmatropic rearrangement.

❶ Oxygen is protonated by the acid.

❷ Water leaves to generate a carbocation.

❹ Loss of a proton gives the final product, a ketone.

❸ₐ This cation is a protonated ketone and is considerably more stable than the initial carbocation because the octet rule is satisfied at all of the atoms.

Figure 22.5

MECHANISM OF THE PINACOL REARRANGEMENT.

Figure 22.6

MECHANISM OF THE
BECKMANN
REARRANGEMENT.

❷ As water leaves, positive charge builds up on the nitrogen. If this continues, a nitrogen with six electrons and a positive charge—the nitrogen analog of a carbocation—would be generated. Because nitrogen is more electronegative than carbon, such a species is much less stable than a carbocation. To avoid forming this highly energetic ion, the phenyl group migrates to the nitrogen as the water leaves.

❸ The product of the rearrangement has the positive charge on the carbon and is stabilized by resonance delocalization of the positive charge onto the nitrogen. The lower resonance structure is the major contributor to the resonance hybrid because it has the octet rule satisfied at both the carbon and the nitrogen. It is quite different from the positive nitrogen with six electrons that would be generated if the rearrangement did not occur.

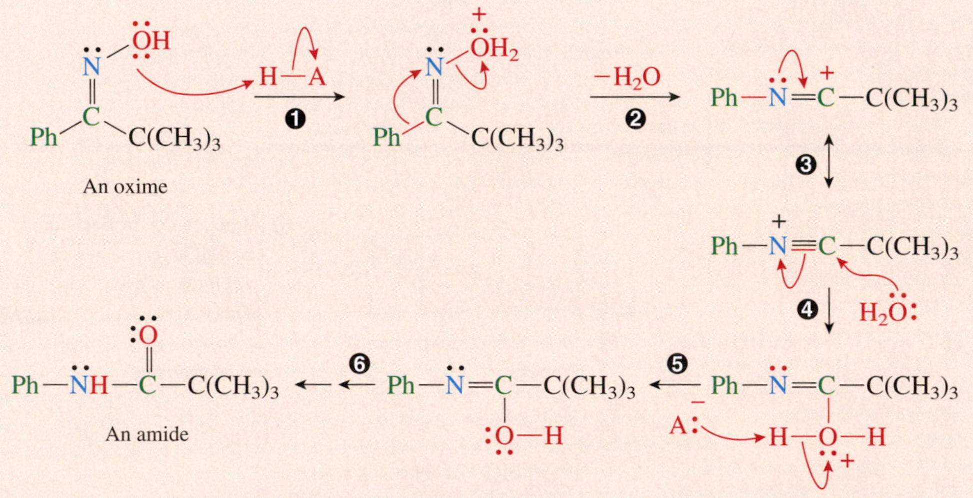

❶ The reaction begins by protonation of the oxygen of the oxime.

❻ Under the reaction conditions, rapid tautomerization gives the final product, an amide (see Chapter 19).

❺ A proton is removed by a base in the reaction mixture. The compound that results is a tautomer of an amide.

❹ Attack by water at the electrophilic carbon produces this ion.

Similar [1,2] migrations involving other electron-deficient atoms are also known. The **Beckmann rearrangement** is a reaction of an oxime to produce an amide:

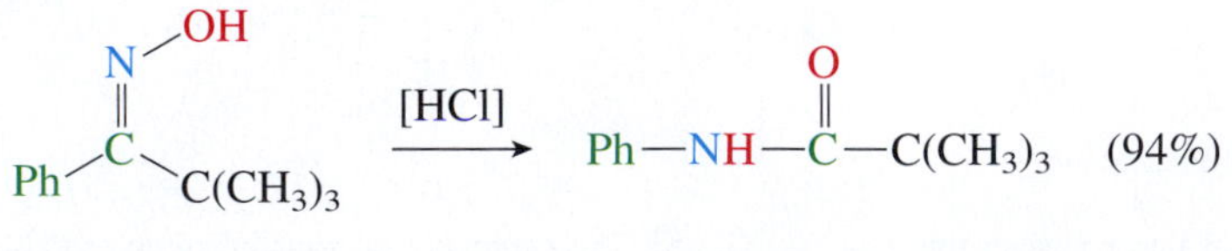

The mechanism for this reaction, shown in Figure 22.6, proceeds via an allowed [1,2] sigmatropic rearrangement to an electron-deficient nitrogen. Another example of the Beckmann rearrangement is provided in the following equation:

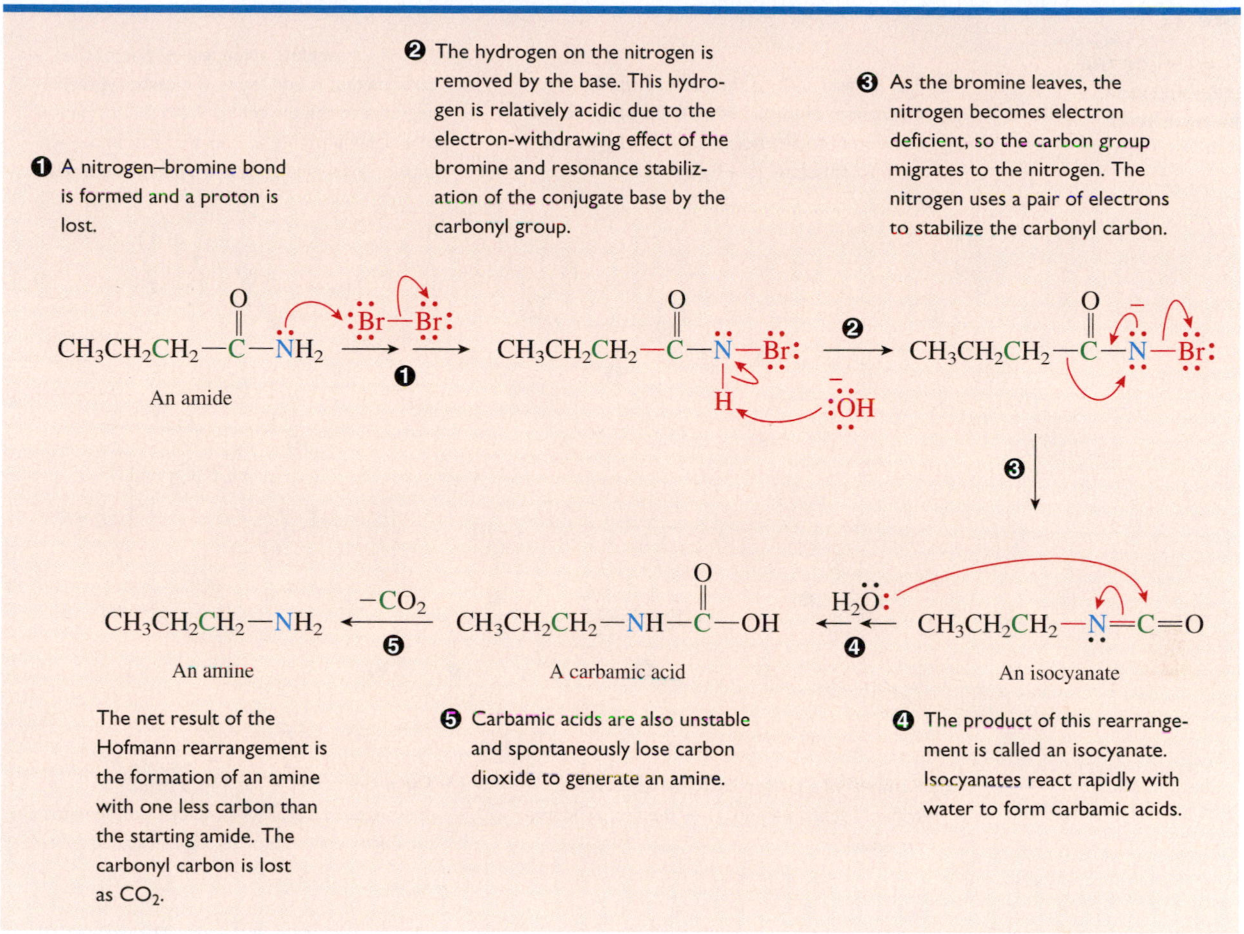
❶ A nitrogen–bromine bond is formed and a proton is lost.
❷ The hydrogen on the nitrogen is removed by the base. This hydrogen is relatively acidic due to the electron-withdrawing effect of the bromine and resonance stabilization of the conjugate base by the carbonyl group.
❸ As the bromine leaves, the nitrogen becomes electron deficient, so the carbon group migrates to the nitrogen. The nitrogen uses a pair of electrons to stabilize the carbonyl carbon.
An amide
A carbamic acid
An amine
An isocyanate
−CO2
The net result of the Hofmann rearrangement is the formation of an amine with one less carbon than the starting amide. The carbonyl carbon is lost as CO2.
❺ Carbamic acids are also unstable and spontaneously lose carbon dioxide to generate an amine.
❹ The product of this rearrangement is called an isocyanate. Isocyanates react rapidly with water to form carbamic acids.

Figure 22.7

MECHANISM OF THE HOFMANN REARRANGEMENT.

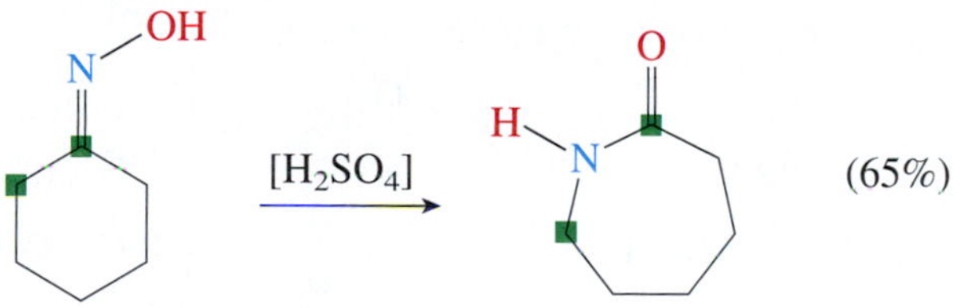
[H2SO4]
(65%)

The **Hofmann rearrangement** also involves a migration to an electron-deficient nitrogen. In this case, an amide is treated with Cl_2 or Br_2 in aqueous base, resulting in the formation of an amine with one less carbon. The original carbonyl carbon is lost as carbon dioxide. The mechanism is shown in Figure 22.7 and an example is provided by the following equation:

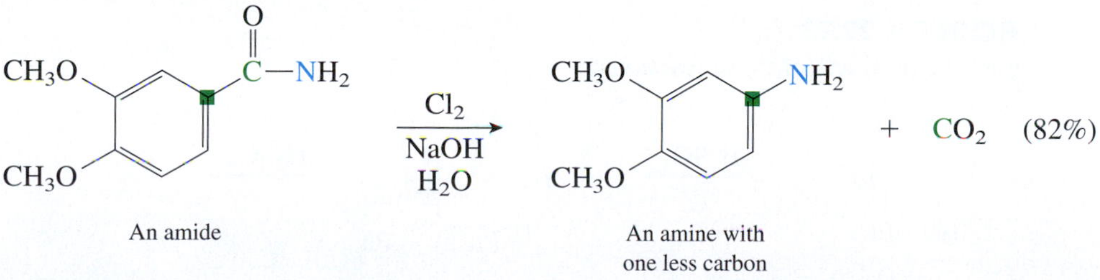
Cl2
NaOH
H2O
+ CO2 (82%)
An amide
An amine with one less carbon

Figure 22.8

MECHANISM OF THE BAEYER-VILLIGER REARRANGEMENT.

❶ The peracid adds to the carbonyl group. The mechanism of this addition is analogous to the addition of an alcohol to a carbonyl group to form a hemiacetal (see Figure 18.5, p. 776).

❷ As trifluoroacetate anion leaves, one of the carbons that is bonded to the carbonyl carbon migrates to the oxygen to avoid the formation of a highly unstable species that has an oxygen with only six electrons and a positive charge.

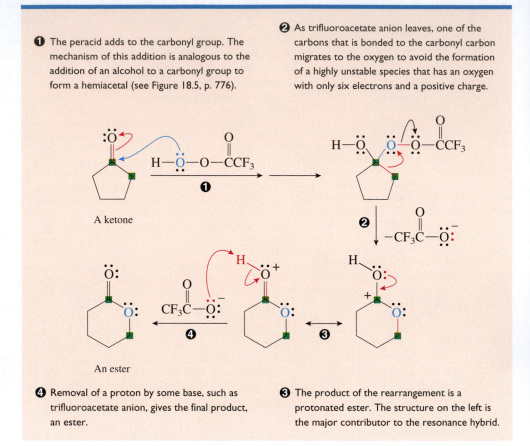

❹ Removal of a proton by some base, such as trifluoroacetate anion, gives the final product, an ester.

❸ The product of the rearrangement is a protonated ester. The structure on the left is the major contributor to the resonance hybrid.

Similar migrations to electron-deficient oxygen also occur. In the **Baeyer-Villiger rearrangement** a ketone is reacted with a peracid. The product is an ester in which one of the alkyl groups of the ketone has migrated to oxygen:

A ketone   A peracid   An ester   (81%)

The mechanism for this reaction, outlined in Figure 22.8, is similar to that for the Beckmann rearrangement. As a group leaves from the oxygen, a carbon group migrates to this electron-deficient oxygen by an allowed [1,2] sigmatropic rearrangement.

PROBLEM 22.23
Show the products of these reactions:

a)
$$\underset{\underset{Et\quad Et}{|\quad\ |}}{\overset{\overset{OH\ OH}{|\quad\ |}}{EtC-CEt}} \xrightarrow{H_2SO_4}$$

b)
$$\underset{Ph}{\overset{N-OH}{\underset{|}{\overset{\|}{C}}}}\underset{Ph}{} \xrightarrow{H_2SO_4}$$

c)

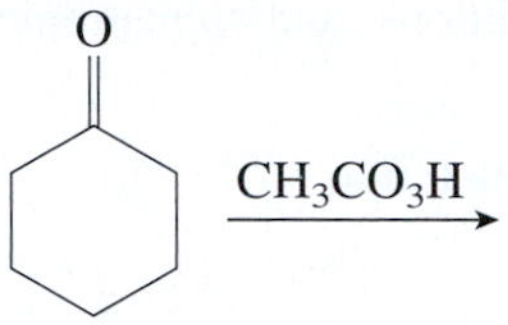

CH_3CO_3H

d)

Br_2
$NaOH$
H_2O

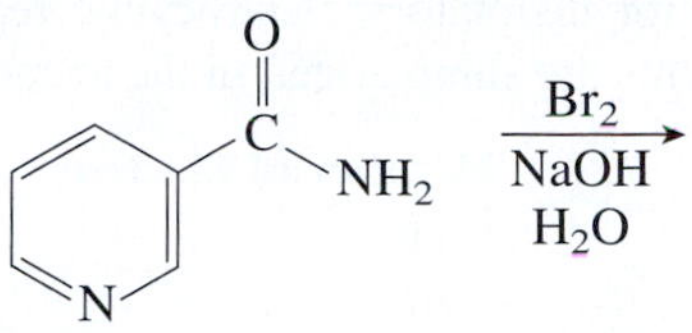

PROBLEM 22.24

The following reaction is known as the Curtius rearrangement. Its mechanism is similar to the Hofmann rearrangement. Show the mechanism of this reaction.

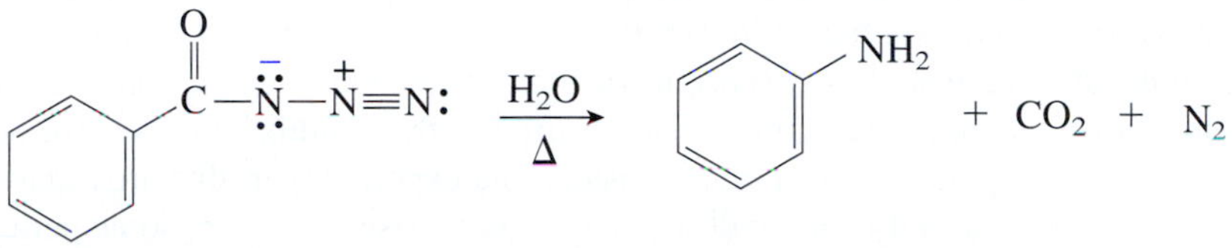

$+ CO_2 + N_2$

Review of Mastery Goals

After completing this chapter, you should be able to:

- Show the energies and nodal properties of the pi MOs of a small conjugated system, whether it is composed of an even or odd number of orbitals. (Problem 22.25)

- Classify reactions as electrocyclic reactions, $[x + y]$ cycloadditions, or $[i,j]$ sigmatropic rearrangements. (Problem 22.30)

- Use the pi MOs to explain whether these reactions are allowed or forbidden. (Problem 22.29)

- Show the products, including stereochemistry, of any of these reactions. (Problems 22.26 and 22.27)

- Show the mechanisms of the pinacol, Beckmann, Hofmann, and Baeyer-Villiger rearrangements. (Problems 22.36, 22.37, 22.38, and 22.39)

- Use the principles of pericyclic reactions to explain the mechanism, selectivity, and stereochemistry of a concerted reaction. (Problems 22.35, 22.40, 22.41, 22.42, 22.43, 22.44, 22.45, 22.46, 22.47, 22.48, and 22.49)

- Use the reactions in syntheses. (Problems 22.28, 22.31, 22.32, and 22.34)

Visual Summary of Key Reactions

The concerted reactions presented in this chapter are called pericyclic reactions. They are divided into electrocyclic reactions, cycloaddition reactions, and sigmatropic rearrangements. Some occur when energy is supplied in the form of heat; others require light energy to occur. Most have strict stereochemical requirements.

Whether a particular pericyclic reaction is allowed or not can be determined by examination of the interaction of the molecular orbitals where the new bonds are forming.

The rules for disrotatory electrocyclic reactions, cycloadditions, and sigmatropic rearrangements are summarized in the accompanying chart:

Number of Electron Pairs	Allowed Reaction
Odd	Thermal
Even	Photochemical

The selection rules for conrotatory electrocyclic reactions are the opposite of those just listed; that is, for a molecule with an even number of electron pairs, conrotation is thermally allowed, and for a molecule with an odd number of electron pairs, conrotation is photochemically allowed.

For many compounds there are several possible allowed pericyclic reactions. On the basis of your limited experience in organic chemistry, it is often difficult to decide which of the allowed reactions will occur for such compounds. In addition, these reactions are allowed in both directions, so the position of the equilibrium may also be important. Even an organic chemist with considerable experience in this area may have difficulty predicting exactly what will happen in every case. However, some pericyclic reactions are more common than others. Table 22.1 summarizes those that are encountered most often.

Table 22.1   Summary of Pericyclic Reactions

Reaction	Comments
Electrocyclic Reactions (Section 22.4)	
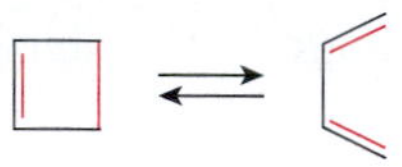	Thermal: conrotation Photochemical: disrotation
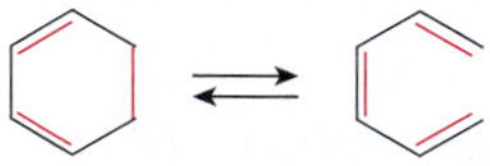	Thermal: disrotation Photochemical: conrotation
	Thermal: conrotation Photochemical: disrotation
Cycloaddition Reactions (Sections 22.7 and 22.6)	
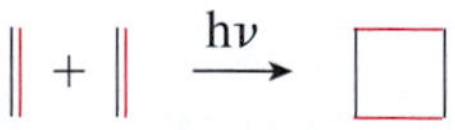	[2 + 2] Cycloaddition, photochemically allowed
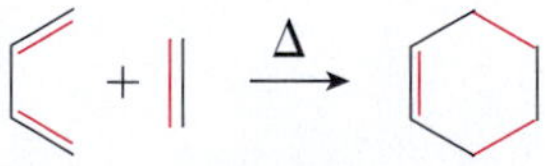	[4 + 2] Cycloaddition, Diels-Alder reaction, thermally allowed
Sigmatropic Rearrangements (Section 22.9)	
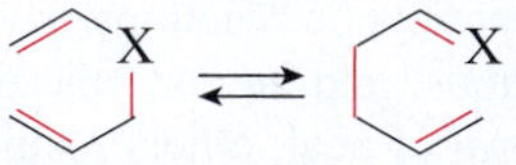	[3,3] Sigmatropic rearrangement, thermally allowed, when X = C, Cope rearrangement, when X = O, Claisen rearrangement

Table 22.1 Summary of Pericyclic Reactions—cont'd

Reaction	Comments
Electron-Deficient Rearrangements (Section 22.10)	
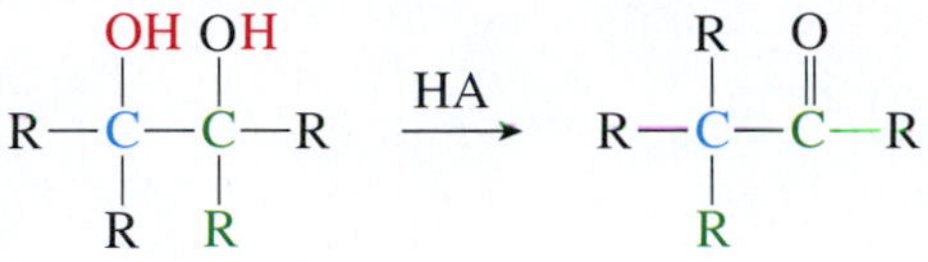	Pinacol rearrangement
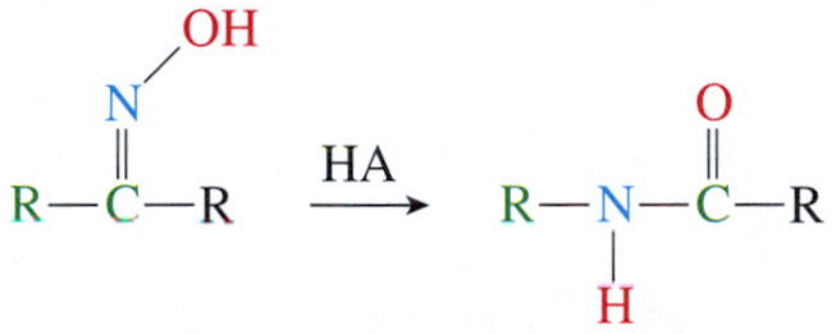	Beckmann rearrangement
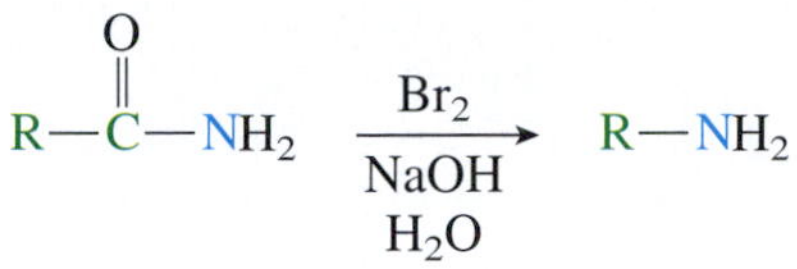	Hofmann rearrangement
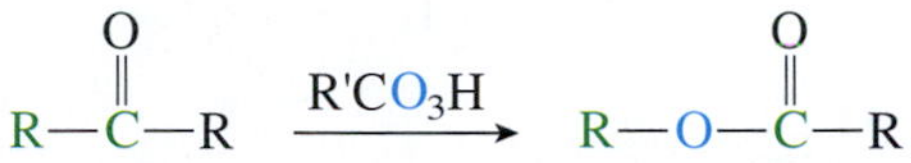	Baeyer-Villiger rearrangement

Additional Problems

ORGANIC Chemistry Now™
Assess your understanding of this chapter's topics with additional quizzing and conceptual-based problems at **http://now.brookscole.com/hornback2**

22.25 Show the energies for the pi MOs of these species. Show the electrons occupying the MOs for each.

a) Lowest-energy excited state of 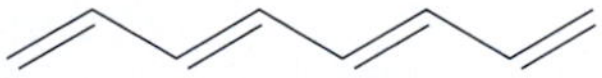

b) Ground state of 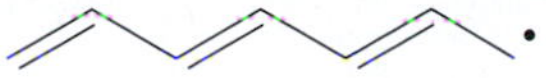

22.26 Show the products of these reactions:

a) 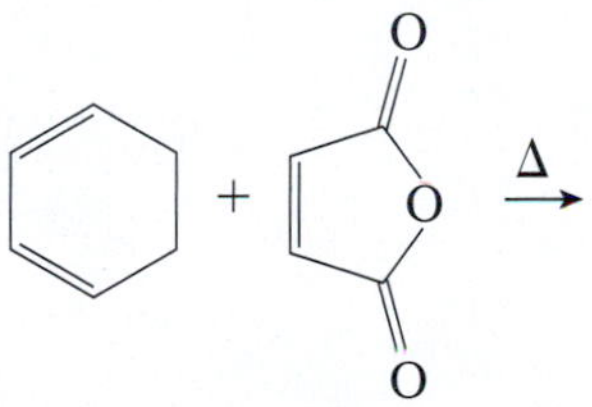

b) 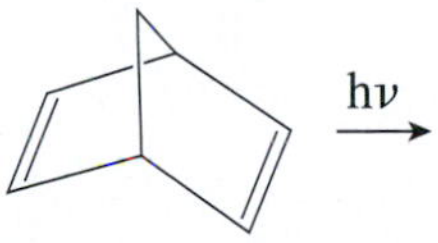

c) 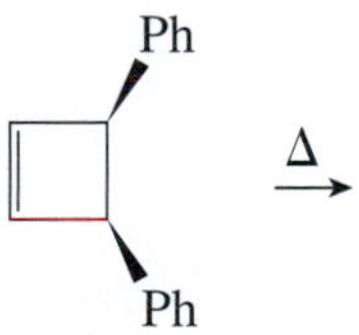

d) 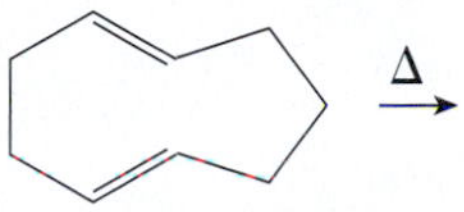

e) 2 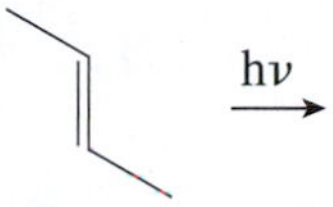

f) 2 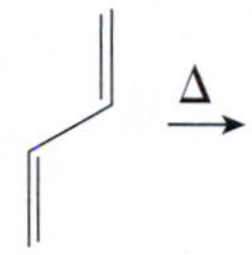

g) 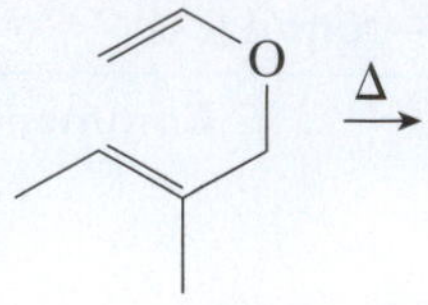   Δ

h) 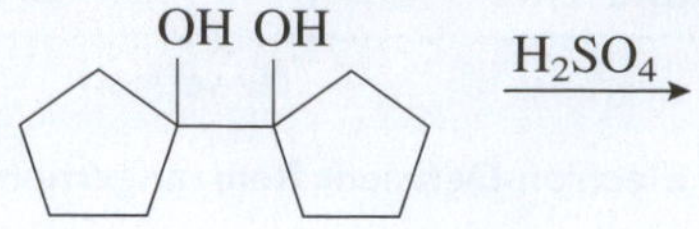   OH OH H$_2$SO$_4$

i)    CF$_3$CO$_3$H

22.27 Show the products of these reactions:

a) 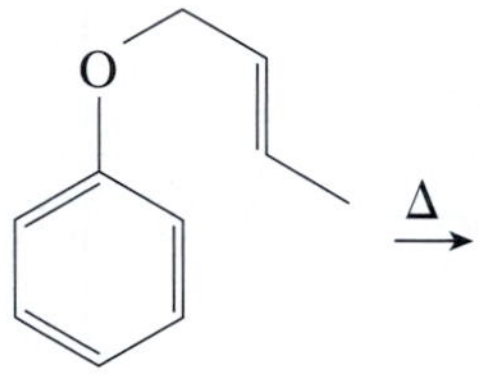   Δ

b) 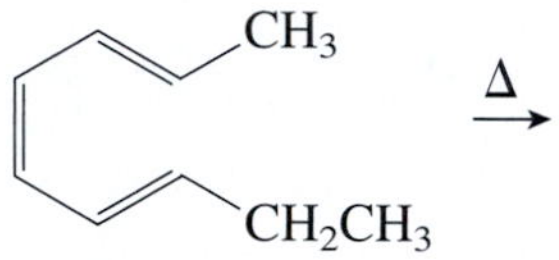   CH$_3$ / CH$_2$CH$_3$ Δ

c) 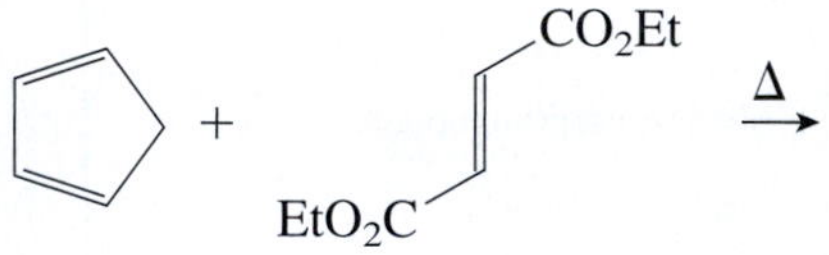   + CO$_2$Et / EtO$_2$C Δ

d) 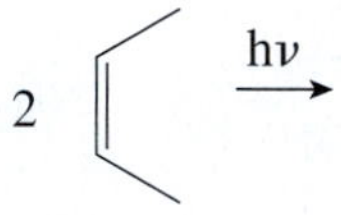 2   hν

e) 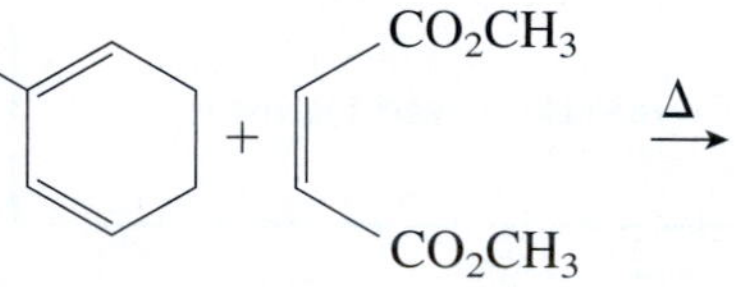   + CO$_2$CH$_3$ / CO$_2$CH$_3$ Δ

f) 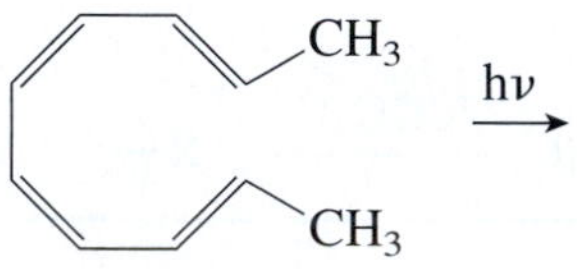   CH$_3$ / CH$_3$ hν

g) 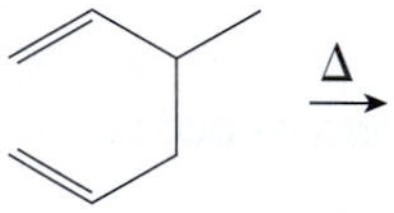   Δ

h) 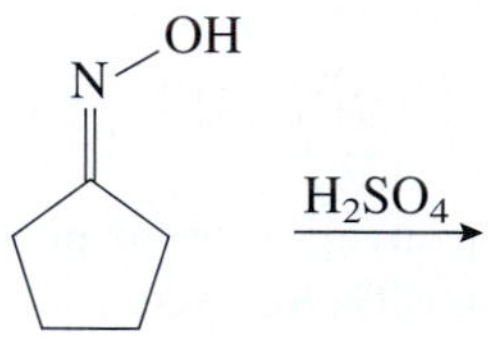   N—OH H$_2$SO$_4$

i) 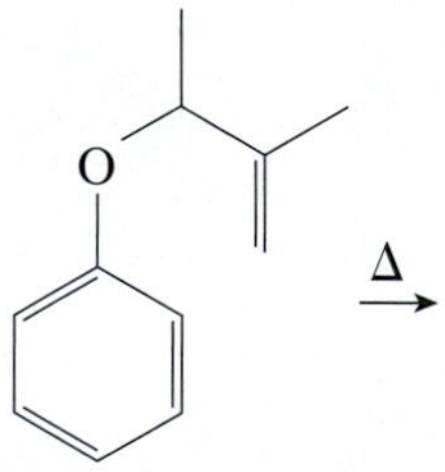   Δ

j) 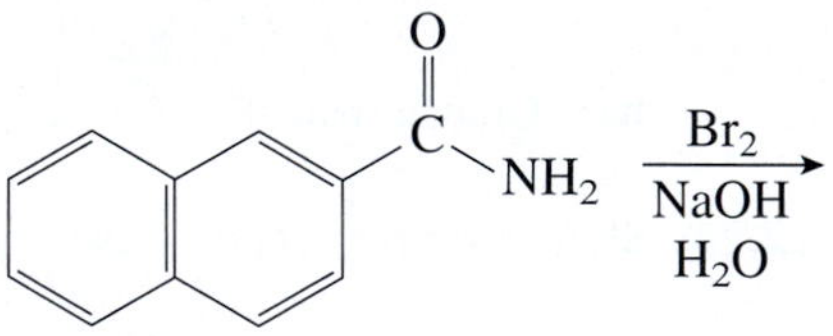   C(=O)—NH$_2$ Br$_2$ / NaOH / H$_2$O

22.28 Show the dienes and dienophiles that could be used to prepare these compounds by Diels-Alder reactions:

a)

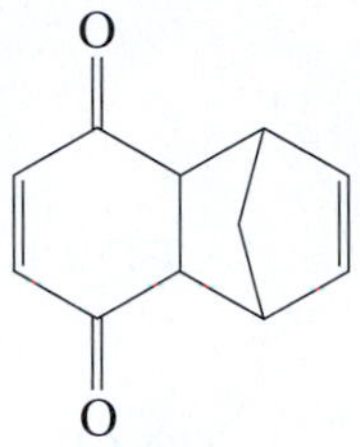

b)

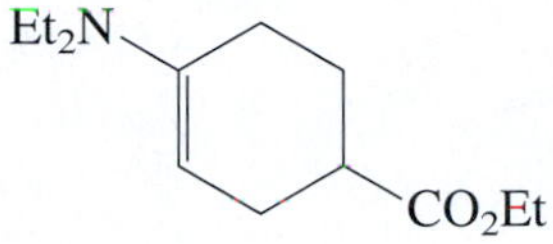

c)

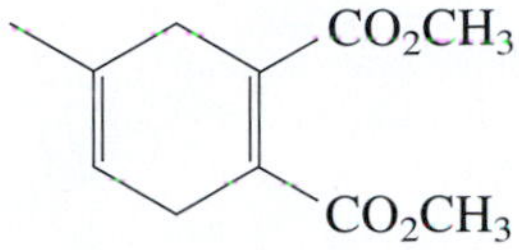

d)

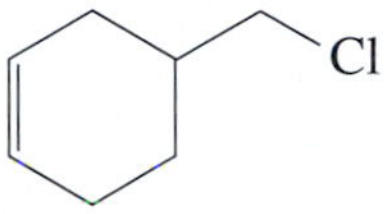

e)

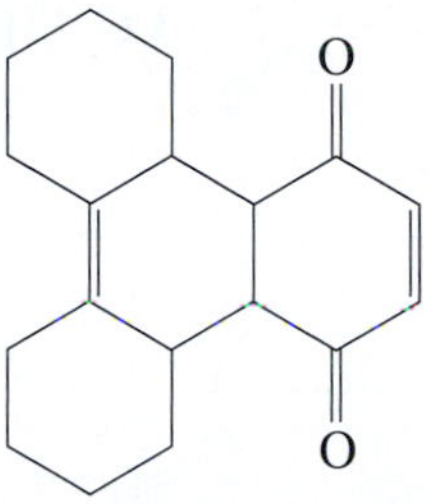

f)

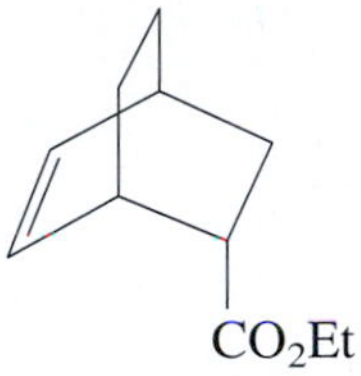

g)

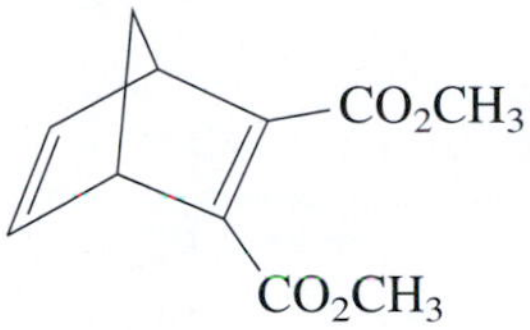

22.29 Use orbital drawings to determine whether these reactions are allowed or forbidden:

a) 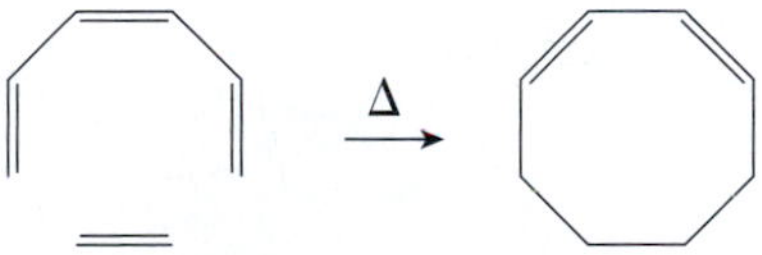 $\xrightarrow{\Delta}$

b) 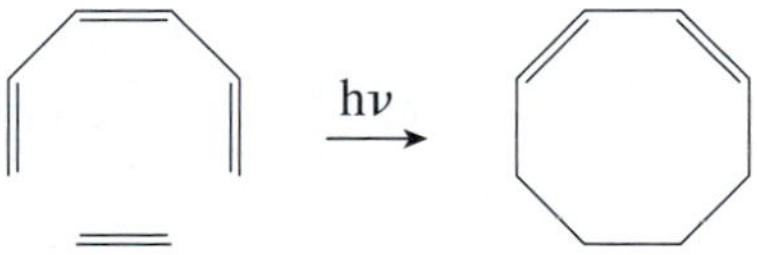 $\xrightarrow{h\nu}$

c) 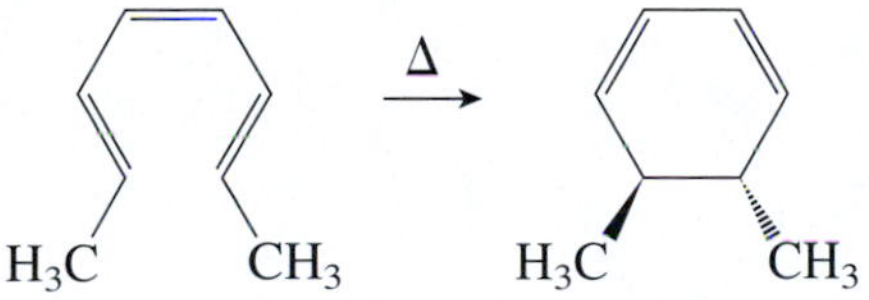 $\xrightarrow{\Delta}$

d) 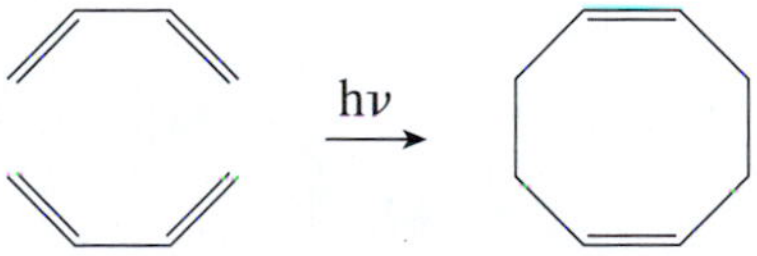 $\xrightarrow{h\nu}$

e) 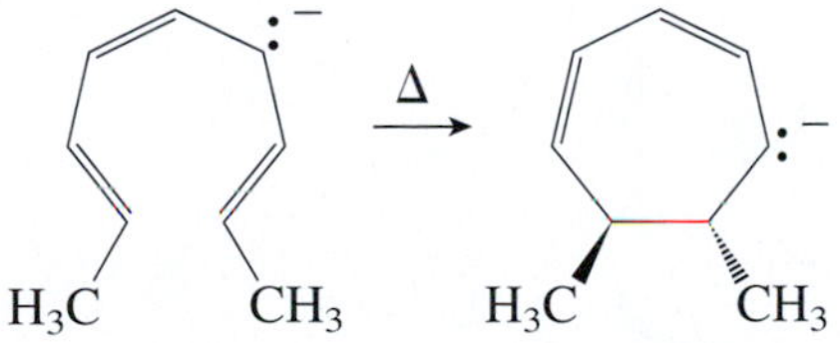 $\xrightarrow{\Delta}$

f) 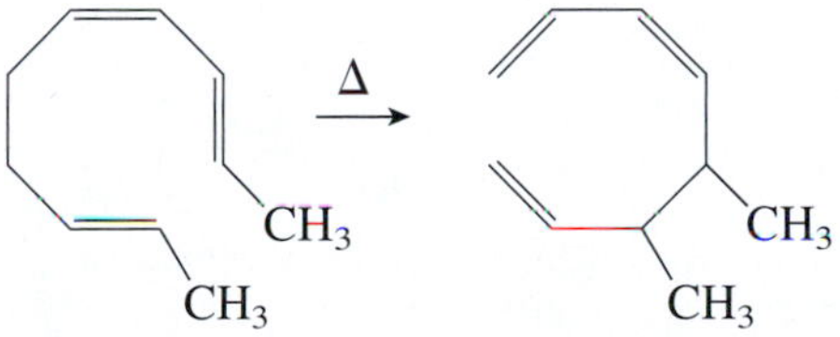 $\xrightarrow{\Delta}$

g) 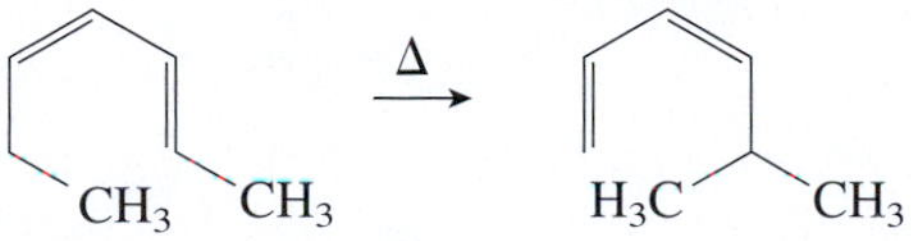 $\xrightarrow{\Delta}$

22.30 Classify these reactions as electrocyclic reactions, $[x + y]$ cycloadditions, or $[i,j]$ sigmatropic rearrangements and explain whether each is allowed thermally or photochemically.

a)

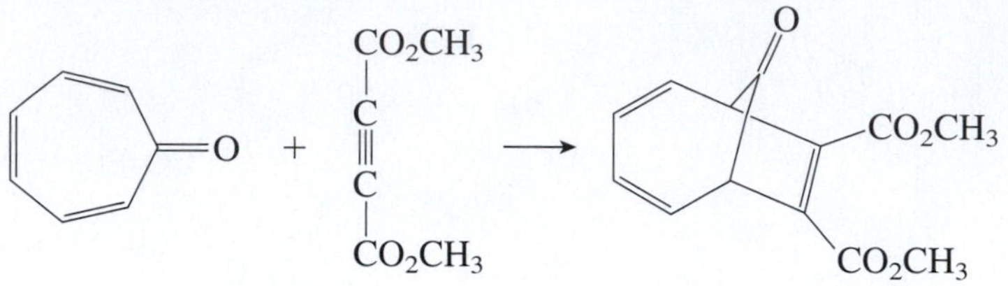

b) 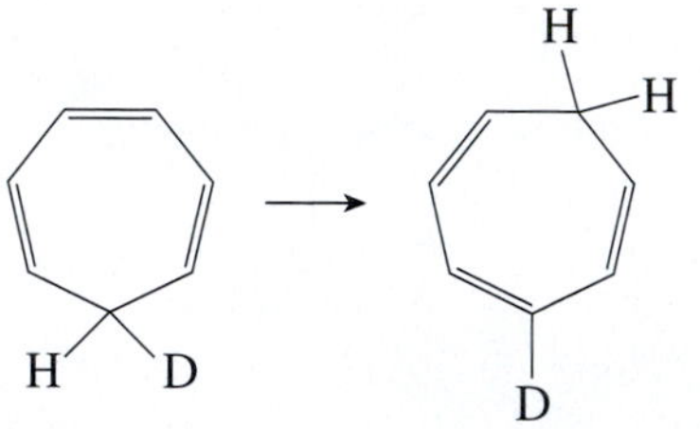   c)

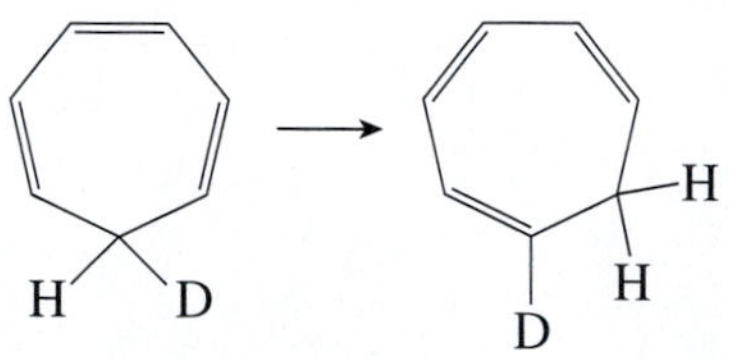

d) 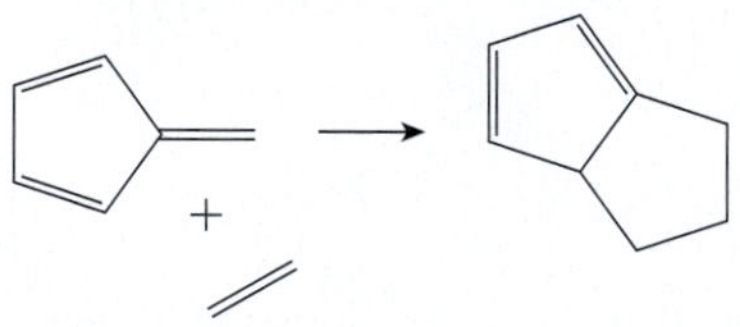   e)

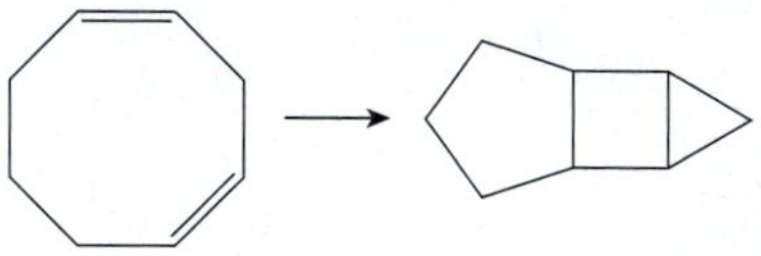

f)

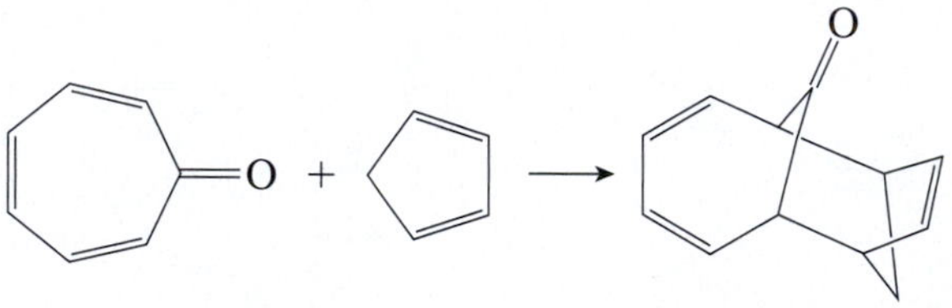

g)

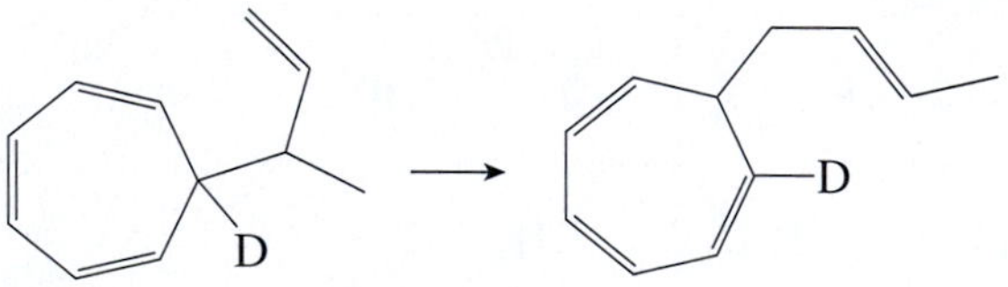

h)

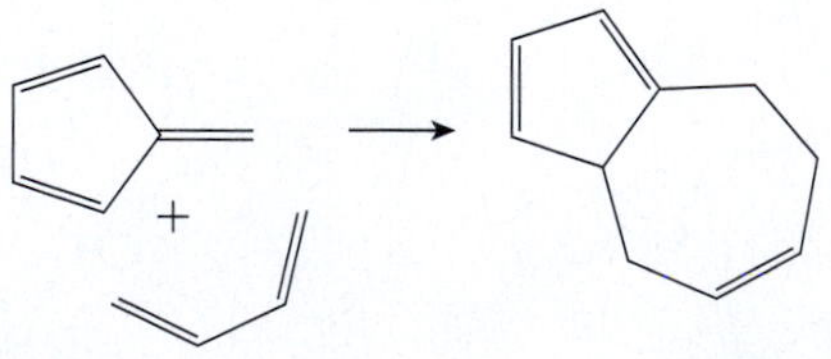

22.31 Show syntheses of these compounds from the indicated starting materials. Reactions from previous chapters may be needed in some cases.

a) 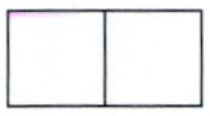 [square structure] from [cyclohexadiene]

b) 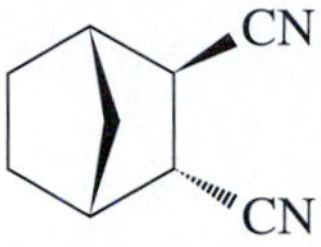 [bicyclic CN structure] from NC—[alkene]—CN

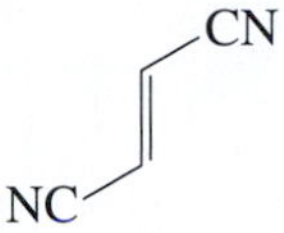

c) 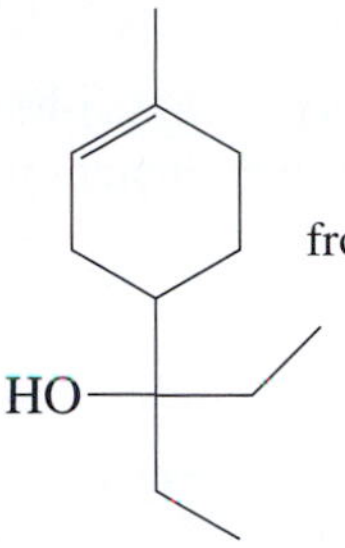 [methylcyclohexene with side chain, HO—] from [acrylate ester]

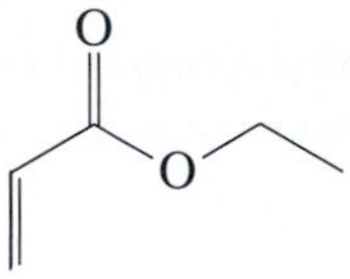

d) 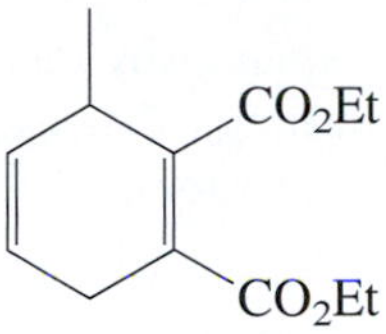 [cyclohexadiene with CO₂Et groups] from compounds without rings

e) 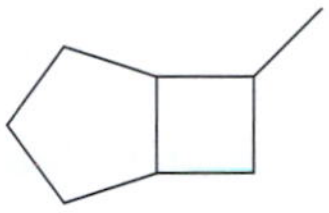 [bicyclic structure] from [cyclopentene]

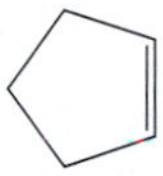

f) 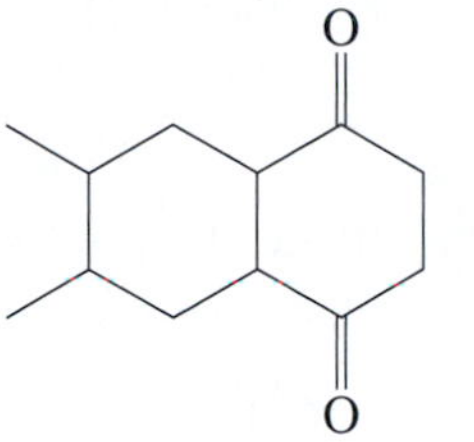 [bicyclic diketone, O at top and bottom] from a compound with one ring

g) 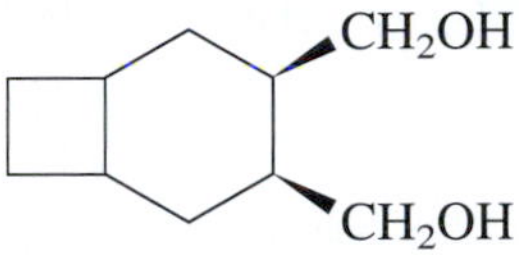 [bicyclic structure with CH₂OH, CH₂OH] from [CO₂CH₃, CO₂CH₃ alkene] and compounds without rings

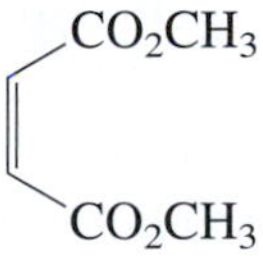

22.32 Suggest a method to convert (2*E*,4*E*)-hexadiene to (2*E*,4*Z*)-hexadiene in a manner so that the stereochemistry is controlled at each step of the process:

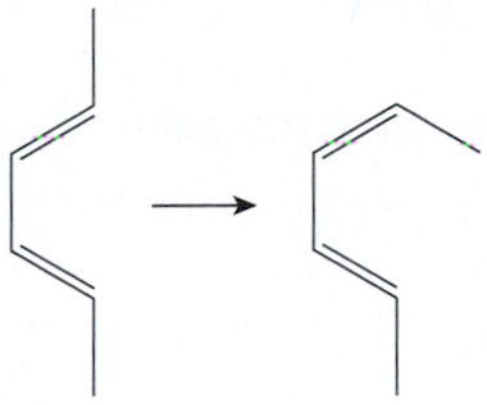

22.33 Explain which compound is more reactive as a diene in the Diels-Alder reaction:

a)  [cyclohexadiene] or [diene]

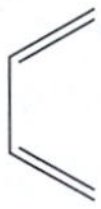

b)  [diene] or [branched diene]

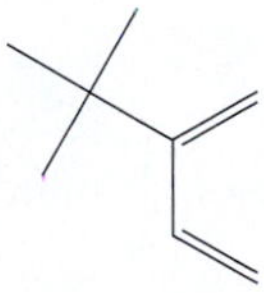

c) 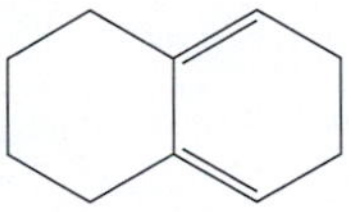 [octahydronaphthalene] or [octahydronaphthalene isomer]

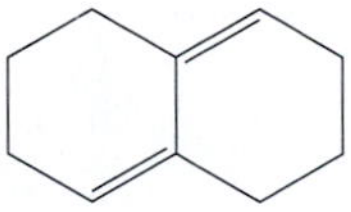

22.34 Suggest a method for the synthesis of *o*-eugenol from *o*-methoxyphenol:

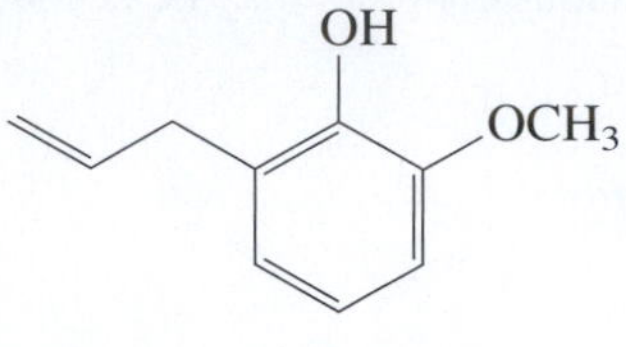

OH

OCH₃

o-Eugenol

22.35 Explain why *trans*-3,4-dimethylcyclobutene produces (2*E*,4*E*)-hexadiene upon heating, whereas (2*Z*,4*Z*)-hexadiene is not formed, even though its formation is an allowed reaction.

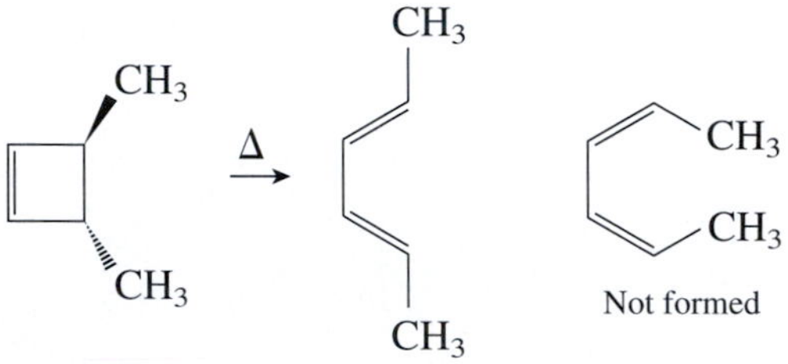

CH₃

CH₃ $\xrightarrow{\Delta}$ CH₃

CH₃

CH₃

CH₃   Not formed

CH₃

22.36 Show all of the steps in the mechanism for this reaction and explain why the aldehyde is formed rather than the ketone:

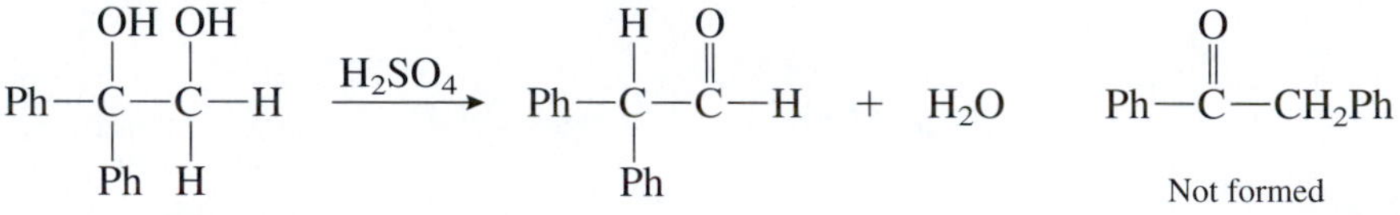

$$\underset{\underset{Ph}{|}\underset{H}{|}}{\overset{\overset{OH}{|}\overset{OH}{|}}{Ph-C-C-H}} \xrightarrow{H_2SO_4} \underset{\underset{Ph}{|}}{\overset{\overset{H}{|}\overset{O}{\|}}{Ph-C-C-H}} + H_2O \qquad \overset{O}{\overset{\|}{Ph-C-CH_2Ph}}$$

Not formed

22.37 Show all of the steps in the mechanism for this reaction:

$$\underset{CH_3CCH_3}{\overset{\overset{N^{OH}}{\|}}{}} \xrightarrow{H_2SO_4} \overset{O}{\overset{\|}{CH_3CNHCH_3}}$$

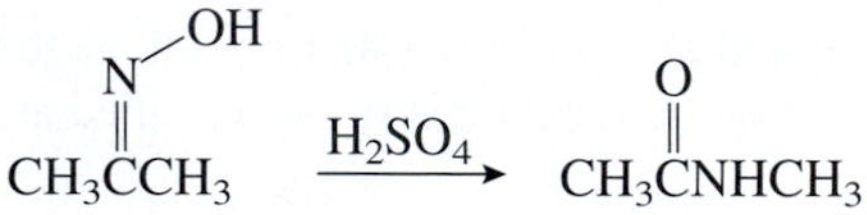

22.38 Show all of the steps in the mechanism for this reaction:

$$\overset{O}{\overset{\|}{CH_3CCH_3}} + \overset{O}{\overset{\|}{CF_3COOH}} \longrightarrow \overset{O}{\overset{\|}{CH_3COCH_3}} + \overset{O}{\overset{\|}{CF_3COH}}$$

22.39 Show all of the steps in the mechanism for this reaction:

$$\overset{O}{\overset{\|}{\underset{NH_2}{C}}} \xrightarrow[\substack{NaOH \\ H_2O}]{Br_2} \qquad NH_2$$

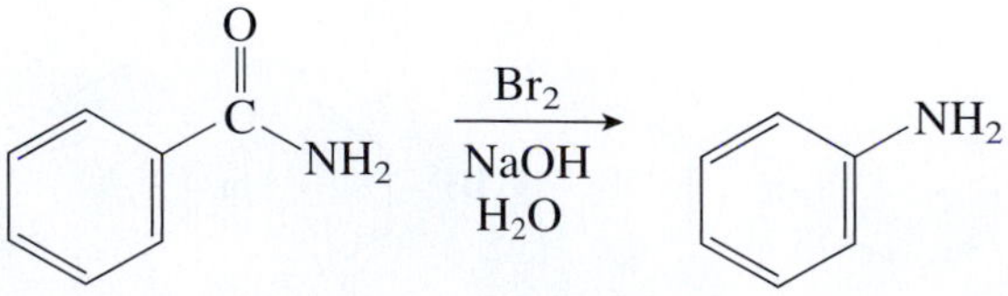

22.40 Attempts to prepare cyclobutadiene usually result in the isolation of this compound. Explain how the formation of this dimer from cyclobutadiene is allowed.

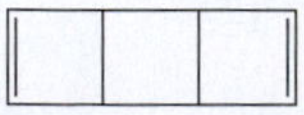

22.41 This reaction has been shown to occur in two thermally allowed steps. Show the structure of the intermediate in the reaction and explain why each step is allowed.

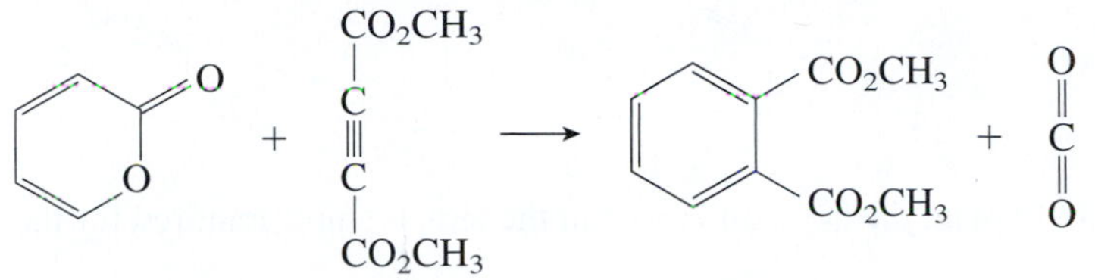

22.42 Suggest a mechanism for this reaction:

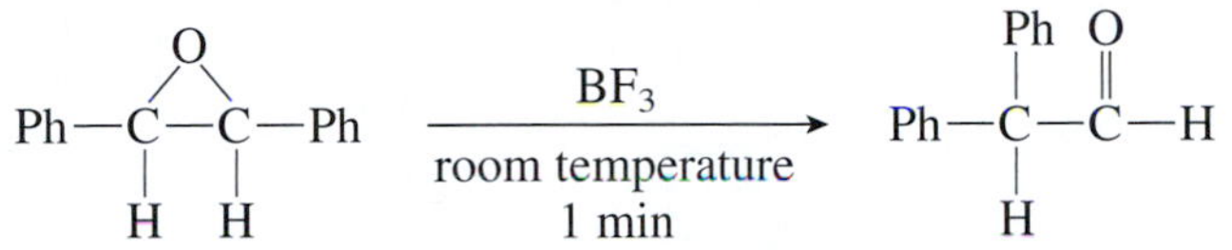

22.43 Suggest a mechanism for this reaction:

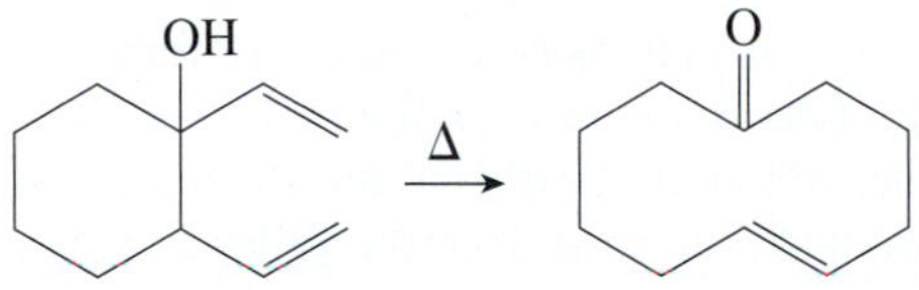

22.44 Explain the stereochemistry of this reaction:

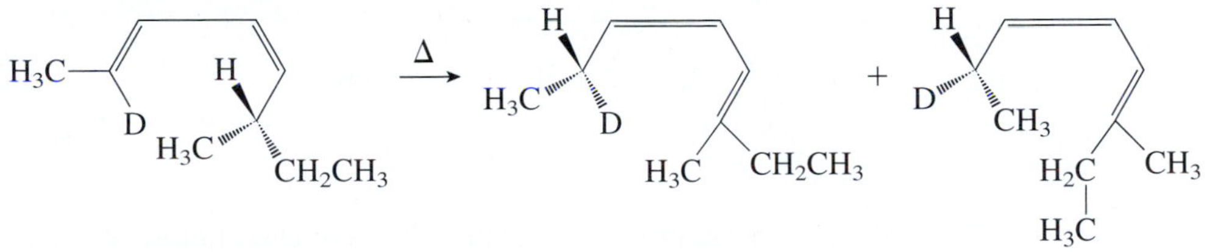

22.45 This reaction has been shown to occur in two thermally allowed steps. Show the structure of the intermediate in the reaction and explain why each step is allowed.

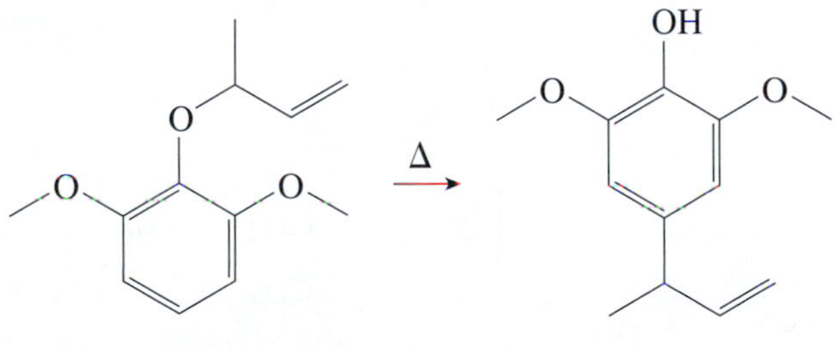

22.46 Suggest a mechanism for this reaction:

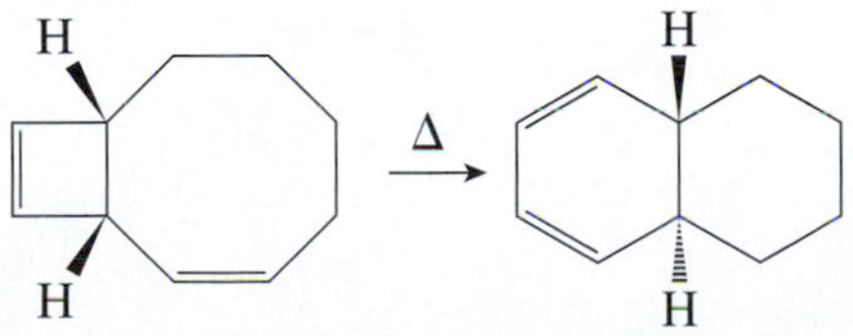

22.47 Explain why this reaction is highly exothermic. Then offer a reason why the reactant is quite stable at room temperature.

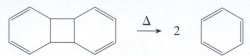

22.48 Explain the large difference in the temperatures required for these reactions.

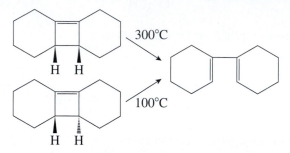

22.49 Cyclopropyl carbocations react rapidly to form allyl carbocations. Explain how this process is allowed. Explain why the allyl carbocation is more stable than the cyclopropyl carbocation. Predict the stereochemistry of the allyl carbocation that is formed from the following *cis*-dimethylcyclopropyl carbocation:

BioLink

22.50 UV radiation causes adjacent thymine bases of DNA to form dimers. The resulting dimers inhibit the normal functioning of DNA and may lead to cancer. Suggest a structure for the thymine dimer that is formed in the following reaction:

22.51 The conversion of chorismate to prephenate, catalyzed by the enzyme chorismate mutase, is involved in the biosynthesis of the amino acids phenylalanine and tyrosine. Classify this pericyclic reaction and explain whether it is thermally allowed or not.

Chorismate   chorismate mutase   Prephenate

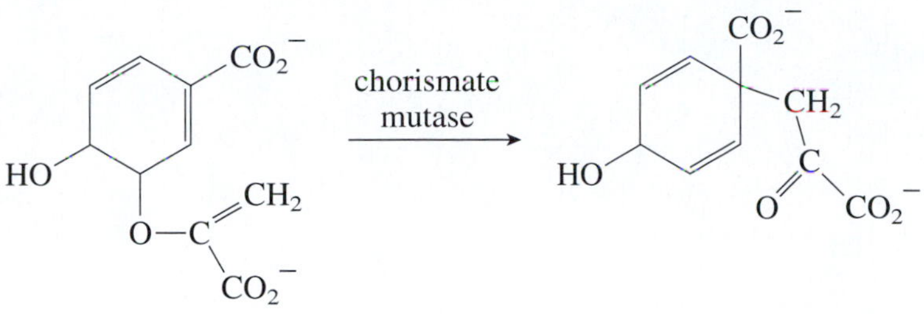

Problems Involving Spectroscopy

22.52 The reaction of anthracene with benzyne gives a product, $C_{20}H_{14}$, that shows only four peaks in its ^{13}C-NMR spectrum. Show the structure of this product and explain its formation.

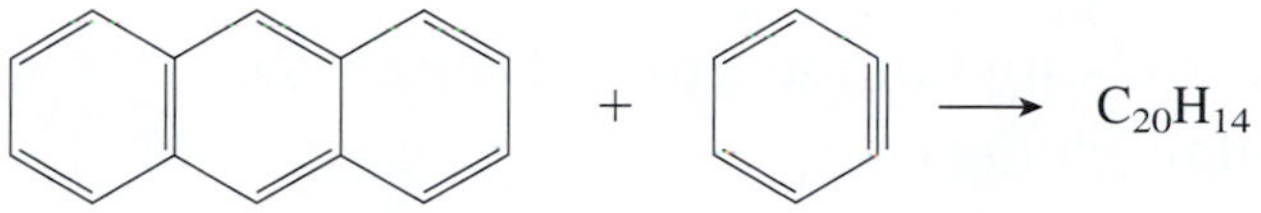

$$+ \quad \longrightarrow \quad C_{20}H_{14}$$

22.53 The mass spectrum of cyclohexene shows a peak at *m/z* 54. Show a structure for this fragment and suggest a mechanism for its formation.

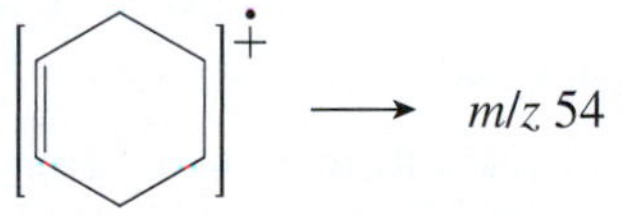

$$\longrightarrow \quad m/z\ 54$$

m/z 82

22.54 With unsymmetrical ketones the Baeyer-Villiger reaction can, in principle, give two products. Usually, one product dominates. The ^1H-NMR spectrum of the product isolated from the Baeyer-Villiger reaction of 3-methyl-2-butanone is shown here. Show the structure of this product.

$$\begin{array}{c} O \\ \| \\ CH_3CHCCH_3 \\ | \\ CH_3 \end{array} \quad \xrightarrow{CF_3CO_3H}$$

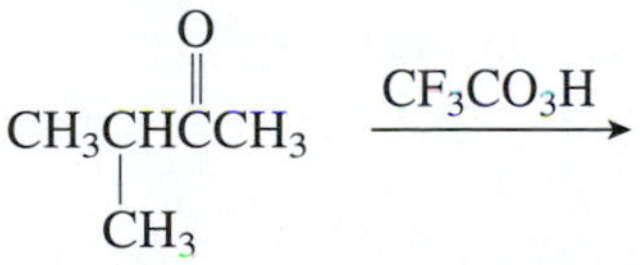

3-Methyl-2-butanone

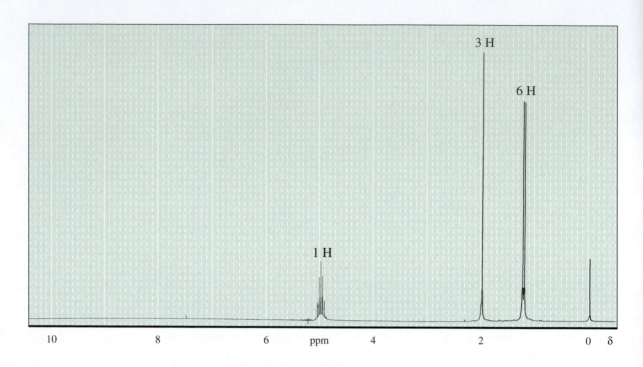

Problems Using Online Three-Dimensional Molecular Models

22.55 Explain whether each of these electrocyclic reactions proceeds by conrotation or disrotation.

22.56 Explain whether each of these electrocyclic reactions is thermally allowed or photochemically allowed.

22.57 Explain which product is formed upon heating the octatriene reactant shown. Explain which product is formed when the octatriene is irradiated with ultraviolet light.

22.58 Show the diene and dienophile that would form this cyclohexene derivative in a Diels-Alder reaction.

22.59 Explain which diene shown is more reactive in the Diels-Alder reaction.

22.60 Explain which product is formed from the Diels-Alder reaction of cyclopentadiene with maleic anhydride.

  Do you need a live tutor for homework problems? Access vMentor at Organic ChemistryNow at **http://now.brookscole.com/hornback2** for one-on-one tutoring from a chemistry expert.

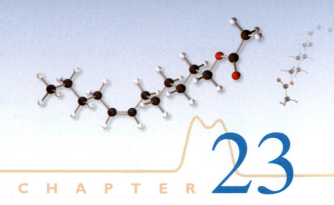

The Synthesis of Organic Compounds

ACH OF THE PRECEDING chapters that presented new reactions contained a number of problems that required using these reactions to synthesize organic compounds. You have probably become at least somewhat comfortable in working this type of problem, even though they are among the more difficult ones in this text. In this chapter the multitude of reactions that have been presented in the previous chapters are applied to the synthesis of more complex compounds. The syntheses here are longer and require somewhat more thought than those done previously, although the general approach and strategies remain the same as those you have already learned.

In the synthesis of complex compounds, which contain more than one functional group, it is often necessary to protect one reactive functional group so that the desired reaction will occur only at the unprotected functional group. Groups that can be used to protect alcohols, aldehydes and ketones, carboxylic acids, and amines during the course of a synthesis are presented first in this chapter. Then the strategy of synthesis is discussed in somewhat more detail. This is followed by the presentation of several syntheses. Tables at the end of this chapter summarize all of the reactions that have been presented in the previous chapters and are of significant synthetic utility. Table 23.2 lists important carbon–carbon bond-forming reactions. Table 23.3 lists all of the reactions, including the carbon–carbon bond-forming reactions of Table 23.2, arranged according to the functional group that is produced in the reaction. Thus, it is possible by consulting Table 23.3 to quickly identify the various reactions that can be used to prepare a particular functional group. Note that this table also constitutes a review of most of the reactions presented in earlier chapters.

MASTERING ORGANIC CHEMISTRY

▶ Using Protecting Groups in Organic Synthesis

▶ Using Retrosynthetic Analysis to Design Syntheses of More Complex Organic Compounds

23.1 PROTECTING GROUPS FOR ALCOHOLS

As discussed in Section 18.9, a protecting group is used to protect one functional group in a complex compound from reacting while a reaction is occurring at another

ORGANIC
Chemistry Now ™
Look for this logo in the chapter and go to OrganicChemistryNow at
http://now.brookscole.com/hornback2 for tutorials, simulations, problems, and molecular models.

functional group in the molecule. For this process to be useful, the functional group must be readily converted to the protecting group. In addition, the protecting group must be stable to the conditions of the desired reaction. Finally, the protected functional group must be easily converted back (deprotected) to the original functional group after the desired reaction has been accomplished.

In the case of a compound containing an alcohol functional group, it is often the acidic hydrogen of the hydroxy group that causes problems when a reaction is attempted at another functional group in the compound. This hydrogen can be replaced by an alkyl group to form an ether, a rather unreactive functional group that is stable to treatment with strongly basic or nucleophilic reagents. However, simple alkyl ethers are quite difficult to cleave, so it is difficult to regenerate the alcohol from a simple alkyl ether. Therefore, if an ether is to be used as a protecting group for an alcohol, it must be specially designed for easy removal after the reaction is complete.

One method that has found widespread use for the protection of an alcohol is reaction with dihydropyran to form a tetrahydropyranyl ether. Once the desired reaction has been accomplished, the protecting group can be removed by treatment with aqueous acid or acid and ethanol. The formation of a tetrahydropyranyl ether and its cleavage are illustrated in the following equation:

Dihydropyran   A tetrahydropyranyl ether (99%)

The mechanism of the formation of the tetrahydropyranyl ether (see Figure 23.1) is an acid-catalyzed addition of the alcohol to the double bond of the dihydropyran and is quite similar to the acid-catalyzed hydration of an alkene described in Section 11.3. Dihydropyran is especially reactive toward such an addition because the oxygen helps stabilize the carbocation that is initially produced in the reaction. The tetrahydropyranyl ether is inert toward bases and nucleophiles and serves to protect the alcohol from reagents with these properties. Although normal ethers are difficult to cleave, a tetrahydropyranyl ether is actually an acetal, and as such, it is readily cleaved under acidic conditions. (The mechanism for this cleavage is the reverse of that for acetal formation, shown in Figure 18.5 on page 776.)

In designing a synthesis, it is useful to have available several protecting groups that are formed under, are stable to, or are cleaved by different conditions so that one can be found that is compatible with the requirements imposed by the particular functional groups and reaction conditions of the proposed synthesis. Another protecting group that is used for alcohols is the methoxymethyl ether. The protecting group is added by reacting the conjugate base of the alcohol with chloromethyl methyl ether in an S_N2 reaction. Like the tetrahydropyranyl ether, this ether is also an acetal and can be readily removed

Figure 23.1

MECHANISM OF THE
FORMATION OF A
TETRAHYDROPYRANYL
ETHER.

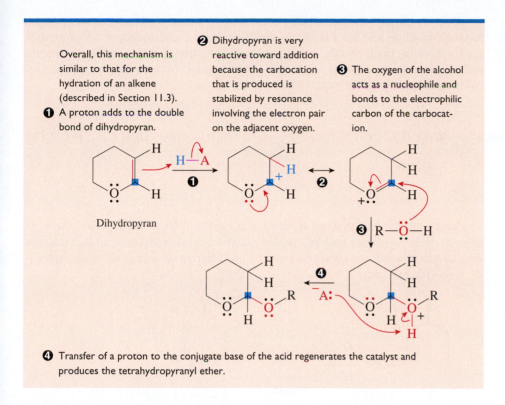

Overall, this mechanism is similar to that for the hydration of an alkene (described in Section 11.3).

❶ A proton adds to the double bond of dihydropyran.

❷ Dihydropyran is very reactive toward addition because the carbocation that is produced is stabilized by resonance involving the electron pair on the adjacent oxygen.

❸ The oxygen of the alcohol acts as a nucleophile and bonds to the electrophilic carbon of the carbocation.

Dihydropyran

❹ Transfer of a proton to the conjugate base of the acid regenerates the catalyst and produces the tetrahydropyranyl ether.

by treatment with aqueous acid. The use of the methoxymethyl ether protecting group is illustrated in the following sequence. In the first step, sodium hydride is used to deprotonate the alcohol. Reaction of this alkoxide ion with chloromethyl methyl ether provides the protected alcohol. Now reactions that would have failed in the presence of the unprotected hydroxy group can be done. After the desired transformations have been accomplished, the protecting group is removed by treatment with acid and water.

1) NaH
2) $ClCH_2OCH_3$

Chloromethyl
methyl ether

(72%)

A methoxymethyl
ether

Several steps

HCl
H_2O

(95%)

Alcohols can also be protected by conversion to a benzyl ether. The ether is stable to base and dilute acid but can be cleaved by treatment with hydrogen and a catalyst. Although most single bonds do not react with hydrogen and a catalyst, benzylic carbon–oxygen single bonds are readily cleaved under these conditions in a process called **hydrogenolysis.** An example is provided by the following equation:

$$\text{PhCH}_2\text{—OCH}_2\text{CH}_2\text{CH}_2\text{CH}_2\text{CH}_3 \xrightarrow[\substack{\text{Pd/C}\\25°\text{C}}]{\text{H}_2 \text{ (1 atm)}} \text{PhCH}_2 + \text{H—OCH}_2\text{CH}_2\text{CH}_2\text{CH}_2\text{CH}_3$$

(100%)

The benzylic carbon–oxygen bond is cleaved.

The benzyl ether is prepared in a similar manner to the methoxymethyl ether, that is, by reaction of the conjugate base of the alcohol with benzyl bromide in an S_N2 reaction. An example of a sequence that employs a benzyl ether protecting group is illustrated in the following sequence:

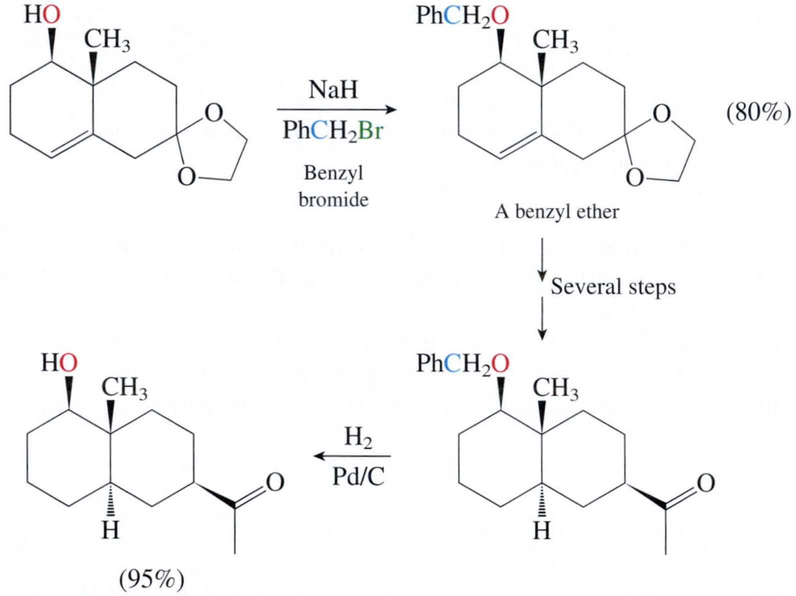

(80%)

A benzyl ether

Several steps

(95%)

Finally, alcohols can also be protected as silyl ethers. For example, the reaction of the alcohol with trimethylsilyl chloride in the presence of triethylamine (to react with the HCl that is produced) produces the trimethylsilyl ether of the alcohol as shown in the following equation. (This reaction is a nucleophilic substitution by the oxygen on the silicon.) The silyl group can be removed in high yield by reaction with fluoride anion.

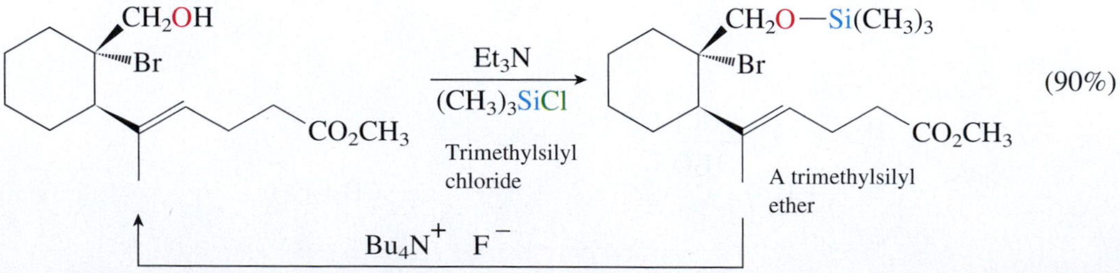

(90%)

A trimethylsilyl ether

PROBLEM 23.1

Show the products of these reactions:

a) 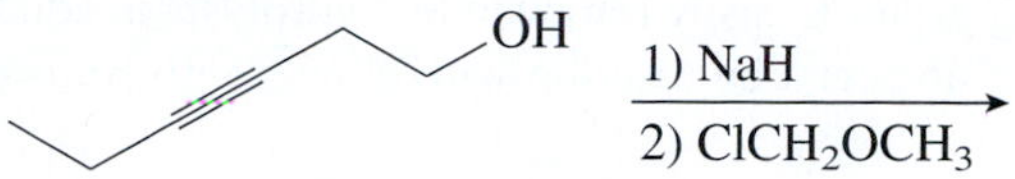 (structure: pent-3-yn-1-ol chain with OH)   1) NaH / 2) ClCH$_2$OCH$_3$

b) 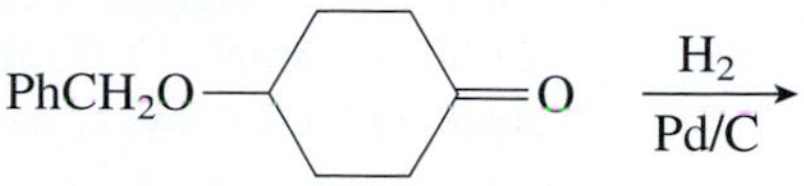 PhCH$_2$O—(cyclohexanone with =O) $\dfrac{H_2}{Pd/C}$

c) (benzaldehyde with CH$_2$OH substituent, CHO group) + (dihydropyran) $\dfrac{}{TsOH}$

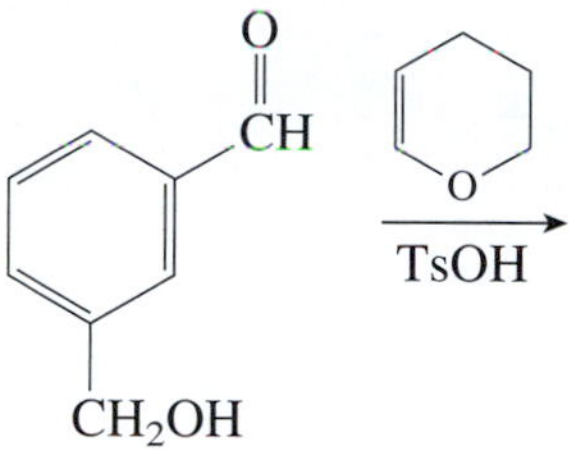

d) 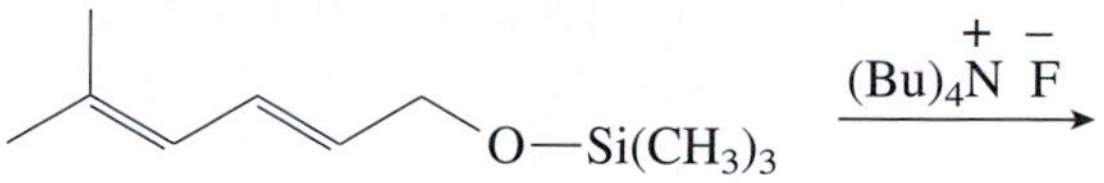 (structure with O—Si(CH$_3$)$_3$) $\overset{+\ \ -}{(Bu)_4N\ \ F}$

e) 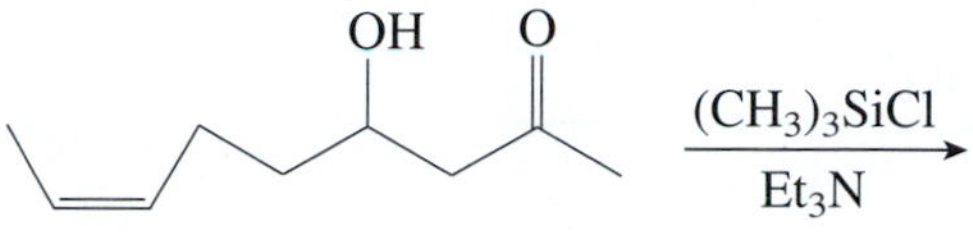 (structure with OH and O ketone) $\dfrac{(CH_3)_3SiCl}{Et_3N}$

f) Ph (chain with O   O acetal) $\dfrac{HCl}{H_2O}$

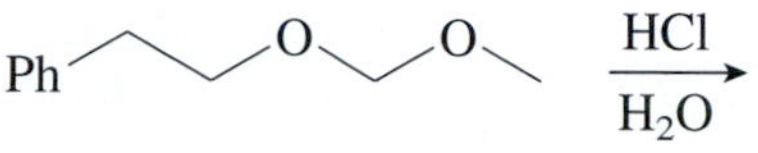

g) 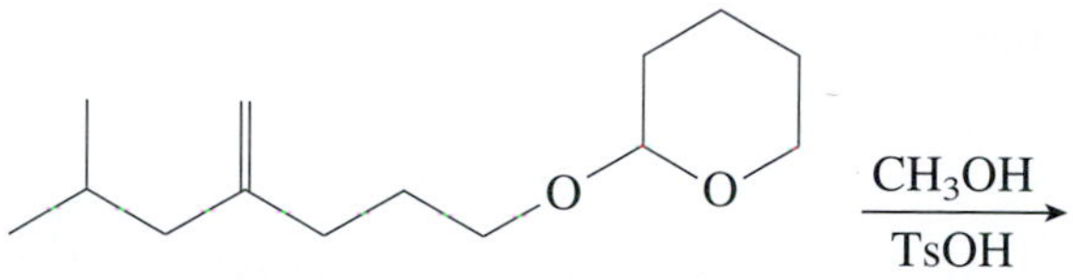 (structure with tetrahydropyranyl O O) $\dfrac{CH_3OH}{TsOH}$

PROBLEM 23.2

Show all of the steps in the mechanism for the hydrolysis of this methoxymethyl ether. What other products are formed in this reaction?

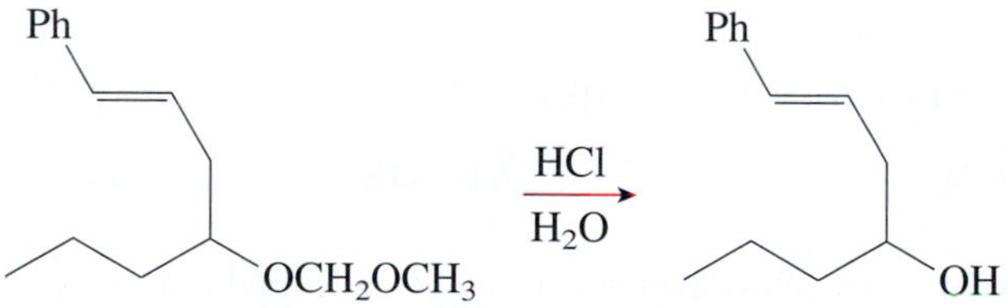

(structure: Ph chain with OCH$_2$OCH$_3$) $\dfrac{HCl}{H_2O}$ (structure: Ph chain with OH)

PROBLEM 23.3

A Grignard reagent cannot be prepared from 4-bromo-1-butanol due to the presence of the acidic hydrogen of the hydroxy group. Show how the use of a protecting group enables this compound to be used to prepare 1,5-hexanediol by a Grignard reaction.

HOCH$_2$CH$_2$CH$_2$CH$_2$Br $\longrightarrow$ HOCH$_2$CH$_2$CH$_2$CH$_2\overset{\displaystyle OH}{\underset{|}{C}}HCH_3$

23.2 PROTECTING GROUPS

FOR ALDEHYDES AND KETONES

Aldehydes and ketones are usually protected by converting them to acetals by reaction with an alcohol in the presence of acid (see Section 18.9). Although many different alcohols could be used, ethylene glycol (1,2-ethanediol) or 1,3-propanediol is most often employed because the intramolecular nature of the addition of the second alcohol group makes the equilibrium more favorable. The acetal, once formed, is stable to basic and nucleophilic reagents. The protecting group is easily removed by hydrolysis in aqueous acid. An example of the formation of an acetal by reaction with ethylene glycol is provided by the following equation:

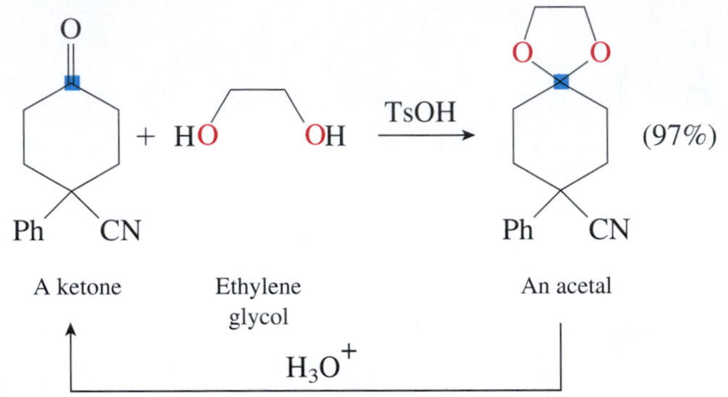

| A ketone | Ethylene glycol | An acetal |

$$H_3O^+$$

PROBLEM 23.4

Show the products of these reactions:

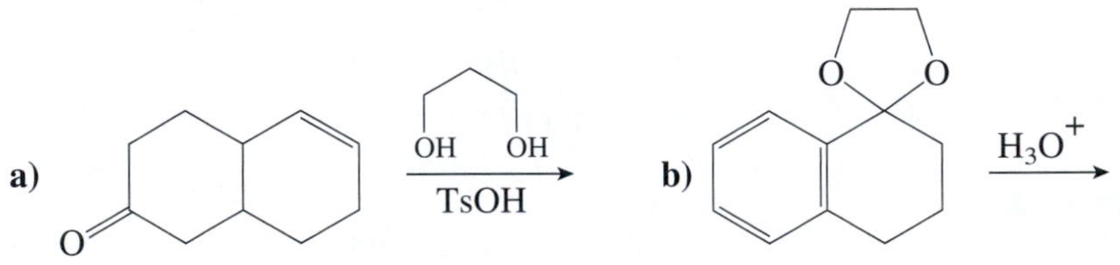

a)

b)

23.3 PROTECTING GROUPS FOR CARBOXYLIC ACIDS

Because the acidic hydrogen is the usual cause of undesired reactivity, carboxylic acids are commonly protected as esters. Most often, simple methyl esters are used. The ester can be prepared via the acyl chloride or by Fischer esterification (see Section 19.4). The ester group can be converted back to the carboxylic acid by hydrolysis under acidic or basic (saponification) conditions (see Section 19.5).

Other esters are sometimes used to protect carboxylic acids, especially when there is a desire to deprotect the acid by using different conditions from those available for methyl esters. Benzyl esters are prepared in the usual manner but can be cleaved by reaction with hydrogen and a catalyst. Again it is the benzylic carbon–oxygen bond that is broken in the hydrogenolysis reaction:

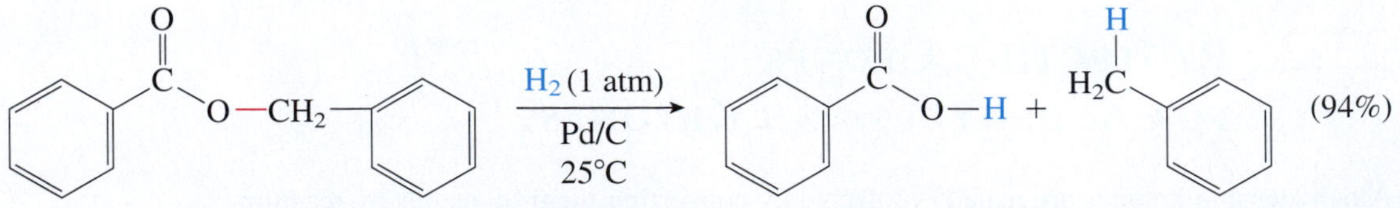

A benzyl ester

t-Butyl esters can be cleaved by reaction with dilute acid under milder conditions than those required to hydrolyze a methyl ester. The reaction follows an S_N1 mechanism, rather than the reverse of the Fischer esterification mechanism, because of the stability of the *t*-butyl carbocation:

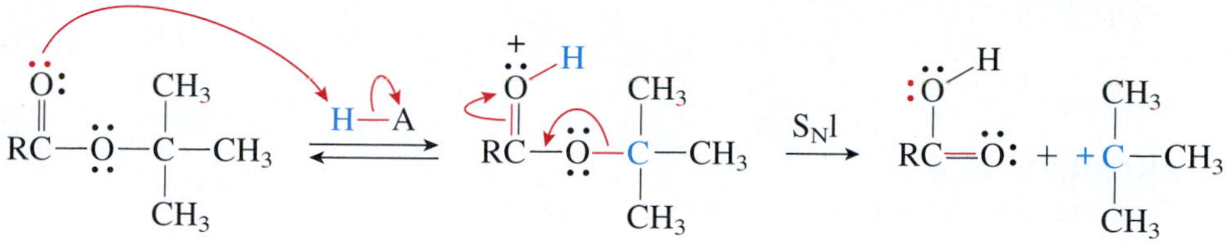

A *t*-butyl ester

The *t*-butyl ester is often prepared by acid-catalyzed addition of the carboxylic acid to isobutylene. The overall protection–deprotection sequence is outlined in the following equation. Note that the *t*-butyl group is removed in the last step without destroying the phosphorus ester or the amide or benzyl ester groups.

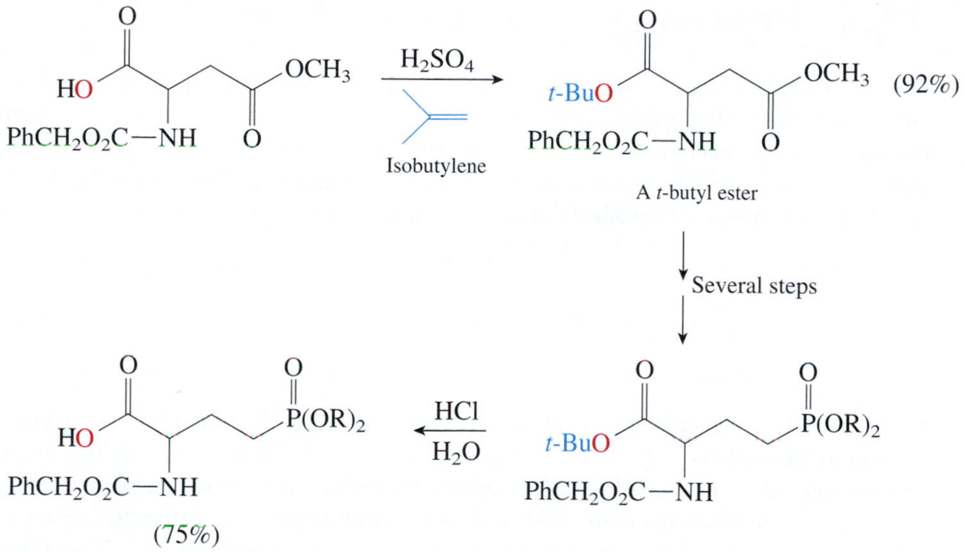

Isobutylene

A *t*-butyl ester

Several steps

(75%)

PROBLEM 23.5
Show the products of these reactions:

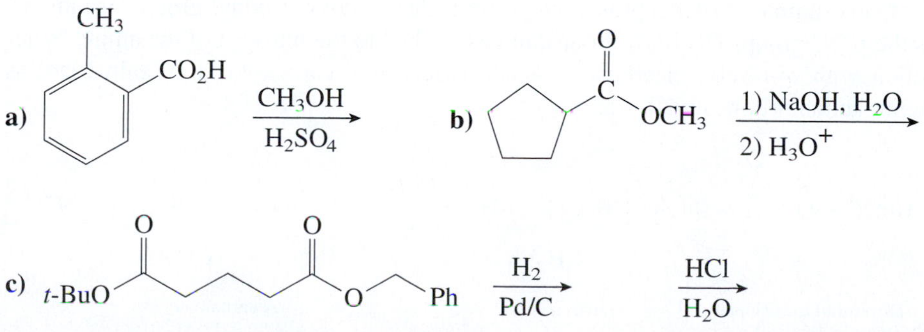

a) [structure with CH₃ and CO₂H groups] $\xrightarrow{\text{CH}_3\text{OH}, \text{H}_2\text{SO}_4}$

b) [structure] $\xrightarrow[\text{2) H}_3\text{O}^+]{\text{1) NaOH, H}_2\text{O}}$

c) *t*-BuO [structure] Ph $\xrightarrow[\text{Pd/C}]{\text{H}_2}$ $\xrightarrow[\text{H}_2\text{O}]{\text{HCl}}$

d)

PROBLEM 23.6
Show the steps in the mechanism for this reaction:

$$CH_3(CH_2)_3\overset{\underset{\displaystyle \|}{O}}{C}OH \ + \ CH_3\overset{\underset{\displaystyle |}{CH_3}}{C}{=}CH_2 \ \xrightarrow{H_2SO_4} \ CH_3(CH_2)_3\overset{\underset{\displaystyle |}{\underset{\displaystyle CH_3}{}}}{\overset{\overset{\displaystyle O}{\|}}{C}}O\overset{\overset{\displaystyle CH_3}{|}}{C}CH_3$$

23.4 PROTECTING GROUPS FOR AMINES

Often it is the basic and nucleophilic unshared pair of electrons on the nitrogen of an amine that causes the problem in a synthesis and needs to be protected. This electron pair can be made much less reactive by converting the amine to an amide. The electron pair on the nitrogen in the amide is delocalized by resonance onto the oxygen of the carbonyl group and is not as available to react as either a base or a nucleophile.

Amine   Protected amine

One example of this strategy has already been presented in Section 17.4, where conversion to an acetamide ($R' = CH_3$) was used to decrease the activating ability of an amino group on a benzene ring in electrophilic aromatic substitution reactions.

An amide is one of the more stable carboxylic acid derivatives, and rather vigorous conditions are required to hydrolyze it to regenerate the unprotected amine. Therefore, several special protecting groups that can more readily be removed have been developed. These groups still employ an amide to deactivate the nitrogen, but they all contain some feature that allows them to be removed under milder conditions. They are especially useful in the synthesis of peptides from amino acids, described in Chapter 26.

One example of such a protecting group is the *t*-butoxycarbonyl group, also known as the BOC group. The BOC group can be attached to the nitrogen of the amine by reaction with di-*t*-butyl dicarbonate, which reacts as a carboxylic acid anhydride, as shown in the following equation:

$$t\text{-BuOC}-$$
$$\overset{\displaystyle O}{\overset{\displaystyle \|}{}}$$
BOC

Di-*t*-Butyl dicarbonate   Phenylalanine   A carbamate

The product contains the **carbamate** functional group, which has characteristics of both an amide and an ester. Thus, the amine is protected as an "amide," but the ester part of the carbamate can be readily removed. In fact, the *t*-butyl group of the carbamate can be removed by treatment with dilute acid. (This is the same S_N1 process that is described in Section 23.3) The resulting carbamic acid is unstable and spontaneously eliminates carbon dioxide, regenerating the amine. An example is provided by the following equation:

$$t\text{-BuOCNH(CH}_2)_6\text{NHCC}{=}\text{CH}_2 \xrightarrow[\text{H}_2\text{O}]{\text{HCl}} \text{H}_2\text{N(CH}_2)_6\text{NHCC}{=}\text{CH}_2 \quad (96\%)$$

with CH_3 substituents

$$\left[\text{HOCNH(CH}_2)_6\text{NHCC}{=}\text{CH}_2\right] \xrightarrow[\text{spontaneous}]{-CO_2}$$

A carbamic acid

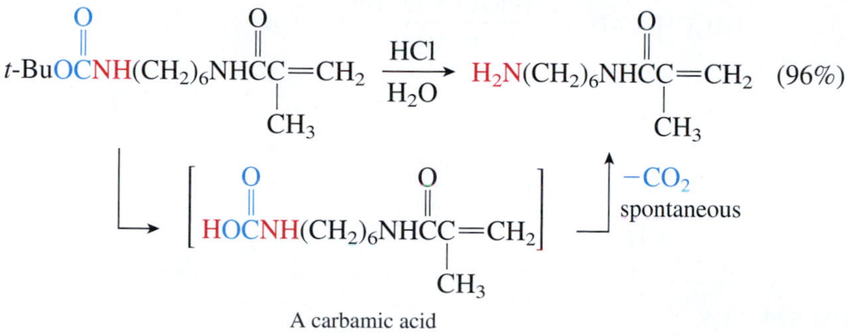

Another protecting group often employed for amines is the **benzyloxycarbonyl** group. The chemistry here is reminiscent of that of the BOC group in that the amine is protected by conversion to a carbamate. This time an acyl chloride derivative, carbobenzoxy chloride, rather than an acid anhydride derivative, is reacted with the amine.

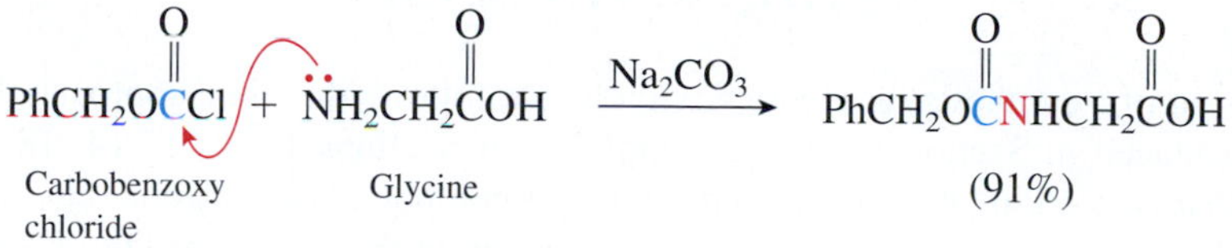

$$\text{PhCH}_2\text{OCCl} + \overset{\cdot\cdot}{\text{N}}\text{H}_2\text{CH}_2\text{COH} \xrightarrow{\text{Na}_2\text{CO}_3} \text{PhCH}_2\text{OCNHCH}_2\text{COH}$$

Carbobenzoxy chloride   Glycine   (91%)

When it is time to remove the benzyloxycarbonyl protective group, the compound is reacted with hydrogen in the presence of a catalyst. The cleavage of the benzyl group by hydrogenolysis (see Section 23.3) is followed by spontaneous decarboxylation of the carbamic acid. An example is provided by the following equation:

$$\text{PhCH}_2\text{OCNHCHCO}_2\text{H} \xrightarrow[\text{Pd/C}]{\text{H}_2} \text{NH}_2\text{CHCO}_2\text{H} \quad (100\%)$$

with CH_2Ph substituents

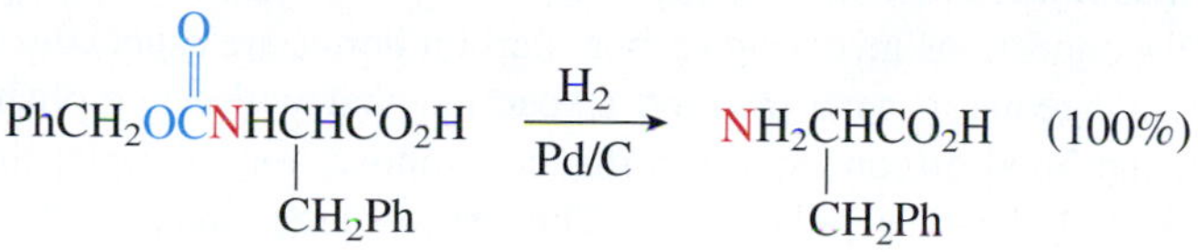

PROBLEM 23.7

BioLink

Show the products of these reactions:

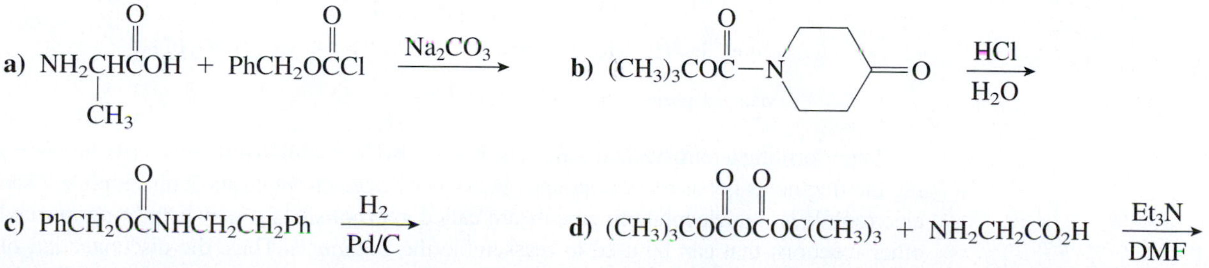

a) $\text{NH}_2\text{CHCOH} + \text{PhCH}_2\text{OCCl} \xrightarrow{\text{Na}_2\text{CO}_3}$ (with CH_3 substituent)

b) $(\text{CH}_3)_3\text{COC}{-}\text{N} \bigcirc {=}\text{O} \xrightarrow[\text{H}_2\text{O}]{\text{HCl}}$

c) $\text{PhCH}_2\text{OCNHCH}_2\text{CH}_2\text{Ph} \xrightarrow[\text{Pd/C}]{\text{H}_2}$

d) $(\text{CH}_3)_3\text{COCOCOC(CH}_3)_3 + \text{NH}_2\text{CH}_2\text{CO}_2\text{H} \xrightarrow[\text{DMF}]{\text{Et}_3\text{N}}$

PROBLEM 23.8

An example of the use of a protecting group for an amine is provided by the following reaction scheme. Show the structures of the missing compounds, **A** and **B,** in this sequence of reactions and explain why the final amide cannot be prepared directly, that is, by reaction of the original amino acid with $SOCl_2$ followed by $(CH_3)_2NH$.

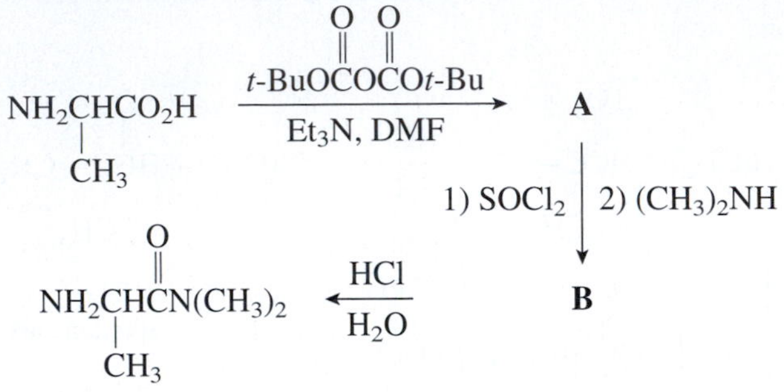

$$NH_2CHCO_2H \quad \xrightarrow[\text{Et}_3\text{N, DMF}]{t\text{-BuOCOCOt-Bu}} \quad \textbf{A}$$
$$\underset{\displaystyle CH_3}{|}$$

$$1)\ SOCl_2 \quad | \quad 2)\ (CH_3)_2NH$$

$$NH_2CHCN(CH_3)_2 \quad \xleftarrow[\text{H}_2\text{O}]{\text{HCl}} \quad \textbf{B}$$
$$\underset{\displaystyle CH_3}{|}$$

PROBLEM 23.9

Explain why the hydrolysis of **B** in problem 23.8 is selective in that the amide bond of the final product is not cleaved.

23.5   RETROSYNTHETIC ANALYSIS

The design of a synthesis for a complex molecule follows the same approach that was first introduced in Section 10.15, and amplified in Sections 11.14, 17.14, 18.11, and 20.11, that is, working backward from the target compound or retrosynthetic analysis. Recall that this process involves envisioning a reaction that can be used to prepare the target compound from a simpler compound. This simpler compound then becomes the target for the next step, and this process is continued until a commercially available compound is reached.

When looking for reactions that can be used to prepare the target compound, we envision **bond disconnections,** or breaking bonds, at or near the functional groups in the compound. Disconnections involving carbon–carbon bonds are especially important. It is usually advantageous to have in mind a reaction that will accomplish the reverse process when the bond disconnection is made. Suppose, for example, that the target compound is 3-methyl-1-phenyl-1-butanol. The Grignard reaction provides an excellent method for the preparation of many alcohols. With this reaction in mind, a disconnection between the phenyl group and the alkyl group, as shown in the following equation, provides fragments that can be reassembled by a Grignard reaction.

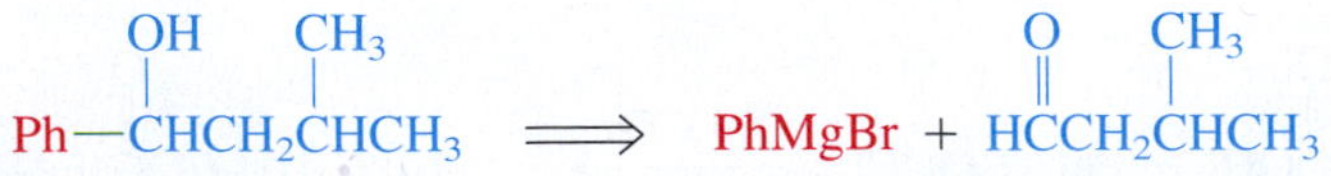

$$Ph\!-\!\underset{\displaystyle OH}{\overset{\displaystyle |}{C}}H\underset{}{CH_2}\underset{\displaystyle CH_3}{\overset{\displaystyle |}{C}}HCH_3 \quad \Longrightarrow \quad PhMgBr\ +\ H\overset{\displaystyle O}{\overset{\displaystyle \|}{C}}CH_2\underset{\displaystyle CH_3}{\overset{\displaystyle |}{C}}HCH_3$$

3-Methyl-1-phenyl-1-butanol

In performing a retrosynthetic analysis, it may also be useful to disconnect a bond, showing the fragments not as real compounds but only as an electrophile and a nucleophile. (The electrophile and nucleophile fragments are called **synthons.**) This may help bring to mind other reactions that can be used to reassemble the fragments. Thus, the disconnection of 3-methyl-1-phenyl-1-butanol can be written as shown in the following equation:

OH   CH₃
| |
(phenyl)CHCH₂CHCH₃

$\Longrightarrow$

Nucleophile   Electrophile

(benzene ring)⁻   OH   CH₃
                  | |
+ HCCH₂CHCH₃
                  +

Synthons

In addition to suggesting the Grignard route shown, this disconnection might bring to mind the fact that aromatic rings also react as nucleophiles in electrophilic aromatic substitution reactions. This suggests that a Friedel-Crafts acylation might be used in this synthesis. (Although the Friedel-Crafts acylation reaction produces a ketone, we recognize that an alcohol or ketone in the target is usually equivalent because either can be converted to the other by oxidation or reduction reactions.)

OH   CH₃
| |
(phenyl)CHCH₂CHCH₃

$\Longrightarrow$

(benzene)   +   O   CH₃
                || |
                ClCCH₂CHCH₃

When a bond is disconnected, two polarities of the fragments are possible. Either fragment can potentially be the nucleophile, while the other is the electrophile. Let's consider the disconnection of the other carbon–carbon bond involving the hydroxy-bearing carbon of 3-methyl-1-phenyl-1-butanol. The disconnection shown in the following equation suggests another Grignard pathway:

OH   CH₃
| |
Ph–CH–CH₂CHCH₃

$\Longrightarrow$

OH   CH₃
| |
Ph–CH + ⁻CH₂CHCH₃
     +

Suggests $\Longrightarrow$

O   CH₃
|| |
Ph–C–H   +   BrMgCH₂CHCH₃

The disconnection with opposite polarity requires a very different approach—the use of a synthon that is nucleophilic at the hydroxy carbon. The acyl anion equivalent described in Section 20.9 provides a nucleophilic carbon with a substituent that can be readily converted to a hydroxy group:

OH   CH₃
| |
PhCH–CH₂CHCH₃

$\Longrightarrow$

OH   CH₃
| |
PhCH + ⁺CH₂CHCH₃

Suggests $\Longrightarrow$

(dithiane ring with S, S)   CH₃
Ph ⁻ |
                        +   Br–CH₂CHCH₃

As an example of this strategy, let's consider the synthesis of 2-methyl-6-methylene-7-octen-4-ol, an insect **pheromone.** (As described in the Focus On box on pages 1025–1026, pheromones are compounds used by animals, especially insects, for communication.) The synthesis of this alcohol, a component of the sex attractant of the male bark beetle, has been described in the literature. Retrosynthetic analysis, similar to that described earlier, suggested the use of a Grignard reaction to prepare this alcohol:

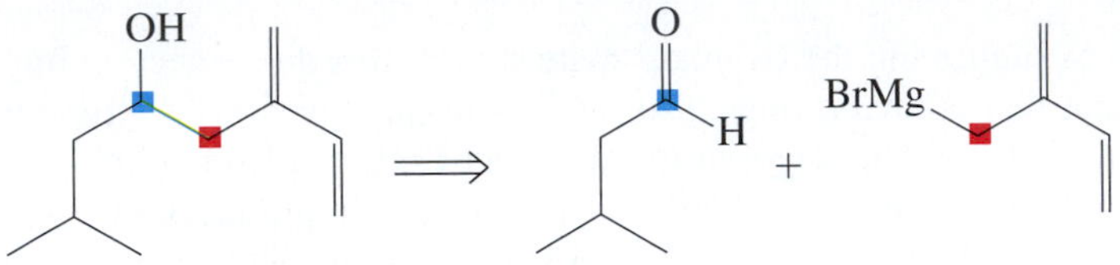

2-Methyl-6-methylene-
7-octen-4-ol

However, the chemists who were conducting this investigation were not able to get this particular Grignard reaction to work. Therefore, a different approach was needed. Again, using analysis similar to that given earlier, they developed a route based on the same disconnection, but with reverse polarity:

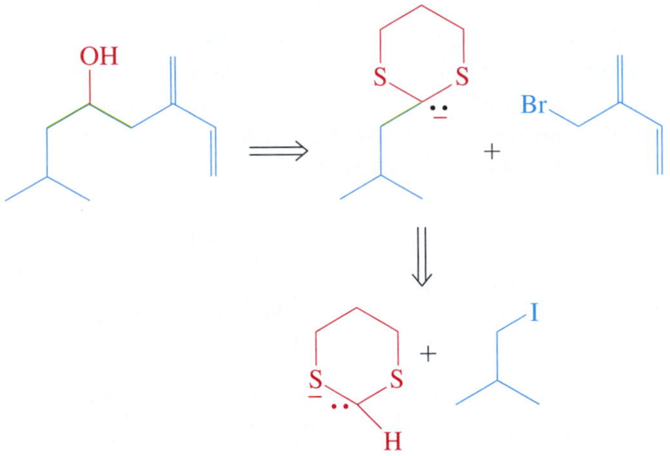

The needed substituted dithiane could itself be prepared by a similar alkylation of the unsubstituted dithiane anion. The synthesis is shown in Figure 23.2.

What makes a good synthesis? The answer to this question depends on many factors. The time required to complete the synthesis and the overall cost of the synthesis are factors whose relative importance often depends on whether the synthesis is being done in an academic research laboratory, an industrial research laboratory, or a factory. How much material is needed—milligrams, grams, kilograms, or tons? Today, environmental concerns are extremely important. Reactions that employ toxic solvents, use large amounts of poisonous heavy metals, or generate toxic waste are avoided as much as possible. In a long synthesis the yield of each step is of critical importance. For example, if each reaction proceeds in 90% yield, a synthesis of 10 steps has an overall yield of only 35% ($0.9^{10} = 0.35$). If each of the steps proceeds in 80% yield, the overall yield falls to 11%.

In designing a synthesis, it is important to have a wide range of reactions available to use. The reactions of most importance in synthesis that have been presented in earlier chapters in this book are summarized in Tables 23.2 and 23.3 (see Sections 23.7 and 23.8). Note that the purpose of Tables 23.2 and 23.3 is not to provide a list of reactions

❶ Butyllithium was used as a strong base to remove one of the weakly acidic hydrogens on the carbon between the sulfurs. The resulting anion was then used as a nucleophile, replacing the iodine in an S_N2 reaction (see Section 20.9).

❷ This process was repeated with a different alkyl halide. The yield in this step was 51%.

❸ The dithioacetal group was removed by hydrolysis with mercuric ion and water in 59% yield.

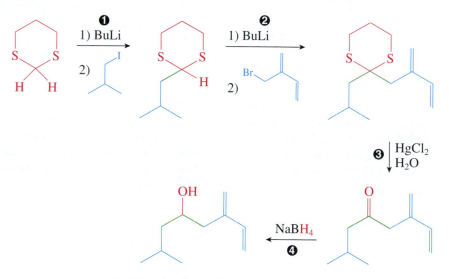

2-Methyl-6-methylene-7-octen-4-ol

❹ To complete the synthesis, the ketone was reduced to an alcohol. This was accomplished in 66% yield using sodium borohydride.

Figure 23.2

THE SYNTHESIS OF 2-METHYL-6-METHYLENE-7-OCTEN-4-OL.

for you to memorize. Rather, use these tables as references when you are designing syntheses. Table 23.2 lists important reactions that form carbon–carbon bonds. Such reactions are often the key to designing a synthesis. Table 23.3 lists all of the reactions, including those of Table 23.2, according to which functional group is formed in the product. Thus, if the target molecule is an alcohol, Table 23.3 can be consulted for a number of different methods that might be employed to prepare that alcohol.

PRACTICE PROBLEM 23.1

Use disconnections of carbon–carbon bonds to generate electrophile and nucleophile synthon fragments that could be used to prepare 3-methyl-4-heptanone. Then show an actual reaction suggested by these synthons.

$$\underset{\substack{|\\CH_3}}{CH_3CH_2CH\overset{\substack{O\\\|}}{C}CH_2CH_2CH_3}$$

3-Methyl-4-heptanone

Solution

Because most reactions occur at a functional group or at the adjacent carbon, this is where most disconnections should be made. This usually still allows a variety of disconnections. As one possibility, let's consider disconnecting the methyl group from the α-carbon of this compound. One of the possible polarities of the synthon fragments is shown:

$$CH_3CH_2CHCCH_2CH_2CH_3 \implies CH_3CH_2\overset{-}{C}HCCH_2CH_2CH_3$$
$$\quad\quad\quad\;\; CH_3 \quad\quad\quad\quad\quad\quad\quad\quad\quad + CH_3$$

Perhaps you recall that a nucleophilic enolate anion can be generated on the carbon adjacent to a carbonyl group (see Chapter 20), so this disconnection looks promising. To accomplish this transformation, we need to add an electrophilic methyl group (a methyl group with a leaving group attached) to the enolate anion. The overall reaction is an alkylation of a ketone described in Section 20.3. Because the ketone is symmetrical, there is no problem with the regiochemistry of the alkylation.

$$CH_3CH_2CH_2CCH_2CH_2CH_3 \xrightarrow[\text{2) } CH_3I]{\text{1) LDA}} CH_3CH_2CHCCH_2CH_2CH_3$$
$$\quad\quad\quad\quad\quad\quad\quad\quad\quad\quad\quad\quad\quad\quad\quad\quad\quad CH_3$$

Note that numerous other disconnections might also prove useful. For example, the bond between the carbonyl carbon and one of the α-carbons can be disconnected with this polarity:

$$CH_3CH_2CH{-}CCH_2CH_2CH_3 \implies CH_3CH_2\overset{-}{C}H \quad \overset{+}{C}CH_2CH_2CH_3$$
$$\quad\;\; CH_3 \;\; O \quad\quad\quad\quad\quad\quad\quad\quad CH_3 \;\; O$$

This suggests one of the ketone syntheses described in Section 19.10. How many other routes can you think of?

PROBLEM 23.10

Use disconnections of carbon–carbon bonds to generate electrophile and nucleophile synthon fragments that could be used to prepare the following compounds. Then show an actual reaction suggested by these synthons that would provide a method for the synthesis of each compound. You may find it useful to consult Table 23.2 at the end of this chapter.

a) $CH_3CH_2CH_2\overset{\overset{\displaystyle OH}{|}}{C}HCH_2\overset{\overset{\displaystyle }{|}}{C}HCH_3$
$\quad\quad\quad\quad\quad\quad\quad\quad CH_3$

b) $PhC{\equiv}CCH_2CH_2CH_2CH_3$

c) $PhCH{=}CHCCH_3$ with carbonyl O

Focus On Biological Chemistry

Pheromones

Insects communicate with chemicals. These chemicals are called pheromones, derived from the Greek words *pherin,* to transfer, and *hormone,* to excite. Perhaps you have seen a line of ants, all following the same invisible trail across a floor or a concrete walkway. One ant has found a food source and has laid down a chemical, a trail pheromone, on its return to the nest. Other ants then follow this chemical trail back to the food source. Insects also use pheromones extensively to attract members of the opposite sex for mating. Pheromones must be volatile enough to readily enter the gas phase so that they can be detected by the insect's antennae. This requires them to be relatively small organic compounds. Some examples of pheromones are provided in the accompanying table.

Pheromones	Source
$CH_3(CH_2)_2$ ⌒⌒ $(CH_2)_9OH$	Bombykol or (10E,12Z)-hexadecadien-1-ol, sex attractant of the female silkworm moth
⌒⌒ $(CH_2)_9\overset{\displaystyle O}{\overset{\|}{C}}H$	Fumilure or 11-tetradecenal, sex attractant of the female spruce budworm; fumilure is a 96:4 ratio of *E:Z* geometrical isomers
(Neocembrene A structure)	Neocembrene A, a termite trail pheromone
$CH_3\overset{\displaystyle O}{\overset{\|}{C}}(CH_2)_5$ ⌒⌒ CO_2H	Queen substance or (*E*)-9-oxo-2-decenoic acid, excreted by the queen bee to prevent the development of other queens
H_3C (methylpyrrole structure) $\overset{\displaystyle}{N}$—$CO_2CH_3$, H	Methyl 4-methylpyrrole-2-carboxylate, trail pheromone of the leaf-cutting ant
$CH_3(CH_2)_3$ ⌒⌒ $(CH_2)_6O\overset{\displaystyle O}{\overset{\|}{C}}CH_3$	(*Z*)-7-Dodecen-1-yl acetate, component of the sex attractant of the female elephant and numerous insects

Because of their small size, insects are able to excrete only minute amounts of a pheromone. However, these compounds are active in extremely small amounts. An insect's antennae can detect one molecule per receptor, and it has been estimated that only

Continued

100 to 1000 molecules are necessary to elicit the appropriate response. Of course, the presence of such a small amount of the pheromone in the insect makes the task of isolating enough of the compound to determine its structure quite difficult. For example, the last two abdominal segments from 78,000 female gypsy moths were needed to isolate enough of the sex pheromone dispalure to allow its structure to be determined.

The structural requirements for a pheromone to be active in a particular species are very strict. Not only must the structure be correct, but the stereochemistry of the pheromone is often of critical importance. For example, the sex attractant of the female silkworm moth is (10*E*,12*Z*)-hexadecadien-1-ol, also known as bombykol. This particular stereoisomer is 10^{12} times more active than the (10*Z*,12*Z*) or (10*E*,12*E*) stereoisomers, which differ only in the configuration about one of the double bonds. Often, a mixture of stereoisomers in a very specific ratio is necessary for maximum activity of the pheromone.

Pheromones are also important in mammals, although many fewer have been identified. As one example, (*Z*)-7-dodecen-1-yl acetate has recently been identified in the urine of female elephants when they are in heat. This ester elicits a specific sexual response in male elephants. Interestingly, this same ester is part of the pheromone mixture that many insects, especially moths, use as a sexual attractant.

One of the major reasons for the interest in insect pheromones is their potential for use to control pests. In one method a large number of traps, baited with small amounts of the sex attractant of the female insect, are used to trap enough males that the breeding of the insects is decreased. In another method that requires fewer traps, a small number of traps are used to monitor the population of the target insect. The best time to apply pesticides can be determined by monitoring these traps. In one case, 10 to 15 applications of a pesticide to control the pink bollworm still resulted in damage to 30% of a cotton crop. This was decreased to almost no damage with only one to two pesticide applications when the ideal times for these applications were determined by the use of traps.

Insect pheromones provide ideal targets for industrial organic synthesis for a number of reasons. First, pheromones cannot be isolated in quantity from natural sources, so they must be synthesized if they are to be used. In addition, they are active in minute quantities, so the synthesis of large amounts is not necessary. Finally, most have relatively simple structures and can be synthesized in a relatively small number of steps. Of course, the stereochemistry of these steps must be carefully controlled.

Nigel Cattlin/Holt Studios International/Photo Researchers, Inc.

23.6   EXAMPLES OF SYNTHESES

This section presents several syntheses to further illustrate retrosynthetic analysis. First, let's consider the synthesis of dispalure, the sex pheromone of the gypsy moth, a serious despoiler of forests.

Dispalure

❶ Sodium amide, a strong base, was used to remove the acidic hydrogen of ethyne. The resulting acetylide ion was used as a nucleophile in an S_N2 substitution (see Section 10.8).

❷ This process was repeated to add the other alkyl group to the other side of the alkyne.

❸ Hydrogenation using the Lindlar catalyst resulted in the formation of the cis-alkene (see Section 11.12).

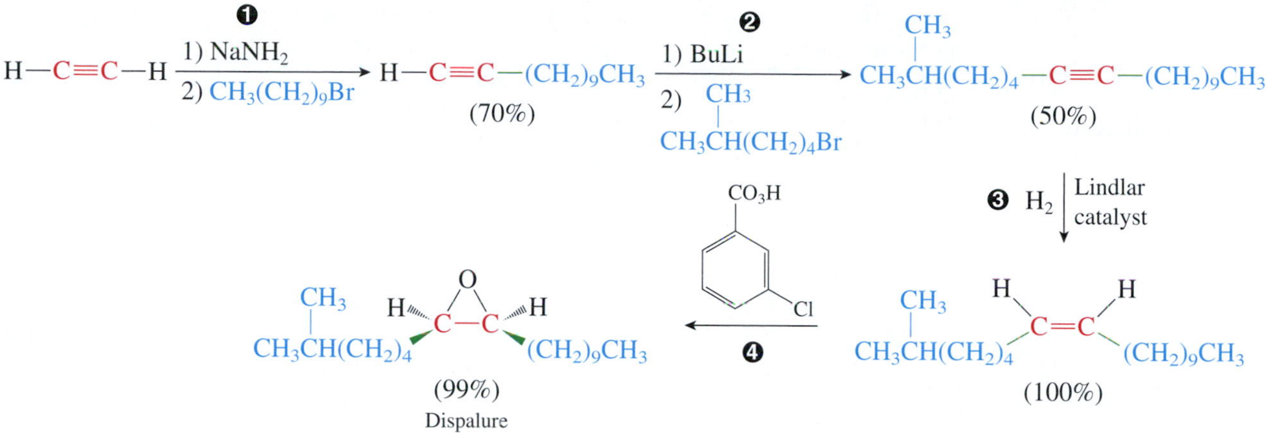

Dispalure
(99%)

(100%)

❹ The synthesis was completed by epoxidation of the alkene with m-chloroperbenzoic acid. This resulted in the formation of the epoxide by a syn addition (see Section 11.9).

Figure 23.3

THE SYNTHESIS OF DISPALURE.

Using retrosynthetic analysis, we recognize that the cis-epoxide can be prepared from the cis-alkene. The cis-alkene can be prepared by catalytic hydrogenation of an alkyne. Finally, substituted alkynes can be prepared by nucleophilic substitution reactions using acetylide ion nucleophiles (see Section 10.8). On the basis of this analysis, the synthesis reported in the literature was accomplished as shown in Figure 23.3.

As another example, let's consider the synthesis of brevicomin, the aggregation pheromone of the Western pine beetle. First, it is necessary to recognize that brevicomin is an acetal and can be disconnected into a diol and a ketone. The diol can be prepared by hydroxylation of an alkene, and the alkene can be prepared by a Wittig reaction:

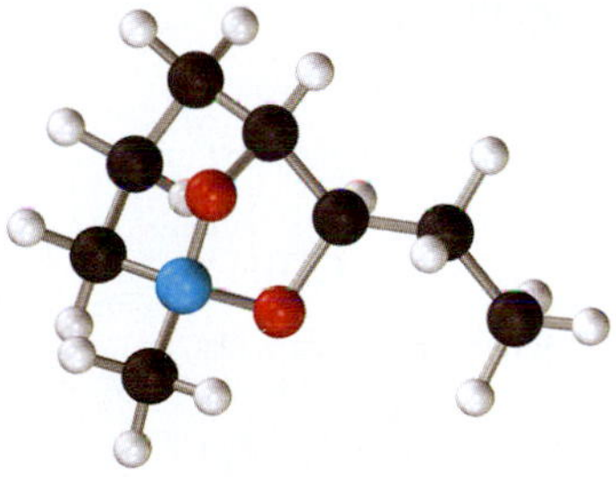

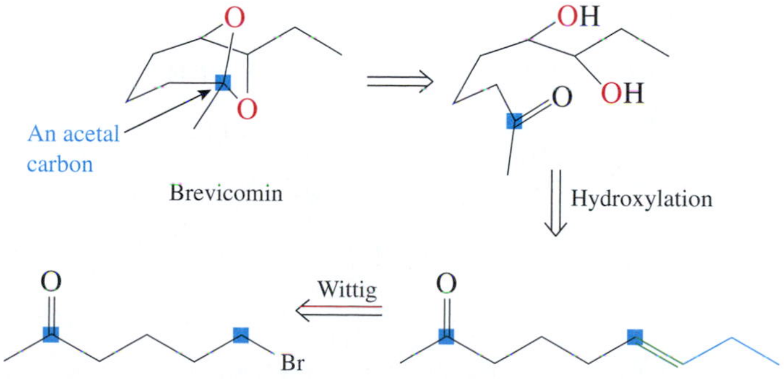

Brevicomin

The reported synthesis was accomplished as shown in Figure 23.4.

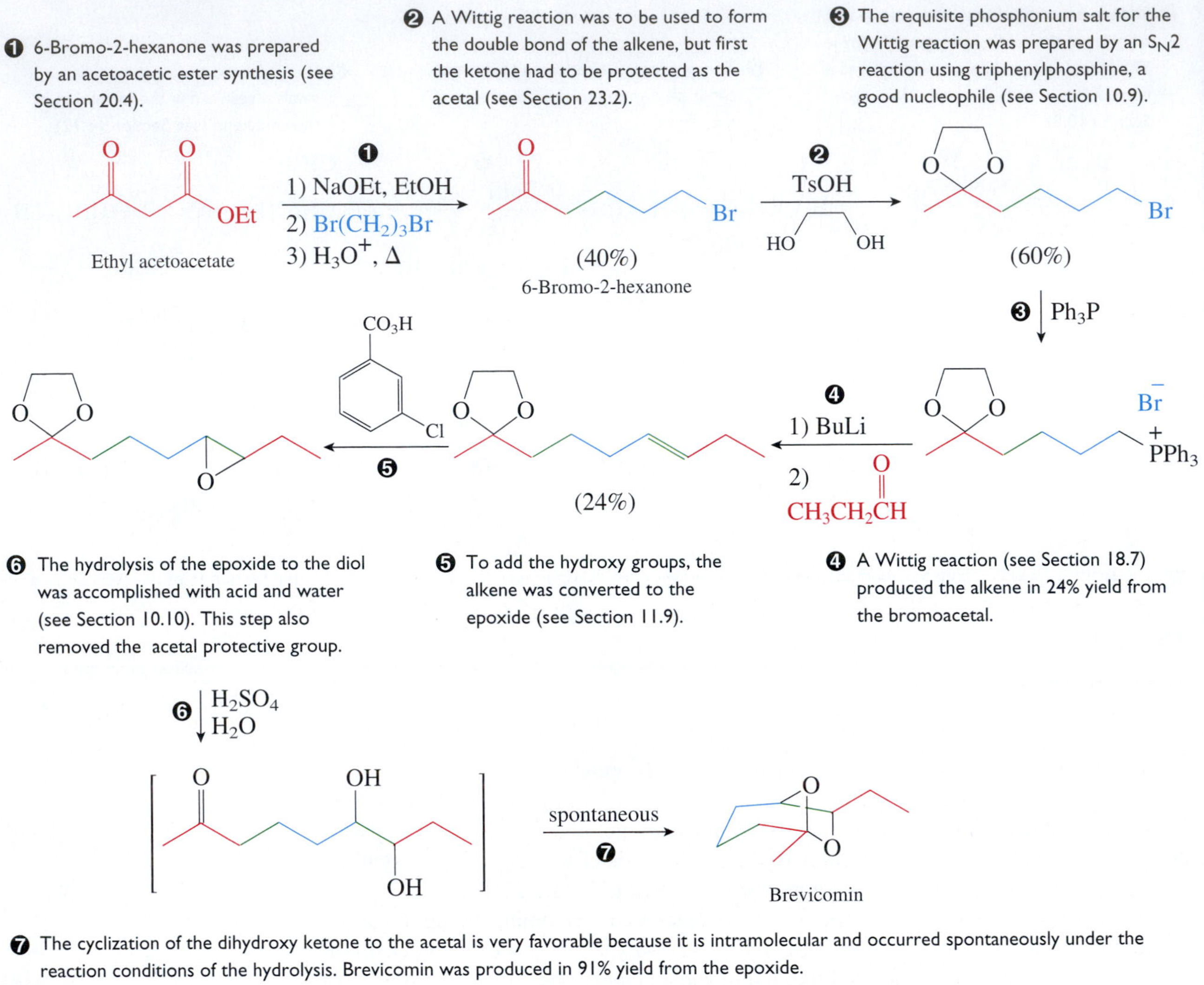

① 6-Bromo-2-hexanone was prepared by an acetoacetic ester synthesis (see Section 20.4).

② A Wittig reaction was to be used to form the double bond of the alkene, but first the ketone had to be protected as the acetal (see Section 23.2).

③ The requisite phosphonium salt for the Wittig reaction was prepared by an S_N2 reaction using triphenylphosphine, a good nucleophile (see Section 10.9).

Ethyl acetoacetate

1) NaOEt, EtOH
2) Br(CH₂)₃Br
3) H₃O⁺, Δ

(40%)
6-Bromo-2-hexanone

TsOH
HO   OH

(60%)

③ Ph₃P

⑥ The hydrolysis of the epoxide to the diol was accomplished with acid and water (see Section 10.10). This step also removed the acetal protective group.

⑤ To add the hydroxy groups, the alkene was converted to the epoxide (see Section 11.9).

④ A Wittig reaction (see Section 18.7) produced the alkene in 24% yield from the bromoacetal.

(24%)

④
1) BuLi
2) CH₃CH₂CH (O)

⑥ H₂SO₄ / H₂O

spontaneous **⑦**

Brevicomin

⑦ The cyclization of the dihydroxy ketone to the acetal is very favorable because it is intramolecular and occurred spontaneously under the reaction conditions of the hydrolysis. Brevicomin was produced in 91% yield from the epoxide.

Figure 23.4

THE SYNTHESIS OF BREVICOMIN.

Next, let's consider the synthesis of oxanamide, a tranquilizer:

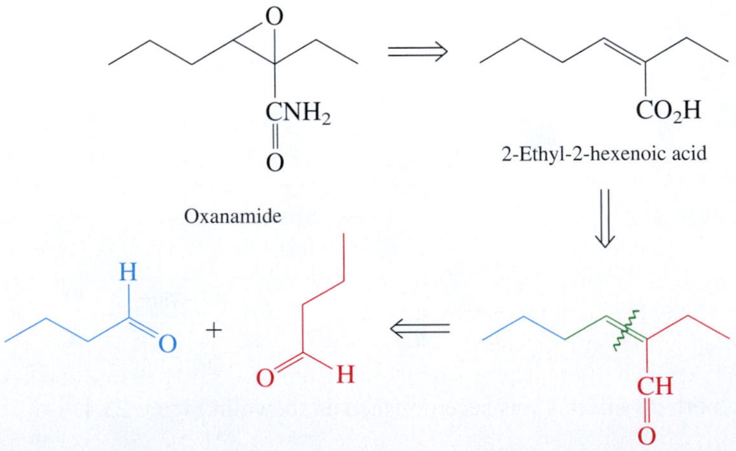

Oxanamide

2-Ethyl-2-hexenoic acid

❶ A simple aldol condensation (see Section 20.5) produced the desired α,β-unsaturated aldehyde in excellent yield.

❷ Oxidation of the aldehyde to a carboxylic acid was accomplished with silver oxide (see Section 10.14).

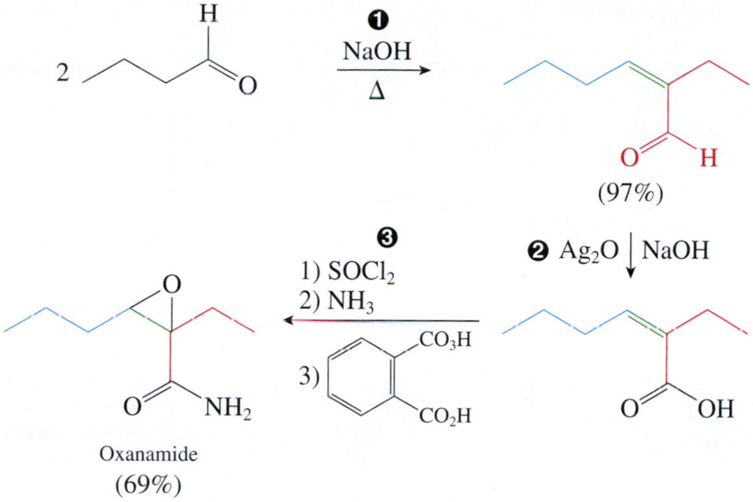

Figure 23.5

THE SYNTHESIS OF OXANAMIDE.

(97%)

Oxanamide
(69%)

❸ The amide was prepared from the acyl chloride (Section 19.6) and then epoxidized (Section 11.9) in 69% overall yield.

The epoxide can be prepared from an alkene and the amide from a carboxylic acid. The new target, 2-ethyl-2-hexenoic acid, has a CC double bond in conjugation with the carbonyl group of the carboxylic acid. Whenever a compound with an α,β-unsaturated carbonyl group is encountered, it is worthwhile to consider the possibility of using an aldol condensation (see Section 20.5) or a related reaction to prepare it. To examine this possibility, the aldehyde that will provide the carboxylic acid upon oxidation is disconnected at the double bond. Because both fragments produced by this disconnection are the same, it is apparent that an aldol condensation of butanal can be employed to prepare this compound. The synthesis was accomplished as shown in Figure 23.5.

23.7 REACTIONS THAT FORM CARBON–CARBON BONDS

We have seen in this and previous chapters that carbon–carbon bond-forming reactions are of critical importance in designing a synthesis of an organic compound. As discussed in Section 23.6, it is often helpful to examine a synthetic target for certain structural features that suggest the use of a particular one of these reactions. Some of the more important disconnections that suggest specific carbon–carbon bond-forming reactions are listed in Table 23.1. (Remember, however, that there are many other ways to prepare these and other target compounds.)

Table 23.2 lists most of the carbon–carbon bond-forming reactions that have been presented in this text. It may prove helpful to consult this table when working synthesis problems. The more familiar you are with these reactions, the easier designing a synthesis will be.

Table 23.1   Important Disconnections and the Reactions They Suggest

Disconnection	Suggested Reaction
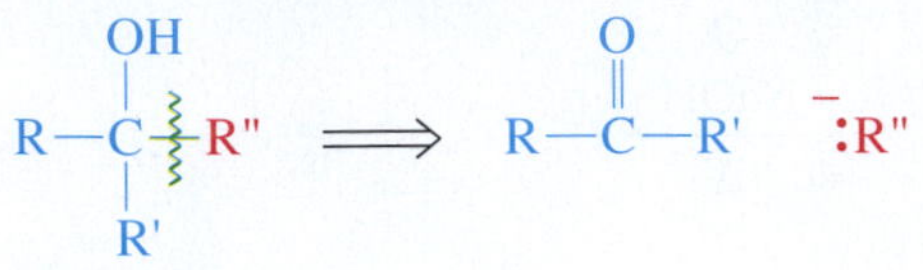	Grignard reaction
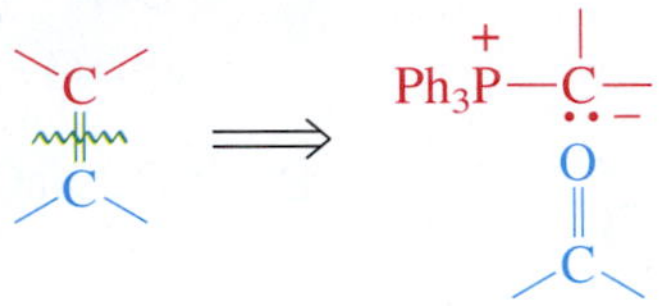	Wittig reaction
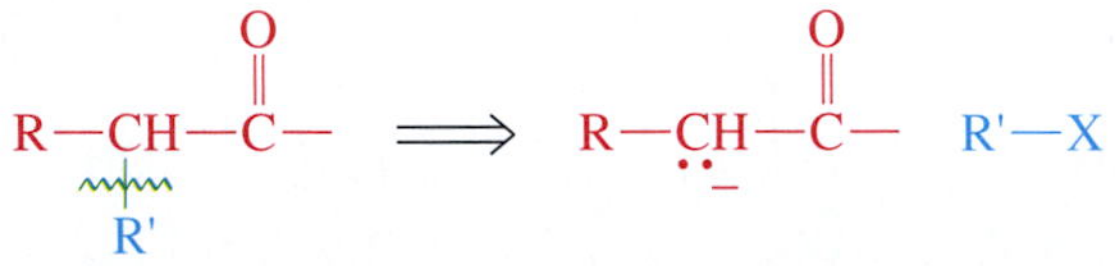	Enolate alkylation, acetoacetic ester synthesis, or malonic ester synthesis
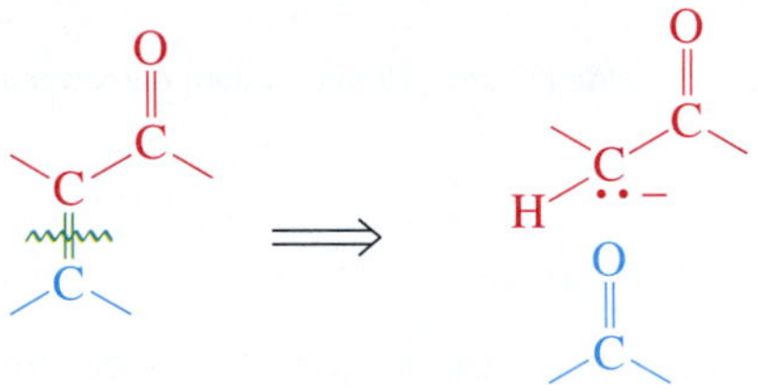	Aldol condensation
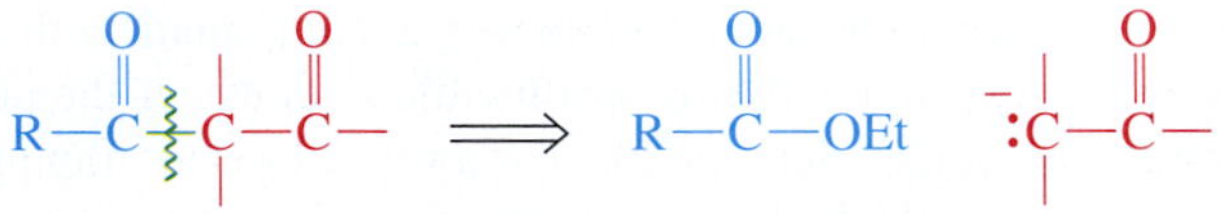	Ester condensation
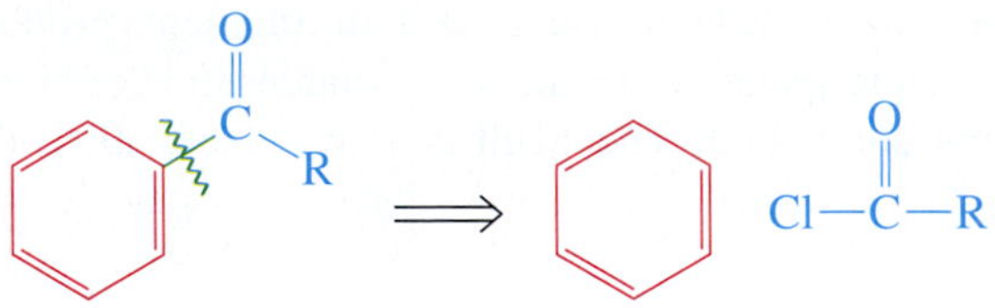	Friedel-Crafts acylation
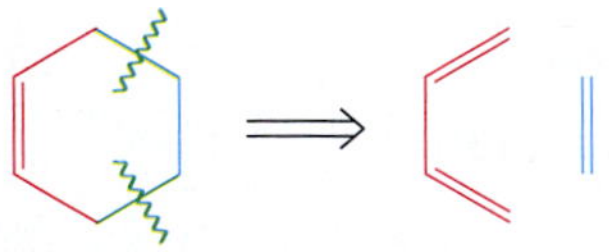	Diels-Alder reaction

23.8   PREPARATION OF FUNCTIONAL GROUPS

Table 23.3 lists most of the reactions presented in preceding chapters that are useful in synthesis, including the reactions of Table 23.2. The reactions are grouped according to the functional group that is present in the product, rather than by their mechanisms. Thus, if an alcohol is the synthetic target, consultation of the appropriate section of the table will reveal a variety of methods that can be used to prepare alcohols.

Text continued on p. 1044.

Table 23.2   Some Important Carbon–Carbon Bond-Forming Reactions

Reaction	Section	Comments
$R-L + {}^-CN \longrightarrow R-CN$	10.8	S_N2, 1° and 2° R OK
$R-L + {}^-\!:C\equiv CR' \longrightarrow R-C\equiv CR'$	10.8	S_N2, 1° R only
[benzene] $\xrightarrow[\text{AlCl}_3]{\text{RX}}$ [benzene]—R	17.7	Friedel-Crafts alkylation, does not work for deactivated rings
[benzene] $\xrightarrow[\text{AlCl}_3]{\text{RCCl}}$ [benzene]—C(=O)—R	17.8	Friedel-Crafts acylation, does not work for strongly deactivated rings
$R-\overset{O}{\overset{\|}{C}}-R' \xrightarrow[\text{2) H}_3\text{O}^+]{\text{1) R''MgX or R''Li}} R-\overset{OH}{\underset{R''}{\overset{\|}{C}}}-R'$	18.6	Grignard or organolithium addition
$R-\overset{O}{\overset{\|}{C}}-R' \xrightarrow{\text{Ph}_3\overset{+}{\text{P}}-\overset{-}{\ddot{\text{C}}}\text{R}''_2} R-\overset{CR''_2}{\overset{\|}{C}}-R'$	18.7	Wittig reaction
[conjugate enone] $\xrightarrow{\text{R}_2\text{CuLi}}$ [1,4-adduct with R]	18.10	Conjugate addition of organocuprate reagents
$R-\overset{O}{\overset{\|}{C}}-OR' \xrightarrow[\text{2) H}_3\text{O}^+]{\text{1) 2 R''MgX or 2 R''Li}} R-\overset{OH}{\underset{R''}{\overset{\|}{C}}}-R''$	19.9	Grignard reagents or organolithium reagents plus esters give tertiary alcohols
$R-\overset{O}{\overset{\|}{C}}-Cl \xrightarrow{\text{R}'_2\text{CuLi}} R-\overset{O}{\overset{\|}{C}}-R'$	19.10	Acyl chlorides plus organocuprate reagents give ketones
$R-C\equiv N \xrightarrow[\text{2) H}_3\text{O}^+]{\text{1) R'MgBr}} R-\overset{O}{\overset{\|}{C}}-R'$	19.10	Nitriles plus Grignard reagents give ketones
$R'-\overset{O}{\overset{\|}{C}}-CH_2R \xrightarrow[\text{2) R''X}]{\text{1) LDA}} R'-\overset{O}{\overset{\|}{C}}-\overset{R''}{\overset{\|}{C}}HR$	20.3	Alkylation of ketones, S_N2, R'' may be 1° (best) or 2°

Continued

Table 23.2 **Some Important Carbon–Carbon Bond-Forming Reactions—cont'd**

Reaction	Section	Comments
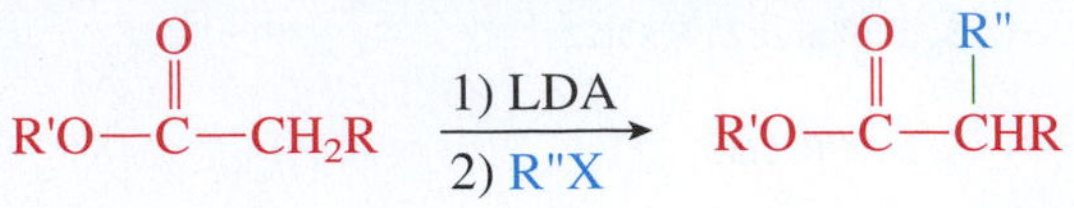 $R'O-C(=O)-CH_2R \xrightarrow[\text{2) R''X}]{\text{1) LDA}} R'O-C(=O)-CHR(R'')$	20.3	Alkylation of esters, S_N2, R″ may be 1° (best) or 2°
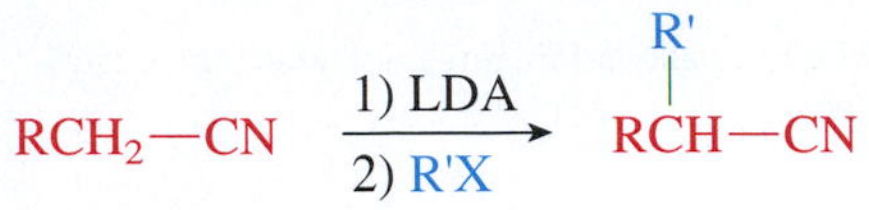 $RCH_2-CN \xrightarrow[\text{2) R'X}]{\text{1) LDA}} RCH(R')-CN$	20.3	Alkylation of nitriles, S_N2, R′ may be 1° (best) or 2°
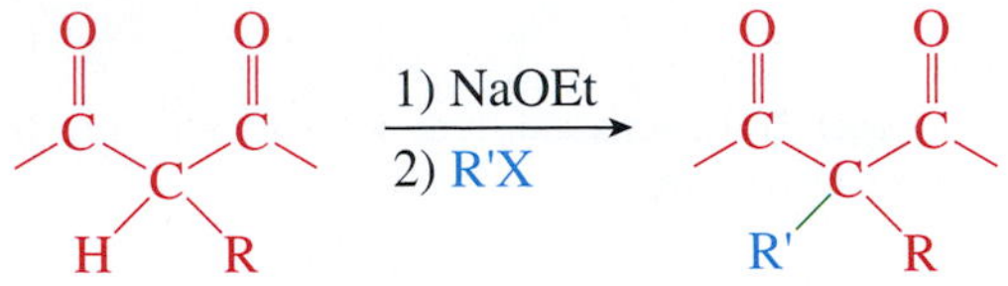 $(O=)C-C(H)(R)-C(=O) \xrightarrow[\text{2) R'X}]{\text{1) NaOEt}} (O=)C-C(R')(R)-C(=O)$	20.4	Alkylation of stabilized anions, acetoacetic ester and malonic ester syntheses
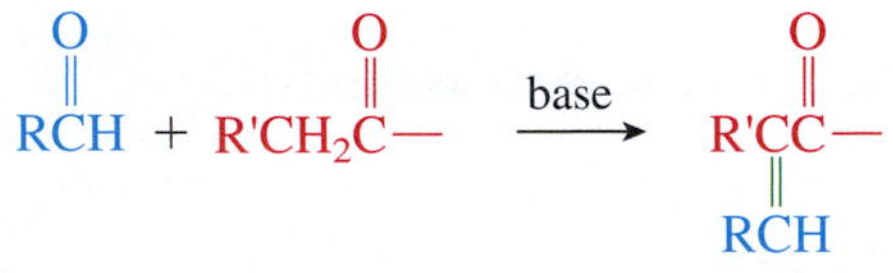 $RCH(=O) + R'CH_2C(=O)- \xrightarrow{\text{base}} R'CC(=O)-(=CHR)$	20.5	Aldol condensation
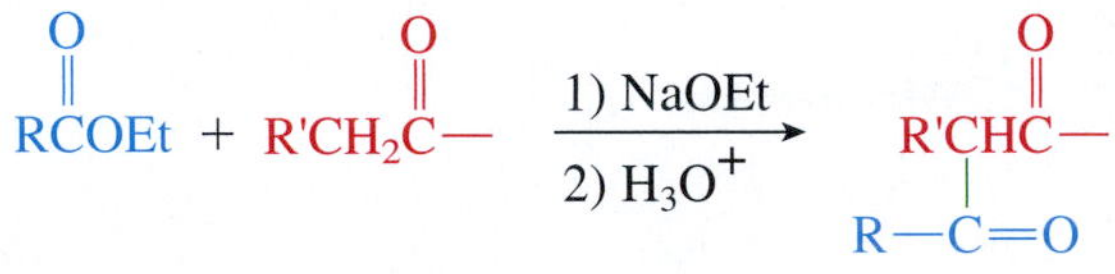 $RCOEt + R'CH_2C(=O)- \xrightarrow[\text{2) H}_3\text{O}^+]{\text{1) NaOEt}} R'CHC(=O)-(R-C=O)$	20.6	Ester condensation
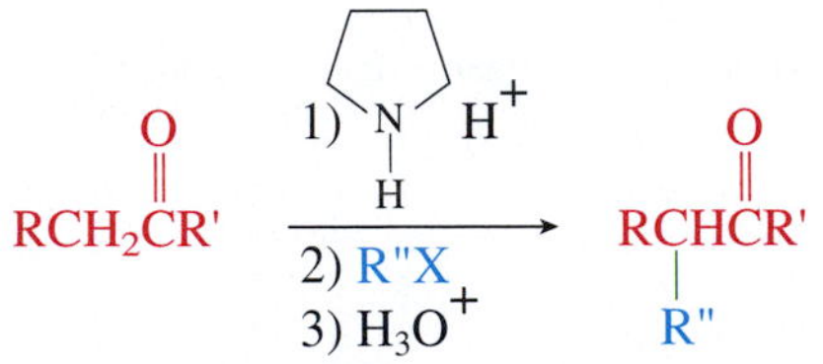 $RCH_2CR'(=O) \xrightarrow[\substack{\text{2) R''X} \\ \text{3) H}_3\text{O}^+}]{\text{1) pyrrolidine } N-H^+} RCHCR'(=O)(R'')$	20.8	Enamine reation, R″X must be very reactive
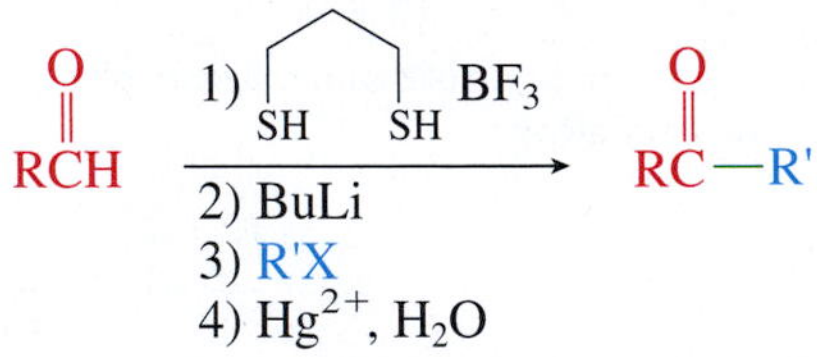 $RCH(=O) \xrightarrow[\substack{\text{2) BuLi} \\ \text{3) R'X} \\ \text{4) Hg}^{2+}, \text{H}_2\text{O}}]{\text{1) (SH)(SH)/BF}_3} RC(=O)-R'$	20.9	Dithiane alkylation, acyl anion equivalent
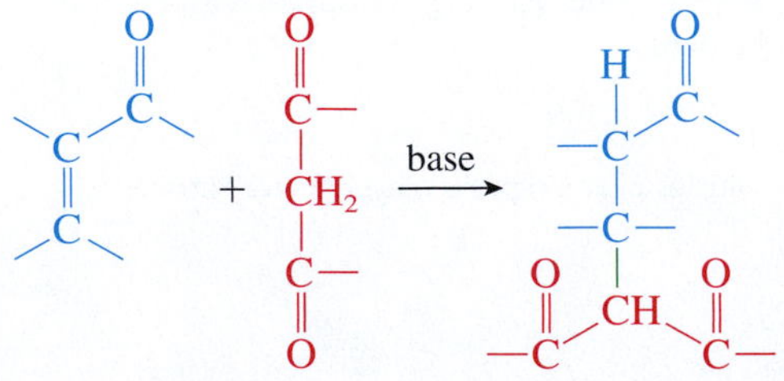 (conjugate addition structures) + $CH_2 \xrightarrow{\text{base}}$ (product)	20.10	Conjugate addition of stabilized anions, Michael reaction
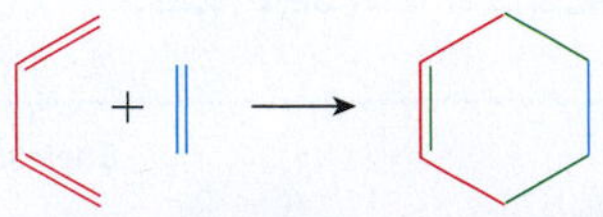 (diene) + (dienophile) → (cyclohexene)	22.6	Diels-Alder reaction

Table 23.3 **Preparation of Functional Groups**

	Reaction	Section	Comments
Alkanes	$R-L \xrightarrow{\text{LiAlH}_4 \text{ or NaBH}_4} R-H$	10.7	S_N2, 1° and 2° OK
	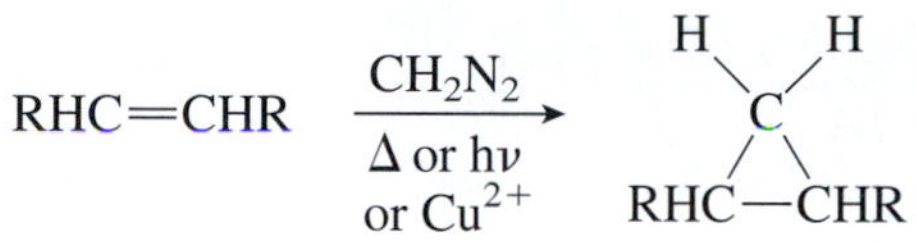 $RHC=CHR \xrightarrow[\Delta \text{ or } h\nu \text{ or Cu}^{2+}]{\text{CH}_2\text{N}_2}$ gives cyclopropane $\begin{smallmatrix}H \quad H \\ C \\ RHC-CHR\end{smallmatrix}$	11.8	Syn addition
	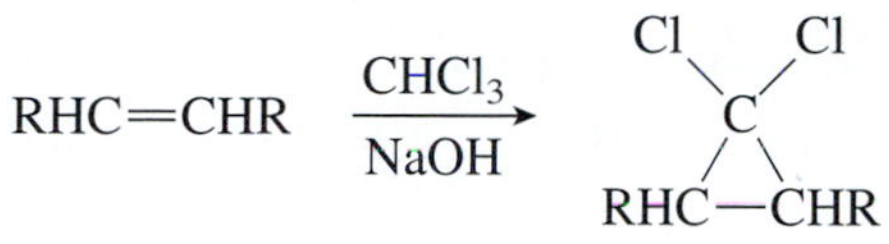 $RHC=CHR \xrightarrow{\text{CHCl}_3 / \text{NaOH}}$ gives cyclopropane $\begin{smallmatrix}Cl \quad Cl \\ C \\ RHC-CHR\end{smallmatrix}$	11.8	Syn addition
	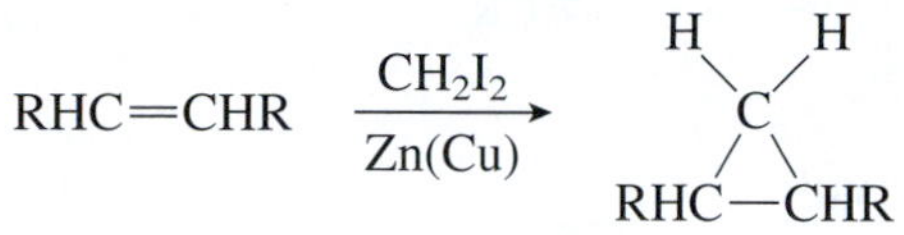 $RHC=CHR \xrightarrow{\text{CH}_2\text{I}_2 / \text{Zn(Cu)}}$ gives cyclopropane $\begin{smallmatrix}H \quad H \\ C \\ RHC-CHR\end{smallmatrix}$	11.8	Simmons-Smith reaction, syn addition
	$\begin{matrix}RHC=CHR \\ \text{or} \\ RC\equiv CR\end{matrix} \xrightarrow{\text{H}_2 / \text{Ni, Pd, or Pt}} RCH_2CH_2R$	11.12	Syn addition
	$R-\overset{O}{\underset{\|}{C}}-R' \xrightarrow[\text{NH}_2\text{NH}_2 \text{ KOH, }\Delta]{\text{Zn(Hg), HCl or}} R-CH_2-R'$	17.13	Clemmensen reduction (H_2/Pd can be used when R = Ar)
		18.8	Wolff-Kishner reduction
	$RBr \xrightarrow{\text{Bu}_3\text{SnH} / [\text{AIBN}]} RH$	21.7	Radical chain mechanism
Alkenes	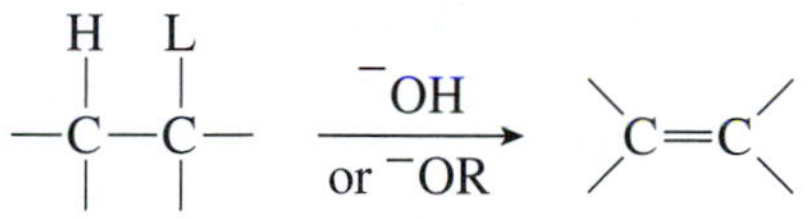 $\begin{smallmatrix}H \quad L \\ -C-C- \end{smallmatrix} \xrightarrow[\text{or } ^-\text{OR}]{^-\text{OH}} \,C=C\,$	10.11	E2, Zaitsev's rule, usually anti elimination
	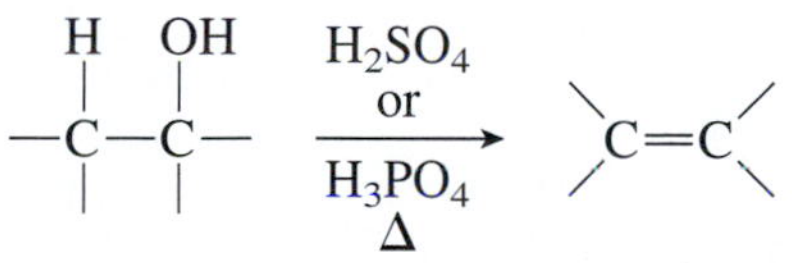 $\begin{smallmatrix}H \quad OH \\ -C-C- \end{smallmatrix} \xrightarrow[\text{H}_3\text{PO}_4 \, \Delta]{\text{H}_2\text{SO}_4 \text{ or}} \,C=C\,$	10.13	Usually E1, Zaitsev's rule
	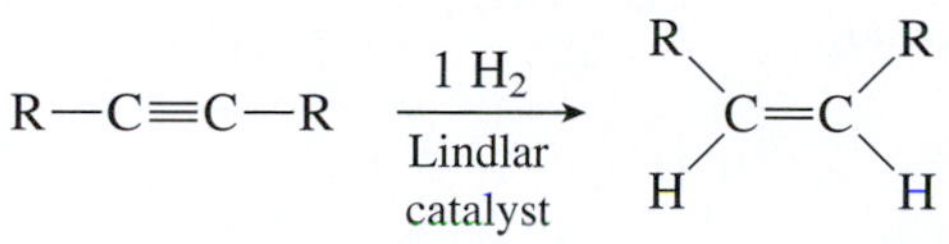 $R-C\equiv C-R \xrightarrow[\text{Lindlar catalyst}]{1\ \text{H}_2} \begin{smallmatrix}R \qquad R \\ C=C \\ H \qquad H\end{smallmatrix}$	11.12	Syn addition

Continued

Table 23.3   Preparation of Functional Groups—cont'd

	Reaction	Section	Comments
Alkenes —cont'd	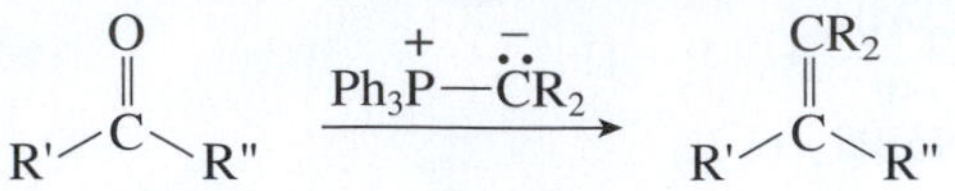 $R'-\underset{\underset{O}{\shortparallel}}{C}-R'' \xrightarrow{\overset{+}{Ph_3P}-\overset{-}{\ddot{C}R_2}} R'-\underset{\underset{CR_2}{\shortparallel}}{C}-R''$	18.7	Wittig reaction
	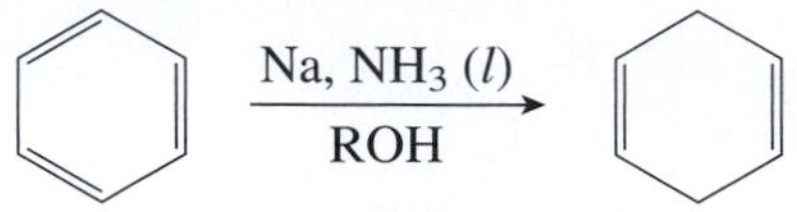 benzene $\xrightarrow[ROH]{Na,\ NH_3\ (l)}$ 1,4-cyclohexadiene	21.10	Birch reduction
	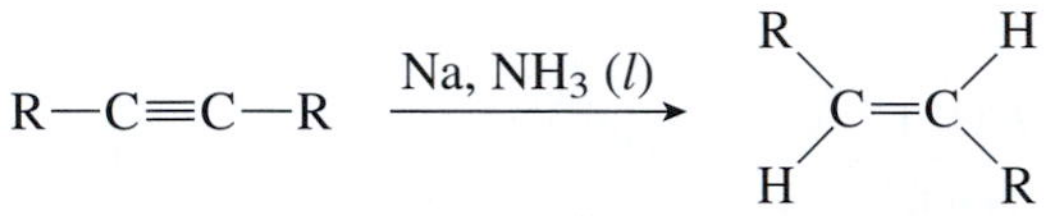 $R-C\equiv C-R \xrightarrow{Na,\ NH_3\ (l)} \underset{H}{\overset{R}{>}}C=C\underset{R}{\overset{H}{<}}$	21.10	Anti addition
Alkynes	$R-L \xrightarrow{\ ^-:C\equiv CR'\ } R-C\equiv CR'$	10.8	S_N2, 1° R only
	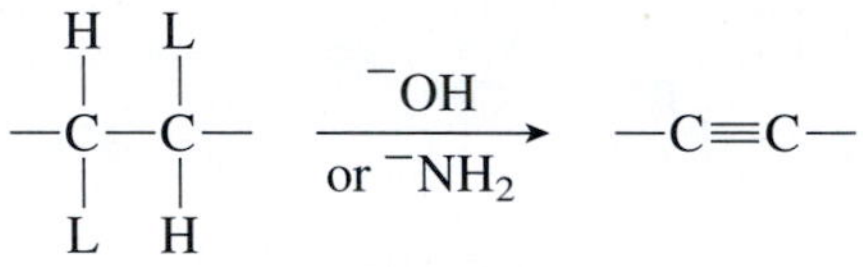 $-\underset{\underset{L}{\overset{H}{\vert}}}{C}-\underset{\underset{H}{\overset{L}{\vert}}}{C}- \xrightarrow[or\ ^-NH_2]{^-OH} -C\equiv C-$	10.12	E2
	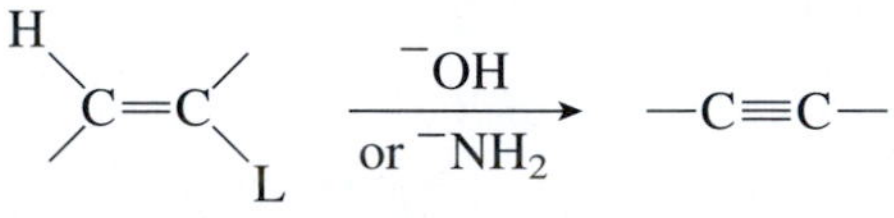 $\underset{\diagdown}{\overset{H}{\diagup}}C=C\underset{L}{\diagdown} \xrightarrow[or\ ^-NH_2]{^-OH} -C\equiv C-$	10.12	E2
Alcohols	$R-X \xrightarrow{^-OH} R-OH$	10.2	S_N2, best with 1° R
	$R-X \xrightarrow{H_2O} R-OH$	10.2	S_N1, OK with 2° and 3° R
	$R-X \xrightarrow[2)\ KOH,\ H_2O]{1)\ CH_3CO_2^-} R-OH$	10.2	S_N2, good with 1° and 2° R, synthetic equivalent of hydroxide ion
	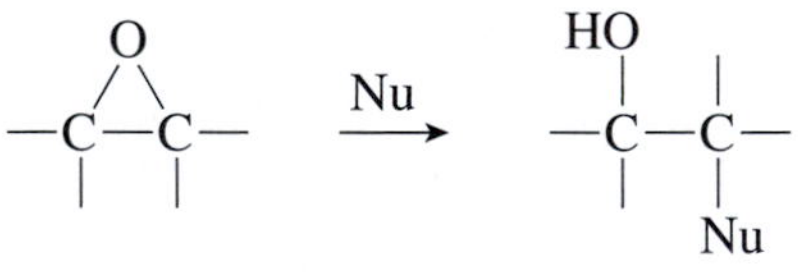 $-\underset{\vert}{\overset{\diagup O \diagdown}{C}}-\underset{\vert}{C}- \xrightarrow{Nu} -\underset{\underset{\vert}{\overset{\overset{HO}{\vert}}{C}}}{}-\underset{\underset{Nu}{\vert}}{C}-$	10.10	Epoxide opening
	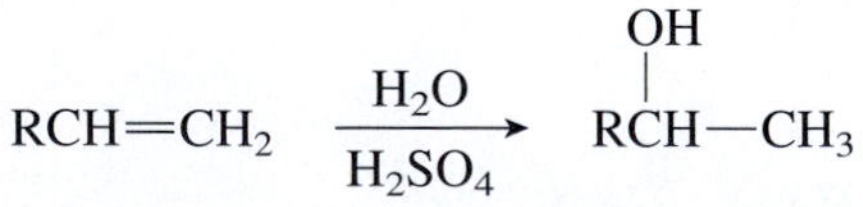 $RCH=CH_2 \xrightarrow[H_2SO_4]{H_2O} RCH-CH_3$ (OH)	11.3	Hydration, low yields, rearrangements may occur, Markovnikov's rule
	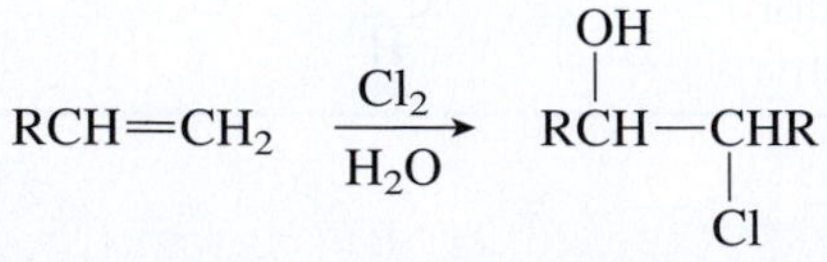 $RCH=CH_2 \xrightarrow[H_2O]{Cl_2} \underset{\underset{Cl}{}}{RCH}-CHR$ (OH)	11.5	Halohydrin formation, anti addition, OH on more substituted carbon

Table 23.3   Preparation of Functional Groups—cont'd

	Reaction	Section	Comments
Alcohols—cont'd	$RCH\!=\!CH_2$ $\xrightarrow[\text{2) NaBH}_4\text{, NaOH}]{\text{1) Hg(O}_2\text{CCH}_3)_2\text{, H}_2\text{O}}$ $RCH\!-\!CH_3$ with OH on RCH	11.6	Oxymercuration–reduction, no rearrangements, Markovnikov's rule
	$RCH\!=\!CH_2$ $\xrightarrow[\text{2) H}_2\text{O}_2\text{, NaOH}]{\text{1) BH}_3\text{, THF}}$ $RCH\!-\!CH_2$ with H, OH	11.7	Hydroboration–oxidation, syn addition, anti-Markovnikov
	$RCH\!=\!CHR$ $\xrightarrow[\text{KMnO}_4\text{, NaOH}]{\text{OsO}_4 \text{ or}}$ $RCH\!-\!CHR$ with OH, OH	11.10	Syn addition
	$R\!-\!\overset{O}{\underset{\|}{C}}\!-\!R'$ $\xrightarrow[\text{2) H}_3\text{O}^+]{\text{1) LiAlH}_4 \text{ or NaBH}_4}$ $R\!-\!\overset{OH}{\underset{H}{C}}\!-\!R'$	18.2	Reduction of aldehydes and ketones
	$R\!-\!\overset{O}{\underset{\|}{C}}\!-\!R'$ $\xrightarrow[\text{2) H}_3\text{O}^+]{\text{1) R''MgX or R''Li}}$ $R\!-\!\overset{OH}{\underset{R''}{C}}\!-\!R'$	18.6	Addition of Grignard and organolithium reagents to aldehydes and ketones
	$R\!-\!\overset{O}{\underset{\|}{C}}\!-\!OR'$ or $R\!-\!\overset{O}{\underset{\|}{C}}\!-\!OH$ $\xrightarrow[\text{2) H}_3\text{O}^+]{\text{1) LiAlH}_4}$ $R\!-\!\overset{OH}{\underset{H}{C}}\!-\!H$	19.7	Reduction of carboxylic acids and esters, LiAlH$_4$ required
	$R\!-\!\overset{O}{\underset{\|}{C}}\!-\!OR'$ $\xrightarrow[\text{2) H}_3\text{O}^+]{\text{1) 2 R''MgX or 2 R''Li}}$ $R\!-\!\overset{OH}{\underset{R''}{C}}\!-\!R''$	19.9	Addition of Grignard and organolithium reagents to esters, adds two identical R″ groups
Ethers	$R\!-\!L$ $\xrightarrow{\,^-\text{OR'}\,}$ $R\!-\!OR'$	10.3	Williamson ether synthesis, S$_N$2, best with 1° R
	$R\!-\!L$ $\xrightarrow{\text{HOR'}}$ $R\!-\!OR'$	10.3	S$_N$1, OK with 2° and 3° R

Continued

Table 23.3   **Preparation of Functional Groups—cont'd**

Reaction	Section	Comments
Epoxides		
$-\overset{\text{OH}}{\underset{\text{X}}{\text{C}-\text{C}}}-$ $\xrightarrow{\text{NaOH}}$ epoxide	11.5	Intramolecular S_N2
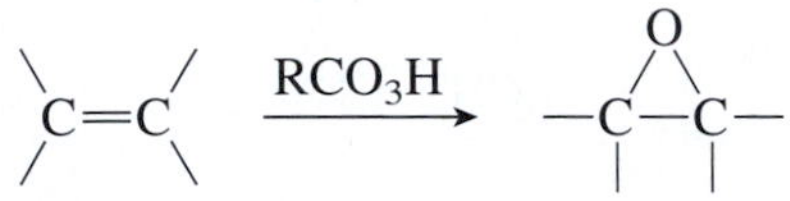 C=C $\xrightarrow{\text{RCO}_3\text{H}}$ epoxide	11.9	Syn addition
Amines R—L $\xrightarrow{\text{NH}_3}$ $\text{R—}\overset{+}{\text{NH}}_3$	10.6	S_N2, multiple alkylation problems
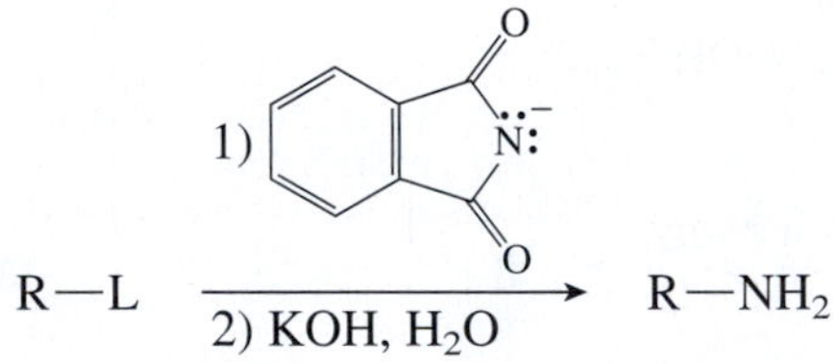 R—L $\xrightarrow[\text{2) KOH, H}_2\text{O}]{\text{1) phthalimide}}$ R—NH_2	10.6	Gabriel synthesis, S_N2, 1° and 2° R OK
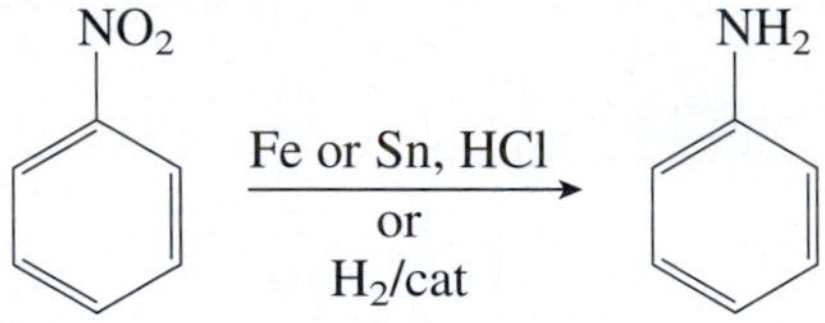 PhNO_2 $\xrightarrow[\text{H}_2/\text{cat}]{\text{Fe or Sn, HCl or}}$ PhNH_2	17.13	Reduction of aromatic nitro compounds
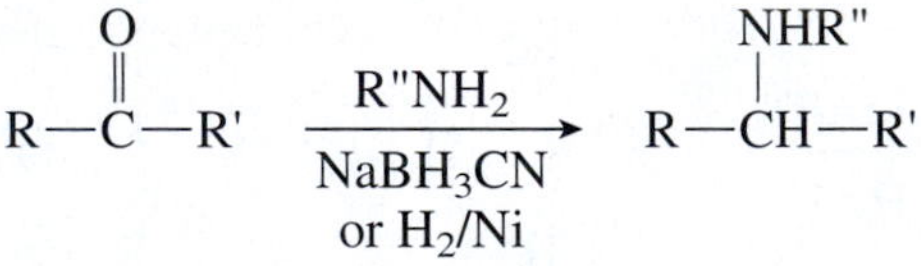 R—CO—R' $\xrightarrow[\text{NaBH}_3\text{CN or H}_2/\text{Ni}]{\text{R''NH}_2}$ R—CH(NHR'')—R'	18.8	Reductive amination
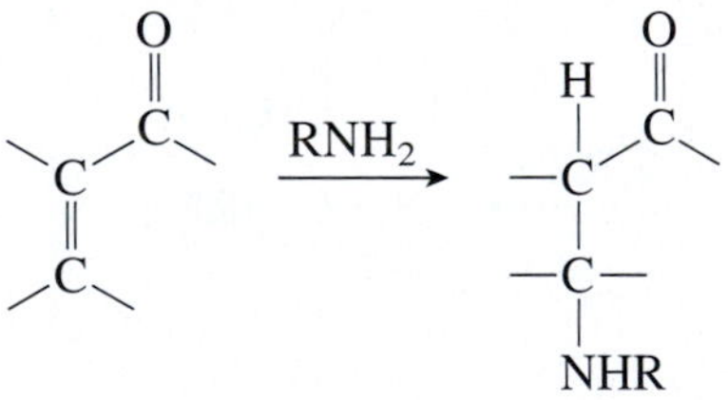 enone $\xrightarrow{\text{RNH}_2}$ conjugate adduct	18.10	Conjugate addition
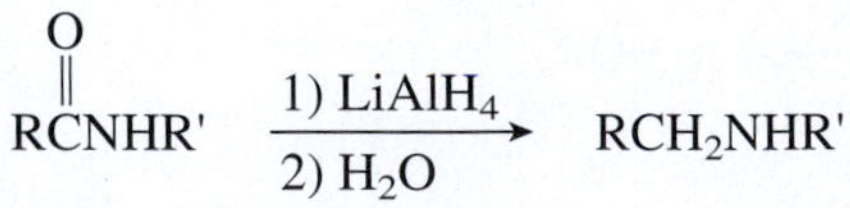 RCNHR' $\xrightarrow[\text{2) H}_2\text{O}]{\text{1) LiAlH}_4}$ $\text{RCH}_2\text{NHR'}$	19.7	Reduction of amides
$\text{R—C}\equiv\text{N}$ $\xrightarrow[\text{2) H}_2\text{O}]{\text{1) LiAlH}_4}$ RCH_2NH_2	19.7	Reduction of nitriles

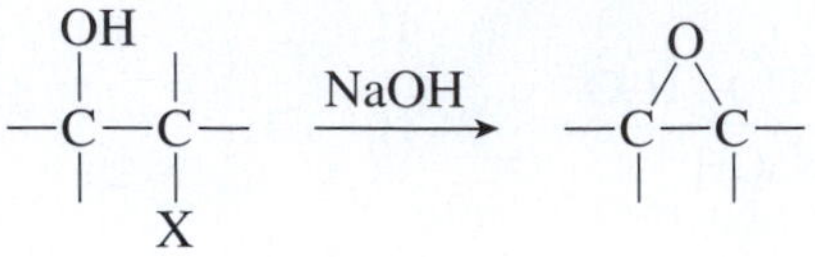

Table 23.3   Preparation of Functional Groups—cont'd

	Reaction	Section	Comments
Halides	$R-OH \xrightarrow{HX} R-X$	10.5	Chlorides require $ZnCl_2$ for 1° and 2° R
	$R-OTs \xrightarrow{X^-} R-X$	10.5	S_N2 conditions
	$R-OH \xrightarrow{SOCl_2} R-Cl$	10.5	Thionyl chloride
	$R-OH \xrightarrow{PBr_3} R-Br$	10.5	Phosphorus tribromide
	$R-OH \xrightarrow{PI_3} R-I$	10.5	Phosphorus triiodide
	$R-CH=CH_2 \xrightarrow{HX} R-\overset{X}{\underset{}{C}}H\overset{H}{\underset{}{C}}H_2$	11.2	Markovnikov's rule
	$RCH=CHR' \xrightarrow[\text{or } Br_2]{Cl_2} RCH\overset{X}{-}CHR'\underset{X}{}$	11.4	Anti addition
	$RCH=CHR' \xrightarrow[H_2O]{X_2} RCH\overset{X}{-}CHR'\underset{OH}{}$	11.5	Halohydrin formation, anti addition, Markovnikov's rule
	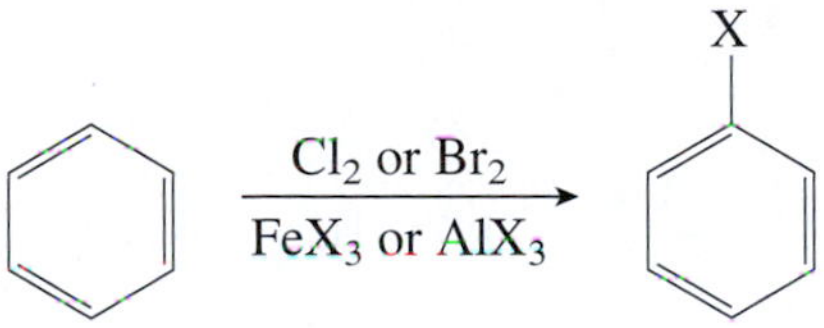 (benzene) $\xrightarrow[FeX_3 \text{ or } AlX_3]{Cl_2 \text{ or } Br_2}$ (aryl-X)	17.5	Electrophilic aromatic substitution
	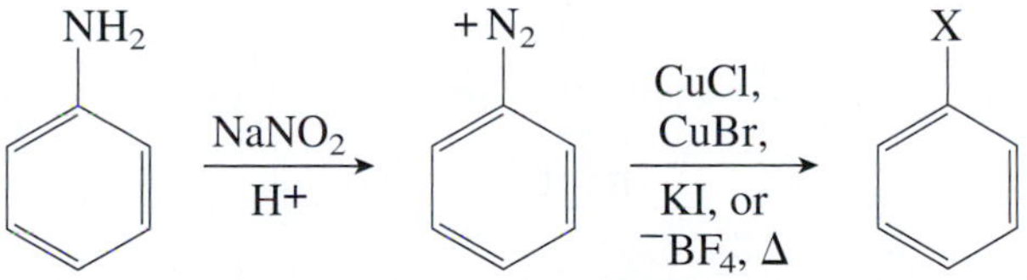 (aniline, NH_2) $\xrightarrow[H+]{NaNO_2}$ (diazonium, $+N_2$) $\xrightarrow[^-BF_4, \Delta]{\substack{CuCl, \\ CuBr, \\ KI, \text{ or}}}$ (aryl-X)	17.10	Substitutions via diazonium ion
	$R\overset{O}{\overset{\|}{C}}CH_3 \xrightarrow[HX]{X_2} R\overset{O}{\overset{\|}{C}}CH_2X$	20.2	Bromination or chlorination

Continued

Table 23.3   Preparation of Functional Groups—cont'd

Reaction	Section	Comments
Halides—cont'd — CH_3(benzene) $\xrightarrow[\text{NBS, initiator}]{\text{Br}_2,\ h\nu\ \text{or}}$ CH_2Br(benzene)	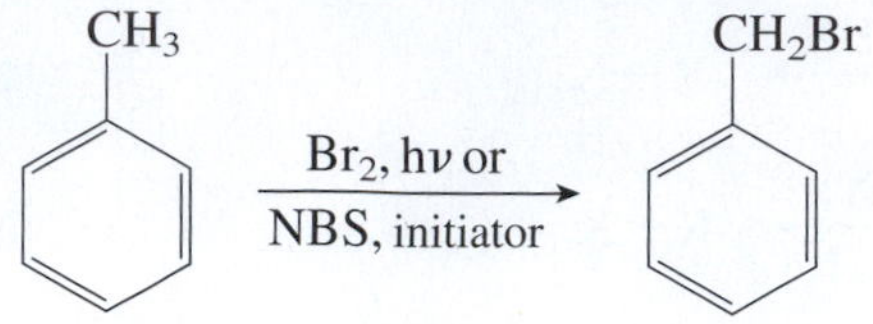 21.6	Used for allylic bromides also, radical chain mechanism
$RCH{=}CH_2 \xrightarrow[\text{initiator}]{\text{HBr}} RCH{-}CH_2$ (H, Br)	21.9	Anti-Markovnikov regiochemistry, HBr only, radical chain mechanism
Aldehydes — $\underset{RCH_2}{\overset{OH}{\mid}} \xrightarrow{CrO_3 \cdot 2\,\text{pyridine}} \underset{RCH}{\overset{O}{\|}}$	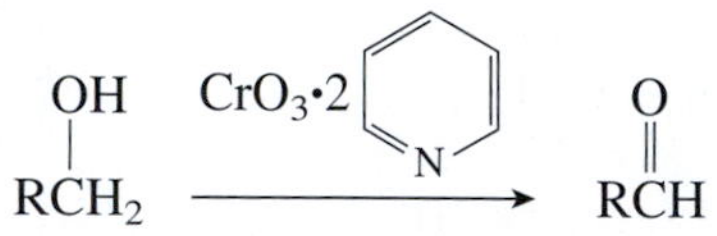 10.14	Oxidation using chromium trioxide–pyridine complex
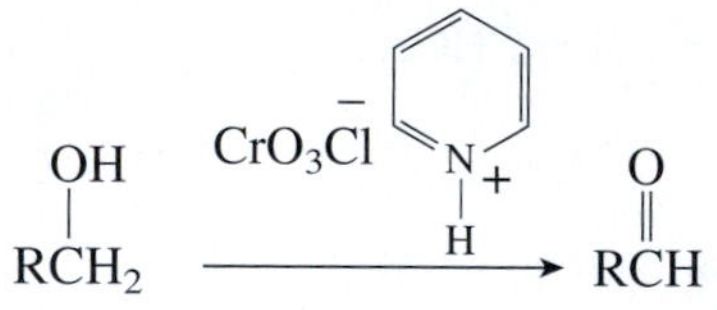 $\underset{RCH_2}{\overset{OH}{\mid}} \xrightarrow{CrO_3Cl^-\ \text{pyridinium}} \underset{RCH}{\overset{O}{\|}}$	10.14	Oxidation using pyridinium chlorochromate (PCC)
$R{-}C{\equiv}C{-}H \xrightarrow[\text{2) H}_2\text{O}_2,\ \text{NaOH}]{\text{1) disiamylborane}} R{-}CH_2CH{=}O$	11.7	Hydroboration of terminal alkynes
$\underset{RC{-}Cl}{\overset{O}{\|}} \xrightarrow[-78°C]{\text{LiAlH(O}t\text{-Bu)}_3} \underset{RC{-}H}{\overset{O}{\|}}$	19.8	Selective reduction of acyl chlorides
$\underset{RC{-}OR}{\overset{O}{\|}} \xrightarrow[\text{2) H}_3\text{O}^+]{\text{1) DIBALH, }-78°C} \underset{RC{-}H}{\overset{O}{\|}}$	19.8	Selective reduction of esters with diisobutylaluminum hydride
$\underset{RCH}{\overset{O}{\|}} + \underset{R'CH_2C{-}H}{\overset{O}{\|}} \xrightarrow{\text{base}} \underset{\underset{RCH}{\|}}{\overset{O}{\overset{\|}{R'CC{-}H}}}$	20.5	Aldol condensation
Ketones — $\underset{RCHR'}{\overset{OH}{\mid}} \xrightarrow[\text{H}_2\text{SO}_4]{\text{Na}_2\text{Cr}_2\text{O}_7} \underset{RCR'}{\overset{O}{\|}}$ or $CrO_3 \cdot 2\,\text{pyridine}$ or $CrO_3Cl^-\ \text{pyridinium}$	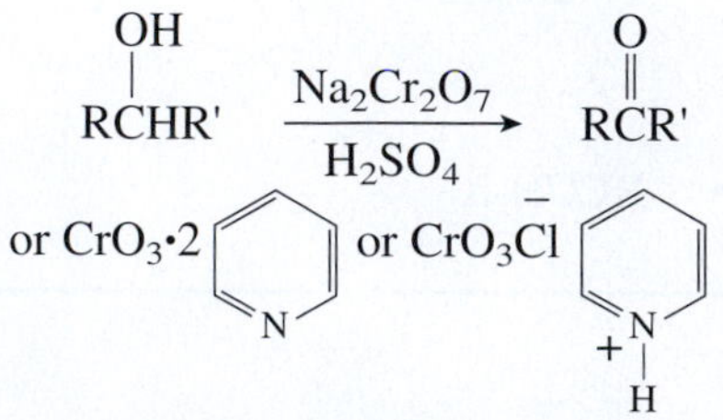 10.14	Oxidation

Table 23.3   Preparation of Functional Groups—cont'd

	Reaction	Section	Comments
Ketones— cont'd	$R-C \equiv C-H \xrightarrow[\substack{H_2SO_4 \\ HgSO_4}]{H_2O} RCCH_3$ (O double bond on C)	11.6	Hydration of alkynes, Markovnikov regiochemistry, alkyne should be terminal or symmetrical
	$R-C \equiv C-R \xrightarrow[\text{2) } H_2O_2, \text{ NaOH}]{\text{1) } BH_3, \text{ THF}} RCCH_2R$ (O double bond)	11.7	Hydroboration of alkynes, alkyne should be symmetrical
	benzene $\xrightarrow[AlCl_3]{RCCl}$ phenyl ketone $\overset{O}{\overset{\|}{C}}-R$	17.8	Friedel-Crafts acylation, does not work when ring has strong deactivating substituents
	$R-\overset{O}{\overset{\|}{C}}-Cl \xrightarrow{R'_2CuLi} R-\overset{O}{\overset{\|}{C}}-R'$	19.10	Addition of organocuprate reagents to acyl chlorides
	$R-C \equiv N \xrightarrow[\text{2) } H_3O^+]{\text{1) } R'MgBr} R-\overset{O}{\overset{\|}{C}}-R'$	19.10	Addition of Grignard reagents to nitriles
	$R'-\overset{O}{\overset{\|}{C}}-CH_2R \xrightarrow[\text{2) } R''X]{\text{1) LDA}} R'-\overset{O}{\overset{\|}{C}}-\overset{R''}{\overset{\|}{C}}HR$	20.3	Alkylation of ketones, S_N2, R'' may be 1° (best) or 2°, regiochemistry must be controlled somehow
	$H_3C \overset{O}{\overset{\|}{C}} CH_2 \overset{O}{\overset{\|}{C}} OEt \xrightarrow[\substack{\text{2) RL} \\ \text{3) KOH, } H_2O \\ \text{4) } H_3O^+, \Delta}]{\text{1) NaOEt}} H_3C \overset{O}{\overset{\|}{C}} CH_2R$	20.4	Acetoacetic ester synthesis, S_N2, R must be 1° or 2°
	$RCH + R'CH_2 \overset{O}{\overset{\|}{C}}-R'' \xrightarrow[\Delta]{\text{base}} R'C \overset{O}{\overset{\|}{C}}-R''$ with $\|$ RCH	20.5	Aldol condensation

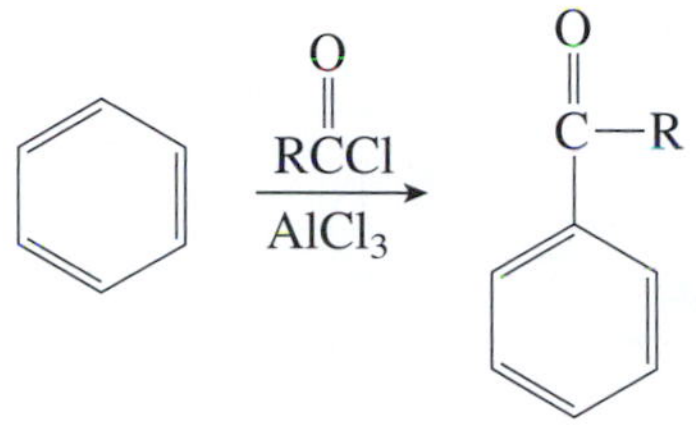

Continued

Table 23.3 Preparation of Functional Groups—cont'd

Reaction	Section	Comments
Ketones—cont'd		
RCH_2CR' (with C=O) $\xrightarrow[\substack{2)\ R''X \\ 3)\ H_3O^+}]{1)\ \text{pyrrolidine } N-H, HA}$ $RCHCR'$ (with C=O and R'')	20.8	Enamine reaction, R''X must be very reactive
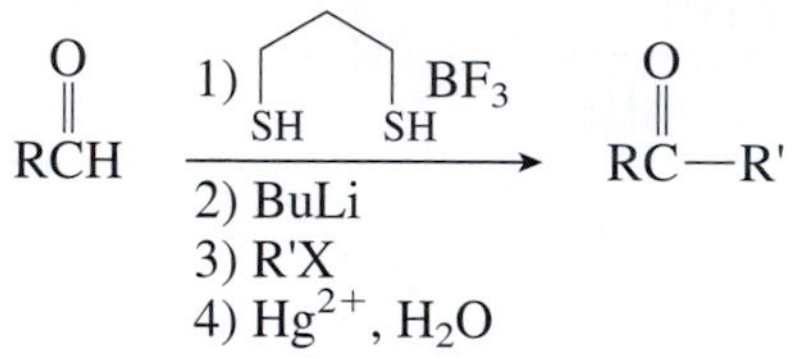 RCH (with C=O) $\xrightarrow[\substack{2)\ BuLi \\ 3)\ R'X \\ 4)\ Hg^{2+},\ H_2O}]{1)\ \text{dithiol SH SH}, BF_3}$ $RC-R'$ (with C=O)	20.9	Dithiane alkylation, acyl anion equivalent, S_N2, R' must be 1° or 2°
Carboxylic Acids		
$R-C-H$ (with C=O) $\xrightarrow[\substack{\text{or} \\ KMnO_4}]{Ag_2O,\ NaOH}$ $R-C-OH$ (with C=O)	10.14	Oxidation of aldehydes
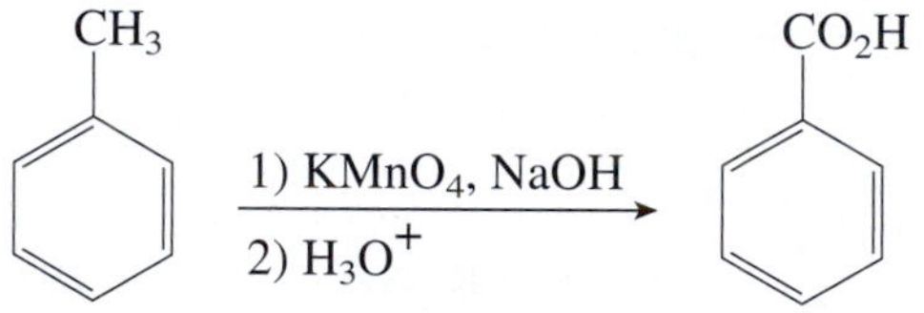 $C_6H_5CH_3$ $\xrightarrow[2)\ H_3O^+]{1)\ KMnO_4,\ NaOH}$ $C_6H_5CO_2H$	17.13	Any alkyl group, except tertiary ones, can be oxidized to the benzoic acid derivative
$RMgX$ or RLi $\xrightarrow[2)\ H_3O^+]{1)\ CO_2}$ $R-CO_2H$	18.6	Reaction of Grignard reagents with carbon dioxide
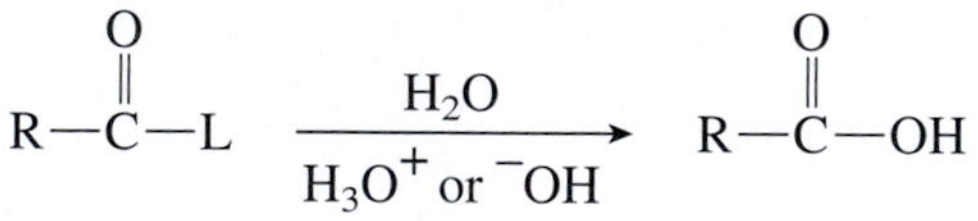 $R-C-L$ (with C=O) $\xrightarrow[H_3O^+\ \text{or}\ {}^-OH]{H_2O}$ $R-C-OH$ (with C=O)	19.5	Any carboxylic acid derivative can be hydrolyzed to a carboxylic acid
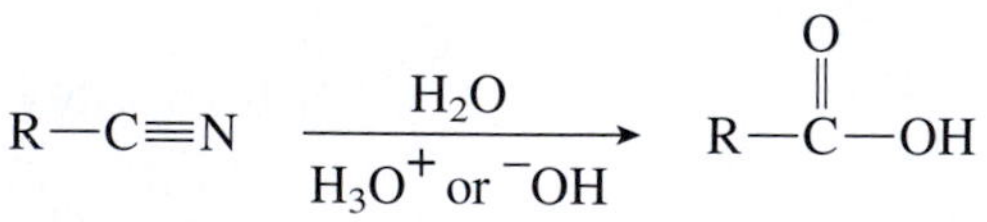 $R-C\equiv N$ $\xrightarrow[H_3O^+\ \text{or}\ {}^-OH]{H_2O}$ $R-C-OH$ (with C=O)	19.5	Nitriles can also be hydrolyzed to carboxylic acids
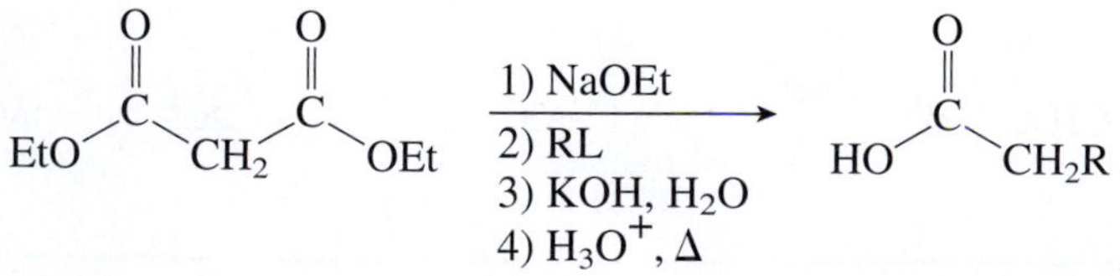 $EtO-C-CH_2-C-OEt$ (diester, both C=O) $\xrightarrow[\substack{2)\ RL \\ 3)\ KOH,\ H_2O \\ 4)\ H_3O^+,\ \Delta}]{1)\ NaOEt}$ $HO-C-CH_2R$ (with C=O)	20.4	Malonic ester synthesis, S_N2, R must be 1° or 2°

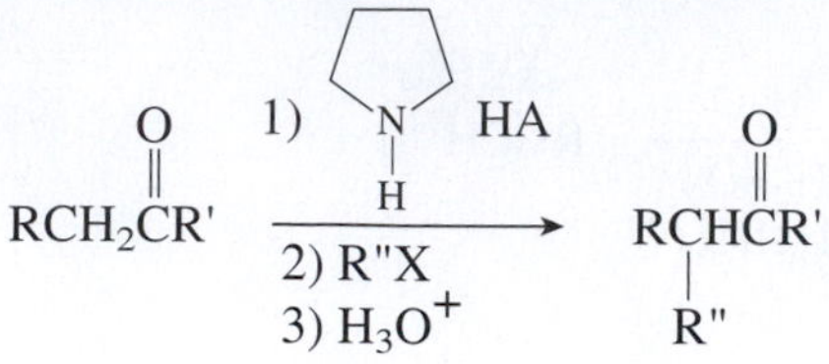

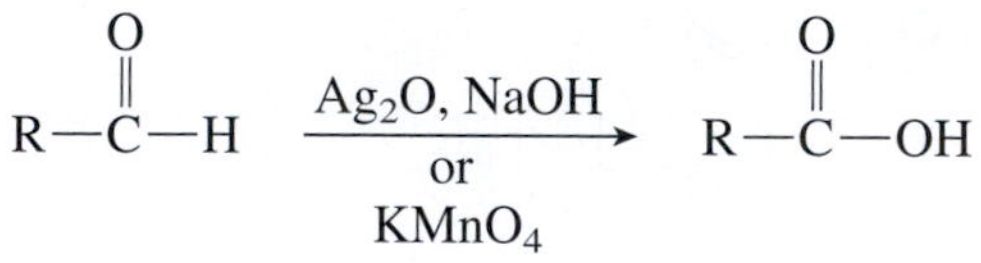

Table 23.3   Preparation of Functional Groups—cont'd

	Reaction	Section	Comments
Acyl Chlorides	$R-\overset{O}{\overset{\|}{C}}-OH \xrightarrow[\text{or}\ PCl_5]{SOCl_2} R-\overset{O}{\overset{\|}{C}}-Cl$	19.2	
Anhydrides	$R-\overset{O}{\overset{\|}{C}}-Cl + HO-\overset{O}{\overset{\|}{C}}-R \longrightarrow R-\overset{O}{\overset{\|}{C}}-O-\overset{O}{\overset{\|}{C}}-R$	19.3	Only symmetrical anhydrides are usually prepared
	$2\ R-\overset{O}{\overset{\|}{C}}-OH \xrightarrow{CH_3-\overset{O}{\overset{\|}{C}}-O-\overset{O}{\overset{\|}{C}}-CH_3} R-\overset{O}{\overset{\|}{C}}-O-\overset{O}{\overset{\|}{C}}-R$	19.3	Anhydride exchange
Amides	$R-\overset{O}{\overset{\|}{C}}-Cl \xrightarrow{NH_2R'} R-\overset{O}{\overset{\|}{C}}-NHR'$	19.6	
	$R-\overset{O}{\overset{\|}{C}}-O-\overset{O}{\overset{\|}{C}}-R \xrightarrow{NH_2R'} R-\overset{O}{\overset{\|}{C}}-NHR'$	19.6	
Esters	$R-L \xrightarrow{^-O-\overset{O}{\overset{\|}{C}}-R'} R-O-\overset{O}{\overset{\|}{C}}-R'$	10.4	S_N2, OK for 1° and 2°
	$R-\overset{O}{\overset{\|}{C}}-Cl + R'OH \longrightarrow R-\overset{O}{\overset{\|}{C}}-OR'$	19.4	
	$R-\overset{O}{\overset{\|}{C}}-O-\overset{O}{\overset{\|}{C}}-R + R'OH \longrightarrow R-\overset{O}{\overset{\|}{C}}-OR'$	19.4	
	$R-\overset{O}{\overset{\|}{C}}-O-H + R'OH \xrightarrow{HA} R-\overset{O}{\overset{\|}{C}}-OR'$	19.4	Fischer esterification
	$R'O-\overset{O}{\overset{\|}{C}}-CH_2R \xrightarrow[\text{2) R''X}]{\text{1) LDA}} R'O-\overset{O}{\overset{\|}{C}}-\overset{R''}{\overset{\|}{C}}HR$	20.3	S_N2, R'' may be 1° (best) or 2°
	$RCOEt + R'CH_2\overset{O}{\overset{\|}{C}}-OEt \xrightarrow[\text{2) H}_3\text{O}^+]{\text{1) NaOEt}} R'\overset{\underset{R-C=O}{}}{C}H\overset{O}{\overset{\|}{C}}-OEt$	20.6	Ester condensation

Continued

Table 23.3 **Preparation of Functional Groups—cont'd**

	Reaction	Section	Comments
Nitriles	$R-L + {}^-CN \longrightarrow R-CN$	10.8	S_N2, OK with 1° and 2°
	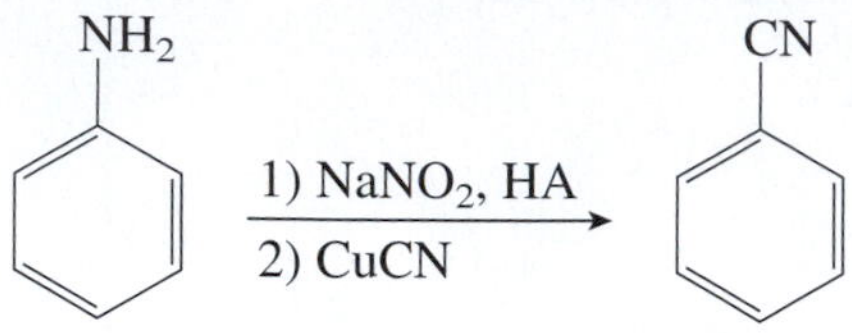 Aniline $\xrightarrow[\text{2) CuCN}]{\text{1) NaNO}_2, \text{HA}}$ Benzonitrile	17.10	Via diazonium ion
	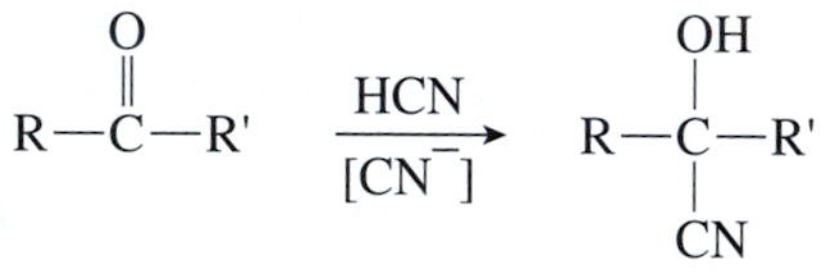 $R-\overset{O}{\underset{}{C}}-R' \xrightarrow[\text{[CN}^-]]{\text{HCN}} R-\overset{OH}{\underset{CN}{C}}-R'$	18.4	Cyanohydrin formation
	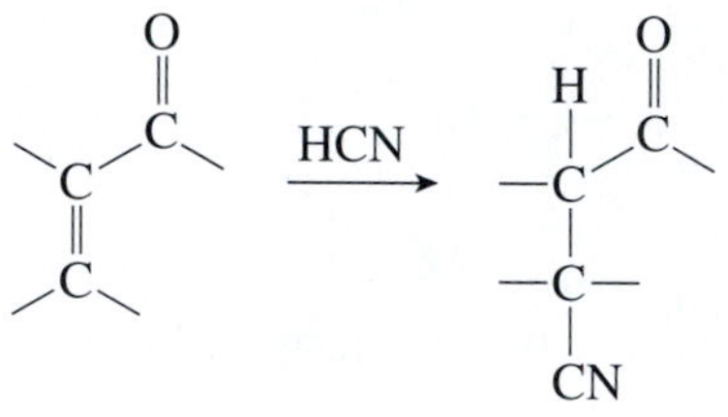 conjugate addition $\xrightarrow{\text{HCN}}$ product	18.10	Conjugate addition
	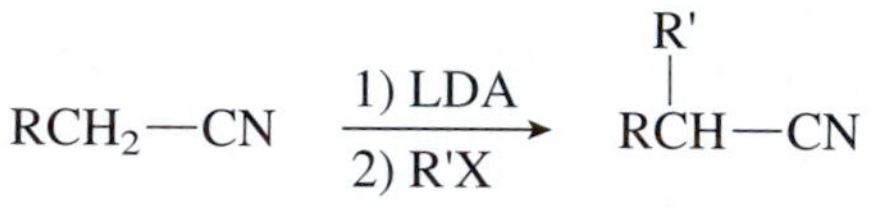 $RCH_2-CN \xrightarrow[\text{2) R'X}]{\text{1) LDA}} R\overset{R'}{\underset{}{C}}H-CN$	20.3	S_N2, R' may be 1° (best) or 2°
Aromatic Compounds	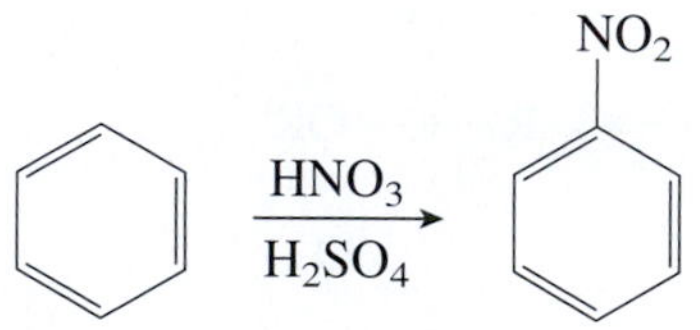 benzene $\xrightarrow[\text{H}_2\text{SO}_4]{\text{HNO}_3}$ nitrobenzene (NO_2)	17.4	Electrophilic aromatic substitution
	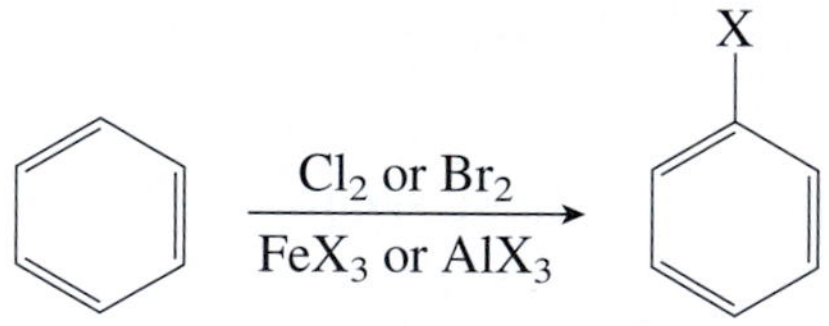 benzene $\xrightarrow[\text{FeX}_3 \text{ or AlX}_3]{\text{Cl}_2 \text{ or Br}_2}$ halobenzene (X)	17.5	Electrophilic aromatic substitution
	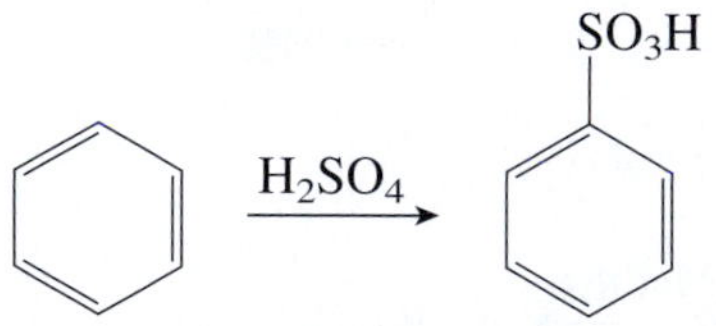 benzene $\xrightarrow{\text{H}_2\text{SO}_4}$ benzenesulfonic acid (SO_3H)	17.6	Electrophilic aromatic substitution, reversible in H_2O/H_2SO_4

Table 23.3   Preparation of Functional Groups—cont'd

	Reaction	Section	Comments
Aromatic Compounds	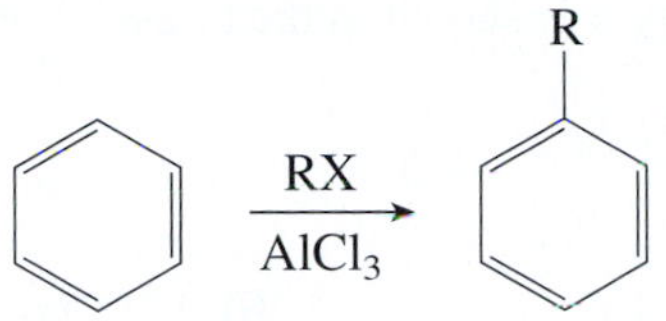 benzene $\xrightarrow[\text{AlCl}_3]{\text{RX}}$ R-substituted benzene	17.7	Friedel-Crafts alkylation, rearrangements, multiple alkylation, fails with deactivated rings
	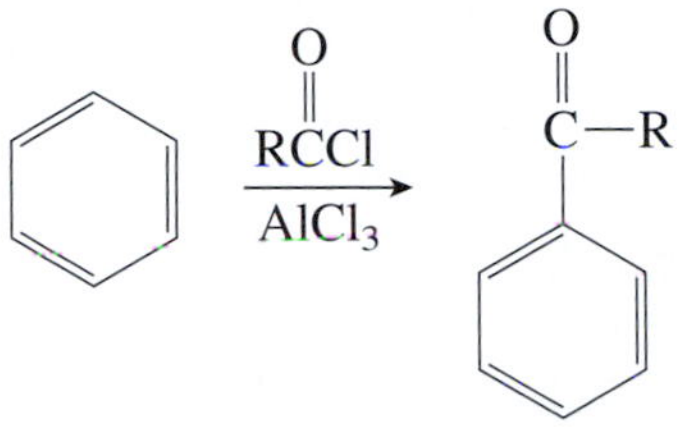 benzene $\xrightarrow[\text{AlCl}_3]{\text{RCCl}, \text{O}}$ phenyl ketone $\overset{O}{C}$-R	17.8	Friedel-Crafts acylation, fails with strongly deactivated rings
	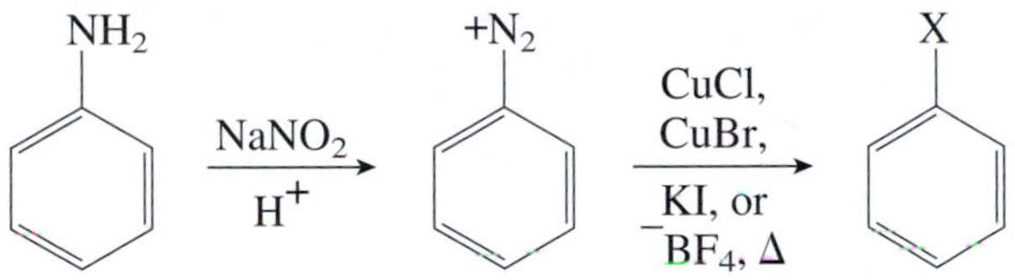 aniline NH_2 $\xrightarrow[\text{H}^+]{\text{NaNO}_2}$ diazonium $^+\text{N}_2$ $\xrightarrow[\overline{\ }\text{BF}_4, \Delta]{\text{CuCl, CuBr, KI, or}}$ X-substituted benzene	17.10	Via diazonium ion
	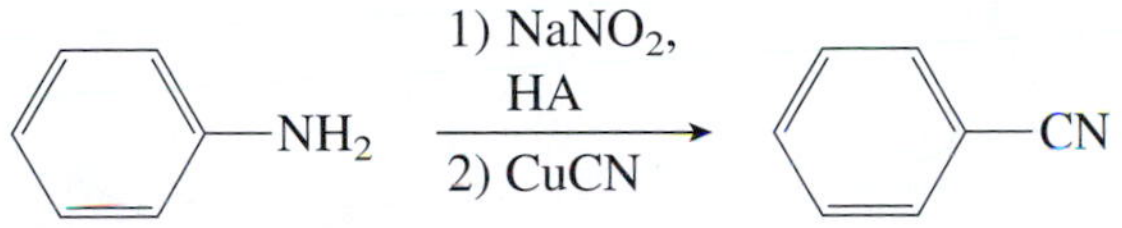 aniline—NH_2 $\xrightarrow[\text{2) CuCN}]{\text{1) NaNO}_2, \text{HA}}$ benzene—CN	17.10	Via diazonium ion
	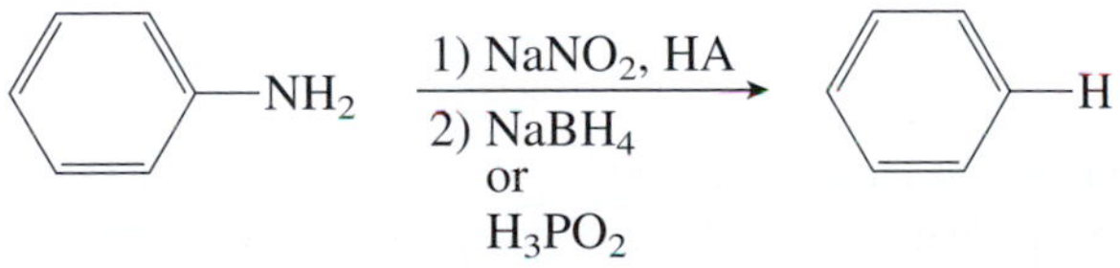 aniline—NH_2 $\xrightarrow[\substack{\text{2) NaBH}_4 \\ \text{or} \\ \text{H}_3\text{PO}_2}]{\text{1) NaNO}_2, \text{HA}}$ benzene—H	17.10	Via diazonium ion
	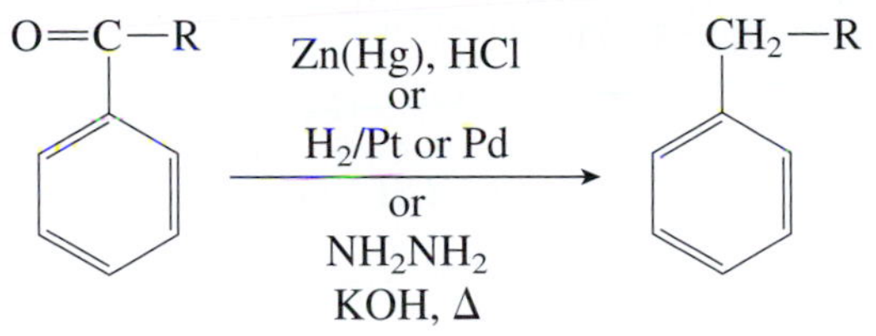 $\text{O}=\text{C}-\text{R}$ phenyl ketone $\xrightarrow[\substack{\text{or} \\ \text{NH}_2\text{NH}_2 \\ \text{KOH}, \Delta}]{\substack{\text{Zn(Hg), HCl} \\ \text{or} \\ \text{H}_2/\text{Pt or Pd}}}$ CH_2-R benzene	17.13 18.8	Clemmensen reduction or hydrogenolysis, or Wolff-Kishner reduction
	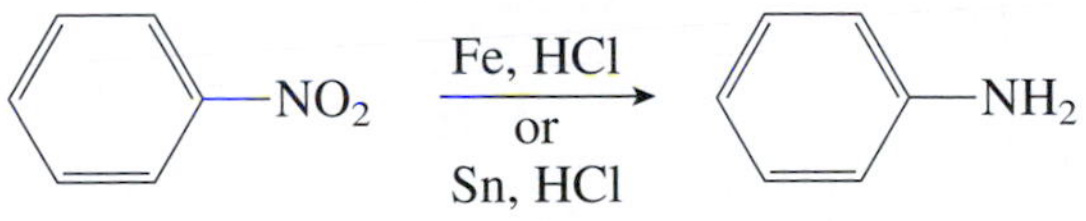 benzene—NO_2 $\xrightarrow[\substack{\text{or} \\ \text{Sn, HCl}}]{\text{Fe, HCl}}$ benzene—NH_2	17.13	Reduction of nitro groups
	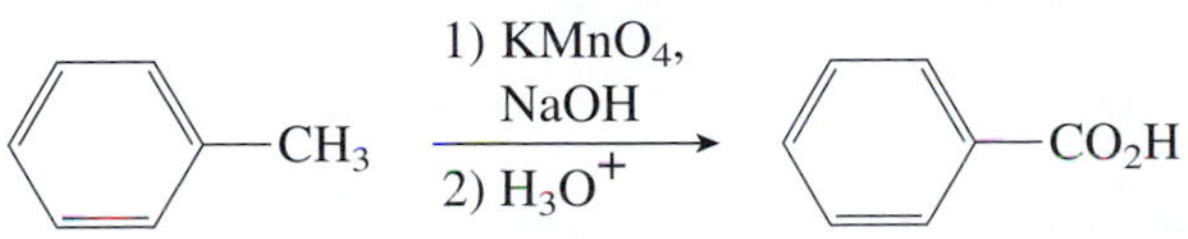 benzene—CH_3 $\xrightarrow[\text{2) H}_3\text{O}^+]{\substack{\text{1) KMnO}_4, \\ \text{NaOH}}}$ benzene—CO_2H	17.13	Oxidation of any alkyl group except tertiary

PROBLEM 23.11

Suggest syntheses of these compounds. Each synthesis should involve the formation of at least two carbon–carbon bonds. Make sure that your synthesis produces the correct stereochemistry where it is shown in the target compound.

a) 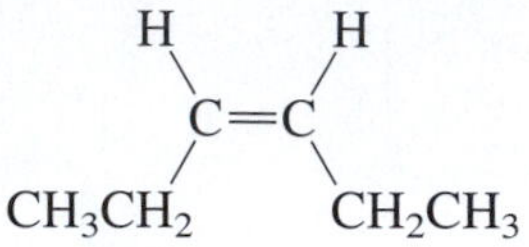 CH₃CH₂ ... CH₂CH₃

b) 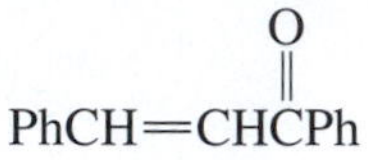 PhCH=CHCPh

c) 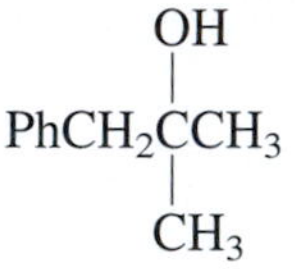 PhCH₂CCH₃ with OH and CH₃

d)

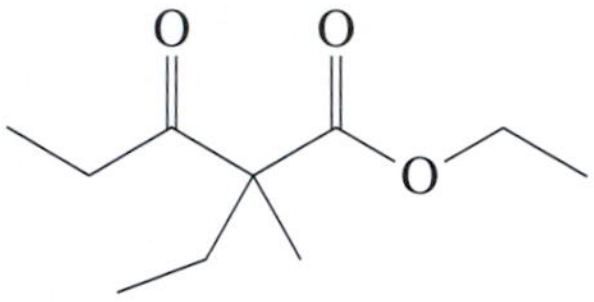

e)

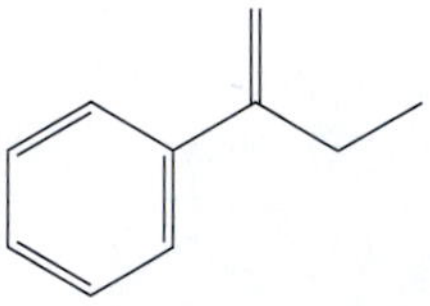

f) 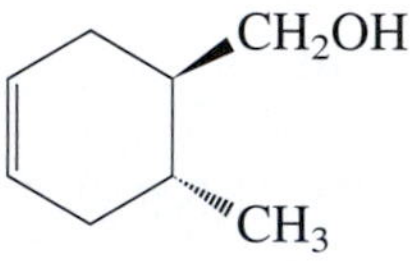 CH₂OH / CH₃

g)

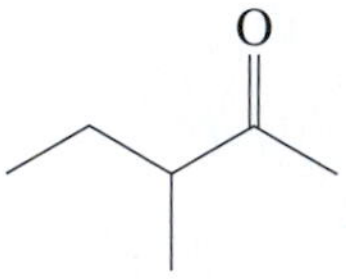

h)

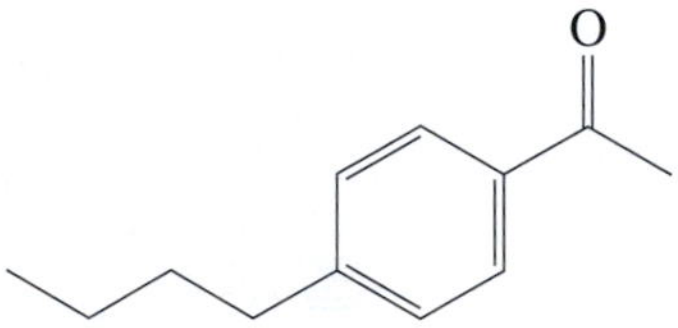

PROBLEM 23.12

Show syntheses of these compounds using methanol, ethanol, and benzene as the only sources for the carbons in the compounds. You may use any solvents or inorganic reagents that you need. Once you have made a compound, you may use it in the synthesis of any subsequent compound.

a) CH₃OCH₂CH₃

b) 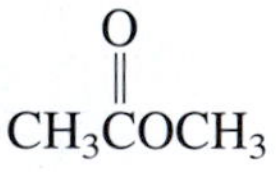 CH₃COCH₃

c)

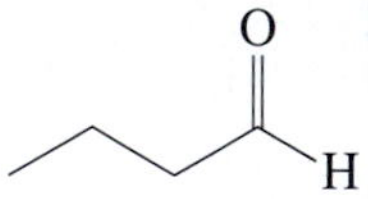

d)

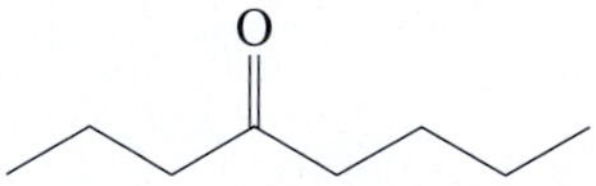

e)

f)

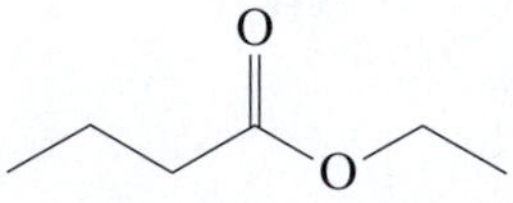

g)

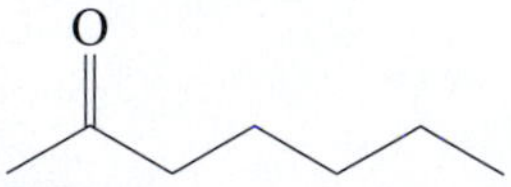

h)

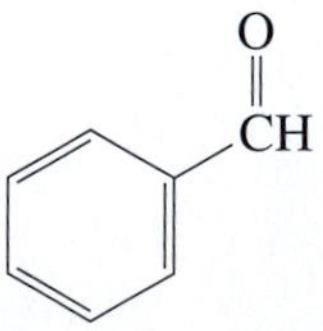

i)

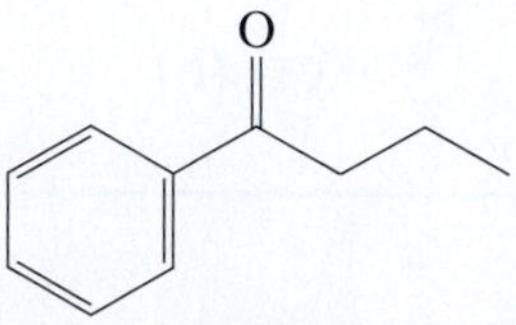

j)

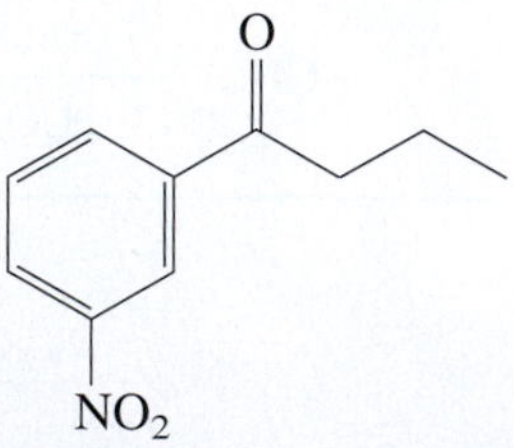

k)

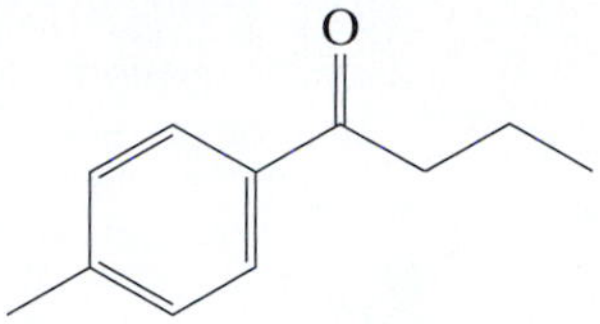

l)

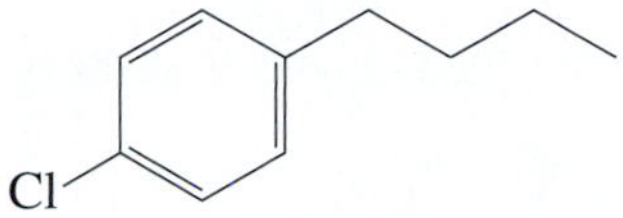

m)

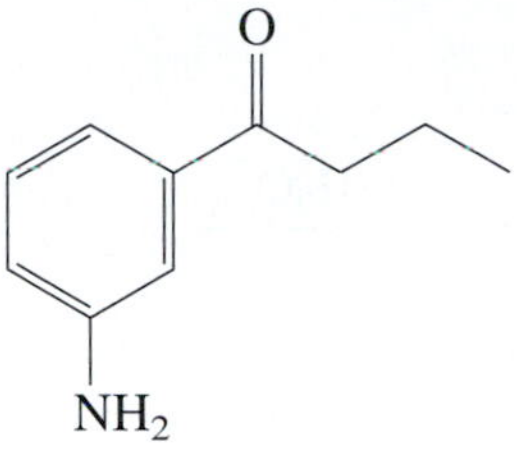

n)

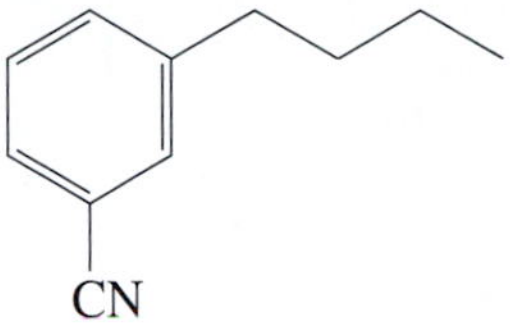

o)

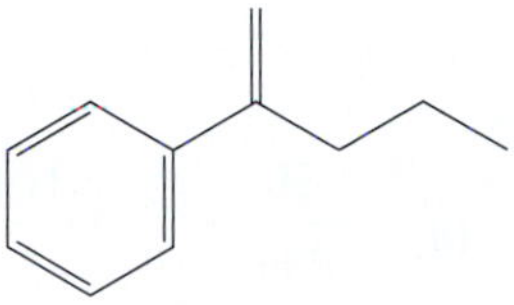

p)

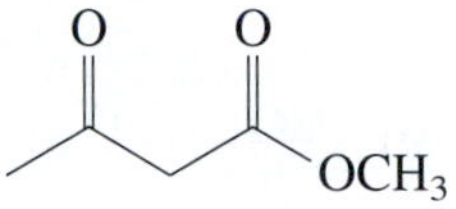

q)

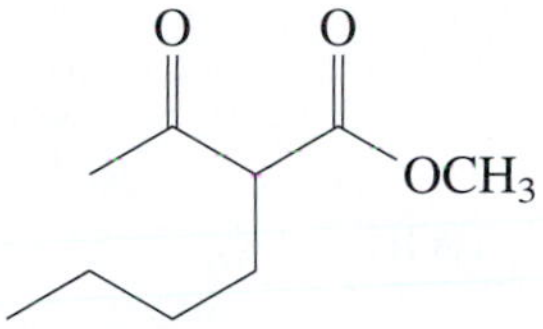

r)

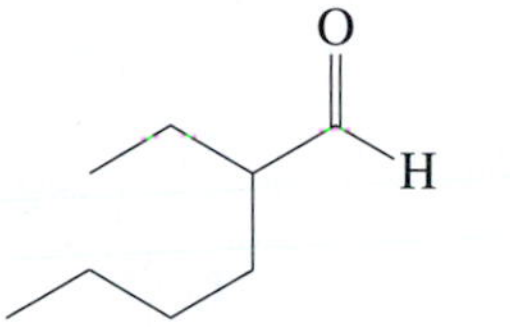

s)

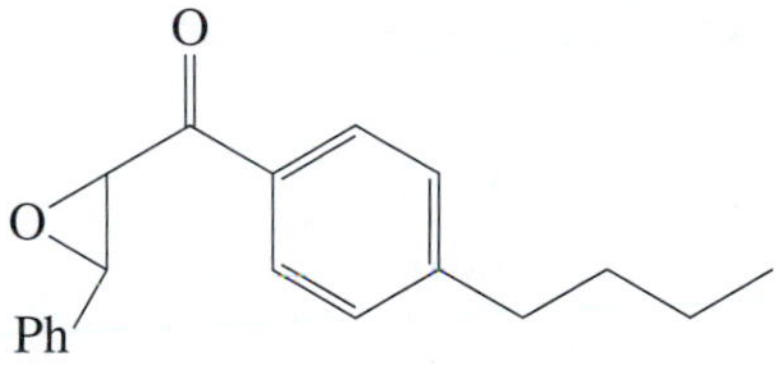

t)

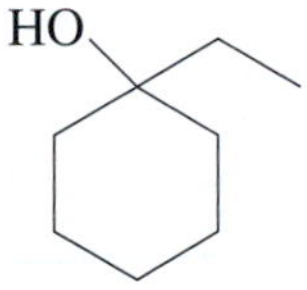

Review of Mastery Goals

After completing this chapter, you should be able to:

■ Recognize when a protecting group is needed and how to use it in the synthesis of an organic compound. (Problem 23.17)

■ Use retrosynthetic analysis to design syntheses of more complex organic compounds. (Problems 23.14, 23.15, 23.16, 23.18, and 23.19)

ORGANIC
Chemistry Now™
Click *Mastery Goal Quiz* to test how well you have met these goals.

Additional Problems

23.13 Show the products of these reactions:

a) 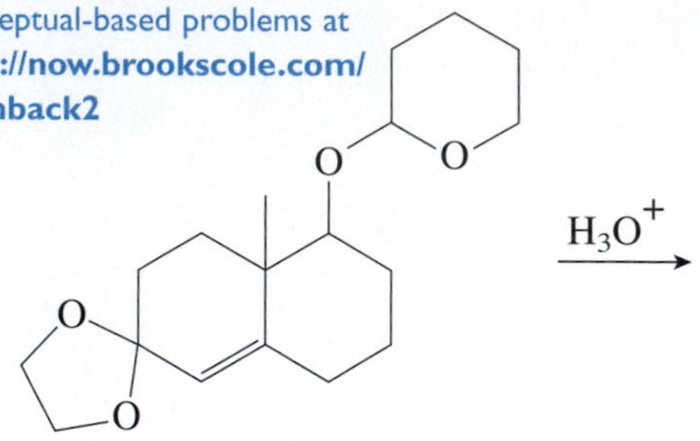

b) 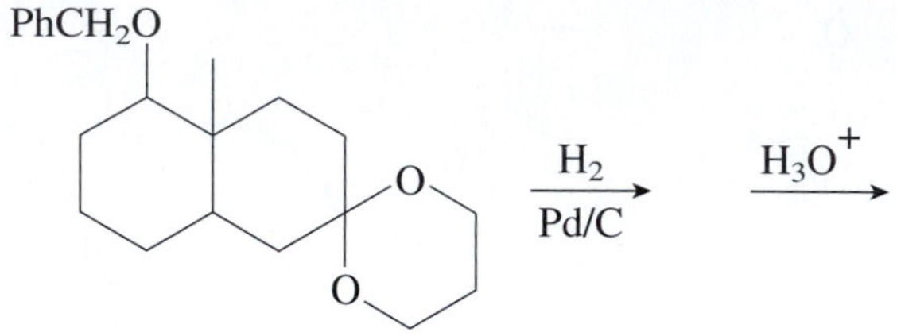

c) 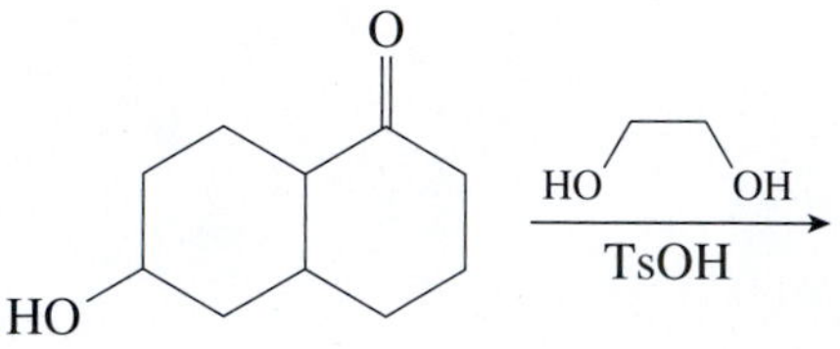

d) 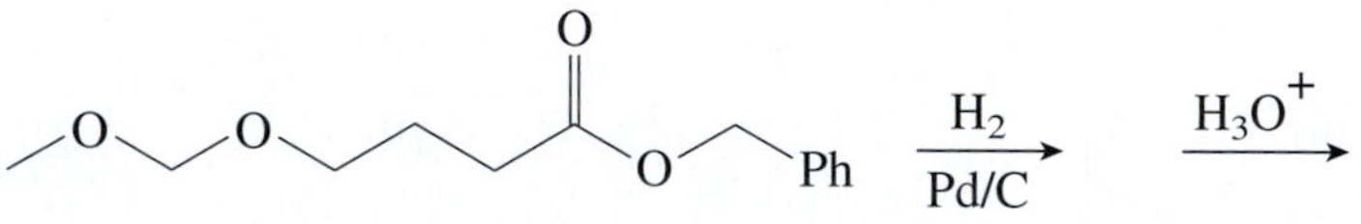

e) 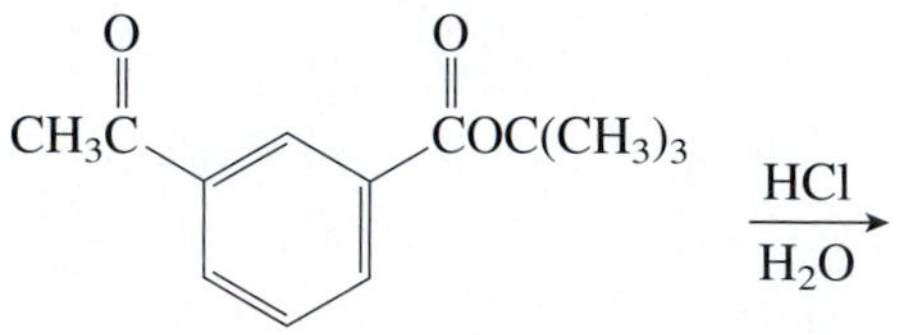

f) 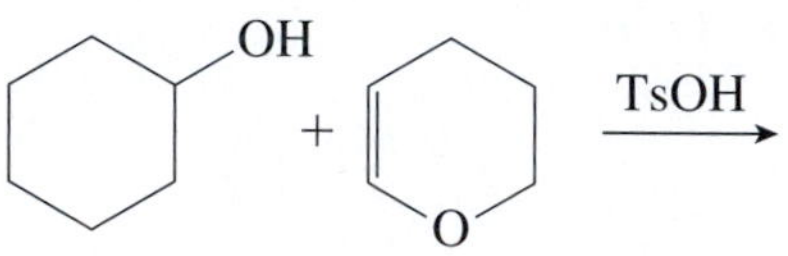

g) 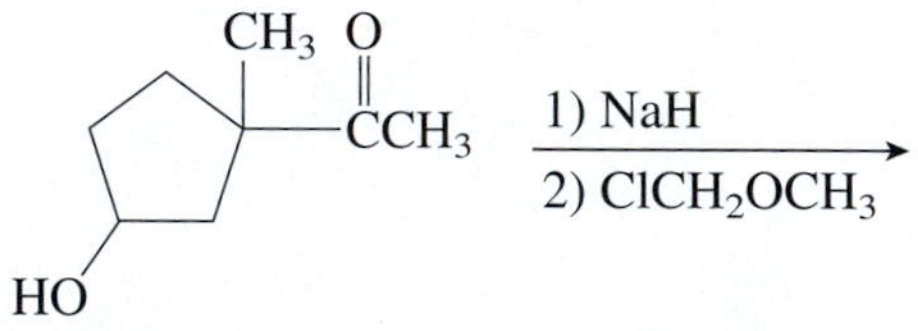

h) 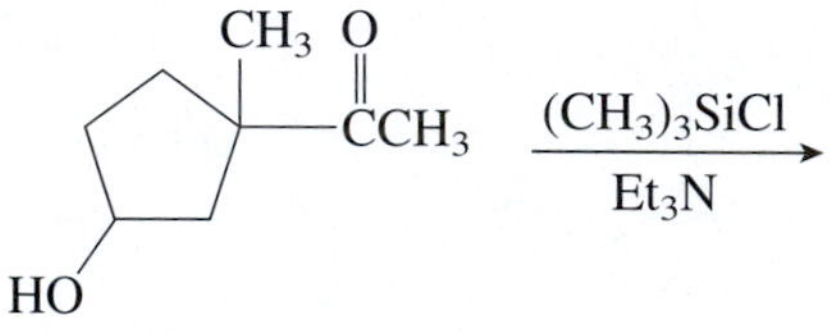

i) 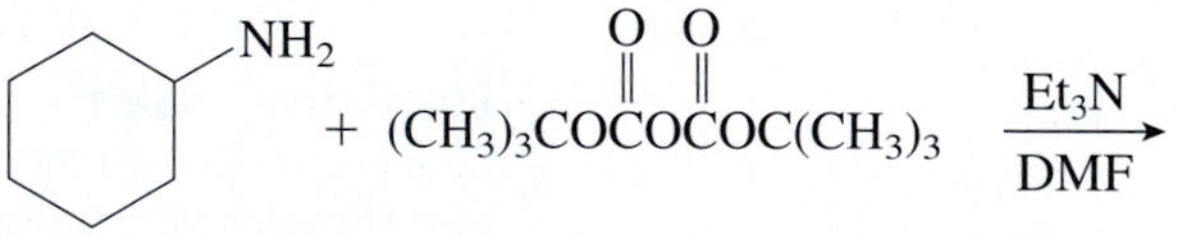

j) 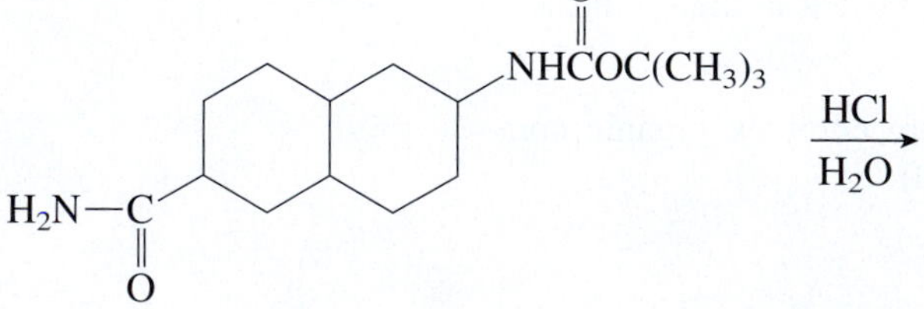

k)

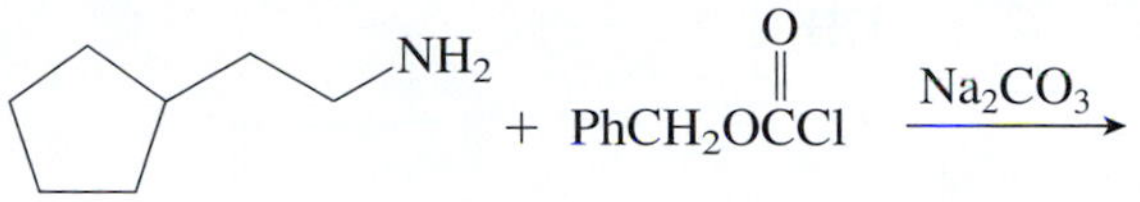

$$\text{(cyclopentyl-CH}_2\text{CH}_2\text{NH}_2) + \text{PhCH}_2\text{OCCl} \xrightarrow{\text{Na}_2\text{CO}_3}$$

l)

$$\xrightarrow{\text{H}_2}{\text{Pd/C}}$$

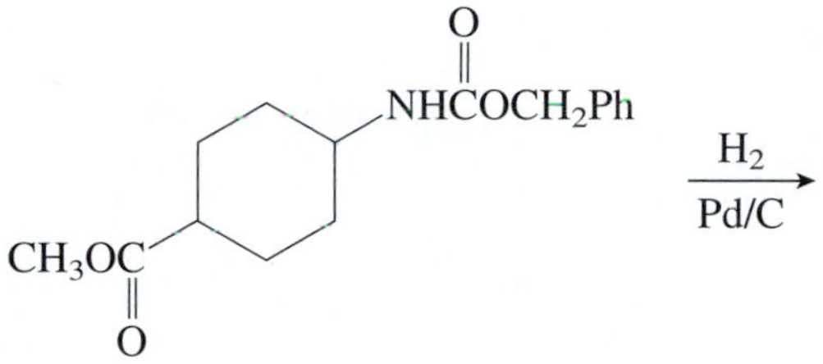

23.14 Use disconnections of carbon–carbon bonds to generate electrophile and nucleophile synthon fragments that could be used to prepare these compounds. Then show an actual reaction suggested by these synthons that would provide a method for the synthesis of each compound. Design at least three different routes to each compound.

a)

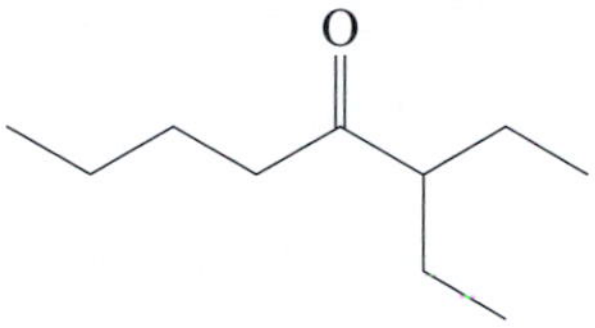

b)

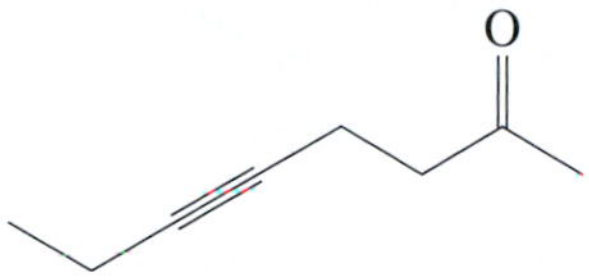

c)

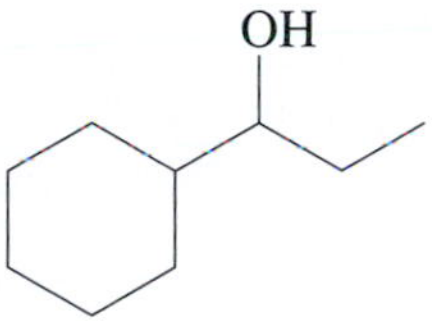

d)

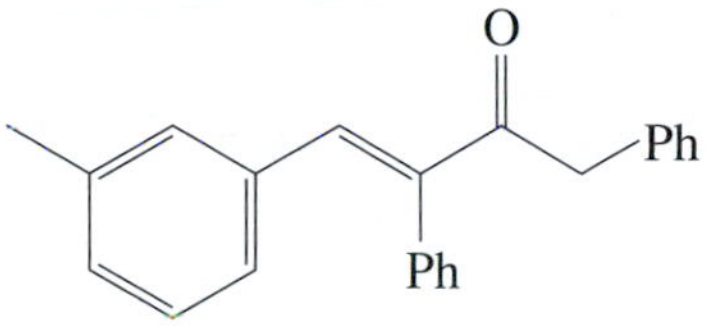

23.15 Show syntheses of these compounds from compounds with the indicated number of carbons:

a)

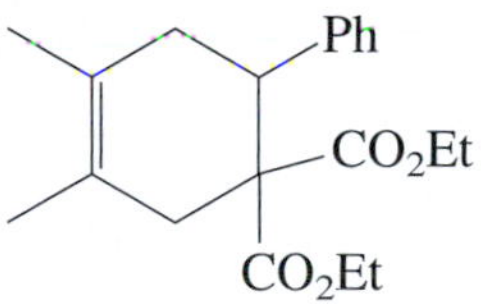

(7 carbons or fewer)

b)

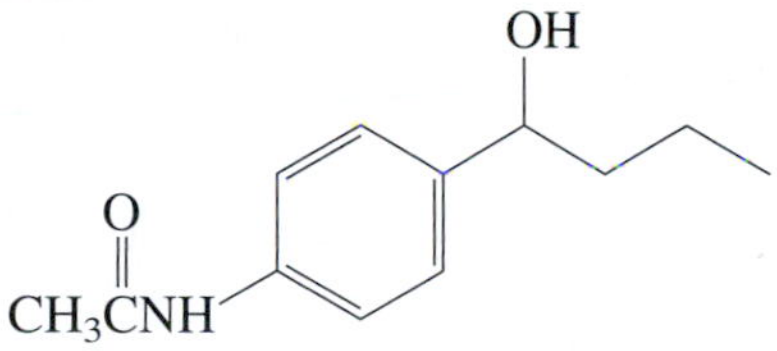

(6 carbons or fewer)

c)

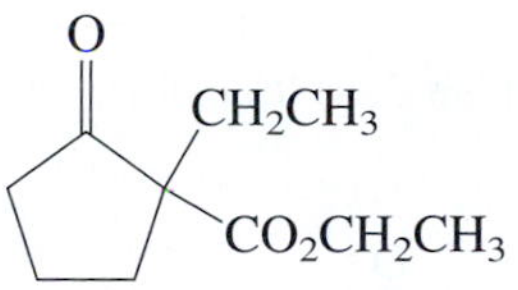

(6 carbons or fewer, no rings)

d)

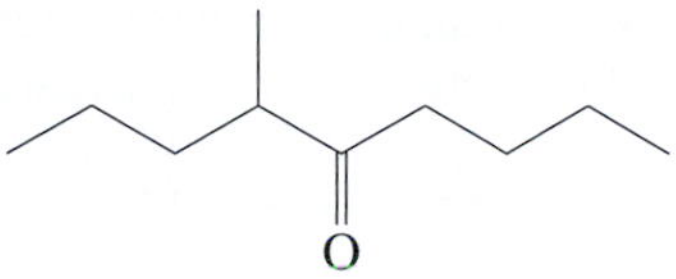

(5 carbons or fewer)

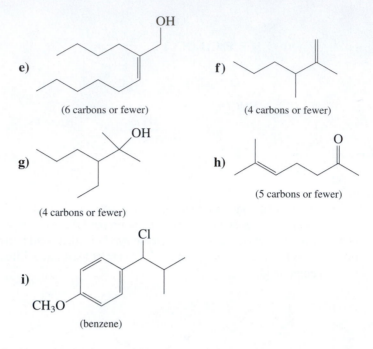

e) (6 carbons or fewer)

f) (4 carbons or fewer)

g) (4 carbons or fewer)

h) (5 carbons or fewer)

i) (benzene)

BioLink

23.16 Show syntheses of these compounds from compounds with the indicated number of carbons:

a) 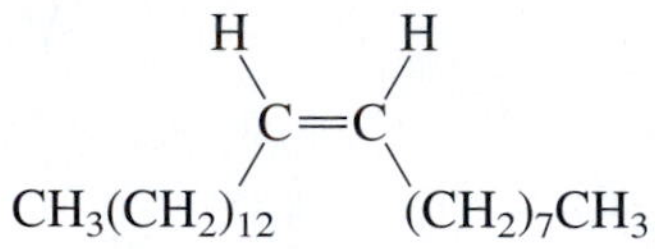 $CH_3(CH_2)_{12}$ ... $(CH_2)_7CH_3$

Muscalure (sex pheromone of the female housefly) (from compounds with 13 carbons or fewer)

b) 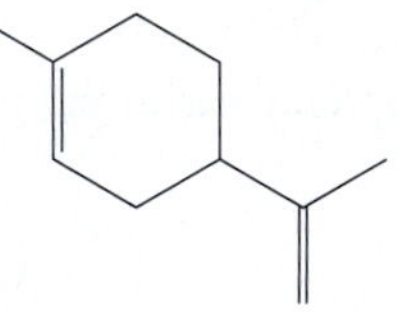 Limonene (constituent of lemon peel) (from compounds with 5 carbons or fewer)

c) Pival (a rat poison) (from compounds with 9 carbons or fewer)

23.17 Show a series of reactions that will accomplish these transformations. The use of protecting groups is necessary.

a)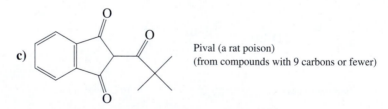

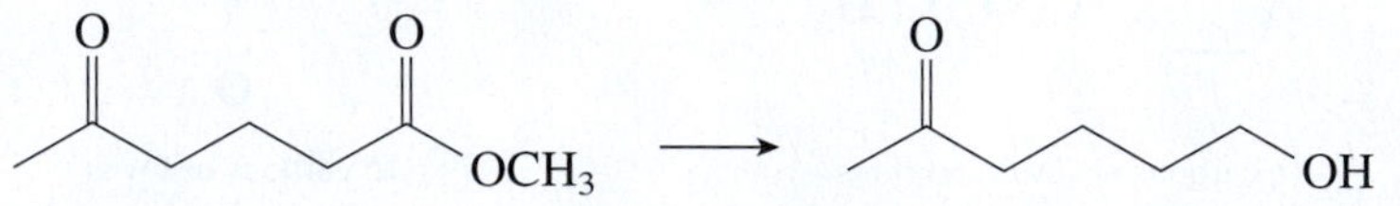

b)

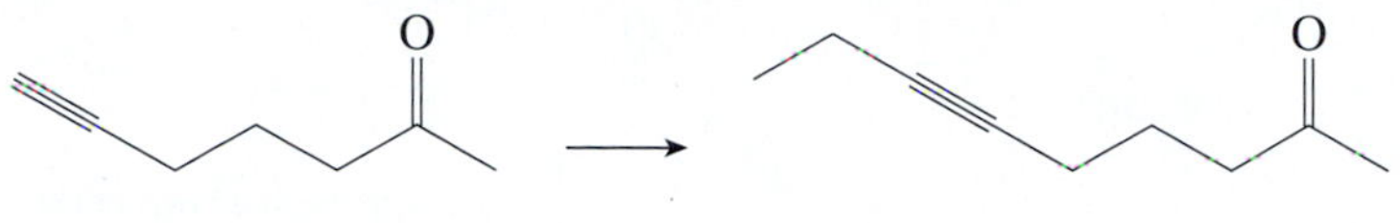

c) 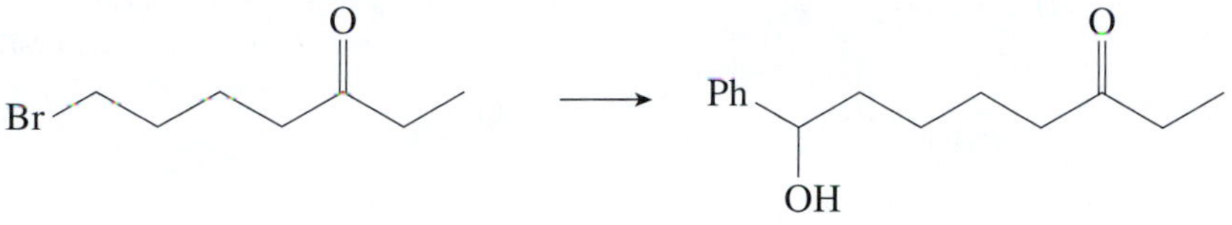 Br~~~~~ → Ph (with OH)

d)

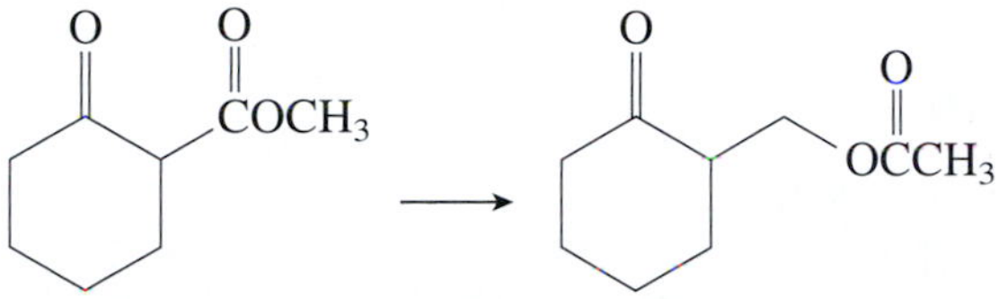

e) HOCH₂C≡CH ⟶ HOCH₂C≡C—CO₂H

f)

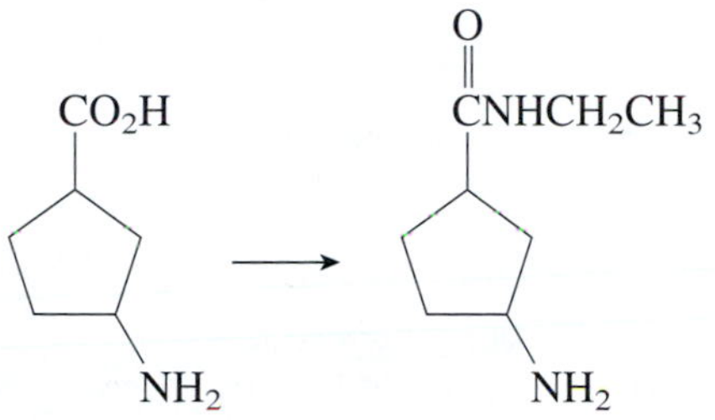

g)

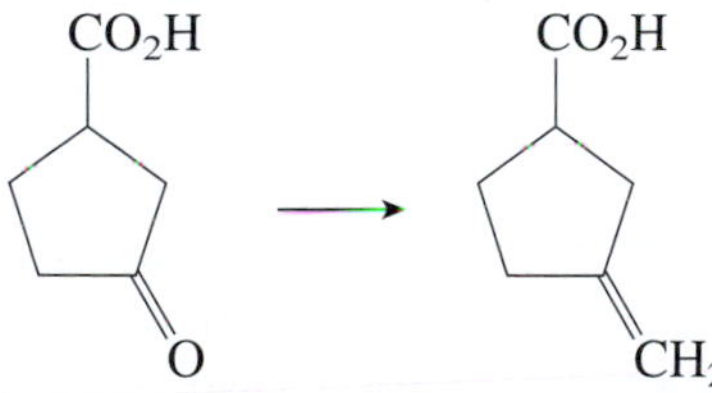

23.18 Show syntheses of these compounds from compounds with the indicated number of carbons:

a) 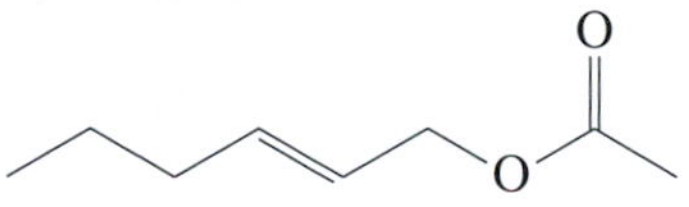 Sex pheromone of Indian water bug (from compounds with 3 carbons or fewer) (ignore stereochemistry)

b) CH₃(CH₂)₃CH=CHCH₂CH₂CH=CH(CH₂)₆OCCH₃

Gossyplure (sex pheromone of pink bollworm moth, a severe cotton pest) (from compounds with 8 carbons or fewer)

c) 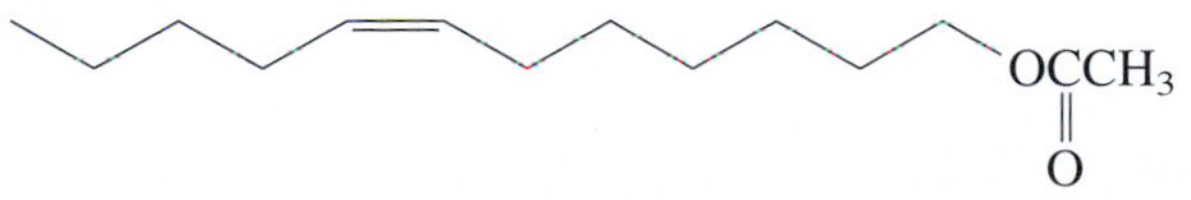 ~~~~~OCCH₃

(Z)-7-Dodecen-1-yl acetate (sex attractant of elephants and moths) (from compounds with 6 carbons or fewer)

d)

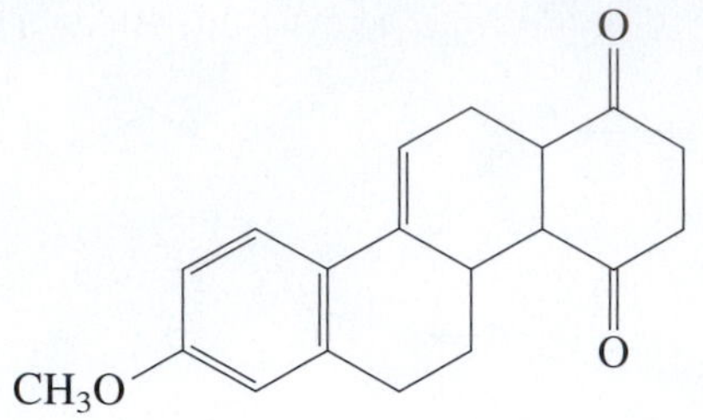

(Used in the synthesis of estrone,
a human sex hormone)
(from compounds with 11 carbons or fewer)

CH₃O

23.19 Show syntheses of these compounds from compounds with the indicated number of carbons:

a)

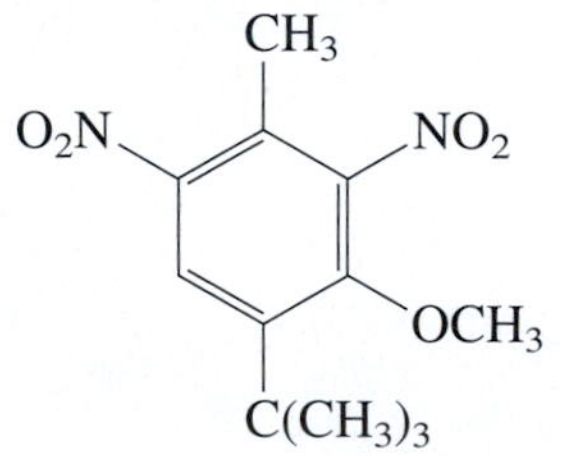

CH₃

O₂N   NO₂

OCH₃

C(CH₃)₃

Musk ambrette, a perfume ingredient
(from compounds with 7 carbons or fewer)

b) 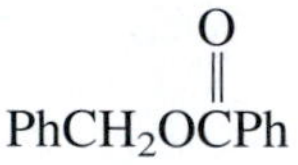 PhCH₂OCPh

O

Perfume ingredient and insect repellent
(from compounds with 6 carbons or fewer)

c)

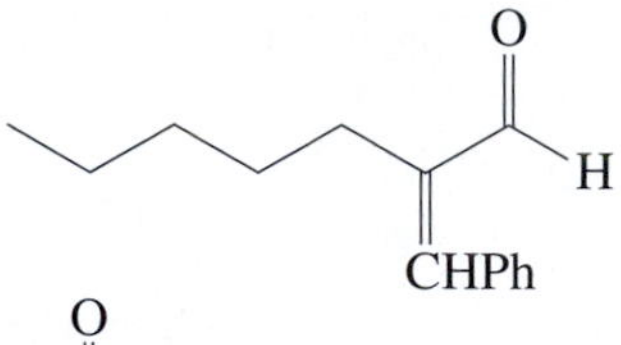

O

H

CHPh

Flosal, a perfume ingredient
(from compounds with 6 carbons or fewer)

d)

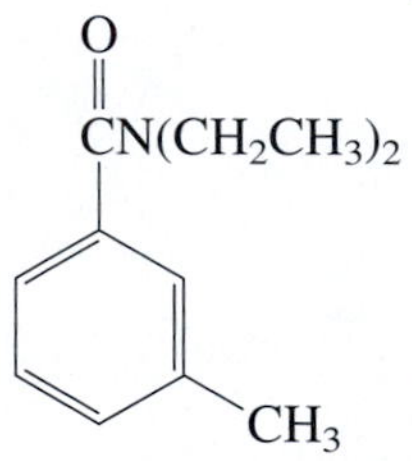

O

CN(CH₂CH₃)₂

CH₃

N,N-Diethyl-m-toluamide, an insect repellent
(from compounds with 7 carbons or fewer)

23.20 Show the steps in the mechanism for the hydrolysis of this tetrahydropyranyl ether:

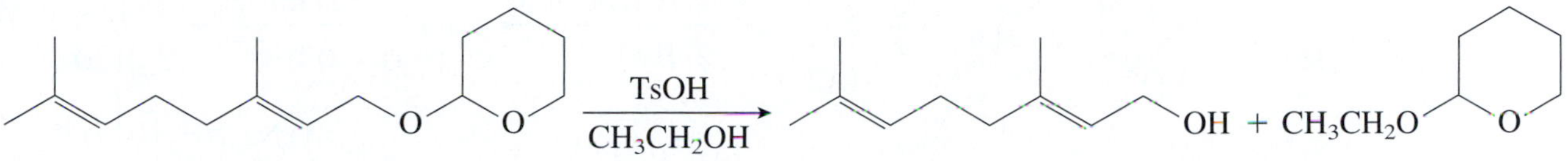

23.21 Show the missing structures in this reaction sequence. This type of problem is sometimes called a road-map problem. Keep both the initial and final structure in mind and don't forget to use the information in the formulas of the missing compounds.

BioLink

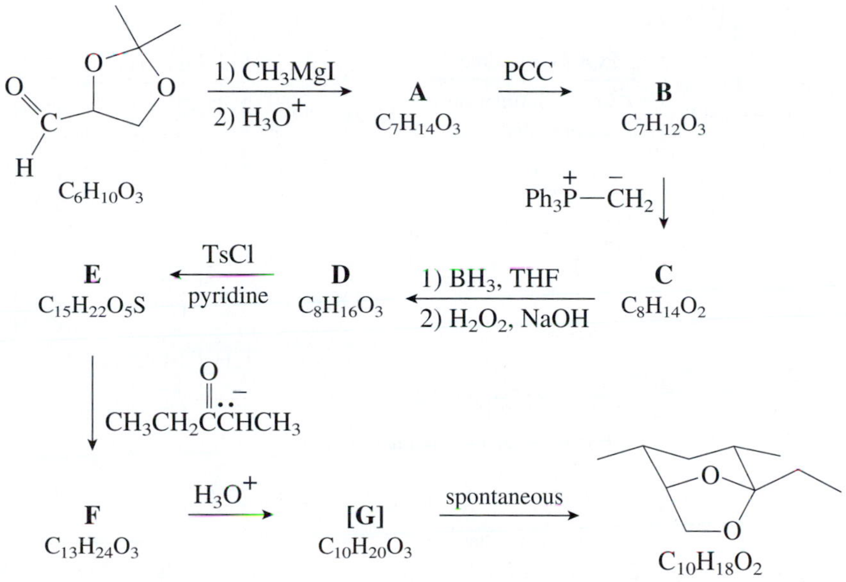

α-Multistriatin, an insect pheromone

23.22 Show the missing structures in this reaction sequence:

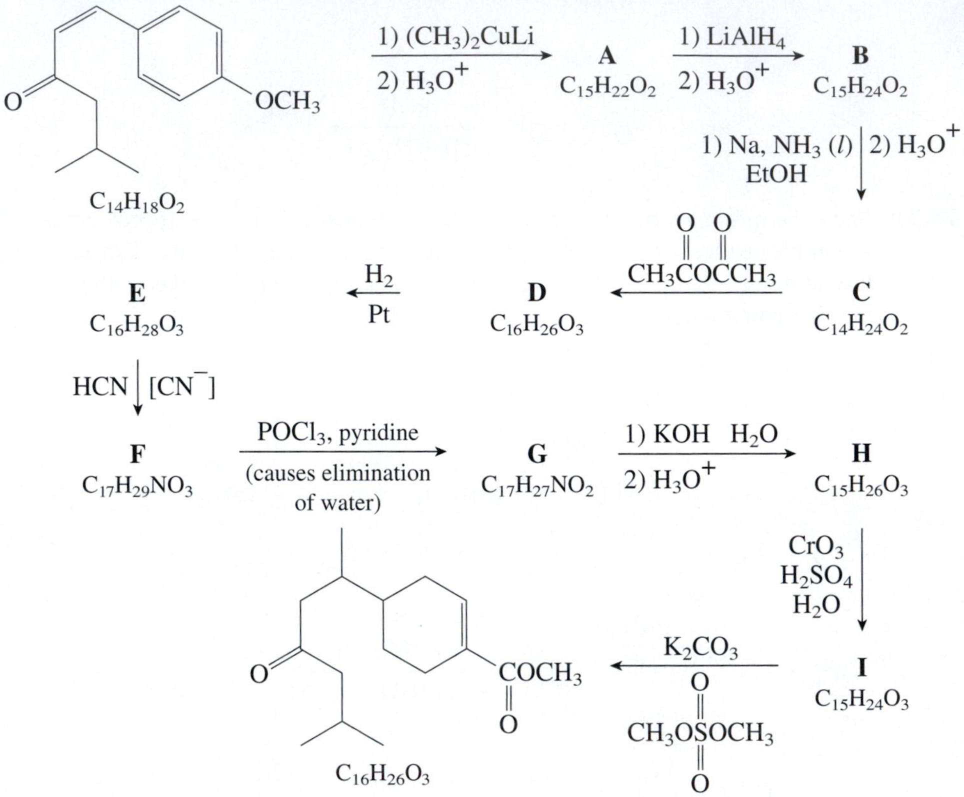

$C_{14}H_{18}O_2$

$\xrightarrow[\text{2) }H_3O^+]{\text{1) }(CH_3)_2CuLi}$ **A** $C_{15}H_{22}O_2$ $\xrightarrow[\text{2) }H_3O^+]{\text{1) LiAlH}_4}$ **B** $C_{15}H_{24}O_2$

1) Na, NH$_3$ (*l*) | 2) H$_3$O$^+$
EtOH

E
$C_{16}H_{28}O_3$ $\xleftarrow[\text{Pt}]{\text{H}_2}$ **D** $C_{16}H_{26}O_3$ $\xleftarrow{\underset{\text{CH}_3\text{COCCH}_3}{\overset{\text{O O}}{}}}$ **C** $C_{14}H_{24}O_2$

HCN $\downarrow$ [CN$^-$]

F
$C_{17}H_{29}NO_3$ $\xrightarrow[\substack{\text{(causes elimination} \\ \text{of water)}}]{\text{POCl}_3\text{, pyridine}}$ **G** $C_{17}H_{27}NO_2$ $\xrightarrow[\text{2) }H_3O^+]{\text{1) KOH   }H_2O}$ **H** $C_{15}H_{26}O_3$

CrO$_3$
H$_2$SO$_4$
H$_2$O

$C_{16}H_{26}O_3$
Juvabione, an insect hormone mimic

$\xleftarrow[\underset{\underset{\text{O}}{\text{CH}_3\text{OSOCH}_3}}{\overset{\text{O}}{}}]{\text{K}_2\text{CO}_3}$ **I** $C_{15}H_{24}O_3$

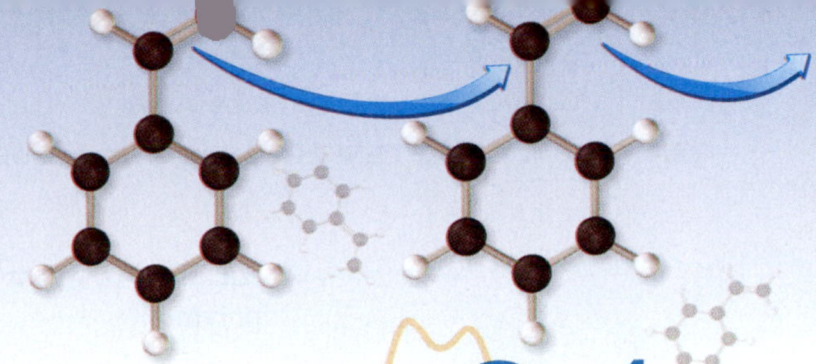

Synthetic Polymers

CHAPTER 24

MASTERING ORGANIC CHEMISTRY

- Identifying the Repeat Unit for Any Addition or Condensation Polymer
- Understanding the Mechanisms for the Formation of Addition Polymers
- Understanding the Structure and Stereochemistry of Polymers
- Understanding How the Structure of a Polymer Affects Its Physical Properties
- Understanding the Chemical Properties of a Polymer

POLYMERS ARE VERY large molecules made up of repeating units. A majority of the compounds produced by the chemical industry are ultimately used to prepare polymers. These human-made or synthetic polymers are the plastics (polyethylene, polystyrene), the adhesives (epoxy glue), the paints (acrylics), and the fibers (polyester, nylon) that we encounter many times each day. It is difficult to picture our lives without these materials. In addition to these synthetic polymers, natural polymers such as wood, rubber, cotton, and wool are all around us. And, of course, life itself depends on polymers such as carbohydrates, proteins, and DNA. This chapter discusses synthetic polymers. Naturally occurring polymers are presented in Chapters 25, 26, and 27.

First, a common method of forming polymers by a radical reaction is discussed. After the structures of the addition polymers made by this method are examined, several other procedures that can be used to prepare these or similar polymers are presented. Next, the effect of the structure of a polymer on its physical properties is discussed. This provides a basis for understanding the properties and uses of a number of other addition polymers. Rubbers (elastomers) are then discussed, followed by condensation polymers and thermosetting polymers. The chapter concludes with a brief examination of the chemical properties of polymers.

24.1 Radical Chain Polymerization

Polymers can be considered to be formed from the bonding of a large number of individual units, called monomers. This can be represented by the following general

ORGANIC Chemistry Now™
Look for this logo in the chapter and go to OrganicChemistryNow at **http://now.brookscole.com/hornback2** for tutorials, simulations, problems, and molecular models.

equation, in which some large number (*n*) of monomers (A) are combined to form a polymer:

$$n \, A \longrightarrow \cdots A-A-A-A-A-A\cdots \quad \text{or} \quad \left(A \right)_n$$

Monomers   Polymer

Because the number of monomers that combines to form an individual polymer molecule is usually very large—hundreds or even thousands—polymers are extremely large and are termed **macromolecules.**

Polymers are often classified as **addition polymers** or **condensation polymers** according to the general mechanism by which they are prepared. Most addition polymers are prepared by the reaction of an alkene monomer as illustrated in the following equation:

$$\text{Initiator} \quad \underset{}{CH_2}{=}\overset{R}{CH} \quad CH_2{=}\overset{R}{CH} \quad CH_2{=}\overset{R}{CH} \quad CH_2{=}\overset{R}{CH} \quad \text{etc.}$$

$$\text{Initiator}-CH_2-\overset{R}{CH}-CH_2-\overset{R}{CH}-CH_2-\overset{R}{CH}-CH_2-\overset{R}{CH}-\text{etc.}$$

In this process, some initiator molecule adds to one carbon of the CC double bond of the monomer to generate a reactive site, such as a radical or a carbocation, at the other carbon. This reactive carbon species then adds to another monomer to produce another reactive carbon species, and the process continues until a large number of monomers have been connected. Another way to represent this reaction shows the repeating unit that is formed when the monomers react to give the polymer:

$$n \quad CH_2{=}\overset{R}{CH} \quad \xrightarrow{\text{initiator}} \quad \left(CH_2-\overset{R}{CH} \right)_n$$

Monomer   Repeat unit
of polymer

These addition polymers are often called **vinyl polymers** because of the vinyl group that is present in the monomers.

Although several variations occur, depending on the reactive intermediate that is present, all addition polymers are formed by chain mechanisms, in which one initiator molecule causes a large number of monomers to react to form one polymer molecule. For this reason these polymers are also known as **chain-growth polymers.** To better understand how this process occurs, let's examine a specific case, the formation of polyethylene by a radical chain mechanism, as shown in the following equation:

$$n \quad CH_2{=}CH_2 \quad \xrightarrow[\text{PhCOOCPh}]{\overset{O \quad O}{\| \quad \|}} \quad \left(CH_2-CH_2 \right)_n$$

Ethylene   Polyethylene

As is the case with other reactions that proceed by a radical chain mechanism (see Chapter 21), this reaction involves three kinds of steps: initiation, propagation, and termination.

Initiation

Recall that the initiation step generates the radicals. In this case the weak oxygen–oxygen bond of the initiator, dibenzoyl peroxide, cleaves to produce two benzoyloxy radicals. A benzoyloxy radical then adds to the CC double bond of an ethylene molecule:

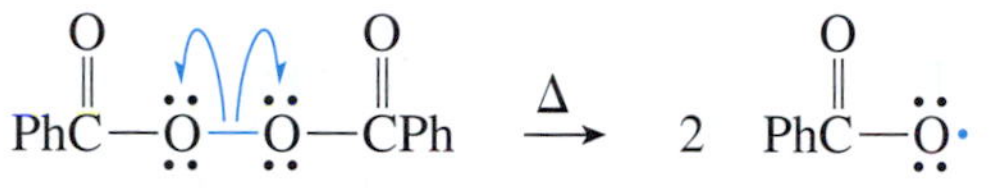

Dibenzoyl peroxide Benzoyloxy radical

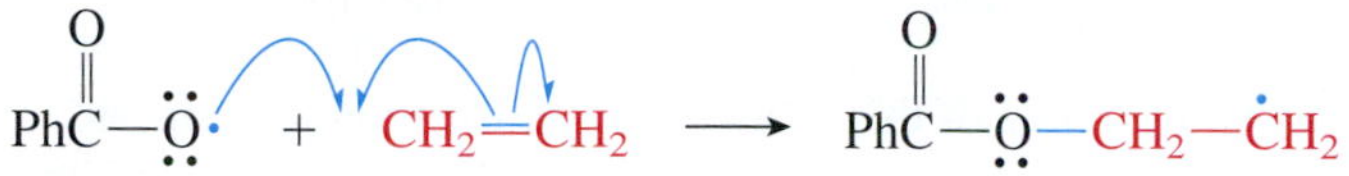

Propagation

In the propagation steps a carbon radical adds to the double bond of a monomer to produce a larger carbon radical. This radical adds to another ethylene to produce an even larger radical, and the process continues until the radical is somehow destroyed.

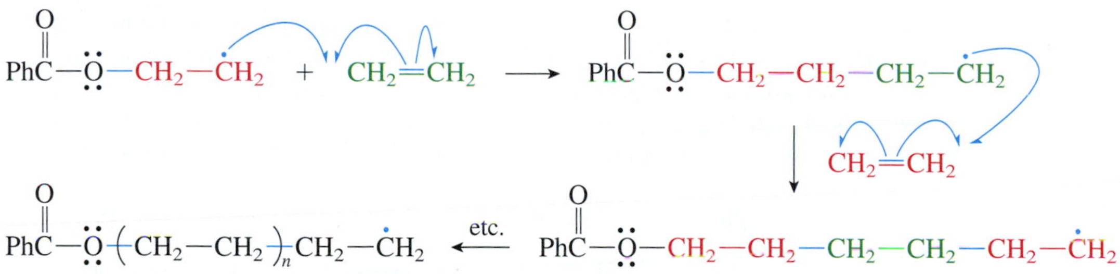

Termination

Various types of termination reactions can occur. Two radicals can couple to form nonradical products:

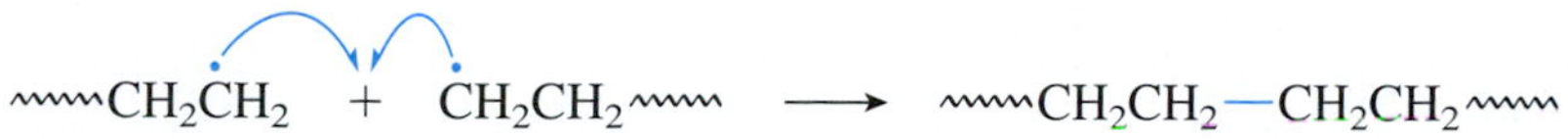

Or one radical can abstract a hydrogen from the carbon adjacent to the radical center of another radical in a disproportionation process:

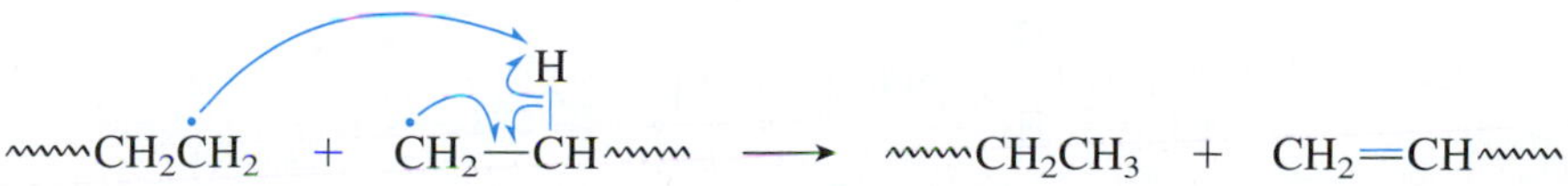

The average length of the polymer depends on the average number of propagation cycles that occur before a termination occurs. As is the case with many radical chain reactions, terminations are relatively rare because the concentration of radicals is extremely low. Therefore, the probability of one radical encountering another is also quite low. This means that a typical polymer molecule is composed of thousands of monomers. For a hydrocarbon polymer such as polyethylene, useful mechanical properties are not present until the polymer contains more than approximately 100 monomer units.

Another polymer that can be produced by radical chain polymerization is polystyrene:

$$n \quad CH_2{=}\underset{\alpha}{\overset{Ph}{CH}} \quad \xrightarrow{[ROOR]} \quad \left(CH_2{-}\overset{Ph}{CH}\right)_n$$
$$\qquad\quad \beta$$

Styrene   Polystyrene

In this case the two carbons of the double bond of the monomer are not identical. The polymer is found to have a regular structure in which the α-carbon of one monomer is always bonded to the β-carbon of the next, as shown in the following partial structure for polystyrene. The polymer is said to be formed by head-to-tail bonding of the monomer units.

$$\text{wwwCH}_2{-}\overset{Ph}{CH}{-}CH_2{-}\overset{Ph}{CH}{-}CH_2{-}\overset{Ph}{CH}{-}CH_2{-}\overset{Ph}{CH}\text{www}$$

Examination of the propagation steps in the mechanism for the formation of polystyrene makes the cause of this regularity readily apparent. Each time a new monomer is added to the growing polymer chain, the new bond is formed to the β-carbon. The resulting radical, with the odd electron located on the α-carbon, is stabilized by resonance involving the phenyl group and the odd electron:

$$\text{wwwCH}_2{-}\overset{Ph}{\underset{\bullet}{CH}} \quad + \quad CH_2{=}\overset{Ph}{CH} \quad \longrightarrow \quad \text{wwwCH}_2{-}\overset{Ph}{CH}{-}CH_2{-}\overset{Ph}{\underset{\bullet}{CH}}$$

As was the case with the radical addition reactions discussed in Chapter 21, the addition always occurs so as to produce the more stable radical. Because most other substituents also stabilize a radical on the carbon to which they are attached, vinyl polymers typically result from head-to-tail coupling of their monomer units.

PROBLEM 24.1

Show all of the steps in the mechanism for the radical polymerization of propylene and explain why the polymer is formed by head-to-tail bonding of the monomer units.

$$n \quad CH_2{=}\overset{CH_3}{CH} \quad \xrightarrow{\left[\overset{O}{\overset{\|}{Ph}}C\overset{O}{\underset{\|}{O}}CPh\right]} \quad \left(CH_2{-}\overset{CH_3}{CH}\right)_n$$

PROBLEM 24.2

Show the repeat unit of the polymers that would result from addition polymerization of these monomers:

a) $CH_2{=}CH{-}Cl$

Vinyl chloride

b) $CH_2{=}CH{-}C{\equiv}N$

Acrylonitrile

c) $CH_2{=}CCl_2$

Vinylidene chloride

PROBLEM 24.3

Show the monomers that could be used to prepare these polymers:

a) $(CH_2{-}CH(OCCH_3{=}O))_n$

Poly(vinyl acetate)

b) $(CF_2{-}CF_2)_n$

Teflon

24.2   STRUCTURES OF POLYMERS

Let's consider the structure of polyethylene in more detail. First, it is important to note that the number of monomers varies widely from one macromolecule to another. The termination steps, which stop the growth of individual polymer chains, occur at random times during the polymerization of those chains. Thus, one reacting chain may terminate early, resulting in a polymer molecule that contains relatively few monomer units, whereas another may terminate much later in the chain process, resulting in a molecule that contains many more monomer units. Although it is possible to exercise some experimental control over the average molecular mass of the polymer—that is, the average number of monomer units in each macromolecule—all synthetic polymers are composed of a variety of individual macromolecules of differing molecular masses.

The structure at one end of a polymer chain depends on the initiator that started that particular chain, and the structure at the other end depends on how that chain terminated. However, most polymers are so large that the ends do not have much effect on their properties. Therefore, the structure of a polymer is usually represented by its repeat unit, and the ends are not specified.

The radical chain mechanism shown in Section 24.1 implies that polyethylene and polystyrene are composed of linear macromolecules—that is, that the carbons of the vinyl groups of the monomers are connected in a straight chain. In fact, two processes can cause individual macromolecules to have branched structures. The first of these occurs when a growing radical chain abstracts a hydrogen from a random position in the interior of another macromolecule. This process, called chain transfer, occurs when the radical does not find another monomer unit to which to add nor another radical

❶ The radical end of a growing polymer chain may occasionally abstract a hydrogen from the interior of another chain. This process is favorable because a primary radical in the reactant is converted to a secondary radical in the product. This terminates the original polymer and forms a reactive radical in the interior of the second chain that can serve as a site to initiate polymerization.

❷ The new radical can add to an ethylene monomer, resulting in a polymer chain growing from the interior of this macromolecule. Although a secondary radical is converted to a primary radical, this step is still exothermic because a weak pi bond is broken and a stronger sigma bond is formed.

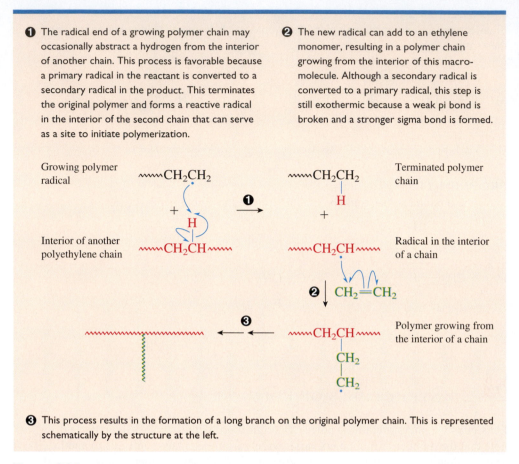

❸ This process results in the formation of a long branch on the original polymer chain. This is represented schematically by the structure at the left.

Figure 24.1

MECHANISM OF THE FORMATION OF A LONG BRANCH IN THE RADICAL POLYMERIZATION OF ETHYLENE.

center so that termination can occur. Chain transfer results in the termination of the original chain but forms a new radical in the interior of the other macromolecule. Polymerization can occur at this new radical site, resulting in the formation of a long branch on the macromolecule. The formation of such a long branch in polyethylene is outlined in Figure 24.1.

A second process results in the formation of shorter branches that contain only four carbons. These result when the radical end of a growing polymer chain reaches back and abstracts a hydrogen from itself. Because the cyclic transition state for this abstraction is most favorable when it contains six atoms, four-carbon butyl group branches are formed. The mechanism for the formation of these butyl branches is outlined in Figure 24.2.

These two branching processes decrease the regularity of the polyethylene macromolecules. Individual polymer chains may have long branches or butyl branches that occur at random positions. As we will see shortly, this irregularity in the structure dramatically affects the physical properties of the polymer.

❶ The radical carbon at the end of a growing polymer chain bends back and abstracts a hydrogen from the interior of the molecule. The most favorable size for the cyclic transition state is six atoms. This process is favorable because the original primary radical is converted to a more stable secondary radical.

❷ Polymerization can continue at the site of the new secondary radical.

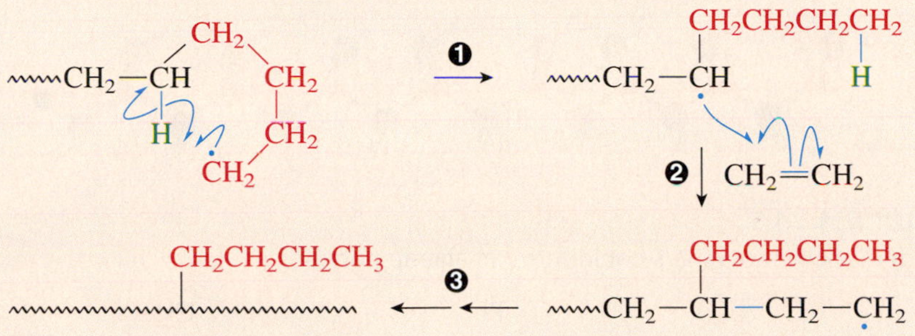

❸ This results in a four-carbon branch on the ultimate macromolecule.

Active Figure 24.2

ORGANIC
Chemistry⚛Now™

MECHANISM OF THE FORMATION OF A BUTYL BRANCH DURING THE POLYMERIZATION OF ETHYLENE. Test yourself on the concepts in this figure at **OrganicChemistryNow.**

PROBLEM 24.4

In addition to the four-carbon branches shown in Figure 24.2, polyethylene has a smaller number of branches containing three carbons. Show the steps in the mechanism for the formation of these three-carbon branches.

Another type of irregularity results if the vinyl monomer that is used to make an addition polymer has two different substituents on one end of the double bond. Propylene (propene), with a hydrogen and a methyl group on one of the vinyl carbons, provides an example. When such a monomer polymerizes, a new stereocenter (an asymmetric carbon chirality center) is created each time a new monomer is added:

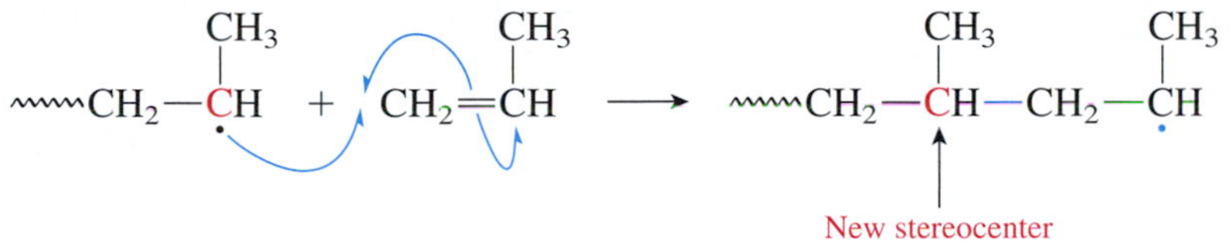

Because there is no preference for one absolute configuration over the other at this new stereocenter, the resulting macromolecule has a random configuration at its many stereocenters and is termed **atactic.** A partial structure for atactic polystyrene is shown in the following diagram:

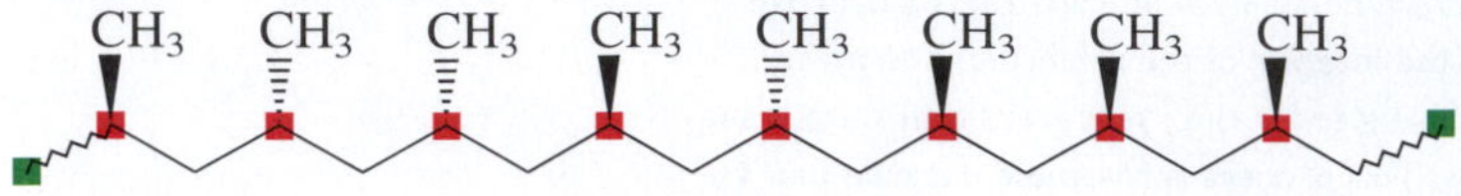

Atactic polypropylene
(random configuration at the carbons bonded to the methyl groups)

PROBLEM 24.5
Which of these monomers could form an atactic polymer?

a) 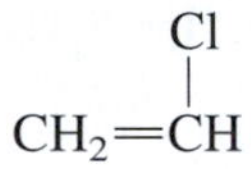 $CH_2{=}CH{-}Cl$ b) 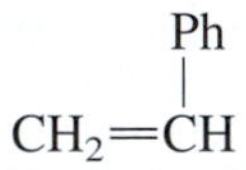 $CH_2{=}CH{-}Ph$ c) 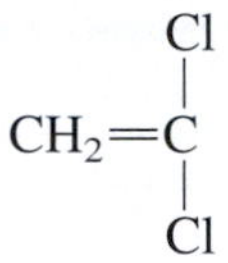 $CH_2{=}C(Cl)_2$

24.3   IONIC POLYMERIZATION

As well as mechanisms involving radical intermediates, addition polymers can be made by mechanisms involving cationic or anionic intermediates. For example, polyisobutylene is made by treating isobutylene with a small amount of boron trifluoride and water:

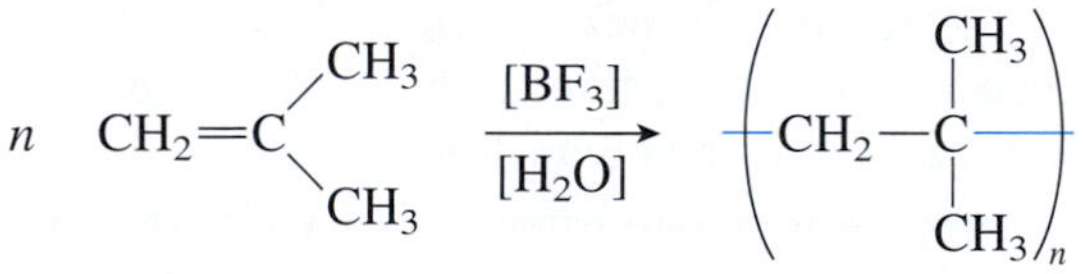

$$ n \ \ CH_2{=}C(CH_3)_2 \xrightarrow[\text{[H}_2\text{O]}]{\text{[BF}_3\text{]}} {+}CH_2{-}C(CH_3)_2{+}_n $$

Isobutylene   Polyisobutylene

The Lewis acid–base complex of water and boron trifluoride donates a proton to an isobutylene monomer to produce a carbocation. This carbocation adds to another isobutylene monomer to produce a larger carbocation, and the process continues, producing the polymer:

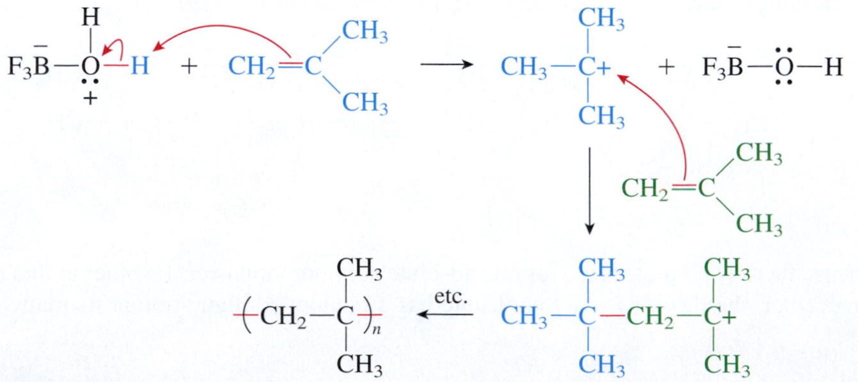

$$ F_3\bar{B}{-}\overset{+}{\underset{..}{O}}{-}H \ + \ CH_2{=}C(CH_3)_2 \longrightarrow CH_3{-}\overset{+}{C}(CH_3)_2 \ + \ F_3\bar{B}{-}\overset{..}{\underset{..}{O}}{-}H $$

$$ {+}CH_2{-}C(CH_3)_2{+}_n \ \xleftarrow{\text{etc.}} \ CH_3{-}C(CH_3)_2{-}CH_2{-}\overset{+}{C}(CH_3)_2 $$

Because each addition occurs so as to form the more stable tertiary carbocation, the monomers are connected in a regular head-to-tail fashion. Cationic polymerization can be used only when one vinyl carbon of the monomer is substituted with groups that can stabilize the intermediate carbocation, as is the case with the two methyl groups of isobutylene.

Addition polymerization can also occur by a mechanism involving anionic intermediates. For example, styrene can be polymerized by the addition of a small amount of sodium amide. In this case the amide anion adds to the double bond to produce a carbanion. This carbanion then adds to another styrene molecule to form a larger carbanion, and the process continues to form polystyrene:

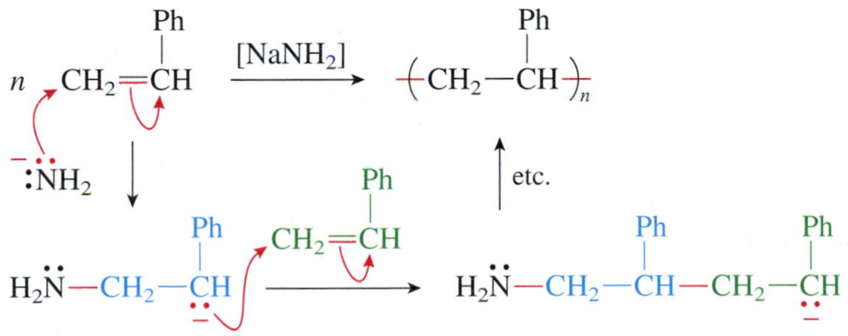

Most addition polymers are prepared from vinyl monomers. However, another type of addition polymer can be formed by ring-opening reactions. For example, the polymerization of ethylene oxide can be accomplished by treatment with a small amount of a nucleophile, such as methoxide ion. The product, a polyether, is formed by a mechanism involving anionic intermediates:

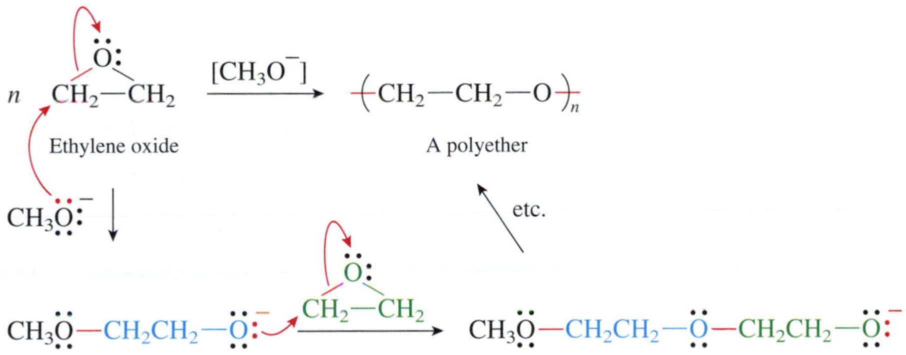

PROBLEM 24.6
Explain why polyethylene cannot be prepared by cationic polymerization whereas polystyrene can.

PROBLEM 24.7
Anionic polymerization is a good method for the preparation of polyacrylonitrile but not polyisobutylene. Explain.

Polyacrylonitrile   Polyisobutylene

Focus On

Super Glue

Super glue is a polymer of methyl cyanoacrylate. Because both the cyano and carbonyl groups of the monomer help stabilize carbanions, this compound is sensitive to polymerization by the anionic mechanism. The tube of glue contains very pure monomer, which does not polymerize until it contacts an initiator. However, contact with any nucleophile causes rapid polymerization. Therefore, when the tube is opened, polymerization is initiated by water in the air, by SiOH groups on a glass surface, by FeOH groups on an iron surface, or by various nucleophiles that are part of the proteins in skin. The adhesion between the polymer and the surface to which it is applied is very strong because the polymer chains are covalently linked to the nucleophiles that are part of the surface!

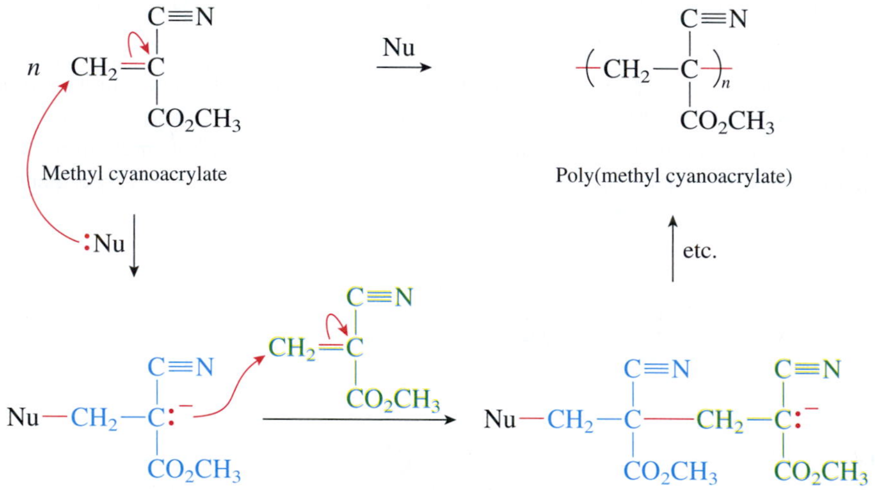

PROBLEM 24.8

In contrast to the formation of a polyether from the reaction of ethylene oxide with a small amount of methoxide ion, a similar reaction using THF does not result in the formation of a polymer. Explain.

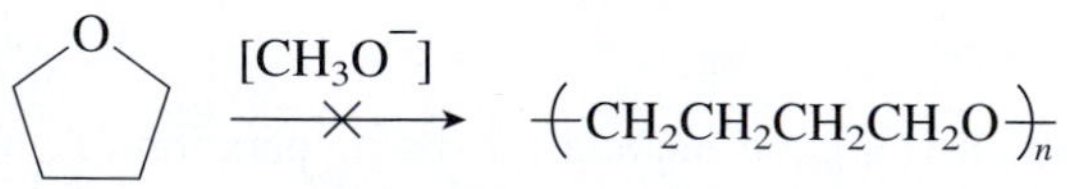

24.4 COORDINATION POLYMERIZATION

Perhaps the most important development in the area of addition polymerization was the discovery of a method for preparing vinyl polymers using metal catalysts. This breakthrough earned Karl Ziegler and Giulio Natta the 1963 Nobel Prize in chemistry

❶ The active catalytic species results from the transfer of an ethyl group from an aluminum to form a sigma bond to a titanium species. Propylene forms a "pi complex" with the titanium by interaction of its pi MOs with a vacant coordination site on the metal.

❷ The ethyl group migrates to one carbon of the double bond of the coordinated propylene, and the other carbon forms a sigma bond to the titanium. This creates a vacant coordination site on the metal, so the process can occur again.

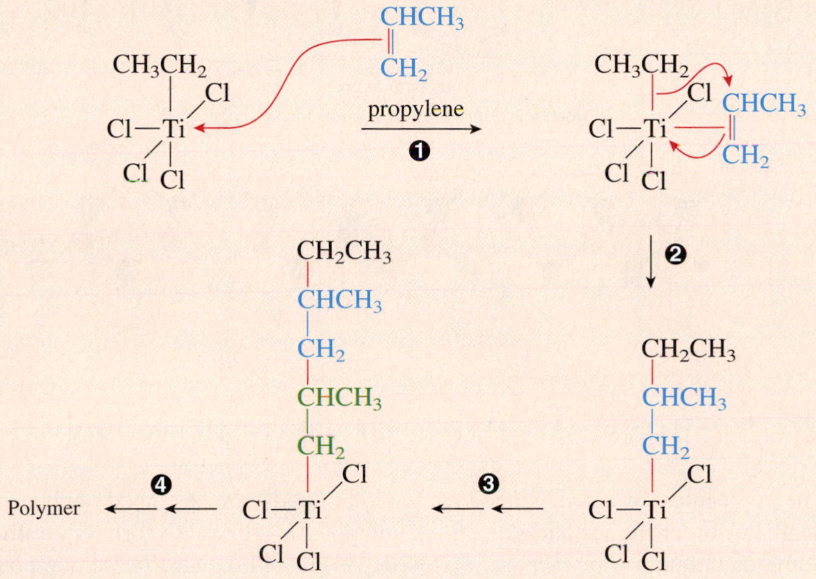

propylene

❹ Continuation of this process produces polypropylene.

❸ The process of steps 1 and 2 is repeated as another propylene coordinates to the titanium, followed by migration of the new, larger alkyl group.

Figure 24.3

MECHANISM OF POLYMERIZATION INVOLVING A METAL COORDINATION CATALYST (SIMPLIFIED VERSION).

and is probably used today to produce a larger amount of vinyl polymers than all other methods combined. Although various catalysts have been used, a typical one uses an organometallic compound, such as triethylaluminum, and a transition metal halide, such as titanium tetrachloride. Although the mechanism is complex, a simplified version that has the correct general features is shown in Figure 24.3 for the polymerization of propylene. Basically, the mechanism proceeds by coordination of a monomer to a titanium that has an alkyl group sigma bonded to it. The alkyl group then migrates to one carbon of the double bond of the coordinated monomer while the other carbon forms a sigma bond to the titanium. This step regenerates the original catalyst, but with a larger alkyl group bonded to it. Additional monomer units are added in a similar fashion.

What makes the method using Ziegler-Natta catalysts so important? The resulting polymers are much more regular than those produced by other methods. For example, polyethylene produced by using a coordination catalyst is linear. It does not have the short or long branches that characterize polyethylene that is produced by a radical initiator. Furthermore, coordination polymerization can be used to prepare stereoregular polymers. For example, polypropylene can be prepared with identical configuration at all of the stereocenters. The resulting **isotactic** polypropylene has very different, and much more useful, properties than the atactic polypropylene that is produced by radical polymerization. Let's see how the structure and regularity of these polymers affect their physical properties.

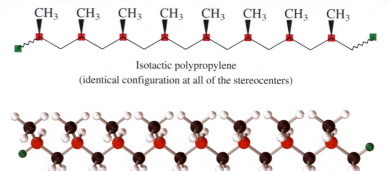

Isotactic polypropylene
(identical configuration at all of the stereocenters)

24.5   PHYSICAL PROPERTIES OF POLYMERS

The molecules of most nonpolymeric compounds are arranged with a very high degree of order in the solid state. Such compounds are said to form **crystalline solids.** Some compounds, however, have no order in the solid state. These compounds form glassy solids with a random arrangement of molecules and are said to be **amorphous.**

The enormous molecules of a polymer such as polyethylene are too long to form a completely crystalline solid. However, many polymers are semicrystalline; that is, they have both crystalline and amorphous regions. For example, solid polyethylene has crystalline regions, called **crystallites,** where the chains are arranged in a very ordered manner, along with amorphous regions that are completely disordered. The part of a polyethylene molecule that is in a crystallite has an anti conformation about each of its carbon–carbon bonds. The resulting zigzag chain can pack well with the zigzag chains of other molecules. A two-dimensional picture of a crystallite can be represented schematically as shown in the following diagram:

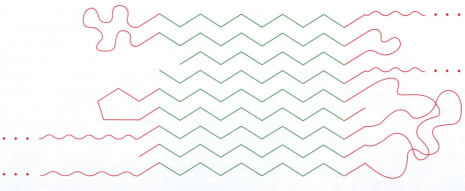

Amorphous region   Crystalline region   Amorphous region

These crystallites vary in size and shape and are much smaller than the crystals of a normal organic compound. An individual crystallite has dimensions on the order of 10^3 Å (100 nm). Recall that the length of most covalent bonds is slightly greater than 1 Å. Therefore, the ordered part of a crystallite extends over a region containing tens to hundreds of bonds. When a chain reaches the boundary of a crystallite, it may bend back and become part of the crystallite again. Other chains wander off into the amorphous region. Some of these remain in the amorphous region; others return to the same crystallite, and some even become part of another crystallite. In general, polymers that have more and larger crystalline regions are stiffer and stronger and are more useful for many applications.

Regularity in the structure of a polymer favors crystallinity because the chains can pack closer together. The head-to-tail bonding of the individual monomer units is one type of regularity that is present in most addition polymers. The presence of branches is an irregularity that decreases the ability of the polymer chain to pack into crystalline regions. Therefore, linear polyethylene prepared by polymerization using a Ziegler-Natta catalyst is more crystalline than the highly branched polyethylene produced by a radical mechanism. In addition, polymers must have a regular stereochemistry if they are to be crystalline. Atactic polymers are completely amorphous. Again, the ability to use coordination polymerization for the preparation of stereoregular polymers is extremely valuable.

Crystallinity is also favored by strong forces between polymer chains. In a nonpolar polymer such as polyethylene, the only forces holding the chains in place are van der Waals forces (see Section 2.5). Although these attractions are relatively weak, there are many of them, so their total force can be quite large. However, polymers with polar groups have stronger intermolecular forces and are more crystalline than nonpolar polymers, other factors being equal.

Most addition polymers are **thermoplastics;** that is, they are hard at room temperature but soften and eventually melt as they are heated. At low temperatures there is very little motion of the molecules and the polymer is glasslike and brittle. As the temperature of the polymer is raised, it passes through its glass transition temperature (T_g). Above T_g, more motion of the chains is possible and the polymer is a rubbery solid. Eventually, the polymer passes through its crystalline melting point (T_m) and melts to form a viscous liquid. Many semicrystalline polymers are most useful at temperatures between T_g and T_m. Both T_g and T_m increase as the crystallinity of the polymer increases and as the strength of the intermolecular forces between the polymer chains increases. The total intermolecular force increases as the length of the polymer chains increases.

In general, polymers have no useful mechanical properties until the chains reach a certain average length. Above this minimum length, the strength increases as the polymer gets longer, but it also becomes more difficult to process the polymer. Therefore, the average molecular mass is usually a compromise, large enough that the polymer has useful mechanical properties but not so large that it cannot be molded, extruded, or drawn into fibers.

PROBLEM 24.9

Explain which of these monomers produces the more crystalline polymer.

a) CH₂=CH(Cl) or CH₂=C(Cl)(Cl)   using radical polymerization

b) CH₂=CH(CH₃) or CH₂=CH(CO₂CH₃)   using radical polymerization

c)

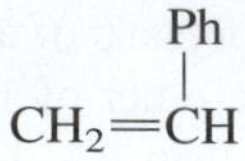

$$\text{CH}_2\text{=CH} \quad \overset{\displaystyle \text{Ph}}{|}$$

using radical polymerization or coordination polymerization

d)

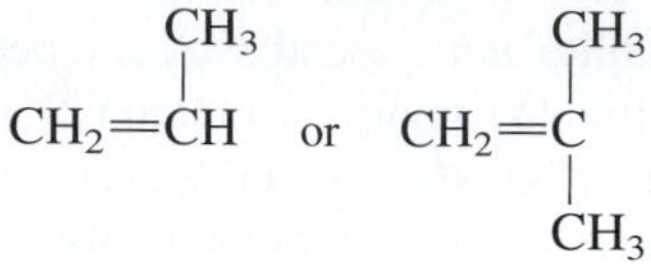

$$\text{CH}_2\text{=CH} \quad \overset{\displaystyle \text{CH}_3}{|} \quad \text{or} \quad \text{CH}_2\text{=C} \quad \overset{\displaystyle \text{CH}_3}{\underset{\displaystyle \text{CH}_3}{|}}$$

using radical polymerization

PROBLEM 24.10

Explain why Teflon is linear and has no stereochemical complications even though it is prepared by radical polymerization.

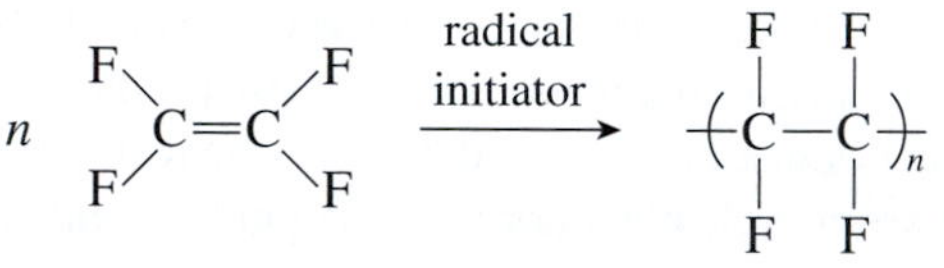

Tetrafluoroethylene   Poly(tetrafluoroethylene)
                                  (Teflon)

24.6 MAJOR THERMOPLASTIC ADDITION POLYMERS

Four thermoplastic addition polymers—polyethylene, poly(vinyl chloride), polypropylene, and polystyrene—comprise the majority of the total amount of polymers manufactured in the United States. In 2002 a total of 33.6 million metric tons of these plastics was produced, distributed as shown in Table 24.1.

Table 24.1   Output of Thermoplastic Polymers

Polymer	Structure	Amount (million metric tons)
Polyethylene	$\{\text{CH}_2\text{—CH}_2\}_n$	16.0
Poly(vinyl chloride)	$\{\text{CH}_2\text{—CH}\}_n$ (Cl)	6.9
Polypropylene	$\{\text{CH}_2\text{—CH}\}_n$ (CH$_3$)	7.7
Polystyrene	$\{\text{CH}_2\text{—CH}\}_n$ (Ph)	3.0

Two types of polyethylene are manufactured: high-density polyethylene (HDPE) and low-density polyethylene (LDPE). HDPE is produced by coordination polymerization using a Ziegler-Natta type catalyst. Its regularity and the absence of branches make it more crystalline, and the resulting closer packing of its chains results in a higher density. It is strong and rigid, with a higher T_m. It is used to make a variety of containers and plastic items, such as bottles, mixing bowls and other kitchen items, and toys. LDPE is produced by radical polymerization and has both long and short branches. As a result, it is more amorphous and less dense because its chains are not packed as closely. It is weaker and less rigid than HDPE. A majority of it is used as film in packaging products, such as garbage bags, household plastic wrap, and the transparent film used to cover trays of meat in the supermarket. Its major advantage is its low cost.

Poly(vinyl chloride), also known as PVC, is prepared by radical polymerization to produce material composed of an average of 10,000 to 24,000 monomer units. It is atactic and therefore amorphous, but it has a relatively high T_g because of the large size of its molecules and its polar carbon–chlorine bonds. It is a rigid material and is used to make pipe, panels, and molded objects. About 68% of PVC is used in the building and construction industry. A more flexible form of PVC is produced by adding a plasticizer such as dioctyl phthalate. This is used to prepare electric wire coatings, film, and simulated leather or "vinyl."

Polypropylene owes its current market success to the development of coordination polymerization. Before 1957 it was not produced commercially because radical polymerization gives an atactic polymer that is amorphous and has poor mechanical properties. Using a coordination catalyst, however, enables the production of an isotactic polymer that is semicrystalline. This material is stiff and hard and has a high tensile strength. Among its many useful products are rope, molded objects, and furniture.

Polystyrene is made by radical polymerization and is atactic and amorphous. Incorporation of small air bubbles produces a foam (Styrofoam) that finds a major use in packaging materials and insulation.

Many other addition polymers are manufactured commercially, although in much smaller amounts than those just described. For example, poly(methyl methacrylate) is prepared by radical polymerization of the methyl ester of methacrylic acid:

Methyl methacrylate   Poly(methyl methacrylate)

Although it is atactic and amorphous, its polar groups cause it to have a relatively high T_g (110°C), and it is rigid and glasslike at normal temperatures. It is used to make products such as Plexiglas and Lucite.

PROBLEM 24.11

Do you think that poly(methyl methacrylate) could be prepared by cationic polymerization? by anionic polymerization? Explain.

24.7   ELASTOMERS

Several general requirements must be met for a polymer to be elastic—that is, to stretch under the application of force but return to its original shape when the force is released. The polymer should be predominantly amorphous so that its T_g is below room temperature. The individual molecules of an amorphous polymer are not in fully extended, anti conformations; instead, they have random, coiled conformations. When a force that pulls on opposite ends of the molecules is applied, the molecules assume an anti conformation about more bonds and thus they become longer; that is, they stretch. Although the stretching tends to arrange the molecules in extended, zigzag conformations that are favorable for crystallization, the overall shapes of the elastomer molecules are such that crystallization does not readily occur. Furthermore, most elastomers are nonpolar, so only weak attractive forces exist between chains. Therefore, when the force is removed, the molecules tend to return to their initial random conformations because these random shapes are favored by entropy (disorder).

In addition, however, it is necessary to somehow keep the chains from slipping past each other entirely as they are stretched. If this is not prevented, the material will simply pull apart rather than stretch and then return to its previous shape when the stretching force is removed. This is accomplished by the formation of **cross-links**—that is, groups that connect separate chains by covalent bonds. If only a few of these cross-links are present, then the molecules are still flexible enough to be stretched, but the cross-links prevent the molecules from separating entirely, and they return to their original shape when the stretching force is removed. The following diagram attempts to illustrate this process for a simplified system of only two polymer chains. In this picture, -L- is used to represent the cross-linking groups. A few cross-links couple the two chains while others connect them to other polymer molecules.

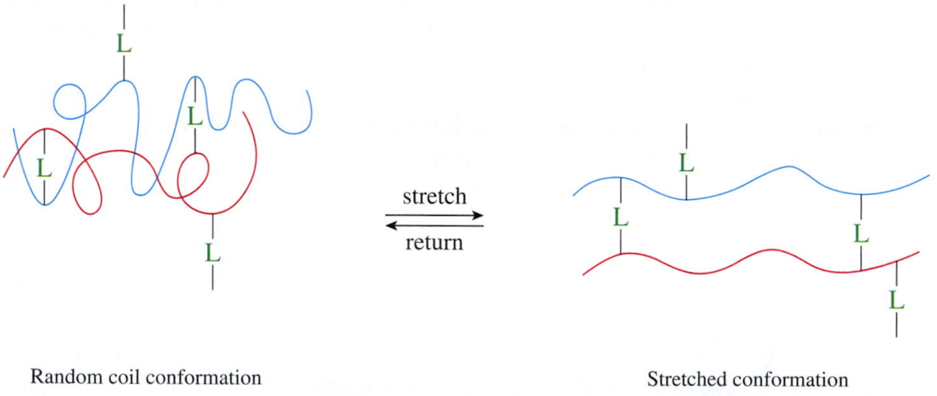

Random coil conformation   Stretched conformation

Natural rubber can be viewed as a polymer formed by addition of isoprene monomers at the ends of the 1,3-diene units (although nature does not prepare it in this manner).

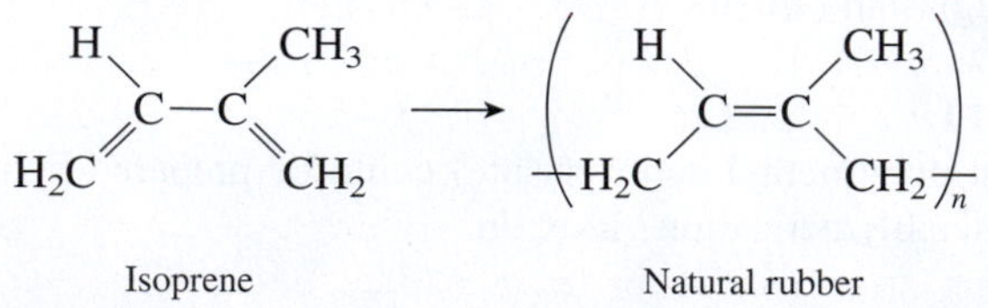

Isoprene   Natural rubber

The double bonds of the rubber molecules all have the cis, or Z, configuration. This causes bends in the chains that make the molecules less able to crystallize. To prevent the molecules from slipping past each other when the rubber is stretched, the molecules are cross-linked by treatment with sulfur in a process called *vulcanization*. Although the exact details of vulcanization are not known, links between different chains are formed by one or two sulfur atoms:

$$
\begin{array}{c}
CH_3 \\
\mid \\
\text{\small www}CH-C=CH-CH_2\text{\small www} \\
\mid \\
S \\
\mid \\
S \quad\quad CH_3 \\
\mid \quad\quad \mid \\
\text{\small www}CH-C=CH-CH_2\text{\small www}
\end{array}
$$

A typical cross-link formed
in vulcanization

Isoprene can be polymerized in the laboratory by a radical chain mechanism. As shown in the following equations, the odd electron of the initially produced radical is delocalized onto both C-2 and C-4 by resonance. Either of these carbons may add to another isoprene monomer to continue the chain reaction. If C-2 adds, the process is called 1,2-addition; if C-4 adds, the process is called 1,4-addition. (This is similar to the addition of electrophiles to conjugated dienes discussed in Section 11.13 and the addition of nucleophiles to α,β-unsaturated carbonyl compounds described in Section 18.10.)

1,2-Addition   1,4-Addition

Radical polymerization results mainly in 1,4-addition of the monomer units, and the double bonds are predominantly in the trans or E configuration. Because its trans double bonds allow its chains to crystallize more readily, this product, known as gutta percha, has a higher T_g than natural rubber and is hard and inelastic at room temperature. Recently, however, a method for the polymerization of isoprene using a coordination catalyst has been developed. The polymer results from 1,4-addition with almost all of its double bonds in the cis configuration and is nearly identical to natural rubber.

Synthetic rubbers now constitute about two thirds of the rubber that is used worldwide. The major synthetic rubber, called styrene-butadiene rubber, or SBR, is a copolymer formed by radical polymerization of a mixture of 25% styrene and 75% 1,3-butadiene. The monomers add in a random sequence. The incorporation of each butadiene monomer occurs predominantly by 1,4-addition, and the double bond is usually trans. However, the random placement of the styrene units and their stereochemical disorder keep the rubber from being too crystalline. The product can be vulcanized in a manner similar to natural rubber and finds a major use in automobile tires.

PROBLEM 24.12
Explain why radicals prefer to add to carbon 1 of isoprene rather than carbons 2, 3, or 4.

PROBLEM 24.13
Show a partial structure for SBR and explain why most of the double bonds have a trans configuration.

24.8   CONDENSATION POLYMERS

Besides addition polymerization, the other general way to prepare polymers is known as **condensation polymerization** or **step growth polymerization.** Much of the pioneering work on condensation polymerization was conducted by Wallace Carothers while he was employed by DuPont. He recognized that many natural polymers are formed from monomers with two reactive functional groups. For example, proteins are polymers of amino acids, which contain both amine and carboxylic acid groups. The formation of amide bonds is used to connect one monomer to another. Carothers's attempts to imitate nature led to a whole industry based on condensation polymerization.

The key to the formation of a condensation polymer is the presence of two reactive functional groups in each monomer. This is illustrated schematically in the following equation, where FG and FG′ represent different functional groups that can react to form a covalent bond.

For example, the functional group represented by FG might be an amine, and the functional group represented by FG′ might be a carboxylic acid. Formation of an amide bond between the amine of one monomer and the carboxylic acid of the other results in the formation of a dimer. Continuation of this process results in polymer formation. Let's examine some specific examples to better understand how this process works.

One important group of condensation polymers is the polyesters. The most important commercial polyester is formed from the reaction of terephthalic acid (a diacid) with ethylene glycol (a diol). This polymerization occurs in a stepwise fashion (hence the name step growth polymerization). First, one carboxylic acid group of a diacid molecule and one hydroxy group of a diol molecule combine to form an ester, with the loss of water. Then a second diol molecule reacts with the unreacted carboxylic group on the other end of the diacid molecule, or a second diacid molecule reacts with the unreacted hydroxy group of the diol. Continuation of this process adds a new monomer unit at each step, ultimately producing a polymer. The resulting polyester is called poly(ethylene terephthalate), or PET.

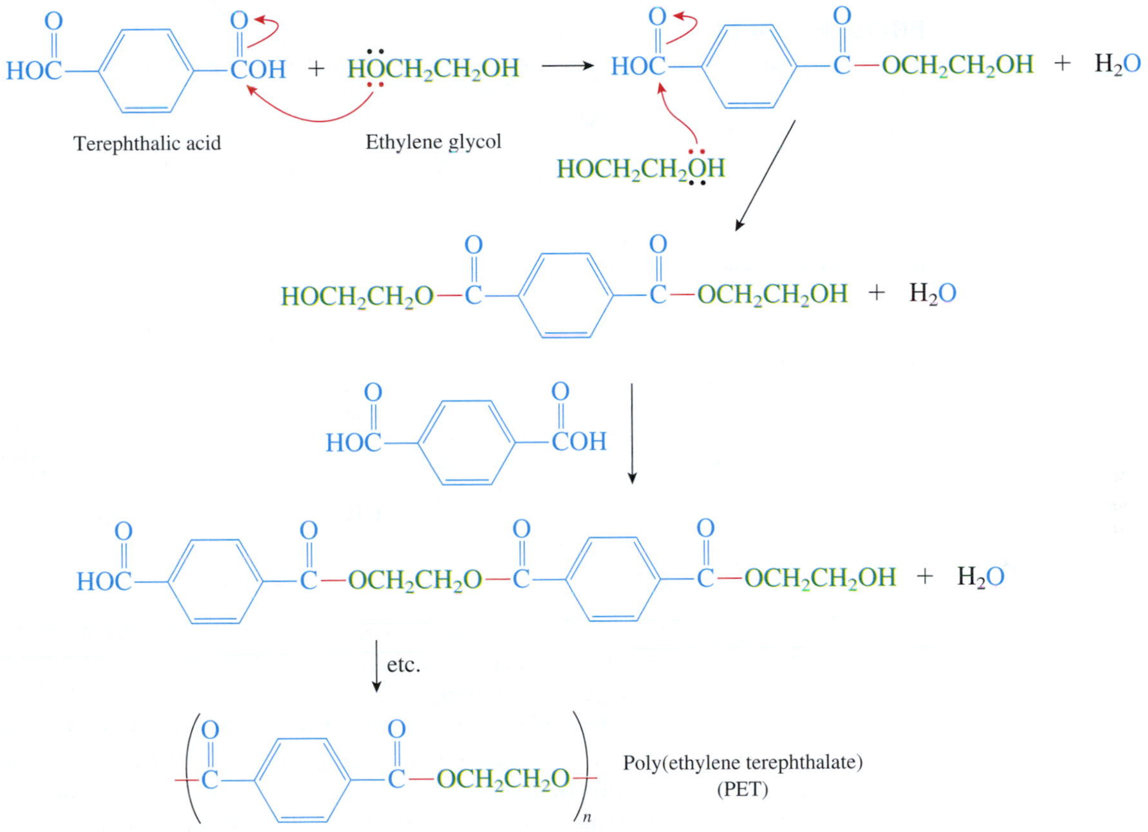

Terephthalic acid   Ethylene glycol

Poly(ethylene terephthalate)
(PET)

Because of the difficulty encountered in removing the water that is produced when an acid reacts with an alcohol, PET is produced commercially by the reaction of dimethyl terephthalate, the dimethyl ester of terephthalic acid, with ethylene glycol in an ester interchange reaction. The methanol that is formed in this reaction is readily removed by distillation.

Dimethyl terephthalate

$+ \quad 2n - 1 \ CH_3OH$

PET has a high degree of crystallinity because it is linear and contains polar ester functional groups. It has a relatively high melting point (270°C). If the molten polymer is drawn through a small hole, fibers are formed. The drawing process extends and orients the molecules that form the fibers, maximizing the dipole–dipole interactions between chains and increasing their crystallinity. This results in a high tensile strength for the fiber. This fiber, known as Dacron or Terylene, is spun into thread that is used to make polyester fabrics or blended with cotton to make permanent press fabrics. PET is also used to make a strong film, called Mylar, and is the major plastic used to make bottles for soft drinks.

PROBLEM 24.14

Show the repeat unit of the polymer formed from adipic acid and ethylene glycol:

$$HO\overset{O}{\overset{\|}{C}}(CH_2)_4\overset{O}{\overset{\|}{C}}OH \quad + \quad HOCH_2CH_2OH \longrightarrow$$

Adipic acid

Another important polyester is the polycarbonate formed by the reaction of bisphenol A with phosgene:

Bisphenol A   Phosgene   A polycarbonate

The resulting polymer, known as Lexan, is useful because of its high-impact strength. It is used in products such as football helmets and motorcycle helmets.

Condensation polymers based on an amide linkage have also found considerable commercial applications. For example, when the salt of adipic acid and hexamethylenediamine is heated to 270°C, a polyamide known as nylon 6,6 is formed. (The first number in the name of the nylon designates the number of carbons in the diamine, six in this case, and the second designates number of carbons in the diacid, also six in this case, that are used to form the nylon.)

$$HO\overset{O}{\overset{\|}{C}}(CH_2)_4\overset{O}{\overset{\|}{C}}OH \quad + \quad H_2N(CH_2)_6NH_2 \longrightarrow \quad {}^-O\overset{O}{\overset{\|}{C}}(CH_2)_4\overset{O}{\overset{\|}{C}}O^- \quad + \quad H_3\overset{+}{N}(CH_2)_6\overset{+}{N}H_3$$

Adipic acid   Hexamethylenediamine

270°C

$$\left(\overset{O}{\overset{\|}{C}}(CH_2)_4\overset{O}{\overset{\|}{C}}-NH(CH_2)_6NH\right)_n$$

Nylon 6,6

This same polyamide can be prepared by reaction of the diacyl chloride of adipic acid with hexamethylenediamine. In Section 19.6 we saw that the conversion of the salt of a carboxylic acid and an amine to an amide could be accomplished by heating but that this method was not often used in the laboratory because of the rather vigorous conditions that must be employed. However, in an industrial setting, cost is a more important factor. Because the reaction starting only with the diacid and the diamine is much less expensive than first converting the diacid to its acyl chloride, the former method is the one used to prepare nylon 6,6.

Because of its polarity and its ability to form hydrogen bonds, nylon 6,6, along with other polyamides, has strong intermolecular forces. It has a high melting point (250°C) and can be drawn into fibers with good tensile strength. Nylon 6,6 was the first synthetic fiber. (In May 1940 the first pairs of nylon stockings went on sale. Within 4 days, nearly the entire stock of 4 million pairs had been sold.) It is also used to make molded objects.

PROBLEM 24.15

Nylon 6,10 is quite rigid and is used in applications such as brush bristles. Show the monomers that are used to prepare this polymer.

$$\left[\overset{\overset{\displaystyle O}{\|}}{C}(CH_2)_8\overset{\overset{\displaystyle O}{\|}}{C}NH(CH_2)_6NH\right]_n$$

Other factors being equal, a polyamide has stronger intermolecular forces than a similar polyester because the polyester cannot form hydrogen bonds. As an illustration of this point, a polyester related to nylon 6,6—that is, one having only alkyl groups connecting the two hydroxy groups and the two carboxylic acid groups in its monomers—has weaker intramolecular forces and does not have useful mechanical properties. Changing to a more rigid diacid, such as terephthalic acid, increases the intramolecular forces and provides a more crystalline and more useful polyester.

Another polyamide is prepared by polymerization of ϵ-caprolactam. This monomer is different from the others that we have seen so far because it has two different functional groups in the same molecule. Therefore, only one type of monomer is needed to form the polymer. When the lactam is heated with a small amount of water, some is hydrolyzed to the amino acid, which forms an internal salt. Heating at 250°C causes formation of an amide bond while simultaneously regenerating water. Continuation of this process provides the polyamide known as nylon 6:

ϵ-Caprolactam   Nylon 6

Use of a more rigid diamine and diacid derivative, such as 1,4-diaminobenzene and terephthaloyl chloride, provides a polyamide, known as Kevlar, with very interesting properties:

1,4-Diaminobenzene   Terephthaloyl chloride   Kevlar

The rigid, rod-shaped molecules of Kevlar are very crystalline and can be used to form strong, stiff, high-strength fibers. Although this polymer is fairly expensive, it is used in high-tech applications such as bulletproof body armor.

Polyurethanes are produced by the reaction between alcohols and isocyanates. Isocyanates have a very reactive functional group that is subject to nucleophilic attack by an alcohol to give a carbamate or urethane:

$$R-N=C=O \quad + \quad H-\ddot{O}-R' \quad \longrightarrow \quad R-N-\overset{\overset{\displaystyle O}{\|}}{C}-O-R'$$

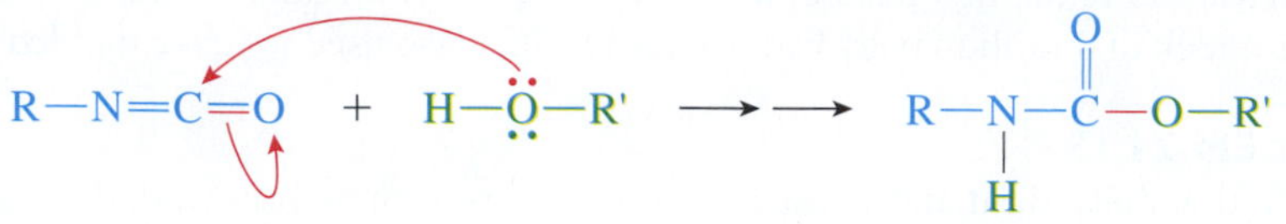

An isocyanate   A carbamate or urethane

Formation of a polyurethane, then, requires the reaction of a dialcohol with a diisocyanate. The dialcohol is often a small polymer, with a molecular mass of 1000 to 2000, that has hydroxy groups at each end. It is prepared by the reaction of a diacid, such as adipic acid, with a slight excess of a diol, such as ethylene glycol. Using a small excess of the diol ensures that the polymer chains are relatively short and have hydroxy groups at the ends:

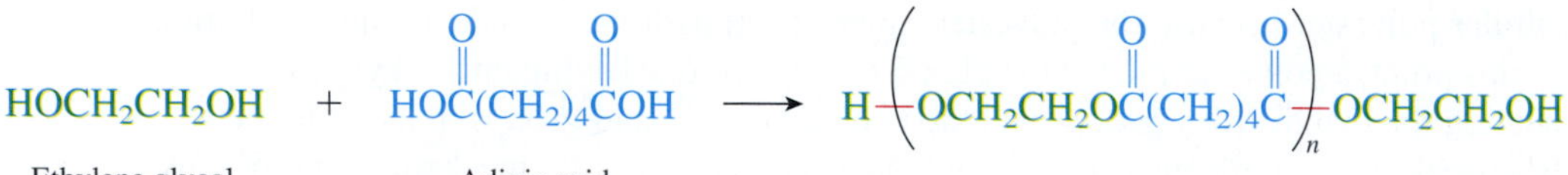

$$HOCH_2CH_2OH \quad + \quad HOC(CH_2)_4COH \quad \longrightarrow \quad H\left(OCH_2CH_2OC(CH_2)_4C\right)_n OCH_2CH_2OH$$

Ethylene glycol   Adipic acid
(slight excess)

This polymeric diol is then reacted with a diisocyanate, such as toluene diisocyanate, to produce the polyurethane:

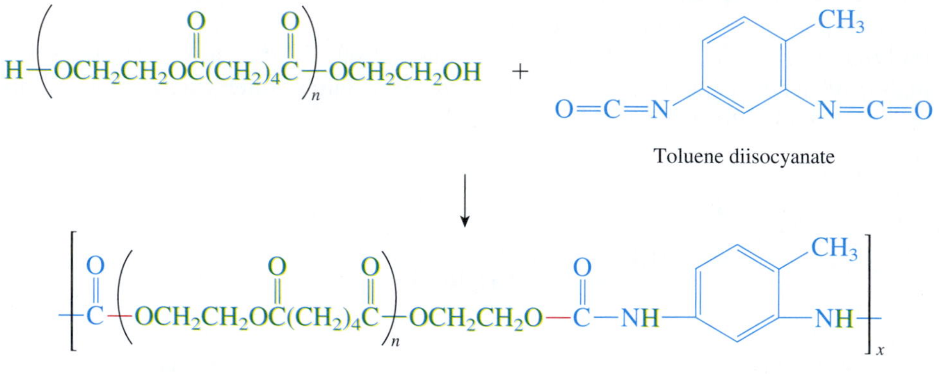

$$H\left(OCH_2CH_2OC(CH_2)_4C\right)_n OCH_2CH_2OH \quad +$$

Toluene diisocyanate

A polyurethane

PROBLEM 24.16

To form high-molecular-mass polyesters, the diol and the diacid must be used in exactly a 1:1 molar ratio. Explain why the use of a slight excess of the diol results in the formation of a relatively short polymer chain.

Polyurethane foams are prepared by including some water when the monomers are mixed. The reaction of an isocyanate with water produces a carbamic acid. As discussed in Section 23.4, a carbamic acid is unstable and spontaneously eliminates carbon dioxide to form an amine. This reaction is illustrated for toluene diisocyanate in the following equation:

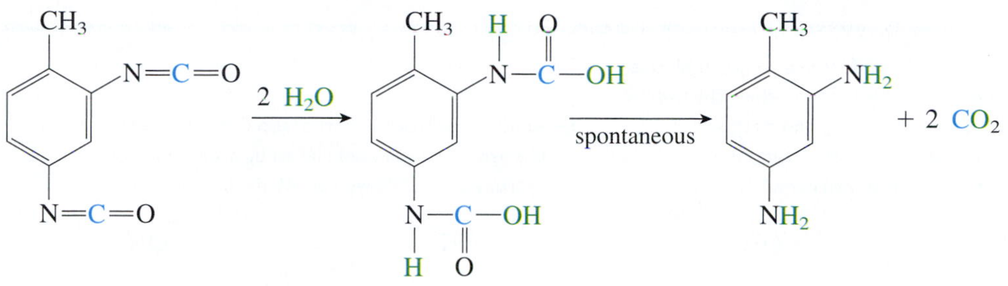

A dicarbamic acid

The carbon dioxide that is produced by the preceding equation is trapped in the polymer as it forms, resulting in a foamy polymer. Polyurethane foam is used to make cushions, pillows, and insulation.

24.9 THERMOSET POLYMERS

In contrast to the thermoplastic polymers that have been discussed so far, thermoset polymers do not soften or melt on heating. Instead, they become larger, harder, and more insoluble because cross-linking becomes very extensive. Essentially, the sample becomes one huge molecule as the polymer cures.

Bakelite, the first synthetic polymer, is an example of a thermoset polymer. It is prepared by the polymerization of phenol and formaldehyde in the presence of an acid. Carbocations produced by protonation of formaldehyde bond to the ortho and para positions of the highly reactive phenol molecules in a Friedel-Crafts alkylation reaction. The benzylic alcohols that are produced in this step react to produce carbocations that then alkylate additional phenol molecules. A mechanism for the first few steps of this polymerization process is shown in Figure 24.4.

Because each phenol has three sites (two ortho and a para) that can be linked by CH_2 groups, the resulting polymer is highly cross-linked. The structure of a part of the polymer can be illustrated as shown in the following diagram:

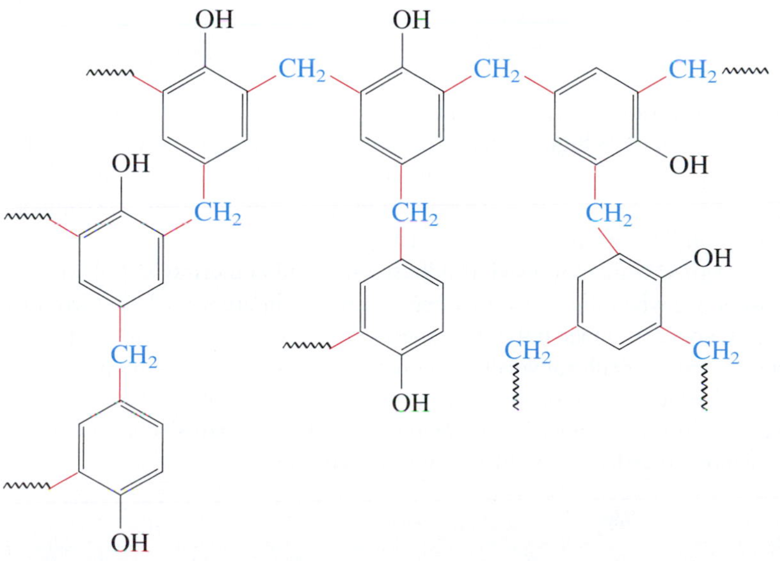

❶ The electrophilic carbon of a protonated formaldehyde molecule adds to phenol in a Friedel-Crafts alkylation reaction. The hydroxy group of phenol is a strong activating group and directs the incoming electrophile to the ortho and para positions. One possibility is the ortho attack that is shown here.

❷ Another electrophilic carbocation can be generated from this hydroxymethylated phenol by protonation of the hydroxy group followed by ❸ the loss of water.

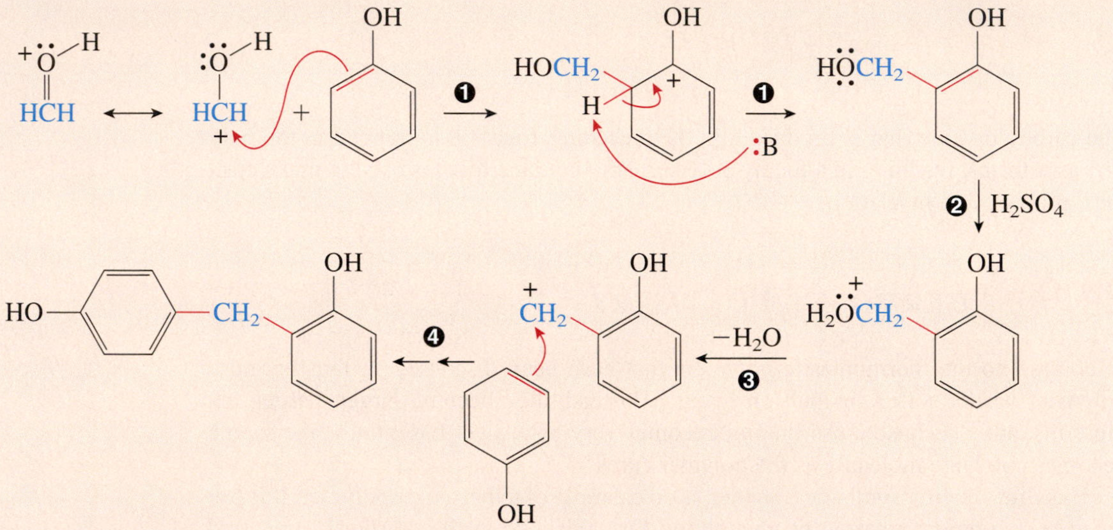

Each phenol nucleus in this molecule has two additional reactive sites (the other ortho and/or para positions) where connections to other phenols can occur by this same mechanism. Ultimately, a highly cross-linked polymer is formed.

❹ The resulting carbocation can alkylate another phenol molecule. Again the new bond can be formed at the ortho or para position.

Figure 24.4

PARTIAL MECHANISM OF THE POLYMERIZATION OF PHENOL AND FORMALDEHYDE.

Some of the phenol units are linked at all three positions, whereas others are linked at only two. Because of its high degree of cross-linking, Bakelite does not melt, burns only with difficulty, and is an excellent electrical insulator. It is molded into a large variety of objects, including billiard balls, automobile distributor caps, instrument control knobs, and handles for cookware and toasters.

Epoxy resins, which are used as adhesives, are also thermoset polymers that form by cross-linking when the two components of the resin are mixed. One component is a low-molecular-mass linear polymer formed by the reaction of the conjugate base of bisphenol A with epichlorohydrin. The nucleophilic oxygens of the phenolate dianion can either displace the chlorine or open the epoxide ring of epichlorohydrin. A slight excess of epichlorohydrin is used to keep these polymer chains short and to ensure that the linear molecules have epoxide groups at their ends.

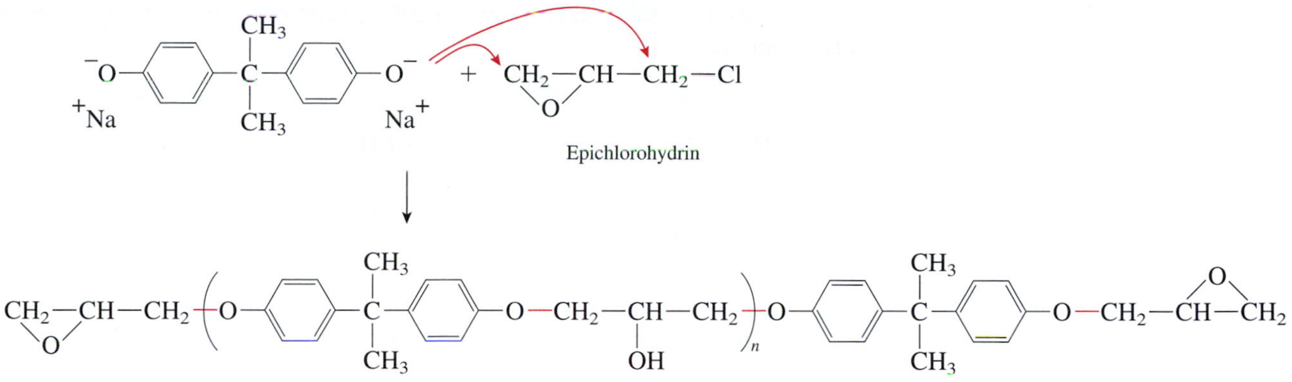

Epichlorohydrin

The second component of the glue is composed of a trifunctional amine such as diethylenetriamine. When the two components are mixed, the three nucleophilic nitrogens of the triamine react with the epoxide groups to form a highly cross-linked polymer, as shown in the following equation:

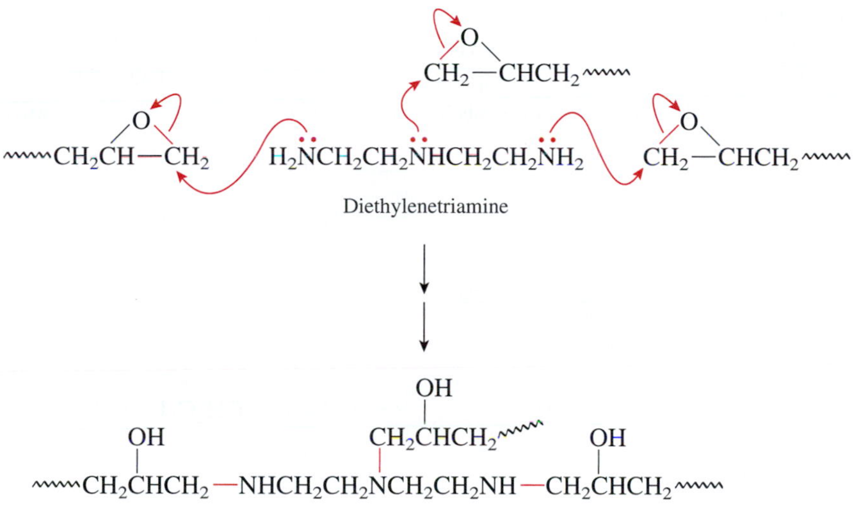

Diethylenetriamine

The reaction is complete in a few hours, and the polymer hardens to a strong adhesive.

24.10 CHEMICAL PROPERTIES OF POLYMERS

Although polymer molecules are enormous in comparison to the other molecules that we have discussed, their chemical reactions are nearly identical to those of smaller molecules. For example, poly(vinyl alcohol) is prepared by saponification of poly(vinyl acetate):

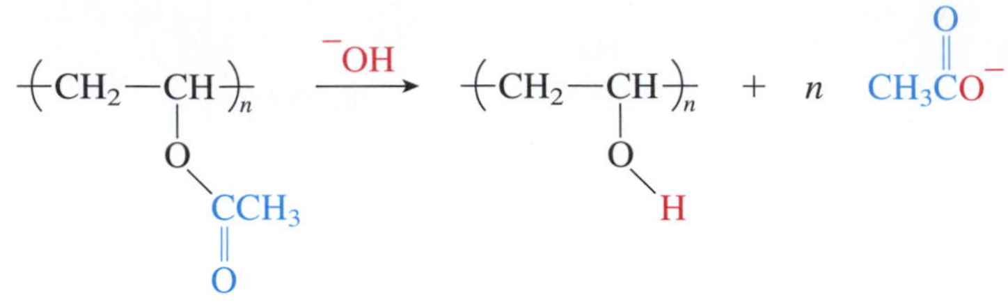

Poly(vinyl acetate)   Poly(vinyl alcohol)

PROBLEM 24.17

Explain why poly(vinyl alcohol) cannot be prepared by a direct polymerization reaction. (*Hint:* Draw the structure of the monomer that would be required.)

As another example, let's consider the chemical reactions that are used to prepare ion exchange resins. These resins are composed of small polymer beads that can be used to exchange one ion for another in aqueous solutions. The polymer that is used in these applications is a cross-linked polystyrene. To make this material, styrene containing a small amount of divinylbenzene is polymerized. Each vinyl group of divinylbenzene can become part of a separate polymer chain, so these groups act as cross-links between chains. The resulting polymer is used in the form of small beads that are completely insoluble in typical solvents.

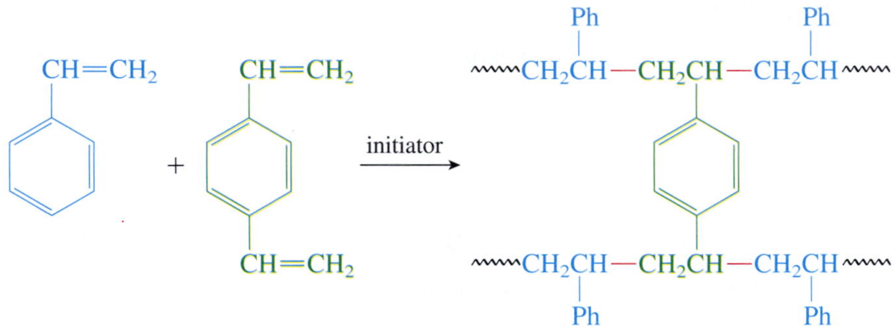

Styrene   Divinylbenzene   Cross-linked polystyrene
98%   2%

The phenyl groups that are on the surface of these beads are accessible to added reagents and react like other aromatic compounds. (Of course, the phenyl groups in the interior of the bead do not usually react because the reagents are seldom able to penetrate the bead and reach them.) For example, the phenyl groups on the surface can be sulfonated by reaction with concentrated sulfuric acid (see Section 17.6).

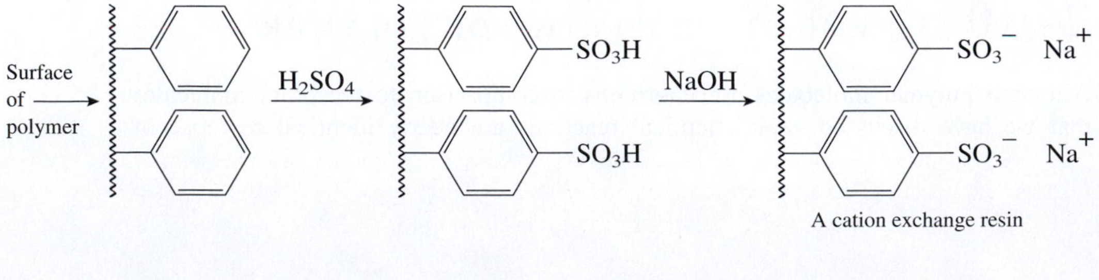

A cation exchange resin

The sulfonated polystyrene beads can be used as a cation exchange resin. For example, if the sulfonated resin is treated with sodium hydroxide, the sodium salt of each of the sulfonic acid groups is formed. This resin can then be employed to exchange sodium cations for other cations in an aqueous solution.

One application of such a cation exchange resin is to prepare soft water for cleaning purposes. Recall from the Focus On box "Synthetic Detergents BHT and BHA" on page 694 that hard water contains ions such as Ca^{2+} and Mg^{2+} that combine with soap molecules to produce a scummy precipitate. If hard water is passed through a cylinder containing the sodium form of an ion exchange resin, Na^+ ions are exchanged for the Ca^{2+} and Mg^{2+} ions. Although the total ionic charge in the water is unchanged, the soft water contains only sodium cations and no longer causes soap molecules to precipitate. After most of the sodium ions in the resin have been used to exchange for the cations in the water, the resin is regenerated by backflushing with a sodium chloride solution to replace the cations with Na^+.

Note that a cation exchange resin has anionic groups attached to the phenyl groups of the polymer. The cations associated with these immobile anions can then be exchanged for other cations. To prepare an anion exchange resin, cationic groups must be attached to the phenyl groups of the resin. This is accomplished by reacting the cross-linked polystyrene with formaldehyde and hydrogen chloride. This first adds hydroxymethyl groups to the phenyl groups by the same mechanism shown in Figure 24.4. Then the hydroxy groups are replaced by chloride in a nucleophilic substitution reaction (see Section 10.5). Finally, the chloromethylated polystyrene is reacted with trimethylamine. This nucleophilic substitution reaction (see Section 10.6) produces a quaternary ammonium salt with an immobile cation and an exchangeable anion, an anion exchange resin.

An anion exchange resin

A combination of a cation exchange resin and an anion exchange resin is used to deionize water. First the water is passed through a cation exchange resin in its acidic form. This exchanges all of the cations in the water for H_3O^+. Then the water is passed through an anion exchange resin in its basic form. This exchanges all of the anions for OH^-, which reacts with the H_3O^+ to form water. On passing through both resins, all of the ions in the water have been removed! This process is represented schematically in the following diagram:

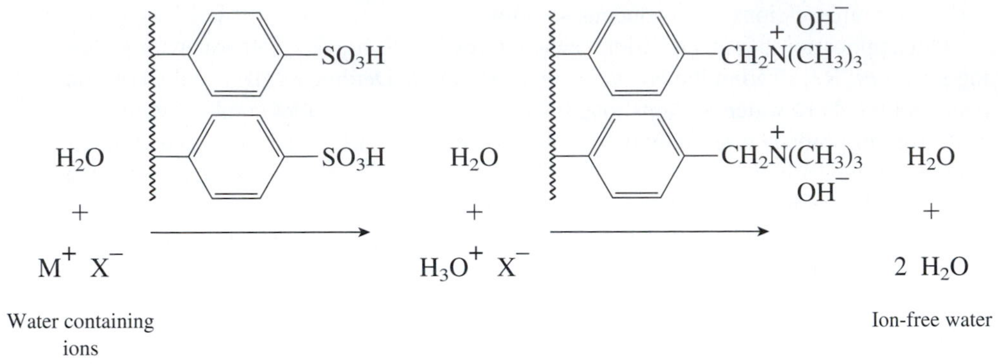

$$\text{H}_2\text{O} + \text{M}^+\ \text{X}^- \longrightarrow \text{H}_2\text{O} + \text{H}_3\text{O}^+\ \text{X}^- \longrightarrow \text{H}_2\text{O} + 2\ \text{H}_2\text{O}$$

Water containing ions

Ion-free water

Focus On

Recycling Plastics

A bench made from recycled plastic.

As people have become more environmentally conscious, the need to recycle many of the materials, such as plastics, that are used in everyday life has been recognized. Although several billion pounds of plastics are currently recycled, this is only a small fraction of the total amount of plastics produced each year. One of the limits encountered in the recycling effort is the expense in collecting the plastic waste. In addition, the waste must be sorted for most recycling processes. For example, a small amount of PVC contaminant in a batch of PET can render the recycled product useless. Currently, such sorting must be done by hand, although there is hope that an automated process based on some type of sensor might be developed.

Once the plastic has been separated into its various types, it can be recycled by reprocessing. More commonly, some recycled polymer is added to virgin polymer, although this may limit the application of the resulting material. For example, poly(vinyl chloride) containing 25% recycled resin is available, but only for applications in which it does not come into contact with food.

In the case of condensation polymers such as PET, it is often possible to reverse the polymerization process and convert the polymer back to its constituent monomers. The monomers can then be purified and used to prepare new polymer. PET is converted back to ethylene glycol and dimethyl terephthalate by heating the polymer with methanol. This is exactly the reverse of the reaction that is used to prepare the polyester:

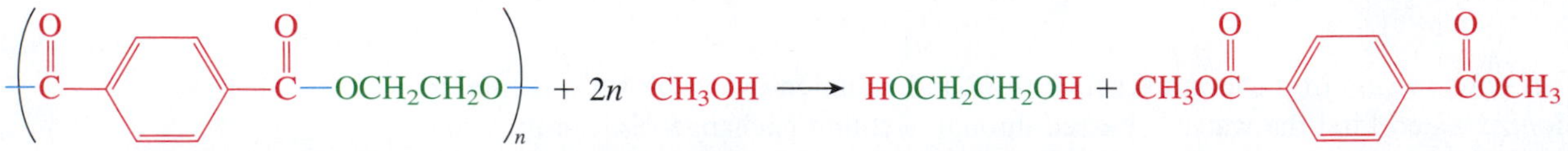

$$\left(\text{—C(=O)—C}_6\text{H}_4\text{—C(=O)—OCH}_2\text{CH}_2\text{O—}\right)_n + 2n\ \text{CH}_3\text{OH} \longrightarrow \text{HOCH}_2\text{CH}_2\text{OH} + \text{CH}_3\text{OC(=O)—C}_6\text{H}_4\text{—C(=O)OCH}_3$$

Unfortunately, there is no apparent way to accomplish the same thing with addition polymers. No one has yet figured out a way to convert polyethylene back to ethylene or polystyrene back to styrene. However, a method has been developed to pyrolyze plastic waste to a chemical feedstock in a process that can be likened to the cracking process that is used in crude oil refining. Although much of the value that was added to the polymer in the manufacturing process is lost, this method has the distinct advantage that a mixture of various plastics can be used.

Finally, there is the option of burning the plastic for fuel. Although much of the value of the plastic is wasted, at least the energy content is recovered. Again, a mixture of various polymers can be used, although there is some concern about burning poly(vinyl chloride) because of its chlorine content.

Review of Mastery Goals

After completing this chapter, you should be able to:

■ Show the repeat unit for any addition or condensation polymer. (Problems 24.18 and 24.19)

■ Write the mechanisms for the formation of addition polymers that are prepared by radical, anionic, or cationic initiation. (Problems 24.20, 24.21, 24.22, 24.26, 24.28, 24.29, 24.30, and 24.32)

■ Discuss the structure and stereochemistry of polymers in terms of both regular and irregular features. (Problems 24.24 and 24.35)

■ Discuss how the physical properties of a polymer are related to its structure. (Problem 24.27)

■ Discuss the chemical properties of a polymer. (Problem 24.31)

Additional Problems

24.18 Show structures for the polymers that could be formed from these monomers and suggest the best method to accomplish each polymerization:

a) $NH_2(CH_2)_6NH_2$ + $HOC(CH_2)_8COH$ (each C=O)

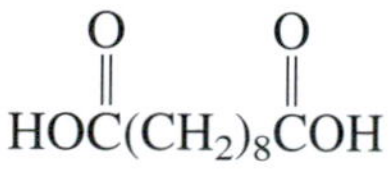

b) 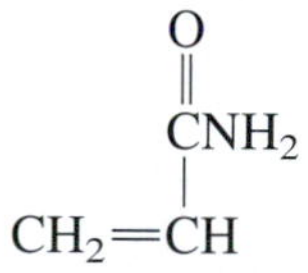 $CH_2{=}CH$ with CNH_2 (C=O)

c) $HO(CH_2)_4OH$ + benzene ring with CO_2CH_3 (top and bottom)

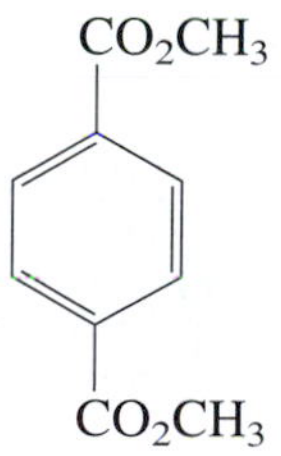

d) 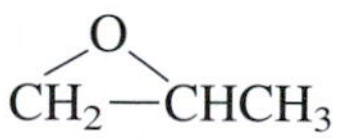 $CH_2{-}CHCH_3$ (epoxide, O)

e) $HO(CH_2)_4OH$ + benzene ring with $N{=}C{=}O$ groups

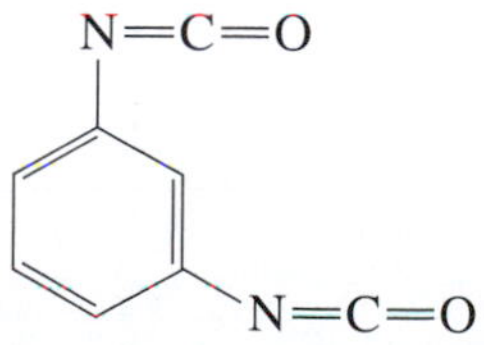

f) 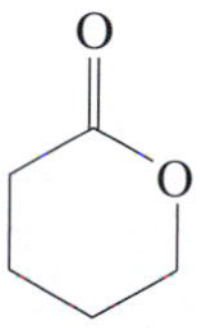 lactone ring (O, C=O)

24.19 Show the monomers that could be used to prepare each of these polymers:

a)

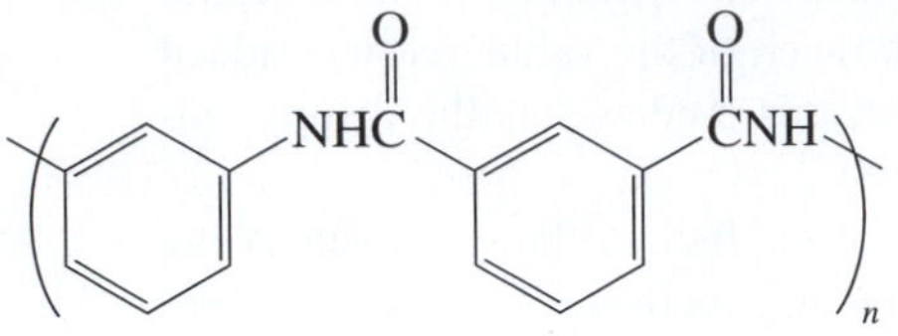

Nomex

b)

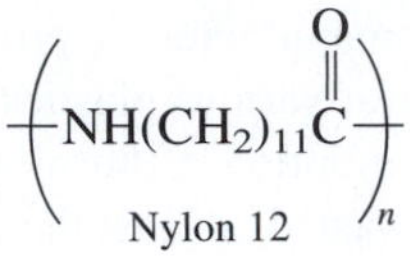

$$\left[NH(CH_2)_{11}\overset{\overset{\displaystyle O}{\|}}{C} \right]_n$$

Nylon 12

c)

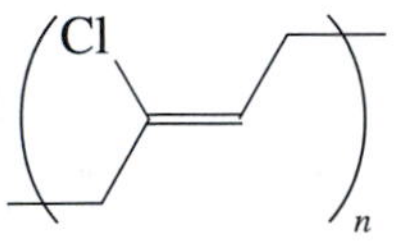

Neoprene

d)

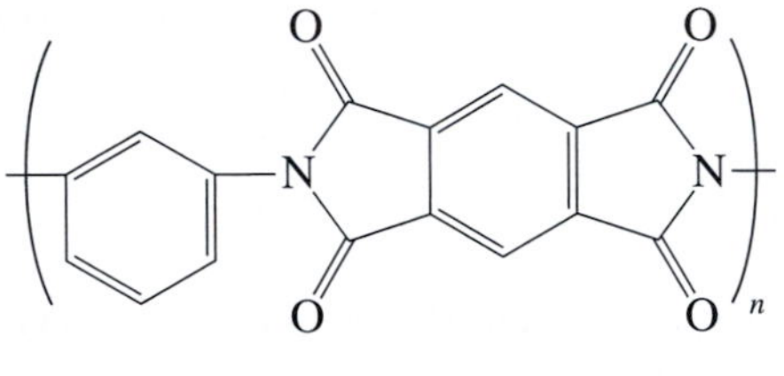

e)

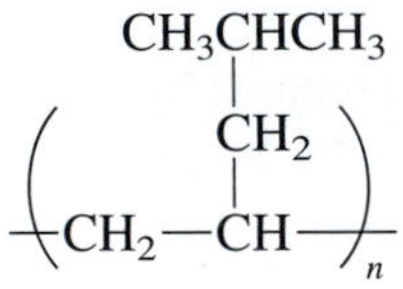

$$\left(CH_2-\underset{\underset{\underset{\underset{CH_3CHCH_3}{|}}{CH_2}}{|}}{CH} \right)_n$$

f)

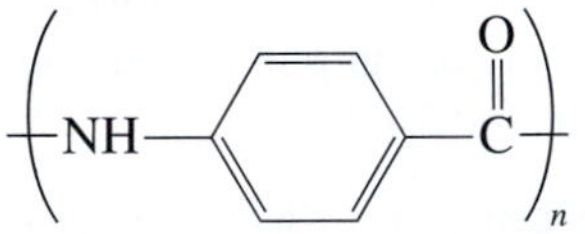

$$\left[NH-\underset{}{\bigcirc}-\overset{\overset{\displaystyle O}{\|}}{C} \right]_n$$

g)

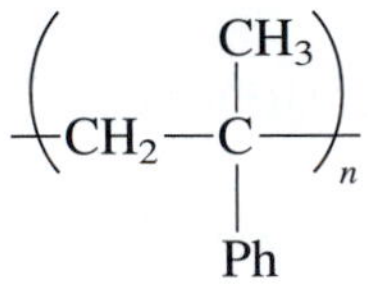

$$\left(CH_2-\underset{\underset{Ph}{|}}{\overset{\overset{CH_3}{|}}{C}} \right)_n$$

h)

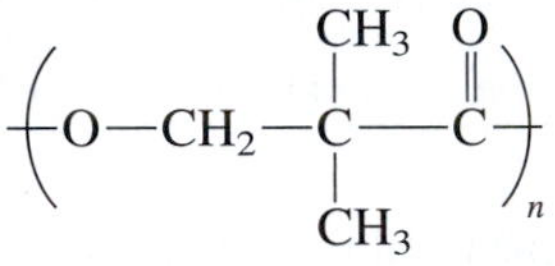

$$\left[O-CH_2-\underset{\underset{CH_3}{|}}{\overset{\overset{CH_3}{|}}{C}}-\overset{\overset{\displaystyle O}{\|}}{C} \right]_n$$

i)

$$\left(CH_2-\underset{\underset{\underset{\underset{O=C-CH_3}{}}{|}}{OCCH_3}}{CH} \right)_n$$

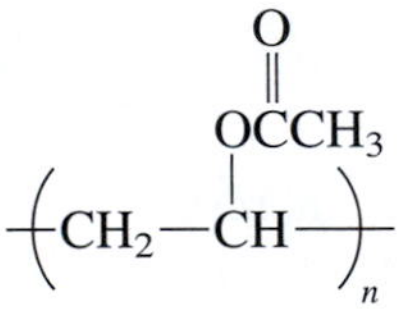

Poly(vinyl acetate)

j)

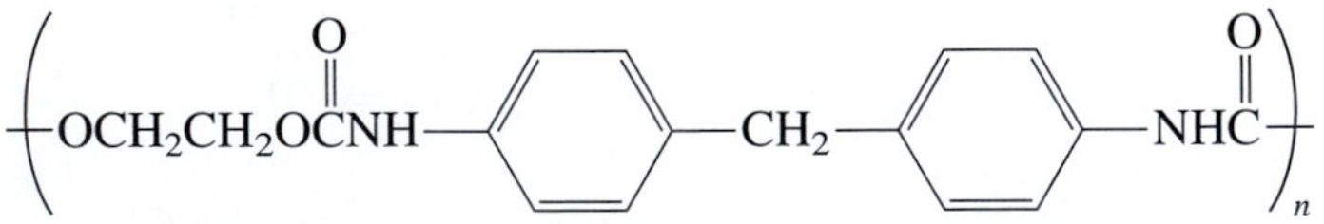

24.20 Show all of the steps in the mechanism for this reaction:

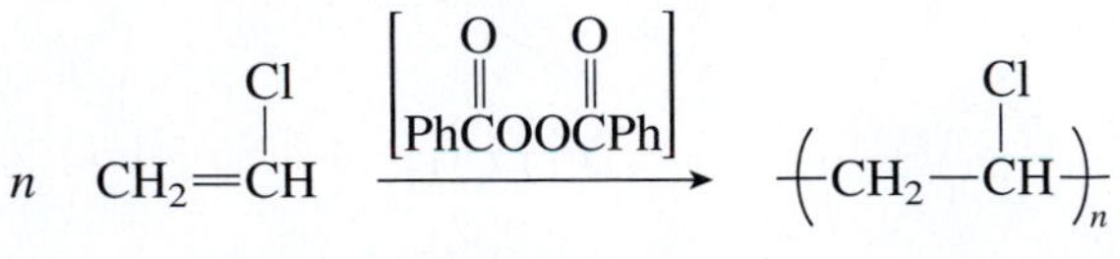

24.21 Show all of the steps in the mechanism for this reaction:

$$n \quad CH_2=\overset{\overset{\displaystyle Ph}{|}}{CH} \xrightarrow[\text{[H}_2\text{O]}]{\text{[BF}_3\text{]}} \left(CH_2-\overset{\overset{\displaystyle Ph}{|}}{CH} \right)_n$$

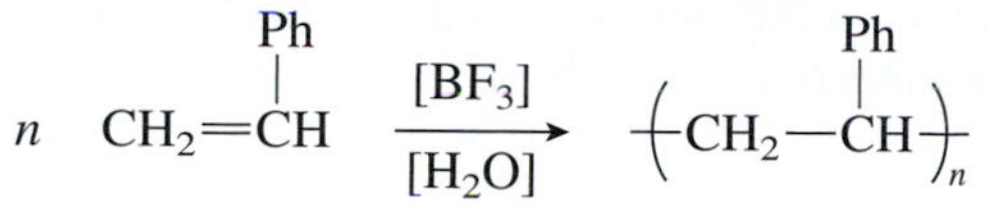

24.22 Show all of the steps in the mechanism for this reaction:

$$n \quad CH_2=\overset{\overset{\displaystyle \overset{\displaystyle O}{\|}}{\underset{}{C}OCH_3}}{CH} \xrightarrow{\text{[NaNH}_2\text{]}} \left(CH_2-\overset{\overset{\displaystyle \overset{\displaystyle O}{\|}}{\underset{}{C}OCH_3}}{CH} \right)_n$$

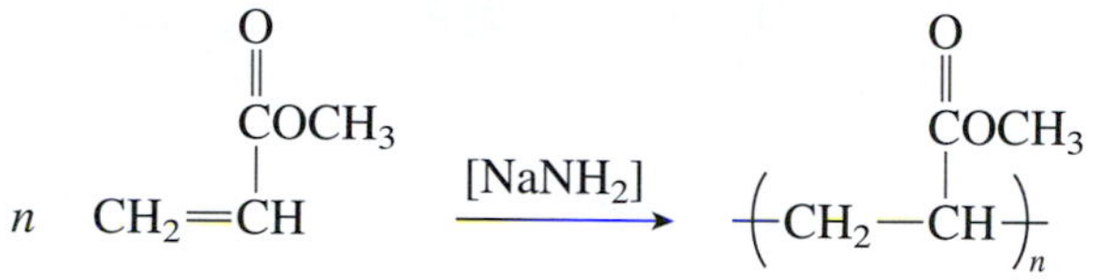

24.23 Show how these polymers could be prepared from the indicated starting materials. More than one step is needed.

a) 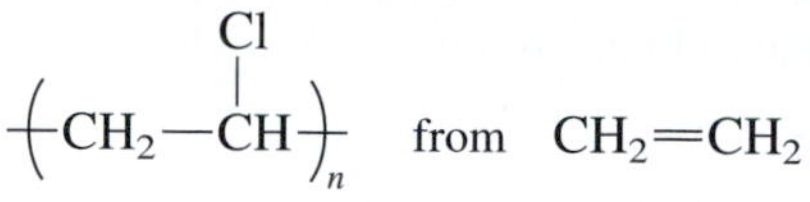 $\left(CH_2-\overset{\overset{\displaystyle Cl}{|}}{CH} \right)_n$ from $CH_2=CH_2$

b) 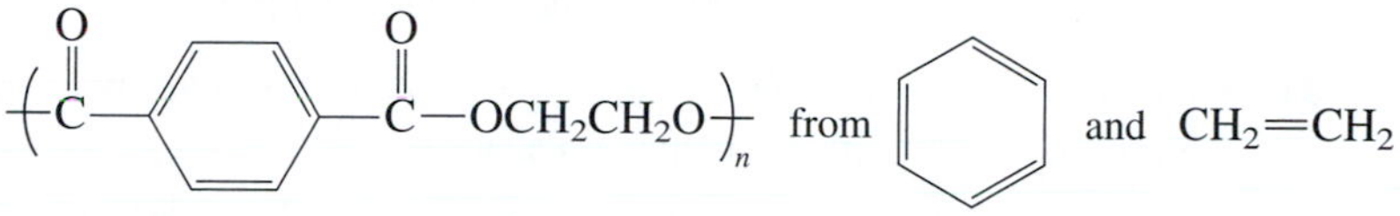 $\left(\overset{\overset{\displaystyle O}{\|}}{C}-\!\!\!\left\langle\right\rangle\!\!\!-\overset{\overset{\displaystyle O}{\|}}{C}-OCH_2CH_2O \right)_n$ from ⬡ and $CH_2=CH_2$

c) $\left(CH_2-\overset{\overset{\displaystyle CN}{|}}{CH} \right)_n$ from $\overset{\displaystyle \;\;\;O}{CH_2-CH_2}$

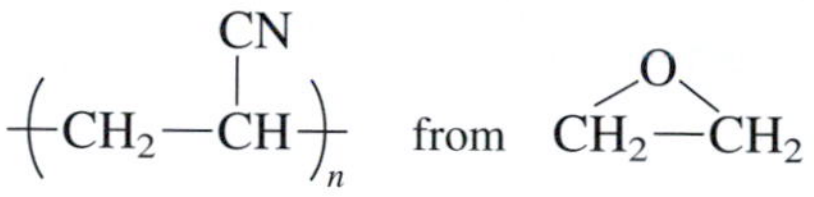

24.24 Show the steps in the mechanism for the formation of a long-chain branch in the polymerization of styrene.

24.25 Show a synthesis of nylon 10 starting from cyclodecanone:

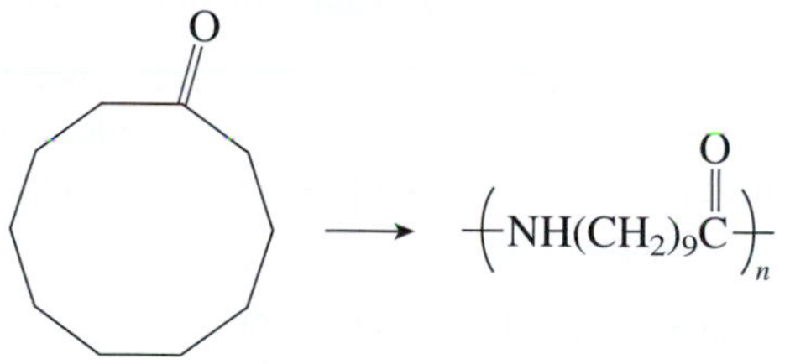

$$\text{(cyclodecanone with =O)} \longrightarrow \left(NH(CH_2)_9\overset{\overset{\displaystyle O}{\|}}{C} \right)_n$$

24.26 Although poly(butylene oxide) cannot be made by anionic polymerization of THF, it can be made by treating THF with acid. Suggest a mechanism for this process.

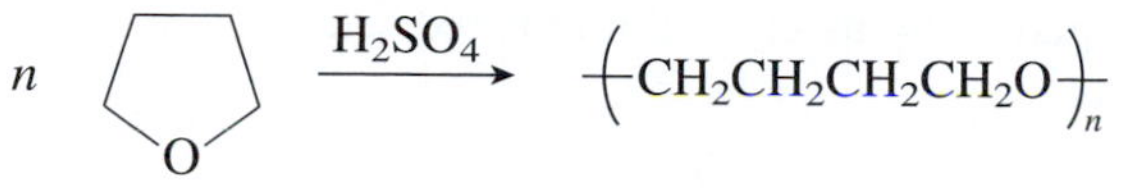

$$n \quad \text{(THF ring with O)} \xrightarrow{H_2SO_4} \left(CH_2CH_2CH_2CH_2O \right)_n$$

THF   Poly(butylene oxide)

24.27 Show a partial structure for the polymer formed from glycerol and phthalic anhydride when they are reacted in a 2:3 ratio. What physical properties do you expect this polymer to have?

$$
\begin{array}{ccc}
\text{OH} & \text{OH} & \text{OH} \\
| & | & | \\
\text{CH}_2\text{—CH—CH}_2
\end{array}
$$

Glycerol   Phthalic anhydride

24.28 Explain why a small amount of BHT (butylated hydroxytoluene) is often added to styrene before it is sold for use in a research laboratory.

24.29 Show the ends of a molecule of polyethylene whose polymerization is initiated by $(PhCO_2)_2$ and terminated by abstraction of a hydrogen atom from another chain.

24.30 Show the steps in the mechanism for the formation of chloromethylated polystyrene by the reaction of polystyrene with formaldehyde and HCl.

24.31 Show a synthesis of an anion exchange resin starting with benzene and any other compounds with six or fewer carbons.

24.32 Show the steps in the mechanism for the conversion of PET back to its monomers using methoxide ion as the catalyst:

$$
\left(\!\!\begin{array}{c}
O \\
\| \\
C
\end{array}\!\!\!\!\raisebox{0pt}{\hspace{1em}}\!\!\!\!\begin{array}{c}
O \\
\| \\
C\text{—OCH}_2\text{CH}_2\text{O}
\end{array}\!\!\right)_{\!\!n} + 2n\ \text{CH}_3\text{OH} \xrightarrow{[\text{CH}_3\text{O}^-]} \text{HOCH}_2\text{CH}_2\text{OH} + \text{CH}_3\overset{O}{\overset{\|}{\text{OC}}}\!\!\!\!\raisebox{0pt}{\hspace{1em}}\!\!\!\!\overset{O}{\overset{\|}{\text{COCH}_3}}
$$

24.33 Contrary to intuition, a rubber band actually shortens when heated. Use the equation $\Delta G = \Delta H - T\Delta S$ to explain this observation.

24.34 Saran, a plastic wrap used for food storage, is a polymer prepared from a mixture of 85% $CH_2{=}CCl_2$ (vinylidene chloride) and 15% $CH_2{=}CHCl$ (vinyl chloride). Why do you suppose that pure vinylidene chloride is not used to make this polymer?

24.35 Show a partial structure for the polymer formed from 1,3-butadiene by radical polymerization. Discuss the possible structural variations in this polymer.

24.36 The crystallinity of polyethylene prepared by coordination polymerization can be decreased in a controlled fashion by the addition of a small amount of another alkene, such as 1-pentene. Show the structure of the resulting polymer and explain why its crystallinity is reduced.

Do you need a live tutor for homework problems? Access vMentor at Organic ChemistryNow at **http://now.brookscole.com/hornback2** for one-on-one tutoring from a chemistry expert.

Carbohydrates

CARBOHYDRATES ARE IMPORTANT, naturally occurring organic compounds. They include simple sugars, or monosaccharides, such as glucose and fructose, and polysaccharides, such as starch and cellulose, which are more complex compounds composed of a number of sugar units. Carbohydrates are one of the initial products of photosynthesis. As such, they serve as the molecules that store the sun's energy for later use in metabolism. In addition, carbohydrate polymers are structural materials used by plants and animals. Even our genetic material, DNA, contains carbohydrate units as part of its polymeric backbone.

This chapter begins with a discussion of the structure and stereochemistry of monosaccharides. Then the formation of cyclic structures from monosaccharides is discussed. This is followed by the presentation of a small number of reactions of these compounds. The classic series of experiments that was used to establish the structure of glucose is presented next. Finally, the structures of disaccharides, polysaccharides, and a few other types of carbohydrate-containing compounds are introduced.

MASTERING ORGANIC CHEMISTRY

▶ Understanding the General Structures of Carbohydrates

▶ Understanding the Stereochemistry of Carbohydrates

▶ Understanding the Cyclization of Monosaccharides to Form Hemiacetal Rings

▶ Predicting the Products of the Common Reactions of Monosaccharides

▶ Understanding Fischer's Structure Proof for Glucose

▶ Understanding the General Structural Features of Disaccharides and Polysaccharides

25.1 STRUCTURES OF CARBOHYDRATES

Many carbohydrates fit the general formula $C_x(H_2O)_x$, so it is apparent how the name originated. Actually, they are polyhydroxy aldehydes or ketones. Glucose, $C_6H_{12}O_6$, is a typical monosaccharide. It is a six-carbon aldehyde with hydroxy groups on all of the other carbons.

ORGANIC
Chemistry Now™
Look for this logo in the chapter and go to OrganicChemistryNow at
http://now.brookscole.com/hornback2 for tutorials, simulations, problems, and molecular models.

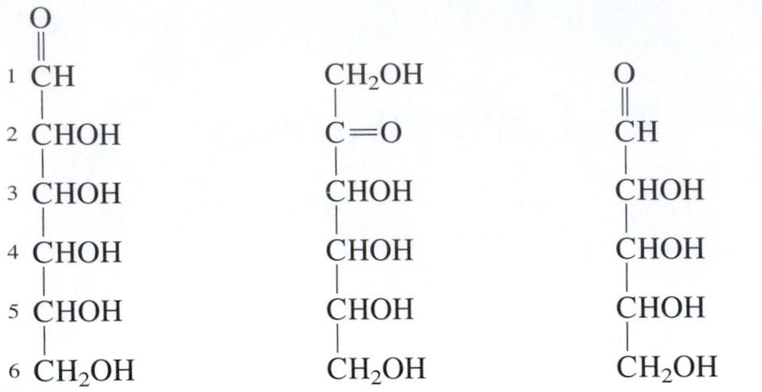

O
‖
1 CH   CH₂OH   O
| | ‖
2 CHOH   C=O   CH
| | |
3 CHOH   CHOH   CHOH
| | |
4 CHOH   CHOH   CHOH
| | |
5 CHOH   CHOH   CHOH
| | |
6 CH₂OH   CH₂OH   CH₂OH

Glucose   Fructose   Ribose
(an aldohexose)   (a 2-ketohexose)   (an aldopentose)

Glucose is an aldohexose, where aldo- indicates that it is an aldehyde, -hex- designates the number of carbons, and -ose is the suffix used for carbohydrates. Some other common monosaccharides are fructose, a 2-ketohexose that is isomeric with glucose, and ribose, an aldopentose that contains one fewer carbons than glucose.

PROBLEM 25.1

Show a structure for ribulose, a 2-ketopentose.

25.2   STEREOCHEMISTRY OF CARBOHYDRATES

As can be seen by examination of its structure, glucose has four stereocenters, C-2, C-3, C-4, and C-5. Therefore, this structure has $2^4 = 16$ stereoisomers. One of these stereoisomers is glucose, and one is the enantiomer of glucose. The other 14 compounds (seven pairs of enantiomers) are diastereomers of glucose.

PROBLEM 25.2

How many stereoisomeric aldopentoses are possible? How many stereoisomeric 2-ketopentoses are possible?

Much of the early research on carbohydrates was done by Emil Fischer, who invented Fischer projections (see Section 7.8) to show the complex stereochemistry of these compounds. Fischer projections have proved quite useful and are still used today. By convention an aldose is drawn with the carbon chain oriented in the vertical direction with the aldehyde carbon on the top. Naturally occurring glucose, designated D-glucose, has the stereochemistry shown in the following Fischer projection (the leftmost structure in the following diagram). Recall that the horizontal bonds in a Fischer projection extend above the plane of the page and the vertical bonds extend below the plane of the page. The middle structure in the diagram shows the horizontal bonds correctly but cannot show all of the vertical bonds projecting below the plane of the page. Although the structure on the right, with all of the carbon–carbon bonds in the plane of the page, perhaps better represents the shape of glucose, Fischer projections are much easier to use.

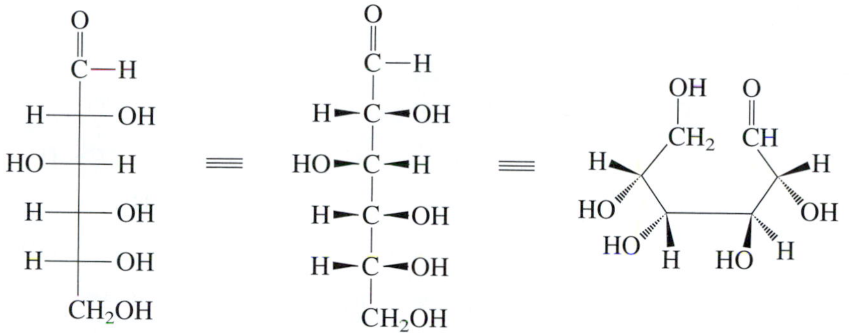

D-Glucose
Fischer projection

L-Glucose is the enantiomer of D-glucose and has inverted configuration at all stereo-centers. D-Galactose is a diastereomer of D-glucose and differs only in the configuration at C-4.

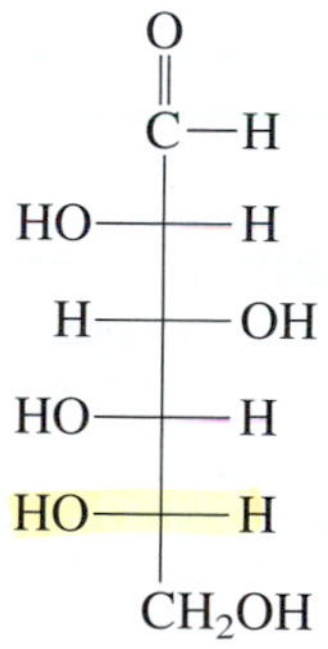

L-Glucose 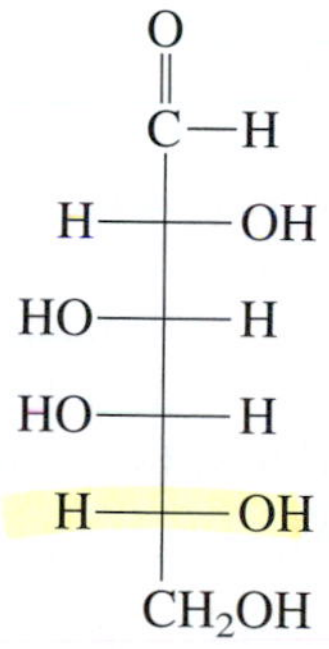   D-Galactose

The configuration of many biological compounds, including sugars and amino acids, is usually designated by employing the letter D or L, rather than R or S. These letters relate the configuration to that of the naturally occurring enantiomer of glyceraldehyde, the simplest sugar, which has the D-configuration. By convention the structure is shown with the carbon chain in the vertical direction and with the most oxidized carbon (the one with the most bonds to oxygen) at the top. When written like this, D-glyceraldehyde has the hydroxy group on the right side of the Fischer projection. Any sugar that has this same absolute stereochemistry at its highest-numbered carbon chirality center (at the carbon chirality center farthest from the top) is also designated as having the D-configuration. Thus, the naturally occurring enantiomer of glucose is D because the configuration at C-5 is the same as the configuration of D-glyceraldehyde at C-2; that is, the hydroxy group is on the right side of the Fischer projection. Likewise, D-ribose has the hydroxy group on C-4 on the right side of the Fischer projection. Most sugars that occur naturally have the D-configuration. Note that D has no relationship to *d*, which designates a compound that is dextrorotatory—that is, one that rotates plane-polarized light in the clockwise or positive direction. Thus, although D-glyceraldehyde and D-glucose are also *d* (or +), D-ribose is levorotatory (*l* or −).

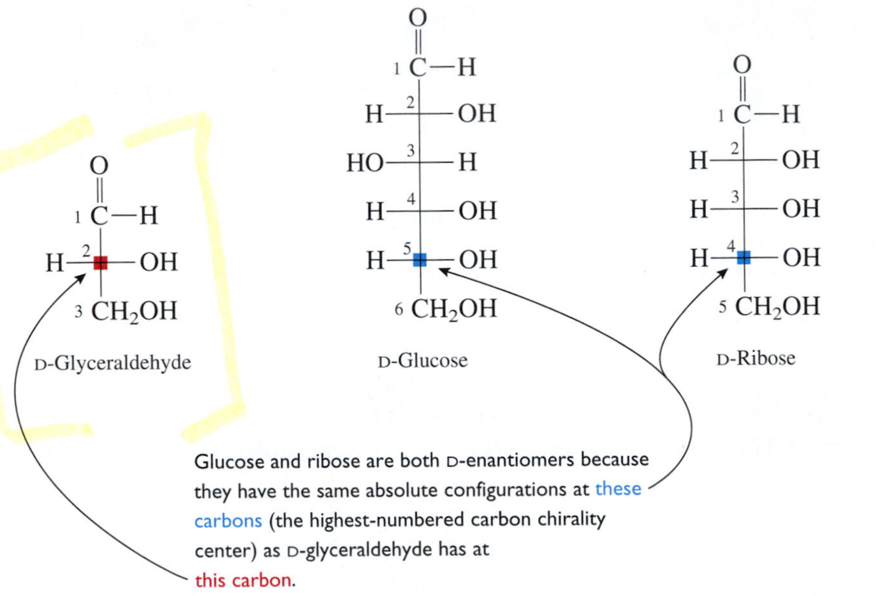

D-Glyceraldehyde

D-Glucose

D-Ribose

Glucose and ribose are both D-enantiomers because they have the same absolute configurations at these carbons (the highest-numbered carbon chirality center) as D-glyceraldehyde has at this carbon.

Figure 25.1 shows all of the possible D-aldoses with three to six carbons. Of these, only D-glucose is commonly found as a monosaccharide in nature. Several others, including D-glyceraldehyde, D-mannose, D-galactose, and D-ribose, are found as part of polysaccharides or in other biological molecules. In addition, the ketoses D-fructose, D-ribulose and D-xylulose (2-ketopentoses with the same configuration at the other carbons as ribose and xylose), and dihydroxyacetone (1,3-dihydroxy-2-propanone) are also common.

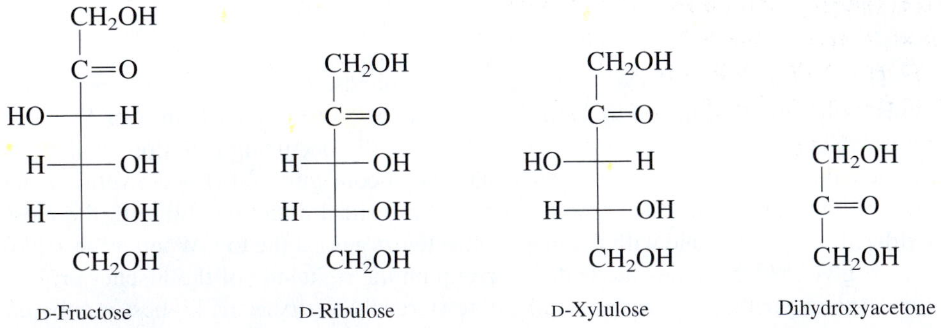

D-Fructose

D-Ribulose

D-Xylulose

Dihydroxyacetone

Ketoses

PROBLEM 25.3

Draw Fischer projections for these monosaccharides:
a) L-Glyceraldehyde   **b)** L-Mannose

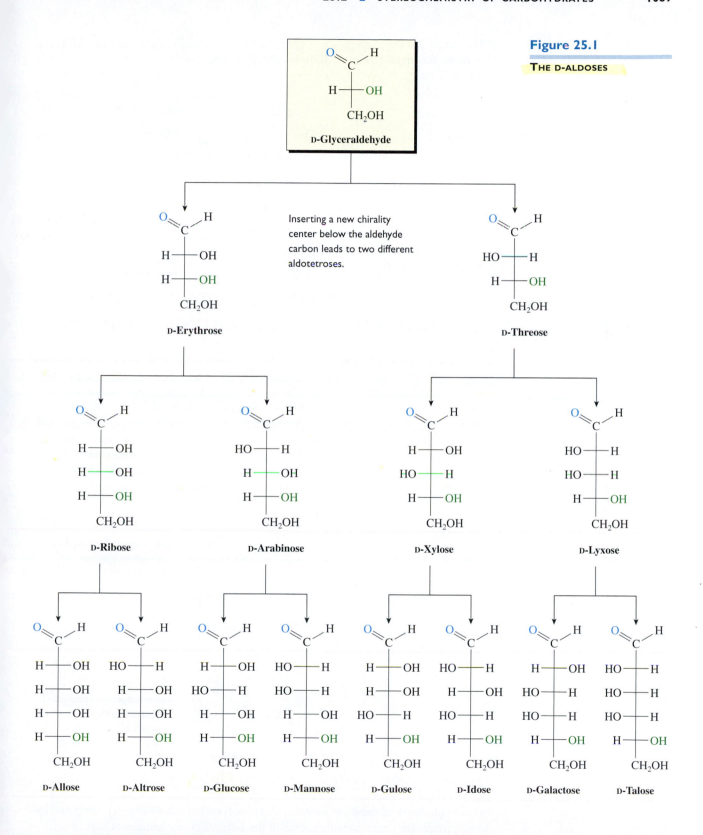

Figure 25.1

THE D-ALDOSES

D-Glyceraldehyde

Inserting a new chirality center below the aldehyde carbon leads to two different aldotetroses.

D-Erythrose

D-Threose

D-Ribose

D-Arabinose

D-Xylose

D-Lyxose

D-Allose

D-Altrose

D-Glucose

D-Mannose

D-Gulose

D-Idose

D-Galactose

D-Talose

PROBLEM 25.4

Determine the identity of each of these carbohydrates:

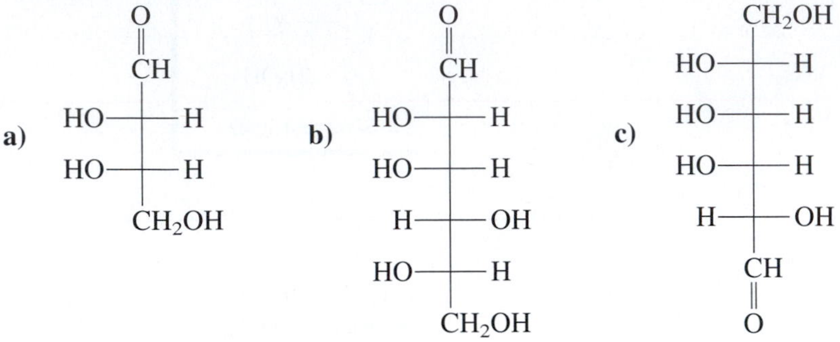

a), b), c) [Fischer projection structures]

(Hint: This must be rotated first.)

25.3   CYCLIZATION OF MONOSACCHARIDES

Up to this point, the structure of glucose has been shown as an aldehyde with hydroxy groups on the other carbons. However, as described in Section 18.9, aldehydes and ketones react with alcohols to form hemiacetals. When this reaction is intermolecular—that is, when the aldehyde group and the alcohol group are in different molecules—the equilibrium is unfavorable and the amount of hemiacetal that is present is very small. However, when the aldehyde group and the alcohol group are contained in the same molecule, as is the case in the second equation that follows, the intramolecular reaction is much more favorable (because of entropy effects; see Sections 8.13 and 18.9) and the hemiacetal is the predominant species present at equilibrium.

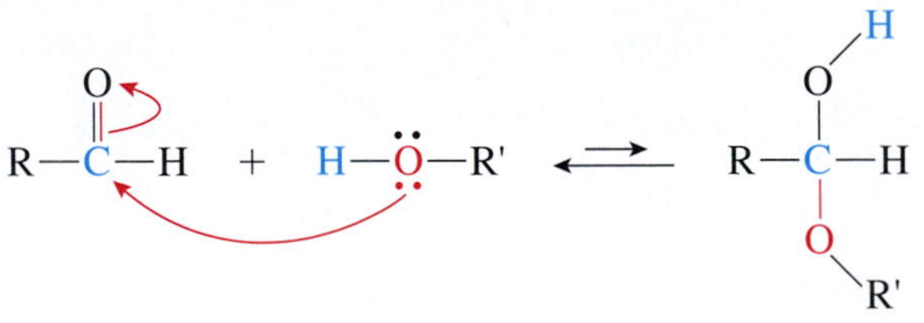

Intermolecular reaction
Equilibrium favors reactants

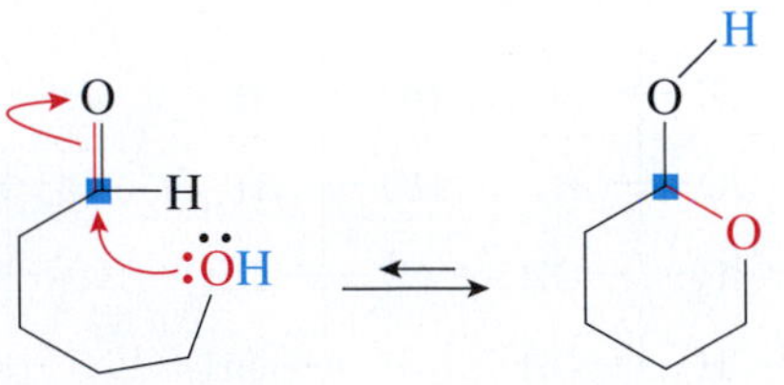

A hemiacetal

Intramolecular reaction
Equilibrium favors products

(6.7%)   (93.3%)
                  A cyclic hemiacetal

Because glucose and the other monosaccharides contain both a carbonyl group and hydroxy groups, they exist predominantly in the form of cyclic hemiacetals.

Let's consider the cyclization of glucose in more detail. There are five different hydroxy groups that might react with the aldehyde group. However, because five- and six-membered rings are much more stable than others, these are the only ring sizes that need be considered. These two possibilities are illustrated for glucose in the following equation:

<0.2%	0.02%	>99.8%
A furanose	D-Glucose	A pyranose

The cyclic hemiacetal that has a five-membered ring is called a **furanose**. This name is derived from that of the five-membered, oxygen-containing heterocyclic compound **furan.** The hemiacetal with a six-membered ring is known as a **pyranose** after the heterocycle **pyran.**

Furan   Pyran

As we saw in Chapter 6, six-membered rings are generally more stable than five-membered rings, primarily because of increased torsional strain in the latter. Therefore, it is not surprising that the pyranose form of a monosaccharide is usually more stable than the furanose form. At equilibrium, glucose exists primarily as the pyranose (>99.8%), with little, if any, furanose (<0.2%) present. There is also a trace amount (0.02%) of the uncyclized aldehyde present. Of course, this equilibrium depends on the structure of the monosaccharide, and some other sugars have larger amounts of the furanose form.

PROBLEM 25.5

Show the steps in the mechanism for the cyclization of the open form of D-glucose to the pyranose form. (Use the acid-catalyzed mechanism of Chapter 18.)

PROBLEM 25.6

At equilibrium, D-fructose exists 67.5% in the pyranose form, 31.5% in the furanose form, and 1% in the open, uncyclized form. Draw the pyranose and furanose forms. Explain why D-fructose has more of the uncyclized form present at equilibrium than does D-glucose.

PROBLEM 25.7

How much pyranose form, furanose form, and uncyclized aldehyde would be present at equilibrium for L-glucose?

PROBLEM 25.8

Show the structure of the hemiacetal formed from D-erythrose.

Let's now address the stereochemistry of the cyclization of D-glucose to a pyranose. Note that carbon 1, the hemiacetal carbon, becomes a new stereocenter when the cyclization occurs. Therefore, two diastereomers of the pyranose, with different configurations at the new stereocenter, are formed when D-glucose cyclizes. Such diastereomers are called **anomers.** The two anomers for the pyranose form of D-glucose are shown in the following equation:

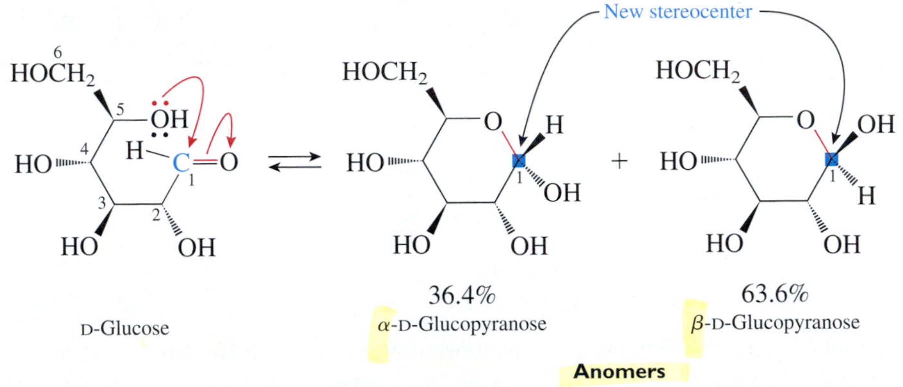

	36.4%	63.6%
D-Glucose	α-D-Glucopyranose	β-D-Glucopyranose

Anomers

By convention, when the sugar is drawn as shown in the preceding equation and the hydroxy group at the new stereocenter projects down, the compound is designated as the α-stereoisomer. When the hydroxy group at the hemiacetal carbon projects up, it is the β-stereoisomer. The full names for these two anomers of glucose are α-D-glucopyranose and β-D-glucopyranose.

PROBLEM 25.9

D-Mannose differs from D-glucose only in its configuration at C-2. Show the formation of α-D-mannopyranose and β-D-mannopyranose from the uncyclized form of D-mannose in the same manner as was done for D-glucose.

Carbohydrate chemists often represent these compounds using **Haworth projections.** In this method the ring is drawn flat and viewed partly from the edge. The bonds to the substituents are shown as coming straight up or straight down from the carbons. Although Haworth projections distort the geometry at the carbons, they are easy to draw and make the stereochemical relationships among the substituents readily apparent.

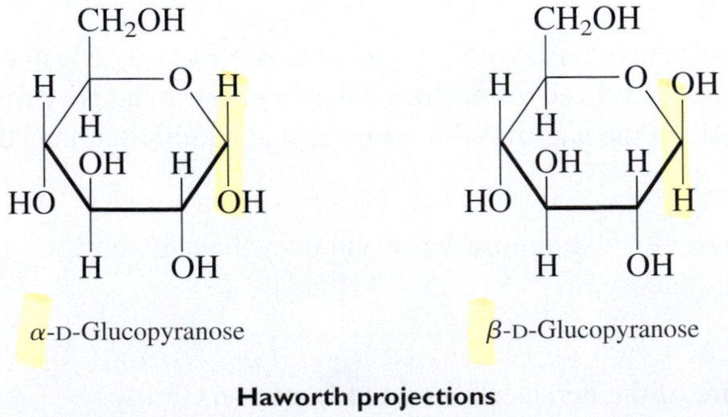

α-D-Glucopyranose   β-D-Glucopyranose

Haworth projections

PROBLEM 25.10

Show Haworth projections for α-D-mannopyranose and its β-anomer. (Remember that D-mannose differs from D-glucose only in its configuration at C-2.)

Haworth projections show the absolute stereochemistries at the various stereocenters of glucose quite well, but the six-membered rings of pyranoses are, of course, not planar. The presence of an oxygen atom in these rings causes only a slight perturbation, so most of the discussion about cyclohexane rings presented in Chapter 6 also applies to pyranose rings. The cyclization of D-glucose to form chair conformations of α- and β-D-glucopyranose is shown in Figure 25.2.

PROBLEM 25.11

Show the chair conformations for α-D-mannopyranose and its β-anomer.

As was the case for cyclohexane derivatives, the chair conformer that has the larger groups equatorial is usually more stable. Therefore, the chair conformers shown in Figure 25.2, which have most or all of the larger substituents equatorial, are more stable than the conformers obtained by ring-flips. The α- and β-anomers differ only in the stereochemistry of the groups at the hemiacetal carbon. In the α-anomer the hydroxy group on this carbon is axial, and in the β-anomer it is equatorial.

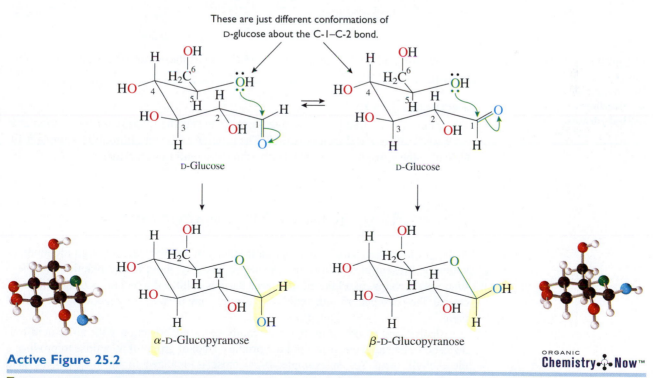

Active Figure 25.2

THE CYCLIZATION OF D-GLUCOSE TO FORM α- AND β-D-GLUCOPYRANOSE. Test yourself on the concepts in this figure at **OrganicChemistryNow.**

Both α-D-glucopyranose and β-D-glucopyranose can be isolated in pure form. Because they are diastereomers, they have different physical properties. For example, the α-stereoisomer has a specific rotation of +112.2, whereas that of the β-isomer is +18.7. However, if either of these pure stereoisomers is dissolved in water, the specific rotation slowly changes, over several hours, to a value of +52.7. This process, termed **mutarotation,** results from the formation of an equilibrium mixture that consists of 36.4% of the α-isomer and 63.6% of the β-isomer. (Of course, the same equilibrium mixture results starting from either of the anomers.) In fact, it is the specific rotation at equilibrium that is used to calculate the equilibrium concentration of the two stereoisomers.

PRACTICE PROBLEM 25.1

Given that the rotation of α-D-glucopyranose is +112.2, the rotation of β-D-glucopyranose is +18.7, and the rotation of an equilibrium mixture of the two anomers is +52.7, calculate the percentages of each anomer present at equilibrium.

Solution

Let x equal the decimal fraction of the α-isomer that is present at equilibrium. Then $1 - x$ equals the decimal fraction of the β-isomer present. The rotations of each must total to +52.7:

$$x(+112.2) + (1 - x)(+18.7) = +52.7$$
$$112.2x + 18.7 - 18.7x = 52.7$$
$$93.5x = 34.0$$
$$x = 0.364$$

Therefore, 36.4% of the α-isomer and 63.6% of the β-isomer are present at equilibrium.

PROBLEM 25.12

D-Mannose exists entirely in pyranose forms. The specific rotation for the α-anomer is +29, and that for the β-isomer is −17. The rotation of the equilibrium mixture is +14. Calculate the percentages of each anomer in the equilibrium mixture.

ORGANIC
Chemistry⚛Now™
Click *Coached Tutorial Problems*
for more practice with
**Cyclizations of
Carbohydrates.**

25.4   REACTIONS OF MONOSACCHARIDES

Although monosaccharides exist predominantly as hemiacetals, enough aldehyde or ketone is present at equilibrium that the sugars give most of the reactions of these functional groups. In addition, monosaccharides exhibit the reactions of alcohols. Of course, the presence of both functional groups may perturb the reactions of either of them.

Alcohol and aldehyde groups are readily oxidized. Reaction with nitric acid oxidizes both the aldehyde group and the primary alcohol group of an aldose to produce a dicarboxylic acid. As an example, D-galactose is oxidized to the dicarboxylic acid known as galactaric acid:

D-Galactose → (HNO₃, 0°C) → Galactaric acid (87%)

PROBLEM 25.13

Although D-galactose rotates plane-polarized light, its oxidation product, galactaric acid, does not. Explain.

PROBLEM 25.14

Identify all of the D-aldopentoses from Figure 25.1 that, on oxidation with nitric acid, give diacids that do not rotate plane-polarized light.

Focus On

The Determination of Anomer Configuration

The determination of the configuration at the anomeric carbon of a cyclic sugar such as D-glucopyranose is a difficult problem. Usually, both stereoisomers must be isolated. Then comparison of their properties to those of compounds of known stereochemistry often enables the correct configuration to be determined.

Nuclear magnetic resonance spectroscopy can be used to answer this question in some cases. It has been demonstrated that the coupling constant between two hydrogens depends on their dihedral angle. This relationship, known as the Karplus correlation, is shown on the next page. Note that the coupling constant is maximum at dihedral angles near 180° and is minimum at angles near 90°. In the case of hydrogens on a six-membered ring, this correlation is especially useful. The dihedral angle for axial hydrogens on adjacent carbons is 180°, resulting in a relatively large coupling constant of 8 to 10 Hz. In contrast, if one hydrogen is axial and one is equatorial, or if both are equatorial, then their dihedral angle is 60°. This results in a much smaller coupling constant, usually about 2 to 3 Hz.

Consider the two anomers α- and β-D-glucopyranose. The peak for the hydrogen on the anomeric carbon, carbon 1, is separated from those for the other hydrogens because of the downfield shift caused by the two electronegative oxygens attached to the carbon. The hydroxy group on carbon 1 of the α-stereoisomer is axial, so the hydrogen on this carbon is equatorial. The hydroxy group on carbon 2 is equatorial, so the

Continued

hydrogen on this carbon is axial. As can be seen from the Newman projections, the dihedral angle between the hydrogens is 60°. The coupling constant is 3 Hz in this case. The β-isomer has the opposite configuration at carbon 1, the anomeric carbon. Therefore, the hydrogen on this carbon is axial. The dihedral angle between this hydrogen and the hydrogen on carbon 2, which is also axial, is 180°. This results in a larger coupling constant of 8 Hz for the β-isomer. This method works for other stereoisomers of glucose also, as long as the hydrogen on carbon 2 is axial.

Karplus Correlation

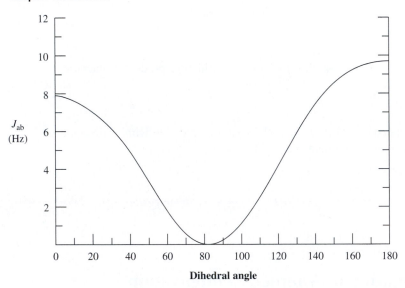

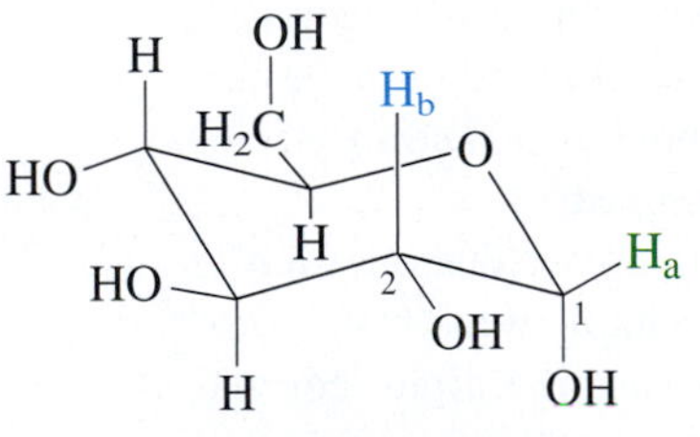

α-D-Glucopyranose

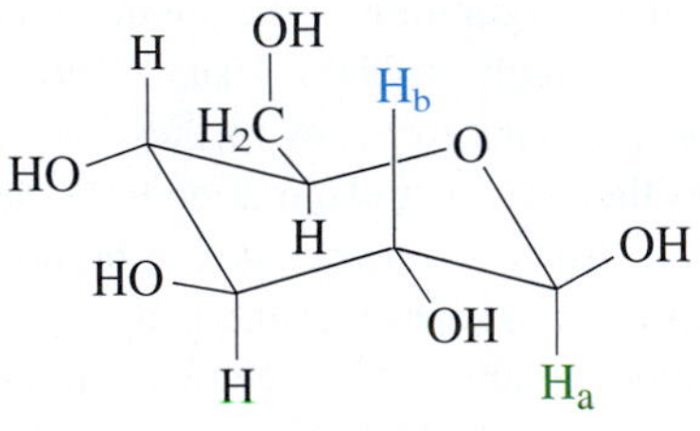

β-D-Glucopyranose

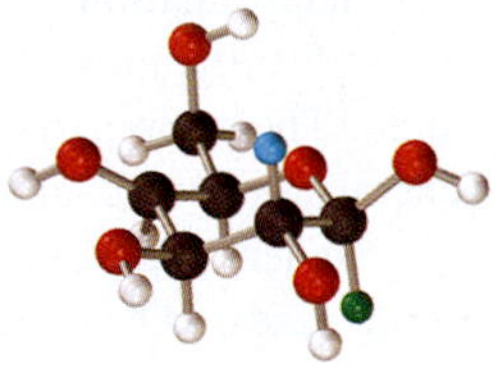

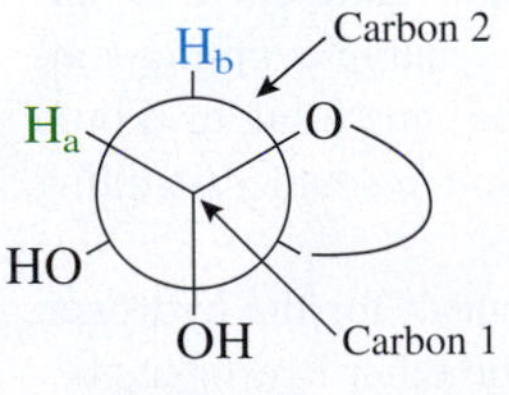

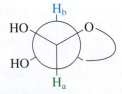

60° dihedral angle
$J_{ab} = 3$ Hz

180° dihedral angle
$J_{ab} = 8$ Hz

It is also possible to selectively oxidize only the aldehyde group with a milder oxidizing agent such as bromine. The oxidation of D-glucose with bromine produces D-gluconic acid in high yield. The reaction of the carboxylic acid group with one of the hydroxy groups to form an intramolecular ester (a lactone) occurs readily. Both five- and six-membered lactone rings can be formed.

(96%)

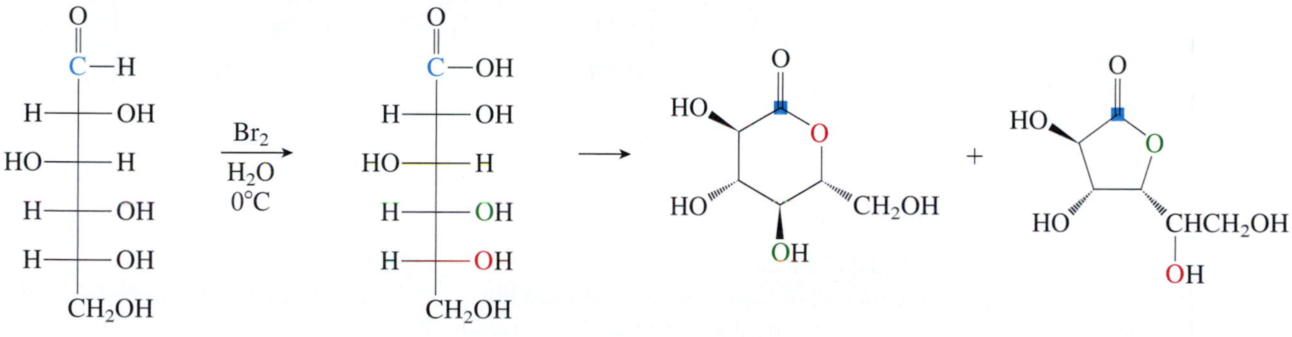

D-Glucose D-Gluconic acid

Oxidation reactions are often used as a test for the presence of an aldehyde group in a carbohydrate. In Tollens' test, the compound is treated with a solution of Ag^+ ion in aqueous ammonia. The aldehyde group is oxidized, and the Ag^+ ion is reduced to metallic silver. The formation of a silver mirror constitutes a positive test. Benedict's test and Fehling's test both employ a solution of Cu^{2+} ion in aqueous base. When the carbohydrate is oxidized, the blue Cu^{2+} ion is reduced to Cu_2O, which forms a brick-red precipitate. (Benedict's reagent has been used in a self-test kit for sugar in the urine caused by diabetes.) Carbohydrates that give a positive test in these reactions are called **reducing sugars** because they reduce the metal ion.

$$\mathrm{RCH{=}O} + \mathrm{Ag^+} \xrightarrow[\ ^{-}\mathrm{OH},\ \mathrm{H_2O}]{\mathrm{NH_3}} \mathrm{RCO_2H} + \text{Other oxidation products} + \underset{\text{Silver mirror}}{\mathrm{Ag}\,(s)}$$

Tollens' test

$$\mathrm{RCH{=}O} + \mathrm{Cu^{2+}} \xrightarrow[\mathrm{H_2O}]{^{-}\mathrm{OH}} \mathrm{RCO_2H} + \text{Other oxidation products} + \underset{\text{Brick-red}}{\mathrm{Cu_2O}\,(s)}$$

Benedict's test and Fehling's test

The aldehyde group can be readily reduced by catalytic hydrogenation or with reagents such as sodium borohydride. The reduction of xylose produces xylitol, which is used as a sweetener in "sugarless" gum:

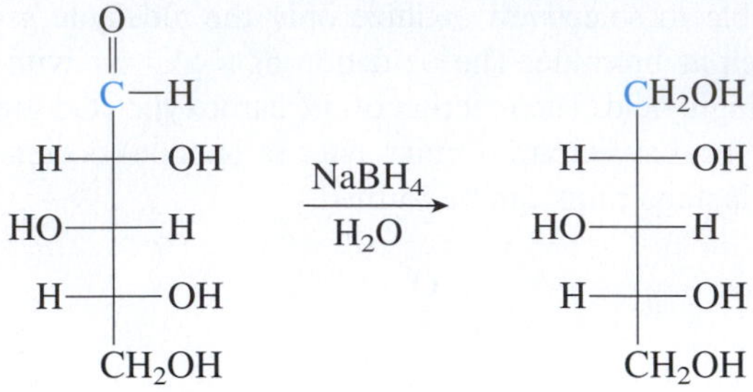

O
‖
C—H
H——OH
HO——H   NaBH₄ / H₂O →
H——OH
CH₂OH

CH₂OH
H——OH
HO——H
H——OH
CH₂OH

D-Xylose   Xylitol

PROBLEM 25.15
Why is xylitol not called D-xylitol?

PROBLEM 25.16
Identify all of the D-aldohexoses from Figure 25.1 that, on reduction with NaBH₄, give products that do not rotate plane-polarized light.

As is the case with other carbonyl compounds, the carbonyl group of a sugar causes any hydrogens on adjacent carbons to be weakly acidic. This provides a mechanism for the isomerization of sugars in basic solution. Thus, D-glucose is isomerized to D-mannose and D-fructose under basic conditions.

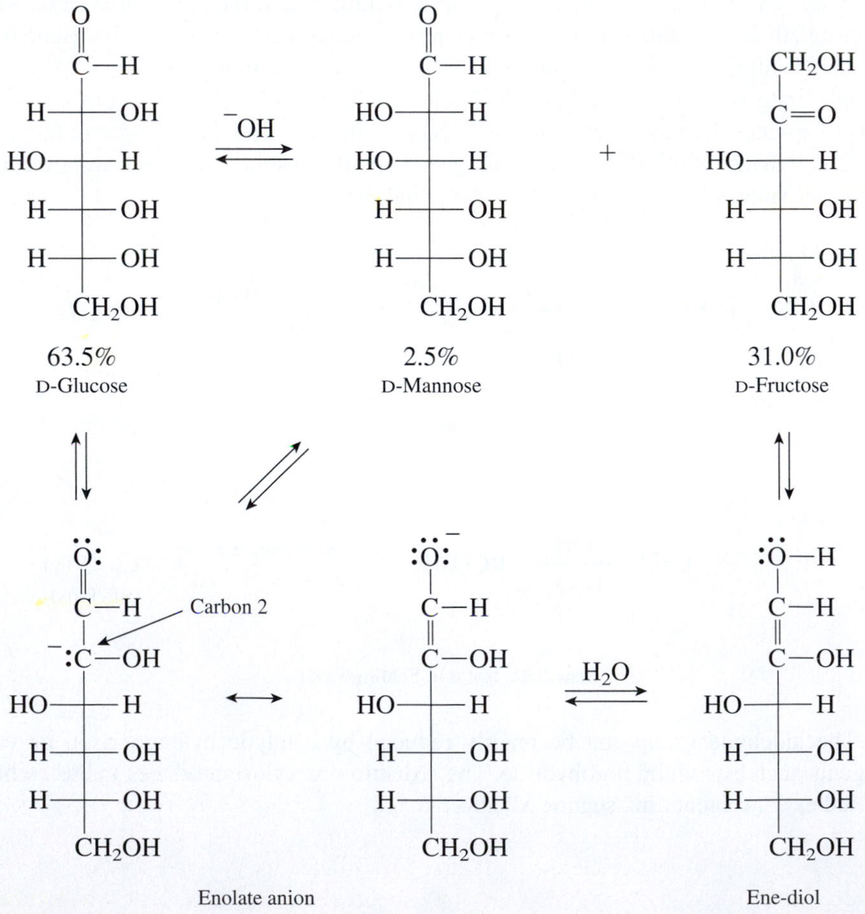

63.5%   2.5%   31.0%
D-Glucose   D-Mannose   D-Fructose

Enolate anion   Ene-diol

Base removes an acidic hydrogen on carbon 2 of glucose to produce an enolate anion. Carbon 2 of the enolate anion is planar and is no longer chiral. Protonation from one side regenerates D-glucose while protonation from the other side produces D-mannose, with the opposite configuration at carbon 2. If, instead, the enolate anion is protonated on oxygen, a double bond with a hydroxy group on each carbon, an ene-diol, is formed. Removal of the proton from the hydroxy group on carbon 2 of the ene-diol gives a new enolate ion. Protonation of this enolate ion on carbon 1 produces D-fructose. This is an equilibrium process, so D-fructose is also isomerized to D-glucose and D-mannose under basic conditions. For this reason, D-fructose is also a reducing sugar even though it does not have an aldehyde group.

The alcohol groups of a monosaccharide can be converted to esters by the methods described in Section 19.4. Thus, the reaction of α-D-glucopyranose with acetic anhydride produces a pentaacetate. Note that the hemiacetal hydroxy group is also esterified in this reaction.

α-D-Glucopyranose

88%

PROBLEM 25.17

If α-D-glucopyranose is reacted with acetic anhydride at 100°C, the major product is the β-isomer of the pentaacetate. Explain how this product is formed.

Perhaps the most important reaction of monosaccharides is the conversion of the hemiacetal to an acetal in the presence of an alcohol and acid. The products are known as **glycosides,** and the bond between the oxygen of the alcohol and the acetal carbon is called a **glycosidic bond.**

(66%)
Methyl-α-D-glucoside

(32%)
Methyl-β-D-glucoside

The mechanism for this reaction is shown in Figure 25.3. It is the same mechanism as that shown in Figure 18.5 on page 776 for the formation of acetals. Because the reaction involves a planar carbocation intermediate, a mixture of stereoisomeric glycosides is formed. This same mixture is produced starting from either α- or β-D-glucopyranose.

Numerous glycosides are found in nature. For example, willow bark contains the glycoside salicin. In ancient Greece, Hippocrates encouraged his patients to chew willow bark

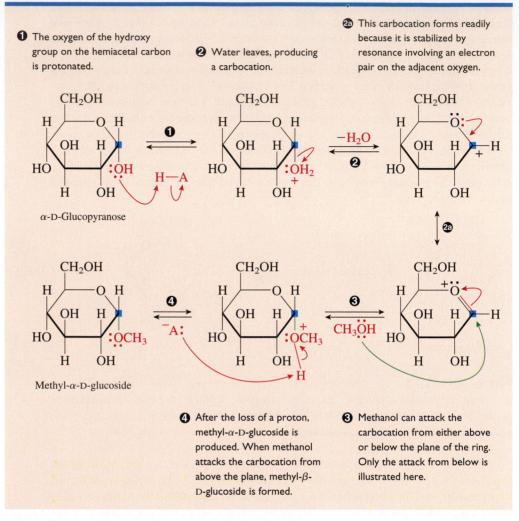

Figure 25.3

MECHANISM OF THE FORMATION OF METHYL-α-D-GLUCOSIDE FROM α-D-GLUCOPYRANOSE. This mechanism is the same as that for the formation of acetals shown in Figure 18.5. It is illustrated here with α-D-glucopyranose but applies also to the β-anomer.

for pain. The alcohol derived from hydrolysis of salicin, called an aglycone, is readily oxidized to salicylic acid, which is very similar to aspirin, so this was an effective remedy.

PROBLEM 25.18

Show all of the steps in the mechanism for the hydrolysis of salicin.

The major reason that the formation of glycosides is so important is that disaccharides and polysaccharides are formed from monosaccharide units held together by glycosidic bonds. The oxygen of a hydroxy group from one sugar is used to form a bond to the acetal carbon of another monosaccharide. This process is discussed in Sections 25.6 and 25.7.

The final reaction to be covered in this section is known as the Kiliani-Fischer synthesis. It is a method that converts an aldose to two diastereomeric aldoses that contain one more carbon than the original sugar. The Kiliani-Fischer synthesis is illustrated in the following reaction sequence, which shows the formation of the aldopentoses D-ribose and D-arabinose from the aldotetrose D-erythrose:

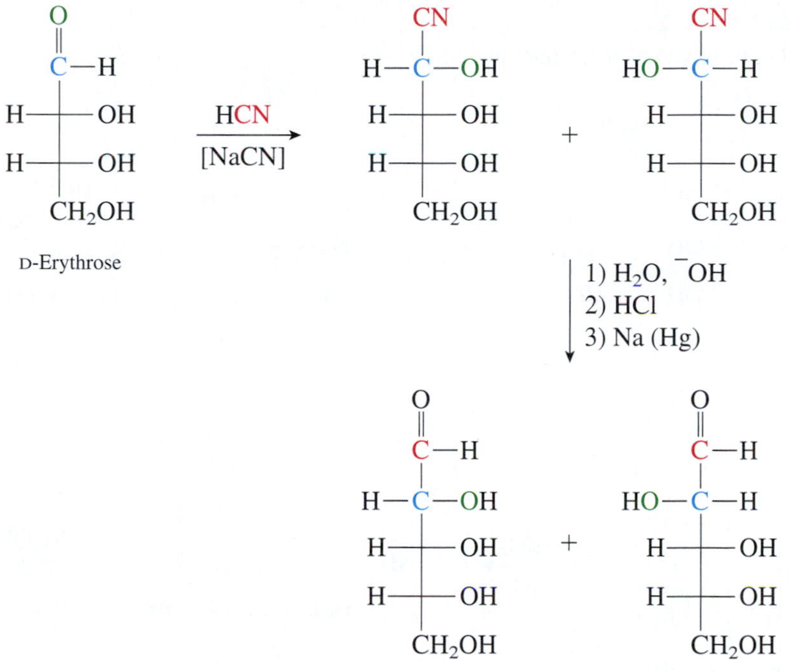

In the Kiliani-Fischer synthesis the starting aldose is first reacted with hydrogen cyanide to form isomeric cyanohydrins (see Section 18.4). This reaction introduces a new stereocenter at the site of the carbonyl carbon of the original aldose. Therefore, the product is a mixture of two diastereomeric cyanohydrins with opposite configurations at the new stereocenter. In the next part of the synthetic sequence, the nitriles are hydrolyzed to carboxylic acids, which spontaneously form lactones. The lactones are reduced to the aldehydes with sodium amalgam. (Today we might choose to accomplish this reduction with some other reagent, such as DIBALH described in Section 19.8.) The overall result is the conversion of the original aldose to two diastereomeric aldoses with one more carbon. The synthesis preserves the configurations of the stereocenters of the original aldose while producing both possible configurations at the new stereocenter, carbon 2 of the aldose products.

PROBLEM 25.19

Show all of the steps in the mechanism for this reaction. What isomeric product is also formed?

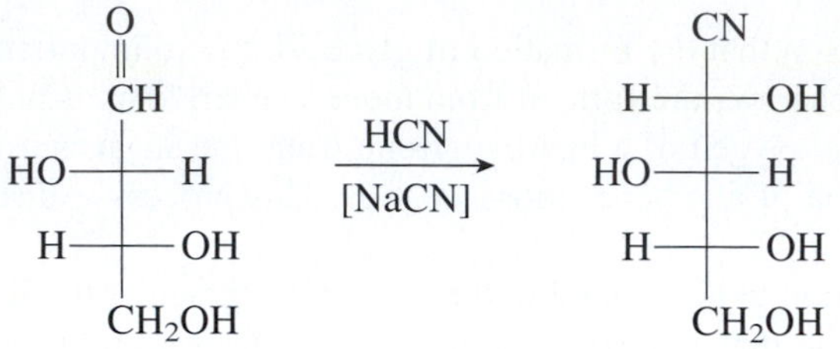

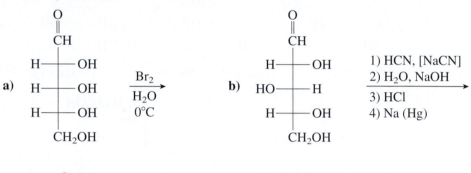

PROBLEM 25.20

Show an equation for a Kiliani-Fischer synthesis starting from D-ribose. What are the names of the monosaccharides produced in this reaction?

ORGANIC
Chemistry·Now™
Click *Coached Tutorial Problems* for more practice with **Reactions of Monosaccharides.**

PROBLEM 25.21

Show the products of these reactions:

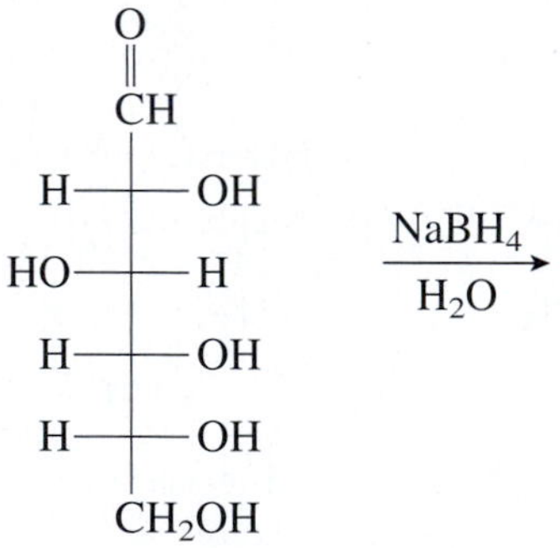

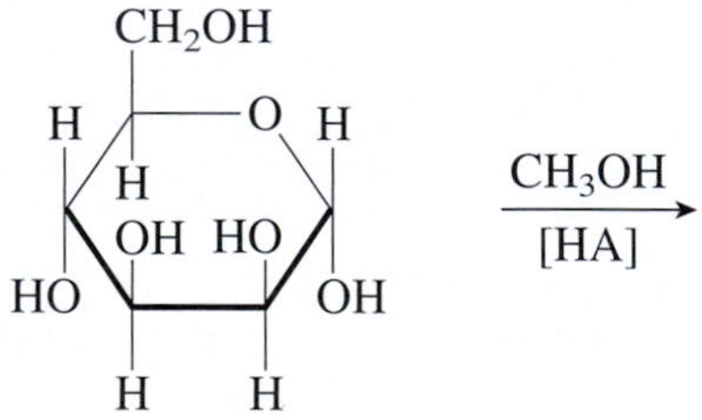

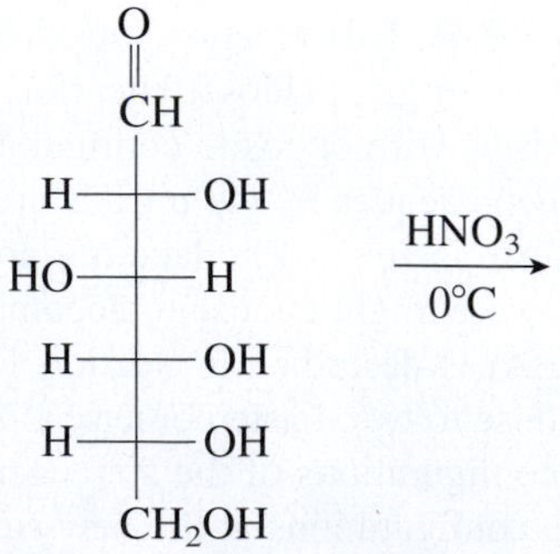

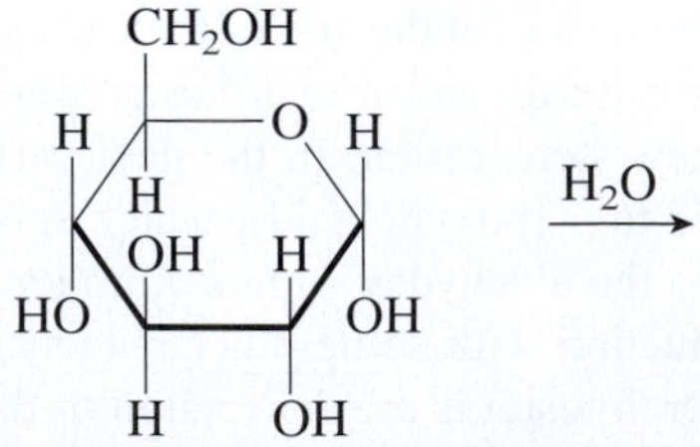

Focus On

Artificial Sweeteners

Roses are red, violets are blue, sugar is sweet, and so are—many other organic compounds. For example, many compounds that have hydroxy groups on adjacent carbons have a sweet taste. The simplest of these is ethylene glycol or 1,2-ethanediol, the main component of antifreeze. Although this diol is sweet, it is also toxic. Every year, numerous cases of poisonings occur when animals, especially dogs, ingest antifreeze that was carelessly discarded.

Xylitol, another polyhydroxy compound, is used as a sweetener in "sugarless" gum. It has approximately the same number of calories per gram as does sucrose and is not a low-calorie sweetener. However, because it does not have a carbonyl group, it is not fermented by bacteria in the mouth and does not promote tooth decay.

In recent years there has been a great search for a safe, calorie-free sweetener to help diabetic people who need to control their sugar intake and, more recently, to meet the demands of health- and diet-conscious consumers. One such sweetener that was marketed in the late 1960s was calcium cyclamate. However, cyclamates were banned in the United States in 1970 because of a suspected link with cancer.

Saccharin is 300 times as sweet as sucrose on a weight basis. Like cyclamate, it is a low-calorie sweetener, not because it has fewer calories than sucrose on a weight basis, but because it is so much sweeter that only a small amount need be used. Saccharin is another artificial sweetener that is suspected of causing cancer in laboratory animals at very high doses. Although the U.S. Food and Drug Administration (FDA) moved to prohibit the use of saccharin in the late 1970s, the ban was blocked by Congress because saccharin was the only artificial sweetener available at that time. Although it has been used for many years now, it has only recently been removed from the government's list of human carcinogens.

The sweetener aspartame was discovered in 1965 and approved by the FDA in 1981. It is the methyl ester of a dipeptide formed from the amino acids aspartic acid and phenylalanine. Because both of these amino acids occur naturally and are part of nearly every protein, there is much less reason to be concerned about the health effects of this compound. Nevertheless, it has been extensively tested. Aspartame is about 180 times sweeter than sucrose, so the amount that is needed to sweeten a can of a soft drink, for example, is so small that it contributes only negligible calories to the diet. In addition, the taste profile of aspartame is much closer to sugar than is that of saccharin. Aspartame, sold under the brand name NutraSweet, has been an enormous financial success. Sucralose (Splenda) is prepared from sucrose by replacing some of the hydroxy groups with chlorines. Its taste closely resembles sucrose, but it is about 600 times sweeter. Acesulfame K (Sunett, Sweet One) is about 200 times sweeter than sucrose. It is quite stable to heat, so it is potentially very useful in baked goods.

As might be expected, the search for an even better artificial sweetener continues. Alitame is a dipeptide formed from aspartic acid and alanine, with an unusual amide at the carboxylate end of the alanine. It is 2000 times as sweet as sucrose—1 pound of alitame has the sweetening power of 1 ton of sucrose! In addition, because an amide bond is more stable than an ester bond, alitame is more stable to hydrolysis than is aspartame. Therefore, alitame keeps its sweetness in aqueous solution better than aspar-

Continued

Some Artificial Sweeteners
Numbers in parentheses indicate how much
sweeter than sucrose these compounds are.

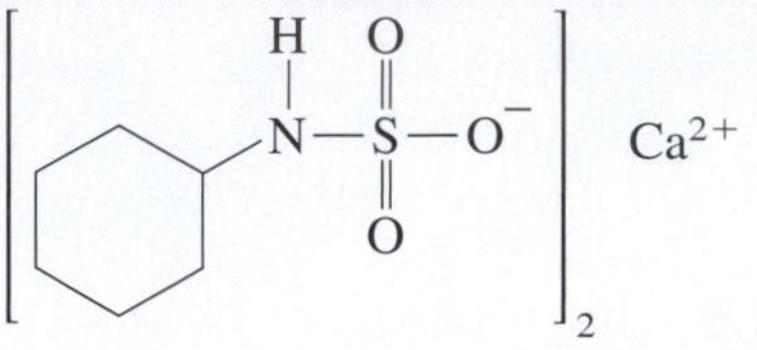

Calcium cyclamate

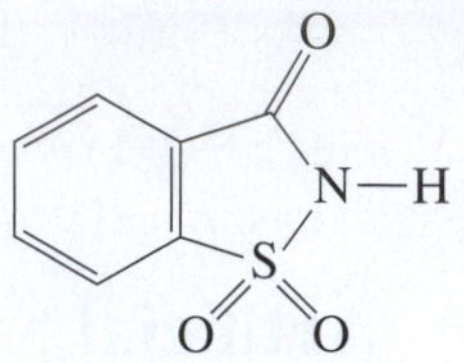

Saccharin
(300×)

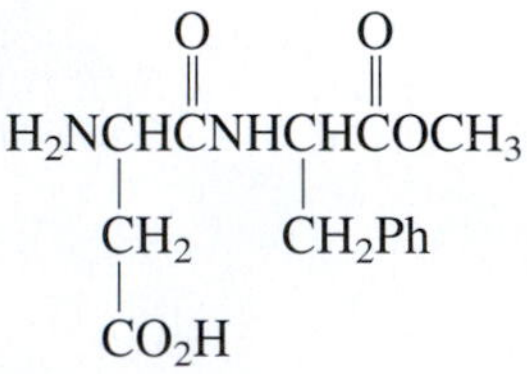

Aspartame
(180×)

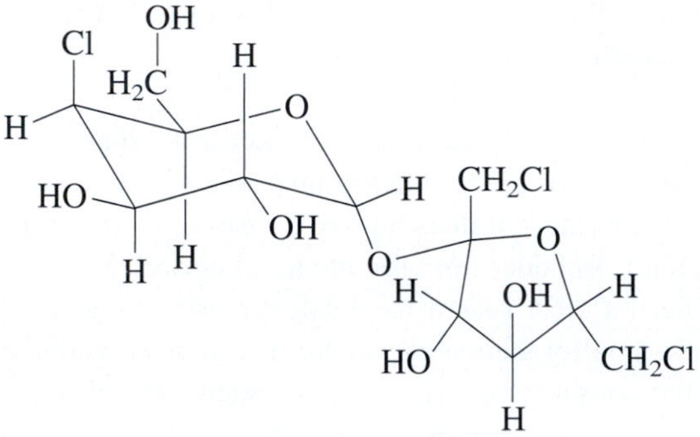

Sucralose
(600×)

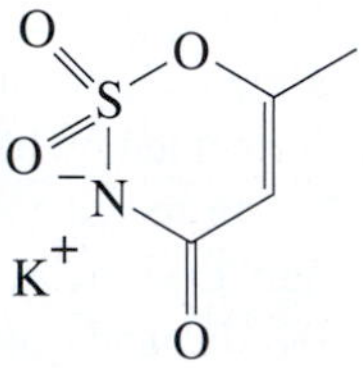

Acesulfame K
(200×)

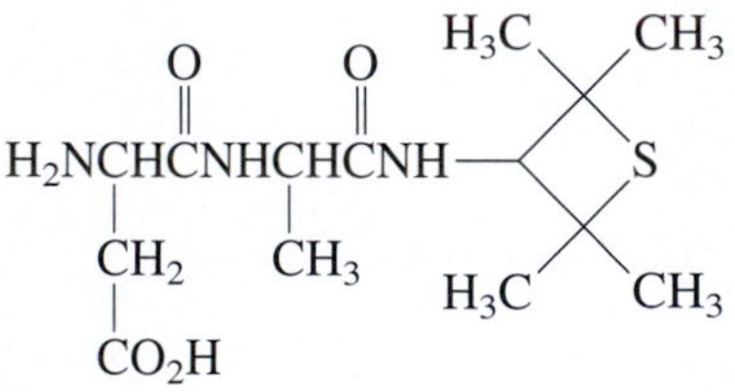

Alitame
(2000×)

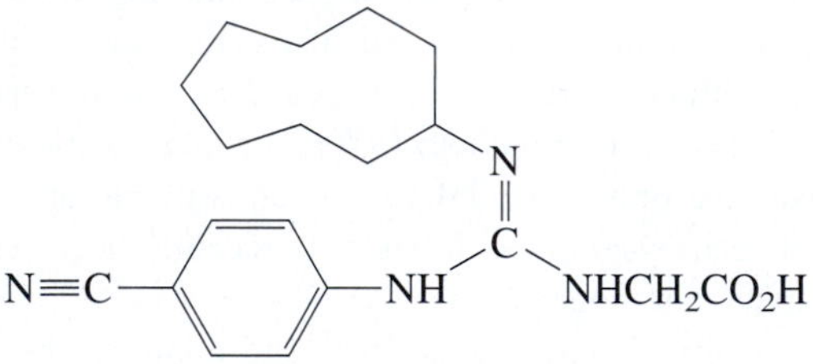

Sucrononic acid
(200,000×)

tame does. Even sweeter compounds have been discovered. For example, sucrononic acid is 200,000 times as sweet as sucrose. Neither alitame nor sucrononic acid has yet been approved for use.

Obtaining approval for the use of a new artificial sweetener is a very expensive proposition and requires considerable expenditure of time and scientific effort. This is absolutely necessary, though, because the number of people that would be exposed to products containing the sweetener is so large. (Of course, the fact that most products would contain only a very small amount of an artificial sweetener somewhat decreases the potential for problems.) In addition, certain groups, such as diabetic people, might have consumption patterns that are much higher than other parts of the population. Although obtaining FDA approval is extremely expensive, we may indeed see the approval of other artificial sweeteners in the future because the financial rewards are potentially so large.

Examination of the structures of these sweet compounds does not reveal any simple pattern for their structural features. Chemists are still attempting to determine what it is that makes a compound sweet. They are also trying to model the taste receptor that is responsible for detecting sweetness. Although some progress has been made in this area, there is still a long way to go.

Because of this difficulty in predicting what features are necessary for a compound to taste sweet, most of the discoveries of artificial sweeteners have been serendipitous. In fact, many of the early discoveries resulted from dangerous laboratory practices that we would not condone today. For example, the sweetness of saccharin was discovered in 1879 by a chemist who spilled some of the compound on his hand. Later, while eating lunch in the laboratory, he noticed the extremely sweet taste. The sweetness of cyclamate was discovered in 1937 by a chemist who tasted it on a cigarette that he had set on the lab bench. And aspartame was found to be sweet by a chemist who got some on his hand and later licked his finger before picking up a piece of paper. This resulted in a billion-dollar-per-year product!

The unsafe nature of these practices needs to be emphasized. Precautions should always be taken to minimize exposure to all laboratory compounds. Hands should be washed whenever a compound is spilled on them. Eating, drinking, and smoking in the laboratory are all extremely unsafe practices and should be forbidden. Even licking a finger to pick up a piece of paper should be avoided.

25.5   FISCHER'S STRUCTURE PROOF FOR GLUCOSE

Consider the problem that Emil Fischer faced in the late 1800s. Using a number of experimental observations, he had determined that glucose was an aldohexose. But which of the 16 possible aldohexoses was it? How could the configuration at each of the stereocenters of glucose be determined experimentally? Fischer realized that it was impossible at that time to determine the absolute configuration of a compound—that is, which enantiomer it was. Therefore, he had no way to determine whether glucose was a D-aldohexose or an L-aldohexose. Because he needed to be able to write something for the structures of the compounds he was working with, he arbitrarily chose the D-configuration for glucose. Although this guess was fortuitously correct, it had no bearing on the experiments and the logic that he used to solve the problem. The experiments established the configuration of carbons 2, 3, and 4 *relative* to that at carbon 5, and on the basis of these experiments, Fischer proved that glucose was either the compound shown in Figure 25.1 or its enantiomer.

Fischer studied carbohydrates for many years and performed numerous experiments. Overall, several subsets of these experiments can be used to establish the structure of glucose. Although the ones described here are not the exact ones that he used in the structure proof that he published, they serve to illustrate his reasoning and are somewhat simpler to follow.

There are eight possible structures for D-glucose (see structures **5** through **12** in Figure 25.4). Its actual structure can be determined by using two reactions, oxidation with nitric acid and the Kiliani-Fischer synthesis, along with the monosaccharides glucose, mannose, and arabinose and one additional sugar. Let's see how this was done.

Experiment 1

Application of the Kiliani-Fischer synthesis to the aldopentose D-arabinose produces aldohexoses D-glucose and D-mannose. Because the Kiliani-Fischer synthesis incorporates the stereocenters of D-arabinose without changes, D-glucose and D-mannose must have the same configurations at carbons 3, 4, and 5 as does D-arabinose at carbons 2, 3, and 4, respectively. Furthermore, D-glucose and D-mannose must differ only in their configuration at carbon 2.

The four possible D-aldopentoses are shown as structures **1** through **4** in Figure 25.4. One of these is D-arabinose. Each of these produces two of the eight possible D-aldohexoses on application of the Kiliani-Fischer synthesis. Note that in each case the configurations of the aldohexoses at carbons 3, 4, and 5 are the same as those of the aldopentose at carbons 2, 3, and 4, respectively. If D-arabinose has structure **1,** then the two aldohexoses produced from it upon Kiliani-Fischer synthesis, **5** and **6,** must be D-glucose and D-mannose. Similarly, if D-arabinose is **2,** then D-glucose and D-mannose must be **7** and **8,** and so forth.

Experiment 2

On oxidation with nitric acid, both D-glucose and D-mannose produce diacids that rotate plane-polarized light. This experiment enables some of the aldohexoses to be eliminated as possibilities for D-glucose and D-mannose. For example, consider the oxidation of aldohexose **5,** shown in the following equation:

5   A diacid

Examination of the structure of the diacid product shows that it has a plane of symmetry. It does not rotate plane-polarized light because it is a meso compound and is not chiral. Therefore, neither D-glucose nor D-mannose can have structure **5.** Because neither can have structure **5,** then, on the basis of experiment 1, the other cannot have structure **6,** nor can D-arabinose have structure **1.**

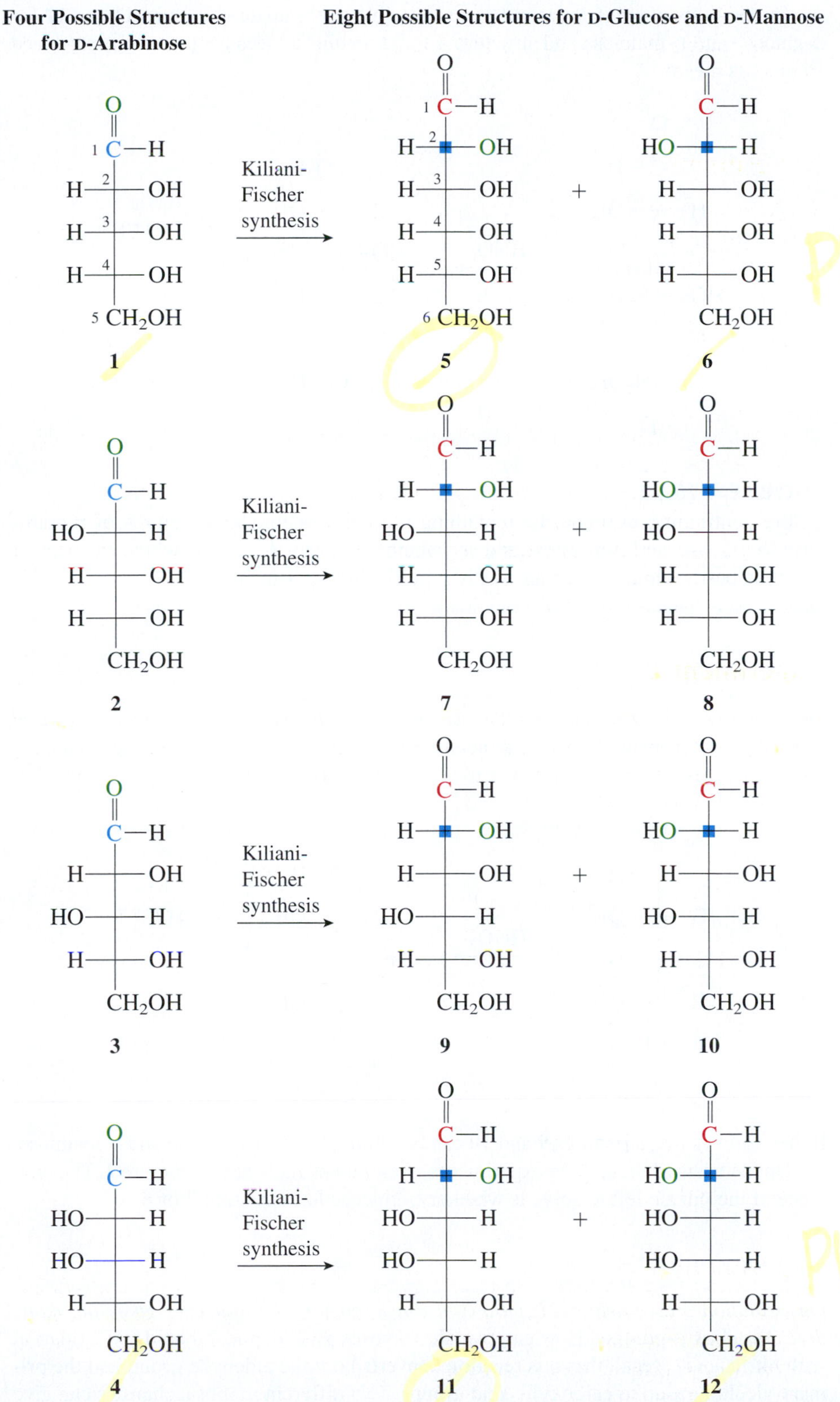

Four Possible Structures for D-Arabinose

Eight Possible Structures for D-Glucose and D-Mannose

Figure 25.4

POSSIBLE STRUCTURES FOR D-ARABINOSE AND ITS KILIANI-FISCHER PRODUCTS, D-GLUCOSE AND D-MANNOSE.

Plane of sym.

Plane of sym

On the basis of similar reasoning, it is possible to eliminate structures **11** and **12** for D-glucose and D-mannose and structure **4** for D-arabinose because the diacid produced from **11** is meso:

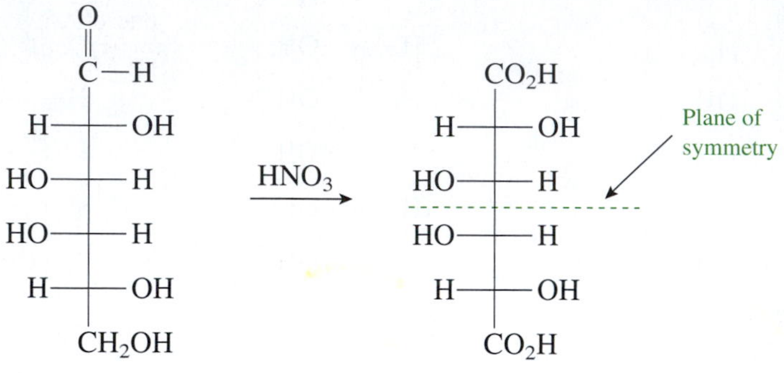

11

PROBLEM 25.22

Before continuing, examine the remaining possibilities for the structures of D-arabinose, D-glucose, and D-mannose and see whether you can determine which experiment might be best to do next. The answer is supplied in Experiment 3.

Experiment 3

On oxidation with nitric acid, D-arabinose produces a diacid that rotates plane-polarized light. This experiment eliminates structure **3** as a possibility for D-arabinose because it produces a meso diacid that does not rotate plane-polarized light:

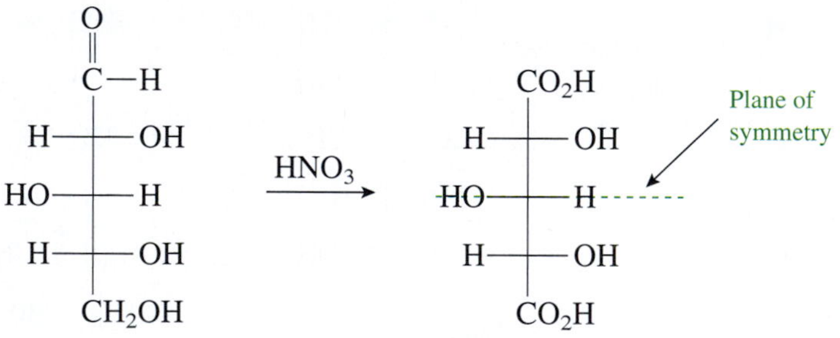

3

It then follows that structures **9** and **10** can be eliminated for D-glucose and D-mannose.

On the basis of these three experiments, D-arabinose must have structure **2.** The only piece of the puzzle left to solve is whether D-glucose has structure **7** or **8.**

Experiment 4

On oxidation with nitric acid, gulose, a diastereomer of D-glucose, gives the same diacid as does D-glucose. How can two aldohexoses give identical diacids on oxidation with nitric acid? Recall that this reaction converts both the aldehyde group and the primary alcohol group to carboxylic acid groups. Two different monosaccharides can give the same product if they have the same configurations at the asymmetric carbons when one is written with its aldehyde group at the top of the structure and its primary alcohol group at the bottom and the other is written with its primary alcohol group at the top and its aldehyde group at the bottom.

This is best seen with an example. Consider aldohexose **7**, one of the structures remaining as a possibility for D-glucose. On oxidation with nitric acid, **7** produces diacid **13**:

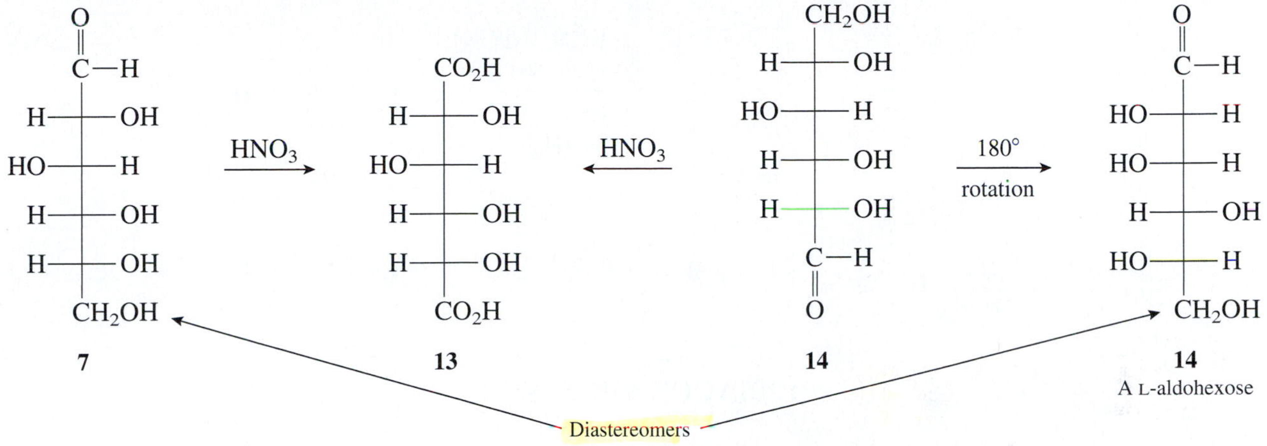

Aldohexose **14**, written with its primary alcohol group at the top of the chain and its aldehyde group at the bottom, also produces diacid **13** on oxidation with nitric acid. A rotation of 180° in the plane of the page (recall that such a rotation of a Fischer projection does not change the configuration of the compound) puts **14** in the more common form with the aldehyde group at the top of the chain. When drawn this way, it is apparent that **14** is an L-aldohexose and is a diastereomer of **7**.

The other possible structure for D-glucose is **8**. Oxidation of **8** gives diacid **15**:

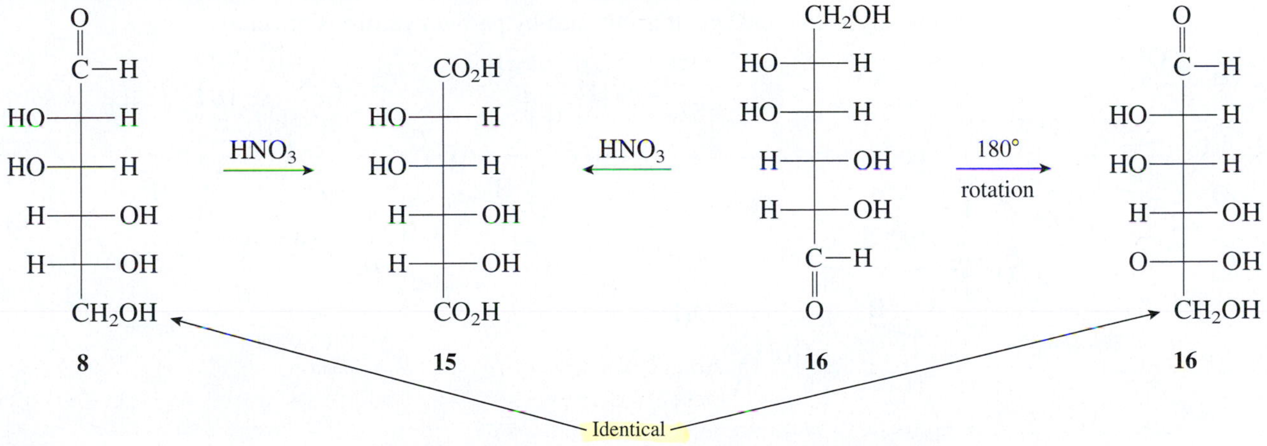

Aldohexose **16** also produces **15** on oxidation with nitric acid. When **16** is rotated 180°, it is found to be identical with **8**. There is only one aldohexose that produces **15** on oxidation!

The structure proof for D-glucose is now complete. If D-glucose were **8**, then the diacid obtained from it on oxidation with nitric acid must be **15**. This is impossible because there is no other aldohexose that produces **15** on oxidation with nitric acid—there is no possible structure for gulose. Therefore, D-glucose must have structure **7**, and D-mannose has structure **8**. Gulose has structure **14** and has the L-configuration.

PROBLEM 25.23

Monosaccharide **A** in the following scheme is a D-aldotetrose. Compound **E** does rotate plane-polarized light, whereas compounds **B** and **F** do not. Show the structures of **A**, **B**, **C**, **D**, **E**, and **F**.

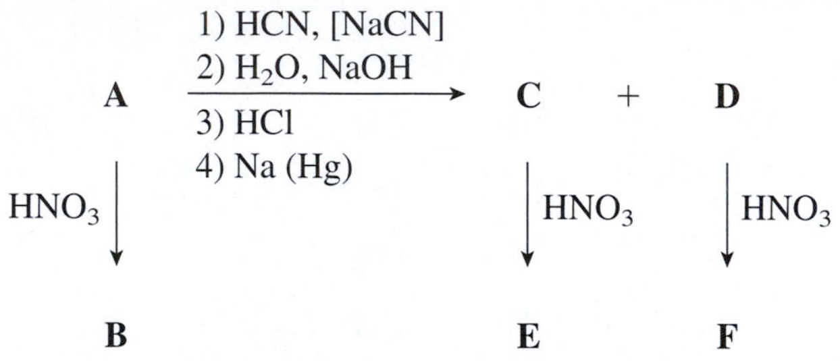

$$
\mathbf{A} \xrightarrow[\substack{\text{3) HCl} \\ \text{4) Na (Hg)}}]{\substack{\text{1) HCN, [NaCN]} \\ \text{2) } H_2O, \text{ NaOH}}} \mathbf{C} \quad + \quad \mathbf{D}
$$

A → (HNO₃) → B

C → (HNO₃) → E

D → (HNO₃) → F

25.6   DISACCHARIDES

Disaccharides are formed from two monosaccharide units that are connected by a glycosidic bond. Different disaccharides result from the combination of different monosaccharides, from the formation of the glycosidic bond using different hydroxy groups on the monosaccharides, or from different stereochemistries of the glycosidic bond. A few of the more important disaccharides are described in the following paragraph.

Maltose is a disaccharide obtained from the partial hydrolysis of starch. It is composed of two molecules of glucose that are linked by a glycosidic bond from the hydroxy group on carbon 4 of one of the glucose units to carbon 1 of the other with α-stereochemistry. Cellobiose is a stereoisomer of maltose with β-stereochemistry at the glycosidic linkage. It is obtained by partial hydrolysis of cellulose.

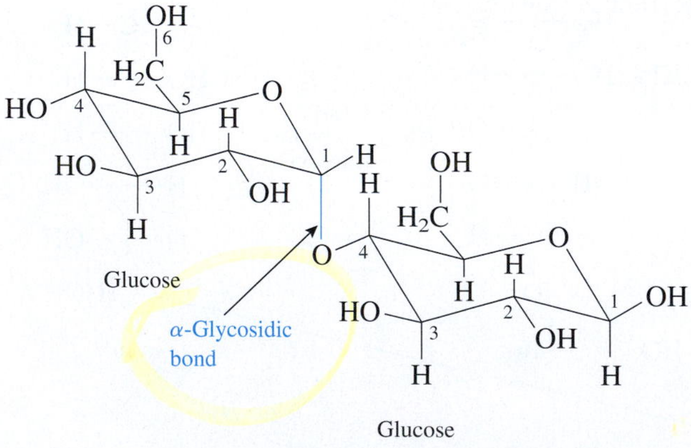

Maltose

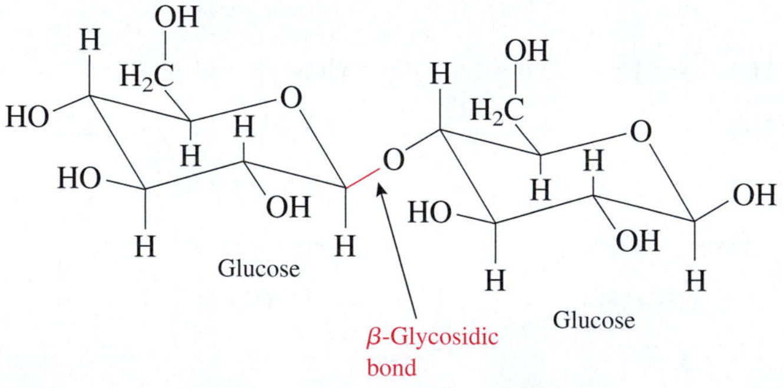

Cellobiose

Lactose, a disaccharide found in milk, is composed of D-galactose and D-glucose, which are joined by a β-glycosidic linkage from the hydroxy group on carbon 4 of glucose to carbon 1 of galactose. Sucrose, also known as table sugar, is the most abundant disaccharide. It is composed of one molecule of D-glucose and one of D-fructose. It differs from the other disaccharides in that its glycosidic bond is formed between carbon 1 (the acetal carbon) of glucose and carbon 2 (the acetal carbon) of fructose. Because of this connection, it has no hemiacetal group but has two acetal groups instead. The glucose is a pyranose ring with an α-glycosidic bond, whereas the fructose is a furanose ring with a β-glycosidic bond:

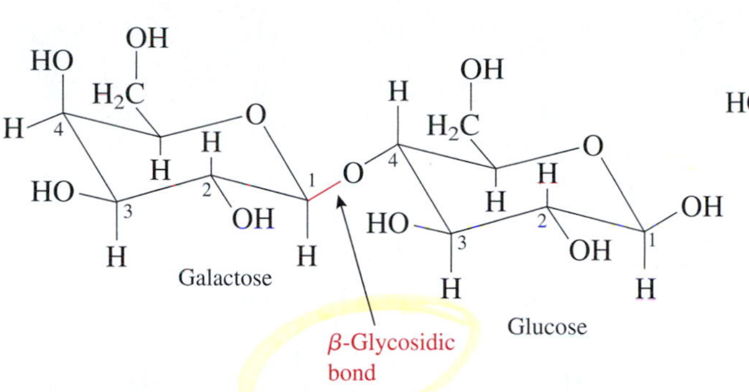

Lactose

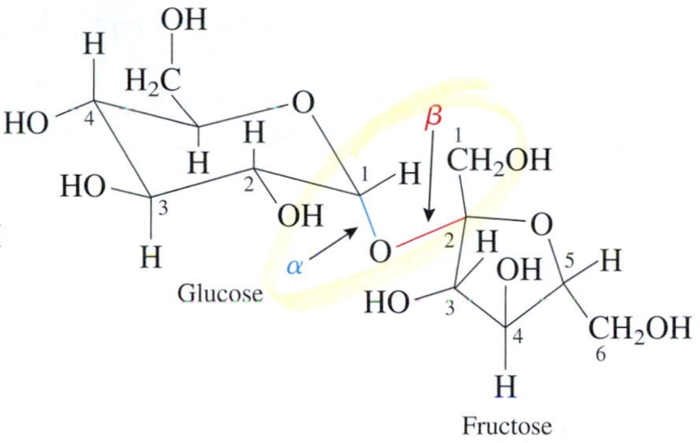

Sucrose

PROBLEM 25.24

Maltose, cellobiose, and lactose are reducing sugars, but sucrose is not. Explain.

25.7 POLYSACCHARIDES

The polysaccharides cellulose, starch, and glycogen are all polymers of D-glucose units linked by glycosidic bonds. Cellulose is the major structural component of the cell walls of plants and is the most abundant organic compound on Earth, accounting for more than half of the carbon in the biosphere. More than 10^{15} kg of cellulose are synthesized and degraded annually. It is a linear polymer containing 2500 to 15,000 D-glucose units linked by β-glycosidic bonds between carbons 1 and 4. As mentioned previously, incomplete hydrolysis of cellulose produces the disaccharide cellobiose, which therefore has the same stereochemistry at its glycosidic bond as does cellulose.

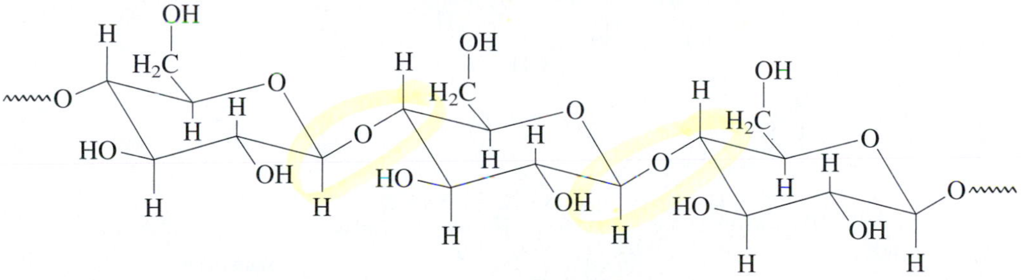

Cellulose

Starch is the food reserve of plants. It consists of two components: amylose and amylopectin. Amylose consists of several thousand glucose units linked by α-glycosidic bonds between carbons 1 and 4 to form a linear polymer. Incomplete hydrolysis of amylose produces the disaccharide maltose, which also has an α-glycosidic bond. Although amylose differs from cellulose only in the stereochemistry of the glycosidic bonds between its glucose monomers, it has a very different shape and quite different physical, chemical, and biological properties. For example, starch can be digested by almost all animals, whereas cellulose can be digested by only a few microorganisms. In the digestive tracts of herbivores, symbiotic microorganisms hydrolyze cellulose to glucose, using enzymes known as cellulases.

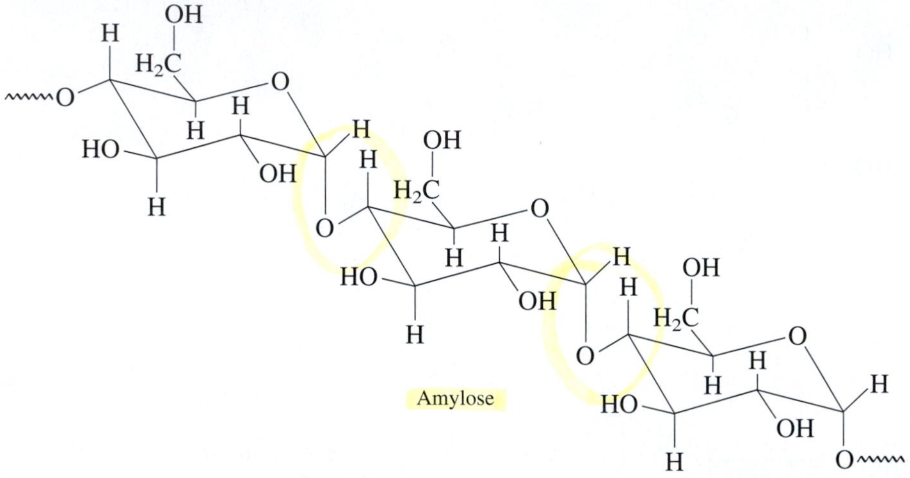

Amylose

The other component of starch, amylopectin, contains up to 10^6 glucose units. It is similar to amylose in that the glucose monomers are connected by α-glycosidic bonds between carbons 1 and 4. It differs, however, in that it has branches that occur every 20 to 30 glucose units. These branches are also amylose-type chains that are connected to the main chain by an α-glycosidic bond from carbon 1 of the branch to carbon 6 of the main chain:

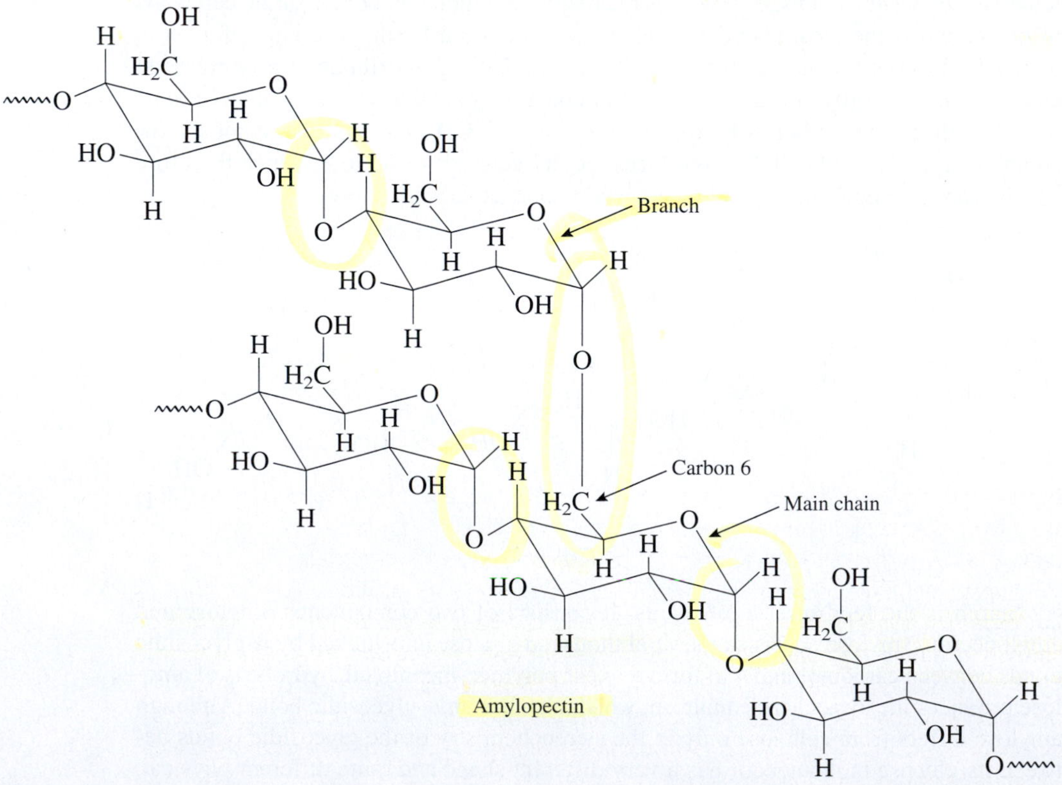

Amylopectin

Branches occur on the branches also, so amylopectin has a treelike structure. Glycogen, the food reserve of animals, has a structure similar to that of amylopectin except that it has branches every 8 to 10 glucose units. The structures of these two polymers are represented schematically in the following drawings:

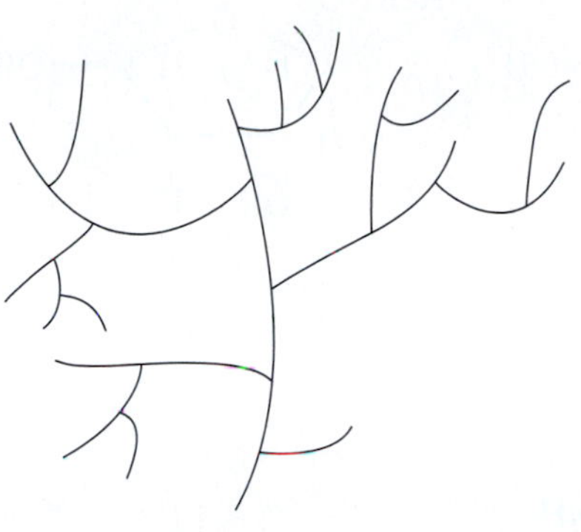

Amylopectin    Glycogen

Chitin, which is the structural component of the exoskeleton of invertebrates such as crustaceans, insects, and spiders, resembles cellulose with the exception that the hydroxy groups on carbon 2 are replaced by acetylamino groups.

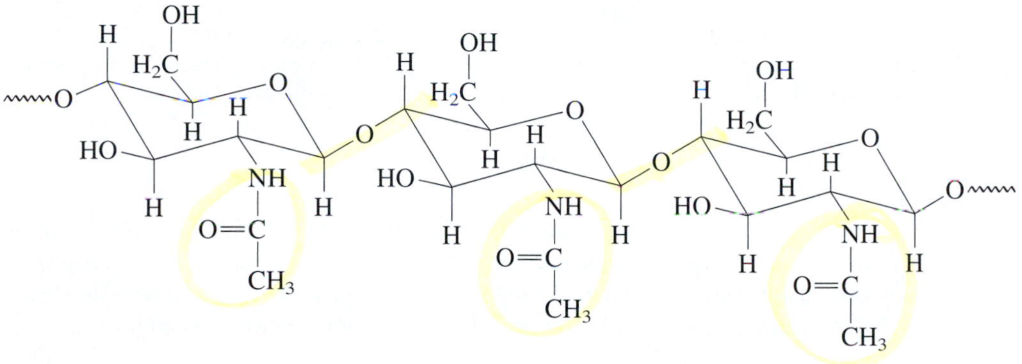

Chitin

25.8   OTHER CARBOHYDRATE-CONTAINING COMPOUNDS

In addition to the sugars already described, many other naturally occurring compounds have related structures or have carbohydrate groups as part of a larger structure. There are too many of these to list completely, so only a few examples are provided here.

Sugars that have an amino group in place of a hydroxy group are widely distributed. For example, the aminosugars D-glucosamine and D-galactosamine are components of chitin and numerous other biologically important polysaccharides. Sugars that have a hydroxy group replaced with a hydrogen, called deoxy sugars, are important also. The monosaccharide D-2-deoxyribose is part of the polymer that stores genetic information, deoxyribose nucleic acid (DNA). And many antibiotics are derived from carbohydrates. Streptomycin is one example.

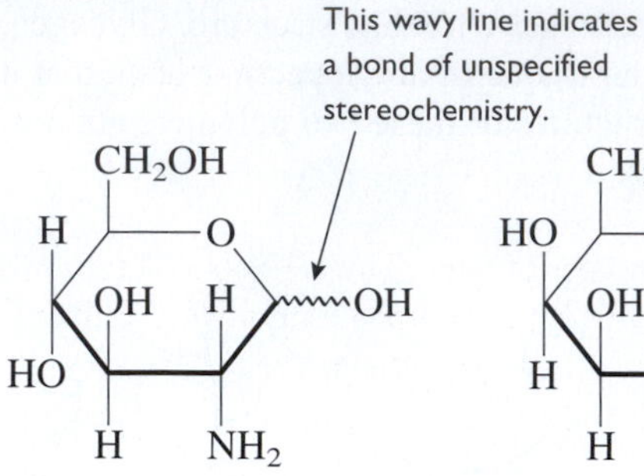

This wavy line indicates
a bond of unspecified
stereochemistry.

D-Glucosamine

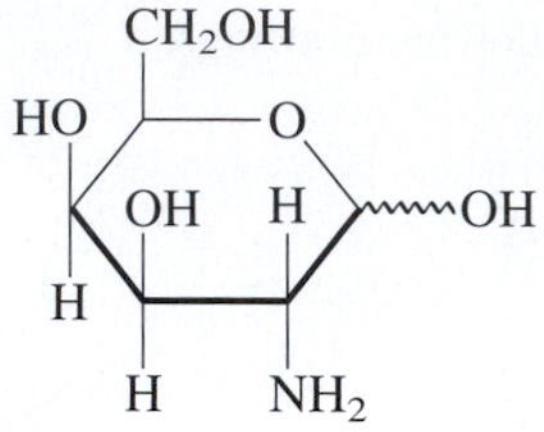

D-Galactosamine

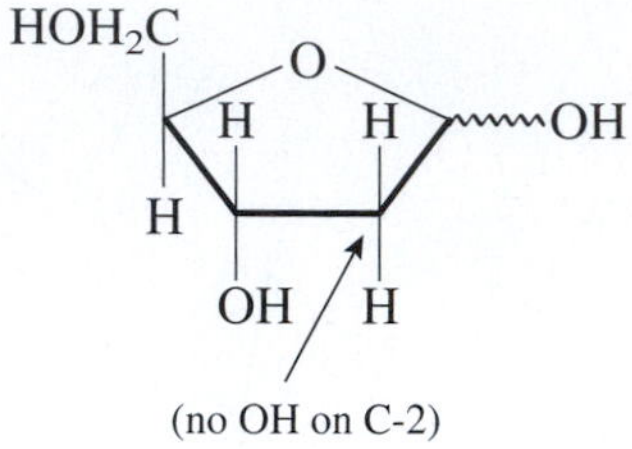

(no OH on C-2)
D-2-Deoxyribose

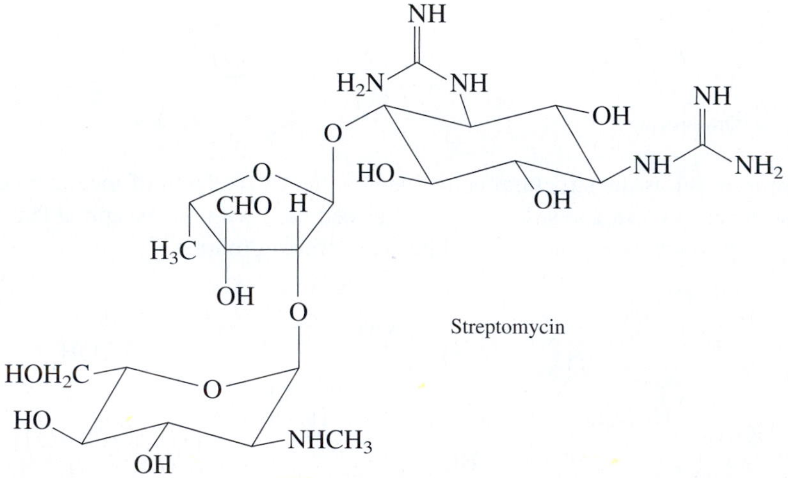

Streptomycin

Digitoxin, a cardiac glycoside that affects the action of the heart, is a complex glycoside formed from three units of the monosaccharide D-digitoxose and the steroidal (see Chapter 28) aglycone digitoxigenin. Small doses of digitoxin reduce the pulse rate, regularize the rhythm of the heart, and strengthen the heartbeat. In larger doses it is a powerful heart poison.

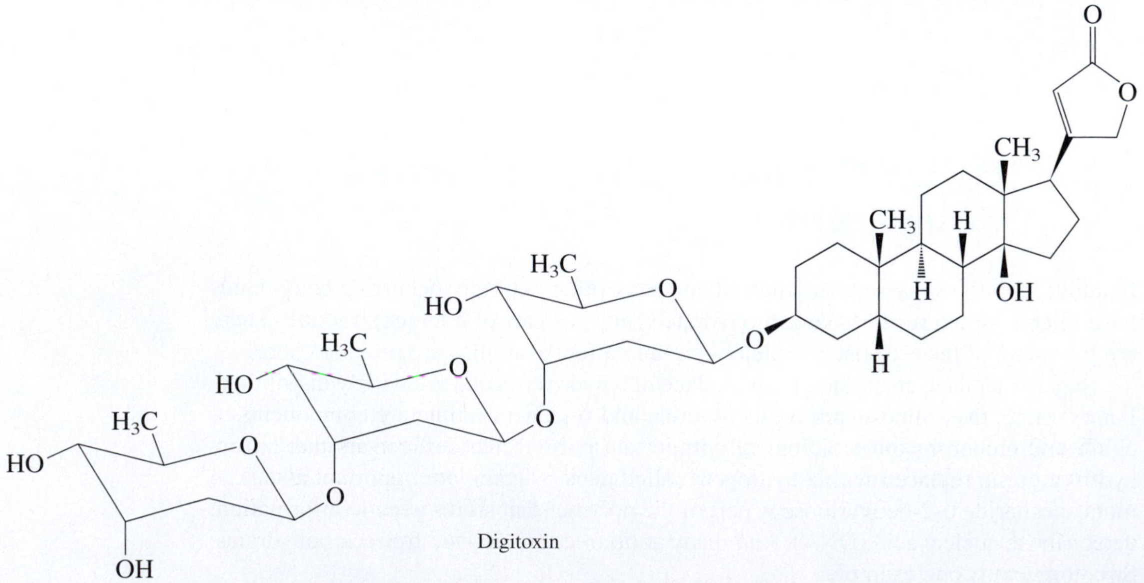

Digitoxin

PROBLEM 25.25

Show the structure of D-digitoxose. What unusual features are present in this monosaccharide?

Many proteins have short carbohydrate chains bonded to them. These **glycoproteins** have their carbohydrate groups linked to the protein chain by a glycosidic bond involving the nitrogen of the amide group of an asparagine amino acid residue or the oxygen of the hydroxy group of a serine or a threonine amino acid residue. The sugar chains are usually branched, so a wide variety of carbohydrate groups are possible. In most cases the sugar groups do not seem to have much effect on the biochemical properties of the protein. Instead, they seem to act as markers so that the proteins can be recognized for a variety of biological processes. An example of an N-acetyl-D-galactosamine attached to a protein through the hydroxy group of a serine amino acid residue is shown in the following structure:

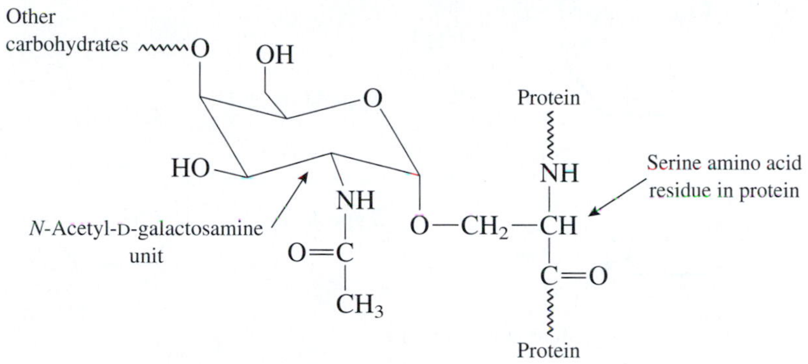

PROBLEM 25.26

Carbohydrates can also attach to proteins through an amino group. A general example is shown in the following equation. Show the steps in the mechanism for this reaction.

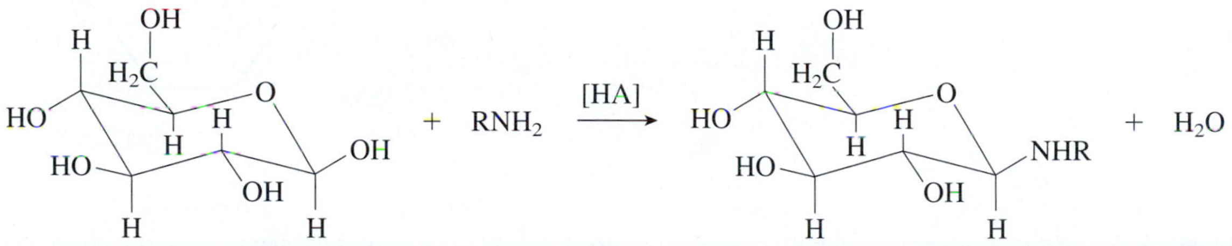

Focus On Biological Chemistry

Blood Groups

Lipids and proteins that are part of cell membranes often are bonded to small polysaccharide groups that project from the surface of the cell. Slight differences in the structures of polysaccharides that are bonded to lipids in red blood cell membranes of humans are responsible for the A, B, and O blood types. Antibodies recognize these groups and cause cells to clump when their surface groups are not the same as those on the cells of the original individual.

Continued

The structures of the ends of the polysaccharide chains of the blood group determinants are shown in the following diagram. For all of the blood types, the end of the polysaccharide chain has the hydroxy group on carbon 4 of an *N*-acetyl-D-glucosamine connected via a glycosidic bond to a D-galactose. The hydroxy group on carbon 2 of the D-galactose is connected to an unusual sugar, L-fucose (L-6-deoxygalactose), by a glycosidic bond. The only difference in the blood groups is the nature of the group, shown as R in the diagram, that is attached to the oxygen on carbon 3 of the D-galactose ring. For Type O, R is a hydrogen; for Type A, R is an *N*-acetyl-D-galactosamine; and for Type B, R is a D-galactose.

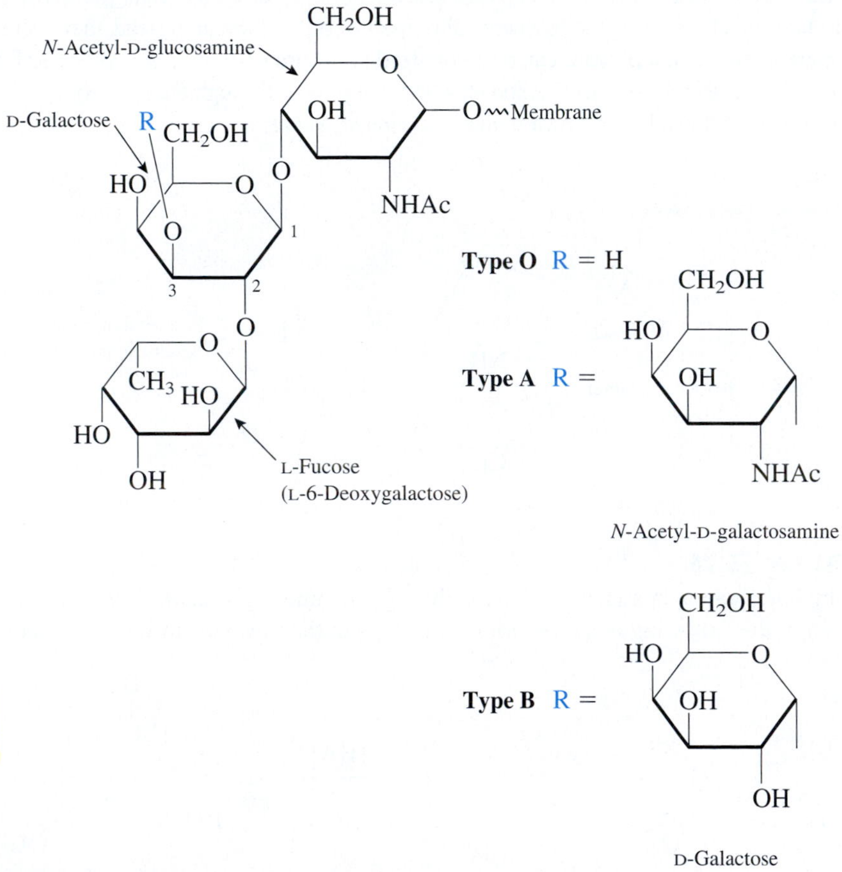

Type O R = H

Type A R =

N-Acetyl-D-galactosamine

Type B R =

D-Galactose

ORGANIC
Chemistry Now™
Click *Mastery Goal Quiz* to test how well you have met these goals.

Review of Mastery Goals

After completing this chapter, you should be able to:

■ Show the general structures for carbohydrates including the variations that occur.

■ Discuss the stereochemistry of carbohydrates, including the use of D or L to designate absolute stereochemistry. (Problem 25.33)

■ Understand the cyclization of monosaccharides to form pyranose and furanose rings. (Problems 25.29, 25.30, 25.31, 25.36, and 25.44)

■ Show the products of the common reactions of monosaccharides that were presented in this chapter: oxidation with nitric acid, oxidation with bromine, reduction with sodium borohydride, esterification, glycoside formation, and the Kiliani-Fischer synthesis. (Problems 25.27 and 25.34)

■ Understand Fischer's structure proof for glucose and apply this type of reasoning to other stereochemical problems. (Problems 25.35, 25.40, 25.42, 25.43, and 25.50)

■ Understand the general structural features of disaccharides and polysaccharides. (Problem 25.47)

Additional Problems

ORGANIC
Chemistry ✦ Now™
Assess your understanding of this chapter's topics with additional quizzing and conceptual-based problems at
http://now.brookscole.com/hornback2

25.27 Show the products of these reactions:

a) 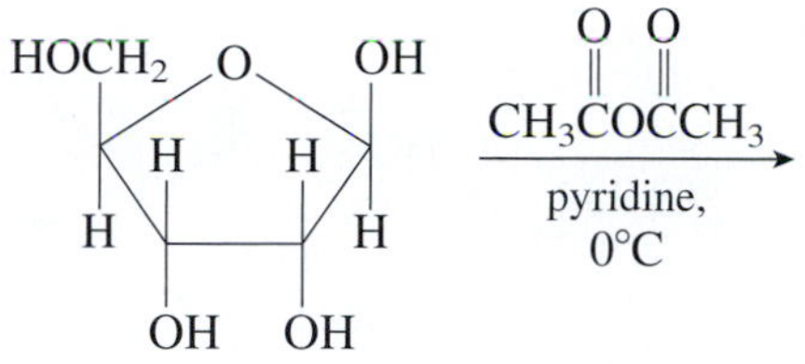 $\xrightarrow[\text{pyridine, } 0°C]{CH_3COCCH_3}$

b) 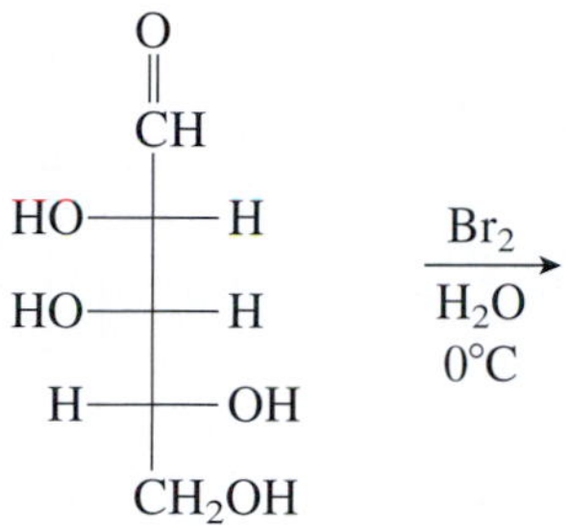 $\xrightarrow[\text{H}_2\text{O} \\ 0°C]{Br_2}$

c) $\xrightarrow[\text{H}_2\text{O}]{NaBH_4}$

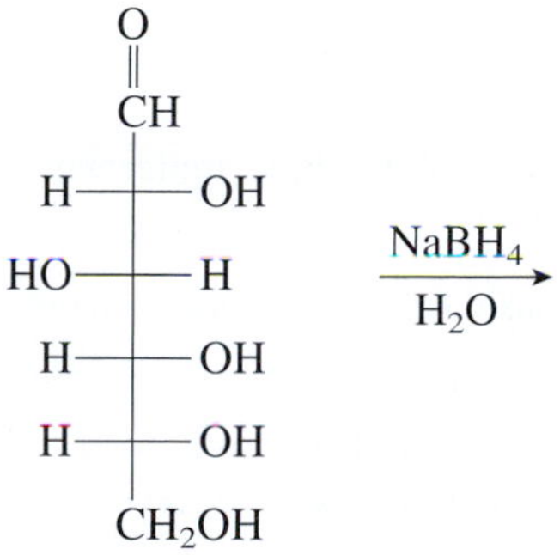

d) 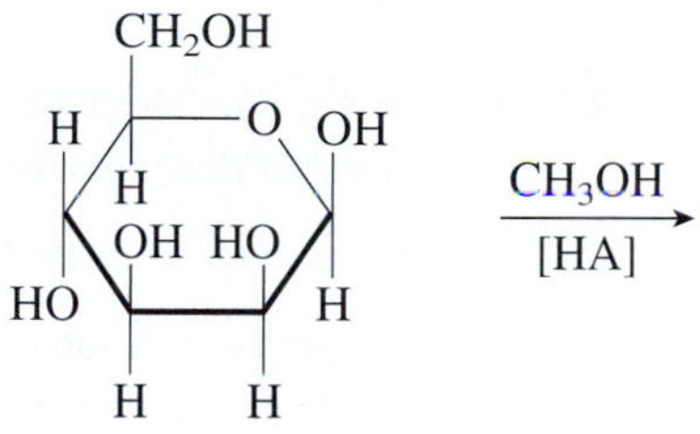 $\xrightarrow[\text{[HA]}]{CH_3OH}$

e) 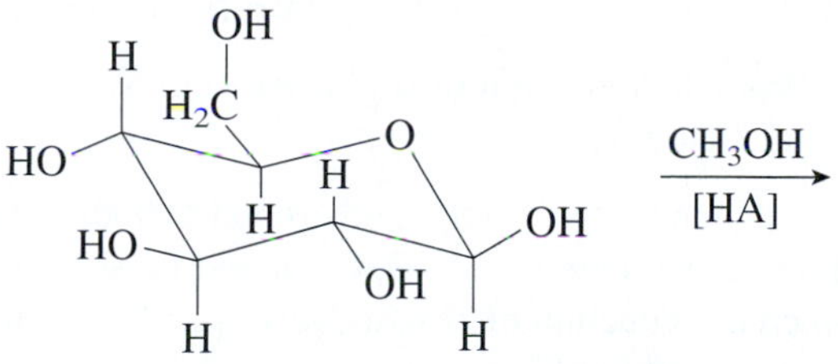 $\xrightarrow[\text{[HA]}]{CH_3OH}$

f) 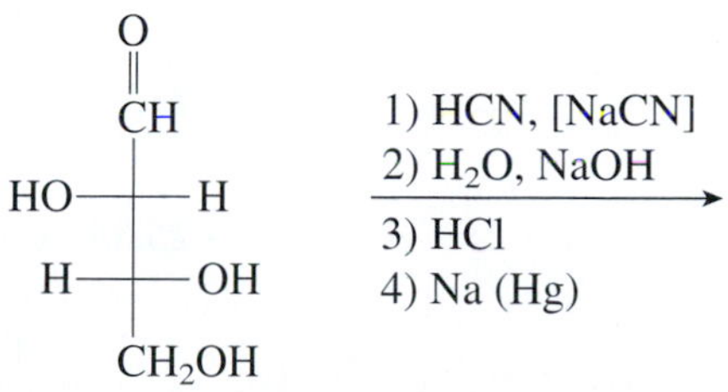 $\xrightarrow{\begin{array}{l}1)\ HCN,\ [NaCN]\\ 2)\ H_2O,\ NaOH\\ 3)\ HCl\\ 4)\ Na\ (Hg)\end{array}}$

g) 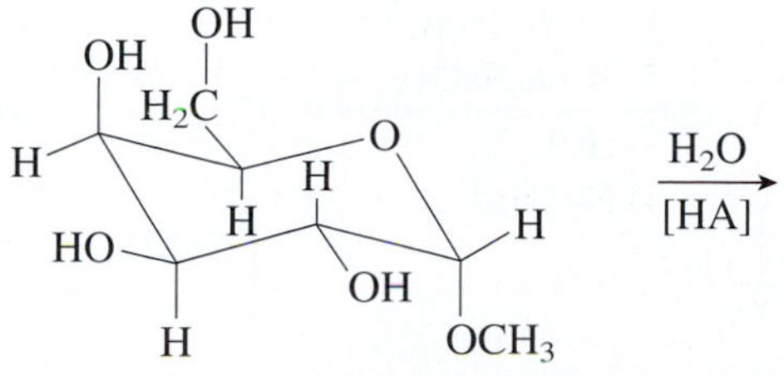 $\xrightarrow[\text{[HA]}]{H_2O}$

25.28 Show a mechanism for this reaction:

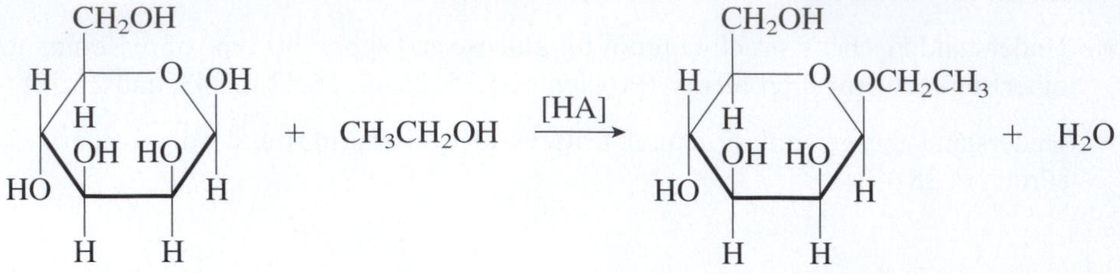

β-Mannose + CH_3CH_2OH $\xrightarrow{[HA]}$ + H_2O

25.29 Show a mechanism for this reaction:

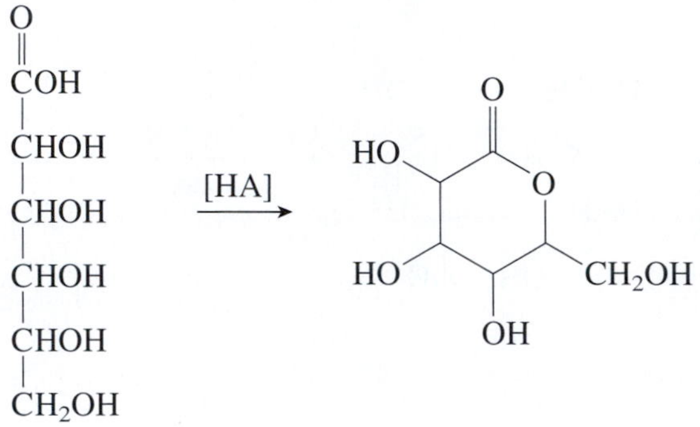

25.30 Explain why methyl-α-D-glucoside does not exhibit mutarotation in basic solution but does in acidic solution.

25.31 Lactose undergoes mutarotation in basic solution but sucrose does not. Explain.

25.32 The specific rotation of α-D-galactopyranose is $+150.7$, and that of the β-anomer is $+52.8$. The rotation of an equilibrium mixture of these two anomers is $+80.2$. Calculate the percentage of each in the equilibrium mixture.

25.33 Draw the chair conformation of β-D-allopyranose. Explain whether you expect this compound to be more or less stable than β-D-glucopyranose.

25.34 How is the product from the reduction of D-glucose with $NaBH_4$ related to that from reduction of D-gulose?

25.35 Carbohydrate **A** is a D-aldotetrose. Compounds **B** and **E** both rotate plane-polarized light. Show the structures of **A, B, C, D,** and **E**. Show the structure of the product formed on reduction of **D** with $NaBH_4$ and explain whether or not it rotates plane-polarized light.

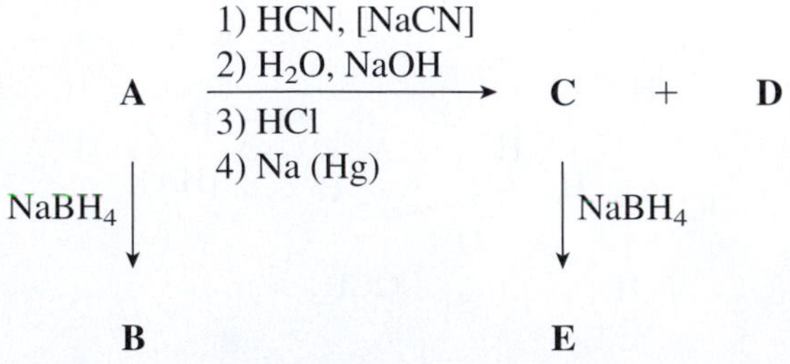

25.36 Show a mechanism for the conversion of D-fructose to a furanose. Do not worry about the stereochemistry of the reaction.

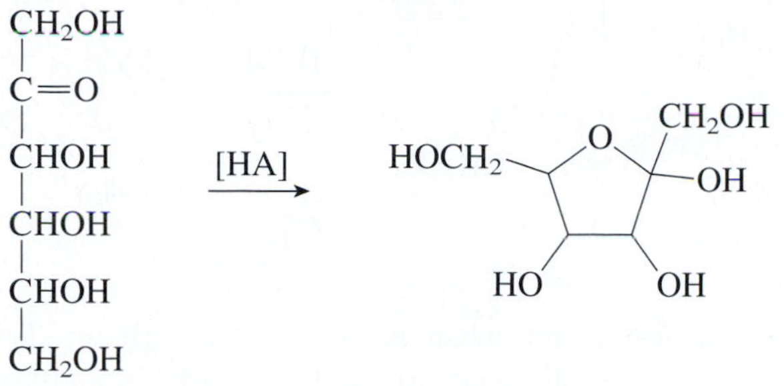

25.37 The interconversion of α-D-glucopyranose and β-D-glucopyranose can occur in aqueous solution without passing through the open aldehyde form. Show a mechanism for this process. (Use acid catalysis.)

25.38 Assign the configuration of each of the stereocenters of these compounds as *R* or *S*.
   a) D-Galactose   **b)** D-Ribose   **c)** L-Xylose

25.39 Show a mechanism for the interconversion of D-allose and D-altrose in aqueous base.

25.40 Explain how the Kiliani-Fischer synthesis could be used to demonstrate that naturally occurring glucose has the D-configuration, that is, that it has the same relative configuration at C-5 as does D-glyceraldehyde at C-2.

25.41 What aldohexose could be formed from D-galactose on treatment with base? Show a mechanism for the formation of this aldohexose.

25.42 A D-aldopentose, **X,** gives a product that rotates plane-polarized light on reaction with HNO_3. Compound **X** can be prepared from aldotetrose **Y** by Kiliani-Fischer synthesis. Reaction of **Y** with HNO_3 gives a product that rotates plane-polarized light. Show the structures of **X** and **Y**.

25.43 Compound **A** is a D-aldopentose. Compound **E** rotates plane-polarized light, whereas compounds **B** and **F** do not. Show the structures of **A, B, C, D, E,** and **F.**

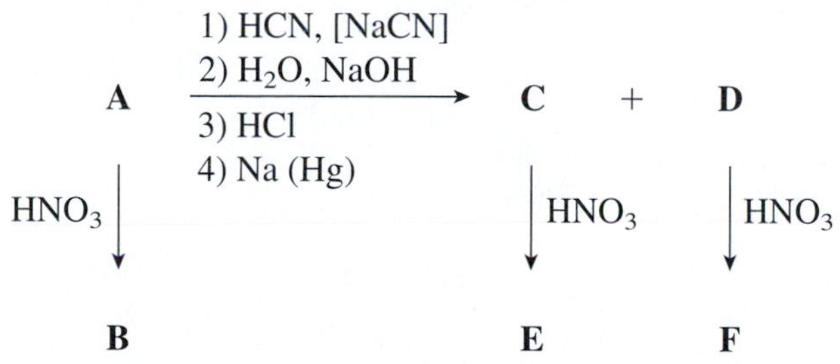

25.44 Show a mechanism for this reaction:

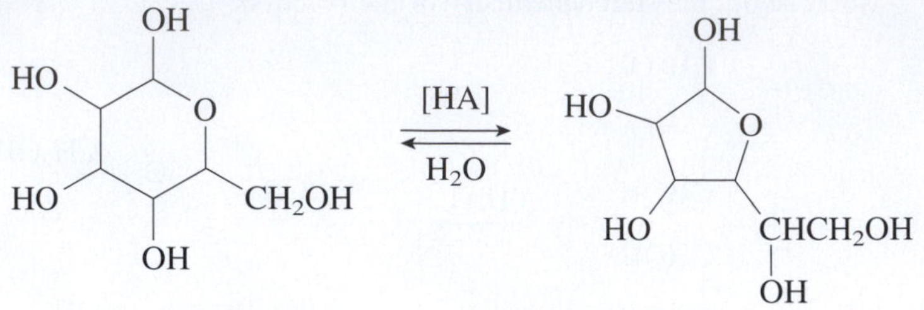

$$\xrightleftharpoons[H_2O]{[HA]}$$

25.45 Dihydroxyacetone phosphate is converted to D-glyceraldehyde 3-phosphate by the enzyme triosephosphate isomerase as part of the glycolytic pathway of metabolism. Show how this interconversion could occur by a base-catalyzed mechanism:

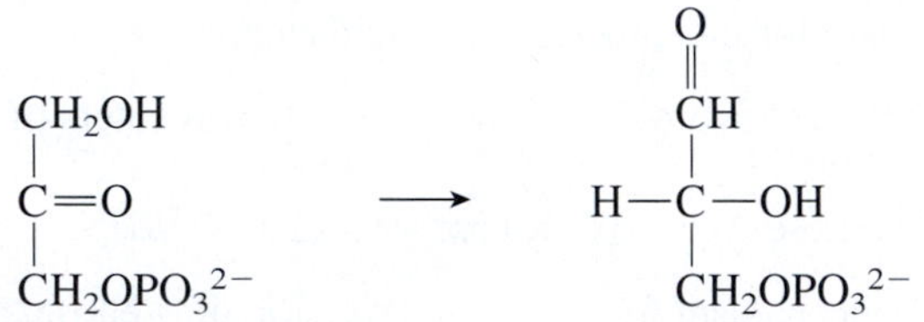

Dihydroxyacetone phosphate

D-Glyceraldehyde 3-phosphate

25.46 Laetrile, found in the seeds of apricots and bitter almonds, has considerable toxicity because it releases hydrogen cyanide on hydrolysis. It has been purported to be useful in the treatment of cancer, but controlled studies have shown no evidence of effectiveness. Show how hydrogen cyanide is produced on hydrolysis of laetrile.

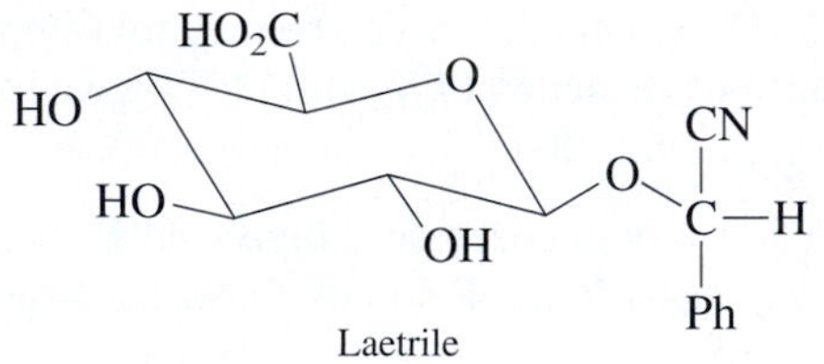

Laetrile

25.47 Trehalose is a disaccharide that is used by insects and some fungi to store energy. What monosaccharides are used to form trehalose? How are these monosaccharides connected? Is trehalose a reducing sugar?

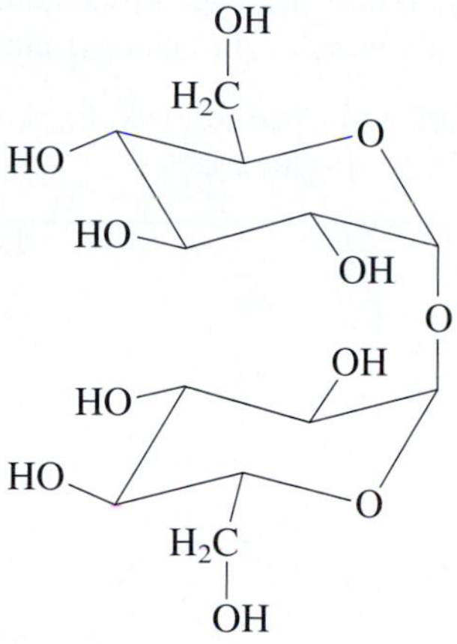

Trehalose

25.48 D-Glyceraldehyde, dihydroxyacetone, or a mixture of these two isomers reacts in the presence of sodium hydroxide to form fructose, along with other products. Show a mechanism for this reaction.

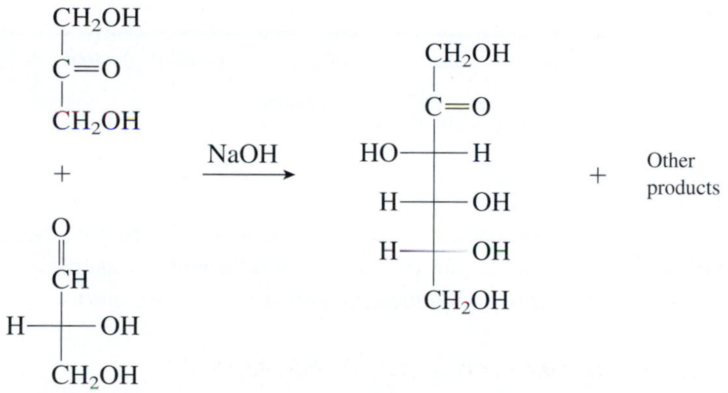

25.49 An example of the Amadori rearrangement is shown in the following equation. Suggest a mechanism for this reaction.

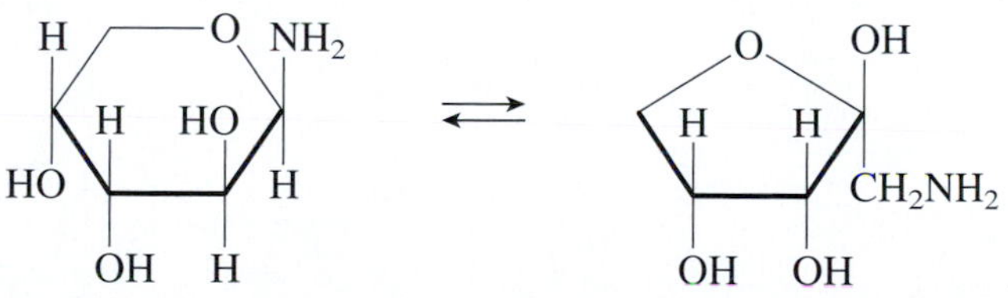

25.50 Two of the D-aldohexoses give the same product upon reduction with NaBH$_4$. Show the structures of these aldohexoses and the reduced product formed from them.

Problems Involving Spectroscopy

25.51 Explain why the NMR method described in the Focus On box "Determination of Anomer Configuration" on page 1095 cannot be used to determine the configurations of the anomers of D-mannopyranose.

25.52 The IR spectrum of D-glucose follows. Explain how this spectrum supports the cyclic structure for D-glucose.

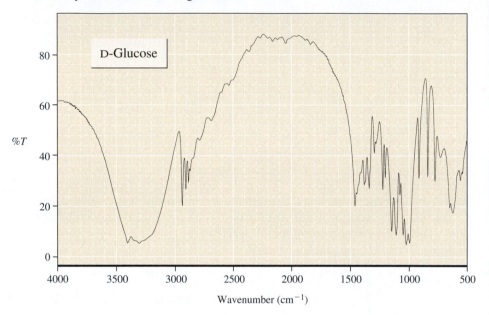

D-Glucose

%T

Wavenumber (cm^{-1})

Amino Acids, Peptides, and Proteins

CHAPTER 26

AMINO ACIDS AND THE polymers that are formed from them, peptides and proteins, are among the most important compounds found in living organisms. They act as structural components, catalysts, hormones, neurotransmitters, and essential nutrients, among other things. This chapter emphasizes the organic chemistry of these important biomolecules. Details of their synthesis and function in living organisms are left for biochemistry textbooks and are only briefly described here.

First the general structure and chemistry of the amino acids is presented. Then several methods that can be used to prepare them in the laboratory are discussed. After an introduction to the structure of peptides and proteins, chemical methods that can be used to determine the amino acid sequence in proteins are presented. Next, the synthesis of peptides in the laboratory is introduced. Finally, the three-dimensional structure of proteins and the mechanism of action of enzymes are briefly addressed.

26.1 AMINO ACIDS

As the name implies, amino acids are carboxylic acids that contain amino groups. The amino acids that are of most importance in nature have the amino group on carbon 2 of the carboxylic acid (the α-carbon), as shown in the following general structure, and are therefore sometimes called α-amino acids:

$$R-\underset{\underset{\displaystyle \alpha\text{-Carbon}}{|}}{\overset{\overset{\displaystyle NH_2}{|}}{CH}}-CO_2H$$

An α-amino acid

Twenty "standard" amino acids commonly occur in nature. They differ in the structure of the side chain that is attached to the α-carbon (the R group in the previous structure). Table 26.1 shows the structures of the standard amino acids at pH 7, along with their names and abbreviations.

Table 26.1   Naturally Occurring Amino Acids

Structure	Name	Abbreviation	pK$_a$ of α-CO$_2$H	pK$_a$ of α-NH$_3{}^+$	pK$_a$ of R Group	pI
Nonpolar Side Chains						
H—CH—CO$_2{}^-$ (⁺NH₃) — Glycine	Glycine	Gly or G	2.3	9.8		6.1
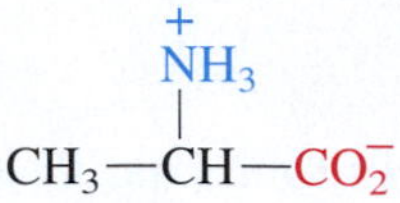 CH$_3$—CH—CO$_2{}^-$ (⁺NH₃)	Alanine	Ala or A	2.3	9.9		6.1
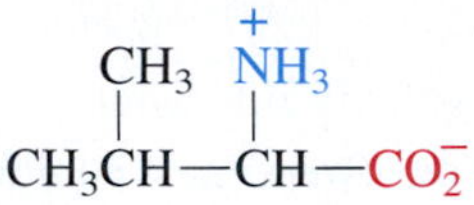 CH$_3$CH—CH—CO$_2{}^-$ (CH₃, ⁺NH₃)	Valine	Val or V	2.3	9.7		6.0
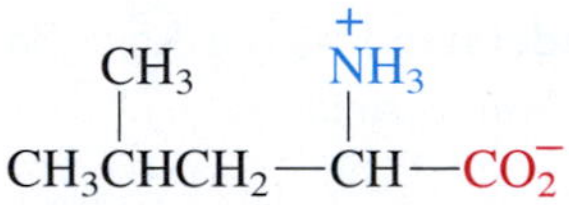 CH$_3$CHCH$_2$—CH—CO$_2{}^-$ (CH₃, ⁺NH₃)	Leucine	Leu or L	2.3	9.7		6.0
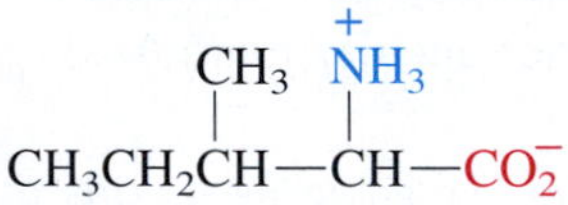 CH$_3$CH$_2$CH—CH—CO$_2{}^-$ (CH₃, ⁺NH₃)	Isoleucine	Ile or I	2.3	9.7		6.0
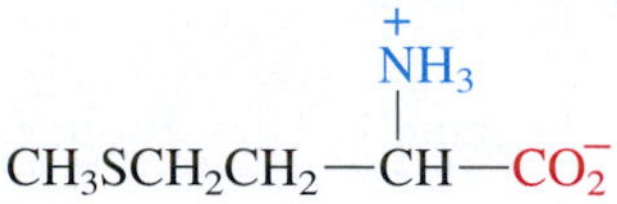 CH$_3$SCH$_2$CH$_2$—CH—CO$_2{}^-$ (⁺NH₃)	Methionine	Met or M	2.1	9.3		5.7
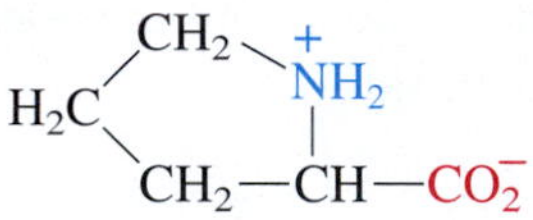 CH$_2$—CH—CO$_2{}^-$ (proline ring, ⁺NH₂)	Proline	Pro or P	2.0	10.6		6.3
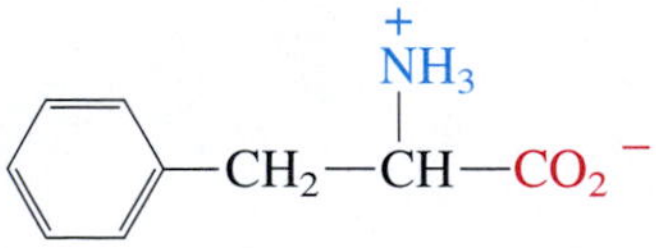 C$_6$H$_5$—CH$_2$—CH—CO$_2{}^-$ (⁺NH₃)	Phenylalanine	Phe or F	2.2	9.3		5.7
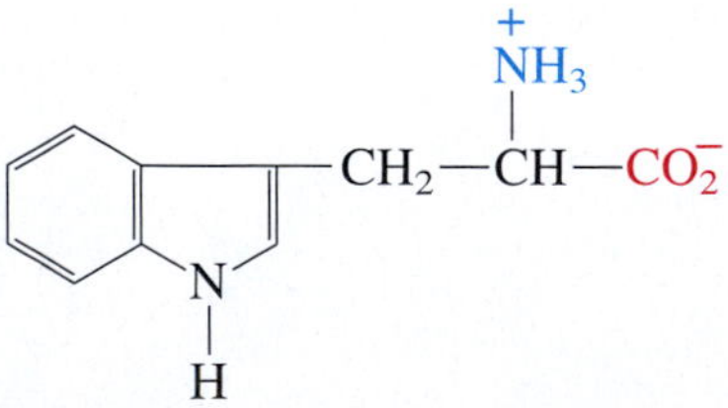 indole—CH$_2$—CH—CO$_2{}^-$ (⁺NH₃)	Tryptophan	Trp or W	2.5	9.4		5.9
Polar Side Chains						
HOCH$_2$—CH—CO$_2{}^-$ (⁺NH₃)	Serine	Ser or S	2.2	9.2		5.7

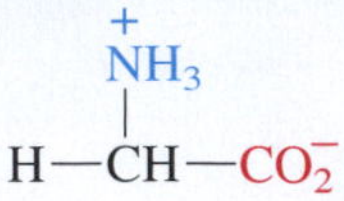

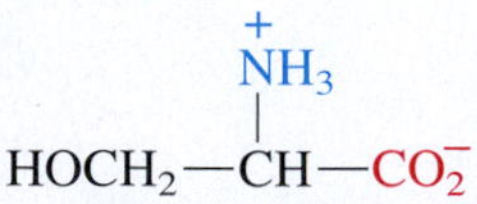

Table 26.1 Naturally Occurring Amino Acids—cont'd

Structure	Name	Abbreviation	pK_a of $\alpha\text{-}CO_2H$	pK_a of $\alpha\text{-}NH_3^+$	pK_a of R Group	pI
Polar Side Chains—cont'd HOCH—CH—CO_2^- with CH$_3$ and $\overset{+}{N}H_3$	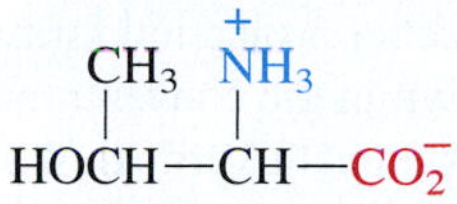 Threonine	Thr or T	2.1	9.1		5.6
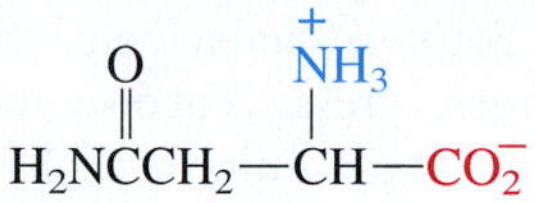 H$_2$NCCH$_2$—CH—CO_2^- with O and $\overset{+}{N}H_3$	Asparagine	Asn or N	2.1	8.7		5.4
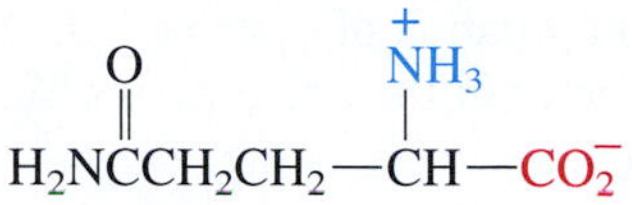 H$_2$NCCH$_2$CH$_2$—CH—CO_2^- with O and $\overset{+}{N}H_3$	Glutamine	Gln or Q	2.2	9.1		5.7
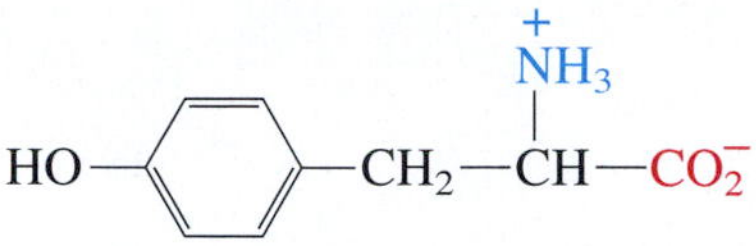 HO—⟨ring⟩—CH$_2$—CH—CO_2^- with $\overset{+}{N}H_3$	Tyrosine	Tyr or Y	2.2	9.2	10.5	5.7
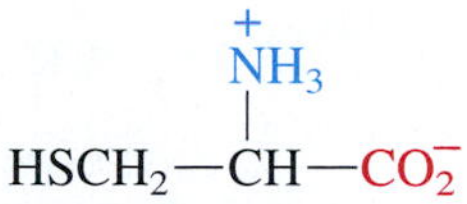 HSCH$_2$—CH—CO_2^- with $\overset{+}{N}H_3$	Cysteine	Cys or C	1.9	10.7	8.4	5.2
Acidic Side Chains $^-$OCCH$_2$—CH—CO_2^- with O and $\overset{+}{N}H_3$	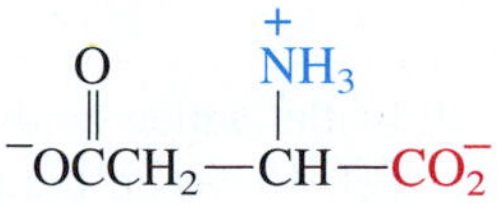 Aspartic acid	Asp or D	2.0	9.9	3.9	3.0
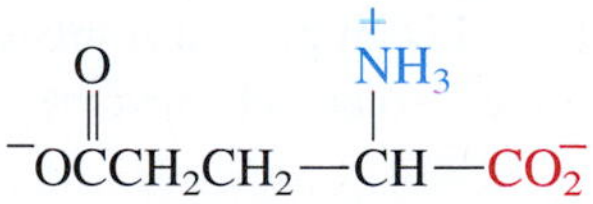 $^-$OCCH$_2$CH$_2$—CH—CO_2^- with O and $\overset{+}{N}H_3$	Glutamic acid	Glu or E	2.1	9.5	4.1	3.1
Basic Side Chains $\overset{+}{H_3N}$CH$_2$CH$_2$CH$_2$CH$_2$—CH—CO_2^- with $\overset{+}{N}H_3$	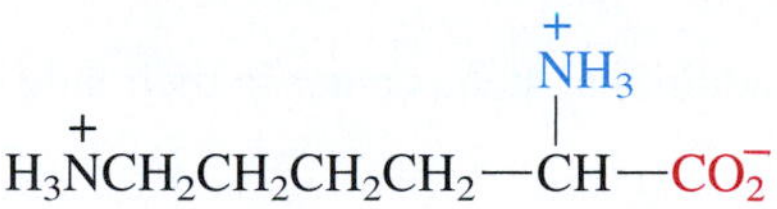 Lysine	Lys or K	2.2	9.1	10.5	9.8
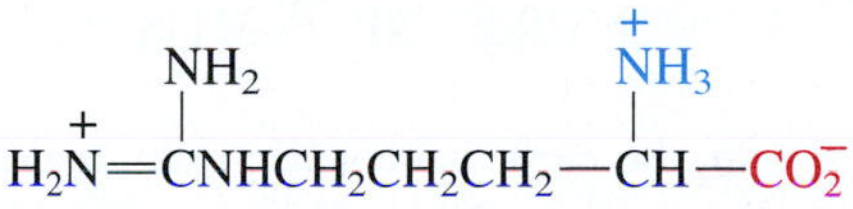 $\overset{+}{H_2N}$=CNHCH$_2$CH$_2$CH$_2$—CH—CO_2^- with NH$_2$ and $\overset{+}{N}H_3$	Arginine	Arg or R	1.8	9.0	12.5	10.8
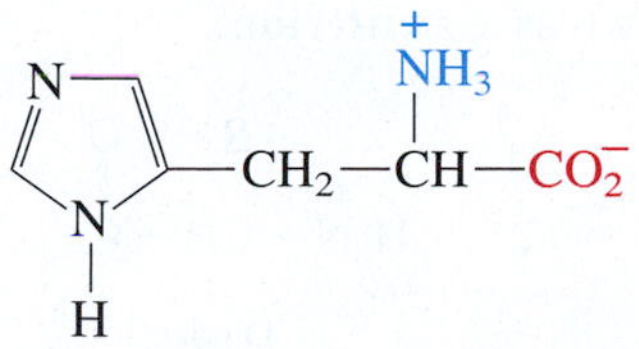 ⟨imidazole ring⟩—CH$_2$—CH—CO_2^- with $\overset{+}{N}H_3$	Histidine	His or H	1.8	9.3	6.0	7.6

In this table the compounds have been grouped according to the polarity of their side chains (nonpolar, polar, acidic, and basic).

All of the amino acids, except the simplest one, glycine (R = H), have four different groups attached to the α-carbon and are, therefore, chiral. In general, only a single enantiomer is found in nature. When the Fischer projection formula for the naturally occurring enantiomer of an amino acid is drawn in the conventional manner, with the carboxylic acid group at the top and the carbon chain vertical, the amino group is on the left. Because D-glyceraldehyde has its hydroxy group on the right, the naturally occurring amino acids belong to the L series. For most amino acids the NH_2 group has the highest priority according to the Cahn-Ingold-Prelog sequence rules, followed in order by the CO_2H group, the R group, and the H. These amino acids have the S absolute configuration. However, in the case of cysteine the CH_2SH group has a higher priority than the CO_2H group; thus, the absolute configuration of cysteine is R. Note that several of the amino acids also contain a stereocenter in their side chains. Again, only one configuration at this carbon is usually found.

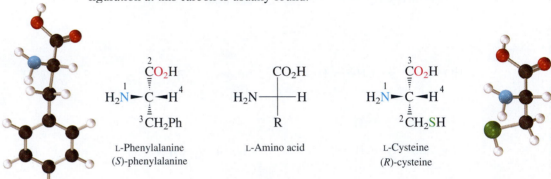

Human beings are able to synthesize 10 of the amino acids, termed *nonessential amino acids,* from other compounds in the diet. However, the other 10 amino acids, termed **essential amino acids,** cannot be synthesized by humans (or are synthesized only in small amounts) and must be obtained from protein sources in the diet. The essential amino acids are arginine, histidine, isoleucine, leucine, lysine, methionine, phenylalanine, threonine, tryptophan, and valine.

PROBLEM 26.1

Which of the amino acids have an additional stereocenter in their side chains?

26.2   ACID–BASE CHEMISTRY OF AMINO ACIDS

Amino acids have both an acidic group, the carboxylic acid group, and a basic group, the amino group. These two functional groups undergo an intramolecular acid base reaction to form a **dipolar ion,** also known as a **zwitterion:**

Dipolar ion
or
zwitterion

A carboxylic acid is a stronger acid ($pK_a \approx 5$) than the conjugate acid of an amine ($pK_a \approx 9$). Therefore, the equilibrium in the preceding reaction favors the dipolar ion, and this is the species that is present in the solid form of an amino acid. Thus, many of the physical properties of amino acids resemble those of salts rather than those of typical organic compounds. They have high melting points (in the vicinity of 300°C), are quite soluble in water, and are rather insoluble in typical organic solvents. Although amino acids exist as dipolar ions, they are often written with their amino and carboxylic acid groups in un-ionized form.

The dipolar ion has both an acidic group, the ammonium cation, and a basic group, the carboxylate anion, so it can act as either an acid or a base. Such compounds are termed **amphoteric.** The species that is present in aqueous solution depends on the pH. In the pH range near neutral the amino acid is present in the form of the dipolar ion. In acidic solution the carboxylate group becomes protonated and the amino acid is present as a cation, whereas in basic solution the ammonium group gives up a proton and the molecule exists as an anion:

$$
\overset{+}{H_3N}-\overset{\underset{|}{R}}{CH}-\overset{\overset{O}{\|}}{C}-OH
\quad\underset{H_3O^+}{\overset{}{\rightleftarrows}}\quad
\overset{+}{H_3N}-\overset{\underset{|}{R}}{CH}-\overset{\overset{O}{\|}}{C}-O^-
\quad\underset{^-OH}{\overset{}{\rightleftarrows}}\quad
H_2N-\overset{\underset{|}{R}}{CH}-\overset{\overset{O}{\|}}{C}-O^-
$$

Cation	Dipolar ion	Anion
pH < 2	pH ≅ 6	pH > 10

The concentration of an acid and its conjugate base in aqueous solution can be calculated from its K_a or pK_a. Recall from Chapter 4 that

$$K_a = \frac{[H_3O^+][A^-]}{[HA]} \quad \text{and} \quad pK_a = -\log K_a$$

Therefore,

$$pK_a = -\log \frac{[H_3O^+][A^-]}{[HA]} = -\log[H_3O^+] - \log \frac{[A^-]}{[HA]}$$

and

$$pK_a = pH - \log \frac{[A^-]}{[HA]}$$

or

$$pH = pK_a + \log \frac{[A^-]}{[HA]} \quad \text{(Henderson-Hasselbach equation)}$$

From the Henderson-Hasselbalch equation, when $pH = pK_a$, then

$$\log \frac{[A^-]}{[HA]} = 0$$

and

$$[HA] = [A^-]$$

Let's use the Henderson-Hasselbalch equation to examine the effect of pH on the concentrations of the various forms of glycine:

$$
\overset{+}{H_3N}CH_2\overset{\overset{O}{\|}}{C}OH
\quad\rightleftarrows\quad
\overset{+}{H_3N}CH_2\overset{\overset{O}{\|}}{C}O^-
\quad\rightleftarrows\quad
H_2NCH_2\overset{\overset{O}{\|}}{C}O^-
$$

Cationic form	Dipolar ion	Anionic form
$pK_{a1} = 2.3$	$pK_{a2} = 9.8$	

Species present

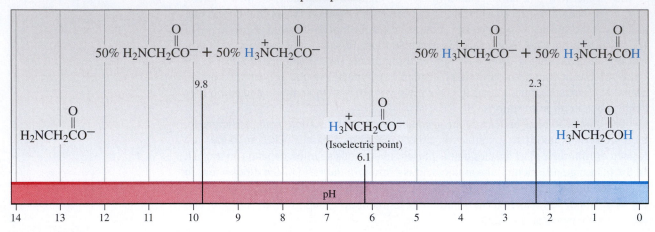

Figure 26.1

FORMS OF GLYCINE PRESENT
AT VARIOUS pHs.

At very low pH, glycine is almost entirely in the cationic form. If the pH is increased, the cationic form gives up a proton from the carboxylic acid group (the carboxylic acid is a stronger acid than the ammonium group), producing the dipolar ion. From the Henderson-Hasselbalch equation the concentration of the cationic form equals the concentration of the dipolar ion when the pH is equal to pK_{a1}—that is, at pH = 2.3. If the pH is increased further, the concentration of the dipolar ion increases until nearly all of the glycine is present in that form. Then the ammonium group begins to give up its acidic proton, producing the anionic form. The concentration of the dipolar ion equals that of the anionic form when the pH is equal to pK_{a2}—that is, at pH = 9.8. At higher pH, glycine is present predominantly in its anionic form. The concentration of the dipolar ion is at a maximum at a pH equal to the average of pK_{a1} and pK_{a2}, or pH = $\frac{1}{2}(2.3 + 9.8) = 6.1$ in the case of glycine. This pH is called the **isoelectric point** (pI) because the amino acid has an overall charge of zero; that is, it is neutral. The forms of glycine that are present at various pHs are illustrated in Figure 26.1.

PROBLEM 26.2

Explain why the carboxylic acid group of the cationic form of glycine ($pK_a = 2.3$) is a stronger acid than the carboxylic acid group of acetic acid ($pK_a = 4.8$).

PROBLEM 26.3

Explain the order of these pK_a values:

$$\overset{+}{H_3}NCH_2\overset{\overset{\displaystyle O}{\|}}{C}OCH_3 \qquad \overset{+}{H_3}NCH_2\overset{\overset{\displaystyle O}{\|}}{C}O^-$$

$$pK_a = 7.8 \qquad\qquad pK_a = 9.8$$

Two important points should be learned from this discussion. First, glycine (and other amino acids) is never present in aqueous solution in a neutral form with uncharged carboxylic acid and amino groups. It is present as a cationic form, a dipolar ion, or an anionic form, depending on the pH. Second, because the pH of most physiological solutions is near 7, which is close to the isoelectric point of glycine, it is commonly present as a dipolar ion in biological fluids.

Table 26.1 lists the pK_a's of the amino acids along with their isoelectric points (pI). Both those with nonpolar side chains and those with polar side chains have pK_a values near those of glycine. Therefore, these all have isoelectric points near 6 and exist predominantly as dipolar ions at neutral pH.

However, some amino acids have an acidic or basic functional group in their side chains. In these cases there is another acidity constant to consider. Let's examine the case of aspartic acid, which has two carboxylic acid groups and an amino group. At low pH, aspartic acid is present in a cationic form:

| Cationic form | Dipolar ion | Anionic form | Dianionic form |

As the pH is increased, the proton is removed first from the carboxylic acid group closest to the ammonium group to generate the dipolar ion, then from the other carboxylic acid group, to generate an anionic form, and finally from the ammonium group to generate a dianionic form. The pI is the average of pK_{a1} and pK_{a2} and equals 3.0. The concentration of the anionic form is a maximum at a pH of 6.9, the average of pK_{a2} and pK_{a3}. Therefore, at neutral pH, aspartic acid is present predominantly in its anionic form. The situation is quite similar for glutamic acid.

The amino acids with basic side chains—lysine, arginine, and histidine—all have pI values greater than 7. In strongly acidic solution they exist in dicationic forms, and all have significant amounts of a cationic form and a dipolar ion present at neutral pH.

PROBLEM 26.4

Explain why the carboxylic acid group in the side chain of aspartic acid ($pK_a = 3.9$) is a weaker acid than the main carboxylic acid group of the amino acid ($pK_a = 2.0$).

PROBLEM 26.5

Explain why the carboxylic acid group in the side chain of glutamic acid is a weaker acid than the carboxylic acid group in the side chain of aspartic acid.

PRACTICE PROBLEM 26.1

Show the structure of the dianion form of tyrosine.

Solution

The pK_a's of the various groups of tyrosine are listed in Table 26.1. Starting from the cationic form that is present in strong acid, first a proton is removed from the carboxylic acid group ($pK_a = 2.2$) to form the dipolar ion. The next proton is removed from the α-NH_3^+ ($pK_a = 9.2$) to give the anionic form. The final proton is removed from the OH group of the side chain ($pK_a = 10.5$) to give the dianionic form:

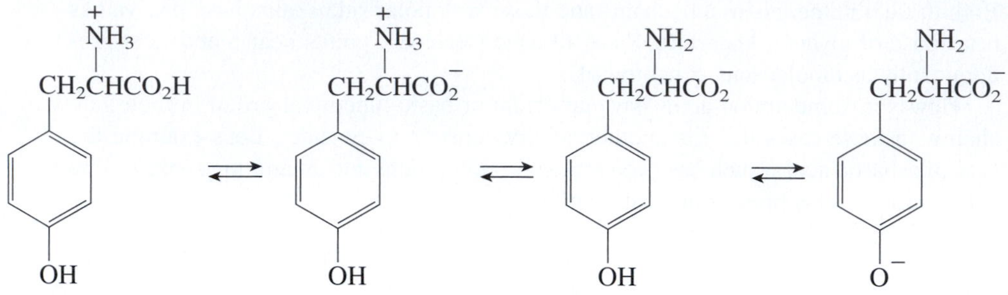

| Cationic form | Dipolar ion | Anionic form | Dianionic form |

PROBLEM 26.6

Show the structures of these species:
a) Dipolar ion of proline
b) Anion form of cysteine (careful)
c) Cation form of arginine

PROBLEM 26.7

a) Show the four differently charged forms of lysine.
b) Construct a diagram like that of Figure 26.1 for lysine.

26.3 CHEMICAL REACTIONS OF AMINO ACIDS

Amino acids exhibit chemical reactions that are typical of both amines and carboxylic acids. For example, the acid can be converted to an ester by the Fischer method. This reaction requires the use of an excess of acid because one equivalent is needed to react with the amino group of the product. As another example, the amine can be converted to an amide by reaction with acetic anhydride. Additional examples are provided by the reactions that are used in the preparation of peptides from amino acids described in Section 26.7:

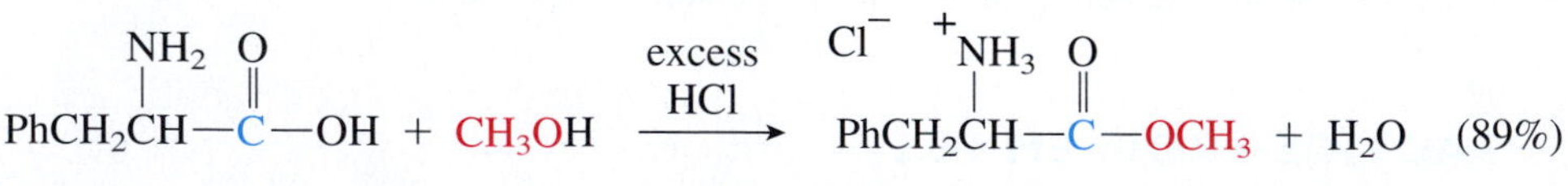

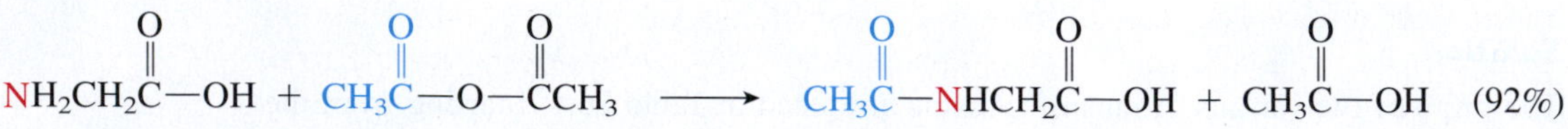

PROBLEM 26.8

Show the products of these reactions:

a) NH₂CH—COH + PhCH₂OH $\xrightarrow{\text{excess HCl}}$

with CH₃ and O groups

b) NH₂CH—COH + PhCCl $\longrightarrow$

with CH₃CHCH₃, CH₂ and O, and O groups

c) NH₂CH—COH + CH₃COCCH₃ $\longrightarrow$

with CH₃ and O, and O O groups

26.4 LABORATORY SYNTHESIS OF AMINO ACIDS

Because of their importance, a number of laboratory methods for the synthesis of amino acids have been developed. In the **Strecker synthesis** an aldehyde is treated with NaCN and NH₄Cl to form an aminonitrile, which is then hydrolyzed to the amino acid:

$$CH_3CH \xrightarrow[\text{NH}_4\text{Cl, H}_2\text{O}]{\text{NaCN}} CH_3CH-C\equiv N \xrightarrow[\text{H}_2\text{O}]{\text{HCl}} CH_3CH-CO_2H \quad (70\%)$$

with O group on first, NH₂ on aminonitrile, NH₂ on Alanine

An aminonitrile

Alanine

$$NH_4Cl \downarrow \qquad \left[CH_3CH \overset{NH}{\underset{}{\parallel}} \right] \xrightarrow{\text{NaCN} \atop \text{H}_2\text{O}}$$

This reaction proceeds by initial reaction of ammonium chloride with the aldehyde to form an imine (see Section 18.8). Then cyanide adds to the imine in a reaction that is exactly analogous to the addition of cyanide to an aldehyde to form a cyanohydrin (see Section 18.4). The final step in the Strecker synthesis is the hydrolysis of the nitrile to a carboxylic acid (see Section 19.5).

In another method the amino group is introduced onto the α-carbon of a carboxylic acid. To accomplish this, an H on the α-carbon is first replaced with a Br, which can then act as a leaving group. The Br is replaced with an amino group by an S_N2 reaction with NH₃ as the nucleophile:

$$CH_3CH_2-COH \xrightarrow[\text{[PBr}_3]}{\text{Br}_2} CH_3CH-COH \xrightarrow[\text{NH}_3]{\text{excess}} CH_3CH-COH \quad (70\%)$$

with O on first, Br and O on second, NH₂ and O on third

The bromine is introduced onto the α-carbon by treating the carboxylic acid with Br₂ and a catalytic amount of PBr₃ in a process known as the **Hell-Volhard-Zelinsky reaction.** This reaction proceeds through an enol intermediate. Because carboxylic acids form enols only with difficulty, a catalytic amount of PBr₃ is added to form a small amount of the acyl bromide, which enolizes more readily than the acid. Addition of bromine to the enol produces an α-bromoacyl bromide (see Section 20.2). This reacts with a molecule of the carboxylic acid in a process that exchanges the Br and OH groups to form the product and another molecule of acyl bromide, which can go through the same cycle. Part of the mechanism for this process is outlined in Figure 26.2.

PROBLEM 26.9

Show all of the steps in the mechanism for the acid-catalyzed enolization of the acyl bromide in the Hell-Volhard-Zelinsky reaction:

$$CH_3CH_2CBr \overset{O}{\parallel} \underset{}{\overset{[HBr]}{\rightleftarrows}} CH_3CH{=}CBr \overset{OH}{\mid}$$

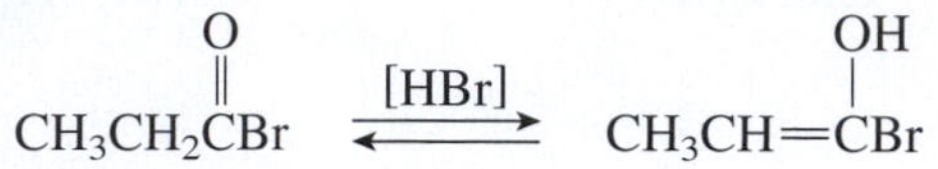

❶ The carboxylic acid reacts with PBr₃ to form an acyl bromide. The mechanism for this part of the reaction is very similar to that described for the formation of an acyl chloride in Section 19.2.

Next the acyl bromide is brominated at the α-carbon. This acid catalyzed halogenation, described in Section 20.2, proceeds through an enol.

❷ The acyl bromide is quite reactive and forms an enol more readily than the carboxylic acid.

❸ As can be seen by the resonance structure, the double bond of the enol is electron rich, so bromine adds rapidly by an electrophilic mechanism.

❹ Loss of a proton produces the α-brominated acyl bromide.

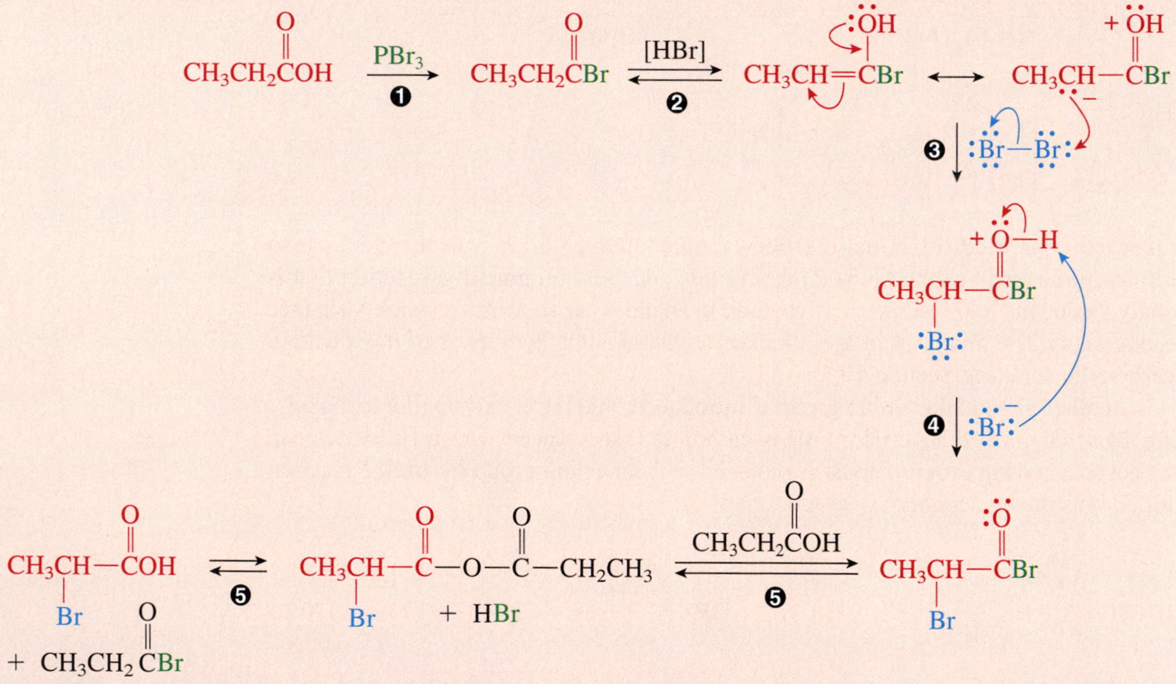

The acyl bromide can reenter the mechanism at step 2 and undergo enolization and bromination. Therefore, only a catalytic amount of PBr₃ is necessary.

❺ The Br of the acyl bromide exchanges with the OH of another molecule of the carboxylic acid to produce the α-brominated acid and an acyl bromide. This exchange reaction proceeds through an anhydride intermediate.

Figure 26.2

PARTIAL MECHANISM OF THE HELL-VOLHARD-ZELINSKY REACTION. Not all of the steps in the mechanism are shown.

A third amino acid synthesis begins with diethyl α-bromomalonate. First the Br is replaced by a protected amino group using the Gabriel synthesis (see Section 10.6). Then the side chain of the amino acid is added by an alkylation reaction that resembles the malonic ester synthesis (see Section 20.4). Hydrolysis of the ester and amide bonds followed by decarboxylation of the diacid produces the amino acid. An example that shows the use of this method to prepare aspartic acid is shown in the following sequence:

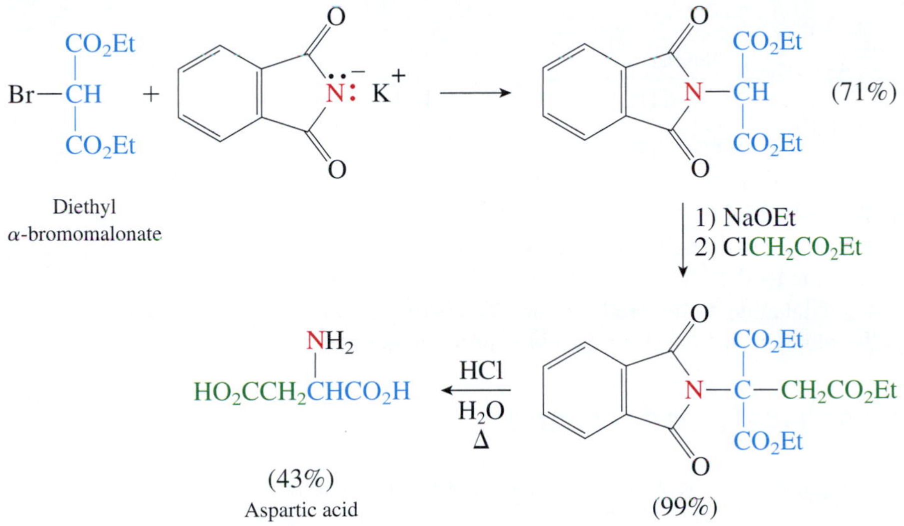

Diethyl α-bromomalonate

(71%)

1) NaOEt
2) ClCH₂CO₂Et

Aspartic acid (43%)

(99%)

A drawback of all of these methods is that they produce racemic amino acids. If the product is to be used in place of a natural amino acid, it must first be resolved. This can be accomplished by the traditional method of preparing and separating diastereomeric salts. Alternatively, nature's help can be enlisted through the use of enzymes. In one method the racemic amino acid is converted to its amide by reaction with acetic anhydride. The racemic amide is then treated with a deacylase enzyme. This enzyme catalyzes the hydrolysis of the amide back to the amino acid. However, the enzyme reacts only with the amide of the naturally occurring L-amino acid. The L-amino acid is easily separated from the unhydrolyzed D-amide. The following equation illustrates the use of this process to resolve methionine:

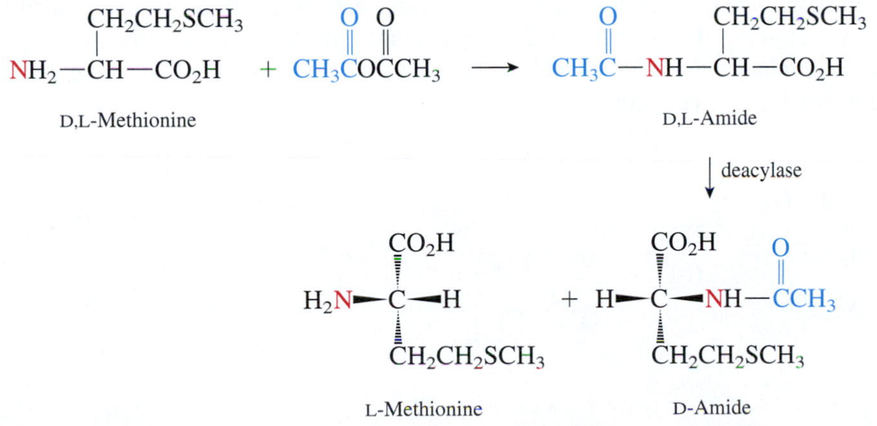

D,L-Methionine

D,L-Amide

deacylase

L-Methionine

D-Amide

PROBLEM 26.10

Show the products of these reactions:

a) 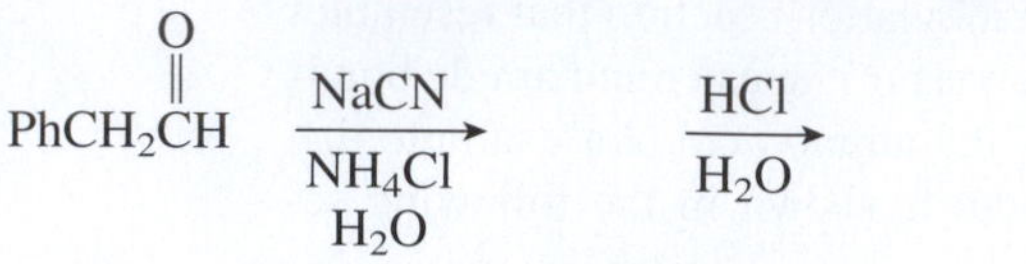 $PhCH_2\overset{O}{\underset{\|}{C}}H$ $\xrightarrow[\substack{NH_4Cl \\ H_2O}]{NaCN}$ $\xrightarrow[H_2O]{HCl}$

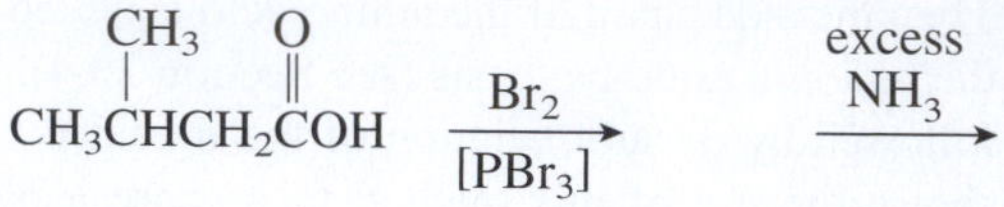

b) $CH_3\overset{CH_3}{\underset{|}{C}}HCH_2\overset{O}{\underset{\|}{C}}OH$ $\xrightarrow[[PBr_3]]{Br_2}$ $\xrightarrow{\substack{excess \\ NH_3}}$

c) Phthalimide $N-\overset{CO_2Et}{\underset{CO_2Et}{\overset{|}{C}H}}$ $\xrightarrow[\substack{2)\ CH_3CHCH_2Br}]{1)\ NaOEt}$ $\xrightarrow[\substack{H_2O \\ \Delta}]{HCl}$

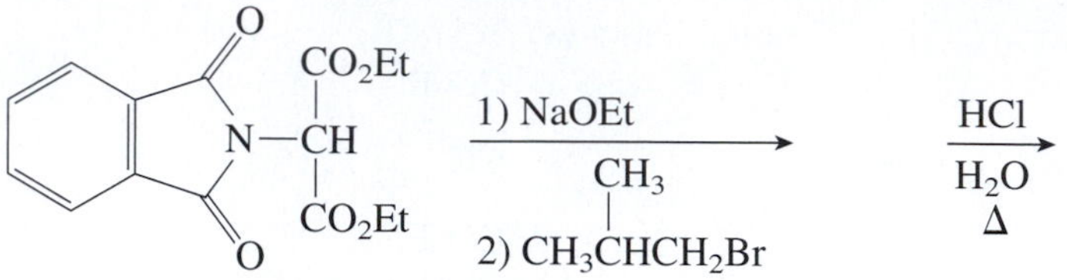

PROBLEM 26.11

Show syntheses of these amino acids:
a) Leucine by the Strecker synthesis
b) Phenylalanine by the Hell-Volhard-Zelinsky reaction
c) Tryptophan starting from diethyl α-bromomalonate

PRACTICE PROBLEM 26.2

Show representative steps in the mechanism for this reaction:

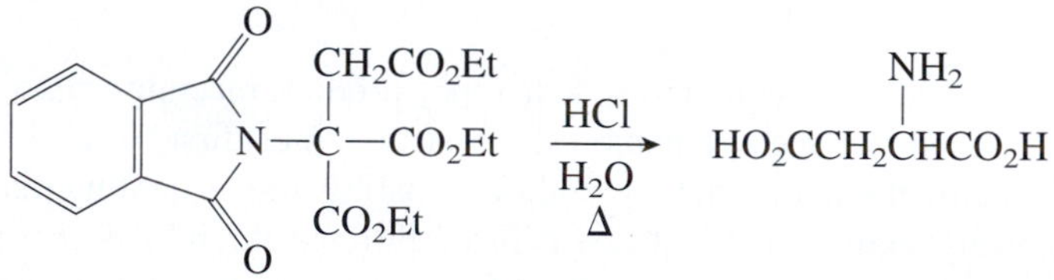

Phthalimide $N-\overset{CH_2CO_2Et}{\underset{CO_2Et}{\overset{|}{C}-CO_2Et}}$ $\xrightarrow[H_2O]{HCl}$ $HO_2CCH_2\overset{NH_2}{\underset{|}{C}HCO_2H}$

Solution

First all three ester bonds and both amide bonds are hydrolyzed to carboxylic acid groups by the aqueous acid. The mechanisms for these reactions are discussed in Section 19.5. The ester hydrolyses follow the exact reverse of the Fischer esterification mechanism shown in Figure 19.3, and the amide hydrolysis occurs by a very similar mechanism. The product of these hydrolysis steps has three carboxylic acid groups and one amino group. Two of these acid groups are attached to the same carbon so that one can be eliminated as carbon dioxide by the cyclic mechanism described in Section 20.4 for the malonic ester synthesis:

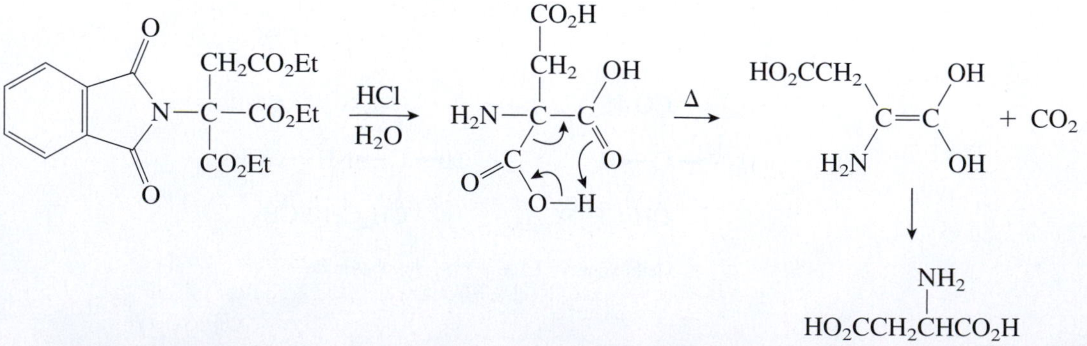

Focus On

Asymmetric Synthesis of Amino Acids

Even if the resolution of an amino acid is relatively easy, the synthesis of a racemic mixture when only one enantiomer is desired is wasteful, because half of the product cannot be used. Recently, considerable effort has been devoted to the development of methods that produce only the desired enantiomer by so-called asymmetric synthesis. As was discussed in Chapter 7, one enantiomer of a chiral product can be produced only in the presence of one enantiomer of another chiral compound. In some asymmetric syntheses a chiral reagent is employed. In others a compound called a chiral auxiliary is attached to the achiral starting material and used to induce the desired stereochemistry into the product. The chiral auxiliary is then removed and recycled.

An example of the use of a chiral reagent to accomplish an asymmetric synthesis of an amino acid is provided in the following equation:

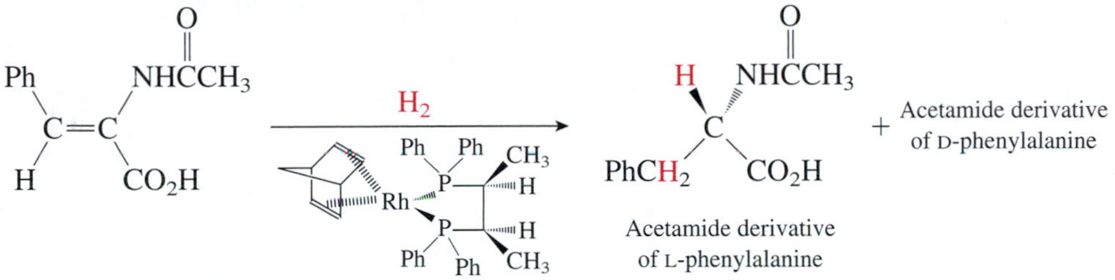

The stereocenter at the α-carbon is introduced by catalytic hydrogenation. To selectively produce one enantiomer of the product, the acetamide of phenylalanine, a chiral catalyst is employed. Rather than using a metal surface as a catalyst, as is common for hydrogenations, a metal complex that is soluble in the reaction solvent is employed. In this particular case the catalyst is rhodium complexed to a bicyclic diene and a chiral phosphorus-containing ligand. Similar to the reaction on the surface of a metal, the reaction occurs by initial coordination of the alkene and hydrogen to the metal atom in place of the bicyclic diene. The chiral phosphorus ligand causes the hydrogen to be transferred to the alkene so as to produce a single enantiomer of the product. A variety of chiral catalysts have been developed, so one can often be found that will accomplish a particular asymmetric hydrogenation. In addition, the use of a chiral catalyst has the advantage that a full equivalent of the expensive chiral reagent is not needed and the catalyst can often be recovered. Another example of this process is provided in the Focus On box on page 449. You might recall that W. S. Knowles shared the 2001 Nobel Prize in chemistry for developing asymmetric hydrogenations like this.

The use of proline methyl ester as a chiral auxiliary in the asymmetric synthesis of alanine is shown on the following page. The idea is to start with 2-oxopropanoic acid (pyruvic acid), which has the correct carbon skeleton, and replace the oxygen on carbon 2 with an amino group and a hydrogen. This must be done in such a manner as to produce only the S-enantiomer of the amino acid, that is, L-alanine. This is accomplished by first attaching a chiral auxiliary, the methyl ester of L-proline, to the acid. In the critical step of the process, the catalytic hydrogenation, the chirality of the

Continued

L-proline is used to induce the proper stereochemistry at the new stereocenter. To put this in terms used in Chapter 7, the α-carbon of proline has S stereochemistry. The new stereocenter generated in hydrogenation, which is the α-carbon of the alanine, could have either R or S stereochemistry. The potential products, with stereochemistries of S,S or S,R, are diastereomers. The hydrogenation occurs preferentially (greater than

The starting material for the synthesis of L-alanine is 2-oxopropanoic acid, also known as pyruvic acid. The carbonyl group at the α-carbon will be replaced with a H and a NH_2 so that only one enantiomer is formed at the new chirality center.

❶ In the first step, an amide is formed by the reaction of the acid with the methyl ester of L-proline, using DCC as the coupling agent (see Section 26.7).

❷ The product is then reacted with ammonia. The ammonia nucleophile attacks the carbonyl carbon of the ester group, resulting in the formation of an amide group. This product is not isolated but spontaneously proceeds to the next step.

❸ An intramolecular nucleophilic attack by the newly introduced nitrogen at the α-carbonyl carbon produces a six-membered ring.

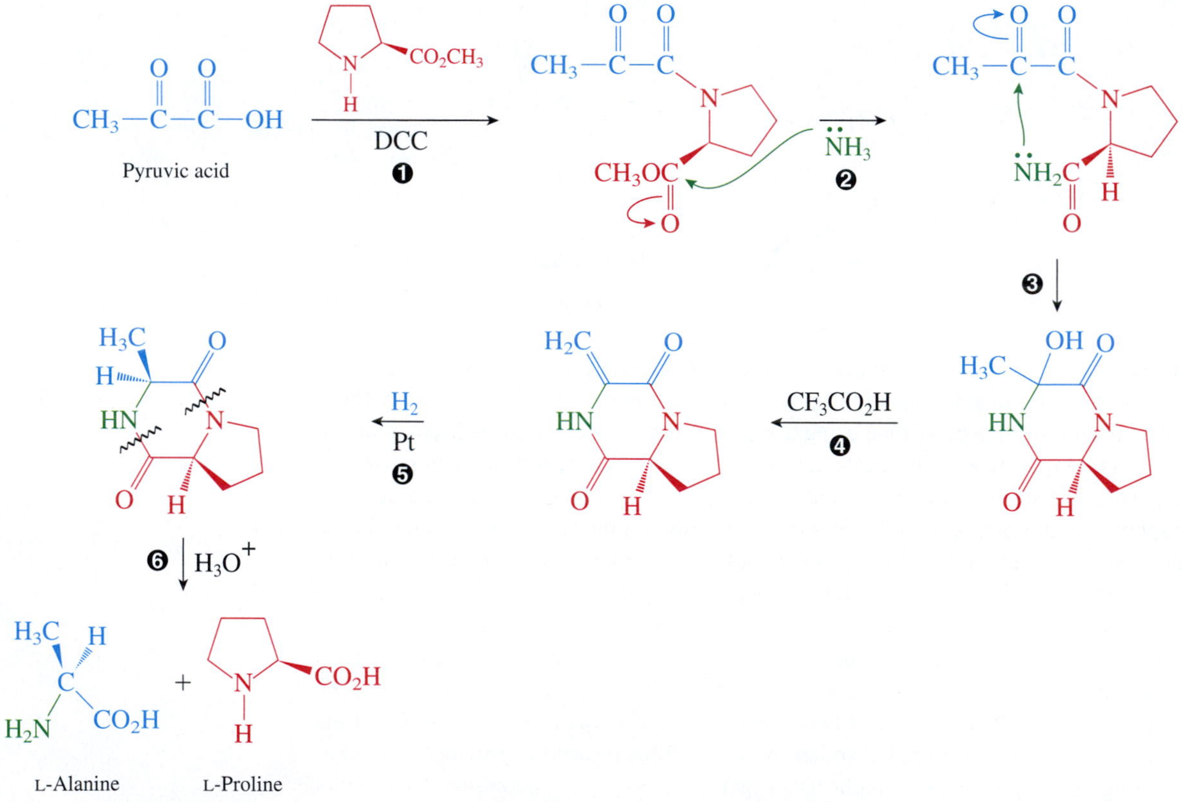

❻ Acid-catalyzed hydrolysis of the two amide bonds produces L-alanine and regenerates the L-proline so that it can be used again.

❺ It is at this stage that the new stereocenter is introduced by catalytic hydrogenation of the double bond. The catalyst prefers to approach from the less hindered bottom side of the molecule, so a single stereoisomer of the product predominates. The yield of this step is quantitative, with more than 90% of the product having the stereochemistry shown.

❹ Trifluoroacetic acid causes the elimination of water to produce a CC double bond.

90%) at the less hindered bottom side of the molecule, so the hydrogen at the α-carbon of the alanine is added cis to the hydrogen at the α-carbon of the proline. This results in the formation of the *S,S*-enantiomer of the product. Hydrolysis produces the *S*-enantiomer of alanine, L-alanine, and regenerates the chiral auxiliary, L-proline, so that it can be used again.

PROBLEM 26.12

What starting material would be used for the synthesis of L-phenylalanine by the method using proline methyl ester as a chiral auxiliary?

26.5 PEPTIDES AND PROTEINS

Because they contain two functional groups, amino acids can react to produce condensation polymers by forming amide bonds. These polymers are called peptides, polypeptides, or proteins. Although there is no universally accepted distinction, the term *protein* is usually reserved for naturally occurring polymers that contain a relatively large number of amino acid units and have molecular masses in the range of a few thousand or larger. The term *peptide* is used for smaller polymers.

As a simple example, the dipeptide formed by the reaction of two glycines has the following structure:

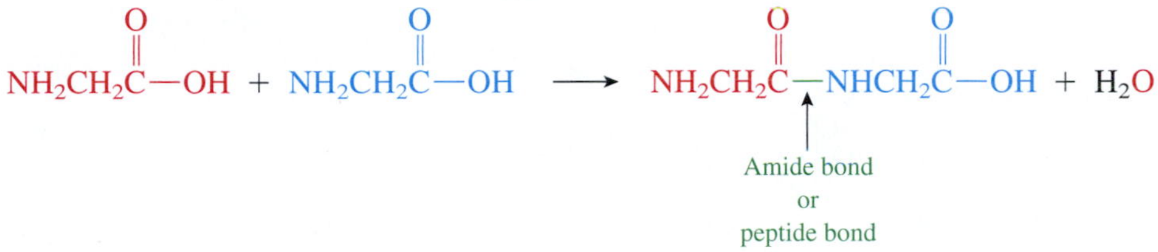

$$\text{NH}_2\text{CH}_2\overset{\overset{\displaystyle O}{\|}}{\text{C}}\!-\!\text{OH} \; + \; \text{NH}_2\text{CH}_2\overset{\overset{\displaystyle O}{\|}}{\text{C}}\!-\!\text{OH} \longrightarrow \text{NH}_2\text{CH}_2\overset{\overset{\displaystyle O}{\|}}{\text{C}}\!-\!\text{NHCH}_2\overset{\overset{\displaystyle O}{\|}}{\text{C}}\!-\!\text{OH} \; + \; \text{H}_2\text{O}$$

Amide bond
or
peptide bond

Biochemists say that the two amino acids are connected by a peptide bond, but, of course, the peptide bond is just an amide bond. A slightly more complex example, the phagocytosis-stimulating tetrapeptide known as tuftsin, derived from four amino acids, can be employed to illustrate some of the conventions that are used in writing the structures of polypeptides and proteins:

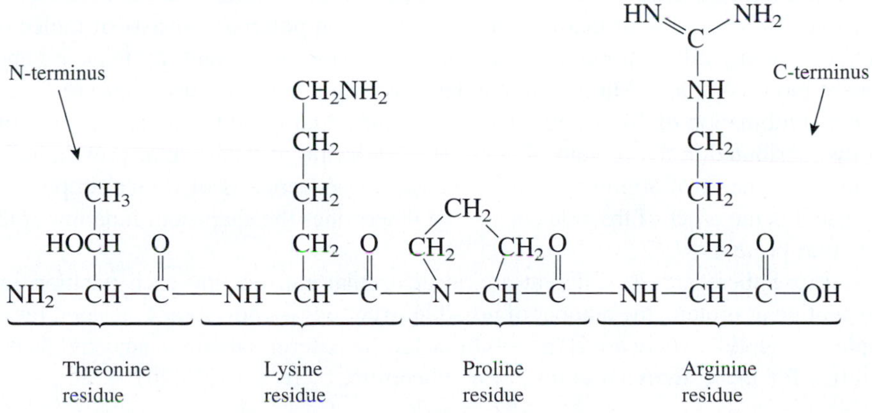

| Threonine residue | Lysine residue | Proline residue | Arginine residue |

Tuftsin

By convention, peptides are written so that the end with the free amino group, called the N-terminus, is on the left and the end with the free carboxyl group, the C-terminus, is on the right. Because it takes considerable space to show the structure of even a small polypeptide like this one, it is common to represent the structures of peptides and proteins by using the three-letter abbreviation for each amino acid (see Table 26.1). Thus tuftsin, with a threonine N-terminal amino acid, followed by lysine, proline, and, finally, arginine as the C-terminal amino acid, is represented as

Thr-Lys-Pro-Arg

Note that the N-terminal amino acid is on the left and the C-terminal amino acid is on the right in this abbreviated representation also. (For very large polypeptides the one-letter codes for the amino acids are used to save even more space.)

PROBLEM 26.13
Draw the complete structure for Phe-Val-Asp.

PROBLEM 26.14
Identify the amino acids in this polypeptide and show its structure using the three-letter abbreviations for the amino acids:

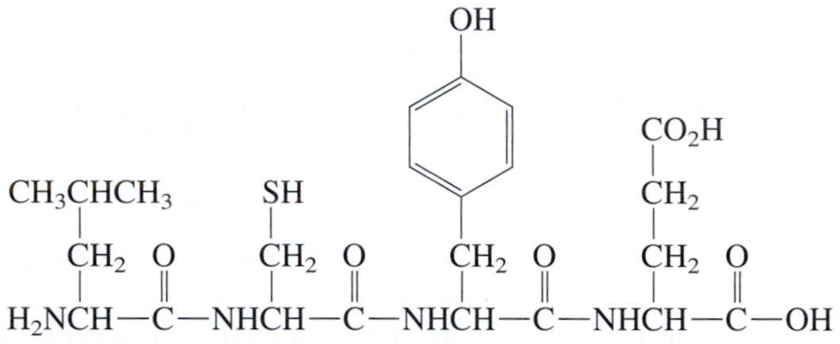

Let's compare proteins to the polymers that were discussed in Chapter 24. One difference is that all the molecules of a particular protein are identical; that is, they have the same molecular mass and contain the same number of amino acids connected in the same sequence. Recall that a typical condensation polymer consists of molecules containing many different numbers of monomers. More important, proteins are enormously more complex than simple condensation polymers because they are formed from a combination of 20 different monomer units. And these monomers are not randomly distributed in the protein. Rather, each molecule of a particular protein has an identical sequence of amino acid units. The exact sequence is of critical importance because it is the order of the side chains that determines the shape and function of that particular protein.

Because there are 20 different amino acids that can occupy each position in a polypeptide or protein, the number of possible structures is enormous. Consider, for example, a dipeptide. There are 20 possibilities for the N-terminal amino acid and 20 possibilities for the C-terminal amino acid. Therefore, there are $(20)(20) = 20^2 = 400$

different dipeptides. The number of possibilities increases rapidly as the number of amino acids in the polymer increases. For a tripeptide there are $20^3 = 8000$ possibilities. And for a polypeptide that contains 100 amino acids (many proteins are considerably larger than this) there are $20^{100} = 1.27 \times 10^{130}$ possibilities. Such large numbers have little meaning for most of us, so let's try to put this number in perspective. It has been estimated that the total number of atoms in the universe is about 10^{80}. The number of possible polypeptides containing only 100 amino acids vastly exceeds the total number of atoms in the entire universe!

The geometry of the amide bond helps determine the overall shape of a peptide or protein. The nitrogen of an amide is sp^2 hybridized, so the electron pair on the nitrogen is in a p orbital that can overlap with the pi bond of the carbonyl group. The nitrogen is planar, and there is considerable double-bond character to the bond connecting it to the carbonyl carbon. In other words, the structure on the right makes an important contribution to the resonance hybrid for an amide:

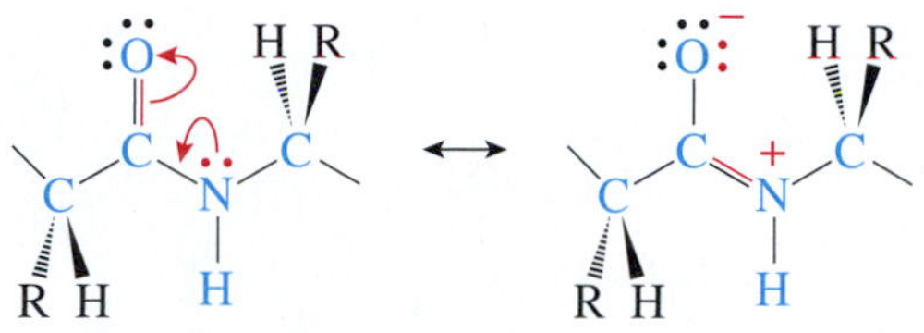

This requires that the carbonyl carbon, the nitrogen, and the two atoms attached to each of them (the α-carbon and the oxygen bonded to the carbonyl carbon and the hydrogen and the other α-carbon bonded to the nitrogen) must all lie in the same plane. The most stable conformation has the bulky α-carbons in a trans relationship about the carbon–nitrogen partial double bond, as shown in the preceding structure.

Another important feature of proteins is due to the presence of cysteine amino acids in the polymer chain. The SH groups of two cysteine residues can react to form a disulfide bond as shown in the following equation:

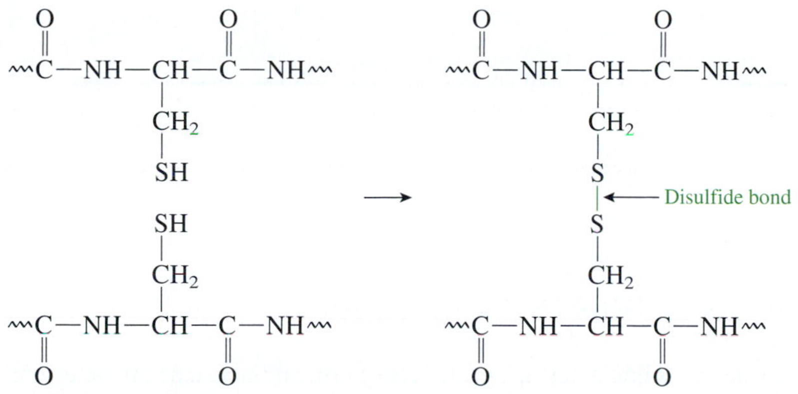

If the cysteines are part of different polypeptide chains, the resulting disulfide bond acts as a cross-link between the chains. If the cysteines are part of the same polypeptide chain, the large ring that is formed helps determine the overall shape of the peptide.

Focus On

NMR Spectra of Amides

Because the bond between the nitrogen and the carbonyl carbon of an amide has considerable double-bond character, rotation about this bond is relatively slow. This slow rotation is evident in the NMR spectra of many amides. For example, the two methyl groups on the nitrogen of *N,N*-dimethylacetamide appear at different chemical shifts. If rotation about the carbon–nitrogen bond were fast, the methyl groups would be equivalent and would appear at the same chemical shift.

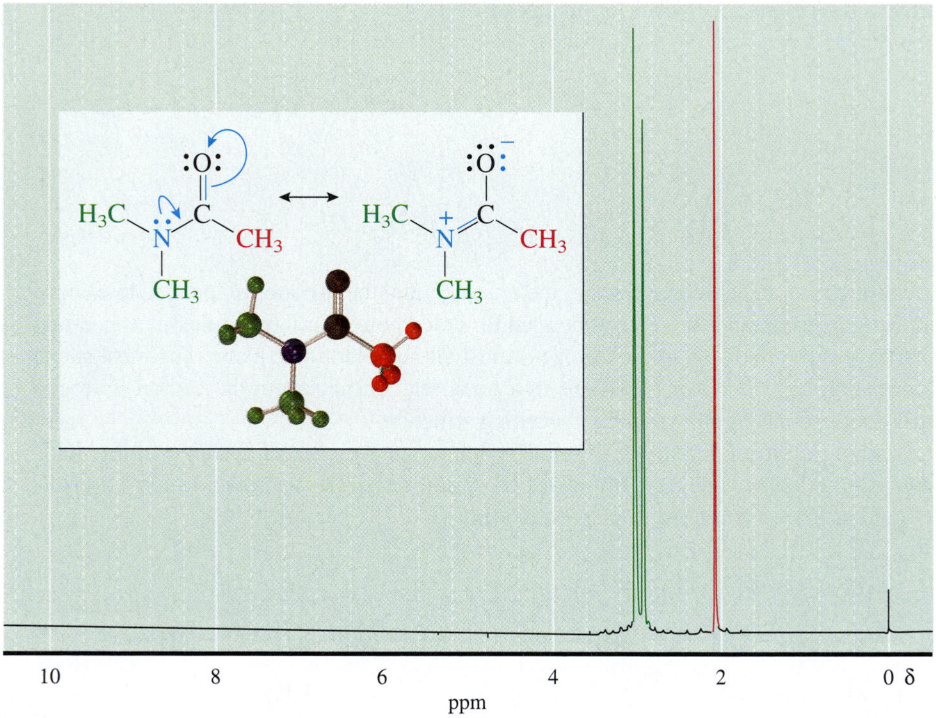

26.6   SEQUENCING PEPTIDES

The sequence of amino acids in a protein is of critical importance in determining the function of that protein. Therefore, considerable effort has been devoted to the development of methods to determine amino acid sequences. The process usually begins with the determination of the relative numbers of each amino acid that are present in the protein. To accomplish this, a sample is completely hydrolyzed to its individual amino acid components by treatment with 6 *M* HCl at 100 to 120°C for

10 to 100 hours. These rather vigorous conditions are needed to completely hydrolyze the protein because amide bonds are rather unreactive. The amino acids are then separated by some type of chromatography, and the relative number of each is determined.

PROBLEM 26.15

Explain why asparagine and glutamine are never found when a peptide is completely hydrolyzed by using HCl, H_2O, and elevated temperatures. What amino acids are found in place of these?

The separation and detection process has been automated. In the original amino acid analyzer, developed by W. H. Stein and S. Moore, who were awarded the 1972 Nobel Prize in chemistry for determining the structure of the enzyme ribonuclease, the amino acids are separated by ion-exchange chromatography. They are then reacted with ninhydrin, and the resulting purple derivatives are detected by visible spectroscopy. In a more modern version the amino acids are reacted with dansyl chloride, and the resulting derivatives are separated by high-performance liquid chromatography. The dansyl group is highly fluorescent, so very small amounts of the dansylated amino acids can be detected. With a modern amino acid analyzer, the complete analysis of a hydrolyzed protein can be done in less than 1 hour. The method is sensitive enough to detect as little as 10^{-12} mol of an amino acid, so only a very small amount of the protein need be hydrolyzed.

Dansyl chloride

Of course, the determination of the number of each kind of amino acid that is present is only a small part of the solution to the structure of a protein. The sequence of the amino acids must also be determined. This is accomplished by taking advantage of the fact that only the N-terminal amino acid has a free NH_2 group that is nucleophilic. All of the other nitrogens are part of amide groups and are not nucleophilic (unless the side chain contains an amino group, as is the case with lysine).

In one method the polypeptide is reacted with Sanger's reagent, 2,4-dinitrofluorobenzene (DNFB). The nucleophilic nitrogen of the N-terminal amino acid displaces the fluorine in a nucleophilic aromatic substitution reaction. (This reaction follows an addition–elimination mechanism; see Section 17.11.) The polypeptide is then hydrolyzed to its individual amino acid components. Because the bond between the nitrogen and the dinitrophenyl group is resistant to hydrolysis, the N-terminal amino acid is labeled and can easily be identified in the hydrolysis mixture:

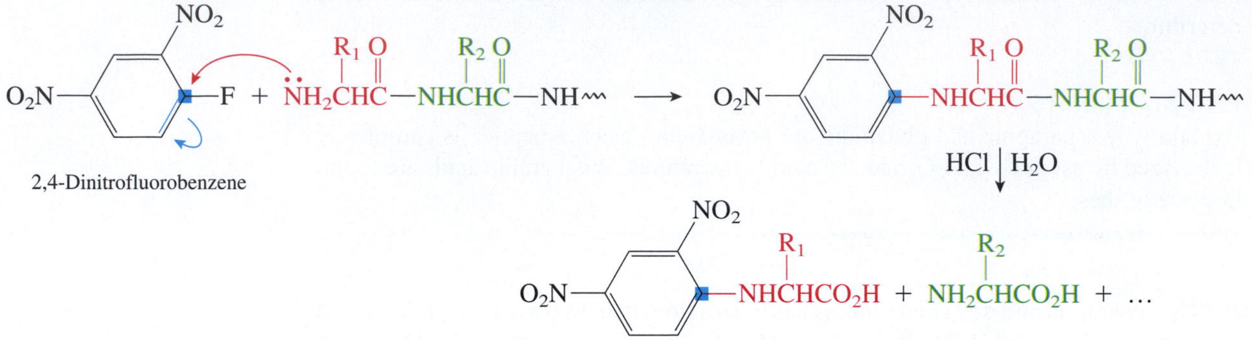

2,4-Dinitrofluorobenzene

In another method the nitrogen of the N-terminal amino acid is reacted with dansyl chloride. The resulting sulfonamide bond is quite resistant to hydrolysis, so the N-terminal amino acid, labeled with the dansyl group, is readily determined after the peptide bonds have been hydrolyzed.

PROBLEM 26.16

Show all of the steps in the mechanism for the reaction of an amino acid with Sanger's reagent and explain why the nitro groups are necessary for the reaction to occur.

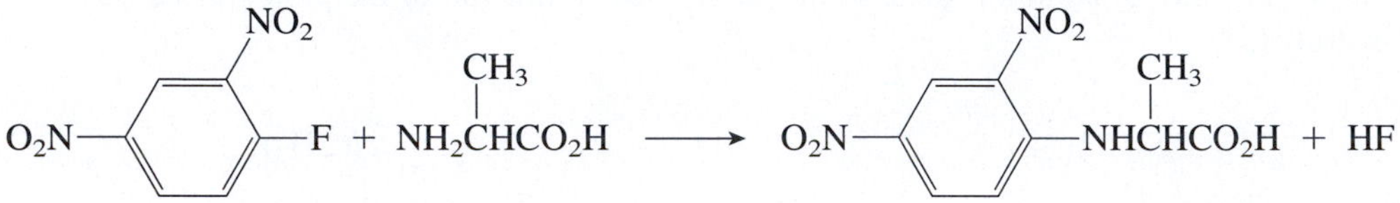

The most useful method of N-terminal analysis is called the **Edman degradation.** This method allows the N-terminal amino acid to be removed and its identity to be determined without hydrolyzing the other peptide bonds. The reaction initially produces a thiazolinone, which is rearranged by aqueous acid to a phenylthiohydantoin for identification by high-performance liquid chromatography:

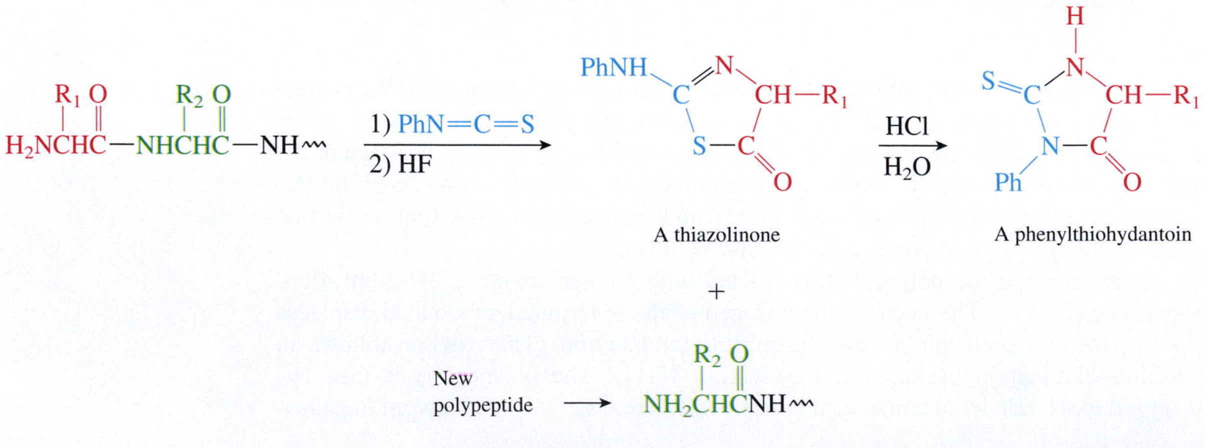

A thiazolinone   A phenylthiohydantoin

+

New polypeptide

The feature that makes the Edman degradation so useful is that the new polypeptide, with one fewer amino acid, can be isolated and submitted to the process again, allowing identification of its N-terminal amino acid. It is possible to continue removing and identifying the N-terminal amino acid for 40 cycles or more before impurities due to incomplete reactions and side reactions build up to the extent that the identification of the last removed amino acid is uncertain. The Edman degradation procedure has also been automated, so it is now possible to sequence a polypeptide from the N-terminal end at the rate of about one amino acid residue per hour.

The critical feature of the Edman degradation is that it allows the N-terminal amino acid to be removed without cleaving any of the other peptide bonds. Let's see how this occurs. The mechanism of the reaction is shown in Figure 26.3. First the nucleophilic nitrogen of the N-terminal amino acid attacks the electrophilic carbon of phenyl isothiocyanate. When anhydrous HF is added in the next step, the sulfur of the thiourea acts as an intramolecular nucleophile and attacks the carbonyl carbon of the closest peptide bond. It is the intramolecular nature of this step and the formation of a five-membered ring that result in the selective cleavage of only the N-terminal amino acid. The mechanism for this part of the reaction is very similar to that for acid-catalyzed hydrolysis of an amide (see Section 19.5). However, because no water is present, only the sulfur is available to act as a nucleophile. The sulfur is ideally positioned for intramolecular attack at the carbonyl carbon of the N-terminal amino acid, so only this amide bond is broken.

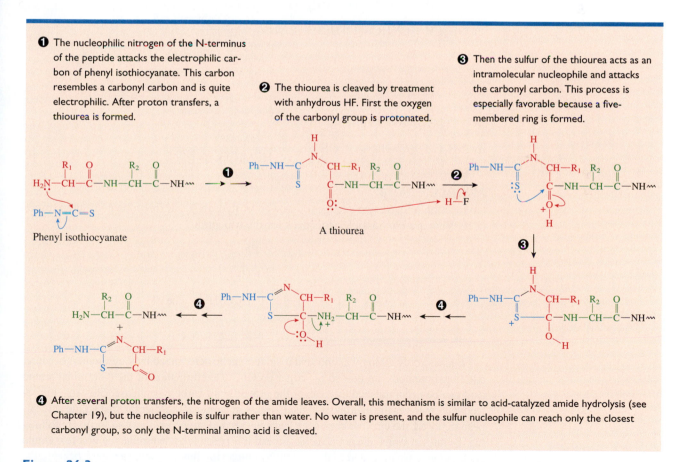

❶ The nucleophilic nitrogen of the N-terminus of the peptide attacks the electrophilic carbon of phenyl isothiocyanate. This carbon resembles a carbonyl carbon and is quite electrophilic. After proton transfers, a thiourea is formed.

❷ The thiourea is cleaved by treatment with anhydrous HF. First the oxygen of the carbonyl group is protonated.

❸ Then the sulfur of the thiourea acts as an intramolecular nucleophile and attacks the carbonyl carbon. This process is especially favorable because a five-membered ring is formed.

Phenyl isothiocyanate

A thiourea

❹ After several proton transfers, the nitrogen of the amide leaves. Overall, this mechanism is similar to acid-catalyzed amide hydrolysis (see Chapter 19), but the nucleophile is sulfur rather than water. No water is present, and the sulfur nucleophile can reach only the closest carbonyl group, so only the N-terminal amino acid is cleaved.

Figure 26.3

PARTIAL MECHANISM OF THE EDMAN DEGRADATION.

PROBLEM 26.17
Show the products of these reactions:

a)

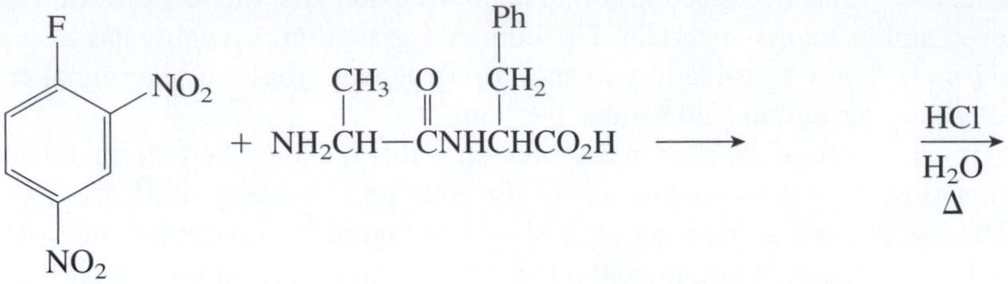

b)

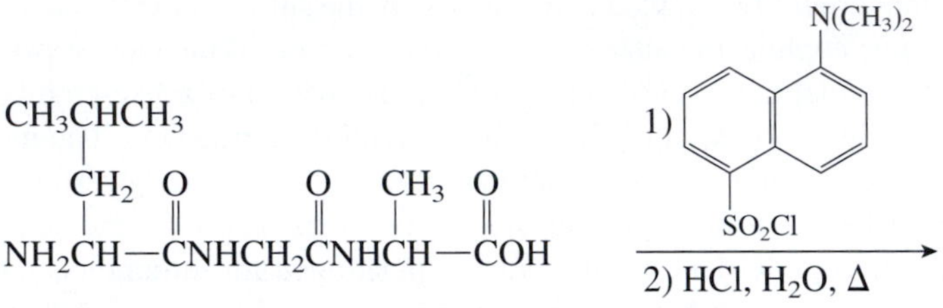

c)

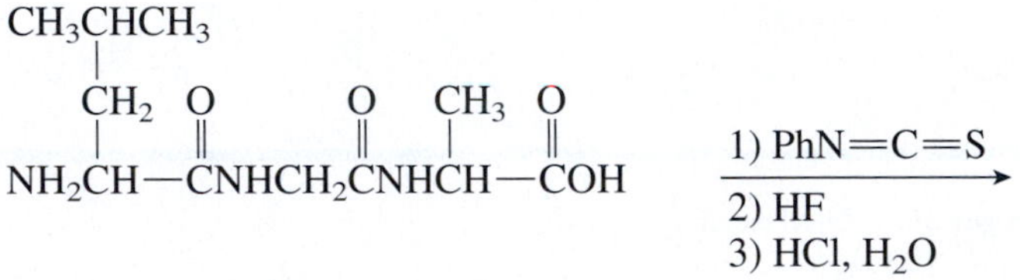

d)

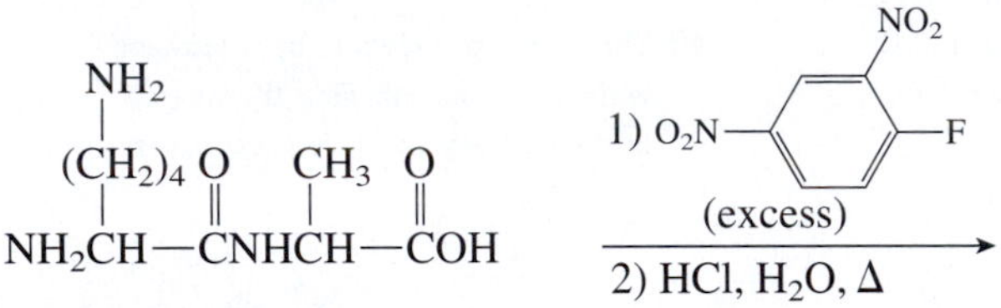

PROBLEM 26.18
On reaction with Sanger's reagent followed by hydrolysis, a tripeptide gives these products. What are the possible sequences for the tripeptide?

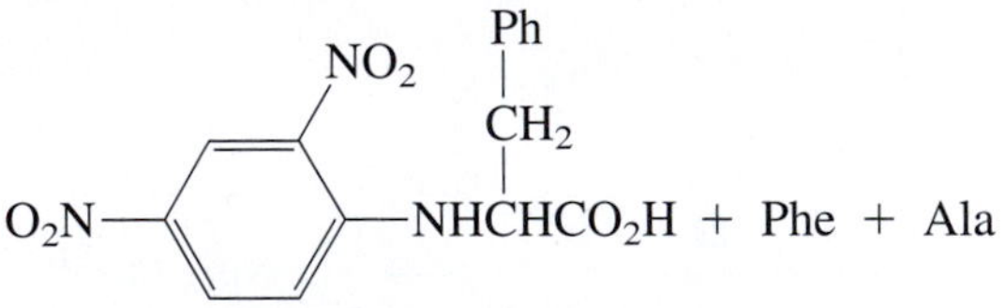

$$O_2N-\text{(ring)}-NHCHCO_2H + Phe + Ala$$

Because only 40 to 60 amino acid residues can be determined by the Edman procedure, additional methods are needed for larger proteins. Determination of the C-terminal amino acid can be accomplished by treating the protein with carboxypeptidase. This enzyme selectively catalyzes the hydrolysis of the C-terminal amino acid. After the first amino acid has been removed, the enzyme begins to cleave the second amino acid, and so forth. By following the rates at which the amino acids appear, it is possible to determine the first few amino acids at the C-terminal end of the protein by employing this enzyme. However, because the enzyme hydrolyzes different peptide bonds at different rates, it is possible to identify only a few amino acids before the reaction mixture becomes too complex.

Most proteins are too large to be completely sequenced by using both N-terminal and C-terminal analysis. In such cases it is necessary to cleave the protein into smaller fragments that can be individually sequenced by using the preceding methods. Although random hydrolytic cleavage can be used, it is more common to use enzymes that cleave the protein in specific positions. For example, the enzyme trypsin cleaves peptides after amino acids with positively charged side chains—that is, after the basic amino acids lysine and arginine. Chymotrypsin cleaves after tyrosine, tryptophan, and phenylalanine, amino acids with aromatic side chains:

Trypsin cleaves here   Chymotrypsin cleaves here

Try-Lys-Ala-Ile-Phe-Val-Asp

Several different cleavages must be done so that the fragments overlap and the amino acid sequence of one can be used to determine how others are connected.

A final problem is to determine which cysteines are connected by disulfide links. Usually, the disulfide bonds are broken by either oxidative cleavage with performic acid or reductive cleavage with 2-mercaptoethanol before initial sequencing is undertaken:

$$
\begin{array}{c}
\text{O} \\
\parallel \\
\sim\sim\text{NHCHC}\sim\sim \\
| \\
\text{CH}_2 \\
| \\
\text{S} \\
| \\
\text{S} \\
| \\
\text{CH}_2 \\
| \\
\sim\sim\text{NHCHC}\sim\sim \\
\parallel \\
\text{O}
\end{array}
\;+\;
\begin{array}{c}
\text{O} \\
\parallel \\
\text{HCOOH} \\
\text{Performic acid}
\end{array}
\;\longrightarrow\;
\begin{array}{c}
\text{O} \\
\parallel \\
\sim\sim\text{NHCHC}\sim\sim \\
| \\
\text{CH}_2 \\
| \\
\text{SO}_3\text{H} \\
+ \\
\text{SO}_3\text{H} \\
| \\
\text{CH}_2 \\
| \\
\sim\sim\text{NHCHC}\sim\sim \\
\parallel \\
\text{O}
\end{array}
$$

$$
\begin{array}{c}
\text{O} \\
\parallel \\
\sim\sim\text{NHCHC}\sim\sim \\
| \\
\text{CH}_2 \\
| \\
\text{S} \\
| \\
\text{S} \\
| \\
\text{CH}_2 \\
| \\
\sim\sim\text{NHCHC}\sim\sim \\
\parallel \\
\text{O}
\end{array}
\;+\; 2\;
\begin{array}{c}
\text{SH} \\
| \\
\text{CH}_2\text{CH}_2\text{OH} \\
\text{2-Mercaptoethanol}
\end{array}
\;\longrightarrow\;
\begin{array}{c}
\text{O} \\
\parallel \\
\sim\sim\text{NHCHC}\sim\sim \\
| \\
\text{CH}_2 \\
| \\
\text{SH} \\
+ \\
\text{SH} \\
| \\
\text{CH}_2 \\
| \\
\sim\sim\text{NHCHC}\sim\sim \\
\parallel \\
\text{O}
\end{array}
\;+\;
\begin{array}{c}
\text{CH}_2\text{CH}_2\text{OH} \\
| \\
\text{S} \\
| \\
\text{S} \\
| \\
\text{CH}_2\text{CH}_2\text{OH}
\end{array}
$$

After the sequence of the individual chains has been determined, a sample of the protein is partially hydrolyzed without cleavage of the disulfide bonds, and the fragments containing the intact disulfide bonds are sequenced.

The first protein to have its sequence determined by these methods was insulin. This was accomplished in 1953 by the research group headed by Frederick Sanger. Sanger won the 1958 Nobel Prize in chemistry for directing this work. (He also shared the 1980 Nobel Prize for his contributions to DNA sequencing, so he is one of the few individuals to have won two Nobel Prizes.) Sanger chose insulin because it is readily available and is relatively small (51 amino acids). To help see the approach used to solve a problem of this type, let's examine a few of the many experiments that were used.

Experiment 1

Insulin was reacted with Sanger's reagent and then completely hydrolyzed. Two amino acids, a glycine and a phenylalanine, were found to be labeled with the 2,4-dinitrophenyl group.

DNP-Gly   DNP-Phe
               + other amino acids

This experiment indicates that there must be two N-terminal amino acids because two amino acids were labeled with the DNP group. Therefore, there must be two polypeptide chains, connected by disulfide linkages. One has glycine as its N-terminus and the other has phenylalanine.

Experiment 2

Insulin was reacted with performic acid to cleave the disulfide bridges. Two polypeptides were isolated from this reaction, polypeptide A with 21 amino acids and polypeptide B with 30 amino acids. Polypeptide B was reacted with Sanger's reagent and then partially hydrolyzed. This produced a very complex mixture containing individual amino acids, dipeptides, tripeptides, and so on. All of the peptides that had the dinitrophenyl group on their N-terminal amino acid were then isolated from this mixture. (This was relatively easy to accomplish because these labeled peptides have very different solubility properties from the unlabeled polypeptides in acidic solution.) The labeled peptides were separated by chromatography, and each component was hydrolyzed. The following results were obtained:

$$\text{Component 1} \xrightarrow[\Delta]{H_3O^+} \text{DNP-Phe} + \text{Val}$$

$$\text{Component 2} \xrightarrow[\Delta]{H_3O^+} \text{DNP-Phe} + \text{Val} + \text{Asp}$$

$$\text{Component 3} \xrightarrow[\Delta]{H_3O^+} \text{DNP-Phe} + \text{Val} + \text{Asp} + \text{Glu}$$

This experiment shows that Phe must be the N-terminal amino acid of peptide B because it is labeled with the DNP group. From hydrolysis of component 1, Val must be attached to Phe; from hydrolysis of component 2, Asp must be attached to Val; and from hydrolysis of component 3, Glu must be attached to Asp. The sequence of the first four amino acids of peptide B, starting from the N-terminus, is Phe-Val-Asp-Glu.

Experiment 3

It was known that there were two Cys residues in polypeptide B. All of the Cys-containing peptides from the hydrolysis mixture were isolated. This was accomplished by taking advantage of the strongly acidic sulfonic acid side chains of the oxidized cysteine residues. These peptides were separated by chromatography. Then each was reacted with Sanger's reagent and hydrolyzed, and its individual amino acids were identified. The following components were identified:

$$\text{Component 1} \quad \xrightarrow[\text{2) H}_3\text{O}^+, \Delta]{\text{1) DNFB}} \quad \text{DNP-Val} + \text{Cys}$$

$$\text{Component 2} \quad \xrightarrow[\text{2) H}_3\text{O}^+, \Delta]{\text{1) DNFB}} \quad \text{DNP-Leu} + \text{Cys}$$

$$\text{Component 3} \quad \xrightarrow[\text{2) H}_3\text{O}^+, \Delta]{\text{1) DNFB}} \quad \text{DNP-Val} + \text{Cys} + \text{Gly}$$

$$\text{Component 4} \quad \xrightarrow[\text{2) H}_3\text{O}^+, \Delta]{\text{1) DNFB}} \quad \text{DNP-Leu} + \text{Cys} + \text{Gly}$$

$$\text{Component 5} \quad \xrightarrow[\text{2) H}_3\text{O}^+, \Delta]{\text{1) DNFB}} \quad \text{DNP-Leu} + \text{Cys} + \text{Val}$$

Component 1 indicates Val-Cys, and component 2 indicates Leu-Cys. (Remember that there are two Cys residues.) Component 3 indicates Val-Cys-Gly because the two Cys residues have Val and Leu attached to the amino end, so Val-Gly-Cys is not possible. Component 4 indicates Leu-Cys-Gly for similar reasons. Component 5 indicates Leu-Val-Cys because Val cannot be attached to the carboxyl end of Cys owing to the information obtained from components 3 and 4. Therefore, two small parts of the sequence are Leu-Val-Cys-Gly and Leu-Cys-Gly.

These experiments constitute only a small part of the project to determine the sequence of insulin. Overall, a large number of scientists worked under Sanger's direction on this project, which took 10 years and 100 g of protein to accomplish. Today, the sequence of such a simple protein could be determined by one experienced technician in only a few days using an automated sequencer employing the Edman method and would require only a few micrograms of sample! Or the sequence might be determined by methods employing mass spectrometry, or it might be read directly from the sequence of the DNA that codes for the protein.

A chain

Gly
|
Ile
|
Val
|
Glu
|
Gln
|
Cys—S——————————→S
| |
Cys—Thr—Ser—Ile—Cys—Ser—Leu—Tyr—Gln—Leu—Glu—Asn—Tyr—Cys—Asn
| |
S   S
| |
S   S
| |
Cys—Gly—Ser—His—Leu—Val—Glu—Ala—Leu—Tyr—Leu—Val—Cys—Gly—Glu
| |
Leu   Arg
| |
His   Gly
| |
Glu   Phe
| |
Asn   Phe
| |
Val   Thr—Lys—Pro—Thr—Tyr
|
Phe

B chain

Intramolecular disulfide bridge

Intermolecular disulfide bridges

Insulin

26.7 LABORATORY SYNTHESIS OF PEPTIDES

Synthesis of polypeptides in the laboratory is much more complicated than the synthesis of the polyamides described in Chapter 24 because more than one monomer must be used and the order of their attachment must be carefully controlled. For example, suppose we want to prepare the dipeptide Gly-Ala. We cannot just heat a mixture of glycine and alanine. Although this would produce some of the desired dipeptide, it would also form a host of other products, including other dipeptides and tripeptides, as shown in the following equation:

$$NH_2CH_2CO_2H + NH_2CHCO_2H \xrightarrow{\Delta} NH_2CH_2C-NHCHCO_2H + NH_2CHC-NHCH_2CO_2H$$

Glycine   Alanine   Gly-Ala   Ala-Gly

+ Gly-Gly + Ala-Ala + Gly-Ala-Gly + Gly-Ala-Ala + ...

The problem with this reaction is that there are too many reactive functional groups: two amino groups and two carboxylic acid groups. To prepare the desired dipeptide in a controlled manner, it is necessary to restrict the reactivity of the compounds so that only the amino group of alanine and the carboxylic acid group of glycine are available to react. As usual, this is accomplished by the use of protecting groups on the functional groups in which reaction is not desired. The synthesis of the dipeptide is performed by first protecting the amino group of glycine and the carboxylic acid group of alanine. To form the amide bond, the carboxylic acid group of the protected glycine is activated by conversion to a more reactive derivative (see Section 19.6). Then the protected alanine is added. In the final step the protecting groups must be removed without breaking the newly formed amide bond.

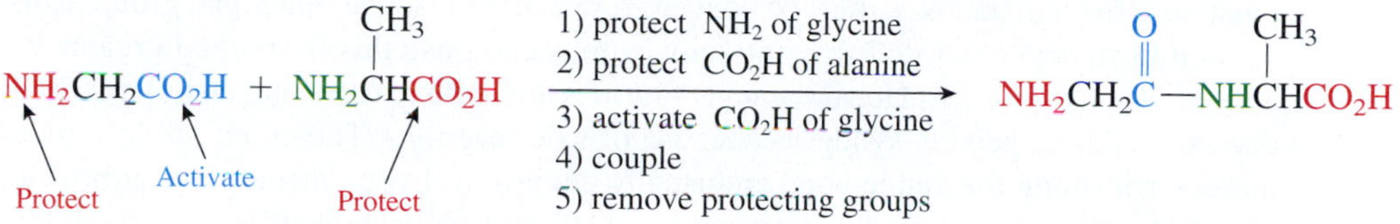

$$NH_2CH_2CO_2H + NH_2\overset{CH_3}{\underset{|}{CH}}CO_2H \xrightarrow{\begin{array}{l}1)\ \text{protect}\ NH_2\ \text{of glycine}\\ 2)\ \text{protect}\ CO_2H\ \text{of alanine}\\ 3)\ \text{activate}\ CO_2H\ \text{of glycine}\\ 4)\ \text{couple}\\ 5)\ \text{remove protecting groups}\end{array}} NH_2CH_2\overset{O}{\overset{||}{C}}-NH\overset{CH_3}{\underset{|}{CH}}CO_2H$$

Protect   Activate   Protect

Recall that an amino group is commonly converted to an amide for protection (see Section 23.4). However, there must be some special feature of the protecting group that allows it to be removed without also cleaving the amide bonds of the peptide. Although a large number of different protecting groups has been developed, the most common ones are the *t*-butoxycarbonyl (BOC) group and the benzyloxycarbonyl group. The BOC group is added by the use of an anhydride and is readily removed by treatment with dilute acid:

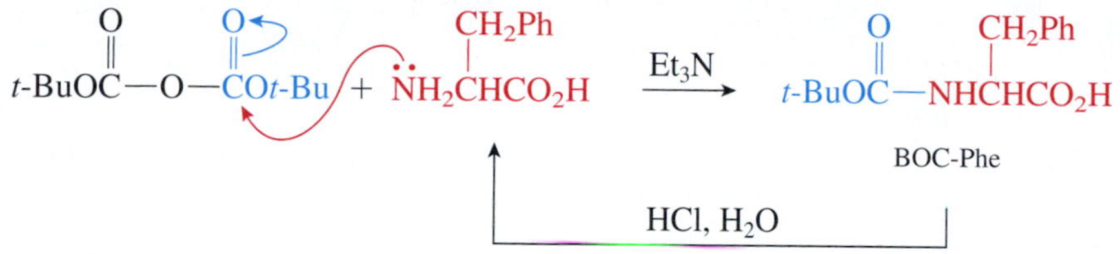

$$t\text{-BuO}\overset{O}{\overset{||}{C}}-O-\overset{O}{\overset{||}{C}}t\text{-Bu} + \ddot{N}H_2\overset{CH_2Ph}{\underset{|}{CH}}CO_2H \xrightarrow{Et_3N} t\text{-BuO}\overset{O}{\overset{||}{C}}-NH\overset{CH_2Ph}{\underset{|}{CH}}CO_2H$$

BOC-Phe

HCl, H₂O

The benzyloxycarbonyl group is added by using an acyl chloride and can be removed by hydrogenolysis:

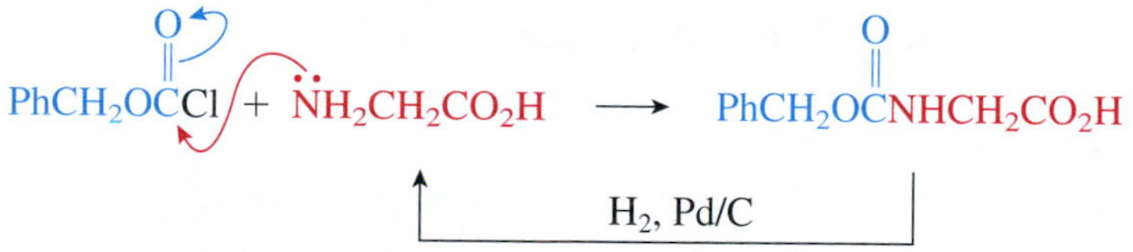

$$PhCH_2O\overset{O}{\overset{||}{C}}Cl + \ddot{N}H_2CH_2CO_2H \longrightarrow PhCH_2O\overset{O}{\overset{||}{C}}NHCH_2CO_2H$$

H₂, Pd/C

PROBLEM 26.19

Show all of the steps in the mechanisms for these reactions:

a) 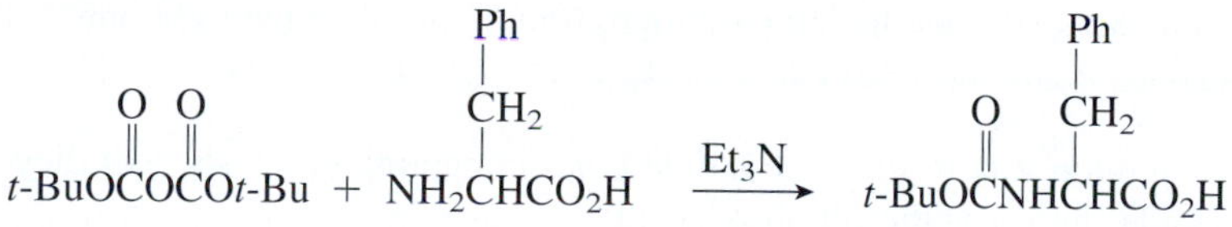 $t\text{-BuO}\overset{O}{\overset{||}{C}}O\overset{O}{\overset{||}{C}}Ot\text{-Bu} + NH_2\overset{Ph}{\underset{|}{\underset{|}{CH_2}}}\overset{}{\underset{|}{CH}}CO_2H \xrightarrow{Et_3N} t\text{-BuO}\overset{O}{\overset{||}{C}}NH\overset{Ph}{\underset{|}{\underset{|}{CH_2}}}\overset{}{\underset{|}{CH}}CO_2H$

b) $t\text{-BuO}\overset{O}{\overset{||}{C}}NH\overset{Ph}{\underset{|}{\underset{|}{CH_2}}}\overset{}{\underset{|}{CH}}CO_2H \xrightarrow[H_2O]{HCl} NH_2\overset{Ph}{\underset{|}{\underset{|}{CH_2}}}\overset{}{\underset{|}{CH}}CO_2H$

c) $PhCH_2O\overset{O}{\overset{||}{C}}Cl + NH_2CH_2CO_2H \longrightarrow PhCH_2O\overset{O}{\overset{||}{C}}NHCH_2CO_2H$

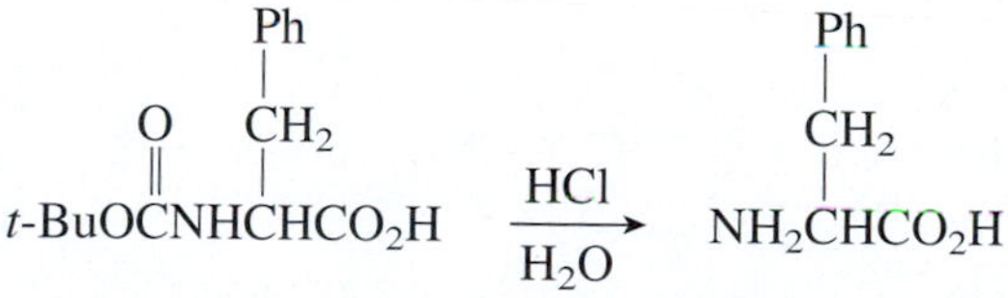

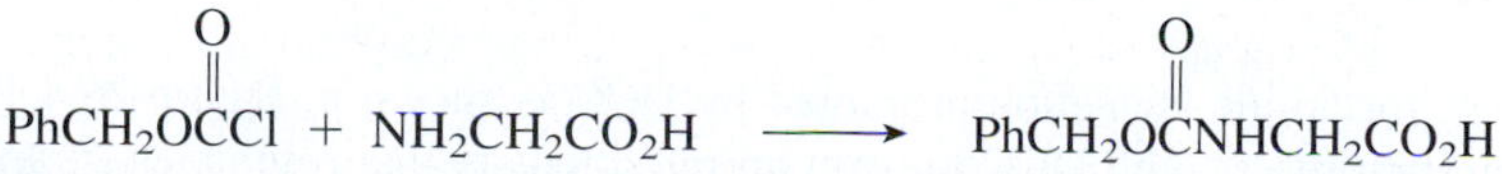

The carboxylic acid group is usually protected as an ester. Because an ester is considerably more reactive than an amide, it is possible to remove this protecting group by hydrolysis without cleaving the amide bond of the peptide.

For a carboxylic acid and an amine to form an amide, the carboxylic acid usually must be activated; that is, it must be converted to a more reactive functional group. Conversion to an acyl chloride is a common way to accomplish this for normal organic reactions (see Chapter 19). However, acyl chlorides are quite reactive and do not give high enough yields in peptide synthesis because of side reactions. Therefore, milder procedures for forming the amide bond are usually employed. In one method the carboxylic acid is reacted with ethyl chloroformate (a half acyl chloride, half ester of carbonic acid) to produce an anhydride. Treatment of this anhydride with an amine results in the formation of an amide:

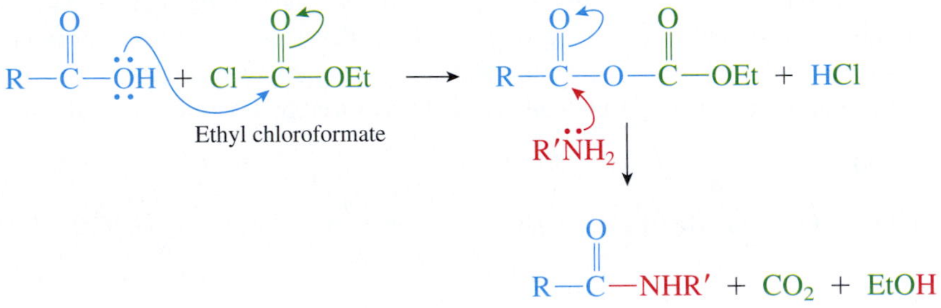

$$R-\overset{O}{\overset{\|}{C}}-\ddot{O}H + Cl-\overset{O}{\overset{\|}{C}}-OEt \longrightarrow R-\overset{O}{\overset{\|}{C}}-O-\overset{O}{\overset{\|}{C}}-OEt + HCl$$

Ethyl chloroformate

$$R'NH_2 \downarrow$$

$$R-\overset{O}{\overset{\|}{C}}-NHR' + CO_2 + EtOH$$

PROBLEM 26.20

Show all of the steps in the mechanisms for these reactions:

a) $\overset{O}{\overset{\|}{RCOH}} + \overset{O}{\overset{\|}{ClCOEt}} \longrightarrow \overset{O\ O}{\overset{\|\ \|}{RCOCOEt}} \quad \overset{O\ O}{\overset{\|\ \|}{RCOCCl}}$

                Formed   Not formed

Explain why the second product is not formed in the preceding reaction.

b) $\overset{O\ O}{\overset{\|\ \|}{RCOCOEt}} + R'NH_2 \longrightarrow \overset{O}{\overset{\|}{RCNHR'}} + CO_2 + EtOH$

Explain why the amine attacks the left carbonyl group rather than the right one.

The most common way to form the amide bond in peptide synthesis uses dicyclohexylcarbodiimide (DCC) as the coupling agent:

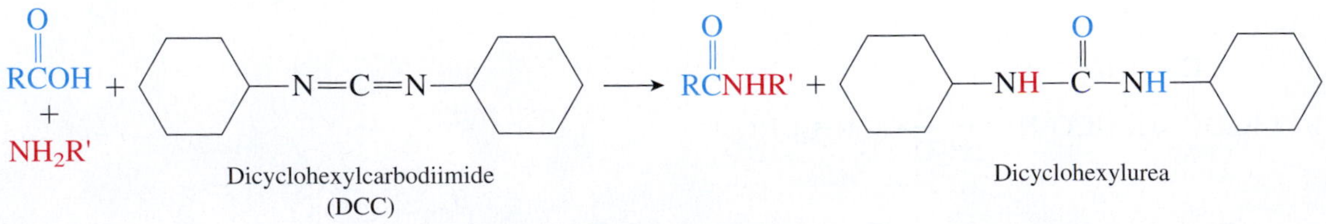

$$\overset{O}{\overset{\|}{RCOH}} + \underset{\text{Dicyclohexylcarbodiimide (DCC)}}{\bigcirc-N{=}C{=}N-\bigcirc} \longrightarrow \overset{O}{\overset{\|}{RCNHR'}} + \underset{\text{Dicyclohexylurea}}{\bigcirc-NH-\overset{O}{\overset{\|}{C}}-NH-\bigcirc}$$

$$+ \ NH_2R'$$

The mechanism for amide formation promoted by DCC is shown in Figure 26.4. The carboxylic acid first reacts with DCC to form an intermediate that resembles an anhydride, but with a carbon–nitrogen double bond in place of one of the carbonyl groups.

① The carbon of DCC is quite electrophilic. The oxygen of the carboxylic acid acts as a nucleophile and adds to the CN double bond in a reaction that is very similar to the addition of nucleophiles to the carbonyl group of an aldehyde or ketone (see Chapter 18).

② Proton transfers complete the addition reaction.

This compound resembles an anhydride, but with a CN double bond rather than a CO double bond.

③ Just like the reaction with an anhydride, the amine nucleophile attacks the carbonyl carbon.

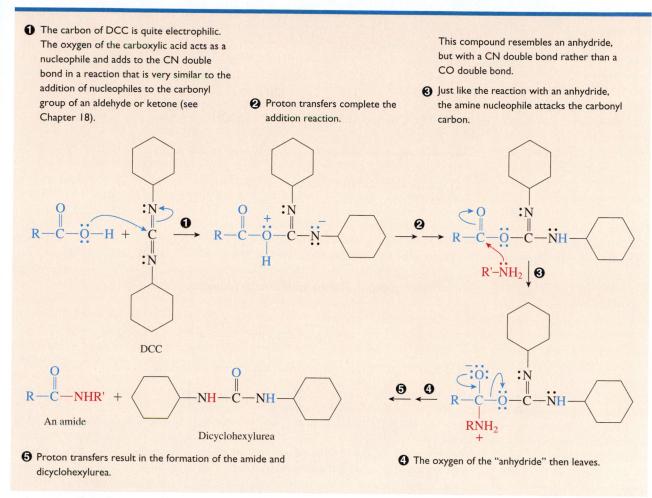

⑤ Proton transfers result in the formation of the amide and dicyclohexylurea.

④ The oxygen of the "anhydride" then leaves.

Active Figure 26.4

ORGANIC
Chemistry⚛Now™

MECHANISM OF AMIDE FORMATION USING DICYCLOHEXYLCARBODIIMIDE. Test yourself on the concepts in this figure at **OrganicChemistryNow.**

When this pseudo-anhydride is reacted with an amine, an amide is produced along with the by-product, dicyclohexylurea. Many other coupling reagents and protecting groups have been developed, but the ones shown here are the most common and illustrate the principles involved.

PROBLEM 26.21

Show the products of these reactions:

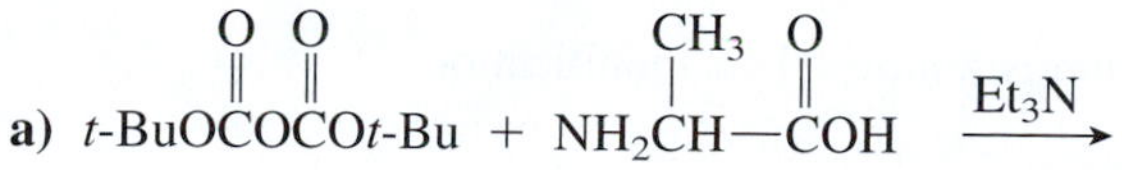

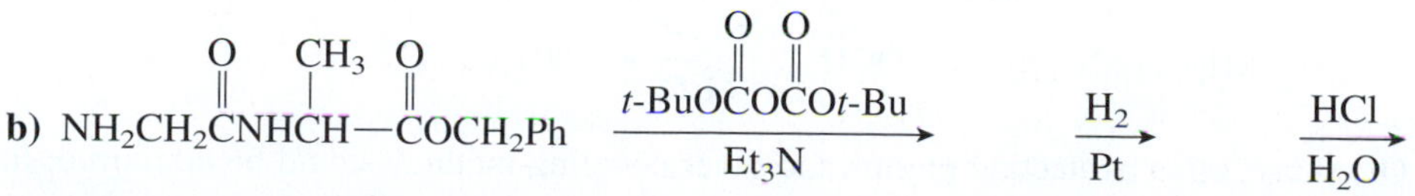

a) *t*-BuOCOCOt-Bu + NH₂CH—COH (with O O and CH₃ O groups) →[Et₃N]

b) NH₂CH₂CNHCH—COCH₂Ph →[t-BuOCOCOt-Bu / Et₃N] →[H₂ / Pt] →[HCl / H₂O]

ORGANIC
Chemistry⚛Now™
Click *Coached Tutorial Problems* for more practice with **Reactions Used in Synthesis of Peptides.**

c) $t\text{-BuOCNHCH}-\overset{\text{O CH}_3\ \text{O}}{\text{COH}}$ $\xrightarrow[\text{2) NH}_2\text{CH}_2\text{CO}_2\text{CH}_3]{\text{1) ClCOEt}}$ $\xrightarrow{\text{HCl} \atop \text{H}_2\text{O}}$ $\xrightarrow{\text{NaOH} \atop \text{H}_2\text{O}}$

d) $\text{PhCH}_2\text{OCNHCH}_2\text{COH} + \text{NH}_2\text{CH}-\text{COCH}_3$ $\xrightarrow{\text{DCC}}$ $\xrightarrow[\text{Pt}]{\text{H}_2}$

Let's consider the synthesis of Gly-Ala. First, the amino group of glycine must be protected. (Today, amino acids can be purchased in BOC protected form.)

$$t\text{-BuOCOCO}t\text{-Bu} + \text{NH}_2\text{CH}_2\text{COH} \longrightarrow t\text{-BuOCNHCH}_2\text{COH}$$

The carboxyl group of alanine must also be protected:

$$\text{NH}_2\text{CHCOH} + \text{CH}_3\text{OH} \xrightarrow{[\text{HA}]} \text{NH}_2\text{CHCOCH}_3$$

Then the carboxyl group of the protected glycine must be activated:

$$t\text{-BuOCNHCH}_2\text{COH} + \text{ClCOEt} \longrightarrow t\text{-BuOCNHCH}_2\text{COCOEt}$$

The next step is the formation of the amide bond by reaction of the activated, N-protected glycine with the carboxyl-protected alanine:

$$t\text{-BuOCNHCH}_2\text{COCOEt} + \text{NH}_2\text{CHCOCH}_3 \longrightarrow t\text{-BuOCNHCH}_2\text{C}-\text{NHCHCOCH}_3$$

To complete the synthesis, it is necessary only to remove the protecting groups. The BOC group is removed by treatment with aqueous acid. This reaction occurs under very mild conditions that do not also hydrolyze the ester group. Note that this dipeptide, deprotected at the N-terminal, can be reacted with another N-protected, carboxyl-activated amino acid to produce a tripeptide. These steps can be repeated to produce a tetrapeptide, and so on.

$$t\text{-BuOCNHCH}_2\text{C}-\text{NHCHCOCH}_3 \xrightarrow[\text{H}_2\text{O}]{\text{HCl}} \text{NH}_2\text{CH}_2\text{C}-\text{NHCHCOCH}_3$$

Finally, the carboxyl protecting group is removed by saponification:

$$\text{NH}_2\text{CH}_2\text{C}-\text{NHCHCOCH}_3 \xrightarrow[\text{H}_2\text{O}]{\text{NaOH}} \text{NH}_2\text{CH}_2\text{C}-\text{NHCHCOH}$$

Of course, other protecting groups and other coupling methods could be employed. In addition, things are a little more complicated because many of the amino acids have functional groups in their side chains that must also be protected. However, the general ideas of the process are provided in the preceding procedure.

PROBLEM 26.22
Show a synthesis of Ala-Phe-Gly-Gly starting from the individual amino acids.

As is apparent from this example, the synthesis of a polypeptide requires numerous steps. At each step, the product must be isolated, and if you have worked in an organic chemistry laboratory, you are certainly aware of how much time and energy are required to isolate and purify a product. This makes a polypeptide synthesis quite tedious. In addition, the mechanical losses that occur in each isolation step contribute to lower yields for the overall process. Motivated by these problems, R. B. Merrifield developed a method, called *solid phase synthesis,* that makes the preparation of a polypeptide much easier. Merrifield was awarded the 1984 Nobel Prize in chemistry for this work. Let's see how it works.

Merrifield's idea was to attach the initial amino acid molecules to the surface of insoluble polymer beads. To isolate the product after the next amino acid has been attached, it is necessary only to collect the beads by filtration and wash them to remove any remaining reagent. Not only is the isolation procedure fast and simple, but mechanical losses are minimized.

The insoluble polymer that is used for solid phase syntheses is the chloromethylated, divinylbenzene cross-linked polystyrene that was described in Section 24.10. The beads of this material have $ClCH_2$ groups bonded to the phenyl groups that are part of the polymer chains on the surface of the beads. As we saw in Section 24.10, functional groups on the surface of such beads exhibit normal chemical reactions. To synthesize a polypeptide, the carboxyl group of the C-terminal amino acid, with its amino group protected, is first bonded to the polymer by an ester linkage. The amino group of this amino acid is deprotected. Then the next amino acid, with its amine group protected, is attached using DCC to promote coupling. This cycle of deprotection of the amino group followed by coupling with another N-protected amino acid is repeated until all of the amino acids of the desired polypeptide have been added. At each step, isolation of the product is accomplished by simply collecting the beads by filtration. Finally, the peptide is cleaved from the polymer bead. An example of a solid phase synthesis is provided in Figure 26.5.

PROBLEM 26.23
Show a synthesis of Ala-Phe-Gly-Gly using the Merrifield solid phase method.

As mentioned earlier, the ease of isolation of the product is the major advantage of the solid phase method. However, this is also a disadvantage because the product, the growing polypeptide attached to the polymer, is never purified until it is finally cleaved from the polymer. Each time a coupling reaction does not proceed in 100% yield, a small amount of the final product will be missing one of the amino acids. These impurities build up with the number of steps in the synthesis and become more and more difficult to remove as the polypeptide becomes larger. For this reason the yield of each step must be as high as possible. Current methods provide yields of 99.5% or better in each step.

Another advantage of the solid phase method is that the steps are repetitive and can be automated. An automatic peptide synthesizer is commercially available. It consists of a flask with a fritted-glass filter at the bottom along with bottles containing solutions of BOC-protected amino acids and the other necessary reagents and wash solutions. The polymer with the first amino acid attached is placed in the flask, and the order of

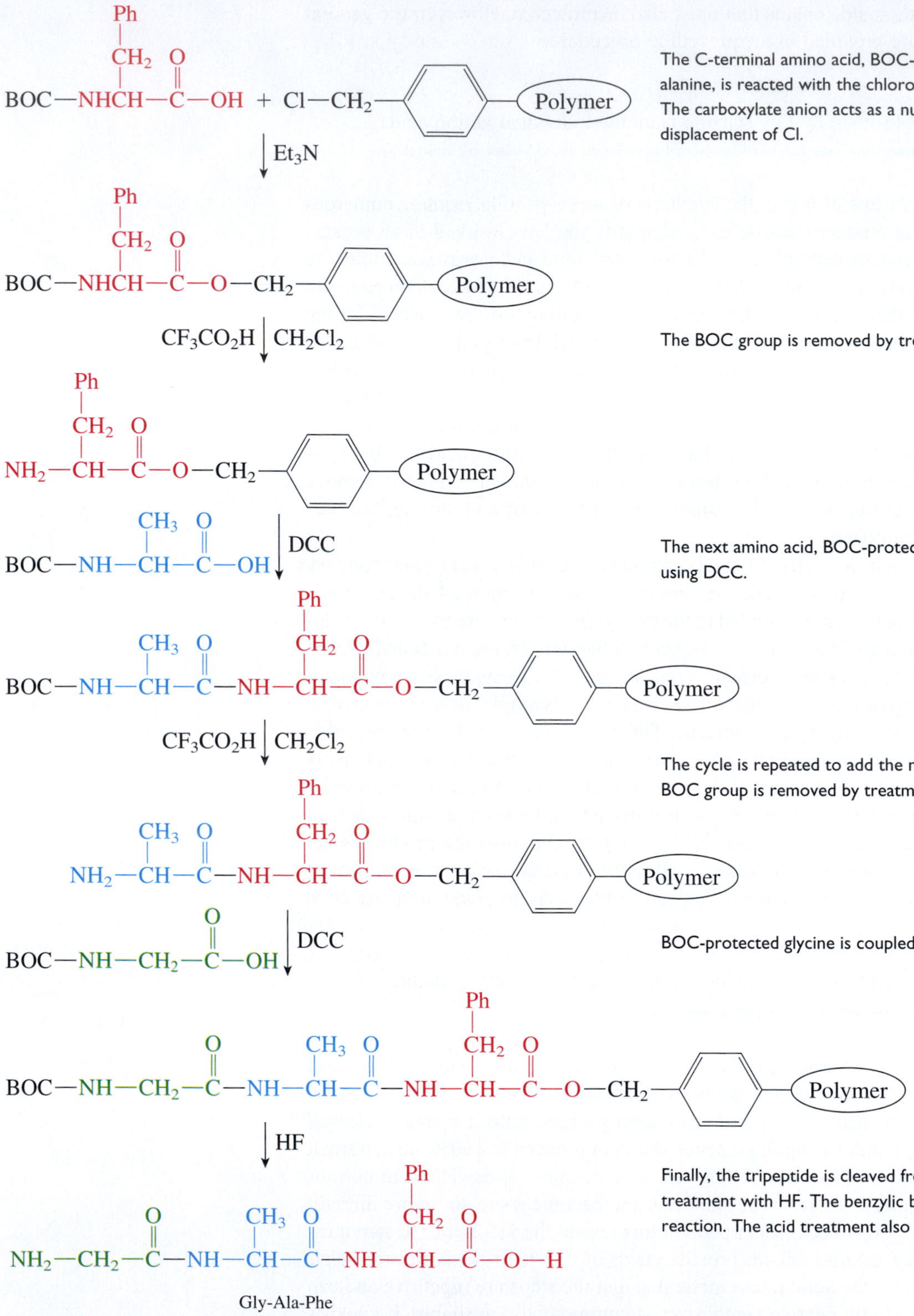

The C-terminal amino acid, BOC-protected phenyl-alanine, is reacted with the chloromethylated polymer. The carboxylate anion acts as a nucleophile in an S_N2 displacement of Cl.

The BOC group is removed by treatment with acid.

The next amino acid, BOC-protected alanine, is coupled using DCC.

The cycle is repeated to add the next amino acid. The BOC group is removed by treatment with acid.

BOC-protected glycine is coupled using DCC.

Finally, the tripeptide is cleaved from the polymer by treatment with HF. The benzylic bond is broken in an S_N2 reaction. The acid treatment also removes the BOC group.

Gly-Ala-Phe

Figure 26.5

THE MERRIFIELD SOLID PHASE SYNTHESIS OF GLY-ALA-PHE.

the amino acids in the desired polypeptide is programmed into the machine. The synthesizer then adds the next amino acid and DCC, reacts the mixture for the appropriate period of time, and removes the reagents by filtration. After the polymer beads have been washed, the next reagent is added and the process continues. The nonapeptide hormone, bradykinin, has been synthesized in 85% yield in 27 hours using such a device. As another example, bovine pancreatic enzyme ribonuclease A, with 124 amino acids, was synthesized in 17% yield in 6 weeks. After purification this material showed 78% of the bioactivity of the natural enzyme. However, this is an example in which the synthesis went particularly well. More commonly, large polypeptides like this contain many impurities that cannot readily be removed and therefore have a bioactivity that is much less than the natural protein. Solid phase synthesis is best for smaller polypeptides, up to about 50 amino acid residues. For larger peptides it is usually best to synthesize fragments of 50 amino acids or less and then couple these by solution phase reactions after they have been purified.

26.8 PROTEIN STRUCTURE

The properties of a protein depend primarily on its three-dimensional structure. The sequence of amino acids in the polypeptide chain is termed its **primary structure.** Its **secondary structure** is the shape of the backbone polypeptide chain. Remember that each amide group is planar, but the chain can have various conformations about the bond between the α-carbon and the nitrogen. The **tertiary structure** is the overall three-dimensional shape of the protein, including the conformations of the side chains.

The primary structure of the protein—that is, the sequence of its amino acids—determines its secondary and tertiary structure. Each protein usually has one shape that has the lowest energy. Individual molecules of the protein naturally and reproducibly fold into this lowest-energy conformation. The interactions that determine the preferred shape include hydrogen bonding involving the hydrogens on the amide nitrogens and the carbonyl groups, disulfide bonds, and interactions of the side chains with other side chains and the aqueous environment of the protein. The protein usually folds so that charged side chains—those of aspartic acid, glutamic acid, lysine, arginine, and histidine—are on the surface of the protein, so they can be solvated by water. Polar side chains also prefer the surface, where they are exposed to the aqueous environment, whereas nonpolar side chains, with their hydrophobic groups, prefer to pack together in the interior of the protein to avoid contact with water.

Two common elements of secondary structure that comprise about 50% of the structure of an average protein are the α-helix and the β-pleated sheet. In an α-helix the polypeptide chain assumes a helical shape that is held in place by hydrogen bonding between the NH group of one amino acid residue and the carbonyl oxygen four residues removed. There are 3.6 amino acid residues per turn of the helix. A simplified picture of an α-helix is shown in Figure 26.6.

In β-pleated sheets the hydrogen bonding occurs between parts of the polypeptide chain separated by larger distances. In one version, the antiparallel β-pleated sheet, the parts of the chain in the sheet run in opposite directions; that is, the N-terminal to C-terminal directions are opposite, as shown in Figure 26.7. In a parallel β-pleated sheet, the N-terminal to C-terminal directions are the same. In either case, when viewed from the edge, the sheet takes on a pleated shape. In an actual protein, α-helixes and β-pleated sheets, connected by loops and turns, pack together to give the overall three-dimensional shape of the protein.

Figure 26.6

PARTIAL STRUCTURE
OF AN α-HELIX.

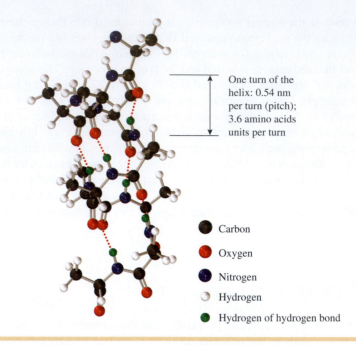

One turn of the
helix: 0.54 nm
per turn (pitch);
3.6 amino acids
units per turn

● Carbon

● Oxygen

● Nitrogen

○ Hydrogen

● Hydrogen of hydrogen bond

Figure 26.7

PARTIAL STRUCTURE OF AN
ANTIPARALLEL β-PLEATED
SHEET.

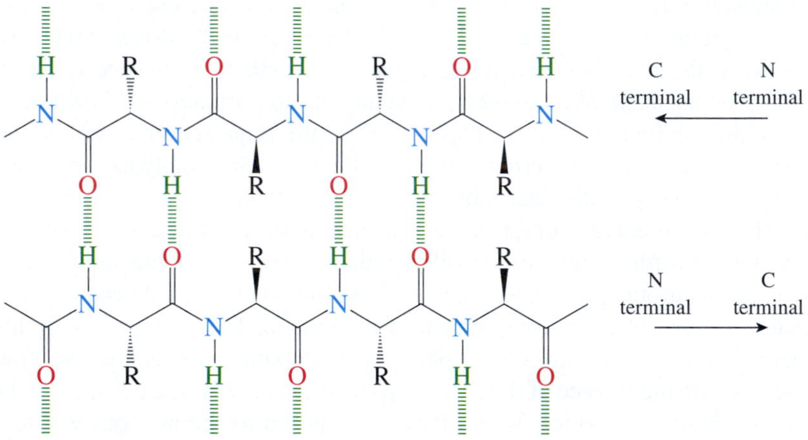

ORGANIC
Chemistry ❖ Now™
Click *Coached Tutorial Problems*
to view an α-**Helix** or a
β-**Sheet**.

26.9   ENZYMES

It was Emil Fischer who first proposed that an enzyme and its substrate (the compound undergoing the enzyme-catalyzed reaction) fit together like a lock and key. The substrate is tightly bound to the enzyme by hydrogen bonding, polar interactions, and hydrophobic interactions in a precise three-dimensional arrangement. In the active site of the enzyme–substrate complex, various functional groups of the side chains of the amino acids comprising the enzyme are held in proximity to the substrate. It is the proximity of these catalytic functional groups to the reactive site of the substrate that causes the rapid rate of catalysis. Essentially, the catalytic reaction resembles the intramolecular reactions that we have seen so many times to proceed at greatly enhanced rates.

As an example, let's consider the mechanism of action of the enzyme chymotrypsin. This enzyme is known as a serine protease because it catalyzes the hydrolysis of

❶ The side chains of a serine, a histidine, and an aspartate residue are all important in the active site. The polypeptide substrate is held in proximity to the OH of the serine. The oxygen of this OH acts as a nucleophile and attacks the carbonyl carbon of the peptide. As this happens, the histidine removes the proton from the OH, making it a better nucleophile. And the aspartate makes the histidine more basic by hydrogen bonding to it.

This is the tetrahedral intermediate shown in the mechanism of Figure 19.1.

❷ As happens in that mechanism, the electrons on the negative oxygen displace the leaving group, the nitrogen in this case. The protonated histidine transfers its proton to the nitrogen to make it a better leaving group.

At this point, the amide bond has been broken and the amine part is free to diffuse out of the active site. The acyl group is bonded to the oxygen of the serine. To complete the catalytic cycle, this ester bond must be hydrolyzed and the OH of the serine must be regenerated.

❸ The amine diffuses away and is replaced by water.

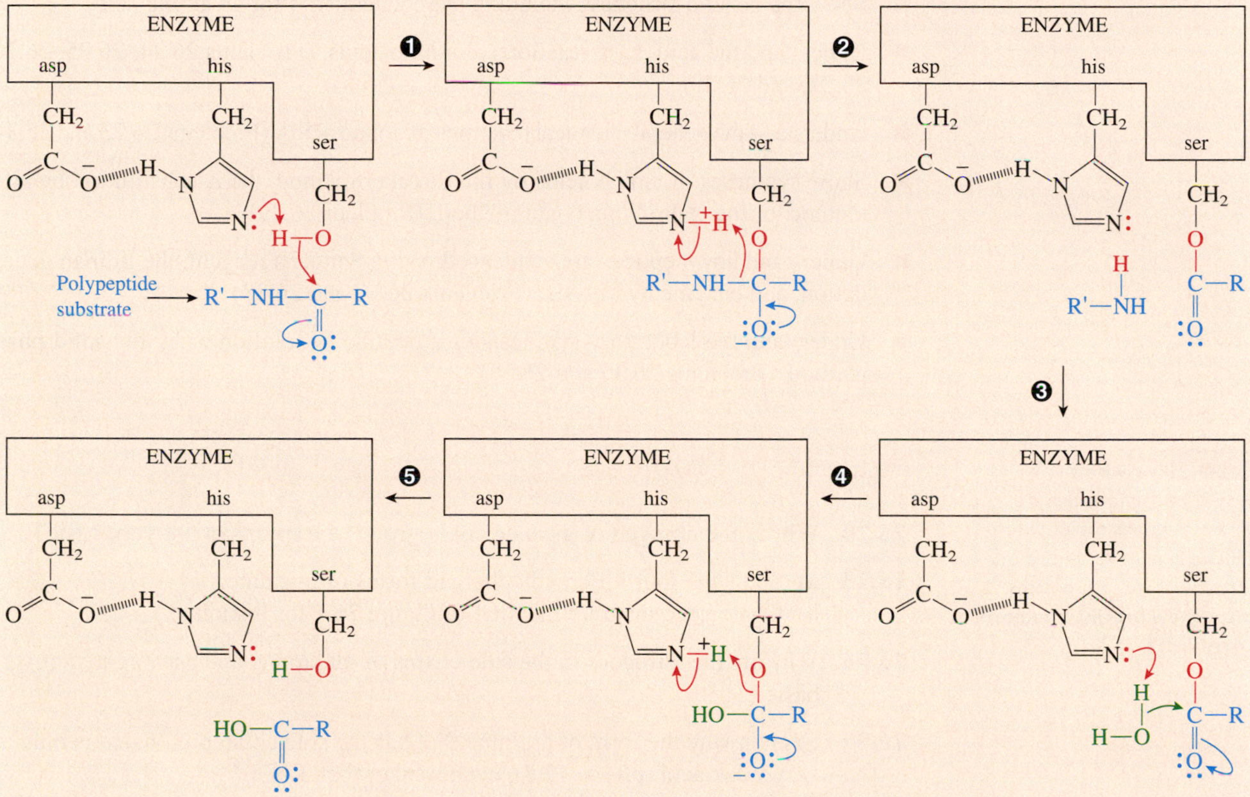

The remaining steps are just the reverse of the preceding steps.

❹ The water acts as a nucleophile, attacking the carbonyl group. Again, the histidine removes the proton from the water as this reaction proceeds.

❺ The serine oxygen acts as a leaving group, generating the free carboxylic acid, which diffuses from the active site. The enzyme is now ready for another catalytic cycle.

Figure 26.8

MECHANISM OF PEPTIDE HYDROLYSIS CATALYZED BY CHYMOTRYPSIN.

polypeptides using the hydroxy group of a serine residue as a nucleophile. Its mechanism is outlined in Figure 26.8. First, the substrate, a polypeptide, is bound to the enzyme active site. (For simplicity this binding is not shown in the mechanism in Figure 26.8, but it is of critical importance because it makes the rest of the steps favorable by making them intramolecular. This binding also gives chymotrypsin its selectivity. Large hydrophobic groups, such as a phenyl group, are bound more tightly at the active site of chymotrypsin, so it preferentially cleaves peptides after amino acid residues with such groups—that is, after phenylalanine, tyrosine, and tryptophan.) The hydrolysis mechanism follows the same steps as shown in Figure 19.1, with the oxygen of the hydroxy group of a serine acting as the nucleophile.

These last two sections have necessarily been brief, but they provide some of the general principles governing the structure of proteins and the action of enzymes. A textbook of biochemistry should be consulted for more details.

Review of Mastery Goals

After completing this chapter, you should be able to:

■ Show the general structure, including stereochemistry, for an amino acid.

■ Understand the acid–base reactions of amino acids. (Problems 26.24, 26.25, 26.26, 26.27, and 26.28)

■ Understand the general chemical reactions of amino acids. (Problems 26.33 and 26.34)

■ Show syntheses of amino acids by the Strecker method, the α-substitution by NH_3 method, or the Gabriel/malonate method. (Problem 26.29)

■ Understand how peptides are sequenced, using Sanger's reagent, the Edman degradation, and enzyme hydrolysis. (Problems 26.37 and 26.38)

■ Understand the laboratory synthesis of a peptide in solution or by the solid phase method. (Problems 26.35 and 26.36)

Additional Problems

26.24 Why is the nitrogen of the side chain group of tryptophan not very basic?

26.25 **a)** Show the four differently charged forms of histidine.
b) Construct a diagram like that of Figure 26.1 for histidine.

26.26 Why are the nitrogens in the side chains of glutamine and asparagine not very basic?

26.27 Explain why the form of arginine that has the side chain protonated is only a very weak acid ($pK_a = 12.5$), even weaker than RNH_3^+.

26.28 Explain which nitrogen of the side chain ring of histidine is protonated in the monocationic form.

26.29 Show syntheses of isoleucine by the Strecker method, the α-substitution by NH_3 method, and the Gabriel/malonate method.

26.30 Show a mechanism for the second part of the Strecker synthesis:

$$\underset{\substack{\displaystyle CH_3CH \\ }}{\overset{\displaystyle NH}{\|}} \quad \xrightarrow[H_2O]{NaCN} \quad \underset{\substack{\displaystyle CH_3CH \\ \displaystyle | \\ \displaystyle C\equiv N}}{\overset{\displaystyle NH_2}{|}}$$

26.31 Show all of the steps in the mechanism for this reaction:

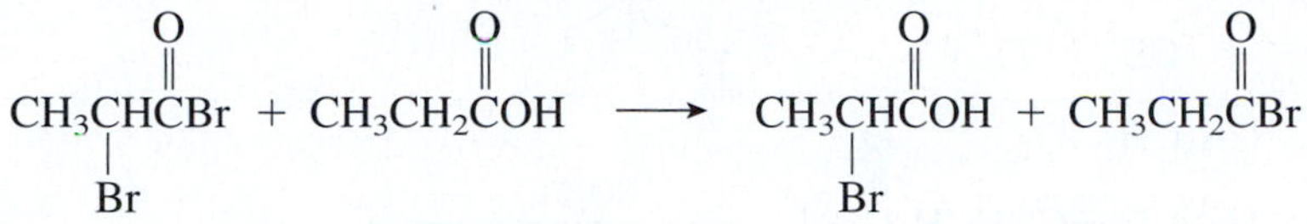

$$\underset{\underset{Br}{|}}{CH_3CHCBr} + CH_3CH_2\overset{O}{\overset{||}{C}}OH \longrightarrow \underset{\underset{Br}{|}}{CH_3CH}\overset{O}{\overset{||}{C}}OH + CH_3CH_2\overset{O}{\overset{||}{C}}Br$$

26.32 Show all of the steps for the mechanism of the first reaction in the Edman degradation:

$$Ph-N=C=S + RNH_2 \longrightarrow PhNH\overset{S}{\overset{||}{C}}NHR$$

26.33 Show the products of these reactions:

a) 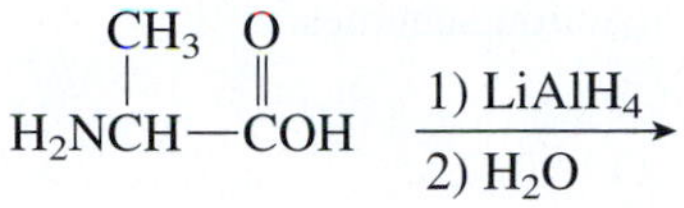 $H_2N\underset{\underset{CH_3}{|}}{CH}-\overset{O}{\overset{||}{C}}OH \xrightarrow[\text{2) H}_2\text{O}]{\text{1) LiAlH}_4}$

b) $H_2N\underset{\underset{CH_2Ph}{|}}{CH}CO_2H$ + [cyclic anhydride structure with two C=O and O] $\longrightarrow$

c) $H_2N\underset{\underset{CH_3}{|}}{CH}-\overset{O}{\overset{||}{C}}OH + CH_3CH_2OH \xrightarrow[\text{HCl}]{\text{excess}}$

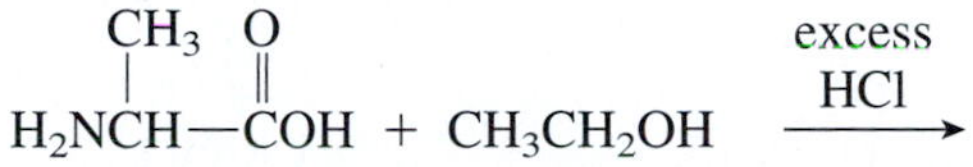

d) 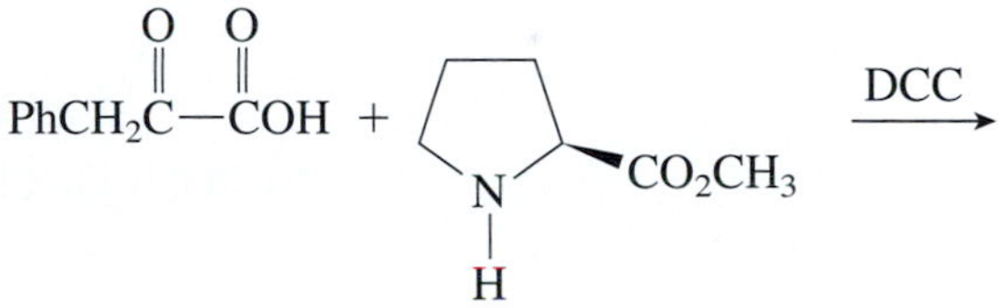 $PhCH_2\overset{O}{\overset{||}{C}}-\overset{O}{\overset{||}{C}}OH$ + [pyrrolidine ring with N–H and CO$_2$CH$_3$ substituent] $\xrightarrow{\text{DCC}}$

26.34 Show the products of these reactions:

a) 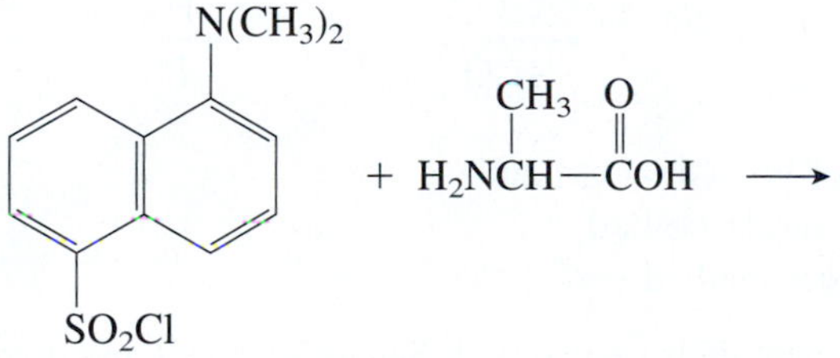 [naphthalene with N(CH$_3$)$_2$ group and SO$_2$Cl group] $+ H_2N\underset{\underset{CH_3}{|}}{CH}-\overset{O}{\overset{||}{C}}OH \longrightarrow$

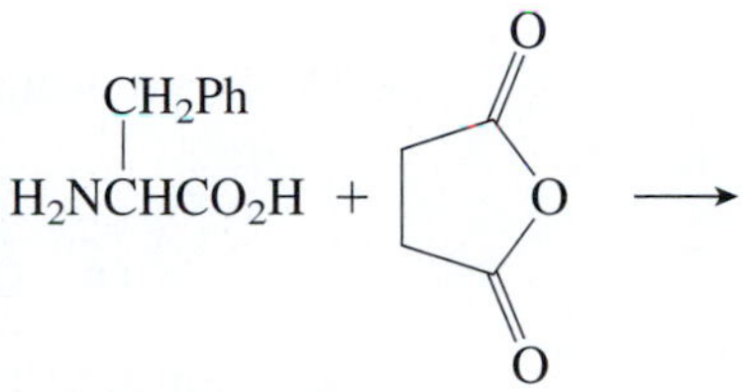

b) 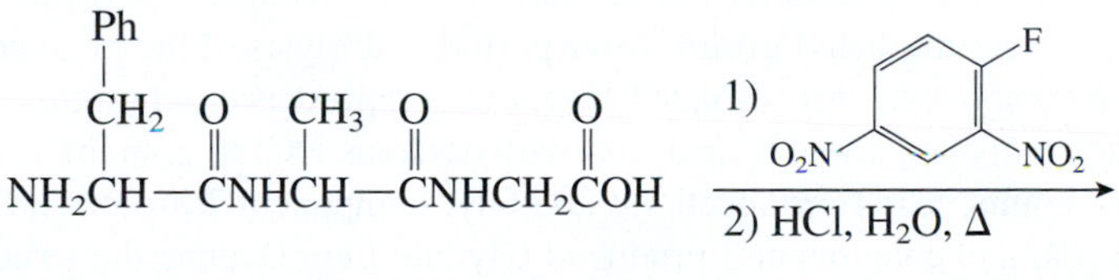 $NH_2\underset{\underset{CH_2}{|}{\underset{|}{Ph}}}{CH}-\overset{O}{\overset{||}{C}}NH\underset{\underset{CH_3}{|}}{CH}-\overset{O}{\overset{||}{C}}NHCH_2\overset{O}{\overset{||}{C}}OH \xrightarrow[\text{2) HCl, H}_2\text{O, }\Delta]{\text{1) [2-fluoro-1,4-dinitrobenzene: F, O}_2\text{N, NO}_2\text{]}}$

c)

$$\underset{\substack{| \\ CH_3CH \\ |}}{CH_3} \quad \underset{\substack{\| \\ }}{O} \quad \underset{\substack{\| \\ }}{O}$$

c) $H_2NCH-CNHCH_2COH$ $\xrightarrow[\text{2) HCl, H}_2\text{O, }\Delta]{\text{1) [naphthalene ring with N(CH}_3)_2\text{ and SO}_2\text{Cl]}}$

d)

$$\underset{\substack{Ph \\ | \\ CH_2 \\ |}}{} \quad \underset{\substack{O \\ \| \\ }}{} \quad \underset{\substack{CH_3 \\ | \\ }}{} \quad \underset{\substack{O \\ \| \\ }}{} \quad \underset{\substack{O \\ \| \\ }}{}$$

d) $NH_2CH-CNHCH-CNHCH_2COH$ $\xrightarrow[\substack{\text{2) HF} \\ \text{3) HCl, H}_2\text{O}}]{\text{1) PhN}=\text{C}=\text{S}}$ $\xrightarrow[\substack{\text{2) HF} \\ \text{3) HCl, H}_2\text{O}}]{\text{1) PhN}=\text{C}=\text{S}}$

26.35 Show the missing products in these reaction schemes:

a)

$$\underset{\substack{Ph \\ | \\ O \quad CH_2 \quad O \\ \| \qquad | \qquad \|}}{}$$

a) $t\text{-BuOCNHCH-COH}$ $\xrightarrow[\text{DCC}]{\underset{\substack{CH_3 \quad O \\ | \qquad \| \\ NH_2CH-COCH_3}}{}}$ $\xrightarrow[\text{H}_2\text{O}]{\text{HCl}}$ $\xrightarrow[\text{H}_2\text{O}]{\text{NaOH}}$

b)

$$\underset{\substack{O \\ \|}}{} \qquad \underset{\substack{O \\ \|}}{}$$

b) $PhCH_2OCCl + NH_2CH_2COH$ $\longrightarrow$ $\xrightarrow[]{\underset{\substack{O \\ \| \\ ClCOEt}}{}}$

$$\xrightarrow[\text{H}_2\text{O}]{\text{NaOH}} \qquad \xleftarrow[\text{Pt}]{\text{H}_2} \qquad \underset{\substack{O \\ \| \\ NH_2CH_2COCH_3 \downarrow}}{}$$

26.36 Show a synthesis of Phe-Gly-Ala-Leu
  a) by the solution phase method.
  b) by the solid phase method.

26.37 An unknown pentapeptide is treated with Sanger's reagent and then hydrolyzed with aqueous HCl to produce Val, Gly, Leu, Ala, and Phe labeled with the dinitrophenyl group. After partial hydrolysis of the pentapeptide, three components were isolated from the complex reaction mixture. Each of these components was hydrolyzed with aqueous HCl to give the following results: component 1 produced Val and Gly; component 2 produced Phe, Leu, and Ala; and component 3 produced Gly and Leu. Deduce the structure of the pentapeptide.

26.38 Bradykinin is a nonapeptide that causes severe pain. It is formed in humans in response to stimuli such as a wasp sting. Complete hydrolysis of bradykinin produces 3 Pro, 2 Phe, 2 Arg, 1 Gly, and 1 Ser. After reaction with Sanger's reagent and complete hydrolysis, an Arg is labeled with the dinitrophenyl group. Reaction with carboxypeptidase first produces Arg, followed by Phe. Partial hydrolysis gives a complex mixture from which five components can be isolated. Treatment of each of these components with Sanger's reagent followed by hydrolysis gives the following results:

Component 1 gives 2 Pro and Arg labeled with a dinitrophenyl group;
Component 2 gives Arg, Phe, and Pro labeled with a dinitrophenyl group;
Component 3 gives Gly and Pro labeled with a dinitrophenyl group;
Component 4 gives Ser, Phe, and Gly labeled with a dinitrophenyl group;
Component 5 gives Ser and Phe labeled with a dinitrophenyl group.

Deduce the sequence of the amino acids of bradykinin.

Nucleotides and Nucleic Acids

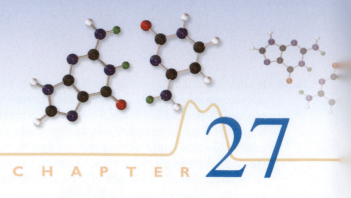

MASTERING ORGANIC CHEMISTRY

▶ Understanding the General Structures of Nucleosides, Nucleotides, DNA, and RNA

▶ Recognizing the Hydrogen Bonding That Occurs between Complementary Bases

▶ Understanding the General Features of Replication, Transcription, and Translation

▶ Understanding How the Sequence of DNA Is Determined

▶ Understanding How DNA Is Synthesized in the Laboratory

T HE NUCLEIC ACIDS, DNA and RNA, store genetic information in living organisms and are responsible for translating this information into the structure of proteins. Because their structure and function are discussed in great detail in modern biochemistry texts, this chapter concentrates on the organic chemical aspects of these important biomolecules.

The chapter begins with a discussion of the structure of nucleosides and nucleotides. Then the structure of the nucleic acids, DNA and RNA, the polymers formed from nucleotide monomers, is presented. The function of these polymers in the replication, transcription, and translation of genetic information is briefly addressed. Next, the organic chemistry involved in determining the sequence of DNA is presented. Finally, the synthesis of small DNA molecules in the laboratory is discussed.

27.1 NUCLEOSIDES AND NUCLEOTIDES

Nucleosides are composed of a sugar, either ribose or 2′-deoxyribose, and a base, a cyclic nitrogen-containing compound. These groups are connected by a bond between one of the nitrogen atoms of the base and the anomeric carbon (acetal carbon) of the sugar (a nitrogen analog of a glycoside bond). The nucleoside cytidine is formed from the sugar ribose and the base cytosine:

A Nucleoside

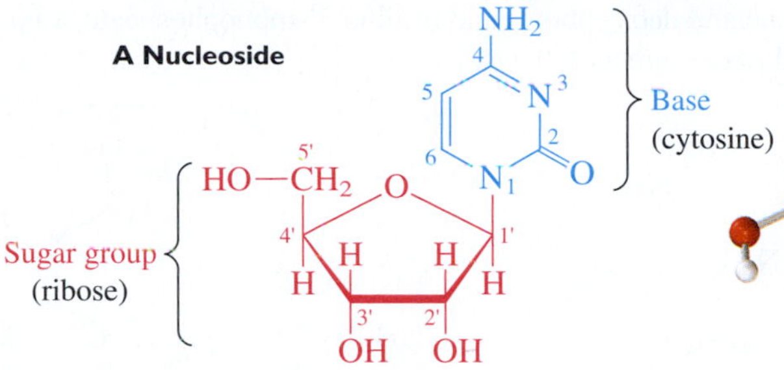

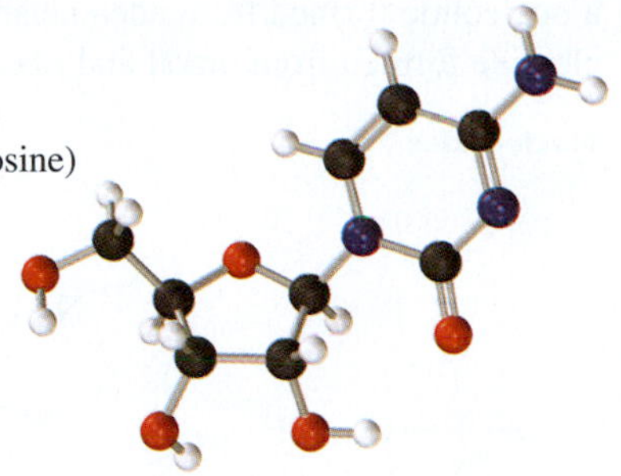

Cytidine

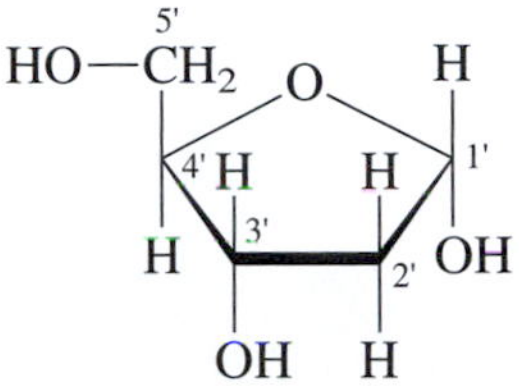

In nucleosides the atoms of the heterocyclic base are numbered 1, 2, and so on, and the positions of the sugar are numbered with primes, that is, $1'$, $2'$, and so on. Based on this numbering system, the base is attached to the $1'$ position of the sugar. Nucleosides may also be formed from the sugar $2'$-deoxyribose, which differs from ribose only in that the hydroxy group at the $2'$ position has been replaced with a hydrogen.

Five nitrogen bases are commonly found in nucleosides. Three of these—cytosine, thymine, and uracil—are based on the pyrimidine ring system. They are attached to the sugar by the nitrogen at position 1. The other two bases—adenine and guanine—are based on the purine ring system. They are attached to the sugar by the nitrogen at position 9:

2'-Deoxyribose

Nitrogen bases

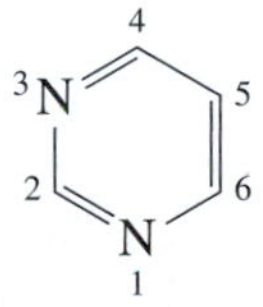

| Pyrimidine | Cytosine | Thymine | Uracil |

| Purine | Adenine | Guanine |

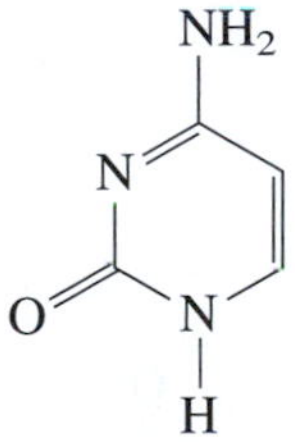

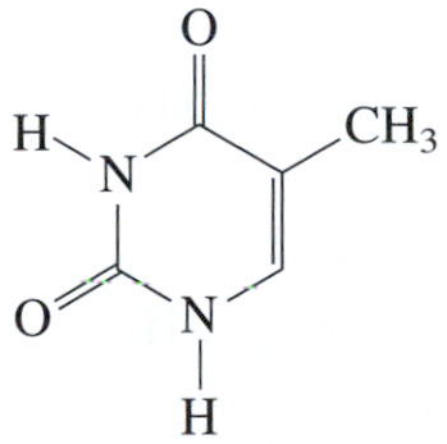

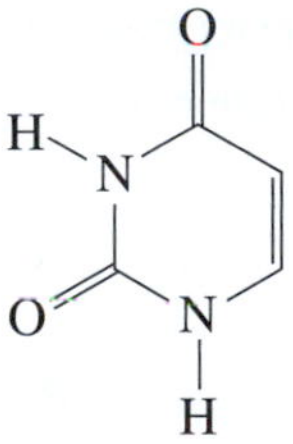

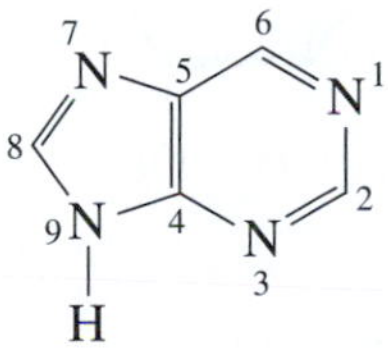

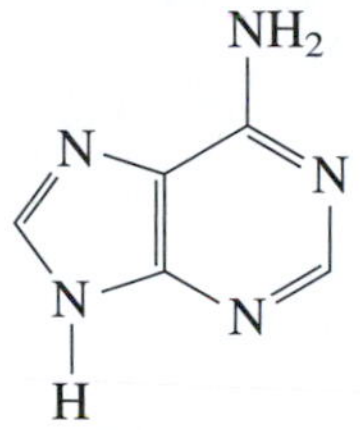

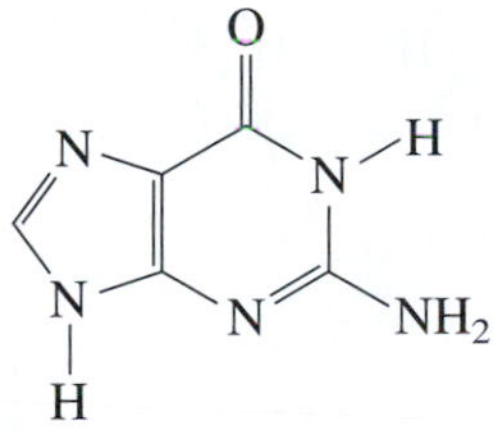

The names of the nucleosides formed from these bases and ribose are cytidine, thymidine, uridine, adenosine, and guanosine. Nucleosides formed from deoxyribose are named using the prefix deoxy-, as in deoxythymidine.

Nucleotides differ from nucleosides in that they have phosphate groups attached to either the $3'$ or $5'$ position of the sugar. The structures of deoxyadenosine $5'$-monophosphate, a nucleotide formed from adenine and deoxyribose, and uridine 3'-monophosphate, a nucleotide formed from uracil and ribose, are as follows:

Nucleotides

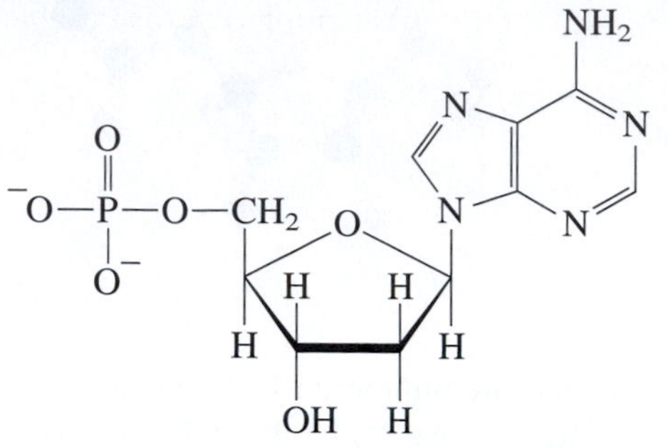

Deoxyadenosine
5'-monophosphate

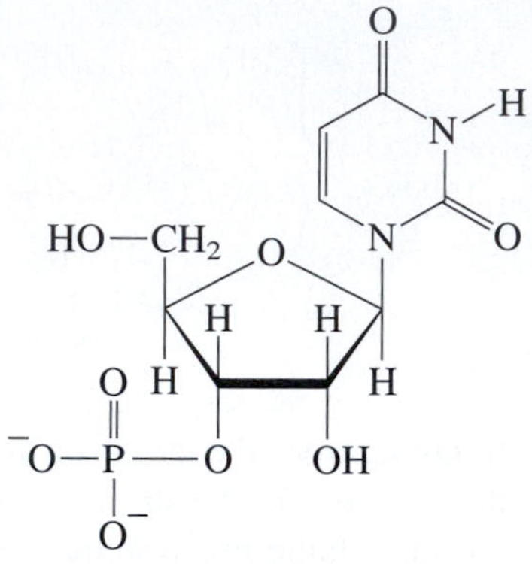

Uridine
3'-monophosphate

PROBLEM 27.1

Draw the structures of these compounds:
a) Thymidine
b) Deoxyguanosine
c) Deoxyuridine 5'-monophosphate
d) Deoxycytidine 3'-monophosphate

PROBLEM 27.2

Provide names for these compounds:

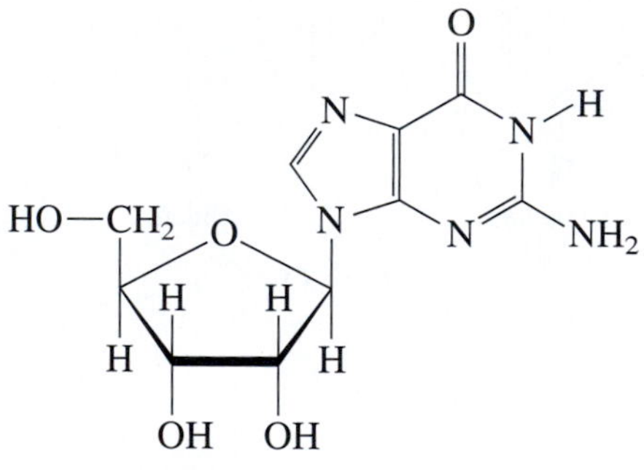

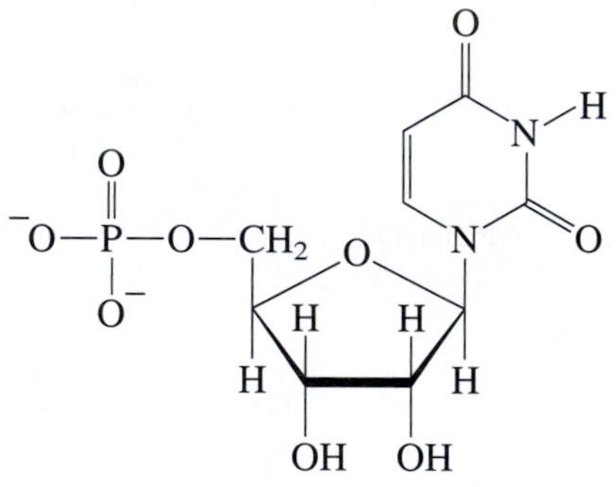

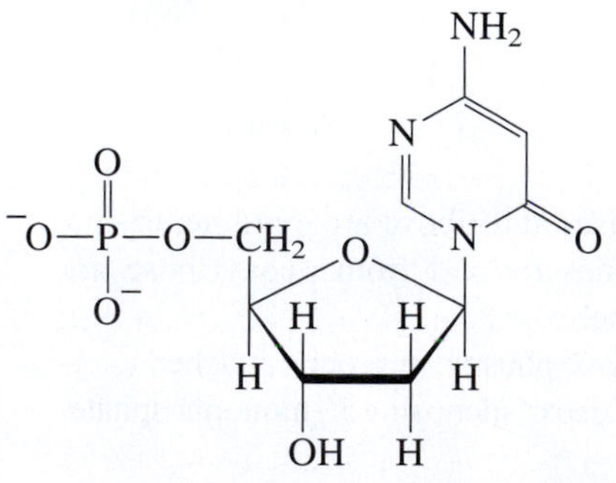

PROBLEM 27.3
Explain why pyrimidine and purine are aromatic compounds.

PROBLEM 27.4
Show a resonance structure for each of these compounds that emphasizes its aromatic character:
a) Guanine
b) Cytosine
c) Thymine

In addition to their role in the formation of DNA and RNA (see Section 27.2), nucleotides have other important biological functions. For example, adenosine triphosphate (ATP) is an important energy carrier in biochemical reactions, and nicotinamide adenine dinucleotide is a coenzyme that is often involved in biochemical oxidation–reduction reactions.

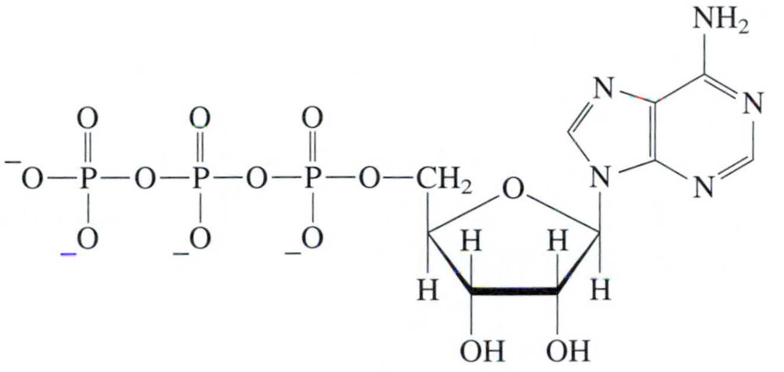

Adenosine triphosphate (ATP)

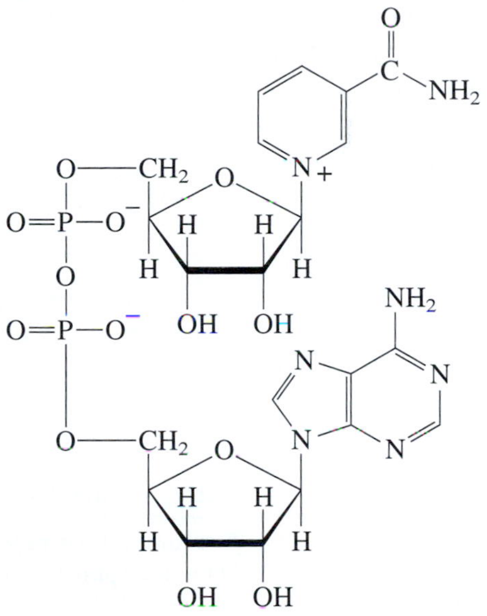

Nicotinamide adenine dinucleotide (NAD$^+$)

27.2 STRUCTURE OF DNA AND RNA

Nucleotides have several functional groups, so they can form condensation-type polymers. The formation of deoxyribonucleic acid (DNA) can be viewed as resulting from the polymerization of deoxyribonucleotides by the formation of a phosphate ester bond between a phosphate group at the 5′ position of one nucleotide and the hydroxy group at the 3′ position of another. A short tetranucleotide is shown in Figure 27.1. It has a backbone of alternating sugar and phosphate groups, with the heterocyclic bases appended to the sugars. DNA is just a much larger version of a

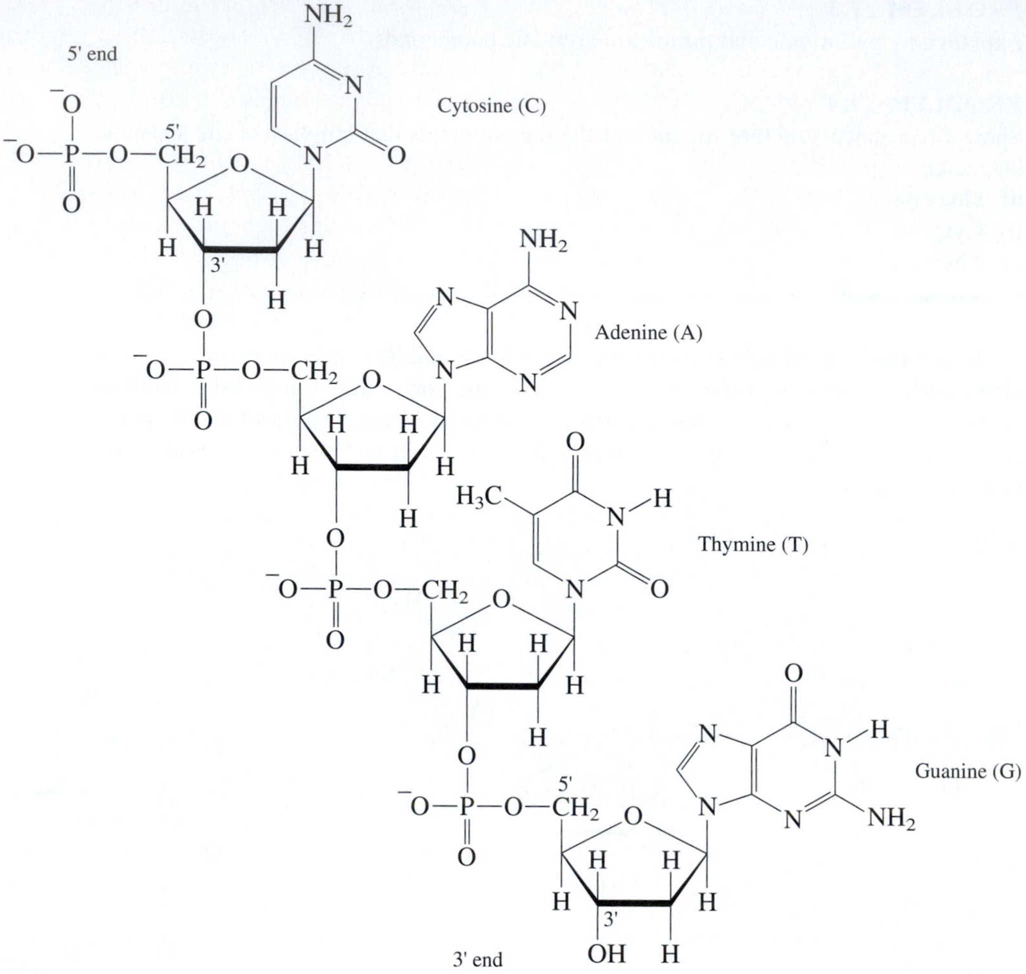

Active Figure 27.1

ORGANIC
Chemistry•ᐧ•Now™

A TETRANUCLEOTIDE WITH THE GENERAL STRUCTURE OF DNA. Test yourself on the concepts in this figure at **OrganicChemistryNow.**

polymer with this general structure. If the negative oxygens of the phosphate groups of DNA were protonated, then each of these groups would be a diester of phosphoric acid. As shown in the following equation, this ester is a relatively strong acid, comparable in strength to phosphoric acid, which has $pK_{a1} = 2.2$. Therefore, nucleic acids are strong acids and are completely ionized at the near neutral pH that occurs in living organisms.

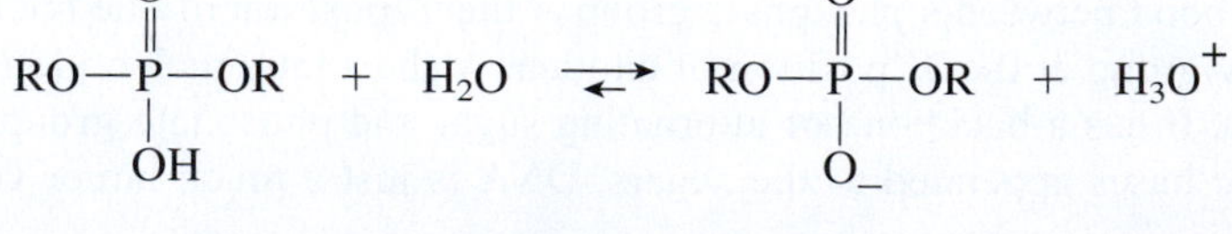

$$RO-\overset{\overset{\displaystyle O}{\|}}{\underset{\underset{\displaystyle OH}{|}}{P}}-OR \ + \ H_2O \ \rightleftharpoons \ RO-\overset{\overset{\displaystyle O}{\|}}{\underset{\underset{\displaystyle O_-}{|}}{P}}-OR \ + \ H_3O^+$$

By the early 1950s, DNA was known to have the general structure shown in Figure 27.1. However, how these molecules are arranged in the cell, how they are used to store genetic information, and how they provide a mechanism for the replication of that genetic information were a mystery until James Watson and Francis Crick solved the puzzle. In 1953, Watson and Crick published a paper that proposed an explanation for the overall structure of DNA and showed that certain features of this structure explained how genetic information was replicated when the cell divided. This paper is certainly one of the most important scientific publications ever and is generally recognized to mark the beginning of modern biochemistry and molecular biology. Watson, Crick, and M. Wilkins were awarded the 1962 Nobel Prize in physiology or medicine for this work. Wilkins obtained the X-ray diffraction patterns that provided some vital information about the structure of DNA.

A number of clues enabled Watson and Crick to make this important discovery. Wilkins's X-ray photographs suggested to Crick, an X-ray crystallographer by background, that DNA was a helical molecule with its aromatic bases arranged in a planar stack. Another clue was found in Chargaff's rules, which state that the number of adenine residues in a particular DNA molecule is always equal to the number of thymine residues and that the number of guanine residues is always equal to the number of cytosine residues. Using this information and building numerous models enabled Watson and Crick to solve the DNA puzzle.

The explanation for Chargaff's rules is provided by the selective hydrogen bonds, termed **complementary base pairing,** that form between adenine and thymine molecules and between guanine and cytosine. Adenine and thymine form two hydrogen bonds with an overall strength of about 10 kcal/mol (42 kJ/mol):

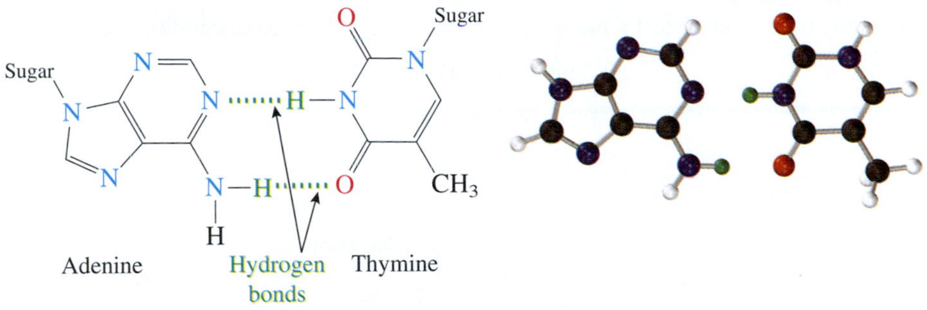

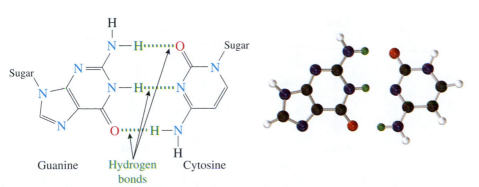

Guanine and cytosine form three hydrogen bonds with an overall strength of about 17 kcal/mol (71 kJ/mol):

ORGANIC
Chemistry ⚛ Now™
Click *Coached Tutorial Problems*
for more experience with
Complementary Base Pairing.

These interactions are strong enough to bind one DNA chain to its complementary chain under most circumstances, thus providing stability for the storage of genetic information. However, on occasions when the chains need to be separated—that is, when they are to be replicated or used to delineate the sequence of amino acids in a protein that is being synthesized—the hydrogen bonds are weak enough to be readily broken.

MODEL BUILDING PROBLEM 27.1

Build models to show the hydrogen bonding that occurs in these base pairs:
a) Adenine and thymine
b) Guanine and cytosine

The model proposed by Watson and Crick for DNA has two polymer chains running in opposite directions and held together by hydrogen bonds between complementary bases. A schematic representation of this structure is shown in Figure 27.2. The left chain has its 5′ end at the top of the page and its 3′ end at the bottom. The right chain has its 5′ end at the bottom of the page and its 3′ end at the top. Note that the bases in the chains are not identical. Instead, they are complementary; that is, wherever one chain has an adenine, the other has a thymine and so forth. Because of the specific hydrogen bonds that form between complementary base pairs, the order of the bases in one chain dictates the order in the other. This means that both chains provide enough information to allow the construction of the other chain. This is used in replication and also helps to ensure that the information in the base sequence is maintained with high fidelity.

PROBLEM 27.5

Show the base sequence of a piece of DNA that is complementary to this piece:

3′ end ATTGCGAGC 5′ end

Figure 27.2

SCHEMATIC REPRESENTATION OF BASE PAIRING BETWEEN DNA CHAINS. A = adenine, T = thymine, G = guanine, C = cytosine.

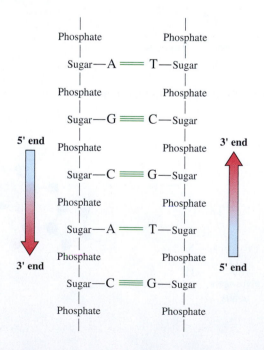

As shown in Figure 27.3, the two chains of this dimeric polymer coil around each other in a helical manner, producing the famous DNA double helix. The distance between the sugar–phosphate backbones of the two chains remains constant, regardless of which bases connect them because a purine base (adenine or guanine) is always hydrogen bonded to a pyrimidine base (thymine or cytosine). The less polar bases are on the inside of the helix, where their interaction with the water molecules of the aqueous environment is minimized, whereas the ionic phosphate groups are on the outer surface of the helix, where they can be readily solvated. In addition, the planes of the bases are stacked one on top of another. The interactions between these stacked bases are important in stabilizing the double helix.

DNA is a truly enormous molecule. Human DNA contains approximately 3 billion base pairs grouped into 23 individual DNA molecules or chromosomes. Each base pair contributes 3.4 Å to the length of a DNA molecule. Therefore, the total length of the DNA in the 23 human chromosomes is approximately $(3.4 \times 10^{-10}\,\text{m})(3 \times 10^9) \cong 1$ m, but it is only 20 Å in diameter. Obviously, DNA is highly coiled in the cell.

The chemical behavior of DNA is primarily what would be expected on the basis of its structure. The sugar–phosphate backbone is held together by phosphate ester bonds. These bonds are not too different from carboxylic ester bonds. Thus, the phosphodiester

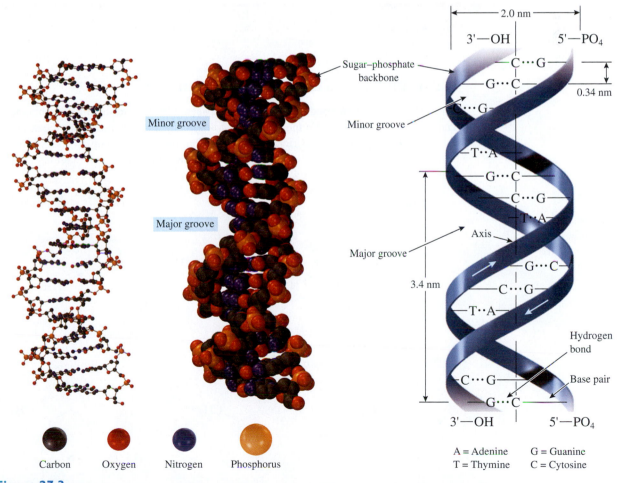

Figure 27.3

MODEL OF THE DNA DOUBLE HELIX. Hydrogens have been omitted for clarity.

linkages can be hydrolyzed in acidic solution. The heterocyclic bases are attached to the ribose or deoxyribose groups by a nitrogen analog of a glycoside bond. Like its oxygen analog, the N-glycosidic bond is also hydrolyzed by aqueous acid.

RNA has the same general structure as DNA with several exceptions. It employs the sugar ribose rather than deoxyribose. It has the base uracil in place of thymine. Note that thymine and uracil are both pyrimidine bases, and both hydrogen bond to adenine in the same manner. Unlike DNA, RNA usually occurs as a single-stranded molecule. Because RNA is used to transfer and translate the information that is stored in DNA the complementary copy is not necessary and is not synthesized by the cell. Finally, RNA is much smaller than DNA.

PROBLEM 27.6

Use drawings to show the hydrogen bonding that occurs in a base pair formed from adenine and uracil.

PROBLEM 27.7

Use drawings to explain why thymine and guanine do not form a base pair that has hydrogen bonds as strong as those between cytosine and guanine.

Focus On

Tautomers of Guanine and Thymine

The specific hydrogen bonding that occurs between adenine and thymine or guanine and cytosine was of crucial importance in the development of the model for the structure of DNA by Watson and Crick. However, at that time, many people believed that guanine existed primarily as the enol tautomer and thymine as the dienol tautomer because both of these structures would be fully aromatic.

Guanine

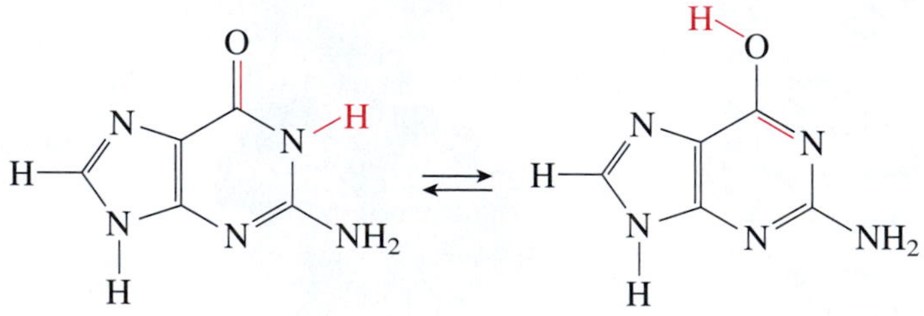

Keto tautomer   Enol tautomer

Thymine

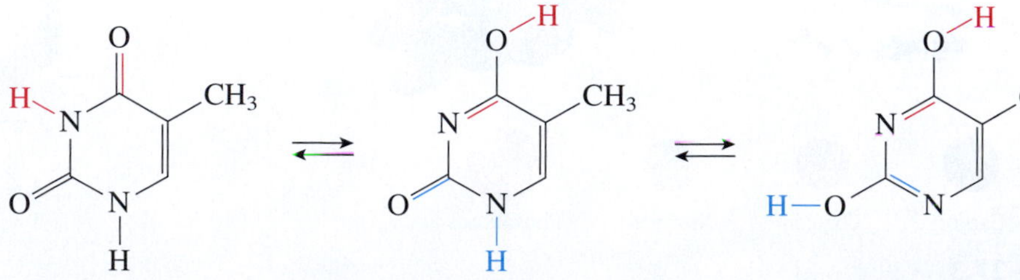

Keto tautomer   Enol tautomer   Dienol tautomer

Actually, because of the strength of the CO double bond and the fact that not much resonance energy is lost, the keto tautomers are more stable. Fortunately, Watson's officemate was an expert on X-ray structures of small organic molecules and knew that the keto isomers were the predominant tautomers. Without this information, Watson and Crick would probably have used the incorrect structures in their model-building studies. Of course, the enol tautomers do not form the same specific hydrogen bonding interactions with the other bases, so this important clue to the structure and function of DNA would have remained hidden. Luck sometimes plays an important role in great discoveries.

27.3   REPLICATION, TRANSCRIPTION, AND TRANSLATION

Genetic information is stored in DNA molecules. The sequence of bases in a particular piece of DNA specifies the sequence of amino acids in a particular protein. Exactly how the DNA is replicated when a cell divides and how the information in the DNA sequence is converted to an amino acid sequence of a protein are of critical importance to the functioning of a living organism and comprise a major part of any modern biochemistry textbook. We have space here only to briefly outline these processes.

The information for the amino acid sequence of a protein is stored in the base sequence of DNA. However, 20 amino acids are found in proteins, whereas only 4 bases occur in DNA. This means that a single base cannot code for an individual amino acid. Likewise, a two-base code, which provides $4 \times 4 = 16$ combinations, still is not large enough to specify 20 different amino acids. The genetic code is actually based on a series of three bases, called a **codon**, which provides $4^3 = 64$ different possibilities. A codon for a particular amino acid is designated by listing the first letters of the three bases that compose it. Thus, one codon for serine is UCA, which designates a base sequence of uracil, cytosine, and adenine. The code is degenerate; that is, most amino acids are specified by two or more codons. For example, the codons CCC, CCU, CCA, and CCG all specify the amino acid proline. In addition, the codons UAA, UAG, and UGA all specify a stop signal; that is, they indicate the end of the protein chain.

The two complementary strands of DNA provide a stable reservoir for genetic information, which must be preserved over the entire lifetime of the organism. Either of the strands, by itself, has all of the genetic information. Therefore, if one strand is damaged, perhaps by random hydrolysis, the information to repair the damage is still present in the complementary strand. For example, suppose a G and a T were lost from the piece of DNA represented by the following schematic structure. When a repair enzyme encounters this damage, the C and A of the complementary strand ensure that the correct bases are inserted in the repair process.

Double strand of DNA   Damaged DNA   Repaired DNA

The double-stranded nature of DNA also provides a method for **replication**, the process whereby DNA is duplicated so that two identical copies are available when a cell divides. In this process, the DNA unwinds, and each strand serves as a template for the synthesis of its complementary strand. When replication is completed, two identical versions of the original DNA helix are present. This process is represented schematically in the following diagram:

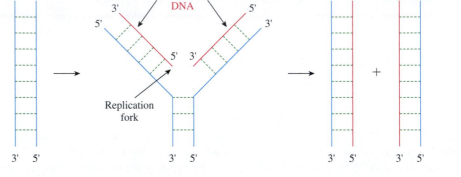

Original DNA   Unwinding and synthesis   Duplicate copies of original DNA

Like all biological processes, replication is controlled by enzymes. Enzymes unwind the DNA double helix so that the individual chains are accessible. Then an enzyme called DNA polymerase attaches a new nucleotide to a deoxyribose in the growing chain of the new DNA strand.

In replication, the nucleotide that is to be attached next is held in place by the DNA polymerase enzyme. This new nucleotide is complementary to the base in the original DNA strand. The reason usually given for the incorporation of the correct nucleotide (the complementary nucleotide) is the favorable hydrogen bonding that occurs between complementary bases. However, recent studies have shown that the size and shape of the base is at least as important. As shown in Figure 27.4, the new nucleotide has a triphosphate group attached to its 5′-hydroxy group. The 3′-hydroxy group on the end of the growing DNA chain reacts with the triphosphate to form a phosphate ester in a reaction that is quite similar to the reaction of an alcohol with a carboxylic acid anhydride to form the ester of a carboxylic acid. Of course, these reactions are under enzymatic control.

PROBLEM 27.8

The mechanism for the attachment of a new nucleotide to a growing DNA chain shown in Figure 27.4 is very similar to the mechanism for the formation of a carboxylic ester from an alcohol and a carboxylic anhydride. Suggest the steps in the mechanism for this reaction:

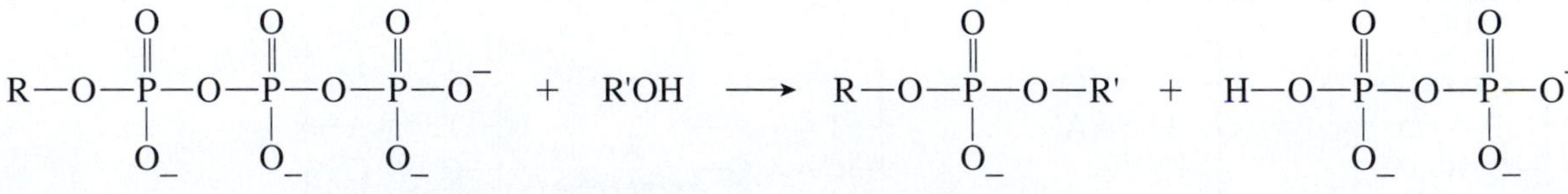

$$R-O-\overset{\overset{\displaystyle O}{\|}}{\underset{\underset{\displaystyle O^-}{|}}{P}}-O-\overset{\overset{\displaystyle O}{\|}}{\underset{\underset{\displaystyle O^-}{|}}{P}}-O-\overset{\overset{\displaystyle O}{\|}}{\underset{\underset{\displaystyle O^-}{|}}{P}}-O^- \;+\; R'OH \;\longrightarrow\; R-O-\overset{\overset{\displaystyle O}{\|}}{\underset{\underset{\displaystyle O^-}{|}}{P}}-O-R' \;+\; H-O-\overset{\overset{\displaystyle O}{\|}}{\underset{\underset{\displaystyle O^-}{|}}{P}}-O-\overset{\overset{\displaystyle O}{\|}}{\underset{\underset{\displaystyle O^-}{|}}{P}}-O^-$$

The transformation of the information in a DNA strand into a protein sequence involves several steps. First an RNA polymer, called **messenger RNA** (mRNA), that is complementary to the DNA is synthesized in much the same manner as the synthesis of new DNA described earlier. This process is called **transcription.** Messenger RNA

Figure 27.4

ATTACHMENT OF A NEW NUCLEOTIDE TO A GROWING DNA CHAIN.

then serves as a template for protein synthesis in a process called **translation.** Individual amino acids are attached to relatively small RNA molecules called **transfer RNA** (tRNA). Each amino acid has its own type of tRNA that has a three-base region, known as the **anticodon,** that is complementary to the codon for that amino acid. The tRNA with the correct amino acid forms three base pairs with the mRNA and brings the amino acid into position for attachment to the growing protein chain. The process for attaching a proline, followed by a phenylalanine, to a growing polypeptide chain is represented schematically in the following diagram:

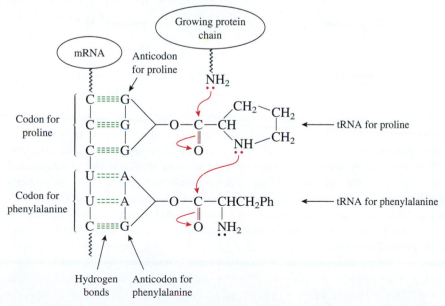

The tRNA for proline, which has the anticodon GGG, hydrogen bonds to the CCC codon for proline in the mRNA and brings its attached proline amino acid into position for attachment to the growing protein chain. Then the tRNA for phenylalanine hydrogen bonds to the codon for phenylalanine in the mRNA and brings its attached phenylalanine into position for attachment to the proline. This process continues until a stop signal is reached.

This has been a necessarily brief overview of the processes of replication, transcription, and translation. If you are interested in learning more, please consult a biochemistry textbook.

Focus On Biological Chemistry

Treatment of AIDS with AZT

Azidothymidine, also known as AZT or zidovudine, was initially prepared in the 1960s with the hope that it might be useful in the treatment of cancer. It is similar to the nucleoside deoxythymidine, except that it has an azido group in place of the 3′-hydroxy group. The reasoning was that this "fraudulent nucleoside" might be incorporated into new DNA chains that were being synthesized by the cancer cells. Because AZT has no 3′-hydroxy group, it would act to terminate the DNA chain and would therefore prevent cell division. Unfortunately, cancer cells recognize AZT as a fake and do not incorporate it.

Azidothymidine
(AZT)

2′,3′-Dideoxycytidine
(DDC)

The AIDS virus is a retrovirus; that is, it stores its genetic information in the form of RNA. On infection it injects its RNA into the target cell and uses an enzyme called reverse transcriptase to synthesize DNA that is complementary to this RNA template. AZT is accepted by reverse transcriptase as a building block for this synthesis and slows or prevents the conversion of the viral RNA information into DNA. By disrupting this process, AZT slows the replication of the virus in the cell. The nucleoside analog 2′,3′-dideoxycytidine (DDC) works in a similar manner. Current treatments, which use a cocktail of two nucleoside analogs and a protease inhibitor, seem to at least slow the progress of the disease.

27.4   SEQUENCING DNA

Determination of the sequence of bases in DNA is accomplished in a manner similar to the determination of the sequence of amino acids in a protein. The problem is somewhat simpler in that there are only four possible bases. However, DNA is much larger than a protein, so there is considerably more information to obtain. First it is necessary to break the enormous DNA molecule into specific fragments of more manageable size. Then the sequence of bases in these fragments must be determined. Several different cleavage processes must be employed to produce the fragments so that one set overlaps the cleavage points of another and can be used to determine how the fragments were connected.

Until recently, sequencing DNA was difficult because no methods were available to cleave the DNA at specific positions. In addition, no reaction analogous to the Edman degradation is available to remove one base at a time for analysis. However, several recent developments have simplified the sequencing of DNA considerably, so it is now easier than the sequencing of a protein. For this reason the sequence of a protein is often determined today by isolating the DNA that codes for that protein and sequencing it. Let's see how DNA sequencing is accomplished.

The discovery of **restriction endonucleases** was of crucial importance in sequencing DNA. These enzymes recognize a specific sequence of four to eight bases in double-stranded DNA and cleave the DNA at a precise point in this sequence. For example, the restriction endonuclease known as *Alu*I cleaves the sequence AGCT between the G and C and the one known as *Pst*I cleaves the sequence CTGCAG between the A and the G. These restriction endonuclease enzymes provide a reproducible way to produce precisely defined fragments of an appropriate size for sequencing.

Cloning techniques have also played an important role in the development of DNA sequencing. Cloning allows many copies of the DNA to be made so that enough material to sequence is available.

Two methods have been developed to sequence the DNA once enough copies of the correct length have been obtained. The Maxam-Gilbert method uses chemical reactions to cleave the DNA at specific bases. The other procedure, called the chain-terminator method, was developed by Frederick Sanger. (Walter Gilbert and Sanger shared the 1980 Nobel Prize in chemistry for developing these sequencing methods. This was Sanger's second Nobel Prize, making him one of the few to have won twice. Recall that he also was awarded one for developing methods to sequence peptides.) The chain-terminator method is used for most sequencing today, so let's examine how this method works.

The chain-terminator method relies on the use of the enzyme DNA polymerase I, which catalyzes the synthesis of DNA. This enzyme makes complementary copies of single-stranded DNA, starting from the 5′ end and proceeding to the 3′ end. First the DNA to be sequenced, termed the template strand, is isolated. To begin the synthesis, the enzyme requires the presence of a small piece of DNA, called a primer, at the 5′ end of the chain to be synthesized. Because the template DNA is prepared by a restriction endonuclease, a few bases at its 3′ end are known. A primer that is complementary to this sequence is prepared by chemical synthesis. The enzyme then adds the bases that are complementary to the template DNA to the 3′-hydroxy group of the primer.

Sequencing is accomplished by incubating the template DNA, the polymerase enzyme, the appropriate primer and the four nucleotides, which are supplied as triphosphates. One of the nucleoside triphosphates is provided in radioactive form, so the newly synthesized DNA is radioactive. Of critical importance, a small amount of the 2′,3′-dideoxynucleoside 5′-triphosphate corresponding to one of the nucleotides is also added. The structures of 2′-deoxyadenosine 5′-triphosphate (dATP) and 2′,3′-dideoxyadenosine 5′-triphosphate (ddATP) are as follows:

2′-Deoxyadenosine 5′-triphosphate
(dATP)

The 3′-hydroxy group is missing. ⟶

2′,3′-Dideoxyadenosine 5′-triphosphate
(ddATP)

Because there is only a small amount of dideoxynucleotide present, it is incorporated only occasionally. However, wherever it is incorporated into the growing piece of DNA, the polymerization is terminated because there is no 3′-hydroxy group to which the next nucleotide can be attached. This results in a series of DNA fragments of different lengths, each terminated at the base that was added as its dideoxy derivative. The synthesis of DNA fragments using dideoxycytosine triphosphate is outlined in Figure 27.5.

To determine the sequence of the template DNA, four versions of the polymerization are run, each with a different dideoxynucleoside triphosphate. The reactions are analyzed using gel electrophoresis. In this technique the mixtures of synthesized fragments from each reaction are placed in separate lanes at the top of a layer of a polyacrylamide gel and a voltage is applied between the top and the bottom. Smaller molecules move faster than larger molecules, so they appear closer to the bottom of the gel

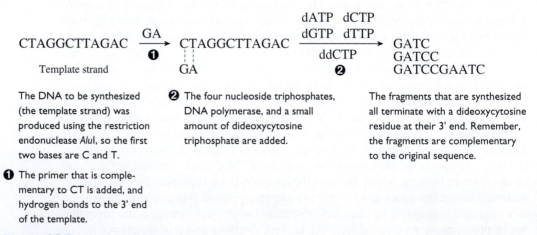

CTAGGCTTAGAC —GA→ ❶ CTAGGCTTAGAC   dATP dCTP / dGTP dTTP / ddCTP —❷→   GATC
                                          GA   GATCC
Template strand   GATCCGAATC

The DNA to be synthesized (the template strand) was produced using the restriction endonuclease *Alu*I, so the first two bases are C and T.

❶ The primer that is complementary to CT is added, and hydrogen bonds to the 3′ end of the template.

❷ The four nucleoside triphosphates, DNA polymerase, and a small amount of dideoxycytosine triphosphate are added.

The fragments that are synthesized all terminate with a dideoxycytosine residue at their 3′ end. Remember, the fragments are complementary to the original sequence.

Figure 27.5

SYNTHESIS OF **DNA** FRAGMENTS USING DIDEOXYCYTOSINE TRIPHOSPHATE IN THE CHAIN-TERMINATOR SEQUENCING METHOD.

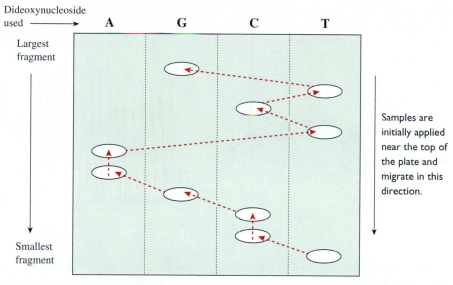

Dideoxynucleoside used ⟶   A   G   C   T

Largest fragment

Samples are initially applied near the top of the plate and migrate in this direction.

Smallest fragment

Figure 27.6

PATTERN OF GEL ELECTROPHORESIS SPOTS OBTAINED FROM SEQUENCING CTAGGCTTAGAC USING THE CHAIN-TERMINATOR SEQUENCING METHOD. The sequence that is complementary to the piece of DNA that is being sequenced is read starting at the bottom of the gel. The first two bases are from the primer (see Figure 27.5), so the smallest fragment terminates in T (the base complementary to A), so the lowest spot appears in the T lane. The sequence of the complementary strand is TCCGAATCTG.

when migration is halted. Once the separation is complete, a piece of film is placed over the gel. The location of the radioactive fragments can be determined by the spots that appear when the film is developed. Because the separation is based solely on size, the spot closest to the bottom of the plate is found in the lane of the reaction using the dideoxynucleoside base complementary to the first base in the template strand. The sequence of the complementary strand can be obtained by simply reading the gel from bottom to top. The electrophoresis gel pattern that would be obtained for the DNA fragment of Figure 27.5 is shown in Figure 27.6. A photograph of the autoradiogram of an actual sequencing gel is shown in Figure 27.7.

PROBLEM 27.9

Show the pattern of the gel electrophoresis spots that would be obtained from sequencing the DNA fragment CTTAGTTGCACCT using the chain-terminator method. The primer is GA.

An automatic sequencing instrument has been developed that uses the chain-terminator method. To avoid the use of radioactive labels, a different color fluorescent dye is attached to the primer in each of the four reactions used to synthesize the DNA fragments. The mixture of fragments from all four reactions is then analyzed using electrophoresis in a single lane. A fluorescent spot appears for each polynucleotide of increasing size. The 3′-terminal base for each spot can be determined by the color of the fluorescence. The detection system is computer controlled, and the acquisition of data is automated. A schematic representation

ATGC

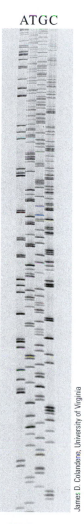

James D. Colandene, University of Virginia

Figure 27.7

AUTORADIOGRAM OF A SEQUENCING GEL.

Figure 27.8

SCHEMATIC DIAGRAM OF AN AUTOMATIC DNA SEQUENCER. Each lane represents a separate sequencing experiment.

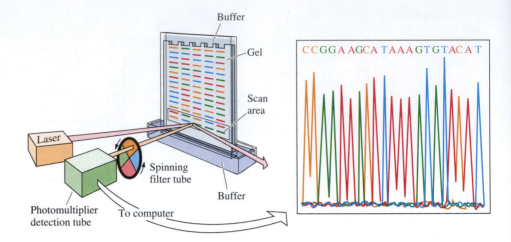

of such an instrument is shown in Figure 27.8. It was the use of numerous automatic sequencers of this type that enabled the determination of the sequence of the entire human genome (2.9 billion base pairs) to be completed in 2001.

27.5   LABORATORY SYNTHESIS OF DNA

The ability to chemically synthesize DNA is quite important. This synthetic DNA is used in cloning, in site-directed mutagenesis (which enables the preparation of a protein with a different amino acid at a single site), and in the diagnosis and prenatal detection of genetic diseases.

The basic strategy of polynucleotide synthesis is similar to that of polypeptide synthesis. Functionality in the nucleotide at which reaction is not desired must be protected. Then the nucleotides must be coupled. Next, the protecting group must be removed so that another cycle of coupling can be accomplished. After the desired cycles of coupling have been performed, any remaining protecting groups must be removed.

Currently, the most widely used synthetic method for DNA is the **phosphoramidite method,** a solid phase procedure that has many similarities to solid phase peptide synthesis. An example of the synthesis of a dinucleotide by this procedure is shown in Figure 27.9. First a protected nucleoside is attached to the surface of a silica particle through its 3′-hydroxy group. The free NH_2 groups of adenine, cytosine, and guanine are protected as amides. The 5′-hydroxy group is protected as a dimethoxytrityl ether. After the nucleoside has been attached to the silica, the trityl protecting group is removed by hydrolysis under S_N1 conditions. This is followed by reaction with a 3′-phosphoramidite derivative of the next nucleoside. The phosphorus in this compound is trivalent and is bonded to an oxygen of deoxyribose and to a diisopropylamino group. The remaining oxygen is protected as a β-cyanoethyl ether. The liberated 5′-hydroxy group displaces the diisopropylamino group, forming a phosphite triester with the desired bond to the phosphorus. Oxidation of the phosphite triester to the phosphotriester with iodine completes the addition of one nucleoside. This cycle of deprotection, coupling, and oxidation is repeated for each nucleoside that is to be added. After the desired number of cycles, the polynucleotide is removed from the silica particle by treatment with aqueous ammonia. This also removes

The first nucleoside is attached to the surface of a silica particle via its 3'-hydroxy group. This also serves to protect the 3'-hydroxy group during subsequent transformations. The NH$_2$ groups of adenine and cytosine are protected as benzamides, and the NH$_2$ group of guanine is protected as an isobutanamide. The 5'-hydroxy group is protected as a dimethoxytrityl ether (DMTr).

❶ Aqueous dichloroacetic acid is used to remove the dimethoxytrityl group in an S$_N$1 reaction.

❷ The unprotected 5'-hydroxy group is then reacted with a 3'-phosphoramidite derivative of the next nucleoside in the presence of tetrazole, which acts as a weak acid catalyst. (These phosphoramidite derivatives are now commercially available.) The diisopropylamino group is displaced by the 5'-hydroxy group, and the phosphorus–oxygen bond is formed.

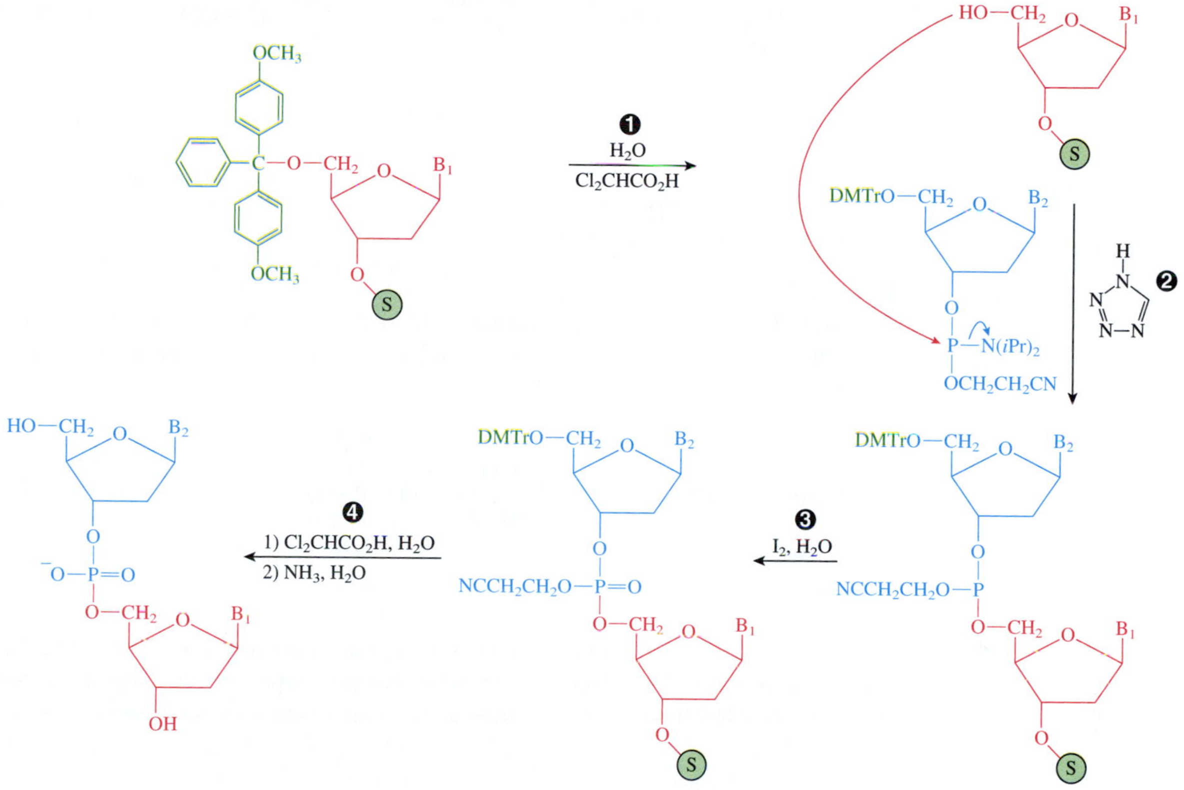

Figure 27.9

SOLID PHASE SYNTHESIS OF A POLYNUCLEOTIDE BY THE PHOSPHORAMIDITE METHOD.

❸ Iodine is used to oxidize the phosphite triester [P(OR)$_3$] to the phosphotriester [O=P(OR)$_3$].

❹ Dichloroacetic acid is used to remove the DMTr protecting group on the 5'-hydroxy group of the second nucleotide. At this point, a third nucleoside can be added and the cycle can be repeated. When all of the desired nucleosides have been added, reaction with aqueous ammonia is employed to remove the polynucleotide from the silica, remove all of the amide protecting groups from the bases, and to remove the β-cyanoethyl protecting group.

the amide protecting groups on the bases and the β-cyanoethyl protecting group. The polynucleotide is then purified by gel electrophoresis or by chromatography.

PROBLEM 27.10

Show a synthesis of this trinucleotide using the phosphoramidite method:

5' end A-T-G 3' end

PROBLEM 27.11

Show the mechanism for the cleavage of the dimethoxytrityl ether. Explain why the methoxy groups accelerate the reaction.

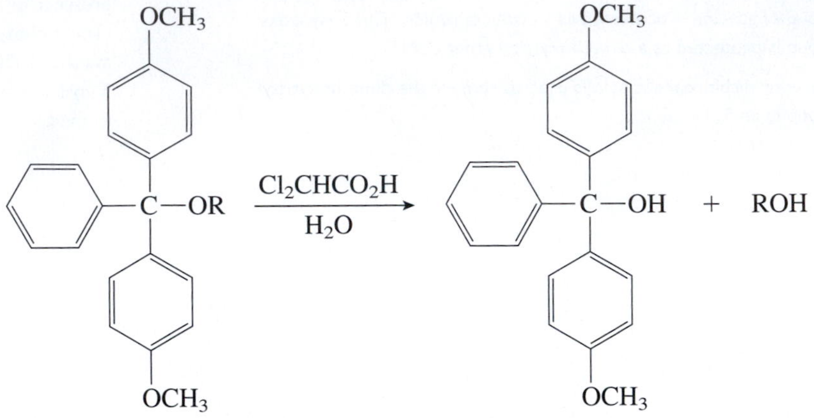

PROBLEM 27.12

Removal of the β-cyanoethyl protecting group occurs according to the following equation. Show a mechanism for this reaction. What is the role of the cyano group in this reaction?

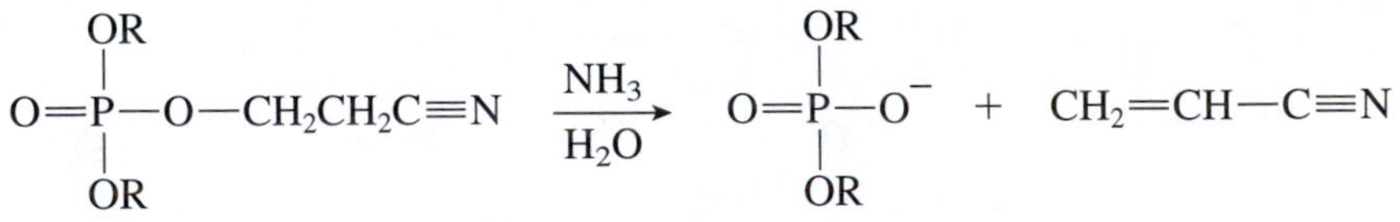

Like the solid phase peptide synthesis, this process has also been automated. Commercial automated synthesizers are available that can prepare polynucleotides containing more than 150 bases with a cycle time of about 10 minutes per base.

ORGANIC Chemistry•Now™

Click *Mastery Goal Quiz* to test how well you have met these goals.

Review of Mastery Goals

After completing this chapter, you should be able to:

- Show the general structures of nucleosides, nucleotides, DNA, and RNA. (Problems 27.13 and 27.14)
- Show the hydrogen bonding that occurs between adenine and thymine or uracil and between guanine and cytosine. (Problems 27.19, 27.24, and 27.25)
- Understand the general features of replication, transcription, and translation.
- Understand the chain-terminator method for determining the sequence of DNA.
- Show a reaction scheme for the synthesis of a polynucleotide. (Problem 27.20)

Additional Problems

27.13 Draw the structures of these compounds:
   a) Adenosine
   b) ddCTP
   c) Deoxythymidine 5′-monophosphate

27.14 Provide names for these compounds:

a)

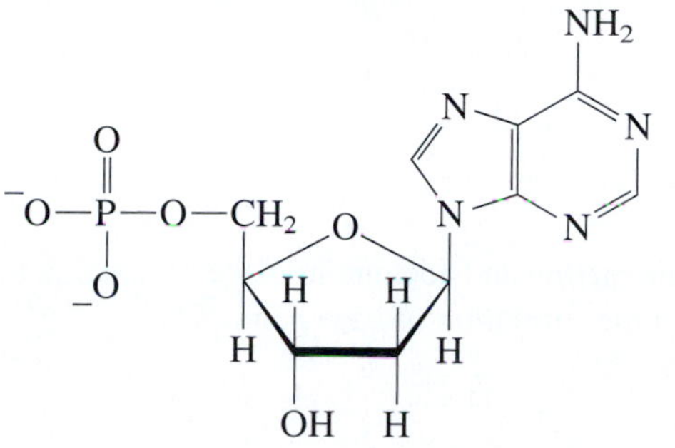

b)

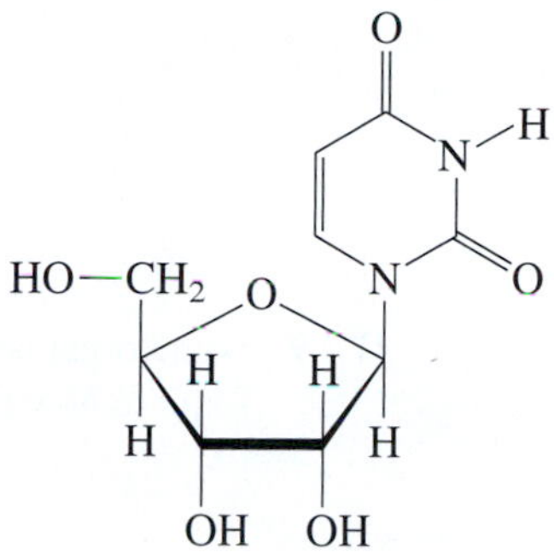

27.15 Explain why cytosine is a much stronger base than thymine or uracil.

27.16 Show the tautomers of uracil.

27.17 Show the sequence of the DNA and of the RNA that is complementary to this piece of DNA:

5′ end GCTTATGC 3′ end

27.18 Show the products of these reactions:

a)

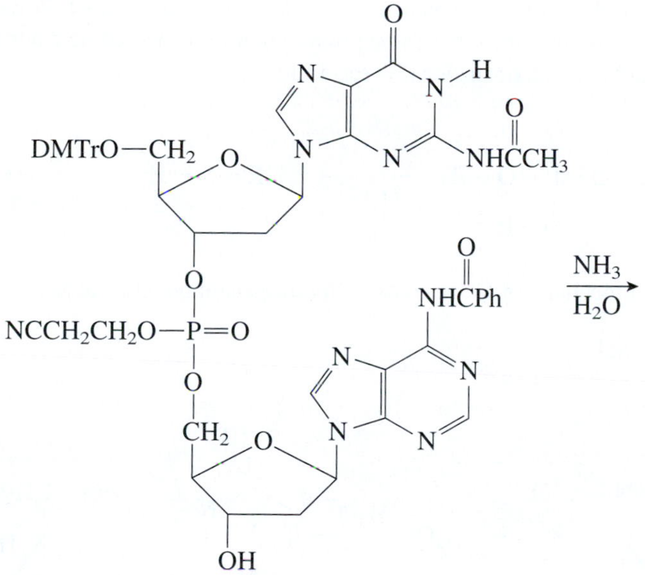

b)

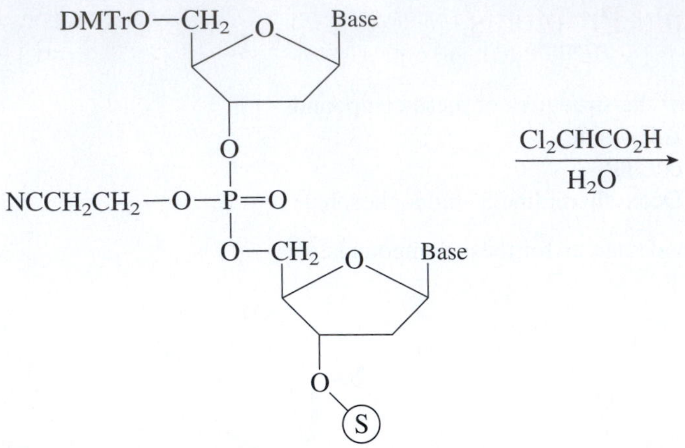

27.19 Some organisms use this methylated adenine in place of adenine in their DNA. Explain how this affects the formation of base pairs.

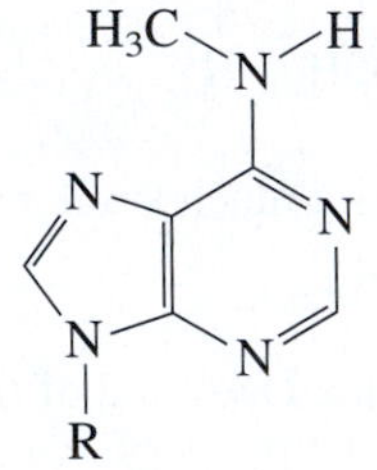

27.20 Show a synthesis of this trinucleotide by the phosphoramidite method:

5′ end GCT 3′ end

27.21 The hydrolysis of a phosphate ester under basic conditions follows a mechanism that is similar to that for saponification of a carboxylic ester. Show the steps in the mechanism for this reaction:

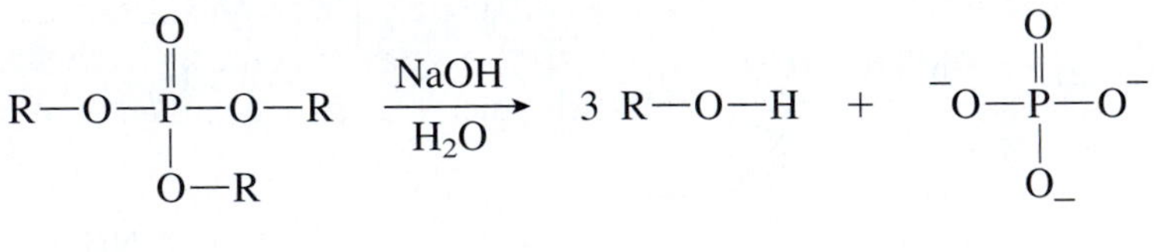

27.22 Show a mechanism for nucleoside hydrolysis in aqueous acid.

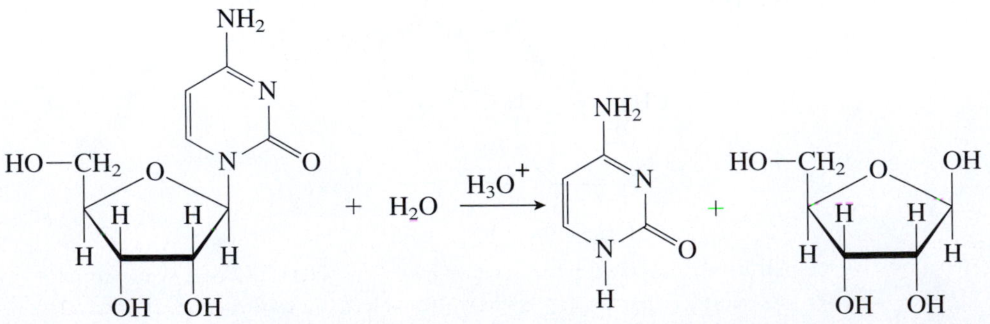

27.23 Explain why this cyclic phosphodiester gives mainly the 3'-monophosphate rather than the 5'-monophosphate when it is treated with aqueous sodium hydroxide:

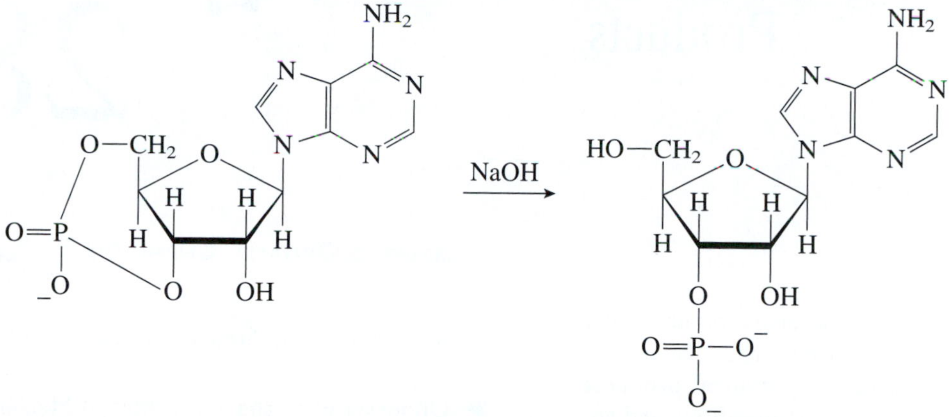

27.24 Some of the nucleophilic atoms in the bases of DNA can be alkylated in an S_N2 reaction with highly reactive electrophiles. For example, the oxygen of guanine can be methylated to form O-methylated guanine. The presence of an O-methylated guanine often results in mutations caused by the incorporation of thymine rather than cytosine during replication. Show how base pairing explains this observation.

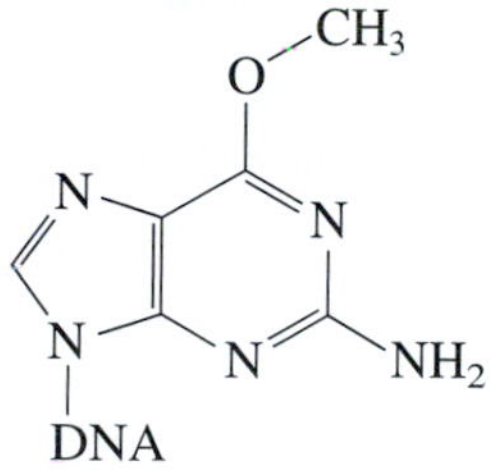

O-Methylated guanine

27.25 Nitrous acid (HNO_2) is a mutagen (a compound that causes mutations). It is thought to operate by changing the structure of some bases in DNA. For example, it is capable of converting adenine to hypoxanthine. Explain how this might cause a mutation.

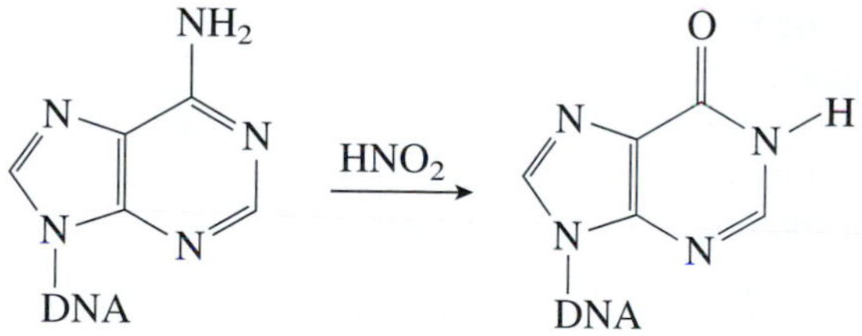

Adenine   Hypoxanthine

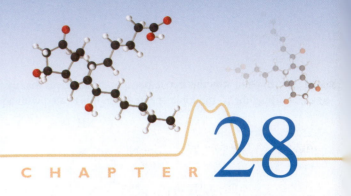

Other Natural Products

THE CARBOHYDRATES, amino acids, proteins, and nucleic acids discussed in Chapters 25, 26, and 27 are sometimes called **primary natural products** because they are found in all types of organisms and are the products of primary metabolism. **Secondary natural products** are usually produced from primary natural product precursors, such as amino acids or acetate ion, and, in general, are less widespread in occurrence. Today, natural product chemistry usually refers to the structure, reactions, and synthesis of these secondary natural products.

Many of these compounds have quite complex structures, often with multiple chirality centers. In addition, many of them are pharmacologically active; that is, they have dramatic physiological effects on any organism that ingests them. For both of these reasons they have always interested organic chemists, and the determination of their structures and their syntheses continue to play an important role in the development of organic chemistry.

This chapter discusses some of the more important natural products: terpenes, steroids, alkaloids, fats, and prostaglandins. (Fats are primary natural products, but it is convenient to include them in this chapter.) The structures and various aspects of their biosynthesis and chemical reactions are presented in subsequent sections. Because entire books have been written on each of these groups of compounds, the coverage here is necessarily incomplete. However, the intent is to present some of the flavor of their chemistry. Many other classes of naturally occurring organic compounds are not included for reasons of space.

MASTERING ORGANIC CHEMISTRY

▶ Recognizing the Structural Features of Terpenes

▶ Understanding the Hypothetical Mechanisms That Form Terpenes

▶ Recognizing Steroids and Prostaglandins and the Mechanisms That Form Them

▶ Recognizing Alkaloids and Fats

28.1 TERPENES

The term **terpene** was originally applied to the compounds obtained from turpentine, an extract from pine trees. Many of the terpenes that were first isolated from

ORGANIC
Chemistry Now™
Look for this logo in the chapter and go to OrganicChemistryNow at
http://now.brookscole.com/hornback2 for tutorials, simulations, problems, and molecular models.

plants had the formula $C_{10}H_{16}$, and it was soon recognized that they could be considered as resulting from the combination of two isoprene units (C_5H_8). Others were found that contain three to six or more isoprene units. (Recall that natural rubber can be viewed as a polymer containing many isoprene units.) Terpenes are now classified as monoterpenes (10 carbons), sesquiterpenes (15 carbons), diterpenes (20 carbons), triterpenes (30 carbons), and so on. Many have various oxygen-containing functional groups. In addition, some do not contain exact multiples of five carbons because carbons have been lost in degradation reactions during their biosynthesis. The structures of several simple terpenes and how they can be considered to be constructed from isoprene units are shown in the following examples. The individual "isoprene units" are shown in different colors.

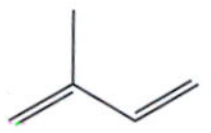

Isoprene

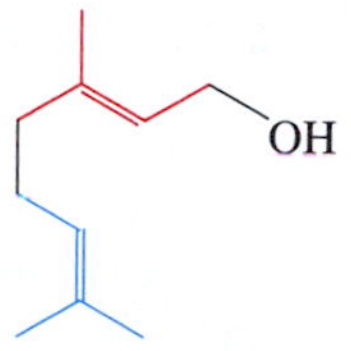

Geraniol
(a monoterpene)

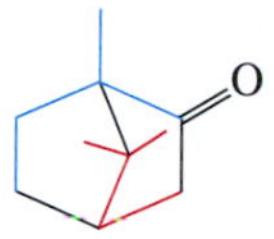

Camphor
(a monoterpene)

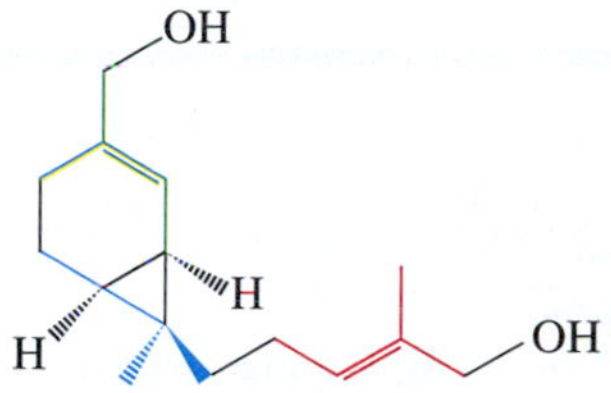

Sirenin
(a sesquiterpene)

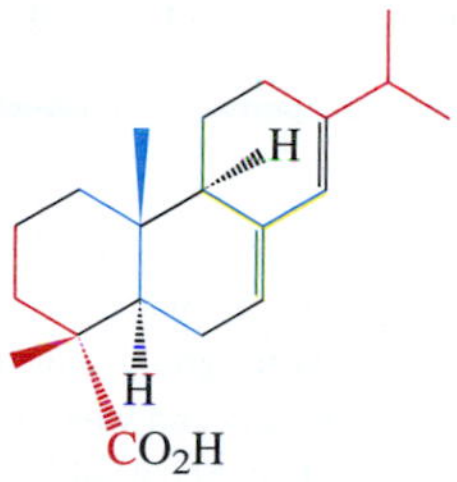

Abietic acid
(a diterpene)

PROBLEM 28.1

Classify these compounds as monoterpenes, diterpenes, and so on, and show the isoprene units that compose them:

a)

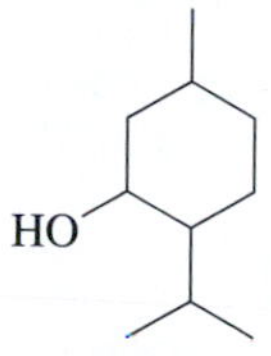

Menthol

b)

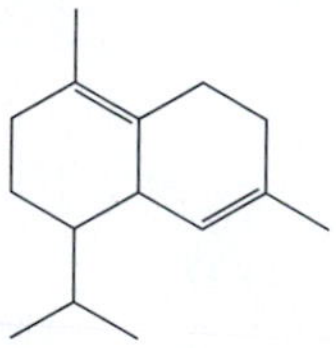

δ-Cadinene

c)

α-Pinene

d)

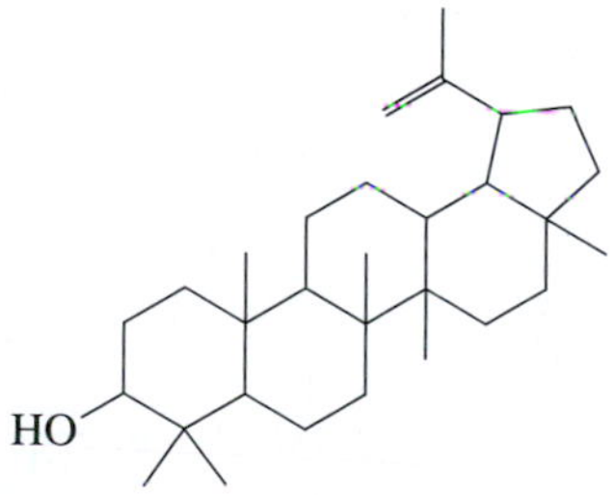

Lupeol

Why do plants synthesize terpenes? In some cases, because of their disagreeable taste, terpenes may act as antifeedants to protect the plant from herbivores. A few are known to be hormones that control plant growth. However, the biological function, if any, of most terpenes is not known. Regardless of their actual purpose, they have served as attractive targets for creative synthetic organic chemists because of their complex structures.

28.2 MONOTERPENES

Monoterpenes, with 10 carbons, can be viewed as resulting from the combination of two isoprene units. Isopentenyl pyrophosphate is the source of these isoprene units in biosynthesis. Dimethylallyl pyrophosphate, which is isomeric with isopentenyl pyrophosphate and is produced from it by the enzyme isopentenyl pyrophosphate isomerase, is also an important intermediate.

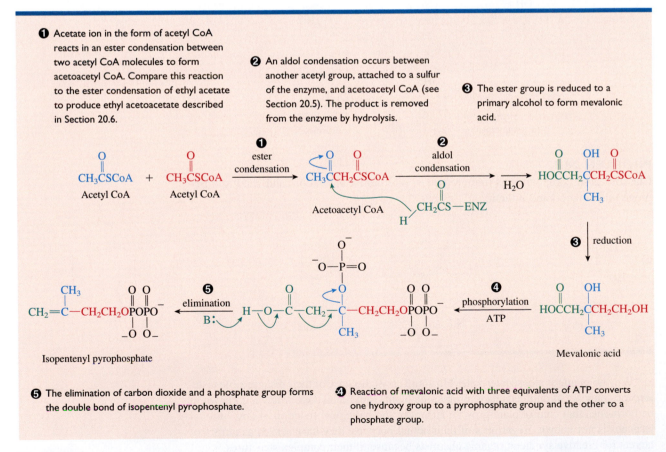

The pathway for the biosynthesis of isopentenyl pyrophosphate is outlined in Figure 28.1. The starting material for this process is represented as acetyl CoA, which

Figure 28.1

THE CONVERSION OF ACETYL COENZYME A TO ISOPENTENYL PYROPHOSPHATE.

has an acetyl group, from acetate ion, attached to a sulfur of coenzyme A. (The structure of coenzyme A need not concern us here.) The first step in the biosynthesis is a reaction between two acetyl CoA molecules to form acetoacetyl CoA. Although this reaction is controlled by an enzyme, it involves only mechanistic steps that are comparable to those that we have seen in earlier chapters. It is an ester condensation, a reaction that we have seen numerous times (see Section 20.6), and its mechanism is quite similar to the one for an ester condensation that does not involve an enzyme. The next step is an aldol condensation, in which acetoacetyl CoA acts as the electrophile and the nucleophile is derived from another acetyl group attached to a sulfur of an enzyme. Hydrolysis of this compound from the enzyme and reduction produces mevalonic acid. This acid, containing six carbons, is converted to isopentenyl pyrophosphate by phosphorylation followed by elimination of carbon dioxide and a phosphate group.

PROBLEM 28.2

Show the mechanism for the ester condensation shown in the first step of the process outlined in Figure 28.1. Assume that the reaction is caused by base and ignore the effect of the enzyme.

PROBLEM 28.3

Show the mechanism for the aldol condensation shown in the second step of the process outlined in Figure 28.1. Assume that the reaction is caused by base and ignore the effect of the enzyme.

The pyrophosphate group (OPP) is a good leaving group. Much of the chemistry involved in the biosynthesis of terpenes can be understood on the basis of the reactions of carbocations that are formed when the pyrophosphate group departs from compounds such as isopentenyl pyrophosphate or dimethylallyl pyrophosphate. Thus, the formation of geranyl pyrophosphate, the parent compound for the monoterpenes, can be viewed as resulting from the addition of the carbocation produced from dimethylallyl pyrophosphate to isopentenyl pyrophosphate as shown in Figure 28.2. However, it is important to remember that these reactions are controlled by enzymes. As such, they do not involve free carbocation intermediates, and they are often more concerted than the "mechanisms" shown in this chapter. Nevertheless, these "mechanisms" serve to show that these enzyme reactions are not magic but follow the same rules presented in previous chapters for carbocation behavior. The enzymes serve to lower the activation barriers for the overall processes and to control the stereochemistry, regiochemistry, and selectivity of each reaction.

PROBLEM 28.4

Show the structure of the species derived from the pyrophosphate leaving group after it has departed and explain why it is a good leaving group.

As shown in Figure 28.2, the allylic carbocation that is produced after pyrophosphate has left from dimethylallyl pyrophosphate adds to the double bond of isopentenyl pyrophosphate to produce a new carbocation containing 10 carbons. Loss of a proton from this carbocation produces geranyl pyrophosphate, which serves as the precursor

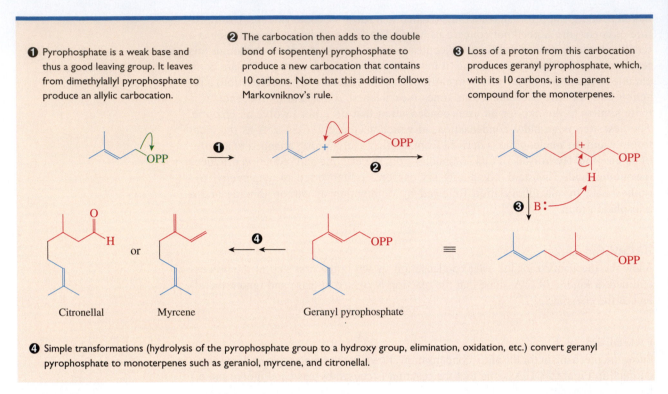

❶ Pyrophosphate is a weak base and thus a good leaving group. It leaves from dimethylallyl pyrophosphate to produce an allylic carbocation.

❷ The carbocation then adds to the double bond of isopentenyl pyrophosphate to produce a new carbocation that contains 10 carbons. Note that this addition follows Markovnikov's rule.

❸ Loss of a proton from this carbocation produces geranyl pyrophosphate, which, with its 10 carbons, is the parent compound for the monoterpenes.

Citronellal   Myrcene   Geranyl pyrophosphate

❹ Simple transformations (hydrolysis of the pyrophosphate group to a hydroxy group, elimination, oxidation, etc.) convert geranyl pyrophosphate to monoterpenes such as geraniol, myrcene, and citronellal.

Figure 28.2

THE CONVERSION OF ISOPENTENYL PYROPHOSPHATE AND DIMETHYLALLYL PYROPHOSPHATE TO GERANYL PYROPHOSPHATE.

for the formation of the monoterpenes. For example, elimination of the pyrophosphate group from geranyl pyrophosphate produces myrcene, and hydrolysis of the OPP group to an alcohol produces geraniol, a terpene obtained from roses and used in many perfumes because of its roselike odor. Reduction of the double bond of geraniol and oxidation of the hydroxy group to an aldehyde produces citronellal, a terpene found in lemon-grass oil that is also used as an ingredient in perfumes and in candles that are used to repel mosquitoes.

PROBLEM 28.5

The conversion of geranyl pyrophosphate to myrcene follows an E1 mechanism. Show the steps in this mechanism.

Isomerization of the double bond of geranyl pyrophosphate from *E* to *Z* produces neryl pyrophosphate. As shown in Figure 28.3, the carbocation that is formed from neryl pyrophosphate can cyclize to a new carbocation that contains a six-membered ring. Nucleophilic addition of water to this carbocation produces α-terpenol, whereas loss of a proton produces limonene, a monoterpene with a lemonlike odor that occurs in citrus fruits. Further transformations lead to other monoterpenes, such as menthol,

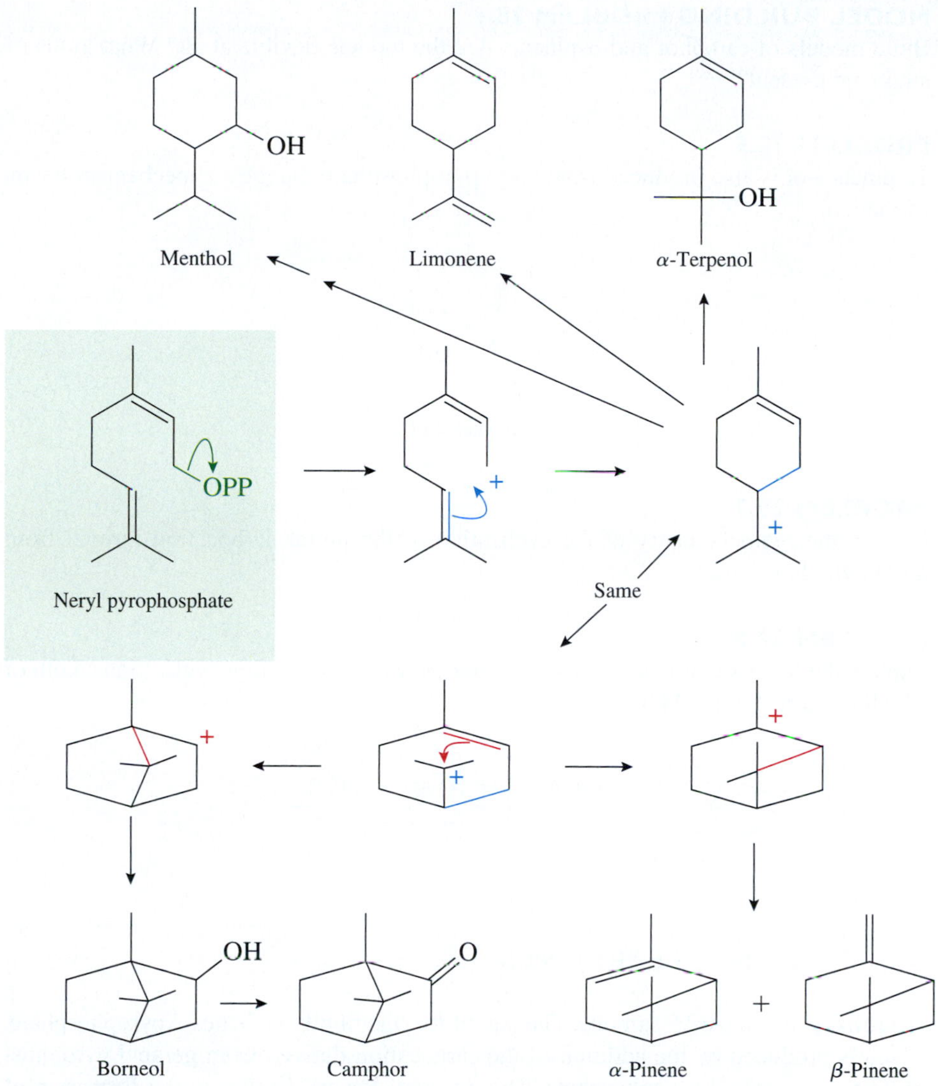

Figure 28.3

THE FORMATION OF SOME CYCLIC AND BICYCLIC MONOTERPENES FROM NERYL
PYROPHOSPHATE.

which is a major component of oil of peppermint and is used as a flavoring for foods,
toothpastes, and mouthwashes.

The cyclic carbocation produced from neryl pyrophosphate can also cyclize a second time by adding to either end of the double bond of the six-membered ring (see Figure 28.3). The two resulting carbocations are said to be bicyclic (having two rings) and
give rise to various bicyclic monoterpenes such as borneol and camphor (which smells
like mothballs and is used in cosmetics, liniment, and anti-itching medications) and α-
and β-pinene, two of the major components of turpentine.

MODEL BUILDING PROBLEM 28.1

Build models of camphor and α-pinene. Are the models flexible at all? What kinds of strain are present?

PROBLEM 28.6

Terpinen-4-ol is also produced from neryl pyrophosphate. Suggest a mechanism for its formation.

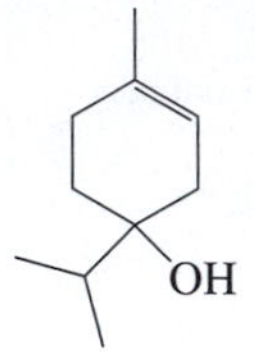

Terpinen-4-ol

PROBLEM 28.7

Explain the regiochemistry of the cyclization of the initial carbocation formed from neryl pyrophosphate.

PROBLEM 28.8

Explain the factors that favor or disfavor each of these carbocation cyclizations, both of which occur in Figure 28.3:

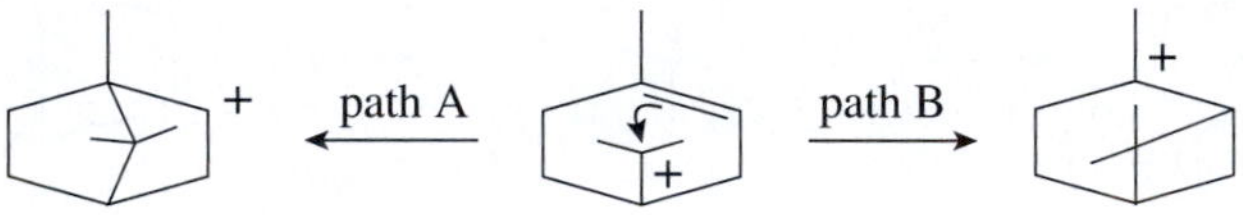

28.3   SESQUITERPENES

Sesquiterpenes have 15 carbons. The parent for this family is farnesyl pyrophosphate, which is produced by the addition of the carbocation derived from geranyl pyrophosphate to isopentenyl pyrophosphate. This reaction is very similar to the formation of geranyl pyrophosphate shown in Figure 28.2.

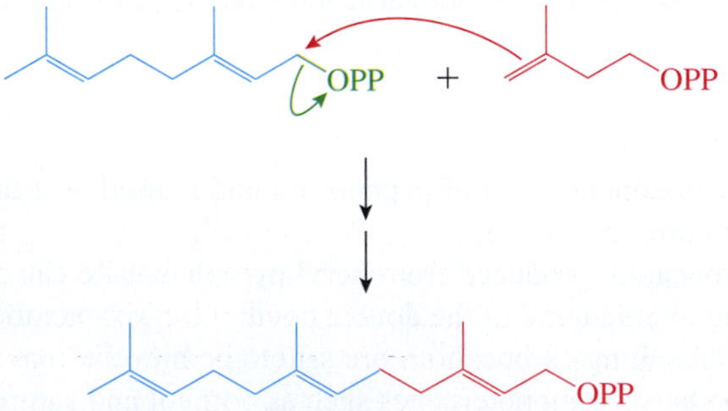

Farnesyl pyrophosphate

In a process that is quite similar to the formation of the cyclic monoterpenes, isomerization about the double bond of farnesyl pyrophosphate, followed by carbocation

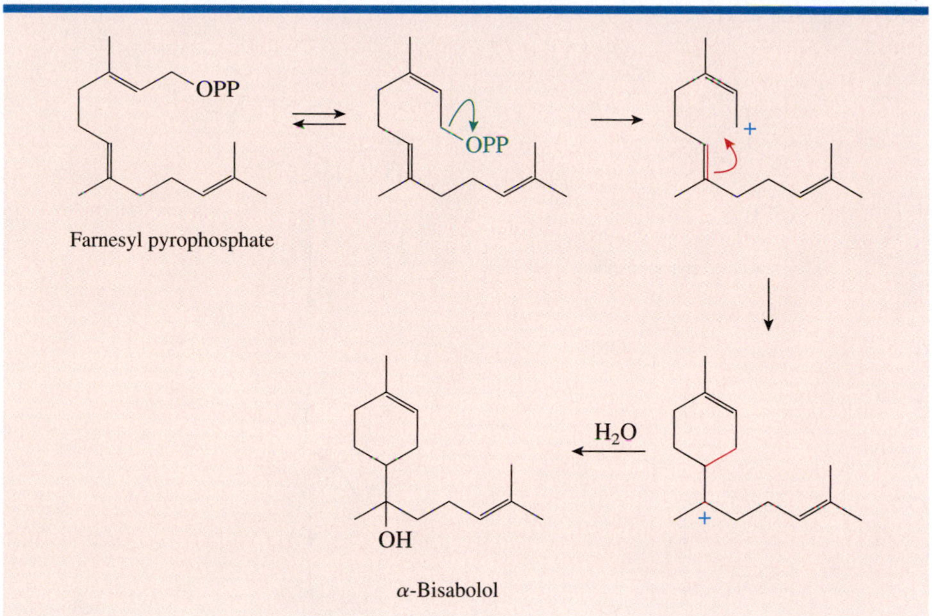

Figure 28.4

FORMATION OF A CYCLIC SESQUITERPENE FROM FARNESYL PYROPHOSPHATE.

formation and cyclization, results in the production of cyclic sesquiterpenes such as α-bisabolol (see Figure 28.4).

PROBLEM 28.9

Explain the regiochemistry of the carbocation cyclization shown in Figure 28.4.

PROBLEM 28.10

Suggest a mechanism for the formation of campherenol from farnesyl pyrophosphate:

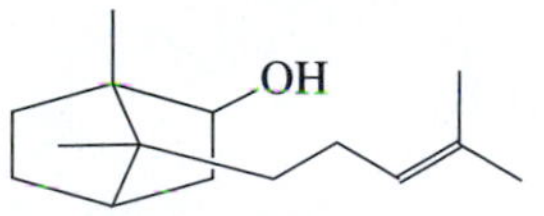

Campherenol

Farnesyl pyrophosphate can also cyclize in a quite different manner that does not involve the pyrophosphate group, as illustrated in Figure 28.5. This process is initiated by protonation of the double bond at the opposite end of the chain from the pyrophosphate group. The resulting carbocation adds to the double bond in the center of the chain, forming a six-membered ring. This is the ring system present in sesquiterpenes such as abscisic acid, a plant hormone that controls the shedding of leaves. The carbocation can also add to the remaining double bond to produce a bicyclic system of two six-membered rings that share one ring bond. This is illustrated in Figure 28.5 by the formation of drimenol. The cyclic structure present in drimenol, with its two fused six-membered rings, is especially important in natural

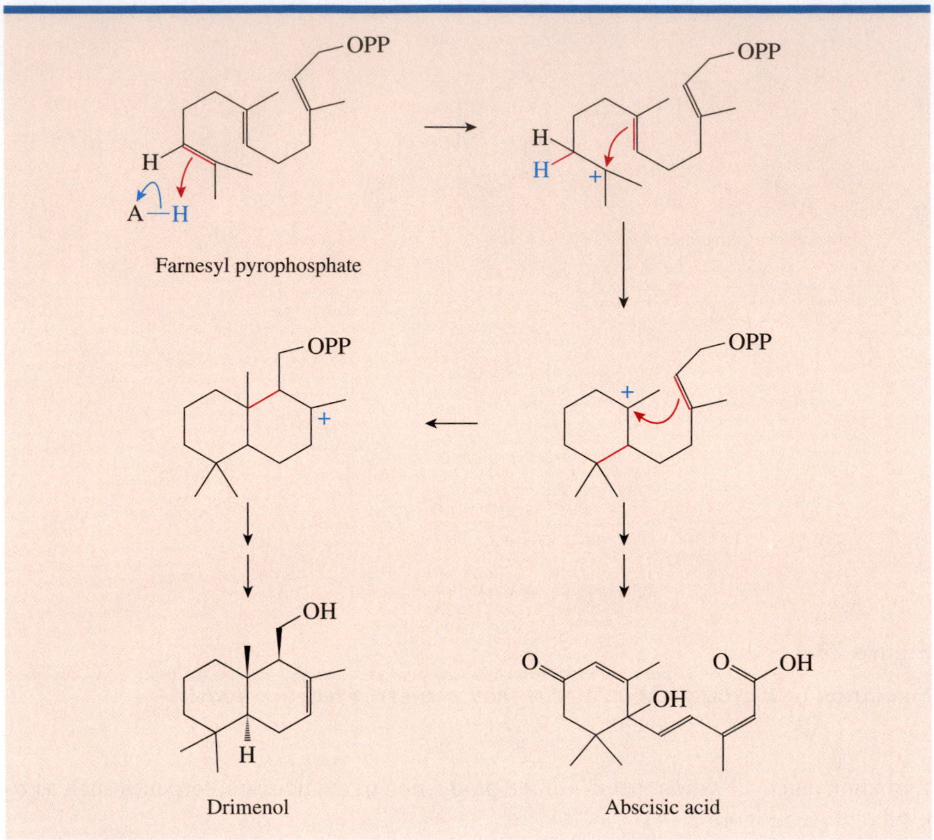

Figure 28.5

AN ALTERNATIVE CYCLIZATION OF FARNESYL PYROPHOSPHATE.

products chemistry. The parent bicyclic ring would be produced by the addition of 10 hydrogens to naphthalene, so it is known as decahydronaphthalene, which is often shortened to decalin.

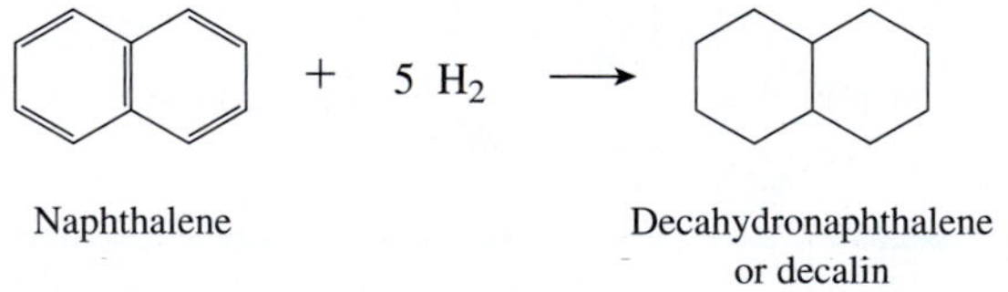

Naphthalene   Decahydronaphthalene
                                    or decalin

Decalin has two saturated six-membered rings fused together; that is, the rings share a common bond. The hydrogens at the ring junctions may be attached on the same side or on opposite sides, so there are two stereoisomers: *cis*-decalin and *trans*-decalin. Neither of these is chiral.

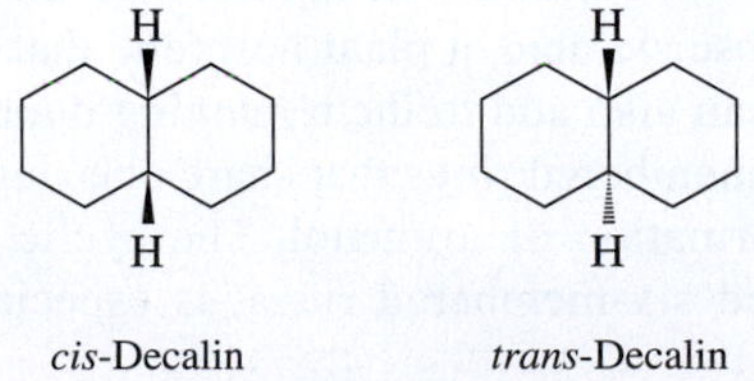

cis-Decalin   *trans*-Decalin

The fused six-membered rings of decalin are still cyclohexane rings, and like most cyclohexane rings, they prefer chair conformations. Analysis of the conformational preferences of these molecules is very similar to that of 1,2-dimethylcyclohexane (see Chapter 6). Recall that *cis*-1,2-dimethylcyclohexane has one methyl group axial and one equatorial and that a ring-flip produces a new chair conformation in which the axial methyl group has become equatorial and vice versa. Likewise, *trans*-1,2-dimethylcyclohexane has two conformers, one with both methyl groups equatorial and one with both groups axial.

In the case of the decalins, both rings assume chair conformations. For *cis*-decalin the carbons of the first ring are attached to an axial position and an equatorial position of the second ring, just like the methyl groups of *cis*-1,2-dimethylcyclohexane. Likewise, the carbons of the second ring are attached to an axial and an equatorial position of the first ring. A ring-flip, which interconverts axial and equatorial positions, can occur.

cis-1,2-Dimethyl-cyclohexane

This bond is equatorial on the red ring.

This bond is axial on the red ring.

cis-Decalin

For *trans*-decalin, both carbons of one ring are attached to equatorial positions of the other. This is true for both rings. The molecule is rigid and cannot undergo a ring-flip because one ring is not large enough to bridge axial positions, which point in opposite directions, on the other. As was the case with the 1,2-dimethylcyclohexanes, *trans*-decalin, with both rings attached equatorially to the other, is more stable than *cis*-decalin by 2.7 kcal/mol (11.3 kJ/mol). Decalin rings, especially those with *trans*-stereochemistry, are an important part of many natural products, including terpenes and steroids (see Section 28.5).

Both of these bonds are equatorial on the red ring.

trans-1,2-Dimethyl-cyclohexane

trans-Decalin

MODEL BUILDING PROBLEM 28.2
Build models of *cis*- and *trans*-decalin and examine the conformational mobility of each.

PROBLEM 28.11
Discuss the regiochemistry of each carbocation reaction shown in Figure 28.5.

PROBLEM 28.12
Suggest a mechanism for the conversion of germacrene D to γ-cadinene:

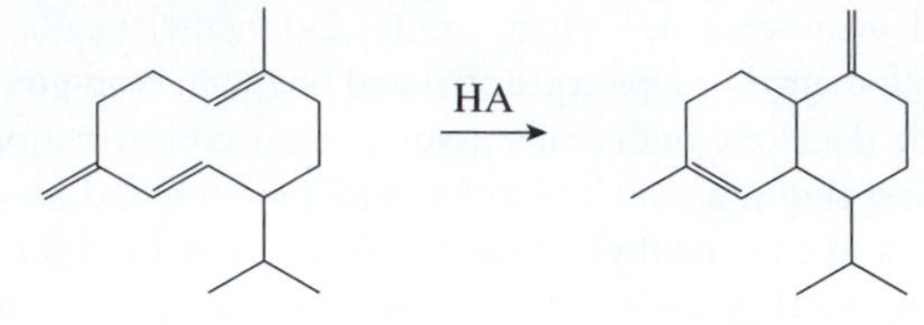

Germacrene D   γ-Cadinene

28.4   LARGER TERPENES

The larger terpenes are formed in the same manner as the monoterpenes and sesquiter-penes. The precursor for the diterpenes is the C_{20} compound geranylgeranyl pyrophos-phate, which is formed from farnesyl pyrophosphate in a manner similar to that shown in Figure 28.2:

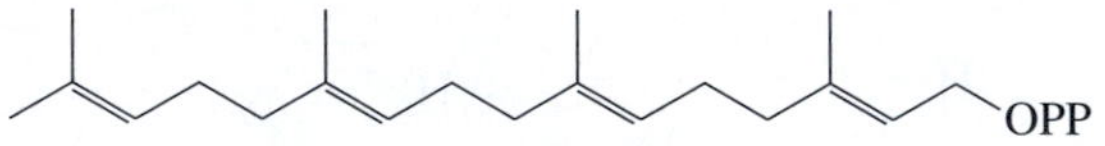

Geranylgeranyl pyrophosphate

Geranylgeranyl pyrophosphate can cyclize in many different ways to produce a wide variety of diterpene ring systems. As one example, cyclization initiated by protonation of the double bond most remote from the pyrophosphate group produces the decalin ring system of labdadienyl pyrophosphate in a reaction that is nearly identical to the cy-clization shown in Figure 28.5. Further cyclization, initiated by departure of the py-rophosphate group and proceeding along the lines of those shown in Figures 28.3 and 28.4, produces the tricyclic diterpene pimaradiene:

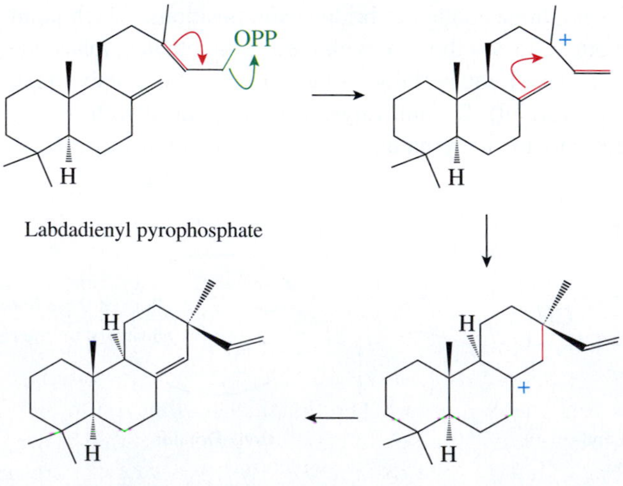

Labdadienyl pyrophosphate

Pimaradiene

PROBLEM 28.13

Suggest a mechanism for the conversion of geranylgeranyl pyrophosphate to labdadi-enyl pyrophosphate.

PROBLEM 28.14

Suggest a mechanism for the conversion of geranyl pyrophosphate to geranylgeranyl pyrophosphate.

Coupling of two C_{15} farnesyl pyrophosphate units produces the C_{30} compound squa-lene, which is the precursor for the triterpenes and also for the steroids (see Section 28.6):

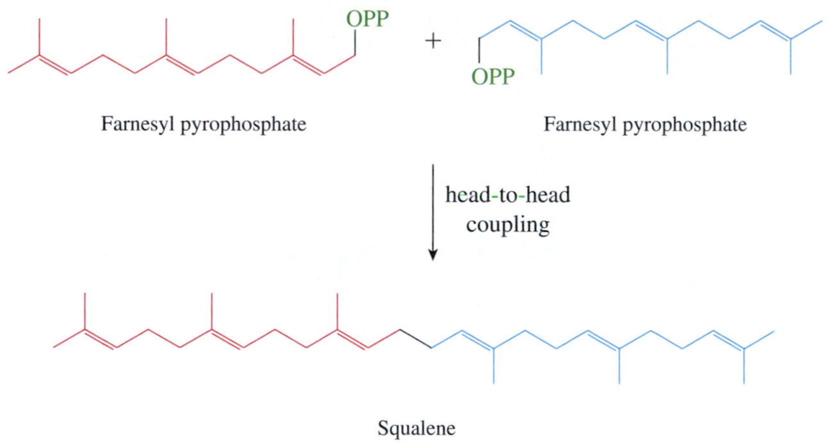

Farnesyl pyrophosphate   Farnesyl pyrophosphate

head-to-head coupling

Squalene

Note that the formation of squalene results from the coupling of both of the farnesyl groups at the carbons attached to the pyrophosphate groups (head-to-head coupling) and that both pyrophosphate groups are lost during this process. The mechanism for this reaction differs dramatically from that shown in Figure 28.2 and is too compli-cated to present here.

A similar coupling of two C_{20} geranylgeranyl pyrophosphate units, followed by dehydrogenation, produces the C_{40} compound, phytoene. This tetraterpene is the pre-cursor for an important group of compounds called carotenes. For example, addi-tional dehydrogenation of phytoene produces lycopene. The long, conjugated system of double bonds of this compound causes it to absorb visible light. Lycopene is the red pigment found in tomatoes. Cyclization of each end of a lycopene molecule, as shown in Figure 28.6, produces β-carotene. This orange compound is responsible for the color of carrots and serves as a light antenna in photosynthesis. The symmetri-cal compound β-carotene is cleaved to two molecules of retinal in the intestine. Reti-nal, also known as vitamin A, is used to form the pigment that absorbs light in the retina of the eye. So carrots really are good for your eyes!

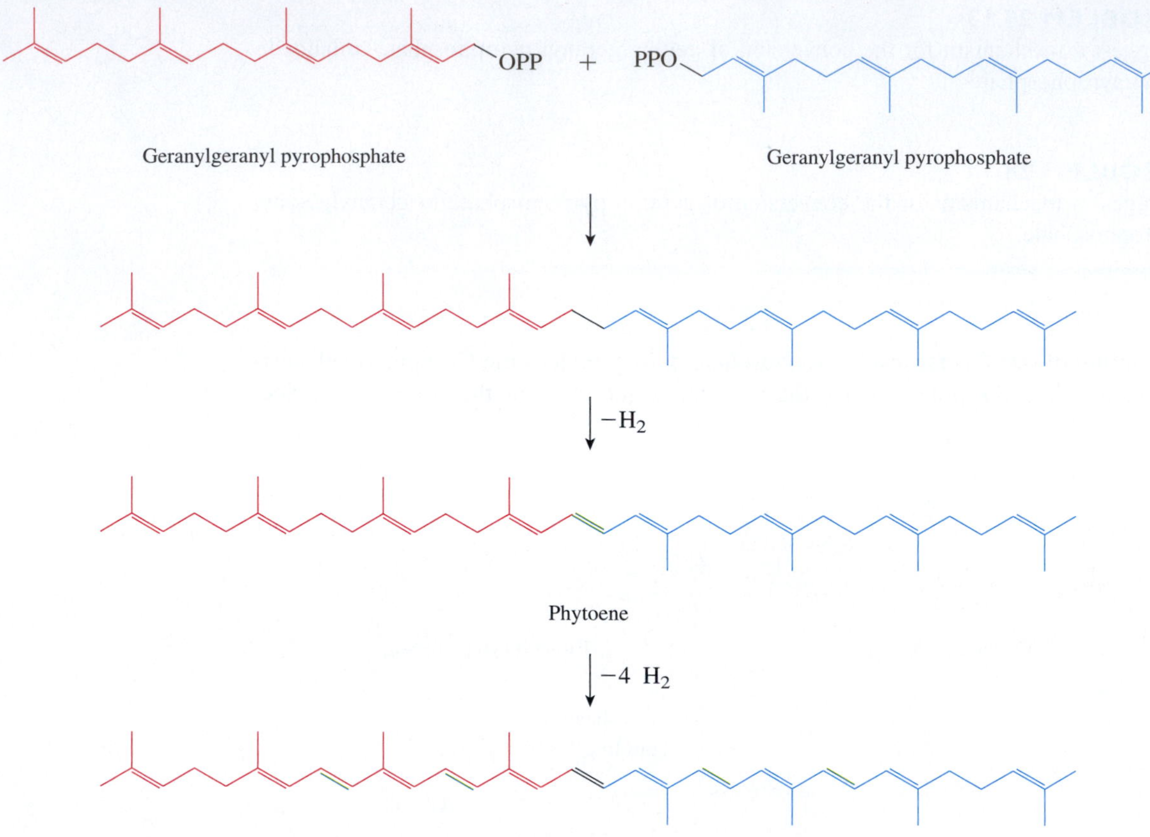
OPP
+
PPO
Geranylgeranyl pyrophosphate
Geranylgeranyl pyrophosphate
−H2
Phytoene
−4 H2
Lycopene

28.5 Steroids

Perhaps the best-known steroid—or at least the most notorious—is **cholesterol.** Like all steroids, it has a tetracyclic ring system consisting of three fused six-membered rings and one five-membered ring. The structure of cholesterol and the standard numbering for the steroid ring system are shown in the following diagram:

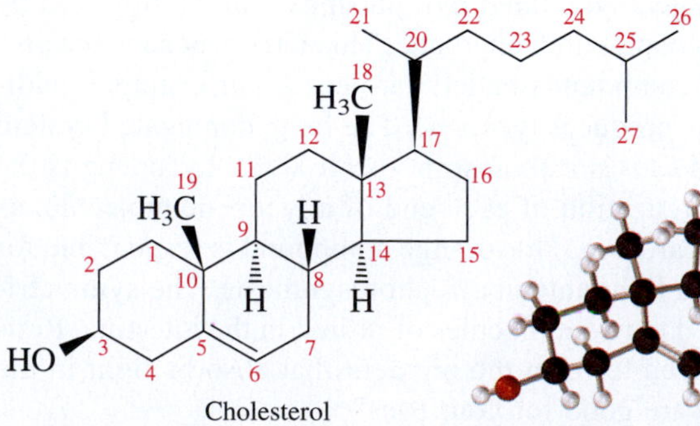
HO
H3C
H3C
H
H
H
1
2
3
4
5
6
7
8
9
10
11
12
13
14
15
16
17
18
19
20
21
22
23
24
25
26
27

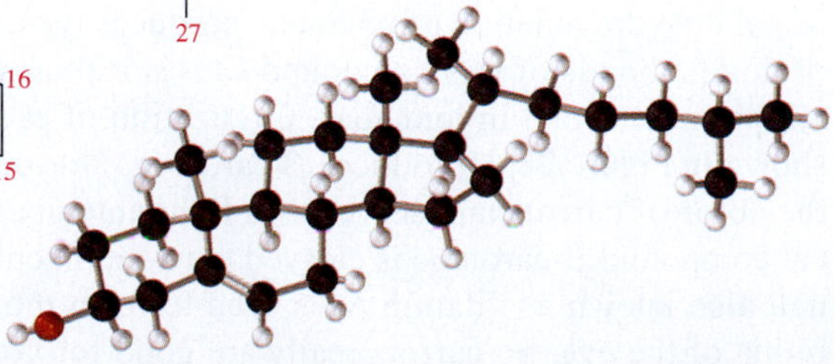

Cholesterol

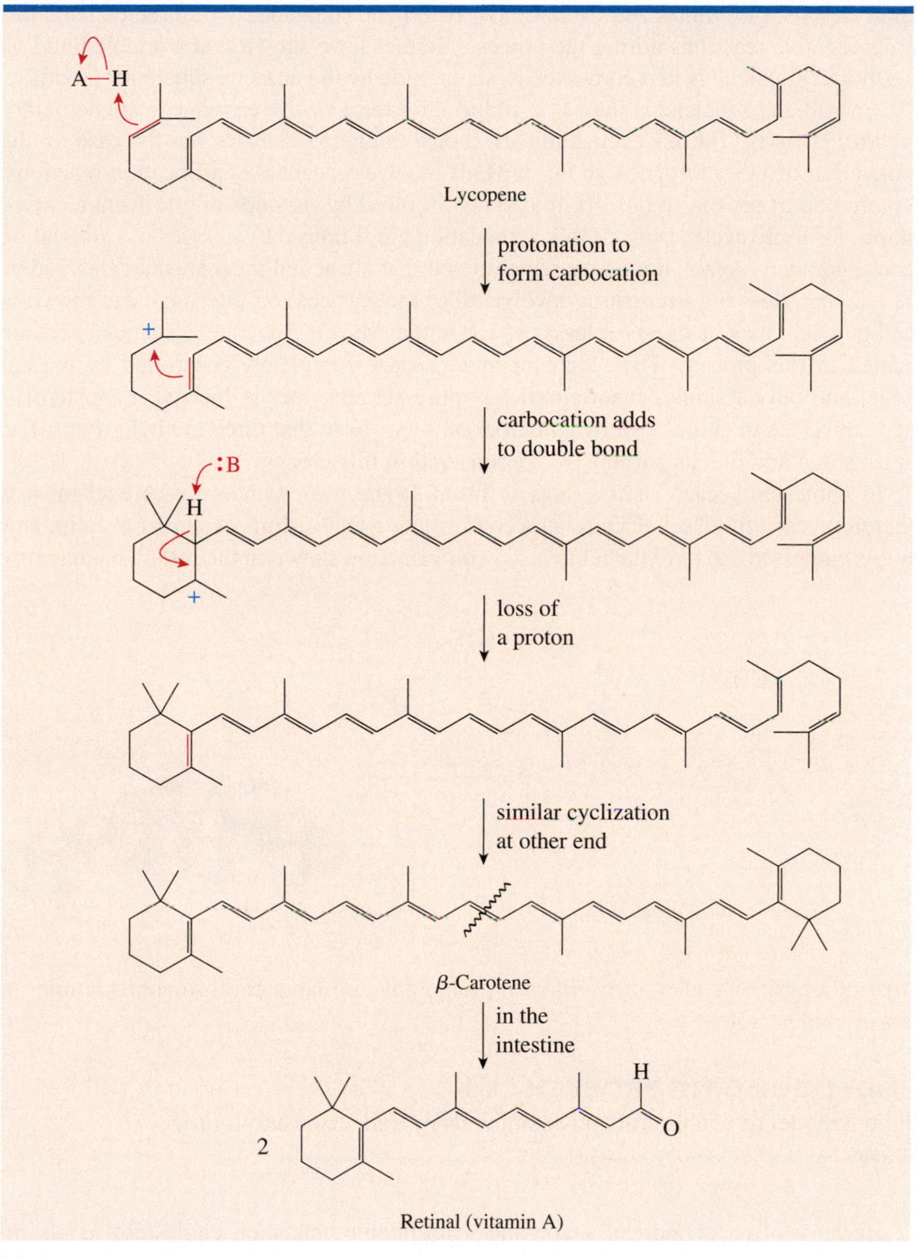

Figure 28.6

THE CONVERSION OF LYCOPENE TO β-CAROTENE AND VITAMIN A.

PROBLEM 28.15

Small amounts of α-carotene are formed along with β-carotene in the process shown in Figure 28.6. Suggest a mechanism for this process and explain why more β-carotene is produced.

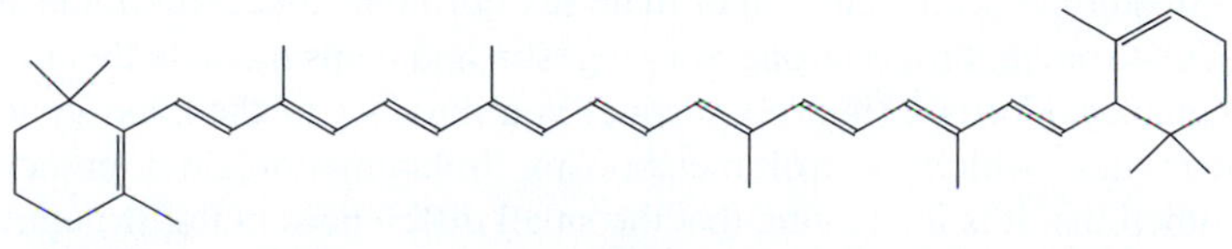

α-Carotene

Cholesterol is biosynthesized from the C_{30} triterpene squalene, with three carbons lost in degradative reactions during the process. Studies have shown that a double bond on one end of squalene is first converted to an epoxide by the enzyme squalene epoxidase. The epoxidized squalene is then cyclized to lanosterol by the enzyme squalene oxido-cyclase. The steps for this cyclization are shown in Figure 28.7. As was the case for the cyclization of other terpenes, these steps all involve reasonable carbocation reactions. Protonation of the oxygen of the epoxide is followed by ring opening to form a carbo-cation. Multiple cyclizations of this carbocation are followed by a series of carbocation rearrangements. Again, it must be remembered that the actual mechanism, catalyzed by the enzyme, does not necessarily involve all of the carbocation intermediates shown in the figure and may be more (or less) concerted than shown. Seven new stereocenters are created in this process. Their stereochemistries are completely controlled by the enzyme, and only a single, enantiomerically pure stereoisomer is formed. Lanosterol is then converted to cholesterol by a number of steps. Note that three methyl groups, two on carbon 4 and one on carbon 14, are removed in this process.

In cholesterol, each of the rings is fused to the next with *trans*-stereochemistry. Therefore, each fused set of rings has a conformation that is similar to *trans*-decalin. This causes cholesterol to have the rather rigid conformation shown in the following diagram:

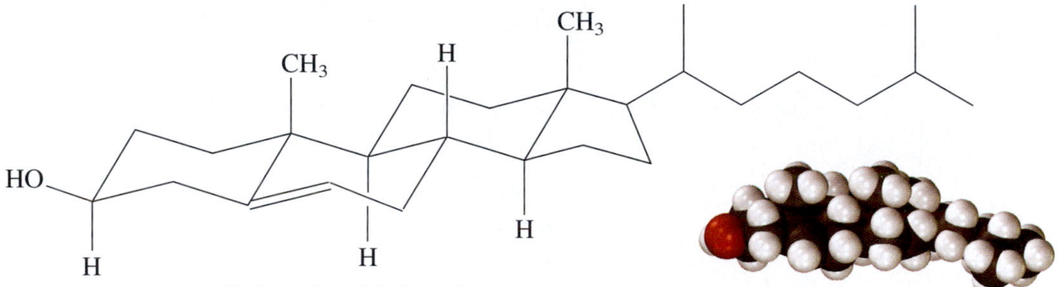

Conformation of cholesterol

Most other steroids also have *trans*-ring junctions and have conformations similar to that of cholesterol.

MODEL BUILDING PROBLEM 28.3

Build a model of cholesterol and examine its overall shape and rigidity.

Because of its cylindrical shape and hydrophobic character, cholesterol is an important component of the membranes of animal cells. Its rigid structure decreases membrane fluidity, but it also inhibits the "crystallization" of fatty acid side chains of the membrane lipids and it acts as a sort of membrane plasticizer.

Other steroids, which are biosynthesized from cholesterol, show a wide variety of hormonal activity. The structures of a few of these steroidal hormones are shown in Figure 28.8. Testosterone is an androgen or male sex hormone, and estradiol is an estrogen or female sex hormone. Progesterone is a progestin and helps mediate the menstrual cycle and pregnancy. Cortisol (hydrocortisone) is a member of the class called adrenocortical hormones, which control metabolism, inflammation, and numerous other biological functions. It is interesting that the small differences in the structures of these compounds cause enormous differences in their physiological activities. For example, loss of the 19-methyl group of the male sex hormone testosterone and the aromatiza-

Squalene oxide is formed by epoxidation of squalene catalyzed by the enzyme squalene epoxidase.

❶ The cyclization is catalyzed by the enzyme squalene oxidocyclase. First the epoxide ring is protonated, and it then opens to form a carbocation.

❷ While the enzyme holds the molecule in a conformation favorable for cyclization, the carbocation adds to the double bond to form a six-membered ring and a new carbocation.

❸ This new carbocation adds to another double bond, and that carbocation adds to another. The enzyme controls both the stereochemistry and the regiochemistry of all of these additions, some of which may be concerted.

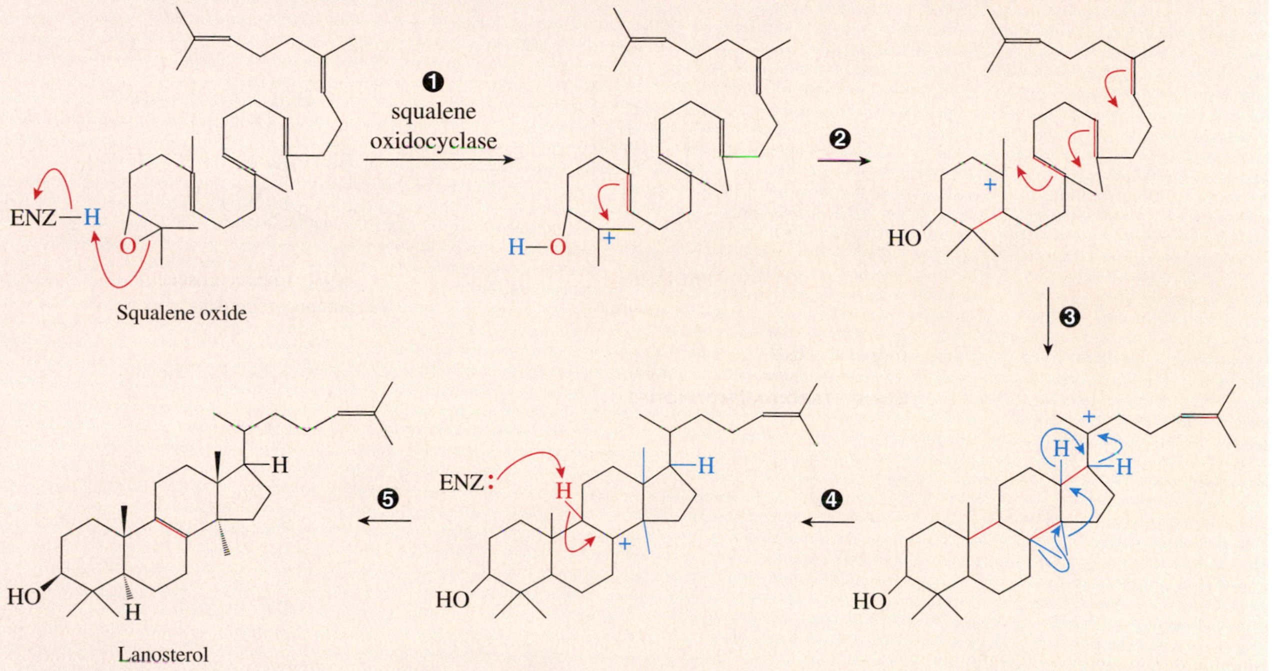

Squalene oxide

Lanosterol

The product, lanosterol, is formed with complete regioselectivity and stereoselectivity.

❺ Finally a proton is removed by the enzyme to form a double bond.

❹ Next a series of carbocation rearrangements occurs. In each case, either a hydrogen or a methyl group moves to the adjacent carbon. These rearrangements may all be concerted.

Active Figure 28.7

ORGANIC
Chemistry ⊷ Now™

THE CYCLIZATION OF SQUALENE OXIDE TO LANOSTEROL. Test yourself on the concepts in this figure at **OrganicChemistryNow.**

tion of one cyclohexane ring produces the female sex hormone estradiol. In fact, estradiol is synthesized from testosterone in females.

PROBLEM 28.16
Explain the regiochemistry of the opening of the epoxide ring that occurs in the first step of the process shown in Figure 28.7.

PROBLEM 28.17
Assume that each cyclization or rearrangement shown in Figure 28.7 produces a discrete carbocation intermediate. Show the structure of each carbocation and comment on the regiochemistry of each step. Do any of the steps proceed with unexpected regiochemistry?

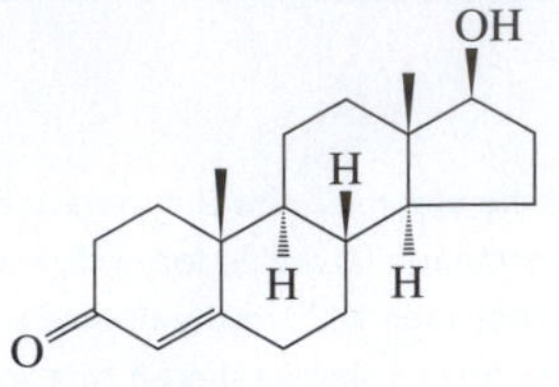

Testosterone
(an androgen)

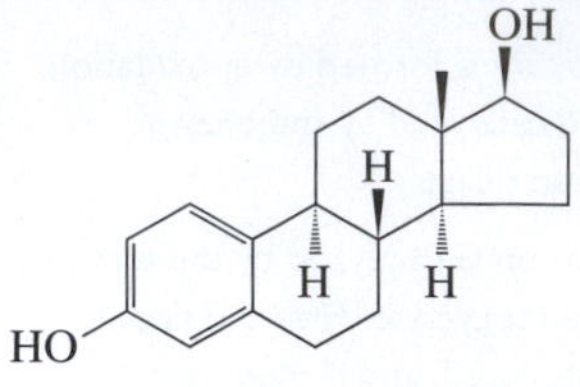

Estradiol
(an estrogen)

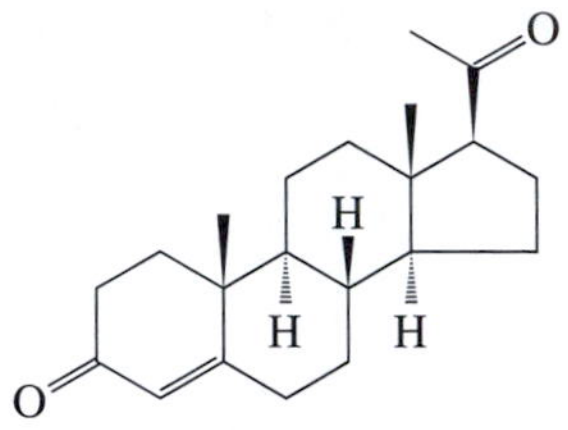

Progesterone
(a progestin)

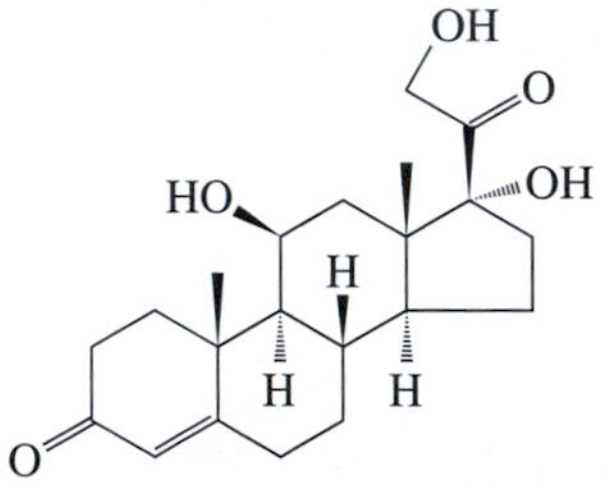

Cortisol (hydrocortisone)
(an adrenocorticoid hormone)

Figure 28.8

SOME STEROIDAL HORMONES.

Focus On

Syntheses That Mimic Nature

Nature often provides excellent suggestions about how to synthesize a compound. After the pathway for the biosynthesis of steroids by cationic cyclization of polyenes was determined, Professor William S. Johnson and coworkers at Stanford University used a very similar reaction to synthesize progesterone. The last part of this synthesis is outlined in the following equations. Alcohol **A** was prepared in 12 steps with an overall yield of 10%. It was then cyclized to form the steroid ring system.

This cyclization of **A** to form **B** is truly amazing. In a single step, three rings are formed, in 71% yield, predominantly with the correct stereochemistry. And this occurs without the aid of enzymes. Although the natural process proceeds in higher yield and is more stereoselective, we see again that the enzyme does not perform magic in the case of these cyclization reactions, but simply makes a reaction that is already favorable the only reaction that occurs.

1 Protonation of the hydroxy group, followed by the loss of water, produces the carbocation that initiates the cyclization.

1a This carbocation is allylic and is stabilized by resonance.

2 The carbocation adds to the double bond to form a new six-membered ring.

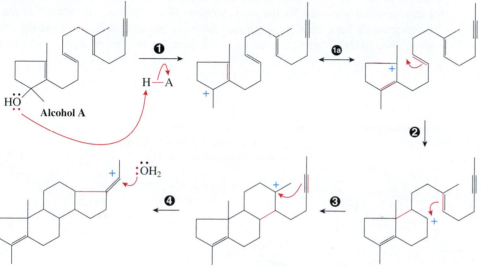

5 Water adds to this final cationic intermediate to form an enol, which then tautomerizes to the final product of the cyclization, ketone **B**.

4 This carbocation adds to the triple bond, forming a five-membered ring and a vinyl cation.

3 This new carbocation adds to the next double bond to form another six-membered ring.

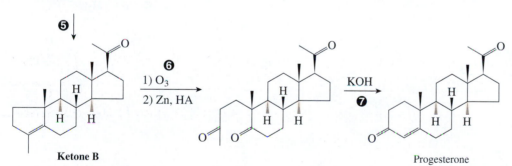

Ketone B → 1) O₃ 2) Zn, HA → KOH → Progesterone

Ketone **B** is formed in 71% yield from **A** in a single step. This step has formed three new rings. Although several stereoisomers are produced, the major product is **B**, which has the correct stereochemistry for conversion to racemic progesterone. To finish the synthesis, it is only necessary to open the five-membered ring and reclose it as a six-membered ring.

6 Ozonolysis (Section 11.11) is used to open the ring.

7 An intramolecular aldol condensation (Section 20.5) produces progesterone in 45% overall yield from **B**.

28.6 SYNTHESIS OF STEROIDS

Steroids are very powerful hormones and therefore are present in animals in only extremely minute concentrations. They are difficult to obtain from natural sources. For example, 4 tons of sow ovaries (from 80,000 pigs) were required to isolate the 12 mg of estrogen that was used for the determination of its structure. Syntheses of most of these hormones have been accomplished from simple starting materials in the laboratory. Although these routes are too long and complex to provide a practical source of

❶ This is an elimination reaction. The oxygen is acetylated to make it a better leaving group. Then an E1 elimination occurs.

❷ The CrO_3 cleaves the double bond. Although this reaction has not been covered in this book, it is quite similar to the ozonolysis reaction (Section 11.11).

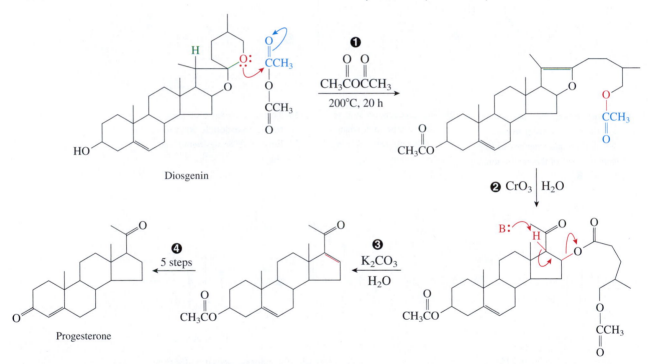

❹ To complete the synthesis of progesterone, it is necessary to selectively reduce one of the double bonds, hydrolyze the acetate group to a hydroxy group, oxidize the hydroxy group to a carbonyl group, and isomerize the remaining double bond into conjugation with the carbonyl group. This is accomplished in five steps.

❸ The ester group is lost by an elimination reaction. The adjacent carbonyl group makes the hydrogen more acidic, so it is more readily removed by the base. This is an example of an E1cb mechanism. (See the Focus On box on page 333.)

Figure 28.9

THE PREPARATION OF PROGESTERONE FROM DIOSGENIN.

the hormones for medicinal uses, many are truly elegant. R. B. Woodward, regarded by many as the greatest organic chemist of the last half of the twentieth century, was awarded the 1965 Nobel Prize in chemistry for his syntheses of steroids and numerous other natural products. (To aid in understanding the stereochemistry of some of the synthetic reactions he was using, he and R. Hoffmann developed the theory that explained the selectivity of pericyclic reactions [see Chapter 22]. Woodward died in 1979, or he would certainly have shared in the Nobel Prize awarded to Hoffmann and Fukui in 1981 and he would have been one of the few scientists to have won two Nobel Prizes.)

The most viable method for obtaining larger amounts of steroid hormones is to start with some readily available natural product with a structure that is similar to a steroid and convert it to the desired compound. Russell Marker, a professor at Pennsylvania State University, developed such a method to prepare progesterone from diosgenin, a material that is readily available from Mexican yams. His synthesis is outlined in Figure 28.9. However, he could not interest a major pharmaceutical company in his process, so in 1944 he founded his own company, Syntex, in Mexico City to develop it.

PROBLEM 28.18
Show a mechanism for the elimination reaction that occurs in the first step of the process shown in Figure 28.9.

PROBLEM 28.19
One way to accomplish the isomerization of a double bond into conjugation with a carbonyl group, one of the reactions needed for the last part of the synthesis of progesterone from diosgenin, is to treat the compound with base as shown in the following equation. Show a mechanism for this reaction.

In the late 1940s, cortisone was the most sought-after steroid because its anti-inflammatory properties appeared to have amazing effects in the treatment of rheumatoid arthritis. The major difficulty in converting a readily available steroid, such as progesterone, to cortisone was the introduction of an oxygen functionality at carbon 11. As we have seen innumerable times, reactions occur at a functional group or sometimes at the carbon adjacent to it. Causing a reaction to occur specifically at an unactivated carbon presents quite a challenge. The problem was solved by a roundabout method that introduced a double bond between carbons 9 and 11. Several syntheses of cortisone were reported in 1951.

However, some chemists at Upjohn found a better method. They were able to introduce the troublesome oxygen at carbon 11 by the microbiological fermentation of progesterone to produce 11-hydroxyprogesterone. The microbe, with its highly specific enzymes, was able to accomplish in a single step what required numerous steps for the synthetic chemist. The completion of the synthesis of cortisone required only nine additional steps.

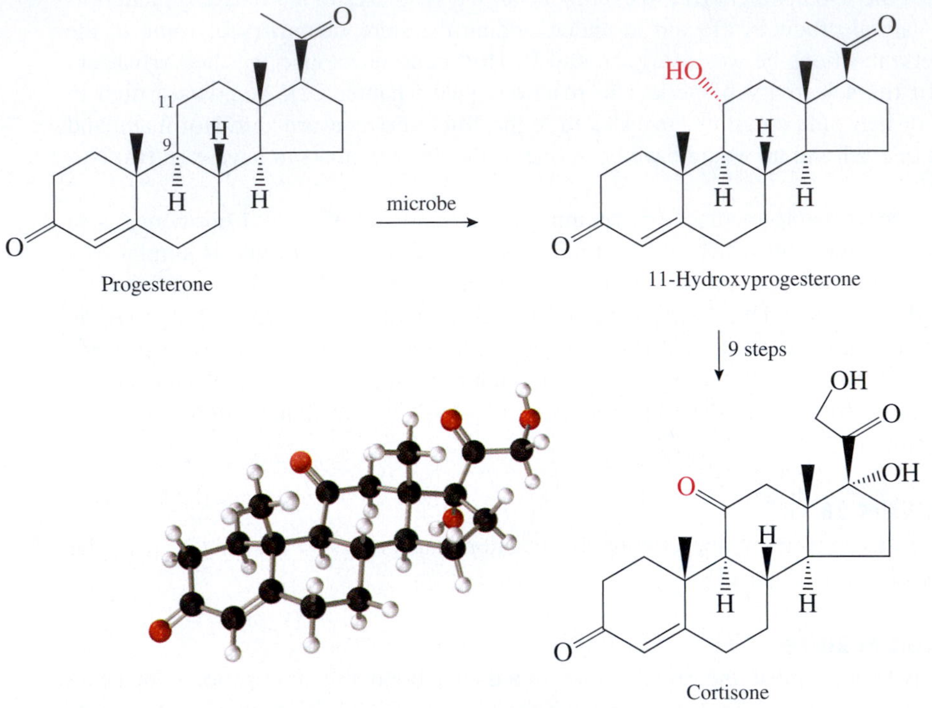

Progesterone

microbe

11-Hydroxyprogesterone

9 steps

Cortisone

Focus On

The Birth Control Pill

Perhaps no single development has influenced today's society more than the birth control pill. This simple, effective, and inexpensive method to limit pregnancy helped bring about both the sexual revolution and women's liberation.

By 1937 it was known that large doses of progesterone inhibited ovulation, and the possibility of its use in birth control was recognized. A major problem, however, was that progesterone displays only weak activity when administered orally. The idea of an injectable contraceptive was not very attractive. Somewhat later, it was discovered that the removal of the methyl group from carbon 10 of progesterone makes it more active. Other researchers discovered that the incorporation of an ethynyl group at position 17 increased the oral activity of these drugs. Carl Djerassi at Syntex decided to combine both of these effects and prepared norethindrone. This compound was found to be an effective oral contraceptive and was quickly patented in 1951.

Norethindrone

Although norethindrone was the first oral contraceptive to be developed, it was not the first to be marketed. This was due mainly to the lack of both biological testing laboratories and marketing expertise at Syntex.

In 1953, G. D. Searle and Co. patented a related compound, norethynodrel. This steroid is also an active contraceptive when taken orally. This fact is not at all surprising to an experienced organic chemist because it is well known that a CC double bond such as the one in norethynodrel is readily isomerized into conjugation with a carbonyl group. This acid-catalyzed isomerization occurs via an enol intermediate, and the equilibrium favors the more stable, conjugated compound. The acidic conditions in the human gastric system are quite sufficient to accomplish the transformation of norethynodrel to norethindrone. It seems that chemical reactions do not care much about patents!

Norethynodrel

Syntex licensed norethindrone to Parke-Davis for development. Eventually, Parke-Davis decided not to market it because of fears about a possible boycott of their other products by groups who were opposed to birth control. Searle had no such concerns and brought its pill to the market in 1960 under the name Enovid. In the meantime, Syntex was forced to find another partner and finally settled on the Ortho division of Johnson & Johnson. The norethindrone pill was first marketed in 1962 under the name Ortho-Novum. By 1965 "the pill" was the most popular form of birth control.

Progestin-based contraceptives work by inducing a state of pseudopregnancy. Therefore, they are administered in a 20-day cycle, followed by a break to allow normal menstruation. Today, most birth control pills use norethindrone or norgestrel, which is similar to norethindrone but has an ethyl group at position 13 rather than a methyl group. In addition, they contain a small amount of estrogen to reduce breakthrough bleeding. Interestingly, the usefulness of estrogen was discovered by accident when one batch of pills was synthesized with a small amount of estrogen as a contaminant.

There has been considerable concern about the health effects of the pill, and many studies have been done. These are extremely powerful compounds and are taken by a large number of healthy women over an extended period, not to cure disease, but to prevent pregnancy. Although there was some evidence for heightened risk of cardiovascular disease in early studies, this risk decreased as the amount of estrogen in the pill was decreased. Today, the amount of estrogen has been reduced from 150 μg per pill for Enovid to 30 to 35 μg per pill. The progestin component has also been reduced, and the pill is a relatively safe method of birth control.

28.7 ALKALOIDS

Alkaloids are a group of nitrogen-containing natural products that occur primarily in higher plants, although they are also found in some fungi, such as mushrooms. The name alkaloid (meaning "alkali-like") is applied because they are amines and thus basic. Their basic character allows them to be readily isolated from their plant source. The plant material is extracted with aqueous acid. This converts the alkaloid to an ammonium cation, which is water soluble. Neutralization of the acidic extract with base causes the alkaloid to precipitate.

Like terpenes, alkaloids have been important in the development of organic chemistry. Some, such as nicotine, have relatively simple structures. Others, such as morphine and strychnine, have very complex structures containing multiple rings. Most are quite physiologically active.

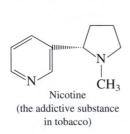

Nicotine
(the addictive substance
in tobacco)

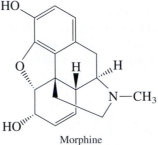

Morphine
(a highly addictive painkiller
obtained from the opium poppy)

Strychnine
(an extremely
poisonous central
nervous system stimulant)

The complex structures of alkaloids, along with their biological activity and relative ease of isolation, have kept the interest of organic chemists over the years. They have provided enormously challenging structure elucidation problems. For example, although strychnine was first isolated in 1818, its complete structure was not determined until 1946. (R. Robinson was awarded the 1947 Nobel Prize in chemistry for the determination of the structure of strychnine and other alkaloids as well as synthetic work in this area.) Of course, soon after this, synthetic chemists accepted the challenge of preparing this complicated compound in the laboratory. The first synthesis, reported by

Woodward in 1953, required 28 steps (not very many when the complexity of the molecule is considered). More recently, strychnine has been prepared, enantiomerically pure, in 20 steps with an overall yield of 3%!

Just as terpenes could be viewed as being formed from isoprene units, alkaloids can be viewed as being derived from amino acids. Four amino acids give rise to important classes of alkaloids. As shown in Table 28.1, the pyrrolidine alkaloids are derived from the amino acid ornithine (not one of the 20 "standard" amino acids), the piperidine alkaloids from lysine, the isoquinoline alkaloids from tyrosine, and the indole alkaloids from tryptophan.

Table 28.1   Some Important Classes of Alkaloids

Amino Acid	Alkaloid Partial Structure	Example	
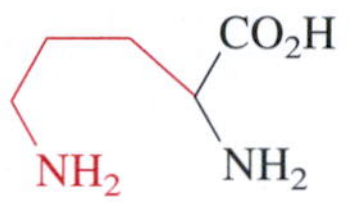 Ornithine	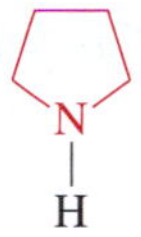 Pyrrolidine	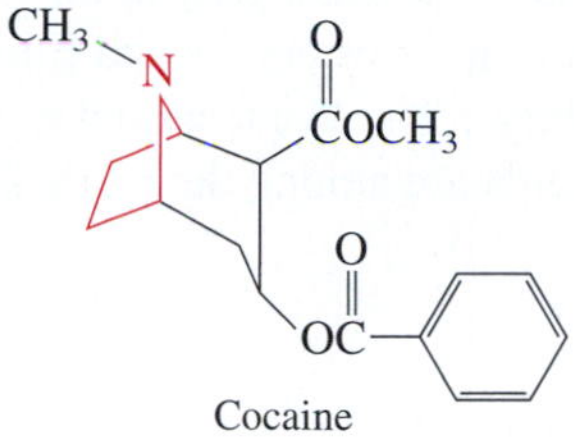 Cocaine	Cocaine is a central nervous system stimulant; it is obtained from the coca plant.
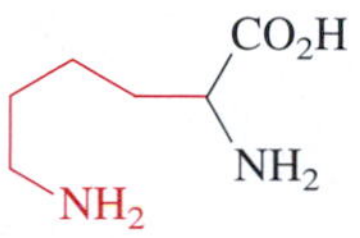 Lysine	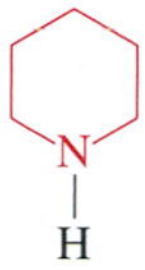 Piperidine	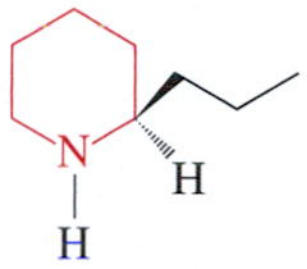 Coniine	Coniine is a major component of poisonous hemlock, which was used to kill Socrates.
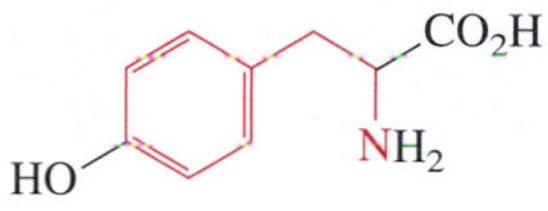 Tyrosine	 Isoquinoline	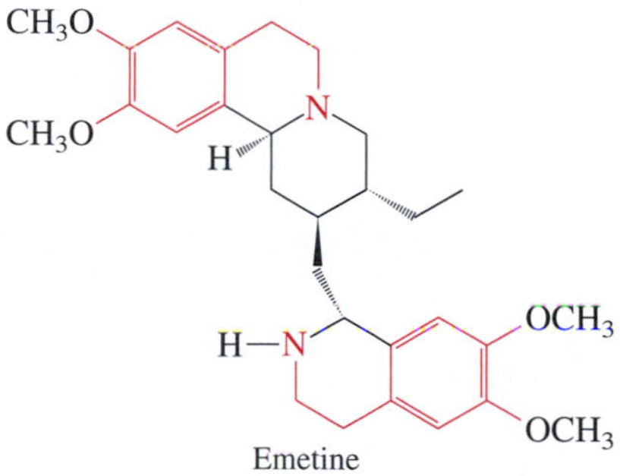 Emetine	Emetine is used to treat amebic dysentery.
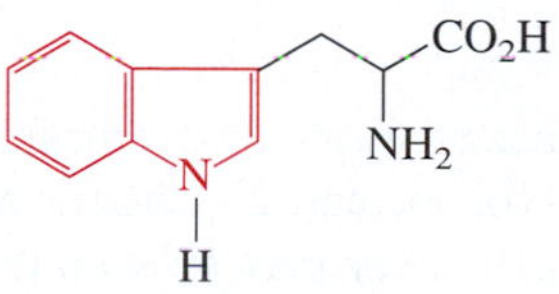 Tryptophan	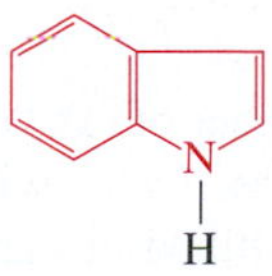 Indole	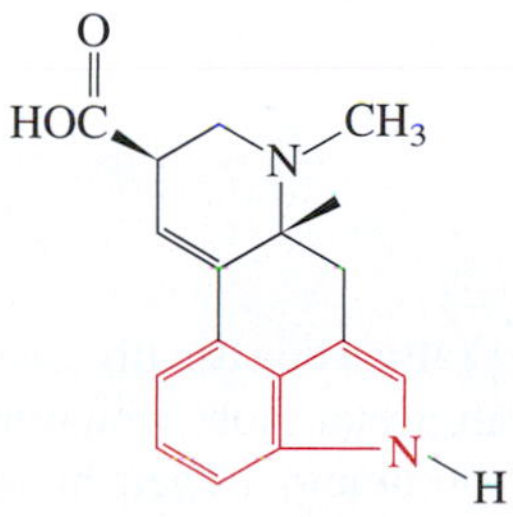 Lysergic acid	Lysergic acid is an ergot alkaloid that is obtained from a fungus that grows on cereal. The diethyl amide is the hallucinogen LSD.

As was the case with terpenes, the function of alkaloids in plants is not known. It has been proposed that they are merely nitrogen-containing waste products of plants, like urea in animals. However, most plants reutilize nitrogen, rather than wasting it. Furthermore, it is difficult to imagine why such complex structures would be needed to store waste nitrogen. Like terpenes, alkaloids have been proposed to serve as protection from herbivores and insects. However, only a few examples of such protection can be demonstrated. Whatever the role of alkaloids is, some 70% to 80% of plants manage to do quite nicely without them.

28.8   FATS AND RELATED COMPOUNDS

As discussed briefly in the Focus On box titled "The Preparation of Soap" in Chapter 19, fats are triesters formed from glycerol and long, linear-chain carboxylic acids, known as fatty acids. The resulting triesters are called **triacylglycerols.** Stearic, palmitic, and oleic acids are among the many different fatty acids that occur in nature:

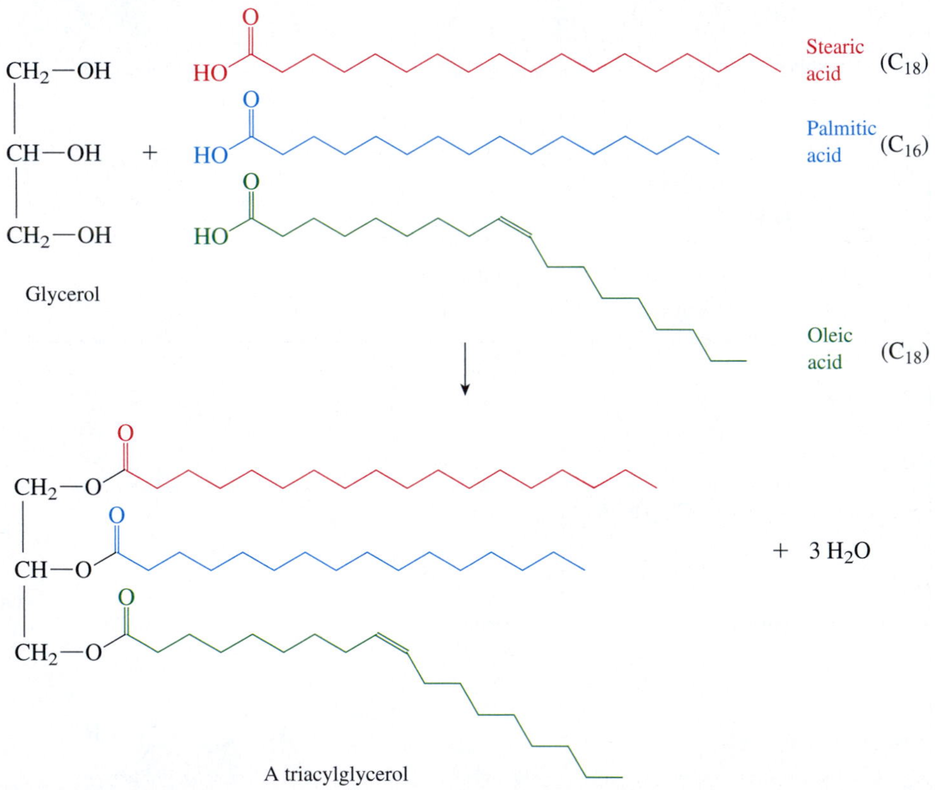

Fatty acids usually have an even number of carbons because they are biosynthesized from acetate ion, which has two carbons. Those with 14, 16, 18, and 20 carbons are most common. Their biosynthesis is outlined in Figure 28.10. They may be saturated, like stearic acid and palmitic acid, or they may have one or more CC double bonds, like

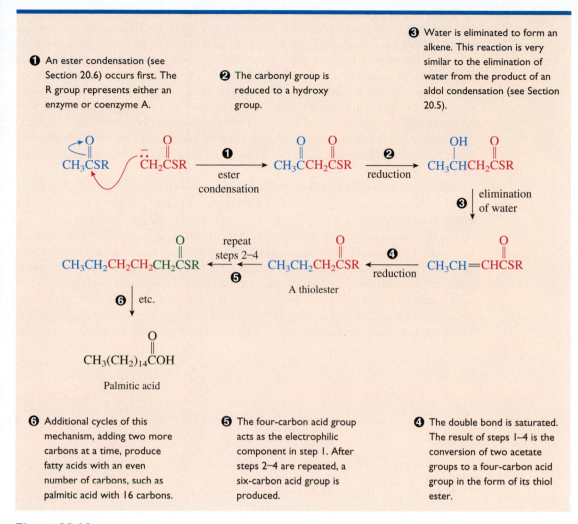

❶ An ester condensation (see Section 20.6) occurs first. The R group represents either an enzyme or coenzyme A.

❷ The carbonyl group is reduced to a hydroxy group.

❸ Water is eliminated to form an alkene. This reaction is very similar to the elimination of water from the product of an aldol condensation (see Section 20.5).

❻ Additional cycles of this mechanism, adding two more carbons at a time, produce fatty acids with an even number of carbons, such as palmitic acid with 16 carbons.

❺ The four-carbon acid group acts as the electrophilic component in step 1. After steps 2–4 are repeated, a six-carbon acid group is produced.

❹ The double bond is saturated. The result of steps 1–4 is the conversion of two acetate groups to a four-carbon acid group in the form of its thiol ester.

Figure 28.10

THE BIOSYNTHESIS OF FATTY ACIDS.

oleic acid. The double bonds invariably have cis geometry, and the kink in the long chain caused by the cis double bond prevents the chain from packing as well into a crystal lattice and decreases the melting point of the lipid. Thus, oils tend to have a larger percentage of unsaturated fatty acids than fats.

PROBLEM 28.20
Show a mechanism for step 1 of Figure 28.10. Ignore the participation of the enzyme.

PROBLEM 28.21
Show a mechanism for step 3 of Figure 28.10. Assume that the reaction is catalyzed by base.

A particular species of plant or animal contains a number of different fatty acid residues in its triacylglycerols. These are randomly distributed, so the individual molecules of a triacylglycerol are not all identical. For example, one molecule may have the structure shown previously, having been formed from one molecule each of stearic, palmitic, and oleic acid. Another may contain two molecules of stearic acid and one of oleic acid. Still another may contain entirely different fatty acids. The average composition of the fatty acid part of triacylglycerols varies with the species of the source. Triacylglycerols from plants usually contain more unsaturated fatty acid residues and have lower melting points than those from animals. Thus, plant triacylglycerols are more likely to be oils, whereas those from animals are fats.

Fats and oils serve as energy reserves for the organism. Because they are in a lower oxidation state than carbohydrates, they provide more energy per gram when they are metabolized (see the Focus On box "Energy Content of Fuels" in Chapter 5).

Glycerophospholipids are an important class of compounds related to fats. They are also triesters of glycerol. However, in this case, two of the ester groups are formed from fatty acids, whereas the third is a phosphate ester that also has an ionic or very polar group. Glycerophospholipids are the major component of biological membranes. Their polar heads project into the aqueous solution, and the nonpolar tails form the bilayer membranes. An example of a glycerophospholipid is shown in the following structure. Again, the fatty acid components, as well as the polar part of the phosphate ester, can vary.

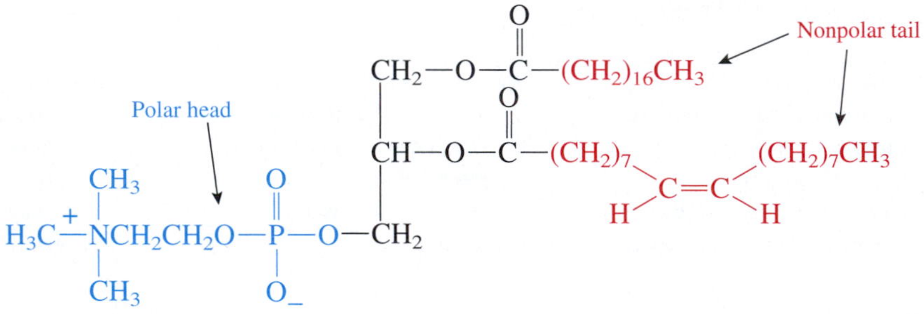

A glycerophospholipid

Focus On

Partially Hydrogenated Vegetable Oil

Margarines are prepared from vegetable oils. However, most people do not like to spread liquid oil on their toast. The presence of cis double bonds in the triacylglycerols causes kinks in the hydrocarbon tails of the fatty acid residues. These kinks prevent the triacylglycerol molecules from packing closely and lower the melting point. To raise the melting point of the oil so that it is a solid at room temperature, some of the double bonds are reduced by catalytic hydrogenation:

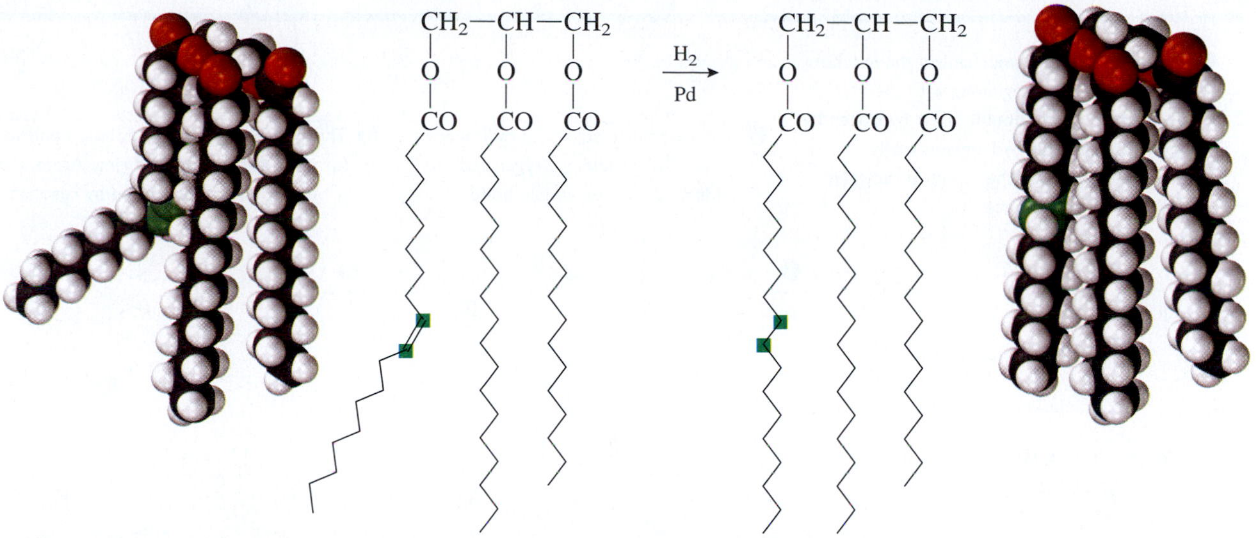

$$CH_2-CH-CH_2$$

As more of the double bonds are saturated, the melting point of the product increases. The degree of hydrogenation is carefully controlled to produce a product with just the right melting point, a partially hydrogenated vegetable oil.

A product with a higher melting point is necessary for consumer acceptance. In addition, triacylglycerides that are more unsaturated tend to spoil more rapidly. This spoilage is due to oxidation caused by radical reactions. (This is an example of the autoxidation process described in Section 21.8 and the Focus On box "Vitamin E and Lipid Autoxidation" on page 937. The hydrogens on the allylic carbons of unsaturated fatty acid residues are more readily abstracted because the resulting radicals are stabilized by resonance, so these compounds oxidize and spoil faster.) However, there is a trade-off, because it has been demonstrated that saturated fats have more deleterious health consequences than unsaturated fats do.

28.9   PROSTAGLANDINS

Prostaglandins are naturally occurring carboxylic acids that are related to the fatty acids. They contain the carbon skeleton of prostanoic acid, with various additional unsaturations and oxygen groups. One example is provided by PGE$_2$:

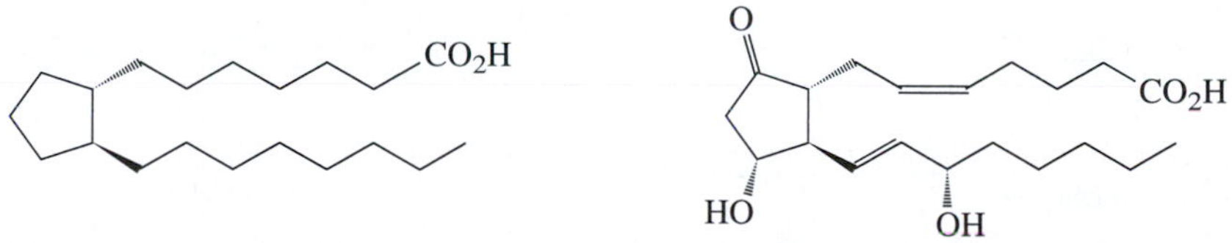

Prostanoic acid   PGE$_2$

Prostaglandins have been found to be involved in a number of important physiological functions, including the inflammatory response, the production of pain and fever, the regulation of blood pressure, the induction of blood clotting, and the induction of labor.

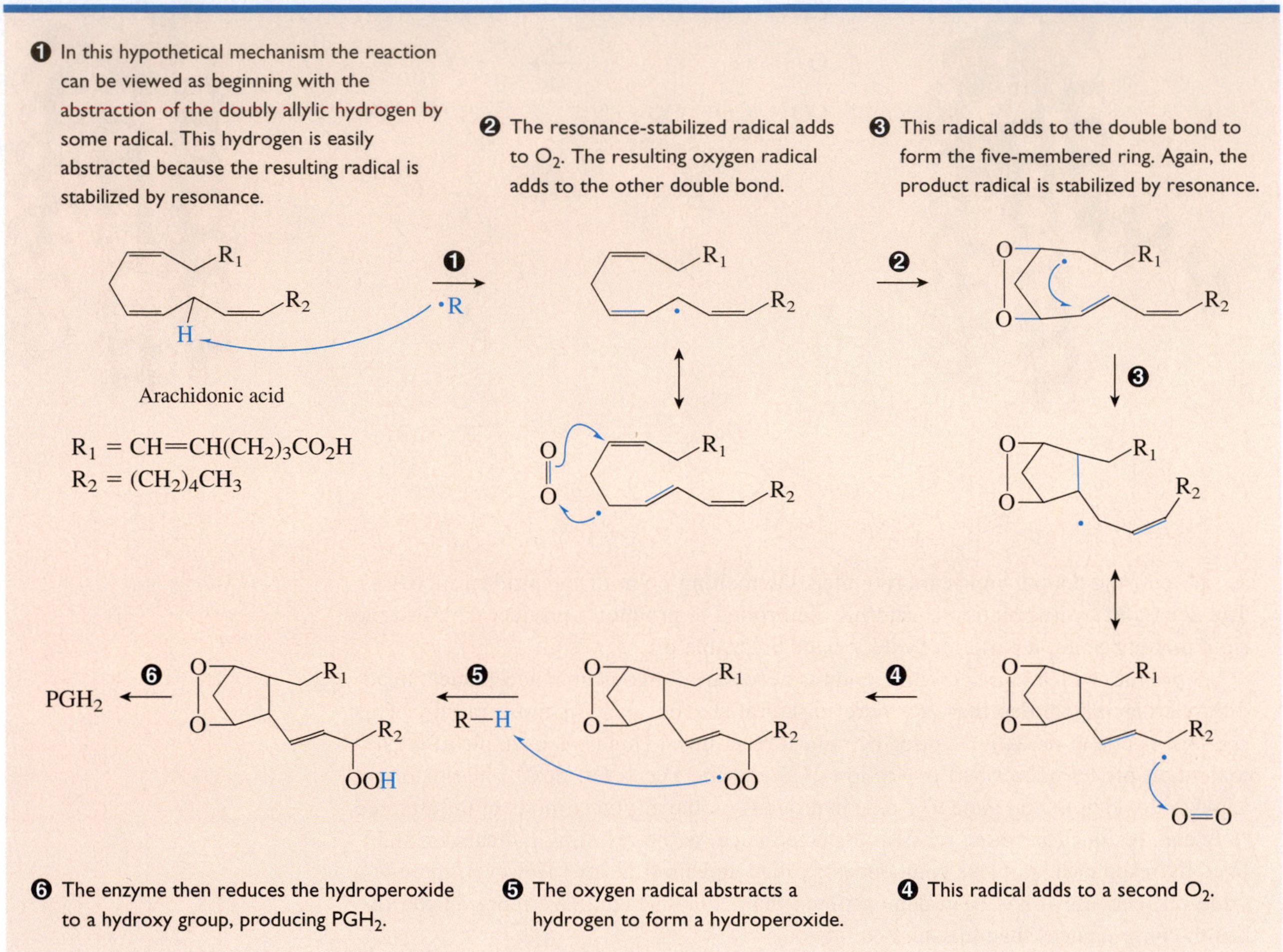

❶ In this hypothetical mechanism the reaction can be viewed as beginning with the abstraction of the doubly allylic hydrogen by some radical. This hydrogen is easily abstracted because the resulting radical is stabilized by resonance.

❷ The resonance-stabilized radical adds to O$_2$. The resulting oxygen radical adds to the other double bond.

❸ This radical adds to the double bond to form the five-membered ring. Again, the product radical is stabilized by resonance.

Arachidonic acid

$R_1 = CH=CH(CH_2)_3CO_2H$
$R_2 = (CH_2)_4CH_3$

❻ The enzyme then reduces the hydroperoxide to a hydroxy group, producing PGH$_2$.

❺ The oxygen radical abstracts a hydrogen to form a hydroperoxide.

❹ This radical adds to a second O$_2$.

Figure 28.11

HYPOTHETICAL "MECHANISM" OF THE FORMATION OF PGH$_2$ FROM ARACHIDONIC ACID.

Prostaglandins are biosynthesized from arachidonic acid, an unsaturated fatty acid containing four double bonds. The enzyme prostaglandin endoperoxide synthase converts arachidonic acid to PGH$_2$, which serves as the precursor for prostaglandins and related compounds. Aspirin exerts its pharmacological effect by inhibiting this enzyme.

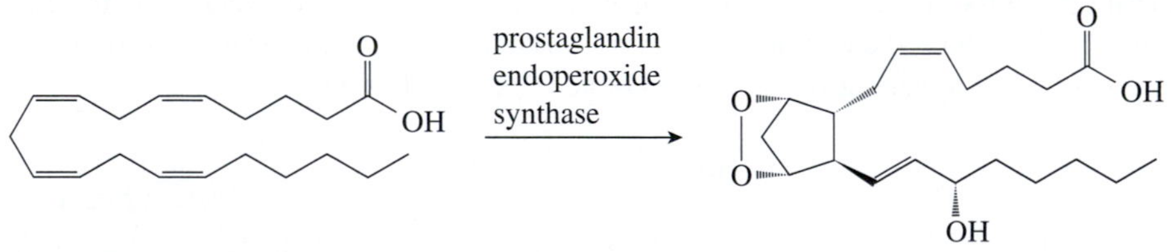

Arachidonic acid

prostaglandin endoperoxide synthase

PGH$_2$

Although the enzyme exerts enormous stereochemical and regiochemical control, the reaction that it catalyzes, like others in this chapter, involves only unexceptional chemical steps. A hypothetical "mechanism" for this process, based on radical chemistry, shows that the steps are, indeed, reasonable (see Figure 28.11). Again, remember that this does not represent the real mechanism, which is certainly more concerted than shown in the figure and probably does not involve any true radical intermediates.

Review of Mastery Goals

After completing this chapter, you should be able to:

■ Recognize the general structural features associated with terpenes and how they can be viewed as being formed from isoprene units. (Problem 28.22)

■ Understand the hypothetical mechanisms by which terpenes are formed. (Problems 28.25, 28.26, 28.27, 28.32, 28.33, and 28.35)

■ Do the same for steroids and prostaglandins. (Problems 28.22, 28.28, 28.29, and 28.31)

■ Recognize the general structural features of alkaloids and fats. (Problem 28.22)

Additional Problems

28.22 Identify each of these compounds as a terpene, steroid, alkaloid, fat, or prostaglandin:

a)

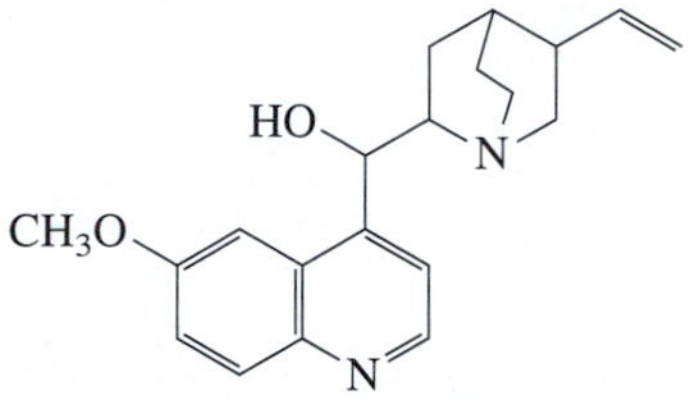

Quinine

b)

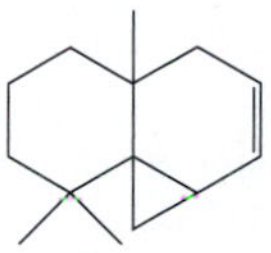

Thujopsene

c)

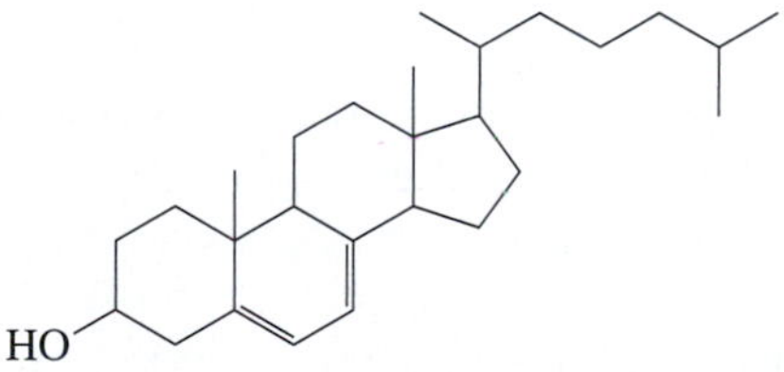

Provitamin D

d)

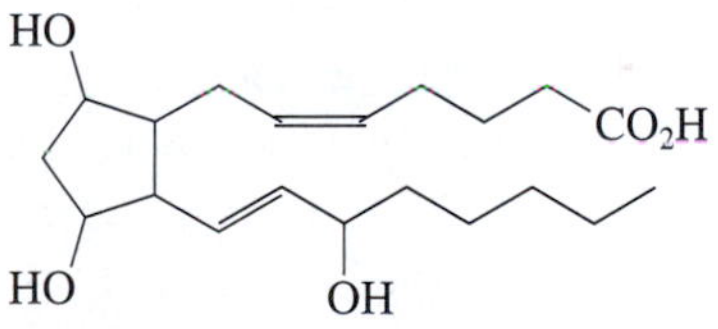

e)

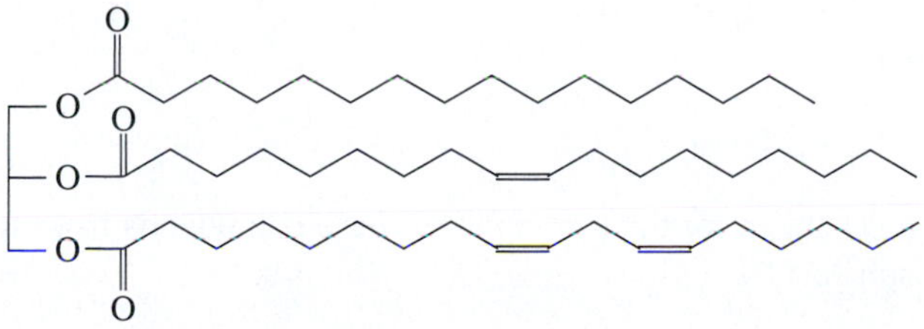

f)

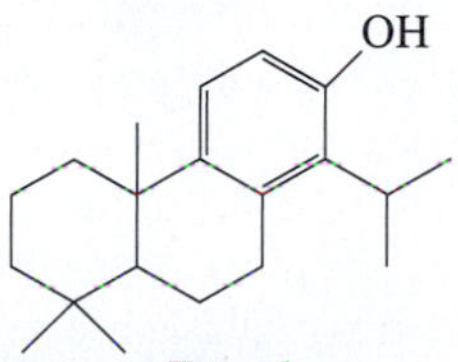

Totarol

28.23 Show the products of these reactions:

a)

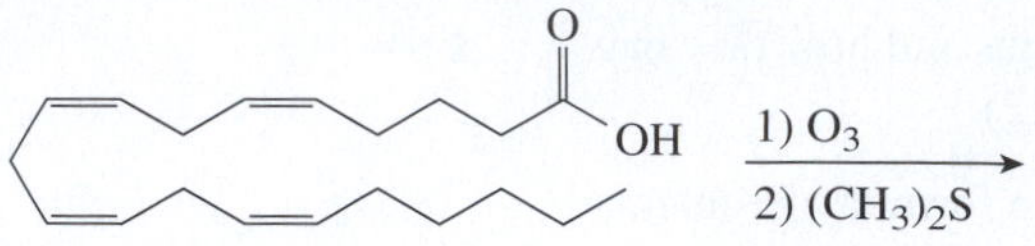

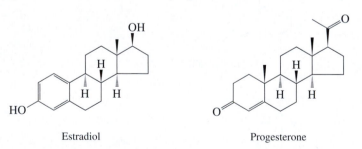

(arachidonic acid structure) $\xrightarrow[\text{2) (CH}_3)_2\text{S}]{\text{1) O}_3}$

b) 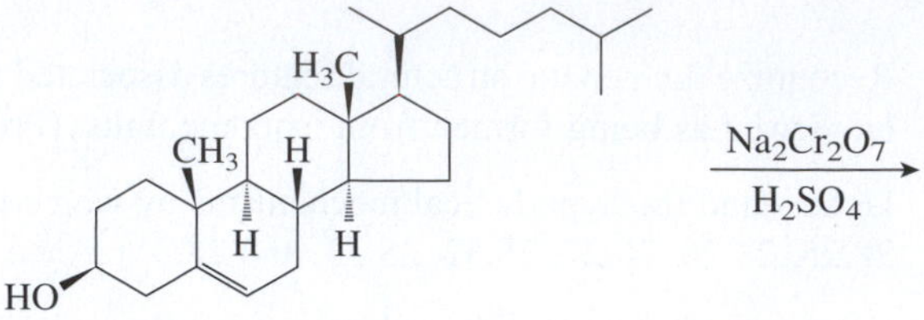 (cholesterol structure with H_3C, CH_3, H, HO) $\xrightarrow[\text{H}_2\text{SO}_4]{\text{Na}_2\text{Cr}_2\text{O}_7}$

c) 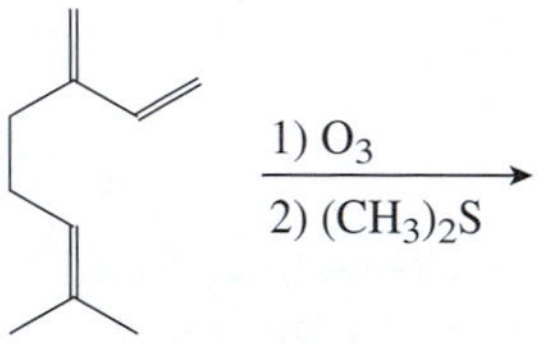 (diene structure) $\xrightarrow[\text{2) (CH}_3)_2\text{S}]{\text{1) O}_3}$

d) 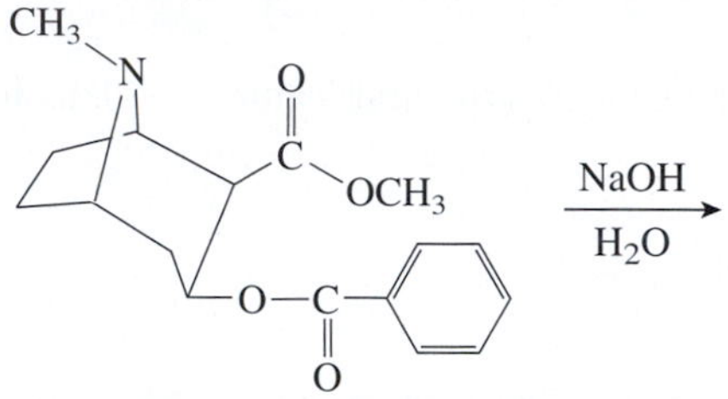 (cocaine structure with CH_3, N, C=O, OCH$_3$, O-C(=O)-phenyl) $\xrightarrow[\text{H}_2\text{O}]{\text{NaOH}}$

e) 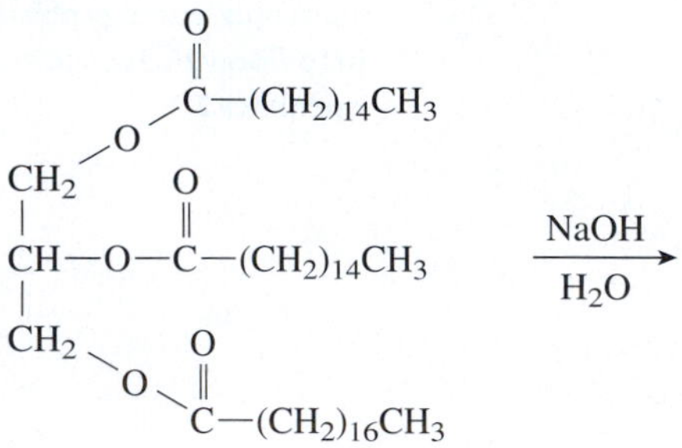 (triglyceride structure)

$$\text{CH}_2-\text{O}-\overset{\overset{\text{O}}{\|}}{\text{C}}-(\text{CH}_2)_{14}\text{CH}_3$$
$$\text{CH}-\text{O}-\overset{\overset{\text{O}}{\|}}{\text{C}}-(\text{CH}_2)_{14}\text{CH}_3$$
$$\text{CH}_2-\text{O}-\overset{\overset{\text{O}}{\|}}{\text{C}}-(\text{CH}_2)_{16}\text{CH}_3$$

$\xrightarrow[\text{H}_2\text{O}]{\text{NaOH}}$

28.24 Describe a method to easily separate estradiol from progesterone by taking advantage of their different chemical properties.

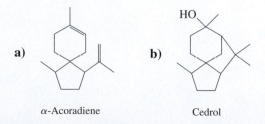

Estradiol   Progesterone

28.25 Show a mechanism for the formation of these compounds from farnesyl pyrophosphate:

a)

α-Acoradiene

b)

Cedrol

28.26 Show a mechanism for the formation of sclareol from geranylgeranyl pyrophosphate.

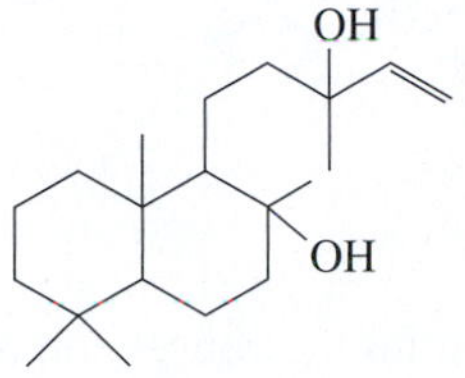

Sclareol

28.27 Show a mechanism for the formation of thujene from neryl pyrophosphate.

Thujene

28.28 Show a mechanism for this cyclization reaction:

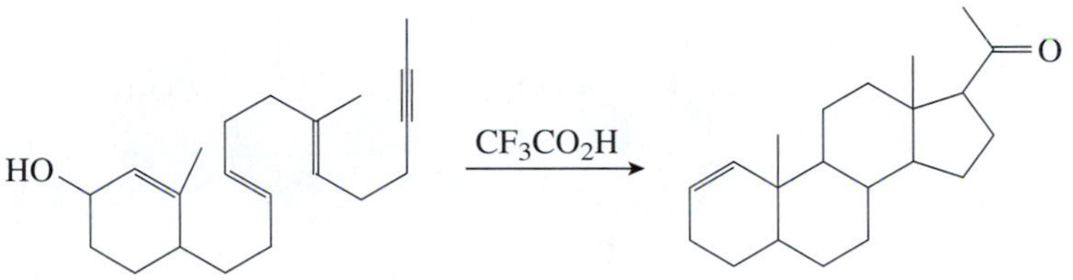

28.29 Show a mechanism for the final step in the synthesis of progesterone shown in the Focus On box "Syntheses That Mimic Nature."

28.30 Discuss the regiochemistry of each step in the conversion of alcohol **A** to ketone **B** in the Focus On box "Syntheses That Mimic Nature."

28.31 Show the steps in the mechanism for the conversion of norethynodrel to norethindrone discussed in the Focus On box "The Birth Control Pill."

28.32 Show a mechanism for the acid-catalyzed conversion of geraniol to α-terpineol and terpin:

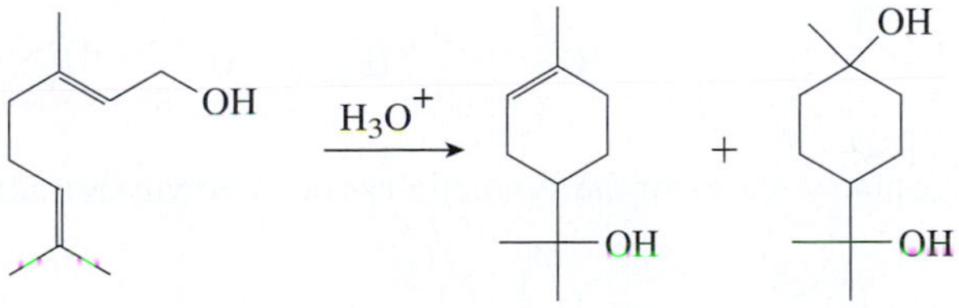

Geraniol α-Terpineol   Terpin

28.33 Suggest a mechanism for the acid-catalyzed conversion of ψ-ionone to β-ionone:

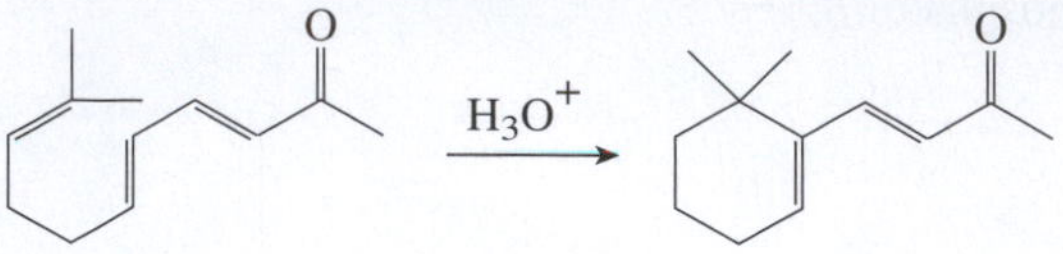

ψ-Ionone   β-Ionone

28.34 Suggest a mechanism for the conversion of limonene to isoprene:

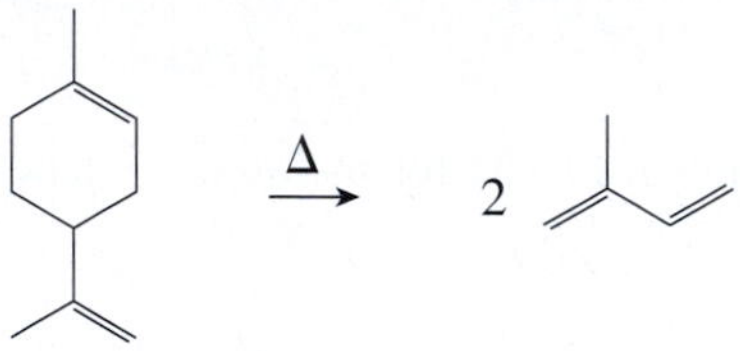

28.35 Suggest a mechanism for the cyclization of **A** and explain why **B** and **C** are unreactive under these conditions.

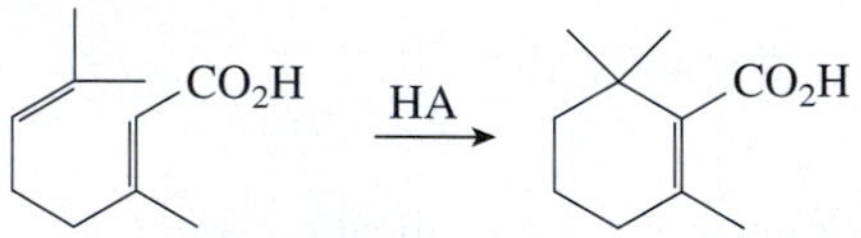

A

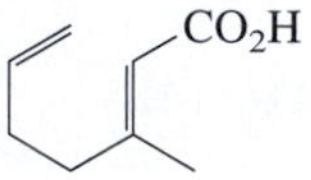

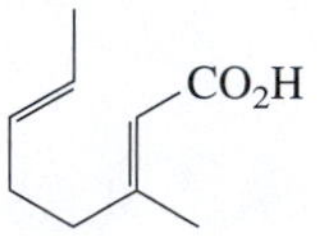

B   **C**

28.36 Show the products of ozonolysis of these terpenes:

a)   b)

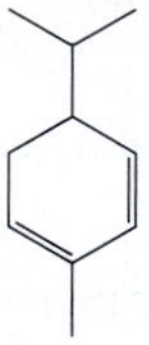

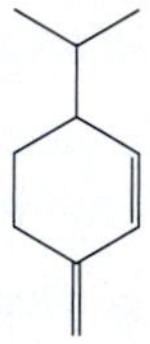

28.37 On ozonolysis, linolenic acid ($C_{18}H_{30}O_2$) gives the products shown in the following equation. Show the structure of linolenic acid.

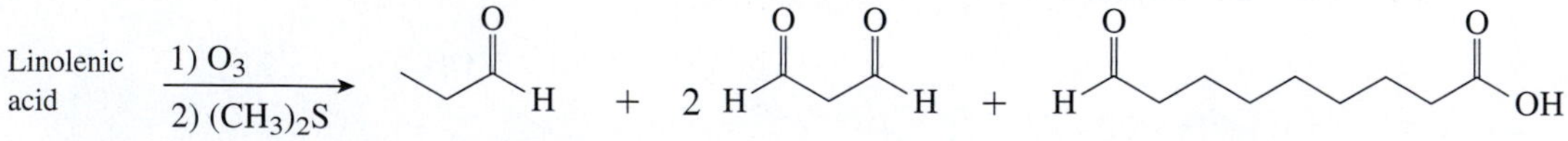

28.38 Suggest a synthesis of the synthetic estrogen ethynylestradiol from estrone.

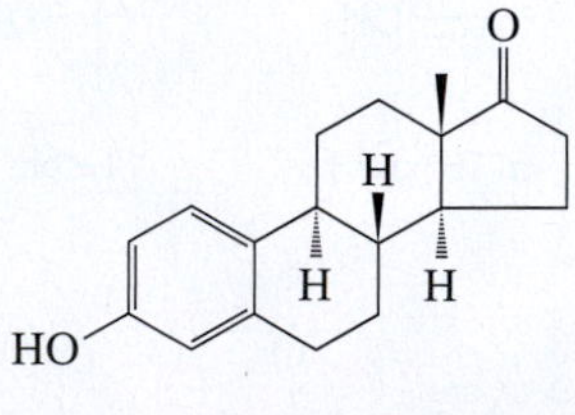

Estrone 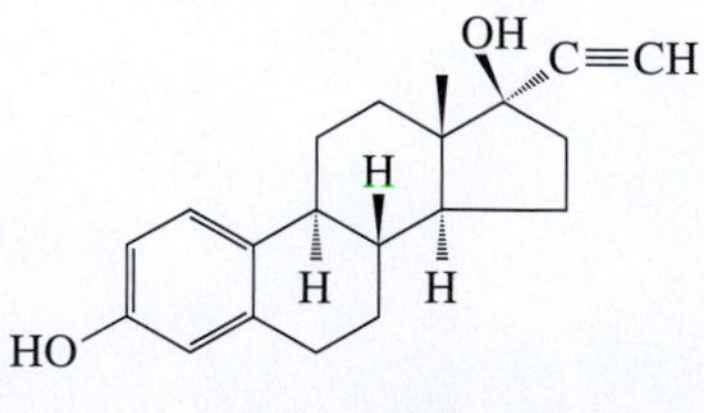   Ethynylestradiol

28.39 Suggest syntheses of oleic acid and stearic acid starting from 1-decyne and 1-bromo-7-chloroheptane.

28.40 Suggest a synthesis of β-santalene from the following ketone. (Do not worry about stereochemistry.)

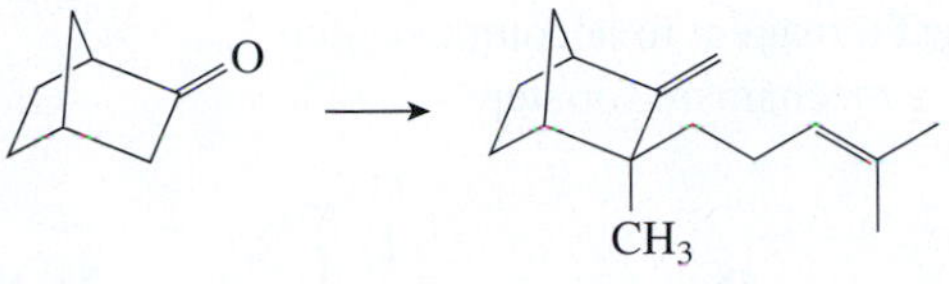

β-Santalene

28.41 The following reaction sequence was used to prepare the starting material for Johnson's synthesis of progesterone described in the Focus On box "Syntheses That Mimic Nature."

a) Show the reagents that are needed to accomplish steps 4, 5, 7, 8, and 9.
b) Show the structures of **A** and **B.**
c) Show the mechanism for step 3.
d) Show the mechanism for step 8.

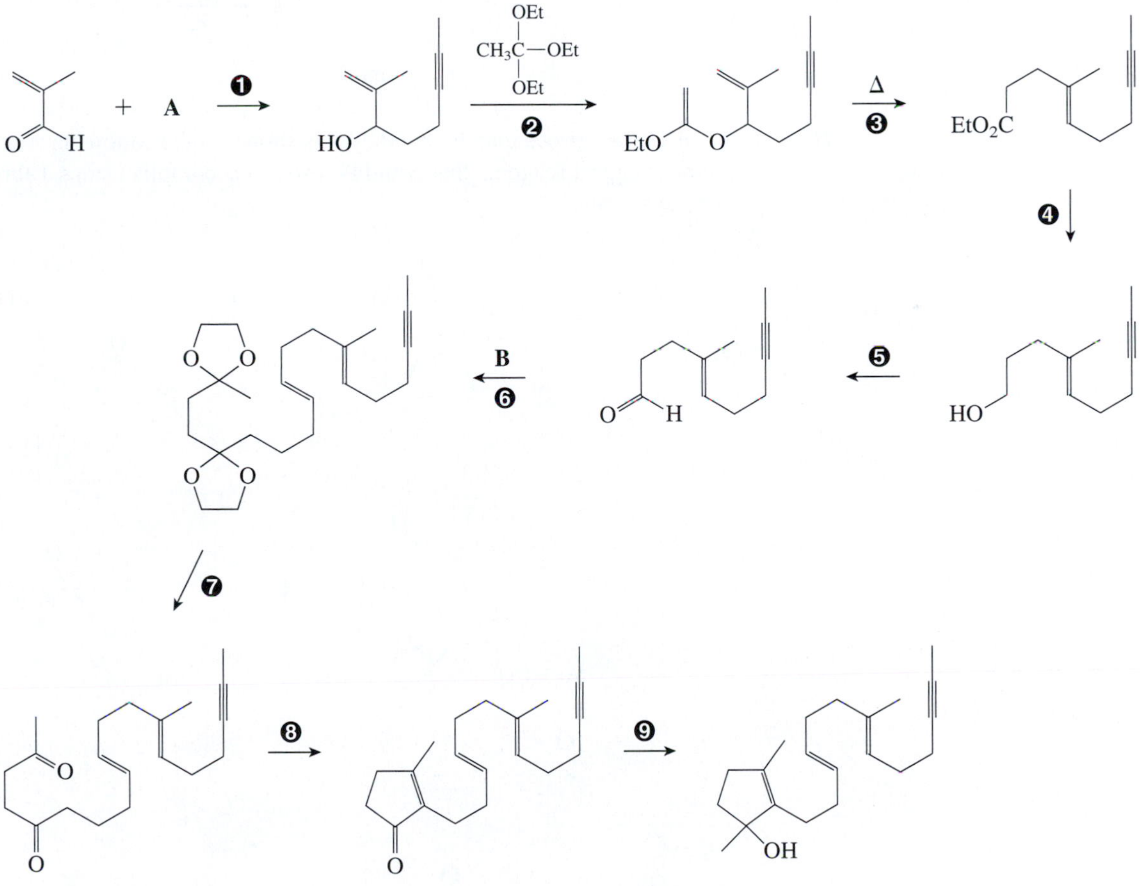

28.42 Part of Woodward's synthesis of cholesterol is shown in the following reaction scheme.
a) Show the structure of **A.** What kind of reaction is this?
b) Show a mechanism for step 2. Explain which of these products you expect to dominate at equilibrium.
c) Suggest a reagent to accomplish step 3.
d) Show a mechanism for step 4.

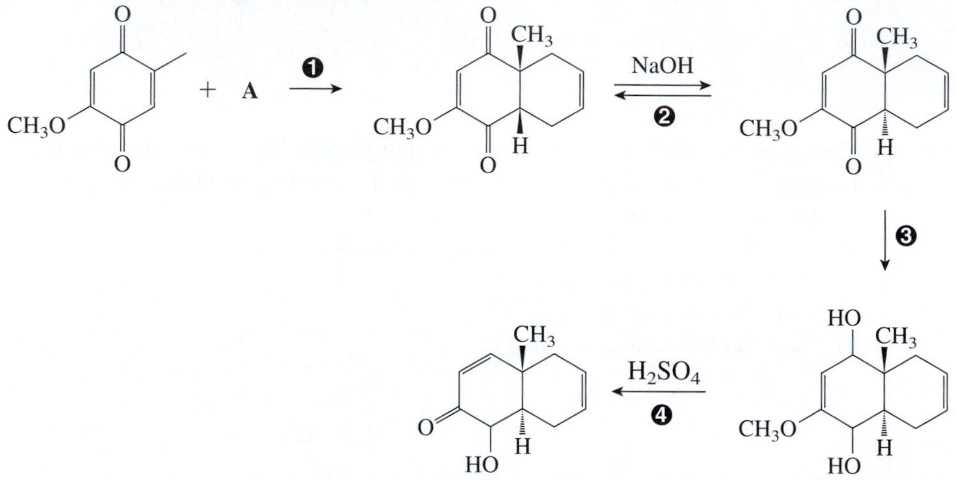

28.43 One of Corey's prostaglandin syntheses is shown in the following reaction scheme. Suggest reagents that could be used to accomplish steps 1 through 11 in this synthesis.

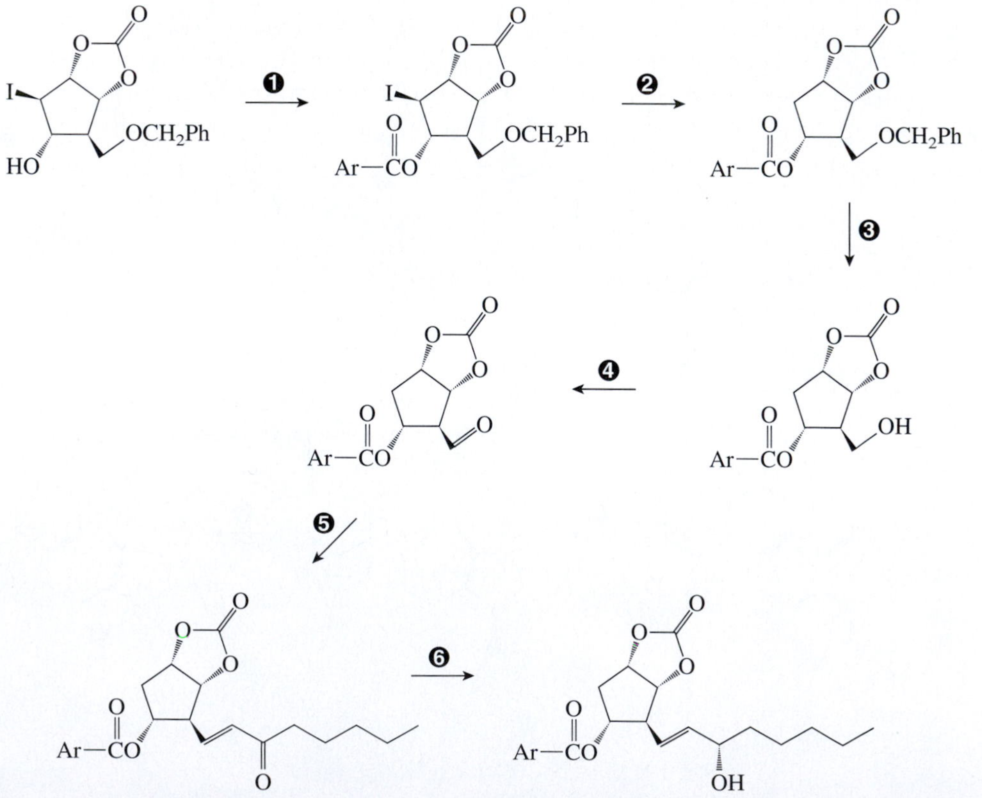

PGF$_{2\alpha}$

Problems Involving Spectroscopy

28.44 The terpene terpinolene, $C_{10}H_{16}$, gives compound **A**, $C_{10}H_{20}$, on reaction with H_2/Pt. **A** shows seven peaks in its ^{13}C-NMR spectrum. The products of ozonolysis of terpinolene are shown in the following equation. Show the structure of terpinolene.

Terpinolene $\xrightarrow[\text{2) } (CH_3)_2S]{\text{1) } O_3}$

Appendix

Answers to selected in-chapter problems are provided in this appendix. Complete answers to all of the problems can be found in the Solutions Manual that accompanies this text.

CHAPTER 1

1.1   a) :B̈r·   b) ·Ca·   c) ·G̈e·

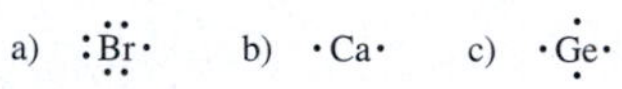

1.2   a) ·Ca· + 2 :C̈l· ⟶ Ca^{2+} + 2 :C̈l:⁻

   b) 2 Na· + ·S̈· ⟶ 2 Na^+ + :S̈:²⁻

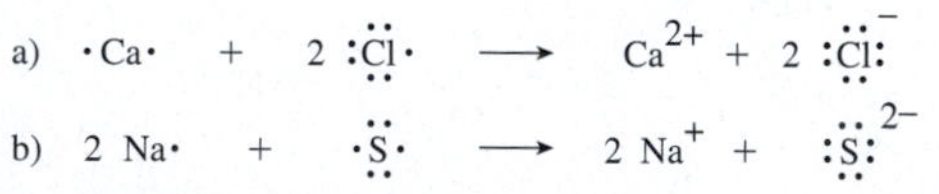

1.3   a) :C̈l:C:C̈l: (with :C̈l: above and :C̈l: below central C)   b) H:B̈r:

1.4   H:S̈:H

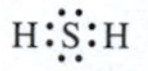

1.5   a) Stable, octet rule satisfied   b) Unstable, 10 electrons on N

1.6   a) H:C:C:C:H (with H H H on top, H H H on bottom)   b) H:C:::C:H   c) H:C::N:H (with H on top)   d) H:N:O:H (with H on top)

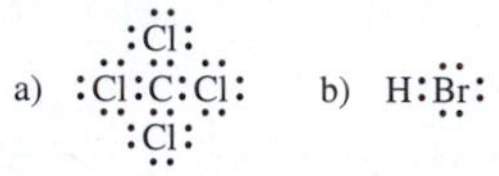

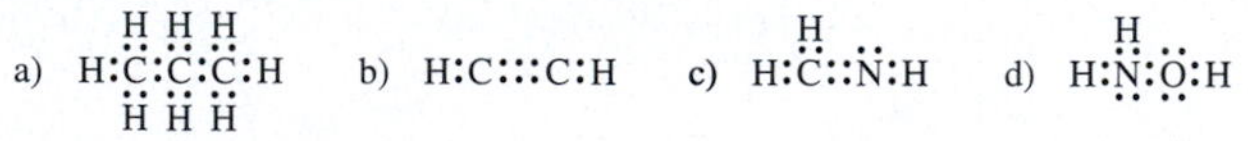

1.7   H:Ö:H + H:C̈l: ⟶ H:Ö:H⁺ (with H below) + :C̈l:⁻

1.8   a)
0 ↘ H   +1
H—C—O—H
   | |
   H   H
Total Charge = TC = +1

   b)
0 ↘ H   −1
H—C—Ö:
   |
   H
TC = −1

   c)
0 ↘ :F̈: −1
F—B—F
   |
   F
TC = −1

   d)
H ↘
   C=N=N̈:
H ↗ ↑ ↑
  0 +1 −1
TC = 0

   e)
+1 ↘   Ö:←0
H—Ö—N
       Ö:←−1
   0
TC = 0

   g)
+1 H   0 0
H—N—C—C   Ö:←0
   | | Ö:←−1
   H   H
TC = 0

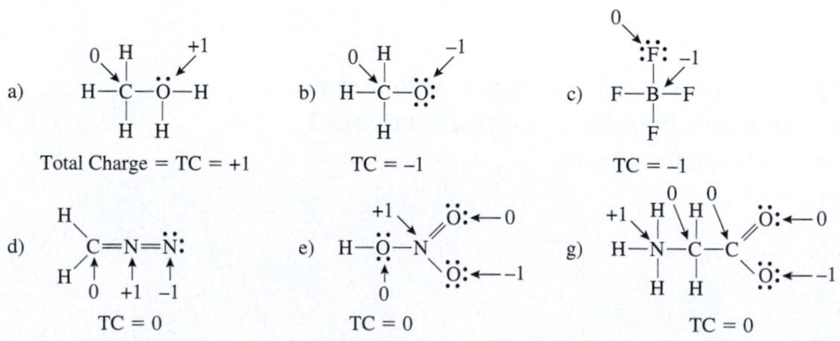

1.9   H—O—Cl more stable because of less formal charges

A-0

1.10

1.11 a) C—N b) O—N c) O—Cl d) C—Cl e) B—O f) C—Mg g) Not polar h) Not polar

1.12 a) Linear b) Trigonal planar c) Tetrahedral

1.13 a) Tetrahedral at C, bent at O (tetrahedral) b) Trigonal planar at C, bent at N (trigonal planar)
  c) Trigonal planar at C, bent at O (trigonal planar)

1.14 a) 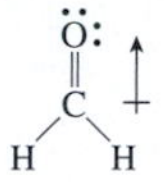 b) 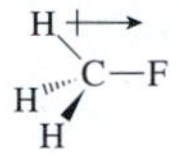 c)

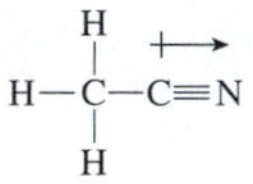

CHAPTER 2

2.1 a) Less stable, octet rule satisfied but charge on C b) Stable, octet rule satisfied c) Very unstable, octet rule not satisfied at N d) and e) Stable, octet rule satisifed

2.2

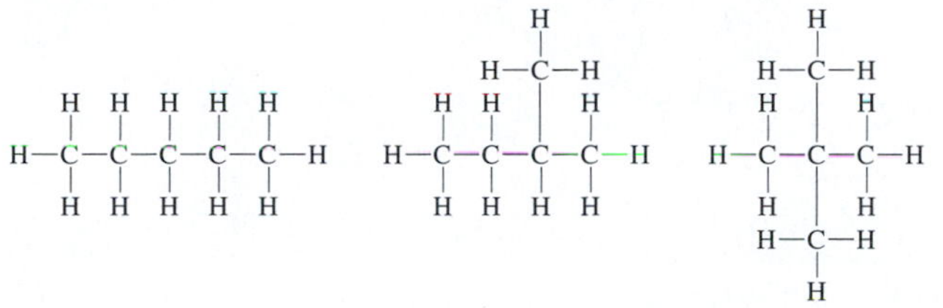

2.3

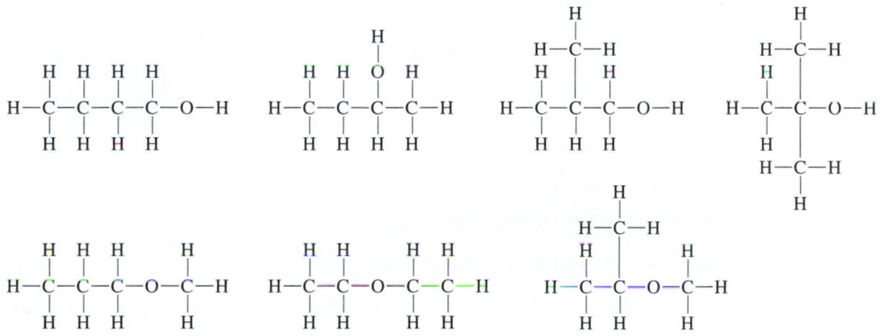

2.4 a) Same b) Same c) Isomers d) Same

2.5 a) DU = 0

  b) DU = 2

  c) DU = 4

2.6 a) 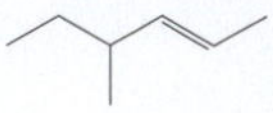 [structure] b) 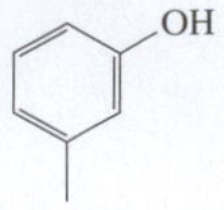 [structure with OH] c) 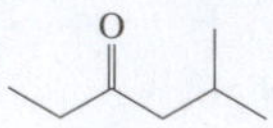 [structure with O] d) 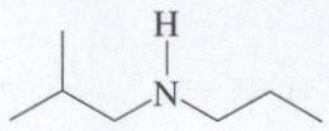 [structure with N—H]

2.8 a) Same b) Same c) Isomers d) Same e) Same f) Isomers

2.9 a) DU = 1 [structures]

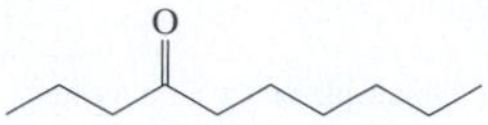

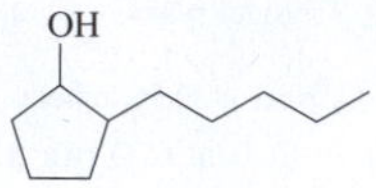

  b) DU = 3 [structures]

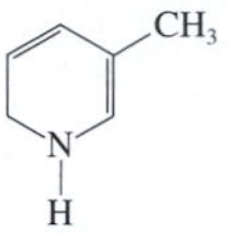

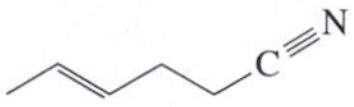

  c) DU = 0 [structures]

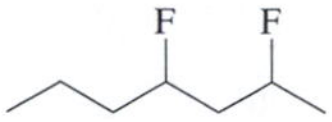

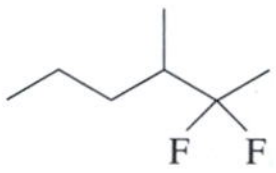

  d) DU = 3 [structures]

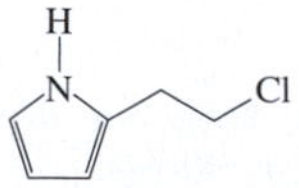

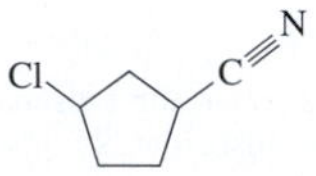

  e) DU = 3 [structures]

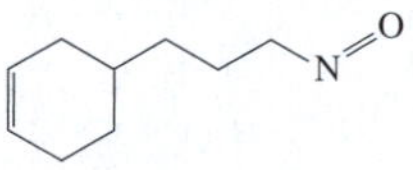

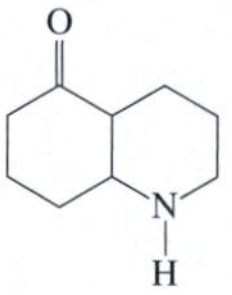

2.10 a) 5 b) 4 c) 3 d) 7 e) 6

2.11 a) London b) Van der Waals c) Ion–ion d) Van der Waals and hydrogen bonding

2.12

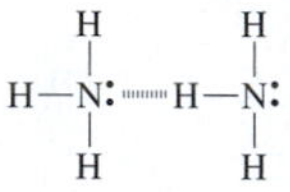

$$H-\overset{H}{\underset{H}{N}}\colon \cdots\cdots H-\overset{H}{\underset{H}{N}}\colon$$

2.13 KBr because it is ionic

2.14 $CH_3CH_2CH_2OH$ because it can hydrogen bond

2.15 $CH_3CH_2CH_2CH_2CO_2H$ because it has a smaller nonpolar part

2.16 a) Ether b) Alcohol c) Carboxylic acid d) Amide e) Ester f) Arene and aldehyde

CHAPTER 3

3.1 [diagram of concentric circles]

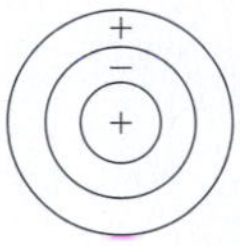

3.2   a)

E
- $3p$ ↑ ↑ __
- $3s$ ↑↓
- $2p$ ↑↓ ↑↓ ↑↓
- $2s$ ↑↓
- $1s$ ↑↓

b)

E
- $3p$ ↑↓ __ __
- $3s$ ↑↓
- $2p$ ↑↓ ↑↓ ↑↓
- $2s$ ↑↓
- $1s$ ↑↓

c)

E
- $3p$ ↑↓ ↑↓ ↑
- $3s$ ↑↓
- $2p$ ↑↓ ↑↓ ↑↓
- $2s$ ↑↓
- $1s$ ↑↓

3.3   a) Excited state   b) Excited state

3.4

E
- ↑↓ $1s_a - 1s_b$
- ↑↓ $1s_a + 1s_b$

3.5

Osp^3 $\sigma Csp^3 + H1s$

H—C—O—C—H with H atoms above and below each C

$\sigma Csp^3 + Osp^3$

3.6   1) $\sigma Csp^2 + H1s$ 2) $\sigma Csp^3 + H1s$ 3) $\sigma Csp^3 + Csp^2$
4) $\sigma Csp^2 + Nsp^2$, $\pi C2p + N2p$ 5) $\sigma Csp^3 + Nsp^2$ 6) $\sigma Csp^3 + H1s$

3.7   a) sp at both
b) one sigma and two pi bonds
c) nonbonding sp AO on N

3.8   1) $\sigma Csp^2 + H1s$ 2) $\sigma Csp^3 + H1s$ 3) $\sigma Csp^3 + Csp^2$
4) $\sigma Csp^2 + Csp^2$, $\pi C2p + C2p$ 5) $\sigma Csp^2 + Csp$
6) $\sigma Csp + Csp$, $2\ \pi C2p + C2p$ 7) $\sigma Csp + H1s$

3.9   a) 1) $\sigma Csp^3 + H1s$ 2) $\sigma Csp^3 + Csp^2$ 3) $\sigma Csp^2 + H1s$ 4) $\sigma Csp^2 + Osp^2$, $\pi C2p + O2p$
b) 1) $\sigma Csp^3 + Csp^2$ 2) $\sigma Csp^2 + Csp^2$, $\pi C2p + C2p$ 3) $\sigma Csp^2 + Csp$
   4) $\sigma Csp + Nsp$, $2\ \pi C2p + N2p$ 5) $\sigma Csp^2 + H1s$
c) 1) $\sigma Csp^2 + Osp^2$, $\pi C2p + O2p$ 2) $\sigma Csp + H1s$ 3) $\sigma Csp + Csp$, $2\ \pi C2p + C2p$
   4) $\sigma Csp^2 + Csp$ 5) $\sigma Csp^3 + Csp^2$ 6) $\sigma Csp^3 + H1s$
d) 1) $\sigma Csp^3 + H1s$ 2) $\sigma Csp^3 + Csp^3$ 3) $\sigma Csp^3 + Nsp^3$ 4) $\sigma Nsp^3 + H1s$

3.10   a) None

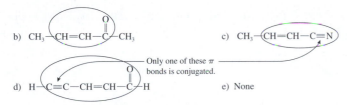

b) CH_3—CH=CH—C(=O)—CH_3

c) CH_3—CH=CH—C≡N

——— Only one of these π bonds is conjugated. ———

d) H—C≡C—CH=CH—C(=O)—H   e) None

3.11   a) 1) sp^3 2) sp^2 3) sp^2 4) sp^2 5) sp^3
b) 1) sp^3 2) sp^2 3) sp^2 4) sp^3
c) 1) sp^2 2) sp^2 3) sp^2 4) sp^3
d) 1) sp^2 2) sp^2

3.12   a) These are not resonance structures because atoms have moved.
b) These are resonance structures.
c) These are not resonance structures because atoms have moved.

3.13 a) Two electrons are missing from the structure on the right.

b) The structure on the right has 5 bonds to C and 2 more electrons.

c) In the structure on the right, the center N has only 6 electrons.

d) The C bonded to O has 5 bonds (10 electrons) in the right structure.

3.14 a) The right structure is less important because it has formal charges and the octet rule is not satisfied at one C.

b) The first two structures are equally important; the last structure is less important because it has more formal charges and the octet rule is not satisfied at the N.

3.15 a) The second structure makes only a minor contribution because it does not satisfy the octet rule. The compound has only a small (but important) amount of resonance stabilization.

b) The first two structures, plus one additional similar structure, contribute equally; the last structure can be neglected. The ion has a large amount of resonance stabilization.

3.16 a)

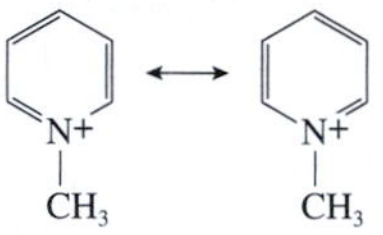

b)

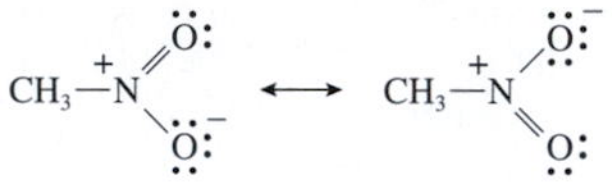

c)

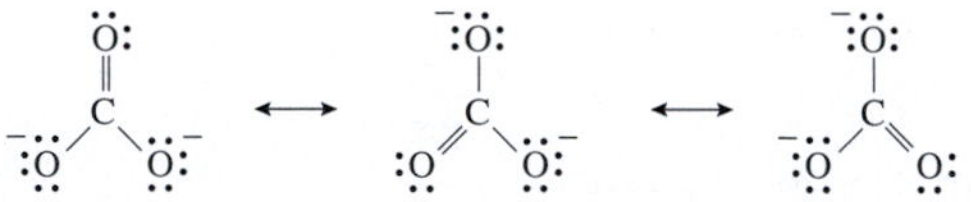

d)

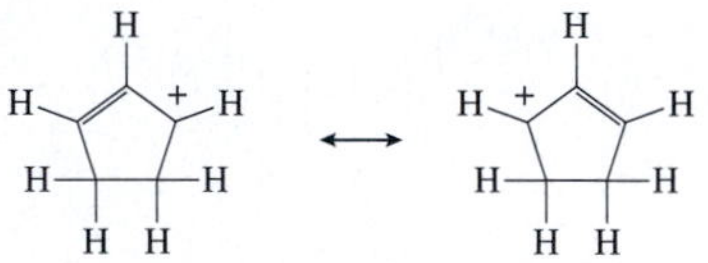

e)

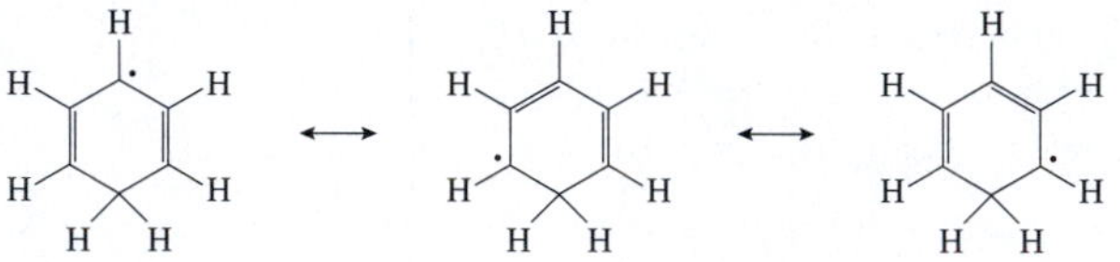

3.17

a)

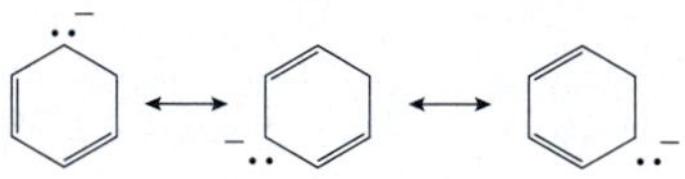

e)

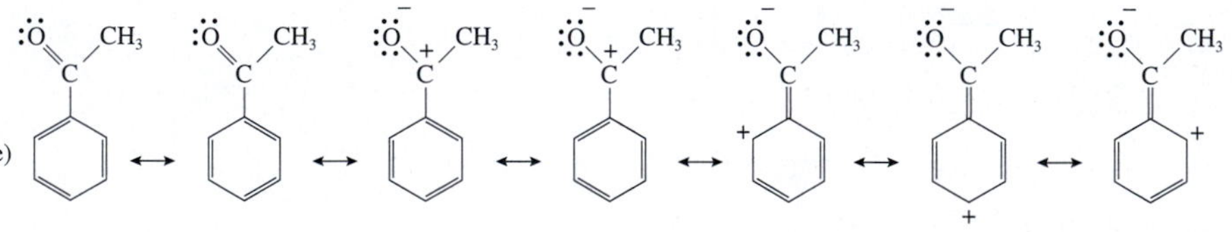

f)

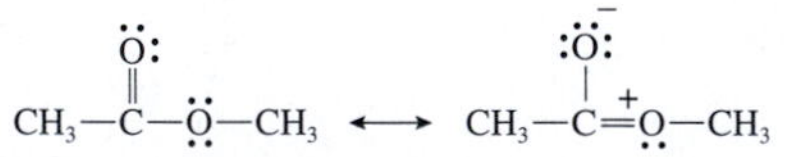

3.18

a)

HCN

— — $2\,\sigma^*$
— ⥮ $2\,\pi^*_{CN}$
⥮ $N_{nonbonding}$
⥮ ⥮ $2\,\pi_{CN}$
⥮ ⥮ $2\,\sigma$

E

b)

$$\overset{O}{\underset{\|}{HCCH_3}}$$

————— $6\,\sigma^*$
π^*_{CO}
⥮ ⥮ $O_{nonbonding}$
⥮ π_{CO}
⥮ ⥮ ⥮ ⥮ ⥮ $6\,\sigma$

E

c)

CH_3NH_2

————— $6\,\sigma^*$
⥮ $N_{nonbonding}$
⥮ ⥮ ⥮ ⥮ ⥮ $6\,\sigma$

E

CHAPTER 4

4.1 a) Acid b) Both c) Acid d) Base e) Both f) Both g) Both

4.2 a) $CH_3—\overset{+}{\underset{H}{\overset{\cdot\cdot}{O}}}—H$ b) $H—\overset{\cdot\cdot}{O}—H$ c) $CH_3—\overset{+}{N}H_3$

4.3 a) $H—\overset{\cdot\cdot}{\underset{\cdot\cdot}{O}}{:}^{-}$ b) $H—\overset{\cdot\cdot}{\underset{H}{O}}{:}^{-}$ c) $H—\overset{-}{\underset{H}{N}}{:}$ d) $H—\overset{H}{\underset{H}{C}}—\overset{H}{\underset{H}{C}}{:}^{-}$

	Base	Acid	Conjugate acid	Conjugate base

4.4 a) $:\overset{-}{N}H_2$ + $H—\overset{\cdot\cdot}{\underset{\cdot\cdot}{O}}—H$ ⇌ $:NH_3$ + $:\overset{\cdot\cdot}{\underset{\cdot\cdot}{O}}—H$

b) $CH_3\overset{\cdot\cdot}{\underset{\cdot\cdot}{O}}{:}^{-}$ + $H—\overset{+}{\underset{\overset{|}{H}}{O}}—H$ ⇌ $CH_3\overset{\cdot\cdot}{\underset{\cdot\cdot}{O}}—H$ + $H—\overset{\cdot\cdot}{\underset{\cdot\cdot}{O}}—H$

4.5 a) Lewis acid b) Lewis base c) Lewis acid d) Lewis base e) Both

4.6 a) $K_a = 1 \times 10^4$ b) $pK_a = 16$ c) $K_a = 1 \times 10^{-38}$ d) $pK_a = -6$

4.7 a) Stronger b) Weaker c) Stronger d) Weaker

4.8 a) Stronger b) Stronger c) Weaker d) Weaker

4.9 a) Favors reactants b) Favors products

4.10 a) Favors products b) Favors products

4.11

a)

Equilibrium favors reactants

$K < 1$
$\Delta G° > 0$

————— Products
↑
$\Delta G°$
↓
————— Reactants

Free energy (G)

b)

Equilibrium favors products

$K > 1$
$\Delta G° < 0$

————— Reactants
↑
$\Delta G°$
↓
————— Products

4.13 a) HCl b) PH_4^+ c) H_2S

4.14 a) HO^- b) CH_3NH^-

4.15 a) $H_2NCH_2CH_2O$—Ⓗ   b) CH_3CH_2O—Ⓗ   c) CH_3S—Ⓗ

4.16 a) CHF_2CO_2H b) CHF_2CO_2H c) $CH_3OCH_2CO_2H$

4.17 $CH_3CH_2C\equiv C$—Ⓗ

4.18

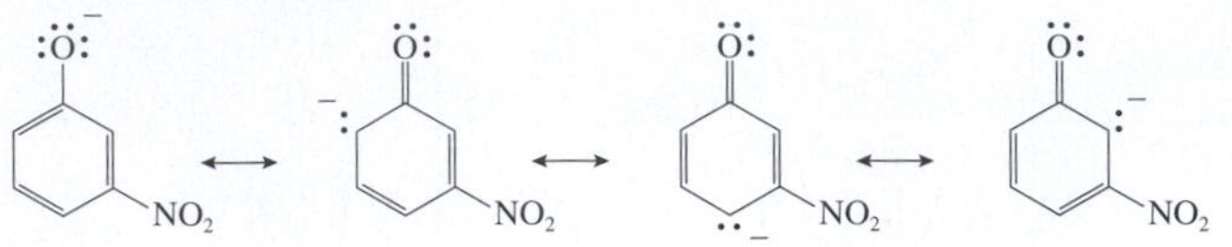

4.19 a) $CH_3CCH_2C\equiv N$ b) 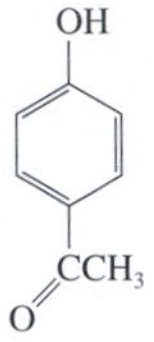   c) 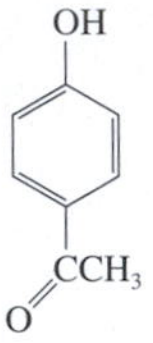   d) CH_3COCH_3 e)

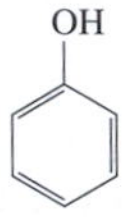

4.20 a) 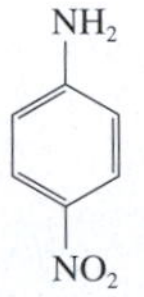   b)

4.21 a) Products   b) Products   c) Reactants

4.22 Answers a) and c) are acceptable because they have no H acidic enough to protonate the anion; b) is not acceptable because the H on the O is too acidic.

CHAPTER 5

5.1 a) 2-Methylpentane   b) 2-Methylpentane   c) 2,4-Dimethylhexane
     d) 5-Ethyl-3-methyl-5-propylnonane   e) 3-Ethyl-2,5-dimethylhexane   f) 3-Ethyl-2,6-dimethylheptane

5.2 a) 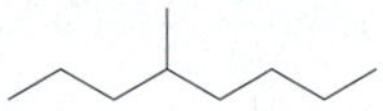   b)

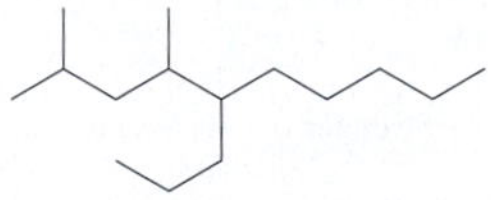

5.3 a) 4-Ethyl-2,2-dimethylhexane   b) 2,2-Dimethylpentane

5.4 a) (2-Methylpentyl)   b) (1-Methylpropyl)   c) (2,2-Dimethylpropyl)

5.5 a) 4-(1-Methylethyl)heptane   b) 5-(1,2-Dimethylpropyl)decane

5.6 a) 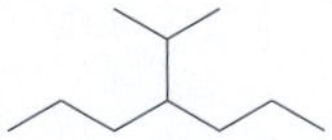   b)

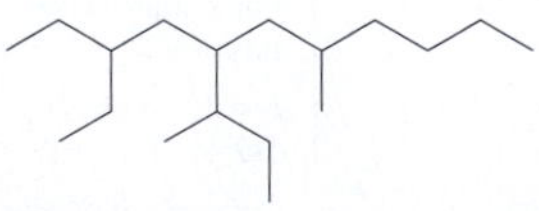

5.7 a) 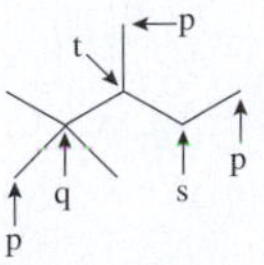   b) 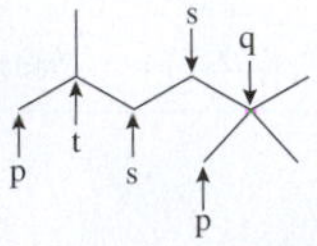   c)

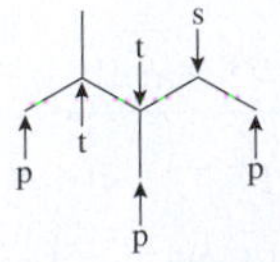

5.8 4-Isopropylheptane

5.9

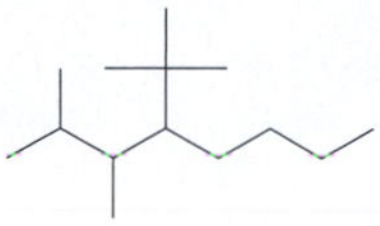

5.10 a) 1,2-Dimethylcyclopentane b) (1-Methylpropyl)cyclohexane

c) 5-Cyclopentyl-2-methylheptane d) 1-Ethyl-3,5-dimethylcyclooctane

5.11 a) 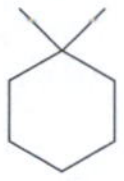 b)

5.12 a) 2,4-Dimethyl-2-hexene b) 2-Methyl-1,3,5-cycloheptatriene c) 3-Ethyl-1,2-dimethylcyclopentene

5.13 a) 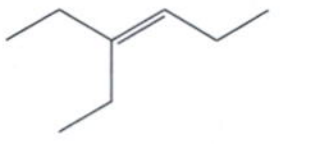 b) c)

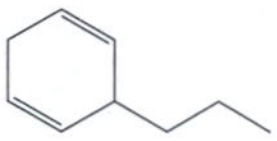

5.14 a) 3-Isopropyl-1-heptyne b) 2-Methylpent-1-en-3-yne c) 3-(2-Methylpropyl)-1,4-hexadiyne

5.15 a) $HC{\equiv}CCH_2CH_2CH_3$ b)

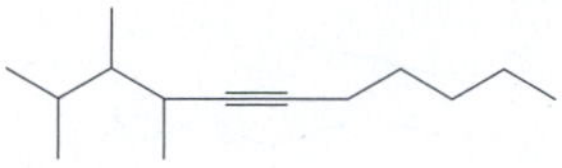

5.16 a) 5-Bromo-2,4,4-trimethylheptane b) 1-Chloro-3-ethyl-1-methylcyclopentane

5.17

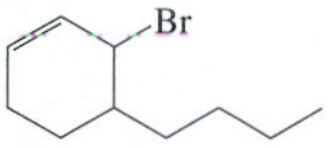

5.18 a) 2-Butanol b) 3-Methyl-3-hexanol c) 3-Cyclopentyl-1-propanol

d) 3-Bromo-3-methylcyclohexanol

5.19 a) 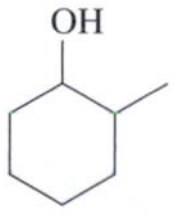 b)

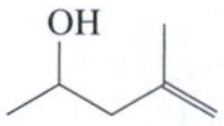

5.20 a) Ethyl methyl ether b) 1-Chloro-3-methoxycyclopentane

5.21 a) Propylamine b) *N*-Ethylcyclopentylamine

5.22 a) *N*,5-Dimethyl-2-hexanamine b) 5-Amino-2-hexanol

5.23 a) 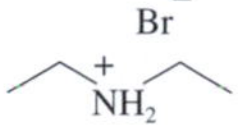 b)

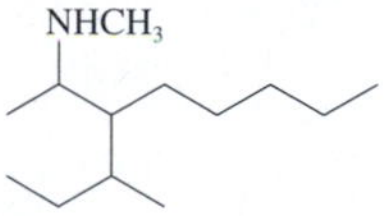

CHAPTER 6

6.1 a) 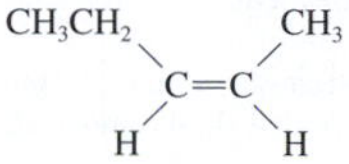 and 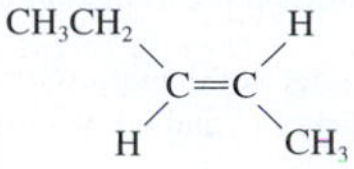 b) None

c) None d) 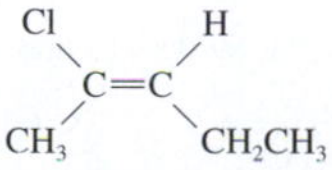 and

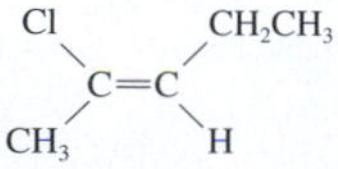

6.2

a) 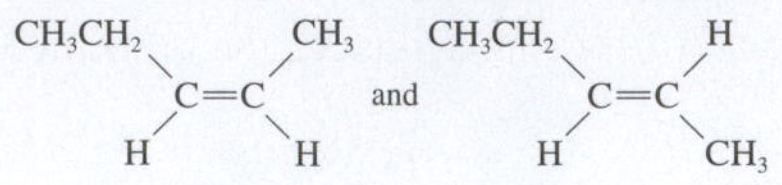 and

b) 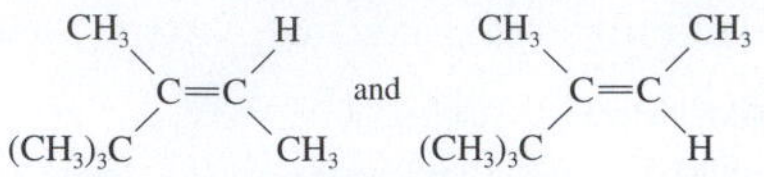 and

In both cases the right isomer is more stable because the larger groups are trans.

6.3   a) —CH_2CH_3 b) —$\overset{\overset{\displaystyle O}{\|}}{C}OH$ c) —$CH_2CH_2CH_2CH_3$ d) —$C\equiv N$ e)

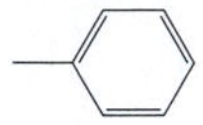

6.4   a) *E*   b) *E*   c) *Z*

6.6

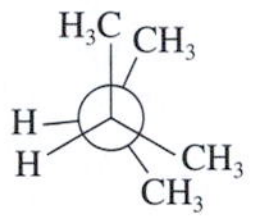

6.7

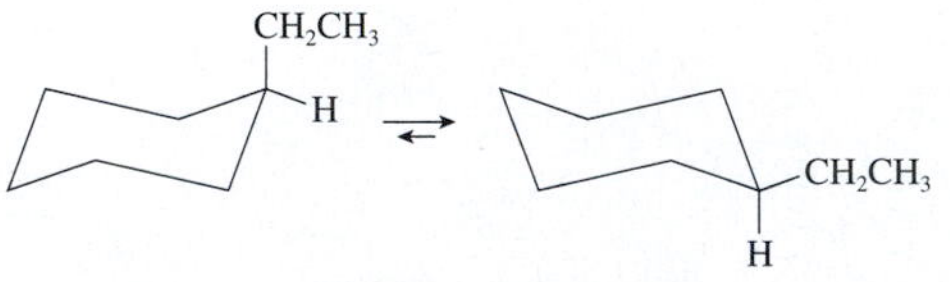

The conformation on the right, with the ethyl group equatorial, is more stable.

6.8   a) 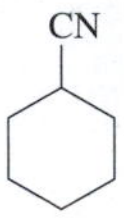   b) 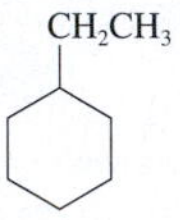   c)

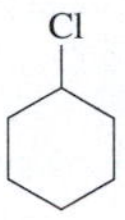

6.10   a) 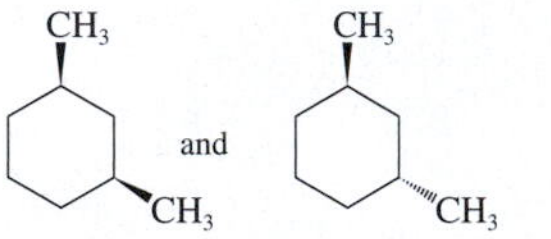 and   b) 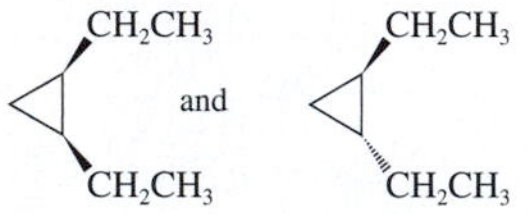 and

c) 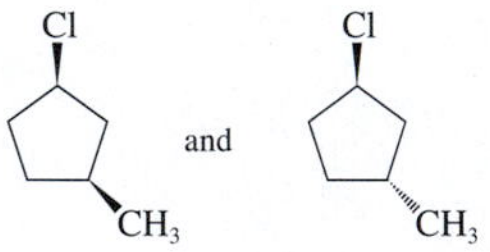 and

6.11

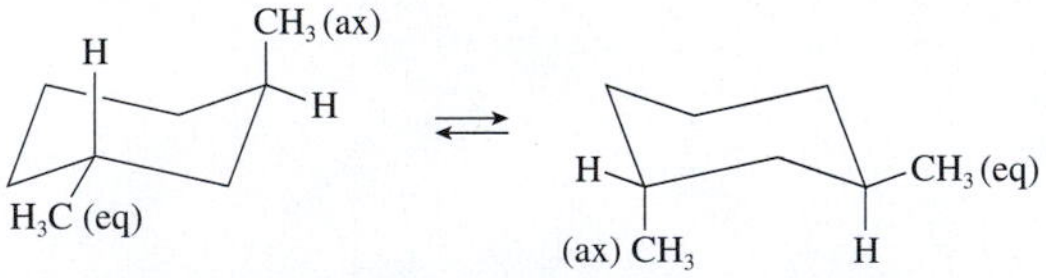

6.12   a) One methyl is axial, and one is equatorial.

b) Both methyls are axial in one conformation, and both are equatorial in the other.

c) The *trans*-isomer is more stable because it has a conformation with both methyls equatorial.

6.13   a) The conformation with both groups equatorial is the more stable conformation of the *cis*-stereoisomer. The conformation with the isopropyl group equatorial and the hydroxy group axial is the more stable conformation of the *trans*-stereoisomer.

b) The *cis*-stereoisomer is more stable than the *trans*-stereoisomer by 0.9 kcal/mol (3.8 kJ/mol).

6.14   a) The methyls are trans; the *t*-Bu is cis to the closer methyl; all are equatorial; the conformation shown is more stable; the stereoisomer shown is most stable.

b) The chlorines are trans; both are axial; the ring-flipped conformation is more stable; the stereoisomer shown is more stable.

c) The groups are trans; the methyl is axial; the phenyl is equatorial; the conformation shown is more stable; the *cis*-stereoisomer is more stable.

CHAPTER 7

7.1   a) Achiral   b) Chiral   c) Chiral   d) Achiral   e) Chiral   f) Chiral

7.2   a) Not   b) Not   c) 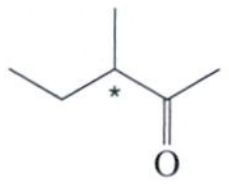   d)

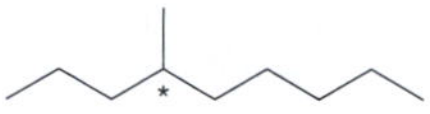

e) 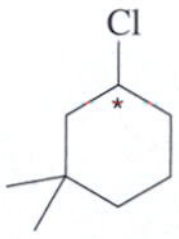   f) Not

7.3   a) Yes   b) Yes   c) No   d) Yes   e) No   f) Yes   g) Yes   h) No

7.4   a) 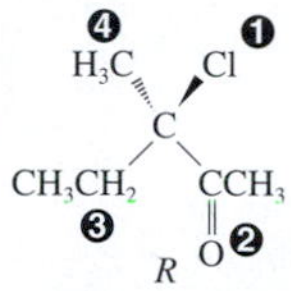   b) 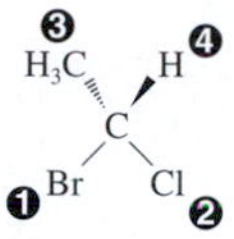   c)

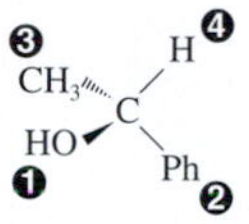

d) 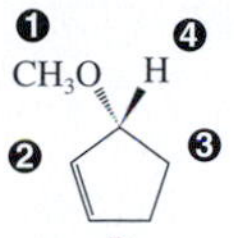   e) 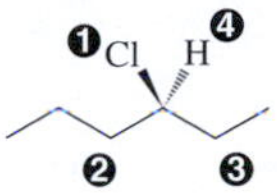   f)

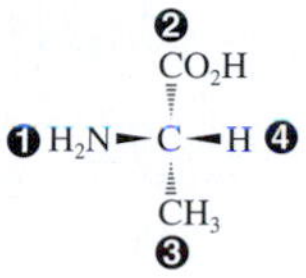

7.5   a) 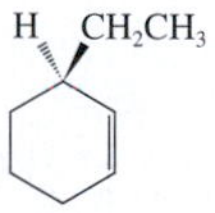   b)

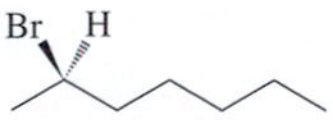

7.6   a) False   b) False   c) True   d) Cannot be determined   e) True

7.7

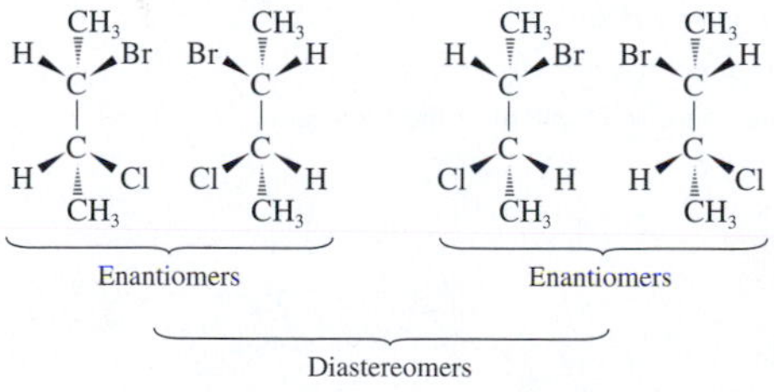

7.8   a) 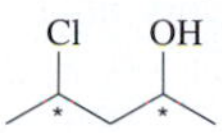 4   b) 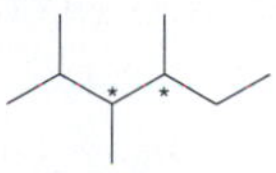 4   c) 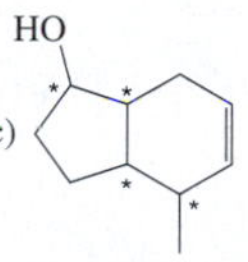 16   d) None

7.9

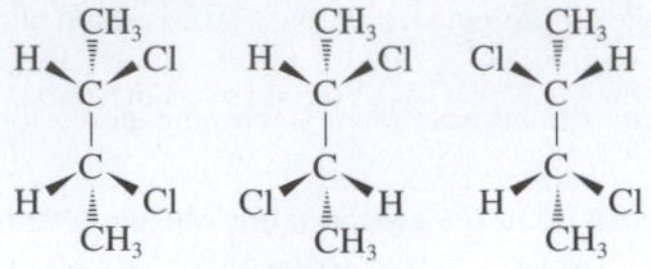

Meso   Rotates   Rotates

7.10

Meso   Rotates   Rotates

7.11   a) Yes   b) No, meso   c) Yes   d) No, meso

7.12   a) and b) and c)

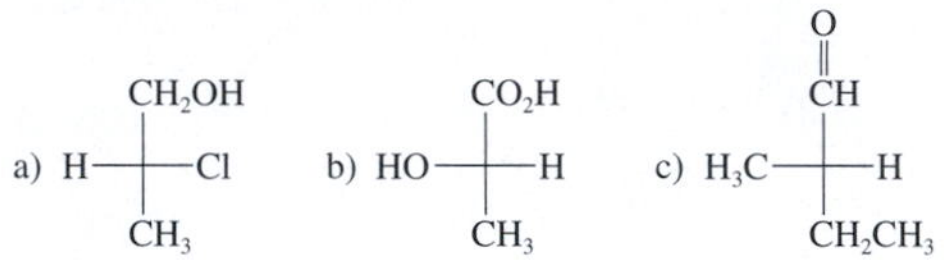

7.13   a) *S*   b) *S*

7.14   a) Chiral   b) Chiral   c) Not chiral   d) Chiral   e) Not chiral   f) Chiral

CHAPTER 8

8.1   a)   b)   c)

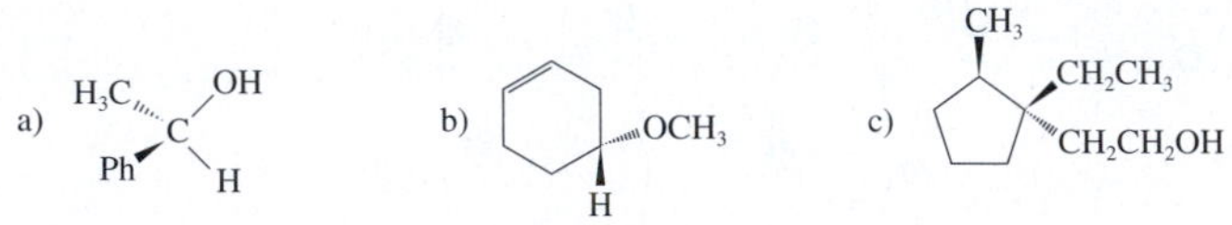

8.2   a) The right compound reacts faster because it has less steric hindrance.

b) The right compound reacts faster because it has less steric hindrance.

c) The left compound reacts faster because of resonance stabilization of the transition state.

8.3

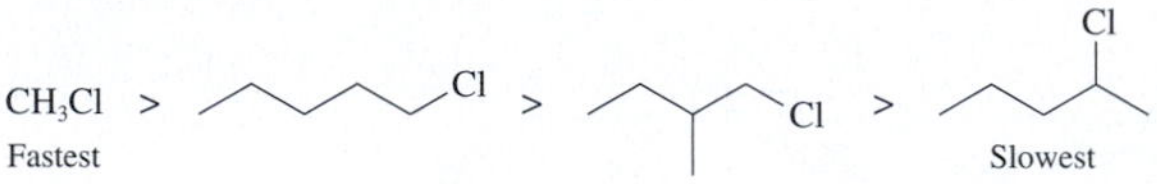

CH_3Cl > ... Fastest > ... > ... Slowest

8.5   Approximately half broken

8.6   a) Left; tertiary carbocation is more stable

b) Left; resonance stabilized carbocation is more stable

c) Left; resonance stabilized carbocation is more stable

d) Right; methoxy group provides extra resonance stabilization of the carbocation

8.7

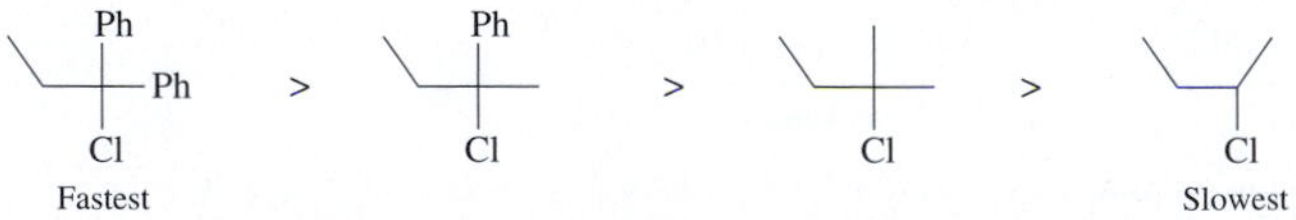

Fastest > ... > ... > Slowest

8.8   a)   b)

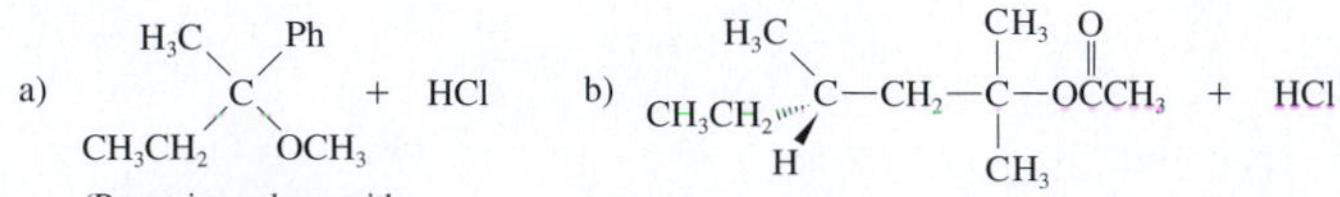

(Racemic, perhaps with some excess inversion)

8.9　a) Primary substrates with a strong nucleophile (hydroxide ion) react by the S_N2 mechanism. The right reaction is faster because mesylate ion is a better leaving group than chloride ion.

　　b) Tertiary substrates react by the S_N1 mechanism. The right reaction is faster because iodide ion is a better leaving group than bromide ion.

8.11　a) Because the leaving group is on a tertiary carbon, the reaction proceeds by an S_N1 mechanism. The reactivity of the nucleophile does not affect the rate of an S_N1 reaction, so both reactions proceed at the same rate.

　　b) Because the leaving group is on a primary carbon, the reaction proceeds by an S_N2 mechanism. The right reaction is faster because the nucleophile is stronger. (Nucleophile strength increases down a column of the periodic table.)

　　c) Because the leaving group is on a primary carbon, the reaction proceeds by an S_N2 mechanism. The left reaction is faster because the nucleophile is stronger. (It is a stronger base.)

8.12　a) S_N2

　　b) S_N1

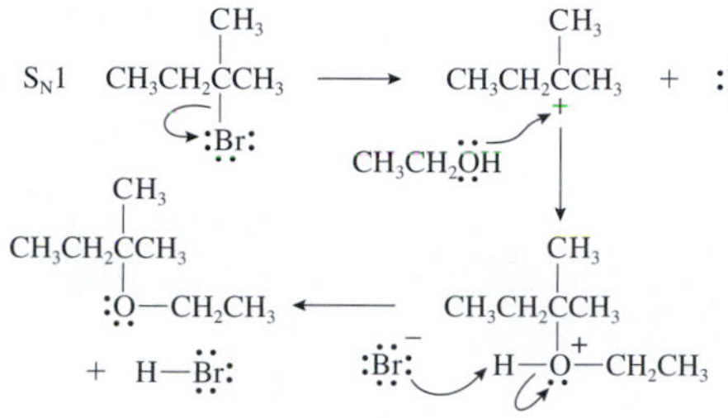

　　c) S_N2

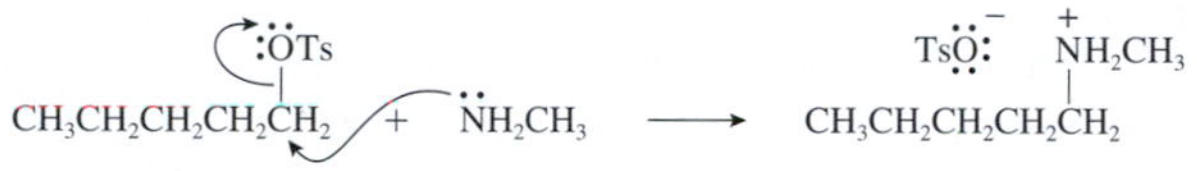

　　d) S_N2

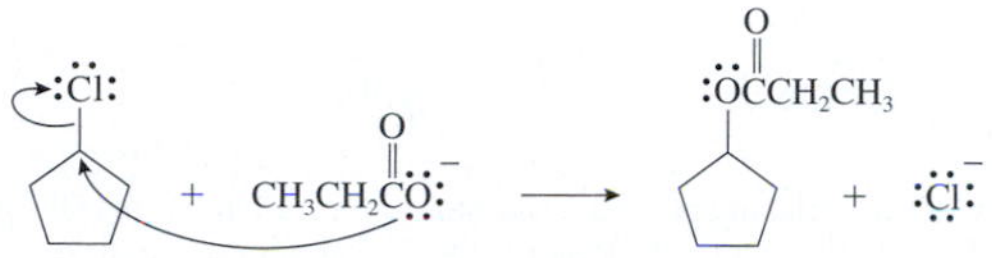

8.15　a) This S_N1 reaction is faster in the more polar solvent, methanol.

　　b) This S_N2 reaction, with a negative nucleophile, is faster in the less polar solvent, pure methanol.

　　c) This S_N2 reaction is faster in the aprotic solvent, DMSO.

8.16　a) Tertiary, so S_N1

　　b) Secondary, with a strong nucleophile and an aprotic solvent, so S_N2

　　c) Methyl substrate (less hindered than primary), so S_N2

　　d) Allylic substrate, weak nucleophile and polar solvent, so S_N1

　　e) Allylic substrate, strong nucleophile and aprotic solvent, so S_N2

　　f) Secondary benzylic substrate, weak nucleophile and polar solvent, so S_N1

8.17　a)

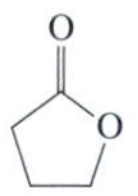

　　b)

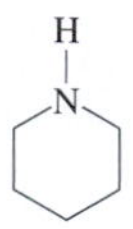

8.18　a)

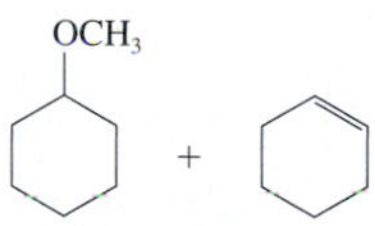

　　b)

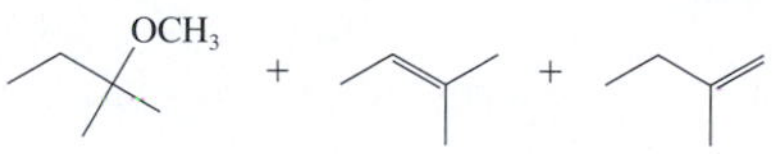

　　c)

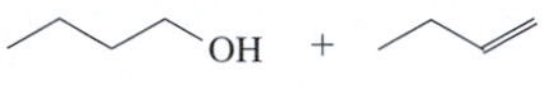

8.19 a) 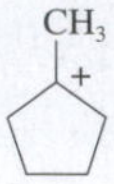 b) 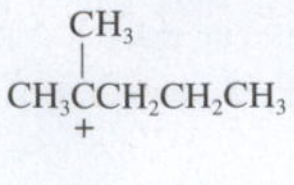 $CH_3\overset{+}{C}CH_2CH_2CH_3$ (with CH_3 substituent) c)

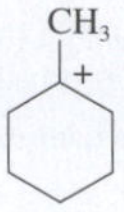

8.20 a) 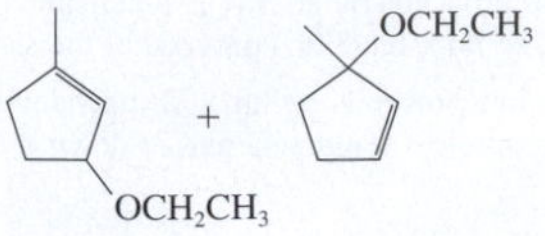 b)

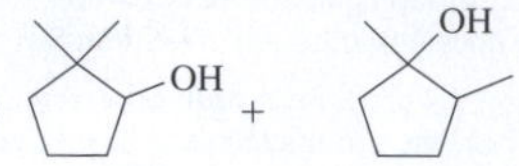

CHAPTER 9

9.1 a) $CH_3CH{=}CHCH_2CH_3$ b)  c)

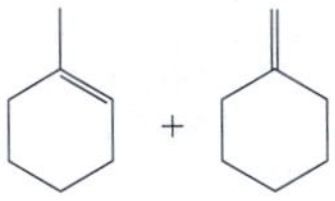

9.2

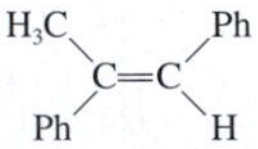

9.3 a) 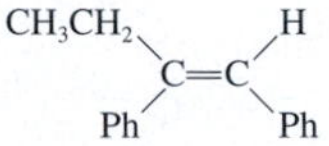 b)

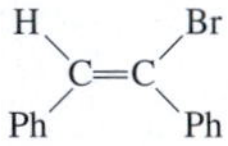

9.5 a) 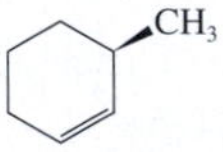 b)

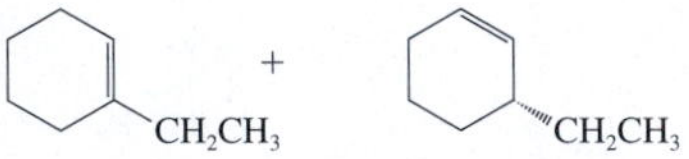

9.6 The bulky *t*-Bu group must be equatorial in both cases. The isomer on the left has the OTs group axial, so it can readily undergo E2 elimination. The isomer on the right has the OTs group equatorial, so it cannot readily react.

9.7 a) 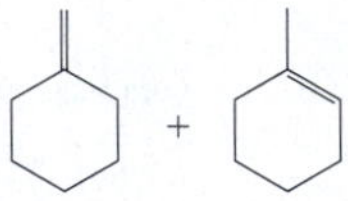 b)  c)

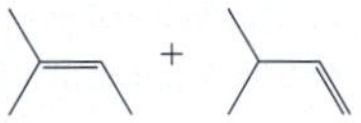

9.8 (*E*)-2-Butene is the major product because it is more stable and the conformation leading to it is more stable.

9.9 a) 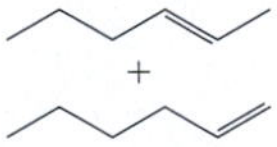 b)

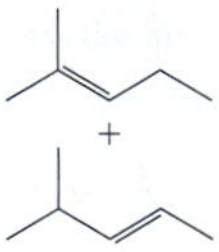

c) 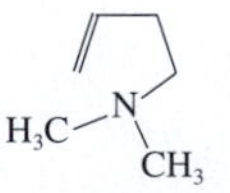 d)

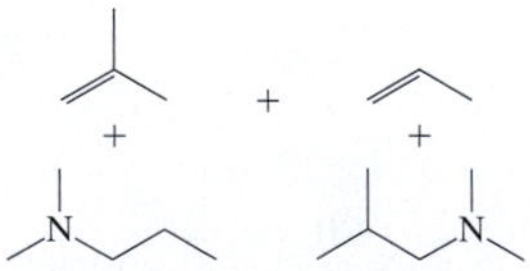

9.10 a) Ph⌒⌒⌒   b) [cyclohexene with Ph]

c) [alkene] + [alkene]   d) Ph⌒⌒
Major

9.11

[Reaction mechanism scheme: protonation of tert-butyl alcohol by H—O—SO₃H giving oxocarbenium, loss of water, and formation of isobutylene CH₂=C(CH₃)CH₃ + H₃O⁺]

9.13 a) Ph—C(—OCH₂CH₃) + [alkene Ph⌒] + [alkene Ph⌒]
Major

b) [cyclohexane with CH₃ and OCH₃ and CH₃] + [cyclohexane with CH₃, OCH₃, CH₃] + [cyclohexene with CH₃, CH₃] + [cyclohexene with CH₃, CH₃]   c) Same as (b)
Major

d) [cyclopentane H₃C OH] + [cyclopentane CH₃ OH] + [cyclopentane CH₃ OH] + [cyclopentene CH₃] + [cyclopentene CH₃] + [methylenecyclopentane CH₂]
+ Ethyl ethers corresponding to the alcohols

9.14 a) [cyclohexenone]   b) Ph⌒⌒ with O and Br

9.15 a) A secondary substrate with a weak base reacts predominately by the S_N2 mechanism. Minor amounts of E2 products may also be formed.

CH₃CH₂CH(OCCH₂CH₃)CH₃ + CH₃CH=CHCH₃ + CH₃CH₂CH=CH₂
Major

b) A secondary substrate with a strong base reacts predominately by the E2 mechanism.

CH₃CH=CHCH₃ + CH₃CH₂CH=CH₂
Major

c) A tertiary substrate in the absence of a strong base reacts by the S_N1 and E1 mechanisms.

Ph—C(CH₃)(CH₃)—OCH₃ + [CH₂=C(Ph)(CH₃)]
Major

9.16 a) The leaving group is on a primary carbon, so an S_N2 reaction occurs.

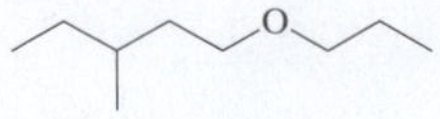

b) A primary substrate with a very hindered base gives predominately E2 elimination.

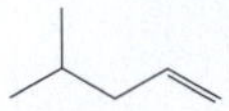

c) A secondary substrate with a strong base gives predominately E2 elimination.

d) A secondary substrate with a weak base that is moderately nucleophilic in an aprotic solvent favors an S_N2 reaction. The reaction occurs with inversion of configuration.

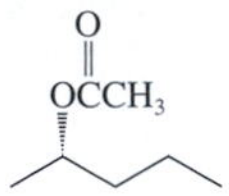

e) A tertiary substrate with a strong base favors E2 elimination.

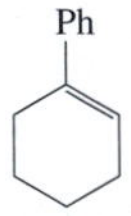

f) A secondary substrate in a polar solvent and in the absence of a strong base or nucleophile favors the S_N1 and E1 reactions.

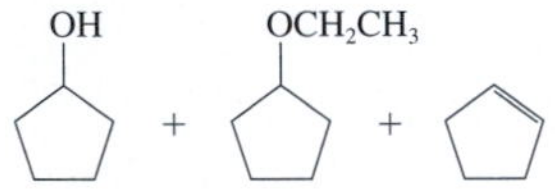

CHAPTER 10

10.1 a) 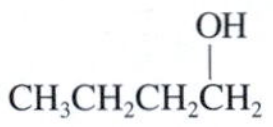 $CH_3CH_2CH_2CH_2$—OH   b) 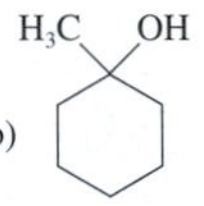   c) 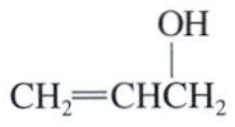 CH_2=$CHCH_2$—OH

10.2 This is an S_N1 reaction, so only the chlorine bonded to the tertiary carbon is replaced.

10.3 Stereochemistry is retained because the CO bond of the product is not broken in the reaction.

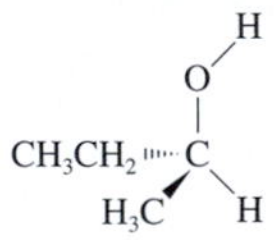

10.4 a)

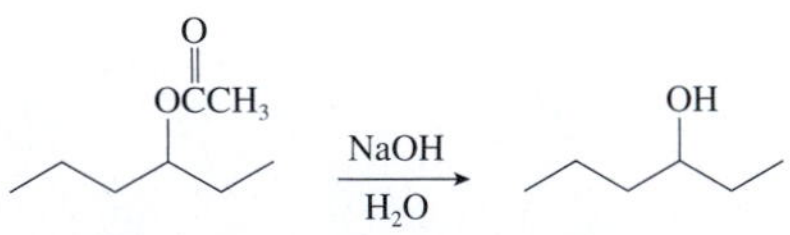

b)

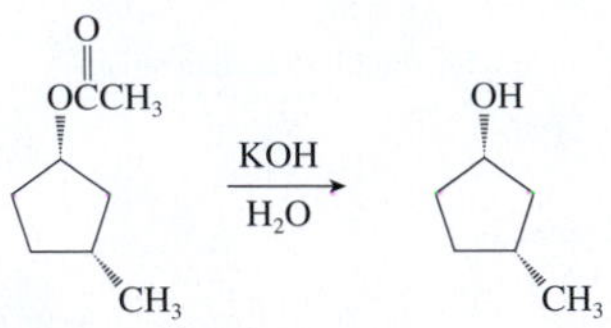

10.5 a) $CH_3CH_2CH_2$ — with OCH$_2$CH$_3$

b)

c)

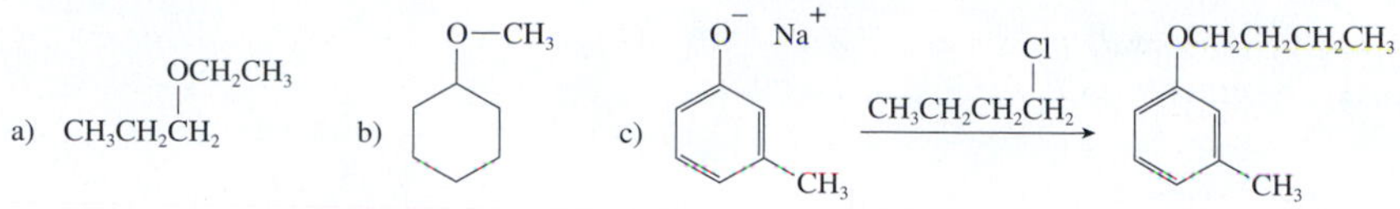

10.7 a) The right route is better because the left route gives primarily E2 elimination.

b) The right route is better because elimination cannot occur.

10.8 a) $CH_3CH_2CH_2CH_2OH$ $\xrightarrow[\text{2) CH}_3\text{I}]{\text{1) Na}}$ $CH_3CH_2CH_2CH_2OCH_3$ $\xleftarrow[\text{2) CH}_3\text{CH}_2\text{CH}_2\text{CH}_2\text{Br}]{\text{1) Na}}$ CH_3OH

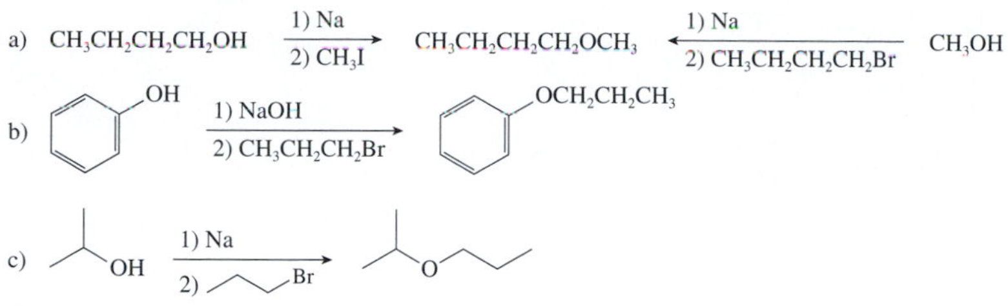

b) $\xrightarrow[\text{2) CH}_3\text{CH}_2\text{CH}_2\text{Br}]{\text{1) NaOH}}$

c) $\xrightarrow[\text{2)}]{\text{1) Na}}$

10.9 a)

b) $CH_3COCH_2CH_3$ with CH_3 ... CH_3

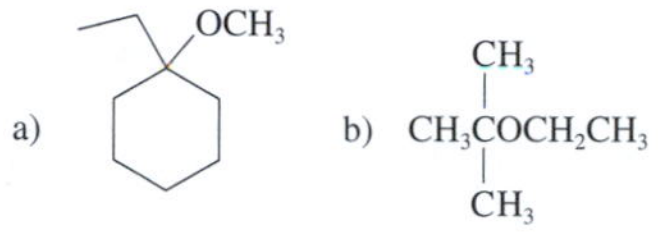

c) $Ph_2CHOCH_2CH_2Cl$

10.12

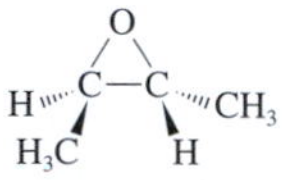

10.14 a)

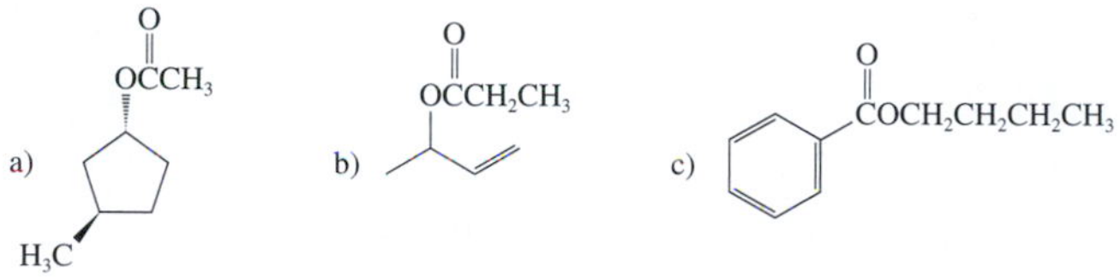

b)

c)

10.16 a)

b)

c)

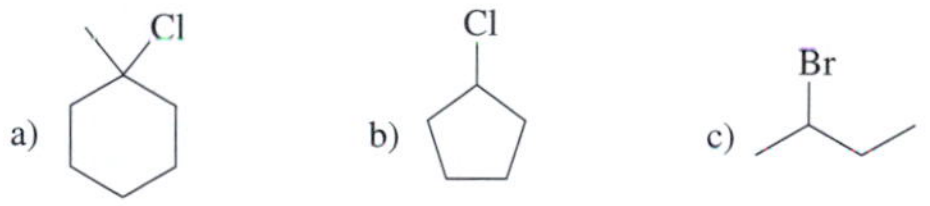

d) $CH_3CH_2CH_2I$

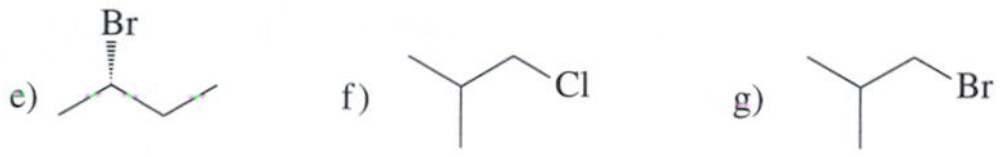

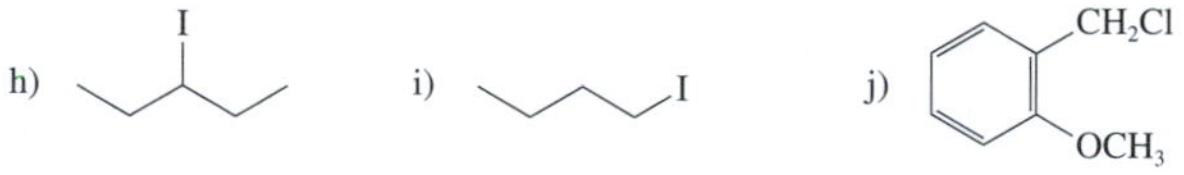

e)

f)

g)

h)

i)

j)

10.17 a) $CH_3\overset{CH_3}{\underset{CH_2CH_3}{C}}OH$ $\xrightarrow{\text{HBr}}$

b) $CH_3\overset{CH_3}{\underset{Ph}{C}}OH$ $\xrightarrow{\text{HCl}}$

c) $\xrightarrow[\text{2) NaI}]{\text{1) TsCl, pyridine}}$

d) $\xrightarrow{\text{PBr}_3}$

e) $\xrightarrow{\text{HBr}}$

f) $CH_3CH_2CH_2CH_2OH$ $\xrightarrow{\text{SOCl}_2}$

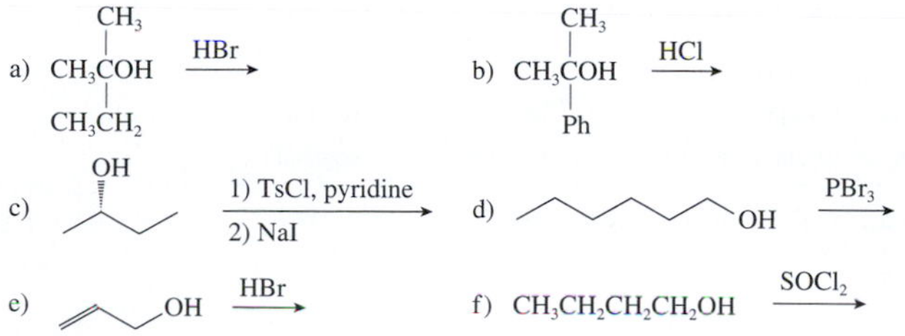

10.18 a) $(CH_3CH_2)_2\overset{\displaystyle H}{\overset{\displaystyle |}{\underset{+}{N}}}CH_2CH_3$ Br^- b) $CH_3CH_2\overset{\displaystyle CH_3}{\overset{\displaystyle |}{\underset{+}{N}}CH_3}$ I^-
CH_3

10.19 a) 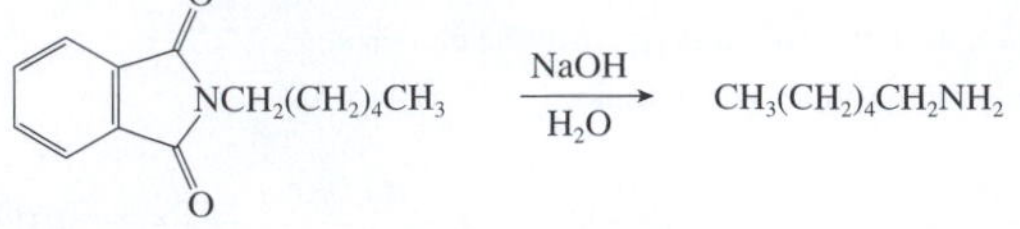 [phthalimide N-hexyl] $\xrightarrow[\text{H}_2\text{O}]{\text{NaOH}}$ $CH_3(CH_2)_4CH_2NH_2$

10.20 [phthalimide NH] $\xrightarrow[\begin{array}{l}\text{2) PhCH}_2\text{CH}_2\text{Br}\\ \text{3) NaOH, H}_2\text{O}\end{array}]{\text{1) KOH}}$ $PhCH_2CH_2NH_2$

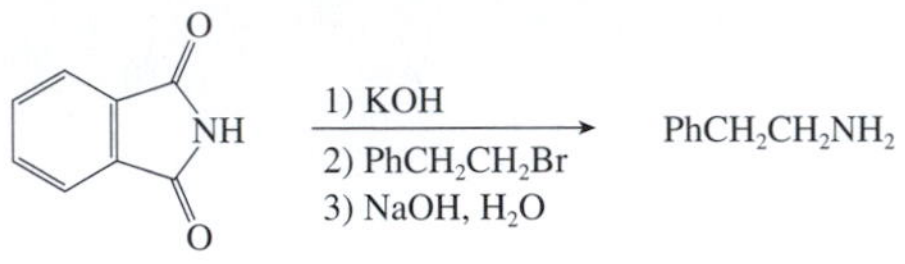

10.21 a)  [heptane skeletal]   b) 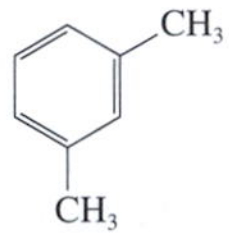 [1,3-dimethylbenzene with CH₃ and CH₃]

10.22 a) $PhCH_2CN$ b) $CH_3C\equiv C-CH_2CH_2CH_3$ c) 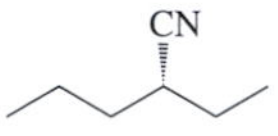 [hexane chain with CN substituent]

d) $HC\equiv CCH_2CH_3$ $\xrightarrow[\text{2) CH}_3\text{I}]{\text{1) NaNH}_2}$ $CH_3C\equiv CCH_2CH_3$

e)  $Cl\diagdown\diagup\diagdown CN$ f)  [cyclopentene] $+$ $HC\equiv CH$

10.23 a) 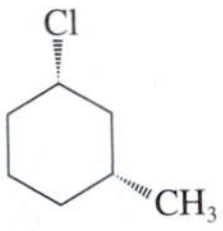 [cyclohexane with Cl up and CH₃ down] $\xrightarrow[\text{DMSO}]{\text{NaCN}}$ b) $BrCH_2CH_2\overset{\displaystyle CH_3}{\overset{\displaystyle |}{CH}}CH_3$ $\xrightarrow[\text{NH}_3(l)]{HC\equiv C:^- \; Na^+}$

c) $CH_3C\equiv CH$ $\xrightarrow[\text{2) PhCH}_2\text{Br}]{\text{1) NaNH}_2}$

10.24 a) 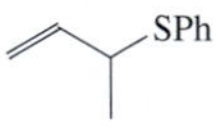 [allyl SPh structure]   b) $Ph_3\overset{+}{P}-CH_2CH_2CH_3$ Br^- c) $CH_3CH_2CH_2CH_2SCH_3$ d) 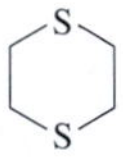 [1,4-dithiane ring with two S]

10.25 a) 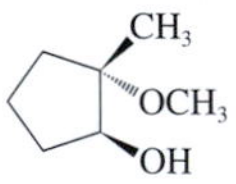 [cyclopentane with CH₃, OCH₃, OH]   b) 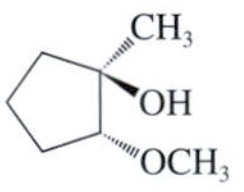 [cyclopentane with CH₃, OH, OCH₃]   c) 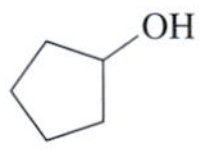 [cyclopentanol with OH]

10.26 a) $CH_3CH_2CH=CH_2$ b)  [cyclohexene]   c)  [pentene structures] Major $+$ [pentene structure] Minor   d) 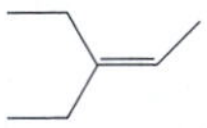 [branched alkene]

10.27 a) No. The alkene shown would be a minor product according to Zaitsev's rule.

b) Yes. The alkene shown would be the major product because it is conjugated.

10.28 The top reaction is better because it can produce only the desired alkene. The bottom reaction would produce 1-pentene also.

10.29 a) 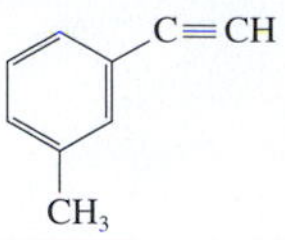   b)

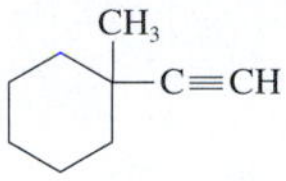

10.30 a) 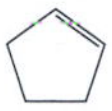   b) 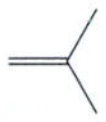   c)

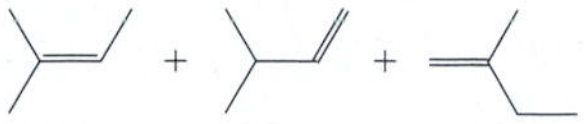

Major   Minor   Minor

10.31 a) 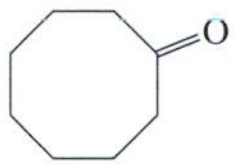   b) 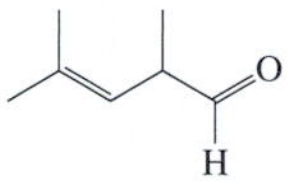   c)

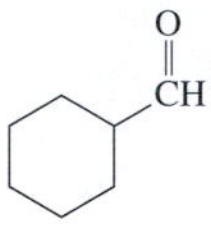

d) 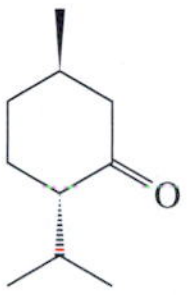   e)

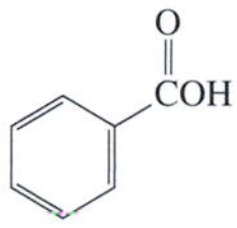

10.32 a) 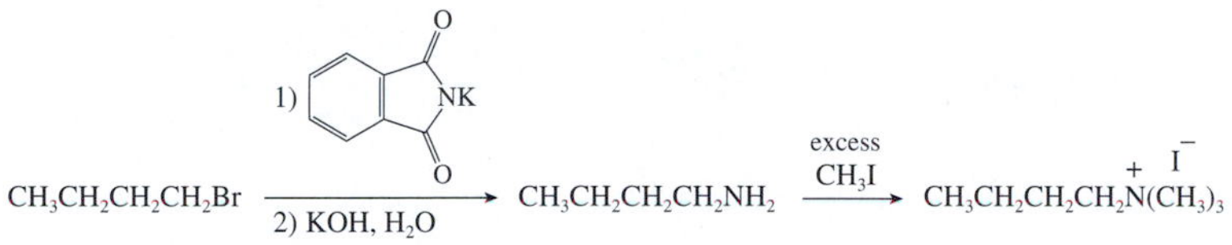 $CH_3CH_2CH_2CH_2Br$ $\xrightarrow[\text{2) KOH, H}_2\text{O}]{\text{1)}}$ $CH_3CH_2CH_2CH_2NH_2$ $\xrightarrow[\text{CH}_3\text{I}]{\text{excess}}$ $CH_3CH_2CH_2CH_2N(CH_3)_3$ $+$ I^-

b) 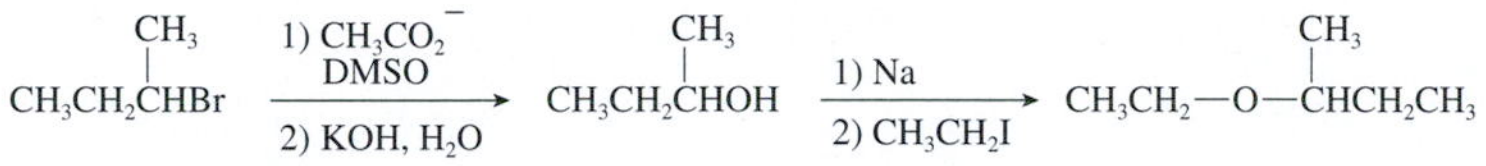 $CH_3CH_2CHBrCH_3$ (CH₃) $\xrightarrow[\text{2) KOH, H}_2\text{O}]{\text{1) CH}_3\text{CO}_2^- \text{ DMSO}}$ $CH_3CH_2CHOHCH_3$ (CH₃) $\xrightarrow[\text{2) CH}_3\text{CH}_2\text{I}]{\text{1) Na}}$ $CH_3CH_2-O-CHCH_2CH_3$ (CH₃)

c) 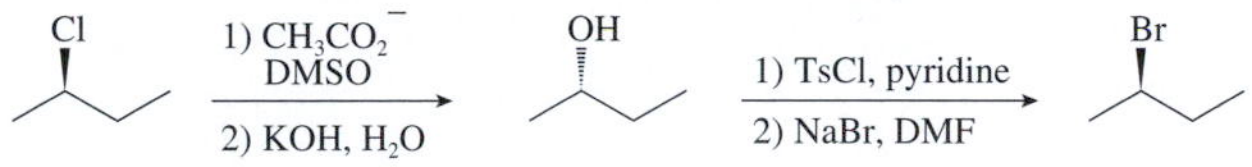 $\xrightarrow[\text{2) KOH, H}_2\text{O}]{\text{1) CH}_3\text{CO}_2^- \text{ DMSO}}$ $\xrightarrow[\text{2) NaBr, DMF}]{\text{1) TsCl, pyridine}}$

d) 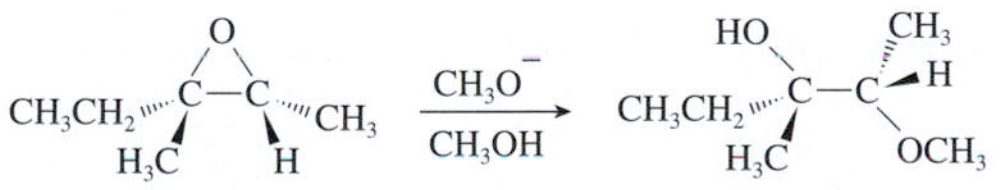 $\xrightarrow[\text{CH}_3\text{OH}]{\text{CH}_3\text{O}^-}$

e) $HC{\equiv}CH$ $\xrightarrow[\text{2) PhCH}_2\text{Cl}]{\text{1) NaNH}_2}$ $PhCH_2C{\equiv}CH$ $\xrightarrow[\text{2) CH}_3\text{I}]{\text{1) NaNH}_2}$ $PhCH_2C{\equiv}CCH_3$

CHAPTER 11

11.1 a) $CH_2{=}CH_2$ $<$ $CH_3CH_2CH{=}CH_2$ $<$ $CH_3CH_2C{=}CH_2$ (CH₃)

Slowest   Fastest

b) $CH_3CH{=}CH_2$ $<$ $<$

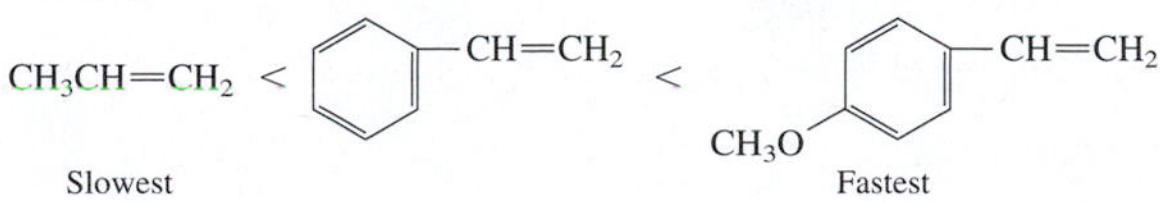

Slowest   Fastest

11.3 a) b) c)

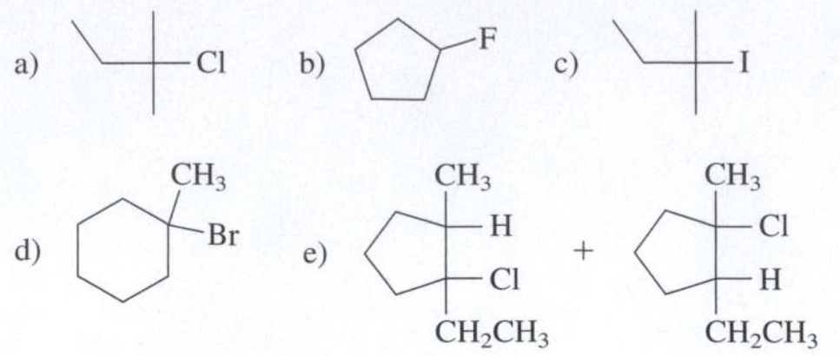

d) e)

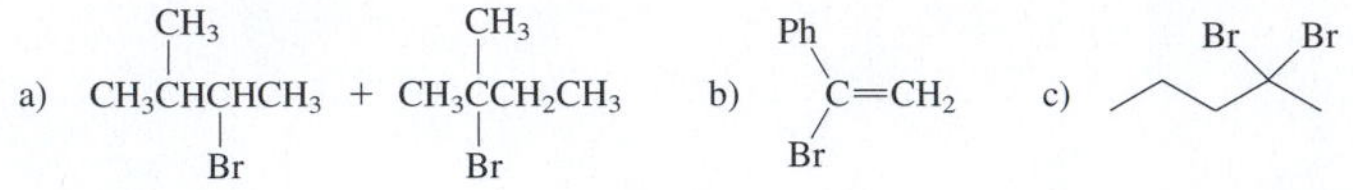

11.4 a) b) c)

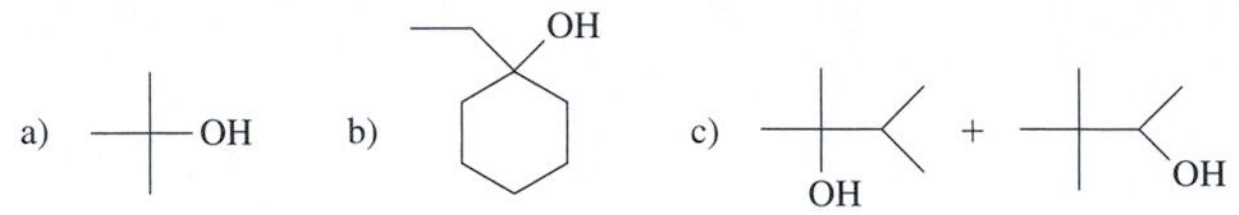

11.5 a) b) c)

11.7 a) $CH_3CH_2CHCH_2$ b) c) d)

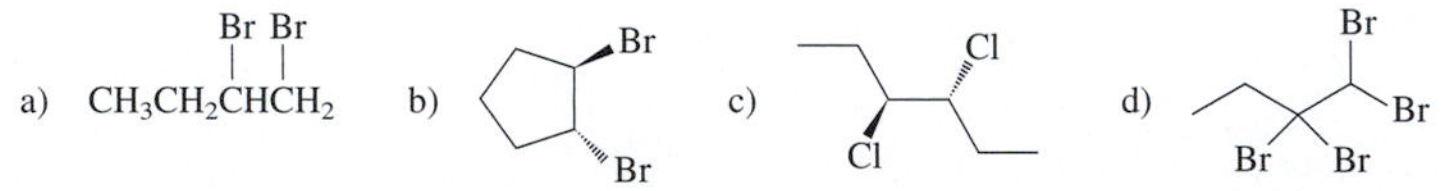

11.8

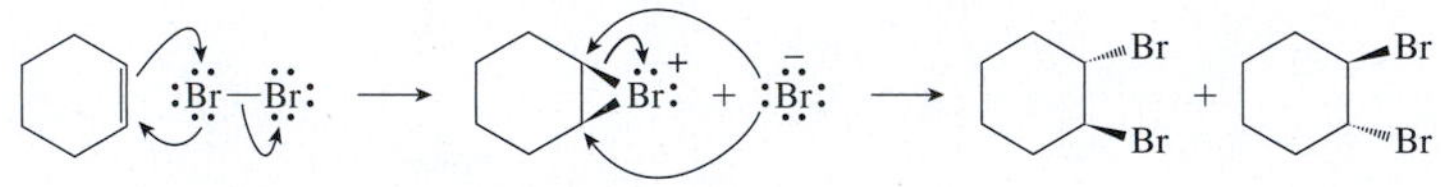

11.9 The right one reacts faster because the methyl group makes the double bond more nucleophilic.

11.10 a) b)

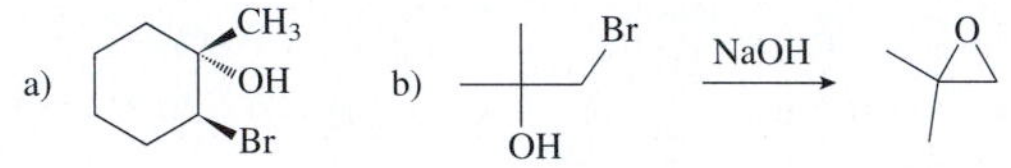

11.13 a) b) c) d)

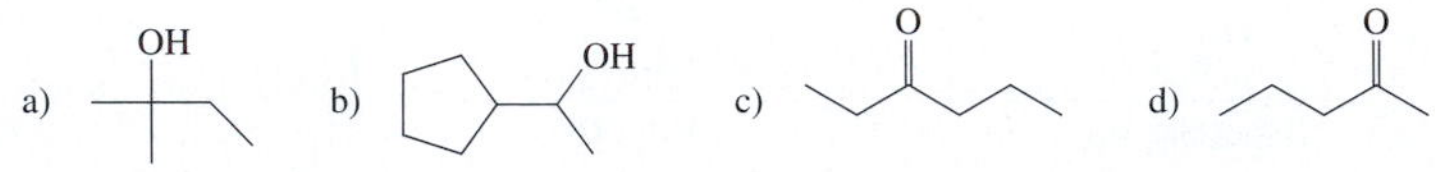

11.14 The left synthesis is better because only 3-hexanone, the desired ketone, is produced. The right reaction also produces 2-hexanone.

11.15 a) b) c)

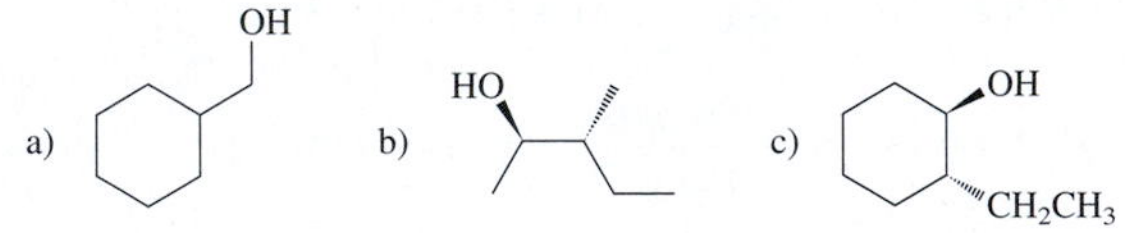

11.16 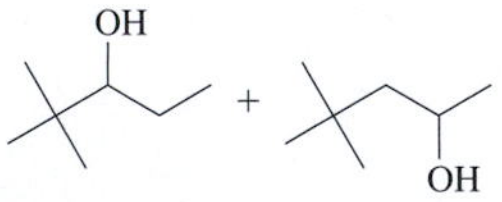   The alcohol on the right is the major product because the boron prefers to bond to the less hindered carbon of the alkene.

11.17 a)

b)

c)

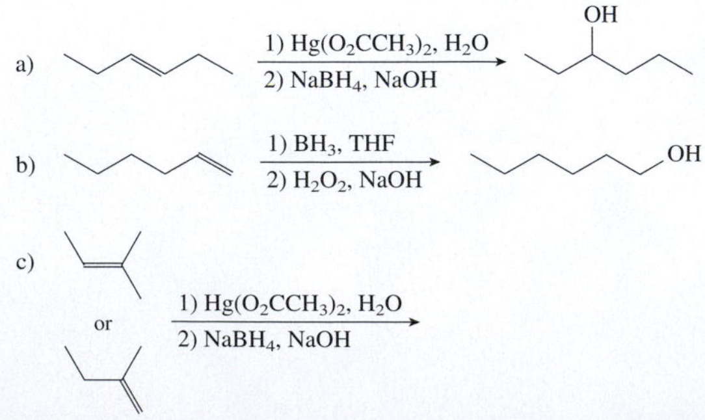

11.18 a) 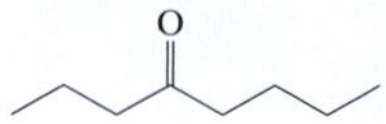 b)

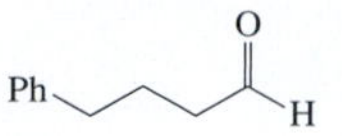

11.19 b)

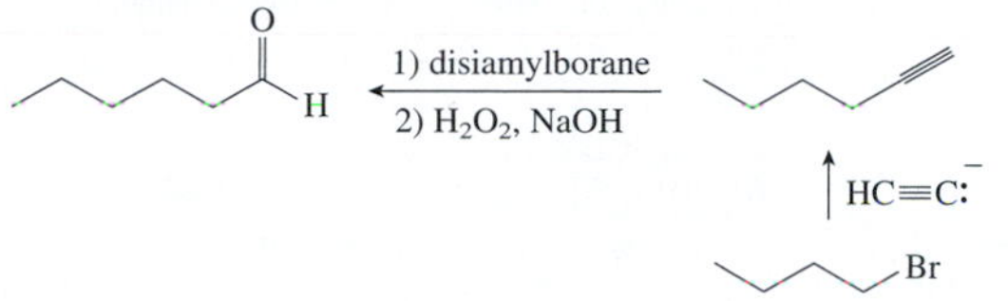

11.22 a) 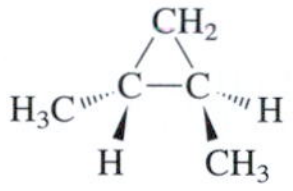 b) 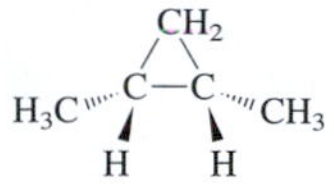 c) 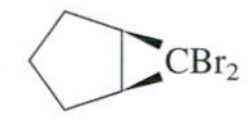 d)

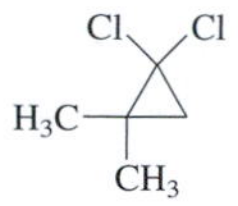

11.23 a) 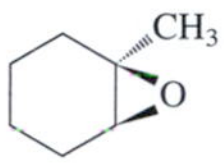 b)

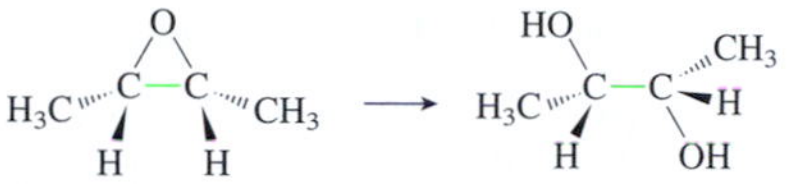

11.24 a) 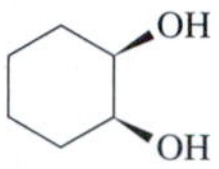 b) 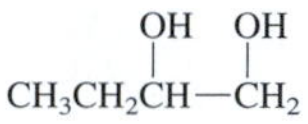 $CH_3CH_2CH-CH_2$ with OH OH c)

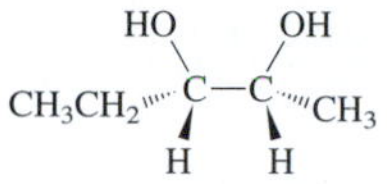

11.25 a) 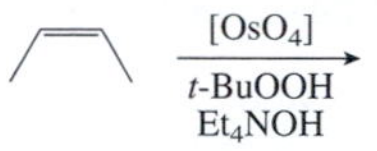 b)

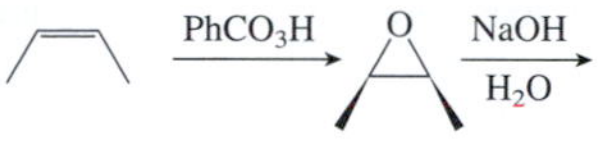

11.26 a) 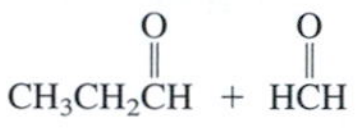 CH_3CH_2CH + HCH b) $2 \ CH_3CCH_3$ c)

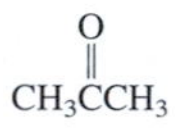

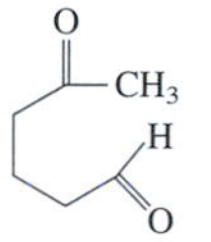

d) 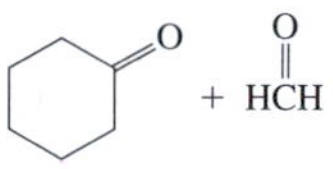 + HCH e) $2 \ H$

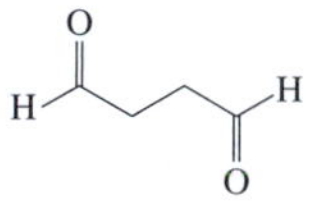

11.27 a) 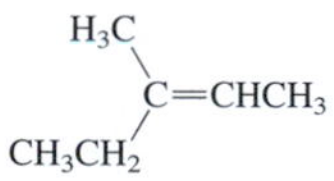 b) 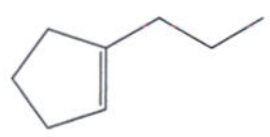 c) 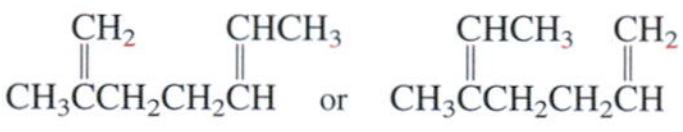 $CH_3CCH_2CH_2CH$ or $CH_3CCH_2CH_2CH$

11.28 a) 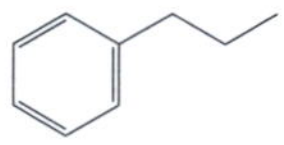 b) 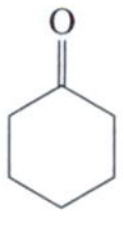 c) $CH_3CH_2CH_2CH_2CH_3$ d) 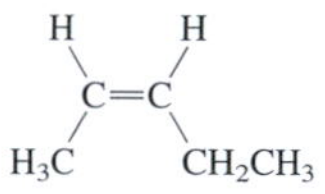 e)

11.30 a) 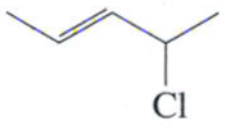 b)

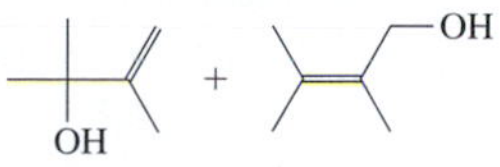

11.31 a)

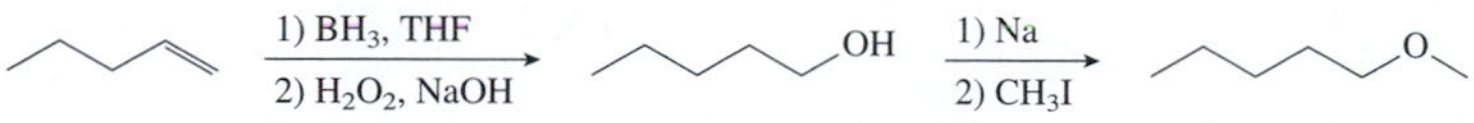

b) 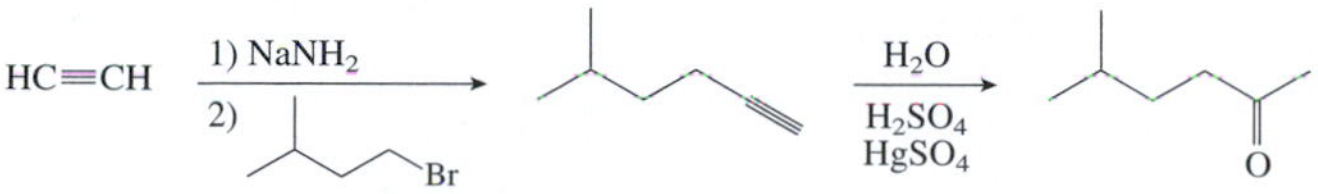 $HC≡CH$

CHAPTER 12

12.1   a) (1-Methylpropyl)benzene   b) *o*-Chlorotoluene   c) 1-Bromo-4-methoxybenzene
d) 4-Butyl-3-chlorotoluene   e) 3-Ethyl-4-phenylcyclohexene   f) 3-Phenyl-1-butanol

12.2   a) 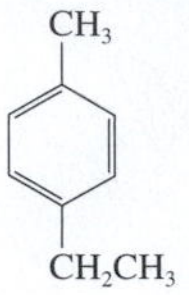   b) 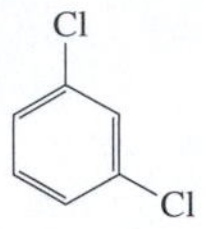   c) 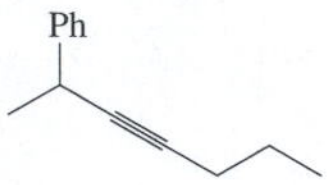
d) 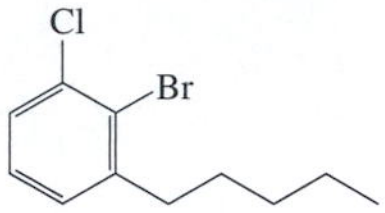   e) 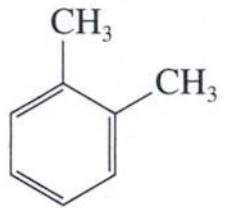   f)

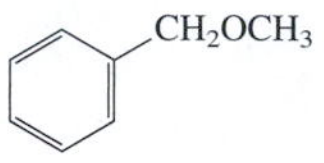

12.3   a) *m*-Propylphenol   b) 3,5-Dibromophenol   c) *p*-Methoxyphenol

12.4   a) 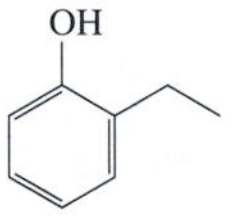   b) 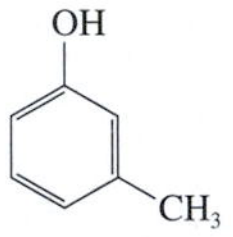   c)

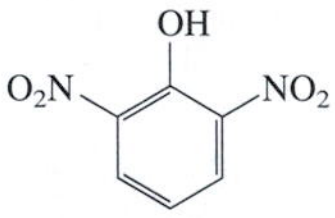

12.7   a) Hexanal   b) 5-Methyl-3-hexanone   c) (Z)-4-Methyl-2-hexenal   d) *p*-Chlorobenzaldehyde
e) 5-(2-Methylpropyl)-3-cycloheptenone   f) 2,4-Pentanedione   g) 3-Phenylcyclopentanone
h) 4-Methylpent-3-en-2-one

12.8   a) 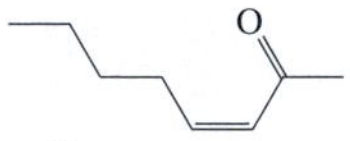   b) 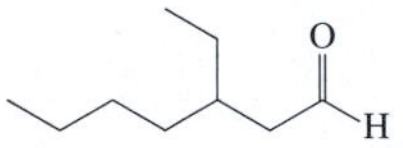   c) CH_2=CHCH=CHCH
d) 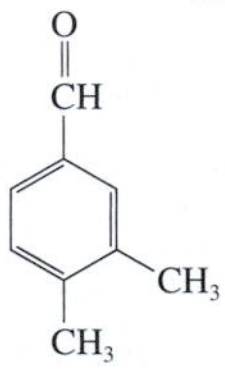   e) $PhCCH_2CH_3$ f)

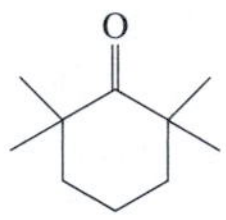

12.11   a) 3-Methylpentanoic acid   b) *p*-Bromobenzoic acid   c) (Z)-4,4-Dichloro-2-pentenoic acid
d) 2-Cyclohexenecarboxylic acid   e) 3-Methyl-4-nitrobenzoic acid   f) 3-Cyclopropylbutanoic acid

12.12   a) 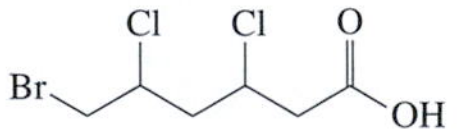   b) 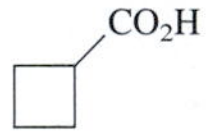   c) 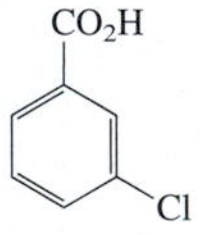   d)

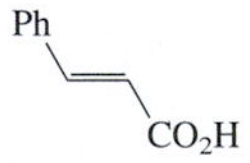

12.14   a) Benzoyl chloride   b) Propanoic anhydride   c) Methyl propanoate
d) Propyl 3-ethylbenzoate   e) Cyclopentyl 4-methylpentanoate   f) *N*-Methyl-3-pentynamide
g) Isopropyl 3-cyanocyclopentanecarboxylate   h) Pentanenitrile   i) Potassium 2-methylpentanoate

12.15

a) CH_3CH_2CCl (with O above)
b) $CH_3CN(CH_3)_2$ (with O above)
c) $CH_3(CH_2)_3COC(CH_2)_3CH_3$ (with O, O above)
d) sodium 4-nitrobenzoate ($CO_2^-Na^+$ on benzene ring with NO_2 para)

e) $CH_3(CH_2)_4CNH_2$ (with O above)
f) $CH_3COCHCH_3$ (with O above, CH_3 branch)
g) $PhCOCH_2Ph$ (with O above)
h) cyclopentyl $COCH_2CH_3$ (with O above)

i) benzene ring with CN and Cl substituents (meta)
j) $CH_3CH_2CH_2CH_2CHCH_2C\equiv N$ (with CH_3 branch)

12.19

a) 4-Hydroxy-2-cyclohexenone   b) 2-Amino-4-methylpentanoic acid
c) 1,4-Butanedioic acid   d) Methyl 4-methyl-5-oxopentanoate
e) 3-Cyanobenzaldehyde   f) 4-Ethyl-3-hydroxy-*N*-methylhexanamide
g) 3-Oxobutanenitrile   h) Isopropyl 3-oxo-4-phenylhexanoate

12.20

a) $HOC(CH_2)_4COH$ (with O, O above)
b) $CH_3CH_2C-COCH_2CH_3$ (with OH, O above; CH_3CH_2 branch)
c) cyclohexyl CH_2CHCH_2OH (with NH_2 branch)

d) $CH_3CH_2CH=CHCH_2CH_2CH-C-OCCH_3$ (with OH, O, CH_3 above; CH_3 branch)
e) $CH_3CH-C-C-N(CH_3)_2$ (with CH_3, O, O above)
f) $PhCH_2CHCO_2H$ (with NH_2 above)

CHAPTER 13

13.1 a) 3.33×10^{-3} cm   b) 9.68×10^{14} s^{-1} c) 0.858 kcal/mol (3.59 kJ/mol)
d) 92.3 kcal/mol (387 kJ/mol)

13.2 Infrared

13.3 3.8×10^{-5} kcal/mol (1.6×10^{-4} kJ/mol)

13.4 a) C—H, lighter atom   b) C≡C, stronger bond   c) C—Cl, lighter atom and stronger bond

13.5 a) OH, 3000 cm^{-1}, very broad; =CH, 3100–3000 cm^{-1}; —CH, 3000–2850 cm^{-1}
b) =CH, 3100–3000 cm^{-1}; —CH, 3000–2850 cm^{-1}
c) NH$_2$, two bands, 3400–3250 cm^{-1}; =CH, 3100–3000 cm^{-1}
d) OH, 3550–3200 cm^{-1}, broad; =CH, 3100–3000 cm^{-1}; —CH, 3000–2850 cm^{-1}
e) NH, one band, 3400–3250 cm^{-1}; —CH, 3000–2850 cm^{-1}
f) —CH, 3000–2850 cm^{-1}; CHO, 2830–2700 cm^{-1}, two bands

13.6 The band in the triple bond region at 2150–2100 cm^{-1} is much stronger for 1-hexyne than it is for the more symmetrical 3-hexyne. In addition, the ≡CH band near 3300 cm^{-1} in the spectrum of 1-hexyne confirms the presence of the triple bond.

13.7 Most hydrocarbons have —CH bonds that absorb in the 3000–2850 cm^{-1} region.

13.8 a) 1745 cm^{-1} b) 1710–1690 cm^{-1} c) 1695–1675 cm^{-1}
d) 1720–1700 cm^{-1} e) 1690–1670 cm^{-1}

13.9 a) The left compound, a ketone, has its carbonyl absorption near 1715 cm^{-1}, whereas the right compound, an aldehyde, has its carbonyl peak near 1730 cm^{-1} and has two bands in the region of 2830–2700 cm^{-1}.
b) The left ester (nonconjugated) has its carbonyl absorption near 1740 cm^{-1}, whereas the right ester (conjugated) has its carbonyl absorption near 1720–1700 cm^{-1}.
c) The carboxylic acid has a very broad band near 3000 cm^{-1}.

13.10 a) 3100–3000, 3000–2850, 1695–1675, 1660–1640

b) 3400–3250 (two bands), 3100–3000, 3000–2850, 1600–1450 (four bands), 900–675

c) 3100–3000, 3000–2850, 1660–1640, 1300–1000

d) 3550–3200 (broad), 3000–2850, 1300–1000

e) 3300, 3000–2850, 2150–2100, 1550, 1380

f) 3100–3000, 3000–2850, 2830–2700 (two bands), 1710–1690, 1600–1450 (four bands), 900–675

13.11 a) 1-Butyne has absorptions at 3300 and 2150–2100 cm^{-1} that are not present in the spectrum of 1-butene.

b) Benzyl alcohol has absorptions at 3100–3000, 1600–1450, and 900–675 cm^{-1} that are not present in the spectrum of *t*-butanol.

c) Benzaldehyde has two bands at 2830–2700 cm^{-1} that are not present in the spectrum of the ketone, acetophenone.

d) The primary amine has two bands in the 3400–3250 cm^{-1} region, whereas the secondary amine has only one.

13.12 a) Ketone, no evidence for C=C

b) Carboxylic acid, no evidence for C=C

c) Conjugated aldehyde, has C=C, possibly aromatic ring

Chapter 14

14.1 a) 7.4 δ b) 1480 Hz on a 200-MHz and 2960 Hz on a 400-MHz instrument c) 7.4 δ

14.2 a) 3 b) 4 c) 5 d) 4 e) 2

14.3
a) $CH_3CCH_2CH_3$
2.2 >2.2 >0.9
b) CH_3CHCH_3
OH ← 2–5
>0.9 >3.3
c) $CH_3COCH_2CH_3$
2.0 >3.7 >0.9
d) CH_3CHCH_2
Cl Cl
>0.9 >3.0
>3.0 (most downfield)

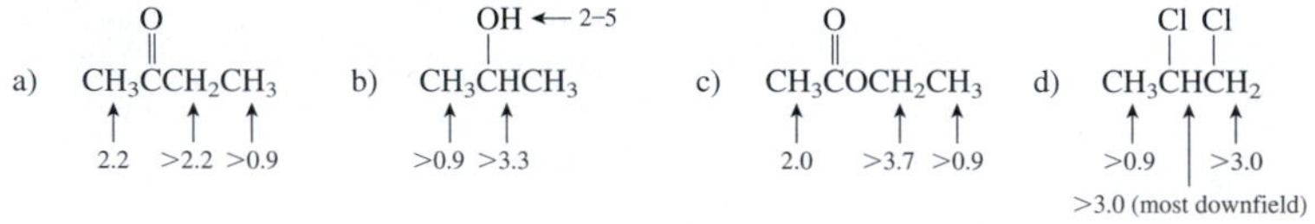

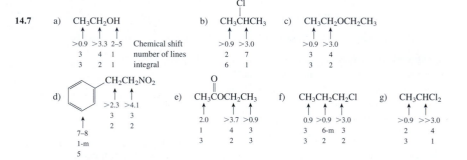

14.4 Triplet

14.6 a) $a = x = 3$ b) $a = 7, x = 2$ c) $a = 3, x = 2$ d) $a = 5, x = 2$ e) $a = 3, m = 6, x = 3$

14.7
a) CH_3CH_2OH
>0.9 >3.3 2–5 Chemical shift
3 4 1 number of lines
3 2 1 integral
b) CH_3CHCH_3
Cl
>0.9 >3.0
2 7
6 1
c) $CH_3CH_2OCH_2CH_3$
>0.9 >3.0
3 4
3 2
d) $CH_2CH_2NO_2$
>2.3 >4.1
3 3
2 2
7–8
1-m
5
e) $CH_3COCH_2CH_3$
2.0 >3.7 >0.9
1 4 3
3 2 3
f) $CH_3CH_2CH_2Cl$
0.9 >0.9 >3.0
3 6-m 3
3 2 2
g) CH_3CHCl_2
>0.9 >>3.0
2 4
3 1

14.8
a) CH_3CCH_3
b) CH_2CH_3
c) CH_3CH
d)
e) $CH_3COCHCH_3$
f) CH_3CH_2CHCl

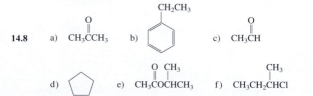

14.9   a) 1   b) 3   c) 6   d) 3

14.10   a) 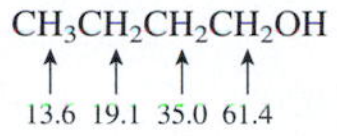 $CH_3CH_2CH_2CH_2OH$
            13.6 19.1 35.0 61.4

   b) 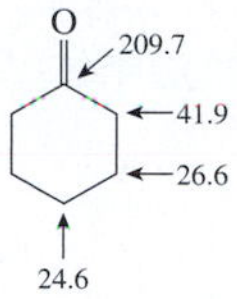
   cyclohexanone with chemical shifts 209.7, 41.9, 26.6, 24.6

14.11   a) 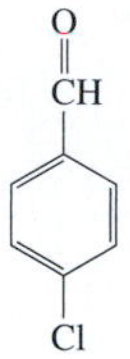
   (para-chlorobenzaldehyde)

   b) 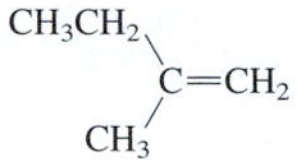
   CH_3CH_2, CH_3, $C{=}CH_2$

   c)
   (cyclohexene)

14.12   a) 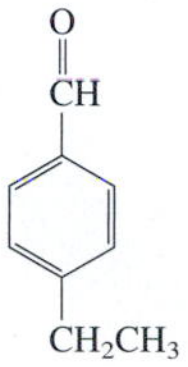
   (para-ethylbenzaldehyde)

   b) $BrCH_2CH_2COCH_2CH_3$

   c) 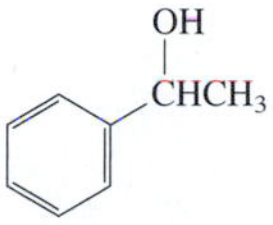
   (1-phenylethanol, $CHCH_3$ with OH)

14.13 $HOCH_2CH_2CH_2CH_2CH_2CH_2CH_2OH$

14.14   a) $CH_3CH_2CHCH_2OH$ with CH_3

   b) 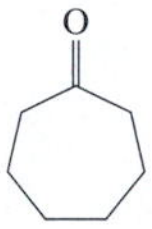
   (cycloheptanone)

   c) 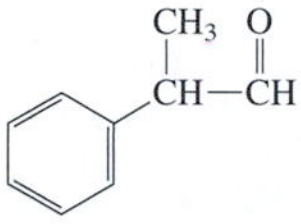
   $CH{-}CH$ with CH_3 and O

CHAPTER 15

15.1   0.349

15.2 2.0×10^4 M^{-1} cm^{-1}

15.3 2.38×10^{-5} M

15.4   n $\longrightarrow \pi^*$

15.5   The absorption at 213 nm is due to a $\pi \longrightarrow \pi^*$ transition, and that at 320 nm is due to a n $\longrightarrow \pi^*$ transition.

15.6   a) The left ketone is conjugated and should absorb at longer wavelength than the right ketone, which is not conjugated.

   b) The right compound is conjugated and should absorb at longer wavelength than the left compound, which is not conjugated.

   c) The left compound has three conjugated double bonds and should absorb at longer wavelength than the right compound, which has only two conjugated double bonds.

   d) Cyclohexanone has an absorption for a n $\longrightarrow \pi^*$ transition in the accessible UV region, whereas the alcohol shows no such absorption.

15.7   Butane = 58.0783; acetone = 58.0419

15.8   Butane = 4.4%; acetone = 3.3%

15.9   a) M : M + 2 : M + 4 = 1 : 0.67 : 0.11 (9:6:1)

   b) M : M + 2 : M + 4 = 1 : 1.33 : 0.33 (3:4:1)

15.10 a) Br is present.   b) N is present.

15.11 a) $CH_3\overset{+}{N}HCH_3$ b)

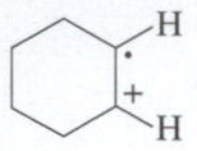

15.12 a) 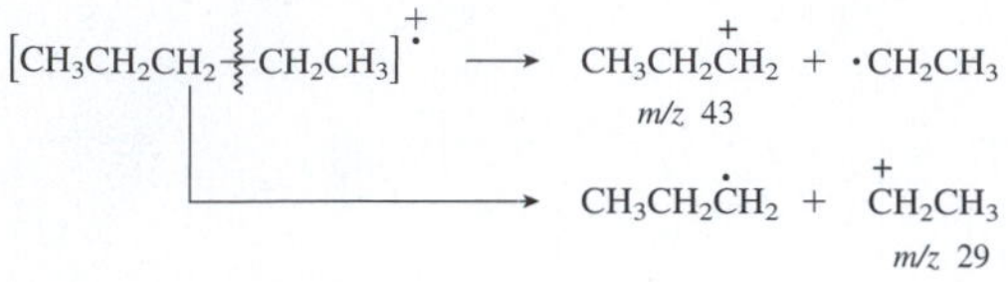 $[CH_3CH_2CH_2 \dashv CH_2CH_3]^{\overset{\cdot}{+}} \longrightarrow CH_3CH_2\overset{+}{C}H_2 + \cdot CH_2CH_3$
$m/z\ 43$

$\longrightarrow CH_3CH_2\overset{\cdot}{C}H_2 + \overset{+}{C}H_2CH_3$
$m/z\ 29$

15.13 a)

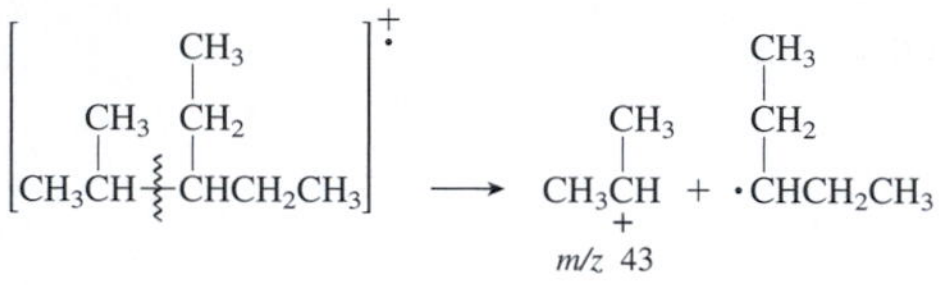

15.14 a) 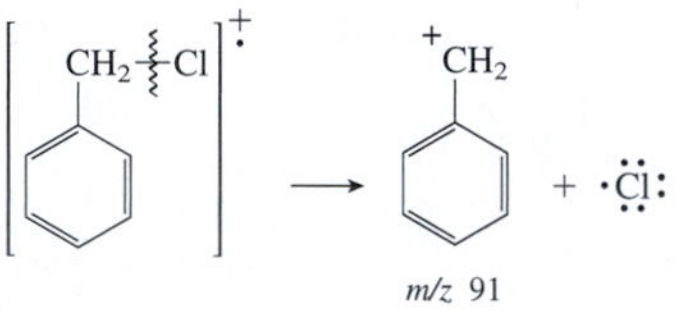
$m/z\ 91$

b) 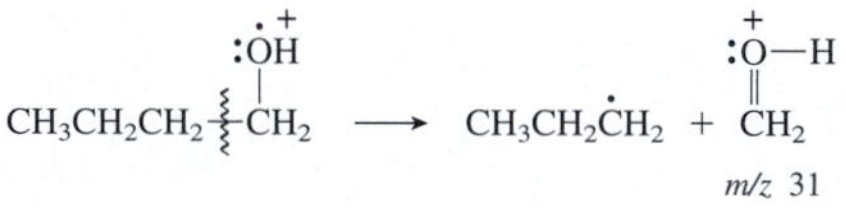
$m/z\ 31$

c)

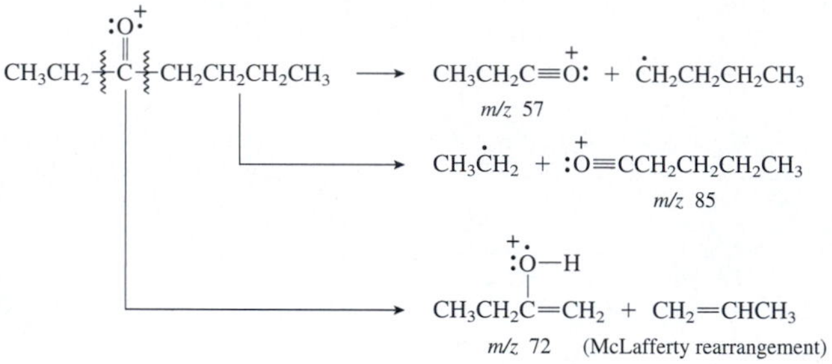

CHAPTER 16

16.1 a) 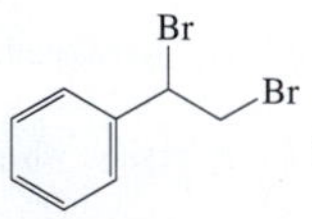   b) 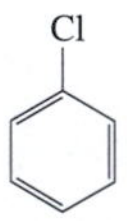   c) 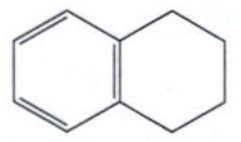   d)

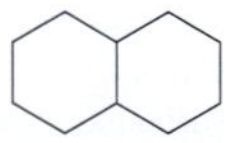

16.2 (5) (28.6 kcal/mol) − 80 kcal/mol = 63 kcal/mol; yes
(5) (120 kJ/mol) − 335 kJ/mol = 265 kJ/mol; yes

16.3

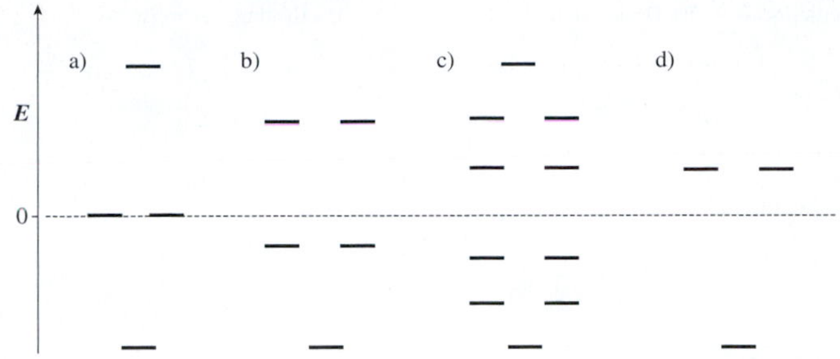

16.7   a) 4 π electrons, so antiaromatic   b) 6 π electrons, so aromatic

   c) 10 π electrons, so aromatic   d) 2 π electrons, so aromatic

16.8   a) 6 π electrons, so aromatic   b) 6 π electrons, so aromatic

   c) 6 π electrons, so aromatic   d) 6 π electrons, so aromatic

   e) 6 π electrons, so aromatic   f) 8 π electrons, so antiaromatic if planar

16.12   a) Left compound because its conjugate base is aromatic

   b) Right compound because conjugate base of left compound is antiaromatic

   c) Right compound because its conjugate base is aromatic

16.13   Left compound because the carbocation that is formed from it is aromatic

CHAPTER 17

17.1

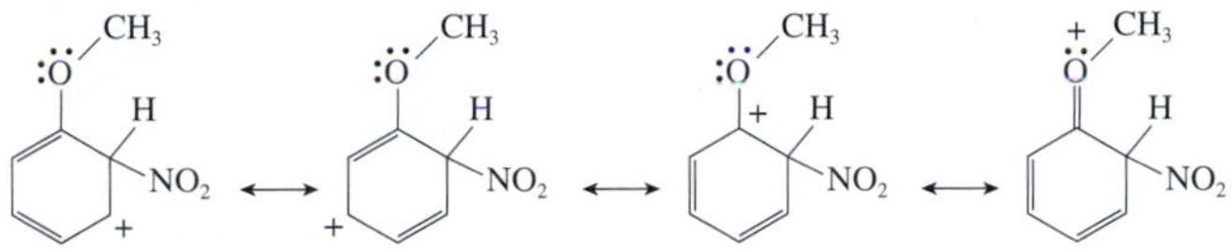

The last structure is especially stable because the octet rule is satisfied at all atoms.

17.2

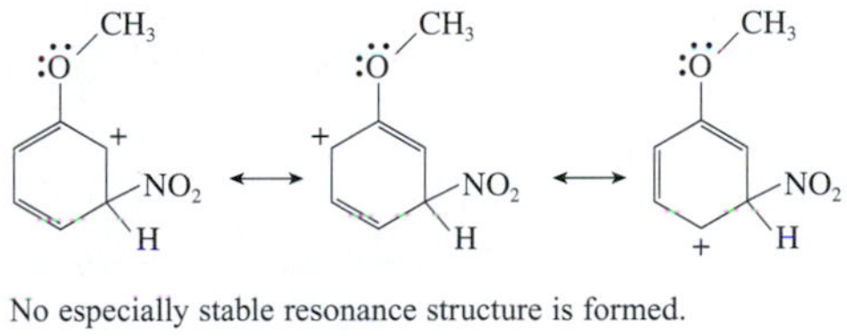

No especially stable resonance structure is formed.

17.3   a) The electron pair on the N can stabilize a positive charge on the adjacent carbon by resonance.

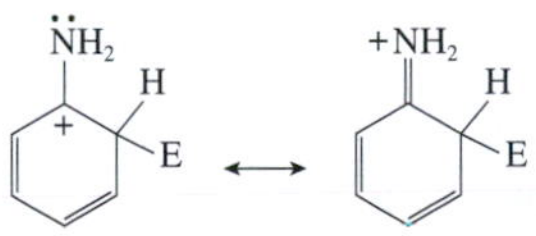

   b) The ethyl group stabilizes a positive charge on the adjacent carbon by hyperconjugation.

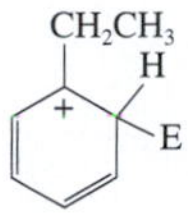

17.4   a) The electronegative fluorines make the CF_3 group electron withdrawing.

   b) The positive nitrogen is an inductive electron-withdrawing group.

17.5 a) Activating, *o*/*p*-director b) Deactivating, *m*-director c) Deactivating, *m*-director

17.6 a) 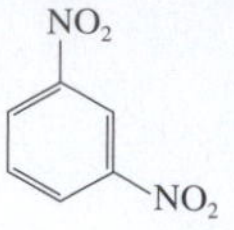 b) 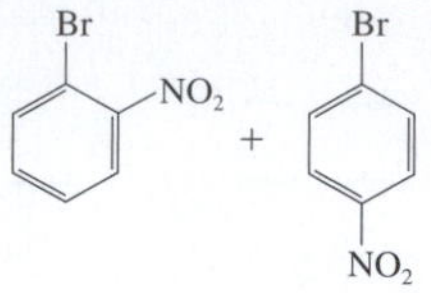

17.7 a) 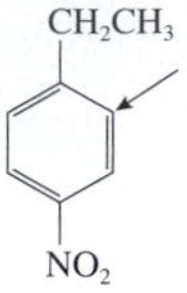 b) 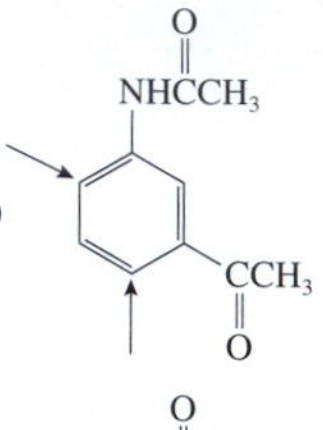 c) 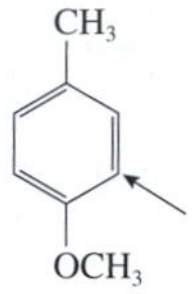

d) 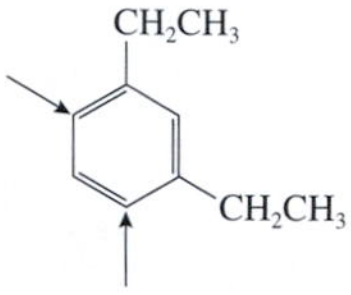 e) 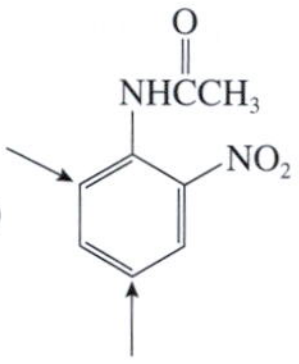 f) 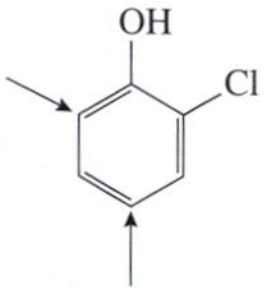

17.8 a) 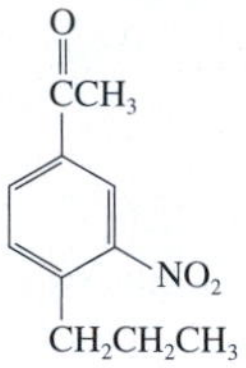 b) 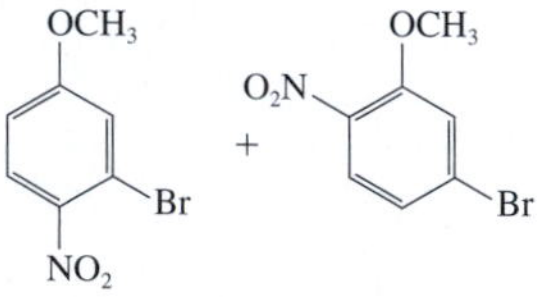

17.9 a) 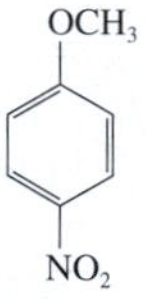
+ Ortho

b) 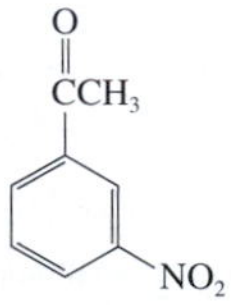

c) 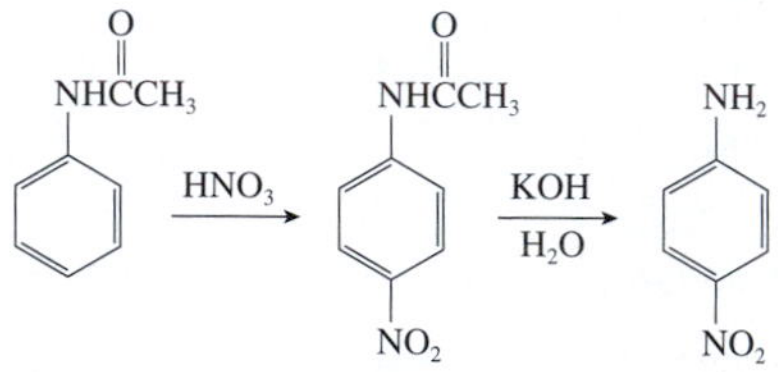
+ Ortho + Ortho

17.11 a) 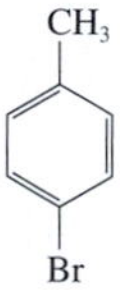 + Ortho b) 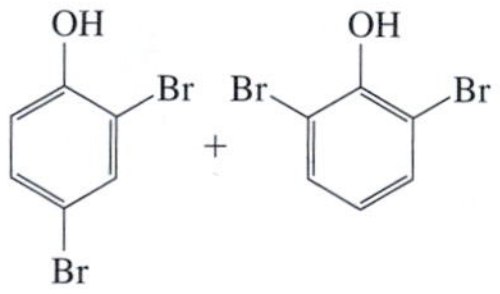 c) 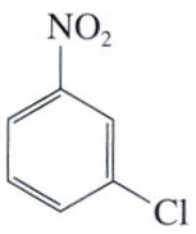

d) 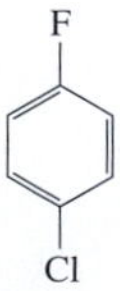 + Ortho e) 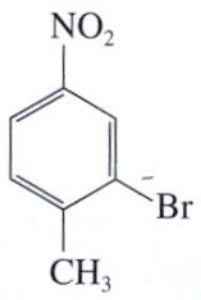 f) 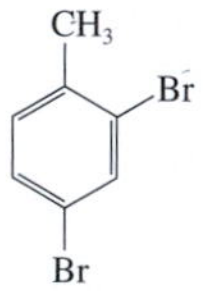

17.13

a)

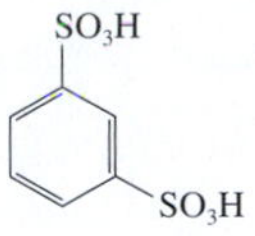

b) 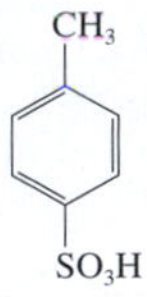 + Ortho

c)

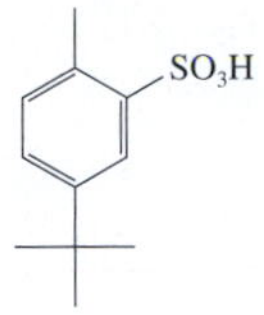

d) 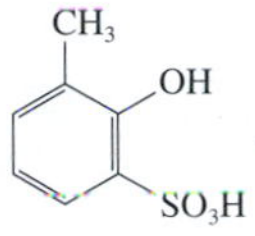 +

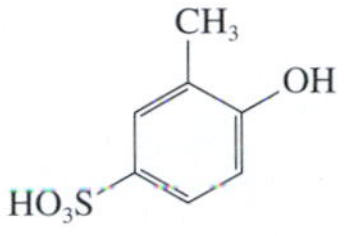

17.15

a) 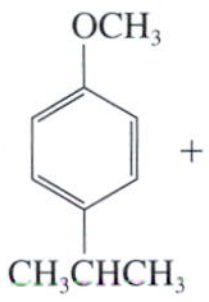 + Ortho

b) 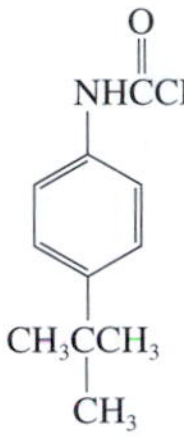 + Ortho

c)

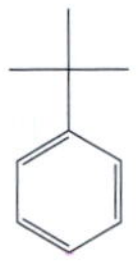

d)

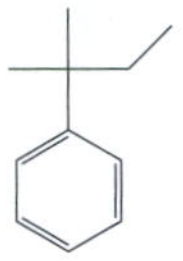

e)

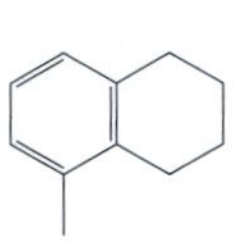

f) No reaction

17.16

a)  + $\xrightarrow{\text{AlCl}_3}$

b) 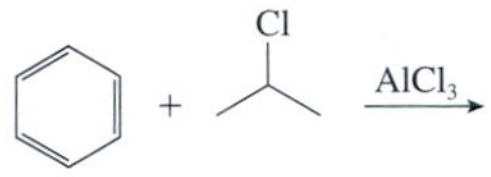 + $\xrightarrow{\text{AlCl}_3}$

17.18 Both groups are weakly activating, but the position ortho to the methyl group is less sterically hindered than the position ortho to the isopropyl group.

17.19

a)

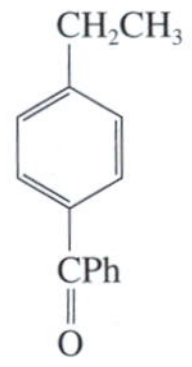

b) No reaction

c)

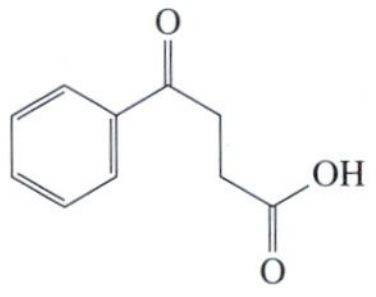

d)

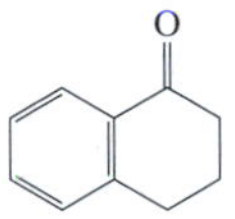

e)

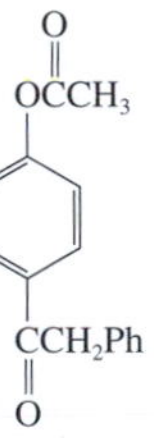

17.20

a) 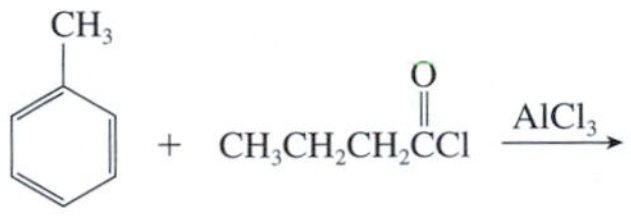 + $CH_3CH_2CH_2\overset{O}{\overset{\|}{C}}Cl$ $\xrightarrow{\text{AlCl}_3}$

b) 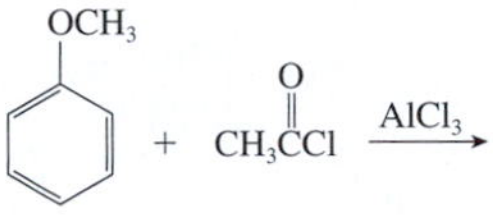 + $CH_3\overset{O}{\overset{\|}{C}}Cl$ $\xrightarrow{\text{AlCl}_3}$

17.22 a) 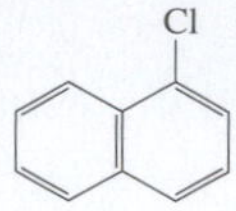 b) 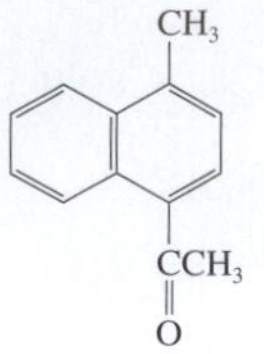 c)

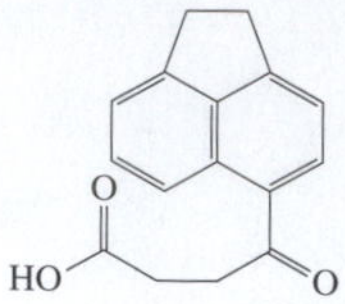

17.23 a) 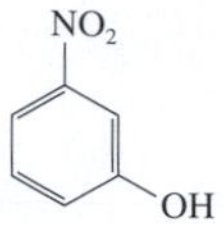 b) 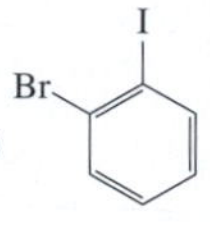 c) 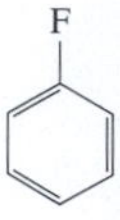 d)

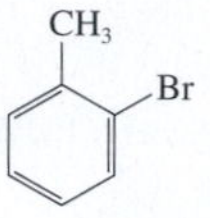

e) 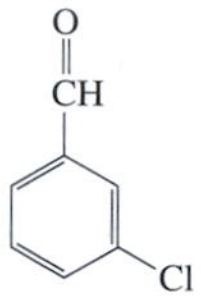 f) 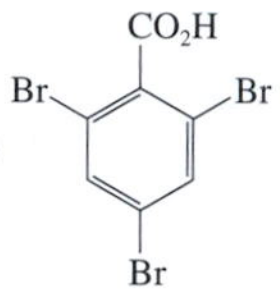 g) 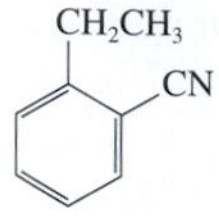 h)

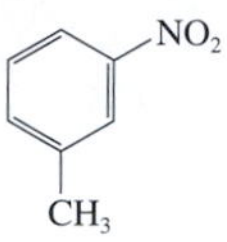

17.25 a) 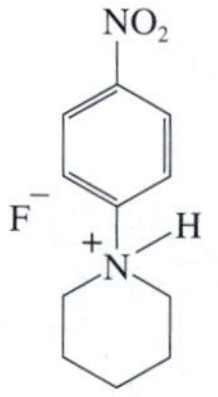 b)

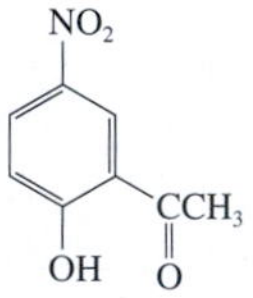

17.26 a) 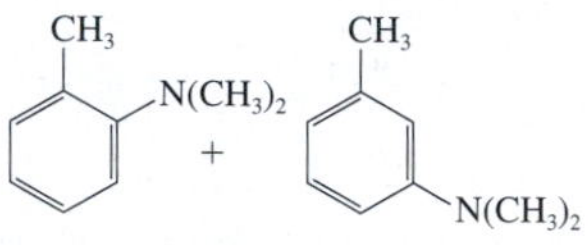 b) 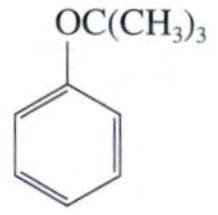 c)

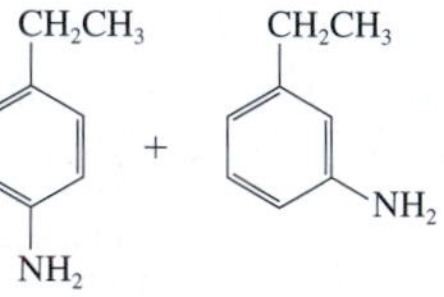

17.27 a) 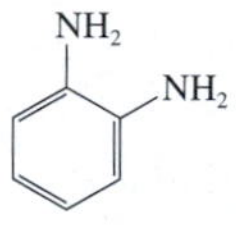 b) 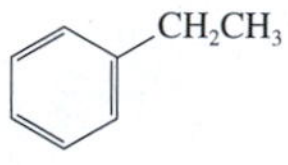 c) 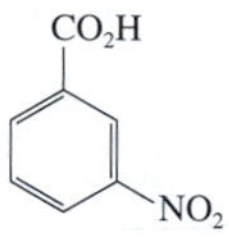 d)

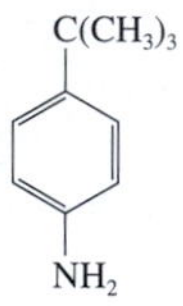

e) 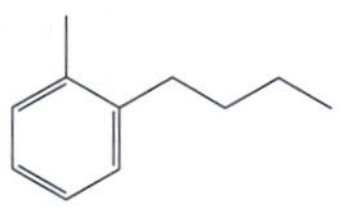 f) 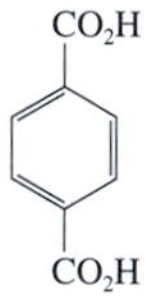 g)

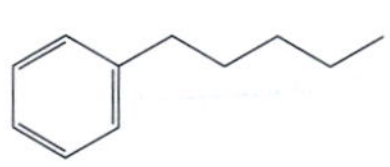

17.28 a)

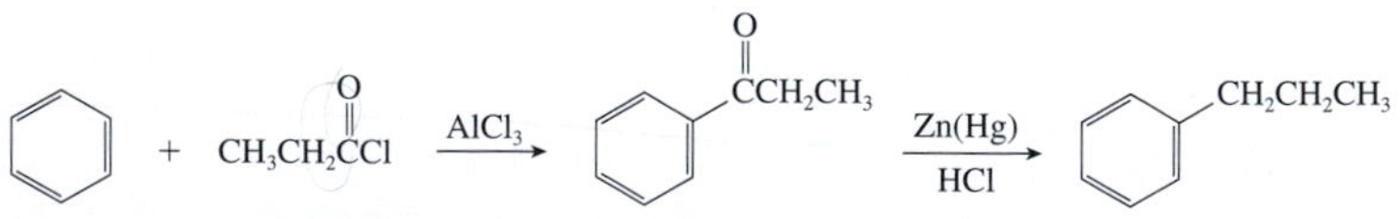

b)

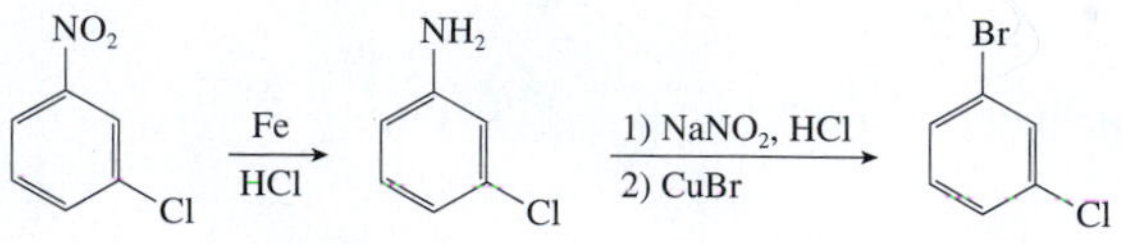

17.29   a)

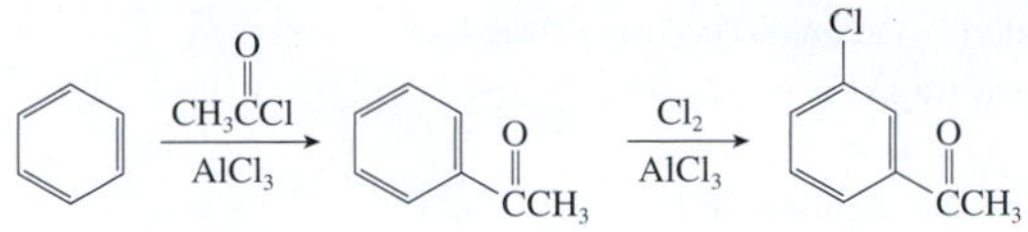

b)

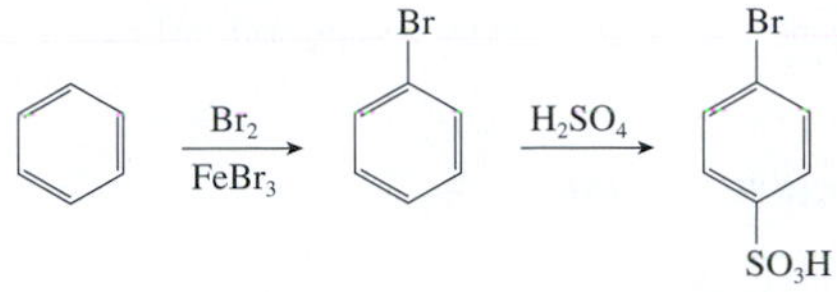

c)

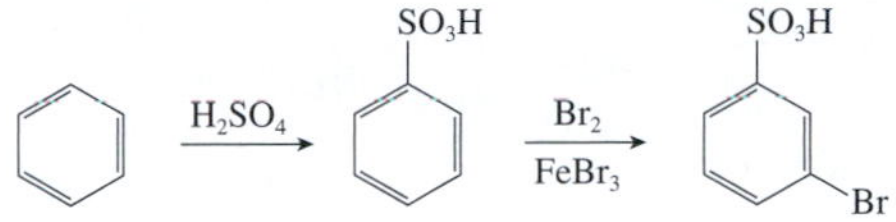

d)

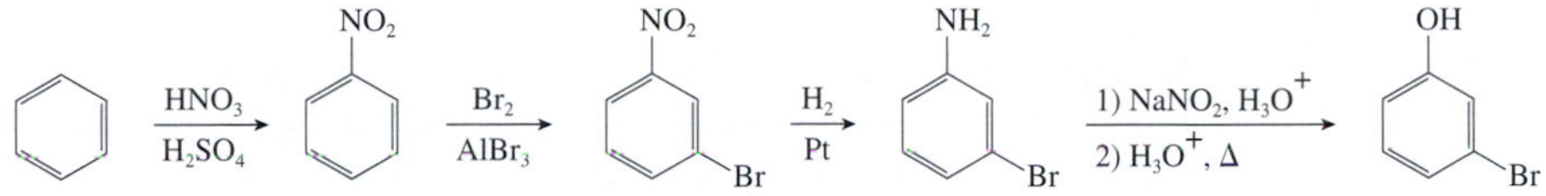

e)

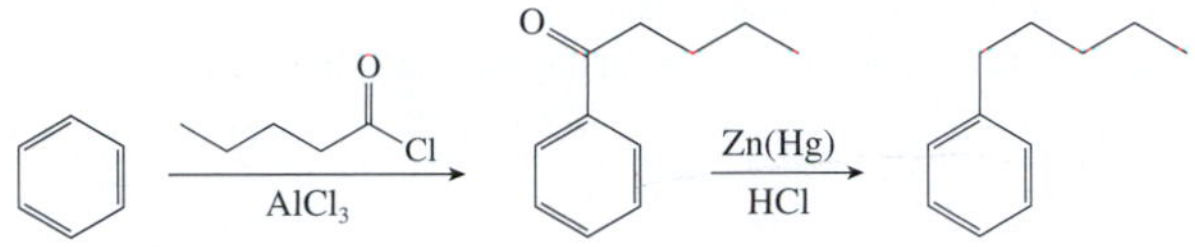

17.30   a)

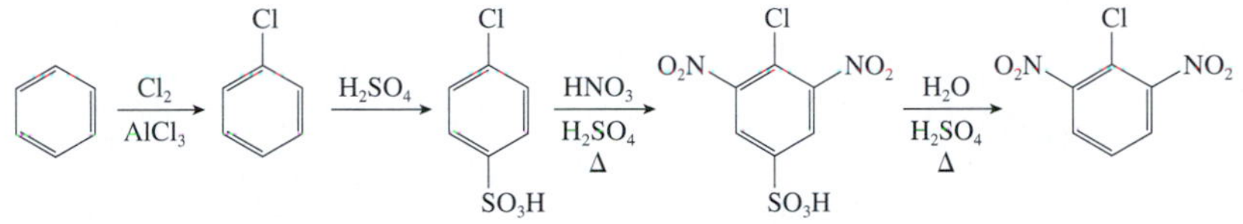

b)

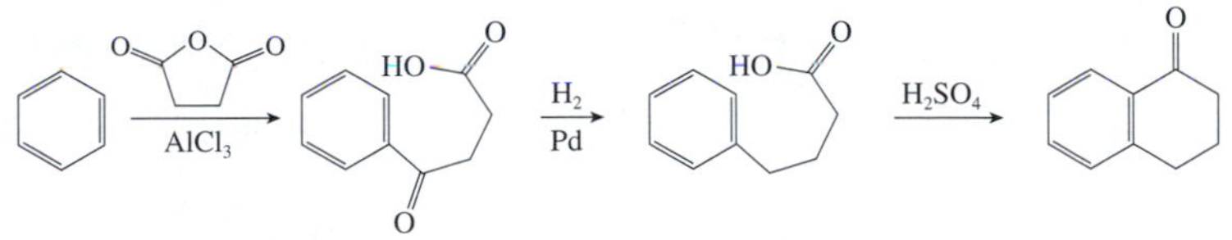

c)

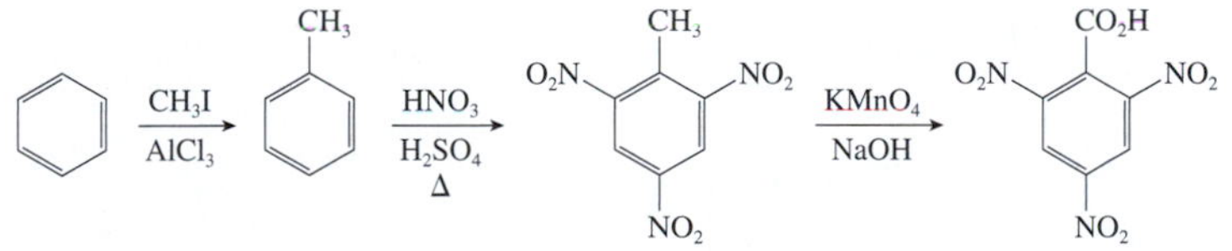

CHAPTER 18

18.2   a)

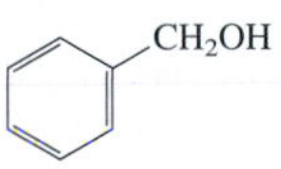

b) $CH_3CH_2\overset{\displaystyle OH}{\underset{|}{C}}HCH_3$

c) $CH_3CH=CHCH_2CH_2\overset{\displaystyle OH}{\underset{|}{C}}H_2$

d)

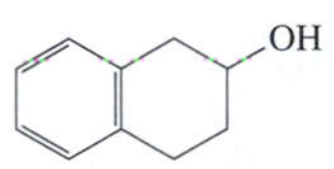

18.4 a) Right compound due to inductive electron-withdrawing effect of F

b) Right compound due to inductive and steric effects

c) Left compound due to less steric hindrance

d) Right compound due to inductive electron-withdrawing effect of Cl

e) Not much difference because the steric effect is too far from the reacting carbonyl carbon

18.5

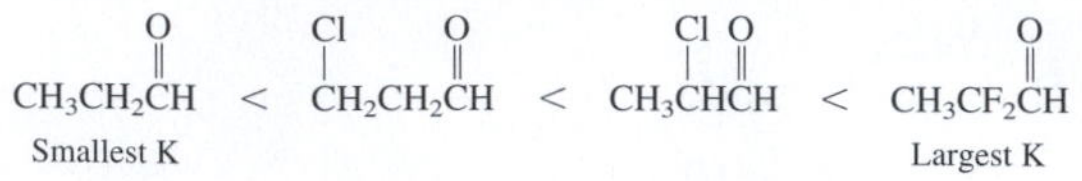

$$\underset{\text{Smallest K}}{CH_3CH_2\overset{O}{\overset{\|}{C}}H} \quad < \quad \overset{Cl}{\underset{}{CH_2}}CH_2\overset{O}{\overset{\|}{C}}H \quad < \quad CH_3\overset{Cl}{\underset{}{CH}}\overset{O}{\overset{\|}{C}}H \quad < \quad \underset{\text{Largest K}}{CH_3CF_2\overset{O}{\overset{\|}{C}}H}$$

18.6 a)   b) $CH_3CH_2\underset{CN}{\overset{OH}{C}}CH_3$ c)

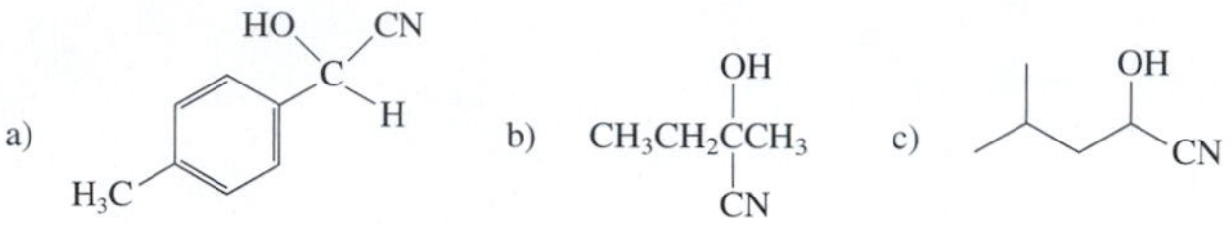

18.7 a) Right compound because aldehydes are less sterically hindered than ketones

b) Left compound because resonance makes the carbonyl carbon of the right compound less electrophilic

c) Left compound because the ethoxy group in right compound donates electrons to the carbonyl group by resonance, making the carbonyl carbon less electrophilic

d) Right compound because the electron-withdrawing nitro group makes its carbonyl carbon more electrophilic

18.8 a) $CH_3CH_2CH_2CH_3$ b) 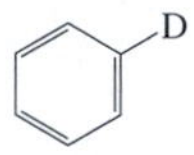   c) $CH_3CH_2C\equiv CMgBr \quad + \quad CH_3CH_3$

18.9 a) 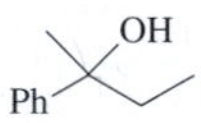   b)    c)

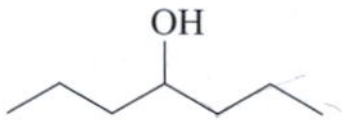

d) 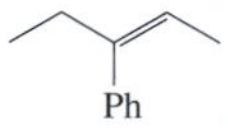   e) 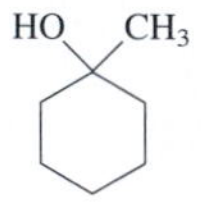   f) 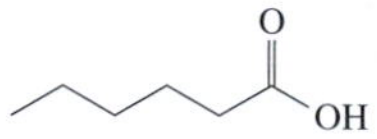   g) 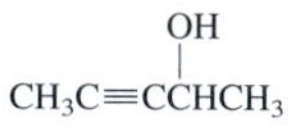 $CH_3C\equiv C\underset{}{\overset{OH}{C}}HCH_3$

18.10 a)

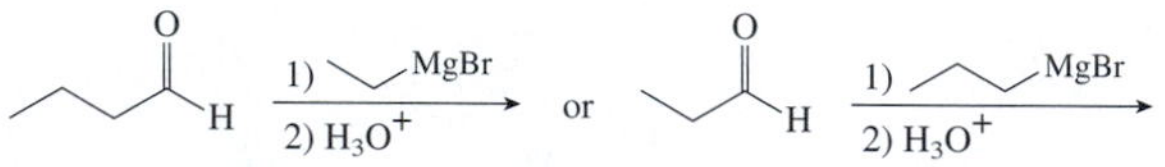

b) 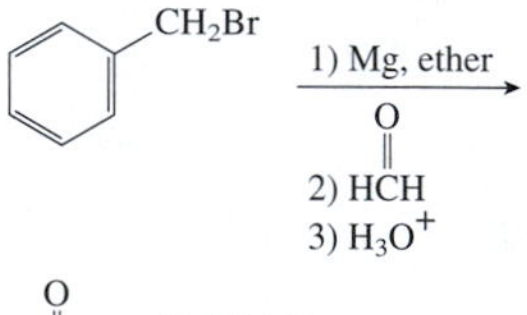   c)

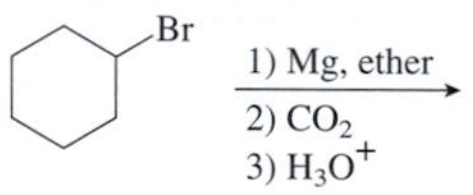

d) 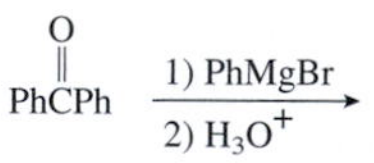 $Ph\overset{O}{\overset{\|}{C}}Ph \xrightarrow[\text{2) } H_3O^+]{\text{1) PhMgBr}}$ e)

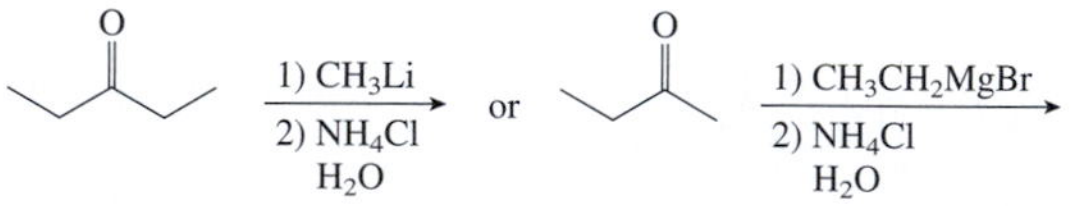

18.11 An acid–base reaction between the Grignard reagent and the hydroxy group on the aldehyde destroys the Grignard reagent.

18.12 $Ph_3P \; + \; PhCH_2Cl \xrightarrow{S_N2} \overset{+}{Ph_3P}-CH_2Ph \quad \overset{-}{Cl}$

18.13 a) PhCH=CHCH$_2$CH$_3$ b)    c) HC≡CCH=CHCH$_2$CH$_2$CH$_3$

d) 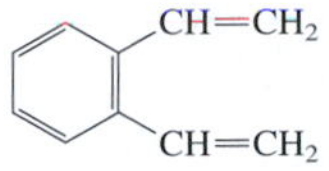   e) PhCH=CHCOEt (with O above C)

18.14 The conjugate base (the ylide) is stabilized by resonance with the phenyl group.

18.15 a) PhCH$_2$CH (O) $\xrightarrow{Ph_3P-CHCH_2CH_3}$ or   CH$_3$CH$_2$CH (O) $\xrightarrow{Ph_3P-CHCH_2Ph}$

b) CH$_3$CH$_2$CH (O) $\xrightarrow{Ph_3P-CHCCH_2CH_3}$ c)

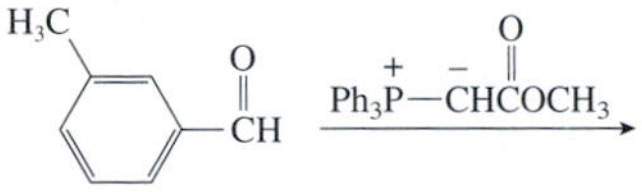

18.16 a)    b) 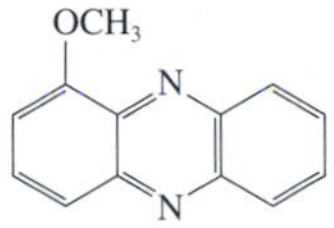   c) CH$_3$CCH$_2$CH$_3$ (with NOH)

d) 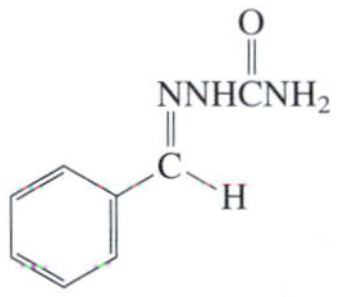   e)

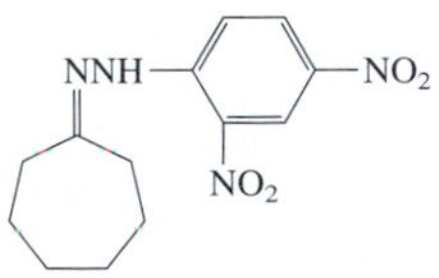

18.18 Formation of the more stable conjugated enamine is preferred.

18.19 a) 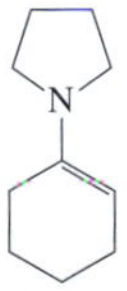   b) 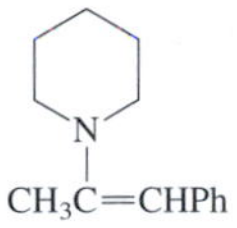 CH$_3$C=CHPh (with piperidine N)

18.20 a) 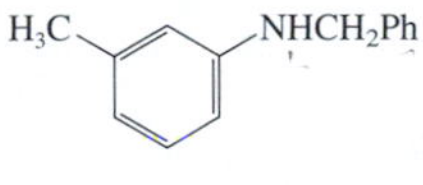   b) 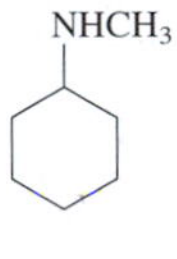   c)

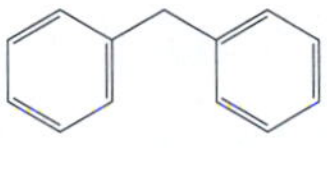

18.21 The right compound gives more cyclic hemiacetal at equilibrium because a five-membered ring is more stable than a four-membered ring.

18.23 a) CH$_3$CH$_2$CH$_2$CH (with CH$_3$CH$_2$O and OCH$_2$CH$_3$)   b) 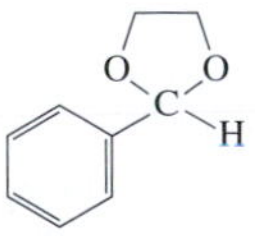   c)

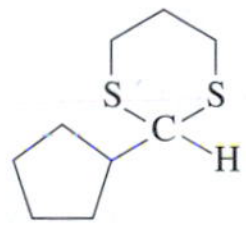

18.24

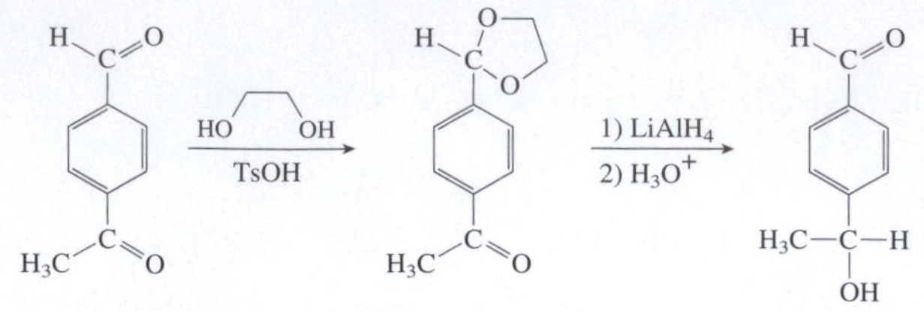

18.25 a) 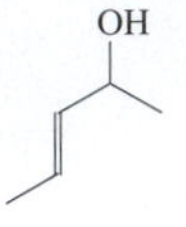 [structure]   b) 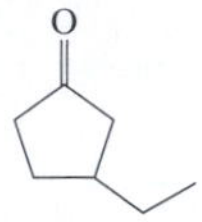 [structure]   c) 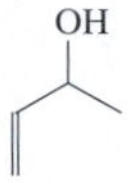 [structure]   d) 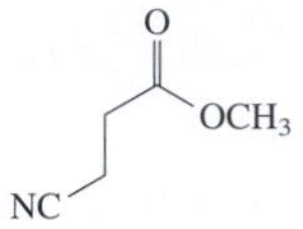 [structure]   e) 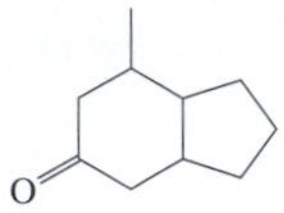 [structure]

18.26 a)

b)

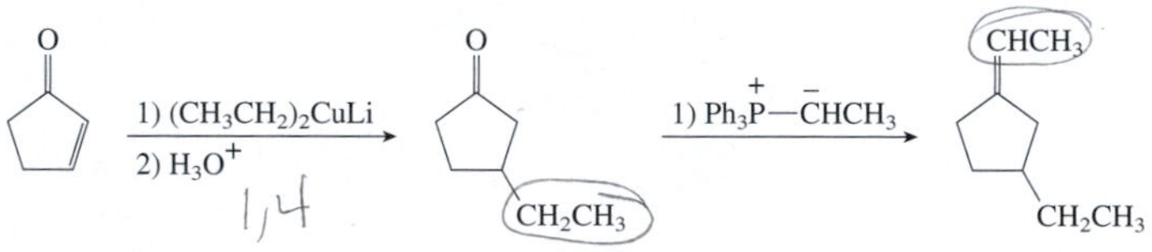

CHAPTER 19

19.1 a) Left ester due to less steric hindrance

b) Left compound due to a more electrophilic carbonyl carbon because of the electron-withdrawing F

c) Left compound due to a more electrophilic carbonyl carbon because of the electron-withdrawing nitro group

19.2 a) Products   b) Reactants   c) Products

19.3 The reaction of the ester is faster because an ester is more reactive than an amide.

19.4 $CH_3CH_2CH_2\overset{O}{\overset{\|}{C}}Cl \ + \ NH_2CH_2CH_2CH_3 \ \longrightarrow$

19.5 a) 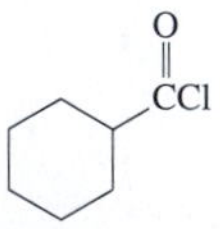 [structure]   b) 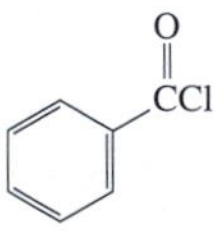 [structure]   c) $CH_3CH_2CH_2\overset{O}{\overset{\|}{C}}Cl$

19.6 a) $CH_3CH_2\overset{O}{\overset{\|}{C}}O\overset{O}{\overset{\|}{C}}CH_2CH_3$ b) 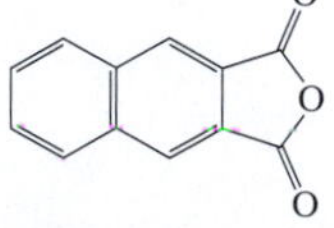 [structure] $+ \ 2 \ CH_3\overset{O}{\overset{\|}{C}}OH$

19.8 The reaction produces aspirin, an ester, rather than the more reactive anhydride.

19.9

a) 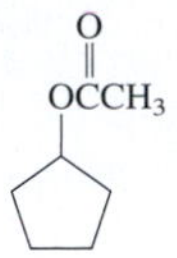 cyclopentyl acetate (OCCH₃ ester on cyclopentane)

b) 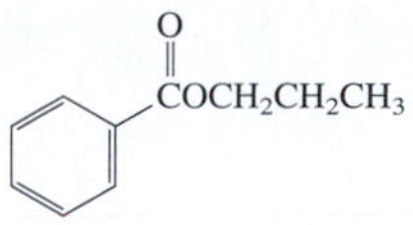 PhCOCH₂CH₂CH₃ (phenyl ketone)

c) $CH_3COCH_2CHCH_3$ (with CH₃) + CH_3COH

d) 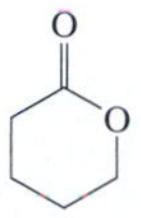 (δ-valerolactone, six-membered lactone)

e) 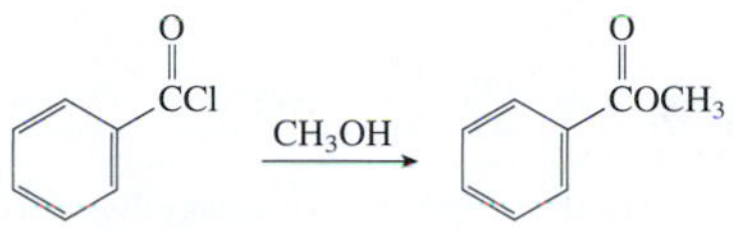 PhCOCl $\xrightarrow{CH_3OH}$ PhCOCH₃ (methyl benzoate)

f) 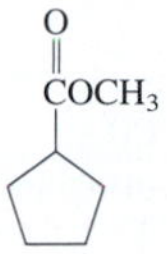 cyclopentyl COCH₃ (acetyl cyclopentane)

19.12

a) 2 CH_3COH

b) $PhCO_2^- + HOCH_2CH_3$

c) $CH_3CO_2H + PhNH_3^+ Cl^-$

d) 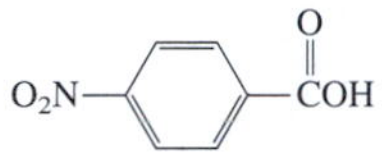 $O_2N-C_6H_4-COH$

e) $CH_3CH_2COH + HOCH_2CH_3$

f) $PhCCH_2CNH_2$ (diketo amide)

g) $CH_3CH_2CHCH_3$ with OH

h) 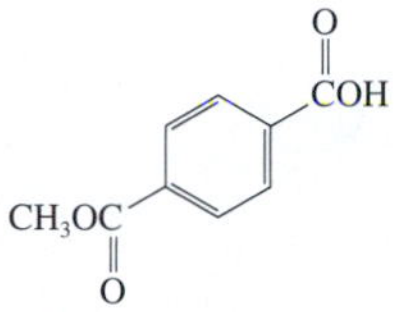 $CH_3OC-C_6H_4-COH$ (terephthalate)

i) 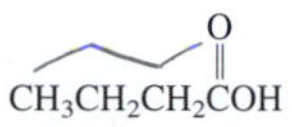 $CH_3CH_2CH_2COH$ (with ether oxygen)

19.13 $PhCH_2Br \xrightarrow[DMSO]{NaCN} PhCH_2C≡N \xrightarrow[\substack{H_2O \\ reflux}]{H_2SO_4} PhCH_2COH$

19.14

a) 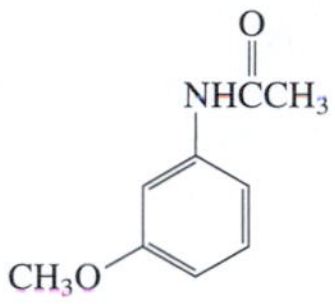 (m-methoxyphenyl)NHCCH₃ , CH₃O substituted

b) $CH_3CH_2CH_2CN(CH_2CH_3)_2$

c) 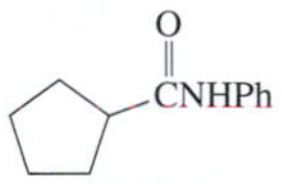 cyclopentyl-CNHPh

d) $PhCNHCH_3$

19.16

a) 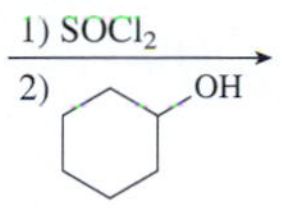 $\xrightarrow[\text{2)} \text{cyclohexyl-OH}]{\text{1) SOCl}_2}$

b) $\xrightarrow[\substack{\text{2) excess} \\ CH_3CH_2CH_2NH_2}]{\text{1) SOCl}_2}$

c) $\xrightarrow[\substack{\text{2) SOCl}_2 \\ \text{3) } CH_3CH_2OH}]{\text{1) H}_3O^+, \Delta}$

19.17

a) $CH_3CH_2CH_2CH_2OH$ + CH_3OH

b) 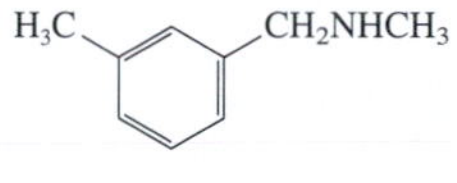 (m-methylphenyl)CH₂NHCH₃

c) 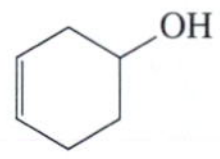 cyclohexenol OH + CH_3CH_2OH

d) 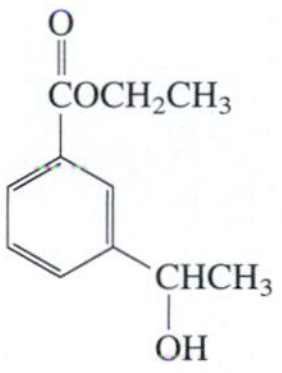 benzene ring with COCH₂CH₃ and CHCH₃ with OH

e)  $\sim\sim NH_2$ (pentylamine)

19.18 If acid were used, the product, an amine, would also be protonated.

19.19 a) b) c) $PhCH_2CH$ (with C=O)

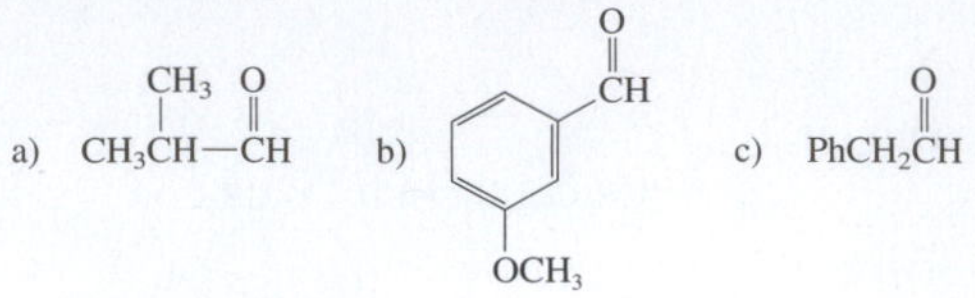

19.20 a) 1) $SOCl_2$ 2) $LiAlH(Ot\text{-}Bu)_3$ $-78°C$
b) 1) $SOCl_2$ 2) $PhNH_2$ 3) $LiAlH_4$ 4) H_2O
c) 1) H_3O^+, Δ 2) $SOCl_2$ 3) $LiAlH(Ot\text{-}Bu)_3$ $-78°C$
d) 1) $LiAlH_4$ 2) H_2O 3) CH_3CCl (with C=O) 4) $LiAlH_4$ 5) H_2O

19.21 a) b) $CH_3CH_2CH_2CCH_3$ with OH and CH_3 c)

19.23 a) b)

19.24 a) CH_3CH_2COH (C=O) $\xrightarrow[H_2SO_4]{CH_3OH}$ $CH_3CH_2COCH_3$ (C=O) $\xrightarrow[2) NH_4Cl, H_2O]{1) 2\ PhMgBr}$ $CH_3CH_2C\text{-}Ph$ with OH and Ph

b) CH_3CH_2COH (C=O) $\xrightarrow{SOCl_2}$ CH_3CH_2CCl (C=O) $\xrightarrow{(H_3C-\bigcirc-)_2CuLi}$ CH_3CH_2C (C=O)$-\bigcirc-CH_3$

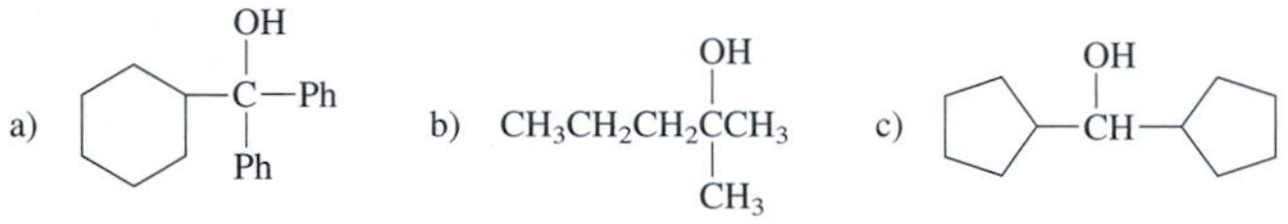

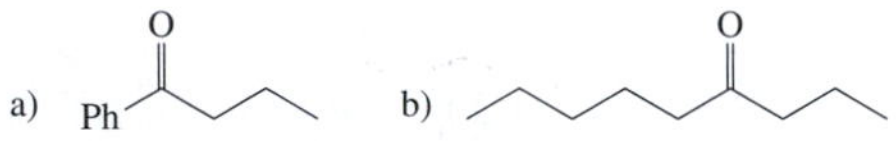

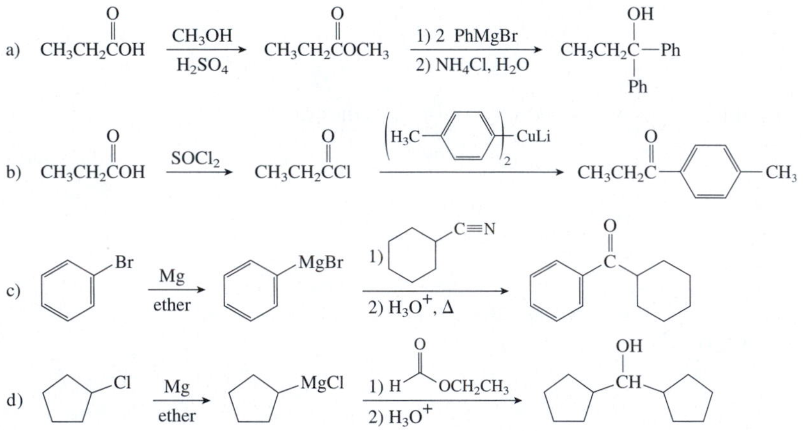

c)

d)

19.25 a) b) c) $CH_3CH_2OPOCH_2CH_3$ with O and OCH_2CH_3

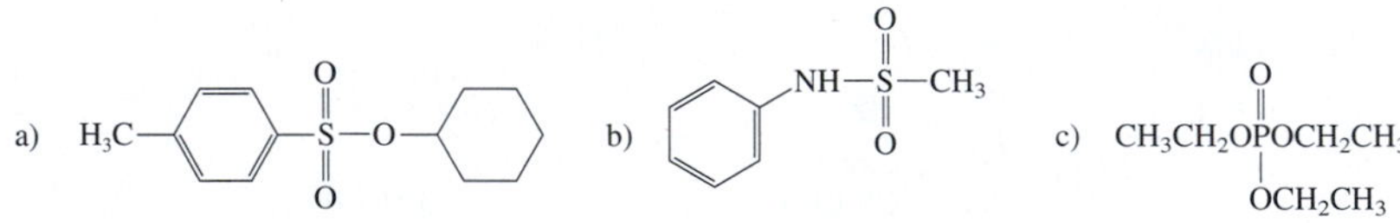

CHAPTER 20

20.1 a) $PhCH=C\text{-}CH_3$ with OH and $PhCH_2C=CH_2$ with OH b) $CH_3CH_2C=CHCH_3$ with OH

This one is more stable because it is conjugated.

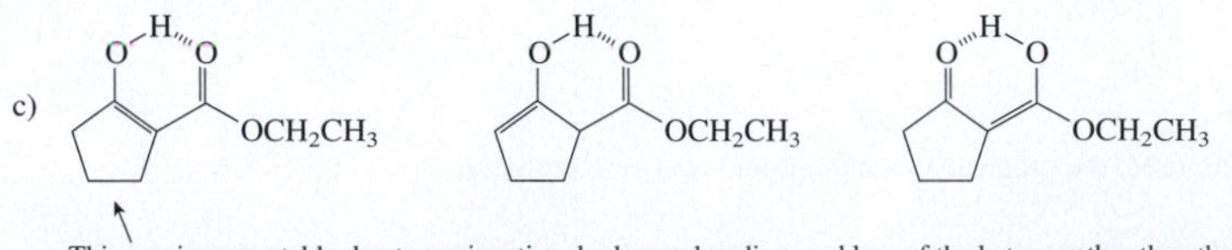

c)

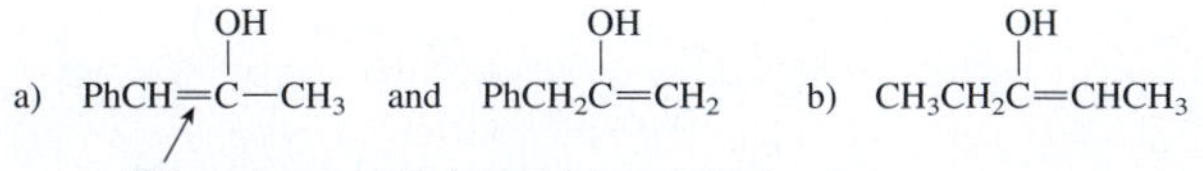

This one is more stable due to conjugation, hydrogen bonding, and loss of the ketone rather than the ester carbonyl group.

20.2   a) 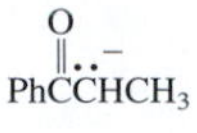   b) 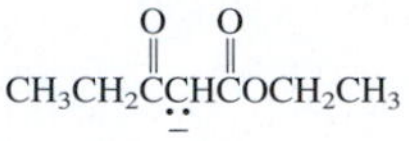   c) 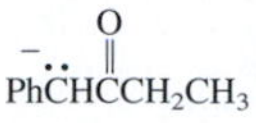   d)

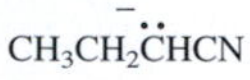

20.3   a) 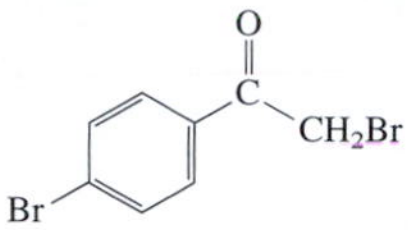   b)

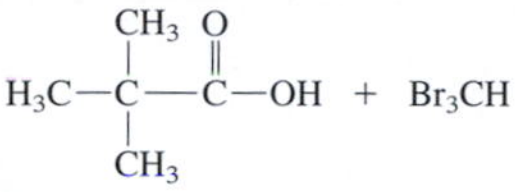

20.4

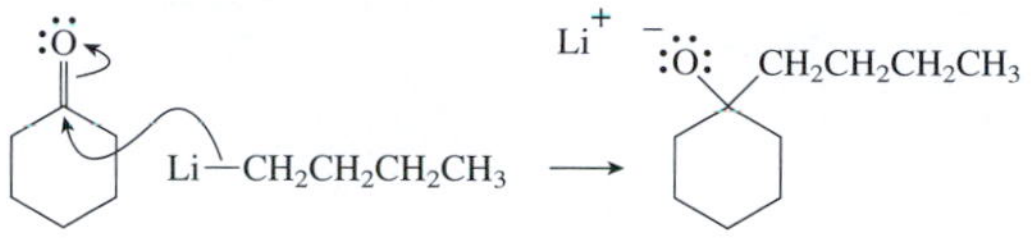

20.5   a) 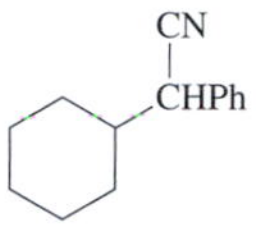   b) 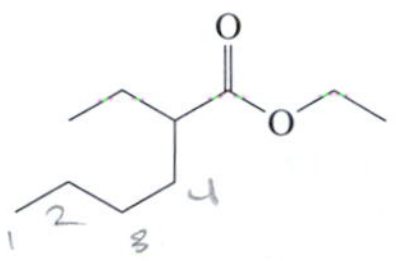   c)

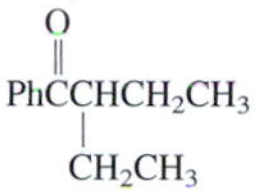

20.6   a)

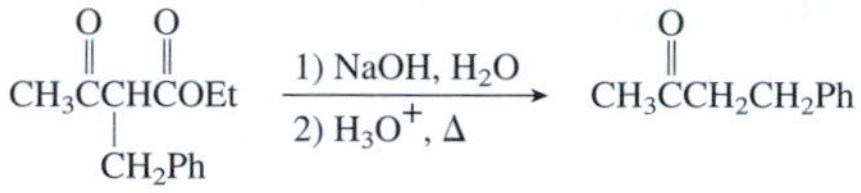

b)

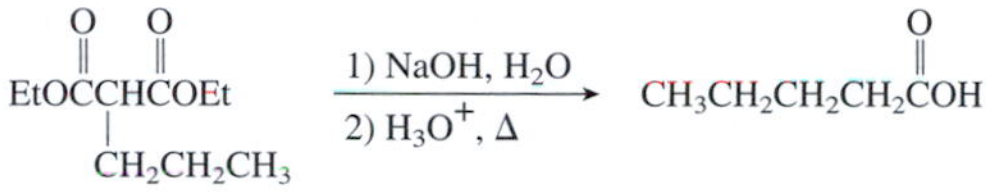

20.7   a)

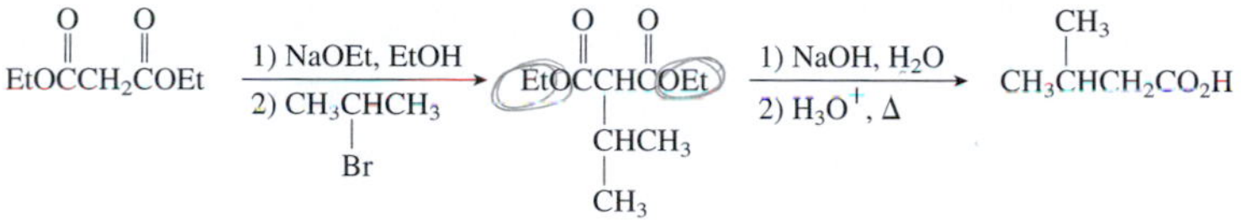

b)

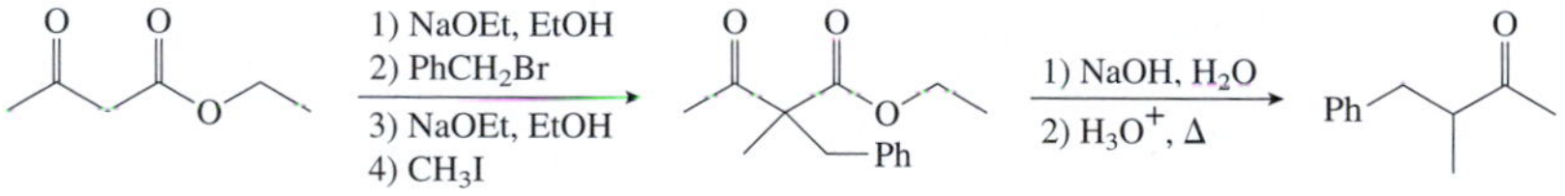

c)

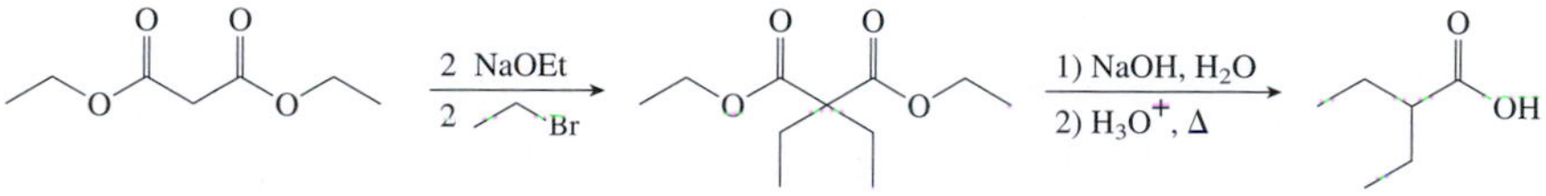

20.8   a) 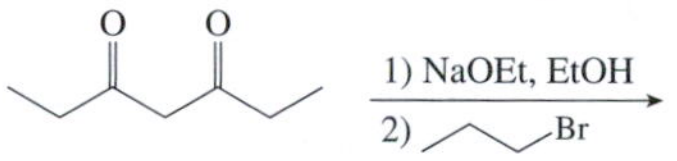   b)

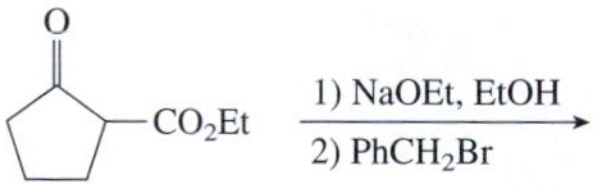

c) 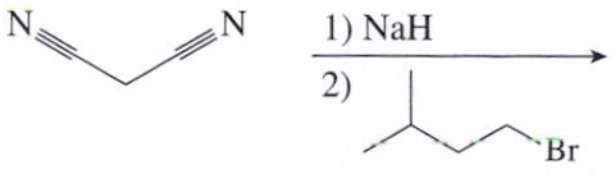   d)

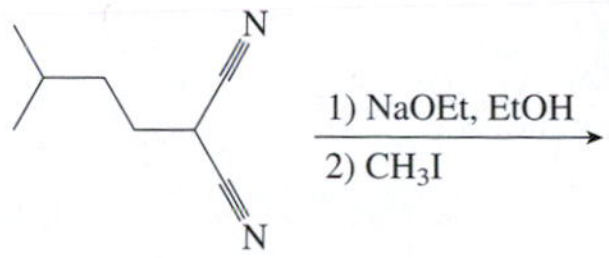

20.9   a) 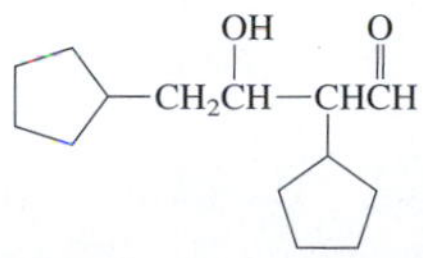   b) 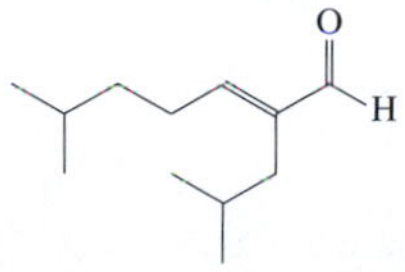   c)

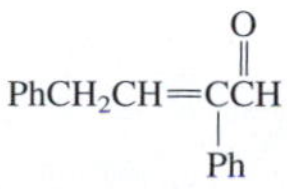

20.11 a) 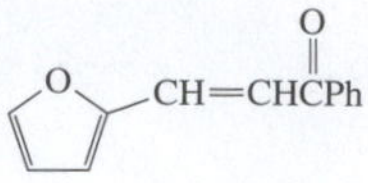 b) 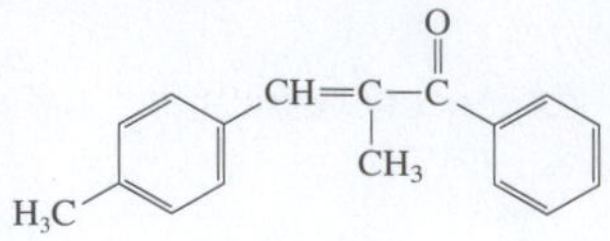 c)

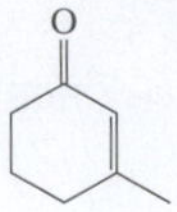

20.13

a) 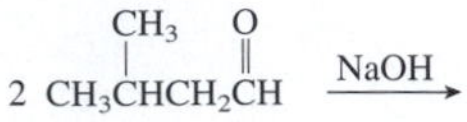 2 CH$_3$CHCH$_2$CH $\xrightarrow{\text{NaOH}}$

b) 2 $+$ CH$_3$CCH$_3$ $\xrightarrow[\Delta]{\text{NaOH}}$

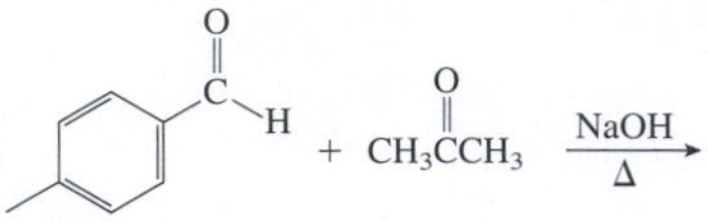

c) 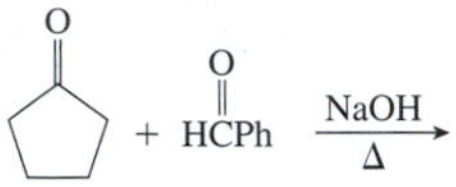 $+$ HCPh $\xrightarrow[\Delta]{\text{NaOH}}$

d) 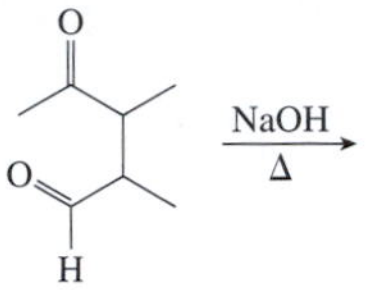 $\xrightarrow[\Delta]{\text{NaOH}}$

20.14

a) 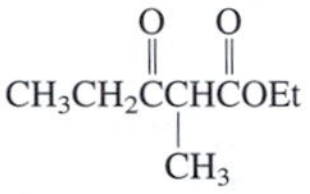 CH$_3$CH$_2$CCHCOEt with CH$_3$

b) 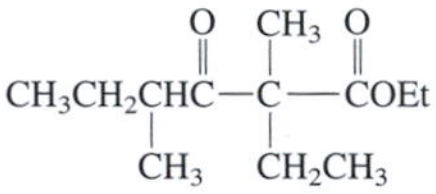 CH$_3$CH$_2$CHC—C—COEt with CH$_3$, CH$_3$, CH$_2$CH$_3$

c) PhCH$_2$CCHCOEt with Ph

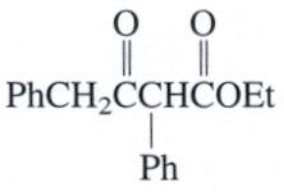

d)

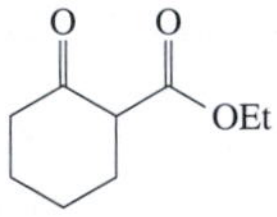

20.15 This product does not form because there is no acidic H between the two carbonyl groups, so the equilibrium driving step cannot occur.

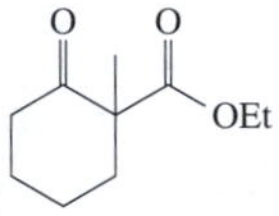

20.16

a) 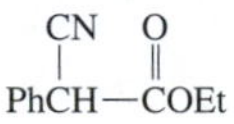 PhCH—COEt with CN

b) 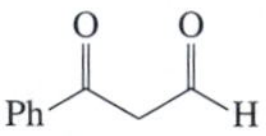 Ph—CH$_2$—CHO (with two C=O)

c)

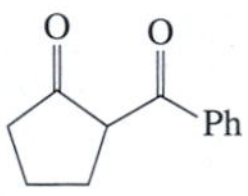

d) 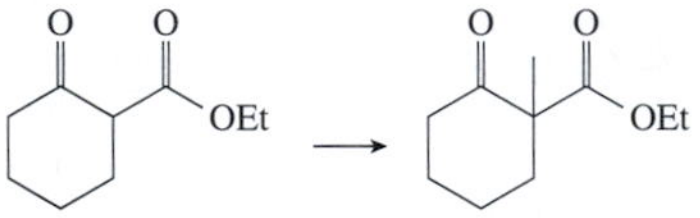 $\longrightarrow$

20.18

a) 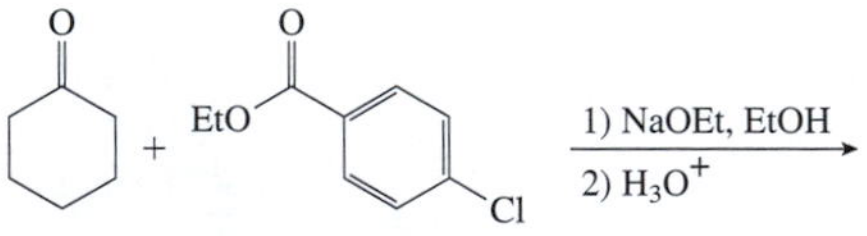 $+$ $\xrightarrow[\text{2) H}_3\text{O}^+]{\text{1) NaOEt, EtOH}}$

b) 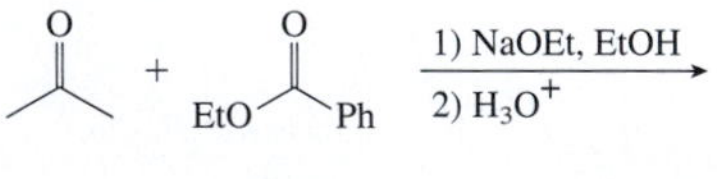 $+$ $\xrightarrow[\text{2) H}_3\text{O}^+]{\text{1) NaOEt, EtOH}}$

c) 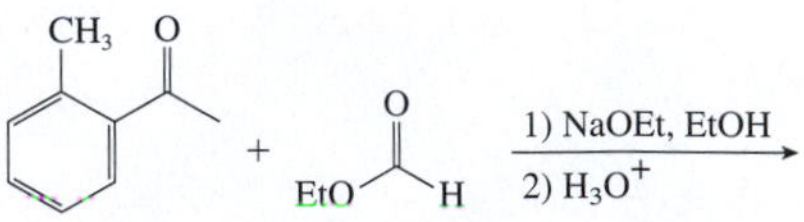 $+$ $\xrightarrow[\text{2) H}_3\text{O}^+]{\text{1) NaOEt, EtOH}}$

20.20

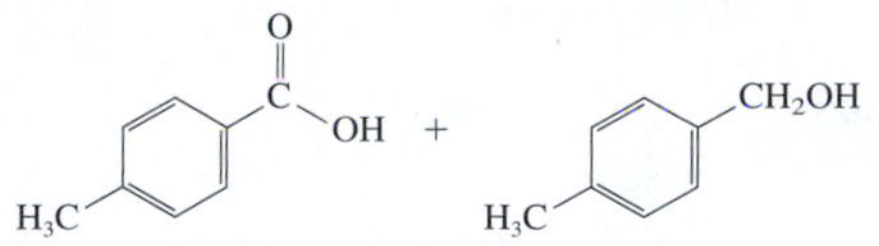
OH +
CH_2OH
H_3C
H_3C

20.22 a) 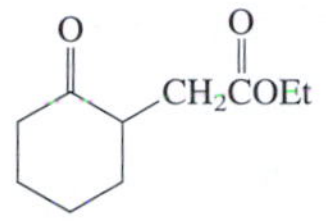
CH_2COEt
b) 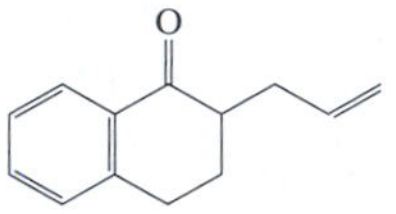
c) 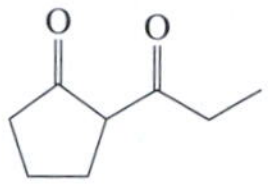

20.23 a) 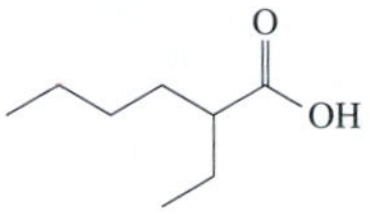
OH
b) 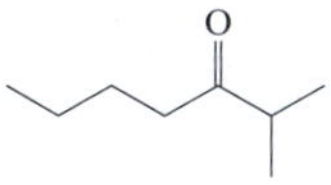

c) 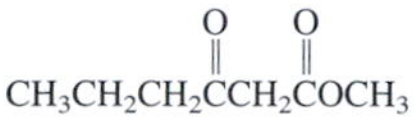
$CH_3CH_2CH_2CCH_2COCH_3$
d) 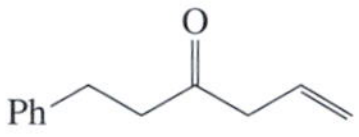
Ph

20.24 a) 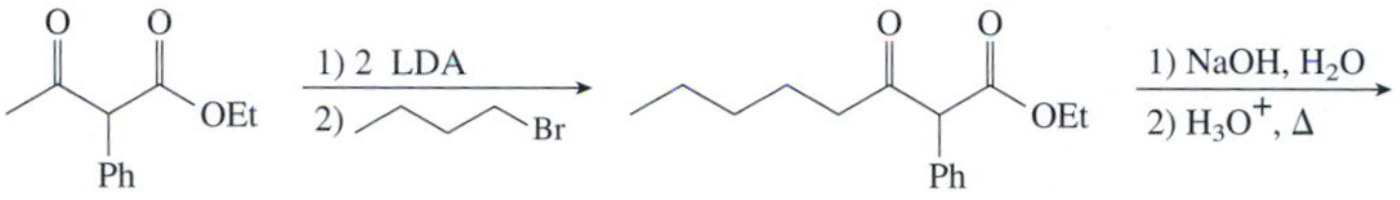
OEt
Ph
1) 2 LDA
2)
Br
OEt
Ph
1) NaOH, H_2O
2) H_3O^+, Δ

b) 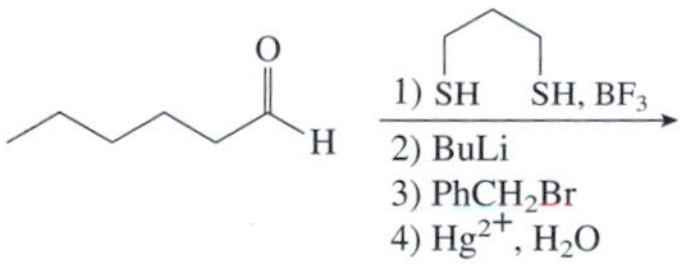
H
1) SH SH, BF_3
2) BuLi
3) $PhCH_2Br$
4) Hg^{2+}, H_2O
c) 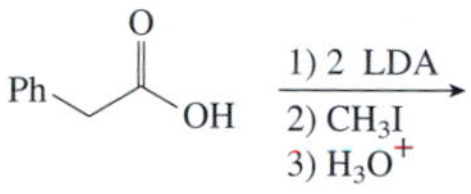
Ph
OH
1) 2 LDA
2) CH_3I
3) H_3O^+

20.25 a) 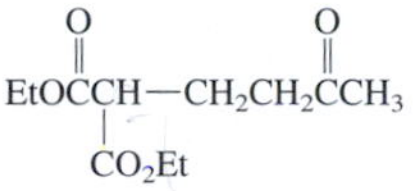
$EtOCCH-CH_2CH_2CCH_3$
CO_2Et
b) 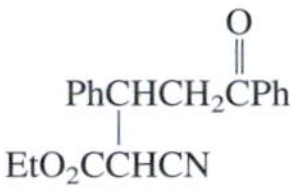
$PhCHCH_2CPh$
EtO_2CCHCN
c) 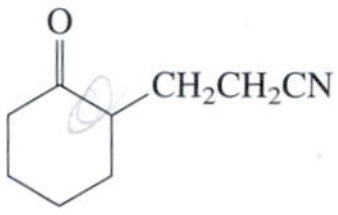
CH_2CH_2CN

20.26 The hydrogen on the α-carbon bonded to the phenyl group is more acidic because the resulting enolate anion is stabilized by resonance with the phenyl group.

20.27

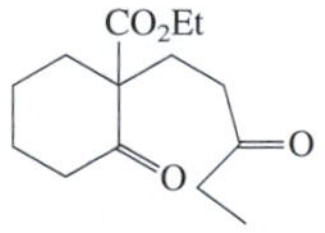
CO_2Et

20.28

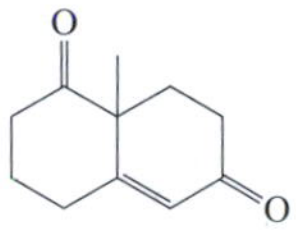

20.29

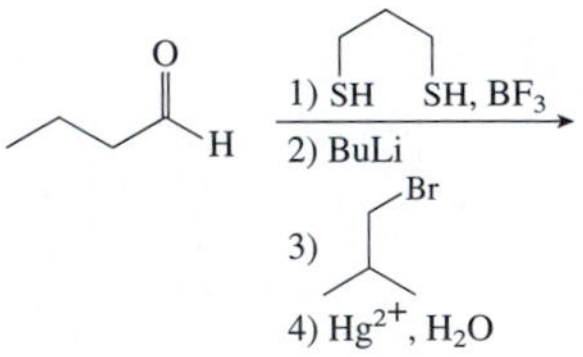

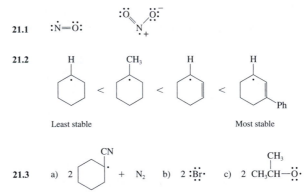

CHAPTER 21

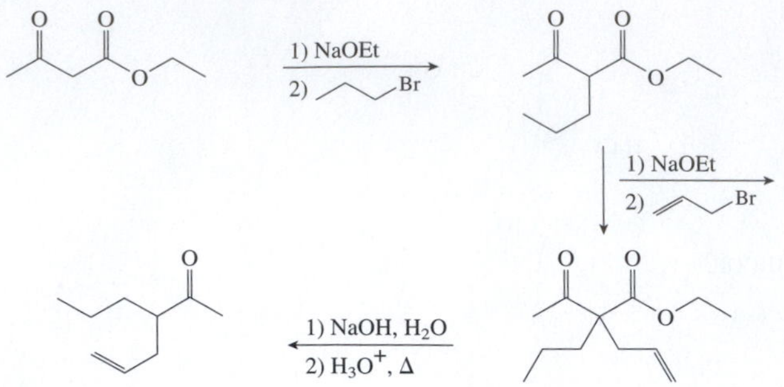

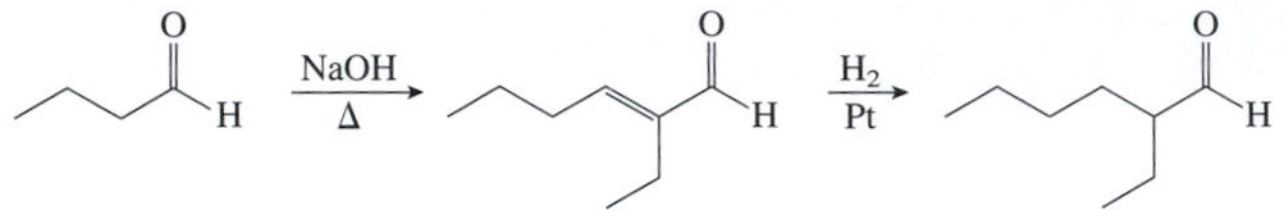

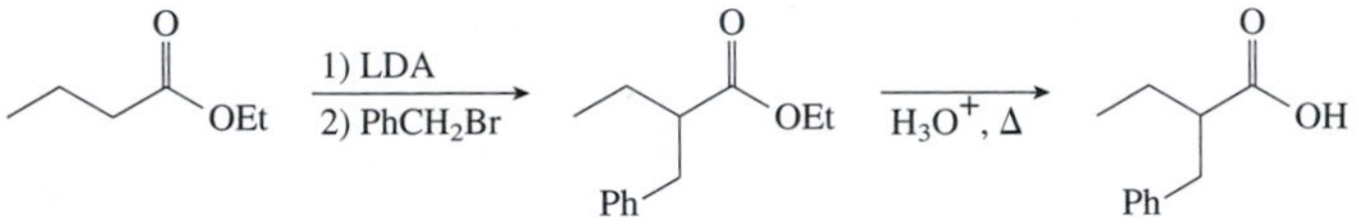

21.1   :N̈=Ö: :Ö̈ Ö:⁻
                          N̈
                          ⁺

21.2

Least stable   Most stable

21.3   a) 2 [cyclohexyl-CN radical] + N₂   b) 2 :B̈r·   c) 2 CH₃CH—Ö· with CH₃

21.4   a) Abstraction of a tertiary hydrogen occurs more readily than abstraction of a secondary hydrogen, so the second reaction is faster.

b) Formation of the resonance stabilized benzylic radical occurs more readily, so the first reaction is faster.

21.5 a) $CH_3\overset{\cdot}{C}HCH_2Br$ b) 2 $CH_3C\overset{\cdot\cdot}{\underset{\cdot\cdot}{O}}\cdot$ c) $CH_3(CH_2)_6CH_3$ + $CH_3CH_2CH_2CH_3$ + $CH_3CH_2CH=CH_2$

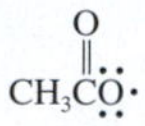

d) 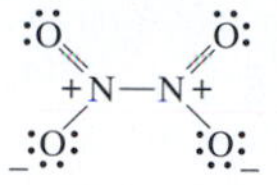 structure   e) 2 $CH_3\overset{CH_3}{\underset{CH_3}{\overset{|}{\underset{|}{C}}}}\cdot$ + N_2

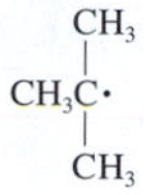

21.7 If the strength of a typical CH bond is taken as 98 kcal/mol (410 kJ/mol) (see Table 2.1), then the abstraction of the hydrogen atom by a fluorine atom is exothermic by 135 − 98 = 37 kcal/mol (155 kJ/mol), whereas a similar abstraction by an iodine atom is endothermic by 98 − 71 = 27 kcal/mol (113 kJ/mol).

21.8 a) 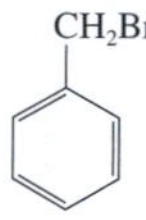 b) 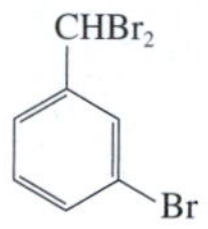 c) 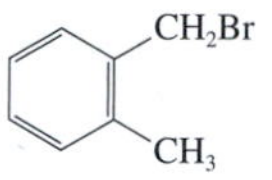 d)

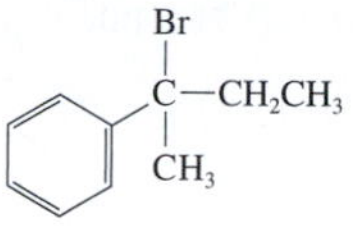

Racemic

e) 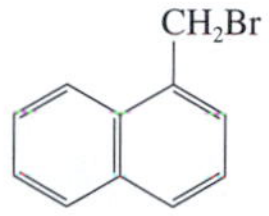 f) 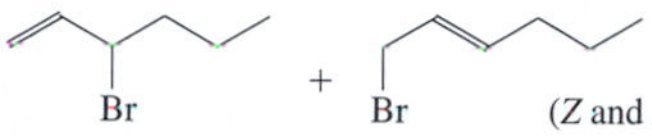 (Z and E)

21.9 The major product is 1-phenyl-3-bromopropene because it is more stable due to conjugation.

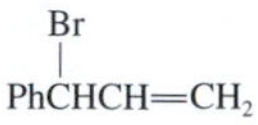

$Ph\overset{Br}{\overset{|}{C}}HCH=CH_2$

21.10 a)    b) $PhCH_2CH_3$

21.11 The radical is stabilized by resonance and its reactions are also slowed by steric hindrance.

21.12 a) 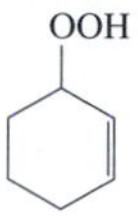 b)

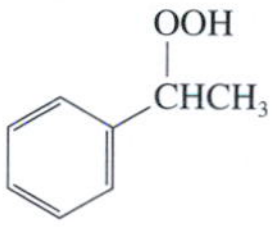

21.14 a) 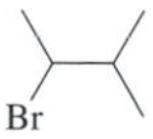 b) 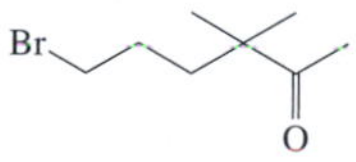 c) 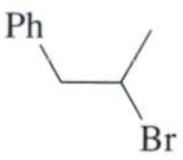 d)

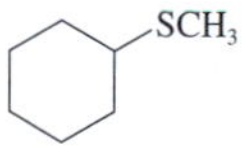

e) 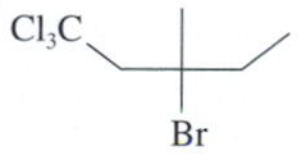 f)

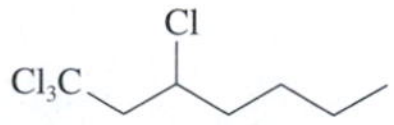

21.15 a) 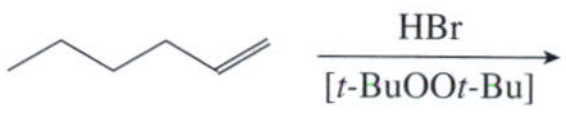 $\xrightarrow[\text{[}t\text{-BuOO}t\text{-Bu]}]{\text{HBr}}$ b) 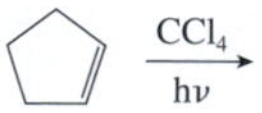 $\xrightarrow[h\nu]{\text{CCl}_4}$ c) 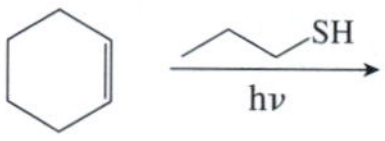 $\xrightarrow[h\nu]{\text{SH}}$

d) 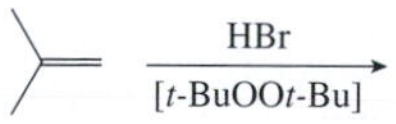 $\xrightarrow[\text{[}t\text{-BuOO}t\text{-Bu]}]{\text{HBr}}$ e) $PhCH=CH_2$ $\xrightarrow[\text{[}t\text{-BuOO}t\text{-Bu]}]{\text{HBr}}$

21.16 a) 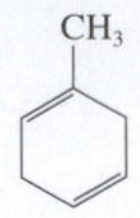 b) 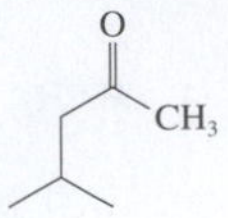 c) 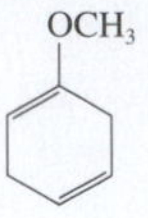 d) 

e) 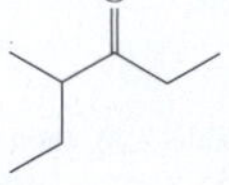 f) 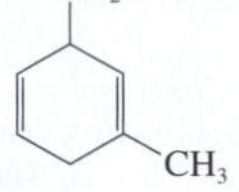 g)  h) 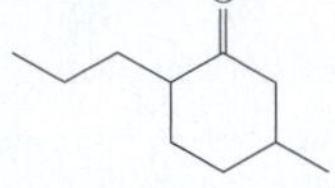

21.17 a) 1) Li, NH_3 (*l*); 2) $PhCH_2Br$ b) 1) Li, NH_3 (*l*); 2) H_3O^+

CHAPTER 22

22.1 a) Conrotation b) Disrotation

22.2

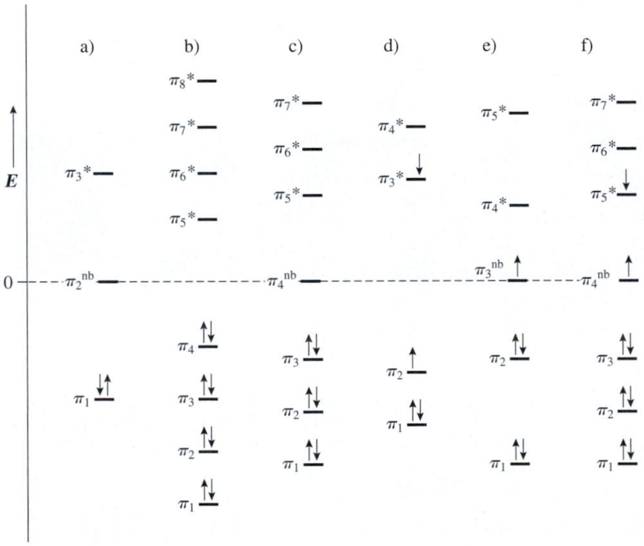

22.3 a) π_3^{nb} 2 nodes  b) π_2 1 node 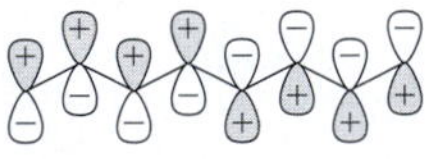

c) π_4^* 3 nodes 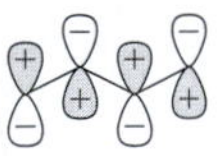 d) π_4^{nb} 3 nodes 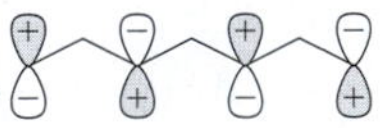 e) π_3 2 nodes 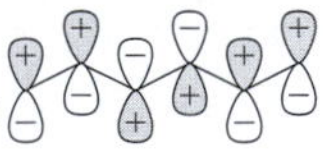

22.4 For the photochemical reaction the HOMO = π_3^*; conrotation is forbidden.

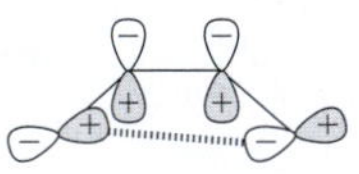

22.5 For the thermal reaction the HOMO = π_3; conrotation is forbidden.

For the photochemical reaction the HOMO = π_4*; disrotation is forbidden.

22.6 a) For the thermal reaction the HOMO = π_1; disrotation is allowed.

b) For the photochemical reaction the HOMO = π_4^{nb}; disrotation is forbidden.

22.7 a) For an odd number of electron pairs (1), disrotation is thermally allowed.

b) For an odd number of electron pairs (3), disrotation is photochemically forbidden.

22.8 a) 2 electron pairs; disrotation; photochemically allowed

b) 3 electron pairs; conrotation; thermally forbidden

c) 2 electron pairs; disrotation; thermally forbidden

d) 4 electron pairs; disrotation; photochemically allowed

22.9 a), b), c), d), e), f), g)

22.11 One overlap is bonding and the other is antibonding, so the reaction is forbidden.

HOMO of one diene π_2

LUMO of the other diene π_3*

22.12 a) A [4 + 4] cycloaddition involves 4 electron pairs and is photochemically allowed.

b) A [6 + 2] cycloaddition involves 4 electron pairs and is photochemically allowed and thermally forbidden.

22.13 The left diene is more reactive because its *s*-cis conformation is less hindered and is therefore present in higher concentration.

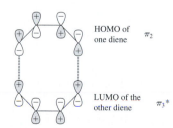

22.14   a) 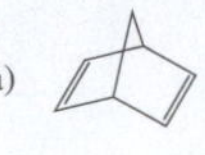   b) 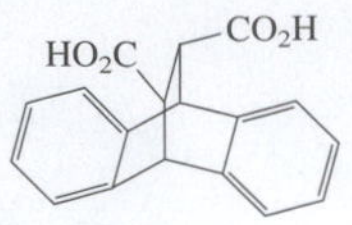   c) 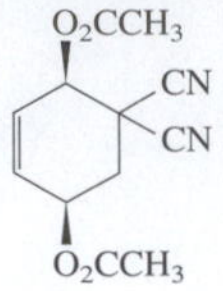   d)

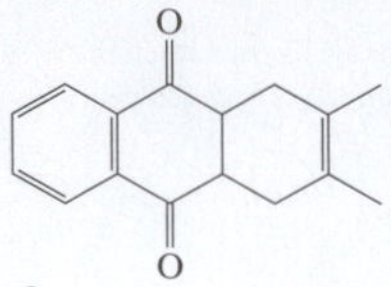

e) 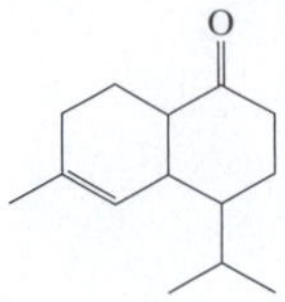   f) 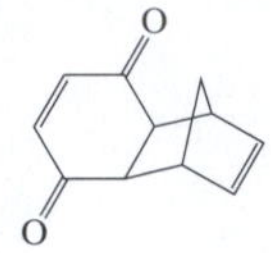   g)

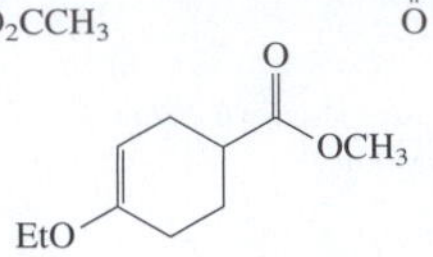

h) 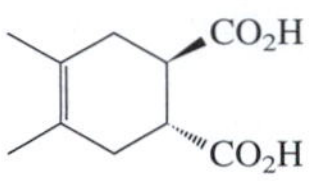   i) 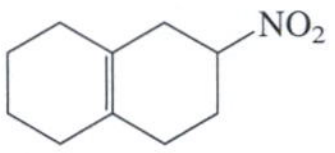   j) 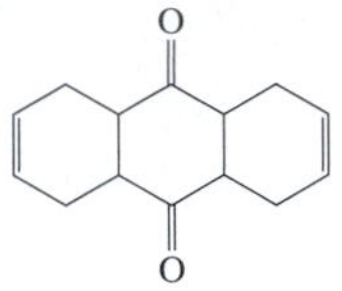   k)

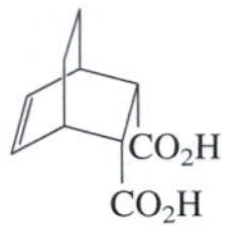

22.15   [10 + 2] cycloaddition; 6 electron pairs; photochemically allowed

22.16   a)    b) 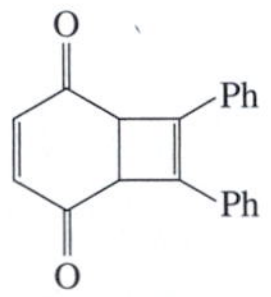   c) 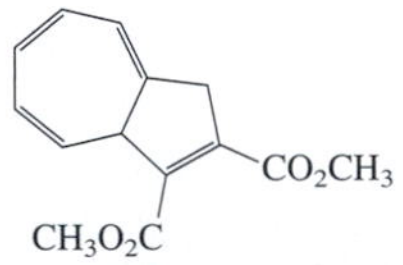   d)

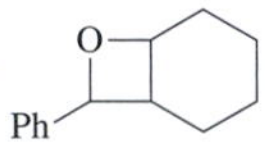

22.17   a) [3,3] Sigmatropic rearrangement   b) [1,7] Sigmatropic rearrangement
c) [3,5] Sigmatropic rearrangement

22.18   Use π_4* of the pentadienyl radical. One overlap is
bonding and one is antibonding, so the reaction is
forbidden.

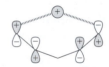

22.20   a) Thermally allowed   b) Photochemically allowed   c) Photochemically allowed

22.21   a) 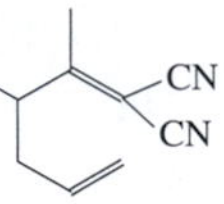   b) 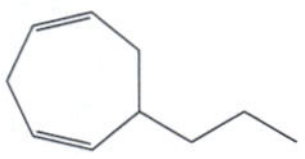   c)

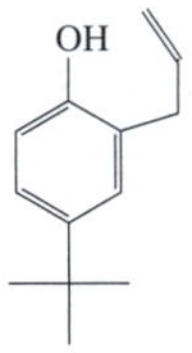

d) 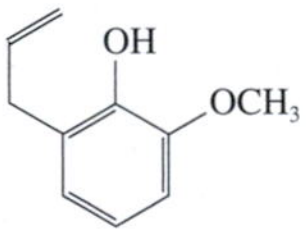   e)

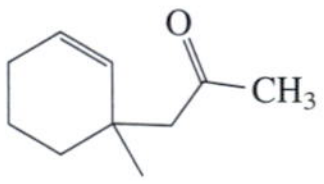

22.22   It is a [3,3] sigmatropic rearrangement and is thermally allowed. The reactant is favored at equilibrium
because it has less angle strain.

22.23   a) Et—C—C—Et   b) PhNHCPh   c) 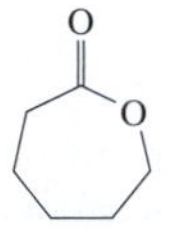   d) 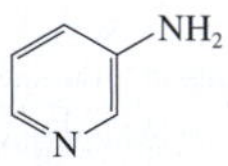
with Et and O substituents

CHAPTER 23

23.1 a) 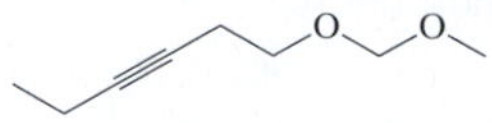   b) PhCH$_3$ +

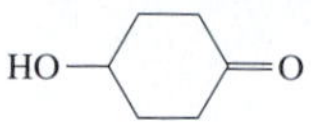

c) 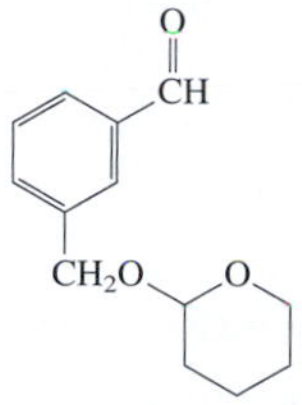   d) 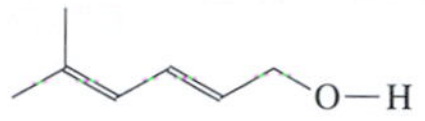   e)

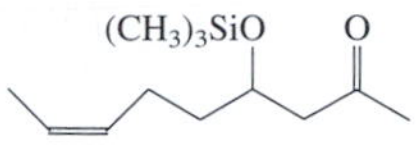

f)  Ph   g)

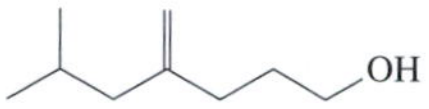

23.3

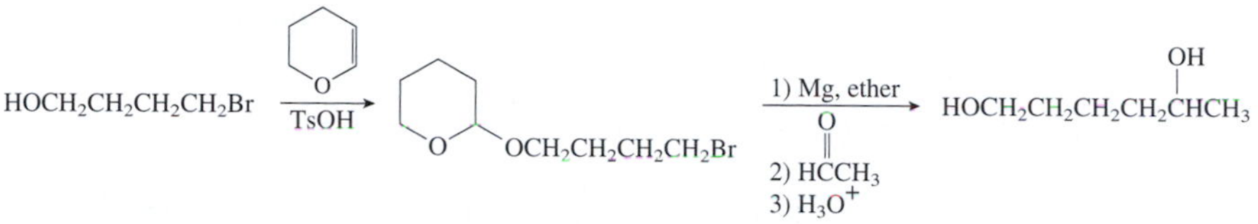

HOCH$_2$CH$_2$CH$_2$CH$_2$Br $\xrightarrow{\text{TsOH}}$ OCH$_2$CH$_2$CH$_2$CH$_2$Br $\xrightarrow[\substack{2)\ \text{HCCH}_3 \\ 3)\ \text{H}_3\text{O}^+}]{\substack{1)\ \text{Mg, ether} \\ \text{O}}}$ HOCH$_2$CH$_2$CH$_2$CH$_2$CHCH$_3$ (OH)

23.4 a) 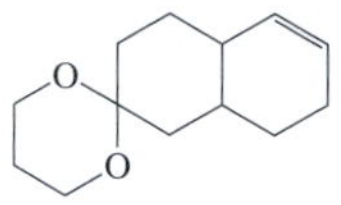   b)

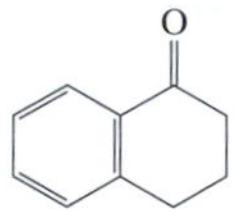

23.5 a) 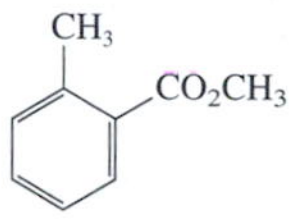   b)

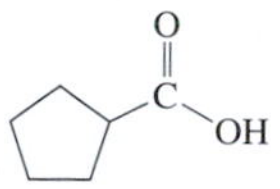

c) 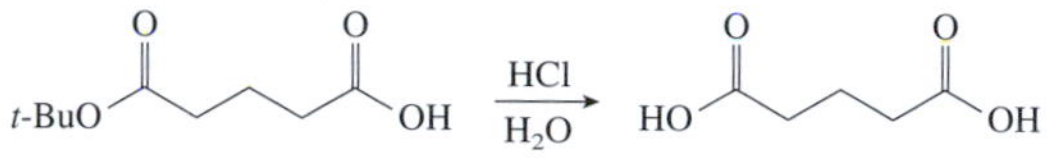 *t*-BuO OH $\xrightarrow[\text{H}_2\text{O}]{\text{HCl}}$ HO OH   d)

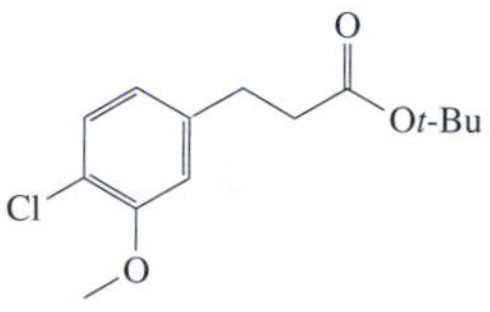

23.7 a) 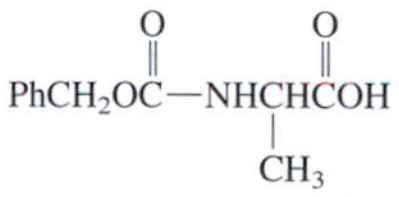 PhCH$_2$OC—NHCHCOH (O, O, CH$_3$)   b)  H—N O + CO$_2$ + (CH$_3$)$_3$COH

c) PhCH$_3$ + NH$_2$CH$_2$CH$_2$Ph + CO$_2$ d)  (CH$_3$)$_3$COCNHCH$_2$CO$_2$H (O)

23.8 *t*-BuOCNHCHCOH (O, O, CH$_3$)   *t*-BuOCNHCHCN(CH$_3$)$_2$ (O, O, CH$_3$)

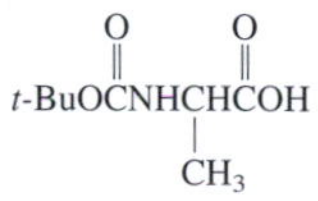

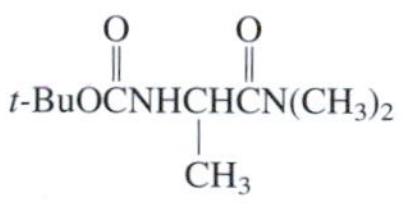

A   **B**

The amide cannot be prepared directly because the unprotected amino group will react with the acyl chloride group.

23.9 The ester bond of the carbamate group is hydrolyzed more rapidly than either amide bond. The resulting carbamic acid then eliminates carbon dioxide to produce the amine. Conditions drastic enough to hydrolyze an amide bond are never employed in this process.

23.10   a) 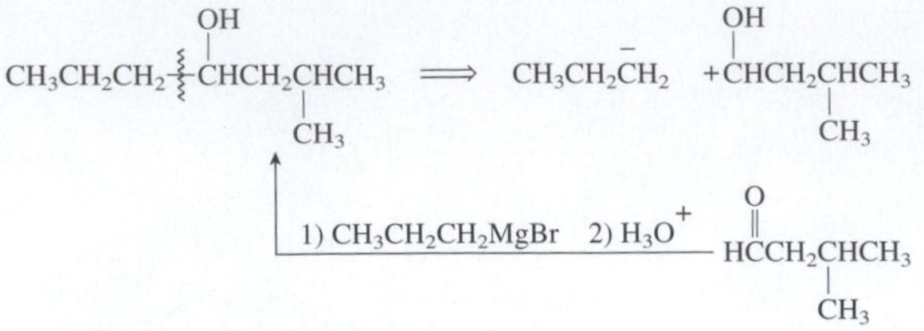 CH₃CH₂CH₂—CHCH₂CHCH₃ ⟹ CH₃CH₂CH₂⁻ + ⁻CHCH₂CHCH₃

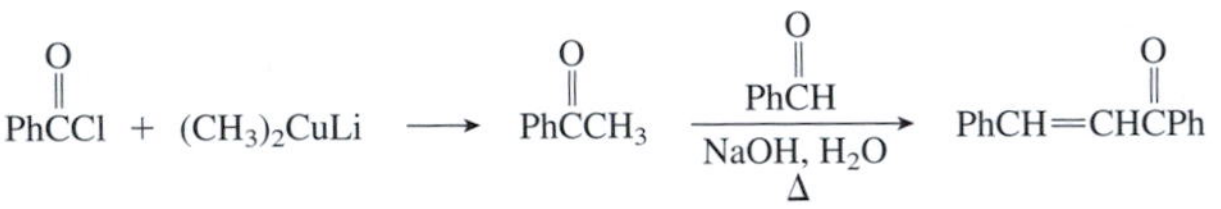

with OH and CH₃ substituents

1) CH₃CH₂CH₂MgBr   2) H₃O⁺

O‖HCCH₂CHCH₃ with CH₃

23.11   a) 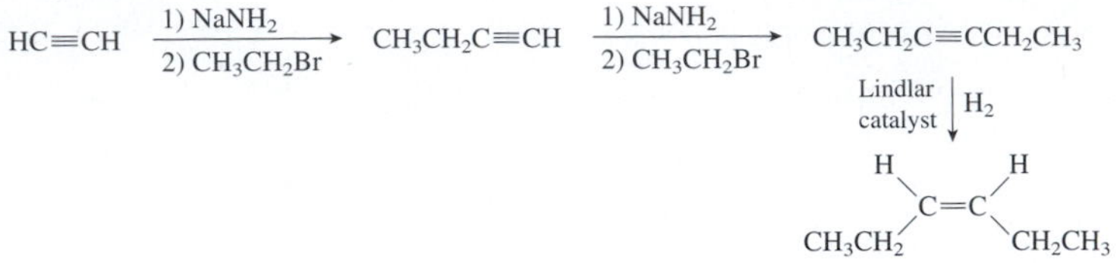 HC≡CH →(1) NaNH₂ 2) CH₃CH₂Br)→ CH₃CH₂C≡CH →(1) NaNH₂ 2) CH₃CH₂Br)→ CH₃CH₂C≡CCH₂CH₃

Lindlar catalyst | H₂

cis-CH₃CH₂, H / C=C / H, CH₂CH₃

b) PhCCl + (CH₃)₂CuLi ⟶ PhCCH₃ →(PhCH, NaOH, H₂O, Δ)→ PhCH=CHCPh

c) PhCH₂Br →(KCN / DMSO)→ PhCH₂CN →(H₃O⁺ / Δ)→ PhCH₂COH →(CH₃OH / H₂SO₄)→ PhCH₂COCH₃

1) 2 CH₃MgI
2) NH₄Cl, H₂O

OH
PhCH₂CCH₃
CH₃

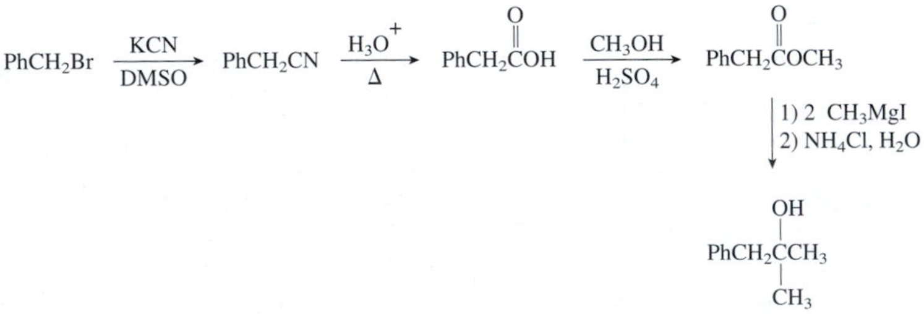

CHAPTER 24

24.2   a) 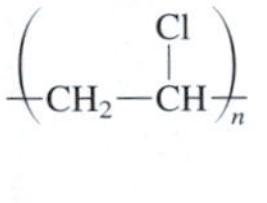 —(CH₂—CH)—ₙ with Cl   b) 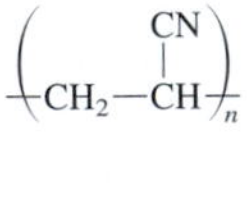 —(CH₂—CH)—ₙ with CN   c) 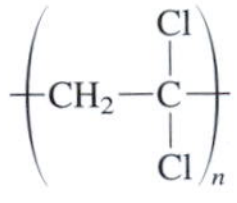 —(CH₂—C)— with Cl, Cl ₙ

24.3   a) 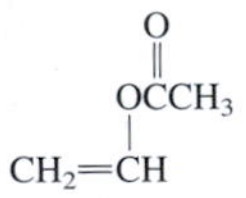 CH₂=CH with OCCH₃ (O)   b) 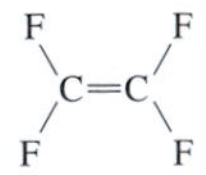 F, F / C=C / F, F

24.5   Both (a) and (b) can form atactic polymers, whereas (c) cannot.

24.6   The primary carbocation that is required as an intermediate in the cationic polymerization of ethylene is too unstable.

24.7   The anionic intermediate formed in the polymerization of acrylonitrile is stabilized by resonance and is readily formed, whereas the anionic intermediate formed in the polymerization of isobutylene is quite unstable and is difficult to form.

24.8   THF is much less reactive toward nucleophiles than is ethylene oxide because no ring strain is relieved when THF reacts.

24.9   a) The alkene on the right produces a more crystalline polymer because it has no chirality centers and is more stereoregular.

        b) The alkene on the right produces a more crystalline polymer because its side chain is more polar.

        c) Coordination polymerization produces a less branched and more stereoregular polymer that is more crystalline.

        d) The alkene on the right produces a more crystalline polymer because it has no chirality centers and is more stereoregular.

24.10   Teflon has no chirality centers, so it has no stereochemical complications; nor does it have any hydrogens that can be abstracted from the interior of the chain, so it is linear.

24.11   Poly(methyl methacrylate) cannot be prepared by cationic polymerization because the carbonyl group destabilizes the intermediate carbocation. It can be produced by anionic polymerization because the carbonyl group stabilizes the carbanion intermediate by resonance.

24.12   Radicals prefer to add to C-1 of isoprene because the resulting radical is the most stable of the four possibilities. It is stabilized by resonance, and the odd electron is on a tertiary carbon in one resonance structure and a primary carbon in the other. In contrast, addition at C-2 or C-3 produces less stable radicals because they have no resonance stabilization. Addition at C-4 produces a resonance stabilized radical, but the odd electron is on a secondary carbon and a primary carbon in the two resonance structures.

24.13

Trans double bonds predominate because they are more stable than cis double bonds.

24.14

24.15

24.16   If the polyester is to be formed from more diol units, then each polymer chain must terminate with a diol unit at each end. To accommodate even a small excess of diol units, there must be many chains, so the chains must be relatively short.

24.17   Direct preparation of poly(vinyl alcohol) would require an enol as the monomer. However, recall from Section 11.6 that most enols cannot be isolated because they spontaneously isomerize to the carbonyl tautomer.

CHAPTER 25

25.1

25.2   Eight aldopentoses; four 2-ketopentoses

25.3   a) 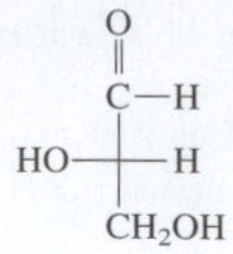   b)

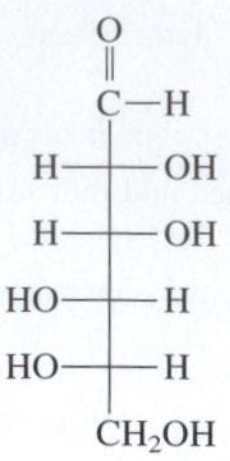

25.4   a) L-Erythrose   b) L-Gulose   c) D-Altrose

25.6

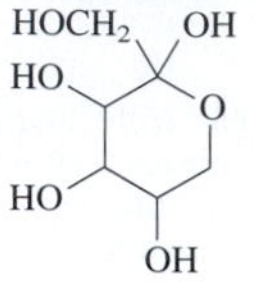

Pyranose 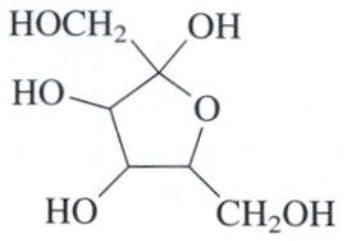   Furanose

D-Fructose has more of its uncyclized form present at equilibrium than does D-glucose because its ketone carbonyl group is less reactive than the aldehyde carbonyl group of glucose.

25.7   Identical to D-glucose

25.8

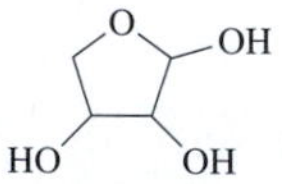

25.9

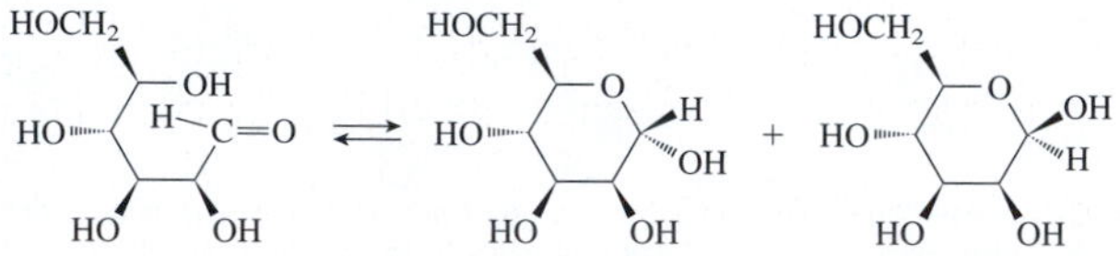

α-D-Mannopyranose   β-D-Mannopyranose

25.10

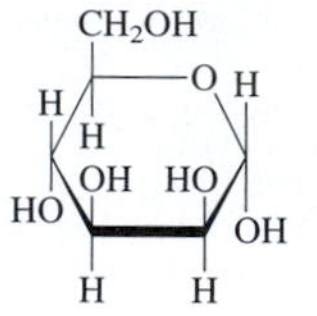

α-D-Mannopyranose 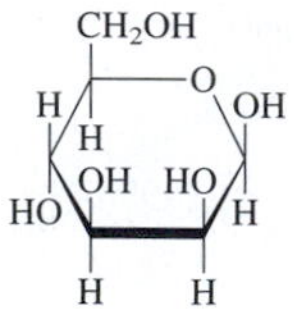   β-D-Mannopyranose

25.11

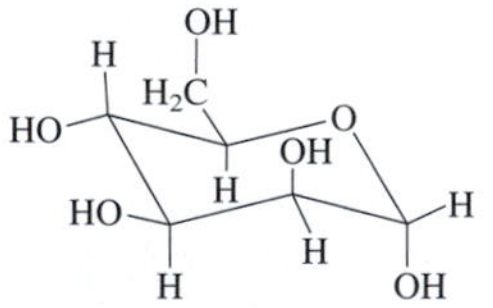

α-D-Mannopyranose 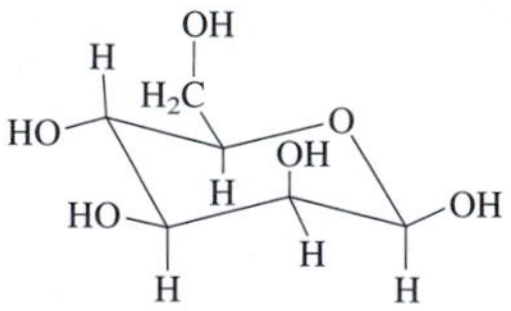   β-D-Mannopyranose

25.12   67.4% α, 32.6% β

25.13   Galactaric acid is a meso compound.

25.14   D-Ribose and D-xylose

25.15   Xylitol is a meso compound.

25.16   D-Allose and D-galactose

25.17 Under these reaction conditions, α-D-glucopyranose undergoes isomerization to the β-isomer.

25.20

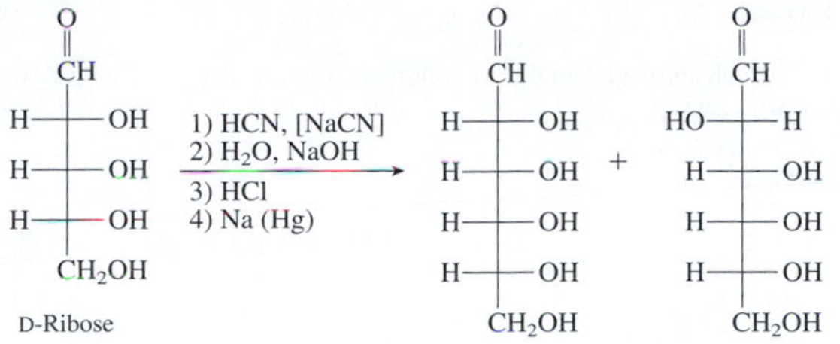

D-Ribose D-Allose D-Altrose

25.21 a)

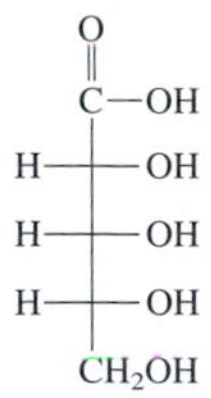

b)

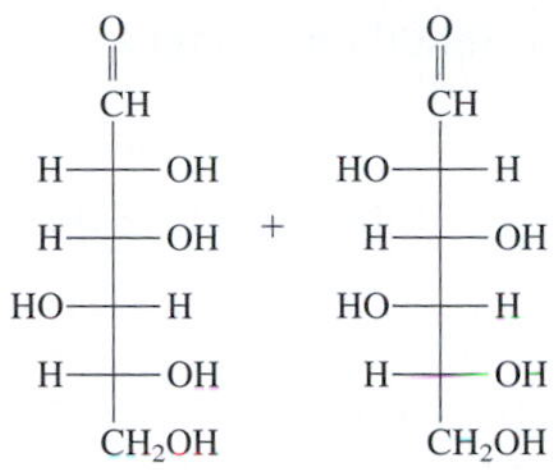

c)

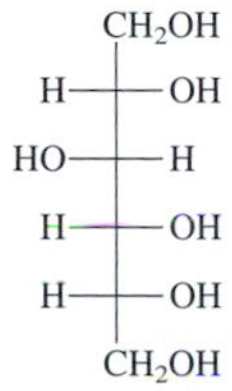

d)

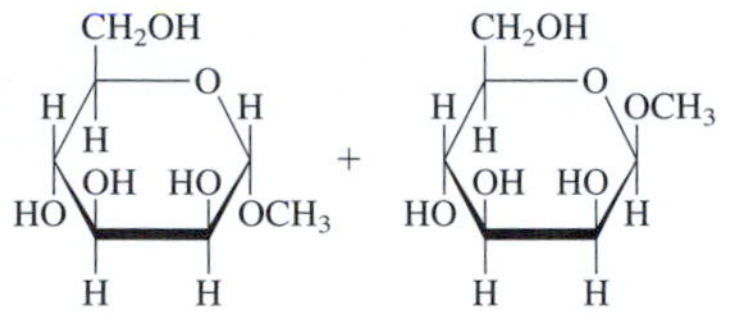

e)

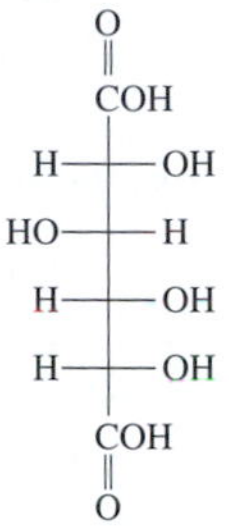

f)

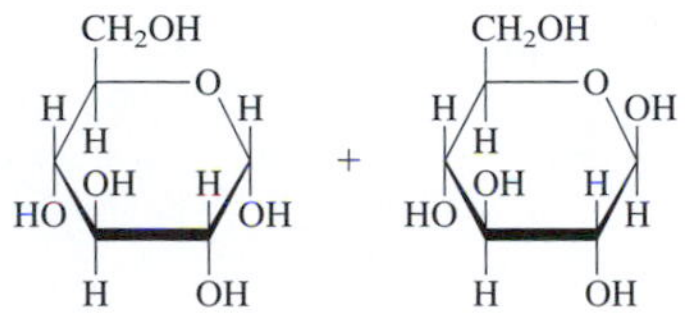

25.24 Maltose and cellobiose both have a hemiacetal group that is in equilibrium with an aldehyde group in aqueous solution. It is the aldehyde group that gives a positive test for a reducing sugar. Sucrose does not have a hemiacetal group, so there is no aldehyde group present at equilibrium in a solution of sucrose.

25.25

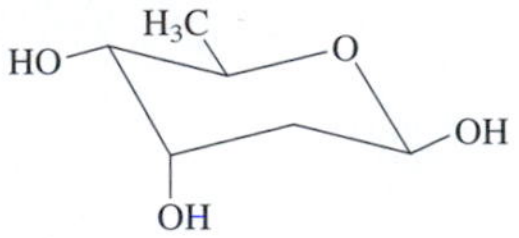

There are no hydroxy groups present on C-2 and C-6.

CHAPTER 26

26.1 Isoleucine and threonine

26.2 The positively charged nitrogen acts as an inductive electron-withdrawing group and therefore makes the carboxylic acid group a stronger acid.

26.3 The ester is the stronger acid because the inductive electron-withdrawing effect of the ester group increases the acid strength of the ammonium group. The carboxylate anion is an electron-donating group and decreases the acid strength of the ammonium group in the amino acid.

26.4 The inductive electron-withdrawing effect of the positively charged nitrogen, which increases acid strength, is stronger at the main carboxylic acid group because the distance separating the groups is smaller. Inductive effects decrease rapidly with distance.

26.5 The carboxylic acid group in the side chain of glutamic acid is farther from the electron-withdrawing positive nitrogen than is the case with aspartic acid.

26.6

a) a pyrrolidine ring with $\overset{+}{N}H_2$ and CO_2^- substituents

b) $^-S-CH_2\overset{\overset{\displaystyle +}{N}H_3}{\underset{}{C}}H-CO_2^-$

c) $H_2N=\overset{\overset{\displaystyle NH_2}{|}}{C}NHCH_2CH_2CH_2\overset{\overset{\displaystyle +}{N}H_3}{\underset{}{C}}HCO_2^-$

26.7

a) $H_3\overset{+}{N}CH_2CH_2CH_2CH_2\overset{\overset{\displaystyle \overset{+}{N}H_3}{|}}{C}HCO_2H$; $H_3\overset{+}{N}CH_2CH_2CH_2CH_2\overset{\overset{\displaystyle \overset{+}{N}H_3}{|}}{C}HCO_2^-$

$H_3\overset{+}{N}CH_2CH_2CH_2CH_2\overset{\overset{\displaystyle NH_2}{|}}{C}HCO_2^-$; $H_2NCH_2CH_2CH_2CH_2\overset{\overset{\displaystyle NH_2}{|}}{C}HCO_2^-$

26.8

a) $\overset{Cl^-}{\underset{}{}}\ \overset{CH_3}{\underset{}{}}\ \overset{O}{\underset{}{}}$ $\overset{+}{N}H_3CH-COCH_2Ph$

b) $\overset{CH_3CHCH_3}{\underset{}{}}$ $\overset{O}{\underset{}{}}\ \overset{CH_2}{\underset{}{}}\ \overset{O}{\underset{}{}}$ $PhC-NHCH-COH$

c) $\overset{O}{\underset{}{}}\ \overset{CH_3}{\underset{}{}}\ \overset{O}{\underset{}{}}$ $CH_3C-NHCH-COH$

26.10

a) $PhCH_2\overset{\overset{\displaystyle NH_2}{|}}{C}HCN \xrightarrow[H_2O]{HCl} PhCH_2\overset{\overset{\displaystyle NH_2}{|}}{C}HCO_2H$

b) $CH_3\overset{\overset{\displaystyle CH_3}{|}}{C}H\overset{\overset{\displaystyle O}{\parallel}}{\underset{\underset{\displaystyle Br}{|}}{C}}HCOH \xrightarrow[NH_3]{excess} CH_3\overset{\overset{\displaystyle CH_3}{|}}{C}H\overset{\overset{\displaystyle O}{\parallel}}{\underset{\underset{\displaystyle NH_2}{|}}{C}}HCOH$

c) phthalimide$-N-\overset{\overset{\displaystyle CO_2Et}{|}}{\underset{\underset{\displaystyle CO_2Et}{|}}{C}}-CH_2\overset{\overset{\displaystyle CH_3}{|}}{C}HCH_3 \xrightarrow[\Delta]{\overset{HCl}{H_2O}} CH_3\overset{\overset{\displaystyle CH_3}{|}}{C}HCH_2\overset{\overset{\displaystyle NH_2}{|}}{C}HCO_2H$

26.11

a) $CH_3\overset{\overset{\displaystyle CH_3}{|}}{C}HCH_2\overset{\overset{\displaystyle O}{\parallel}}{C}H \xrightarrow[\substack{NH_4Cl,\ H_2O \\ 2)\ HCl,\ H_2O}]{1)\ NaCN} CH_3\overset{\overset{\displaystyle CH_3}{|}}{C}HCH_2\overset{\overset{\displaystyle NH_2}{|}}{C}HCO_2H$

b) $PhCH_2CH_2-\overset{\overset{\displaystyle O}{\parallel}}{C}OH \xrightarrow[2)\ excess\ NH_3]{1)\ Br_2,\ [PBr_3]} PhCH_2\overset{\overset{\displaystyle NH_2}{|}}{C}H-\overset{\overset{\displaystyle O}{\parallel}}{C}OH$

26.13

$NH_2\overset{\overset{\displaystyle \overset{\displaystyle Ph}{|}}{\overset{\displaystyle CH_2}{|}}}{C}H-\overset{\overset{\displaystyle O}{\parallel}}{C}-NH\overset{\overset{\displaystyle \overset{\displaystyle CH_3}{|}}{\overset{\displaystyle CH_3CH}{}}}{C}H-\overset{\overset{\displaystyle O}{\parallel}}{C}-NH\overset{\overset{\displaystyle \overset{\displaystyle CO_2H}{|}}{\overset{\displaystyle CH_2}{|}}}{C}H-\overset{\overset{\displaystyle O}{\parallel}}{C}OH$

26.14 Leu-Cys-Tyr-Glu

26.15 The amide groups in the side chains of asparagine and glutamine residues in a peptide are cleaved under the same conditions that hydrolyze the amide bonds of the peptide. Therefore, aspartic acid and glutamic acid are isolated instead.

26.17 a) 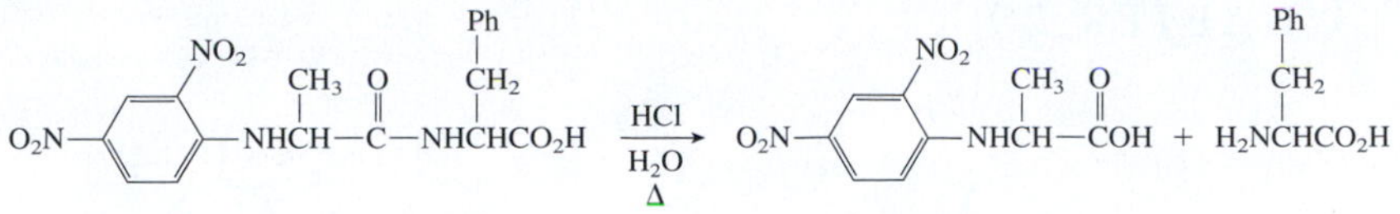 O_2N —(ring, NO_2)— $NHCH(CH_3)$ —$C(O)$— $NHCHCO_2H$ (CH$_2$Ph) $\xrightarrow[\Delta]{HCl\ H_2O}$ O_2N —(ring, NO_2)— $NHCH(CH_3)COH$(O) + H_2NCHCO_2H(CH$_2$Ph)

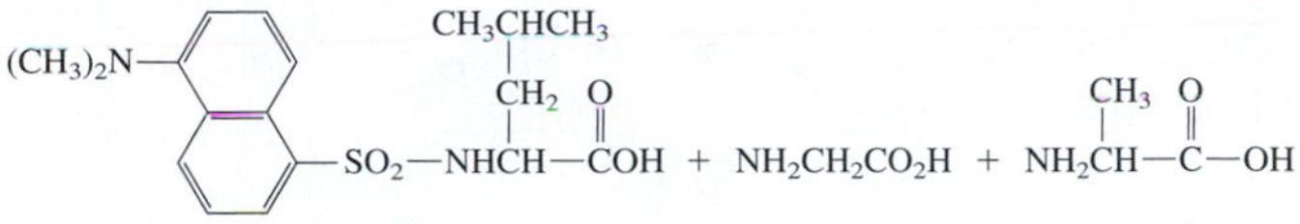

b) $(CH_3)_2N$—(naphthalene)—SO_2—$NHCH(CH_2CH(CH_3)CH_3)COH$(O) + $NH_2CH_2CO_2H$ + $NH_2CH(CH_3)C(O)OH$

c) (thiohydantoin ring: S=, N–H, N–Ph, =O, CH_2CHCH_3/CH_3) + $NH_2CH_2C(O)NHCH(CH_3)CO_2H$

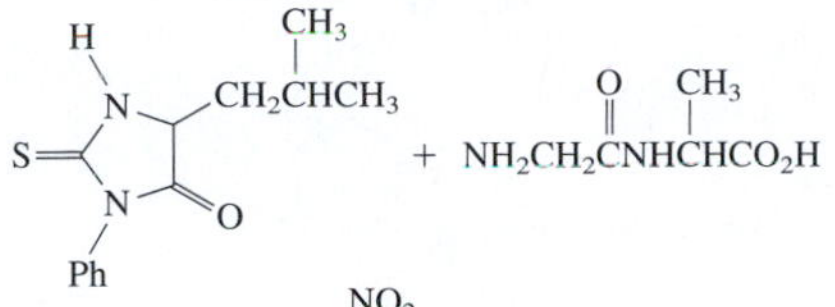

d) 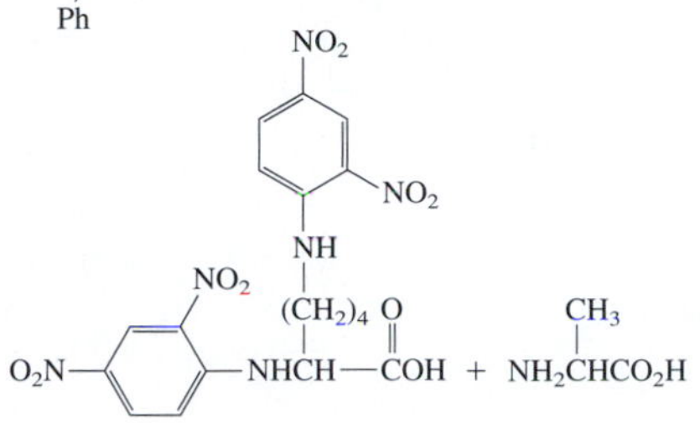 O_2N—(ring, NO_2)—$NHCH$—COH(O), side chain $(CH_2)_4$–NH–(ring $2NO_2$) + $NH_2CH(CH_3)CO_2H$

26.18 Phe-Phe-Ala or Phe-Ala-Phe

26.21 a)  t-BuOC(O)NHCH(CH$_3$)CO$_2$H

b) 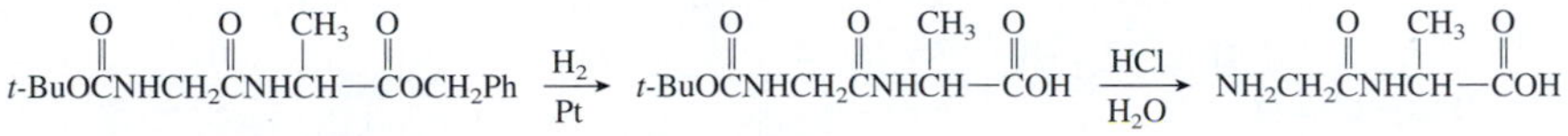 t-BuOC(O)NHCH$_2$C(O)NHCH(CH$_3$)C(O)OCH$_2$Ph $\xrightarrow{H_2/Pt}$ t-BuOC(O)NHCH$_2$C(O)NHCH(CH$_3$)C(O)OH $\xrightarrow{HCl/H_2O}$ $NH_2CH_2C(O)NHCH(CH_3)C(O)OH$

c) t-BuOC(O)NHCH(CH$_3$)C(O)NHCH$_2$C(O)OCH$_3$ $\xrightarrow{HCl/H_2O}$ $NH_2CH(CH_3)C(O)NHCH_2C(O)OCH_3$ $\xrightarrow{NaOH/H_2O}$ $NH_2CH(CH_3)C(O)NHCH_2C(O)OH$

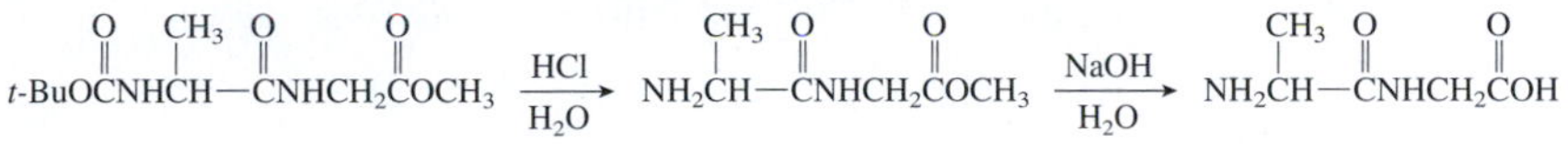

d) 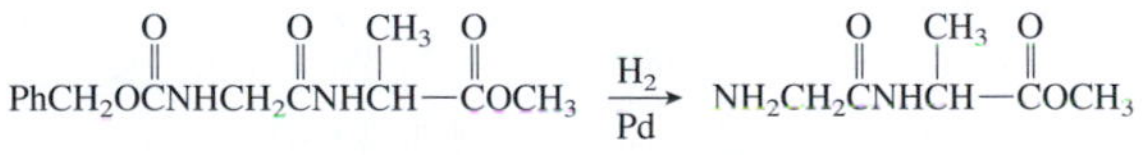 PhCH$_2$OC(O)NHCH$_2$C(O)NHCH(CH$_3$)C(O)OCH$_3$ $\xrightarrow{H_2/Pd}$ $NH_2CH_2C(O)NHCH(CH_3)C(O)OCH_3$

CHAPTER 27

27.1

a)

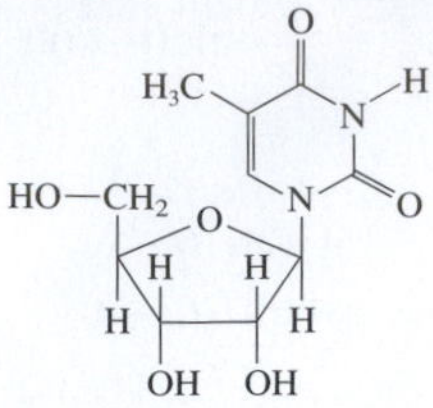

b)

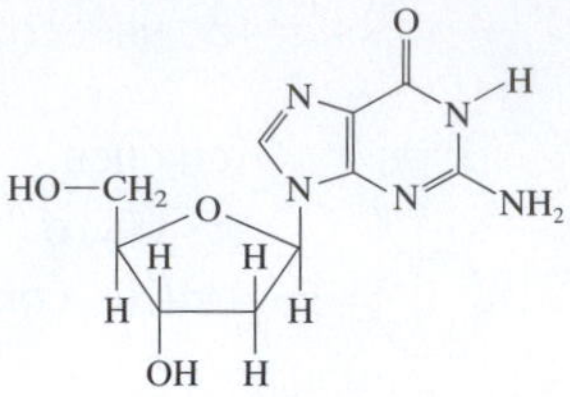

c)

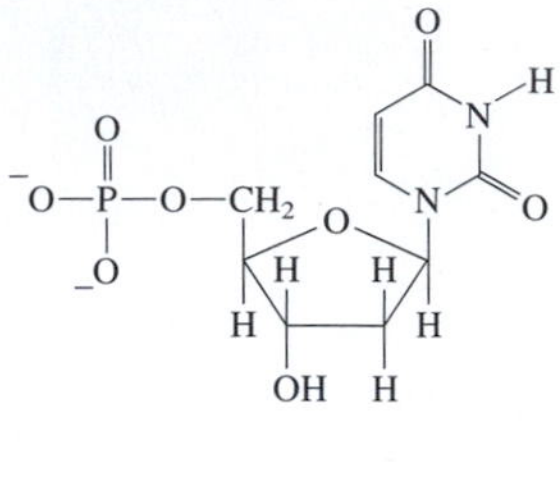

d)

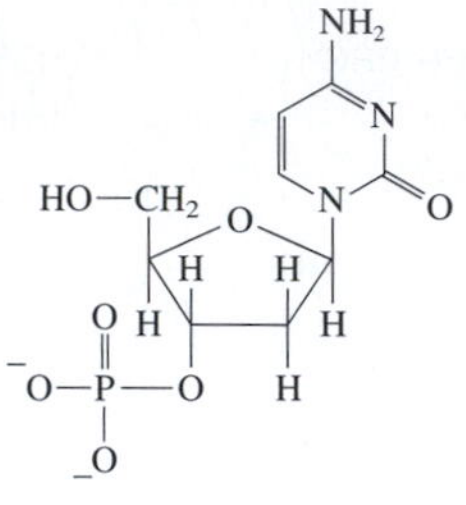

27.2   a) Guanosine   b) Uridine 5'-monophosphate

       c) Deoxycytidine 5'-monophosphate

27.3   Pyrimidine has six electrons in its cyclic pi system and therefore fits Huckel's rule. The unshared electrons on the nitrogens are not part of the pi system. The six-membered ring of purine is aromatic for the same reason as pyrimidine. The five-membered ring is also aromatic. Like imidazole (see Chapter 16), it has a total of six electrons in its cyclic pi system. The electrons on N-9 are part of the pi system, whereas the electrons on N-7 are not.

27.4   a)

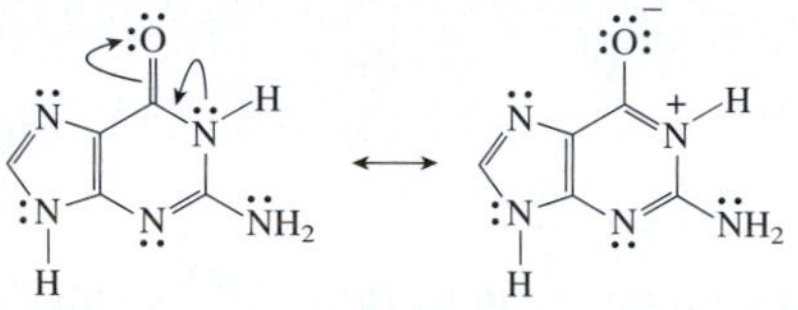

b)

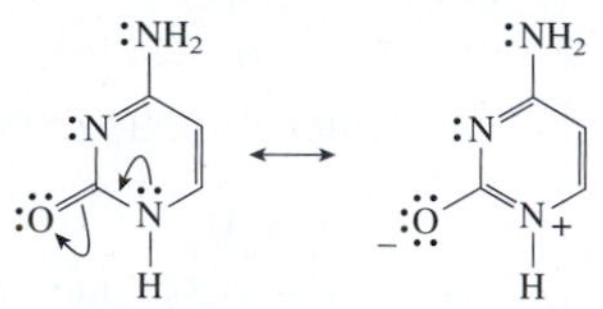

c)

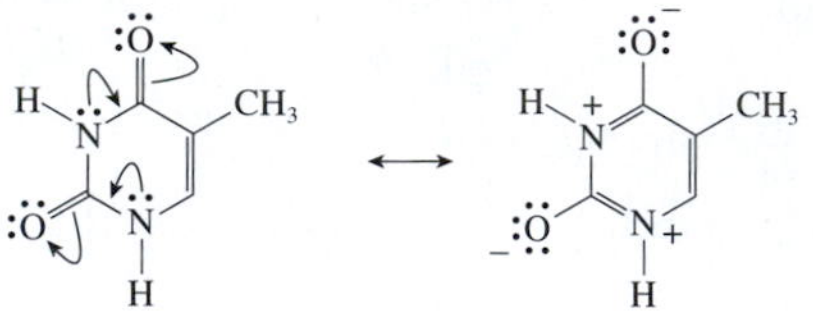

27.5   5' end TAACGCTCG 3' end

27.6

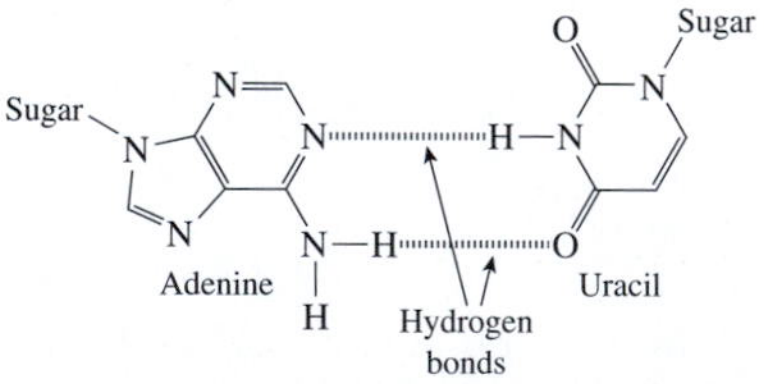

27.7

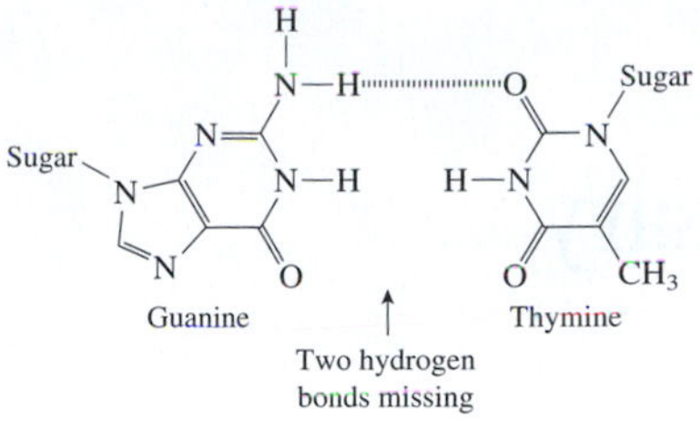

Guanine   Thymine

↑
Two hydrogen
bonds missing

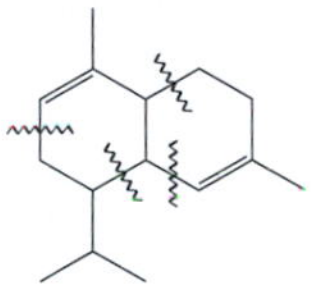

27.9

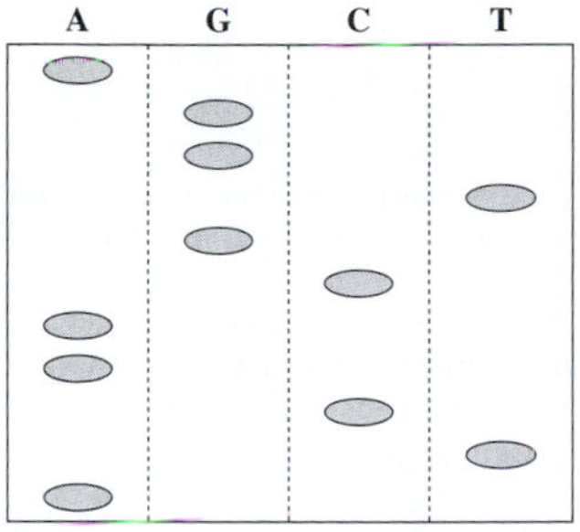

CHAPTER 28

28.1   a)

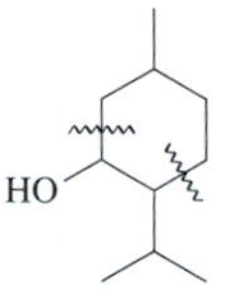

Monoterpene   Sesquiterpene

c) 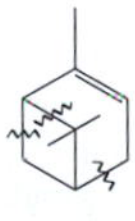   d)

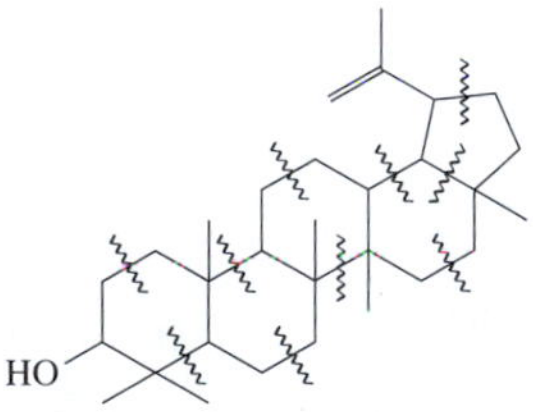

Monoterpene   Triterpene

28.4

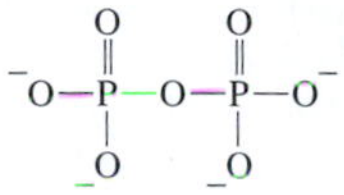

This is a good leaving group because it is a weak base.

28.7   The cyclization occurs so as to form the more stable tertiary carbocation.

28.8   Path A is disfavored because it produces a secondary carbocation. It is favored because the five-membered rings that are formed have lower ring strain. Path B is favored because it produces a tertiary carbocation. It is disfavored due to the strain in the four-membered ring of the product.

28.9   The cyclization occurs so as to form the more stable tertiary carbocation.

28.11 Both cyclizations occur so as to form the more stable tertiary carbocations.

28.16 The epoxide opens so as to form the more stable tertiary carbocation.

Glossary

Absolute configuration (Section 7.3): The actual three-dimensional arrangement of groups around a chirality center.

Acetal (Section 18.9): The product of the addition of two equivalents of an alcohol to an aldehyde or a ketone; $R_2C(OR')_2$.

Acetylide anion (Section 10.8): A carbon nucleophile generated by treating 1-alkynes with a very strong base, such as sodium amide; $RC{\equiv}C{:}^-$.

Achiral molecule (Section 7.1): A molecule that is superimposable on its mirror image.

Acidity constant (K_a) (Section 4.2): An equilibrium constant for the reaction of an acid with water as a base; used as a measure of the strength of an acid. For a general acid, HA, the equation for K_a is

$$K_a = \frac{[A^-][H_3O^+]}{[HA]}$$

Activation energy (Section 4.3): The energy required to surmount the energy barrier separating reactants and products.

Acyl group (Section 12.5): A carbonyl group with an attached alkyl group.

Addition polymer or **chain-growth polymer** (Section 24.1): A polymer formed by a chain mechanism, where one initiator molecule causes a large number of monomers to react to form one polymer molecule.

Addition reaction (Chapter 10): A reaction that results in the addition of two groups to opposite ends of a multiple bond.

Alkaloid (Section 28.7): A nitrogen-containing natural product that occurs primarily in higher plants and in some fungi, such as mushrooms.

Alkyl group (Section 2.7): The part of a compound that has carbons that are only singly bonded to other carbons and hydrogens.

Alkylation reaction (Section 20.3): A reaction that results in the attachment of an alkyl group to the reactant.

Allowed reaction (Section 22.2): A pericyclic reaction that is energetically favorable because the electrons in the occupied MOs do not increase in energy.

Allyl group (Section 5.5): The $CH_2{=}CHCH_2{-}$ group.

Alpha-carbon (α-carbon) (Section 12.3): The carbon adjacent to a functional group; most commonly used to refer to the carbon adjacent to a carbonyl group.

Amorphous solid (Section 24.5): A solid in which the individual molecules have a random arrangement; a glassy solid with no order in the arrangement of its molecules.

Amphoteric compound (Section 26.2): A compound that can act as either an acid or a base.

Angle strain (Section 6.4): Destabilization, usually found in compounds having three- or four-membered rings, that occurs when the orbitals of a bond do not point directly at each other, so the amount of overlap is decreased.

Annulene (Section 16.10): A name sometimes given to rings that contain alternating single and double bonds in a single Lewis structure.

Anomers (Section 25.3): Diastereomers that are formed by cyclization of a carbohydrate with differing configurations at the new stereocenter.

Anti addition (Section 10.2): The addition of groups to opposite faces of a double bond.

Anti conformation (Section 6.3): Conformation in which the dihedral angle between two groups on adjacent atoms is 180°.

Anti elimination (Section 9.3): Elimination of groups from a conformation in which the dihedral angle between them is 180°. This is the preferred geometry in the E2 reaction.

Antiaromatic compound (Section 16.5): A compound that is destabilized because of the presence of conjugated cycle of p orbitals containing $4n$ electrons.

Antibonding MO (Section 3.2): A MO that is higher in energy than the AOs that combine to form it.

Anti-periplanar bonds (Section 9.3): Bonds with a dihedral angle of 180°.

Aprotic solvent (Section 8.11): A solvent that does not have a hydrogen bonded to nitrogen or oxygen and therefore cannot hydrogen bond.

Arenium ion (Section 17.1): The carbocation formed by addition of an electrophile to a benzene derivative.

Arenium ion

Aromatic compound (Sections 16.1 and 16.5): A compound that is especially stable because of the presence of conjugated cycle of p orbitals containing $4n + 2$ electrons.

Atactic polymer (Section 24.2): A polymer with random configurations at its many stereocenters.

Atomic orbital (Section 3.1): The region about the nucleus of an atom where, if the orbital contains an electron, the probability of finding that electron is very high.

Axial bond (Section 6.5): In the chair conformation of cyclohexane, a bond that is parallel to the axis of the ring.

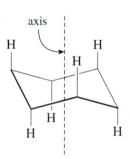

Axial CH bonds

Axial strain energy (Section 6.7): The amount of destabilization caused by a group in the axial position in the chair conformation of cyclohexane.

Base ion (Section 15.4): The most abundant ion in the mass spectrum.

Benzyl group (Section 12.1): The $PhCH_2{-}$ group.

Benzyne (Section 17.12): A highly reactive intermediate that has a benzene ring with a formal triple bond.

Benzyne

Boat conformation (Section 6.5): The boat-shaped conformation of cyclohexane that has no angle strain but does have some steric strain and some torsional strain.

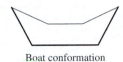

Boat conformation

Bond disconnections (Section 23.5): The imaginary process of breaking bonds at or near the functional groups in a compound during the process of retrosynthetic analysis.

Bond dissociation energy (Section 2.2): The amount of energy that must be added in the gas phase to break the bond in a homolytic manner.

Bonding MO (Section 3.2): A MO that is lower in energy than the AOs that combine to form it.

Bromonium ion (Section 11.4): A three-membered ring containing a positively charged bromine atom; an intermediate formed in the addition of bromine to an alkene.

Bronsted-Lowry acid (Section 4.1): A proton donor.

Bronsted-Lowry base (Section 4.1): A proton acceptor.

Cahn-Ingold-Prelog sequence rules (Section 6.2): Rules that are used to assign priorities to groups attached to a stereocenter so that the configuration of the compound can be designated.

Carbanion (Section 2.1): A carbon with three bonds, an unshared pair of electrons, and a negative charge.

Carbene (Section 11.8): A reactive species having a carbon with only two bonds and an unshared pair of electrons.

Carbenoid (Section 11.8): An organometallic species that reacts like carbene.

Carbocation (Sections 2.1 and 8.6): A carbon with three bonds and a positive charge.

Carbohydrate (Section 25.1): Naturally occurring compounds, often with the general formula $C_x(H_2O)_x$, that are polyhydroxy aldehydes or ketones or derivatives formed from these; includes sugars and starches.

Carbonyl group (Section 12.3): A CO double bond.

Carboxy group (Section 12.4): A carbonyl group with an attached hydroxy group.

Chain reaction (Section 21.6): A reaction in which a reactive intermediate, such as a radical, reacts with a normal molecule, ultimately generating a molecule of product and regenerating the original reactive intermediate, which then causes another reaction cycle to occur. A chain reaction involves three types of steps: initiation, propagation, and termination.

Chain-growth polymer. See *Addition polymer.*

Chair conformation (Section 6.5): The chair-shaped conformation of cyclohexane that has no angle strain and has no torsional strain either be-cause it is perfectly staggered about all the CC bonds. It is strain free.

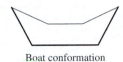

Chair conformation

Chemical shift (Sections 14.1 and 14.3): The position of an absorption on the *x*-axis in an NMR spectrum; provides information about the local environment of the atom that is responsible for the absorption.

Chirality center (Section 7.2): A carbon or other tetrahedral atom bonded to four different groups; a type of stereocenter.

Chiral molecule (Section 7.1): A molecule that is not superimposable on its mirror image.

Chromophore (Section 15.2): The part of a molecule that is responsible for the absorption of ultraviolet or visible light.

Cis–trans isomers (Section 6.1): Stereoisomers that differ in the placement of groups on one side or the other of a double bond.

Codon (Section 27.3): A series of three bases in a nucleic acid polymer that specifies a particular amino acid.

Concerted reaction (Section 8.3): A reaction that occurs in one step.

Condensation polymer or step growth polymer (Section 24.8): A polymer formed from monomers with two reactive functional groups by normal reactions, such as ester formation, between the functional groups.

Configuration (Section 6.2): The three-dimensional arrangement of groups about a stereocenter in a molecule.

Conformation (Section 6.3): A shape that a molecule can assume by rotation about a single bond.

Conformational analysis (Section 6.3): Analysis of the energies of the various conformations of a compound.

Conjugate acid (Section 4.1): The acid formed by protonation of a base in an acid–base reaction.

Conjugate addition or 1,4-addition (Section 18.10): The addition of a nucleophile to the β-carbon of an α,β-unsaturated carbonyl compound.

Conjugate base (Section 4.1): The base formed by the loss of a proton from an acid in an acid–base reaction.

Conjugated molecule (Section 3.6): A molecule that has a series of three or more overlapping parallel *p* orbitals on adjacent atoms.

Connectivity (Section 1.7): The arrangement of bonded atoms in a structure.

Conrotation (Section 22.1): Rotation of orbitals in the same direction in an electrocyclic reaction.

Constitutional isomers (Section 1.7): Compounds that have the same molecular formula but a different arrangement (connectivity) of bonded atoms.

Coupling constant (J_{ax}) (Section 14.4): The separation between two adjacent peaks in a group of peaks that results from coupling in an NMR spectrum.

Covalent bonding (Section 1.4): Bonding that results from atoms sharing electrons in order to arrive at the same number of electrons as a noble gas.

Cross-link (Section 24.7): A group that connects separate polymer chains by covalent bonds.

Crystalline solid (Section 24.5): A solid in which the individual molecules are arranged with a very high degree of order.

Cycloaddition reaction (Section 22.5): A pericyclic reaction in which two molecules react to form two new sigma bonds between the end atoms of their pi systems, resulting in the formation of a ring.

Degenerate orbitals (Section 3.1): Orbitals with the same energy.

Degree of unsaturation (DU) (Section 2.4): The total number of multiple bonds plus rings in a compound. The DU is calculated by subtracting the actual number of hydrogens in a compound from the maximum number of hydrogens and dividing the result by 2.

Delocalized MO (Sections 3.2 and 3.6): A MO that extends around more than two atoms.

DEPT-NMR (distortionless enhancement by polarization transfer) (Section 14.9): A technique used in ¹³C-NMR that allows the number of hydrogens attached to each carbon to be determined.

Dextrorotatory, (*d*) or (+) (Section 7.4): Clockwise rotation of plane-polarized light.

Diastereomers (Section 7.5): Non–mirror-image stereoisomers.

1,3-Diaxial interaction (Section 6.7): An interaction that destabilizes two axial groups on the same face of a cyclohexane ring because of steric crowding between them.

Dienophile (Section 22.6): The species, most often an alkene or alkyne, that acts as the two-electron component in a Diels-Alder reaction.

Dihedral angle (Section 6.3): The angle between a marker group on the front atom and one on the back atom in the Newman projection.

Dipolar ion or zwitterion (Section 26.2): A neutral compound containing a covalently linked cation and anion.

Dipole moment (μ) (Section 1.9): The product of the amount of charge separation in a molecule times the distance of the charge separation.

Disaccharide (Section 25.6): A carbohydrate formed from two monosaccharide units that are connected by a glycosidic bond.

Disrotation (Section 22.1): Rotation of orbitals in opposite directions in an electrocyclic reaction.

Downfield (Section 14.3): A chemical shift at a higher delta value in the NMR spectrum.

E (Section 6.2): Letter used to designate the isomer of an alkene that has the high priority groups on opposite sides of the double bond.

E1 reaction or **unimolecular elimination reaction** (Section 9.5): An elimination reaction that occurs in two steps through a carbocation intermediate.

E2 reaction or **bimolecular elimination reaction** (Section 9.2): An elimination reaction that follows a concerted mechanism in which a base removes a proton simultaneously with the departure of the leaving group.

Eclipsed conformation (Section 6.3): Conformation in which the bonds on one atom are directly in line with the bonds on the adjacent atom when a Newman projection is viewed.

Eclipsed conformation

Electrocyclic reaction (Section 22.3): A pericyclic reaction that forms a sigma bond between the end atoms of a series of conjugated pi bonds within a molecule.

Electromagnetic spectrum (Section 13.3): The range of electromagnetic radiation (light), from very energetic cosmic rays to low-energy radio waves.

Electronegativity (Section 1.9): Electron-attracting ability of an atom.

Electronic transition (Section 15.1): The excitation of an electron from an occupied MO to an antibonding MO.

Electrophile (Section 8.1): An electron-poor species that seeks an electron-rich site; similar to a Lewis acid.

Electrophilic addition reaction (Section 11.1): A reaction that results in the addition of two groups, an electrophile and a nucleophile, to the carbons of a CC double or triple bond.

Electrophilic aromatic substitution reaction (Section 17.1): A reaction in which an electrophile is substituted for a hydrogen on an aromatic ring.

Elimination reaction (Section 8.14, Chapter 9): A reaction in which groups (most commonly a proton and a leaving group) are lost from adjacent atoms, resulting in the formation of a double bond.

1,1-Elimination (Section 11.8): An elimination of two groups from the same carbon to produce a carbene; also called an α-elimination.

1,2-Elimination (Section 9.1): An elimination of two groups from adjacent atoms to produce a pi bond; also called a β-elimination.

Enantiomers (Section 7.1): Nonsuperimposable mirror-image stereoisomers.

Endergonic reaction (Section 8.3): A reaction for which $\Delta G°$ is positive; the products are higher in energy than the reactants and the reaction favors the reactants at equilibrium.

Endothermic reaction (Section 8.3): A reaction for which the enthalpy change ($\Delta H°$) is positive.

Enol (Section 11.6): A compound with a hydroxy group attached to a CC double bond.

Enolate ion (Chapter 20): A carbanion adjacent to a carbonyl group.

Enthalpy change (ΔH) (Section 2.6): The heat of reaction; $\Delta H > 0$ for an endothermic reaction and $\Delta H < 0$ for an exothermic reaction.

Entropy change (ΔS) (Section 2.6): Entropy is a measure of disorder. Processes that increase the disorder in a system ($\Delta S > 0$) are favored.

Epoxide (oxirane) (Section 10.10): A three-membered cyclic ether.

Equatorial bond (Section 6.5): In the chair conformation of cyclohexane, a bond that projects outward from the "equator" of the ring.

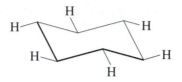

Equatorial CH bonds

Equilibrium arrows (Section 1.8): Two arrows that are used to represent an equilibrium reaction where the reactants and products are interconverting. ⇌

Excited state (Section 3.1): Any state that is higher in energy than the ground state.

Exergonic reaction (Section 8.3): A reaction for which $\Delta G°$ is negative; the products are lower in energy than the reactants and the reaction favors the products at equilibrium.

Exothermic reaction (Section 8.3): A reaction for which the enthalpy change ($\Delta H°$) is negative.

Fat (Section 28.8): A triester formed from glycerol and long, linear-chain carboxylic acids, known as fatty acids.

Fingerprint region (Section 13.9): The region below 1500 cm^{-1} in the infrared spectrum that contains numerous absorptions due to single bond stretches and a variety of bending vibrations. Because of the large number of bands, this is the region where comparison of the spectrum of an unknown to that of a known compound can establish the identity of the unknown.

Fischer projection (Section 7.8): A two-dimensional drawing of a chiral molecule in which the chirality center is represented as a cross with the atom at its center. Although the four bonds to the chirality center are shown in the plane of the page, the horizontal bonds project above the plane of the page and the vertical bonds project behind the page.

Forbidden reaction (Section 22.2): A pericyclic reaction that is energetically unfavorable because the electrons in the occupied MOs increase in energy.

Formal charge (Section 1.7): The approximate charge that is present on an atom in a covalent structure. It is the number of valence electrons in the neutral atom before any bonding (this is the same as the group number of the atom), minus the number of unshared electrons on the atom in the structure of interest, and also minus one-half the number of shared electrons on that atom.

Free energy of activation, $\Delta G^{\ddagger}$ (Sections 4.4 and 8.3): The energy difference between the transition state and the reactants.

Free energy change ($\Delta G°$) (Section 4.2): The difference in energy between the products and reactants. It is related to the enthalpy change, the entropy change, and the equilibrium constant for a reaction by the equations $\Delta G° = \Delta H° - T\Delta S° = -RT \ln K$. For a reaction to be spontaneous, K must be greater than 1 and $\Delta G°$ must be negative.

Frequency (ν) (Section 13.1): The number of wave cycles that pass a point in a second.

Frontier orbitals (Section 22.2): The highest occupied molecular orbital (HOMO) and the lowest unoccupied molecular orbital (LUMO).

Functional group (Section 2.7): The group of atoms and bonds that determines the chemical behavior of a compound.

Furanose (Section 25.3): A cyclic form of a carbohydrate that has a five-membered ring.

Gauche conformation (Section 6.3): Conformation in which the dihedral angle between two groups on adjacent atoms is 60°.

Geminal hydrogens (Section 14.4): Hydrogens bonded to the same carbon.

Glycoside (Section 25.4): An acetal formed by the reaction of a carbohydrate hemiacetal with an alcohol in the presence of acid.

Grignard reagent (organomagnesium halide) (Section 18.5): An organometallic species with a carbon–magnesium bond; RMgX.

Ground state (Section 3.1): The lowest energy state.

Halogens (Section 1.5): The elements of group 7A of the periodic table: F, Cl, Br, and I.

Hammond postulate (Section 8.6): The structure of the transition state for a reaction step is more similar to that of the species (reactant or product of that step) to which it is closer in energy.

Heteroatom (Section 12.5): An atom other than C and H.

Heterolytic bond cleavage (Section 21.1): A bond-breaking process in which both electrons of that bond remain with one of the atoms; ions are produced.

Highest occupied MO (HOMO) (Sections 15.1 and 22.2): The highest-energy orbital that is occupied by electrons.

Hofmann's rule (Section 9.4): The major product of an elimination reaction has fewer alkyl groups bonded to the carbons of the double bond (the less highly substituted product). The Hofmann elimination follows this rule.

Homolytic bond cleavage (Section 21.1): A bond-breaking process in which one electron of the bond remains with each of the atoms; radicals are produced.

Hückel's rule (Section 16.5): A cyclic, fully conjugated, planar molecule with $4n + 2$ pi electrons (n = any integer) is aromatic.

Hund's rule (Section 3.1): Electrons first occupy degenerate orbitals singly, with the same (parallel) spins until they all contain one electron.

Hydride ion (Sections 8.14 and 10.7): A hydrogen with a pair of electrons and a negative charge; H:⁻.

Hydrocarbon (Sections 2.3 and 5.1): A compound made up of only carbon and hydrogen.

Hydrogen bond (Section 2.5): A relatively strong attraction of a hydrogen bonded to an electronegative atom (O, N, or F) to another electronegative atom (O, N, or F).

Hydrogenolysis (Section 23.1): A reaction in which a single bond is broken by hydrogen and a catalyst.

Hydrolysis reaction (Section 10.2): A reaction in which water is both the nucleophile and solvent.

Hydrophilic compound (Section 2.6): A compound that has a favorable interaction with water because of its polar nature.

Hydrophobic compound (Section 2.6): A compound that has an unfavorable interaction with water because of its nonpolar nature.

Hyperconjugation (Section 8.7): A stabilizing interaction of a sigma bonding MO with an empty p orbital on an adjacent atom.

Inductive effect (Section 4.5): The effect of a dipole in a molecule on a reaction elsewhere in that molecule.

Infrared (IR) spectroscopy (Section 13.4): A type of spectroscopy that employs infrared light. IR spectroscopy uses transitions between vibrational energy levels to provide information about the functional groups that are present in a compound.

Integral (Section 14.1): The area under a group of peaks in a ¹H-NMR spectrum; proportional to the number of hydrogens that produce that group of peaks.

Intermolecular reaction (Section 8.13): A reaction that involves two separate molecules.

Intramolecular reaction (Section 8.13): A reaction in which both reacting groups are part of the same molecule.

Inversion of configuration (Section 8.4): The stereochemical result of a reaction in which the product has the opposite relative configuration to the reactant.

Ionic bonding (Section 1.3): Bonding that results when atoms gain or lose electrons to form ions with the same number of electrons as one of the noble gases.

Isoelectric point (pI) (Section 26.2): The pH at which an amino acid has an overall charge of zero.

Isotactic polymer (Section 24.4): A polymer with identical configuration at all its stereocenters.

Leaving group (Section 8.1): The group that is replaced in a substitution reaction.

Levorotatory, (l) or (−) (Section 7.4): Counterclockwise rotation of plane-polarized light.

Lewis acid (Section 4.1): An electron pair acceptor.

Lewis base (Section 4.1): An electron pair donor.

Lewis structures (Section 1.5): Structures that show all the electrons in the valence shells of the atoms as dots.

London force (Section 2.5): The attraction between nonpolar molecules that results from the interaction of an instantaneous dipole with an induced dipole.

Lowest unoccupied MO (LUMO) (Sections 15.1 and 22.2): The lowest-energy empty MO.

Macromolecule (Section 24.1): A very large molecule. This term is often applied to polymer molecules, which are composed of a very large number of monomers (hundreds or even thousands) and therefore are extremely large.

Markovnikov's rule (Section 11.2): In addition reactions of HX to alkenes, the H bonds to the carbon with more hydrogens (fewer alkyl substituents) and the X bonds to the carbon with fewer hydrogens (more alkyl substituents). A more modern version of Markovnikov's rule, based on mechanistic reasoning, is that the electrophile adds so as to form the more stable carbocation.

Mass spectrometry (Section 15.4): An instrumental technique used to measure the molecular mass of a compound; provides information about the structure of the compound from the masses of fragments that are produced from the compound.

meso-Stereoisomer (Section 7.5): A compound that contains chirality centers but is not chiral because it has the same chirality centers, but with opposite configurations, placed symmetrically in the compound, so it has an internal plane of symmetry.

Methylene (Section 11.8): The simplest carbene (carbon with only two bonds and an unshared pair of electrons); CH_2.

Methylene group (Section 5.3): The CH_2 group.

Molecular ion (M⁺⋅) (Section 15.4): A radical cation produced in a mass spectrometer that has the same mass as the original molecule but has one less electron.

Molecular orbital (MO) (Section 3.2): An electron orbital that extends around more than one atom.

Monomer (Section 24.1): One of the individual units that are bonded together to form a polymer.

Monosaccharide (Section 25.1): A carbohydrate formed from one sugar unit.

Multiplicity (Sections 14.1 and 14.4): The number of peaks in each absorption in a ¹H-NMR spectrum; provides information about the other hydrogens that are near the hydrogen(s) that produces the peaks.

Newman projection (Section 6.3): An end-on view of a bond so that one atom of the bond is directly behind the other. By convention, the front atom is represented by the intersection of its bonds and the back atom is represented by a circle.

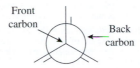

Newman projection

Node (Section 3.1): A region of an orbital where the value of the wave function equals zero, and therefore the electron density is zero.

Nonbonding AO (Section 3.3): An atomic orbital that is not involved in bonding.

Nonpolar bond (Section 1.9): A bond between atoms of similar electronegativities in which the electrons are shared nearly equally.

Nuclear magnetic resonance (NMR) spectroscopy (Section 14.1): A type of spectroscopy that uses transitions between the energy states of certain nuclei when they are in a magnetic field to supply information about the hydrocarbon part of a compound. There are two NMR techniques that are of most use to organic chemists: proton magnetic resonance (¹H-NMR) spectroscopy, which provides information about the hydrogens in a compound, and carbon-13 magnetic resonance spectroscopy (¹³C-NMR), which provides information about the carbons in a compound.

Nucleophile (Section 8.1): An electron-rich species that seeks an electron-poor site; similar to a Lewis base.

Nucleoside (Section 27.1): A component of a nucleic acid containing a sugar, either ribose or 2′-deoxyribose, and a base, a cyclic nitrogen-containing compound.

Nucleotide (Section 27.1): A nucleoside that has a phosphate group attached to the sugar.

Octet rule (Section 1.5): Atoms of the second period of the periodic table need eight electrons in their outer shell to be most stable.

Optically active (Section 7.4): A compound that rotates the plane of polarization of plane-polarized light.

Organometallic compound (Section 18.5): A compound with a covalent carbon–metal bond. Because the metal is less electronegative than carbon, the bond is polarized in the opposite direction to that found in most organic compounds; that is, the negative end of the dipole is on the carbon and the positive end is on the metal. These often make useful carbon nucleophiles.

Oxidation reaction (Section 10.14): A reaction that results in an increase in oxygen content of the compound and/or a decrease in hydrogen content.

Pauli exclusion principle (Section 3.1): No two electrons can have all four quantum numbers the same.

Peptide (Section 26.5): A term used for a polymer of amino acids that is smaller than a protein.

Percarboxylic acid (Section 11.9): A carboxylic acid that has an extra oxygen and an oxygen–oxygen bond; RCO_3H.

Pericyclic reaction (Section 22.1): A concerted reaction that proceeds through a cyclic transition state in which two or more bonds are made and/or broken.

Phenyl group (Section 12.1): A benzene ring group; Ph—.

Pheromone (Section 23.5): A compound used by animals, especially insects, for communication.

Photolysis reaction (Section 22.1): A reaction caused by light.

Pi (π) bond (Section 3.4): A bond formed by overlap of parallel *p* orbitals.

Plane of symmetry (Section 7.2): An imaginary plane that passes through the center of an object so that one-half of the object is the mirror image of the other half; also called a mirror plane.

Plane-polarized light (Section 7.4): A light beam in which the electromagnetic field of all the waves oscillate in a single plane.

Polar bond (Section 1.9): A bond with charge separation because the atoms involved have different electronegativities.

Polarimeter (Section 7.4): An instrument used to determine the amount by which a compound rotates the plane of polarization of plane-polarized light.

Polymer (Chapter 24): A very large molecule made up of repeating units.

Polysaccharide (Section 25.7): A polymeric carbohydrate formed from monosaccharides that are linked by glycosidic bonds.

Primary carbon (Section 5.3): A carbon that is bonded to one other carbon.

Primary natural products (Chapter 28): Naturally occurring compounds that are found in all types of organisms and are the products of primary metabolism; includes carbohydrates, amino acids, proteins, and nucleic acids.

Product development control (Section 9.4): More of the lower-energy product being formed in a reaction because the stability of the products results in a somewhat more stable transition state.

Prostaglandins (Section 28.9): A group of naturally occurring carboxylic acids that are related to the fatty acids and have been found to be involved in a number of important physiological functions, including the inflammatory response, the production of pain and fever, the regulation of blood pressure, the induction of blood clotting, and the induction of labor.

Protecting group (Sections 18.9 and 23.1): A group that is used to protect one functional group in a complex compound from reacting while a reaction is occurring at another functional group in the molecule.

Protein (Section 26.5): A term usually reserved for a naturally occurring polymer that contains a relatively large number of amino acid units and has a molecular mass in the range of a few thousand or larger.

Protic solvent (Section 8.11): A solvent that has a hydrogen bonded to nitrogen or oxygen. Water, alcohols, and carboxylic acids are examples of protic solvents.

Pyranose (Section 25.3): A cyclic form of a carbohydrate that has a six-membered ring.

Quaternary carbon (Section 5.3): A carbon that is bonded to four carbons.

***R* configuration** (Section 7.3): Term used to designate the configuration of a carbon chirality center. When viewed so that the bond from the chirality center to the group with the lowest priority is pointed directly away from the viewer, the direction of the cycle of the remaining groups, proceeding in decreasing order of priority, is clockwise.

Racemic mixture (Section 7.4): An equal mixture of enantiomers. Also called a racemate.

Racemization (Section 8.4): The stereochemical result of a reaction in which complete randomization of stereochemistry has occurred in the product (50% inversion and 50% retention).

Radical (Section 2.1 and Chapter 21): A species with an odd number of electrons.

Radical anion (Section 21.10): A species that has both an odd number of electrons and a negative charge.

Radical cation (Section 15.4): A species with both an odd number of electrons and a positive charge.

Rate-limiting step or **rate-determining step** (Section 8.6): The step in a mechanism that has the highest energy transition state and therefore determines the rate of the reaction.

Reaction mechanism (Sections 4.3 and 8.2): The individual steps in a reaction that show how the nuclei and the electrons move, how the bonds change as the reaction proceeds, and the order in which the bonds are made and broken.

Reactive intermediate (Section 8.6): A high-energy, reactive species, such as a carbocation, that is formed along a reaction pathway. Under most conditions it has a very short lifetime.

Reduction reaction (Section 10.14): A reaction that results in a decrease in oxygen content of the compound and/or an increase in hydrogen content.

Regiochemistry (Section 9.4): The result of a reaction that can produce two or more structural isomers.

Regioselective reaction (Section 11.2): A reaction that produces predominantly one possible orientation (regiochemistry) in a reaction but does form some of the product with the other orientation.

Regiospecific reaction (Section 11.2): A reaction that produces only one of two possible orientations (regiochemistries) in a reaction.

Relative configuration (Section 7.3): The configuration of a chiral center relative to that of a chiral center in another molecule.

Resolution (Section 7.7): The process of separating the enantiomers of a racemic mixture.

Resonance arrow (Sections 1.8 and 3.6): The double-headed arrow (⟷) used to show that structures are resonance structures.

Resonance hybrid (Section 1.8): The actual structure of a compound that is represented by two or more resonance structures. It is a blend of the extremes represented by the various resonance structures.

Resonance stabilization energy (Sections 1.8 and 3.7): The extra stabilization of a compound because of resonance.

Resonance structures (Section 3.7): Structures for a compound that differ only in the position of multiple bonds and unshared electrons.

Retention of configuration (Section 8.4): The stereochemical result of a reaction in which the product has the same relative configuration as the reactant.

Retrosynthetic analysis (Section 10.15): The process of designing a synthesis by working backward from the target compound.

Retrosynthetic arrow (Section 11.14): An arrow used in the development of a synthetic scheme that points from the target to the reactant from which it can be prepared.

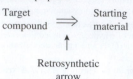

Ring-flip (Section 6.5): The conversion of one chair conformation of cyclohexane to another in a process that changes all axial bonds to equatorial and vice versa.

***S* configuration** (Section 7.3): Term used to designate the configuration of a carbon chirality center. When viewed so that the bond from the chirality center to the group with the lowest priority is pointed directly away from the viewer, the direction of the cycle of the remaining groups, proceeding in decreasing order of priority, is counterclockwise.

Saponification (Section 19.5): The hydrolysis of an ester under basic conditions.

Saturated compound (Section 5.5): A compound that does not contain multiple bonds.

Secondary carbon (Section 5.3): A carbon that is bonded to two carbons.

Secondary natural products (Chapter 28): Natural products that are usually produced from primary natural product precursors, such as amino acids or acetate ion, and, in general, are less widespread in occurrence than primary natural products.

Sigma (σ) bond (Section 3.2): A bond formed by overlap of orbitals that point directly toward each other so that its MO is symmetric about the internuclear axis.

Sigmatropic rearrangement (Section 22.8): The intramolecular migration of a group along a conjugated pi system.

S$_N$1 reaction or **unimolecular nucleophilic substitution reaction** (Section 8.6): A reaction in which the nucleophile replaces the leaving group at an *sp*^{3}-hybridized carbon in a two-step mechanism that proceeds through a carbocation intermediate.

S$_N$2 reaction or **bimolecular nucleophilic substitution reaction** (Section 8.3): A reaction in which the nucleophile replaces the leaving group at an *sp*^{3}-hybridized carbon in a one-step mechanism.

Solid phase synthesis (Section 26.7): A synthetic method in which a compound is attached to the surface of insoluble polymer beads. After a reaction is run on the compound, it is only necessary to collect the beads by filtration and wash them to remove any remaining reagent and isolate the product. Not only is the isolation procedure fast and simple but mechanical losses are minimized. After a number of reactions are run, the final product is cleaved from the polymer and isolated.

Solvolysis reaction (Section 9.7): A reaction in which the solvent acts as both the nucleophile and the solvent.

sp-Hybridized AOs (Section 3.5): The two AOs formed from combining one *s* orbital and one *p* orbital. These AOs are used to form sigma bonds by atoms with linear geometry.

sp²-Hybridized AOs (Section 3.4): The three AOs that are formed from combining one *s* orbital and two *p* orbitals. These AOs are used to form sigma bonds by atoms with trigonal planar geometry.

sp³-Hybridized AOs (Section 3.3): The four AOs that are formed from combining one *s* orbital and three *p* orbitals. These AOs are used to form sigma bonds by atoms with tetrahedral geometry.

Specific rotation ([α]) (Section 7.4): The magnitude of rotation of plane-polarized light by a chiral compound at a concentration of 1 g/mL and a path length of 1 dm.

Spectroscopy (Chapter 13): The study of the interaction of electromagnetic radiation (light) with molecules.

Spectrum (Section 13.2): A plot of the amount of light that is absorbed versus the frequency or wavelength of the light.

Spin coupling (Section 14.5): The interaction of the small magnetic fields of nearby nuclei in an NMR experiment that results in the splitting of the absorption into multiple peaks.

Staggered conformation (Section 6.3): Conformation in which the bonds on one atom bisect the angle between the bonds on the adjacent atom when a Newman projection is viewed.

Staggered conformation

Step growth polymer. See *Condensation polymer.*

Stereocenter or **stereogenic atom** (Section 6.1): An atom at which the interchange of two groups produces a stereoisomer.

Stereochemistry (Chapter 6): The three-dimensional structure of a molecule.

Stereoelectronic requirement (Section 4.3): The orientation required for the orbitals of the reactants in order for the reaction to occur.

Stereoisomers (Chapter 6): Isomers that have the same bonds or connectivity but a different three-dimensional orientation of these bonds.

Stereoselective reaction (Section 11.4): A reaction that produces predominantly one stereoisomer in a reaction but does form some of the other stereoisomer.

Stereospecific reaction (Section 11.4): A reaction that produces only one of two possible stereoisomers in a reaction.

Steric strain (Section 6.1): Strain caused by repulsion between nonbonded atoms that results in distortion of bond lengths and/or bond angles.

Steroid (Section 28.5): A naturally occurring compound that has a tetracyclic ring system con-

sisting of three fused six-membered rings and one five-membered ring.

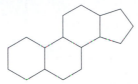

Steroid ring system

Strain (Section 6.1): Any factor that destabilizes a molecule by forcing it to deviate from its optimum bonding geometry.

Structural isomers: See *Constitutional isomers.*

Substitution reaction (Chapter 8): A reaction in which one group bonded to carbon is replaced by another.

Syn addition (Section 11.7): The addition of groups to the same face of a double bond.

Syn-periplanar bonds (Section 9.3): Bonds with a dihedral angle of 0°.

Synthetic equivalent (Section 10.2): A species that is used in place of another reagent in a synthesis. A synthetic equivalent gives the same final product, usually in better yield, as the other reagent, although more steps are required.

Synthons (Section 23.5): Electrophilic and nucleophilic fragments generated by bond disconnections during retrosynthetic analysis.

Tautomers (Section 11.6): Structural isomers that differ only in the position of a hydrogen and a pi bond.

Terpene (Section 28.1): A compound that occurs naturally in plants and can be considered as resulting from the combination of isoprene units.

Tertiary carbon (Section 5.3): A carbon that is bonded to three carbons.

Tetrahedral intermediate (Section 18.1): The intermediate formed in a substitution reaction at a carbonyl carbon, in which a nucleophile bonds to the carbonyl carbon, resulting in the formation of an *sp³*-hybridized carbon in the intermediate.

Thermolysis (pyrolysis) reaction (Section 22.1): A reaction caused by heat.

Thermoplastic polymer (Section 24.5): A polymer that is hard at room temperature but softens and eventually melts as it is heated.

Thermoset polymer (Section 24.9): A polymer that does not soften or melt on heating but becomes larger, harder, and more insoluble because cross-linking becomes more extensive.

Torsional strain (Section 6.3): Destabilization resulting from eclipsed bonds.

Transannular strain (Section 6.6): A type of strain that occurs in larger rings because of steric crowding of atoms on the opposite sides of the ring. Also called cross-ring strain.

Trans-diaxial elimination (Section 9.3): Elimination of axial groups on adjacent atoms on a cyclohexane ring. This is the preferred geometry for the E2 reaction with cyclohexane derivatives.

Transition state (Section 8.3): The structure of the complex at the maximum on the energy versus reaction progress curve and in which all the requirements for a reaction have been met.

Ultraviolet-visible (UV) spectroscopy (Section 15.1): A type of spectroscopy that employs ultraviolet or visible light; UV-visible spectroscopy uses transitions between electronic energy levels to provide information about the conjugated part of a compound.

Umpolung (polarity reversal) (Section 20.9): The formal reversal of the normal polarity of a functional group in synthetic reactions.

Unsaturated compound (Section 5.5): A compound that contains multiple bonds.

Upfield (Section 14.3): A chemical shift at a lower delta value in the NMR spectrum.

Valence bond theory (Section 3.2): A bonding model that uses overlap of AOs on the two bonded atoms to form MOs that are localized around these two atoms.

Valence electrons (Section 1.2): The electrons in the outermost electron shell.

Valence shell electron pair repulsion theory (VSEPR) (Section 1.10): A theory that is used to predict geometry based on the principle that pairs of electrons in the valence shell of an atom repel each other and try to stay as far apart as possible.

van der Waals forces (Section 2.5): The attraction between molecules that results from dipole–dipole interactions, dipole-induced dipole interactions, and instantaneous dipole-induced dipole (London force) interactions.

Vicinal hydrogens (Section 14.4): Hydrogens bonded to adjacent atoms.

Vinyl group (Section 5.5): The $CH_2{=}CH{-}$ group.

Vinyl polymer (Section 24.1): An addition polymer that is prepared from a monomer that contains a vinyl group.

Wavelength (λ) (Section 13.1): The distance of one complete cycle of the light wave; the distance between successive crests or troughs of the wave.

Wavenumber (Section 13.4): The reciprocal of wavelength (units are cm^{-1}); used in infrared spectroscopy.

Ylide (Section 18.7): A carbanion that is bonded to a positive phosphorus group.

Z (Section 6.2): Letter used to designate the isomer of an alkene that has the high priority groups on the same side of the double bond.

Zaitsev's rule (Section 9.4): The major product of an elimination reaction is the alkene with more alkyl groups on the carbons of the double bond (the more highly substituted product). Most E1 and E2 reactions follow this rule.

Zwitterion: See *Dipolar ion.*

Index

Important Absorption Bands in the Infrared Spectral Region

Position (cm⁻¹)	Group	Comments
3550–3200	$-$O$-$H	Strong intensity, very broad band
3400–3250	$-$N$-$H	Weaker intensity and less broad than O$-$H; NH$_2$ shows two bands, NH shows one
3300	$\equiv$C$-$H	Sharp, C is sp hybridized
3100–3000	$=$C$-$H	C is sp^2 hybridized
3000–2850	$-$C$-$H	C is sp^3 hybridized; 3000 cm⁻¹ is a convenient dividing line between this type of C$-$H bond and the preceding type
2830–2700	$-\overset{\overset{\displaystyle O}{\|\|}}{C}-$H	Two bands
2260–2200	$-$C$\equiv$N	Medium intensity
2150–2100	$-$C$\equiv$C$-$	Weak intensity
1820–1650	$-\overset{\overset{\displaystyle O}{\|\|}}{C}-$	Strong intensity, exact position depends on substituents; see Table 13.1
1660–1640	C=C	Often weak intensity
1600–1450	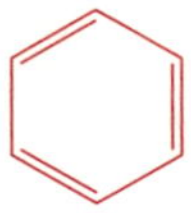 (benzene ring)	Four bands of variable intensity
1550 and 1380	$-$NO$_2$	Two strong intensity bands
1300–1000	$-$C$-$O$-$	Strong intensity
900–675	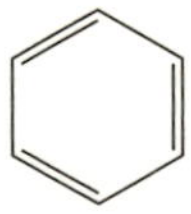 (benzene ring)	Strong intensity